ALS WARNING LABELS

will be found on the back cover.

RADIOACTIVE MATERIALS

POISONOUS MATERIAL

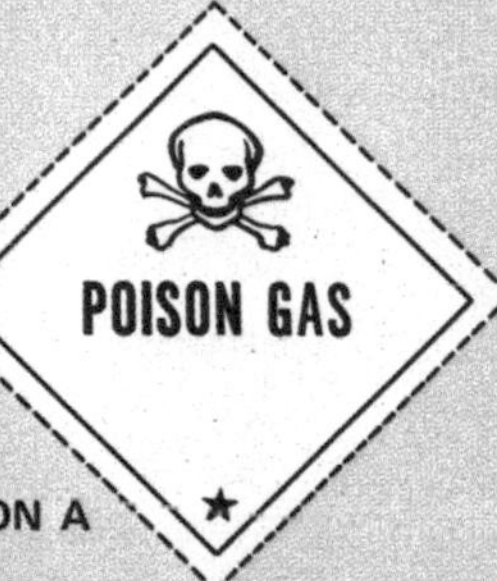

ON A

POISON B

IRRITATING MATERIAL

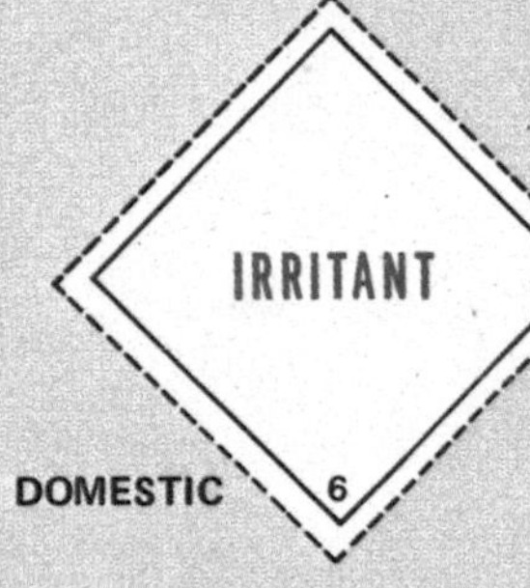

DOMESTIC

EXPORT

ETIOLOGIC AGENT

ETIOLOGIC AGENTS
BIOMEDICAL MATERIAL
IN CASE OF DAMAGE OR LEAKAGE
NOTIFY DIRECTOR CDC
ATLANTA, GEORGIA
404/633-5313

BUNG

CAUTION — Unscrew This Bung SLOWLY

Do not unscrew entirely until all interior pressure has escaped through the loosened threads.

REMOVE BUNG IN OPEN AIR. Keep all open flame lights and fires away. Enclosed Electric Lights are safe.

EMPTY

EMPTY

IR SHIPMENTS

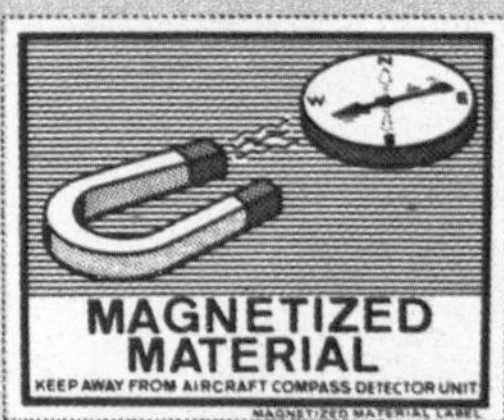

MAGNETIZED MATERIAL

NOTE

1. DOT hazardous materials warning labels indicate the hazard(s) of the material(s) in the package. These labels may or may not indicate the proper DOT hazard class of material.

2. LABEL SIZES, COLORS AND BORDERS—All diamond shaped labels illustrated above must be at least 4 x 4 inches with a solid line border 1/4 inch from each edge. Bung label: 5 x 3 inches; Empty label: 6 x 6 inches (min.); Etiologic agent: 4 x 2 inches; Cargo Aircraft Only: 4-5/16 x 4-3/4 inches; Magnetized material label: 3-9/16 x 4-5/16 inches. Each label must be affixed to a background of contrasting color or must have dotted or solid line outer border. Colors for labels must be as specified in Sec. 172.407 and in Appendix A, Part 172.

Dangerous Properties of Industrial Materials

Dangerous Properties of Industrial Materials

Fifth Edition

N. IRVING SAX

Assisted by:

Marilyn C. Bracken/Robert D. Bruce/William F. Durham/Benjamin Feiner/
Edward G. Fitzgerald/Joseph J. Fitzgerald/Barbara J. Goldsmith/John H. Harley/
Robert Herrick/Richard J. Lewis/James R. Mahoney/John F. Schmutz/
E. June Thompson/Elizabeth K. Weisburger/David Gordon Wilson

VNR VAN NOSTRAND REINHOLD COMPANY
NEW YORK CINCINNATI ATLANTA DALLAS SAN FRANCISCO
LONDON TORONTO MELBOURNE

Van Nostrand Reinhold Company Regional Offices:
New York Cincinnati Chicago Millbrae Dallas

Van Nostrand Reinhold Company International Offices:
London Toronto Melbourne

Library of Congress Catalog Card Number: 78-20812
ISBN: 0-442-27373-8

Manufactured in the United States of America

Published by Van Nostrand Reinhold Company
135 West 50th Street, New York, N.Y. 10020

Published simultaneously in Canada by Van Nostrand Reinhold Ltd.

15 14 13 12 11 10 9 8 7 6 5 4 3 2

Library of Congress Cataloging in Publication Data

Sax, Newton Irving.
Dangerous properties of industrial materials.

First published in 1951 under title: Handbook of dangerous materials.
Includes bibliographical references.
1. Hazardous substances. I. Bruce, Robert D.
II. Title.
T33.3.H3S3 1979 604'.7 78-20812
ISBN 0-442-27373-8

To Murray

PREFACE

This new, completely revised, fifth edition of *Dangerous Properties of Industrial Materials* provides a single source for quick, up-to-date, concise, hazard-analysis information for nearly 15,000 common industrial laboratory materials. The information is set forth mainly in Section 12, "General Chemicals," which expedites the retrieval of data as follows:

- Direct alphabetical listing of over 16,000 entries.
- Physical constants, synonym(s), description, and flammability and explosion data as available for each entry.
- The whole point of this book is to promote safety by providing hazard information about many materials. To that end, we have now incorporated into more than 10,000 entries some basic toxicologic information, including route of exposure, test animal(s), and quantity of exposure. THR is included to put into words the importance of the toxicological data. In some cases where there are no numerical data, a THR will still report carcinogenicity, mutagenicity, and teratogenicity as well as an opinion of the hazard of a given entry. Any material with an LD_{50} of 500 mg/kg or less is considered HIGHLY toxic; 500 to 7500 mg/kg, is labeled MOD toxic; 7500 mg/kg to 15,000 mg/kg is considered LOW toxic; above 15,000 mg/kg is labeled NONE. Concise definitions of these toxicity levels are given with the abbreviations, located just before Section 12. As to fire hazards, a material with a flash point of 100°F or less is dangerously flammable. From 100° to 200°F, the flammability is moderate, and above 200°F the flammability is listed as low.
- In the countermeasures portion of each entry, fire-extinguishment materials are listed. Also listed for the benefit of warehouse and transportation workers, and firefighters are the disaster hazards, which are not always obvious, of involving a listed entry in a fierce consuming fire, an earthquake (or some other source of shock), or a flood. For instance (flood) water and high temperatures can form steam, which may adversely react with many materials to liberate heat, explosive gases, and toxic fumes. We have listed more than 1000 chemical incompatibilities to forewarn chemical users, storers, or transporters of possible problems.
- The acute toxicity data, THR, and chemical incompatibilities have all been referenced and annotated to permit further study.
- More than 2500 new compounds have been added; larger, more readable type and better spacing were used. To avoid repetition, a counter-measures directory is included to permit the reader to rapidly find, in the 11 sections of text, detailed information about any problem concerning a given material. Thus, if the reader is interested in the TLVs, the entire ACGIH listing for 1977 is included with Section 1.
- Section 2 includes detailed ventilation procedures, sample calculations, and methods for determining the amount and kind of ventilation required. The safe handling of carcinogens is included. Section 3 is a dissertation on the effects and engineering control aspects of noise.
- Section 4 details the interaction of industrial and community environmental pollution with present and forthcoming legislation. Section 5, in two parts, discusses the theoretical and practical aspects of ionizing radiation, its production, and control. Maximum permissible levels tables for environmental contamination are included as well as a great deal of information about the use of short-lived isotopes in medical and industrial research. Environmental contamination, including waste disposal and control, as well as procedures for personnel protection are discussed.
- Section 6 deals with the enormously increasing problems of solid-waste production and disposal. Section 7 discusses the theory of fire and its control, and methods of storing, handling, and using flammable and explosive materials. Section 8 is devoted to the identification and control of environmental carcinogens.
- Section 9 addresses itself to toxicological problems of industrial materials from a theoretical standpoint. Section 10A is an up-to-date summary of the regulations of industrial chemicals. Section 10B discusses chemical regulations from the standpoint of industry. Section 11 goes into the labeling and indentification of industrial materials for storage and shipment purposes.

Each section is written by an expert in his field, and each contains a bibliography for further study by the reader. As before, no effort has been spared to make this edition useful and relevant.

N. Irving Sax
Boca Raton, Florida

ACKNOWLEDGMENT

I wish to unstintingly thank my wife Paula, without whose tireless dedication to the Fifth Edition, this three-year project could not have been completed.

I also wish to thank Lucille McKevitt and Mimi Schlesinger for clerical and proofreading assistance.

My gratitude to Alberta W. Gordon, Managing Editor, and to Ashak M. Rawji, Senior Editor of the Van Nostrand Reinhold Company for their professional assistance with this book.

N. IRVING SAX

CONTENTS

Section 1

The Historical Perspective

Robert Herrick

THE HISTORICAL PERSPECTIVE

The protection of the health of workers from the hazards of their occupation is of relatively recent concern. More advances in science and legislation have been made in the past 65 years than in all prior time. An understanding of the social, political, and technical factors which brought us to the present provide an insight into the future of occupational safety and health.

Historically, the goal of occupational medicine is the promotion of optimal health, productivity, and social adjustments for the worker. Attaining this goal requires identification of the factors which deprive him of these conditions, the definition of possible solutions, and the implementation of such corrective action as is necessary.

As the sciences of medical diagnosis and engineering control advance, new steps forward will become possible. The present rate of growth and complexity of our industrial society creates new hazards at a rapid rate, so it can safely be said that the next decades will bring even more significant advances in raising the control of worker health closer to a science than an art.

The earliest of man's occupational injuries were probably the cuts, bruises, and eye injuries he suffered as he chipped out his first primitive rock tools. This same hunter might also have contracted anthrax from skinning his game, but that was a secondary hazard.

At this stage in human development, just staying alive was such a significant occupational problem that probably no one paid any attention to chronic or secondary problems which might develop in time. Life was so simple that there was no concern over specialized occupational hazards; indeed, there were few such in any case.

As agrarian societies developed, specialized occupations became identified and no longer was each individual responsible for meeting all the needs of himself and his family. As the processes for winning metals from their ores and working with these metals developed, separate workers identified as craftsmen were freed from their primary work of food production. Metallurgists were, in fact, the first class of craftsmen which came to be clearly identified as industrial workers.

Early history of mankind clearly points out the reality of social stratification by occupation, with individuals identified in one distinct group or another. As civilization developed, the practice of slavery also developed and provided a cheap source of power needed to keep civilization in motion. The supply of slaves seemed inexhaustible and, as slavery spread, the value and respect for the individual which we today hold so high nearly died out. This social stratification coupled with the moral philosophy of the day led to strange conflicting (by today's standards) social values. Occupational medicine did not yet exist, although there were many occupations known to lead to specific diseases. Since mining and heavy labor were done by slaves, and it was their ills that resulted from the occupation in question, these ills were not considered medical problems. Ancient physicians were concerned only with the diseases of the ruling, i.e., powerful or able-to-pay classes, almost totally ignoring the craftsman and the laborer. Thus the medical writing of Hippocrates, which exemplified this attitude, dealt almost exclusively with the health of citizens rather than workers.

Ancient Egyptian, Greek, and Roman societies placed very high values on physical and moral perfection but their contempt for craftsmanship, i.e., handwork or manual labor, was so great that in many cities it was illegal for even an ordinary citizen to ply a trade. This prejudice against manual labor and the trades severely limited the interest in and the study of occupational diseases. Yet the fruits, the products produced by these laboring classes, were highly valued to the point that commerce demanded that they be even further developed.

One of the earliest known writings to include specific references to occupational medicine was *Historia Naturalis*, an encyclopedia of the natural sciences written by Pliny the Elder in the first century A.D. He recommended that a protective mask consisting of a crude bladder be placed over the face of the worker to prevent inhalation of the dust from cinnabar (mercury ore) grinding and lead fumes from lead smelting or working.

The Middle Ages, although a time of change in social systems, accomplished little to advance occupational medicine. Feudalism was the characteristic social order in Europe and the great landlord and his serfs lived nearly independent existences.

As the Renaissance approached, peasants and workers were attracted to the walled and crowded Medieval towns. New classes of merchants, free artisans, and laborers developed as did the guild system. The rigidly organized guilds regulated nearly every aspect of their members lives, but on the other hand, the same guilds also provided assistance (a sort of health insurance) to members who became ill and in time developed lasting prestige through the high standards of workmanship which guild members maintained.

The Middle Ages were, from a public health standpoint, characterized by epidemic diseases, truly terrible by today's living standards, and a total lack of knowledge of public hygiene. Even in the midst of these enormous problems, specific occupational diseases can be noted, isolated, and identified. For instance, the beautiful illuminated Medieval manuscript books contained illustrations drawn with metallic paints and some say that Medieval scribes developed lead and mercury poisoning as a result of moistening their metal pigment paint brushes on their tongues, in a manner similar to the workers who painted radium dials on watches in the early twentieth century. Also "Grinder's Disease" was identified as a hazard to those workers who, in sharpening knives and swords, worked close to sandstone wheels.

The specialized use of metals in industry led to significant problems. Paracelsus wrote of many of the hazards of metallurgy and mining in *Von der Bergsucht und anderen Bergkrankheiten* (ca. 1530). His description of mercury poisoning and his warnings of the hazards of toxic metals brought about an initial awareness of these problems. In France, for example, the hazardous nature of the vapors from mercury distillation were well enough acknowledged that workers were allowed only one month of employment at this trade in a year.

Georgius Agricola's *De Re Metallica*, published in 1556, was a classic volume on metallurgy. The 12 sections of the book describe nearly every facet of mining, smelting, and refining. The last section of the book describes the diseases and accidents prevalent among miners and the means for their prevention. Agricola was concerned with ventilation in mining, considering lack of fresh air as one of its greatest hazards. He proposed several ventilating devices as possible solutions to this hazard.

Bernardino Ramazzini is to occupational medicine what Galileo and Newton are to physical science. He developed an entirely new concept of occupational health and hygiene. *De Morbis Artificum Diatribe*, published in 1700, is a classic in the field. He examined the conditions of work and the diseases due to most of the occupations of his time. In addition to describing the diseases, Ramazzini proposed preventive measures which, although ignored for several centuries, were basically sound.

Among the early occupations wherein specific industrial diseases were identified were gilding, marble cutting, grinding, pyrite roasting, dyeing, mercury smelting, silvering of mirrors, and glass blowing.

The eighteenth century and the Industrial Revolution created a completely new set of occupational hazards. The social status of the worker was still very low, with children and women mainly being exposed to long hours in hazardous trades. In particular, English coal miners lived a precarious life and as a result of trying to improve this situation several English physicians made specific contributions which led to improved working conditions. Sir George Baker, for instance, traced lead poisoning to "Devonshire colic" to the practice or lining cider presses and vats with lead. This logically led to the abandonment of lead in the cider industry.

Percivall Pott identified the first occupational cancer in 1775. This was scrotal cancer caused by the soot to which chimney sweeps were exposed (see Section 8 and *Soot* in Section 12). Charles Thackrah devoted his life to the study and prevention of occupational hazards, and his treatise on occupational hazards and occupational medicine in the early 1800's was the first book of its kind published in England. It played an important part in stimulating factory and health legislation.

Thomas Beddoes founded the Pneumatic Institution in the late 1700s for the study of inhalation therapy in treating respiratory disease. Beddoes identified several occupations which made workers more susceptible to tuberculosis, including stone and metal grinding. Sir Humphry Davy studied at the Pneumatic Institution and is remembered for his study of mine explosions and his development of the miner's safety lamp.

In the United States an act was passed in Boston in 1726 preventing the use of lead in the distilling of rum and other liquors. Thomas Cadwalader published the first medical monograph in the United States, dealing with lead colic and lead palsy. Benjamin Franklin published the book in 1745.

The nineteenth century was a turning point for occupational medicine. The advent of the industrial society and the absence of social legislation make this the low point in society's concern for the worker and his health. The excesses of dangerous occupations and hazardous employment brought about a new social consciousness in Western society. The trade union movement gave organization and leadership to large groups of workers who had heretofore been without a voice. Concurrently, certain industrialists realized that worker well-being was in their economic self-interest. Combined with the genuine humanitarianism and utopian socialism, goals espoused by increasingly larger portions of the population, a reform movement was generated which has lasted well into the twentieth century.

During 1910 to 1912, modern occupational medicine

was born. Physicians wrote about occupational diseases, the public was becoming aware of occupational hazards, and long needed legislation was finally being enacted. The work of Dr. Alice Hamilton at this point was outstanding, as she applied the science of pathology to occupational health. Her work in the area of lead poisoning is particularly significant.

From this point in time, occupational medicine, industrial hygiene, and industrial nursing became identified as specific occupations within a social system that now placed a high value on the individual and therefore his health and well-being. Advances in science and technology began to be applied to the protection of the worker and the emphasis shifted from diagnosis and treatment after the fact of occupational disease to the much more desirable prevention and control of hazardous situations so that the exposures leading to ill health would be recognized in advance and avoided.

B. LEGISLATION

With the exception of the banning in Boston of lead as a processing material for foodstuffs, the English factory and health laws were the first major worker health-related legislation. The early twentieth century is then, for all practical purposes, the beginning of modern industrial health legislation.

In 1898 the U.S. Supreme Court made the first broad statement that the health of the laborer as a producer is considered to be as much a public benefit as the health of the consumer. The protection of the laborer thus became a public purpose. However, it was some years before this basic court decision was to become a daily reality. The State of Massachusetts was one of the first to adopt progressive legislation. In 1905 Massachusetts employed health inspectors to investigate occupational dangers. By 1909, 21 states had enacted laws which attempted to regulate industrial working conditions in one way or another. Some of these dealt with ventilation, others with dust control, and still others with workshop sanitation. Unfortunately, enforcement lagged significantly behind the intent of the regulations. In many cases inspectors were political appointees untrained in the technical aspects of occupational health and unappreciative of the elements of the problems with which they dealt.

The accomplishments of Dr. Alice Hamilton as a member of a special Illinois commission centered on her correlation of 35,000 cases of lead poisoning which had occurred in American industry from 1908 to 1914. As a result, several states adopted "lead laws" in 1913 and 1914. The regulatory and enforcement powers for this legislation in New Jersey, Ohio, and Pennsylvania were vested in the Department of Labor at the state level. This was the origin of the splitting of public health functions between health and labor agencies which has continued through the years to the present time.

Workmen's compensation laws were enacted by about half of the states in 1913. Occupational diseases were included as compensable under these acts in the 1930's; but the greatest impact of this inclusion was that industry gained an economic awareness of the importance of the disciplines of industrial hygiene.

The Social Security Act of 1935 gave an economic boost to many health programs, including industrial hygiene. The period of 1936 to 1939 was the most important in the development of industrial hygiene as a distinct profession.

In the early 1940s, the U.S. Public Health Service provided professional industrial hygienists to serve in wartime munitions plants. Then in 1945 most of these personnel were transferred to state agencies on a loan basis to pursue their professions.

The Walsh-Healy Act, covering Federal requirements for industries engaged in government contract work, was the first piece of broad legislation which included numerical occupational health standards. The threshold limit values (TLV's) of the American Council of Governmental Industrial Hygienists (ACGIH) were included in the health requirements under this Act.

The most significant individual piece of legislation is the Occupational Safety and Health Act of 1970. When considered with the Federal Coal Mine Health and Safety Act of 1969, nearly every occupation and industry in the United States comes under Federal regulation. The Coal Mine Health and Safety Act is very specific in setting environmental levels for coal mine dust with health and safety being regarded as separate considerations.

The Occupational Safety and Health Act has been hailed as a particularly important piece of legislation because of its thoroughness in approaching the entire scope of occupational safety and health. The administration of the Act is the joint responsibility of the Department of Labor and the Department of Health, Education, and Welfare. The agencies involved are the Occupational Safety and Health Administration (OSHA) and the National Institute for Occupational Safety and Health (NIOSH), respectively.

The Department of Labor now has primary responsibility for determining priorities, setting standards, enforcement, operating a national record keeping and reporting system, providing employer/employee education, approving state plans, and awarding state grants. The Department of Health, Education, and Welfare has a supportive role in nearly all of these activities. HEW has specific responsibilities under the Act including health and safety research, industry-wide studies, hazard evaluations, toxicity determinations, and publication of an annually compiled list of toxic substances.[3,103]

The Occupational Safety and Health Act is specific in defining "consensus standards" as the principal mechanism for the adoption of specific regulations. Consensus standards are those which had been adopted by various technical societies and were generally accepted as good practice guidelines by professionals, industry, and regulatory agencies. Most of the consensus standards initially adopted as law concerned safety, but there were included, as well, several industrial hygiene standards. Among these were the Threshold Limit Values (TLV's) of the ACGIH and the ventilation standards of the American National Standards Institute (ANSI).

Most of the health-related standards will be promulgated in future years. The process will be one of standard setting by specific chemical, e.g., carbon monoxide, or physical hazard, e.g., heat stress. NIOSH will prepare a "criteria document" and present its recommendations by a panel of experts and consultation with affected parties; OSHA will promulgate and enforce the standard.

The development of a "criteria document" is a lengthy, in-depth procedure. All available toxicological data are collected and interpreted. Simple identification of a hazard is only the first step. The "criteria document" includes recommended procedures for measurement of the stress agent of interest, not only in terms of sampling and analysis but in terms of biological monitoring when appropriate. The source(s) of the agent are examined and the options for engineering control are discussed in detail. Personal protective measures are also examined in detail. Good practice recommendations are included. The net result is that the "criteria document" becomes a complete manual on the degree of hazard, procedures for monitoring, and control measures to be taken to reduce hazard to acceptable limits for the chemical studied. Even after OSHA standards are set and enforced, the "criteria document" remains as a primary reference for the working industrial hygienist.

These useful and carefully considered procedures will result in the promulgation of specific industrial hygiene standards on a national level. More of the enforcement of these standards will be on a state level as the states increase their budget and staff capabilities. While it is conceivable that the numerical standards adopted by a state could be more stringent than those of OSHA for certain compounds, it is not likely to happen, mainly because the Federal procedure described above is very scientific and detailed, with many checkpoints and opportunity for interested parties to make their influence felt before the standard is promulgated.

On-site inspections which could be made by Federal, state or local compliance officers will likely be handled at the lowest practical jurisdictional level. It is to the advantage of all concerned to have inspections performed by a local compliance officer who is familiar with the individual plant or process and who will be aware of and will be able to act upon any significant changes over a period of years. His observations can verify the benefits of process changes or added engineering controls. Likewise, he can be aware of the use or non-use of personal protective equipment by employees.

Certain hazards, such as carbon monoxide and noise, can be monitored by continuous instrumentation. The law allows for requiring such data from the employer when the need is justified and reliable instrumentation is available. Since this means considerable outlays of money on the part of a plant operator, it would normally not be required in any but critical situations. As research identifies these critical situations and new instrumentation is developed to provide monitoring data, it can be expected that more use will be made of this procedure for verifying the safety of working conditions.

The Occupational Safety and Health Act and its administration during the first few years of its existence have come under heavy attack because of certain alleged inconsistencies. However, as with any new piece of legislation, particularly with such far-reaching effects, these inconsistencies are being shaken out. The basic principles of the Act relating to occupational health are sound, and it is expected that this legislation will be the pattern to be followed for many years.

TSCA

The Federal Government in the United States has taken a broad approach to the control of potentially dangerous industrial materials. The Clean Air and Clean Water Acts include provisions that allow the Federal Government to regulate against the discharge of toxic or hazardous materials into the waterways or air. Under the Clean Air Act, for example, specific regulations have been promulgated to define allowable emissions of such materials as asbestos and mercury. Both of these Acts have provisions that allow the government to set emission standards as a means of attaining appropriately low levels of hazardous materials in the ambient air and waterways of the nation.

The Department of Transportation has many specific regulations concerning the transport of hazardous materials. These are described in detail in Section II of this book.

Effective January 1, 1977, significant new legislation became effective. The Toxic Substances Control Act (TSCA) will have a major impact on all persons involved in the development, manufacture, and use of potentially toxic substances. In its broadest context, TSCA calls for the screening of new chemical substances prior to their introduction into the market place. Statistics used in the development of this legislation refer to some 30,000 or more chemical substances presently in use, with an additional 1000 new substances introduced annually. Once the baseline data are generated on the substances in use, the TSCA procedures will provide scientific information concerning the problems posed by the presence of these chemicals in the environment. The overall goal of TSCA is to protect human health and the environment from unreasonable risks.

Highlights of TSCA include defining the responsibilities for initial testing requirements on the manufacturer of that compound. As a first step in the implementation of the act, every manufacturer of chemical compounds is required to submit an inventory of chemicals currently being produced. Once this initial inventory of existing chemicals is completed, any new compounds proposed for introduction into the market place will require a pre-market notification procedure. Specific information required for different types of compounds will be developed. Testing requirements will vary depending upon the generic category of the new chemical. It will be significant to industry that new chemical compounds cannot be introduced into the market place until approval has been granted by TSCA.

TSCA legislation is intended to prevent the type of problems that have not been dealt with by other legislation. Examples used during the development of the legislation included polychlorinated biphenyls (PCB's) and chlorofluorocarbons. These two classes of compounds have been in general use for many years, with specific problems identified only recently. PCB's in the nation's waterways have caused contamination of some fish species to such a degree that the fish is no longer suitable for human consumption. In addition, it is a highly persistent chemical, and the problems that it causes will endure for several years, even if no more PBC is added to the environment. The chlorofluorocarbons, on the other hand, have been identified as dangerous only because there is a presumed stratospheric chemical reaction that might change the quantity and quality of solar energy reaching the earth's surface. These compounds have been verified by toxicity tests to be quite safe from an industrial hygiene standpoint.

The charter defined by TSCA is both broad and ambitious. If the intent of the legislation can be carried out with a high degree of technical competence it should be a major step forward in the protection of both the people and the environment in general.

C. GENERAL PRINCIPLES OF INDUSTRIAL HYGIENE

Goals

The overall goal of the industrial hygienist is to protect workers from the health hazards of their working environment. The working industrial hygienist recognizes that there will nearly always be toxic and potentially hazardous materials to be processed or handled, and that some degree of physical stress such as noise or heat or radiation will be encountered as well.

Within this broad scope, or overall goal, is contained the need for a wide range of professional skills and disciplines. The industrial hygienist must be sensitive to the cost-effectiveness of proposed measures to control the situations with which he works. It is his function to evaluate the degree of a hazard and the applicable regulatory requirements and balance the requirements against the costs of reducing exposures by means of a number of options. These can include engineering controls, process isolation in space or time, substitution of process materials, personnel protection, personnel rotation, and, in extreme cases, discontinuance of a process. However, before we can productively discuss implementation of the industrial hygienist's goals we must clarify the terms *toxicity* and *hazard*.

Toxicity and Hazard

The toxicity of a material is a property of that material, and can only be described by its effects upon a living organism. As a general principle, toxic materials are toxic to all living things, while the individual susceptibility of each species to the toxic material varies, as do the susceptibilities of the individual members within a species. Within a species, susceptibility varies with age, sex, state of health, rate of dosage, diet, etc. Therefore, toxicity data, to be useful, must include all of the above and must be stated in terms of the specific test animal or organism used, the routes of administration of the toxic matter, and the time of exposure. Also a sufficient number of test animals or organisms must have been tested to rule out some of the variables listed above and to make the data statistically significant; for further discussion see Sections 8 and 9. For purposes of this section, suffice it to state that all materials are more or less toxic. Those of any concern to this section are the ones which are or could be damaging to humans in amounts that ordinarily occur or might occur in the living environment or in the industrial or medical or scientific environments. Here is where hazard enters the picture. For instance, it is conceivable for a material to have a high "toxicity" rating but a low "hazard" rating. This is because the hazard rating or "hazard" associated with a material is simply a measure of the *likelihood* of damage to humans working with or studying the material in question. Thus while highly toxic materials such as arsenic and mercury or beryllium compounds are generally considered very hazardous as well, it is usually possible for the industrial hygienist to apply his professional know-how to, say, an industrial process so as to reduce its hazard to the people who work with it. The toxicity of the materials remain constant; he cannot change that.

Therefore, when highly toxic matrials are included in an industrial process the competent industrial hygienist realizes that the chance for damage is great and he must do what he can to reduce the hazard or likelihood of the occurrence of damage from such materials.

Of course, other hazards, such as flammability and explosibility, though completely separate from toxicity, are still part of the industrial hygienist's job and must be controlled for the well-being of the workers and even the physical equipment.

A knowledge of toxicity and the ability to evaluate hazards are basic to the industrial hygienist's work performance. When evaluating an existing situation the goal is to evaluate the probability that a hazardous situation does or might exist. In new situations, such as a proposed raw material change in a process or the design of new process facilities, the goal is ample controls to be applied as necessary to reduce the probability of hazardous exposures to an acceptably low level.

Specialty Skills

The field of industrial hygiene encompasses a wide spectrum of specialty areas. Of particular importance is the fact that the industrial hygienist must be aware of the physiological effects of a multitude of potentially hazardous compounds. Other aspects are the identification and quantification of exposure levels, application of realistic methods of evaluating the working environment, and selecting options for control or lessening of exposures. The basic disciplines are chemistry, toxicology, physics, and engineering. Few other professions cover such a broad spectrum of professional disciplines.

The industrial hygiene *chemist* has a broad knowledge of both organic and inorganic chemistry, especially those aspects dealing with the analysis of trace quantities of materials. Prior to about 1950 nearly all analytical work was done by wet-chemical methods. The advent of automated analytical techniques has provided new tools to the industrial hygiene chemist. Atomic absorption spectrometry, for example, is useful when analyzing for the heavy metals such as lead. The old reliable dithizone technique for lead analysis is quite sensitive but it is time-consuming and tricky in that it requires tight quality control over the reagents and laboratory practices involved. The automated determination of lead in biological specimens still requires extreme laboratory cleanliness to prevent contamination of samples, but the analysis time has been reduced from hours per sample to minutes by the new techniques. There is also voltammetry, polarography, infrared and ultraviolet spectroscopy, emission spectroscopy, mass spectrometry, gas and liquid chromatography, nuclear activation, and many combinations of all these.

The chemist is concerned also with the actual collection as well as the analysis of samples. The field industrial hygienist who collects samples at work stations must work closely with the industrial hygiene chemist to assure that the sample which is taken is of adequate size to allow accurate determination of the contaminant in question, that the sampling equipment and reagents are appropriate for the task at hand, and that the presence of significant interferences identified by the chemist is properly noted at the time of collection of the sample.

In the past it was the chemist who commonly interpreted toxicological data, but more and more there is a role for the industrial hygiene *toxicologist* as a specialist in industrial hygiene. In any case, some knowledge of toxicology is essential for all industrial hygienists and this knowledge should be based, in part, upon actual toxicological research work. The specialist in toxicology or the industrial hygiene toxicologist must understand not only the application of toxicological information to problems at hand but the procedures used to determine the toxicology of various compounds; whether the toxicity be acute or chronic; whether the significant route of exposure is via skin absorption, inhalation, or ingestion; and the possible synergistic effects of exposure to a number of compounds. He may himself perform these evaluations or supervise their performance in biological testing laboratories.

The industrial hygiene *physicist* generally works in the areas of radiation, heat stress, and noise (see also Sections 3 and 5). These are classed as physical rather than chemical hazards and their evaluation and control require the application of sound physical principles. Developing approaches to the control of heat and noise problems are particularly challenging parts of their work.

The principal function of the industrial hygiene *engineer* is to evaluate potentially hazardous situations and develop viable solutions for the control of exposures. Many innovative solutions to industrial problems have been developed by industrial hygiene engineers. Ventilation engineering, for example, is an important aspect of industrial hygiene engineering (see Section 2).

The melding of all these factors, with an overview that reaches from the occupational physician to the plant manager, is the scope of the industrial hygienist. Usually, no single individual is an expert in all aspects, but many are capable of meeting their acknowledged goal of protecting employees. An abilty to adapt training and experience to new situations and to seek out expert assistance when required is important for a really functional team in this small but viable profession.

Toxicology for Industrial Hygienists

A basic knowledge of toxicology is one of the first skills needed by the industrial hygienist. Since this is covered in some detail in Section 9, and in a special sense in Section 8, the following paragraphs of this section will simply outline the areas of particular interest to the industrial hygienist.

Toxicity data are presented in many forms. The most common is the LD_{50} or the lethal dose of a material for 50% of a population of specified laboratory animals. This can be a single dose in mg/kg of weight of experimental animal to determine acute toxicity or a long-term repeated dosage to determine chronic effects. The National Institute of Occupational Safety and Health (NIOSH) has started to publish an annual Toxic Substance List, with LD (lethal dose) data on many thousands of compounds. It should become a valuable basic reference.

Each route of entry into the body—skin, via absorption; lungs, via inhalation; and intestines, via ingestion—has a special membrane wall which protects the body from absorption of some classes of chemicals. The skin, for example, is nearly impervious to hydroxy, carboxy, and ionized molecules while hydrocarbons, fats, and esters as well as some organometallics pass through with relative ease.

Some compounds have great mobility within the body. Benzidene base (see Section 12) placed on the skin can be detected in the urine within 20–30 minutes. Benzidene hydrochloride or sulfate, being of different molecular structure, do not penetrate the skin.

Certain critical organs of the body have definite protective screens. A brain-blood barrier exists which prevents many ions from reaching the brain tissue even though they get into the blood. Inorganic lead encephalitis is hardly ever found in adults as a result of this barrier, but it is often found in young children because in them this barrier is not fully completed and the young are often exposed to lead.

The liver and kidneys perform the vital function of cleansing from the body most of what toxic substances get into it, but these "shock" organs can be damaged in the process. The effect of mercury, for instance, can be most severe on the kidneys, which, incidentally, do the work of separating it from body fluids. The liver, it appears, usually takes the brunt of the load of blood purification, and carbon tetrachloride exposures may thus lead to cirrhosis as the liver may thereby be overtaxed. See Section 9.

Acute exposures, strangely enough, put additional stress on the liver and kidneys in a different manner. The first effect of acute poisoning is usually a severe stress similar to shock. Circulation and respiration are threatened. The brain's body supply is, however, preferentially maintained by blood shunting, and the liver and kidneys are among the first organs to have their blood supply cut off. While this also protects these organs from chemical damage, long periods of time in this condition can cause them to atrophy or otherwise be damaged.

In acute poisoning cases, therefore, treating the victim for shock is liable to save his life even without knowing the type of poison involved.

The route of entry into the body sometimes changes the whole toxicity mechanism and broad toxicological generalizations from a smattering of information may be quite dangerous. Trichlorethylene, for instance, acts as a systemic poison by ingestion while acute inhalation exposure mainly causes anesthesia. Two modes of entry; two completely different effects. This explains the advice against causing vomiting for the victim of an ingestion of solvents because vomiting can put some of the solvent into the lungs and the victim may then be subjected to both effects at the same time.

The industrial hygienist needs knowledge of this type because in doing his job, he is often near the site of an emergency. Even more, he may be the one whose name comes first to the mind of the person calling for aid because of his demonstrated knowledge and interest in the health of the employees.

Certain other specific toxic agents should be noted because of their wide exposure and/or high hazard. Hemoglobin has a preferential affinity for carbon monoxide which is about 200 times that of its affinity for O_2. A CO concentration in the air of 1000 ppm (0.1%), therefore, competes for hemoglobin on an equal basis with O_2 at its normal level of 20.4%. Carbon monoxide thus is very hazardous because it can be found in so many areas; because it is colorless, odorless, and tasteless; and because it is insidious, i.e., in acute exposures, collapse comes almost without warning and the victim is left in the high-exposure area in an unconscious state. Even at concentrations which do no more than cause an occasional headache, continuous or chronic exposures over a period of years can so starve the brain for O_2 that scar tissue develops and permanent brain damage ensues.

Hydrogen sulfide (H_2S), the rotten-egg gas, is particularly dangerous. Its toxic action is via an enzyme inhibition which causes respiratory paralysis. One breath of a very high concentration can be fatal, making it one of the most highly toxic materials found commonly in industry. It is particularly hazardous because its odor is strong and disagreeable at safe levels, but the more dangerous, high concentrations cause olfactory fatigue and there is soon no odor sensation at all. The rule of thumb in H_2S exposure area is: "If you smelled it before but can't smell it now, get out while you still can."

There are three routes by which a toxic agent can enter the body: ingestion, inhalation, and skin absorption. The industrial hygienist must think of each of these possible mechanisms.

Ingestion, or swallowing of industrial products, reagents, and solvents, is sometimes encountered in industry where workers eat or even smoke without washing; or where foods or candy, gum, or cigarettes are stored near where these materials are used. Some cases of lead and mercury poisoning have been traced to this route of entry into the body.

Considerable ingestion toxicological data obtained from work on experimental animals is usually available for compounds encountered in industry. This is the quickest and easiest test used by toxicologists when evaluating new compounds. Also, FDA requirements demand this data for nearly every material which might end up in consumer hands. Thus, it is likely that some data will be available (though in many cases far from *readily* available) for the use of the industrial hygienist for most situations which he might encounter.

Skin absorption is a more insidious or hazardous route of entry into the body than ingestion because it is not easily noticed and there is so much handling of products and chemicals in industrial activities which give rise to it. Dermatitis from skin contact with materials is the most common industrial disease (see Section 9). Oil and solvent dermatitis cases are more common than acid or caustic burns simply because workers have a healthy respect for the acute damage caused by strong chemicals but chronic exposure to oils and solvents defat the skin, making it prone to cracking and subsequent infection, and this does not command the same respect. This example also points up the basic difference between toxicity and hazard as discussed above.

Inhalation is an even more general hazard than ingestion or skin absorption because no visible physical contact between the worker and the compound in question is required. Airborne particulates, gases, vapors, and mists travel freely with air currents in the work area, so potential inhalation hazards must be evaluated carefully even though the worker is unaware of any exposure.

The difference between acute and chronic toxicity must also be clearly understood. Acute toxicity refers to the effects of a single exposure or close-together-in-time series of exposures to the agent in question. Chronic toxicity refers to the effects of many exposures repeated over a long period of time.

Another important factor is physiological response to the exposure in question. Exposures to ionizing radiation must be considered in terms of a lifetime cumulative dose since there is no proof of complete recovery from each individual exposure. Chemical compounds which display the same characteristic are termed radiomimetic, i.e., their effect mimics that of radiation (see also Sections 5 and 9). Another broad grouping of compounds includes those with only a temporary effect and with complete recovery after exposure, such as the anesthetic class of hydrocarbons.

The effects of physical agents such as heat and noise are defined somewhat differently than the chemical toxicity discussed above, but the same principles apply.

The mechanism of the Occupational Safety and Health Legislation in the United States involves a detailed process for setting standards. Basic to this process is the issuance of criteria documents by NIOSH. These documents contain a compilation and interpretation of appropriate toxicological or human-effect data as justification for the recommended standards. They are a "must" in the library or every industrial hygienist concerned with the agents under consideration.

Survey Techniques

All of the industrial hygienist's training and skills are converted from theory to practice during surveys conducted at the job site. Whether the survey is a quick walk-through inspection or a detailed study of a particular work operation, many of the same principles apply.

A starting point for a survey should be the medical history of the employees in the area. Is there evidence of excessive illness or absenteeism? Are working conditions recognized as so severe that short hours are the rule? Has a pattern developed in workman's compensation claims? These questions are properly addressed to the medical or personnel department, if available.

The next step is to gain an understanding of the process; the actual raw materials, intermediates, and products which knowledge coupled with a knowledge of toxicity allows the industrial hygienist to frequently indentify critical areas before even seeing the work site.

Upon entering the work area the first impression is very important. General housekeeping and cleanliness have a definite effect on the probability of hazardous conditions. Messy areas increase the chances for dermatitis due to the vapors of solvent spills and increase other safety problems as well.

The physical layout of the work area is another important aspect which can easily be noted. Certain operations should not be located close to each other. An example is welding, which must not take place where there are or may be halogenated solvent vapors, as from a degreaser because the intense heat from arc or flame welding can pyrolyze chlorinated hydrocarbons to form, among other things, the highly toxic phosgene gas.

Once a potentially hazardous condition has been identified, the task of the industrial hygienist is to protect the workers from the hazards. Two general approaches that have proved to be effective are engineering controls and materials substitution, both of which are discussed in detail in Section 2.

Industrial hygiene engineers are well versed in the techniques used for process enclosure and ventilation. These are the most widely used techniques and have proved to be effective for many years. The recent awareness of the high cost and limited availability of energy will tend to discourage general ventilation as a control technique. In this approach the number of air changes per unit time is calculated so as to reduce the concentration of the contaminant material to an acceptably low level. It is implicit in this approach that the system relies on 100% makeup air, with 100% exhaust. In climate controlled spaces this is an energy intensive process. The development of design information for the construction of enclosures and hoods is progressing as a science. Those versed in these techniques are generally aware that the capture of a contaminant material in an air stream may require some type of air cleaning device to prevent an air pollution problem at the expense of solving an industrial hygiene problem. Likewise, if wet-type air cleaning devices are used, the potential for generation of a water pollution problem must be considered.

The only positive way to solve a potential exposure problem is to eliminate the problem. No matter how well designed the engineering system to remove the contaminants from a process area, there is always the possibility of an equipment malfunction resulting in high exposure to workers. There are many successful instances where material substitution has proved to be a workable and desirable solution to an industrial hygiene problem. One of the more recent situations concerns vinyl chloride. Polyvinyl chloride (PVC) is a heavily used industrial resin. When the carcinoma problems resulting from long exposure to vinyl chloride monomer were identified, the manufacturers of PVC resin rose to the challenge. For many years there was no reason for serious concern over the residual vinyl chloride monomer in PVC. Users of the resin, in operations such as plastic molding, became legitimately concerned over exposure of their personnel to vinyl chloride. By reformulating their product, modifying their process, and maintaining strong quality control, the manufacturers have reduced the residual vinyl chloride monomer in PVC resin to a level that does not present a potential exposure problem to those handling it.

There are many instances where a process formulation has been modified to eliminate a compound that contributes to an undesirable industrial hygiene exposure potential. For many years, certain types of pipe insulation were formulated using asbestos fibers to provide material strength in combination with high temperature integrity. Rather than continually being plagued with the problems of respirable dust control in both the manufacture and use of these products, the product formulation has been changed and other fibrous materials substituted for the asbestos.

In the above examples, significant changes were required in the manufacture of the products. It would be naive to believe that such changes were made smoothly or easily. It is the task of the industrial hygienist to thoroughly evaluate the alternatives available to solve specific problems. This includes an analysis of technical and cost alternatives, both of which are legitimate concerns. In a broad view, the health professional has a responsibility to both the profession and to his employer to be innovative in reaching an acceptable resolution to such problems. It is his goal to be able to resolve the situation before it becomes a problem in terms of health damage to the unsuspecting workers.

Once it is established that a potential hazard is unavoidable the industrial hygienist utilizes his sampling and analytical skills to determine the personal exposure levels and then he must determine if such are acceptable. The TLV's are his primary reference in this regard (see below and Section 9), since they define the concentration levels of chemical compounds and physical agents below which the average healthy worker will suffer no demonstrably damaging effects. These levels are, whenever possible, determined at the breathing zone of the employee for an inhalation hazard, rather than simply at the work station.

For measuring noise exposures, use a simple and relatively inexpensive sound-level meter calibrated for A scale slow-response (the A scale is an electronically defined broad frequency band range which approximates the response of the human ear). See Section 3.

The TLV's define the maximum average daily noise dosage to which employees may be exposed in terms of dbA, decibels measured on the A scale. A slightly different instrument is required for measurement of impact noise, for which there is also a TLV.

Personal noise protective devices, i.e., earplugs and earmuffs, are frequently found in areas where noise levels approach or exceed the TLV criteria. Also, employee work duties can be rotated between noisy and relatively quiet work stations so that the allowable daily noise dosage is not exceeded.

For chemical hazards some type of air sampling and analysis is required. The quickest or spot-check method uses detector tubes. These are transparent tubes containing chemicals which change color or extend a stain in proportion to the amount of a specified compound drawn through the tube by means of a hand-held air pump. The tubes may be specific for one compound (NO_2 CO) or for a category of compounds (olefins). Their accuracy is a subject of continual improvement by manufacturers, with ±25% as the NIOSH goal. Detector tubes must be recognized as a screening technique rather than a sophisticated sampling/analysis technique. If this is understood when they are used they will provide much useful information and can be a powerful field tool.

The details of sophisticated sampling and analysis techniques are beyond the scope of this volume. Since they are in a continual state of improvement, the reader is encouraged to scan current literature and follow OSHA and NIOSH recommendations as a best course of action when a specific need arises. The Analytical Guides published by the American Industrial Hygiene Association are good primary reference for analytical information.

The ventilation of process equipment is an important aspect of exposure control. Hood and other dust equipment design are complex in some cases, simple in others. An initial survey of a ventilated process should always include an inspection of the fan while it is in operation. It is surprising to find that many systems are operating well below their design capacity because the fan is rotating in reverse.

The industrial hygienist must, above all other considerations, be guided by common sense. He must realize that first impressions have to be tempered by repeat visits to clarify and justify preliminary conclusions; that sampling must be conducted over extended time periods in borderline cases; that working conditions can become radically different under different weather conditions; that acute exposure hazards can develop from so-called "freak" conditions but which might be predicted when the situation is thoroughly studied; that several occupational diseases such as silicosis may show no symptoms for several years after exposure, so reliance for employee health must be placed on thorough survey procedures.

In terms of employee protection from hazards, the substitution of nonhazardous for hazardous materials is certainly the most foolproof. The next choice is engineering control to remove the potential hazard from the work station. The least satisfactory, unfortunately often the first one considered, is the stop-gap device of personal protective devices.

The basic fault of personal protection, whether it be protective clothing, ear protectors, or respirators, is that its successful use is dependent on the cooperation of the employee. Every manner of personal protective device is for some reason considered to be at least an encumbrance or perhaps even an "insult to manhood" by a certain segment of the workforce. In addition, there is a continuing problem with proper fit of the devices even with cooperative employees. Personal protection must never be considered anything but a temporary expedient which is to be used only until more positive methods of control are applied (see Section 2).

Detection of several hazardous exposures are amenable to the use of biological monitoring or bioassay to provide evidence of excessive exposures. In lead workers, for example, blood lead levels can identify those workers who have excessive lead input into their bodies. While not generally a legally acceptable proof of compliance with regulations, biological monitoring can be extremely valuable in protecting the health of the occasional worker whose exposure is excessive in spite of all efforts to protect him. After all, if there is no untoward exposure of a worker to lead or mercury it is difficult to explain abnormally high tissue concentrations as observed via bioassay. This technique is therefore an excellent screening device.

These survey technique comments, while not comprehensive, define the philosophy of the working industrial hygienist. He is a pragmatist with a substantial background in theory and an understanding of the physical stresses of the workplace. The industrial hygienist serves as a valuable part of the industrial economy by identifying potential hazards and effecting solutions *before* worker health is impaired.

D. THRESHOLD LIMIT VALUES

For many years the ACGIH has published an annual list of TLV's. The 1977 edition is reprinted below in its entirety because TLV's are a prime reference for industrial hygienists and not only are the numerical values important to occupational safety and health, but at least of equal importance are the explanations and limitations of these values as detailed in the text which accompanies the list.

The introduction to the TLV's states that they "represent conditions under which it is believed that nearly all workers may be repeatedly exposed day after day without adverse effect."

Since the TLV's are presented in terms of numerical standards, they suffer from common misinterpretations. The most common error is to interpret these values as being a precise definition of a hazard/nonhazard situation. This is not the case. The TLV's have been established on the basis of the best information available and are subject to annual revision. They are not intended to define values below which all individuals will be protected and above which damage will occur to all; it is predictable that some hypersensitives may suffer adverse effects even at exposure levels well below those defined by the TLV. The TLV's must be considered simply as good-practice guidelines and as such have been accepted as official U.S. government standards for industrial air.

Another common misinterpretation concerning the TLV's is the lack of understanding that they are mostly long-term average concentration values. When instantaneous grab-sampling techniques are used to determine the concentration of a specific substance in workroom air, it is to be expected that in many cases, even though a TLV is actually being met, there will be instantaneous concentration excursions above the TLV number. This is specifically covered in Appendix D of the TLV's, which discusses permissible excursions for time-weighted average (TWA) limits. It is clearly pointed out in Appendix D that the limiting excursions quoted are to be considered "rules of thumb." When the numerical value of the TLV is between 0 and 1 the rule of thumb excursion factor is that instantaneous values should not be more than 3 times the limit. When the numerical value of the TLV is between 100 and 1000 the rule of thumb for the excursion factor is that instantaneous concentrations should not exceed the TLV by more than a factor of 1.25. Intermediate factors are quoted for TLV's between 1 and 10 and 100.

Certain of the TLV's are listed as "C" or ceiling values rather than long-term averages. For these compounds, e.g., certain acid mists, the short-term (acute) effects of excursion concentrations are so undesirable that they outweigh the long-term (chronic) effects and careful protection of workers depends upon maintaining concentrations below the ceiling value at all times, i.e., no excursions over the TLV are permitted.

The selection of numerical values in working areas to compare with TLV's is based on the most reliable available sampling and analysis of workroom air. In almost all cases this means the use of sophisticated sampling and analytical techniques. The widespread use of detector tubes for the estimation of the concentration of several contaminants in workroom air has proven both a curse and a blessing. The blessing comes from the fact that the tubes are direct reading, relatively inexpensive, and convenient to use even by unskilled people. The "curse" comes from the use of them by personnel who frequently do not understand the principle of short-term excursions and the fact that the detector tubes are, at best, accurate to ±25%. It is not necessary that the user of detector tubes be a qualified industrial hygienist. It is, however, important that the individual who interprets such readings have a solid understanding of both the principles of TLV and the inherent shortcomings of the method of measurement he uses.

The TLV for mineral dusts is dependent upon the free silica content of the material in question. There are two techniques for evaluating these exposures: "total" airborne dust and "respirable" airborne dust. Both can be valid measurements yet the numerical limits applicable to each are different because the sampling techniques are not the same. The "total" dust technique has been used for many more years than the "respirable dust" technique, and so more data based on it are available.

Determination of "total" dust is based on the use of impinger-collected samples and requires somewhat detailed and lengthy laboratory analysis using microscopic techniques to interpret the data. The impinger itself was developed about 1930 for the purpose of evaluating the dust content of industrial air. The device draws dusty air at high velocity through a nozzle which is submerged in a liquid. Dust particles larger than about 0.8 microns mean diameter, impact on a collecting surface and are thus removed from the airstream. The liquid continually washes the impaction plate and retains the particles so they are not swept back into the airstream. For analytical purposes an aliquot of the liquid is transferred to a dust-counting cell, whose depth is carefully controlled. After a period of time, long enough for the particles to settle to the bottom of the cell, the particles are optically counted using a light-field microscope at 100× magnification. The results of the microscopic count are calculated to millions of particles per unit volume of air sampled.

The "respirable" dust sampling technique is a gravimetric procedure based on the increase in weight of a filter through which a known volume of air containing size-classified dust has passed. The size classification is accomplished via a small "cyclone" which, at a specified air flow rate is known to pass dust particles in approximately the same size range as the human respiratory system. The size selector passes 90% of 2-micron particles and 0% of 10-micron particles. The stated size is the aerodynamic diameter; that is, the diameter of a spherical particle with a d = 1.00. Since industrial particles can vary widely in their d (generally in the range of 2.0 to 3.5) an appropriate correction must be made in the rate of air flow through the size classification device in order to duplicate the size characteristics as required by the TLV's.

Regardless of the sampling technique used, the TLV depends upon the per cent of quartz in the respirable fraction. When using the respirable-mass sampler the determination is straightforward in that the quartz content can be determined from the sample filters themselves. The analytical methods for free silica are quite accurate, with a minimum detectable mass of 0.01 mg of silica. When using the impinger technique it is inconvenient to recover the impinger-collected particles in most cases so other techniques are used for the collection of the dust sample to be analyzed for free silica. One standard technique is to take a "rafter" sample. This is a sample of the float dust which has settled in a relatively inaccessible elevated location in the work area, such as on the rafters. It is more accurate to elute any bulk sample of dust to a size classification matching that of the respiratory retention curve than to perform the free silica analysis of the bulk sample. Several studies have shown that the concentration of free silica in different size fractions or bulk material can be radically different. If this is not considered, the results of the survey can be misleading at best and very possibly in gross error.

Many industrial situations are encountered where a variety of potentially hazardous compounds are in concurrent use. Appendix C of the TLV's describes the procedures to be used for interpreting such situations. In its simplest form these calculations amount to the summation of individual compound as a per cent of their TLV's.

In summary, the TLV's provide a useful *guideline* to the evaluation of industrial health exposures. As with almost every powerful tool they are subject to misapplication when the user shortcuts the underlying principles and tries to lift portions out of context.

It cannot be overstated: *read the directions;* understand how the TLV's were developed. Then you will be able to use them effectively to protect the worker.

The following listing of Threshold Limit Values is reprinted with permission from the American Conference of Governmental Industrial Hygienists.

BIBLIOGRAPHY

Alden, J. L. and Kane, J. M. *Design of Industrial Exhaust Systems*. Industrial Press, New York, 1970.

Anon. *Compliance Operations Manual—January 1972*. U.S. Department of Labor Publication No. OSHA-2006, 1972.

Anon. *Industrial Ventilation*. American Conference of Governmental Industrial Hygienists, Cincinnati, 1972.

Anon. *Inspection Survey Guide*. U.S. Department of Labor Bulletin No. 326, 1970.

Anon. *Legislative History of the Occupational Safety and Health Act of 1970*. Subcommittee on Labor of the Committee on Labor and Public Welfare, United States Senate. U.S. Government Printing Office, Washington, 1971.

Anon. *Occupational Health in Transition*. Public Health Reports, Vol. 7, No. 11, November 1964.

Doyle, Henry N. *Notebook on Industrial Hygiene for Engineers*. College of Mines and Mineral Industries, University of Utah, 1951.

Felton, J. S., et al. *Man, Medicine, and Work—Historic Events in Occupational Medicine*. U.S. Public Health Service Publication No. 1044, April 1964.

Hemeon, W. C. L. *Plant and Process Ventilation*. Industrial Press, New York, 1963.

Patty, F. A. (ed.). *Industrial Hygiene and Toxicology*. Second Revised Edition, Volumes I and II. John Wiley and Sons, New York, 1958, 1963.

Powell, C. H. and Hosey, A. D. (eds.). *The Industrial Environment—Its Evaluation and Control: Syllabus*. U.S. Public Health Service Publication No. 614, 1965.

Trasko, V. M. *Occupational Health and Safety Legislation—A Compilation of State Laws and Regulations*. U.S. Public Health Service Publication No. 357, Revised 1970.

TLVs®
Threshold Limit Values for Chemical Substances and Physical Agents in the Workroom Environment with Intended Changes for 1978

PRICE EACH

1–49	$1.50
50–199	1.25
200–999	1.10
1000–4999	.75
5000 or more	.65

Documentation of the Threshold Limit Values for Substances in Workroom Air.© A separate companion piece to the Chemical TLVs is issued by ACGIH under this title. This publication gives the pertinent scientific information and data with reference to literature sources that were used to base each limit. Each documentation also contains a statement defining the type of response against which the limit is safeguarding the worker. For a better understanding of the TLVs it is essential that the Documentation be consulted when the TLVs are being used.

Information concerning the availability of copies of the Documentation of the Threshold Limit Values for Substances in Workroom Air should be directed to the Executive Secretary, ACGIH (third edition, fourth printing, 1978, $20.00).

TLVs®
Threshold Limit Values for Chemical Substances in Workroom Air Adopted by ACGIH for 1978

1978 TLV AIRBORNE CONTAMINANTS COMMITTEE

Hervey B. Elkins, Ph.D., Chairman
Charles E. Adkins
Mary O. Amdur, Ph.D.
Hector P. Blejer, M.D.
Paul E. Caplan, P.E., M.P.H.
Paul Gross, M.D.
James W. Hammond
John W. Knauber, M.P.H.
Jesse Lieberman, P.E.
Trent R. Lewis, Ph.D.

Keith R. Long, Ph.D.
Frederick T. McDermott
Floyd A. Madsen
Walter W. Melvin, Jr., M.D., Sc.D.
Leonard D. Pagnotto, Secretary
Ronald S. Ratney
Meier Schneider, P.E., C.I.H.
Richard D. Stewart, M.D.
Herbert E.Stokinger, Ph.D.
Vera F. Thomas, Ph.D.
William D. Wagner
Elizabeth K. Weisburger, Ph.D.
David H. Wegman, M.D.

CONSULTANTS

E. Mastromatteo, M.D.
James F. Morgan
Marshall Steinberg, Ph. D.
Theodore R. Torkelson, Sc.D.
Ralph C. Wands
Mitchell R. Zavon, M.D.

Any comments or questions regarding these limits should be addressed to:

Executive Secretary
American Conference of Governmental
Industrial Hygienists
P.O. Box 1937
Cincinnati OH 45201: (513) 825-0312

PREFACE
CHEMICAL CONTAMINANTS

Threshold limit values refer to airborne concentrations of substances and represent conditions under which it is believed that nearly all workers may be repeatedly exposed day after day without adverse effect. Because of wide variation in individual susceptibility, however, a small percentage of workers may experience discomfort from some substances at concentrations at or below the threshold limit; a smaller percentage may be affected more seriously by aggravation of a pre-existing condition or by development of an occupational illness.

Tests are available (J. Occup. Med. 15: 564, 1973; Ann. N.Y. Acad. Sci., *151, Art. 2:* 968, 1968) that may be used to detect those individuals hypersusceptible to a variety of industrial chemicals (respiratory irritants, hemolytic chemicals, organic isocyanates, carbon disulfide).

Three categories of Threshold Limit Values (TLVs) are specified herein, as follows:

a) Threshold Limit Value-Time Weighted Average (TLV-TWA) — the time-weighted average concentration for a normal 8-hour workday or 40-hour workweek, to which nearly all workers may be repeatedly exposed, day after day, without adverse effect.

b) Threshold Limit Value-Short Term Exposure Limit (TLV-STEL) — the maximal concentration to which workers can be exposed for a period up to 15 minutes continuously without suffering from 1) irritation, 2) chronic or irreversible tissue change, or 3) narcosis of sufficient degree to increase accident proneness, impair self-rescue, or materially reduce work efficiency, provided that no more than four excursions per day are permitted, with at least 60 minutes between exposure periods, and provided that the daily TLV-TWA also is not exceeded. The STEL should be considered a maximal allowable concentration, or ceiling, not to be exceeded at any time during the 15-minute excursion period. STELs are based on one or more of the following criteria: (1) Adopted TLVs including those with a "C" or "ceiling" limit. (2) TWA-TLV Excursion Factors listed in Appendix D. (3) Pennsylvania Short-Term Limits for Exposure to Airborne Contaminants (Penna. Dept. of Hlth., Chapter 4, Art. 432, Rev. Jan. 25, 1968). (4) OSHA Occupational Safety and Health Standards, 40 FR 23073, May 28, 1975. (5) NIOSH criteria for recommended standards for occupational exposure to specific substances. The TWA-STEL should not be used as engineering design criterion or considered as an emergency exposure level (EEL).

c) Threshold Limit Value-Ceiling (TLV-C) — the concentration that should not be exceeded even instantaneously.

For some substances, e.g., irritant gases, only one category, the TLV-Ceiling, may be relevant. For other substances, either two or three categories may be relevant, depending upon their physiologic action. It is important to observe that if any one of these three TLVs is exceeded, a potential hazard from that substance is presumed to exist.

The TLV-TWA should be used as guides in the control of health hazards and should not be used as fine lines between safe and dangerous concentrations.

Time-weighted averages permit excursions above the limit provided they are compensated by equivalent excursions below the limit during the workday. In some instances it may be permissible to calculate the average concentration for a workweek rather than for a workday. The degree of permissible excursion is related to the magnitude of the threshold limit value of a particular substance as given in Appendix D. The relationship between threshold limit and permissible excursion is a rule of thumb and in certain cases may not apply. The amount by which threshold limits may be exceeded for short periods without injury to health depends upon a number of factors such as the nature of the contaminant, whether very high concentrations — even for short periods — produce acute poisoning, whether the effects are cumulative, the frequency with which high concentrations occur, and the duration of such periods. All-factors must be taken into consideration in arriving at a decision as to whether a hazardous condition exists.

Threshold limits are based on the best available information from industrial experience, from experimental human and animal studies, and, when possible, from a combination of the three. The basis on which the values are established may differ from substance to substance; protection against impairment of health may be a guiding factor for some, whereas reasonable freedom from irritation, narcosis, nuisance or other forms of stress may form the basis for others.

The amount and nature of the information available for establishing a TLV varies from substance to substance; consequently, the precision of the estimated TLV is also subject to variation and the latest *Documentation* should be consulted in order to assess the extent of the data available for a given substance.

The committee holds to the opinion that limits based on physical irritation should be considered no less binding than those based on physical impairment. There is increasing evidence that physical irritation may initiate, promote or accelerate physical impairment through interaction with other chemical or biologic agents.

In spite of the fact that serious injury is not believed likely as a result of exposure to the threshold limit concentrations, the best practice is to maintain concentrations of all atmospheric contaminants as low as is practical.

These limits are intended for use in the practice of industrial hygiene and should be interpreted and applied only by a person trained in this discipline. They are not intended for use, or for modification for use, (1) as a relative index of hazard or toxicity, (2) in the evaluation or control of community air pollution nuisances, (3) in estimating the toxic potential of continuous, uninterrupted exposures or other extended work periods, (4) as proof or disproof of an existing disease or physical condition, or (5) for adoption by countries whose working conditions differ from those in the United States of America and where substances and processes differ.

Ceiling vs Time-Weighted Average Limits. Although the time-weighted average concentration provides the most satisfactory, practical way of monitoring airborne agents for compliance with the limits, there are certain substances for which it is inappropriate. In the latter group are substances which are predominantly fast acting and whose threshold limit is more appropriately based on this particular response. Substances with this type of response are best controlled by a ceiling "C" limit that should not be exceeded. It is implicit in these definitions that the manner of sampling to determine noncompliance with the limits for each group must differ; a single brief sample, that is applicable to a "C" limit, is not appropriate to the time-weighted limit; here, a sufficient number of samples are needed to permit a time-weighted average concentration throughout a complete cycle of operations or throughout the work shift.

Whereas the ceiling limit places a definite boundary which concentrations should not be permitted to exceed, the time-weighted average limit requires an explicit limit to the excursions that are permissible above the listed values. The magnitude of these excursions may be pegged to the magnitude of the threshold limit by an appropriate factor shown in Appendix D. It should be noted that the same factors are used by the Committee in determining the magnitude of the value of the STELs, or whether to include or exclude a substance for a "C" listing.

"Skin" Notation. Listed substances followed by the designation "Skin" refer to the potential contribution to the overall exposure by the cutaneous route including mucous membranes and eye, either by airborne, or more particularly, by direct contact with the substance. Vehicles can alter skin absorption. This attention-calling designation is intended to suggest appropriate measures for the prevention of cutaneous absorption so that the threshold limit is not invalidated.

Mixtures. Special consideration should be given also to the application of the TLVs in assessing the health hazards which may be associated with exposure to mixtures of two or more substances. A brief discussion of basic considerations involved in developing threshold limit values for mixtures, and methods for their development, amplified by specific examples are given in Appendix C.

Nuisance Particulates. In contrast to fibrogenic dusts which cause scar tissue to be formed in lungs when inhaled in excessive amounts, so-called "nuisance" dusts have a long history of little adverse effect on lungs and do not produce significant organic disease or toxic effect when exposures are kept under reasonable control. The nuisance dusts have also been called (biologically) "inert" dusts, but the latter term is inappropriate to the extent that there is no dust which does not evoke some cellular response in the lung when inhaled in sufficient amount. However, the lung-tissue reaction caused by inhalation of nuisance dusts has the following characteristics: (1) The architecture of the air spaces remains intact. (2) Collagen (scar tissue) is not formed to a significant extent. (3) The tissue reaction is potentially reversible.

Excessive concentrations of nuisance dusts in the workroom air may seriously reduce visibility, may cause unpleasant deposits in the eyes, ears and nasal passages (Portland Cement dust), or cause injury to the skin or mucous membranes by chemical or mechanical action per se or by the rigorous skin cleansing procedures necessary for their removal.

A threshold limit of 10 mg/m³, or 30 mppcf, of total dust < 1% quartz, or, 5 mg/m³ respirable dust is recommended for substances in these categories and for which no specific threshold limits have been assigned. This limit, for a normal workday, does not apply to brief exposures at higher concentrations. Neither does it apply to those substances which may cause physiologic impairment at lower concentrations but for which a threshold limit has not yet been adopted. Some nuisance particulates are given in Appendix E.

Simple Asphyxiants — "Inert" Gases or Vapors. A number of gases and vapors, when present in high concentrations in air, act primarily as simple asphyxiants without other significant physiologic effects. A TLV may not be recommended for each simple asphyxiant because the limiting factor is the available oxygen. The minimal oxygen content should be 18 percent by volume under normal atmospheric pressure (equivalent to a partial pressure, pO_2 of 135 mm Hg). Atmospheres deficient in O_2 do not provide adequate warning and most simple asphyxiants are odorless. Several simple asphyxiants present an explosion hazard. Account should be taken of this factor in limiting the concentration of the asphyxiant. Specific examples are listed in Appendix F.

Physical Factors. It is recognized that such physical factors as heat, ultraviolet and ionizing radiation, humidity, abnormal pressure (altitude) and the like may place added stress on the body so that the effects from exposure at a threshold limit may be altered. Most of these stresses act adversely to increase the toxic response of a substance. Although most threshold limits have built-in safety factors to guard against adverse effects to moderate deviations from normal environments, the safety factors of most substances are not of such a magnitude as to take care of gross deviations. For example, continuous work at temperatures above 90°F, or overtime extending the workweek more than 25%, might be considered gross deviations. In such instances judgment must be exercised in the proper adjustments of the Threshold Limit Values.

Biologic Limit Values (BLVs). Other means exist and may be necessary for monitoring worker exposure other than reliance on the Threshold Limit Values for industrial air, namely, the Biologic Limit Values. These values represent limiting amounts of substances (or their effects) to which the worker may be exposed without hazard to health or well-being as determined in his tissues and fluids or in his exhaled breath. The biologic measurements on which the BLVs are based can furnish two kinds of information useful in the control of worker exposure: (1) measure of the individual worker's over-all exposure; (2) measure of the worker's individual and characteristic response. Measurements of response furnish a superior estimate of the physiologic status of the worker, and may be made of (a) changes in amount of some critical biochemical constituent, (b) changes in activity of a critical enzyme, (c) changes in some physiologic function. Measurement of exposure may be made by (1) determining in blood, urine, hair, nails, in body tissues and fluids, the amount of substance to which the worker was exposed; (2) determination of the amount of the metabolite(s) of the substance in tissues and fluids; (3) determination of the amount of the substance in the exhaled breath. The biologic limits may be used as an adjunct to the TLVs for air, or in place of them. The BLVs, and their associated procedures for determining compliance with them, should thus be regarded as an effective means of providing health surveillance of the worker.

Unlisted Substances. Many substances present or handled in industrial processes do not appear on the TLV list. In a number of instances the material is rarely present as a particulate, vapor or other airborne contaminant, and a TLV is not necessary. In other cases sufficient information to warrant development of a TLV, even on a tentative basis, is not available to the Committee. Other substances, of low toxicity, could be included in Appendix E pertaining to nuisance particulates. This list (as well as Appendix F) is not meant to be all inclusive; the substances serve only as examples.

In addition there are some substances of not inconsiderable toxicity, which have been omitted primarily because only a limited number of workers (e.g., employees of a single plant) are known to have potential exposure to possibly harmful concentrations.

"Notice of Intent." At the beginning of each year, proposed actions of the Committee for the forthcoming year are issued in the form of a "Notice of Intended Changes." This Notice provides not only an opportunity for comment, but solicits suggestions of substances to be added to the list. The suggestions should be accompanied by substantiating evidence. The list of Intended Changes follows the Adopted Values in the TLV booklet. Values listed in parenthesis in the "Adopted" list are to be used during the period in which a proposed change for that Value is listed in the Notice of Intended Changes.

Legal Status. By publication in the Federal Register (Vol. 36, No. 105, May 29, 1971) the Threshold Limit Values for 1968 were made official federal standards for industrial air. Since 1971, new standards for certain of these substances have been promulgated by OSHA.

Reprint Permission. This publication may be reprinted provided that written permission is obtained from the Secretary-Treasurer of the Conference and that it be published in its entirety.

	ADOPTED VALUES		TENTATIVE VALUES	
	TWA		STEL	
Substance	ppm[a]	mg/m^3[b]	ppm[a]	mg/m^3[b]
Abate	—	10	—	20
Acetaldehyde	100	180	150	270
Acetic acid	10	25	15	37
C Acetic anhydride	5	20	—	—
Acetone	1,000	2,400	1,250	3,000
Acetonitrile	40	70	60	105
Acetylene	F	—	—	—
Acetylene dichloride, see 1, 2-Dichloroethylene	200	790	250	1,000
Acetylene tetrabromide	1	14	1.5	18
Acrolein	0.1	0.25	0.3	0.8
Acrylamide — Skin	—	0.3	—	0.6
** Acrylonitrile — Skin	(20)	(45)	(30)	(65)
Aldrin — Skin	—	0.25	—	0.75
Allyl alcohol — Skin	2	5	4	10
Allyl chloride	1	3	2	6
Allyl glycidyl ether (AGE) — Skin	5	22	10	44
Allyl propyl disulfide	2	12	3	18
Aluminum oxide	—	E	—	20
4-Aminodiphenyl — Skin	—	A1b	—	A1b
2-Aminoethanol, see Ethanolamine	3	8	6	15
2-Aminopyridine	0.5	2	2	4
Ammonia	25	18	35	27
Ammonium chloride-fume	—	10	—	20
Ammonium sulfamate (Ammate)	—	10	—	20
n-Amyl acetate	100	530	150	800
sec-Amyl acetate	125	670	150	800
** Aniline — Skin	(5)	(19)	—	—
Anisidine (o-, p-isomers) — Skin	0.1	0.5	—	—
** Antimony & Compounds (as Sb)	—	(0.5)	—	—
* Antimony trioxide, handling and use (as Sb)	—	0.5	—	—
** Antimony trioxide production (as Sb)	—	(0.5, A2)	—	—
ANTU (α-Naphthyl thiourea)	—	0.3	—	0.9
Argon	F	—	F	F
** Arsenic & compounds (as As)	—	(0.5)	—	—
** Arsenic trioxide production (as As)	—	(A1a)	—	—
Arsine	0.05	0.2	—	—
** Asbestos (all forms)	—	(A1a)	—	(A1a)
Asphalt (petroleum) fumes	—	5	—	10
* Atrazine	—	10	—	—
Azinphos-methyl — Skin	—	0.2	—	0.6
Barium (soluble compounds), as Ba	—	0.5	—	—
Baygon (propoxur)	—	0.5	—	2
Benzene	10, A2	30, A2	—	—
Benzidine production — Skin	—	A1b	—	A1b
p-Benzoquinone, see Quinone	0.1	0.4	0.3	2
Benzoyl peroxide	—	5	—	—
Benz(a)pyrene	—	A2	—	A2
Benzyl chloride	1	5	—	—
Beryllium	—	0.002	—	0.025
Biphenyl	0.2	1.5	0.6	4

Capital letters refer to Appendices.
Footnotes (a thru h) see Page 22.
**See Notices of Intended Changes.
*1978 Addition.

	ADOPTED VALUES		TENTATIVE VALUES	
	TWA		STEL	
Substance	ppm[a]	mg/m³[b]	ppm[a]	mg/m³[b]
C Bisphenol A, see Diglycidal ether (DGE)	0.5	3.0	—	—
Bismuth telluride	—	10	—	20
Bismuth telluride, Se-doped	—	5	—	10
Borates, tetra, sodium salts,				
Anhydrous	—	1	—	—
Decahydrate	—	5	—	—
Pentahydrate	—	1	—	—
Boron oxide	—	10	—	20
Boron tribromide	1	10	3	30
C Boron trifluoride	1	3	—	—
Bromine	0.1	0.7	0.3	2
Bromine pentafluoride	0.1	0.7	0.3	2
Bromochloromethane/ chlorobromomethane	200	1,050	250	1,300
Bromoform — Skin	0.5	5	—	—
Butadiene (1, 3-butadiene)	1,000	2,200	1,250	2,750
Butane	600	1,430	750	1,780
Butanethiol, see Butyl mercaptan	0.5	1.5	—	—
2-Butanone	200	590	300	885
2-Butoxyethanol (Butyl Cellosolve) — Skin	50	240	150	720
n-Butyl acetate	150	710	200	950
sec-Butyl acetate	200	950	250	1,190
tert-Butyl acetate	200	950	250	1,190
* Butyl acrylate	10	55	—	—
C n-Butyl alcohol — Skin	50	150	—	—
sec-Butyl alcohol	150	450	—	—
tert-Butyl alcohol	100	300	150	450
C Butylamine — Skin	5	15	—	—
C tert-Butyl chromate (as CrO_3) — Skin	—	0.1	—	—
n-Butyl glycidyl ether (BGE)	50	270	—	—
n-Butyl lactate	5	25	—	—
Butyl mercaptan	0.5	1.5	—	—
p-tert-Butyltoluene	10	60	20	120
Cadmium, dust & salts (as Cd)	—	0.05	—	0.2
C Cadmium oxide fume (as Cd)	—	0.05	—	—
** Cadmium oxide production (as Cd)	—	(A2)	—	—
Calcium carbonate/ marble	—	E	—	20
Calcium arsenate (as As)	—	1	—	—
Calcium cyanamide	—	0.5	—	1
* Calcium hydroxide	—	5	—	—
* Calcium oxide	—	2	—	—
Camphor, synthetic	2	12	3	18
Caprolactam				
Dust	—	1	—	3
Vapor	5	20	10	40
Captafol (Difolatan®) — Skin	—	0.1	—	—
Captan	—	5	—	15
Carbaryl (Sevin®)	—	5	—	10
Carbofuran (Furadan®)	—	0.1	—	—
Carbon black	—	3.5	—	7
Carbon dioxide	5,000	9,000	15,000	18,000
** Carbon disulfide — Skin	(20)	(60)	(30)	(90)

	ADOPTED VALUES		TENTATIVE VALUES	
	TWA		STEL	
Substance	ppm[a]	mg/m³[b]	ppm[a]	mg/m³[b]
Carbon monoxide	50	55	400	440
Carbon tetrabromide	0.1	1.4	0.3	4
Carbon tetrachloride — Skin	10	65	20	130
* Carbonyl chloride (Phosgene)	0.1	0.4	—	—
* Carbonyl fluoride	5	15	—	—
Catechol (Pyrocatechol)	5	20	—	—
Cellulose (paper fiber)	—	E	—	20
Cesium hydroxide	—	2	—	—
Chlordane — Skin	—	0.5	—	2
Chlorinated camphene — Skin	—	0.5	—	1
Chlorinated diphenyl oxide	—	0.5	—	2
Chlorine	1	3	3	9
Chlorine dioxide	0.1	0.3	0.3	0.9
C Chlorine trifluoride	0.1	0.4	—	—
C Chloroacetaldehyde	1	3	—	—
α-Chloroacetophenone (Phenacyl chloride)	0.05	0.3	—	—
Chlorobenzene (Monochlorobenzene)	75	350	—	—
o-Chlorobenzylidene malonoitrile — Skin	0.05	0.4	—	—
Chlorobromomethane/ Bromochloromethane	200	1,050	250	1,300
2-Chloro-1, 3-butadiene, see β Chloroprene — Skin	25	90	35	125
Chlorodifluoromethane	1,000	3,500	1,250	4,375
Chlorodiphenyl (42% Chlorine) — Skin	—	1	—	2
Chlorodiphenyl (54% Chlorine) — Skin	—	0.5	—	1
1-Chloro, 2, 3-epoxy-propane (Epichlorhydrin) — Skin	5	20	10	40
C 2-Chloroethanol (Ethylene chlorohydrin) — Skin	1	3	—	—
** Chloroethylene (Vinyl chloride)	(A1c)	—	(A1c)	—
* Chloroform (Trichloromethane)	10, A2	50, A2	—	—
bis-Chloromethyl ether	0.001	A1a	—	A1a
1-Chloro-1-nitro-propane	20	100	—	—
Chloropicrin	0.1	0.7	0.3	2
** β-Chloroprene — Skin	(25)	(90)	(35)	(135)
Chlorpyrifos (Dursban®) — Skin	—	0.2	—	0.6
o-Chlorostyrene	50	285	75	430
o-Chlorotoluene — Skin	50	250	75	375
2-Chloro-6-(trichloromethyl pyridine (N-Serve®)	—	10	—	20
Chromates, certain insoluble forms	—	0.05, A1a	—	A1a
Chromic acid and Chromates, (as Cr)	—	0.05	—	—
* Chromite ore processing (chromate), as Cr	—	0.05, A1a	—	—
Chromium, Sol. chromic, chromous salts (as Cr)	—	0.5	—	—

Capital letters refer to Appendices.
Footnotes (a thru h) see Page 22.
**See Notice of Intended Changes.
*1978 Addition.

Substance	ADOPTED VALUES TWA ppm[a]	ADOPTED VALUES TWA mg/m³[b]	TENTATIVE VALUES STEL ppm[a]	TENTATIVE VALUES STEL mg/m³[b]
Clopidol (Coyden®)	—	10	—	20
Coal tar pitch volatiles (See Particulate polycyclic aromatic hydrocarbons)	—	0.2, A1a	—	A1a
** Cobalt metal, dust and fume (as Co)	—	(0.1)	—	—
Copper fume	—	0.2	—	—
Dusts & Mists (as Cu)	—	1	—	2
Corundum (Al_2O_3)	—	E	—	E
Cotton dust, raw	—	0.2[D]	—	0.6
Crag® herbicide	—	10	—	20
Cresol, all isomers — Skin	5	22	—	—
Crotonaldehyde	2	6	6	18
Crufomate®	—	5	—	20
Cumene — Skin	50	245	75	365
Cyanamide	—	2	—	—
Cyanide, as CN — Skin	—	5	—	—
Cyanogen	10	20	—	—
Cyclohexane	300	1,050	375	1,300
Cyclohexanol	50	200	—	—
Cyclohexanone	50	200	—	—
Cyclohexene	300	1,015	—	—
Cyclohexylamine — Skin	10	40	—	—
Cyclopentadiene	75	200	150	400
2, 4-D (2, 4-Diphenoxy-acetic acid)	—	10	—	20
DDT (Dichlorodiphenyl-trichloroethane)	—	1	—	3
DDVP, see Dichlorvos — Skin	0.1	1	0.3	3
Decaborane — Skin	0.05	0.3	0.15	0.9
Demeton® — Skin	0.01	0.1	0.03	0.3
Diacetone alcohol (4-hydroxy-4-methyl-2-pentanone)	50	240	75	360
1, 2-Diaminoethane, see Ethylenediamine	10	25	—	—
Diazinon — Skin	—	0.1	—	0.3
Diazomethane	0.2	0.4	—	—
Diborane	0.1	0.1	—	—
** 1, 2-Dibromoethane (Ethylene dibromide) — Skin	(20)	(155)	(30)	(230)
Dibrom®	—	3	—	6
2-N-Dibutylaminoethanol — Skin	2	14	4	28
Dibutyl phosphate	1	5	2	10
Dibutyl phthalate	—	5	—	10
C Dichloracetylene	0.1	0.4	—	—
C o-Dichlorobenzene	50	300	—	—
p-Dichlorobenzene	75	450	110	675
Dichlorobenzidine — Skin	—	A2	—	A2
Dichlorodifluoromethane	1,000	4,950	1,250	6,200
1, 3-Dichloro-5, 5-dimethyl hydantoin	—	0.2	—	0.4
1, 1-Dichloroethane	200	810	250	1,010
** 1, 2-Dichloroethane	(50)	(200)	(75)	(300)
1, 2-Dichloroethylene	200	790	250	1,000
Dichloroethyl ether — Skin	5	30	10	60
** Dichloromethane, see Methylene chloride	(200)	(700)	(250)	(870)
** Dichloromonofluoro-methane	(1,000)	(4,200)	—	—

Capital letters refer to Appendices.
[D] See Page 23.
Footnotes (a thru g) see Page 22.

Substance	ADOPTED VALUES TWA ppm[a]	ADOPTED VALUES TWA mg/m³[b]	TENTATIVE VALUES STEL ppm[a]	TENTATIVE VALUES STEL mg/m³[b]
C 1, 1-Dichloro-1-nitroethane	10	60	—	—
1, 2-Dichloropropane, see Propylene dichloride	75	350	110	510
Dichlorotetrafluoro-ethane	1,000	7,000	1,250	8,750
Dichlorvos (DDVP) — Skin	0.1	1	0.3	3
Dicrotophos (Bidrin®) — Skin	—	0.25	—	—
Dicyclopentadiene	5	30	—	—
Dicyclopentadienyl iron	—	10	—	20
Dieldrin — Skin	—	0.25	—	0.75
Diethylamine	25	75	—	—
Diethylaminoethanol — Skin	10	50	—	—
Diethylene triamine — Skin	1	4	—	—
Diethyl ether, see Ethyl ether	400	1,200	500	1,500
Diethyl phthalate	—	5	—	10
Difluorodibromomethane	100	860	150	1,290
C Diglycidyl ether (DGE)	0.5	3	—	—
Dihydroxybenzene, see Hydroquinone	—	2	—	4
Diisobutyl ketone	25	150	—	—
Diisopropylamine — Skin	5	20	—	—
Dimethoxymethane, see Methylal	1,000	3,100	1,250	3,875
Dimethyl acetamide — Skin	10	35	15	50
* Dimethyl carbamyl chloride	A2	A2	—	—
Dimethylamine	10	18	—	—
Dimethylaminobenzene, see Xylidene — Skin	5	25	10	50
Dimethylaniline (N, N-Dimethylaniline) — Skin	5	25	10	50
Dimethylbenzene, see Xylene — Skin	100	435	150	650
Dimethyl-1, 2-dibromo-2-dichloroethyl phosphate, see Dibrom	—	3	—	6
Dimethylformamide — Skin	10	30	20	60
2, 6-Dimethyl-4-heptanone, see Diisobutyl ketone	25	150	—	—
1, 1-Dimethylhydrazine — Skin	0.5	1	1	2
Dimethylphthalate	—	5	—	10
C Dimethyl sulfate — Skin	0.1, A2	0.5, A2	—	—
Dinitrobenzene (all isomers) — Skin	0.15	1	0.5	3
Dinitro-o-cresol — Skin	—	0.2	—	0.6
3, 5-Dinitro-o-toluamide (Zoalene®)	—	5	—	10
Dinitrotoluene — Skin	—	1.5	—	5
Dioxane, tech. grade — Skin	50	180	—	—
Dioxathion (Delnav®) — Skin	—	0.2	—	—
Diphenyl, see Biphenyl	0.2	1.5	0.6	4
Diphenylamine	—	10	—	20

*1978 Addition.
**See Notice of Intended Changes.

Substance	ADOPTED VALUES TWA ppm[a]	mg/m³[b]	TENTATIVE VALUES STEL ppm[a]	mg/m³[b]
C Diphenylmethane diisocyanate, see Methylene bisphenyl isocyanate (MDI)	0.02	0.2	—	—
Dipropylene glycol methyl ether — Skin	100	600	150	900
Diquat	—	0.5	—	1
Di-sec, octyl phthalate (Di-2-ethylhexyl-phthalate)	—	5	—	10
Disulfiram	—	2	—	5
Disyston — Skin	—	0.1	—	0.3
2, 6-Ditert. butyl-p-cresol	—	10	—	20
Diuron	—	10	—	—
Dyfonate	—	0.1	—	—
Emery	—	E	—	20
Endosulfan (Thiodan®) — Skin	—	0.1	—	0.3
Endrin — Skin	—	0.1	—	0.3
** Epichlorhydrin — Skin	(5)	(20)	(10)	(40)
EPN — Skin	—	0.5	—	2
1, 2-Epoxypropane, see Propylene oxide	100	240	150	360
2, 3-Epoxy-1-propanol, see Glycidol	50	150	75	225
Ethane	F	—	F	—
Ethanethiol, see Ethyl mercaptan	0.5	1	2	3
Ethanolamine	3	8	6	15
Ethion (Nialate®) — Skin	—	0.4	—	—
2-Ethoxyethanol — Skin	100	370	150	560
2-Ethoxyethyl acetate (Cellosolve acetate) — Skin	100	540	150	810
Ethyl acetate	400	1,400	—	—
Ethyl acrylate — Skin	25	100	—	—
Ethyl alcohol (Ethanol)	1,000	1,900	—	—
Ethylamine	10	18	—	—
Ethyl sec-amyl ketone (4-Methyl-3-heptanone)	25	130	—	—
Ethyl benzene	100	435	125	545
Ethyl bromide	200	890	250	1,110
Ethylbutyl ketone (3-Heptanone)	50	230	75	345
Ethyl chloride	1,000	2,600	1,250	3,250
Ethyl ether	400	1,200	500	1,500
Ethyl formate	100	300	150	450
Ethyl mercaptan	0.5	1	2	3
** Ethyl sillicate	(100)	(850)	—	—
Ethylene	F	—	F	—
C Ethylene chlorohydrin — Skin	1	3	—	—
Ethylenediamine	10	25	—	—
** Ethylene dibromide, see 1, 2-Dibromoethane	(20)	(155)	(30)	(230)
** Ethylene dichloride, see 1, 2-Dichloroethane	(50)	(200)	(75)	(300)
Ethylene glycol, Particulate	—	10	—	20
Vapor	100	250	125	325
C Ethylene glycol dinitrate and/or Nitroglycerin — Skin	0.2[d]	2	—	—
Ethylene glycol monomethyl ether acetate (Methyl cellosolve acetate) — Skin	25	120	35	170
Ethylene oxide	50	90	75	135
Ethylenimine — Skin	0.5	1	—	—
Ethylidene chloride, see 1, 1-Dichloroethane	200	810	250	1,010
C Ethylidene norbornene	5	25	—	—
N-Ethylmorpholine — Skin	20	94	—	—
Fensulfothion (Dasanit)	—	0.1	—	—
Ferbam	—	10	—	20
Ferrovanadium dust	—	1	—	0.3
Fluoride (as F)	—	2.5	—	—
Fluorine	1	2	2	4
Fluorotrichloromethane	1,000	5,600	1,250	7,000
C Formaldehyde	2	3	—	—
Formamide	20	30	30	45
Formic acid	5	9	—	—
Furfural — Skin	5	20	15	60
Furfuryl alcohol — Skin	5	20	10	40
Gasoline	—	B2	—	B2
Germanium tetrahydride	0.2	0.6	0.6	1.8
Glass, fibrous[e] or dust	—	10	—	—
** C Glutaraldehyde, activated or unactivated	—	(0.25)	—	—
Glycerin mist	—	E	—	E
Glycidol (2, 3-Epoxy-1-propanol)	50	150	75	225
Glycol monoethyl ether, see 2-Ethoxyethanol — Skin	100	370	150	560
Graphite (Synthetic)	—	E	—	—
Guthion®, see Azinphos-methyl — Skin	—	0.2	—	0.6
Gypsum	—	E	—	20
Hafnium	—	0.5	—	1.5
Helium	F	—	F	—
Heptachlor — Skin	—	0.5	—	2
Heptane (n-Heptane)	400	1,600	500	2,000
Hexachlorocyclopentadiene	0.01	0.1	0.03	0.3
Hexachloroethane — Skin	1	10	3	30
Hexachloronaphthalene — Skin	—	0.2	—	0.6
Hexafluoroacetone	0.1	0.7	0.3	2
Hexane (n-hexane)	100	360	125	450
* Hexamethyl phosphoramide — Skin	A2	A2	—	—
2-Hexanone, see Methyl butyl ketone — Skin	25	100	40	165
Hexone (Methyl isobutyl ketone) — Skin	100	410	125	510
sec-Hexyl acetate	50	300	—	—
C Hexylene glycol	25	125	—	—
Hydrazine — Skin	0.1, A2	0.1, A2	—	—
Hydrogen	F	—	F	—
Hydrogenated terphenyls	0.5	5	—	—
Hydrogen bromide	3	10	—	—
C Hydrogen chloride	5	7	—	—

Capital letters refer to Appendices.
Footnotes (a thru g) see Page 22.
*1978 Addition.
**See Notice of Intended Changes.

Substance	ADOPTED VALUES TWA ppm[a]	mg/m³[b]	TENTATIVE VALUES STEL ppm[a]	mg/m³[b]
** Hydrogen cyanide — Skin	(10)	(11)	(15)	(16)
Hydrogen fluoride	3	2	—	—
Hydrogen peroxide	1	1.5	2	3
Hydrogen selenide	0.05	0.2	—	—
Hydrogen sulfide	10	15	15	27
Hydroquinone	—	2	—	4
Indene	10	45	15	70
Indium & Compounds (as In)	—	0.1	—	0.3
C Iodine	0.1	1	—	—
Iodoform	0.6	10	1	20
Iron oxide fume	B3	5	—	10
Iron pentacarbonyl	0.01	0.08	—	—
Iron salts, soluble (as Fe)	—	1	—	2
Isoamyl acetate	100	525	125	655
Isoamyl alcohol	100	360	125	450
Isobutyl acetate	150	700	187	875
Isobutyl alcohol	50	150	75	225
C Isophorone	5	25	—	—
Isophorone diisocyanate — Skin	0.01	0.09	—	—
Isopropyl acetate	250	950	310	1,185
Isopropyl alcohol — Skin	400	980	500	1,225
Isopropylamine	5	12	10	24
Isopropyl ether	250	1,050	310	1,320
Isopropyl glycidyl ether (IGE)	50	240	75	360
Kaolin	—	E	—	20
Ketene	0.5	0.9	1.5	3
Lead, inorg., fumes & dusts (as Pb)	—	0.15	—	0.45
Lead arsenate (as Pb)	—	0.15	—	0.45
Lead chromate (as Cr)	—	0.05, A2	—	—
Limestone	—	E	—	20
Lindane — Skin	—	0.5	—	1.5
Lithium hydride	—	0.025	—	—
L.P.G. (Liquified petroleum gas)	1,000	1,800	1,250	2,250
Magnesite	—	E	—	20
Magnesium oxide fume (as Mg)	—	10	—	—
Malathion — Skin	—	10	—	—
Maleic anhydride	0.25	1	—	—
C Manganese & Compounds (as Mn)	—	5	—	—
Manganese cyclopentadienyl tricarbonyl (as Mn) — Skin	—	0.1	—	0.3
* Manganese Tetroxide	—	1	—	—
Marble/calcium carbonate	—	E	—	20
Mercury (Alkyl compounds) — Skin, (as Hg)	0.001	0.01	0.003	0.03
Mercury (All forms except alkyl), as Hg	—	0.05	—	0.15
Mesityl oxide	25	100	—	—
Methane	F	—	F	—
Methanethiol, see Methyl mercaptan	0.5	1	—	—
Methomyl (Lannate®) — Skin	—	2.5	—	—
Methoxychlor	—	10	—	—
2-Methoxyethanol — Skin (Methyl cellosolve)	25	80	35	120
Methyl acetate	200	610	250	760
Methyl acetylene (propyne)	1,000	1,650	1,250	2,040
Methyl acetylene-propadiene mixture (MAPP)	1,000	1,800	1,250	2,250
Methyl acrylate — Skin	10	35	—	—
Methylacrylonitrile — Skin	1	3	2	6
Methylal (dimethoxymethane)	1,000	3,100	1,250	3,875
Methyl alcohol (methanol) — Skin	200	260	250	310
Methylamine	10	12	—	—
Methyl amyl alcohol, see Methyl isobutyl carbinol — Skin	25	100	40	160
Methyl 2-cyanoacrylate	2	8	4	16
Methyl n-amyl ketone (2-Heptanone)	100	465	150	710
Methyl bromide — Skin	15	60	—	—
Methyl butyl ketone, see 2-Hexanone — Skin	25	100	40	165
Methyl cellosolve — Skin see 2-Methoxyethanol	25	80	35	120
Methyl cellosolve acetate — Skin, see Ethylene glycol monomethyl ether acetate	25	120	35	170
Methyl chloride	100	210	125	260
Methyl chloroform (1, 1, 1-Trichloroethane)	350	1,900	450	2,380
Methylcyclohexane	400	1,600	500	2,000
Methylcyclohexanol	50	235	75	350
o-Methycyclohexanone — Skin	50	230	75	345
Methylcyclopentadienyl manganese tricarbonyl (as Mn) — Skin	0.1	0.2	0.3	0.6
Methyl demeton — Skin	—	0.5	—	1.5
C Methylene bisphenyl isocyanate (MDI)	0.02	0.2	—	—
** Methylene chloride (dichloromethane)	(200)	(700)	(250)	(870)
4, 4′-Methylene bis (2-chloraniline) — Skin	0.02, A2	—	A2	—
C Methylene bis (4-cyclohexylisocyanate)	0.01	0.11	—	—
Methyl ethyl ketone (MEK), see 2-Butanone	200	590	300	885
C Methyl ethyl ketone peroxide	0.2	1.5	—	—
Methyl formate	100	250	150	375
Methyl iodide — Skin	5	28	10	56
Methyl isoamyl ketone	100	475	150	710
Methyl isobutyl carbinol — Skin	25	100	40	165
Methyl isobutyl ketone, see Hexone — Skin	100	410	125	510

Capital Letters refer to Appendices.
*1978 Addition.
Footnotes (a thru g) see Page 22.
*See Notice of Intended Changes.

Substance	ADOPTED VALUES TWA ppm[a]	mg/m^3 [b]	TENTATIVE VALUES STEL ppm[a]	mg/m^3 [b]
Methyl isocyanate — Skin	0.02	0.05	—	—
Methyl mercaptan	0.5	1	—	—
Methyl methacrylate	100	410	125	510
Methyl parathion — Skin	—	0.2	—	0.6
Methyl propyl ketone, see 2-Pentanone	200	700	250	875
C Methyl silicate	5	30	—	—
C α-Methyl styrene	100	480	—	—
Molybdenum (as Mo)				
Soluble compounds	—	5	—	10
Insoluble compounds	—	10	—	20
Monocrotophos (Azodrin®)	—	0.25	—	—
Monomethyl aniline — Skin	2	9	4	18
C Monomethyl hydrazine — Skin	0.2	0.35	—	—
Morpholine — Skin	20	70	30	105
Naphthalene	10	50	15	75
β-Naphthylamine	—	A1b	—	A1b
Neon	F	—	F	—
Nickel carbonyl	0.05	0.35	—	—
Nickel metal	—	1	—	—
Nickel, soluble compounds (as Ni)	—	0.1	—	0.3
Nickel sulfide roasting, fume & dust (as Ni)	—	1, A1a	—	—
Nicotine — Skin	—	0.5	—	1.5
Nitric acid	2	5	4	10
Nitric oxide	25	30	35	45
p-Nitroaniline — Skin	1	6	2	12
Nitrobenzene — Skin	1	5	2	10
p-Nitrochlorobenzene — Skin	—	1	—	2
4-Nitrodiphenyl	—	A1b	—	A1b
Nitroethane	100	310	150	465
C Nitrogen dioxide	5	9	—	—
Nitrogen trifluoride	10	29	15	45
C Nitroglycerin — Skin	0.2[d]	2	—	—
Nitromethane	100	250	150	375
1-Nitropropane	25	90	35	135
** 2-Nitropropane	(25)	(90)	—	—
N-Nitrosodimethylamine (dimethylnitrosoamine) — Skin	—	A2	—	A2
Nitrotoluene — Skin	5	30	10	60
Nitrotrichloromethane, see Chloropicrin	0.1	0.7	0.3	2
Nonane	200	1,050	250	1,300
Octachloronaphthalene — Skin	—	0.1	—	0.3
Octane	300	1,450	375	1,800
Oil mist, mineral	—	5[f]	—	10
Osmium tetroxide (as Os)	0.0002	0.002	0.0006	0.006
Oxalic acid	—	1	—	2
Oxygen difluoride	0.05	0.1	0.15	0.3
Ozone	0.1	0.2	0.3	0.6
Paraffin wax fume	—	2	—	6
* Paraquat, respirable sizes	—	0.1	—	—
Parathion — Skin	—	0.1	—	0.3
Particulate polycyclic aromatic hydrocarbons (PPAH), as benzene solubles	—	0.2, A1a	—	A1a

Substance	ADOPTED VALUES TWA ppm[a]	mg/m^3 [b]	TENTATIVE VALUES STEL ppm[a]	mg/m^3 [b]
Pentaborane	0.005	0.01	0.015	0.03
Pentachloronaphthalene	—	0.5	—	2
Pentachlorophenol — Skin	—	0.5	—	1.5
Pentaerythritol	—	E	—	20
Pentane	600	1,800	750	2,250
2-Pentanone	200	700	250	875
Perchloroethylene — Skin	100	670	150	1,000
Perchloromethyl mercaptan	0.1	0.8	—	—
Perchloryl fluoride	3	14	6	28
Phenol — Skin	5	19	10	38
Phenothiazine — Skin	—	5	—	10
p-Phenylene diamine — Skin	—	0.1	—	—
Phenyl ether (vapor)	1	7	2	14
Phenyl ether-Diphenyl mixture (vapor)	1	7	2	14
Phenylethylene, see Styrene, monomer	100	420	125	525
Phenyl glycidyl ether (PGE)	10	60	15	90
* Phenyl mercaptan	0.5	2	—	—
Phenylhydrazine — Skin	5	22	10	44
C Phenylphosphine	0.05	0.25	—	—
Phorate (Thimet®) — Skin	—	0.05	—	0.2
Phosdrin (Mevinphos®) — Skin	0.01	0.1	0.03	0.3
* Phosgene (carbonyl chloride)	0.1	0.4	—	—
Phosphine	0.3	0.4	1	1
Phosphoric acid	—	1	—	3
Phosphorus (yellow)	—	0.1	—	0.3
** Phosphorus pentachloride	—	(1)	—	(3)
Phosphorus pentasulfide	—	1	—	3
Phosphorus trichloride	0.5	3	—	—
Phthalic anhydride	1	6	4	24
m-Phthalodinitrile	—	5	—	—
Picloram (Tordon®)	—	10	—	20
Picric acid — Skin	—	0.1	—	0.3
Pival® (2-Pivalyl-1, 3-indandione)	—	0.1	—	0.3
Plaster of Paris	—	E	—	20
Platinum (Soluble salts) as Pt	—	0.002	—	—
Polychlorobiphenyls, see Chlorodiphenyls — Skin	—	—	—	—
Polytetrafluoroethylene decomposition products	—	B1	—	B1
C Potassium hydroxide	—	2	—	—
Propane	F	—	F	—
β-Propiolactone	—	A2	—	A2
Propargyl alcohol — Skin	1	2	3	6
n-Propyl acetate	200	840	250	1,050
Propyl alcohol — Skin	200	500	250	625
n-Propyl nitrate	25	105	40	470
Propylene	F	—	F	—
Propylene dichloride (1, 2-Dichloropropane)	75	350	110	510
*C Propylene glycol dinitrate — Skin	0.2	2	—	—

Capital letters refer to Appendices.
Footnotes (a thru g) see Page 22.
*1978 Addition.
**See Notice of Intended Changes.

Substance	ADOPTED VALUES TWA ppm[a]	mg/m^3 [b]	TENTATIVE VALUES STEL ppm[a]	mg/m^3 [b]
Propylene glycol monomethyl ether	100	360	150	540
Propylene imine — Skin	2	5	—	—
Propylene oxide	100	240	150	360
Propyne, see Methyl acetylene	1,000	1,650	1,250	2,040
Pyrethrum	—	5	—	10
Pyridine	5	15	10	30
Quinone	0.1	0.4	0.3	2
RDX — Skin	—	1.5	—	3
Resorcinol	10	45	20	90
Rhodium, Metal fume and dusts (as Rh)	—	0.1	—	0.3
Soluble salts (as Rh)	—	0.001	—	0.003
Ronnel	—	10	—	—
Rosin core solder pyrolysis products (as formaldehyde)	—	0.1	—	0.3
Rotenone (commercial)	—	5	—	10
Rouge	—	E	—	20
Rubber solvent (Naphtha)	400	1,600	—	—
Selenium compounds (as Se)	—	0.2	—	—
Selenium hexafluoride, as Se	0.05	0.4	0.05	0.4
Sevin® (see Carbaryl)	—	5	—	10
Silane (see Silicon tetrahydride)	0.5	7	1	2
Silicon	—	E	—	20
Silicon carbide	—	E	—	20
Silicon tetrahydride (Silane)	0.5	0.7	1	2
** Silver, metal and soluble compounds, as Ag	—	(0.01)	—	(0.03)
C Sodium azide	0.1	0.3	—	—
Sodium fluoroacetate (1080) — Skin	—	0.05	—	0.15
C Sodium hydroxide	—	2	—	—
Starch	—	E	—	20
Stibine	0.1	0.5	0.3	1.5
Stoddard solvent	100	575	125	720
Strychnine	—	0.15	—	0.45
Styrene, monomer (Phenylethylene)	100	420	125	525
C Subtilisins (Proteolytic enzymes as 100% pure crystalline enzyme)	—	0.00006[n]	—	—
Sucrose	—	E	—	20
** Sulfur dioxide	(5)	(13)	—	—
Sulfur hexafluoride	1,000	6,000	1,250	7,500
Sulfuric acid	—	1	—	—
Sulfur monochloride	1	6	3	18
Sulfur pentafluoride	0.025	0.25	0.075	0.75
Sulfur tetrafluoride	0.1	0.4	0.3	1
Sulfuryl fluoride	5	20	10	40
Systox, see Demeton® — Skin	0.01	0.1	0.03	0.3
2, 4, 5-T	—	10	—	20
Tantalum	—	5	—	10
TEDP — Skin	—	0.2	—	0.6
Teflon® decomposition products	—	B1	—	B1
Tellurium & compounds (as Te)	—	0.1	—	—
Tellurium hexafluoride, as Te	0.02	0.2	—	—
TEPP — Skin	0.004	0.05	0.01	0.2
**C Terphenyls	(1)	(9)	—	—
1, 1, 1, 2-Tetrachloro-2, 2-difluoroethane	500	4,170	625	5,210
1, 1, 2, 2-Tetrachloro-1, 2-difluoroethane	500	4,170	625	5,210
1, 1, 2, 2-Tetrachloroethane — Skin	5	35	10	70
Tetrachloroethylene, see Perchloroethylene — Skin	100	670	150	1,000
Tetrachloromethane, see Carbon tetrachloride — Skin	10	65	20	130
Tetrachloronaphthalene	—	2	—	4
Tetraethyl lead (as Pb) — Skin	—	0.100[g]	—	0.3
Tetrahydrofuran	200	590	250	735
Tetramethyl lead (as Pb) — Skin	—	0.150[g]	—	0.5
Tetramethyl succinonitrile — Skin	0.5	3	2	9
Tetranitromethane	1	8	—	—
Tetryl (2, 4, 6-trinitrophenyl-methylnitramine) — Skin	—	1.5	—	3.0
Thallium, soluble compounds (as Tl) — Skin	—	0.1	—	—
4, 4′-Thiobis (6-tert. butyl-m-cresol)	—	10	—	20
* Thioglycolic acid	1	5	—	—
Thiram®	—	5	—	10
Tin, inorganic compounds, except SnH_4 and SnO_2 (as Sn)	—	2	—	4
Tin, organic compounds (as Sn) — Skin	—	0.1	—	0.2
Tin oxide (as Sn)	—	E	—	20
Titanium dioxide (as Ti)	—	E	—	20
Toluene (toluol) — Skin	100	375	150	560
**C Toluene-2, 4-diisocyanate (TDI)	(0.02)	(0.14)	—	—
o-Toluidine	5	22	10	44
Toxaphene, see Chlorinated camphene — Skin	—	0.5	—	2.0
Tributyl phosphate	—	5	—	5
*C 1, 2, 4-Trichlorobenzene	5	40	—	—
1, 1, 1-Trichloroethane, see Methyl chloroform	350	1,900	440	2,380
1, 1, 2-Trichloroethane — Skin	10	45	20	90
Trichloroethylene	100	535	150	800
* Trichloromethane, see Chloroform	10, A2	50, A2	—	—
Trichloronaphthalene	—	5	—	10
1, 2, 3-Trichloropropane	50	300	75	450
1, 1, 2-Trichloro 1, 2, 2-trifluoroethane	1,000	7,600	1,250	9,500
Triethylamine	25	100	40	160
Tricyclohexyltin hydroxide (Plictran®)	—	5	—	10

Capital letters refer to Appendices.
Footnotes (a thru g) see Page 22.
*1978 Addition.
**See Notice of Intended Changes.

Substance	ADOPTED VALUES TWA ppm[a)]	ADOPTED VALUES TWA mg/m³[b)]	TENTATIVE VALUES STEL ppm[a)]	TENTATIVE VALUES STEL mg/m³[b)]
Trifluoromonobromo-methane	1,000	6,100	1,200	7,300
Trimethyl benzene	25	125	35	170
2, 4, 6-Trinitrophenol, see Picric acid — Skin	—	0.1	—	0.3
2, 4, 6-Trinitrophenyl-methylnitramine, see Tetryl — Skin	—	1.5	—	3.0
*C 2, 4, 6-Trinitrotoluene (TNT)	—	0.5	—	—
Triorthocresyl phosphate	—	0.1	—	0.3
Triphenyl phosphate	—	3	—	6
Tungsten & compounds, as W				
Soluble	—	1	—	3
Insoluble	—	5	—	10
Turpentine	100	560	150	840
Uranium (natural) soluble & insoluble compounds, as U	—	0.2	—	0.6
Vanadium (V_2O_5), as V				
Dust	—	0.5	—	1.5
C Fume	—	0.05	—	—
* Valeraldehyde	50	175	—	—
Vinyl acetate	10	30	20	60
Vinyl benzene, see Styrene	100	420	150	630
** Vinyl bromide	(250)	(1,100)	—	—
** Vinyl chloride	(A1c)	—	(A1c)	—
Vinyl cyanide, see Acrylonitrile — Skin	20	45	30	65
Vinyl cyclohexene dioxide	10	60	—	—
Vinylidene chloride	10	40	20	80
Vinyl toluene	100	480	150	720
Warfarin	—	0.1	—	0.3
Welding fumes (NOC)†	—	5, B3	—	B3
Wood dust (nonallergenic)	—	5	—	10
Xylene (o-, m-, p-isomers) — Skin	100	435	150	655
C m-Xylene α, α′-diamine	—	0.1	—	—
Xylidene — Skin	5	25	10	50
Yttrium	—	1	—	3
Zinc chloride fume	—	1	—	2
Zinc chromate (as Cr)	—	0.05, A2	—	—
Zinc oxide fume	—	5	—	10
Zinc stearate	—	E	—	20
Zirconium compounds (as Zr)	—	5	—	10

Capital letters refer to Appendices.
†(NOC) Not Otherwise Classified.
*1978 Addition.
**See Notice of Intended Changes.

a) Parts of vapor or gas per million parts of contaminated air by volume at 25°C and 760 mm. Hg. pressure.
b) Approximate milligrams of substance per cubic meter of air.
d) An atmospheric concentration of not more than 0.02 ppm, or personal protection may be necessary to avoid headache for intermittent exposure.
e) $< 7\ \mu m$ in diameter.
f) As sampled by method that does not collect vapor.
g) For control of general room air, biologic monitoring is essential for personnel control.

Radioactivity: For permissible concentrations of radioisotopes in air, see U.S. Department of Commerce, National Bureau of Standards Handbook 69, "Maximum Permissible Body Burdens and Maximum Permissible Concentrations of Radionuclides in Air and in Water for Occupational Exposure," June 5, 1959; Addendum 1, August 1963 (NCRP Report No. 22). Also, see U.S. Department of Commerce National Bureau of Standards, Handbook 59, "Permissible Dose from External Sources of Ionizing Radiation," September 24, 1954, an addendum of April 15, 1958. A report, Basic Radiation Protection Criteria, published by the National Committee on Radiation Protection, revises and modernizes the concept of the NCRP standards of 1954, 1957 and 1958; obtainable as NCRP Rept. No. 39, 7910 Woodmont Ave., Washington, D.C. 20014.

MINERAL DUSTS

Substance

SILICA, SiO_2

Crystalline

Quartz — TLV in mppcf[h)]:

$$\frac{300^{i)}}{\%\ \text{quartz} + 10}$$

TLV for respirable dust in mg/m³:

$$\frac{10\ \text{mg/m}^{3j)}}{\%\ \text{Respirable quartz} + 2}$$

TLV for "total dust," respirable and nonrespirable:

$$\frac{30\ \text{mg/m}^3}{\%\ \text{quartz} + 3}$$

Cristobalite .... Use one-half the value calculated from the count or mass formulae for quartz.

Tridymite — Use one-half the value calculated from formulae for quartz.

Silica, fused — Use quartz formulae.

Tripoli — Use respirable[o)] mass quartz formula

***Amorphous* ....(20 mppcf[h)])

SILICATES (< 1% quartz)

**Asbestos, all forms	(5 fibers/cc> 5μm in length[m)]; A1a)
Mica	20 mppcf
Mineral wool fiber	10 mg/m³
Perlite	30 mppcf
Portland Cement	30 mppcf
Soapstone	20 mppcf
Talc (nonasbestiform)	20 mppcf

**Talc (fibrous), use Asbestos limit.
**Tremolite, see Asbestos.

COAL DUST

2 mg/m³ (respirable dust fraction < 5% quartz). If > 5% quartz, use respirable mass formula.

NUISANCE PARTICULATES

(see Appendix E)

30 mppcf or 10 mg/m³[k)]

of total dust < 1% quartz, or, 5 mg/m³ respirable dust.

Conversion factors:
mppcf × 35.3 = Million particles per cubic meter
= particles per cc

**See Notice of Intended Changes.
See Footnotes Page 23.

h) Millions of particles per cubic foot of air, based on impinger samples counted by light-field technics.

i) The percentage of quartz in the formula is the amount determined from airborne samples, except in those instances in which other methods have been shown to be applicable.

j) Both concentration and percent quartz for the application of this limit are to be determined from the fraction passing a size-selector with the following characteristics:

Aerodynamic Diameter (μm) (unit density sphere)	% passing selector
≂ 2	90
2.5	75
3.5	50
5.0	25
10	0

k) containing <1% quartz; if quartz content > 1%, use formulae for quartz.

l) Lint-free dust as measured by the vertical-elutriator, cotton-dust sampler described in the Transactions of the National Conference on Cotton Dust, J. R. Lynch, pg. 33, May 2, 1970.

m) As determined by the membrane filter method at 400–450X magnification (4 mm objective) phase contrast illumination.

n) Based on "high volume" sampling.

o) "Respirable" dust as defined by the British Medical Research Council Criteria (1) and as sampled by a device producing equivalent results (2).

(1) Hatch, T. E. and Gross, P., Pulmonary Deposition and Retention of Inhaled Aerosols, p. 149. Academic Press, New York, New York, 1964.

(2) Interim Guide for Respirable Mass Sampling, AIHA Aerosol Technology Committee, AHIA J. *31:* 2, 1970, p. 133.

NOTICE OF INTENDED CHANGES (for 1978)

These substances, with their corresponding values, comprise those for which either a limit has been proposed for the first time, or for which a change in the "Adopted" listing has been proposed. In both cases, the proposed limits should be considered trial limits that will remain in the listing for a period of at least two years. If, after two years no evidence comes to light that questions the appropriateness of the values herein, the values will be reconsidered for the "Adopted" list. Documentation is available for each of these substances.

Substance	TWA ppm[a]	TWA mg/m³[b]	STEL ppm[a]	STEL mg/m³[b]
† Acetylsalicylic acid (Aspirin)	—	5	—	—
† Acrylonitrile	A1c	A1c	—	—
Aluminum metal and oxide	—	10	—	20
Aluminum pyro powders	—	5	—	—
Aluminum welding fumes	—	5	—	—
Aluminum, soluble salts	—	2	—	—
Aluminum, alkyls (NOC)*	—	2	—	—
3-Amino 1, 2, 4-triazole	A2	A2	—	—
† Aniline and homologues — Skin	2	10	5	20

Capital letters refer to Appendices.
†1978 Revision or Addition.
*Not otherwise classified (NOC).

Substance	TWA ppm[a]	TWA mg/m³[b]	STEL ppm[a]	STEL mg/m³[b]
Antimony, soluble salts (as Sb)	—	2	—	—
Antimony trioxide production	—	—, A2	—	—
Arsenic (soluble), as As	—	0.2	—	—
Arsenic trioxide production	—	—, A1a	—	—
† Baytex	—	0.1	—	0.3
Benomyl	—	10	—	15
Bromacil	—	10	—	20
† o-sec Butylphenol — Skin	5	—	—	—
Cadmium oxide production	—	—, A2	—	—
† Carbon disulfide	10	30	—	—
† Chloroacetyl chloride	0.05	0.2	—	—
† Chloromethyl methyl ether	A1b	A1b	—	—
† β-Chloroprene — Skin	10	45	—	—
Cobalt metal, dust & fume (as Co)	—	0.05	—	0.1
†C Cyanogen chloride	0.3	0.6	—	—
Cyclopentane	300	850	450	1,000
† Dalapon	1	—	—	—
† 1, 2-Dibromoethane — Skin	A1c	A1c	—	—
† Dichloromonofluoromethane	10	40	—	—
† Dichloropropene	1	5	10	50
† Diethanolamine	3	15	—	—
† Divinyl benzene	10	50	—	—
† Epichlorhydrine — Skin	2	10	5	20
Ethyl silicate	10	85	30	250
† Ethylene dibromide, see 1, 2-Dibromoethane	A1c	A1c	—	—
† Ethylene dichloride, see 1, 2-Dichloroethane	10	40	15	60
C Glutaraldehyde	0.2	0.8	—	—
Hexachlorobutadiene	A2	A2	—	—
†C Hydrogen cyanide — Skin	10	10	—	—
† 2-Hydroxypropyl acrylate — Skin	0.5	3	—	—
† N-Isopropylaniline — Skin	2	10	5	20
Manganese fume (as Mn)	—	1	—	3
† Methylene chloride (dichloromethane)	100	360	500	1,700
† 4, 4-Methylene dianiline	0.1	0.8	0.5	4
†C 2-Nitropropane	25, A2	90, A2	—	—
Phenyl-beta-naphthylamine	A2	A2	—	—
† Phosphorous pentachloride	0.1	1	—	—
† Propionic acid	10	30	15	45
† Silver, metal	—	0.1	—	—
† Sodium bisulfite	—	5	—	—
† Sodium metabisulfite	—	5	—	—
† Sulfur dioxide	2	5	5	15
†C Terphenyls	0.5	5	—	—
† Tetrasodium pyrophosphate	—	5	—	—
† Toluene-2, 4-diisocyanate (TDI)	0.002	0.015	0.005	0.035
† Trichloroacetic acid	—	1	—	—
† Vinyl bromide	5, A2	20, A2	—	—
† Vinyl chloride	5, A1a	10, A1a	—	—
VM & P Naphtha	300	1,350	400	1,800

Capital letters refer to Appendices.
†1978 Addition.

NOTICE OF INTENDED CHANGES
MINERAL DUSTS

Substance	TLV
† Asbestos	
Amosite	0.5 fiber/cc, A1a
Chrysotile	2 fibers/cc, A1a
Crocidolite	0.2 fiber/cc, A1a
Tremolite	0.5 fiber/cc, A1a
Other forms	2 fibers/cc, A1a
Diatomaceous earth, natural	1.5 mg/m³, Respirable dust
Silica, amorphous	5 mg/m³, Total dust (all sampled sizes)
	2 mg/m³, Respirable dust (< 5 μm)
† Talc (fibrous)	0.5 fiber/cc

APPENDIX A
CARCINOGENS

The Committee lists below those substances in industrial use that have proven carcinogenic in man, or have induced cancer in animals under appropriate experimental conditions. Present listing of those substances carcinogenic for man takes three forms: Those for which a TLV has been assigned (1a), those for which environmental conditions have not been sufficiently defined to assign a TLV (1b), and (1c), those whose reassignment of a TLV is awaiting more definitive data, and hence should be treated as a 1b carcinogen.

A1a. *Human Carcinogens*. Substances, or substances associated with industrial processes, recognized to have carcinogenic or cocarcinogenic potential, with an assigned TLV:

	TLV
** Arsenic trioxide production	(As_2O_3, 0.05 mg/m³ as As) (SO_2, C 5.0 ppm) (Sb_2O_3, 0.5 mg/m³ (as Sb))
** Asbestos, all forms*	(5 fibers/cc, > 5 μm in length)
bis (Chloromethyl) ether	0.001 ppm
Chromite ore processing (chromate)	0.05 mg/m³ (as Cr)
Nickel sulfide roasting, fume & dust	1.0 mg/m³ (as Ni)
Particulate Polycyclic Aromatic Hydrocarbons (PPAH)	0.2 mg/m³, as benzene solubles
Vinyl Chloride	5 ppm

A1b. *Human Carcinogens*. Substances, or substances associated with industrial processes, recognized to have carcinogenic potential without an assigned TLV:

Chloromethyl methyl ether
4-Aminodiphenyl (p-Xenylamine)
Benzidine production
beta-Naphthylamine
4-Nitrodiphenyl

A1c. *Human Carcinogens*. Substances with recognized carcinogenic potential awaiting reassignment of TLV pending further data acquisition:

Acrylonitrile
1, 2-Dibromoethane (Ethylene dibromide)

For the substances in 1b or 1c, no exposure or contact by any route — respiratory, skin or oral, as detected by the most sensitive methods — shall be permitted.

"No exposure or contact" means hermitizing the process or operation by the best practicable engineering methods. The worker should be properly equipped to insure virtually no contact with the carcinogen.

A2. *Industrial Substances Suspect of Carcinogenic Potential for MAN*. Chemical substances or substances associated with industrial processes, which are suspect of inducing cancer, based on either (1) limited epidemiologic evidence, exclusive of clinical reports of single cases, or (2) demonstration of carcinogenesis in one or more animal species by appropriate methods.

Substance	TLV
3-Amino 1, 2, 4-triazole	
** Antimony trioxide production*	(0.5 mg/m³)
Benzene	10 ppm
Benz(a)pyrene	——
Beryllium	2.0 μg/m³
** Cadmium oxide production	(0.05 mg/m³)
Chloroform	10 ppm
Chromates of lead and zinc (as Cr)	0.05 mg/m³
3, 3′-Dichlorobenzidine	——
Dimethylcarbamyl chloride	——
1, 1-Dimethyl hydrazine	0.5 ppm
Dimethyl sulfate — Skin	0.1 ppm
Epichlorhydrin	5 ppm
Hexachlorobutadiene	——
Hexamethyl phosphoramide — Skin	——
Hydrazine	0.1 ppm
Lead chromate	0.05 mg/m³
4, 4′-Methylene bis (2-chloroaniline) — Skin	0.02 ppm
Monomethyl hydrazine	0.2 ppm
C 2-Nitropropane	25 ppm
Nitrosamines	——
Phenyl-beta-naphthylamine	——
Propane sultone	——
beta-Propiolactone	——
Vinyl cyclohexene dioxide	10 ppm
Zinc chromate (as Cr)	0.05 mg/m³

For the above, worker exposure by all routes should be carefully controlled to levels consistent with the animal and human experience data (see Documentation), including those substances with a listed TLV.

*Cigarette smoking can enhance the incidence of respiratory cancers from this and others of these substances or processes.

**See Notice of Intended Changes

†1978 Addition.

*Cigarette smoking can enhance the incidence of respiratory cancers from this or others of these substances or processes.

A3. *Guidelines for the Classification of Experimental ANIMAL Carcinogens*. The following guidelines are offered in the present state of knowledge as an aid in classifying substances in the occupational environment found to be carcinogenic in experimental animals. A need was felt by the Threshold Limits Committee for such a classification in order to take the first step in developing an appropriate TLV for occupational exposure.

Determination of Approximate Threshold of Response Requirement. In order to determine in which category to classify an experimental carcinogen for the purpose of assigning an industrial air limit (TLV), an approximate threshold of neoplastic response must be determined. Because of practical experimental difficulties, a precisely defined threshold cannot be attained. For the purposes of standard-setting, this is of little moment, as an appropriate risk, or safety, factor can be applied to the approximate threshold, the magnitude of which is dependent on the degree of potency of the carcinogenic response.

To obtain the best 'practical' threshold of neoplastic response, dosage decrements should be less than logarithmic. This becomes particularly important at levels greater than 10 ppm (or corresponding mg/m^3). Accordingly, after a range-finding determination has been made by logarithmic decreases, two additional dosage levels are required within the levels of "effect" and "no effect" to approximate the true threshold of neoplastic response.

The second step should attempt to establish a metabolic relationship between animal and man for the particular substance found carcinogenic in animals. If the metabolic pathways are found comparable, the substance should be classed highly suspect as a carcinogen for man. If no such relation is found, the substance should remain listed as an experimental animal carcinogen until evidence to the contrary is found.

Proposed Classification of Experimental Animal Carcinogens. Substances occurring in the occupational environment found carcinogenic for animals may be grouped into three classes, those of high, intermediate and low potency. In evaluating the incidence of animal cancers, significant incidence of cancer is defined as a neoplastic response which represents, in the judgment of the Committee, a significant excess of cancers above that occurring in negative controls.

EXCEPTIONS: No substance is to be considered an occupational carcinogen of any practical significance which reacts by the respiratory route at or above 1000 mg/m^3 for the mouse, 2000 mg/m^3 for the rat; by the dermal route, at or above 1500 mg/kg for the mouse, 3000 mg/kg for the rat; by the gastrointestinal route at or above 500 mg/kg/d for a lifetime, equivalent to about 100 g T.D. for the rat, 10g T.D. for the mouse.

These dosage limitations exclude such substances as dioxane and trichlorethylene from consideration as carcinogens.

Examples: Dioxane — rats, hepatocellular and nasal tumors from 1015 mg/kg/d, oral

Trichloroethylene — female mice, tumors (30/98 @ 900 mg/kg/d), oral

A3a. *INDUSTRIAL SUBSTANCES OF HIGH CARCINOGENIC POTENCY IN EXPERIMENTAL ANIMALS*

1. A substance to qualify as a carcinogen of high potency must fulfill one of the three following conditions in two animal species:

1a. *Respiratory*. Elicit cancer from (1) dosages below 1 mg/m^3 (or equivalent ppm) via the respiratory tract in 6- 7-hour daily repeated inhalation exposures throughout lifetime; or (2) from a single intratracheally administered dose not exceeding 1 mg of particulate, or liquid, per 100 ml or less of animal minute respiratory volume;

Examples: bis-Chloromethyl ether, malignant tumors, rats, @ 0.47 mg/m^3 (0.1 ppm) in 2 years;

Hexamethyl phosphoramide, nasal squamous cell carcinoma, rats, @ 0.05 ppm, in 13 months

OR

1b. *Dermal*. Elicit cancer within 20 weeks by skin-painting, twice weekly at 2 mg/kg body weight or less per application for a total dose equal to or less than 1.5 mg, in a biologically inert vehicle;

Examples: 7, 12-Dimethylbenz(a)anthracene — skin tumors @ 0.12-0.8 mg T.D. in four weeks

Benz(a)pyrene, mice 12 μg, 3X/wk for 18 mos. T.D. 2.6 mg, 90.9% skin tumors

OR

1c. *Gastrointestinal*. Elicit cancer by daily intake via the gastrointestinal tract, within six months, with a six-month holding period, at a dosage below 1 mg/kg body weight per day; total dose, rat, $\leq$ 50 mg; mouse, $\leq$ 3.5 mg;

Examples: 7, 12-Dimethylbenz(a)anthracene — mammary tumors from 10 mg 1X

3-Methylcholanthrene — Tumors @ 3 sites from 8 mg in 89 weeks

Benz(a)pyrene, mice, 3.9% leukemias, from 30 mg T.D. 198 days

2. Elicit cancer by all three routes in at least two animal species at dose levels prescribed for high or intermediate potency.

A3b. *INDUSTRIAL SUBSTANCES OF INTERMEDIATE CARCINOGENIC POTENCY IN EXPERIMENTAL ANIMALS*

To qualify as a carcinogen of intermediate potency, a substance should elicit cancer in two animal species at dosages intermediate between those described in A3a and A3c by two routes of administration.

Example: Carbamic acid Ethyl Ester
Dermal, mammary tumors, mice, 100%, 63 weeks, 500–1400 mg T.D.
Gastrointestinal, various type tumors, mice 42 weeks, 320 mg T.D.

Gastrointestinal, various type tumors, rats, 60 weeks, 110–930 mg T.D.

A3c. *INDUSTRIAL SUBSTANCES OF LOW CARCINOGENIC POTENCY IN EXPERIMENTAL ANIMALS*

To qualify as a carcinogen of low potency, a substance should elicit cancer in one animal species by any *one* of three routes of administration at the following prescribed dosages and conditions:

1a. *Respiratory*. Elicit cancer from (1) dosages greater than 10 mg/m^3 (or equivalent ppm) via the respiratory tract in 6- 7-hour, daily repeated inhalation exposures, for 12 months' exposure and 12 months' observation period; or (2) from intratracheally administered dosages totaling more than 10 mg of particulate or liquid per 100 ml or more of animal minute respiratory volume;

Examples: Beryl (beryllium aluminum silicate) malig. lung tumors, rats, @ 15 mg/m^3 @ 17 months

Benzidine, var. tumors, rats, 10–20 mg/m^3 @ $>$ 13 mos.

OR

1b. *Dermal*. Elicit cancer by skin-painting of mice in twice weekly dosages of $>$ 10 mg/kg body weight in a biologically inert vehicle for at least 75 weeks, i.e., $\geq$ 1.5g T.D.

Examples: Shale tar, mouse, 0.1 ml $\times$ 50 – g T.D. 59/60 skin tumors

Arsenic trioxide, man, dose unknown, but estimated to be high

1c. *Gastrointestinal*. Elicit cancer from daily oral dosages of 50 mg/kg/day or greater during the lifetime of the animal.

APPENDIX B
SUBSTANCES OF VARIABLE COMPOSITION

B1 *Polytetrafluoroethylene* decomposition products*. Thermal decomposition of the fluorocarbon chain in air leads to the formation of oxidized products containing carbon, fluorine and oxygen. Because these products decompose in part by hydrolysis in alkaline solution, they can be quantitatively determined in air as fluoride to provide an index of exposure. No TLV is recommended pending determination of the toxicity of the products, but air concentrations should be minimal.

B2 *Gasoline*. The composition of gasoline varies greatly and thus a single TLV for all types of these materials is no longer applicable. In general, the aromatic hydrocarbon content will determine what TLV applies. Consequently the content of benzene, other aromatics and additives should be determined to arrive at the appropriate TLV (Elkins, et al. A.I.H.A.J. *24*:99, 1963); Runion, ibid. *36*, 338, 1975).

B3 *Welding Fumes — Total Particulate (NOC)***
TLV, 5 mg/m^3

Welding fumes cannot be classified simply. The composition and quantity of both are dependent on the alloy being welded and the process and electrodes used. Reliable analysis of fumes cannot be made without considering the nature of the welding process and system being examined; reactive metals and alloys such as aluminum and titanium are arc-welded in a protective, inert atmosphere such as argon. These arcs create relatively little fume, but an intense radiation which can produce ozone. Similar processes are used to arc-weld steels, also creating a relatively low level of fumes. Ferrous alloys also are arc-welded in oxidizing environments which generate considerable fume, and can produce carbon monoxide instead of ozone. Such fumes generally are composed of discreet particles of amorphous slags containing iron, manganese, silicon and other metallic constituents depending on the alloy system involved. Chromium and nickel compounds are found in fumes when stainless steels are arc-welded. Some coated and flux-cored electrodes are formulated with fluorides and the fumes associated with them can contain significantly more fluorides than oxides. Because of the above factors, arc-welding fumes frequently must be tested for individual constituents which are likely to be present to determine whether specific TLV's are exceeded. Conclusions based on total fume concentration are generally adequate if no toxic elements are present in welding rod, metal, or metal coating and conditions are not conducive to the formation of toxic gases.

Most welding, even with primitive ventilation, does not produce exposures inside the welding helmet above 5 mg/m^3. That which does, should be controlled.

APPENDIX C
MIXTURES

C.1 THRESHOLD LIMIT VALUES FOR MIXTURES

When two or more hazardous substances are present, their combined effect, rather than that of either individually, should be given primary consideration. In the absence of information to the contrary, the effects of the different hazards should be considered as additive. That is, if the sum of the following fractions,

$$\frac{C_1}{T_1} + \frac{C_2}{T_2} + \ldots \frac{C_n}{T_n}$$

exceeds unity, then the threshold limit of the mixture should be considered as being exceeded. C_1 indicates the observed atmospheric concentration, and T_1 the corresponding threshold limit (See Example 1A.a. and 1A.c.).

Exceptions to the above rule may be made when there is a good reason to believe that the chief effects of the different harmful substances are not in fact additive, but *independent* as when purely local effects on different organs of the body are produced by the various components of the mixture. In such cases the threshold limit ordinarily is exceeded only when at least one member of the series $\left(\frac{C_1}{T_1} + \text{or} + \frac{C_2}{T_2} \text{etc.}\right)$ itself has a value exceeding unity (See Example 1A.c.).

*Trade Names: Algoflon, Fluon, Halon, Teflon, Tetran

**Not otherwise classified (NOC).

Antagonistic action or potentiation may occur with some combinations of atmospheric contaminants. Such cases at present must be determined individually. Potentiating or antagonistic agents are not necessarily harmful by themselves. Potentiating effects of exposure to such agents by routes other than that of inhalation is also possible, e.g. imbibed alcohol and inhaled narcotic (trichloroethylene). Potentiation is characteristically exhibited at high concentrations, less probably at low.

When a given operation or process characteristically emits a number of harmful dusts, fumes, vapors or gases, it will frequently be only feasible to attempt to evaluate the hazard by measurement of a single substance. In such cases, the threshold limit used for this substance should be reduced by a suitable factor, the magnitude of which will depend on the number, toxicity and relative quantity of the other contaminants ordinarily present.

Examples of processes which are typically associated with two or more harmful atmospheric contaminants are welding, automobile repair, blasting, painting, lacquering, certain foundry operations, diesel exhauts, etc.

C.1A Examples of THRESHOLD LIMIT VALUES FOR MIXTURES

The following formulae apply only when the components in a mixture have similar toxicologic effects; they should not be used for mixtures with widely differing reactivities, e.g. hydrogen cyanide & sulfur dioxide. In such case the formula for Independent Effects (1A.c.) should be used.

1A.a. General case, where air is analyzed for each component:

a. *Additive effects. (Note: It is essential that the atmosphere be analyzed both qualitatively and quantitatively for each component present, in order to evaluate compliance or noncompliance with this calculated TLV.)*

$$\frac{C_1}{T_1} + \frac{C_2}{T_2} + \frac{C_3}{T_3} + \ldots = 1$$

Example No. 1A.a.: Air contains 400 ppm of acetone (TLV = 1000 ppm) 150 ppm of secbutyl acetate (TLV = 200 ppm) and 100 ppm of 2-butanone (TLV = 200 ppm)

Atmospheric concentration of mixture = 400 + 150 + 100 = 650 ppm of mixture

$$\frac{400}{1000} + \frac{150}{200} + \frac{100}{200} = 0.4 + 0.75 + 0.5 = 1.65$$

Threshold Limit is exceeded.

1A.b. Special case when the source of contaminant is a liquid mixture and the atmospheric composition is *assumed* to be similar to that of the original material; e.g. on a time-weighted average exposure basis, all of the liquid (solvent) mixture eventually evaporates.

Additive effects (approximate solution)

1. The percent composition (by weight) of the liquid mixture is known, the TLVs of the constituents must be listed in mg/m³.

(Note: In order to evaluate compliance with this TLV, field sampling instruments should be calibrated, in the laboratory, for response to this specific quantitative and qualitative air-vapor mixture, and also to fractional concentrations of this mixture; e.g., 1/2 the TLV; 1/10 the TLV; 2 × the TLV; 10 × the TLV; etc.)

TLV of mixture =

$$\frac{1}{\frac{f_a}{TLV_a} + \frac{f_b}{TLV_b} + \frac{f_c}{TLV_c} + \ldots \frac{f_n}{TLV_n}}$$

Example No. 1: Liquid contains (by weight)

50% heptane: TLV = 400 ppm or 1600 mg/m³
1 mg/m³ ≡ 0.25 ppm

30% methyl chloroform: TLV = 350 ppm or 1900 mg/m³
1 mg/m³ ≡ 0.28 ppm

20% perchloroethylene: TLV = 100 ppm or 670 mg/m³
1 mg/m³ ≡ 0.15 ppm

$$\text{TLV of Mixture} = \frac{1}{\frac{0.5}{1600} + \frac{0.3}{1900} + \frac{0.2}{670}}$$

$$= \frac{1}{0.00031 + 0.00016 + 0.00030}$$

$$= \frac{1}{0.00077} = 1300 \text{ mg/m}^3$$

of this mixture
50% or (1300) (0.5) = 650 mg/m³ is heptane
30% or (1300) (0.3) = 390 mg/m³ is methyl chloroform
20% or (1300) (0.2) = 260 mg/m³ is perchloroethylene

These values can be converted to ppm as follows:

heptane: 650 mg/m³ × 0.25 = 162 ppm
methyl chloroform: 390 mg/m³ × 0.18 = 70 ppm
perchloroethylene: 260 mg/m³ × 0.15 = 39 ppm

TLV of mixture = 162 + 70 + 39 = 271 ppm, or 1300 mg/m³

1A.c. *Independent effects*.
Air contains 0.15 mg/m³ of lead (TLV, 0.15) and 0.7 mg/m³ of sulfuric acid (TLV, 1).

$$\frac{0.15}{0.15} = 1; \quad \frac{0.7}{1} = 0.7$$

Threshold limit is not exceeded.

1B. TLV for Mixtures of Mineral Dusts.

For mixtures of biologically active mineral dusts the general formula for mixtures may be used.
For mixture containing 80% nonasbestiform talc and 20% quartz, the TLV for 100% of the mixture is given by:

$$TLV = \frac{1}{\frac{0.8}{20} + \frac{0.2}{2.7}} = 9 \text{ mppcf}$$

TLV of nonasbestiform talc (pure) = 20 mppcf
TLV of quartz (pure) =

$$\frac{300}{100 + 10} = \frac{300}{110} = 2.7 \text{ mppcf}$$

Essentially the same result will be obtained if the limit of the more (most) toxic component is used provided the effects are additive. In the above example the limit for 20% quartz is 10 mppcf.

For another mixture of 25% quartz, 25% amorphous silica and 50% talc:

25% quartz — TLV (pure) = 2.7 mppcf
25% amorphous silica — TLV (pure) = mppcf
50% talc TLV (pure) = 20 mppcf

$$TLV = \frac{1}{\frac{0.25}{2.7} + \frac{0.25}{20} + \frac{0.5}{20}} = 8 \text{ mppcf}$$

The limit for 25% quartz approximates 9 mppcf.

APPENDIX D
PERMISSIBLE EXCURSIONS FOR TIME-WEIGHTED AVERAGE (TWA) LIMITS

The Excursion TLV Factor in the Table automatically defines the magnitude of the permissible excursion above the limit for those substances not given a "C" designation; i.e., the TWA limits. Examples in the Table show that nitrobenzene, the TLV for which is 1 ppm, should never be allowed to exceed 3 ppm. Similarly, carbon tetrachloride, TLV = 10 ppm, should never be allowed to exceed 20 ppm. By contrast, those substances with a "C" designation are not subject to the excursion factor and must be kept at or below the TLV ceiling.

These limiting excursions are to be considered to provide a "rule-of-thumb" guidance for listed substances generally, and may not provide the most appropriate excursion for a particular substance e.g., the permissible excursion for CO is 400 ppm for 15 minutes.

For appropriate excursions for 142 substances consult Pa. Rules & Regs., Chap. 4, Art. 432, and "Acceptable Concentrations," ANSI.

Substance	TLV	Excursion Factor	Max. Conc. Permitted for short time
Nitrobenzene	1	3	2
Carbon tetrachloride	10	2	20
Trimethyl benzene	25	1.5	35
Acetone	1000	1.25	1250
Boron trifluoride	C 1	—	1
Butylamine	C 5	—	5

EXCURSION FACTORS

For all substances not bearing C notation

		Excursion Factor	
TLV> 0–1	(ppm or mg/m³),		= 3
TLV> 1–10	"	"	= 2
TLV> 10–100	"	"	= 1.5
TLV> 100–1000	"	"	= 1.25

The number of times the excursion above the TLV is permitted is governed by conformity with the Time-Weighted Average TLV.

INTERPRETATION OF MEASURED PEAK CONCENTRATIONS

With increasing use of rapid, direct-reading analytical instruments for airborne contaminants in the work area, the question of interpretation of essentially "instantaneous" peaks arises. Although no general statement can be made covering all occupational substances, the following guidelines should prove helpful, assuming peak excursions conform to time-weighted average TLV as stated above.

The toxicologic importance of momentary peak concentrations depends on whether the substance is fast or slow acting. If slow acting, as for quartz, lead, or carbon monoxide, momentary peaks are of no toxicologic concern provided, of course, they are not astronomic. On the other hand, fast-acting substances that rapidly produce disabling narcosis, e.g., H_2S, or intolerable irritation or asphyxiation, NH_3, SO_2, CO_2, or initiate sensitization — the organic isocyanates, even "instantaneous" peaks appreciably above the permissible excursion, should not be permitted, unless information exists to the contrary. Other more specific excursions will be developed in the future.

APPENDIX E
Some Nuisance Particulates[p)]

TLV, 30 mppcf or 10mg/m³ of total dust < 1% quartz, or, 5 mg/m³ respirable dust

Aluminum oxide	Limestone
Calcium carbonate	Magnesite
Calcium silicate	Marble
Cellulose (paper fiber)	Mineral Wool Fiber
Portland Cement	Pentaerythritol
Corundum (Al_2O_3)	Plaster of Paris
Emery	Rouge
Glycerin Mist	Silicon
Graphite (synthetic)	Silicon Carbide
Gypsum	Starch
Vegetable oil mists (except castor, cashew nut, or similar irritant oils)	Sucrose
	Tin Oxide
	Titanium Dioxide
	Zinc Stearate
Kaolin	Zinc oxide dust

p) When toxic impurities are not present, e.g. quartz < 1%.

APPENDIX F
Some Simple Asphyxiants[q)]

Acetylene	Hydrogen
Argon	Methane
Butane	Neon
Ethane	Propane
Ethylene	Propylene
Helium	

q) As defined on pg. 14.

APPENDIX G

Calculations for Conversion of Particle Count Concentration (by Standard Light Field — Midget Impinger

Techniques), in mppcf, to Respirable Mass Concentration (by Respirable Sampler) in mg/m³.†

1. In 1967, Jacobsen and Tomb,† derived an empirical relationship of 5.6 mppcf to 1 milligram of respirable dust per cubic meter of air, based on 23 sets of samples, mostly coal dust. The following calculation results in an equivalence of 6.37 mppcf to 1 mg/m³ of respirable dust. Thus, an approximate ratio of 6 mppcf to 1 mg/m³ of respirable dust is suggested for conversion of TLVs from a count to a mass basis when the density and mass median diameter have not been determined.

2. Basic assumptions:
   a) Average density for silica containing dusts $\approx$ 2.5 gms/cm³ (2500 mg/cm³). Pulmonary significant dust densities may vary from 1.2 gm/cm³ for coal dust to 3.1 gms/cm³ for Portland Cement. Silica densities vary from 2.2 (amorphous) to 2.3 (cristobalite and tridymite) to 2.5 (alpha-quartz.) gms per cm³.
   b) The mass median diameter (mmd) of particles collected in midget impinger samplers and counted by the standard light field technique, *and* collected in a respirable sampler is approximately 1.5 μm or 1.5×10^{-4} cm. This assumption is, of course, quite arbitrary since the mmd of all dust clouds is quite variable, depending on many independent parameters, such as source of dust, age of dust cloud, meteorological conditions, etc.

3. Calculation:
   a) vol. per particle: $4/3\ \pi\ r^3$; $r = 0.75 \times 10^{-4}$ cm
   $= 4/3 \cdot \pi \cdot (0.75 \times 10^{-4})^3$
   $= 1.77 \times 10^{-12}$ cm³
   b) wt. per particle = vol. × density
   $= 1.77 \times 10^{-12}$ cm³ $\times 2.5 \times 10^3$ mg/cm³
   $= 4.425 \times 10^{-9}$ mg/particle
   c) 1 particle/ft.³ = 35.5 part./m³
   (since 35.5 cu ft = 1 cu m.)
   10^6 part./ft³ = mppcf = 35.5×10^6 part./m³
   wt. of 1 mppcf $\equiv 35.5 \times 10^6$ part./m³ $\times$ 4.425×10^{-9} mg/part.
   1 mppcf $\equiv$ 0.157 mg/m³
   or
   6.37 mppcf $\equiv$ 1 mg/m³
   or approximately 6 mpccf $\equiv$ 1 mg/m³.

4. Equivalent TLVs in mppcf and mg/m³ (respirable mass) for Mineral Dusts.

Substance	Threshold Limit Value		
	Count mppcf	Resp. Mass mg/m³	Total Mass* mg/m³
Silica (SiO_2)			
Amorphous	20	(3)**	(6)
Cristobalite	1.5	0.05	0.15
Fused silica	3	0.1	0.3
Quartz	3	0.1	0.3
Tridymite	1.5	0.05	0.15
Coal Dust	(12)	2	(4)
Diatomaceous earth, natural	—	1.5	—
Graphite	15	(2.5)	(5)
Mica	20	(3)	(6)
Mineral wool fiber	—	(5)	10
Nuisance particulates	30	(5)	10
Perlite	30	(5)	(10)
Portland Cement	30	(5)	(10)
Soapstone	20	(3)	(6)
Talc (nonasbestiform)	20	(3)	(6)
Tripoli	(3)	0.1	(0.3)

*Unless otherwise specified, respirable mass is presumed to equal approximately 50% of total mass.
**All values in parentheses ( ) represent newly calculated values based on equivalence of 6 mppcf ≡ 1 mg/m³ respirable mass and respirable mass ≡ 50% total mass.

†"Relationship Between Gravimetric Respirable Dust Concentration and Midget Impinger Number Concentration," by Murray Jacobson and T. F. Tomb, AIHAJ, 28: Nov.–Dec. 1967.

TLVs®
Threshold Limit Values for Physical Agents Adopted by ACGIH for 1978

1978 TLV PHYSICAL AGENTS COMMITTEE

Herbert H. Jones, Central Missouri State University, Chairman
Peter A. Breysse, University of Washington
Gerald V. Coles, Dept. of Public Health — Australia
Thomas Cummings, Dept. of Health — Ontario, Canada
Irving H. Davis, Dept. of Health — Michigan
Ronald D. Dobbin, NIOSH
LCDR Joseph J. Drozd, USN
Dr. Allan P. Heins, OSHA
LCDR Richard Johnson, USN
LTC. George S. Kush, USAF
Edward J. Largent, OSHA

William E. Murray, NIOSH
Dr. Wordie H. Parr, NIOSH
David H. Sliney, USAEHA
LTC Robert T. Wangemann, USAEHA
Thomas K. Wilkinson, USPHS-NIH

Any comments or questions regarding these limits should be addressed to:

Executive Secretary
P.O. Box 1937
Cincinnati, Ohio 45201

PREFACE

PHYSICAL AGENTS

These threshold limit values refer to levels of physical agents and represent conditions under which it is believed that nearly all workers may be repeatedly exposed day after day without adverse effect. Because of wide variations in individual susceptibility, exposure of an occasional individual at, or even below, the threshold limit may not prevent annoyance, aggravation of a pre-existing condition, or physiological damage.

These threshold limits are based on the best available information from industrial experience, from experimental human and animal studies, and when possible, from a combination of the three.

These limits are intended for use in the practice of industrial hygiene and should be interpreted and applied only by a person trained in this discipline. They are not intended for use, or for modification for use, (1) in the evaluation or control of the levels of physical agents in the community, (2) as proof or disproof of an existing physical disability, or (3) for adoption by countries whose working conditions differ from those in the United States of America.

These values are reviewed annually by the Committee on Threshold Limits for Physical Agents for revisions or additions, as further information becomes available.

Notice of Intent — At the beginning of each year, proposed actions of the Committee for the forthcoming year are issued in the form of a "Notice of Intent." This notice provides not only an opportunity for comment, but solicits suggestions of physical agents to be added to the list. The suggestions should be accompanied by substantiating evidence.

As Legislative Code — The Conference recognizes that the Threshold Limit Values may be adopted in legislative codes and regulations. If so used, the intent of the concepts contained in the Preface should be maintained and provisions should be made to keep the list current.

Reprint Permission — This publication may be reprinted provided that written permission is obtained from the Executive Secretary of the Conference and that this Preface be published in its entirety along with the Threshold Limit Values.

THRESHOLD LIMIT VALUES

HEAT STRESS

These Threshold Limit Values refer to heat stress conditions under which it is believed that nearly all workers may be repeatedly exposed without adverse health effects. The TLVs shown in Table 1 are based on the assumption that nearly all acclimatized, fully clothed workers with adequate water and salt intake should be able to function effectively under the given working conditions without exceeding a deep body temperature of 38°C (WHO technical report series #412, 1969 *Health Factors Involved in Working Under Conditions of Heat Stress*; F. N. Dukes-Dobos and A. Henschel: *"Development of Permissible Heat Exposure Limits for Occupational Work."* ASHRAE Journal, Vol. 15: No. 9, September 1973, pp. 57–62.)

Since measurement of deep body temperature is impractical for monitoring the workers' heat load, the measurement of environmental factors is required which most nearly correlate with deep body temperature and other physiological responses to heat. At the present time Wet Bulb-Globe Temperature Index (WBGT) is the simplest and most suitable technique to measure the environmental factors. WBGT values are calculated by the following equations:

1. Outdoors with solar load:
   $WBGT = 0.7WB + 0.2GT + 0.1\,DB$
2. Indoors or Outdoors with no solar load:
   $WBGT = 0.7WB + 0.3GT$

where:

WBGT = Wet Bulb-Globe Temperature Index
WB = Natural Wet-Bulb Temperature
DB = Dry-Bulb Temperature
GT = Globe Thermometer Temperature

The determination of WBGT requires the use of a black globe thermometer, a natural (static) wet-bulb thermometer, and a dry-bulb thermometer.

TABLE 1

Permissible Heat Exposure Threshold Limit Values
(Values are given in °C. WBGT)

	Work Load		
Work — Rest Regimen	Light	Moderate	Heavy
Continuous work	30.0	26.7	25.0
75% Work — 25% Rest, Each hour	30.6	28.0	25.9
50% Work — 50% Rest, Each hour	31.4	29.4	27.9
25% Work — 75% Rest, Each hour	32.2	31.1	30.0

Higher heat exposures than shown in Table 1 are permissible if the workers have been undergoing medical surveillance and it has been established that they are more tolerant to work in heat than the average worker. Workers should not be permitted to continue their work when their deep body temperature exceeds 38.0°C.

EVALUATION AND CONTROL

I. *Measurement of the Environment*

The instruments required are a dry-bulb, a natural wet-bulb, a globe thermometer, and a stand. The measurement of the environmental factors shall be performed as follows:

A. The range of the dry and the natural wet bulb thermometer shall be −5°C to 50°C with an accuracy of ±0.5°C. The dry bulb thermometer must be shielded from the sun and the other radiant surfaces of the environment without restricting the airflow around the bulb. The wick of the natural wet-bulb thermometer shall be kept wet with distilled water for at least 1/2 hour before the temperature reading is made. It is not enough to immerse the other end of the wick into a reservoir of distilled water and wait until the whole wick becomes wet by capillarity. The wick shall be wetted by direct application of water from a syringe 1/2 hour before each reading. The wick shall extend over the bulb of the thermometer, covering the stem about one additional bulb length. The wick should always be clean and new wicks should be washed before using.

B. A globe thermometer, consisting of a 15 cm. (6-inch) diameter hollow copper sphere, painted on the outside with a matte black finish or equivalent shall be used. The bulb or sensor of a thermometer (range −5°C to 100°C with an accuracy of ±0.5°C) must be fixed in the center of the sphere. The globe thermometer shall be exposed at least 25 minutes before it is read.

C. A stand shall be used to suspend the three thermometers so that they do not restrict free air flow around the bulbs, and the wet-bulb and globe thermometers are not shaded.

D. It is permissible to use any other type of temperature sensor that gives identical reading as that of a mercury thermometer under the same conditions.

E. The thermometers must be so placed that the readings are representative of the condition where the men work or rest, respectively.

The methodology outlined above is more fully explained in the following publications:

1. "Prevention of Heat Casualties in Marine Corps Recruits, 1955–1960, with Comparative Incidence Rates and Climatic Heat Stresses in other Training Categories," by Captain David Minard, MC, USN, Research Report No. 4, Contract No. MR005.01–0001.01, Naval Medical Research Institute, Bethesda, Maryland, 21 February 1961.

2. "Heat Casualties in the Navy and Marine Corps, 1959–1962, with Appendices on the Field Use of the Wet Bulb-Globe Temperature Index," by Captain David Minard, MC, USN, and R. L. O'Brien, HMC, USN. Research Report No. 7, Contract No. MR 005.01–0001.01, Naval Medical Research Institute, Bethesda, Maryland, 12 March 1964.

3. Minard, D.: Prevention of Heat Casualties in Marine Corps Recruits. Military Medicine *126(4)*: 261–272, 1961.

II. *Work Load Categories*

Heat produced by the body and the environmental heat together determine the total heat load. Therefore, if work is to be performed under hot environmental conditions, the workload category of each job shall be established and the heat exposure limit pertinent to the work load evaluated against the applicable standard in order to protect the worker from exposure beyond the permissible limit.

A. The work load category may be established by ranking each job into light, medium, and heavy categories on the basis of type of operation. Where the work load is ranked into one of said three categories, i.e.

(1) light work (up to 200 Kcal/hr or 800 Btu/hr): e.g., sitting or standing to control machines, performing light hand or arm work,

(2) moderate work (200–350 Kcal/hr or 800–1400 Btu/hr): e.g., walking about with moderate lifting and pushing,

(3) heavy work (350–500 Kcal/hr or 1400–2000 Btu/hr): e.g., pick and shovel work,

the permissible heat exposure limit for that work load shall be determined from Table 1.

B. The ranking of the job may be performed either by measuring the worker's metabolic rate while performing his job or by estimating his metabolic rate by the use of the scheme shown in Table 2. Tables available in the literature listed below and in other publications as well may also be utilized. When this method is used the permissible heat exposure limit can be determined by Figure 1.

1. Per-Olaf Astrand and Kaare Rodahl: "Textbook of Work Physiology" McGraw-Hill Book Company, New York, San Francisco, 1970.

2. "Ergonomies Guide to Assessment of Metabolic and Cardiac Costs of Physical Work." Amer. Ind. Hyg. Assoc. J. *32:* 560, 1971.

3. Energy Requirements for Physical Work. Purdue Farm Cardiac Project. Agricultural Experiment Station. Research Progress Report No. 30, 1961.

4. J. V. G. A. Durnin and R. Passmore: "Energy, Work and Leisure." Heinemann Educational Books, Ltd., London, 1967.

TABLE 2

Assessment of Work Load

Average values of metabolic rate during different activities.

A. Body position and movement	Kcal./min.
Sitting	0.3
Standing	0.6
Walking	2.0–3.0
Walking up hill	add 0.8 per meter (yard) rise

B. Type of Work		Average Kcal./min.	Range Kcal./min.
Hand work			
	light	0.4	0.2–1.2
	heavy	0.9	
Work with one arm			
	light	1.0	0.7–2.5
	heavy	1.8	
Work with both arms			
	light	1.5	1.0–3.5
	heavy	2.5	
Work with body			
	light	3.5	2.5–15.0
	moderate	5.0	
	heavy	7.0	
	very heavy	9.0	

Light hand work: writing, hand knitting

Heavy hand work: typewriting

Heavy work with one arm: hammering in nails (shoemaker, upholsterer)

Light work with two arms: filing metal, planing wood, raking of a garden

Moderate work with the body: cleaning a floor, beating a carpet

Heavy work with the body: railroad track laying, digging, barking trees

Sample Calculation: Using a heavy hand tool on an assembly line

A. Walking along	2.0 Kcal./min.
B. Intermediate value between heavy work with two arms and light work with the body	3.0 Kcal./min.
	5.0 Kcal./min.
C. Add for basal metabolism	1.0 Kcal./min.
Total	6.0 Kcal./min.

Adapted from Lehmann, G. E., A. Muller and H. Spitzer: Der Kalorienbedarf bei gewerblicher Arbeit. Arbeitsphysiol. *14:* 166, 1950.

III. *Work-Rest Regimen*

The permissible exposure limits specified in Table 1 and Figure 1 are based on the assumption that the WBGT value of the resting place is the same or very close to that of the work place. Where the WBGT of the work area is different from that of the rest area a time-weighted average value should be used for both environmental and metabolic heat. When time-weighted average values are used the appropriate curve on Figure 1 is the solid line labeled "continuous."

The time-weighted average metabolic rate (M) shall be determined by the equation:

$$\text{Av. M} = \frac{(M_1) \times (t_1) + (M_2) \times (t_2) + \ldots + (M_n) \times (t_n)}{(t_1) + (t_2) + \ldots + (t_n)}$$

Where M_1 M_2, M_n are estimated or measured metabolic rates for the various activities of the worker during the total time period. t_1, t_2, t_n are the elapsed times in minutes spent at the corresponding metabolic rate as determined by a time study.

The time-weighted average WBGT shall be determined by the equation:

$$\text{Av. WBGT} = \frac{(\text{WBGT}_1) \times (t_1) + (\text{WBGT}_2) \times t_2 + \ldots + (\text{WBGT}_n) \times (t_n)}{(t_1) + (t_2) + \ldots + (t_n)}$$

where WBGT_1, WBGT_2, WBGT_n are calculated values of WBGT for the various work and rest areas occupied during total time periods. t_1, t_2, t_n are the elapsed times in minutes spent in the corresponding areas which are determined by a time study. Where exposure to hot environmental conditions is continuous for several hours or the entire work day, the time-weighted averages shall be calculated as hourly time-weighted average i.e., $t_1 + t_2 + \ldots t_n = 60$ minutes. Where the exposure is intermittent, the time-weighted averages shall be calculated as two-hour time-weighted averages, i.e., $t_1 + t_2 + \ldots t_n = 120$ minutes.

The permissible exposure limits for continuous work are applicable where there is a work-rest regimen of a 5-day work week and an 8-hour work day with a short morning and afternoon break (approximately 15 minutes) and a longer lunch break (approximately 30 minutes). Higher exposure limits are permitted if additional resting time is allowed. All breaks, including unscheduled pauses and administrative or operational waiting periods during work may be counted as rest time when additional rest allowance must be given because of high environmental temperatures.

It is a common experience that when the work on a job is self-paced, the workers will spontaneously limit their hourly work load to 30–50% of their maximum physical performance capacity. They do this either by setting an appropriate work speed or by interspersing unscheduled breaks. Thus the daily average of the workers' metabolic rate seldom exceeds 330 kcal/hr. However, within an 8-hour work shift there may be periods where the workers' hourly average metabolic rate will be higher.

IV. *Water and Salt Supplementation*

During the hot season or when the worker is exposed to artificially generated heat, drinking water shall be made available to the workers in such a way that they are stimulated to frequently drink small amounts, i.e., one cup every 15–20 minutes (about 150 ml or 1/4 pint).

The water shall be kept reasonably cool (10°–15°C or 50.0°–60.0°F) and shall be placed close to the workplace so that the worker can reach it without abandoning the work area.

The workers should be encouraged to salt their food abundantly during the hot season and particularly during hot spells. If the workers are unacclimatized, salted drinking water shall be made available in a concentration of 0.1% (1g NaCl to 1.0 liter or 1 level tablespoon of salt to 15 quarts of water). The added salt shall be complete-

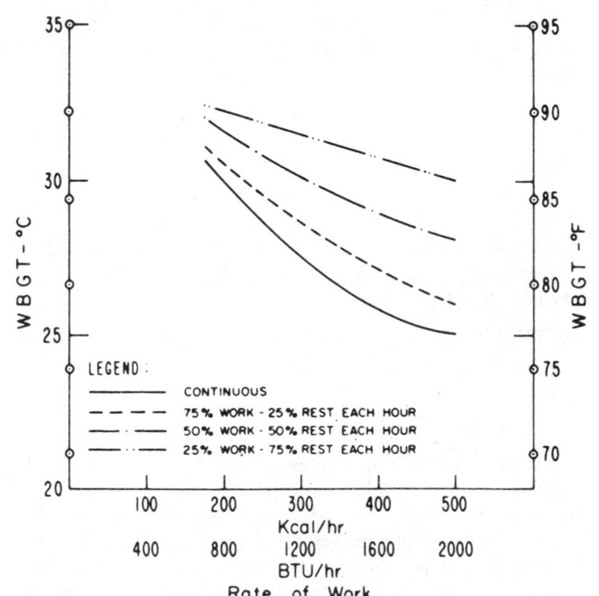

Figure 1 — Permissible Heat Exposure Threshold Limit Value

ly dissolved before the water is distributed, and the water shall be kept reasonably cool.

V. *Other Considerations*

A. Clothing: The permissible heat exposure TLVs are valid for light summer clothing as customarily worn by workers when working under hot environmental conditions. If special cothing is required for performing a particular job and this clothing is heavier or it impedes sweat evaporation or has higher insulation value, the worker's heat tolerance is reduced, and the permissible heat exposure limits indicated in Table 1 and Figure 1 are not applicable. For each job category where special clothing is required, the permissible heat exposure limit shall be established by an expert.

B. Acclimatization and Fitness: The recommended heat stress TLVs are valid for acclimated workers who are physically fit.

IONIZING RADIATION

The recommendations of the National Council on Radiation Protection and Measurements (NCRP) are suggested as threshold limit values to which nearly all workers may be exposed without adverse effects. The limits should be used as guides for the control of exposure and should not be regarded as fine lines between safe and dangerous levels. The underlying philosophy of radiation protection is to keep all exposures as low as reasonably achievable.

The two basic reference documents are as follows:

a. "Basic Radiation Protection Criteria," NCRP Report No. 39, issued January 15, 1971.

b. "Maximum Permissible Body Burdens and Maximum Permissible Concentrations of Radionuclides in Air and in Water for Occupational Exposure," US Department of Commerce, National Bureau of Standards Handbook 69, issued June 5, 1959, with Addendum 1 issued August 1963. Available as NCRP Report No. 22.

The above documents, as well as information on numerous other NCRP Reports addressing specific subjects in ionizing radiation protection are available from: NCRP Publications, 7910 Woodmont Ave., Washington, DC 20014.

LASERS

The threshold limit values are for exposure to laser radiation under conditions to which nearly all workers may be exposed without adverse effects. The values should be used as guides in the control of exposures and should not be regarded as fine lines between safe and dangerous levels. They are based on the best available information from experimental studies.

Limiting Apertures

The TLVs expressed as radiant exposure or irradiance in this section may be averaged over an aperture of 1 mm except for TLVs for the eye in the spectral range of 400–1400 nm, which should be averaged over a 7 mm limiting aperture (pupil); and except for all TLVs for wavelengths between 0.1–1 mm where the limiting aperture is 10 mm. No modification of the TLVs is permitted for pupil sizes less than 7 mm.

The TLVs for "extended sources" apply to sources which subtend an angle greater than α (Table 5) which varies with exposure time. This angle is *not* the beam divergence of the source.

Correction Factors A and B (C_A and C_B)

The TLVs for ocular exposure in Tables 3 and 4 are to be used as given for all wavelength ranges. The TLVs for wavelengths between 700 nm and 1049 nm are to be increased by a uniformly extrapolated factor (C_A) as shown in Figure 2. Between 1049 nm and 1400 nm, the TLV has been increased by a factor (C_A) of five. For certain exposure times at wavelengths between 550 nm and 700 nm, correction factor (C_B) must be applied.

The TLVs for skin exposure are given in Table 6. The TLVs are to be increased by a factor (C_A) as shown in Figure 2 for wavelengths between 700 nm and 1400 nm. To aid in the determination of TLVs for exposure durations requiring calculations of fractional powers Figures 3, 4, 5 and 6 may be used.

Repetitively Pulsed Lasers

Since there are few experimental data for multiple pulses, caution must be used in the evaluation of such exposures. The protection standards for irradiance or radiant exposure in multiple pulse trains have the following limitations:

(1) The exposure from any single pulse in the train is limited to the protection standard for a single comparable pulse.

(2) The average irradiance for a group of pulses is limited to the protection standard as given in Tables 3, 4, or 6 of a single pulse of the same duration as the entire pulse group.

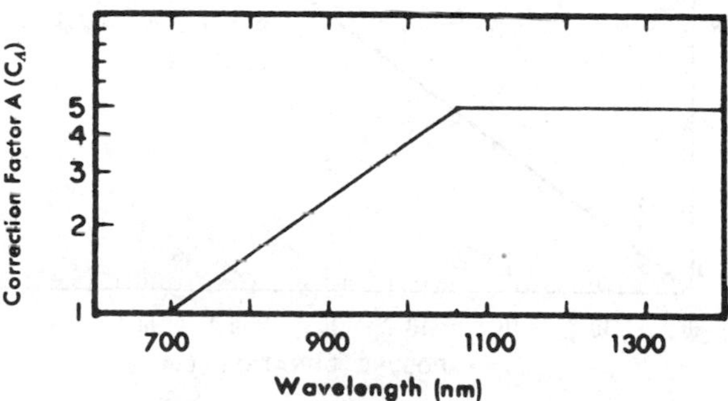

Figure 2 — TLV correction factor for λ = 700 – 1400 nm*

*For λ = 700 – 1049 nm, $C_A = 10^{[0.002(\lambda - 700)]}$
For λ = 1050 – 1400 nm, C_A = 5

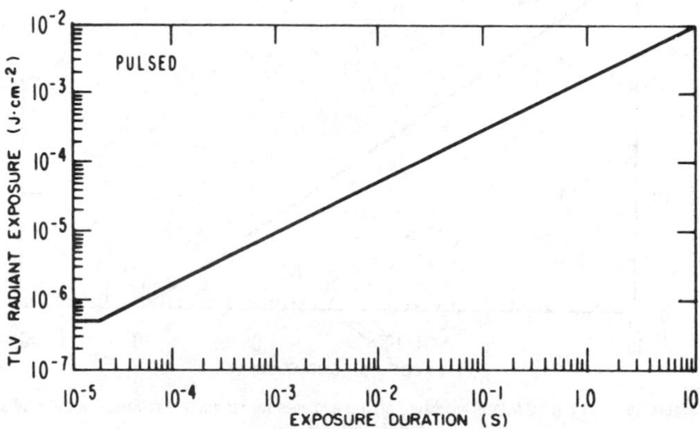

Figure 3a — TLV for intrabeam (direct) viewing of laser beam (400–700 nm).

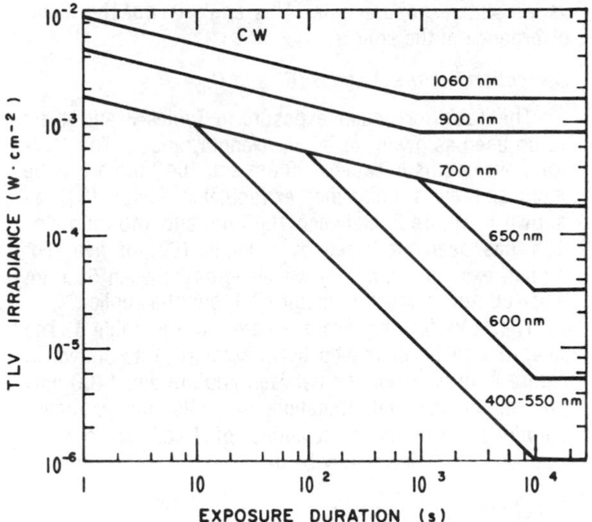

Figure 3b

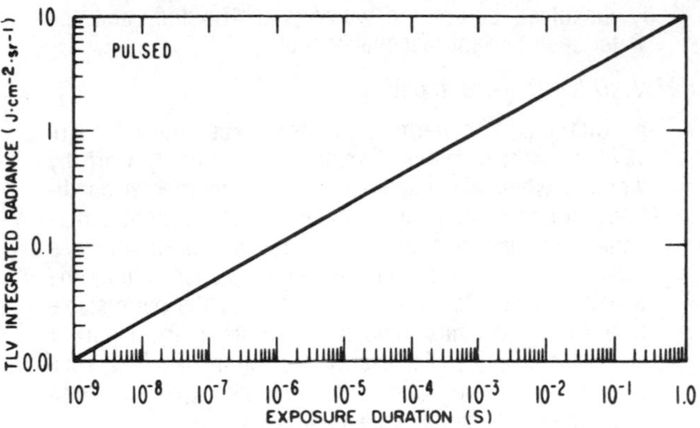

Figure 5a — TLV for extended sources or diffuse reflections of laser radiation (400–700 nm).

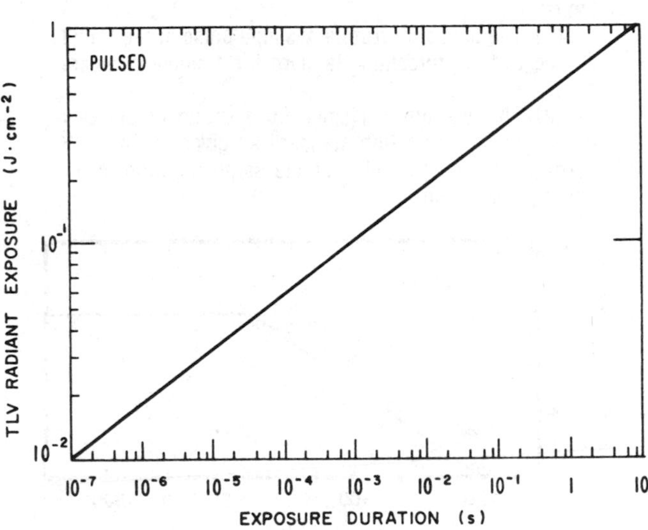

Figure 4a

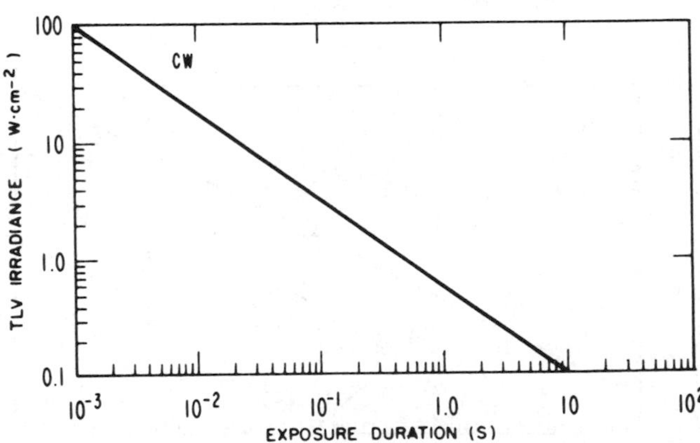

Figure 4b — TLV for CW laser exposure of skin and eyes for far-infrared radiation (wavelengths greater than 1.4 μm).

TABLE 3

Threshold Limit Value for Direct Ocular Exposures (Intrabeam Viewing) from a Laser Beam

Spectral Region	Wave Length	Exposure Time, (t) Seconds	TLV
UVC	200 nm to 280 nm	10^{-3} to 3×10^4	3 mJ • cm^{-2}
UVB	280 nm to 302 nm	"	3 "
	303 nm	"	4 "
	304 nm	"	6 "
	305 nm	"	10 "
	306 nm	"	16 "
	307 nm	"	25 "
	308 nm	"	40 "
	309 nm	"	63 "
	310 nm	"	100 "
	311 nm	"	160 "
	312 nm	"	250 "
	313 nm	"	400 "
	314 nm	"	630 "
UVA	315 nm to 400 nm	10^{-9} to 10	$.56\, t^{1/4}$ J • cm^{-2}
	" "	10 to 10^3	1.0 J • cm^{-2}
	" "	10^3 to 3×10^4	1.0 mW • cm^{-2}
Light	400 nm to 700 nm	10^{-9} to 1.8×10^{-5}	5×10^{-7} J • cm^{-2}
	400 nm to 700 nm	1.8×10^{-5} to 10	1.8 $(t/\sqrt[4]{t})$ mJ • cm^{-2}
	400 nm to 549 nm	10 to 10^4	10 mJ • cm^{-2}
	550 nm to 700 nm	10 to T_1	1.8 $(t/\sqrt[4]{t})$ mJ • cm^{-2}
	550 nm to 700 nm	T_1 to 10^4	10 C_B mJ • cm^{-2}
	400 nm to 700 nm	10^4 to 3×10^4	C_B μW • cm^{-2}
IR-A	700 nm to 1049 nm	10^{-9} to 1.8×10^{-5}	5 $C_A \times 10^{-7}$ J • cm^{-2}
	700 nm to 1049 nm	1.8×10^{-5} to 10^3	1.8 C_A $(t/\sqrt[4]{t})$ mJ • cm^{-2}
	1050 nm to 1400 nm	10^{-9} to 10^{-4}	5×10^{-6} J • cm^{-2}
	1050 nm to 1400 nm	10^{-4} to 10^3	9$(t/\sqrt[4]{t})$ mJ • cm^{-2}
	700 nm to 1400 nm	10^3 to 3×10^4	320 C_A μW • cm^{-2}
IR-B & C	1.4 μm to 10^3 μm	10^{-9} to 10^{-7}	10^{-2} J • cm^2
	" "	10^{-7} to 10	0.56 $\sqrt[4]{t}$ J • cm^{-2}
	" "	10 to 3×10^4	0.1 W • cm^{-2}

C_A – See Fig. 2.
C_B = 1 for λ = 400 to 549 nm; $C_B = 10^{[0.015 (\lambda - 550)]}$ for λ = 550 to 700 nm.
T_1 = 10 s for λ = 400 to 549 nm; $T_1 = 10 \times 10^{[0.02 (\lambda - 550)]}$ for λ = 550 to 700 n.

TABLE 4

Threshold Limit Values for Viewing a Diffuse Reflection of a Laser Beam or an Extended Source Laser

Spectral Region	Wave Length	Exposure Time, (t) Seconds	TLV
UV	200 nm to 400 nm	10^{-3} to 3×10^4	Same as Table 3
Light	400 nm to 700 nm	10^{-9} to 10	10 $\sqrt[3]{t}$ J • cm^{-2} • sr^{-1}
	400 nm to 549 nm	10 to 10^4	21 J • cm^{-2} • sr^{-1}
	550 nm to 700 nm	10 to T_1	3.83 $(t/\sqrt[4]{t})$ J • cm^{-2} • sr^{-1}
	550 nm to 700 nm	T_1 to 10^4	21/C_B J cm^{-2} • sr^{-1}
	400 nm to 700 nm	10^4 to 3×10^4	2.1/$C_B \times 10^{-3}$ W • cm^{-2} • sr^{-1}
IR-A	700 nm to 1400 nm	10^{-9} to 10	10 C_A $\sqrt[3]{t}$ J • cm^{-2} • sr^{-1}
	700 nm to 1400 nm	10 to 10^3	3.83 C_A $(t/\sqrt[4]{t})$ J • cm^{-2} • sr^{-1}
	700 nm to 1400 nm	10^3 to 3×10^4	0.64 C_A W • cm^{-2} • sr^{-1}
IR-B & C	1.4 μm to 1 mm	10^{-9} to 3×10^4	Same as Table 3

C_A, C_B, and T_1 are the same as in footnote to Table 3.

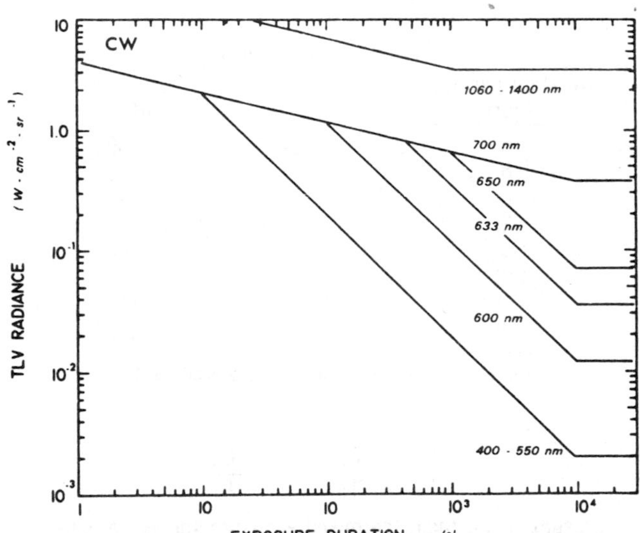

Figure 5b — TLV for intrabeam (direct) viewing of CW laser beam (400–1400 nm).

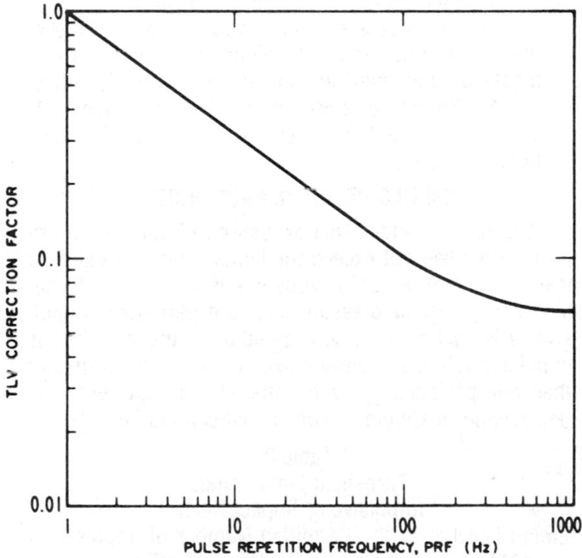

Figure 6 — Multiplicative correction factor for repetitively pulsed lasers having pulse durations less than 10^{-5} second. TLV for a single pulse of the pulse train is multiplied by the above correction factor. Correction factor for PRF greater than 1000 H_z is 0.06.

(3) When the Instantaneous Pulse Repetition Frequency (PRF) of any pulses within a train exceeds one, the protection standard applicable to each pulse is reduced as shown in Figure 6 for pulse durations less than 10^{-5} second. For pulses of greater duration, the following formula should be followed:

$$\text{Standard}\begin{pmatrix}\text{single pulse}\\ \text{in train}\end{pmatrix} = \frac{\text{Standard (pulse } n\tau)}{n}$$

where:

n = number of pulses in train

τ = duration of a single pulse in the train

Standard ($n\tau$) – protection standard of one pulse having a duration equal to $n\tau$ seconds.

TABLE 5

Limiting Angle to Extended Source Which May Be Used for Applying Extended Source TLVs

Exposure Duration(s)	Angle α (mrad)
10^{-9}	8.0
10^{-8}	5.4
10^{-7}	3.7
10^{-6}	2.5
10^{-5}	1.7
10^{-4}	2.2
10^{-3}	3.6
10^{-2}	5.7
10^{-1}	9.2
1.0	15
10	24
10^{2}	24
10^{3}	24
10^{4}	24

TABLE 6

Threshold Limit Value for Skin Exposure from a Laser Beam

Spectral Region	Wave Length	Exposure Time, (t) Seconds	TLV
UV	200 nm to 400 nm	10^{-3} to 3×10^{4}	Same as Table 3
Light &	400 nm to 1400 nm	10^{-9} to 10^{-7}	$2\,C_A \times 10^{-2}\,J \bullet cm^{-2}$
IR-A	" "	10^{-7} to 10	$1.1\,C_A \sqrt[4]{t}\,J \bullet cm^{-2}$
IR-B & C	1.4 μm to 1 mm	10^{-9} to 3×10^{4}	Same as Table 3

C_A = 1.0 for λ = 400–700 nm; see Figure 2 Laser TLV list for greater wavelength values.

NOTE: To aid in the determination of TLV's for exposure durations requiring calculations of fractional powers Figures 3, 4, 5 and 6 may be used.

MICROWAVES

These Threshold Limit Values refer to microwave energy in the frequency range of 300 MHz to 300 GHz and represent conditions under which it is believed that nearly all workers may be repeatedly exposed without adverse effect.

Under conditions of moderate to severe heat stress, the recommended values may need to be reduced.* Therefore, these values should be used as guides in the control of exposure to microwave energy and should not be regarded as a fine line between safe and dangerous levels.

Recommended Values:

The Threshold Limit Value for occupational exposure to microwave energy, where power density or field intensity is known and exposure time is controlled, is as follows:

1. For exposure to continuous wave (CW) sources, the power density level shall not exceed 10 milliwatts per square centimeter (mW/cm^2) for continuous exposure, and the total exposure time shall be limited to an 8-hour workday. This power density is approximately equivalent to a free-space electric field

*Mumford, W.W., "Heat Stress Due to R. F. Radiation," Proceedings of IEEE, Vol. 57, No. 2, Feb. 1969, pp. 171–178.

strength of 200 volts-per-meter rms (V/m) and a free-space magnetic field strength of 0.5 ampere-per-meter rms (A/m).

2. Exposures to CW power density levels greater than 10 mW/cm² are permissible up to a maximum of 25 mW/cm² based upon an average energy density of 1 milliwatt-hour per square centimeter (mWh/cm²) averaged over any 0.1 hour period. For example, at 25 mW/cm², the permissible exposure duration is approximately 2.4 minutes in any 0.1 hour period.
3. For repetitively pulsed microwave sources, the average field strength or power density is calculated by multiplying the peak-pulse value by the duty cycle. The duty cycle is equal to the pulse duration in seconds times the pulse repetition rate in Hertz. Exposure during an 8-hour workday shall not exceed the following values which are averaged over any 0.1 hour period:

Power Density	10 mW/cm²
Energy Density	1 mWh/cm²
Mean Squared Electric Field Strength	40,000 V²/m²
Mean Squared Magnetic Field Strength	0.25 A²/m²

4. Exposure is not permissible in CW or repetitively pulsed fields with an average power density in excess of 25 mW/cm² or approximate equivalent free-space field strengths of 300 V/m or 0.75 A/m.

NOISE

These threshold limit values refer to sound pressure levels and durations of exposure that represent conditions under which it is believed that nearly all workers may be repeatedly exposed without adverse effect on their ability to hear and understand normal speech. The medical profession has defined hearing impairment as an average hearing threshold level in excess of 25 decibels (ANSI-S3.6-1969) at 500, 1000, and 2000 Hz, and the limits which are given have been established to prevent a hearing loss in excess of this level. The values should be used as guides in the control of noise exposure and, due to individual susceptibility, should not be regarded as fine lines between safe and dangerous levels.

It should be recognized that the application of the TLV for noise will not protect all workers from the adverse effects of noise exposure. A hearing conservation program with audiometric testing is necessary when workers are exposed to noise at or above the TLV levels.

Continuous or Intermittent

The sound level shall be determined by a sound level meter, conforming as a minimum to the requirements of the American National Standard Specification for Sound Level Meters, S1.4 (1971) Type S2A, and set to use the A-weighted network with slow meter response. Duration of exposure shall not exceed that shown in Table 7.

These values apply to total duration of exposure per working day regardless of whether this is one continuous exposure or a number of short-term exposures but does not apply to impact or impulsive type of noise.

When the daily noise exposure is composed of two or more periods of noise exposure of different levels, their combined effect should be considered, rather than the individual effect of each. If the sum of the following fractions:

Table 7
Threshold Limit Values

Duration per day Hours	Sound Level dBA[a]
16	80
8	85
4	90
2	95
1	100
1/2	105
1/4	110
1/8	115*

*No exposure to continuous or intermittent in excess of 115 dBA

$$\frac{C_1}{T_1} + \frac{C_2}{T_2} + \ldots \frac{C_n}{T_n}$$

exceeds unity, then, the mixed exposure should be considered to exceed the threshold limit value, C_1 indicates the total duration of exposure at a specific noise level, and T_1 indicates the total duration of exposure permitted at that level. All on-the-job noise exposures of 80 dBA or greater shall be used in the above calculations.

a) Sound level in decibels as measured on a sound level meter, conforming as a minimum to the requirements of the American National Standard Specification for Sound Level Meters, S1.4 (1971) Type S2A, and set to use the A-weighted network with slow meter response.

IMPULSIVE OR IMPACT NOISE

It is recommended that exposure to impulsive or impact noise shall not exceed the limits listed in Table 8 or taken from Figure 7. No exposures in excess of 140 decibels peak sound pressure level are permitted. Impulsive or impact noise is considered to be those variations in noise levels that involve maxima at intervals of greater than one per second. Where the intervals are less than one second, it should be considered continuous.

Table 8
Threshold Limit Values
Impulsive or Impact Noise

Sound Level dB**	Permitted Number of Impulses or Impacts per day
140	100
130	1000
120	10,000

**Decibels peak sound pressure level.

ULTRAVIOLET RADIATION*

These threshold limit values refer to ultraviolet radiation in the spectral region between 200 and 400 nm and represent conditions under which it is believed that nearly all workers may be repeatedly exposed without adverse effect. These values for exposure of the eye or the skin apply to ultraviolet radiation from arcs, gas, and vapor discharges, fluorescent, and incandescent sources, and solar radiation, but do not apply to ultraviolet lasers.* These values do not apply to ultraviolet radiation exposure of photosensitive individuals or of indi-

*See Laser TLVs.

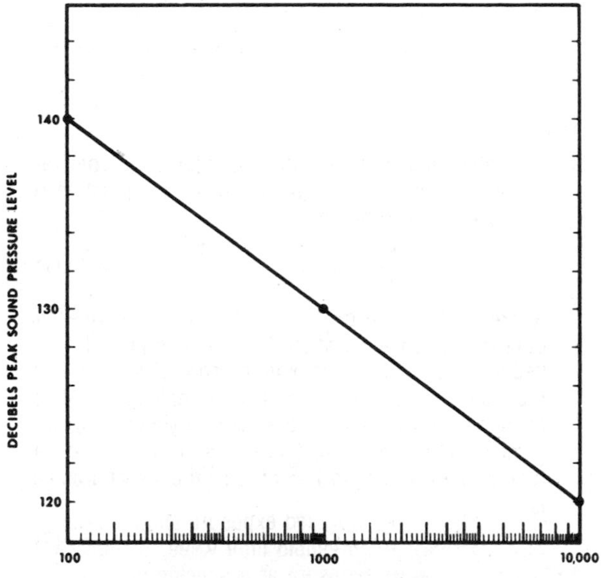

Figure 7 — Threshold Limit Values for Impulse/Impact Noise.

viduals concomitantly exposed to photosensitizing agents (Fitzpatrick, et al., eds., Sunlight and Man, Univ. Tokyo Press, Tokyo, Japan, 1974). These values should be used as guides in the control of exposure to continuous sources where the exposure duration shall not be less than 0.1 sec.

These values should be used as guides in the control of exposure to ultraviolet sources and should not be regarded as a fine line between safe and dangerous levels.

Recommended Values:

The threshold limit value for occupational exposure to ultraviolet radiation incident upon skin or eye where irradiance values are known and exposure time is controlled are as follows:

1. For the near ultraviolet spectral region (320 to 400 nm) total irradiance incident upon the unprotected skin or eye should not exceed 1 mw/cm² for periods greater than 10^3 seconds (approximately 16 minutes) and for exposure times less than 10^3 seconds should not exceed one J/cm².
2. For the actinic ultraviolet spectral region (200 — 315 nm), radiant exposure incident upon the unprotected skin or eye should not exceed the values given in Table 9 within an 8-hour period.
3. To determine the effective irradiance of a broadband source weighted against the peak of the spectral effectiveness curve (270 nm), the following weighting formula should be used:

$$E_{eff} = \sum E_\lambda S_\lambda \Delta\lambda$$

where:

E_{eff} = effective irradiance relative to a monochromatic source at 270 nm in W/cm² (J/s/cm²)

E_λ = spectral irradiance in W/cm²/nm

S_λ = relative spectral effectiveness (unitless)

$\Delta\lambda$ = band width in nanometers

TABLE 9

Relative Spectral Effectiveness by Wavelength

Wavelength (nm)	TLV (mJ/cm²)**	Relative Spectral Effectiveness S_λ
200	100	0.03
210	40	0.075
220	25	0.12
230	16	0.19
240	10	0.30
250	7.0	0.43
254	6.0	0.5
260	4.6	0.65
270	3.0	1.0
280	3.4	0.88
290	4.7	0.64
300	10	0.30
305	50	0.06
310	200	0.015
315	1000	0.003

**1 m J/cm² = 10^{-3} J/cm²

4. Permissible exposure time in seconds for exposure to actinic ultraviolet radiation incident upon the unprotected skin or eye may be computed by dividing 0.003 J/cm² by E_{eff} in W/cm². The exposure time may also be determined using Table 10 which provides exposure times corresponding to effective irradiances in μW/cm².

TABLE 10

Permissible Ultraviolet Exposures

Duration of Exposure Per Day	Effective Irradiance, E_{eff} (μW/cm²)***
8 hrs.	0.1
4 hrs.	0.2
2 hrs.	0.4
1 hr.	0.8
30 min.	1.7
15 min.	3.3
10 min.	5
5 min.	10
1 min.	50
30 sec.	100
10 sec.	300
1 sec.	3,000
0.5 sec.	6,000
0.1 sec.	30,000

***1μW/cm² = 10^{-6} W/cm²

All the preceding TLVs for ultraviolet energy apply to sources which subtend an angle less than 80°. Sources which subtend a greater angle need to be measured only over an angle of 80°.

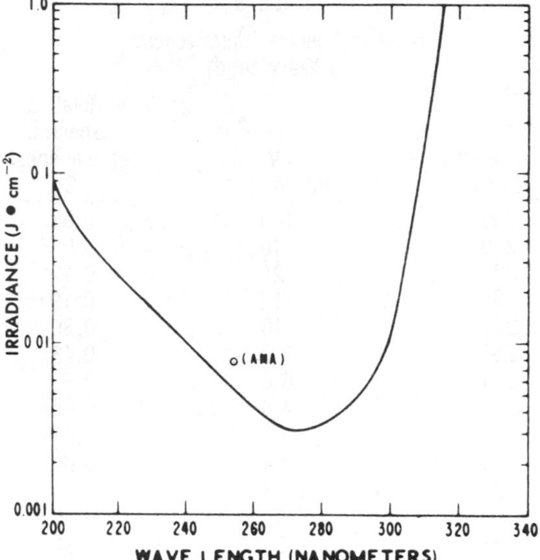

Figure 8 — Threshold Limit Values for Ultraviolet Radiation

Conditioned (tanned) individuals can tolerate skin exposure in excess of the TLV without erythemal effects. However, such conditioning may not protect persons against skin cancer.

NOTICE OF INTENDED CHANGES (for 1978)

These physical agents, with their corresponding values, comprise those for which either a limit has been proposed for the first time, or for which a change in the "Adopted" listing has been proposed. In both cases, the proposed limits should be considered trial limits that will remain in the listing for a period of at least one year. If after one year no evidence comes to light that questions the appropriateness of the values herein the values will be reconsidered for the "Adopted" list.

NOTICE OF INTENT TO ESTABLISH THRESHOLD LIMIT VALUES

LIGHT AND NEAR-INFRARED RADIATION

These Threshold Limit Values refer to visible and near-infrared radiation in the wavelength range of 400 nm to 1400 nm and represent conditions under which it is believed that nearly all workers may be exposed without adverse effect. These values should be used as guides in the control of exposure to light and should not be regarded as a fine line between safe and dangerous levels.

Recommended Values:

The Threshold Limit Value for occupational exposure to broad-band light and near-infrared radiation for the eye apply to exposure in any eight-hour workday and require knowledge of the spectral radiance (L_λ) and total irradiance (E) of the source as measured at the position(s) of the eye of the worker. Such detailed spectral data of a white light source is generally only required if the luminance of the source exceeds 1 cd cm^{-2}. At luminances less than this value the TLV would not be exceeded.

The TLV's are:

1. To protect against retinal thermal injury, the spectral radiance of the lamp weighted against the function R (Table 11) should not exceed:

$$\sum_{400}^{1400} L_\lambda R_\lambda \Delta\lambda \leq 1/\alpha t^{1/2} \quad (1)^*$$

where L_λ is in W cm^{-2} sr^{-1} and t is the viewing duration (or pulse duration if the lamp is pulsed) limited to 1 μs to 10 s, and α is the angular subtense of the source in radians. If the lamp is oblong, α refers to the longest dimension that can be viewed. For instance, at a viewing distance r = 100 cm from a tubular lamp of length l = 50 cm, the viewing angle is:

$$\alpha = l/r = 50/100 = 0.5 \text{ rad} \quad (2)$$

2. To protect against retinal photochemical injury from chronic blue-light exposure the integrated spectral radiance of the lamp weighted against the blue-light hazard function B_λ (Table 11) should not exceed:

$$\sum_{400}^{1400} L_\lambda t B_\lambda \Delta\lambda \leq 100 \text{ Jcm}^{-2} \text{ sr}^{-1} \ (t \leq 10^4 \text{s}) \quad (3a)$$

$$\sum_{400}^{1400} L_\lambda B_\lambda \Delta\lambda \leq 10^{-2} \text{ Wcm}^{-2} \text{ sr}^{-1} \ (t > 10^4 \text{s}) \quad (3b)$$

For a source radiance L which exceeds 2 mW cm^{-2} sr^{-1} in the blue spectral region, the permissible exposure duration t_{max} in seconds is simply:

$$t_{max} = 100 \text{ J cm}^{-2} \text{ sr}^{-1}/L \text{ (blue)} \quad (4)$$

The latter limits are greater than the maximum permissible exposure limits for 440 nm laser radiation (see Laser TLV) because a 2–3 mm pupil is assumed rather than a 7 mm pupil for the Laser TLV.

3. *Infrared Radiation:* To avoid possible delayed effects upon the lens of the eye (cataractogenesis), the infrared radiation ($\lambda > 770$ nm) should be limited to 10 $mWcm^{-2}$. For an infrared heat lamp or any near-infrared source where a strong visual stimulus is absent, the near infrared (770–1400 nm) radiance as viewed by the eye should be limited to:

$$\sum_{770}^{1400} L_\lambda \Delta\lambda = 0.6/\alpha \quad (5)^*$$

for extended duration viewing conditions. This limit is based upon a 7 mm pupil diameter.

AIRBORNE UPPER SONIC AND ULTRASONIC ACOUSTIC RADIATION

These threshold limit values refer to sound pressure levels that represent conditions under which it is believed that nearly all workers may be repeatedly exposed without adverse effect. The values listed in Table 15 should be used as guides in the control of noise expo-

*Formulae (1) and (2) are empirical and are not, strictly speaking, dimensionally correct. To make the formulae dimensionally correct, one would have to insert a dimensional correction factor k in the right hand numerator in each formula. For formula (1) this would be k_1 = 1 W • rad • $s^{1/2}/(cm^2 \bullet sr)$, and for formula (5) k_2 = 1 W • rad/(cm^2 • sr).

sure and, due to individual susceptibility, should not be regarded as fine lines between safe and dangerous levels. The levels for the third octave bands centered below 20 kHz are below those which cause subjective effects. Those levels for 1/3 octaves above 20 kHz are for prevention of possible hearing losses from subharmonics of these frequencies.

TABLE 11

SPECTRAL WEIGHTING FUNCTIONS FOR ASSESSING RETINAL HAZARDS FROM BROAD — BAND OPTICAL SOURCES

Wavelength (nm)	Blue-Light Hazard Function B_λ	Burn Hazard Function R_λ
400	0.10	1.0
405	0.20	2.0
410	0.40	4.0
415	0.80	8.0
420	0.90	9.0
425	0.95	9.5
430	0.98	9.8
435	1.0	10
440	1.0	10
445	0.97	9.7
450	0.94	9.4
455	0.90	9.0
460	0.80	8.0
465	0.70	7.0
470	0.62	6.2
475	0.55	5.5
480	0.45	4.5
485	0.40	4.0
490	0.22	2.2
495	0.16	1.6
500–600	$10^{[(450-\lambda)/50]}$	1.0
600–700	0.001	1.0
700–1049	0.001	$10^{[(700-\lambda)/505]}$
1060–1400	0.001	0.2

TABLE 12

Permissible Ultrasound Exposure Levels

Mid-Frequency of Third-Octave Band kHz	One-Third Octave — Band Level in dB reference 0.0002 dynes/cm^2
10	80
12.5	80
16	80
20	105
25	110
31.5	115
40	115
50	115

PHYSICAL AGENTS UNDER STUDY

These agents comprise those which the Physical Agents Committee of ACGIH proposes to study during this year to determine the feasibility of establishing proposed TLVs in 1979. Comments and suggestions, accompanied by substantitive evidence, are solicited.

1. *Radiofrequency Radiation*. Specifically, that portion of the spectrum from 10 MHz to 100 MHz.
2. *Extremely Low Frequency (ELF) Radiation*. Specifically, that portion of the spectrum from 0 to 300 Hz.
3. *Magnetic Fields*. Both pulsed and continuous.
4. *Laser Radiation*. Specifically ultraviolet radiation for pulsed exposures, and repetitively pulsed light and infrared-A laser exposures.
5. *Ultrasonic Energy*. Specifically, acoustic energy at frequencies above 10 kHz.
6. *Vibration*. Segmental and whole-body.
7. *Cold Stress*.
8. *Pressure Variations*.

The American Conference of Governmental Industrial Hygienists was organized in 1938 by a group of governmental industrial hygienists who desired a medium for the free exchange of ideas, experiences and the promotion of standards and techniques in industrial health. The Conference is not an official Government Agency.

It is an organization devoted to the development of administrative and technical aspects of worker health protection. The association has contributed substantially to the development and improvement of official industrial health services to industry and labor. The committees on Industrial Ventilation and Threshold Limit Values are recognized throughout the world for their expertise and contributions to industrial hygiene.

Membership is limited to professional personnel in governmental agencies or educational institutions engaged in occupational safety and health programs. The more than 1800 members from across the United States and around the world give the organization an international scope.

SECTION 2
Industrial Air Contaminant Control

Benjamin Feiner
Environmental Consultant
New York, New York
Former Head, Environmental Control Unit, Engineering Section
Division of Industrial Hygiene
New York State Labor Department

When materials are so used industrially that air contaminants are created, generated, or released in concentrations which may injure the health of workers, the usual method of providing protection is by means of ventilation, usually local exhaust. There are, however, other methods of protection which should be investigated, even before consideration of ventilation control.

CONTROL BY METHODS OTHER THAN VENTILATION

The basic principle of industrial hygiene is to prevent or minimize exposure of workers to dangerous materials (see Section 1).

While ventilation control is the most widely used method of achieving this protection, a number of other, simpler procedures are available. These can often actually effect considerable operational economies as well as reduce or eliminate the cost of ventilation, and are frequently the most practical method of providing such protection.

Design into Plant or Process

It is becoming increasingly common for plant and design engineers to consult with the industrial hygiene engineer at the design stage of a new plant or process. Consideration of industrial hygiene control principles at this point can eliminate or simplify the costly exhaust ventilation which would otherwise be necessary. Automation and automatic operations which require few if any workers are examples of this principle. Other examples include the design of continuous, enclosed chemical processes, underground trenches for ventilation ductwork, and grouping of hazardous operations so as to localize control, such as central shakeout stations, etc.

Process Change

A simple change in a process can often reduce contaminant dispersion and sometimes improve production efficiency, too. The following are examples of decreasing contaminant dispersal or otherwise reducing exposure: metal joining by crimping instead of soldering; a change in temperature, speed, or pressure of a chemical reaction; automatic electrostatic paint spraying instead of manual compressed air paint spraying; mechanical continuous hopper charging instead of manual batch charging.

A change in the physical condition or container specifications of raw materials received by a plant for further processing may be salutary. Thus, use of pelletized or briquetted materials that are ordinarily dusty, such as carbon powder, may drastically reduce atmospheric dust contamination at several stages in a process. Batch charging of materials that are slightly wetted or contained in paper bags rather than in a dry bulk state may eliminate or reduce the need for control in storage bins and batch mixers.

Substitution

An often effective and usually inexpensive method of control is the substitution in a process of nontoxic or less toxic materials for highly toxic ones. The classic examples of substitution as a control measure include replacing of mercury used in the processing of fur into hatter's felt with non-mercury compounds; replacement of white lead in paint pigments by zinc, barium or titanium oxides; the use of mixtures of paraffin HC instead of benzene (benzol) in the rubber products industries; and the almost complete disappearance of the hazardous carbon tetrachloride from industry by substitution of much less toxic chlorinated HC. We may also cite the use of steel shot instead of sand for abrasive blasting, synthetic rather than sandstone grinding wheels, and nonsilica parting compounds in foundry molding operations.

Frequently, such substitution carries with it a bonus in the form of an operational improvement. For example, the replacement of benzene by toluene in the manufacture of self-sealing aircraft gasoline tanks during World War II resulted in elimination of the troublesome problem of too rapid drying of the benzene between the various steps of the process.

Substitution often requires a good selling job to overcome the reluctance of production engineers to change the status quo. Fortunately, it can usually be demonstrated that substitution can result in an overall increase in economy.

Isolation in Time or Space

Many operations which do not readily lend themselves to ventilation control because of their nature or extent may generate contaminants in such quantities as to be able to permeate an entire workroom or building and expose all the workers to a hazard, although only a few of them are actually engaged in the operations. In such instances, an attempt should be made to perform the operation so that only those workers immediately concerned with it need be within its influence.

One such method is isolation in time or space. In a foundry without central shakeout this operation may be performed after the regular shift has gone for the day. The three or four shakeout workers can be provided with suitable respirators for the 1 or 2 hours during which they are exposed to the silica dust. Blasting in mines at the end of or between shifts and housekeeping procedures or plant painting at night are other examples of work that can be scheduled so as to minimize the number of workers exposed to a hazard.

An operation can also be isolated in space. The ordinary furnace is a prime example of such isolation. Complete enclosure of a sand blast operation with an air line respirator for the worker within the enclosure is another instance. Some operations require complete enclosure and remote control so that nobody is exposed, as in many processes involving nuclear radiation.

Enclosing a dangerous operation or locating one or more dangerous operations together in a separate room or building not only sharply reduces the number of workers exposed but greatly simplifies the necessary control procedures. Plating tanks, lead melting pots, paint dipping operations, and similar processes, when located in a separate room and grouped together, can usually be provided with efficient and relatively inexpensive local exhaust systems. Where continuous supervision of such operations by a worker is not necessary, only general ventilation may be required to prevent escape of contaminants into the main workroom. If necessary, the exposed worker can in such a case be provided with a respirator for use during his brief period of exposure.

Segregation of Personnel

An opposite approach, where contaminant-producing operations must be carried out over a large area, is to segregate the worker from the operation. The crane operator in a large foundry, bulk material storage building, or cement clinker shed can be provided with a completely enclosed cab ventilated under positive pressure to keep contaminants out. In automatic stone crushing, grinding, and conveying processes, where only periodic or emergency attendance is required by an operator, small ventilated rooms strategically located within the large workroom can be occupied by the workers during the major part of the work day. If tempered air is supplied to these rooms, the necessity of heating a very large building can frequently also be avoided.

Local Suppression of Contaminants

Mechanical or physical prevention or lessening of contaminant release can reduce or eliminate the need for ventilation control. The wetting of dust with water or other liquids is one of the oldest methods of control and may be very effective if properly used. Wet drilling in mines and quarries; water sprays at blasting, crushing, and conveying operations, and foundry shakeout; wet grinding and machining—all these can reduce dust dispersion if properly applied. The application of water must be designed to blanket the dust source completely. The particles must be thoroughly wetted by means of high pressure sprays, wetting agents, deluge sprays, or other procedures as indicated. Means must be provided for containment or continuous removal and disposal of the wetted dust. Baffles can be used around an operation releasing dust at high speed. Considerable mist and vapor suppression can be accomplished by flotation of plastic foam or nontoxic, nonmiscible liquids with low vapor pressure on the top of a large, high-vapor-pressure liquid surface which is releasing a contaminating vapor.

Housekeeping

The conventional need for and advantages of good housekeeping are familiar to everyone. Less commonly appreciated is the role of good industrial housekeeping as a positive factor in contaminant control. Dust or other contaminants which fall or settle onto the floor, workbenches, machines, walls, rafters, and ledges may become airborne again by the action of ambient air currents, drafts, vibration, and normal plant activity.

Constant good housekeeping by vacuum cleaning or wet washing is necessary to remove these materials continually and to prevent their becoming an additional source of air contamination. This is frequently true of gross materials as well as dusts. Leadpot dross accumulating on the floor may be reduced to airborne lead particles by the abrading action of being stepped on continuously.

In many industries, air recontamination resulting from inadequate housekeeping may be as significant a source of contaminant as the operations themselves. Lead storage battery manufacture, high silica stone crushing and screening, and mercury thermometer manufacture are examples of the need for proper housekeeping practice as a means of contaminant control.

Design is an important factor. Light-colored walls, good illumination, smooth, impervious, coved floors, and careful workspace design simplify and increase the incentive for proper housekeeping.

Respiratory Protection

Principles. Personal respiratory protection is one of the classical methods of control in industrial hygiene. However, the importance of this technique is often misunderstood. All the other methods of control, when properly designed and applied, can provide adequate continuous protection of a worker against harmful contaminants under normal working conditions. Personal respiratory protection, however, finds its chief usefulness as an emergency or short-term means of protection. It should be used as a primary protective device for normal operations only when no other method of control is possible. At times, it may be useful as an adjunct to exhaust ventilation or other control measures (see Section 1).

Respirators are thus normally emergency devices. They are needed when it is necessary to enter a highly contaminated atmosphere for a short time for rescue or emergency repair work; as a means of escape from a suddenly highly contaminated atmosphere; for a periodic, short-term inspection, maintenance, or repair of equipment located in a contaminated atmosphere; and for normal operations in conjunction with other control measures where the contaminant is so toxic that a single control measure such as ventilation cannot safely be relied on.

However, proper selection and use of respirators are becoming increasingly important in the United States since the promulgation of the Occupational Safety and Health Administration (OSHA) in 1970. OSHA permits and/or requires respiratory protection for workers when engineering controls are not feasible; during the period when engineering controls are being implemented; where the state of the art is such that engineering controls alone will not provide sufficient control at a particular operation; in emergency and rescue operations; and for maintenance, inspection, and other short term non-routine exposures.

A respirator must be designed and selected for the particular environment in which it is to be used. The type of contaminant, its probable maximum concentration, the possibility of oxygen deficiency, the useful life of the respirator, the escape routes available—all these and other factors must be considered in selecting a respirator for emergency use, for periodic use, or for standby purposes. Where these factors are not known with certainty, the device providing the widest spectrum of protection must be used.

Respirators must fit well, if possible without discomfort, should permit breathing without undue effort, should not interfere with vision, and should permit complete freedom of movement where danger may otherwise result. Respirators should be cleaned after each use and sterilized frequently, assigned to individual workers, and stored in dust-free cabinets readily available for use; where necessary, they should be available both in the workroom for escape and just outside the workroom in a safe atmosphere for rescue. Careful records should be kept of length of time in use, the remaining oxygen or air supply or condition of purifying element, and they should be renewed when necessary.

Atmospheric Hazards. The type and degree of hazard to be encountered will govern the kind of respiratory protective device to be used. Several classifications may be described to assist in proper selection of a device:

(1) Oxygen deficiency.
(2) Gases or vapors immediately dangerous to life.
(3) Gases or vapors not immediately dangerous to life.
(4) Particulates.
(5) Combination of particulates and gases or vapors immediately dangerous to life.
(6) Combination of particulates and gases or vapors not immediately dangerous to life.
(7) Possible skin irritation or absorption.
(8) Eye irritation.
(9) Poor inherent warning properties.

Where the degree of hazard is not known, it must be assumed to be immediately dangerous to life.

Oxygen Deficiency. Normal air contains approximately 21% oxygen by volume. This oxygen content may be reduced by such factors as dilution or displacement of oxygen by other gases and loss of oxygen by reaction with other substances or by absorption in certain materials. These conditions are usually found in such confined spaces as storage bins, silos, tanks, sewers, wells, mines, and ships' holds, and may also exist in burning rooms and in closed areas where natural oxidation of materials can occur.

Atmospheres containing 16% or less of oxygen may cause serious injury or death to people breathing them, depending on the actual concentration, length of exposure, and physical activity of the exposed persons. In such atmospheres, the respiratory protective device must be of a type which supplies fresh air or oxygen to the wearer. If the atmosphere is such that the wearer cannot safely escape, without respiratory protection, from the remotest location he is likely to be in to an uncontaminated location, the device chosen must be a self-contained type.

Gas or Vapors. Gases and vapors may be classified as toxic or inert. Toxic gases and vapors may cause injury or death depending on the concentration present. Inert gases can displace oxygen and are dangerous only under the conditions described above.

Where gases are so toxic that even in low concentrations they may be immediately dangerous to life, the respiratory protective device must be chosen to provide positive, reliable control for the conditions to be encountered.

Particulates. Particulates may assume the form of dispersions of solids, such as dusts, fumes, and smokes; liquids, such as mists and fogs; and in combinations of the two such as mists created by sprays of suspensions of dusts and paints. With rare exceptions, such as the organic phosphorus insecticides and, possibly, massive concentrations of the more highly toxic metals, particulate contaminants are not immediately dangerous to life. Mechanical filtration of inspired air is the usual protection method used.

Particulates may be classified as toxic, pneumoconiosis producing, and nonspecific or nuisance. The design of a respirator depends on the type of dust to be encountered. Toxic particulates, such as lead, flouride, and phosphorus dusts, enter the blood stream directly from the lungs and cause systemic poisoning. Pneumoconiosis-producing dusts (silica, asbestos) remain in the lungs and cause localized fibrotic diseases. Nuisance dusts (flour, wool, wood) may do either, but usually do not produce local or systemic effects. However, they may be irritating or allergenic and, in massive concentrations, may cause debility by their physical presence in the lungs.

Other types of particulates include carcinogens, the injurious effects of which are not usually manifest for many years, and biohazards, such as bacteria and viruses, which can cause illness almost immediately.

Combination of Particulates and Gases. Special respiratory devices designed to protect against mixtures of more than one type of contaminant are available. These should be selected to provide protection against the maximum expected concentration of each contaminant.

Skin Absorption. Exposure to substances which may cause injury by contact with or absorption through the skin requires protection of the exposed surface in addition to respiratory protection. Full-facepiece respirators, head covering or helmets, or supplied air suits may be required to afford skin protection.

Eye Irritation. When eye irritation occurs during normal work exposure, full-facepiece respirators are necessary as a minimum.

Poor Warning Properties. Air purifying respirators are suitable only for those vapors and gases with adequate warning properties (odor, eye or respiratory irritation) so that exhaustion of the absorption material, poor fit, or other malfunction can be detected by the wearer. If the concentration at which the warning effects are detectable is appreciably greater than the permissible limits, air purifying respirators should not be used. The anesthetic action of some chemicals on the olfactory nerves must also be taken into consideration.

Types of Respirators

There are two principal types of respiratory protective devices, i.e., oxygen or air supply and air purifying. There are, in turn, several classifications within each type. Table 2.1 lists all the common types and describes their principal limitations and necessary precautions to be observed when they are used.

(A) Supplied Air or Oxygen Respirators. (1) Self-contained. These are completely independent of the atmosphere surrounding them. They are entirely self-contained and can be worn by the user, may be used in any atmosphere, and permit complete freedom of movement. All self-contained units are designed for specific hours (length of time) of use. There are three principal types,

(a) Recirculating compressed oxygen. Compressed oxygen in a cylinder carried by the user is made available for breathing to a breathing bag through a suitable tube connected to a mouthpiece or facepiece by means of check and pressure regulating valves. Exhaled breath passes through another check valve to a canister where carbon dioxide is absorbed and the breath is returned to the breathing bag through a cooler. In the event of any failure of this system, it can be bypassed by means of a manually operated valve which admits oxygen directly to the breathing bag.

(b) Demand compressed air or oxygen. Compressed oxygen or air, contained in a cylinder worn by the user, is supplied through pressure-reducing valves to the facepiece only when the wearer inhales, in quantities governed by his breathing. There is no recirculation and the exhaled breath is directed to the atmosphere.

(c) Self-generating oxygen (recirculating). In this device, the exhaled breath is directed to a chemical canister which simultaneously releases oxygen by

TABLE 2.1. Respiratory Protective Equipment: Applications, Limitations, Precautions

Atmosphere	Recommended Type of Respirator	Applications and Limitations	Precautions
All particulates, gases, vapors, oxygen deficiency	*Self-Contained* Recirculating compressed oxygen Demand compressed air or oxygen Self-generating oxygen	Use in any atmosphere, allows freedom of movement, allows worker to leave atmosphere by any route; limited time of use, careful training required for proper use.	Wearer should be in good physical condition, throughly trained, assure plentiful supply of air or oxygen in tank, check for proper and tight fit, use with life line, leave at once if an odor is detected, do not remove until out into respirable air.
	Supply Air Hose mask with blower Hose mask without blower Air-line respirator	Unlimited time of use, use in any atmosphere (except air line not to be used in oxygen deficient or immediately dangerous atmosphere), not to be used where worker cannot escape unharmed without protection, must exit by entrance route, 150′ maximum from exit (75′ hose mask without blower), limits freedom of movement.	Place inlet in respirable air location, adjust fit and air lines properly, test before entering dangerous atmosphere, use life line, protect air line or hose from sharp edges or falling objects, leave at once if air flow is interrupted, do not remove until in respirable air, (air-line respirators—ensure a clean supply of air free from dust, oil and carbon monoxide).
Particulates alone	*Mechanical Filter* Special filter respirators	Allows freedom of movement, not to be used in excessively dusty atmospheres or in oxygen deficient atmospheres or in atmospheres containing gases or vapors, not to be used for abrasive blasting, relatively difficult to breathe, limited time of use.	Use clean filter and change when plugged, ensure good fit and good operating conditions, leave at once if difficulty in breathing increases significantly.
Gases and vapors alone	*Chemical Absorbers* Universal gas mask Special canister gas mask Special cart-tridge respirator	Allows freedom of movement, do not use in atmospheres deficient in oxygen or containing excessive contaminants (above 2% with gas masks, above 1000 ppm with cartridge type), used for limited time and specific contaminant only, (cartridge respirators not to be used in atmospheres immediately dangerous to life), relatively difficult to breathe, limited time of use.	Adjust properly, insure good tight fit, check operating condition, always use fresh canister or cartridge at start of use if possible, enter atmosphere cautiously, whenever odor is detected leave at once. Leave at once also if difficulty in breathing increases significantly.
Combination of particulates and gases and vapors	*Chemical-Mechanical Filters* Gas mask with filter Filter respirator with chemical cartridge	See mechanical filters and chemical absorbers.	See mechanical filters and chemical absorbers.

the action of the moisture in the breath on the chemicals and absorbs carbon dioxide from the breath. The breath containing the released oxygen enters the breathing bag for inhalation and the exhaled breath repeats the cycle.

(2) Hose type. These respirators supply air to the wearer from an uncontaminated source and are therefore independent of the workroom atmosphere.

(a) Hose mask (with or without blower). A large-diameter hose connected to a motor or hand-operated air mover located in a clean atmosphere supplies air to a full facepiece worn by the user. The hose diameter is sufficiently large so that the wearer can inhale clean air even if the blower is not operating. This type of equipment is frequently used without a blower. In this event, the inlet end is anchored in a respirable atmosphere location and may be provided with a screen to filter out coarse dust.

(b) Air-line respirators. These may consist of a full or half-mask facepiece or a head-covering helmet or hood to which air is supplied, by means of suitable reducing valves, from a source of compressed air, usually an air compressor, although a compressed air tank may be used. They may be of the continuous flow type or the demand type which is governed by the wearer's breathing. The latter is used only with a facepiece. The abrasive blasting respirator is a special type of air-line respirator, with a hood designed to be resistant against abrasion and to protect the head, face, and neck of the wearer against abrasive particles.

The air compressor should be equipped with the necessary safety and standby devices. The air supply should be from a noncontaminated source, protected by in-line absorbents and filters. When oil lubricated compressors are used they should have high temperature and carbon monoxide alarms.

(B) Air-purifying Respirators. In these, gaseous contaminants are removed from otherwise respirable air by absorption or chemical reaction and particulate contaminants are removed by mechanical filtration. They cannot be used in an oxygen-deficient atmosphere and must not be used in atmospheres containing contaminants in concentrations higher than those for which they were designed.

(1) Chemical respirators. The inspired air is drawn over suitable chemicals, where gaseous contaminants are removed before inhalation. Breath is exhaled to the atmosphere. The respirator is usually designed for combinations of contaminants.

(a) Gas masks. These consist of a full facepiece attached to a canister containing suitable contaminant-removing chemicals, and may be used for emergency purposes under certain conditions. The Universal Gas-Mask Canister protects against a number of contaminants including carbon monoxide.

(b) Chemical cartridge respirators. These consist of a half-mask facepiece attached to one or more cartridges containing suitable air purifying chemicals. They are for nonemergency use only and are usually designed for a single gas or vapor or single classes of gas or vapor.

(c) Self-rescue respirators. These are similar to the chemical cartridge type. They are not used routinely but carried on the person for use to escape to a safe atmosphere in an emergency or catastrophic situation.

(2) Mechanical filters. These are similar to the chemical cartridge respirators except that the purifying chemicals in the cartridge are replaced by filters, usually a felt pad. Filter respirators are designed to remove a specific single contaminant or class of particulate contaminant but are also available for several different particulate contaminants. Certain dusts, such as mercury compounds, may have a vapor pressure so high that when caught on a filter, inspired air passing over them may introduce toxic vapors into the air being breathed. Respirators for these may require a chemical cartridge in series with the filter.

A special, powered, air-purifying mechanical filter has recently been developed for the protection of coke-oven workers against tarry particulate contaminants. It is being adapted, by selection of suitable filtering media, for a whole spectrum of particulates.

A filter cartridge, connected to a battery-operated motor blower, is worn by the worker, usually on a belt around his waist. The filtered air is discharged through a hose to a half-mask facepiece, full facepiece, hood, or helmet, in a manner similar to the conventional air-line respirator.

Usual design calls for the filter and air hose to have a resistance below 1 in. of water and a minimum flow rate of 4 cfm. The battery should be rechargeable and be capable of providing flow for an entire workshift. The filter characteristics should permit operation for a similar period of time without plugging and without undue increase in resistance.

Although this equipment combines some features of the air-line respirator with those of the mechanical filter, it cannot be used in atmospheres containing harmful concentrations of gases or vapors or oxygen-deficient atmospheres. Its principal advantage lies in the fact that it is much more comfortable to wear than the conventional filter respirator; it does not impede natural respiration; and there are no long air hoses to interfere with movement and create a safety hazard.

(3) Combination respirators for gases and particulates. These are a combination of one unit, for simultaneous protection against gas, vapor, and particulates. They consist of gas mask canisters or chemical cartridges with mechanical filters in series, so that inspired air passes first through the filter and then over the chemical granules.

(4) The single use or disposable filter respirator is a recent development. The filter (resin-impregnated natural wool fiber, synthetic fiber, felt or fabric) is an integral part of the facepiece or may comprise the entire facepiece. Proper fit is essential with these respirators. They provide protection against pneumoconiosis- and fibrosis-producing dusts and could probably be adapted for mists and more toxic dusts.

Respirator Testing and Approval

The U.S. Bureau of Mines formerly had sole responsibility for inspecting, testing, and issuing approvals for respirators. At present, the Mining Enforcement and Safety Administration of the U.S. Department of the Interior (MESA) and the National Institute of Occupational Safety and Health of the U.S. Department of Health, Education and Welfare (NIOSH) test and approve respirators jointly under the provisions of Title 30, Part 11 of the Code of Federal Regulations.

Selection of Respirators

A number of the OSHA standards in Part 1910, Subpart Z, Toxic and Hazardous Substances, contain specific requirements for respirator selection and use. In addition, some of the criteria documents issued by NIOSH (criteria for recommended standards for occupational exposure) for a number of chemicals contain specific recommendations for respirator selection and use for protection against exposure to these chemicals. These official requirements and recommendations should be followed. In addition to these specific standards, OSHA Part 1910:134 lists general requirements.

Tables 2.2a and 2.2b give a basis for the selection of a respiratory device according to the degree of toxicity and expected concentration of the contaminant against which protection is required and for which no official standard has been issued. Officially approved devices should be used when available. Devices that have not yet been offically approved, such as air supplied suits, should be selected on the basis of tests for a specific applicability by a recognized organization.

A number of other factors must be considered before selecting a respirator.

1. Type of contaminant. In addition to the toxicity and expected concentration of the contaminant, both the physical form and identity of the contaminant must be known so that the proper respirator can be selected.

2. Period of required protection. The maximum time to be spent in the contaminated air must be known, so that a device which will function properly for that period can be selected. Self-contained or air-purifying respirators have limited periods of use. Hose-type respirators have relatively unlimited usefulness.

3. Availability of safe atmosphere Use of hose-type respirators in contaminated areas is limited to a distance equal to the maximum amount of hose which can be used as well as the need to enter and leave the area by the same route. Other conditions will usually necessitate self-contained apparatus or, in less severe situations, air-purifying respirators.

4. Activity of wearer. Certain types of activity, such as traveling over a large area, climbing, manipulating equipment, etc., preclude the use of hose-type respirators. Hard physical labor not only increases the discomfort of some types of respiratory equipment but also increases the breathing rate three or four times over that of rest conditions. This will result in a much more rapid depletion of the air or oxygen supply in a self-contained respiratory device and likewise much more rapid overwhelming of a canister, cartridge, or filter.

5. Ease of use. All users of respiratory devices require careful instruction in proper selection, fitting, and operation of the devices. Some are much more

TABLE 2.2a. Selection of Respirators for Emergency or Short-Term Use on the Basis of Hazard and Expected Concentration (Gases and Vapors)

Toxicity	Expected Concentrations of Gases or Vapors			
	Two to five times TLV or up to 1000 ppm	Five to ten times TLV or 1000–5000 ppm	Above ten times TLV or 5000–20,000 ppm	Oxygen deficiency, emergency or above 20,000 ppm
Low	No respirator, or chemical cartridge needed	Canister gas mask	Canister gas mask or air-line respirator	Self-contained air or oxygen
Moderate	Chemical cartridge	Canister gas mask or air-line respirator	Air-line or self-contained air or oxygen	Self-contained air or oxygen
High	Canister gas mask	Air-line respirator	Self-contained air or oxygen	Self-contained air or oxygen

NOTES:
(1) TLV refers to the Threshold Limit Values for a number of substances published by the American Conference of Governmental Industrial Hygienists (see Section 1 and section 12).
(2) See Sections 1 and 2 for a discussion of toxicity ratings and their relation to TLV.
(3) When unavoidable conditions necessitate using respirators for longer periods (above 1 hour), use equipment in a higher protective category than shown above.
(4) Subject to limitiations (Table 2.1), hose-type respirators may be used in place of air line.

TABLE 2.2b. Selection of Respirators for Emergency or Short-Term Use on the Basis of Hazard and Expected Concentration (Particulates)

Toxicity	Expected Concentrations of Particulate Matter (Dusts, Fumes and Mists)			
	Two to five times TLV	Five to twenty times TLV	Above twenty times TLV	Oxygen deficient, emergency, highly corrosive
Low	Respirator not usually needed	Filter	Filter or air-line respirator	Where exposure is to extremely corrosive dusts or to dusts in an oxygen deficient atmosphere, a self-contained air or oxygen respirator must be used.
Moderate or High (toxicity no greater than lead)	Filter	Filter or air-line respirator	Air-line or self-contained air or oxygen	
Extremely High (toxicity greater than lead)	Filter or air-line respirator	Air-line respirator	Self-contained air or oxygen	

NOTES:
(1) TLV refers to the Threshold Limit Values for a number of substances published by the American Conference of Governmental Industrial Hygienists (Sections 1, 12).
(2) See Sections 1 and 2 for a discussion of toxicity ratings and their relation to TLV's.
(3) Expected concentrations of particulate matter have been shown only as multiples of the threshold limit values. Where these values are not available, the following concentrations may be used as a guide:

	Mineral Dusts	Other Dusts, Fumes, and Mists
2 to 5 (TLV)	up to 50 mppcf*	Up to 0.5 milligrams per cubic meter
5 to 20 (TLV)	50 to 1000 mppcf*	0.5 to 10 milligrams per cubic meter
Above 20 (TLV)	above 1000 mppcf*	Above 10 milligrams per cubic meter

(4) When unavoidable conditions necessitate using respirators for longer periods (above 1 hour), use equipment in a higher protective category than shown above.
(5) Subject to limitations (Table 2.1), hose-type respirators may be used in place of air line.

* Mppcf = millions of particles per cubic foot.

complicated than others, and this may be a governing factor in the selection of one device over another, assuming equal protection is provided by each. All users or potential users of devices should be required to familiarize themselves completely with their principles, use, operation, and emergency procedures.

Care of Respirators

Maintenance. Wherever possible, care of respiratory devices should be the full-time function of a specially trained operator. Careful records should be kept of time of use of each device so that the air or oxygen cylinders are recharged when necessary and canisters, cartridges, and filters are replaced before exhaustion or plugging. Worn or broken parts should quickly be repaired or replaced.

Inspection. All devices, whether for routine or emergency use, should be inspected periodically to ensure that they will operate properly when needed. Rubber parts should be checked for deterioration, metal parts for corrosion, and plastic or glass parts for cracks or breaks.

Cleaning. Equipment should be cleaned, sterilized, and dried after each use to prevent the spread of infection and to ensure continuing efficient operation. Recommendations of the manufacturer of the device for proper cleaning and sterilization methods should be followed.

Storage. Respirators should be stored in clean compartments protected against humidity, extremes of temperature, and sunlight. Nonemergency respirators may be centrally stored. Equipment for emergency use should be stored in suitable compartments within work areas where the emergency may occur, for escape purposes, as well as outside the potentially dangerous areas for use in entering the area for rescue or repair purposes. Replacement canisters, cartridges, and filters should be similarly stored.

CONTROL BY VENTILATION

The design of exhaust ventilation usually depends on several factors, such as the physical state of the contaminant, i.e., dust, fume, smoke, mist, gas, or vapor; the manner in which it is generated; the velocity and direction with which it is released to the atmosphere; and its relative toxicity.

Section 1 dealt with the methods by which relative toxicity is determined and discussed two methods for describing it quantitatively. The concept of maximum allowable concentration (mac), now more commonly described as threshold limit values (TLV), expresses toxicity in numerical values. The units may be parts of the contaminant per million parts of air (ppm) for gases and vapors, milligrams of contaminant per cubic meter of air (mg/m^3) for solids and mists, and millions of respirable particles per cubic foot of air (mppcf) for mineral dusts. Some mineral dust TLV's are now also expressed in terms of mg of dust per cubic meter of air. Recent studies have shown that the TLV for asbestos dust is more properly expressed in terms of number of fibers above 5 microns in length per milliliter of air.

These values, when used with understanding, give a semiquantitative index of concentrations which a normal worker can safely tolerate for an eight-hour daily exposure for an indefinite length of time at the rate of 5 days/week. The manner in which these values have been derived, their limitations, and the precautions with which they must be used have already been described in Section 1. These considerations are stressed here because, as we shall see later, the TLV data are sometimes used in ventilation design calculations. They will also influence selection of ventilation type and must be carefully interpreted and never used as absolute values. A list of TLV's is published annually by the ACGIH (Section 1).

The other method of delineating relative toxicity (described in Sections 1 and 9) is by the more general concept of slightly toxic, moderately toxic, and highly toxic. Wherever toxicity information is available on the materials listed in Section 12 it has been classified into categories of slightly, moderately, or highly toxic. Substances of unknown toxicity should usually be treated as highly toxic, until evidence to the contrary becomes available. The substances for which numerical TLV's have been established may also be classified as slightly, moderately, and highly toxic in accordance with Table 2.3.

A reliable determination of the health hazard created by an industrial process and the need for control requires careful evaluation by a trained industrial hygienist. Chemical analyses of air are usually necessary. The concentration of contaminant thus measured, its toxicity rating, and an evaluation of such pertinent factors as rate of work, length of workday, time of year, housekeeping, and working conditions will determine the need for and extent of ventilation control.

TABLE 2.3. Threshold Limit Values

Toxicity	ppm	mg/m^3	mppcf
Slight	over 500	over 0.5	50
Moderate	101–500	0.101–0.5	20
High	0–100	0–0.1	5

In general, it may be stated that an operation or process releasing highly toxic air contaminants almost always requires control; an operation releasing moderately toxic contaminants usually requires control; one releasing slightly toxic contaminants occasionally requires control. In all cases, one of the most important factors in determining the need for control will be the toxicity and concentration of contaminants created—because the more toxic the contaminant, the lower the atmospheric concentration which can be tolerated without injury and without the need for control.

The ventilation design data which appear later in this section have been related to the toxicity ratings in Section 12. Ventilation control is applicable to all airborne substances which are shown in Section 12 as having a potential systemic toxicity upon entering the body by inhalation. In addition, those substances which, when airborne, can cause local injury to the mucous membranes or skin are also amenable to control by ventilation and will be so treated.

The basic ventilation methods are local exhaust ventilation and dilution or general ventilation.

Local Exhaust Ventilation

In local exhaust ventilation, the airborne contaminant is removed or captured from the environment at or as close as possible to the source.

It is the method of choice wherever contaminants are released from a sharply defined and narrowly circumscribed operation. Its basic advantages are that contaminant control can be positive and efficient, exhaust air volumes are relatively low, and the cost of make-up air is kept at a minimum. Its disadvantages are the high initial cost of installation; the elaborate system of hoods and piping frequently required, which use valuable plant space; and relatively high power requirements.

The basic principles and elements of local exhaust must be understood before design can be attempted. These fall into definite categories.

Basic Principles. Air will flow from one point to another when there exists a difference in "head" or pressure between two points. In nature, such flow is random, as the flow of air from a high-pressure weather system to a low-pressure one.

A local exhaust system is a mechanical device for creating a pressure differential between two points so as to direct air flow. At one end of a system the fan creates pressure in a hood or orifice lower than the air just outside by means of piping connecting the hood and the fan. The air which flows into the hood because of the pressure difference travels to the fan and is then conveyed from the fan by means of discharge piping to a desired location.

This series of events performs a number of functions. The hood, which is located around, at, or near a source of atmospheric contamination, prevents the escape of the contaminant, captures it, or acts as a curtain between the contaminant and the workroom. The piping through which the air is moved by the fan serves both to pull air through the hood and to convey away the contaminant captured by the hood. The contaminant-laden air reaching the fan may be discharged by the fan to the outdoors untreated, to the outdoors after being cleaned of contaminant by means of an air cleaner, or in certain instances, back to the workroom after being cleaned (see discussion of recirculation later in this section).

Hoods. The function of an exhaust system is to protect the worker from exposure to potentially hazardous contaminants created in the workroom. As such, the heart of the system is the hood or orifice. Design of a system begins with the hood, which is usually a compromise between the ideal and the practical. Hood design is governed by a consideration of the physical state of the contaminant to be controlled as well as such details as temperature of process, velocity and direction of contaminant trajectory, nature of operation, degree and closeness of attention required by the operation, and extent of control needed. In simple terms, the designer will first mentally enclose the entire source of contaminant and then make such openings or changes in it as are dictated by the operation. He will depend upon air flow and velocity to make up for the openings or design compromises.

Hood types include complete enclosure; partial enclosure; shaped enclosure; lateral, rear, or overhang canopy; and free hanging. The complete enclosure is typified by the abrasive blasting room where the worker is inside the room but is completely protected by an air-supplied helmet. The function of the enclosure is to protect not the operator but other workers in the vicinity. The standard spray booth is a partial enclosure where the worker sprays objects inside the enclosure through an open front or from within the enclosure. Here the flow of air through the frontal opening acts as a curtain between the point of spray or contaminant release and the workroom. It provides a capture velocity between the worker and source.

An example of a shaped enclosure is the ordinary grinding-wheel hood which assumes the contours of the wheel with a minimum-sized working opening through which, ideally, a velocity of air high enough to capture the fine particles carried around the wheel is maintained. In such hoods, the pipe take-off is set tangential to the point of operation so that the heavy particles, projected with high velocity, travel directly into the pipe. Inner baffles are often used to decrease particle velocity and prevent its escape.

Open surface tanks are usually provided with lateral slots, vertical or partially overhanging rear hoods, or rarely, overhead canopy hoods with one or more sides. When properly designed, these hoods rely on a number of factors to provide good control. There are, for instance, a capturing effect, a control effect produced by bringing a mass of air down to the tank and into the hood, and a dilution effect, whereby that portion of the contaminant which is not captured is diluted to safe levels.

A common free-hanging hood is the welding-fume hood consisting of a flexible pipe with a flanged orifice which can easily be located near the point of weld. Here the contaminant is captured as it is released and carried into the hood before it becomes dispersed and thus difficult to capture.

The design of hood shapes is governed by two inherent characteristics of air entering an opening under suction: (1) air will enter the opening from all directions, and (2) in general the velocity of this air at distances from the opening follows an inverse square law, with some exceptions to be noted later. Thus, if we assume an ideal point source of suction, air will travel to the point equally from all directions, and the equal velocity contour at any distance from the point will have the shape of a sphere whose radius equals the distance. Round, square, or essentially round or square openings are practical counterparts of the ideal point source of suction. They will have essentially spherical velocity contours, and the velocity at a distance from the opening will largely follow the inverse-square law. Suitable baffling or flanging of the hood opening can limit the directions from which air will flow into the hood and thus create air velocities at a selected point higher than with an unflanged hood.

Similarly, one can postulate an ideal line source of suction. Flow into the line, which will come from all directions perpendicular to the line, will have a cylindrical contour. The slot is the practical counterpart of the ideal line source. Experimental data indicate that flow into a slot will be such that for practical purposes velocity at a distance from the slot will vary inversely as the distance rather than as the square of the distance. Here also, flanging can increase capture efficiency.

Rectangular hoods with large width-to-length ratios are essentially similar to round or square openings. As the hood shape changes from square to increasingly narrow rectangular, the characteristics of air flow will vary gradually from those of square to those of slot hoods. Since the characteristics of air flow into these

TABLE 2.4a. Range of Capture Velocities

Condition of Dispersion of Contaminant	Examples	Capture Velocity, fpm
Released with no velocity into quiet air	Evaporation from tanks; degreasing, etc.	50–100
Released at low velocity into moderately still air	Spray booths; intermittent container filling; low speed conveyor transfers; welding; plating; pickling	100–200
Active generation into zone of rapid air motion	Spray painting in shallow booths; barrel filling; conveyor loading; crushers	200–500
Released at high initial velocity into zone of very rapid air motion	Grinding; abrasive blasting, tumbling	500–2000

transition hoods are complex, it may be arbitrarily assumed that rectangular hoods with width-to-length ratios greater than 0.3 are square hoods, and hoods with ratios 0.3 or less are slot hoods.

These considerations are utilized in the derivation of hood equations and in the design of hoods to the end that the hoods provide a capture velocity at the point of contaminant release with the least possible air flow. Typical capture velocities are shown in Table 2.4a.

Piping. Piping in an exhaust system serves the dual function of acting as a link between the fan and hood, so that the desired air quantity is drawn into the hood, and of conveying the contaminant-laden air to a desired location. Pipe construction, including elbows, fittings, tapers, and joints, requires due consideration for aerodynamic principles so that resistance to flow is kept at a minimum. Material and gauge of piping will depend on the length and diameter of the pipe, the pressure of the system, and the abrasive or corrosive nature of the contaminant being conveyed. Design pipe velocities, and therefore diameters, are a function of the minimum required transport velocity for a particular contaminant. Since a truly airborne contaminant is part of the air stream, there is no minimum pipe velocity requirement for such contaminants and pipe diameters may be selected on the basis of compromise. Low-velocity (large-diameter) pipe calls for less fan horsepower but more metal and plant space. High-velocity (small-diameter) pipe requires less metal and space but more horsepower. For airborne contaminants (gases and vapors, most fumes, and mists) a pipe velocity of 2000 fpm has been found to be optimum. For particulate matter, pipe velocities required for transport will range from 2500–5000 fpm, with usual velocities in the 3000–4000 fpm range, depending on the density, size, and shape of the contaminant particles. Typical transport velocities are shown in Table 2.4b.

TABLE 2.4.b Average Transport Velocities for Dust

Material	Minimum Transport Velocity, fpm
Very fine, light dusts	2500
Fine, dry dusts and powders	3000
Average industrial dusts	3500
Coarse dusts	4000–4500
Heavy or moist dust loading	4500 and up

A very simple, but useful equation of flow, $Q = AV$, is basic to exhaust system calculations. Q is rate of flow in cubic feet/minute (cfm) of air flowing through a certain area A in square feet, usually a hood opening or pipe, at a velocity V in linear feet/minute (fpm). If any two terms are known or specified, the third can be obtained by calculation. In this manner, hood or slot face area, pipe diameter, and pipe or hood air flow or velocity can be determined.

Fans. Numerous fan types are available to the exhaust system designer, but as a rule the individual exhaust system limits the type of fan which may be selected. The propeller fan, usually with a cast metal blade, is used principally for spray booths and similar applications for movement of large volumes of air at relatively low pressures. The axial (vane or tube) fan is similar, but it can generate higher pressures and is used where higher pressures must be maintained and where an in-line fan is advantageous.

There are three principal types of centrifugal fans. Each has its own application, although there is some overlapping of characteristics. The forward curve (commonly known as "squirrel cage" or "sirocco") fan is a relatively high volume, quiet fan with low space requirements; it is used principally to move air at low pressure with some applications for fume or mist handling. The radial (paddle wheel) fan is a pressure exhauster used for dust-handling systems. The backward curved fan is an efficient, non-overloading fan used to move clean air or air containing gases and vapors at moderate pressures, although it can be used for dust systems when located on the clean side of a dust collector.

Selection of a fan is governed by the material to be handled, need for flexibility, location, space limitations, noise generated, and necessary efficiency. Although selection is usually done from the manufacturer's certified multirating tables, the fan laws, which describe the relationship between cfm, pressure, horsepower, speed, and size in fan systems can be used. This permits selection of a fan, evaluation of the adequacy of an existing fan for a new system, and determination of the effect of a change in conditions in an existing system.

Air Cleaning. Air cleaning, when required, is an integral part of an exhaust system and must be designed to be compatible with all the other components of the system. The nature and amount of contaminants and required degree of air cleaning will dictate the design and size of air cleaner to be used. This in turn will influence the type and horsepower of the fan to be used (see *Air-Cleaning* later in this section).

Make-up Air. One of the most important and least understood considerations in exhaust system design is the necessity to supply tempered make-up air to the workroom to replace the air removed by the exhaust system. All too frequently this need is overlooked and natural infiltration is relied on. In some instances natural infiltration may be acceptable, particularly where the exhaust volumes are low and plant heating capacity is adequate. However, this method usually produces adverse effects, particularly in a new "tight" building which will not permit enough make-up air to enter by infiltration. This will increase the resistance to be overcome by the exhaust fan and thus reduce the exhaust volume and the degree of control provided by the exhaust hoods. Other adverse effects include negation of the action of natural draft stacks with introduction of contaminants into the workroom; and creation of cold drafts, particularly in undesirable locations. Such drafts can cause workers to shut off exhaust fans and can result in inefficient or inadequate heating. High-velocity cross drafts may interfere with the operation of exhaust hoods and can reintroduce settled contaminants into the workroom air.

A properly designed make-up air system should supply tempered air approximately equal in volume to the air exhausted. A slight excess of supply over exhaust is usually desirable for most efficient operation. In some cases, an excess of exhaust is required to prevent contaminants from entering adjacent workrooms. Make-up air can be provided by means of a central-heated supply system, with filtered intakes drawing air from uncontaminated outside areas, and ductwork terminating in suitable grills or registers. This technique is preferable since the supply air can be directed to desired locations to achieve such incidental benefits as spot cooling or heating and dilution of airborne contaminants near their point of origin. However, make-up can also be provided by means of fresh air supply unit heaters properly located in windows, exterior wall, or on the roof. These systems cannot be carefully designed, but they are flexible and eliminate the need for expensive ductwork. In some instances, untempered make-up air can be brought directly to the vicinity of an exhausted operation when there are no workers in that vicinity. It is frequently possible to design supply systems to provide comfort ventilation at the living zone of the plant (below the 8–10′ level) and thus eliminate the need for heating air in unoccupied zones in high ceilinged workrooms. Where possible, the supply air should be zoned, from clean to increasingly contaminated areas, to achieve a dilution effect (see Dilution Ventilation later in this section).

ENERGY CONSERVATION AND EXHAUST SYSTEM DESIGN

However, tempering (heating or cooling) make-up air, which is then discharged to the outdoors by the exhaust system, is becoming increasingly unacceptable in light of the energy shortage. Fortunately, there are measures that can be taken to minimize or avoid these unproductive energy expenditures.

I. Energy Conservation by Recirculation

Recirculation is a process whereby the air discharged by an exhaust system is completely or partially returned to the workplace after suitable cleaning, eliminating the need for costly make-up air systems or making it possible to use much smaller air supply systems. Recirculation has always been attractive to management because of the savings in both the capital costs of installing a tempered supply system and the fuel costs. Recent developments have accelerated the trend toward recirculation. Many workrooms are now equipped with air conditioning in the summer and heat in the winter. The new and more sophisticated technologies frequently require the maintenance of workroom air to rigid temperature, humidity, and dust conditions. Continuous discharge of expensively conditioned air imposes a severe economic burden. Furthermore, the increasing emphasis on air pollution control requires air cleaning devices where none was needed before or the use of expensive high efficiency air cleaners where cheaper collectors were previously acceptable. Under conditions where recirculation is permissible, it may be economically feasible to provide the high efficiency air cleaning required for recirculation where air pollution requirements would permit low efficiency cleaners with discharge to the outdoors. In some cases recirculation with proper air cleaning is virtually unavoidable, as, for example, when operations are so located that it is practically impossible to run a discharge pipe to the outdoors. In any event, future energy conservation regulations may make recirculation mandatory in many instances.

Many states and governmental agencies having applicable jurisdiction require prior approval before instituting recirculation. Such prior approval encourages recirculation designed into a new system, where it can be of optimum efficiency, rather than an add-on feature. Recirculation from systems handling carcinogens or other highly toxic contaminants is rarely acceptable because very few air cleaners are efficient or reliable enough to give the required degree of uninterrupted cleaning. Even where such air cleaners are available, recirculation from systems handling highly toxic contaminants is contraindicated because of the severe potential hazard created by improper operation and maintenance of air cleaners or by a sudden massive collector failure. Systems handling moderately toxic contaminants may be recirculated if compelling reasons exist, if favorable conditions are present, and if proper safeguards are provided. However, recirculation from systems handling contaminants of little toxicity is becoming increasingly acceptable.

Fan systems handling solid particulates are the most likely candidates for recirculation since appropriate high efficiency, reliable air cleaners are available. Collectors are also available for such fine particles as fume and smoke but they require considerable energy to operate. In addition, operations that generate fume or smoke often produce toxic gases as well. Recirculation from systems handling gases or vapors is usually not acceptable since air cleaners for these contaminants are, by and large, not efficient or reliable enough, and most cleaners in this category can be saturated or overwhelmed with a sharp decrease in efficiency. In any event, these systems lend themselves to heat recovery by other means, as will be discussed later.

Other factors which will influence the decision to recirculate, assuming it is not mandated by energy needs, include adverse employee attitudes and the likelihood of ongoing reduction in TLV's. The latter may make systems designed for one set of TLV's unacceptable for lower values unless a safety factor is incorporated into the original design.

A. Design Features. There are a number of desired or required features for recirculation systems:

1. High efficiency air cleaning must be available so that the concentration of the contaminant in the return air is a fraction (preferably no greater than 10%) of the TLV. In addition, the average concentration in the workroom should be equal to or less than the TLV, except that where transient high concentrations are present or inadequate mixing of the room air exists, the average concentration should be no greater than 25% of the TLV. It should be noted here that collector efficiency is usually expressed in terms of percent by weight. While this is acceptable for systems discharging to the outdoors, recirculating systems require specifications and performances in terms of concentration of the contaminant in the return air. Selection of a collector must also take into consideration a possible future increase in the amount of contaminants to be cleaned because of a change in process, product, or the raw material.

2. A number of collector types are available. Fabric filters are preferred for particulate recirculating systems. Of the other types of particulate air cleaners, venturi scrubbers introduce humidity to the workroom, dynamic collectors are not efficient enough and electrostatic precipitators, unless carefully designed, may discharge ozone to the workroom.

Gases and vapors can be removed from the air stream by incineration (thermal or catalytic), absorption or adsorption. Incineration can introduce toxic products of combustion to the workroom; absorption (scrubbing) is usually not efficient enough and the return air may be close to saturation; adsorption (as on charcoal beds) may be contaminated by particulates, collection of mixtures may be selective, and saturation with contaminants may be rapid, resulting in loss of retentivity. Accordingly, recirculation from gas and vapor exhaust systems is not considered safe at present. If it is used, highly reliable, accurate and specific monitors in a fail-safe system must be provided.

It should be stressed that the above discussion applies only to recirculating exhaust systems. All of the air cleaning types described above have applications in standard exhaust systems discharging to the outdoors. They are described in greater detail under "Air Cleaning Devices."

3. Suitable safeguards should be provided in the event of cleaner failure. These may include the following:

(a) Redundant air cleaners by two identical cleaners situated parallel in the system. The failure of one cleaner as determined by monitoring (discussed below) results in automatic switching of the air flow to the other cleaner.

(b) Multiple-stage cleaning with a high efficiency collector at the end of the system would provide temporary protection in the event one of the intermediate stages failed. The high efficiency collector could be smaller in a multiple-stage system than if it were the only collector. However, the installation and power costs of such a system are not usually warranted unless required by the physical state and highly toxic nature of the contaminants.

(c) By-pass for discharge of the air to the outside in the event of cleaner failure. Special provision for make-up air to replace the air temporarily discharged outdoors is not usually necessary. Infiltration would normally be satisfactory during the relatively brief period of malfunction.

4. Other design features:

(a) Wherever possible, recirculating systems should be located in large workrooms with good natural or mechanical ventilation so that odors, fumes or traces of contaminants that returned to the workroom can be effectively diluted. The air should be recirculated in such a way as to provide good mixing to achieve this dilution. Therefore, the workroom to which the air is returned should, by virtue of its size and natural exfiltration factor or existing mechanical ventilation to the outdoors, be such that the prevailing dilution ventilation (for odors, fine dust, etc.) in CFM, would be at least 20% of the recirculated air volume. If this is not feasible, recirculation should be less than 100%, by a suitable factor.

(b) It would be desirable to utilize the by-pass described above to discharge air outdoors at times when the workroom is not heated (or air conditioned) so that general comfort ventilation is enhanced. This can be accomplished by a thermostatically actuated motorized damper.

(c) Recirculating air registers or ducts should not subject workers to drafts or high velocity air noise, which is usually in the injurious high frequency range.

B. Monitoring of Recirculating Systems. As recirculation becomes more prevalent with the likelihood that it may be used with gas and vapor systems, monitoring will become increasingly necessary to insure that toxic concentrations of contaminants are not returned to the workroom. An exception to this would be inert or low toxicity dust systems for which monitoring is not usually necessary since a cleaner failure would be detected long before injury resulted.

1. Duct monitors, where available, should be specific, accurate, sensitive and continuously reliable. They should automatically actuate the required safety measures when excessive concentrations are sensed. Such measures could consist of shunting the recirculating air to a redundant air cleaner, directing it outdoors through a by-pass, sounding an audible alarm, or shutting the operation down, provided this would not continue to generate contaminants.

2. Other monitoring or surveillance techniques:

(a) Many odorous and irritating contaminants can be detected by the senses at concentrations below the TLV. This can be used as a recirculation monitoring technique. Caution must be observed since these agents may desensitize workers after a period of exposure and excessive concentrations can go undetected.

(b) A carefully designed program of periodic air tests in the workroom and breathing zone can serve to monitor recirculation, often at lower cost and equivalent or greater efficiency than a duct monitor.

(c) Visual observation of dust in the return air or observing the opacity of the return air stream, possibly with a modified Ringelman chart, may be satisfactory in some instances.

(d) A pressure differential gauge across a filter will detect a gross leak or a tear in the cloth. For very toxic dusts, the gauge should be sensitive enough to detect even minute pressure changes caused by a slight leak.

C. Maintenance. An integral part of recirculation design must include setting up a periodic exhaust system maintenance schedule at least as thorough as that for machinery and process equipment. Inadequately cleaned collectors, fan belt slippage, fan wear, an increase in collector loading—all may result in operation at reduced efficiency for long periods of time without detection. Monitoring techniques are available to alert such a condition.

II. Heat Conservation by Heat Recovery

When recirculation is undesirable or impractical, considerable heat may be recovered from the exhaust air by a number of techniques, most of which utilize the reclaimed heat to temper the entering make-up air. In these applications, the exhaust and make-up air ducts are contiguous or in close proximity to each other.

A. The heat wheel is a device in which a large wheel, with specially constructed blades, rotates slowly within a partitioned housing. The air discharged by the exhaust system enters one side of the partition and gives up heat to the blades, before discharging to the outdoors. At the same time, make-up air enters the other side of the partition and absorbs heat from the blades before entering the make-up air duct system. An analogous process takes place during the summer months in an air-cooled workroom. The efficiency of this heat transfer may be as high as 60% to 70%. (This device is not usually acceptable for dust handling systems unless it is specially designed since particulate contaminants can be picked up by the blades from the discharge air and be reintroduced to the entering make-up stream.)

B. A similar device incorporates heat exchange coils situated in the discharge and make-up ducts. During the heating season, the discharge air heats the fluid in the coils, which circulates by convection to the coils within the make-up duct and gives up heat to the entering air duct. A reverse process takes place during the air cooling season. There can be no transfer of contaminants as with the heat wheel.

C. The duct work may be designed so that heat is conducted from a rectangular discharge duct to a contiguous rectangular make-up air duct. Concentric ducts may also be used, with the make-up air duct within the discharge duct. Rectangular or round ducts may be used with the latter configuration.

III. Energy Conservation by Engineering Design

A. It is frequently possible to design processes or redesign existing processes so that exhaust control can be accomplished with a minimum exhaust and make-up air CFM.

B. Hoods should be carefully engineered with optimum shapes and maximum enclosures, so that exhaust CFM's are as low as possible consistent with good control.

C. Other design techniques are available to minimize exhaust volumes:

1. Low-volume high-velocity systems, with special nozzles, orifices and pressure piping, provide very high velocity (up to 16,000 FPM) an inch or two from the contaminant source so that control is afforded with low CFM. However, these systems develop extremely high static pressures and in some instances the power required to operate them may negate the saving in heat resulting from the low CFM. They are particularly useful when they replace conventional hoods.

2. Open surface tanks are used in many industries and conventional exhaust systems require high exhaust volumes to provide control. A number of exhaust reduction techniques are available for these systems.

(a) In push-pull systems, compressed air jets are directed from one side of a tank toward an exhaust slot at the other side, resulting in up to a 50% saving of CFM. However, this design is not recommended for operations involving lowering and raising of large flat surfaces in and out of the tank since the effectiveness of the jets would be negated.

(b) Systems exhausting large numbers of adjacent tanks may be provided with exhaust rates at lower than normal amounts because of the dilution effect provided by the large volumes of air exhausted and the additional control effect created by the mass of air moving down toward the tanks and into the exhaust openings.

(c) Plastic chips, foam or other surfactants floated on top of the liquid surface reduce the amount of escaping contaminants and thus the required CFM exhaust.

D. Frequently systems can be designed so that untempered make-up air can be supplied to the immediate vicinity of the exhaust system. The supplied air laboratory fume hood is an example of this principle.

E. Dilution exhaust should be used only when a local system is not feasible. It should be carefully engineered so that only the air needed for control is exhausted. These systems are frequently overdesigned because equipment and installation costs are low.

F. In large workrooms with few workers, it may be feasible to provide enclosed heated work stations for the employees, leaving the workroom unheated.

G. Split systems with interlocking dampers may be provided for intermittently used machines so that only part of the system is exhausted at any one time. When used for dust systems, design must ensure transport velocity at all times. When toxic contaminants are involved, it is advisable to provide interlocks so that a machine or process cannot be activated unless the required air is moving in the exhaust duct for that machine.

Intermittently used systems which cannot be recirculated may achieve an energy saving by interlocking the exhaust and make-up air systems so that they are actuated by a single starter.

System Testing. Since exhaust system design involves empirical as well as theoretical considerations, it is necessary that all systems be tested after installation to ensure their effectiveness. Such tests are made to ensure that the design exhaust air quantity is being maintained into the hoods and that this design air quantity is actually providing the necessary control. This latter point is particularly important where toxic materials are to be controlled.

A swinging-vane anemometer (velometer) will suffice to determine the air velocity into a simple hood with a clearly defined opening. Usually velocity measurements by means of a Pitot tube in the duct behind an exhaust hood are necessary. Other physical measurements which are made to evaluate an exhaust system will include static pressure or throat suction at all hoods, static pressure at both sides of the fan and collector, fan rpm, and brake horsepower.

Determination of the degree of control provided by a design air quantity can usually only be made by chemical analyses for the contaminant at the hood and in the worker's breathing zone. Chemical analyses of air are almost always required also for testing a dilution ventilation system. Such tests are also frequently made at the inlet and outlet of a collector to determine its efficiency.

Exhaust System Maintenance. Proper maintenance is vital because systems will decrease in efficiency for a number of reasons. An inefficient exhaust system may be worse than none at all since it provides a false sense of security. Obvious maintenance procedures include routine inspection to replace broken or corroded hoods, ducts, fan blades, and collector elements, as well as routine cleaning of hoods, ducts, and collectors. Constant surveillance to warn against gradual diminution of air flow is the critical element in maintenance.

Fortunately, such surveillance is relatively simple. After a system has been designed, installed, and tested, the hood suction and the fan inlet suction measured when the system is new and performing at maximum efficiency are accurate indices of the proper quantity of air flow into each hood and the system as a whole. (Systems with cloth dust arresters should be checked in this manner after several weeks of use.) A pressure tap at the fan inlet pipe and at each hood, connected to suitable pressure-indicating devices, will reveal when the air flow decreases significantly. The pressure created by proper air flows can be posted at each gauge, which should be designed to give a visual or audible indication of air flow diminution. In some instances, automatic devices are used to shut down operations if the air flow drops too much.

In summary, a local exhaust system is a mechanical air handling and cleaning system with a number of interrelated components. The inadequacy of any one of these components because of improper design or the failure of any one of these components because of poor maintenance will result in a loss of efficiency and the possible introduction of a health hazard.

Specific Hood Design Considerations. Local exhaust ventilation comprises a more or less elaborately engineered system containing a number of interrelated components. In usual practice, design of such systems is an engineering procedure involving the utilization of aerodynamic principles, mechanical engineering fundamentals, knowledge of the behavior of particles in an air stream, as well as industrial toxicology.

It does not fall within the scope of this book to treat these matters thoroughly or in detail, nor do space limitations permit an exhaustive discussion. Complete treatments are available in numerous published works, a number of which have been listed at the end of this section. When possible, exhaust systems should be designed by an industrial hygiene engineer, or the standard books in the field should be consulted as a complement to the design features to be described below.

However, industrial hygiene engineering is a specialized profession, with very few private consultants available to industry. There is a need in work of this kind, for a simplified approach to local exhaust hood and dilution ventilation design which can be used by the plant or production engineer. Management will not very often have industrial hygiene engineering personnel; there will frequently be the need for emergency installation of exhaust ventilation; and there may not always be the time or personnel for careful study and application of the engineering data in the standard works.

Therefore a simplified approach is offered to meet this pressing need. Required cfm or fpm for basic hood types under average conditions have been indicated and modifying factors for unusual or varying conditions are given. The values used are derived from a consensus of published data tempered by many years of experience. It is felt that intelligent application of the material to follow will result in sufficient data for adequate control. No specific information on design and selection of piping, fans, and collectors will be given, since a wealth of information is available in the various sources of information listed at the end of this section.

The hoods to be described are basic and can be applied, as shown in the figures, to many operations. They can be modified, adapted or even combined in many ways, limited only by the designer's imagination, to meet a particular situation.

As has been previously noted, the hood is the heart of the local exhaust system, and dictates the design of all its other elements. Several considerations bear repeating, i.e., the hood should enclose the operation as much as possible or be as close to it as possible; it should permit air passage with a minimum of extraneous restrictions and obstructions; baffles and flanges should be provided to concentrate flow of air in the desired areas; and the hood cfm should provide a velocity high enough to capture or contain the contaminant. (See also Table 2.4a.)

Special attention must be given to ensure uniformity of flow into the hood face or slot. For a simple hood slot, a gradually tapered transformation from the face to the duct inlet will suffice to provide uniform face velocities. Long hoods can have several transformations. Where a slot is part of a manifold, internal vanes can provide uniform air velocity into the slot. In most cases, uniformity of distribution is designed by providing a restriction at the slot, or within the hood, so that the velocity (and therefore resistance to air flow) in the restriction will be much higher than in the hood face. When this is done, air flow through the restriction will distribute itself uniformly through the restriction, and consequently through the hood opening.

In certain instances, a slot acts as both the restriction and the hood. A simple illustration of this principle is the spray booth line of baffles, where the velocity into the spaces between the baffle plates is much higher than the velocity in the section of the booth behind the baffles. Here, air velocity through the baffle plates, and into the booth face will be uniform. Other examples of providing for distribution include oversized, low-velocity manifolds behind narrow, high-velocity slots and plenums with perforated plates.

Theoretically, the air flow required to control a contaminant using a specific hood depends in most cases only on such physical considerations as temperature, method of contaminant generation, velocity and trajectory of propulsion of contaminant, and distance of hood face to point of contaminant release. However, hood cfm design data must show increasing air flow requirements with increasing toxicity of contaminant. This is so because very few exhaust hoods are 100% effective in controlling all the contaminant released at an operation, and control for the more toxic contaminants which have much lower TLV's must be greater than for the less toxic substances.

The basic hood types include free-hanging plain (square or round) openings, free-hanging slot openings, enclosure with frontal opening, lateral exhaust at a tank or table top, down-draft and canopy hoods. Much data on hood design and cfm requirements for specific operations is available (see references), and the discussion to follow must of necessity be general. The design data for each type of hood are shown in Figs. 2.1 through 2.6.

The enclosing-hood (Fig 2.1) method is preferred whenever operations permit its use. It is the simplest to build, usually does not interfere with operations, and can provide the most efficient control with minimum air flow rates. The enclosure principle finds many applications in local exhaust ventilation, including spray booths, melting pots, cabinet blasting, tunnel drying, belt transfer, mulling, and screening. Most of the spe-

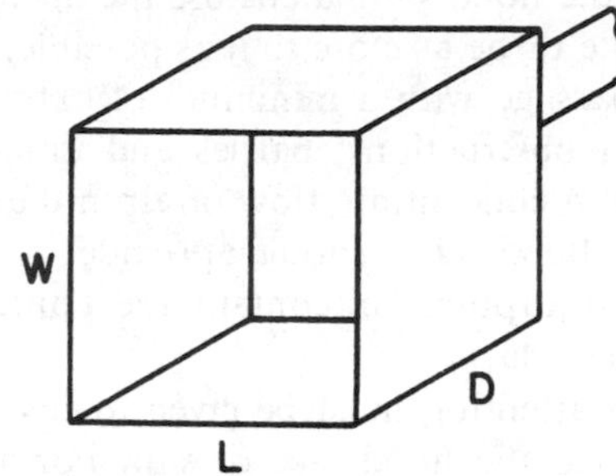

$$cfm = (L)(W)(V)(a)$$

where
L and W are open face dimensions in feet
V is required control velocity at the face in fpm
D is depth from face to rear or line of baffles
a is a factor reflecting temperature of operation and rate of evolution.

V* values for different depths (D)**

Toxicity	Shallow	Average	Deep
Slight	125	75	50
Moderate	150	100	75
High	200	150	125

Rate of Evolution and Temperature Factor (a)

Temperature	Rate	a
Up to 150 °F	Low	1.0
151–300 °F	Moderate	1.25
301–600 °F	High	1.50
601 °F and up	—	2.0

NOTES:
* For extremely toxic and some radioactive substances, V may have to be 200 or 250 for average depths and increased proportionately for shallow hoods. Where a contaminant is released within the hood at a high velocity, particularly toward the opening, the selected V should also be increased.
** The booth should completely contain the operation and allow working room. Depth to the rear or the line or baffles should be as large as possible. Expressing depth as a function of W or L (whichever is the larger), depths from 65% to 85% of W or L may be considered average. Depths up to 65% may be considered shallow and depths above 85% may be considered deep.

Figure 2.1. Enclosing hood or booth.

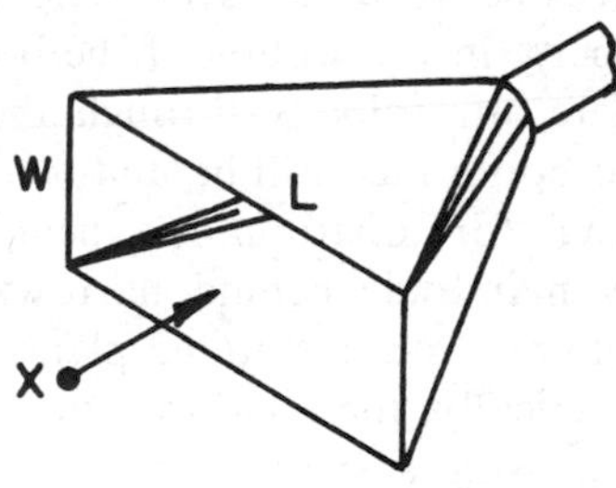

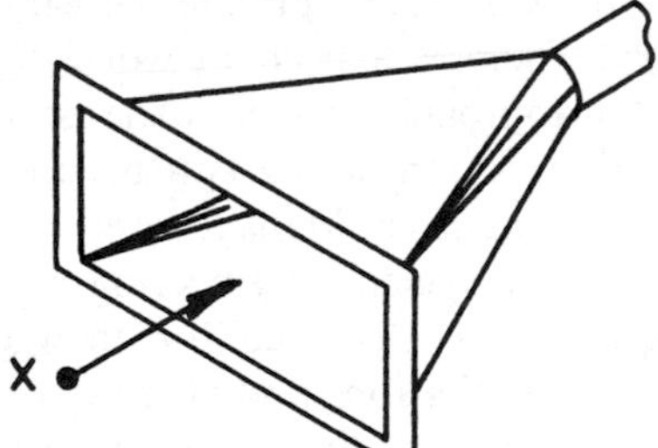

$$cfm = (V)(10X^2 + A)^*$$

where V is required velocity (in fpm) at distance X (in feet) from hood face of area A (in square feet).

V values for different rates of contaminant release**

Toxicity	No Upward Velocity	Moderate Upward Velocity	Active Generation	Violent*** Generation
Slight	50	75	125	500
Moderate	100	150	250	1000
High	150	225	400	1500
For flanged hoods or hoods on table tops the cfm may be decreased by 25%.				

NOTES:
* This formula may be used for openings with W/L ratios above 0.3 as well as square or round openings. This type of hood should not be used where significant cross drafts may be present unless side shields are provided.
** Where reliable capture data are available for a particular operation and contaminant, these data should be used in place of the V values shown above.
*** This type of hood is not recommended for violently generated contaminants.

Figure 2.2. Free-hanging plain openings.

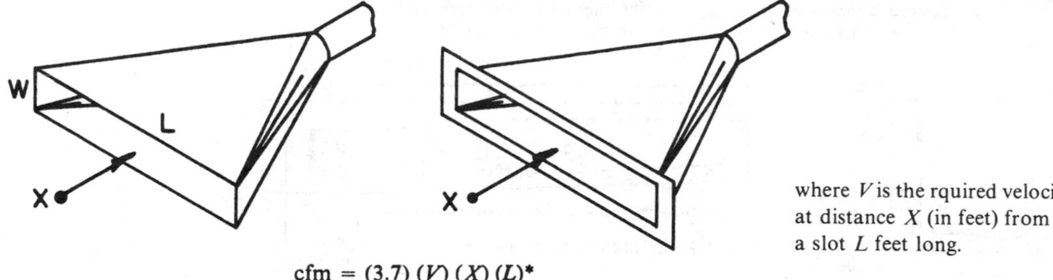

where V is the rquired velocity (in fpm) at distance X (in feet) from the face of a slot L feet long.

$$\text{cfm} = (3.7)(V)(X)(L)^{*}$$

V values for different rates of contaminant release**

Toxicity	No Upward Velocity	Moderate Upward Velocity	Active Generation	Violent*** Generation
Slight	50	75	125	500
Moderate	100	150	250	1000
High	150	225	400	1500

For flanged hoods or hoods on table tops the cfm may be decreased by 25%.

NOTES:
*This formula may be used for openings with W/L ratios of 0.3 or less. This type of hood should not be used where significant cross drafts are present unless side shields are provided.
**Where reliable capture data are available for a particular operation and contaminant, these data should be used in place of the V values shown above.
***This type of hood is not recommended for violently generated contaminants.

Figure 2.3. Free-hanging slot openings.

cially shaped hoods for such equipment as woodworking machines and buffers are adaptions of the enclosure principle, taking advantage of the characteristics of each machine.

Free-hanging plain openings (Fig 2.2) are essentially round or square openings or rectangular openings with width-to-length ratios greater than 0.3. They are used where the contaminant is derived from a point source or a very small area and where operations do not permit enclosure or encumbrance by hoods. They may be permanently mounted in position or may be attached to a flexible pipe for adjustment as operations require. Applications will include arc welding, soldering, tail-pipe exhaust, and high-toxicity, high-precision milling and machining. The latter use requires very careful design and operation.

The basic formula in Fig. 2.2 is derived from Dalla Valle. The formula indicates that the cfm value needed to maintain a required velocity varies as the square of the distance of the hood face from the contaminant source. For this reason, the hood must be kept as close as possible to the contaminant source. If close placement is not possible, use of this type of hood may become prohibitively expensive in terms of required cfm.

Flanging the hood greatly increases its efficiency and should be used whenever possible.

The free-hanging slot opening (Fig 2.3) is a rectangular hood similar to the free-hanging round or square hood which is used in similar circumstances where the zone of contaminant generation requires a long narrow source of suction. The hood may be considered a slot when the ratio of width to length is equal to or less than 0.3.

The formula developed by the late Professor L. Silverman, of Harvard University School of Public Health, indicates that required cfm varies as the distance from the contaminant, and close placement, while important, is not as critical as with the round or square hood. Again, flanging increases efficiency.

Lateral tank or table ventilation (Fig. 2.4) is a modification of the free-hanging slot which finds widespread use in local exhaust ventilation, where operations make enclosing hoods impractical. This design utilizes slots along one or preferably two sides of the tank or table, although rear hoods along one side of the tank or table may also be used. Wherever possible the slot or hood should be located along the long side or sides.

A basic ventilation design has evolved for this type of exhaust in which required ventilation is expressed in terms of cfm per square foot of tank or table top surface. As previously noted in the general discussion of local exhaust ventilation, this method of control has three features—a capture effect, a mass control effect, and a dilution effect (see discussion of Dilution later in this section) for the uncaptured contaminants.

These hoods can be utilized in an operation where the contaminants are released at or immediately above a flat surface. Tank or table top operations, plating and all associated tank operations, degreasing, paint dipping, rubber cementing, and air drying are among the operations amenable to this design. The tank or table width-to-length ratio is an important factor in determining cfm since it reflects the distance over which the air flow must provide control, and should be as small as possible. Figure 2.4 shows basic rates for tanks with a width-length (W/L) ratio of 0.5 with modifying factors for other ratios and for varying conditions.

Downdraft hoods (Fig. 2.5) consist of a grille top

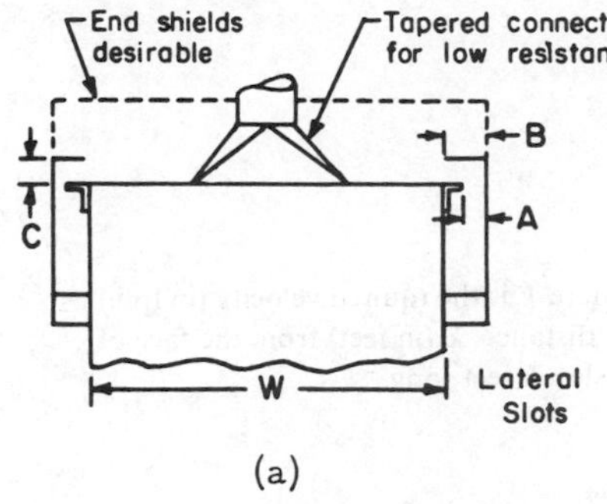

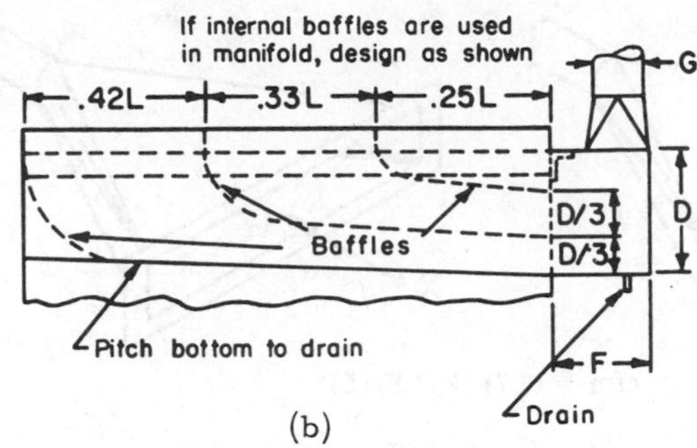

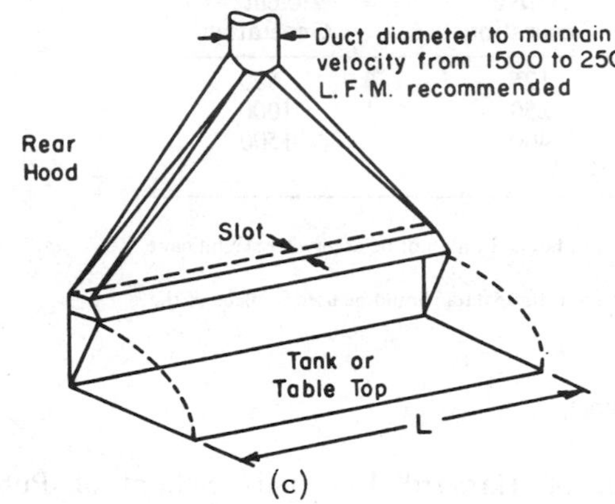

where

Q is cfm per square foot of tank or table surface
W and L are open surface of tank or table top dimensions (in feet)
K is a correction factor for different width-length (W/L) ratios

$$\text{cfm} = (W)(L)(Q)(K)$$

Q values for lateral exhaust at tanks or tables where W/L is 0.5

Toxicity	Flash Point	Low Rate of Evolution, cfm	Moderate Rate of Evolution, cfm	High Rate of Evolution, cfm
Slight	over 200 °F	50–75	75–100	100–150
Moderate	100–200 °F	75–100	100–150	125–175
High	below 100 °F	100–150	125–175	150–200

The lower figure in each pair is for tanks or tables against a wall or baffle; the higher figure for free standing tanks or tables. The hood shown in "c" also serves as a baffle.

Correction factors (K) for different tank W/L ratios

W/L	K	W/L	K
0.1	0.70	0.6	1.05
0.2	0.80	0.7	1.10
0.3	0.90	0.8	1.20
0.4	0.95	0.9	1.30
0.5	1.0	1.0	1.40

Notes to Figure 2.4:

(1) Parts (a) and (b) show end and side views of a tank with lateral slots. Balanced flow through slot C can be achieved by a high velocity inner slot (A), large manifold cross-section area (B × D), or internal baffle vanes as in (b). Slot A width = approximately 1/2 slot C. Use vanes for manifold greater than 6 feet.

(2) Rear hood shown in (c) may be used when the front or ends of the tank must be unencumbered. Balance can be secured by 60° transformation pieces (1 for each 5 feet of hood length) or high-velocity inner slot.

(3) When slots on 2 opposite sides are provided, W is halved for the purpose of determining the correction factor. Side shields should be provided when cross drafts are present. Use 2 slots for tanks more than 2′ wide.

(4) Rate of evolution depends on temperature, gassing, evaporation rate and agitation. At normal room temperature, quiescent operations will have a low rate.

(5) Slots can normally be sized for velocities from 1500–2000 fpm. Rear hood faces can normally be sized for velocities from 200–600 fpm.

(6) Flash points of flammable liquids are also a factor in this type of ventilation design. Both the toxicity and the flash point of such a liquid should be considered and the higher Q value used.

(7) Vapor phase degreasers with condensing coils constitute a special case. Poorly operated or improperly designed degreasers will have a moderate or high rate of evolution. Good degreaser operation will have a low rate of evolution.

Figure 2.4. Lateral ventilation.

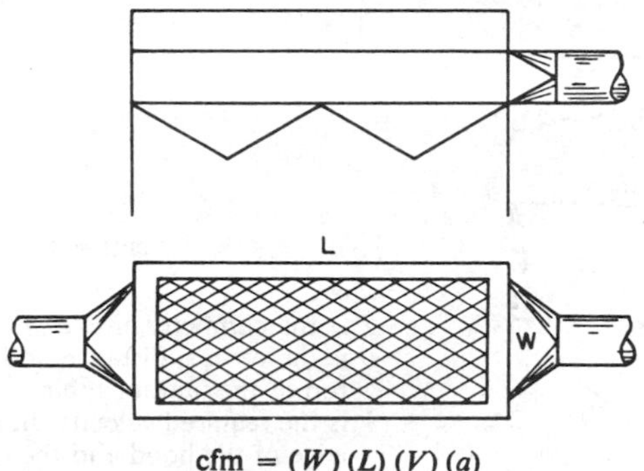

cfm = (*W*) (*L*) (*V*) (*a*)

where *V* is required velocity through grille (based on gross open area)*
a is a temperature correction factor.

V values at various conditions of room cross drafts**

	Cross Drafts			
Toxicity	Low 0 to 35 fpm Draft Velocity	Moderate 35 to 75 fpm Draft Velocity	High 75 to 150 fpm Draft Velocity	Strong*** 150 to 300 fpm Draft Velocity
Slight	100	125	250	350
Moderate	125	150	300	400
High	150	200	350	500

Temperature Correction Factor (*a*)

Temperature of Operations	*a*
Up to 100°F	1.0
100 to 150 °F	1.25
150 to 300 °F	1.50
Above 300 °F	2.00

Notes:
* For a required velocity above the grille use the formula in Figure 2.2.
** Baffles or side shields on one or more sides should be used wherever possible. These will permit selection of a *V* value from a lower cross draft category.
*** This type of hood is not recommended for highly toxic contaminants, very hot processes, or, when unshielded, in locations with strong cross drafts.

Figure 2.5. Downdraft hoods.

table or platform above a hopper or plenum through which air is exhausted. They are used where the operation must be unencumbered on all sides or where it is desirable to have a downward flow of air past the source of contaminant. Examples of operations where these hoods may be used include welding, soldering, metallizing, shakeout, portable tool chipping and grinding, spray painting, and solvent drying. Their efficiency is adversely affected by cross drafts and thermal updrafts, and they are usually used where no other hood type is feasible.

The familiar canopy hood (Fig 2.6) should never be used where the operations require the worker to place his head under the hood since he will thus be exposed to the contaminant-laden air. Nor is its use advised in locations where cross drafts are prevalent. Where only infrequent tending of the operation is needed, the duct to which the hood is attached can be provided with a telescoping joint so that the hood can be lowered very close to the tank or table to overcome the effect of cross drafts.

Canopy hoods are often used at hot processes to take advantage of the thermal updraft. A frequently overlooked feature of such use is the secondary air induced into the area under the hood by the heated air rising into it. The fan must be selected to accommodate this secondary air as well as the design air quantity. Otherwise, air will spill out of the top of the hood. Design of canopy hoods for hot processes is fully discussed in Hemeon, *Plant and Process Ventilation*.

If carefully designed, canopy hoods can be used in a natural draft system sometimes more successfully than in a fan system. Of critical importance in such use are the shape and size of the hood, adequate duct diameter to move the air at low velocities, sufficient length of vertical pipe to create a thermal head and, most important, a positive make-up air supply. In workrooms without adequate make-up air, during winter condi-

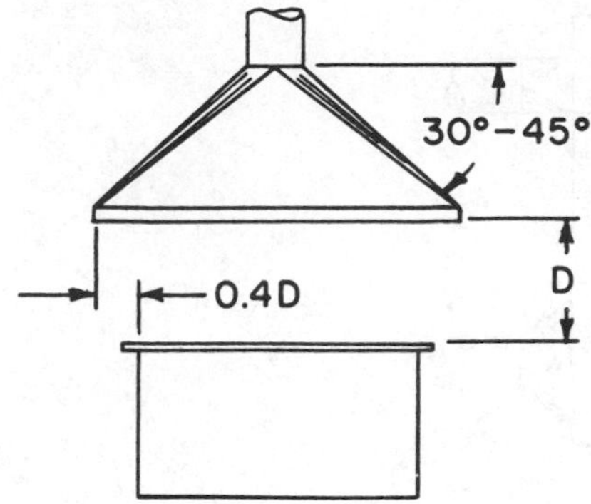

cfm = (1.4) (*P*) (*D*) (*V*)

where
P is the perimeter, in feet, of the hood base
D is the vertical distance, in feet, from the hood base to the top of the tank or table
V is the required velocity, in fpm, into the area between the edge of the hood and the edge of the table.

V values at various conditions of room drafts (unbaffled hood)

Toxicity	No Cross Drafts	Slight Cross Drafts	Moderate Cross Drafts
Slight	75	125	175
Moderate	125	175	225
High	175	225	275

See Figure 2.5 for definition of cross draft categories.
See text for a discussion of canopies for hot processes.

Effect of baffles on required V values

Baffles on two opposite sides	—decrease *V* by 25
Baffles on two adjacent sides	—decrease *V* by 50
Baffles on three sides	—see Figure 2.1, enclosed hoods

V cannot be less than 50 fpm.

Note:
Unbaffled canopy hoods should not be used in strong drafts. Hoods with baffles on two adjacent sides may be used in strong drafts, selecting *V* values from the table above under moderate cross drafts. In all cases where baffles are provided at canopy hoods, they should be perpendicular to the prevailing cross draft direction and on the lee side.

Figure 2.6. Canopy hoods.

tions air flow through natural draft systems may reverse and flood the workroom with contaminants. Wherever possible, canopy hoods should be provided with one or more sides.

Dilution Ventilation. Dilution ventilation consists of general ventilation of a workroom so designed that the contaminants released into the atmosphere are continuously diluted by the introduction of uncontaminated air to levels to which a worker can safely be exposed for 8 hours a day. It is usually applied to the control of contaminants released over such a large area or in such a manner that local exhaust ventilation is impossible, impractical, or prohibitively expensive. It should almost never be used where local exhaust ventilation is feasible or where highly toxic air contaminants are involved. It is most successfully used to control the vapors evaporated from liquids such as solvents or thinners. Operations at which dilution ventilation may be used include dip or roller coating and air drying and cementing.

Dilution ventilation must be designed and tailored to meet a particular situation. Design factors will include rate of contaminant release as well as physical constants of the contaminant, toxicity, and workroom conditions. The random placement of one or two exhaust fans in the wall will usually be inadequate. Similarly, the obsolete concept of air changes/hour cannot be used because it cannot be related to contaminant generation rate. Another obvious drawback to the air change method of ventilation design is that it is a function of room volume. For example, an operation involving volatile solvents may evaporate one pint of acetone/minute and general room ventilation at 12 air changes/hour might be specified. In a workroom 100×100×20 feet, this is equivalent to 40,000 cfm. Identical conditions may exist in a workroom 20×30×10 feet. Twelve air changes an hour in this workroom will require 1200 cfm. The former cfm would be excessive, the latter, much too low for adequate control.

The principal disadvantage of dilution ventilation lies in the high exhaust rates usually required, which result in the need for heating and tempering large volumes of make-up air for efficient operation. The principal advantages are its simplicity, low original cost, and low power requirements.

Dilution ventilation is used primarily to control

vapors and gases which mix readily and uniformly with the workroom air and whose rate of evolution is relatively constant and can be readily determined. It is rarely applied to the control of particulate matter such as dust, fumes, and mists since these do not mix uniformly and their rate of evolution is usually impossible to ascertain. In some instances, empirical data for fume or mist control by dilution are available. One such case is black iron or mild steel welding where cfm dilution rates have been set up on the basis of number of welders in a workroom and size of welding rod used.

Calculation of Dilution Ventilation Rates. One method of calculating cfm dilution rates involves a knowledge of prevailing conditions during normal operations. If tests can be made to determine the actual concentration of contaminant and the rate of natural infiltration of air into the workroom during existing conditions, the required dilution ventilation will be found by the following formula:

$$\text{Dilution (CFM)} = \frac{\text{Prevailing Concentration (ppm)} \times \text{Infiltration (cfm)} \times K}{\text{TLV (ppm)}} \qquad (2.1)$$

where TLV = the threshold limit value
K = a factor of safety (see Table 2.5)

However, these data are difficult to obtain and are subject to error, and the cfm obtained cannot be extrapolated to a future change in conditions which may require lower or higher ventilation rates.

A more rational method is to use data on rate of contaminant release to determine rate of exhaust necessary to continuously dilute the contaminant to safe levels, where

d = density of parent liquid (Section 12)
mol. wt. = molecular weight of parent liquid (Section 12)
TLV = threshold limit value of the vapor in ppm (Section 1, Table 1.3)
K = factor of safety (see Table 2.5)

The formula is:

$$\text{cfm} = \frac{(403)(d)(10^6)(\text{pints of liquid evaporated/min})(K)}{(\text{mol. wt.})(\text{TLV})} \qquad (2.2)$$

Where the evolution rate is expressed in pounds/minute, the formula becomes:

$$\text{cfm} = \frac{(387)(10^6)(\text{lb of liquid evaporated/min})(K)}{(\text{mol. wt.})(\text{TLV})} \qquad (2.3)$$

Equation (2.2) or (2.3) above are only applicable to volatile organic liquids. Dilution ventilation for gases can also be designed, but determination of the rate of generation is much more difficult than with liquids. If the rate can be determined with reasonable accuracy, it should be expressed in terms of pounds/minute, and equation (2.3) should be used. Table 2.6 describes a simplified method of calculating dilution ventilation for substances for which TLV's have not been established using the general formulas and safety factor data shown.

Rate of Contaminant Generation. The amount of liquid evaporated per minute must be determined as accurately as possible. Several methods for such determinations are available. Usually process control data are the most reliable sources. For example, in a coating operation, the amount of solids applied to the product can be determined, and this information, coupled with production rate and coating material composition data, can be used to calculate pints of volatile material released per minute. Often, operating and supervisory personnel can supply fairly reliable information on rate of use of volatile materials. Purchasing department data over a sufficiently large period of time can also be utilized to determine the average amount of volatile material per month. This can be translated into pints evaporated per working minute. Where an operation is performed for only part of a day, the time during which the vapor is being released is used to determine the rate of contaminant generation.

Effect of Distribution and Make-up Air. One of the most important factors in the design of dilution ventilation, and the one most frequently overlooked, is the need to supply sufficient and properly directed make-up air to replace the air exhausted. Ideally, the make-up air will be positively introduced and distributed in such a manner that the source of contaminant is situated between the worker and the exhaust outlet, the exhaust is located as close as possible to the source of contamination, all the air will pass through the zone of contamination, and the contaminant will be diluted to the design levels as soon as it is generated. Under such conditions, the K factor in the formulas could be 1, since no factor of safety would be required. However, ideal distribution conditions can never be assumed and actual conditions will vary considerably.

Average conditions may be assumed to be those

where the toxicity of the contaminant is moderate; the exhaust fan or fans are placed reasonably close to the operation releasing vapors; make-up is by infiltration through doors, windows, and walls, so that a reasonable amount of dilution occurs as the contaminant is generated; and the worker is not too near or in the zone of concentrated contaminant release. Under such circumstances, the average *K* factor may be taken as 4 so that the average concentration in the workroom is $\frac{1}{4}$ of the TLV and the concentration in any part of the workroom will probably not be higher than the TLV.

However, average distribution conditions are probably as elusive as ideal conditions and the need arises for a more rational basis for selecting a factor of safety to reflect actual conditions, whether poor, average, good, or excellent. Poor distribution, such as may be caused by short circuiting the make-up air, inadequacy of infiltration as a source of make-up air, location of the worker directly downstream from or in the evolving vapors, will require a higher factor of safety than 4. A lower factor may be used with good distribution, which will usually consist of a positive mechanical make-up air supply system with properly placed supply and exhaust fans. Excellent distribution will be provided by properly located exhaust fans with a mechanical supply system directing tempered air in the optimum manner by means of diffusers, ductwork, or a perforated plenum. This will make possible a futher reduction in safety factor. The influence of distribution on safety factors is shown in Table 2.5. Typical distribution conditions are shown in Fig. 2.7.

Effect of Toxicity. The TLV or relative toxicity of the substance being controlled enters into the design directly since it is part of the calculation formula. Where the TLV (in ppm) of a substance has been promulgated by a recognized authority such as the ACGIH or the American Standards Association, it is recommended that dilution ventilation rates be calculated using this TLV value in equation (2.2) and applying the safety factors shown in Table 2.5. For other substances, the general equations in Table 2.6, as well as the *K* factors in Table 2.5 should be used.

The general dilution ventilation equations in Table 2.6 have been derived from equation (2.2) using arbitrary values of 400 ppm as the TLV for slightly toxic substances, and 200 ppm for moderately toxic substances, and 50 ppm for highly toxic substances, as shown in Sections 1 and 12.

The relative toxicity enters into the design calculations indirectly in another manner. Theoretically the dilution cfm calculation already reflects the relative toxicity since TLV appears in the formula. However, a local high concentration of vapor caused by poor distribution or other reasons is much less tolerable with a highly toxic substance than with a slightly toxic substance. The factor of safety must reflect this situation. Accordingly, assuming average distribution conditions, basic safety factors (*K*) of 4, 5, and 8 have been assigned to slightly, moderately and highly toxic substances respectively. Thus the lower air volumes required by a well-engineered exhaust-supply system will afford considerable savings in heating costs.

(Note: Table 2-7 lists a number of common organic liquids, the TLV's and cfm values required to dilute the vapor released by 1 pint to the TLV. These can be multiplied by the appropriate *K* factor in Table 2.5 and the rate of evaporation in pints/minute to obtained required dilution cfm values.)

Mixtures. The above considerations apply to vapors released by a single substance. Frequently, dilution ventilation must be applied to vapors released by a mixture of volatile liquids, such as the volatile portion of the average lacquer. The systemic effect of inhalation of mixtures of solvent vapors cannot be readily predicted or evaluated. If more knowledge were avail-

TABLE 2.5. *K* factors at different distribution conditions (see Fig. 2.7)

Toxicity	Poor	Average	Good	Excellent
Slight	7	4	3	2
Moderate	8	5	4	3
High	11	8	7	6

TABLE 2.6. Dilution cfm Formula

Slightly toxic substances	$\text{cfm} = \dfrac{(1 \times 10^6)(d)(\text{pints evaporated per min})(K)}{\text{mol wt}}$
Moderately toxic substances	$\text{cfm} = \dfrac{(2 \times 10^6)(d)(\text{pints evaporated per min})(K)}{\text{mol wt}}$
Highly toxic substances	$\text{cfm} = \dfrac{(8 \times 10^6)(d)(\text{pints evaporated per min})(K)}{\text{mol wt}}$

Note: d = specific gravity of liquid (section 12, density)
mol wt = molecular weight of liquid (Section 12)

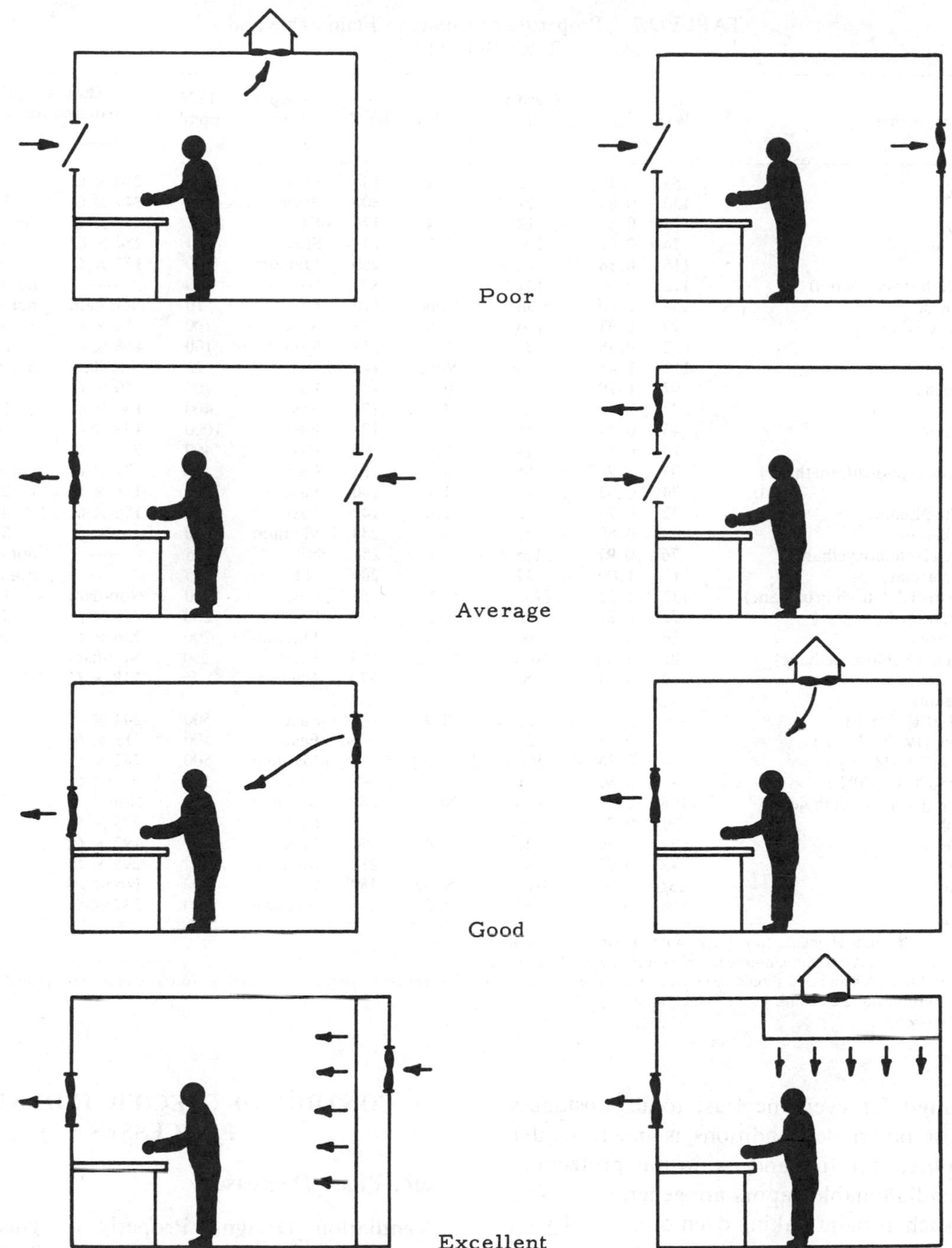

Figure 2.7. Typical distribution conditions. Courtesy *Industrial Ventilation Manual*, 7th edition, American Conference of Governmental Industrial Hygienists.

able, a rational method for designing dilution ventilation for such mixtures could be formulated. In the absence of such information, dilution ventilation rates should be calculated for each component of the mixture and the cfm values obtained added together to give a total cfm rate. A method which is occasionally used to afford greater safety is to assume that the entire volume of mixture being evaporated consists of the component with the highest cfm rate per pint and to calculate cfm accordingly.

Fire and Explosion Safety

Dilution ventilation for health protection will always provide sufficient protection against flammable vapor fire and explosions since the health protection ventila-

TABLE 2.7. Properties of Common Flammable and Toxic Solvents[a]

Name	Mol. Wt.	d	Flash P °F	lel. %/Vol.	bp °F	Evap. Rate	TLV ppm[b]	Cu ft air/pint solvent required for dilution below	
								lel[c]	TLV[d]
Acetone	58	0.79	0	2.2	134	Fast	1000	210 × C	5,500 × K
n-Amyl acetate	130	0.88	77	1.1	300	Slow	100	244 × C	27,200 × K
Benzol (benzene)	78	0.88	12	1.4	176	Fast	10	290 × C	not recommended
n-Butanol (butyl alcohol)	74	0.81	100	1.7	243	Slow	50	254 × C	88, 200 × K
n-Butyl acetate	116	0.88	72	1.7	260	Medium	150	177 × C	24,000 × K
Butyl cellosolve (2-butoxyethanol)	118	0.90	141	—	340	Nil	50	——	not recommended
Carbon tetrachloride	154	1.60	None	None	170	Fast	10	Non-flam	not recommended
Cellosolve (2-ethoxyethanol)	90	0.93	104	2.6	275	Slow	100	156 × C	41, 600 × K
Cellosolve acetate	132	0.98	124	1.7	313	Slow	100	168 × C	20,970 × K
Chloroform	119	1.48	None	None	142	Fast	25	Non-flam	not recommended
1,2-Dichloroethylene	97	1.29	43	9.7	141	Fast	200	50 × C	26,900 × K
Ethyl acetate	88	0.90	24	2.2	171	Fast	400	184 × C	10,300 × K
Ethyl alcohol (ethanol)	46	0.79	55	3.4	173	Fast	1000	193 × C	6,900 × K
Ethyl ether	74	0.71	49	1.9	95	Fast	400	205 × C	9,630 × K
Ethylene dichloride (1,2-dichloroethane)	99	1.26	56	6.2	181	Fast	50	77 × C	not recommended
Methyl acetate	74	0.92	14	4.1	140	Fast	200	118 × C	25,000 × K
Methyl alcohol (methanol)	32	0.79	52	6.0	147	Fast	200	156 × C	49,100 × K
Methyl isobutyl ketone	100	0.80	73	—	244	Medium	100	——	32,300 × K
Methyl cellosolve (2-methoxyethanol)	76	0.97	105	—	255	Nil	25	——	not recommended
Methyl cellosolve acetate	118	1.00	132	—	289	Nil	25	——	not recommended
Methyl chloroform (1,1,1-trichloroethane)	133	1.33	None	None	165	Fast	350	Non-flam	11,520 × K
Methyl ethyl ketone (2-butanone)	72	0.81	30	1.8	176	Fast	200	250 × C	22,750 × K
Methyl propyl ketone	86	0.82	60	1.6	216	Medium	200	236 × C	19,000 × K
Methylene chloride (dichloromethane)	85	1.34	None	None	104	Fast	200	Non-flam	31, 875 × K
Monochlorobenzene	113	1.11	85	1.8	270	Medium	75	218 × C	5,290 × K
Naphtha (petroleum)									
Low boiling (140°F–206°F)	—	0.70	50	0.9	—	Fast	500	344 × C	7,000 × K
Medium boiling (196°F–250°F)	—	0.73	50	to	—	Fast	500	313 × C	6,000 × K
High Boiling (217°F–288°F)	—	0.76	100	1.3	—	Medium	500	282 × C	5,400 × K
Safety Solvent (300°F–400°F)	—	0.80	100–110		—	Slow	500	250 × C	4,600 × K
Perchloroethylene (tetrachloroethylene)	166	1.62	None	None	249	Medium	100	Non-flam	39,600 × K
Isopropyl alcohol	60	0.79	53	2.5	181	Fast	400	206 × C	13,200 × K
Isopropyl acetate	102	0.89	40	1.8	194	Fast	250	192 × C	14, 000 × K
Toluol (toluene)	92	0.87	40	1.3	232	Medium	100	295 × C	38, 000 × K
Trichloroethylene	131	1.47	None	None	189	Fast	100	Non-flam	45,000 × K
Xylol (xylene)	106	0.88	63	1.0	291	Medium	100	332 × C	33,000 × K

[a]Courtesy, Division of Industrial Hygiene, New York State Department of Labor.
[b]1977 Threshold Limit Values, American Conference of Governmental Hygienists.
The latest TLV list and OSHA standards should be consulted. If either of these show different values they should be used. Note also that a change in TLV from the above value will result in a proportionate change in the last column.
[c]See "ovens" in this section.
[d]See Table 2.6.

tion rate required for even the least toxic substances under the most favorable conditions is much greater than that required for fire and explosion protection. However, when flammable vapors are generated within an enclosure, such as paint baking oven or an air drying cabinet, where workers are not exposed to vapors, dilution ventilation of the enclosure is still required to dilute the vapors to below the lower explosive limit. The formula in this instance is:

$$\text{cfm} = \frac{(403)(d)(\text{pints evaporated/min})(100)(C)}{(\text{mol. wt.})(\text{lel})(B)} \qquad (2.4)$$

where lel = lower explosive limit in % (Section 12)
C = Factor of safety
B = lel correction factor

(Note: See discussion of flammable atmosphere ovens.)

CONTROL OF SPECIFIC INDUSTRIAL PROCESSES

Vapor Phase Degreasers

Ventilation Design. Properly designed, operated, and located degeasers do not normally require local exhaust ventilation if adequate, draft-free general ventilation is available to the area. However, the usual degreasing operation is such that local exhaust ventilation is required to control the vapors released.

Design data will be found in Fig. 2.4. The most commonly used degreasing solvents—trichlorethylene and perchlorethylene—fall into the high-toxicity class (Hazard Rating 3; see Section 12). The "Freons" which have recently been introduced for degreasing are mostly of low toxicity (Hazard Rating 1, Section 12).

The rate of solvent vapor evolution requires individual subjective evaluation. Any conditions of design, operation, or location which cause excessive "dragout" of solvent vapors will create a high rate of evolution. Average conditions will result in a low or moderate rate. Where operations are such that the work is not dry upon removal from the tank, secondary ventilation, as in Figs. 2.4c or 2.5, may be required to control the vapors evaporated from the work subsequent to its removal from the tank.

The most commonly used hood for degreasing is the "lateral slot hood" shown in Fig. 2.4. The outer slot C should be sized for velocities between 400 and 600 fpm to minimize loss of solvent by dragout of vapors.

Equipment Design. Thermostatic control is required to prevent the temperature of the liquid from rising in excess of 20°F above its boiling point, in the vapor phase to prevent the rise of the vapor level above the condenser, and in the cooling water to maintain condensing temperatures between room temperature and 110°F.

The size of the tank should be such that the horizontal cross-section area of the work basket or loading rack is less than two-thirds of the horizontal cross-section area of the tank.

The height of the tank should be such that during normal operations the distance between the top of the vapor zone and the top of the tank is at least 15 inches and not less than two-thirds of the tank width.

The combustion chambers of gas- or oil-heated tanks shall be enclosed except for necessary openings and independently vented to the outdoors by means of a corrosion-proof pipe. Natural draft pipes shall have draft diverters.

Tight-fitting tank covers shall be provided and kept closed whenever the degreaser is idle. The covers shall not interfere with the exhaust hood when in an open position.

Location. Wherever possible, the room in which a degreaser is located shall have a volume of at least 20,000 cubic feet. Degreasers located in smaller rooms shall have local exhaust ventilation, even if not otherwise deemed necessary.

Ample clearance shall be provided on all sides.

Degreasers should not be located near pits in the floor, corner pockets, or dead air spaces. Pits containing large degreasers shall be large enough for a worker to enter conveniently for cleanup and maintenance, and should be provided with mechanical ventilation.

Degreasers should be located away from other exhaust systems, in strong cross drafts or in areas with ambient air velocities in excess of 50 fpm. When such a location is unavoidable, provide a shield or baffle upstream.

Locate degreasers away from open flames, electric heaters, welding torches, or other sources of ignition or excessively hot surfaces.

Operation. The operator should be taught to follow the manufacturer's instructions.

If possible provide a mechanical loading hoist with a maximum vertical speed of 11 fpm. If not, the work should be introduced and removed as slowly as possible and hooks or stops should be provided to permit convenient holding of the work in the boiling liquid and vapor zone without the necessity of the worker bending over the tank.

Use only the designated solvent containing suitable inhibitors.

Remove adhering aluminum dust or chips from parts before degreasing.

For small parts, use only metal racks, wire, or perforated metal baskets. Place the work so as to facilitate draining and to prevent cupping or retention of liquid. Do not load beyond design capacity.

Hold the work in the vapor zone until drainage and condensation of liquid on the work stops and the work is completely dry. Hold above vapor zone briefly before removal from the tank.

When removal of wet work is unavoidable, provide a ventilated drying station as described above in Ventilation Design.

When spray nozzles are used, always operate with the nozzle below the vapor level.

Maintenance and Cleaning. Inspect all piping, valves, pumps, thermostats, etc., periodically in accordance with the manufacturer's instructions.

Clean tanks and pits frequently.

(1) Allow tank to reach room temperature, drain contents completely, open all cleanout doors, shut off and lock all power switches, and ventilate thoroughly.

(2) Provide the worker with suitable protective clothing, air line respirator and, if it is necessary for him to enter the tank or pit, a safety harness. A trained worker should be stationed nearby to render immediate assistance in the event of an emergency. All ventilation should be operated continuously during the cleaning process.

(3) Store collected sludge and residue in a covered container until it can be disposed of safely.

Air Contaminant Emissions. The nature of vapor degreasing is such that the concentration of contaminant discharged by the exhaust system is usually rarely sufficiently high to constitute an air pollution problem. In addition, the high cost of most degreasing solvents acts as an economic self-limiting factor. A degreasing exhaust system will create excessive air contamination only when operating conditions are very poor. Steps should be taken to improve operations, not only to avoid the installation of expensive air-cleaning equipment, but also to prevent the loss of costly solvent. (Although trichlorethylene, a common degreasing fluid, has not been considered an air pollution problem because of the relatively small amount of emissions from the average degreaser, it has been incriminated as a photochemical oxidant source and allowable emissions have been sharply reduced).

Potential solvent emissions of a system can be estimated from its operational losses which consist of liquid solvent entrapped in the sludge at the bottom of the tank, evaporation to the workroom, and vapors captured by the exhaust hood and discharged to the outdoors. If we assume the 25% of the losses are accounted for by the first two factors, 75% of losses will be emitted as environmental contaminants. (See also Table 2.15.)

Air Cleaning. The only practical air-cleaning methods for vapors released by degreasing are adsorption and condensation. Combustion of trichlorethylene is not practical because of the high temperature required and the hydrochloric acid produced which is more objectional than the original contaminant (see Industrial Air Cleaning—Gases and Vapors).

Spray Coating (Fig. 2.8)

I. Design Data (see also Fig. 2.1).

(a) Both worker and operator inside of booth:

W = Width of widest work plus 6 ft.
H = Height of tallest work plus 3 ft.
C = Depth of deepest work plus at least 3-ft clearance between work and baffle line and between work and face of booth.
V = 100–200 fpm through booth cross section.
cfm = $W \times H \times V$.

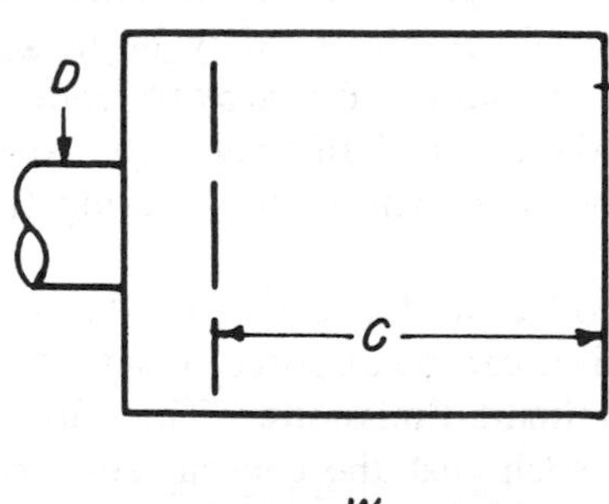

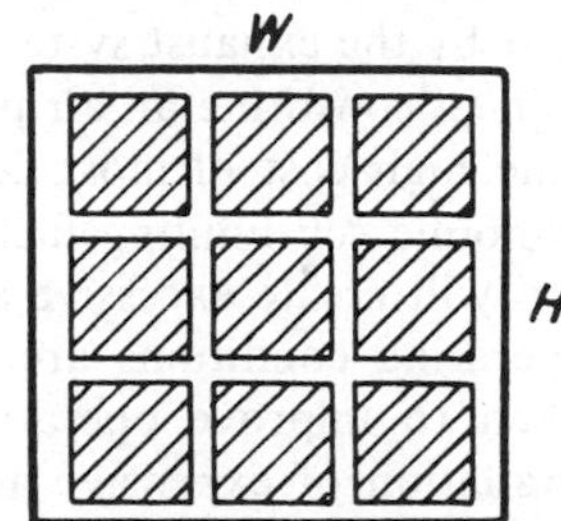

Figure 2.8. Spray booth.

(b) Work inside booth—operator in front of booth:

W = width of widest work to be sprayed plus 12 in.
H = Depth of deepest work plus 1 ft rear and 1.5 ft front clearnace. Depth shall be at least equal to three-quarters of the heigth or width, whichever is greater.
V = 100–200 fpm through face area and all other openings.
cfm = $W \times H$ (minus any approved face obstructions down from booth roof or up from floor) $\times V$

(c) 100 fpm minimum velocity: 150–200 fpm is required velocity for shallow booths, spraying toxic material (Hazard Rating 3, Section 12), presence of strong cross drafts, unavoidable rebound because of size and shape of work.

(d) Baffles—open area between baffle plates from 25–50% of cross-sectional booth area. Distance between line of baffles and bottom of booth from 3 to 5 times the distance between baffles and the sides and top of booth.

(e) Where worker is outside of the booth, reduce face area by baffling the frontal opening as much as possible.

II. Fan and Piping

(a) Cast, nonferrous metal-bladed propeller fan, belt-driven, motor outside of duct or air stream, belt covered, all moving parts electrically grounded. For systems with high resistance, use tube axial or centrifugal (backward curved or radial blade) fans. Forward curved centrifugal fans (squirrel cage or sirocco) should not be used.

(b) Locate fan as close as possible to the point of discharge to the atmosphere.

(c) Select fan and pipe diameter for approximately 2000 fpm (except where long runs dictate smaller diameters and higher velocities).

(d) Pipe seams and joints should be of airtight construction. Airtight construction mandatory for pipes under pressure inside of the workroom or building.

(e) Interlocking of fan with spray gun by means of solenoid desirable.

(f) No dampers, blast gates, bird screens in ductwork.

III. Construction and Installation

(a) Gauge of metal from 18–22 depending on size. Standard air tight construction.

(b) Inner booth walls of galvanized sheet steel or equivalent fireproof, smooth, hard, impervious construction. Floor of nonsparking, nonflammable material. No pipes, cabinets, beams or other encumbrances in booth.

(c) Windows, if used, shall be of $\frac{1}{4}$-inch safety glass or wired glass, maximum dimensions 18 × 54 inches.

(d) Locate booth away from strong drafts and at least 20 ft from open flames, sparks, welding, or other sources of ignition.

(e) Ground all metal parts, paint containers, paint hose, etc.

(f) Sprinklers to be protected against paint deposition by a film of light grease or lightweight paper bags. Provide fire extinguisher.

(g) No storage or combustible construction within 3 feet on all sides.

(h) For heavy, continuous spraying in booths without water wash or paint filters, provide flanged pipe-joints for convenient cleaning. Flanges every fourth joint recommended in all cases.

IV. Electrical

(a) Fan motors not to be located in any hood, booth, or duct. Fan belts to be effectively grounded and completely enclosed.

(b) Wiring within a booth and within 10 ft of booth opening to be in sealed metal conduit with explosion-proof fittings (UL approved).

(c) Lamps are not permitted in a booth when they may be subject to accumulations of flammable residue.

(d) Lamps within a booth, when permitted, and within 10 ft of booth opening to be totally enclosed, protected against breakage by location or guards and provided with sockets with nonmetallic shells.

(e) When such lamps are behind transparent panels in the walls or ceiling of a booth, the panels shall comply with Item III(c), shall effectively isolate the interior of the booth from the lamps, and shall be separated from the lamps by an air space.

(f) Switches, controls, and all other electrtical equipment within a booth or within 10 feet of the booth opening shall be explosion proof (UL approved).

V. Operations

(a) Locate the work as deep in the booth as possible. Position work and operate gun so that spray is directed at an angle upstream toward the rear of the booth. Spray gun to be held at optimum distance from work so that proper coverage is achieved with minimum overspray and rebound.

(b) Use turntable when possible. Turntable to be rotatable 360° while supporting the largest piece of work.

(c) When in the booth do not spray toward the front or perpendicular to the air stream. If such spraying is unavoidable, and in all cases where toxic materials are sprayed inside the booth, use approved respirator.

(d) Alternate spraying of materials, the combination of which may result in spontaneous ignition, should be avoided. Where it must be done, clean booth, fan, and piping before each change of material.

(e) Separate booths must be provided for exclusive spraying of hydrogen peroxide, perchlorates, and other strong oxidizers.

VI. Special Types and Applications

(a) Downdraft (see also Fig. 2.5): average conditions, 100 fpm minimum velocity into gross downdraft area; other conditions, see Fig. 2.5; avoid cross drafts; provide side shields if possible; grates to be nonferrous metal; projected area of work to be no greater than 67% gross area of grate.

(b) Conveyorized booth: Area of side conveyor openings as small as possible; openings as far from front as possible; side opening area to be included with face area for cfm calculations. For automatic painting, interlock so that paint flow ceases if conveyor stops or if fire starts.

(c) Electrostatic Spraying.

(1) Location: Electrostatic apparatus and devices used in connection with paint spraying, dipping, or other coating operations shall have transformers and power packs located outside spray booths or enclosures and areas subject to paint deposit.

(2) Supports: Electrodes and electrostatic atomizing heads, shall be rigidly suported and permanently installed and shall be effectively insulated from ground.

(3) Clearance: Space of at least twice the sparking distance shall be maintained between articles being finished and electrodes, electrostatic atomizing heads, or conductors.

(4) Articles on conveyors shall be so arranged and supported as to maintain the required clearances.

(5) Electrostatic apparatus shall be equipped with automatic controls arranged to stop the entire operation if the ventilation of the spraying or drying area falls below a velocity of 100 fpm, if an article on the conveyor projects into the required clearance, or if the conveyor stops.

(6) Guarding and isolation of process: The electrical field and all parts of the equipment carrying high potential shall be located, guarded, and fenced off to provide safe isolation of the process. Guards and fences shall be of conducting material and shall be grounded.

(7) Signs: A suitable sign stating the sparking distance shall be posted conspicuously near the assembly. Signs designating the process zone as dangerous by reason of fire and accident hazard shall be posted.

(8) Insulators: Insulators shall meet test standards for the potential used and shall be maintained clean and dry.

(d) Airless spraying (hot, steam, hydrostatic): Locate heaters outside of booth; interlock heater with exhaust fan; provide suitable hose, valves and fittings for high pressure applications; locate equipment, including paint hose, so that in the event of a rupture flammable material will not be discharged into an ignition source.

(e) Multiple booth systems: With a single fan, balanced air flow should be provided by system design and not with blast gates. If individual fans are used, all fan motors to be interlocked so that all must operate together.

VII. Maintenance

(a) Clean booth fan, and piping frequently. Solvents, if used, must be nonflammable and nontoxic (Hazard Rating 1).

(b) Interior walls of booth can be coated with nonflammable strippable or water soluble compound.

(c) Use nonferrous cleaning tools only. If deposits are dry, wet them before cleaning.

(d) Wet and clean booths, fans, and piping before repair, dismantling, or alteration.

(e) Check fan belts and pulleys periodically.

(f) Store cleaning rags, residues, loaded filters, etc., in covered containers, preferably under water, and dispose of daily.

VIII. Air Contaminant Emissions

(a) Minimize overspray by proper spraying techniques.

(b) Minimize overspray by electrostatic or airless spraying.

(c) Potential emissions will vary widely with operations. See Table 2.15 (Emissions).

IX. Air Cleaning

(a) Water wash: Interlock water pump with fan; maintain water filter, pump, and nozzles in clean condition; dispose of slurry in wet condition.

(b) Dry paint filters: Do not use for materials or combinations of materials which are susceptible to spontaneous ignition; provide pressure-actuated interlock between filters and spray guns so that the latter cannot operate when the booth ventilation velocity falls below the design amount (replace filters at this point); provide access doors and sprinklers in plenum behind the filters.

Cleaning Efficiency

	Volatiles (%)	Nonvolatiles (%)
Water wash	0–10	50–90
Filters	0	90
Baffles	0	Indeterminate

(c) Baffles: Air distributing baffles provide an incidental air cleaning effect by impingement; cleaning can be improved by counter-current flanges on baffle plates and by using triple row of staggered, flanged baffles.

Flammable Atmosphere Ovens

The flammable atmosphere oven is one of the most prevalent pieces of equipment found in industry. It is used to evaporate flammable liquids from objects coated or impregnated with a variety of materials as well as for any process requiring evaporation of volatile flammable liquids. Two hazards must be controlled: fire and explosion may result from any source of ignition in the presence of flammable concentrations of vapor, and escape of toxic vapors into the workroom may create a health hazard.

Common Industrial Ovens. There are two main types of oven: the batch type in which the work is loaded, heated for a period of time, and removed; and the conveyor type through which the work passes continuously entering freshly coated through one opening, and emerging dry through a second opening, or the same opening.

Ovens are usually arranged with either a recirculating system in which a fixed portion of the heated vapor-laden air is continuously recirculated through the oven and the balance exhausted to the outside, or nonrecirculating with continuous exhaust of air from the oven. In direct fired ovens, the source of heat may be in contact with the air within the oven. In indirect ovens, an intermediate heat exchanger is used to convey heat from the source to the oven atmosphere. Gas, oil, steam, or electricity are the heat sources. Indirect fired gas or oil ovens require independent venting of the combustion chamber to remove the products of combustion. In all other cases, the ventilation aspects of oven design are concerned only with removal of vapors, which in the case of direct fired gas or oil heated ovens are intimately mingled with the products of combustion.

Ventilation. Oven ventilation has two functions. The first is to provide safety by diluting the evolving vapor concentration within all parts of the oven to well below the lel (see Fire and Safety Dilution Ventilation this Section and also Section 7). This ventilation is particularly important at the peak evaporation rates which occur during the early part of a batch cycle or near the entrance end of a continuous oven. Also, the ventilation must provide protection for the worker by maintaining a controlling velocity into all operating openings of the oven to prevent escape of vapor into the workroom. Ventilation rates are calculated on the basis of both criteria, and the higher rate is used. Economical oven design will minimize both the amount of solvent introduced into the oven as well as the area of the oven opening (the term "solvent" denotes any volatile flammable liquid).

(*a*) *Safety Ventilation:*

(1) Determination of solvent quantity evaporated into the oven. Before ventilation design can be undertaken, the amount and nature of the solvents evaporated into the oven by the drying process must be known, since safety ventilation rates will depend on these factors.

The amount and composition of solvent present in the coating or impregnating material must be determined, taking into account the amount of thinner added to the original material. This information can usually be obtained from the supplier. In the average dipping process, the coating material may contain up to 50% by volume of solvent. In spray coating, the amount may be as high as 85%. In fabric coating, the material will usually contain less solvent. A typical baking paint or enamel will contain petroleum naphtha; some formulations contain toluene or xylene as well. Some processes require a coating material containing mixtures of the alphatic and aromatic HC, various esters, ketones, and alcohols. Thus, fairly accurate information on the ingredients present is important.

A simple and widely used method of estimating the quantity of solvent evaporated is to utilize plant records of coating material used per specified period of time and number of oven batches or hours of oven operation in the same period. If the solvent percentage is known, a simple calculation will yield volume of solvent per batch or per hour of oven operation. Another method of determining solvent use involves weighing the solids deposited per unit of surface area and calculating back to the original coating material. The increased accuracy of this method is usually not sufficient to warrant the trouble involved unless the information is also desired for operational purposes.

The difference between the solvent in the coating material as applied and that on the work entering the oven may arise from a number of conditions, as follows:

- Overspray: This will depend to a large extent on the spraying technique and the size and shape of the work being sprayed. With good technique, the overspray for large flat surfaces may be as low as 10%, i.e., 10% of the material leaving the spray gun does not reach the work. For small or open-surfaced objects, or with poor technique, the overspray may reach 40–80%.
- Dipping: In dipping, especially with low-viscosity substances, there will be some material dripping from the object as it is raised from the tank. Some of this will drip back on to the drainboard or floor before reaching the oven. Evaporation from the tank is also a factor.
- Evaporation: There is an inevitable evaporative loss of solvent between the time the article is coated and the time it reaches the oven. This will depend on the coating process, the ambient room air temperature and velocity, distance from oven, speed of conveyor travel, amount of incidental or operational preliminary air drying, thickness of coat, evaporation rate, and other factors.

These pre-oven solvent evaporation losses are so difficult to estimate with accuracy that they are frequently ignored. Where they may be large enough to drastically affect the required oven heat capacity and ventilation rates, they can be considered, but extreme precaution must be observed because to overestimate the pre-oven solvent losses in the design of oven ventilation may introduce a serious fire and explosion hazard.

(2) Ventilation rates. The internal oven ventilation rate required to dilute the solvent vapor evaporated within the oven can be calculated from the following formula based on Avogadro's number at 1 atmosphere and 70° F:

$$\text{cfm (at oven conditions)} = \frac{(403)(d)(\text{pints/min})(100)\dfrac{(T + 460)}{(530)}(C)}{(\text{mol. wt.})(\text{lel})(B)} \qquad (2.5)$$

where d is the specific gravity of the solvents.

For batch ovens, pints of solvent evaporated into the oven is determined by dividing the amount of solvent evaporated per batch by the time of the batch. For continuous ovens, it is the amount of solvent entering the oven per minute.

The quantity T is the oven temperature in °F. The T factor in the equation reflects the volume of expanded air at the elevated temperature of the oven.

The factor of safety C is designed to maintain at least $\frac{1}{4}$ of lel at all times. For continuous ovens $C = 4$. For batch ovens, where the average evaporation rate in pints/minute will not reflect the higher and usually indeterminate peak evaporation rate, the C factor is taken at 10 to ensure the maintenance of $\frac{1}{4}$ of the lel at the peak rate.

"Mol. wt." is the molecular weight of the solvent; lel is the lower explosive limit in per cent of the solvent vapor at standard conditions.

The factor B reflects the decrease in lel at elevated temperatures. It may be ignored for temperatures up to 250° F. For the usual oven temperatures between 250 and 500° F it is assumed to equal 0.7.

The formula refers to single substances. Since the lel of a particular mixture of solvent vapors can not usually be calculated accurately and is difficult to determine experimentally, it is normal practice to regard a mixture as consisting entirely of that component requiring the largest amount of dilution air/pint and to determine the air quantity on that basis (see Table 2.7 for constants and dilution air/pint data for a number of common solvents).

(*b*) *Health hazard ventilation.* A ventilation rate from 50–100 cfm/ft^2 of oven opening will be adequate to prevent vapors from escaping into the workroom. The total cfm thus obtained, multiplied by the temperature correction factor, will be the hygienic ventilation rate. The required ventilation will then be this rate or the safety ventilation rate, whichever is greater.

In most batch ovens the latter will be greater. However, in many continuous ovens, the entrance and exist openings may be so large that the hygienic ventilation rate will be greater and may result in a cfm so excessive as to make it prohibitively expensive to maintain the required oven temperature. To overcome this, "air seals" may be used.

In the simplest and safest form of the air seal, the ends of the oven are provided with independently ventilated vestibules or canopy hoods to control the contaminated air which may spill out of the oven which is ventilated at the safety ventilation rate. Alternate designs make use of one fan to provide both oven safety ventilation and an "air curtain" at the oven opening by cycling part of the air handled by the fan into a vestibule and back into the oven. The latter technique has the operational advantage of preheating the work and the air entering the oven. However, unless carefully designed, it will not provide adequate control of vapors escaping from the oven and, more importantly, the internal oven exhaust feature of the system may not provide adequate dilution ventilation in all parts of the oven.

Equipment design features are as follows:

(1) Location of make-up air inlets, recirculating vents, exhaust outlets to ensure continuous ventilation in all parts of the oven, especially near the flame or heating element.

(2) Oven heating system (except steam) to be located at least 20 feet from operations and/or storage areas involving flammable materials (Section 7). Floor surface of oven to be of incombustible material and to extend at least 1 ft beyond oven outline.

(3) Oven to be constructed of noncombustible material throughout. Expansion joints to be provided in oven framing. Oven interiors to have smooth surfaces.

(4) Explosion relief vents (1 square foot area per 15 cubic feet of oven volume) to be provided, preferably in top. Inner panel edges of the vents to be maintained free of condensate.

(5) Discharge ducts not to be connected to other ventilating systems, chimney, or flue. Ducts to be of noncombustible material and free of obstructions and screens.

(6) Two-inch clearance or $\frac{3}{4}$-inch insulation to be provided for ducts passing through combustible walls, floors, and roofs. Duct external surface temperature not to exceed 160° F at these points.

(7) Products of combustion from indirect fired ovens to be exhausted or vented to the outside, preferably by means of an independent system.

Safeguards and interlocks include the following:

(1) Exhaust fan to be interlocked with heat supply by means of air flow switch so that heat supply is cut off unless fan is maintaining required air flow. Switch to be installed preferably on suction side of fan.

(2) Recirculating fans to be interlocked with heat supply be means of air-flow switch so that heat supply is cut off unless fans are operating. Rotational or electrical interlock recommended as additional safety measure.

(3) Conveyors to be interlocked with oven exhaust and recirculating fans so that conveyors cannot operate unless fans are on.

(4) Switch for fans to include time-delay relay to secure proper preventilation. If necessary, oven doors to have limit-switch interlock to prevent time-delay from operating unless oven doors are wide open.

(5) Air supply or exhaust volume control dampers to be properly adjusted and permanently fixed to maintain proper air supply. Means to be provided to prevent full closure at any time.

(6) Excess temperature limit switch with manual reset to be provided in order to cut off heat supply when temperature exceeds safe limit.

(7) Pressure switch with manual reset (gas and oil ovens) to be installed in fuel line to shut off fuel supply when pressure is insufficient.

(8) Combustion (flame) failure safeguards to be provided to shut off fuel supply in the event of flame failure (gas and oil ovens).

Operating features are as follows:

Ovens must be designed to meet a particular set of conditions. Many fires and explosions occur as a result of changing these conditions without making the necessary changes in the oven ventilation and safety features. The worst conditions to be encoutered in a particular oven operation must be used as a design basis, and a suitable placard or warning notice placed on the oven outlining these conditions.

Any change in these conditions will cause a change in amount or rate of solvent evaporated into the oven.

In batch ovens, particularly in production shops, the maximum amount of solvent per batch, once it has been determined, will not materially change unless there is a change in product or process. If the oven has been designed for the maximum and is so operated, no problems will arise.

In continuous ovens, however, a number of conditions such as conveyor or web speed, doctor blade adjustment, composition of coating material, etc., can be readily and frequently changed during the operation. The general principle of designing for maximum conditions is thus particularly important.

Continuous ovens, particularly those used in conjunction with cloth coating and roller coating operations, lend themselves to a design feature which has operational as well as safety advantages.

It is possible to install continuous vapor concentration indicators and controls calibrated in terms of %-lel. The sampling probes are located near the entrance to the oven at the zone of highest vapor concentrations. It is common practice to set the control to actuate an alarm or turn the heat off when the concentration at this point exceeds 40% of the lel. This is usually equivalent to an average concentration below 25% of the lel.

In many instances, these indicators are being used as an additional safeguard since the conveyor speed can be adjusted and maintained at such a rate that 40% of the lel is not reached. However, it must be borne in mind that these devices are very sensitive and may be inaccurate. They must be carefully calibrated and adjusted, zeroed with fresh air every day, serviced frequently, and always used as a check on the ventilation but not as a sole means of determining safe operations (coating materials containing silicones and other substances may poison the indicator and render it useless).

Any of the following changes in design and operating conditions may make the design safety ventilation inadequate and require a recalculation of the adequacy of the ventilation (conversely, a change may introduce conditions which will permit a decrease in ventilation with a resultant saving in heat or time):

(1) Elevation in temperature.
(2) Increase in amount of coated work in a batch, or /unit time.
(3) Increase in thickness of coating.
(4) Increase in conveyor speed.
(5) Change in size or surface area of work.
(6) Change in composition or nature of solvent or coating material.
(7) Change in physical shape of the work. A coated object with a high surface-area-to-mass ratio may result in a much higher peak evaporation rate and higher average rate of evaporation than one with a low surface-area-to-mass ratio.
(8) Change in physical nature of work. Solvent evaporation from a coated impervious surface will be much greater initially and come to completion in a much shorter time than from a coated or impregnated porous surface.

Maintenance. The interior of ovens and ductwork must periodically be cleaned to remove dripped paint, deposits, and condensates. Nonferrous scraping tools must be used, and if necessary, ductwork should be taken apart and scraped, steamed or, in unusual circumstances, burned clean in a safe location.

All controls, probes, switches, etc., must be examined and cleaned. Explosion vents must be carefully cleaned and loosened to prevent locking by polymerization of condensed fume at the joints.

Controls, switches, interlocks, conveyors, etc., must periodically be checked to ensure proper operation.

Air Contaminant Emissions. The solvent evaporation rates which are used to detemine required ventilation will also be the potential emission rates. (See also Table 2.15)

Air Cleaning. Ovens and their ventilation features are carefully designed to ensure operational efficiency and safety. If air cleaning is required, extreme caution must be taken to ensure that the air cleaner does not interfere with the oven or introduce excessive back-pressure to reduce the air flow and introduce an explosion hazard. Similarly, the air cleaner itself may constitute a fire or explosion hazard unless properly designed, located, and operated. Additional interlocks and alarms for the operation of the air cleaner may be advisable.

The usual air cleaning methods consist of adsorption on activated charcoal, incineration with after-burners, or condensation. Local ordinances may prohibit 1 or more of the above techniques because of the fire hazard.

Air Cleaning Efficiency. Equipment can usually be designed to provide any desired efficiency. However, the usual criteria may not always be valid for oven emissions. Depending on the binders, drying oils, or plasticizers used in the coating formulations, fumes may be created in very low concentrations which will nevertheless have a high pollution potential because of their odor and visibility. In such cases, cleaning efficiency higher than othewise needed may be required.

Laboratory Hoods

Although laboratory hoods are, in most cases, purchased as a package from any number of suppliers, and are rarely custom fabricated, a brief discussion of hood design features and good installation practices will assist in intelligent purchase and proper installation of laboratory hood systems.

General Principles:

Installation. Laboratory hoods should be located away from doors, windows, walkways, and any other sources of drafts since these may interfere with the exhaust to the point where contaminants can escape from the hood into the room. Properly tempered make-up air should be furnished to replace the air exhausted. It is usually desirable to maintain a slight negative pressure in the room by means of an excess of exhaust over supply air. In some cases it may be necessary to maintain positive pressure when it is essential to exclude outside, possibly contaminated air. When the supply is an integral part of the building ventilation design, it is desirable to arrange for movement of supply air from clean areas to more contaminated ones.

The discharge duct should extend above the turbulent zone of the roof (see Section 4) to prevent downflow of contaminated air and it should be at least 10′ away from any building inlet air openings. A standard weather cap or elbow should not be used since these also result in downflow of air with possible reentry. A powerhouse-type or no-loss sleeve drain stack, which permits upblast of air, is preferable. The fan should be located on the roof so that the ductwork within the building is under negative pressure and contaminants cannot leak into the room or building.

Laboratory hood principles. The basic principle is to provide a capture or controlling velocity at the face so that contaminants generated within the hood cannot escape. This velocity, to be effective, will depend on a number of factors such as location of the hood within the room; the thermal load; use and location of equipment within the hood; characteristics of contaminants generated; rate, velocity and direction of contaminant generation; and toxicity or hazard of contaminant.

A velocity of 100 fpm into the fully opened face of the hood will be adequate in a majority of applications. Less, i.e., 75 fpm, may be acceptable for routine control laboratory hoods and more, 150 to 200 fpm, may be required for high hazard or radioactive operations. (The higher figure for the latter is necessary because a much greater degree of control is required.)

The cfm required to maintain a design velocity into a particular sized hood may be reduced by using horizontally sliding doors rather than a vertically sliding sash. The number of such doors will depend on the operations and the size of the hood. Two such doors will result in a maximum sized opening of 50% of the face. Three doors will give a maximum opening of 67% of the face.

If possible, the hood face should have shaped or airfoil edges to reduce turbulence. The bottom foil edge should be mounted about 2″ above the bottom of the bench (or floor) so that some air always enters the hood. (Ordinary hoods should have a sash stop 2″ above the bottom of the hood for similar reasons.)

When excessive velocities within the hood are contraindicated because of the operations being performed, a bypass damper or atmospheric damper at the top may be used. As the door of the hood is lowered, air enters through the dampered opening in proportional amounts so that uniform face velocity is maintained regardless of face opening.

The interior of the hood and ductwork, as well as the fan, should be as corrosion resistant as the comtemplated operations to be performed dictate. Hoods may be constructed of natural or synthetic stone, asbestos-cement, stainless steel, and plastic, or they may be lined or coated with suitable corrosion-resistant material. Similarly, ducts and fans can be fabricated of stainless steel or plastic or they can be suitable coated.

Pipe velocities in te 2000-fpm range are recommended, although velocities of 3000 fpm or higher may be required to scour condensate from the ductwork or to convey particulate contaminants.

Baffling is required to ensure uniform face velocity. This will usually consist of a back baffle with a slot opening at the bottom, one at the top, and frequently one in the middle. Slot widths should be adjustable and set to maintain slot velocities from 1000 to 1500 fpm.

Specific Hood Types:

General Utility Bench Hoods. (See Figs. 2.9a and 2.9b.) The requirements for these are as indicated above. Face velocity should be 100 fpm as a rule but frequently 75 fpm is adequate.

Floor Hoods. The requirements are similar to those above, except tht uniform face velocity will be more difficult to achieve because of the face opening which is usually from 5 to 6 feet high. However, uniform velocity over the entire face, especially at the bottom, is particularly important since operations in these hoods can frequently result in leaks or spillage of volatile contaminant to the bottom. The face height will depend on the size of the equipment to be used in the hood and the need for personnel to enter the enclosure. If they can be used safely, horizontally sliding doors as described above can be used to reduce the cfm requirements which might otherwise be excessive.

Supplied Air (Auxiliary Air) Hoods. (See Fig. 2.9c.) These hoods are usually installed in air conditioned laboratories and are designed to reduce the amount of costly, conditioned air exhausted to the outdoors from the room. In these hoods, a portion of the air exhausted from the hood is delivered to the hood usually from an outdoor, uncontaminated source, heated during the winter months, if necessary. This supply air should be introduced directly into the room so that it enters the hood face along with that portion of the exhaust air which comes from the room itself. This can be accomplished by means of either a diffuser or ceiling grille directly above the hood or, as is more common, by means of a special commercially available auxiliary air hood which has an integral supply fan bringing in outside air which discharges into the room through a grille at the top front protion of the hood. Auxiliary air hoods are usually adjusted so that 50% of the total air volume exhausted by the hood is supplied air and 50% comes

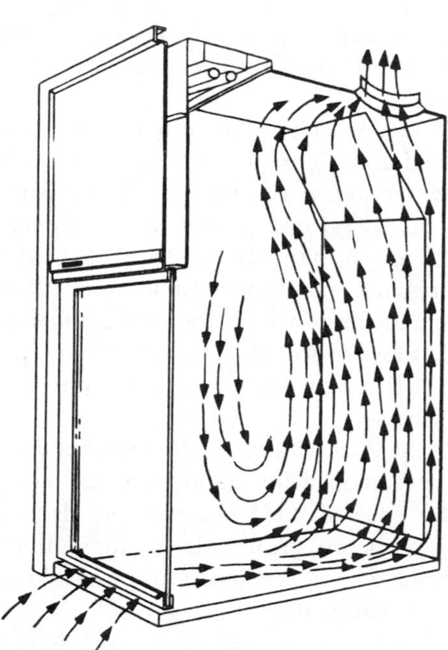

Figure 2.9a. Standard utility laboratory hood (sash closed).

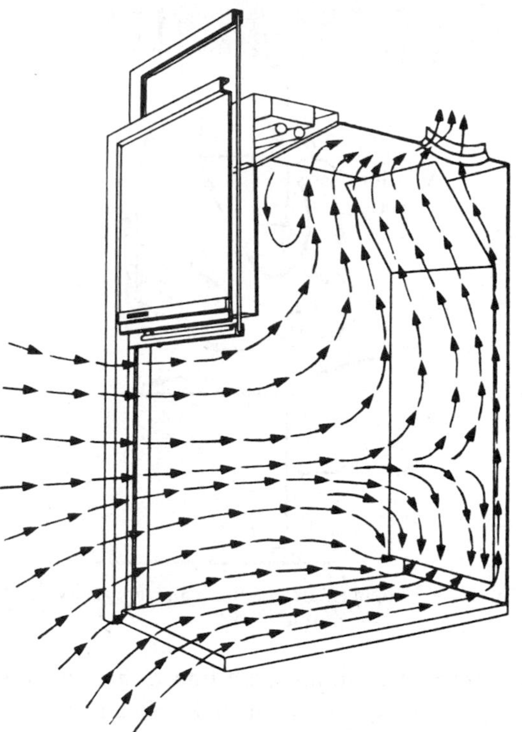

Figure 2.9a (cont.). Standard utility laboratory hood (sash open).

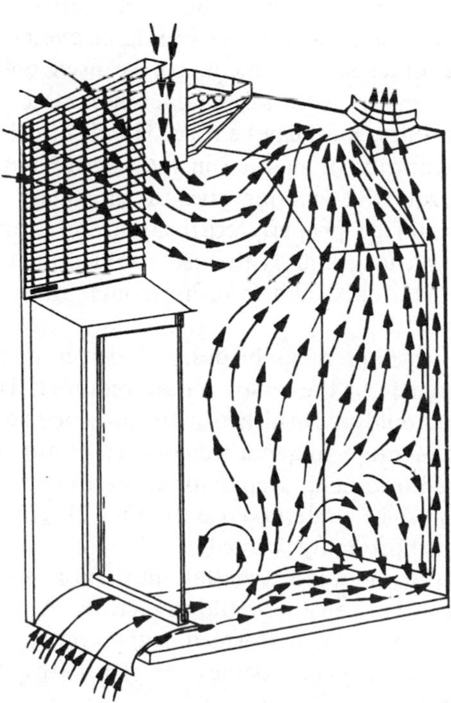

Figure 2.9b. Standard by-pass laboratory hood (sash closed).

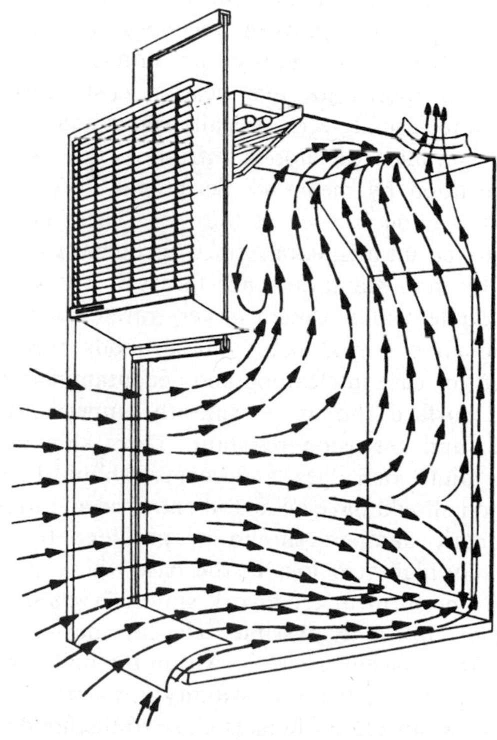

Figure 2.9b. (cont.). Standard by-pass laboratory hood (sash open).

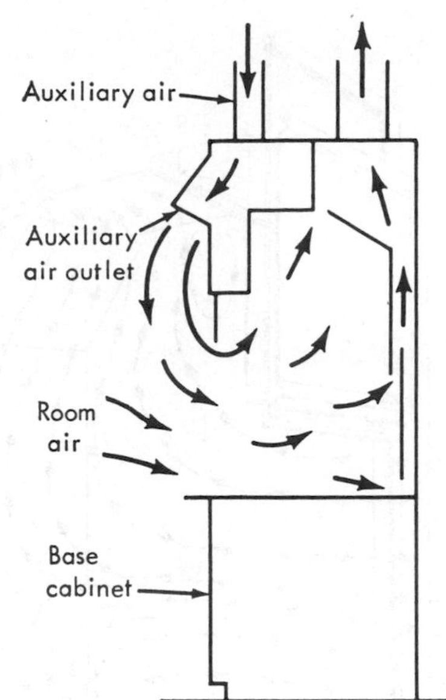

Figure 2.9c. Supplied air (auxiliary air) laboratory hood.

directly from the room. These figures may be modified to as high as 65% supply air and 35% air directly from the room.

If the air-conditioned laboratory is situated in an otherwise non-air-conditioned building, the supply air may be taken from an uncontaminated source within the building. This will usually eliminate the need for special heating provisions for the supply air.

Before considering a supplied-air hood, it is desirable to weigh the savings in air-conditioning costs against the cost of supply-air ductwork, fans, and heaters. (Installation of supply air hoods in non-air-conditioned rooms to save heating costs during the winter is not usually economically warranted).

Some supplied air hoods introduce the supply air into the hood by means of peripheral slots or grilles within the hood face or at the rear. These have been found to be unsatisfactory since turbulence is created within the hood, and entrance face velocity is too low for adequate control except in very rare circumstances.

Radioactive Use Hoods. These hoods must reflect the need for ease in cleaning and decontamination. The interior surface should be smooth, impervious, continuous, and corrosion resistant. There should be no cracks, joints, or ledges, and corners should be coved. The hood should have an airfoil entrance and bypass or atmospheric dampers should be provided to prevent excessively high velocities at the base where the operations are performed. Ductwork should be as short and as free of elbows as possible. The fan must be on the roof to keep the entire system within the building under negative pressure. If the possibility of particulate emissions exists, absolute filters as close to the hood as possible are indicated, with a pressure gauge to signal the need for filter replacement. Roughing pre-filters will increase the viable life of the absolute filters. Filters must be replaced and disposed of under the usual radioactive or high hazard handling precautions. If radioactive gases or vapors may be emitted, air cleaning in the form of activated charcoal or special scrubbers may be necessary.

Required face velocity is 100 fpm minimum, with 150 fpm preferred and higher when conditions dictate.

Biological Organism Use Hood. Design features are essentially as described above for radioactive use hoods with an absolute filter. If virulent viruses or other pathogenic organisms 0.3 microns or less in size may become airborne within the hood, absolute filters may not be able to provide adequate filtration and incineration of the air stream beyond the absolute filters would be necessary.

High Hazard Use Hoods. The requirements are essentially the same as for radioactive use hoods. The necessity for ease of hood cleaning and decontamination may not be so great. Air cleaning, if required, would depend on the composition, physical state, and toxicity of the emissions.

Where an explosive as well as a health hazard exist, hoods require special precautions.

Perchloric Acid Use Hoods. A severe explosion hazard is created when perchloric acid comes in contact with any organic substance. Accordingly, all perchloric acid operations must be performed in a specially constructed hood. The air-borne health hazard is not great and a face velocity of 100 fpm is usually sufficient.

As with radioactive use hoods, the interior must be impervious, smooth, continuous, and corrosion resistant. Cracks, joints, and ledges should be avoided. Hard soapstone or asbestos-cement can be used only if the joints are made with an inert, nonorganic cement. Special stainless steel can be used. Hoods of other types of metal can not be painted in the interior but should have a protective coating such as porcelain enamel. Ducts should be properly corrosion-proof and should be short, vertical, and easily disassembled for cleaning. Elbows, manifolds, and long horizontal runs must be avoided.

Fans with nonferrous blades, belt-driven by a motor located outside of the airstream are required. The hood and duct should be furnished with spray nozzles to permit daily or more frequent wash-down of any accumulated perchloric acid. Repair or inspection of exhaust systems must be undertaken only after all components have been thoroughly washed.

Contact with organic matter must be avoided. The hood must be kept scrupulously clean. Any heating operations must be performed with electric or steam heating devices. Gas flames or oil baths may be dangerous.

Glove Boxes. (See Fig. 2.9d.) When operations permit their use, glove boxes are the method of choice

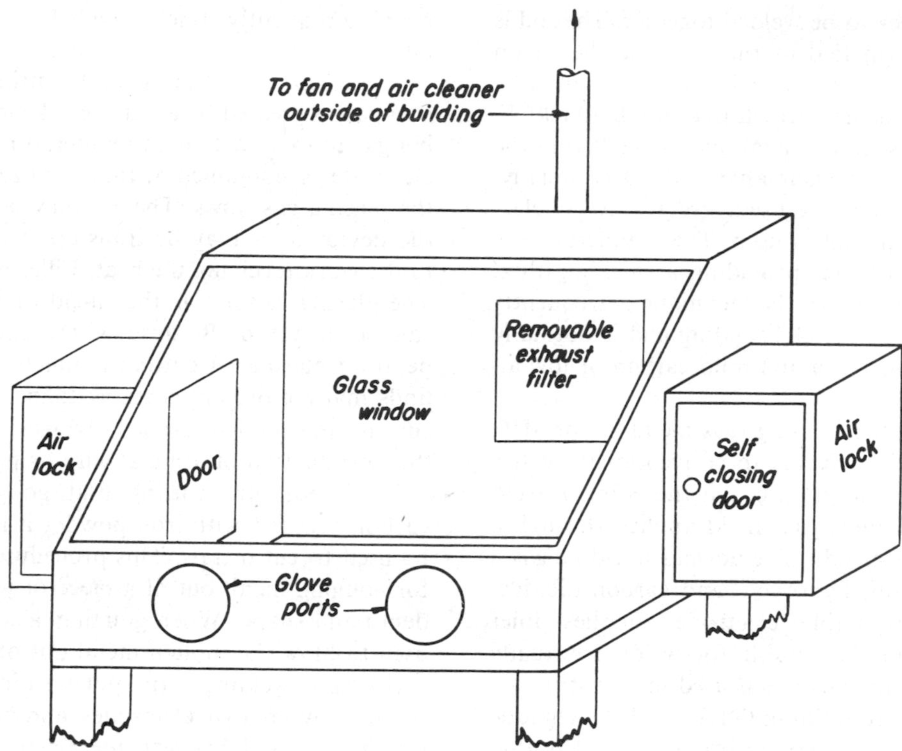

Figure 2.9d. Glove box.

for radioactive use, high-hazard and biological organism use hoods. They consist essentially of an airtight box, with tightly connected gauntlets for interior manipulation; a supply and exhaust system (filtered if necessary); and an access door for introduction and removal of equipment and materials. The door should preferably be a double door with an air lock and the ventilation should be designed to maintain at least 50 fpm into the access door opening.

This equipment provides much more positive control of contaminants from the operations at a much lower initial and operating cost.

Clean Hoods. The "clean" hood is a laboratory hood specially designed to simulate cleanroom conditions so that work requiring such an environment can be performed within the hood while it is situated in an ordinary room.

Exhaust is downdraft through a perforated plate or a perforated slot on all 4 sides. Air is supplied by a blower at the top through an absolute filter which serves to clean the supply air and at the same time provide downdraft laminar air flow. The balance between supply and exhaust is designed to provide a slight excess of supply over exhaust. This balance is usually accomplished by adjustment of the excess supply air so that it provides a curtain of downflow air just outside of the hood face.

The operations to be performed, particularly if they involve contaminants, must be located well within the hood so that the contaminants are not released within the influence of the air curtain and discharged into the room. It is usual practice to mark the area in the hood within which operations can be performed.

Air Contaminant Emissions. With the exception of radioactive and biologic organism use hood, laboratory hoods exhaust systems rarely discharge emissions in quantities requiring air cleaning. Air cleaning for the former has been described above.

Welding and Flame Cutting

Welding is a technique of joining metal surfaces together by fusing them into each other, sometimes with the addition of a molten filler metal to enhance the process. In most welding techniques it is necessary to shield the point of weld to exclude oxygen and ensure proper joining of the metals. The method with which this shielding is accomplished distinguishes the welding technique and is also an important factor governing the amount and nature of the contaminant emitted by the process. A variety of welding techniques are now available to meet the increasingly complicated needs of industry. A number of these are briefly described, along with contaminants emitted and methods of controlling them.

1. Types of welding.

(a) Electric arc. In this conventional welding technique, an electric arc is struck between a metal electrode (rod) and the surfaces to be welded together. The rod is usually consumable (melted by the arc), and the fusion of the surfaces, along with the rod, is brought about by the temperature of the arc, which may reach 10,000° F. In some applications, filler metal may be melted in the weld, along with the consumable rod. Occasionally, nonconsumable rods may be used. Frequently, an electrode coated with minerals is used. These minerals fuse and decompose in the arc, providing a shielding flux. When coated rods are used, the technique is frequently known as metal-arc or "stick" welding and is normally used for welding black iron and mild carbon or low alloy steels.

(b) Inert gas shielded welding (gas metal-arc or MIG welding). This is comparable to conventional electric arc welding except that the arc and the point of weld are blanketed by an inert gas shield applied through a nozzle around the electrode. The normal shielding gases include helium, argon, hydrogen, and carbon dioxide. A continuous consumable electrode supplies filler metal. This technique is suitable for welding virtually any ferrous and nonferrous metal used in industry.

(c) Gas tungsten arc welding (TIG). This technique is identical to MIG welding described above except that a nonconsumable tungsten electrode is used.

(d) Flux-cored arc welding. This differs from conventional electric arc welding only in that the flux is present as a core within a hollow consumable elctrode rather than on the surface of the electrode. Inert gas shielding may also be used, in which event the process is similar to MIG welding.

(e) Stud welding. Stud welding is performed by an arc drawn between a metal stud and the part to which it is to be attached. When the surfaces to be joined are sufficiently heated, they are brought together under pressure. Shielding is usually accomplished by means of a ceramic ferrule surrounding the stud or with an inert gas shield.

(f) Submerged arc welding. Submerged arc welding is performed by an arc between a bare, consumable metal electrode and the work. Shielding consists of granual fusible flux, usually composed of fluorides, which blankets the work. Slag also forms a protective layer.

(g) Electroslag welding. Is similar to submerged arc welding. A blanket of flux, electrically melted by the flow of current from the consumable electrode to the work melts the electrode and the surface of the work. Shielding is furnished by the pool of molten flux and slag which blankets the work.

(h) Oxyacetylene gas welding. A hot (up to 7500° F) oxyacetylene gas flame can be used for welding by melting surfaces of work to be welded, with or without the addition of a filler metal. This technique may also be used for soldering and hard soldering or brazing of metal. It also finds applications in flame-cutting of metal. Frequently, finely divided iron powder may be introduced into the flame to enhance the cutting process.

(i) Plasma jet welding and cutting. Plasma (ionized gas) is created by an arc-gas device capable of heating gas to extremely high temperatures (60,000° F). The electrode is contained within a nozzle through which the plasma gas flows. The arc may be contained within the device or it may be transferred from the electrode to the work receiving the heat. Filler metal may be used. The plasma can act as the shield or inert gas shielding can be provided. Because of the extremely high temperature plasma jet cutting is much more efficient and finds many more applications than oxyacetylene flame cutting. Plasma jets can also be used for metallizing by the introduction of a metal wire or powder into the jet.

(j) Carbon arc cutting and gouging. An electric carbon arc, fed with iron powder if necessary, can also be used to cut metals. This procedure is also adaptable for gouging metal out of a piece of work to the desired depth and shape. When gouging, a compressed air jet is used to blow the molten metal out of the gouged area.

(k) Spot welding. In spot welding, electric current flows between two electrodes and the pieces of work which are fixed between the electrodes. The resultant resistance heating is great enough to fuse the pieces together at a small spot.

(l) Coated metal welding. Any of the above techniques may be used to weld or cut coated metal. Usual coatings include zinc (galvanized), plated metal, paint or lacquer, and plastic. The contaminants released by the fusion or decomposition of the coating material may be more toxic than the contaminants which would normally be generated by welding or cutting the bare metal. For this reason, it is frequently advisable to mechanically remove the coating for 3 or 4 inches on each side of the desired location of the weld or cut.

2. *Contaminants.* Welding operations cannot be characterized precisely since they may be carried out under widely different conditions; with various voltages, amperages, and arc lengths; and on a broad spectrum of metals. The amount and nature of contaminants and the potential health hazard at a particular welding operation will depend on the many variables which can influence the impact on the environment. Thus, a specific operation cannot be considered implicitly hazardous or nonhazardous; the conditions must be evaluated each time. It is true, however, that certain welding techniques have a much greater potential for hazard than others.

Welding in a confined space, such as the interior of a tank or ship's hold, will have a much greater health hazard potential than the identical operation performed in a large workroom with good natural or mechanical general ventialtion. Even ordinarily innocuous operations must be considered suspect under confined space conditions.

Ordinary electric arc welding in unconfined spaces will not usually generate O_3 or NO_x in toxicologically significant concentrations. Since this method is largely used for black iron or mild steel welding, the only significant fumes produced normally are nontoxic concentrations of FeO fumes or fumes released by the decomposition and fusion of the flux when a coated rod or flux is used.

In general, inert gas shielded welding and plasma jet welding and cutting operations can produce excessive concentrations of O_3 and NO_x. This results from the oxidation of the N_2 in the air to NO_x and the conversion of O_2 to O_3 by the intense heat of the arc and the ultraviolet radiation produced by the inert gas shielded arc. The action of the latter on the O_2 in the air converts it to O_3. Although ordinary non-gas shielded electric arc welding also produces ultraviolet light, the frequency and intensity of the light are not sufficient to produce NO_x or O_3 in significant concentrations. Depending on the process, the influence of the light created by the arc may extend to a radius of 18″ from the arc.

Carbon monoxide may be produced by the decomposition of the carbon dioxide when it is used as a shielding gas. Partial decomposition of organic compounds in the vicinity of the arc may also create carbon monoxide. A hazard which is frequently overlooked is the formation of $COCl_2$ gas by the action of the ultraviolet light or heat upon trichlorethylene vapor in the air. This can result even at a considerable distance from the arc and significant concentration of $COCl_2$ or chlorinated hydrocarbon intermediates can be produced even when the trichlorethylene vapor concentrations are low.

Fumes are generated by the melting of the metal being welded and the fusion of the flux, when one is used. When the flux contains flourine, as many of them do, the concentration of fluorine fumes generated may be excessive. The metal fume hazard may be minimal as from mild steel welding; or very great when welding such highly toxic metals as lead or beryllium.

Normally, spot welding does not produce contaminants in concentrations requiring control. However, large-scale automatic welding, such as in automobile body assembly, can produce excessive concentrations. As previously discussed, the contaminants from coated metal welding depend on the coating. The contaminants created by plasma jet and oxyacetylene welding and cutting are similar to those from conventional welding but usually of a higher order of magnitude.

3. *Ventilation Control:*

A. UNCONFINED SPACES

1. General (dilution) ventilation: The use of general ventilation is usually acceptable only for ordinary welding in unconfined spaces with uncoated rods or rods coated with a non-F-flux. Average conditions will require 2000 cfm/welder. The following table may also be used:

Welding Rod Diameter	cfm per Welder
5/32″	1000
3/16″	1500
1/4″	3500
3/8″	4500

For design considerations, see "Dilution Ventilation," this section.

2. Local exhaust ventilation. The most common type of local exhaust is the free-hanging hood which can be positioned close to the arc and can be readily moved as the work progresses. (See Figs. 2.2 and 2.10). This hood is provided with a flexible duct and is either counter-balanced or equipped with a magnetic or other type of holding device to permit ready placement of the hood as close as possible to the arc. Ventilation is designed to provide a velocity of 100 fpm at the point of the arc towards the hood. The required cfm at a particular distance from the hood opening can be calculated from the formula for free hanging plain openings (Fig. 2.2), using 100 as the V value. The following empirical table may also be used with the standard welding hood with 8″ × 3″ face:

Distance from arc to hood	cfm (flanged hood)	cfm (unflanged hood)	Branch Diameter
4″ to 6″	150	200	3-3½″
6″ to 8″	275	350	4-4½″
8″ to 10″	425	550	5-5½″
10″ to 12″	600	750	6-6½″

Enclosing downdraft or side hoods may also be used (see illustration in this section). The side hood in this instance would be a version of the free-hanging plain opening (Fig. 2.2), fixed in a permanent position at an angle close to the welding operation. This hood is applicable for production welding operations where the type and extent of the operations do not vary. With the side hood, design velocity at the arc is 100 fpm as previously indicated for the free-hanging, plan opening. Minimum face velocity for the enclosing hood is 100 fpm. The downdraft hood may also be used and is commonly the method of choice for flame-cutting operations. Design for the downdraft hood is 100 cfm/square foot of gross grate area, except that for plasma-jet welding or cutting minimum design would be 150 cfm/square foot, with higher quantities frequently necessary for the high-temperature plasma operations or for high-toxicity operations. (Similar considerations apply to oxyacetylene operations.)

B. CONFINED SPACES

For ordinary arc welding in a confined space it is usual to provide each welder with an approved air-line respirator or supplied-air helmet or to ventilate the confined space at a rate of 1000 cfm/welder in the space. If other nonwelding workers are present in the same confined space, the ventilation should always be provided in addition to the respiratory protection for the welders. It is frequently desirable to ventilate the confined space by blowing in the required cfm from an uncontaminated source. The jet action of the blowing air will create turbulent conditions which will provide better mixing and dilution of the contaminants in all parts of the space. The diluted contaminants, which will have low concentrations, exit from the space under slight positive pressure.

At all welding or cutting operations in a confined space which are likely to produce contaminants in harmful concentrations, each welder should be provided with respiratory protection as described above and the space ventilated at a rate of 1000 cfm/welder. If nonwelding workers are also present in the space, they must also be provided with an approved air-line respirator or a supplied-air helmet.

C. AIR CLEANING

Most welding operations do not require cleaning of the exhaust air before discharge to the outdoors. However, flame cutting, especially when done with a plasma jet, usually requires air cleaning because of the large amounts of metal fumes generated. Cloth arresters, high-energy scrubbers, or absolute filters may be used.

Brazing and Soldering

Brazing and soldering procedures for joining metals are analogous to welding in that liquid (molten) metal is applied to the surface of metals to be joined. Brazing refers to processes using Cu-Zu, Cu-Sn, Ag-Cd, and a number of other alloys as the filler metal (braze), depending on the metals to be joined. In soldering, the filler metals consist of low melting alloys such as Pb-Sn and Pb-Sn-Cd with melting points ranging from 150° F to 700° F. The lower the melting point alloys are classified as soft solders; the high melting ones constitute hard solders.

Brazing. Most brazing processes are defined by the method of heating and melting the brazing metals.

(a) Dip brazing. The cleaned parts, which have been preheated, are immersed, along with the braze, into a molten flux bath for a specified period of time. The braze melts and flows into the joint.

(b) Furnace brazing. Cleaning parts, assembled along with the braze, are placed in a furnace and heated to the brazing temperature. Normally, furnace brazing is performed without flux in an inert (Ar) atmosphere, reducing (H_2) atmosphere, or in a vacuum furnace.

(c) Induction brazing. The metals to be joined, with pre-placed braze, are subjected to an induction coil field. The resistance of the metal to the induced current provides localized heating and melting of the braze. Induction brazing may be done in air, or vacuum, or in an inert atmosphere.

(d) Resistance brazing. This is similar to induction brazing except that the metals to be joined and the braze are placed between two electrodes and a high-amperage, low-voltage current is passed through them in a process similar to spot welding. Heating and melting of the braze metal results from the resistance of the metal to the current.

(e) Torch brazing. The area of the metal to be brazed is heated by a gas torch flame or a plasma jet. Fluxes are normally used and braze may be pre-placed before heating or may be fed into the joint at the time of heating.

Soldering. In most soldering processes, the work is cleaned with a flux and heated to temperature prior to application of the lead-containing solder alloy. The solder, which may also be heated, melts and flows into the joint.

Examples of these typical soldering processes include torch or flame soldering, electric- or gas-heated iron soldering, and electrical-resistance, induction, and hot-plate soldering.

A relatively new, conveyorized soldering technique called wave soldering, is a special, widely used process in the electronics industry for the fabrication of circuit boards. The board is conveyed, face down, over a tank containing an alcoholic solution of flux; over a hot plate for drying of the alcohol; and finally over a bath of molten solder. Both the flux solution and the molten solder are impinged to the under side of the board by a mechanically or air produced "wave" in the liquid.

Special soldering techniques which are not normally of industrial hygiene significance include friction soldering and ultrasonic soldering.

Contaminants. As with welding, the nature and amount of contaminant and the potential health hazard at any brazing or soldering operation will depend on many variables. However, airborne fumes from such metals as lead or cadmium or from fluorine fluxes can be present in harmful concentrations. In addition, furnace brazing in hydrogen atmospheres can creat serious explosion hazards unless proper safeguards and interlocks are provided.

Ventilation Control. When required, ventilation control of brazing and soldering operations is identical to the local exhaust ventilation described above under "Welding."

Air Cleaning. Air cleaning is very rarely required. If necessary, the techniques would be identical to those described for Welding.

Plating, Metal Finishing and Cleaning, and Allied Processes

Table 2.8 lists the more common plating-room and allied open surface tank operations. The usual contaminants released to the atmosphere, their hazard rating, usual rates of evolution, and recommended hood types are given (see "Hoods" in this section). Special operations may involve contaminants and rates differing from those shown, and the required ventilation in such cases would have to be modified upward or downward accordingly.

Since most of these operations require relatively unencumbered access to the surface of the tank, the hood of choice is usually one of those shown in Fig. 2.4. For certain operations with a very high rate of evolution, such as boiling, caustic etching of aluminum, or where contaminant continues to be released from the work as it is raised from the tank, as in HNO_3 bright dipping of Cu, hoods shown in Fig. 2.1 or 2.4c will provide better control.

Since virtually all the contaminants listed are highly soluble in water, air cleaning, where required, can readily be accomplished by a simple wet scrubber. For contaminants in the form of mists, as in chromic acid plating, special mist collectors requiring little or no water are available.

See also "Industrial Air Cleaning and Emissions," later in this section, and "Hazard Ratings," Section 12.

Miscellaneous Operations

Table 2.9 lists some of the more common industrial processes for which exhaust ventilation is usually required. Hood and air cleaning data will be found elsewhere in this section.

CONTROL OF WORKPLACE CARCINOGENS

The incrimination of an increasing number of industrial chemicals as potential carcinogens requires special attention because of the special hazard of these materials. A single exposure of the body to even trace amounts of a carcinogen may result in permanent damage to the replicative mechanism of the exposed cells. Such damage can remain localized and latent for many years until malignancy is triggered. Ongoing exposures create additional nonreversible and ultimately proliferating damage so that the effect is cumulative. Removal of an affected worker from exposure to carcinogens will not mitigate the damage already done nor prevent it from increasing. This is in contrast to exposures to a noncarcinogenic toxic chemical where the effect is directly and immediately related to the severity of the exposure and in most cases no further injury results after removal from exposure. (A more complete discussion of industrial carcinogens will be found in Section 8.)

Ideally, the most desirable precaution would be to prevent exposure. To accomplish this with certainty would require discontinuance of carcinogen manufacture and use. In some instances this drastic measure might be required but usually control measures should be directed at attaining the least possible exposure. Since the lowest feasible exposures will vary with different carcinogens and processes, evaluation of the risk/benefit ratio and cost effectiveness of the different controls available must be considered.

Thus, all the tools of industrial hygiene must be brought into play to reduce exposure to the lowest feasible level. Where possible, inherent design should be utilized to prevent or minimize exposure before ameliorating engineering, work practice, and respiratory protection techniques are applied. When threshold and ceiling limits have been established, these should be maintained by engineering controls where feasible, supplemented when necessary by the other control techniques. Exposures to carcinogens for which limits have not been established should be as close to zero as possible, utilizing all available techniques.

Controls

The function of controls is to protect against inhalation of airborne carcinogenic dusts, fibers, mists, gases and fumes, as well as against skin contact, absorption, and ingestion.

(Specific regulations for the control of a number of carcinogens on an individual basis have been promulgated by OSHA as Part 1910, Subpart Z. A draft of control standards applying to all carcinogens is now being prepared by OSHA. They will establish criteria for the identification of carcinogens; classify them into different risk categories, and issue regulations establishing the degree of control required for each category. The discussion to follow is designed to supplement and complement the official standards.)

Plant, Process and Equipment Design

Equipment and process design and layout procedures should include consideration of ease in maintenance and repair. Equipment and piping, etc., should be fabricated of stainless steel or other corrosion-resistant

TABLE 2.8. Plating, Metal Finishing and Cleaning, and Allied Processes

Operation	Contaminant	Toxic or Flammable Rating	Rate of Evolution	Hood Type	Comments
Chrome plating	Chromic acid mist	High	High	4	Hydrogen and oxygen released—possible explosion hazard with foam blanket
Cyanide plating					
(a) Brass-bronze	Cyanide and alkali mist	Slight	Low	4	
(b) Zinc	Cyanide and alkali mist	Slight	Low	4	
(c) Copper (other than conventional)	Cyanide and alkali mist	Slight	Low	4	
(d) Gold, silver, cadmium, conventional copper	Cyanide and alkali mist	Slight	Nil–Low	4	Local exhaust not usually necessary
Strike (copper and silver)	Cyanide mist	Slight	Moderate	4	
Acid plating					
(a) Copper	Sulfuric acid mist	Slight	Nil–Low	4	Local exhaust not usually necessary
(b) Nickel (hydrofluoric acid)	Hydrogen fluoride gas	High	Low–moderate	4	Magnesium base metal
(c) Tin (halide bath)	Halide mists	Slight	Moderate	4	
(d) Zinc (chloride bath)	Chloride mists	Slight	Low	4	
(e) Nickel (sulfate and chloride)	—	—	—	—	Local exhaust not usually necessary, otherwise see Zinc (chloride bath) above
(f) Tin and zinc (sulfate)	—	—	—	—	
Fluoborate plating					
(a) Lead	Fluoborates, hydrogen fluoride gas	High	Low	4	
(b) Lead-tin, nickel, tin cadmium, copper, zinc	Fluoborate mist	Moderate	Low–moderate	4	
Electroless copper	Formaldehyde gas	High	High	4	
Anodizing					
(a) Chromic acid	Chromic acid mist	High	High	4	
(b) Sulfuric acid	Sulfuric acid mist	Slight	Moderate–high	4	
Electropolish					
(a) Brass, bronze, copper	Acid mists, arsine	High	Low	4	High toxic rating because of arsine possibility
(b) All other	Acid mists, arsine	High	Moderate	4	High toxic rating because of arsine possibility
Etching					
(a) Aluminum	Alkali mists	Slight	High	4	
(b) Copper	Hydrogen chloride gas	High	Moderate	4	
Surface treatment					
(a) Parkerize, bonderize	Steam, inorganic salts	Nil or high	High	1,4	Hood No. 1 preferable, high toxic rating if carbon monoxide in products of combustion

(b) Alkaline oxidizing	Alkali mists	Slight	High	1,4	"Black Magic," "Ebanol," etc.
(c) Stainless steel descaling	Nitric, sulfuric, hydrochloric acids	High	Moderate-high	1,4	Hood No. 1 or 4(c) recommended
Pickling					
(a) Aluminum-nitric acid	Oxides of nitrogen (gas)	High	Moderate	1,4	Hood No. 1 or 4(c) recommended
(b) Aluminum-chromic acid	Chromic acid mists	High	Low	1,4	
(c) Aluminum-sodium hydroxide	Alkali mists	Slight	Moderate-high	1,4	
(d) Copper-sulfuric acid	Sulfuric acid mists	Slight	Low-moderate	1,4	
(e) Iron, steel, nickel (hydrochloric, sulfuric acid)	Hydrogen chloride gas and sulfuric acid mist	High	Moderate-high	1,4	
(f) Stainless steel-nitric acid	Oxides of nitrogen (gas)	High	Moderate	1,4	Passivation, immunization, hood No. 1 or 4(c) recommended
Acid dipping					
(a) Bright dip-copper, brass, bronze, aluminum	Oxides of nitrogen, sulfuric acid mist	High	Moderate-high	1,4	Hood No. 1 or 4(c) recommended
(b) Zinc-hydrochloric acid	Hydrogen chloride gas	High	Low	1,4	
Stripping agents					
(a) Hydrochloric acid	Hydrogen chloride gas	High	Low-moderate	1,4	These items refer to the stripping of a plated coating from a base metal. The usual stripping agents and rates of evolution are shown. For mixtures of the agents, the component with the more stringent requirement is the basis of design.
(b) Sulfuric acid	Sulfuric acid mists	Slight	Low-moderate	1,4	
(c) Nitric acid	Oxides of nitrogen	High	Moderate-high	1,4	
(d) Hydrofluoric acid	Hydrogen fluoride gas	High	Low-moderate	1,4	
(e) Chromic acid	Chromic acid mists	High	Low-moderate	1,4	
(f) Sodium hydroxide	Alkali mists	Slight	Low-moderate	1,4	
(g) Ammonium hydroxide	Ammonia gas	High	Low-moderate	1,4	
Metal cleaning					
(a) Vapor degreasing	See Degreasing (this section)				
(b) Alkaline cleaning	Alkali mist	Slight	Moderate-high	1,4	The lower rate of evolution is for cold baths, the higher for heated or air agitated baths
(c) Emulsion cleaning (petroleum and coal tar solvents)	Solvent vapors	Slight-moderate	Low-high	1,4	
(d) Emulsion cleaning (chlorinated hydrocarbons)	Chlorinated hydrocarbon vapor	Moderate-high	Moderate-high	1,4	

TABLE 2.9. Miscellaneous Operations

Operation	Contaminant	Toxic or Flammable Rating	Rate of Evolution	Hood Type	Comments	Type of Air Cleaner (see Industrial Air Cleaners)
Dipping and drying						
(a) Paint and enamel						
(1) Dipping	Vapors of petroleum naphtha, and possibly xylene	Slight–High	Low–High	1,4	Hood No. 1 or 4(c) recommended	Adsorber or Condenser
(2) Draining		Slight–High	Low–High	1,4(c)	See also Dilution Ventilation, this section	"
(3) Air drying		Slight–High	Low–High	1,4(c)		
(b) Lacquer						
(1) Dipping	Vapors of naphtha, esters, aromatics	High	High	1,4	Hood No. 1 or 4(c) Recommended	"
(2) Draining		High	High	1,4(c)	See also Dilution Ventilation, this section	"
(3) Air drying		High	High	1,4(c)		
Flow coating						
(a) Doctor blade	Vapors of naphtha and xylene	Slight–High	Moderate–High	1,3,6	Hood No. 3 may consist of the entrance to a drying tunnel. Hood No. 6 as close as possible to operation.	"
(b) Deluge	Vapors of naphtha and xylene	Slight–High	Moderate–High	1,4(c)	Hood No. 1 to enclose the operation as much as possible. Ventilation must also take into account explosive concentrations. See also Dilution Ventilation, Fire Safety.	"
Melting	Metal fume, gases	Slight–High	High	1,6	Hood No. 6 as close as possible to pot. Products of combustion to be vented to hood or independently.	Cloth filters, electrostatic precipitator, high efficiency scrubbers
Toxic metal machining	Metal dust	High	Moderate–High	1,2	Hood No. 2 to be shaped to suit operation, if possible 4 in. maximum from tool.	Cloth filters, electrostatic precipitator, absolute filter.
Portable grinding, chipping, snagging	Metal, abrasive and possibly silica dust	Slight–High	High	1,2,5		High efficiency cyclone, cloth filter
Garages						
(a) Local tailpipe exhaust	Carbon monoxide, aldehydes, etc.	High	High	Special	Hood or open pipe end to fit over tailpipe. CFM per Branch Vehicles up to 100 hp—100cfm Vehicles above 100 hp—200cfm Diesels—400 cfm	Not usually needed
(b) General ventilation	Carbon monoxide, aldehydes, etc.	High	Varies		10,000 cfm per running auto 20,000 cfm per running truck 200 cfm/hp per running diesel	
Epoxy mixing, potting, etc.	Vapors of amines, styrene, etc.	High	Low–Moderate	1,2	Amine vapors can cause contact dermatitis, Hood No. 1 preferable. Ventilation must also maintain required flow at operator's hands.	Not usually needed

material, readily disassembled for cleaning. Smooth piping with a minimum of bends should be used. Floors should be well drained, smooth, and coated with an impervious material such as epoxy or polyurethane paint. Completely enclosed reactors and kettles and similar equipment, remotely controlled and automatic processes, and chemical or physical modification of the carcinogen are examples of control by design.

Relief and breather vents must discharge to an exhaust system or air cleaner and not into the workroom. All carcinogenic operations must be performed in separate rooms or buildings used exclusively for these operations and ventilated to maintain negative pressure. Designation of such work places as regulated areas into which only properly protected workers are permitted to enter will limit the number of workers exposed.

Substitution

Substitution of a non-carcinogenic chemical, when possible, will obviously eliminate exposure. However, this approach must be used with extreme caution. Substitution, especially of a highly toxic non-carcinogen or one of unknown toxicity, may create a false sense of security and lessen attention to strict controls. In addition, substitution may result in a mixture of non-carcinogens which is carcinogenic.

Enclosure and Isolation

Particularly hazardous processes within a regulated area can be completely enclosed and ventilated at negative presssure, and operated by remote controls or by a fully protected worker who enters the enclosure only briefly when necessary. Conversely, the worker can be located in a small, enclosed room or booth, supplied with clean air under positive pressure, within the regulated area. Provision of completely covering air supply suits fills a similar function.

Other techniques for reducing and limiting exposure that may be useful include scheduling operations at times when no other work is being done and rotation of workers between carcinogenic and non-carcinogenic operations.

Monitoring

Monitoring of carcinogens for which sensitive, specific and reliable instrumentation or analytic procedures are available can perform a number of important functions.

The average daily exposure (8 hour time weighted average) of a worker can be determined by use of a personal monitor. This consists of a battery operated pump, attached to the belt and connected by hose to a filter or adsorbent tube pinned to the worker's shirt so that the inlet is in the breathing zone. It is worn during the entire course of a work day and then sent to a laboratory for quantification or analysis. The results may be subject to error due to inaccurate sampling air flow measurement, improper placing of the sampling device, unauthorized manipulation of the sampler, or contamination.

Ceiling concentrations can be monitored by recording instruments located at operations most likely to produce high concentrations. They can also be determined by personal monitoring of selected workers for selected 15-minute periods.

Area monitoring is useful to afford information on day-to-day variations of concentrations, overall exposures, and a sudden change in concentrations. By judicious selection of sampling locations and times, area monitors can be used to calculate 8 hour time weighted averages rather accurately. They can also be used to estimate exposures of workers who must enter a regulated area occasionally. Area monitors may be read visually or by means of a computer printout or pen recorder. Long-term air samples may be taken for subsequent laboratory analysis as another technique for area monitoring.

Sampling may be continuous or periodic. Sequential monitoring, with probes at different locations, can monitor an entire plant with one system. Monitors can be designed to signal an audible or visual alarm and to automatically shut down equipment or actuate stand-by control equipment.

In addition to routine area or personal monitoring, special surveillance techniques may be advisable. Pressure monitors in exhaust system ducts can signal a decrease in efficiency and need for service or replacement of air cleaning devices. Spot tests of work spaces at the end of a shift can highlight problem areas of contamination, signal diminution of control, or evaluate adequacy of clean-up measures.

Monitoring should always be performed for a suitable period of time after a change in production, process or control measures which may result in new, additional, or higher exposures.

Protective Clothing and Personal Hygiene

Protective clothing may include coveralls, hats or head coverings, workshoes or shoe coverings, gloves or gauntlets, and goggles or face shields. The amount of body protection will depend on the degree and nature of exposure. Clothes should be fabricated of launderable, fine weave, minimum porosity fabric or plastic-impregnated throw-away paper.

Change rooms, with suitable lockers, should be provided so that prior to entering a regulated area, the worker enters a "clean" room for removal and storage of street clothing. He then proceeds to a connecting "dirty" room where protective clothing is put on before he exits directly to the regulated area. At the end of the shift, (and of each exit from a regulated area), he enters the "dirty" room (preferably through a vacuum dusting vestibule where he can remove as much contaminant as possible). Work clothing is placed in properly labeled waste bins, plastic bags or laundry containers. (Work clothing must not be cleaned by shaking or blowing with compressed air.) A shower should be provided between the rooms so the worker can shower before entering the clean room. Suitable clothes washers and driers may be located in the "dirty" change room so that employees may put their work clothing into the washer as soon as they discard them. A maintenance worker can be assigned the task of laundering the clothing. (The washer waste water must be properly treated and disposed of.) Alternatively, clothing in sealed, labeled containers may be sent to a special laundry along with proper precautionary and procedural information.

Throw away clothing should be treated as hazardous waste.

Lunch rooms must be readily accessible to regulated area employees and have washing facilities at the entrance. Vacuum dedusting and removal of work clothing before entry to a lunch room is desirable to minimize ingestion. Lunch rooms should be ventilated to provide positive pressure. Eating, drinking or smoking must be prohibited in the workroom.

Housekeeping and Maintenance

All process equipment requires proper cleaning and maintenance but kettles, reactors, stills, and similar equipment pose a special problem since they frequently require removal of hardened residues and slag. Manual tool scraping of residues, if unavoidable, must be performed by a worker completely protected by clothing, hand, skin and face protection, and air supplied respirators. Preferable techniques include melting of the residues for subsequent disposal by complete incineration or as a hazardous waste, and solvent treatment or steam cleaning to loosen and dissolve residues which can then be removed with a minimum of contact.

Exhaust systems require maintenance to clean surface accumulations and remove settled material or condensed material from ducts as well as contaminant collected by the air cleaning device.

All equipment should be completely decontaminated before service or repair work is undertaken, either at the site or in a repair shop.

All housekeeping, maintenance, and laundry workers require clothing, protective equipment, and respirators to the extent consonant to the expected exposure.

Dry sweeping or compressed air cleaning of floors and work spaces should be forbidden. Deluge wet cleaning should be preceded by a fine spray shower to lay the dust. Dry cleaning of floors, work spaces, and equipment must be done by vacuum systems with high efficiency filters. Discharge must be to an existing exhaust system or to HEPA filters. In some cases, venting to the outdoors without secondary cleaning may be acceptable. Specailly shaped nozzles should be provided for efficient cleaning of all surfaces.

All sweepings, debris, throw-away protective clothing, plugged respirator filters, discarded air cleaner elements, collected material from air cleaner dust hoppers, etc., must be immediately placed into covered, labeled containers or sealed impervious bags. Disposal must be by incineration at temperatures high enough to insure complete combustion; or burial at on-site or special locations with precautions against unauthorized access, release of contaminant to the atmosphere, or to ground water or bodies of water. As an alternative, an adequately alerted hazardous waste disposal contractor may be engaged to dispose of the wastes. Liquid wastes must be treated or decontaminated before discharge. (Incineration techniques are available for disposal of hazardous liquids.)

Decontamination and Emergency Procedures

An emergency, such as a ruptured pipe or other piece of equipment, resulting in massive or continuous discharge of carcinogens into the workroom, requires that the workers assigned to shut down and decontamination procedures be fully protected. This includes all protective clothing and a full face or helmet air supply respirator. In many situations, an air supply suit would be highly desirable.

The contaminated area and equipment should be cleaned by all available, relevant techniques, including vacuum cleaning and scrubbing with a suitable reagent. After cleaning, the area should be tested for residual contamination and the process repeated as often as necessary until tests reveal no further contamination. Any spills or leaks which occur during normal operations should be similarly cleaned at once.

Clean-up workers should leave the regulated area in a manner similar to that of the production workers, namely, through the "dirty" change room, discarding their contaminated work clothing, showering, and then entering the "clean" room.

Respiratory Protection

Respirators are required as an adjunct method of control when local exhaust ventilation, engineering controls, and work practices cannot reduce exposure to acceptable limits; during the time prior to installation of exhaust ventilation and other controls; for housekeeping, maintenance, repair and decontamination workers in conjunction with proper work practices and protective equipment; and during emergency procedures.

Selection, use, and care of respirators are discussed earlier in this section. In all cases the respirator type recommended for highly toxic exposures should be used; and a full face or helmet respirator should be used when the face may come into contact with carcinogenic materials.

Respirators should be cleaned after each use. All adhering materials should be removed from the respirator prior to cleaning and protection of the worker engaged in the cleaning operation must be considered. (The assignment of one properly trained person to be responsible for the maintenance, cleaning and storage of respirators would minimize exposures and insure a ready supply of cleaned, functioning respirators.) Filters and adsorbent canisters should be replaced daily and more often if necessary.

Local Exhaust Ventilation

Exposure to carcinogens can exist at many different operations in a large variety of industries. Accordingly, since local exhaust ventilation design parameters are related to the machine, process or operation creating airborne contaminants, discussion of local exhaust ventilation control of work place carcinogens must be largely general in any treatment of this kind.

A potential for carcinogen exposure exists at four points of activity; primary manufacture of carcinogenic chemicals, carcinogenic intermediates, or use of carcinogenic raw materials (viz. synthesis of vinyl chloride monomer and its subsequent polymerization into polyvinyl chloride); creation of carcinogens as a by-product, impurity, or intermediate in a process (coke oven emissions created during the destructive distillation of coal in the production of coke); use of carcinogens, or materials containing a carcinogen as an impurity in the preparation of resins used in the finished product such as polyvinyl chloride handbags; and laboratory research involving carcinogens.

Design of ventilation control of carcinogens will utilize the principles and hood design details (with an added factor of safety and higher CFM design bases) shown under "Control by Ventilation" (this section). The specifications in the *Industrial Ventilation Manual*, published by the American Conference of Governmental Hygienists (Reference 1) is an excellent source of ventilation data for many different types of operations.

Chemical Process Industries

With few possible exceptions, such as use of asbestos, and pesticide manufacture or formulation, the most significant exposure to carcinogens will be in the chemical process and related industries where they are manufactured, are present as intermediates, or are used as a raw material. There are many actual potential sources of contamination in these industries and they must be provided with local exhaust ventilation.

Peripheral, slot, or enclosing hoods, modified as necessary, can be provided at many types of process equipment such as pumps, valves and seals; kettle and reactor loading and unloading; drum, barrel, bag and container filling; conveyors, elevators and screens; blenders, mills and solution tanks; filter presses and tray driers. In most cases, minimum ventilation rates should be 150 CFM per square foot of enclosure opening or area of surface of tank or opening to be controlled. Filter presses which require manual manipulation are difficult to control. Enclosure and ventilation plus respiratory and personal protection of the worker may not be practical or adequate. Replacement of the press by centrifuges or rotary or vacuum driers may be required. Similarly, tray drying and dumping and subsequent handling of the dried product may be a significant source of exposure, even with exhaust ventilation. Use of fluid bed or spray driers would be preferable.

Special ventilation designs are available for vacuum ovens, stills, pressure relief and breather vents and box car loading. Glove boxes are useful for handling, transfer and weighing of small amounts.

Typical Hood Details for the Chemical Process Industry

The following sketches show recommended hood details for the more common processes and equipment. (Courtesy of Mr. W. V. Andresen, Director, Occupational Health Services, American Cyanamid Company). They can be adapted for other equipment with similar exposures. (Figs. 2.9e–2.9z, 2.9zz)

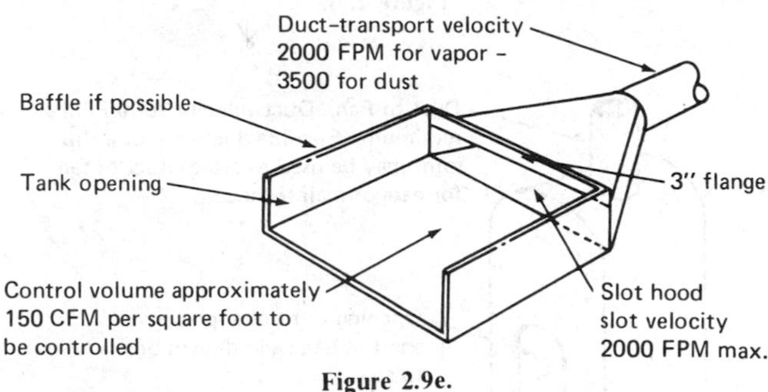

Figure 2.9e.

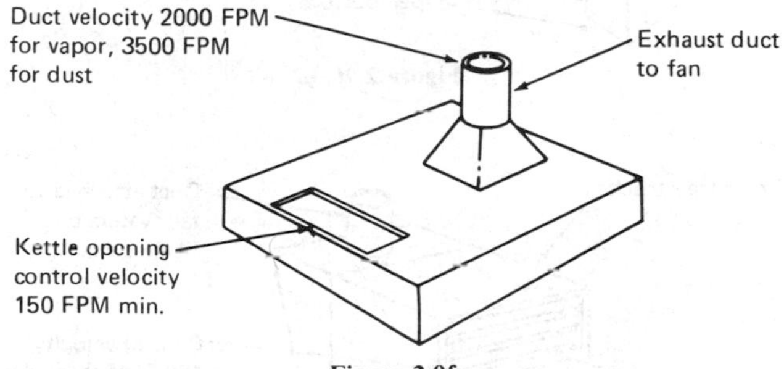

Figure 2.9f.

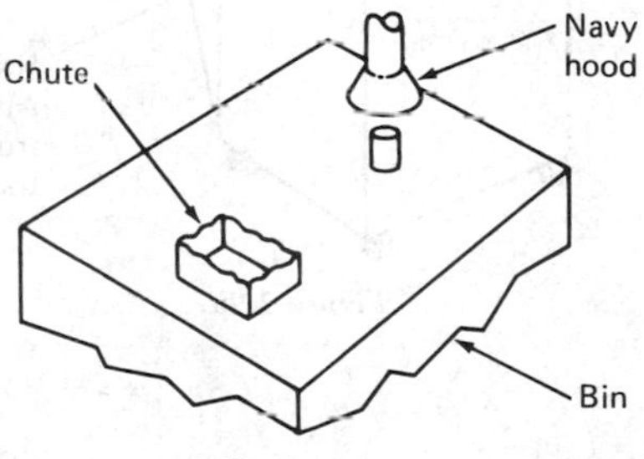

Figure 2.9g.

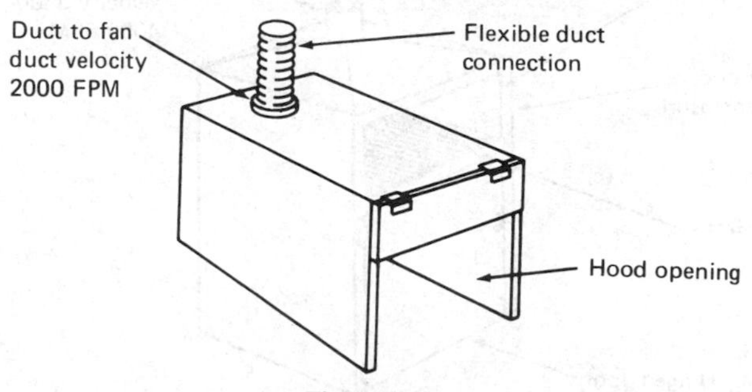

Figure 2.9h.

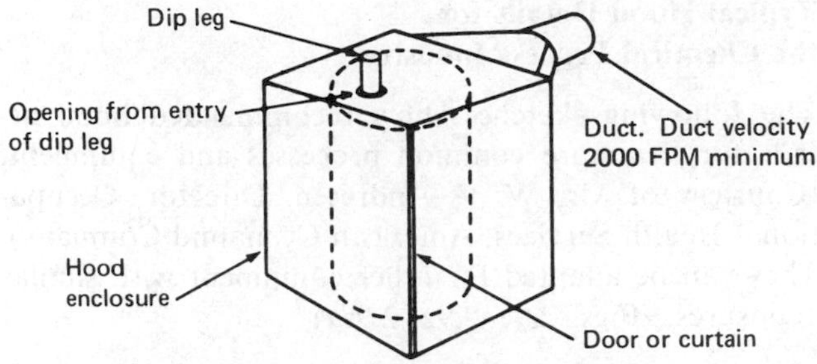

Figure 2.9i.

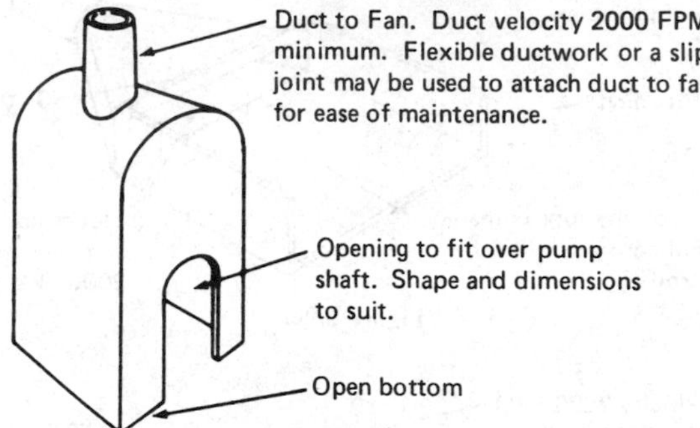

Figure 2.9j.

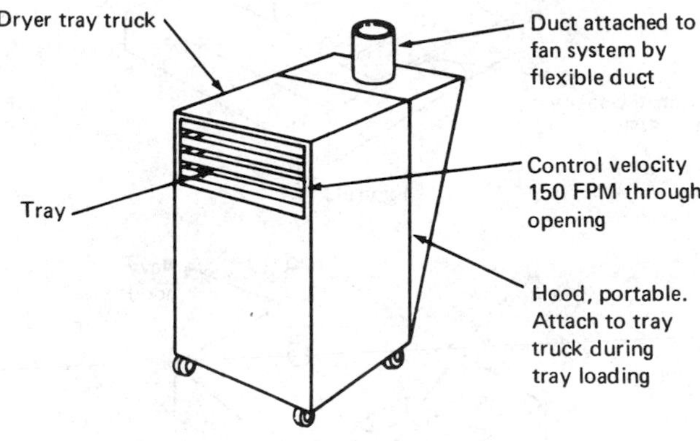

Figure 2.9k.

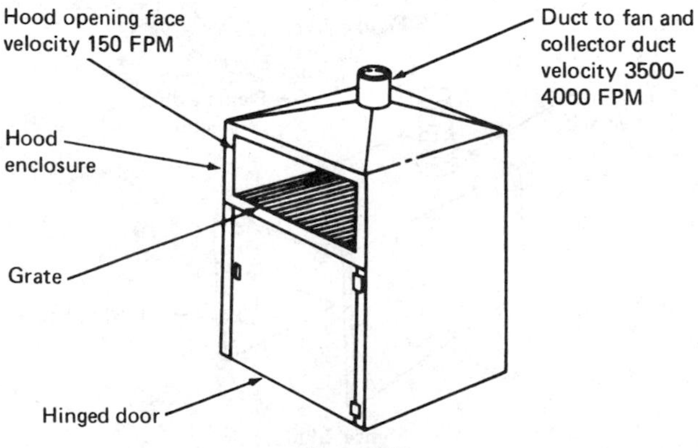

Figure 2.9l.

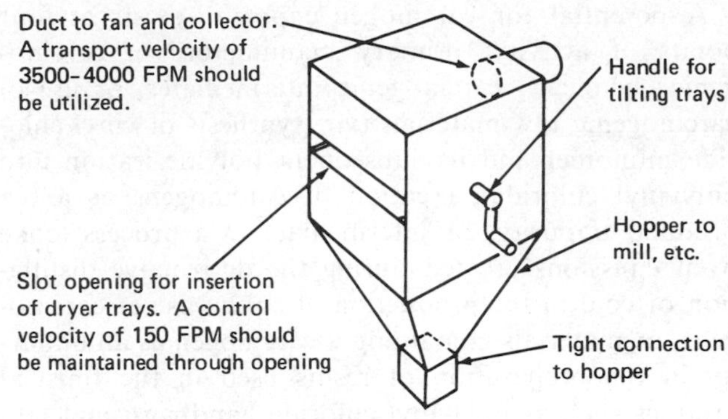

Figure 2.9m.

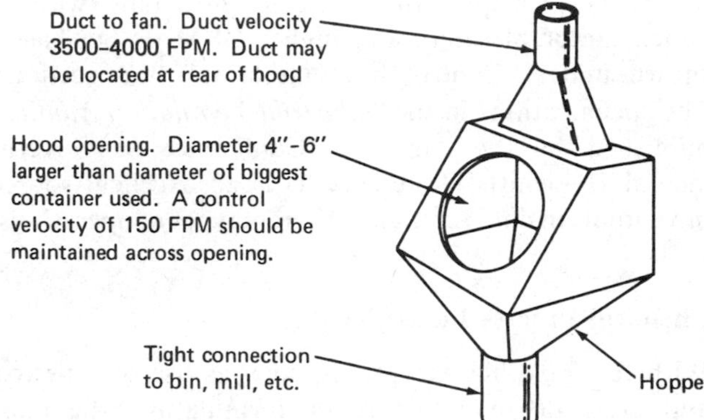

Figure 2.9n.

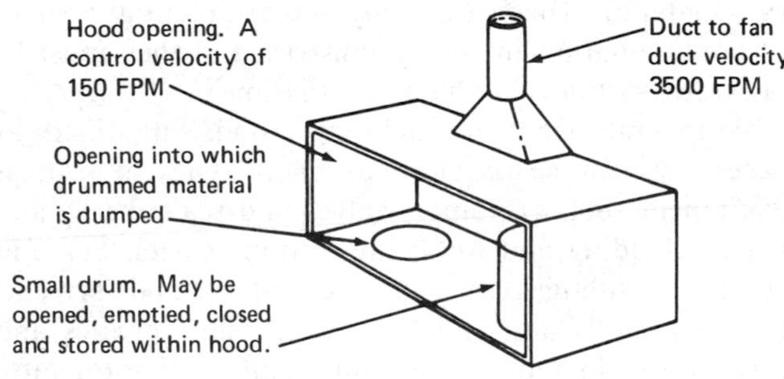

Figure 2.9o.

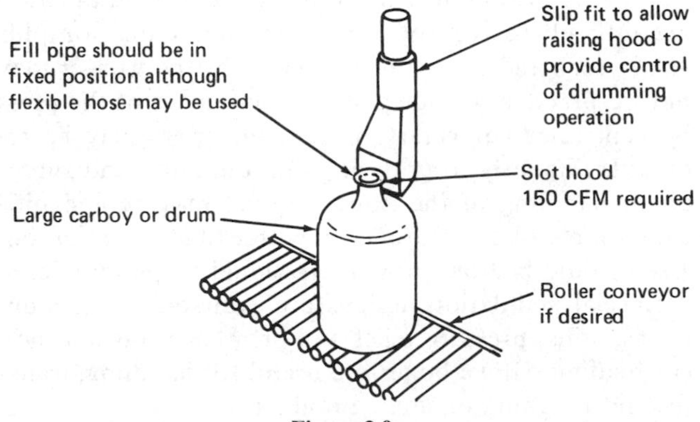

Figure 2.9p.

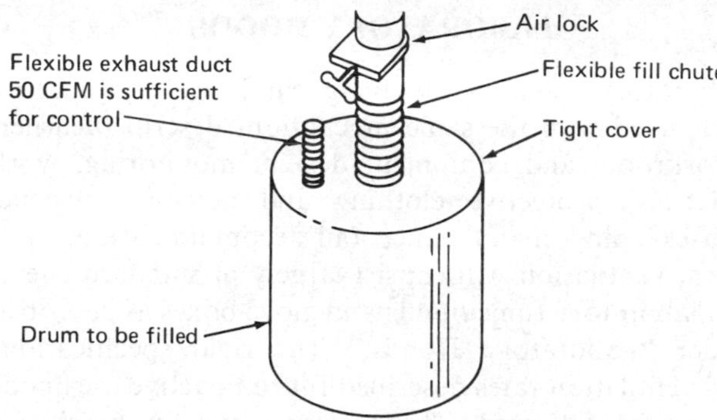

Figure 2.9q.

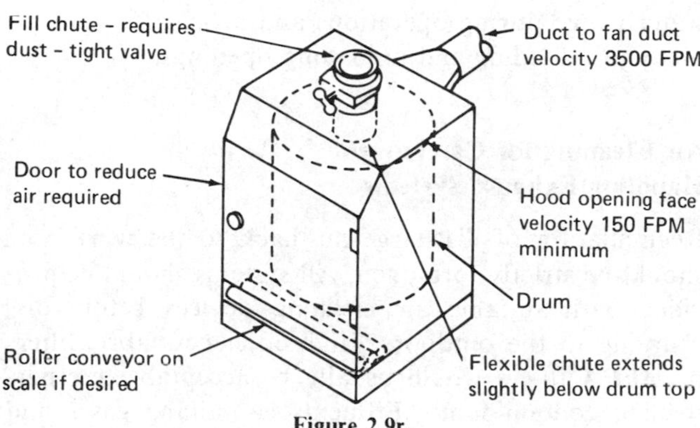

Figure 2.9r.

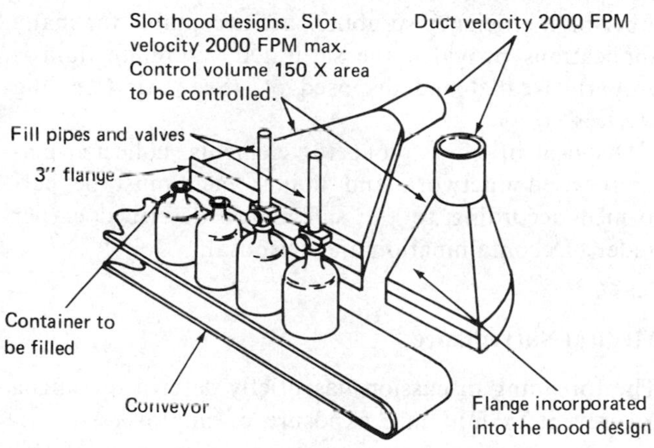

Figure 2.9s.

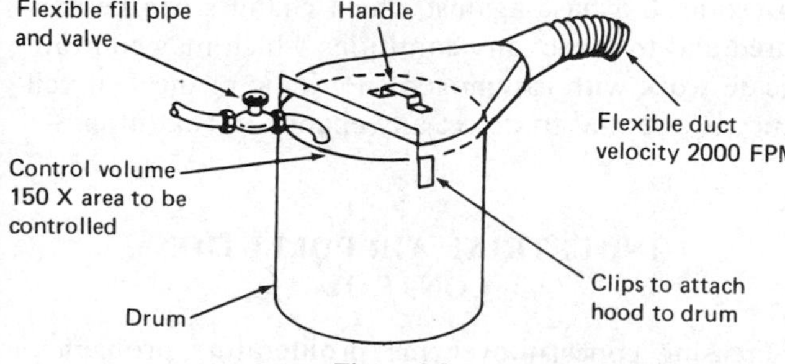

Figure 2.9t.

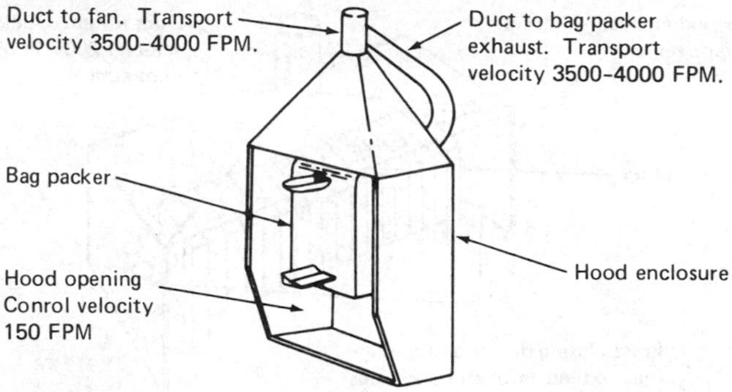

Typical hood enclosure for the control of dust during bag packing from belt packers. A drawer may be provided to contain spillage.

Figure 2.9u.

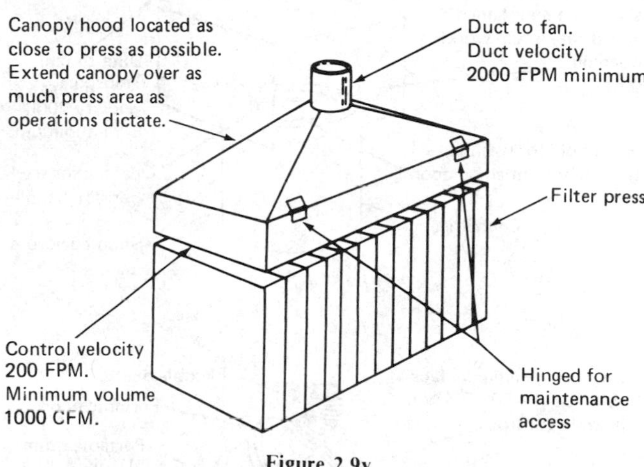

Figure 2.9v.

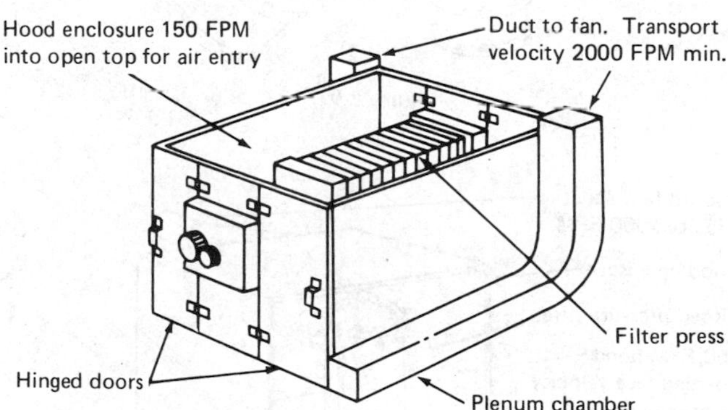

Open top downdraft enclosure for the control of vapors during filter press operation. Control air at a minimum of 150 FPM is drawn through top. Top may be closed and air drawn through upper sides. Operator cleans and sets up press inside of enclosure.

Figure 2.9w.

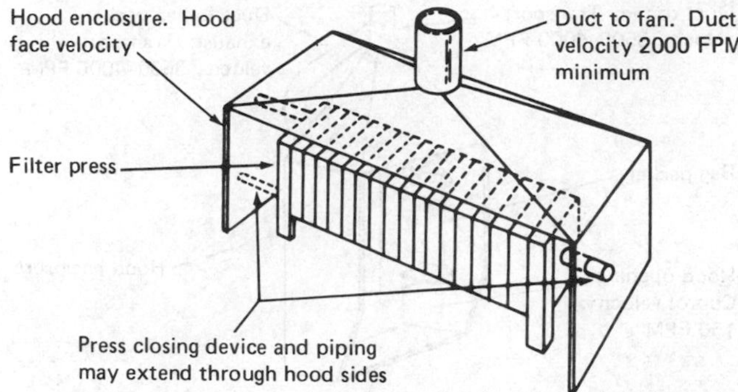

Hood enclosure for filter pressed up to 24" square. Filter press enclosed within hood. All filter press operations performed by one operator within hood. Curtain may be installed across hood face.

Figure 2.9x.

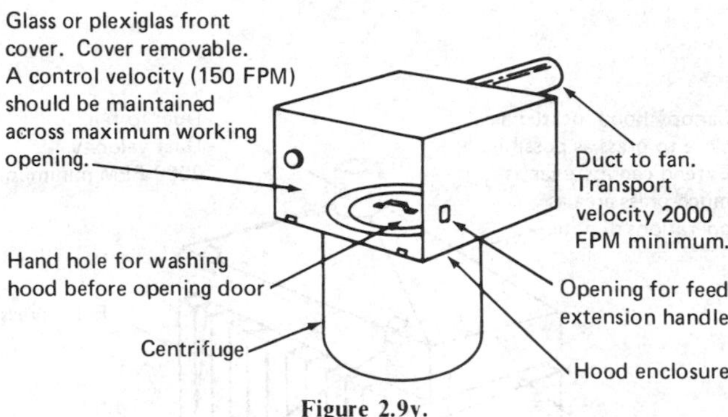

Figure 2.9y.

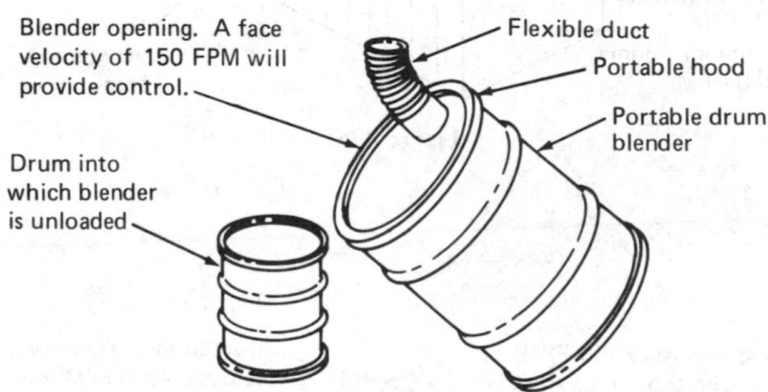

Figure 2.9z.

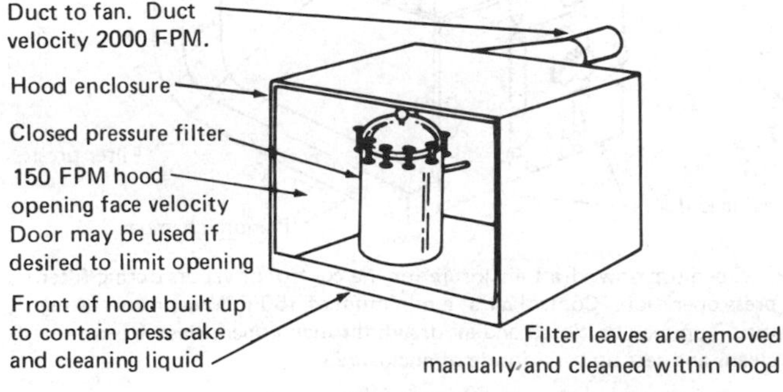

Figure 2.9zz.

LABORATORY HOODS

Laboratory research involving carcinogens should be performed with the same precautions described earlier. (Workroom and equipment design, monitoring, work practices, protective clothing and personal hygiene, housekeeping, maintenance and decontamination.)

Local ventilation will consist largely of standard chemical laboratory fume hoods and glove boxes as described under "Laboratory Hoods." The rigid specifications and ventilation rates described for radioactive use hoods should be followed. The specifications for perchloric acid hoods may also be applicable. Glove boxes should be exhausted to maintain a negative pressure of 0.5 inch water gauge during operations and at least 50 fpm velocity into loading and unloading openings.

Air Cleaning for Carcinogen Handling Exhaust Systems

Recirculation of discharge air back to the workroom should be strictly forbidden. All systems should be provided with suitable air cleaning devices before discharging to the outdoors. High efficiency fabric filters or HEPA filters would usually be acceptable for particulate contaminants. Effluents containing gases and vapors must be cleaned by incineration or adsorption on charcoal filters or other efficient sorbent. Venturi or other high efficiency scrubbers are acceptable for many applications provided the saturated scrubbing fluid is properly treated and disposed of. (See "Air Cleaning Devices".)

Disposal of filters, collector elements, collected material, used ductwork and hoods, etc., must be performed according to the safeguards described earlier under "Decontamination and Disposal."

Medical Surveillance

The foregoing discussion has briefly described controls to prevent or minimize exposure of employees to carcinogens. Medical surveillance is a vital part of this control. Before an assignment to work in a regulated area, employees should be given physical examinations to establish a base against which changes can be measured and to detect any contitions which may contraindicate work with carcinogens. Periodic medical surveillance is essential to detect any change in condition.

INDUSTRIAL AIR POLLUTION CONTROL

Increasing concern over the proliferating problem of environmental pollution decreases the options available to management in carrying out its responsibility for the control of industrially produced air contaminants.

Air pollution control rules and ambient air quality standards for the environment are becoming more stringent, and now the design of industrial exhaust systems must always include consideration of immediate or possible future need for suitable air cleaning equipment to control emissions of air contaminants.

The atmosphere is a natural resource and can no longer be considered a sink into which unlimited quantities of airborne contaminants may be discharged without deleterious effects upon the public health. Careful plant siting remote from settled areas can no longer be relied upon as a totally adequate substitute for expensive air cleaning devices and procedures since many such locations have become populated, often as a result of the plant's presence.

Many industrial exhaust systems which now discharge into the atmosphere, without treatment, gases, fumes, and mists from such operations as plating, paint dipping, and die casting, will ultimately be required to include air cleaning. Also, in many places where dust collectors have long been required to control particulate emissions, the collection efficiency will have to be greater than now acceptable. Specific emission and air quality standards have been written into air pollution control rules and the air cleaner selected will have to meet these standards.

However, the economic burden of providing high-efficiency air cleaning can be mitigated. As discussed previously, exhaust systems with air cleaners handling contaminants of little or no toxicity (Hazard Rating 0 or low, Section 12) can frequently be designed to return the cleaned air to the workroom (see "Energy Conservation and Exhaust System Design"). This eliminates the need for a costly tempered make-up air system to replace the air discharged to the outdoors, and results in a considerable saving in heating, air conditioning, and expensive ductwork installation.

Material collected by an air cleaning system can have economic value as in cement manufacture, where it can be reintroduced into the process; in agricultural lime manufacture, where it can be sold as a high-quality product; in jewelry buffing, where it can be reclaimed for its intrinsic value; or, in the case of collected wood dust, where it can be utilized as fuel.

Since the cost and size of air cleaners are almost directly proportional to the quantity of air to be cleaned and the concentration of contaminants, both of these parameters should be minimized. Careful design of hoods (q.v.) can minimize the amount of exhaust air required to provide adequate in-plant control and reduce the amount of contaminant introduced into the system. Low-velocity takeoffs will diminish the amount of contaminant collected by the exhaust pipe. Hoods can also act as inertial settling chambers to keep heavier particles from getting into the air stream.

Contaminant generation can in many cases be reduced by the selection of appropriate equipment, a change in operations, or improved working technique. Correct paint spraying procedures limit overspray and, concomitantly, air contaminants. Electrostatic or airless spraying minimizes overspray; and paint dipping or flow coating, instead of spraying, will reduce airborne particulates to zero. Floatation of plastic chips or surfactant foam on the surface of plating baths will act as a mist suppressant and frequently eliminate the need for a collector. A properly designed and operated degreaser (*see* Vapor Phase Degreasers) will require no air cleaning and perhaps no exhaust system at all. Hydroblast which eliminates particulate generation can replace dry abrasive blasting.

The cost of air cleaning can also be minimized by providing separate exhaust systems for groups of machines or operations generating similar contaminants. Thus it is not advisable to combine into a single exhaust system operations which generate particulates requiring air cleaning and processes which generate gases or vapors. The latter may not require air cleaning or may require more expensive air cleaners. Using separate systems, especially if only one will need an air cleaner, will result in lower air cleaning costs. Similar considerations apply to such installations as woodworking shops where sanders create dust requiring high-efficiency air cleaning, and planers and jointers create chips for which less efficient cleaners are adequate.

Another approach for particulates is to provide two different collectors in series. The primary colletor, usually an inexpensive, low-pressure cyclone, removes most of the larger particles. The remaining large particles and all the fine particles are collected by a high-efficiency cleaner which can be much smaller and less frequently overloaded than if no primary collector were used.

Control by Atmospheric Dispersion and Dilution

Although this method of control (see also Section 4) is encountering increasing disfavor and opposition, its use will continue in certain applications. The use of a stack to dilute and disperse contaminants is one of the oldest and until recently most widely used techniques for control of air pollution. In theory, a contaminated air stream emitted from a stack will be diluted to acceptable levels by the time it reaches ground level. Most industrial applications of this technique consist merely of an exhaust pipe discharging 6 or 10 feet above the roof of the building. The use of a standard weather cap is contraindicated since it will direct the contaminated air downward. Preferably, the means of weather protection should permit discharge of the air directly upward, adding to the effective stack height. The usual pragmatic measure of efficiency of this method is the number of complaints elicited from residents in the environs. However, ground-level concentrations can be measured.

A number of factors influence the efficiency of the atmospheric dispersion and dilution technique. These include local topography, micrometeorology, height, distribution, and configuration of buildings, and other structures in the vicinity, all of the weather variables, rate and nature of contaminant emissions, height and cross-sectional area of stack, and stack gas temperature and flow rate.

A number of empirical equations have been derived, utilizing the above factors, for the prediction of ground-level concentrations at various distances and cardinal points from the source. These can be used to design stacks against a desired ground-level concentration. However, they yield only approximations and are highly dependent on accurate but not readily accessible long-term meteorological data at the source.

Present-day control standards severly limit dispersion and dilution as an acceptable technique. It cannot be used in locations which have stable atmospheres or frequent temperature inversions since turbulent mixing is required. Also and more seriously, air pollution is becoming so ubiquitous that the concept of acceptable ground-level concentrations at a certain distance from a specific emitter is rapidly losing validity. Many separate sources may each be contributing an "acceptable" ground-level concentration to the same point, making the total effect intolerable.

As a result of this developing problem, the concept of ambient air quality standards has emerged. Accordingly, discharge from a particular emitter will be evaluated not only in terms of its contaminant concentration but for its effect on ambient air quality and impact on the environment as well.

Accordingly, industrial air pollution control is coming more and more to mean industrial air cleaning.

Air Cleaning

It is not the function of this section to supply technical design information but rather to discuss general principles and design features as a guide to the selection of suitable air cleaning equipment. It is seldom feasible or desirable for the plant to build its own equipment since all types of equipment are now commercially available. Manufacturers of such equipment can furnish valuable guidance and also guarantee specific performance standards to meet requirements of the law.

Air Cleaning Design Factors:

(1) Cleaning efficiency. The most important factor is required air cleaning efficiency, and this is dictated by existing or probable future air pollution control rules of the agency having applicable jurisdiction. Efficiency is usually expressed as per cent by weight of contaminant removed from the air stream by the air cleaner.

The criteria for determining required efficiency are usually expressed in terms of the nature of the contaminant and potential emission rate (pounds/hour) before air cleaning. Contaminants with high air pollution potential (toxicity, nuisance or odor factor, property damage potential) require higher cleaning efficiency. High potential emission rates also require higher efficiencies.

As previously indicated, high-efficiency cleaning can usually be most economically provided by a two-stage cleaning cycle, with a lower-efficiency collector preceding the higher-efficiency one. The latter can then be much smaller than if only one high-efficiency collector was provided. The overall cost of such a two-cycle system is usually less than a single high-efficiency collector. However, a two-cycle system may be contraindicated by the lack of space since the primary collector will usually be large.

Although a low-efficiency collector may be accept-

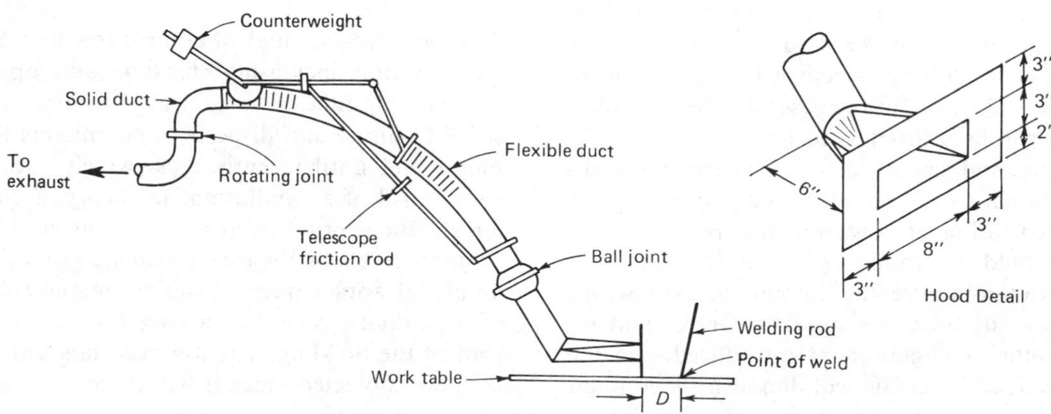

Figure 2.10. Welding hood.

able at the time of the original installation, it is usually advisable to plan for the possible need for higher cleaning efficiency at a later date and initially provide excess fan and motor capacity and reserve space for a second collector.

(2) Air stream characteristics:

(a) Temperature: High-temperature air filters may require special filtering cloth media and special construction. Consideration should be given to the possible economic advantage of pre-cooling the air before it enters the collector, permitting the use of standard equipment.

(b) Steam or moisture: Excessive moisture content of the air to be filtered may result in condensation, packing of the contaminants, and plugging of filter fabric or other components of a dry collector.

(3) Contaminant characteristics:

(a) Corrosive or oxidizing contaminants may require neutralizing wet cleaners or corrosion-resistant construction.

(b) Wax impregnating, buffing, rubber manufacture and other processes involving waxy or greasy contaminants can quickly plug cloth arrestors and some dynamic air cleaners and may require special equipment.

(c) Abrasive materials require wet or specially designed collectors to prevent excessive wear.

(d) Particle size and shape and size distribution will dictate the efficiency of the collector with the smallest particles usually the determining factor.

(e) Combustibility of the contaminant may require explosion-proof construction, venting, and fire protection.

(4) Collected Material Disposal:

(a) Dry collectors may be continuously or intermittently unloaded through special valves or gates to conveyors or containers. Manual unloading should be directly into bags or containers. In all cases, the unloading should be conducted in such a manner that a secondary air pollution problem does not arise and defeat the purpose of the air cleaner. When dust bins are not too large, they may be lined with heavy-duty plastic bags which can be tied before removal.

(b) Wet collectors can be continuously discharged into a sewer, settling basins, etc., or can periodically be unloaded in the form of a slurry. Wherever possible, wastes should be clarified or treated before discharge. Water pollution regulations may circumscribe the manner in which wet wastes may be disposed. In many areas, such requirements limit the use of wet collectors or make necessary costly treatment.

Air Cleaning Devices

Dust Collectors, Cloth Filters (see Figs. 2.11 and 2.12 and Table 2.10).

One of the most efficient and widely used media for removing dry particulates from an air stream is the cloth filter, also known as the dust arrester or bag filter. Practically any fibrous, woven, or felted material can be used as the filtering agent.

The usual design takes the form of tubes, screens, envelopes, or mats. Dust-laden air enters one side of the fabric and emerges through the other side as cleaned air to be discharged out of doors or, in some cases, recirculated back to the workroom. In most commercial designs, the fabric is mounted in a suitable framework enclosed in a housing. Air enters from below, usually via a plenum large enough to permit settling of the larger particles by gravity before they reach the filter zone. A manual or motorized shaking of the fabric can dislodge adhering aggregates. Mat filters can either be shaken, replaced, or cleaned by other means. The dust falls into a hopper for subsequent removal.

The cleaning process is not simple filtration since the pores of the medium are usually much larger than the particles to be collected. When new, much dust will pass through the filter until a bed of deposited dust is built up on the fabric. The process is complicated and not clearly understood. It probably involves impingement of the duct particles on the fibers as well as deposition under the influence of settling, Brownian motion, and static electricty created by the flowing air. The dust mat will rapidly build up on the medium, and it is this mat, rather than the fabric, which acts as the filtering agent. A permanent dust base will be created within the pores which will not be dislodged by shaking, so that cleaning efficiency remains high.

The most commonly used fabric medium for normal applications is cotton or wool sateen or felt. Operating temperature is usually the determining factor in the selection of fabric. Cotton may be used for temperatures up to 180°F. Wool is acceptable for temperatures of 200°F, and synthetics such as acrylics are available for temperatures up to 350°F and higher. Glass fiber and asbestos fabrics have been used for applications up to 650°F, but the fragility of the fibers results in failure after a number of shaking cycles.

The size of the filter, i.e., the area of filter medium through which a given stream of air passes,will affect the resistance to the air flow and therefore the required fan horsepower. In designing a cloth filter, the optimum size will usually be a compromise between available space, initial cost, and required horsepower. In some instances, the nature of the pollutants being exhausted or their toxicity will be the governing factor. The usual practice is to select a filter velocity of 3 linear fpm (cfm to cloth area ratio of 3 to 1).

Normal resistance of the filter to flow ranges from 2 to 6 inches of water, the usual resistance for a filter with a ratio of 3 to 1 being about 3 inches. Ratios higher than 3 to 1 will make possible smaller units, but the resultant high resistance will require more horsepower and the lower filtering area will necessitate more frequent shaking. A less recognized consequence is greater var-

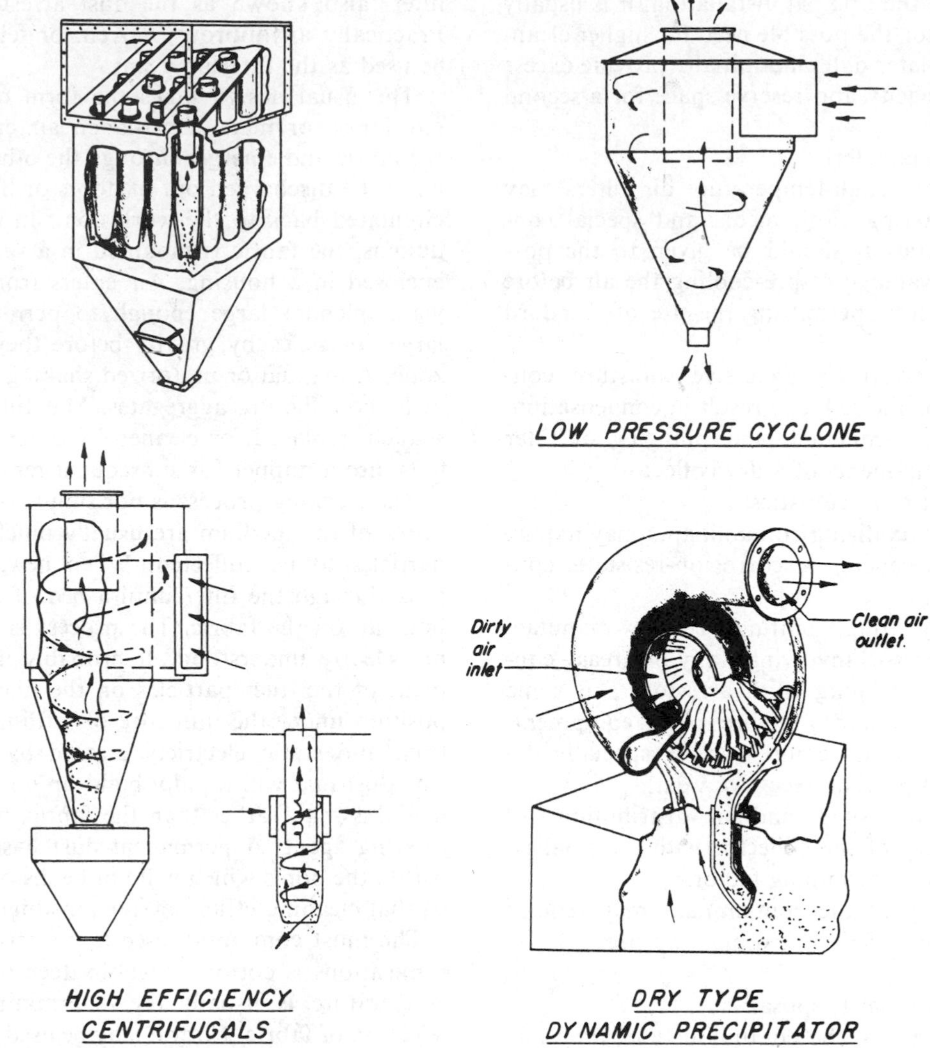

Figure 2.11a. Dry type centrifugal collectors.

iation in cfm developed by the fan when smaller units are used, since cfm will decrease as the resistance increases. With systems involving the control of highly toxic substances, where minimum flow rates at the hood must be exceeded at all times, lower cloth ratios are always used.

An important feature of the design of filters is the shaking mechanism and cycle. The fabric must be cleansed of adhering dust sufficiently often and thoroughly so that the rate of exhaust does not fall below the required minimum. The shaking mechanism must vigorously rap, shake, and flex the fabric to dislodge the dust which may adhere very firmly. The shaking cycle must, as indicated, be frequent enough to ensure continuous maintenance of the design cfm. For many normal applications, shaking can be performed periodically at normal process shutdowns such as lunch time or at the end of the day. An automatic interlock can be provided which will operate the shaking mechanism for a predetermined period each time the fan motor is turned off.

When abnormally heavy dust loading exists, it may be necessary to shut down the process during a shift in order to permit shaking. However, it is usually not advisable and frequently not safe to shut down operations involving the control of highly toxic, flammable, or explosive contaminants. Similarly, it may not be feasible to shut down operations performed on a 24-hour-a-day basis. In such situations, continuous cleaning must be provided.

One method of providing continuous cleaning is to use a three-compartment cleaner, each compartment being independent of the others and provided with a separate shaking mechanism. In this design, each compartment is successively and automatically removed from air cleaning service for shaking. Although only

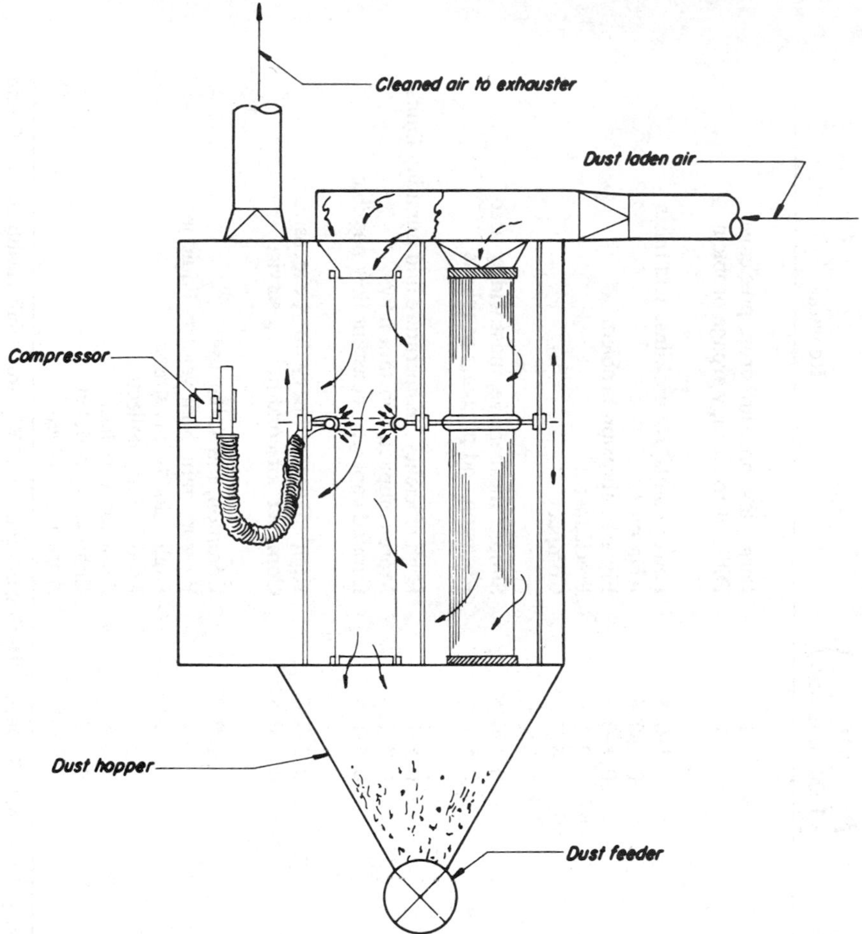

Figure 2.11b. Reverse jet type of continuous fabric arrester.

two-thirds of the cloth area is available at any one time for filtration, the cloth ratio can be the same as for a conventional arrester since the disadvantage of utilizing only two-thirds of the cloth is overcome by the more frequent cleaning.

A modification of the standard cloth arrester was developed a number of years ago and is now available commercially as a reverse jet arrester. It may consist of the standard cloth tubes or a single, large-diameter bag, usually of wool felt. The bag or tubes are circumscribed by a pipe ring which travels up and down the surface continuously. Jets of compressed air are continuously blown through the fabric, contercurrent to the flow of cleaned air out of the cloth, through a slot or orifices. The material caked on the cloth is continuously dislodged from the surface and, since it is agglomerated, is not reentrained in the air stream but falls to the collecting hopper. Continuous cleaning prevents excessive build up of resistance, making possible much smaller units with cloth ratios from 5–10 times greater than with standard arresters. The resistance, after initial build-up, is essentially constant, resulting in essentially constant cfm at the hoods and making possible more efficient fan selection and operation.

There are a number of other bag cleaning techniques including air shaking, wind whip, "bubble' cleaning, jet pulse, reverse air flexing, reverse flow of ambient air, repressuring, and sonic cleaning. Each technique has its place, depending on the fabric used for the bags. In some cases, depending on the nature and amount of dust loading and the design of the cleaning process, higher air-to-cloth ratios may be used.

There are many special types of filters, including dry paint filters, mats or panels of treated paper or cloth, and dry or oiled wire mesh, which have applications for light dust loading or special conditions. The "absolute" or HEPA (high efficiency particulate air filter) filter is a special deep-bed fibrous filter with extremely high efficiency for dust and fume particulates as fine as 0.3μ. These are particularly useful for extremely toxic materials such as beryllium dust, radioactive particulates, and virulent biologicals. Their special advantage lies not only in high efficiency but also in the fact that, when loaded, the complete filter unit can be discarded, usually

TABLE 2.10. Approximate Characteristics of Dust and Mist Collection Equipment[a]

Equipment Type	Smallest Particle Collected (μ)[b]	Pressure Drop (in. H_2O)	Power Used[c] $\left(\frac{kW}{1000 \text{ cu ft/min}}\right)$	Remarks
Settling chambers				
(1) Simple	40	0.1–0.5	0.1	Large, low pressure drop, precleaner
(2) Multiple tray	10	0.1–0.5	0.1	Difficult to clean, warpage problem
Inertal separators				
(1) Baffle chamber	20	0.5–1.5	0.1–0.5	Power plants, rotary kilns, acid mists
(2) Orifice impaction	2	1–3	0.2–0.6	Acid mists
(3) Louver type	10	0.3–1	0.1–0.2	Fly ash, abrasion problem
(4) Gas reversal	40	0.1–0.4	0.1	Precleaner
(5) Rotating impeller	5	—	0.5–2	Compact
Cyclones				
(1) Single	15	0.5–3	0.1–0.6	Simple, inexpensive, most widely used
(2) Multiple	5	2–10	0.5–2	Abrasion and plugging problems
Filters				
(1) Tubular	<0.1	2–6	0.5–1.5	High efficiency, temperature and humidity limits
(2) Reverse jet	<0.1	2–6	0.7–1.5	More compact, constant flow
(3) Envelope	<0.1	2–6	0.5–1.5	Limited capacity, constant flow possible
Electrical precipitators				
(1) One-stage	<0.1	0.1–0.5	0.2–0.6	High efficiency, heavy duty, expensive
(2) Two-stage	<0.1	0.1–0.3	0.2–0.4	Compact, air-conditioning service
Scrubbers				
(1) Spray tower	10	0.1–0.5	0.1–0.2	Common, low water use
(2) Jet	2	—	2–10	Pressure gain, high velocity liquid jet
(3) Venturi	1	10–15	2–10	High velocity gas stream
(4) Cyclonic	5	2–8	0.6–2	Modified dry collector
(5) Inertial	2	2–15	0.8–8	Abrasion problem
(6) Packed	5	0.5–10	0.6–2	Channeling problem
(7) Rotating impeller	2	—	2–10	Abrasion problem

[a] From "Dust Collector Review," David G. Stephan, Air Pollution Engineering Research, U.S. Dept. of Health, Education and Welfare, Public Health Service, Cincinnati, Ohio, 1960.
[b] With 90–95% efficiency by weight.
[c] Includes pressure loss, water pumping, electrical energy.
Note: See Figure 2.12 for cost estimates.

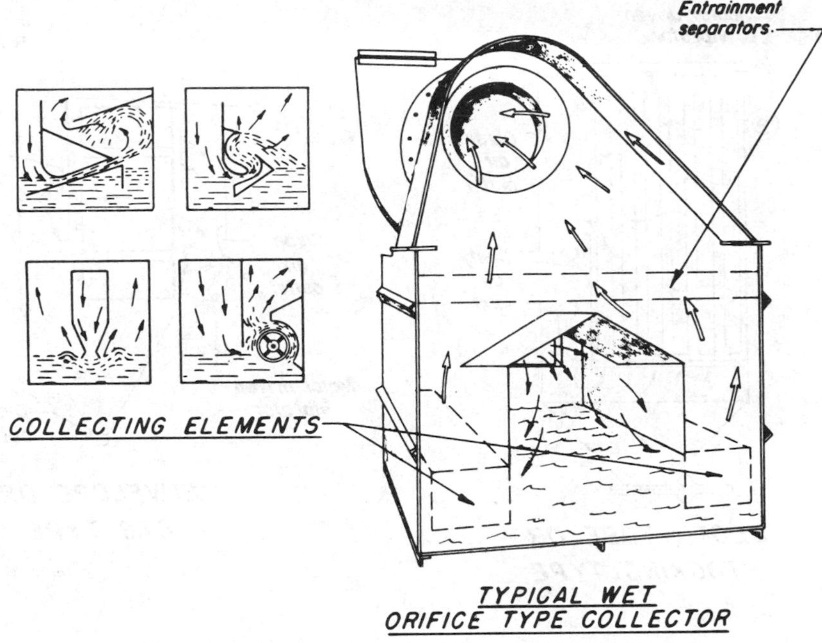

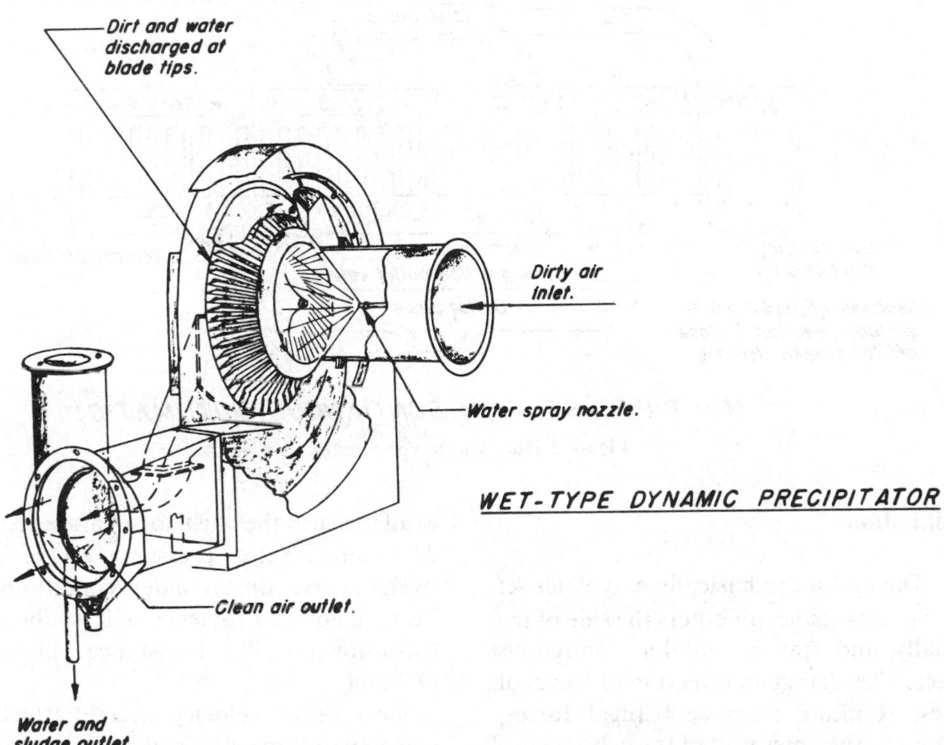

Figure 2.11c. Orifice type collector.

by companies specializing in hazardous wastes disposal. Disposal of hazardous dusts collected by conventional filters can create a secondary air pollution problem as well as an industrial hygiene problem.

Filters have wide application and can be used for almost all type of dusts and fumes in the fine particle range with efficiencies above 99%. Use of less efficient, low-pressure pre-cleaners will reduce the required size of the more expensive arrester and will increase its useful life by removing the larger, more abrasive particles and by making possible shorter and less frequent cleaning cycles. Tables 2.11 and 2.12 list recommended filtering ratios for various dusts in conventional and reverse-jet cloth dust arresters.

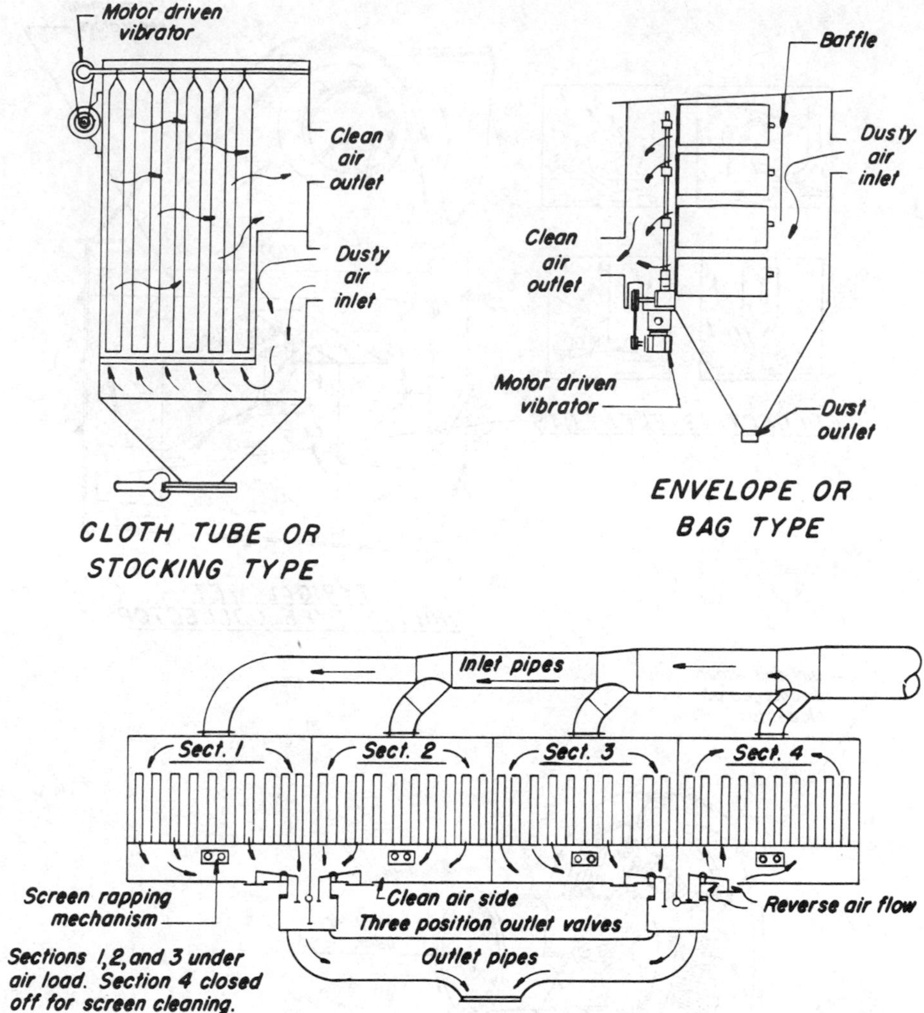

Figure 2-11d. Cloth type arresters.

Dynamic Precipitation.

(1) Cyclones. The cyclone is basically a cylinder set on top of a cone. The dust-laden air enters the side of the cylinder tangentially and travels radially, contiguous to its inside surface. The change in direction of travel of the dust particles, resulting from centrifugal forces, causes them to move to the inner wall of the cylinder in a zone of lower air velocity. Here they are precipitated from the air stream and fall to a dust hopper at the bottom of the cone by the action of gravity. The cleaned air loses velocity as it travels downward in a helical motion until is reverses direction and is discharged axially out the top of the cylinder through a concentric oulet pipe.

Efficiency of collection is a function of entrance and radial velocities within the cylinder and is therefore directly proportional to pressure drop or resistance. Length of cone is also a factor since it increases the time available for the dust to separate from the air stream. The standard low-pressure cyclone is useful for relatively coarse dusts under conditions of moderate to heavy loading. Efficiency will be about 75–80% for particles down to 40μ. Resistance will be about 1–3 inches of water.

Since radial velocity directly affects efficiency, for a given air volume efficiency will increase with decrease in cylinder diameter. Modifications of the dimensions of the standard cyclone which have been developed empirically over the years have resulted in the so-called high-pressure or high-efficiency cyclone. Performance can be further improved by using a number of very small cyclones in parallel instead of a single unit. Multiple units with efficiencies up to 90% for particles down to 10μ can thus be obtained with pressure drops ranging from 2–10 inches.

Principle advantages of the cyclone include its sim-

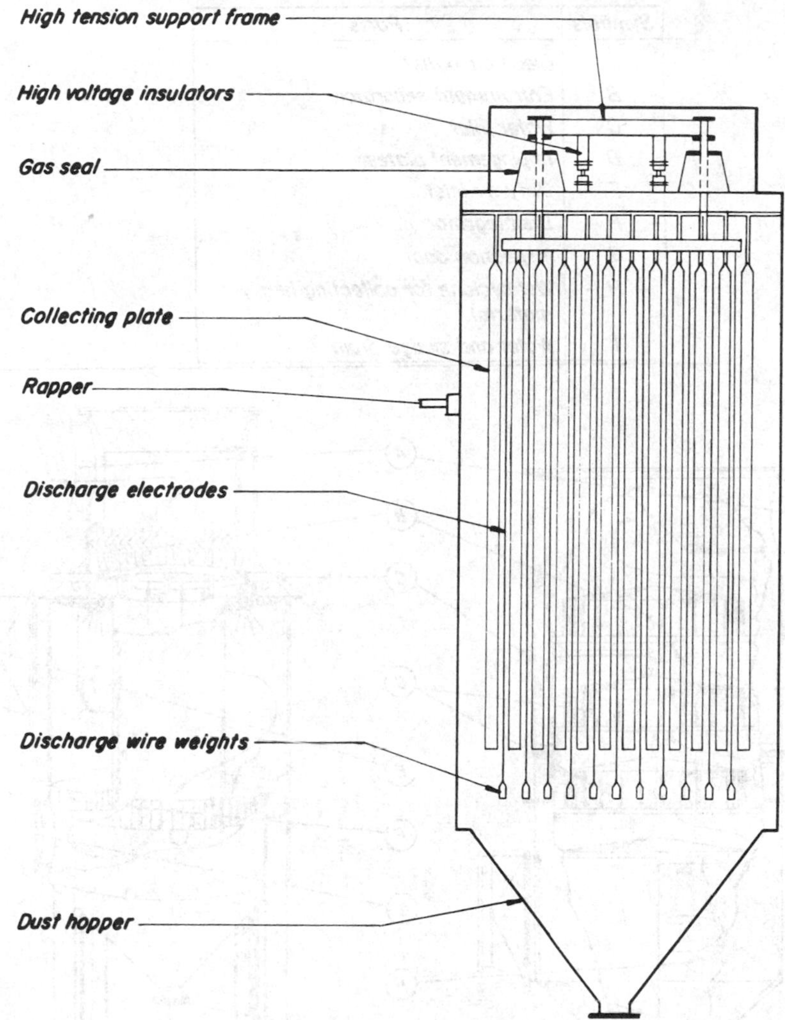

Figure 2.11e. High-voltage electrostatic precipitator.

plicity, lack of moving parts, and constant air flow. (Table 2.13 lists applications of centrifugal collectors.)

(2) Dry dynamic precipitator. This collector uses a specially shaped impeller to precipitate dust from the air stream by centrifugal and dynamic forces. The dust travels along the impeller blade surfaces and is projected into a separate dust chamber in the impeller housing from which it is transported through an opening to a hopper below. Collection efficiency and resistance are similar to the multiple cyclone units described above. The principal advantage of this equipment is that the impeller serves as both the air mover and the dust collector, resulting in a lower initial cost and space requirement than for a separate fan and cyclone with similar characteristics. However, it cannot be used for large air quantities and is not suitable for fibrous or adhesive dusts or for very large particles.

(3) Louver collectors. These consist of a series of louvers or plates set an angle to the air stream. Separation results from the centrifugal forces created by the rapid and frequent change in direction of air flow as the air passes over the louver surfaces. Decreases in the spaces between the louver surfaces results in an increase in efficiency but also increases the possibility of plugging by deposition of dust in the spaces. Characteristics and advantages are similar to those of the cyclone, but the louver collector is not as adaptable for large volumes of air because of its space requirements. The usual resistance will be between 0.3 and 1.0 inches of water.

(4) Inertial separators. These are large settling chambers in which dust settles from the air stream by gravity. The pressure drop is very low, and the settling chamber will have some limited usefulness where space is available and the contaminant consists principally of large, heavy particles. Chambers as long as possible with cross sections as large as possible will result in low particle velocity with maximum possible distance of travel to permit the particles to reach terminal settling velocity before the air is discharged outdoors.

Inertial separators can also be provided with suitably

Symbols	*Parts*
A	*Clean air outlet.*
B	*Entrainment separator.*
C	*Water inlet.*
D	*Impingement plates.*
E	*Dirty air inlet.*
F	*Disintegrator.*
G	*Inspection door.*
H	*Wet cyclone for collecting heavy material.*
I	*Water and sludge drain.*

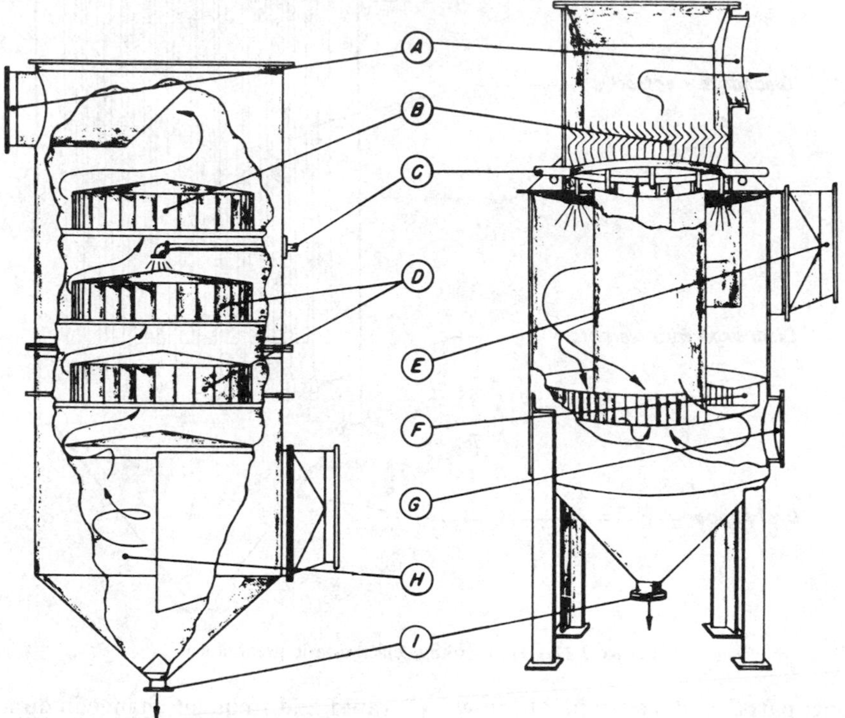

Figure 2.11f. Wet centrifugal collectors.

arranged baffle plates so that there is a continuous change of direction of air flow. The resultant lowering of particle velocity plus an impingement factor increases the rate of settling. Such an arrangement is particularly suited for removal of mists from an air stream.

The settling chamber principle has effectively been used in the low-velocity plenum main where the "cleaned air" is either discharged directly to the outdoors or into a suitable collector. The latter can usually be quite small because of the ultimately very light dust loading. A bonus advantage of such equipment is that branch pipes can be added, removed, or relocated at will without affecting the dynamic balance of the system.

The dust which settles in the plenum may be removed by a conveyor or other method. In some cases, the dust settles into a hopper or hoppers under the plenum from which it is conveyed by a conventional, small, high-velocity system to a suitable collector. See Fig. 2.12.

Wet Collectors. The basic principle of the wet collector is to provide intimate contact between the collecting liquid and the contaminant in the air to be cleaned. When the contaminant is particulate, the wetting of the particle increases its mass and it becomes relatively easy to remove from the air stream. Wet collection is also suitable for gases or vapors which are either soluble in, or react with, the collecting medium. The intimate mixture of the air and liquid provides a large interface area between the two media so that solution or reaction of the contaminant is enhanced. There are many designs of wet collectors with a wide range of pressure drops and efficiencies. As in all cleaners, the former varies directly with the latter.

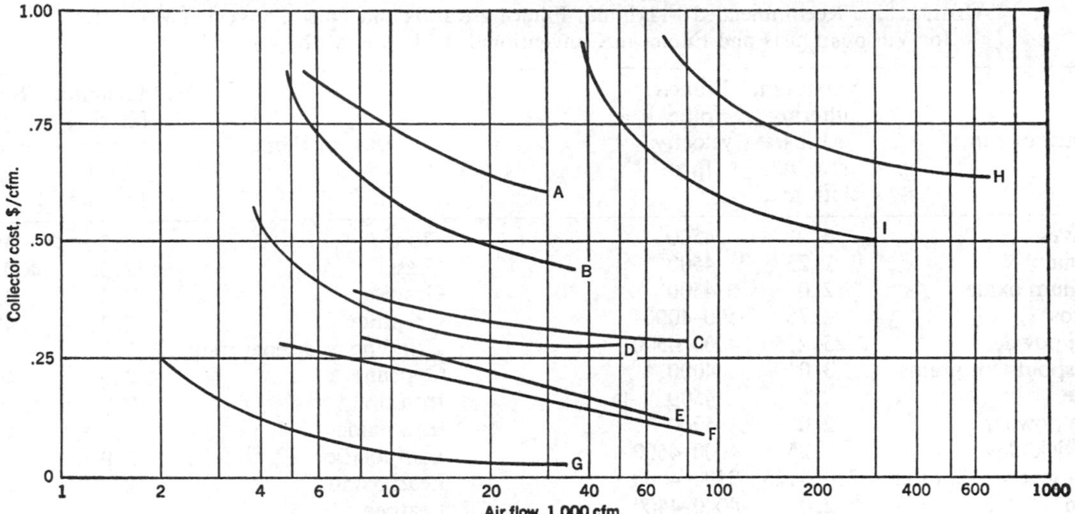

A—High temperature fabric collector (continuous duty)
B—Reverse jet fabric collector (continuous duty)
C—Wet collector (maximum cost range)
D—Intermittant duty fabric collector
E—High efficiency centrifugal collector
F—Wet collector (minimum cost range)
G—Low pressure drop cyclone (maximum cost range)
H—High voltage precipitators
I—High voltage precipitators (minimum cost range)

NOTE 1: Cost based on collector section only. Cost does not include ducting, water requirement, power requirement or exhausters (unless exhaust is integral part of secondary air circuit.)

NOTE 2: Cost of continuous duty sectional fabric collector approaches cost of reverse jet continuous duty collector.

NOTE 3: Costs of electrostatic precipitators will vary with the contact time and the electrical equipment required. Costs shown are for fly ash installations when high velocities of 300 to 600 fpm are usual. Precipitators for metallurgical fumes, etc. will be considerably higher in cost per cfm.

Figure 2.12. Cost estimates of air-cleaning equipment (1968). Courtesy *Industrial Ventilation Manual*, 11th edition, American Conference of Governmental Hygienists. An inflation correction factor should be applied.

The principal advantages of wet collectors are the maintenance of a constant air flow, applicability to corrosive or hot air streams, suitability for both particulate and nonparticulate contaminants, and the absence of a secondary air contamination during disposal of collected contaminants.

Among the disadvantages are the creation or required use of corrosive liquids in the unit (increasing construction and maintenance costs), elaborate installations frequently required, the need for a reliable and plentiful water supply, and disposal problems which may require expensive pretreatment. Suitable protection against freezing must be provided if located outdoors.

Efficiency for particle collection can be as high as 98% With proper design and operation, essentially 100% efficiency can be obtained for gases and vapors.

The basic wet collectors include the following:

(1) Spray chamber. This is perhaps the simplest wet collector. It consists of sprays, usually in combination with scrubber plates, through which the contaminated air flows. Efficiency will be governed by the pressure, number, and location of the spray nozzles as well as cross-sectional area. The standard dry cyclone can also be converted to a wet centrifugal collector by suitable placed nozzles within the cyclone to provide both centrifugal and wetting action, with the latter enhancing the former.

Water may be distributed by nozzles, an elevated reservoir, or the air stream itself. Pressure drop will vary from 2–6inches. Water consumption is usually from 3–5 gal/1000 cfm of air.

(2) Wet dynamic. The dry dynamic precipitator previously described can be modified by the addition of spray nozzles at the inlet designed to maintain a film of water on the impeller surfaces. This increases the efficiency of collection retention of the particles on the im-

TABLE 2.11. Recommended Maximum Filtering Ratios and Dust Conveying Velocities for Various Dusts and Fumes in Conventional Baghouses with Woven Fabrics

Dust or fumes	Maximum filtering ratios, cfm/ft² cloth area	Branch pipe velocity, fpm	Dust or Fumes	Maximum filtering ratios, cfm/ft² cloth area	Branch pipe velocity, fpm
Abrasives	3.0	4500	Flour	2.5	3500
Alumina	2.25	4500	Glass	2.5	4000–4500
Aluminum oxide	2.0	4500	Granite	2.5	4500
Asbestos	2.75	3500–4000	Graphite	2.0	4500
Baking powder	2.25–2.50	4000–4500	Grinding and separating	2.25	4000
Batch spouts for grains	3.0	4000	Gypsum	2.5	4000
Bauxite	2.5	4500	Iron ore	2.0	4500–5000
Bronze powder	2.0	5000	Iron oxide	2.0	4500
Brunswick clay	2.25	4000–4500	Lampblack	2.0	4500
Buffing wheel operations	3.0–3.25	3500–4000	Lead oxide	2.25	4500
Carbon	2.0	4000–4500	Leather	3.5	3500
Cement crushing and grinding	1.5	4500	Lime	2.0	4000
Cement kiln (wet process)	1.5	4000–4500	Limestone	2.75	4500
Ceramics	2.5	4000–4500	Manganese	2.25	5000
Charcoal	2.25	4500	Marbel	3.0	4500
Chocolate	2.25	4000	Mica	2.25	4000
Chrome ore	2.5	5000	Oyster shell	3.0	4500
Clay	2.25	4000–4500	Packing machines	2.75	4000
Cleanser	2.25	4000	Paint pigments	2.0	4000
Cocoa	2.25	4000	Paper	3.5	3500
Coke	2.25	4000–4500	Plastics	2.5	4500
Conveying	2.5	4000	Quartz	2.75	4500
Cork	3.0	3000–3500	Rock	3.25	4500
Cosmetics	2.0	4000	Sanding machines	3.25	4500
Cotton	3.5	3500	Silica	2.75	4500
Feeds and grain	3.25	3500	Soap	2.25	3500
Feldspar	2.5	4000–4500	Soapstone	2.25	4000
Fertilizer (bagging)	2.4	4000	Starch	2.25	3500
Fertilizer (cooler, dryer)	2.0	4500	Sugar	2.25	4000
Flint	2.5	4500	Talc	2.25	4000
			Tobacco	3.5	3500
			Wood	3.5	3500

(From *Control Techniques for Particulate Air Contaminants.* National Air Pollution Control Administration Publ. #AP-51).

peller blade surface, as well as the disposal of the material. Water consumption will be from ½–1 gal/1000 cfm of air.

(3) Packed towers. These are used extensively by the chemical industry for process purposes, but they can also function as air cleaners. Essentially they consist of shells of steel, ceramic, plastic, or other suitable material packed with such materials as irregularly shaped ceramic saddles or rings, or granular materials such as sand, coke, and gravel. Fibrous packing such as glass or steel wool may also be used. Water, at the usual rate of about 10 gal/1000 cfm air is introduced at the top and contaminated air at the bottom. Efficiency depends on period of contact between the air and the wetted packing so that air velocity through the unit should be as low as possible and the surface area and depth of packing as great as possible. Pressure drops may vary widely depending on design but the usual range is from 1.5–3.5 inches. The simplicity of design of these towers minimizes maintenance problems and makes them particularly adaptable for hot or corrosive air streams. Heavy dust loading will, however, require cleaning or replacement of the packing.

(4) Wet orifice collectors. In these, the passage of the contaminated air at a high velocity through an orifice or baffled opening located above a water reservior creates a water curtain with considerable turbulence. The contaminants are in this manner thoroughly wetted and retained in the water reservoir. The lack of ledges, obstructions, or moving parts makes the design features of orifice collectors fairly simple. This, plus the presence of water, makes orifice collectors particulary adaptable for explosive, pyrophoric, sticky, or fibrous contaminants. The pressure drop varies from 2.5–6 inches, and higher. Facilities must be provided for separating en-

TABLE 2.12. Recommended Maximum Filtering Ratios and Fabric for Dust and Fume Collection in Reverse-Jet Baghouses

Material or Operation	Fabric	Filtering ratios, cfm/ft²	Material or Operation	Fabric	Filtering ratios, cfm/ft²
Aluminum oxide	Napped cotton	11	Mica	Napped cotton	11
Bauxite	Cotton sateen	10	Paint pigments	Cotton sateen	10
Carbon, calcined	Napped cotton, wool felt	8[a]	Phenolic molding powders.	Cotton sateen	10
Carbon, green	Orlon felt	7	Polyvinyl chloride (PVC)	Wool felt	10[a]
Carbon, banbury mixer.	Wool felt	8	Refractory brick sizing (after firing).	Napped cotton	12
Cement, raw	Cotton sateen	9	Sandblasting	Napped cotton, wool felt	6–8[a]
Cement, finished	Cotton sateen	10	Silicon carbide	Cotton sateen	9–11
Cement, milling	Cotton sateen	8	Soap and detergent powder.	Dacron felt, orlon felt	12[a]
Chrome, (ferro) crushing.	Cotton sateen	10	Soy bean	Cotton sateen	14
Clay, green	Cotton sateen	10	Starch	Cotton sateen	10
Clay, virtified silicious.	Cotton sateen	12	Sugar	Cotton sateen, wool felt	10[a]
Enamel, (porcelain)	Napped cotton	12	Talc	Cotton sateen	11
Flour	Cotton sateen	14[a]	Tantalum fluoride	Orlon felt	6[a]
Grain	Wool felt, cotton sateen	16	Tobacco	Cotton sateen	12
Graphite	Wool felt	7[a]	Wood flour	Cotton sateen	10
Gypsum	Cotton sateen, orlon felt	10	Wood sawing operations.	Cotton sateen	12
Lead oxide fume	Orlon felt, wool felt	8[a]	Zinc, metallic	Orlon felt, dacron felt	11
Lime	Napped cotton	10	Zinc oxide	Orlon felt	8[a]
Limestone (crushing)	Cotton sateen	10	Zirconium oxide	Orlon felt	8
Metallurgical fumes	Orlon felt, wool felt	10[a]			

[a]Decrease 1 cfm/ft^2 if dust concentration is high or particle size is small.
(From *Control Techniques for Particulate Air Contaminants* National Air Pollution Control Administration Publ. #AP-51).

trained moisture from the effluent before discharge. Maintenance of a constant water level in the reservoir is critical for the creation of the proper water curtain.

(5) High-efficiency wet collectors. As the size of the water droplets created by a wet collector approaches the size of the smallest particles to be collected, the probability of contact, and therefore collection, increases. For this reason, the fog filter and the venturi scrubber are among the most efficient of wet collectors since their action results in the dispersion of very fine water particles.

Fog filters consist of many small, high-pressure nozzles located in centrifugal tower collectors. Water pressures of 250–600 psi at flow rates of 5–10 gal/1000 cfm of air are used. The very fine orifices at high pressures require special nozzle construction. The water may have to be pre-filtered to prevent plugging, and for this same reason, recirculation of the water is not practical.

In the venturi scrubber, the contaminated air is passed through a venturi throat at very high velocity. Water, which is simultaneously fed into the throat, is atomized by the air into a very fine fog. The intimate mixing of the contaminant and water in the turbulence created by the venturi results in rapid wetting and collection. Pressure lossed may be as high as 12–50 inches, but efficiencies may run as high as 99% in the submicron particle size range.

Electrostatic Precipitators. The basic principle involves the creation of a strong unidirectional electrostatic field between two electrodes. The dust particles in the contaminated air passing between the electrodes are electrically charged and attracted toward an oppositely charged collecting electrode where they are deposited.

There are two general types of electrostatic precipitators:

(1) The high-potential, single-stage precipitator in which ionization and collection are simultaneous through-out the unit. These are used for the collection of particulates (dust, fume, mist) in industrial exhaust systems where loading can be high and will generate from 50,000–75,000 volts.

(2) The low-potential, two-stage precipitator is divided into a pre-ionizing section followed by a nonionizing collection section. It is used for general air cleaning or industrial air cleaning where the concentration of par-

TABLE 2.13. Applications of Centrifugal Collectors

Operation or Process	Air Contaminant	Type of Air Cleaning Equipment	Collector Efficiency, wt %
Crushing, pulverizing, mixing, screening			
Alfalfa feed mill	Alfalfa dust	Cyclone, settling chamber	85
Barley feed mill	Barley flour dust	Cyclone	85
Wheat air cleaner	Chaff	Cyclone	85
Drying, baking			
Catalyst regenerator (petroleum)	Catalyst dust	Cyclone, ESP	95
Detergent powder spray drier	Detergent powder	Cyclone	85
Orange pulp feed drier	Pulp dust	Cyclone	85
Sand drying kiln	Silica dust	Cyclone	78
Sand and gravel drying	Silica dust	Inertial collector	50
Stone drying kiln	Silica dust	Cyclone	86
Mixing fluids			
Asphalt mixing	Sand and gravel dust	Cyclone	50–86
Bituminous concrete mixing	Sand and stone dust	Cyclone, scrubber	95
Polishing, buffing, grinding, chipping			
Grinding (aluminum)	Aluminum dust	Cyclone	89
Grinding (iron)	Iron scale and sand	Cyclone	56
Grinding (machine shop)	Dust	Impeller collector	91
Surface coating			
Rubber dusting	Fluffy zinc stearate	Impeller collector	78–88
Surface treatment—physical			
Abrasive cleaning	Talc dust	Cyclone	93
Abrasive stick trimming and shaping	Silicon carbide and alumina dust	2 parallel cyclones	51
Casting cleaning with metal shot, sandblasting and tumbling	Metallic and silica dust	Impeller collector	97–99+
Foundry tumbling	Dust	Impeller collector	99
Truing and shaping abrasive products	Silicon carbide and alumina dust	Cyclone	58
Woodworking, including plastics rubber, paper board			
Milling planing	Wood dust and chips	Cyclone	97

(From *Control Techniques for Particulate Air Pollutants* National Air Pollution Control Administration Publ. #AP-51).

ticulates is low. The potential ranges from 12,000–15,000 volts.

The design of these units will depend on air flow rate, characteristics of the particles and the air stream, and required efficiency. Although efficiencies approaching 100% can be obtained, it is not usually practical to do so since the cost rises very sharply as this figure is approached. Such factors as local air pollution laws, available space, reclaim value of collected material, etc., will affect the design.

The major advantages of the electrostatic precipitator include constant air flow rate and low pressure drop. Among its disadvantages are high initial and operating costs, unsuitability for flammable or explosive atmospheres, frequent necessity to condition the entering air stream, and the need for primary separators when high dust loading (more than 25 grains/ft^3 of air) may be encountered.

Gas and Vapor Cleaning (see Fig. 2.14). As indicated above, since atmospheric control of gases and vapors by dispersion and dilution is becoming less acceptable, air cleaners for these contaminants, which were formerly seldom required, will in many cases ultimately become mandatory. Therefore, the problem of design of air cleaning for gases and vapors is rapidly attracting much more attention than in the past, particularly for relatively small installations.

The nature of gases (the term gases will be used to signify both gases and vapors) is such that their removal from an air stream is limited to four basic methods: absorption by solution and/or reaction in a suitable liquid medium, adsorption on the surface of a solid material,

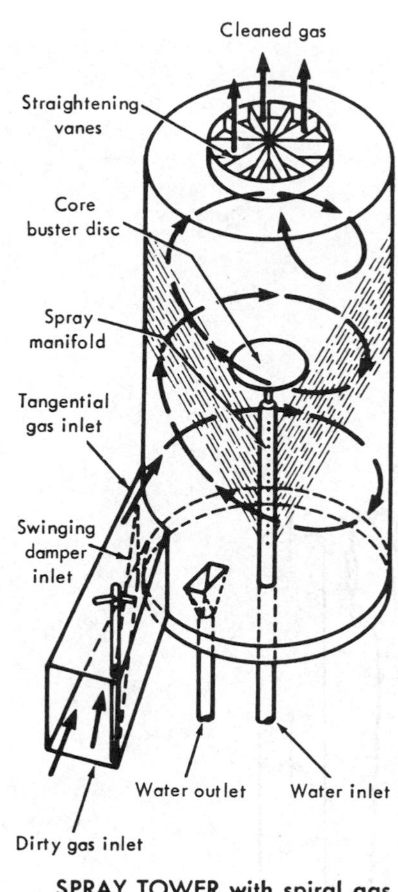

SPRAY TOWER with spiral gas motion.

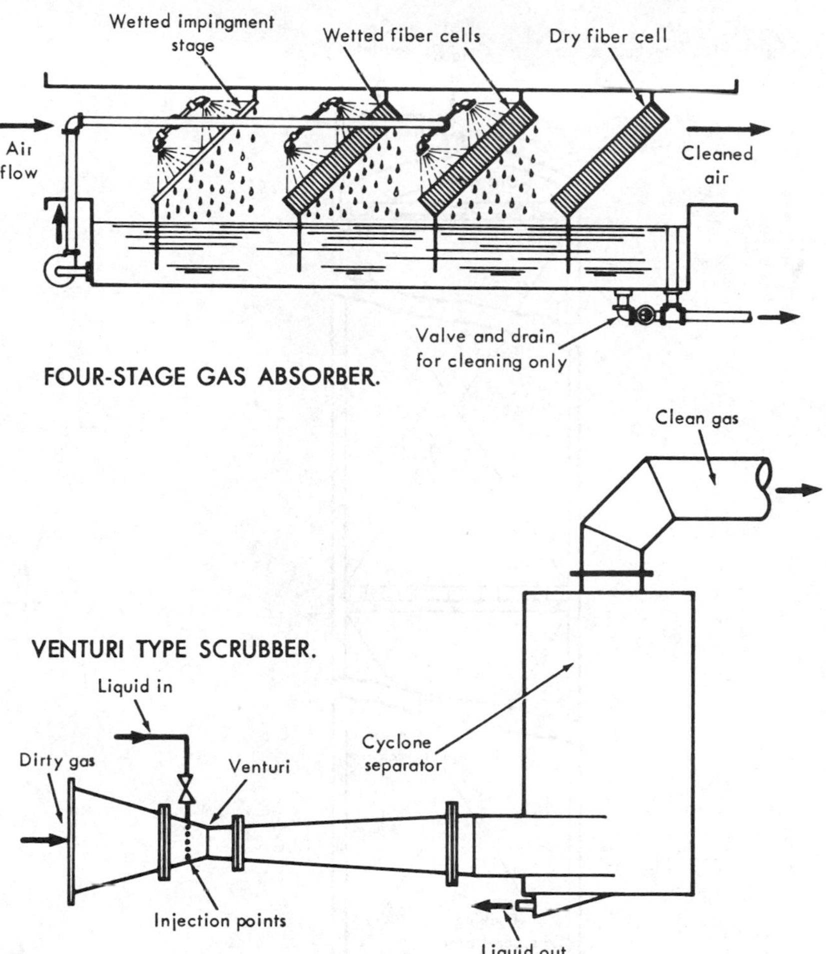

Figure 2.13a. Wet scrubbers.

combustion to yield innocuous products, and condensation to the liquid state. Proper design can provide any desired degree of efficiency up to virtually 100%. Resistance to air flow may be as high as 20–30 inches of static pressure.

(1) Absorption. This is normally accomplished by any of the wet collectors previously described for particulates. (Dynamic scrubbers, however, serve no useful purpose in gas collection except when the dynamic features of the design also increase the contact between the contaminant and the scrubbing medium.)

Since gaseous molecules will not be physically entrained and retained by a liquid, the medium must be one in which the contaminant is highly soluble or with which it will react. In many instances water alone will be satisfactory. In other cases, caustic soda or other chemicals or reactants must be added to the water. The scrubber should be designed to provide an intimate mixture of contaminated air and cleaning liquid for sufficient duration and magnitude of contact so that maximum solution or reaction takes place. The medium is usually continually or partially recirculated, particularly when water is in short supply, when disposal of liquid is a problem, or when chemical solutions are used as the medium. Arrangements usually are made for continuous or periodic replenishment to prevent the medium from becoming saturated or vitiated with a resultant decrease in absorbing efficiency. Mist entrainment devices are usually necessary in the discharge stack.

If the contaminant is collected in a viable condition, it can frequently be recovered for sale or reintroduction into the process. Otherwise, disposal of waste liquid must be to a sewer, stream, ground, or sump, with due consideration for local or state rules. Frequently, dilution, clarification, or chemical treatment is required before disposal. The necessity for costly treatment of waste liquid or the unavailability of a reliable water supply may contraindicate the use of wet collectors.

(2) Adsorption. Adsorption of gases on the surface of a solid is possible because of the existence of available molecular binding energy on the surface of the adsorbent. Maximum adsorption of gases from an air

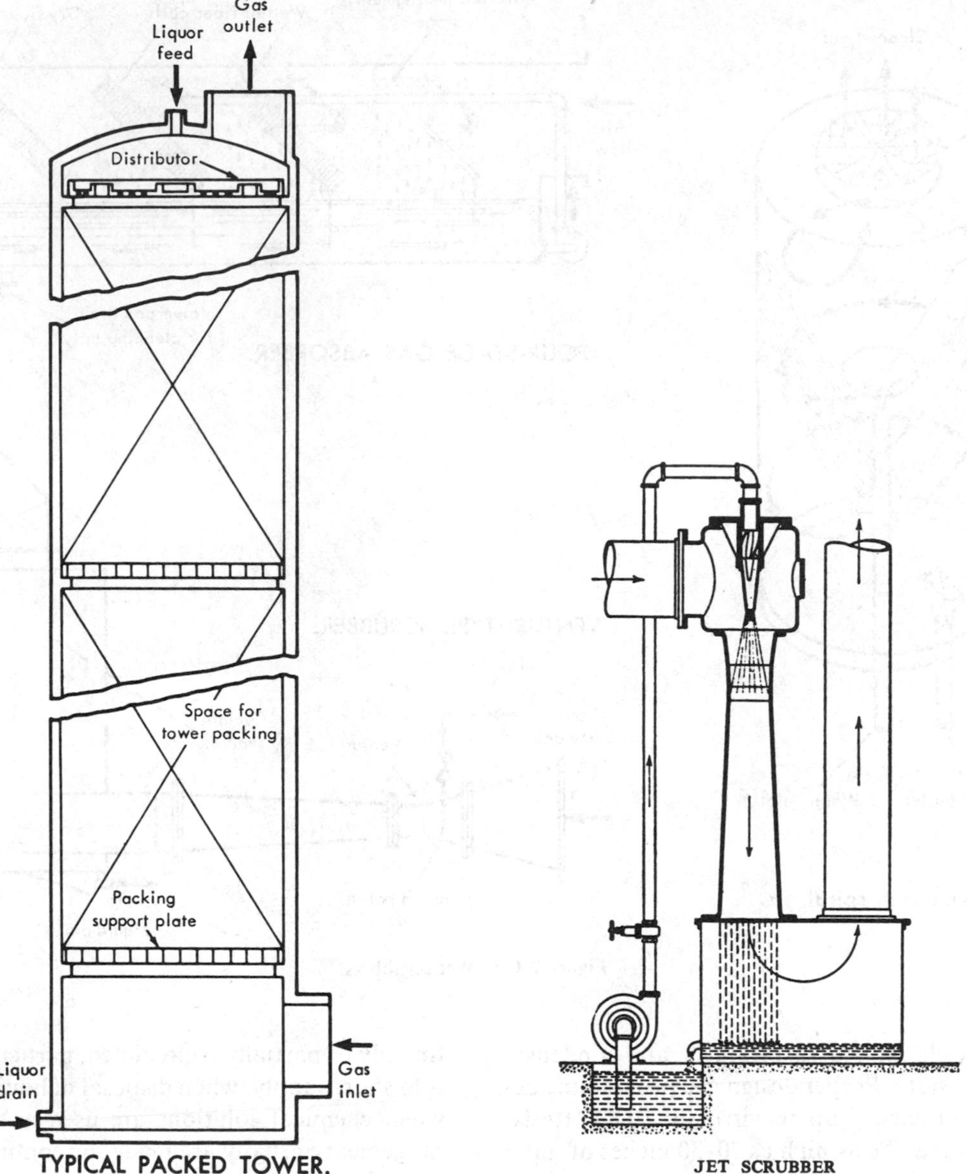

Figure 2.13b. Wet collectors.

stream occurs when there is a high concentration of gas, a large adsorbing surface (finely divided material) absence of interfering substances, low temperature, and favorable characteristics of gas molecule with respect to the molecular structure of the adsorbent (shape, size, and polarity of the respective molecules). All factors being equal, efficiency and resistance to air flow will bear a direct relationship to fineness and shape of the adsorbing particles and the thickness of the bed of adsorbent.

The usual adsorbents consist of activated charcoal or silica gel, although a number of substances can be used. The efficiency can frequently be increased by impregnation of the adsorbent with a chemical agent which converts the pollutant to a harmless or much more adsorbable material. The adsorbent can also be coated onto an inert carrier. This is frequently done when the adsorbent may be an expensive material which cannot be reactivavated for re-use after it is saturated.

The adsorbent is placed in a perforated cannister, flat bed, or other suitable container through which the contaminated air can be directed. The units are arranged in series and parallel so an optimum surface area and depth of bed is obtained. The design basis will provide for maximum contact between the contaminant and the adsorbent, sufficient capacity for the desired service life, resistance to air flow within the capacity of the air mover, uniform distribution over the adsorbent bed with avoidance of channeling, pretreatment of the air if nec-

TABLE 2.14. Typical Industrial Application of Wet Scrubbers

Scrubber Type	Typical Application
Spray chambers	Dust cleaning, electroplating, phosphate fertilizer, kraft paper, smoke abatement
Spray tower	Precooler, blast furnace gas
Centrifugal	Spray dryers, calciners, crushers, classifiers, fluid bed processes, kraft paper, fly ash
Impingement plate	Cupolas, driers, kilns, fertilizer, flue gas
Venturi	
Venturi throat Flooded disk Multiple jet	Pulverized coal, abrasives, rotary kilns, foundries, flue gas, cupola gas, fertilizers, lime kilns, roasting, titanium dioxide processing, odor control, oxygen steel making, coke oven gas, fly ash
Venturi jet	Fertilizer manufacture, odor control, smoke control
Vertical venturi	Pulverized coal, abrasive manufacture
Packed bed	
Fixed	Fertilizer manufacturing, plating, acid pickling
Flooded	Acid vapors, aluminum inoculation, foundries, asphalt plants, atomic wastes, carbon black, ceramic frit, chlorine tail gas, pigment manufacture, cupola gas, driers, ferrite, fertilizer
Fluid (floating) ball	Kraft paper, basic oxygen steel, fertilizer, aluminum ore reduction, aluminum foundries, fly ash, asphalt manufacturing
Self-induced spray	Coal mining, ore mining, explosive dusts, air conditioning, incinerators
Mechanically-induced spray	Iron foundry, cupolas, smoke, chemical fume control, paint spray
Disintegrator	Blast furnace gas
Centrifugal fan Inline fan	Metal mining, coal processing, foundry, food, pharmaceuticals
Wetted filters	Electroplating, acid pickling, air conditioning, light dust
Dust, mist eliminators	
Fiber filters	Sulfuric, phosphoric, and nitric acid mists; moisture separators; household ventilation; radioactive and toxic dusts, oil mists
Wire mesh	Sulfuric, phosphoric, and nitric acid mists; distillation and absorption
Baffles	Coke quenching, kraft paper manufacture, plating
Packed beds	Sulfuric and phosphoric acid manufacture, electroplating, spray towers

(From *Control Techniques for Particulate Air Pollutants*, National Air Pollution Control Administration Publ. #AP-51.)

essary to remove particulates and other interfering substances, and provision for renewing or replacing the adsorbent periodically.

The collected material may be removed from the adsorbent by suitable means (this also regenerates the adsorbent for further use); the saturated adsorbent may be discarded (highly toxic or radioactive contaminants may require special handling); or the contaminant may be oxidized or combusted on the surface of the adsorbent.

(3) Combustion. When a gas or vapor can be oxidized to an innocuous substance (as with most HC), combustion is frequently used as a method for removing such contaminants from the air stream. In some instances, where suitable liquid or solid collecting media are not available or practical for a particular contaminant, combustion may be the only feasible method. Design of a system for maximum conbustion requires bringing the oxygen into intimate contact with the gas molecules at adequate temperatures for a sufficient duration. This will require a combination of high temperature and turbulent mixing in a properly designed combustion chamber.

When the contaminant is present in the air in concentrations between the lel (lower explosive limit) and the uel (upper explosive limit), provision of a source of ignition will initiate combustion which will then be self-sustaining (see Section 7). In those rare instances where the contaminant is present in concentrations above the uel, provision of additional dilution air to bring the level to below the uel will make this process possible. However, in most industrial applications, the concentration of gases in the effluent will be below the lel so that the air stream must be heated and maintained at the autogenous temperature (temperature at which organic gases and vapors will burn regardless of their concentration) for sufficient duration and with adequate turbulent mixing to complete combustion. This is usually accomplished by means of direct fired burners in a furnace or similar chamber.

Direct fired incineration presents an explosion hazard

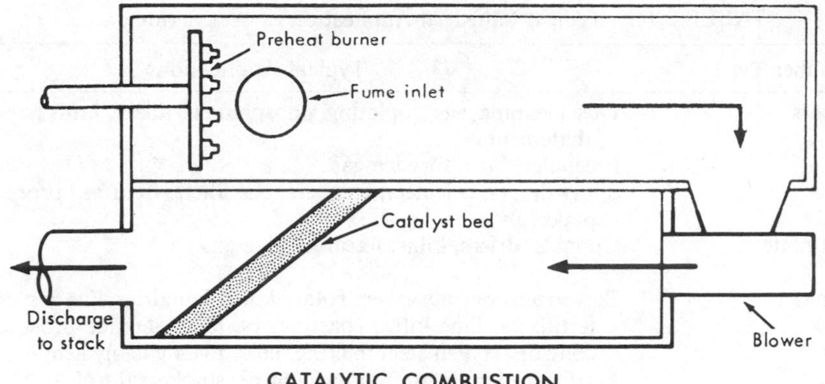

CATALYTIC COMBUSTION.

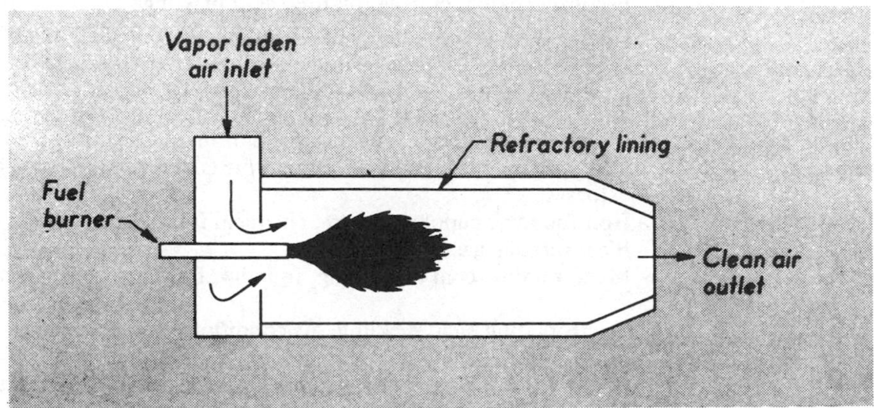

INCINERATION.

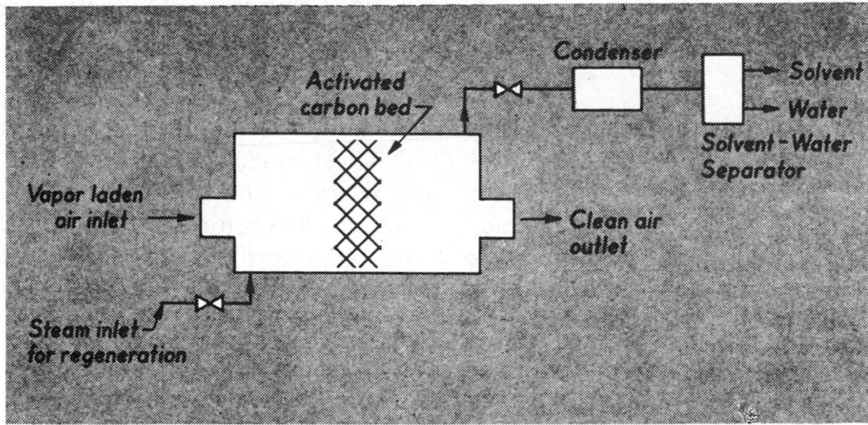

ADSORPTION.

Figure 2.14. Gas and vapor cleaning. Courtesty *Heating, Piping, and Air Conditioning*, December 1959.

unless all necessary precautions are followed. These include such standard safety features as pre-purge, ignition and temperature controls, limit switches, fuel pressure controls, etc. See, "Ovens" this section.

Catalysts are now available which will initiate self-sustaining combustion on their surface, even of dilute concentrations of contaminant, at temperatures much lower than than the autogenous temperature. Catalytic combustion is not suitable when certain metallic vapors or other inorganics which may poison the catalyst are present in the air stream. Pre-filtration may also be necessary to remove particulates from the air.

(4) Condensation. Vapors can be removed from effluent by condensation to the liquid state by standard

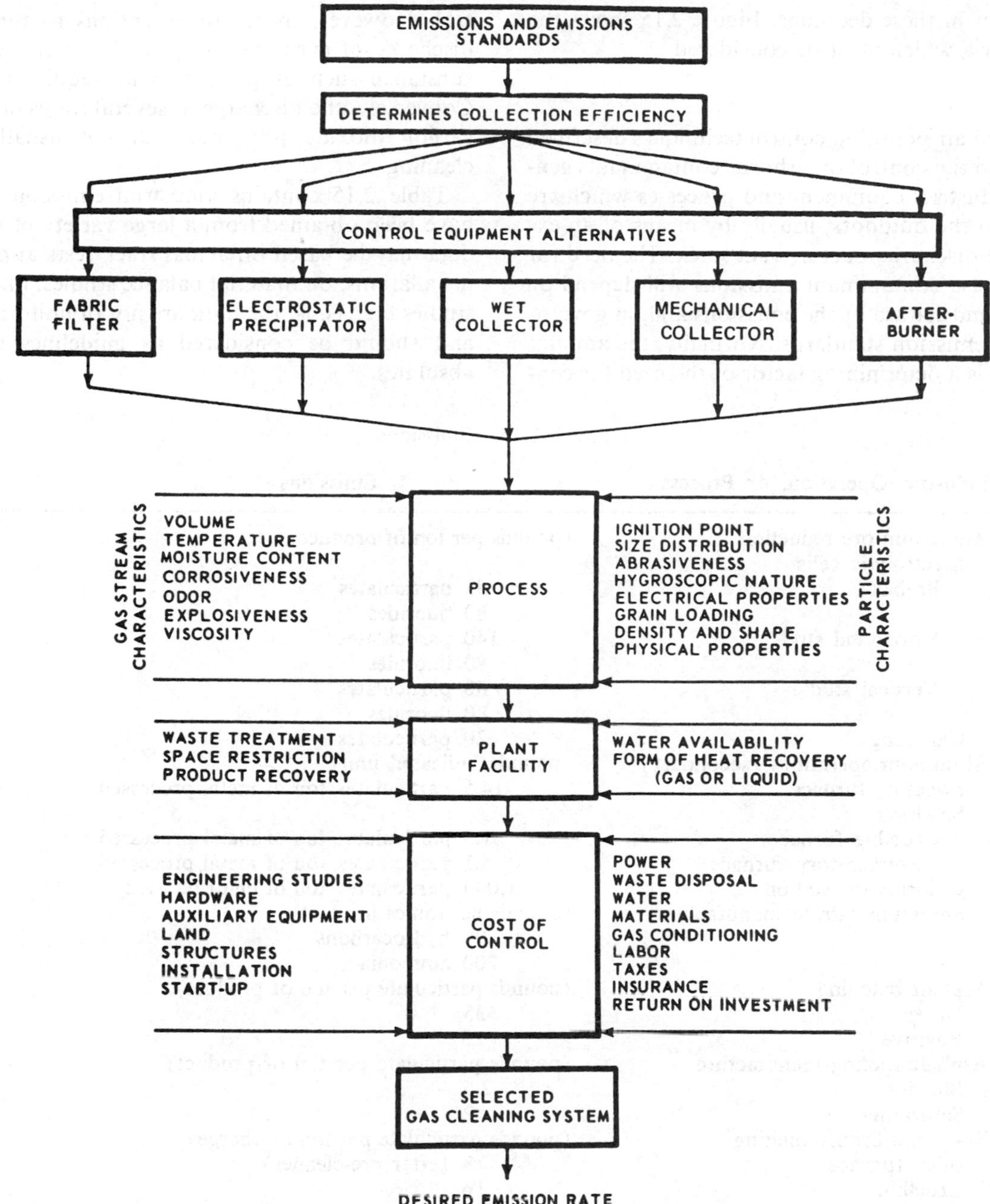

Figure 2.15. Criteria for selection of air cleaning equipment. From "Control Techniques for Particulate Air Pollutants," National Air Pollution Control Administration Publ. No. AP-51, 1970.

techniques or refrigeration. The liquid is collected in suitable containers for subsequent disposal. This technique is quite expensive and is usually economically feasible only for substances with a relatively high boiling point, where low air volumes are to be treated, where the concentration of the contaminant in the air is relatively high, or where the collected material can be reused or sold.

(5) Masking. Masking is a control method of very limited usefulness. Low concentrations of contaminants which are suficiently odorous to be objectionable but which are otherwise innocuous may be treated by injection of masking agents into the air stream before discharge to the outdoors. These agents, which should not create a secondary air pollution problem, are designed to mask the objectionable odor or neutralize it.

Criteria

Many factors must be carefully considered before a decision is made on the need for aircleaning equipment and on the type of equipment to be used. In many cases the important, or even the controlling factors may be subjective. However, in all cases a number of objective factors will enter in these decisions. Figure 2.15 shows the various criteria which must be considered.

Emissions

The industrial air pollution control techniques described above refer to the control of airborne contaminants generated by industrial equipment and processes which are discharged to the outdoors, usually by means of an exhaust system discharge duct or vent stack. The need for control of these contaminant emissions will depend on the amount and nature of the emissions and on governmental legal emission standards. Normally, the amount of emissions is a determining factor of the need for control. However, there are exceptions to this rule. The discharge of even several g per hour of a malodorous substance such as pyridine may require air cleaning. Conversely, the discharge of several thousand pounds of carbon dioxide per hour will not usually need air cleaning.

Table 2.15 contains industrial emission data which have been obtained from a large variety of sources. the data may be based on actual stack tests at one or more installations, on material balance studies, on production studies etc. Thus, the data are not of uniform reliability and should be considered as guidelines rather than absolutes.

TABLE 2.15. Emissions

Industry, Operation, or Process	Emissions
Aluminum ore reduction	(pounds per ton of product)
Electrolytic cells	
Prebake	55 particulates
	80 fluorides
Horizontal stud	140 particulates
	80 fluorides
Vertical stud	80 particulates
	80 fluorides
Calcining	20 particulates
Aluminum operations (secondary)	(pounds/indicated unit)
Sweating furnace	14.5 particulates/ton of metal processed
Smelting	
Crucible furnace	1.9 particulates/ton of metal processed
Reverberatory furnace	4.3 particulates/ton of metal processed
Chlorination station	1,000 particulates/ton of chlorine used
Ammonium nitrate manufacture	(pounds per ton of product)
	90 hydrocarbons
	200 ammonia
Asphalt batching	(pounds particulate per ton of product)
Dryer	35
Fugitive	10
Asphalt roofing manufacture	(pounds particulate per ton of product)
Blowing	2.5
Saturating	2
Brass and bronze melting	(pounds particulate per ton of charge)
Blast furnace	18 (after pre-cleaner)
Crucible	16
Cupola	73
Rotary furnace	60
Reverberatory furnace	70
Electric induction furance	2
Brick manufacture	(pounds per ton of bricks)
Raw material	
Drying and grinding	73 particulates
Storage	34 particulates
Curing and firing	1 nitrogen oxides
	1 fluorides
Buffing and polishing	0.01 grains/cubic foot of exhaust air
Calcium carbide manufacture	(pounds per ton of product)
Furnace	19 particulates
	3 sulfur oxides
Coke dryer	2 particulates
	3 sulfur oxides
Room vents	26 particulates
	18 acetylene gas

TABLE 2.15. Emissions (*Continued*)

Industry, Operation, or Process	Emissions
Carbon black manufacture	(pounds per ton of product)
Channel process	2,300 particulates
	33,500 carbon monoxide
	11,500 hydrocarbons
Furnace process	220 particulates
	5,000 carbon monoxide
	1,100 hydrocarbons
Cement manufacture	(pounds particulates per barrel of cement)
Dry process	
Kilns	55
Dryers, grinders, etc.	20
Wet process	
Kilns	35
Dryers, grinders, etc.	6
Ceramic clay	(pounds particulates per ton of input)
Drying	70
Grinding	76
Storage	34
Charcoal manufacture	(pounds per ton of product)
	400 particulates (tar, oil)
	320 carbon monoxide
	100 hydrocarbons
	152 methanol
	232 acetic acid
	60 other gases (formaldehyde, etc.)
Chemical milling	(see electroplating)
Chlor-alkali plant	(pounds chlorine per 100 tons of product)
Liquefaction blow gases	
Diaphragm cell	6,000
Mercury cell	10,000
Water absorber	500
Chlorine tank car loading vents	450
Chlorine storage tank loading vents	1,200
Cell brine air blowing	500
Coffee roasting	(pounds per ton of coffee beans)
Roaster	6 particulates
	0.1 nitrogen oxides
	0.2 aldehydes
	0.9 organic acids
Stoner and cooler	1.5 particulates
Spray drier	1.5 particulates
Coke (metallurgical)	(pounds per ton of coal charged)
By-product coking	0.7 particulates
	2.0 sulfur dioxide
	0.42 carbon monoxide
	1.4 hydrocarbon
	0.02 nitrogen oxides
	0.06 ammonia
Beehive ovens	200 particulates
	1 carbon monoxide
	8 hydrocarbons
	2 ammonia
Concrete batching	5 lb particulates/ton process weight
	or
	0.2 lb particulates/cubic yard concrete
Copper smelting (primary)	(pounds per ton of concentrated ore)
Roasting	45 particulates
	60 sulfur oxides
Smelting	20 particulates
	320 sulfur oxides
Converting	60 particulates
	870 sulfur oxides
Refining	10 particulates

TABLE 2.15. Emissions (*Continued*)

Industry, Operation, or Process	Emissions
Cotton ginning	(pounds particulates per bale of cotton)
Unloading fan	5
Cleaner	0.3
Stick and burr	0.2
Miscellaneous	1.5
Degreasing (vapor phase)	0.5 lb/sq ft of tank surface/hr. or 1 gallon/hr/ton of metal cleaned
Dry cleaning	(pounds of solvent vapor per ton of clothes)
Petroleum solvents	305
Synthetic solvents	210
Explosives	(pounds per ton of product)
TNT	
Nitration reactors	160 nitrogen oxides
Sulfuric acid regenerators	18 sulfur oxides
Nitrocellulose	
Reactor pots	12 nitrogen oxides
Sulfuric acid concentrators	29 nitrogen oxides 65 sulfur oxides
Fermentation	(pounds per ton of product)
Grain handling	3 particulates
Drying, etc.	5 particulates
Aging (whiskey)	10 hydrocarbons
Ferroalloy production	(pounds particulates/ton of specified product)
Open furnace	
50% FeSi	200
75% FeSi	315
90% FeSi	565
Silicon metal	625
Silicomanganese	195
Semi-covered furnace	
Ferromanganese	45
Fibers (synthetic)	(pounds per ton of product)
Viscose rayon	55 carbon disulfide 6 hydrogen sulfide
Nylon	7 hydrocarbons 15 oil mist or vapor
Dacron	7 oil mist or vapor
Fiber glass manufacture	(pounds particulates per ton of input)
Furnace	2
Forming line	50
Curing oven	7
Fish meal processing	(pounds per ton of product)
Stale fish cookers	3.5 trimethylamine 0.2 hydrogen sulfide
Fresh fish cookers	0.3 trimethylamine 0.01 hydrogen sulfide
Frit manufacture	(pounds per ton of charge)
Rotary furnace	16 particulates 5 fluorides
Galvanizing (see zinc smelting) (secondary)	
Glass melting	(pounds per ton of product) 2 particulates K fluorides $K = (\%\text{ fluoride in charge}) \times (4)$
Grain handling	(pounds particulates per ton of product)
Elevators	
Shipping or receiving	1–5
Transferring, conveying, etc.	2–3
Screening and cleaning	5–8
Drying	6–7
	(The lower figures are for terminal elevators. The higher ones are for country elevators.)

TABLE 2.15. Emissions (*Continued*)

Industry, Operation, or Process	Emissions
Grain processing	From 0.5 to 7 pounds of particulates per ton of product depending on the grain and process.
Grinding (grinding wheels)	
Aluminum	0.7 grains/cubic foot of exhaust air
Iron	0.15 grains/cubic foot of exhaust air
Gypsum manufacture	(pounds of particulates per ton of throughput)
Raw material dryer	40
Primary grinder	1
Calciner	90
Conveying	1
Heat treat (oil quench)	Oil mist and vapors equivalent to 10% of the makeup oil weight.
Hydrochloric acid manufacture	
By-product hydrogen chloride	3 lb hydrogen chloride per ton of product
Hydrofluoric acid manufacture	
Rotary kiln	50 lb fluorides per ton of acid
Fluorspar grinding and drying	20 lb particulates per ton of fluorspar
Incineration (auto body)	(pounds per car)
	2 particulates
	2.5 carbon monoxide
	1 organics
Incineration (wire)	225 particulates per ton of combustibles
Iron production (primary)	(pounds per ton of metal produced)
Blast furnace	
Ore charge	110 particulates
Agglomerates charge	40 particulates
Coke ovens (see Coke, metallurgical)	
Sintering	
Windbox	20 particulates
	3 sulfur dioxide
	22 particulates
Discharge	44 carbon monoxide per ton finished sinter
Iron foundry (grey iron)	(pounds per ton of metal charged)
Cupola	17 particulates
	145 carbon monoxide
Reverberatory furnace	2 particulates
Electric induction furnace	1.5 particulates
Furnace pre-heat	1 particulates
Charge weighing	4 grains/cubic foot of exhaust air
Shakeout	5 grains/cubic foot of exhaust air
Abrasive cleaning	6 grains/cubic foot of exhaust air
Grinding	10–12 grains/cubic foot of exhaust air
Sand screening, mixing, etc	3–5 grains/cubic foot of exhaust air
Core baking	6 lb gas/gallon of core oil
Lead smelting (primary)	
Blast furnace	75 lb particulates per ton of ore concentrate
Reverberatory furnace	12 lb particulates per ton of ore concentrate
Sintering and sinter crushing	30 lb particulates per ton of sinter
	660 lb sulfur oxides per ton of sinter (Overall plant emissions of sulfur oxides)
Lead smelting (secondary)	(pounds per ton of metal processed)
Pot furnace	0.8 particulates
Reverberatory furnace	130 particulates
	85 sulfur oxides
Blast (cupola) furnace	190 particulates
	90 sulfur oxides
Rotary reverberatory furnace	70 particulates
Lime manufacture	(ponds of particulates per ton of lime processed)
Primary crushing	31
Secondary crushing	2
Vertical kiln calcining	8
Rotary kiln calcining	200
Magnesium smelting	
Pot furnace	4 lb particulates per ton of metal processed

TABLE 2.15. Emissions (*Continued*)

Industry, Operation, or Process	Emissions
Meat smoking	(pounds per ton of meat processed) 0.3 particulates 0.6 carbon monoxide 0.4 organics
Mineral wool manufacture	(pounds particulates per ton of charge)
Cupola	22
Reverberatory furnace	5
Blow chamber	17
Curing oven	4
Cooler	2
Nitrate fertilizer manufacture	(pounds per ton of product)
Neutralizer	2 ammonia
Prilling tower	0.9 particulates
Granulator	0.4 particulates 0.9 nitrogen oxides 0.5 ammonia
Drying and cooling	7 particulates 3 nitrogen oxides 1.3 ammonia
Nitric acid manufacture	(pounds nitrogen oxides per ton of product)
Ammonia-oxidation	
old plant	57
new plant	5
Nitric acid concentrators	
old plant	5
new plant	0.2
Ovens (paint drying) (See also the discussion of flammable atmosphere ovens under "Control of Specific Industrial Processes.")	Emissions from these ovens depend on the painting process and the amount of volatiles evaporated into the air before the coated work enters the oven. The oven emissions are shown below as a percentage of the total volatiles in the coating material used by the process. The lower emission figures in each category will reflect appreciable prior air drying; the higher ones indicate a lower quantity of volatiles evaporated prior to entry into the oven. (Although the incompletely combusted drying oils and plasticizers constitute an insignificant emission from the weight standpoint, their irritating and unpleasant odors may constitute a significant problem requiring air cleaning.)

Coating Operation	*Oven Emissions* (% of total paint volatiles)
Spraying (normal)	10 to 30%
Spraying (large, flat pieces)	20 to 40%
Dip and flow coat	30 to 60%
Roller coat	50 to 80%

Industry, Operation, or Process	Emissions	
Paint manufacture (see also varnish)	2 lb particulate per ton of pigment 30 lb hydrocarbons per ton of product	
Painting (surface coating)	(pounds of hydrocarbon per ton of coating material)	
Paint	1,120	when coated work is subsequently dried in an oven, some of these emissions will be oven emissions. (See "Ovens")
Varnish and shellac	1,000	
Lacquer	1,540	
Enamel	840	
Primer (zinc chromate)	1,320	

Paper (Kraft process) (pounds per ton air-dried unbleached pulp)

	Part.	SO_2	CO	H_2S	Organic sulfides
Blow tank accumulator				0.1	3.0
Washers and screens				0.02	0.2
Multiple-effect evaporators				0.5	0.4
Recovery boilers	151	5.0	60	12	0.9
Smelt dissolving tank	2			0.03	0.04
Lime kilns	45		10	1.0	0.6

TABLE 2.15. Emissions (*Continued*)

Industry, Operation, or Process	Emissions
Condenser	0.01 0.5
Calciner	72
Perlite furnaces	21 lb particulates per ton of charge
Petroleum refineries	Petroleum refinery emissions vary very widely both quantitatively and qualitatively. The data do not lend themselves to inclusion in a general table of this type. See the references.
Phosphate fertilizer manufacture	(pounds per ton of product)
Normal superphosphate	
Grinding, drying	9 particulates
Main stack	0.15 fluorides
Triple superphosphate	
Run-of-pile	0.03 fluorides
Granular	0.10 fluorides
Diammonium phosphate	
Dryer, cooler	80 particulates
Ammoniator-granulator	2 particulates 0.04 fluorides
Phosphoric acid	(pounds per ton of product)
Wet process	
Reactor	18 fluorides
Gypsum pond	1 fluorides
Condenser	20 fluorides
Thermal process	1–6 particulates depending on the degree of air cleaning which is present in all thermal process plants
Phthalic anhydride (overall plant)	32 lb organics per ton of product
Pipe coating (asphalt or coal tar)	10 lb particulates per ton of coating material
Plastics manufacture	(pounds per ton of product)
Polyvinyl chloride	35 particulates 17 gases and vapors
Polypropylene	3 particulates 0.7 gases and vapors
General	5–10 particulates
Plating, anodizing, etching, pickling, metal cleaning, etc.	Emissions depend on rate of evolution of contaminants from the tank surface. (See Rate of Evolution column in Table 2.8) Emissions can be expressed as a percentage of the weight of make-up chemicals added.

Rate of Evolution	% Loss
High	5%
Moderate	3
Low	2
Nil	0

Industry, Operation, or Process	Emissions
Polishing and buffing	0.005–0.01 grains per cubic foot of exhaust air
Printing ink manufacture	
Vehicle cooking	(pounds emitted/unit indicated)
General	120 gaseous organics per ton of product
Oils	40 gaseous organics per ton of product
Oleoresinous	150 gaseous organics per ton of product
Alkyds	160 gaseous organics per ton of product
Pigment mixing	2 particulates per ton of pigment
Refractories (castable)	(pounds particulates per ton of feed material)
Raw material dryer	30
Raw material crushing	120
Melting	50
Curing oven	0.2
Molding and shakeout	25
Rubber compounding	0.5 lb particulates per hour of operation
Soaps and detergents	

TABLE 2.15. Emissions (*Continued*)

Industry, Operation, or Process	Emissions
Soap manufacture	Odor is the only significant emission
Detergent (spray dryer)	90 lb particulates per ton of product
Sodium carbonate manufacture	
Ammonia recovery	7 lb ammonia per ton of product
Conveying, etc.	6 lb particulates per ton of product
Soldering	0.5 lb particulates per 100 lb solder used
Starch manufacturing	8 lb particulates per ton of product
Steel foundry	(pounds of particulates per ton of metal)
Melting	
Electric arc	13
Open-hearth	11
Open-hearth oxygen lanced	10
Electric induction	0.1
Other (see iron foundry)	
Steel production (primary)	(pounds of particulates per ton of metal)
Open-hearth furnace	
Oxygen lance	22
No oxygen lance	12
Basic oxygen furnace	46
Electric-arc furnace	
Oxygen lance	11
No oxygen lance	7
Scarfing	20
Stone quarrying and crushing	(pounds particulates per ton of raw material)
Primary crushing	0.5
Secondary crushing and screening	1.5
Tertiary crushing and screening	6
Recrushing and screening	5 (per ton of throughput)
Fines mill	6
Screening, conveying, handling	2
Storage pile	10
Sulfuric acid manufacture	40 lb sulfur dioxide per ton sulfuric acid
Surface coating (see Painting)	
Varnish manufacture	(pounds hydrocarbon per ton of product)
Bodying oil	40
Oleoresinous	150
Alkyd	160
Acrylic	20
Welding	2 lb particulates per 100 lb welding rods
Zinc smelting (primary)	(pounds particulates per ton of ore concentrate)
Roasting (multiple hearth)	120
	1,100 sulfur oxides (gas)
Sintering	90
Horizontal retorts	8
Vertical retorts	100
Electrolytic process	3
Zinc smelting (secondary)	(pounds particulates per ton of metal)
Retort reduction	47
Horizontal muffle	45
Pot furnace	0.1
Kettle sweat furnace	
Clean metallic scrap	Negligible
General metallic scrap	11
Residual scrap	25
Reverberatory sweat furnace	
Clean metallic scrap	Negligible
General metallic scrap	13
Residual scrap	32
Galvanizing kettles	5
Calcining kiln	89

INFORMATION SOURCES

There are many governmental agencies in the fields of industrial hygiene and air pollution control which are available for consultation and advice. Most of them publish data sheets and informational brochures. A large majority of the state and local agencies have promulgated codes, rules, and standards which must be followed. Nearly all the states as well as Puerto Rico have both industrial hygiene and air pollution control divisions, either as autonomous agencies or within a larger department. Many counties and cities also have established similar agencies on a local level.

In the federal government, the Occupational Safety and Health Administration (OSHA) within the Labor Department and the National Institute of Occupational Safety and Health (NIOSH), within the Department of Health, Education, and Welfare (HEW), have the principal responsibility in the field of industrial hygiene. Specialized agencies such as the Bureau of Mines, Department of Energy, and the NASA may also be sources of information within the sphere of their specialized interests. The EPA is the federal agency principally concerned with air pollution control.

The principal publications are the *Journal of the APCA* and the *Journal of the AIHA*, although there are many other journals which publish relevant material. Many trade associations are also excellent sources of information.

BIBLIOGRAPHY

1. *Industrial Ventilation*, 14th edition. A Manual on Recommended Practice, Committee on Industrial Ventilation, P.O. Box 16153 Lansing, Michigan, 1976.
2. Alden, J. L. *Design of Industrial Exhaust Systems*. New York, The Industrial Press, 1970.
3. Brandt, A. D. *Industrial Health Engineering*. New York, John Wiley & Sons, 1947.
4. Hemeon, W. C. L. *Plant and Process Ventilation*. New York, Industrial Press, 2nd ed., 1965.
5. Drinker, P. and Hatch, T. *Industrial Dust*. New York, McGraw-Hill Book Co., 1954.
6. Patty, F. A., editor. *Industrial Hygiene and Toxicology*, Vol. 1. New York, Interscience Publishers, 1958.
7. Dalla Valle, J. M. *Exhaust Hoods*. New York, Industrial Press, 1952.
8. Dalla Valle, J. M. *The Industrial Environment and Its Control*. New York, Pitman, 1948.
9. McCabe, L. C. *Air Pollution*. New York, McGraw-Hill Book Co., 1952.
10. Mallette, F. S. *Problems and Control of Air Pollution*. New York, Reinhold Publishing Corp., 1955.
11. Magill, P. L. *Air Pollution Handbook*. New York, McGraw-Hill Book Co., 1956.
12. *Handbook on Air Cleaning*. U.S. Atomic Energy Commission, Washington, D.C., 1952.
13. *Heating, Ventilating and Air Conditioning Guide*. Am. Soc. of Heating Air Cond. Engineers, New York, 1958.
14. *Engineering Manual for Control of In-Plant Environment*. American Foundrymen's Society, Des Plaines, Ill., 1956.
15. *Air Pollution Abatement Manual*, Manufacturing Chemists' Association, Washington, D.C.
16. *Handbook of Industrial Loss Prevention*. Factory Mutual Engineering Division, New York, McGraw-Hill Book Co., 1959.
17. *Control Techniques for Particulate Air Pollutants*. U.S. Department of Health, Education and Welfare, Public Health Service, Environmental Health Sevice, National Air Pollution Control Administration Publication AP-51, 1969.
18. *Power*, January 1971.
19. *Michigan's Occupational Health*. Michigan Department of Public Health, Lansing, Michigan. Vol. 5, No. 3 and Vol. 16, No. 2.
20. J. H. Greenberg and H. E. Conover. "Iron Foundry Emissions," *Foundry*, April 1971.
21. *Compilation of Air Pollutant Emission Factors*. U.S. Environmental Protection Agency, Office of Air Programs, Publication AP-42, Research Triangle Park, North Carolina, April, 1973.
22. *Proceedings: Third National Conference on Air Pollution*, December 1966. Public Health Service Publication 1649. National Center for Air Pollution Control.
23. *Air Pollution Engineering Manual*. Public Health Service Publication #999-AP-40. 1973, National Center for Air Pollution Control.
24. Stern, Arthur C., editor. *Air Pollution*, Vol. II and III, Academic Press, New York, 1976.
25. *Welding, Brazing, and Soldering*. Publication TSP69-10264, National Aeronautics and Space Administration, Office of Technology Utilization, Washington, D.C.
26. New York State Industrial Code, Rule 12, "Control of Air Contaminants."
27. New York State Industrial Code, Rule 18, "Exhaust Systems."
28. Kehoe, J. W. and Ferry, J. J. *Possible Health Hazards Associated with Welding and Cutting Processes*. Industrial Hygiene Foundation, Pittsburgh, Pa. 1954.
29. Thorpe, M. L. "The Plasma Jet and Its Uses." *Research/Development*, Vol. 11, No. 1, January, 1960.
30. Brief, R. S., Church, F. W., and Hendricks, N. V. "Design and Selection of Laboratory Hoods," *Air Engineering*, Detroit, Michigan, 1963.
31. Peterson, J. E. and Peay, J. A. "Laboratory Fume Hoods," *Air Conditioning, Heating and Ventilating*, May 1963.
32. Dinman, B. "The Nature of Occupational Cancer," Charles C. Thomas, Springfield, Ill., 1974.
33. Rappaport, S. M. and Campbell, E. E., "The Interpretation and Application of OSHA Carcinogen Standards for Laboratory Operations," American Industrial Hygiene Assoc. Journal, Vol. 37, No. 12, Dec, 1976.
34. "National Cancer Institute Safety Standards for Research Involving Chemical Carcinogens," Office of Research Safety, National Cancer Institute, 1976.
35. "Recirculation of Exhaust Air," HEW Publication No. (NIOSH) 76–186, Feb. 1976.
36. "A Guide to Industrial Respiratory Protection," HEW Publication No. (NIOSH) 76–189, June 1976.

SECTION 3

Industrial Noise: Effects and Controls

Robert D. Bruce
Bolt Beranek and Newman Inc.
Cambridge, Massachusetts

ABSTRACT

The physiological and psychological effects of noise on people have been under study for some time, and general guidelines have been available for allowable noise in industry; however, the focus of attention on the environment has increased both public and private concern about the high noise levels in industry. Meanwhile, continuing progress in the understanding of noise generation and propagation has resulted in the development of effective noise control treatments and techniques.

In this section, appropriate noise criteria will be compared with typical noise levels in several industries, and noise control measures will be discussed in the framework of case histories. In addition, several studies will be summarized that estimate the cost of complying with occupational noise standards.

INTRODUCTION

As usually defined, *noise* is any unwanted sound. Under this definition, noise may vary from the tick of a clock to teenagers' music, to the roar of a jet engine. In industry, noise may come from compressors, pumps, turbines, motors, extruders, and hundreds of other machines.

The selection of appropriate noise criteria for industry depends on a knowledge of the effects of noise on people, as well as on the activities in which they are engaged. The first part of this section is a discussion of how the human sensory system is affected by noise and what criteria are important in industrial situations, both for the worker within the plant and the surrounding community.[1] The second part gives case studies illustrating noise levels encountered in various industries and includes methods for controlling that noise.[2] The third part presents a brief discussion of how much each industry would have to spend in order to comply with existing federal regulations governing exposure to occupational noise.

THE HUMAN AUDITORY SYSTEM

The sounds we hear are small pressure variations in the atmosphere which strike the eardrum and cause it to vibrate. These vibrations activate nerves which in turn send signals to the brain. It should be kept in mind that the ear is but one component of the human auditory system, which also comprises the neural pathways that transmit sensory information and the cognitive aspects of information processing. Thus, our concern is not limited to the mechanics of the effects of noise on the ears, but includes the impact of noise on the entire range of human behavior.

The ear can operate effectively over a range in sound pressures of more than one million to one, from 0 to 120 dB. To help visualize this range, try to imagine, as an analogy, a postal scale or balance that is delicate enough to weigh accurately an airmail letter (the 0 dB level) and an 10-lb sack (the 44 dB level) routinely without loss of sensitivity. The balance would also need to weigh a 30-ton tractor trailer (120 dB) without damage. Exposing the ear to gun shots or large punch press impulses (140 to 160 dB) would be somewhat like dropping a Navy destroyer or battleship on a delicate balance and still expecting it to work. As is evident from this analogy, the ear has tremendous ability to accept severe strain and to recuperate. Even with the advancements made in electronic instrumentation, there is no device that can be compared with the human ear in sensitivity and dynamic range: it is an extremely valuable instrument that needs protecting.

Although the ear is capable of functioning over an extremely wide range of levels, it does not work identically at all levels. For example, pitch, the subjective dimension associated with the frequency content of an acoustic signal, is known to be a function of level, such that the pitch of a pure tone can vary with the intensity of the tone. Other differences in the ear's performance dependent upon level are protective reflexes, that operate at high levels but not at low levels, and distortion, which increases with level.

One aspect, in particular, of the ear's functioning that is not generally appreciated by the engineering community is its performance in detecting very low-level signals. The basic problem seems to stem from a misunderstanding of what is meant by a sensory threshold. A sensory threshold is *not* a level on a continuum of physical energy such that stimulation in excess of that level will invariably yield sensation. Instead, a sensory threshold is a statistical concept.

The correct interpretation of a sensory threshold is an amount of physical energy associated with a given probability of sensation. In most cases, a sensory threshold is defined as a physical level that gives rise to a sensation 50% of the time. Consequently, when it is said that the "threshold of hearing" at 1000 Hz is 0 dB re 20 micro-newtons/m^2, the statement should not be interpreted as implying that no lower level signals can be heard. The statement, in fact, implies that average subjects can hear this level half the time. Thus, in some engineering applications, when it is necessary to be very precise in specifying that a given noise source will be "inaudible," one must understand what is meant by "inaudibility." The range of frequencies to which people are sensitive exceeds the range of frequencies to which the ear is sensitive. The reason is that the body possesses multiple receptors for very low frequency acoustic signals, especially vibration. One does not need ears, for example, to detect vibration-induced motion of limbs or other body parts. Furthermore, sensitivity to acoustic signals of different frequencies is a relative matter. The statement that people can hear acoustic signals in the range of 20 to 20,000 Hz is incomplete; given a powerful enough acoustic source, the eardrum can be made to vibrate and to produce sensations of some sort at almost any frequency. In short, what is meant by the statement that people can hear in the range of 20 to 20,000 Hz is that people's differential sensitivity to signals in this range, generated by familiar acoustic sources, is readily measurable.

Few adults in mechanized societies can hear well in the octave above 10,000 Hz; many suffer from varying degrees of presbycusis "old age hearing" or sociocusis (noise induced hearing loss). Nonetheless, constant attention is focused on high-frequency sensitivity. The manufacturers of high fidelity equipment and popular accounts of ultrasonic death rays seem responsible for most of this attention. Suffice it to say that there are no known menaces to health or happiness caused by exposure to infra- or ultra-sound in the everyday environment.

The functioning of the ear in the temporal domain is still somewhat controversial. Most attention has focused on potential differences in the way the ear responds to steady-state sounds and the way it responds to very short duration or impulsive sounds. The question of whether or not the ear has a time constant, and what its value might be, has not yet been resolved. Some researchers suggest that the ear integrates energy up to about 0.1–0.2

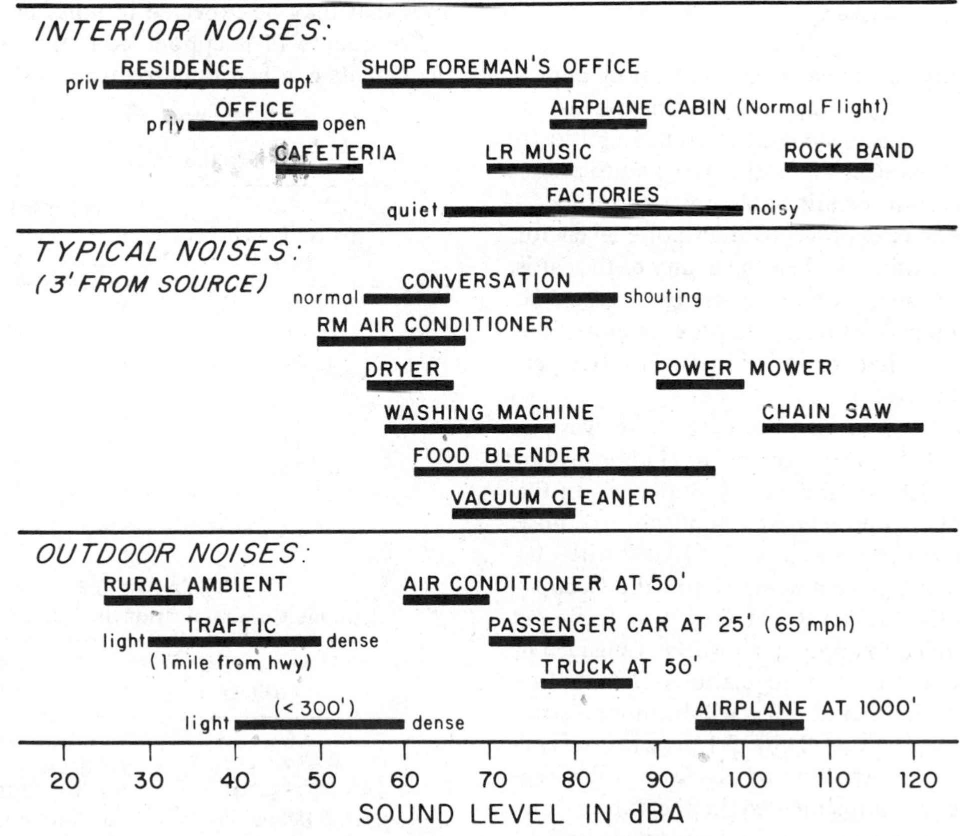

Figure 3.1. Examples of A-weighted sound levels.

sec; others suggest that the ear acts as a simple energy detector, continuously integrating energy over time. The implications of the latter research are that the noisiness of impulsive sounds increases 3 dB for every doubling of duration.

Thus far, the sorts of acoustic signals to which people are sensitive have been discussed in general terms. The effects of such signals upon individuals, giving special attention to criteria for hearing damage, speech interference, sleep interference, task interference, and annoyance will be discussed next.

HOW NOISE AFFECTS PEOPLE

Many of the effects of noise are level dependent; that is, the magnitude of the effect varies with the level of the noise. One convenient scale for expressing the magnitude of noise is the "A" scale. This is a physical scale that was designed to approximate human frequency sensitivity by a weighting procedure which gives relative emphasis to the components of sound in the mid- and high-frequency range (500–4000 Hz). The unit of measurement on the A scale is the decibel and is distinguished by the notation "dB(A)." Some representative values of A-weighted sound levels for common sounds are given in Fig. 3.1.

Hearing Damage

The great versatility of the ear is often not fully appreciated until one's hearing is impaired. During the last 20 years, much research has been conducted in an effort to understand the mechanism of hearing loss and to establish a suitable criterion for allowable noise exposure.[3]

Whenever the ear is exposed to high noise levels for extended periods of time, the hearing acuity of the ear is affected. If the exposure to noise levels in excess of 90 dB(A), is on a regular basis (e.g., 8 hr/day for more than 10 years), it is likely that the individual will suffer permanent hearing damage.

In 1969, a criterion for exposure to noise was introduced by the U.S. Department of Labor as an amendment to the Health and Safety Standards of the Walsh-Healy Public Contracts Act, applicable to those companies having contracts exceeding $10,000 with any agency of the federal government. Under the Occupational Safety and Health Act (OSHA) of 1970, this same criterion was extended to apply to all workers engaged in interstate commerce. Future regulations for limiting worker exposure to industrial noise will be more restrictive that existing regulations. Table 3.1 presents the current OSHA regulation and one of several possible regulations likely to be implemented in the future. The OSHA standards, where applied, will effectively protect approximately 85% of the exposed population from permanent hearing damage due to industrial noise.

TABLE 3.1.

Maximum Exposure Time per Day, hr	Current Regulation Noise Level dB(A)	Possible Regulation Noise Level dB(A)
8	90	85
6	92	87
4	95	90
3	97	92
2	100	95
$1\frac{1}{2}$	102	97
1	105	100
$\frac{1}{2}$	110	105
$\frac{1}{4}$	115	110

Speech Interference

Hearing damage is not the only criterion by which to judge excessive noise; it is also essential to consider the ability of people to communicate, both by telephone and in person. Therefore, criteria have been developed relating the existing noise environment to the ability of the typical individual to communicate in spaces that are likely to be somewhat noisy. Table 3.2 may be used as a guide in estimating the limitations in voice communications that may be expected in noisy environments.

The quality of telephone communication in such environments can be estimated from Table 3.3.

TABLE 3.2. Maximum Speaker-Listener Distance.

Noise Level dB(A)	Voice Effort	
	Normal	Very Loud
55	10′	35′
61	5′	18′
69	2′	7′
75	1′	4′
81	$\frac{1}{2}$′	2′
87	—	1′
93	—	$\frac{1}{2}$′

TABLE 3.3.
Telephone Communication In Noise Environments.

Noise Level dB(A)	Quality of Communication
Above 85	Unsatisfactory
70–85	Difficult
55–70	Slightly Difficult
Below 55	Satisfactory

Task Interference

Task interference, or the effects of noise on job performance, is not well understood. To date, research in this area has produced numerous conflicting and inconclusive reports. In general, it seems safe to say that high-intensity, aperiodic, intermittent noises impede efficient work to a greater extent than low-intensity, steady-state noise. A number of studies has failed to uncover any effects of noise on performance, and a few even report paradoxical improvements in performance attributable to noise exposure.

It appears obvious that the complexity of the task with which noise interferes plays a major role in determining how much the noise affects performance. Anecdotal evidence about the effects of unexpected noise on heart surgeons and diamond cutters abounds, but no one has been able to produce a dosage-effect relationship for task interference in general. Thus, no criteria have yet been established for task interference due to noise exposure. As far as industrial noise control is concerned, each suspected case of interference must be evaluated on its own merits by experienced professionals.

Sleep Interference

The two principal ways in which noise exposure interferes with sleep are to delay the onset of sleep and to shift sleep "stages." Numerous studies are available on the sleep-delaying and stage-shifting effects of noise exposure. Although there is frequently broad agreement among studies, detailed agreement is lacking. Discrepancies among outcomes of similar studies are attributable to incomparable control conditions, differences in experimental design, and the host of individual differences which beset sleep research.

For example, it is universally observed that the initial time required for subjects to fall asleep increases monotonically with exposure to increasing noise levels. Unfortunately, different studies produce estimates of the sleep-delaying effects of noise that are more than 35 dB apart. Thus, two studies report delays in onset of sleep from 20 to 90 minutes, corresponding to exposure to continuous noise at levels of 35 dB(A) and 50 dB(A), respectively. Other studies, however, report that subjects can fall asleep in as little as 12 minutes despite exposure to noise levels of 70 dB(A). Furthermore, prolonged exposure to high noise levels can produce tinnitus (ringing in the ears), which has been claimed to delay the onset of sleep. In other words, aftereffects of noise, even in the absence of any noise exposure at bedtime, can impede sleep. It is also claimed in the literature that levels as low as 35 dB(A) can either induce a shift from deep to light sleep or awaken certain people. Pronounced differences in sensitivity to noise during sleep have been observed as a function of age.

An absolute criterion for noise exposure levels in sleeping quarters is obviously unjustifiable on the basis of extant research. Two possible levels to bear in mind when considering sleep interference effects are 40 and 70 dB(A). As a rule of thumb, levels of 40 dB(A) have been known to awaken approximately 25% of the sleeping population. As another approximation, about 50% of the population may be awakened by noise exposure of levels of 70 dB(A), while about 50% of the population will find some difficulty in falling asleep when exposed to such levels.

Annoyance to Individuals

One effect of noise that does not seem to depend strongly on level is annoyance. Under some circumstances, a dripping water faucet or chalk squeak on a blackboard can be as annoying as a jackhammer. Furthermore, annoyance frequently accompanies other effects of noise on people, such as speech interference or sleep interference. Thus, there are no generally accepted criteria for noise levels associated with annoyance. Perhaps the most reasonable approach to establishing criteria for noise levels that do not annoy people is to ensure that the environments are not sufficiently noisy to violate other criteria (as presented above) for effects of noise on people. Even this approach has its limitations, and it will not be until research currently in progress and planned for the future has been completed that adequate criteria for annoyance can be established.

Annoyance to the Community

A consideration in establishing criteria for allowable noise, particularly from industrial plants, is the response of the residents of a neighboring community. That communities are becoming more sensitive to intrusive noise is evidenced by an increasing number of legal actions relating to noise, as well as a growing number of community ordinances for industrial noise. These limits are often given in terms of a frequency analysis of the noise rather than by the single-number A-weighted sound level that we have used in describing hearing damage and speech interference criteria. The acoustic data are often presented in 9 octave bands that indicate the distribution of the sound energy over the range of frequencies, extending from low-frequency "rumbles" to high-frequency "hisses."

One obvious criterion for allowable plant noise would be the community noise ordinance. A sampling of existing community ordinances for industrial noise is presented in Fig. 3.2. However, the noise ordinance is not the only important criterion. For example, if the erection of a new plant, or the installation of additional equipment at an existing facility, can be expected to increase the noise in the community over the ambient noise

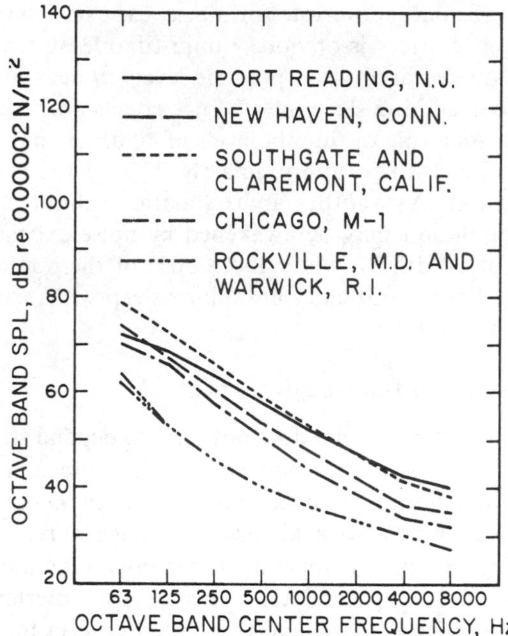

Figure 3.2. Examples of community noise ordinances.

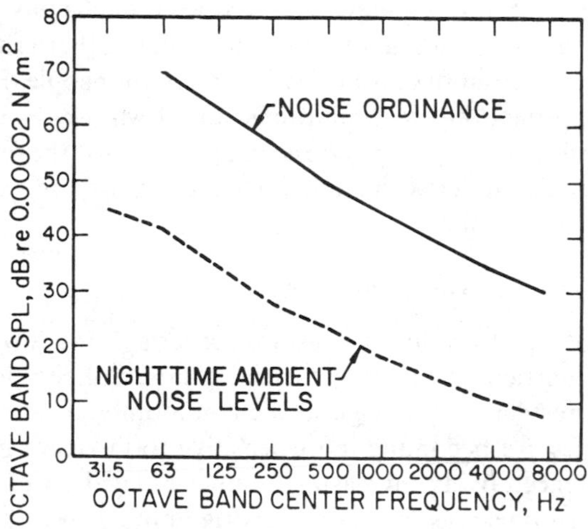

Figure 3.3. Case history of community annoyance.

levels (even though the total noise level may be still less than prescribed in the town ordinance), it is likely that residents in the community will find the noise objectionable. This principle is illustrated in Fig. 3.3, which presents ambient noise levels measured in a community near a proposed site for an industrial complex. Also presented is the noise ordinance for the industrial zone of that community. A comparison of the existing noise levels (ambient) with the allowable limits will illustrate the problem. If the noise of a new industrial facility were to increase the noise levels in a community from the nighttime ambient levels shown in Fig. 3.3 to the noise ordinance levels, this 25-dB increase in noise level would be subjectively perceived as a four- or five-fold increase in loudness. Under such circumstances it is likely that neighbors would complain.

TABLE 3.4.

Increase in Octave Band Noise Levels, dB	Anticipated Response of Community
0–5	Acceptable
5–10	Sporadic complaints
10–15	Complaints threaten legal action
15–20	Legal Action

Unless there are specific socioeconomic-political considerations (e.g., the individuals living in the area are employed by the plant), a rough estimate of the community's response to an increase in noise level can be obtained in Table 3.4.

If there are pure tones in the frequency spectrum, or if the noise is impulsive in character, serious complaints may exist even when the increase in octave band noise level is less than 5 dB. On the other hand, if the noise exists for a small fraction of the time or if it occurs only during the daytime hours, then the increase in noise levels could be greater than shown for the same indicated response.

Thus far, we have dealt with how the human auditory system responds to noise, and have discussed the known effects of noise on people. These effects include hearing damage, sleep and task interference, and individual and community annoyance.

Although much in known, there still remains a continuing need for refinement of our knowledge of the effects of noise on people.

The remainer of this section will deal specifically with noise problems encountered in industrial environments and the cost of noise control to protect the hearing of workers.

REPRESENTATIVE NOISE LEVELS IN INDUSTRY

Figure 3.4 shows a comparison of the current OSHA regulation with noise levels found in several industries. In many of the work areas shown, the noise levels exceed the OSHA 8-hr standard. For example, employee activity in the vicinity of the steam valves should be limited to less than 8 hours/day for purposes of hearing conservation. Employee exposure to the noise of the boiler feed pumps should be less than 4 or 5 hr/day. Daily exposure to the punch press noise should be less than 1 hr, and less than half an hour to the noise in the wood chipping room. While not all similar work areas in industry are necessarily as noisy as the examples given here, such levels are not uncommon and represent serious hearing damage risk to employees having many years of exposure.

In many plant areas, voice communication is extremely difficult, if not impossible. This difficulty is usually resolved by retreating to a quieter area of the

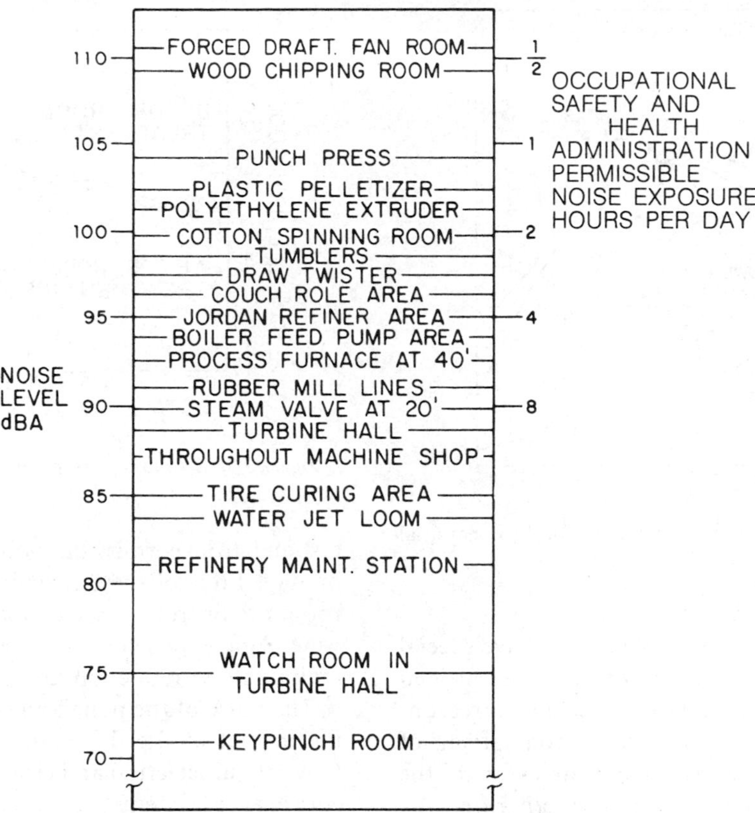

Figure 3.4. Comparison of industrial noise levels with current Occupational Safety and Health Administration Standard.

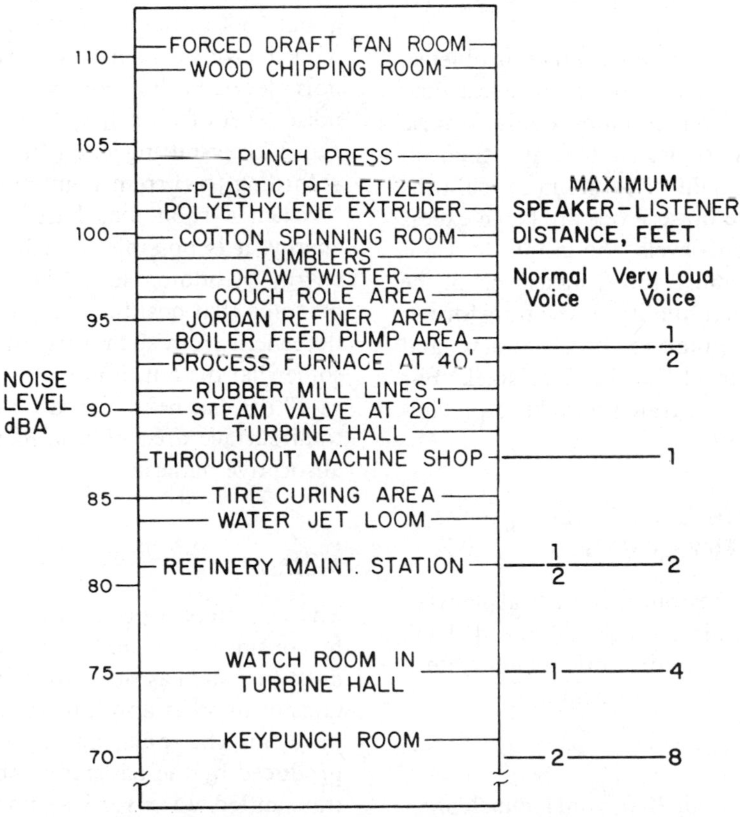

Figure 3.5. Comparison of noise levels with speech interference criteria.

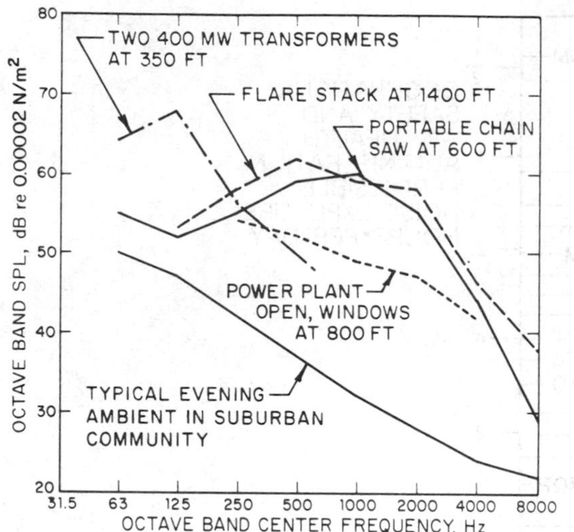

Figure 3.6. Examples of noise in communities due to industrial operations.

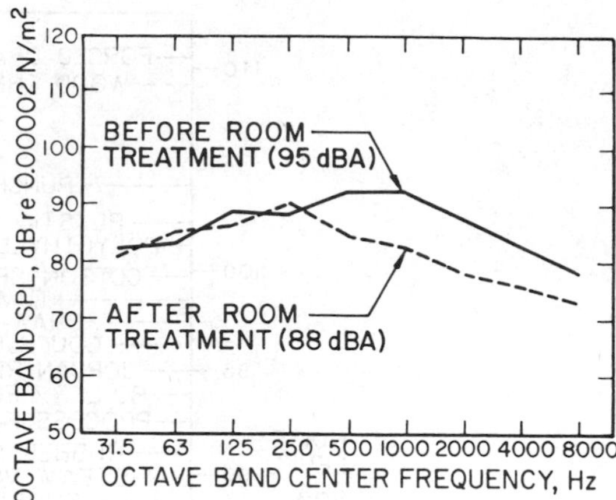

Figure 3.7. Case history of noise control using room absorption.

plant. However, many plants simply do not have special areas convenient to the work station that are shielded from noise where conversation can be held with reasonable speech effort. Figure 3.5 shows a comparison of noise levels found in several plant areas with the generally accepted speech interference criterion. Although these levels are not necessarily hearing damage risk levels, communication in areas such as tire curing rooms and spinning rooms may require shouting over distances of more than a few feet.

To illustrate potential community noise problems, Fig. 3.6 shows the relationship between noise levels measured at various distances from common outdoor noise sources and the somewhat typical, albeit hypothetical, ambient noise levels that might be found in the suburbs in the evening. While these noise levels are far in excess of the hypothetical ambient levels, the graph does not fully describe some of the more annoying aspects of the noise, for example, the tonal quality to the transformer noise, the starting and stopping of the chain saw, and the fluctuations in noise levels of the flare stack. Far less severe situations than illustrated here have precipitated widespread complaints.

CASE HISTORIES OF NOISE CONTROL TREATMENTS

As examples of what can be done in industrial plants to control noise, several case histories are discussed. Each of these cases provides an opportunity to illustrate a particular type of noise control treatment.

Absorptive Room Treatment

The noise levels measured about 30 ft from a punch press operation, illustrated by the solid curve in Fig. 3.7, correspond to a permissible noise exposure of less than 4 hr/day. Four of these machines were located in a large room of concrete construction. A number of employees were being exposed to these levels for sustained periods, making noise reduction desirable.

The noise of the punch presses is caused principally by the impact of the die stamping out parts. Frequently, however, air ejection and cleaning mechanisms, as well as conveyance systems for finished parts, generate noise levels comparable to the impact noise.

As a noise control measure, sound-absorbing baffles were hung from the ceiling on 3-ft center spacings, and about 75% of the wall behind the machines was covered with fiber-glass board. As a result of this treatment, the noise levels in the room were reduced to an A-weighted noise level of 88 dB(A) (shown by the bottom curve in Fig. 3.7) permitting a daily noise exposure exceeding 8 hr according to current regulations.

When a room has hard surfaces that reflect sound energy, it is possible to reduce the amount of reflected energy by adding sound-absorbing materials. It should be noted that positions very near the machine receive little benefit from such a treatment; however, at positions somewhat distant, a reduction in noise level of as much as 10 dB can be achieved by covering 25 to 35% of the total surface area of a hard room with high-efficiency absorptive material.

Mufflers

The comunity aspects of noise problems are illustrated by considering the exhaust of a large flat-bladed induced draft fan, such as those used in air pollution control systems or in other applications where corrosion or dirt is a problem. The tones, and in some cases their harmonics, produced by the passage of the fan blades are generally transmitted up a vertical stack (perhaps 100 to 200 ft high); and the noise is radiated in all directions to neigh-

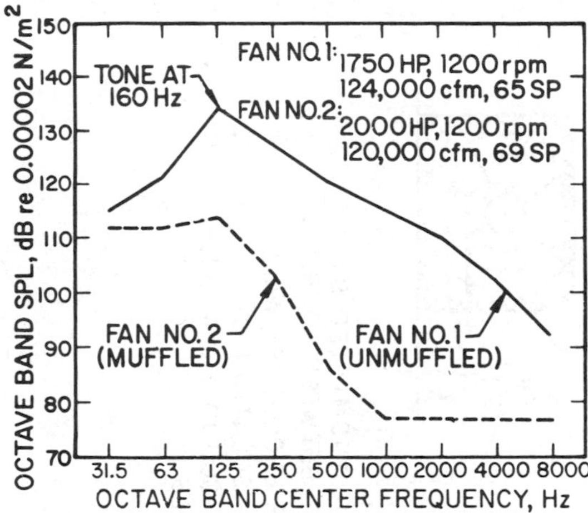

Figure 3.8. Case history of noise control using mufflers.

bors. Complaints of noise generated by fans have been registered by residents several miles away.

The top curve in Fig. 3.8 presents the noise levels measured at the top of a stack connected to an unmuffled 108-in.-diameter fan rated at 1750 hp.[5] A tone at 160 Hz is shown by the sharp peak in the spectrum in the 125-Hz octave band. The lower curve shows the reduced noise levels at the top of the stack for a similar fan with suitable muffling. With this treatment, the fan noise in the community was reduced to acceptable levels.

Mufflers are of two general types. Reactive mufflers, like the exhaust mufflers for reciprocating engines, achieve sound attenuation by reflecting the incident energy. Dissipative mufflers, like the one in this example, use absorptive materials to convert sound energy into heat. Many styles and sizes of both reactive and dissipative mufflers are commercially available; others can be custom-designed. Lining ducts with acoustical absorbing material can be effective if the material is sufficiently thick and absorptive.

Enclosures

Textile mills often house many machines in very large rooms; the noise levels throughout these rooms are typically 90–100 dB(A). The need for noise control is evident since workers spend 8 hrs a day in this environment.

While it was clear that the high-speed belt and spindles of a tube-drive ring twister[5] were the major noise sources, total enclosure of the bobbins, spindles, and spindle drives would have prohibited proper machine operation. It was decided that partial enclosures for the lower spindles, drive belt, and idlers would be acceptable. An enclosure was constructed of 16-gauge sheet metal, lined internally with a damping material and acoustically absorptive foam. Small penetrations were cut in the enclosure, where required, to accommodate moving parts. The enclosure panels were mounted to the frame through rubber washers.

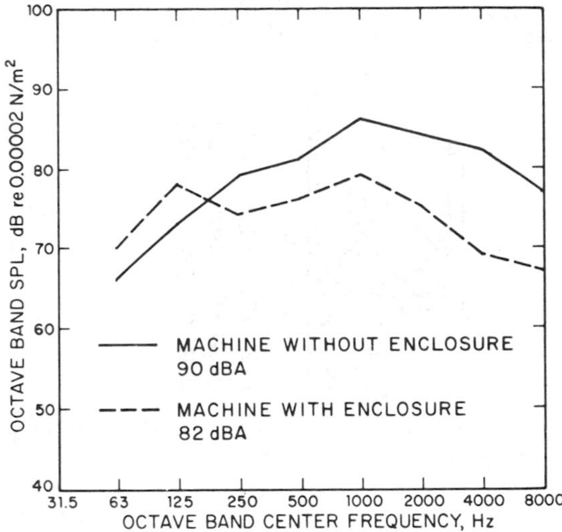

Figure 3.9. Case history of noise control for textile ring twister using an enclosure.

Figure 3.9 shows the octave band sound pressure level analysis measured 12 in. from the spindles before and after installation of the partial enclosures. The noise reduction achieved was about 8 dB.

Octave band noise levels measured in the aisle near a plastic pelletizer are shown by the upper curve in Fig. 3.10. The high-speed rotation of the cutting blades generates a siren-like sound whose fundamental frequency is the blade passage frequency of the cutting head (the number of blades × the rotation speed of the cutting head). The 99-dB(A) noise level limited permissible exposure to less than 3 hr/day.

A thick-walled, acoustically-lined housing was designed to enclose the cutter and its drive mechanism; acoustically lined openings were provided for the entry of the plastic strands and for the exit of the pelletized stock. The noise levels in the aisle after treatment are shown by the lower line—a nose reduction of approximately 10 dB.

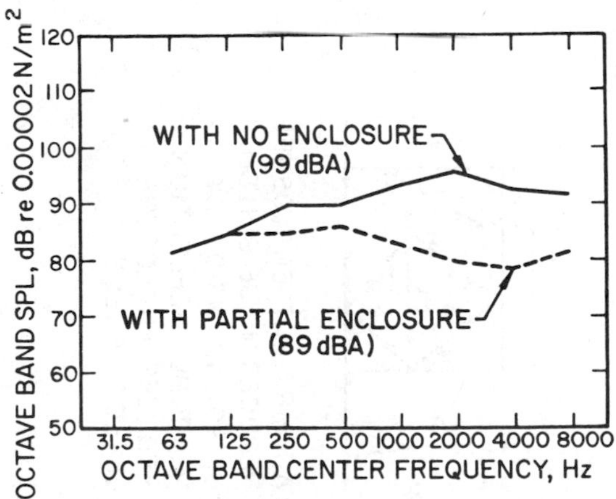

Figure 3.10. Case history of noise control for plastic pelletizer using an enclosure.

AN AIRTIGHT ENCLOSURE
REDUCES NOISE. BUILD UP
OF SOUND INSIDE BY
MULTIPLE REFLECTION
REDUCES ENCLOSURE
EFFECTIVENESS.

ACOUSTICAL ABSORPTION
LINING REDUCES INTERNAL
REFLECTIONS. MAXIMUM
EFFECTIVENESS OF ENCLO-
SURE RESULTS.

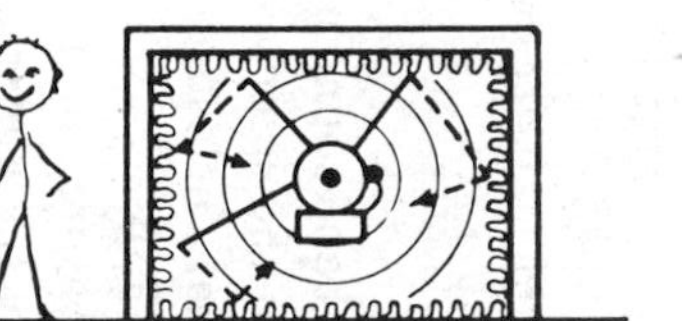

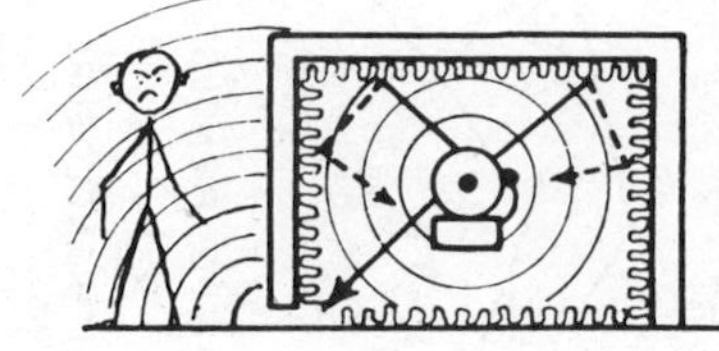

AN ENCLOSURE OF POROUS
SOUND ABSORBING MATERIAL
IS A LEAKY ENCLOSURE AND
IS OF LITTLE IF ANY VALUE.

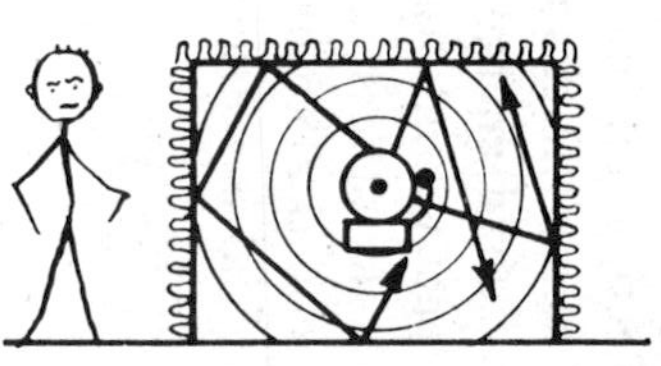

ACOUSTICAL ABSORBING
MATERIAL OUTSIDE THE
ENCLOSURE DOES NOT
IMPROVE ISOLATION.

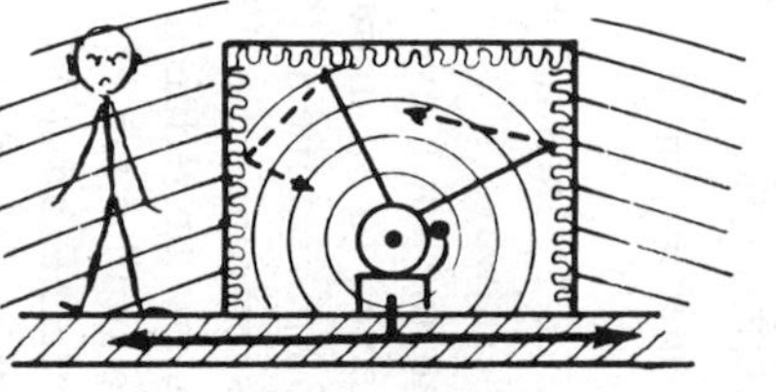

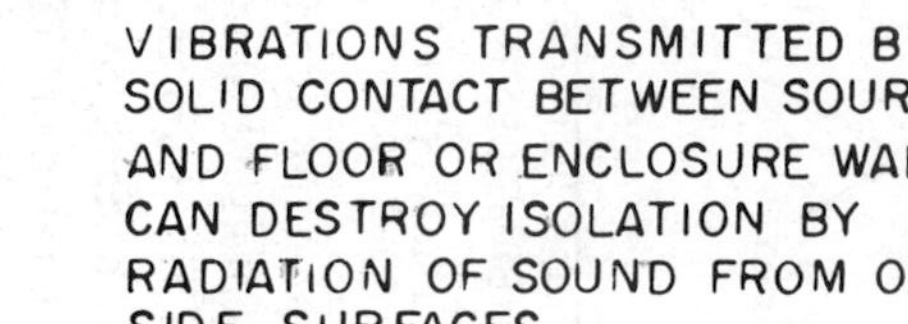

VIBRATIONS TRANSMITTED BY
SOLID CONTACT BETWEEN SOURCE
AND FLOOR OR ENCLOSURE WALL
CAN DESTROY ISOLATION BY
RADIATION OF SOUND FROM OUT-
SIDE SURFACES.

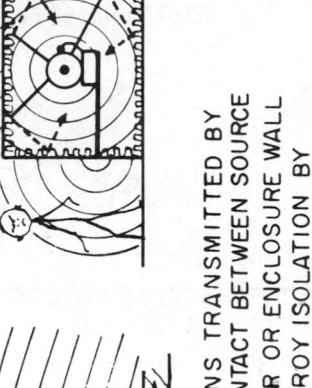

Figure 3.11. Airborne sound isolation by enclosures.

TABLE 3.5.

Percentage of Open Area in Enclosure	Maximum Average Noise Reduction
50	3 dB
25	6 dB
10	10 dB
1	20 dB

Figure 3.11 illustrates airborne sound isolation by enclosures. In general, the enclosures offering the most noise reduction are airtight constructions of heavy materials and isolated from the vibrations of the machine. Enclosures are commonly made from any nonporous construction material such as metal, acrylics, plywood, or masonry. The sound-absorbing material placed inside the enclosure in the example above prevented a reverberant build-up of sound within the cavity which would have reduced the effectiveness of the enclosure. The damping material was fastened directly to the metal panels to prevent strong resonant vibrations of the panels. Although provision must be made for material feed, product output, and preservation of the operating environment of the machine (e.g., cooling), openings should be kept to a minimum.

How complete the enclosure must be depends on how much noise reduction is necessary and on the location of the openings in relation to the points where noise reduction is required. Table 3.5 provides a means of estimating the percent of allowable unmuffled openings in an enclosure surface if a given level of noise reduction is to be achieved.[6]

In other instances, it may prove more cost-effective to provide an enclosure, or booth, for the machine operator rather than for the machine. This approach is common even when the noise levels do not represent a potential hearing damage risk, since a well-designed booth will permit better telephone usage, conversation among personnel, and will, in general, yield a more satisfactory working environment.

Lagging

The cutting edges of the Jordan refiner used in the pulp and paper industry for shredding pulp generate prominent tones that lie in the 250-Hz and 500-Hz octave bands. The octave band noise measured at a 3-ft distance from the untreated machine is shown by the upper curve in Fig. 3.12. After the lagging treatment was applied, the noise levels were reduced to those shown by the lower curve—a reduction of some 8 dB. At the operator's bench, the reduced noise levels were below the 8-hr hearing damage risk criterion.

The most effective lagging treatment consists of a sandwich composition—heavy, impervious, outer layer

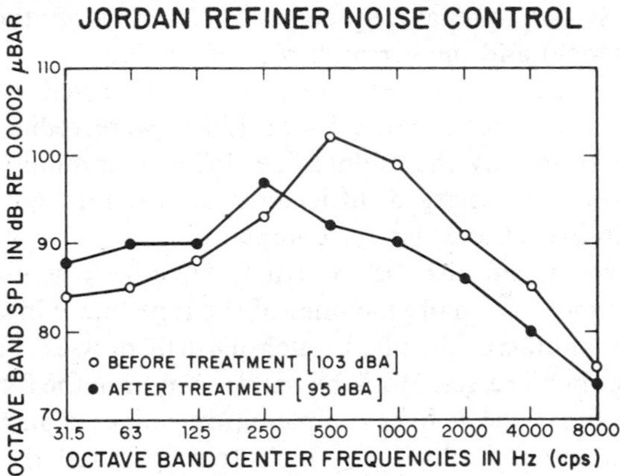

Figure 3.12. Case history of noise control: Jordan refiner lagging treatment.

separated from the machine or pipe wall by an inner layer of a material having good vibration isolation and sound absorption properties, such as glass fiber or open-cell polyurethane foam sheet.

Barriers

If large enough, a wall, building, or other type of solid structure can serve as a partial barrier to sound and thus provide a moderate amount of noise reduction.

The top line in Fig. 3.13 shows the octave band noise level at 150 ft produced by 2400-MW transformers.[7] The bottom line shows the ambient noise levels in a community 1200 ft away. Complaints from the residents were prompted by the transformer tones which could be heard to modulate, as shown by the vertical lines, above and below the ambient noise. Transformers commonly gen-

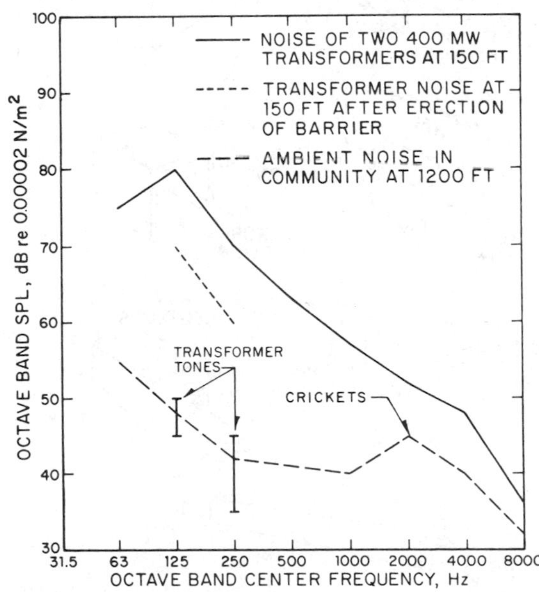

Figure 3.13. Case history of noise control using a barrier.

erate strong tones at frequencies of 120 Hz (twice the line frequency) and integer multiples of 120 Hz.

After a large concrete block wall was erected around the transformers, the levels at 150 ft were reduced to those shown by the middle line. In the community, the tones were submerged sufficiently into the ambient noise to eliminate the cause for complaint.

Barriers can also be extremely effective where there are numerous small machines of the type found in a typical manufacturing plant, such as drill presses, lathes, and punch presses. Here the barrier can be in the form of a Plexiglas sheet, for example, and it might replace the typical open-mesh safety shield that is frequently used on production equipment. Examples of some barriers are shown in Fig. 3.14.

If the barrier is to be effective, it must weigh at least 1 lb/ft^2 for high-frequency noise. It should be large enough to provide an acoustical shadow that extends several feet on all sides of the operator. Barriers can shield an operator from the direct sound field by directing the energy away from him, but, of course, they will not provide shielding from the reverberant sound field or from the direct sound from other machines.

Partial-height walls may be used to break up large spaces into smaller rooms and provide surfaces for absorptive material as well. To be effective in dividing the room, the barrier should extend several feet above and below typical head height and be used with an absorptive ceiling treatment. Effective partial-height barriers can provide as much as 10 dB of isolation from the nearby sources that they shield.

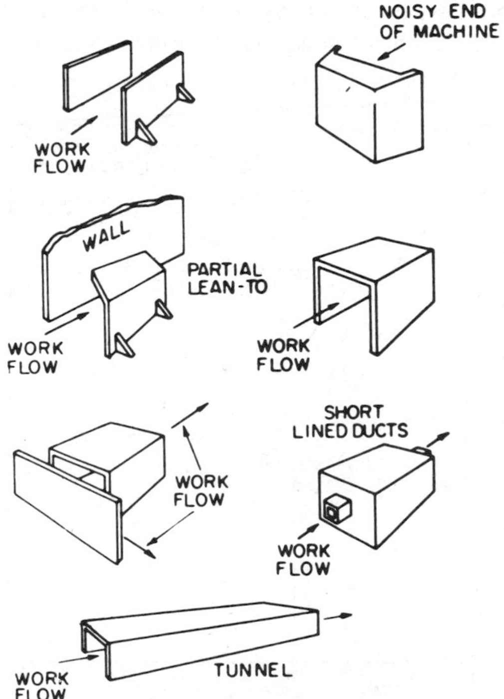

Figure 3.14. Examples of barriers.

If a barrier is not much larger than the noise source itself, its effectiveness is hardly more than visual.

Vibration Isolation

Vibration isolation is always an important consideration in noise control, because solid structures such as floors, walls, cover plates, and enclosures can act as sounding boards for the vibration source.

A vane-axial fan in a high-pressure air handling system produced, in a classroom below, the noise levels shown by the upper curve in Fig. 3.15. The design goal noise levels for the classroom were defined by the smooth curve labeled "NC-30." After a diligent search, the problem was found to be caused by inadequate vibration isolation between the fan intake bell and the plenum wall. The tone produced by the fan blades was transmitted to the concrete plenum wall, down the wall, into the floor slab, and finally into the classroom. Suitable vibration isolation reduced the intrusive fan noise to that shown by the lowest curve.

As shown in Fig. 3.16, vibration isolation generally requires a spring or a rubber isolator that accommodates for vibration at one end and remains relatively stationary at the other. Such an isolator is effective over a limited frequency range—for very low frequencies, the vibrator isolator appears hard, and vibratory forces pass through it with little loss. At high frequencies, the isolator may function well, but much noise may be radiated directly from the vibrating mechanism because of the small wavelengths of high-frequency sound. In the intermediate region, care must be taken to avoid an isolator which might create a resonance between the vibratory mechanism and its support, possibly causing an increase in noise radiation and, in the extreme case, destruction or damage to the isolated mechanism. Often, a combination of springs and high-frequency isolators may be

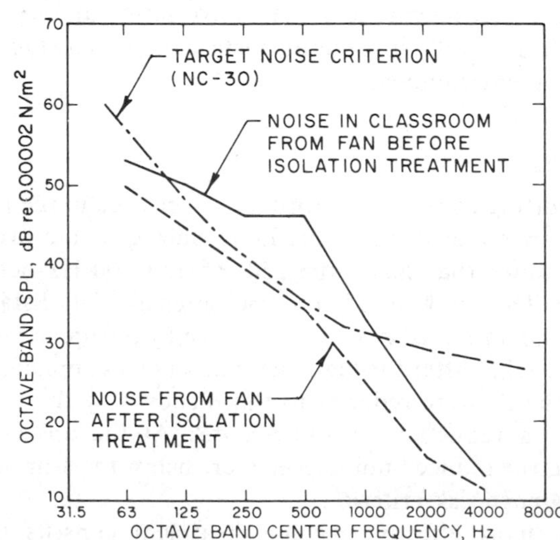

Figure 3.15. Case history of noise control using vibration isolation.

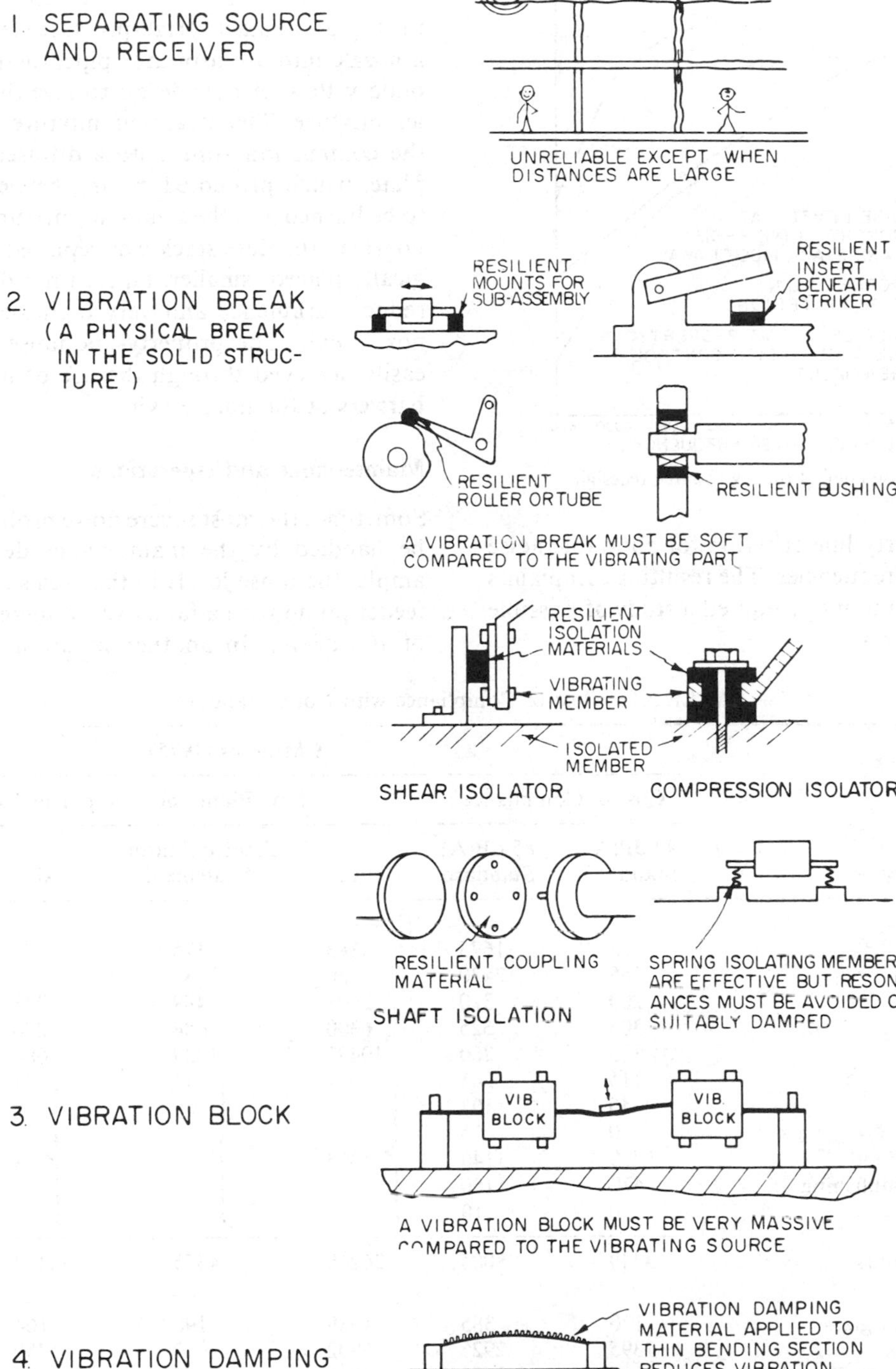

Figure 3.16. Vibration isolation systems.

needed to isolate vibrations throughout a broad range of frequencies. In any case, a vibration-isolated system must have no solid ties bridging the isolated and unisolated members. Even one solid connection can nullify a good vibration-isolation system exactly analogous to the way that a single wire connection can short-circuit a good electrical insulation.

Source Redesign

Sometimes a noise problem is so severe that unless a noise control treatment is especially effective, a modification of the source itself is required. The more common noise control measures can then be applied as secondary treatments to achieve the noise reduction desired.

The flare stack noise shown in Fig. 3.17 was some 20

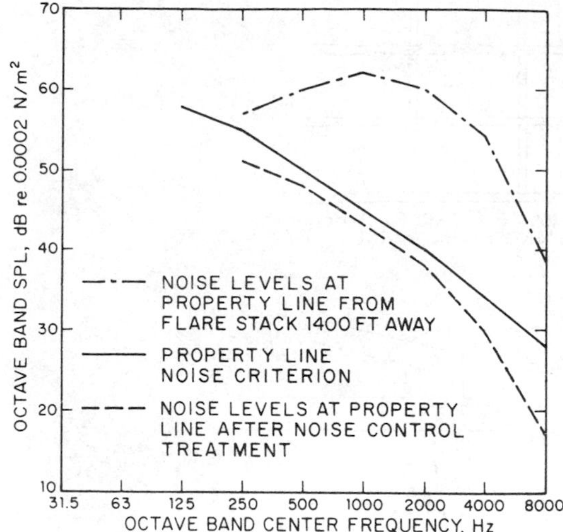

Figure 3.17. Noise control through source redesign.

dB above the property-line criterion for allowable noise in the mid- to high-frequencies. The resulting complaints from a nearby community prompted a study of possible noise control measures.

In order to burn excess HC gases smokelessly, a steam-air mixture was introduced into the combustion zone at the top of the stack. High-pressure steam exhausted from a nozzle into an inspirator pipe, thereby inducing a secondary flow of outside air to give the necessary steam-air mixture. The steam-air mixture then discharged at the combustion zone onto a diffuser, a perforated flat plate, which promoted mixing between the waste gases to be burned and the steam-air mixture. The single steam nozzle in the flare stack was replaced with several strategically placed, smaller, tapered nozzles which served to reduce turbulence and very high local velocities at the nozzle exit. The property-line noise criterion was then easily achieved through the use of mufflers and partial barriers at the nozzle exit.

Maintenance and Operations

Sometimes the most severe noise problems are matters to be handled by the maintenance department. For example, the noise levels in the area surrounding a boiler-feeder pump with a faulty speed increaser were in excess of 106 dB(A). In another situation, the noise problem

TABLE 3.6. Cost of Compliance with Noise Standards

	$ Millions (1975)						
	Cost of Compliance[1]		New Plant and Equipment Expenditures[2]				
Industry	90 dB(A) Standard	85 dB(A) Standard	Total	Total Pollution Abatement	Air	Water	Solid Waste
Non-durable Goods							
SIC 20 Food	575	1675	3383	175	71	92	12
22 Textile	1155	2470	680	31	15	15	1
26 Paper	200	310	2908	489	273	189	27
28 Chemicals	305	625	6300	684	250	394	40
29 Petroleum	175	260	10497	1239	684	483	72
30 Rubber	115	245	1037	41	25	14	2
21 Tobacco	45	105	[4]	[4]	[4]	[4]	[4]
23 Apparel	0	15					
24 Lumber and Wood	700	1140	1463	41	14	22	6
27 Printing and Publishing	470	1150					
31 Leather	0	10					
Total Non-durable Goods	3740	8005	26268	4475	2494	1737	245
Durable Goods							
SIC 32 Stone, clay, and glass	170	385	1389	198	164	31	3
33 Primary Metals	1395	2925	5892	1012	750	221	41
35 Machinery, except Electrical	2185	2620	4736	83	40	37	6
36 Electrical Machinery	145	370	2327	136	34	93	9
37 Transportation	670	1050	3387	116	51	50	15
25 Furniture and Fixture	360	445	[3]	[3]	[3]	[3]	[3]
34 Fabricated Metal Products	1305	1560	4315	229	122	97	10
Other	—	—					
Total Durable Goods	6230	9555	22046	1175	1161	529	85
Total Manufacturing	9970	17560	48314	4475	2494	1737	245

1. Bruce, R. D. "Economic Impact Analysis of the Proposed Noise Control Regulation." Bolt Beranek and Newman Inc. Report No. 3246, April 1976.
2. Survey of Current Business, July 1976, Vol. 56, No. 7, p. 14, U.S. Department of Commerce, Bureau of Economic Analysis, Washington, D.C.
3. Total Expenditures for SICs 25, 34, and other are aggregated.
4. Expenditures for SICs 21, 23, 24, 27, and 31 are aggregated.

could be reduced through adjustments in normal operations. The mid- and high-frequency noise levels 550 ft south of a power plant could be reduced about 6 dB when the southside windows were closed. Recognizing the noise source is usually the difficult part in cases like these. A little detective work can sometimes uncover the source of a problem and its solution, too.

Cost of Noise Control

During the past five years, the feasibility of noise control and the cost of providing noise control to comply with current OSHA standards have been debated at technical meetings, public hearings, and in courts of law. Although there are specific machines and/or operations for which current technology does *not* provide a solution, e.g., forging hammers, fly-shuttle looms, and chipping hammers, in general noise control is technically feasible. The basic problem preventing industry from compliance with the existing noise regulation is cost.

Since the passage of the Occupational Safety and Health Act of 1970, Bolt Beranek and Newman Inc. (BBN) has assisted several hundred firms in developing plans for complying with the noise standard. For many of these firms, we developed estimates of the cost of compliance with the noise standard. Using this base data, we have developed estimates for most of the 3-digit industries in the manufacturing sector to comply with the existing noise standard. In addition, we have developed estimates of the cost for compliance with a more stringent standard, allowing 8-hr exposure to an 85-dB(A) noise level. These estimates are summarized in Table 3.6. Also presented in this table are the 1975 expenditures of the manufacturing industry for total new plant and equipment and the expenditure for air, water, solid waste, and total pollution abatement. Currently, the industries that comprise nondurable goods are spending about 10% of their expenditures for new plant and equipment on pollution abatement; durable goods industries are expending about 8%. Compliance with either the 90-dB(A) or the 85-dB(A) standard would require each group to spend anywhere from 14 to 43% of their plant and equipment expenditures if they were to comply within 1 year. Clearly, industry will be unable to respond this quickly. It is anticipated that it will take some industries at least 10 years to comply. With their expenditures spread out over 10 years, the maximum expenditure required in a given year would be 2 to 7%, depending upon what standard is chosen.

Although noise control can be costly, it does not have to be excessive. However, it is imperative that compliance plans be developed and discussed with OSHA personnel, so that an equitable schedule of compliance can be established. Clearly, cost and its economic impact on a company are important ingredients in how quality engineering controls are installed.

CONCLUSION

Noise levels in many work areas in industry exceed the OSHA criteria; levels outside the plants may exceed community noise criteria. Recent evidence suggests that regulations may become more restrictive as people become increasingly sensitive to noise. Thus, greater concern about plant and community noise may be expected in the near future.

Fortunately, there exist considerable literature and practical experience related to industrial noise control problems.[8,9,10,11,12] The challenge is to select those particular types of noise control measures that are appropriate for each specific situation and to design efficient, balanced, practical, and cost-effective treatments for noise control.

GLOSSSARY*

Acceleration—Acceleration occurs whenever a velocity changes with time. This change may be a change in speed, as for example, in an apple falling off a tree; it may be a change in direction, as when an automobile rounds a curve at constant speed; or it may be a combination of both types of change.

Acoustic, Acoustical—These two qualifying adjectives can be confusing and in fact are often misused. The qualifying adjective acoustic is used when the term which it modifies designates something which has the properties, dimensions, or physical characteristics associated with sound waves. The adjective acoustical, on the other hand, is used when the term being qualified does not innately contain some property, dimension, or physical characteristic which is intimately associated with sound. Thus, we speak of an acoustic impedance, but we speak of the Acoustical Society of America.

Acoustics—Acoustics is the science of sound, including its production, its transmission, and any effects that it may cause.

Hearing Loss (Hearing Level) (Hearing Threshold Level)—The hearing loss of an ear at a specified frequency is the amount, in decibels, by which the threshold of audibility for that ear exceeds a standard audiometric threshold. Hearing loss and deafness are both legitimate qualitative terms for describing the medical condition of a moderate or severe impairment of hearing, respectively. Hearing level, however, should only be used to designate a quantitative measure of the deviation of the hearing threshold from a prescribed standard.

*Taken from *Glossary of Terms Frequently Used in Acoustics*, American Institute of Physics.

Noise—Noise is any undesired sound. By extension, noise is any unwanted disturbance within a useful frequency band, such as an undesired electric wave in a transmission channel or device. It is generally assumed that noise is an erratic, intermittent, or statistically radom oscillation.

Noise Level—Noise level is simply the reading of a standard sound level meter when the excitation is that of the noise.

Sound Level—The expression "sound level" has a very particular technical meaning. There exists in this country a line of instruments whose performances are specified in an American Standard, and these are called American Standard Sound Level Meters for the Measurement of Noise and Other Sounds. Now, each of these meters is equipped with three different electrical networks inside, which serve the purpose of weighting the frequency components in different manners. These networks are called Weightings A, B, or C. Now, a sound level is simply the reading of such a meter when subjected to a given sound. The level has meaning, of course, only if the particular weighting used at the time of reading the meter is specified.

Sound-Level Meter—A sound-level meter is an instrument including a microphone, an amplifier, an output meter, and frequency weighting networks for the measurement of noise and sound levels in a specified manner. These instruments are constructed and adjusted to agree with specifications laid down under an American Standard.

Sound Pressure—The sound pressure at a point in a medium is the instantaneous pressure at that point in the presence of a sound wave, minus the static pressure at that point.

Vibration—Vibration is an oscillation wherein the quantity that is being observed is specifically associated with a mechanical system.

Wavelength—The notation of wavelength is quite analogous to the distance between the crests of successive waves on the ocean. By extension, the idea can be generalized to mean a comparison between analogous points on any two successive waves in any kind of a disturbance.

REFERENCES

1. Bruce, R. D., and S. A. Fidell. "Noise and Noise Levels Affecting the Human Sensory System." Text of a talk delivered at the International Pollution Engineering Exposition and Congress for Government and Industry, Cleveland, Ohio, 5 December 1972.
2. Shadley, J. R. and R. D. Bruce. "Industrial Noise Control Treatments." IEEE-IGA 1971 Annual Meeting, Cleveland, Ohio, 18 October 1971.
3. Kryter, K. D., W. D. Ward, J. D. Miller, and D. H. Eldredge. "Hazardous Exposure to Intermittent and Steady-State Noise." *JASA*, Vol. 39, No. 3, March 1966. Much of the pioneering work in the area of hearing loss has been performed by the National Academy of Science—National Research Council Committee on Hearing, Bioacoustics, and Biomechanics (CHABA). The CHABA working group, after reviewing all available information, defined hearing loss as a permanent threshold shift of: 10 dB at 1000 Hz, and below, 15 dB at 2000 Hz, and 20 dB at 3000 Hz and above. These frequencies and levels were selected to provide protection in the frequency range associated with speech communication. Using the above definition of hearing loss, it is possible to estimate the percentage of the population that is then protected by any particular criterion.
4. "Health and Safety Standards." *U.S. Department of Labor, Federal Register*, Vol. 34, No. 96, May 29, 1969 (corrections, July 1, 1969). With this criterion, it is estimated that after 40 years of exposure, 20% of the exposed population will suffer hearing loss in excess of that normally experienced by their age group.
5. Bruce, R. D. and N. F. Gubitose. "Noise Control for a Textile Machine." *Sound and Vibration*, Vol. 5, No. 5, May 1971.
6. Holmer, C. I. "Reducing Machine Noise with Proper Damping and Acoustical Treatment." *Techniques of Plant Engineering and Maintenance* (*Proc.*), Vol. 23, 1972, Session 29b.
7. Hoover, R. M. Lecture at Westinghouse School for Environmental Management, Fort Collins, Colorado, 16 July 1971.
8. Beranek, L. L. *Noise Reduction.* New York, McGraw-Hill Book Co., 1960.
9. Harris, C. M. *Handbook of Noise Control.* New York, McGraw-Hill Book Co., 1957.
10. Peterson, A. P. G. and E. E. Gross, Jr. *Handbook of Noise Measurements.* General Radio Company, West Concord, Massachusetts, 1967.
11. Beranek. L. L. *Noise and Vibration Control.* New York, McGraw-Hill Book Co., 1971.
12. Miller, L. N. Lecture Notes on "Noise and Vibration Control for Mechanical and Electrical Equipment in Buildings" and "Noise in Manufacturing Plants" Bolt Beranek and Newman Inc. 1971.

SECTION 4

Air Pollution Control Requirements for Industrial, Commercial and Public Facilities

James R. Mahoney, Ph.D.
President
ERT International, Inc. (ERTI)
Concord, Massachusetts

and

Barbara J. Goldsmith, M.C.P.
Manager, Office of Environmental Policy
Environmental Research & Technology, Inc. (ERT)
Concord, Massachusetts

The Clean Air Act Amendments of 1977 (P.L. 95–95) were signed into law on August 7, 1977. These Amendments impose a wide range of new responsibilities upon the operators of stationary source facilities and substantially change the conditions for obtaining permits for new and expanded plants. Compliance with the Amendments will require development, interpretation, and submission of several kinds of technical data and will generally increase the lead time for obtaining required permits to construct new or expanded facilities.

This chapter discusses two sets of provisions contained in the 1977 Amendments: Prevention of Significant Deterioration (PSD) of air quality and improvement of air quality in Non-Attainment (NA) areas. These provisions impact all new, and most of the existing, major emitting facilities in the U.S.

The Amendments establish detailed rules for the prevention of significant deterioration of air quality in regions presently in compliance with the National Ambient Air Quality Standards (NAAQS), and specific procedures for air quality improvement in all remaining areas, identified as non-attainment regions. The 1977 Act establishes specific PSD provisions for sulfur dioxide (SO_2) and total suspended particulate matter (TSP), and requires EPA to promulgate similar provisions for the other criteria pollutants (nitrogen oxides, carbon monoxide, and oxidants) within two years. The SO_2 and TSP provisions divide all PSD regions into three categories and require that the maximum increases in SO_2 and TSP concentrations throughout such regions not exceed specified increment values, which are small percentages of the related NAAQS concentrations for these pollutants. The Act establishes very stringent requirements for new facilities to be located in NA areas; such facilities must be controlled to the level of the "lowest achievable emission rate," and any emission of the controlled pollutants at such new facilities must be balanced by emission reductions at other sites within the NA area. Taken together, the PSD and NA provisions of the 1977 Act significantly influence the strategics for the development and siting of industrial and energy-related facilities in the country:

- These provisions can have the effect of substantially limiting the size of individual facilities and the total number of sites potentially suitable for development.
- Small scale facilities are generally favored, because of the small emission limits required at any individual site.
- Substantially more stringent source emission limits will generally be required for sites in regions of rough terrain, because the effluent can occasionally impact on the elevated terrain.
- Uncertainties and delays in the development of major new facilities may occur as a result of the requirements for use and interpretation of air pollution models and monitoring data in the permit application and review process.
- The difficulties in the permitting process, and the relatively higher costs of required emission controls for new facilities, may in some cases, favor the retention and continued use of older industrial plants.

The language of the 1977 Clean Air Act Amendments is complex. In some cases, the Act:

Portions of the material presented in this chapter originally appeared in *Environmental Science & Technology*, Vol. 12, No. 2, pages 144–149, February 1978, © 1978 American Chemical Society.

- contains multiple definitions of the same term which are applicable to the different provisions in the Act to which sources may be subject.
- stipulates interim rules for both PSD and NA until State Implementation Plans (SIP's) are revised to reflect the requirements of the new law, and
- exempts some sources in certain cases from PSD requirements.

While this chapter covers the principal requirements for stationary source operators wishing to expand or modify existing facilities or construct new facilities in PSD or NA areas, a careful review of the law will generally be required in individual cases, and the technical advice of air quality experts should be obtained.

DESIGNATION OF PSD AND NA AREAS

The Clean Air Act of 1970 divided the country into Air Quality Control Regions (AQCR's) for regulatory purposes. At that time, a goal was established that by 1975, air quality in each AQCR would be in compliance with primary (designed to protect public health) National Ambient Air Quality Standards for each of the criterion pollutants: sulfur dioxide, suspended particulate matter, nitrogen oxides (NO_x), carbon monoxide (CO), and photochemical oxidants. Subsequently, the deadline for attainment of primary NAAQS in some areas was extended to 1977. Secondary (designed to protect public welfare) NAAQS were to be met within a reasonable time, but most states adopted the same attainment date as for primary standards. The mechanism for attaining and maintaining NAAQS is the State Implementation Plan prepared by each state. The 1977 Amendments to the Clean Air Act extended the latest allowable date for attainment of the primary standards to December 31, 1982 when it became apparent that attainment would not generally be achieved by the earlier date. For areas with severe photochemical oxidant or carbon monoxide non-attainment conditions, an additional five-year extension to December 31, 1987 is possible under very limited conditions.

In December 1977, every state was required to submit to EPA a listing of the attainment status of its Air Quality Control Regions for each of the five pollutants, for which a National Ambient Air Quality Standard (NAAQS) has been promulgated. Areas shown to have air quality better than the NAAQS for SO_2 and TSP (based on air pollution monitoring or modeling data), were designated as Prevention of Significant Deterioration (attainment) areas for these pollutants; where measured or estimated air quality was worse than the NAAQS, non-attainment areas were designated. Areas for which there is insufficient information to determine whether the standards have been met were initially classified as PSD areas. A state-by-state designation of attainment status was subsequently published by EPA in March 1978. The compliance status designations throughout the country for four of the criteria pollutants are shown in Table 4.1. Entire AQCR's may be designated as attainment or non-attainment areas or such regions may be subdivided by the state.

The designations promulgated by EPA have very important influence upon the development of industries within each state. Because attainment/non-attainment designations are made on a pollutant specific basis, an industry desiring to construct in a particular area of the country may be subject to PSD rules for one pollutant (for example, sulfur dioxide) and NA rules for another pollutant (for example, nitrogen oxides).

PSD

Prevention of Significant Deterioration provisions are generally extensions of similar regulations promulgated by EPA in December 1974. Three classes of clean air areas are established for which maximum allowable increases in pollution levels over baseline measurements of air quality are specified. (The PSD provisions presently pertain to TSP and SO_2 only. However, by August 1979 EPA must promulgate PSD regulations for the remaining criteria pollutants. If a national standard is promulgated for any additional pollutant species in the future, EPA must promulgate PSD regulations for such new pollutants within two years following promulgation of the standard.) The provisions require that the maximum increase in SO_2 and TSP concentrations throughout these regions not exceed the specified increment limits, which are small percentages of the related NAAQS concentrations for these pollutants. The classes and their allowable air quality increment limitations established in the 1977 Amendments are shown in Table 4.2. It is very important to note that the increment limits must be attained for all future times; therefore individual new facilities will not generally be permitted to "use up" the increment. Instead, they will be required to limit emissions so as to "use" only a part of the increment in each case. (This division of the increment limits may occur in particular relative to energy-producing facilities, because of the need to accommodate future capacity expansions at individual sites.)

Facilities commencing construction between June 1, 1975 and March 1, 1978 are reviewed for a permit according to EPA PSD regulations promulgated in 1974. The term "commenced construction" means having all permits and having begun a program of continuously preparing a site, or having entered into binding agreements for construction equipment, and the like.

Facilities commencing construction after March 1, 1978 are to be reviewed according to the PSD require-

TABLE 4.1. Attainment Status Designations by State.[a]

State	TSP	SO_2	OX	NO_2
Alabama	NA,PSD[b]	NA,PSD	NA,PSD	PSD
Alaska	PSD	PSD	PSD	PSD
Arizona	NA,PSD	NA,PSD	NA,PSD	PSD
Arkansas	NA,PSD	PSD	NA,PSD	PSD
California	NA,PSD	NA,PSD	NA,PSD	NA,PSD
Colorado	NA,PSD	PSD	NA,PSD	NA,PSD
Connecticut	NA,PSD	PSD	NA	PSD
Delaware	PSD	PSD	NA,PSD	PSD
District of Columbia	NA,PSD	PSD	NA	PSD
Florida	NA,PSD	NA,PSD	NA,PSD	PSD
Georgia	NA,PSD	PSD	NA,PSD	PSD
Hawaii	PSD	PSD	PSD	PSD
Idaho	NA,PSD	NA,PSD	PSD	PSD
Illinois	NA,PSD	NA,PSD	NA,PSD	NA,PSD
Indiana	NA,PSD	NA,PSD	NA,PSD	PSD
Iowa	NA,PSD	NA,PSD	NA,PSD	PSD
Kansas	NA,PSD	PSD	NA,PSD	PSD
Kentucky	NA,PSD	NA,PSD	NA,PSD	PSD
Louisiana	PSD	PSD	NA,PSD	PSD
Maine	NA,PSD	NA,PSD	NA,PSD	PSD
Maryland	NA,PSD	PSD	NA,PSD	PSD
Massachusetts	NA,PSD	PSD	NA	PSD
Michigan	NA,PSD	NA,PSD	NA,PSD	PSD
Minnesota	NA,PSD	NA,PSD	NA	PSD
Mississippi	NA,PSD	PSD	PSD	PSD
Missouri	NA,PSD	NA,PSD	NA,PSD	PSD
Montana	NA,PSD	NA,PSD	NA,PSD	PSD
Nebraska	NA,PSD	PSD	NA,PSD	PSD
Nevada	NA,PSD	NA,PSD	NA,PSD	PSD
New Hampshire	NA,PSd	NA,PSD	NA,PSD	PSD
New Jersey	NA,PSD	PSD	NA	PSD
New Mexico	NA,PSD	NA,PSD	NA,PSD	PSD
New York	NA,PSD	NA,PSD	NA	PSD
North Carolina	NA,PSD	PSD	NA,PSD	PSD
North Dakota	PSD	PSD	NA,PSD	PSD
Ohio	NA,PSD	NA,PSD	NA,PSD	PSD
Oklahoma	NA,PSD	NA,PSD	NA,PSD	PSD
Oregon	NA,PSD	PSD	NA,PSD	PSD
Pennsylvania	NA,PSD	NA,PSD	NA	PSD
Rhode Island	NA,PSD	PSD	NA	PSD
South Carolina	NA,PSD	PSD	NA,PSD	PSD
South Dakota	NA,PSD	PSD	PSD	PSD
Tennessee	NA,PSD	NA,PSD	NA,PSD	PSD
Texas	NA,PSD	PSD	NA,PSD	PSD
Utah	NA,PSD	NA,PSD	NA,PSD	PSD
Vermont	NA,PSD	PSD	NA	PSD
Virginia	PSD	PSD	NA,PSD	PSD
Washington	NA,PSD	NA,PSD	NA,PSD	PSD
West Virginia	NA,PSD	NA,PSD	NA,PSD	PSD
Wisconsin	NA,PSD	NA,PSD	NA,PSD	PSD
Wyoming	NA,PSD	PSD	PSD	PSD

[a]Based on information in the *Fed. Register*, Vol. 43, No. 43, pp. 8962–9059, dated March 3, 1978. Areas which do not meet primary or secondary NAAQS are classified as Non-attainment (NA) areas; areas which have air quality better than NAAQS or cannot be classified because of insufficient data are classified as Prevention of Significant Deterioration (PSD) areas. Source: ERT

[b]Indicates that some parts of the state are classified NA while other parts are classified PSD.

TABLE 4.2. PSD Permitted Increments ($\mu g/m^3$).[a]

	Class I	Class II	Class III	NAAQS
SO_2				
Annual	2	20	40	80
24-hour	8	91	182	365
3-hour	25	512	700	1300(s)
TSP				
Annual	5	19	37	75 60(s)
24-hour	10	37	75	260 150(s)

[a]All 24-hour and 3-hour values may be exceeded once per year.
(s) indicates a secondary standard.

ments of the 1977 Amendments. For these latter facilities the new PSD regulations are in effect until State Implementation Plans are revised to incorporate the PSD requirements of the 1977 Amendments. States are required to complete appropriate SIP revisions by March 19, 1979.

The EPA significant deterioration regulations apply to all areas of the country not exceeding the NAAQS for SO_2 and particulate matter in 1974. For consistency, the EPA regulations have been interpreted to include: the Class I area designations, stack height limitations, allowable increment limitations and rules for area redesignations legislated in the 1977 Amendments. Immediately effective changes to PSD requirements are found in the *Fed. Regist.*, Volume 42, No. 212, pages 57459–57462, dated November 3, 1977.

Facilities that had permit applications pending when the Amendments passed or had received preconstruction permits under the EPA PSD regulations but did not commence construction prior to August 7, 1977, are required to undergo a new PSD review based upon the immediately effective modifications described above.

CLASS I AREAS

Immediately upon passage of the 1977 Amendments, the following PSD areas were designated as Class I areas:

- all international parks
- national wilderness areas greater than 5000 acres in size
- national memorial parks greater than 5000 acres in size
- national parks greater than 6000 acres and in existence as of August 7, 1977.

Areas redesignated as Class I areas under the EPA regulations are also classified as Class I areas under the new Act. (Under EPA regulations, all areas of the country were initially designated as Class II areas; the only area redesignated to Class I is the Northern Cheyenne Indian Reservation in Montana.) At present, 158 areas throughout the country are designated as "mandatory"

Class I areas. All other PSD areas of the country are initially classified as Class II.

A state may redesignate any area to Class I. States are also permitted to redesignate certain areas to Class III except the following areas greater than 10,000 acres in size: present national monuments, primitive areas, recreation areas, wild and scenic rivers, wildlife refuges, lakeshores and seashores, and future national parks and wilderness areas. Redesignation of an area to Class III is a complicated process requiring approval by the governor, public notices and hearings, consultation with the state legislature and approval by a majority of potentially affected local residents.

Detailed analysis is required prior to public hearing (in the area to be redesignated and in any area that may be affected by the redesignation), including health, environmental, economic, social and energy impacts of the proposal. Redesignation of areas within Indian reservations may only be done by the applicable Indian governing body.

The EPA Administrator may disapprove a proposed redesignation only if the redesignation does not meet the procedural requirements of the PSD provisions. If federal lands are included in the proposed redesignation area, the Federal Land Manager is to submit recommendations on the proposal, but the state's decision, if it differs, is binding. EPA may be requested to resolve disputes on proposed redesignations.

INCREMENT LIMITATIONS

Major emitting facilities commencing construction after August 7, 1977 must comply with the PSD numerical increment limitations established in the 1977 Amendments. A "major emitting facility" is defined as one from 28 specifically named categories of industry having emissions of more than 100 tons/year of any pollutant, *and* any other source with emissions of more than 250 tons/year of any pollutant. A listing of the 28 designated industries appears in Table 4.3.

TABLE 4.3. Major Emitting Facilities Subject to PSD Review.

Power plants (> 250 million Btu/hr)
Specific sources (> 100 tons/year any pollutant)
- power plants
- coal cleaning plants
- kraft pulp mills
- Portland Cement plants
- primary zinc smelters
- iron and steel mill plants
- primary aluminum ore reduction plants
- primary copper smelters
- municipal incinerators > 250/tons/day
- hydrofluoric acid plants
- sulfuric acid plants
- petroleum refineries
- lime plants
- phosphate rock processing plants
- coke oven batteries
- sulfur recovery plants
- carbon black plants (furnace process)
- primary lead smelters
- fuel conversion plants
- sintering plants
- secondary metal production facilities
- chemical process plants
- fossil-fuel boilers > 250 million Btu/hr
- petroleum storage and transfer facilities > 300,000 bbls
- taconite ore processing facilities
- glass fiber processing plants
- charcoal production facilities

Any other source (> 250 tons/year any pollutant)

The short-term (3 and 24 hour) SO_2 and TSP increments in any Class I, II or III area (shown in Table 4.2) may be exceeded once per year. The increment limitations represent allowable concentrations above baseline measurements of air quality. "Baseline air quality" is defined as the concentration level of actual emissions on August 7, 1977 in an area subject to the PSD rules. Determination of baseline air quality is to be assessed using air quality models, available air quality data from EPA or a state air pollution control agency and using monitoring data which the permit applicant may be required to submit. The baseline air quality concentration also includes all approved allowable emissions from any major emitting facility which commences construction before 6 January 1975 but which has not begun operation by August 7, 1977. SO_2 and particulate emissions from any major emitting facility that commenced construction after 6 January 1975 cannot be included in the baseline but must be counted against the maximum allowable increment limitaion for the applicable PSD area.

The governor of a state can exempt the following cases in determining compliance with the allowable PSD increments:

- increases in ambient concentrations owing to fuel conversion orders
- conversion from natural gases to coal resulting from a curtailment of supplies
- increases in particulate concentrations resulting from construction or other temporary emission-related activities
- effect of sources outside the U.S.

Discounting the contribution of pollutants from coal conversion projects (conversion of a fuel-burning facility to the use of a coal) will be allowed for a maximum of five years from the date of exemption approval.

In order for an existing or proposed major emitting facility to be modified or constructed in any of the three areas, the facility must:

- have obtained a permit (preceeded by an extensive analysis of impacts and public hearing)
- not exceed or cause exceedance of the applicable increases in pollutant concentrations allowed
- agree to conduct whatever monitoring is necessary in any area potentially affected by emissions from the source and use the Best Available Control Technology (BACT).

BACT is to be determined on a case-by-case basis for each pollutant subject to regulation under the Clean Air Act and must represent an emission limitation based on the maximum degree of reduction (taking into account energy, environmental and economic impacts) which the permitting authority determines is achievable for a specific facility. All sources in the country, regardless of where the plant is located, are also subject to New Source Performance Standards (NSPS), expressed as emission requirements for the applicable source category.

Foremost among the questions of interpretation is the issue of the permitted one excess per year. Based upon preliminary EPA guidance materials, it appears that the increments may be exceeded once per year at each location (receptor site) surrounding a proposed facility, rather than permitting only a single excess anywhere in the region surrounding the facility. Another important issue is the "sharing" of PSD increments among adjacent or nearby sources. The increment limits must be attained for all future times; therefore, individual new facilities will not generally be permitted to "use up" the increment. Generally the sharing of the PSD increments will result in lower permitted emission rates for each facility than would be the case if a facility were evaluated independently of its potential future industrial neighbors.

A related issue to be resolved is the separation distance that will be required between facilities before the impacts of each facility can be considered independently rather than additively. The production capacity and size of a facility may have to be reduced, or stringent controls required, depending on interpretation of these and other issues in regulatory practice.

These issues may become even more pronounced when PSD rules are promulgated for the other criteria pollutants. EPA is required to promulgate PSD regulations governing carbon monoxide, oxides of nitrogen and photochemical oxidants by August 1979. Because of the complexity of chemically reactive pollutants, the control of such pollutants in relation to PSD will likely preclude the use of the air quality increment concept employed with SO_2 and TSP.

EXCEPTIONS—CLASS I AREAS

The protection of "air quality-related values" is a key factor in determining whether a facility may be granted a permit when it would impact a Class I area. If the Federal Land Manager finds that emissions from a proposed facility would have an adverse impact on the "air quality-related values" of the land area (even if allowable Class I increments would not be exceeded) a permit cannot be issued unless the source can demonstrate that no adverse impact would occur.

If, however, the emissions from the facility would not adversely impact these values but would exceed the Class I increments, the state may issue a permit which would allow the facility to comply with less stringent air quality increments. (Even in this case the maximum increments are the Class II values, except for the three hour SO_2 increment limit that is not to exceed 325 $\mu g/m^3$).

If a permit is denied in this procedure, the facility owner may demonstrate to the governor after public hearing that the facility cannot be constructed in a Class II area so as to comply with the increments in a neighboring Class I area. The governor may then permit the facility to exceed the short-term Class I SO_2 increments on 18 days a year if the Federal Land Manager concurs.

If the Federal Land Manager disagrees with the proposed variance, the final decision is made by the President, based on the "national interest." If a variance is granted, the emission of SO_2 from the source together with all other sources, may not exceed the maximum allowable increases over baseline concentrations on these 18 days as shown in Table 4.4.

In all cases, the facility owner/operator must still demonstrate that the variance will not adversely affect the air quality-related values of the Class I area. No similar variance is allowed for the increment limit in Class II or Class III areas.

VISIBILITY

It should be noted that visibility is the only "air quality-related value" that is specifically cited in the 1977

TABLE 4.4. SO_2 Increment Limits for 18 Exceeding Days in a Class I Region.

Period of Exposure	High Terrain Areas	Low Terrain Areas
	(900′ above stack base)	
24-hour maximum	62 $\mu g/m^3$	36 $\mu g/m^3$
3-hour maximum	221 $\mu g/m^3$	130 $\mu g/m^3$

Note that a 3-hour and 24-hour exceedance occurring on a single day or two or more 3-hour violations occurring on a single day constitutes one exceedance day.

Amendments. Although it is likely that potential adverse impact on visibility will act as the primary "trigger mechanism" governing variances from the stated Class I increment limits, other "values" might include odor, vegetative stress, climatic change and so on. In February 1978 the Secretary of the Interior determined that visibility is an important value in 153 of the 158 mandatory Class I Federal areas.

Visibility protection has emerged as a major issue in the 1977 Amendments, and EPA is required to promulgate regulations on visibility protection by August 1979 to enforce a national goal of visibility protection. Existing sources in operation for less than 15 years may be required to retrofit control technology if they are shown to contribute to visibility impairment.

NON-ATTAINMENT

Construction permits for major new and modified stationary sources in non-attainment areas will be issued in accordance with EPA's December 1976 Interpretive Ruling on emission offsets until July 1, 1979, unless the state obtains a waiver to this policy. (States may be granted a waiver if they have a detailed emission inventory of sources, an enforceable permit program, and have a plan for reducing overall emissions.)

Major sources subject to this ruling include any sources that emit, or have potential to emit, 100 tons a year of any pollutant, including any source of fugitive emissions. The emission offset policy requires that a new or modified source must meet the following conditions before a permit can be granted.

- The plant must obtain an "emissions offset"—emissions from existing sources in the region must be reduced in an amount more than sufficient to offset the new plant's emissions.
- The plant must attain an emission limitation that is the Lowest Achievable Emission Rate (LAER).
- The emission offsets obtained must produce a "positive net air quality benefit" resulting in "reasonable further progress" toward attainment of the applicable standard.
- All the applicant's existing facilities within the AQCR must be in compliance with applicable emission limitations or standards.

Emission offsets are determined on a case-by-case and pound-per-pound basis with reductions required to exceed new emissions when all facilities are operating at maximum load. In limited circumstances "banking" of unused offset credits may be permitted until July 1979; credit may also be obtained, in limited conditions, by the permanent shutdown of an existing source.

After June 30, 1979 each state must have revised its SIP to provide for attainment of primary ambient air quality standards by December 31, 1982 before a major new source may locate in a non-attainment area. For areas with photochemical oxidant or carbon monoxide non-attainment conditions an additional five-year extension is available to December 31, 1987.

Since new sources must be included in the SIP it will generally be to the advantage of new source operators to communicate development plans to the applicable state permitting authorities as soon as they are known.

DATA, MODELING AND MONITORING

Owners of all existing stationary sources of pollution have several specific responsibilities for compliance under the 1977 Amendments. Generally, compliance will require conformance with several provisions of the SIP and will require periodic reporting of such conformance to the state agencies.

In regions designated as non-attainment areas, even those industries presently in compliance will frequently have new requirements for additional controls imposed to bring about compliance in the area. The long time scales required for planned changes or facility modifications under the provisions of the 1977 Amendments make it imperative that periodic short- and long-term planning of compliance with the Amendments be carried out by all major industrial operators.

Owners of planned new facilities, or facilities undergoing substantial modification, will face all the requirements imposed upon existing facilities. These owners will also have requirements for development and submission of the necessary permit applications.

Supporting data for such applications will include specific evaluation of the PSD or non-attainment area status of the new operation and will include reports of baseline monitoring at the proposed site of the new activities. Time schedules required for the development of all the supporting data will be lengthy in many cases, and will require detailed advanced planning for new facilities.

The review and permitting process under the PSD provisions involves the use of air quality modeling for estimating the impact of facilities not yet built. Therefore, in these cases, the model formulation, data used, and interpretation of the model results will be direct determinants of site suitability and emission limits required for industrial sites.

In particular, since the short-term PSD increments can be exceeded only once a year, models must be used to evaluate the meteorological or topographical conditions in a region that has the highest expected concentrations of pollutants.

The use of special models necessary to accommodate unique terrain or meteorological features of any area may be allowed by the EPA Regional Administrator following notice and public hearing. In the case of non-attainment plan development, models are also used to evaluate the air quality benefits to result from various emission reduction alternatives, including evaluation of allocation of emission reductions among multiple sources in a region. There is significant uncertainty, and variability in regulatory practice, concerning the choice of specific air quality models, and their associated input data, for site evaluation studies. The use of models to evaluate compliance with the PSD provisions can create a circumstance where evaluation of the impact of a single air pollutant (SO_2) during a few "worst hours" per year can act as an absolute veto on the environmental suitability of a proposed site regardless of other environmental, economic or social factors. Even if a site were best suited relative to other environmental considerations, it could be prohibited from development if the specific increment limits were not attained as required. The question of attainment of these limits is not determined by typical or even occasional meteorological conditions, but by very unusual conditions which occur only a few days each year at most locations in the country. The presence or absence of these conditions in a specific meteorological record can control the selection of a site which might otherwise be most suitable in terms of environmental considerations. These considerations can cause significant delays and uncertainties in the review of candidate sites for industrial and energy facility development.

By February 7, 1978 EPA was required to promulgate regulations specifying requirements for the evaluation facilities proposed in PSD areas, including the models to be used. The time sequence of this and other major EPA rulemaking activities that are required under the 1977 Amendments are shown in Table 4.5.

Beginning in August 1978, air quality monitoring may be required as part of the permit process for proposed facilities in PSD areas. This requirement can be waived or shortened if the state determines that adequate data already exists or a shorter period will be adequate to document baseline conditions.

For new facilities and modifications, the requirement to submit one year of baseline monitoring data as part of a permit application causes a substantial lengthening of the total time required to achieve issuance of a permit since actual field measurements will be required in the location of every alternative site being considered for a particular facility. Furthermore, it is not known in how many areas sufficient meteorological and air quality data already exist in the vicinity of sites that will be proposed for development or whether these data would be suitable for use.

Even with maximum efficiency achieved in the planning and implementation of a monitoring program, the time required to achieve a permit approval will likely be as long as two years, and may be as long as $3\frac{1}{2}$ years in some cases. Table 6 summarizes time estimates for various steps in the permit sequence. Consideration of this time schedule suggests that an early start of baseline monitoring may be appropriate for many facility owners, in advance of the development of specific plans for facility expansion or modifications.

In many cases, operators of individual stationary source facilities will find it useful to arrange for audits, or periodic reviews, of conditions of compliance with the 1977 Amendments. Such audits can be used as the basis for development of long-range planning for compliance with the Amendments. Generally, an environmental audit will include a detailed initial survey, together with provisions of periodic updates of the survey information. A standard reporting format will assist in routine evaluation of the audit data and will facilitate the evaluation of trends of impact associated with emission from the facility.

Compliance with the 1977 Amendments will require

TABLE 4.5. Principal EPA Rulemaking Activities Affecting Stationary Sources.

Action	Statutory Deadline
• Listing of AQCR's by attainment status	February 1978
• Short-term NO_2 standard	February 1978
• Stack height regulations (limiting credit to $2\frac{1}{2}$ times the height of the source)	February 1978
• Analysis requirements in PSD areas	February 1978
• Revised NSPS for fossil boilers	August 1978
• Ozone protection regulations	November 1979
• PSD regulations for other criteria pollutants	August 1979
• Visibility protection regulations	November 1979
• NSPS for stationary sources	August 1979–August 1982[a]

[a]EPA is required to list major stationary source categories for NSPS by August 1979 and promulgate NSPS for 25%, 75% and 100% of list by August 1981, 1982 and 1983, respectively.

TABLE 4.6. Estimates of Time Required for Permit Issuance. (Influenced by Requirement for Monitoring)

Activity	Time Required (months)	Cumulative (months)
Specify monitoring required	1–2	1–2
Select vendors and contractors	1–3	2–5
Procure and install equipment	1–4	3–9
One-year baseline monitoring	12	15–21
Complete data analysis and modeling	1–4	16–25
(Permit application complete)		
Request special model, with agency hearing and review (if necessary)	2–6	18–31
Hearings on application and final agency review	3–12	21–43

careful planning of control strategies in many cases. For example, in a non-attainment area facility owners should generally plan to schedule the implementation of any emission reduction technology so as to use the associated credit at a time related to the schedule for facility modification or expansion. Similarly, the development of new types of control technology and the establishment of cogeneration capacity are favored by Congress in the 1977 Amendments. Therefore, options for such developments should be considered as part of the long-range plans for emission control at every major stationary source.

SECTION 5A

Radiation Hazards

John H. Harley, Ph.D.
Director
U.S. Department of Energy
New York

The field of atomic energy and the applications of radioactive materials has expanded very rapidly, particularly since about 1960. A further expansion in the application of nuclear power is predicted for the 1980s. This will be accompanied by a lesser increase in the other applications of radioactivity, so that it appears that the amount of radioactive material in use will be very large.

The rapid growth of the industry means that a larger number of people now has a possible occupational exposure to radiation and that the control of radioactive materials must be exercised on a large scale to minimize population exposure. Essentially all of the scientific and industrial applications of radiation are subject to regulations by federal or local governments. These regulations, however, only set the required performance standards and do not specify the means of controlling exposure. This section is intended to lay down the principles and describe the techniques of radiation protection.

The increased interest of scientists and the public in the condition of our environment has led to a great deal of public concern over possible radioactive releases from power reactors and other sources. This degree of public concern has brought about greater restrictions and requirements for greater involvement of industry in the radiation environment. This has led to the recommendation here of a conservative approach in the control of radiation hazards. Our best scientific information leads to the conclusion that the present standards of protection for occupationally exposed workers and the public are adequate to protect them from measurable effects. This may not be a sufficient response to present concern with the environment and it will be evident in this section that there is a marked difference between having no measurable effect on the surroundings and in not allowing measurable radiation to exist in these surroundings.

To safely operate a plant or laboratory it is necessary to have an understanding of the nature of the radiation, its measurement, and the possible effect of overexposure to it. With this knowledge methods of exposure evaluation and control may be set up. This section provides the background information required by the user of small amounts of radioactivity or of low power x-ray equipment. In each case, however, there has been an attempt to indicate what additional precautions will be necessary if greater amounts of radiation are involved.

Radiation protection is the prevention of illness or injury from overexposure to x-rays and nuclear radiation. Nuclear radiation and x-rays are invisible, and except in the case of acute overexposure such as might occur in an atomic explosion, the effects do not appear for some time. In the case of radium poisoning, for example, delayed effects still were being discovered over 40 years after the exposure incurred in the 1920s. At the present time, there is no cure for the various effects produced, so all our efforts must be concentrated on prevention.

When we consider radiation exposure, we must think of that radiation which is added to the normal background radiation to which man has been exposed for hundreds of generations. This background radiation from the naturally occurring radioisotopes and from cosmic rays is the basic level of exposure and cannot be reduced.

The theoretical ideal of radiation control would be maintenance of exposure at background level, but this is often not possible. Economic considerations become increasingly important as exposure is restricted to lower and lower levels. On the other hand, while maximum permissible levels are considered safe based on past experience, designing for exposure at just these levels is not necessarily a guarantee of good radiation protection. Frequently, by careful planning, exposures can be reduced to considerably below the permissible level without extensive changes in facilities or without additional costs. In such cases, the reductions should be made. If further reduction of radiation levels requires large changes in equipment or would be very expensive, there must be a compromise between the desired and the practical levels.

The experience of radiologists, radiographers, and

isotope users has provided a reasonable basis for adopting a program for personnel protection. This experience has resulted in a series of maximum permissible levels for occupational exposure to external radiation, for inhalation or ingestion of isotopes, and for laboratory or plant waste products disposal. The accepted levels contain a reasonable factor of safety. However, needless exposure to any potential hazard is foolish and every reasonable effort should be made to minimize such exposure.

Fortunately, exposures can be measured and can be controlled so that permissible levels are not exceeded, and thus preventable damage from radiation can be avoided.

A radiation protection program must consist of two parts—evaluation and control. *Evaluation* consists of the measurements required to estimate possible radiation exposures and the necessary interpretation of these measurements. *Control* consists of taking measures required to reduce radiation exposure. A good control program minimizes necessary exposures and eliminates needless exposures. A continuing evaluation program will indicate the effectiveness of the controls and will give a continuous record of all exposures. This section will emphasize these two facets of a realistic radiation protection program.

REGULATIONS

Radiation exposure is almost always subject to some form of regulation. These regulations usually cover the purchase, possession, use, and disposal of radioactive material and the use of x-ray equipment.

The federal government is responsible for most radioactive material, although it is relinquishing this authority to the individual states as they request such authority and show that they are competent to handle radiation problems. Some natural radioactive materials and x-ray equipment usage are not covered by federal regulation.

The federal government has classified radioactive materials into three groups; source materials including U and Th, special nuclear materials including Pu and U that is enriched in the ^{233}U or ^{235}U isotopes, and by-product material including any other radioactive material resulting from the production or utilization of special nuclear material. The federal regulations have been set forth in the *Code of Federal Regulations* (CFR), Title 10, Chapter 1. The parts of this chapter of interest here are indicated below.*

Part 19 Notices to Workers
Part 20 Standards for Protection against Radiation
Part 30 Licensing of By-product Material
Part 34 Licenses for Radiography and Radiation Safety Requirements for Radiographic Operations
Part 40 Licensing of Source Material
Part 50 Licensing of Production and Utilization Facilities
Part 70 Special Nuclear Material
Part 71 Packaging of Radioactive Material for Transport

*Copies of the NRC Rules and Regulations applying to work with radioactive materials are available on subscription from the Superintendent of Documents, Government Printing Office, Washington, D.C., 20402.

The Nuclear Regulatory Commission now has the federal responsibility for radiation exposures from by-product material, source material, and special nuclear material. They are the agency that draws up the Code of Federal Regulations as to radiation exposure.

The granting of a license by the NRC or an agreement state requires that the equipment and facilities are adequate to protect health and minimize danger to life or property and that the applicant is qualified by training and experience to use the material in such a way as to protect health and minimize danger to life or property.

Licensees are generally required to maintain records on the receipt, transfer, and disposal of radioactive materials and are subject to inspection of the facility and of the required records.

The federal government or the states issue specific licenses for purchase, possession, use, or disposal of source material, special nuclear material, or by-product material. The licensee must show that a suitable radiation protection program is set up before obtaining the license and he is subject to inspection for compliance with the terms of the license. Violations of the radiation protection requirements subject the licensee to possible revocation of the license.

Certain small quantities of various radioisotopes are available without license. The quantities are considered to be harmless under any conditions of usage and are available for experimental work. These exempt quantities will be discussed later and are listed in Table 5A.7.

While the CFR applies specifically to licensees, the same standards are applied to operations under government contracts. Both groups are subject to inspection but the basic responsibility for compliance with either regulations or standards lies with the operator. Thus, he must have an adequate radiation protection program.

In most states, x-rays and naturally occurring radioactive materials are included under a code of either health or labor regulations. These codes of radiation protection standards usually apply to medical installations as well as those of an industrial nature. In general, there has been an attempt to standardize codes developed by the states to follow the federal regulations where they are applicable.

NATURE OF RADIATION

Radiation is a form of energy and, as such, can be put to use for a variety of purposes. As with other forms of energy, it can be dangerous when uncontrolled. To control radiation intelligently, it is necessary to understand its nature and then proceed to the practical aspects of measurement, evaluation, and control.

In the broad sense, radiation includes light, radio waves, cosmic rays, and many other forms, but our immediate interest is limited to nuclear radiation and x-rays. These two forms represent controllable industrial or laboratory hazards that are utilized for the very properties that make them dangerous.

The fundamental properties of radiation are mass, electrical charge, and energy. While all the types of radiation that we are interested in have energies within a rather narrow range, they differ markedly in the properties of mass and charge. The radiations all originate within the atom, and their similarities and differences are characteristic of their origin.

Atomic Structure

The scientist's picture of the structure of matter has changed radically in the last century. Knowledge has progessed from the concept of the atom as an indivisible unit of a chemical element through the planetary atom consisting of a dense central nucleus surrounded by electrons moving in orbits around it, to the modern quantum theory. The latter theory is invaluable in theoretical physics for describing the behavior of atoms, but it has almost completely eliminated any simply physical picture of atomic structure. For the moment, however, we may consider the atom as a positively charged nucleus surrounded by a negatively charged electron cloud.

For many years, every atom of each chemical element was considered to have a particular atomic mass, known as its atomic weight, which was characteristic of that element. In practice, the weights of individual atoms were not known, and the atomic weights used represented the average of a large number of atoms. These atomic weights were used in calculating the proportions of various substances entering chemical reactions and were considered to be among the most basic physical properties of atoms.

In studying atomic structure, however, it became apparent that the average atomic weights did not have as great a regularity as might have been expected. Actually, the irregular nature of these atomic weights stems from the fact that most of the elements consist of mixtures of atoms having different atomic weights. The range of these differences is small, being only one to a few atomic mass units. The individual atoms of an element having a particular atomic weight make up an isotope of the element. The naturally occurring mixture of the various isotopes produces the average atomic weight, which is the one used for chemical calculations.

It is possible to have more than one form of a particular isotope. The two forms differ in their energy content, and one is usually indicated as a metastable state. The most general term applied to the isotopes of all elements in all forms is *nuclide*, and *radionuclide* is a proper term for the radioactive form of any element having a particular energy state and atomic weight. Common usage, however, has accepted the term *radioisotope* as an exact equivalent and it will be used here.

We can describe the properties of the atoms, their components, and their radiations in physical units; the mass in atomic mass units (amu), which are $\frac{1}{12}$ the mass of the most abundant isotope of carbon; the charge in units of the charge on a single electron; and the energy in terms of the energy acquired by an electron when it is accelerated by a potential of 1 volt. This latter quantity, the electron volt (eV) is so small that we commonly speak of thousands (keV) or millions (MeV) of electron volts. The absolute energy of an atom or its components is of little interest, but the energy changes involved in the production of radiation are of primary interest to us.

Within the nucleus are protons with a mass of 1 amu and a positive charge of 1 unit, and neutrons with a mass of 1 amu and no electrical charge. Outside the nucleus are electrons (extra-nuclear electrons) with a mass of about $\frac{1}{1800}$ amu and a negative charge of 1 unit. The characteristics of an atom are due to the number of protons and neutrons which make up the nucleus. Since the atom is electrically neutral, the number of positively charged protons in the nucleus and negatively charged electrons outside the nucleus must be equal.

The number of protons in the nucleus determines the particular chemical "species" or element that the atom belongs to, while its mass is determined by the number of protons and neutrons in the nucleus. We may completely describe any atom in terms of its mass and its elemental species.

The different chemical species are designated by the element symbol or by the atomic number (the number of protons in the nucleus). In the scientific shorthand used, ${}^{12}_{6}C$ describes an atom of carbon with an atomic number of 6 and a mass of 12. A different atomic number indicates a different element, but isotopes of the same element can have different masses. Our example, carbon, exists as three isotopes of mass 12, 13, and 14. These would be represented as ${}^{12}_{6}C$, ${}^{13}_{6}C$, and ${}^{14}_{6}C$, respectively.

Actually, since the element symbol also designates the atomic number, the more usual terminology is ${}^{12}C$, ${}^{13}C$, ${}^{14}C$ (or C-12, C-13, and C-14).

The chemical nature of an element is almost independent of the particular isotope, so that in a chemical reaction all of the isotopes follow the same reaction pattern. Even the isotopes of hydrogen, where the masses are 1, 2, and 3, show only slight differences in chemical reactivity. In cases of heavier elements where the relative differences in mass between isotopes are smaller, variations in chemical behavior usually cannot be observed. If it is necessary to separate the isotopes of an element, physical properties highly dependent on mass must be used. Some of those which have proved practicable are the deflection of ions in a magnetic field, variations in diffusion rate of gases, and thermal diffusion. Isotopes separated in this way can hardly be distinguished chemically but may be distinguished by instruments in which measurement is based upon mass—for example, the mass spectrograph or the optical spectrograph.

Over 300 stable isotopes of the elements are now known. The total number of isotopes has been increased by the production of artificially radioactive isotopes by particle acclerators and by atomic fission. Up to the year 1965, over 1100 unstable isotopes had been identified, and a few more are discovered each year.

A large fraction of the naturally existing isotopes are stable, but some of the natural isotopes and most of the manmade isotopes disintegrate spontaneously. The process of disintegration (also called radioactive decay) produces nuclear radiation, and the emitting isotopes (nuclides) are called radioisotopes (radionuclides). The most common emissions are α, β, and γ radiation. Each type will be discussed, together with the nuclear transformation which accompanies it.

Nuclear Radiation

The radiations with which we are concerned are x-rays and the α, β, γ emissions from radioactive isotopes. There are installations where exposure to neutrons or accelerator-produced particles is possible. These should also be under competent safety supervision, but such a discussion is outside the scope of this section.

X-rays and γ radiation are electromagnetic; i.e., their properties are similar to UV rays or visible light, only they are much more penetrating. Beta and alpha radiations are particulate; they are extremely small particles moving at high velocity.

All types of radiation share the property of losing energy by absorption in passing through matter. In any given case the degree of absorption depends upon the type of radiation, but all types are absorbed to some extent. Also, the process of absorption in some cases results in ionization, i.e., the removal of an extra-nuclear electron from an atom of the absorbing material. It is this process of ionization that both produces damage in tissue and allows us to design instruments for detection and measurement.

Since the properties of the various radiations determine not only the protective measures needed but also the methods of measurement, a brief general description of the different radiations will be given at this point.

The α particle is emitted by many of the heavy artificial and naturally radioactive elements. The α particle is the same as the nucleus of the helium atom, with a mass of 4 amu and a positive charge of 2. A typical α decay could be expressed

$$^{226}_{90}\mathrm{Ra} \longleftarrow {}^{4}_{2}\mathrm{He} + {}^{222}_{88}\mathrm{Rn}$$

where Ra is transformed to Rn by α emission. The relatively high mass of the α particle means that for a given energy, the velocity is relatively low. The exact relationship is shown by the kinetic energy relationship for mass and velocity:

$$E = \tfrac{1}{2} mv^2$$

where

E = kinetic energy
m = mass
v = velocity

The heavy, slow-moving, highly charged α particle has a great opportunity to interact with an absorber. Thus, on the average it dissipates its energy in a short path length (e.g., most α particles are completely absorbed by a few centimeters of air, or less than 0.005 mm of Al). When all its kinetic energy has been lost the α particle will take up two electrons from its surroundings and become an electrically neutral atom of He. Most of the energies of α particles given off by the different emitters lie in the range of 4–8 MeV but all the particles emitted from a single isotope will have nearly the same energy.

Beta emission is a property of both heavy and light radioactive elements. The β particle is an electron that possesses kinetic energy because of the speed with which it is emitted from the nucleus. In representing β emission, the change in mass of the atom is negligible, but the loss of the unit charge causes a change in atomic number; e.g.,

$$^{14}_{6}\mathrm{C} \longrightarrow e + {}^{14}_{7}\mathrm{N}$$

The energies of the β particles given off by a single emitting isotope cover a range from 0 MeV to a maximum value characteristic of the isotope. Their average energy is approximately $\frac{1}{3}$ of the maximum value. The maximum energies of different isotopes lie in the range of 0–4 MeV. The velocities of the more energetic β particles approach the speed of light, they may be computed from the kinetic energy formula. The range of β particles in air may be more than a meter for energetic β particles, and such particles will penetrate several millimeters of aluminum or plastic.

Gamma emission is a secondary process following rapidly after certain α or β decays. The emission of the latter particles may leave the nucleus in an unstable state, and the excess energy is released in the form of γ radiation. Since γ radiation has neither mass nor charge, there is no change in the atomic number or atomic weight of the emitting isotope.

The energies of γ rays are mostly in the range of 0–2 MeV, and the radiation from a single isotope will be made up of one or more groups of rays each having a single value of energy. Gamma rays can be very penetrating. Absorbing material does not completely stop them, but only reduces their intensity. For example, the γ radiation from a radium source is reduced to 50% of its incident value by 0.5 inch of lead. One inch would pass 25%; 1.5 inches, 12.5%; and so on. Therefore, we speak of the half-thickness of a given absorber material for a particular γ emitter rather than its range.

Most of the known radioisotopes emit α or β particles, sometimes accompanied by γ rays. Three other, less common processes—positron emission, electron capture, and internal conversion—should, however, be considered.

The *positron* is similar to a β particle but has a positive charge. A typical reaction would be

$$^{30}_{15}\text{P} \longrightarrow e^{+} + ^{30}_{14}\text{Si}$$

All the effects of β radiation are present, and in addition, 0.51 MeV γ rays are always emitted. These result when the positron loses its kinetic energy by absorption and combines with a normal electron. The two particles are annihilated and their mass is completely transformed into two γ photons with an energy of 0.51 MeV.

The reaction for *electron capture* is similar to positron emission. For example:

$$^{64}_{29}\text{Cu} + e \longrightarrow ^{64}_{28}\text{Ni}$$

The electron is one of the extra-nuclear electrons from the original atom. Most commonly it is one of the inner electrons, and as the inner electron is replaced by an outer electron, x-rays are emitted. Whenever the nucleus is left in an unstable state after capture, γ rays are also emitted.

When a nucleus emits γ radiation, it is possible that some of the γ rays may be absorbed by one of the extra-nuclear electrons in the atom. If this occurs, the electron may be ejected from the atom with considerable kinetic energy. This process is known as *internal conversion* and may be exemplified by

$$^{60}_{27}\text{Co} \longrightarrow ^{60}_{27}\text{Co}^{+} + e^{-}$$

The electron leaves a charged Co^{+} which must combine with another electron to resume electrical neutrality. (Note that since the electron involved is extra-nuclear, there is no change in atomic number.)

Internal conversion electrons may be distinguished from β particles in that they have a single energy rather than a range from zero to a maximum. Theoretically, they they could appear with any γ emission. Possibly they do, but where the γ ray energy is greater than 0.5 MeV, the probability of internal conversion is usually very low. With lower energy γ rays, the fraction of γ rays causing internal conversion may become quite large.

All the disintegration or decay processes described are the result of energy changes, which eventually end up in the production of a stable nucleus. The excess energy of the unstable nucleus is released by one or more of these processes according to characteristic rates. These decay rates are discussed later in this section.

X-Radiation

The production of x-rays is not a spontaneous process. Also, in contrast to nuclear radiation, it is not the nucleus that is involved but the extra-nuclear electrons. These electrons may exist in many different energy states or levels. If this level is unstable, the electron will return to a lower energy level, and the atom will emit the energy in the form of radiation. Depending on the quantity of energy involved, the radiation may appear as visible light, UV radiation, or x-rays. X-rays have the highest energy, and are consequently the most penetrating and present the greatest hazard, although any of the others may cause damage to the skin.

The excitation energy for the production of x-rays is derived from the acceleration of a stream of electrons by a high electric potential. The stream of electrons emitted by the heated cathode is accelerated and strikes a target. Most of these electrons have their energy dissipated as heat in the target, but a fraction of them excite an extra-nuclear electron in an atom of the target material and produce x-rays.

Formerly, x-rays were considered to have lower energy than γ rays, but the advent of multimillion-volt x-ray installations has brought about an overlapping of their energy ranges. The true distinction lies not in the energy but rather in the origin. Gamma radiation comes spontaneously from nuclear transformations, while x-rays are produced by excitation of extra-nuclear electrons. Like γ radiation, x-rays have no mass or charge, only energy. The energies of the x-radiation produced by a tube depend on the target material and the acceleration voltage. Any target gives a continuous band of x-rays with peak intensities at energies characteristic of the target material.

Their maximum energy cannot be greater than the energy of the accelerating potential, and it is customary to describe x-rays in terms of the voltage used for acceleration, for example, 75 or 200 keV. Actually, only a portion of the x-radiation produced would have energies of 75 or 200 keV and a considerable quantity of softer (lower energy) x-rays are always emitted. The intensity of an x-ray source is usually expressed not in terms of a number of x-rays of a particular energy produced but in terms of the electric current flowing in the anode or target circuit. The description of x-radiation in terms of the instrument characteristics of voltage, current, and target material may be roughly translated into the more fundamental physical units as will be shown later. Hazards and protective measures in this system will be discussed later.

The Natural Radioisotopes

The majority of the naturally-occurring radioisotopes are members of three series of elements. These series all have a long-lived radioisotope as the first member, pass

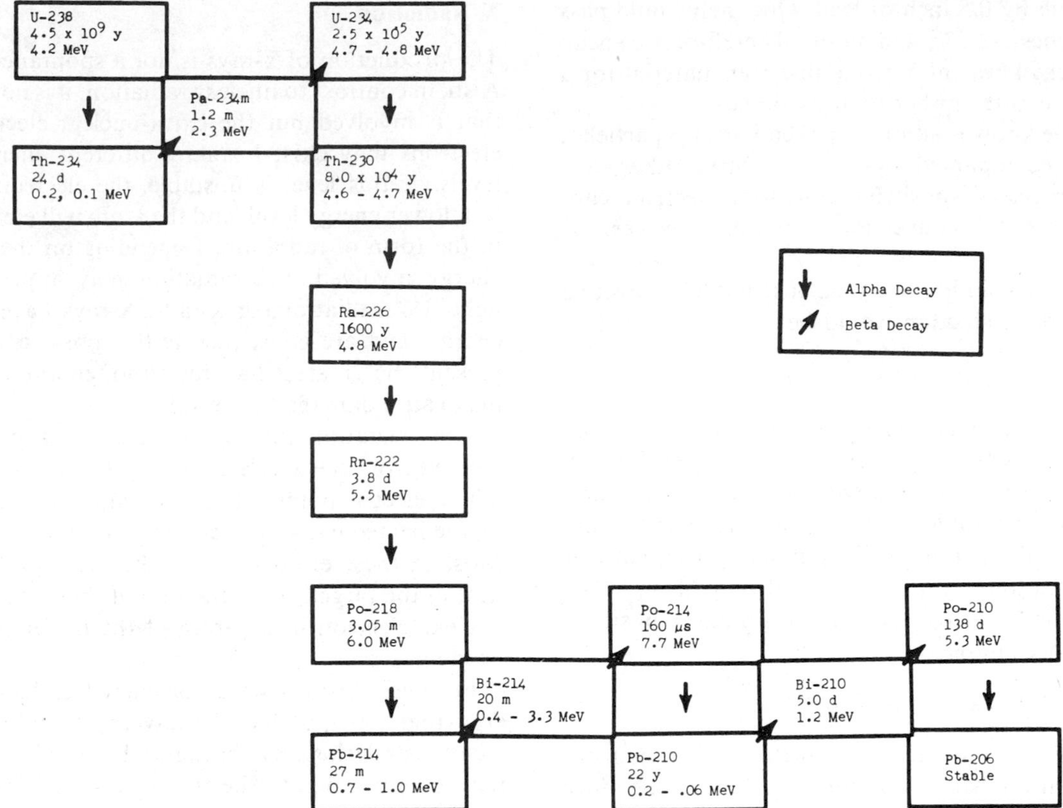

Figure 5A.1. Principal decay scheme of the uranium series.

through a sequence of radioactive disintegrations involving α, β, and γ emissions, and end up as a stable isotope of lead. The principal decay schemes for the three natural series are shown in Figs. 5A.1, 5A.2 and 5A.3.

Other important, naturally-occurring radioisotopes are ^{3}H, ^{14}C, and ^{40}K. These undergo simple decay, yielding stable isotopes. ^{3}H and ^{14}C are not long-lived on the scale of uranium and thorium. They are always present, however, because they are continually being produced in the atmosphere by cosmic ray bombardment.

Artificial Radioisotopes

Many manmade isotopes are now available. They include not only new isotopes of the natural elements, but entirely new elements as well. The new elements are those having an atomic number greater than that of U (92) and are called the transuranic elements. The most familiar one is plutonium produced by neutron irradiation of $^{238}_{92}U$ in a reactor.

The neutron bombardment of ^{238}U produces another radioactive series, the Np series which decays in a fashion similar to the natural series described above.

The nuclear reactor is designed to sustain a controlled fission process, where a fissionable element is split into two lighter elements. For example:

$$^{235}_{92}U + ^{1}_{0}n \longrightarrow ^{95}_{38}Sr + ^{139}_{54}Xe + 2^{1}_{0}n$$

The reaction produces about 200 MeV of energy as well as 1–3 neutrons. These neutrons in turn cause fission in more atoms of the fissionable material (fuel). This chain reaction starts either by injecting neutrons or by spontaneous fission of a few atoms of fuel and is controlled by absorbing neutrons in excess of those required to maintain the reaction. The reactor is thus a source of a high flux of neutrons for irradiation.

The products of fission that may be separated chemically from the spent fuel consist of over 30 elements. The original isotopes are unstable and undergo a series of 1–6 β decays before a stable isotope is formed. The average fission series consists of three or four members, so as many as 100 isotopes may be present in a sample of mixed fission products. The different elements may be separated chemically, but this process does not separate isotopes of the same element.

The major sources today of individual artificial radioisotopes are neutron irradiation (in a reactor) or bombardment in an accelerator such as a cyclotron with protons or other charged particles. A typical transformation obtainable in a reactor, using the notation previously given, would be:

$$^{31}_{15}P + ^{1}_{0}n \longrightarrow ^{32}_{15}P$$

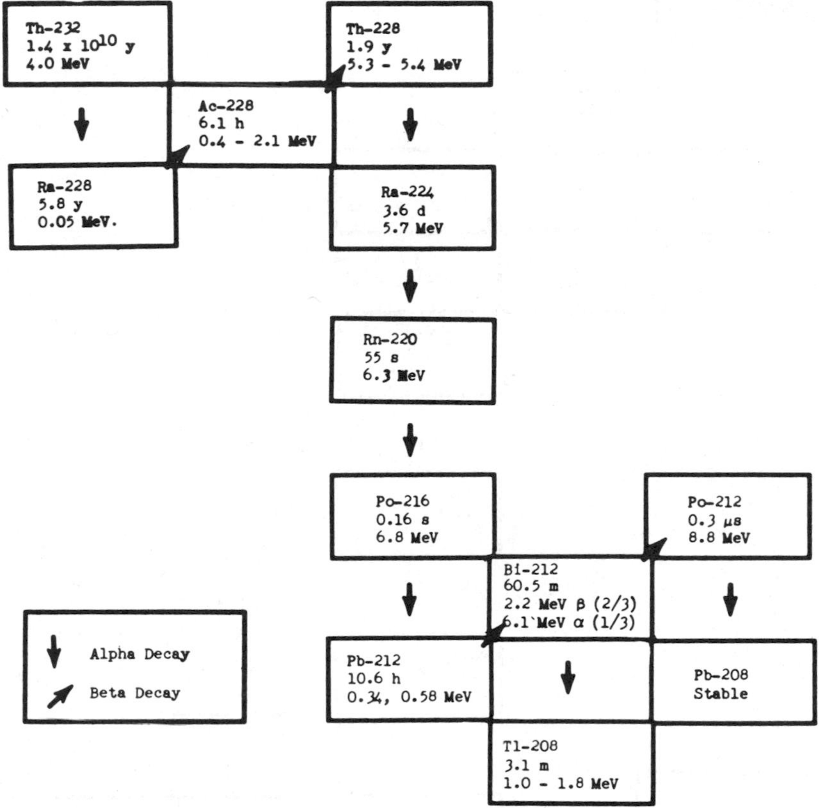

Figure 5A.2. Principal decay scheme of the thorium series.

producing an unstable phosphorus isotope which decays by emission:

$$^{32}_{15}P \longrightarrow ^{32}_{16}S + e$$

A typical cyclotron-produced reaction would be

$$^{7}_{3}Li + ^{1}_{1}H \longrightarrow ^{7}_{4}Be + ^{1}_{0}n$$

producing a beryllium isotope which decays by electron capture:

$$^{7}_{4}Be + e \longrightarrow ^{7}_{3}Li$$

These processes of irradiation and bombardment as well as the chemical separation of specific elements from fission products and from natural series produce radioisotopes of practically every element. The use of these isotopes in the laboratory, in medicine, and in industry has brought many benefits, and while most of them are hazardous materials, they can be handled safely.

The radioisotopes covered by federal regulations and their radiation properties are given under each element in Section 12.

Radioactive Decay

Radioactive isotopes may be characterized by the rate at which they disintegrate. All isotopes follow the same law of disintegration but the rates differ. This law states that a fixed fraction of the number of atoms present disintegrates in a unit time. The fraction, which differs for different isotopes, is known as the disintegration constant and is a fundamental constant for each radioisotope. Another way of expressing the same thought is that for each isotope there is a period of time during which half of the atoms initially present will disintegrate. This time is defined as the half-life of the radioisotope.* Thus, if a given number of atoms is present at a particular instant, half of them will have decayed after one half-life, $\frac{3}{4}$ after two half-lives, and so on. Mathematically the number of atoms remaining at any given time may be expressed

$$N = N_0 e^{-\lambda t}$$

where

N_0 = the original number of atoms
λ = the disintegration constant
t = the time

(t and λ must be expressed in consistent units).

The half-life relationship may be arrived at from the above equation as

$$T_{0.5} = 0.693/\lambda$$

*All values for half-life and energy in this section and in Section 12 were selected from Lederer, C. M., Hollander, J. M., and Perlman, I., *Table of Isotopes* (6th edition). Wiley, New York, 1967.

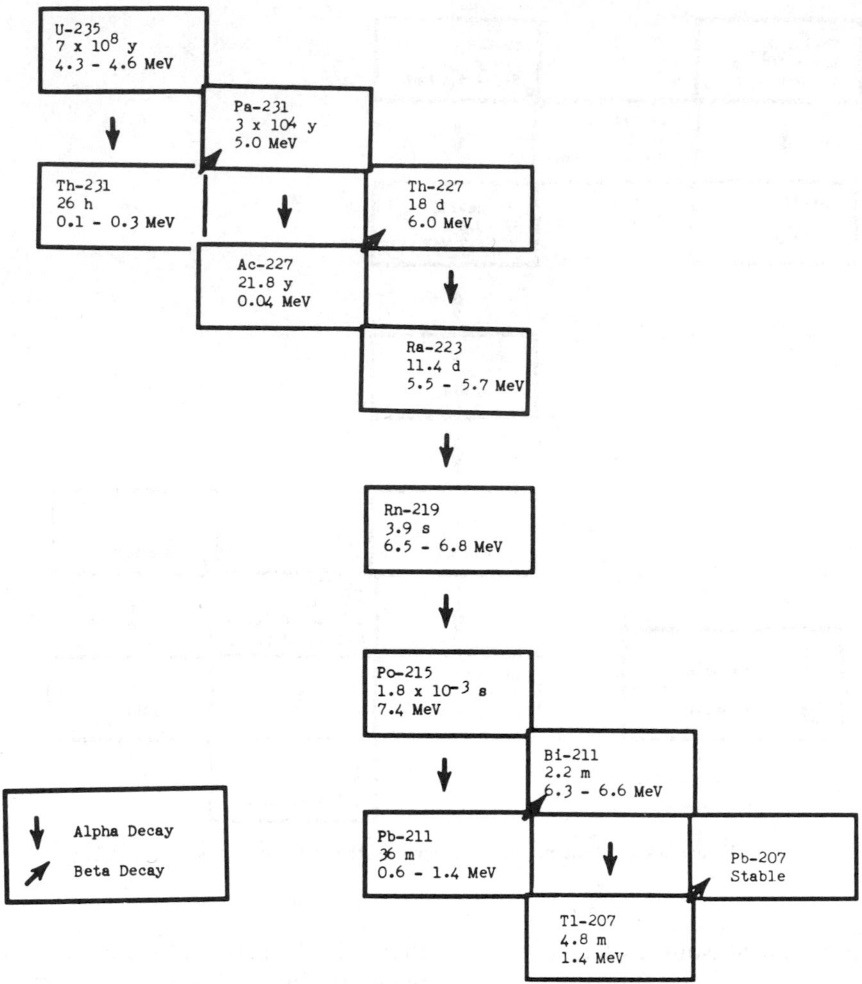

Figure 5A.3. Principal decay scheme of the actinium series.

This simple relationship between half-life and disintegration constant is quite useful since λ is the basic unit used by the physicist, the $T_{0.5}$ is the common unit. The actual half-lives of the known isotopes range from less than 10^{-6} second to more than 10^{17} seconds. The half-life for each radionuclide is constant since the rate of radioactive decay is independent of physical variables such as temperature, pressure, or concentration.

Radioactive Series

As indicated in Figs. 5A.1, 5A.2, and 5A.3 the parent of a natural radioactive series undergoes a series of disintegrations before reaching a stable form. Shorter series frequently occur in the fission products where the so-called fission product chains usually have 3 or 4 members. The importance of this stepwise disintegration is that the radiation effect may be greater than from the parent alone or the chemical differences arising as decay proceeds along the chain may be of metabolic significance. These factors cannot be discussed here, but it is necessary to keep in mind that such decay series exist and that they may require consideration in hazard evaluation.

Probably the most significant chains are those following ^{222}Rn and ^{220}Rn. Radon is a gaseous element produced by the decay of radium. As a gas it may emanate from the solid material containing the Ra and be distributed in the air of the workroom or the atmosphere. When the Rn decays, the daughter products are solid and these, in turn, decay to additional solid members of the chain. Chemically, the principal elements are Po, Pb, and Bi. On formation, the atoms generally attach themselves to small aerosol particles and their hazard is considered to be by inhalation. The shorter-lived daughters are always readily measurable on air filters while ^{210}Pb and ^{210}Po are found in all vegetation, in foods, and in man. They are distributed by a process of natural fallout that carries them to the soil and to plant surfaces.

Environmental Radiation

Radiation in the environment comes from both natural and manmade sources. The natural external radiation includes γ radiation from K, the members of the Th and U series present in the ground, and cosmic radiation from outer space. Internally, the body contains small amounts of Ra, U, and the daughter products of Rn, as well as K, ^{14}C, and minor radioactive elements. The manmade contribution is chiefly radioactive fallout from nuclear weapons tests. The contributions from nuclear accidents or from waste disposal into the environment have so far not been a measureable contribution to worldwide exposure.

If we consider the possible radiation hazards from a single source such as a reactor, we must differentiate sharply between normal operations and accidents. In the course of normal operation, the releases should be minimal and, depending on the type of reactor, would consist of tritium, ^{41}A, ^{85}Kr, and possibly traces of fission products, particularly volatile radioisotopes of intermediate half-life such as ^{131}I. These are normally monitored directly at the source, the point of highest concentration. Reactor accidents have an extremely small probability, and reactors are designed so that most of the expected accidents will not release significant amounts of radioactivity into the environment. If an accident were to occur, the expected release would probably be the more volatile fission products such as the gases and radioiodine.

While accidents are extremely unlikely, it is necessary to guard against the gradual build-up of manmade radioactivity in the environment from chronic low-level releases. This could eventually result in problems comparable to those presently encountered in air and water pollution by nonradioactive materials.

Absorption of Radiation

The single property shared by all forms of radiation is energy. Any radiation is a form of energy, and it is the dissipation of this energy in matter that causes biological damage or that allows us to detect and measure radiation. X-rays and γ rays are electromagnetic energy; the α and β particles possess kinetic energy. While these two types of energy are quite different, the effect they produce in matter is much the same.

In passing through matter, nuclear and x-radiations lose their energy principally by causing ionization in the absorbing material. In this process, neutral atoms of the absorber are ionized, yielding an electron and a positively charged ion. The electrons may acquire sufficient kinetic energy so that they cause further ionization in the absorber. The original ion-electron pairs are spoken of as primary ionization, and those produced by the electrons as secondary ionization. Each ion pair requires the same amount of energy for production; the excess energy imparted in any ionizing event appears as kinetic energy of the electron which then causes secondary ionization. Thus a certain amount of energy when dissipated in an absorber will produce a certain number of ion pairs. This is true whether the energy is introduced as x-rays or as nuclear radiation.

The average energy required to produce an ion pair in air is 34 eV for either primary or secondary ionization. The total energy of any particle or ray will be expended in producing these ion pairs, so that the number of ion pairs resulting from any radiation may be calculated.

When the ions and electrons recombine, the ionization energy will appear in the absorber as heat. In fact, one method of determining the total energy produced by strong sources is by measurement with a calorimeter, an instrument designed to determine the amount of heat produced in a process.

Particulate radiation, α and β, has a finite range in an absorber and this range is dependent on the energy of the particle. One method of determining the approximate energy of α or β radiation is to determine the amount of absorber which just absorbs the radiation completely.

Electromagnetic radiation, x-rays and γ rays does not have a finite range but is reduced in intensity as described earlier. The half-thickness or amount of absorber required to reduce the radiation intensity to one-half is a characteristic of electromagnetic radiation dependent on the energy of the rays. Determination of the half-thickness may be used to measure the energy of x- and γ radiation.

RADIATION UNITS

In relating the measured values of isotope radiation to the predicted or actual biological effects produced by this radiation, two distinct types of units have arisen. The first, based on the number of radioactive disintegrations per unit time, is a measure of the quantity of the isotope present. This is, in effect, a unit of radiation flux, whether measured at the source or at a considerable distance. The second type of unit is one of radiation dose, measured in terms of the quantity of radiation that is absorbed, and is a more direct measure of the possible biological effect.

The relationship between flux and dose units for radioisotopes is not simple and cannot be readily calculated with any degree of accuracy. Such a relationship can be determined empirically for an individual isotope but is only valid under conditions which are identical with those used in determining the relationship.

A similar relationship of flux and dose exists for x-rays. This section will define and explain the various systems of units in common usage for radiation intensity, energy, and dose.

Units of Radiation Intensity

The units of radiation flux differ for x-ray and isotope radiation. However, the basic principle is that radiation flux measures the strength of the source. In the case of x-rays, this can only be expressed in terms of the amount of energy dissipated by the x-ray tube. The common units are milliamperes (ma) of plate current flowing in the tube. This energy includes both that emitted as x-radiation and that dissipated as heat within the tube. However, if the operating voltage, plate current, and efficiency of the x-ray tube are known, the radiation flux may be calculated; i.e., the product of voltage and current expresses the wattage dissipated, and the product of this wattage dissipated, and the product of this wattage and efficiency would give the wattage of x-rays produced.

The energy of the individual x-rays is determined by the target material in the x-ray tube and the operating voltage. Flux, which is a measure of the number of emitted rays, must include the other factors mentioned. However, although the number of x-rays emitted by a source is the most meaningful description, it is not a common characteristic for describing flux.

In the case of isotopes, the source flux (activity) means the number of radioactive disintegrations occurring in the source per unit time. As in the case of x-rays, this is the total of the radiation emitted and dissipated within the source itself. The actual emission can only be determined by measurement and the correction factor obtained would include both the geometrical effects and the absorption of the radiation within the source.

The practical units for radioisotopes are disintegrations per minute (dpm) or per second (dps). More intense sources may be expressed in terms of the curie, which is equivalent to 2.2×10^{12} dpm. Smaller units such as the millicurie (mCi), microcurie (μCi) and picocurie (pCi) are 10^{-3}, 10^{-6}, and 10^{-12} curie, respectively. Again it should be emphasized that these units express only the source activity or the quantity of radioisotope present and not the radiation dose which this quantity would produce.

The new International Standard (SI) system of units is being introduced over a period of years. The SI unit of radioactivity is the bequerel (Bq), equal to 1 dps. but the curie should remain the common unit for many years.

Units of Energy

The energy of any radiation is expressed in terms of electron volts. An electron volt (eV) is the energy which would be acquired by an electron accelerated by an electrical potential of one volt. In terms of the more common units of physics, the electron volt is 1.60×10^{-12} erg. This unit is so small that energies are usually given in thousands of electron volts (keV) or milions of electron volts (MeV). The introduction of high-powered accelerators has also brought billion electron volts (BeV) into the physicist's vocabulary. This latter unit is also called the giga electron volt (GeV) to avoid confusion between the U.S. (10^9) and European (10^{12}) billion.

The usual range of energies for the common radiations are tabulated below:

Radiation	Energy
X-rays	0–3 MeV
Alpha particles (α)	4–8 MeV
Beta particles (β)	0–4 MeV
Gamma rays (γ)	0–2 MeV

These ranges include most of the radiations from the common isotopes and x-ray equipment, but higher energies of each radiation do exist.

The energies shown are for single particles or rays. The total energy rate for emission of a source would be the product of the individual energy and the number of particles or rays per unit time. This product would give a different measure of source activity (radiation flux), as, for example, million electron volts per minute. The SI unit of energy is the joule (J). As with the other SI units, the joule will not replace the more common units of radiation energy for many years.

Units of Radiation Dose

Radiation dose is a measure of the amount of energy that is absorbed by the material being irradiated. *This is the only quantity which can be related to biological effect*. The instruments and measurements for health protection therefore should be designed in terms of radiation dose. The basic quantity is the amount of energy liberated in a unit mass of material and the unit is the roentgen (R). This is a purely arbitrary unit which has been set up as the quantity of x- or γ radiation such that the associated ionization per cubic centimeter of air at standard conditions will produce 1 electrostatic unit (esu) of electricity of either sign. A more useful definition is that the roentgen is the quantity of x- or γ radiation which results in the absorption of 83.4 ergs/g of air.

This unit was set up at a time when only the dosage from x- or γ radiation was considered to be important and the equivalent absorption for tissue was approximately the same as that for air. This is not true for β and α radiation.

In addition, the biological effects of a given quantity of absorbed energy were supposed to be higher for α particles, neutrons, and protons than for the others. The factor for α radiation was set at 20 (later 10, now 20 again) and a new unit was devised to incorporate this factor. For α radiation in particular the Dose Equivalent (rem) is defined as the quantity of radiation which, when absorbed in tissue, produces an effect equivalent to the absorption of 1R of x- or γ radiation. The factor of 20 is called the *quality factor* (QF) and is used as a multiplying factor. For example, if measurement in air of α radiation showed 1R, the possible biological effect would be more closely approximated if the result were given as 20 rem. The QF for β radiation is 1.

The International Commission on Radiation Units (ICRU) in 1953 adopted the rad as the unit for absorbed dose. This unit is defined as the equivalent of 100 ergs/g absorbed in the material of interest. Since this quantity does not include any factor for QF, the type of radiation involved must be specified. For soft tissues, the numerical value for rads is sufficiently close to the numerical value of R for protective purposes.

By applying these and various later terms, it was hoped to develop a system for expressing dosages which would be additive when exposure to several types of radiation had occurred. The SI unit, not yet in common use is the gray (Gy) which is equal to 1 joule per kilogram. For the present, the rad is the accepted unit of absorbed dose and the rem the accepted unit of dose equivalent.

Two distinctions must be kept in mind. The first is that measurements are ordinarily made in terms of dose rate, i.e., R (rems, rads)/hour, while chronic biological effects are assessed in terms of integrated dose, i.e., total R (rems, rads). The second distinction is that the R or rad is defined in terms of energy absorption per unit weight or volume. Thus, a total body dose of 100 mR would indicate a much greater total quantity of radiation absorbed than a 100 mR dose to a hand or an arm. No description of dose is complete unless it tells the amount of material or portion of the body involved.

The instruments and measurements for health protection should be designed in terms of radiation dose. This is rather complex in execution since it is difficult to produce a detector which absorbs radiation in the same way as air or tissue. A dose measurement made with a simple instrument therefore must be carefully interpreted to yield valid estimates of dose to man.

While the units of intensity and energy of radiation are readily understood, the units of dose are somewhat more difficult to put into perspective. For this reason, a comparison will be made here of various doses.

The permissible levels for occupational radiation exposure which have been adopted are based on the value of a maximum of 5 rem/yr to the whole body.* With our present knowledge, it is believed that this level can be absorbed by a man for a working lifetime without the appearance of any sign of body damage. An idea of the magnitude of the permissible level can be obtained by a study of Tables 5A.1, 5A.2, and 5A.3, which show various radiation dosages and effects.

Table 5A.1, taken from the NCRP report on natural background radiation,[1] cites the normal radiation levels from natural sources to which everyone is exposed. Table 5A.2 from the U.N. Scientific Committee of the Effects of Atomic Radiation[2] shows the radiation exposure received in various diagnostic x-rays. Table 5A.3 shows the effects that can be expected from various large instantaneous doses of gamma radiation.[3]

*permissible levels are covered in detail in a later part of this section.

TABLE 5A.1. Average Dose Equivalent Rates from Natural Background in the United States (mrem)

Type	Gonads	Lung	Bone Marrow
Cosmic radiation	28	28	28
External terrestrial	26	26	26
Inhaled radionuclides	—	100	—
Radionuclides in the body	28	25	25
Rounded total	80	180	80

TABLE 5A.2. Gonad Dose from Diagnostic X-rays

Barium enema	600 mrad
Pelvis	300
Obstetrical abdomen	300
Mass survey chest	2
Dental	½

EFFECTS OF RADIATION

The effects of radiation are largely based on the ionization produced when the energy of the radiation is absorbed in matter. While the physical and even some of the biological mechanisms are understood, many of the biological mechanisms and the ultimate radiation damage in humans are not.

Physical Effects

Since the same physical processes are involved in the absorption of radiation, whether the absorber is air, metal, or tissue, we must first consider the general action of radiation when it is absorbed in any material.

The primary effect of the absorption of radiation in matter is that of ionization. Atoms of the absorber are given sufficient energy so that they ionize, i.e., an electron is removed from the atom leaving a positively charged residue or ion. This ionization takes place in any substance which absorbs radiation. Very roughly, if the amount of absorber in the path of the radiation is expressed in terms of the weight per unit area, the degree of absorption is independent of the material.

For both scientific and practical purposes the common units are mass/unit area of absorber. This quantity is obtained by multiplying the thickness of absorber, usually expressed in centimeters (cm), by the density, expressed in grams per cubic centimeter (g/cm^3). This product will be in units of grams per square centimeter (g/cm^2). For example, if the β radiation from a source

TABLE 5A.3. Summary of Effects Resulting from Acute Whole Body External Exposure to Radiation

0–25 rems	25–100 rems	100–200 rems	200–300 rems	300–600 rems	600 or more
No detectable clinical effects. Delayed effects may occur.	Slight transient reductions in lymphocytes and neutrophils. Disabling sickness not common, exposed individuals should be able to proceed with usual duties. Delayed effects possible, but serious effects on average individual very improbable.	Nausea and fatigue, with possible vomiting about 125 rems. Reduction in lymphocytes and neutrophils with delayed recovery. Delayed effects may shorten life expectancy in the order of 1 percent.	Nausea and vomiting on first day. Latent period up to 2 weeks or perhaps longer. Following latent period symptoms appear but are not severe: loss of appetite, general malaise, sore throat, pallor, petechiae, diarrhea, moderate emaciation. Recovery likely in about 3 months unless complicated by poor previous health, superimposed injuries or infections.	Nausea, vomiting and diarrhea in first few hours. Latent period with no definite symptoms, perhaps as long as 1 week. Epilation, loss of appetite, general malaise, and fever during second week, followed by hemorrhage, purpura, petechiae, inflammation of mouth and throat, diarrhea, and emaciation in the third week. Some deaths in 2 to 6 weeks. Possible eventual death to 50 percent of the exposed individuals for about 450 rems.	Nausea, vomiting and diarrhea in first few hours. Short latent period with no definite symptoms in some cases during first week. Diarrhea, hemorrhage, purpura, inflammation of mouth and throat, fever toward end of first week. Rapid emaciation and death as early as the second week with possible eventual death of up to 100 percent of exposed individuals.

would be completely absorbed by 1.1 g/cm^2 of aluminum it would also be absorbed by 1.1 g/cm^2 of air. However, the absorber thickness would be only about 0.25 inch for the aluminum and over 30 feet for air.

The result of ionization is merely a conversion of the radiation energy into another form of energy within the absorber, and it is these secondary effects which are of the greatest importance in radiation protection work.

The primary effects of ionization and the distribution of this ionization over various path lengths in different absorbers have been mentioned previously. The different types of radiation also show different degrees of absorption and these differences also are biologically significant. Alpha particles are heavy, slow moving, and expend their energy in a relatively short path. They are, therefore, spoken of as showing high specific ionization, i.e., a large number of ions are formed per unit length of path in the absorber. Gamma and x-radiations, on the other hand, require a great thickness of absorber for complete absorption. Gamma rays and x-rays have a low specific ionization, i.e., the ionization is spread out over the relatively long path required for complete absorption. Beta particles are intermediate in their specific ionization.

Biological Effects

The biological effects of radiation are considered here only in sufficient detail to be of assistance in problems of radiation protection. Some of the information is also required for an understanding of the concepts that have gone into the formulation of permissible levels.

X-rays or γ rays, because of their penetrating nature, may dissipate only a fraction of their energy in passing through the body. This is particularly true of high-energy rays. The energy dissipated is, of course, the absorbed dose delivered to the body or portion of the body.

Radioisotopes, in contrast, may present a further hazard when the material is taken into the body where it irradiates the tissues or organs internally. The most serious effects from this standpoint are produced by the α emitters such as Ra, U, and Pu. They are particularly marked because α emitters outside the body expend their energy either in penetrating the clothing or the dead cells of the epidermis; usually the radiation cannot penetrate to living cells. Once they are taken into the body via ingestion or inhalation, this same property of short-range and high specific ionization increases their relative effect considerably. Emitters located in a small section of tissue will irradiate that small section very heavily.

Beta emitters can be both an internal and an external hazard. The range of most external β radiation is great enough that the outer tissues, at least, will be penetrated. The most common external effects have been radiation burns and malignancies of the skin. Internally, they may produce a considerable effect. Their specific ionization is high although not as great as that for α radiation.

The preceding paragraphs have emphasized the ionization effects, particularly specific ionization. Many secondary effects can be caused by the ionization process. It may disrupt molecules, it may destroy body cells, or the energy may merely appear in final form as heat released within the absorber. Depending on the location of the absorbing atom within the molecule, the ionization may or may not disrupt the molecule. If this molecule is in a critical place within the cell, the cell, its function, or its ability to reproduce itself may be destroyed. Many of these processes are reversible; that is, damage caused by molecule disruption or cell destruction can be reversed by the usual reparative mechanism of the body. This is confirmed by experimental data which show that a fixed total dose spread out over a period of weeks produces a smaller effect than the same dose delivered in a few minutes. However, in the case of a large acute dose or continued chronic overexposure, there is the possibility that non-reversible damage will occur.

Another type of cell change which is possible is that the regulative functions of a tissue may be destroyed. In this case a carcinoma (cancer) may be produced. Although the mechanism is not fully understood, there is direct evidence that continued insult to a tissue may produce this result. The high rates of leukemia among radiologists, bone cancer among Ra dial painters, and lung cancer among miners of the Czechoslovakian, German, and U.S. uranium mines all point to radiation as the causative agent. This irreversible damage in chronic radiation exposure was apparently cumulative and the cumulative effects led to the illnesses.

Internal Emitters. The biological effects of radiation from radioisotopes in the body are complicated by several factors. In any determination of radiation effects, whether in working populations or in animal experiments, the following factors must be considered: (1) the location of specific isotopes in the body, and (2) the relative sensitivity of different tissues to radiation.

The general effects of external radiation have been previously described but there are certain modifications in the consideration of radiation from internal sources. The first is that different elements tend to localize in different organs of the body, e.g., calcium or strontium in bone, iron in the red blood cells, and iodine in the thyroid. This is true for any material which is metabolized following either inhalation or ingestion. Of course, many not readily soluble substances will remain in the lungs for long periods after inhalation. This means that the total amount of such a radioactive material is not distributing its dose uniformly but rather is concentrating its effect on a relatively small fraction of the body.

Most of the heavy metals tend to be deposited in the bone structure. After deposition, there is usually a continuous excretion of the isotope which gradually reduces the amount present. The excretion rate of such materials has been considered to follow much the same pattern as the radioactive decay of an isotope. The time required by the body to eliminate one-half the total quantity it contains is thus referred to as "biological half-life." Most of the experimental data on excretion seem to fit a power function which is the resultant of a number of exponentials rather than a simple exponential function, but the concept of biological half-life is still used in deriving permissible levels.

Such body deposits may depend on many physiological factors both in the process of deposition and of excrecretion. For many years a high calcium diet was recommended for radium workers, as it was supposed that a large excess of calcium entering the body would reduce the amount of Ra deposition. Actually, the relative radium deposition is a function of the ratio of radium to calcium in the blood stream. Unless the calcium level of the blood is maintained at a very high value there will still be deposition of radium. The increase in the blood calcium required to cut the radium deposition by even a factor of three would be impossible to attain.

Besides the bone structure, common sites of deposition are the lungs and lymph nodes for inhaled particles, and specific organs for certain isotopes, such as the thyroid for iodine and spleen for iron.

A second consideration is that certain organs or tissues are more radiosensitive than others. The membranes lining the bronchi are supposedly quite sensitive to radiation and this is the primary site of many lung cancers attributed to inhaled radioactive material. The spleen is also sensitive to radiation and relatively small doses have produced more irreversible damage in that organ than in other parts of the body.

The organ most likely to be damaged because of the combined effects of concentration and radiosensitivity is known as the critical organ for a particular isotope. In general, any cell in the process of division (mitosis) is radiosensitive and for that reason a person is more sensitive to radiation during his growing period than as an adult.

Radiation Injury. The effects of radiation are nonspecific; i.e., other agents or diseases can cause the same damage. For example, it is impossible to distinguish between radiation-induced anemia and normally incident anemia. Other possible effects such as lung cancer, leukemia, and bone cancer present similar difficulties.

In any case, where the effects of radiation are being studied, conclusions can only be drawn on the basis of incidence of a particular type of damage above that normally occurring in a comparable population. If tabulations are made of incidence in a particular group, such as chemical operators exposed to radiation in a process plant, the radiation effects can only be evaluated by a statistically valid comparison with a similar group exposed to the same chemical hazards but not the radiation.

Different people exhibit differing degrees of sensitivity to toxic agents (see Section 9). This is also true of radiation effects. Many of the early injuries to radium dial painters occured with relatively small amounts of radium deposited in the body. Other workers with many times as much radium deposited in their bodies showed no injury even after many years.

The direct effects of ionizing radiation with which we are concerned are those on the cells or tissues exposed to the radiation. The five principal damaging effects:

(a) Superficial injuries such as skin damage or erythema
(b) General effects on the body, particularly the blood-forming organs, and non-specific shortening of life span.
(c) Induction of cancer
(d) Miscellaneous effects such as cataracts or impaired fertility
(e) Genetic effects

All of the effects mentioned are possible with excessive exposure to ionizing radiation. The first four, the somatic effects, were the basis for setting the original permissible dose levels for occupational exposure.

Not only the total dose, but the rate at which it is delivered will influence the nature of the effects. For example, about 250 R would be necessary to give a superficial radiation burn from x-rays in a single dose to the skin. The lifetime dose permitted under present standards is about 250 R to the whole body spread over 50 years, and this should produce no measureable effects.

All of the experience resulting in human injury has been gained from large doses of radiation. The effects of low doses of radiation are still not clear. Many scientists believe that all toxic materials including radiation require a certain minimum threshold dose before effects can appear. This is possibly true of somatic effects, i.e., injury to the individual receiving the dose, although this is not conceded by all. In the case of genetic effects, it is considered that a linear relationship exists—a very small dose of radiation will produce a very small effect. It is not possible to demonstrate this linear hypothesis since the expected effects could only be shown on a statistical basis by using large populations. The hypothesis cannot be disproved, and it is presently accepted as a conservative assumption in evaluating radiation hazards. This has led to the general concept of comparative risk as a basis for deciding the permissible amounts of radiation to large populations.

PERMISSIBLE LEVELS

A Threshold Limit Value (TLV) is a recommendation on exposure to a toxic material or a hazardous condition (see Sections 1, 2, and 9). In chemical exposures, it is clear that there is a threshold below which no damage occurs. In the case of radiation, this is not fully accepted, and the maximum (mpl's) permissible levels of exposure are considered as levels which constitute an acceptable risk.

The initial experience with internal radionuclides came from the radium dial painters who tipped their brushes with their tongues. The dosage estimated in those with apparent damage led to the selection of the permissible body burden, 0.1 μCi of radium and this in turn was used to develop comparable standards for other radionuclides in the body, particularly α emitters. In the same way, the early experience of radiologists led to the acceptance of permissible levels for x-rays. The first permissible levels used were in terms of the erythema dose—the dose causing reddening of the skin. This obviously did not allow for the delayed affects of radiation and the permissible level has been reduced considerably since this early value was used.

An mpl can be set up in two ways: One is on the basis of the largest quantity of radiation or radioactive material which is known to have caused no damage to the exposed individual; the other is the smallest amount that is known to have caused damage to exposed individuals. Fixing permissible levels could be done best if both the maximum and minimum figures were known. Our experience has been that the maximum value is usually obtained from known exposures of working populations, while the minimum values are obtained from animal experimentation.

Of course, in the cases of x-rays and ingested radium, the minimum values also come from human experiences. The weakness in evaluating human exposure is that the hazard is not recognized until some clinical evidence of damage appears, and at that time, satisfactory records of exposure are often not available. Animal experimentation is unsatisfactory because there must be a final extrapolation from animal experience to an estimate of what might be expected in humans. However, in spite of the failings of the available methods of determining mpl's, the values which have been set can serve as a standard for operation.

The approach to setting maximum levels in the case of radiation and radioisotopes has been more unified than the approach to setting levels for other toxic materials. The large proportion of work with radioactive isotopes in the period of development either was a direct government function or was under private contract with government agencies. This has meant that a centralized group has been studying the common problems appearing in all fields where radiation is a possible hazard. In addition, recognition by scientists that such a hazard existed has resulted in considerable group and community activity in evaluating hazards.

The reductions in permissible levels that have occurred

since the concept first arose have been based on a change in the type of damage that was considered to be limiting. The reductions in 1958 were largely based on a change from consideration of somatic damage to the individual to possible genetic damage to the population at large. This also involved the knowledge that an increasingly larger proportion of the population as a whole was receiving some exposure to radiation either occupationally or from other man-made sources.

Since 1970 there has also been considerable pressure from scientists and the public for a further reduction in permissible levels. This has generally been proposed on the basis that the effects of low levels of radiation over a lifetime are not sufficiently well known and that caution would indicate that the level should be as low as possible. The authorities have yielded to this pressure to some degree although there is certainly more known about the effects of radiation than the effects of any other natural material or contaminant in our environment.

ICRP

The International Commission on Radiological Protection (ICRP) standards are set as a guide for radiation protection. They define the objectives of radiation protection as the prevention of acute radiation effects and the limiting of the risks of late effects to an acceptable level.

The basic Maximum Permissible Dose for occupational exposure is set at 5 rem/year to the whole body. As will be shown later for other recommendations higher doses are allowed for portions of the body.

It should be remembered that the Maximum Permissible Doses or the Dose Limits are for irradiation other than natural background or medical uses. The ICRP has recommended that measures should be taken to reduce medical exposure when this is possible without loss of information. On the other hand, no such exposure is included in evaluating occupational or population doses from scientific and industrial radiation.

Based on the considerations previously described, a set of permissible amounts in the body (body burdens) and of permissible concentrations of radioactive materials in air and water has been established. The detailed data on distribution of materials in the body and permissible levels for various organs may be found in Publication 2 of the ICRP.[4]

In contrast to previous sets of permissible levels, limits are not given for the rate at which exposure may be accumulated within the annual limit.

The maximum permissible levels set by the ICRP and other groups are for 40 hours exposure weekly over a working lifetime. This must be remembered in comparing single acute exposures with the ICRP values. These acute exposures should be compared with some integrated value over some reasonable period of time such as 13 weeks or even a calendar year. This, of course, is for cases where the exposure will not be repeated.

NCRP

The national authority in the field at present is the National Council on Radiation Protection and Measurements (NCRP). This committee was organized in its present form in 1946 and, in cooperation with other committees and the ICRP, has set the present national working standards for radiation protection. The Council has been organized into subcommittees of specialists in different fields, and these subcommittees have published several reports on different types of radiation hazards. Many current reports are listed in the bibliography and additional ones will no doubt appear from time to time, both on new subjects and on modifications of the current material.

In general, the NCRP has adopted the same levels[5] as the ICRP since there is considerable overlapping of membership. While the recommendations of these two groups do not have the force of law in themselves, they have been adopted in drafting both federal and state codes on radiation protection. In 1971 the NCRP issued their report No. 39 on Basic Radiation Protection Criteria. This upheld their previous standards even though they had heard a number of arguments for reducing the permissible levels. They found that the evidence did not warrant such a decrease.

FRC

Since so many agencies of the U.S. Government are concerned with radiation and radiation protection, the Federal Radiation Council (FRC) was formed in 1959, to provide a federal policy on human radiation exposure. The FRC relied quite heavily on the recommendations of the NCRP and ICRP in its deliberations. It avoided the use of the terms "maximum permissible level" and "maximum permissible concentration" and adopted the following:[6]

> (1) Radiation Protection Guide (RPG) is the radiation dose which should not be exceeded without careful consideration of the reasons for doing so; every effort should be made to encourage the maintenance of radiation doses as far below this guide as practicable.
>
> (2) Radioactivity Concentration Guide (RCG) is the concentration of radioactive matter in the environment which it is determined will result in whole body or organ doses equal to the Radiation Protection Guide.

These guides were expected to provide the target levels for operation of government facilities and licenses. Table 5A.4 gives the RPG values for occupational exposures and for the general population.

Only a few RCG values have been issued, and the permissible concentrations for air and water given in Table 5A.5 are those shown in Part 20 of the CFR, Chapter 1. These are reprinted from the Federal Register as revised to mid-1977.

The exposures considered by the NCRP and ICRP are theoretically under control, i.e., they are recevied under industrial conditions and there is some control of the source. The FRC also attempted to develop guides for uncontrolled situations such as those arising from nuclear weapons tests or nuclear accidents. The principle invoked was the balancing of risk against benefit or the comparative risk of receiving a radiation dose as against the possible consequences of actions taken to avoid the dose. The Federal Radiation Council was dropped in 1970 and its basic functions were largely taken over by the EPA. This agency has not addressed itself to the same questions considered by the FRC but has looked at a number of practical limited situations.

It must be stressed that the permissible levels set by the ICRP, NCRP, and other groups are for continuous exposure over a working life-time at the occupational levels or continuous exposure for a lifetime at the non-occupational levels. Measurements in the working environment or of the surroundings will always show peak values, and some of these may exceed the recommended levels. This is not necessarily serious. It does show that a problem exists, but if the levels are merely a few times the permissible, corrections can be instituted in an orderly manner and not on a crash basis. If the levels are hundreds of times the permissible, then immediate action is desirable but it is still not expected that any individual would suffer radiation damage as long as the exposure is continued for only a day or two. This warning on the use of maximum permissible levels may seem unnecessary but experience has shown that this is a major source of misunderstanding.

The NRC limits the exposure of the workers on the basis of a calendar quarter. The limits in terms of rems/calendar quarter are:

(1) Whole body; head and trunk; active blood forming organs; lens of eyes; or gonads ... $1\frac{1}{4}$
(2) Hands and forearms; feet and ankles ..... $18\frac{3}{4}$
(3) Skin of whole body .... $7\frac{1}{2}$

The exposure of minors below the age of 18 is limited to 10% of above limits and their exposure to airborne material should be limited to concentrations as shown in Table 5A.5 using the values listed in Table II rather than those in Table I allowed for adults.

An additional restriction is also encouraged in the case of the case of women workers in the early stages of pregnancy. The NCRP has recommended that a woman limit her possible exposure to 0.5 rem during the period of pregnancy. This is lower than that allowed other workers. While it is not legal to remove a possibly pregnant woman from radiation exposure it is required that she be informed as to the possible dangers. The required information is available as NRC Regulatory Guide 8.13. Exposures to individuals in unrestricted areas are limited to 0.5 rem/y. Additional restrictions would be a maximum dose rate of 2 mrem/h or 100 mrem/wk.

TABLE 5A.4. Radiation Protection Guides

Type of Exposure	Condition	Dose[a] (rem)
Radiation worker		
(a) Whole body, head and trunk, active blood forming organs, gonads, or lens of eye	Accumulated dose 13 weeks	5 times number of years beyond age 18 3
(b) Skin of whole body and thyroid	Year 13 weeks	30 10
(c) Hands and forearms, feet and ankles	Year 13 weeks	75 25
(d) Bone	Body burden	0.1 μg of ^{226}Ra or its biological equivalent
(e) Other organs	Year 13 weeks	15 5
Population		
(a) Individual[b]	Year	0.5 (whole body)
(b) Average[b]	30 years	5 (gonads)

[a] Minor variations here from certain other recommendations are not considered significant in light of present uncertainties.

[b] The distinction between individual and average values is necessary since under some conditions only measurements of average values will be possible. The assumption is made here that the individual values will not exceed 3 times the average.

TABLE 5A.5. Concentrations in Air and Water above Natural Background
Table I: In restricted areas (occupational)
Table II: Outside restricted areas

Element-Z	Isotope[a]		Table I		Table II	
			Column 1 Air (μCi/ml)	Column 2 Water (μCi/ml)	Column 1 Air (μCi/ml)	Column 2 Water (μCi/ml)
Actinium-89	^{227}Ac	S	2×10^{-12}	6×10^{-5}	8×10^{-14}	2×10^{-6}
		I	3×10^{-11}	9×10^{-3}	9×10^{-13}	3×10^{-4}
	^{228}Ac	S	8×10^{-8}	3×10^{-3}	3×10^{-9}	9×10^{-5}
		I	2×10^{-8}	3×10^{-3}	6×10^{-10}	9×10^{-5}
Americium-95	^{241}Am	S	6×10^{-12}	1×10^{-4}	2×10^{-13}	4×10^{-6}
		I	1×10^{-10}	8×10^{-4}	4×10^{-12}	2×10^{-5}
	^{242m}Am	S	6×10^{-12}	1×10^{-4}	2×10^{-13}	4×10^{-6}
		I	3×10^{-10}	3×10^{-3}	9×10^{-12}	9×10^{-5}
	^{242}Am	S	4×10^{-8}	4×10^{-3}	1×10^{-9}	1×10^{-4}
		I	5×10^{-8}	4×10^{-3}	2×10^{-9}	1×10^{-4}
	^{243}Am	S	6×10^{-12}	1×10^{-4}	2×10^{-13}	4×10^{-6}
		I	1×10^{-10}	8×10^{-4}	4×10^{-12}	3×10^{-5}
	^{244}Am	S	4×10^{-6}	1×10^{-1}	1×10^{-7}	5×10^{-3}
		I	2×10^{-5}	1×10^{-1}	8×10^{-7}	5×10^{-3}
Antimony-51	^{122}Sb	S	2×10^{-7}	8×10^{-4}	6×10^{-9}	3×10^{-5}
		I	1×10^{-7}	8×10^{-4}	5×10^{-9}	3×10^{-5}
	^{124}Sb	S	2×10^{-7}	7×10^{-4}	5×10^{-9}	2×10^{-5}
		I	2×10^{-8}	7×10^{-4}	7×10^{-10}	2×10^{-5}
	^{125}Sb	S	5×10^{-7}	3×10^{-3}	2×10^{-8}	1×10^{-4}
		I	3×10^{-8}	3×10^{-3}	9×10^{-10}	1×10^{-4}
Argon-18	^{37}A	Sub[b]	6×10^{-3}	—	1×10^{-4}	—
	^{41}A	Sub	2×10^{-6}	—	4×10^{-8}	—
Arsenic-33	^{73}As	S	2×10^{-6}	1×10^{-2}	7×10^{-8}	5×10^{-4}
		I	4×10^{-7}	1×10^{-2}	1×10^{-8}	5×10^{-4}
	^{74}As	S	3×10^{-7}	2×10^{-3}	1×10^{-8}	5×10^{-5}
		I	1×10^{-7}	2×10^{-3}	4×10^{-9}	5×10^{-5}
	^{76}As	S	1×10^{-7}	6×10^{-4}	4×10^{-9}	2×10^{-5}
		I	1×10^{-7}	6×10^{-4}	3×10^{-9}	2×10^{-5}
	^{77}As	S	5×10^{-7}	2×10^{-3}	2×10^{-8}	8×10^{-5}
		I	4×10^{-7}	2×10^{-3}	1×10^{-8}	8×10^{-5}
Astatine-85	^{211}At	S	7×10^{-9}	5×10^{-5}	2×10^{-10}	2×10^{-6}
		I	3×10^{-8}	2×10^{-3}	1×10^{-9}	7×10^{-5}
Barium-56	^{131}Ba	S	1×10^{-6}	5×10^{-3}	4×10^{-8}	2×10^{-4}
		I	4×10^{-7}	5×10^{-3}	1×10^{-8}	2×10^{-4}
	^{140}Ba	S	1×10^{-7}	8×10^{-4}	4×10^{-9}	3×10^{-5}
		I	4×10^{-8}	7×10^{-4}	1×10^{-9}	2×10^{-5}
Berkelium-97	^{249}Bk	S	9×10^{-10}	2×10^{-2}	3×10^{-11}	6×10^{-4}
		I	1×10^{-7}	2×10^{-2}	4×10^{-9}	6×10^{-4}
	^{250}Bk	S	1×10^{-7}	6×10^{-3}	5×10^{-9}	2×10^{-4}
		I	1×10^{-6}	6×10^{-3}	4×10^{-8}	2×10^{-4}
Beryllium-4	^{7}Be	S	6×10^{-6}	5×10^{-2}	2×10^{-7}	2×10^{-3}
		I	1×10^{-6}	5×10^{-2}	4×10^{-8}	2×10^{-3}
Bismuth-83	^{206}Bi	S	2×10^{-7}	1×10^{-3}	6×10^{-9}	4×10^{-5}
		I	1×10^{-7}	1×10^{-3}	5×10^{-9}	4×10^{-5}
	^{207}Bi	S	2×10^{-7}	2×10^{-3}	6×10^{-9}	6×10^{-5}
		I	1×10^{-8}	2×10^{-3}	5×10^{-10}	6×10^{-5}
	^{210}Bi	S	6×10^{-9}	1×10^{-3}	2×10^{-10}	4×10^{-5}
		I	6×10^{-9}	1×10^{-3}	2×10^{-10}	4×10^{-5}
	^{212}Bi	S	1×10^{-7}	1×10^{-2}	3×10^{-9}	4×10^{-4}
		I	2×10^{-7}	1×10^{-2}	7×10^{-9}	4×10^{-4}
Bromine-35	^{82}Br	S	1×10^{-6}	8×10^{-3}	4×10^{-8}	3×10^{-4}
		I	2×10^{-7}	1×10^{-3}	6×10^{-9}	4×10^{-5}
Cadmium-48	^{109}Cd	S	5×10^{-8}	5×10^{-3}	2×10^{-9}	2×10^{-4}
		I	7×10^{-8}	5×10^{-3}	3×10^{-9}	2×10^{-4}
	^{115m}Cd	S	4×10^{-8}	7×10^{-4}	1×10^{-9}	3×10^{-5}
		I	4×10^{-8}	7×10^{-4}	1×10^{-9}	3×10^{-5}
	^{115}Cd	S	2×10^{-7}	1×10^{-3}	8×10^{-9}	3×10^{-5}
		I	2×10^{-7}	1×10^{-3}	6×10^{-9}	4×10^{-5}

See footnotes at end of table.

TABLE 5A.5. (*Continued*)

Element-Z	Isotope[a]		Table I Column 1 Air (μCi/ml)	Table I Column 2 Water (μCi/ml)	Table II Column 1 Air (μCi/ml)	Table II Column 2 Water (μCi/ml)
Calcium-20	^{45}Ca	S	3×10^{-8}	3×10^{-4}	1×10^{-9}	9×10^{-6}
		I	1×10^{-7}	5×10^{-3}	4×10^{-9}	2×10^{-4}
	^{47}Ca	S	2×10^{-7}	1×10^{-3}	6×10^{-9}	5×10^{-5}
		I	2×10^{-7}	1×10^{-3}	6×10^{-9}	3×10^{-5}
Californium-98	^{249}Cf	S	2×10^{-12}	1×10^{-4}	5×10^{-14}	4×10^{-6}
		I	1×10^{-10}	7×10^{-4}	3×10^{-12}	2×10^{-5}
	^{250}Cf	S	5×10^{-12}	4×10^{-4}	2×10^{-13}	1×10^{-5}
		I	1×10^{-10}	7×10^{-4}	3×10^{-12}	3×10^{-5}
	^{251}Cf	S	2×10^{-12}	1×10^{-4}	6×10^{-14}	4×10^{-6}
		I	1×10^{-10}	8×10^{-4}	3×10^{-12}	3×10^{-5}
	^{252}Cf	S	2×10^{-11}	7×10^{-4}	7×10^{-13}	2×10^{-5}
		I	1×10^{-10}	7×10^{-4}	4×10^{-12}	2×10^{-5}
Californium-98	^{253}Cf	S	8×10^{-10}	4×10^{-3}	3×10^{-11}	1×10^{-4}
		I	8×10^{-10}	4×10^{-3}	3×10^{-11}	1×10^{-4}
	^{254}Cf	S	5×10^{-12}	4×10^{-6}	2×10^{-13}	1×10^{-7}
		I	5×10^{-12}	4×10^{-6}	2×10^{-13}	1×10^{-7}
Carbon-6	^{14}C	S	4×10^{-6}	2×10^{-2}	1×10^{-7}	8×10^{-4}
	(CO_2)	Sub	5×10^{-5}	—	1×10^{-6}	—
Cerium-58	^{141}Ce	S	4×10^{-7}	3×10^{-3}	2×10^{-8}	9×10^{-5}
		I	2×10^{-7}	3×10^{-3}	5×10^{-9}	9×10^{-5}
	^{143}Ce	S	3×10^{-7}	1×10^{-3}	9×10^{-9}	4×10^{-5}
		I	2×10^{-7}	1×10^{-3}	7×10^{-9}	4×10^{-5}
	^{144}Ce	S	1×10^{-8}	3×10^{-4}	3×10^{-10}	1×10^{-5}
		I	6×10^{-9}	3×10^{-4}	2×10^{-10}	1×10^{-5}
Cesium-55	^{131}Cs	S	1×10^{-5}	7×10^{-2}	4×10^{-7}	2×10^{-3}
		I	3×10^{-6}	3×10^{-2}	1×10^{-7}	9×10^{-4}
	^{134m}Cs	S	4×10^{-5}	2×10^{-1}	1×10^{-6}	6×10^{-3}
		I	6×10^{-6}	3×10^{-2}	2×10^{-7}	1×10^{-3}
	^{134}Cs	S	4×10^{-8}	3×10^{-4}	1×10^{-9}	9×10^{-6}
		I	1×10^{-8}	1×10^{-3}	4×10^{-10}	4×10^{-5}
	^{135}Cs	S	5×10^{-7}	3×10^{-3}	2×10^{-8}	1×10^{-4}
		I	9×10^{-8}	7×10^{-3}	3×10^{-9}	2×10^{-4}
	^{136}Cs	S	4×10^{-7}	2×10^{-3}	1×10^{-8}	9×10^{-5}
		I	2×10^{-7}	2×10^{-3}	6×10^{-9}	6×10^{-5}
	^{137}Cs	S	6×10^{-8}	4×10^{-4}	2×10^{-9}	2×10^{-5}
		I	1×10^{-8}	1×10^{-3}	5×10^{-10}	4×10^{-5}
Chlorine-17	^{36}Cl	S	4×10^{-7}	2×10^{-3}	1×10^{-8}	8×10^{-5}
		I	2×10^{-8}	2×10^{-3}	8×10^{-10}	6×10^{-5}
	^{38}Cl	S	3×10^{-6}	1×10^{-2}	9×10^{-8}	4×10^{-4}
		I	2×10^{-6}	1×10^{-2}	7×10^{-8}	4×10^{-4}
Chromium-24	^{51}Cr	S	1×10^{-5}	5×10^{-2}	4×10^{-7}	2×10^{-3}
		I	2×10^{-6}	5×10^{-2}	8×10^{-8}	2×10^{-3}
Cobalt-27	^{57}Co	S	3×10^{-6}	2×10^{-2}	1×10^{-7}	5×10^{-4}
		I	2×10^{-7}	1×10^{-2}	6×10^{-9}	4×10^{-4}
	^{58m}Co	S	2×10^{-5}	8×10^{-2}	6×10^{-7}	3×10^{-3}
		I	9×10^{-6}	6×10^{-2}	3×10^{-7}	2×10^{-3}
	^{58}Co	S	8×10^{-7}	4×10^{-3}	3×10^{-8}	1×10^{-4}
		I	5×10^{-8}	3×10^{-3}	2×10^{-9}	9×10^{-5}
	^{60}Co	S	3×10^{-7}	1×10^{-3}	1×10^{-8}	5×10^{-5}
		I	9×10^{-9}	1×10^{-3}	3×10^{-10}	3×10^{-5}
Copper-29	^{64}Cu	S	2×10^{-6}	1×10^{-2}	7×10^{-8}	3×10^{-4}
		I	1×10^{-6}	6×10^{-3}	4×10^{-8}	2×10^{-4}
Curium-96	^{242}Cm	S	1×10^{-10}	7×10^{-4}	4×10^{-12}	2×10^{-5}
		I	2×10^{-10}	7×10^{-4}	6×10^{-12}	3×10^{-5}
	^{243}Cm	S	6×10^{-12}	1×10^{-4}	2×10^{-13}	5×10^{-6}
		I	1×10^{-10}	7×10^{-4}	3×10^{-12}	2×10^{-5}
	$^{244\alpha}$Cm	S	9×10^{-12}	2×10^{-4}	3×10^{-13}	7×10^{-6}
		I	1×10^{-10}	8×10^{-4}	3×10^{-12}	3×10^{-5}

See footnotes at end of table.

TABLE 5A.5. *(Continued)*

Element-Z	Isotope[a]	Table I Column 1 Air (μCi/ml)	Table I Column 2 Water (μCi/ml)	Table II Column 1 Air (μCi/ml)	Table II Column 2 Water (μCi/ml)
Curium-96 continued	^{245}Cm S	5×10^{-12}	1×10^{-4}	2×10^{-13}	4×10^{-6}
	I	1×10^{-10}	8×10^{-4}	4×10^{-12}	3×10^{-5}
	^{246}Cm S	5×10^{-12}	1×10^{-4}	2×10^{-13}	4×10^{-6}
	I	1×10^{-10}	8×10^{-4}	4×10^{-12}	3×10^{-5}
	^{247}Cm S	5×10^{-12}	1×10^{-4}	2×10^{-13}	4×10^{-6}
	I	1×10^{-10}	6×10^{-4}	4×10^{-12}	2×10^{-5}
	^{248}Cm S	6×10^{-13}	1×10^{-5}	2×10^{-14}	4×10^{-7}
	I	1×10^{-11}	4×10^{-5}	4×10^{-13}	1×10^{-6}
	^{249}Cm S	1×10^{-5}	6×10^{-2}	4×10^{-7}	2×10^{-3}
	I	1×10^{-5}	6×10^{-2}	4×10^{-7}	2×10^{-3}
Dysprosium-66	^{165}Dy S	3×10^{-6}	1×10^{-2}	9×10^{-8}	4×10^{-4}
	I	2×10^{-6}	1×10^{-2}	7×10^{-8}	4×10^{-4}
	^{166}Dy S	2×10^{-7}	1×10^{-3}	8×10^{-9}	4×10^{-5}
	I	2×10^{-7}	1×10^{-3}	7×10^{-9}	4×10^{-5}
Einsteinium-99	^{253}Es S	8×10^{-10}	7×10^{-4}	3×10^{-11}	2×10^{-5}
	I	6×10^{-10}	7×10^{-4}	2×10^{-11}	2×10^{-5}
	^{254m}Es S	5×10^{-9}	5×10^{-4}	2×10^{-10}	2×10^{-5}
	I	6×10^{-9}	5×10^{-4}	2×10^{-10}	2×10^{-5}
	^{254}Es S	2×10^{-11}	4×10^{-4}	6×10^{-13}	1×10^{-5}
	I	1×10^{-10}	4×10^{-4}	4×10^{-12}	1×10^{-5}
	^{255}Es S	5×10^{-10}	8×10^{-4}	2×10^{-11}	3×10^{-5}
	I	4×10^{-10}	8×10^{-4}	1×10^{-11}	3×10^{-5}
Erbium-68	^{169}Er S	6×10^{-7}	3×10^{-3}	2×10^{-8}	9×10^{-5}
	I	4×10^{-7}	3×10^{-3}	1×10^{-8}	9×10^{-5}
	^{171}Er S	7×10^{-7}	3×10^{-3}	2×10^{-8}	1×10^{-4}
	I	6×10^{-7}	3×10^{-3}	2×10^{-8}	1×10^{-4}
Europium-63	^{152}Eu S	4×10^{-7}	2×10^{-3}	1×10^{-8}	6×10^{-5}
	(T/2 = 9.2 hr) I	3×10^{-7}	2×10^{-3}	1×10^{-8}	6×10^{-5}
	^{152}Eu S	1×10^{-8}	2×10^{-3}	4×10^{-10}	8×10^{-5}
	(T/2 = 13 yr) I	2×10^{-8}	2×10^{-3}	6×10^{-10}	8×10^{-5}
	^{154}Eu S	4×10^{-9}	6×10^{-4}	1×10^{-10}	2×10^{-5}
	I	7×10^{-9}	6×10^{-4}	2×10^{-10}	2×10^{-5}
	^{155}Eu S	9×10^{-8}	6×10^{-3}	3×10^{-9}	2×10^{-4}
	I	7×10^{-8}	6×10^{-3}	3×10^{-9}	2×10^{-4}
Fermium-100	^{254}Fm S	6×10^{-8}	4×10^{-3}	2×10^{-9}	1×10^{-4}
	I	7×10^{-8}	4×10^{-3}	2×10^{-9}	1×10^{-4}
	^{255}Fm S	2×10^{-8}	1×10^{-3}	6×10^{-10}	3×10^{-5}
	I	1×10^{-8}	1×10^{-3}	4×10^{-10}	3×10^{-5}
	^{256}Fm S	3×10^{-9}	3×10^{-5}	1×10^{-10}	9×10^{-7}
	I	2×10^{-9}	3×10^{-5}	6×10^{-11}	9×10^{-7}
Fluorine-9	^{18}F S	5×10^{-6}	2×10^{-2}	2×10^{-7}	8×10^{-4}
	I	3×10^{-6}	1×10^{-2}	9×10^{-8}	5×10^{-4}
Gadolinium-64	^{153}Gd S	2×10^{-7}	6×10^{-3}	8×10^{-9}	2×10^{-4}
	I	9×10^{-8}	6×10^{-3}	3×10^{-9}	2×10^{-4}
	^{159}Gd S	5×10^{-7}	2×10^{-3}	2×10^{-8}	8×10^{-5}
	I	4×10^{-7}	2×10^{-3}	1×10^{-8}	8×10^{-5}
Gallium-31	^{72}Ga S	2×10^{-7}	1×10^{-3}	8×10^{-9}	4×10^{-5}
	I	2×10^{-7}	1×10^{-3}	6×10^{-9}	4×10^{-5}
Germanium-32	^{71}Ge S	1×10^{-5}	5×10^{-2}	4×10^{-7}	2×10^{-3}
	I	6×10^{-6}	5×10^{-2}	2×10^{-7}	2×10^{-3}
Gold-79	^{196}Au S	1×10^{-6}	5×10^{-3}	4×10^{-8}	2×10^{-4}
	I	6×10^{-7}	4×10^{-3}	2×10^{-8}	1×10^{-4}
	^{198}Au S	3×10^{-7}	2×10^{-3}	1×10^{-8}	5×10^{-5}
	I	2×10^{-7}	1×10^{-3}	8×10^{-9}	5×10^{-5}
	^{199}Au S	1×10^{-6}	5×10^{-3}	4×10^{-8}	2×10^{-4}
	I	8×10^{-7}	4×10^{-3}	3×10^{-8}	2×10^{-4}
Hafnium-72	^{181}Hf S	4×10^{-8}	2×10^{-3}	1×10^{-9}	7×10^{-5}
	I	7×10^{-8}	2×10^{-3}	3×10^{-9}	7×10^{-5}

See footnotes at end of table.

TABLE 5A.5. (*Continued*)

Element-Z	Isotope[a]	Table I		Table II	
		Column 1 Air (μCi/ml)	Column 2 Water (μCi/ml)	Column 1 Air (μCi/ml)	Column 2 Water (μCi/ml)
Holmium-67	^{166}Ho S	2×10^{-7}	9×10^{-4}	7×10^{-9}	3×10^{-5}
	I	2×10^{-7}	9×10^{-4}	6×10^{-9}	3×10^{-5}
Hydrogen-1	^{3}H S	5×10^{-6}	1×10^{-1}	2×10^{-7}	3×10^{-3}
	I	5×10^{-6}	1×10^{-1}	2×10^{-7}	3×10^{-3}
	Sub	2×10^{-3}	—	4×10^{-5}	—
Indium-49	^{113m}In S	8×10^{-6}	4×10^{-2}	3×10^{-7}	1×10^{-3}
	I	7×10^{-6}	4×10^{-2}	2×10^{-7}	1×10^{-3}
	^{114m}In S	1×10^{-7}	5×10^{-4}	4×10^{-9}	2×10^{-5}
	I	2×10^{-8}	5×10^{-4}	7×10^{-10}	2×10^{-5}
	^{115m}In S	2×10^{-6}	1×10^{-2}	8×10^{-8}	4×10^{-4}
	I	2×10^{-6}	1×10^{-2}	6×10^{-8}	4×10^{-4}
	^{115}In S	2×10^{-7}	3×10^{-3}	9×10^{-9}	9×10^{-5}
	I	3×10^{-8}	3×10^{-3}	1×10^{-9}	9×10^{-5}
Iodine-53	^{125}I S	5×10^{-9}	4×10^{-5}	8×10^{-11}	2×10^{-7}
	I	2×10^{-7}	6×10^{-3}	6×10^{-9}	2×10^{-4}
	^{126}I S	8×10^{-9}	5×10^{-5}	9×10^{-11}	3×10^{-7}
	I	3×10^{-7}	3×10^{-3}	1×10^{-8}	9×10^{-5}
	^{129}I S	2×10^{-9}	1×10^{-5}	2×10^{-11}	6×10^{-8}
	I	7×10^{-8}	6×10^{-3}	2×10^{-9}	2×10^{-4}
	^{131}I S	9×10^{-9}	6×10^{-5}	1×10^{-10}	3×10^{-7}
	I	3×10^{-7}	2×10^{-3}	1×10^{-8}	6×10^{-5}
	^{132}I S	2×10^{-7}	2×10^{-3}	3×10^{-9}	8×10^{-6}
	I	9×10^{-7}	5×10^{-3}	3×10^{-8}	2×10^{-4}
	^{133}I S	3×10^{-8}	2×10^{-4}	4×10^{-10}	1×10^{-6}
	I	2×10^{-7}	1×10^{-3}	7×10^{-9}	4×10^{-5}
	^{134}I S	5×10^{-7}	4×10^{-3}	6×10^{-9}	2×10^{-5}
	I	3×10^{-6}	2×10^{-2}	1×10^{-7}	6×10^{-4}
	^{135}I S	1×10^{-7}	7×10^{-4}	1×10^{-9}	4×10^{-6}
	I	4×10^{-7}	2×10^{-3}	1×10^{-8}	7×10^{-5}
Iridium-77	^{190}Ir S	1×10^{-6}	6×10^{-3}	4×10^{-8}	2×10^{-4}
	I	4×10^{-7}	5×10^{-3}	1×10^{-8}	2×10^{-4}
	^{192}Ir S	1×10^{-7}	1×10^{-3}	4×10^{-9}	4×10^{-5}
	I	3×10^{-8}	1×10^{-3}	9×10^{-10}	4×10^{-5}
	^{194}Ir S	2×10^{-7}	1×10^{-3}	8×10^{-9}	3×10^{-5}
	I	2×10^{-7}	9×10^{-4}	5×10^{-9}	3×10^{-5}
Iron-26	^{55}Fe S	9×10^{-7}	2×10^{-2}	3×10^{-8}	8×10^{-4}
	I	1×10^{-6}	7×10^{-2}	3×10^{-8}	2×10^{-3}
	^{59}Fe S	1×10^{-7}	2×10^{-3}	5×10^{-9}	6×10^{-5}
	I	5×10^{-9}	2×10^{-3}	2×10^{-9}	5×10^{-5}
Krypton-36[b]	^{85m}Kr Sub	6×10^{-6}	—	1×10^{-7}	—
	^{85}Kr Sub	1×10^{-5}	—	3×10^{-7}	—
	^{87}Kr Sub	1×10^{-6}	—	2×10^{-8}	—
Krypton-36	^{88}Kr Sub	1×10^{-6}	—	2×10^{-8}	—
Lanthanum-57	^{140}La S	2×10^{-7}	7×10^{-4}	5×10^{-9}	2×10^{-5}
	I	1×10^{-7}	7×10^{-4}	4×10^{-9}	2×10^{-5}
Lead-82	^{203}Pb S	3×10^{-6}	1×10^{-2}	9×10^{-8}	4×10^{-4}
	I	2×10^{-6}	1×10^{-2}	6×10^{-8}	4×10^{-4}
	^{210}Pb S	1×10^{-10}	4×10^{-6}	4×10^{-12}	1×10^{-7}
	I	2×10^{-10}	5×10^{-3}	8×10^{-12}	2×10^{-4}
	^{212}Pb S	2×10^{-8}	6×10^{-4}	6×10^{-10}	2×10^{-5}
	I	2×10^{-8}	5×10^{-4}	7×10^{-10}	2×10^{-5}
Lutetium-71	^{177}Lu S	6×10^{-7}	3×10^{-3}	2×10^{-8}	1×10^{-4}
	I	5×10^{-7}	3×10^{-3}	2×10^{-8}	1×10^{-4}
Manganese-25	^{52}Mn S	2×10^{-7}	1×10^{-3}	7×10^{-9}	3×10^{-5}
	I	1×10^{-7}	9×10^{-4}	5×10^{-9}	3×10^{-5}
	^{54}Mn S	4×10^{-7}	4×10^{-3}	1×10^{-9}	1×10^{-4}
	I	4×10^{-8}	3×10^{-3}	1×10^{-9}	1×10^{-4}
	^{56}Mn S	8×10^{-7}	4×10^{-3}	3×10^{-8}	1×10^{-4}
	I	5×10^{-7}	3×10^{-3}	2×10^{-8}	1×10^{-4}

See footnotes at end of table.

TABLE 5A.5. *(Continued)*

Element-Z	Isotope[a]	Table I Column 1 Air (μCi/ml)	Table I Column 2 Water (μCi/ml)	Table II Column 1 Air (μCi/ml)	Table II Column 2 Water (μCi/ml)
Mercury-80	^{197m}Hg S	7×10^{-7}	6×10^{-3}	3×10^{-8}	2×10^{-4}
	I	8×10^{-7}	5×10^{-3}	3×10^{-8}	2×10^{-4}
	^{197}Hg S	1×10^{-6}	9×10^{-3}	4×10^{-8}	3×10^{-4}
	I	3×10^{-6}	1×10^{-2}	9×10^{-8}	5×10^{-4}
	^{203}Hg S	7×10^{-8}	5×10^{-4}	2×10^{-9}	2×10^{-5}
	I	1×10^{-7}	3×10^{-3}	4×10^{-9}	1×10^{-4}
Molybdenum-42	^{99}Mo S	7×10^{-7}	5×10^{-3}	3×10^{-8}	2×10^{-4}
	I	2×10^{-7}	1×10^{-3}	7×10^{-9}	4×10^{-5}
Neodymium-60	^{144}Nd S	8×10^{-11}	2×10^{-3}	3×10^{-12}	7×10^{-5}
	I	3×10^{-10}	2×10^{-3}	1×10^{-11}	8×10^{-5}
	^{147}Nd S	4×10^{-7}	2×10^{-3}	1×10^{-8}	6×10^{-5}
	I	2×10^{-7}	2×10^{-3}	8×10^{-9}	6×10^{-5}
	^{149}Nd S	2×10^{-6}	8×10^{-3}	6×10^{-8}	3×10^{-4}
	I	1×10^{-6}	8×10^{-3}	5×10^{-8}	3×10^{-4}
Neptunium-93	^{237}Np S	4×10^{-12}	9×10^{-5}	1×10^{-13}	3×10^{-6}
	I	1×10^{-10}	9×10^{-4}	4×10^{-12}	3×10^{-5}
	^{239}Np S	8×10^{-7}	4×10^{-3}	3×10^{-8}	1×10^{-4}
	I	7×10^{-7}	4×10^{-3}	2×10^{-8}	1×10^{-4}
Nickel-28	^{59}Ni S	5×10^{-7}	6×10^{-3}	2×10^{-8}	2×10^{-4}
	I	8×10^{-7}	6×10^{-2}	3×10^{-8}	2×10^{-3}
	^{63}Ni S	6×10^{-8}	8×10^{-4}	2×10^{-9}	3×10^{-5}
	I	3×10^{-7}	2×10^{-2}	1×10^{-8}	7×10^{-4}
	^{65}Ni S	9×10^{-7}	4×10^{-3}	3×10^{-8}	1×10^{-4}
	I	5×10^{-7}	3×10^{-3}	2×10^{-8}	1×10^{-4}
Niobium-41 (Columbium)	^{93m}Nb S	1×10^{-7}	1×10^{-2}	4×10^{-9}	4×10^{-4}
	I	2×10^{-7}	1×10^{-2}	5×10^{-9}	4×10^{-4}
	^{95}Nb S	5×10^{-7}	3×10^{-3}	2×10^{-8}	1×10^{-4}
	I	1×10^{-7}	3×10^{-3}	3×10^{-9}	1×10^{-4}
	^{97}Nb S	6×10^{-6}	3×10^{-2}	2×10^{-7}	9×10^{-4}
	I	5×10^{-6}	3×10^{-2}	2×10^{-7}	9×10^{-4}
Osmium-76	^{185}Os S	5×10^{-7}	2×10^{-3}	2×10^{-8}	7×10^{-5}
	I	5×10^{-8}	2×10^{-3}	2×10^{-9}	7×10^{-5}
	^{191m}Os S	2×10^{-5}	7×10^{-2}	6×10^{-7}	3×10^{-3}
	I	9×10^{-6}	7×10^{-2}	3×10^{-7}	2×10^{-3}
	^{191}Os S	1×10^{-6}	5×10^{-3}	4×10^{-8}	2×10^{-4}
	I	4×10^{-7}	5×10^{-3}	1×10^{-8}	2×10^{-4}
	^{193}Os S	4×10^{-7}	2×10^{-3}	1×10^{-8}	6×10^{-5}
	I	3×10^{-7}	2×10^{-3}	9×10^{-9}	5×10^{-5}
Palladium-46	^{103}Pd S	1×10^{-6}	1×10^{-2}	5×10^{-8}	3×10^{-4}
	I	7×10^{-7}	8×10^{-3}	3×10^{-8}	3×10^{-4}
	^{109}Pd S	6×10^{-7}	3×10^{-3}	2×10^{-8}	9×10^{-5}
	I	4×10^{-7}	2×10^{-3}	1×10^{-8}	7×10^{-5}
Phosphorus-15	^{32}P S	7×10^{-8}	5×10^{-4}	2×10^{-9}	2×10^{-5}
	I	8×10^{-8}	7×10^{-4}	3×10^{-9}	2×10^{-5}
Platinum-78	^{191}Pt S	8×10^{-7}	4×10^{-3}	3×10^{-8}	1×10^{-4}
	I	6×10^{-7}	3×10^{-3}	2×10^{-8}	1×10^{-4}
	^{193m}Pt S	7×10^{-6}	3×10^{-2}	2×10^{-7}	1×10^{-3}
	I	5×10^{-6}	3×10^{-2}	2×10^{-7}	1×10^{-3}
	^{197m}Pt S	6×10^{-6}	3×10^{-2}	2×10^{-7}	1×10^{-3}
	I	5×10^{-6}	3×10^{-2}	2×10^{-7}	9×10^{-4}
	^{197}Pt S	8×10^{-7}	4×10^{-3}	3×10^{-8}	1×10^{-4}
	I	6×10^{-7}	3×10^{-3}	2×10^{-8}	1×10^{-4}
Plutonium-94	^{238}Pu S	2×10^{-12}	1×10^{-4}	7×10^{-14}	5×10^{-6}
	I	3×10^{-11}	8×10^{-4}	1×10^{-12}	3×10^{-5}
	^{239}Pu S	2×10^{-12}	1×10^{-4}	6×10^{-14}	5×10^{-6}
	I	4×10^{-11}	8×10^{-4}	1×10^{-12}	3×10^{-5}
	^{240}Pu S	2×10^{-12}	1×10^{-4}	6×10^{-14}	5×10^{-6}
	I	4×10^{-11}	8×10^{-4}	1×10^{-12}	3×10^{-5}

See footnotes at end of table.

TABLE 5A.5. (*Continued*)

Element-Z	Isotope[a]	Table I Column 1 Air (μCi/ml)	Table I Column 2 Water (μCi/ml)	Table II Column 1 Air (μCi/ml)	Table II Column 2 Water (μCi/ml)
Plutonium-94 continued	^{241}Pu S	9×10^{-11}	7×10^{-3}	3×10^{-12}	2×10^{-4}
	I	4×10^{-8}	4×10^{-2}	1×10^{-9}	1×10^{-3}
	^{242}Pu S	2×10^{-12}	1×10^{-4}	6×10^{-14}	5×10^{-6}
	I	4×10^{-11}	9×10^{-4}	1×10^{-12}	3×10^{-5}
	^{243}Pu S	2×10^{-6}	1×10^{-2}	6×10^{-8}	3×10^{-4}
	I	2×10^{-6}	1×10^{-2}	8×10^{-8}	3×10^{-4}
	^{244}Pu S	2×10^{-12}	1×10^{-4}	6×10^{-14}	4×10^{-6}
	I	3×10^{-11}	3×10^{-4}	1×10^{-12}	1×10^{-5}
Polonium-84	^{210}Po S	5×10^{-10}	2×10^{-5}	2×10^{-11}	7×10^{-7}
	I	2×10^{-10}	8×10^{-4}	7×10^{-12}	3×10^{-5}
Potassium-19	^{42}K S	2×10^{-6}	9×10^{-3}	7×10^{-8}	3×10^{-4}
	I	1×10^{-7}	6×10^{-4}	4×10^{-9}	2×10^{-5}
Praseodymium-59	^{142}Pr S	2×10^{-7}	9×10^{-4}	7×10^{-9}	3×10^{-5}
	I	2×10^{-7}	9×10^{-4}	5×10^{-9}	3×10^{-5}
	^{143}Pr S	3×10^{-7}	1×10^{-3}	1×10^{-8}	5×10^{-5}
	I	2×10^{-7}	1×10^{-3}	6×10^{-9}	5×10^{-5}
Promethium-61	^{147}Pm S	6×10^{-8}	6×10^{-3}	2×10^{-9}	2×10^{-4}
	I	1×10^{-7}	6×10^{-3}	3×10^{-9}	2×10^{-4}
	^{149}Pm S	3×10^{-7}	1×10^{-3}	1×10^{-8}	4×10^{-5}
	I	2×10^{-7}	1×10^{-3}	8×10^{-9}	4×10^{-5}
Protoactinium-91	^{230}Pa S	2×10^{-9}	7×10^{-3}	6×10^{-11}	2×10^{-4}
	I	8×10^{-10}	7×10^{-3}	3×10^{-11}	2×10^{-4}
	^{231}Pa S	1×10^{-12}	3×10^{-5}	4×10^{-14}	9×10^{-7}
	I	1×10^{-10}	8×10^{-4}	4×10^{-12}	2×10^{-5}
	^{233}Pa S	6×10^{-7}	4×10^{-3}	2×10^{-8}	1×10^{-4}
	I	2×10^{-7}	3×10^{-3}	6×10^{-9}	1×10^{-4}
Radium-88	^{223}Ra S	2×10^{-9}	2×10^{-5}	6×10^{-11}	7×10^{-7}
	I	2×10^{-10}	1×10^{-4}	8×10^{-12}	4×10^{-6}
	^{224}Ra S	5×10^{-9}	7×10^{-5}	2×10^{-10}	2×10^{-6}
	I	7×10^{-10}	2×10^{-4}	2×10^{-11}	5×10^{-6}
	^{226}Ra S	3×10^{-11}	4×10^{-7}	3×10^{-12}	3×10^{-8}
	I	5×10^{-11}	9×10^{-4}	2×10^{-12}	3×10^{-5}
	^{228}Ra S	7×10^{-11}	8×10^{-7}	2×10^{-12}	3×10^{-8}
	I	4×10^{-11}	7×10^{-4}	1×10^{-12}	3×10^{-5}
Radon-86	^{220}Rn S	3×10^{-7}	—	1×10^{-8}	—
	I	—	—	—	—
	^{222}Rn S	1×10^{-7}	—	3×10^{-9}	—
Rhenium-75	^{183}Re S	3×10^{-6}	2×10^{-2}	9×10^{-8}	6×10^{-4}
	I	2×10^{-7}	8×10^{-3}	5×10^{-9}	3×10^{-4}
	^{186}Re S	6×10^{-7}	3×10^{-3}	2×10^{-8}	9×10^{-5}
	I	2×10^{-7}	1×10^{-3}	8×10^{-9}	5×10^{-5}
	^{187}Re S	9×10^{-6}	7×10^{-2}	3×10^{-7}	3×10^{-3}
	I	5×10^{-7}	4×10^{-2}	2×10^{-8}	2×10^{-3}
	^{188}Re S	4×10^{-7}	2×10^{-3}	1×10^{-8}	6×10^{-5}
	I	2×10^{-7}	9×10^{-4}	6×10^{-9}	3×10^{-5}
Rhodium-45	^{103m}Rh S	8×10^{-5}	4×10^{-1}	3×10^{-6}	1×10^{-2}
	I	6×10^{-5}	3×10^{-1}	2×10^{-6}	1×10^{-2}
	^{105}Rh S	8×10^{-7}	4×10^{-3}	3×10^{-8}	1×10^{-4}
	I	5×10^{-7}	3×10^{-3}	2×10^{-8}	1×10^{-4}
Rubidium-37	^{86}Rb S	3×10^{-7}	2×10^{-3}	1×10^{-8}	7×10^{-5}
	I	7×10^{-8}	7×10^{-4}	2×10^{-9}	2×10^{-5}
	^{87}Rb S	5×10^{-7}	3×10^{-3}	2×10^{-8}	1×10^{-4}
	I	7×10^{-8}	5×10^{-3}	2×10^{-9}	2×10^{-4}
Ruthenium-44	^{97}Ru S	2×10^{-6}	1×10^{-2}	8×10^{-8}	4×10^{-4}
	I	2×10^{-6}	1×10^{-2}	6×10^{-8}	3×10^{-4}
	^{103}Ru S	5×10^{-7}	2×10^{-3}	2×10^{-8}	8×10^{-5}
	I	8×10^{-8}	2×10^{-3}	3×10^{-9}	8×10^{-5}

See footnotes at end of table.

TABLE 5A.5. *(Continued)*

Element-Z	Isotope[a]	Table I Column 1 Air (μCi/ml)	Table I Column 2 Water (μCi/ml)	Table II Column 1 Air (μCi/ml)	Table II Column 2 Water (μCi/ml)
Ruthenium-44 continued	^{105}Ru S	7×10^{-7}	3×10^{-3}	2×10^{-8}	1×10^{-4}
	I	5×10^{-7}	3×10^{-3}	2×10^{-8}	1×10^{-4}
	^{106}Ru S	8×10^{-8}	4×10^{-4}	3×10^{-9}	1×10^{-5}
	I	6×10^{-9}	3×10^{-4}	2×10^{-10}	1×10^{-5}
Samarium-62	^{147}Sm S	7×10^{-11}	2×10^{-3}	2×10^{-12}	6×10^{-5}
	I	3×10^{-10}	2×10^{-3}	9×10^{-12}	7×10^{-5}
	^{151}Sm S	6×10^{-8}	1×10^{-2}	2×10^{-9}	4×10^{-4}
	I	1×10^{-7}	1×10^{-2}	5×10^{-9}	4×10^{-4}
	^{153}Sm S	5×10^{-7}	2×10^{-3}	2×10^{-8}	8×10^{-5}
	I	4×10^{-7}	2×10^{-3}	1×10^{-8}	8×10^{-5}
Scandium-21	^{46}Sc S	2×10^{-7}	1×10^{-3}	8×10^{-9}	4×10^{-5}
	I	2×10^{-8}	1×10^{-3}	8×10^{-10}	4×10^{-5}
	^{47}Sc S	6×10^{-7}	3×10^{-3}	2×10^{-8}	9×10^{-5}
	I	5×10^{-7}	3×10^{-3}	2×10^{-8}	9×10^{-5}
	^{48}Sc S	2×10^{-7}	8×10^{-4}	6×10^{-9}	3×10^{-5}
	I	1×10^{-7}	8×10^{-4}	5×10^{-9}	3×10^{-5}
Selenium-34	^{75}Se S	1×10^{-6}	9×10^{-3}	4×10^{-8}	3×10^{-4}
	I	1×10^{-7}	8×10^{-3}	4×10^{-9}	3×10^{-4}
Silicon-14	^{31}Si S	6×10^{-6}	3×10^{-2}	2×10^{-7}	9×10^{-4}
	I	1×10^{-6}	6×10^{-3}	3×10^{-8}	2×10^{-4}
Silver-47	^{105}Ag S	6×10^{-7}	3×10^{-3}	2×10^{-8}	1×10^{-4}
	I	8×10^{-8}	3×10^{-3}	3×10^{-9}	1×10^{-4}
	^{110m}Ag S	2×10^{-7}	9×10^{-4}	7×10^{-9}	3×10^{-5}
	I	1×10^{-8}	9×10^{-4}	3×10^{-10}	3×10^{-5}
	^{111}Ag S	3×10^{-7}	1×10^{-3}	1×10^{-8}	4×10^{-5}
	I	2×10^{-7}	1×10^{-3}	8×10^{-9}	4×10^{-5}
Sodium-11	^{22}Na S	2×10^{-7}	1×10^{-3}	6×10^{-9}	4×10^{-5}
	I	9×10^{-9}	9×10^{-4}	3×10^{-10}	3×10^{-5}
	^{24}Na S	1×10^{-6}	6×10^{-3}	4×10^{-8}	2×10^{-4}
	I	$i \times 10^{-7}$	8×10^{-4}	5×10^{-9}	3×10^{-5}
Strontium-38	^{85m}Sr S	4×10^{-5}	2×10^{-1}	1×10^{-6}	7×10^{-3}
	I	3×10^{-5}	2×10^{-1}	1×10^{-6}	7×10^{-3}
	^{85}Sr S	2×10^{-7}	3×10^{-3}	8×10^{-9}	1×10^{-4}
	I	1×10^{-7}	5×10^{-3}	4×10^{-9}	2×10^{-4}
	^{89}Sr S	3×10^{-8}	3×10^{-4}	3×10^{-10}	3×10^{-6}
	I	4×10^{-8}	8×10^{-4}	1×10^{-9}	3×10^{-5}
	^{90}Sr S	1×10^{-9}	1×10^{-5}	3×10^{-11}	3×10^{-7}
	I	5×10^{-9}	1×10^{-3}	2×10^{-10}	4×10^{-5}
	^{91}Sr S	4×10^{-7}	2×10^{-3}	2×10^{-8}	7×10^{-5}
	I	3×10^{-7}	1×10^{-3}	9×10^{-9}	5×10^{-5}
	^{92}Sr S	4×10^{-7}	2×10^{-3}	2×10^{-8}	7×10^{-5}
	I	3×10^{-7}	2×10^{-3}	1×10^{-8}	6×10^{-5}
Sulfur-16	^{35}S S	3×10^{-7}	2×10^{-3}	9×10^{-9}	6×10^{-5}
	I	3×10^{-7}	8×10^{-3}	9×10^{-9}	3×10^{-4}
Tantalum-73	^{182}Ta S	4×10^{-8}	1×10^{-3}	1×10^{-9}	4×10^{-5}
	I	2×10^{-8}	1×10^{-3}	7×10^{-10}	4×10^{-5}
Technetium-43	^{96m}Tc S	8×10^{-5}	4×10^{-1}	3×10^{-6}	1×10^{-2}
	I	3×10^{-5}	3×10^{-1}	1×10^{-6}	1×10^{-2}
	^{96}Tc S	6×10^{-7}	3×10^{-3}	2×10^{-8}	1×10^{-4}
	I	2×10^{-7}	1×10^{-3}	8×10^{-9}	5×10^{-5}
	^{97m}Tc S	2×10^{-6}	1×10^{-2}	8×10^{-8}	4×10^{-4}
	I	2×10^{-7}	5×10^{-3}	5×10^{-9}	2×10^{-4}
	^{97}Tc S	1×10^{-5}	5×10^{-2}	4×10^{-7}	2×10^{-3}
	I	3×10^{-7}	2×10^{-2}	1×10^{-8}	8×10^{-4}
	^{99m}Tc S	4×10^{-5}	2×10^{-1}	1×10^{-6}	6×10^{-3}
	I	1×10^{-5}	8×10^{-2}	5×10^{-7}	3×10^{-3}
	^{99}Tc S	2×10^{-6}	1×10^{-2}	7×10^{-8}	3×10^{-4}
	I	6×10^{-8}	5×10^{-3}	2×10^{-9}	2×10^{-4}

See footnotes at end of table.

TABLE 5A.5. *(Continued)*

Element-Z	Isotope[a]	Table I Column 1 Air (μCi/ml)	Table I Column 2 Water (μCi/ml)	Table II Column 1 Air (μCi/ml)	Table II Column 2 Water (μCi/ml)
Tellurium-52	^{125m}Te S	4×10^{-7}	5×10^{-3}	1×10^{-8}	2×10^{-4}
	I	1×10^{-7}	3×10^{-3}	4×10^{-9}	1×10^{-4}
	^{127m}Te S	1×10^{-7}	2×10^{-3}	5×10^{-9}	6×10^{-5}
	I	4×10^{-8}	2×10^{-3}	1×10^{-9}	5×10^{-5}
	^{127}Te S	2×10^{-6}.	8×10^{-3}	6×10^{-8}	3×10^{-4}
	I	9×10^{-7}	5×10^{-3}	3×10^{-8}	2×10^{-4}
	^{129m}Te S	8×10^{-8}	1×10^{-3}	3×10^{-9}	3×10^{-5}
	I	3×10^{-8}	6×10^{-4}	1×10^{-9}	2×10^{-5}
	^{129}Te S	5×10^{-6}	2×10^{-2}	2×10^{-7}	8×10^{-4}
	I	4×10^{-6}	2×10^{-2}	1×10^{-7}	8×10^{-4}
	^{131m}Te S	4×10^{-7}	2×10^{-3}	1×10^{-8}	6×10^{-5}
	I	2×10^{-7}	1×10^{-3}	6×10^{-9}	4×10^{-5}
	^{132}Te S	2×10^{-7}	9×10^{-4}	7×10^{-9}	3×10^{-5}
	I	1×10^{-7}	6×10^{-4}	4×10^{-9}	2×10^{-5}
Terbium-65	^{160}Tb S	1×10^{-7}	1×10^{-3}	3×10^{-9}	4×10^{-5}
	I	3×10^{-8}	1×10^{-3}	1×10^{-9}	4×10^{-5}
Thallium-81	^{200}Tl S	3×10^{-6}	1×10^{-2}	9×10^{-8}	4×10^{-4}
	I	1×10^{-6}	7×10^{-3}	4×10^{-8}	2×10^{-4}
	^{201}Tl S	2×10^{-6}	9×10^{-3}	7×10^{-8}	3×10^{-4}
	I	9×10^{-7}	5×10^{-3}	3×10^{-8}	2×10^{-4}
	^{202}Tl S	8×10^{-7}	4×10^{-3}	3×10^{-8}	1×10^{-4}
	I	2×10^{-7}	2×10^{-3}	8×10^{-9}	7×10^{-5}
	^{204}Tl S	6×10^{-7}	3×10^{-3}	2×10^{-8}	1×10^{-4}
	I	3×10^{-8}	2×10^{-3}	9×10^{-10}	6×10^{-5}
Thorium-90	^{228}Th S	9×10^{-12}	2×10^{-4}	3×10^{-13}	7×10^{-6}
	I	6×10^{-12}	4×10^{-4}	2×10^{-13}	1×10^{-5}
	^{230}Th S	2×10^{-12}	5×10^{-5}	8×10^{-14}	2×10^{-5}
	I	1×10^{-11}	9×10^{-4}	3×10^{-13}	3×10^{-5}
	^{232}Th S	3×10^{-11}	5×10^{-5}	1×10^{-12}	2×10^{-6}
	I	3×10^{-11}	1×10^{-3}	1×10^{-12}	4×10^{-5}
	Th (natural) S	3×10^{-11}	3×10^{-5}	1×10^{-12}	1×10^{-6}
	I	3×10^{-11}	3×10^{-4}	1×10^{-12}	1×10^{-5}
	^{234}Th S	6×10^{-8}	5×10^{-4}	2×10^{-9}	2×10^{-5}
	I	3×10^{-8}	5×10^{-4}	1×10^{-9}	2×10^{-5}
Thulium-69	^{170}Tm S	4×10^{-8}	1×10^{-3}	1×10^{-9}	5×10^{-5}
	I	3×10^{-8}	1×10^{-3}	1×10^{-9}	5×10^{-5}
	^{171}Tm S	1×10^{-7}	1×10^{-2}	4×10^{-9}	5×10^{-4}
	I	2×10^{-7}	1×10^{-2}	8×10^{-9}	5×10^{-4}
Tin-50	^{113}Sn S	4×10^{-7}	2×10^{-3}	1×10^{-8}	9×10^{-5}
	I	5×10^{-8}	2×10^{-3}	2×10^{-9}	8×10^{-5}
	^{125}Sn S	1×10^{-7}	5×10^{-4}	4×10^{-9}	2×10^{-5}
	I	8×10^{-8}	5×10^{-4}	3×10^{-9}	2×10^{-5}
Tungsten-74 (Wolfram)	^{181}W S	2×10^{-6}	1×10^{-2}	8×10^{-8}	4×10^{-4}
	I	1×10^{-7}	1×10^{-2}	4×10^{-9}	3×10^{-4}
	^{185}W S	8×10^{-7}	4×10^{-3}	3×10^{-8}	1×10^{-4}
	I	1×10^{-7}	3×10^{-3}	4×10^{-9}	1×10^{-4}
	^{187}W S	4×10^{-7}	2×10^{-3}	2×10^{-8}	7×10^{-5}
	I	3×10^{-7}	2×10^{-3}	1×10^{-8}	6×10^{-5}
Uranium-92	^{230}U S	3×10^{-10}	1×10^{-4}	1×10^{-11}	5×10^{-6}
	I	1×10^{-10}	1×10^{-4}	4×10^{-12}	5×10^{-6}
	^{232}U S	1×10^{-10}	8×10^{-4}	3×10^{-12}	3×10^{-5}
	I	3×10^{-11}	8×10^{-4}	9×10^{-13}	3×10^{-5}
	^{233}U S	5×10^{-10}	9×10^{-4}	2×10^{-11}	3×10^{-5}
	I	1×10^{-10}	9×10^{-4}	4×10^{-12}	3×10^{-5}
	^{234}U S	6×10^{-10}	9×10^{-4}	2×10^{-11}	3×10^{-5}
	I	1×10^{-10}	9×10^{-4}	4×10^{-12}	3×10^{-5}
	^{235}U S	5×10^{-10}	8×10^{-4}	2×10^{-11}	3×10^{-5}
	I	1×10^{-10}	8×10^{-4}	4×10^{-12}	3×10^{-5}

See footnotes at end of table.

TABLE 5A.5. (*Continued*)

Element-Z	Isotope[a]	Table I		Table II	
		Column 1 Air (μCi/ml)	Column 2 Water (μCi/ml)	Column 1 Air (μCi/ml)	Column 2 Water (μCi/ml)
Uranium-92 continued	^{236}U S	6×10^{-10}	1×10^{-3}	2×10^{-11}	3×10^{-5}
	I	1×10^{-10}	1×10^{-3}	4×10^{-12}	3×10^{-5}
	^{238}U S	7×10^{-11}	1×10^{-3}	3×10^{-12}	4×10^{-5}
	I	1×10^{-10}	1×10^{-3}	5×10^{-12}	4×10^{-5}
	^{240}U S	2×10^{-7}	1×10^{-3}	8×10^{-9}	3×10^{-5}
	I	2×10^{-7}	1×10^{-3}	6×10^{-9}	3×10^{-5}
	U (natural) S	7×10^{-11}	5×10^{-4}	3×10^{-12}	2×10^{-5}
	I	6×10^{-11}	5×10^{-4}	2×10^{-12}	2×10^{-5}
Vanadium-23	^{48}V S	2×10^{-7}	9×10^{-4}	6×10^{-9}	3×10^{-5}
	I	6×10^{-8}	8×10^{-4}	2×10^{-9}	3×10^{-5}
Xenon-54	^{131m}Xe Sub	2×10^{-5}	—	4×10^{-7}	—
	^{133}Xe Sub	1×10^{-5}	—	3×10^{-7}	—
	^{133m}Xe Sub	1×10^{-6}	—	3×10^{-7}	—
	^{135}Xe Sub	4×10^{-6}	—	1×10^{-7}	—
Ytterbium-70	^{175}Yb S	7×10^{-7}	3×10^{-3}	2×10^{-8}	1×10^{-4}
	I	6×10^{-7}	3×10^{-3}	2×10^{-8}	1×10^{-4}
Yttrium-39	^{90}Y S	1×10^{-7}	6×10^{-4}	4×10^{-9}	2×10^{-5}
	I	1×10^{-7}	6×10^{-4}	3×10^{-9}	2×10^{-5}
	^{91m}Y S	2×10^{-5}	1×10^{-1}	8×10^{-7}	3×10^{-3}
	I	2×10^{-5}	1×10^{-1}	6×10^{-7}	3×10^{-3}
	^{91}Y S	4×10^{-8}	8×10^{-4}	1×10^{-9}	3×10^{-5}
	I	3×10^{-8}	8×10^{-4}	1×10^{-9}	3×10^{-5}
	^{92}Y S	4×10^{-7}	2×10^{-3}	1×10^{-8}	6×10^{-5}
	I	3×10^{-7}	2×10^{-3}	1×10^{-8}	6×10^{-5}
	^{93}Y S	2×10^{-7}	8×10^{-4}	6×10^{-9}	3×10^{-5}
	I	1×10^{-7}	8×10^{-4}	5×10^{-9}	3×10^{-5}
Zinc-30	^{65}Zn S	1×10^{-7}	3×10^{-3}	4×10^{-9}	1×10^{-4}
	I	6×10^{-8}	5×10^{-3}	2×10^{-9}	2×10^{-4}
	^{69m}Zn S	4×10^{-7}	2×10^{-3}	1×10^{-8}	7×10^{-5}
	I	3×10^{-7}	2×10^{-3}	1×10^{-8}	6×10^{-5}
	^{69}Zn S	7×10^{-6}	5×10^{-2}	2×10^{-7}	2×10^{-3}
	I	9×10^{-6}	5×10^{-2}	3×10^{-7}	2×10^{-3}
Zirconium-40	^{93}Zr S	1×10^{-7}	2×10^{-2}	4×10^{-9}	8×10^{-4}
	I	3×10^{-7}	2×10^{-2}	1×10^{-8}	8×10^{-4}
	^{95}Zr S	1×10^{-7}	2×10^{-3}	4×10^{-9}	6×10^{-5}
	I	3×10^{-8}	2×10^{-3}	1×10^{-9}	6×10^{-5}
	^{97}Zr S	1×10^{-7}	5×10^{-4}	4×10^{-9}	2×10^{-5}
	I	9×10^{-8}	5×10^{-4}	3×10^{-9}	2×10^{-5}
Any single radionuclide not listed above with decay mode other than alpha emission or spontaneous fission and with radioactive half-life less than 2 hr	Sub	1×10^{-6}	—	3×10^{-8}	—
Any single radionuclide not listed above with decay mode other than alpha emission or spontaneous fission and with radioactive half-life greater than 2 hr		3×10^{-9}	9×10^{-5}	1×10^{-10}	3×10^{-6}
Any single radionuclide not listed		6×10^{-13}	4×10^{-7}	2×10^{-14}	3×10^{-8}

See footnotes at end of table.

TABLE 5A.5. (*Continued*)

Element-Z	Isotope[a]	Table I Column 1 Air (μCi/ml)	Table I Column 2 Water (μCi/ml)	Table II Column 1 Air (μCi/ml)	Table II Column 2 Water (μCi/ml)
above, which decays by alpha emission or spontaneous fission					
Element and Isotope					
If it is known that ^{90}Sr, ^{125}I, ^{126}I, ^{129}I, ^{131}I, (^{133}I, Table II only), ^{210}Pb, ^{210}Po, ^{211}At, ^{223}Ra, ^{224}Ra, ^{226}Ra, ^{227}Ac, ^{228}Ra, ^{230}Th, ^{231}Pa, ^{232}Th, Th (natural), ^{248}Cm, ^{254}Cf, and ^{256}Fm are not present		9×10^{-5}		3×10^{-6}	
If it is known that ^{90}Sr, ^{125}I, ^{126}I, ^{129}I, (^{131}I, ^{133}I, Table II only), ^{210}Pb, ^{210}Po, ^{223}Ra, ^{226}Ra, ^{228}Ra, ^{231}Pa, Th (natural), ^{248}Cm, ^{254}Cf, and ^{256}Fm are not present		6×10^{-5}		2×10^{-6}	
If it is known that ^{90}Sr, ^{129}I, (^{125}I, ^{126}I, ^{131}I, Table II only), ^{210}Pb, ^{226}Ra, ^{228}Ra, ^{248}Cm, and ^{254}Cf are not present		2×10^{-5}		6×10^{-7}	
If it is known that (^{129}I, Table II only), ^{226}Ra, and ^{228}Ra are not present		3×10^{-6}		1×10^{-7}	
If it is known that alpha emitters and ^{90}Sr, ^{129}I, ^{210}Pb, ^{227}Ac, ^{228}Ra, ^{230}Pa, ^{241}Pu, and ^{249}Bk are not present		3×10^{-9}		1×10^{-10}	
If it is known that alpha emitters and ^{210}Pb, ^{227}Ac, ^{228}Ra, and ^{241}Pu are not present		3×10^{-10}		1×10^{-11}	
If it is known that alpha emitters and ^{227}Ac are not present		3×10^{-11}		1×10^{-12}	
If it is known that ^{227}Ac, ^{230}Th, ^{231}Pa, ^{238}Pu, ^{239}Pu, ^{240}Pu, ^{242}Pu, ^{244}Pu, ^{248}Cm, ^{249}Cf, and ^{251}Cf are not present		3×10^{-12}		1×10^{-13}	

[a] Soluble (S); Insoluble (I).

[b] "Sub" means that values given are for submersion in hemispherical infinite cloud of airborne material.

Note: In any case where there is a mixture in air or water of more than one radionuclide, the limiting values should be determined as follows:

(1) If the identity and concentration of each radionuclide in the mixture is known, the limiting vales should be derived as follows: Determine, for each radionuclide in the mixture, the ratio between the quantity present in the mixture and the limit otherwise established for the specific radionuclide when not in a mixture. The sum of such ratios for all the radionuclides in the mixture may not exceed "1" (i.e., "unity").

Example. If radionuclides A, B, and C are present in concentrations C_A, C_B, and C_C, and if the applicable MPC's are MPC_A and MPC_B, and MPC_C, respectively, then the concentrations shall be limited so that the following relationship exists:

$$\frac{C_A}{MPC_A} + \frac{C_B}{MPC_B} + \frac{C_C}{MPC_C} \leqslant 1$$

(2) If either the identity or the concentration of any radionuclide in the mixture is not known, the limiting values shall be:

(a) For purposes of Table I, Col. 1 = 6×10^{-13}
(b) For purposes of Table I, Col. 2 = 4×10^{-7}
(c) For purposes of Table II, Col. 1 = 2×10^{-14}
(d) For purposes of Table II, Col. 2 = 3×10^{-8}

(3) If any of the conditions specified below are met, the corresponding values specified below may be used in lieu of those specified in paragraph (2) above.

(a) If the identity of each radionuclide in the mixture is known but the concentration of one or more of the radionuclides in the mixture is not known, the concentration limit for the mixture is the limit specified for the radionuclide in the mixture having the lowest concentration limit: or

(b) If the identity of each radionuclide in the mixture is not known, but it is known that certain radionulcides specified are not present in the mixture, the concentration limit of the mixture is the lowest concentration limit specified for any radionuclide which is not known to be absent from the mixture: or

(4) If the mixture of radionuclides consists of uranium and its daughter products in ore dust prior to chemical processing of the uranium ore, the values specified below may be used in lieu of those determined in accordance with paragraph (1) above or those specified in paragraphs (2) and (3) above.

(a) For purposes of Table I, Col. 1 = 1×10^{-10} μCi/ml gross α activity; or 2.5×10^{-11} μCi/ml natural uranium; or 75 μg/m^3 of natural uranium in air.

(b) For purposes of Table II, Col. 1 = 3×10^{-12} μCi/ml gross α activity; or 8×10^{-13} μCi/ml natural uranium; or 3 $\mu g/m^3$ of natural uranium in air.

(5) For purposes of this note a radionuclide may be considered as not present in a mixture of (a) the ratio of the concentration of that radionuclide in the mixture (C_A) to the concentration limit for that radionuclide specified (MPC_A) does not exceed 1/10 (i.e., $C_A/MPC_A \leqslant 1/10$) and (b) the sum of such ratios for all the radionuclides considered as not present in the mixture does not exceed 0.25, i.e.,

$$\frac{C_A}{MPC_A} + \frac{C_B}{MPC_B} + \ldots \leqslant 0.25$$

The Nuclear Regulatory Commission states that radiation exposure should be maintained as low as is reasonably achievable, commonly referred to as the ALARA principle. Thus, while the limits of exposure shown above are allowed, they are not considered as satisfactory performance. NRC Regulatory Guide 8.10 explains that management should be committed to minimizing exposures and that the personnel responsible for radiation protection should be continually vigilant for ways to reduce exposures.

It should be clearly understood that the maximum permissible levels have been set up for radiation protection purposes. They do follow the latest proven scientific information as closely as possible; however, the various constants and metabolic factors that have proved to be quite acceptable in radiological health calculations are not satisfactory for more exact scientific purposes.

Radiation Measurement

The measurement of radiation for the purposes of protection requires both the proper instruments and the proper techniques of using them. This section must necessarily concentrate on the second of these requirements.

There are two major types of instruments for measurement of radiation and radioactivity. These are the field survey instruments used for measurements on the spot and the laboratory instruments which require that a sample be brought back into the laboratory. The latter, of course, are usually more accurate, but field equipment is adequate for the accuracy required in radiation surveys.

Laboratory and field survey equipment for the measurement of radiation are based on three types of detectors. In the first type, the ionization produced in a gas is collected to give an electrical pulse current which can be measured. These gas ionization instruments include the ionization chamber, the proportional counter, and the Geiger counter. The second type is based on conversion of radiation energy into a light pulse by a scintillator. The light pulse is then reconverted to an electrical pulse or current by means of a photomultiplier tube. Various scintillating materials, such as zinc sulfide for α particles and sodium iodide for γ rays have been widely applied. The third type is based on the electric current induced in relatively small silicon or germanium diodes when irradiated. These solid-state detectors are quite expensive and are generally part of a sophisticated set of equipment for determining α or γ spectra. This system allows qualitative identification and quantitative analysis of a large number of nuclides simultaneously. It is beyond the scope of this section to go into the theory and operation of counters, but references are appended in which such information is available.

A valuable tool for determination of the body burden of a γ emitting isotope is the whole-body counter. This instrument usually consists of a large sodium iodide scintillation detector in a shielded room with a multichannel analyzer as auxiliary equipment. Such a spectrometer system allows qualitative and quantitative estimates of nuclides in samples as large as a human body. This is important since a normal man contains about 150 g of K which is naturally radioactive and can mask small amounts of other radioisotopes if qualitative identification is not carried out.

For body burdens of γ emitters approaching mpl's, a simpler, more portable shield can be used. Several of these have been reported in the literature and their use does allow measurements to be made without the costly shielded room.

While no details of instruments are given in this section, it is desirable to point out certain precautions that must be observed in their operation.

(1) Instruments must be maintained in good repair and calibrated at intervals to assure the desired accuracy.

(2) The instrument used must be selected so that it is sensitive to the radiation to be measured. For example, the standard Geiger survey meter does not respond well to α or weak β radiation.

(3) As a corollary, the instrument calibrations must be carried out with isotopes having the same or similar energy to those being measured.

(4) Personnel handling or using radiation instruments must have an understanding of their characteristics and of the possible interpretation of readings obtained.

Survey instruments should be calibrated in terms of the particular radiation which will be present in the actual operations. However, it is still standard practice to calibrate such instruments in terms of the γ radiation from a Ra or ^{60}Co source. Ra and Co sources are prepared in terms of Rhm units, i.e., R/hr at a distance of 1 meter. Other dose rates may be obtained by varying the distance between the source and the instruments.

Table 5A.6 gives the Rhm values[7] for 1 Ci of various isotopic sources, illustrating the possible differences between sources. A useful approximation is $\text{Rhm} = 0.6\, CiE$, where Ci is the number of curies and E is the energy of the radiation measured in MeV.

TABLE 5A.6. Calculated Gamma Radiation Levels[a] for 1 Curie of Some Radioisotopes

Isotope	Half-life	Roentgens per Hour at 1 Foot	Roentgens per Hour at 1 Meter
^{22}Na	2.6 yr	14.1	1.31
^{24}Na	15 hr	20.6	1.92
^{52}Mn	5.7 days	20.8	1.93
^{54}Mn	314 days	5.22	0.485
^{59}Fe	45 days	7.0	0.651
^{58}Co	71 days	6.02	0.560
^{60}Co	5.3 yr	14.3	1.3
^{64}Cu	12.8 hr	4.52	0.42
^{65}Zn	245 days	3.22	0.30
^{76}As	26.5 hr	5.81	0.54
^{82}Br	36 hr	16.1	1.50
^{128}I	25 min	0.194	0.018
^{130}I	12.5 hr	13.5	1.25
^{131}I	8.1 days	2.54	0.236
^{137}Cs	30 yr	3.4	0.32
^{170}Tm	127 days	0.04	0.004
^{192}Ir	74 days	5.4	0.5
^{198}Au	2.7 days	2.5	0.23
Radium in equilibrium (0.5 mm Pt filtration)		9.04	0.84

[a]For an unshielded point source except as noted. The values shown have been modified from the original.

Radium calibration sources in themselves present an additional hazard above that of their γ radiation. The powdered radium salts are enclosed in platinum or a glass and platinum capsule. If the capsule should be ruptured, a considerable quantity of α-emitting material may be released. Cobalt sources, on the other hand, are solid metallic material and are not subject to rupture. Their only disadvantage is the change in their activity with time—the ^{60}Co decays with a half-life of 5.3 years.

Personnel Monitoring

Licensees under the NRC are required to provide appropriate personnel monitoring equipment under the following circumstances.

(1) Any individual entering a restricted area who is likely to receive in excess of 25% of the permissible dose in any calendar quarter.

(2) Any individual under age 18 who is likely to receive an excess of 5% of the permissible dose in any calendar quarter.

(3) Any individual who enters a high radiation area which is defined as an area where a major portion of the body could receive in excess of 100 mrem in any 1 hour.

Photographic film has been the accepted standard for evaluation of personnel exposure to low radiation levels. Special monitoring films are readily available in convenient package form. Naturally, the degree of blackening of the film must be interpreted by means of a microphotometer or densitometer in terms of radiation dose. This is done by calibrating each batch of film with known amounts of radiation.

The photographic film is not a perfect dosimeter, and considerable effort has been devoted to the development of glass dosimeters and thermoluminescent dosimeters (TLD) for personnel monitoring. The sensitivity of the latter is somewhat better than film, and it has been finding increasingly wide use. In principle, calcium fluoride or lithium fluoride are phosphors which store energy when exposed to ionizing radiation. When the phosphor is heated to several hundred degrees Centigrade, this energy is released in the form of visible light and can be measured with a photometer.

The film or TLD packet can be merely pinned to the work clothes, or if only penetrating radiation is to be encountered, it can be carried in the pocket. However, the common method of wearing it is to enclose the packet in a holder (badge) for convenience. This badge, allows identification of the film with the individual, it holds the film or TLD in the proper place wth respect to the body and by bcing worn outside the clothing it can be made more responsive to less-penetrating radiation such as β particles. Unless an installation employs a large number of workers with possible radiation exposure, the expense of setting up and operating a badge service, including the calibrating and processing, is very high. Fortunately, several reliable organizations offer a badge service for smaller installations at a reasonable cost.

The film badge and thermoluminescent dosimeters are limited by the requirement that they be processed before a measure of radiation exposure can be obtained. This is perfectly satisfactory where the personnel monitoring is done only as a matter of record and there is no danger of high-level exposures. If such exposures are possible, a pocket ionization chamber (dosimeter) in addition to the film is desirable. Pocket chambers are available in various ranges of total radiation dose and are made either as direct-reading dosimeters with their own internal scale or as simple chambers which must be inserted in a "reader" to obtain the dose measurement. All of these operate by charging the chamber before use. Exposure to radiation causes ionization within the chamber and allows a portion of this charge to leak off. After wearing, the residual charge is measured and the amount of leakage related to the radiation exposure. The internal scale or the reader scale is calibrated directly in terms of dose received. Dosimeters can be read daily or hourly to follow the exposure level if the circumstances require. However, they are chiefly sensitive to γ or x-radiation, since the walls of the chamber will effectively cut out α and most β particles.

The recommended monitoring period, e.g., the time of wearing a film badge or TLD, is usually 1 month in the case of predictable low-level exposure. If the operator can show that the maximum exposure is less than one-fourth of the permissible occupational level, individual personnel monitoring is not required.

Radiation Surveys

The general area survey made with a portable survey meter is most useful for locating sources of radiation and evaluating possible hazard, but is also valuable for the immediate determination of the effectiveness of corrective measures or of cleanup procedures. When used for exposure evaluation, particular care should be taken to make measurements with the detector probe in the locations occupied by the operator's body, as well as measurements throughout the entire working area.

Area surveys are carried out with a portable dose rate survey meter sensitive to the radiation present. General area surveys for β and γ radiation are best performed by holding the detector away from the body at waist level and tracing a systematic path through the area, noting any rise in radiation level and marking these spots either directly with chalk or on a diagram. Radiation levels above the permissible level require cleaning or supplementary shielding and resurveying until the level is reduced to acceptable limits. Levels markedly above background but below the permissible level may be attacked in a more leisurely manner.

Surface contamination from α or β emitters can also be surveyed by moving a suitable detector slowly over a systematic path covering the floor, walls, and working surfaces. Areas indicating contamination well above background may be checked with a wipe test as described in the discussion on surface contamination.

Permissible radiation levels are set up based on the dose received above that due to the radiation that is normally present. This background radiation is variable and must be considered when surveying work areas. Background level can best be measured with the actual instrument used for the survey in a spot well away from the work area. Readings on survey instruments ranging from 0.01–0.02 mR/hr are quite common and the background may be even higher in geographical areas where large deposits of radioactive elements are found.

Any instruments used for area surveys should be sensitive to the least penetrating radiation present. Survey meters are worthless unless they are maintained in proper condition and are calibrated frequently enough to assure the accuracy of the readings. Small "button" sources are available from the instrument manufacturers for frequent rough checks of operating conditions, but a calibration of the instrument on all scales should be made 3 or 4 times a year. Inaccurate or uncalibrated instruments can be dangerous if they give low readings in the presence of high radiation fields.

On the other hand, incorrect high readings may cause considerable expense in trying to remedy a situation which does not actually exist. Calibration services are available from instrument manufacturers and from consulting laboratories in the field of radiation protection. If several instruments are in use, the cost of calibrating sources and their shielding will be negligible compared with the convenience of being able to run calibrations without having the instruments out of service for extended periods.

Surface Contamination

The contamination of working surfaces (or other portions of the working area) with radioactive material is referred to as surface contamination. If the isotopes are γ emitters, the contamination is a source of external radiation. The magnitude of this radiation can best be evaluated with survey equipment. Surface contamination measured with survey instruments is not a good measure of the hazard for α or β emitters. Alpha emitters in particular become critical when they are airborne and available for inhalation or ingestion. Several methods of measuring the amount of surface activity that can become airborne have been devised, none of which is completely satisfactory.

The most widely accepted method is to wipe a fixed area with a piece of filter paper and measure the activity taken up by the paper. The results are expressed in terms of disintegrations/minute/100 cm^2 or similar units. Since the correlation between the amount that can be wiped off with a paper and the amount that can become airborne or can be picked up on the hands is not good, this cannot be considered very meaningful. Low but positive results from a wipe test should not lead to ignoring a removable surface contamination, as it is always desirable to keep working areas as free as possible from radioactivity. However, the cleaning of such surfaces need not be performed on an emergency basis.

Airborne Radioactivity

Almost all operations involving radioisotopes may result in some of the activity becoming airborne. This may arise from the dusting of dry materials, spraying from solutions, or the liberation of radioactive gaseous products. It is common practice in both laboratory and plant processes to carry out all operations in enclosed hoods or with good general ventilation.

The high toxicity of small amounts of some radioisotopes when taken into the body make the collection and evaluation of airborne radioactivity a critical part of most radiation protection systems. Solid or liquid materials can be determined by drawing known quantities of contaminated air through a filter and measuring the activity collected on the filter. This can be an extremely rapid and accurate method of analysis.

Two types of hazard evaluation are possible. One is the continuous sampling of air-borne material in the general work area for an overall picture of the probable exposure. The other is the sampling of air in the operator's breathing zone during the time that the actual process which may give rise to airborne activity is being carried out.

Gaseous isotopes, of course, will not be collected on a filter. (The natural Rn isotopes can be estimated, however, by the collection of their solid daughter products on a filter.) Other radioactive gases must be absorbed in a gas washer or an impinger and the resulting solution treated to convert the activity to a form suitable for counting. Even this will not work with some of the inert gases such as krypton and argon and more elaborate methods are needed for their measurement. However, their production and measurement are very specialized problems encountered only in reactor operation and fuel reprocessing.

Bioassay

Measurement of airborne dust or other activity is difficult on a continuous basis, particularly if it is desirable to monitor the exposure of each individual. Some evaluation is possible by measurements on the individual himself (bioassay). For example, analysis for the amount of an isotope excreted in the urine of an individual may furnish an indication of relative exposure. While the conversion of the concentration of isotope in the urine to the body burden or the concentration of that isotope in the air breathed is not always possible, the excretion does constitute a measure of the relative intake and can be valuable as a secondary means of exposure evaluation. No permissible levels of urine excretion have been set up, since in addition to the usual variables found in biological systems, others such as diet and fluid intake can cause marked fluctuations in the urinary concentration of an isotope. Urine analyses, therefore, are only valuable when a continuing sequence is maintained. Methods of analysis have been recommended by a joint panel of the WHO, the IAEA, and the FAO.[8]

One specific bioassay method is accepted for radium exposure evaluation. That is the measurement of the radon concentration in the expired air. Radon is one of the rare gases and an immediate daughter of radium. A portion of the radon formed by radium decay in the body is transported to the lungs and appears in the exhaled breath. The value of 1 pCi of Rn/liter has been set as an indication of 0.1 μg of Ra in the body. This latter value is the maximum permissible amount based on the experience of the radium dial painters. The actual measurement of breath radon content requires specialized apparatus.

Bioassay samples are taken at various intervals depending on the degree of exposure. Slightly exposed personnel may be sampled yearly while those engaged in actual processing may be sampled quarterly. In a few cases where there is exposure to highly toxic substances such as plutonium, urine analysis may be performed more frequently.

Total Exposure

The overall evaluation of plant personnel exposure to radiation must include all possible factors. For example, if there is total body external irradiation, radiation from inhaled or ingested isotopes and surface irradiation of the hands when handling isotopes, it is necessary that the sum of these exposures does not exceed the maximum permissible level. Such a summation can be a complicated calculation. However, a reasonable approximation can be made, relatively simply, by taking the percentage of permissible level represented by each exposure and summing them up. If they are less than 100% the total exposure should be acceptable.

Overexposures are handled according to their severity. Where a weekly dose is excessive, it should be possible to maintain the 13-week summation below the permissible level by reducing the subsequent exposure, or if necessary by removing the individual entirely from possible exposure. Any exposure greater than that which can be treated in this way is serious enough so that the NRC regulations as well as any applicable state or local codes will require notification* of such an occurrence, the reason for it, and procedures instituted to prevent recurrence. Whenever overexposure occurs, the source must be located by a thorough survey. The source must be removed and adequate safeguards set up to prevent recurrence. It should be noted here, however, that no system of rules and safeguards can completely protect careless or ignorant individuals. Therefore, considerable stress must be placed on training the worker in safe practices.

Waste Monitoring

Since a radiation protection program includes prevention of exposure of those outside the plant as well as plant personnel, it is necessary to monitor plant effluents and wastes before they can be disposed of through the usual channels. The levels of activity permitted in waste materials are quite low and direct measurement with survey instruments is not usually accurate enough to be adequate. In most cases, only a very few isotopes can possibly be present. Thus samples of the waste material can be put through relatively simple chemical separations to isolate a particular isotope from the bulk of the material and to prepare it in a form sutiable for counting. A few simple cases may be handled by merely concentrating the isotope, for example, by evaporation of

*Requirements for notification of the NRC are detailed in Part 20 of Title 10, Chapter 1, of the *Code of Federal Regulations*.

waste solutions or by ashing of combustible solid wastes. Where the isotope is a γ emitter, the concentration procedures may be perfectly adequate, but for β emitters, particularly those of low energy, the sample bulk may absorb a large fraction of the emited particles. Then it is necessary to resort to some form of chemical separation. The chemical methods required for such analyses are described in the literature. The sampling process is frequently much more complex, particularly where the wastes are not homogeneous. For this reason, many installations where only small amounts of wastes are accumulated do not bother with sampling or analysis, but treat all such material as contaminated and dispose of it by appropriate methods.

RESPONSIBILITY

The specific responsibilities of the licensee handling isotopes (by-product material), source materials, or radiographic sources are set forth in the CFR, and similar responsibilities apply to government contractors. Medical x-ray and nuclear medical operations are usually covered in state and local codes. All regulations place the responsibility for the control of radiation exposure squarely on the management of the installation who must design facilities and plan operations so that overexposures do not occur.

The responsibility of the supervisor of work with x-ray equipment or with radioisotopes is perhaps even greater than that of a supervisor in charge of dangerous chemical operations. The immediate effects of most chemical accidents are fully appreciated and the chemical agent can usually be neutralized or destroyed to prevent further exposure. This is not true for radiation or radiation emitters, and the lack of immediate effect makes supervisory control more difficult.

In considering the handling of reasonably long-lived isotopes, it must be realized that responsibility involves being certain both that the workers who handle the material are not overexposed and that none of the material escapes to endanger others.

All instruction designed to minimize the radiation exposure of plant workers should originate with the technically qualified man in charge. Most of the attention must be directed to giving the operating supervisors a complete picture of the nature of the hazards, the ways in which the resulting exposures can be minimized, and the value of the monitoring and control procedures to be used. Depending on the size of the organization and the quality of the personnel, this same instruction may be given to all workers or else the complete responsibility may be given to the operating supervisor. In any case, the workers should know that they are being exposed to radiation, the steps being taken to protect them from overexposure, and the safety regulations which are made necessary by the nature of radiation. It is sometimes difficult to teach safe practices without causing excessive alarm in some people, but the fear of alarm cannot be used to justify withholding such information. For example, a licensee for radiography must describe an adequate training program of:

(a) Initial training
(b) Periodic training
(c) On-the-job training
(d) Means to be used by the licensee to determine the radiographer's assistant's knowledge of, understanding of, and ability to comply with the NRC regulations and licensing requirements and the operating and emergency procedures of the licensee

The paragraphs below will explain the responsibility of supervisory personnel to the various groups under their control, as well as to those outside the plant. Management, operational supervisors and technical personnel must all play a part in minimizing radiation exposure.

Workers

The most immediate responsibility of management is to the workers directly concerned with radiation. Responsible persons must see that the exposure of the workers is continuously evaluated, that records of their exposure are maintained, and that no continued overexposure is allowed. With a group of nontechnical personnel this responsibility can best be delegated to the operating supervisor. From his other duties, he is familiar with the operations and personnel and is able to issue direct orders. It is not necessary that he be completely trained in the technical aspects of radiation protection, but rather that he understand the measurements made and the meaning of the numbers obtained. In addition, there must be someone available to give technical advice on radiation protection problems. While a large organization may have its own radiation safety officer (RSO) or similarly qualified person, the smaller organizations may have to depend on consultants in the field. In either case, the direct day-to-day supervision can best be handled through the operational supervisor or a radiation safety supervisor. In technical groups such as laboratories or hospitals, there is still need for someone to have charge of records and to perform the duties of an operational supervisor. This person must have the authority to issue direct orders concerning exposure. Almost all of the recorded cases of serious radiation overexposure have been to technical personnel who knew and ignored the possible consequence. Management is still responsible for the actions of these employees. To protect the workers and themselves, they must set up an adequate supervisory system and follow the records to see that the program is being carried out. Those working directly with radiation will have the highest potential exposure and must be given the most attention in a radiation protection program. This must include exposure evaluation, exposure records, and job training to minimize hazards.

All other employees may be split into two classifications: those who enter the work area for relatively short periods, such as maintenance workers, and those who do not enter the area. In an operation involving low radiation levels, their exposure can be considered negligible if that of the direct workers is well below the maximum permissible levels. In any case, casual visits to the work area should be minimized in length and cut to a very small number by proper planning. If the radiation level in the work area is higher, it becomes necessary to include anyone entering the work area in the complete control program of personnel monitoring and exposure records. In a large organization this can become a matter of considerable expense, but again, by careful planning, the number of these casual visitors can be kept to a minimum.

The casual worker in the operating area has the added disadvantage that he is not fully acquainted with the operations or the hazards involved and may do something which would expose him to more radiation than someone familiar with the operations. The supervisor for the area is responsible for the safety of these part-time workers and must either supervise their activities or instruct them in the safety procedures of the area. Employees outside the work area must receive only a minimum exposure and this is the responsibility of the technically qualified person in charge. Generally, minimizing such exposure is accomplished in the original design of the installation, but periodic checks of the operation of control measures must be made. Personnel monitoring should not be required as any design which would allow a measurable exposure outside the work area is entirely unsatisfactory, and no operations should be carried out in such an area.

Environment

Minimizing the hazards of exposure to persons in the neighborhood of the plant is the responsibility of the qualified technical man in charge. Direct radiation should not be a hazard: The problems are rather those of radioisotope waste disposal. Periodic checks of contamination of the environment must be made, and if the level of activity handled in the installation increases, continuous monitoring may be required. The permissible levels of exposure for plant personnel are not satisfactory for those living or working in the neighborhood. All plant workers must realize the hazards involved in their jobs and, in accepting a job, must also accept the hazards. This is true in any field and is not unique for those engaged in work with radiation. People outside the plant are not aware of the hazards and have not entered into any agreement with the company. Therefore, every effort should be made to see that their possible exposure is far below that allowed for plant personnel. The limitation of exposure for the population has been set at 5 rem/generation, or 170 mrem/year. The NRC principle of "as low as reasonably achievable" (ALARA) appears to be interpreted as 10 mrem/yr for exposure in the neighborhood of nuclear facilities.

RADIATION CONTROL

A radiation protection program has two facets—the continuous evaluation of exposure and the reduction of exposure by any applicable control measure. The fact that all exposures are maintained below maximum permissible levels is an indication that the control procedures are working, but since any unnecessary exposure is foolish, the possibility of maintaining values as low as practicable should be continuously kept in mind.

A major factor in control is the proper training of operating personnel. It is part of supervisory responsibility that every worker know what he is working with, what the hazards are, and what measures are being taken to secure his safety. He must be trained in safe techniques and know what to do in case of any accident. And he must be made to realize that observance of safety rules and personnel monitoring requirements are just as much a part of the job as the actual operations performed.

At the present time, the use of most radio-isotopes requires licensing through the NRC or the states. This means that a certain degree of control is exercised in the allocation of isotopes. Applications are reviewed, and applicants are required to describe the purpose for which the isotopes are to be used and the facilities and equipment available for radiation protection. The applicant agrees to keep records of the receipt, use, and distribution of the isotopes, and further agrees not to use them for purposes other than those described or even to redistribute the materials to others without authorization.

NRC licensees, government contractors and other users of radiation sources are subject to various types of inspection. It has been pointed out, however, that these inspections are not intended to control radiation, but merely to determine compliance with regulations. Responsibility for control rests with the operator.

There are many laboratories working with μCi amounts of various radioisotopes which do not require any special handling or special equipment. The NRC has issued a list of exempt quantities of various radioisotopes. This list is reproduced in Table 5A.7. The quantities are exempted based on the concept that this amount of the given isotope cannot cause any damage under any conditions of use or misuse. On this basis, these quantities indicate whether or not the laboratory should take any precautions in handling.

TABLE 5A.7. Quantities of Radioisotopes Exempt from License Requirements

Material	Microcuries	Material	Microcuries
^{122}Sb	100	^{134}I	10
^{124}Sb	10	^{135}I	10
^{125}Sb	10	^{192}Ir	10
^{73}As	100	^{194}Ir	100
^{74}As	10	^{55}Fe	100
^{76}As	10	^{59}Fe	10
^{77}As	100	^{85}Kr	100
^{131}Ba	10	^{87}Kr	10
^{140}Ba	10	^{140}La	10
^{210}Bi	1	^{177}Lu	100
^{82}Br	10	^{52}Mn	10
^{109}Cd	10	^{54}Mn	10
^{115m}Cd	10	^{56}Mn	10
^{115}Cd	100	^{197}Hg	100
^{45}Ca	10	^{197m}Hg	100
^{47}Ca	10	^{203}Hg	10
^{14}C	100	^{99}Mo	100
^{141}Ce	100	^{147}Nd	100
^{143}Ce	100	^{149}Nd	100
^{144}Ce	1	^{59}Ni	100
^{131}Cs	1,000	^{63}Ni	10
^{134m}Cs	100	^{65}Ni	100
^{134}Cs	1	^{93m}Nb	10
^{135}Cs	10	^{95}Nb	10
^{136}Cs	10	^{97}Nb	10
^{137}Cs	10	^{185}Os	10
^{36}Cl	10	^{191m}Os	100
^{38}Cl	10	^{191}Os	100
^{51}Cr	1,000	^{193}Os	100
^{58m}Co	10	^{103}Pd	100
^{58}Co	10	^{109}Pd	100
^{60}Co	1	^{32}P	10
^{64}Cu	100	^{191}Pt	100
^{165}Dy	10	^{193m}Pt	100
^{166}Dy	100	^{193}Pt	100
^{169}Er	100	^{197m}Pt	100
^{171}Er	100	^{107}Pt	100
^{152}Eu(9.2 h)	100	^{210}Po	0.1
^{152}Eu(13 y)	1	^{42}K	10
^{154}Eu	1	^{142}Pr	100
^{155}Eu	10	^{143}Pr	100
^{18}F	1,000	^{147}Pm	10
^{153}Gd	10	^{149}Pm	10
^{159}Gd	100	^{186}Re	100
^{72}Ga	10	^{188}Re	100
^{71}Ge	100	^{103m}Rh	100
^{198}Au	100	^{105}Rh	100
^{199}Au	100	^{86}Rb	10
^{181}Hf	10	^{87}Rb	10
^{166}Ho	100	^{97}Ru	100
^{3}H	1,000	^{103}Ru	10
^{113m}In	100	^{105}Ru	10
^{114m}In	10	^{106}Ru	1
^{115m}In	100	^{151}Sm	10
^{115}In	10	^{153}Sm	100
^{125}I	1	^{46}Sc	10
^{126}I	1	^{47}Sc	100
^{129}I	.1	^{48}Sc	10
^{131}I	1	^{75}Se	10
^{132}I	10	^{31}Si	100
^{133}I	1	^{105}Ag	10

TABLE 5A.7 (*Continued*)

Material	Microcuries	Material	Microcuries
^{110m}Ag	1	^{204}Tl	10
^{111}Ag	100	^{170}Tm	10
^{24}Na	10	^{171}Tm	10
^{85}Sr	10	^{113}Sn	10
^{89}Sr	1	^{125}Sn	10
^{90}Sr	0.1	^{181}W	10
^{91}Sr	10	^{185}W	10
^{92}Sr	10	^{187}W	100
^{35}S	100	^{48}V	10
^{182}Ta	10	^{131m}Xe	1,000
^{96}Tc	10	^{133}Xe	100
^{97m}Tc	100	^{135}Xe	100
^{97}Tc	100	^{175}Yb	100
^{99m}Tc	100	^{90}Y	10
^{99}Tc	10	^{91}Y	10
^{125m}Te	10	^{92}Y	100
^{127m}Te	10	^{93}Y	100
^{127}Te	100	^{65}Zn	10
^{129m}Te	10	^{69m}Zn	100
^{129}Te	100	^{69}Zn	1,000
^{131m}Te	10	^{93}Zr	10
^{132}Te	10	^{95}Zr	10
^{160}Tb	10	^{97}Zr	10
^{200}Tl	100	Any by-product not listed above other than alpha-emitting by-product material	0.1
^{201}Tl	100		
^{202}Tl	100		

The measurement and evaluation of exposure have been discussed previously; the following discussion will introduce the various methods of reducing radiation exposure.

Reducing External Exposure

The three methods of reducing external radiation exposure are: (1) increasing shielding; (2) decreasing the time of exposure; and (3) increasing the distance between worker and source. In the usual x-ray or isotope installation the shielding is part of the original design, but modifications of equipment or changes in operating technique may require shielding changes. Such shielding is subject to the same criteria as that in the original design and will be covered later. The actual physical setup may vary, however, if only temporary shielding is required. Temporary shielding is set up with interlocking lead bricks or steel plates of the proper thickness. Such shields can be readily assembled or stored when no longer needed. If these temporary shielding materials are made easily available in the work area they will be used more often.

Reducing the time of exposure is not merely a matter of calculating the permissible time in a certain area which will keep the exposure just at the maximum permissible level. Exposure time should be reduced to the minimum consistent with sound economical operation. This can be facilitated by careful planning of procedures. In a practical sense this means that isotopes should be kept in shielded containers when not in actual use, x-ray workers should not enter any area where measurable scattering from x-ray equipment occurs except when absolutely necessary, and all those not directly concerned with an immediate operation should be excluded from the irradiated area. Again, these steps are not taken merely to maintain exposures below maximum permissible level but to keep overall exposure to a minimum.

One method of reducing exposure time which may be applicable in some instances is rotation of personnel. The operators or technicians may be trained to interchange jobs, and the higher-exposure operations rotated through the largest possible group.

The factor of distance is probably the most valuable means of controlling radiation exposure. This has been emphasized indirectly in the preceding paragraph in that those not directly concerned with the operation should be excluded from the work area and also that operating personnel should stay away from radiation sources when their presence is not required.

Radiation intensity decreases according to the inverse square law, i.e., doubling the distance drops the exposure rate to $\frac{1}{4}$; tripling the distance, to 1/9, and so on. This rapid fall-off with distance can be of greatest assistance when properly used. It is certainly the cheapest barrier to overexposure.

Actual work areas where the possible exposure approaches permissible levels should be posted with signs indicating the level of radioactivity, and whenever possible, temporary barriers should be set up to keep those not working directly with the radiation source at a safe distance.

Federal regulations require posting of radiation areas (greater than 5 mrem/hr), high radiation areas (greater than 100 mrem/hr) and airborne radioactivity areas (greater than 25 per cent of the maximum permissible air concentration).

When handling isotopes, various personnel protection equipment may be required. Depending on the activity level and the physical state of the active material, coveralls, gloves, or respirators may be required. This will be discussed further in connection with decontamination. In cases where heavy β exposure is possible, it is sometimes advisable that the workers wear heavy-lensed goggles to reduce eye exposure.

Decontamination

Isotope spills may occur, and prompt decontamination is necessary to minimize exposure from them. This is not only required for the reduction of direct radiation but to prevent spilled material from becoming airborne or being picked up by contact. The supervisor and preferably the workers as well should know the proper means of cleanup for any possible isotope spill in the operations which they are performing. When an accident occurs, the operator should first move to a safe distance and call for assistance. The greatest difficulty with spills in the past has been that the workers, scientifically trained or not, try to clean up by themselves and spread contamination throughout the laboratory or plant.

The first operation in decontamination is to check the personnel to see if their clothing or bodies have become contaminated. This should be done by the supervisor or technically qualified person. If the workers are contaminated, their clothing should be removed and their bodies checked for radioactivity. Any active areas should be washed with plenty of water, then soap and water, and finally scrubbed with a soft brush. Decontaminating agents are not always satisfactory for application on skin.[9] The following decontamination procedures are taken from *Medical Aspects of Radiation Accidents*.[3]

USEFUL DETERGENT PREPARATIONS FOR DECONTAMINATION OF WOUNDS

Aqueous:

1. Soap and water.
2. Abrasive soap and water.
3. Commercial detergent (10%) Tide, Dreft, Alconox, Hemo Sol.
4. Complexing agent (1% citric acid solution).
5. Chelating agent (1% versene solution) with or without detergent.

Waterless:

1. Corn meal and commercial powdered detergent in equal parts made into a water paste and used without additional water. Scrubbing with brush. Removal with cotton or soft tissues.
2. Waterless mechanics' hand cream used without additional water. Scrubbing with brush. Removal with cotton or soft tissues.
3. Homogenized cream of 8% carboxymethyl cellulose, 3% commercial powdered detergent, 1% versene and 88% distilled water used without additional water. Scrubbing with brush. Removal with cotton or soft tissues.

The procedure for washing contaminated skin and hands is as follows:

1. The skin area should be monitored by a suitable survey meter if immediately available. If a survey instrument is not immediately available, and a person is suspected of being contaminated, wash immediately. Do not delay for want of a monitoring instrument. All readings should be recorded. Wipes of contaminated areas may be useful. These can be made with filter paper of appropriate size to fit in available gas flow counters for α or β activity. Usually a one-inch-diameter disk is adequate. These wipes can also be counted by β end-window counters, or can be placed in test tubes and counted in γ scintillation well counters. Cotton tipped swabs should be used for detection of activity in nares, ear canals, mouth, and other orifices as needed. If possible, clean areas should be covered with plastic drapes or towels to prevent spread of radioactive materials.
2. Wash thoroughly for 2–3 minutes by the clock. Use tepid (not hot) water and a mild detergent or soap. Cover the entire contaminated surface with a good lather. Rinse off completely with running water. Repeat at least 3 times. Do not use abrasive or highly alkaline soaps or powders. The skin area should be surveyed between washes.
3. If the above procedure does not remove all dirt and contamination, scrub the hands for a period of 8–10 minutes by the clock using liquid or cake soap, hand brush, and tepid water. Light pressure should be exerted on the brush. One should not press so hard that the bristles are bent out of shape. There should be at least three complete changes of soap and water. All surfaces should be covered with a minimum of four brush strokes. A convenient routine is to start by scrubbing one thumb, being certain to brush all surfaces, and then proceed to the web space between the thumb and index finger, and similarly over the other fingers. Next scrub palm and dorsal surface. Each nail and

cuticle must be similarly scrubbed. The hand should then be thoroughly rinsed and checked by a survey meter. An emollient cream should then be applied. The brush and towels should be discarded into a "radioactive" waste can or bag suitably labeled. If contamination is localized to a specific part, masking above the contaminated area with plastic material and cellophane tape will prevent the spread of the contaminant.

4. Where water is limited in quantity, waterless cleansers can be used although they tend to defat the skin. Thus a protective barrier may be removed and percutaneous absorption may be increased.
5. Contamination of the eyes, nose, and mouth can be handled only by copious amounts of water at first. Isotonic irrigants should be obtained as soon as possible. Eye cups should not be used.
6. Fission product removal using titanium dioxide. Titanium dioxide may be used as a paste or slurry made by shaking the powder into the wet palm of the hand until a good paste is formed. Run tap water over the hands continually so that the paste is kept wet, and apply this lather thoroughly to all hand surfaces, especially around the fingernails, for a minimum time of two minutes. Rinse off thoroughly with lukewarm water and follow by a thorough washing with soap and water and a hand brush. If any of the paste is left under the nails after washing, it will form a rather hard cake which is difficult to remove.

Note: A hand decontamination kit should be maintained in each washroom associated with laboratories in which work is done with radioisotopes.

If a significant* quantity of radioactive material has been ingested, the treatment should follow that given for general chemical poisoning such as dosing with an emetic. Introduction of significant quantities of radioisotopes into a puncture wound or cut requires immediate action. The injury should be washed with large amounts of water and treatment similar to that for snake bite given immediately afterward. The treatments in this paragraph should be under the direction of a physician.

The problems of skin contamination, ingestion, and entry through wounds have been mentioned. However, it must be remembered that these are treatments indicated where possible toxic quantities of material are present. If small quantities of isotopes are being handled, there is no point in using drastic means of first aid when an accident occurs. It is the responsibility of the operational supervisor to know whether an accident can be a source of hazard and what procedures should be followed in the event of an accident.

*Several times the permissible body burden.

The reason for immediate action in area decontamination is to prevent the spread of radioactive material. The return of the area to operating condition is usually not a critical factor and the cleanup can proceed in an orderly, planned fashion. All cleaning should be from the outer part of the area working in toward the center of contamination. The operation should be monitored for personnel overexposure and to check that contamination is not being spread. In most cases a preliminary washing with water and detergent will remove most of the isotope. After that, smaller areas may be cleaned up with decontaminating agents suitable for the isotope and the surface to be cleaned. Table 5A.8 lists decontamination methods recommended by the Federal Civil Defense Agency.[10] These are quite old, but still valid.

The maximum level of β and γ radiation which is frequently recommended is 1.0 mrad/hr for large surfaces. While this may seem to be a low level it is not impossible to attain, and maintenance of contamination below this value will keep the general level of the work area within satisfactory limits.

The decontamination procedure should be followed with survey instruments and the working area should not be reopened for operations without the approval of the qualified technical person in charge. If the quantity of isotope involved in the accident is at the toxic level, the cleanup crew should wear gloves and other protective clothing. Any significant airborne contamination makes necessary the use of respirators. When the cleanup procedure is completed, the bodies of the crew should be monitored, after removal of their protective clothing, and any radioactive material on the body removed as described above.

Ventilation

Ventilation is needed only when dealing with isotopes that may become airborne. This can happen much more readily than is commonly imagined, since transfer of dry solids, boiling of liquids, or other normal processes frequently give rise to airborne contamination. Such operations should be carried out in well- ventilated areas only, and monitoring of the level of airborne activity must be a part of the protection program. This is true only if the total amount of isotope handled is possibly toxic; it is not required for lower levels of activity.

General ventilation requirements for the handling of toxic materials are covered in Section 2. The general requirements apply as well to handling radioactive materials. The only additional factor to consider is that hoods for use with isotopes emitting penetrating radiation may also require shielding. This shielding may disturb the air flow characteristics of the hood, and measurements of air flow should be made before any operations are carried out to ensure adequate ventilation.

TABLE 5A.8. Decontamination Methods

Surface	Method	Advantages	Disadvantages
Paint	Water	Most practical method of gross decontamination from a distance. Contamination reduced by approximately 50%.	Protection needed from contaminated spray. Runoff must be controlled. Water under high pressure should not be used on a surface covered with contaminated dust.
	Steam (with detergent if available)	Most practical method for decontaminating large horizontal, vertical, and overhead surfaces. Contamination reduced by approximately 90%.	Same as for water.
	Soapless detergents	Where effective, reduces activity to safe level in 1 or 2 applications.	Mild action.
	Complexing agents:[a] oxalates, carbonates, citrates[b]	Holds contamination in solution. Contamination on unweathered surfaces reduced by approximately 75% in 4 min. Easily stored, nontoxic, noncorrosive.	Requires application from 5–30 min for effectiveness. Has little penetrating power; hence of small value on weathered surfaces.
	Organic solvents	Quick dissolving action makes solvents useful on vertical and overhead surfaces.	Toxic and flammable. Requires good ventilation and fire precautions.
	Caustics[c]	Minimum contact with contaminated surface. Contamination reduced almost 100%.	Applicable only on horizontal surfaces. Personnel hazard. Not to be used on aluminum or magnesium.
	Abrasion (wet sandblasting)	Complete removal of surface and contamination. Feasible for large-scale operations.	Contaminated sand spread over large area. Method too harsh for many surfaces.
Metal	Water	Contamination reduced by approximately 50%.	Same as for painted surfaces.
	Detergents	Removal of oil or grease films.	Same as for painted surfaces.
	Organic solvents	Stripping of grease	Same as for painted surfaces.
	Complexing agents:[a] oxalates, carbonates, citrates[b]	Holds contamination in solution.	Difficult to keep in place on any but horizontal surfaces. Limited value on weathered porous surfaces.
	Inorganic acids	Fast, complete decontamination.	Good ventilation required; acid fumes toxic to personnel. Possibility of excessive corrosion. Acid mixture cannot be safely heated.
	Acid mixtures	Action of weak acid. Reduces contamination of unweathered surfaces.	Same as for inorganic acids.
	Abrasion (buffers, grinders)	Useful for detailed cleaning.	Follow-up procedure required to pick up powdered contamination.
	Abrasion (wet sand blasting)	Same as for painted surfaces.	Same as for painted surfaces.
Concrete	Abrasion (vacuum blasting)	Direct removal of contaminated dust.	Contamination of equipment.
	Vacuum cleaning	Same as for vacuum blasting on concrete.	Same as for vacuum blasting on concrete.

TABLE 5A.8. Decontamination Methods (*Continued*)

Surface	Method	Advantages	Disadvantages
	Flame cleaning	Only method of trapping contamination on surface.	Slow and painstaking. Fire and airborne radiation hazard is great.
Brick	Same as for concrete	Same as for concrete.	Same as for concrete.
Asphalt	Abrasion	No direct contact with surface; contamination may be reduced to safe level.	Residual contamination fixed into asphalt. If road is subject to further contamination, may require recovering.
Wood	Flame cleaning	Same as for flame cleaning on concrete.	Same as for flame cleaning on concrete.

[a]Another complexing agent would be a solution of 1% ethylenediaminetetraacetic acid ("Versene," "Sequestrene") in water.
[b]Oxalates, carbonates or citrates may be the ammonium or sodium salts.
[c]Caustics should preferably be trisodium phosphate or sodium sesquicarbonate rather than hydroxides since they are less hazardous.

Almost all operations involving small amounts of isotopes can be carried out in a totally enclosed hood or dry box. The dry box is set up so that the worker performs all operations through gloved ports. The ventilation intake is at the ends of the box, and the outlet at the rear or top is connected to suitable ducts. Such a totally enclosed system offers a great margin of safety while a small amount of practice allows normal operations to be carried out with facility.

Personnel Contamination

The hazard from radioisotopes taken internally has been described earlier in this section. Actually, effective prevention of isotope ingestion depends on the personal habits of the worker. The common preliminary to ingestion is transfer of activity from contaminated surfaces or objects to the mouth by way of contaminated hands, food, etc., when eating or smoking. It is an absolute requirement for laboratories handling toxic amounts of unsealed isotopes that edible materials should not be allowed into the work area. The general spread of contamination by hands to personal clothing and personal items such as pocketbooks can only be stopped by frequent hand cleaning, particularly before leaving the work area. This cannot readily be a supervised process, but whenever dangerous quantities of isotopes are handled, the workers must be explicitly instructed in the requirements of personal cleanliness and good housekeeping.

The requirement for protective clothing in isotope handling depends on the level of activity handled. At any level of activity, it is desirable that street clothing should not be worn in the plant or laboratory. Nor are laboratory coats good protection for prevention of contamination. Actually, when levels above tracer quantities are handled in such a way that contamination can occur, a change room should be provided with separate lockers for street and work clothing and with adquate shower and scrubbing facilities. (At this same level, the laundering of work clothing should be done at the site. Contaminated clothing should not be sent to commercial laundries.)

Rubber or plastic gloves are a very good means of preventing the spread of radioactivity by the hands. Some delicate manual operations may require surgical gloves, but whenever possible cloth-lined heavy gloves should be worn. These should never be removed from the work area until ready for disposal. More specialized equipment such as rubbers, overshoes, or shoe covers are rarely required, unless serious floor contamination has occurred.

Reevaluation

A radiation control program is almost useless unless it is continuously reevaluated. Many changes can take place without being apparent. Shields may develop cracks or other openings, spills of radioisotopes may cause contamination in an unexpected part of the work area, ventilation blocks may occur, or some change in operating technique may present a new source of radiation. Therefore, a reevaluation of the control system must be made periodically. A portion of this is done through the personnel monitoring program, but surveys with portable instruments can be of great value in detecting sources of radiation before they appear in the results of personnel monitoring. This concept is true of many types of health protection. No control program can be considered completely static and free from the possibility of breakdown.

DESIGN OF INSTALLATIONS

The design of the installation where radiation exposure is possible is probably the most important single factor in maintaining low exposure levels. The addition of radiation control measures to an operating laboratory or plant is always more expensive and usually less effective than when control measures are built into the original design. The various factors of responsibility must be included when designing a work area, and a complete knowledge of the processes to be carried out in the area must be available to the designer. Redesign may be necessary if the quantity of isotopes or the x-ray energy is increased and this will add very considerably to the cost. Suitable protection for the foreseeable future should be part of the initial plan.

The overall design must be viewed from the standpoint of exposure reduction in isotope and x-ray installations. Adequate shielding and electrical safeguards are the primary considerations for x-rays. In isotope laboratories such additional factors as storage space, materials of construction, and ventilation are also important. In both cases every effort should be made to use distance as a method of exposure reduction.

No other working areas should be laid out near a high-level radiation source. Partitioning may be designed to act not only as shielding but as a physical barrier to prevent passage of workers through or near areas of high radiation exposure. Wherever possible, separate buildings or wings should be adapted for isotope or x-ray operations.

The primary attention is given to the working area, called a "restricted area" in federal regulations and a "controlled area" by the NCRP. The NCRP definition is "a defined area in which the occupational exposure of personnel to radiation or to radioactive materials is under the supervision of an individual in charge of radiation protection. (This implies that a 'controlled area' is one that requires control of access, occupancy, and working coniditions for radiation protection purposes)."

Installations where radiation is a hazard may be classed according to whether the source of radiation is electrical, such as x-ray equipment, or nuclear, such as radioisotopes. Under the classification of x-ray operations we have medical units for diagnosis and therapy and industrial units for testing of products as well as laboratory equipment for x-ray diffraction and other scientific measurements. The isotope field may be divided into laboratory installations using isotopes for experimental purposes and industrial operations such as radiography. The large-scale utilization of low-priced isotopes, such as mixed fission products, for sterilization of foods or drugs or similar processes is still awaiting approval by the FDA.

One factor in laboratory design which lies outside the field of protection is the requirement that radiation measuring equipment be installed in areas which have low radiation background and which are not easily subject to contamination. Counters and other equipment are efficient only when the ratio of background to the desired activity is maintained at a low level. This requires isolation or shielding of the measuring area beyond the requirements for eliminating health hazards.

Medical X-Ray Installations

The medical x-ray technician and radiologist have given less consideration in the past to their safety precautions than anyone else with hazardous exposure. Most of our knowledge leading to the setting up of permissible levels came from overexposures incurred in the medical use of x-rays. It is also probable that many patients received excessive exposures in the course of diagnosis. X-ray therapy is always a calculated risk, where the possible damage from the high x-ray dose is balanced against the possible effects of the disease.

Medical x-ray facilities do not come under federal regulations but may come under state and local codes. Both the ICRP and the NCRP have issued recommendations on measures to protect the staff of such facilties and the references should be consulted for details. This section will deal only with the general features of design.

The basic criterion for x-ray operation is that the radiologist or technician not receive more than the maximum permissible dose and reduction of dose rate can best be accomplished by proper design of the installation. The two main sources of exposure are the useful beam from the x-ray tube and radiation scattered from the patient and the four walls and ceiling of the room. It is not possible to shield the useful beam, although 1 mm of aluminum should be used in it as a minimum filter to reduce the low-energy x-rays in the useful beam which are of little value anyway in radiography yet represent considerable biological hazard. The useful beam, of course, must also be prevented from leaving the room and exposing those outside the x-ray installation. The requirements for shielding vary for beams of different energies and intensities as does the distance between source and shield.

As was previously discussed, x-rays and γ rays cannot be completely blocked, but only reduced in intensity by shielding. Thickness of shielding is expressed as half value (HVL) or tenth value (TVL) layers, the thickness required to reduce the radiation intensity to half or 1/10 of the initial value. Figure 5A.4 shows the HVL's in lead and in concrete for radiation of various energies. Figure 5A.5 shows the number of HVL's required for various degrees of reduction.

The usual description of an x-ray source is in terms of the applied voltage and the target current. These parameters may be converted approximately to dose units by the graph shown in Fig. 5A.6.[11] This gives the R/minute/milliampere at various high voltages with various filters built into the x-ray equipment.

Figures 5A.4, 5A.5, and 5A.6 may be used to calculate whether existing shielding is adequate for approximate design work. From the characteristics of the source, Fig. 5A.6 may be used to obtain the R/milliampere-minute at

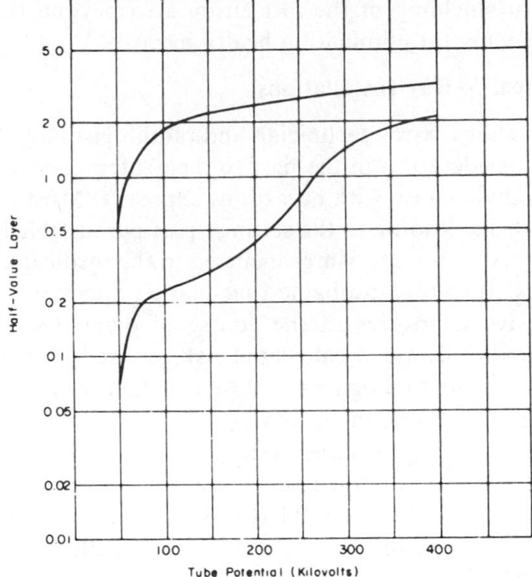

Figure 5A.4. Approximate HVL's for x-rays at high filtration for various tube potentials: (a) centimeters of concrete (upper curve); (b) millimeters of lead (lower curve).

1 m. This may be converted to the required distance by the inverse square law to give the dose rate at the point of interest. Figure 5A.5 will then give the number of HVL's needed to reduce this to the permissible level, and Fig. 5A.4 will give the thickness of the HVL layer.

For example, a wall between the radiation area and an adjoining office is to be shielded. The source is a 150 kV machine with 2 mm Al filtration and operates at 5 ma, 16 min/day (average), 5 days a week. The wall is 3 m from the source.

From Fig. 5A.6, the dose rate is about 2.1 R/mA-min, so (2.1) (5) (16) (5) = 840 R/week at 1 m. At 3 m this would be

$$840/9 = 93 \text{ R/week}$$

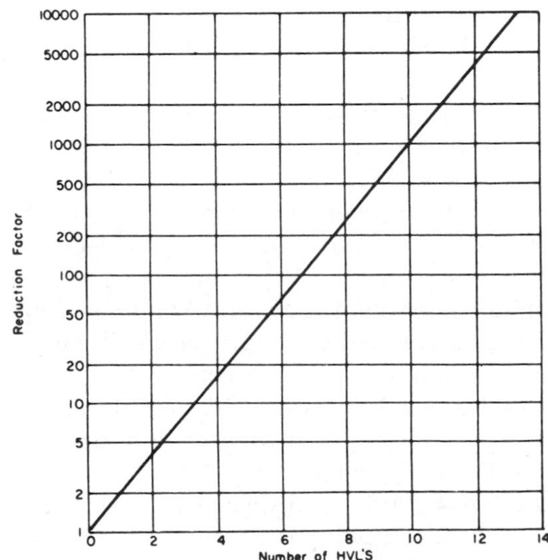

Figure 5A.5. Reduction factor as a function of number of HVL's.

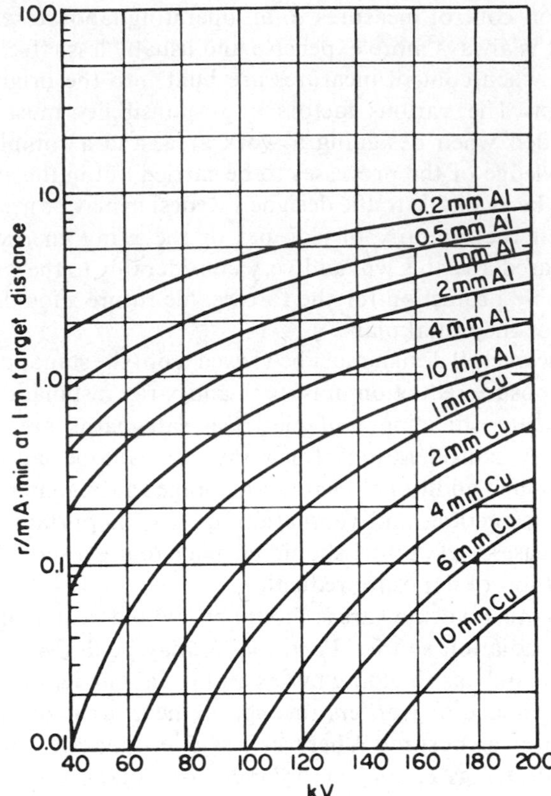

Figure 5A.6. Approximate radiation output in roentgens per milliampere-minute at a distance of 1 m, measured in air, of the primary radiation from a tube with tungsten target and total filtrations from 0.2 mm Al to 10 mm Cu at 40–200 kV constant potential.

This would require a reduction of

$$93/0.01 = 9300$$

or 13 HVL's (Fig. 5A.5). From Fig. 5A.4, the HVL for 150 kV x-rays in lead is 0.3 mm and the total shield would be 3.9 mm to attain 0.01 R/week.

Since not all areas being shielded have the same occupancy or are exposed to the beam for the same time, the occupancy factor (T) and use factor (U) are applied to compensate. These have been defined by the ICRP[11] and are shown in Table 5A.9. This can reduce the beam shielding required.

It is apparent that extending the primary barrier to the floor, ceiling, and all walls would be extremely expensive, both for the shielding material itself and for suitable building construction to support such a weight of shielding. Therefore, it is more economical to mechanically limit the direction in which the beam can be pointed. If this is done, a primary barrier is required to cover only the field which can be intercepted by the useful beam.

The primary barriers described for shielding do not completely cover the protection requirements. Since the tube housing gives off some radiation and since some low-energy x-rays are produced by scattering from the subject, additional protection is required.

TABLE 5A.9. Occupancy Factors

Full occupancy ($T = 1$)	Control space, offices, corridors and waiting space large enough to hold desks, darkrooms, workrooms and shops, nurse stations, rest and lounge rooms routinely used by occupationally exposed personnel, living quarters, children's play areas, occupied space in adjoining buildings.
Partial occupancy ($T = 1/4$)	Corridors too narrow for desks, utility rooms, rest and lounge rooms not used routinely by occupationally exposed personnel, wards and patients' rooms, elevators using operators, unattended parking lots.
Occasional occupancy ($T = 1/16$)	Closets too small for future occupancy, toilets not used routinely by occupationally exposed personnel, stairways, automatic elevators, sidewalks, streets.
Use Factors	
Full use ($U = 1$)	Floors of radiation rooms except dental installations, doors, wall and ceiling areas of radiation rooms routinely exposed to the useful beam.
Partial use ($U = 1/4$)	Doors and wall areas of radiation rooms not routinely exposed to the useful beam, floors of dental installations.
Occasional use ($U = 1/16$)	Ceiling areas of radiation rooms not routinely exposed to the useful beam.

All walls out of the direct beam and the floor and ceiling must have suitable secondary barriers to absorb scattered radiation plus leakage radiation through the tube shield if the area on the other side of the shield is inhabited. These secondary barriers are computed on the basis that the scattered radiation is 0.1% of the primary beam at a distance of 1 m from the scatterer for x-rays up to 500 kV.

Such calculations are adequate for determining whether barriers installed are probably adequate for the particular application. In designing a new installation it may be desirable for economic reasons to follow the the more detailed calculations given in the original references.[11,12]

All of the computations are intended to yield adequate barriers for the conditions described. It must be pointed out, however, that the final test is a measurement of the performance of these barriers under working conditions and that the criterion is the protection of the operators.

The operator of medical x-ray equipment might possibly receive an excessive dose from scattered radiation. This may be reduced somewhat by wearing protective clothing, but this method is certainly not reliable. It is much better to include a shielded booth for the operator in the original design. A proper design would require that the operator be in the booth in order for the x-ray equipment to operate. Thus, after placement of the patient, he would enter the booth and begin the exposure. While automatic timers are desirable, they may fail and the presence of an operator at all times will ensure that the patient does not get an excessive dose. The patient may be observed through lead glass or heavy-liquid-filled windows.

The installation described is one step away from the so-called totally protected system. Such a system includes electrical interlocks so that it is impossible for the x-ray unit to be operated unless everyone is out of the x-ray room with the exception of the patient. Such an installation is most desirable where heavy doses are administered for therapeutic purposes. The less rigid design, requiring only that the operator be at the control panel behind a shield before radiation can begin, is suitable for diagnostic x-ray.

Industrial X-Ray Installations

Industrial radiography requirements have brought about the development of higher voltage x-ray equipment. The need for seeking flaws in large castings as well as similar nondestructive tests requires extremely penetrating, high-energy radiation. In addition, radiographic equipment is in operation a larger fraction of the time than ordinary medical equipment. Therefore, even more care is required in industrial work than in medical work. One advantage of industrial design is that the direct beam can usually be absolutely fixed in a vertical downward position on the lowest floor of the building and the need for primary shielding can be eliminated.

There is no excuse for failure of any industrial installation to use a totally protected system. The operator is dealing with inanimate objects which can be set immovably in place and the exposure must be controlled to produce a usable film. Therefore, the operator should be in a booth where he can control the total exposure without being exposed to either primary or secondary radiation. If the objects to be radiographed are changed by any other workers, there must also be an interlock and signal system so that even a careless x-ray operator cannot start the exposure unless everyone is clear of the area.

The high-voltage industrial installations usually rely on concrete shielding rather than lead. The high mechanical strength and simple construction of concrete barriers more than compensates for the greater thickness required in comparison to lead.

Gamma Ray Radiography

Isotopic sources for industrial or medical radiography are more closely allied to x-ray equipment than to isotopes as used for laboratory purposes. The source must be a γ emitter of high intensity and usually of high energy. When installed, it is protected by heavy shielding to give a useful beam very close in properties to the useful beam from x-ray equipment and is completely encapsulated to prevent escape of the isotope. All of the design criteria for x-ray installation would apply to the use of γ sources for radiography. The Federal requirements for licensing the use of radiographic sources and the radiation safety requirements for operations are given in Part 34 of Title 10, *Code of Federal Regulations*.

The common isotopes in radiography include ^{60}Co, ^{137}Cs, and ^{192}Ir. In addition, others such as ^{182}Ta, ^{169}Yb, ^{153}Eu, and ^{170}Tm are being used. Radium sources are still found in older installations. Radiographic sources are similar to x-ray equipment in their shielding requirements. As for x-rays, the design dose rates are 0.1 R/week in the working area and 0.01 R/week in uncontrolled areas.[12]

Values may be obtained for concrete shielding of ^{226}Ra, ^{60}Co, and ^{137}Cs from Figs. 5A.7, 5A.8 and 5A.9. Charts for iron and lead shielding plus requirements for secondary shielding of leakage and scattered radiation are given in NCRP Report #49.[13]

As an example a 5 Ci ^{137}Cs source (0.66 MeV) is to be shielded from the work area. From Table 5A.6, the unshielded dose rate is 1.6 R/hr @ 1 m. From Figure 5A.9, the required barrier is 20 inches of concrete. For reduction of exposure to nonoccupational levels, an additional TVL should be added. This can be estimated roughly as 6 inches by noting that a source 1/10 the activity would require only 14 inches of shielding. Thus the TVL is about 20 minus 14.

Figures 5A.7, 5A.8 and 5A.9 also indicate the value of distance in reducing exposure and shielding requirements. For the ^{137}Cs source noted, the shield would only need to be 10 inches thick if the area to be shielded were 6 m away rather than one meter. Distance is always the cheapest way of reducing exposure.

Isotope Installations

Operations with isotopes that are not encapsulated still may require shielding in the same way as the radiographic sources described above. They do offer additional hazards, however, over those found with encapsulated sources.

First, materials may become airborne, so the external radiation from an isotopic source may be negligible compared with the hazard from its inhalation or ingestion.

Secondly, appreciable amounts of the radioisotope may be left on working surfaces or equipment surfaces during the operations. This material in turn may be generally spread about by later work. Good design, therefore, must include not only proper shielding for minimizing exposure to direct radiation but proper ventilation, proper methods of handling, containers where necessary, and means for ready decontamination of working surfaces and equipment. Shielding from γ radiation may follow the design outlined for x-radiation of the same energy. The actual intensity is known from the quantity

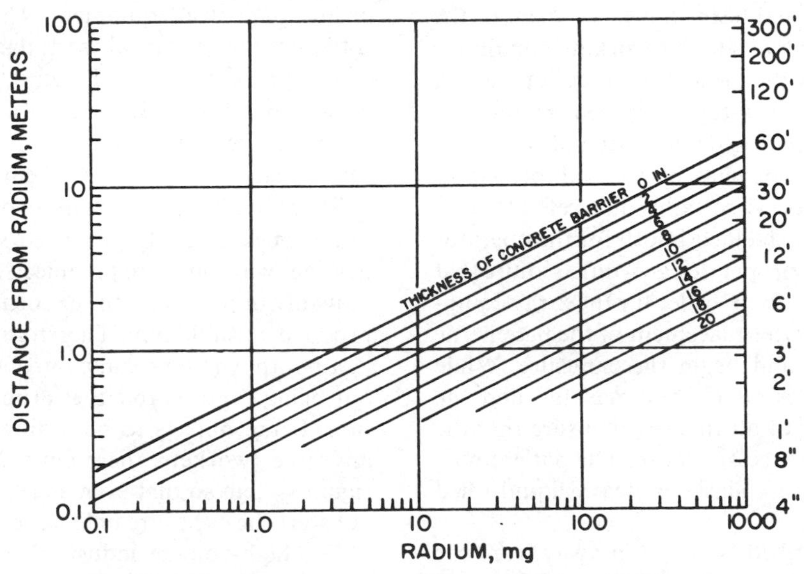

Figure 5A.7. Relation between amount of radium, distance, and shielding for controlled areas.

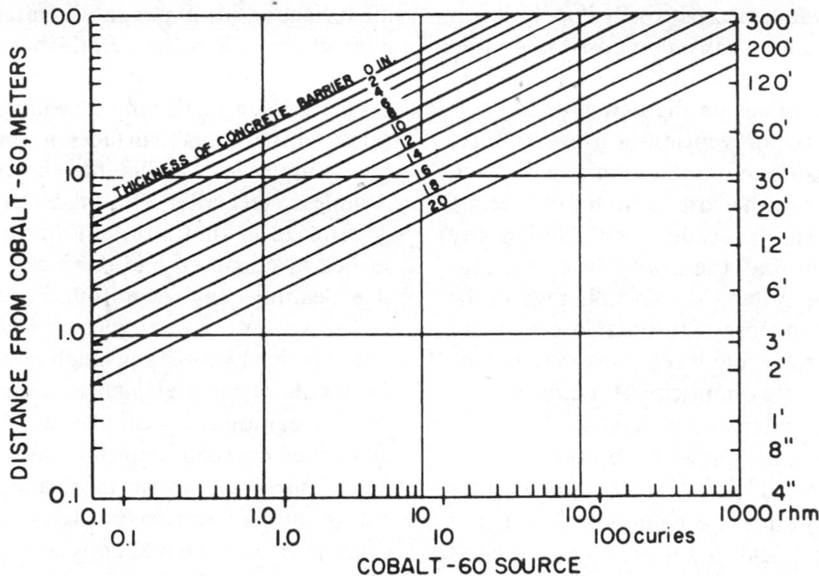

Figure 5A.8. Relation between rhm, distance, and shielding for controlled areas.

of isotope available, and the shielding requirements may be obtained by calculation. Again, the Rhm values of Table 5A.6 will be useful.

Laboratory operations with isotopes, unless they are of a routine nature, are usually shielded by interlocking lead bricks. This is the most flexible system for providing shielding in experimental work. The quantities of radioisotopes involved are usually such that primary shielding alone is sufficient. Of course, this cannot be assumed, but should be checked with survey instruments just as any operation would be checked. Also, it must not be assumed that bench tops or table tops offer good shielding for radiation directed downward.

Industrial tracer operations with isotopes follow those common in the laboratory. The major differences are that ordinarily larger quantities of isotopes are required and that the personnel with possible exposure to radiation are not technically trained. Therefore, it is more critical that they be given proper instruction and protection. Such operations are so varied that only generalized statements may be made. The installation must be designed around the process to be carried out. Protec-

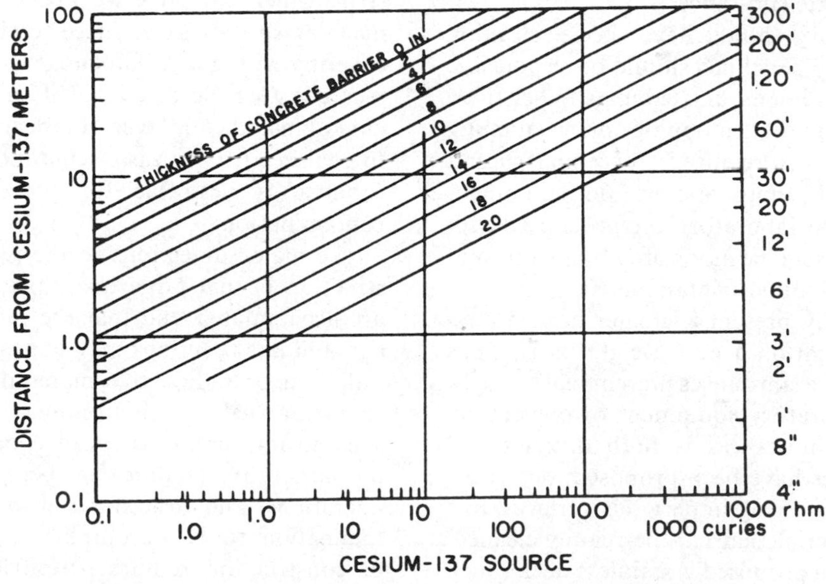

Figure 5A.9. Relation between rhm, distance, and shielding for controlled areas.

tive measures can be made less expensively if included in the original design. This is usually not possible as it is more likely that equipment previously used for normal operations would be modified for the new process.

The basic criterion is the prevention of overexposure. If the concentration of isotope in the material is known, the possible dose rate from the total amount of isotope can be calculated. The safe procedure is to design any required shielding so that if all the isotope were concentrated at any point in the system, the shielding would be sufficient to maintain the dose rate outside the shield below the maximum permissible level. This may not be possible with extremely large quantities of radioisotopes. Where the equipment is extensive it may not be necessary, particularly when dealing with solutions which may be readily homogenized. In such a case the amount of isotope possibly present in any area may be considered as a point source and the shielding for the area calculated on this basis.

With β emitters, the shielding from direct radiation is less of a problem, and thick ($\frac{1}{2}$ inch or greater) transparent plastic barriers will give sufficient protection. This simplifies any laboratory manipulative operations, because the source is always in sight. Alpha emitters should require no shielding whatsoever. Both of these emitters, however, as well as γ emitters, do require consideration of the inhalation, ingestion, and contamination problems in their design. The problem of ventilation for any hazard is described in Section 2, and the criteria for handling various toxic materials apply equally well to radioactive isotopes. The direct ingestion problem can only be avoided by habits of personal caution and cleanliness and wearing of suitable protective clothing. It is from the viewpoint of decontaminability that the design of the laboratory equipment for radioactive materials differs from that for other toxic substances.

An isotope laboratory should have special cleaning services. The laboratory personnel should be responsible for cleaning all benches, hoods, and cabinets. When floor cleaning is performed, wet mopping or vacuuming through suitable filters is required. Sweeping should not be allowed. All pails, mops, and cleaning equipment should be assigned to the laboratory and never used elsewhere. In this way they can be monitored frequently and discarded or cleaned if found contaminated.

Radioisotopes usually present a greater hazard than any other toxic material on a pure weight basis. Thus smaller amounts of these substances present real hazards, and the design of laboratory equipment to prevent accumulation of these substances is both difficult and costly. Two approaches have been proposed; one is the type of construction involving impervious surfaces, free of seams, using a material that can be readily cleaned. The material most often proposed is stainless steel, which is expensive to buy and to fabricate. The second approach utilizes normal laboratory equipment and covers all surfaces with expendable materials that can be stripped off and discarded. Both methods have their merits and will be discussed in some detail.

One fallacy in the initial concept of stainless steel or other "impervious" surfaces is the belief that they are truly impervious. This has been shown to be false. Stainless steel after one vigorous cleaning is found to deteriorate in that more and more material may be absorbed or adsorbed and retained on the surface. Successive cleanings have been found to become more difficult and to require more vigorous methods of decontamination. Glass always shows high adsorptive characteristics for traces of most elements, and even the drastic methods of cleaning chemical glassware do not always remove all of the adsorbed material. On the other hand, "impervious" surfaces can be fabricated to be free of cracks, seams, and crevices so that large quantities of an isotope cannot become permanently lodged in the structure.

The removable, expendable covering for laboratory surfaces is probably less expensive than the impervious construction. In a typical operation, the bench surfaces may be divided into areas with stainless or plastic trays. These in turn are covered with an adsorbent paper with a waterproof backing. In case of spills the paper is discarded as radioactive waste and the trays taken up and cleaned as vigorously as required. Floors are made of rubber or asphalt tile which can be removed and replaced in small areas at a time when or if they become contaminated. Several strippable plastic coatings have been developed for covering vertical surfaces subject to contamination. These resemble paints in texture and application except that they do not form a firm bond on the surface to which they are applied. If they become contaminated, they can be stripped off and replaced.

The replaceable material concept can be carried one step further. Movable workbenches of the type used in machine shops can be fitted to chemical procedures by covering with glass, linoleum, or "Transite." The coverings can then be discarded if they become excessively contaminated, and even the benches are not expensive to replace. In any case, separation of work tables and cabinets is desirable for preventing the spread of contamination.

The major objections to the replaceable type of construction is that large quantities of radioactive wastes are accumulated and that the removal and disposal of contaminated sections may expose the workers to more radiation or radioactive material than a cleaning procedure carried out on an impervious surface. However, such an installation is cheaper, particularly when existing laboratory facilities are being converted to isotope operations, and procedures now exist to allow decontamination to be accomplished safely. Both methods of construction require periodic monitoring to determine when decontamination is required.

The preceding paragraphs have discussed the aspects

of three factors—shielding, decontaminability, and ventilation—with respect to design. A fourth factor which must be considered in isotope installations is that of liquid waste disposal. This will be covered later but it must also be considered in the design of an installation.

Sinks and drains for carrying off highly radioactive solutions cannot be fed directly into the normal channels. Such installations require two entirely separate waste systems; one for inactive material going directly to the regular sewer system and one for radioactive material which is led into tanks for holdup or processing prior to disposal. In extreme cases, these waste lines may also require shielding to prevent overexposure to workers in the area. In any case, they should be designed to minimize accumulation of isotopes in the lines. Such accumulations can be hazardous when making repairs on high-level disposal systems.

ISOTOPE STORAGE, HANDLING, AND SHIPPING

Exposures found in shipping and storage of isotopes are distinguished from those occurring in actual operations because they generally involve an entirely different group of personnel. This group is usually untrained, unaware of the hazards, and must operate solely on the basis of directions, without allowance for personal judgment. For this reason DOT, CG, USPS, and IATA have set up definite regulations for the shipping of radioisotopes. The storage problems within any installation must be met by similar rules set up to fit the activity level and exposure time involved.

Since the storage and handling of isotopes includes many of the problems of storing and handling dangerous chemicals, the general safety rules for these must be complied with and will be discussed here in a general way. More details will be given on the specific problems involved in storing and handling radioisotopes.

Storage

Storage facilities designs are dictated by the type of emitter and the quantity being stored. Laboratory quantities of radioisotopes may not even require a separate storage area but are usually kept away from counting areas to prevent an increased counter background from stray radiation or contamination. Alpha or beta emitters are usually quite well shielded by the bottle or other container that they are shipped in and additional shielding is best provided by a plastic container having walls that are about 1 cm thick. These containers can be more useful if they are designed to be liquid-tight and thus to retain liquids if the inner container becomes broken. Additional protection is provided by filling with vermiculite or other absorbing material. Loose-fitting, heavy caps provide sufficient shielding and ease of handling. Laboratory quantities of γ emitters are best stored in a glass or plastic inner container held within a solid lead block of suitable thickness. Such storage blocks are commercially available or may be readily fabricated. The block should have a heavy, loose-fitting cap to provide sufficient shielding from the top and to allow ready removal. Radioisotope solutions, whether in tracer quantities or large amounts, should preferably be contained in an unbreakable plastic bottle. Polyethylene bottles resist the action of most chemicals and can stand accidental dropping without breaking. With large amounts of radioactivity, a secondary waterproof container packed with absorbent material should always surround the primary bottle before it is placed in the storage shield. This will prevent contamination of the shield if leakage of the primary container occurs. Radioisotope containers should always be labelled carefully in accordance with good practice, listing the isotope, its form, the amount, and the date when that amount was present. Federal regulations require licensees to label each container holding more than exempt quantites and to apply a suitable hazard warning.

In many cases the lead bricks described for temporary shielding may serve to protect a γ emitter for the period of storage. Shields prepared from lead bricks should only be considered for temporary operations as they do not give complete protection from all directions unless an elaborate structure is erected.

Whenever larger than laboratory quantities of isotopes must be stored, it is advantageous to set up a separate storage area. This will minimize the shielding requirements, if the area is isolated from work areas and normal traffic. Frequently a basement is most adaptable, as no shielding need be provided for the floor and most of the wall area. With α or β emitters, mere isolation of the source from the work areas or traffic is sufficient and no shielding would be required. With high-level γ emitters it is desirable to lower the source into steel tanks or cylinders embedded in concrete beneath the cellar or basement floor. The cylinder or tank acts as a container and does provide some shielding from radiation coming through the walls at an angle which would allow it to reach a populated area. The most important shielding is on the top, which is the direct radiation field. A lead or steel cap for the tank or cylinder can be fitted with a ring-bolt and raised or lowered by means of a chain hoist. Ventilation is required, particularly if radium or thorium sources are kept in the storage room.

Any open industrial storage installation should meet the requirements that no exposure above the permissible level is possible with all sources in their storage containers and capped. While this requirement may seem excessive, since personnel ordinarily do not operate within the storage area, it is desirable because the workers involved are frequently untrained. For the very few installations using multicurie quantities of isotopes, a time and distance requirement may be set up and higher dose rates allowed. The storage area in such an installation should be locked and should only be opened by a responsible supervisor. When entering such an area the handling crew should be under the control of the responsible supervisor who should remain outside the doorway to the storage area. The doors, although locked from the outside, should be capable of being readily opened from the inside to prevent anyone from locking himself into the room. The general instructions for layout of a completely protected radiation source, given above, should apply to the storage area for multicurie lots of isotopes.

Handling

As in all operations, the handling of radioisotopes must be set up to prevent excessive exposure to personnel and in fact to prevent any unnecessary exposure. The operational techniques, whether for the plant or laboratory, must be carefully planned to minimize direct radiation exposure and inhalation or ingestion of radioactive materials. After planning any operation, it is advisable to make one or more test runs of the complete procedure using inactive materials. It will be found that this will reduce exposure markedly by familiarizing personnel with the techniques so that they may be performed more rapidly. In addition, it is frequently found that such tests will show how process improvements may be made. The advantage is even more marked where unfamiliar apparatus is required or where there is a necessity for working around or over shields where normal body movement is not possible.

During usual handling operations on either a laboratory or plant scale, the greatest exposure is likely to occur during transfer of the radioisotopes. Transfer is involved in preparation of solutions, weighing or measuring quantities into new containers, or loading reaction facilities. Regardless of the type of emitter or its physical state, the operations are best carried out in a well-ventilated hood or glove box. This is most important with finely divided solids or other materials which may become airborne. The hood interior should be readily decontaminated to remove the normal spillage which occurs during transfer operations.

The shielding of actual operations depends entirely on the type of emission produced by the isotope. Alpha emitters require no shielding, beta emitters are best shielded with transparent plastic, while strong gamma sources may require lead or steel. If the operation requires heavy shielding, remote control of the apparatus must also generally be provided. This would not necessarily be a complete mechanical robot system but might merely be an extension of all controls outside the shield. Remote control equipment, such as tongs or solution—transfer devices are commercially available, which allow handling at a distance. The usual β-emitter operations would take place behind a plastic shield with cutouts for arms, so that the operator can reach inside to carry on the process. Where high activity is present or if possible hand contamination is a problem, the operator may wear long gauntlet gloves where the cuffs are sealed to the shield. Such complete units, called either glove boxes or dry boxes, are available commercially and are very convenient for carrying out small-scale laboratory or pilot plant operations.

Shipping

Radioisotope shipments in the U.S. are controlled by regulations of the Department of Transportation and the U.S. Postal Service. These regulations are designed both to prevent any excessive exposure to shipping personnel and to prevent damage to shipped material such as photographic films which are sensitive to radiation.

Since the limiting factor of most regulations is the dose rate at the surface of the package, it has become customary to increase the size of the outer packing container rather than to increase the weight of shielding. This is usually more effective if packing and shipping costs are considered. In other words, a 1-inch cube source might have to be contained in a 4-inch cube of lead for adequate shielding. The same reduction in surface dose rate (for high-energy γ) could be obtained by placing the original source in a corrugated carton 1 foot in each dimension. The latter method would be much less expensive. However, the shipping carton must be strong enough to survive handling and the source must be fixed in the center of the carton so that it cannot shift to one of the carton faces.

RADIOISOTOPE WASTE

All operations involving radioisotopes produce radioactive material which is no longer suitable for use. Such material must be handled properly and this is a major responsibility of those in charge of the operations. The actual method of management depends on the activity level, the type of waste, and the radiochemical properties of the isotopes. The two major methods of management are dilution for environmental disposal and containment, and these will be discussed in relation to the considerations mentioned and the quantity and type of inert waste material associated with the radioisotope.

As the total quantity of isotopes being handled in the world increases, more and more emphasis is being placed on containment, and this method should always be given primary consideration. The adoption of a conservative attitude to waste management has several advantages. First, in some cases, little is actually known about the process of concentration whereby plant and animal life may collect appreciable amounts of an isotope. Second, poor public relations may result from the exposure of large off-site populations to any measurable radioactivity. Third, local public health officials may yield to public appeal and set further restrictions on disposal if levels near the maximum permissible are found.

Most of the federal regulations for work with radioisotopes are based on the performance of the radiation protection system. In the case of disposal of radioactive waste, they also require prior approval of the method to be used. Systems involving incineration or disposal at sea are likely to receive particularly close scrutiny. In essence, approval is required for any disposal except that of small amounts to the sanitary sewers.

Activity Level

For the purposes of this discussion, a low activity level for gases and liquids will be defined as one in which the concentration can be reduced below the maximum permissible level by reasonable dilution. Other considerations may dictate the use of a containment procedure, but this definition will assist in waste disposal planning. High activity levels are those that cannot be treated by moderate dilution.

Radiochemical Properties

The important properties of waste for disposal relate to its toxicity, such as the type and energy of the radiation and the half-life and relative biological hazard of the isotope.

The probable fate of the waste will regulate the relative amounts of these materials that can be disposed of safely through normal channels. Alpha emitters are all heavy metals and tend to be deposited in the bone. Thus they must be kept out of food or drinking water and must be prevented from becoming airborne. However, they may be stored in large quantities without hazard, so they are not ordinarily released into the normal disposal channels. Long-lived β emitters such as ^{90}Sr are subject to the same considerations. Storage of γ emitters requires isolation or even shielding to prevent exposure to possible external radiation hazard.

Half-life is an important consideration, as short-lived emitters may be economically stored until their activity has decayed to a negligible value. Also, the biological hazard from the short-lived materials is considerably less than for long-lived emitters. As the half-life becomes longer the cost of storing large quantities for the required period of years becomes appreciable. Safe methods are therefore sought for low cost disposal of these materials by burial or retrievable surface storage.

Waste Management

As mentioned earlier, it is possible to dispose of radioactive wastes after suitable dilution in some cases. In others it is necessary to contain them and to store the material. Dilution is the process of mixing the waste with sufficient inert material to reduce the concentration of activity below the permissible levels. An example would be the flushing of microcurie amounts of an isotope into the regular sewer system. Containment is the storage of waste under conditions that keep it isolated from the environment to prevent direct radiation exposure of populations.

The three forms of waste material produced in isotope operations are solid, liquid, and airborne. In most contaminated solid materials, the activity per pound of waste is relatively low. Such material presents a considerable storage problem and efforts are usually made to reduce its bulk by incineration or baling. Liquid wastes of low-level activity are also bulky and must be concentrated by evaporation or the activity must be removed by some process such as precipitation, flocculation, or ion exchange. Airborne wastes may involve either radioactive gases as in reactor effluents, or airborne dusts such as hood exhausts. The containment and storage of waste gases is impossible, and reduction in contamination must be made by collecting the radioactive material by absorption, scrubbing, or removal or particles by filtration. Contaminated gases must be diluted to satisfactory levels before release. Since many of the noble gas fission products have short half-lives, even a brief hold-up is effective. Present regulations require measurement of concentrations at the point of release even though the stack effluent may be greatly diluted before reaching an inhabited area. The incineration process for solid wastes suffers from the requirement that considerable care must be taken to remove radioactivity from the flue gases of the incinerator. Entrainment of solid material by the hot gases can carry appreciable amounts of material out the stack, thereby making some form of stack-gas cleaning necessary. As noted above, it is necessary to obtain approval before incineration may be used. Many installations with small quantities of waste, of course, find it possible to handle their material without bulk reduction.

Liquid wastes are ordinarily solutions or suspensions of radioactive materials in a large bulk of solvent. A simple method of bulk reduction is evaporation, but the amounts of heat required for dilute solutions may make this an expensive operation in spite of its simplicity. As in the case of incineration, the exit gases or vapors from the evaporator must be cleaned before discharge to the atmosphere. The amount of liquid entrained in vapor coming off a boiling solution can represent several per cent of the total liquid volume. This can only be removed at considerable cost as any filters in the system must be heated to prevent excessive condensation.

An isotope in solution or suspension can frequently be concentrated by precipitation with specific materials or by general collection with a flocculating agent. The alum process used for water purification will scavenge many different isotopes and yield a resulting liquid which may be suitable for disposal through the normal sewer system. Other methods such as the precipitation of calcium phosphate have given effective scavenging from aqueous solutions. Any precipitation or flocculation process requires filtration to remove the concentrated activity and the carrying material. Sedimentation processes are not sufficiently efficient, and a large volume of liquid waste rquires a sand bed or other large filter. Small volumes may be treated with standard laboratory equipment. Setting aside a small area for such processes will reduce possible contamination of other laboratory areas.

Considerable research has been done on the removal of isotopes from liquids by ion exchange. Both the ion exchange resins and the montmorillonite clays have been tested as scavenging agents. These have the advantage for small installations that they can be run efficiently on an intermittent basis. When their capacity has been reached, as indicated by activity passing through the system, the resin or clay can be treated as a compact solid waste for disposal. The clays have the advantage that high temperature baking gives a massive product which is readily handled.

Any intermittent handling process for liquid wastes requires that adequate storage tank capacity be available to contain the waste accumulated in the period between disposals. Naturally, with short-lived isotopes a system of hold-up tanks may be the only requirement for waste disposal. The liquids are stored until a test shows that the activity level is below the limit permissible for disposal into the normal sewer system. Long-lived isotopes may merely be held until processing facilities are ready to accept the waste.

Since the usual laboratory problem encountered is the disposal of small amounts of waste, the requirements of Part 20 are paraphrased here.

No licensee shall discharge licensed material into a sanitary sewerage system unless:

(a) It is readily soluble or dispersible in water.

(b) The quantity of any licensed or other radioactive material released into the system by the licensee in any single day does not exceed the larger of:

  (1) the quantity which, if diluted by the average daily quantity of sewage released into the sewer by the licensee, will result in an average concentration equal to limits specified in Table 5A.5, column 2, or

  (2) daily release of 10 times the quantity of such material specified in Table 5A.7.

(c) The quantity of any licensed or other radioactive material released in any one month, if diluted by the average monthly quantity of water released by the licensee, will not result in an average concentration exceeding the limits specified in Table 5A.5, column 2.

(d) The gross quantity of licensed and other radioactive material released into the sewerage system by the licensee does not exceed 1 Ci/yr.

Excreta from individuals undergoing medical diagnosis or therapy with radioactive material shall be exempt from any limitations.

Airborne wastes may be concentrated by filtration of particulate matter or by absorption or scrubbing of radioactive gases. These processes yield solid or liquid material as a concentrate, and the wastes are disposed of as described above.

For the smaller isotope user, there are waste disposal services in several large cities which have been authorized to collect, store, and dispose of isotopes.

Equipment which has become contaminated cannot be disposed of simply by dumping, as it may appear in a critical location later as second-hand equipment. Frequently the cheapest method of disposal is to decontaminate the article sufficiently so that it can be returned to the normal channels of commerce.

MEDICAL PROGRAMS AND RECORDS

As radioisotope and x-ray utilization in industry increases it becomes much more important that adequate knowledge of any significant exposure of those employed in these fields is maintained. The employer must protect himself and his employees having the possibility of significant radiation exposure with a medical program and must maintain complete medical and radiation exposure records on each employee. These will be invaluable if any question of radiation damage occurs and in addition will furnish future information for the scientific evaluation of long-term exposure to low radiation levels.

The World Health Organization (WHO) in its technical report #196 has outlined the requirements for medical supervision in radiation work.[14] This is a clear description of the program required including the role of the physician. The ICRP merely indicates that the medical surveillance should be essentially the same as in general industrial medical practice.

At the present time there is only a limited number of medical tests available for radiation protection. Most exposure information is still obtained from personnel and area monitoring. Any radiation protection program is a failure if clinical evidence of radiation damage appears. Such a statement is not always true for other toxic materials, as clinical evidence sometimes may appear and a course of treatment may be indicated which can restore the worker to complete health. In case of radiation, the first clinical signs appear after considerable nonreversible damage has occurred. Thus, medical tests are not as much a part of a protection program as they are a confirmation that some acute overexposure has occurred.

A complete blood count was often recommended as a check on radiation exposure, but is not in favor with medical men interested in radiation protection. The most marked change is ordinarily seen in the leucocyte count and more specifically in the lymphocytes. The normal variation in various cell counts obscures the small variation caused by low-level exposure. When a positive indication is given by blood count it means that the subject has received definite overexposure approaching the acute stage. A systematic following of each individual could possibly give trends indicating chronic lower-level exposure, but such data have not been available on industrial or laboratory populations in the past.

One useful test is based on testing the chromosome changes that occur as a result of relatively low doses of radiation, perhaps as few as 10 rads. This involves measuring chromosome abnormalities in certain types of blood cells after suitable culture. Such a procedure may be helpful in evaluating the dose received in accidents.

One of the most valuable functions of a medical staff in a radiation protection program is in the selection of employees. Those candidates showing symptoms of normal disease which may also be attributable later to radiation should not go into a job requiring radiation exposure. This will obviously protect the employer and, less obviously, the candidate, as radiation may aggravate an existing condition. A typical example would be preemployment candidates showing signs of anemia.

The examination for selection of employees should be as complete as the periodic examination. In addition, it is desirable to obtain the entire past radiation history of the candidate, particularly if he has incurred industrial radiation exposure. An exposure history should not be grounds for turning down the candidate, but a knowledge of previous exposure is desirable and should be included in the employee's record.

If the medical examiner has some idea of the operations, he may be more useful in selecting candidates. Certain positions require specific mental characteristics as well as physical features, particularly where failure of the individual may endanger others. This has been a familiar industrial problem for many years and is not peculiar to jobs with a possible radiation exposure.

In particular, the medical examiner should be familiar with the details of metabolism and the permissible levels of any isotopes handled in the plant or laboratory under his control. This is required not for the treatment of overexposure but rather for the recognition of the hazards which may exist.

A complete set of records should be maintained for each employee. This should include the results of each medical examination, history of illnesses, exposure data, and the results of any special tests such as urinary excretion of isotopes or breath-Rn measurements. Such records should be reviewed at each periodic physical examination.

The exact form of medical records is subject to individual preference. Monitoring records for NRC licensees should be kept on form NRC-5 or in a form containing all of the information required by that form. This includes records of personnel monitoring with personnel dosimeters, the results of radiation surveys with instruments, and any disposal of radioactive materials to the environment. The regulations contain requirements for notification of overexposures and excessive levels and concentrations. In addition, licensees are required to notify employees of their radiation exposure on an annual basis. In the case of licensees, they must not only maintain records of each employee's cumulative exposure, but must obtain the previously accumulated occupational dose from former employers.

It is worthwhile for the person in charge of radiation protection to review the cumulative records of personnel monitoring at least semiannually. This will indicate which employees and which operations are showing the greatest exposure. Such knowledge may enable him to correct minor difficulties and reduce exposure appreciably. This again is part of the concept that maintenance of exposure below the permissible levels is not a sign of a perfect protection program and that efforts should be directed toward reducing all exposure to the minimum.

REFERENCES

1. Natural Background Radiation in the United States. NCRP Report No. 45 Washington, 1975.
2. Ionizing Radiation: Levels and Effects. U.N. Scientific Committee on the Effects of Atomic Radiation, United Nations, New York 1972.
3. E. I. Saenger (ed.). Medical Aspects of Radiation Accidents. U. S. Atomic Energy Commission, 1963.
4. Permissible Dose of Internal Radiation. ICRP Publication 2, New York, Pergamon Press, 1959.
5. Maximum Permissible Body Burdens and Maximum Permissible Concentrations of Radionuclides in Air and in Water for Occupational Experience. National Bureau of Standards Handbook 69, U.S. Government Printing Office, Washington, 1959.
6. Background Material for the Development of Radiation Protection Standards. Federal Radiation Council, Report No. 1, U.S. Government Printing Office, Washington 1960.
7. Nuclear Science Series, Preliminary Report. National Research Council Publication 251, 1951.
8. Methods of Radiochemical Analysis. World Health Organization, Geneva, 1966.
9. Control and Removal of Radioactive Contamination in Laboratories. NCRP Report No. 8, Washington 1951.
10. Radiological Decontamination in Civil Defense. Federal Civil Defense Agency, TM-11-6.
11. Protection Against X-rays up to Energies of 3 MeV and Beta and Gamma Rays from Sealed Sources. International Commission on Radiological Protection. ICRP Publication 3, New York, 1960.
12. Medical X-ray and Gamma Ray Protection for Energies up to 10 MeV. NCRP Report No. 33, Washington, 1968.

13. Structural Shielding Design and Evaluation for Medical Use of X-Rays and Gamma Rays up to 10 MeV. NCRP Report No. 49, Washington, 1976.
14. Medical Supervision in Radiation Work. World Health Organization, Technical Report Series No. 196, Geneva, 1960.

Selected Publications of the ICRP (Available from Pergamon Press, New York City.)

Pub. 5. The Handling and Disposal of Radioactive Materials in Hospitals and Medical Research Establishments (1964).
Pub. 7. Principles of Environmental Monitoring Related to the Handling of Radioactive Materials (1965).
Pub. 8. The Evaluation of Risks from Radiation (1965).
Pub. 9. Recommendations of the ICRP (1965).
Pub. 10. Evaluation of Radiation Doses to Body Tissues from Internal Contamination due to Occupational Exposure (1968).
Pub. 12. General Principles of Monitoring for Radiation Protection of Workers (1968).
Pub. 15. Protection Against Ionizing Radiation from External Sources (1969).
Pub. 21. Data for Protection Against Ionizing Radiation from External Sources (1971).
Pub. 22. Implications of Commission Recommendations that Doses be kept as Low as Readily Achievable (1973).
Pub. 26. Recommendations of the ICRP (1977).

Selected Reports of the NCRP (Available from NCRP Publications, P.O. Box 30175, Washington, D.C. 20014.)

No. 28. A Manual of Radioactivity Procedures (1961).
No. 30. Safe Handling of Radioactive Materials (1964).
No. 32. Radiation Protection in Educational Institutions (1966).
No. 35. Dental X-Ray Protection (1970).
No. 48. Radiation Protection for Medical and Allied Health Personnel (1976).
No. 50. Environmental Radiation Measurements (1976).
No. 53. Review of NCRP Radiation Dose Limit for Embryo and Fetus in Occupationally-Exposed Women (1977)

Selected Reports in the IAEA Safety Series (Available from Unipub, P.O. Box 433, Murray Hill Station, New York, N.Y. 10016.)

No. 1. Safe Handling of Radionuclides (1973 edition).
No. 2. Safe Handling of Radioisotopes: Health Physics Addendum.
No. 3. Safe Handling of Radioisotopes: Medical Addendum.
No. 6. Regulations for the Safe Transport of Radioactive Materials (1973 revised edition).
No. 12. The Management of Radioactive Wastes Produced by Radioisotope Users.
No. 13. The Provision of Radiological Protection Services.
No. 14. The Basic Requirements for Personnel Monitoring.
No. 16. Manual on Environmental Monitoring in Normal Operation.
No. 19. The Management of Radioactive Wastes Produced by Radioisotope Users: Technical Addendum.
No. 22. Respirators and Protective Clothing.
No. 24. Basic Factors for the Treatment and Disposal of Radioactive Wastes.
No. 25. Medical Supervision of Radiation Workers.
No. 30. Manual on Safety Aspects of the Design and Equipment of Hot Laboratories.
No. 38. Radiation Protection Procedures.
No. 40. Safe Use of Radioactive Tracers in Industrial Processes.
No. 41. Objectives and Design of Environmental Monitoring Programmes for Radioactive Contaminants.

SECTION 5B

Large Radiation Sources Applications & Safeguards

J. J. Fitzgerald, *President*
E. G. Fitzgerald, *Operations Manager*
Cambridge Nuclear Corporation
Billerica, Massachusetts

INTRODUCTION

Problems constitute part of the price paid for progress, particularly in the field of nuclear energy. Thus, our requirements for the industrial usage of radioisotope heat and power and penetrating radiation sources are going to require the highest degree of radiation protection and control. Safety must be an integral part of the construction and application of such sources.

The magnitude of the potential associated environmental problems and the methods to prevent their occurrence will be reviewed here. We shall concern ourselves also with the benefits derived from the use of these isotope sources. Society must weigh the benefits derived from the use of an inherently dangerous product versus the risks of its use. In some instances, it is clear that the benefits are substantial and the risks negligible, whereas in other instances, it is not easy to make an objective evaluation.

MAJOR SOURCES OF RADIATION POLLUTION

The major sources of radiation that have or will have an increasing impact on environmental pollution will be discussed in this chapter. They involve medical, space, terrestrial, radiation processing, and power applications. The potential problems and solutions associated with radioactive waste disposal and the transportation of radioactive materials will also be reviewed.

MEDICAL APPLICATIONS

In medicine, radioisotopes are being used as diagnostic tools in the treatment of disease and as sources of energy. As diagnostic tools, small or tracer quantities of radioisotopes may be attached to or incorporated in chemicals or compounds that, when injected into the blood of a patient, will concentrate in specific organs. The radiation from these radioactively labeled compounds permit the external (to the body) measurement of normal or abnormal functions of the organs in question. These radioisotope products are known as "In-vivo" radiopharmaceutical products since they are taken into the body as opposed to other diagnostic products that are injected into test tubes and are called "In-vitro" radiopharmaceutical products. Such diagnostic products are the tracer quantities of radioisotopes that are used to label proteins, hormones or antigens to determine the quantities of such antigens as insulin, growth hormone, etc., in the patients extracted blood or excreted urine. Solid relatively large sources of 60-Co applied externally to the body are used to irradiate and destroy cancerous tissue. Solid large sources of radioisotopes may be implanted within the body as a source of energy for a cardiac pacemaker and for an artificial heart pump. Most of the above categories will be covered briefly after a discussion on the production of radioisotopes to be used in nuclear medicine.

Radioisotope Production for Nuclear Medicine

There are two major sources of radioisotope production for nuclear medicine; the nuclear reactor and the particle accelerator. The first isotopes used in nuclear medicine were produced in accelerators. Both sources of radioisotope production are used today. For in-vivo radiopharmaceutical operations, the most used radioisotope is Tc^{99m} which is the decay product of Mo^{99}. Technetium-99m is produced by the exposure of Mo^{98} to reactor neutrons or as a fission product from the fissioning of U-235 in a nuclear reactor. In both cases, Mo^{99} (half-life of 67 hours) is formed and decays with the emission of a beta particle to the meta-stable state of Technetium-99 that has a 6-hour half-life. Tc^{99m} is a very interesting radioisotope and has excellent characteristics as a tracer radioisotope for in-vivo diagnostic purposes. Since it is in its meta-stable state there is no particulate emission or absorption of particulate radiation in the body. Tc^{99m} emits a photon with energy of approximately 140 Kev that is excellent for external scanning of the organ that is being diagnosed. The half-life is 6 hours which is long enough for daily use and for daily extraction from the Mo-99 source which is prepared weekly. The half-life is yet short enough to present only a relatively small radiation exposure to the patient. From a chemical viewpoint, technetium can be prepared to take many valences and thus be tagged to a number of compounds that concentrate in specific organs which must be scanned and traced. The radioactive technetium product can be provided as a molybdenum-technetium generator or as sodium pertechnetate ($NaTcO_4$). The Mo-Tc generator permits the daily extraction of Tc from Mo over a period of approximately one week. Each week a new generator must be purchased since the parent product Mo-99 has a 67-hour half-life. Purchasing of the ready-to-use $NaTcO_4$ must be done on a daily basis since Tc-99m has a half-life of six hours.

For in-vitro radiopharmaceutical operations, the most widely used radioisotope is ^{125}I (60 day half-life). The characteristics of ^{125}I for in-vitro diagnostic operations are also very desirable. The half-life is long enough to provide relatively stable products for at least one month with respect to the radioactive product. Since the radioactive product is not injected into the patient's body, but mixed with the extracted blood or urine of a patient and in a test tube, there is, therefore no dose to the patient and consequently the longer half-life product presents no problem from a patient-dose viewpoint. The production of ^{125}I in nuclear reactors is made with ^{124}Xe as the target material as shown below:

$$_{54}Xe^{124} + {}_0n^1 \longrightarrow {}_{54}Xe^{125} \xrightarrow{EC\ (17\ hr)} {}_{53}I^{125}\ (EC\ 60\ days)$$

In-Vivo Applications

At present there are more than 2000 hospitals in the United States with facilities for the practive of nuclear medicine, the fastest growing branch of medicine. Nuclear medicine may be defined as the scientific and clinical discipline concerned with diagnostic, therapeutic (exclusive of sealed radiation sources), and investigative uses of radionuclides. Nuclear medicine departments of licensed hospitals use more than 10^6 radiopharmaceutical packages of pyrogen-free, sterile doses of relatively pure radioisotopes per year for application to patients to measure certain of their organ functions. Radiopharmaceuticals thus make it possible for the physician to diagnose the ills of many patients that he would not otherwise be able to do. A few examples of radiopharmaceutical applications are presented below:[1]

Brain Scanning via Tc-99m. Radioisotope scanning of the brain following injection of $NaTcO_4$ using Tc-99m is used to detect primary and metastatic brain tumors, strokes, subdural hematomas, or other organic brain lesions; see "normal" brain scan (Fig. 5B.1). Abnormal conditions in the brain are indicated by post injection traces of radioactive material in the area of the brain such that a breakdown of the brain-blood barrier must have occurred (see Fig. 5B.2).

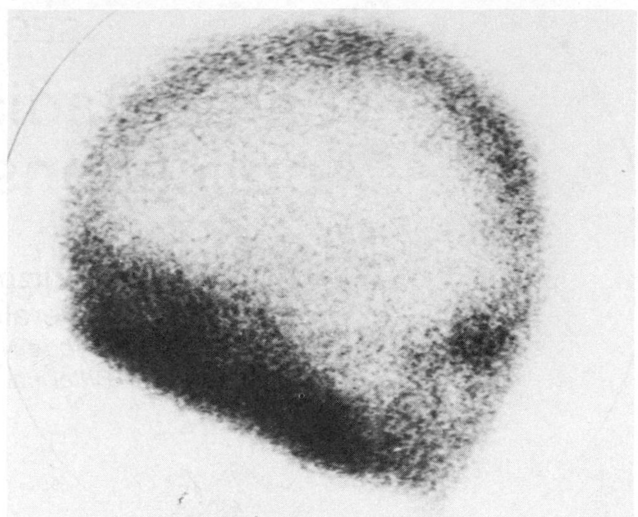

Fig. 5B.1. Scan of normal brain.

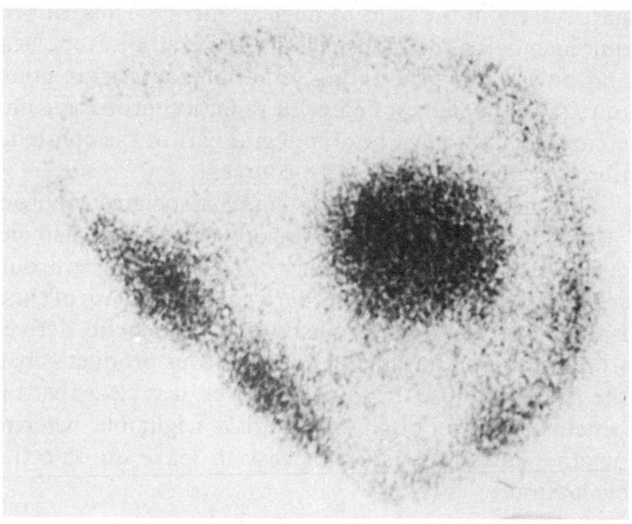

Fig. 5B.2. Scan of abnormal brain (ependymoma).

Lung Scanning via Tc-99m and Xe-133. Radioisotope scanning of the lungs is applied to detect carcinoma, asthma, emphysema, TB, and pneumonia. The injection of Tc-99m labeled macroaggregates of serum albumin into the patient's blood stream enables the physician to study the patient's lungs (see Figs. 5B.3 and 5B.4). Xenon-133, a radioactive gas, may also be used for lung ventilation function studies to provide a means of diagnosing bronchitis, emphysema, and such chronic diseases (Fig. 5B.5).

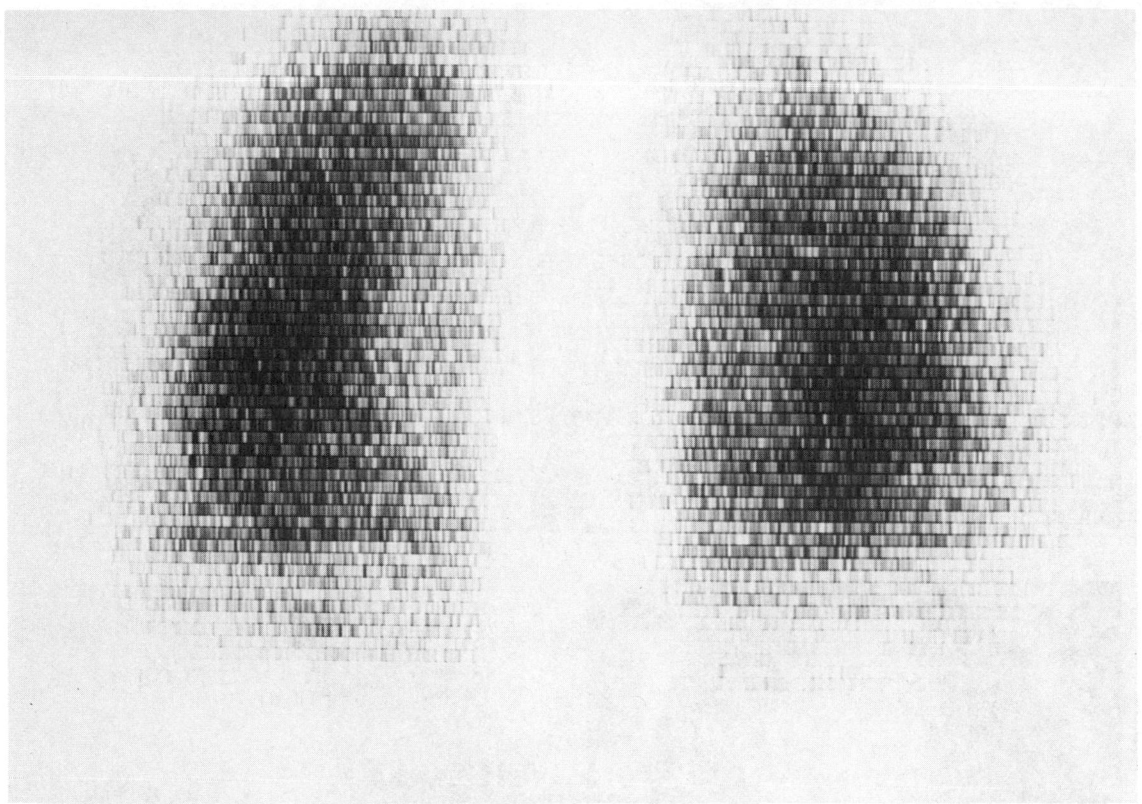

Fig. 5B.3. Scan of normal lung.

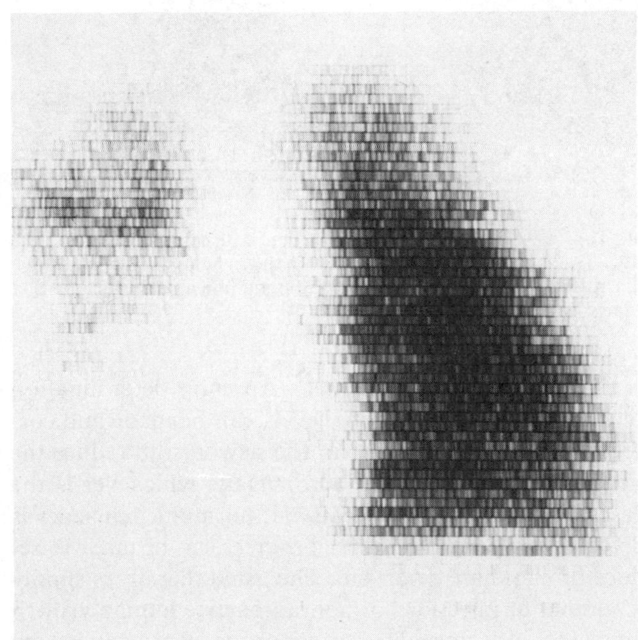

Fig. 5B.4. Scan of abnormal lung (bilateral pulmonary emboli).

Bone Scanning via F-18, Tc-99m, Sr-85 and Ca-47. A bone scan following the injection of F-18 or a polyphosphate compound tagged with Tc-99m may show evidence of metastatic disease months before an x-ray analysis could indicate such an abnormality (see Fig. 5B.6). Other bone scanning isotopes, i.e., Sr-85, and Ca-47, are also used, but the first two mentioned have the greatest promise for the future because they reduce population dose and provide effective bone disease analysis.

In-Vitro Applications

Closely allied with the in-vivo application of radioisotope labeled compounds is the field of radioimmunoassay (RIA) involving the use of radioisotope labeled antigen or in-vitro (in glass or in test tubes) radiopharmaceuticals. In this process, a small sample of the circulating blood or patient's urine is added to a test tube together with a small known amount of labeled antigen (for example, insulin or growth hormone) and a small and known amount of a specific antibody that would bind the corresponding labeled and unlabeled antigen. This process is very sensitive and capable of measuring picogram amount of antigens (proteins, hormones, steroids, etc.) in circulating blood. It is a highly specific process using highly specific antibody which bind essentially only the corresponding antigens, for example, a glucagon antibody may be produced and used that is approximately 100% specific for pancreatic glucagon and essentially ineffective (less than 0.1%) for binding of insulin or gut glucagon. In this in-vitro process, there is no exposure to the patient since the activity used in the

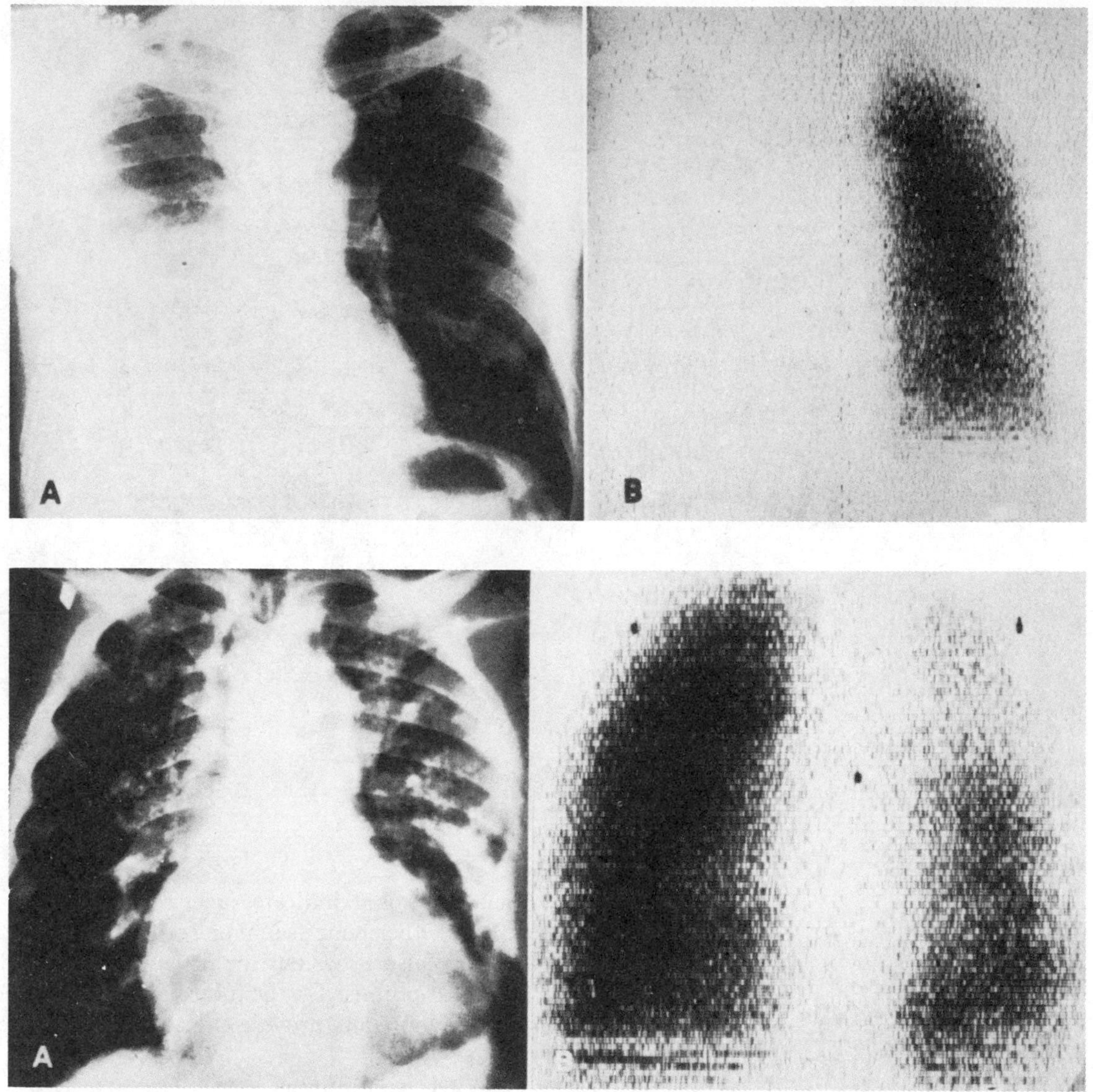

Fig. 5B.5. Radioactive xenon in diagnosis of pulmonary embolism. Top: A—Chest radiograph of a patient with carcinoma of right lung. B—Lung scan in the same patient. There is essentially no pulmonary arterial blood flow to the right lung. The oxygen uptake by the right lung was zero as measured by bronchospirometry. Bottom: A—Chest radiograph in a patient with pulmonary embolism and infarction. B—Lung scan in the same patient.

analysis is added to the test tube. A few examples of the applications and the associated benefits will be discussed.

Angiotensin. A pressor hormone, is formed in the blood-stream and causes a powerful vasoconstriction of the arteries, thereby causing a rise in the blood pressure. Two important disease conditions can be diagnosed by abnormal levels of angiotensin: (1) Unilateral renal disease (URD), which is indicated by the pressure of high angiotensin levels in the blood causing high blood pressure as the result of hypersecretion of renin in the kidney (see Fig. 5B.7), and (2) Conn's syndrome, diagnosed by the presence of low angiotensin levels, because it is usually related to an aldosterone-secreting tumor in the adrenal gland area and aldosterone is known to inhibit renin or angiotensin formation.

Human Growth Hormone. An early determination of human growth hormone levels can be made and corrections quickly instituted in the newborn to reduce the problems of dwarfism or acromegaly, whichever is the case. This is important because, if hormone deficiency is detected early in a newborn, progress can be made to reduce or eliminate dwarfism. The usual therapy is simply a number of injections of nonradioactive human growth hormone. Acromegaly is a condition characterized by an abnormal increase in the number of cells in normal arrangement in the extremities of the skeleton (the nose, jaws, fingers, and toes). The usual therapy for acromegaly is proton bombardment of the pituitary gland (see HGH extraction form pituitary in Fig. 5B.8). The effects of the bombardments can than be successively measured by RIA.

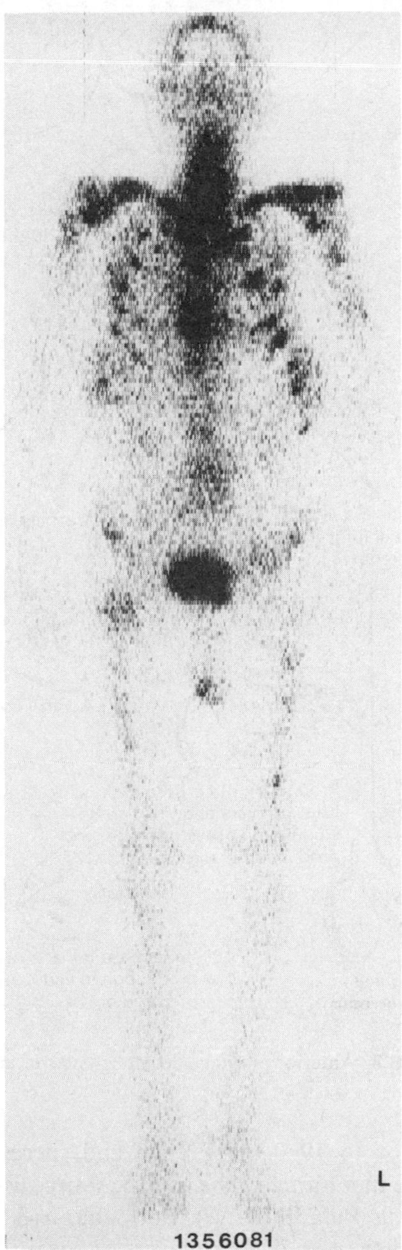

Fig. 5B.6a. Abnormal bone scan (metastases).

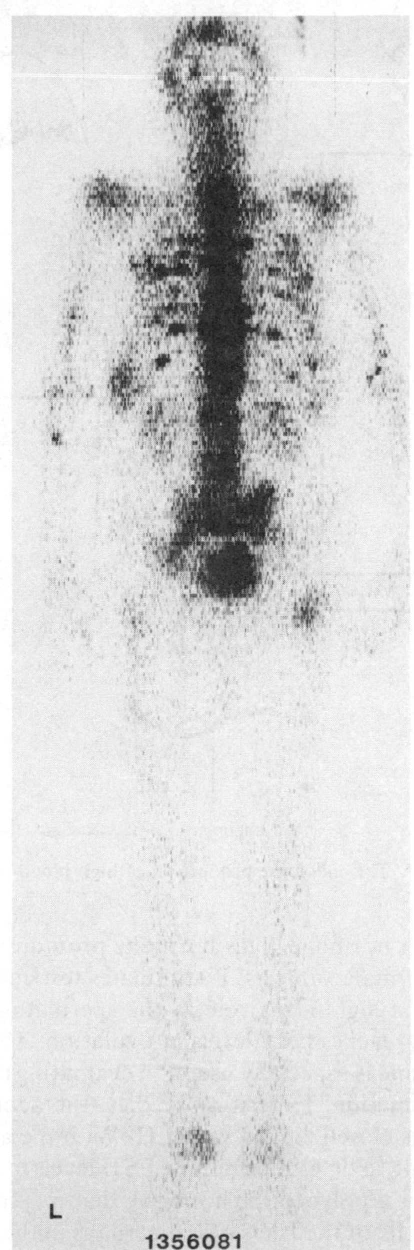

Fig. 5B.6b. Abnormal bone scan (CA prostate metastases).

Luteinizing Hormone (LH). LH is a protein hormone synthesized in the basophilic cells of pituitary glands. It exerts action on both male and female gonadal tissue. In the female, it activates the maturation of the graafian follicle, ovulation, and development of the corpora lutea. Estrogen and progesterone secretion are also stimulated by this hormone (see Fig. 5B.8). In the male, LH stimulates testosterone production by the testis, thereby maintaining spermatogenesis and the development of vas deferens, prostate, and seminal vesicles. LH level is of particular importance in cases of infertility. RIA is being used to measure the LH level in the body. Either urine or serum samples may be used, although the serum sample method is preferable.

Thyroid Stimulating Hormone (TSH). The anterior lobe of the pituitary gland secretes TSH, also called thyrotropin (see Fig. 5B.8). It is a mucoprotein and its presence increases both thyroid growth and general metabolic activity. Iodine uptake and the breakdown of thyroglobulin results from TSH with the ultimate release of thyroid hormone. Clinically, TSH is used to differentiate primary hypothyroidism (myxedema) from pituitary insufficiency. Bioassay has been the usual method of determining the level of TSH. Now, with the availability of RIA, more precise and highly sensitive assays are performed.

Follicle Stimulating Hormone (FSH). Hypothalamic releasing factor mediates the secretion of follicle

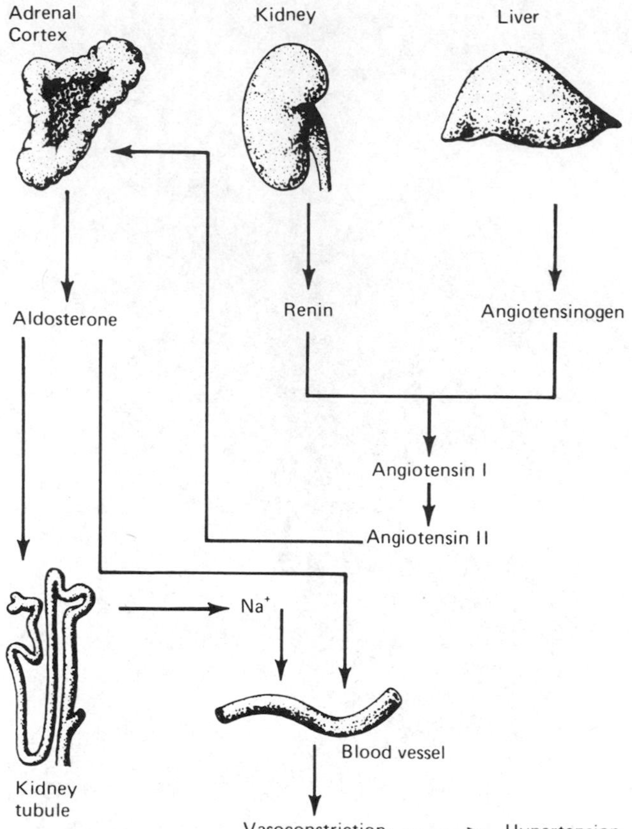

Fig. 5B.7. The endocrine production of high blood pressure.

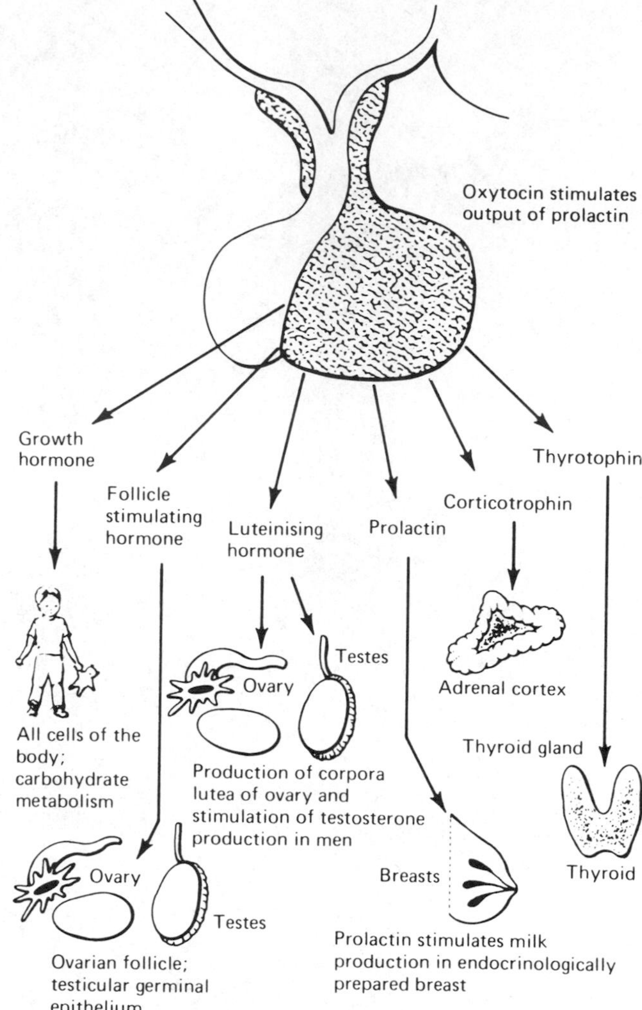

Fig. 5B.8. Anterior pituitary hormones sites of action.

stimulating hormone. This hormone promotes follicular growth. In male subjects, it stimulates testicular growth and plays a significant role in the spermatogenesis. In females, it reaches peak levels at ovulation. The assay of this hormone is especially useful in evaluating the general gonadal function. In fertility studies, the action of this hormone is closely linked with LH. With the aid of RIA it is possible to determine minute FSH levels in the blood.

Insulin is a polypeptide hormone that is secreted from the beta-cells of the Islets of Langerhans in the pancreas. The function of insulin is to maintain the blood glucose at a constant level. Insulin affects the permeability of cell membranes to glucose thereby regulating the flow of glucose out of the blood. In high blood glucose levels, insulin is present in relatively high concentrations in normal patients, and drops dramatically as suger levels decrease. Therefore, glucose acts as a negative feedback for insulin production. Abnormal insulin values are seen in certain pathological states, most notably Diabetes Mellitus.

Isotope-Powered Pacemakers

Isotope-powered cardiac pacemakers are a reality today. In 1972, after years of research and development, the first such unit was inserted into a human in the United States. Prior to 1970 the French had successfully implanted one in a human. The isotope source used in both instances was Pu-238 (89 yr). Both countries used solid sources and the two or more layers of encapsulation for the sake of safety. A pacemaker source would normally contain approximately 3 Ci of an α emitter and yield approximately 0.1 watt (thermal) of energy.

The natural cardiac pacemaker is a small portion of tissue located in the attrium of the heart. In operation, impulses are sent from the pacemaker to essentially every muscle cell of the ventricle, causing them to contract together. If one has a complete heart blockage or symptoms of Adam Stokes disease, it is because the "Bundle of His" loses its conductivity and the atrial impulses cannot get through the ventricle. However, the heart does not stop completely under these conditions because the ventricles have their own natural pacemakers. Normally, the ventricular cells synchronize with the signal from the atrium, but in the absence of the atrial signal, the ven-

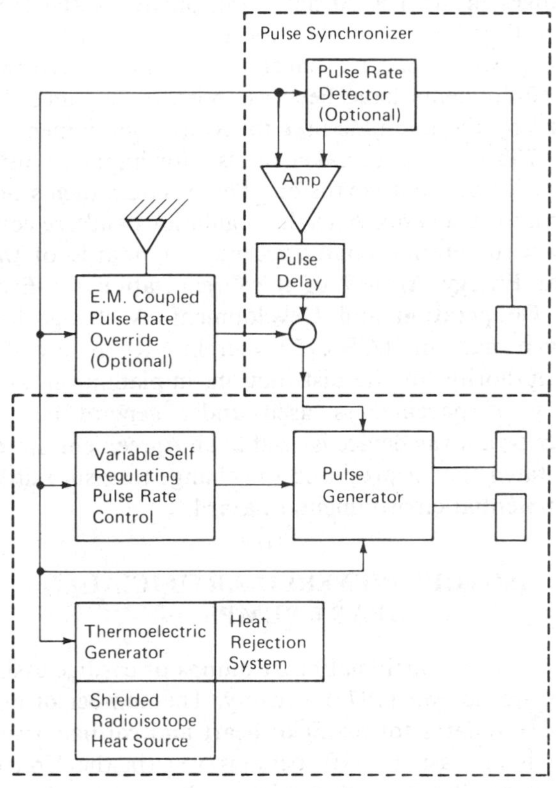

Fig. 5B.9. Asynchronous radioisotope-powered cardiac pacemaker.

tricular cells revert to their own natural rate which is 20 to 35 beats/minute for a human being. Thus, the attrium continues to contract at a rate of 72 beats/minute while the ventricles beat at a completely asynchronous rate. This is clearly indicated in an EKG which shows the first peak in the EKG graph coinciding with the contraction of the atrium at the normal rate and the QRST complex occurring at tandem.

In the radioisotope-powered artificial pacemaker, a small electronic pulse generator delivers a pulse directly to the ventricles at a preset rate with an amplitude of 4 to 8 volts and a duration of 1.5 msec. The ventricular cells then synchronize on the signal. Although the above contraction rate is still asynchronous, patients are able to adapt rapidly to the constant heart beat. Pacemakers have also been built that actually synchronize with the P wave.

Until recently artificial cardiac pacemakers were powered by energy cells or batteries. With the development of compact long-lived radioisotope power units, a new source of power suitable for pacemakers became available. The major in-use source of failures now appears to be electrode-lead breakage causing a loss of electrical contact in the heart. While battery-operated pacemakers whose leads do not break can operate continuously for about 2 years, an isotope source unit with a thermoelectric generator and no leads to worry about can oper-

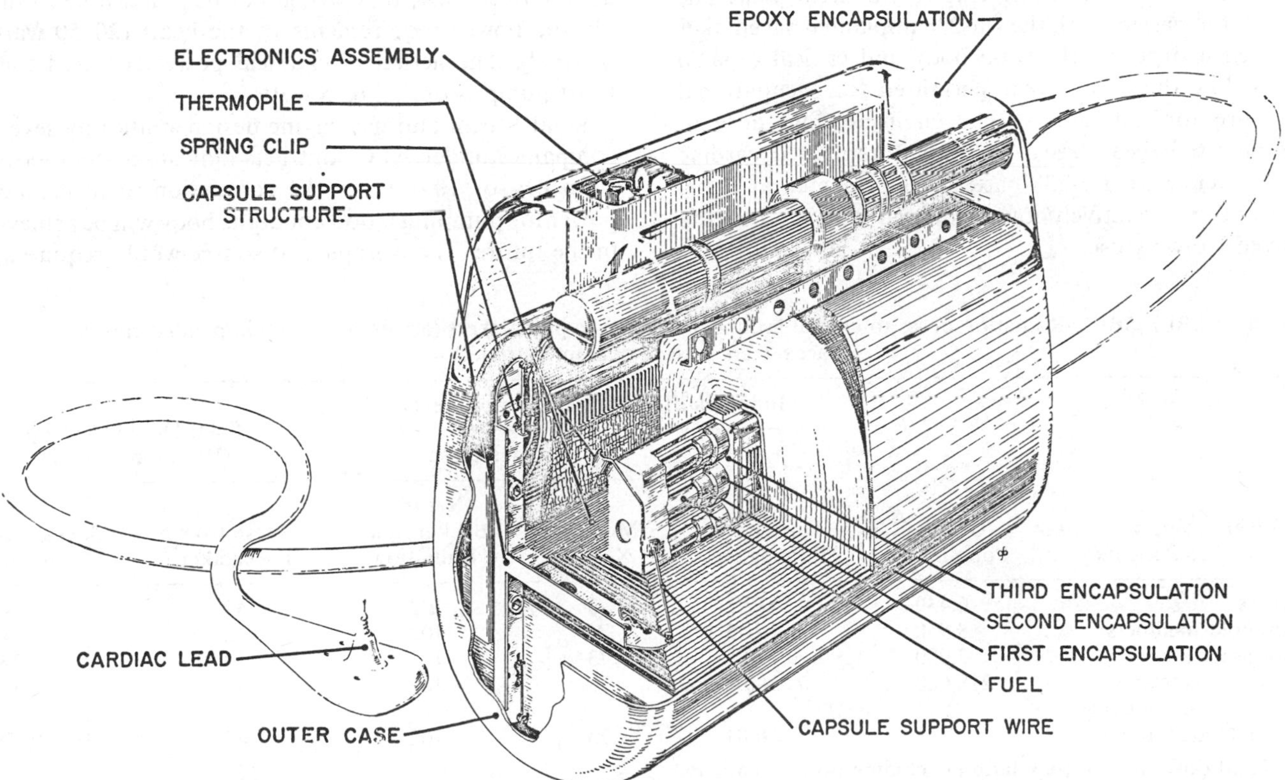

Fig. 5B.10. Pacemaker (Numec).

ate up to 10 years or more without requiring additional surgery to change power supplies (see Figs 5B.9 and 5B.10). This is particularly important to the elderly patient.

Since the first implantation of an isotope-powered pacemaker in France and the United States, there have been many more such successful implants. Since it has been estimated that in the United States alone between 10,000 and 50,000 persons/year require pacemakers, it is reasonable to assume that tens of thousands of people will soon be using implanted isotope-powered pacemakers at any one time.

The benefits derived from these pacemakers are substantial, whereas the possible risk associated with the violation of the integrity of the sources is very small. These sources are well encapsulated and are exhaustively tested to assure proper containment.

As to the environment, although as many as 10^5 units containing 3 Ci of activity each, or up to 3×10^5 Ci total, might be moving about on the surface of the earth, the actual contamination of the environment from leakage should be nearly zero.

Pacemakers powered with Pu-238 sources can now be licensed on a routine basis, based on the relatively low environment impact and the high benefits received by the use of Pu-238. As shown in Table 5B.1, the dose to individual spouses is up to 7.5 mrem/year, while nonfamily associates may receive up to 0.2 mrem/year. The dose to the patient with the Pu-238 implanted pacemaker received a dose (to the total body and critical organs) well below the 5 rem/year permitted for occupational exposure for individuals. The environmental dose of 128 man-rem/year based on 10,000 implanted cardiac pacemakers with Pu-238 batteries is presented in Table 5B.1 and it is relatively low compared with the benefits derived from its use.

Changes in 10CFR-70 have been made by the U.S. Nuclear Regulatory Commission to simplify the use of Pu-238 powered cardiac pacemakers. Safety performance requirements have been established that must be satisfied by the manufacturer to assure containment of the Pu-238 in the pacemaker in tests for impact, crush, fire, cremation, and corrosion. These requirements are set forth in Appendix A in the regulations and are consistent with internationally accepted standards of the Nuclear Energy Agency of the Organization for Economic Cooperation and Development. A change has also been made in 10CFR150 wherein NRC is now the single authority for the distribution, implantation, and recovery of pacemakers used under general license whether or not the device is used in an Agreement State. NRC stated that it proposed the change to assure little or no potential environmental hazard.

ISOTOPE-POWERED ARTIFICIAL HEART PUMPS

Isotope-powered artificial heart pumps or cardiac assist devices are not yet (1977) a reality. The number of potential candidates for artificial heart and cardiac assist devices is at least 2×10^5 persons/year in the United States alone. But, while batteries can be used for the artificial pacemaker because its power level requirements are relatively low, they would not be practicable to supply the power requirements of the heart (20–50 watts, thermal). The actual mechanical power required for a heart pump is only 2 to 5 watts.

Small steam and gas engine design studies by several companies under AEC contracts indicate that a thermal efficiency of 10–15% in the conversion to mechanical power for pumping blood about the body will be achieved in the future. The isotope heat source would require ap-

TABLE 5B.1 Radiation doses to critical groups from cardiac pacemakers (assuming 10,000 implanted cardiac pacemakers with plutonium batteries)

		Individual Dose (mrem/person/year)			Total Dose to Group (Person-rem/year)	
			Average Dose			
Relationship to Pacemaker Patients	Group Population	Dose from Pacemaker[a]	Medical X-rays	Natural Background Radiation	Dose from Pacemaker[b]	Natural Background Radiation
Spouses	6,430	5–7.5	73	102	42	646
Household members	8,950	1–1.5	73	102	12	912
Work associates[c]	72,000	0.1–0.2	73	102	10.5	7,344
Nonwork associates[c]	218,000	0.05–0.1	73	102	14.5	22,378
Total in U.S. populace not included above		<<0.01	73	102	49	21,400,000
Total dose to U.S. population excluding dose to patients[d]					128	

[a]Dose will vary depending upon the plutonium content, fuel characteristics, and shielding effects of a particular pacemaker model.
[b]Integrated dose using 4 Ci of plutonium which is the average amount of plutonium used in any battery.
[c]A patient is predicted to associate with about 30 persons during his daily activities.
[d]U.S. population of 210,000,000.

proximately 30 Ci of α emission per thermal watt, 300 Ci of an energetic β emitter per thermal watt, or more than 3000 Ci of a weak β emitter per thermal watt. Thus, the total heat source of an artificial heart might contain from 1000 Ci of an α emitter to 100,000 Ci of a weak β emitter. Since the artificial heart pump source is approximately 300 times more powerful than the pacemaker power source, much more concern must be given to its radiation protection and control problems. In addition, the number of persons requiring these power units in the future is potentially ten times greater than those that will require pacemakers. To be safe to use, the leak rate from the source which has several layers of encapsulation must be vanishingly low to prevent deposition of substantial amounts of activity in the body of the patient and thus into the environment.

Only a few of all the available radioisotopes can be considered as isotopes of choice for powering the artificial heart. They are generally considered to be Tm-171, Pm-147, and Pu-238 each has advantages and disadvantages. The dose-rate and the integrated dose to the body of a wearer through a shielded source for Pu-238 may be too high for wide-spread use or perhaps the weight of an adequate shield in the case of Pu-238 may be too great for popular use. Earlier, it was believed that Tm-171 has only limited use because of its relatively short half-life (1.9 yr) and possible difficulties in production or cost. A simple new method of hypodermic injection or removal of radioactive power source has been developed by Fitzgerald to eliminate the problem of half-life altogether. This would enable a surgeon to add or extract radioisotope power without additional surgery. Thus, the useful life of an isotope rather than its physical half-life may better describe the operational life of the power unit. This method puts back into the hands of the physician the flexibility to increase or decrease the pumping power of the patient's heart-assist device without additional surgery. In addition, the physician can install the cardiac assist device without its isotope charge as a stand-by in case it is ever needed by a patient who may be in the early stages of a heart problem. A diagram of the double-barreled needle and the method of injection and extraction are shown in Fig. 5B.11 As indicated by the diagram the outer barrel, which alone contacts the skin and never goes beyond the lower level of the decontaminating middle zone, is not removed until the sonic-probe cleaning process indicates that there is no contamination in the barrel of the hypodermic. Then if a patient with a standby unit needs to use it, a simple hypodermic injection of liquid radioisotope, which can be done in a physician's office, starts the heart-assisting action.

Since large quantities of radioactivity will be used in large numbers of potential patients, considerable effort must be made to assume adequate safeguards. Detailed analyses have been made of the internal and external hazards to patients. Analyses of the environmental hazards indicate that the artificial heart pump and cardiac-assist device can be designed for safe use (see Figs. 5B.12 a,b).[4]

It is evident that isotopes can be put to their best use in saving or prolonging life. It is not immediately evident that these sources can be functionally designed in an almost inherently safe manner, although detailed analyses makes it clear that they can be designed to operate and function safely even under very adverse conditions.

SPACE APPLICATIONS[5,6,7]

Power requirements for space vehicles and associated equipment are being satisfied in part by the use of radioisotope-powered generators. The advantages are their high power density, long operating life, and reliability. Some of the uses of isotope-powered generators in space and on the moon will be discussed. The basic environ-

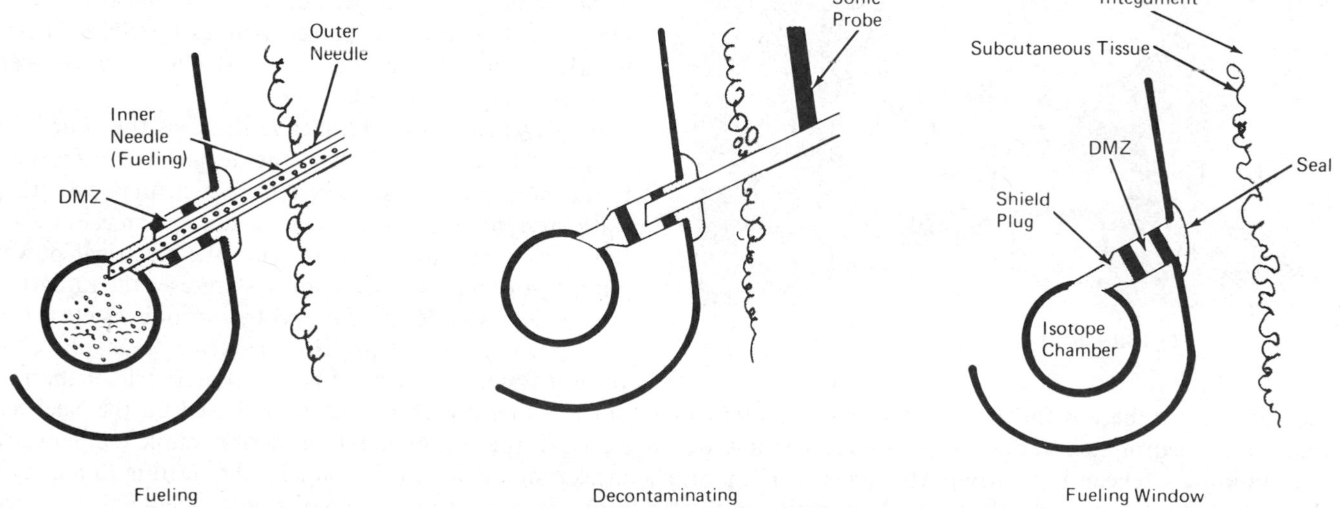

Fig. 5B.11. Transcutaneous fueling and defueling of implanted nuclear-powered devices.

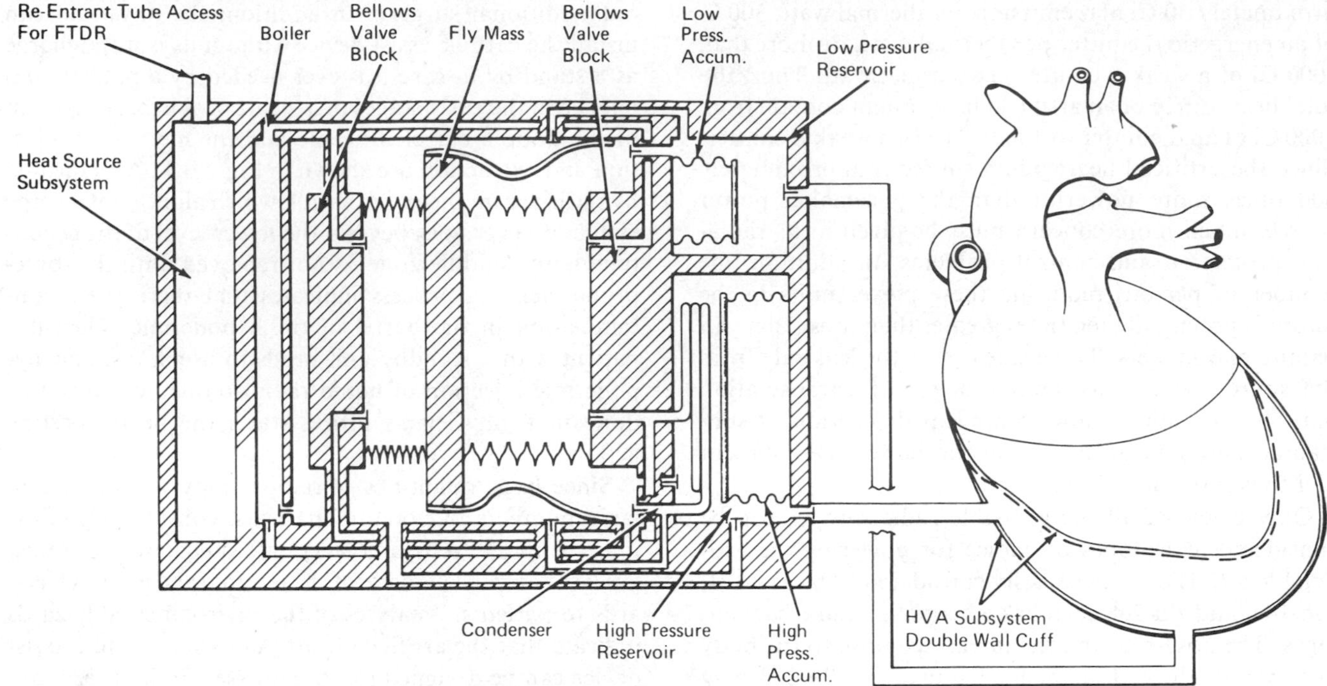

Fig. 5B.12a. Conceptual CAPE system schematic.

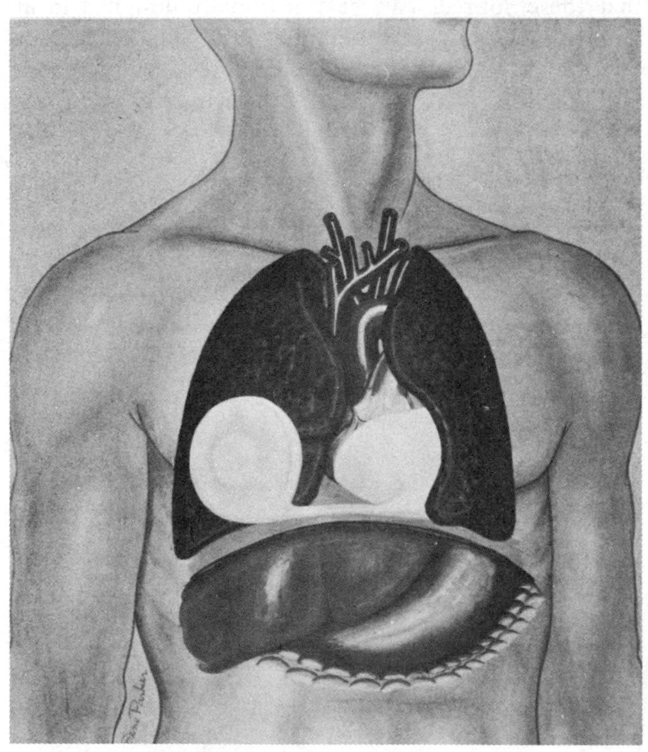

Fig. 5B.12b. Heart assist pump unit.

mental problem here is the ability of a manufacturer to make an encapsulated radioisotope source capable of withstanding the heat of reentry of the space vehicle into the earth's atmosphere, and the force of impaction of the vehicle on the surface of the earth. The source encapsulation must be able to remain intact through all of that to prevent dispersion of the source radioactivity over the surface of the earth or the sea. (See Fig. 5B.13.)

ISOTOPE POWERED SATELLITE POWER UNITS

The ability to convert the energy of radioactive decay into mechanical and electrical energy has led to the development of small nuclear-power supplies for use in locations and environments where conventional sources of power are either not feasible or less than satisfactory. Under AEC contracts, power sources for SNAP (Systems for Nuclear Auxiliary Power) devices have been and are being developed.

In SNAP devices, radioactive decay is the source of heat, but that is only half of the device. The other half is the power conversion system, which may be either static or dynamic. A static conversion system uses radiation per se or directly to provide the electrical power output. A dynamic conversion system uses the radiation energy as a source of heat and then produces electrical energy from the heat perhaps via steam. Dynamic conversion systems may be either thermoelectric or thermionic. A thermoelectric system is based on the Seebeck effect, or that which makes a thermocouple temperature measuring device work, namely, the heating of a circuit juncture composed of two dissimilar metals to produce an electric current which flows along a circuit. (See Fig.

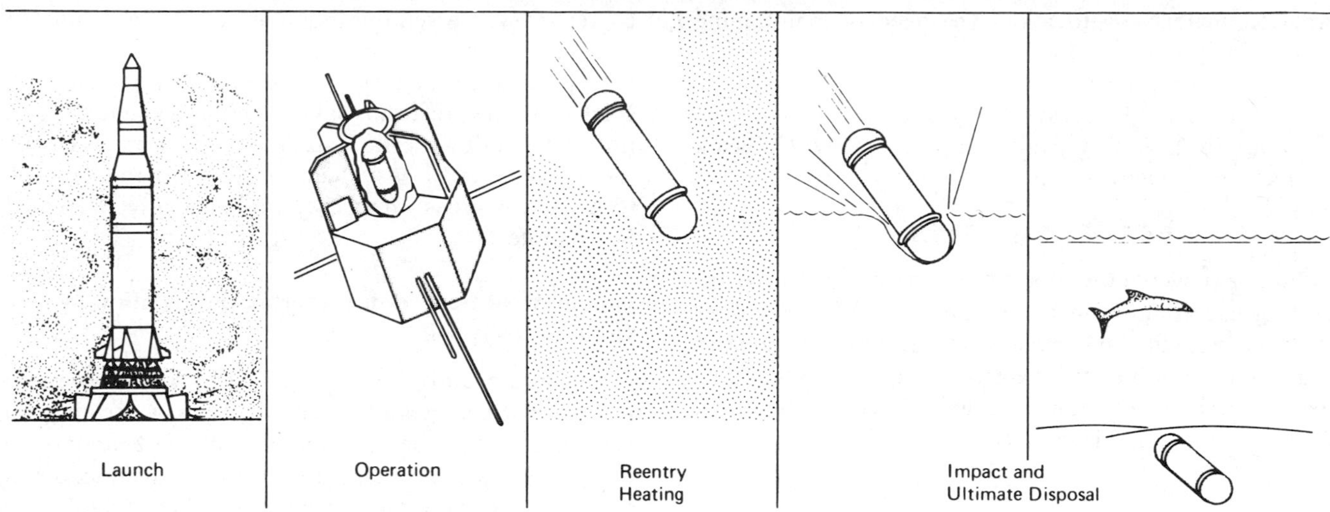

Fig. 5B.13. Isotope fuel systems operating environments. (From Newby, "The USAEC space nuclear electric power program—The objectives and status," ANS meeting, June 1968, Toronto, Canada.)

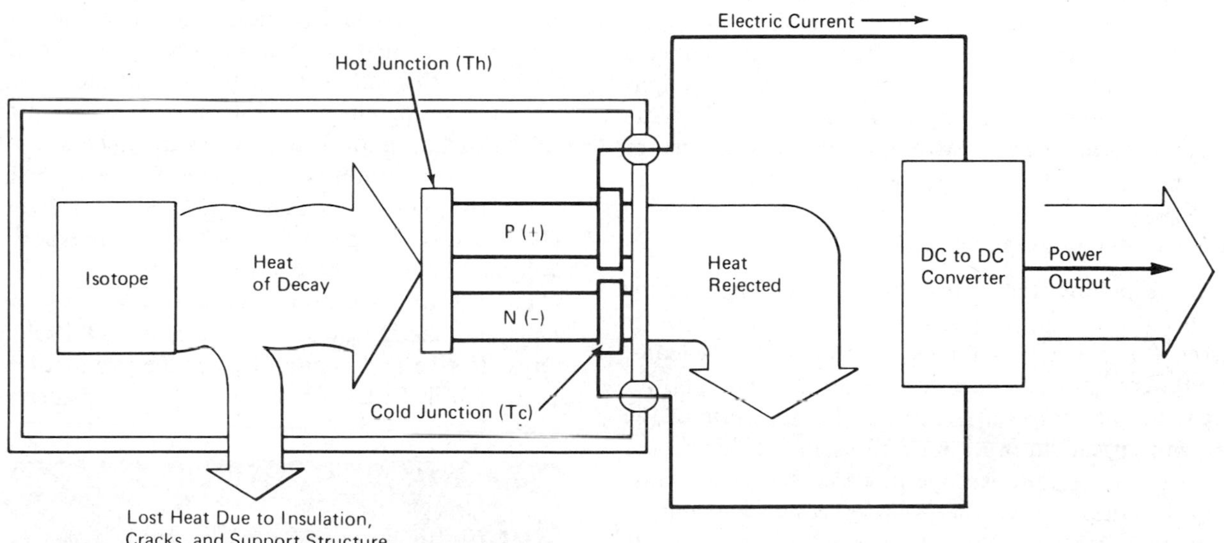

Fig. 5B.14. Thermal power heat flow schematic.

5B.14.*) A thermionic system resembles an ordinary vacuum tube in operation in that the radiation heat source acts as a cathode from which electrons are boiled off to be captured by an anode of different material after they have passed through a vacuum gap separating the two electrodes.

Radioisotope-powered units in space applications may be used in scientific satellites, communication satellites, and in military satellites. Scientific satellites include weather satellites, orbiting geophysical observatory satellites, orbiting solar observatory satellites, and interplanetary monitoring platform satellites. Communication satellites are most certain to increase in number in the future to provide better telephone and television coverage. Military satellites may be used to provide scientific data and communications as indicated above, in addition they may be operated to provide navigational aid and surveillance. With increased space operations, the use of radioisotope-powered units will increase. Substantial quantities of relatively long-lived radioisotopes may be used to provide the necessary electrical requirements. The most popular radioisotope presently considered or used in these operations is Pu-238 (89 yr). To protect the environs the integrity of those radioisotopes sources must be maintained over a long period of time since the half-life of the material is 89 yr. A solution to the problem of maintaining the integrity of the encap-

*David L. Purdy, Nuclear Materials and Equipment Corporation, Apollo, Pa. Design of Isotopically Powered Thermoelectric Generators, ANS meeting, June 1968, Toronto, Canada.

sulation material is to avoid the need of maintaining source integrity for a long time or to adopt the use of shorter-lived radioisotopes such as Tm-170 (127 d). The activity of Tm-170 decays by a factor of 8 each year whereas Pu-238 (89 yr) takes approximately 270 years to decay by a factor of 8.

Isotope-Powered Lunar-Based Units

A Pu-238 powered thermoelectric generator has been operating successfully on the moon since 1970. The first such unit provided 63 watts of electrical power and later units will undoubtedly have greater capacity. Such units provide reliable power over a long period of time and need not pose significant safety problems.

TERRESTRIAL APPLICATIONS

Radioisotope-powered energy generators have also been proposed for a variety of terrestrial applications. A few such applications will be discussed. Radioisotope-powered units become attractive in comparison to other forms of energy producers when the power requirements are, in general, less than 1000 watts, although designs have been proposed for requirements up to 10^4 watts and for remote operations requiring a high degree of reliability.

Isotope-Powered Oceanographic Units

Radioisotope-powered units for marine applications have been used and proposed for further use as power sources for navigational aids, marine radio beacons, sonar beacons, automatic weather stations, and as power sources for offshore oil well units. Perhaps one of the most widespread future applications might be the use of air-droppable radioisotope-powered buoys to transmit information on weather and sea conditions and, in conjunction with a navigation satellite, to provide world wide knowledge of weather conditions. Such a program would involve a large number of radioisotope sources and therefore substantial amount of radioisotope material. Such a program could be environmentally safe to conduct if Tm-170, for instance, was the radioactive source of power.

Isotope-Powered Airplane Navigational Aids

There is a need for remotely located aircraft navigational aids. Commercial airlines pioneering new routes over undeveloped lands as well as business aircraft operating in areas remote from company landing fields, i.e., at mineral recovery or plantation sites, could use what is known as a vector omnirange (VOR) navigational aid using a radioisotope source as a power unit. Radioisotope power enables one to construct a small integrated power unit that can be transported by helicopter to even a remote mountain top. Such a unit located in a remote area could continuously and for a long time provide the necessary directional information for aviators in the area, thus saving lives and property. Central and South America and Africa are prime areas for the use of VOR's, particularly since such units are safe to use in the areas where they would usually be located, i.e., remote.

Isotope Heat Sources for Thermal Stimulation of High-Viscosity Oil

A new and promising use of the decay heat from radioisotopes, which would otherwise be a waste disposal problem, is the thermal stimulation of secondary oil deposits in sand and clay at depths relatively closer to the earth's surface than those in oil shale. The known deposits that may be thermally tapped by the decay heat of such radioisotopes have been placed at 400 to 500 billion barrels of oil. (See Fig. 5B.15.)

The basic problem of extracting high-viscosity oil is one of efficiently and economically depositing heat into the crude oil matrix. The present primary method of extracting this oil is by a process of steam injection wherein one produces high-temperature steam (550° F) on the surface and pipes it down the injection well.

A second method is called *fire flooding*. In this process, air is fed into the deposit through an injection tube. The crude is ignited and its heat of combustion is used to heat the rest of the deposit.

The use of an isotope source properly encapsulated which is placed downhole in the deposit itself can be very effective and economical in the production of extractable oil. In the final analysis, it seems certain that

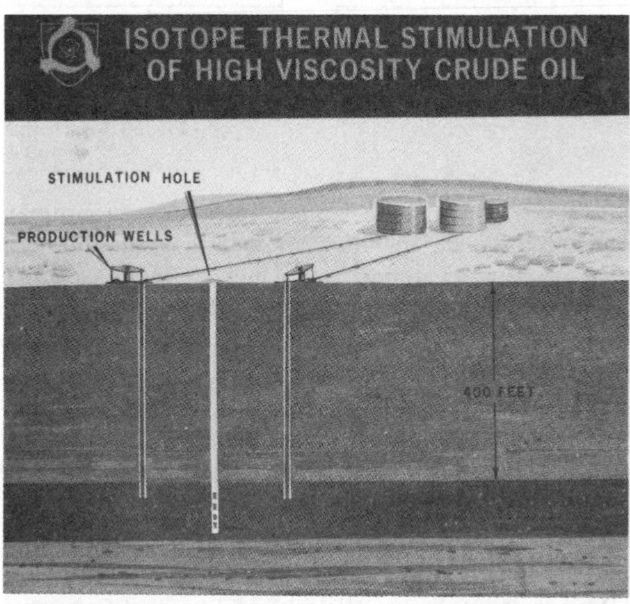

Fig. 5B.15. Thermal stimulation of high-viscosity oil by radioisotope heat sources.

the use of long-lived waste materials from reprocessing plants properly encapsulated would be better placed down-hole in an oil deposit than left to decay in a salt mine where no economic return is possible. Other nuclear sources may be economically and efficiently used both up-hole and down-hole as reliable sources of energy.

RADIATION PROCESSING OPERATIONS

Perhaps one of the largest future uses of radioisotopes may be in the field of process radiation. This includes the industrial use of relatively large sources of Co-60 or Cs-137 to produce, via irradiation, chemical changes in materials, to improve the physical characteristics of materials, or to destroy the bacteria in food (pasteurize it).

Large radioisotope sources are valuable to the chemical industry because chemical reactions can be initiated or accelerated thereby and the yield of a desired product increased often by the application of ionizing radiation. The Dow Chemical Company developed the first commercial-scale operation for the production of ethyl bromide using Co-60. In this process C_2H_4 and liquid HBr are fed into the bottom of a reaction vessel and allowed to flow upward around a core containing a Co-60 source. A radiation-catalyzed reaction occurs and ethylene goes into solution as ethyl bromide. The outstanding advantage of the new process is that it produces a nearly 100% pure material without side products or chemical contaminants. The most recent additions to the list of chemical products being produced by Co-60 irradiation is a polyethylene oxide and an unnamed flocculating agent.

Wood-plastic irradiation illustrates the use of ionizing radiation to change the characteristics of a material. A new family of irradiated wood plastics has been developed that can be made fire resistant and to have greater strength and more resistance to blows, scratches, warping, and swelling. In this process, the wood is impregnated with a liquid plastic monomer. The monomer is polymerized by Co-60 irradiation and results in an improved final product. Current research sponsored by the AEC has been concentrated on the irradiation of concrete-plastic combinations. Compressive and tensile strengths of such concrete polymers are up to four times higher than in ordinary "untreated concrete," and resistance to adverse chemical environments is also vastly improved. It has been reported that these materials are being developed in a variety of compositions to be evaluated for use as housing panels, bridge decking, undersea submersibles, tunnel linings, desalination plant vessels, drainage pipes, and minewall shoring. A glass-polymer combination is also being investigated. In this material crushed (waste) discarded bottles are mixed with methyl methacrylate and are irradiated by Co-60 to yield a product with compressive strengths up to four times that of ordinary concrete. This program may result in a solution to the growing waste disposal problem of discarded glass bottles.

The use of radiation to preserve food is an interesting and important possibility since over 25% of all the food produced in the world is lost by deterioration. The effect of radiation in relation to the dose applied is shown in Fig. 5B.16. The earlier efforts of the AEC are shown in Table 5B.2 and the highlights of foreign research are listed in Table 5B.3. More recently, the AEC has focused its attention on the irradiation and preservation of strawberries, papayas, and haddock. It should be noted that because 50% of the people of the world are either underfed or suffering from malnutrition, the preservation of food by radioisotope irradiation may be a powerful social, economic, and political instrument in the next decade. The quantity of radioisotopes required to accomplish even a few of the above goals would be substantial. The risks could be kept small since all of the sources could be solid and have several layers of encapsulation. The benefits, however, would be substantial.

In November 1977 a symposium held under the auspices of the International Atomic Energy Agency at Wageningen, Netherlands, discussed the wholesomeness of irradiated foods; the use of irradiation to control sprouting, pest infestation, spoilage and foód-borne diseases, and dosimetry. The conclusions drawn from the symposium were that the potentialities of the process justify intensification of further work on the technological application and practical introduction of radiation for the preservation of a number of foods. Furthermore, it was concluded that there was continued need for international collaboration. As of 1977, twenty-six food items have received limited or unrestricted clearance for human consumption in individual member states of the IAEA.

POWER APPLICATIONS

Substantial sources of electrical energy or power can be obtained by the use of nuclear fission reactors, breeder fission reactors and fusion reactors. These power sources will be discussed in more detail with respect to relative safety or hazards, magnitude of available energy, and in comparison to other sources of energy.

Fission Reactors

The nuclear fission reactor operates on the basis that its fuel, Uranium-235, undergoes fissioning spontaneously into fragments when it captures a slowly moving neutron. The splitting of the U-235 atom is accomplished by the release of appoximately 200 million electron volts per fission and the release of approximately 2.5 neutrons

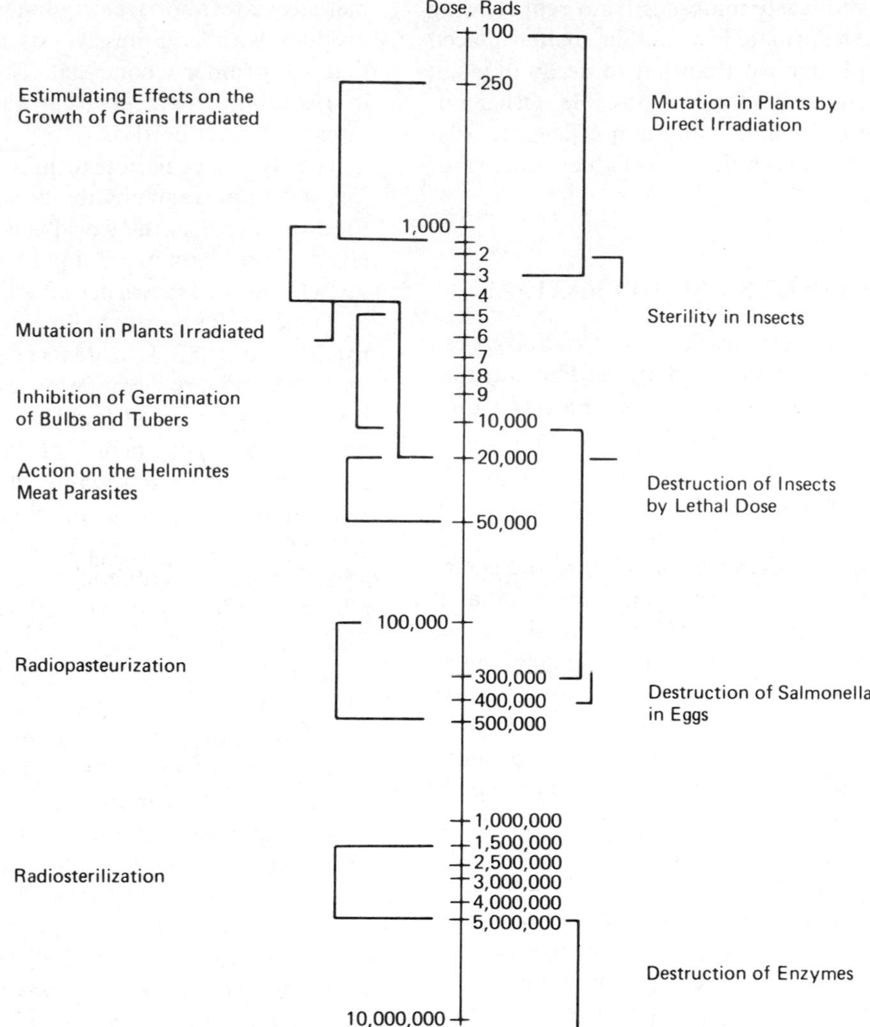

Fig. 5B.16. Effect of radiation in relation to the dose applied. (After Ready-to-process food, *Atomics*, **18**(3):18, 1965.)

per fission. These excess neutrons provided the means to maintain the fissioning of additional U-235 atoms. The energy released from the fissioning of 1 pound of U-235 is approximately equivalent to that produced from the burning of 3 million pounds of coal or 20,000 barrels of oil.

A schematic drawing of a nuclear reactor is shown in Fig. 5B.17. The fuel elements are generally enriched U-235 and the control rods may be made of neutron-absorbing materials such as silver, indium, cadmium, boron, etc. The two most frequently used systems are the boiling water reactor (BWR) and the pressurized water reactor (PWR) shown in Fig. 5B.18.

In naturally occurring Uranium, there is only one atom of U-235 in approximately 140 atoms of Uranium. The U-235 used in reactors is usually enriched to approximately 3.5% of the total uranium content. It should be noted that bomb material contains usually more than 90% of U-235 in the total uranium content. The nuclear reactor is therefore not a nuclear bomb. It should be noted, as shown in Table 5B.4 that industrial reactors which contain less than 5% enriched U-235 cannot be made into explosive devices. Reactors are designed with some inherent safety features and with some noninherent but rather built-in safety features. Although it is

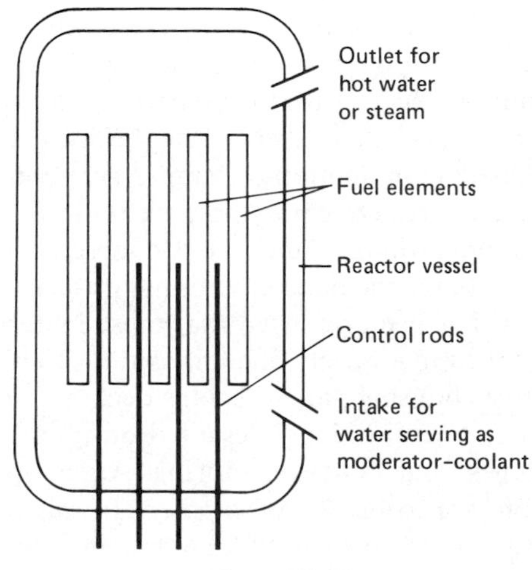

Figure 5B.17

TABLE 5B.2. Irradiators in AEC Food Programs.

Type	Purpose and Status	Co-60 Gamma Source	Capacity	Cost
Research irradiators, located at a number of educational institutions*	Immediate availability for research support; have been operating several years	30,000 curies, double plaque, pool storage	75 lb/hr at 1 megarad	$35,000 plus source
Mobile gamma irradiator, truck-mounted	Demonstration of fruit processing feasibility, economic determinations: was scheduled to begin operating in mid-1965	125,000 curies, four-pass quadrant irradiation	1000 lb/hr at 200,000 rads	$225,000 plus source
Grain-products irradiator, USDA Entomological Research Center Savannah, Ga.	Bulk-grain or packaged-product disinfestation; was scheduled to begin operating in mid-1965	35,000 curies; stored in shipping cask, or extended to radiation positions for grain or packages	5000 lb/hr bulk, or 2800 lb/hr packaged product, either at 25,000 rads	$200,000 plus source
Marine Products Development Irradiator, USDI Technological Laboratories, Gloucester, Mass.	Semicommercial seafood irradiation; operating in 1965	250,000 curies, pool storage, four-pass quadrant irradiation	1000 lb/hr at 500,000 rads	$500,000 plus source
On-board ship irradiators, two transportable 18-ton self-contained units	Research support, operating in 1965	30,000 curies, continuous liquid irradiation	150 lb/hr at 100,000 rads	$36,000 plus source
Tropical fruit irradiator	Funds allotted in 1965; completion expected in spring 1967			

*Universities of California (Davis), Florida, Hawaii, Michigan, North Carolina, North Dakota, and Washington and the Massachusetts Institute of Technology.

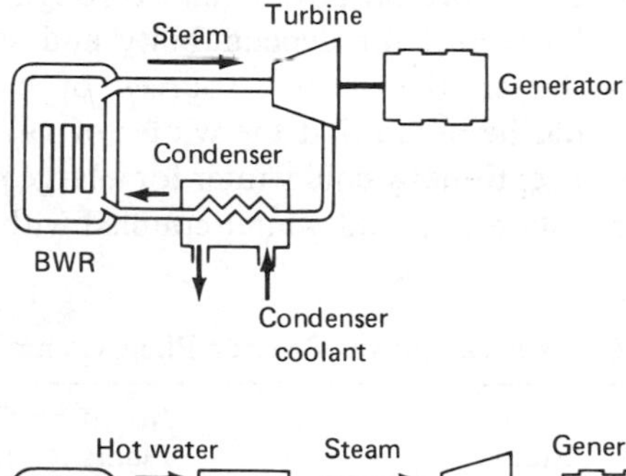

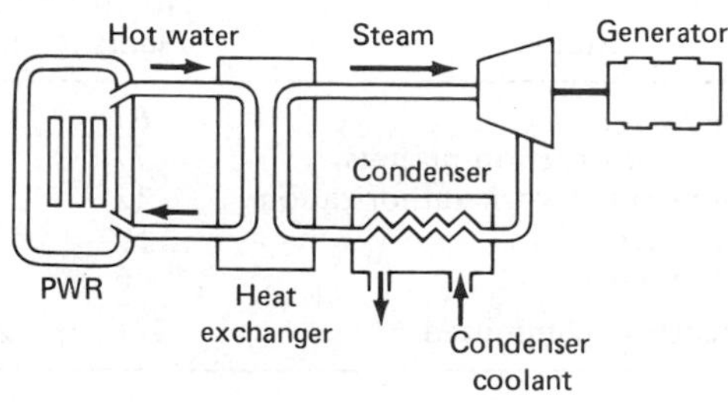

Figure 5B.18

possible to have a significant or serious reactor accident, the history of reactor operations has been good and is a reflection of the inherent and built-in safety features. All power sources of their nature have associated safety risks. An example of the comparative mining risks associated with coal and uranium is presented in Table 5B.5.

Before we discuss breeder reactors and their value in the generation of electrical energy, it may be desirable to compare the utilization of nuclear fuel and oil in the production of electrical power.

At present, nuclear reactors with power levels of approximately 1000 Megawatts (electrical) are being built which could save approximately 40,000 barrels of oil per day or enough oil to fuel cars, buses, trains, planes, and ships in a city of 1 million people. In 1974, we imported approximately 7 million barrels of oil per day and by 1977 that number was increased to approximately 9 million barrels of oil per day. Thirty 1000 Mw(e) reactors

TABLE 5B.3 Highlights of Major Foreign Research in Food Irradiation[3]

Country	Foods and Areas of Study
Austria	Fruit juices, wines, beer
Australia	Control of Queensland fruit fly
Belgium	Strawberries, vegetables, grain, flour, powdered eggs, gluten, disinfestation of grain and grist
Denmark	Consumer tests planned on meats; potato sprouting; strawberries, fruits, and vegetables
France	Tomatoes, chicken, fermented milk, grain, economic study on irradiation of African produce (disinfestation of fish, sterilization of meat)
Germany	Basic studies of dosimetry in foodstuffs and effects of radiation on microbes, packaging, potato sprouting, fruits, and fish
Greece	Radioresistance of yeasts and molds in fruit; disinfestation of figs; extension of shelf life of Valencia oranges
Holland	Pilot plant for strawberry research planned
India	Mangos, sapotas, guavas, green tomatoes, limes, ginger, garlic, and beans
Israel	Irradiation of plastic-coated oranges
Italy	Inactivation of foot and mouth virus in imported beef
Norway	Marine products, beef, potatoes, fruit, and vegetables
Pakistan	Disinfestation of leguminous pulses
Poland	Potato sprouting, fruit, barley, and malt; mushrooms, cellulose, milk, meat; bilberries in polyethylene; development of potato decoctions for forage
Republic of South Africa	Apples, grapes, peaches, and juices
Sweden	Wholesomeness of potatoes; fresh and smoked fish
United Kingdom	Wholesomeness of wheat, eggs; studies of *Salmonella* in horse meat used in pet food, fish, strawberries
USSR	Potato sprouting; irradiation of hen eggs in hatchability and egg-production studies; extension of shelf life of several foods, disinfestation

TABLE 5B.4. Quantities of Fissionable Materials Required to Make an Atomic Bomb[a(8)]

Fissionable Material	As Metal (kg)[b]	As Oxide (kg)[b]
Weapons-grade plutonium	4	about 6
Commercial plutonium	8	about 10
Fully enriched (93%) uranium-235	17	about 20
20% enriched uranium-235	250	about 375
10% enriched uranium-235	1000	about 1500
5% enriched uranium-235 (LWR fuel)	Cannot be made into an explosive device.	

[a]Quantities are minimum amounts assuming an effective neutron reflector.
[b]1 kg = 2.2 pounds.

TABLE 5B.5. Safety Risks Associated with Mining for Source of Power.[9]

Source of Power	Deaths/10^9MWH[a]	Disability Days/10^6MWH[a]	Lung Deaths/10^9MWH[a]
Coal	189	1,545	1,000
Uranium	2	157	20

[a]Megawatt hours.

would provide enough power to preserve all the oil in the entire Alaskan oil field (10 billion barrels) over a 40-year period. Approximately, two hundred and twenty nuclear reactors operating at a power level of 1000 Mw(e) would provide sufficient electrical power to eliminate the import of 9 million barrels of oil per day, restore our economy and make our balance of payments strongly positive. Moreover, oil shipment pollution problems would be considerably reduced.

In 1977 there were 67 nuclear reactors in operation in the United States with a capacity of approximately 48,000 Megawatts (electrical). Table 5B.6 lists the additional commitments to nuclear power in the United States as of July 1977.

In 1975 nuclear reactors produced approximately 8% of the electrical power in the United States and represented a savings of approximately 500,000 barrels of oil per day. In 1977 nuclear power provided approximately 9% of the electrical power in the United States. It should be noted that during the "exceptionally cold winter" of 1976–1977, when the coal banks were frozen and the rivers so ice packed that the oil barges could not move, and when natural gas supplies were very low, nuclear power averaged approximately 15% of the electrical power. In Connecticut, at that time, nuclear power supplied 50% of the electrical power, in Chicago and northern Illinois it provided 47%, and in New York City it accounted for 32%.[10] Nuclear power came through when it was needed. Its record for dependability and safety is considerably better than other sources of electrical power. It should be noted that the winter of 1977–1978 was another exceptionally cold winter for a large portion of the United States. A cold winter coupled with a long

TABLE 5B.6. Nuclear Power Reactor Plant Commitments.

Status[a]	No. of Plants	Capacity in Mwe.
Plants operating	67	47,568
Plants with construction permits	77	82,424
Plants with limited work authorization	12	12,950
Plants on order	59	67,597
Letters of intent	6	5,970
Total reactors committed	221	216,509

[a]July 1977.

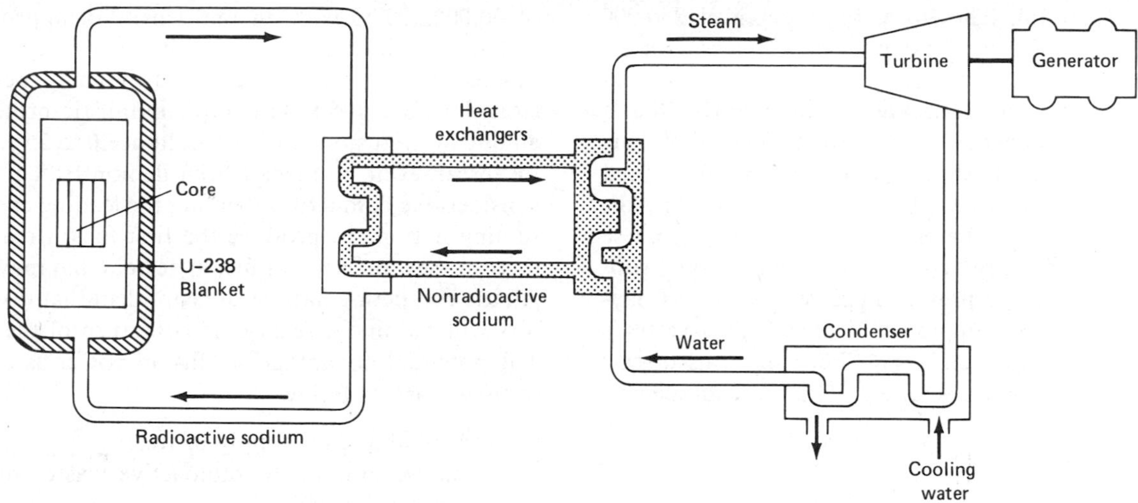

Figure 5B.19

coal miner's strike, emphasizes the need for the use of nuclear energy to supplement our energy requirements.

Breeder Reactors

A breeder reactor is a nuclear reactor designed to producc both power and new fuel at the same time. When fissionable uranium or plutonium is burned (fissioned) in a reactor, some of the nonfissionable but fertile uranium or thorium also in the reactor is converted to plutonium or uranium-233. In breeder reactors, the amount of new fuel produced exceeds the amount of fuel burned in the reactor. A schematic diagram of a fast breeder reactor is shown in Fig. 5B.19 wherein the blanket of Uranium-238 is converted to Pu-239 (a fissionable material) at a faster rate than the burning up of U-235 in the core. The breeder reactor is necessary to make maximum utilization of uranium and thorium fuel materials. Several countries, at the present time, have demonstration (breeder reactor) plants in operation. Because of the production of Pu-239 in breeder reactors, and the possible proliferation of Pu-239 which could be used to manufacture an atomic bomb, a difference in opinion exists in the United States (1977) as to the future of the breeder reactor program. This problem involves ethical, moral, political, financial, and scientific aspects.

In this author's mind the resolution of this problem is clear with respect to all the above mentioned aspects. We should, and I believe we must, develop an effective and safe breeder reactor program. Control of the Pu-239 would be vital to the program. The comparative estimated life of the several non-nuclear energy sources, as well as nuclear energy sources, are presented in Fig. 5B.20. As shown in Fig. 5B.21, the breeder reactor source would provide sufficient power for approximately 4000 years. Fusion, which will be discussed in the following section, will provide the greatest source of power in the future.[8]

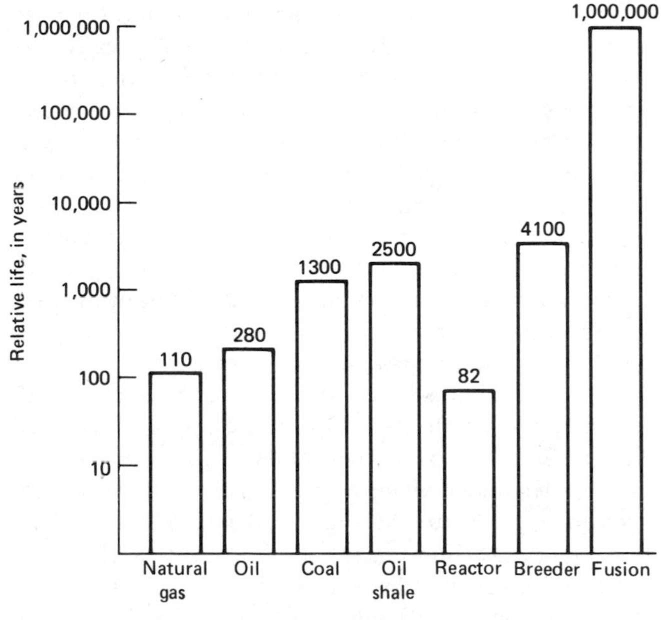

Figure 5B.20

It is estimated that there are at least 250,000 tons of U-238 stockpiled in the United States as a by-product from the enrichment of U-235. This material is stored at diffusion plants where the enrichment process is carried on. The U-238 could be transformed into a valuable energy source (Pu-239) by neutron bombardment in a breeder reactor. The above stockpiled material has the energy equivalent of approximately a trillion tons of coal and could provide for all of our electrical energy needs for hundreds of years. The energy from this stockpiled material is equivalent to more than 1000 years of coal mined at the 1976 rates, or more than 400 years of oil at the 1976 rate of consumption. One pound of U-238 converted to one pound of Pu-239 is equivalent to more than

5000 barrels of oil. Therefore, the stockpiled 250,000 tons of U-238 is the energy equivalent of more than 2.5 trillion barrels of oil in storage. It is anticipated that the United States will continue to stockpile U-238 at a rate equivalent in energy to more than 1000 billion barrels of oil per year. It should be noted that the United States uses approximately 6 billion barrels of oil per year, and it imports approximately 50% of its total usage. Use of the stockpiled energy could relieve our energy problem and eliminate the present negative balance of payments due to the large import of oil. The United States is by nature an energy rich nation both in terms of raw materials and human resources. They should be put together.

Fusion Reactor

Fusion occurs when light weight atoms such as hydrogen are heated to very high temperatures (approximately 100 million °F) so as to fuse the lighter atoms together to make heavier ones. In this process, theoretically, the energy released is well in excess of the energy used to fuse the atoms. The released energy can then be used to power a steam generator. Since the abundance of hydrogen (and deuterium, heavy hydrogen) is much greater than the abundance of uranium and plutonium, the resultant available energy from fusion is considerably greater as shown in Fig. 5B.20. One may calculate that 1 gram of heavy water will theoretically yield 3 megawatt days power, whereas 1 gram of uranium-235 has experimentally yielded approximately 1 megawatt day. The above calculation is based on the use of 5 atoms of deuterium (heavy hydrogen) to produce approximately 25 Mev or 15×10^{23} Mev/gram of deuterium while 1 atom of U-235 fissioned yields approximately 200 Mev or approximately 5×10^{23} Mev/gram of U-235 (fissioned).

A practical fusion reactor is not a reality at this time. Nuclear reactors and breeder reactors have been demonstrated. Reactors without breeder operations have long histories of operations. Breeder reactor operations are in their early stages of demonstration or commercial operations. There is no question of the breeder reactor ability to produce more fuel than used, but more experience with these operations is required to make their operations more economical and efficient. In the case of the fusion reactor, although the basic principles have been demonstrated, one has not yet liberated more energy from the reactor than has been consumed. The problem to date has been the inability to hold the hot plasma (highly ionized gas) together long enough for the light weight nuclei to fuse together and to release more energy than it takes to heat the plasma and to operate the associated system. The goal for a break-even fusion reaction set forth as the "Lawson Criteria" is to maintain the very hot plasma of fuel material in a stable condition for about 0.1 to 1.0 seconds at a minimum temperature of 50,000,000°C with an ion density of approximately 100 trillion ions per cubic centimeters. It is now estimated that a break-even situation may be demonstrated before 1980. The most optimistic projection of an operating fusion reactor would be 20 to 25 years after the break-even point had been demonstrated. There is considerable competition among the leading energy consuming nations to produce the first fusion reactor and there are several types of fusion devices that may be used. All the competing nations are confident that fusion reactors will become a reality after the turn of the century. The potential advantages of fusion power as compared to fission are as follows:

1. More abundant supply of fuel.
2. Smaller amount of radioactive waste per unit of power generated.
3. Higher energy efficiency.
4. Safer operations (a fusion reactor cannot go supercritical).

WASTE DISPOSAL PROBLEMS AND SOLUTIONS

Waste Generated from Nuclear Reactor Operations

A 1000 Megawatt (electrical) reactor produces a few pounds of fission product nuclear waste per day whereas a coal-fired plant of 1000 MW(e) produces 30 pounds of ashes per second. A coal-fired 1000 MW(e) plant also releases approximately 600 lb of carbon dioxide/second; approximately 10 lb of sulfur dioxide per second and substantial quantities of nitrogen oxides. The quantities of nuclear wastes are relatively small but they involve wastes with relatively long half-lives which require long-term storage.

One solution to the problem of handling high-level waste from nuclear reactors involves the incorporation of the waste in glass cylinders approximately 300 cm in length and 30 cm in diameter. It has been estimated that one years waste from a 1000 MW(e) reactor will involved the use of 10 glass canisters. These canisters, shown in Fig. 5B.21 would be stored in underground salt mines as illustrated in Fig. 5B.22. Since the canisters would be buried 10 meters distance from each other, one years waste from a 1000 MW(e) reactor would require 1000 square meters. Therefore, an all-nuclear U.S. electrical-power system would involve approximately 400–1000 MW(e) reactor plants which would require an area of less than 0.5 square kilometer per year. Thousands of square miles of suitable salt formation are potentially available in the United States for storage of high-level waste. Therefore, there exists potential reactor waste storage area for thousands of years at high levels of reactor operations.

In 1977 scientists at Catholic University of America developed a new radwaste glassification process that

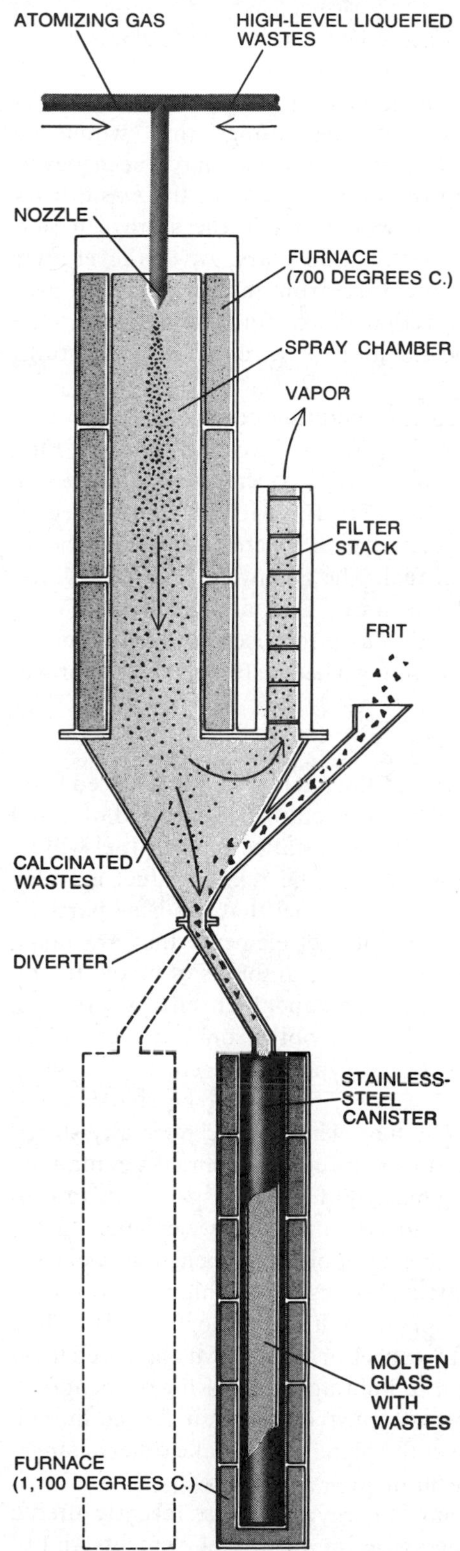

Figure 5B.21. Current plan for handling high-level radioactive wastes calls for their incorporation into glass cylinders about 300 centimeters long and 30 centimeters in diameter. In the single-step solidification process depicted here the liquid high-level waste is first converted into a fine powder inside a calcining chamber (*top*), then mixed with glass-making frit (*middle*) and finally melted into a block of glass within the thick stainless-steel canister in which it will eventually be stored (*bottom*). When canister is full, flow is switched by a diverter valve into a new canister (*broken outline*); hence the process is continuous.[11]

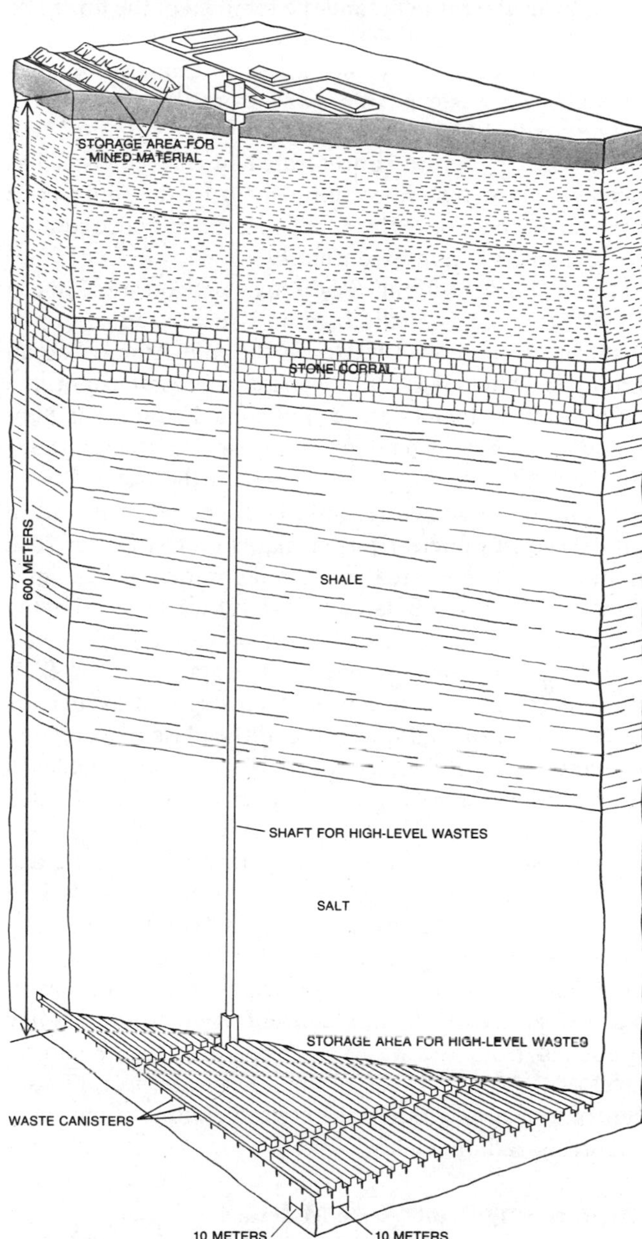

Figure 5B.22. Deep underground burial is at present the method favored by most nuclear power experts in the U.S. for the long-term storage of high-level radioactive wastes. In this idealized diagram of a proposed Federally operated respository in southeastern New Mexico the waste canisters are shown emplaced at a depth of 600 meters in a geologically stable salt formation. In order to dissipate the heat from the canisters they would be buried about 10 meters apart; thus each canister would occupy an area of about 100 square meters. On this basis the total high-level wastes generated annually by an all-nuclear U.S. electric-power system (assuming roughly 400 1,000-megawatt plants) would occupy an area of less than half a square kilometer.[12]

reportedly yields a more chemically durable glass to provide greater containment of radioisotopes. The glass has a chemical durability (resistance to heat and corrosion) 1000 times greater than previously available materials used in radioactive waste storage. The new process involves the isolation of high silica glass from silica by heat treatment, and the removal of the low silica by dissolution in acid. The waste is inserted in the spaces previously occupied by the low silica that has been removed. The waste is kept away from the surface by a process called *profiling*. The mass is then collapsed to a solid with the waste concentrated in the center. It has been reported that safe confinement of the radioactive waste can be achieved well beyond the periods required for Cesium-137 and Strontium-90.

In 1977 The National Radiological Protection Board (NRPB) of the United Kingdom issued a report on the disposal of high level waste on the mid-Atlantic ocean floor. It concluded "No overriding reason connected with the radiological protection consideration has been identified which would preclude the disposal of suitably conditioned high-level waste on the ocean floor." The analysis assumed that all the high-level waste from all the world's nuclear power plants up to the year 2000 is dumped in one area of 35 miles radius in the mid-Atlantic. The accumulated high-level waste is estimated to be in the form of 72,000 vitrified cylinders of 3-meter length and 0.5 meter diameter. The stainless steel cladding of the glass is assumed to survive in seawater for only a few years and fracturing of the glass is postulated to reduce all the cylinders to fist-sized lumps. The leach rate of the borosilicate glass is estimated to be such that all the vitrified waste would be dissolved in a period of 3500 years. A multitude of pathways from the dissolved waste at ocean depths to man were considered. In the worst case, it was assumed that in the future man will be harvesting plankton and fish from the deep ocean for consumption. Under these conservative assumptions, it was found that the highest concentration of any actinide predicted from the waste is comparable to the natural level of Ra-226 and that all the other waste isotope concentrations are less than the concentrations of naturally occurring radionuclides.

Reprocessing Plant Waste Materials

The spent fuel from reactor operations is stored in a pool of water before being shipped to a reprocessing plant where the fissionable uranium and plutonium are separated from the fission products or potential waste materials. Present (White House) plans are for the long-term storage of the spent fuel and the elimination of reprocessing wherein plutonium is separated and accumulated. Future long-term nuclear operations and the development of breeder reactors require separations reprocessing operations. Most, if not all, other nuclear-power plant countries with reprocessing capabilities are continuing to use their facilities for reprocessing.

Nuclear waste from the reprocessing of spent fuel elements will contain Sr-90 and Cs-137 as mixed fission products and less than 0.5% of plutonium from spent fuel. The limit on the period of storage will be set by the Pu-239 content in the mixed fission product waste material. In general, the storage time has been taken as 10 half-lives of the longest lived isotope. With Pu-239 (half-life 24,000 years), the storage time would be 240,000 years. Conceptual processes have been developed for stripping the plutonium from the waste from 0.5 to 0.005% which would enable the storage period to be reduced to 600 to 1000 years. At this level after 600 to 1000 years, the waste from the mixed fission products containing less than 0.005% of plutonium from the spent fuel will be no more hazardous than a natural uranium mine.

There have been no operating commercial reprocessing plant in the United States since 1972. Nuclear Fuel Services in West Valley, N.Y., operated a reprocessing facility from 1966 to 1971. The Morris, Illinois, reprocessing facility of the General Electric Co. is now being used to store spent fuel. The Barnwell, South Carolina, facility of Allied Chemical, Gulf and Shell Oil Co. is still under construction as a reprocessing facility and is in the process of obtaining a license for spent fuel storage.

Spent Fuel Storage

Approximately $\frac{1}{4}$ to $\frac{1}{3}$ of the nuclear fuel is replaced from a power reactor each year such that in a period of 3 to 4 years all of the previously installed nuclear fuel will be removed from the reactor. The removed fuel is called "spent fuel"; that is, nuclear fuel that has been partially used or spent. The spent fuel elements that are taken from the reactor contain large amounts of mixed fission products emitting neutrons and high energy gammas. These fuel elements are hot both from a nuclear and a thermal or non-nuclear viewpoint. Therefore, after being removed from the reactor, they must be shielded and cooled. These spent fuel elements are normally stored in a pool of water in spent fuel storage racks containing neutron-absorbing material with such spacing as to provide safe storage and cooling for the elements. In the past, it was desirable to cool these spent fuel elements for a period of 100 to 120 days before shipment to a spent fuel reprocessing plant. With the most recent White House directive to cancel or slow down the breeder reactor program and to hold up reprocessing of spent fuel, reactor power plants must either plan for additional spent fuel storage at the plant site or make other arrangements for long-term or permanent storage.

The Department of Energy (DOE) has recently offered to provide storage space for spent fuel elements and to assume complete responsibility for its safe handling and disposal thereafter. This new spent fuel storage policy has been also extended to foreign country reactor owners on a "limited basis" as part of the American non-proliferation policy to induce selected nations to store their spent fuel at secure sites in this country.

Under the new policy, utilities will transport their

spent fuel to a government center and pay a one time storage fee. The utilities will not receive any credit for the energy value of plutonium or the unburned uranium-235 in the delivered spent fuel. If, however, the fuel elements are reprocessed at a later date, the utilities could receive a partial refund. The Department of Energy may use federally owned storage facilities or possibly privately owned facilities such as the General Electric facility at Morris, Illinois, or the Allied General Nuclear Services facility at Barnwell, South Carolina. In 1977 approximately 2000 metric tons of spent fuel were stored at reactor sites in the United States. The storage volume is expected to increase at the rate of 1000 tons per year. Present spent fuel storage facilities may be adequate at best to the mid-1980s. Therefore, either additional storage facilities will be needed or reprocessing will be resumed to handle accumulated spent fuel elements.

TRANSPORTATION OF RADIOACTIVE MATERIALS

Transportation of Medical Isotopes

Medical isotopes as indicated earlier in this chapter are generally small amounts of short half-life materials used for diagnostic purposes. An estimated 600,000 packages of radioactive materials were shipped in the United States during 1974. Over 75% of these shipments were associated with three materials—Molybdenum-99, Technetium-99m and radioiodines (^{123}I, ^{125}I and ^{131}I). A summary of standard shipments for the nuclear industry as of July 1, 1975 is given in Table 5B.7. The majority of these shipments involve medical isotopes. A summary of radiological impacts of current shipments is presented in Table 5B.8. The annual dose from the radioisotopes shipped in 1974–1975 was approximately 8000 person-rem. The Nuclear Regulatory Commission analyses of these exposures determined that the average exposure was approximately 0.5 rem/year. A statistical increase of approximately one latent cancer fatality per year in the United States from the normal transportation of radioactive materials has been estimated. This compares to the existing rate of 300,000 cancer fatalities per year from all other causes.

Analyses of exposures to passengers on planes carrying radioisotopes have been made and the NRC has indicated the acceptance of doses to individuals from normal shipments on aircraft of 340 mrem/year (maximum) and 60 mrem/year (average). The EPA issued recommendations to the FAA in December 1974 for a dose rate limit of 0.5 mrem/hour at sea level (42 mrem/year).

Transportation of Spent Nuclear Fuel Elements

The safety record for the transportation of spent nuclear fuel elements has been excellent. There has never been an accident that has resulted in a death, an injury, or an overexposure to radiation. Figure 5B.23 illustrates a spent fuel shipping cask and truck for overland transportation to a fuel reprocessing plant. The determination of potential accident rates has been placed at 2.5 accidents per million vehicle miles for either trucks or rail modes of transportation.[10]

The packaging requirements of spent fuel elements are such that most accidents will fail to release radioactive materials into the environs. The AEC experience from

TABLE 5B.7. Standard shipments for the nuclear industry.[12]

Shipment Class	Avg. Ci/Pkg.	Avg. Transport Indices/Pkg.	Total Packages/Year		Transport Mode Split—% of Total			
			1975	1985	Pass. air and van	Cargo air and van	Truck	Rail
^{99m}Tc	0.28	0.1	1.14×10^5	2.96×10^5	100	—	—	—
Iodine (small)	0.012	0.8	1.91×10^5	4.95×10^5	52.2	37.5	10	—
Iodine (large)	56.7	1.6	1.00×10^3	2.59×10^3	—	100	—	—
^{99}Mo (small)	0.60	2.4	1.01×10^5	2.62×10^5	32.5	32.6	34.9	—
^{99}Mo (large)	80	3.6	7.20×10^3	1.87×10^4	—	100	—	—
[a]RAPH (small)	0.042	0.5	1.33×10^5	3.45×10^5	52.5	37.5	10	—
[a]RAPH (large)	74	1.7	2.70×10^3	7.00×10^3	—	100	—	—
^{192}Ir (source)	100	5.0	6.6×10^3	1.71×10^4	—	—	100	—
^{192}Ir (raw material)	5000	10.0	1.30×10^2	3.37×10^2	—	100	—	—
Spent Fuel: SF Kr	2000 (Equiv.)	0	370	3.6×10^3	—	—	14.2	85.8
Spent Fuel: SF I	2 (Equiv.)	0	370	3.6×10^3	—	—	14.2	85.8
Spent Fuel: SF FP	800	100.0 (Equiv.)	370	3.6×10^3	—	—	14.2	85.8
PuO_2	3628	5.0	370	925	—	35	65	—

[a]All radiopharmaceutical other than molybdenum-99, technetium-99m, and iodine (including the isotopes ^{123}I, ^{125}I, and ^{131}I).

TABLE 5B.8 Radiological impacts of current shipments.[12]

Standard Shipment	Annual-Pop. Dose-Normal Trans. Dose (Person-rem)	Annual-Pop. Dose-Normal Trans. Latent Cancer Fatalities	Annual-Pop. Dose due to Accidents Organ	Annual-Pop. Dose due to Accidents Latent Cancer Fatalities
^{99m}Tc	171	0.021	Lung	8.0×10^{-10}
Iodine (small)	2779	.338	Thyroid	8.5×10^{-7}
Iodine (large)	16	.002	Thyroid	8.5×10^{-7}
^{99}Mo (small)	4240	.55	Lung	7.3×10^{-7}
^{99}Mo (large)	299	.036	Lung	2.7×10^{-7}
[a]RAPH (small)	1217	.148	Thyroid	2.1×10^{-6}
[a]RAPH (large)	46	.006	Thyroid	3.0×10^{-6}
^{192}Ir (source)	248	.030	Whole body	3.4×10^{-13}
^{192}Ir (raw material)	5	.001	Whole body	6.5×10^{-13}
Spent fuel	535	.065	Thyroid, lung, bone marrow	7.3×10^{-8}
PuO_2	33	.004	Lung, bone	5.5×10^{-4}
TOTAL	9589 (9600)	1.166 (1.2)		5.58×10^{-4} (5.6×10^{-4})

[a]All radiopharmaceutical other than molybdenum-99, technetium-99m, and iodine (including the isotopes ^{123}I, ^{125}I, and ^{131}I).

TABLE 5B.9 Environmental impact of transportation of fuel and waste to and from one lightwater-cooled nuclear power reactor.[a(12)]

Normal Conditions of Transport

	Environmental Impact
Heat (per irradiated fuel cask in transit)	250,000 Btu/h
Weight (governed by Federal or State restrictions)	73,000 lb per truck; 100 tons per cask per rail car
Traffic density	
Truck	less than 1 per day
Rail	less than 3 per month

Exposed population	Estimated number of persons exposed	Range of doses to exposed individuals[b] (per reactor-y)	Cumulative dose to exposed population (per reactor-y)[c]
Transportation workers	200	0.0 to 300 mrem	4 person-rem
General public			
Onlookers	1,100	0.003 to 1.3 mrem	3 person-rem
Along route	600,000	0.0001 to 0.06 mrem	

Accidents in Transport

	Environmental Risk
Radiological effects	Small[d]
Common (nonradiological) causes	1 fatal injury in 100 reactor-years; 1 nonfatal injury in 10 reactor-years; $475 property damage per reactor-year.

[a]Data supporting this table are given in the Commission's "Environmental Survey Transportation of Radioactive Materials to and from Nuclear Power Plants," WASH-1238, December 1972.

[b]The Federal Radiation Council has recommended that the radiation doses from all sources of radiation other than natural background and medical exposures should be limited to 5,000 millirem per year for individuals as a result of occupational exposure and should be limited to 500 millirem per year for individuals in the general population. The dose to individuals due to average natural background radiation is about 130 millirem per year.

[c]Person-rem is an expression for the summation of whole body doses to individuals in a group. Thus, if each member of a population group of 1,000 people were to receive a dose of 0.001 rem (1 millirem), or if 2 people were to receive a dose of 0.5 rem (500 millirem) each, the total of person-rem dose in each case would be 1 person-rem.

[d]Although the environmental risk of radiological effects stemming from transportation accidents is currently incapable of being numerically quantified, the risk remains small regardless of whether it is being applied to a single reactor or a multi-reactor site.

Figure 5B.23. The roadmaster special spent fuel transportation system. This heavy-duty performer carries six High Temperature Gas-cooled Reactor (HTGR) spent fuel elements, and meets all gross vehicle weight limitations for unrestricted highway travel.

1949 through 1967 has been used to estimate that in 2.5% of the accidents, a fraction of the radioactive material being transported will be released into the environment. An analysis of the potential transportation risks in the year 2000 has been made by C. Starr et al. Assuming the above frequency of accidents with 2.5% of the accidents involving the maximum credible accident (all fission gases in the shipping containers plenum are released), the number of serious injuries was less than one in 1000 years for the projected fuel logistics requirements. It is clear from these analyses that the transportation of spent fuel elements does not add measurably to the total public health risks of the production of power.[14,15]

With the growth of nuclear reactor operations, there will be a corresponding growth in the volume of transportation of spent fuel from reactor sites to reprocessing plants. A large reactor with a power level of 1100 MWe will have approximately 100 MTU (100 metric tons of uranium). The pressurized water reactor (PWR) fuel element contains approximately 500 g or 0.5 MTU, while a boiling water reactor (BWR) fuel element contains approximately 250 g or 0.25 MTU. In 1970, the shipment of spent fuel amounted to approximately 50 MTU, whereas it is estimated that in 1980 it will reach a level of 3000 MTU/year. Each year on the average $\frac{1}{5}$ to $\frac{1}{3}$ of the reactor fuel must be taken from the reactor, retained in a water cooling system for approximately 150 days and then transferred to a reprocessing facility. A single PWR element, even after 150 days cooling, will contain 2,000,000 Ci, of which 5000 Ci is in gaseous form as shown in Tables 5B.10, 5B.11 and 5B.12.

RADIATION PROTECTION AND CONTROL

To enable us to proceed safely with the use of nuclear fuels and power-reactors and their by-products, adequate radiation protection and control of the operations must be exercised to protect personnel, plant, and the environment. Each facility must adhere to federal, state, and/or local regulations with respect to the use, handling, transportation, and disposal of radioactive materials. Operating procedures and working standards must be designed to assure that operations will not lead to greater than maximum permissible levels (mpl's), or in most cases, to certain fractions thereof. The basic functions and responsibilities of a radiation protection and

TABLE 5B.10. Radioactivity of Irradiated Fuel (curies per metric ton of uranium).[16]

	Cooling Period (in days)			
	90	150	365	3650
Fission Products	6.19×10^6	4.39×10^6	2.22×10^6	3.17×10^5
Actinides (Pu, Cm, Am, etc.)	1.42×10^5	1.36×10^5	1.24×10^5	
Total	6.33×10^6	4.53×10^6	2.34×10^6	

TABLE 5B.11. Predominant Fission Products in Gaseous Form Included in Radioactivity of Irradiated Fuel (curies per metric ton of uranium).[16]

	Cooling Period (in days)			
	90	150	365	3650
Krypton-85	1.13×10^4	1.12×10^4	1.08×10^4	6.05×10^3
Xenon-131m	1.06×10^2	3.27	1.08×10^{-5}	
Iodine-131	3.81×10^2	2.17	1.98×10^{-8}	

TABLE 5B.12 Thermal Energy in Irradiated Fuel (watts per metric ton of uranium)

	Cooling Period (in days)			
	90	150	365	3650
Thermal Energy	2.71×10^4	2.01×10^4	1.04×10^4	1.06×10^3

control program are personnel monitoring, in-plant monitoring, and environmental monitoring.

Methods of Protection

Methods of radiation protection from internal and external sources of radiation with respect to the person exposed are shown in Fig. 5B.24. If you increase the shielding or mass of material in front of a person, reduce the time of his exposure, or increase the distance between the person exposed and the source, you will reduce the radiation dose to his body. Although medical treatment prior to or following exposure can afford some protection, such is not recommended as a substitute for the time, distance, and shielding methods of protection.[19,20] Protection from ingestion or inhalation of internal sources can be achieved by proper ventilation in work areas and by the use of protective clothing, including a respirator. Chemical and medical treatment can be used to accelerate removal of absorbed activity from the body, but at this time, prevention is much more effective.

The National Radiological Protection Board (NRPB) of the United Kingdom has received encouraging reports on the use of a new drug called Puchel that is a derivative of DTPA, a chelating agent capable of removing heavy metals such as Pu from the bone. This new drug appears to be considerably more effective in removing Pu from the body than DTPA. In animal tests, it was shown that when Puchel and DTPA were administered three days after the ingestion of plutonium, Puchel removed 35% of the Pu in a period of 24 hours whereas DTPA removed only 4%. It should be noted that this work is in the development phase and has not yet been reported to have been used on humans, but the results are encouraging. DTPA, however, has been recently used effectively to reduce AM-241 that was deposited in the bones of exposed persons.

Time
Distance
Shielding
External
Medical Treatment
Radiation Protection
Ventilation
Internal
Chemical Treatment
Respirator
Clothing

Figure 5B.24. Methods of radiation protection.

Methods of Control

Methods of radiation control from internal and external sources with respect to the person exposed are shown in Fig. 5B.25. Film badges and personnel dosimeters are a means for measuring the exposure to an individual and a method to control his future and cumulative dose. Ra-

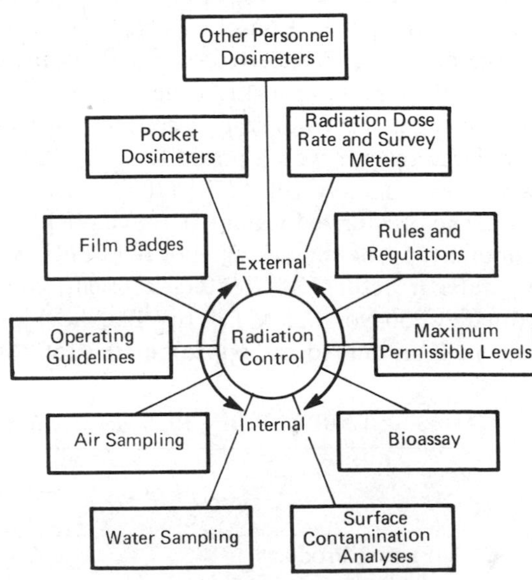

Figure 5B.25. Methods of radiation control.

diation survey-meters can detect sources of radiation and contamination to reduce and control both the external and internal exposures. Air-sampling units are used to collect, for measurement, the airborne radioactive material, and water samples can be taken to determine the concentration of radioactivity in such water before releasing it from a hold-up tank to final discharge. Urinalyses provide an indication of the species and roughly the amount of radioactive material that may be deposited in a person's body. All these measurements and concentrations are compared to the mpl's and to the working levels established as guides. As the levels approach the mpl or the so-called working level, steps must be taken to reduce the radiation dose of the contamination producing sources of these levels and to thus provide more radiation protection and control.

SUMMARY OF RADIATION SOURCE PROBLEMS AND SAFEGUARDS

The benefits and potential risks have been presented for the use of or the exposure to a variety of nuclear sources. A summary of the benefits and risks associated with the use of the major sources are presented in Table 5B.13,

TABLE 5B.13. Summary of Magnitude and Impact of Nuclear Sources.

	Unit of Use	Population Dose* In man-rem yr in Year 2000	Benefit vs. Risk Comments
MEDICAL APPLICATIONS			
Diagnostic uses			
In vivo	1 mCi	5×10^6	Substantial life-saving diagnoses to a very large population with essentially no risk or an insignificant one
In vitro	1 μCi		
Pacemakers	0.1 watt (th)	1×10^3	Risks are small under the circumstances
Artificial heart	30 watt (th)	2×10^7 (Pu-238) 4×10^5 (Tm-171)	Life-saving benefits to a relatively small population; the risks are very small. Very significant benefits to a large population with a small risk if the proper isotope is used.
SPACE APPLICATIONS			
Isotope-powered satellite			
Scientific	50 to 500 Watt (e)	Neg.	With appropriate encapsulation and safety precautions as stated in this report, the risks are very small and the benefits can be substantial
Communication	50 to 500 Watt (e)	Neg.	
Military	50 to 500 Watt (e)	Neg.	
Lunar-based power units	50 to 300 Watt (e)	Neg.	
TERRESTRIAL APPLICATIONS			
Oceanographic units			
Navigational lights	5 to 10 Watt (e)	Neg.	With effective encapsulation and the use of air droppable short-lived radioisotopes the risks can be made insignificant while the benefit can be life saving and substantial
Marine radio beacon	5 to 25 Watt (e)	Neg.	
Sonar beacon	1 to 10 Watt (e)	Neg.	
Automatic weather stn.	5 to 10 Watt (e)	Neg.	
Air-droppable buoys	5 to 10 Watt (e)	Neg.	
Land-based remote area units:			
VOR for air navigation	1000 to 5000 Watt (e)	Neg.	No significant risks are expected from well-encapsulated sources and remote operated units. The benefits can be quite substantial
Wintering-Over Sta.	3000 to 10,000 Watt (e)		
Remote area microwave stations	3000 to 10,000 Watt (e)		
Isotope heat sources secondary oil recovery	Mw-yr	Neg.	Control against ground and water contamination must be exercised in order to keep the risks at an insignificant level. The benefits from oil recovery and the effective use and control of waste materials can be significant
INDUSTRIAL APPLICATIONS			
Food processing			
Sterilization of meat	10^6 rad/lb.	Neg.	With the use of solid and properly encapsulated sources, the risks should be negligible while the possibility of using the technology to provide food for undernourished people would be significant
Pasteurization of fish	10^5 rad/lb.	Neg.	
Disinfestation of grain	10^4 rad/lb.	Neg.	
Reduction of sprouting	10^3 rad/lb.	Neg.	

TABLE 5B.13. *(Continued)*

	Unit of Use	Population Dose* In man-rem yr in Year 2000	Benefit vs. Risk Comments
Radiation Processing			
Sterilization of medical supplies	10^6 rad/lb.	Neg.	Similar to Food Processing above with respect to the risks. The benefits are more economic than social except for the cold sterilization of medical supplies. These irridiations can be of great benefit to man
Wood-plastic irradiation	10^6 rad/lb.	Neg.	
Concrete-polymer irradiation	10^6 rad/lb.	Neg.	
Chemical production	10^4 rad/lb.	Neg.	
REACTOR OPERATIONS			
Spent fuel shipping	N.A.	1×10^4	Offshore siting of reactor power plants together with improvements in contamination of potential effluents should keep the risks at a relatively low level. The use of the tuned platform or its equivalent will aid in reducing the risks. The need for energy and the associated benefits are evident
Reactor operations effluent	N.A.	6×10^4	
Natural radiation	100 mrem/yr/person	42×10^6	
Fallout radiation	6 mrem/yr/person	2×10^6	

together with an estimate of their future impact on the environment (year 2000) and a comparison of the natural radiation and fallout impacts.[19] In essence, the risks can be substantially reduced by applying presently known technology and new developments in each of these sources, as indicated in Table 5B.13. The use of shorter half-life radioisotopes and non-particulate emitters such as Tc-99^m will reduce exposures to patients. The effective control of encapsulated sources must be maintained and improved. The analysis indicates that substantial medical benefits can be obtained by the use of nuclear sources for diagnostic, therapeutic, and power uses. The use of nuclear fuel for reactor operations has its potential problems but they are relatively small in comparison to other sources of energy. With advances in safety technology we can further reduce the potential exposures from these sources of nuclear radiation.

REFERENCES

1. Industrial Pollution, Sax, N. I., Environment Impact of Large Radiation Sources, Fitzgerald, J. J. New York: Van Nostrand Reinhold Co.
2. Human Hormone, Greene, Raymond, World University Library.
3. Design of Isotopically Powered Thermoelectric Generators, Purdy, D. L. Nuclear Materials and Equipment Corporation, Apollo, PA, ANS Meeting, June 1968, Toronto, Canada.
4. Physical Aspects of the Nuclear Powered Artificial Heart System, Fitzgerald, J. J., Texas Medical Association Annual Meeting, May 1972.
5. The U.S. AEC Space Nuclear Electric Power Program—The Objectives and Status, Newby, ANS Meeting, June 1968, Toronto, Canada.
6. Radioisotopic Power Generation, Corliss, W. R., and Harvey, D. G., Prentice-Hall, Inc., Englewood Cliffs, New Jersey, 1964.
7. Factors Affecting Future of Isotope Power, Fitzgerald, J. J. and Mayo, K. E., ANS Meeting, June 1968, Toronto, Canada.
8. Nuclear Power and the Environment, Questions and Answers, American Nuclear Society, April 1976.
9. The Health Hazards of Not Going Nuclear, Beckmann, P. The Goldem Press, Boulder, Colorado, 1976.
10. The NRC and Its Regulatory Role, Kennedy, R. T., U.S. NRC Kansas City Kiwanis Club, Kansas City, Missouri, April 21, 1977.
11. The Disposal of Radioactive Wastes from Fission Reactors, Cohen, B. L., *Scientific American*, Vol. 236, No. 6, June 1977.
12. Radiological Quality of the Environment in the U.S.A., 1977 U.S. Environmental Protection Agency (EPA 520/1-77-009), Washington, D.C., September 1977.
13. Safeguards—A Regulatory View, Kennedy, R. T. U.S. NRC Institute of Nuclear Materials Management, Washington, D.C., June 29, 1977.
14. Spent Fuel Transportation Risks, Yadigorogly, G., et al., *Nuclear News*, page 71, November 1972.
15. A comparison of Public Health Risks: Nuclear vs. Oil-Fired Power Plants, Starr C. et al., *Nuclear News*, page 37, October 1972.
16. Siting of Fuel Reprocessing Plants and Waste Management Facilities, ORNL-4451, July 1970.
17. Applied Radiation Protection and Control, Fitzgerald, J. J., Gordon Breach Publishing Co., New York, 1969.
18. Mathematical Theory of Radiation Dosimetry, Fitzgerald, J. J. et al., Gordon Breach Publishing Co., New York, 1968.
19. Estimates of Ionizing Radiation Doses in the U.S., 1960–2000 U.S. Environmental Protection Agency (ORP/CSD 72-1) Washington, D.C., August 1972.

SECTION 6

Health Hazards of Solid-Waste Treatment

David Gordon Wilson
Massachusetts Institute of Technology
Cambridge, Massachusetts

INTRODUCTION

Because solid waste may potentially contain any of the solid materials found in nature and in addition many of the man-made materials, they constitute the most heterogeneous collection of substances possible. Laws and ordinances may prohibit certain materials from being put into the solid-wastes stream, but these regulations are no guarantee that occasionally prohibited substances might not appear. Domestic refuse has, with alarming frequency, been found to contain, for instance, live ammunition or explosives, highly toxic substances, or disease cultures.

Industrial wastes from any one location usually display a lesser degree of heterogeneity than that of domestic refuse, but the variations of compositions and properties from one operation to another and one type of industry to another is very large. Someone charged with responsibility for solid-waste treatment may, therefore, need to refer from time to time to the main listing in this handbook, Section 12, for information about any of the specific dangerous substances.

The major constituents of some representative samples of residential wastes are listed in Table 6.1.

The dangerous properties listed elsewhere in the handbook are principally those of pure substances or compounds. These materials may be present in solid wastes; however, there are many other dangerous possibilities resulting from mixtures of substances which individually may not be hazardous; from the properties of a large mass being dangerous whereas the properties of small samples may be considered safe; from the form in which wastes appear; or from the manner in which they are handled and subsequently treated.

The scale of the potential dangers in solid wastes, previously regarded as a comparatively innocuous although unaesthetic form of pollution, has only recently been realized, and consequently data are extremely scanty. The treatment here will be limited principally to an organized discussion of potential dangers in the belief that the categorization used will form a useful framework for the compilation of data as collected.

Solid wastes may be hazardous as generated; they may subsequently combine to produce secondary hazards; or they may be hazardous by reason of their location or the method of processing used. These categories will be used in the following discussion.

A decision as to whether or not a particular type of waste should be regarded as hazardous may be made with the EPA flow chart shown in Fig. 6.1.

INITIALLY HAZARDOUS WASTES

Solid wastes which are dangerous at the point of discard, such as explosive or incendiary devices, or toxic substances, need little discussion here because they constitute the most obvious category of hazard. The listing in Section 12 provides a good guide to many of these potential dangers.

A related but distinct category includes the broad range of wastes which, while not toxic, have the potential of causing injury. A plateglass window becomes solid waste at the moment it is broken, and at that point it has obvious injury potential for men and animals. Machine-shop turnings belong in the same category, with additional dangers because they are often hot enough to cause burns or alternatively may be contaminated with cutting oils which may cause adverse physiological reactions if the skin is punctured by such turnings.

A wide variety of substances can cause harm or disfigurement either in general or to sensitive individuals. Acids and alkalis are universal problems; though not strictly solid wastes, they appear along with many other liquid substances in discarded containers. Pesticide containers have caused intoxication and death after being played with by children.* Minute quantities of foaming or hardening agents used for plastic resins cause acute reactions in some individuals. In some cases,

Material for this section is condensed and revised from the Handbook of Solid-Waste Management (Van Nostrand Reinhold, 1977).

*H. P. Wolfe, et al., "Health hazards of discarded pesticide containers," *Archives of Environmental Health*, 3:45–51 (Nov. 1961).

TABLE 6.1. Percentage by Weight of Various Materials in Municipal Refuse (dry basis).

Category	Quad City New Jersey	Cincinnati, Ohio	San Diego, California	Batelle estimate	E.P.A.
Paper	47	42	46.1	55	45
Glass	5	8	8.3	9	10
Metals	9	9	7.7	9	11
Plastic	2	2	0.3	1	5
Wood	4	3	7.5	4	2
Food	31	28	0.8	14	17
Yard	–	6	21.1	5	4
Cloth	–	1	3.5	–	3
Miscellaneous	2	1	4.7	3	3

John M. Malarkey, "Design of a mechanical-handling system for an automatic refuse-reclamation plant," Massachusetts Institute of Technology, thesis, 1973.

death may occur in otherwise fit individuals within 30 minutes of exposure to, for instance, penicillin.

Medical and Institutional Solid Wastes

A somewhat related category of solid wastes which are dangerous at the point of disposal is that which is potentially disease-producing. Although organisms which can transmit disease are organic and will live on organic refuse, the quantity of organic material necessary to maintain a culture may be an invisibly small coating on a plate or container. Or the culture might be maintained at the tip of a hypodermic needle. The careless discarding of hypodermic needles after use into wastebaskets has unfortunately frequently resulted in janitors becoming infected when they use their hands to clean out the residue from a waste container.

A German study on illnesses caused by refuse* stated that the practice of over 70% of medical laboratories and institutes of allowing their wastes to be collected along with regular domestic wastes is particularly dangerous. All medical wastes should be contained in special, closed, plastic bags and collected and disposed of separately.

Half the outbreaks of salmonellosis in the United States were found to occur in hospitals.** Regular fecal examination of food handlers in hospitals and institutions and action to eliminate salmonellae from animal feeds is recommended.

Pathogenic bacteria can be and are present in very high concentrations in hospital solid wastes. The highest coliform counts were found by Trigg*** to occur in wastes from the intensive-care, pediatric, and psychiatric wards.

Hospital refuse chutes were found by Armstrong to be potentially transmitters of airborne bacteria.†. The use of closed plastic bags in refuse chutes is especially important, and an air-removal and purifying system to remove the smallest particles which can penetrate the respiratory tract is necessary.

The use of disposable needles, knives, swabs, diapers, examination gowns, thermometers, transfusion fittings, bed pads, sheets, plates and utensils, and so forth, in hospitals and institutions requires special handling if injuries and cross-infections are to be avoided. A particular danger arises if a contaminated non-woven item is laundered along with washable linen. The nonwoven item will probably disintegrate and spread all over other clothing, thus potentially spreading infection over the whole institution.‡

AIRBORNE PATHOGENS ASSOCIATED WITH SOLID-WASTES HANDLING

Air samples taken from municipal incinerators and a municipal compost plant in the United States showed that the potential health hazard from the pathogens associated with the dust is significant. The dust was found to carry a large number of microorganisms including pathogens of intestinal and respiratory-tract origin. The degree of hazard depends on the concentration of the dust and on the types of organisms carried.

Table 6.2 shows microbial characteristics of representative environments and of the ambient air in various

*Goettsching, H., "Wie konnen pathogene Keime in den Hausmuell gelangen?," *Städtehygiene*, 23(3): 183–188 (1972); "Krankheiten durch muell—Vorschlage zur abhilfe," *Städtehygiene*, 23(10): 230–236 (1972).
**Edwards, Phillip W. "Salmonellosis—an expanding problem" in proceedings of the national conference on salmonellosis, P.H.S. Publication No. 1262, Washington, D.C., Mar. 1965.
***Trigg, Jere A., "Microbial examination of hospital solid wastes," M.S. thesis, West Virginia University, Morgantown, W.VA. 1971.

†Armstrong, David Harold, "Hospital refuse-chute sanitation," M.S. thesis, West Virginia University, Morgantown, W.VA. 1969.
‡Lane, Thomas M., "Trends in the use of disposables in health-care facilities," in Salkowski, ed., "Disposal of single-use items from health-care facilities; report of the second national conference," National Sanitation Foundation, Ann Arbor, Michigan, Monograph No. 8, Sept. 1970.

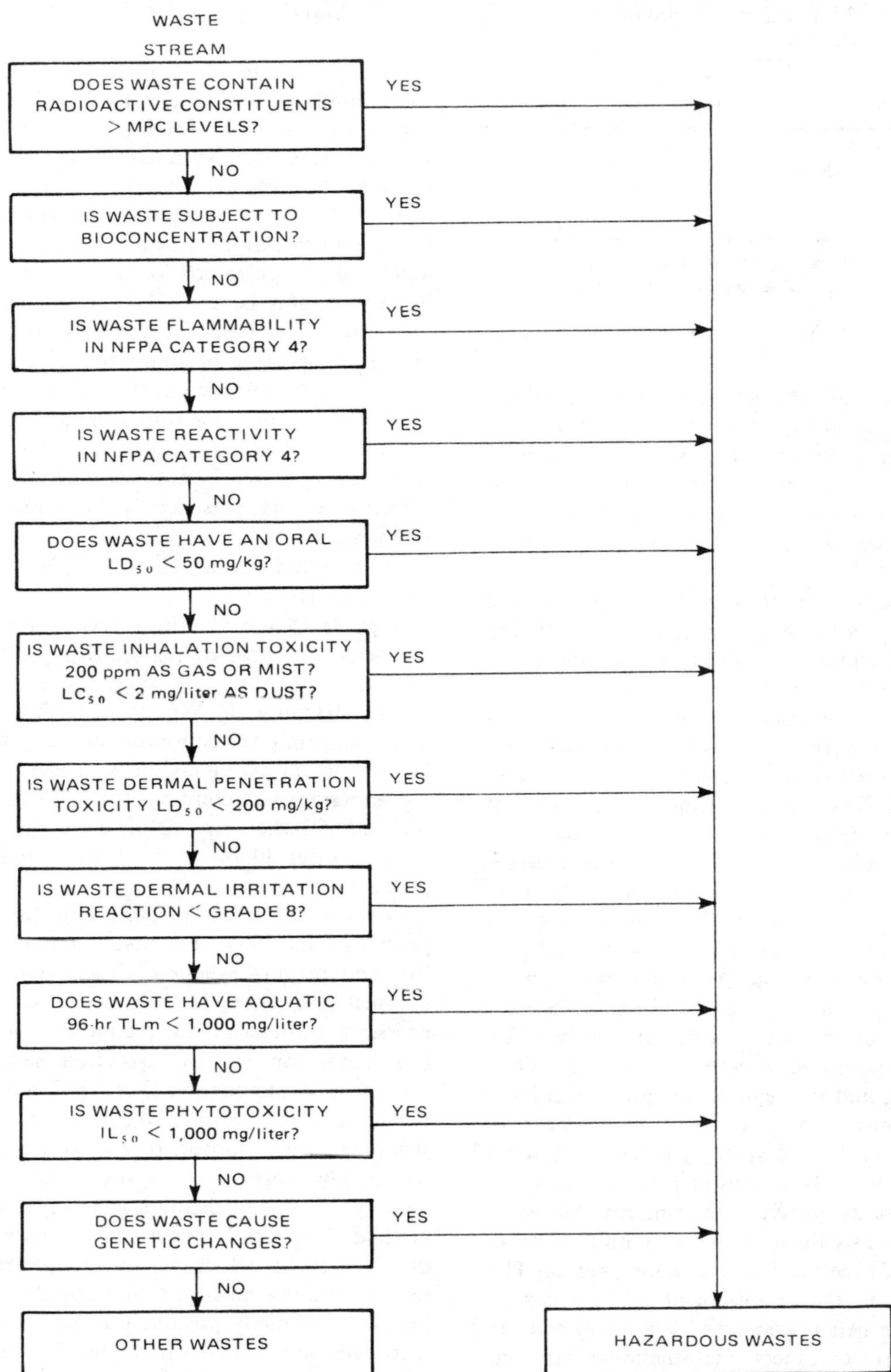
WASTE STREAM
DOES WASTE CONTAIN RADIOACTIVE CONSTITUENTS > MPC LEVELS?
YES
NO
IS WASTE SUBJECT TO BIOCONCENTRATION?
YES
NO
IS WASTE FLAMMABILITY IN NFPA CATEGORY 4?
YES
NO
IS WASTE REACTIVITY IN NFPA CATEGORY 4?
YES
NO
DOES WASTE HAVE AN ORAL LD50 < 50 mg/kg?
YES
NO
IS WASTE INHALATION TOXICITY 200 ppm AS GAS OR MIST? LC50 < 2 mg/liter AS DUST?
YES
NO
IS WASTE DERMAL PENETRATION TOXICITY LD50 < 200 mg/kg?
YES
NO
IS WASTE DERMAL IRRITATION REACTION < GRADE 8?
YES
NO
DOES WASTE HAVE AQUATIC 96-hr TLm < 1,000 mg/liter?
YES
NO
IS WASTE PHYTOTOXICITY IL50 < 1,000 mg/liter?
YES
NO
DOES WASTE CAUSE GENETIC CHANGES?
YES
NO
OTHER WASTES
HAZARDOUS WASTES

Figure 6.1. Flowchart of the hazardous-waste screening model. (Source: Office of solid-waste-management programs, E.P.A. "Report to Congress—disposal of hazardous wastes," U.S. E.P.A. report SW-155, 1974.)

locations in municipal incinerators. The dumping-floor values can be considered to be representative of dumping floors of other types of solid-waste-handling plants; the microbial levels might be somewhat lower for incinerators because, in good designs, the combustion air is extracted from the dumping area to reduce dust, smell, and pathogen levels.

Fungus spores can also cause disease (mycosis) through inhalation and passage to the throat and lungs (see Table 6.10).

TABLE 6.2. Representative Environmental Microbial Levels.

Environment	Total microbial levels (colonies/cu ft air)
Country air	56
General offices and schools	95
City streets	72
Factories	113

Mirdza L. Peterson, "Pathogens associated with solid-waste processing," U.S. E.P.A., report SW-49r, 1971.

Asbestos Dust

The high incidence of lung and other cancers among asbestos workers, including those in the demolition industry, has been recognized comparatively recently. Stringent regulations are now in force in many countries to control one form of asbestos—the blue crocidolite—which has been found to be prinicipally responsible.

The British Asbestosis Research Council and the British Scrap Federation have kindly given permission for the following summaries of their guides and codes of practice to be made here.*

Those engaged in the manufacture of asbestos products and those who both use and work with such products are mainly aware of the health hazards in question and take every precaution to safeguard the health of employees. However, the seriousness of both the health and legal implications involved in dealing with asbestos materials is perhaps not so fully recognized by the scrap industry. The industry is nevertheless vitally involved, because its members have so often to handle and process scrap materials which are contaminated with asbestos thermal lagging, e.g., pipes, boilers, tanks, etc., which are received into their yards, and in addition often have to deal with such materials on demolition/dismantling sites, and the regulations quite specifically cover all operations which give rise to the liberation of asbestos-containing dust to such an extent as may result in the emission of dangerous quantities of dust.

The health hazards involved are primarily the risk of contracting asbestosis through the inhalation of heavy concentrations of asbestos dust over a long period. This sometimes leads to the development of lung cancer, particularly if the person concerned is a heavy smoker. Another rare form of cancer, mesothelioma, affecting the chest or stomach wall, sometimes occurs chiefly from exposure to certain types of "blue" asbestos.

Types of Asbestos.** The main types of asbestos used in the past have been:

Amosite	Color—grey/brown
Anthophyllite	Color—white
Chrysotile	Color—white
Crocidolite	Color—lavender blue

Crocidolite is the most dangerous type of asbestos, and while it is now very little used in industry it was widely used in the past in particular for its acid-resistant qualities. It is therefore likely to be found in acid-fume areas such as chemical plants and gas works, and it is also found in power stations. Total emphasis upon the ability to recognize crociodolite by color alone cannot be relied upon because it is sometimes found mixed with the other types of asbestos. Although the significant proportion of admixture of crocidolite may be as low as 1 to 2%, even in these proportions it could be highly hazardous if the dust is inhaled. Unless the original owners of asbestos lagged materials can certify that such lagging is free from crocidolite, the only sure identification of its presence is by x-ray diffraction of samples.

The threshold limit value (TLV) for asbestos dust is listed in Section 1.

A guide to the identification of possible sources of the origin of asbestos materials is given in Table 6.3, while Table 6.4 lists nonhazardous wastes.

The Stripping of Asbestos Lagging. The removal of old lagging gives cause for serious concern since this is generally an extremely dusty process, and the following paragraphs attempt to deal with the problems associated with the stripping of such lagging in the varying circumstances likely to be encountered by the scrap processor.

Preparations Prior to Stripping Old Lagging. As previously indicated, the first requirement is to establish the type(s) of asbestos in the lagging that has to be stripped and where no record of this exists it will be necessary to determine if it contains "blue" asbestos. If this is not apparent by inspection, samples should be taken from a number of different points for examination. It will then be necessary to decide upon the method of stripping to be adopted, that is, wet or dry, which will be largely governed by site conditions obtaining, although the wet method is always preferable. The method chosen will have a bearing upon the amount of dust generated, which will in turn determine the protective clothing and type of respiratory equipment to be used. Normally positive-pressure respiratory equipment will be required where "blue" asbestos is present. In other cases a simpler form of approved dust respirator may suffice.

Protective Screens. Where stripping is to be carried out in occupied premises, screens or tents made from plastic sheets erected around the plant may be sufficient to reduce the asbestos dust level in adjacent working areas to an acceptable level.

The Wet Method of Stripping. Thorough water saturation of asbestos lagging wherever possible prior to

*The British government and demolition industry have in general been ahead of their U.S. counterparts in asbestos regulations.
**See also Section 12.

TABLE 6.3. Identification of Possible Sources of Origin of Asbestos Material.

Industry	Asbestos Usage
Electricity generating	Thermal lagging and delagging (Special risk from 'blue' asbestos).
Steel	Thermal lagging and delagging.
Heavy engineering	Furnace insulation.
Locomotive building	Heat and sound insulation. (Special risk from 'blue' asbestos used in the past.)
Railroad-car building	Heat and sound insulation. (Special risk from 'blue' asbestos used in the past.)
Boiler making	Heat insulation.
Paper making	Manufacture of filter papers and grinding of rollers.
Motor-vehicle repair	Repairs to brake and clutch parts.
Building trades	Asbestos cement sheets and insulation boards, asbestos spraying.
Electrical insulation	Insulation systems.
Chemical plants	Thermal insulation. (Special risk from acid-resistant 'blue' asbestos.)
Gas works	Thermal insulation. (Special risk from acid-resistant 'blue' asbestos.)

Other Sources of Asbestos Lagged Materials

Dockyards
Shipbuilding
Shiprepairing
Installation of plant in heavy industries, e.g. steel.
Large building projects–industrial and domestic–or insulation and repair of heating apparatus, e.g., schools.

TABLE 6.4. Nonhazardous Asbestos Wastes.

Linoleum Floor tiles Rubber Paints Plastics Adhesives Roofing compounds	Used as a filler and locked in with the other components.
Motor assembly	Grinding in assembly of brake and clutch parts.
Light engineering	Asbestos washers and gaskets.

stripping can greatly reduce the dust generated. Each job is likely to require its own technique; for example, nonabsorbent surfaces will need to be punctured to permit water to be introduced into the insulation. This operation may be by a hollow probe drilled along its length to allow water to penetrate into the insulation, such probes in parallel being coupled to a suitable water supply. Alternatively, a fine, low-pressure water supply may be used in such a way that dust does not arise from the spray hitting the surface of the lagging. Spraying in this manner is not as effective as total saturation in controlling the dust and the arrangements may need to be augmented by portable exhaust ventilation equipment and the use of respirators.

On removal, the lagging should not be allowed to fall but should be placed immediately in suitable plastic bags (200-gauge polyethylene double-sealed at bottom has been found satisfactory) for removal from site and subsequent disposal by burying. The slurry created should not be permitted to dry out but should be removed from all resting places while still wet. Special additional precautions are required when the waste contains "blue" asbestos.

Removal of Lagging in Dry Conditions. It. is not always possible to adopt a "wet" method of stripping and consequently these are circumstances in which old lagging must be removed in a dry condition. Heavy bags of saturated asbestos may be too cumbersome to carry through restricted openings and up steep stairways, e.g., from factory basements or in ship-breaking, and full advantage must therefore be taken of all other precautions.

Dust extraction must be as effective as possible and in the selection of tools for removing the lagging preference should be given to those producing least dust. When lagging has to be removed in a dry condition, the delaggers may have to be provided with the highest standard of protective equipment—heavy-duty impermeable overalls with head coverings and air-line breathing apparatus connected to a supply of clean compressed air or positive-pressure power respirators irrespective of the type of asbestos present.

In addition to the delaggers, any other workers in the vicinity of the stripping operations and exposed to the dust should wear protective equipment. The area over which the precaution is needed should be roped off and signposted. If possible, delagging operations should be carried out at times when other workers are not in the vicinity.

All asbestos waste must be shipped in impermeable dust-tight bags and broken or damaged bags should not be used.

Part-empty bags should be securely tied between working periods.

Bags containing crocidolite waste must be marked "BLUE ASBESTOS—DO NOT INHALE DUST," in large legible lettering.

Landfill Disposal of Asbestos Wastes. The British Asbestos Research Council has laid down a code of practice for landfill disposal, given below in part.

Asbestos waste, whether loose or in sealed bags or other sealed receptacles, should be disposed of in such a way that no dust is emitted into the air during transport, in the act of dumping, or after dumping. Where the waste is derived from blue asbestos, the receptacles containing it should be clearly marked to show this. Waste which is adequately contained in impermeable bags or other suitably sealed receptacles does not require special transport.

Special containers can be obtained which, when filled, are removed by specially fitted road vehicles direct to a landfill. Whatever type of container is used, steps should be taken to prevent emission of dust from containers when in use.

Wet waste should be transported to the disposal area in sludge tankers or similar vehicles. After use, all vehicles and containers should be cleared of loose fiber or dust by vacuum cleaning or other dustless method.

All deposited waste should be covered with at least 9 in. of consolidated earth or other suitable material capable of forming a seal to prevent subsequent interference and dispersal of dust. Where other dry wastes are disposed of at the same site, those wastes may well include materials which can be used to provide the required cover. Covering should be done promptly (immediately in the case of waste derived from blue asbestos). No waste should be left uncovered at the end of a working day.

The disposal of wet-waste on a dry-waste landfill is acceptable, provided that the quantities are not excessive. It should be remembered that, even if waste is wet when dumped, it must still be covered so that dust cannot escape from the dried-out material.

HAZARDOUS INDUSTRIAL WASTES

Estimated quantities of hazardous industrial wastes, for the year 1970, by geographic region of the United

TABLE 6.5. Estimated Industrial Hazardous Waste Generation by Region* in Tons Per Year (1970).†

Region	Inorganics in aqueous		Organics in aqueous		Organics		Sludges,‡ slurries, solids		Total		Percent of total
	Tons	Metric tons	Tons	Metric tons	Tons	Metric tons	Tons	Metric tons	Tons	Metric tons	
New England	95,000	86,000	170,000	154,000	33,000	30,000	6,000	5,450	304,000	275,450	3.1
Mid Atlantic	1,000,000	907,200	1,100,000	1,000,000	105,000	90,600	55,000	50,000	2,260,000	2,047,800	22.9
East North Central	1,300,000	1,180,000	850,000	770,000	145,000	132,000	90,000	81,600	2,385,000	2,163,600	24.2
West North Central	65,000	59,000	260,000	236,000	49,500	45,000	18,500	16,800	393,000	350,800	4.0
South Atlantic	230,000	208,500	600,000	545,000	75,000	68,000	80,000	72,600	985,000	894,100	10.0
East South Central	90,000	81,700	385,000	350,000	44,000	40,000	9,500	8,600	528,000	480,300	5.4
West South Central	320,000	290,000	1,450,000	1,315,000	180,000	163,000	39,000	35,400	1,989,000	1,803,400	20.2
West (Pacific)	120,000	109,000	550,000	500,000	113,000	103,000	30,500	27,770	813,500	739,770	8.3
Mountain	125,000	113,500	5,000	4,540	50,000	45,400	11,500	10,400	191,500	173,840	1.9
Totals	3,345,000	3,034,900	5,370,000	4,874,540	794,500	717,000	340,000	308,620	9,849,500	8,929,060	100.0

*Refers to Bureau of Census regions, as defined in Appendix B.
†Source: EPA Contract No. 68-01-0762.
‡Predominantly inorganic.
Office of solid-waste-management programs, E.P.A. "Report to congress-disposal of hazardous wastes," U.S. E.P.A., report SW-115, 1974.

TABLE 6.6 Representative Hazardous Substances Within Industrial Waste Stream.

Industry	Hazardous Substances										
	As	Cd	Chlorinated Hydrocarbons*	Cr	Cu	Cyanides	Pb	Hg	Miscellaneous Organics†	Se	Zn
Mining and metallurgy	√	√		√	√	√	√	√		√	√
Paint and dye		√		√	√	√	√	√	√	√	
Pesticide	√		√			√	√	√	√		√
Electrical and electronic			√		√	√	√	√		√	
Printing and duplicating	√			√	√		√		√	√	
Electroplating and metal finishing		√		√	√	√					√
Chemical manufacturing			√	√	√			√	√		
Explosives	√				√		√	√	√		
Rubber and plastics			√			√		√	√		√
Battery		√					√	√			√
Pharmaceutical	√							√	√		
Textile				√	√				√		
Petroleum and coal	√		√				√				
Pulp and paper								√	√		
Leather				√					√		

*Including polychlorinated biphenyls.
†For example, acrolein, chloropicrin, dimethyl sulfate, dinitrobenzene, dinitrophenol, nitroaniline, and pentachlorophenol.
Office of solid-waste-management programs, E.P.A. "Report to congress—disposal of hazardous wastes," U.S. E.P.A., report SW-115, 1974.

States, are shown in Table 6.5. These figures are only approximate guides. Not only are reporting methods extremely crude at this time, but the definition of what constitutes a hazardous waste is imprecise and continually changing. Liquid and slurry wastes are given as well as solid wastes. Waste can be changed from one form to another with relative ease, so that to list solid wastes alone would be insufficient.

Table 6.6 lists some representative hazardous substances in industrial wastes, in broad categories, by industry. Summary data for nonradioactive wastes are shown in Table 6.7 again for liquid-as-discharged wastes as well as for solids and slurries. Some currently available processes for treatment and disposal of nonradioactive hazardous wastes are listed in Table 6.8, with costs shown in Table 6.9.

A promising method, though still in the experimental stage, for the destruction of toxic chemicals has been found at M.I.T. during an investigation of electron irradiation as a treatment for sewage sludge prior to on-land disposal. Substances such as 3,4,2′PCB, monochloro PCB, and the pesticide Monuron, dissolved in water to the limits of saturation, were totally destroyed by electron irradiation at dose levels as low as 10 kilorads. PCB in 0.5% soap solution, which is a model for the lipid-water distribution of sludge, showed virtually complete destruction of the PCB at 400 kilorads.

The following notes on some particularly troublesome metals and organic wastes are meant to be illustrative rather than all-inclusive of the health hazards that can result from incorrect handling of as-generated wastes.

Inorganic Materials*

Lead.* Lead poisoning leads to brain damage, especially in children, and interferes with the ability of people of all ages to resist infection. Lead is dissolved out of lead water-supply pipes to enter the diet through the drinking water. Children obtain additional lead from eating paint chips in old lead-painted buildings. The disposal of metallic lead and demolition debris in landfills therefore poses a potential hazard if the leachate is allowed to pass into ground-water feeding a water-supply system.

Mercury.* Metallic mercury is so valuable that it is unlikely to appear as such in solid wastes to any extent. Organic and inorganic compounds of mercury are widely used, however, and can appear in industrial and agricultural wastes. All mercury compounds are highly toxic to man and animals, causing severe mental and motor symptoms, neurological symptons, kidney damage, gastrointestinal and pulmonary symptoms, and genetic damage.

*See also Section 12.

Beryllium.* Beryllium is a low-density metal used in many industries, prinicpally in alloy form, and as a rocket fuel in the pure form. When ingested as atmospheric particles it can cause severe disease involving respiratory-tract irritation, dermatitis and conjunctivitis and possibly involves other organs. Mortality is still 35% from Be disease in its chronic form. Whether long-term, low-level exposure can lead to cancers is still uncertain.

Manganese.* Manganese is used prinicipally as an alloying element in steels. Inhalation of manganese oxides may cause chronic manganese poisoning or manganic pneumonia.

Organic Materials

Benzopyrene and Benzofluoranthene.* 3,4-benzopyrene and 3,4-benzofluoranthene are formed when organic substances are heated above 500°C for incineration and other processes. These hydrocarbons are carcinogenic to animals. The substances have been found in soils, incorporated from sediments from flue emissions, and in root vegetables (potatoes, carrots, radishes, etc.) grown in such soils.

DEHP.* Di-2-ethylhexyl phthalate is used as a plasticizer for polyvinylchloride plastic packaging. It is toxic to animals and man. It is suspected that deaths in patients receiving blood transfusions from blood stored in PVC bags were due to this plasticizer being transferred from the bag to the blood. Handling and disposal of quantities of waste DEHP must be carried out with stringent precautions.

PBB-Polybrominated biphenyls.* They were made (11-million lb in Michigan from 1970–1974) as flame-retardants for fibers and thermoplastics (radios, TVs, etc.), and are still (1977) made for export. The inclusion, in error, of 4000 lb of PBB in cattle feed in Michigan led to an environmental disaster. Tens of thousands of head of cattle and swine and a million head of chickens were slaughtered. Some were consumed as food. PBB is extremely toxic, more so than PCBs (see below) causing liver, kidney, and thyroid damage, in addition to severely affecting the nervous and reproductive systems of experimental animals. It is a teratogenic agent. It is extremely persistent, and bioaccumulates to 20,000–30,000 times the ambient level.

PCB's.* Polychlorinated biphenyls are still used for electrical-equipment cooling systems and were formerly used for no-carbon copy paper. A small proportion recycled into combination paper board used for packaging breakfast cereals permitted transfer of PCB to the foodstuffs and caused measureable levels in human fat. PCB's or an associated contaminant, chlorinated dibenzofuran, cause liver malfunction and damage, severe skin eruptions and hydropericardium.

TABLE 6.7. Summary Data for Nonradioactive Wastes.

Waste stream title	Standard Industrial Code	Fractions by Geographic Area									Volume (lb/yr)	Remarks
		NE	MA	ENC	WNC	SA	ESC	WSC	M	W		
Aqueous inorganic:												
Chromate wastes from textile dyeing	22	0.101	0.178	0.034	0.005	0.568	0.034	0.014	0.006	0.060	2×10^7 maximum	Fertilizers
Chlorine production brine sludges	2812	.02	.11	.10		.19	.22	.24	—	.12	1×10^8	
Potassium chromate production wastes	2819	.19	.06	.015	.005	.60	.10	.01	.01	.01	1×10^6	
Cellulose ester production wastes	2821	.10	.21	.21	.16	.14	.07	.10		.02	5×10^7	
Intermediate agricultural product wastes (nitric acid)	287	.005	.075	.145	.074	.299	.207	.090	.046	.058	2×10^5	
Production works from ammonium sulfate	2873			.040	—		—	—	.96	—	1×10^3	
Copper- and lead-bearing petroleum refinery wastes	291	.001	.102	.175	.056	.019	.031	.417	.039	.160	8×10^8	
Chrome tanning liquor	31	.22	.29	.29	.03	.086	.05	.004	—	.03	2×10^7	
Mirror production wastes	3231	.09	.25	.23	.01	.28	.10	.04	—	—	9×10^6	Metalworking
Cold finishing wastes	331	.03	.34	.43	.01	.07	.02	.05	.04	.01	5×10^9	
Consolidated steel plant wastes	331	.02	.33	.42	.02	.09	.02	.03	.05	.02	5×10^8	
Stainless steel pickling liquor	3312	.050	.259	.404	.026	.068	.055	.044	.028	.067	5×10^7	
Brass mill wastes	333	.04	.29	.01	.25	.01	.04	.04	.13	.19	5×10^7	
Metal finishing wastes:	33	.115	.179	.379	.046	.050	.015	.036	.011	.169	4×10^7	Cyanide solution
Aluminum anodizing bath with drag out		.115	.179	.379	.046	.050	.015	.036	.011	.169	8×10^6	Metal sludges
Brass plating wastes		.115	.179	.379	.046	.050	.015	.036	.011	.169	Not available	
Cadmium plating wastes		.131	.285	.321	.045	.049	.023	.036	.007	.103	1×10^6	
Chrome plating wastes		.115	.179	.379	.046	.050	.015	.036	.011	.169	Not available	
Cyanide copper plating wastes		.115	.179	.379	.046	.050	.015	.036	.011	.169	2×10^6	
Finishing effluents		.115	.179	.379	.046	.050	.015	.036	.011	.169	Not available	
Metal cleaning wastes		.115	.179	.379	.046	.050	.015	.036	.011	.169	Not available	
Plating preparation wastes		.115	.179	.379	.046	.050	.015	.036	.011	.169	Not available	
Silver plating wastes		.115	.179	.379	.046	.050	.015	.036	.011	.169	Not available	
Zinc plating wastes		.115	.179	.379	.046	.050	.015	.036	.011	.169	Not available	
Metal finishing chromic acid	34	.244	.198	.149	.095	.081	.032	.031	.041	.031	4.4×10^7	
Graphic arts and photography wastes	3555	.06	.19	.20	.08	.15	.06	.09	.13	.04	4×10^3	
Electronic circuitry manufacturing wastes	36	.143	.342	.170	.037	.053	.019	.032	.039	.165	5×10^5	
Aircraft plating wastes	372	.123	.158	.117	.093	.057	.013	.095	.019	.325	4×10^3	
Cooling tower blowdown	—	.005	.150	.170	.060	—	.58		.035	—	2×10^7	As chromate
Subtotal											7×10^9	
Organic:												
Cosynthesis methanol production wastes	2818	—	—	.05	—	.05	—	.90	—	—	1×10^6	Sludge
Formaldehyde production wastes	2818	—	—	.02	—	.05	—	.93	—	—	8×10^5	Sludge
n-Butane dehydrogenation butadiene production wastes	2818	—	—	.03		—		.02		.05	3×10^5	Sludge
Rubber manufacturing wastes	2822	—	.07	.14	—	.11	.11	.50	—	.07	1×10^6	
Benzoic herbicide wastes (DOD)	2879	.168	.130	.009	—	.447	—	—	—	.246	3×10^3	
Chlorinated aliphatic herbicide wastes (DOD)	2879	.196	.062	.027	—	.649	—	.010	.057	—	5×10^3	
Phenyl urea herbicide wastes (DOD)	2879	.539	.059	—	—	.343	.059	—	—	—	2×10^3	
Halogenated aliphatic hydrocarbon fumigant wastes (DOD)	2879	1.0	—	—	—	—	—	—	—	—	2×10^2	
Organophosphate pesticide wastes (DOD)	2879	.0007	.014	.010	—	.033	—	—	.014	.929	1×10^5	
Phenoxy herbicide wastes	2879	.0002	.0001	.0007	—	.0008	.849	.149	.0002	.0004	8×10^6	
Carbonate pesticide manufacturing (DOD)	2879	—	—	—	.006	.848	—	—	—	.145	3×10^2	
Polychlorinated hydrocarbon pesticide wastes (DOD)	2879	.097	.142	.018	.003	.096	.004	.033	.017	.591	1×10^5	
Miscellaneous organic pesticide manufacturing waste (DOD)	2879	—	.026	.012	.002	.257	—	—	—	.702	3×10^4	
Contaminated and waste industrial propellants and explosives	2892	—	—	—	—	—	—	—	.344	.655	3×10^5	
Contaminants and waste from primary explosives production	2892	—	.096	.001	.898	—	.001	—	.003	.001	4×10^6	
Nitrocellulose base propellant contaminated waste	2892	—	.041	—	.457	.492	—	—	.009	—	9×10^6	

High explosive contaminated wastes	2892	–	.005	.094	.394	.397	.027	.004	.023	.012	1×10^7	
Incendiary contaminated wastes	2892	–	–	–	–	–	–	1.0	–	–	6×10^5	
Production of nitroglycerin	2892	–	–	–	–	.42	.19	–	–	.39	7×10^5	
Solid waste from old primers and detonators	2892	–	.005	.430	.454	.001	.006	–	.084	.014	3×10^5	
Wastes from production of nitrocellulose propellants and smokeless powder	2892	–	.060	.046	.387	.477	–	.006	.025	–	6×10^6	
Waste high explosives	2892	.002	.006	.346	.174	.218	.104	.127	.010	.001	1×10^7	
Waste incendiaries	2892	–	.014	.002	.002	–	–	.718	.009	.255	8×10^5	
Waste nitrocellulose and smokeless powder	2892	–	–	–	–	–	–	–	.406	.594	2×10^6	
Waste nitroglycerin	2892	–	–	.01	–	.50	.22	–	.004	.266	5×10^5	
Nonutility polychlorinated biphenyl wastes	2899	.037	.221	.372	.153	.040	.041	.057	.009	.072	8×10^6	
Gasoline blending wastes	2911	.006	.086	.159	.055	.025	.025	.477	.033	.134	4×10^8	
Reclaimers residues	2992	.040	.120	.205	.081	.135	.082	.139	.044	.155	3×10^8	
Coke plant raw waste	3312	.02	.33	.41	.01	.07	.02	.06	.06	.02	8×10^7	
Military arsenical wastes	9711	–	.002	.001	–	.015	.031	.001	.024	.926	3×10^6	
Outdated or contaminated tear gas	9711	–	.138	.189	–	.022	.044	.252	.144	.209	3×10^8	
Subtotal											1×10^9	
Aqueous organic:												
Dimethyl sulfate production wastes	2611	–	(‡)	–	–	(‡)	–	–	–	–	2×10^5	Still bottoms
Acetaldehyde via ethylene oxidation	281	.015	.170	.156	.047	.156	.111	.265	.020	.060	8×10^7	
Residue from manufacture of ethylene dichloride/vinyl chloride	2821	–	.021	.015	–	.163	.171	.533	–	.117	2×10^7	
Nitrobenzene from rubber industry wastes	2822	–	.07	.14	–	.11	.11	.50	–	.07	5×10^9	Probably too dilute to be of concern
Drug manufacturing wastes	283	.056	.348	.183	.089	.100	.033	.060	.011	.115	5×10^9	Probably too dilute to be of concern
Chlorinated hydrocarbon pesticide production wastes	2879	.115	.148	.136	.073	.141	.057	.093	.054	.183	2×10^8	
Miscellaneous organic herbicide production wastes	2879	.076	.135	.124	.080	.156	.062	.108	.059	.200	4×10^8	
Organo-phosphate pesticide production wastes	2879	.115	.148	.136	.073	.141	.057	.093	.054	.183	6×10^7	
Organic pesticide production wastes	2879	.115	.148	.136	.073	.141	.057	.093	.054	.183	3×10^8	
Phenoxy herbicide production wastes	2879	.076	.135	.124	.080	.156	.062	.108	.059	.200	4×10^7	
Subtotal											1×10^{10}	
Solid, slurry, or sludge:												
Recovered arsenic from refinery flues (stored)	1021	–	–	–	–	–		–	–	1.00	4×10^7	Tacoma, Wash.
Sodium dichromate production wastes	2819	–	.150	.243	–	.437	–	.170	–	–	3×10^8	
Solvent-based paint sludge	285	.044	.243	.269	.072	.103	.041	.069	.012	.147	4×10^7	
Water-based paint sludge	285	.044	.243	.269	.072	.103	.041	.069	.012	.147	3×10^7	
Tetraethyl and tetramethyl lead production wastes	2869	–	–	–	–	–	–	.63	–	.37	3×10^5	
Urea production wastes	2873	–	.05	.09	.18	.09	.15	.29	–	.14	2×10^5	Dry basis
Benzoic herbicide contaminated containers	2879	–	–	.655	.154	.006	.017	–	.009	.160	2×10^4	
Calcium arsenate contaminated containers	2879	.C3	.02	.08	.07	.16	.16	.35	.09	.03	6×10^3	
Carbonate pesticide contaminated containers	2879	.0003	.016	.382	.070	.022	.108	.321	.020	.060	5×10^4	
Chlorinated aliphatic pesticide contaminated containers	2879	.381	–	.076	.418	–	.105	.010	–	.010	1×10^4	
Dinitro pesticide contaminated containers	2879	.496	.168	.023	.017	.228	–	.003	.006	.165	2×10^4	
Lead arsenate contaminated containers	2879	.03	.02	.08	.07	.17	.17	.35	.03	.08	1×10^4	
Mercury fungicide contaminated containers	2879	.02	.03	.04	.03	.28	.32	.05	.01	.22	5×10^2	
Miscellaneous organic insecticide contaminated containers	2879	.148	.084	.054	.039	.197	.143	.148	.017	.170	4×10^4	
Organic arsenic contaminated containers	2879	–	.007	–	–	.011	.764	.218	–	–	5×10^3	
Organic fungicide contaminated containers	2879	.048	.125	.047	.028	.441	.001	.036	.007	.266	8×10^4	
Organophosphorus contaminated containers	2879	.043	.050	.018	.125	.139	.192	.175	.049	.208	1×10^5	
Phenoxy contaminated containers	2879	.035	.033	.196	.321	.031	.030	.067	.141	.146	2×10^5	
Phenyl urea contaminated containers	2879	.106	.085	.106	.033	.106	.424	.042	.003	.095	9×10^3	
Polychlorinated hydrocarbon contaminated containers	2879	017	.107	.019	.138	.306	.211	.133	.024	.044	2×10^5	

TABLE 6.7. (*Continued*)

Waste Stream Title	Standard Industrial Code	Fractions by Geographic Area NE	MA	ENC	WNC	SA	ESC	WSC	M	W	Volume (lb/yr)	Remarks
Triazine contaminated containers	2879	.147	.121	.320	.372	.013	.003	.011	.002	.011	6×10^4	
Miscellaneous organic pesticide contaminated containers	2879	.014	.162	.385	.068	.162	.123	.041	.014	.034	1×10^4	
Petroleum refining still bottoms	2911	.006	.086	.159	.055	.025	.025	.477	.033	.134	2×10^4	
Petroleum waste brine sludges	2911	.002	.06	.09	.011	.12	.10	.55	.022	.045	4×10^6	
Iron manufacturing waste sludge	331	.05	.05	.56	.02	.12	.03	.09	.05	.03	6×10^6	
Arsenic trioxide from smelting industry	333	—	.03	.015	.07	.005	.01	.10	.70	.07	2×10^7	
Selenium production wastes	3339	—	.75	—	—	—	—	—	.25	—	2×10^4	
Duplicating equipment manufacturing wastes	3555	—	1.00	—			—	—	—	—	7×10^5	Upstate New York
Refrigeration equipment manufacturing wastes	3585	.013	.232	.408	.096	.040	.069	.086	.011	.045	2×10^8	
Battery manufacturing waste sludge	3691	.117	.043	—	.118	.117	—	—	.118	—	5×10^7	
Arsenic trichloride recovered from coal	49	.05	.23	.07	.05	.33	.25	—	.07	—	6×10^6	
Military paris green (stored)	9711	—		1.00	—	—		—	—	—	3×10^4	
Stored military mercury compounds	9711	.47	—		.51	—		—		.02	2×10^2	
Subtotal											7×10^8	
Aqueous inorganic (insufficient quantity or distribution data):												
Zinc ore roasting acid wash	1031					Not available					Not available	
Mercury ore extraction wastes	1092	—	—	—	—	—	—	—	.28	.72	Not available	
Cadmium ore extraction wastes	1099					Not available					2×10^8	
Mercury bearing textile wastes	22					Not available					Not available	
Wastes from pulp and paper industry	26	.11	.11	.19	.04	.23	.08	.10	.03	.11	Negligible	
Cadmium-selenium pigment wastes	28					Not available					Not available	
Waste or contaminated perchloric acid	28					Not available					Negligible	
Arsine production wastes	2813	(‡)	(‡)	(‡)	—	—	(‡)	(‡)	—	(‡)	1×10^4	
Borane production wastes	2813	—	1.0	—	—	—	—	—	—	—	Negligible	
Nickel carbonyl production wastes	2813	—	1.0	—	—	—	—	—	—	—	Negligible	
Waste bromine pentafluoride	2813	—	—	—	—	—	—	1.0	—	—	Negligible	
Waste chlorine pentafluoride	2813	—	—	(‡)	—	—	—	(‡)	—	—	Negligible	
Waste chlorine trifluoride	2813	—	—	(‡)	—	—	—	(‡)	—	—	Negligible	
Chromate wastes from pigments and dyes	2816	.015	.170	.156	.047	.156	.111	.265	.020	.060	1×10^{11}(§)	
Arsenic wastes from purification of phosphoric acid	2819	.015	.170	.156	.047	.156	.111	.265	.020	.060	Negligible	
Contaminated fluorine	2819	.007	.101	.166	.075	.147	.207	.147	.054	.096	Negligible	
Cyanide production wastes	2819	.007	.101	.166	.075	.147	.207	.147	.054	.096	2×10^{11}(§)	
Waste from manufacture of mercuric cyanide	2819	—	1.0	—	—	—	—	—	—	—	Negligible	
Waste from production of barium salts	2819	.007	.101	.166	.075	.147	.207	.147	.054	.096	Negligible	
Urethane manufacturing wastes	2821	.046	.121	.101	.018	.404	.182	.101	—	.027	2×10^{12}(§)	
Wastes from polycarbonate polymer production	2821	.046	.121	.101	.018	.404	.182	.101	—	.027	2×10^{12}(§)	
Pharmaceutical arsenic wastes	283	.056	.348	.183	.089	.100	.033	.060	.011	.115	Negligible	
Pharmaceutical mercurial wastes	283	.056	.348	.183	.089	.100	.033	.060	.011	.115	Negligible	
Wood preservative wastes	2865	.007	.029	.117	.060	.267	.141	.174	.042	.162	2×10^{11}(§)	
Contaminated antimony pentafluoride	2869	—	(‡)	(‡)	—	—	—	(‡)	—	—	Negligible	
Contaminated antimony trifluoride	2869	—	(‡)	(‡)	—	—	—	(‡)	—	—	Negligible	
Hydrazine production wastes	2869	(‡)	(‡)	—	—	(‡)	—	(‡)	—	—	Not available	
Agricultural chemical production wastes	287	.005	.075	.145	.074	.299	.207	.090	.046	.058	2×10^{12}(§)	
Agricultural pesticide arsenic wastes	2879	.005	.075	.145	.074	.299	.207	.090	.046	.058	2×10^{12}(§)	
Mercuric fungicide production wastes	2879	.005	.075	.145	.074	.299	.207	.090	.046	.058	2×10^{12}(§)	
Pesticide arsenate wastes	2879	.005	.075	.145	.074	.299	.207	.090	.046	.058	2×10^{12}(§)	
Pesticide arsenic wastes	2879	.005	.075	.145	.074	.299	.207	.090	.046	.058	2×10^{12}(§)	
1080 production wastes and contaminated lots	2879	—	(‡)		—	(‡)	1.0	—	—	—	Negligible	Production of sodium fluor-acetate
Wastes from pesticide-herbicide manufacture (arsenites)	2879	.005	.075	.145	.074	.299	.207	.090	.046	.058	2×10^{12}(§)	
Electrical fuse manufacturing wastes	2899	.037	.221	.372	.153	.039	.040	.057	.009	.072	2×10^{12}(§)	

Beryllium salt production wastes	3339	–	1.0	–	–	–	–	–		–	Negligible	
Thallium production wastes	3339	–	–	–	–	–	–	1.0	(‡)	–	Negligible	Small amount in Colorado
Rotogravure printing plate wastes	3555	.105	.446	.320	.051	.019	–	.028	–	.031	1×10^{12}(§)	
Computer manufacturing wastes	3573	.143	.342	.170	.037	.053	.019	.032	.039	.165	1×10^{12}(§)	
Electronic tube production wastes	367	.143	.342	.170	.037	.053	.019	.032	.039	.165	Not available	
Magnetic tape production wastes	3679	.171	.336	.165	.120	.077	.016	.060	–	.060	Not available	
Battery manufacturing wastes	3691	.030	.236	.289	.111	.103	.029	.056	.012	.134	1.1×10^{12}	
Mercury cell battery wastes	3692	.060	.138	.556	.049	.074	.019	.017	(¶)	(¶)	1.1×10^{12}(§)	
Railroad engine cleaning	40					Not available					Not available	
Arsenic wastes from transportation industry	40					Not available					Not available	
Military cadmium wastes from plating	9711					Not available					Not available	
Military sodium chromate	9711					Not available					Not available	
Subtotal											2×10^{8}	
Organic (insufficient quantity or distribution data):												
Spent wood-preserving liquors	2491	.007	.029	.117	.060	.267	.141	.174	.042	.162	2×10^{12}(§)	
Off-specification "agent orange" defoliant	9711					Not available					Not available	
Paint stripping wastes, Vance Air Force Base, Oklahoma	9711	–	–		–	–		1.0	–	–	Not available	
Subtotal											Not available	
Aqueous organic (insufficient quantity or distribution data):												
Synthetic fiber production wastes	2824	.046	.121	.101	.018	.404	.182	.101		.027	2×10^{12}(§)	
Dye manufacturing wastes	2865	.015	.170	.156	.047	.156	.111	.265	.020	.060	1×10^{11}(§)	
Nitrile pesticide wastes	2879	.005	.075	.145	.074	.299	.207	.090	.046	.058	2×10^{12}(§)	
Organic arsenicals from production of cacodylates	2879	–	.200	.800		–	–	–	–	–	Not available	
Torpedo process wastes	2879	–	–	–	–	–	1.0		–	–	Negligible	
Utilities and electrical station waste	49					Not available					3×10^{7}	
Wastes from production of chloropicrin	9711	–	(‡)	–	–	–	–	(‡)	–	(‡)	Negligible	
Subtotal											3×10^{7}	
Solid, slurry, or sludge (insufficient quantity or distribution data):												
Wastes from seed industry	011	.017	.088	.371	.213	.053	.060	.081	.023	.094	Not available	
Contaminated orchard soil	0175	.05	–	.15	–	.33	–	.35	.03	.09	Unknown	
Old or contaminated thallium and thallium sulfate rodenticide	2879	.005	.075	.145	.074	.299	.207	.090	.046	.058	Not available	
Highly contaminated soil	9711	–	–		–	–	–		1.00	–	3×10^{9} (not included in total)	Stored at Rocky Mountain Arsenal
Spent filter media from military operations	9711					Not available					Not available	
Waste chemicals from military	9711					Not available					3×10^{5}	
Explosives from military ordnance	9711					Not available					4×10^{8}	
Drugs and contraband seized by customs	–					Not available					Not available	
Etiological materials from commercial production	–					Not available					3×10	
Subtotal											4×10^{8}	
Total											2×10^{10}	

*This is an updated version of the table that appeared in the first edition of this report.

†NE = New England: Connecticut, Maine, Massachusetts, New Hampshire, Rhode Island, and Vermont; MA = Mid Atlantic: New Jersey, New York, and Pennsylvania; ENC = East North Central: Illinois, Indiana, Michigan, Ohio, and Wisconsin; WNC = West North Central: Iowa, Kansas, Minnesota, Missouri, Nebraska, North Dakota, and South Dakota; SA = South Atlantic: Delaware, District of Columbia, Florida, Georgia, Maryland, North Carolina, South Carolina, Virginia, and West Virginia; ESC = East South Central: Alabama, Kentucky, Mississippi, and Tennessee; WSC = West South Central: Arkansas, Louisiana, Oklahoma, and Texas; M = Mountain: Arizona, Colorado, Idaho, Montana, Nevada, New Mexico, Utah, and Wyoming; W = West (Pacific): Alaska, California, Hawaii, Oregon, and Washington.

‡Exists but quantity is unknown.

§Total liquid discharge for the larger 3-digit standard industrial code category.

¶Percentage for the Mountain and Pacific areas combined is 0.087.

Office of solid-waste-management programs, E.P.A. "Report to congress–disposal of hazardous wastes," U.S.E.P.A., report SW-115, 1974.

TABLE 6.8. Currently Available Hazardous Waste-Treatment and Disposal Processes.*

Process	Functions Performed†	Types of Waste‡	Forms of Waste§	Resource Recovery Capability
Physical treatment:				
Carbon sorption	VR, Se	1, 3, 4, 5	L, G	Yes
Dialysis	VR, Se	1, 2, 3, 4	L	Yes
Electrodialysis	VR, Se	1, 2, 3, 4, 6	L	Yes
Evaporation	VR, Se	1, 2, 5	L	Yes
Filtration	VR, Se	1, 2, 3, 4, 5	L, G	Yes
Flocculation/settling	VR, Se	1, 2, 3, 4, 5	L	Yes
Reverse osmosis	VR, Se	1, 2, 4, 6	L	Yes
Ammonia stripping	VR, Se	1, 2, 3, 4	L	Yes
Chemical treatment:				
Calcination	VR	1, 2, 5	L	
Ion exchange	VR, Se, De	1, 2, 3, 4, 5	L	Yes
Neutralization	De	1, 2, 3, 4	L	Yes
Oxidation	De	1, 2, 3, 4	L	
Precipitation	VR, Se	1, 2, 3, 4, 5	L	Yes
Reduction	De	1, 2	L	
Thermal treatment:				
Pyrolysis	VR, De	3, 4, 6	S, L, G	Yes
Incineration	De, Di	3, 5, 6, 7, 8	S, L, G	Yes
Biological treatment:				
Activated sludges	De	3	L	No
Aerated lagoons	De	3	L	No
Waste stabilization ponds	De	3	L	No
Trickling filters	De	3	L	No
Disposal/storage:				
Deep-well injection	Di	1, 2, 3, 4, 6, 7	L	No
Detonation	Di	6, 8	S, L, G	No
Engineered storage	St	1, 2, 3, 4, 5, 6, 7, 8	S, L, G	No
Land burial	Di	1, 2, 3, 4, 5, 6, 7, 8	S, L	No
Ocean dumping	Di	1, 2, 3, 4, 7, 8	S, L, G	No

*Sources: EPA Contract Nos. 68-03-0089, 68-01-0762, and 68-01-0556.

†Functions: VR, volume reduction; Se, separation; De, detoxification; Di, disposal; and St, storage.

‡Waste types: 1, inorganic chemical without heavy metals; 2, inorganic chemical with heavy metals; 3, organic chemical without heavy metals; 4, organic chemical with heavy metals; 5, radiological; 6, biological; 7, flammable; and 8, explosive.

§Waste forms: S, solid; L, liquid; and G, gas.

Office of solid-waste-management programs, E.P.A. "Report to congress–disposal of hazardous wastes," U.S. E.P.A., report SW-115, 1974.

TABLE 6.9. Costs of Representative Hazardous-Waste-Treatment Processes.*†

Process	Capacity 1,000 gal/day	Capacity 1,000 liters/day	Capital Costs‡ ($1,000)	Operating Costs§ $/1,000 gal	Operating Costs§ $/1,000 liters
Chemical oxidation of cyanide wastes	25	94.8	400	68	18
Chemical reduction of chromium wastes	42	159	340	29	7.65
Neutralization/precipitation	120	452	3,000	50	13.20
Liquid-solid separation	120	452	9,000	40	10.60
Carbon sorption	120	452	910	7	1.85
Evaporation	120	452	510	10	2.64
Incineration	¶74	**67	4,900	††95	‡‡105

*Source: EPA Contract No. 68-01-0762.

†Data correspond to a typical medium-size treatment and disposal facility capable of processing approximately 150,000 tons (136,000 metric tons) per year or 600 tons (545 metric tons) per day.

‡Capital costs include land, buildings, and complete processing and auxiliary facilities.

§Operating costs include neutralization chemicals, labor, utilities, maintenance, amortization charges (7 percent interest), insurance, taxes, and administrative expenses.

¶Tons per day.

**Metric tons per day.

††Dollars per ton.

‡‡Dollars per metric ton.

Office of solid-waste-management programs, E.P.A. "Report to congress–disposal of hazardous wastes," U.S. E.P.A., report SW-115, 1974.

Fiber Polymers and Polyurethane Foams. The open burning of some fiber polymers, such as Orlon, and foamed plastics, such as polyurethane, can lead to the formation of potentially toxic concentrations of hydrogen cyanide.

HEALTH AND INJURY TO SANITATION WORKERS

A survey conducted for the Public Health Service in 1966 showed that, among organizations maintaining surveillance on sanitation workers' occupational injuries, the average accident frequency rate of 156 disabling injuries per million man-hours was $4\frac{1}{4}$ times that of the highest rate (36.71 for coal mining) of any major industry reported by the National Safety Council for 1965.

It was found, however, that many organizations believed that there was no particular problem with hazards to sanitation workers. One reason for this generally false belief is that people who work in solid-waste handling are often undereducated, with little bargaining power, in largely transient employment, and they often quit when they are injured or become ill.

Illness data reported in the same survey indicated that sanitation workers suffered average or below-average illness rates for industry in general. Other scattered reports state that "refuse-removal workers (in Germany) have been found to be astonishingly immune against infectious diseases, while street cleaners often fall ill with toxoplasmosis." Another disease to which street cleaners and others are exposed is cutaneous larva migrans, a form of dermatitis characterized by linear lesions marking the migratory paths of larval parasites. As an occupational disease, cutaneous larva migrans is most frequently seen in plumbers, electricians, and construction workers who are required to be in contact with damp or wet soil contaminated with feces of cats or dogs. The only practical preventive measure is avoiding contact with infective soil or water.

Occupational mycoses—infections from fungi—also affect workers who handle certain types of wastes (Table 6.10). Histo-plasmosis can be contracted most readily by contact with chicken feces. Cryptococcosis is a fungus disease carried in pigeon feces, and can afflict demolition workers particularly. Some deaths have been attributed to this mycosis.

TABLE 6.10. Fungus Diseases (Mycoses) Associated with Solid Wastes.

Coccidioidomycosis
Sporotrichosis
Histoplasmosis

SECONDARY HAZARDS FROM SOLID WASTE

After materials are discarded they become solid wastes and normally pass first to a container, either immediately or via a refuse chute; then the wastes are normally transported by vehicle either in the original container or after transfer to a larger container; then there may be a processing stage such as incineration, compaction, shredding, or reclamation; and so-called "final" disposal (if not by recycling) must by definition be back to the environment in solid, liquid, or gaseous form.

During this general sequence of events, non-toxic wastes may become poisonous; sterile wastes may become capable of transmitting disease; nonexplosive wastes may cause explosions; and wastes which apparently have no injury-causing potential may in fact produce injury and even death. The mechanisms involved, and steps which can be taken to combat the problems, are discussed below.

Public-Health Hazards

Within a matter of hours in a warm temperature, sterile organic matter, such as cooked meat, can become a potentially lethal source of toxic or disease-producing organisms. The organisms do not have to be originally present in the host material because the environment is normally well provided with spores, bacteria, viruses, insects, vermin and other vectors awaiting a favorable site on which to multiply.

The mere presence of these cultures of potentially disease-producing organisms in solid wastes is not enough to cause a major health hazard. The blame for disease transmittal must be placed particularly on the flies (Table 6.11), mosquitoes (Table 6.12) and rodents (Table 6.13). It has been estimated that 90% of the urban house-fly population breeds in the contents of

TABLE 6.11. Fly-Borne Diseases.

Typhoid
Bacillary dysentery
Amoebic dysentery
Diarrheas
Asiatic cholera
Helminth/infections (worm)
Myiasis
Loiasis
Onchocerciasis
Ozzard's filariasis
Leishmaniasis
African sleeping sickness (trypanosomiasis)
Yaws
Tularemia
Bartonellosis
Cararrhal conjunctivitis
Sandfly fever

TABLE 6.12. Mosquito-Borne Diseases.

Dengue
Encephalitis
Filariasis
Malaria
Yellow fever
Tularemia
Lymphocytic choriomeningitis
Melioidosis
Rift-Valley fever

open trash barrels. Rats also rely on readily available scraps of food such as those which are found in open trash barrels. Refuse overflowing from or carelessly disposed around containers outside restaurants or supermarkets or the use of containers with ill-fitting or bent closures are equally responsible. One remedy is obvious: refuse containing organic materials should be transferred to closed containers with zero spillage so far as is possible.

There are alternative forms of treatment which might be practicable in certain cases. One is compaction. Reasonably high-density compaction (perhaps defined as that producing a density of about 40 lb/cu ft or 1000 lb/cu yd) eliminates most of the crevices and voids in which insects prefer to lay their eggs; greatly reduces the surface area available; and squeezes out many of the juices so leaving the surface dry, or perhaps at most uniformly moist without the local areas of high organic concentration which are necessary for optimum hatching of fly larvae.

A second approach, less desirable, is to process refuse within two days of it having been discarded, so that fly larvae do not have time to emerge (2–3 days being required). Rats can still feed if containment is not required, and there is no presently available method of treatment other than denying rodents access to food.

A third approach seems contradictory to the first in that it involves shredding solid wastes. Shredding increases the surface area and decreases the density. In these circumstances an interesting transformation takes place in organic wastes. If shredded refuse is not kept in closed containers, and if the volume in any one place is not such as to give a depth of refuse greater than perhaps three feet, sufficient air can reach all parts of the mass so that aerobic rather than anaerobic bacteria grow vigorously. These bacteria work many times faster than the anaerobic variety and give off combustion heat as well as being thermophilic. Carefully controlled tests have shown that insects are unable to hatch on aerobically decomposing refuse, and the material is also unattractive to rats and other vermin. The rate of air movement possible within such a mass is small, and fires, if set, tend to smolder rather than to burn. As most fire deaths are due to smoke inhalation rather than to flame-produced injuries, the shredding of refuse could be regarded as an acceptable way to reduce many of the secondary hazards of solid wastes so long as the shredded refuse is not stored within a building or other enclosure in which people may be trapped.

TABLE 6.13. Rodent-Borne Diseases.

Echinostomiasis
Hemorrhagic septicemia
Histoplasmosis
Lymphocytic choriomeningitis
Plague
Rat-bite fever
Rat-mite dermatitis
Rat-tapeworm infection
Rocky Mountain spotted fever
Salivary-gland virus infection
Salmonellosis
Schistosomiasis
Bilharziasis
Sporotrichosis
Swine erysipelas
Trichinosis
Leptospirosis
Leishmaniasis
Relapsing fever
Tularemia
Rickettsial pox
Murine ryphus

Fire and Explosions

There is an extreme danger of fire in most categories of untreated solid wastes simply because a large proportion of the material is paper products, wood or plastics. The potential for these materials to be ignited by carelessly disposed-of smoking materials is high. However, whenever organic materials are amassed together there is the danger that these and less flammable materials may undergo spontaneous combustion.

Spontaneous Combustion

Spontaneous combustion occurs when organic materials decompose, usually initially by action of aerobic (oxygen-loving) bacteria and the heat released is not all able to escape so that the temperature of the mass rises. Thermal insulation is usually provided by the neighboring mass of nonreacting material, so that in an analogy to a nuclear reaction, a critical mass of organic material has to be present before heat transfer outwards from the reaction zone is sufficiently reduced for the temperature to rise and for high-temperature burning to commence.

Spontaneous combustion has long been recognized to be a problem in agricultural materials such as hay; in coal; and in other finely divided fuels such as woodchips and sawdust. A recent example emphasizing the need to avoid large masses of any materials which have combustion potential is the fusing together of a pile of steel turnings which were awaiting reclamation in an open-air yard. The turnings were wet with cutting oil. The plant supervisor maintained that the inside of the pile became red hot and subsequently cooled down to form a solid mass. The resulting pyramidical block was too thick to be cut with torches and all other attempts to reduce it to moveable pieces failed.

There do not appear to be scientifically useful data to give guidance for avoiding spontaneous combustion in all types of solid wastes. Common-sense solutions are to simply avoid storing solid wastes in warm moist conditions for several days when the smallest dimension of the waste mass is greater than 6 to 10 ft (2 to 3 m). Smaller quantities, or very dry, or very wet conditions, all appear to be inhibitors of spontaneous combustion.

Explosions

While it is obvious that explosive materials, such as live ammunition, are very liable to be detonated during handling and processing of mixed solid wastes, it is less well known that combinations of, or large quantities of, apparently innocuous materials can also cause explosions in certain circumstances. Incinerator explosions, for instance, have resulted from the charging of large masses of plastic wastes. Presumably the combination of gaseous pyrolysis products with the available oxygen reached explosive proportions. The gas produced by anaerobic decomposition of organic waste can also lead in certain circumstances to explosive mixtures, and this is discussed below.

Gas Production

The principal gaseous product of anáerobic decomposition of organic wastes is methane, CH_4, although other gases such as carbon dioxide, CO_2, nitrogen, N_2, and hydrogen sulfide, H_2S, are also present. Wastes buried in landfills undergo predominantly anaerobic decomposition, and the gases normally take the shortest or easiest route to the surface. Occasionally the presence of fissures in or surrounding the filled land, possibly together with the presence of a surface barrier, such as a roadway or parking area, will cause the evolved gases to travel large distances horizontally. In some cases these gases have found their way into basements of houses and buildings, and deaths and injuries have resulted from asphyxiation, poisoning, or from the detonation of explosions of air-methane mixtures.

Sanitary-landfill regulations require that ducts be incorporated into landfills to lead the evolved gases to be discharged at a safe location. Building codes for structures near new or old landfills normally require particularly stringent standards for the construction of impermeable basements and for sealed and ventilated underground services.

Leachate Production

During decomposition, organic wastes normally produce a liquid which is termed a "leachate." In landfills, some of this liquid results from the pressure applied by the fill above squeezing out juices from, for instance, garbage. Another principal component of leachate is rain water which, during its passage through the landfill, dissolves a large range of materials. Leachate may emerge from a landfill with a biological oxygen demand of over 20,000 mg/liter, which is about 100 times stronger than raw sewage. Sanitary-landfill regulations call for the capture of all leachate and for its treatment by normal sewage-treatment methods or the equivalent. Studies have shown that the leachate concentration from a landfill might still be significant after 20 years.

DANGER POINTS IN SOLID-WASTE HANDLING

In this section the progress of typical solid waste will be followed through typical sequences of locations and operations and particular hazards will be identified.

Wastebaskets

For typical residential, institutional, and in some cases commercial refuse, the wastebasket is the start of the journey to final disposal. Wastebaskets are usually dry and reasonably clean and are emptied every one or two days, so that problems of fly breeding and multiplication of undesirable bacteria are usually small. The principal hazards are fires originating from wastebaskets and from personal injury to the personnel emptying wastebaskets.

Paper products predominate in wastebaskets. The contents are therfore usually highly flammable. As wastebaskets are frequently used as receptacles for ashes from cigarettes, cigars, and pipes, and frequently for the butts and pipe residues, the contents of ash trays, and the matches and spent cigarette-lighter cartridges, there has to be only a very low incidence of any of these materials being in a smoldering condition for wastebasket fires to be comparatively frequent. Wastebaskets are often placed half hidden behind curtains and drapes and by upholstered furniture so that the fire danger is typically extremely high.

Some obvious methods of control are to limit smoking within buildings; to provide a large-capacity, floor-standing ash tray near each wastebasket; or to use one of the many types of self-closing waste containers.

Danger to personnel emptying wastebaskets comes

predominantly from thoughtlessly discarded dangerous materials. Examples encountered within the author's institution are half-filled bottles containing strong acids; and used hypodermic needles from animal ex-experiments.

Two avenues of correction suggest themselves in these cases. One is public education to the dangers to personnel of dangerous substances and objects in the waste stream, together with education of the service personnel to empty wastebaskets directly into receiving containers without the use of hands. The second approach is to promulgate regulations which make the person who discards prohibited materials into the regular waste stream legally liable for damages resulting from these actions. The second approach is in fact a variant of the first in that education is usually more rapidly assimilated when there are strong incentives to learn.

Refuse Chutes

A frequently heard complaint against architects is that the handling of solid wastes is seldom considered in the design stage of buildings, large or small. Occasionally, however, refuse chutes are incorporated, and these undoubtedly greatly reduce handling costs and thereby reduce some of the dangers involved in solid-waste handling.

Some other dangers are introduced, such as the extreme fire danger if a refuse chute is designed without appropriate traps to prevent the chute acting as a chimney and if the materials and construction are insufficiently fire-resistant. Building codes in most cities require that fire traps be used. A secondary danger which refuse chutes introduce is that the anonymity in the disposal of refuse which they permit encourages people to dispose of all manner of prohibited, dangerous, or incriminating wastes.

Compactors

Compactors are designed to withstand the high loads and pressures imposed on wastes. Accordingly, they are able to withstand and contain minor explosions from, for instance, vapor-propellant bottles. Fires are normally extinguished by compaction. However, fires are comparatively frequent in collector-compaction trucks after, for instance, some hot ashes might be loaded toward the beginning of a collection run, before the vehicle contents have become compressed to any extent. When fire inside a truck body is confirmed, standing instructions to the crew usually are to dump the burning load in the middle of the road, insuring first, of course, that other traffic is alerted and rerouted. A fire department can more easily put out burning refuse than exploding gasoline.

Other dangers from compactors come from the loading process when loading is carried out by hand. In most stationary compactors there are interlocks to insure that the compaction device, usually a ram and sometimes a screw, is closed off before operation. Accidents to loaders of refuse-collection trucks are fairly frequent due to imploding T.V. tubes, or the sudden movement of pieces of steel when these are pinched half in and half out of the mechanism. When workers are on an incentive system, protective devices are often removed in the interest of speed, and good management is required to educate what is usually a transient work force on the dangers involved.

Containers

The term "container" is usually reserved for enclosures with volumes larger than those of wastebaskets or ash barrels: containers normally hold from 1 to 20 cu yd. However, containers share many of the dangers of wastebaskets and refuse barrels: fire, dangerous materials, and the encouragement of insects and vermin are examples. In addition, the storage of large volumes of refuse in a compact mass encourages biological degradation which can lead to spontaneous combustion given the right combinations of moisture, warmth, insulation, and fuel availability. Some control measures have been suggested above.

Incinerators

Small incinerators as manufactured for apartment houses and commercial establishments have been banned in many areas because of the difficulty of reducing pollutant emissions from the stack. Larger incinerators (50 tons/day to 3000 tons/day) will normally have a discharge or tipping floor; a refuse pit or horizontal refuse conveyors; usually an over-head crane; one or more furnaces; ash-handling equipment; gas-cleaning equipment; and a stack. Such a plant contains all the dangerous characteristics of an industrial processing plant involving the mechanical handling of extremely varied materials, together with some danger of explosions resulting from intrinsically explosive materials or from a high concentration of plastics. Poisonous gases are given off during the combustion of some plastics, notably from the urethanes and the vinyls, but because furnaces are normally maintained under negative pressure the dangers to personnel should be small.

Shredders

Solid wastes are increasingly being comminuted, the most popular device being a hammermill. Single machines have capacities from one-half ton per hour to fifty tons per hour. No unusual incidence of danger from shredders has been reported, although fires and explosions are occasionally experienced. The robust construction of shredders enables them to contain all but the most violent explosions.

Transfer Stations

In its irreducible minimum, a transfer station is a location where solid wastes are transferred from, usually, the collection vehicles to long-haul vehicles without any additional processing. The long-haul vehicles are usually large-capacity highway trucks, sometimes with trailers; sometimes railroad cars; and occasionally barges. Frequently, processing by compaction, or comminution is included as part of a transfer-station operation. Danger points are those associated with the unit operations as discussed above.

Reclamation Plants

Reclamation is practiced in a crude way at many town dumps by authorized and unauthorized personnel picking over refuse for possible valuable items. This has been long known as an extremely dangerous practice because of the intrinsically unsafe nature of the surface of a dump; because of the difficulty of extracting desired items from their resting position in the dump; and from the danger of being run down by vehicles while they discharge their loads. Dumps and landfills should always be adequately fenced and policed to prevent scavenging and the reasons for these restrictions should be clearly stated.

Some degree of mechanization was introduced into reclamation in the early decades of the present century by loading refuse onto traveling belts which then pass by so-called "picking stations" where individual personnel have responsibility for removing items in particular categories (newspapers, corrugated cardboard bottles and so forth). There are obvious dangers involved, and in present-day conditions it has been found to be difficult to attract any but an extremely transient and poorly educated work force.

A new generation of automated reclamation plants is being developed and in one or two cases such plants are in actual operation. These are generally marked by a lack of human contact with refuse and by the use of sophisticated control systems and protective devices. The dangers involved should therefore be small.

Landfills

Dangers from improper landfilling procedures have been discussed above, and fall into the categories of fires; gas production leading to possible fires, asphyxiation and explosions in buildings; leachate production leading to possible pollution of aquifers; rat and insect infestation, leading to spread of disease; and scavenging leading to a probability of injuries. Of these, the production of gases and leachates, and the spread of underground fires, which can persist for 20 years and more, are the most troublesome. While easy to prevent during landfilling by following one of several approved techniques, subsequent prevention of these conditions after poorer landfilling methods have been used is extremely difficult, if not impossible. Even though, for instance, modern infrared measurements may locate with considerable accuracy the area where an underground fire is smoldering, such a fire is virtually impossible to extinguish with certainty. The cavern left by a smoldering fire may subsequently cause a sudden collapse of the surface, with obvious dangers to people and to any buildings which have been erected without sufficient design precautions. The disposal of sewage sludge brings particular problems, discussed below.

POTENTIAL HEALTH HAZARDS OF SEWAGE-SLUDGE COMPOSTING AND LAND DISPOSAL

Recent findings at the University of Toronto indicate "alarming amounts of cadmium and lead are present in sewage sludge used as fertilizers for food crops. These metals may be assimilated by crops in amounts harmful to human and animal health. Medical evidence shows a correlation between cadmium and hypertensive, cardiovascular, and respiratory problems. Lead can cause brain damage or death."

Recent German research has also found that there has been "no decrease in the number of pathogens in raw sewage over the last few years, and even increased incidence of hepatitis, salmonelloses and bovine cysticercoses has been noticed."

There is much disagreement as to what extent high-quality aerobic composting kills pathogenic bacteria present in the feed material (particularly in sewage sludge, if used, and household refuse). One view, given in a 1974 German report, is that "after the temperature has stayed for one day between 30 and 45° C the temperature may further rise to 70° C . . . after one week all pathogenic germs are killed. . . . " Another German report states: "The temperature of 65 to 70° C at which composting processes usually take place is not sufficiently high to kill all pathogenic germs, viruses, bacteria and bacilli found in garbage and sewage sludge. But studies of the decomposition process in composting plants revealed that microbial substances with antibiotic effects can form which kill even heat-resistance forms of pathogenic germs." These antibiotic substances apparently work effectively after the initial composting process,usually taking 3 to 7 days, during the so-called "curing" process. During this process, com-

TABLE 6.14. Diseases Associated with Human Fecal Waste.

Bacterial infections
Typhoid fever
Paratyphoid fevers
Cholera
Shigellosis (bacillary dysentery)
Viral infections
Poliomyelitis
Coxsackie infection
Infectious hepatitis
(very many other enteric viruses exist)
Protozoal infections
Entamoeba histolytica
Helminthiasis
Fish tapeworm
Beef tapeworm
Pork tapeworm
Pinworm
Roundworm
Whipworm
Hookworm

post is stored in piles or windrows, often in the open, for perhaps 3 months. Formerly this was regarded as a subsidiary process, but with this finding it takes on primary significance.

When sewage sludge is used for field irrigation there is no curing process. Enteroviruses (poliomyelitis, ECHO, Coxsackie) have been found to be taken up by the soil and to contaminate vegetables grown in them (see Tables 6.14 and 6.15).

Parasites can also be spread from sewage. Hookworm infection, prevalent over much of the world, has been reduced to a negligible point in most of Europe and the United States. However, tests reported in 1971 by the E.P.A. showed two of five samples of marketed compost contained helminth ova and larvae. This finding has great significance for the use of such composts on areas where people might walk barefoot. The larvae penetrate the skin, pass through the liver and lodge in lungs. The pulmonary migration may cause symptoms ranging from mild, dry cough to severe dyspnoea, cyanosis, etc. Intestinal ascariasis appears to aggravate malnutrition and may cause dysentery. Hookworm infection causes an anemia syndrome. Strongyloides infection has a mild form characterized by catarrhal enteritis, with mucous secretions, intermediate forms characterized by edematous enteritis, and serious forms characterized by ulcerous enteritis. The VLM (visceral larva migrans) syndrome, like massive trichuriasis and ascariasis, is seen in dirt-eating children aged one to four years. The principal symptoms are eosinophilia, hepatomegaly, hyperglobulinemia, fever and pneumonitis. Larval toxocariasis may cause damage to the eye.

TABLE 6.15. Diseases Associated with Animal Fecal Waste.

Salmonellae
Infection from pig intestinal protozoan (Balantidium coli)
Helminthiasis
Pig ascaris
Animal tapeworms
Hydatid worm

As mentioned earlier, high-energy electron irradiation could overcome many of these problems.

HAZARDS FROM SEA DISPOSAL OF SOLID WASTES

Hard data about the harmful effects of dumping a vast range of solid wastes and sludges at sea are unavailable, except for a few specific cases. These cases, involving DDT, oil, mercury, and bacteria from sewage, alarmed and alerted the public and regulatory agencies, so that the dumping of some hazardous or possibly hazardous materials has been outlawed. For instance, since 1967 radioactive materials may not be dumped in United States waters, and not only the dumping but the use, except in special circumstances, of DDT have been outlawed. The discharge or dumping of other potentially toxic or injurious materials have been reduced by regulations or legislation.

While these regulations do not presently have international recognition, international rules have been proposed and will eventually be adopted. Meanwhile, because the United States is the world's major industrial power, regulation of its hazardous wastes dumped at sea should both accomplish a major reduction in global pollution and should set an example for the international community to follow.

Metals

The principal metals which may offer potential hazards are arsenic, cadmium, chromium, cobalt, copper, iron, lead, manganese, mercury, nickel, silver and zinc.

Mercury discharging from a chemical plant in Japan was taken up by the fish caught locally and extensively used for food. Deaths and serious disablement followed.

Sewage sludge contains metals in a quantity which can be toxic to marine life when the deposits are allowed to build up rather than to disperse. Some of the metals, such as lead and copper, dissolve out of water-supply systems; some, such as iron and lead, come from highway runoff into combined sewer systems; mercury may run into a sewer system after use as a garden fungicide; other metals such as cadmium, silver and zinc are discharged, sometimes illegally, into sewers by industry.

Inorganics

Inorganic materials strictly include the metals. However, the term is more usually taken to include such materials as dredging spoils, sulfuric-acid sludges, glass from domestic refuse and demolition debris.

Organics

Wastes from the synthetic-chemical industry are particularly complex, and new compounds are discovered and join the waste stream faster than they can be tested for long-term hazard-potential. It was the discovery in Berkeley, California that the decline of the brown pelican was due to the effects of DDT on the shells of the eggs, and later that measurable concentrations of DDT could be detected in Antartic penguins, which catalyzed public concern over pesticide use and the discharge of organic chemicals.

Biological Wastes

Three components of the sludge from human sewage cause concern. Disease organisms, particularly infectious hepatitis, may contaminate shellfish which may reinfect people. Metals dissolved from plumbing systems and from other sources may be toxic to marine growth and may concentrate in shellfish and other diet fish (e.g., mercury in swordfish). The nutrients may cause the growth of undesirable organisms.

This third concern, the possibility of excess nutrients being present, is quite different from the problem which arises in inland waters. The ocean is starved of nutrients, and any excess is purely a local phenomenon. Dispersal is therefore an acceptable solution.

SECTION 7
Industrial Fire Protection

N. Irving Sax
Consultant, Industrial Hazards
Boca Raton, Florida

WHAT IS FIRE?

Fire, combustion, or burning require four things: (1) a fuel (any oxidizable material); (2) oxygen (usually air); (3) a certain temperature (heat); and finally (4) free radical reactions. Recent research on the kinetics and chemistry of combustion indicates that the union of oxygen and the fuel is not direct but via a series of steps wherein the actual reactions occur between the oxygen and free radicals emitted by the heated fuel at the point of ignition. These free radical reactions also give rise to the visible flames and the evolution of heat. Certain techniques of fire extinguishment involve interfering with these free radical reactions via the use of various extinguishers or certain dry chemical mixtures. Usual methods of extinguishment involve manipulating the fuel, oxygen, or temperature. If any one of these three constituents is not present in the proper proportions or degree, no fire will occur. If a fire exists and even one of them is sufficiently altered, the fire will go out.

CLASSIFICATION OF FIRES

Class A Fires. These are fires in ordinary combustible materials where the quenching and cooling effects of quantities of water or solutions containing large percentages of water are of first importance. Ordinary combustible materials tend to produce glowing embers after burning, and these must be quenched to prevent rekindling. For extinguishment of such fires, use water (or solutions which are mainly water), foam, multipurpose dry chemical, or loaded stream-type extinguishers, which should be marked by an A against a background which, if colored, shall be green.

Class B Fires. These are fires in flammable liquids (oils, gasoline, solvents, etc.), where a blanketing or smothering effect is essential to put the fire out. This effect keeps oxygen away from the fuel, and can be obtained with CO_2, dry chemical (essentially sodium bicarbonate), foam, a vaporizing liquid type of extinguishing agent, loaded stream, or multipurpose dry chemical. Water is most effective when used as a fine spray or mist. Water spray, CO_2, and dry chemical extinguishers for these types of fires are marked by a B on a background (if colored) of red; multipurpose dry chemical or vaporizing liquid is satisfactory.

Class C Fires. These are fires in energized electrical equipment, where the use of a non-conducting extinguishing agent is essential.

Note: CO_2 extinguishers with a horn are not acceptable. Extinguishers for Class C fires should be marked with a C against (if colored) a blue background.

Class D Fires. These are fires in combustible metals such as Li, Mg, Ti, Zn, Na, and K. Extinguishment of such fires poses a special problem and requires materials and techniques approved for use on local specific combustible metal fires. Since mishandling such a fire can cause explosions and spread of area involved, combustible metals installations should be provided with the proper extinguishment materials, generally dry powder, sand, graphite, etc. Such an extinguisher should be marked by a five-pointed star containing the letter D. If colored, the star shall be yellow.

Definition of Terms

The meanings of fire control terms as used in this book are as follows:

Flash Point (flash p). This is the lowest temperature at which a liquid will give off enough flammable vapor at or near its surface such that in an intimate mixture with air and a spark or flame, it ignites. The flash point of liquids is usually determined by the Standard Method of Test for Flash Point with the Tag Closed Cup (TCC) or (CC) Tester (ASTM D56-52, available from the ASTM, 1916 Race St., Philadelphia, Pa.). This method is also the standard of the American Standards Association. The Interstate Commerce Commission uses the Tag Open Cup (TOC) Tester giving results 5–10°F higher (less flammable). Other methods frequently used are Cleveland Open Cup (COC) and Pensky-Martens (PM). The closed cup flash point value is usually several degrees lower (more flammable) than the open cup, as the test in the former case is made on a saturated vapor-air mixture, whereas in the latter case the vapor has free access to air and thus is slightly less concentrated. For this reason, open cup values more nearly simulate actual conditions.

Fire Point (fire p). This is the lowest temperature at which a mixture of air and vapor continues to burn in an open container when ignited. It is usually above the flash point. Where the flash point is available, only it is given; if it is not, the fire point may be given. It is at least as significant as the flash point as an indication of the fire hazard of a material.

Autoignition Temperature (autoign. temp). This is the temperature at which a material (solid, liquid, or gas) will self-ignite and sustain combustion in the absence of a spark or flame (ASTM Designation D286-36). This value is influenced by the size, shape, and material of the heated surface, the rate of heating (in the case of a solid), and other factors.

Vapor Density (vap d:). This value expresses the ratio of the density of a vapor to the density of air. The vapors of most flammable liquids are heavier than air, thus they can readily flow into low areas, excavations, and similar localities. Hence, ventilating outlets in a plant should be located near ground level. For combustible gases and vapors which are lighter than air, ventilating outlets should be near the ceiling.

Melting Point (mp). This is the temperature at which the solid and liquid forms of a substance exist in equilibrium. This value indicates at what temperature flammable materials that are solid at room temperature may become flammable liquids.

Boiling Point (bp). This is the temperature at which a continuous flow of vapor bubbles occurs in a liquid being heated in an open container. The boiling point may be taken as an indication of the volatility of a material. Thus, in the case of a flammable liquid, bp can be a direct measure of the hazard involved in its use.

Formula. In the event of a lack of information regarding a material, its formula can give a clue to its fire hazard. For instance, all materials composed solely of carbon and hydrogen are combustible and in some degree, flammable. If they are liquids with a low boiling point, they can be assumed to be fire hazards.

Underwriter's Laboratories Classification (ulc). This is a standard classification for grading the relative fire hazard of flammable liquids against the following standards:

Ether class	100
Gasoline class	90–100
Ethyl alcohol class	60–70
Kerosene class	30–40
Paraffin oil class	10–20

Where this value is known it is an excellent measure of the relative fire hazard of a flammable liquid. Unfortunately, it is available in only a few instances.

Susceptibilty to Spontaneous Heating. Many materials combine with atmospheric oxygen at ordinary temperatures and liberate heat. If the heat is evolved faster than it is dissipated due to poor housekeeping, a fire can start, particularly in the presence of easily ignited waste, etc. ["Factory Mutual Modified Mackey Method," Industrial and Engineering Chemistry, (March 1927)].

Explosive Range or Flammability Limits. These values expressed in per cent by volume of fuel vapor in air are the ranges of concentration over which a particular vapor or gas mixture with air will burn when ignited. If a mixture within its explosive range of concentration is ignited, flame propagation will occur. This range will be indicated by *lel* for lower explosive limit or *uel* for upper explosive limit. The values given, unless otherwise indicated, are for normal ambient conditions of temperature and pressure.

Flammable Liquids (NFPA Standard #30)	
Class I	flash p $<100°$ F
Class II	flash p $\geqq 100°$ F but $<140°$ F
Class III	flash p $\geqq 140°$ F

FIRE PROTECTION

The two main aspects of fire protection are prevention and loss limitation.

Prevention

Fire prevention is an inseparable requirement of fire safety. Since, in order for a fire to start, all three necessary constituents—fuel, oxygen, heat—must be represented, effective fire prevention simply boils down to manipulation of these constituents so that a fire cannot start. For instance, where a flammable liquid such as acetone is used out in an open work shop, two of the needed constituents are immediately present, i.e., the fuel and a supply of oxygen. Now the only thing lacking to start a fire is heat. Thus, by referring to acetone in Section 12 it is found that the flash p is 0° F, which means that at any temperature above 0° F, acetone can evolve enough vapor to form a flammable mixture with air which will catch fire if exposed to a spark, flame, or other source of ignition. Thus, strictly from the standpoint of fire prevention in an installation using acetone, the following avenues of protection are open:

(1) The working (ambient) temperature must be kept below 0° F, or

(2) The supply of atmospheric oxygen must be cut off, or

(3) Sources of ignition, such as flames, glowing cigars and cigarettes, sparks, etc., must be eliminated from the area, or

(4) The area must be ventilated so that even though the acetone gives off enough vapor to form a flammable mixture with air, the vapor will be drawn out of the area by means of fume exhaust equipment as rapidly as it is evolved, thus preventing the build-up of dangerous concentrations of vapors.

Naturally, since conditions (1) and (2) above are relatively difficult to attain on an industrial scale, conditions (3) and (4) are the ones most likely to be used.

Furthermore, although total removal of any one of the necessary conditions for a fire will absolutely prevent its occurrence, such stringent restrictions on industrial operations are seldom economically feasible. Industrial materials are, however, studied with a view to ascertaining just how much leeway there is, so that a compromise between absolute fire prevention and economy of operation may be reached. It is for this reason that, while we know how to prevent fires, they still do start, and why loss limitation is such an important part of industrial fire protection.

Below is some discussion of the three essentials of fire.

Oxygen. Although under certain unusual circumstances it is possible to produce combustion-like chemical reactions with materials such as chlorine or sulfur, it is safe to say that nearly all combustion requires the presence of oxygen. Also the higher the concentration of oxygen in an atmosphere, the more rapidly will burning proceed. Industrially, it is difficult to manipulate the oxygen concentration in a working area, particularly since a concentration of oxygen far enough below normal to keep fires from starting would also be too low to support human life.

When industry found it necessary to work with materials so sensitive to oxygen that they would catch fire at ordinary temperatures merely upon being exposed to air, it has found it possible to isolate such materials from air, either in a vacuum chamber or in a chamber filled with an inert atmosphere, such as argon, helium, or nitrogen. In Section 12 the materials which require such isolation are so noted.

Heat. As a necessary component of fire, this is often manipulated to render an industrial set-up safe from fire. The most difficult aspect of controlling the heat component of a fire is the easily overlooked fact that to start a fire it is often necessary to heat to a sufficient degree only a very small quantity of fuel and oxygen mixture. Then, since fires are by definition exothermic, the very small fire started by a tiny heat source supplies to its surroundings more heat than it absorbs, thus enabling it to ignite more fuel and oxygen mixture, and so on, until very quickly there is more heat available than is needed to propagate a large fire. The heat may be provided by various sources of ignition, such as high environmental (ambient) temperatures, hot surfaces, mechanical friction, sparks, or open flame.

A discussion of this aspect of fire prevention should include most sources of ignition.

(1) Open flames. At or near a flammable liquid installation it is necessary to check for such sources as burners, matches, lamps, welding torches, lighting torches, lanterns, small furnaces, and the possibility of broken gas or oil lines becoming flaming torches. Ample isolation may often be obtained by means of partitions. In this respect the partition should be substantial enough to contain the fire, if one starts, while the sprinklers or other fire-fighting apparatus put it out. Fire-resistant construction (brick or concrete walls) is generally recommended.

(2) Electrical sources (electric power supply and generating equipment, heating equipment, and lighting equipment). The provisions of the National Electrical Code (see NFPA No. 70; ASA C1-1965) are the recognized standards and these should be carefully observed in installing electrical equipment in hazardous locations.

(3) Overheating (excessive temperatures at points requiring heat). Such processes should be kept out of combustible buildings and closely supervised. The use of automatic temperature controls and high-temperature limit switches is recommended, although supervision is still important.

(4) Hot surfaces. The incomplete immersion of hot metal in quenching baths, the contact of flammable vapors and hot combustion chambers, hot dryers, ovens, boilers, ducts, and steam lines all are frequent causes of flammable vapor fires. Care should be taken that material whose autoignition temp is lower than the temperature sometimes reached by operating equipment be kept at a safe distance from such equipment. This equipment should be carefully supervised and maintained to prevent accidental overheating, etc.

(5) Spontaneous ignition. Many fires are caused by spontaneous heating of materials, accelerated by external heat from processes such as dryers, ovens, ducts, impregnating or steam lines adjacent to piles of waste materials. Sometimes the accumulated heat in a closed, unventilated warehouse will be sufficient to accelerate oxidation to the point of an actual fire. Wherever flammable liquids are handled, particularly those which are known to be liable to spontaneous heating, it is important to pay particular attention to housekeeping and ventilation. Fires are almost sure to follow neglect of these matters. All equipment and buildings should be kept free of deposits and accumulations of wiping rags, waste materials, oil mops, etc.

(6) Sparks, etc. Sparks from mechanical tools and equipment, hot ashes from smoking, unprotected extension lights, boilers and furnaces, and backfire from gasoline engines are all potential causes of fire. Smoking should be prohibited in areas where flammable liquids are stored or used in the open. All equipment in such areas should be maintained in first class condition.

Wherever possible, sparkproof or nonsparking tools and materials should be used.

(7) Static electricity.[6] This is due to electrical charges generated on the surface of a material by friction, such as calendaring, printing, and the like. Many fires are caused in the rubber and paper industries by this means. Most of these occur during the months when humidity is relatively low, and artificial heat is used. Maintaining a relative humidity of from 40–50% in rooms where flammable liquids are used will greatly reduce the chance of static sparks. Electrical grounding, static discharge devices, etc., should be mandatory, and all flammable liquid tanks, piping, and equipment should be so interconnected and grounded that the chances for static sparks are minimized. In all this type of equipment, belts should be eliminated and direct or chain drives used wherever possible. If belt drives must be used, the belt speed should be kept below 150 ft/min or a special belt dressing should be used which will reduce the possibility of the formation of a static spark.

(8) Friction.[4] Many fires are caused by mechanical (heat producing) friction, e.g., from fan impellers rubbing on casings, poorly lubricated fan bearings, grinding processes, and machining, etc. Fans and other equipment should be frequently inspected and maintained in the best possible condition. Other processes known to generate a good deal of heat due to friction should be well separated from locations where flammable materials are stored or used.

Fuel. Combustion takes place most readily between oxygen and a fuel in its vapor or other finely divided state. Solids are most easily ignited when reduced to powders or vaporized by the application of heat, but except in a few cases the temperatures required for the vaporization of solids are well above normal ambient temperatures.

Liquids present a different case. Some liquids will give off dangerous quantities of flammable vapors well below normal room temperature (the vapor pressure of a liquid is a measure of this effect); others do so at points only slightly above room temperature, and still others at much higher temperatures. It is apparent that the temperature at which a liquid evolves vapors which can form flammable mixtures with air is a measure of its hazard potential. This is indicated by the flash point. These flammable liquids can be categorized as:

(1) High fire hazards—flash p $< 100°$F

(2) Moderate fire hazards—flash p $> 100°$F but $<$ $200°$F

(3) Slight fire hazard—flash p $> 200°$F.

It should be understood, however, that practically all organic materials will burn if exposed to sufficiently high temperatures. The ratings given above are merely an indication of the risk involved in handling or storing them.

It is important to confine the flammable liquid while it is in use. Safety cans should be used for transporting small quantities of flammable liquids about a working area, as well as for storage at a bench. Whenever possible, closed systems should be used to prevent the spread of fumes, etc. In the event of a fire, it is imperative to prevent spreading of the fire. Hence all tanks should have trapped overflow drains leading to a safe place. Dikes must be used to contain the overflow of burning liquid; otherwise fires could easily spread over large areas, trapping personnel, and causing great damage. The principle behind this form of protection is to contain the fire at all costs. Installations of flammable liquids in upper stories should be made only in such a fashion that burning liquid will be prevented from flowing down stairwells, pipe openings, cracks in walls, etc., by means of waterproof floors, dikes, overflow pipes, etc.[11]

A vital point in flammable liquids safety is the prevention of the accumulation of explosive concentrations of vapors in closed-off areas (see Section 2 for a full discussion of ventilation control for toxic vapors). Wherever either moderately or highly flammable liquids are used or stored, ventilation is a very important consideration.

The amount needed, whether natural or mechanical (fans and blowers), depends upon the materials and the conditions involved. No dependence should be placed upon the odor of the material as a warning, because some flammable vapors are heavy and tend to settle and because smell is deceptive. The safe procedure is continual testing with an explosion or flammable vapor indicator.

It is vitally necessary that a complete program for the handling, transfer, and use of flammable liquids be set up and maintained. This program should start when the process is initially under construction. Where flammable liquids are called for in the original write-up of a process, the first question is to determine whether the flammable material can be replaced by a nonflammable one. If the question of cost arises it should be remembered that to the possibly low cost of the flammable liquid should be added the cost of special protection needed to use it safely as well as its effect upon the insurance rate. It may well be that in the final analysis, the cost of a flammable material is not as favorable as it seemed at first. However, there are many flammable materials in constant use for which there are no substitutes, but even they can be safely handled if proper precautions are taken.

Besides flammable liquid fires, the results of which can be somewhat mitigated by effective loss limitation techniques, there are two more types of disaster, protection from which is nearly entirely dependent upon prevention.

Dust explosions. Practically any combustible, when in the form of dust and mixed with air in the proper proportion, will burn so rapidly as to cause a severe explosion if ignited by heat, a spark, or flame. Ignorance of this fact has led to many disasters involving plastics, grain, flour, coal dust, and metal powders all constitute hazards in this regard. Explosions have been known to occur in plants handling fertilizers, wood dust, powdered milk, soap powder, paper dust, cocoa, spices, cork, sulfur, hard rubber dust, leather dust, and many other products. For the prevention of dust explosions good housekeeping is of the utmost importance. All equipment must be made dust-tight and kept so. Explosion vents should lead outdoors to a safe location, and the vent ducts themselves should be strong enough to withstand the force of the explosion. Vacuum cleaning is superior to sweeping. The use of compressed air to blow dust off equipment and thus create dust clouds should be forbidden!

Ledges, exposed piping, beams, etc., in the ceiling should be kept free from accumulation of dust. Where a dusty operation is to be installed in a location where there is piping and projections overhead, it is often erroneously considered satisfactory to install a smooth ceiling below the piping and other projections. This does not eliminate the hazard and may intensify it; unless the ceiling is extremely well designed and installed, dust will penetrate it and settle not only on the piping but on the upper side of the ceiling itself. Then a shock may be sufficient to fill the entire false space with a combustible dust cloud which a spark may set off. If piping cannot be relocated or eliminated, it would be better to leave it exposed and provide for a regular cleaning program.

It was for reasons such as the above that a starch dust plant recently constructed in the southwest, where weather conditions are moderate the year round, has been built entirely of open construction so that there is no confinement of the force of any explosion and the constant flow of air through the plant provides little opportunity for dust layers to build up.

As in the case of flammable liquid fires and explosions, the control of dust explosions is based upon prevention of ignition and secondarily limitation of damage in the event ignition does occur.

To prevent ignition, open flames, smoking and cutting or welding are prohibited until the area is made dust free. Electrical wiring should be of the type suitable for a dusty atmosphere and static electricity, too, must be eliminated. Highly dangerous materials of this sort are handled most satisfactorily in enclosed systems in which suitable inert gases are introduced into the system to replace the air normally present. This precaution is particularly applicable to the field of powder metallurgy. The kind of inert gas used must be chosen on the basis of its suitability for the operation in question.

See as applicable the following codes and standards:

NFPA Nos.	10	80	211
	14	91	622
	31	92	653
	54	93	66
	58	101	657
	60	601	13
	61A	655	656
	61B	77M	77
	61C	69	661
	62	70	220
	63	86A	82
	64	654	
	68	78	
ASA No. 2	12.16		
	12.3		
	12.5		
	12.7		

which cover the hazards of sulfur, aluminum, magnesium, titanium, zinc, pulverized fuel, starch manufacture, grain elevators, flow and feed mills, pulverized sugar and cocoa, coal preparations, plastics, powdered spices, ground confections, woodworking and woodflow plus venting, conveying, and handling of dusty materials.

Salt-bath explosions.[12] The third type of disaster in which after-the-fact protection is much less important than prevention is the molten salt-bath explosion. There have been serious disasters involving such baths, because personnel involved on both the management and the operating level failed to appreciate the potential hazards of the situation. Due to mechanical failure or human failure, or a combination of both, molten salt baths have been allowed to explode. The hazards of molten salt-baths may be summarized as follows:

(1) Violent generation of steam due to water introduced as "carry-over" on a piece of work from a preliminary cleansing or quenching bath, condensation on overhead service piping, leaky roofs, and operation of automatic sprinklers, as well as contact with liquid foods placed on ledges near the baths for "warming-up" by workmen.

(2) Sudden and explosive expansion of air occluded in blow-holes of castings and that trapped in tubes, closed piping, or hollow metal work when immersed in molten baths without pre-warming.

(3) Violent and uncontrollable chemical reactions between nitrate baths, and carbonaceous materials such as oils, soot, graphite, and cyanide carry-over from adjacent carburizing baths.

(4) Vigorous and explosive reaction between overheated nitrate baths and aluminum alloys.

(5) Explosive reaction between normally heated nitrate baths and carelessly introduced magnesium alloys.

(6) Thermit-like reaction between aluminum alloy articles lost in a bath and the iron oxide sludge blanketing and insulating the bottom of bath container.

(7) Structural failure of bath container while in operation under conditions tending to lower the normal durability; reaction between metal of bath container and nitrate due to localized overheating.

(8) Failure of temperature controls, with consequent overheating of nitrate bath.

(9) Storage and handling of bulk supply of sodium nitrate, and careless disposal and storage of waste nitrate without regard to active properties of the salt.

(10) Accidental or uninstructed setting of temperature control above safe operating limits.

The precautions for safe operation of molten salt baths are summarized as follows:

(a) Guard against the introduction of any extraneous matter.

(b) Protect completely from overheating by automatic control and temperature readings taken at regular intervals.

(c) Isolate the operation as far as practicable.

(d) Instruct all personnel thoroughly and completely in regular and emergency procedures.

From the foregoing it can be clearly seen that the handling of flammable liquids, flammable dust or molten salt-baths are the three chief operations in which prevention is the most important phase of fire protection. In each of these cases, whatever action is taken after the fact and whatever physical protection is provided can only furnish some degree of mitigation of loss and it is often insufficient to prevent a large scale disaster.

Loss Limitation[10]

The other aspect of a realistic fire protection program is limitation of loss due to a fire which includes a provision for the prompt discovery and equally prompt extinguishment of the fire (see NFPA No. 27). It is certain that everyone has at one time or another wondered why a particularly destructive fire was allowed to happen, when supposedly a great deal of effort is constantly being devoted to the prevention of such a fire.

It may be that too often the prevention aspect of fire protection has been the total or nearly total effort at protection, with the result that when even a small fire starts it has a good chance to become a calamity. However, even the utmost vigilance would have been of no avail in guarding against some of the fires which are on record, and once the fire has started its cause is immaterial. The cause of the loss is much more important, and the facts which determine that are the physical conditions and those measures which have or have not been taken to limit the extension of the fire.[8]

One of the means of preventing the extension of fire is to segregate hazardous processes and storage into separate buildings. But even where hazardous processes are not involved, the concentration of too much value in one fire area must be guarded against. This is best accomplished by the erection of separate buildings, adequately spaced, which in turn presents problems of maintenance and operation. Suppose a major plant has an operation involving flammable liquids. Such processes as spray painting or dipping use tremendous quantities of flammable solvent; extraction processes and the manufacture of products of which flammable solvents are major constituents are typical examples of processes which require subdivision. The flammable liquid operations and all its appurtenances must be physically separated from the rest of the plant for maximum safety. Care should be taken that the separate buildings are no larger than production efficiency demands; in other words, only as much should be put under one roof as is necessary to be in one building. Where space or production requirements preclude separate buildings, the area in which flammable liquids are handled must be physically separated from the rest of the plant by approved fire walls, or, where these are not practicable, by water-curtain type sprinklers.

Subdivision of one large risk into smaller fire areas may also be accomplished by means of fire walls which stop the spread of fire from one area to another. To accomplish this, the wall must be carried through the roof and either go through the side walls for a distance of at least 36 inches or turn back on both ends for a distance of several feet to provide a barrier around which the fire cannot travel.

Many otherwise sound fire walls have failed because holes were made in the walls to permit the passage of pipes, conduit, etc., and then never properly closed. Every hole in a fire wall must be sealed at the time such work is being done. The weakest point in a fire wall is the fire door provided to permit access from one section to another. At any given time a high percentage of such doors are found upon inspection to be useless as fire barriers. They must be test operated regularly. The chief deficiencies are missing fusible links, damage to the doors by materials handling equipment, which damage would prevent their operation, or blocking by material left in the doorway so that the door cannot close.

Proper maintenance includes regular inspection, physical guarding to prevent damage, the painting of "keep clear" lines on the floor, and a constant program of education. Wherever practicable, such doors should be closed at night to ensure that they will be closed in the event of fire. Where the use or occupancy of the building has changed and fire doors are not needed any longer, the openings should be bricked up to the same thickness as the original walls. Even a pair of fire doors, one on each side of the opening is less resistant to the passage of fire than the fire wall in which they are installed. In normal operation, because of the aisle space leading to the opening, the heat on the door should be less than elsewhere in the building. If the doors are closed and contents are piled against the doors, fire may be transmitted from one side of the wall to the other. In all cases, the doors provided should be of the type approved for the opening in the wall. In many cases, doors approved for use only on vertical enclosures such as stairways are installed in fire walls and will not serve their intended purposes in event of fire. If openings are necessary in fire walls for the passage of conveyors and no type of door is practical, then the openings should be specially protected by hooded automatic sprinkler heads directly over the opening of each side of the wall. A fire wall should be thought of as a dam which any small leak can cause to fail.

The spread of fire from floor to floor in a building is prevented by the proper enclosure of vertical openings such as stairways, elevators, shafts, and process openings through the floor. The question of stairways deserves particular comment because in so many cases, self-closing stairway doors are found to be wedged open to permit easy passage from floor to floor. Such examples of poor management entirely negate the cost of closing the stairway off by providing the doors in the first place.

One of the primary reasons for enclosing stairways is to permit the passage without injury of personnel from upper floors to the street level past the floor which is on fire. If the stairway doors are wedged open this may be impossible. The stairways can immediately become choked with hot air, smoke, and gases from the stairwell.

Provision of fusible link arrangements to close such doors is not very satisfactory because fumes and smoke will pass through without operating the link. In at least one laboratory the problem of stairway doors being wedged open versus the desirability of having them closed in a hurry has been solved by providing for electrical latches. These latches hold all the doors open, but connected to the fire alarm system is a relay which causes all the electric locks to release when a fire alarm is sounded, thus closing the doors.

A further loss-limiting device which is useful where flammable dusts and vapors are used is the explosion vent. It is important to install explosion vents in areas where flammable liquids or dusts are used because of the possibility of great damage due to explosive ignition of such mixtures and air. Therefore, on a practical basis, properly designed explosion vents are a suitable safeguard, as they reduce the chances of destruction indoors by allowing the force of the explosion to be transmitted outdoors.

In order to relieve the pressures produced by explosions in vapor and air mixtures, a vent area as large as 1 square foot for every 10 cubic feet of room volume would be necessary. However, it is unlikely that more than a fraction of the total volume of a room will at any time be within the explosive range. Therefore, for a small room with a floor area of about 200 square feet, the venting area would be at least 1 square foot for each 30 cubic feet of room volume. For larger areas, this proportion may not be obtainable, but in no case should the vent area be less than 1 square foot for each 50 cubic feet of volume.

Approved explosion venting windows are available. Also skylights, roof hatches, or light windows hinged at the top and carefully installed to swing outward under even slight pressure can be useful. Under some conditions, doors equipped with releasing latches may be utilized as vents.[4]

Furthermore, where the conservation of heat in a plant is important, and the walls of the building are otherwise of strong construction, a section of exterior wall may be built of light wood, hollow tile, or some other material which is relatively weak compared to the rest of the building so that in case of an explosion, these sections will give first.

It is important that snow and ice be kept from collecting on explosion vents so that they can operate freely in case of an explosion.

FIRE DISCOVERY AND TURNING IN THE ALARM[7,10]

With the exception of such occurrences as flammable liquid explosions, dust explosions, and molten salt-bath explosions, most fires are quite small at first and can be readily extinguished if discovered early. Although no one would dispute the truth of this statement, it is amazing how many organizations completely ignore the fact and make no provision for the prompt discovery and sounding of an alarm, in the event of a fire.

The greatest single cause of large fires is delayed alarms. This does not necessarily mean delayed discovery; the files are replete with examples of people who discover fires and take entirely the wrong action because they had not been properly trained.

The detection of fire is too often left to chance. If it is left to the chance passer-by it can be assumed that no alarm will be turned in until flames are actually leaping from the front of the building, at which point a heavy fire loss has already taken place. The best solution to this problem is an automatic fire alarm system connected to an office manned by responsible personnel on an around-the-clock basis. In larger installations this may be the plant's own security office; in small installations, the office of a private company specializing in this service. In small cities or rural locations this service can often terminate at the police station or fire house. Such an alarm system, if properly installed and maintained, assures that fires will be discovered and reported promptly in their incipient stages. Automatic sprinkler systems with provision for an alarm can serve the same purpose.

Most automatic fire alarm systems are of two general types; one measures the rate at which the temperature of the air rises; the other sounds an alarm when the temperature reaches a certain predetermined point. Each has its merits for a particular installation, and some types of systems combine both features. Any such system should bear the approval of Factory Mutual or Underwriter's Laboratories.

Under some circumstances, such as the storage of furs or documents, or as in electrical equipment rooms, a slow-burning, smoldering type of fire may be expected. In such a case, neither the fixed temperature nor the temperature-rate-of-rise device may function rapidly enough to transmit an alarm before serious damage has been done. A photoelectric cell device might be used here to trigger the alarm when traces of smoke pass through the light beam. Such a system also starts all necessary action, such as shutting down fans, operating dampers, and transmitting the alarm. This type of protection is a "must" in the modern windowless factory which depends for its ventilation entirely upon a mechanical air-conditioning system into which the introduction of smoke at any point can cause it to be transmitted throughout the area, with consequent panic and possible loss of life.

The most ancient and still most common form of fire protection is the watchman. He is an extremely important employee. In many cases the entire physical property of an organization together with its business future is entrusted to his care over half the time. However, he is rarely chosen with these responsibilities in mind. Just as rarely is an effort made to see to it that the watchman is schooled to react automatically to take proper action if the emergency against which he was employed does occur. Too often the position is used for the semi-retirement of the superannuated employee.

Even when the watchman is hired outright, the pay scale is often so low as to attract only the completely unfit. There are many instances of watchmen for valuable properties who cannot write or read English and in truth can hardly speak it, who do not know how to use the dial telephone, who do not know how to turn in a fire alarm—in short, who do more to present a hazard to a property than to protect it.

There are three considerations in watchman protection which cannot be be overlooked: (a) selection of personnel; (b) training; and (c) supervision. The watchman should be neither too young nor too old. If too young he may lack judgment and in addition is very likely to be bored with the job. If too old he may lack the physical stamina which the job requires. He must be thoroughly and carefully trained in the proper action to be taken in the event of any emergency in the plant.

Particular stress should be laid upon obtaining assistance for the watchman before attempting to fight any fire or deal with any other dangerous emergency. Many concerns use a "watchman daily report form," which also contains an illustrated paragraph of instruction on some phase of the watchman's duties. If the watchman's service is to be performed effectively, regular routes must be laid out and means provided to determine that the watchman is or has been in a required location at the required time. The watchman's clock is the most familiar example of such a device, but it has a serious limitation—no check can be made until the next day.

There is no instantaneous assurance that the rounds are being made in accordance with instructions, or in the case of a single watchman in the plant, there is no provision for investigation should the watchman be injured, become ill, or leaves his post for any other reason.

The best type of protection is that provided by electric watchman-reporting-stations; most often they are incorporated in fire alarm boxes whereby the watchman sends a signal to the supervisory office and indicates his location at the same time. If the watchman does not report at each location within the time specified, an immediate investigation can be made. In a large plant, this service would probably terminate at the plant guard office; in the smaller plant it would be provided by a company specializing in fire protection.

Even when the plant is in operation, the question of sounding the alarm must not be left to chance. The fire protection organization composed of plant personnel should provide for a specific person (with a number of alternates) to turn in the alarm, and emphasis should be laid at all times on the matter of doing this promptly.

No fire department objects to being called to trivial fires. Over 95% of the responses of all fire departments, large and small, are to fires which could be classed as trivial. A high proportion of the other 5% could have be kept in the trivial class had the fire department been called immediately when the fire was discovered. Delayed alarms cause fire disasters (see also NFPA Nos. 101 and 601).

EXTINGUISHMENT OF FIRE

To extinguish a fire it is necessary to remove or sufficiently lower the concentration of only one of the three requirements mentioned previously.

Normally, the removal of fuel is not considered as a practical fire-extinguishing method; however, in such cases as multiple flammable liquid tanks which are interconnected, it is often entirely practicable to pump the flammable contents of a tank which is burning at its surface to another tank, and thus extinguishing the fire by leaving the burning tank nearly empty. In the main, however, fire extinguishment operates by either cooling the burning material below its ignition temperature or cutting off its supply of oxygen. Water, foam, carbon dioxide, bicarbonate of soda (dry chemical), multipurpose dry chemical, loaded stream, and halogenated hydrocarbon comprise the most widely used fire-extinguishing media. Burning metals require special extinguishing media and techniques.

The chief problem is to determine which medium to use and in what manner. For all the media there is at least one method of application varying both in scale and in type of operation, i.e., manual or automatic, for single spot delivery or general coverage.

Water

The oldest, cheapest, and most commonly used fire-extinguishing agent is water. As an example of the use of bulk or "coarse" water, an extremely effective fire extinguisher for small blazes is a pail of water. However, it is inadvisable to use bulk water on a fire involving oils, gasoline, paints, or solvents; since they may be lighter than water, they will tend to be physically dispersed by the impact of a stream. As a result, the burning material will be scattered about the area and spread the fire. Moreover, no blanketing action is possible, as the oil or solvent may float on the water. Water should never be used to extinguish fires of metals which may react violently with it to cause an explosion.

Special fire pails are available for protective purposes; they have rounded bottoms which make them unsuitable for general service. The use of 55-gallon used oil drums full of water equipped with 2 or 3 fire pails is practical for yard storage where the material is not piled too high, as well as for the early stages of construction projects before permanent water supplies are installed. Calcium chloride can be added to the water to prevent freezing during cold weather.

Water is usually most effective when applied in the form of fine droplets or spray. This has a blanketing action, and avoids the difficulty of impact scattering of materials lighter than water. Proper fire extinguishers should always be at hand in areas where flammable solids or liquids are used or stored. Modern sprinkler systems are the most efficient means of supplying such a spray over a wide area.

Fire Extinguishers[3]

A standard 2.5-gallon soda-acid fire extinguisher provides 2.5 gallons of water. When the extinguisher is inverted, a small bottle of sulfuric acid is overturned into the water in which bicarbonate of soda has been dissolved. The pressure of CO_2 gas so formed expels the water. Such an extinguisher requires annual recharge and must be protected from freezing. The soda-acid extinguisher is rapidly being supplanted by a type in which the CO_2 gas is provided by a small, high-pressure cylinder which can be punctured when the extinguisher is put into operation. The water is then driven out by the gas. This type of extinguisher can be winterized by the addition of antifreeze chemicals if desired; in addition, some manufacturers provide an antifreeze additive which also increases the extinguishing effectiveness of the unit. The maintenance required is an annual check on the weight of the CO_2 cylinder.

Automatic Sprinkler Systems

A properly designed, installed, and maintained automatic sprinkler system is by far the most efficient fire protection device yet produced. It is no exaggeration to state that American industry could not exist as it does today without automatic sprinklers. NFPA statistics for over 50 years show that automatic sprinklers extinguished or held in check 96 per cent of the fires in sprinklered areas. Of the failures, about 1 per cent were due to water being shut off in the sprinkler system or the water supply being defective. Another 1% were due to the fact that the system could not get into operation fast enough because the fire started in an unprotected portion of the building and had gained tremendous headway by the time it entered the protected section. Obstructions to distribution of the water from the sprinklers, which in effect rendered at least portions of the building unsprinklered, also caused some failures.

Objection is sometimes made to installing automatic sprinklers for fear that premature operation would cause excessive water damage. The possibility of a sprinkler head misoperating from defects in manufacture is negligible—only 1 chance in 8,400,000 per year. Premature operation from other causes can be easily prevented: from overheating and corrosion by using special sprinkler heads; from mechanical damage by installing wire guards; from freezing by providing antifreeze solutions or by use of dry-pipe systems.

A hose stream delivers the equivalent of a dozen sprinklers, whereas 5 heads or less extinguish 73 per cent of all reported fires where sprinklers operate. Sprinkler piping is tested for a pressure of at least 200 lb/in^2 before the system is accepted. No such tests apply to ordinary water piping, and the possibility of damage from a burst water pipe is therefore much greater in the case of ordinary domestic service. In addition, no sprinkler system is complete without an alarm device which notifies responsible authorities when any water flows in the system—a feature entirely lacking in domestic and process piping.

An automatic sprinkler system is essentially an arrangement of pipes throughout the protected area provided at suitable intervals with sprinkler heads equipped with fusible (wood's metal) links to release the water when the temperature at the head reaches the desired point.

Wet-type automatic sprinkler systems consist of such an array of piping together with an alarm check valve which prevents back flow from the system to the water supply and provides a means whereby a local central station alarm can be received in the event that there is a flow of water in the system due to a fire or accident.

In a low building where water pressure and volume are good, the water supply may come solely from the street main. In other cases, it may be necessary to provide an elevated supply of water so that the water will be automatically available under gravity pressure. A fire department connection is provided so that the fire department can pump into the system and increase the volume of water available in the event of a serious fire. The maintenance is simple but the utmost care must be taken to make sure that there is water in the system at all times.

The equipment is designed to facilitate inspection. The outside valve leading from the street main is almost invariably a post indicator valve, i.e., a valve brought up above the surface of the ground with an indicator as part of the valve, which shows whether the valve is open or shut. Interior valves are of the outside stem and yoke type; i.e., the valve stem protrudes through the valve yoke and is visible for its full length when the valve is open. If the stem cannot be seen the valve is closed.

Seals and padlock arrangements are often used to prevent unauthorized tampering with sprinkler valves. A tag system including notification of responsible authority should be used in connection with routine work to make sure that the valves are reopened when the work is completed. If extensive alterations are necessary, they should be carefully planned so that the maximum portion of the system can be plugged off and so be kept in operation until the alterations and additions are ready to be tied back into the main system. Every effort should be made to put the system back in service to at least give partial coverage during nights and over weekends during the period of repair. There are many instances of fires getting headway in plants which had an excellent sprinkler system which was out of service for repairs at the critical moment.

The ultimate in protection of sprinkler valves against tampering, whether accidental or malicious, is an electrical alarm device, attached to the valve, which will transmit a signal to a central headquarters when the valve is closed.

When alteration or new construction is planned for a sprinklered building, a part of the alteration cost must be provision for the necessary changes, extensions, or additions to the sprinkler system. An unsprinklered mezzanine constructed outside can become so heavily involved in a fire that by the time the fire reaches the sprinklered area, the system is overpowered.

In areas subject to freezing, a dry-pipe automatic sprinkler system may be installed. In such a system, air under pressure in the piping keeps the water from entering the unheated area through the dry pipe valve where a differential type of valve seat enables a relatively low air pressure to hold back a much higher pressure of water. Such a system requires a dependable supply of compressed air to make up any losses in air pressure due to air leakage in the system. If the air pressure fails, the system "goes wet," and an alarm will be received. In nonfreezing weather the system may be allowed to remain wet for a short period of time before it is drained and the dry-pipe system restored. A word of caution, however: once the system has "gone wet," no further alarm will be received from a flow of water due to the operation of a sprinkler head. During freezing weather, of course, immediate action must be taken to drain the system and restore the valve to prevent freezing. Large dry-pipe systems are provided with exhaust valves to vent the air in the system quickly and thus speed up the water discharge.

When settling of a building occurs, the pitch of the piping must be carefully checked to see that all sections of the system will drain back to the valve and that no pockets are created where water may remain and cause the pipe to burst during freezing weather. A dry-pipe system necessarily operates more slowly than a wet-pipe system. In the event that the situation changes and a permanent heating system is installed in the building, the sprinkler system should be converted to a wet-pipe system. Another practice that should be guarded against is the "pumping up" of the air pressure on the sprinkler system to an excessively high point to prevent the accidental tripping of the dry-pipe valve because of leakage of air. By filling this system with many times the normal volume of air, a situation is created where in any emergency it may take several minutes for air to exhaust itself from the system and finally permit the flow of water from a sprinkler head. The proper way to correct air leakage is to inspect the piping, to locate and eliminate the air leaks.

Modern warehousing may present difficult new problems, such as the high piling of stock now permitted by means of mechanical handling equipment; the palletizing of material providing for horizontal and vertical spaces through the pile; and the storage of such materials as rubber tires.

Most sprinkler heads now in use project the water upward in a very fine spray; the spray patterns from the various sprinkler heads join just below the ceiling to provide a barrier against the transmission of heat to the building proper. It is highly effective as a heat absorber, which is, after all, the primary purpose for using water in fire extinguishment. These new heads are by no means a cure-all, particularly for an inherently unsatisfactory situation, but they will definitely provide a higher degree of protection than has heretofore been possible in the type of occupancies discussed above (NFPA No. 13).

Fire Hoses

Since the use of water to extinguish uncontrolled fires is very important, fire hoses—the means by which most of the water used to extinguish uncontrolled fires is brought to them and then applied—require some discussion.

Hose lines fall into interior and exterior categories. By tradition hose is defined by its trade diameter, proof pressure (600–800 psi), jacketing, and any special construction features. Thus 1″ to 3.5″ diameters are commonly used. Rubber hose of 1.5″ diameter with rubber-covered woven jacketing is often used for interior protection to minimize damage; the larger sizes are commonly used on exteriors. Jacketing can be single or double depending upon use. And depending upon usage and materials of construction, a program of testing the hoses must be set up and adhered to. For full discussion of hoses and applicable standards see NFPA Nos. 196, 197, 198, 14.

When the type of hazard demands heavy-duty outside fire protection and the organization of a thoroughly trained brigade (see NFPA No. 27), serious consideration should be given to the provision of fewer strategically located hose reels and the elimination of hose houses. For the cost of several hose houses a lightweight truck can be purchased and readily adapted as a hose and equipment carrier, thus providing greater flexibility and ease of maintenance at lower cost.

Foam

Another important type of fire-extinguishment material is foam. Most foam installations are small enough to be portable and very effective on fires covering a relatively small area. There are a few installations, however, which have installed permanent piping to bring foam to bear on relatively large areas of acute fire hazard. Foam, sometimes referred to as alcohol foam, is particularly important in the case of liquids which are lighter than water and which would float on the surface of the water and spread out, thereby carrying the fire to points beyond which it originated. Chemically this foam is often produced by bubbling CO_2 or some other gas through a liquid containing a foam producing chemical. The hard resilient bubbles of foam thus formed coat a burning surface with what is essentially the gas used to bubble through the liquid in the first place, and if this gas is carbon dioxide it effectively smothers the fire by preventing access of oxygen.

Other Agents

Still other techniques for fire extinguishment are the use of CO_2, dry chemical, and halogenated HC such as carbon tetrachloride or methyl bromide. Here a portable extinguisher of CO_2 or dry chemical type is applicable to relatively small fires. Large-scale protection would require permanently piped CO_2 systems so that enough gas can be brought to bear on the burning mass to smother it effectively. As to the use of halogenated HC, it must be remembered that when the most commonly used members of this group are heated to high temperatures they decompose and evolve highly toxic decomposition products. In the case of carbon tetrachloride, one of the decomposition products is the well-known poison, phosgene. However, in the case of a fire where the ventilation is relatively good, it may be advisable to put up with the comparatively small amount of phosgene formed in order to put out the fire quickly and effectively. An ideal example of where not to use halogenated HC types of fire extinguishers is in public conveyances where people might be trapped, such as a subway train without adequate ventilation. The use of halogenated HC fire extinguishers should be restricted to relatively small fires in well-ventilated areas.

Gas Fires

The extinguishment of gas fires may be carried out with CO_2, dry chemical, or in some cases, with water supply. But it must be borne in mind that if the gas fire is put out and the gas supply is not shut off, the gas will continue to pour out, ultimately filling a large room or building with a flammable or even explosive mixture, which upon ignition could cause more damage than the original burning gas. The best method of handling gas fires is to stop the flow of gas. If this cannot be done at once, it is often wise to simply spray the surrounding area and surfaces with water to keep them cool and thus prevent their ignition until the supply of gas is exhausted or a valve controlling the flow can be closed.

Burning Metals

Fires caused by burning metals are very difficult to extinguish and cannot be handled in the ordinary manner. For instance, to spray water upon burning metal might cause an explosion which would spatter flaming particles of metal to great distances. Usually the best way to attack burning metal is with specially formulated dry-type fire-extinguishing agents. For instance, when one is planning to use a metal in a form in which it might readily become ignited, it is wise to discuss the situation with the manufacturer or supplier of this metal and obtain from him explicit instructions for the storage, handling, and fire extinguishment of the metal. This is particularly applicable to the use of sodium, potassium, lithium, zinc, uranium, thorium, and magnesium. For instance, it has been found that ordinary sand, even when dry, is a very poor material for extinguishing metal fires; it may react with the hot metal and add more heat to an already intense fire. Often salt (sodium chloride), sodium bicarbonate, graphite, magnesium carbonate, magnesium oxide, or mixtures of all of these materials have been found effective; in every case the supplier of this material will know how it must be handled. Water should never be applied to burning metals (see NFPA Nos. 48, 481, 482M, 651, 652).

INTANGIBLE LOSSES DUE TO FIRE

While ordinary fire insurance may pay the dollar losses incurred in a fire, and insurance coverage may also be provided for the loss of profit during the shutdown period, what protection is there against the loss of market and the fact that customers unable to be supplied may turn to a competitor?

When it seems that all the normal precautions have been taken and that there is a generally good level of protection throughout a plant, it is necessary to consider the entire operation and determine if there are bottlenecks where a relatively minor loss could result in a severe one by a consequent shutdown in production. This may be within the organization or elsewhere in the production chain. Power supply, for instance, is an extremely critical item. Inquiry should be made of the local utility to determine just how dependable it is. Should the local utility have a serious breakdown in its own facilities, arrangements should be made for possible connections to a utility grid in order to continue an uninterrupted flow of power.

Many areas experience periodic shortages of natural gas, during which industrial organizations must conserve its use in order that more may be available for home heating. If such fuel is used, arrangements should be made for the emergency use of liquefied petroleum gas. Safety should be provided for in advance; if precautions are improvised at the last minute, unnecessary risks may be taken. As to utility arrangements, transformers and other substation equipment should be adequately guarded against damage or destruction from fire in the plant.

Another very important source of serious loss results from the destruction of records. The old iron safe in the treasurer's office does not represent adequate protection for the vital records it contains, should there be a destructive fire. The Underwriter's Laboratories and the Safe Manufacturers Association test and label safes and cabinets for their fire resistance. Very often we find that while some documents easily recognized as vital are well protected, those considered to be just ordinary "business records" are housed in sheet metal filing cabinets which are no more protection than wood in the event of a severe fire. The loss of Accounts Receivable records can be disastrous, and other records, even though not directly concerned with the collection of payments and accounts, may require reconstruction in the event of loss. Also, an accurate proof of loss is necessary for the prompt and proper payment of insurance claims. If the records supporting such a proof of loss have been destroyed, it may be necessary to resort to litigation or accept a lesser figure because of lack of proof.

Engineering drawings of the plant and its equipment may also be priceless, especially when they relate to underground or otherwise concealed installations.

Provision of fire-safe record cabinets is in itself not enough; precaution must be taken to close the files and record containers in the event of a fire. Normally, in case of fire, employees would immediately evacuate the area without any attention to this important detail, which would render the entire installed protection useless during the working day. The production chain should further be carefully surveyed for vital bottlenecks which might be subjected to fire loss due to the type of operation or to their exposure to other fire-producing potentials. The protection provided should be thoroughly considered to determine whether or not it is adequate under the circumstances. If adequate protection for one reason of another is unduly costly, a study should be undertaken to determine where else the work could be carried on if the operations were wiped out. In some cases it may be cheaper to assume the risk, because in the event of a fire the work could be contracted out to other organizations in the vicinity with similar facilities. Or it may be found that such an arrangement would be impractical and that the loss of the spray painting and detearing steps, for example, would completely shut down production; continuity of production might then best be protected by rearranging the area and subdividing it with fire walls so that only a portion of the facilities would be destroyed at any time.

Having examined the production bottlenecks in a plant and determined the best way with which each one must be dealt to prevent it from shutting down production, it is necessary to examine the situation with respect to suppliers. It is quite a shock to receive word that the XYZ Corporation has burned to the ground, and then realize that it was the sole supplier of an important subassembly of one's production and that there is only a single week's inventory on hand. When subcontracts for any portion of production are undertaken, it is necessary to consider the subcontractor's plant in the same light as one's own. There is no reason why this matter cannot be brought out during the contract negotiations, but even prior to this time, a discreet survey of the situation can be made in connection with the ordinary inspection of his facilities. If the situation looks dangerous, the least precaution that should be taken is to divide up the subcontract and allow a portion of it to another contractor, thus spreading the risk. Or one might advance his delivery schedule to provide himself with a larger inventory.

Then there is the question of storage. Many industrial organizations produce the year round at a relatively steady rate to meet the requirements of a seasonal market. This requires the utilization of storage space often not available at one's own plant. And in too many cases storage space is leased with little or no inquiry into the fire safety of the situation, with the result that months of production effort can go up in smoke and a market may be lost forever. Here again, the basic principles of subdivision should be employed to the fullest extent practicable, even though this may involve some inconvenience and expense (see NFPA Nos. 231, 213A, 232).

STORAGE AND HANDLING OF HAZARDOUS MATERIALS[5,13,2,12]

Due to the wide range of materials which have to be stored in factories, laboratories and warehouses, good storage practice has become very important and is steadily becoming more so. Economically, it is probably not feasible to store each item of inventory in an environment which is ideally safe. It is therefore necessary to group items together for storage in such a way that the best use is made of available space. To effect a compromise between safety and the competitive demands of an industrial economy, a source of information is necessary. The description of each material listed in Section 12, for instance, contains as much of this information as was available, judged to be reliable and useful for the purposes of this book.

The disastrous effects of neglecting the physical and chemical properties of stored materials are fires, explosions, emission of toxic gases, vapors, dusts, and radiation, and various combinations of these effects.

In order to make it possible to discuss most of the materials with which we are concerned, we shall group the many thousands of individual items into categories as follows:

(1) Explosives

This classification includes materials which under certain conditions of temperature, shock, or chemical action can decompose rapidly to evolve either large volumes of gas or so much heat that the surrounding air is forced to expand very rapidly; in either case, an explosion results. These materials are particularly dangerous when involved in shock or heat-producing disaster conditions. Appendix 1 to this section discusses in detail the storage, handling, and disposal of explosives.

(2) Flammable Materials

Practically all combusition takes place between oxygen and a fuel in its vapor or some other finely divided state. Thus, in order to start a solid to burning it is necessary to heat at least a portion of it to a point where enough vapor is evolved to catch fire. Since at a given temperature liquids are generally at a higher vapor pressure than solids, flammable liquids are usually very easy to ignite. Flammable dusts are about as easy to ignite as vapors or gases.

A further, quantitative difference exists between propagation rates of solid, liquid, and gas fires. Fires of solids propagate slowly, fires of liquids propagate relatively rapidly, and gas, vapor, or dust fires propagate so rapidly they often seem to explode.

Section 12 lists, in alphabetical order, thousands of materials with varying degrees of flammability. In each case where sufficiently reliable data are available the degree of fire hazard of the material is indicated.

Materials which ignite easily under normal industrial conditions are considered to be dangerous fire hazards, e.g., pyrophoric substances such as finely divided metals, hydrides of B, P, liquids with flash points of 100°F or lower, and flammable gases. Such materials must be stored in places that are cool enough to prevent accidental ignition in the event that vapors of the fuel mix with the air. It is important to provide adequate ventilation in the storage space, so that normal leakage of such vapors from containers will be diluted enough to prevent a spark from igniting them. A further precaution is to locate the storage area well away from areas of fire hazard, for example, where torch cutting of metals is to be performed, etc. Highly flammable, i.e., reducing or easily oxidized materials must be kept apart from powerful oxidizing agents, materials which are susceptible to spontaneous heating, explosives, or materials which react with air or moisture to evolve heat.

Ample fire-fighting equipment, either automatic or manual, must be readily available for emergency use. The storage area must be posted to prevent smoking or striking of matches. Bare filament heaters or other sources of ignition must be kept away. The area must be electrically grounded and periodically inspected or equipped with automatic smoke or fire detection equipment.

All combustible materials are dangerous in a fire and/or explosion, both because heat will cause them to catch fire and add to the smoke and fume hazard, and because small explosions have been known to scatter clouds of flammable dusts which in turn have caused still greater explosions than the one which gave rise to them (see Appendix 3 to this section).

(3) Oxidizing Agents

Oxidizing agents are sources of oxygen, one of the necessary components of a fire. Normal air with its 20% O_2 content is the primary source. However, many other materials are so constituted chemically that they can supply oxygen to a reaction, even in the absence of air. Some of these oxygen suppliers require heat before they will yield oxygen; others evolve significant amounts of it at room temperature. When containers of oxidizing materials are damaged, for example, by the forces unleashed during a disaster, their contents can mix with the contents of other damaged containers and possibly start fires and explosions. This is a strong argument in favor of separate storage over mixed storage. The following classes of compounds are noted for their ability to supply oxygen: organic and inorganic peroxides, oxides, permanganates, perrhenates, chlorates, perchlorates, persulfates, organic and inorganic nitrates, iodates, periodates, bromates, perselenates, perbromates, chromates, dichromates, ozone and perborates.

In Section 12 each material which is an oxidizing agent is so rated. It is important to know this when planning storage.

In general, it is unsafe to store oxidizers close to liquids of low flash point. In fact, even slightly flammable materials should be isolated from oxidizers. For instance, ordinary glycerol and potassium permanganate, when simply blended at room temperature for a few minutes, react violently to produce a hot fire.

As a general rule, it is wise to keep all flammables away from an area where oxidizing agents are stored. This storage area should be kept cool and ventilated, and should be fireproof. Normal fire fighting equipment is less useful here, since the blanketing or smothering effect of fire extinguishers is less effective because the oxidizers supply their own oxygen. With these materials it is best to keep fuel away.

(4) Water-Sensitive Fire and Explosion Hazards

These are materials which react with water, steam, or water solutions to evolve heat, flammable gases, or explosive gases. Examples of such materials are: lithium, sodium, potassium, calcium, cobalt, rubidium alloys and amalgams of the above, hydrides, nitrides, sulfides, carbides, borides, silicides, tellurides, selenides, arsenides, phosphides, acid anhydrides, and concentrated acids or alkalies.

Of the above list, the first eight are materials which react exothermally with moisture and evolve hydrogen (see Section 12); the next nine (nitrides through phosphides) react rapidly with moisture to evolve volatile, flammable, sometimes spontaneously flammable and/or explosive hydrides. Acid anhydrides and concentrated acids or alkalies react with moisture to evolve heat. Such materials must be stored in well-ventilated, cool, dry areas. Because many of these materials are also flammable, it is essential that *no* automatic sprinkler system be used in a storage area which houses them; in fact, such an area should have no water coming to it at all. Heating may be electrical or with hot, dry air. The building must be waterproof, located on high ground, and separated from other storage. This building should conform to that required for storage of hydrogen.

Particular attention must be paid to the following: pocketing of light gases under the roof, introduction of sources of ignition, periodic inspection, automatic detection, and alarm due to dangerous concentrations of flammable gases.

Since many disasters cause wetting of extensive areas either by flooding (if that is the nature of the disaster), by breaking of water pipes and sprinkler systems, or possibly by damaging the warehouse so that rain can get in, it is important that the items listed above receive special storage consideration for protection against disasters (see Appendix 3 to this section).

(5) Fire and Explosion Hazards of Acid and Acid Fume Sensitive Materials

These are materials which react with acid and acid fumes to evolve heat, hydrogen, and flammable and/or explosive gases. Examples are: Li through phosphides in the foregoing list, concentrated alkalies, metals (including structural alloys), As, Se, Te, and cyanides. Therefore, it follows that acids should not be stored in close proximity to these materials. If, however, such storage is contemplated, the following precautions are called for: the area must be kept cool; it must be ventilated and periodically inspected; sources of ignition must be kept away; construction as for hydrogen storage is required (see NFPA No. 567).

In fact, keeping in mind that acid and/or acid fumes can attack structural alloys and evolve hydrogen, acid storage might as well be in ventilated wooden sheds. Or if metal is used in construction, it should be painted or otherwise rendered immune to attack by acid. If hydrogen can be evolved, the building should be so constructed that possible hydrogen pockets are eliminated.

Disaster conditions of high temperatures, much vibration, and/or flooding can bring acids or acid fumes into contact with Li, Na, K, Ca, and Rb; unless a good deal of thought has gone into the problem of storing such materials far enough apart to avoid this, fires and explosions may result (see Appendix 3 to this section).

(6) Compressed Gases

Tanks of compressed gases should be stored upright and chained or otherwise securely attached to some substantial support to minimize the chance of falling over and breaking or straining the valve or other part of the tank. The tank storage area should be kept cool, out of the direct rays of the sun, away from hot pipes, in a ventilated area where care has been taken in construction so that pocketing of gases which may escape from their containers will be kept to a minimum.

The building which houses the tanks should be fireproof. It should provide some means (such as sprinkler system) of keeping the tanks cool in case of external or internal fire. The reason for concern about overheating is that this will raise the pressure of the gas in the tank to where the safety disk or a valve might rupture, thus releasing a significant volume of, say, hydrogen or carbon monoxide which might then catch fire and possibly explode.

While it is true that today the handling of compressed air tanks is relatively safe, due in great part to engineering improvements and standardization of components, certain precautions are still advisable. Aside from upright, cool storage, care must be taken to keep from damaging tanks in handling; valves must be operated carefully and kept in good condition. Avoid drastic changes in temperature; do not hammer valve cocks; keep valve cover on whenever possible; do not interchange reduction gauges from one gas to another without checking for compatibility; discourage tampering with tanks in any way; put tank storage in charge of personnel competent to handle it. For further information on compressed gas tank storage consult the Compressed Gas Association of New York City, your gas supplier, or your local fire department.

Disaster conditions including high ambient temperatures due to nearby fires, shock and vibration due to nearby explosions, etc., are particularly trying for compressed gas storage. Great care should be exercised in the storage of large quantities of compressed gases, particularly in areas of high population density (see Appendix 3 to this section).

(7) Toxic Hazards

These are materials which under either normal conditions or disaster conditions, or both, can be dangerous to living things around them. Thus carbon tetrachloride, for example, if stored in a poorly ventilated place under standard conditions, can evolve enough vapor to render the storage area toxic. Under disaster conditions of high temperature, if carbon tetrachloride is decomposed it can form significant quantities of the highly toxic phosgene.

Since it is nearly impossible to seal containers perfectly, it is always to be expected that some of whatever volatile materials are stored will escape into the atmosphere of the storage area. Likewise, air, atmospheric moisture, and carbon dioxide will come into contact with the contents of imperfectly sealed containers. Some initially well-sealed containers will build up enough internal pressure to break a seal or even burst the sealed unit.

Materials which are toxic because of their radioactivity are also included in this category. Such materials are dangrous if allowed to become airborne, or to be ingested or inhaled into the body. Thus, under normal storage conditions, small amounts of radioactive materials may escape their containers and contaminate the atmosphere; disaster conditions such as high temperatures, severe shocks, floods, or combinations of these can burst containers and volatilize or scatter by the force of an explosion or spread about, by means of flood waters, much larger quantities of radioactive materials. Furthermore, it must be borne in mind that sources of radiation enclosed in lead shielding can become dangerous in a disaster because conditions can be such as to melt the shield or volatilize a radioactive material (see also Sections 5 and 8, and Appendix 3 to this section).

In general, materials which are toxic as stored or which can decompose into toxic components due to contact with heat, moisture, acids, or acid fumes, should be stored in a cool, well-ventilated place, out of the direct rays of the sun, away from areas of high fire hazard, and should be periodically inspected and monitored. Incompatible materials should be isolated from each other.

(8) Corrosive Materials

Corrosive materials include acids, acid anhydrides, and alkalies. Such materials often destroy their containers and get into the atmosphere of a storage area; some are volatile, others react violently with moisture. Acid fumes react to evolve toxic fumes with sulfides, sulfites, cyanides, arsenides, tellurides, phosphides, borides, silicides carbides, flourides, selenides; they liberate hydrogen upon contact with metals and hydrides. Alkalies may liberate hydrogen upon contact with aluminum, etc. Acid mists or fumes corrode structural materials and equipment and are toxic to personnel. Such materials should be kept cool, but well above freezing points. Acetic acid, for instance, can freeze in an unheated room and, due to differential expansion, crack a glass container. There should be sufficient ventilation to prevent accumulation of fumes; there must also be regular inspection. Containers of corrosive materials should be carefully handled, kept closed, and labeled. All exposed metal in the vicinity of such storage should be painted and checked for weakening by corrosion. Corrosive materials should be isolated from materials noted above (cyanides, sulfides, etc.), reaction with which can produce highly toxic fumes.

It should be added that strong acids and alkalies will cause serious burns and eye damage to personnel, and hence adequate protection in the form of gloves, aprons, goggles, etc., should be worn when handling them into and out of storage areas (see Section 2).

Some examples of materials which warrant special care in storage are: alkali metals, Ca, acid anhydrides, concentrated acids, concentrated alkalies, As, As compounds, Be, Be compounds, borides, phosphides, P, silicides, cyanides, nitrides, nitrates, nitrites, S, sulfides, Te, Te compounds, Se, Se compounds, carbides, halogenated HC, Hg, Hg compounds, Pb, Pb compunds, Cd, Cd compounds, etc.

Storage of significant quantities of any of the above materials should be brought to the attention of the safety director of the plant owning the material, the local fire department, and the local Civil Defense headquarters.

The term "significant quantity" is difficult to define. In an unventilated room even one pound is significant, while in well-ventilated storage areas it may take as much as 100–1000 pounds to be significant, particularly to a safety director and a fire department. Ton lots and over are of concern to Civil Defense authorities, as well as to fire departments and safety directors. In Section 12 of this book, the disaster hazard potential of many materials is discussed. Often indicated are the most toxic capabilities of the materials and the disaster conditions which might release them. Naturally, the decomposition of complex materials yields a number of products, and it is beyond the scope of this book to list them all. However, when at least one toxic material is known to be formed, it is listed. Often the product listed is not the one formed in the greatest abundance. An effort has been made to list at least one product of toxicological significance.

Appendix 3 to this section discusses in some detail the the problem of planning in advance for the time when a combination of circumstances approximating a disaster might occur in an industrial installation.

A disaster can occur in a plant from without, due to flood, fire, bombing, or earthquake, or from within, due to fire or explosion.

Whatever the cause, the effcts are much the same and the information in Appendix 3 will be found to apply.

HANDLING OF TOXIC AND CORROSIVE MATERIALS

When loading, transporting, or unloading such materials, HANDLE CONTAINERS CAREFULLY! Damaged drums may leak and, in an enclosed space may cause dangerous concentrations of vapors or dusts.

IF A LEAKY CONTAINER IS DISCOVERED, put on protective clothing (see below). After thoroughly airing out the space, enter and turn the leaking container on its side or end to stop the leak. Block or rope off the contaminated area and post with DANGER signs. Then report the leak immediately to your safety department or to the supplier. Wire or phone and report the following information:

(1) What is leaking (see drum label)?
(2) Where is the leaky container now located?
(3) Where can you be reached for instructions (gave phone number or address for telegram)?
(4) Has anything or anybody been contaminated by the spilled material? If any persons are contaminated, call a local doctor.

Finally, be sure that the affected area will be properly guarded in your absence by a responsible person until necesary instructions are received.

Before you open a container or handle a toxic or corrosive material in any way PUT ON COMPLETE PROTECTIVE CLOTHING. Because your eyes could be harmed, you should wear goggles (or a plastic face shield). Since you could be poisoned by breathing vapors or getting poison into your mouth, you should wear an effective respirator. Since many poisons can be absorbed through the skin, you should wear a clean cap, clean rubber gloves (rubber can absorb dangerous amounts of poison after prolonged use), rubber boots, and coveralls which enclose your neck.

Drum Handling

HANDLE DRUMS CAREFULLY! Don't cause leaks. Store and use drums in a cool, well-ventilated area. Do not store drums near steam pipes, boilers, or other sources of heat.

To Open Drums. Set the drum with the bung end up under a ventilating hood. Unscrew the bung slowly to gradually release any internal pressure. Then fit the valve in place of the top bung. Be sure it is tight. Now tip the drum on its side with the side bung up and fit another valve in place. Be sure that both connections are leakproof. Support the drum in its cradle with one valve uppermost.

To Empty Drum. Connect the upper valve to an open-air vent pipe. Connect the other valve to a closed system storage or processing unit. Watch for leaky connections and correct immediately if found. IF LIQUID SPILLS, absorb spillage with dry clay or sawdust. Bury or burn these sweepings. (Avoid the smoke. It may carry poisonous vapors.) Scrub the spill area with soda ash and soap. Flush this thoroughly and repeat the operation as often as necessary to destroy or remove as much of the poison as possible. Don't track spilled poisons around.

To Discard Empty Drums Safely. NEVER USE CONTAMINATED DRUMS FOR ANYTHING ELSE, no matter how clean they seem to be. Empty drums must be decontaminated and disposed of as follows: (1) Thoroughly scrub out the open drum with hot water and soda ash or 5% caustic solution; (2) flush and wash again until as much as possible of the poison has been removed; (3) to be sure drums are not reused, perforate sides, top, and bottom before discarding; (4) for greater safety, burn out the perforated drum (avoid the smoke) before finally dumping it in a safe area. Rain water may rinse traces of poison out of a drum onto soil or into drainage streams where the poison could eventually reach animals, fish, or even humans. Take no chances of this in drum disposal. Perform all operations carefully and with thoroughness.

APPENDIX 1

Explosives are materials, chemical compounds, or devices, whose primary usefulness is to function by explosion. That is, by the rapid release of gases or energy. Relatively great, if short-lived, pressures are brought to bear upon the surroundings.

There are many materials, mixtures, compounds, or devices which can be made to explode, (see Section 12 under disaster conditions), but in this appendix, we shall discuss materials which are designed and formulated so that they can be made to explode when and with the force desired, safe to store and transport and safe to use from the standpoint of toxicity of decomposition products. Nuclear explosives, explosive dusts, vapor, and gases are specifically excluded.

Commercial explosives which are the subject of these paragraphs fall into four classes:

Class A Explosives. Also called "high" explosive or detonating explosives and includes dynamite, nitroglycerine, picric acid, lead azide, mercury fulminate, black powder, blasting caps, and detonating primes.

Class B Explosives. These materials are flammable and include such products as propellant explosives, photographic flash powders, and some special fireworks.

Class C Explosives. These include manufactured combinations of Class A and Class B explosives in restricted concentrations and quantities.

Forbidden Explosives are those not acceptable for transportation by regulations of DOT, including but not limited to:

Liquid nitroglycerin.

Dynamite (except gelatin dynamite) containing $>$ 60% of liquid explosive ingredient. Dynamite having an unsatisfactory absorbent or one that permits leakage of a liquid explosive ingredient under any conditions liable to exist during storage.

Nitrocellulose in a dry and uncompressed condition in quantity greater than 10 lb net weight in one package.

Fulminate of mercury in a dry condition and fulminate of all other metals in any condition except as a component of manufactured articles not hereinafter forbidden.

Explosive compositions that ignite spontaneously or undergo marked decomposition rendering the products or their use more hazardous when subjected for 48 consecutive hours or less to a temperature of 167°F (75°C).

Explosives containing an ammonium salt and a chlorate.

New explosives until approved by DOT, except that a permit may be granted for transportation and possession for laboratory examination of such explosives when under development by responsible research organizations.

Explosives not packed or marked in accordance with the requirements of DOT.

Explosives condemned by DOT.

Explosives are shock sensitive to varying degrees (see Explosives, High, in Section 12, for exact data). Explosive manufacturers as well as other organizations have issued numerous recommendations as to safe and effective storage. The Institute of Makers of Explosives has published a pamphlet "Standard Storage Magazines Recommended by the Institute of Makers of Explosives" and has also issued the "American Table of Distances, Specifiying Distances to be Maintained between Magazines for Explosives and Inhabitated Buildings, Public Railways, and Public Highways." In addition, the Institute has issued a pamphlet entitled "Suggested State Law Compiled by Institute of Makers of Explosives," which has been the foundation of the law of at least some states with respect to explosives.

Those in close contact with explosives are of the opinion that conditions of storage largely determine not only the safety but the efficiency that may be expected when explosives are used. Extremes of heat or cold, of air dryness, or of moisture in storage, the roughness or carefulness with which handled, the length of time stored, the length of time out of their original container before used—these and many other considerations have a vital influence on the behavior of a commercial explosive. The manufacturers issue pamphlets in connection with their products, and many excellent suggestions are made with respect to the safe and efficient transportation and storage as well as use of explosives; a particularly informative pamphlet entitled "Safety in Handling and Use of Explosives" was issued by the Institute of Makers of Explosives under the auspices of representative explosives manufacturing organizations of the United States.

The Bureau of Mines of the U.S. Department of Interior has issued much information about the testing, transportation, storage and use of explosives in the mining and allied industries, but more especially with respect to coal mining. The information has been published in numerous pamphlets (Bulletins, Technical Papers, Reports of Investigations, and Information Circulars), many of which are now out of print. Other publications on explosives can be obtained from the Superintendent of Documents, Washington, D.C.; reports of Investigations can be obtained from the Bureau of Mines, U.S. Department of Interior, Washington, D.C. Safety in the use of explosives is treated in publications of the National Safety Council, the American Standards Association, as well as in various other technical journals and NFPA No. 495.

Location and Construction of Magazines

Magazines should be situated far enough away from other buildings or vital structures so that in case of an explosion in the magazine the least possible damage will be done to buildings or persons in the surrounding area.

Recommendations as to Surface Magazines

(1) Surface magazines should be well ventilated.

(2) Natural light or light by portable electric storage lamps or by protected electrical systems, or flood lights from the outside of the magazine should be provided.

(3) Floors should be made of wood or other non-sparking material and have no metal exposed.

(4) Magazines should be grounded if constructed of steel or covered with sheet iron.

Precautions Against Theft

Explosives must be protected against theft by storing them in magazines constructed and locked as required by regulations. Additional protection may be secured in appropriate cases by using:

(1) Mortise locks wherever practicable. When padlocks are used they should be protected by steel hoods or shields placed around the locks to prevent tampering. The hood or shield may be made from pipe or of steel large enough in diameter to enclose and protect the lock and hasp and at least 4 inches in length, securely fastened to the door by welding or by lugs that are riveted or bolted or fastened by some other secure method.

(2) Fences.

(3) Floodlights.

(4) Gas bombs.

(5) Alarm systems, such as the "electric eye," etc.

(6) Additional guards or inspection by watchmen.

Precautions Against Fires

In order to prevent fires the following precautions should be taken:

(1) The magazine should be kept clean.

(2) Wastepaper, sawdust, used empty boxes and containers, and other combustible material should not be left to accumulate in or around any magazine.

(3) All surrounding area for 25 feet, and preferably 50 feet in all directions, should be kept free of rubbish, dry grass, or other materials of a combustible nature, and where feasible, the area should be covered with a material to prevent the growth of grass, weeds, and brush.

(4) Smoking, open lights, or other flame, or carrying of matches and smoker's articles into or around any magazine should be prohibited.

(5) Black powder and high explosives should be stored in separate magazines.

General Rules

The following rules, although not directly applicable to magazine construction, do apply to the safe operation of explosives and detonator magazines.

No denotators, tools, or other materials should be stored in a magazine containing explosives.

No high—or low—explosives or blasting device heaters should be stored in a detonator magazine.

Lighting in the magazine should be natural or by permissible lights; if a magazine is electrically lighted, the lamps should be of the vaporproof type, the switch should be outside the building, and the wiring should be in conduit.

Only tools made of wood or other nonmetallic material should be used in opening cases of explosives.

Explosives should be stored so that the oldest stock is used first.

Cases of explosives should be stored topside up, in other words, so that cartridges are lying flat. However, they should be turned at regular intervals as this will help to prevent their deterioration.

Preferably cases of explosives should not be piled in stacks more than 6 feet high.

Cases containing explosives should always be lifted and set down carefully; never slide one over on another or drop from one level to another or otherwise mishandle them.

Only authorized persons should be allowed in the magazine. One person should be made responsible for the operation and be held responsible and accountable for the contents of the magazine.

Safety with Explosives

Safe use of explosives is the result of planning and doing the thing right. The user must remember that he is dealing with a powerful force, and that various devices and methods have been prepared to enable him to direct this force. He should realize that this force, if misdirected, may seriously and permanently injure him and his fellow-workers.

Destruction of Explosives. Explosives to be destroyed may be fresh material from damaged packages, or material that has deteriorated either from natural aging or from improper storage to the point where it is unfit for use.

Deteriorated explosives may be more dangerous to handle than explosives in good condition. When there is any question about the safety of the undertaking, a representative of the manufacturer of the particular lot of explosives should be consulted, or a request for assistance should be made to an authorized representative of the Bureau of Mines or to someone else known to have had the necessary experience. This is especially true if large quantities of explosives must be destroyed.

Most explosives, except detonators, are best destroyed

by burning. The hazards of an explosion is always present, even under the most favorable conditions, so it is of prime importance to select a site where no damage will be done, either to persons or property, if the explosives detonate. This means a safe distance from any structure, railroad, or highway and from any place where one or more persons may even accidentally be exposed to danger, including that from flying fragments.

During the destruction of any type of explosives the possibility of preignition should be prevented by eliminating smoking and open lights. Only one type of explosive should be destroyed at a time, and the utmost care should be taken to see that no detonators are accidentally included in explosives to be destroyed by burning.

High explosives should never be burned in cases or in deep piles. Dynamites, especially permissible gelatins, become increasingly sensitive when overheated before ignition. Quantities of dynamite to be burned should not exceed 100 pounds of regular dynamite or 10 pounds of permissible gelatin. Local conditions may limit destruction to much smaller amounts, but when more than these maximum quantities must be destroyed, a new space should be selected for each lot, as it is not safe to place explosives on ground heated by the preceding burning.

No attempt should be made to return to the site as long as any flame or smoke can be observed.

As soon as all dynamite has been burned, it is believed to be a good practice to plow the ground, as the residue may contain salts said to be attractive to livestock, which if eaten can produce toxic symptoms.

Dynamite

When properly stored, dynamite shoud remain in good condition for a long time. It usually deteriorates rapidly if improperly stored or handled. The most common signs of deterioration are discoloration, leakiness, hardness or excessive softness, or the formation of crystals on the outside of the wrapper. Frequently a combination of two or more of these signs can be noted.

Many persons believe that the crystals are nitroglycerin and that they are especially dangerous, actually, the crystals are salts that have exuded through the wrapper, whereas nitroglycerin is an oily liquid at normal temperatures, their presence on the outside of the wrapper or on the container shows that the dynamite has deteriorated to some degree and is therefore extremely dangerous.

Care should be exercised in handling deteriorated explosives, whether loose or in containers. Most persons experience undesirable effects, especially headaches of varying degrees of severity, by absorption of nitroglycerin through the skin when handling leaky or loose dynamite, and some persons are so sensitive as to have headaches after working over loose dynamite for only a short time, even without touching it. Therefore, if leaky or loose dynamite must be handled, gloves should be worn and then burned as often they become impregnated with nitroglycerin.

Some dynamites are rather difficult to ignite, especially when wet, so it is best to prepare a bed of dry, combustible material, such as excelsior, wood shavings, or sawdust; to maintain combustion it is sometimes necessary before igniting the pile or bed to pour a little kerosene over the dynamite and the fuel bed. The area of the bed should be such that the dynamite can be arranged in a single layer if sticks or part sticks are being destroyed, or arranged not to exceed 2 inches in thickness if loose dynamite is to be burned. The bed should be long and narrow rather than square or circular.

It is often recommended that each stick of dynamite be slit and the loose material scattered on the fuel bed, but considering the extra hazard to the operator, it seems preferable to deposit whole sticks of the more common sizes on the fuel bed as carefully as possible without slitting. If the cartridges are of large diameter, such as those often used in quarry blasting, the loose material should be spread; "free running" (loose) dynamite may be so spread, but never in a thickness exceeding 2 inches.

When the bed has been formed and the dynamite deposited on it, a train of paper or similar readily ignitable material should be laid to it, preferable on the downward side, and the explosives ignited thus. The train should be long enough to permit the operator to reach a safe place.

Dynamite should ordinarily burn quietly, with a bluish flame. If solid pieces remain, as sometimes happens, especially if the dynamite is wet, it is dangerous to poke about the debris or attempt to handle the pieces for reburning until it is certain that they are cool. The containers should be burned separately.

Detonators

Blasting caps, electric blasting caps, and delay electric blasting caps which have so deteriorated from age or improper storage that they are unfit for use should be destroyed. These devices should also be destroyed if they have ever been under water as, for example, during a flood, regardless of whether or not they have been subsequently dried out. In some cases, the shells of caps that have been wet and then dried will show signs of corrosion. Such caps may be very dangerous to handle, and it is recommended that they not be disturbed until a representative of the manufacturer has had an opportunity to pass on them. The method most generally used for destroying detonators is to explode them under some confinement as described below. Detonators should not be thrown into small bodies of water such as rivers, creeks, ponds, or wells.

If possible, it is advisable to explode ordinary

(fuse) blasting caps in the original container with the cover removed. Otherwise, they should be prepared for blasting as follows: They should be placed in a small box or bag; a hole should be dug in the ground, preferably in dry sand, at least one foot deep; the container should be placed in the bottom of the hole and primed with one cartridge of dynamite and a good electric blasting cap or ordinary cap and fuse; the caps and the primed cartridge should be carefully covered with paper and then with dry sand or fine dirt and fired from a safe distance. It is recommended that not more than 100 caps be destroyed at a time and that the ground around the shots be thoroughly examined after the shot to make certain that no unexploded caps remain. The same hole should not be used for successive shots.

To destroy electric blasting caps or delay electric blasting caps, it is necessary first to cut the wires off about one inch from the top of the cap, preferably with a pair of tin snips. No attempt should be made to cut the wires from more than one cap at a time. Not more than 100 caps should be placed in a box or paper bag, primed with a cartridge of dynamite and a good electric blasting cap, buried under paper and sand or dirt, and exploded as described above. The same precautions mentioned above should be observed.

Blasting caps should never be destroyed by placing them in a hole which is to be shot, especially by dropping them into well drill holes. Many bad accidents have occurred in this way.

Electric Squibs and Delay Electric Squibs

These devices should be destroyed by the same procedure as used for electric blasting caps.

Safety Fuse

This material may be disposed of very satisfactorily by burning in a bonfire.

Primacord

The preferred method of destroying Primacord is by burning. It should not be burned on the spool but should be strung out in parallel lines $\frac{1}{2}$ inch or more apart on paper or dry straw.

Removal of Nitroglycerin from Magazine Floors

Contaminated floors should be scrubbed well with a stiff broom, hard brush, or mop, using an ample volume of a solution in the proportion of 1.5 quarts of water, 3.5 quarts of denatured alcohol, 1 quart of acetone, and 1 pound of sodium sulfide (60% commercial). The liquid should be used freely to decompose the nitroglycerin thoroughly. If the magazine floor is covered with "Ruberoid" or any material impervious to nitroglycerin, this portion of the floor should be swept thoroughly with dry sawdust and the sweepings taken to a safe distance from the magazine and destroyed by burning.

It was formerly considered good practice to cover explosives magazine floors with rubber or similar material, but the best practice now is generally believed to be a tight, smooth wood floor, free from exposed nails or other metal fastenings, so as to eliminate holes and ragged edges that sooner or later result from excessive wear. Floor coverings often hide floor stains and, when worn, render proper cleaning difficult or impossible unless the whole covering is removed. With the tight wood floor now considered best, cleaning is more certain and, if necessary, parts of the floor can be carefully removed and replaced.

Conclusion

Manufacturers frequently employ other methods of destroying explosives and blasting supplies than those described above, but when employed these should be used only under direction of the manufacturer's representative.

As indicated previously, the information given here is not for the purpose of encouraging persons unfamiliar with the proper methods of destroying explosives to undertake such work unassisted. It is realized, however, that expert advice and assistance are not always available and it is believed that destruction of unwanted explosives can be accomplished with less hazard by following the foregoing suggestions (see also Explosives; Explosives, High; Explosives, Low in Sections 6 and 12). See also bibliography below.

APPENDIX 2 TANKS FOR STORAGE OF FLAMMABLE LIQUIDS

The contents of storage tanks have played a part in some very serious fires. There are three types of tank storage (a) underground; (b) aboveground, outside building; (c) aboveground, inside building.

Underground Tanks

Plans for underground storage tanks designed for flammable liquids should be discussed with local building groups, national regulatory groups, and an insurance carrier. The following is a check list of items to be considered:

(a) The optimum size and position of the buried tank, e.g., whether it shall be vertical or horizontal.

(b) Location of the tank with respect to the terrain in the general area which is to be considered, the location of buildings, cellars, pits, etc., in the same area. The possibility of corrosion is an important factor.

(c) Adequate anchorage for the tank, keeping in mind the possibility of flood, heavy rainfall, or other events and what effect these might have on the installation.

(d) Sufficient covering for the tank; this will improve the safety of the installation and help with the problem of anchorage.

(e) In selecting a location in which to bury a tank, cognizance must be taken of the possibilities for corrosion, to prolong the life of the tank. It is important to determine if the region chosen is reasonably free of corrosive effluents from nearby plants, corrosive cinderfill, or possibly corrosive groundwater. If corrosion is considered likely, tanks should be painted with at least one coat of red lead in linseed oil primer and then one coat of asphalt or coal tar base paint. Other formulas may be equally effective.

(f) It must be remembered that the equivalence of a below-ground location may be obtained with a partially buried tank. In such a situation the exposed part of the tank may be covered with a dirt fill and a concrete cap as well as a concrete retaining wall for anchorage.

Aboveground Tanks

Aboveground tanks can be a source of extreme danger, if not planned carefully. For instance, rupture of an aboveground tank or a leak in such a tank at a point below the liquid level may very easily lead to a serious fire. Therefore, such tanks should be located on ground sloping away from the main buildings and plant utility installations. Protection against fire spread is provided on level ground by tank spacing, proper drainage facilities, and adequate dikes. In hilly terrain provisions must be made for safe drainage past installations at lower levels. The following considerations must be kept in mind when planning a flammable-liquid storage tank above ground:

(a) The possibility of damage to nearby buildings or tanks.

(b) The amount of flammable material contained in the unit under consideration.

(c) The values involved in the unit under consideration and those adjoining.

(d) The burning and ignition characteristics of the materials involved.

(e) The provision of adequate room for fire-fighting operations and access for fire-fighting equipment.

(f) Dike capacity of such an installation: In the ideal case the dike capacity should equal the total tank capacity inside the dike area. Where the flammable liquid involved is such as may be subject to boil-over, i.e., fuel oil, etc., the diking capacity must be 1.5 times the total liquid capacity of the tanks in the dike area.

(g) Adequate drainage arrangements must be made to prevent the accumulation of water in the dike areas.

Tank Construction

The following points must be considered in making sure that tank construction is adequate:

(a) In the case of horizontal steel tanks, it should have an Underwriters' label indicating that it meets definite construction standards.

(b) All horizontal tanks must be given a hydrostatic strength and leakage test before installation is complete.

(c) In the case of aboveground tanks, they must be constructed upon fire-resisting supports such as brick or reinforced concrete.

(d) Vertical steel tanks should be constructed in accordance with the American Petroleum Institute's specification for standard tanks.

(e) A leakage test is recommended for vertical tanks.

(f) Provision must be made for adequate venting facilities and flame arrestors.

Tank Connections and Fittings

(a) Connections to horizontal tanks.

(b) The use of steel flanges or wrought iron couplings to make pipe connections.

(c) Provision for manholes.

(d) Provision for shut-off valves.

(e) The use of heating equipment.

(f) The use of electrical equipment with the attendant necessities for prevention of sparks, i.e., grounding, electrical bonding, etc.

Fire Protection

The protection required for tanks containing flammable liquids is determined by the tank size, type, location, exposure to or from buildings or other tanks, value of contents, flash point of the liquid stored, and the probability of the interrruption to production by loss of tank contents.

Conditioning and Use of Old Tanks

Failure to take proper precautions during the cleaning and repair of used tanks can result in serious explosion hazards. To avoid such hazards the following precautions should be observed:

(a) Tanks which have contained flammable liquids should be purged with steam or filled with water and allowed to drain, or combinations of these methods may be employed.

(b) Use a portable flammable vapor indicator to determine that vapor concentrations inside such tanks have been reduced to safe levels.

(c) Remove all remaining scale and sludge from inside such tanks with nonferrous scrapers. The use of detergents to clean such tanks, while helpful in actually removing the residues, should be undertaken with care, because, for example, caustic solutions promote decomposition of nitrocellulose residues, and the reaction thus started will move with explosive violence under the proper conditions.

(d) Take the additional precaution of filling the tank with water or inert gas such as CO_2, if cutting or welding torches must be used on it.

(e) Proper supervision of the workers and sufficient ventilation is essential inside a tank.

(f) While such reconditioning or repair is in progress, it is wise to move any readily flammable materials from the neighborhood. See also NFPA No. 30.

APPENDIX 3—INDUSTRIAL DISASTER CONTROL

The function of disaster control in industry is taking on increased importance. Such factors as industrial plant decentralization and the advent of newer and more powerful weapons of war have made it necessary for a plant to be as self-sufficient as possible, particularly during times of emergency. All that disaster planning involves is an extension of routine fire and emergency programs to cope with disaster.

The first requirement of a disaster control program is common sense. No highly specialized knowledge or backlog of experience is necessary. The essence of disaster planning is thorough analysis of the existing plant facilities, and observance of how they may be affected by a disaster. Unless a plant is in the zone of total destruction of an atom bomb, or some similar destructive agency, an adequate disaster control plan is important. Although it is possible to protect against an atomic detonation, the protection necessary is so elaborate and so costly that it may not be economically feasible.

Model Disaster Control Program

Make a complete survey of existing functions and facilities, including possible emergency facilities. It is necessary to determine what is done under routine conditions and what is to be done when these conditions are disrupted, e.g., during the emergencies created by fires, hurricanes, earthquakes, release of toxic materials, bombs, etc.

Determine the key production point in the plant—the departments and processes which, if interrupted, would result in a complete production stoppage. Such key locations should have the best possible protection. A plan, to be successful, must be sufficiently broad in scope to be adaptable to any emergency. Specific operational details, of course, need not be included, since these will depend upon the actual nature of the emergency. The following factors, however, must be taken into consideration: executive authority and duties; handling the alarm; emergency headquarters; operations of the police force; operations of maintenance force; operations of fire department; operations of public information and industrial relations group; operation of monitoring group for toxic inhalants and radioactivity.

Executive Authority and Duties

A person responsible for coordinating the operation of all emergency groups must be designated in advance to serve as plant emergency director. Provisions must be made for succession of command. There must be no confusion as to who is in charge. The plant manager is usually a good choice during office hours, and the general shift foreman for nights and holidays. It is recognized that someone with authority must be available during emergencies.

The plant emergency director shall coordinate the efforts of the emergency groups which shall include: determination of seriousness of emergency condition; sounding general alarm throughout plant; directing operation of all plant forces from emergency headquarters; promptly summoning any outside aid which may be deemed necessary. (In calling outside assistance, the plant emergency director may secure excellent advice from the heads of his own forces, e.g., the plant fire chief might be consulted on the necessity of summoning outside fire assistance.) Aid of various types should be promptly summoned. If necessary it can be turned back if not needed upon arrival. Delay can be disastrous; inform the press as to requirements of public interest in an attempt to avoid promoting misinformation; attempt to maintain a log of events to serve as a record.

Handling the Alarm

Upon discovering an emergency, the individual shall:

For a fire or explosion, pull the nearest fire alarm box or telephone fire department. (The importance of indoctrinating all personnel in advance as to how to transmit a fire alarm cannot be overemphasized.)

For other emergencies, telephone for ambulance, police patrol, emergency repair, medical, safety, or decontamination aid as may be necessary.

For notification, the plant emergency director should be notified immediately after completing the call for assistance. The supervisor shall, immediately upon learning of an emergency in his area, proceed to the scene and see that the appropriate notifications as required above have been made. He will direct the forces at hand in initiating appropriate action and then report to the plant emergency director. An emergency procedure should be established for the telephone switchboard operators to screen telephone calls, restricting the use of lines to official calls.

When an emergency is reported, a general alarm should be sounded alerting the entire plant; the method of sounding the alarm depends on the equipment available at the plant. In such an eventuality, all personnel except those initiating on-the-spot action or assigned to some specific function, should evacuate their respective buildings, closing all doors but not locking them, and assemble in front of their buildings to await further orders. All radioactive, highly toxic, or corrosive materials should be locked up.

Emergency Headquarters (see NFPA 232, 231)

Primary and secondary emergency plant headquarters from which operations will be directed should be established in advance. The communications room in the police or guard headquarters is usually a good location for the primary headquarters. A secondary headquarters is necessary in the event that the primary location is involved in a disaster; if both the primary and secondary locations are knocked out, an alternate will have to be set up on the spot by the executive authority. Depending upon the plant, the secondary or alternate headquarters might be established at the fire headquarters, plant management office, safety director's office, etc., or in the field. A portable sign reading "Emergency Headquarters" should be provided at the appropriate location. (Red lettering on a white background is recommended.) Red electric lanterns should be used at night in emergencies to designate the headquarters. Both primary and secondary headquarters should be equipped with: telephone facilities, both interplant and outside; radio (if available); map showing fire mains, fire alarm boxes, fire hydrants, general piping layouts and similar pertinent data; telephone numbers (plant and home) and home addresses of key personnel, together with information as to the time and route required for each person to report to the plant in response to an emergency call; same information as above for all other personnel; specific information as to how to contact and/or obtain the following services: fire department, police department, medical services, ambulances, public utilities, and other public services, the means of contacting nearby military establishments if assistance is needed, e.g., air transportation.

Operation of Police Force

The chief of the police force is responsible for maintaining the material and data required at primary and secondary emergency headquarters. In case of emergency, the senior police officer on the premises shall: maintain liaison with the plant emergency director; alert guards on duty; maintain order at all scenes of emergency; assist plant emergency director in placing calls for outside assistance and recalling off-duty personnel; escort emergency vehicles and personnel returning to the plant to appropriate locations; if equipped with radio, dispatch radio patrol cars to locations in order to maintain communications, particularly in the event of telephone failure (if both telephone and radio fail, police will have to handle communications within plant and with outside agencies by messenger); stand by to evacuate entire plant if necessary; if the emergency requires outside personnel the police should guide them to the emergency headquarters where the director will direct their admittance and disposition.

Operation of Maintenance Staff

The maintenance staff initiates emergency action if large scale plant emergency groups are not maintained; e.g., in the event a plant does not have a full-time fire department, the maintenance personnel organized and trained as a fire brigade should immediately combat a fire and attempt to prevent it from spreading pending arrival of the outside fire department.

Maintenance personnel should be organized by the chief of maintenance into appropriate squads to handle electrical repairs, emergency lighting, rescue, maintenance of power, water supply, salvage operations to minimize water damage, transportation, etc. The senior maintenance man should maintain liaison with the plant emergency director and direct the operations of his various divisions. Maintenance personnel should be kept informed of hazardous materials in the plant and how to cope with them.

Operation of Fire Department (see NFPA No. 4A)

The cooperation between the plant and municipal fire departments should be outlined in detail on a mutual aid agreement plan. Regardless of how an alarm is received, the fire department immediately responds in accordance with the prearranged plan. Upon arrival at the scene, the senior fire officer present is in charge of all fire department apparatus and personnel. he maintains liason with the plant emergency director and with the monitors as concerns radiological and toxic chemicals safety.

An extra supply of rubber boots, clothing, helmets, and and fire hose is desirable where regular equipment may become contaminated.

Fire department personnel should be kept informed of all hazardous materials in a plant and how to cope with same (see NFPA Nos. 4C and 6).

Operation of the Medical Department

The chief of the medical department shall be responsible for first aid training and for the procurement, maintenance, and availability of appropriate medical supplies.

The senior medical man present at the time of an emergency shall: maintain liaison with plant emergency director; care for injured; transport casualties to municipal hospitals as may be necessary; arrange for appropriate ambulances and for other vehicles to serve as ambulances in case of catastrophe; when casualities are sent to outside hospitals, maintain liaison with the hospital personnel as to the nature of the injuries and whether or not radioactivity is a factor; coordinate efforts of Red Cross units which may be called in.

Depending upon the outside hospital facilities available, the chief of the medical department may want to have mimeographed copies of medical data available relative to contamination control for distribution to municipal hospital medical personnel in times of emergencies. Immediately after a disaster the medical division, in cooperation with industrial hygiene personnel, shall evaluate the degree of severity of exposures. Special clinical and laboratory examinations of blood and urine should then be performed on all individuals who may have received significant exposure. Follow-up examinations will have to be made at the descretion of the medical department.

Operation of Public Information and Industrial Relations Group

Public information releases should generally be prepared by the plant emergency director or some other authoritative individual designated in advance. It is recommended that these be in writing, if time permits, to prevent misinterpretation. A list of editors and radio newsmen should be prepared well in advance, so that these individuals can be contacted as quickly as possible and informed that a written release statement is being prepared with authoritative information relative to the emergency.

The senior industrial relations representative, the safety director, and the senior insurance department representative shall maintain liaison with the plant emergency director. The insurance representative shall, in in addition, obtain names, addresses, and severity of injury to all casualties; when ordered by plant emergency director, notify the families of the casualties as to the nature of the injury and location of injured; prepare necessary reports for insurance carrier.

Operations of Monitoring Group for Toxic Inhalants and Radioactivity

When an emergency occurs this group shall assemble at its base and immediately contact the plant emergency director, who shall instruct the monitoring group leader as to the areas to be monitored and the contaminants or radiations to be measured. While maintaining constant contact with the plant emergency director both for direction and to report monitoring results, this group has the following responsibilities:

In accordance with a prearranged plan, to have personnel evacuated from contaminated or high-radiation-level areas; to assist in the control of exposures to radiation and toxic materials by firemen or other emergency workers; to monitor possible contaminated personnel in order to control the spreading of contamination as well as to be able to advise personal decontamination procedures; to help in decontamination of any contaminated areas; to make checks upon possible environmental contamination, e.g., fall-out, radiation, contamination of water, air or food, livestock, houses, etc. Since hastily given numbers are often misunderstood, it is a prime responsibility of a monitor to see to it that his data go only to the plant emergency director to use as he sees fit.

While the foregoing discussion does not provide a complete plan for controlling plant disasters, it is hoped that it will serve as a guide and provide stimulus for thought and planning relative to continuity of production and safeguarding of personnel during actual emergencies.

Special Reference

National Fire Codes, Vol. 12, 1977, Recommended Practices and Manuals. NFPA,

BIBLIOGRAPHY

1. National Fire Codes, Vol. 12, 1977; National Fire Protection Association, Boston, Mass.
2. Manual of Hazardous Chemical Reactions, 1975, 5th ed. NFPA.
3. Installation of Sprinkler Systems, 1976, NFPA #13.
4. Explosion Venting 1974. NFPA #68.
5. Fire Protection for Laboratories Using Chemicals, 1975, NFPA #45.
6. Static Electricity, 1972, NFPA #77.
7. Laboratories in Health Related Institutions, 1973, NFPA #56C (ANSI MD2-1-1975; 8/21/75) NFPA.
8. Fire Hazards of Materials 1975; NFPA #704.
9. Portable Fire Extinguishers 1975; NFPA #10 (ANS Z 112.1-1976. 2/27/76).
10. Organization of Industrial Fire Loss Prevention 1974, NFPA #6.
11. Flammable and Combustible Liquids Code - 1976 NFPA #30.
12. Hazardous Chemicals Data. 1975, NFPA #49.
13. Flash Point Index of Trade Name Liquids, 1972, NFPA #325A.
14. Flammable Liquids, Gases, Volatile Solids-1969. NFPA #325M.

GENERAL BIBLIOGRAPHY

Hartmann, Irving. "The Explosibility of Titanium, Zirconium, Thorium, Uranium and Their Hydrides." Bureau of Mines Report #3202, NYO-1562.

Bararas, G. D., et al. "Zirconium Metal Powder Precautionary Handling Suggestions." *J. Am. Chem. Soc.*, **70**, 877 (1948).

Birchall, James D. The Classification of Fire Hazards and Extinction Methods. London, Ernest Benn Ltd. 1954.

"Fire and Explosion Hazards of Thermal Insectidal Fogging," National Board Fire Underwriters Research Report #9.

"Safety in the Mining Industry." BM 481, Bureau of Mines Washington, D.C.

"Investigations on the Explosibility of Ammonium Nitrate." Bureau of Mines Report of Investigations 4994, August 1953.

"Potential Hazards in Molten Salt Baths for Heat Treatment of Metals." National Board Fire Underwriters Research Report #2, 1954.

"Fire Hazards and Safeguards for Metal Working Industries." Technical Survey No. 2, National Board of Fire Underwriters, 1954.

"Flammable Liquid Pumping Equipment." Factory Mutual Bulletin of Loss Prevention Nos. 13, 24. Associated Factory Mutual Fire Insurance Companies, December 1949.

"Explosives Drivers' Handbook." OP 2239, Department of the Navy, Bureau of Ordnance, March 29, 1956.

"Storage, Handling, and Use of Flammable Liquids." National Board of Fire Underwriters No. 30, July 1956.

"Dangerous Chemicals Code." 1951 Edition, Bureau of Fire Prevention, City of Los Angeles Fire Department. Parker and Co. Los Angeles, Calif.

Marshall Sittig, *Sodium, Its Manufacture, Properties, Uses.* Reinhold Publishing Corp., 1956.

Robinson, Clark Shove. *Explosions, Their Anatomy and Destructiveness.* McGraw-Hill Book Co., New York, 1944.

Blasters' Handbook. 12th Edition, E. I. duPont de Nemours and Co., Inc. Wilmington, Delaware.

"Storage Tanks for Flammable Liquids." Factory Mutual Bulletin of Loss Prevention Nos. 13, 23. Associated Factory Mutual Fire Insurance Companies, December 1940.

"Ventilation and Operation of Open Surface Tanks." American Standard Safety Code No. Z9, 1-1951, American Standards Association, Inc. New York.

Babcock, Chester I. "Ammonium Nitrate—Behavior in Fires." *NFPA Quarterly*, January 1960.

Guise, A. B. "The Chemical Aspects of Fire Extinguishment." *NFPA Quarterly*, April 1960.

Nickerson, M. H. "Palletized Storage Fire Problems." *NFPA Quarterly*, July 1957.

Betz, G. M. "Organic Peroxides Storage and Handling." *NFPA Quarterly*, July 1962.

"Handling Cryogenic Fluids." *NFPA Quarterly*, July 1960.

Le Vine, R. Y. "Electrical Equipment in Chemical Atmospheres." *NFPA Quarterly*, April 1964.

Bahme, C. W. "Our Protection for Chemicals." *NFPA International*, 1961.

John, H. Perry (ed.). *Chemical Engineers' Handbook.* McGraw-Hill Book Co. New York, 1964.

G. G. Hawley, (ed.). *Condensed Chemical Dictionary.* Van Nostrand Reinhold Co., New York, 1977.

R. C. Weast (ed.). *Handbook of Chemistry and Physics.* Chemical Rubber Co. Cleveland, 1972–73.

Los Angeles Fire Prevention Code, Divisions 70–76 inc. Bureau of Fire Prevention, Los Angeles Fire Department, Los Angeles, Calif.

Paul G. Stecker (ed.). *Merck Index of Chemical and Drugs.* Merck and Co., Rahway, N.J., 1976.

Ethel Browning. *Toxicity of Industrial Organic Solvents.* Chemical Publishing Co., Inc., New York, 1953.

Threshold Limit Values. American Conference of Governmental Industrial Hygienists, Cleveland, Ohio.

Melvin A. Cook. *The Science of High Explosives.* Reinhold Publishing Co., New York, 1958.

National Fire Protection Assoc., Boston, Mass. *Fire Protection Guide on Hazardous Materials*, 1966; NFPA Contg. NFPA Nos. 325A, 325M, 49, 491M, 704M. Volumes 1–10 incl. of National Fire Codes for 1966–1967.

Instituion of Flame Reactions: A Preliminary Investigation of the Role of Ions and Electrons. N.B.S. Report 6588, 1959; E. C. Creitz.

J. Taylor. "Effect of Various Chemicals, More Particularly Sodium Bicarbonate, on the Ignition of Cellulose, Coal Dust and Activated Charcoal." *F. Pollol. Fuel* **27,** 77–78, 1948.

D. J. Price, H. H. Brown, R. Hylton and R. E. Roethe. *Dust Explosions, Causes and Methods of Prevention.* NFPA 1923.

I. Hartmann. "Dust Explosions in Coal Mines and Industry." *Scientific Monthly* **79,** 97–108, 1954.

G. Long. "Preventing Aluminium Powder Dust Cloud Explosions." *Ind. Eng. Chem.* **53,** 823–825, 1961.

SECTION 8

Industrial and Environmental Cancer Risks

Elizabeth K. Weisburger, Ph.D.
Chief, Laboratory of Carcinogen Metabolism
National Cancer Institute
Bethesda, Maryland

Since the previous publication of this material,[1] substantial changes have occurred with respect to occupational cancer. One has been the adoption by the Occupational Safety and Health Administration (OSHA), a branch of the Department of Labor, of a list of regulated carcinogens. These compounds can be used only under certain specific guidelines including: dilution of any mixture containing the compounds to a concentration of 1% or less, the use of adequate safety equipment to minimize exposure, the need to maintain facilities for showering at the end of the work day, and the necessity to eliminate entry by unauthorized persons in areas where any of these carcinogens are used. Although some of the compounds on the OSHA list are important industrially, others are generally research chemicals.[2] Legal action, however, has resulted in deletion of the standards as applied to laboratory uses of these chemicals.

More importantly, the Toxic Substances Control Act became law in 1976.[3] This Act gave the Environmental Protection Agency (EPA) the authority to control the entry of toxic or carcinogenic substances into the environment. EPA also was specifically ordered to request manufacturers to provide lists of all chemicals made by them for sale and to furnish all available material on the toxic, mutagenic, or other biological properties of such compounds. In addition, the Act stated that any compound introduced for commercial or industrial purposes must first be tested to insure that its use will not constitute a hazard. Compounds proposed or synthesized for research or developmental purposes were exempted from these restrictions.

The Toxic Substances Act does not specifically state that a two-year study in animals is required but that the proper studies be done. Standards for testing are to be developed by EPA.

The adoption of the above-mentioned regulations has changed the climate regarding toxicity studies of new environmental or commercial chemicals. Previously such tests were desirable in order to avoid unfortunate consequences; presently such tests are necessary due to regulation.

In addition, the National Institute of Occupational Safety and Health (NIOSH) has taken a more active role in pinpointing areas where various chemicals may present a hazard to human health. NIOSH also recommends to OSHA the standards or levels on allowable exposures for individual compounds. The background documentation for setting these standards is found in the criteria documents prepared by NIOSH (Table 1). Furthermore, NIOSH issues "Current Intelligence Bulletins" to alert those concerned to possible human hazards. These have included statements on such diverse materials as polychlorinated biphenyls, nitrosamines in cutting fluids, 2-nitropropane, N-phenyl-2-naphthylamine, 2,4-diaminoanisole, and others.

Within the past two or three years, the costs of doing a long-term bioassay have increased at least two- to three-fold. The adoption by the Food and Drug Administration (FDA) of a Code of Good Laboratory Practice for animal studies may lead to further increases in cost in qualifying institutions. Consequently, it is almost too costly for a medium or small-sized manufacturing establishment to fund a study on the long-term toxicity or carcinogenicity of any chemicals proposed for commercial development. Therefore many short-term tests are being developed to predict which compounds would more likely be carcinogenic. A prominent one is a test for mutagenicity in various strains of the bacterium *Salmonella typhimurium*.[4] The tests generally involve addition of the compound to a culture dish which is seeded with one of these bacterial strains. The medium is deficient or lacking in histidine; therefore no bacterial growth occurs unless mutation by the test chemical yields a form of bacteria which does not require histidine. A count of the bacterial colonies can therefore be used as a measure of the mutagenicity of the test compound.

Even though this technique is rapid and much less expensive than doing a full scale bioassay in animals,

TABLE 8.1. NIOSH Criteria Documents on Chemicals and Recommendations for Occupational Health Standards.

Substance	Current OSHA Environmental Standard	NIOSH Recommendation for Environmental Exposure Limit	Health Effect Considered
	Documents Submitted		
Acetylene	2500 ppm (10% of lower explosive limit)	No exposure in excess of 2500 ppm	Indirect asphyxia
Acrylamide	0.3 mg/cu m, 8 hr TWA (skin)	0.3 mg/cu m TWA[a]	Skin, eye, nervous system effects
Acrylonitrile	20 ppm, 8 hr TWA	Not greater than 4 ppm by recommended method	Lung and bowel cancer
Alkanes (C5-C8)	Pentane: 1000 ppm, 8 hr TWA, *n*-hexane, *n*-heptane, octane: 500 ppm, 8 hr TWA	350 mg/cu m TWA; Mixtures to be not greater than 350 mg/cu m TWA; 1800 mg/cu m ceiling singly or mixtures (15 min)	Skin and nervous system effects
Allyl chloride	1 ppm, 8 hr TWA	1 ppm TWA; 3 ppm ceiling (15 min)	Liver, kidney, lung effects
Ammonia	50 ppm; 8 hr TWA	50 ppm ceiling (5 min)	Airway irritation
Arsenic inorganic	0.5 mg/As/cu m TWA	2 μg As/cu m ceiling (15 min)	Dermatitis, lung and lymphatic cancer
Asbestos	2,000,000 fibers/cu m 8 hr TWA 10,000,000 fibers/cu m ceiling	100,000 fibers/cu m over 5 microns TWA; 500,000 fibers/cu m over 5 μ ceiling (15 min)	Asbestosis, lung cancer
Asphalt fumes	2.5 mg/cu m	5 mg/cu m ceiling (15 min)	Eye and respiratory irritation
Benzene	10 ppm, 8 hr TWA; 25 ppm acceptable ceiling; 50 ppm maximum ceiling (10 min)	1 ppm ceiling (60 min)	Blood changes including leukemia
Benzoyl peroxide	5 mg/cu m, 8 hr TWA	5 mg/cu m TWA	Airway and eye irritation, skin effects
Beryllium	2 μg/cu m, 8 hr TWA 5 μg/cu m acceptable ceiling; 25 μg/cu m maximum ceiling (30 min)	.5 μg/cu m (130 min)	Lung cancer
Boron trifluoride	1 ppm ceiling	None recommended	Respiratory system effects
Cadmium	0.1 mg/cu m, 8 hr TWA; 0.3 mg/cu m ceiling (fume; erroneously published as 3 mg/cu m) 0.2 mg/cu m, 8 hr TWA; 0.6 mg/cu m ceiling (dust)	40 μg Cd/cu m TWA; 200 μg Cd/cu m ceiling (15 min)	Lung and kidney effects
Carbaryl	5 mg/cu m, 8 hr TWA	5 mg/cu m TWA	Nervous and reproductive system effects
Carbon dioxide	5000 ppm, 8 hr TWA	10,000 ppm TWA; 30,000 ppm ceiling (10 min)	Respiratory effects
Carbon disulfide	20 ppm, 8 hr TWA; 30 ppm acceptable ceiling; 100 ppm maximum ceiling	1 ppm TWA; 10 ppm ceiling (15 min)	Heart, nervous and reproductive system effects
Carbon monoxide	50 ppm, 8 hr TWA	35 ppm TWA 200 ppm ceiling	Heart effects
Carbon tetrachloride	10 ppm, 8 hr TWA 25 ppm acceptable ceiling; 200 ppm maximum ceiling (5 min in 4 hr)	2 ppm ceiling (60 min)	Liver cancer
Chlorine	1 ppm, 8 hr TWA	0.5 ppm ceiling (15 min)	Eye/airway irritation

[a]TWA = Time weighted average based on up to a 10 hour exposure unless otherwise noted.

TABLE 8.1. NIOSH Criteria Documents on Chemicals and Recommendations for Occupational Health Standards. (*Continued*)

Substance	Current OSHA Environmental Standard	NIOSH Recommendation for Environmental Exposure Limit	Health Effect Considered
	Documents Submitted		
Chloroform	50 ppm ceiling	2 ppm ceiling (60 min)	Liver or kidney tumors and central nervous system effects
Chloroprene	25 ppm, 8 hr TWA	1 ppm ceiling (15 min)	Reproductive effects; potential for cancer
Chromic acid	1 mg/10 cu m ceiling	0.05 mg Cr03/cu m TWA 0.1 mg Cr03/cu m ceiling (15 min)	Nasal ulceration
Chromium (VI)	100 μg/cu m ceiling.	1 μg/cu m for carcinogen Cr (VI); 25 μg/cu m TWA for other Cr (VI); 50 μg/cu m ceiling (15 min)	Lung cancer, skin ulcers, lung irritation
Coal tar products	None	0.1 mg/cu m TWA, (cyclohexane-extractable fraction)	Lung and skin cancer
Coke oven emissions	150 μg/cu m TWA	Work practices to minimize exposure to emissions	Lung cancer
Cotton dust	1 mg/cu m (raw-cotton dust)	0.2 mg/cu m lint-free cotton dust	Pulmonary disease (byssinosis)
Cyanide, hydrogen and cyanide salts	10 ppm, 8 hr TWA (alkali cyanides) cyanide 5 mg CN/cu m (skin)	5 mg CN/cu m ceiling (10 min)	Thyroid, blood, respiratory system effects
Decomposition products of fluorocarbon	None	None recommended	Lung effects, polymer fume fever
Dibromochloropropane	None	10 ppb ceiling (30 min)	Sterility; renal and liver effects
Dioxane	100 ppm, 8 hr TWA (skin)	1 ppm ceiling (30 min)	Liver and kidney effects; cancer
Epichlorohydrin	5 ppm, 8 hr TWA (20 mg/cu m)	2 mg/cu in TWA; 19 mg/cu m ceiling (15 min)	Skin, kidney, liver and respiratory system effects
Ethylene dibromide	20 ppm, 8 hr TWA 30 ppm acceptable ceiling; 50 ppm maximum peak (5 min)	1 mg/cu m ceiling (15 min)	Damage to skin, eyes, heart, liver, spleen, respiratory and central nervous systems. Potential for cancer and mutagenesis
Ethylene dichloride	50 ppm, 8 hr TWA 100 ppm acceptable ceiling; 200 ppm maximum ceiling (5 min in 3 hr)	5 ppm TWA 15 ppm ceiling (15 min)	Nervous system, respiratory, heart, liver effects
Fibrous glass	15 mg/cu m total dust; 5 mg/cu m respirable fraction (nuisance dust)	3,000,000 fibers/cu m TWA (fibers $<$ 3.5 μ diameter and $>$ 10 μ length); 5 mg/cu m TWA (total fibrous glass)	Eye and skin and airway effects
Fluorides, inorganic	2.5 mg/cu m, 8 hr TWA	2.5 mg F/cu m TWA	Kidney and bone effects
Formaldehyde	3 ppm, 8 hr TWA; 5 ppm acceptable ceiling; 10 ppm maximum ceiling (30 min)	1.2 mg/cu m ceiling (30 min)	Irritation, lung effects
Hydrogen fluoride	3 ppm, 8 hr TWA	2.5 mg F/cu m TWA; 5.0 mg/cu m ceiling (15-min, fluoride ion)	Skin/eye/airway irritation; bone effects
Hydrogen sulfide	20 ppm acceptable ceiling; 50 ppm maximum ceiling (10 min)	15 mg/cu m ceiling (10 min)	Irritation; severe acute effects, nervous and respiratory systems

TABLE 8.1. NIOSH Criteria Documents on Chemicals and Recommendations for Occupational Health Standards. (*Continued*)

Substance	Current OSHA Environmental Standard	NIOSH Recommendation for Environmental Exposure Limit	Health Effect Considered
	Documents Submitted		
Isopropyl alcohol	400 ppm, 8 hr TWA	400 ppm TWA, 800 ppm ceiling (15 min)	Mucous membrane irritation; possible cancer threat in manufacturing process
Kepone	None	1 μg/cu m ceiling (15 min)	Nervous system effects; liver cancer
Lead, inorganic	0.2 mg/cu m, 8 hr TWA	Less than 100 μg/cu m	Kidney, blood, and nervous system effects
Malathion	15 mg/cu m, 8 hr TWA	15 mg/cu m TWA	Nervous system effects
Mercury, inorganic	0.1 mg/cu m ceiling	0.05 mg/cu m TWA	Central nervous system and mental effects
Methyl alcohol	200 ppm TWA	200 ppm TWA; 800 ppm ceiling (15 min)	Blindness; metabolic acidosis
Methyl parathion	None	0.2 mg/cu m TWA	Nervous system effects
Methylene chloride	500 ppm, 8 hr TWA; 1000 ppm acceptable ceiling; 2000 ppm maximum (15 min in 2 hr)	75 ppm TWA; 500 ppm ceiling, (15 min) TWA to be lowered in presence of carbon monoxide	Central nervous system effects; carbon monoxide toxicity
Nickel, inorganic and compounds	1 mg/cu m, 8 hr TWA	15 μg Ni/cu m TWA	Skin effects; lung and nasal cancer
Nitric acid	2 ppm, 8 hr TWA	2 ppm TWA	Dental erosion, nasal/lung irritation
Nitrogen, oxides	NO 2: 5 ppm, 8 hr TWA	NO 2: 1 ppm ceiling	Airway effects
Organotin compounds	0.1 mg tin/cu m 8 hr TWA	0.1 mg tin/cu m TWA	Eye, skin, liver, nervous system, and heart effects
Parathion	0.11 mg/cu m TWA	0.05 mg/cu m TWA	Nervous system effects
Phenol	5 ppm, 8 hr TWA (skin)	20 mg/cu m TWA; 60 mg/cu m ceiling (15 min)	Skin, eye, CNS, liver, and kidney effects
Phosgene	0.1 ppm, 8 hr TWA	0.1 ppm TWA; 0.2 ppm ceiling (15 min)	Airway effects
Polychlorinated biphenyls	1 mg/cu m, 8 hr TWA (42% chlorine); 0.5 mg/cu m, 8 hr TWA (54% chlorine)	1 μg/cu m TWA	Cancer; skin, liver and reproductive effects
Refined petroleum solvents	500 ppm, 8 hr TWA (solvent)	350 mg/cu m TWA; (15 min)	Skin, lung, and nerve irritation
Silica, crystalline	250/%SiO + 5 in mppcf, or 10 mg/cu m/%SiO + 2 (respirable quartz)	50 μg/cu m TWA, respirable free silica	Chronic lung disease (Silicosis)
Sodium hydroxide	2 mg/cu m, 8 hr TWA	2 mg/cu m ceiling (15 min)	Airway irritation
Sulfur dioxide	5 ppm, 8 hr TWA	0.5 ppm TWA	Respiratory effects
Sulfuric acid	1 mg/cu m, 8 hr TWA	1 mg/cu m TWA	Pulmonary irritation
1,1,2,2-Tetrachloroethane	5 ppm, 8 hr TWA (skin)	1 ppm TWA	Liver, gastrointestinal, and nervous system effects
Tetrachloroethylene	100 ppm, 8 hr TWA; 200 ppm acceptable maximum ceiling; 300 ppm maximum ceiling (5 min in 3 hr)	50 ppm TWA; 100 ppm ceiling (15 min)	Nervous system, heart, respiratory, liver effects
Toluene	200 ppm, 8 hr TWA; 300 ppm acceptable ceiling; 500 ppm maximum ceiling (10 min)	100 ppm TWA; 200 ppm ceiling (10 min)	Central nervous system depressant

TABLE 8.1. NIOSH Criteria Documents on Chemicals and Recommendations for Occupational Health Standards. (*Continued*)

Substance	Current OSHA Environmental Standard	NIOSH Recommendation for Environmental Exposure Limit	Health Effect Considered
	Documents Submitted		
Toluene diisocyanate	0.02 ppm ceiling	0.005 ppm TWA; 0.02 ceiling (20 min)	Airway effects
1,1,1-Trichloroethane	350 ppm, 8 hr TWA	350 ppm ceiling (15 min)	Nervous system, liver, and heart effects
Trichloroethylene	100 ppm, 8 hr TWA; 200 ppm acceptable ceiling; 300 ppm maximum ceiling; (5 min in any 2 hr)	100 ppm TWA; 150 ppm ceiling (10 min)	Central nervous system depressant
Tungsten and cemented tungsten carbide	None	Insoluble tungsten: 5 mg/cu m TWA; soluble tungsten: 1 mg/cu m TWA; Dust of cemented tungsten carbide (> 2% cobalt): 0.1 mg cobalt/cu m TWA: Dust of cemented tungsten carbide (> 0.3% nickel): 15 micrograms nickel/cu m TWA	Lung and skin effects
Vanadium	Vanadium pentoxide (dust): 0.5 mg/cu m ceiling; (fume) 0.1 mg/cu m ceiling; Ferrovanadium: 1 mg/cu m, 8 hr TWA	Vanadium compounds; 0.05 mg/cu m ceiling (15 min); metallic vanadium and vanadium carbide: 1 mg/cu m TWA	Eye, skin, and lung effects
Vinyl chloride	1 ppm, 8 hr TWA; 5 ppm ceiling, (15 min sample)	Minimum detectable level; 1 ppm ceiling (15 min)	Liver cancer
Waste anesthetic gases and vapors	None for substances when used as anesthetic agents	2 ppm ceiling (halogenated anesthetic agents) (1 hr); 25 ppm TWA during periods of use (nitrous oxide)	Reproductive effects and audiovisual performance decrements
Xylene	100 ppm, 8 hr TWA	100 ppm TWA; 200 ppm ceiling (10 min)	Central nervous system depressant; airway irritation
Zinc oxide	5 mg/cu m, 8 hr TWA	5 mg/cu m TWA, 15 mg/cu m ceiling (15 min)	Metal fume fever

Documents in Preparation or Proposed

1978	1979	1980	1981
Antimony compounds	Aliphatic primary monoamines	Aliphatic di- and polyamines	Alkanolamines
Benzyl chloride	Brominated aromatics	Aromatic amines	Aluminum compounds
Carbon black	Chlorinated benzenes	Brominated aliphatics	Cement manufacturing and use
Coal gasification	Coal liquifaction	Dichloropropane	Chlorophenols
Cresol and cresylic acid	Cobalt compounds	Diphenyl	Epoxides
Dinitro-*o*-cresol	Flame retardants	Dyeing and finishing textiles	Glycol ethers Glycols
Glycidyl ethers	Fluorocarbons	Hexachlorobutadine	Hexachlorocyclopentadiene
Hydrazines	Furfuryl alcohol	Hexachloroethane	Hexafluoroacetone
Hydroquinone	Methyl chloride	Hydrogen chloride	Inorganic azides
Ketones	Nitrobenzenes	Manufacture of nonmetallic pigments and dyes	Inorganic chromium (non hexavalent)

TABLE 8.1. NIOSH Criteria Documents on Chemicals and Recommendations for Occupational Health Standards. (*Continued*)

Documents in Preparation or Proposed			
1978	1979	1980	1981
Mercaptans	Nitrotoluenes	Monochloracetic acid	Iron compounds
Manufacture and formulation of pesticides	Organo arsenic compounds	Monochloroethane	Lubricants and greases
Nitriles	Organo mercury compounds	Pentachloroethane	Manganese compounds
Nitroglycerine and EGD	Oxalic acid	Pulp and paper mills	Methyl ethyl ketone peroxide
Oil mists	Plastics and resins manufacturing	Secondary aliphatic monoamines	Naphthalene
o-Tolidine	Synthetic rubber manufacturing	Terphenyl	Nitrophenols
Organo-isocyanates	Talc	Tertiary aliphatic monoamines	Organic anhydrides
Ozone	Wood dust	Tetrahydrofuran	Organophosphates (excluding pesticides and fire-retardants)
Paint and allied products manufacturing		Trichloropropane	Phthalates
Styrene		Wood preserving	Tire manufacturing
Vinyl acetate			Vinyl cyclohexene dioxide
Vinyl compounds			

certain deficiencies exist. For one, a fair number of compounds are not mutagenic even though they are strong carcinogens in animals.[5] These are usually compounds which require metabolic activation to demonstrate their carcinogenicity. To overcome this deficiency, liver microsomal fractions from untreated rats or rats which have been previously treated (induced) with a polychlorinated biphenyl (Aroclor 1254) are often added to the culture to provide metabolic activation. A further difficulty is the extrapolation of the effects in bacteria with those in mammals.[6] Mammals have many detoxication pathways and immunological defense systems which may not be available to bacteria and other lower organisms. For these reasons some effort has been made to develop short-term bioassays by transformation of various mammalian cells such as Chinese hamster cells, mouse lymphoma cells or other cells in culture. Another test that has been proposed is determination of unscheduled DNA repair in cell systems, a measure of DNA damage by the test compounds, based on the premise that interaction with the DNA is one of the attributes of a chemical carcinogen. Still other short-term tests entail investigation of chromosomal aberrations, of teratogenic effects, of reaction with nucleic acids, or of cytological alterations in target cells.[7] A systematic evaluation of some of these tests and others has been made, using both carcinogens and non-carcinogens.[8] Although mutagenesis, DNA repair synthesis and cell transformation offer the most promise, much effort is still needed to develop these systems to a practical, reliable stage. However, they can be used as prescreens to give indications of whether further developmental work on a series of compounds might be justified.

With the increased emphasis on testing of compounds for long-term toxicity or other deleterious effects, various Government agencies have attempted to systematize the protocols they consider necessary for bioassay experiments. For example, the National Cancer Institute (USA) and Health and Welfare (Canada) have each issued a technical report on guidelines for bioassay of carcinogens.[9,10] Other Government agencies may follow suit. Consultation with the proper authorities to determine whether a test protocol is suitable and meets their criteria is necessary for any individual or concern which wishes to introduce a new compound.

TOXICITY TESTING

In the previous edition a section on long-term toxicity tests covered methods of administration of compounds, additional influences such as the species or strain of animal, age, sex, vehicle, diet and other factors which can influence the outcome of a bioassay experiment.[1] Additional reviews on the conduct of long-term bioassays and the effect of genetic or sex differences on the response to carcinogens have appeared.[11-14] Furthermore, there has been increased emphasis on administration of compounds by inhalation, leading to a new specialty in the field of toxicology. Inhalation is the most expensive route for testing due to the expensive apparatus required, including inhalation chambers, monitoring instruments, and pollution control equipment or scrubbers for the effluent from the chambers.[12,15,16] There also has been renewed interest in dietary factors which influence the effect of chemical carcinogens.[17-19] The matter of cocarcinogenesis or promoting agents, also discussed previously,[1] has become the subject of much research aimed at determining the mechanism of these effects.[20]

Extensive reviews on the structure-activity relationships of polycyclic hydrocarbons, polycyclic heteroaromatic compounds, aromatic amines and aminoazo compounds have become available.[21,22]

Cancer in Man

Substantiating an epidemiological study on members of the American Chemical Society (cf. 1), a study on graduates of the Swedish Royal Institute of Technology also indicated that cancer deaths are higher in chemists.[23] The total malignant neoplasms and lymphomas were particularly increased over the expected number. Hodgkins disease and tumors of the urinary system were also higher than usual in the Swedish chemists. The earlier study did not find such increases in the American chemists (Table 2).

TABLE 8.2. Comparison of Death Rates in American and Swedish Chemists.

	American[a]		Swedish	
	Observed	Expected	Observed	Expected
Cause of Death	Numbers			
Malignant neoplasms	444	354	22	13.0
Digestive system	143	126	6	4.5
Pancreas	36	22		
Lymphatic/ hematopoietic	94	59	6	1.7
Hodgkins disease			3	0.3
Respiratory system	74	64	1	2.4
Urinary tract	10	10	3	1.0
Other	109	84	6	3.4
Diabetes	21	26		
Respiratory diseases	30	50	3	1.8
Liver cirrhosis	17	41		
Digestive diseases			2	4.0
Heart disease	995	971		
Vascular lesions—nervous system	117	134		
Circulatory			14	20.7
Ischemic			7	15.2
Accidents and suicides	252	248	9	18.0
Other	276	328	8	9.5
Total	2152	2152	58	67.0

[a]For males, aged 20–64 years at death.

The higher cancer rate in chemists may be caused by exposure to small amounts of many different chemicals. This complicates finding whether specific chemicals may have been responsible. However, for chemical production workers, who are more likely exposed to large amounts of specific chemicals for defined periods, it is easier to determine whether such exposure led to development of cancer. If the cancers induced are of an unusual type, the correlation with occupational exposure is made more readily. Such was the case among workers in industries where polymerization of *vinyl chloride* was a major process.[24,25] These findings led to the realization that vinyl chloride was a carcinogen in man; studies in several species of animals have fully substantiated the carcinogenicity of this high-volume industrial intermediate.[26,27] Therefore OSHA regulates vinyl chloride as a human carcinogen[2] and has set 1 ppm as a limit for occupational exposure.

One epidemiological study of pliofilm workers who had a sustained exposure to *benzene* reported an increased leukemia incidence in these workers;[28] a result disputed by others.[29] In any event OSHA has established an emergency standard of 1 ppm for benzene in the workplace.

A "Current Intelligence Bulletin" from NIOSH stated that an epidemiological survey of workers in an *acrylonitrile* polymerization operation showed they had an increased risk from lung and intestinal cancer. Furthermore, administration of acrylonitrile to rats, either by inhalation or by ingestion in drinking water, led to the development of various types of tumors (central nervous system, ear duct, mammary gland, stomach, and Zymbal gland). NIOSH thus concluded that it would be prudent to handle acrylonitrile as if it were a human carcinogen.[30]

Additional bulletins from NIOSH have implicated as human carcinogens lead chromate, polychlorinated biphenyls, and possibly beryllium. The NIOSH conclusion on beryllium has been challenged.[31]

Chemicals Which Are Carcinogens in Animals

Due to the extensive bioassay program of the National Cancer Institute (NCI), results on long-term tests of several hundred pesticides, industrial intermediates, drugs, and other environmental substances are now available. These reports can be obtained from the Office of Cancer Communications, NCI, Bethesda, Md. 20014, as announced in the *Federal Register*. In due course, synopses of the results from these NCI sponsored tests plus all others reported in the scientific literature are available in the "Survey of Compounds Which Have Been Tested for Carcinogenic Activity,"[32] which is published periodically by NCI. Other sources of data on carcinogenicity tests of various compounds are the various Monographs on the Evaluation of Carcinogenic Risk of Chemicals to Man, published by the International Agency for Research on Cancer (IARC). These monographs summarize various test data and furnish an evaluation both of the effects in animals and the risk to humans.[33] IARC also publishes at intervals an information bulletin on chemicals being tested for carcinogenicity in laboratories all over the world. Therefore, these various sources should be consulted for data on any specific chemicals of interest.

As an example of some recent results from the NCI bioassays, one class of compounds will be discussed in some detail. The NCI tests indicated that various aliphatic halogenated hydrocarbons are carcinogenic in animals, in addition to carbon tetrachloride which was mentioned previously.[1]

Chloroform induced a high incidence of liver tumors in a hybrid strain of mice which generally has a relatively low incidence of such tumors. It also caused kidney tumors in male Osborne-Mendel rats and thyroid tumors in female rats after oral administration.[34] On this basis the FDA has ordered that chloroform no longer be used in such products as cough medicines, mouth washes, or toothpaste. However, exposure of the general population to chloroform continues, for it occurs frequently in potable water as a result of the reaction of humic material in raw water with chlorine during the purification process.[35]

Trichloroethylene, a solvent for degreasing, drycleaning, dissolving fats, greases, waxes, for decaffeinating coffee, or as a general anesthetic in some situations, led to liver tumors in mice when given orally in a corn oil solution. Rats were not affected significantly by trichloroethylene.[36] Because of the effect in mice, NIOSH has recommended to the Department of Labor that occupational exposure to trichloroethylene be limited. There is a recent claim that the effect of trichloroethylene was due to the stabilizers in the technical material.[37] However, the experimental evidence was based on the high mutagenic activity of the stabilizers and the low mutagenicity of pure trichloroethylene. As discussed previously, there is no absolute correlation between mutagenicity and carcinogenicity. The positive carcinogenic action in mice therefore takes precedence over the negative effect in bacteria.

Tetrachloroethylene, a widely used drycleaning and degreasing solvent caused tumors in mice similar to the effect of trichloroethylene.[38]

In the same series of experiments it was found that *ethylene dibromide* (1,2-dibromoethane), used as a gasoline additive or as a fumigant for grain supplies, induced a high incidence of squamous cell carcinomas of the stomach in both rats and mice.[38,39]

Another compound of this series, the nematocide *1,2-dibromo-3-chloropropane*, caused stomach tumors in both rats and mice and mammary tumors in female rats.[39,40] Reports that 1,2-dibromo-3-chloropropane also induced sterility in workers employed in a manufacturing operation led to an emergency temporary standard of 10 ppb for exposure to this compound. Epidemiological studies of such populations would be indicated.

The following compounds have each been the subject of a "Current Intelligence Bulletin" from NIOSH.

2-Nitropropane, a solvent for industrial coatings, printing inks and adhesives, apparently caused liver cancer in rats exposed to 200 ppm by inhalation over a 6 month period.[41]

Dimethylcarbamoyl chloride, a specialty chemical, was previously reported to cause skin or subcutaneous tumors in mice after injection (cf 1). A new study indicated that squamous cell cancer of the nose occurred in rats which had inhaled 1 ppm of the compound for 200 days.[42]

Hexamethylphosphoramide is a useful solvent both for certain synthetic fibers and for synthetic organic reactions. This compound led to nasal cancers, which penetrated the brain, in rats inhaling the compound for 8 months or longer at levels of 400 ppb or higher.[43]

An additional report from NIOSH indicates that *epichlorohydrin*, widely used to produce synthetic glycerol, epoxy resins, and elastomers, also led to a small number of nasal cancers in rats which had inhaled a level of 100 ppm. Subcutaneous injection of epichlorohydrin led to local malignant tumors in mice.[44]

Another compound implicated as a carcinogen is *hexachlorobutadiene*, the major component of the tarry waste from manufacture of tri- and tetrachloroethylene or carbon tetrachloride. Rats ingesting 20 mg/kg/day developed kidney tumors.[45]

Although it was not definitely carcinogenic in tests in mice and dogs, the antioxidant *N-phenyl-2-naphthylamine* was dephenylated by humans and dogs to 2-naphthylamine, a known human bladder carcinogen.[46,47] Thus, NIOSH recommends finding alternatives for this antioxidant or minimizing human exposure.

The intermediate and solvent, *quinoline*, caused various types of liver tumors in rats when fed in the diet at levels of 0.05–0.25 for 4–10 months.[48] Exposure to quinoline may be self-limiting because of the odor, but precautions to avoid human contact should be followed.

Nitrosamines

Other cancer risks to workers may occur because of the formation of carcinogens *in situ* from industrial precursors. For example, nitrosamines, up to a level of 3%, can be formed in the cutting fluids used in machining operations. The probable cause is the reaction of nitrite, added as an antioxidant, and the di- or triethanolamine often employed in cutting fluids.[49] Alternatives to nitrite as an antioxidant or antirusting agent in cutting fluids should be investigated to avoid this exposure. Furthermore, pesticides or herbicides which are formulated as diethanolamine salts may contain traces of nitrosamines,[50] through a similar reaction as in cutting fluids—diethanolamine with nitrite or nitrite esters used as antirusting agents. However, the levels of nitrosamines found in pesticides generally have been rather low.

Synthetic chemicals may not be the only ones involved in possible cancer risks. A report on the high cancer rate in workers from a factory processing oilseeds implicated aflatoxin contamination of the oilseeds and factory dust as the cause.[51] However, the number of workers was relatively small and further studies would be needed to confirm this conclusion.

Cancer in the General Population

In the general population there is usually no outstanding exposure to any one suspected carcinogen as may be the case for industrial workers but to very low levels of numerous environmental carcinogens. Environmentally many polynuclear aromatic hydrocarbons, (PAH), including known carcinogens as benzo[a]-pyrene are present in soots, combustion products, in soils, marine sediments, and in roasted or broiled foods.[52–54] In some populations which consume large amounts of heavily smoked fish or meat PAH may be responsible for their somewhat higher cancer incidence.

Tobacco smoking is another personal habit which leads to exposure to various types of carcinogens. Tobacco smoke contains numerous polycyclic aromatic hydrocarbons, many of which are carcinogenic in animals. It also contains other known carcinogens such as vinyl chloride, hydrazine, aromatic amines, and various nitrosamines.[55] Unburned tobacco itself contains at least four nitrosamines, one of which is a liver carcinogen.[56] Thus, the higher cancer incidence in those who smoke to excess might be explained on the basis of continued exposure to many different carcinogens.[57]

Aromatic amines occur not only in tobacco smoke but also in consumer products. Some of the components of hair and fur dyes may cause cancer in animals.[58] The excessive use of phenacetin, an aromatic amide and a common ingredient of over-the-counter analgesics, is associated with increased renal cancer incidence.[59] Laboratory studies have shown that N-hydroxyphenacetin, one of the metabolites of phenacetin, is probably responsible.[60]

Other commonly used drugs, besides those employed in the chemotherapeutic treatment of cancer, may present a cancer risk after long-continued use. Diphenylhydantoin, anabolic steroids, some oral contraceptives, and reserpine are examples.[61]

Consumption of ethanol in various beverages increases the risk of cancer in humans. Animal studies indicate that the congeners, rather than the ethanol itself, may be responsible.[62] The problem is complex, however, due to many other complicating factors which include use of tobacco and nutritional deficiencies.[63]

However, it is considered that dietary exposure to either low levels of various carcinogens or to factors which enhance or promote the action of naturally occurring carcinogens may represent one of the greatest risks to the general population. Statistics on cancer rates in various countries substantiate this premise. For example, there is an association with high fat levels in the diet and excess risk of breast cancer in many populations.[64–66] Colon cancer is linked with a high meat-high fat diet.[67] Ingestion of aflatoxin or related mycotoxins in the diet is often associated with an increased risk from liver cancer.[68] However, aflatoxin contamination of cereal grains is not only a problem in under-developed countries. Many food crops in the more developed countries may be contaminated with aflatoxin.[69] The aflatoxin problem is usually greater during years when extensive insect damage of cereal grain crops occurs. However, FDA, other agencies, and food manufacturers check carefully to avoid aflatoxin or similar compounds in peanuts, corn, and the like.

Other naturally occurring compounds are carcinogenic in animals when consumed in high doses.[70] These would include safrole, estragole, and β-asarone, all constituents of essential oils. Even constituents of certain mushrooms have come under question.[71]

The problem of endogenous formation of carcinogens is of utmost importance, particularly in the case of nitrosamines. Nitrosamines can be formed *in vivo* from secondary or tertiary amines which are present in many foods and drugs and nitrite which can be formed endogenously from nitrate.[72–74] Although some bacteria in the intestinal tract can synthesize nitrosamines, other bacteria can degrade these compounds. Nitrosamines are also formed during the cooking of meats, especially bacon, which have been cured with nitrate or nitrite, but ascorbic acid and other compounds can inhibit this reaction.[75,76]

Nitrosamines may be formed in other consumer products besides foods. The presence of di- and triethanolamine in cosmetic products apparently have led to formation of nitrosamines, analogous to the situation in cutting fluids.[77]

Another endogenous carcinogen can be formed during fermentation of foods or beverages. Ethyl carbamate, a lung carcinogen in mice and a weak or non-effective compound in rats, may be formed from carbamyl phosphate and ethanol in beverages and foods such as ale, beer, wine, yogurt, olives, bread, etc. Except for sake the levels are generally very low and probably pose no risk to the general population.[78]

The examples discussed are only a sampling of those which may be partly responsible for the background level of cancer in people. Furthermore as the population ages due to improved nutrition and control of infectious diseases, the likelihood of development of cancer increases. However, the mammalian organism is equipped with various detoxification and repair systems. These usually can cope with exposure to limited amounts of toxic materials. Moderation in lifestyle and avoidance of overwhelming exposure to toxic substances aid the natural defense mechanisms in protecting against cancer.

Chemoprevention of Cancer

There have been promising developments in the chemoprevention of cancer in laboratory animals. Several commonly used antioxidants and other naturally occurring substances can inhibit the action of carcinogens.[79] Retinoids, synthetic derivatives of vitamin A, decrease the effect of certain potent carcinogens.[80] Future research in this area may eventually allow application of these results in man.

Acknowledgment

The secretarial assistance of F. M. Williams is gratefully acknowledged.

BIBLIOGRAPHY

1. Weisburger, E. K. Industrial cancer risks. In: *Dangerous Properties of Industrial Materials*, 4th ed. Sax, N. I. (ed.), pp. 274–288, New York, Van Nostrand Reinhold, 1975.
2. Lassiter, D. V. Occupational carcinogens. In: *Environmental Cancer*, *Advances in Modern Toxicology*, Vol. 3, Kraybill, H. F. and Mehlman, M. A. (eds.), pp. 63–86, Washington, Hemisphere, 1977.
3. Public Law 94-469, 94th Congress, October 11, 1976. An Act to regulate commerce and protect human health and the environment by requiring testing and necessary use restrictions on certain chemical substances, and for other purposes.
4. Ames, B. N., McCann, J., and Yamasaki, E. Methods for detecting carcinogens and mutagens with the Salmonella/mammalian-microsome mutagenicity test. *Mutation Res.* **31,** 347, 1975.
5. McCann, J., Choi, E., and Ames, B. N. Detection of carcinogens as mutagens in the Salmonella/microsome test: Assay of 300 chemicals. *Proc. Natl. Acad. Sci. USA* **72,** 5135, 1975.
6. Dybas, R. A., Hite, M. and Flamm, W. G. Detecting mutagens—Correlation between mutagenicity and carcinogenicity of chemicals. In: *Annual Reports in Medicinal Chemistry*, Vol. 12, Clarke, F. H. (ed.), pp. 234–248, New York, Academic Press, 1977.
7. Stoltz, D. R., Poirier, L. A., Irving, C. C., Stich, H. F., Weisburger, J. H., and Grice, H. C. Evaluation of short-term tests for carcinogenicity. *Toxicol. Appl. Pharmacol.* **29,** 157, 1974.
8. Purchase, I. F. H., Longstaff, E., Ashby, J., Styles, J. A., Anderson, D., Lefevre, P. A., and Westwood, F. R. Evaluation of six short term tests for detecting organic chemical carcinogens and recommendations for their use. *Nature* **264,** 624, 1976.
9. Sontag, J. M., Page, N. P., and Saffiotti, U. *Guidelines for Carcinogen Bioassay in Small Rodents*, *National Cancer Institute*, Department of Health, Education and Welfare, DHEW Publication No. (NIH) 76-801, Bethesda, Maryland, February 1976.
10. Health and Welfare Canada. The testing of chemicals for carcinogenicity, mutagenicity, and teratogenicity. September 1973.
11. Weisburger, J. H. Bioassays and tests for chemical carcinogens. In: *Chemical Carcinogens*, Searle, C. E. (ed.). pp. 1–23, Amer. Chem. Soc. Monograph 173, Washington, 1976.
12. Page, N. P. Concepts of a bioassay program in environmental carcinogenesis. In: *Environmental Cancer*, *Advances in Modern Toxicology*, Vol. 3, Kraybill, H. F. and Mehlman, M. A. (eds.), pp. 87–171, Washington, Hemisphere, 1977.
13. Toh, Y. C. Physiological and biochemical reviews of sex differences and carcinogenesis with particular reference to the liver. *Advances Cancer Res.* **18,** 155, 1973.
14. Thorgeirsson, S. S. and Nebert, D. W. The Ah locus and the metabolism of chemical carcinogens and other foreign compounds. *Advances Cancer Res.* **25,** 149, 1977.
15. Karbe, E. and Park, J. F. (eds.). *Experimental Lung Cancer: Carcinogenesis and Bioassays*, New York, Springer, 1974.
16. Cassarett, L. J. Toxicology of the respiratory system. In: *Toxicology. The Basic Science of Poisons*. Cassarett, L. J. and Doull, J. (eds.), pp. 201–224, New York, Macmillan, 1975.
17. Clayson, D. B. Nutrition and experimental carcinogenesis: A review. *Cancer Res.* **35,** 3292, 1975.
18. Wattenberg, L. W., Loub, W. D., Lam, L. K., and Speier, J. L. Dietary constituents altering the response to chemical carcinogens. *Fed. Proc.* **35,** 1327, 1976.
19. Sporn, M. B., Dunlop, N. M., Newton, D. L., and Smith, J. M. Prevention of chemical carcinogenesis by vitamin A and its synthetic analogs (retinoids). *Fed. Proc.* **35,** 1332, 1976.
20. Slaga, T. J., Boutwell, R. K., and Sivak, A. (eds.). Mechanisms of Tumor Promotion and Cocarcinogenesis; Vol. 2 of *Carcinogenesis: A Comprehensive Survey*. New York, Raven Press, 1978.
21. Arcos, J. C. and Argus, M. F. *Chemical Induction of Cancer*, Vol. IIA, 387 pp., New York, Academic Press, 1974.
22. Arcos, J. C. and Argus, M. F. *Chemical Induction of Cancer*, Vol. IIB, 379 pp., New York, Academic Press, 1974.
23. Rawls, R. L. Cancer death rate higher for chemists. *Chem. and Eng. News*, p. 17, June 28, 1976.
24. Heath, C. W., Jr., Falk, H., and Creech, J. L., Jr. Characteristics of cases of angiosarcoma of the liver among vinyl chloride workers in the United States. *Ann. N. Y. Acad. Sci.* **246,** 231, 1975.

25. Byren, D. and Holmberg, B. Two possible cases of angiosarcoma of the liver in a group of Swedish vinyl chloride-polyvinyl cloride workers. *Ann. N. Y. Acad. Sci.* **246,** 249, 1975.
26. Maltoni, C. and Lefemine, G. Carcinogenicity bioassays of vinyl chloride: Current results. *Ann. N. Y. Acad. Sci.* **246,** 195, 1975.
27. Keplinger, M. L., Goode, J. W., Gordon, D. E., and Calandra, J. C. Interim results of exposure of rats, hamsters, and mice to vinyl chloride. *Ann. N. Y. Acad. Sci.* **246,** 219, 1975.
28. Infante, P. F., Rinsky, R. A., Wagoner, J. K., and Young, R. J. Leukaemia in benzene workers. *Lancet* **ii,** 76, July 9, 1977; also *Lancet* **ii,** 868, October 22, 1977.
29. Tabershaw, I. R. and Lamm, S. H. Benzene and leukaemia. Lancet ii, p. 867, October 22, 1977.
30. NIOSH. Current intelligence bulletin: Acrylonitrile, July 1, 1977.
31. Shapley, D. Occupational cancer: Government challenged in beryllium proceeding. *Science* **198,** 898, 1977.
32. Public Health Service Publication #149: *Survey of Compounds Which Have Been Tested for Carcinogenic Activity.* 1951; Supplement 1, 1957; Supplement 2, 1969; 1961–1967; 1968–1969; 1970–1971; 1972–1973.
33. International Agency for Research on Cancer. IARC Monographs on the *Evaluation of Carcinogenic Risk of Chemicals to Man.* Vol. 13, 14 and 15, Lyon, France, 1977.
34. National Cancer Institute. Report on the carcinogenesis bioassay of chloroform. March 1, 1976.
35. Kraybill, H. F. Global distribution of carcinogenic pollutants in water. *Ann. N. Y. Acad. Sci.* **298,** 80, 1977.
36. National Cancer Institute, Carcinogenesis bioassay of trichloroethylene. DHEW Publication No. (NIH) 76-802, 1976.
37. Henschler, D., Eder, E., Neudecker, T., and Metzler, M. Carcinogenicity of trichloroethylene: Fact or artifact? *Arch. Toxicol.* **37,** 233, 1977.
38. National Cancer Institute. Bioassay of tetrachloroethylene for possible carcinogenicity. DHEW Publication No. (NIH) 77-813, 1977.
39. Olson, W. A., Habermann, R. T., Weisburger, E. K., Ward, J. M., and Weisburger, J. H. Induction of stomach cancer in rats and mice with halogenated aliphatic fumigants. *Jour. Natl. Cancer Inst.* **51,** 1993, 1973.
40. Weisburger, E. K. Carcinogenicity studies on halogenated hydrocarbons. *Environ. Health Perspective* **21,** 7, 1977.
41. NIOSH. Current intelligence bulletin: 2-Nitropropane. April 25, 1977.
42. NIOSH. Current intelligence bulletin: Errata dimethylcarbamoyl chloride. July 7, 1976.
43. NIOSH. Background information on hexamethylphosphoric triamide. October 24, 1975.
44. Van Duuren, B. L., Goldschmidt, B. M., Katz, C., Seidman, I., and Paul, J. S. Carcinogenic activity of alkylating agents. *Jour. Natl. Cancer Inst.* **53,** 695, 1974.
45. NIOSH. Hexachlorobutadiene (HCBD). Renal tubular cancer in laboratory rats. September 7, 1976.
46. NIOSH. Current intelligence bulletin: Metabolic precursors of a known human carcinogen, beta-naphthylamine. December 17, 1976.
47. Batten, P. L. and Hathaway, D. E. Dephenylation of N-phenyl-2-naphthylamine in dogs and its possible oncogenic implications. *Brit. Jour. Cancer* **35,** 342, 1977.
48. Hirao, K., Shinohara, Y., Tsuda, H., Fukushima, S., Takahashi, M., and Ito, N. Carcinogenic activity of quinoline on rat liver. *Cancer Res.* **36,** 329, 1976.
49. Fan, T. Y., Morrison, J., Rounbehler, D. P., Ross, R., Fine, D. H., Miles, W., and Sen, N. P.: N-Nitrosodiethanolamine in synthetic cutting fluids: a part-per-hundred impurity. *Science* **196,** 70, 1977.
50. Ross, R. D., Morrison, J., Rounbehler, D. P., Fan, S., and Fine, D. H. N-Nitroso compound impurities in herbicide formulations. *Jour. Agric. Food Chem.* **25,** 1416, 1977.
51. van Nieuwenhuize, J. P., Herber, R. F. M., de Bruin, A., Meyer, P. B., and Duba, W. C. Aflatoxinen: Epidemiologish onderzoek naar carcinogeniteit bij langdurige 'low level' expositie van een fabriekspopulatie. *T. soc. Geneesk.* **51,** 706; 717; 754, 1973.
52. National Academy of Sciences. *Particulate Polycyclic Organic Matter.* 361 pp., Washington, D.C., 1972.
53. Pike, M. C., Gordon, R. J., Henderson, B. E., Menck, H. R., and SooHoo, J. Air pollution. In: *Persons at High Risk of Cancer.* Fraumeni, J. F., Jr. (ed.), pp. 225–238, New York, Academic Press, 1975.
54. Blumer, M. Polycyclic aromatic compounds in nature. *Scientific American* **234,** 35, 1976.
55. Hoffman, D., Schmeltz, I., Hecht, S. S., and Wynder, E. L. Tobacco carcinogenesis. In: *Polycyclic Hydrocarbons and Cancer, Chemistry, Molecular Biology and Environment.* T'so, P. O. P. and Gelboin, H. V. (eds.), New York, Academic Press, 1978. (*In press*).
56. Schmeltz, I. and Hoffmann, D. Nitrogen-containing compounds in tobacco and tobacco smoke. *Chem. Rev.* **77,** 295, 1977.
57. Hammond, E. C. Tobacco. In: *Persons at High Risk of Cancer.* Fraumeni, J. F., Jr. (ed.), pp. 131–137, New York, Academic Press, 1975.
58. NIOSH. Current intelligence bulletin 19: 2,4-Diaminoanisole (4-methoxy-m-phenylenediamine) in hair and fur dyes. January 13, 1978.
59. IARC Monograph on the *Evaluation of Carcinogenic Risk of Chemicals to Man.* Some miscellaneous pharmaceutical substances. Vol. 13, p. 141.
60. Calder, I. C., Goss, D. E., Williams, P. J., Funder, C. C., Green, C. R., Ham, K. N., and Tange, J. D. Neoplasia in the rat induced by N-hydroxyphenacetin, a metabolite of phenacetin. *Pathology* **8,** 1, 1976.
61. Hoover, R. and Fraumeni, J. F., Jr. Drugs. In: *Persons at High Risk of Cancer*, Fraumeni, J. F., Jr., (ed.), pp. 185–198, New York, Academic Press, 1975.
62. Rothman, J. J. Alcohol. In: *Persons at High Risk of Cancer.* Fraumeni, J. F., Jr. (ed.), pp. 139–148, New York, Academic Press, 1975.
63. Vitale, J. J. and Gottlieb, L. S. Alcohol and alcohol-related deficiences as carcinogens. *Cancer Res.* **35,** 3336, 1975.
64. de Waard, F. Breast cancer incidence and nutritional status with particular reference to body weight and height. *Cancer Res.* **35,** 3351, 1975.
65. Carroll, K. K. Experimental evidence of dietary factors and hormone-dependent cancers. *Cancer Res.* **35,** 3374, 1975.

66. Berg, J. W. Can nutrition explain the pattern of international epidemiology of hormone-dependent cancers? *Cancer Res.* **35,** 3345, 1975.
67. Hill, M. J. Metabolic epidemiology of dietary factors in large bowel cancer. *Cancer Res.* **35,** 3398, 1975.
68. Wogan, G. N. Naturally occurring carcinogens. In: *The Physiopathology of Cancer, Vol. 1. Biology and Biochemistry*. Homburger, F. (ed.), pp. 64–109 Basel, Karger, 1974.
69. Lillehoj, E. B., Fennel, D. I., and Kwolek, W. F. *Aspergillus flavus* and aflatoxin in Iowa corn before harvest. *Science* **193,** 495, 1976.
70. Miller, J. A. and Miller, E. C. Carcinogens occurring naturally in foods. *Fed. Proc.* **35,** 1316, 1976.
71. Toth, B. and Nagel, D. Tumors induced in mice by N-methyl-N-formylhydrazine of the false morel *Gyromitra esculenta*. *Jour. Natl. Cancer Inst.* **60,** 201, 1978.
72. Mirvish, S. S. Formation of N-nitroso compounds: Chemistry, kinetics, and in vivo occurrence. *Toxicol. Appl. Pharmacol.* **31,** 325, 1975.
73. Wogan, G. N. and Tannenbaum, S. R. Environmental N-nitroso compounds: Implications for public health. *Toxicol. Appl. Pharmacol.* **31,** 375, 1975.
74. Issenberg, P. Nitrite, nitrosamines and cancer, *Fed. Proc.* **35,** 1322, 1976.
75. Weisburger, J. H. and Raineri, R. Assessment of human exposure and response to N-nitroso compounds: A new view on the etiology of digestive tract cancers. *Toxicol. Appl. Pharmacol.* **31,** 369, 1975.
76. Sen, N. P., Donaldson, B., Seaman, S., Iyengar, J. R., and Miles, W. F. Inhibition of nitrosamine formation in fried bacon by propyl gallate and L-ascorbyl palmitate. *Jour. Agric. Food Chem.* **24,** 397, 1976.
77. Fan, T. Y., Goff, U., Song, L., Fine, D. H., Arsenault, G. P., and Biemann, K. N-Nitrosodiethanolamine in cosmetics, lotions and shampoos. *Fd. Cosmet. Toxicol.* **15,** 423, 1977.
78. Ough, C. S. Ethylcarbamate in fermented beverages and foods. I. Naturally occurring ethylcarbamate. *Jour. Agric. Food Chem.* **24,** 323, 1974.
79. Wattenberg, L. W. Inhibition of chemical carcinogenesis. *Jour. Natl. Cancer Inst.* **60,** 11, 1978.
80. Sporn, M. B., Squire, R. A., Brown, C. C., Smith, J. M., Wenk, M. L., and Springer, S. 13-*cis*-Retinoic acid: Inhibition of bladder carcinogenesis in the rat. *Science* **195,** 487, 1977.

SECTION 9

Toxicology

William F. Durham, Ph.D.
Director, Pesticides and Toxic Substances Effects Laboratory
National Environmental Research Center
Environmental Protection Agency
Research Triangle Park, NC

Toxicology may be defined as the science of poisons, their effects, antidotes, and detection.

In the modern-day world with our constant exposure to foreign chemicals in the form of food additives, pesticides, fertilizers, cosmetics, therapeutic drugs, and air pollutants, it is important to have some knowledge of toxicology. Basic to the science of toxicology is the fact that chemical compounds vary in their hazard to man, domestic animals, and other living members of the environment.

Thus, while practically all materials can be toxic under the proper conditions, there is a classification of materials of high toxicity and high hazard under normal conditions of use and encounter in small quantities which are called poisons. The final section of this book discusses the toxic qualities of a large number of commonly-encountered chemical materials. See also Section 1.

In addition, the dosage or amount of the chemical required to produce harm is important. The numbers used to express the toxic level constitute the quantitative aspect of toxicology.

TOXICITY RATINGS

In Section 12 in this book, the following system of toxicity ratings is used to indicate the relative hazard:

U = Unknown

This rating has been assigned to chemicals for which insufficient toxicity data were available to enable a valid assessment of hazard to be made. These compounds usually are in one of the following categories:

(a) No toxicity information could be found in the literature and none was known to the authors.

(b) Limited information based on animal experiments was available but in the opinion of the authors this information could not be applied to human exposures. In some cases this information is mentioned so that the reader may know that some experimental work has been done.

(c) Published toxicity data were felt by the authors to be of questionable validity.

NONE = No Toxicity

This designation is given to materials which fall into one of the following categories:

(a) Materials which cause no harm under any conditions of normal use.

(b) Materials which produce toxic effects on humans only under the most unusual conditions or by overwhelming dosage.

LOW = Slight Toxicity

(a) *Acute local.* Materials which on single exposures lasting seconds, minutes, or hours cause only slight effects on the skin or mucous membranes regardless of the extent of the exposure.

(b) *Acute systemic.* Materials which can be absorbed into the body by inhalation, ingestion (oral) or through the skin (dermal) and which produce only slight effects following single exposures lasting seconds, minutes, or hours, or following ingestion of a single dose, regardless of the quantity absorbed or the extent of exposure.

(c) *Chronic local.* Materials which on continuous or repeated exposures extending over periods of days, months, or years can cause only slight and usually reversible harm to the skin or mucous membranes. The extent of exposure may be great or small.

(d) *Chronic systemic.* Materials which can be absorbed into the body by inhalation, ingestion, or through the skin and which produce only slightly usually reversible effects following continuous or repeated exposures extending over days, months, or years. The extent of the exposure may be great or small.

In general, those substances classified as having "slight toxicity" produce changes in the human body which are readily reversible and which will disappear following termination of exposure, either with or without medical intervention.

MOD = Moderate Toxicity

(a) *Acute local.* Materials which on single exposure lasting seconds, minutes, or hours cause moderate effects on the skin or mucous membranes. These effects may be the result of intense exposure for a matter of seconds or moderate exposure for a matter of hours.

(b) *Acute systemic.* Materials which can be absorbed into the body by inhalation, ingestion, or through the skin and which produce moderate effects following single exposures lasting seconds, minutes, or hours, or following ingestion of a single dose.

(c) *Chronic local.* Materials which on continuous or repeated exposures extending over periods of days, months, or years cause moderate harm to the skin or mucous membranes.

(d) *Chronic systemic.* Materials which can be absorbed into the body by inhalation, ingestion, or through the skin and which produce moderate effects following continuous or repeated exposures extending over periods of days, months, or years.

Those substances classified as MOD or having "moderate toxicity" may produce irreversible as well as reversible changes in the human body. These changes are not of such severity as to threaten life or produce serious physical impairment.

HIGH = Severe Toxicity

(a) *Acute local.* Materials which on single exposure lasting seconds or minutes cause injury to skin or mucous membranes of sufficient severity to threaten life or to cause permanent physical impairment or disfigurement.

(b) *Acute systemic.* Material which can be absorbed into the body by inhalation, ingestion, or through the skin and which can cause injury of sufficient severity to threaten life following a single exposure lasting seconds, minutes, or hours, or following ingestion of a single dose.

(c) *Chronic local.* Materials which on continuous or repeated exposures extending over periods of days, months, or years can cause injury to skin or mucous membranes of sufficient severity to threaten life or cause permanent impairment, disfigurement, or irreversible change.

(d) *Chronic systemic.* Materials which can be absorbed into the body by inhalation, ingestion or through the skin and which can cause death or serious physical impairment following continuous or repeated exposures to small amounts extending over periods of days, months, or years.

TOXICITY AND HAZARD

Toxicity is the ability of a chemical to produce injury once it reaches a susceptible site in or on the body. Hazard is the probability that injury will be caused by the circumstances of use.

EXPOSURES

(1) Acute exposure: This term is used to mean "of short duration." Applied to materials which are inhaled or absorbed through the skin, it refers to a single exposure of a duration measured in seconds, minutes, or hours. As applied to materials which are ingested, it refers generally to a single quantity or dose.

(2) Subacute exposure: This refers to exposures of intermediate duration, i.e., between acute and chronic. Generally speaking, subacute exposures include durations up to about 90 days.

(3) Chronic exposure: This term will be used in contrast to "acute" and means "of long duration." As applied to materials which are inhaled or absorbed through the skin, it refers to prolonged or repeated exposures of a duration measured in days, months, or years. As applied to materials which are ingested, it refers to repeated doses over a period of days, months, or years. The term "chronic" will not refer to severity of symptoms but will carry the implication of exposures or doses which would be relatively harmless unless extended or repeated over long periods of time (days, months, or years).

EFFECTS

It is important to differentiate between acute and chronic *exposure* and acute and chronic *effects*. Although the expression "chronic toxicity" is sometimes used to indicate the result of repeated exposure to a chemical or to ionizing radiation, it would be much clearer if "chronic toxicity" were equated with chronic illness resulting from these agents without any commitment regarding the duration of exposure. The fact is that some compounds have a strong tendency to produce chronic illness even though the exposure may be acute (i.e., only a single dose). Such compounds include the heavy metals and most carcinogens. Of course, their tendency to produce chronic sickness is accentuated if they are absorbed in repeated doses. At the opposite extreme are compounds such as cyanide with which it is virtually impossible to produce chronic illness—even though a single excessive dose may produce acute poisoning and rapid death. Most compounds lie somewhere between the two extremes. Ordinary alcoholic intoxication is acute, but years of excessive drinking and the accompanying malnutrition can produce chronic organic disease that remains even though no more alcohol is consumed.

The words acute and chronic applied to illness have nothing to do with severity but only with the duration and character of illness. The common cold, intoxication from social drinking, plague, and parathion poisoning are all acute illnesses. The first two are mild, the last two potentially fatal; all are brief with little tissue reaction. Pulmonary tuberculosis and lead poisoning are almost always chronic diseases. They are characterized by a prolonged course and by pathological changes in tissue that reflect continuing injury and perhaps ineffectual repair.

Toxic effects may also be subdivided on the basis of site of action:

(1) Local effect: This term means that the action takes place at the point or area of contact. The site may be skin, mucous membranes of the eyes, nose, mouth, throat, or anywhere along the respiratory or gastrointestinal system depending on the body area exposed. Absorption does not necessarily occur.

(2) Systemic effect: This term refers to a site of action other than the point of contact and presupposes that absorption has taken place. It is possible, however, for toxic agents to be absorbed through a channel (skin, lungs, or intestinal canal) and produce later manifestations on one of those channels which are not a result of the original direct contact. Thus it is possible for some agents to produce harmful effects on a single organ or tissue as a result of both "local" and "systemic" actions.

DOSAGE

The single most important factor in determining whether or not illness will occur as the result of exposure to a specific chemical compound is dosage. The dosage concept leads to the conclusion that no chemical compound is completely safe and that none is entirely harmful. The same idea was expressed by Paracelsus (1493–1541), who wrote "All things are poisons, for there is nothing without poisonous qualities. It is only the dose which makes a thing a poison." See also Section 1.

Compounds vary tremendously in their toxicity (see also Section 1). Usually harmless and even essential substances, such as water and salt, may cause illness or death if consumed in sufficient amount. Even compounds recognized as poisons may differ in toxicity by a factor of at least ten billion. The most toxic material presently known is botulinum toxin, which is more poisonous by several orders of magnitude than any other compound known. It is interesting to note that this is not a synthetic compound but is a material of biological origin.

In comparing the toxicity of different compounds, it is convenient to have a standardized notation for describing the toxic level. The most commonly used notation is the median lethal dose or LD_{50}, which is a statistical estimate of the dosage necessary to kill 50 per cent of an infinite population of the test animals. The LD_{50} is usually expressed in terms of the weight of poison per unit of body weight, most often as mg of chemical/kg of animal (mg/kg). This permits a meaningful comparison of the susceptibility of animals of different species regardless of their size.

The LD_{50} is a special case of a more general measure of effect, the median effective dosage (ED_{50}). The ED_{50} is the dosage necessary to produce any specified effect in 50 per cent of the test animals. The effect may be anything that can be observed. It may be a specified degree of inhibition of an enzyme. It may be the production of a tumor. These measures of effect (i.e., ED_{50} or LD_{50}) for a particular compound have meaning only if experimental condition are defined, including the species, age, and sex of experimental animal, the number of doses and the route of administration.

In addition to the identity of the compound involved and dosage, there are a number of less important factors which play a role in determining whether or not illness may result from any particular exposure episode.

Among the other factors which should be considered are:

1. Route of exposure: It is important to learn whether the toxicant was ingested, inhaled, spilled on the skin, or, as most often happens, exposure was caused by some combination of these.
2. Type of formulation or state of dispersion of the toxicant.
3. Temperature.
4. Humidity.
5. Physiologic condition of the subject.
6. Interaction of the toxicant with other chemicals and drugs. Interaction of chemicals may occur either in the external environment (such as when exposure to a mixture of two industrial chemicals occurs), or inside the patient (such as when previous dosage of a therapeutic drug may alter response to another foreign chemical). It is now recognized that the latter type of reaction occurs with some frequency as a result of stimulation of the drug-metabolizing enzyme system of the liver (microsomal enzymes) by various chemicals and drugs.

TOXIC LEVELS

A method of expressing toxic levels for different compounds is the use of Threshold Limit Values (TLV), formerly known as Maximum Allowable Concentrations (MAC) (see Section 1).

It should be noted that the LD_{50} represents an experimentally-derived value while the TLV is arbitrarily set on the basis of known experimental and other available data. The acute oral LD_{50} value is most useful if one is dealing with an ingestion case while the TLV is more pertinent for industrial and occupational exposure restrictions.

In the United States, TLV's have received wide acceptance. The ACGIH has now set TLV's for about 700 chemical compounds which are listed in Section 1, refer to airborne concentrations and represent conditions under which it is believed that nearly all workers may be repeatedly exposed for an 8-hour day, 5 days a week, without harmful effect.

Since there are wide variations in individual susceptibility, a small percentage of workers may experience some discomfort and a smaller fraction may be affected more seriously by substances at concentrations at or near the TLV. Sensitivity testing should be carried out on new industrial workers being placed in jobs involving significant exposure to potentially toxic chemicals so that those persons who are hypersusceptible may be screened out.

With some few exceptions, the TLV's are time-weighted average concentrations. Thus, limited temporary excursions above the limit may be permitted provided they are compensated for by equivalent excursions below the limit during a work day.

In spite of the fact that no serious injury is expected to occur as a result of exposure to TLV concentrations, the best industrial hygiene practice is to maintain concentrations of all atmospheric contaminants in workroom air, at levels as low as practicable.

Closely related to TLV's are the so-called acceptable concentration standards promulgated by the American Standards Association. According to the ASA, these standards are designed to prevent "(1) undesirable changes in body structure or biochemistry; (2) undesirable functional reactions that may have no discernible effects on health; (3) irritation or other adverse sensory effects."

For gases and vapors the TLV is usually expressed in parts per million (ppm), that is parts of the gas or vapor per million parts of air. For fumes or mists, and for some dusts, the TLV is usually listed as milligrams per cubic meter, (mg/m^3). It is possible to make conversions between these two sets of units (i.e., ppm and mg/m^3) by means of the following formula:

$$ppm = \frac{mg/m^3 \times 24.45}{\text{molecular weight of compound}}$$

For some dusts, particularly those containing silica, the TLV is usually expressed as millions of particles per cubic foot of air (mppcf).

Classes of Toxic Substances

Toxic or harmful substances encountered in industry or in some instances released into the environment (such as pesticides) may be classified in various ways. A simple and useful classification is given below, together with definitions adopted by the ASA. See also Sections 1 and 2.

Dusts. Solid particles generated by handling, crushing, grinding, rapid impact, detonation, and decrepitation of organic or inorganic materials such as rocks, ore, metal, coal, wood, grain, etc. Dusts do not tend to flocculate except under electrostatic forces; they do not diffuse in air, but settle under the influence of gravity.

Fumes. Solid particles generated by condensation from the gaseous state, generally after volatilization from molten metals, etc., and often accompanied by a chemical reaction such as oxidation. Fumes flocculate and sometimes coalesce.

Mists. Suspended liquid droplets generated by condensation from the gaseous to the liquid state or by breaking up a liquid into a dispersed state, such as by splashing, foaming, and atomizing.

Vapors. The gaseous form of substances which are normally in the solid or liquid state and which can be changed to these states by either increasing the pressure or decreasing the temperature alone. Vapors diffuse.

Gases. Normally formless fluids which occupy the space of enclosure and which can be changed to the liquid or solid state only by the combined effect of increased pressure and decreased temperature. Gases diffuse.

This classification does not include the obvious categories of solids and liquids which may be harmful, nor does it encompass physical agents. The latter, strictly speaking, cannot be considered "substances." Living agents, such as bacteria, molds, and other parasites comprise another group of "substances" that would appear in a comprehensive classification of industrial health hazards.

ROUTES OF ABSORPTION

In the physiological sense, a material is said to have been absorbed only when it has gained entry into the blood stream and consequently is carried to all parts of the body. Something which is swallowed and which is later excreted more or less unchanged in the feces has not necessarily been absorbed, even though it may have remained within the gastrointestinal tract for hours or even days.

Although there are a number of ways by which an individual may be exposed to a poisonous chemical, the three most important and most usually encountered routes are: oral, dermal, and respiratory. In accidental poisoning, as is usually the case with children, ingestion or oral exposure is most common. In industrial situations, exposure is usually either respiratory or dermal or a combination of the two.

Absorption Through the Skin = Dermal route

In general, chemicals are absorbed more slowly and less completely through the skin than from the gastrointestinal tract or the lungs, and thus toxic effects are less likely to occur when a toxicant is spilled on the skin than when it is ingested or inhaled. Fortunately, many compounds are just not absorbed to a significant degree by the skin. For example, botulinum toxin, the substance which produces botulism, is extremely toxic to man when taken orally, but is relatively harmless when applied to the skin. The principal reason for this difference is that the botulinum toxin molecule is very large compared to most chemicals, and is therefore not readily absorbed by the skin. On the other hand, when it enters the digestive tract, the toxin is immediately absorbed and transported to the susceptible sites of action. Thus, the chemical and physical properties of the toxicant determine which routes of exposure will produce what toxic effects. There is greater magnitude of difference between compounds with regard to skin absorption than for other common exposure routes.

There are some few compounds which are more toxic when spilled on the skin than when swallowed. Usually these are compounds which are rapidly detoxified by the liver. The detoxication site is important since materials absorbed by the gastrointestinal tract are carried directly to the liver and only subsequently are distributed to the rest of the body. Materials absorbed through the skin or the lungs are distributed by the blood to the rest of the body at the same instant that a portion passes to the liver.

One of the problems with compounds which present a skin absorption hazard is that it is frequently difficult to convince a layman, such as a farmer or an industrial worker, that a chemical can cause poisoning merely by spilling some on his skin.

It is recognized that skin absorption may be a significant factor in occupational mercury poisoning as well as in a number of other industrial diseases. In the case of metals other than mercury, however, entry through the skin is relatively unimportant except for some organometallic compounds such as lead tetraethyl.

Skin absorption attains its greatest importance in connection with the organic solvents. It is generally recognized that significant quantities of these compounds may enter the body through the skin either as a result of direct accidental contamination or indirectly when the material has been spilled on the clothing. An additional source of exposure is found in the fairly common practice of using industrial solvents for removing grease and dirt from the hands and arms, in other words, for washing purposes. This procedure, incidentally, is a fruitful source of dermatitis.

There are a number of other important environmental and industrial chemicals which are also absorbed to a significant degree through the skin. Some of the organic phosphorus pesticides, including parathion, are notable in this regard.

Gastrointestinal Absorption = Oral route

In accidental poisoning of children, oral ingestion is the most common route of exposure. In the past it has been common practice to attribute certain cases of occupational poisoning to unclean habits on the part of the victim, particularly failure to wash his hands before eating. There is no doubt that some toxic materials used industrially can be absorbed through the intestinal tract, but it is now generally believed that with certain notable exceptions this portal of entry is of minor importance. One outstanding exception is the case of the radium dial painters who followed the practice of "pointing" their brushes between their lips, thus ingesting lethal quantities of radioactive material. Accidental swallowing of harmful amounts of poisonous compounds in single large doses has also been known to occur. In general it can be said that intestinal absorption of industrial poisons is of minor importance and that the "dirty hands" theory of poisoning has been pretty well discredited. Recent experimental studies of pesticide workers have shown that eating and smoking without washing the hands, while not recommended procedure, are not major sources of exposure.

Absorption through the Lungs = Inhal route

The inhalation of contaminated air is by far the most important means by which occupational poisons gain entry into the body. It seems safe to estimated that at least 90 per cent of all industrial poisoning (exclusive of dermatitis) can be attributed to absorption of poisons through the lungs. Harmful substances may be suspended in the air in the form of dust, fume, mist, or vapor, and may be mixed with the respired air in the case of true gases. Since an individual under conditions of moderate exertion will breathe about 10 cubic meters of air in the course of an ordinary 8-hour working day, it is readily understood that any poisonous material present in the respired air offers a serious threat to health.

Fortunately, all foreign matter which is inhaled is not necessarily absorbed into the blood. A certain amount, particularly that which is in a very finely divided state, will be immediately exhaled. Another portion of respired particulate matter is trapped by the mucus which lines the air passages and is subsequently brought up in the sputum. In this connection it might be mentioned that some of the sputum may be consciously or unconsciously swallowed, thus affording an opportunity for intestinal absorption. Other particles are taken up by "scavenger cells" following which they may enter the blood stream or may be deposited in various tissues or organs. True gases will pass directly from the lungs into the blood in the same manner as the oxygen in inspired air. Because of the fact that a great majority of the known industrial poisons may at some time be present as atmospheric contaminants and thus constitute a potential threat to health, programs directed toward the prevention of occupational poisoning generally place major emphasis on ventilation control (section 2) to reduce the hazard.

In section 12 of this book the following (mainly experimental) routes of exposure are included under, "acute tox data," to indicate how the various data were obtained;

im = intra muscular
ic = intra cerebral
id = intra duodenal
imp = via implantation
in = intra dermal
ip = intra peritoneal
ipl = intra pleural
ir = intra renal
it = intra tracheal
iv = intra venous
ivg = intra vaginal
rec = via rectum
sc = via subcutaneous
oral = via ingestion
dermal = via skin contact
inhal = via inhalation

STORAGE AND EXCRETION

Some toxic substances can be retained or stored in the body for indefinite periods of time, being excreted but slowly over periods of months or years. Lead, for example, is stored primarily in the bones and mercury principally in the kidneys. Smaller amounts may be stored in other organs or tissues. Particulate matter, when inhaled, can be phagocytosed and remain in regional lymph nodes where it may have little effect, as in the case of coal dust, or may produce pathological lung changes as in the case of SiO_2 and beryllium.

The excretion of toxic agents takes place through the same channels as does absorption, namely, lungs, intestines, and skin, but the kidneys (urine) are the main excretory organs for many substances. Sweat, saliva, milk, and other body fluids may participate to a small extent in the excretory process. Gases and volatile vapors are commonly excreted via the lungs and breath. This fact can sometimes be used as a measure of earlier absorption.

Many organic compounds are not excreted unchanged, but pass through what is known as biotransformation. The processes by which this occurs are also considered "detoxication mechanisms." The resulting new compounds, or metabolites, can subsequently be found in the urine and are used as evidence of absorption of the present substance.

INDIVIDUAL SUSCEPTIBILITY

The term "individual susceptibility" has long been used to express the well-known fact that under conditions of like exposure to potentially harmful substances there may be a marked variability in the manner in which individuals will respond to a given exposure. Some may show no evidence of intoxication whatsoever; others may show signs of mild poisoning, while still others may become severely or even fatally poisoned. Comparatively little is known about the factors responsible for this variability. It is believed that differences in the anatomical structure of noses may be concerned with different degrees of efficiency in filtering out harmful dusts from the inspired air. Previous infections of the lungs, particularly tuberculosis, are known to enhance susceptibility to silicosis. Most industrial toxicologists believe that obesity is an important predisposing factor among persons who are subject to occupational exposure to organic solvents and related compounds. Age and sex are also believed to play a part and a previous illness may be significant. Liver enzyme activity is known to be important in determing the fate of many compounds.

Other possible factors relating to individual susceptibility are even less understood than those just mentioned. It has been suggested that different rates of working speed, resulting in variations in respiratory rate, in depth of respiration, and in pulse rate may play a part. The action of the cilia, those tiny hairs present in the cells which line the air passages, may have some importance. The permeability of the lungs may influence absorption and the efficiency of the kidneys may govern the rate at which toxic materials are excreted, but the underlying nature of these possible variations is not known.

There is considerable literature purporting to show that nutritional factors may have something to do with susceptibility to occupational poisoning. Most of the published material is rather unscientific and unconvincing, but a few reports strongly suggest that there actually exists a relationship between the nature of the diet and susceptibility to poisoning. There is as yet no substantial evidence that the addition of vitamin concentrates, milk, or special foods have any protective value, but when diets are deficient in some of the essential nutritional elements it appears that poisoning is more likely to occur. There is considerable evidence that indulgence in alcoholic beverages will significantly increase the possibility of occurrence of occupational poisoning, particularly from organic solvents.

ACUTE AND CHRONIC EFFECTS

Industrial toxicology is generally concerned with the effects of low-grade sublethal exposures which are continued over a period of months or years. It is, of course, true that toxicological problems are frequently presented as a result of accidents which create sudden massive exposures to overwhelming concentrations of toxic compounds. The acute poisoning which results may cause unconsciousness, shock, or collapse, severe inflammation of the lungs, or even sudden death. An understanding of the nature of the action of the offending agent may be of great value in the treatment of acute poisoning but in some instances the only application of toxicological knowledge will be to establish the cause of death.

The detection of minute amounts of toxic agents in the atmosphere and in body fluids (blood and urine) and the recognition of the effects of exposure to small quantities of poisons are among the principal jobs of the industrial toxicologist. The manifestations of chronic poisoning are often so subtle that the keenest judgment is required to detect and interpret them. The most refined techniques of analytical chemistry and of clinical pathology are called into play, involving studies of the working environment and of exposed individuals (see also Section 1).

In order to demonstrate that chronic industrial poisoning has taken place or is a possibility it must be shown that an offending agent is present in significant concentrations, that it has been absorbed, and that it has produced in the exposed subject, disturbances compatible with poisoning by the suspected substance. Significant concentrations are ordinarily expressed in terms of TLV's. Absorption of a substance may be proved by demonstrating its presence in the blood or urine in concentrations above those found in nonexposed persons, or by finding certain metabolic products of the substance in the excreta. To prove that disturbances have occurred in an exposed subject may require the application of all the diagnostic procedures used in medicine, including a medical history, physical examination, blood counts, urinalysis, x-ray studies and other measures.

A few of the more widely used industrial chemicals, notably lead and benzene, will produce changes in the blood in the very early stages of poisoning. Other chemicals, particularly chlorinated hydrocarbons, give no such early evidence of their presence. Heavy metals such as mercury and lead produce their chronic harmful effects through what is known as "cumulative action." This means that over a period of time the material which is absorbed is only partially excreted and that increasing amounts accumulate in the body, which is to say that the toxicant is coming in faster than it is being excreted. Eventually the quantity in the body, becomes great enough to cause physiologic disturbances. Volatile compounds do not accumulate in the body but probably produce their chronic toxic effects by causing a series of small insults to one or more of the vital organs.

SITE OF ACTION OF POISONS

Brief mention has already been made of the fact that different poisons act on different parts of the body. Many substances can produce a local or direct action upon the skin. The fumes and mists arising from strong acids, some of the war-gases, and many other chemicals have a direct irritating effect on the eyes, nose, throat, and lower air passages. If they reach the lungs they may set up a severe inflammatory reaction called chemical pneumonitis. These local effects are of greatest importance in connection with acute poisoning. More important to the industrial toxicologist are the so-called systemic effects.

Systemic or indirect effects occur when a toxic substance has been absorbed into the blood stream and distributed throughout the body. Some materials, such as arsenic, when absorbed in toxic amounts, may cause disturbances in several parts of the body: blood, nervous systems, liver, kidneys, and skin. Benzene, on the other hand, appears to significantly affect only one organ, namely, the blood-forming bone marrow. Carbon monoxide causes asphyxia by preventing the hemoglobin of the blood from carrying out its normal function of transporting oxygen from the lungs to the tissues of the body. Although oxygen starvation occurs equally in all parts of the body, brain tissue is most sensitive; consequently the earliest manifestations are those due to damage to the brain. An understanding of what organ or organs can be damaged, and the nature and manifestations of the damage caused by various compounds, is among the more important functions of the industrial toxicologist.

At the cellular level, toxic agents may act on the cell surface or within the cell, depending on "receptors" or binding sites. A familiar example is the affinity of arsenic and mercury for sulfhydryl (S–H) groups in biological material.

ABSORPTION AND POISONING

As mentioned above, with the exception of external irritants, toxic substances generally must be absorbed into the body and distributed through the body by means of the blood stream in order for poisoning to occur. In other words, poisoning ordinarily does not occur without absorption. On the other hand, absorption does not necessarily or always result in poisoning. The human body is provided with an elaborate system of protective mechanisms and is able to tolerate to an amazing degree, the presence of many toxic materials. Some foreign materials are excreted unchanged through the urine and feces. Toxic gases, following absorption, may be excreted through the lungs. Some chemical compounds go through processes of metabolism and are excreted in an altered form. Some of these processes are known as detoxication (or detoxification) mechanisms. In some instances the biotransformation products in a detoxication process may be more toxic than the original substance, for example, the conversion of parathion to paraoxon.

PREVENTION

First Aid

Emergency Treatment of Acute Poisoning. Acute poisoning may be the result of entry into the body of large or concentrated doses of a poison through

(a) Breathing (inhalation),
(b) Swallowing (ingestion),
(c) Skin absorption,
(d) Injection (hypodermic or intravenous entry).

It is obvious that the route of entry will influence the type of emergency treatment.

In every case of poisoning, summon medical assistance immediately. The names and telephone numbers of one or more on-call physicians, the nearest hospital, and ambulance service should be posted near appropriate telephones.

If the police department, fire department, or utility company maintains an emergency service, its telephone numbers should also be posted.

Every industrial establishment, no matter how small, should have at least one person on duty at all times who is trained and designated to take charge in the event of an emergency due to poisoning. This individual should be conversant with the emergency handling of the particular situations that may arise.

Improper first aid may be more harmful than none at all.

Although prompt action is always important, there are relatively few situations in which a delay of seconds or minutes will have a significant bearing on the outcome.

When possible, a sample of the suspected poison, or the container from which it came, should be preserved for the guidance of the treating physician, the police, or the medical examiner (coroner).

General Procedures

(A) Inhalation:

(1) Remove victim from contaminated area. Rescuers should be properly protected or provided with life lines.

(2) Keep victim warm (not hot) and quiet. Lying flat is usually the best position.

(3) If breathing has stopped, give artificial respiration.

(4) Administer oxygen, if it is available.

(5) Keep breathing passages open. Examine mouth for false teeth and chewing gum and if present, remove them.

(B) Ingestion:

(1) Attempt to empty the stomach by causing vomiting by use of an emetic. This should be done even if a period of several hours has passed since the poison was swallowed. Exceptions: Corrosive chemicals such as strong acids or caustic alkalies; victim having convulsions; victim unconscious.

(2) Dilute the poison by administering fluids in any of the following forms:

(a) Plain tap water: 3–4 glasses.
(b) Soapy water: 2–3 glasses.
(c) Table salt in warm water: one tablespoon to an ordinary 8-ounce tumbler.
(d) Milk: 3–4 glasses.

If these fluids are vomited, which is desirable, the dose may be repeated several times.

(3) Give the victim a "universal antidote" i.e. a mixture of powdered burnt toast (charcoal), strong tea, and milk of magnesia. This will absorb and neutralize many poisons. (One piece of toast and 4 tablespoons of milk of magnesia in a cup of strong tea.)

(C) Skin Contact:

(1) Dilute the contaminating substance with large amounts of water. This is best done in a shower, but may also be done with a hose, buckets or other means. The water should be lukewarm if possible.

(2) Remove contaminated clothing. Those assisting the victim should protect their own skin with gloves, if available.

(3) Chemical burns of the eye should be treated with large amounts of water or with a weak solution of bicarbonate of soda (a level teaspoonful of bicarbonate to 1 quart of warm, clean water).

(D) Injected Poisons:

(1) Absorption may be delayed by the application of a tourniquet about the point of injection if this is in the arm or leg. The tourniquet should not be so tight that it impairs the flow of arterial blood.

(2) Excretion can be hastened by administering large doses of water or other fluid.

(3) Attempt to determine the nature of the poison so that the proper antidote can be given.

ALLERGY AND HYPERSENSITIVITY

One of the most difficult toxic manifestations of chemical substances with which the toxicologist has to deal is allergy or hypersensitivity.

Allergy is estimated to be a health problem for from 10 to 30 per cent of the population. Generally, an allergic disease is not fatal. However it may be disabling. In industry, the problem of lost worktime due to allergic response from on-the-job contact with chemical or other allergies is a perplexing one and is complicated by off-the-job contacts over which the industrial hygienist has little control.

Any substance which produces an allergic reaction is called an allergen. Almost any substance can be allergenic to an allergy-predisposed individual. An allergen affects only the susceptible individual because only the allergic person exhibits an abnormal response to the particular allergen. Usually the substance is otherwise harmless; in fact, some usually wholesome foods, such as milk, eggs, or wheat, may be specific causes of allergic disease.

Allergy is related to a specific substance. Thus, if one is allergic to an animal, he may not be allergic to all animals, or to all furs, or to all chemicals. This specificity can often permit a worker sensitized to one type of chemical to work with another material. However, an allergy-prone person can become cross-sensitized; i.e., having become allergic to one chemical he subsequently develops allergy to related compounds.

Although the general belief is to the contrary, the allergic reaction is, in general, a quantitative one. Even though a very minute quantity of an allergen may precipitate a reaction. In general, the amount of chemical required to produce an allergic response is less than that needed to bring about a systemic toxic response. Slight contact with an allergen produces light symptoms, while contact with a large amount of the same material may produce a much more severe response. It is important to understand the quantitative nature of the allergic response in order to appeciate that simply decreasing the magnitude of an individual's exposure to an offending substance may eliminate his allergic symptoms. Protective measures to guard personnel against allergens are the same as for toxic substances, generally, and are discussed in Sections 1, 2, and 8.

Time is also important in the build-up of an allergy. It usually requires long contact with a substance for an individual to become allergic to that material. The nature of the allergic response depends to some extent on the route of exposure. Thus, since industrial exposure is primarily through the skin or respiratory tract, symptomatology most often involves the skin, sinuses, nasal or respiratory system, and eyes. Current research indicates that other immunologic mechanisms may also be important in the physiologic reaction to exposure to foreign chemicals. Thus, some compounds may act to suppress the production of antibodies and other response mechanisms important in resistance to bacterial or viral infections.

SECTION 10A*

Chemical Substances Legislation

Marilyn C. Bracken, Ph.D.
Office of Toxic Substances
Environmental Protection Agency
and
E. June Thompson
Engineering & Sciences Directorate
Consumer Product Safety Commission

INTRODUCTION

Chemical substances legislation has steadily evolved during this Century, from initial recognition of the need to prevent people from being poisoned by chemicals in processed foods and medicinal products, to a new awareness of the need for protecting the entire ecosystem from the adverse effects of chemicals.

Prior to the 1970s, the legislation to control the use of chemical substances had been enacted with the objective of protecting individuals from toxic exposure to food and cosmetic additives, drugs, pesticides, and certain hazardous substances in consumer products intended for household use.

Manufacturers of food additives, drugs, and pesticides were required to provide evidence of the safety and effectiveness of the proposed chemical substance or chemical products.

Although there were no premarket clearance requirements for household chemicals such as cleaning agents, manufacturers were required to label their products with appropriate warning statements as to the specific hazard(s) which might be associated with the use of a given product. The label was required to be readily legible and to include precautionary information.

However, the chemical substances legislation enacted during the earlier decades of this Century did not adequately address the industrial and environmental exposures to the thousands of chemicals in production, nor did they offer opportunity to evaluate potential risks associated with the hundreds of new chemicals that are introduced into the marketplace each year.

Most of the early legislation addressed acute toxicity hazards and not the adverse effects resulting from chronic, long-term exposure to chemicals.

*This section was written by Marilyn C. Bracken and E. June Thompson in their private capacity. No official support or endorsement by EPA or CPSC is intended or should be inferred.

In his statements supporting passage of the Toxic Substances Control Act (TSCA), the Honorable Russell E. Train, former Administrator of EPA,[1] offered the following observation:

> We live surrounded by growing amounts of new and displaced chemical substances, several of which pose hazards to human health or the environment even in minute quantities. Some 2 million chemical compounds have been identified and many thousands of new chemicals are discovered each year. Most new compounds are laboratory curiosities, but thousands of chemical compounds are already in commercial use and several hundred new chemicals are introduced into such use each year.
>
> Many of these chemicals ultimately are discharged into our water, air, and soil systems. After the substances enter the environment, they may be diluted or concentrated by physical forces, and they may undergo chemical changes, including combination with other chemicals, that affect their toxicity. The substances may be picked up by living organisms which may further change and either store or eliminate them. The results of the interactions between living organisms and chemical species are often unpredictable, but such interactions may produce materials or concentrations that are more dangerous than that of the initial pollutants.

Mr. Train further noted that current and past controls over chemical substances pertain to only a small percentage of those chemicals which ultimately find their way into the environment, and that existing controls ". . . suffer from the limited focus of their authority . . .".

Most early chemical substances legislation did not provide adequate authority for identifying potential hazards beforehand, nor for preventing long-term adverse effects. For example, polychlorinated biphenyls (PCB's) were first introduced into commerce in the 1930s, but the hazard potential was not recognized until the late 1960s. Mr. Train cited an incident in Japan involving accidental exposure to high levels of PCB's in rice contaminated from improper handling of the chemicals, which produced a variety of adverse effects including skin disorders, blindness, and damage to the liver among the exposed population. He went on to note that PCB's are extremely stable, slowly degradable organic chemicals and, given their former widespread use, have been found in significant levels in the soil, water, and several species of wildlife, as well as in human tissue samples. Continuing, Mr. Train stated "Tests with PCB's have shown that concentrations of 0.1 ppm were fatal to juvenile pink shrimp after 48-hour exposure and the same concentration stopped oyster shell growth in 96 hours and that . . . "the full range and effects of these chemicals are yet to be determined."

Previous legislation allowed regulatory agencies to react, albeit on a somewhat limited basis, once adverse effects resulting from exposure to hazardous chemicals were recognized. Thus, when early reports of the carcinogenic effects associated with industrial exposure to vinyl chloride monomer (VCM) became known, all the regulatory agencies with jurisdictional responsibility over release of and products containing VCM quickly acted to promulgate regulations which either banned the use of VCM in commercial products, or which minimized industrial and environmental exposures.

Mr. Train concluded: "Our awareness of environmental threats, our ability to screen and test substances for adverse effects, and our capabilities for monitoring and predicting, although inadequate, are now sufficiently developed that we need no longer remain in a purely reactive posture with respect to chemical hazards. We need no longer be limited to repairing damage after it has been done; nor should we allow our population to be used as a laboratory for discovering adverse health effects. There is no longer any valid reason for continued failure to develop and exercise reasonable controls over toxic substances in the environment."

Regulation of chemical substances is a dynamic process, reflecting changing "state-of-knowledge." The emphasis of recent regulations is prevention of adverse effects through the many varied mechanisms available to regulatory agencies under the statutes, including labeling requirements, safety packaging, requirements for protective work clothing and equipment in the workplace, mandatory emission control mechanisms, as well as provisions requiring evidence of safety and efficacy of many chemical substances prior to their being marketed.

There is increasing emphasis on the need to develop predictive capabilities, as well as on prevention, which will enable regulatory agencies to act prospectively before a hazard has been demonstrated.

Various carcinogen policies are now being developed or have recently been proposed[2] which can form a basis for promulgation of regulations (or for setting emergency temporary standards) once a chemical substance is classified as a potential human carcinogen based on criteria developed in the carcinogen policies, such as evidence the carcinogenic effect of a substance in experimental animal studies (See Sections 1, 2, 8.)

The most potentially effective preventive regulations will be those promulgated under the new Toxic Substances Control Act (TSCA) which not only provides for premarket testing and use restrictions on certain chemical substances, but also for pre-manufacture regulation as well.

CHEMICAL SUBSTANCES LEGISLATION PRIOR TO 1970

In the early 1900s, legislators and medical scientists recognized the need to protect individuals from exposure to deleterious substances increasingly used in or contaminating processed food products and drugs and other medications, as well as from harmful effects associated with the burgeoning use of pesticides.

This important early legislation included the Food and Drug Act of 1906 (replaced in 1938 by the Federal Food, Drug and Cosmetic Act), the Federal Insecticide Act of 1910 (repealed in 1947 and replaced by the Federal Insecticide, Fungicide and Rodenticide Act) FIFRA and the Federal Hazardous Substances Labeling Act of 1961 (amended in 1966 as the Federal Hazardous Substances Act).

Early legislation was also enacted which reflected concern for water pollution. In 1948, laws were enacted for the purpose of restoring and preserving the national water resources. In 1961, Congress created the Water Pollution Control Program administered by the Department of Health, Education and Welfare (DHEW). This program was further implemented by passage of the Water Quality Act of 1965 and the Clean Waters Restoration Act of 1966. The functions of the Water Pollution Control Program were transferred to the Secretary of the Interior in 1966, and the administration of the Acts was subsequently transferred to the Environmental Protection Agency in 1970.

Other important legislation was enacted to require the labeling of hazardous chemical substances in transportation. These labeling requirements are discussed in Sections 11, 1, 2, 10B.

The Federal Food, Drug and Cosmetic Act

The present Federal Food, Drug and Cosmetic Act was enacted in 1938. It replaced the earlier Food and Drug Law of 1906 which was the first major Federal chemical substances legislation, retaining and extending most of the original provisions.

Although fraud in food and medicinals is an age-old problem, it did not assume alarming proportions until the industrial 19th Century when manufacturers, their ethical standards dulled by the impersonality and remoteness of the ultimate consumer, debased their goods in the struggle to survive economically.[3]

During the latter part of the 19th Century, an industrial society had, to a great extent, succeeded or replaced the agricultural society. With this transition, larger proportions of the populace became dependent upon packaged food supplies. Improved transportation facilities, as well as processing and packaging advancements, permitted increased amounts of foods to be distributed over wide areas. Competition in product development and for sales profits led to debasement and/or mislabeling of foods and other products.[4]

The abuses ranged from fraud to the use of poisonous preservatives—from the attempted sale of mixtures of glucose, flavoring and hayseed as "raspberry jam" to the use of toxic chemical preservatives, such as formaldehyde.[3]

Between 1887–1893, Dr. Harvey Wiley, Chief of the Division of Chemistry of the Department of Agriculture, provided scientific documentation of the need for Federal food and drug legislation, in the form of serial publications entitled "Bulletin 13."[5]

Dr. Wiley and his group of volunteers called the "poison squad" were dedicated to alerting the public to the dangers of preservatives[3] and other chemical hazards.

The initiative behind proposed food and drug regulation was two-fold:

1. Concern for public health, both at the Federal level and by local medical practitioners.
2. Concern for the economic interest of honest businessmen and producers by the Federal Government. In lobbying for the Law, producers sought protection from fraud, including dairymen who sought protection from oleomargarine producers. The Act provided for protecting the public from injurious chemicals used as food additives and preservatives, and also from similar hazards of medicinal products. The purity of drugs was at issue, as was a growing concern about the increased number of patent remedies being offered, many of which contained dangerous, habit-forming ingredients.[3]

While the 1906 Food and Drug Law did much to improve the quality and representation of foods and to ban the use of a number of injurious ingredients and preservatives, its limitations soon became apparent.

The 1906 Law made few positive labeling requirements (i.e., requiring little information about composition or ingredients). In fact, the misbranding provisions were essentially negative in character, simply classing a food or drug as "misbranded" if the label bore any false or misleading representations.[4]

In 1911, the Supreme Court ruled that the labeling provision could only address the identity of the drug product and could not be used to restrict claims about the product's curative properties.[3]

In response to the ruling, Congress passed an amendment (the "Sherley Amendment" of 1912) which prohibited false and fraudulent curative or therapeutic claims on the label. However, the law now required statements to be shown as both false *and* fraudulent, a matter extremely difficult to prove, since establishing fraud involves proving an intent to deceive.[6]

The limitations imposed by the "false and fraudulent" provisions of the "Sherley Amendment" were subsequently revised to become "false *or* fraudulent" under the Federal Food, Drug and Cosmetic Act of 1938.[4]

One of the greatest weaknesses of the 1906 Food and Drug Law was the nature of its administration. The Bureau of Chemistry within the Department of Agriculture had responsibility for enforcement of the Law to protect consumers; while the Department's primary mission was to promote the production and distribution of food.[3]

The Bureau of Chemistry became the Food, Drug and Insecticide Administration in 1927. In 1941, President Franklin D. Roosevelt transferred the Food and Drug Administration to the Federal Security Agency under Reorganization Plan IV.

The Food and Drug Administration became part of the Department of Health, Education and Welfare (DHEW) when, in 1952 by an Act of Congress, the Federal Security Agency was subsumed by the new Department.[5]

In an attempt to improve the 1906 Food and Drug Law, legislation to enact a new law was introduced in 1933 but Congress took no action until the "Elixir of Sulfanilamide" disaster of 1937 which pointed out one of the serious weaknesses of the 1906 Law—lack of premarket safety testing requirements for new drugs—and provided an impetus for the passage of the Federal Food, Drug and Cosmetic Act of 1938 (FFDCA).[6]

The "Elixir of Sulfanilamide" tragedy resulted in more than 107 deaths and called attention to the need for additional legislation to protect the public from similar disasters.

Sulfanilamide is not readily soluble in ordinary solvents, but is soluble in diethylene glycol. So, sulfanilamide in diethylene glycol solution was marketed without its toxicity having been investigated. Diethylene glycol was subsequently demonstrated to be highly toxic. Even a preliminary animal toxicty test would have revealed the hazard of this mixture and would have precluded its use.

Without legal authority to take action on Elixir of Sulfanilamide as a "dangerous drug," the FDA could only recall it on the basis that it was "misbranded," inasmuch as it was not an "elixir" (i.e., not an alcoholic solution).[7]

Enactment of the Federal Food, Drug and Cosmetic Act of 1938 (FFDCA) now required that all "new drugs" be shown to be safe for use *prior to marketing*.

The 1938 Act also brought cosmetics under regulation for the first time. An 1897 proposed bill, introduced by the Association of Official Agricultural Chemists, included cosmetics in the definition of "drugs," but due to trade-offs, was dropped in the passage of the 1906 Law.[3]

The FFDCA has been amended many times over the years to correct deficiencies and improve interpretation and enforcement of the regulations. More importantly, many of the amendments have greatly strengthened the 1938 Act.

The pre-market principle (one of the more important aspects of the Act) has been expanded by amendments to include (in addition to safety of drugs).

- Pesticide tolerances—under Pesticides Amendments of 1954.
- Safety of food additives—under the Food Additives Amendment of 1958.
- Safety of color additives—under the Color Additives Amendment of 1960.
- Effectiveness (as well as safety) requirements of New Drugs—under the Drug Amendments of 1962.
- Therapeutic devices—1976.

Perhaps the most important of the Drug Amendments of 1962 is that provision which shifted the burden from the Government (to disprove safety or efficacy) to the manufacturer (to prove safety *and* efficacy). The efficacy provisions also authorized the FDA to retroactively apply tests of effectiveness to every product subject to New Drug provisions of the 1938 Act. Therefore, every new drug introduced between 1938 and 1962 was subjected to efficacy review.

Probably the most controversial provision in FFDC legislation is the "Delaney Clause" of the Food Additives Amendment of 1958, to wit:

> no additive shall be deemed to be safe if it is found to induce cancer when ingested by man or animal, or if it is found after tests which are appropriate for the evaluation of the safety of food additives, to induce cancer in man or animal.[8]

The FDA recently stayed a ban on saccharin that would have been mandatory under this provision (based on Canadian studies) pending further investigation. The Congress included an 18-month delay of any ban on saccharin as a rider to the HEW supplemental appropriation for FY 78.

There have been many court challenges of FDA regulatory actions. The tendency of the courts has been to apply a very broad interpretation of the authority given the Secretary under Section 701 (2) "to promulgate regulations for the efficient enforcement of this act."[9]

Liberal construction of the Act by the Courts has made the statute a strong and effective weapon, perhaps more potent than Congress had originally contemplated when they passed the Act in 1938.[9] Several court decisions have strengthened administration and enforcement, and expanded the purview of the Act.

The Food and Drug Law, the earliest Federal chemical substances legislation, has not only become a strong mechanism for protection of the consuming public, but has also provided the impetus for significant advancements in the fields of foods and drug technology.

Early Pesticide Legislation

As early as 1910, the use of insecticides had reached a high level of public concern and resulted in Federal legislation to protect users and the general public from this class of potentially harmful chemical substances.[10]

The Federal Insecticide Act of 1910 prevented the manufacture, sale or transportation of adulterated or misbranded insecticides and fungicides and authorized the Department of Agriculture to regulate the sale of these substances through labeling requirements.

With even greater use and variety of chemical pesticides over the years, the Federal Government determined that more comprehensive legislation was needed. In 1947, the Insecticide Act of 1910 was repealed and replaced by the Federal Insecticide, Fungicide and Rodenticide Act (FIFRA), administered by the Department of Agriculture.

The 1947 Act required:[11]

- Registration of economic poisons (chemical pesticides) with the Department of Agriculture prior to sale in interstate commerce.
- Warning labeling.
- Coloring (or discoloring) of powdered insecticides to prevent them from being mistaken for foodstuffs.
- Use instructions on the labels.
- Shipment and holdings of pesticides reports filed with the Department of Agriculture.

The Act was amended several times over the years following enactment in 1947 and in 1970, with the establishment of the Environmental Protection Agency (EPA), administration of FIFRA was transferred from the Department of Agriculture to EPA.

The Federal Hazardous Substances Act

Poisoning from household chemicals, including lye and other caustics, was early recognized by the Food and Drug Administration (FDA) as a health problem.[12] The Federal Caustic Poisons Act of 1927 required warning labeling for 12 chemicals which were responsible for many acute poisonings.

However, the FDA, public health experts, and various chemical and manufacturing associations recognized the need for cautionary labeling of many potentially dangerous household substances which were not subject to provisions of existing statutes.

The Federal Hazardous Substances Labeling Act (FHSLA) was passed in 1960 and superseded the Caustic Poisons Act except for products subject to FDCA. The implementing regulations were published in the *Federal Register* on August 12, 1961.[13]

The FHSLA, administered by FDA, contained broad new responsibilities for protection of the public from hazards associated with a myriad of chemical substances that may be found in and around the home. However, the FHSLA specifically excluded substances subject to the Federal Food, Drug and Cosmetic Act and economic poisons subject to the Federal Insecticide, Fungicide and Rodenticide Act (FIFRA).

Included under FHSLA jurisdiction were household cleaning agents, rust and stain removers, waxes and polishes, home construction and repair substances (such as soldering solutions), paints, solvents and adhesives and automotive products including brake fluids and antifreeze solutions, among other household chemical substances.[14]

The new Act greatly expanded the concept of "poisons" for which warning labeling was required to include a broad definition of "hazardous substances" that became subject to labeling provisions under FHSLA:

"Hazardous substances," as defined, mean a substance, or mixture of substances which:

1. Possesses certain hazardous properties: it must be toxic, an irritant, or corrosive, or a strong sensitizer, or it must be flammable or combustible or generate pressure through decomposition, heat or other means.
2. It must be capable of producing substantial injury or illness under reasonably foreseeable conditions of use, including reasonably foreseeable ingestion by children.

The Act further defined these "hazards" and specified testing parameters for evaluating several hazards, including eye and dermal irritation test procedures.

Specific labeling requirements included:

- Use of signal words commensurate with the hazard (e.g., Warning, Danger)
- Hazardous ingredient(s) identification.
- Manufacturer identification.
- Injury prevention instructions.
- Antidote or first aid instructions.
- Other necessary statements (e.g., Keep out of reach of children).[14]

The Federal Hazardous Substances Labeling Act was just that—a labeling law. Regardless of the hazard that might be associated with its use, if a product were legally labeled under provisions of FHSLA, there was no way of prohibiting its sale to consumers.

Following enactment of FHSLA, the need for a "banning" mechanism was soon illustrated by the report of serious personal injuries associated with products that were labeled in accordance with legal requirements under FHSLA. The only recourse FDA had to remove dangerous products from the marketplace was by instituting multiple seizure provisions. An importer/distributor of a firework device called "cracker balls" which resembled gum balls (and which children attempted to eat) challenged the FDA seizure provisions and took FDA to court. The court ruled against FDA's contention that these "cracker balls" could not be adequately labeled. As a result of the court's decision, the seized firework devices were released and distributed under revised labeling.[15]

The Child Protection Act, enacted in 1966, added the necessary power to ban products so dangerous that cautionary labeling would be insufficient protection. This Act also retitled the FHSLA as the Federal Hazardous Substances Act (FHSA). The "cracker ball" hazard of cut and burned mouths was discussed in the legislative history of this Act.

Subsequently, the Child Protection and Toy Safety Act of 1969 added the power to regulate (including ban as imminent hazards) toys and other articles intended for use by children which present an electrical, mechanical or thermal hazard.

Under Section 2 (q) (1) of the amended Act, the definition of a "banned hazardous substance" is "(A) any toy, or other article intended for use by children, which is a hazardous substance, or which bears or contains a hazardous substance in such manner as to be susceptible of access by a child to whom such toy or other article is entrusted; or (B) any hazardous substance intended, or packaged in a form suitable, for use in the household, which the Commissioner, by regulation, classifies as a 'banned hazardous substance' on the basis of a finding that, notwithstanding such cautionary labeling as is or may be required under this Act for that substance, the degree or nature of the hazard involved . . . is such that the objective of the protection of the public health and safety can be adequately served only by keeping such substance . . . out of channels of interstate commerce . . ."

CHEMICAL SUBSTANCES LEGISLATION IN THE 1970's

During the decades preceding 1970, the focus of most chemical substances legislation had been directed toward the use of chemicals in certain products such as pesticides, food and drugs, and certain household chemical products. The manufacture, industrial exposure, and ultimate release of many chemical substances into the environment had not been adequately regulated by earlier statutes.

Legislators, public health scientists, and industry alike recognized the need for additional regulation of toxic chemical substances. Thus, after much compromise, the first major Federal occupational safety and health legislation was enacted in 1970 and, in the same year, the U.S. EPA was established to administer existing environmental laws and to propose enactment of new protective legislation as needed. Many of the previously existing environmental laws have been substantially amended, expanded, and strengthened. In addition, legislation has been enacted which contains wide-range provisions for the regulation of toxic substances which had not been adequately regulated by other laws.

In 1973 a new independent Commission was established by the Consumer Product Safety Act, and under provisions of this Act, other consumer protection laws were transferred to the Commission for more unified administration.

Occupational Safety and Health Act of 1970

Industrial and public health experts, both in industry and in government, have long expressed concern for the need for specialized occupational health programs to protect workers exposed to chemicals.

Voluntary safety labeling standards were proposed and widely utilized by the developing chemical industry. The Surgeon General's Agreements for Labeling certain solvents and Manual L-1 developed by the Labeling and Precautionary Information Committee of the Manufacturing Chemists' Association were examples of early voluntary labeling standards.[14]

In addition, the American Conference of Governmental and Industrial Hygienists developed standards, such as recommended Threshold Limit Values (TLV's) which have been widely utilized for limiting exposures to noxious gases and dusts in industrial environments.

But these were all voluntary standards which had limited effect in protecting workers from harmful exposures to chemicals in the workplace.

Prior to 1970, the Federal role in occupational safety and health had been severely limited.[16]

The Walsh-Healey Public Contracts Acts set standards for occupational safety, but the coverage was limited to workers employed on Federal contracts for supplies and equipment.

For many decades, governmental responsibility for occupational safety rested mainly with the states. However, state laws varied widely. Some States were fearful that strict standards would place them at a competitive disadvantage with other states in attracting and holding industry.

The Occupational Safety and Health Act of 1970, then, represents the most important occupational safety legislation yet passed. The purpose of this Act is to ensure that workers will be able to work in surroundings that are as free as possible from hazards and toxic pollutants.

The Occupational Safety and Health Act of 1970 authorizes the Secretary of Labor to set standards, to assure safe and healthful working conditions for workers by authorizing enforcement of the standards developed under the Act, to assist and encourage states to participate in efforts to assure such working conditions and to provide for research, information, education, and training in the field of occupational safety and health.

This Act superseded the Walsh-Healy Public Contracts Act and the Services Contract Act of 1965 which set standards for protection of employees of contractors with the Federal government.

The Act required, under Sec. 5(a) that each employer:

(1) Shall furnish to each employee, employment and a safe place of employment which are free from recognized hazards that are causing or are likely to cause, death or serious physical harm to his employees;

(2) Shall comply with . . . "standards promulgated under this Act" and that each employee—

  (b) Shall comply with occupational safety and health standards and all rules, regulations and orders issued pursuant to this Act which are applicable to his own actions and conduct.

The Secretary of Labor, under Sec. 6 is authorized to promulgate occupational safety or health standards, with provisions for public participation in the rulemaking process.

The Act also makes the following provisions:

(1) Provides for an independent Occupational Safety & Health Review Commission to adjudicate enforcement cases and to set penalties for violations, and provides judicial review of Commission decisions in the Court of Appeals.

(2) Established the National Institute of Occupational Safety and Health Administration (NIOSH) in HEW to conduct research in occupational safety and health areas and to develop and establish recommended safety and health standards.

(3) Provides for a 15-member Commission to study State workmen's compensation laws and for submission of a report to Congress at the end of two years.

(4) Provides for administrative closing of worksites for 72 hours in cases where an inspector determines an imminent danger exists.

(5) Provides civil penalties for violations and criminal penalties for willful violations resulting in the death of an employee.

Poison Prevention Packaging Act of 1970

Prior to May 1973, the Poison Prevention Packaging Act of 1970 (PPPA) was administered by FDA. Authority to administer the Act was transferred to the Consumer Product Safety Commission upon its activation in 1973. The regulations which appeared at 21 CFR Part 295 of the Code of Federal Regulations were recodified in Title 16, Chapter II, Part 1700–Poison Prevention Packaging.

The Commission, by authority of this Act, may establish standards for special packaging of any household substance that the Commission finds possesses a degree of hazard, by reason of its availability to children, such that special packaging is required to protect children from serious personal injury resulting from handling, using, or ingesting such substances.

Household substances subject to the provisions of PPPA include (1) products covered by FHSA, (2) food, drugs and cosmetics, and (3) fuels in portable containers. Economic poisons, or pesticides, formerly subject to PPPA, have been excluded from it by Sec. 3(a) of the Consumer Product Safety Commission Improvements Act of 1976, Pub. L. 94-284.

Section 6 of the PPPA provides for the formation of a Technical Advisory Committee composed of members from regulated industries, scientists and medical practitioners, consumers, and packaging manufacturers, as well as government representatives. Prior to establishment of safety packaging requirements, the Commission must consult with the Technical Advisory Committee. However, this requirement specifies only that the Committee must be consulted and not that Committee approval must be obtained for setting standards.

In order to propose safety packaging under provisions of PPPA, certain findings are required:

(1) Hazard. The degree, or nature of the hazard to children under 5 years of age must be such that special packaging is required for protection from serious personal injury or illness resulting from handling, using, or ingesting the chemical substance.

(2) Feasibility of Packaging. It must be technically possible to produce, or manufacture, the needed safety packaging on a mass production basis and the packaging, while complying with the safety requirements, must not interfere with the intended use or storage of the product.

The PPPA does not contain its own enforcement provisions. Instead, under Sec. 30 of the Consumer Product Safety Act, the Commission enforces PPPA packaging standards through provisions of the FHSA and the Food, Drug and Cosmetic Act. Household substances covered by the FHSA or food, drugs, and cosmetics covered by the Federal Food, Drug and Cosmetic Act (for which safety packaging standards have been promulgated), are "misbranded" within the meaning of these two acts if not packaged in accordance with provisions of PPPA.

The Consumer Product Safety Act

The Consumer Product Safety Act became law on October 27, 1972, and the Consumer Product Safety Commission, charged with its administration, was formally activated on May 14, 1973.

The stated purposes for the Consumer Product Safety Act (CPSA) are:

- To protect the public against unreasonable risks of injury associated with consumer products.
- To assist consumers in evaluating the comparative safety of consumer products.
- To develop uniform safety standards for consumer products and to minimize conflicting state and local regulations.
- To promote research and investigations into the causes and prevention of product-related deaths, illness, and injuries.

Chemical substances or mixtures in consumer products would be subject to regulation under CPSA (as amended on May 11, 1976) under the following conditions:

Section 30(d)—"a risk of injury which is associated with a consumer product and which could be eliminated or reduced to a sufficient extent by action under the Federal Hazardous Substances Act or the Poison Prevention Packaging Act of 1970 . . . may be regulated under this Act only if the Commission by rule finds that it is in the public interest to regulate such risk of injury under this Act. Such a rule shall identify the risk of injury proposed to be regulated under this Act and shall be promulgated in accordance with Section 553 of Title 5, U.S. Code . . ."

The Commission recently decided under this Section that it was in the public interest to propose the ban of consumer patching compounds and emberizing materials containing respirable free-form asbestos under the CPSA. Although CPSC had been petitioned under FHSA, and the risk of injury could have been eliminated or reduced to a sufficient extent under the FHSA, the Commission concluded, in part, that the rule-making proceedings under FHSA would probably have been lengthy and resource-consuming and that those proceedings could have made it more difficult for interested persons to participate. In accordance with Section 8 of the CPSA, a period of time was provided wherein any person could submit written comments on the proposed banning action for consideration by the Commission. In addition, an opportunity for oral presentation of data, views, or arguments was provided. During these proceedings, any additional information or data related to the nature or degree of the hazard associated with the affected products could be brought to the Commission's attention for consideration prior to the promulgation of the final rule.[17]

Following the public participation period, the final banning notice was published in the *Federal Register* on December 15, 1977 and included responses to all written comments which had been received. There was an immediate effective date for the ban of asbestos-containing emberizing materials and the final ban on the sale or distribution of patching compounds containing intentionally added asbestos at any point in commerce is effective on June 15, 1978.[18]

Similar public interest findings have been made under CPSA for other consumer products, including highly flammable contact adhesives and certain paints and coatings containing lead.

The Consumer Product Safety Act empowers the Commission to take a wide range of actions to eliminate, or to reduce risks of injury associated with consumer products. Provisions for actions range from assisting in voluntary standards development to enactment of mandatory standards, to product removal or recall, to banning. The Commission is also empowered to seek civil and criminal penalties for noncompliance with enacted standards.

ENVIRONMENTAL PROTECTION AGENCY ACTS

Probably the most wide-ranging impact on chemical substances legislation in the 1970s has been the establishment of the Environmental Protection Agency (EPA) under Presidential Reorganization Plan No. 3 of 1970, which was intended to consolidate existing environmental legislation by transferring functions administered under several existing Acts, administered by five different departments and independent agencies.[19]

Thus, water quality responsibilities were transferred from the Department of Interior and from the Department of Health, Education and Welfare (DHEW). Air pollution control activities and solid waste management were also transferred from DHEW.

Authority to regulate pesticides was also transferred to EPA from the Department of Agriculture (registration), from the Federal Food and Drug Administration, DHEW (tolerance levels for pesticides), and from the Department of Interior (pesticides research program).

In addition, functions of the Council on Environmental Quality pertaining to ecological systems under the National Environmental Policy Act of 1969 were also transferred to EPA, as were functions of the Atomic Energy Commission under the Atomic Energy Act of 1954 for establishing environmental standards for the protection of the environment from radioactive material.

Since the establishment of EPA in 1970, much of the previously existing legislation has been amended, expanded, and/or rewritten. In addition, significant new legislation has been enacted which empowers EPA to effectively restrict or prevent toxic chemical substances from entering the environment.

The Clean Air Act (as Amended 1977)

This Act provides EPA with authority to control air pollution by a variety of means.

Some of the major provisions of the Act are:

- *National Ambient Air Quality Standards.* Under Sections 108 and 109, EPA can set primary and secondary national ambient standards for certain air pollutants (referred to as "criteria pollutants") resulting from mobile or stationary sources. Control is effected by state action under State Implementation Plans. The Office of Air Quality Planning and Standards is responsible for recommending the designation of criteria pollutants, for coordinating the development of criteria and control techniques documents, for setting primary and secondary standards levels, and for providing guidance for the development of State Implementation Plans. There are six "criteria" pollutants: (1) particulate matter, (2) sulfur oxides, (3) carbon monoxide, (4) nitrogen dioxide, (5) photochemical oxidants, and (6) hydrocarbons. EPA has also proposed an air quality standard for lead. The complexity and comprehensiveness of National Ambient Air Quality Standards effectively limit the number of pollutants which can be listed under this section.
- *Standards of Performance for New Stationary Sources.* Section 111 requires EPA to set emission standards for new or modified sources which cause or significantly contribute to air pollution. The standard reflects the best control (taking cost, energy, and other factors into consideration) which EPA determines has been adequately demonstrated. Within 120 days after a source category is listed, the Administrator must propose performance standards for the category. Promulgation of the standard must follow within 90 days. For both new and modified sources, the standard is effective as of date of proposal.

EPA may set a new source performance standard for limiting emissions of both "criteria pollutants" and "non-criteria" pollutants (those for which no National Ambient Air Quality Standards have been established). When a New Stationary Source standard for a non-criteria pollutant is set, the states must, within nine months following notification of the availability of final EPA guidelines, submit a plan to EPA containing regulations for control of all their existing sources for this pollutant [Section 111 (d)]. Generally existing sources must be in compliance within three years after EPA approves the State plan. Regulations have been promulgated under Section 111 for:

- Acid mist (H_2SO_4)
- Carbon monoxide (CO)
- Fluorides
- Hydrocarbons
- Nitrogen oxides (NO_2)
- Particulates
- Sulfur oxides (SO_2)
- Total reduced sulfur (TRS)

• *National Emission Standards for Hazardous Air Pollutants.* Section 112 of the Act provides for control of non-criteria pollutants which may cause or contribute to an increase in mortality, or an increase in serious irreversible, or incapacitating reversible, illness. For a pollutant considered hazardous, EPA establishes emission standards which provide an ample margin of safety to protect public health. Although the Act does not explicitly provide for consideration of cost or availability of demonstrated control technology in determining allowable emissions, these have been considered to some extent in practice.

These regulations are effective upon promulgation for new sources and effective 90 days after promulgation for existing sources. The Act allows for waivers of compliance in certain situations for up to two years.

Hazardous emission standards have been promulgated for mercury, asbestos, beryllium, and vinyl chloride. Benzene has been listed under Sec. 112 and regulations are under development for significant benzene sources.

• *Mobile Sources.* Section 202 provides for emission standards for any air pollutant coming from a motor vehicle if the pollutant may be harmful to the public health or welfare. This section includes the mandated reduction of CO, HC, and NO_x emissions for light-duty vehicles. Emission standards have also been set for light-duty trucks, light diesel engines, and heavy-duty gasoline and diesel trucks.

In Sec., 231, EPA is directed to establish emission standards for aircraft. After consultation with the Department of Transportation (DOT), EPA has issued emission standards for propeller, turbo-prop, and jet aircraft. These standards will be enforced by DOT.

• *Regulations of Fuels and Fuel Additives.* Section 211 provides for the registration of any fuel or fuel additive. The EPA Administrator may require that the manufacturer of any fuel notify him as to the commercial name of any additive, the concentration, purpose and chemical composition of the additive. EPA may also require the manufacturer to conduct tests to determine the possible health effects of any additive or the emissions resulting from its use.

If an additive may endanger the public health, EPA may prohibit or control its use or sale. It is under this authority that EPA is requiring a phase-down in the lead content of gasoline. Also, if an additive significantly impairs the action of any existing or proposed emission control device, EPA may prohibit or control its use. Under this latter authority EPA has required that at least one grade of lead-free gasoline be available at all gas stations (over a certain size specified by EPA guidelines).

• *Emergency Powers.* Under Sec. 303, where state officials have not acted, EPA is authorized to bring suit to control emission of air pollutants which pose an imminent and substantial endangerment to the public health. This provision is mainly applicable to control of criteria pollutants during air inversions. Its applicability to control toxics has not been tested.

• *Grants for Support of Air Pollution Planning and Control Programs.* Under Sec. 105, provisions are made for grants to state and local agencies for planning, developing, and maintaining air pollution control programs, including implementation of National Ambient Air Quality Standards.

• *Use of Authorities to Control Toxic Pollutants.* Current Office of Air Quality Planning and Standards programs which control toxic air pollutants include (1) indirect regulation through National Ambient Air Quality Standards and New Stationary Source Performance Standards for Photochemical Oxidant/Hydrocarbon/Particulate Matter and (2) direct regulation by National Ambient Air Quality Standards (lead), 111(d), (Sulfuric acid mist, fluorides, reduced sulfur), National Emission Standards for Hazardous Air Pollutants for beryllium, asbestos, mercury, and vinyl chloride.

• *Oxidant National Ambient Air Quality Standards.* The major control strategy for meeting these standards is control of volatile organic chemicals. Mobile source controls significantly reduce emissions of POM and other potentially toxic organics. State implementation plans must include sufficient volatile organic chemicals stationary source controls to attain the standard by a stated date, and in the meantime to make reasonable annual progress toward meeting the standard. EPA prepares guideline documents for the states defining reasonably available control technology for major sources of volatile organics. To date, guidelines have not explicitly recommended consideration of toxicity of volatile organics in applying specified control technology. However, in EPA policy for the control of volatile organics, some low oxidant producing (non-reactive) solvents were removed from control exemption because of their toxicity.

• *Regulation of Emissions from the Synthetic Organic Chemical Industry.* The objective of this program is to reduce existing levels and to minimize growth in emissions of photochemical oxidants and organic chemicals. Both Sec. 111 (New Source Performance Standards) and Sec. 112 (Hazardous Standards) will provide the mechanisms for control which may include generic process and equipment specifications, as well as emission limitations. This program is established on the basis of the need to limit emissions of hydrocarbons (for which National Ambient Air Quality Standards exist), but EPA is also aware of the benefits of strict control for volatile organic compounds which may be toxic.

• *Particulate National Ambient Air Quality Standards.* Control of emissions of total suspended particulates has resulted in reduction of urban ambient levels of a number of toxic trace elements (V, Mn, Ni) and some organic particulates (BaP, and POM).

• *Section B, Title I - Ozone Protection.* This section provides for the study of the cumulative effects on the stratospheric ozone layer of:

1. Halocarbons released into the ambient air.
2. Release of other sources of chlorine into the air.
3. Uses of bromine compounds.
4. Emissions of aircraft and aircraft propulsion systems.

The Clean Water Act (Formerly the Federal Water Pollution Control Act [FWPCA])

The need for comprehensive, consolidated water pollution legislation was demonstrated during the Hearings before the 92nd Congress. Subsequently, passage of the Federal Water Control Amendments of 1972 represented a comprehensive revision of previously existing water pollution control laws (administration of which had been transferred to EPA in 1970) and became the Federal Water Pollution Control Act (FWPCA).

Some of the major provisions of the Act, as amended in 1972, are:[20]

1. Standards for direct discharges into waters.
2. Standards for discharges into publicly owned treatment works.
3. Provisions for responding to, preventing, and penalizing spills of oil and hazardous substances.
4. Regulation of vessel sewage.
5. Regulation of disposal of dredged material.

Other provisions of the Act include financial assistance for sewage treatment plant construction, and provisions for Federal support for research and demonstration projects and Federal support for state water pollution control programs.

The Act established separate regulatory schemes for two classes of point source discharge:

(1) Direct Discharge into the Navigable Waters. Direct dischargers are subject to dual requirements: (a) Effluent standards and (b) water quality standards.

(*a*) *Effluent Standards.* Effluent standards are limitations upon dischargers on the amounts of pollutants that may be discharged, and are based on the capability of pollution control technology. Existing dischargers (other than publicly owned, or municipal, treatment works) are required to adopt "best practicable control technology currently available" by July 1977 and "best available technology (BAT) economically achievable" by July 1983. (These statutory definitions of "technologies" are defined in EPA regulations, or in the absence of regulations, by permit issuers). EPA is additionally required to issue effluent standards for new sources (defined as those, the construction of which commences after a proposed standard for the source is published) which are to be based on "the best available demonstrated control technology, processes, operating methods, or other alternatives." Publicly owned treatment works must meet "secondary treatment" by July 1, 1977, and "best practicable waste treatment technology by July, 1983.

EPA is also required to establish special effluent standards for toxic pollutants bases, primarily, on environmental effects of the pollutants.

(*b*) *Water Quality Standards.* Water quality standards are based on technical data on the minimum requirements necessary to sustain various uses of water, including its use in recreation activities such as swimming, propagation of aquatic life, and public water supply. The standards consist of rules, or criteria, defining a required quality for the ambient water such as a requirement for at least 5 ppm dissolved oxygen, no more than 1000 fecal coliform bacteria/100 ml water, etc.

State water quality standards are submitted to EPA for approval. If EPA fails to approve a state-adopted water quality standard, it may promulgate Federal standards for the particular state. Water quality standards may vary depending on uses of certain waters and individual state preferences, although EPA has made some attempt, through its approval authority, to impose a certain degree of uniformity in water quality standards.

Enforcement. The basic mechanism for enforcement of Effluent and Water Quality Standards is through the issuance of discharge permits which define maximum levels of permissible discharge (if any). Permits must require compliance with both effluent standards and water quality standards. If statutory criteria are met, States may be authorized to issue Federal permits (although they are subject to EPA veto). Various administrative and judicial remedies are established for violations of the permits.

(2) Discharges to Publicy-Owned Treatment Plants. The second regulatory scheme requires EPA to promulgate pretreatment standards designed to prevent the discharge of any pollutant through publicly owned treatment works which "interferes with, passes through, or otherwise is incompatible with such works." A discharger into a municipal, or other publicly owned treatment plant, is not required to obtain a permit under this Act, although a permit may be required by state or municipal law. However, the treatment plants must have permits for discharges into the water and holders of such required permits must notify EPA (or the state, depending on which issued the permit) of any substantial change in the volume or character of pollutants introduced into the plant or treatment works. If a treatment works permit is violated, an action may be brought to restrict or prohibit any new tie-ins to the treatment works.

In addition, users of publicly owned treatment works approved for Federal financing after 1973 are required to pay a "proportionate share" of the operation and maintenance costs. Industrial users will be required to pay a portion of construction costs.

(3) Spills of Oil and Hazardous Substances. EPA is required to designate a list of Hazardous Substances and to define what constitutes a harmful discharge of both oil and each hazardous substance. A list of hazardous substances, other than oil, "harmful quantity" and rate of penalty were published March 13, 1978, in the *Federal Register*, Part II—Water Programs (vol. 43, no. 49, p. 10474–10508), along with a proposal to add an additional twenty-eight substances to the designation list of hazardous substances.

Any person in charge of a vessel or facility from which there is a harmful discharge of oil or hazardous substance is required to notify EPA or the Coast Guard immediately so that the Government may quickly act to clean up the discharge or mitigate its effects. The discharger is responsible for cleanup costs (with certain exceptions and limitations). Regulations are also to be promulgated requiring potential dischargers to have equipment for spill prevention and containment. Those who discharge "harmful quantities" or fail to report spills are subject to stringent penalties.

(4) Vessel Sewage. EPA is required to issue regulations defining a standard of performance for "marine sanitation devices" and the Coast Guard is required to establish equipment requirements which conform to the performance standards. These standards are applicable to both existing and new vessels within certain time limitations after promulgation of the regulations.

(5) Dredged and Fill Material. The Act establishes a separate permit system for the discharge of dredged or fill material at specified disposal sites. The permits are to be issued by the Army Corps of Engineers, subject to EPA guidelines. EPA is authorized to deny or restrict the use of any particular area as a disposal site.

Clean Water Act of 1977

The Clean Water Act of 1977 was signed into law on December 28, 1977. The 1977 Amendments (P.L. 95-217) preserved the basic structure of the Federal Water Pollution Control Act, but it provided extensions of the 1977 and 1983 deadlines for dischargers and strengthened provisions dealing with toxic pollutants.

Among the changes, under the 1977 Amendments, have been those made in requirements for control of industrial discharges and certain extensions granted for compliance with discharge standards.

Some of the provisions of the 1977 Amendments are:[21]

- Companies or plants not having achieved compliance with the 1977 requirements have been granted limited deadline extensions. For plants discharging directly into navigable waters, the maximum extension is 21 months (or until April 1979). For companies planning to tie into municipal systems, extensions are authorized up to July 1983.
- The deadline for achieving standards for "Best available technology (BAT)" control (previously scheduled for compliance by July 1983) is extended to July 1984.

The 1977 Amendments also modify requirements for control of conventional pollutants and, at the same time, increase the emphasis on control of toxic pollutants. EPA is directed to develop BAT effluent standards covering 21 industries and 65 pollutants identified in the *Natural Resources Defense Council vs Train* 1976 Consent Decree, entered into by EPA.[21]

Other provisions under the 1977 Amendments are:

- Variances are permitted from requirements for control of non-conventional pollutants under specified conditions.
- Extension of the deadline for achieving BAT requirements for plants installing certain types of innovative technology.
- Expansion of regulatory framework by authorizing EPA to issue "best management practices" regulations to control certain parts of plant operations to restrict the runoff of toxic and hazardous materials.
- Modification of pretreatment requirements to provide dischargers credit for removal of toxic pollutants by municipal sewage treatment plants.

The Federal Insecticide, Fungicide and Rodenticide Act (FIFRA)

In 1970, with establishment of the Environmental Protection Agency (EPA), administration of the Federal Insecticide, Fungicide and Rodenticide Act (FIFRA) of 1947, as amended was transferred from the Department of Agriculture to EPA. FIFRA requires the registration, or pre-market clearance, of pesticides in the United States.

Functions of several Federal agencies dealing with pesticides were also transferred to EPA at that time, including responsibility for establishing tolerances for pesticide residues that might remain in or on harvested food or feed crops (under the Pesticide Amendments to the Federal Food, Drug and Cosmetic Act) which was tranferred from FDA to EPA. Enforcement of tolerances remained the responsibility of FDA.

In 1972 Congress enacted the Federal Environmental Pesticide Control Act which substantially amended FIFRA.

The basic registration and labeling authorities of the original FIFRA were retained, while greatly expanding EPA's regulatory authority under the 1972 Act.

Some major provisions of the amended Act are:[22]

- Re-registration of all currently registered pesticides to ensure conformance to today's safety standards.
- Authority to classify pesticides for "restricted" or "general" use. "Restricted" pesticides may be used only by, or under the supervision of, certified applicators or under other conditions deemed appropriate by EPA to protect man and the environment.
- Enforcement provisions against misuse of any registered pesticide.

This is the first time that the Agency received authority to prosecute for deviation from registered labels.

Other provisions of the Act include:

- Registration of all pesticide products distributed in *intra*state commerce, as well as in interstate commerce.
- Authority to set Federal standards for certification of applicators of pesticides classified for "restricted" use. State certification programs, based on Federal standards, must be submitted to EPA for approval.
- Federal assistance to the states to enforce provisions of the law and help develop and administer applicator certification programs.
- Registration of all establishments producing pesticides with EPA, and reporting of information on amounts and types of pesticides manufactured, distributed, and sold. An EPA agent may enter and inspect establishments where pesticides are held for distribution or sale and take samples of finished products.
- Authority to develop procedures and regulations for storage and/or disposal of pesticides. EPA is also required to accept at convenient locations for disposal, a pesticide for which registration is suspended and then cancelled.
- Authority to undertake research on pesticides, particularly biologically integrated alternatives, to issue experimental use permits and to monitor pesticide use and presence in the environment.
- The EPA Administrator may issue a "stop sale, use, and removal" order when it appears a pesticide violates the law, or its registration has been suspended or finally cancelled. Pesticides may also be seized if they violate the law.
- EPA can impose civil and criminal sanctions for violations of the law.

Still other provisions of the 1972 law included:

- Indemnification is authorized for certain owners of pesticides, registrations of which are suspended, and than cancelled.
- Applicants for registration of a pesticide may retain proprietary rights to certain test data submitted to EPA in support of the application. However, the law creates a mandatory licensing system for the data; a second applicant may pay reasonable compensation to the original applicant.
- States can be authorized to issue limited registrations for pesticides intended for special local needs.

The basis for EPA's regulatory decisions under FIFRA as to the registration and use classification, or suspension and cancellation is the determination of whether a pesticide can perform its intended function without "unreasonable adverse effects" on the environment at the time it is registered.[23] "Unreasonable adverse effects" are defined as " . . . any unreasonable risk to man or the environment, taking into account the economic, social and environmental costs and benefits of the use of any pesticide."

This basic mandate means that EPA must balance the risks of pesticide use against their benefits to Society in reaching decisions to allow or deny use.

Risk/benefit balancing is at the heart of several regulatory processes developed under FIFRA:

Rebuttable Presumption Against Registration (*RPAR*) is a public process wherein if EPA determines that use of a pesticide may present substantial risks to man or the environment, EPA invites the registrant, the public and/or other concerned groups to rebut the risk and submit any available benefit information. EPA then evaluates all submitted data and decides whether to restrict, cancel, or permit continued uses as registered, of subject pesticides.

Cancellation. Federal registration of a pesticide product may be cancelled by EPA if the Agency determines that its use generally causes unreasonable adverse effects to man or to the environment. The manufacturer of the pesticide subject to cancellation of registration can appeal the cancellation action within 30 days. During the appeal proceedings, the manufacturer can continue to market his product until all administrative procedures have been completed and a final order has been issued on the appeal.

Suspension. EPA has authority to suspend Federal registration of a pesticide product if a finding is made that its continued use during the time required for cancellation or change in classsification proceedings, constitutes an imminent hazard to public health.

Simultaneously, EPA must file a notice of intent to cancel the registration or to change the "use" classification of a pesticide. The manufacturer may, within 5 days after receipt of notification, request an "expedited hearing" as to whether or not an imminent hazard exists.

The public may also petition EPA for cancellation or suspension of a pesticide registration.

The 1975 amendment to the Act adds procedural provisions for participation of a Scientific Advisory Panel, the U.S. Department of Agriculture and House and Senate Agriculture oversight committees in certain EPA decisions on pesticides.[23]

Proposed final pesticide regulations and cancellation actions are to be forwarded to the Scientific Advisory Panel, to the Agricultural committees and to the Secretary of Agriculture for comments. The response of the Panel and USDA are to be published, along with EPA's response, in the *Federal Register*. EPA is additionally required to prepare and submit an agricultural impact assessment to the Secretary of the Department of Agriculture prior to a pesticide cancellation action, which is also published in the *Federal Register*.

It is likely that significant new provisions will be added in upcoming amendments to FIFRA.

Resource Conservation and Recovery Act of 1976 (RCRA)

The purpose of this Act is "to provide technical and financial assistance for the development of management plans and facilities for the recovery of energy and other resources from discarded materials and for the safe disposal of discarded materials, and to regulate the management of hazardous waste."

The RCRA is an amendment of the Solid Waste Disposal Act and is administered by the Environmental Protection Agency.

Among the findings made by Congress in support of the need for solid wastes management and recovery were:

• The problem of solid waste disposal is national in scope, necessitating Federal financial and technical assistance in the development and application of new methods and processes to reduce the amount of waste and unsalvageable materials and to provide for appropriate disposal practices.

• Environmentally unsound practices for the disposal of solid waste have created greater amounts of air and water pollution including ground water and present a danger to human health.

• Much of the discarded solid waste is recoverable and/or could become the source of additional fuel.

• *Hazardous Waste Regulatory Program.* A hazardous waste regulatory program is established under Subtitle C of the Resource Conservation and Recovery Act of 1976 (RCRA). The hazardous waste regulatory program is established in two stages: First, under provisions contained in Subtitle C, EPA must publish criteria for determining the characteristics of "hazardous waste" (Sec. 3001[a]). EPA must promulgate hazardous waste regulations which identify the characteristics and list specific hazardous wastes.

Within 90 days after promulgation of these regulations anyone generating or transporting such hazardous wastes, or owning or operating a treatment, storage, or disposal facility for them, must file with EPA a notification of such activity (Sec. 3010[a]).

• *Federal Programs.* Under Sections 3002, 3003, and 3004, EPA is required to promulgate such standards as may be "necessary to protect human health and the environment . . ." applicable to generators of hazardous waste, transporters, and owners and operators of treatment, storage, and disposal facilities. These standards take effect six months after their promulgation.

The standards for generators and transporters differ from those applicable to owners and operators of treatment facilities:

(a) The standards applicable to generators and transporters will require labeling, appropriate containers, a "manifest" system, and record keeping. The standards will require that hazardous waste be transported to treatment, storage, or disposal facilities having permits.

(b) For owners and operators of treatment, storage, and disposal facilities, EPA is required (under Sections 3004 and 3005) to establish performance standards and a permit system for treatment, storage, and disposal facilities. Existing facilities which have made a permit application and have filed the required notification of their activities in the handling of hazardous waste will be treated as having been issued a permit until the final EPA disposition of the application is made, unless the applicant has failed to provide information needed for the EPA permit (Sec. 3005[e]).

• *State Programs.* Under Sec. 3006(a), EPA is required to promulgate guidelines to assist states in the development of state hazardous waste programs. If EPA finds that the following three conditions are met, a State is authorized to implement its own program "in lieu of" the Federal program:

- Equivalence to the Federal program.
- Consistency with Federal or state programs applicable in other states.
- Adequate compliance with enforcement provisions.

Subject to approval, state regulations for generators and transporters and for owners and operators of treatment, storage, and disposal facilities will replace Federal regulations and permits. States with hazardous waste programs already in existence by July 1978 may apply to EPA for temporary (24 months) authorization which must be granted if the state's program is "substantially equivalent" to the Federal program (3006[e]).

States may not impose less stringent requirements than Federal regulations (except during such time that Federal regulations are postponed or enjoined).

• *Federal Enforcement.* The Act prohibits violation of the regulations and permits issued under both the Federal Hazardous Waste Program and the Federally approved state programs.

Provision for Federal enforcement authorized EPA to issue compliance orders, as well as to impose civil and/or criminal penalties, which can include fines and injunctions for certain violations.

EPA compliance orders or civil actions must be preceded by a notice of violation. If a violation is not corrected within 30 days, the Administrator of EPA may pursue either civil remedy.

Any person who knowingly transports a listed hazardous waste to a non-permitted facility, or disposes of a hazardous waste without a permit, is subject to certain criminal penalties.

• *Information-Gathering.* All person handling hazardous wastes are required under RCRA to provide EPA or state authorized agents access to records and to permit on-site inspections, including sample collections. The Act grants public access to such information unless it is entitled to confidential treatment (Sec. 3007).

• *State or Regional Solid Waste Plans.* Under Subtitle D of RCRA, Federal technical and financial assistance is authorized to states (or regional authorities) for planning and developing solid waste disposal methods which are environmentally sound, which maximize resource utilization and which encourage resource conservation.

EPA is required to publish guidelines for the identification of appropriate units for planning regional solid management, and guidelines for development and implementation of state solid waste plans. In order for a state plan to be approved (among other requirements), it must contain a timetable for closing or upgrading existing dumps, and require that solid waste be utilized for resource recovery, or otherwise be disposed of, in an "environmentally sound manner."

EPA must publish criteria to distinguish between "Open Dumps" and "Sanitary Landfills" and the Administrator of EPA is directed to make an inventory of all open dumps in the United States (Sec. 4005[b]). The Act "prohibits" all open dumps except those operating under compliance schedules not to extend beyond 5 years from the date of inventory.

• *Research, Development, Demonstration and Information Programs and Grants.* Under the Act, EPA is given broad authority to conduct and coordinate research, demonstration, and information studies (among other related activities) with the objectives of developing new methods for collecting and disposing of solid wastes, as well as for processing and recovering useful materials and energy from solid waste.

The actual promotion of new technology and development of markets is assigned to the Secretary of Commerce.

Other provisions of RCRA are:

- Federal facilities, including Federal procurement agencies, must comply with solid waste management provisions under RCRA.
- EPA is to consult with other Federal agencies in the investigation of economic, social, and environmental consequences of resource conservation.
- RCRA provides for citizen petitions and civil actions.
- EPA is to coordinate provisions under RCRA with other programs.
- EPA may sue to restrain an "imminent and substantial endangerment to health or the environment" (Sec. 7003).[24]

Resource conservation and recovery of waste materials are the ultimate goals under RCRA. However, for the present, EPA's objectives under the Act are to promulgate strong regulations for disposal and control of hazardous solid waste and to encourage the phasing-out of open-dumping.

Safe Drinking Water Act, As Amended

This Act, administered by EPA, amended the Public Health Service Act in 1974 under a new title, "Title XIV—Safety of Public Water Systems," and is the basis for EPA's Public Water System supervision program. This Act was amended in November 1977.

Under Sec. 1412, the Administrator of EPA is required to promulgate National primary drinking water regulations applicable to public water systems in each State (with certain exceptions) for protection of public health, and to promulgate National secondary drinking water regulations applicable to public water systems, specifying maximum contaminant levels for protecting the public welfare.

Under Sec. 1412(e) of the Safe Drinking Water Act, the Administrator of EPA is required to arrange for a study by the National Academy of Sciences (NAS) of known or suspected contaminants in drinking water. This study serves as a basis for the Administrator's promulgation of primary drinking water regulations. The NAS study was completed in 1977.

Sec. 1413 of the Act provides that a state has primary enforcement responsibility for public water systems if it demonstrates that it satisfies the conditions specified in this Section and in regulations promulgated thereunder. In establishing an effective drinking water program, EPA and the states have worked together in developing state programs which meet the minimal requirements.

For those states not accepting primary enforcement responsibility, EPA is authorized to administer and enforce the Federal regulations in those states.

The Safe Drinking Water Act also empowers EPA, under Part C, Secs. 1421 and 1422 to designate states and to establish regulations for state underground injection control programs designed to protect underground sources of drinking water. Such regulations must set forth minimal requirements for effective programs to prevent underground injection which endangers drinking water sources.

Upon application by the state and approval by the Administrator, a state may be granted primary enforcement responsibility for underground water sources if it satisfies the requirements of Section 1422 and regulations promulgated thereunder. EPA is authorized to enforce applicable underground injection control requirements where a state fails to take appropriate action or during any period which a state does not have primary enforcement responsibility.

Under Section 1424, the Administrator may designate for protection an area where one acquifer is the sole or principal drinking water source for the area and which, if contaminated, would create a significant hazard to public health.

Finally, Section 1431 of the Act authorizes the Administrator to take such action as he deems necessary to protect the public health in any situation where a contaminant is present or is likely to enter a public water system and present an imminent and substantial endangerment to the health of persons.

Toxic Substances Control Act

After several years' legislative effort in Congress, a compromise Toxic Substances Control bill was passed by both houses in September 1976 and was enacted on October 11, 1976, as the "Toxic Substances Control Act" (TSCA).

The stated purpose of the Act is "to regulate commerce and protect human health and the environment by requiring testing and necessary use restrictions on certain chemical substances, and for other purposes."

A finding was made by Congress that:

- Humans and the environment are being exposed each year to a large number of chemical substances.
- Among those chemicals being developed and produced, some may cause or contribute to an unreasonable risk of injury to health or the environment.
- Effective regulation of such substances in interstate commerce necessitates the regulation in *intra*state commerce as well.

Unlike most other laws, the Toxic Substances Control Act provides EPA with the mechanism to control chemical substances at the point of manufacture rather than at the point of release.

Provisions of the Law. Unlike most other laws regulating chemical substances, the Toxic Substances Control Act (TSCA) provides the authority to control chemical substances at the point of manufacture rather than at the point of release. TSCA permits EPA to act *before* harmful substances threaten human health or the environment.

TSCA provides EPA with comprehensive and flexible authority to gather certain kinds of basic information on chemicals, to identify harmful substances, and to control those toxic chemicals whose risks of injury to public health and the environment outweigh their benefits to society and the economy. The reach of the law is extremely broad. It encompasses the estimated 70,000 chemical substances manufactured for commercial purposes and several million research and development chemicals.

It makes the entire chemical industry subject to comprehensive Federal regulation for the first time. EPA's authority under TSCA is extended into virtually every facet of industry—product development, testing, manufacturing, processing, distribution, use, and disposal. TSCA specifically exempts pesticides, tobacco, nuclear material, firearms and ammunition, food additives, drugs and cosmetics since these products are currently regulated under other laws.

Some of the major provisions under TSCA are:

- TESTING

Under Sec. 4 (Testing of Chemical Substances and Mixtures), the EPA Administrator may require manufacturers or processors to conduct tests on chemicals. Testing may be directed to evaluate the characteristics of the chemical, including carcinogenic, mutagenic and teratogenic effects. Testing standards must be promulgated by regulation. These standards will be as consistent as possible with those already developed by EPA under the Federal Insecticide, Fungicide and Rodenticide Act (FIFRA).

Sec. 4 also provides for an interagency committee of experts to make recommendations to the Administrator concerning chemicals that should be tested. The Committee, designated as the Interagency Testing Committee (ITC), made its first report to the Administrator in October 1977 and its second report in April 1978. Within one year after submission of the respective reports, EPA must either initiate testing requirements for these recommended chemicals or explain its reasons for not doing so.

- PREMANUFACTURE NOTIFICATION

Sec. 5 of the Act (Manufacturing and Processing Notices) requires that manufacturers of new chemical substances must give the Administrator 90 days notice before the manufacture of the chemicals. A "new" chemical substance is defined as any substance which is not listed on an inventory of existing chemicals which is expected to be published by the Administrator in early 1979.

The Administrator may designate a use of an existing

chemical compound as a "significant new use" based on the extent to which a use changes the type or form of exposure to human beings or to the environment. Any person who intends to manufacture or process a chemical for such a significant new use must also report 90 days before manufacturing the chemical for that use.

The premanufacturing notification process will be initiated by EPA in 1979, 30 days after publication of the inventory of existing chemical substances.

• REGULATION OF CHEMICAL SUBSTANCES AND MIXTURES:

Under Sec. 6 (Regulation of Hazardous Chemical Substances and Mixtures), the Administrator of EPA may prohibit or limit the manufacture, processing, distribution in commerce, use, or disposal of a chemical substance or mixture if he finds that these activities or any combination of them presents or will present an unreasonable risk of injury to health or to the environment. Sec. 6 also provides EPA the authority to require labeling for a chemical or any article containing the chemical.

Under Sec. 6(e) the Agency has initiated rulemaking dealing with polychlorinated biphenyls (PCB's). Marking and disposal rules were promulgated in February 1978 and regulations implementing the statutory ban on manufacture and use in any way other than in a totally enclosed manner were proposed in June 1978.

Also under Sec. 6(c), the Agency has issued temporary rules under which interested persons may be compensated for certain expenses related to participation in rulemaking proceedings. The temporary rules establish a pilot program related specifically to the rulemaking on PCB's.

Regulations on aerosol uses of chlorofluorocarbons were also promulgated in March 1978.

• IMMINENT HAZARDS

Under Sec. 7 (Imminent Hazards), the Administrator may ask a court to require whatever action that may be necessary to protect against serious or widespread injury to health or the environment for those chemicals that present an imminent and unreasonable risk.

• REPORTING REQUIREMENTS

Sec. 8 (Reporting and Retention of Information), provides the Administrator of EPA with the authority to propose general rules for reporting production, use, exposure and other types of information needed for determinations about human and environmental risks. The Agency also is authorized to require submission of health and safety studies and to require maintenance of records of significant adverse reactions to chemicals.

The first rule, under Sec.(a), issued in December 1977, concerned the inventory reporting regulations. This rule will provide the nation's first comprehensive list of commercial chemical substances and, in addition, requires information on where the chemicals are manufactured and in what quantities.

EPA published final rules in July 1978 under Sec. 8(d) to require industry to submit results of relevant health and safety testing already performed on chemicals included in the Interagency Testing Committee's first report (October 1977). Under Sec. 8(e) chemical manufacturers, processors and distributors are required to furnish information which reasonably supports the conclusion that a chemical substance or mixture presents a substantial risk of injury to health or the environment. Policy guidance on Sec. 8(e), reporting of substantial risk information, was published in March 1978.

• OTHER FEDERAL LAWS

Under Sec. 9 (Relationship to Other Federal Laws), the EPA Administrator may determine that a chemical risk may be prevented or sufficiently reduced by action under an existing Federal law administered by another agency. If such a determination is made, the Administrator will submit a report requesting the agency administering the other law to determine whether the risk exists and if the agency's action would sufficiently reduce the risk. If the agency to which such a report was made finds no risk or takes action directed to the risk, the Administrator may not take any action under Sec. 6 or 7 with respect to such risk.

The Administrator is also directed to use other laws administered by EPA to protect against unreasonable risks, such as the Clean Water or Air Acts unless the Administrator determined that it is in the public interest to protect against such risks under TSCA.

• DISCLOSURE OF DATA

Under Sec. 14 (Disclosure of Data), confidential business information such as trade secrets will be protected from disclosure by the Administrator.

All health and safety information on chemicals in commerce submitted under the Act is subject to disclosure. Exemptions to disclosure of health and safety data are authorized by Sec. 14(b) to protect process or formulary information.

• INDUSTRY ASSISTANCE

Under Sec. 26(d), EPA has established an Industry Assistance Office (IAO) to furnish the chemical industry information on implementation of the Act and on requirements applicable to manufacturers and processors. From its inception, the IAO has been providing technical and other non-financial assistance to industry on TSCA requirements, compliance measures, and agency policy.

Other Provisions of the Law:

- Research, development, collection, dissemination and utilization of data
- Authority to designate establishment inspections

and to subpoena witnesses and testimony for the purposes of carrying out this Act

- Regulation of a chemical produced for export that presents an unreasonable risk to health or to the environment of the U.S. and to require imported chemical substances to comply with any existing U.S. rules promulgated under TSCA

Establishment of Priorities. Given the enormity of existing chemical substances potentially subject to TSCA regulations, plus many additional substances possibly in development, EPA has developed an early strategy for implementing some of the regulatory provisions and has established several objectives which include:

- Define methods for assigning priorities to chemical substances for investigation and regulation. Systematic selection of chemical substances for investigation, or for regulatory action is needed. Criteria for selection of chemicals include two factors of risk: (1) toxicity and (2) exposure. Thus, a chemical that is toxic or that is suspected to cause adverse chronic health effects and for which there is reasonable levels of exposure in the population would receive a high (or relatively high) priority for selection. On the other hand, a toxic substance for which there is little exposure to humans or to the environment would be given a lower priority for selection

Chemical substances that may produce adverse chronic health effects will be given higher priority than acute toxins. Most acutely toxic chemicals are already addressed by other Acts, while chronic hazards have often escaped identification and regulation (e.g., PCB's).

EPA will emphasize those chemicals whose effects are either irreversible or slowly reversible and debilitating: e.g., oncogenic, mutagenic, teratogenic, and neurotoxic effects. With respect to cardiovascular, respiratory, immunological, dermatological, and reproductive effects, EPA will determine priorities based on the severity and irreversibility of the effects.

High priority will also be given to environmental effects of substance that are widely dispersed in the environment and either indirectly threaten human health, effect commercially important species, or significantly disrupt ecosystems.

FEDERAL REGULATORY AGENCY LIAISON

From the early 1900s when the public decried the fact that there was little or no Federal control of hazardous chemicals, the emphasis has shifted to concern by industry and other groups that chemical substances may be overregulated by the Federal Government.

In an address to the Drug, Chemical and Allied Trades Association in June 1976, Hon. Russell Train, then Administrator of the U.S. Environmental Protection Agency, cited an advertisement in *Chemical Week*, "WARNING—TSCA (Toxic Substances Control Act) Can Be Hazardous to The Well-Being of this Country."[26]

Congress recognized this concern for the "well-being of industry," as well as for the public "well-being" in the several years' deliberations that preceeded passage of the Toxic Substances Control legislation. Congress, in enacting the Toxic Substances Control Act, recognized that such authority must reflect industry concern that excessive costs and unnecessary burdens must be avoided.

Thus, one of the primary objectives of the new law emphasizes that while TSCA is to protect the public and the environment from unreasonable effects of toxic substances, it is not to unnecessarily burden industry, nor is it to inhibit the growth and development of new products.

In his address to the Drug, Chemical and Allied Trades Association (ibid) Mr. Train pointed out that industry can no better afford these so-called "chemical incidents" than can the public. Not only are such "incidents" costly to the public in terms of human injury and harm to the environment, but they are also significant financial burdens to industry—costs that could possibly be avoided by effective controls.

Mr. Train stated: ". . . I do not think any of us imagines that there is, or is ever going to be, a real world without real risks. But there is no reason why we cannot—there is, in fact, every reason why we must—take sensible steps to exercise some intelligent and effective control over the risks that we ourselves create."

With passage of TSCA, along with previously existing statutes, there is now chemical substances legislation which provide the several Federal regulatory agencies with authority to regulate chemicals from the point of manufacture to the point of ultimate disposal.

With regulations being promulgated under authority of more than a dozen Acts, the major Federal regulatory agencies recognized the need to coordinate their regulatory activities to avoid overlapping or conflicting actions, as well as the need to conserve resources whenever possible.

In 1977 the four major agencies joined in an historic new initiative in interagency cooperation, seeking ways to coordinate compliance and enforcement activities, as well as to reduce the burden on the regulated industries.

The U.S. Consumer Product Safety Commission (CPSC), the U.S. Environmental Protection Agency (EPA), the Food and Drug Administration (FDA) and the Occupational Safety and Health Administration (OSHA) have agreed to work together as the Interagency Regulatory Liaison Group (IRLG) to improve the public health through:

- Information sharing.
- Avoiding duplication of effort.
- Development of consistent regulatory policy.

On October 11, 1977, the IRLG published in the *Federal Register* an Interagency Agreement Relating to Regulation of Toxic and Hazardous Substances (42 FR 54856). To implement this Agreement, eight work groups were established and on February 17, 1978, the IRLG published in the *Federal Register* a Notice of IRLG Provisional Work Plans of the eight work groups and the schedule of public meetings to discuss these plans.[27]

The objectives of the IRLG work groups are to develop common, consistent, or compatible practices in areas of activity common to the four Agencies. The eight work groups are:

(1) *Compliance and enforcement*—with objectives to coordinate compliance and enforcement efforts in the field, including cooperative application of compliance sanctions.

(2) *Education and communications*—to coordinate public education activities, to coordinate communications materials, including preparation of a general information booklet on Toxic Substances and the IRLG and films/audiovisual materials.

(3) *Epidemiology*—coordination of information on epidemiological programs and resource personnel.

(4) *Information exchange*—proposed development of a coordinated chemical substances information network and coordination of information-gathering/identification projects.

(5) *Regulatory development*—coordinated interagency regulatory action of 24 items/toxic chemicals,* and implementation of an interagency alert system designed to inform the various agencies of an upcoming regulatory action by one of the agencies.

(6) Research planning—(A) A collaborative review of existing agency projects to eliminate duplication, maximize effectiveness and to identify gaps in needed research, (B) Joint planning and coordination of budget proposals for selected mutual interest programs, and (C) Development of a comprehensive action plan for utilizing nonregulatory research programs in regulatory decision-making.

(7) *Risk assessment*—development of procedures and criteria for uniform application by the four Agencies for the purposes of characterizing and quantifying human health risks associated with certain chemicals.

(8) *Testing standards and guidelines*—development of testing guidelines in cooperation with experts and with international agencies such as OECD and WHO. Also, development of criteria for interpretations, quality assurance and other policies relating to the testing of toxic and hazardous substances.

The Agencies have already cooperated in issuance of regulatory proposals and in investigation of toxic hazards of common concern.

The Food and Drug Administration (FDA), EPA and CPSC have cooperated in developing proposals to regulate nonessential chlorofluorocarbons as aerosol propellants.[28] Banning has been proposed under statutes administered by FDA and EPA. CPSC has taken no independent action to ban because the regulation proposed by EPA under the Toxic Substances Control Act would cover the use of fully halogenated chlorofluoroalkanes as aerosol propellants in consumer products.

Recently, three of the Agencies (CPSC, EPA and FDA) have cooperated in the evaluation of glassware having external decorations which contain lead and cadmium. While their statutory authorities differ, the three Agencies share responsibility for the evaluation of the potential toxicity of these substances. A special government task force will evaluate a joint agency report, a trade association's recommendations and any comments received from a recently held public meeting and will make recommendations to the three Agencies for action, if needed.[29]

Undoubtedly many more cooperative efforts will be forthcoming from this new interagency initiative which will benefit the regulators, the regulated trade, and the public.

*See matrix below.

Matrix

Chemical	Hazard	EPA	CPSC	FDA	OSHA
Acrylonitrile	• Systemic toxin • Cancer • Toxic effects of leaded polymer	Hazard assessment being contracted. Report due 5/78. Air risk assessment by CAG 2/78. Decision whether to regulate under CAA 4/78.	Research to investigate the extent and nature of hazard.	Hearing on proposal (bottles) held 6/77. Proposal (other than bottles) published 3/11/77. Final Commissioner decision on bottles 9/77. Options being considered on other uses in strategy document.	Emergency temporary standard and proposed standard published 1/17/78. Public hearing scheduled for 3/21/78.
Arsenic	• Systemic toxin • Lung cancer	Cu smelter control regs. sched for proposal 2/79. Fungi-, Insecti-, and Herbicides Pre-RPAR Review 9/77.		FDA is interested in receiving pertinent data. Certain compounds of arsenic are used in animal drugs.	Current standard 0.5 mg/m^3 Preparing final standard Coordination with EPA on pesticides.

Matrix (*Continued*)

Chemical	Hazard	EPA	CPSC	FDA	OSHA
Asbestos	• Cancer • Asbestosis	EPA analyzing sources and uses of asbestos, monitoring various asbestos sources, looking into asbestos in fireproofing and home insulation hoping to propose new regs. under Clean Air Act re: asbestos in existing buildings. Advisory Panel on Asbestos	Ban on consumer patching compounds and artificial emberizing material (embers and ash) containing respirable, free form asbestos published in FR 12/15/77 (42 FR 63354). Identification and evaluation of other consumer products with asbestos due April.	FDA contracted for method to identify asbestos in food and for animal feed study to determine hazards of ingestion.	Current standard 2 fibers/cc. Hearing to be scheduled. Issues include permissible exposure limit, feasibility and cost of proposal, separate construction standard.
Benzene	• Leukemogen • Blood dyscrasias	Listed as HAP in 6/8/77. F.R. Decision on sources to be controlled scheduled for 1978. Health risk analysis completed, preparing final draft due April.	Research to study extent and nature of hazard.	FDA interested in receiving pertinent data. Permitted in adhesives not in contact with food.	Final standard issued 02/10/78. Separate proceeding on benzene in gasoline—special EPA OSHA Task Force
Beryllium	• Berylliosis • Cancer	OWPS—BAT; 311 NESHAPS several sources—S112 April 1973.			Current standard 2 mg/m^3. Hearings on proposal completed. Analyzing record of rulemaking.
Cadmium	• Systemic toxin • Cancer	A. Determine need to regulate air sources by 8/78. Contact: Richard Johnson 629-5355 B. Health Impacts of air sources. C. Carcinogenic Assessment Group Eval. of Cd. 8/22/77. To assess significance of atmospheric levels. D. Prepare hazard assessment summary (12/77), source analysis (1/78), environmental impacts of control options. (6/79) E. RPAR Procedure F.R.—10/26/77	Determination of need to limit Cd in ceramic pigments in decorative frits for glassware (with EPA and FDA).	Investigating chronic effects in rats, levels of cadmium in food supply, entry into food supply thru use of sludge (with EPA, USDA), levels of CD in shellfish.	Current standard 0.1 mg/m^3 (fume); 0.2 mg/m^3 (dust). NIOSH Criteria Document received 8/24/76. OSHA Advance Notice of Rulemaking 1/28/77.
Chlorinated solvents Trichoroethylene (TCE); Perchloroethylene (PCE); Methyl Chloroform; chloroform	• CNS disorders • Liver dysfunction • Possible carcinogenesis	Hazard assessment Phase I draft reports due 9/77. Phase II reports 8/78. Contact: S. Nasaleki 755-6956. OWS plans proposal for CHCl3 in drinking water. Air risk assessment for PCE due from CAG 3/78. Decision to regulate PCE under CAA 5/78.	Research to study extent and nature of hazard (PCE). Proposal drafted to add additional warnings on paint strippers containing methylene chloride for people with heart and lung problems.	Final rule revoking us of TCE has been drafted. Proposed rule revoking use of PCE has been drafted. CHCl3 final published 6/29/76 (drug and cosmetic uses). Analyzing record on indirect food additive uses.	Proposal for TCE published 10/20/75. Proposals for PCE and methyl chloroform being developed. CHCl3—NIOSH Criteria Document Rec'd 9/11/78.
Chlorofluorocarbons (CFC)		Proposed regulation 5/77 on aerosols. Promulgate 02/78. Phase II nonaerosal proposal 6/78. Promulgate 12/78.	CPSC/FOA/EPA Interagency Activity Labeling and Reporting Regulation will become effective on 02/78.	CPSC/FDA/EPA Interagency Activity. Proposal 5/3/77. Final approx. 04/78.	

Matrix (*Continued*)

Chemical	Hazard	EPA	CPSC	FDA	OSHA
Chromates	• Irritation and ulceration of skin; perforation of nasal spetum • Cancer	Drinking Water Std. 1976; Water Qual. Std. 1976 currently being revised; Some effluent guidelines, BAT revisions will include additional limitation; Hazardous substance (S311, FWPCA)		FDA is interested in pertinent data. Use is permitted as colorant in rubber products contacting food.	Current standard 1 mg/m^3 Proposed standard under development.
Coke oven emissions	• Lung disease • Cancer	Determination whether to list as "new source" or "hazardous" pollutant under CAA or neither by 7/78 after public hearing. First set of controls to be proposed 10/78. Water programs assessing need to propose effluent guidelines.			OSHA Final Standard issued 10/22/76.
DES	• Physiological effects; carcinogen			New labelling requirements in effect. Proposing to ban DES as animal growth stimulant.	Subject of OSHA compliance inspections. No permanent standard.
Dibromochloropropane	• Reduced sperm counts • Cancer	RPAR handled by OSPR. Proposed cancellation of pesticide registration.		Monitoring of food for DBCP residues by Office of Technology. Office of Compliance has prepared enforcement guidelines.	OSHA permanent standard issued 03/15/78.
Ethylene dibromide	• Cancer	RPAR issued 12/14/77. Rebuttal period closed 04/03/77. Office of Pesticide Programs and Office of Toxic Substances reviewing risks.		FDA is interested in receiving pertinent data. Tolerances established for use as a grain fumigant.	NIOSH Criteria Document rec'd 8/77. Request for comments and information published in 3/17/78 FR
Ethylene oxide	• Local toxicity, burns, mutagencity	EPA to take action on RPAR.		Preparing to publish Federal Register announcement of proposed maximum residue limits on surgical instruments and maximum levels of exposure.	Current standard 50 ppm. (NIOSH has issued special hazard review.)
Lead	• Lead intoxication and poisoning • Subclinical effects	Ambient air standard proposed 12/77. Promulgation expected 6/78. Revised national primary drinking water regulations. In progress. Criteria document for lead in ambient waters. Final due by 6/78 Effluent guidelines for lead discharge from 21 industrial categories. Scheduled for 1978–1981. Implementation of RCRA (resource recovery). Regulation of lead in solid waste by 6/79.	Ban of paints containing lead in excess of .06% C.	Proposed tolerance for lead in evaporated milk published 12/6/74. New proposal on temporary action level to be completed other regulatory options to be explored.	Current standard 0.2 mg/m^3 Preparing final regulation Special issues of susceptible groups, medical removal protection.

Matrix (*Continued*)

Chemical	Hazard	EPA	CPSC	FDA	OSHA
		Study of environmental and economic impacts of limitation of lead and its compounds.			
Mercury	• Mercurialism	Study to begin 3/77. NESHAP's under CAA—sec. 112, April 1973.		Proposal published 12/6/74. Final expected 4 QTR 78. Analyzing record of rulemaking.	NIOSH Criteria Document received 8/13/73. Advanced Notice of Proposed Rulemaking OSHA exposure standard of 0.1 mg/m^3 published 1/18/74.
Nitrosamines	• Cancer	FIFRA, Air, Effluent Guidelines, Water	Candidate for the development of a monograph.	Notice published 9/2/77 for info on poultry products. Exposure in other products to be considered.	
Ozone		Revising Criteria Document on ozone; completion of review with respect to air 2/78; Publication of proposal on limits of Trihalonethane in H_3O, alternatives to chlorine—12/77.		Notice of filing of Petition 6/25/75. Petition filed by American bottled water association. Contact:	OSHA exposure standard of 0.1 ppm. Proposed Std. 10/8/75. 40 FR 47309 NIOSH Crit. Doc. to be transmitted to OSHA fy 78
PBB's	• Environmental contamination	Proposed regulation by 9/78.	Used as a flame retardant; CPSC is cooperating with EPA.	Petition to establish tolerance denied 8/12/77.	Subject of OSHA Compliance inspections. No permanent standard.
PCB's	• Environmental contaminant	Final regulation on marking/disposal will be promulgated by 2/78. Draft ban regulation will be proposed in early 1978.	Studying substitute dielectric material.	Proposal published 4/1/77. Analyzing comments. Final expected I QTR 78.	Subject of OSHA Compliance Inspections. Regulated by OSHA as chlorodiphenyl (54% and 42%)
Radiation, ionizing		Determination whether to list radioactive materials as "criteria" or "hazardous" or neither by 8/78. Office of Radiation Programs	Reviewing question of consumer exposure to ionizing radiation.	Guidelines under development on radionuclides/irraiated food. Notice for comments on guidelines published 5/7/76. Bureau of Radiological Health.	OSHA standards at 29 CFR 1910.96. Revisions under consideration.
Waste disposal to the food chain	• Environmental contamination	Solid Waste Disposal Regs. to be promulgated in Mid 78. Guidelines and bulletin under development.			
Sulfur dioxide	• Respiratory irritant • Environmental damage	EPA Criteria Pollutant. Criteria Revision scheduled 5/80.		Generally recognized as safe (GRAS). Status currently under review.	Current standard 5 ppm Analyzing record of rulemaking. Preparing final standard.
Vinyl chloride (VC); Polyvinyl chloride (PVC).	• System toxicity • Cancer	VC Proposal 6/2/77. Comment period extended to 9/23/77. Final rule due SPR 1978.	Reissuance of final standard classifying as "banned hazardous substance" self-press pressurized household substances containing vinyl chloride monomer.	Proposal published 9/3/75. Analyzing record of rulemaking in light of acrylonitrile decision.	OSHA final VC standard published 10/4/74.

Source: Regulation Development Work Group—IRLG.

Matrix (*Continued*)

Chemical Substances Legislation		
Agency/Act	Code of Federal Regulations (CFR)	Jurisdiction
Consumer Product Saftey Commission (CPSC)	16 CFR	
• Consumer Product Safety Act (CPSA)	Part 1100–1499	Chemicals in consumer products IF public interest finding by Commission
• Fed. Hazardous Substances Act (FHSA)	Part 1500–	Household chemical substances
• Poison Prevention Pkg. Act (PPPA)	Part 1700–	Safety Pkg.—Chem. substs. toxic to children under 5 yrs. of age
Environmental Protection Agency (EPA)	40 CFR	
• Clean Air Act	Parts 50–55, 60–69	Toxic emissions from stationary sources and moving vehicles
• Clean Water Act (Formerly Fed. Water Pollution Control Act)	Parts 100–140, 300–, 400–460	Discharge of Effluents incl. Haz. Substances into water
• Fed. Insecticide, Fungicide & Rodenticide Act (FIFRA)	Part 162–180	Regulation and control of pesticides
• Resource Conservation & Recovery Act (RCRA) (Solid Wastes Disposal Act)	Part 240–255	Generation, transportation disposal & reclamation of solid wastes
• Safe Drinking Water Act	Part 141–149	Chem. contaminants of primary & secondary drinking water supply
• Toxic Substances Control Act (TSCA)	Part 700–799	Reg. of existing environ. toxics and control of Toxic substs. at point of mfgr.
Food & Drug Administration (FDA)	21 CFR	
• Fed. Food, Drug & Cosmetic Act	Part 8–9	Color additives & color certification
	Part 10–199	Food & food products
	Part 200–299	Drugs: general
	Part 300–499	Drugs for human use
	Part 500–599	Animal Drugs, Feeds & related products
	Part 600–699	Biologics
	Part 700–799	Cosmetics
	Part 800–899	Medical devices
	Part 1000–1099	Radiological Health
	(eg. Pt. 1020)	(eg. Ionizing rad.)
• Misc. Acts: (Fed. Import Milk Act, Import. Tea Act, etc.)	Part 1200–1299	Imported milk products, tea, etc.
Occupational Safety & Health Admin. (OSHA)	29 CFR	
• Occup. Safety & Health Act of 1970	Part. 1900–1919	Industrial use of, and worker exposure to, chemical substs.

BIBLIOGRAPHY

1. Train, Russell E.: "Statements of Hon. Russell Train, Chairman, Council on Environmental Quality": Congressional Record, 71-179, 0-72-pt.1-6, pp. 63–76, Oct. 1971 and 79-875-72-6, pp. 65–79, Nov. 1971.
2. Dept. of Labor, Occupational Safety and Health Administration: "Identification, Classification and Regulation of Toxic Substances Posing a Potential Occupational Carcinogenic Risk." *Federal Register*, Part IV, **42** (192), 54148-54247 (Oct. 4, 1977).
3. Anderson, O. E.: "Pioneer Statute: The Pure Food and Drugs Act of 1906," Emory University Law School, J. Public Law: The Government and the Consumer: Evolution of Food and Drug Laws., Vol. 13, No. 1, pp. 189–196, 1964.
4. Roe, R. S.: "The Food and Drugs Act—Past, Present, and Future." The Impact of the Food and Drug Administration on Our Society., Welch, H. and Marti-Ibanez, eds., MD Publications, Inc., New York, 1956.
5. Janssen, W. F.: "Introduction: Annual Reports 1950–1974," Food and Drug Administration, pp. 1–xv, June 1, 1976.
6. Lehnhard, M. N.: "A Legislative History of the Federal Food, Drug and Cosmetic Act," Congressional Research Service, TX 501 B, 73-147 ED, pp. 1–28, Oct. 1973.
7. King, E. A.: "New Drugs." The Impact of the Food and Drug Administration on Our Society, Welch, H. and Marti-Ibanez, eds., MD Publications, Inc., New York, 1956.
8. Janssen, W. F.: "FDA Since 1938: The Major Trends and Developments," Emory University, *J. Public Law*, **13** (1) 205-220 (1964).
9. Kleinfeld, V. A.: "Introduction," Federal Food, Drug

& Cosmetic Act. *Judicial Record*, 1969–74, published by Food and Drug Law Institute, Washington, D.C., 1976.

10. Federal Environmental Pesticide Control Act of 1971: Report No. 92-511, 92nd Congress, 1st Session, pp. 1–13, Sept. 1971.
11. U.S. Environmental Protection Agency: "The Federal Environmental Pesticide Control Act of 1972. Highlights," Jan. 1973.
12. Kerlan, I.: "Public Health Activities of the Food & Drug Administration," The Impact of the Food & Drug Administration on Our Society, p. 120–122, Welch, H. and Marti-Ibanez, eds., MD Publications, Inc., New York, 1956.
13. Harvey, J. L.: "The Federal Hazardous Substances Labeling Act." *Food, Drug, Cosmetic Law Jour.*, **16,** (12) 708–714 (Dec. 1961).
14. Ligon, E. W.: "Federal Hazardous Substances Labeling Act," *Arch. Environ. Health*, **10,** 596–598 (Apr. 1965).
15. Miller, D. C.: "Child Protection Under the Federal Hazardous Substances Act," FDA Papers, May 1967.
16. Committee on Labor & Public Welfare, U.S. Senate: "Legislative History of the Occupational Safety & Health Act of 1970" (S.2193, P.L. 91-596), June 1971.
17. Hehir, R. M., Bayard, S. P., and Thompson, E. J.: "CPSC Regulation of Non-Occupational Exposure to Asbestos in Consumer Products," presented July 20, 1977 at the NBS/OSHA Workshop on Asbestos and to be publ. in its Proceedings.
18. Ehrlich, A. M., Bayard, S. P., and Thompson, E. J.: "Consumer Protection and the Consumer Product Safety Commission," presented Dec. 7, 1977 at the Soc. on Environ. & Occup. Health Symposium on Occupational Exposure to Fibrous and Particulate Dust and Their Extension into the Environment, to be publ. in its Proceedings.
19. *Federal Register*, Vol. 35, No. 494, Oct. 6, 1970.
20. Zenner, R.: "The Federal Laws of Water Pollution Control," in *Federal Environmental Law* (E. Dolgin and T. Guilbert, ed., 1974).
21. Quarles, J. R.: "Impact of the 1977 Clean Water Act Amendments on Industrial Dischargers," Environment Reporter, Monograph No. 25, Vol. 8, No. 46, The Bureau of National Affairs, Mar. 18, 1978.
22. Environmental Protection Agency: "Pesticides and the Law," Environmental Facts, publ. by EPA, Washington, D.C., Dec. 1974.
23. Quarles, J. R.: "Statement Before the Committee on Agriculture," House of Representatives, Jan. 26, 1977.
24. Environmental Protection Agency—Office of General Counsel. Memo dated Oct. 27, 1976—M. Wright, Atty., Air Quality & Noise Control Div. to G. W. Frick, General Counsel.
25. Environmental Protection Agency—Toxic Substances Control Act (TSCA) Strategy, Jan. 23, 1978.
26. Train, R. E.: "Controlling the Risks We Create: The Need for Toxic Substances Control Legislation." Remarks prepared for delivery before the Spring Luncheon—Drug, Chemical and Allied Trades Assoc., New York, June 17, 1976.
27. CPSC, EPA, FDA, OSHA: "Interagency Regulatory Liaison Group: Notice of IRLG Work Plans and Public Meetings," *Federal Register*, Part VII, Vol. 43, No. 34, pp. 7174–7198, Feb. 17, 1978.
28. U.S. Dept. of Health, Education & Welfare: *HEW News*, P78-13, Mar. 15, 1978.
29. U.S. Dept. of Health, Education & Welfare: *HEW News*, P78-7, Jan. 31, 1978.

SECTION 10B

An Industrial Response to Chronic Health Hazards

John F. Schmutz
Assistant General Counsel
E. I. Du Pont de Nemours & Company
Wilmington, Delaware

Chronic health hazards are of concern to industry and the Nation. Twenty-five years ago, two or three chemicals were known to cause cancer in man. Today, OSHA regulates 20 as potential carcinogens, and NIOSH has a list of 2240 suspected carcinogens. We must move aggressively to deal with the issue—and we are.

The Nation is now at the same stage with regard to chronic health hazards as it was with respect to air and water pollution control five to ten years ago. Key laws have been enacted, and we are in the process of developing policies and the bases for regulations necessary to establish the statutory authority. From a policy viewpoint, we can draw from prior experiences with the environment in charting a course which focuses our effort, conserves our resources, and makes positive progress toward a cleaner and more healthful environment.

The following discussion provides a perspective to chronic health hazards; discusses the critical issues, particularly acceptable risk; and provides a suggestion as to how we might deal with those issues.

FIVE FACTS ABOUT CHRONIC HEALTH HAZARDS

Because it is so personal to each of us, the question of chronic illness, particularly cancer, is one which is difficult to view in perspective. The deep concern for those stricken leads us to react emotionally rather than rationally. To manage chronic health hazards effectively, they must be looked at via a reasoned approach, based on available facts. Five facts which help frame the chronic health hazards issue with respect to chemicals are set forth below.

1. Chemicals are not necessarily "good" or "bad," whether "man-made" or natural. Almost everything in nature involves chemicals. To change from man-made materials to naturally based products does not avoid chemicals. Food is as much an organic chemical as a plastic sheet or a solvent. Acetic acid, for example, is made in large volumes synthetically and also found naturally in vinegar. In its testimony before Congress on June 23, 1977, the Environmental Defense Fund identified 59 toxic environmental pollutants from the manufacture of paper; 13 of these are also found naturally occurring in foods or are normal metabolites of substances found in such foods.

2. Chemical carcinogens and other chemical chronic health hazards are not necessarily man-made. Asbestos, which helped initiate the focus on chronic health hazards, is a chemical and a naturally occurring carcinogen. Peanuts and grains frequently contain traces of aflatoxin, a potent carcinogen formed by a common mold. See sections 8, 12.

3. Many chemicals essential for health in small quantities are highly toxic in larger quantities. We cannot live without small amounts of zinc, silicon, nickel, tin, vanadium, potassium, and many other elements which also display severe acute and chronic toxicity in larger amounts.[1] For example, nickel and selenium in some forms are carcinogens.

4. The incidence of cancer is actually, despite some statistics, not rapidly increasing. If the effects of cigarette smoking are excluded and statistics are age adjusted, cancer incidence and cancer deaths per unit population have remained about *constant* over the last twenty-five years.

5. The statement that 80 to 90% of cancers are environmentally caused does not mean that 80 to 90% of cancers are caused by industry. Environment in this sense includes not only the air we breathe and water we drink, but our diet and all elements of our lifestyle, at home and elsewhere, on and off the job.

"Environment," therefore does, *not* equal "industry."

The major causes of environmental cancer are smoking and diet. The American Cancer Society estimates that up to 80% of *all lung cancer* or up to 22% of *all* cancers among males is caused by cigarette smoking. Furthermore, the National Cancer Institute estimates that diet, even excluding food additives and contaminants, may contribute to as much as 50% of all cancers. Reliable experts estimate that 5% or less of cancer is industrially related. That number includes known hazards such as asbestos, β-naphthylamine and others now well under control. See Sections 12, 8.

In summary, health hazards are a national problem resulting from a wide variety of causes of which industrial chemicals are only a part. Because of the intense human suffering involved, industrial chronic health hazards are a major concern to industry. However, industrial health hazards are far from being the principal cause of such chronic illnesses. Why then the focus on industry?

NEED TO DEFINE ISSUES

Cancer has become an immediate, readily identifiable and emotional issue in all our lives. All of us have lost friends and relatives from it. Also, tragic events in the last several years have focused on several instances where cancer or other serious chronic illnesses have been caused by industry-related chemicals. This has led to a tendency to assume that if an industrial chemical is present in the environment, it is harmful. Emotionally, we are prone to consider a man-made chemical found in the environment as more dangerous than a naturally occurring chemical found there in the same or even larger quantities.

In addition, enormous strides in the detection of chemicals in our environment also have helped focus attention on the many chemicals to which workers and the public have been exposed over long periods. During the last five years, our techniques for detecting chemicals have improved greatly. In water, for example, accuracy of detection has progressed a thousandfold, from parts per million to parts per billion—from drops per 100 gallons to drops per *100 thousand* gallons.

The rapid evolution in toxicology and epidemiology in pinpointing hazards over the last twenty years also have made it difficult to segregate the events of today from the events of many years ago. Cancers or other chronic illnesses diagnosed today often are the result of exposure many years ago when hazards were unknown and so industrial practices were far different.

Judging the work practices of twenty years ago by the increased knowledge and more informed standards of today, is often not constructive. One of Du Pont's experiences in this regard illustrates the point.

More than twenty years ago, the Du Pont Company established a system to collect morbidity and mortality data on its employees. Its purpose was to provide one check among many on the exposure of our workers to chronic health hazards. In the last several years, the cancer registry part of this epidemiology program has been examined in-depth by government, actually as one of the few available to so examine. In voluntarily submitting such data, we ourselves have pointed out certain deficiencies in our registry—deficiencies not even unique to our system. Yet, focus from some sources was on the shortcomings of our system and has, unfortunately, masked the fact that it is a unique and pioneering effort and a contribution to the health of our employees. There must be, it seems, an emotional conditioning for critics to focus on the negative, rather than on the positive aspects of the system.

There also is an element of frustration caused by the very nature of chronic hazards. In most situations we are quickly able to see improvement in response to our positive actions. Fish return to the cleaner water. People can swim again in waters previously polluted. There is reduced smog and air pollution. But with chronic hazards it is quite different. Exposures of many years ago may cause illnesses today and possibly will for years to come, so that the results of today's controls are not yet apparent, and may not be for twenty or more years.

That frustration of not being able to see improvement simply increases the pressure to "do something" and feeds a desire to mandate uniform technology-forcing control at "absolutely" safe levels. However, that approach will not solve the problem of past exposures to chronic hazards nor is the theory in the national interest.

Because of the emotionalism and pressures described above, there is a critical need to bring objectivity to focus on the issue of chronic health hazards and acceptable levels of risk.

Zero risk or zero exposure is neither technically possible, nor, given its consequences, desirable. For example, it is not possible to reduce to zero, worker exposure to a gaseous chemical used in a plant, nor in products made from that material. Industrial exposure could, however, be reduced to a level so low that the probability of causing cancer would be nearly nil. There may, in fact, be no hazard either from minute quantities of residual monomer in a polymer product, but we could not reduce the quantity of monomer to zero.

The delicious aroma of your Thanksgiving turkey is in part caused by acrolein, a highly toxic material. That delectable picnic-grilled meat contains benzopyrene, a carcinogen. Are we going to stop eating turkeys for Thanksgiving or using outdoor grills? Are we going to ban peanuts? Those are some of the consequences of zero risk.

In speaking of an acceptable level of risk, I am not talking about those situations in which it is known that the level and duration of exposure would be likely to cause some people to contract a chronic illness. What I am talking about is accepting a level of risk at which it is unlikely that anyone will become chronically ill but at which one *cannot* prove whether *or not* there is one hundred percent safety. It means living with reasonable assurance of safety and acceptable uncertainty.

Congress has accepted that uncertainty. The report by the Committee on Interstate and Foreign Commerce of the House of Representatives on the Toxic Substances Control Act states:

> The Committee has limited the Administrator to taking action only against unreasonable risks because to do otherwise assumes that a risk-free society is attainable, an assumption the Committee does not make.[2]

Note that many years ago the concept of acceptable risk was adopted in handling radiation. The risk of X-rays is generally accepted. The nuclear power industry is regulated on that premise.

The concepts of zero exposure and zero risk ignore another point—the many positive benefits of chemical products. They ignore the fact that those products have become essential to our health and safety as well as our comfort and convenience. They ignore the many jobs made possible by some chemical products. They ignore the many socially beneficial results obtained with taxes generated by the manufacture of products made from them.

Food provides an example of those benefits. Because of synthetic fibers, millions of acres otherwise required to produce natural fibers are available for food cultivation. Because of agricultural chemicals and pesticides, yields per acre are increased manyfold. Refrigerants and preservatives reduce spoilage and permit us to have healthful fruits and vegetables year-round.

SETTING AN ACCEPTABLE LEVEL OF RISK

How is an acceptable level of risk set?

The starting point is the fact that the effect of carcinogens and other chronic health hazards is related to dose and period of exposure. Stockinger and others have suggested that below a certain dose, natural mechanisms may effectively "neutralize" carcinogens and prevent tumor formation.[3] Even above that dose, tumors that result may require a period far exceeding a normal lifetime to show. One accepted relationship for carcinogens is that of Druckrey that $Dt^n = k$, where D is the total accumulated dose, t is the time to the first appearance of tumors, and n and k are constants related to the particular carcinogen.[4] Note that dose itself involves concentration times duration of exposure.

Considering level and duration of exposure, those conditions should be found under which no effect is observed. Then a substantial margin of safety should be applied by further reducing the level of exposure. Let me add, however, that a necessary element of this approach is full disclosure of all known significant hazards to all those accepting the risk.

It follows from the concept of acceptable risk, that use of a rigid guideline of control to the lowest level feasible is inappropriate, except as an interim measure. With a potent carcinogen, for which a no-effect level has not yet been established, the lowest level feasible may present an unacceptable risk if it is continued over a prolonged period. In such case, use of the carcinogen should be discontinued unless a no-effect level can be established. In other cases, control to the lowest level feasible may unnecessarily waste jobs as well as capital, energy, and other resources.

Duration of exposure, as well as level, is important. Control to a lowest feasible level while further data are gathered may provide reasonable assurance of safety and acceptable temporary uncertainty. Prolonged exposure at that level may later be shown to be unacceptable.

One can well sympathize with the regulators' frustration at extended proceedings and the delay involved in a product-by-product approach. Factors such as oncogenicity vs. carcinogenicity and screening tests are subject to policy decisions. Many aspects of recent proposals for regulation of chronic hazards in the workplace are sound. But, ultimately, hazards must be evaluated on a product-by-product basis. Furthermore, controls mandated, that is, whether administrative controls, engineering controls or personal protective equipment are to be used, should vary from case to case. In water pollution, the promulgation of effluent guidelines has confirmed the administrative feasibility of the case-by-case approach. The NPDES permits to tens of thousands of individual plants confirm it.

Just as product-by-product evaluation of risk is essential if there is to be socially effective regulation, so, too, is the setting of priorities for efforts. The Nation does not have the laboratories or the toxicologists to test the chronic effects of all chemicals immediately, nor would such effort be desirable. The efforts of the Advisory Committee, established under the Toxic Substances Control Act to help set priorities, are to be commended. Selectivity also should be applied to all aspects of the regulation of chronic hazards. For example, broad production of detailed data as now required by TSCA inventory reporting regulations can only obscure the critical issues and delay meaningful analysis by the Government.

WHEN A HAZARD IS DISCOVERED

What might be a typical response when a hazard is discovered? When the determination is made that there is a chronic health hazard in an existing process, all those who may need to act on that information should be promptly notified. This includes employees, appropriate government agencies, and, in many cases, customers, other producers, and the media. Programs to measure and monitor exposure levels also should be reviewed and, if necessary, supplemented. Review of medical histories of exposed employees and other epidemiological data also help to define the hazard.

Based on all available data, an acceptable level of risk is determined. If insufficient data are available to fix that level, long-term, further toxicological testing may be necessary.

When the hazard is defined and an acceptable level of risk set, at least short-term programs and practices should be reviewed to determine if additional measures are needed to safeguard the health of employees. These might include revising work practices; specifying additional protective equipment; or modifying or redesigning equipment, process changes or other improved engineering controls. Appropriate medical surveillance procedures should also be instituted. See Section 2.

Specific actions will vary with each particular case. In some cases with materials of low potency, only minor changes in operations may be needed to guard the safety of employees. In other cases, major changes may be indicated. In a few cases, it may be necessary to halt production or sales until operations which protect employees and the public can be assured.

The following case history illustrates the above as well as the complexity of the carcinogen issue from an industrial viewpoint. A promising new product, Kevlar® aramid fiber, is made using hexamethyl phosphoramide (HMPA) as a polymer solvent in an early step. That chemical is and has been used widely as a solvent in laboratories for many years. No human cancer has been attributed to it.

From its earliest use, because of its acute toxic hazards, Du Pont treated HMPA as a potentially hazardous chemical and controlled exposure to it to very low levels. Based on early toxicological screens, rat inhalation tests were also begun, but before the tests could be run, special techniques to handle the low exposure levels had to be developed. Then, eight months into the rat tests, cancer emerged at levels as low as 400 parts per billion (ppb) and after thirteen months, at 50 ppb.

Although there was no evidence of cancer-related problems with employees, the first step was to quickly and systematically disclose findings to employees and government agencies to which such findings would be helpful. Since the chemical was widely used in research, Du Pont also sought disclosure in major publications.

Concurrently, as the test data developed, Du Pont lowered the permissible airborne exposure limit to 25 *ppb*, then to 5 and, within one year, to *0.5 ppb*. Initially, there was no method to detect such small quantities, the equivalent of about one drop per 10 million gallons. Therefore, research and engineering people had to devise a test method. In the plant, a combination of engineering controls and personal protection was used, the latter because of the freedom of manipulation required of employees for some of the fiber operations.

Simultaneously, new animal tests were initiated to substantiate a no-effect level.

Although the search for an alternate was begun immediately, many man-years of research have been required to find an alternate solvent and another 18–24 months will be required to modify the equipment and process to fully eliminate the use of HMPA.

The foregoing provides a perspective on chronic hazards from an industrial viewpoint and an example of how one company has dealt with one new discovery. How should such hazards be regulated?

REGULATION OF CHRONIC HEALTH HAZARDS

The condition precedent to sound regulation is an adequate and balanced legislative base. With the passage of the Occupational Safety and Health Act in 1970 and the Toxic Substances Control Act in 1977, that base has been provided. Now success of implementation will depend critically on balanced regulation.

An Interagency Regulatory Liaison Group, representing CPSC, EPA, FDA and OSHA, has been formed to coordinate approaches to such issues as testing, risk assessment, regulation and enforcement. Another group called the Toxic Substances Strategy Committee has been formed by the Council on Environmental Quality to develop policies and coordinate the federal regulatory approach to toxic chemicals. These groups can do much to help keep that regulation within the bounds of congressional intent by assuring a stable, coordinated approach.

To the regulators, five guidelines in drafting chronic hazards regulations are suggested:

1. Data within federal agencies should be reviewed to determine where additional data acquisition or development may be necessary. Where gaps exist, EPA can selectively require submission of information from the private sector and develop testing requirements. Appropriate selection criteria would include the magnitude and routes of exposure, the extent of existing data, and chemical properties. This approach would focus on those chemicals which may present unreasonable risks and permit meaningful utilization of existing testing resources to evaluate priority needs.

Nonselectivity could be counterproductive and serve merely to slow down review of major high risk areas of concern. Thus, recommendations for testing by categories of chemical substances threaten priority needs by requiring extensive and unnecessary preemption of limited testing resources.

2. There should be full disclosure of all significant hazards to all those who may need to act on that information, including employees, customers, the Government, and others. Existing authority under TSCA provides an adequate statutory basis to require industry to notify the Government of pertinent health and safety data.

3. Workplace concentrations of suspected animal or human carcinogens should be promptly but temporarily limited to the lowest feasible level considering length of exposure, the physical form of the material, its concentration, and existing toxicological data. The level of control should take into consideration existing data and apply a safety factor. For example, in some cases the acceptable airborne concentration might be one-tenth of the no-observed effect concentration in a suitable animal study.

The level and duration of control, not the means thereof, should be of primary concern. The means of control should be a practical combination of engineering controls to the extent technically and economically feasible augmented by administrative controls and personal protective equipment as necessary.

4. An acceptable risk level should be defined as quickly as possible. Acceptable risk should be based on a suitable reduction from the no-observed effect level in animals, epidemiology studies, or their equivalent. Functionally, the acceptable risk level would be a level at which it would be reasonable to predict that *no one* would be likely to get cancer or other chronic illness from exposure to the chemical. In making this determination, it must be recognized that one could not prove, nor is it possible to ever prove, that some uniquely sensitive person could not get cancer in a situation considered as safe. See Section 8.

5. There should be control to the acceptable exposure level. If extended testing is required to determine that level, we should continue control to the lowest level feasible. If an acceptable risk level cannot be determined and the product cannot be made and used safely, then the operation should be discontinued.

It will be a challenge for industry, labor, the Government and the public to work together objectively on chronic hazards. But the stakes are high and we must do it. Our mistakes and successes will not be measured for many years. In the interim—while we are hard at work finding and reducing the hazards—good judgment, objectivity, and an acceptance of the fact that life cannot be made risk-free must tide us over.

REFERENCES

1. Thorn, G. W., *et al., Principles of Internal Medicine*. p. 466, 8th ed., 1977.
2. H. R. Rep. No. 94-1341, 94th Congress, 2nd Session, p. 15, 1976.
3. Stockinger, H. E., "The Case for Carcinogen TLV's Continues Strong," *Occupational Safety and Health*, p. 54, March/April 1977.
4. Druckrey, H., Quantitative Aspects of Chemical Carcinogenesis. *Potential Carcinogenic Hazards from Drugs*. R. Truhaut, ed. UICC Monograph Series 7:60–78. Springer-Verlag. New York, 1967.

SECTION 11*

Labeling and Identification of Hazardous Materials

Richard J. Lewis, Sr.
Division of Technical Services
National Institute for Occupational Safety and Health
Cincinnati, Ohio

Spillage, accidental discharge, improper storage, or use of hazardous materials can cause unexpected human contact or create potentially dangerous situations. Adverse consequences can be minimized if prompt corrective action is taken based on the nature of the danger presented by the material. Safety and health experts have developed procedures for controlling hazards and treating persons who have been exposed. Corrective action depends on knowing the nature of the material and its toxic or hazardous properties. To supply this information where it is most needed, various governmental agencies and private associations have developed rules, systems, and guidelines for imparting concise hazard information and control instructions at the point of storage or use of hazardous materials. This section provides an overview of labeling and identification systems in current use, arranged by the organization that controls the system. Each system is discussed in terms of its:

(1) Originator.
(2) Legal standing.
(3) Extent of coverage.
(4) Format and application.

These systems undergo regular modification in response to the introduction of products containing new hazardous materials, new applications of old materials, refinement in medical and safety knowledge and analytical testing procedures, and the impetus of environmental and consumer advocate groups. Since many of the regulations governing labeling systems are undergoing rapid expansion or extensive revision, this Section will only attempt to provide a general discussion of the scope and content of the various labeling systems. As a result of this constant change, most publications, including this one, will contain some out-of date material. The brief descriptions are but introductions and not full explanations. This is intentional, since accurate application of any labeling system requires careful study of the most current rules. Direct contact with the sponsoring organization is highly recommended.

It is generally agreed that proper application of the more complex and detailed systems is a difficult and frustrating task. The inexperienced labeler is well advised to contact the appropriate agency or association for basic information and resolution of conflicts before generating a label. Some regulations provide for substantial penalties for improper labeling of a hazardous material. Those involved in enforcing a labeling system generally realize the complexity of the matter, and have available numerous information packets which try to explain and clarify the labeling rules and systems.

The terms *label* and *labeling* can refer to a specific device containing printed or graphic matter and attached to a container of a hazardous material. *Labeling* can also refer to placards, signs, and information sheets which contain precautionary information about hazardous materials. The latter, broader definition is used in this Section.

The users of labeling systems can be grouped into three categories: (1) manufacturers, (2) shippers, and (3) employers. These categories reflect the scope of the systems. There is usually considerable overlapping since manufacturers must ship their products and inform their employees of in-plant hazards. Because each system is designed with a specific purpose, there can be conflicts. Some materials may require several separate labels to comply with all applicable regulations.

The *manufacturer* of a material is responsible for complying with all applicable laws and regulations dealing with his product. The same product may be classified as hazardous for inclusion in consumer goods and require one type of label, and nonhazardous or hazardous in a different way for industrial use, thus requiring different labeling. In addition to complying with

*This section was written by R. J. Lewis, Sr. in his private capacity. No official support or endorsement by NIOSH or DHEW is intended or should be inferred.

the law, or where no law is applicable, the manufacturer may consider labeling certain products in accordance with industry or trade association guidelines. Customers are demanding increasingly detailed hazard information to satisfy internal safety needs.

The *shipper*, who may be the manufacturer or intermediate handler of the material, must comply with the extensive shipping regulations of the Department of Transportation. International air shipments must meet the requirements of the destination country and comply with international agreements on labeling arranged under the auspices of the United Nations. International air shipments must comply with the rules of the International Air Transport Association (IATA).

The *employer* must comply with the requirements of the Occupational Safety and Health Administration (OSHA), the rules of the Department of Energy for manmade radioactive materials, and the Environmental Protection Agency's pesticide regulations.

Table 11.1 lists the major agencies or associations dealing with the labeling of hazardous materials and the coverage intended by each system.

State and local regulations exist for a variety of hazardous materials. Lead and other metals in paint and finishing products, flammable, and reactive materials are controlled to a variable degree in different jurisdictions. The rules and extent of coverage are far from uniform thus requiring a review of each state and local code for applicability to a particular material.

TABLE 11.1. Classification of Groups Active in the Labeling of Hazardous Materials in Terms of Users and Area of Coverage.

User of Labels	Area of Coverage	Groups Regulating the System
Manufacturers	Consumer products	Consumer Product Safety Commission Food and Drug Administration Environmental Protection Agency Manufacturing Chemists Association Local ordinances
	Nonconsumer	Occupational Safety and Health Administration Environmental Protection Agency Department of Energy Manufacturing Chemists Association National Fire Protection Association American National Standards Institute Local ordinances
Shippers	All shipments by common carrier	Department of Transportation International Air Transport Association United Nations
Employers	Materials to which employees may be exposed	Occupational Safety and Health Administration Environmental Protection Agency Department of Energy Local ordinances

FEDERAL REGULATORY AGENCIES

Information on proposed changes in rules and regulations as well as the final notices of changes are published in the FR, the daily newspaper of the U.S. Government. These rules are the detailed implementation of the general laws enacted by Congress giving an agency authority to regulate in a certain area. These rules are compiled annually in the CFR. Each agency has one or more volumes of the CFR assigned to it, i.e., 10 CFR contains AEC regulations. The FR and CFR are available in most large public libraries. These publications can be consulted for detailed, exact definitions and interpretations of the rules.

These publications are in legal language and often difficult to read. Each agency publishes pamphlets, interpretations, and informational flyers concerning its regulations in an effort to clarify and explain the requirements. The address of each group discussed below is listed at the end of this Section.

Department of Transportation—Office of Hazardous Materials Operations (DOT-OHM)

The DOT regulates shipment by common carrier in all modes, rail, truck, barge, and air. Its regulations are found in 14, 46, and 49 CFR. These rules are specific and detailed. The regulations indicate the proper shipping container, the information that must appear on the shipping documents, and the form and content of the labels and placards. The labels are slightly different for the various modes of shipment. The present system requires the labeling of shipping containers, and the placarding of vehicles with a diamond-shaped design. Examples are given on the inside covers of this book. The designs indicate toxic, flammable, explosive, etiologic, and corrosive liquids, gases, and solids. Radioactive materials are controlled in cooperation with the Department of Energy.

The basic documentation other than labels and placards is the information required on shipping papers. General requirements for various modes follow.

HAZARDOUS MATERIALS TRANSPORTATION GUIDE FOR HAZARDOUS MATERIALS SHIPPING PAPERS

The following information has been abstracted from the Code of Federal Regulations, Title 49, CFR, Parts 100–199. (Refer to Subpart C.)

1 SHIPPING PAPER DEFINED (Sec. 171.8)—A shipping order, bill of lading, manifest, or other shipping document serving a similar purpose and containing the information required by Sec. 172.202 and 172.203.

2 REQUIREMENTS FOR ALL SHIPPERS (Sec. 172.200)—Except as otherwise provided in this subpart, each person who offers a hazardous material for transportation *shall* describe the hazardous material on the shipping paper (Sec. 172.200(a)).

3 DESCRIPTION OF HAZARDOUS MATERIALS ON SHIPPING PAPERS (Sec. 172.202)

(a) Each description of a hazardous material on the shipping paper must include—
 (1) The proper shipping name prescribed for the material as required by Sec. 172.101 (Hazardous Materials Table).
 (2) The class prescribed for the material as required by Sec. 172.101. When the words of the proper shipping name are identical (excluding the entry "n.o.s.") with the words of the class, the inclusion of the class is not required.
 (3) Except for empty packagings, the total quantity (by weight, volume, or as otherwise appropriate) of the hazardous material covered by the description.

(b) The basic description specified in (a)(1) and (a)(2) above *must* be shown in sequence except that the technical name of the material may be entered between the proper shipping name and the class. For example: "Gasoline, Flammable liquid"; or "Flammable solid, n.o.s."; or "Corrosive liquid, n.o.s. (caprylyl chloride) corrosive material."

(c) The total quantity of the material covered by one description must appear before or after, or both before and after, the description required and authorized by this subpart.
 (1) Abbreviations may be used to specify the type of packaging and weight or volume. For example: 40 cyl. Nitrogen, Non-flammable Gas-800 pounds; 1 box Cement, liquid n.o.s., Flammable liquid, 25 pounds.
 (2) The type of packaging may be entered in any appropriate manner.

4 GENERAL ENTRIES REQUIRED ON SHIPPING PAPERS (Sec. 172.201)

(a) *Contents*—When a description of hazardous material is required to be included on a shipping paper, that description must conform to the following requirements:
 (1) When a hazardous material and a material not subject to the requirements of this subchapter are described on the same shipping paper, the hazardous material description entries required by Sec. 172.202 and those additional entries that may be required by Sec. 172.203.
  (i) Must be entered first or (See Figure 11.3).
  (ii) Must be entered in a color that clearly contrasts with any description of a material not subject to the requirements of this subchapter on the shipping paper except that a description on a reproduction of a shipping paper may be highlighted, rather than printed, in a contrasting color. (The provisions of this paragraph apply only to the basic description required by Sec. 172.203(a)(1) and (2).) or (See Figure 11.2)
  (iii) Must be identified by the entry of an "X" placed before the proper shipping name in a column captioned "HM." (See Figure 3).
 (2) The required shipping description on a shipping paper and all copies thereof used for transportation purposes, must be legible and printed (manually or mechanically) in English.
 (3) Unless it is specifically authorized or required in this subpart, the required shipping description *may not* contain any code or abbreviation.
 (4) A shipping paper may contain additional information concerning the material provided the information is not inconsistent with the required basic description. Unless otherwise permitted or required by this subpart, additional information must be placed after the basic description required by Sec. 172.202(a)(1) and (a)(2).
  (i) When appropriate, the entry "IMCO" may be entered immediately following the class in the basic description.
  (ii) For a material meeting the definition of more than one hazard class, the additional hazard class or classes entered may be entered after the hazard class in the basic description.

(b) *Name of Shipper*—A shipping paper for a shipment *by water* must contain the name of the shipper.

5 ADDITIONAL DESCRIPTION REQUIREMENTS (Sec. 172.203) (ALL MODES)

(a) *Exemptions*—Each shipping paper issued in connection with a shipment made under an exemption must bear the notation "DOT-E" *followed by the exemption number* assigned and so located that the notation is clearly associated with the description to which the exemption applies.

(b) *Limited quantities*—Descriptions for materials defined as "Limited quantities" . . . must include the words "Limited Quantities" or "Ltd. QTY." following the basic description.

(c) *Blasting caps*—Description for a shipment of blasting caps must have an entry stating the number of caps in the shipment, either *before or after* the basic description.

(d) *Radioactive material*—For additional descriptions for radioactive material, refer to Sec. 172.203(d).

(e) *Empty packagings*—For other than a tank car, the description on the shipping paper for an empty packaging containing the residue of a hazardous material may contain the word(s) "EMPTY": or "EMPTY: Last contained . . ." followed by the name of the hazardous material last contained in the packaging. This entry may be before or after the basic description. For empty tank cars, see Sec. 174.25(c) of this subchapter.

NOTE: Additional description requirements will be found in this handout for all modes.

<u>6 EXCEPTIONS ORM-A, B, C and D (Sec. 172.200 (b))</u>—This subpart does not apply to any material that is—

(a) An ORM-A, B or C unless it is offered or intended for transportation by air or water when it is subject to the regulations pertaining to transportation by air or water as specified in Sec. 172.101 (Hazardous Materials Table); or

(b) An ORM-D unless it is offered or intended for transportation by air.

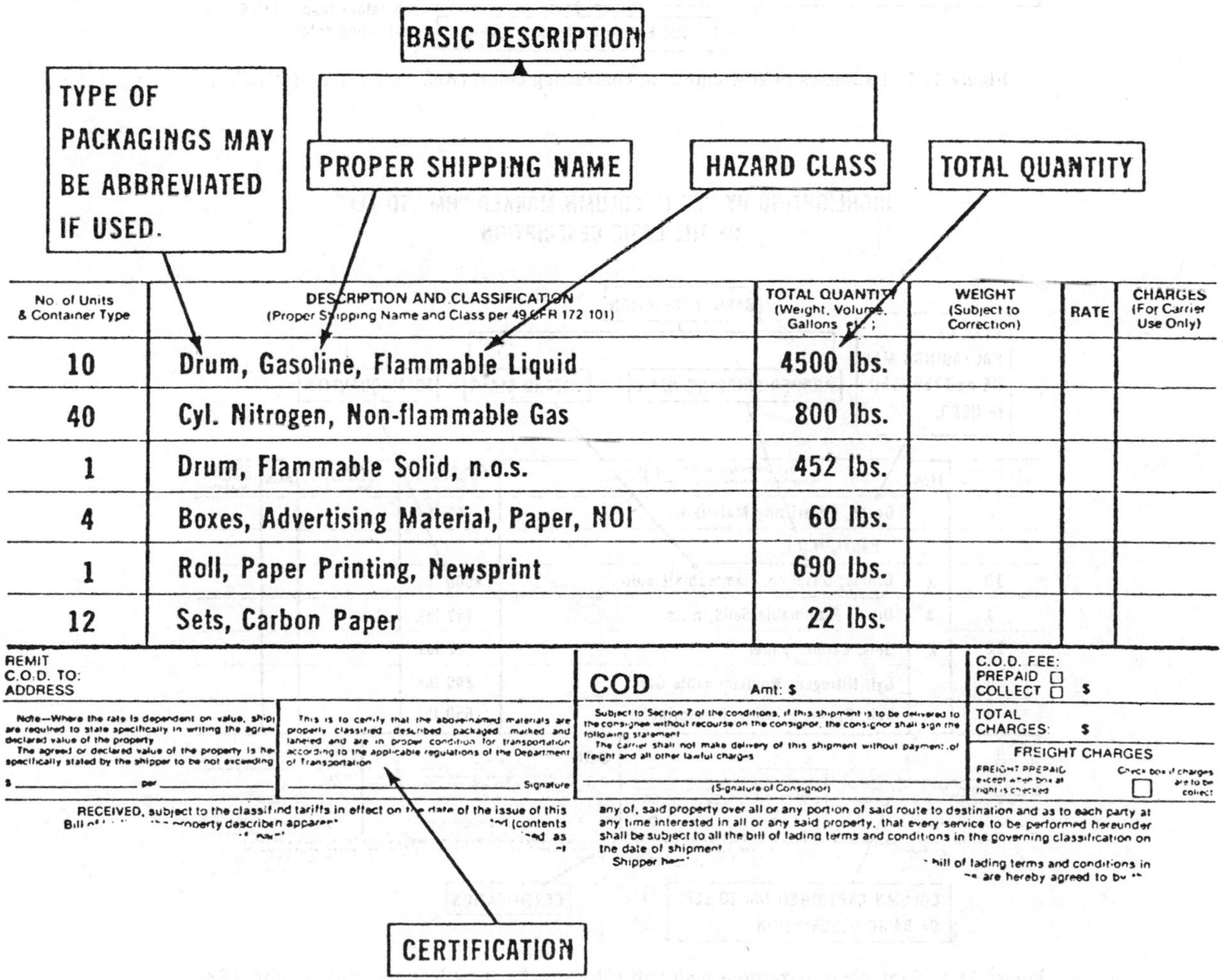

No. of Units & Container Type	DESCRIPTION AND CLASSIFICATION (Proper Shipping Name and Class per 49 CFR 172 101)	TOTAL QUANTITY (Weight, Volume, Gallons etc.)	WEIGHT (Subject to Correction)	RATE	CHARGES (For Carrier Use Only)
10	Drum, Gasoline, Flammable Liquid	4500 lbs.			
40	Cyl. Nitrogen, Non-flammable Gas	800 lbs.			
1	Drum, Flammable Solid, n.o.s.	452 lbs.			
4	Boxes, Advertising Material, Paper, NOI	60 lbs.			
1	Roll, Paper Printing, Newsprint	690 lbs.			
12	Sets, Carbon Paper	22 lbs.			

REMIT C.O.D. TO: ADDRESS

COD Amt: $

C.O.D. FEE: PREPAID ☐ COLLECT ☐ $

Note—Where the rate is dependent on value, shippers are required to state specifically in writing the agreed or declared value of the property. The agreed or declared value of the property is hereby specifically stated by the shipper to be not exceeding $ ______ per ______

This is to certify that the above-named materials are properly classified, described, packaged, marked and labeled and are in proper condition for transportation according to the applicable regulations of the Department of Transportation. ______ Signature

Subject to Section 7 of the conditions, if this shipment is to be delivered to the consignee without recourse on the consignor, the consignor shall sign the following statement: The carrier shall not make delivery of this shipment without payment of freight and all other lawful charges. ______ (Signature of Consignor)

TOTAL CHARGES: $

FREIGHT CHARGES — FREIGHT PREPAID except when box at right is checked — Check box if charges are to be collect ☐

RECEIVED, subject to the classified tariffs in effect on the date of the issue of this Bill of Lading, the property described apparent ... (contents ... as ... any of, said property over all or any portion of said route to destination and as to each party at any time interested in all or any said property, that every service to be performed hereunder shall be subject to all the bill of lading terms and conditions in the governing classification on the date of shipment. Shipper here... bill of lading terms and conditions in ... are hereby agreed to by ...

Figure 11.1. Example of entering hazardous materials first. (Ref. Sec. 172.201(a)(1)(i).)

HIGHLIGHTING BY COLOR-HAZARDOUS MATERIALS ENTRIES IN COLOR CONTRASTING WITH NON-HAZARDOUS MATERIALS ENTRIES· ONE EXAMPLE OF HIGHLIGHTING USING THIS OPTION.

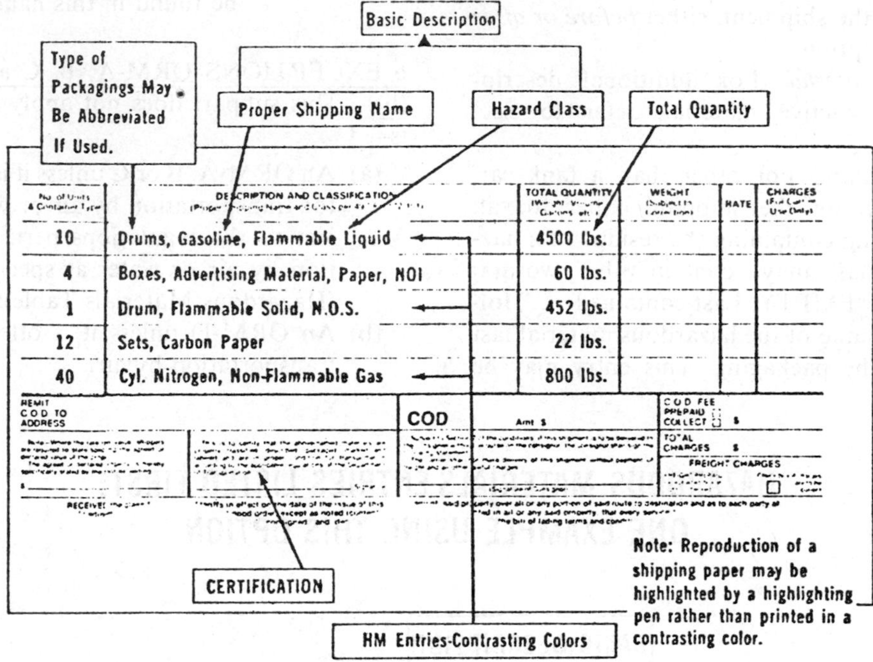

Figure 11.2. Examples of HM entries in contrasting color. (Ref. Sec. 172.201(a)(1)(ii).)

HIGHLIGHTING BY "X" IN COLUMN MARKED "HM" TO LEFT OF THE BASIC DESCRIPTION

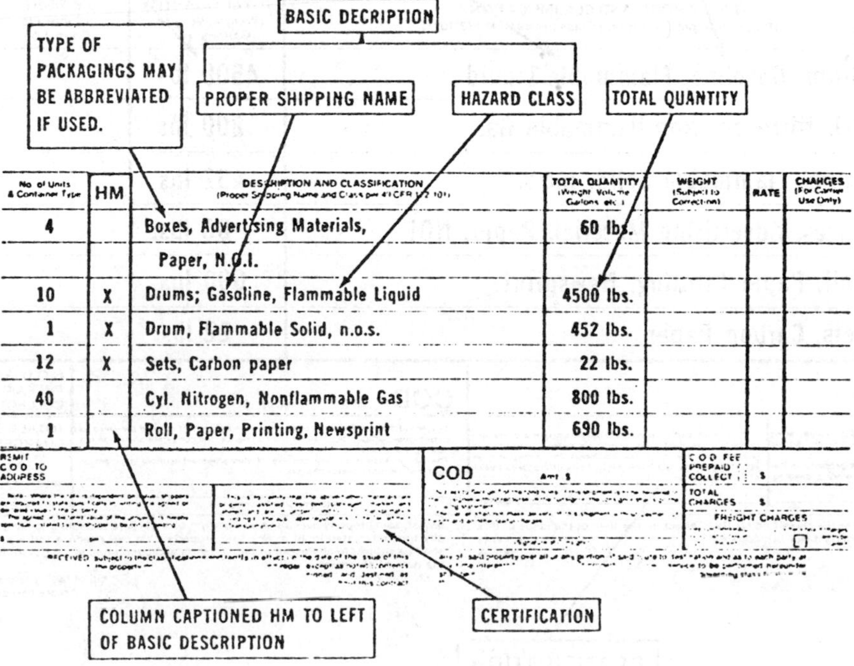

Figure 11.3. Example of hazardous materials entries prefixed by X in the HM column. (Ref. Sec. 172.201(a)(1)(iii).)

TRANSPORTATION BY RAIL

1 SHIPPING PAPERS (Sec. 174.24)

(a) Except as provided in paragraph (b) of this section, no person may accept for transportation by rail any hazardous material which is subject to this subchapter unless he has received a shipping paper prepared in a manner specified in Sec. 172.200. In addition, the shipping paper must include a certificate, if required by Sec. 172.204. However, no member of the train crew of a train transporting the hazardous material is required to have a shippers certificate on the shipping paper in his possession if the original shipping paper containing the certificate is in the originating carriers possession.

(b) This subpart does not apply to materials classed as ORM-A, B, C or D.

2 ADDITIONAL DESCRIPTION FOR SHIPPING PAPERS (Sec. 172.203(g))

(a) The shipping paper for a rail car containing a hazardous material must contain the notation "Placarded" followed by the name of the placard required for the rail car.

(b) The shipping paper for each specification DOT 112A or 114A tank car (without head shields) containing a flammable compressed gas must contain the notation, "DOT 112A" or "DOT 114A," as appropriate, and either "Must be handled in accordance with FRA E.O. No. 5" or "Shove to rest per E.O. No. 5."

NOTE: For additional details, refer to Part 174.

TRANSPORTATION BY AIR

1 SHIPPING PAPERS ABOARD AIRCRAFT (Sec. 175.35)—A copy of the shipping papers required by Sec. 175.30(a)(2) must accompany the shipment it covers during transportation aboard an aircraft.

NOTE: The documents required (shipping papers and notification of pilot in command) may be combined into one document if it is given to the pilot-in-command before departure of the aircraft. (Sec. 175.35(b)

2 NOTIFICATION OF PILOT IN COMMAND (Sec. 175.33)—The operator of the aircraft shall give the pilot-in-command the following information in writing before takeoff (Sec. 175.35):

(1) Description of hazardous materials on shipping papers (Sec. 172.202 and 172.203);

(2) Location of the hazardous material in the aircraft; and

(3) The results of the inspection requirements by Sec. 175.30(b).

NOTE: For additional details, refer to Part 175.

TRANSPORTATION BY WATER

1 SHIPPING PAPERS (Sec. 176.24)—A carrier may not transport a hazardous material by vessel unless the material is properly described on the shipping paper in the manner prescribed in Part 172.

2 CERTIFICATE (Sec. 176.27)

(a) A carrier may not transport a hazardous material by vessel unless he has received a certificate prepared in accordance with Sec. 172.204.

(b) In the case of an *import or export shipment* of hazardous materials which will not be transported by rail, highway, or air, the shipper *may certify* on the bill of lading or other shipping paper that the hazardous material is properly classed, described, marked, packaged, and labeled according to Part 172 or in accordance with the requirements of the IMCO Code. (See Sec. 171.12)

3 DANGEROUS CARGO MANIFEST (Sec. 176.30)

(a) The master of a vessel transporting hazardous materials or his authorized representative shall prepare a dangerous cargo manifest, list, or stowage plan. This document may not include a material which is not subject to the requirements of 49CFR or the IMCO Code. This document must be kept in a designated holder on or near the vessel's bridge. (See Sec. 176.30 for details)

4 EXEMPTIONS (Sec. 176.31)—If a hazardous material is being transported by vessel under the authority of an exemption and a copy of the exemption is required to be on board vessel, it must be kept with the dangerous cargo manifest.

NOTE: For additional details, refer to Part 176.

5 ADDITIONAL DESCRIPTION FOR SHIPPING PAPERS (Sec. 172.203(i))

(a) Each shipment by water must have the following additional shipping paper entries:

(1) Identification of the type of packages such as barrels, drums, cylinders, and boxes.

(2) The number of each type of packages including those in freight container or on a pallet, and

(3) The gross weight of each type of package or the individual gross weight of each package.

(b) The shipping papers for a hazardous material offered for transportation by water to any country outside the United States must have in parenthesis the *technical name* of the material *following the proper shipping name* when the material is described by a "n.o.s." entry in Sec. 172.101 (Hazardous Materials Table). For example: Corrosive liquid, n.o.s. (caprylyl chloride), Corrosive material. However, for a mixture, only the technical name of any hazardous material giving the mixture its hazardous properties *must be identified.*

TRANSPORTATION BY HIGHWAY

1 SHIPPING PAPERS (Sec. 177.817)

(a) *General*—A carrier may not transport a hazardous material unless it is accompanied by a shipping paper that is prepared in accordance with Sec. 172.201, 172.202, and 172.203 of this subchapter.

(b) *Shipper's certification*—An initial carrier may not accept hazardous materials offered for transportation unless the shipping paper describing the material includes a shipper's certification which meets the requirements in Sec. 172.204 of this subchapter. The certification is not required for shipments to be transported entirely by private carriage and for bulk shipments to be transported in a cargo tank supplied by the carrier. (Sec. 177.817(b).)

(c) *Interlining with carriers by rail*—A motor carrier shall *mark on the* shipping paper required by this section, if it offers or delivers a freight container or transport vehicle to a rail carrier for further transportation: (Sec. 177.817(c))

(1) A description of the freight container or transport vehicle; and

(2) The *kind of placard* affixed to the freight container or transport vehicle.

(d) This subpart does not apply to materials classed as an ORM-A, B, C or D.

(e) *Shipping paper accessibility-accident or inspection*—A driver of a motor vehicle containing hazardous material, and each carrier using such a vehicle, shall ensure that the shipping paper required . . . is readily available to, and recognizable by, authorities in the event of accident or inspection. (See Sec. 177.817(e) for details)

2 ADDITIONAL DESCRIPTION FOR SHIPPING PAPERS (Sec. 172.203(h))—Following the basic description for a hazardous material in a specification MC 330 or MC 331 cargo tank made of quenched and tempered steel, there must be entered for—

(1) Anhydrous ammonia. (See Sec. 172.203(h)(1))

(2) Liquefied petroleum gas. (See Sec. 172.203 (h)(2))

3 EXEMPTIONS (Sec. 172.203(a))

SHIPPER'S CERTIFICATION ALL MODES (Sec. 172.204)

(a) *General*—Except as provided in paragraph (b) *Exceptions* and (c) *Transportation by air, each person* who offers a hazardous material for transportation shall certify that the material offered for transportation in accordance with this subchapter *by printing* (manually or mechanically) the following statement on the shipping paper containing the required descriptions:

This is to certify that the above-named materials are properly classified, described, packaged, marked and labeled, and are in proper condition for transportation according to the applicable regulations of the Department of Transportation.

NOTE: Preprinted certificates complying with 49CFR 173.430(a) in effect on June 30, 1976, *may be used through June 30, 1979.* After June 30, 1979, use of the certificate required by Sec. 172.200 is *mandatory.*

(b) *Exceptions*—No certification is required for hazardous material offered for transportation by highway that is transported—

(1) In a cargo tank supplied by the carrier, or

(2) By the shipper as a private carrier except for a hazardous material that is to be reshipped or transferred from one carrier to another.

(c) *Air transportation*

(1) *General*—Certification containing the following language may be used in place of the certification required by paragraph (a) above.

I hereby certify that the contents of this consignment are fully and accurately described above by proper shipping name and are classified, packed, marked and labeled, and in proper condition for carriage by air according to applicable national governmental regulations.

(2) *Duplicate certificate*—Each person who offers a hazardous material to an *aircraft operator for transportation by air* shall provide two copies of the certification required in this section. (Sec. 175.30)

(3) *Passenger and cargo aircraft*—Each person who offers for transportation by air a hazardous material authorized for air transportation shall add to the certification required in this section the following statement:

This shipment is within the limitations prescribed for passenger aircraft/cargo-only aircraft (delete nonapplicable).

(4) *Radioactive material*—Each person who offers any radioactive material for transportation aboard a passenger-carrying aircraft *shall sign* (mechanically or manually) a printed certificate stating that the shipment contains radioactive material intended for use in, or

incident to, research, medical diagnosis or treatment. This requirement does not apply to materials excepted under the provisions of Sec. 175.10(a)(6) and (a)(8) of this subchapter.

(d) *Signature*—The certifications required by Sec. 172.204(a) or (c) above must be legibly signed (mechanically or manually) by a principal, officer, partner, or employee of the shipper or his agent. (Sec. 172.204(d))

NOTE: This handout is designed as a training aid for shippers and carriers of hazardous materials. It does not relieve persons from complying with the Department of Transportation Hazardous Materials Regulations. Final authority for use of shipping papers are found in Title 49, CFR, Parts 100-199.

Department of Energy (DOE)

The DOE regulates manmade radioactive materials wherever they are used in the United States. The regulations are found in 10CFR. The universal radiation symbol in magenta and yellow is the standard label or placard. Depending on the class and intensity of radiation, the signal words DANGER HIGH RADIATION or DANGER RADIATION are used. Additional information on radiation type and concentration of radioactive element or isotope, and warnings restricting access are required for certain applications.

Most radioactive materials are licensed for use and their handling is strictly monitored. As a result, correct labels and instructions for use will be supplied in most cases at the time of first use of the materials through the licensing procedure.

Consumer Product Safety Commission (CPSC)

The CPSC has responsibilities for labeling hazardous materials in consumer goods under the Hazardous Substances Act and the Poison Prevention Packaging Act. Its regulations are found in 21CFR. This still relatively new agency has so far considered labeling on a case by case basis and does not have a system for general labeling of consumer chemicals. Increasing activity in the area of toys containing hazardous chemicals can be expected.

The definitions of hazard as viewed by CPSC are consumer and especially child-oriented and may differ considerably from the definitions used for regulating shipping or industrial use of materials. The wording on labels generally follow those of ANSI Z129.1.

Food and Drug Administration (FDA)

The FDA regulates all food additives, drugs, and cosmetic products under the Federal Food, Drug, and Cosmetic Act. Its regulations are found in 21CFR. Drugs require full disclosure of biological effects and extensive testing before public distribution begins. Since all drugs by their nature have biological effects, their labels are very complex and individual to each formulation.

Cosmetics require general labeling as detailed in 21CFR. Disclosure of ingredients is a recent change in the regulations. In addition, claims made for the product, since they affect use of materials by consumers for long periods, are monitored by FDA. The compliance branch of the nearest FDA office should be contacted for latest regulations and answers to queries. Food additives follow a similar procedure. All substances regulated by FDA are highly sensitive in terms of labeling requirements since they directly affect the health and nutrition of consumers.

Environmental Protection Agency (EPA)

The Office of Toxic Substances handles premarked clearance, registration, reregistration, and certification of pesticide applicators under the Federal Insecticide, Fungicide, and Rodenticide Act. The regulations are contained in 40CFR. Extensive information must be submitted to EPA in the process of registration of a pesticide. A very detailed container label results from the registration process. Specific instructions on handling and permitted applications are included on the label in addition to the customary health and safety warnings. The EPA registration number must appear on the label.

The EPA has many duties under the Toxic Substances Control Act (TSCA). All chemicals in use in the United States must be reported to the Office of Toxic Substances of EPA. New uses for previously registered chemicals, and all new chemicals proposed for commercial distribution must be reported to EPA for premarket clearance. Many other activities related to hazardous material control are mandated by TSCA which will have an impact on future labeling activities. A computer-based Chemical Information System (CIS) is being expanded to contain data on materials from over 100 government data collections. The CIS will be the source for data on EPA regulated materials and for information on the regulations of other federal agencies. The system is available on a commercial fee basis. The CIS should be of considerable benefit to anyone striving to accurately comply with a labeling system by supplying data needed to classify a material.

A major outgrowth of TSCA is increased cooperation among EPA, FDA, OSHA, and the CPSC. These regulatory agencies are attempting to coordinate many aspects of their activities. Coordination should ultimately benefit those subject to the regulations by minimizing any conflict between the regulations and reducing duplication of effort in data collection and review. If additional agencies such as DOE and DOT join in this cooperative effort, unified labeling of hazardous materials could result, thereby simplifying the job of the labeler and increasing the comprehension of the end user of the information contained on the label.

No specific labeling requirements have been developed under TSCA, but the possibility exists that restrictions on new materials, or new uses for old materials may generate the need for some labeling requirements.

Occupational Safety and Health Administration (OSHA)

The OSHA of the Department of Labor has responsibility for the labeling of hazardous industrial substances with respect to the protection of the worker. The OSHA regulations are found in 29CFR. Under the Longshoremen's and Harbor Worker's Compensation Act regulations were established (29 CFR Part 1501, 1502, and 1503) requiring the completion of a Material Safety Data Sheet (MSDS) for each product used by workers which meet certain specific definitions of hazard. These MSDS contain basic data on the hazard and instructions for storage and use of the materials in question. Separate labeling of containers is not required.

The standards for asbestos, 13 carcinogens, and vinyl chloride require specific labels and workplace signs. The National Institute for Occupational Safety and Health of the Department of Health Education and Welfare sent recommendations to OSHA for "An Identification System of Occupationally Hazardous Materials." OSHA has developed several draft standards based on these recommendations. The final standard, when promulgated, will probably contain requirements for container labels, material safety data sheets, and employee education. There may be some requirements for signs or posting of information at the worksite relating to the materials used at that worksite.

U.S. GOVERNMENT PURCHASES

The Department of Defense and the General Services Administration have established Federal Standard 313 A. This standard is beginning to be used by federal procurement agents by incorporating it into government procurement contracts for hazardous materials. The standard requires specified label statements on the containers and the completion of a MSDS similar to the OSHA form. Detailed instructions are supplied with the purchase contract.

LABELING SYSTEMS

There is no agreement on the design or content of a universal system for labeling hazardous materials. A single system would be desirable to facilitate instant recognition of danger whether found in the home, in industry, or while in transit. The six systems discussed below were selected because of their acceptance and extensive use. Each consists of a set of criteria defining what materials are hazardous and to what degree, rules for selection of information to be conveyed, and a format for displaying the information. The complex criteria and rules reflect the variety of materials found in commerce. The displays are quite variable ranging from a simple picture to a detailed fact sheet. The intended audience is expected to have at least some background of understanding of hazard potential and in some cases detailed knowledge and the availability of back-up reference sources are assumed.

Each system is described in sufficient detail to facilitate recognition of the system's output. Space limitations prevent inclusion of sufficient details to allow full preparation of a label at this point. Therefore, the original sources must be checked to ensure availability of up-to-date information.

(1) American National Standards Institute (ANSI)

The Labels and Precautionary Information (LAPI) Committee of the Manufacturing Chemists Association has developed Manual L-1 entitled (Guide to Precautionary Labeling of Hazardous Chemicals). This guide was prepared for use by chemical manufacturers to identify hazards associated with their products. It has found wide use within industry and for consumer product labeling. A modified version of this Manual has been adopted by the ANSI as standard Z129.1-1976.

The labels generated by this system are designed not for protection against specific *chemicals*, but against specific *hazards* or combinations of hazards of a product as a whole. This approach emphasizes the fact that a precautionary label should be based upon hazards that the final product is known to possess. The Guide includes definitions of hazards in terms of toxicity, flammability, corrosivity, and reactivity, and the applicable test for each class. Hazards are classified as "Danger," "Warning," and "Caution."

Tables listing useful precautionary statements and model labels for various combinations of hazards are provided. The final output is intended to be placed on the container holding the hazardous material. Figure 11.4 illustrates several suggested hazard labels. Most of the common combinations are found in the Guide.

(2) The United Nations Labeling System

The U.N. system is designed to ease recognition of hazards by persons of all nations. It seeks to overcome language barriers by giving a pictorial representation of the hazard. This system, therefore, conveys a mini-

HAZARD:

Strong Sensitizer, Lungs

WARNING! MAY CAUSE ALLERGIC RESPIRATORY REACTION

Avoid breathing (dust, mist, vapor, gas). *
Keep container closed.
Use with adequate ventilation.

**Select applicable word or words in parentheses.*

HAZARD:

Physiologically Inert Vapor or Gas

CAUTION! (VAPOR) (GAS) REDUCES OXYGEN AVAILABLE FOR BREATHING*

Keep container closed.
Use with adequate ventilation.
Do not enter storage areas unless adequately ventilated.

**Select applicable word or words in parentheses.*

HAZARD:

Extremely Flammable Liquid

DANGER! EXTREMELY FLAMMABLE

Keep away from heat, sparks, and open flame.
Keep container closed.
Use with adequate ventilation.

See Section 7 for selection of appropriate fire-extinguishing statement.

HAZARD:

Flammable Liquid

WARNING! FLAMMABLE

Keep away from heat, sparks, and open flame.
Keep container closed.
Use with adequate ventilation.

See Section 7 for selection of appropriate fire-extinguishing statement.

HAZARDS:

Highly Toxic by Inhalation

Toxic by Ingestion

Toxic by Absorption

DANGER! MAY BE FATAL IF INHALED HARMFUL IF SWALLOWED OR ABSORBED THROUGH SKIN

Do not breathe (dust, vapor, mist, gas).*
Avoid contact with eyes, skin, and clothing.
Keep container closed.
Use only with adequate ventilation.
Wash thoroughly after handling.

POISON

Call a Physician

FIRST AID: If inhaled, remove to fresh air. If not breathing give artificial respiration, preferably mouth-to-mouth. If breathing is difficult, give oxygen.
If swallowed, induce vomiting by sticking finger down throat or by giving soapy or strong salty water to drink. Repeat until vomit is clear. Never give anything by mouth to an unconscious person.
In case of contact, immediately flush eyes or skin with plenty of water for at least 15 minutes while removing contaminated clothing and shoes. Wash clothing before re-use. (Discard contaminated shoes.)

**Select applicable word or words in parentheses.*

Figure 11.4

mum of information. The U.N. system uses a diamond shape with graphics identical to those used by DOT (illustrated on the inside front and back cover of this book). Figure 11.5 indicates the numbers frequently displayed on U.N. type labels.

(3) The Department of Transportation

The DOT system is specified in detail in 49CFR. Each chemical substance in the various forms and concentrations found in interstate commerce is specified by hazard class and label requirements. Figure 11.6 indicates general rules for using the system and refers to specific sections of DOT regulations where exact specifications are listed. These specifications are quite lengthy and constantly being revised by publication of changes in the Federal Register. Specifications for hazard tests and definitions of hazardous materials are also contained in 49CFR.

Various placards have been required over the years for different modes of shipment. These symbols were used as labels on outside shipping containers and as placards affixed to the sides of trucks, railroad cars, and barges. The older labels have been replaced by those shown on the end papers.

(4) NFPA 704

The National Fire Protection Association's "Committee on Fire Hazards of Materials 704," has developed a numerical rating system for identification of hazardous materials with respect to fire hazard. Since fire hazard potential includes contact with the material, in case of spills incident to a fire, the system is generally applicable to labeling of hazardous materials. The definitions are generally subjective and general rather than specific and only flash point tests are specified (see Section 7).

The system allows rating of hazard on a scale of 0 (nonhazardous) to 4 (extremely hazardous) in the categories of health, fire, and reactivity. The result is a four pointed diamond within a diamond symbol containing the numerical ratings. Figure 11.7 illustrates the symbol and the rating criteria.

Like the UN system, 704M is independent of lan-

HAZARDOUS CLASS NUMBERS ARE FREQUENTLY DISPLAYED ON UNITED NATIONS TYPE LABELS. THESE NUMBERS CAN BE OF CONSIDERABLE ASSISTANCE IN IDENTIFYING HAZARDOUS MATERIALS SHIPMENTS. THESE U.N. CLASSES ARE AS FOLLOWS:

Hazard Class Number	Description
1	EXPLOSIVES—Class A, B and C explosives
2	NON-FLAMMABLE AND FLAMMABLE GASES
3	FLAMMABLE LIQUIDS
4	FLAMMABLE SOLIDS (Readily combustible) SPONTANEOUSLY COMBUSTIBLE SUBSTANCES WATER REACTIVE SUBSTANCES
5	OXIDIZING MATERIALS (Oxidizing materials and/or organic peroxides)
6	POISONOUS MATERIALS (Class A, B, and C poisonous or toxic substances)
7	RADIOACTIVE MATERIALS—White I, Yellow II or Yellow III
8	CORROSIVE MATERIALS—Acids, corrosive liquids or solids and alkaline caustic liquids
9	MISCELLANEOUS HAZARDOUS MATERIALS—These are materials which during transport present a danger not covered by other classes. No specific label authorized.

NOTE: Uusually only one label will be applied to a package. A second label is normally applied to a package only when a significant secondary hazard exists. The secondary hazard label usually will not carry a class number.

Figure 11.5

guage but it requires prior learning of the code for understanding. It is in use in some foreign countries and by many fire departments and chemical manufacturers for identification of tanks and bulk storage of hazardous materials. Once memorized, the symbol can convey warnings at some distance in three categories of possible danger.

(5) Material Safety Data Sheets (MSDS)

The Department of Labor regulations under Public Law 85-742 for "Ship Repairing, Shipbuilding, and Shipbreaking" require employers whose employees may encounter hazardous materials to have on hand a completed Material Safety Data Sheet (MSDS). DOL Form LSB 00S-4 (Fig. 11.8) is the basic design from which many essentially similar forms have been developed for specific applications. The MSDS contains detailed information related to the possible hazardous nature of a product. It is an information sheet rather than a label, but often contains wording suitable for use on labels and placards.

The MSDS is only required for the ship construction industry, but is widely used by manufacturers to inform their customers of product safety information, and by employers as part of their programs to train employees in safe work practices. The use of a format of this type will probably increase with time as more demands are made for full disclosure of harmful properties of materials.

The definitions of a hazardous material and instructions for completing the various sections of the MSDS are found in 29CFR Parts 1501, 1502, 1503. Since these are not mandatory for non-ship construction areas, the MSDS and its contents can be modified to suit the application.

(6) IATA Restricted Articles Regulations

The International Air Transport Association (IATA) publishes a detailed handbook of regulations for the international air shipment of hazardous materials. All member air-carriers conform to these regulations. The labels and markings employ the UN pictograph in the upper portion of a diamond shaped symbol with specific words in the lower half, which relate to the nature of the hazard and indicate emergency precautionary measures.

These regulations, like those of DOT, are regularly modified. The most current edition of the regulations must be consulted for specific details in any specific case.

GENERAL GUIDELINES ON USE OF LABELS

NOTE: These guidelines do not include all of the DOT hazardous materials warning label requirements. Complete detailed requirements are found in Title 49, Code of Federal Regulations, Part 172.

1. Each person who offers a hazardous material for shipment must label the package containing the material, if required, with the appropriate label(s). (Sec. 172.400 (a))

2. Labels may be affixed to packages even though not required by the regulations provided each label represents a hazard of the material in the package. (Sec. 172.401)

3. Exceptions to the labeling requirements for limited quantities of certain hazardous materials are specified in the regulations.

4. The number appearing at the bottom corner of some labels represent the UN and IMCO hazard class number. These are permitted, but not required, by DOT regulations. (Sec. 172.407 (g))

5. Label(s), when required, must be affixed to or printed on the surface of the package near the marked proper shipping name. (Sec. 172.406(a))

6. When two or more different warning labels are required, they must be displayed next to each other. (Sec. 172.406(c))

7. When two or more packages containing compatible hazardous materials are packaged within the same overpack, the outside container must be labeled as required for each class of material contained therein. (Sec. 172.404(b))

8. Packages containing a sample of a hazardous material other than an explosive must be labeled in accordance with the requirements of Sec. 172.402(h). (For Explosives, see Title 49, CFR, Part 173, Subpart C)

9. A material classed as an Explosive A, Poison A, or Radioactive material, that also meets the definition of another hazard class, must be labeled as required for each class (Sec. 172.402 (a))

10. Packages containing Radioactive material, that also meets the definition of one or more additional hazards, must be labeled as a Radioactive material and for each additional hazard on opposite sides of the package. (Sec. 172.403(e) and (f))

11. A material classed as an Oxidizer, Flammable solid, or Flammable liquid, that also meets the definition of a Poison B, must be labeled POISON, in addition to the hazard class label. (Sec. 172.402 (a) (3))

12. A material classed as a Flammable solid, that also meets the definition of a water reactive material, must have both FLAMMABLE SOLID and DANGEROUS WHEN WET labels affixed. (Sec. 172.402(a) (4))

13. For OXYGEN, the word "OXYGEN" may be used in place of the word "OXIDIZER" on the OXIDIZER label. (Sec. 172.405(a)) For foreign shipments, the NON-FLAMMABLE GAS label may also be required.

14. For CHLORINE, a CHLORINE label may be used in place of the NON-FLAMMABLE GAS and POISON labels. (Sec. 172.405(b)) For foreign shipments, the NON-FLAMMABLE GAS label may also be required.

15. Each person who offers a package containing a hazardous material for air transport that is authorized only on Cargo Aircraft, shall affix to the package a CARGO AIRCRAFT ONLY label. (Sec. 172.402 (b))

Figure 11.6

DISCUSSION OF LABELING

A manufacturer, shipper, or employer wishing to comply with all regulations relating to proper labeling of a hazardous material is faced with a complex series of rules, definitions, symbols, suggested text, and methods of implementation. Hopefully, the preceding sections will direct the labeler to the correct agency or system which applies in each case. At least this way he will come in contact with experts in this complex field who can assist in developing the necessary information or contacts in the area of interest.

A major area of confusion lies with the definition of a hazardous material. The various systems and regulations take divergent approaches to definition, and especially to relative rating of degree of hazard. This situation partly results from the lack of agreement on suitable testing procedures. Within the "labeling community" violent arguments abound on the details to be included in a warning, the leeway given to the user in constructing the label, and the response expected in the event of misuse. As a result, constant revision is presently the hallmark of a labeling system.

The question of toxicity suffers from a lack of adequate experimental data and the difficulty of projecting the results of animal studies to human exposure. In addition, the chronic exposure of employees or consumers to low concentrations produces different effects and calls for varying warnings from those related to massive contact in the case of spills or disasters.

System for the
IDENTIFICATION OF
THE FIRE HAZARDS OF MATERIALS

NFPA 704M

NATIONAL FIRE PROTECTION ASSOCIATION

A W in the bottom space of the diamond alerts fire fighting personnel to the possible hazard in use of water. The violence of the reaction with water is indicated by the degree number in the REACTIVITY category.

For detailed information on this identification system see "Identification of the Fire Hazards of Materials, NFPA 704M"

For the assignment of degrees for individual chemicals see: Hazardous Chemical Data (over 250 chemicals) NFPA No. 49 • Fire Hazard Properties of Flammable Liquids, Gases, and Volatile Solids (over 1500 chemicals) NFPA No. 325M • Standard for Fumigation (over 20 fumigants) NFPA No. 57

IDENTIFICATION OF HEALTH HAZARD

Health

In general, health hazard in fire fighting is that of a single exposure which may vary from a few seconds up to an hour. The physical exertion demanded in fire fighting or other emergency conditions may be expected to intensify the effects of any exposure. Only hazards arising out of an inherent property of the material are considered. The following explanation is based upon protective equipment normally used by fire fighters.

4 Materials too dangerous to health to expose fire fighters. A few whiffs of the vapor could cause death or the vapor or liquid could be fatal on penetrating the fire fighter's normal full protective clothing. The normal full protective clothing and breathing apparatus available to the average fire department will not provide adequate protection against inhalation or skin contact with these materials.

3 Materials extremely hazardous to health but areas may be entered with extreme care. Full protective clothing, including self-contained breathing apparatus, coat, pants, gloves, boots, and bands around legs, arms and waist should be provided. No skin surface should be exposed.

2 Materials hazardous to health, but areas may be entered freely with full-faced mask self-contained breathing apparatus which provides eye protection.

1 Materials only slightly hazardous to health. It may be desirable to wear self-contained breathing apparatus.

0 Materials which on exposure under fire conditions would offer no hazard beyond that of ordinary combustible material.

Flammability

Susceptibility to burning is the basis for assigning degrees within this category. The method of attacking the fire is influenced by this susceptibility factor.

4 Very flammable gases or very volatile flammable liquids. Shut off flow and keep cooling water streams on exposed tanks or containers.

3 Materials which can be ignited under almost all normal temperature conditions. Water may be ineffective because of the low flash point.

Figure 11.7

2 Materials which must be moderately heated before ignition will occur. Water spray may be used to extinguish the fire because the material can be cooled below its flash point.

1 Materials that must be preheated before ignition can occur. Water may cause frothing if it gets below the surface of the liquid and turns to steam. However, water fog gently applied to the surface will cause a frothing which will extinguish the fire.

0 Materials that will not burn.

Reactivity (Stability)

The assignment of degrees in the reactivity category is based upon the susceptibility of materials to release energy either by themselves or in combination with water. Fire exposure was one of the factors considered along with conditions of shock and pressure.

4 Materials which (in themselves) are readily capable of detonation or of explosive decomposition or explosive reaction at normal temperatures and pressures. Includes materials which are sensitive to mechanical or localized thermal shock. If a chemical with this hazard rating is in an advanced or massive fire, the area should be evacuated.

3 Materials which (in themselves) are capable of detonation or of explosive decomposition or of explosive reaction but which require a strong initiating source or which must be heated under confinement before initiation. Includes materials which are sensitive to thermal or mechanical shock at elevated temperatures and pressures or which react explosively with water without requiring heat or confinement. Fire fighting should be done from an explosive resistant location.

2 Materials which (in themselves) are normally unstable and readily undergo violent chemical change but do not detonate. Includes materials which can undergo chemical change with rapid release of energy at normal temperatures and pressures or which can undergo violent chemical change at elevated temperatures and pressures. Also includes those materials which may react violently with water or which may form potentially explosive mixtures with water. In advance or massive fires, fire fighting should be done from a safe distance or from a protected location.

1 Materials which (in themselves) are normally stable but which may become unstable at elevated temperatures and pressures or which may react with water with some release of energy but not violently. Caution must be used in approaching the fire and applying water.

0 Materials which (in themselves) are normally stable even under fire exposure conditions and which are not reactive with water. Normal fire fighting procedures may be used.

Figure 11.7 (Continued)

Except for flammable liquids, there are no accepted tests for flammability. The combustion products of many materials present a greater danger than the materials themselves. Useful caution statements, and where to place them to maximize availability during confusing fire situations remain unresolved. It is hard to read a burning label!

Reactive and unstable products also suffer from a lack of suitable definitions and adequate tests. Details of storage compatibility, and reactivity with common substances can be extremely complex. For many materials this information is vital for safe use or storage. Much remains to be done within the labeling system's structures and in the development of testing procedures.

Developing a label is often an arduous and time-consuming task. The rules of any selected system will probably be unclear on some point, the required data will be unavailable or in the wrong form, the system selected or required will demand too much or too little information in the view of the labeler. Judgment is called for in the absence of hard data. The resolution of conflicts often requires contact with the originators of the labeling system for both clarification and the granting of wavers or modifications to suit the facts.

The understanding of language is a variable thing. Where options are available, the labeler should test the proposed label for clarity. Avoid words with multiple meanings. Instructions such as "Dispose of in the usual manner" or "Use with adequate ventilation" are not helpful if the "usual manner" is to dump it back of the shop or "adequate" is thought to mean opening a small window halfway.

The responsibility for labeling and the penalty for mislabeling fall heavily on the labeler. The federal agencies can levy expensive penalties for failure to abide by the rules. Consumer, environmental, and third-party suits are on the increase in response to the rise in the number and complexity of products containing hazardous materials. These trends will continue to place the burden for proper disclosure of harmful properties on the seller, shipper, or employer.

WHERE TO GO FOR INFORMATION

If you need definitive information on labeling, write to the following addresses for information on the laws, current rules and pamphlets explaining the rules. In addition, trade associations and similar groups often know the requirements for specific areas of interest, and should be consulted for assistance.

Shipping

Office of Hazardous Materials
U.S. Department of Transportation
Washington, D.C. 20590

Extensive informational materials are available. The OHM publishes the *OHM Monthly Newsletter* which

Form Approved
Budget Bureau No. 44-R1387
Approval Expires April 30, 1971

Form No. LSB-OOS-4
May 1969

U.S. DEPARTMENT OF LABOR

WAGE AND LABOR STANDARDS ADMINISTRATION

Bureau of Labor Standards

MATERIAL SAFETY DATA SHEET

SECTION I

MANUFACTURER'S NAME	EMERGENCY TELEPHONE NO.
ADDRESS *(Number, Street, City, State, and ZIP Code)*	
CHEMICAL NAME AND SYNONYMS	TRADE NAME AND SYNONYMS
CHEMICAL FAMILY	FORMULA

SECTION II HAZARDOUS INGREDIENTS

PAINTS, PRESERVATIVES, & SOLVENTS	%	TLV (Units)	ALLOYS AND METALLIC COATINGS	%	TLV (Units)
PIGMENTS			BASE METAL		
CATALYST			ALLOYS		
VEHICLE			METALLIC COATINGS		
SOLVENTS			FILLER METAL PLUS COATING OR CORE FLUX		
ADDITIVES			OTHERS		
OTHERS					

HAZARDOUS MIXTURES OF OTHER LIQUIDS, SOLIDS, OR GASES	%	TLV (Units)

SECTION III PHYSICAL DATA

BOILING POINT (°F.)		SPECIFIC GRAVITY (H_2O=1)	
VAPOR PRESSURE (mm Hg.)		PERCENT VOLATILE BY VOLUME (%)	
VAPOR DENSITY (AIR=1)		EVAPORATION RATE (=1)	
SOLUBILITY IN WATER			
APPEARANCE AND ODOR			

SECTION IV FIRE AND EXPLOSION HAZARD DATA

FLASH POINT (Method used)	FLAMMABLE LIMITS	Lel	Uel
EXTINGUISHING MEDIA			
SPECIAL FIRE FIGHTING PROCEDURES			
UNUSUAL FIRE AND EXPLOSION HAZARDS			

Figure 11.8

THRESHOLD LIMIT VALUE
EFFECTS OF OVEREXPOSURE
EMERGENCY AND FIRST AID PROCEDURES

SECTION VI REACTIVITY DATA

STABILITY	UNSTABLE		CONDITIONS TO AVOID
	STABLE		

INCOMPATABILITY *(Materials to avoid)*
HAZARDOUS DECOMPOSITION PRODUCTS

HAZARDOUS POLYMERIZATION	MAY OCCUR		CONDITIONS TO AVOID
	WILL NOT OCCUR		

SECTION VII SPILL OR LEAK PROCEDURES

STEPS TO BE TAKEN IN CASE MATERIAL IS RELEASED OR SPILLED
WASTE DISPOSAL METHOD

SECTION VIII SPECIAL PROTECTION INFORMATION

RESPIRATORY PROTECTION *(Specify type)*		
VENTILATION	LOCAL EXHAUST	SPECIAL
	MECHANICAL *(General)*	OTHER
PROTECTIVE GLOVES		EYE PROTECTION
OTHER PROTECTIVE EQUIPMENT		

SECTION IX SPECIAL PRECAUTIONS

PRECAUTIONS TO BE TAKEN IN HANDLING AND STORING
OTHER PRECAUTIONS

Figure 11.8 (Continued)

announces changes in the rules. It is available free from the above address.

International Air Transport Association (IATA)
1155 Mansfield Street
Montreal 113
Quebec, Canada

IATA publishes the "IATA Restricted Articles Regulations" which apply to all air shipments and cover in detail international air shipping requirements.

Manufacturing

Bureau of Biomedical Sciences
Consumer Product Safety Commission
5401 Westbard Avenue
Bethesda, Maryland 20207

Department of Energy
Washington, D.C. 20545

Office of Hazardous Materials
Environmental Protection Agency
Waterside Mall
401 M St., SW
Washington, D.C. 20460

Food and Drug Administration
Contact the Compliance Branch of the nearest regional FDA office.

American National Standards Institute
1430 Broadway
New York, New York 10018

The ANSI publishes two standards of interest to labeling activities: Z48.1-1954(R1971) entitled "Marking of Portable Compressed Gas Containers to Identify the Material Contained," and Z129.1-1976 entitled "Precautionary Labeling of Hazardous Industrial Chemicals."

National Fire Protection Association
470 Atlantic Avenue
Boston, Mass 02110

The 704 identification system is described in a pamphlet entitled "Fire Hazard of Materials 1975" NFPA No. 704. This and other pamphlets on rating of hazardous materials, testing methods, and flammability data are available from NFPA.

Employers

Office of Health Standards
Occupational Safety and Health Administration
3rd and Constitution Avenue, N.W.
Washington, D.C. 20210

SECTION 12

GENERAL CHEMICALS

COUNTERMEASURES AND ABBREVIATIONS

Countermeasures Directory

For a listing and discussion of ACGIH TLV's for chemical substances and physical agents in the workroom environment with intended changes for 1979 see **Section 1.**

For industrial air contaminant control, i.e., ventilation, process change, substitution, filtration, isolation in time or space, personnel segregation, local suppression of contaminants, housekeeping, respiratory protection, personnel protection, hoods, glove boxes, carcinogen control systems, tabular data, diagrams and sample calculations see **Section 2.**

For industrial noise—effects and controls, definitions, the human auditory system, how noise effects people, hearing damage, speech interference, task interference, sleep interference, annoyance, representative noise levels in industry, case histories of noise control treatments, mufflers, enclosures, lagging, barriers, vibration isolation, source redesign, costs of noise control, maintenance and operations, and bibliography see **Section 3.**

For air pollution control requirements for industrial, commercial and public facilities via legislation for Class I areas, increment limitations, exceptions, visibility, non-attainment, data, modeling and monitoring, tabular data on permit issuance time, EPA rule-making activities for stationary sources, SO_2 increment limits in a Class A region, major emitting facilities subject to PSD review, PSD permitted increments, and attainment status designations by state see **Section 4.**

For radiation sources, hazards, controls, i.e., regulations, nature of radiation, decay schemes, environmental radiation, radiation units, natural and medical dosage, effects, permissible levels, tabular protection guides, monitoring, bioassay, responsibility, controls, storage, handling and shipping of radioisotopes, and radioisotope waste problems, and controls see **Section 5A.**

For environmental impact of industrial radiation, i.e., major sources, medical applications, space exploration uses, terrestrial uses, industrial processing uses, power production uses, problems of waste, transportation, protection and control see **Section 5B.**

For health hazards of solid waste treatment—initially hazardous, airborne pathogens, asbestos, analysis, sanitation workers, fungal diseases, secondary hazards (flies, mosquitos, rodents, fires and explosions, spontaneous combustion, gas production, leachates), danger points in solid-waste handling, potential hazards and sea disposal see **Section 6.**

For industrial fire protection—classification of fires, fire protection, loss limitation, fire discovery and alarm, extinguishment, intangible fire losses, storage and handling of hazardous materials, explosives, storage and handling, flammable liquid storage in tanks and industrial disaster control see **Section 7.**

For industrial and environmental cancer risks—new carcinogens legislation, testing procedures, toxicity testing, human cancer, animal carcinogens, cancer in the general population, chemo prevention of cancer, bibliography, Table of NIOSH criteria documents, schedule of NIOSH documents, and comparison of death rates in chemists see **Section 8.**

For toxicology—toxicity ratings, types of toxicity, dosage, classes of toxic materials, routes of absorption, storage and excretion, individual susceptibility, acute and chronic effects, sites of action of poisons, absorption and poisoning, prevention (first aid), allergy and hypersensitivity see **Section 9.**

For Chemical substances regulation including legislation prior to 1970, i.e., the Federal Food, Drug and Cosmetic Act, the Delaney Clause, Early Pesticide Legislation, The Federal Hazardous Substances Act (including The Federal Hazardous Substances Labeling Act), The Caustic Poisons Act and Chemical Substances Legislation in the 1970's such as OSHA (including NIOSH); The Poison Prevention Packaging Act of 1970; the Consumer Product Safety Act of 1973; The Environmental Protection Agency Acts; The Clean Air Act (as amended in 1977); The Clean Water Act; The Federal Insecticide, Fungicide and Rodenticide Act (FIFRA); The Resource Conservation and Recovery Act of 1976 (RCRA); The Safe Drinking Water Act, as amended; The Toxic Substances Control Act (TSCA); The Federal Regulatory Agency Liason to include a MATRIX for fast reference to the regulations which cover some 28 dangerous chemicals; plus a listing of Federal Legislation, Agencies involved and jurisdiction; plus a 29 item list of references see **Section 10A.**

For industrial response to chronic health hazards—five facts about chronic health hazards, need to define issues, setting an acceptable level of risk, discovering a hazard, regulation of chronic health hazards see **Section 10B.**

For labeling and certification of hazardous materials in terms of originator, legal status, extent of coverage, format and application, federal regulatory agencies, hazardous materials transportation guides, rail transport, air transport, water transport, highway transport, radioactives, FDA, EPA, TSCA, OSHA, labeling systems of ANSI, UN, DOT, general labeling guidelines, NFPA, IATA see **Section 11.**

ABBREVIATIONS

ACGIH	American Conference of Governmental Industrial Hygienists
Acute tox data	Acute toxicity data
anh	anhydrous
ASA	American Standards Association
ASTM	American Society for Testing Materials
atm	atmospheres
at wt	atomic weight
autoign temp	autoignition temperature
BeV	billions of electron volts
bp	boiling point
carc	carcinogen
(CC)	closed cup
cc	cubic centimeters
cfm	cubic feet per minute
CG	Coast Guard
CNS	central nervous system
(COC)	Cleveland open cup
conc	concentration(s)
contg	containing
CTD	chronic toxicity data
°	degrees Celsius
d	density
decomp	decomposes or decomposition
dermal	skin absorption mode of exposure
DOT	Department of Transportation
exper	experimental
expl	explosive, explodes
°F	degrees Fahrenheit
flam	flammable
flash p	flash point
fc	foot candle
fp	freezing point
fpm	feet per minute
ft	feet
g	grams
GI	gastrointestinal
gpm	gallons per minute
HC	hydrocarbon(s)
HIGH	capable of causing death or permanent injury due to the exposures of normal use; incapacitating and poisonous. Requires special handling.
hr(s)	hour(s)
IATA	International Air Transport Association
IARC	International Agency for Research on Cancer
ic	intracerebral
id	intraduodenal
inhal	inhalation mode of exposure
im	intramuscular
imp	implantation
in	intradermal
ip	intraperitoneal
ipl	intrapleural
ir	intrarenal
irr	irritant, irritation
it	intratracheal
iv	intravenous
ivg	intravaginal
°K	degree Kelvin
KeV	thousands of electron volts
Kg (kg)	kilograms
Km (km)	kilometers
Kw (kw)	kilowatt(s)
Kwh (kwh)	kilowatt hours
LC	lethal concentration

LC_{50}	lethal concentration to 50% of a specified population
LC_{LO}	lowest published *lethal* concentration
LD	lethal dose
LD_{50}	lethal dose to 50% of a specified population
lel	lower explosive limit
LOW	causes readily reversible tissue changes which disappear after exposure stops. Causes some discomfort
lpm	liters per minute
m-	meta
M^3, m^3	cubic meters
MCA	Manufacturing Chemists Association
mg	milligrams
ml	milliliters
MLD	minimum lethal dose
mm	millimeters of mercury
MOD	may cause reversible or irreversible changes to exposed tissue, not permanent injury or death. Can cause considerable discomfort
mp	melting point
mppcf	millions of particles per cubic foot
mu mem	mucous membrane(s)
μg	micrograms
n-	normal
neo	causes formation of neoplasm(s) i.e., non-metastasizing abnormal or new growth(s).
NONE	no harm via exposures of normal use; harmful only due to overwhelming dose or unusual conditions
NO_x	oxides of N
o-	ortho
(OC)	open cup
oral	ingestion mode of exposure
p-	para
pa	parenteral mode of exposure
PO_x	oxides of P
ppb	parts per billion
pphm	parts per hundred million
ppm	parts per million
rec	rectal mode of exposure
recog	recognized
(S)	upon review IARC classifies it as a susp carc
sec-	secondary
SO_x	oxides of S
sc	subcutaneous
spont	spontaneous(ly)
spont htg	spontaneous heating
susp, (S)	suspected
sym	symmetrical
syn	synonym
(TCC)	Tag closed cup
TC_{LO}	lowest published toxic concentration
TD	toxic dose
tert-	tertiary
THR	summary tox statement; acute unless otherwise indicated
TLV	threshold limit values
(TOC)	Tag open cup
tox	toxic(ity)
U or uk	unknown, insufficient data or experience recorded or available to permit a statement
uel	upper explosive limit
ulc	Underwriters' Laboratory Classification
uns-	unsymmetrical
vap d	vapor density
vap press:	vapor pressure
>	greater than
<	less than
Δ	via heat or heating
1°	primary
α	alpha
β	beta
γ	gamma
(+)	upon review IARC classifies it a carcinogen
(−)	upon review IARC classifies it not a carcinogen
(±)	upon review IARC states insufficient data to classify
$\longrightarrow$	yields or causes

RADIOLOGIC ABBREVIATIONS

cpm	counts per minute
d	days
dpm	disintegrations per minute
e	electrical
ec	electron capture
ev	electron volts
fCi	femto curies
h	hours
m	minutes
mCi	millicuries
Ci	curies
MeV	millions of electron volts
MT	metric tons
MTU	metric tons of Uranium
Mw	megawatts
μCi	micro curies
nCi	nano curies
pCi	pico curies
s	seconds
t	thermal
$T\frac{1}{2}$	half-life
y	years

TOXICITY SCALE

NONE = No harm via exposures of normal use; harmful only due to overwhelming dose or unusual conditions.

LOW = Causes readily reversible tissue changes which disappear after exposure stops; causes some discomfort.

MOD = May cause reversible or irreversible changes to exposed tissue, not permanent injury or death; can cause considerable discomfort.

HIGH = Capable of causing death or permanent injury due to the exposures of normal use; incapacitating and poisonous; requires special handling.

UNKNOWN (uk) (u) = Insufficient data or experience recorded or available to permit a statement.

a

"666." See benzene hexachloride.

"1068." See chlordane.

"1080." See sodium fluoroacetate.

"1081." See fluoroacetamide.

AAMX. See acetoacet-*m*-xylidide.

ABALYN. See methyl abietate.

ABATE. Syn: *0,0,0′,0′-tetramethyl-0,0′-thiodi-p-phenylene phosphorothioate*. White crystals. $((CH_3O)_2PSOC_6H_4)S$, mw: 466.1, mp: 30°.
Acute tox data: Oral LD_{50} (rat) = 145 mg/kg; dermal LD_{50} (rat) = 202 mg/kg; oral LD_{50} (pigeon) = 27 mg/kg [*3*]
THR = HIGH via oral, dermal and inhal routes. A cholinesterase inhibitor, insecticide.

ABIETIC ACID. Syns: *abietinic acid, sylvic acid*. Yellow powder. $C_{20}H_{30}O_2$, mw: 302.4, mp: 172–175°.
THR = MOD via oral route.
Fire Hazard: Slight.
Explosion Hazard: Slight, as dust.

ABIETIC ACID, ETHYL ESTER. See ethyl abietate.

ABIETINIC ACID. See abietic acid.

ABITOL. See hydroabietyl alcohol.

"A" BLASTING POWDER. See explosives, low.

ABRIN. See ricin.

ABSINTHIUM. Syn: *wormwood*. Dried leaves and flowering tops of *Artemisia absinthium L.*
THR = MOD via oral route. An allergen. Habitual users develop "absinthism," with tremors, vertigo, vomiting and hallucinations. May cause contact dermatitis.
Fire Hazard: Slight.

AC 3810. $C_{20}H_{27}O_3N_5$, mw: 385.5.
THR = HIGH via oral route. An exper teratogen. [*3*]

AC 3092. $C_{20}H_{23}O_3HCl$, mw: 347.7.
THR = HIGH via oral route. An exper teratogen. [*3*]

ACACIA GUM. Syn: *gum arabic*. Spheres or "tears." mw: 240,000, d: 1.35–1.49.
THR = Weak allergen. Local contact may cause contact dermatitis. Inhalation as dust may cause respiratory symptoms such as asthma, watery nose and eyes, cough and wheezing. Hives, eczema and angioedema may also result from inhalation or ingestion. A stabilizer food additive and is a migrating substance from packaging materials. [*109*]
Fire Hazard: Slight.

ACACIA WOOD. See sawdust.

ACANTHITE. See silver sulfide.

ACCELLERENE. See *p*-nitroso dimethyl aniline.

4,10-ACE-1,2-BENZANTHRACENE. Syn: *5,6-dihydrobenzene-(e)aceanthrylene*. $C_{20}H_{14}$, mw: 254.3.
THR = A deadly test-animal poison. An exper neo. [*3*]

8,9-ACE-1,2-BENZANTHRACENE. Syn: *1,2-dihydrobenz(l)-aceanthrylene*. $C_{20}H_{14}$, mw: 254.3.
THR = An exper carc to mice via sc route. [*3*]

ACE NAPHTHANTHRACENE. Syn: *3,4-dimethylene-1,2-benzanthracene*. $C_{20}H_{14}$, mw: 254.3.
THR = HIGH to animals. An exper carc. [*3*]

ACENAPHTHENE. Syn: *1,8-ethylene naphthalene*. White, elongated crystals. $C_{10}H_6(CH_2)_2$, mw: 154.2, mp: 95°, bp: 277.5°, d: 1.024 @ 99°/4°, vap. press.: 10 mm @ 131.2°, vap. d: 5.32.
THR = Irr to skin and mu mem. An exper neo. [*3*] May cause acute vomiting if swallowed in large quantities.
Fire Hazard: Slight.

ACERDOL. See calcium permanganate and manganese compounds.

ACETAL. Syns: *1,1,-diethoxy ethane, acetaldehyde diethylacetal, ethylidene diethyl ether*. Colorless, volatile liquid, agreeable odor, nutty after-taste. $C_6H_{14}O_2$, mw: 118.17, bp: 102.7°, flash p: −5°F (CC), lel = 1.65%, uel = 10.4%, d: 0.831; autoign. temp.: 446°F, vap. press: 10 mm @ 8.0°, vap. d: 4.08, mp: −100°.
Acute tox data: Oral LD_{50} (rat) = 4600 mg/kg; inhal LC_{LO} (rat) = 4000 ppm for 4 hrs. [*3*]
THR = MOD via oral and inhal routes. No industrial intoxication known. It is narcotic and more toxic than paraldehyde.
Fire Hazard: Dangerous, when exposed to heat or flame; can react vigorously with oxidizing materials.
Spontaneous Heating: No.
Explosion Hazard: Mod, when exposed to flame. Old samples have been known to explode upon heating.
Disaster Hazard: Dangerous from fire and explosion.
To Fight Fire: CO_2, alcohol foam, dry chemical.

ACETALDEHYDE. Syns: *acetic aldehyde, ethyl aldehyde*. Colorless, fuming liquid; pungent, fruity odor. C_2H_4O, mw: 44.05, mp: −123.5°, bp: 20.8°, lel = 4.0%, uel = 57%, flash p: −36°F (CC), d: 0.7827 @ 20°/20°, autoign. temp.: 347°F, vap. d: 1.52.
Acute tox data: Oral LD_{50} (rat) = 1930 mg/kg. inhal LC_{LO} (rat) = 4000 ppm for 4 hrs, sc LD_{50} (mouse) = 560 mg/kg. [*3*]
THR = MOD via oral, inhal and sc routes. [*24*] A local irr, CNS narcotic. A synthetic flavoring substance and adjuvant. A common air contaminant. [*2*], [*109*]
Fire Hazard: Dangerous when exposed to heat or flame; can react vigorously with acid anhydrides, alcohols, ketones, phenols, NH_3, HCN, H_2S, halogens, P, isocyanates and strong alkalis. [*19*]
Spontaneous Heating: No
Explosion Hazard: Severe when vapors exposed to flame.
Disaster Hazard: Highly dangerous due to fire and explosion hazard.
To Fight Fire: CO_2, dry chemical, alcohol foam.

***m*-ACETALDEHYDE.** See metaldehyde.

***p*-ACETALDEHYDE.** See paraldehyde.

ACETALDEHYDE AMMONIA. See aldehyde ammonia.

ACETALDEHYDE CYANOHYDRIN. See lactonitrile.

ACETALDEHYDE DIETHYL ACETAL. See acetal.

7-ACETALDEHYDE-12-METHYL BENZ(a)ANTHRACENE. An exper carc. [*23*]

ACETALDEHYDE OXIME. Syns: *acetaldoxime, ethylidene hydroxylamine*. A crystalline material, very sol in water, alcohol, ether. C_2H_5NO, mw: 59.1, mp: 46.5° mp: 12°B, d: 0.966, bp: 114.5°.
THR = HIGH via ip route. [*3*]

ACETALDEHYDE SODIUM BISULFITE. White crystals decomp by acids, sol in water, insol in alcohol. $C_2H_5O_2SO_2Na \cdot \frac{1}{2}H_2O$, mw: 157.
THR = MOD irr.
Disaster Hazard: Dangerous; when heated to decomp, emits highly toxic fumes.

ACETALDOL. See aldol.

ACETALDOXIME. See acetaldehyde oxime.

ACETAMIDE. Syn: *acetic acid amide*. Colorless, crystals; mousy odor. CH_3CONH_2, mw: 59.07, mp: 81°, bp: 221.2°, d: 1.159 @ 20°/4°, vap. press: 1 mm @ 65.0°.
THR = MILD irr, LOW toxicity. An exper (+) carc. [*3, 23*]
Disaster Hazard: Dangerous. See also cyanides.

ACETAMIDINE HYDROCHLORIDE. Syns: *ethanamidine hydrochloride, α-amino-α-iminoethane hydrochloride*. Long, somewhat deliquescent prisms when crystallized from ethanol. $C_2H_6N_2 \cdot HCl$, mw: 94.6, mp: 164°, sol in water and alcohols.
THR = MOD also MOD irr.
Disaster Hazard: Dangerous; see chlorides.

***p*-ACETAMIDO BENZALDEHYDE.** $C_9H_9O_2N$, mw: 163.
THR = LOW to test animals.

2-ACETAMIDO(TERT-BUTYL)-6-PROPYL ISOTHIOURACIL. Acute tox data; ip LD_{50} (mouse) = 200 mg/kg. [*3*]
THR = HIGH via ip route to test animals.

2-ACETAMIDO-6H-DIBENZO(*b,d*)PYRAN-6-ONE. $C_{15}H_{10}O_3N$, mw: 252.3.
THR = An exper carc. [*3*]

2-ACETAMIDO FLUORENE. Syns: *N-fluoren-2-yl acetamide, 2-acetyl amino fluorene*. A powder. $C_{15}H_{13}O$, mw: 223.3.
THR = HIGH to animals. An exper carc, neo and teratogen. [*3, 23*] An insecticide.
Disaster Hazard: Dangerous; see also cyanides.

4-ACETAMIDO-3(5-NITRO-2-FURYL)-6H-1,2,4-OXADIAZINE. $C_6H_8O_5N_4$, mw: 252.2.
THR = An exper carc. [*3, 23*]

5-*p*-ACETAMIDO PHENYLAZO-8-QUINOLINAL HYDROCHLORIDE. Acute tox data: ip LD_{LO} (mouse) = 200 mg/kg. [*3*]
THR = HIGH via ip route.

ACETAMIDO PHENYL THIOACETATE. $C_{10}H_{10}O_2SN$, mw: 208.3. Acute tox data: ip LD_{50} (mouse) = 100 mg/kg. [*3*]
THR = HIGH via ip route.

ACETAMINE YELLOW CG. Syn: *acetate yellow fast G*. $C_{15}H_{15}O_2N_3$, mw: 269.3.
THR = An allergen. An exper (±) carc. [*3, 4*]

ACETAMINOPHEN. Syn: *4′-hydroxy acetanilide*. A crystalline material. $C_8H_9NO_2$, mw: 151.2, d: 1.293 @ 21°/4°, mp: 170°.
Acute tox data: Oral LD_{50} (mouse) = 338 mg/kg; oral LD_{50} (rat) = 2400 mg/kg. [*3*] Has been implicated in aplastic anemia.
THR = HIGH to MOD via oral route.

ACETANILIDE. Syns: *N-phenyl acetamide, antifebrin*. White, shining crystalline scales. C_8H_9ON, mw: 135.16, mp: 113.5°, bp: 303.8°, flash p: 345°F (OC), d: 1.2105 @ 4°/4°, autoign. temp.: 1004°F, vap. press: 1 mm @ 114.0°, vap. d: 4.65.
Acute tox data: Oral LD_{50} (rat) = 800 mg/kg; oral LD_{50} (mouse) = 1210 mg/kg. [*3*]

THR = MOD via oral and ip routes. In experimental poisoning, a variety of effects have been reported, depending on the animal or tissue used for testing. Acute poisoning in humans may occur if several grams are taken by mouth. Cyanosis is a prominent finding in both acute and chronic poisoning. Habitual use of acetanilide for relief of headache has resulted in damage to the blood-forming organs, with anemia and cyanosis. The latter is due to the presence of methemoglobin. Sulfhemoglobin has also been demonstrated in such cases. It may also cause contact dermatitis and inhalation or ingestion may cause an eczematous eruption of the skin.

Fire Hazard: Slight, when exposed to heat or flame.

Spontaneous Heating: No.

To Fight Fire: Water, CO_2, foam, mist, dry chemical.

Disaster Hazard: Dangerous; see aniline.

ACETARSONE. See acetphenarsine.

ACETATE FIBER. See cellulose acetate.

ACETATE YELLOW FAST G. See acetamine yellow CG.

ACETAZOLAMIDE. See N-(5-sulfamoyl-1,3,4-thiadiazol-2-yl)acetamide.

ACETIC ACID. Syns: *methane carboxylic acid, vinegar acid, ethanoic acid*. Clear, colorless liquid, pungent odor. CH_3COOH, mw: 60.05, mp: 16.7°, bp: 118.1°, flash p: 109°F (CC), lel = 5.4%, uel = 16.0% @ 212°F, d: 1.049 @ 20°/4°, autoign. temp.: 869°F, vap. press: 11.4 mm @ 20°, vap. d: 2.07.

Acute tox data: Inhal TD_{Lo} (human) = 816 ppm for 3 min, highly irr, oral LD_{50} (rat) = 3310 mg/kg; dermal LD_{50} (rabbit) = 1060 mg/kg. [*3*]

THR = MOD via inhal, oral and dermal routes. Caustic, irr, can cause burns, lachrymation and conjunctivitis. It attacks the skin easily and can cause dermatitis and ulcers. Inhalation causes irr of mu mem. In case of contact with skin, eyes or clothing, immediately flush skin or eyes with plenty of water and remove all contaminated clothing. If swallowed give magnesia, chalk or whiting in water. It is a miscellaneous and/or general purpose food additive which can migrate to food from packaging materials. A common air contaminant. [*109*]

Fire Hazard: MOD, when exposed to heat or flame; can react vigorously with oxidizing materials.

Spontaneous Heating: No.

Particularly dangerous in contact with chromic acid, sodium peroxide, nitric acid, acetaldehyde, 2-amino-ethanol, NH_4NO_3, BrF_5, ClF_3, chlorosulfonic acid, (O_3 + diallyl methyl carbinol), ethylene diamine, ethylene imine, H_2O_2, (HNO_3 + acetone), oleum, $HClO_4$, permanganates, P(OCN), PCl, KOH, NaOH, *n*-xylene. [*19*]

Explosion Hazard: MOD, when exposed to flame.

Disaster Hazard: Dangerous; when heated to decomp, emits toxic fumes.

To Fight Fire: CO_2, dry chemical, alcohol foam, foam, mist.

ACETIC ACID AMIDE. See acetamide.

ACETIC ACID SECONDARY BUTYL ESTER. See sec-butylacetate.

ACETIC ACID DIMETHYLAMIDE. See N,N-dimethyl acetamide.

ACETIC ACID, GLACIAL. See acetic acid.

ACETIC ALDEHYDE. See acetaldehyde.

ACETIC ANHYDRIDE. Syns: *acetyl oxide, acetic oxic, ethanoic anhydride*. Colorless, very mobile, strongly refractive liquid; very strong acetic odor. $(CH_3CO)_2O$, mw: 102.09, mp: −73.1°, bp: 140°, flash p: 129°F (CC), d: 1.082 @ 20°/4°, lel = 2.9%, uel = 10.3%, autoign. temp.: 734°F, vap. press: 10 mm @ 36.0°, vap. d: 3.52.

Acute tox data: Oral LD_{50} (rat) = 1780 mg/kg; dermal LD_{50} (rabbit) = 3000 mg/kg; inhal LC_{Lo} (rat) = 1000 ppm for 4 hrs. [*3*]

THR = MOD via oral, dermal and inhal routes. An irr, corrosive on contact with tissue, particularly the eyes and upper respiratory tract. Systemic effects can be avoided by heeding the warning of its presence—coughing and a burning sensation in nose and throat. Ingestion causes a burning pain in the stomach followed by nausea and vomiting. If not removed from skin at once it will cause a reddening which may be followed by wrinkling, whitening and peeling. Continued contact with skin may cause dermatitis. It is especially hazardous to the eyes, with delayed action causing eye burns.

Fire Hazard: Mod, when exposed to heat or flame.

Spontaneous Heating: No.

Explosion Hazard: Mod, when exposed to flame. Can react violently with 2-aminoethanol, aniline, boric acid, chlorosulfonic acid, CrO_3, (CrO_3 + acetic acid), ethylene diamine, ethylene imine, glycerol, HCl, HNO_3, oleum, $HClO_4$, H_2O_2, HF, permanganates, NaOH, Na_2O_2, H_2SO_4, water, N_2O_2. [*19*]

Disaster Hazard: Dangerous; when heated to decomp, it emits toxic fumes; can react vigorously with oxidizing materials, will react violently on contact with water or steam.

To Fight Fire: CO_2, dry chemical, water mist, alcohol foam.

ACETIC ETHER. See ethyl acetate.

ACETIC OXIDE. See acetic anhydride.

ACETIN. See glyceryl monoacetate.

ACETO ACETANILIDE. Syns: *α-ketobutyranilide*, *α-acetyl acetanalide*. White crystalline solid. $C_{10}H_{11}O_2N$, mw: 177.2, mp: 85°, bp: decomp, flash p: 365°F (COC), d: 1.260 @ 20°, vap. press: 0.01 mm @ 20°.
Acute tox data: ip LD_{LO} (mouse) = 300 mg/kg. [*3*]
THR = HIGH via ip route. A weak allergen. See also acetanilide.
Fire Hazard: Slight, when exposed to heat or flame.
Disaster Hazard: Dangerous; see aniline and cyanides.
To Fight Fire: Alcohol foam, water mist, CO_2, dry chemical.

ACETOACET-*o*-ANISIDINE. Crystals. $C_{11}H_{13}O_3N$, mw: 207.3 mp: 86.6°, flash p: 325°F (OC), d: 1.132 @ 86.6°/20°, vap. d: 7.0.
Acute tox data: Acute oral LD_{50} (rat) = 2290 mg/kg. [*3*]
THR = MOD via oral route.
Fire Hazard: Slight, when exposed to heat or flame; can react with oxidizing materials.
To Fight Fire: CO_2, mist, dry chemicals.

ACETOACET-*o*-CHLORANILIDE. Crystals. $NC_{10}H_{10}O_2Cl$, mw: 211.65, mp: 107°, bp: decomp, flash p: 350°F (COC), d: 1.438 @ 20°, vap. press: 0.01 mm @ 20°, vap. d: 7.31.
Acute tox data: ip LD_{LO} (mouse) = 500 mg/kg. [*3*]
THR = HIGH via ip route. See also acetanilide.
Fire Hazard: Slight, when exposed to heat or flame.
Disaster Hazard: Dangerous; see aniline and cyanides; can react vigorously with oxidizing materials.
To Fight Fire: Water, foam, CO_2, water mist, dry chemical.

ACETOACET-*p*-CHLORANILIDE. Crystals. $CH_3COCH_2CONHC_6H_4Cl$, mw: 211.65, mp: 133°, bp: decomp, flash p: 320°F (CC), d: 1.348 @ 20°, vap. press: $<$0.011 mm @ 20°, vap. d: 7.31.
Acute tox data: ip LD (mouse) = 500 mg/kg. [*3*]
THR = HIGH via ip route.
Fire Hazard: Slight, when exposed to heat or flame.
Disaster Hazard: Dangerous; see aniline, phosgene and cyanides; can react vigorously with oxidizing materials.
To Fight Fire: Water, foam, CO_2, dry chemical.

ACETOACETIC ACID. Syn: *acetyl acetic acid*. Colorless syrup. $C_4H_6O_3$, mw: 102.1, bp: $<$ 100° (decomp), mp: 36.4°
THR = Details U. An irr.
Fire Hazard: Slight; when heated, emits acrid fumes; reacts with oxidizing materials.

ACETOACETIC ESTER. See ethyl acetoacetate.

ACETOACET-*p*-PHENETIDIDE. Crystals. $C_{12}H_{15}O_3N$, mw: 221.25, mp: 108.5°, bp: decomp, flash p: 325°F (OC), d: 1.220 @ 20°, vap. press: 0.02 mm @ 20°, vap. d: 7.63.
THR = Details U. See acetanilide.
Fire Hazard: Slight, when exposed to heat or flame; can react with oxidizing materials.
To Fight Fire: Water, foam, CO_2, water spray or mist, dry chemical (multi-purpose), dry chemical.

ACETOACET-*o*-TOLUIDIDE. Crystals. $C_{11}H_{13}O_2N$, mw: 191.22, mp: 106°, bp: decomp, d: 1.300 @ 20°, vap. press: 0.01 mm @ 20°, flash p: 320°F (COC).
Acute tox data: Oral LD_{50} (rat) = 1600 mg/kg; ip LD_{50} (rat) = 800 mg/kg. [*3*]
THR = MOD via oral and ip routes.

ACETOACET-*m*-XYLIDIDE. Syn: *AAMX*. White to light yellow crystalline solid, sol in water to 0.5% @ 25°. $C_{12}H_{15}O_2N$, mw: 205, mp: 89–90°, d: 1.238, flash p: 340°F (OC).
THR = U.
Fire Hazard: Combustible.
To Fight Fire: Alcohol foam, water spray or mist, dry chemical (multi-purpose).

ACETOGUANAMINE. Crystals. $CH_3(C_3N_3(NH_2)_2$, mw: 125.15, mp: 270° (Decomp), d: 1.44.
THR = U. See amines.
Fire Hazard: Slight, when exposed to heat or flame; can react with oxidizing materials.

ACETOHYDROXAMIC ACID. $C_2H_5O_2N$, mw: 75.1.
THR = An exper teratogen. [*3*]

ACETOIN. Syns: *acetyl methyl carbinol-3-hydroxy-2-butanone*, *dimethylketol*. Slightly yellow liquid or crystalline solid, sol in alcohol, miscible with water. $C_4H_8O_2$, mw: 88, d: 1.016, bp: 147–148°, mp: 15°.
THR = LOW. A synthetic flavoring substance and adjuvant. [*109*]

ACETOL. Syns: *acetyl salicylic acid*, *aspirin*. White crystals. $C_9H_8O_4$, mw: 180.2, mp: 135°, d: 1.35.
Acute tox data: Oral LD_{50} (rat) = 1000 mg/kg.
THR = MOD via oral route. An allergen. A 10 g dose to an adult may be fatal. Contact, inhal or ingestion can cause asthma, sneezing, irr and watering of eye and nose as well as hives and eczema. Has been implicated in aplastic anemia. An exper teratogen. [*3*]
Fire Hazard: Slight, when exposed to heat or flame.

ACETONANYL. See 2,2,4-trimethyl-1,2-dihydroquinoline.

ACETONE. Syns: *dimethyl ketone*, *ketone propane*, *propanone*. Colorless liquid, fragrant mintlike odor. CH_3COCH_3, mw: 58.08, mp: −94.6° bp: 56.48°, ulc = 90, flash p: 0°F (CC), lel = 2.6%, uel = 12.8%, d:

0.7972 @ 15°, autoign. temp. (color): 869°F, vap. press: 400 mm @ 39.5°, vap. d: 2.00.

Acute tox data: Oral LD_{50} (rat) = 9750 mg/kg; dermal LD_{50} (rabbit) = 20,000 mg/kg; ip LD_{50} (mouse) = 1297 mg/kg; inhal TC_{LO} (human) = 500 ppm ⟶ eye symptoms. [3]

THR = MOD via oral, ip and inhal routes; VERY LOW via dermal route. Acetone is narcotic in high conc. In industry, no injurious effects from its use have been reported, other than the occurrence of skin irr resulting from its de-fatting action, or headache from prolonged inhal. A food additive permitted in food for human consumption. A common air contaminant. [109]

Fire Hazard: Dangerous, when exposed to heat or flame or oxidizers. Can react violently with ($CHCl_3$ + a base), CrO, $Cr(OCl)_2$, (nitric + acetic acid), (nitric + sulfuric acid), NOCl, nitrosyl perchlorate, nitryl perchlorate, permonosulfuric acid, potassium tert-butoxide, NaOBr, (sulfuric acid + potassium dichromate), (thio-diglycol + hydrogen peroxide), trichloromelamine. [19]

Explosion Hazard: Mod, when vapor is exposed to flame.

Disaster Hazard: Dangerous, due to fire and explosion hazard, can react vigorously with oxidizing materials.

To Fight Fire: CO_2, dry chemical, alcohol foam.

ACETONE CHLOROFORM. See chlorobutanol.

ACETONE CYANHYDRIN. Syn: *α-hydroxy isobutyro nitrile*. CH_3COCH_3HCN, mw: 85.10, mp: −20°, bp: 82° @ 23 mm, d: 0.932 @ 19°, autoign. temp.: 1270°F, flash p: 165°F, vap. d: 2.93.

Acute tox data: Oral LD_{50} (rat) = 17 mg/kg; inhal LD_{50} (mouse) = 575 ppm for 2 hrs. [3]

THR = HIGH via oral and inhal routes. This material readily decomp to HCN and acetone. It should not be stored for long periods and it should be kept cool. Contact with H_2SO_4 can explode. [19] See also hydrocyanic acid and acetone.

Fire Hazard: Slight, when exposed to heat or flame.

Disaster Hazard: Dangerous; see cyanides.

To Fight Fire: CO_2, dry chemical, alcohol foam.

ACETONE DICARBOXYLIC ACID. Syn: *β-ketoglutaric acid*. White crystals. $C_5H_6O_5$, mw: 146.10, mp: 138° (decomp).

THR = LOW. An insecticide.

Fire Hazard: Slight; when heated, emits acrid fumes.

ACETONE DICHLORIDE. See 2,2-dichloropropane.

ACETONE OIL. (a) Standard: light, lemon-yellow. (b) Refined: almost water white. (c) Heavy: dark, orange-yellow. bp: (a) 75°–160°, (c) 80°–225°, d: (a) 0.826–0.830, (b) 0.812, (c) 0.885–0.865.

THR = U.

Fire Hazard: Dangerous; when exposed to heat or flame.

Explosion Hazard: Mod, when exposed to flame.

Disaster Hazard: Dangerous; can react vigorously with oxidizing materials.

To Fight Fire: CO_2, dry chemical.

ACETONE PEROXIDE. Liquid. The trimeric form is crystalline. mp: 97°.

THR = U. See also peroxides, organic.

Fire Hazard: Mod by spont chemical reaction; can react vigorously with reducing materials.

Explosion Hazard: The trimeric form is shock-sensitive and static-electricity-sensitive and may detonate.

ACETONE SEMICARBAZONE. $(CH_3)_2CNNHCONH_2$, mw: 115.1.

Acute tox data: iv LD_{50} (mouse) = 90 mg/kg. [3]

THR = HIGH via ip route.

Disaster Hazard: Mod dangerous; when heated to decomp, it emits toxic fumes.

ACETONITRILE. See methyl cyanide.

ACETONYL ACETONE. Syns: *hexanedione-2,5, 1,2-diacetyl ethane*. Colorless liquid. Gradually turns yellow. $C_6H_{10}O_2$, mw: 114.14, mp: −9°, bp: 188°, flash p: 174°F (CC), d: 0.970 @ 20°/4°, autoign. temp.: 920°, vap. d: 3.94.

Acute tox data: Oral LD_{50} (rat) = 2700 mg/kg; inhal LC_{LO} (rat) = 200 ppm for 4 hrs. [3]

THR = MOD via oral and HIGH via inhal routes. Can irr the eyes.

Fire Hazard: Mod, when exposed to heat or flame; can react with oxidizing materials.

Spontaneous Heating: No.

To Fight Fire: CO_2, dry chemical, (multi-purpose dry chemical), water spray or mist, alcohol foam.

3-(α-ACETONYL BENZYL)-4-HYDROXY COUMARIN. See warfarin.

3-α-ACETONYL FURFURYL-4-HYDROXY COUMARIN. Syn: *coumafuryl*. White powder; practically insol in water, sol in alcohols. $C_{19}H_{14}O$, mw: 298.3, mp: 124°.

Acute tox data: Oral LD_{50} (rat) = 25 mg/kg; oral LD_{50} (mouse) = 15 mg/kg. [3]

THR = HIGH via oral and inhal routes. See also warfarin.

This rodenticide is almost always used mixed with bait preparations unpalatable to humans and it resembles warfarin in action. However, in case of accidental ingestion, induce vomiting until fluid is clear. Administer vitamin K (oral or intravenous) in large doses. Call a physician immediately. It is highly toxic.

7-ACETONYL-12-METHYL BENZ(*a*)ANTHRACENE. An exper carc. [*23*]

***p*-ACETOPHENETIDE.** See acetophenetidine.

ACETOPHENETIDINE. Syns: *p-acetophenetide, phenacetin.* Glistening, crystalline powder. $C_{10}H_{13}O_2N$, mw: 179.21, mp: 135°.

Acute tox data: Oral LD_{50} (rat) = 1650 mg/kg; oral LD_{50} (mouse) = 1220 mg/kg; oral TD_{LO} (human) = 1000 mg/kg ⟶ CNS symptoms. [*3*]

THR = MOD via oral and inhal routes. An exper carc via oral route. [*23, 103*] Although less toxic than acetanilide, the symptoms from it are nearly identical: Weakness, dizziness, depression, collapse and very evident cyanosis. Chronic effects consist of loss of weight, insomnia, shortness of breath, weakness and often aplastic anemia.

Disaster Hazard: Mod dangerous; when heated to decomp, it emits toxic fumes.

ACETOPHENONE. See phenyl methyl ketone.

ACETOPHENONE OXIME. $C_6H_3O(NOH)CH_3$, mw: 139.2.

THR = U. An insecticide.

Disaster Hazard: Dangerous; see cyanides.

ACETOTHIOPHINE. See thioacetamide.

***p*-ACETO TOLUIDIDE.** Syn: *p-tolylacetamide.* Crystals. $C_9H_{11}ON$, mw: 149.1, bp: 307°, flash p: 335°F (CC), d: 1.212, vap. d: 5.14, mp: 153°.

Acute tox data: Oral LD_{50} (mouse) = 980 mg/kg. [*3*]

THR = MOD via oral route. See also aceto phenetidine and acetanilide.

Fire Hazard: Slight, when exposed to heat or flame; can react with oxidizing materials.

To Fight Fire: Water, foam, CO_2, dry chemical.

N-ACETOXY-4-ACETAMIDO BIPHENYL. $C_{16}H_{15}O_3N$, mw: 269.3.

THR = An exper neo. [*3*]

N-ACETOXY-2-ACETAMIDO FLUORENE. $C_{17}H_{13}O_3N$, mw: 279.3.

THR = An exper neo. [*3*]

N-ACETOXY-3-ACETAMIDO FLUORENE.

THR = An exper neo and carc. [*3*]

N-ACETOXY-4-ACETAMIDO STILBENE. $C_{18}H_{17}O_3N$, mw: 295.4.

THR = HIGH to animals. An exper neo. [*3*]

***trans*-N-ACETOXY-4-ACETYLAMINO STILBENE.** $C_{18}H_{17}O_3N$, mw: 295.4.

THR = An exper carc. [*3*]

N-ACETOXY-4-BIPHENYL ACETAMIDE. $C_{15}H_{15}O_3N$, mw: 257.2.

THR = An exper carc. [*23*]

17-α-ACETOXY-6-DEHYDRO-6-METHYL PROGESTERONE. $C_{24}H_{32}O_4$, mw: 384.6.

THR = An exper neo. [*3*]

N-ACETOXY-4-FLUORENYL ACETAMIDE. $C_{17}H_{13}O_3N$, mw: 279.3.

THR = An exper neo. [*3, 23*]

ACETOXY ISOSUCCINO DINITRILE.

THR = HIGH by animal exper. See also nitrites.

Disaster Hazard: Dangerous. When heated to decomp, it emits highly toxic fumes of nitriles.

ACETOXY MERCURI BENZO THIOPHENE.

THR = HIGH. See mercury compounds, organic.

Disaster Hazard: Dangerous; see mercury and oxides of sulfur.

10-ACETOXY METHYL-1,2-BENZANTHRACENE. $C_{21}H_{17}O_2$, mw: 301.3.

THR = An exper carc and neo. [*3*]

6-ACETOXY METHYLBENZO(*a*)PYRENE. $C_{23}H_{16}O_2$, mw: 324.4.

THR = An exper carc. [*3*]

ACETOXY METHYL METHYL NITROSAMINE. $C_4H_8O_3N_2$, mw: 132.1.

THR = HIGH via oral to rats. An exper carc. [*3*]

N-ACETOXY-2-PHENATHRYL ACETAMIDE. $C_{18}H_{15}O_3N$, mw: 293.3.

THR = An exper carc. [*23*]

N-ACETOXY-4-PHENANTHRYL ACETAMIDE. An exper carc. [*23*]

1-ACETOXY SAFROLE. $C_{12}H_{12}O_4$, mw: 220.2.

THR = An exper neo. [*3, 23*]

3-ACETOXY XANTHINE. $C_7H_5O_4N_4$, mw: 209.2.

THR = An exper neo. [*3, 23*]

ACETOZONE. See acetyl benzoylperoxide.

ACETPHENARSINE. Syns: *acetarsone, stovarsol.* A crystalline material, slightly water sol. $C_8H_{10}NO_5As$, mw: 275.1, decomp @ 240°–250°.

Acute tox data: Oral LD_{50} (mouse) = 4 mg/kg. [*3*]

THR = HIGH via oral route. See arsenic compounds.

ACETULAN. Pale yellow liquid. Acetylated lanolin fractions. d: 0.867 @ 25°.

THR = Limited information indicates virtually no tox.

Fire Hazard: Slight when exposed to heat or flame. Can react with oxidizers.

To Fight Fire: Foam, CO_2, mist, water spray, dry chemical.

α-ACETYL ACETANILIDE. See acetoacetanilide.

ACETYL ACETIC ACID. See acetoacetic acid.

ACETYL ACETONATE OF CHROMIUM. See chromium 2,4-pentanedione derivative.

ACETYL ACETONATE OF COPPER. See copper 2,4-pentanedione derivative.

ACETYL ACETONE. See pentanedione-2,4.

ACETYL AMINO AZOTOLUENE. A red powder. $C_{16}H_{17}N_3O$, mw: 267.3, mp: 186°.
THR = A (S) carc. [*14*]

***p*-ACETYL AMINO BENZALDEHYDE THIOSEMICARBAZONE.** See "thiosemicarbazone."

2-ACETYL AMINO DIBENZO THIOPHENE. See N-2-dibenzothienyl acetamide.

2-ACETYLAMINO-9,10-DIHYDROPHENANTHRENE. $C_{16}H_{16}ON$, mw: 238.3.
THR = LOW to rats via oral route. An exper carc. [*3*]

3-ACETYLAMINO FLUORANTHENE. Syn: *N-3-fluoranthenylacetamide*. $C_{18}H_3ON$, mw: 259.3.
THR = LOW to animals via oral route. An exper carc. [*3*]

2-ACETYL AMINO FLUORENE. See acetamido fluorene.

2-ACETYL AMINO PYRENE. See N-pyren-2-yl-acetamide.

ACETYL BENZENE. See phenyl methyl ketone.

1-ACETYL BENZO(*a*)PHENANTHRENE.
THR = An exper carc. [*23*]

5-ACETYL BENZO(*c*)PHENANTHRENE.
THR = An exper carc. [*23*]

ACETYL BENZOYL ACONINE. See aconitine.

ACETYL BENZOYL PEROXIDE. Syn: *acetozone*. White crystals. $C_6H_5COOOCOCH_3$, mw: 180.2, mp: 36°–37°, bp: 130° @ 19 mm.
THR = HIGH irr; HIGH via oral or inhal routes. This material is a powerful oxidizing agent which is corrosive to skin and mu mem.
Fire Hazard: Mod, by spont chemical reaction.
Explosion Hazard: Mod, when shocked or exposed to heat.
Disaster Hazard: Dangerous; shock or heat will cause detonation with evolution of toxic fumes; will react with water or steam to produce heat; can react vigorously with reducing materials.
To Fight Fire: CO_2 or dry chemical.

ACETYL BENZOYL PEROXIDE, SOLUTION. See acetyl benzoyl peroxide.

ACETYL BENZYL PEROXIDE, DRY. A powder.
THR = U. See acetyl benzoyl peroxide and peroxides, organic.
Fire Hazard: Mod, by spont chemical reaction. A powerful oxidizing agent.
Explosion Hazard: Mod, when shocked or exposed to heat.
Disaster Hazard: Dangerous; shock will cause detonation with evolution of toxic fumes; will react with water and steam to produce heat; can react vigorously with reducing materials.

ACETYL BENZYL PEROXIDE, WET.
THR = U. See also peroxides, organic.
Fire Hazard: Mod, by spont chemical reaction.
Disaster Hazard: Mod dangerous; when heated to decomp, emits toxic fumes; can react vigorously with reducing materials.

ACETYL BROMIDE. Syn: *ethanoyl bromide*. Colorless fuming liquid; turns yellow in air. CH_3COBr, mw: 122.96, mp: −96.5°, bp: 76.7°, d: 1.52 @ 9.5°/4°.
THR = HIGH via contact, oral and inhal routes. This material readily hydrolyzes. See also hydrobromic acid, acetic acid.
Explosion Hazard: Great, by spont chemical reaction; decomp violently upon contact with moisture or alcohols.
Disaster Hazard: Dangerous; when heated to decomp, emits highly corrosive and toxic fumes of carbonyl bromide and bromine; will react with water or steam to produce heat.
To Fight Fire: Dry chemical, CO_2.

α-ACETYL BUTYROLACTONE. Syn: *α-acetyl-γ-hydroxybutyric acid-γ-lactone*. Liquid, fruity odor, sol in water. $C_6H_8O_3$, mw: 128.1, d: 1.187 @ 20°/20°, bp: 143° @ 30 mm.
THR = MOD via irr, oral and inhal routes. Avoid prolonged contact with skin or any contact with eyes or mu mem.
Fire Hazard: Slightly hazardous when exposed to heat or fire. Can react with oxidizers.
To Fight Fire: Water, foam, CO_2, dry chemical.

ACETYL CHLORIDE. Syn: *ethanoyl chloride*. Colorless, fuming liquid. CH_3COCl, mw: 78.50, mp: −112°, bp: 51°–52°, flash p: 40°F (CC), autoign. temp.: 734°F, d: 1.1051 @ 20°/4°, vap. d: 2.70.
THR = HIGH via irr, oral and inhal routes. It readily hydrolyzes to form HCl and acetic acid. See also hydrochloric acid, acetic acid.
Fire Hazard: Dangerous; when exposed to heat or flame, it reacts violently with water.
Spontaneous Heating: No.
Explosion Hazard: By spont chemical reaction, with contact with water, dimethyl sulfoxide or ethanol.
Disaster Hazard: Dangerous; when heated to decomp,

emits highly toxic fumes of phosgene; will react with water or steam to produce heat and toxic or corrosive fumes.

To Fight Fire: CO_2, or dry chemical.

ACETYLENE. Syns: *ethyne*, *ethine*. Colorless gas, garlic-like odor. Flammable. HC≡CH, mw: 26.04, bp: −84.0° (sublimes), lel = 2.5%, uel = 82%, mp: −81.8°, flash p: 0°F (CC), d: 1.173 g/liter @ 0°, autoign. temp.: 581°F, vap. press: 40 atm @ 16.8°, vap. d: 0.91.

THR = When mixed with O_2 in proportions of 40% or more, acetylene acts as a narcotic and has been used in anesthesia. Acetylene acts as a simple asphyxiant by diluting the O_2 in the air to a level which will not support life. However, the presence of impurities in commercial acetylene may result in the production of symptoms before an asphyxiant contribution is reached. Thus: 10% in air ⟶ slight intox; 20% ⟶ staggering gait; 30% ⟶ general incoordination; 33% ⟶ unconsciousness in 7 min; up to 80% ⟶ complete anesthesia, increased blood pressure, narcosis and stimulated respiration. See Table XII-5. [*96*]

Dizziness, headache, mild gastric symptoms, and (in high conc) semi-asphyxia and brief loss of consciousness have all been reported. In general industrial practice, however, acetylene does not constitute a serious hazard. See argon for discussion of simple asphyxiants.

Fire Hazard: Very dangerous, when exposed to heat, flame or oxidizers.

Spontaneous Heating: No.

Explosion Hazard: Mod, when exposed to heat or flame or by spont chemical reaction. At high pressures and even moderate tempreatures, acetylene has been known to decomp explosively. Forms explosive compounds with copper, brass, copper salts, copper carbide, Co, Hg, Hg salts, K, Ag and Ag salts, RbH, CsH, halogens, HNO_3, NaH. [*19*] See acetylides.

Disaster Hazard: Dangerous; when ignited it burns with an intensely hot flame; can react vigorously with oxidizing materials.

To Fight Fire: CO_2, water spray, or dry chemical. Stop flow of gas.

ACETYLENE BLACK. See soot.

ACETYLENE CHLORIDE. Syn: *chloroethyne*. A gas. CHCCl, mw: 60.5, bp: −31°, vap. d: 2.0.

THR = U. Probably has anesthetic properties if inhaled. See also chlorinated HC, aliphatic.

Fire Hazard: Dangerous, by spont chemical reaction.

Spontaneous Heating: Spont flam in air.

Explosion Hazard: Severe, when shocked or exposed to heat.

Disaster Hazard: Dangerous; shock will explode it; when heated to decomp, it emits highly toxic fumes of phosgene; can react vigorously with oxidizing materials.

cis-**ACETYLENE DICHLORIDE.** See cis-dichloro ethylene.

trans-**ACETYLENE DICHLORIDE.** See trans-dichloro ethylene.

ACETYLENE DIUREINE. Syn: *glycol uril*. White, needle-like crystals. NHCONHCNCHNHCONH, mw: 142.12, d: 1.599.

THR = U.

Fire Hazard: Slight.

ACETYLENE TETRABROMIDE. Syn: *tetrabromo ethane*. Colorless to yellow liquid. $CHBr_2CHBr_2$, mw: 345.70, bp: 151° @ 54 mm, fp: −1°, d: 2.9638 @ 20°/4°, autoign. temp.: 635°F.

Acute tox data: Oral LD_{50} (rabbit) = 400 mg/kg, oral LD_{50} (guinea pig) = 400 mg/kg. [*3*]

THR = HIGH via oral and inhal routes. It is irr and narcotic.

Fire Hazard: Low.

To Fight Fire: Water foam, fog, CO_2, dry chemical.

Disaster Hazard: Dangerous; when heated, it emits highly toxic fumes of carbonyl bromide.

ACETYLENE TETRACHLORIDE. Syn: *tetrachloroethane-1,1,2,2*. Heavy, colorless, mobile liquid, chloroform-like odor. $CHCl_2CHCl_2$, mw: 167.86, mp: −43.8°, bp: 146.4°, d: 1.600 @ 20°/4°.

Acute tox data: Oral LD_{50} (rat) = 200 mg/kg; inhal LC_{LO} (rat) = 1000 ppm for 4 hrs; dermal LD_{50} (rabbit) = 3990 mg/kg. [*3*]

THR = HIGH via oral and inhal routes; MOD via dermal route. This is generally considered the most toxic of the common chlorinated HC. It has a fairly strong irritant action on mu mem of the eyes and upper respiratory tract; a conc of 3 ppm produces a detectable odor. There is thus an initial warning effect. Its narcotic action is stronger than that of chloroform, but because of its low volatility, narcosis is less severe and much less common in industrial poisoning than in the case of other chlorinated HC. The toxic action of this material is chiefly on the liver, where it produces acute yellow atrophy and cirrhosis. Fatty degeneration of the kidneys and heart, hemorrhage into the lungs and serous membranes, and edema of the brain have also been found in fatal cases. Some reports indicate a toxic action on the CNS, with changes in the brain and in the peripheral nerves. The effect on the blood is one of hemolysis, with appearnce of young cells in the circulation and a monocytosis. Due to its solvent action on the natural skin oils, dermatitis is not uncommon (Section 9).

The initial symptoms resulting from exposure to the vapor are lacrimation, salivation and irritation of the nose and throat. Continued exposure to high concentrations results in restlessness, dizziness, nausea and vomiting and narcosis. The latter, however, is rare in industry. More commonly, the exposure is less severe, and the complaints are vague and referable to the digestive and nervous systems. The patient's complaints gradually progress to serious illness, with development of a toxic jaundice, liver tenderness, etc., and possibly albuminuria and edema. With serious liver damage the jaundice increases and toxic symptoms appear, with somnolence, delirium, convulsions and coma usually preceding death.

This material is considered to be a very severe industrial hazard and its use has been restricted or even forbidden in certain countries.

Explosion Hazard: Reacts violently with N_2O_4, 2,4-dinitrophenyl disulfide [*19*] and contact with sodium or potassium. When heated in contact with solid potassium hydroxide, a spont flam gas is evolved. Any water can cause appreciable hydrolysis even at room temp. and both hydrolysis and oxidation become comparatively rapid above 110°.

Disaster Hazard: Dangerous; when heated, emits highly toxic decomp products.

N-ACETYL ETHANOLAMINE. See hydroxy ethyl acetamide.

N-ACETYLETHYLCARBAMATE. $C_5H_9O_3N$, mw: 131.2.

THR = An exper neo. [*3*]

ACETYLETHYLENE IMINE. C_4H_7ON, mw: 85.1.

THR = HIGH. An exper neo. [*3, 23*]

N-ACETYL ETHYL NITROSO UREA. $C_5H_9O_3N_3$, mw: 159.2.

THR = An exper carc. [*23*]

ACETYL ETHYL TETRAMETHYL TETRALIN. Syn: *AETT.* White crystals.

Acute tox data: TD_{LO} (rats) = 3 mg/kg. [*112*]

THR = Exposure causes blue coloration of internal organs. It is slowly metabolized and excreted via feces. Exposure ⟶ CNS effects i.e. hyperexcitability, tremors, lack of coordination, hunched backs and loss of weight. Symptoms persist for 90 days after exposure stops. Severity of symptoms seem proportional to length of exposure. It is freely absorbed via human skin. It is HIGHly toxic.

ACETYL GLYCOLIC ACID ETHYL ESTER. Liquid. $C_6H_{10}O_4$, mw: 146, bp: 184–189°, flash p: 349°F, d: 1.094.

THR = MOD via oral and inhal routes.

Fire Hazard: Slight, when exposed to heat or flame; can react vigorously with oxidizing materials.

ACETYL HYDRAZINE. $C_2H_6ON_2$, mw: 74.1.

Acute tox data: ip LD_{50} (mouse) = 153 mg/kg. [*3*]

THR = HIGH via ip route. Possible liver damage from chronic exposure. Large doses cause haemolysis. See also phenyl hydrazine.

ACETYL HYDROPEROXIDE. See peracetic acid (40% solution).

ACETYLIDES

THR = See individual compounds.

Fire Hazard: U.

Explosion Hazard: Severe, when shocked or exposed to heat. Acetylides are very sensitive to shock, friction and heat. They explode readily and are one of the few commercial explosives which contain no O_2 or N_2. Their explosion produces no gas but simply is an effect of the large amount of heat, instantaneously produced. Acetylides are used for detonating compositions, or in combination with lead azide in detonating rivets where the acetylides reduce the flash point of the more insensitive azides. They are in a class with the fulminates and the azides as primary detonants.

Because these materials are so sensitive to shock and temperature, they must be handled with extreme care. They must be kept cool, and if they are to be stored, should be kept wet. (See fulminates for suggested precautions in storage and handling of acetylides). Metal powders, such as finely divided Cu or Ag should not be stored or kept with acetylene or acetylides, because it is possible for them to react with these metal powders to form very sensitive acetylides which, while they are not dangerous in themselves, can cause enough of a flash to ignite a possibly explosive mixture of gases and thus cause an explosion in a warehouse or storage area (For destruction of acetylides, see Section 7). Examples of commercially used acetylides are silver acetylide and copper acetylide.

See acetylene.

N-ACETYL IMIDAZOLE. $C_5H_6ON_2$, mw: 110.1.

THR = An exper neo. [*3*]

ACETYL IODIDE. Syn: *ethanoyl iodide*. Brown, transparent, fuming liquid, CH_3COI, mw: 169.96, bp: 108°, d: 2.067 @ 20°/4°.

THR = HIGH via oral, inhal routes; irr to eyes, skin and mu mem. [*3*]

Disaster Hazard: Dangerous, see iodides; will react with water or steam to produce toxic or corrosive fumes.

1-ACETYL-2-ISONICOTINOYL HYDRAZINE. $C_8H_9O_2N_3$, mw: 179.2.
THR = An exper carc. [*3*]

ACETYL KETENE. See diketene.

ACETYL METHYL CARBINOL. See acetoin.

3-ACETYL-1-METHYL-1-NITROSO UREA. $C_4H_7O_3N_3$, mw: 145.1.
THR = An exper carc via oral route. [*3, 23*]

2-ACETYL-3-METHYL TIOPHEN-4-ONE. $C_7H_8O_2S$, mw: 156.2.
THR = Unstable. Heat can explode it. [*19*]

3-ACETYL-6-METHYL-1,2-PYRAN-2,4(3H)DIONE. See dehydroacetic acid.

N-ACETYL MORPHOLINE. Liquid.
$CH_3CONCH_2CH_2OCH_2CH_2$, mw: 129.16, mp: 14°, bp: decomp, flash p: 235°F (OC), d: 1.1164, vap. press: 0.02 mm @ 20°, vap. d: 4.46.
Acute tox data: pa LD_{LO} (mouse) = 2400 mg/kg. [*3*]
THR = MOD via pa route. See also morpholine.
Fire Hazard: Slight, when exposed to heat or flame; can react vigorously with oxidizing materials.
To Fight Fire: Alcohol foam.

N-ACETYL-2-NAPHTHYL HYDROXYLAMINE. $C_{12}H_{11}O_2N$, mw: 201.2.
THR = An exper neo. [*3*]
To Fight Fire: Water, alcohol foam, mist, water spray, CO_2, dry chemical.

ACETYL NITRATE. Colorless, fuming, mobile liquid. CH_3COONO_2, mw: 105.1, bp: 22° @ 70 mm, d: 1.24 @ 15°/4°.
THR = HIGH tox irr and corrosive material.
Explosion Hazard: Very dangerous. Must NOT be strongly heated (> 60°). Must be stored in the dark. HgO explodes it; sometimes glass surfaces explode it.

ACETYL (*p*-NITROPHENYL)SULFANILAMIDE. Syn: *sulfanitran*. $C_{14}H_{13}N_3O_5S$, mw: 335, mp: 260°–261°.
THR = U. A food additive permitted in the feed and drinking water of animals and/or for the treatment of food producing animals. [*109*]
Disaster Hazard: Dangerous; when heated to decomp, it can give off highly toxic fumes.

ACETYL OXIDE. See acetic anhydride.

11-ACETYL-17-OXO-16,17-DIHYDRO-15H-CYCLO PENTA(*a*)-PHENANTHRENE.
THR = An exper carc. [*23*]

ACETYL PEROXIDE. Syns: *ethanol peroxide, diacetyl peroxide*. Solid or colorless crystals, or liquid. $(CH_3CO)_2O_2$, mw: 118.1, mp: 30°, bp: 63° @ 21 mm, d: 1.18.
Acute tox data: TD_{LO} (mouse) = 280 mg/kg. [*3*]
THR = HIGH irr to eyes, skin and mu mem via oral and inhal routes. An exper neo. [*3, 23*]
Fire Hazard: Dangerous, by spont chemical reaction. A powerful oxidizing agent; can cause ignition of organic materials on contact.
Explosion Hazard: Severe, when shocked or exposed to heat. It may explode spont, possibly when more than 24 hrs old; it should be used up as soon as prepared.
Disaster Hazard: Highly dangerous; shock will explode it; it will react with water or stream to produce heat; can react vigorously with reducing materials; can emit toxic fumes on contact with acid or acid fumes.
To Fight Fire: CO_2, dry chemical.
Storage and Handling: Must be kept below 27° and not warmed over 30°. Do not add to hot materials. Do not add accelerator to this material. Store in original container with vented cap. Avoid bodily contact. See peroxides, organic, also Section 7. This material is nearly always stored and handled as a 25% solution in an inert solvent. See also acetyl peroxide 25% solution (in dimethyl phthalate).

ACETYL PEROXIDE 25% SOLUTION (IN DIMETHYL PHTHALATE). Syn: *diacetyl peroxide solution*. Crystal clear liquid. mp: −7°, flash p: 113°F (OC), d: 1.18 @ 20°.
Acute tox data: ip LD_{LO} (mouse) = 150 mg/kg. [*3*]
THR = HIGH via ip route. See acetyl peroxide, also peroxides, organic.
Fire Hazard: Mod when exposed to heat or flame, or by spont chemical reaction. An oxidizing agent.
Disaster Hazard: Mod dangerous; when heated to decomp, emits toxic fumes; can react vigorously with oxidizing materials.
To Fight Fire: Foam, CO_2.

ACETYL PHENOL. See phenyl acetate.

ACETYL-*p*-PHENYLENE DIAMINE. See *p*-amino acetanilide.

1-ACETYL-2-PHENYL HYDRAZINE. $C_8H_{10}ON_2$, mw: 150.2, mp: 129°.
THR = U.
Disaster Hazard: Mod dangerous; when heated to decomp may emit toxic fumes.

1-ACETYL-2,2-PICOLINOYL HYDRAZINE. $C_8H_9O_2N_3$, mw: 179.2.
THR = HIGH. An exper carc. [*3*]

ACETYL SALICYLIC ACID. See acetol.

ACETYL TRIBUTYL CITRATE. Odorless powder. $C_{20}H_{34}O_8$, mw: 402.5, bp: 172–174° @ 1 mm, flash p: 400° F (COC), d: 1.048 @ 25°/25°.
THR = LOW via irr, oral and inhal routes.
Disaster Hazard: Slight; when heated, it emits acrid fumes.

ACETYL TRIETHYL CITRATE. $C_{12}H_{20}O_7$, mw: 276.3.
Acute tox data: Oral LD (rat) = 5 g/kg. [*3*]
THR = MOD via oral route.

ACETYL TRI-2-ETHYL HEXYL CITRATE. Liquid. $C_{32}H_{58}O_8$, mw: 570.8, bp: 225° @ 1 mm, flash p: 430° F (COC), d: 0.983 @ 25°/25°.
THR = U. Probably LOW.
Fire Hazard: Slight, when exposed to heat or flame; can react with oxidizing materials.

ACID BARIUM OXALATE. See barium binoxalate.

ACID BRILLIANT GREEN BS. $C_{27}H_{26}O_7N_2S_2$, mw: 577.7.
THR = An exper carc. [*3*]

ACID BUTYL PHOSPHATE. Syns: *n-butyl acid phosphate*, *n-butyl phosphoric acid*. Water white liquid, sol in alcohol, acetone, and toluene, insol in water, petroleum and naphtha. $C_4H_{10}O_4P$, mw: 153.1, d: 1.120–1.125 @ 25°/40°, flash p: 230° F (COC).
THR = A toxic and corrosive, combustible liquid. See phosphoric acid.
Fire Hazard: Slight when exposed to heat or flame.
Disaster Hazard: Dangerous; when heated to decomp, emits highly toxic fumes. See phosphoric acid.

ACID CALCIUM PHOSPHATE. See calcium phosphate, monobasic.

ACID CARBOYS, EMPTY.
Warning: These containers may contain conc vapors or even some liquid acid remaining from their original contents. Therefore, they can give rise to all the hazards of their original contents.

ACID ETHYL SULFATE. See ethyl sulfuric acid.

ACID FAST VIOLET 5 BN. $C_{39}H_{41}O_6N_3S_2$, mw: 736.
THR = An exper carc [*3*] in rats via oral, sc routes. [*108*]

ACID GREEN B. $C_{37}H_{36}O_6N_2S_2$, mw: 690.8; a dark green powder.
THR = An exper carc [*3*] in rats via oral, sc routes. [*108*]

ACID LEATHER BLUE R. $C_{26}H_{19}O_{10}S_3N_3$, mw: 698.6.
THR = HIGH.

ACIDS, LIQUID, N.O.S. See corrosive liquids.

ACID, SPENT, or **ACID, SLUDGE.** See nitric acid, sulfuric acid.

ACONITE. Syns: *monkshood*, *aconitum*, *wolfsbane*. The dried tuberous root of Aconitum Napellus, composed of several alkaloids, the chief one being aconitine.
THR = HIGH via all routes. The poisonous alkaloid can be absorbed through the skin sufficiently to cause death. Usually causes pupils to dilate, but may cause them to be contracted. A lethal dose causes diminution in force and frequency of pulse, a cold and clammy skin and a tingling and numbness of the mouth, face and throat. Somewhat larger doses cause burning of the throat and stomach, increased salivation, nausea, retching and vomiting, grinding of the teeth and extension of the numbness and tingling to other parts of the body. There is difficulty in swallowing and speaking and pain in eyes and head. The fatal period is from 8 minutes to 3 or 4 hours.
Fire Hazard: Slight, when exposed to heat or flame.
Disaster Hazard: Dangerous; when heated to decomposition, it emits highly toxic fumes.
Treatment and Antidotes: Wash stomach with tannic acid or powdered charcoal. Heart stimulants, such as strong coffee or caffeine. Artificial respiration and oxygen if necessary. Keep patient warm. Call a physician.

ACONITIC ACID. Syn: *1,2,3-propene tricarboxylic acid*. White crystalline powder. $C_6H_6O_6$, mw: 174.11, mp: decomp.
THR = U. A synthetic flavoring substance and adjuvant. [*109*]

ACONITINE. Syn: *acetyl benzoyl aconine*. White crystalline alkaloid; feeble bitter taste. $C_{34}H_{49}NO_{11}$, mw: 645.7, mp: 204°.
Acute tox data: sc LD_{50} (mouse) = 0.295 mg/kg. [*3*] Oral LD_{50} (mouse) = 1 mg/kg; iv LD_{50} (mouse) = 0.166 mg/kg; ip LD_{50} (mouse) = 0.328 mg/kg. [*2*]
THR = HIGH via all routes. Both the amorphous and crystalline forms are about equally highly toxic. This intensely poisonous alkaloid can be absorbed through the skin to cause death. Ingestion in small quantities may affect the eyes and cause blindness. Its toxicity is about equal to that of aconite.
Disaster Hazard: Dangerous; when heated to decomposition, it emits highly toxic fumes.

ACONITINE HYDROBROMIDE. Hexagonal crys-

tals. $C_{34}H_{47}NO.HBr$, mw: 726.4, mp: 200°–207°. [*2*]
Acute tox data: iv LD_{50} (mouse) = 0.232 mg/kg. [*3*]
THR = HIGH via all routes. A very powerful poison.

ACONITINE HYDROCHLORIDE. Crystals. $C_{34}H_{47}NO.HCl$, mw: 672, mp: 151°.
THR = A very powerful poison. See also aconitine hydrobromide. [*2*]

ACONITINE NITRATE. Crystals. $C_{34}H_{47}NO_{11}HNO_3 \cdot 5H_2O$, mw: 798.8, mp: 200° (decomp).
THR = Very poisonous. See aconitine.
Fire Hazard: Mod, by spont chemical reaction; an oxidizing agent.
Disaster Hazard: Dangerous; when heated, emits highly toxic fumes of oxides of N_2, can react vigorously with reducing materials.

ACONITUM. See aconite.

ACONITUM FEROX. Syns: *bish*, *visha*, *Indian aconite*.
THR = Most powerful poison of the aconites. See also aconite.
Disaster Hazard: Dangerous; when heated to decomp, it emits highly tox fumes.

ACRIDINE. Small colorless needles. $C_{13}H_9N$, mw: 179.21, mp: 110.5°, bp: 346°, d: 1.005 @ 19.7°/4°, vap. press: 1 mm @ 129.4°.
Acute tox data: Oral LD_{50} (rat) = 2000 mg/kg; oral LD_{50} (mouse) = 500 mg/kg; sc LD_{50} (mouse) = 400 mg/kg; iv LD_{50} (dog) = 90 mg/kg; iv LD_{50} (rabbit) = 100 mg/kg. [*3*]
THR = MOD–HIGH via oral; HIGH via sc, iv routes. Occasional injuries stem from its industrial use either as a solid or as a vapor. Is strongly irritating to the skin and mucous surfaces. Upon inhalation it causes sneezing, itching or even violent burning of the skin; sometimes even inflammatory swelling. It is regarded as the effective irritant in tar and creosote or pitch, etc., which can sensitize the skin to light.
Disaster Hazard: Moderately dangerous; when heated to decomposition, emits toxic fumes.

ACRIDINE MUSTARD ICR 191. $C_{19}H_{21}ON_3Cl_2$, mw: 378.2.
THR = An exper carc. [*23*]

ACRIDINE ORANGE. Syn: *3,6-bis(dimethyl amino) acridine*. $C_{17}H_{19}N_3$, mw: 265.4, mp: 182°.
THR = An exper neo [*3*, *23*] and mutagen. [*108*]

ACRIDINE RED. $C_{15}H_{14}ON_2HCl$, mw: 274.8.
THR = An exper carc. [*3*]

ACRIFLAVIN. $C_{14}H_{14}N_3Cl$, mw: 259.8.
THR = HIGH. An exper mutagen. [*3*]

ACROLEIC ACID. See acrylic acid.

ACROLEIN. Syns: *propenal*, *acrylic aldehyde*, *allyl aldehyde*; *acraldehyde*. Colorless or yellowish liquid, disagreeable choking odor. CH_2CHCHO, mw: 56.06, mp: −87.7°, bp: 52.5°, flash p: <0°F, d: 0.841 @ 20°/4°, autoign. temp.: unstable (455°F), lel = 2.8%, uel = 31%, vap. d: 1.94.
Acute tox data: Oral LD_{50} (rat) = 46 mg/kg; dermal LD_{50} (rabbit) = 562 mg/kg. [*3*]
THR = HIGH via oral and MOD via dermal routes. Due to its extreme lachrymatory effect it serves as its own warning agent. Industry records one fatality ascribed to exposure to it formed by the heat of welding in an enclosed space. It affects particularly the membranes of the eyes and respiratory tract. It is a weak sensitizer.
Fire Hazard: Dangerous, when exposed to heat, flame or oxidizers.
Spontaneous Heating: No.
Explosion Hazard: Can react violently with acids, alkalis, amines, SO_2, thiourea, metal salts, oxidants, (light + heat). [*19*]
Disaster Hazard: Dangerous; when heated to decomp, emits highly tox fumes; can react vigorously with oxidizing materials.
To Fight Fire: CO_2, dry chemical or alcohol foam.

ACROLEIN DIMER. Syn: *2-formyl-3,4-dihydro-2H-pyran*. Liquid, sol in water.
OCH:CHCH$_2$CH$_2$CHCHCHO (ring), mw: 112, d: 1.0775 (20°), bp: 151.3°, fp: −100°, flash p: 118°F (OC).
Acute tox data: Oral LD_{50} (rat) = 4920 mg/kg. [*3*]
THR = MOD via oral route.
Fire Hazard: Mod flam when exposed to heat, flame or powerful oxidizing agents.
To Fight Fire: Alcohol foam and multipurpose dry chemical.

ACRYLAMIDE. White crystalline solid.
$CH_2CHCONH_2$, mw: 71.08, mp: 84.5 ± 0.3°, bp: 125° @ 25 mm, d: 1.122 @ 30°, vap. press: 1.6 mm @ 84.5°, vap. d: 2.45.
Acute tox data: Oral LD_{50} (rat) = 170 mg/kg; dermal LD (rabbit) = 1000 mg/kg. [*3*,*97*]
THR = HIGH via oral and MOD via dermal routes. Intoxication from it has caused a peripheral neuropathy, erythema and peeling of the palms. [*42*] In industry, intoxication from it is mainly via dermal, next via inhal and last via oral routes. Time of onset varied from 1–24 mos to 8 yrs. Symptoms were, via dermal route, a numbness, tingling and touch tenderness. In a couple of weeks, coldness of extremities; later, excessive sweating, bluish-red and peeling of palms, [*97*] marked fatigue and limb-weakness. This material is dangerous because it can

be absorbed through the unbroken skin. From animal experiments it would seem that its effect is toxic upon the CNS. Adult rats fed an average of 30 mg/kg of this material for 14 days were all partially paralyzed and had reduced their food consumption by 50%.

Disaster Hazard: Mod, when heated to decomp, emits acrid fumes.

ACRYLAMIDE-ACRYLIC ACID RESIN. Produced by the polymerization of acrylamide with partial hydrolysis; or by copolymerization of acrylamide and acrylic acid.

THR = U. A food additive permitted in food for human consumption. [*109*]

ACRYLIC ACID. Syns: *propene acid, acroleic acid.* Liquid, acrid odor. C_2H_3COOH, mw: 72.06, mp: 14°, bp: 141°, d: 1.062, vap. press: 10 mm @ 39.9°, flash p: 130°F (OC), vap. d: 2.45.

Acute tox data: Oral LD_{50} (rat) = 340 mg/kg. [*3*] An intense irr, dermal LD_{50} (rabbit) = 280 mg/kg. [*3*]

THR = HIGH via oral and dermal routes. An exper teratogen. [*3*]

Fire Hazard: Mod, when exposed to heat or flame.

Explosion Hazard: Mod by spont chemical reaction; it polymerizes violently upon contact with amines, NH_3, oleum, chlorosulfonic acid. [*19*]

Disaster Hazard: Dangerous; when heated to decomp, emits highly toxic fumes; it can react with oxidizing materials.

To Fight Fire: Alcohol foam.

ACRYLIC ACID METHYL ESTER. See methyl acrylate.

ACRYLIC ALDEHYDE. See acrolein.

ACRYLONITRILE. Syns: *propene nitrile, vinyl cyanide.* Colorless, mobile liquid; mild odor. CH_2CHCN, 53.06, mp: −82°, bp: 77.3°, fp: −83°, flash p: 30°F (TCC), lel = 3.1%, uel = 17%, d: 0.806 @ 20°/4°, autoign. temp.: 898°F, vap. press: 100 mm @ 22.8°, vap. d: 1.83.

THR = HIGH. See cyanides and nitriles. Acrylonitrile closely resembles hydrocyanic acid in its toxic action. By inhibiting the respiratory enzymes of tissue, it renders the tissue cells incapable of oxygen absorption. Poisoning is acute; there is little evidence of cumulative action on repeated exposure.

Acute tox data: Oral LD_{50} (rat) = 82 mg/kg; dermal LD_{50} (rabbit) = 280 mg/kg. [*3*]

THR = HIGH via oral and dermal routes. An exper carc. [*3*] (S) human carc. [*3, 14, 40*] Exposure to low conc is followed by flushing of the face and increased salivation; further exposure results in irritation of the eyes, photophobia, irritation of the nose, deepened respiration, and, if exposure continues, shallow respiration, nausea, vomiting, weakness, an oppressive feeling in the chest, and occasionally headache and diarrhea are other complaints. Several cases of mild jaundice accompanied by mild anemia and leucocytosis have been reported. Urinalysis is generally negative, except for an increase in bile pigment. Serum and bile thiocyanates are raised. See also hydrocyanic acid.

Fire Hazard: Dangerous when exposed to heat, flame or oxidizers.

Explosion Hazard: Mod when exposed to flame, can react violently with strong acids, amines, strong alkalis, Br_2, 1,2,3,4-tetrahydrocarbazole. [*19*]

Disaster Hazard: Dangerous; see cyanides; can react vigorously with oxidizing materials.

To Fight Fire: CO_2, dry chemical or alcohol foam.

2-ACRYLOXY ETHYL DIMETHYL SULFONIUM METHYL SULFATE.

THR = U. Limited animal exper suggest MOD.

Disaster Hazard: Dangerous; when heated to decomp, emits highly toxic fumes.

ACTIDIONE. See cycloheximide.

ACTINIC RADIATION. Outdoor workers such as fishermen, sailors, soldiers and farmers show a high incidence of skin cancer. The commonest acute manifestation of actinic radiation effects on skin is sunburn.

ACTINIUM. Brownish-red granular powder. Ac, at wt: 227, mp: 1050°, bp: 3200° (±300°), d: 10.1.

Radiation Hazard: Section 5A, Table 5A.5. Natural isotope ^{227}Ac, $T\frac{1}{2}$ = 22y. Decays to radioactive ^{227}Th via β-emission of 0.05 meV. Natural isotope ^{228}Ac (mesothorium$_2$). $T\frac{1}{2}$ = 6.1 h. Decays to radioactive ^{228}Th via β-emission of 0.45–2.2 MeV. Also emits γ's of 0.06–0.13 meV.

ACTINOMYCIN A.

THR = An exper carc. [*23*]

ACTINOMYCIN D. $C_{62}H_{86}O_{16}N_{12}$, mw: 1255.6.

THR = An exper (+) carc, neo and teratogen. See Plant and Fungal Products. Also HIGH via oral route. [*3, 7, 23*] An antibiotic, used medically as an anti-neoplastic agent.

ACTINOMYCIN L.

THR = An exper (+) carc and neo. [*3, 7, 23*]

ACTINOMYCIN S.

THR = An exper (+) carc. [*3, 7, 23*]

ACTIVATED CARBON. Syn: *charcoal, activated.* Black amorphous mass. C, at wt: 12.01, mp: 3500°, bp: 4000°, d: 3.51.

THR = Pure C is non-toxic. However, activated C, usually made from organic matter, is frequently found associated with small amounts of irr and possibly toxic impurities. Avoid wetting and subsequent drying in storage. Store with ventilation. See also coal tar.

Fire Hazard: Slight, when exposed to heat or flame.

Spontaneous Heating: Yes.

Explosion Hazard: Slight, when dust is exposed to flame. See dust explosions.

To Fight Fire: Water.

ADAMSITE. See diphenylamine chloroarsine.

ADC RHODAMINE B. $C_{28}H_{31}O_3N_2 \cdot Cl$, mw: 478.1, mp: 165°.

THR = An exper neo [*3*] in rats via sc route. [*108*]

ADDITIONAL INFLUENCES ON TOXICITY, PARTICULARLY CARCINOGENICITY. Section 8.

ADENINE. $C_5H_5N_5$, mw: 135.2.

THR = An exper teratogen. [*3*]

ADENINE-1-OXIDE. $C_5H_5ON_5$, mw: 151.2.

THR = An exper neo. [*3, 23*]

ADHESIVE PLASTER. Syn: *adhesive tape.*

THR = The reaction may be of allergic nature, manifested as itching and burning of the skin with redness and blister formation.

Fire Hazard: Slight; when heated, emits acrid fumes.

ADHESIVE TAPE. See adhesive plaster.

ADHESIVES, N.O.S. See cement, liquid, n.o.s.

ADIPIC ACID. Syns: *hexane dioic acid, 1,4-butane dicarboxylic acid.* Fine white crystals or powder. $C_6H_{10}O_4$, mw: 146.14, mp: 152°, bp: 337°, flash p: 385°F (CC), d: 1.360 @ 25°/4°, vap. press: 1 mm @ 159.5°, vap. d: 5.04, autoign. temp.: 788°F.

Acute tox data: Oral LD_{50} (mouse) = 1900 mg/kg. [*3*]

THR = MOD via oral route. A general purpose food additive. [*109*]

Fire Hazard: Slight, when exposed to heat or flame; can react with oxidizing materials.

To Fight Fire: Water, foam, CO_2, dry chemical.

ADIPONITRILE. Syns: *1,4-dicyanobutane, tetramethylenecyanide.* Water white liquid, practically odorless. $CN(CH_2)_4CN$, mw: 108.14, mp: 2.3°, bp: 295°, flash p: 199.4°F (OC), d: 0.965 @ 20°/4°, vap. d: 3.73.

Acute tox data: ip LD_{50} (mouse) = 40 mg/kg; sc LD_{50} (guinea pig) = 50 mg/kg; oral LD_{50} (rat) = 300 mg/kg. [*3*]

THR = HIGH via oral, inhal routes. It is toxic since the nitrile group will behave as a cyanide when ingested or absorbed in the body. It produces disturbances of the respiration and circulation, irritation of the stomach and intestines, and loss of weight. Its low vapor pressure at room temperature makes exposure to harmful concentrations of its vapors unlikely if handled with reasonable care in well ventilated areas. Combustion products of adiponitrile may contain hydrocyanic acid. Accordingly, fires involving this material may be hazardous from a toxicological standpoint. See also hydrocyanic acid and nitriles.

Fire Hazard: Mod, when exposed to heat or flame.

Disaster Hazard: Dangerous; when heated to decomposition, it emits highly toxic fumes; can react with oxidizing materials.

To Fight Fire: Foam, CO_2, dry chemical.

"ADRENALINE." Syn: *epinephrine.* Light brown or nearly white crystals. $C_9H_{13}O_3N$, mw: 183.3, mp: 211–212°.

Acute tox data: Oral LD_{50} (mouse) = 50 mg/kg; iv LD_{50} (rat) = 1 m/kg. [*3*]

THR = HIGH via oral and iv routes. May cause pallor, tremor, anxiety, nervousness, rapid, forceful pulse, rise in blood pressure and temperature, rapid breathing, dilation of the pupils. Allergic skin reactions have been known to result from contact causing contact dermatitis.

Fire Hazard: Slight; when heated.

Treatment and Antidotes: Guanine or atropine may be helpful. Usually the symptoms are of short duration and clear up spont.

ADRIAMYCIN. $C_{25}H_{27}O_{11}N.HCl$, mw: 554.

THR = HIGH to humans via iv route. An exper (±) neo. [*3, 7*] Causes baldness, stomatitis and bone-marrow aplasia. Also reports of fatal cardiac disturbance.

AEROSOL IB. See diisobutyl sodium sulfosuccinate.

AFLATOXICOL.

THR = An exper carc. [*23*]

AFLATOXIN. Mixture of B1 and G1.

THR = An exper mutagen, neo and carc. [*3*]

AFLATOXIN B1. Toxic metabolite of the fungus Aspergillus flavus. A crystalline material. $C_{17}H_{12}O_6$, mw: 312.3, mp: 268°.

Acute tox data: Oral LD_{50} (dogs) = 0.5 mg/kg.

THR = An exper carc, teratogen, neo. [*3, 23*] EXTREMELY tox via oral route.

AFLATOXIN B2. $C_{17}H_{14}O_6$, mw: 314.3.

THR = HIGH via oral route. An exper carc. [*3, 23*]

AFLATOXIN G1. $C_{17}H_{12}O_7$, mw: 328.3.

THR = Deadly oral poison to rats. An exper carc and (S) human carc. [*3, 23, 103*]

AFLATOXIN G2. $C_{17}H_{14}O_7$, mw: 330.3.

THR = Deadly oral poison (animal). An exper carc. [*3, 23, 103*]

AFLATOXIN B2a.
THR = An exper cars. [*23*]

AFLATOXIN G 2a.
THR = An exper carc. [*23*]

AFLATOXIN GM 1.
THR = An exper carc. [*23*]

AFLATOXIN GM 2a.
THR = An exper carc. [*23*]

AFLATOXIN M2.
THR = An exper carc. [*23*]

AFLATOXIN P.
THR = An exper carc. [*23*]

AFLATOXIN Q.
THR = An exper carc. [*23*]

AFLATOXIN B1,2,3-DICHLORIDE. $C_{17}H_{12}O_6Cl_2$, mw: 383.2.
THR = An exper neo and carc. [*3*]

AFLATOXIN M1. $C_{17}H_{12}O_7$, mw: 328.3.
THR = An exper (S) neo, and carc. [*3, 9, 23*]

AFRIDOL BLUE. $C_{32}H_{18}O_{14}N_6S_4Na_4$, mw: 1001.7.
THR = An exper teratogen. [*3*]

AGALMATOLITE. See pyrophyllite.

AGAR. See agar-agar.

AGAR-AGAR. Syns: *agar*, *gelose*, *Japanese*, *Bengal*, *Ceylon or Chinese isinglass or gelatin*, *macassar gum*. Unground-in thin, translucent membranous pieces; ground-pale, buff powder, sol in boiling water, insol in cold water and organic solvents.
THR = LOW. A stabilizer food additive. [*109*]

AGE. See allyl glycidyl ether.

AGERITE WHITE. See *p*-benzyl hydroquinone.

AGGLUTININ. See ricin.

AIRBORNE RADIOACTIVITY. See Sections 5A and 5B.

AIR CLEANING DEVICES. See Section 2.

AIR CONTAMINATING INDUSTRIAL EMISSIONS. See Section 2.

AIR (LIQUID AND COMPRESSED). Bluish, mobile liquid. $O_2 + N_2$, bp: −189° (liq), flash p: none, autoign. temp.: none.
THR = Liquid air at atmospheric pressure exists at so low a temperature that contact with it will destroy tissue. Personnel exposed to compressed air may develop caisson disease (the bends, the chokes) if decompression is too rapid.
Explosion Hazard: Mod when containers under pressure are shocked or exposed to heat or flame or flam materials; i.e., ethyl ether, hydrocarbons or charcoal, which have been in contact with liquid air may explode very easily. Ordinary oxidation is greatly accelerated in compressed air.
Disaster Hazard: Mod dangerous; can react vigorously with reducing materials.

AIREDALE BLUE D. $C_{34}H_{28}O_{16}N_6S_4$, mw: 996.9.
THR = An exper teratogen. [*3*]

AIREDALE BLUE FFD.
THR = An exper teratogen. [*3*]

AIRPLANE FLARES OR AEROPLANE FLARES. See aeroplane flares.

AIR POLLUTANTS AS CARCINOGENS. See Section 8.

AKLOMIDE. See 2-chloro-4-nitrobenzamide.

ALABANDITE. See manganous sulfide.

ALAMOSITE. See lead-*m*-silicate.

ALANAP. See *n*-1-naphthyl phthalamic acid.

ALANINE. Syns: *α-alanine*, *α-amino propionic acid*, *2-amino propanoic acid*. A naturally occurring nonessential amino acid; colorless crystals, sol in water, slightly sol in alcohol, insol in ether, occurs in *dl*, *l* and *d* forms, the data here refer to the l and *dl* forms. $C_3H_7O_2N$, mw: 89, mp: (*dl*) 295° (decomp) (*l*) 297° (decomp).
THR = U. A nutrient and dietary supplement food additive. [*109*]

α-ALANINE. See alanine.

ALCOHOL, DENATURED. Syn: *denatured spirits*. Liquid. Composed of alcohol and denaturants.
THR = This will depend upon the alcohol in question (generally ethyl alcohol) plus the denaturants used. Methyl alcohol is a common denaturant.
Fire Hazard: Dangerous; can react vigorously with oxidizing materials.
Explosion Hazard: Mod; see ethyl alcohol.

ALCOHOLS, N.O.S. (See specific compound).
A generic term applied to a series of compounds, the simplest of which has the general formula $C_nH_{2n+1}OH$.
THR = No general statement can be made as to the toxicity of alcohols due to wide differences in toxicity effects. When the term "alcohol" is used alone it usually applies to ethyl alcohol (C_2H_5OH), which is relatively nontoxic when compared with methyl and some other alcohols.
Fire Hazard: Dangerous when exposed to heat or flame. Can react violently in contact with (H_2O + H_2SO_4), HOCl, Cl_2, isocyanates, $LiAl_4$, N_2O_4, $HClO_4$, H_2SO_5 (Caro's acid), $Ba(ClO_4)_2$, $(CH_2)_2O$, acetaldehyde, diethyl aluminum bromide, hexamethylene diisocyanate, triisobutyl aluminum. [*2, 19*]

ALCOHOL, TERTIARY. See specific compound.

ALDEHYDE AMMONIA. Syn: *acetaldehyde ammonia.* White crystalline solid. $CH_3CH(NH_2)OH$, mw: 61.08, bp: 110°, mp: 97°.

THR = MOD irr to skin, eyes and mu mem via oral and inhal routes.

Fire Hazard: Mod, when exposed to heat or flame; readily decomp into acetaldehyde and ammonia when heated.

Explosion Hazard: Mod, when exposed to heat or flame.

Explosive Range: See ammonia.

Disaster Hazard: Mod dangerous when heated to decomp, emits toxic fumes, can react with oxidizing materials.

ALDEHYDE C-8. See caprylaldehyde.

ALDEHYDE C-16. See ethyl methyl phenyl glycidate.

ALDEHYDES. See also specific compounds. All the aldehydes possess anesthetic properties, but this is obscured by their highly irr action on the eyes and mu mem of the respiratory tract. The lower aldehydes, very sol in water, act chiefly on the eyes and tissues of the upper respiratory tract. The higher aldehydes, less sol in water, tend to penetrate more deeply into the respiratory system and may affect the lungs. The above toxicity hazard rating is more accurate for the lower molecular weight aldehydes. Some higher aldehydes and also the aromatic aldehydes may exhibit much lower toxicity.

ALDEHYDINE. See 5-ethyl-2-methyl pyridine.

ALDERLIN. $C_{15}H_{19}ON$, mw: 229.3.

Acute tox data: Oral LD_{50} (rat) = 900 mg/kg; iv LD_{50} (rat) = 50 mg/kg; ip LD_{50} (mice) = 124 mg/kg. [*3*]

THR = HIGH via ip and iv routes; MOD via oral route. An exper carc to mice. [*3*]

ALDOL. Syns: *acetaldol, 3-butanolal, oxybutyricaldehyde.* Clear, white to yellow syrupy liquid. $C_4H_8O_2$, mw: 88.10, bp: 83° @ 20 mm, flash p: 150°F (OC), d: 1.11, autoign. temp.: 482°F, vap. d: 3.04.

THR = MOD via oral route.

Fire Hazard: Mod, when exposed to heat or flame; decomp into crotonaldehyde and water when heated. See crotonaldehyde.

Spontaneous Heating: No.

Disaster Hazard: Dangerous; when heated to decomp, emits highly toxic fumes of crotonaldehyde; can react with oxidizing materials.

To Fight Fire: Water, alcohol foam, mist or water spray, CO_2, dry chemical.

ALDOMET. See methyl dopa.

ALDRIN. Syns: *1,2,3,4,10,10-hexachloro-1,4,4a,5,8,8a-hexahydro-1,4,5,8-dimethanonaphthalene, octalene, compound 118.* Crystals, insol in water, sol in aromatics, esters, ketones, paraffins and halogenated solvents. $C_{12}H_8Cl_6$, mw: 365, mp: 104°–105°.

Acute tox data: Oral LD_{50} (rat) = 55 mg/kg, dermal LD_{50} (rat) = > 200 mg/kg. [*94*]

THR = HIGH via oral, dermal and CNS routes. Ingestion, inhal or absorption of this material into the body can cause irritability, convulsions and depression in from 1 to 5 hrs. Continued exposure causes liver damage. An exper carc. [*3, 12, 102*] See chlorinated HC.

Disaster Hazard: Dangerous; see chlorides.

ALFALFA MEAL.

THR = A mild sensitizer, which when inhaled may cause asthma, running nose, sneezing, coughing and tearing eyes. Contact with skin may cause contact dermatitis.

Fire Hazard: Mod, when exposed to heat or flame; by spont chemical reaction.

Spontaneous Heating: Yes. Avoid moisture content extremes. Fires may smolder for 72 hrs before becoming noticeable.

ALIPHATIC AMINES. See fatty amines.

ALIPHATIC AND AROMATIC EPOXIDES.

THR = Exper carc of skin, lung, blood-forming tissues. [*14*]

ALIPHATIC AND AROMATIC EPOXIDE MONOMERS AND POLYMERS.

THR = Exper carc. See aliphatic and aromatic epoxides. [*14*]

ALIZARIN. Syn: *1,2-dihydroxy anthraquinone.* Orange-red crystals. $C_{16}H_8O_4$, mw: 240.2, bp: 430° (sublimes), mp: 289°.

THR = MILD allergen.

Fire Hazard: Slight, when exposed to heat or flame; can react with oxidizing materials.

ALKALIES. (see also specific compounds.)

A term loosely applied to the hydroxides and carbonates of the alkali metals and alkaline earth metals, as well as the bicarbonate and hydroxide of ammonium. They can neutralize acids, change the color of indicators and impart a soapy taste and feel to aqueous solutions.

THR = Variable. The alkalies, as a group, constitute the commonest causes of occupational dermatitis. They act on the skin as primary irr. Alkaline solutions soften and dissolve the keratin layer, and the skin becomes white, soggy, wrinkled and macerated. Repeated exposure frequently results in the development of chronic eczematous skin conditions. The stronger caustics may produce chemical burns which are often deep and slow in healing. Systemically, the only alkali presenting any hazard is ammonia.

ALKALIES, N.O.S. See sodium hydroxide.

ALKALI METALS.
THR = Vigorous reactions with BF_3, (hydrazine + NH_3), maleic anhydride. [*19*] See also specific metals.

ALKALI METAL SALTS.
THR = Violent reaction with Zr. [*19*]

ALKALINE CAUSTIC LIQUID, N.O.S. Syn: *caustic alkali solution.*
THR = HIGH irr via oral and inhal routes.

ALKALINE CORROSIVE BATTERY FLUID or **ELECTROLYTE (ALKALINE), CORROSIVE BATTERY FLUID.** See alkaline caustic liquid, n.o.s.

ALKALINE CORROSIVE LIQUID, N.O.S. See alkaline caustic liquids, n.o.s.

ALKALOID, POISONS, ALSO THEIR SALTS, LIQUID, N.O.S., OR ALKALOIDS, POISONS, AND THEIR SALTS, SOLID, N.O.S. Syn: *alkaloids.*
THR = Practically all of the alkaloid salts are poisonous. Some of them also cause allergic symptoms in humans, such as contact dermatitis or asthma. See specific alkaloid salt.
Disaster Hazard: Dangerous; when heated to decomp, they emit highly toxic fumes.

ALKANES. Syns: *n-pentane + 2 isomers, n-hexane + 4 isomers, n-heptane + 8 isomers, n-octane + 17 isomers.* All colorless neutral liquids with light aromatic odors. See also individual alkanes as listed.
THR = Hexane has been often described as causing neuropathy upon chronic exposure. This may also be a property of other single or mixtures of alkanes.
Acute effects: transient CNS depression. The following symptoms are associated with exposure to *n*-hexane: numbness, coldness, redness and roughening of skin, muscular weakness, hypoactive reflexes, dysesthesia, emaciation, blurred vision, hyperactive reflexes, muscular atrophy, loss of sense of smell, face numbness, etc. Dermal alkane exposure: erythema, hyperemia, swelling and pigmentation. No topical anesthesia after 5 hrs. Dermal absorption is not reported. See also individual alkanes as listed. [*98*]

ALKANE SULFONIC ACID, MIXED. Liquid, mp: −40°, bp: 120°, d: 1.38.
THR = U; probably MOD; when heated to decomp, it emits high tox fumes of oxides of S.

ALKARGEN. See cacodylic acid.

ALKRON. See parathion.

ALKYL ALUMINUM HALIDES. See pyrophoric liquid, n.o.s.

ALKYL ARYL POLYETHYLENE GLYCOL ETHER. Liquid. bp: 150°, flash p: 440°F, d: 1.106.
THR = U; see glycols.
Fire Hazard: Slight, when exposed to heat or flame; can react with oxidizing materials.
To Fight Fire: Water foam, CO_2, water spray or mist, dry chemical.

ALKYL AZIDES.
THR = Very unstable. [*19*]

ALKYL BENZENES.
THR = Very reactive with F_2. [*19*]

ALKYL DIMETHYL BENZYL AMMONIUM CHLORIDE. Syn: *BTC.* Clear, mobile liquid. $C_9H_{13}NRCl$, mw: 365, d: 0.9884 @ 20° (50%).
THR = LOW via oral and inhal routes. An irr.
Fire Hazard: Slight.

ALKYL DISULFIDES. Syns: *methyl, ethyl, n-propyl, n-butyl, isobutyl and isoamyl disulfides.*
THR = Limited information, based mainly on animal exper, suggests that these compounds are dangerous and may cause hemolytic anemia. They may also produce allergic dermatitis.

ALKYL ISOTHIOUREA SALTS.
THR = Highly reactive with Cl_2. [*19*]

ALKYL MERCURIACETATES. See mercury compounds, organic.

ALKYL-N-PHENYL CARBAMATES.
THR = (S) carc. See carbamates.

ALKYL PHENYL POLYETHYLENE GLYCOL ETHER. Liquid mp: −5°, d: 1.0643 @ 20°/20°, autoign. temp.: 590°.
THR = MILD irr, MOD via oral route, LOW via dermal and inhal routes. Water solutions of less than 1% of this material have irr properties comparable to soap. Before this material is used in food or drug preparations a complete physiological evaluation of it in its intended use should be made. See also glycols.
Fire Hazard: Slight, when exposed to heat or flame; can react with oxidizing materials.
To Fight Fire: Water, foam, CO_2, dry chemical.

ALKYL PHOSPHINES.
THR = Very reactive with Cl_2. [*19*]

ALLENE. Syns: *dimethylene methane, propadiene.* Colorless, unstable, flam gas, sweet odor. $H_2C{:}C{:}CH_2$, mw: 40.06, d: 1.787, mp: −146°, bp: −32°, lel = 2.1%.
THR = U. Probably anesthetic.

Fire Hazard: Dangerous when exposed to heat, flame or powerful oxidizers.

Explosion Hazard: Mod, when exposed to flame or is compressed >2 atm. [*19*]

To Fight Fire: Stop flow of gas.

ALLETHRIN. Syns: *dl-2-allyl-4-hydroxy-3-methyl-2-cyclopenten-1-one ester of cis and trans-dl-chrysanthemum monocarboxylic acid.* Viscous liquid. $C_{19}H_{26}O_3$, mw: 302.4.

Acute tox data: Oral LD_{50} (mice) = 480 mg/kg.

THR = HIGH via oral and MOD via inhal routes. A MOD allergen. An insecticide. It can cause liver and kidney damage by all routes of entry into the body. Lung congestion may occur due to exposure. Local contact may cause contact dermatitis. Inhal may cause asthma, coughing, wheezing, running nose and eyes.

Fire Hazard: Slight.

***d-trans* ALLETHRIN.** Syn: *dl-2-allyl-4-hydroxy-3-methyl-2-cyclopenten-1-one of d-trans-chrysanthemum monocarboxylic acid.*

Acute tox data: Oral LD_{50} (rat) = 860 mg/kg. [*3*]

THR = MOD to mammals. An insecticide. See also allethrin.

ALLICIN. Colorless oily liquid, sharp garlic odor. $C_6H_{10}OS_2$, mw: 162.3, d: 1.112 @20°/4°.

THR = MOD irr, MOD via oral and inhal routes.

Disaster Hazard: Slightly dangerous; when heated to decomp, it emits toxic fumes.

ALLOXANTIN. Syn: *uroxin.* Crystalline powder, on exposure to air turns red; yellow at 225°. $C_8H_6N_4O_8 \cdot 2H_2O$, mw: 322.19.

THR = MOD via oral route. On a chronic basis caused disturbed carbohydrate metabolism leading to diabetes.

Disaster Hazard: Mod dangerous; when heated to decomp, it emits toxic fumes.

ALLSPICE. Syn: *pimenta.* Composition: eugenol, etc.

THR = A weak sensitizer which may cause dermatitis on local contact.

Fire Hazard: Slight.

ALLTOX. See toxaphene.

ALLYL ACETATE. Syn: *2-propenyl methanoate.* Liquid. $C_5H_8O_2$, mw: 100.1, vap. d: 3.45, d: 0.928, bp: 104°.

Acute tox data: Oral LD_{50} (rat) = 130 mg/kg; dermal LD_{50} (rabbit) = 1100 mg/kg; inhal LC_{50} (rat) = 1000 ppm for 1 hr. [*3*]

THR = HIGH via oral and inhal routes, MOD via dermal route.

Fire Hazard: Slight; can react with oxidizing materials.

ALLYL ACETONE. Syn: *5-hexen-2-one.* Colorless liquid. C_6H_9O, mw: 98.1, bp: 129.5°, d: 0.841 @20°/20°, vap. d: 3.39.

THR = U. Allyl compounds are generally toxic.

Fire Hazard: Mod; when exposed to heat or flame, can react with oxidizing material.

To Fight Fire: Foam, mist, CO_2, dry chemical.

ALLYL ALCOHOL. Syn: *2-propen-1-ol, vinyl carbinol.* Limpid liquid, pungent odor, sol in water, alcohol and ether. C_3H_6O, mw: 58.08, mp: −50°, bp: 96°–97°, lel = 2.5%, uel = 18%, flash p: 70°F (CC), d: 0.854 @20°/4°, autoign. temp.: 713°F, vap. press: 10 mm @ 10.5°, vap. d: 2.00.

Acute tox data: 25 ppm has caused marked human inhal tox. Inhal LC_{50} (rat) = 165 ppm for 4 hrs; oral LD_{50} (mouse) = 96 mg/kg; dermal LD_{50} (rabbit) = 54 mg/kg. [*3*]

THR = HIGH via dermal, oral and inhal routes. Animal exper have resulted in marked irr of skin, eyes and respiratory tract, as well as damage to kidneys and liver. Human exposure causes irr of eyes, mu mem and skin. Systemic poisoning is possible but has not been reported.

Fire Hazard: Dangerous when exposed to heat or flame.

Spontaneous Heating: No.

Explosion Hazard: Mod when exposed to flame, also in presence of CCl_4; a heat-unstable mixture, also presence of chlorosulfonic acid or HNO_3, H_2SO_4, oleum, NaOH, diallyl phosphite, PCl_3, tri-*n*-bromomelamine. [*19*]

Disaster Hazard: Dangerous; when heated, emits toxic fumes; can react vigorously with oxidizing materials.

To Fight Fire: CO_2, alcohol foam, dry chemical.

ALLYL ALDEHYDE. See acrolein.

ALLYL AMINE. Syn: *2-propenyl amine.* Colorless liquid, burning taste, sharp odor. C_3H_7N, mw: 57.09, bp: 55.2°, d: 0.761 @20°/4°, flash p: −20°F, autoign. temp.: 705°F, vap. d: 2.00, lel = 2.2%, uel = 22%.

Acute tox data: Oral LD_{50} (rat) = 106 mg/kg; inhal LD_{50} (rat) = 286 ppm for 4 hrs; dermal LD_{50} (rabbit) = 35 mg/kg; inhal TC_{LO} (man) = 5 ppm for 5 min ⟶ pulmonary effects. [*3*]

THR = HIGH via oral, inhal and dermal routes, HIGH to human via inhal route. In animal exper, inhal produced irr of nose and mouth, congestion of the eyes, irregular respiration, cyanosis, excitement, convulsions, coma and death. Extraordinary precautions against fumes are advised.

Fire Hazard: Mod, when exposed to heat or flame.

Disaster Hazard: Dangerous; when heated to decomp,

emits toxic fumes; can react with oxidizing materials.
To Fight Fire: Alcohol foam, CO_2, dry chemical.

ALLYL AMINE ANHYDROUS. See allyl amine.

ALLYL BENZENE SULFONATE. $C_9H_{10}O_3S$, mw: 198.2.
THR = Can explode on distillation. [*19*]

ALLYL BROMIDE. Colorless liquid, pungent odor. C_3H_5Br, mw: 121.0, mp: −119°, bp: 71.3°, flash p: 30° F, d: 1.3980 @ 20°/4°, autoign. temp.: 563° F, vap. d: 4.17, lel = 4.4%, uel = 7.3%.
THR = MOD via oral, inhal and dermal routes. See also allyl chloride.
Fire Hazard: Dangerous, when exposed to heat or flame; can react vigorously with oxidizing materials.
Disaster Hazard: Dangerous; see bromides.
To Fight Fire: Alcohol foam, water spray or mist, CO_2, dry chemical.

ALLYL CARBAMATE. $C_4H_7O_2N$, mw: 101.1.
THR = An exper neo. [*3*]

ALLYL CHLORIDE. Syn: *3-chloropropene*. Colorless liquid. C_3H_5Cl, mw: 76.53, mp: −136.4°, bp: 44.6°, flash p: −25° F, lel = 2.9%, uel = 11.2%, d: 0.938 @ 20°/4°, autoign. temp.: 905° F, vap. d: 2.64.
Acute tox data: Dermal LD_{50} (rabbit) = 2200 mg/kg; inhal LC_{50} (human) = 3000 ppm; oral LD_{50} (rat) = 64 mg/kg. [*3*] Inhal (mice) = 129 ppm for 1 hr ⟶ great pulmonary damage, liver damage and slight kidney changes, 100 mg/l for ½ hr (inhal) ⟶ 100% lethal to rats; 50 mg/l for 1.25 hrs (inhal) ⟶ 100% lethal to rats; 20 mg/l for 2 hrs (inhal) ⟶ 100% lethal to rats; 10 mg/l for 3 hrs (inhal) ⟶ 100% lethal to rats; 1 mg/l for 8 hrs (inhal) ⟶ 100% lethal to rats. Using guinca pigs to 100% lethality; 50 mg/l need 3/4 hrs of exposure; 10 mg/l needed 2 hrs of exposure; 1 mg/l needed 4 hrs of exposure. [*99*] The vapors of allyl chloride are quite irritating to the eyes, nose and throat and contact of the liquid with the skin, in addition to local vasoconstriction and numbness, may lead to rapid absorption and distribution through the body. If remedial measures are not taken promptly, such contact may result in burns and internal injuries. inhalation may cause headache, dizziness, and in high conc, loss of consciousness; however, even in low conc, its odor and irritating effects furnish warning of its presence in most cases. Conc of the vapors high enough to cause serious effects, including damage to the lungs, especially on repeated exposure, may not be intolerable. Consequently, the warning characteristics should never be disregarded. In general, precautions should be taken at all times to avoid spillage and accumulation of noticeable conc of the vapors in the atmosphere. Acute exposure in exper animals has resulted in marked inflammation of lungs, irritation of skin and swelling of the kidneys. Chronically exposed animals have shown degenerative changes in the liver and kidneys. Reported human exposures have been principally cases of irr of eyes, skin and respiratory tract, sometimes accompanied by aches and pains in the bones. Liver and kidney injury is possible.
Storage and Handling: Keep cool, away from heat sources. Maintain good ventilation. Work in a fume hood or with closed system if possible; otherwise, use enough ventilation so that the odor of allyl chloride does not persist. If it should be necessary to enter an area in which the odor of allyl chloride is at all noticeable, use a gas mask equipped with an "organic vapor" canister. Do not disregard the warning odor or eye irritation of allyl chloride.
Fire Hazard: High, when exposed to heat or flame.
Explosion Hazard: Mod when exposed to flame; can react vigorously with HNO_3, H_2SO_4, ethylene imine, ethylene diamine, chlorosulfonic acid, oleum, NaOH, (benzene or toluene to −70° in presence of trichlorotriethyl dialuminum or ethyl aluminum dichloride or diethyl aluminum chloride). [*19*]
Disaster Hazard: Dangerous; see chlorides.
To Fight Fire: CO_2, alcohol foam, dry chemical.

ALLYL CHLOROCARBONATE. Syn: *Allyl chloroformate*. Liquid. $C_4H_5O_2Cl$, mw: 120.5, bp: 106°–114°, flash p: 88° F (CC), d: 1.14, vap. d: 4.2.
Acute tox data: Inhal LC_{LO} (mouse) = 2000 mg/m³. [*3*]
THR = MOD via oral routc.
Fire Hazard: High when exposed to heat, open flame (or sparks) or powerful oxidizers.
To Fight Fire: Alcohol foam, spray or mist, dry chemical.
Disaster Hazard: Dangerous; when heated to decomp, emits highly toxic fumes of phosgene; can react with oxidizing matcrials.

ALLYL CHLOROFORMATE. See allyl chlorocarbonate.

ALLYL CHLOROPHENYL CARBONATE.
THR = U. An herbicide. Allyl compounds are generally toxic.
Disaster Hazard: Dangerous; when heated to decomp, it emits highly toxic fumes of phosgene.

ALLYL CYANIDE. Syn: *vinyl acetonitrile*. Colorless liquid, onion-like odor. CH_2CHCH_2CN, mw: 67.1, bp: 116°–119°, d: 0.8341 @ 20°/4°, mp: −87°.
Acute tox data: Oral LD_{50} (rat) = 115 mg/kg; inhal

LC_{LO} (guinea pig) = 2500 mg/m^3 for 4 hrs; dermal LD_{LO} (rabbit) = 1410 mg/kg. [*3*]
THR = HIGH via oral and MOD via inhal and dermal routes.
Disaster Hazard: Dangerous; emits highly toxic fumes when heated to decomp or on contact with acids or acid fumes.
To Fight Fire: Alcohol foam, mist.

ALLYL DIGLYCOL CARBONATE. Liquid. bp: 162°, flash p: 378°F (OC), d: 1.14.
Acute tox data: ip LD_{30} (mouse) = 270 mg/kg. [*3*]
THR = The allyl compounds are generally toxic.
Fire Hazard: Slight, when exposed to heat or flame; can react with oxidizing material.
To Fight Fire: Water mist or spray, foam, CO_2, dry chemical.

ALLYL DIMETHYL ARSINE. Syn: *dimethyl allyl arsine.* $C_5H_{11}As$, mw: 146.
THR = HIGH. See arsenic and compounds. Ignites spont in air. [*19*]

ALLYL DISULFIDE. See allyl sulfide.

ALLYLENE. Syns: *propyne, methyl acetylene.* Gas. CH_3CCH, mw: 40.06, mp: −104°, lel = 1.7%, bp: −23.3°, vap. press: 3876 mm @ 20°, d: 1.787 g/l @ 0°, vap. d: 1.38.
THR = This compound is a simple anesthetic and in high conc is an asphyxiant.
Fire Hazard: Dangerous, when exposed to heat or flame.
Explosion Hazard: Mod, when exposed to flame. Can decomp explosively. [*19*]
Disaster Hazard: Mod dangerous; can react vigorously with oxidizing materials.
To Fight Fire: Stop flow of gas.

ALLYL-3,4-EPOXY-6-METHYL CYCLOHEXANE CARBOXYLATE. $C_{11}H_{16}O_3$, mw: 196.3.
Acute tox data: Oral LD_{50} (rat) = 500 mg/kg; dermal LD_{50} (rabbit) = 2830 mg/kg. [*3*]
THR = HIGH via oral and MOD via dermal routes.

ALLYL-9,10-EPOXY STEARATE. $C_{21}H_{38}O_3$, mw: 338.6.
Acute tox data: Oral LD_{50} (rat) = 1198 mg/kg. [*3*]
THR = MOD via oral route.

ALLYL ESTER THIOCYANIC ACID. See allyl sulfocyanide.

17-ALLYLESTR-4-EN-17-β-OL. $C_{21}H_{32}O$, mw: 300.5.
THR = An exper teratogen. [*3*]

ALLYL ETHER. Syn: *diallyl ether.* Liquid, odor of radishes. $C_6H_{10}O$, mw: 98.1, bp: 94.3°, d: 0.805, vap. d: 3.38, flash p: 20°F (OC).
Acute tox data: Oral LD_{50} (rat) = 320 mg/kg. [*3*]
THR = HIGH via oral route.
Fire Hazard: Mod dangerous; can react with oxidizing materials.
To Fight Fire: Alcohol foam.

ALLYL FLUORIDE. Syn: *3-fluoropropene.* Colorless gas. C_3H_5F, mw: 60.07, bp: −10°.
THR = HIGH via oral and inhal routes. A strong irr.
Disaster Hazard: Dangerous; when heated to decomp, emits highly toxic fumes of fluorides; will react with water or steam to produce toxic and corrosive fumes.

ALLYL FORMATE. Syn: *2-propenyl methanoate.* Liquid, slightly water sol, sol in organic solvents. $C_4H_6O_2$, mw: 86.1, d: 0.948 @ 18°/4°, bp: 83°.
Acute tox data: Oral LD_{50} (rat) = 124 mg/kg; dermal LD_{50} (guinea pig) = 1000 mg/kg. [*3*]
THR = HIGH via oral; MOD via dermal routes.

ALLYL GLYCIDYL ETHER. Syns: *1-allyloxy-2,3-epoxypropane, AGE.* $C_6H_{10}O_2$, mw: 114.14, bp: 153.9°, fp: −100° (forms glass), flash p: 135°F (OC), d: 0.9698 @ 20°/4°, vap. press: 21.59 mm @ 60°, vap. d: 3.94.
Acute tox data: Oral LD_{50} (rat) = 922 mg/kg; inhal LC_{50} (mouse) = 270 ppm for 4 hrs; dermal LD_{50} (rabbit) = 2550 mg/kg. [*3*]
THR = HIGH via inhal route. Can cause CNS depression and pulmonary edema.
Fire Hazard: Mod, when exposed to heat or flame; can react with oxidizing materials.
To Fight Fire: Foam, CO_2, dry chemical.

ALLYL HOMOLOG OF CINERIN I. See allethrin.

ALLYL HYDRAZINE HYDROCHLORIDE. $C_3H_8N_2 \cdot HCl$, mw: 108.6.
THR = An exper carc to mice via oral route. [*103*]

ALLYLIDENE DIACETATE. Liquid. $C_7H_{10}O_4$, mw: 158.15, mp: −36.6°, bp: 107° @ 50 mm, flash p: 180°F (OC), d: 1.0749 @ 20°/20°, vap. d: 5.46.
Acute tox data: Oral LD_{50} (rat) = 250 mg/kg; dermal LD_{50} (rabbit) = 320 mg/kg; oral LD_{50} (mouse) = 38 mg/kg; inhal LC_{LO} (mouse) = 853 ppm for 15 min. [*3*]
THR = HIGH via oral, dermal and inhal routes.
Fire Hazard: Mod, when exposed to heat or flame; can react with oxidizing materials.
To Fight Fire: Water may be used to blanket the fire. Foam, CO_2, dry chemical.

ALLYL IODIDE. Syn: *3-iodo propene.* Yellow liquid. pungent odor. C_3H_5I, mw: 168.0, mp: −99°, bp: 103.1°, d: 1.825 @ 20°/4°, vap. d: 5.8.

THR = Powerful irr. HIGH via oral and inhal routes.
Fire Hazard: Mod, when exposed to heat or flame.
Disaster Hazard: Dangerous; when heated to decomp, it emits highly toxic fumes of iodine and iodides; can react with oxidizing materials.
To Fight Fire: Water, foam, CO_2, dry chemical.

ALLYL ISOPROPYL ACETYL CARBAMIDE.
THR = U. Reported as causing purpura due to depression of blood platelets.

ALLYL ISOTHIOCYANATE. Syn: *allyl mustard oil.* Colorless to pale yellow liquid; irr odor. C_4H_5NS, mw: 99.15, mp: −80°, bp: 150.7°, flash p: 115°F, d: 1.013–1.016 @ 25°/25°, vap. press: 10 mm @ 38.3°, vap. d: 3.41.
Acute tox data: Oral LD_{50} (rat) = 148 mg/kg; ip LD_{LO} (mouse) = 4 mg/kg. [*3*]
THR = HIGH via oral, ip routes. A fumigant. Mild sensitizer and allergen. Local contact may cause contact dermatitis. Inhal may cause asthma, watery eyes and sneezing.
Fire Hazard: Mod, when exposed to heat or flame or oxidizers.
Disaster Hazard: Dangerous; when heated to decomp or on contact with acid or acid fumes, emits highly toxic fumes of cyanides; can react with oxidizing materials.
To Fight Fire: Foam, CO_2, dry chemical.

ALLYL MALEATE. See diallyl maleate.

ALLYL MERCAPTAN. *Syns: allyl thiol-2-propene-1-thiol.* Water white liquid with a strong garlic odor which darkens on standing. CH_2CHCH_2SH, mw: 74.1, d: 0.925 @ 23°/4°, bp: 68°.
THR = HIGH, and HIGH irr via inhal and ingestion route. Irr to skin and mu mem.
Disaster Hazard: Dangerous; when decomp, emits highly toxic fumes.
Fire Hazard: Dangerous.
To Fight Fire: Water mist or spray, alcohol foam, CO_2, or dry chemical.

ALLYL METHANE SULFONATE. $C_4H_8O_3S$, mw: 136.2.
THR = An exper neo. [*3, 23*]

4-ALLYL-2-METHOXY PHENOL. See eugenol.

4-ALLYL-1,2-METHYLENE DIOXYBENZENE. See safrol.

ALLYL MUSTARD OIL. See allyl isothiocyanate.

1-ALLYLOXY-2,3-EPOXYPROPANE. See allyl glyceryl ether.

ALLYL PHOSPHATE. $C_9H_{15}O_4P$, mw: 218.2.
Acute tox data: iv LD_{50} (mice) = 71 mg/kg. [*3*]
THR = HIGH via iv route. An exper neo via sc route. [*3*] Can explode on distillation. [*19*]

ALLYL PHOSPHITE. Syn: *diallyl phosphite.* $C_6H_{11}O_3P$, mw: 162.1.
THR = No information. When vacuum distilled in a stream of CO_2, explosions occur. [*19*]

ALLYL PROPENYL. See 1,4-hexadiene.

ALLYL PROPYL DISULFIDE. Syn: *onion oil.* Liquid, pungent odor. $C_6H_{12}S_2$, mw: 148.16.
Acute tox data: Inhal TC_{LO} (human) = 3 ppm ⟶ irr. [*3*]
THR = Powerful irr. HIGH via inhal and oral routes. This material is particularly irr to the eyes and respiratory passages.
Fire Hazard: Mod.
Disaster Hazard: Dangerous; when heated to decomp, emits highly toxic fumes of oxides of sulfur; can react with oxidizing materials.
To Fight Fire: Foam, CO_2, dry chemical.

ALLYL SUCCINIC ANHYDRIDE. $C_7H_8O_3$, mw: 140.2.
Acute tox data: Oral LD_{50} (rat) = 1070 mg/kg; dermal LD_{50} (rabbit) = 320 mg/kg. [*3*]
THR = MOD via oral and HIGH via dermal routes.

ALLYL SULFIDE. Syns: *allyl disulfide, diallyl disulfide, garlic oil.* A colorless liquid; garlic odor. $C_3H_5S_2$, mw: 114.20, mp: −83°, bp: 139°, d: 0.888, vap. d: 3.90.
Acute tox data: ip LD_{LO} (mouse) = 512 mg/kg. [*3*]
THR = MOD via ip route.
Disaster Hazard: Dangerous; see S compounds.

ALLYL SULFOCARBAMIDE. See allyl thiourea.

ALLYL SULFOCYANIDE. Syn: *allyl ester thiocyanic acid.* An oil. C_4H_5NS, mw: 99.2. bp: 161°, d: 1.056 @ 15°.
Acute tox data: sc LD_{LO} (rabbit) = 12 mg/kg. [*3*]
THR = HIGH via sc route. The allyl compounds are generally toxic.
Disaster Hazard: Dangerous; see cyanides and S compounds.

ALLYL THIOUREA. Syns: *allyl sulfocarbamide, allyl thiocarbamide.* White crystalline solid, slight garlic odor, bitter taste. $C_4H_8N_2S$, mw: 116.1, mp: 72°–74°, d: 1.22.
Acute tox data: Oral LD_{50} (rat) = 200 mg/kg; ip LD_{50} (rat) = 500 mg/kg. [*3*]
THR = HIGH via oral and ip routes. A powerful allergen. Contact eczema due to sensitization in humans has been reported.
Disaster Hazard: Dangerous; see S compounds.

ALLYL TRICHLORIDE. See trichloro propene.

ALLYL TRICHLORO SILANE. Colorless liquid, pungent, irr odor. $C_3H_5SiCl_3$, mw: 175.5, bp: 117.5°, d: 1.217 (27°), flash p: 95°F (COC).
THR = Irr, corrosive toxic material. See also silanes.
Fire Hazard: Mod.
Disaster Hazard: Dangerous; see chlorides.

ALLYL TRISULFIDE. Syn: *diallyl trisulfide.* Liquid. $C_6H_{10}S_3$, mw: 178.3, bp: 140°, d: 1.085 @ 15°.
THR = U. The allyl compounds are generally toxic.
Disaster Hazard: Dangerous; see S compounds.

4-ALLYL VERATROLE. See methyl eugenol.

ALLYL VINYL ETHER. Syn: *vinyl allyl ether.* Very slightly sol in water. $CH_2{:}CHOCH_2CH_2O(CH_2)_3CH_3$, mw: 144, d: 0.8, bp: 67°, flash p: <68°F (OC).
Acute tox data: Oral LD_{50} (rat) = 550 mg/kg; inhal LC_{LO} (rat) = 8000 ppm for 4 hrs. [*3*]
THR = MOD via oral and inhal routes.
Fire Hazard: Dangerous; see ethers.
To Fight Fire: Water may be ineffective. Use alcohol foam, dry chemical, mist.

ALMOND OIL. Syns: *almond oil expressed*, almond oil sweet. Fixed, non-drying oil; oily liquid. Composition: oleic, linoleic, myristic, palmitic acids. d: 0.910–0.915 @ 25°/25°.
THR = A weak sensitizer. Contact dermatitis may result from local contact.
Fire Hazard: Slight, when exposed to heat or flame.
To Fight Fire: Use alcohol foam, dry chemical, water, mist.

ALMOND OIL, BITTER. Colorless oil which turns to yellow; bitter almond odor. Compositions: Chief known constituents are benzaldehyde, hydrocyanic acid, benzaldehyde cyanhydrin. bp: 179°, d: 1.045–1.070 @ 15°.
THR = U. Depends upon purity of sample. An allergen. It can be quite toxic if it has not been separated from its hydrogen cyanide. Weak sensitizer; may cause a contact dermatitis.
Fire Hazard: Slight, when exposed to heat or flame.
Disaster Hazard: Dangerous; see cyanides.

ALMOND OIL EXPRESSED. See almond oil.

ALMOND OIL SWEET. See almond oil.

ALODAN. See 1,2,3,4,7,7-hexachloro-5,6-bis-(chloromethyl)-2-norbornene.

ALOPERIDIN. $C_{21}H_{23}O_2NClF$, mw: 375.9.
THR = An exper teratogen. [*3*] HIGH via oral route.

ALPEROX C. See lauroyl peroxide.

ALPHA RAYS. Particulate radiation emitted by certain radioactive isotopes. Alpha rays consist of heavy charged particles (helium nuclei) moving at high velocity.
Radiation Hazard: A recog carc.

ALPHASOL IB. See diisobutyl sodium sulfosuccinate.

ALROSEPT MBC. See 1-tridecyl-2-benzyl-2-hydroxyimidazolium chloride.

ALTAITE. See lead telluride.

ALUM. Syn: *potasssium aluminum sulfate.* Colorless crystals. $KAl(SO_4)_2 \cdot 12H_2O$, mw: 474.39, mp: 92.5°, d: 1.725.
THR = MILD irr and allergen, used in styptic pencils. A general purpose food additive, it may migrate to food from packaging materials. A weak sensitizer. Local contact may cause contact dermatitis.

ALUM, AMMONIA. See aluminum ammonium sulfate.

ALUMINA. Syn: *aluminum oxide.* White powder. Al_2O_3, mw: 101.94, mp: 2050°, bp: 2977°, d: 3.5–4.0, vap. press: 1 mm @ 2158°.
THR = There has been some record of lung damage due to the inhal of finely divided aluminum oxide particles. However, this effect (known as Shaver's disease) is complicated by the presence in the inhaled air of silica and oxides of iron. A nuisance particulate. An exper neo via ipl route. [*103*]

ALUMINA TRIHYDRATE. See aluminum hydroxide.

ALUMINUM. A silvery ductile metal. Al, atwt: 26.97, mp: 660°, bp: 2056°, d: 2.702, vap. press 1 mm @ 1284°.
THR = Aluminum is not generally regarded as an industrial poison. Inhal of finely divided aluminum powder has been reported as a cause of pulmonary fibrosis.
Fire Hazard of Dust: Mod, when exposed to heat or flame or by chemical reaction.
Spontaneous Heating: No.
Explosion Hazard of Dust: Mod, when exposed to heat or flame or on contact with powerful oxidizers such as chlorates, bromates, iodates, peroxides, perchlorates, nitrates, nitrites, oxides, performates, persulfates, halogens, NO_x, melted sulfates, SO_2, (trichloroethylene + HCl), (Na_2O_2 + CO_2), SCl_2, $COCl_2$, PCl_3, AgCl, O_2 compressed or liquid, (Pd + Δ), NOCl, (Nb oxide + S), chloro and/or fluoro methanes and ethanes, ICl, (Mn + air), CH_3Br, CH_3Cl, (fluoro-chloro lubricants + pressure), (Mg + $KClO_4$), propylene dichloride, Na_2C_2, Na_2CO_3, NaOH. [*19*]
To Fight Fire: Special mixtures of dry chemical.

ALUMINUM ACETATE. Amorphous white powder. $Al(C_2H_3O_2)_3$, mw: 204.1, mp: decomp.

THR = Weak sensitizer. Local contact may cause contact dermatitis. A MILD irr.

ALUMINUM ACETOACETONATE. See aluminum acetate.

ALUMINUM ALKYLS. See pyrophoric liquids.

ALUMINUM AMINO BOROHYDRIDES.
THR = Spont flam in air and decomp by water. [*19*]

ALUMINUM AMMONIUM SULFATE. Syns: *alum ammonia, ammonium alum*. Colorless crystals, odorless, sol in water, glycerine; insol in alcohol.
$Al_2(SO_4)_3(NH_4)_2SO_4 \cdot 24H_2O$, mw: 906, d: 1.645, mp: 94.5°, bp: loses $20H_2O$ @ 120°.
THR = A mild astringent used as a general purpose food additive. Irr if inhal or ingested.
Disaster Hazard: Dangerous; see sulfates.

ALUMINUM-*o*-ARSENATE. White powder. $AlAsO_4 \cdot 8H_2O$, mw: 310, d: 3.011.
THR = See arsenic compounds.

ALUMINUM ARSENIDE. A solid. AlAs, mw: 101.9.
THR = See arsenic compounds and arsine.

ALUMINUM BENZOATE. Crystalline powder, very slightly sol in H_2O. $Al(C_7H_5O_2)_3$, mw: 390.3.
THR = U. Limited data indicate LOW toxicity. See also aluminum.

ALUMINUM BORIDE. Powder.
THR = See aluminum, also boron compounds (hydrides).

ALUMINUM BOROFORMATE. White lustrous scales, freely sol in water and alcohol. Composition: *ca* 33% Al_2O_3, 20% H_3BO_3, 15% formic acid, 32% H_2O.
THR = See aluminum, boron and formates.

ALUMINUM BOROHYDRIDE. Syn: *aluminum tetrahydroborate*. Liquid. AlB_3H_{12}, mw: 71.53, bp: 44.5°, mp: −64.5°, vap. press: 400 mm @ 28.1°.
THR = U. See also hydrides and boron compounds.
Fire Hazard: Dangerous by spont chemical reaction; ignites spont in air, particularly in moist air.
Explosion Hazard: Explodes in O_2, at temperatures as low as 20°. An explosive range of 5% to 90%. [*19*] Evolves H_2 on contact with water.
Disaster Hazard: Mod dangerous; will react with water or steam to produce heat, H_2 or toxic fumes; can react vigorously with oxidizing materials; can emit toxic fumes on contact with acid or acid fumes.
To Fight Fire: CO_2, dry chemical.

ALUMINUM BOROTANNATE. Syn: *cutal*. Light brown powder, insol in water, sol in dilute tartaric acid solution.
THR = See boron compounds and tannic acid.

ALUMINUM BROMATE. Crystals. $Al(BrO_3)_3 \cdot 9H_2O$, mw: 572.9, mp: 62.3°, bp: decomp.
THR = U. See also bromates.
Fire Hazard: Dangerous; see bromates.
Disaster Hazard: See bromates.

ALUMINUM BROMIDE. White to yellowish-red lumps. $AlBr_3$, mw: 266.7, mp: 97.5°, bp: 263.3° @ 748 mm, d: 3.2, vap. press: 1 mm @ 81.3°.
THR = Toxic and corrosive material. See also bromides.
Disaster Hazard: See bromides. Mixtures with Na or K explode violently on impact. [*19*]

ALUMINUM CALCIUM SILICATE.
THR = U, probably LOW. An anti-caking agent food additive. [*109*]

ALUMINUM CARBIDE. Yellow crystal or powder, hygroscopic. Cl_4C_3, mw: 143.91, mp: 2100°, bp: decomp >2200°, d: 2.36.
THR = Dust can cause pulmonary irr. Decomp by water. Incandesces in contact with $KMnO_4$ or PbO_2. [*19*]

ALUMINUM CHLORATE. Colorless deliquescent crystals. $Al(ClO_3)_3$, mw: 277.4, mp: decomp.
THR = U. See chlorates.
Fire Hazard: Mod, by spont chemical reaction; a powerful oxidizer; may ignite upon contact with combustibles.
Explosion Hazard: Mod, when shocked, exposed to heat or by spont chemical reaction with reducing agents. When contaminated may become sensitized.
Disaster Hazard: Dangerous; shock or heat will explode it. See chlorides and chlorates.

ALUMINUM CHLORIDE. Yellowish-white granular crystals. $AlCl_3$, mw: 133.35, mp: 190° @ 1900 mm, bp: 183° @ 752 mm, d: 2.44 @ 25°, vap. press: 1 mm @ 100.0°.
Acute tox data: Oral LD_{50} (rat) = 3.7 g/kg. [*3*]
THR = MOD via oral route. The dust is an oral and inhal skin irr. Old containers can explode on opening; may catalyze violent reactions with (allyl chloride + H_2O), ethylene, (ethylene oxide + acid or base), (nitrobenzene + 5% phenol), (nitromethane + organic matter), OF_2, (perchloryl fluoride + C_6H_6), Na, K, water. [*19*] See chlorides and hydrochloric acid.
Disaster Hazard: Dangerous. See hydrochloric acid; will react with water or steam to produce heat and toxic or corrosive fumes.

ALUMINUM CHLOROHYDROXIDE COMPLEX. See chlorhydrol.

ALUMINUM DEXTRAN.
THR = An exper carc. [*3*]

ALUMINUM DIBORIDE. Hexagonal crystals. AlB_2, mw: 48.6, d: 3.19.
THR = U. See also boron compounds.
Fire Hazard: See boron hydrides.
Explosion Hazard: See boron hydrides.
Disaster Hazard: Dangerous; will react with water or steam, acid or acid fumes; can emit toxic fumes.

ALUMINUM DICHROMATE. Solid. $Al_2(Cr_2O_7)_3$, mw: 702.
THR = See chromium compounds.
Fire Hazard: See chromates.
Disaster Hazard: Mod; reacts with reducing materials.

ALUMINUM DIETHYL MONOCHLORIDE. See diethyl aluminum chloride.

ALUMINUM DROSS, DRY OR WET. See aluminum.

ALUMINUM DUST. See aluminum.

ALUMINUM ETHOXIDE. See aluminum ethylate.

ALUMINUM ETHYLATE. Syn: *aluminum ethoxide*. Liquid. Decomp by H_2O. $Al(OC_2H_5)_3$, mw: 162.15, bp: 200° @ 6–8 mm, mp: 140°.
THR = HIGH irr to skin, eyes and mu mem via inhal route. See also organo metals.

ALUMINUM FERROCYANIDE. Brown powder. $Al_4((Fe(CN)_6))_3 \cdot 17\ H_2O$, mw: 1050.1.
THR = See ferrocyanides.
Disaster Hazard: Dangerous; see ferrocyanides.

ALUMINUM FERROSILICON. See components.

ALUMINUM FLUORIDE. Syn: *aluminum trifluoride*. Colorless hex crystals. AlF_3, mw: 84, mp: 1291°, subl @ 1260°, d: 2.88, vap. press: 1 mm @ 1238°, bp: 1537°.
Acute tox data: Oral LD_{LO} (guinea pig) = 600 mg/kg; sc LD_{LO} (guinea pig) = 3000 mg/kg. [*3*]
THR = MOD via oral and sc routes. See fluorides. When impacted, reacts violently in contact with sodium or potassium. [*19*]

ALUMINUM FLUOROACETIC ACID. $C_6H_6O_6F_3Al$, mw: 258.1.
Acute tox data: Oral LD_{LO} (rat) = 10 mg/kg. [*3*]
THR = HIGH via oral route. [*3*] See also fluorides.

ALUMINUM FORMATE. White powder. $Al(CHO_2)_3$, mw: 162.0.
THR = U. See also formic acid.

ALUMINUM FORMOACETATE. White powder. $Al(OH)(OOCH)(OOCCH_3)$, mw: 148.1.
THR = Slight, an irr.
Disaster Hazard: Slight.

ALUMINUM HEXAFLUOSILICATE. Syns: *topaz, aluminum silicofluoride*. White powder. $Al_2(SiF_6)_3$, mw: 480.2. Occurs as a sol nonahydrate.
Acute tox data: Oral LD_{LO} (guinea pig) = 5000 mg/kg; sc LD_{LO} (guinea pig) = 4000 mg/kg. [*3*]
THR = MOD via oral and sc routes.

ALUMINUM HYDRATE. See aluminum hydroxide.

ALUMINUM HYDRIDE. Colorless powder. AlH_3, mw: 30.0.
THR = Hydrides of some metals (such as AsH_3) are extremely toxic; little is known about AlH_3. See also hydrides.
Fire Hazard: Spont flam in air or O_2. Evolves H_2 upon contact with moisture.
Explosion Hazard: Severe, by chemical reaction wherein H_2 gas is produced, also in contact with dimethyl ether contaminated by CO_2. [*19*]
Disaster Hazard: Mod dangerous; will react with water or steam to produce heat and H_2; reacts with oxidizing materials. On contact with acid or acid fumes, it can emit toxic fumes.

ALUMINUM HYDROXIDE. Syns: *alumina trihydrate, aluminum hydrate, hydrated alumina, hydrated aluminum oxide*. White crystalline powder, balls or granules, insol in water, sol in mineral acids and caustic soda. $Al(OH)_3$, mw: 78, d: 2.42.
THR = Probably LOW. A substance migrating to food [*109*] from packaging materials. When coprecipitated with bismuth hydroxide and reduced by H_2, it is violently flam in air. [*19*]

ALUMINUM HYPOPHOSPHITE. $Al(H_2PO_2)_3$, mw: 222.1.
THR = Liberates highly toxic PH_3 @ 220°, PH_3 is spont flam. [*19*]

ALUMINUM IODIDE. White leaflets. AlI_3, mw: 407.7, mp: 191°, bp: 360°, d: 3.98 @ 25°, vap. press: 1 mm @ 178.0° (sublimes).
THR = see iodides.
Disaster Hazard: Dangerous. See iodides. Reacts violently with water.

ALUMINUM LACTATE. White-yellowish powder. $Al(C_3H_5O_3)_3$, mw: 294.2.
THR = Probably LOW. A MILD irr. See also aluminum compounds.
Fire Hazard: Slight.

ALUMINUM LIQUID OR PAINT. Silvery liquid. Al in volatile vehicle, flash p: 100° F or lower.
THR = See aluminum and vehicle involved.
Fire Hazard: Dangerous, when exposed to heat or flame.
Explosion Hazard: Mod.
Disaster Hazard: Mod dangerous; when heated to decomp emits toxic fumes; can react vigorously with oxidizing materials.
To Fight Fire: Water, foam, CO_2, dry chemical.

ALUMINUM, METALLIC, POWDER. See aluminum.

ALUMINUM METHYL. Syn: *trimethyl aluminum.* Colorless liquid. $Al(CH_3)_3$, mw: 72.07, bp: 130°, mp: 0°.

THR = Related alkyl aluminum compounds show strong irr properties and HIGH toxicity.

Fire Hazard: Dangerous, by spont chemical reaction with air; also reacts violently with water.

Explosion Hazard: Mod by chemical reaction with air, explodes on contact with water. Violent reaction with halogenated hydrocarbons. [*19*]

Disaster Hazard: Dangerous; when heated to decomp, emits toxic fumes; will explode on contact with moisture. Reacts vigorously with oxidizing materials.

To Fight Fire: Do not use water, foam or halogenated extinguishing agents. Use dry chemical.

ALUMINUM MONOPALMITATE. See aluminum palmitate.

ALUMINUM NICOTINATE. Syn: *nicalex.* A powder. $C_{18}H_{12}AlN_3O_6$, mw: 393.3.

THR = Can cause GI upset, pruritis, flushing. A food additive permitted in food for human consumption. [*109*]

ALUMINUM NITRATE. White crystals. $Al(NO_3)_3 \cdot 9H_2O$, mw: 375.14, bp: decomp @ 150°, mp: 70°.

Acute tox data: Oral LD_{50} (rat) = 264 mg/kg. [*3*]

THR = HIGH via oral route. See also nitrates.

Fire Hazard: See nitrates.

Disaster Hazard: See nitrates.

ALUMINUM NITRIDE. White or colorless crystals. AlN, mw: 41, mp: 2200°, bp: sublimes @ 2000°, d: 3.26.

THR =See nitrides and ammonia.

Disaster Hazard: Mod; will react with water or steam to produce toxic or corrosive fumes.

ALUMINUM OLEATE. Yellowish-white viscous mass, insol in water, sol in alcohol, benzene, ether, oil and turpentine. $Al(C_{18}H_{33}O_2)_3$, mw: 87.1.

THR = A metal soap. Probably astringent. A substance migrating to food from packaging materials. [*109*]

ALUMINUM-1-PHENOL-4-SULFONATE. Reddish-white powder. $Al(C_6H_5O_4S)_3$, mw: 546.5.

THR = MOD irr to skin, eyes and mu mem. An allergen.

Disaster Hazard: Dangerous; see phenol and sulfur compounds.

ALUMINUM PHENOXIDE. Gray-white powder or crystalline mass. $Al(C_6H_5O)_3$, mw: 306.3, mp: 265° (decomp), d: 1.23.

THR = MOD irr to skin, eyes and mu mem via oral and inhal routes.

Disaster Hazard: Dangerous; see phenol.

ALUMINUM PHOSPHIDE. Dark gray or dark yellow crystals.

AlP, mw: 57.96, d: 2.85 @ 25°/4°.

Acute tox data: Inhal LC_{LO} (rat) = 1 ppm. [*3*]

THR = An insecticide and a fumigant; releases phosphine. See phosphine. A poison. HIGH via oral and inhal routes. A food additive permitted in the food and drinking water of animals and/or for the treatment of food-producing animals. [*109*]

Disaster Hazard: Dangerous; in contact with water or steam it yields PH_3, which is spont flam in air. See phosphine.

ALUMINUM PICRATE. A solid. $Al(C_6H_2O(NO_2)_3)_3$, mw: 711.3.

THR = A powerful allergen. A poison. See also picric acid.

Fire Hazard: Dangerous, by chemical reaction with reducing materials; a powerful oxidizer.

Explosion Hazard: Severe, when shocked or exposed to heat. See also explosives (high).

Disaster Hazard: Highly dangerous; explodes upon shock; when heated to decomp, it emits Highly toxic fumes of NO_x and explodes; can react vigorously with reducing materials.

ALUMINUM POWDER. See aluminum.

ALUMINUM PROPOXIDE. White crystals. $Al(C_3H_7O)_3$, mw: 204.2, mp: 106°, bp: 248° @ 14 mm, d: 1.058 @20°/0°.

THR = U. See also aluminum.

Fire Hazard: Slight; a combustible material.

ALUMINUM RESINATE. Brown mass. $Al(C_{44}H_{63}O_5)_3$, mw: 2042.8.

THR = U. Possibly an irr.

Fire Hazard: Slight, when exposed to heat or flame.

ALUMINUM RICINOLEATE. Plastic solid, yellow-brown to dark brown. $Al((CO_2(CH_2)_7CHCHCH_2CHOH(CH_2)_5CH_3))_3$, mw: 912, mp: 95°.

THR = See aluminum and ricin.

Fire Hazard: Slight.

ALUMINUM SILICATE. Al_2SiO_5, mw: 162.1.

THR = An exper neo via ipl route. [*103*]

ALUMINUM SODIUM SULFATE. Syn: *sodium aluminum sulfate.* Colorless crystals. $NaAl(SO_4)_2 \cdot 12H_2O$, mw: 458.29, mp: 61°, d: 1.675.

THR = A weak sensitizer. A general purpose food additive. [*109*] Local contact may cause contact dermatitis. An irr.

ALUMINUM SULFATE, DRY. Syn: *cake alum.* White powder. $Al_2(SO_4)_3$, mw: 342.1 mp: decomp. @ 770°, d: 2.71.
THR = This material hydrolyzes readily to form some sulfuric acid which acts as a tissue irr, particularly to the lungs. See sulfuric acid. A general purpose food additive. It is at least MOD via inhal and oral routes. [*109*]

ALUMINUM SULFIDE. Yellowish-gray lumps; odor of hydrogen sulfide. Al_2S_3, mw: 150.12, mp: 1100°, bp: subl @ 1550°, d: 2.02 @ 13°F.
THR = HIGH. See sulfides.
Fire Hazard: See sulfides.
Explosion Hazard: See sulfides.
Disaster Hazard: See sulfides.

ALUMINUM SULFOCARBOLATE. See aluminum-1-phenol-4-sulfonate.

ALUMINUM TARTRATE. White, odorless granules; slowly sol in cold water, readily sol in hot water, sol in ammonia. $Al_2(C_4H_4O_6)_3$, mw: 498.16.
THR = See aluminum and tartaric acid.

ALUMINUM TETRA AZIDO BORATE. $Al[(B(N_3)_4)]_3$, mw: 563.7.
THR = A shock sensitive explosive. [*19*]

ALUMINUM TETRAHYDRO BORATE. See aluminum borohydride.

ALUMINUM THALLIUM SULFATE. Cubic, octagonal, colorless crystals. $AlTl(SO_4)_2 \cdot 12H_2O$, mw: 639.7, mp: 91°, d: 2.32 @ 20°/4°.
THR = HIGH. See thallium compounds.
Disaster Hazard: See thallium compounds and sulfates.

ALUMINUM TRIETHYL. See triethyl aluminum.

ALUMINUM TRIMETHYL. See aluminum methyl.

ALUMINUM TRIPROPYL. Syn: *tripropyl aluminum.* Liquid. $Al(C_3H_7)_3$, mw: 156.24.
THR = U. Related alkyl aluminum compounds have HIGH toxicity. See also diisobutyl aluminum chloride.
Fire Hazard: Dangerous, by spont chemical reaction with air; can react vigorously with halogenated hydrocarbons. [*19*]
Explosion Hazard: Hydrolyzes to evolve flam vapor.
To Fight Fire: Do not use water, foam or halogenated extinguishing agents. Use dry chemical or a special powder extinguisher.

"ALVAR." A solid polyvinyl acetal resin.
THR = Local contact may cause a contact dermatitis. An allergen and MILD irr.
Fire Hazard: Slight, when exposed to heat or flame; can react with oxidizing materials.

ALYPIN. White crystalline powder. $C_{16}H_{26}O_2N_2$, mw: 278.4.
THR = An allergen. MOD to HIGH via sc, iv, ip routes. Poisonings have occurred from small doses but are very unusual.
Disaster Hazard: Mod dangerous; when heated to decomp, emits toxic fumes.

AMATOL. A high explosive. Composition: NH_4NO_3: 80% and TNT: 20%, d: 1.47.
THR = MOD via irr, oral and inhal routes. An allergen. Local contact may cause contact dermatitis.
Fire Hazard: High, as result of spont chemical reaction. A powerful oxidizing mixture.
Explosion Hazard: High, due to shock, spont chemical reaction or exposure to flame.
Disaster Hazard: Highly dangerous; emits highly toxic fumes. Can react vigorously with oxidizing materials.

AMERICAN COUNCIL OF GOVERNMENTAL INDUSTRIAL HYGIENISTS (ACGIH). Section 1.

AMERICAN NATIONAL STANDARDS INSTITUTE (ANSI). Section 1.

AMERICAN WORM SEED OIL. See oil of chenopodium.

AMBER OIL. Syn: *oil of amber.* Pale yellow to brown oil, volatile. Composition: Phenols and terpenes, d: 0.85–0.92.
THR = A weak sensitizer. Local contact may cause contact dermatitis. A MILD irr.
Fire Hazard: Slight; when exposed to heat, emits acrid fumes.

AMBIENT AIR QUALITY STANDARDS. Section 4.

AMBOCHLORIN. Syn: *chloraminophen.* $C_{14}H_{19}O_2NCl_2$, mp: 64°–67°, fine white crystals.
THR = An exper (+) teratogen, neo and carc. [*3, 8*]

AMERICIUM. A silvery, somewhat malleable metal. Am, mp: 994°, bp: 2607°, d: 13.67 @ 20°.
THR = HIGH toxicity, bone-seeking radioactive element.
Radiation Hazard: See Table 5A.5 for listing of permissible levels. Artificial isotope ^{241}Am, $T_{\frac{1}{2}} = 460y$. Decays to radioactive ^{237}Np via α's of 5.4–5.5 MeV.
Fire Hazard: See powdered metals.
Disaster Hazard: In a disaster this highly toxic radioactive material can be disseminated over a wide area, causing a long-lived inhal hazard which is difficult to remove from surfaces or from a body once it enters.

AMETYCIN. $C_{15}H_{18}O_5N_4$, mw: 334.4, mp: > 360°. Blue-violet crystals.
THR = HIGH via oral route to rats and mice. An exper (+) teratogen, carc, and neo. [*3, 7*]

AMIBEN. Syn: *3-amino-2,5-dichlorobenzoic acid* $C_7H_3O_2(Cl)_2(NH_2)$, mw: 206.
Acute tox data: Oral LD_{50} (rat) = 3500 mg/kg.
THR = MOD via oral route. An herbicide.
Disaster Hazard: When heated to decomp, it emits highly toxic fumes. See also chlorides.

AMIDES. Organic compounds containing the structural group -$CONH_2$, and closely related to the organic acids with the grouping COOH. Common examples are: acetamide, CH_3CONH_2, and urea, $CO(NH_2)_2$.
THR = Most of the saturated amides have LOW toxicity, but the unsaturated and N-substituted amides are frequently irr, and may be absorbed via the skin. In animal exper the latter two classes have caused injury to the liver, kidney and brain.

AMIDOL. Syn: *diamino phenol*. Grayish-white crystals. $C_6H_8N_2O$, mw: 124.1.
THR = MILD irr and allergen.
Disaster Hazard: Dangerous; see chlorides.

"AMINE 220." Liquid. $C_{17}H_{33}\overline{CNC_2H_4N}C_2H_4OH$, mw: 350, bp: 235° @ 1 mm, flash p: 465°F (OC), d: 0.9300 @ 20°/20°, vap. d: 12.1.
Acute tox data: Oral LD_{50} (rat) = 3130 mg/kg. [*3*]
THR = MOD via oral route.
Fire Hazard: Slight, when exposed to heat or flame; can react with oxidizing materials.
To Fight Fire: Foam, CO_2, dry chemical.

AMINE PERCHLORATES.
THR = React violently above 215°. [*19*]

AMINES. See also specific compounds.
A large group of organic compounds, containing nitrogen and considered as derived from ammonia (NH_3) by replacement of one or more H atoms by an organic radical.
THR – Variable; some are HIGH in toxicity, others only slightly. Many are skin irr and some are sensitizers. See aromatic amines and also fatty amines. Amines are common air contaminants.

***p*-AMINOACETANILIDE.** Syn: *acetyl-p-phenylene diamine*. White to reddish crystals. $NH_2C_6H_4NHCOCH_3$, mw: 150.2, mp: 164°.
Acute tox data: Oral LD_{50} (rat) = 3350 mg/kg. [*3*]
THR = MOD via oral route.
Disaster Hazard: Mod dangerous; when heated to decomp, emits toxic fumes.

AMINO ACETIC ACID. See glycine.

AMINOACETO NITRILE BISULFATE.
THR = An exper teratogen. [*3*]

***m*-AMINO ACETOPHENONE.** Syn: *m-aminoacetyl benzene*. Yellow oily liquid. $CH_3COC_6H_4NH_2$, mw: 135.2, bp: 250°–252° (slight decomp).
Acute tox data: Oral LD_{50} (rat) = 1870 mg/kg; ip LD_{50} (mouse) = 465 mg/kg. [*3*]
THR = HIGH via ip and MOD via oral routes.

***p*-AMINOACETOPHENONE.**
Acute tox data: ip LD_{LO} (mouse) = 300 mg/kg. [*3*]
THR = HIGH via ip route.

***o*-AMINOACETOPHENONE.**
Acute tox data: sc TD_{LO} (mouse) = 2000 mg/kg. [*3*]
THR = MOD via sc route. An exper carc on a chronic basis. [*3*]
Fire Hazard: Slightly dangerous.

***o*-AMINOACETYL BENZENE.** See *o*-aminoaceto phenone.

***α*-AMINO ANTHROQUINONE.** Syn: *1-amino anthroquinone*. Ruby red crystals, insol in water, sol in alcohol, benzene, chloroform, ether, glacial acetic acid, hydrochloric acid. $C_6H_4(CO_2)C_6H_3NH_2$ (tricyclic), mw: 223.22 mp: 253°, bp: sublimes.
THR = Has produced anemia and degenerative changes in liver and kidneys of exper animals. A sensitizer. An exper neo. [*3*]

2-AMINO-4-ARSENOSOPHENOL HYDROCHLORIDE. $C_6H_6O_2NAs \cdot HCl$ mw: 235.
THR = HIGH via oral route. An exper teratogen. [*3*] See arsenic.

AMINOARSON. See carbasone.

***p*-AMINO AZOBENZENE.** Syn: *p-phenylazoaniline*. Yellow crystals. $C_6H_5NNC_6H_4NH_2$, mw: 197.2, bp: 360°, mp: 128°.
THR = An exper (+) carc. [*3, 23, 4*]
Fire Hazard: U.
Disaster Hazard: Mod dangerous; when heated to decomp, emits toxic fumes.

***o*-AMINOAZOTOLUENE.** Syn: *2-amino-5-azotoluene, solvent yellow 3, toluazotoluidine*. Reddish brown to yellow crystals, sol in alcohol, ether, oils and fats, slightly sol in water. $CH_3C_6H_4N_2C_6H_3NH_2CH_3$, mw: 225.3, mp: 102°.
Acute tox data: Oral LD_{50} (rat) = 1500 mg/kg. [*3*]
THR = MOD via oral route. On a chronic basis, oral, sc, ip ⟶ exper (+) carc. [*3, 23*]
Fire Hazard: U.
Disaster Hazard: Mod dangerous; when heated to decomp, emits toxic fumes.

2-AMINO-2,5-AZOTOLUENE.
THR = An exper carc. [*3, 23*]

4-AMINO-2,3-AZOTOLUENE.
THR = An exper carc. [*3*, *23*]

4-AMINO-4,2-AZOTOLUENE.
THR = An exper carc. [*3*,*23*]

4-AMINO-4,3-AZOTOLUENE.
THR = An exper (+) carc. [*3*, *4*]

10-AMINOBENZ(*a*)ACRIDINE. $C_{17}H_{12}N_2$, mw: 244.3.
Acute tox data: ip LD_{LO} (mouse) = 200 mg/kg. [*3*]
THR = HIGH via ip route. An exper neo. [*3*]

AMINO BENZENE. See aniline.

o-**AMINOBENZENE SULFONIC ACID.** See sulfanilic acid.

2-AMINO BENZENETHIOL. Syn: *2-amino phenyl mercaptan*. Liquid. $NH_2C_6H_4SH$, mw: 125.2, mp: 23°, bp: 227.2°, flash p: 175°F, d: 1.168, vap. d: 4.3.
Acute tox data: Oral LD_{LO} (rat) = 500 mg/kg; ip LD_{LO} (mouse) = 25 mg/kg. [*3*]
THR = MOD via oral and HIGH via ip routes.
Fire Hazard: Mod, when exposed to heat or flame.
Disaster Hazard: Dangerous; see sulfur compounds. Can react with oxidizing materials.
To Fight Fire: Water, foam, CO_2, mist or spray, dry chemical.

m-**AMINOBENZOIC ACID.**
THR = An exper neo. [*3*]

o-**AMINOBENZOIC ACID.** See anthranilic acid.

p-**AMINO BENZOIC ACID.** Syns: *PABA*, *paraminol*. Yellowish to red crystals. $NH_2C_6H_4COOH$, mw: 137.1, mp: 187°.
Acute tox data: Oral LD_{50} (mouse) = 2850 mg/kg; oral LD_{50} (rabbit) = 1830 mg/kg; iv LD_{50} (rabbit) = 2000 mg/kg. [*3*, *108*]
THR = MOD via oral and iv routes. Large doses by mouth can cause nausea, vomiting, skin rash, methemoglobinemia and possibly toxic hepatitis.
Fire Hazard: Slight.

2-AMINO BENZOTHIAZOLE. Crystals. $C_6H_4SCNH_2N$, mw: 150.2, mp: 132°.
Acute tox data: ip LD_{50} (mouse) = 200 mg/kg; iv LD_{50} (mouse) = 126 mg/kg. [*3*]
THR = HIGH via ip and iv routes.
Disaster Hazard: Dangerous; when heated to decomp, emits highly toxic fumes.

m-**AMINO BENZOTRIFLUORIDE.** Colorless liquid with aniline-like odor. $H_2NC_6H_4CF_3$, mw: 161.13, mp: 3°, bp: 189°, d: 1.303 @ 15.5°/15.5°, vap. d: 5.56.
THR = U, probably toxic. See fluorides.
Disaster Hazard: Dangerous; see fluorides.

p-**AMINOBENZOYL-α-DI-*n*-BUTYLAMINO PROPANOL SULFATE.** See butyn.

p-**AMINO BIPHENYL.** See *p*-amino diphenyl.

4-AMINO-(1,1′-BIPHENYL)-3-OL. $C_{12}H_{11}ON$, mw: 185.2.
THR = An exper carc. [*3*]

4′-AMINO-4-BIPHENYLOL.
THR = An exper carc. [*3*]

1-AMINOBUTANE. See butylamine.

2-AMINOBUTANE. See *sec*-butylamine.

2-AMINO-1-BUTANOL. Water-white liquid. $CH_3CH_2CHNH_2 \cdot CH_2OH$, mw: 89.14, mp: −2°, bp: 178°, flash p: 165°F (OC), d: 0.944 @ 20°/20°, vap. d: 3.06.
Acute tox data: Oral LD_{50} (mouse) = 2330 mg/kg; ip LD_{LO} (mouse) = 250 mg/kg. [*3*]
THR = MOD via oral and ip routes.
Fire Hazard: Mod; when exposed to heat or flame, can react with oxidizing materials.
Spontaneous Heating: No.
To Fight Fire: Water spray, alcohol foam, dry chemical.

1-AMINO-4-tert-BUTYL BENZINE. See *p-tert*-butyl aniline.

AMINOCYCLOHEXANE. See cyclohexyl amine.

AMINO DIACETIC ACID. See imino diacetic acid.

9-AMINO-1,2,5,6-DIBENZ ANTHRACENE. $C_{20}H_{13}N$, mw: 267.3.
THR = An exper neo. [*3*]

14-AMINO DIBENZ(a,h)ANTHRACENE. $C_{22}H_{15}N$, mw: 293.4.
THR = An exper carc. [*3*]

3-AMINO-2,5-DICHLOROBENZOIC ACID. See amiben.

p-**AMINO DIMETHYL ANILINE.** See dimethyl-*p*-phenylene diamine.

AMINO DIMETHYL BENZENE. See xylidene.

2-AMINO-4,6-DIMETHYL PYRIDINE. Crystals. $C_7H_{10}N_2$, mw: 122.17, bp: 235.3°, mp: 68.5°.
THR = U. See also pyridine.
Disaster Hazard: Slight; on decomp, emits toxic fumes.

2-AMINO-4,6-DINITROPHENOL. See picramic acid.

p-**AMINODIPHENYL.** Syns: *xenylamine; p-biphenylamine; 4-aminobiphenyl*. Colorless crystals. $C_6H_5C_6H_4NH_2$, mw: 169.2, mp. 53°, bp: 302°, d: 1.160 @ 20°/20°, autoign. temp.: 842°F.
Acute tox data: Oral LD_{50} (rat) = 500 mg/kg; oral LD_{LO} (dog) = 25 mg/kg. [*3*]

THR = HIGH via oral and inhal routes. An irr. Has caused bladder cancer in humans and exper animals. Effects resemble those of benzidine. See benzidine. [*3*, *4*, *14*, *23*]
Disaster Hazard: Mod; when strongly heated, emits toxic fumes.
Fire Hazard: Low to mod when exposed to heat, flames, (sparks) or powerful oxidizers.
To Fight Fire: Water spray, mist, dry chemical.

4-AMINODIPHENYL AMINE. $C_{12}H_{12}N_{12}$, mw: 184.3.
Acute tox data: Oral LD_{50} (rat) = 464 mg/kg; oral LD_{50} (mouse) = 464 mg/kg. [*3*]
THR = HIGH via oral route. An exper carc. [*23*]

AMINO DITHIOFORMIC ACID. See dithiocarbamic acid.

AMINOETHANE. See ethylamine.

2-AMINO ETHANE SULFONIC ACID. Syn: *taurine*. Large rods, sol in water, insol in alcohol.
$NH_2CH_2CH_2SO_3H$, mw: 125.14, decomp @ 300°.
Acute tox data: ip LD_{LO} (mouse) = 64 mg/kg. [*3*]
THR = HIGH via ip route.
Disaster Hazard: Dangerous; when heated to decomp, emits highly toxic fumes of SO_x.

2-AMINOETHANOL. See monoethanolamine.

1-AMINO-4-ETHOXYBENZENE. See *p-phenetidine*.

1-[(2-AMINOETHYL)AMINO]-2-PROPANOL. Syn: *n-(2-hydroxy propyl)ethylene diamine*. Viscous liquid, odor of ammonia. $C_5H_{14}N_2O$, mw: 118.2, bp: 112° @ 10 mm, d: 0.984 @ 25°/4°.
THR = U. An epoxy resin curing agent. Probably a skin irr and sensitizer. See also ethylene diamine.

***o*-AMINOETHYL BENZENE.** See *o*-ethylaniline.

2-AMINO ETHYL ETHANOL AMINE. Syn: *(2-hydroxyethyl) ethylene diamine*. Colorless liquid.
$NH_2CH_2CH_2NHCH_2CH_2OH$, mw: 104.16, bp: 243.7°, flash p: 216°F, d: 1.0304 @ 20°/20°, autoign. temp.: 695°F, vap. press: < 0.01 mm @ 20°, vap. d: 3.59.
Acute tox data: Oral LD_{50} (rat) = 3000 mg/kg; dermal LD_{50} (guinea pig) = 1800 mg/kg. [*3*]
THR = MOD via oral and dermal routes.
Fire Hazard: Low.
To Fight Fire: Alcohol foam, mist, dry chemical.

1-AMINOETHYL-2-HEPTADECYL GLYOXALIDINE.
THR = U. A plant fungicide.
Disaster Hazard: Slight; when exposed to heat, emits acrid fumes.

3-(2-AMINOETHYL)-INDOL-5-OL. $C_{10}H_{12}ON_2$, mw: 176.24.
THR = HIGH. An exper teratogen. [*3*]

3-(2-AMINOETHYL)-INDOL-5-OL + SULFATE CREATININE.
THR = An exper teratogen. [*3*]

***n*-AMINO ETHYL MORPHOLINE.** Liquid.
$C_2H_4OC_2H_4NC_2H_4NH_2$, mw: 130.2, mp: 25.6°, bp: 204.2°, flash p: 347°F (OC), d: 0.9915 @ 20°/20°, vap. d: 4.49.
Acute tox data: Oral LD_{50} (rat) = 3000 mg/kg; dermal LD_{50} (guinea pig) = 300 mg/kg; sc LD_{50} (mouse) = 2145 mg/kg. [*3*]
THR = HIGH via dermal and MOD via oral and sc routes.
Fire Hazard: Mod; when exposed to heat or flame, can react with oxidizing materials.
To Fight Fire: Alcohol foam, dry chemical.

AMINOETHYL PIPERAZINE. Light colored liquid.
$H_2NCH_2CH_2N(CH_2)_4NH$, mw: 129.2, d: 0.9852 @ 20°/20°, mp: −19°, bp: 220.4°, flash p: 200°F (OC), vap. d: 4.4.
Acute tox data: Oral LD_{50} (rat) = 2140 mg/kg; dermal LD_{50} (rabbit) = 880 mg/kg. [*3*]
THR = MOD via oral and dermal routes.
Fire Hazard: Mod via exposure to heat, flame or sparks, powerful oxidizers.
To Fight Fire: Alcohol foam.

2-AMINO-2-ETHYL-1,3-PROPANEDIOL. Colorless liquid or solid. $CH_2OHC(C_2H_5)NH_2CH_2OH$, mw: 119.16, mp: 38°, bp: 152° @ 10 mm, d: 1.099 @ 20°/20°, vap. d: 4.11.
THR = U. Probably LOW.
Fire Hazard: Slight; when exposed to heat or flame, can react with oxidizing materials.

2-AMINOETHYL SULFAMATE.
THR = U. Limited animal exper suggest LOW.
Disaster Hazard: Dangerous; see sulfamates.

2-AMINO-4-(ETHYLTHIO)-BUTYRIC ACID
$C_6H_{13}O_2NS$, mw: 163.3.
THR = An exper teratogen and carc. [*3*]

2-AMINO-N-FLUOREN-2-YL ACETAMIDE.
$C_{15}H_{14}ON_2$, mw: 238.3.
THR = MOD to animals. An exper carc. [*3*, *23*]

4-AMINO-4-FLUORO DIPHENYL. $C_{12}H_{10}NF$, mw: 187.2.
THR = An exper carc. [*3*, *23*]

4-AMINO FOLIC ACID. See aminopterin.

α-AMINOGLUTARIC ACID. See glutamic acid.

AMINO GUANIDINE NITRATE. $CH_6N_4HNO_3$, mw: 137.1.
THR = Can explode during vacuum evaporation. [*19*]

AMINO-4-GUANIDO VALERIC ACID. See argenine.

1-AMINO HEPTANE. See heptyl amine.

7-AMINOHEPTANOIC ACID ISOPROPYL ESTER. THR = U. Limited animal exper suggest MOD toxicity and irr.

6-AMINO HEXANOIC ACID. $C_6H_{13}O_2N$, mw: 131.2. THR = An exper teratogen. [*3*]

***p*-AMINOHYDROGENATED CARDANOL.** See 4-amino-3-pentadecyl phenol.

2′-AMINO-3′-HYDROXY ACETOPHENONE. $C_8H_9O_2N$, mw: 151.2.
THR = HIGH to mice. An exper neo. [*3*]

2-AMINO-3-HYDROXY BENZOIC ACID. $C_7H_7O_3N$, mw: 153.2.
THR = An exper neo and carc. [*3*]

1-AMINO-5-HYDROXY BENZOTHIAZOLE. Crystals. $(NH_2)(OH)(C_6H_3)SCN$, mw: 166.2.
THR = U. Probably toxic.
To Fight Fire: Alcohol foam, CO_2, dry chemical.

α-AMINO-β-HYDROXYBUTYRIC ACID. See threonine.

8-AMINO-7-HYDROXY-3,6-NAPHTHALEN DISULFONIC ACID. $C_{10}H_7O_7NS_2Na_2$, mw: 363.3.
THR = An exper teratogen. [*3*]

α-AMINO-β-*p*-HYDROXY PHENYL PROPIONIC ACID. See tyrosine.

α-AMINO-β-HYDROXY PROPIONIC ACID. See serine.

5-AMINO IMIDAZOLE-4-CARBOXAMIDE. $C_4H_6ON_4$, mw: 126.1.
THR = An exper neo. [*3*]

α-AMINO-β-IMIDO AZOLEPROPIONIC ACID. See histidine.

1,2-AMINO-3-INDOLEPROPIONIC ACID. See tryptophane.

2-AMINO ISOBUTANE. See *tert*-butylamine.

α-AMINO ISOCAPROIC ACID. See leucine.

α-AMINO ISOVALERIC ACID. See valine.

AMINO METHANE. See monomethyl amine.

4-AMINO-4′-METHOXY-3-BIPHENYLOL. $C_{13}H_{13}O_2N$, mw: 215.3.
THR = An exper carc. [*3*]

1-AMINO-2-METHOXY-4-NITROBENZENE. Syns: *5-nitro-2-aminanisole, 2-methoxy-4-nitroaniline*. $C_6H_3(NO_2)(OCH_3)(NH_2)$, mw: 168.2.
Acute tox data: Oral LD_{50} (rat) = 704 mg/kg. [*3*]
THR = MOD via oral route.
Fire Hazard: See nitrates.
Disaster Hazard: See nitrates.

AMINO-β-METHYLAMINO-PROPIONIC ACID. $C_4H_{10}N_2$, mw: 118.2.
Acute tox data: ip LD_{LO} (rat) = 840 mg/kg; ip LD_{LO} (mouse) = 1680 mg/kg. [*3*]
THR = MOD via ip route. A neuro toxin. An exper carc. [*23*]

2-AMINO-4-(12-METHYL-7-BENZ(*a*)ANTHRYL METHYL)THIOBUTYRIC ACID. $C_{24}H_{23}O_2NS$, mw: 389.3.
THR = An exper carc to mice via iv route. [*103*]

2-AMINO-4-METHYL PENTANE. See 1,3-dimethyl butylamine.

2-AMINO-3-METHYL PENTANOIC ACID. See isoleucine.

1-AMINO-2-METHYL PROPANE. See isobutyl amine.

2-AMINO-2-METHYL-1,3-PROPANEDIOL. Clear liquid. $CH_2OHC(CH_3)NH_2CH_2OH$. mw: 105.2, mp: 110° bp: 151° @ 10 mm, vap. d: 3.63.
Acute tox data: Oral LD_{LO} (rabbit) = 1500 mg/kg. [*3*]
THR = MOD via oral route.
Fire Hazard: Slight; when exposed to heat or flame, can react with oxidizing materials.

2-AMINO-2-METHYL-1-PROPANOL. Syns: *AMP, isobutanolamine*. Colorless liquid or crystals. $CH_3CH_2NH_2CCH_2OH$, mw: 89.14, mp: 30°–31°, bp: 165°, flash p: 153°F (Tag OC) d: 0.934 @ 20°/20° vap. d: 3.04.
Acute tox data: Oral LD_{LO} (rabbit) = 1000 mg/kg. [*3*]
THR = MOD via oral route.
Fire Hazard: Mod; when exposed to heat or flame, can react with oxidizing materials.
To Fight Fire: Alcohol foam, dry chemical, mist or spray.

2-AMINO-3-METHYL PYRIDINE. Liquid. $NC_5H_3(CH_3)NH_2$, mw: 108.2, bp: 221.1°, vap. d: 3.73.
Acute tox data: Oral LD_{50} (rat) = 446 mg/kg; oral LD_{50} (wild birds) = 2 mg/kg. [*3*]
THR = HIGH via oral route.
Fire Hazard: Slight.

2-AMINO-4-METHYL PYRIDINE. Crystals. mp: 99°, bp: 230.9°, vap. d: 3.73.
Acute tox data: sc LD_{50} (mouse) = 100 mg/kg; iv LD_{50} (mouse) = 39 mg/kg. [*3*]
THR = HIGH via sc and iv routes.
Fire Hazard: Slight.

2-AMINO-5-METHYL PYRIDINE. Crystals; mp: 76.6°, bp: 227.1°, vap. d: 3.73.
THR = U. See also 2- amino-4-methyl pyridine.
Fire Hazard: Slight.

Disaster Hazard: On decomp, emits toxic fumes; reacts with oxidizing materials.

2-AMINO-6-METHYL PYRIDINE. mp: 43.7°, bp: 214.4°, vap. d: 3.73.
THR = U. See 2-amino-4-methyl pyridine.

2-AMINO-4-(METHYLTHIO)BUTYRIC ACID. See methionine.

α-AMINO-β-METHYL VALERIC ACID. See isoleucine.

α-AMINO-γ-METHYL VALERIC ACID. See leucine.

3-AMINO-2-NAPHTHOIC ACID. Syns: *3-aminonaphthalene-2-carboxylic acid, 3-aminoiso naphthoic acid.* Yellow scales from dilute alcohol, ether. $C_{11}H_9NO_2$, mw: 187.19, mp: 214°.
Acute tox data: Oral LD_{50} (mouse) = 1600 mg/kg. [*3*]
THR = MOD via oral route.

2-AMINO-1-NAPHTHOL HYDROCHLORIDE. $C_{10}H_9ON \cdot HCl$, mw: 195.7.
THR = An exper neo. [*3*]

12-AMINO-6-NAPHTHOL HYDROCHLORIDE.
THR = An exper neo. [*3*]

6-AMINO NICOTINAMIDE. $C_6H_7ON_3$, mw: 137.2.
THR = An exper teratogen. [*3*]

1-AMINO-4-NITROBENZENE. See *p*-nitroaniline.

4-AMINO-4'-NITROBIPHENYL. $C_{12}H_{10}O_2N_2$, mw: 214.2.
THR = An exper neo. [*3*]

2-AMINO-5-(5-NITRO-2-FURYL)-1,3,4-OXADIAZOLE. $C_6H_4O_4N_4$, mw: 196.1.
THR = An exper carc. [*3*]

2-AMINO-5-(5-NITRO-2-FURYL)-1,3,4-THIADIAZOLE. $C_6H_4O_3N_4S$, mw: 212.2.
THR = An exper (+) carc. [*3, 1*]

2-AMINO-4-(5-NITRO-2-FURYL)THIAZOLE. $C_7H_4O_3N_3S$, mw: 210.2.
THR = An exper neo. [*3*]

2-AMINO-4-(*p*-NITROPHENYL)THIAZOLE. $C_9H_6O_2N_3S$, mw: 220.2.
THR = An exper carc. [*3*]

2-AMINO-5-NITROTHIAZOLE. $C_3H_4O_2N_3$, mw: 146.2.
THR = An exper carc. [*3*]

1-AMINO-4-NONANE. See *n*-nonylamine.

1-AMINO OCTANE. See octyl amine.

6-AMINO PENICILLANIC ACID. $C_8H_{12}O_3N_2S$, mw: 216.3.
THR = An exper neo. [*3*]

4-AMINO-3-PENTADECYL PHENOL. Syn: *p-amino hydrogenated cardanol.* Tan colored, crystalline solid. $C_{15}H_{31}C_6H_3(NH_2)OH$, mw: 319.5, mp: 99–101°, bp: 225°-230° @ 1 mm.
THR = U.
Fire Hazard: Slight; when exposed to heat or flame, can react with oxidizing materials.

1-AMINOPENTANE. See amylamine.

2-AMINOPENTANE. See *sec*-amylamine.

2-AMINOPHENETOLE. See *o*-phenetidine.

4-AMINOPHENETOLE. See *p*-phenetidine.

***p*-AMINOPHENOL.** Syns: *rodinol, ursal P, p-hydroxy aniline.* Colorless crystals, slightly sol in water, alcohol and ether, insol in chloroform. $NH_2C_6H_4OH$, mw: 109.1, mp: 189.6°-190.2°, bp: 284° (decomp).
Acute tox data: Oral LD_{50} (rat) = 375 mg/kg; sc LD_{LO} (mouse) = 470 mg/kg. [*3*]
THR = HIGH via oral and sc routes. An allergen. This material resembles *p*-phenylene diamine. It has reportedly been the cause of contact dermatitis and bronchial asthma; can cause methemoglobinemia with cyanosis.
Disaster Hazard: Mod dangerous; when heated to decomp, emits toxic fumes.

***m*-AMINOPHENOL.**
Acute tox data: Oral LD_{50} (rat) = 1660 mg/kg; ip LD_{50} (mouse) = 150 mg/kg. [*3*]
THR = HIGH via ip; MOD via oral routes.

***o*-AMINOPHENOL.**
Acute tox data: ip LD_{LO} (mouse) = 200 mg/kg; sc LD_{LO} (cat) = 37 mg/kg. [*3*]
THR = HIGH via sc and ip routes.

2-AMINOPHENOL ACETIC ACID, ISOAMYL ESTER. $C_6H_3(OH)NH_2CH_2COOC_5H_{11}$, mw: 237.
THR = A powerful allergen. Human contact eczema has been reported.

AMINO PHENYL ARSINE ACID. See arsanilic acid.

4-(*p*-AMINOPHENYLAZO)-6-METHYL-*o*-CRESOL. $C_{14}H_{15}ON_3$, mw: 241.3.
THR = An exper carc. [*3*]

***p*-((*p*-AMINO PHENYL)AZO)PHENOL.** $C_{12}H_9ON_3$, mw: 211.2.
THR = An exper carc. [*3*]

AMINOPHENYL BORIC ACID. White crystals, slightly sol in water. $(NH_2C_6H_4)B(OH)_2$, mw: 137.0.
THR = U. See boron compounds.

***p*-AMINOPHENYL CADMIUM DILACTATE.**
THR = HIGH. See cadmium compounds.
Disaster Hazard: Dangerous; when heated to decomp, emits highly toxic fumes.

4-(o-AMINOPHENYL)-2-HYDRAZINO THIAZOLE. $C_9H_{10}N_4S$, mw: 206.3.
THR = An exper carc. [*3, 23*]

2-AMINOPHENYL MERCAPTAN. See 2-aminobenzenethiol.

AMINO PHENYL MERCURIC ACETATE. Colorless crystals, insol in water. $O_2Hg(C_6H_4)(NH_2)(C_2H_3)$, mw: 351.8, mp: 167°.
THR = HIGH. See mercury compounds, organic.
Disaster Hazard: Dangerous; see mercury compounds, organic.

p-AMINOPHENYL MERCURIC LACTATE.
THR = HIGH. See mercury compounds, organic.
Disaster Hazard: See mercury compounds, organic.

m-AMINOPHENYL METHYL CARBINOL. Solid, sol in water. $NH_2C_6H_4CH(OH)CH_3$, mw: 137, d: 1.12, bp: 217.3° @ 100 mm, fp: 66.4°, flash p: 315°F (OC).
THR = U.
Fire Hazard: Slight.
To Fight Fire: Alcohol foam.

2-AMINO-5-PHENYL-2-OXAZOLIN-4-ONE.
THR = An exper carc. [*23*]

α-AMINO-β-PHENYLPROPIONIC ACID. See phenyl alanine.

1-AMINOPROPANE. See *n*-propyl amine.

2-AMINOPROPANE. See isopropylamine.

2-AMINOPROPANOIC ACID. See alanine.

1-AMINO-2-PROPANOL. Syn: *MIPA, 2-hydroxypropylamine, isopropanolamine.* Liquid, slight ammonia odor, sol in water. $CH_3CH(OH)CH_2NH_3$, mw: 75.1, d: 0.969, mp: 1.4°, flash p: 171°F, vap. d: 2.6.
Acute tox data: Oral LD_{50} (rat) = 4260 mg/kg; dermal LD_{50} (rabbit) = 1640 mg/kg. [*3*]
THR = MOD via oral and dermal routes.
Fire Hazard: Mod via exposure to heat, flame, sparks, powerful oxidizers.
To Fight Fire: Alcohol foam.

3-AMINOPROPANOL. Colorless liquid, fishy odor. $H_2NCH_2CH_2CH_2OH$, mw: 75.11, bp: 168° @ 500 mm, flash p: >175°F (TOC), fp: 12.4°, d: 0.9786 @ 30°, vap. press: 2.1 mm @ 60°, vap. d: 2.59.
Acute tox data: Oral LD_{LO} (rat) = 2830 mg/kg; dermal LD_{50} (rabbit) = 1250 mg/kg. [*3*]
THR = MOD via dermal and oral routes. See also amines. A mod strong base which, from animal exper, would seem to be a non-specific irr.
Fire Hazard: Mod, when exposed to heat or flame; can react with oxidizing materials.
To Fight Fire: Foam, CO_2, dry chemical.

α-AMINOPROPIONIC ACID. See alanine.

β-AMINOPROPIONITRILE. See 3-aminopropionitrile.

3-AMINOPROPIONITRILE. Syn: *β-aminopropionitrile.* Liquid with amine odor. $H_2NCH_2CH_2CN$, mw: 70.1, bp: 185°.
THR = Nitriles generally have cyanide-like effects. See hydrocyanic acid and cyanides. An exper teratogen. [*3*]
Fire Hazard: See cyanides.
Explosion Hazard: See cyanides.
Disaster Hazard: See cyanides.

3-AMINO PROPIONITRILE FUMARATE.
Acute tox data: Oral LD_{LO} (rat) = 800 mg/kg. [*3*]
THR = MOD via oral route. An exper teratogen. [*3*]

p-AMINOPROPIOPHENONE. Syn: *PAPP.* Yellow needle-like crystals, sol in water and alcohol. $C_9H_{11}NO$, mw: 149.2, mp: 140°.
Acute tox data: ip LD_{50} (mouse) = 223 mg/kg; oral LD_{50} (dog) = 7.5 mg/kg; oral LD_{50} (mouse) = 86 mg/kg. [*3*]
THR = HIGH via oral, and ip routes. Large doses can cause cyanosis.
Disaster Hazard: Dangerous, when strongly heated, emits toxic fumes.

N-(3-AMINOPROPYL)CYCLOHEXYLAMINE. Sol in water. $C_6H_{11}NHC_3H_6NH_2$, mw: 156, d: 0.9, vap. d: 5.4, flash p: 175°F (OC).
THR = U.
Fire Hazard: Mod, via heat, flame, spark or oxidizers.
To Fight Fire: Alcohol foam.

N-AMINOPROPYL MORPHOLINE. Liquid. $C_2H_4OC_2H_4NC_3H_5NH_2$, mw: 144.21, mp: −15°, bp: 224.7°, flash p: 220°F (OC), d: 0.9872 @ 20°/20°, vap. press: 0.06 mm @ 20°, vap. d: 4.97.
Acute tox data: Oral LD_{50} (rat) = 3560 mg/kg; dermal LD_{50} (rabbit) = 1230 mg/kg. [*3*]
THR = MOD via oral and dermal routes.
Fire Hazard: Low when exposed to heat or flame; can react with oxidizing materials.
To Fight Fire: Alcohol foam, dry chemical.

AMINOPTERIN. Syn: *4-aminofolic acid.* Yellow needles, sol in sodium hydroxide solution. $C_{19}H_{20}N_8O_5$, mw: 440.4.
Acute tox data: Oral LD_{LO} (women) = 0.12 mg/kg ⟶ blood problems; ip LD_{50} (rat) = 3.4 mg/kg; dermal TD_{LO} (mouse) = 6 mg/kg ⟶ neo; oral LD_{LO} (rat) = 2.5 mg/kg. [*3*]
THR = VERY HIGH via oral, ip, dermal routes. An exper teratogen and neo. [*3*] An anti-metabolite sometimes used as a rodenticide. Can produce bone marrow depression.

Disaster Hazard: Dangerous; when strongly heated can emit highly toxic fumes.

2-AMINO PURINE-6-THIOL. $C_5H_5N_5S$, mw: 167.2.
THR = An exper teratogen. [*3*]

2-AMINOPYRIDINE. Syn: *α-aminopyridine*. White powder or crystals. $C_5H_6N_2$, mw: 94.11, mp: 58.1, bp: 210.6°.
Acute tox data: Inhal TC_{LO} (human) = 5 ppm for 5 hrs ⟶ CNS effects; ip LD_{50} (mouse) = 35 mg/kg; sc LD_{50} (mouse) = 70 mg/kg; iv LD_{50} (mouse) = 23 mg/kg. [*3*]
THR = HIGH via inhal, ip, sc and iv routes. A convulsant poison with effects resembling strychnine. In humans the symptoms are headache, weakness, collapse and epileptiform convulsions.
Disaster Hazard: Dangerous; when heated to decomp, emits highly toxic fumes.

3-AMINOPYRIDINE.
Acute tox data: ip LD_{50} (mouse) = 28 mg/kg; sc LD_{50} (mouse) = 30 mg/kg; iv LD_{50} (mouse) = 24 mg/kg. [*3*]
THR = HIGH via ip, sc and iv routes.

4-AMINOPYRIDINE.
Acute tox data: Oral LD_{50} (rat) = 21 mg/kg; ip LD_{50} (rat) = 6.5 mg/kg; sc LD_{50} (mouse) 5 mg/kg; oral LD_{50} (wild birds) = 4 mg/kg; im LD_{50} (wild birds) = 2.4 mg/kg. [*3*]
THR = HIGH via oral, ip, sc and im routes.

AMINO PYRIDINO TRIBROMO GOLD. Black crystals, sol in water. $(NH_2)(C_5H_4N)AuBr_3$, mw: 53.1, mp: 106° (decomp).
THR = U. See gold compounds.
Disaster Hazard: Dangerous; see bromides.

AMINOPYRINE. Syn: *dimethyl amino antipyrine*. Colorless leaflets, somewhat sol in water. $C_3ON_3(CH_3)_4C_6H_5$, mw: 231.29, mp: 107°–109°.
Acute tox data: Oral LD_{50} (rat) = 1070 mg/kg; ip LD_{50} (mouse) = 250 mg/kg. [*3*]
THR = MOD via oral and HIGH via ip routes. An allergen and irr. In sensitive individuals, bone marrow depression can occur, resulting particularly in leucopenia. Also can cause drug rash. Has been implicated in development of aplastic anemia. Mixed with $NaNO_2$ (1:1) it is an exper carc. [*3*]
Disaster Hazard: Mod dangerous; when heated to decomp, emits toxic fumes.

AMINO PYRINE SODIUM SULFONATE.
$C_{13}H_{17}O_4N_3SNa$, mw: 334.4.
THR = An exper teratogen. [*3*]

4-AMINO QUINOLINE-1-OXIDE.
THR = An exper carc. [*23*]

***p*-AMINO SALICYLIC ACID.** Syns: *deapasil*, *PAS*. Minute crystals (from alcohol), sol in dilute nitric acid and dilute sodium hydroxide. $C_7H_7NO_3$, mw: 153.13, mp: 150°.
Acute tox data: Oral LD_{50} (mouse) = 4000 mg/kg; sc LD_{50} (mouse) = 4000 mg/kg. [*3*]
THR = MOD via oral and sc routes. Intolerance or sensitivity may result in vomiting, diarrhea, skin rash and disturbed blood formation. In severe cases there is disturbed electrolyte balance, jaundice and anaphylactoid reactions.
Fire Hazard: Slight.

AMINO STILBENES.
THR = A recog carc. See aromatic amines. [*14*]

AMINOSUCCINIC ACID. See aspartic acid.

2-AMINO-5-SULFANILYL THIAZOLE. Syns: *promizole*, *thiazosulfone*. Fine needles from alcohol, sol in water. $C_9H_9N_3O_2S_2$, mw: 255.32, mp: 220°.
THR = LOW via oral and inhal routes. Possible chronic effects. Can cause a transient anemia. Over a period of time, thyroid enlargement may occur. Large doses can cause hemolysis, hematuria, cyanosis, gastrointestinal and CNS disturbances.
Disaster Hazard: Dangerous; see sulfur compounds.

4-AMINO-1,3,4-THIADIAZOLE HYDROCHLORIDE. $C_2H_3N_3S \cdot HCl$, mw: 137.6.
THR = An exper teratogen. [*3*]

2-AMINOTHIAZOLE. Light brown crystals. $C_3H_2NSNH_2$, mw: 100.14, mp: 90°, bp: decomp.
Acute tox data: Oral LD_{50} (rat) = 480 mg/kg. [*3*]
THR = HIGH via oral and MOD via inhal routes. Believed to cause nausea, headache and general weakness after exposure to from 3–100 mg/m^3. Ignites spont upon warming to 100° for $3\frac{1}{2}$ hrs. [*19*]
Disaster Hazard: Dangerous; see sulfur compounds.

α-AMINO-β-THIOLPROPIONIC ACID. See cysteine.

AMINOTOLUENE. See benzylamine.

***p*-(4-AMINO-*m*-TOLYL)AZO BENZENE SULFONAMIDE.** $C_{13}H_{14}O_2N_4$, mw: 290.4.
THR = An exper carc. [*3*]

3-AMINO-1,2,4-TRIAZOLE. Syns: *amizol*, *amitrole*, *aminotriazole*. Crystalline, sol in water, alcohol and chloroform. $C_2H_4N_4$, mw: 84.1, mp: 159°.
Acute tox data: Oral LD_{50} (rat) = 1100 mg/kg. [*3*]
THR = MOD via oral route. An exper (+) carc, neo. [*3,14,23,1*]
Disaster Hazard: Dangerous; when strongly heated, emits highly toxic fumes.

AMINO URACIL MUSTARD. $C_8H_{11}O_2N_3Cl_2$, mw: 252.1.

Acute tox data: Oral LD_{50} (rat) = 7.5 mg/kg; ip LD_{50} (rat) = 3.7 mg/kg. [*3*]
THR = HIGH via oral and ip routes. An exper (+) carc and teratogen via ip route. [*3,8,23*]

1-AMINO-4,5-XYLENOL. $C_8H_{11}ON$, mw: 137.2.
THR = An exper carc. [*3*]

AMIODOXYL BENZOATE. Syn: *arthrytin, oxoate.* White, odorless, slightly bitter, crystalline powder. $C_7H_8INO_4$, mw: 297.06.
Acute tox data: Oral LD_{50} (rat) = 2500 mg/kg. [*94*]
THR = MOD via oral route. It is a non-selective, systemic herbicide.
Disaster Hazard: Mod dangerous; when heated to decomp, emits toxic fumes.

AMITROLE. See 3-amino-1,2,4-triazole.

AMIZOL. See 3-amino-1,2,4-triazole.

AMMATE. See ammonium sulfamate.

AMMONAL.
THR = U, probably toxic.
Fire Hazard: A powerful oxidizer.
Explosion Hazard: High, when shocked or exposed to heat.
Disaster Hazard: Highly dangerous; it can explode when shocked or heated. When heated to decomp, emits toxic fumes; can react vigorously with reducing material.

AMMONIA-*d*$_3$. Syn: *trideuterio ammonia.* Gas. ND_3, mw: 20.05, mp: −74°, bp: −33.4°.
THR = See ammonia, anhydrous.
Fire Hazard: Mod, when exposed to flame or heat.
Explosion Hazard: Mod, when exposed to flame.
Disaster Hazard: Mod dangerous; reacts vigorously with oxidizing materials.

AMMONIA, ANHYDROUS. Syn: *ammonia gas.* Colorless gas, extremely pungent odor, liquified by compression. NH_3, mw: 17.03, mp: −77.7°, bp: −33.35°, lel = 16%, uel = 25%, d: 0.771 g/liter @ 0°, 0.817 g/liter @ −79°, autoign. temp.: 1204°F, vap. press: 10 atm @ 25.7°, vap. d: 0.6.
Acute tox data: Inhal LC_{LO} (human) = 10,000 ppm for 3 hrs; inhal TC_{LO} (human) = 20 ppm ⟶ irr; inhal LC_{50} (mouse) = 4837 ppm for 1 hr; inhal LC_{50} (cat) = 1066 ppm for 1 hr; inhal LC_{50} (rabbit) = 1066 ppm for 1 hr. [*3*]
THR = HIGH via inhal route. A powerful irr. Irr to eyes and mu mem of respiratory tract. Signs and symptoms of exposure are irr of the eyes, conjunctivitis, swelling of the eyelids, irr of the nose and throat, coughing, dyspnoea and vomiting. Irr of the skin may be experienced, especially if it is moist. Corneal ulcers have been reported following splashing of ammonia water in the eye. A common air contaminant.
Fire Hazard: Low, because it is difficult to ignite, when exposed to heat or flame.
Spontaneous Heating: No. Requires high conc in air before it catches fire.
Explosion Hazard: Mod, when exposed to flame. Forms explosive compounds in contact with Ag, acetaldehyde, acrolein, B, BI_3, halogens, BrF_5, $HClO_3$, ClO, ClF_3, chlorites, chlorosilane, CrO_3, chromyl chloride, (ethylene dichloride + liquid ammonia), ethylene oxide, Au, hexachloromelamine, (hydrazine + alkali metals), HBr, HOCl, $Mg(ClO_4)_2$, Hg, HNO_3, NO_2, N_2O_4, NCl_3, NF_3, nitryl chloride, OF_2, P_2O_5, P_2O_3, picric acid, (K + AsH_3), (K + PH_3), (K + $NaNO_2$), $KClO_3$, potassium ferricyanide, potassium mecuric cyanide, AgCl, (Na + CO), Sb, S, SCl_2, tellurium hydropentachloride, trichloromelamine. [*19*]
Disaster Hazard: Mod dangerous; when exposed to heat, emits toxic fumes.
To Fight Fire: Stop flow of gas.

AMMONIA AQUA. See ammonium hydroxide.

AMMONIACAL COPPER CARBONATE.
THR = U. A fungicide. See copper compounds.

AMMONIA GAS. See ammonia anhydrous.

AMMONIA SOLUTIONS CONTAINING FREE AMMONIA (NH_3), PRESSURIZED: AMMONIA SOLUTIONS CONTAINING 10% OR MORE FREE AMMONIA (NH_3), NON-PRESSURIZED: AMMONIA SOLUTIONS CONTAINING 10% AMMONIA (NH_3). See ammonia.

AMMONIUM ACETATE. $C_2H_7NO_2$, mw: 77.1, d: 1.07, mp: 114°. Crystals.
Acute tox data: iv LD (mice) = 90 mg of NH_4^+/kg. [*2*]
THR = HIGH via iv route. Decomp on contact with NaOCl. [*19*]

AMMONIUM ACID CARBONATE. See ammonium bicarbonate.

AMMONIUM ACID PHOSPHATE. See ammonium phosphate, monobasic.

AMMONIUM ALGINATE. Syn: *ammonium polymannurate.* Filamentous, grainy, granular or powdered, colorless or slightly yellow; slowly sol in water, insol in alcohol. $(C_6H_7O_6 \cdot NH_4)_n$, mw: 32,000–250,000.
THR = U. Stabilizer food additive. [*109*]

AMMONIUM ALUM. See aluminum ammonium sulfate.

AMMONIUM ARSENATE. White powder or crystal. $2NH_3 \cdot AsH_3O_4$, mw: 176, mp: decomp to yield HN_3.
THR = HIGH; see arsenic compounds.
Disaster Hazard: See arsenic compounds.

AMMONIUM-*m*-ARSENITE. White powder. NH_4AsO_2, mw: 125.
THR = HIGH; see arsenic compounds.

AMMONIUM AZIDE. Colorless plates. NH_4N_3, mw: 60.1. mp: 160°, bp: explodes, d: 1.346, vap. press: 1 mm @ 59.2° (sublimes).
THR = HIGH via oral and inhal routes.
Fire Hazard: Mod.
Explosion Hazard: Mod when heated.

AMMONIUM BENZENE SULFONATE. Crystals. $NH_4C_6H_5SO_3$, mw: 175.2, mp: 271–275° (decomp), d: 1.342.
THR = U. Probably toxic.
Disaster Hazard: Dangerous; see sulfonates.

AMMONIUM BIBORATE. Syn: *ammonium tetraborate*. Colorless, tetragonal crystals. $(NH_4)_2B_4O_7 \cdot 4H_2O$, mw: 263.4, mp: decomp.
THR = U. An herbicide. See also boron compounds.

AMMONIUM BICAMPHORATE. Syns: *acid ammonium camphorate, ammonium camphorate*. Crystalline powder; freely sol in water. $NH_4HC_{10}H_{14}O_4 \cdot 3H_2O$, mw: 271.31.
THR = See camphor.

AMMONIUM BICARBONATE. Syns: *ammonium acid carbonate, ammonium hydrogen carbonate*. White crystals, sol in water, insol in alcohol. NH_4HCO_3, mw: 79, d: 1.586, mp: 108°.
Acute tox data: iv LD_{50} (mouse) = 245 mg/kg. [*3*]
THR = HIGH via iv route. A general purpose food additive. [*109*]

AMMONIUM BICHROMATE. Syn: *ammonium dichromate*. Bright orange red needles. $(NH_4)_2Cr_2O_7$, mw: 252.10, mp: decomp before it melts, d: 2.15 @ 25°.
Acute tox data: sc LD_{LO} (guinea pig) = 30 mg/kg. [*3*]
THR = HIGH via sc, oral, inhal and contact routes. If swallowed it causes prompt vomiting, but if retained, may lead to kidney injury and ulceration of stomach. Chrome ulcers or sores of skin are well known, as is perforation of the nasal septum from chronic exposure to chrome salts. Prolonged inhal of dust can cause asthmatic symptoms. See chromium compounds.
Fire Hazard: Mod; reacts with reducing agents. An oxidizer. Decomp around 200°. Flam.

AMMONIUM BIFLUORIDE, SOLID or SOLUTION. Syn: *ammonium hydrogen fluoride*. White crystals. NH_4FHF, mw: 57.05, d: 1.21 @ 12°/12° (liquid), mp: 124.6°.
THR = HIGH via all routes. See also fluorides.
Disaster Hazard: See fluorides.

AMMONIUM BIMALATE. Syn: *acid ammonium malate*. Crystals, sol in 3 parts water, slightly sol in alcohol. $NH_4HC_4H_4O_5$, mw: 151.02, d: 1.51, mp: 161°.
THR = MOD via irr, oral and inhal routes.

AMMONIUM BINOXALATE. Colorless crystals. $NH_4HC_2O_4 \cdot H_2O$, mw: 125.08, mp: decomp, d: 1.556.
THR = See oxalates.

AMMONIUM BIPHOSPHATE. See ammonium phosphate, monobasic.

AMMONIUM BISULFATE. See ammonium hydrogen sulfate.

AMMONIUM BISULFITE. Syn: *ammonium hydrogen sulfite*. White crystals. NH_4HSO_3, mw: 99.1, mp: decomp.
THR = See bisulfites.
Disaster Hazard: See bisulfites.

AMMONIUM BITARTRATE. Syn: *acid ammonium tartrate*. White crystals, sol in water, acids and alkalies, insol in alcohol. $(NH_4)HC_4H_4O_6$, mw: 167, d: 1.636.
THR = See tartaric acid.

AMMONIUM BORATE. See ammonium biborate.

AMMONIUM BOROFLUORIDE.
THR = A strong irr. See also fluorides.
Disaster Hazard: Dangerous. See fluorides.

AMMONIUM BROMATE. Colorless crystals. Very water sol. NH_4BrO_3, mw: 145.96, mp: explodes.
THR = See bromates.
Fire Hazard: See bromates.
Explosion Hazard: Severe.
Disaster Hazard: Dangerous; see bromates.

AMMONIUM BROMIDE. Colorless, cubic, slightly hygroscopic crystals. NH_4Br, mw: 98.0, mp: sublimes @ 452°, bp: 235° (in vacuo), d: 2.429, vap. press: 1 mm @ 198.3°.
THR = See bromides. Can react violently with BrF_3, IF_7, K. [*19*]
Disaster Hazard: Dangerous; see bromides.

AMMONIUM BROMOPLATINATE. Red-brown, cubic crystals. $(NH_4)_2PtBr_6$, mw: 710.8, mp: 145° (decomp), d: 4.625.
THR = U. See also bromides and platinum compounds.

AMMONIUM BROMOSELENATE. Red octagonal crystals. $(NH_4)SeBr_6$, mw: 594.5, d: 3.326.
THR = HIGH. See selenium compounds and bromides.
Disaster Hazard: Dangerous; see selenium compounds and bromides.

AMMONIUM BROMOSTANNATE. Colorless crystals. $(NH_4)_2SnBr_6$, mw: 634.3, mp: decomp, d: 3.50.
THR = U. See also bromides and tin compounds.
Disaster Hazard: Dangerous; see bromides.

AMMONIUM CADMIUM BROMIDE. See cadmium ammonium bromide.

AMMONIUM CADMIUM CHLORIDE. Crystals. $4NH_4Cl \cdot CdCl_2$, mw: 397.3, d: 2.01.
THR = See cadmium compounds.
Disaster Hazard: See cadmium compounds and chlorides.

AMMONIUM CALCIUM ARSENATE. Colorless crystals. $NH_4CaAsO_4 \cdot 6H_2O$, mw: 305.1, mp: 140° (decomp), d: 1.905.
THR = See arsenic compounds.
Disaster Hazard: See arsenic compounds.

AMMONIUM CARBAMATE. White, crystalline rhombic powder, sol in water and alcohol, ammonia odor. $NH_4CO_2NH_2$, mw: 78, sublimates at 60°.
Acute tox data: iv LD_{50} (rat) = 39 mg/kg; iv LD_{50} (mouse) = 77 mg/kg. [*3*]
THR = HIGH via iv route. See also carbamates.
Disaster Hazard: Dangerous; see carbamates.

AMMONIUM CARBAZOTATE. See ammonium picrate.

AMMONIUM CARBONATE. Syn: *crystal ammonia.* Colorless crystalline plates. $(NH_4)_2CO_3$, mw: 96.09.
Acute tox data: sc LD_{50} (mouse) = 96 mg/kg. [*3*]
THR = HIGH via iv route. MILD irr. A general food additive. [*109*] Can decomp on contact with NaOCl.

AMMONIUM CARBOXY METHYL CELLULOSE.
THR = An exper neo. [*3*]

AMMONIUM CHLORATE. White crystal or mass. NH_4ClO_3, mw: 101.5.
THR = U. See also chlorates.
Fire Hazard: Mod due to spont chemical reaction with reducing agents. A powerful oxidizer. When contaminated by combustibles, may ignite.
Explosion Hazard: High due to shock, chemical reaction or exposure to heat. When contaminated it is very sensitive. It can be detonated.
Disaster Hazard: Highly dangerous; can explode when shocked or exposed to heat; when heated to decomp, emits highly toxic fumes; can react vigorously with reducing material.

AMMONIUM CHLORIDE. Syn: *sal ammonia.* White crystals. NH_4Cl, mw: 53.50, bp: 520°, mp: 337.8°, d: 1.520, vap. press: 1 mm @ 160.4° (sublimes).
Acute tox data: Oral LD_{50} (rat) = 1650 mg/kg; im LD_{50} (rat) = 30 mg/kg; ip LD_{50} (mouse) = 1300 mg/kg. [*3*]
THR = HIGH via im; MOD via oral and ip routes. A substance migrating to food from packaging materials. [*109*] Large doses can cause nausea, vomiting and acidosis. The fume is toxic by inhal. Can react violently with NH_4NO_3, BrF_3, IF_7, $KClO_3$. [*19*]

AMMONIUM CHLOROAURATE. Yellow crystals. NH_4AuCl_4, mw: 357.1.
THR = See gold compounds.

AMMONIUM CHLOROGALLATE. White crystals. NH_4GaCl_4. mw: 229.6, mp: 275°.
THR = See gallium compounds and chlorides.

AMMONIUM CHLOROIRIDATE. Red-black crystals. $(NH_4)_2IrCl_3$, mw: 441.9, mp: decomp, d: 2.856.
THR = See iridium compounds and chlorides.

AMMONIUM CHLOROIRIDITE. Greenish-brown solid. $(NH_4)_3IrCl_3 \cdot 1\frac{1}{2}H_2O$, mw: 487.
THR = See iridium compounds and chlorides.

AMMONIUM CHLOROOSMATE. Crystals. $(NH_4)_2OsCl_6$, mw: 439, d: 2.93.
THR = See osmium compounds and chlorides.

AMMONIUM CHLOROPALLADATE. Red-brown crystals. $(NH_4)_2PdCl_6$, mw: 355.5, mp: decomp, d: 2.418.
THR = See palladium compounds and chlorides.

AMMONIUM CHLOROPALLADITE. Olive-green crystals. $(NH_4)_2PdCl_4$, mw: 284.6, mp: decomp, d: 2.17.
THR = See palladium compounds and chlorides.

AMMONIUM CHLOROPLATINATE. Syn: *platinic ammonium chloride.* Cubic, yellow crystals. $(NH_4)_2PtCl_6$, mw: 443.9, mp. decomp, d: 3.065.
Acute tox data: Inhal TC_{LO} (human) = 0.0009 mg/min ⟶ pulmonary symptoms. [*3*]
THR = HIGH via inhal route. See platinum compounds and chlorides.

AMMONIUM CHLOROPLATINITE. Syn: *platinous ammonium chloride.* Red crystals. $(NH_4)_2PtCl_4$, mw: 373.1, mp: decomp, d: 2.936.
THR = See platinum compounds and chlorides.

AMMONIUM CHLOROPLUMBATE. Yellow crystals. $(NH_4)_2PbCl_6$, mw: 456. mp: 120° (decomp), d: 2.925.
THR = See lead compounds and chlorides.

AMMONIUM CHLOROSTANNATE. White crystals. $(NH_4)_2SnCl_6$, mw: 367.5, mp: decomp, d: 2.4.
THR = See tin compounds and chlorides.

AMMONIUM CHROMATE. Yellow, crystalline material. $(NH_4)_2CrO_4$, mw: 152.1, mp: decomp, d: 1.866.
THR = HIGH. See chromium compounds.
Fire Hazard: See chromium compounds.
Explosion Hazard: Slight, when shocked or heated.
Disaster Hazard: Mod dangerous; shock or heat will explode it; can react with reducing agents.

AMMONIUM CHROME ALUM. See ammonium chromic sulfate.

AMMONIUM CHROMIC SULFATE. Syn: *ammonium chrome alum.* Green or violet crystals. $NH_4Cr(SO_4)_2 \cdot 12H_2O$, mw: 478.4, mp: 94° ($-9H_2O$ @ 100°), d: 1.720.
THR = HIGH. See chromium compounds.
Disaster Hazard: See sulfates.

AMMONIUM COBALTOUS-*o*-PHOSPHATE. Violet crystals or powder. $NH_4CoPO_4 \cdot H_2O$, mw: 190.
THR = See cobalt compounds.

AMMONIUM COBALTOUS SULFATE. Red crystals. $(NH_4)_2SO_4 \cdot CoSO_4 \cdot 6H_2O$, mw: 395.3, d: 1.902.
THR = See cobalt compounds and sulfates.

AMMONIUM COPPER ARSENITE. See arsenic compounds.

AMMONIUM CUPRIC CHLORIDE. Blue crystals. $2NH_4Cl \cdot CuCl_2 \cdot 2H_2O$, mw: 277.5, mp: decomp @ 110°, d: 1.993.
THR = See copper compounds and chlorides.

AMMONIUM CUPRIC SULFATE. Blue violet crystals. $(NH_4)_2SO_4 \cdot CuSO_4$, mw: 291.8.
THR = See copper compounds and sulfates.

AMMONIUM CUPROUS IODIDE. Crystals. $NH_4I \cdot CuI \cdot H_2O$, mw: 353.4.
THR = See copper compounds and iodides.

AMMONIUM CYANATE. White crystals. NH_4OCN, mw: 60.1, mp: decomp @ 60°.
THR = See cyanates.

AMMONIUM CYANIDE. Solid, white powder or crystal. NH_4CN, mw: 44.1, mp: 36°, bp: sublimes @ 40°, d: 1.002 g/liter @ 100°, vap. press: 400 ppm @ 20.5°.
THR = See cyanides.
Fire Hazard: Mod, when exposed to heat or flame. At 35°, a flam gas is evolved.
Explosion Hazard: Mod. At 35°, a flam gas is liberated; see hydrogen cyanide.
Disaster Hazard: See cyanides.

AMMONIUM CYANOAURATE. Colorless plates. $NH_4Au(CN)_4 \cdot H_2O$, mw: 337.3, mp: decomp @ 200°.
THR = See cyanides and gold compounds.

AMMONIUM CYANOAURITE. Colorless crystals. $NH_4Au(CN)_2$, mw: 267.3, mp: decomp @ 100°.
THR = See cyanides and gold compounds.

AMMONIUM CYANOPLATINITE. Yellow crystals. $(NH_4)_2Pt(CN)_4 \cdot H_2O$, mw: 353.4.
THR = See cyanides and platinum compounds.

AMMONIUM DECABORATE. See ammonium pentaborate.

AMMONIUM DICHROMATE. See ammonium bichromate.

AMMONIUM DICYANOGUANIDINE. Crystals. $NH_4(NH_3CN(NCN)CN)$, mw: 126.2, mp: > 300°.
THR = See cyanides.

AMMONIUM DIFLUOPHOSPHATE. Colorless crystals. $NH_4PO_2F_2$, mw: 119, mp: 213°.
THR = See fluorides and phosphates.

AMMONIUM DIHYDROGEN ARSENATE. Colorless crystals. $NH_4H_2AsO_4$, mw: 159, mp: decomp to lose NH_3, d: 2.311.
THR = See arsenic compounds.

AMMONIUM DIHYDROGEN PHOSPHATE. See ammonium phosphate monobasic.

AMMONIUM DINITRO-*o-sec*-BUTYL PHENATE.
THR = U. An herbicide.
Disaster Hazard: Dangerous; when heated to decomp, emits highly toxic fumes.

AMMONIUM DINITRO-*o*-CRESOL. Crystals. $C_7H_9O_5O_3$, mw: 215.2.
THR = HIGH irr via oral and inhal routes.
Fire and Explosion Hazard: See nitrates.
Disaster Hazard: Dangerous; see nitrates.

AMMONIUM DINITRO-*o*-CRESOLATE.
THR = U. An herbicide. See also cresols.
Fire Hazard: Dangerous. A powerful oxidizer.
Explosion Hazard: U.
Disaster Hazard: Dangerous. A powerful oxidizer. When heated to decomp, emits highly toxic fumes.

AMMONIUM DITHIOCARBAMATE. Syn: *ammonium sulfocarbamate.* Yellow, lustrous, almost odorless crystals when fresh; sol in water, decomp in air. NH_2CSSNH_4, mw: 110.19.
THR = U. See sulfocarbamates.

AMMONIUM EMBELATE. Grayish-violet powder. $(NH_4)_2C_{18}H_{26}O_4$, mw: 342.47.
THR = LOW via oral and inhal route. A MILD irr. If inhal, may cause violent and prolonged sneezing.

AMMONIUM ETHYL SULFATE. Syn: *ammonium sulfethylate.* Hygroscopic crystals, freely sol in water and alcohol. $NH_4C_2H_5SO_4$, mw: 143.17, mp: 99°.
THR = U. Probably an irr.
Disaster Hazard: Dangerous; see sulfates.

AMMONIUM FERRICYANIDE. Red crystals. $(NH_4)_3Fe(CN)_6$, mw: 266.1, mp: decomp.
THR = See ferricyanides.

AMMONIUM FERROCYANIDE. Yellow to blue crystals. $(NH_4)_4Fe(CN)_6 \cdot 3H_2O$, mw: 338.2, mp: decomp.
THR = See ferrocyanides.

AMMONIUM FLUOANTIMONITE. Colorless crystals. $(NH_4)_2SbF_6$, mw: 252.8, mp: sublimes with decomp.
THR = See fluorides and antimony compounds.

AMMONIUM FLUOBORATE. Crystals. NH_4BF_4, mw: 104.9, mp: sublimes, d: 1.851 @ 15°.
THR = See fluorides and boron compounds.

AMMONIUM FLUOGALLATE. White crystals. $(NH_4)_2GeF_6$, mw: 222.7, d: 2.564 @ 25°/25°.
THR = See fluorides and germanium compounds.

AMMONIUM FLUORIDE. White crystals. NH_4F, mw: 37.04, mp: subl, d: 1.009 @ 25°.
Acute tox data: ip LD_{50} (rat) = 32 mg/kg; oral LD_{50} (guinea pig) = 150 mg/kg. [*3*]
THR = HIGH via ip and oral routes.

AMMONIUM FLUOSILICATE. Syns: *cryptohalite, ammonium silico fluoride*. $(NH_4)_2SiF_6$, mw: 178.1, mp: subl, d: 2.01.
Acute tox data: Oral LD_{LO} (rat) = 100 mg/kg; oral LD_{50} (guinea pig) = 150 mg/kg. [*3*]
THR = HIGH via oral route. See fluosilicates and fluorides.

AMMONIUM FLUOSULFONATE. Colorless needles. NH_4SO_3F, mw: 117.1, mp: 244.7.
THR = See fluosulfonates.

AMMONIUM FLUOTITANATE. Crystals. $(NH_4)_2TiF_6$, mw: 198, mp: decomp.
THR = See fluorides.

AMMONIUM FLUOZIRCONATE. Crystals. $(NH_4)_2ZrF_6$, mw: 241.3, d: 1.154.
THR = See fluorides.

AMMONIUM FORMATE. White, deliquescent crystals. NH_4COOH, mw: 63.1, mp: 116°, bp: decomp @ 180°, d: 1.280.
Acute tox data: Oral LD_{50} (mouse) = 2250 mg/kg; iv LD_{50} (mouse) = 410 mg/kg. [*3*]
THR = MOD via oral and HIGH via iv routes.

AMMONIUM GLUTAMATE. See monoammonium glutamate.

AMMONIUM HEXAFLUOPHOSPHATE. Syn: *ammonium phosphorus hexafluoride*. Colorless crystals. NH_4PF_6, mw: 163, mp: decomp, d: 2.180.
THR = See fluorides and phosphates.

AMMONIUM HEXAFLUOPHOSPHATE FLUORIDE. White crystals or powder. $NH_4PF_6NH_4F$, mw: 200.06, mp: no melting, subl @ about 140°.
THR = See fluorides and ammonium hexafluophosphate.

AMMONIUM HEXANITRO COBALTATE. $(NH_4)_3Co \cdot (NO_2)_6$, mw: 389.
THR = See cobalt compounds and nitrites. Explodes @ 200°. Impact sensitive. [*19*]

AMMONIUM HYDRATE. See ammonium hydroxide.

AMMONIUM HYDROGEN CARBONATE. See ammonium bicarbonate.

AMMONIUM HYDROGEN FLUORIDE. See ammonium bifluoride.

AMMONIUM HYDROGEN SELENATE. Crystals. NH_4HSeO_4, mw: 162.0, mp: decomp, d: 2.162.
THR = See selenium compounds.

AMMONIUM HYDROGEN SULFATE. White, rhombic crystals, sol in water, insol in acetone. NH_4HSO_4, mw: 115.11, mp: 146.9°, d: 1.78.
Acute tox data: Oral LD_{50} (rat) = 3000 mg/kg. [*3*]
THR = MOD via oral route.
Disaster Hazard: Dangerous, when heated to decomp, emits highly toxic fumes and sulfuric acid and sulfur oxides.

AMMONIUM HYDROGEN SULFIDE. See ammonium sulfhydrate.

AMMONIUM HYDROGEN SULFITE. See ammonium bisulfite.

AMMONIUM HYDROSULFIDE. See ammonium sulfhydrate.

AMMONIUM HYDROXIDE. Syns: *aqua ammonium, water of ammonia, aqua ammonia, ammonium hydrate*. Colorless liquid. NH_4OH, mw: 35.05, mp: −77°.
Acute tox data: Oral LD_{LO} (human) = 43 mg/kg; inhal LC_{LO} (human) = 5000 ppm; oral LD_{50} (rat) = 350 mg/kg. [*3*]
THR = HIGH via oral and inhal routes. A general purpose food additive which migrates to food from packaging materials. [*109*]
Fire Hazard: Slight; when heated, emits toxic fumes; can react from mod to violently with acrolein, acrylic acid, chlorosulfonic acid, dimethyl sulfate, halogens, (Au + aqua regia), HCl, HF, HNO_3, oleum, β-propiolactone, propylene oxide, $AgNO_3$, Ag_2O, (Ag_2O + C_2H_5OH), $AgMnO_4$, H_2SO_4. [*19*]
Disaster Hazard: Dangerous; emits irr fumes, and liquid can inflict burns. Use with adequate ventilation.

AMMONIUM HYPOPHOSPHITE. White granules. H_6NO_2P, mw: 83.
THR = When heated it can liberate highly toxic and flam PH_3. [2] See phosphine.

AMMONIUM IODIDE. Colorless, hygroscopic crystals. NH_4I, mw: 145, mp: subl @ 551°, bp: 220° (vacuo), d: 2.514 @ 25°, vap. press: 1 mm @ 210.9°.
THR = Can react violently with BrF_3, IF_7, K. [*19*]

AMMONIUM MAGNESIUM ARSENATE. Colorless crystals. $NH_4MgAsO_4 \cdot 6H_2O$, mw: 289.4, mp: decomp, d: 1.932 @ 15°.
THR = See arsenic compounds.

AMMONIUM MAGNESIUM CHROMATE. Yellow crystals. $(NH_4)_2CrO_4 \cdot MgCrO_4 \cdot 6H_2O$, mw: 400.5, mp. decomp, d: 1.84.
THR = See Chromium compounds.
Fire Hazard: Mod, as a result of chemical reaction with reducing agents. An oxidizer.
Disaster Hazard: Mod dangerous; when heated, can explode.

AMMONIUM MOLYBDATE. Colorless or slightly greenish or yellowish crystals. $(NH_4)_6Mo_7O_{24} \cdot 4H_2O$, mw: 1236.0, mp: $-H_2O$ @ 90°, bp: decomp @ 190°, d: 2.398.
Acute tox data: Oral LD_{50} (rat) = 333 mg/kg. [*3*]
THR = HIGH via oral, inhal routes. An irr. No cases of human poisoning have been reported. Animal exper indicate relatively LOW systemic tox but MOD severe local irr of skin, eyes and mu mem. Large doses have produced kidney damage in exper animals. See molybdenum compounds.

AMMONIUM MOLYBDO TELLURATE. Colorless crystals. $(NH_4)_6(TeMo_6O_{24}) \cdot 7H_2O$, mw: 1321.7, mp: 550° (decomp), d: 2.78.
THR = See tellurium compounds.

AMMONIUM MONOHYDROGEN ARSENATE. White crystals or powder $(NH_4)_2HAsO_4$, mw: 176, mp: decomp, d: 1.989.
THR = See arsenic compounds.

AMMONIUM MONOSULFIDE. See ammonium sulfide.

AMMONIUM NICKEL CHLORIDE. Green crystals. $NH_4Cl \cdot NiCl_2 \cdot 6H_2O$, mw: 291.2, d: 1.654.
THR = See nickel compounds and chlorides.

AMMONIUM NICKEL SULFATE. Syn: *double nickel salt*. Black to green crystals. $(NH_4)_2SO_4 \cdot NiSO_4 \cdot 6H_2O$, mw: 395, d: 1.923.
THR = See nickel compounds and sulfates.

AMMONIUM NITRATE. Colorless crystals. NH_4NO_3, mw: 80.05, mp: 169.6°, bp: 210° @ 11 mm, d: 1.725 @ 25°.
THR = LOW via irr; allergen. There have been reports of faintness and low blood pressure in workers exposed. These symptoms could be due to nitrites present as impurities. See also nitrates.
Fire Hazard: See nitrates; can ignite when mixed with acetic acid. [*19*]
To Fight Fire: Use water in large amounts. It is important that the mass of materials be kept cool and that burning be extinguished promptly. Ventilate well.
Explosion Hazard: May explode under confinement and high temperatures. Explosions have occurred in ships' holds, etc. There have been warehouse fires that did not detonate. See also nitrates. This material explodes more readily if contaminated, and must be kept cool and unconfined. Can react violently or explode when mixed with powdered metals, (NH_4Cl + heat), (C + heat), chlorides, organic matter, P, (K + $(NH_4)_2SO_4$), NaOCl, $NaClO_4$, (NaK + $(NH_4)_2SO_4$), S. [*19*] See also explosives, high.
Disaster Hazard: Dangerous; heat and confinement may explode it; when heated to decomp, emits highly toxic fumes of oxides of nitrogen; can react vigorously with reducing materials.

AMMONIUM NITRATE, FERTILIZER. See ammonium nitrate.

AMMONIUM NITRIDO OSMATE. NH_4OsNO_3, mw: 270.3.
THR = Explodes @ 150°. [*19*] See also osmium.

AMMONIUM NITRITE. White to yellow crystals. NH_4NO_2, mw: 64, mp: explodes @ 60°–70°, bp: subl. 30° in vacuo, d: 1.69.
THR = See nitrites.
Fire Hazard: See nitrites.
Explosion Hazard: Severe, when shocked or exposed to heat.
Disaster Hazard: See nitrites.

AMMONIUM OXALATE. Colorless crystals. $(NH_4)_2C_2O_4 \cdot H_2O$, mw: 142.12, mp: decomp, d: 1.50.
THR = See oxalates. Can react violently with (NaOCl + ammonium acetate). [*19*]

AMMONIUM PENTABORATE. Syn: *ammonium decaborate*. White solid. $NH_4B_5O_8 \cdot 4H_2O$, mw: 272.20.
THR = See boron compounds.

AMMONIUM PERCHLORATE. White crystals. NH_4ClO_4, mw: 117.50, mp: decomp, d: 1.95.
THR = See perchlorates.
Fire Hazard: MOD, when exposed to heat or flame or by spont chemical reaction with reducing materials. A very powerful oxidizer. Ignites violently with combustibles.
Explosion Hazard: Severe, decomp @ 130° and explodes @ 380°. When contaminated by powdered carbon, ferrocene, S, organic matter, powdered metals it becomes impact sensitive. [*19*] See also perchlorates.
Disaster Hazard: See perchlorates and explosives, high.

AMMONIUM PERCHROMATE. See ammonium peroxychromate.

AMMONIUM-*m*-PERIODATE. Colorless crystals. NH_4IO_4, mw: 209, mp: explodes, d: 3.056.

THR = Can become very unstable and explode on contact. See iodates.

AMMONIUM PERMANGANATE. Crystalline solid. NH_4MnO_4, mw: 137.0, mp: explodes, d: 2.208 @ 10°.

THR = See manganese compounds.

Fire Hazard: Mod, by chemical reaction with reducing agents. A powerful oxidizer.

Explosion Hazard: High, when shocked or warmed to 60°. Can be exploded by percussion.

Disaster Hazard: Mod dangerous shock and heat will explode it; when heated to decomp, emits toxic fumes; can react with reducing material.

AMMONIUM PEROXYBORATE. White crystals. $NH_4BO_3 \cdot \frac{1}{2}H_2O$, mw: 85.9, mp: decomp.

THR = See boron compounds.

Fire Hazard: Slight, by chemical reaction with reducing agents. An oxidizer.

AMMONIUM PEROXY CHROMATE. Red-brown crystals. $(NH_4)_3CrO_8$, mw: 234.1, mp: decomp @ 40°, bp: explodes @ 50°.

THR = See chromium compounds.

Fire Hazard: Mod, by chemical reaction with reducing agents. A powerful oxidizer.

Explosion Hazard: Mod, when heated.

Disaster Hazard: Mod dangerous; when heated to decomp emits toxic fumes and may explode.

AMMONIUM PEROXY DISULFATE. See ammonium persulfate.

AMMONIUM PERRHENATE. White plates. NH_4ReO_4, mw: 268.2, mp: decomp, d: 3.97.

THR = U.

Fire Hazard: Mod, by chemical reaction with reducing agents.

AMMONIUM PERSULFATE. Syn: *ammonium peroxydisulfate*. White crystals. $(NH_4)_2S_2O_8$, mw: 228.20, mp: decomp @ 120°, d: 1.982.

Acute tox data: Oral LD_{50} (rat) = 820 mg/kg. [*3*]

THR = MOD via oral route.

Fire Hazard: Mod by chemical reaction with reducing agents. A powerful oxidizer.

Explosion Hazard: Mod, oxygen released quietly in a fire, probably at a low temperature. When mixed with Na_2O_2 and heated and/or crushed, can explode. Also can explode when mixed with (powdered Al and H_2). [*19*]

Disaster Hazard: Dangerous, see sulfates. Can react vigorously with reducing agents.

AMMONIUM PHOSPHATE, DIBASIC. Syns: *ammonium phosphate secondary*, *diammonium hydrogen phosphate*, *diammonium phosphate*, *DAP*. White crystals or powder, sol in water, insol in alcohol. $(NH_4)_2HPO_4$, mw: 132, d: 1.619, mp: 155° (decomp).

THR = U. A general purpose food additive which migrates to food from packaging materials. [*109*]

Disaster Hazard: See phosphates.

AMMONIUM PHOSPHATE, MONOBASIC. Syns: *ammonium acid phosphate*, *ammonium biphosphate*, *ammonium dihydrogen phosphate*, *ammonium phosphate primary*. Brilliant white crystals or powder, mod sol in water. $NH_4H_2PO_4$, mw: 115, d: 1.803 @ 19°, mp: 190°.

THR = U. A general purpose food additive. [*109*] Mixed with NaOCl the NaOCl decomp. [*19*]

Disaster Hazard: Dangerous; see phosphates.

AMMONIUM PHOSPHATE PRIMARY. See ammonium phosphate, monobasic.

AMMONIUM PHOSPHATE, SECONDARY. See ammonium phosphate, dibasic.

AMMONIUM PHOSPHIDE.

Acute tox data: Inhal LC_{LO} (rat) = 580 ppm for 1 hr, Inhal LC_{LO} (guinea pig) = 288 ppm for 2 hrs. [*3*]

THR = HIGH via inhal and oral routes. See phosphine.

AMMONIUM PHOSPHORUS HEXAFLUORIDE. See ammonium hexafluophosphate.

AMMONIUM PICRATE. Syns: *ammonium carbazotate*, *ammonium picronitrate*. Yellow crystals. $NH_4C_6H_2N_3O_7$, mw: 246.14, mp: decomp, bp: explodes @ 423°, d: 1.719.

THR = An allergen. MOD irr to skin, eyes and mu mem. See also picric acid, nitrates.

Fire Hazard: Mod, by spont chemical reaction. A powerful oxidizer.

Explosion Hazard: High, when shocked or exposed to heat or flame, particularly if contaminated by metals. See also explosives, high.

Disaster Hazard: Highly dangerous; will explode when shocked; when heated to decomp, emits highly toxic fumes of NO_x, etc; can react vigorously with reducing materials.

AMMONIUM PICRATE, WET. Syn: *ammonium carbazotate*.

THR = See ammonium picrate.

Fire Hazard: Mod by chemical reaction with reducing agents. An oxidizer.

Explosion Hazard: Mod, when heated.

Disaster Hazard: Dangerous; when heated to decomp, emits toxic fumes of oxides of nitrogen, etc., and explodes.

AMMONIUM PICRONITRATE. See ammonium picrate.

AMMONIUM POLYMANNURATE. See ammonium alginate.

AMMONIUM POLYSULFIDE. See ammonium sulfide.

AMMONIUM POTASSIUM SELENIDE. NH_4SeK, mp: 136.1.
THR = An exper (+) neo and carc. [*3*] There is question as to exper carc. [*9*]

AMMONIUM SACCHARIN. White crystals or a white crystalline powder; freely sol in water. $C_7H_8N_2O_3S$, mw: 200.
THR = A non-nutritive sweetener food additive. See also saccharin. [*109*]

AMMONIUM SELENATE. Colorless crystals. $(NH_4)_2SeO_4$, mw: 179.04, mp: decomp, d: 2.194 @ 20°/4°.
THR = See selenium compounds.

AMMONIUM SELENIDE. White crystals. $(NH_4)_2Se$, mw: 115.0, mp: decomp.
THR = See selenium compounds.
Fire Hazard: See hydrogen selenide.
Disaster Hazard: Dangerous; when heated to decomp, or on contact with acid or acid fumes, it emits highly toxic fumes of selenium and will react with water or steam to produce toxic and flam vapors.

AMMONIUM SELENITE. Colorless or slightly reddish crystals. $(NH_4)_2SeO_3$, mw: 163.
THR = See selenium compounds.

AMMONIUM SILICOFLUORIDE. See ammonium fluosilicate.

AMMONIUM SULFAMATE. Syn: *ammate*. Deliquescent crystalline material (white crystalline solid). $NH_4OSO_2NH_2$, mw: 114.1, bp: 160° (decomp), mp: 131°.
Acute tox data: Oral LD_{50} (rat) = 1600 mg/kg; ip LD_{50} (rat) = 800 mg/kg. [*3*]
THR = MOD via oral and ip routes.
Explosion Hazard: Slight, when exposed to heat or by spont chemical reaction (hydrolysis); in a hot acid sol this material can undergo spont hydrolysis, liberating much heat.
Disaster Hazard: Dangerous; see sulfonates.

AMMONIUM SULFATE. Brownish-gray to white crystals. $(NH_4)_2SO_4$, mw: 132.09, mp: > 280° (decomp), d: 1.77.
Acute tox data: Oral LD_{50} (rat) = 3000 mg/kg [*3*]
THR = MOD via oral route. A general purpose food additive. [*109*]
Disaster Hazard: Dangerous. Can react violently when mixed with $(K + NH_4NO_3)$, $KClO_3$, KNO_2, $(NaK + NH_4NO_3)$. [*19*]

AMMONIUM SULFHYDRATE. Syns: *ammonium hydrosulfide, ammonium hydrogen sulfide*. Powder or crystals. NH_4HS, mw: 51.11, mp: 118° (150 atm), d: 1.17, vap. press: 400 mm @ 21.8°.
Acute tox data: Oral LD_{LO} (mouse) = 80 mg/kg; dermal LD_{LO} (mouse) = 2457 mg/kg; ip LD_{LO} (mouse) = 10 mg/kg. [*3*]
THR = HIGH via oral and ip routes; MOD via dermal route. HIGH irr. Penetrates skin readily.
Fire Hazard: See sulfides.
Disaster Hazard: See sulfides.
Explosion Hazard: See sulfides.

AMMONIUM SULFIDE. Syn: *ammonium polysulfide*. Yellow, hygroscopic crystals. $(NH_4)_2S$, mw: 68.2, mp: decomp.
THR = HIGH via oral and dermal routes of exposure. See sulfides. Evolves H_2S on contact with acid or acid fumes. Fatal poisoning has been reported from use in hair waving lotion.
Fire Hazard: See sulfides.
Explosion Hazard: See sulfides.
Disaster Hazard: See sulfides.

AMMONIUM SULFITE. Colorless crystals. $(NH_4)_2SO_3 \cdot H_2O$, mw: 134.16, mp: 60°–70° (decomp), bp: subl @ 150°, d: 1.41 @ 25°.
THR = See sulfites.

AMMONIUM SULFOCYANATE. See ammonium thiocyanate.

AMMONIUM TELLURATE. White powder. $(NH_4)_2TeO_4$, mw: 227.7, mp: decomp, d: 3.01 @ 25°.
THR = See tellurium compounds.

AMMONIUM TETRABORATE. See ammonium biborate.

AMMONIUM TETRACHLORO CUPRATE. $(NH_4)_2CuCl_4$, mw: 241.4.
THR = U. Can react violently with Na or K. [*19*]

AMMONIUM TETRACHLORO ZINCATE. White, thin shiny platelets. Hygroscopic and water sol. $ZnCl_2 \cdot 2NH_4Cl$, mw: 243.3, mp: 150° (approx), d: 1.879.
THR = Effects are those of components: Zinc chloride and ammonium chloride, both of which are tox and irr and are described under appropriate headings.
Disaster Hazard: Dangerous; see chlorides.

AMMONIUM TETRACHROMATE. $(NH_4)_2Cr_4O_{13}$, mw: 452.1.
THR = See chromium compounds. Decomp @ 175°. [*19*] A powerful oxidizer.

AMMONIUM TETRAPEROXY CHROMATE. $(NH_4)_3CrO_8$, mw: 234.1.
THR = See chromium compounds. Contact with

H_2SO_4 can explode. Also it is unstable and can explode from shock or 50° Δ. [*19*]

AMMONIUM THALLIUM CHLORIDE. Colorless crystals. $3NH_4Cl \cdot TlCl_3 \cdot 2H_2O$, mw: 507.3, d: 2.39.
THR = See thallium compounds and chlorides.

AMMONIUM THIOANTIMONATE. Yellow crystals. $(NH_4)_3SbS_4 \cdot 4H_2O$, mw: 376.2, mp: decomp.
THR = See antimony and sulfur.
Disaster Hazard: Dangerous; when heated to decomp or on contact with acid or acid fumes, emits highly toxic fumes of SO_x and Sb.

AMMONIUM THIOCYANATE. Syn: *ammonium sulfocyanate*. Colorless solid or deliquescent crystals. NH_4SCN, mw: 76.1, mp: 149.6°, bp: decomp @ 170°, d: 1.305.
Acute tox data: Oral LD_{LO} (mouse) = 330 mg/kg; ip LD_{LO} (mouse) = 500 mg/kg. [*3*]
THR = HIGH via oral and ip routes. An herbicide. Mixtures with $Pb(NO_3)_2$ can explode. [*19*]
Disaster Hazard: See thiocyanates.

AMMONIUM THIOGLYCOLATE. Colorless liquid, strong skunk-like odor. $HSCH_2COONH_4$, mw: 109.1.
Acute tox data: ip LD_{LO} (mouse) = 100 mg/kg. [*3*]
THR = HIGH via ip, oral and inhal routes. A strong allergen. Emits hydrogen sulfide. Upon local contact it can cause a contact dermatitis.
Disaster Hazard: Dangerous; when heated to decomp or on contact with acid or acid fumes, emits highly toxic fumes of sulfides.

AMMONIUM THIOSULFATE. White, monoclinic crystals. $(NH_4)_2S_2O_3$, mw: 148.2, d: 1.679, mp: 150° (decomp).
THR = See thiosulfates. Mixed with NaClO, can explode. [*19*]
Disaster Hazard: Dangerous; see sulfates.

AMMONIUM TRICHLOROACETATE. Colorless crystals. $NH_4O_2CCCl_3$, mw: 180.6.
THR = Powerful irr; HIGH via oral and inhal routes. A (S) carc. [*14*] An herbicide.
Disaster Hazard: Dangerous; when heated to decomp or on contact with acid or acid fumes, emits toxic fumes; will react with water or steam to produce toxic or corrosive fumes.

AMMONIUM TRICHROMATE. $(NH_4)_2Cr_3O_{10}$, mw: 350.
THR = See chromium compounds. It detonates at about 190°. [*19*]

AMMONIUM TRIIODIDE. Brown crystals. NH_4I_3, mw: 398.8, d: 3.749.
THR = See iodides.

AMMONIUM URANIUM FLUORIDE. See uranium ammonium fluoride.

AMMONIUM URANYL CARBONATE. Yellow crystals. $2(NH_4)_2CO_3 \cdot UO_2CO_3 \cdot 2H_2O$, mw: 558.3, mp: decomp @ 100°, d: 2.773.
THR= See uranium compounds, sol.
Radiation Hazard: See uranium.

AMMONIUM URANYL PENTAFLUORIDE. Crystals. $(NH_4)_3UO_2F_5$, mw: 419.2, mp: subl, d: 3.186.
THR = See uranium compounds, sol and fluorides.
Radiation Hazard: See uranium.
Disaster Hazard: Dangerous; see fluorides.

AMMONIUM-*m*-VANADATE. Colorless to yellow crystals. NH_4VO_3, mw: 117, mp: 200° (decomp), d: 2.326.
THR = See vanadium compounds.

AMMONIUM ZINC SULFATE. White crystals. $(NH_4)_2SO_4 \cdot ZnSO_4 \cdot 6H_2O$, mw: 401.7, mp: decomp, d: 1.931.
THR = See zinc compounds and sulfates.

AMMONIUM ZIRCONYL CARBONATE. $(NH_4)_3ZrOH(CO_3)_2 \cdot 2H_2O$, mw: 302.
THR = U. See zirconium compounds.

AMMONOBASIC MERCURIC CHLORIDE. See mercury compounds, inorganic.

AMOBARBITAL. Syns: *5-ethyl-5-isoamyl barbituric acid, amytal*. Slightly bitter crystals. $C_{11}H_{18}N_2O_3$, mw: 226.27.
Acute tox data: Oral LD_{50} (rats) = 575 mg/kg. [*3*]
THR = MOD via oral route. A depressant. Large doses cause poor judgment, emotional instability and at times a toxic psychosis. [*20*] Neurological signs may include nystagmus, dysarthria and ataxia. A potentially fatal dose = 5g, although recovery reported from 33g. Continued use of this material may lead to habituation.

AMOSITE. Syn: *brown asbestos*.
THR = An exper (S) carc. [*3, 23, 6*]. See also asbestos.

AMP. See 2-amino-2-methyl-1-propanol.

AMPHETAMINE. See "benzedrine."

AMPHIBOLE.
THR = An exper (S) carc via ipl route. [*103, 3, 6*] See asbestos.

AMPROLIUM. Syn: *1-((4-amino-2-propyl-5-pyrimidyl) methyl)-2-picolinium hydrochloride*.
THR = U. A food additive permitted in the feed and drinking water of animals and/or for the treatment of food-producing animals. A food additive permitted in food for human consumption. [*109*]
Disaster Hazard: Dangerous; see chlorides.

"AMSCO" NAPHTHAS AND SOLVENTS. Liquid petroleum hydrocarbons. Composition: 40–90% aromatic hydrocarbons.
THR = Variable. See specific petroleum hydrocarbon.
Fire Hazard: Variable.
Explosion Hazard: Variable.

AMSINCKA INTERMEDIA. Syn: *Tarweed.*
THR = An exper carc via oral route. [*103*]

AMYGDALIC ACID. See mandelic acid.

N-AMYL ACETATE. Syns: *amyl acetic ether, pear oil.* Colorless liquid, pear or banana-like odor. $CH_3COO(CH_2)_4CH_3$, mw: 130.18, mp: −78.5°, bp: 148° @ 737 mm, ulc: 55–60, lel = 1.1%, uel = 7.5%, flash p: 77°F (CC), d: 0.879 @ 20°/20°, autoign. temp.: 714°F, vap. d: 4.5.
Acute tox data: Oral LD_{50} (rabbit) = 7400 mg/kg; inhal TC_{LO} (human) = 188 ppm for 30 min ⟶ symptoms of irr. [*3*]
THR = MOD to LOW via oral and inhal route. An irr. Chronic tox is of a low order. When inhaled in high conc, amyl acetate is irr to the mu mem; it also possesses a narcotic effect, and from animal exper, it appears to be more toxic than butyl acetate. A conc of 1,000 ppm, breathed for half an hour, has caused headache, fatigue, oppression in the chest, and irr of the eyes and mu mem of the nose and throat, with excessive salivation. 5,000 ppm produces deep narcosis in cats in 30 min. Symptoms are burning of the eyes, lacrimation, headache, irr and dryness of the throat, "dopiness," fatigue, and occasionally vague nervousness. See also esters, amyl alcohol, acetic acid.
Fire Hazard: Dangerous, when exposed to heat or flame; when heated, it emits acrid fumes; can react with oxidizing materials.
Spontaneous Heating: No.
Explosion Hazard: Mod, when exposed to flame.
To Fight Fire: Alcohol foam, dry chemical.

***sec*-AMYL ACETATE.** Colorless liquid. $CH_3CO_2C_5H_{11}$, mw: 130.1, bp: 120°, flash p: 89°F (CC), d: 0.862–0.866 @ 20°/20°, vap. d: 4.48.
Acute tox data: Inhal TC_{LO} (human) = 200 ppm; inhal TC_{LO} (human) = 5000 ppm; inhal LC_{LO} (guinea pig) = 10,000 ppm for 5 hrs. [*3*]
THR = MOD via inhal route.
Fire Hazard: Dangerous, when exposed to heat or flame; can react with oxidizing materials.
To Fight Fire: Alcohol foam, dry chemical.

AMYL ACETATE ISO-. See isoamyl acetate.

AMYL ACETIC ETHER. See amyl acetate.

AMYL ALCOHOL. Syns: *1-pentanol, pentasol, primary amyl alcohol.* Clear liquid. $C_5H_{12}O$, mw: 88.1, mp: −79°, bp: 137.8°, flash p: 91°F (CC), d: 0.8168 @ 20°/20°, ulc: 40, lel = 1.2%, uel = 10% @ 212°F, autoign. temp.: 572°F, vap. press: 1 mm @ 13.6°, 10 mm @ 44.9°, vap. d: 3.04.
Acute tox data: Oral LD_{50} (rat) = 3030 mg/kg; dermal LD_{50} (rabbit) = 4490 mg/kg; oral LD_{50} (mouse) = 200 mg/kg. [*3*]
THR = MOD via oral, dermal routes. According to animal exper, the amyl alcohols are about four times as toxic as ethyl alcohol. However, because of their low volatility and their low solubility in the body fluids, they are absorbed slowly and only to a small extent. There is little definite evidence that their use in industry has resulted in poisoning. The vapor may, however, be irr to the eyes and the upper respiratory tract. See also alcohols. Ingestion can cause headache, nausea, vomiting, delirium and methemoglobin formation.
Fire Hazard: Dangerous, when exposed to heat or flame; can react with oxidizing materials.
Spontaneous Heating: No.
Explosion Hazard: Mod, when exposed to flame. Hydrogen trisulfide (H_2S_3) can explode it. [*19*]
To Fight Fire: Alcohol foam, dry chemical.

***n*-AMYL ALCOHOL.** See 1-pentanol.

***sec-n*-AMYL ALCOHOL.** Syns: *methyl propyl carbinol, 2-pentanol, diethyl carbinol.* Colorless liquid. $C_5H_{12}O$, mw: 88.2, bp: 116°, flash p: 94°F (CC), ulc: 40–45, uel = 9.0%, lel = 1.2%, fp: −50°, d: 0.8169 @ 20°/20°, autoign. temp.: 650°–725°F, vap. d: 3.04.
Acute tox data: Oral LD_{50} (rat) = 1470 mg/kg; ip LD_{LO} (rat) = 2130 mg/kg; oral LD_{LO} (rabbit) = 3500 mg/kg. [*3*]
THR = MOD via oral and ip routes.
Fire Hazard: Mod, when exposed to heat or flame; can react with oxidizing materials.
To Fight Fire: Alcohol foam, dry chemical.

AMYL ALCOHOL, COMMERCIAL. See isoamyl alcohol.

AMYL ALCOHOL, PRIMARY, ACTIVE. See 2-methyl-1-butanol.

***tert*-N-AMYL ALCOHOL, REFINED.** Syns: *dimethyl ethyl carbinol, 2-methyl-2-butanol.* Colorless liquid. $C_4H_9O(CH_3)$, mw: 88.15, mp: −11.9°, bp: 101.8°, flash p: 105°F (CC), d: 0.809, autoign. temp.: 819°F, vap. press: 10 mm @ 17.2°, lel = 1.2%, uel = 9%, vap. d: 3.03.
Acute tox data: Oral LD_{50} (rat) = 1000 mg/kg; ip LD_{LO} (rat) = 1530 mg/kg; [*3*] dermal LD_{50} (rabbit) = 2520 mg/kg. [*2*]
THR = MOD via oral, dermal and ip routes.

Fire Hazard: Mod, when exposed to heat or flame; can react with oxidizing materials.
To Fight Fire: Alcohol foam, dry chemical.

AMYL ALCOHOL, SYNTHETIC. See amyl alcohol.

AMYLAMINE. Syn: *1-amino pentane*. Water-white liquid $CH_3(CH_2)_4NH_2$, mw: 87.16, mp: −55°, bp: 104°, flash p: 45°F (OC), d: 0.7614 @ 20°/4°, vap. d: 3.01, lel = 2.2%, uel = 22%.
Acute tox data: Oral LD_{50} (rat) = 470 mg/kg; inhal LC_{LO} (rat) = 2000 ppm for 4 hrs; dermal LD_{50} (rabbit) = 1120 mg/kg. [*3*]
THR = MOD via oral, inhal and dermal routes. A strong irr. See also amines.
Fire Hazard: Dangerous, when exposed to heat or flame; can react with oxidizing materials.
To Fight Fire: Alcohol foam, dry chemical.

sec-**AMYLAMINE.** Syns: *2-aminopentane, methyl propyl carbinylamine*. $CH_3(CH_2)_2CH(CH_3)NH_2$, mw: 87.16, d: 0.7, vap. d: 3.0, bp: 93°, flash p: 20°F.
THR = See also amylamine.
Fire Hazard: Dangerous, when exposed to heat or flame.
To Fight Fire: Alcohol foam.

AMYLAMINE LAURATE. Liquid. flash p: 150°F, d: 0.88.
THR = U. See amines.
Fire Hazard: mod, when exposed to heat or flame; can react with oxidizing materials.
To Fight Fire: Water, mist or spray, CO_2, dry chemical.

AMYLAMINE OLEATE. Liquid. flash p: 220°F, d: 0.89.
THR = U. See amines.
Fire Hazard: Mod, when exposed to heat or flame; can react with oxidizing materials.
To Fight Fire: Water, mist or spray, CO_2, dry chemical.

AMYLAMINE STEARATE. Liquid. flash p: 160°F, d: 0.88.
THR = U. See amines.
Fire Hazard: Mod, when exposed to heat or flame; can react with oxidizing materials.
To Fight Fire: Water, mist or spray, CO_2, dry chemical.

p-tert-**AMYLANILINE.** $(C_2H_5)(CH_2)_2CC_6H_4NH_2$, mw: 161, d: 0.9, bp: 260°, flash p: 215°F.
THR = U. See also aniline.
Fire Hazard: Low, when exposed to heat or flame.
To Fight Fire: Dry chemical, spray or mist.

AMYL AZIDE. $C_5H_{11}N_3$, mw: 113.
THR = MOD via irr, oral and inhal routes. Narcotic in high conc. Can cause a fall in blood pressure.

5-N-AMYL-1,2-BENZANTHRACENE. $C_{23}H_{22}$, mw: 298.5.
THR = An exper carc. [*3*].

AMYL BENZENE. Syn: *n-pentyl benzene*. Liquid. $C_6H_5C_5H_{11}$, mw: 148.24, bp: 202.2°, mp: −78.3°, flash p: 150°F (OC), d: 0.8627 @ 20°/4°, vap. d: 5.11.
Fire Hazard: Mod, when exposed to heat or flame; reacts with oxidizing materials.
To Fight Fire: Foam, dry chemical.

sec-**AMYLBENZENE.**
Acute tox data: Oral LD_{LO} (rat) = 5000 mg/kg. [*3*]
THR = MOD via oral and inhal routes.

tert-**AMYLBENZENE.**
Acute tox data: Oral LD_{LO} (rat) = 5000 mg/kg. [*3*]
THR = MOD vial oral and inhal routes.

AMYL BENZOATE. Clear liquid. $C_{12}H_{15}O_2$, mw: 192.3, bp: 260°, flash p: 230°F, d: 0.98.
THR = U. See also esters, amyl alcohol, benzoic acid.
Fire Hazard: Mod, when exposed to heat or flame; reacts with oxidizing materials.
To Fight Fire: CO_2, dry chemical.

N-AMYL BENZYL CYCLOHEXYLAMINE. Liquid. $C_{18}H_{29}N$, mw: 259.5.
Acute tox data: Oral LD_{LO} (rat) = 2000 mg/kg; sc LD_{LO} (rat) = 2000 mg/kg; oral LD_{LO} (rabbit) = 2000 mg/kg. [*3*]
THR = MOD via oral and sc routes.
Fire Hazard: Mod flam liquid; can react with oxidizing materials.

AMYL BENZYL ETHER. Syns: *benzyl isoamyl ether, "gardenia oxide."* Liquid. $C_{12}H_{18}O$, mw: 178.3, bp: 235°, d: 0.965 @ 15.5°/15.5°.
THR = U. See also ethers.
Fire Hazard: Mod, when exposed to heat or flame; can react with oxidizing materials.
To Fight Fire: Foam, CO_2, dry chemical.

AMYL BIPHENYL. Liquid. $C_{17}H_{20}$, mw: 224.3, mp: −60°, bp: 305°–337°, flash p: 300°F, d: 0.958 @ 20°/20°, vap. d: 7.73.
Acute tox data: Oral LD_{LO} (rat) = 5000 mg/kg. [*3*]
THR = MOD via oral and inhal routes.
Fire Hazard: Slight, when exposed to heat or flame.
Disaster Hazard: Mod dangerous, when heated to decomp, emits toxic fumes; can react with oxidizing materials.
To Fight Fire: Foam, CO_2, dry chemical.

AMYL BORIC ACID. Colorless crystals, water sol. $(C_5H_{11})B(OH)_2$, mw: 116.0, mp: 94° (decomp).
THR = See boron compounds.

d-**AMYL BROMIDE.** Colorless liquid. $CH_3(CH_2)_4Br$,

mw: 151.1, bp: 120°, flash p: 90°F, fp: < −30°, d: 1.211 @ 25°/25°.

Acute tox data: ip LD_{LO} (mouse) = 150 mg/kg. [*3*]

THR = HIGH via ip route. Probably a local irr and narcotic in high conc. See also chlorinated HC, aliphatic. Causes liver damage.

Fire Hazard: High, when exposed to heat or flame.

Disaster Hazard: Dangerous, see bromides. Can react with oxidizing materials.

To Fight Fire: Alcohol foam, water mist or spray, dry chemical.

***tert*-AMYL CARBAMATE.** Crystals, camphor odor. $CH_3CH_2CH_2CH_2CH_2CO_2NH_2$, mw: 131.2, mp: 94.5°.

THR = U. See carbamates.

Fire Hazard: Slight, when heated.

AMYL CARBINOL. See *n*-hexyl alcohol.

AMYL CHLORIDE. Syn: *1-chloropentane*. Water white liquid with a sweet odor. $CH_3(CH_2)_3CH_2Cl$, mw: 106.60, mp: −99°, bp: 108.2°, flash p: 54°F (OC), d: 0.883 @ 20°/4°, autoign. temp.: 500°F, vap. d: 3.67, lel = 1.4%, uel = 8.6%.

THR = U. See also chlorinated HC, aliphatic.

Fire Hazard: Dangerous, when exposed to heat or flame.

Spontaneous Heating: No.

Explosion Hazard: Mod.

Disaster Hazard: Dangerous; when heated to decomp, emits highly toxic fumes of phosgene; can react with oxidizing materials.

To Fight Fire: Foam, CO_2, dry chemical.

***tert*-AMYL CHLORIDE.** Liquid. $CH_3CH_2CCl(CH_3)CH_3$, mw: 106.6, bp: 87°, d: 1.407, autoign. temp.: 653°F, vap. d: 3.67, lel – 1.5%, ucl = 7.4%.

THR = U. See also chlorinated HC, aliphatic.

Fire Hazard: Mod, when exposed to heat or flame.

Disaster Hazard: Dangerous; see chlorides. Can react with oxidizing materials.

To Fight Fire: Water may be used as a blanket. Foam, water mist or spray, dry chemical.

AMYL CHLORIDE, MIXED. Straw to deep purple liquid. $C_5H_{11}Cl$, mw: 106.6, bp: 85°–109°, flash p: 38°F (OC), d: 0.88 @ 20°, vap. d: 3.67.

THR = U. See chlorinated HC, aliphatic. See individual amyl chlorides.

Fire Hazard: Dangerous; see chlorides. Can react vigorously with oxidizing materials.

To Fight Fire: Foam, CO_2, dry chemical.

AMYL CHLORONAPHTHALENE. Liquid. $C_5H_{11}ClC_{10}H_6$, mw: 232.8, bp: 241°, flash p: 295°F, d: 1.07.

THR = U. See also chlorinated naphthalene.

Fire Hazard: Slight, when exposed to heat or flame.

Disaster Hazard: Dangerous; see chlorides. Can react with oxidizing materials.

To Fight Fire: Water, foam, CO_2, water mist or spray, dry chemical.

***p-tert*-AMYL-*o*-CRESOL.** $C_5H_{11}C_6H_3OHCH_3$, mw: 178.3, bp: 258°, flash p: 240°F, d: 0.97.

THR = HIGH via oral and inhal routes. See also 6-*n*-amyl-*m*-cresol.

Fire Hazard: Mod, when exposed to heat or flame.

Disaster Hazard: Dangerous; when heated, emits highly toxic fumes; can react vigorously with oxidizing materials.

6-*n*-AMYL-*m*-CRESOL.

Acute tox data: LD_{50} (rat) = 1500 mg/kg. [*3*]

THR = MOD via oral, and inhal routes.

4-*tert*-AMYL CYCLOHEXANOL. Liquid. $C_5H_{11}C_6H_4OH$, mw: 164.3, bp: 245°, flash p: 212°F, d: 0.91.

THR = MOD. See also cyclohexanol.

Fire Hazard: Mod, when exposed to heat or flame; can react with oxidizing materials.

To Fight Fire: Foam, CO_2, dry chemical.

AMYL DICHLOROSILANE. Liquid. $C_5H_{11}SiHCl_2$, mw: 171.2, bp: 235°.

THR = See chlorosilanes.

α-*n*-AMYLENE. Syns: *propylethylene, 2-methyl butene-2, 1-pentene*. Liquid, highly disagreeable odor. $CH_3(CH_2)_2CHCH_2$, mw: 70.13, mp: −124°, bp: 30.1°, lel = 1.6%, uel = 8.7%, flash p: 0°F (OC), d: 0.643, autoign. temp.: 527°F, vap. d: 2.42.

THR = MOD via oral and inhal routes. Narcotic in high conc. A simple asphyxiant.

Fire Hazard: Very dangerous, when exposed to heat, flame or powerful oxidizers.

Disaster Hazard: When heated to decomp, emits toxic fumes.

Spontaneous Heating: No.

Explosion Hazard: Mod, when exposed to flame.

To Fight Fire: Alcohol foam, spray, mist, dry chemical.

AMYLENE HYDRATE. See *tert*-amyl alcohol.

AMYLENES, MIXED. Water white liquid. C_5H_{10}, mw: 70.58, bp: 32.2°, flash p: 0°F, d: 0.66 @ 20°.

THR = MOD. See also α-*n*-amylene.

Fire Hazard: Dangerous, when exposed to heat or flame; reacts with oxidizing materials.

To Fight Fire: Foam, CO_2, dry chemical.

AMYL ETHER. Syns: *amyl oxide, diamyl ether*. Liquid. $((CH_3(CH_2)_3CH_2)_2)O$, mw: 158.3, mp: −69.3°, bp: 187°, flash p: 135°F (OC), d: 0.783 @ 20°/4°, vap. d: 5.46, autoign. temp.: 340°F.

THR = U. See also ethers.
Fire Hazard: Mod, when exposed to heat or flame; reacts with oxidizing materials. See ethers.
To Fight Fire: Alcohol foam, dry chemical.

AMYL FORMATE. Clear liquid. $HCOO(CH_2)_4CH_3$, mw: 116.2, mp: −73.5°, bp: 130.4°, flash p: 80°F, d: 0.893 @ 15°/4°.
THR = MOD irr via oral route of exposure.
Fire Hazard: Dangerous, when exposed to heat or flame; can react vigorously with oxidizing materials.
To Fight Fire: Foam, CO_2, dry chemical.

AMYL HYDRIDE. See *n*-pentane.

AMYL LACTATE. Colorless liquid. $CH_3CH(OH)COO(CH_2)_4CH_3$, mw: 160.2, bp: 210°, flash p: 175°F, d: 0.960 @ 20°.
THR = LOW irr via oral and inhal routes. See also esters.
Fire Hazard: Mod, when exposed to heat or flame; can react with oxidizing materials.
To Fight Fire: Foam, CO_2, dry chemical.

AMYL LAURATE. $C_5H_{11}O_2C(CH_2)_{10}CH_3$, mw: 270.44, bp: 290°, flash p: 300°F, d: 0.86.
THR = U. See also esters. May de-fat skin and lead to dermatitis.
Fire Hazard: Slight, when exposed to heat or flame; can react with oxidizing materials.
To Fight Fire: CO_2, dry chemical.

AMYL MALEATE. $C_7H_{24}O_4$, mw: 172.2, flash p: 270°F, bp: 270°–315°, insol in water.
THR = U. See esters.
Fire Hazard: Low.
To Fight Fire: Dry chemical, water spray or mist.

AMYL MERCAPTAN. Syn: *1-pentanethiol.* Water white to yellow liquid. $CH_3(CH_2)_4SH$, mw: 104.21, d: 0.857 @ 20°, bp: 123. 64°, flash p: 65°F, vap. press: 13.8 mm @ 25°, vap. d: 3.59.
Acute tox data: Inhal LC_{LO} (rat) = 2000 ppm, 4 hrs. [*3*]
THR = MOD irr via inhal and oral routes. An allergen. See also mercaptans. A weak sensitizer. Local contact may cause contact dermatitis.
Fire Hazard: Dangerous, when exposed to heat or flame.
Disaster Hazard: Dangerous; see mercaptans. Reacts with oxidizing materials.
To Fight Fire: Foam, CO_2, dry chemical.

AMYL MERCAPTANS (MIXED). $C_5H_{11}SH$, mw: 105, bp: 80°–120°, d: 0.8, flash p: 65°F (OC).
THR = U. See also amyl mercaptan.
Fire Hazard: Dangerous if exposed to heat or flame.
Disaster Hazard: Dangerous; see mercaptans.
To Fight Fire: Dry chemical, water mist, alcohol foam.

AMYL-2-METHYL ALCOHOL. Liquid. $CH_3(CH_2)_2CH(CH_3)CH_2OH$, mw: 102.2, bp: 130°, flash p: 114°F (CC), d: 0.804, vap. d: 3.52.
Acute tox data: Oral LD_{50} (rat) = 1410 mg/kg; dermal LD_{50} (rabbit) = 3560 mg/kg. [*3*]
THR = MOD via oral and dermal routes.
Fire Hazard: Mod, when exposed to heat or flame; can react with oxidizing materials.
To Fight Fire: CO_2, dry chemical.

AMYL METHYL CARBINOL. See 2-heptanol.

AMYL-4-METHYL ALCOHOL.
Acute tox data: Oral LD_{50} (rat) = 2600 mg/kg; Inhal LC_{LO} (rat) = 2000 ppm, for 4 hrs; ip LD_{50} (mouse) = 812 mg/kg; dermal LD_{50} (rabbit) = 3560 mg/kg. [*3*]
THR = MOD via oral, inhal, ip and dermal routes.

AMYL METHYL KETONE. See 2-heptanol.

N-AMYL METHYL KETONE. Liquid. $CH_3CO(CH_2)_4CH_3$, mw: 114.2, bp: 150°, flash p: 120°F (OC), d: 0.817, autoign. temp.: 991°F, vap. d: 3.94.
Acute tox data: Oral LD_{50} (rat) = 1670 mg/kg; oral LD_{50} (mouse) = 730 mg/kg; dermal LD_{50} (rabbit) = 13000 mg/kg; inhal LC_{LO} (rat) = 4000 ppm, 4 hrs. [*3*]
THR = MOD via oral, inhal; LOW via dermal routes.
Fire Hazard: Mod, when exposed to heat or flame; can react with oxidizing materials.
To Fight Fire: Foam, CO_2, dry chemical.

AMYL NAPHTHALENE. Liquid. $C_{10}H_7C_5H_{11}$, mw: 198.3, mp: −30°, bp: 288°, flash p: 255°F (OC), d: 0.973, vap. d: 6.86.
THR = U. See also naphthalene.
Fire Hazard: Slight, when exposed to heat or flame; can react with oxidizing materials.
To Fight Fire: Foam, CO_2, dry chemical.

AMYL-β-NAPHTHOL. Liquid. $C_5H_{11}C_{10}H_7O$, mw: 214.3, bp: 308°, flash p: 325°F, d: 0.91.
THR = U. See also β-naphthol.
Fire Hazard: Slight, when exposed to heat or flame.
Disaster Hazard: Mod dangerous; when heated to decomp, emits toxic fumes; can react with oxidizing materials.
To Fight Fire: Foam, CO_2, dry chemical.

AMYL NITRATE (MIXED ISOMERS). Liquid. $C_5H_{11}NO_3$, mw: 133.15, bp: 145°, flash p: 125°F (OC), d: 0.99.
Acute tox data: Inhal LC_{LO} (rat) = 3593 ppm; inhal LC_{LO} (mouse) = 1374 ppm; inhal LC_{LO} (rabbit) = 1703 ppm. [*3*]
THR = MOD via inhal route.
Fire Hazard: Mod, when exposed to heat or flame or by spont chemical reaction. An oxidizing agent.

Explosion Hazard: See nitrates.
Disaster Hazard: See nitrates.
To Fight Fire: Foam, dry chemical.

AMYL NITRITE. Syn: *isoamyl nitrite.* Clear yellowish liquid, peculiar ethereal fruity odor and pungent aromatic taste. $CH_3(CH_2)_4NO_2$, mw: 117.15, bp: 96°–99°, d: 0.8528 @ 20°/4°, autoign. temp.: 408°F, vap. d: 4.0.
THR = MOD via oral and inhal routes. Causes flushing of skin, rapid pulse, headache and fall in blood pressure.
Fire Hazard: Mod, when exposed to heat or flame or by spont chemical reaction. An oxidizing material. Vapors explode when heated. [*19*]
Disaster Hazard: Dangerous: see nitrates. Can react with oxidizing or reducing materials.
To Fight Fire: Alcohol foam.

AMYL OCTYLAMINE. Liquid. $C_5H_{11}NHC_8H_{17}$, mw: 199.3.
THR = U. An insecticide. See also amines.
Fire Hazard: Mod flam material; can react with oxidizing materials.

AMYLOFORM. Soft white powder. A formaldehyde-carbohydrate condensation product.
THR = An allergen. Local contact may cause contact dermatitis.
Fire Hazard: Slight, when exposed to heat or flame; can react with oxidizing materials.

AMYL OLEATE. Liquid.
$CH_3(CH_2)_7CH:CH(CH_2)_7COOC_5H_{11}$, mw: 352.6, bp: 200°, flash p: 366°F, d: 0.86.
THR = MOD irr via oral and inhal routes.
Fire Hazard: Slight, when exposed to heat or flame; can react with oxidizing materials.
To Fight Fire: CO_2, dry chemical.

AMYL OXALATE. $(COOC_5H_{11})_2$, mw: 230, d: 1.0, bp: 240°–225°, flash p: 245°F.
THR = HIGH. See also oxalates.
Fire Hazard: Low, when exposed to heat, flame or powerful oxidizers.
To Fight Fire: Dry chemical, CO_2, water mist.

AMYL OXIDE. See amyl ether.

***o*-AMYL PHENOL.** Liquid. $C_5H_{11}C_6H_4OH$, mw: 164.24, bp: 342°, flash p: 219°F (OC), d: 0.96–0.97, vap. d: 5.66.
THR = No data. See also *p-tert*-amyl phenol.
Fire Hazard: Slight, when exposed to heat or flame.
Disaster Hazard: Dangerous; when heated to decomp, emits toxic fumes; can react vigorously with oxidizing materials.
To Fight Fire: Foam, CO_2, dry chemical.

***o-sec*-AMYL PHENOL.** Clear, straw colored liquid; very slightly sol in water; sol in oil and organic solvents. $C_6H_{11}C_6H_4OH$, mw: 164.24, d: (30/30°) 0.955–0.971, initial bp: >235°, final bp: <250.0°, flash p: 200°F.
THR = No data. See also *p-tert*-amyl phenol.
Fire Hazard: Mod, when exposed to heat or flame.
To Fight Fire: Foam, fog, dry chemical, water mist or spray, multi-purpose dry chemical.

***o-tert*-AMYL PHENOL.** Pale yellow liquid; slightly sol in water, sol in oil and organic solvents.
$(CH_3)_2(C_2H_5)CC_6H_4OH$, mw: 164.24, d: 0.96–0.97 @ 30°, initial bp: not < 233°, final bp: not > 245°, flash p: 219°.
THR = U. See also *p-tert*-amyl phenol.
Fire Hazard: Mod, when exposed to heat or flame.
To Fight Fire: Water, foam.

***p-sec*-AMYL PHENOL.** $C_5H_{11}C_6H_4OH$, mw: 164.24, d: <1.0, bp: 482°–516°F, flash p: 270°F.
THR = U. See also *p-tert*-amyl phenol.
Fire Hazard: Slight when exposed to heat or flame.
To Fight Fire: Dry chemical, water mist, CO_2.

***p-tert*-AMYL PHENOL.** Syns: *p-(dimethyl propyl)*-phenol, *pentaphen*. Colorless needles.
$CH_3CH_2C(CH_3)_2C_6H_4OH$, mw: 164.2, bp: 250°, mp: 92°–93°, flash p: 232°F (OC).
Acute tox data: Oral LD_{50} (rat) = 3080 mg/kg; dermal LD_{50} (rabbit) = 2000 mg/kg. [*3*]
THR = MOD via oral and dermal routes.
Fire Hazard: Slightly flam material.
Disaster Hazard: Mod dangerous; when heated to decomp, emits toxic fumes; can react with oxidizing materials.
To Fight Fire: Dry chemical, water mist, CO_2.

***p-tert*-AMYL PHENOXY ETHANOL.** Liquid.
$C_{13}H_{20}O_2$, mw: 212.3, bp: 291°, flash p: 282°F, d: 1.01.
THR = No data. See also alcohols.
Fire Hazard: Low, when exposed to heat or flame.
Disaster Hazard: Mod dangerous; when heated to decomp, emits toxic fumes; can react with oxidizing materials.
To Fight Fire: CO_2, dry chemical, water mist.

***p-tert*-AMYL PHENOXY ETHYL LAURATE.** Liquid. $(CH_2)_{10}C_{15}H_{22}O_3$, mw: 390.6, bp: 240°, flash p: 410°F, d: 0.94.
THR = No data. See also esters.
Fire Hazard: Low, when exposed to heat or flame.
To Fight Fire: Dry chemical, water mist, CO_2.

***p-tert*-AMYL PHENYL ACETATE.** $C_{13}H_{18}O_2$, mw: 206.2, flash p: 240°F, bp: 265°, d: 0.99.
THR = No data, probably low. See also esters.

Fire Hazard: Low, when exposed to heat or flame; can react with oxidizing materials.
Explosion Hazard: U. See also esters.
To Fight Fire: CO_2, dry chemical or water mist.

***p-tert*-AMYL PHENYL-*n*-AMYL ETHER.** Liquid $C_{16}H_{27}O_2$, mw: 251.4, bp: 285°, flash p: 260°F, d: 0.9.
THR = MOD irr via oral route.
Fire Hazard: Slight, when exposed to heat or flame; can react with oxidizing materials.
To Fight Fire: Foam, CO_2, dry chemical.

***p-tert*-AMYL PHENYL BUTYL ETHER.** $C_{15}H_{24}O$, mw: 220, d: 0.9, bp: 263°–265°, flash p: 275°F.
THR = No data. See also ethers.
Fire Hazard: Slight, when exposed to heat or flame.
To Fight Fire: Dry chemical, water mist, CO_2.

AMYL PHENYL ETHER. Liquid. $C_{11}H_{17}O_2$, mw: 181.3, flash p: 185°F, d: 0.92, vap. d: 5.7.
Acute tox data: Oral LD_{50} (mouse) = 2200 mg/kg. [*3*]
THR = MOD via oral route. See also ethers.
Fire Hazard: Mod, when exposed to heat or flame; can react with oxidizing materials.
To Fight Fire: Foam, dry chemical.

AMYL PHENYL METHYL ETHER. Liquid. $C_{12}H_{19}O_2$, mw: 195.3, bp: 240°, flash p: 210°F, d: 0.94.
THR = No data. See also ethers.
Fire Hazard: Low, when exposed to heat or flame; can react with oxidizing materials.
To Fight Fire: Foam, CO_2, dry chemical.

AMYL PHTHALATE. See di-*n*-amyl phthalate.

AMYL PROPIONATE. Syn: *pentyl propanoate*. Stable, colorless, apple-like odor, liquid. $(CH_2)_4C_4H_8O_2$, mw: 144.21, mp: −73.1°, bp: 106.2°, flash p: 106°F (OC), d: 0.8761 @ 15°/4°, vap. press: 10 mm @ 46.3°, vap. d: 5.0, autoign. temp.: 712°F.
THR = MOD irr via oral and inhal routes.
Fire Hazard: Mod, when exposed to heat or flame; can react with oxidizing materials.
To Fight Fire: Foam, CO_2, dry chemical.

AMYL SALICYLATE. Liquid. $C_{11}H_{16}O_3$, mw: 208.3, bp: 266.5°, flash p: 270°F (CC), d: 1.065, vap. d: 7.17.
Acute tox data: iv LD_{50} (dog) = 500 mg/kg. [*3*]
THR = HIGH via iv route. Mild sensitizer. Local contact may cause contact dermatitis.
Fire Hazard: Low, when exposed to heat or flame; can react with oxidizing materials.
To Fight Fire: CO_2, dry chemical, water mist.

AMYL SILICATE. Liquid. $Si(OC_5H_9)_4$, mw: 368.6, mp: 148° @ 3mm.
THR = No data. Probably MOD irr to skin, eyes and mu mem.
Fire Hazard: Slight.

AMYL STEARATE. Liquid. $(CH_2)_{16}C_7H_{14}O_2$, mw: 354.60, bp: 359.5°, flash p: 368°F (OC), d: 0.860.
THR = No data. Probably MOD irr.
Fire Hazard: Slight, when exposed to heat or flame; can react with oxidizing materials.
Spontaneous Heating: No.
To Fight Fire: CO_2, dry chemical, water mist.

AMYL SULFIDE. Syn: *diamyl sulfide*. Liquid. $(C_5H_{11})_2S$, mw: 174.3, bp: 170°–180°, flash p: 185°F (OC), d: 0.85–0.91.
THR = No data. At least MOD irr via oral and inhal routes.
Fire Hazard: Mod, when exposed to heat or flame.
Disaster Hazard: Dangerous; see sulfides. Can react vigorously with oxidizing materials.
To Fight Fire: Foam, CO_2, dry chemical.

AMYL SULFIDES (MIXED). $C_5H_{11}S$, mw: 103, d: 0.9, bp: 170°–180°, flash p: 185°F (OC).
THR = No data.
Fire Hazard: Mod, when exposed to heat or flame.
Disaster Hazard: Dangerous; see sulfur compounds.
To Fight Fire: Foam, fog, dry chemical.

AMYL TETRATHIO-*o*-STANNATE. Solid. $(C_5H_{11}S)_4Sn$, mw: 531.
THR = U. See tin compounds.
Disaster Hazard: Dangerous; see sulfur compounds.

***tert*-AMYL TETRATHIO-*o*-STANNATE.** Solid. $[CH_3CH_2C(CH_3)_2S]_4Sn$, mw: 531.5, mp: 44°.
THR = U. See tin compounds.
Disaster Hazard: Dangerous; see sulfur compounds.

AMYL TOLUENE. Liquid. $C_{12}H_{18}$, mw: 162.25, bp: 205°–210°, flash p: 180°F (OC), d: 0.87, vap. d: 5.6.
THR = U. See also toluene.
Fire Hazard: Mod, when exposed to heat or flame; can react with oxidizing materials.
To Fight Fire: Foam, CO_2, dry chemical.

AMYL TOLYL ETHER. Liquid. $C_{12}H_{18}O$, mw: 178.3, bp: 227°, flash p: 196°F, d: 0.91.
THR = U. See also ethers.
Fire Hazard: Mod, when exposed to heat or flame; can react with oxidizing materials.
To Fight Fire: Foam, CO_2, dry chemical.

AMYL TRICHLOROSILANE. Liquid. $C_5H_{11}SiCl_3$, mw: 205.6.
Acute tox data: Oral LD_{50} (rat) = 2340 mg/kg; inhal LC_{LO} (rat) = 2000 ppm, 4 hrs; dermal LD_{50} (rabbit) = 1500 mg/kg. [*3*]
THR = MOD via oral, inhal and dermal routes.
Disaster Hazard: See chlorosilanes.

AMYL TRIETHOXY SILANE. Liquid.

$C_5H_{11}Si(OC_2H_5)_3$, mw: 234.4, bp: 198°, d: 0.889 @ 20°.

Acute tox data: Oral LD_{50} (rat) = 20,000 mg/kg; dermal LD_{50} (rabbit) = 7130 mg/kg. [*3*]

THR = LOW via oral and dermal routes.

Fire Hazard: Mod flam material; reacts with oxidizing materials.

AMYL TRIMETHOXY SILANE. $C_8H_{20}O_3Si$, mw: 192.4.

Acute tox data: Oral LD_{50} (rat) = 4920 mg/kg [*3*]

THR = MOD via oral route.

AMYL TRIPHENYL GERMANIUM. Colorless crystals, insol in water, sol in organic solvents. $Ge(C_5H_{11})(C_6H_5)_3$, mw: 375.0, mp: 43°.

THR = U. See germanium compounds.

AMYL XYLYL ETHER. Liquid. $(CH_3)_2C_{11}H_{14}O$, mw: 192.29. bp: 250°–260°, flash p: 205°F (OC), d: 0.907.

THR = U. See also ethers.

Fire Hazard: Low, when exposed to heat or flame; can react with oxidizing materials.

To Fight Fire: Foam, CO_2, dry chemical.

AMYTAL. See amobarbital.

ANABASINE. Syns: *neonicotine, nicotine isomer*. Colorless, oily liquid, miscible with water. $C_{10}H_{14}N_2$, mw: 162.2, bp: 105° @ 22mm, d: 1.048 @ 20°/20°, fp: 9°.

Acute tox data: Oral LD_{LO} (rat) = 10 mg/kg; oral LD_{LO} (dog) = 50 mg/kg; iv LD_{LO} (dog) = 3 mg/kg; dermal LD_{LO} (guinea pig) = 290 mg/kg. [*3*]

THR = HIGH via oral, iv and dermal routes. Causes increased salivation, mental confusion, dizziness, disturbed hearing and vision, nausea, vomiting. In severe cases there is unconsciousness with convulsions.

Disaster Hazard: Dangerous; when heated to decomp, emits highly toxic fumes of NO_x and cyanides.

ANACARDIC ACID. Syns: *cashew nut shell oil, o-pentadecadienyl salicylic acid*. Crystals. $C_{22}H_{32}O_3$, mw: 344.5, mp: 36°.

THR = A powerful irr and allergen.

Fire Hazard: Slight.

ANASADOL. See salicylamide.

ANCIENT USE OF LEAD AND MERCURY. See Section 1.

ANDROST-4-ENE-3,17-DIONE. $C_{19}H_{26}O_2$, mw: 286.5.

THR = An exper teratogen and neo. [*3*]

Δ-ANDROSTEN-17(γ)-OL-3-ONE. See testosterone.

ANESTHESIA ETHER. See ethyl ether.

ANESTHESIN. Syn: *ethyl-p-aminobenzoate*. Crystals. $C_9H_{11}NO_2$, mw: 165.2, mp: 88°–90°.

THR = Used medicinally. See also esters, ethyl alcohol and *p*-amino benzoic acid.

Acute tox data: Oral LD_{50} (wild birds) = 56 mg/kg; uk LD_{50} (mouse) = 216 mg/kg. [*3*]

THR = HIGH via oral route. A mild sensitizer. Local contact may cause contact dermatitis.

Disaster Hazard: Mod dangerous, when heated to decomp, emits highly toxic fumes.

ANETHOLE. Syns: *p-propenyl anisole, anise camphor, p-methoxy propenyl benzene*. White crystals, anise oil odor, sol in 8 volumes of 80% alcohol, almost immiscible with water. $CH_3CH{:}CHC_6H_4OCH_3$, mw: 148, d: (*cis*) 0.9878 @ 20°/4°, (*trans*) 0.9883 @ 20°/4°, mp: (*cis*) −22.5°, (*trans*) 21.4°, bp: (*cis*) 79° @ 2.3 mm, (*trans*) 81° @ 2.3 mm.

Acute tox data: Oral LD_{50} (rat) = 2090 mg/kg; oral LD_{50} (mouse) = 3050 mg/kg. [*3*]

THR = MOD via oral route.

ANETHOLE-7-ANGELYL HELIOTRIDINE.

THR = An exper carc. [*23*]

ANETHOLE, *cis*.

Acute tox data: Oral LD_{50} (rat) = 150 mg/kg; ip LD_{50} (rat) = 93 mg/kg. [*3*]

THR = HIGH via oral and ip routes.

ANGIOTONIN

THR = An exper teratogen. [*3*]

ANGLISITE. See lead sulfate.

ANG-STERANTHRENE. $C_{23}H_{18}$, mw: 294.4.

THR = An exper carc. [*3*]

ANHALONIDINE. White crystals. $C_{12}H_{17}NO_3$, mw: 223.3, mp: 161°.

THR = HIGH via oral and inhal routes. A derivative of mescaline.

Disaster Hazard: Dangerous; when heated to decomp, emits highly toxic fumes.

ANILINE. Syns: *phenylamine, aminobenzene, aniline oil*. Colorless, oily liquid, characteristic odor. $C_6H_5NH_2$, mw: 93.12, bp: 184.4°, lel = 1.3%, ulc: 20–25, flash p: 158°F (CC), fp: −6.2°, d: 1.02 @ 20°/4°, autoign. temp.: 1139°F, vap. press: 1 mm @ 34.8°, vap. d: 3.22.

Acute tox data: Oral LD_{LO} (human) = 350 mg/kg; oral LD_{50} (rat) = 440 mg/kg; dermal LD_{50} (rat) = 1400 mg/kg; inhal LC_{LO} (rat) = 250 ppm, 4 hrs; dermal LD_{50} (guinea pig) = 1290 mg/kg; iv LD_{50} (rabbit) = 64 mg/kg. [*3*]

THR = HIGH via oral, iv and inhal routes. LOW via dermal routes. An exper (±) carc. [*3, 23, 10*] The most important action of aniline on the body is the formation of methemoglobin, with the resulting anoxemia and depression of the central nervous system. Some investigators believe that aniline may also have a direct toxic action, resulting in a fall in blood pressure and cardiac arrhythmia. In acute exposures, which usually result from spilling the liquid on the skin and clothes, but which may also follow inhal of the vapor given off when aniline is heated, the signs are of methemoglobinaemia and anoxemia. In less acute exposure which has been prolonged over some weeks or months, there is usually hemolysis of the red blood cells, followed by stimulation of the bone marrow and attempts at regeneration. The red cells may show stippling; immature cells may be present. The white blood cells usually show little change either in number or morphology. The liver may be affected, with production of jaundice. The urine is frequently dark brown or wine colored, and may contain hemoglobin, hematoporphyrin, and in some cases, excretion products of aniline, such as *p*-amino-phenol. Long continued employment in the manufacture of aniline dyes has been associated with the development of papillomatous growths of the bladder, some of which became malignant. Aniline itself has not been proven to be a carc, but the intermediates benzedine and naphthylamines have been incriminated. See α- and β-naphthylamine. A common air contaminant. Mild sensitizer. Local contact may cause contact dermatitis. A poison. Mod to violent reactions caused by mixing with acetic anhydride, chlorosulfonic acid, hexachloro melamine, HNO_3, (HNO_3 + N_2O_4 + H_2SO_4), (nitrobenzene + glycerine), oleum, O_3, (HCHO + $HClO_4$), perchromates, performic acid, K_2O_2, β-propiolactone, $AgClO_4$, Na_2O_2, H_2SO_4, trichloromelamine. [*19*]

Fire Hazard: Mod, when exposed to heat or flame.

Spontaneous Heating: No.

Disaster Hazard: Dangerous; when heated to decomp, emits highly toxic fumes; can react vigorously with oxidizing materials.

To Fight Fire: Alcohol foam, CO_2, dry chemical.

ANILINE ACETATE. Colorless liquid. $C_8H_{11}O_2N$, mw: 153.2, d: 1.071.

THR = See aniline.

ANILINE ANTIMONYL TARTRATE. Syn: *antimonyl aniline tartrate.* White crystals. $C_{10}H_{11}O_7 \cdot Sb$, mw: 380.

THR = See aniline and antimony compounds.

ANILINE CAMPHORATE. Yellow crystals. $(C_6H_5NH_2)_2C_{10}H_{16}O_4$, mw: 386.5.

THR = See aniline.

ANILINE CHLORIDE. See aniline hydrochloride.

ANILINE DYES.

THR = Variable. The finished dyes are generally very much less toxic than many of the intermediates occurring or used in the manufacture of the dyes. Some of the aniline dyes cause local irr effects to the eyes, mu mem and skin; the basic dyes are believed to be more irr than the acid dyes. Allergic responses to aniline dyes have been known to occur. See also specific compounds.

Disaster Hazard: Mod dangerous; when heated to decomp, they emit toxic fumes.

ANILINE FLUORIDE. Syn: *aniline hydrofluoride.* Crystalline powder. $C_6H_5NH_2 \cdot HF$, mw: 113.1.

THR = See aniline and fluorides.

ANILINE GREEN. See malachite green.

ANILINE HYDROBROMIDE. Syn: *aniline bromide.* White to reddish crystals. $C_6H_5NH_2 \cdot HBr$, mw: 174.0, mp: 286°.

THR = See aniline and bromides.

ANILINE HYDROCHLORIDE. Syn: *aniline chloride.* Crystals. $C_6H_5NH_2 \cdot HCl$, mw: 129.59, vap. d: 4.46, d: 1.22, mp: 198°, bp: 245°, flash p: 380°F (OC).

THR = See aniline.

Fire Hazard: Slight, when exposed to heat or flame.

Spontaneous Heating: No.

Disaster Hazard: Dangerous; when heated to decomp, or on contact with acid or acid fumes, emits highly toxic fumes of aniline and chlorine compounds; can react vigorously with oxidizing materials.

To Fight Fire: Water, CO_2, water mist or spray, dry chemical.

ANILINE HYDROFLUORIDE. See aniline fluoride.

ANILINE MUSTARD. Syn: *N,N-di(2-chloroethyl)-aniline.* Crystals. $C_{10}H_{13}NCl_2$, mw: 218.1, mp: 45°.

THR = An exper carc. [*23*]

ANILINE OIL. See aniline.

ANILINE OIL DRUMS, EMPTY.

THR = See aniline.

Fire Hazard: Slight, when exposed to heat or flame. Such drums may fill with vapors and under the proper conditions ignite.

Disaster Hazard: Dangerous, depending on the number involved; when heated, they emit highly toxic fumes of aniline.

ANILINE SODIUM SULFONATE. Syns: *metanilic acid, sodium salt.* Tan flakes. $C_6H_4(NH_2)(NaSO_3)$, mw: 195.

THR = HIGH via oral and inhal routes of exposure. A mild sensitizer. Local contact may cause contact dermatitis. See also aniline and sulfonates.

Disaster Hazard: Dangerous; when heated to decomp or on contact with acid or acid fumes, emits highly toxic fumes of aniline and SO_x.

ANILINE TRIBROMIDE. See 2,4,6-tribromoaniline.

ANILINO BENZENE. See diphenylamine.

2-ANILINO ETHANOL. Syns: *β-anilinoethanol, ethoxyaniline, β-hydroxy ethylaniline.* $C_8H_{11}ON$, mw: 137, d: 1.1, bp: 268°, flash p: 305°F (OC).

Acute tox data: Oral LD_{50} (rat) = 2230 mg/kg; dermal LD_{50} (rabbit) = 63 mg/kg; iv LD_{LO} (dog) = 165 mg/kg. [*3*]

THR = HIGH via dermal and iv routes; MOD via oral route.

Fire Hazard: Slight, when exposed to heat or flame.

To Fight Fire: Dry chemical, water mist.

β-ANILINOETHANOL ETHOXY ANILINE. See 2-anilino ethanol.

ANILINO PHENOL. See *p*-hydroxy diphenylamine.

ANILITE. A high explosive mixture composed of liquid NO_2 and carbon disulfide or gasoline. It is extremely sensitive to shock.

***o*-ANISALDEHYDE.** See *o*-methoxy benzaldehyde.

ANISE CAMPHOR. See anethole.

ANISEED OIL. Colorless or pale yellow liquid. Composition: 80–90% anethol, methyl chavicol, anisaldehyde. d: 0.978–0.988 @ 25°/25°.

Acute tox data: Oral LD_{50} (rat) = 2250–2570 mg/kg. [*3*]

THR = MOD via oral route. A weak sensitizer. Local contact may cause contact dermatitis.

Fire Hazard: Slight.

ANISIC ACID. Syn: *p-methoxy benzoic acid.* Needle-like crystals, sol in alcohol, chloroform, ether. $CH_3OC_6H_4COOH$, mw: 152.1, bp: 277°, d: 1.385, mp: 184°.

Acute tox data: ip LD_{LO} (mouse) = 250 mg/kg. [*3*]

THR = HIGH via ip route.

Fire Hazard: Slight, when exposed to heat or flame; it can react with oxidizing materials.

***o*-ANISIC ALDEHYDE.** See *o*-methoxy benzaldehyde.

***o*-ANISIDINE.** Syn: *o-methoxyaniline.* Yellowish liquid. C_7H_9ON, mw: 123.15, mp: 5.2°, bp: 225°, d: 1.108 @ 26°.

THR = No data. See also *p*-anisidine.

Disaster Hazard: Dangerous, when heated to decomp, emits toxic fumes.

***p*-ANISIDINE.** Syn: *p-methoxyaniline.* Crystals. C_7H_9ON, mw: 123.15, mp: 57°, bp: 246°, d: 1.071 @ 55°/4°, vap. d: 4.28.

Acute tox data: Oral LD_{50} (rat) = 1400 mg/kg; ip LD_{50} (rat) = 1400 mg/kg. [*3*]

THR = MOD via oral and ip routes. Mild sensitizer. Local contact may cause a contact dermatitis. See also aniline.

Disaster Hazard: Dangerous; when heated to decomp, evolves toxic fumes.

ANISOLE. Syn: *phenyl methyl ether.* Mobile liquid, clear straw color. $C_6H_5OCH_3$, mw: 108.13, vap. d: 3.72, mp: −37.3°, bp: 153.8°, flash p: 125°F (COC), d: 0.996 @ 18°/4°, vap. press: 10 mm @ 42.2°, autoign. temp.: 887°F.

Acute tox data: Oral LD_{50} (rat) = 3700 mg/kg; oral LD_{50} (mouse) = 2800 mg/kg. [*3*]

THR = MOD via oral route.

Fire Hazard: Mod, when exposed to heat or flame; can react with oxidizing materials.

To Fight Fire: Foam, CO_2, dry chemical.

ANISOYL CHLORIDE. Syn: *p-anisyl chloride.* Needle-like crystals, insol in water, sol in ether and acetone. $CH_3OC_6H_4COCl$, mw: 170.6, mp: 22°, bp: 262°–263°, (slight decomp).

THR = HIGH irr via oral and inhal routes. Evolves HCl by hydrolysis. See also hydrochloric acid.

Disaster Hazard: Dangerous; see chlorides.

Explosion Hazard: Can explode spont at room temp.

ANISYL ACETONE. Syn: *4-(p-methoxyphenyl)-2-butanone.* Colorless to pale yellow liquid. $CH_3OC_6H_4C_2COCH_3$, mw: 178.2, mp: 8°.

THR = An insect attractant. Tox U, probably LOW.

Fire Hazard: Combustible.

ANISYL ALCOHOL. $C_8H_{10}O_2$, mw: 138.2.

Acute tox data: Oral LD_{50} (rat) = 1200 mg/kg; oral LD_{50} (mouse) = 1600 mg/kg. [*3*]

THR = MOD via oral route.

ANISYL BORIC ACID. Syn: *o-methoxy phenyl boric acid.* White crystals. $CH_3OC_6H_4B(OH)_2$, mw: 152.0.

THR = See boron compounds.

***p*-ANISYL CHLORIDE.** See anisoyl chloride.

ANNOYANCE TO INDIVIDUALS FROM NOISE. Section 3.

ANNOYANCE TO COMMUNITY FROM NOISE. Section 3.

ANNUM.

THR = A (S) carc. See plant and fungal products. [*14*]

ANOGON. See mercury compounds and iodine.

ANSOL M. See alcohol, denatured.

ANTHANTHRENE. A polycyclic HC found in air pollution studies.

ANTHION. See potassium persulfate.

ANTHOPHYLITE. See asbestos.

ANTHRA(9,1,2-*cde*)BENZO(*h*)CINNOLINE.
$C_{22}H_{12}N_2$, mw: 304.4.
THR = An exper neo. [*3*]

ANTHRACENE. Syns: *p-naphthalene, green oil, Anthracene oil.* Colorless crystals, violet fluorescence. $C_6H_4(CH)_2C_6H_4$, mw: 178.22, mp: 217°, bp: 345°, lel = 0.6%, flash p: 250°F (CC), d: 1.24 @ 27°/4°, autoign. temp.: 1004°F, vap. press: 1 mm @ 145.0°, (sublimes), vap. d: 6.15.
THR = Allergen and mild irr. A recog carc of skin, hands, forearms and scrotum. [*14*] An exper carc of the bladder. [*3, 14*]
Fire Hazard: Low, when exposed to heat or flame; reacts with oxidizing materials.
Spontaneous Heating: No.
Explosion Hazard: MOD, when exposed to flame, $Ca(OCl)_2$, chromic acid. [*19*]
To Fight Fire: Water, foam, CO_2, water spray or mist, dry chemical.

1-ANTHRACENE AMIDE. $C_{14}H_{11}N$, mw: 193.3.
THR = An exper neo. [*3*]

2-ANTHRACENE AMIDE.
THR = An exper carc. [*3, 23*]

ANTHRACENE OIL. See anthracene.

1,8,9-ANTHRACENTRIOL. See anthralin.

ANTHRACITE PARTICLES. Syn: *coal dust.* Black powder or dust.
THR = Depends upon SiO_2 content. See appendix for mineral dusts in the ACGIH-THRESHOLD LIMIT VALUE.
Fire Hazard: Mod, when exposed to heat or flame or by chemical reaction with oxidizers.
Explosion Hazard: Slight, when exposed to flame.

ANTHRALIN. Syn: *1,8-dihydroxy anthranol.* Yellow, crystals, insol in water, sol in chloroform, acetone and benzene. $C_{14}H_{10}O_3$, mw: 226.2, mp: 176°–181°.
THR = MOD irr via oral and inhal routes. Locally it can cause folliculitis of skin. Absorption may result in kidney injury and intestinal disturbances. An exper carc. [*23, 3*]
Fire Hazard: Slight, when heated.

9-ANTHRAMINE.
THR = An exper carc. [*23*]

ANTHRANILIC ACID. Syn: *o-amino benzoic acid.* Needle-like crystals. $C_6H_4(NH_2)COOH$, mw: 137.1, mp: 146°, bp: subl, d: 1.412 @ 20°.
Acute tox data: Oral LD_{50} (rat) = 4620 mg/kg. [*3*]
THR = MOD via oral route. An exper neo [*3*] in mice via imp. [*108*]
Fire Hazard: Slight.

ANTHRAQUINONE. Yellow crystals.
$C_6H_4(CO)_2C_6H_4$, mw: 208.20, mp: 286°, bp: 376.9°, flash p: 365°F (CC), d: 1.438, vap. press: 1 mm @ 190.0°, vap. d: 7.16.
THR = A mild allergen. LOW via oral route. An exper neo. [*3*]
Fire Hazard: Slight, when exposed to heat or flame.
Spontaneous Heating: No.
To Fight Fire: Water, foam, CO_2, water spray or mist, dry chemical.

ANTHRAROBIN. Syn: *3,4-dihydroxy anthranol.* Yellow-brown powder. $C_{14}H_{10}O_3$, mw: 226.22, mp: 208°.
THR = Mild allergen. Local contact may cause contact dermatitis.
Fire Hazard: Slight.

ANTIDOTES AND TREATMENT FOR POISONING. Section 9.

ANTIFEBRIN. See acetanilide.

ANTIFREEZE COMPOUND, LIQUID. See specific components.

ANTIMONIC CHLORIDE. See antimony pentachloride.

ANTIMONOUS BROMIDE. See antimony tribromide.

ANTIMONOUS CHLORIDE. See antimony trichloride.

ANTIMONY. Syns: *antimony regulus, stibium.* Silvery or gray lustrous metal. Sb, atwt: 121.76 mp: 630°, bp: 1380°, d: 6.684 @ 25°, vap. press: 1 mm @ 886°.
Acute tox data: Oral LD_{50} (rat) = 100 mg/kg; ip LD_{LO} (rat) = 100 mg/kg. [*3*]
THR = HIGH via oral and ip routes. See antimony compounds.
Radiation Hazard: Artificial isotope ^{124}Sb, $T_{\frac{1}{2}}$ = 60d, decays to stable ^{124}Te via β's of 0.24–2.3 MeV emits γ's of 0.60–2.09 MeV. Artificial isotope ^{125}Sb, $T_{\frac{1}{2}}$ = 2.7*y*, decays to stable ^{125}Te via β's of 0.12–0.61 MeV emits γ's of 0.04–0.67 MeV. For permissible levels see Table 5A.5.
Fire Hazard: Mod in the forms of dust and vapor, when exposed to heat or flame. See also powdered metals.
Explosion Hazard: Mod, in the form of dust when exposed to flame. It can react mod to violently with NH_4NO_3, halogens, BrF_3, BrN_3, $HClO_3$, ClO, ClF_3, HNO_3, KNO_3, $KMnO_4$, K_2O_2, $NaNO_3$. [*19*]
Disaster Hazard: Mod dangerous; when heated or on contact with acid, emits toxic fumes.

ANTIMONY ARSENATE. Heavy, white powder. Composition: Sb_2O_3 + 20% H_4AsO_7. See arsenic and antimony compounds.

ANTIMONY ARSENITE. Fine white powder. Composition: Sb_2O_3 + As_2O_3. See arsenic and antimony compounds.

ANTIMONY BROMIDE. See antimony tribromide.

ANTIMONY CHLORIDE. See antimony pentachloride or antimony trichloride.

ANTIMONY COMPOUNDS.
THR = Most Sb compounds are highly toxic via oral, inhal and ip routes. See also antimony. Because of the association with lead and arsenic in industry, it is often difficult to assess the toxicity of antimony and its compounds. Animals exposed to fumes of Sb oxide have developed pneumonitis, fatty degeneration of the liver, a decreased leucocyte count affecting in particular the polymorphonuclears, and damage to the heart muscle. In humans, complaints referable to the nervous system have been reported. In assessing human cases, however, the possibility of lead or arsenical poisoning must always be borne in mind. Locally antimony compounds are irr to the skin and mu mem. Signs and symptoms may include irr and eczematous eruption of the skin, inflammation of the mu mem of the nose and throat, metallic taste and stomatitis, gastrointestinal upset, with vomiting and diarrhea, and various nervous complaints, such as irritability, sleeplessness, fatigue, dizziness and muscular and neuralgic pains. See also specific compounds. (Sb^{+++} + hot $HClO_3$) can form an explosive mixture. [*19*]

ANTIMONY DIOXYSULFATE. White powder. $Sb_2O_2SO_4$, mw: 371.6, d: 4.89.
THR = See antimony compounds and sulfates.

ANTIMONY ETHOXIDE. Syn: *triethyl antimonite.* Colorless liquid, decomp upon contact with moisture. $Sb(C_2H_5O)_3$, mw: 256.9, d: 1.524 @ 17°, bp: 95° @ 11 mm.
THR = HIGH. See antimony compounds.
Disaster Hazard: See antimony compounds and ethers.

ANTIMONY FLUORIDE. See antimony pentafluoride or antimony trifluoride.

ANTIMONY GLANCE. See antimony trisulfide.

ANTIMONY HYDRIDE. Syns: *stibine, hydrogen antimonide.* Colorless gas. SbH_3, mw: 124.78, mp: −88°, bp: −17°, d: 5.30 g/liter @ 0°.
Acute tox data: Inhal LC_{Lo} (mouse) = 100 ppm, 20 min, iv LD_{50} (rabbit) = 8 mg/kg. [*3*]
THR = HIGH via oral, inhal, iv routes. See antimony compounds.
Fire Hazard: Mod, when exposed to flame.
Explosion Hazard: Violent reaction with Cl_2, HNO_3, O_3, NH_3. [*19*]
Disaster Hazard: Dangerous; when heated to decomp, emits highly toxic fumes; can react vigorously with oxidizing materials.

ANTIMONY IODIDE. See antimony tri-iodide or antimony pentaiodide.

ANTIMONY LACTATE. Tan colored mass, sol in water. $Sb(C_3H_5O_3)_2$, mw: 300.
THR = See antimony compounds.
Disaster Hazard: See antimony compounds.

ANTIMONYL ANILINE TARTRATE. See aniline antimonyl tartrate.

ANTIMONYL PERCHLORATE. $SbOClO_4$, mw: 237.2.
THR = See antimony compounds. Decomp @ 60°. [*19*]

ANTIMONYL PYROGALLOL. $HOC_6H_3O(SbOH)O$, mw: 262.9.
THR = An allergen. See also antimony compounds.
Disaster Hazard: See antimony.

ANTIMONY-α-MERCAPTO ACETAMIDE. Syn: *antimony thioglycolamide.* White crystals. $Sb(C_2H_4NOS)_3$, mw: 392.1, mp: 139°.
THR = See antimony compounds.
Disaster Hazard: Dangerous. See antimony and SO_x.

ANTIMONY ORE.
The raw material from which antimony is extracted. A recog carc. [*14*] See also arsenic compounds.

ANTIMONY OXIDES. See antimony trioxide, antimony tetraoxide and antimony pentaoxide.

ANTIMONY OXYCHLORIDE. Syns: *antimonyl chloride, powder of algroath.* White, amorphous powder. SbOCl, mw: 173.22, mp: decomp @ 170°.
THR = See antimony compounds and chlorides. Can react violently with BrF_3. [*19*]

ANTIMONY PENTACHLORIDE. Syns: *antimonic chloride, antimony perchloride.* Reddish-yellow, oily liquid, offensive odor. $SbCl_5$, mw: 299.05, mp: 2.8°, bp: 79° @ 22 mm, d: (liq.) 2.336, vap. press: 1 mm @ 22.7°.
Acute tox data: Oral LD_{50} (rat) = 1115 mg/kg; oral LD_{50} (guinea pig) = 900 mg/kg. [*3*]
THR = MOD via oral route. See also antimony trichloride.

ANTIMONY PENTAFLUORIDE. Syn: *antimony fluoride.* Oily, colorless liquid. Very reactive. SbF_5, mw: 216.76, mp: 7.0°, bp: 149.5°, d: (liq) 2.99 @ 23°.
THR = See fluorides and antimony compounds. Strong reaction with phosphates.

ANTIMONY PENTAIODIDE. Syn: *antimony iodide.* Brown solid. SbI_5, mw: 756.3, mp: 79°, bp: 400.6°.
THR = See antimony compounds and iodides.

ANTIMONY PENTASULFIDE. Syn: *antimony sulfide.* Orange-yellow powder. Sb_2S_5, mw: 403.82, mp: (decomp), d: 4.120.
Acute tox data: ip LD_{50} (rat) = 1500 mg/kg. [3]
THR = MOD via ip route. See also antimony and sulfides.
Fire Hazard: Mod, when exposed to heat or by chemical reaction with powerful oxidizer.
Explosion Hazard: Mod, when shocked or by spont chemical reaction in contact with powerful oxidizers.
Disaster Hazard: Dangerous; when heated to decomp or on contact with acid or acid fumes, emits highly toxic fumes of oxides of sulfur and antimony; will react with water or steam to produce toxic and flam vapors; can react vigorously with oxidizers, i.e., $Ag(ClO_3)_2$, $HClO_3$, ClO_2, $Mg(ClO_3)_2$,TlO, $Zn(ClO_3)_2$. [19]
To Fight Fire: Water.

ANTIMONY PENTOXIDE. Yellow powder. Sb_2O_5, mw: 323.5, mp: −O @ 380°, $-O_2$ @ 930°, d: 3.78.
Acute tox data: ip LD_{LO} (rat) = 4000 mg/kg. [3]
THR = MOD via ip route. See also antimony compounds.
Fire Hazard: Mod, by chemical reaction with reducing agents. An oxidizer.
Disaster Hazard: Dangerous, see antimony.

ANTIMONY PERCHLORIDE. See antimony pentachloride.

ANTIMONY POTASSIUM OXALATE. Syn: *potassium-antimony oxalate*; $K_3Sb(C_2O_4)_3$, mw: 503.1.
THR = See antimony compounds and oxalates.
Disaster Hazard: Dangerous; see antimony compounds.

ANTIMONY POTASSIUM TARTRATE. Syns: *tartar emetic, potassium and antimony tartrate.* Colorless crystals. White powder. $KSbC_4H_4O_7 \cdot \frac{1}{2}H_2O$, mw: 333.94, d: 2.607, mp: $-\frac{1}{2}H_2O$ @ 100°.
Acute tox data: Oral LD_{50} (rat) = 115 mg/kg; oral LD_{50} (mouse) = 600 mg/kg; ip LD_{50} (mouse) = 50 mg/kg; sc LD_{50} (mouse) = 55 mg/kg; iv LD_{50} (mouse) = 65 mg/kg. [3]
THR = HIGH via oral, ip and sc routes. This compound is used medicinally but the therapeutic dose is close to the toxic dose. It can cause cough, metallic taste, salivation, nausea and diarrhea. Skin rash may also occur. Large doses can cause severe damage to the liver.
Disaster Hazard: See antimony compounds.

ANTIMONY REGULUS. See antimony.

ANTIMONY SALT. Syn: *de Haens salt.* White crystals. Composition: Mixture of antimony trifluoride and sodium fluoride or ammonium sulfate.
THR = See antimony compounds and fluorides.

ANTIMONY SODIUM TARTRATE. $C_4H_4O_7NaSb$, mw: 308.8.
Acute tox data: ip LD_{50} (mouse) = 60 mg/kg; sc LD_{50} (mouse) = 48 mg/kg; iv LD_{50} (mouse) = 25 mg/kg. [3]
THR = HIGH via ip, sc and iv routes.

ANTIMONY SULFATE. Syn: *antimony trisulfate.* White powder. $Sb_2(SO_4)_3$, mw: 531.70, mp: decomp, d: 3.625 @ 4°.
Acute tox data: ip LD_{LO} (rat) = 1000 mg/kg. [3]
THR = MOD via ip route. See antimony compounds and sulfates.

ANTIMONY SULFIDE. See antimony pentasulfide or antimony trisulfide.

ANTIMONY, SULFURATED. Syn: *kermes mineral.* Reddish-brown, odorless, tasteless powder. Composition: 50–60% Sb.
THR = See antimony compounds and sulfides.

ANTIMONY-*d*-TARTRATE. White crystals. $Sb_2(C_4H_4O_6)_3 \cdot 6H_2O$, mw: 795.8.
Acute tox data: im LD_{50} (rabbit) = 90 mg/kg. [3]
THR = HIGH via im route.

ANTIMONY TELLURIDE. See antimony tritelluride.

ANTIMONY TETROXIDE. Syn: *antimony oxide.* White powder. Sb_2O_4, mw: 307.5, mp: −0 @ 930°, d: 5.82.
THR = See antimony compounds.

ANTIMONY THIOGLYCOLAMIDE. See antimony -α-mercaptoacetamide.

ANTIMONY TRIBROMIDE. Syn: *antimonous bromide.* Yellow, deliquescent, crystalline mass. $SbBr_3$, mw: 361.51, mp: 96.6°, bp: 280°, d: 4.148, vap. press: 1 mm @ 93.9°.
THR = See antimony compounds and bromides. Can react violently with Na, K. [19]

ANTIMONY TRICHLORIDE. Syns: *antimonous chloride, butter of antimony.* Colorless, transparent crystalline mass. $SbCl_3$, mw: 228.13, mp: 73.4°, bp: 283.0°, d: 3.140 @ 25°, vap. press: 1 mm @ 49.2° (subl).
Acute tox data: Inhal TC_{LO} (human) = 73 mg/kg⟶ pulmonary problems and GI problems; oral LD_{50} (rat) = 675 mg/kg. [3]
THR = HIGH via inhal; MOD via oral route of exposure. Reacts vigorously with moisture, generating heat and hydrogen chloride gas (high irr) which

can cause pulmonary edema when inhal. Systemic effects can be caused by the antimony components. See antimony compounds.

Disaster Hazard: See antimony compounds and HCl. Can react violently with Al, K, Na.

ANTIMONY TRIETHYL. Syn: *triethyl stibine*. Liquid, water insol. $Sb(C_2H_5)_3$, mw: 209.0, d: 1.324 @ 16°, mp: −29°, bp: 159.5°.

THR = Alkyl metallics of this type are often highly toxic and corrosive. See also antimony compounds.

Fire Hazard: Dangerous, by chemical reaction. Spont flam in air. Explodes in water, CCl_4, halogenated hydrocarbons, dimethyl formamide, triethyl borine and air. [*19*]

Disaster Hazard: Dangerous; when heated to decomp, burns and emits highly toxic fumes of antimony; can react vigorously with oxidizing materials.

ANTIMONY TRIFLUORIDE. Syn: *antimony fluoride*. Octagonal crystals. SbF_3, mw: 178.8, mp: 292°, bp: 319° (subl), d: 4.379 @ 20.9°.

Acute tox data: sc LD_{50} (mouse) = 23 mg/kg; oral LD_{LO} (guinea pig) = 100 mg/kg. [*3*]

THR = HIGH via oral and sc routes. See also antimony compounds and fluorides.

ANTIMONY TRIIODIDE. Syn: *antimony iodide*. Red to yellow crystals. SbI_3, mw: 502.5, mp: 170°, bp: 401°, d: 4.768 @ 22°, vap. press: 1 mm @ 163.6°. Reacts violently with Na, K.

ANTIMONY TRIMETHYL. Syn: *trimethyl stibine*. Liquid, slightly sol in water. $Sb(CH_3)_3$, mw: 166.9, bp: 80.6°, d: 1.523 @ 15°.

Acute tox data: sc LD_{LO} (cat) = 1370 mg/kg. [*3*]

THR = MOD via sc route. See antimony triethyl and antimony compounds.

Fire Hazard: Dangerous by chemical reaction. Spont flam in air, and explodes in water. Reacts violently with halogenated hydrocarbons.

Disaster Hazard: Dangerous; when heated to decomp, emits highly toxic fumes of antimony; can react vigorously with oxidizing materials.

ANTIMONY TRIOXIDE. White, odorless, tasteless, crystalline powder. Sb_2O_3, mw: 291.52, mp: 656°, bp: 1550° (subl), d: 5.2, vap. press: 1 mm @ 574°.

THR = See antimony compounds.

ANTIMONY TRIPHENYL. $C_{18}H_{15}Sb$, mw: 353.1.

Acute tox data: Oral LD_{50} (rat) = 187 mg/kg; ip LD_{50} (rat) = 168 mg/kg. [*3*]

THR = HIGH via ip, and oral routes. Upon heating it burns in air. Reacts violently with BrF_3. [*19*]

ANTIMONY TRIPROPYL. $Sb(C_3H_7)_3$, mw: 250.9.

THR = See antimony compounds. Spont flam in air and explodes in water. Reacts violently with halogenated hydrocarbons. [*19*]

ANTIMONY TRISELENIDE. Gray powder. Sb_2Se_3, mw: 480.40, mp: 611°.

THR = See antimony and selenium compounds.

Disaster Hazard: Dangerous, See antimony and selenium; can react vigorously with oxidizing materials.

ANTIMONY TRISULFIDE. Syn: *antimony sulfide*. Red to black crystals. Sb_2S_3, mw: 339.7, mp: 550°, d: 4.64, bp: ca. 1150°.

Acute tox data: ip LD_{LO} (rat) = 1000 mg/kg; inhal TC_{LO} (human) = 0.58 mg/kg; blood and GI problems. [*3*]

THR = MOD via ip; HIGH via inhal routes. See antimony compounds and sulfides.

Fire Hazard: Mod, by spont chemical reaction in contact with strong oxidizers.

Explosion Hazard: Mod, by spont chemical reaction in contact with chlorates, perchlorates, ClO, thallic oxide; burns in air. [*19*]

Disaster Hazard: Dangerous; when heated to decomp or on contact with acid or acid fumes, emits highly toxic fumes of oxides of sulfur and antimony; will react with water or steam to produce toxic and flam vapors; can react vigorously with oxidizing materials.

ANTIMONY TRISULFATE. See antimony sulfate.

ANTIMONY TRITELLURIDE. Syn: *antimony telluride*. Gray powder. Sb_2Te_3, mw: 626.4, mp: 629°, d: 6.50 @ 13°.

THR = No data. Probably HIGH. See also antimony and tellurium.

Fire Hazard: Mod, by chemical reaction in contact with strong oxidizers.

Explosion Hazard: Mod, by chemical reaction in contact with chlorates and perchlorates.

Disaster Hazard: Dangerous; when heated to decomp or on contact with acid or acid fumes, emits highly toxic fumes of antimony and tellurium; will react with water or steam to produce toxic and flam vapors; can react vigorously with oxidizing materials.

ANTIMONY TRIVINYL. $Sb(C_2H_2)_3$, mw: 200.

THR = See antimony compounds. Spont flam in air. [*19*]

ANTIMONY YELLOW. See lead antimonate.

ANTIPYRINE. Syn: *phenozone*. Fine white, crystalline powder. $C_{11}H_{12}N_2O$, mw: 188.23, mp: 113°, bp: 319° @ 174 mm, d: 1.19.

Acute tox data: Oral LD_{50} (rat) = 1800 mg/kg; ip LD_{50} (mouse) = 750 mg/kg; sc LD_{LO} (rat) = 1570 mg/kg. [*3*]

THR = MOD via oral, ip and sc routes. An exper carc. [*23*]

Disaster Hazard: Mod dangerous; when heated to decomp emits toxic fumes.

ANTISEPTIC OIL. Syn: *hydroquinone in oil.* Liquid.

THR = A weak sensitizer. Local contact may cause contact dermatitis. See also hydroquinone.

Fire Hazard: A combustible material.

ANTOXYLIC ACID. See arsanilic acid.

ANTU. See naphthyl thiourea.

APACHE COAL POWDERS. See explosives, high.

APHAMIDE. Syn: *N,N-ethylene-bis(1-aziridinyl)-N-methyl phosphinic amide.*

THR = A chemosterilant. No tox data.

Disaster Hazard: Dangerous, when heated to decomp emits highly toxic fumes.

APHOLATE. $C_{12}H_{24}N_9P_3$, mw: 387.4.

THR = HIGH via oral route. An exper (±) carc. [*3, 8, 23*] and teratogen.

APHOXIDE. See Tepa.

APHRODINE. See yohimbine.

APOATROPINE. Syn: *atropamine.* Prismatic crystals, sol in alcohol, ether, chloroform, benzene, carbon tetrachloride, insol in water. $C_{17}H_{21}NO_2$, mw: 271.4 mp: 62°.

Acute tox data: Oral LD_{50} (mouse) = 160 mg/kg; ip LD_{50} (mouse) = 14 mg/kg. [*3*]

THR = HIGH via ip and oral routes. A poisonous alkaloid which causes a marked conjunctivitis. Large doses cause death by respiratory failure. See also atropine.

APOATROPINE HYDROCHLORIDE. See atropine.

APOATROPINE SULFATE. See atropine.

APOCHOLIC ACID. $C_{24}H_{38}O_4$, mw: 390.6.

THR = An exper neo. [*3*]

APOCODEINE. White crystalline solid. $C_{18}H_{19}NO_2$, mw: 281.34, mp: 124°.

THR = A poisonous alkaloid. See also codeine. Weak sensitizer. Local contact may cause contact dermatitis. HIGH via oral and inhal routes.

Disaster Hazard: Dangerous; when heated to decomp, emits highly toxic fumes.

APOCODEINE HYDROCHLORIDE.

See apocodeine.

APOMORPHINE. White crystalline alkaloid.

$C_{17}H_{17}NO_2$ mw: 267.4, mp: 195° (decomp).

Acute tox data: ip LD_{50} (mouse) = 160 mg/kg; iv LD_{LO} (dog) = 80 mg/kg. [*3*]

THR = HIGH via oral and iv routes. Small doses cause depression while large doses produce excitement, convulsions and death. A powerful emetic. A weak sensitizer. Local contact may cause contact dermatitis. Mild symptoms require no treatment, but stimulants such as coffee or caffeine may be given. For severe cases, chloral hydrate, chloroform or ether. Call a physician.

Disaster Hazard: Dangerous; when heated to decomp, emits highly toxic fumes.

APOMORPHINE HYDROCHLORIDE. See apomorphine.

APPLE ACID. See malic acid.

AQUA AMMONIUM. See ammonium hydroxide.

AQUACIDE. $C_{12}H_{12}N_2Br_2$, mw: 344.1.

THR = HIGH via oral route. An exper neo. [*3*]

AQUA FORTIS. See nitric acid.

AQUAMARINE. See beryl.

AQUAPHOR.

THR = An allergen. Weak sensitizer. Local contact may cause contact dermatitis.

AQUA REGIA. Syn: *nitro hydrochloric acid.* Fuming yellow, corrosive, suffocating volatile liquid. Composition: (USP): 18 cc HNO_3 + 82 cc HCl.

THR = HIGH corrosive. A very powerful irr via oral, inhal and dermal routes to eyes and mu mem.

Fire Hazard: Mod by chemical reaction with easily oxidized materials. A powerful oxidizer.

Disaster Hazard: Dangerous; when heated to decomp emits highly toxic fumes of nitrosyl chloride; can react vigorously with reducing materials.

1-β-D-ARABINO FURANOSYL-5-FLUOROCYROSINE. $C_9H_{12}O_5N_3F$, mw: 261.2.

THR = An exper teratogen. [*3*]

ARABINOGALACTIN.

A water sol polysaccharide extracted from the timber of western larch trees. Dry, light tan-colored powder. mw: 72,000–92,000.

THR = No data. Probably LOW. A food additive permitted in food for human consumption. [*109*]

ARAGONITE. See calcium carbonate.

ARALKONIUM CHLORIDE. Syn: *dynaltone.* $C_{21}H_{36}Cl_3N$, mw: 408.9.

THR = MOD irr to skin, eyes and mu mem. A quaternary ammonium compound. See alrosept.

Disaster Hazard: Dangerous. See chlorides.

ARAMITE. See sulfurous acid-2-(*o-tert*-butyl phenoxy)-1-methyl ethyl-2-chloroethyl ester.

ARASAN. See *bis*(dimethylthio carbamyl)disulfide.

ARATHANE. See dinocap.

ARECA NUT. Syn: *betel.* Mottled brown, with fawn color.

THR = An exper neo to mice via sc route. [*103*]
Fire Hazard: Slight, when heated.

ARECOLINE. Syn: *methyl 1,2,5,6-tetrahydro-1-methylnicotinate*. Oily liquid. $C_8H_{13}NO_2$, mw: 155.2, bp: 209°.
Acute tox data: sc LD_{LO} (mouse) = 65 mg/kg. [*3*]
THR = HIGH via sc, oral and inhal routes. An exper carc. [*23*]
Fire Hazard: Combustible liquid.
Disaster Hazard: Dangerous, when heated to decomp, emits highly toxic fumes; can react with oxidizing materials.

ARESKAP. See monobutyl phenyl phenol sodium sulfate.

ARESKET. See monobutyl diphenyl sodium monosulfonate.

ARGENTUM. See silver.

ARGININE. Syns: *guanidine amino valeric acid, amino-4-guanidovaleric acid*. An essential amino acid for rats; occurs naturally in the *l* (+) form. Data here refer to *l* and *dl* forms. Prisms from water. $NHC(NH_2)NH(CH_2)_3CH(NH_2)COOH$. mw: 174, dehydrates @ 105°, decomp @ 244°.
THR = LOW via oral route. A dietary supplement food additive. [*109*]

ARGON. Colorless, inert gas. A, atwt: 39.94, mp: −189.2°, bp: −185.7°, d: 1.784 g/liter @ 0°, 1.40 @ −186°, 1.65 @ −233°.
THR = Classified as a simple asphyxiant gas. Gases of this type have no specific toxicity effect, but they act by excluding O_2 from the lungs. The effect of simple asphyxiant gases is proportional to the extent to which they diminish the amount (partial pressure) of O_2 in the air that is breathed. The oxygen may be diminished to $\frac{2}{3}$ of its normal percentage in air before appreciable symptoms develop, and this in turn requires the presence of a simple asphyxiant in a conc of 33% in the mixture of air and gas. When the simple asphyxiant reaches a conc of 50%, marked symptoms can be produced. A conc of 75% is fatal in a matter of minutes. The first symptoms produced by simple asphyxiant gases such as argon are rapid respirations and air hunger. Mental alertness is diminished and muscular coordination is impaired. Later, judgment becomes faulty and all sensations are depressed. Emotional instability often results and fatigue occurs rapidly. As the asphyxia progresses, there may be nausea and vomiting, prostration and loss of consciousness, and finally, convulsions, deep coma and death.
Radiation Hazard: For permissible levels see Section 5, Table 5A.5. Artificial isotope ^{41}A, $T_{\frac{1}{2}}$ = 1.8h, decays to stable ^{41}K via β's of 1.2 MeV, emits γ's of 1.29 MeV. Although ^{41}A has a short half-life, it may present a slight hazard as an effluent from nuclear reactors where it is formed by neutron bombardment of ^{40}A, the most abundant natural isotope.

ARGYROL. Syn: *mild silver protein*. Brown to black crystals. Composition: 19–23% silver.
THR = See silver. An allergen. Continued application can cause argyria. See also silver compounds. Local contact may cause contact dermatitis.

ARICYL. See disodium acetoarsenate.

ARICYL ACID. Syn: *arsonoacetic acid*. Crystals. $C_2H_5AsO_5$, mw: 184.1.
THR = See arsenic compounds.

ARISTODERM. $C_{24}H_{31}O_6$, mw: 415.6.
THR = An exper teratogen. [*3*]

ARNICA. Syns: *Wolfsbane, mountain tobacco*. An alcoholic infusion.
THR = MOD irr and allergen. HIGH via oral and inhal routes. Overdose can be fatal. It can cause gastroenteritis, nervous disturbances and collapse. Local contact may cause contact dermatitis.
Fire Hazard: Slight, when exposed to heat or flame; can react with oxidizing materials.

"AROCHLORS." See chlorinated diphenyls.

AROMATIC AMINES.
Amines which contain one or more rings of unsaturated or cyclic HC such as benzene. There are a vast number of such amines. The term is largely due to the characteristic odor. Many of these aromatic amines are recognized as carc to the human bladder, ureter and renal pelvis and (S) carc to the intestines, lung, liver and prostate. See amines.

AROMATIC SPIRITS OF AMMONIA. Syn: *spirit of Hartshorn*. Colorless liquid, suffocating odor of ammonia. Composition: 10% by weight of NH_3 in alcohol.
THR = See ammonia.
Fire Hazard: Dangerous, in its usual form (solution of NH_3, in alcohol).
Explosion Hazard: Mod, in its usual form.
Disaster Hazard: Mod dangerous; when heated, emits toxic fumes of ammonia; can react with oxidizing materials.

ARSACETIN (SODIUM SALT). Syn: *sodum acetyl arsanilate*. White crystalline powder odorless, tasteless. $C_8H_9O_4NaNAs + 4H_2O$, mw: 353.
Acute tox data: sc LD_{LO} (rat) = 550 mg/kg. [*3*]
THR = HIGH via sc and oral routes. See arsenic compounds.

ARSANILIC ACID. Syns: *antoxylic acid, amino-*

phenylarsine acid. White, crystalline powder. $NH_2C_6H_4AsO(OH)_2$, mw: 217.0, mp: 232°.
Acute tox data: Oral LD_{50} (rat) = 216 mg/kg. [*3*]
THR = HIGH via oral route. See also arsenic. A grasshopper bait; a food additive permitted in the feed and drinking water of animals and/or for the treatment of food-producing animals. [*109*] See arsenic compounds and aniline. A recog carc. [*14*]
Fire Hazard: Mod. Decomp by heat to yield flam vapors.
Disaster Hazard: Dangerous; when heated to decomp or on contact with acid or acid fumes, emits highly toxic fumes of aniline and arsenic.

ARSENATE OF IRON, FERRIC. See ferric arsenate.

ARSENATE OF IRON, FERROUS. See ferrous arsenate.

ARSENIC (see also arsenic vapor). Silvery to black, brittle, crystalline and amorphous metalloid. As_4. mw: 299.64, mp: 814° @ 36 atm, bp: subl @ 615°, d: black crystals 5.724 @ 14°; black amor 4.7, vap. press: 1 mm @ 372° (sublimes).
Acute tox data: im LD_{LO} (rat) = 25 mg/kg; sc LD_{LO} (rabbit) = 300 mg/kg. [*3*]
THR = HIGH via im and sc routes. A poison. An exper (±) carc. [*3, 6*] Some human carc implication as well. [*6, 23, 95*] Used as a food additive in food for human consumption. [*109*] See arsenic compounds.
Radiation Hazard: For permissible levels see Table 5A.5, Section 5. Artificial isotope ^{74}As, $T_{\frac{1}{2}}$ = 18d. Decays to stable ^{74}Ge via ec and via positrons of 0.91 MeV (26%), 1.53 MeV (4%). Also decays to stable ^{74}Se via β's of 0.72 MeV (14%), 1.35 MeV (18%). Emits γ's of 0.60 and 0.63 MeV.
Fire Hazard: Mod in the form of dust when exposed to heat or flame or by chemical reaction with powerful oxidizers such as bromates, chlorates, iodates, peroxides, Li, NCl_3, KNO_3, $KMnO_4$, Rb_2C_2, $AgNO_3$, NOCl, IF_5, CrO_3, ClF_3, ClO, BrF_3, BrF_5, BrN_3, RbC ≡ CH, CsC ≡ CH. [*19*]
Explosion Hazard: Slight in the form of dust when exposed to flame.
Disaster Hazard: Dangerous; when heated or on contact with acid or acid fumes, emits highly toxic fumes; can react vigorously on contact with oxidizing materials.

***m*-ARSENIC ACID.** White crystals. $HAsO_3$, mw: 123.9, mp: decomp.
THR = See arsenic compounds.

***o*-ARSENIC ACID.** Syn: *true arsenic acid.* White, translucent crystals. $H_3AsO_4 \cdot \frac{1}{2}H_2O$, mw: 150.9, mp: 35.5°, bp: $-H_2O$ @ 160°, d: 2.0–2.5.
Acute tox data: Oral LD_{50} (rat) = 48 mg/kg; iv LD_{50} (rabbit) = 8 mg/kg. [*3*]
THR = HIGH via oral and iv routes. See arsenic compounds.

ARSENIC ACID, LIQUID. See arsenic acid.

ARSENIC ACID, SOLID. See arsenic acid.

ARSENICAL BABBITT. A bearing metal. Composition: Up to 3% As.
THR = See arsenic compounds.

ARSENICAL COMPOUNDS OR MIXTURES, N.O.S. LIQUID.
THR = See arsenic compounds.

ARSENICAL DIP, LIQUID. Syn: *sheep dip.*
THR = See arsenic compounds.
Disaster Hazard: See arsenic compounds.

ARSENICAL DUST. See arsenic.

ARSENICAL FLUE DUST. See arsenic.

ARSENICAL MIXTURE OR COMPOUNDS, N.O.S. SOLID. See arsenic compounds.

ARSENIC BISULFIDE. Syn: *arsenic sulfide, realgar.* Red-brown crystals. As_2S_2, mw: 214, bp: 565°, mp: β = 307°, d: α = 3.506 @ 19°, β = 3.254 @ 19°.
THR = See arsenic compounds and sulfides.
Fire Hazard: Mod, in the form of dust when exposed to heat or flame.
Explosion Hazard: When intimately mixed with powerful oxidizers such as Cl_2, KNO_3 chlorates.
Disaster Hazard: Dangerous, see SO_x and arsenic compounds, it will react with water or steam to produce toxic and flam vapors; can react vigorously with oxidizing materials.

ARSENIC BROMIDE. Syns: *arsenic tribromide, arsenous bromide.* Yellowish-white crystals. $AsBr_3$, mw: 314.7, mp: 32.8°, bp: 220.0°, d: 3.54 @ 25°, vap. press: 1 mm @ 41.8°.
THR = See arsenic compounds and bromides.

ARSENIC CHLORIDE. See arsenic pentachloride.

ARSENIC COMPOUNDS. Syn: *arsenicals.* Used as insecticides, herbicides, silvicides, defoliants, desiccants and rodenticides. Poisoning from arsenic compounds may be acute or chronic. Acute poisoning usually results from swallowing arsenic compounds; chronic poisoning from either swallowing or inhal. Acute allergic reactions to arsenic compounds used in medical therapy have been fairly common. The type and severity of reaction depending upon the compound of arsenic. Inorganic arsenicals are more toxic than organics. Trivalent is more toxic than pentavalent. [*89*]

Acute arsenic poisoning (from ingestion) results in marked irritation of the stomach and intestines with nausea, vomiting and diarrhea. In severe cases the vomitus and stools are bloody and the patient goes into collapse and shock with weak, rapid pulse, cold sweats, coma and death.

Chronic arsenic poisoning, whether through ingestion or inhalation, may manifest itself in many different ways. There may be disturbances of the digestive system such as loss of appetite, cramps, nausea, constipation or diarrhea. Liver damage may occur, resulting in jaundice. Disturbances of the blood, kidneys and nervous system are not infrequent. Arsenic can cause a variety of skin abnormalities including itching, pigmentation and even cancerous changes. A characteristic of arsenic poisoning is the great variety of symptoms that can be produced. A recog carc of the skin, lungs, liver. An exper carc of the mouth, esophagus, larynx, bladder and para nasal sinus. [*14, 3, 23, 95, 89*]

In treating acute poisoning from ingestion BAL (dimercaptol) is of questionable effectiveness for acute and chronic poisoning with trivalent arsenicals, such as As trioxide, arsine and arsenites. It is of no value for pentavalent arsenicals, such as cacodylic acid, methanearsonic acid, sodium, cacodylate, MSMA, DSMA, arsanilic acid, arsenic acid, and arsenates. Vomiting and gastric lavage are the preferred emergency treatments for acute arsenical poisoning. Modern medical treatment of arsenical poisoning uses exchange transfusion and dialysis (A. E. De Palma, *J. Occup. Med.*, Vol. 11, 582–587 (1969). Note: Arsenic compounds are common air contaminants.

Disaster Hazard: Dangerous; when heated to decomp, or for metallic arsenic on contact with acids or acid fumes, or when water solutions of arsenicals are in contact with active metals such as Fe, Al, Zn, emits highly toxic fumes of arsenic.

ARSENIC COPPER. See copper arsenide.

ARSENIC DIETHYL. Syns: *ethyl cacodyl, tetraethyldiarsine*. Liquid or oil. $[As(C_2H_5)_2]_2$, mw: 266.2, bp: 185°–190°, d: about 1.

THR = See arsenic compounds.

Fire Hazard: Dangerous, by spont chemical reaction. A spont flam liquid.

Disaster Hazard: Dangerous; see arsenic; can react vigorously with oxidizing materials.

ARSENIC DIIODIDE. Syn: *arsenic iodide*. Red crystals. AsI_2, mw: 328.8, mp: decomp @ 136°.

THR = See arsenic compounds and iodides.

ARSENIC DIMETHYL. Syn: *tetramethyl diarsyl*. Colorless to yellow oily liquid. $[As(CH_3)_2]_2$, mw: 210.0, mp: −6°, bp: 186°, d: 1.15.

THR = HIGH via inhal and oral routes. See also arsenic.

Fire Hazard: Mod flam liquid.

Disaster Hazard: Dangerous; see arsenic.

ARSENIC DISULFIDE. See arsenic bisulfide.

ARSENIC FLUORIDE. See arsenic pentafluoride or arsenic trifluoride.

ARSENIC HEMISELENIDE. Black crystals with metallic luster. As_2Se, mw: 228.78.

THR = See arsenic and selenium compounds.

Disaster Hazard: Dangerous; see arsenic compounds and selenium compounds; can react vigorously with oxidizing materials.

ARSENIC HYDRIDE. See arsine.

ARSENIC IODIDE. See arsenic diiodide.

ARSENIC OXIDES. See arsenic trioxide or arsenic pentoxide.

ARSENIC OXYCHLORIDE. Brown crystals. AsOCl, mw: 126.4, bp: decomp.

THR = See arsenic compounds and chlorides.

ARSENIC PENTACHLORIDE. Colorless liquid. $AsCl_5$, mw: 252.2, mp: −40° (approx).

THR = See arsenic compounds and HCl.

ARSENIC PENTAFLUORIDE. Syn: *arsenic fluoride*. Colorless gas. AsF_5, mw: 169.9, mp: −80°, bp: −53°, d: 7.71 g/liter.

THR = See arsenic compounds and fluorides.

ARSENIC PENTASELENIDE. Black, brittle solid with a metallic luster. As_2Se_5, mw: 544.62.

THR = See arsenic and selenium compounds.

ARSENIC PENTASULFIDE. Brownish-yellow, glassy, amorphous, highly refractive mass. As_2S_5, mw: 310.2, mp: 500° (subl).

THR = See arsenic compounds and sulfides.

Fire Hazard: A flam material.

Explosion Hazard: See arsenic bisulfide.

Disaster Hazard: Dangerous; see SO_x and arsenic compounds; will react with steam or water to produce toxic and corrosive fumes; can react vigorously with oxidizing materials.

ARSENIC PENTAOXIDE. Syn: *arsenic oxide*. White, amorphous solid, Deliquescent. As_2O_5, mw: 229.8, mp: 315° (decomp), d: 4.32.

Acute tox data: Oral LD_{50} (rat) = 8 mg/kg; oral LD_{50} (mouse) = 55 mg/kg; iv LD_{Lo} (rabbit) = 6 mg/kg. [*3*]

THR = HIGH via oral and iv routes. An exper (+) carc [*3, 6*] Reacts vigorously with Rb_2C_2. [*19*]

Disaster Hazard: See arsenic compounds.

ARSENIC PHOSPHIDE. Brown to red powder. AsP, mw: 105.9, mp: subl with decomp.

THR = See arsenic compounds and phosphine.
Fire Hazard: Mod by spont chemical reaction. Phosphine is liberated upon contact with moisture.
Explosion Hazard: U.
Disaster Hazard: Dangerous, see phosphorus and arsenic compounds; will react with water or steam to produce toxic and flam vapors; can react vigorously with oxidizing materials.

ARSENIC, SOLID. See arsenic.

ARSENIC SULFIDE, (POWDER). See arsenic trisulfide, arsenic bisulfide and arsenic pentasulfide.

ARSENIC TRIBROMIDE. See arsenic bromide.

ARSENIC TRICHLORIDE. Clear, almost colorless to pale yellow corrosive oily liquid, or needle-like crystals. $AsCl_3$, mw: 181.3, mp: −8.5°, bp: 130.2°, d: (liq) 2.163 @ 14°/4°, vap. press: 10 mm @ 23.5°, vap. d: 6.25.
Acute tox data: Inhal LC_{LO} (mouse) = 2500 mg/m^3, 10 min; inhal LC_{LO} (cat) = 100 mg/m^3, 1 hr. [*3*]
THR = HIGH via inhal and oral routes. Can produce strong explosions upon impact when mixed with Na, K, Al. [*19*]

ARSENIC TRIETHYL. See triethyl arsenic.

ARSENIC TRIFLUORIDE. Syn: *arsenic fluoride*. Oily liquid. AsF_3, mw: 131.9, mp: −8.5°, bp: 63° @ 752 mm, d: (liq) 2.666, vap. press: 100 mm @ 13.2°, 400 mm @ 41.5°.
THR = See arsenic compounds and fluorides. Reacts violently with P_2O_3. [*19*]

ARSENIC TRIMETHYL. See trimethyl arsenic.

ARSENIC TRIIODIDE. Orange-red crystals. AsI_3, mw: 455.6, mp: 140.9°, bp: 400°, d: 4.688 @ 25°/4°.
THR = HIGH. See arsenic compounds and iodides. Can form a shock sensitive explosive with Na, K. [*19*]

ARSENIC TRIOXIDE. Syn: *white arsenic*. White, odorless, tasteless, amorphous powder. As_2O_3, mw: 197.8, mp: 315° (subl), d: (arsenalite) 3.865 @ 25°; (claudedite) 4.15; (amorphous) 4.09.
Acute tox data: Oral LD_{50} (man) = 1.43 mg/kg; oral LD_{50} (rat) = 20 mg/kg; sc LD_{LO} (rat) = 15 mg/kg; inhal TC_{LO} (man) = 0.7 mg/m^3 ⟶ carc. [*3*]
THR = HIGH via oral and sc routes. A rodenticide. See arsenic compounds. An exper human carc. [*3, 6*] Reacts vigorously with Rb_2C_2, ClF_3, F_2, Hg, OF_2, $NaClO_3$. [*19*]
Disaster Hazard: See arsenic compounds.

ARSENIC TRIPHENYL. See triphenyl arsenic.

ARSENIC TRISELENIDE. Syn: *arsenious selenide*. Brown crystals. As_2Se_3, mw: 386.7, mp: 360°, d: 4.75.
THR = See arsenic and selenium compounds.

ARSENIC TRISULFIDE. Syns: *arsenic sulfide, orpiment*. Yellow or red crystals. As_2S_3, mw: 246.0, mp: 300°, bp: 707°, d: 3.43.
THR = HIGH. See arsenic compounds and sulfides. See arsenic bisulfide. Reacts violently with H_2O_2, (KNO_3 + S). [*19*]
Disaster Hazard: Dangerous; when heated to decomp or on contact with acid or acid fumes, emits highly toxic fumes of sulfur and arsenic; will react with water or steam to produce toxic and flam vapors; can react vigorously on contact with oxidizing materials.

ARSENIC TRISILYL. See trisilyl arsine.

ARSENIC VAPOR. As, atwt: 75.
THR = See arsenic compounds.
Fire Hazard: Mod by chemical reaction with oxidizers.
Disaster Hazard: See arsenic compounds.

ARSENIOUS SELENIDE. See arsenic triselenide.

ARSENIURETTED HYDROGEN. See arsine.

ARSENOACETIC ACID. Yellow crystals, insol in water. $(AsCH_2COOH)_2$, mw: 267.9, mp: >260°.
THR = See arsenic compounds.

ARSENOBENZENE. White crystals, insol in water, sol in benzene. $C_6H_5As_2C_6H_5$, mw: 304.0, mp: 212°.
THR = See arsenic compounds.

ARSENOUS ACID, SOLID. See arsenic trioxide.

ARSENOUS BROMIDE. See arsenic bromide.

ARSENOUS AND MERCURIC IODIDE SOLUTION, LIQUID.
THR = See arsenic and mercury compounds.
Disaster Hazard: See arsenic and mercury compounds.

ARSINE. Syns: *arsenic hydride, arseniuretted hydrogen*. Colorless gas, mild garlic odor. AsH_3, mw: 78, mp: −116°, bp: −55°, d: 3.484 g/liter, vap. d: 2.66.
Acute tox data: Inhal TC_{LO} (human) = 3 ppm; inhal TC_{LO} (human) = 3ppm; inhal TC_{LO} (human) = 25 ppm, 30 min; inhal LC_{LO} (human) = 500 mg/kg. [*3*]
THR = HIGH via inhal route.

The toxicity of arsine is due to its hemolytic action. On entering the blood stream it combines with the hemoglobin of the red blood cells; gradually the arsenic in this hemoglobin-arsenic complex is oxidized and the oxidation process is accompanied by hemolysis of the cell. The resulting anemia is responsible for the production of many of the symptoms accompanying arsine poisoning; other symptoms result from the hemolysis itself, and occur during the excretion of the hemoglobin. Hemoglobin and its degradation products are commonly found in the urine. Less commonly whole blood may be passed. Occasionally, the renal tubules may be plugged by debris, with resultant suppression of urine. Jaundice, which may be severe, is a common result of the hemolysis. Frequently there is edema of the lungs, which may be accompanied by cyanosis. Kidney damage is common in patients surviving acute effects of the gas. A recognized carc. [*14*]

Signs of poisoning usually develop within several hours of exposure. Headache, dizziness, nausea and vomiting, epigastric pain and weakness occur early, followed by tea-colored urine, or bloody urine in the more severe cases. Some time later, albumen, and casts may appear in the urine, or, in serious cases, there may be suppression of urine. Jaundice and tenderness over the liver may appear about the same time. Blood examination shows an anemia which may be marked. In fatal cases, the patient may develop delirium, followed by coma and death. During the acute stage of poisoning and for some weeks after, arsenic may be demonstrated in the urine. See also arsenic and arsenic compounds.

Fire Hazard: Moderate, when exposed to flame.

Explosion Hazard: MOD, when exposed to Cl_2, HNO_3, ($K + NH_3$) or open flame. [*19*]

Disaster Hazard: Dangerous, extremely toxic. More toxic than its oxidation product; when heated to decomp, emits highly toxic fumes; can react vigorously with oxidizing materials.

ARSONOACETIC ACID. See aricyl acid.

ARSPHENAMINE. Syn: *3-diamino-4-dihydroxy-1-arsenobenzene hydrochloride, Ehrlich 606, salvarsan.* Light yellow hygroscopic powder. $C_{12}H_{12}As_2N_2O_2 \cdot 2HCl \cdot 2H_2O$, mw: 475.0.

Acute tox data: iv LD_{LO} (rat) = 100 mg/kg. [*3*]

THR = HIGH via iv route. Implicated in the development of aplastic anemia. See also arsenic.

ARSYSODILA. See sodium cacodylate.

ARTHRYTIN. See amiodoxyl benzoate.

ARTIFICIAL ALMOND OIL. See benzaldehyde.

ARTIFICIAL GUM. See dextrin.

ASBESTOS PARTICLES. Syns: *asbestos dust, amosite, amphibole.*

THR = MOD via inhal route. A recog (+) carc. [*3, 6, 102*] The essential lesion produced by asbestos dust is a diffuse fibrosis which probably begins as a "collar" about the terminal bronchioles. Usually, at least 4 to 7 years of exposure are required before a serious degree of fibrosis results. There is apparently less predisposition to tuberculosis than is the case with silicosis. Prolonged inhal can cause cancer of the lung, pleura and peritoneum, and has exper produced cancers of the peritoneum, intestine, bronchus and oropharynx. [*80, 3, 6, 23*] Clinically, the most striking sign is shortness of breath of gradually increasing intensity, often associated with a dry cough. In the early stages physical signs are absent or slight; in the later stages rales may be heard, and in long standing cases there is frequently clubbing of the fingers. In early stages of the disease the chest x-rays reveal a groundglass or granular change, chiefly in the lower lung fields; as the condition progresses the heart outline becomes "shaggy," and irregular patches of mottled shadowing may be seen. "Asbestos bodies" may be found in the sputum. At autopsy, the pleurae are thickened and adherent and thick subpleural fibrous plaques are often present. Where the disease is far advanced there are usually large areas of fibrosis, with emphysematous changes in the apices and bases. The alveolar walls are thickened, and the characteristic "asbestos bodies" are found. A common air contaminant. [*45*]

ASBESTOS, BLUE. See asbestos particles.

ASBESTOS, BROWN (AMOSITE). See asbestos particles.

ASBESTOS, WHITE. See asbestos particles.

ASCARIDOLE. Syn: *ascarisin.* Unstable liquid. $C_{10}H_{16}O_2$, mw: 168.2, mp: 3.3°, bp: 40° @ 2 mm; 115° @ 15 mm, d: 1.011 @ 13°/15°.

THR = HIGH oral systemic. An exper neo. [*3*] See oil of chenopodium and peroxides, organic.

Fire Hazard: Mod, by spont chemical reaction. An oxidizer.

Explosion Hazard: Explodes when heated above 130° or when exposed to organic acids.

Disaster Hazard: Dangerous; when heated, emits toxic fumes and may explode; reacts with reducing materials.

ASCARISIN. See ascaridole.

ASCORBIC ACID. Syns: *l-ascorbic acid, vitamin C.* White crystals, sol in water, slightly sol in alcohol, insol in ether, chloroform, benzene, petroleum ether, oils and fats. $OCOCOH:COHCHCHOHCH_2OH$, mw: 176, mp: 192°.

Acute tox data: iv LD_{50} (mouse) = 518 mg/kg. [*3*]

THR = MOD via iv route. A chemical preservative food additive and a dietary supplement food additive. [*109*]

ASCORBYL PALMITATE. A white or yellowish white powder, citrus-like odor, sol in alcohol, animal and vegetable oils, slightly sol in water. $C_{22}H_{38}O_7$, mw: 414, mp: 116°–117°.

THR = No data. Probably LOW to MOD. A chemical preservative food additive. [*109*]

ASIATICOCIDE.
A glycoside terpene from plant *Centella Asiatica.*
THR = An exper neo. [*3*]

ASPARAGIC ACID. See aspartic acid.

ASPARAGINIC ACID. See aspartic acid.

ASPARTIC ACID. Syns: *asparaginic acid, asparagic acid, aminosuccinic acid.* A naturally occurring non-essential amino acid. The common form is *l*-aspartic acid. The data below refer to both *l* and *dl* forms. Colorless crystals, sol in water, insol in alcohol and ether. $COOHCH_2CH(NH_2)COOH$, mw: 133, mp: (*dl*) 278°–280° (decomp), d: 1.663 (12°/12°), mp: (*l*) 251°,
THR = U. Probably LOW to MOD. A dietary supplement food additive. [*109*]

ASPHALT. Syns: *bitumen, petroleum pitch.* Black or dark brown mass. bp: <470°, flash p: 400 +°F (CC), d: 0.95–1.1, autoign. temp.: 905°F.
THR = MOD irr. May contain carc components.
Fire Hazard: Slight, when exposed to heat or flame or F_2.
Spontaneous Heating: No.
To Fight Fire: Foam, CO_2, dry chemical.

ASPHALT, CUTBACK. A liquid petroleum product, solubility of residue from distillation in carbon tetrachloride = 99.5%, flash p: < 50°F.
THR = Contains carc. [*14, 23*]
Fire Hazard: Dangerous, when exposed to heat or flame.
To Fight Fire: Dry chemical, water mist, fog.

ASPHALT, LIQUID MEDIUM CURING. Flash p: 100°F (OC), (minimum) grades MC-0 and MC-1; 150°F (OC) (minimum) grades MC-2 through MC-5.
THR = Contains carc. [*14, 23*]
Fire Hazard: Mod when exposed to heat or flame.
To Fight Fire: Dry chemical, fog, water mist.

ASPHALT, LIQUID RAPID CURING. Flash p: 80°F (OC) minimum (grades RC-0 through RC-5).
THR = An exper carc. [*14*]
Fire Hazard: Dangerous, when exposed to heat or flame.
To Fight Fire: Dry chemical, fog, water mist.

ASPHALT, LIQUID-SLOW CURING. Flash p: 150°F+ (OC), grades SC-0 + SC-1; 175°F+ (OC) grades SC-2; 200°F+ (OC) grades SC-3; 225°F+ (OC) grades SC-4; 250°F+ (OC) grades SC-5.
THR = An exper carc. [*14*]
Fire Hazard: Mod, when exposed to heat or flame (grades SC-0 ⟶ SC-4); slight (grade SC-5).
To Fight Fire: Dry chemical, fog, water mist.

ASPIDIUM. Syn: *male fern.*
THR = LOW irr and allergen, MOD via oral route.
Fire Hazard: Slight.

ASPIRIN. See acetol.

ASTATINE. A highly radioactive and therefore toxic member of the halogen family of chemical elements. At, atwt: (210), mp: 302°, bp: 337°. Somewhat more metallic than iodine and collects in the thyroid, as does iodine. All isotopes are radioactive and have short $T_{\frac{1}{2}}$. Occurs naturally in small quantities.
Radiation Hazard: For permissible levels see Table 5A.5. Section 5A.

"ATABRINE." Syn: *quinacrine.* Bright yellow crystals. $C_{23}H_{30}ClN_3O$, mw: 400, mp: decomp @ 248°–250°.
THR = HIGH via oral route to chickens. HIGH via SC to mice. Has been implicated in aplastic anemia. [*3*]
Disaster Hazard: Dangerous.

ATMOSPHERIC HAZARDS REQUIRING RESPIRATORY PROTECTION. Section 2.

ATOXYL. See sodium arsanilate.

ATRAZINE. See 2-chloro-4-ethylamino-6-isopropylamino-5-triazine.

ATROPAMINE. See apoatropine.

ATROPIC ACID. See α-phenyl acrylic acid.

ATROPINE. Syn: *daturine.* Colorless, crystalline alkaloid $C_{17}H_{23}NO_3$, mw: 289.4, mp: 115.5°; subl @ 118°.
Acute tox data: Oral TD_{LO} (human) = 0.1 mg/kg ⟶ CNS effects. Oral LD_{LO} (human) = 0.07 mg/kg; ip LD_{50} (rat) = 280 mg/kg; scLD_{50} (rat) = 250 mg/kg; oral LD_{50} (mouse) = 400 mg/kg. [*3*]
THR = HIGH via oral, ip and sc routes.
Fire Hazard: Slight.

ATROPINE METHYL BROMIDE. See atropine.

ATROPINE METHYL NITRATE. See atropine.

ATROPINE SULFATE. See atropine.

AURAMINE. See 4,4′-(imidocarbonyl)*bis*(N,N-dimethyl aniline).

AURIC BROMIDE. Gray powder, brown crystals. $AuBr_3$, mw: 436.7, mp: 160°.
THR = See gold compounds and bromides.

AURIC CHLORIDE. Syn: *gold chloride.* Claret red crystals. $AuCl_3$, mw: 303.3, mp: 254° (decomp), bp: subl 265°, d: 3.9.
Acute tox data: Sc LD_{LO} (mouse) = 1500 mg/kg. [*3*]
THR = MOD via sc route. See gold compounds and chlorides.

AURIC CYANIDE. See cyanoauric acid.

AURIC HYDROGEN NITRATE. Syn: *gold nitrate.* Yellow crystals. $AuH(NO_3)_4 \cdot 3H_2O$, mw: 500.29, mp: 72°, (decomp), d: 2.84.
THR = See gold compounds and nitrates.

AURIC IODIDE. Syn: *gold iodide*. Dark green crystals. AuI_3, mw: 577.96.
THR = See gold compounds and iodides.

AURIC OXIDE. Syn: *gold oxide*. Brown-black powder. Au_2O_3, mw: 441.9, mp: $-O_2$ @ 160°, bp: $-3O_2$ @ 250°.
THR = See gold compounds.

AURIC SULFIDE. Syn: *gold sulfide*. Brown-black powder. Au_2S_3, mw: 490.60, mp: decomp @ 197°, d: 8.754.
THR = See gold compounds and sulfides.

AUROTHIOGLUCOSE. $C_6H_{12}O_5S \cdot Au$, mw: 393.2.
THR = An exper neo. [*3*]

AUROUS BROMIDE. Yellowish-gray mass or crystalline powder. AuBr, mw: 276.9, mp: decomp @ 115°, d: 7.9.
THR = See gold compounds and bromides.

AUROUS CHLORIDE. Yellow crystals. AuCl, mw: 232.4, mp: decomp to $AuCl_3$ @ 170°, bp: decomp 289.5°.
THR = See gold compounds and chlorides.

AUROUS CYANIDE. Syn: *gold cyanide*. Light yellow, crystalline powder. AuCN, mw: 223.0, mp: decomp, d: 7.12 @ 25°.
THR = See cyanides and gold compounds.

AUROUS IODIDE. Syn: *gold iodide*. Greenish-yellow powder. AuI, mw: 324.12, mp: decomp 120°, d: 8.25.
THR = See gold compounds and iodides.

AUROUS OXIDE. Gray-violet crystals. Au_2O, mw: 410.4, mp: −0 @ 205°, d: 3.6.
THR = See gold compounds.

AUROUS SULFIDE. Brown-black powder. Au_2S, mw: 426.0, mp: decomp 240°.
THR = See gold compounds and sulfides.

AUSTEN RED DIAMONDS. See explosives, high.

AUSTINITE. See calcium zinc arsenate monohydrate.

AUTARITE. See calcium iodate.

AUTOIGNITION TEMPERATURE. Defined in Section 7.

AUTOMATIC SPRINKLER SYSTEM. Section 7.

AUTUNITE. Syns: *calco uranite*, *lime uranite*. Bright yellow mineral of pearly luster. $Ca(UO_2)_2P_2O_8 \cdot 8H_2O$,
THR = See uranium compounds, insol.
Radiation Hazard: See uranium.

AVADEX. See 2,3-dichlorallyl diisopropyl thiocarbamate.

AVERTIN. See tribromo ethanol.

AVOGADRITE. See potassium fluoborate.

Az. See azodicarbonamide.

1-AZABENZO(*a*)PYRENE.
THR = An exper carc. [*23*]

5-AZACYTIDINE. $C_8H_{12}O_5N_4$, mw: 244.2.
THR = An exper neo. [*3*]

8-AZAGUANINE. $C_4H_4ON_6$, mw: 152.1.
THR = An exper teratogen. [*3*]

2-AZAHYPO XANTHINE. $C_4H_3ON_5$, mw: 137.1.
THR = An exper neo. [*3*]

AZELAIC ACID. Syn: *nonanedioic acid*. Powder. $HOOC(CH_2)_7COOH$, mw: 188.22, mp: 106.5°, bp: 265° @ 50 mm, d: 1.038 @ 110°, vap. press: 1 mm @ 178.3°.
THR = No data. Probably LOW. Closely related to glutaric acid and adipic acid.
Fire Hazard: Slight; when exposed to heat or flame, can react with oxidizing materials.

AZIDES. (See also specific compound).
THR = Variable. Many azides cause fall in blood pressure and some inhibit enzyme action, thus resembling nitrites and cyanides. An azide is a compound of H or a metal and the monovalent $-N_3$ radical. The azides as a group are one of the few commercially produced explosives that contain no oxygen. Hydrogen azide or azoic acid and its sodium salt are sol in water. All of its salts and the acid are unstable and decompose explosively; although lead azide, which is one of the most important azides, is not very sensitive and is used as a detonating agent, in which respect it is more efficient, weight for weight, than fulminate. Azides can be used wherever fulminate may be used; in detonators and priming compositions as well. Their only drawbacks are their high ignition temperature (detonators) and relative insensitiveness to friction (priming compositions). However, by compounding them with other materials, this can be overcome. Mercury azide, however, is more sensitive than lead azide; in fact it is more sensitive then mercury fulminate. When packed in bulk, azide should contain not less than 20% water and in this wet condition should be placed in bags of 4 oz or heavier duck. Each bag contains approximately 25 lb, dry weight, of lead azide. In each bag and over the azide is placed a cap of the same cloth and of the diameter of the bag, and the bag is tied securely. Five of these bags are placed in a larger bag of 4 oz or heavier duck and this large bag is tied securely. The large bag should contain not more than 150 lb, dry weight of lead azide. This bag is placed in the center of a container complying with ICC specification 5 or 5B (pertaining to metal barrels or drums), 17H (pertaining to metal drums,

single trip) or 10B (pertaining to wooden barrels or kegs). The container is lined with a heavy close fitting bag of jute or other suitable bag material of equal strength. The large duck bag is completely surrounded within the jute bag by not less than 3 in. of well-packed sawdust, saturated with water and the jute bag is closed by securely sewing to prevent escape of sawdust. The outer container, drum or barrel is water-tight. When considered necessary to prevent freezing in shipment, or storage, an alternative method of packing is used. The lead azide is wet with not less than 20% of its weight of a solution of denatured ethy alcohol in water containing not less than 33% by weight of ethyl alcohol. The 25 lb duck bags are placed in a strong waterproof bag instead of a duck bag and the drum or barrel is filled with sawdust saturated with the same 33% by weight solution of ethyl alcohol. Marking should be in accordance with U.S. Standard Specification for marking shipment no. 100-2. In addition, each barrel or cask is marked plainly "INITIATING EXPLOSIVE—DANGEROUS DO NOT STORE OR LOAD WITH ANY HIGH EXPLOSIVE." For further information concerning surface storage and the destruction of excess azide explosives, see Section 7.

Lead and mercury azides are the best examples of the azide type of explosive. Further uses of azide are in loading of fuse detonators and in the manufacture of priming compositions. Azides are used as substitutes for mercury fulminate. On account of its high temperature of ignition, lead azide is not easily ignited by the "spit" of a safety fuse. However, the addition of lead styphnate for instance, which is more readily ignited, overcomes the difficulty. It is also used in explosive rivets with the addition of silver acetylide or tetracine to lower its ignition temp.

Disaster Hazard: Dangerous; shock and heat will explode it; when heated to decomp, emits highly toxic fumes; if exposed to CS_2, it forms violently explosive salts. Organic azides are made sensitive by metal salts or traces of strong acid. [*19*]

AZIDO ALKANES. Compounds of -NNN are very unstable.

4-AZIDO-N,N-DIETHYL ANILINE. NNN · $C_{10}H_{14}N$, mw: 190.1.
THR = U, but probably HIGH. This compound is very unstable.

AZIDO FLUORINE. See fluorine azide.

3-AZIDO-s-TRIAZOLE. $C_2H_2N_6$, mw: 110.
THR = No data. Very unstable to heat. [*19*]

AZIMETHYLENE. See diazomethyanyl.

AZIMINO BENZENE. See 1,2,3-benzotriazole.

AZINPHOS ETHYL. See ethyl guthion.

AZINPHOS METHYL. See guthion.

AZIRIDINE. See ethylene imine.

1-AZIRIDINE ETHANOL. Syn: *3-hydroxy-1-ethyl aziridine.* C_4H_9ON, mw: 87.1.
THR = An exper (+) neo and carc. [*3, 8, 23*]

1-AZIRIDINYL PHOSPHINE OXIDE (TRIS). See tris (1-aziridinyl) phosphine oxide solution.

***p*-AZOANILINE.** See *p*-diamino azobenzene.

AZOBENZENE. Syns: *azobenzol, azobenzide.* Solid, orange-red crystals. $C_6H_5NNC_6H_5$, mw: 182.2, mp: 68°, d: 1.203 @ 20°/4°, vap. press: 1 mm @ 103.5°, bp: 293°.
Acute tox data: Oral LD_{50} (rat) = 1000 mg/kg. [*3*]
THR = MOD via oral route. An insecticide. sc (rat) = 17000 mg/kg. An exper neo and carc. [*3, 4*]
Disaster Hazard: Dangerous. When strongly heated, emits highly toxic fumes.

AZOBENZIDE. See azobenzene.

AZOBENZOL. See azobenzene.

1,1-AZOBIS FORMAMIDE. See azodicarbonamide.

2,2-AZOBIS ISOBUTYRONITRILE. White powder, sol in many organic solvents and in vinyl monomers, insol in water. $(CH_3)_2C(CN)NNC(CN)(CH_3)_2$, mw: 164, mp: 105° (decomp).
Acute tox data: Oral LD_{50} (mouse) = 700 mg/kg; ip LD_{50} (mouse) = 25 mg/kg. [*3*]
THR = HIGH via ip and MOD via oral routes. See also nitriles.

AZOCHLORAMIDE. Syn: *chlorozodin.* Bright yellow crystals. $C_2H_4Cl_2N_6$, mw: 183.0, mp: explodes at 155°.
THR = MILD allergen. No other data.
Explosive Hazard: Severe when shocked or exposed to heat; explodes at 155° or more, particularly in the presence of metals.
Disaster Hazard: Mod dangerous; when heated to decomp, emits toxic fumes and may explode.

AZODICARBONAMIDE. Syns: *Az, 1,1'-azobis formamide.* Yellow powder, insol in common solvents, sol in dimethyl sulfoxide.
THR = A food additive permitted in food for human consumption. [*109*]

"AZODRIN." Syn: *monocrotophos.* A reddish-brown solid, mild ester odor. $C_6H_{14}O_5NP$, mw: 211.2, bp: 125°.
Acute tox data: Oral LD_{50} (rat) = 21 mg/kg; dermal LD_{50} (rat) = 112 mg/kg; oral LD_{50} (wild birds) = 1.6 mg/kg; dermal LD_{50} (rabbit) = 354 mg/kg. [*3*]

THR = HIGH via oral and dermal routes of exposure.
Fire Hazard: Dangerously flam.
Disaster Hazard: When heated to decomp, evolves highly toxic fumes.

AZO DYES. Are often found to be carc. [*14*]

AZOETHANE. $C_4H_{10}N_2$, mw: 86.2.
THR = An exper trans placental carc. [*3*, *23*]

AZOMIDE. See hydrazoic acid.

AZOLE. See pyrrole.

AZO MUSTARD. $C_{16}H_{17}N_3Cl_2$, mw: 322.2.
THR = An exper carc. [*23*]

1,1′-AZONAPHTHALENE. $C_{20}H_{14}N_2$, mw: 282.4.
THR = An exper carc. [*3*]

2′,2-AZONAPHTHALENE. $C_{20}H_{14}N_2$, mw: 282.4.
THR = An exper carc. [*3*]

AZOMINE BLUE 2B. $C_{32}H_{24}O_{14}N_6S_4$, mw: 936.8.
THR = An exper carc and teratogen. [*3*]

AZOTIC ACID. See nitric acid.

2,3-AZOTOLUENE. $C_{14}H_{14}N_2$, mw: 210.3.
THR = An exper neo. [*3*]

AZOXY BENZENE. Syn: *azoxybenzide*. Pale yellow crystals. $C_{12}H_{10}N_2O$, mw: 198.2, mp: 36°, d: 1.159 @ 26.4°.
Acute tox data: Oral LD_{50} (rat) = 620 mg/kg; dermal LD_{50} (rabbit) = 1090 mg/kg. [*3*]
THR = MOD via oral and dermal routes. An exper carc. [*3*] An insecticide.
Fire Hazard: Slightly dangerous, on decomp, emits toxic fumes.

AZOXYBENZIDE. See azoxybenzene.

AZOXY ETHANE. $C_4H_{10}ON_2$, mw: 102.1.
THR = An exper trans placental carc. [*3*, *23*]

AZOXY METHANE. $C_2H_6ON_2$, mw: 74.1.
THR = An exper trans placental carc and teratogen. [*3*, *23*]

AZULENO(5,6,7-*cd*)PHENALENE. $C_{20}H_{12}$, mw: 252.3.
THR = An exper neo. [*3*]

b

BACITRACIN. An antibiotic; white to pale buff, hygroscopic powder, odorless or slight odor, freely sol in water, sol in alcohol, methanol and glacial acetic acid, insol in acetone, chloroform and ether. $C_{66}H_{103}N_{17}O_{16}S$, mw: 1411.
THR = U. A food additive permitted in feed and drinking water of animals. Also a food additive permitted in food for human consumption. [*109*]

BACITRACIN METHYLENE DISALICYLATE. White to gray-brown powder, slight unpleasant odor, sol in water, pyridine, ethanol; less sol in acetone, ether, chloroform, benzene.
THR = U. A food additive permitted in feed and drinking water of animals and/or for the treatment of food producing animals. Also a food additive permitted in food for human consumption. [*109*]

BACTERIAL CATALASE. *Derived from micrococcus lysodeikticus* by a pure culture fermentation process.
THR = U. A food additive permitted in food for human consumption. [*109*]

BADDELYTE. See zirconium dioxide.

BAGASSE DUST.
THR = A nuisance dust; inhal can cause bronchial asthma, sneezing, rhinorrhea, pneumonia etc. See also cotton dust.
Fire Hazard: Mod, when exposed to heat or flame.
Explosion Hazard: Mod, when exposed to flame. See dust explosions.

BAGS, NITRATE OF SODA, EMPTY AND UNWASHED. See also sodium nitrate.

BAKELITE. See polyvinyl chloride.

BAKER, SIR GEORGE. See Section 1.

BAKING POWDER.
THR = A mild allergen and a mild irr.

BAKING SODA. See sodium bicarbonate.

BAL. See 2,3-dimercapto-1-propanol.

BALATA. Dried juice of the bully tree, mimusops balata. Resembles gutta percha.
THR = A MILD irr and allergen.
Fire Hazard: Slight, when exposed to heat or flame.

BALM OF GILEAD. See canada balsam.

BALSAM COPAIBA. See copaiba.

BALSAM OF PERU. Syn: *peruvian balsam.* Dark brown viscid liquid, vanilla odor.
THR = A MILD allergen.
Fire Hazard: Slight, when heated.

BANANA OIL. See iso amyl acetate.

BANANA PEEL OIL.
THR = A MILD allergen.
Fire Hazard: Yes.

BANOMITE. Syn: *benzoyl chloride (2,4, 6-dichlorophenyl) hydrazone.* $C_{13}H_8Cl_4N_2$, mw: 334. White to yellow crystals almost insol in water, sol in organic solvents, mp: 98°.
Acute tox data: Oral LD_{50} (rat) = 389 mg/kg; dermal LD_{50} (rabbit) = > 10 mg/kg. [*3*]
THR = HIGH via oral and dermal routes. An acaricide. Skin irr and allergen.
Disaster Hazard: Dangerous; when heated to decomp, evolves highly toxic fumes.

BANVEL D. See 3,6-dichloro-*o*-anisic acid.

BANVEL T. See 3,5,6-trichloro-*o*-anisic acid.

BAP. Syn: *N-nitroso-bis(2-acetoxypropyl)amine.*
THR = An exper carc. [*3*].

BARBAN. See carbyne.

BARBITURATES. Syn: *Derivatives of barbituric acid; i.e., barbital, barbitone, barbital sodium.*
THR = MOD by ingestion. Large doses cause marked depression (sometimes preceded by excitation), prolonged coma and death. Allergic skin reactions may occur from contact. Has been implicated in development of aplastic anemia. A truly habit forming drug.
Fire Hazard: Slight, when heated.

BARBITURIC ACID. Syn: *malonylurea.* Crystals or white to yellow-white powder. $C_4H_4O_3N_2$, mw: 128.1; mp: 245°, bp: 260° (decomp).
THR = MOD irr to skin, eyes and mu mem. An allergen. Has no hypnotic properties. [*3*]
Fire Hazard: Slight.

BARITE. See barium sulfate.

BARIUM. Silver-white, slightly lustrous, somewhat malleable metal. Ba, atwt: 137.36, mp: 725°, bp: 1640°, d: 3.5 @ 20°, vap. press: 10 mm @ 1049°.
THR = See barium compounds (sol).
Radiation Hazard: For permissible levels see Table 5A.5. Artificial isotope ^{131}Ba: $T_{\frac{1}{2}}$ = 12 d, decays to ^{131}Cs via ec, emits γ's of 0.5 MeV and x-rays. Artificial isotope ^{140}Ba: $T_{\frac{1}{2}}$ = 12.8 d, decays to

^{140}La via β's of 0.48–1.02 MeV, emits γ's of 0.03–0.54 Mev.

Fire Hazard: Dangerous and explosive in form of dust when exposed to heat or flame or by chemical reaction. Reacts violently with acids, CCl_4, $C_2Cl_3F_3$, $C_2H_2FCl_3$, C_2Cl_4, C_2HCl_3 and water. [*19*]

BARIUM BENZENE SULFONATE. White nacreous leaflets. $Ba(C_6H_5 \cdot SO_3)_2 \cdot H_2O$, mw: 469.71.

THR = See barium compounds, sol and sulfonates.

BARIUM BENZOATE. White nacreous leaflets. $Ba(C_7H_5O_2)_2 \cdot 2H_2O$, mw: 415.61, mp: $-2H_2O$ @ 100°.

THR = Deadly poison! See barium compounds (sol).

BARIUM BICHROMATE. See barium dichromate.

BARIUM BINOXALATE. Syn: *acid barium oxalate*. Crystals. $Ba(HC_2O_4)_2 \cdot 2H_2O$, mw: 351.45, mp: $-H_2O$ @ 80°.

THR = See barium compounds (sol) and oxalates.

BARIUM BINOXIDE. See barium peroxide.

BARIUM BOROWOLFRAMATE. Syn: *barium borotungstate*. Crystals, effloresce rapidly in air. $2BaO \cdot 9WO_3 \cdot B_2O_3 \cdot 18H_2O$, mw: 2787.93.

THR = See barium compounds (sol).

BARIUM BROMATE. White crystals or crystalline powder. $Ba(BrO_3)_2 \cdot H_2O$, mw: 411.21, mp: decomp 260°, d: 3.99 @ 18°.

THR = See barium compounds (sol).

Fire Hazard: Mod by chemical reaction with easily oxidized materials. Violent reaction with Al, As, C, Cu, metal sulfides, organic matter, P, S. [*19*]

Disaster Hazard: Dangerous; see bromides; can react with reducing materials.

BARIUM BROMIDE. Colorless crystals. $BaBr_2$, mw: 297.09, mp: 847°, d: 4.781 @ 24°.

THR = See barium compounds (sol) and bromides.

BARIUM ACETATE. White crystals. $Ba(C_2H_3O_2)_2 \cdot H_2O$, mw: 273.46, mp: $-H_2O$ @ 150°, d: 2.19.

THR = See barium compounds (sol).

BARIUM ALLOYS, NON-PYROPHORIC.

THR = Reacts rapidly with acids, water.

BARIUM ALLOYS, PYROPHORIC.

THR = Reacts rapidly with acids, water.

BARIUM AMIDE. Gray-white crystals. $Ba(NH_2)_2$, mw: 169.41, mp: 280°.

THR = See barium compounds (sol).

BARIUM AMYL SULFATE. White crystals, fatty feel. $Ba(C_5H_{11}SO_4)_2 \cdot H_2O$; mw: 507.80.

THR = See barium compounds (sol) and sulfates.

BARIUM ANTIMONYL TARTRATE. Crystals. $BaSb_2(C_2H_4O_6)_4$, mw: 973.2.

THR = See antimony and barium compounds.

BARIUM-*o*-ARSENATE. Black crystals. $Ba_3(AsO_4)_2$, mw: 689.90, mp: 280°.

THR = See arsenic compounds and barium compounds (sol).

BARIUM ARSENIDE. Brown crystals. Ba_3As_2, mw: 561.90, d: 4.1 @ 15°.

THR = See arsenic and barium compounds (sol).

BARIUM AZIDE. Monoclinic prisms. $Ba(N_3)_2$, mw: 221.41, mp: $-N_2$ @ about 120°, bp: explodes, d: 2.936.

THR = See barium compounds (sol) and azides.

Explosion Hazard: Mod when shocked or exposed to heat. Around 275°, spont flam in air. Very unstable. [*19*]

Disaster Hazard: Dangerous; shock and heat will explode it.

BARIUM BROMIDE FLUORIDE. Platelets. $BaBr_2 \cdot BaF_2$; mw: 472.55, d: 4.96 @ 18°.

THR = See fluorides and barium compounds (sol) and bromides.

BARIUM BROMOPLATINATE. Monoclinic crystals. $BaPtBr_6 \cdot 10H_2O$, mw: 992.25, d: 3.71.

THR = See barium compounds (sol) and bromides.

BARIUM BUTYRATE. Crystals. $Ba(C_4H_7O_2)_2 \cdot 2H_2O$, mw: 347.59.

THR = See barium compounds (sol).

BARIUM CARBIDE. Gray crystals. BaC_2, mw: 161.4, d: 3.75.

THR = See barium compounds (sol).

Fire Hazard: Mod, by chemical reaction with moisture to form acetylene.

Explosion Hazard: Mod; evolves acetylene upon contact with moisture. Can react violently with Se, S, H_2O. [*19*]

Disaster Hazard: Dangerous; will react with water or steam to produce flam vapors.

To Fight Fire: CO_2, dry chemical.

BARIUM CARBONATE. White powder. $BaCO_3$, mw: 197.37, mp: 1740 @ 90 atm, bp: decomp, d: 4.43.

THR = See barium compounds (sol).

BARIUM CHLORATE. Colorless prisms or white powder. $Ba(ClO_3)_2 \cdot H_2O$, mw: 322.29, mp: $-H_2O$ @ 414°, d: 3.18.

THR = See barium compounds (sol).

Fire Hazard: See chlorates.

Explosion Hazard: See chlorates. Can react violently with Al, As, C, charcoal, Cu, MnO_2, metal sulfides, S_4N_4, organic matter, P,S. [*19*]

Disaster Hazard: See chlorates.

BARIUM CHLORIDE. Colorless flat crystals. $BaCl_2$, mw: 208.27, mp: transition @ 925° to cubic crystals, bp: 1560°, d: 3.856 @ 24°.

THR = HIGH. See barium compounds (sol). Reacts violently with BrF_3 and 2-furan percarboxylic acid. [*19*]

BARIUM CHLOROPLATINATE. Rhombic orange-yellow crystals. $BaPtCl_6 \cdot 6H_2O$, mw: 653.2, mp: $-5H_2O$ @ 70°, d: 2.868.
THR = See barium compounds (sol) and chlorides.

BARIUM CHLOROPLATINITE. Crystals. $BaPtCl_4 \cdot 3H_2O$, mw: 528.4, d: 2.868, mp: $-3H_2O$ @ 150°.
THR = See barium compounds (sol) and chlorides.

BARIUM CHROMATE. Heavy yellow crystalline powder. $BaCrO_4$, mw: 253.37, d: 4.498 @ 15°.
THR = See barium compounds (sol) and chromium compounds. A recog carc. [*14*]
Fire Hazard: See chromates.
Disaster Hazard: Dangerous; reacts vigorously with reducing materials.

BARIUM CITRATE. White powder. $Ba_3(C_6H_5O_7)_2 \cdot 7H_2O$ @ 150°.
THR = See barium compounds (sol).

BARIUM COMPOUNDS (SOLUBLE).
The soluble barium salts, such as the chloride and sulfide, are poisonous when taken by mouth. The insoluble sulfate used in radiography is not acutely toxic. See also barium sulfate. Few cases of industrial systemic poisoning have been reported, but one investigator describes a fatal case of poisoning attributed to barium oxide, the symptoms being severe abdominal pain with vomiting, dyspnoea, rapid pulse, paralysis of the arm and leg, and eventually cyanosis and death. The same investigator produced paralysis in animals with barium oxide and carbonate. The usual result of exposure to the sulfide, oxide and carbonate is irr of the eyes, nose and throat, and of the skin, producing dermatitis. The salts mentioned are somewhat caustic.

BARIUM CYANIDE. White crystalline powder. $Ba(CN)_2$, mw: 189.40.
THR = HIGH. See also cyanides and barium compounds (sol).

BARIUM CYANOPLATINITE. (a) monoclinic yellow crystals; (b) rhombic crystals. $BaPt(CN)_4 \cdot 4H_2O$, mw: 508.6, mp: $-2H_2O$ @ 100°, d: (a) 2.076, (b) 2.085.
THR = HIGH. See barium compounds (sol) and cyanides.

BARIUM DICHROMATE. Syn: *barium bichromate.* Brownish-red crystalline masses. $BaCr_2O$, mw: 353.38.
THR = HIGH. See barium compounds (sol) and chromium compounds. Some chromates are recog carc. [*14, 3*]
Fire Hazard: Mod, by chemical reaction with easily oxidized materials; can react vigorously with reducing materials. A powerful oxidizer.

BARIUM DIOXIDE. See barium peroxide.

BARIUM DIPHENYLAMINE SULFONATE. Crystals. $Ba(C_6H_5 \cdot NH \cdot C_6H_4SO_3)_2$, mw: 633.91.
THR = See barium compounds (sol).
Disaster Hazard: Dangerous; see sulfonates.

BARIUM DI-*o*-PHOSPHATE. Rhombic white crystals. $BaHPO_4$, mw: 233.35, d: 4.165 @ 15°.
THR = See barium compounds (sol) and phosphates.

BARIUM DITHIONATE. Syn: *barium hyposulfate.* Rhombic or monoclinic colorless crystals. $BaS_2O_6 \cdot 2H_2O$, mw: 333.52, mp: 120° (dec) d: 4.536 @ 13.5°.
THR = See barium compounds (sol).
Disaster Hazard: Dangerous, see sulfates.

BARIUM DIURANATE. See uranium barium oxide.

BARIUM ETHYLSULFATE. Syn: *barium sulfovinate.* Colorless crystals, white lustrous leaf. $Ba(C_2H_5SO_4)_2 \cdot 2H_2O$, mw: 423.63.
THR = See barium compounds (sol).
Disaster Hazard: Dangerous, see sulfates.

BARIUM FERROCYANIDE. Yellow crystals. $Ba_2Fe(CN)_6 \cdot 6H_2O$, mw: 594.77, mp: $-H_2O$ @ 40°.
THR = See barium compounds (sol) and ferrocyanides.

BARIUM FLUOGALLATE. White crystals. $Ba_3(GaF_6)_2 \cdot H_2O$, mw: 797.54, mp: $-\frac{1}{2}H_2O$ @ 110°, $-\frac{1}{2}H_2O$ @ 230°, d: 4.06.
THR = See barium compounds (sol) and fluogallates.

BARIUM FLUORIDE. White powder. BaF_2, mw: 175.36, mp: 1280°, bp: 2137°, d: 4.89.
THR = See fluorides and barium compounds (sol).

BARIUM FLUORIDE CHLORIDE. Tetragonal crystals. $BaCl_2 \cdot BaF_2$, mw: 383.63, mp: 1008°, d: 4.51 @ 18°.
THR = See barium compounds (sol), fluorides and iodides.

BARIUM FLUORIDE IODIDE. Plates. $BaF_2 \cdot BaI_2$, mw: 566.56, d: 5.21 @ 18°.
THR = See barium compounds, sol, fluorides and iodides.

BARIUM FLUOSILICATE. Syn: *barium silicofluoride.* White crystalline powder. $BaSiF_6$, mw: 279.42, d: 4.29 @ 21°/4°, mp: 300° (dec).
THR = An insecticide. See barium compounds (sol) and fluosilicates.
Disaster Hazard: See fluosilicates.

BARIUM FORMATE. Crystals. $Ba(COOH)_2$, mw: 227.40, d: 3.21, mp: decomp.
THR = See barium compounds (sol).

BARIUM-*d*-GLUCONATE. Prisms or rhombic leaf-

lets. Ba $(C_6H_{11}O_7)_2 \cdot 3H_2O$, mw: 581.71, mp: $-3H_2O$ @ 100°, 120° decomp.
THR = See barium compounds (sol).

BARIUM HEXABORIDE. Cubic black metallic crystals. BaB_6, mw: 202.28, mp: 2270°, d: 4.36 @ 15°.
THR = See barium compounds (sol) and boron.

BARIUM HYDRATE. See barium hydroxide.

BARIUM HYDRIDE. Gray crystals or lumps. BaH_2, mw: 139.38, mp: decomp 675°, bp: 1400°, d: 4.21 @ 0°. Rapidly decomp by water and acids.
THR = See barium compounds (sol) and hydrides. In powder form it ignites spont in air, and reacts vigorously with water. [*19*]

BARIUM HYDROGEN PYROTELLURATE. Voluminous precipitate, yellow when hot, white when cold. $Ba(HTe_2O_7)_2 \cdot H_2O$, mw: 891.83.
THR = See barium compounds (sol) and tellurium compounds.

BARIUM HYDROSULFIDE. Rhombic yellow crystals. $Ba(HS)_2 \cdot 4H_2O$, mw: 275.57, mp: decomp @ 50°.
THR = See barium compounds (sol) and sulfides.

BARIUM HYDROXIDE. Syns: *barium hydrate, caustic baryta.* White powder. $Ba(OH)_2 \cdot 8H_2O$, mw: 315.51, mp: 78°, bp: $-8H_2O$ @ 780°, d: 2.18 @ 16°, anhy: 4.50.
THR = See barium compounds (sol).

BARIUM HYPOCHLORITE. Colorless crystals. $Ba(ClO)_2 \cdot 2H_2O$, mw: 276.31, mp: decomp.
THR = See barium compounds (sol) and hypochlorites.

BARIUM HYPOPHOSPHATE. Needle-like crystals. $BaPO_3$, mw: 216.34.
THR = See barium compounds (sol) and hypophosphates.

BARIUM HYPOPHOSPHITE. Crystalline powder. $Ba(H_2PO_2)_2 \cdot H_2O$, mw: 285.38, mp: decomp, d: 2.90 @ 17°.
THR = See barium compounds (sol) and hypophosphites. Reacts strongly with $KClO_3$.

BARIUM HYPOSULFATE. See barium dithionate.

BARIUM HYPOSULFITE. See barium thiosulfate.

BARIUM IODATE. White crystalline powder. $Ba(IO_3)_2$, mw: 487.20, mp: decomp, d: 4.998.
THR = See barium compounds (sol) and iodates. Reacts strongly with Al, As, C, Cu, metal sulfides, organic matter, P, S. [*19*]

BARIUM IODIDE. Colorless crystals. BaI_2, mw: 391.19, mp: 740°, d: 5.15 @ 25°/4°.
THR = See barium compounds (sol) and iodides.

BARIUM LACTATE. White powder. $Ba(CH_3 \cdot CH \cdot OH \cdot COO)_2$, mw: 315.50.
THR = See barium compounds (sol).

BARIUM LAURATE. White, leaflet crystals. $Ba(C_{12}H_{23}O_2)_2$, mw: 535.97, mp: 260°.
THR = See barium compounds (sol).

BARIUM MALATE. White powder. $BaC_4H_4O_5$, mw: 269.43.
THR = See barium compounds (sol).

BARIUM MALONATE. White powder. $BaC_3H_2O_4 \cdot H_2O$, mw: 257.42.
THR = See barium compounds (sol).

BARIUM MANGANATE. Syns: *manganese green, cassels green.* $BaMnO_4$, mw: 256.29, d: 4.85.
THR = See barium compounds (sol) and manganese compounds.

BARIUM MERCURY BROMIDE. See mercuric barium bromide.

BARIUM METHYLSULFATE. Efflorescent crystals. $Ba(CH_3SO_4)_2 \cdot 2H_2O$, mw: 395.59.
THR = See barium compounds (sol) and sulfates.

BARIUM MOLYBDATE. White powder, tetragonal crystals. $BaMoO_4$, mw: 297.31, d: 4.65, mp: 1480°.
THR = See barium compounds (sol).

BARIUM MONO-HYDROGEN-*o*-ARSENATE. Rhombic or monoclinic colorless crystals. $BaHAsO_4 \cdot H_2O$, mw: 295.29, mp: $-H_2O$ @ 150°, d: 1.93 @ 15°.
THR = See barium compounds (sol) and arsenic compounds.

BARIUM MONO-*o*-PHOSPHATE. Triclinic crystals. $BaH_4(PO_4)_2$, mw: 331.35, d: 2.9 @ 4°.
THR = See barium compounds (sol) and phosphates.

BARIUM MONOSULFIDE. Syn: *barium sulfide.* Cubic colorless crystals. BaS, mw: 169.43, d: 4.25 @ 15°, mp: 1200°.
THR = See barium compounds (sol) and sulfides.
Fire Hazard: Mod, by spont chemical reaction; air, moisture or acid fumes may cause it to ignite.
Explosion Hazard: See sulfides. Also reacts violently with ClO, PbO_2, $KClO_3$, KNO_3. [*19*]
Disaster Hazard: See sulfides.
To Fight Fire: CO_2, dry chemical.

BARIUM MONOXIDE. See barium oxide.

BARIUM MYRISTATE. White crystalline powder. $Ba(C_{14}H_{27}O_2)_2$, mw: 592.07.
THR = See barium compounds (sol).

BARIUM NITRATE. Syn: *nitrobarite.* Lustrous crystals. $Ba(NO_3)_2$, mw: 261.38, mp: 592°, bp: decomp, d: 3.24 @ 23°.
THR = See barium compounds (sol) and nitrates. Can react rapidly with ($Mg + BaO_2 + Zn$). [*19*]

BARIUM NITRIDE. Colorless crystals. Ba_3N_2, mw: 440.10, bp: 1000° (vacuo), d: 4.783 @ 25°/4°.
THR = See barium compounds (sol).
Fire Hazard: Mod, by spont chemical reaction with water to liberate flam vapor; see also ammonia.
Explosion and Disaster Hazard: Dangerous; explodes upon heating and by spont chemical reaction to liberate vapor which can form explosive mixtures with air. See also ammonia.

BARIUM NITRITE. White crystals. $Ba(NO_2)_2$, mw: 229.38, mp: decomp @ 217°, d: 3.23 @ 23°.
THR = See barium compounds (sol) and nitrites.

BARIUM OLEATE. Yellowish-white granular masses. $Ba(C_{18}H_{33}O_2)_2$, mw: 700.25.
THR = See barium compounds (sol).

BARIUM OSMIAMATE. $Ba(OsNO_3)_2$, mw: 641.7.
THR = Probably HIGH. See barium compounds, osmium compounds and nitrates. Detonates at 150°. [*19*]

BARIUM OXALATE. White crystalline powder. BaC_2O_4, mw: 225.38, d: 2.658, mp: 400° (decomp).
THR = See barium compounds (sol) and oxalates.

BARIUM OXIDE. Syns: *barium monoxide, barium protoxide.* White to yellowish-white powder. BaO, mw: 153.36, mp: 1923°, bp: 2000° (approx), d: 5.72.
THR = See barium compounds (sol).
Fire Hazard: Slight, by spont chemical reaction; produces heat on contact with water or steam. Can react violently with H_2S, hydroxylamine, N_2O_4. [*19*]

BARIUM PALMITATE. White crystalline powder. $Ba(C_{16}H_{31}O_2)_2$, mw: 648.19, mp: decomp.
THR = See barium compounds (sol).

BARIUM PERCHLORATE. Colorless crystals, $Ba(ClO_4)_2 \cdot 3H_2O$, mw: 390.4, mp: decomp @ 400°, d: 2.74.
THR = See barium compounds (sol) and perchlorates. Heating (reflux) with an alcohol yields highly explosive product. [*19*]

BARIUM PERMANGANATE. Brownish-violet crystals. $Ba(MnO_4)_2$, mw: 375.22, d: 3.77, mp: 200° (decomp).
THR = See barium compounds (sol) and manganese compounds.

BARIUM PEROXIDE. Syns: *barium binoxide, barium dioxide.* Grayish-white powder. BaO_2, mw: 169.36, mp: 450°, bp: −0 @ 800°, d: 4.96.
THR = See barium compounds (sol) and peroxides. Reaction with hydroxylamine sol causes flaming. Mixture of $(Mg + Zn + Ba(NO_3)_2)$ has caused an explosion. Mixing with organic matter can ignite or explode. [*19*]

BARIUM PEROXYDISULFATE. Monoclinic white crystals. $BaS_2O_8 \cdot 4H_2O$, mw: 401.57, mp: decomp.
THR = See barium compounds (sol).

BARIUM PHENOLSULFONATE. Syns: *barium sulfophenylate, barium sulfocarbolate.* White odorless powder. $C_{12}H_{12}BaO_9S_2$, mw: 501.71.
THR = See barium compounds (sol) and phenosulfonates.

BARIUM PHOSPHATE, DIBASIC. Syn: *secondary barium phosphate.* Crystalline powder. $BaHPO_4$, mw: 233.35, mp: 410° @ 710 mm (decomp), d: 4.17 @ 15°.
THR = See barium compounds (sol) and phosphates.

BARIUM PHOSPHITE. Crystalline powder. $BaHPO_3$, mw: 217.35.
THR = See barium compounds (sol) and phosphites.

BARIUM PROPIONATE. White powder. $C_6H_{10}BaO_4 \cdot H_2O$, mw: 301.52, mp: 300° decomp.
THR = See barium compounds (sol).

BARIUM PROTOXIDE. See barium oxide.

BARIUM PYROPHOSPHATE. Rhombic white crystals. $Ba_2P_2O_7$, mw: 448.68, d: 3.9 @ 20°.
THR = See barium compounds (sol) and phosphates.

BARIUM PYROVANADATE. White crystals. $Ba_2V_2O_7$, mw: 488.62, mp: 863°.
THR = See barium compounds (sol) and vanadium compounds.

BARIUM RHODANIDE. $BaC_6H_4O_2N_2S_4$, mw: 401.6.
THR = See barium compounds. Mixture with $NaNO_3$ may explode. [*19*]

BARIUM RICINOLEATE. Fine white powder. $Ba(C_{18}H_{34}O_3)_2$, mw: 734.3, mp: 116°, d: 1.21 @ 25°/25°.
THR = See barium compounds (sol).

BARIUM SALICYLATE. White needles. $Ba(C_7H_9O_3)_2 \cdot H_2O$, mw: 429.60.
THR = An allergen. HIGH toxicity. See also barium compounds (sol).

BARIUM SELENATE. Orthorhombic crystals. $BaSeO_4$, mw: 280.32, mp: decomp, d: 4.75.
THR = See barium compounds (sol) and selenium compounds.

BARIUM-*m*-SILICATE. Rhombic, colorless crystals. $BaSiO_3$, mw: 213.42, mp: 1604°, d: 4.399.
THR = See barium compounds (sol).

BARIUM SILICOFLUORIDE. See barium fluosilicate.

BARIUM STEARATE. White powder. $Ba(C_{18}H_{35}O_2)_2$, mw: 704.1.
THR = See barium compounds (sol).

BARIUM SUCCINATE. Crystalline powder. $BaC_4H_4O_4$, mw: 253.43.
THR = See barium compounds (sol).

BARIUM SULFATE. Syns: *blanc fixe*, *barite*. White, heavy, odorless powder, not sol in water or dilute acids. $BaSO_4$, mw: 233.4, d: 4.50 @ 15°, mp: 1580°.
THR = Insol salt used as an opaque medium in radiography. Sol impurities can lead to toxic reactions. See barium compounds (sol). Heating with aluminum can produce an explosion. An exper neo to rats via ipl route. [*103*]

BARIUM SULFHYDRATE. Yellow crystals. $Ba(SH)_2$, mw: 203.51.
THR = See barium compounds (sol) and sulfides.

BARIUM SULFIDE. See barium monosulfide.

BARIUM SULFITE. White powder. $BaSO_3$, mw: 217.42, mp: decomp.
THR = See barium compounds (sol) and sulfites.

BARIUM SULFOCARBOLATE. See barium phenosulfonate.

BARIUM SULFOCYANATE. Syn: *barium sulfocyanide*. White crystals. $Ba(SCN)_2 \cdot 2H_2O$, mw: 289.55.
THR = See also barium compounds (sol) and thiocyanates.

BARIUM SULFOCYANIDE. See barium sulfocyanates.

BARIUM SULFOPHENYLATE. See barium phenolsulfonate.

BARIUM SULFOVINATE. See barium ethylsulfate.

BARIUM TARTRATE. White granular powder. $BaC_4H_4O_6 \cdot H_2O$, mw: 303.45, d: 2.980 @ 20.8°.
THR = See barium compounds (sol).

BARIUM TELLURATE. White crystals. $BaTeO_4 \cdot 3H_2O$, mw: 383.02, mp: decomp > 200°, d: 4.2 @ 200°.
THR = See barium compounds (sol) and tellurium compounds.

BARIUM TETRASULFIDE. Rhombic red or yellow crystals or powder. BaS_4, mw: 265.63, mp: decomp @ 200°, d: 2.988.
THR = See barium compounds and sulfides.

BARIUM THIOCYANATE. See barium sulfocyanate.

BARIUM THIOSULFATE. Syn: *barium hyposulfite*. Crystalline powder. BaS_2O_3, mw: 249.49, mp: decomp.
THR = See barium compounds (sol) and sulfates.

BARIUM TITANATE. Syn: *barium-m-titanate*. Light gray-buff powder (also exists in five crystal modifications), insol in water and alkalies, slightly sol in dilute acids, sol in concentrated sulfuric acid and hydrosulfuric acid. $BaTiO_3$, mw: 233.26, mp: 3010°F, d: 5.95.
THR = Animal exper show LOW toxicity. See also titanium compounds.

BARIUM-*o*-TRIPHOSPHATE. Cubic white crystals. $Ba_3(PO_4)_2$, mw: 602.04, d: 4.1 @ 16°.
THR = See barium compounds (sol) and phosphates.

BARIUM TRISULFIDE. Yellow green crystals. BaS_3, mw: 233.54, mp: 554° (decomp).
THR = See barium compounds (sol) and sulfides.

BARIUM TUNGSTATE. Tetragonal, colorless crystals. $BaWO_4$, mw: 385.28, d: 5.04.
THR = See barium compounds (sol).

BARIUM-*m*-TUNGSTATE. Rhombic crystals. $Ba_3(H_2W_{12}O_{40}) \cdot 27H_2O$, mw: 3747.57, d: 4.30.
THR = See barium compounds (sol).

BARIUM ZIRCONATE. Light gray-buff powder, insol in water and alkalies, slightly sol in acid. $BaZrO_3$, mw: 276, d: 5.52, mp: 2510°.
THR = Animal exper show LOW oral toxicity and interstitial pneumonitis on inhal. See also zirconium compounds.

BARLEY OIL.
THR = MILD allergen.
Fire Hazard: Slight when heated.

BARON. See erbon.

BARRIERS as used in Noise Engineering is discussed in Section 3.

BARTHRIN. Syn: *6-chloropiperonyl chrysanthemumate*. $C_{18}H_{21}O_{14}Cl$, mw: 336.8.
THR = A pyrethrin-like material. Toxicity is MOD on an acute basis. [*3*]

BAS F. See potassium ammonium nitrate.

BASIC ANTIMONY CHLORIDE. See antimony oxychloride.

BASIC COPPER ZEOLITES.
THR = See copper compounds.

BASIC PRINCIPLES of personnel protection are discussed in Sections 1 and 2.

BASORA CORRA. Aqueous extract from the root of the plant.
THR = An exper neo to rats via sc route. [*103*]

BATTERY ACID. See sulfuric acid.

BATTERY CHARGER WITH ELECTROLYTE (ACID) OR ALKALINE CORROSIVE LIQUID. See sulfuric acid and/or sodium hydroxide (whichever is present).

BAUXITE. See alumina.

BAYBERRY OIL.
THR = A MILD allergen.
Fire Hazard: Slight when heated.

BAYGON. Syn: *o-isopropoxy phenyl methyl carbamate*. A white to tan crystalline solid, slightly sol in

water, sol in all polar organic solvents. $C_{11}H_{15}O_3N$, mw: 209.2.
Acute tox data: Oral LD_{50} (rat) = 83 mg/kg; iv LD_{50} (rat) = 11 mg/kg; dermal LD_{50} (rat) = 800 mg/kg; im LD_{50} (rat) = 53 mg/kg; oral LD_{50} (wild birds) = 4 mg/kg; oral LD_{50} (duck) = 9.6 mg/kg. [*3*]
THR = HIGH via oral, iv, im and MOD via dermal routes. A carbamate insecticide. See also carbamates.
Disaster Hazard: Dangerous. When heated to decomp, emits highly toxic fumes.

BAYTEX. See o,o-dimethyl-*o*-(4-(methylthio)-*m*-tolyl)-phosphorothioate.

***p-tert*-BBA.** See *p-tert* butylbenzoic acid.

BBC. See bromobenzylcyanide.

B BLASTING POWDER. See explosives, low.

BEDDOES, THOMAS. Section 1.

BEEF FAT OIL.
THR = MILD allergen.
Fire Hazard: Slight, when heated.

BEESWAX. Syn: *yellow beeswax.* Yellow to brownish-yellow, soft to brittle wax. Mp: 62°–65°, d: 0.95–0.96.
THR = MILD allergen. A general purpose food additive. [*109*]
Fire Hazard: Slight, when heated.

BEESWAX, WHITE. Syns: *white wax, cera alba.* Beeswax bleached by sunlight or oxidizing agents. Yellowish white solid, translucent in thin layer. Other properties are those of beeswax.
THR = U. A general purpose food additive. [*109*]

"BEETLE." Urea-formaldehyde condensation product.
THR = A MILD irr by inhalation. An allergen. Inhal of dust may cause allergic response or irr of lungs.
Fire Hazard: Slight.

BELLADONNA. Syn: *deadly nightshade, from which the alkaloids, atropine and belladonnine, are derived.*
THR = A deadly poison. See also hyoscyamine and atropine. Local contact may cause a contact dermatitis.

BENAZOLIN. Syn: *4-chloro-2-oxobenzothiazolin-3-ylactic acid.* A white crystalline solid. $C_9H_6O_3NCIS$, mw: 243.6, mp: 193°.
Acute tox data: Oral LD_{50} (rat) = 3000 mg/kg. An herbicide.
THR = MOD via oral route.
Disaster Hazard: See chlorides and sulfur compounds.

BENTONITE. Syn: *colloidal clay.* A clay containing appreciable amounts of the clay mineral montmorillonite; light yellow or green, cream, pink, gray to black, plastic, insol in water and common organic solvents.
THR = U. A general purpose food additive. [*109*]

BENTRANIL. Syns: *2-phenyl-3,1-benzoxazinone-4.* White solid, slightly water sol, sol in ethanol benzol. $C_{14}H_9O_2N$, mw: 223.2, mp: 124°.
Acute tox data: Oral LD_{50} (rat) = 1600 mg/kg. A post-emergence herbicide. [*3*]
THR = MOD via oral route.

BENZ(a)ACEPHENANTHRYLENE. $C_{20}H_{12}$, mw: 252.3, mp: 168°.
THR = An exper (+) carc. [*3, 11*] Some implication as human carc as well. [*11*]

BENZ(c)ACRIDINE. $C_{17}H_{11}N$, mw: 229.3, mp: 108°.
THR = An exper (+) carc. [*3, 23, 12*] Some implication as human carc as well. [*12*]

BENZ(c)ACRIDINE-7-CARBONITRILE. $C_{18}H_{10}N_2$, mw: 254.3.
THR = An exper neo. [*3*]

BENZ(c)ACRIDINE CARBOXALDEHYDE. $C_{17}H_{11}ON$, mw: 245.3.
THR = An exper neo. [*3*]

α-(BENZ(c)ACRIDINE-7-YL)-N-*p*-(DIMETHYL AMINO)PHENYL NITRONE. $C_{26}H_{21}ON_3$, mw: 391.5.
THR = An exper neo to mice via sc route. [*103*]

BENZAHEX. See benzene hexachloride.

BENZAL CHLORIDE. Syns: *benzyl dichloride, benzylidine chloride.* Very refractive liquid. $C_6H_5CHCl_2$, mw: 161.03, mp: −16°, bp: 214°, d: 1.29.
THR = Details unknown. A strong irr and lachrymator. High conc cause CNS depression.
Disaster Hazard: Dangerous; see chlorides.

BENZALDEHYDE. Syns: *benzoic aldehyde, artificial almond oil.* Refractive liquid. C_6H_5CHO, mw: 106.12, mp: −26°, bp: 179°, flash p: 148°F, d: 1.050 @ 15°/4°, autoign. temp.: 377°F, vap. press: 1 mm @ 26.2°, vap. d: 3.65.
Acute tox data: Oral LD_{50} (rat) = 1300 mg/kg; sc LD_{LO} (rat) = 5000 mg/kg; oral LD_{50} (guinea pig) = 1000 mg/kg. [*3*]
THR = MOD via oral and sc routes. An allergen. A synthetic flavoring substance and adjuvant. [*109*] Acts as a feeble local anesthetic. See aldehydes. Local contact may cause contact dermatitis. Causes CNS depression in small doses and convulsions in larger doses.
Fire Hazard: Mod when exposed to heat or flame; reacts with oxidizing materials such as peroxyformic acid.
Spont Heating: No.

To Fight Fire: Water may be used as a blanket, alcohol, foam, dry chemical.

7-BENZALDEHYDE(a)ANTHRACENE.
THR = An exper carc. [*23*]

BENZALDEHYDE CYANHYDRIN. Syn: *mandelonitrile*. Yellow viscous liquid. $C_6H_5CH(OH)CN$, mw: 133.14, mp: −10°, bp: 170° decomp, d: 1.124.
THR = Details unknown, but probably HIGH. See also cyanides and nitriles.
Disaster Hazard: See cyanides.

BENZALKONIUM CHLORIDE. Syn: *zephiran chloride*. White or yellowish-white powder, aromatic odor, very bitter taste. Alkyl dimethyl benzylammonium chlorides.
Acute tox data: Oral LD_{50} (frogs) = 30 mg/kg.
THR = HIGH via oral route. Also an irr. A bactericide and fungicide.
Disaster Hazard: Dangerous; see chlorides.

BENZALMALONONITRILE. Crystals.
$C_6H_5CH_2CH(CN)_2$, mw: 156.2.
THR = HIGH. See cyanides.
Disaster Hazard: See cyanides.

BENZAL PURINE MUSTARD.
THR = An exper carc. [*23*]

BENZ(a)ANTHRACENE. Colorless leaflets or plates. $C_{18}H_{12}$, mw: 228.3, bp: 400°, mp: 160°.
THR = HIGH via many routes. An exper (+) carc. [*23, 3, 11*] Found in oil, wax, smoke, food, drugs.

BENZ(a)ANTHRACEN-7-ACETIC ACID, METHYL ESTER. $C_{21}H_{16}O_2$, mw: 300.4.
THR = An exper carc. [*3*]

BENZ(a)ANTHRACENE-7-ACETONITRILE.
$C_{20}H_{13}N$, mw: 267.3.
THR = An exper neo. [*3*]

BENZ(a)ANTHRACENE-7-AMINE. $C_{18}H_{13}N$, mw: 243.3.
THR = An exper neo. [*3*]

BENZ(a)ANTHRACENE-8-AMINE.
THR = An exper neo. [*3*]

BENZ(a)ANTHRACENE-7-CARBOXALDEHYDE.
$C_{19}H_{12}O$, mw: 256.3.
THR = An exper carc. [*3*]

BENZ(a)ANTHRACENE-7-12-DIMETHANOL.
$C_{20}H_{18}O_2$, mw: 290.4.
THR = An exper neo. [*3*]

BENZ(a)ANTHRACENE-7,12-DIMETHANOL DIACETATE. $C_{24}H_{20}O_4$, mw: 372.4.
THR = An exper neo. [*3*]

BENZ(a)ANTHRACENE-7-ETHANOL. $C_{20}H_{16}O$, mw: 272.4.
THR = An exper neo and carc. [*3*]

5,6-BENZ(a)ANTHRACENE OXIDE.
THR = An exper carc. [*23*]

BENZ(a)ANTHRACENE-7-THIOL. $C_{18}H_{12}S$, mw: 230.4.
THR = An exper neo. [*3*]

BENZ(a)ANTHRACENE-7-YL-TRICHLOROMETHYL KETONE. $C_{20}H_{11}OCl_3$, mw: 373.7.
THR = HIGH. An exper neo. [*3*]

BENZANTHRONE. Pale yellow needles. $C_{17}H_{10}O$, mw: 230.25, mp: 174°, vap. press: 1 mm @ 225.0°.
THR = U.
Fire Hazard: Slight, when heated.

1,2-BENZANTHRYL-3-CARBAMIDO ACETIC ACID. $C_{21}H_{16}O_3N_2$, mw: 344.4.
THR = An exper neo. [*3*]

1,2-BENZANTHRYL-10-CARBAMIDO ACETIC ACID.
THR = An exper neo. [*3*]

1,2-BENZANTHRYL-10-METHYL MERCAPTAN.
$C_{19}H_{15}S$, mw: 275.4.
THR = An exper neo. [*3*]

2-BENZAZINE. See isoquinoline.

3,4-BENZCHRYSENE. See picene.

"BENZEDRINE." Syn: *amphetamine*. Liquid. $C_9H_{13}N$, mw: 135.20, bp: 200°, flash p: < 212°F, (OC), d: 0.931, vap. d: 4.65.
Acute tox data: sc LD_{50} (rat) = 180 mg/kg; oral LD_{50} (mouse) = 22 mg/kg; ip LD_{50} (mouse) = 18 mg/kg; iv LD_{50} (mouse) = 15 mg/kg. [*3*]
THR = HIGH via oral, sc, ip and iv routes of exposure. A stimulant. Overdoses cause hyperactivity, restlessness, insomnia, rapid pulse, rise in blood pressure, dilated pupils, dryness of the throat.
Fire Hazard: Low when exposed to heat or flame; can react with oxidizing materials.
Disaster Hazard: Dangerous upon exposure to heat or flame.
Treatment and Antidotes: Evacuation of stomach if taken by mouth; sedatives. Call a physician.
To Fight Fire: CO_2, dry chemical, alcohol foam, water mist, fog.

BENZENE. Syns: *benzol*, *phenyl hydride*, *coal naphtha*. Clear colorless liquid. C_6H_6, mw: 78.11, mp: 5.51°, bp: 80.093°–80.094°, flash p: 12°F (CC), d: 0.8794 @ 20°, autoign. temp.: 1044°F, lel: 1.3%, uel: 7.1%, vap. press: 100 mm @ 26.1°, vap. d: 2.77, ulc: 95–100.
THR = Poisoning occurs most commonly through inhal of the vapor, though benzene can penetrate the skin, and poison in that way. Locally, benzene has a comparatively strong irr effect, producing

erythema and burning, and, in more severe cases, edema and even blistering. Exposure to high conc of the vapor (3000 ppm or higher) may result from failure of equipment or spillage. Such exposure, while rare in industry, may result in acute poisoning, characterized by the narcotic action of benzene on the CNS. The anesthetic action of benzene is similar to that of other anesthetic gases, consisting of a preliminary stage of excitation followed by depression and, if exposure is continued, death through respiratory failure. The chronic, rather than the acute form, of benzene poisoning is important in industry. It is a recog leukemogen. [*14, 3, 1, 102*] There is no specific blood picture occurring in cases of chronic benzol poisoning. The bone marrow may be hypoplastic, normal, or hyperplastic, the changes reflected in the peripheral blood. Anemia, leucopenia, macrocytosis, reticulocytosis, thromocytopenia, high color index, and prolonged bleeding time may be present. Cases of myeloid leukemia have been reported. For the supervision of the worker, repeated blood examinations are necessary, including hemoglobin determinations, white and red cell counts and differential smears. Where a worker shows a progressive drop in either red or white cells, or where the white count remains below 5,000 per cu mm or the red count below 4.0 million per cu mm, on two successive monthly examinations, he should be immediately removed from exposure. Following absorption of benzene, elimination is chiefly through the lungs, when fresh air is breathed. The portion that is absorbed is oxidized, and the oxidation products are combined with sulfuric and glycuronic acids and eliminated in the urine. This may be used as a diagnostic sign. Benzene has a definite cumulative action, and exposure to relatively high conc is not serious from the point of view of causing damage to the blood-forming system, provided the exposure is not repeated. On the other hand, daily exposure to conc of 100 ppm or less will usually cause damage if continued over a protracted period of time. In acute poisoning, the worker becomes confused and dizzy, complains of tightening of the leg muscles and of pressure over the forehead, then passes into a stage of excitement. If allowed to remain in exposure, he quickly becomes stupefied and lapses into coma. In non-fatal cases, recovery is usually complete and no permanent disability occurs. In chronic poisoning the onset is slow, with the symptoms vague; fatigue, headache, dizziness, nausea and loss of appetite, loss of weight and weakness are common complaints in early cases. Later, pallor, nosebleeds, bleeding gums, menorrhagia, petechiae and purpura may develop. There is great individual variation in the signs and symptoms of chronic benzene poisoning. Benzene is a common air contaminant.

Fire Hazard: Dangerous, when exposed to heat or flame; can react vigorously with oxidizing materials, such as BrF_5, Cl_2, CrO_3, O_2NClO_4, O_2, O_3, perchlorates, ($AlCl_3$ + $FClO_4$), (H_2SO_4 + permanganates), K_2O_2, ($AgClO_4$ + acetic acid), Na_2O_2, [*19*]

Spont Heating: No.

Explosion Hazard: Mod, when its vapors are exposed to flame. Use with adequate ventilation.

Disaster Hazard: Dangerous, highly flam.

To Fight Fire: Foam, CO_2, dry chemical.

BENZENE ARSONIC ACID. See phenyl arsonic acid.

BENZENE CARBONYL CHLORIDE. See benzoyl chloride.

1,3-BENZENEDIOL. See resorcinol.

BENZENE DIAZOANILIDE. See β-diazoamidobenzol.

BENZENE DIAZONIUM-2-CARBOXYLATE HYDROCHLORIDE. $C_7H_4O_2N_2 \cdot HCl$, mw: 184.6.

THR = No toxicity data. Very unstable in dry state.

BENZENE DIAZONIUM CHLORIDE. See diazobenzene chloride.

BENZENE DIAZONIUM CHROMATE. See diazobenzol chromate.

BENZENE DIAZONIUM NITRATE. See diazobenzene nitrate.

1,3-BENZENE DICARBONITRILE. See *o*-dicyanobenzene.

BENZENE DICARBOXYLIC ACID. See phthalic acid.

1,2-BENZENE DIOL. See pyrocatechol.

1,4-BENZENE DIOL. See hydroquinone.

BENZENE-1,3-DIPHENYL. See *m*-terphenyl.

BENZENE-1,4-DIPHENYL. See *p*-terphenyl.

***m*-BENZENEDISULFONIC ACID.** Syn: *MBDSA*. Gray crystalline hygroscopic powder. $C_6H_6O_6S_2$, mw: 238.23.

THR = HIGH irr to skin, eyes and mu mem. In solutions it forms an extremely corrosive liquid.

Disaster Hazard: Dangerous; see sulfonates.

BENZENE HEXACHLORIDE. See hexachlorocyclohexane.

BENZENE PHOSPHONIC ACID. Colorless crystals. $C_6H_5PO(OH)_2$, mw: 158.1, mp: 165°, d: 1.475.

THR = U.

BENZENE PHOSPHORUS DICHLORIDE. Syn: *phenyl dichlorophosphine*. Colorless liquid. $C_6H_5PCl_2$, mw: 179, mp: −55°, bp: 224.6°, d: 1.319 @ 20°/20°, vap. d: 6.2.
THR = HIGH irr to skin, eyes and mu mem and via oral and inhal routes.
Disaster Hazard: See hydrochloric acid.

BENZENE PHOSPHORUS OXYDICHLORIDE. Syn: *phenyl dichlorophosphine oxide*. Colorless liquid, faint fruity odor. $C_6H_5POCl_2$, mw: 195, bp: 258°, d: 1.375 @ 20°/20°, vap. d: 6.7.
THR = U. HIGH irr.
Disaster Hazard: See chlorides and phosphates.

BENZENE PHOSPHORUS THIODICHLORIDE. Syn: *phenyl dichlorophosphine sulfide*. A colorless liquid which fumes a little in air. $C_6H_5PSCl_2$, mw: 211, bp: 205° @ 130 mm, d: 1.376 @ 20°/13°.
THR = U.
Disaster Hazard: Dangerous; see chlorides, phosphates and sulfates; will react with water or steam to produce toxic and flam vapors.

BENZENE-*o*-NITRO SULFENYL CHLORIDE. Yellow crystals. $O_2NC_5H_4SCl$, mw: 189.6, bp: decomp @ 170° (explosively), mp: 73°–74°.
THR = U.
Explosion Hazard: Mod, when heated.
Disaster Hazard: Dangerous; see sulfates and chlorides; can react vigorously with oxidizing materials.

BENZENE SULFONANILIDE. $C_{12}H_{11}O_2NS$, mw: 233.3
THR = An exper neo. [*3*]

BENZENE SULFONIC ACID. Deliquescent plates or tablets. $C_6H_5SO_3H \cdot 1\frac{1}{2}H_2O$, mw: 185.2, mp: 43°–44°.
Acute tox data: Oral LD_{50} (rat) = 890 mg/kg; oral LD_{50} (wild birds) = 75 mg/kg. [*3*]
THR = HIGH to MOD via oral and probably inhal routes.
Disaster Hazard: Dangerous; see sulfates.

BENZENE SULFONYL CHLORIDE. $C_6H_5SO_2Cl$, mw: 176.5.
THR = MOD acute oral toxicity to rats. [*3*] Reacts vigorously with dimethyl sulfoxide, methyl formamide. [*19*]

BENZENE SULFONYL FLUORIDE. Clear liquid. $C_6H_5SO_2F$, mw: 160, bp: 209°, fp: −5°, flash p: 196°F, d: 1.329, vap. press: 8 mm @ 80°, vap. d: 5.52.
THR = It appears to be of HIGH toxicity by vapor inhalation. Possibly slightly irr to skin.
Fire Hazard: Mod when exposed to heat or flame.
Disaster Hazard: Dangerous. See fluorides, chlorides and sulfates; can react vigorously with oxidizing materials.
To Fight Fire: Water, foam, CO_2, water spray or mist, dry chemical.

1,2,4,5-BENZENE TETRACARBOXYLIC ACID. See pyromellitic acid.

BENZENE TETRACHLORIDE. See 1,2,4,5-tetrachlorobenzene.

BENZENE THIOL. See phenyl mercaptan.

1,2,3-BENZENE TRIOL. See phloroglucinol.

BENZETHONIUM CHLORIDE. Syn: *hyamine 1622*. Sol crystals. $C_{27}H_{42}O_2N \cdot Cl$, mw: 448.2.
Acute tox data: Oral LD_{50} (rat) = 368 mg/kg; sc LD_{50} (rat) = 119 mg/kg; ip LD_{LO} (mice) = 8 mg/kg. [*103*, *95*] Est LD (man) = 1–3g.
THR = HIGH via oral, sc and ip routes. Ingestion may cause nausea, vomiting, collapse, convulsions or coma. [*2*] An exper neo via sc route. [*103*]
Disaster Hazard: Dangerous; see chlorides.

BENZHYDROL. Syns: *benzohydrol*, *diphenyl carbinol*. Needle-like crystals. $(C_6H_5)_2$ CHOH, mw: 184.2, mp: 68°–69°, bp: 301.0° @ 50 mm, vap. press: 1 mm @ 110°.
THR = U. An insecticide.
Fire Hazard: Slight, when heated.

BENZIDINE. Syns: *benzidine base*, *p-diaminodiphenyl*. Grayish-yellow crystalline powder; white or slightly reddish crystals, powders or leaf. $NH_2C_6H_4C_6H_4NH_2$, mw: 184.23, mp: 127.5°–128.7° @ 740 mm, bp: 401.7°, d: 1.250 @ 20°/4°.
Acute tox data: Oral LD_{50} (rat) = 309 mg/kg. [*3*]
THR = HIGH via oral and probably inhal and dermal routes. Can cause damage to blood, including hemolysis and bone marrow depression. On ingestion causes nausea and vomiting which may be followed by liver and kidney damage. A recog carc. [*3*, *9*, *23*] See also aromatic amines. Any exposure is considered extremely hazardous.
Disaster Hazard: Dangerous; when heated to decomp, emits highly toxic fumes.
Treatment and Antidotes: In case of contact, immediately wash skin with plenty of soap and water and flush eyes with plenty of water for at least 15 minutes. Secure medical attention at once.

BENZIDINE BASE. See benzidine.

BENZIDINE + 4-BIPHENYLAMINE + 4-NITROBIPHENYL + 2-NAPHTHYLAMINE (1:1:1:1)
THR = An exper carc. [*3*]

3,3′-BENZIDINE DICARBOXYLIC ACID. $C_{14}H_{12}O_4N_2$, mw: 272.3.
THR = An exper neo. [*3*, *23*]

3,3′-BENZIDINE DICARBOXYLIC ACID, DISODIUM SALT. See 3,3′-benzidinedicarboxylic acid.

BENZIDINE SULFATE. $C_{12}H_{12}N_2 \cdot SO_4$; mw: 280.3.
THR = An exper carc. [*3*]

BENZIL. Syn: *dibenzoyl.* Yellow crystals. $C_6H_5COCOC_6H_5$, mw: 210.2, mp: 95°, bp: 346°–348°, d: 1.23 @ 15°/4°, vap. press: 1 mm @ 128.4°.
THR = Details U. An insecticide.
Fire Hazard: Slight when heated.

BENZILIC ACID. Syn: *diphenylglycolic acid.* Light tan powder. $C_6H_5C(OH)(COOH)C_6H_5$, mw: 228.24, mp: 146°–149°, bp: 180° decomp.
THR = U.
Fire Hazard: Slight, when exposed to heat or flame; can react with oxidizing materials.

BENZIMIDAZOLE. Syn: *n',n'-methenyl-o-phenylene diamine.* Tabular crystals sol in alcohol, sparingly sol in water. $C_7H_6N_2$, mw: 118.1, mp: 170.5°, bp: > 360°.
THR = U. Limited animal exper suggests MOD tox.
Disaster Hazard: Dangerous; when heated to decomp, emits highly toxic fumes.

BENZ IMIDAZOLE MUSTARD.
THR = An exper carc. [*23*]

BENZ(e)INDENO(1,2-*b*)INDOLE. $C_{19}H_{13}N$, mw: 255.3.
THR = An exper neo. [*3*]

BENZINE. See petroleum spirits and naphtha, VM and P.

BENZINOFORM. See carbon tetrachloride.

1,2-BENZISOTHIAZOL-3(2H)ONE-1,1-DIOXIDE. Syn: *saccharine.* $C_7H_5O_3NS$, mw: 183.2.
THR = An exper carc and neo. [*3*]

BENZOATE OF SODA. See sodium benzoate.

BENZO(f,1)BENZOTHIENO(3,2-b)QUINOLINE. $C_{19}H_{11}NS$, mw: 285.4.
THR = An exper neo. [*3*]

BENZO (h,1)BENZOTHIENO(3,2-b)QUINOLINE.
THR = An exper neo. [*3*]

BENZO(f)BENZO(2,3)THIENO(3,2-b)QUINOLINE.
THR = An exper carc. [*23*]

BENZO(h)BENZO(2,3)THIENO(3,2-b)QUINOLINE.
THR = An exper carc. [*23*]

BENZO(e)(1)BENZOTHIOPYRANO-(4,3-b)INDOLE. $C_{19}H_{11}NS$, mw: 285.4.
THR = An exper carc. [*3*] to mice via sc route.

BENZOCAINE. See anesthesin.

11H-BENZO(a)CARBAZOLE. $C_{16}H_{11}N$, mw: 217.3.
THR = An exper carc. [*3, 23*] to mice via dermal route.

7H-BENZO(g)(a)-CARBOLINE.
THR = An exper carc. [*23*]

BENZO(b)CHRYSENE. $C_{22}H_{14}$; mw: 278.4.
THR = An exper neo. [*3, 23*]

BENZO(c)CHRYSENE. Syn: *1,2:5,6-dibenz phenanthrene.*
THR = An exper neo. [*3, 23*]

3,4-(BENZO COUMORINYL)ACETAMIDE.
THR = An exper carc. [*23*]

BENZO(a)CYCLOPENT(h)ANTHRACENE. $C_{21}H_6$; mw: 268.4.
THR = An exper carc. [*3*]

BENZO(de)CYCLOPENT(a)ANTHRACENE. $C_{20}H_{12}$; mw: 252.3.
THR = An exper neo. [*3*]

BENZOFLEX. See ethylene glycol dibenzoate.

BENZO(b)FLUORANTHENE.
THR = An exper carc. [*23*]

BENZO(i)FLUORANTHENE. $C_{20}H_{12}$; mw: 252.3.
THR = An exper neo and carc. [*3, 23*]

BENZO(j)FLUORANTHENE.
THR = An exper carc. [*23*]

BENZO(k)FLUORANTHENE.
THR = An exper carc [*3, 23*] and neo to mice via dermal and sc routes.

***p*-(5-BENZOFURYLAZO)-N,N-DIMETHYL ANILINE.** $C_{16}H_{15}ON_3$; mw: 265.3.
THR = An exper carc. [*3*]

***p*-(6-BENZOFURYLAZO)-N,N-DIMETHYL ANILINE.** $C_{16}H_{15}ON_3$; mw: 265.3.
THR = An exper neo. [*3*]

***p*-(7-BENZOFURYLAZO)-N,N-DIMETHYL ANILINE.**
THR = An exper carc. [*3*]

BENZOGUANAMINE. Crystals. $C_6H_5(C_3N_3)(NH_2)_2$, mw: 187.2, mp: 227°, d: 1.4.
THR = U.
Fire Hazard: Slight.

BENZOHYDROL. See benzhydrol.

BENZOIC ACID. Syn: *phenyl formic acid.* White powder. C_6H_5COOH, mw: 122.12, mp: 121.7°, bp: 249°, flash p: 250°F (CC), d: 1.316, autoign. temp.: 1060°F, vap. press: 1 mm @ 96.0° (sublimes), vap. d: 4.21.
Acute tox data: Oral LD_{50} (rat) = 3040 mg/kg; dermal TD_{LO} (human) = 6 mg/kg. [*3*]
THR = MOD via oral and HIGH via dermal routes. A chemical preservative food additive. [*109*]
Fire Hazard: Slight, when exposed to heat or flame; can react with oxidizing materials.
To Fight Fire: Water, CO_2, water spray or mist, dry chemical.

BENZOIC ACID ANHYDRIDE. See benzoic anhydride.

BENZOIC ACID-α-METHYLBENZYL ESTER. See α-methylbenzyl benzoic acid.

BENZOIC ALDEHYDE. See benzaldehyde.

BENZOIC ANHYDRIDE. Syn: *benzoic acid anhydride*. Crystals. $(C_6H_5CO)_2O$, mw: 226.2, mp: 42°, bp: 360°, d: 1.1989 @ 15°/4°, vap. press: 1 mm @ 135.6°.
THR = A MILD irr and allergen.
Fire Hazard: Slight, when heated.

BENZOL. See benzene.

BENZOL DILUENT. Flash p: −25°F, autoign. temp.: 450°F (these values will vary depending on the manufacturer).
THR = U.
Fire Hazard: Dangerous, when exposed to heat or flame or powerful oxidants.
To Fight Fire: Alcohol foam, water mist, fog, dry chemical.

BENZO(a)NAPHTHO(2,1,8-hi j)NAPHTHACENE.
THR = An exper carc. [*23*]

BENZO(a)NAPHTHO(8,1,2-cde)NAPHTHACENE. $C_{28}H_{16}$, mw: 352.4.
THR = An exper neo. [*3, 23*]

BENZO(h)NAPHTHO(1,2,f)QUINOLINE. $C_{21}H_{13}N$, mw: 279.4.
THR = An exper neo. [*3, 23*]

BENZO NITRILE. Syn: *phenyl cyanide*. Transparent, colorless oil, almond-like odor. C_6H_5CN, mw: 103.1, d: 1.246 @ 20°/4°, bp: 191°, d: 1.0102 @ 15°/15°F (OC), mp: −12.8°.
THR = HIGH. See nitriles.

BENZO(r,s,t)PENTAPHENE. Green-yellow needles. $C_{24}H_{14}$, mw: 302.4, mp: 280°–282°.
THR = An exper (+) neo and carc. [*3, 11, 23*]

BENZO(r,s,t)PENTAPHENE-5-CARBOXALDEHYDE. $C_{25}H_{14}O$, mw: 330.4.
THR = An exper neo. [*3*]

BENZO(ghi)PERYLENE.
THR = An exper carc. [*23*]

BENZO(a)PHENALENO(1,9-hi)ACRIDINE. $C_{27}H_{15}N$, mw: 353.4.
THR = An exper neo. [*3*]

BENZO(h)PHENALENO(1,9-6c)ACRIDINE.
THR = An exper neo. [*3*]

BENZO(d,e,f)PHENANTHRENE. See pyrene.

BENZO(c)PHENANTHRENE. $C_{18}H_{12}$, mw: 228.3.
THR = An exper carc. [*3, 23*]

BENZO(c)PHENANTHRENE-8-CARBOXALDEHYDE. $C_{19}H_{12}O$, mw: 256.3.
THR = An exper neo. [*3*]

5-BENZO(c)PHENANTHRYL METHYL KETONE. $C_{12}H_{14}O$, mw: 270.3.
THR = An exper carc. [*3*]

BENZO PHENONE. Syn: *phenyl ketone, diphenyl ketone*. Rhombic white crystals, persistent rose-like odor. $C_6H_5COC_6H_5$, mw: 182.21, mp (α): 49°, mp (β): 26°, mp (γ): 47°, bp: 305.4°, d (α): 1.0976 @ 50°/50°, d (β): 1.108 @ 23°/40°, vap. press: 1 mm @ 108.2.
THR = Details U. See also ketones.
Fire Hazard: Slight, when heated; can react with oxidiing materials.

BENZO PYRENE. See benzo(a)pyrene.

BENZO(a)PYRENE. Yellow crystals insol in water, sol in benzene, toluene, xylene. $C_{20}H_{12}$, mw: 252.3, mp: 179°, bp: 312° @ 10 mm.
THR = HIGH. An exper (+) carc, [*3, 11, 23*] neo and mutagen. A common contaminant of air, water, food, smoke.

BENZO(a)PYRENE-6-CARBOXALDEHYDE. $C_{21}H_{12}O$, mw: 280.3.
THR = An exper neo and carc. [*3*]

BENZO(a)PYRENE-6-CARBOXALDEHYDE THIO SEMICARBAZONE. $C_{22}H_{15}N_3S$, mw: 353.5.
THR = An exper carc. [*3*]

BENZO(a)PYRENE-4,5-EPOXIDE. $C_{20}H_{14}O$, mw: 270.2.
THR = An exper neo to mice via dermal route. [*103*]

BENZO(a)PYRENE-7,8-EPOXIDE. $C_{20}H_{14}O$, mw: 270.2.
THR = An exper neo to mice via dermal route. [*103*]

BENZO(a)PYRENE-6-METHANOL. $C_{21}H_{14}O$, mw: 282.4.
THR = An exper neo and carc. [*3*]

BENZO(a)PYRENE-4,5-OXIDE. $C_{20}H_{12}O$, mw: 268.3.
THR = An exper neo. [*3*]

BENZO(a)PYRENE-7,8-OXIDE.
THR = An exper carc. [*3*]

BENZO(a)PYREN-6-OL. $C_{20}H_{12}O$, mw: 268.3.
THR = An exper neo. [*3*] An exper neo to mice via sc and in routes. [*103*]

7H-BENZO(a)PYRIDO(3,2-g)CARBAZOLE. $C_{19}H_{12}N_2$, mw: 268.3.
THR = An exper neo. [*3, 23*]

7H-BENZO(c)PYRIDO(2,3-g)CARBAZOLE.
THR = An exper neo. [*3, 23*]

7H-BENZO(c)PYRIDO(3,2-g)CARBAZOLE. THR = An exper neo. [*3*]

7H-BENZO(g)PYRIDO(2,3-a)CARBAZOLE. THR = An exper carc. [*23*]

7H-BENZO(g)PYRIDO(3,2-a)CARBAZOLE. THR = An exper carc. [*23*]

13H-BENZO(a)PYRIDO(3,2-i)CARBAZOLE. THR = An exper carc. [*3, 23*]

13H-BENZO(g)PYRIDO(2,3-a)CARBAZOLE. THR = An exper carc and neo. [*3*]

13H-BENZO(g)PYRIDO(3,2-a)CARBAZOLE. THR = An exper carc. [*3*]

11H-BENZO(g)PYRIDO(4,3-b)INDOLE. $C_{15}H_{10}N_2$, mw: 218.3.
THR = An exper neo. [*3*]

1,2-BENZOPYRONE. See coumarin.

BENZOQUINONE. See quinone.

BENZOSULFINIDE. See saccharin.

BENZOTHIAZOLE. Liquid, odor of quinoline, slightly water sol. C_7H_5NS, mw: 135.2, d: 1.246 @ 20°/4°, bp: 228° @ 765 mm.
THR = U. No data available on human toxicity. iv LD_{50} for mice has been found to be 100 mg/kg. This suggests HIGH toxicity.
Disaster Hazard: Dangerous; see sulfides, cyanides.

2-BENZOTHIAZOLETHIOL. See mercaptobenzothiazol.

2-BENZOTHIAZYL DISULFIDE. Cream to light yellow powder. $(C_6H_4SNCS)_2$, mw: 332.5, mp: 175°, d: 1.5.
THR = U.
Disaster Hazard: See sulfur compounds.

2-BENZOTHIAZOLYL-N-MORPHOLINO SULFIDE. $C_{11}H_{12}ON_2S_2$, mw: 252.4.
Acute tox data: Oral LD_{50} (mouse) = 1870 mg/kg; ip LD_{LO} (mouse) = 100 mg/kg.
THR = MOD via oral and HIGH via ip routes. An exper neo via sc route. [*3*]

1,2,3-BENZOTRIAZOLE. Syn: *aziminobenzene*. Needle-like crystals. C_6H_4NHNN, mw: 119.1, mp: 100°, bp: 204° @ 15 mm.
Acute tox data: Oral LD_{LO} (rat) = 500 mg/kg; ip LD_{50} (mouse) = 1000 mg/kg; iv LD_{50} (mouse) = 238 mg/kg. [*3*]
THR = MOD via ip; HIGH via iv; HIGH to MOD via oral routes.
Disaster Hazard: Has been known to detonate during vacuum distillation. [*19*] Moderately dangerous; when heated to decomp, emits toxic fumes.

BENZOTRICHLORIDE. Syns: *a-trichlorotoluene, phenyl chloroform*. Clear, colorless to yellowish liquid; penetrating odor. $C_6H_5CCl_3$, mw: 195.46, mp: −5°, bp: 221°, d: 1.38 @ 15.5°/15.5°, vap. d: 6.77.
THR = Can cause local irr of skin and mu mem. Large doses can cause CNS depression. Inhalation of 125 ppm for 4 hrs was lethal to rats. HIGH via inhal route.
Disaster Hazard: See HCl.

BENZOTRIFLUORIDE. Syn: *trifluoromethylbenzene*. Water white liquid, aromatic odor. $C_6H_5CF_3$, mw: 146.1, mp: −29.1°, bp: 104°, flash p: 54°F (CC), d: 1.197 @ 15.5°/15.5°, vap. d: 5.04, vap. press: 11 mm @ 0°.
Acute tox data: In animal exper this has caused CNS depression. Ip LD (mouse) = 100 mg/kg. [*3*]
THR = HIGH via ip and oral routes as well.
Fire Hazard: Dangerous, when exposed to heat or flame or oxidants.
Explosion Hazard: U.
Disaster Hazard: Dangerous; see fluorides; can react vigorously with oxidizing materials.
To Fight Fire: Water to blanket fire; foam, CO_2, water spray, mist, dry chemical.

***trans*-β-BENZOYL ACRYLIC ACID.** See β-benzoyl acrylic acid.

β-BENZOYL ACRYLIC ACID. Syn: *trans-β-benzoyl acrylic acid*. Straw-yellow crystals. $C_6H_5COCHCHCOOH$, mw: 176.2, mp: 99.0°.
THR = U. See also acrylic acid.
Fire Hazard: Slight, when heated.

BENZOYL CHLORIDE. Syn: *benzene carbonyl chloride*. Colorless, fuming pungent liquid, decomp in water. C_6H_5COCl, mw: 140.5, mp: −0.5°, bp: 197°, flash p: 162°F (CC), d: 1.2187 @ 15°/15°, vap. press: 1 mm @ 32.1°, vap. d: 4.88.
THR = Powerful irr. HIGH irr to skin, eyes and mu mem and via oral and inhal routes. Possible implications as a carc. [*102*]
Fire Hazard: Mod, when exposed to heat or flame.
Disaster Hazard: Vigorous reaction with dimethyl sulfoxide, $(NaN_3 + KOH)$. [*19*]
Disaster Hazard: Dangerous; see chlorides. Will react with water or steam to produce heat and toxic and corrosive fumes; can react vigorously with oxidizing materials.
To Fight Fire: Alcohol foam, CO_2, dry chemical.

BENZOYL HYDRAZINE. $C_7H_8ON_2$, mw: 136.2.
THR = An exper carc. [*3*]

BENZOYL HYDROPEROXIDE. See perbenzoic acid.

N-BENZOYLOXY-N-ETHYL-4-AMINO AZOBENZENE. $C_{21}H_{19}O_2N_3$, mw: 345.4.
THR = An exper neo. [*3*]

N-BENZOYLOXY-4-ETHYL-N-METHYL-4-AMINOAZOBENZENE. $C_{22}H_{21}O_2N_3$, mw: 359.5.
THR = An exper neo. [*3*]

N-BENZOYLOXY FLUORENYL ACETAMIDE.
THR = An exper carc. [*23*]

N-BENZOYLOXY-4-METHYL-N-METHYL-4-AMINOAZOBENZENE. $C_{21}H_{19}O_2N_3$, mw: 345.4.
THR = An exper neo. [*3*]

N-BENZOYLOXY-N-METHYL-4-AMINOAZOBENZENE. $C_{20}H_{17}O_2N_3$, mw: 331.4.
THR = An exper carc. [*3*]

6-BENZOYLOXY METHYL BENZO(*a*)PYRENE. $C_{28}H_{18}O_2$, mw: 386.5.
THR = An exper carc. [*3*]

BENZOYL PEROXIDE, DRY. Syn: *lucidol.* White, granular, tasteless, odorless, insol in water, sol in benzene, acetone, chloroform. $(C_6H_5CO)_2O_2$, mw: 242.22, mp: 103°–105° (decomp), bp: decomp explosively, autoign. temp.: 176°F.
Acute tox data: Inhal TC_{LO} (human) = 12 mg/m^3 ⟶ pulmonary symptoms; dermal TD_{LO} (mouse) = 25,000 mg/kg ⟶ exper neo. [*3*] Oral LD_{50} (rat) = 950 mg/kg; ip LD_{50} (albino mice) = 167 mg/kg. [*35*]
THR = HIGH via ip and inhal routes. MOD via oral and LOW via dermal routes. Can cause dermatitis and asthmatic effects, testicular atrophy and vasodilation. [*35*]
Fire Hazard: Mod, by spont chemical reaction in contact with reducing agents; a powerful oxidizer.
Explosion Hazard: Mod, when heated to above melting point or when overheated under confinement. Reacts violently in contact with various organic or inorganic acids, alcohols, amines, metallic naphthenates, as well as with polymerization accelerators; i.e., dimethylaniline, $(CCl_4 + C_2H_4)$ and methyl methacrylate. [*19*] Reacts violently with acids, amines, alcohols, metallic naphthenates, as well as other easily oxidized materials. Decomp ⟶ dense white smoke of benzoic acid, phenyl benzoate, terphenyls, biphenyls, benzene and carbon dioxide. [*35*]
Disaster Hazard: Dangerous; may explode spont.
To Fight Fire: Water spray, foam.
Storage and Handling: All precautions must be taken to guard against fire and explosion hazards. Keep in a cool place; out of the direct rays of the sun; away from sparks, open flames and other sources of heat; away from shock, rough handling, friction from grinding, etc. Isolated storage is required; keep away from possible contact with acids, alcohols, ethers or other reducing agents or polymerization catalysts such as dimethylaniline. Complete instructions on storage and handling available from manufacturer.

BENZOYL PEROXIDE, WET. A paste or wetted granular material containing at least 30% water. Autoign. temp. 176°F.
THR = See benzoyl peroxide, dry.
Fire Hazard: Mod, by chemical reaction with reducing agents; a powerful oxidizer.
Disaster Hazard: See benzoyl peroxide, dry. Mixed with a large surplus of water (i.e., 30%), this material is relatively safe. It is most dangerous when it contains very little water (1% or less).
To Fight Fire: Water, foam or spray.
Storage and Handling: Care must be taken to prevent drying out of wet material. See benzoyl peroxide, dry.

***p*-BENZOYL PHENOXY ACETIC ACID.** $C_{15}H_{12}O_4$, mw: 256.3.
Acute tox data: Ip LD_{LO} (mouse) = 128 mg/kg. [*3*]
THR = Probably HIGH via ip route.

BENZ(a)PHENANTHRENE. See chrysene.

1,2-BENZPYRENE. Colorless crystals. $C_{20}H_{12}$, mw: 252.3, mp: 179°, bp: 492°.
THR = An exper (S) carc and neo. [*3, 11, 102*]

3,4-BENZPYRENE. See benzo(*a*)pyrene.

1,2-BENZPYRENE PICRATE. $C_{20}H_{12}N_3O_7C_6H_3$, mw: 481.4.
THR = An exper carc. [*3*]

2,3-BENZPYROL. See indole.

BENZYL ACETATE. Liquid. $CH_3COOCH_2C_6H_5$, mw: 150.17, mp: −51.5°, bp: 213.5°, flash p: 216°F (CC), d: 1.06, autoign. temp.: 862°F, vap. press: 1 mm @ 45°, vap. d: 5.1.
Acute tox data: Oral LD_{50} (rat) = 2490 mg/kg; inhal TC_{LO} (human) = 15 ppm.
THR = HIGH via inhal and MOD via oral route. An irr.
Fire Hazard: Slight, when exposed to heat or flame; can react with oxidizing materials.
Spont Heating: No.
To Fight Fire: Alcohol foam, CO_2.

BENZYL ALCOHOL. Syns: *hydroxytoluene, phenyl carbinol.* Water white liquid, faint aromatic odor. $C_6H_5CH_2OH$, mw: 108.13, mp: −15.3°, bp: 205.7°, flash p: 213°F (CC), d: 1.050 @ 15°/15°, autoign. temp.: 817°F, vap. press: 1 mm @ 58.0°, vap. d: 3.72.
Acute tox data: Oral LD_{50} (wild birds) = 100 mg/kg;

inhal LD_{50} (rat) = 1000 ppm for 8 hrs; oral LD_{50} (mice) = 1580 mg/kg. [*3*]

THR = MOD to HIGH via oral route depending upon species; MOD via inhal route.

Fire Hazard: Slight, when exposed to heat or flame; can react with oxidizing materials and acids.

Spont Heating: No.

To Fight Fire: Alcohol foam, CO_2, dry chemical.

BENZYLAMINE. Syn: *aminotoluene*. Strongly alkaline liquid, miscible with water, alcohol and ether. $C_6H_5 \cdot CH_2 \cdot NH_2$, mw: 107.2, d: 0.983 @ 19°/4°, bp: 185°.

THR = HIGHLY irr to skin, eyes and mu mem. [*2*] See amines.

Disaster Hazard: Mod dangerous; when heated to decomp, emits toxic fumes.

BENZYL BENZENE. See diphenyl methane.

BENZYL BENZOATE. Liquid. $C_6H_5COOCH_2C_6H_5$, mw: 212.2, mp. 21°, bp: 324°, flash p: 298°F (CC), d: 1.114, vap. d: 7.3, autoign. temp.: 898°F.

Acute tox data: Oral LD_{50} (rat) = 1700 mg/kg; oral LD_{50} (guinea pig) = 1000 mg/kg. [*3*]

THR = MOD via oral route. No data on chronic effects.

Fire Hazard: Slight, when exposed to heat or flame; can react with oxidizing materials.

To Fight Fire: CO_2, water spray or mist, dry chemical.

BENZYL BROMIDE. Syn: *α-bromotoluene*. Clear refractive liquid, pleasant odor, lachrymator, insol in water. $C_6H_5CH_2Br$, mw: 171.04, mp: −4.0°, bp: 198°, d: 1.438 @ 22°/0°, vap. d: 5.8.

THR = Intensely irr to skin, eyes and mu mem. Large doses cause CNS depression.

Disaster Hazard: Dangerous; see bromides.

BENZYL BUTYL PHTHALATE. See butyl benzyl phthalate.

3-BENZYL-4-CARBAMOYL METHYL SYDNONE. $C_{11}H_{11}O_3N_3$, mw: 233.3.

Acute tox data: Oral LD_{50} (rat) = 4450 mg/kg. [*3*]

THR = MOD via oral route. An exper carc [*3*] in rats via oral route.

BENZYL CARBINOL. See phenethyl alcohol.

BENZYL "CELLOSOLVE." Syn: *ethylene glycol monobenzyl ether*. Water white liquid, faint rose-like odor. $C_6H_5CH_2OCH_2CH_2OH$, mw: 152.19, mp: −75°, bp: 256°, flash p: 265°F (OC), d: 1.068, autoign. temp.: 665°F, vap. d: 5.25.

Acute tox data: Oral LD_{50} (rat) = 1190 mg/kg. [*3*]

THR = MOD via oral route. MILD irr. See also glycols.

Fire Hazard: Slight, when exposed to heat or flame; can react with oxidizing materials.

Spont Heating: No.

To Fight Fire: CO_2, dry chemical.

BENZYL CHLORIDE. Syn: *α-chlorotoluene*. Colorless liquid, very refractive, irr, unpleasant odor. $C_6H_5CH_2Cl$, mw: 126.58, mp: −43°, bp: 179°, lel: 1.1%, flash p: 153°F, d: 1.1026 @ 18°/4°, autoign. temp.: 1085°F, vap. d: 4.36.

Acute tox data: Oral LD_{50} (rat) = 1231 mg/kg; inhal LD_{50} (rat) = 150 ppm for 2 hrs.

THR = HIGH irr via inhal route; MOD irr via oral route.

Fire Hazard: Mod, when exposed to heat or flame.

Spont Heating: No.

Explosion Hazard: Mod, when exposed to flame. The decomp can reach explosive violence in presence of metals such as iron.

Disaster Hazard: Dangerous; see chlorides. Will react with water or steam to produce toxic and corrosive fumes; can react vigorously with oxidizing materials.

To Fight Fire: CO_2, dry chemical, water to blanket fire.

BENZYL CHLOROCARBONATE. See benzyl chloroformate.

2-(N-BENZYL-2-CHLOROETHYLAMINO)-1-PHENOXY PROPANE HYDROCHLORIDE. $C_{18}H_{22}ONCl \cdot HCl$, mw: 340.3, mp: 39°.

THR = An exper (+) carc. [*3, 8*]

BENZYL CHLOROFORMATE. Syn: *benzyl chlorocarbonate*. Colorless to pale yellow liquid, odor of phosgene. $ClCOOC_6H_5CH_2$, mw: 170.6.

THR = A MOD powerful irr. MOD via oral and inhal routes.

Disaster Hazard: Dangerous; see chlorides; will react with water or steam to produce heat, toxic and corrosive fumes.

***o*-BENZYL-*p*-CHLOROPHENOL.** Syns: *santophen, p-chlorophene*. Nearly colorless flakes. $C_{13}H_{11}OCl$, mw: 218.6, mp: 49°, bp: 175° @ 5 mm, d: 1.2 @ 55°/25°.

Acute tox data: Oral LD_{50} (rat) = 1700 mg/kg. [*3*]

THR = MOD via oral route and a MILD irr.

Disaster Hazard: Dangerous; see chlorides.

BENZYL CINNAMATE. Syn: *cinnamein*. White crystals, aromatic odor. $C_9H_7O_2C_7H_7$, mw: 238.27, mp: 39°, bp: 350.0°, vap. press: 1 mm @ 173.8°.

Acute tox data: Oral LD_{50} (guinea pig) = 3760 mg/kg. [*3*]

THR = A MILD allergen. MOD via oral route.

BENZYL CYANIDE. See phenyl acetonitrile.

BENZYL DICHLORIDE. See benzal chloride.

BENZYL DIETHYLAMINE. Liquid. $C_6H_5CH_2(C_2H_5)_2$, mw: 163.26, bp: 207°–216°, flash p: 170°F (OC), d: 0.89, vap. d: 5.6.
THR = Details U. See also amines.
Fire Hazard: Mod when exposed to heat or flame; can react with oxidizing materials.
Spont Heating: No.
To Fight Fire: Foam, CO_2, dry chemical.

BENZYL ETHER. See dibenzyl ether.

BENZYL ETHYL ETHER. Syn: *ethoxy methyl benzene*. Oily liquid, aromatic odor, insol in water, miscible with alcohol and ether. $C_6H_5CH_2OC_2H_5$, mw: 136.29, d: 0.949, bp: 186°.
THR = Probably narcotic in HIGH conc. May also be an irr.

BENZYL FORMATE. $C_8H_8O_2$, mw: 136.2.
Acute tox data: Oral LD_{50} (rat) = 1700 mg/kg; dermal LD_{50} (rabbit) = 2000 mg/kg. [*3*]
THR = MOD via oral and dermal routes. Probably narcotic in high conc.

BENZYL HYDRAZINE DIHYDROCHLORIDE. $C_7H_{10}N_2 \cdot 2HCl$, mw: 195.0.
THR = An exper neo to mice via oral route. [*103*]

BENZYL HYDROQUINONE. $C_{13}H_{12}O_2$, mw: 200.3.
THR = LOW acute toxicity. An exper neo. [*3*]

1,1′-BENZYLIDINE BIS((2-METHOXY-*p*-PHENYLENE)AZO)DI-2-NAPHTHOL. $C_{38}H_{32}O_4N_4$, mw: 608.7.
THR = An exper neo. [*3*] Very LOW acute toxicity.

BENZYLIDINE CHLORIDE. See benzal chloride.

BENZYL IODIDE. Syn: *α-iodotoluene*. Colorless crystals. $C_6H_5CH_2I$, mw: 218.05, mp: 24°, bp: 93° @ 10 mm, d: 1.733 @ 25°/4°.
THR = Probably MOD irr to skin, eyes and mu mem.
Disaster Hazard: Dangerous, see iodides.

BENZYL ISOAMYL ETHER. See amyl benzyl ether.

BENZYL MERCAPTAN. Syns: *α-toluene thiol, phenyl methane thiol*. A water white, mobile liquid, strong odor. C_7H_8S, mw: 124.2, bp: 194.8°, flash p: 158°F (CC), d: 1.058 @ 20°, vap. d: 4.28.
Acute tox data: Oral LD_{50} (rat) = 493 mg/kg. [*3*]
THR = HIGH via oral route. An exper carc. [*3*] See also mercaptans and benzene.
Fire Hazard: Mod, when exposed to heat or flame.
Disaster Hazard: Dangerous; when heated to decomp and on contact with acid or acid fumes, emits highly toxic fumes; can react vigorously with oxidizing materials.
To Fight Fire: Foam, CO_2, dry chemical, water spray, mist, fog.

BENZYL METHYL ETHER. Syn: *methyl benzyl ether*. Colorless liquid, insol in water, sol in alcohol, ether. $C_6H_5CH_2OCH_3$, mw: 122.16, d: 0.987, bp: 174°.
THR = Probably has narcotic properties typical of ether.

1-BENZYL-2-METHYL HYDRAZINE. $C_8H_{12}N_2$, mw: 136.2.
THR = An exper transplacental teratogen and carc. [*3, 23*] HIGH acute toxicity.

BENZYL MUSTARD OIL. See benzyl thiocyanate.

BENZYL NICOTINIUM CHLORIDE. Crystals. $C_{10}H_{14}N_2Cl(C_6H_5CH_2Cl)$, mw: 288.8.
THR = HIGH via oral, inhal and dermal routes.
Disaster Hazard: Dangerous; see chlorides.

***p*-BENZYL OXYPHENOL.** See benzyl hydroquinone.

BENZYL PENICILLIN. $C_{16}H_{18}O_4N_2S$, mw: 334.4.
Acute tox data: Iv LD_{50} (mice) = 329 mg/kg. [*3*]
THR = HIGH via iv route. An exper carc [*3, 23*] via sc route.

BENZYL PENICILLINIC ACID, SODIUM SALT. $C_{16}H_{17}O_4N_2S \cdot Na$, mw: 356.4.
THR = An exper carc. [*3*]

***o*-BENZYL PHENOL.** Syn: *(2-hydroxydiphenyl) methane*. Crystals or liquid. $C_6H_5CH_2C_6H_4OH$, mw: 184.2, mp: 20°, bp: 154°–156° @ 10 mm.
THR = Toxic, but details U. See also phenol.
Disaster Hazard: Dangerous; when heated to decomp, emits highly toxic fumes.

***p*-BENZYL PHENOL.** Syn: *(4-hydroxydiphenyl)methane*. Crystals, colorless to slightly pink powder, faintly pleasant phenolic odor. $C_6H_5CH_2C_6H_4OH$, mw: 184.23, mp: 84°, bp: 322°, d: 1.10 @ 25°/25°.
THR = Toxic, but details U. See also phenol.
Disaster Hazard: Dangerous; when heated to decomp, emits highly toxic fumes.

BENZYL PYRIDINE. $C_6H_5CH_2C_5H_4N$, mw: 169.2.
Acute tox data: Oral LD_{50} (wild birds) = 18 mg/kg. [*3*]
THR = HIGH via oral route.
Fire Hazard: A mod flam material.
Disaster Hazard: Dangerous; when heated to decomp, emits highly toxic fumes; can react vigorously with oxidizing materials.

BENZYL SALICYLATE. Thick liquid, pleasant odor. $HOC_6H_4CO_2CH_2C_6H_5$, mw: 228.2, bp: 208° @ 26 mm, d: 1.175 @ 20°.
Acute tox data: Oral LD_{50} (rat) = 2227 mg/kg. [*3*]
THR = MOD via oral route. See also benzyl alcohol and salicylic acid.
Fire Hazard: Slight when exposed to heat or flame; can react with oxidizing materials.

BENZYL SILANE. $C_7H_{10}Si$, mw: 122.2.
THR = No toxicity data. Spont flam in air.

BENZYL SODIUM. C_7H_7Na, mw: 114.1.
THR = No toxicity data. Spont flam in air.

BENZYL THIOCYANATE. Syn: *benzyl mustard oil.* Orange-red crystalline solid. $C_6H_5CH_2CNS$, mw: 149.2, mp: 41°, bp: 230°, d: 1.125.
THR = Intensely irr. See also thiocyanates.
Fire Hazard: Mod via heat, flame, oxidizers.
To Fight Fire: Water, spray, foam, dry chemical.

BENZYL TRIMETHYL AMMONIUM CHLORIDE. Thick liquid. $C_6H_5CH_2N(CH_3)_3Cl$, mw: 185.7, bp: > 135° (decomp), fp: <−50° (61% sol), d: 1.07 @ 20°/20° (61% sol).
THR = U.
Fire Hazard: Slight, when exposed to heat or flame; can react with oxidizing materials.

N-BENZYL-N,N,N-TRIMETHYL AMMONIUM HEXAFLUOPHOSPHATE. Crystals. $C_6H_5CH_2N(CH_3)_3PF_6$, mw: 295, mp: 160°.
THR = U. See also fluorides and phosphates.
Disaster Hazard: Dangerous; when heated to decomp, or on contact with acid or acid fumes, emits highly toxic fumes.

BENZYL TRIMETHYL AMMONIUM HYDROXIDE. Solid. $C_6H_5CH_2N(CH_3)_3OH$, mw: 150.3.
THR = MOD irr to eyes and mu mem and via oral route exposure.
Fire Hazard: Mod when exposed to heat; can react with oxidizing materials; products of heat decomp are flam.

BENZYL TRIPHENYL GERMANIUM. Colorless crystals, insol in water. Sol in organic solvents. $(CH_2C_6H_5)(C_6H_5)_3Ge$, mw: 395.0, mp: 83.0°.
THR = Details U. See germanium compounds.

BENZYLYT. $C_{18}H_{22}ONCl$, mw: 303.9.
Acute tox data: Oral LD_{50} (rat) = 2500 mg/kg. [*3*]
THR = MOD via oral route. An exper (±) carc. [*3, 8*]

BEP.
THR = An exper neo [*3, 23*] and exper carc to rats via sc route.

BERBERINE. White to yellow crystals. $C_{20}H_{19}NO_5$, mw: 353.36, mp (anhy): 145°.
Acute tox data: ip LD_{50} (mice) = 24.3 mg/kg. [*3*]
THR = HIGH via ip route. An alkaloid poison. In toxic doses it lowers the temperature, increases peristalsis, and causes death by central paralysis. Should carry a poison label. Should never be ingested without the advice of a physician. Should not be handled excessively, since it may be absorbed through the skin and have a toxic effect upon the body.
Disaster Hazard: Dangerous; when heated to decomp, emits highly toxic fumes.

BERBERINE COMPOUNDS. See berberine.

BERGAMOT OIL. Yellow-green liquid, agreeable odor.
Composition: 1-linalyl acetate, 1-linalool, d-limonene, dipentene, bergaptene, d: 0.875–0.880 @ 25°/25°.
THR = A MILD allergen.
Fire Hazard: Slight.

BERGIUS HYDROGENATED COAL OILS. See mineral oils.
THR = A recog carc. [*14*]

BERGIUS OIL TAR. See coal tar and pitch, also petroleum asphalt.
THR = A recog carc. [*14*]

BERKELIUM. A silvery metal, sol in dilute mineral acids; d: approx. 14. Bk.
THR = HIGH in radioactivity and therefore HIGH in toxicity. Bone-seeking actinide element.
Radiation Hazard: For permissible levels see Table 5A.5. Artificial isotope ^{249}Bk, $T\frac{1}{2}$ = 314d, decays to ^{249}Cf via β's of 0.12 MeV.

BERLIN GREEN. See ferricyanide.

BERTRANDITE. $Be_4Si_2H_{10}O_9H_2O$, mw: 264.3.
THR = An exper (+) carc and neo. [*3, 9*] See also beryllium.

BERYL. Green, blue, yellow or white crystals. $3BeO \cdot Al_2O_3 \cdot 6SiO_2$, mw: 537.5, d: 2.63–2.91.
THR = An exper (+) neo and carc. [*3, 9*] See also beryllium.

BERYLLIUM. Syn: *glucinum.* A grayish-white, hard light metal. Be, atwt: 9.013, mp: 1278°, bp: 2970°, d: 1.85.
THR = HIGH. See beryllium compounds. An exper (+) neo and carc. [*3, 9, 23, 95*] A common air contaminant.
Radiation Hazard: For permissible levels see Table 5A.5. Artificial and natural isotope ^7Be, $T\frac{1}{2}$ = 53d, decays to stable ^7Li via ec. Emits γ's of 0.48 MeV and x-rays. Artificial isotope ^{10}Be, $T\frac{1}{2}$ = 2.7×10^6y, decays to stable ^{10}B via β's of 0.56 MeV.
Fire Hazard: Mod, in the form of dust or powder, or when exposed to flame or by spont chemical reaction. Mixed with CCl_4 or C_2HCl_3, it will flash or spark on impact. Reacts with Li, P. [*19*]
Explosion Hazard: Slight, in the form of powder or dust.

BERYLLIUM ACETATE. Plates. $Be(C_2H_3O_2)_2$, mw: 127.10, mp: decomp @ 300°.
THR = See beryllium compounds.

BERYLLIUM ACETYLACETONATE. Crystals. Practically insol in water. Freely sol in alcohol, acetone, ether, benzene, carbon disulfide and other organic solvents. $C_{10}H_{14}BeO_4$, mw: 207.23, mp: 108°, bp: 270°.
THR = See beryllium compounds.

BERYLLIUM ALGINATE.
THR = See beryllium compounds.

BERYLLIUM ALLOYS OF ALUMINUM.
THR = See beryllium compounds.

BERYLLIUM ALLOYS OF COPPER.
THR = See beryllium compounds.

BERYLLIUM ALLOYS OF MAGNESIUM. See beryllium compounds.

BERYLLIUM ALLOYS OF NICKEL. See beryllium compounds and nickel compounds.

BERYLLIUM ALLOYS OF STEEL. See beryllium compounds.

BERYLLIUM ALUMINATE. Syn: *chrysoberyl.* Rhombic crystals. $BeAl_2O_4$, mw: 126.95, d: 3.76.
THR = See beryllium compounds.

BERYLLIUM ALUMINUM SILICATE. Syn: *euclase.* Monoclinic crystals. $Be_2Al_2(SiO_4)_2(OH)_2$, mw: 290.12, d: 3.1.
THR = See beryllium compounds.

BERYLLIUM BENZENE SULFONATE. Monoclinic crystals. $Be(C_6H_5O_3S)_2$, mw: 323.35.
THR = See beryllium compounds.
Disaster Hazard: Dangerous; when heated to decomp, emits highly toxic fumes of beryllium oxide and SO_x.

BERYLLIUM-*o*-BORATE, BASIC. Syn: *hambergite.* Rhombic crystals. $Be_2(OH)BO_3$, mw: 93.85, d: 2.35.
THR = See beryllium compounds.

BERYLLIUM BOROCARBIDE.
THR = See beryllium compounds.

BERYLLIUM BOROHYDRIDE. Solid. $Be(BH_4)_2$, mw: 38.72, mp: 123°, sublimes @ 91°, vap. press: 10 mm @ 28.1°.
THR = See beryllium compounds, boron compounds and hydrides. Spont flam in air. $(O_2 + H_2O)$ can explode it.

BERYLLIUM BROMIDE. White deliquescent needles. $BeBr_2$, mw: 168.85, mp: 490°, sublimes @ 474°, bp: 520°, d: 3.465 @ 25°.
THR = See beryllium compounds and bromides.

BERYLLIUM BUTYRATE, BASIC. Crystals. $Be_4O(C_4H_7O_2)_6$, mw: 574.6, bp: 239 @ 19 mm.
THR = See beryllium compounds.

BERYLLIUM CARBIDE. Hexagonal yellow crystals. Be_2C, mw: 30.0, mp: > 2100° decomp, d: 1.90 @ 15°.
THR = See beryllium compounds.
Fire Hazard: Details U. See also carbides.
Explosion Hazard: See carbides.
Disaster Hazard: Dangerous; will react with water or steam to produce flam vapors and highly toxic fumes of beryllium oxides.

BERYLLIUM CARBONATE, BASIC. White powder. $BeCO_3 + Be(OH)_2$, mw: 112.05.
THR = See beryllium compounds.

BERYLLIUM CHLORIDE. Colorless deliquescent needles. $BeCl_2$, mw: 79.93, mp: 440°, bp: 520°, d: 1.899 @ 25°, vap. press: 1 mm @ 291° (sublimes).

BERYLLIUM COMPOUNDS.
THR = The extraction of Be from its ore is attended by exposure to acid salts of the metal, particularly the fluoride (BeF_2), the ammonium fluoride and the sulfate ($BeSO_4$) and also to beryllium oxide (BeO), and hydroxide [$Be(OH)_2$]. Exposure to the oxide also occurs in the casting of Be alloys and in operations with beryllia ceramics. In the manufacture of fluorescent powders, lamps and sign tubes there may be exposure to Be carbonate and to more complex salts, such as ZnMnBe silicate. Even alloys of low Be content have been shown to be dangerous. Be compounds can enter the body through inhalation of the dusts and fumes and they may act locally on the skin. Exposure to Be compounds encountered in the extraction of the metal or its oxide from the ore, particularly the halide salts, has been attended, in certain individuals, by the development of dermatitis of an edematous and papulovesicular type (Section 9), chronic skin ulcers, rhinitis, nasopharyngitis, epistaxis, bronchitis and in severe cases, by the development of an acute pneumonitis, with cough, scanty sputum, low-grade fever, rales, dyspnea and substernal pain. Radiographs show diffuse haziness throughout both lungs, followed by the appearance of soft, ill-defined opacities. The condition occurs while the worker is exposed, sometimes within 1 or 2 months of starting work, and recovery occurs within 2 months, as a rule, though radiographic changes sometimes persist for longer periods. Certain investigators have reported occasional failure of complete resolution, followed by fibrosis. In severe cases of pneumonitis the patient may die. Necropsies have revealed diffuse pulmonary edema, hemorrhagic extravasation, large numbers of plasma cells and a relative absence of polymorphonuclear infiltration. On the basis of experimental work with animals, certain investigators are of the opin-

ion that the acute upper and lower respiratory effects are due chiefly to the acid radical present in the dust or fume, but this view has little support.

A delayed form of lung disease, characterized by the occurrence of granulomatous areas in the lung tissue, has been reported in workers manufacturing fluorescent powders, lamps and sign tubes, casting beryllium master alloys, and in the production of beryllium from beryl ore. Symptoms can start during exposure, but they might be delayed up to 5 years or more after leaving work. The commonest symptoms are coughing, shortness of breath, loss of appetite, loss of weight, and fatigue. Rales are usually present in the bases and axillae, and the red cell count is frequently elevated. Cyanosis is common and the pulse and respiratory rates are often increased. Radiographically, three stages of the disease are described: (1) a diffuse, uniform granular shadowing extending throughout both lung fields; (2) a diffuse reticular pattern on the granular background; (3) the appearance of distinct nodules scattered through the lungs, with some enlargement and blurring of the hilar shadows. The intensity of the shadowing is usually greater in the middle third of the lung fields. The prognosis is poor. Clinical improvement may occur gradually over a period of several years, but there appears to be little tendency for the radiographic shadowing to clear. In certain cases, the disease has progressed gradually for some months or years, with death resulting from respiratory and cardiac failure. In several instances necropsies have shown the presence of a diffuse fibrosis with coarse strands of hyalinized collagen between the alveoli and, in some places, replacing them. The hyalinized areas contained granulomatous foci, the alveolar walls are thickened and fibrosed, the blood vessels being engorged and dilated. In some cases the hilar lymph nodes show granulomatous change and fibrosis. Granulomatous change has also been noted in the liver and hyaline fibrosis in the spleen. Two cases of delayed lung disease not coming to autopsy have presented papular lesions on the dorsum of the hands; on the biopsy these showed "sacroid-like" lesions with central necrosis.

Several cases have been reported in which localized granulomatous lesions developed following penetrating wounds caused by splinters of glass from broken fluorescent light tubes. Several weeks or months following the accident, swellings were noted in the injured areas and excision revealed granulomatous tumors, which in one case was shown to contain beryllium. Several cases of beryllium granuloma have been reported in persons residing near processing plants and in families of beryllium workers Be and its compounds are considered as exper carc of the lungs and bones.

There is no specific treatment, but temporary remissions have been produced by ACTH and cortisone.

BERYLLIUM-COPPER ALLOY. A metallic alloy. Be_xCu_y.
THR = Cases of berylliosis have been reported from exposure to so called low beryllium alloys. See beryllium compounds.

BERYLLIUM DIISOPROPYL. $Be(C_3H_7)_2$, mw: 95.2.
THR = Probably HIGH. See beryllium compounds. Explodes on contact with water. [*19*]

BERYLLIUM ETHYL. See diethyl beryllium.

BERYLLIUM ETHYLENE DIAMINE CHLORIDE. Crystals, practically insol in water. $Be(NH_2CH_2CH_2NH_2)_2$, mw: 200.13.
THR = See beryllium compounds.

BERYLLIUM FLUORIDE. Amorphous, colorless mass. BeF_2, mw: 47.02, mp: 800°, d: 1.986 @ 25°.
THR = See beryllium compounds and fluorides.

BERYLLIUM FORMATE. Crystals. $(Be(COOH)_2)_x$, mw: $(99.1)_x$, mp: decomp @ 150°.
THR = See beryllium compounds.
Disaster Hazard: Dangerous; when heated to decomp, emits highly toxic fumes of beryllium oxide.

BERYLLIUM HYDRIDE. White solid, reacts with water, dilute acids, methanol. BeH_2, mw: 11.0.
THR = See beryllium compounds and hydrides.
Fire Hazard: This material liberates hydrogen rapidly when heated to 220°; also reacts with water, dilute acids and methanol to liberate hydrogen; see hydrogen and hydrides.

BERYLLIUM HYDROGEN PHOSPHATE. $BeHPO_4$, mw: 105.
THR = An exper (+) neo and carc. [*3, 9*] See beryllium.

BERYLLIUM HYDROXIDE. Amorphous powder or crystals. $Be(OH)_2$, mw: 43.04, mp: decomp @ 138°, d (cr): 1.909.
THR = See beryllium compounds.

BERYLLIUM IODIDE. Colorless needles. BeI_2, mw: 262.85, mp: 510°, bp: 590°, d: 4.325 @ 25°, vap. press: 1 mm @ 283°.
THR = See beryllium compounds.

BERYLLIUM MANGANESE ZINC SILICATE. $BeMnZnSiO_4$, mw: 221.4.
THR = An exper (+) carc. [*3*] See beryllium.

BERYLLIUM METAL POWDER OR FLAKE. See beryllium.

BERYLLIUM NITRATE. White, yellowish crystals, deliquescent. $Be(NO_3)_2 \cdot 3H_2O$, mw: 187.08, mp: 60° bp: decomp @ 100°–200°.
THR = See beryllium compounds and nitrates.

BERYLLIUM NITRIDE. Cubic colorless crystals. Be_3N_2, mw: 55.06, mp: 2200° ± 100°, bp: decomp @ 2240°.
THR = See beryllium compounds and nitrides.

BERYLLIUM OXALATE. Rhombic crystals. $BeC_2O_4 \cdot 3H_2O$, mw: 151.08, mp: $-2H_2O$ @ 100°, $-3H_2O$ @ 220°, bp: decomp @ 350°.
THR = See beryllium compounds and oxalates.

BERYLLIUM OXIDE. Syn: *bromellete*. White amorphous powder. BeO, mw: 25.0, mp: 2530° ± 30°, bp: 3900° (approx), d: 3.025.
THR = An exper (+) carc and neo [*3*, *9*] See beryllium. Reacts explosively with (Mg + heat). [*19*]

BERYLLIUM-2,4-PENTANEDIONE DERIVATIVE. Monoclinic white crystals. $Be(C_5H_7O_2)_2$, mw: 207.23, mp: 108°, bp: 270°, d: 1.168 @ 4°.
THR = See beryllium compounds.

BERYLLIUM PERCHLORATE. Very hygroscopic crystals, sol in water: 148.6 g/100 ml. $Be(ClO_4)_2 \cdot 4H_2O$, mw: 279.49.
THR = See beryllium compounds and perchlorates.

BERYLLIUM-*m*-PHOSPHATE. Crystals. $Be(PO_3)_2$, mw: 167.0.
THR = See beryllium compounds and phosphates. An exper (+) carc and neo. [*3*]

BERYLLIUM-*o*-PHOSPHATE. Crystals. $Be_3(PO_4)_2 \cdot 3H_2O$, mw: 271.05, mp: $-H_2O$ @ 100°.
THR = See beryllium compounds and phosphates.

BERYLLIUM POTASSIUM FLUORIDE. Syn: *glucinum potassium fluoride*. White crystal masses. $BeF_2 \cdot (KF)_2$, mw: 163.30.
THR = See beryllium compounds and fluorides.

BERYLLIUM POTASSIUM SULFATE. Colorless, brilliant crystals, sparingly sol in water, sol in concentrated K_2SO_4 solution, insol in alcohol. $K_2Be(SO_4)_2 \cdot 2H_2O$, mw: 315.37.
THR = See beryllium compounds and sulfates.

BERYLLIUM PROPIONATE ACETATE, BASIC. Crystals. $Be_4O(C_2H_3O_2)_3(C_3H_5O_2)_3$, mw: 448.39, mp: 127°, bp: 330°.
THR = See beryllium compounds.

BERYLLIUM SELENATE. Orthorhombic crystals, freely sol in water. $BeSeO_4 \cdot 4H_2O$, mw: 224.04, d: 2.03.
THR = See beryllium compounds and selenium.

BERYLLIUM-*o*-SILICATE. Syn: *phenazite*. Colorless crystals. Be_2SiO_4, mw: 110.1, d: 3.0.
THR = An exper (±) carc and neo. [*3*, *9*] See also beryllium compounds.

BERYLLIUM SODIUM FLUORIDE. Syn: *glucinum sodium fluoride*. White crystalline mass. $BeF_2 \cdot (NaF)_2$, mw: 131.10.
THR = See beryllium compounds and fluorides.

BERYLLIUM STEARATE. White waxy crystals. $Be(C_{18}H_{35}O_2)_2$, mw: 575.94, mp: 45°.
THR = See beryllium compounds.

BERYLLIUM SULFATE. Crystals. $BeSO_4$, mw: 105.08, mp: 550°–600° (decomp), d: 2.443.
THR = An exper (+) carc. [*3*] See beryllium compounds and sulfates.

BERYLLIUM SULFATE TETRAHYDRATE. $BeSO_4 \cdot 4H_2O$, mw: 177.2.
THR = An exper (+) carc. [*3*, *9*] HIGH acute toxicity also.

BERYLLIUM SULFIDE. Solid crystalline mass. BeS, mw: 41.08, d: 2.36.
THR = See beryllium compounds and sulfides.

BERYLLIUM ZINC SILICATE. Crystalline solid. $BeZn(SiO_4)$, mw: 166.5.
THR = An exper (+) carc. [*3*, *9*]. See also beryllium compounds.

BERNARDINO RAMAZZINI AND OCCUPATIONAL MEDICINE. Section 1.

BETA RAYS. Particulate radiation emitted by certain radioactive isotopes. Beta rays consist of electrons moving at high velocity.
Radiation Hazard: Recog cause of cancer. See Radiation, Ionizing. See Section 5 for complete discussion.

BETEL. See areca nut.

BETEL NUT QUID. Chewing of betel nuts a susp carc. [*14*] See also plant and fungal carc.

BFE. See bromo trifluoroethylene.

BGE. See *n*-butyl glycidyl ether.

BHA. See butylated hydroxyanisole.

BHC. See benzene hexachloride.

BHT. See di-*tert*-butyl-*p*-cresol.

4′,4‴-BIACETANILIDE. $C_{16}H_{16}O_2N_2$; mw: 268.3; mp: 329°.
THR = An exper carc [*3*, *23*, *108*] in rats via oral, sc, ip routes.

BIBLIOGRAPHY FOR INDUSTRIAL CANCER RISKS. See Section 8.

BIBLIOGRAPHY FOR GENERAL CHEMICALS SECTION. See Section 12.

BIBLIOGRAPHY FOR INDUSTRIAL FIRE PROTECTION AND STORAGE; HANDLING OF INDUSTRIAL HAZARDS. See Section 7.

BIBLIOGRAPHY FOR INDUSTRIAL HYGIENE. See Sections 1 and 12.

BIBLIOGRAPHY FOR INDUSTRIAL SOLID WASTE PROBLEMS. See Section 6.

BIBLIOGRAPHY FOR INDUSTRIAL NOISE. See Section 3.

BIBLIOGRAPHY FOR TOXICOLOGY. See Section 9, 12.

BIBLIOGRAPHY FOR RADIATION HAZARDS AND CONTROLS. See Section 5A.

BIBLIOGRAPHY FOR RADIOLOGICAL SOURCES AND SAFEGUARDS. See Section 5B.

BIBLIOGRAPHY FOR SHIPPING REGULATIONS. See Section 11.

BIBLIOGRAPHY FOR INDUSTRIAL AIR CONTAMINANT CONTROL, RESPIRATORY PROTECTION AND PERSONAL HYGIENE. See Section 2.

BIBLIOGRAPHY FOR METEOROLOGICAL PROBLEMS IN INDUSTRY. See Section 4.

BICALCIUM PHOSPHATE. See calcium phosphate, dibasic.

BICHROMATE OF SODA. See sodium dichromate.

***cis*-BICYCLO(2,2,1)-5-HEPTANE-2,3-DICARBOXYLIC ACID DIMETHYL ESTER.** See dimethyl carbate.

BICYCLOHEXYL. Syn: *dicyclohexyl.* Colorless oil, pleasant odor. $C_{12}H_{22}$, mw: 166.3, mp: 2°, bp: 240°, flash p: 165°F, d: 0.883 @ 25°/15.6°, autoign. temp.: 471°F, vap. d: 5.73, lel = 0.7% @ 212°F, uel = 5.1% @ 302°F.
THR = U. See also cyclohexanol.
Fire Hazard: Mod, when exposed to heat or flame; can react with oxidizing materials.
To Fight Fire: Alcohol foam, foam, CO_2, dry chemical.

BICYCLOHEPTADIENE DIBROMIDE. $(C_7H_9Br)_2$, mw: 346.
THR = HIGH irr to skin, eyes, mu mem and via oral and inhal routes. Human exposure has resulted in severe dermatitis, asthma, and injury to blood-forming organs.
Disaster Hazard: Dangerous. See bromides.

BIDRIN. See 3-hydroxy-*n*,*n*-dimethyl-*cis*-crotonamide-dimethyl phosphate.

BI-(ETHYL MERCURIC) PERTHIOCYANATE. See mercury compounds, organic, and thiocyanates.

BIFORMYL. See glyoxal.

BIG REDS. See explosives, high.

BIMETHYL. See ethane.

BINAPACRYL See dimethyl-1-carbomethoxy-1-propenyl-2-phosphate.

(1,1′-BINAPHTHALENE)-2,2′-DIAMINE. $C_{20}H_{16}N_2$, mw: 284.4.
THR = An exper carc and neo. [*3*]

(1,2-BINAPHTHALENE)-1,2-DIAMINE.
THR = An exper neo. [*3*]

BIOASSAY APPLICATIONS TO RADIATION SAFETY. See Section 5A.

BIOLOGICAL EFFECTS OF RADIATION. See Section 5A.

BIOTIN. Syns: *vitamin H, hexahydro-2-oxo-1H-thieno-[3,4-d]imidazole-4-pentanoic acid.* Biotin is frequently referred to as a member of the Vitamin B complex. White crystals, sol in water and alcohol, insol in petroleum, ether and chloroform. $C_{10}H_{16}N_2O_3S$, mw: 244, mp: 230°–232°.
THR = U. A nutrient and/or dietary supplement food additive. [*109*]

BIPHENYL. See diphenyl.

2-BIPHENYL ACETAMIDE.
THR = An exper carc. [*23*]

3-BIPHENYL ACETAMIDE.
THR = An exper carc. [*23*]

4-BIPHENYL ACETAMIDE.
THR = An exper carc. [*23*]

4-BIPHENYLACETHYDROXAMIC ACID.
THR = An exper carc. [*23*]

***p*-BIPHENYLAMINE.** See *p*-amino diphenyl.

2,4′-BIPHENYLDIAMINE. Syns: *diphenyline, 2,4′-diphenyldianiline, 2,4′-diaminodiphenyl.* Needles, very slightly sol in alcohol and ether. $(C_6H_4)_2(NH_2)_2$, mw: 184.23, mp: 54.5°, bp: 363°.
Acute tox data: Oral LD_{50} (rat) = 311 mg/kg. [*3*]
THR = HIGH via oral route. An exper neo. [*3, 108*] in dogs via oral route.

4-BIPHENYL DIMETHYL AMINE.
THR = An exper carc. [*23*]

2-BIPHENYLYL DIPHENYL PHOSPHATE. Clear mobile liquid. $C_{24}H_{19}O_4P$, mw: 402.37, mp $< 0°$, bp: 250°–285° @ 5 mm, flash p: 437°F, d: 1.2 @ 60°, vap. d: 13.8.
THR = U.
Fire Hazard: Slight, when exposed to heat or flame.
Disaster Hazard: Dangerous; see phosphates; can react with oxidizing materials.
To Fight Fire: Water, foam, CO_2, water spray or mist, dry chemical.

***o*-BIPHENYLENE METHANE.** See fluorene.

4-BIPHENYL HYDROXYLANINE.
THR = An exper carc. [*23*]

BIPHENYL MERCURY. Crystals, not easily dissolved in ordinary solvents. $Hg(C_6H_5C_6H_4)_2$, mw: 507.0, mp: 216°.
THR = See mercury compounds, organic.

3,3′,4,4′-BIPHENYLTETRAMINE. $C_{12}H_{14}N_4$, mw: 214.3.
THR = An exper neo. [*3*]

N-4-BIPHENYLYL ACETOHYDROXAMIC ACID. $C_{14}H_{13}O_2N$, mw: 227.3.
THR = An exper carc. [*3*]

N-4-BIPHENYLYL BENZENE SULFONAMIDE. $C_{12}H_{15}O_2N$, mw: 237.3.
THR = An exper neo. [*3*]

N-4-BIPHENYLYL BENZOHYDROXAMIC ACID. $C_{19}H_{14}ON$, mw: 372.3.
THR = An exper carc. [*3*]

N-4-BIPHENYLYL-N-HYDROXY BENZENE SULFONAMIDE. $C_{18}H_{11}O_3NS$, mw: 321.4.
THR = An exper carc. [*3*]

1-(4-BIPHENYLYLOXY)-2-PROPANOL. See propylene glycol-4-biphenylyl ether.

2,2′-BIPYRIDINE. See α,α-dipyridyl.

BIRCH TAR OIL. Brown liquid; leather-like odor. D: 0.886–0.950.
THR = MOD irr to eyes and mu mem. A MILD allergen.
Fire Hazard: Slight, when exposed to heat or flame; can react with oxidizing materials.

BIRCH WOOD DUST. A typical wood dust.
THR = See saw dust.

BIS-4-ACETAMINO-PHENYL SELENIUM DIHYDROXIDE. $C_8H_{11}O_3NSe$; mw: 248.2.
THR = An exper neo. [*3*]

BIS-ACETYL ACETONE GERMANIUM DIBROMIDE. Colorless, tiny crystals.
$[CH(CCH_3O)_2]_2GeBr_2$; mw: 430.7, mp: 226°.
THR = Details U. See germanium compounds.
Disaster Hazard: Dangerous. See bromides.

BIS-ACETYL ACETONE GERMANIUM DICHLORIDE. Colorless crystals. $GeCl_2 \cdot CH(CCH_3O_2)_2$; mw: 341.7, mp: 240° (decomp).
THR = Details U. See germanium compounds.
Disaster Hazard: Dangerous; see chlorides.

BIS(4-AMINO-3-CHLOROPHENYL)ETHER. $C_{12}H_{10}ON_2Cl_2$; mw: 269.1.
THR = An exper carc. [*3*] See also ethers.

BIS(2-AMINO-1-NAPHTHYL)SODIUM PHOSPHATE. $C_{20}H_{17}O_4N_2P \cdot Na$; mw: 403.4.
THR = An exper neo. [*3*]

BIS(*o*-AMINO PHENYL)DISULFIDE.
Acute tox data: ip LD_{50} (mouse) = 50 mg/kg. [*3*]
THR = HIGH via ip route.
Disaster Hazard: Dangerous; see sulfides.

BIS(3-AMINO PROPYL)AMINE. $NH(C_3H_6NH_2)_2$, mw: 131.1.
Acute tox data: A corrosive material. Oral LD_{50} (rat) = 810 mg/kg. [*3*]
THR = MOD via oral exposure.

N,N-BIS(3-AMINO PROPYL)METHYL AMINE. Liquid, completely miscible in water. $CH_3N(C_3H_6NH_2)_2$, mw: 145, d: 0.9307 @ 20°/20°, bp: 240.6°, fp: −29.6°, flash p: 220°F.
Acute tox data: Oral LD_{50} (rat) = 1540 mg/kg. [*3*]
THR = MOD via oral route.
Fire Hazard: Mod, when exposed to heat or flame.
To Fight Fire: Foam, fog, dry chemical.

BIS(1-AZIRIDINYL) MORPHOLINE PHOSPHINE SULFIDE.
THR = An exper carc. [*23*]

1,3-BIS(2-BENZOTHIAZOLYL MERCAPTOMETHYL)UREA. See di-(2-benzothiazylthiomethyl) urea.

2,2-BIS(*p*-BROMOPHENYL)-1,1,1-TRICHLOROETHANE. $C_{14}H_9Br_2Cl_3$, mw: 443.43.
Acute tox data: LD (mouse) = 100 mg/kg. [*3*]
THR = HIGH via oral route. The bromine analog of DDT. See DDT.
Disaster Hazard: Dangerous, see chlorides.

BIS(2-BUTYL OCTYL)SODIUM SULFOSUCCINATE. White waxy pellets.
$(C_{12}H_{25}O_2C)_2CH_2CHSO_3Na$, mw: 556, mp: >200°, d: 1.0 @ 25°.
THR = Probably LOW.
Disaster Hazard: Dangerous; see sulfonates.

2,2-BIS(*tert*-BUTYLPEROXY)BUTANE. Syn: *22-PB*. Liquid. $(CH_3)_3COOC(CH_2)_3OOC(CH_3)_3$, mw: 220.3, mp: −10° to −26°, flash p: 84°F (OC), d: 0.8655 @ 20°/4°, vap. d: 7.59.
THR = An irr. See also peroxides, organic.
Fire Hazard: Dangerous, it burns when exposed to heat or flame or by spont chemical reaction with reducing agents. An oxidizer as well as a flam material.
To Fight Fire: Foam, CO_2, dry chemical.

BIS(2-*p*-*tert*-BUTYL PHENOXYETHYL)SODIUM SULFOSUCCINATE. White powder.

$(C_{12}H_{17}OCO_2)_2CH_2SCHO_3Na$, mw: 540, mp: 130°–140°, d: 1.0 @ 25°.
THR = U.
Disaster Hazard: Dangerous, see sulfonates.

BIS(*p*-tert-BUTYL PHENYL)PHENYL PHOSPHATE. Clear viscous liquid. $C_{26}H_{31}O_4P$, mw: 438.49, bp: 260°–275° @ 5 mm, fp: 0°, flash p: 482°F, d: 1.11 @ 25°/25°, vap. d: 15.1.
THR = U.
Fire Hazard: Slight, when exposed to heat or flame.
Disaster Hazard: Dangerous, see phosphates; can react with oxidizing materials.
To Fight Fire: CO_2, dry chemical.

BIS[2-(2-CHLOROETHOXY)ETHYL]ETHER. $C_8H_{16}O_3Cl_2$, mw: 231.1, flash p: > 250°F.
THR = MOD.
Fire Hazard: LOW.
To Fight Fire: Alcohol foam, dry chemical.

BIS(2-CHLOROETHYL)-2,3-DIMETHYOXY ANILINE. $C_{12}H_{17}O_2NCl_2$, mw: 278.2.
THR = An exper carc. [*3*]

BIS(1-CHLOROETHYL)ETHER. $C_4H_8OCl_2$, mw: 143.
THR = An exper neo. [*3*]

BIS-β-CHLOROETHYL ETHER. See 2,2′-dichloroethyl ether.

N,N-BIS(2-CHLOROETHYL)-2-NAPHTHYLAMINE. See naphthyl amine mustard.

1,3-BIS(2-CHLOROETHYL)-1-NITROSO UREA. $C_5H_9O_2N_3Cl_2$, mw: 214.1.
THR = HIGH via oral. An exper teratogen. [*3*]

BIS(5-CHLORO-2-HYDROXYPHENYL)METHANE. Crystals, nearly insol in water. $C_{13}H_{10}O_2Cl_2$, mw: 269.3, mp: 178°, vap. press: 10^{-4} mm @ 100°.
Acute tox data: Oral LD_{50} (rat) = 2690 mg/kg; iv LD_{50} (rat) 17 mg/kg. [*3*]
THR = HIGH via iv route and MOD via oral route. A MOD irr to skin. Can cause cramps and diarrhea. Possibly similar to DDT. See DDT.
Disaster Hazard: Dangerous, see chlorides.

1,4-BIS(CHLOROMETHOXYMETHYL)BENZENE. $C_{20}H_{12}O_2Cl_2$, mw: 235.1.
THR = An exper (+) carc via dermal or sc route. [*3, 81*] Exper produced malignant tumors at sites of application.

BIS-1,2-(CHLOROMETHOXY)ETHANE. Viscous liquid. $C_4H_8O_2Cl_2$, mw: 159.0, bp: 99°–100° @ 22 mm, d: 1.2879 @ 14°/15°.
THR = An exper (+) neo and carc via dermal, sc and ip routes. [*3, 81*]

9,10-BIS(CHLOROMETHYL)ANTHRACENE. Syn: *ICR 450*. $C_{16}H_{12}Cl_2$, mw: 275.2.
THR = An exper carc to mice via iv route. [*103*]

BIS(CHLOROMETHYL)ETHER. See dichloromethyl ether.

3,3-BIS-CHLOROMETHYL OXYCYCLOBUTANE.
THR = Probably a strong irr to skin, eyes and mu mem. Animal exper show irr and narcotic effects.
Disaster Hazard: Dangerous; see chlorides.

1,3-BIS(CHLOROMETHYL)-1,1,3,3-TETRAMETHYL DISILAZANE. $C_6H_{16}NCl_2Si_2$, mw: 229.3.
THR = An exper neo. [*3*]

BIS(*p*-CHLOROPHENOXY METHANE). Syn: *di(4-chlorophenoxy) methane, neotran*. White crystals. $C_{13}H_{10}Cl_2O_2$, mw: 269.1, mp: 70°, bp: 189°–194° @ 6 mm.
Acute tox data: Oral LD_{50} (rat) = 5800 mg/kg. [*3*]
THR = MOD via oral route. An insecticide.
Disaster Hazard: Dangerous; see chlorides.

2,2-BIS(*p*-CHLOROPHENYL)-1,1-DICHLOROETHANE. See DDD.

1,1-BIS(*p*-CHLOROPHENYL)-2-NITROBUTANE. See dilan.

1,1-BIS(*p*-CHLOROPHENYL)-2-NITROPROPANE. See dilan.

BIS(*o*-CHLOROPHENYL)PHENYL PHOSPHATE. Clear pale straw-colored mobile liquid. $C_{18}H_{13}O_4Cl_2P$, mw: 395.2, mp: < 0°, bp: 255°–275° @ 5 mm, flash p: > 437°F, d: 1.34 @ 25°/25°, vap. d: 13.6.
THR = U.
Fire Hazard: Slight, when exposed to heat or flame.
Disaster Hazard: Dangerous; see chlorides and phosphates; can react with oxidizing materials.
To Fight Fire: Water, foam, CO_2, water spray or mist, dry chemical.

2,2-BIS(*p*-CHLOROPHENYL)1,1,1-TRICHLOROETHANE. See DDT.

1,1-BIS(*p*-CHLOROPHENYL)-2,2,2-TRICHLOROETHANOL. Syn: *kelthane*.
Acute tox data: Oral LD_{50} (rat) = 800 mg/kg (tech). [*3*]
THR = MOD via oral route. See also DDT.
Disaster Hazard: Dangerous; see chlorides.

BISCYCLO PENTADIENYL MANGANESE. See manganocene.

BIS(DIETHYLTHIOCARBAMYL)DISULFIDE. Syn: *tetraethylthiuram disulfide*. Yellow-white crystals. $C_{10}H_{20}N_2S_4$, mp: 70°–72°, mw: 296.5.
Acute tox data: Oral LD_{50} (rabbit) = 2050 mg/kg. [*3*]
THR = MOD via oral route. An exper (+) carc and neo. [*12, 3*] Toxic when accompanied by

ingestion of alcohol. See also bis(dimethyl thiocarbamyl)disulfide.
Disaster Hazard: Dangerous; see SO_x.

BIS DEHYDRODO ISYNOLIC ACID-7-METHYL-ETHER. $C_{19}H_{22}O_3$, mw: 298.4.
THR = An exper carc. [*3*] See also ethers.

BIS DEHYDRODO ISYNOLIC ACID-7-METHYL-ESTER.
THR = An exper neo. [*3*]

3,6-BIS(DIMETHYLAMINO)ACRIDINE. See acridine orange.

3,6-BIS(DIMETHYL AMINO)ACRIDINE HYDROCHLORIDE.
THR = An exper carc. [*23*]

BIS(1,3-DIMETHYLBUTYL)AMINE. Liquid. $C_{12}H_{26}NH$, mw: 185.2, bp: 190°–200°, d: 0.775 @ 20°/20°, vap. d: 6.38.
THR = Details U. See also amines.
Fire Hazard: Slight.

BISDIMETHYL GLYOXIME COBALTO CHLORIDE. Pale green crystals, sol in water. $Co(C_8H_{15}O_4N_4Cl_2)$, mw: 361.1.
THR = Details U. See cobalt compounds.
Disaster Hazard: Dangerous; see chlorides.

BIS(1,2-DIMETHYL PROPYL)BORINE. Syn: *di-sec-isoamyl borane.*
THR = See boron hydrides.

BIS(DIMETHYL STIBINE)OXIDE. $C_4H_{12}OSb_2$, mw: 319.6.
THR = No data. See antimony. Spont flam in air.

BIS(DIMETHYLTHIOCARBAMYL)DISULFIDE. Syns: *tetramethylthiuram disulfide*, *disulfiram*, *TTD*, *thiram*. Crystals, insol in water, sol in alcohol, ether, acetone, chloroform. $C_6H_{12}N_2S_4$, mw: 296.6, mp: 156°, d: 1.30, bp: 129° @ 20 mm.
Acute tox data: Oral LD_{50} (rat) = 560 mg/kg; oral LD_{50} (rabbit) = 210 mg/kg. [*3*] Dermal LD_{50} (rat) = 2000 mg/kg. [*12*]
THR = MOD via dermal route. HIGH via oral route. A MILD allergen and irr. An exper teratogen. [*3*] Acute poisoning in exper animals produced liver and kidney injury and also brain damage. In the presence of alcohol this compound produces violent nausea, vomiting and collapse. A fungicide.
Disaster Hazard: Dangerous; see NO_x and SO_x.

BIS(2,3-EPOXY-2-METHYL PROPYL)ETHER $H_2COC(CH_3)CH_2OH_2C(CH_3)COCH_2$, mw: 158.
Acute tox data: Oral LD_{50} (rat) = 1680 mg/kg; dermal LD_{50} (rabbit) = 1250 mg/kg. [*3*]
THR = MOD via oral and dermal routes.

2,2′-BIS(2,3-EPOXY PROPYL)-N-*tert*-BUTYL DIPROPYLAMINE. $C_{16}H_{31}O_4N$, mw: 301.5.
THR = An exper neo. [*3*]

2,3-BIS(2,3-EPOXY PROPOXY)-1,4-DIOXANE. $C_{10}H_{16}O_6$, mw: 232.3.
Acute tox data: Oral LD_{50} (rat) = 1070 mg/kg; dermal LD_{50} (rabbit) = 1590 mg/kg. [*3*]
THR = MOD via oral and dermal routes.

N,N-BIS-2-(2,3-EPOXY PROPOXY)ETHOXY ANILINE. $C_{16}H_{23}O_6N$, mw: 325.4.
THR = An exper neo. [*3*]

N,N-BIS[2-(2-EPOXY PROPYOXY)ETHYL]ANILINE. $C_{16}H_{23}O_4N$, mw: 293.4.
THR = An exper carc. [*3*]

1,2-BIS(3-ETHOXY CARBONYL-2-THIOUREIDO) BENZENE. See thiophanate.

1,3-BIS-ETHYL AMINOBUTANE. Liquid. $(C_2H_5)_2C_4H_7NH_2$, mw: 129.2, bp: 180°, flash p: 115°F (OC), d: 0.81, vap. d: 4.44.
THR = U.
Fire Hazard: Mod, when exposed to heat or flame; can react with oxidizing materials.
To Fight Fire: Alcohol foam, dry chemical.

BIS(ETHYL AMINO)SILOXENE. $C_8H_{24}O_3N_2Si_3$, mw: 280.5.
THR = No data. Spont flam in air. [*19*]

2,5-BIS(ETHYLENIMINO)-*p*-BENZOQUINONE. Gray needles. $C_{16}H_{22}O_6N_2$, mw: 338.4; mp: 89°.
THR = An exper carc. [*3, 8*]

1,3-BIS(ETHYLENE IMINO SULFONYL)PROPANE. See BEP.

BIS(2-ETHYL HEXYL)AZELATE. $C_{25}H_{48}O_4$, mw: 412.7.
Acute tox data: Iv LD_{50} (rat) = 1060 mg/kg; iv LD_{50} (rabbit) = 640 mg/kg. [*3*]
THR = MOD via ip and iv routes.

BIS(2-ETHYL HEXYL)-2-ETHYL HEXYL PHOSPHONATE. Colorless liquid, very mild odor. $C_8H_{17}P(O)(OC_8H_{17})_2$, mw: 418.6, bp: 161° @ 0.25 mm, flash p: 419°F (COC), d: 0.908 @ 20°/4°, vap. d: 14.4.
THR = U.
Fire Hazard: Slight, when exposed to heat or flame.
Disaster Hazard: Dangerous; see phosphates; can react with oxidizing materials.
To Fight Fire: Foam, CO_2, dry chemical.

BIS(2-ETHYL HEXYL)SUCCINATE. Slightly water sol. $C_{20}H_{38}O_4$, mw: 342.2, d: 0.9, bp: 258° @ 50 mm; flash p: 315°F.
THR = LOW. See also esters.

Fire Hazard: Low.
To Fight Fire: Alcohol foam.

2,2-BIS(*p*-ETHYL PHENYL)-1,1-DICHLOROETHANE. $C_{18}H_{20}Cl_2$, mw: 307.3.
THR = An exper carc. [*3*]

2,2′-BIS(ETHYL SULFONYL)BUTANE. $C_8H_{18}O_4S_2$, mw: 242.4.
THR = An exper neo. [*3*] via dermal route.

2,2-BIS-(*p*-FLUOROPHENYL)-1,1,1-TRICHLOROETHANE. See DFDT.

1,1-BIS(*p*-FLUOROPHENYL)-2-PROPYNYL CYCLOHEPTANE CARBAMIC ACID. $C_{23}H_{23}O_2NF_2$, mw: 383.5.
Acute tox data: Oral LD_{50} (mice) = 405 mg/kg; ip LD_{50} (mice) = 318 mg/kg. [*3*]
THR = An exper carc. [*3, 12*]

1,1-BIS(4-FLUOROPHENYL)-2-PROPYNYL-N-CYCLOOCTYL CARBAMATE. $C_{24}H_{25}O_2NF_2$, mw: 398.
THR = An exper carc. [*3, 12*]

1,4-BIS(4-FLUOROPHENYL)-2-PROPYNYL-N-CYCLOOCTYL CARBAMATE.
THR = An exper carc. [*12*]

2,3-BIS(GLYCIDYLOXY)-1,4-DIOXANE. See 2,3-bis(2,3-epoxy propoxyl)-1,4-dioxane.

2,2-BIS-*p*-(2,3-GLYCIDYLOXY)PHENYL. $C_{21}H_{24}O_4$, mw: 340.5.
Acute tox data: Oral LD_{50} (rat) = 11000 mg/kg. [*3*]
THR = LOW via oral route.

BISH. See aconitum ferox.

BIS-(2-HYDROXY-5-CHLOROPHENYL)SULFIDE. $(C_6H_3OHCl)_2S$, mw: 287.2.
Acute tox data: Ip LD_{LO} (mouse) = 250 mg/kg. [*3*]
THR = HIGH via ip route. A fungicide. See chlorinated phenols.
Fire Hazard: U.
Disaster Hazard: Dangerous; see chlorides and oxides of sulfur.

BIS-HYDROXYCOUMARIN. Syn: *dicoumarol.* Very small crystals, slight pleasant odor, bitter taste, sol in alkali. $C_{19}H_{12}O_6$, mw: 336.3, mp: 287°–293°.
Acute tox data: Oral LD_{50} (rat) = 710 mg/kg; ip LD_{50} (rat) = 95 mg/kg; iv LD_{50} (rat) = 52 mg/kg. [*3*]
THR = HIGH–MOD oral and HIGH via iv and ip routes. An anticoagulant. Excessive doses can cause hemorrhages.
Antidote: Vitamin K.

BIS-(2-HYDROXY-3,5-DICHLORO PHENYL)SULFIDE. See bithionol.

4-BIS-(2-HYDROXY ETHYL)AMINO-2-(5-NITRO-2-FURYL)QUINAZOLINE. $C_{16}H_{16}O_5N_4$, mw: 344.4.
THR = An exper carc to rats via oral route. [*103*]

4-BIS-(2-HYDROXY ETHYL)AMINO-2-(5-NITRO-2-THIENYL)QUINAZOLINE. $C_{16}H_{16}O_4N_4S$, mw: 360.5.
THR = An exper oral carc to rats. [*103*]

BIS(2-HYDROXY ETHYL)DITHIOCARBAMATE, POTASSIUM SALT. $C_5H_{10}NS_2 \cdot K$, mw: 219.4.
THR = An exper (+) carc via oral route ⟶ liver cell tumors. [*3, 12*]

3,3-BIS-(*p*-HYDROXY PHENYL)PHTHALIDE. See phenolphthalein.

2,2-BIS(4-HYDROXY PHENYL)PROPANE. Syns: *bisphenol A, p,p′-isopropylidene diphenol.* White flakes, mild phenolic odor, insol in water, sol in alcohol and dilute alkalies, slightly sol in CCl_4. $C_{15}H_{16}O_2$, mw: 228.
Acute tox data: Oral LD_{50} (mammal) = 6500 mg/kg; ip LD_{50} (mouse) = 150 mg/kg. [*3*]
THR = MOD via oral and HIGH via ip routes.

BIS(ISOPROPYL AMIDO)FLUOROPHOSPHATE. Syns: *isopestox, mipafox.* $C_6H_{16}ON_2PF$, mw: 182.2.
Acute tox data: Oral LD_{50} (rabbit) = 100 mg/kg; ip LD_{50} (rat) = 90 mg/kg. [*3*]
THR = HIGH via oral and ip routes.
Disaster Hazard: Very dangerous; see fluorides and phosphates. See also parathion.

BISMARCK BROWN 331.
THR = A MILD allergen.

3,4-BIS(METHOXY)BENZYL CHLORIDE. See 3,4-dimethoxy benzyl chloride.

2,2-BIS(*p*-METHOXY PHENYL)-1,1,1-TRICHLOROETHANE. See methoxychlor.

BIS(α-METHYL BENZYL)AMINE. See α-methyl benzylamine.

BIS-α-METHYL BENZYL ETHER. See methyl benzyl ether.

N,N′-BIS(1-METHYL HEPTYL)ETHYLENE DIAMINE. Insol in water.
$HC(CH_3)(C_6H_{13})NHCH_2(CH_2NHCH)(CH_3)(C_6H_{13})$, mw: 284, d: 0.8, bp: 218° @ 43 mm, flash p: 400°F.
Acute tox data: Oral LD_{50} (rat) = 2400 mg/kg; dermal LD_{50} (rabbit) = 1800 mg/kg. [*3*]
THR = MOD via oral and dermal routes.
Fire Hazard: Slight, when exposed to heat or flame.
To Fight Fire: Dry chemical, water fog, mist.

BISMUTH. Hexagonal silver-white or reddish metallic crystals. Bi, atwt: 209.00, mp: 271.3°, bp: 1420°–1560°, d: 9.80, vap. press: 1 mm @ 1021°.
THR = See bismuth compounds.

Radiation Hazard: For permissible levels see Table 5A.5. Natural isotope ^{210}Bi (RaE), $T\frac{1}{2}$ = 5.0 d, decays to ^{210}Po via β's of 1.16 MeV. Natural isotope ^{212}Bi, $T\frac{1}{2}$ = 60.5 m (Th-C), decays to ^{208}Ti via α's of 6.05 and 6.09 MeV (36%). Also decays to ^{212}Po via β's of 0.45 to 2.3 MeV.

Fire Hazard: Mod, when exposed to flame and by chemical reaction with [$Bi(OH)_3$ + $Al(OH)_3$] co-precipitated and H_2 reduced $\longrightarrow$ spont flam product. Vigorous reaction with NH_4NO_3, $HClO_3$, Cl_2, IF_5, HNO_3, $HClO_4$. [*19*]

Disaster Hazard: Mod dangerous; can react with acid or acid fumes to emit toxic fumes.

BISMUTH ACETATE. White crystals. $Bi(C_2H_3O_2)_3$, mw: 386.13, mp: decomp.

THR = See bismuth compounds.

BISMUTH (COLLOIDAL SOLUTION).

THR = Details U. See also bismuth compounds.

BISMUTH-*o*-ARSENATE. Monoclinic crystals. $BiAsO_4$, mw: 347.91, d: 7.14.

THR = See arsenic compounds.

BISMUTH BENZOATE. White powder. $Bi(C_7H_5O_2)_3$, mw: 572.33.

THR = Scc bismuth compounds.

Fire Hazard: Slight, when heated.

BISMUTH CARBONATE, BASIC. Syns: *bismuth oxycarbonate, bismuthospherite.* White powder. $Bi_2O_2CO_3$, mw: 510.10, mp: decomp, d: 6.86.

THR = See bismuth compounds.

BISMUTH CITRATE. White crystals. $BiC_6H_5O_7$, mw: 398.10, mp: decomp, d: 3.458.

THR = Scc bismuth compounds.

BISMUTH COMPOUNDS.

THR = Bi and its salts can cause kidney damage, although the degree of such damage is usually mild. Large doses can be fatal. Industrially it is considered one of the less toxic of the heavy metals, although intoxication has occurred from its use in medicine. The similarity between the pharmacologic and toxic behavior of Pb and Bi has been pointed out in the literature. Like Pb, Bi may be liberated from tissue deposits during periods of acidosis. Serious and sometimes fatal poisoning may occur from the injection of large doses into closed cavities and from extensive application to burns. Death of animals from bismuth nephritis following injections of soluble salts occurs within several hours to 24 days, the time being generally inversely proportional to the dose, and it appears to be in the order of 5 to 10 times higher than the dose by slow intravenous injection for rabbits. It is stated that the administration of Bi should be stopped when gingivitis appears, for otherwise serious ulcerative stomatitis is likely to result. Other toxic results may develop, such as malaise, albuminuria, diarrhea, skin reactions, and sometimes serous exodermatitis. Industrial Bi poisoning has not been reported, although bismuth absorbed in industrial cases may complicate a diagnosis of plumbism, since the dark line in the gums, which is often present in lead poisoning, is also produced by bismuth. All Bi compounds do not have equal toxicity. See individual entries.

Treatment and Antidotes: Personnel showing some of the symptoms noted above which might indicate that they were absorbing too much bismuth into the body should be removed from exposure as soon as possible. Get medical advice. Personnel should be cautioned against careless handling of these materials.

BISMUTH DICHROMATE, BASIC. Yellow or red crystals. $(BiO)_2Cr_2O_7$, mw: 666.

THR = See chromium compounds and bismuth compounds.

Fire Hazard: Mod, by chemical reaction with easily combustible materials; can react with reducing materials; a powerful oxidizer.

BISMUTH DIMETHYL DITHIO CARBAMATE. $C_9H_{18}N_3S_6 \cdot Bi$, mw: 569.6.

THR = An exper neo. [*3*]

BISMUTH ETHYL CHLORIDE. Powder. $BiHC_2H_5Cl$, mw: 274.5.

THR = See bismuth compounds.

Fire Hazard: Dangerous, by spont chemical reaction; also spont flam in air.

Explosion Hazard: U.

Disaster Hazard: Dangerous; See chlorides; can react with oxidizing materials.

BISMUTH FLUORIDE. See bismuth trifluoride.

BISMUTH GALLATE, BASIC. Syns: *subgallate, dermatol* (*commercial*). Yellow amorphous crystals. $Bi(OH)_2C_7H_5O_5$ (approx), mw: 412.13, mp: decomp.

THR = See bismuth compounds.

BISMUTH GLANCE. See bismuth trisulfide.

BISMUTH HEPTADIENE CARBOXYLATE.

THR = See bismuth compounds.

Fire Hazard: Slight, when heated.

BISMUTH HEPTOXIDE.

THR = See bismuth compounds.

Fire Hazard: Mod, by chemical reaction with easily oxidized materials; can react with reducing materials; a powerful oxidizer.

BISMUTH HYDROXIDE. White amorphous powder. $Bi(OH)_3$, mw: 260.02, $-H_2O$ @ 100°, $-\frac{1}{2}$ H_2O @ 400°, decomp @ 415°.
THR = See bismuth compounds.

BISMUTH ACID. $HBiO_3$, mw: 258.
THR = No data. See bismuth compounds. Reacts violently with 40% HF. [*19*]

***m*-BISMUTH ACID.** See bismuth pentoxide.

BISMUTHINE. See hydrogen bismuthide.

BISMUTH IODATE. White crystals. $Bi(IO_3)_3$, mw: 733.76.
THR = See bismuth compounds and iodates.
Fire Hazard: Mod, by spont chemical reaction with easily oxidized materials; a powerful oxidizer.

BISMUTH-*dl*-LACTATE. Prisms. $Bi(C_6H_9O_6) \cdot 7H_2O$, mw: 512.25.
THR = See bismuth compounds.
Fire Hazard: Slight, when heated.

BISMUTH MOLYBDATE. Tetragonal needles. $Bi_2(MoO_4)_3$, mw: 897.85, mp: 643°, d: 6.07.
THR = See bismuth compounds and molybdenum compounds.

BISMUTH MONOSULFIDE. Gray crystals. BiS, mw: 241.07, mp: 685°, d: 7.7.
THR = See bismuth compounds and sulfides.
Fire Hazard: See sulfides.
Explosion Hazard: See sulfides.
Disaster Hazard: Dangerous; see oxides of sulfur; will react with water or steam to produce toxic and flam vapors.

BISMUTHNITE. See bismuth trisulfide.

BISMUTH NITRATE. Triclinic, colorless, slightly hygroscopic crystals, $Bi(NO_3)_3 \cdot 5H_2O$, mw: 485.10, bp: $-5H_2O$ @ 80°, d: 2.83, mp: 30° (decomp).
THR = See bismuth compounds and nitrates.

BISMUTH OXALATE. White powder. $Bi_2(C_2O_4)_3$, mw: 682.06.
THR = See bismuth compounds and oxalates.

BISMUTH OXYBROMIDE. Colorless crystals or white powder. BiOBr, mw: 304.92, d: 8.08.
THR = See bismuth compounds and bromides.

BISMUTH OXYCARBONATE. See bismuth carbonate, basic.

BISMUTH OXYCHLORIDE. White lustrous crystalline powder. BiOCl, mw: 260.46, mp: decomp @ red heat, d: 7.72.
THR = See bismuth compounds and chlorides.

BISMUTH OXYFLUORIDE. White crystals or powder. BiOF, mw: 244.00, mp: decomp @ red heat, d: 7.5.
THR = See bismuth compounds and fluorides.

BISMUTH OXYIODIDE. Red powder. BiOI, mw: 351.92, mp: decomp @ red heat, d: 7.92.
THR = See bismuth compounds and iodides.

BISMUTH OXYIODOGALLATE. Grayish-green, odorless, tasteless powder. $C_6H_2(OH)_3COO(BiOHI)$, mw: 522.0.
THR = See bismuth compounds and iodides.

BISMUTH OXYNITRATE. Hexagonal plates or white powder. $BiONO_3 \cdot H_2O$, mw: 305.02, mp: decomp @ 260°, d: 4.928 @ 15°.
THR = See bismuth compounds and nitrates.

BISMUTH PENTAFLUORIDE. Crystals which react violently with water and petrolatum above 50°. BiF_5, mw: 304, sublimes @ 550°.
THR = HIGH irr via oral and inhal routes. HIGH irr. Decomp readily on contact with moisture to yield O_2 and bismuth trifluoride. See fluorides and ozone.
Disaster Hazard: Very dangerous. When heated to decomp, emits highly toxic fluoride fumes. In contact with moisture, acids, etc., it reacts violently, liberating much heat and ozone.

BISMUTH PENTOXIDE. Syn: *m-bismuthic acid.* Brown or dark red crystals. Bi_2O_5, mw: 498.00, $-O$ @ 150°, $-O_2$ @ 357°, d: 5.10.
THR = See bismuth compounds.
Fire Hazard: Slight, when exposed to heat and oxidizing materials or by chemical reaction with reducing agents; when heated, evolves O_2. Reacts violently with BrF_3.

BISMUTH-*o*-PHOSPHATE. Monoclinic white crystals. $BiPO_4$, mw: 303.98, mp: decomp, d: 6.323 @ 15°.
THR = See bismuth compounds and phosphates.

BISMUTH PROPIONATE, BASIC. White powder, faint odor of propionic acid. $BiOC_3H_5O_2$, mw: 298.07.
THR = See bismuth compounds.
Fire Hazard: Slight, when heated.

BISMUTH SUBCARBONATE. White, odorless, tasteless powder, insol in water or alcohol. $(BiO)_2CO_3 \cdot \frac{1}{2}H_2O$, mw: 519, d: 6.86.
THR = See bismuth compounds. Has been used as an opaque medium in x-ray work.

BISMUTH SUBNITRATE. Syn: *magistery of bismuth.* White, heavy powder. $4BiNO_3(OH)_2 \cdot BiO(OH)$, mw: 1462.11, mp: decomp @ 260°, d: 4.928.
THR = See bismuth compounds and nitrates.

BISMUTH SUBSALICYLATE. See bismuth salicylate, basic.

BISMUTH SULFATE. White needles. $Bi_2(SO_4)_3$, mw: 706.20, mp: decomp, d: 5.08 @ 15°.
THR = See bismuth compounds and sulfates.

BISMUTH TARTRATE. White powder. $Bi_2(C_4H_4O_6)_3 \cdot 6H_2O$, mw: 970.31, $-3H_2O$ @ 105°, d: 2.595 @ 25°.
THR = See bismuth compounds.

BISMUTH TELLURATE. Syn: *montanite*. Powder or crystals. $Bi_2TeO_6 \cdot 2H_2O$, mw: 677.64, d: 3.79.
THR = See bismuth and tellurium compounds.

BISMUTH TELLURIDE. See bismuth tritelluride.

BISMUTH TELLURIDE-SE DOPED.
THR = See compounds of bismuth, tellurium, selenium.

BISMUTH TETRACHLORIDE. Colorless crystals. $BiCl_4$, mw: 350.83, mp: 225°.
THR = See bismuth compounds and hydrochloric acid.

BISMUTH TETRAOXIDE. Heavy yellowish-brown powder. Bi_2O_4, mw: 482, mp: 305°, d: 5.6.
THR = See bismuth compounds.
Fire Hazard: Slight, by chemical reaction with reducing agents; an oxidizer.

BISMUTH TITANATE. $BiTiO_3$, mw: 305.
THR = U. Animal exper show LOW. See also bismuth and titanium compounds.

BISMUTH TRIBROMIDE. Yellow crystalline deliquescent powder. $BiBr_3$, mw: 448.75, mp: 218°, bp: 461°, d: 5.7, vap. press: 10 mm @ 282°.
THR = See bismuth compounds, bromides and hydrobromic acid. Reacts violently with Na, K. [*19*]

BISMUTH TRICHLORIDE. White, deliquescent crystals. $BiCl_3$, mw: 315.37, mp: 230°, bp: 447°, d: 4.75, vap. press: 10 mm @ 264°.
THR = See bismuth compounds and hydrochloric acid. Reacts violently with Na, K. [*19*]

BISMUTH TRIETHYL. See triethyl bismuthine.

BISMUTH TRIFLUORIDE. Syn: *bismuth fluoride*. White cubic crystals, insol in water. BiF_3, mw: 266, d: 8.3, mp: 725°–730°.
THR = See fluorides and bismuth compounds.

BISMUTH TRIIODIDE. Hexagonal reddish, brown-gray-blue crystals. BiI_3, mw: 589.76, mp: 439°, bp: decomp @ 500°, d: 5.7.
THR = See bismuth compounds and iodides. Reacts violently with Na, K. [*19*]

BISMUTH TRIOXIDE. Rhombic yellow crystals. Bi_2O_3, mw: 466.00, mp: 820°, bp: 1890°, d: 8.9.
THR = See bismuth compounds.

BISMUTH TRIPHENYL. See triphenyl bismuthine.

BISMUTH TRISELENIDE. Syn: *guanajuatite*. Rhombic black crystals. Bi_2Se_3, mw: 654.88, mp: 710°, bp: decomp, d: 6.82.
THR = See bismuth and selenium compounds.
Fire Hazard: Mod, by chemical reaction with powerful oxidizers. On contact with moisture a flam gas is evolved.
Explosion Hazard: Slight, by chemical reaction with powerful oxidizers and moisture.
Disaster Hazard: See selenium compounds.

BISMUTH TRISULFIDE. Syns: *bismuthnite*, *bismuth glance*. Rhombic black crystals. Bi_2S_3, mw: 514.20, mp: 685° decomp, d: 7.39.
THR = See bismuth compounds and sulfides.

BISMUTH TRITELLURIDE. Syns: *tetradymite*, *bismuth telluride*. Gray crystals. Bi_2Te_3, mw: 800.83, mp: 573°, d: 7.7.
THR = See bismuth and tellurium compounds.
Fire Hazard: Mod by spont chemical reaction with powerful oxidizers; reacts with moisture to evolve a gas.
Explosion Hazard: Slight, by chemical reaction with powerful oxidizers; reacts with moisture.
Disaster Hazard: See tellurium compounds.

BISMUTH TRIVINYL. See trivinyl bismuthine.

BIS(5-OXY-2,8-DITHIOOCTANE)GERMANIUM. Colorless crystals, sol in benzene. $[(SCH_2CH_2)_2O]_2Ge$, mw: 345.1, mp: 159.5°.
THR = Details U. See germanium compounds.
Disaster Hazard: Dangerous; see sulfates.

BISPHENOL GLYCIDYL ETHER. See 2,2-bis-*p*-(2,3-glycidyl)phenyl propane.

BIS-PROPIONYL ACETONE GERMANIUM DICHLORIDE. White crystals.
$GeCl_2[CHC(C_2H_5)CH_2O_2]_2$, mw: 369.8, mp: 129°.
THR = U. See germanium compounds.
Disaster Hazard: Dangerous; see chlorides.

BIS-TRIBENZYL GERMANYL SULFIDE. Colorless crystals, sol in alcohol. $SGe_2(C_6H_5CH_2)_6$, mw: 724, mp: 124°.
THR = Details U. See germanium compounds.
Disaster Hazard: Dangerous; see sulfur compounds.

BIS-TRIBIPHENYLYL GERMANYL SULFIDE. Colorless crystals, sol in organic solvents.
$[(C_6H_5 \cdot C_6H_4)_3Ge]_2S$, mw: 1096.4, mp: 238°.
THR = See bis-tribenzyl germanyl sulfide.

BIS-TRICHLOROGERMANYL METHANE. Colorless liquid, hydrolyzed by water. $CH_2(GeCl_3)_2$, mw: 372, bp: 110° @ 18 mm.
THR = See bis-propionyl acetone germanium dichloride.

BIS-TRICYCLOHEXYL GERMANIUM DISULFIDE. Colorless crystals, insol in water.
$[(C_6H_{11})_3Ge]_2S_2$, mw: 708.2, mp: 88°.
THR = See bis-tribenzyl germanyl sulfide.

BIS(TRIDECYL)SODIUM SULFOSUCCINATE. White waxy pellets. $(C_{13}H_{27}CO_2)_2CH_2SCHO_3Na$, mw: 584, mp: 70°–75°, d: 1.0 @ 25°.
THR = U.
Fire Hazard: Slight.

BIS-TRIETHYL GERMANYL SULFIDE. Colorless oily liquid, sol in organic solvents. $[(C_2H_5)_3Ge]_2S$, mw: 351.6, bp: 150° @ 12 mm.
THR = See bis-tribenzyl germanyl sulfide.

BIS(TRIFLUOROMETHYL)BENZENE. See xylene hexafluoride.

BIS(TRIFLUOROMETHYL)CHLORO PHOSPHINE. $(CF_3)_2PCl$, mw: 204.5.
THR = No data. Probably HIGH. Spont flam in air. [*19*]

BIS(TRIFLUOROMETHYL)CYANO PHOSPHINE. $(CF_3)_2PCN$, mw: 195.1.
THR = No data. Probably HIGH. Spont flam in air. [*19*]

BIS(TRIFLUOROMETHYL)PHOSPHINE. $(CF_3)_2PH$, mw: 170.
THR = Probably HIGH. Spont flam in air. [*19*]

2,2-BIS-(2,2,2-TRIMETHYL ETHYL)OXIRANE. $C_{12}H_{24}O$, mw: 184.4.
Acute tox data: Oral LD_{50} (rat) = 6690 mg/kg; dermal LD_{50} (rabbit) = 14000 mg/kg. [*3*]
THR = MOD via oral and very LOW via dermal routes.

BIS-(2,2,4-TRIMETHYL PENTANEDIOL MONOISOBUTYRATE)DIGLYCOLLATE. $C_{28}H_{27}O_9$, mw: 507, d: 1.1, bp: 337°, flash p: 383°F (OC).
THR = U.
Fire Hazard: Slight, when exposed to heat or flame.
To Fight Fire: Dry chemical, water mist, fog.

BIS(TRIMETHYL SILYL)ACETAMIDE. $C_8H_{21}ONSi_2$, mw: 203.5.
Acute tox data: ip LD_{LO} (mice) = 750 mg/kg. [*103*]
THR = MOD via ip route. An exper neo via ip route. [*103*]

BIS-TRIPHENYL GERMANYL SULFIDE. Colorless crystals, sol in organic solvents. $[(C_6H_5)Ge]_2S$, mw: 640, mp: 138°.
THR = See bis-tribenzyl germanyl sulfide.

BIS(TRIPHENYL SILYL)CHROMATE.
Acute tox data: Oral LD_{50} (rat) = 3360 mg/kg; dermal LD_{50} (rabbit) = 710 mg/kg. [*3*]
THR = MOD via oral and dermal routes.

BIS-TRITOLYL GERMANYL SULFIDE. Colorless crystals, sol in organic solvents. $[(C_6H_4CH_3)_3Ge]_2S$, mw: 724, mp: 157°.
THR = See bis-tribenzyl germanyl sulfide.

BISULFITE DRY POWDER SODA. See sodium bisulfite.

BISULFITE OF SODIUM. See sodium bisulfite.

BISULFITES.
THR = See individual compounds.
Disaster Hazard: Dangerous; when heated to decomp, or on contact with acid or acid fumes, emits highly toxic fumes; emits highly toxic fumes of sulfur dioxide; will react with water or steam to produce toxic and corrosive fumes.

BITHIONOL. Syns: *2,2-thiobis(4,6-dichlorophenol)*, *bis (2-hydroxy-3,5-dichlorophenyl)sulfide*. White crystalline powder, very faint phenolic odor. $(C_6H_2Cl_2OH)_2S$, mw: 356.1, mp: 187°–188°, d: 1.61 @ 25°, vap. press: 1.1×10^{-9} mm @ 37°.
Acute tox data: Oral LD_{50} (rat) = 7 mg/kg; oral LD_{50} (mouse) = 900 mg/kg. [*3*]
THR = Species dependent HIGH or MOD via oral routes. An exper neo. [*3*] A food additive permitted in feed and drinking water of animals and/or for the treatment of food-producing animals. Also a food additive permitted in food for human consumption. [*109*]
Disaster Hazard: Dangerous; when heated to decomp or on contact with acid or acid fumes, emits highly toxic fumes of oxides of sulfur.

BITHIONOL, DISODIUM.
THR = An exper neo. [*3*]

(*m,o'*-BITOLYL)-4-AMINE.
THR = An exper carc. [*3*]

3,3-BITOLYLENE-4,4'-DIISOCYANATE. Small, pale yellowish flakes. $[C_6H_3(OCN)(CH_3)]_2$, mw: 264.3, fp: 69.6°, d: 1.197 @ 80°/4°.
THR = Details U, but probably an irr. See also 2,4-tolylene diisocyanate.
Disaster Hazard: Mod dangerous; when heated to decomp, it emits toxic fumes.

BITTER APPLE. See colocynth.

BITTER CUCUMBER. See colocynth.

BITTER WOOD TREE. See quassia.

BITUMEN. See petroleum asphalt.

BLACK BLASTING POWDER. See explosives, low.

BLACK EBONY WOOD. A wood.
THR = A MILD irr and allergen.
Fire Hazard: Slight, in the form of dust when exposed to heat or flame; can react with oxidizing materials.

BLACK IRON OXIDE. Syn: *Magnetite*. Fe_3O_4, mw: 231.6.
THR = An exper neo. [*3*]

BLACK LEAD. See graphite.

BLACK POWDER. See charcoal, sulfur, potassium or sodium nitrate.

BLACK POWDER IGNITERS WITH EMPTY CARTRIDGE BAGS. See explosives, low.

BLACK PRECIPITATE. See mercurous nitrate, ammoniated.

BLACK WAX.
THR = A recog carc. [*14*] See also paraffin and petroleum waxes.

BLADAFUME. See tetraethyl dithiopyrophosphate.

BLADEX. Syns: *2-(4-chloro-6-ethylamino-s-triazine-2-ylamino)-2-methyl propionitrile.* A white, crystalline material. $C_{11}H_{13}N_6Cl$, mw: 264.6, mp: 167°.
Acute tox data: Oral LD_{50} (rat) = 334 mg/kg; acute dermal LD_{50} (rat = > 1200 mg/kg (for 20% tech).
THR = HIGH. A selective herbicide.
Disaster Hazard: High temp. can decomp it to yield highly toxic fumes.

BLANC FIXE. See barium sulfate.

BLAST FURNACE GAS. See carbon monoxide.

BLACK FURNACE TAR.
THR = A recog carc. [*14*] See also coal tar and pitch.

BLASTING CAPS, 1000 OR LESS. See explosives, high.

BLASTING CAPS, MORE THAN 1000. See explosives, high.

BLASTING CAPS WITH METAL CLAD MILD DETONATING FUSE, 1000 OR LESS. See explosives, high.

BLASTING CAPS WITH METAL CLAD MILD DETONATING FUSE, MORE THAN 1000. See explosives, high.

BLASTING CAPS WITH SAFETY FUSE, 1000 OR LESS CAPS. See explosives, high.

BLASTING CAPS WITH SAFETY FUSE, MORE THAN 1000. See explosives, high.

BLASTING GELATIN. See explosives, high.

BLASTING OIL. See nitroglycerine.

BLASTING POWDERS. See explosives, low.

BLAU GAS. A gas. Comp: typical gas will analyze 51.9% illuminants; 0.1% carbon monoxide; 2.7% hydrogen; 44.1% methane; 1.2% nitrogen, etc.
THR = Variable depending upon comp, CO, methane, etc.
Fire Hazard: Dangerous, when exposed to flame.
Explosion Hazard: Slight, when exposed to flame.
Disaster Hazard: Mod dangerous; when heated, it burns and can react with oxidizing materials.

BLEACHING FOOD ADDITIVES. [*109*]

BLEACHING POWDER. Syns: *calcium hypochlorite, chlorinated lime, calcium chloride hypochlorite.* White powder. $CaCl(ClO) \cdot 4H_2O$, mw: 199.0.
THR = HIGH via oral and inhal routes. Can cause severe irr of skin and mu mem and emit fumes capable of causing pulmonary edema.
Fire Hazard: Mod by chemical reaction with combustible materials; i.e., anthracene, C, charcoal, C_2H_5OH, glycerol, grease, oil, mercaptans, methyl carbitol, nitromethane, organic matter, organic sulfides, phenol, 1-propanethiol, propylmercaptan, S, turpentine. Can explode with CCl_4, amines. [*19*] A powerful oxidizer. Deflagration occurs in contact with combustible substances.
Explosion Hazard: Mod, in its solid form when heated.
Explosive Range: When heated suddenly above 212°F.
Disaster Hazard: Dangerous, when heated to decomp or on contact with acid or acid fumes, emits highly toxic fumes and explodes; will react with water or steam to produce toxic and corrosive fumes; can react vigorously with reducing materials.

BLISTERING BEETLES. See cantharides.

BLISTERING FLIES. See cantharides.

BLOWN ASPHALT.
THR = A recog carc. [*14*] See also petroleum asphalt.

BLUE LEAD. See lead sulfate.

BLUE OIL.
THR = A recog carc. [*14*] See also mineral oils.

BLUE STONE. See copper sulfate.

BLUE VITRIOL. See copper sulfate.

BOILER COMPOUND LIQUID.
THR = A caustic solution. See sodium hydroxide.

BOLIDEN SALT. See chromated zinc arsenate.

BONE ASH. See calcium phosphate, tribasic.

BONE BLACK. See soot.

BONE DUST. Comp: Phosphates, silicates of calcium, magnesium, etc.
THR = A nuisance dust.

BONE PITCH.
THR = Susp carc. [*14*]

BOOSTERS (EXPLOSIVE). See explosives, high.

BORACIC ACID. See boric acid.

BORANES. See boron hydrides.

BORAX. Syns: *tincal, tinkal.* White-gray, bluish- or greenish-white streak, vitreous or dull luster. $Na_2B_4O_7 \cdot 10H_2O$, mw: 381.44, mp: 75°, $-8H_2O$ @ 60°, $-10H_2O$ @ 200°, d: 1.69–1.72.
Acute tox data: Oral LD_{50} (rat) = 2660 mg/kg; oral

LD_{LO} (human) = 214 mg/kg; oral LD_{LO} (infant) = 1700 mg/kg. [*3*]
THR = MOD via oral route. Reacts violently with Zr. An herbicide. See boron compounds.

BORAZINE. See borazole.

BORAZOLE. Colorless liquid. $B_3N_3H_6$, mw: 80.53, mp: −58°, bp: 53°, d: 0.824 @ 0°.
THR = HIGH irr to skin, eyes and mu mem. See also boron compounds.
Fire Hazard: Dangerous, by chemical reaction to produce flam or even spont flam gases. Hydrolyzes in water to evolve boron hydrides.
Explosion Hazard: Details U. See also boron hydrides.
Disaster Hazard: Dangerous; when heated to decomp, emits toxic fumes; upon reaction with water, can evolve toxic and flam gases.

BORDEAUX, ARSENITES, LIQUID. See arsenic and copper compounds.

BORDEAUX, ARSENITES, SOLID. See arsenic and copper compounds.

BORDEAUX MIXTURE. See copper compounds.

BORIC ACETIC ANHYDRIDE. See boron acetate.

BORIC ACID. Syns: *boracic acid*, *o-boric acid*. White crystals or powder. H_3BO_3, mw: 61.84, mp: 185° (decomp), − $1\frac{1}{2}H_2O$ @ 300°, d: 1.435 @ 15°.
THR = An exper neo. [*3*] HIGH via oral; MOD via iv routes. Affects skin, and CNS. See also boron compounds. Can explode with K, $(CH_3CO)_2O$. [*19*]

BORIC ACID OINTMENT. See boric acid.

BORINE CARBONYL. Colorless, unstable gas. BH_3CO, mw: 41.85, mp: −137.0°, bp: −64°.
THR = See CO, boron compounds and boron hydrides.
Fire Hazard: Dangerous, when exposed to heat or flame or by chemical reaction. See also boron hydrides and CO, which are readily evolved by this material upon contact with heat or moisture.
Explosion Hazard: See boron hydrides and CO.
Disaster Hazard: Dangerous; when heated to decomp, emits highly toxic fumes and will react with water or steam to produce toxic and flam vapors.

BORINOAMINOBORINE. Colorless liquid. B_2H_7N, mw: 42.70, mp: −66.5°, bp: 76.2°, vap. d: 14.7.
THR = See boron compounds.
Fire Hazard: It may evolve boron hydrides upon reaction with water or acids.
Disaster Hazard: Dangerous; when heated to decomp it can emit toxic and flam decomp products.

BORNEO CAMPHOR. See *d*-borneol.

***d*-BORNEOL.** Syns: *α-camphanol*, *borneo camphor*. Hexagonal crystals, peppery odor and burning taste. $C_{10}H_{17}OH$, mw: 154.24, mp: 208°, bp: 212°, flash p: 150°F, d: 1.01 @ 20°/4°, vap. d: 5.31.
Acute tox data: Oral LD_{LO} (rabbit) = 2000 mg/kg. [*3*]
THR = MOD via oral route and MILD irr. Can cause nausea, vomiting, mental disturbances and convulsions.
Fire Hazard: Mod, when exposed to heat or flame; can react with oxidizing material.
To Fight Fire: Water, CO_2, water spray, dry chemical.

BORNEOL ACETATE. See bornyl acetate.

BORNEOL THIOCYANOACETATE. See bornyl thiocyanoacetate.

BORNYL ACETATE. Syn: *borneol acetate*. Crystals, very slightly sol in water, sol in alcohol and ether. $C_{10}H_{17} \cdot OCOCH_3$, mw: 196.28, mp: 29°, bp: 225°–226°.
THR = U. See borneol.

BORNYL THIOCYANOACETATE. Syn: *borneol thiocyanoacetate*. Crystals. $C_{10}H_{17}OOCCH_2CNS$, mw: 253.4.
THR = U. An insecticide. See also isobornyl thiocyanoacetate.
Disaster Hazard: Dangerous; when heated to decomp or on contact with acid or acid fumes, emits highly toxic fumes of cyanide.

BOROBUTANE. See dihydrotetraborane.

BOROCAINE. Syn: *procain borate*. Crystals. $C_{13}H_{20}N_2O_2 \cdot 5HBO_2$, mw: 455.5, mp: 165°–166°.
THR = MOD via irr to skin, eyes and mu mem. Can cause an eczematoid dermatitis upon contact with skin.
Fire Hazard: Slight, when heated.

BOROETHANE. See diborane.

BORON. Monoclinic crystals, yellow or brown amorphous powder. B, atwt: 10.82, mp: 2300°, bp: 2550°, d: 3.33 @ 20°.
Acute tox data: Oral LD_{50} (mouse) = 2000 mg/kg. [*3*]
THR = MOD via oral route. See boron compounds.
Fire Hazard: Mod, in the form of dust when exposed to air or by chemical reaction with NH_3, Br_2, BrF_3, Cs_2C_2, Cl_2, CuO, F_2, HIO_3, PbO_2, HNO_3, NO, NOF, N_2O, $KClO_3$, KNO_3, Rb_2C_2, AgF, S. [*19*]
Explosion Hazard: An explosion hazard in the form of dust, which ignites on contact with air. See also powdered metals.

BORON ACETATE. Syn: *boric acetic anhydride*. White to cream, hygroscopic crystals. $(CH_3CO_2)_3B$, mw: 188.0, mp: 135°–144°.
THR = See boron compounds and acetic acid.

BORON ARSENOTRIBROMIDE. BAs_2Br_3, mw: 400.4.

THR = No data. See boron, arsenic and bromides. Reacts strongly with air or O_2. [*19*]

BORON BROMIDE. See boron tribromide.

BORON BROMIDE DIIODIDE. Colorless liquid. $BBrI_2$, mw: 344.58, bp: 180°.

THR = No data. See bromides and iodides. Reacts violently with water.

Disaster Hazard: Dangerous; see bromides and iodides.

BORON BROMIDE PENTAHYDRIDE. Colorless gas. B_2H_5Br, mw: 106.60, mp: −104°, bp: 10°.

THR = U. This material can evolve a boron hydride.

Disaster Hazard: Dangerous; when heated to decomp, emits highly toxic fumes of boron hydrides; will react with water or steam to produce toxic and corrosive fumes; can react vigorously on contact with oxidizing materials.

BORON CARBIDE. Black crystals. B_4C, mw: 55.29, mp: 2450°, bp: > 3500°, d: 2.50.

THR = See nuisance dusts.

BORON CHLORIDE. See boron trichloride.

BORON CHLORIDE PENTAHYDRIDE. Colorless gas, highly unstable. B_2H_5Cl, mw: 62.4, bp: −78° @ 18 mm.

THR = See boron compounds and hydrochloric acid.

Fire Hazard: See boron hydrides.

Explosion Hazard: See boron hydrides.

Disaster Hazard: Dangerous; when heated to decomp, emits highly toxic fumes of boron hydrides and hydrochloric acid; will react with water or steam to produce toxic and corrosive fumes; can react vigorously with oxidizing material.

BORON COMPOUNDS.

THR = HIGH and therefore considered an industrial poison. Used in medicine as sodium borate, boric acid or borax, which is a common cleaner. Fatal poisoning of children has been caused in some instances by the accidental substitution of boric acid for powdered milk. The medical literature reveals many instances of accidental poisoning due to boric acid, oral ingestion of borates or boric acid, and presumably absorption of boric acid from wounds and burns. The fatal dose of orally ingested boric acid for an adult is somewhat more than 15 or 20 g and for an infant from 5 to 6 g; i.e. LD (man) = 200 mg/kg.

Boron is one of a group of elements, such as Pb, Mn, As, which affects the CNS. Boron poisoning causes depression of the circulation, persistent vomiting and diarrhea, followed by profound shock and coma. The temp becomes sub-normal and a scarletina-form rash may cover the entire body. Boric acid intoxication can come about from absorbing toxic quantities from ointments applied to burned areas or to wounds involving loss or damage to such areas of skin, but it is not absorbed from intact skin. When a 5% boric acid solution is used to irrigate body cavities most of the boric acid is absorbed by the tissues. Continuous irrigation of the body cavities with solutions containing B can be dangerous.

Treatment and Antidotes: Large intravenous doses of isotonic salt solution and plasma have been shown to act as an antidote. Care should be observed in applying ointments and dressings which contain boron over large areas of the body where the skin has been destroyed. It can be absorbed by the body in this way with the toxic effects noted above. Containers of boric acid should be plainly labeled and should differ radically from those which contain powdered milk, particularly in institutions such as hospitals. The careless use of borax as a skin cleaner should be discouraged as well as the continuous irrigation of body cavities with solutions containing boron.

BORON DECAHYDRIDE. See boron hydrides.

BORON DIBROMIDE IODIDE. Colorless liquid. BBr_2I, mw: 297.57, bp: 125°, vap. d: 10.3.

THR = See boron compounds, bromides and iodides.

Disaster Hazard: Dangerous; see bromides and iodides; will react with water or steam to produce toxic and corrosive fumes.

BORON FLUORIDE. See boron trifluoride.

BORON FLUORIDE ETHERATE. See boron trifluoride etherate.

BORON HEXAHYDRIDE. See boron hydrides.

BORON HYDRIDES. Syn: *diborane; dihydrotetra borane; pentaborane (stable); pentaborane (unstable); hexaborane; decaborane.*

THR = HIGH irr to skin, eyes and mu mem; comparable with chlorine, fluorine, arsine and phosgene. The liquids cause local inflam, blisters, redness and swelling. HIGH via dermal route. [*20*] Experimental studies in which dogs were exposed to diborane by inhalation resulted in severe irr of the lungs and pulmonary edema. Injuries to CNS, liver and kidneys have also been produced in experimental animals. Similar observations have been reported in humans resulting at times in a reaction resembling metal fume fever. Human exposure to pentaborane has produced signs of severe central nervous system irritation such as drowsiness, dizziness, visual disturbances, muscle twitching and in severe cases painful muscle spasm.

Fire Hazard: Dangerous, when exposed to heat or flame or by chemical reaction. Pentaborane (stable) is spont flam in air. On contact with moisture, hydrogen is usually evolved (Section 7).

Explosion Hazard: When exposed to heat, flame, air or HNO_3, O_2. Diborane reacts explosively with Cl_2. Other boron hydrides evolve H_2 upon contact with moisture or can propagate a flame rapidly enough to cause an explosion.

Disaster Hazard: Severe. Heat can cause these materials to decomp violently or at least to evolve H_2; these materials react with water or steam to evolve hydrogen. Powerful oxidizing agents such as chlorine gas, etc., can react violently with boron hydrides.

BORON METHYL. See boron trimethyl.

BORON MONOXIDE. See boron compounds.

BORON NITRIDE. Hexagonal, white crystals. BN, mw: 24.83, mp: 3000° (sublimes), d: 2.20.

THR = U. This material is nearly insol and therefore possibly a nuisance dust (see also boron compounds). Added to molten NaO_2, it incandesces. [*19*]

BORON OXIDE. Vitreous, colorless crystals. B_2O_3, mw: 69.64, mp: 450° (approx), bp: 1860°, d: 2.46.

THR = Animal exper suggest relatively LOW. An herbicide. See boron compounds. Mixed with CaO and put into fused $CaCl_2$, the mixture incandesces. [*19*]

BORON PENTASULFIDE. White crystals. B_2S_5, mw: 181.97, mp: 390°, d: 1.85.

THR = See boron compounds and sulfides.

Fire Hazard: See sulfides.

Disaster Hazard: See sulfides.

BORON PHOSPHATE. A white powder. BPO_4, mw: 105.8, mp: > 1200°, d: 2.52.

THR = See boron compounds.

BORON PHOSPHIDE. Maroon powder. BP, mw: 41.80, mp: 200°.

THR = See boron compounds and phosphides.

Fire Hazard: Ignites @ 200°. See also phosphides. Reacts violently with HNO_3. Deflagrates with fused alkali nitrates. [*19*]

Explosion Hazard: See phosphides.

Disaster Hazard: See phosphides.

BORON PHOSPHODIIODIDE. BPI_2, mw: 295.6.

THR = No data. Reacts violently with Cl_2, Mg, Hg. [*19*]

BORON TRIAZIDE. $B(N_3)_3$, mw: 137.5.

THR = No data. Very unstable.

BORON TRIBROMIDE. Syn: *boron bromide*. Colorless, fuming liquid. BBr_3, mw: 250.57, mp: −45°, bp: 91.7°, d: 2.650 @ 0°, vap. press: 40 mm @ 14.0°, 100 mm @ 33.5°.

THR = An irr to skin, eyes and mu mem. See boron compounds and hydrobromic acid.

Disaster Hazard: Dangerous; when heated to decomp, can explode; emits toxic fumes of bromides and will react with water or steam to produce toxic and corrosive fumes and possibly explode. Reacts strongly with K, Na.

BORON TRI-*n*-BUTYL. See tributylborane.

BORON TRICHLORIDE. Syn: *boron chloride*. Colorless, fuming liquid. Pungent, irr odor. BCl_3, mw: 117.19, mp: −107°, bp: 12.5°, d: 1.434 @ 0°, vap. press: 1 atm @ 12.7°, vap. d: 4.03.

THR = HIGH. An irr to skin, eyes and mu mem. See boron compounds and hydrochloric acid.

Disaster Hazard: Dangerous; when heated to decomp, emits toxic fumes of chlorides; will react with water or steam to produce heat, toxic and corrosive fumes. Reacts strongly with NO_2, PH_3, grease, organic matter, O_2. [*19*]

BORON TRICYCLOHEXYL. See tricyclohexyl borine.

BORON TRIETHYL. Syns: *boron ethyl, triethyl borine, triethyl boron*. Colorless liquid. $B(C_2H_5)_3$, mw: 98.0, mp: −93°, d: 0.6961 @ 23°.

Acute tox data: Oral LD_{50} (rat) = 235 mg/kg; inhal LC_{50} (rat) = 2800 mg/m^3. [*3*]

THR = HIGH via oral, MOD via inhal routes. Animal exper show HIGH vapor and MOD leading to pulmonary irr and convulsions.

Fire Hazard: Dangerous; by spont chemical reaction with oxidizers. Spont flam in air.

Explosion Hazard: Reacts violently with (triethyl aluminum + air). [*19*]

Disaster Hazard: Highly dangerous; when heated to decomp or upon contact with air, emits toxic fumes; will react with water or steam to produce toxic and flam vapors; can react vigorously with oxidizing materials.

To Fight Fire: Do not use halogenated extinguishing agents.

BORON TRIFLUORIDE. Syn: *boron fluoride*. Colorless gas. Pungent, irr odor. BF_3, mw: 67.82, mp: −126.8°, bp: −99.9°, d: 2.99 g/liter.

Acute tox data: Inhal LC_{50} (mouse) = 3460 mg/kg for 2 hrs; inhal LC_{50} (guinea pig) = 109 mg/m^3 for 4 hrs.

THR = HIGH to MOD via inhal route. [*49*] A strong irr; reacts vigorously with alkali metals, alkaline earth metals, CaO. [*19*]

Disaster Hazard: Dangerous; when heated to decomp or upon contact with water or steam, will produce toxic and corrosive fumes.

BORON TRIFLUORIDE-ACETIC ACID COMPLEX. See boron, acetic acid and fluorides.

BORON TRIFLUORIDE ETHERATE. Comp: Boron trifluoride, ether. $CH_3CH_2O(BF_3)CH_2CH_3$, mw: 142, d: 1.1, bp: 126°, flash p: 147°F (OC).
THR = See boron compounds, fluorides and ether.
Fire Hazard: Mod; see ether. Reacts violently with $LiAlH_4$. [*19*]
Explosion Hazard: See ether.
Disaster Hazard: Dangerous; see fluorides; will react with water or steam to produce toxic, corrosive and flam vapors; can react vigorously with oxidizing materials.
To Fight Fire: Dry chemical, CO_2, fog or mist.

BORON TRIFLUORIDE MONOETHYLAMINE. $C_2H_7NBF_3$, mw: 122.9.
Acute tox data: Ip LD_{LO} (mouse) = 42 mg/kg. [*3*]
THR = HIGH via ip route.

BORON TRIIODIDE. Colorless, hygroscopic plates. BI_3, mw: 391.58, mp: 43°, bp: 210°, d: 3.35 @ 50°.
THR = See boron compounds and iodides. Reacts vigorously with NH_3, ethers, carbohydrates, P, POCl. [*19*]
Disaster Hazard: Dangerous, see iodides.

BORON TRIMETHYL. Syns: *boron methyl, trimethyl borine*. Colorless gas. $B(CH_3)_3$, mw: 55.9, mp: −161.5°, bp: −20°, d: 1.91 g/liter (gas), 0.625 (solid @ −100°).
THR = HIGH. See boron compounds.
Fire Hazard: Dangerous, when exposed to flame or by chemical reaction with oxidizing agents. Spont flam gas.
Explosion Hazard: U.
Disaster Hazard: Highly dangerous; can react vigorously with oxidizing materials.

BORON TRIOXIDE. See boron oxide.

BORON TRIPHENYL. Syn: *triphenyl boron*. Colorless crystals decomp by water. $(C_6H_5)_3B$, mw: 242.1, mp: 136°, bp: 203° @ 15 mm.
THR = U. See phenol and boron compounds.

BORON TRIPHENYL AMMINE. Colorless crystals sol in alcohol. $(C_6H_5)_3BNH_3$, mw: 259.2, mp: 216° (decomp).
THR = U. See phenols and boron compounds.

BORON TRIPROPYL. See tripropyl borane.

BORON TRISELENIDE. Yellow gray powder. B_2Se_3, mw: 258.52.
THR = See boron compounds and selenium compounds.
Disaster Hazard: See selenium compounds.

BORON TRISULFIDE. White crystals. B_2S_3, mw: 117.84, mp: 310°, d: 1.55.
THR = See boron compounds and sulfides. Very strong reaction with Cl_2.

BOROPHOSPHORIC ACID. White crystals. $BPO_4 \cdot H_2O$, mw: 123.8, mp: decomp, d: 1.873.
THR = See boron compounds and phosphoric acid.

BOROTUNGSTIC ACID. Tetragonal, colorless crystals. $H_5BW_{12}O_{40} \cdot 3OH_2O$, mw: 3403.39, mp: 45°–51°, d: 3.0.
THR = See boron compounds and tungsten compounds.

BOTTLE GAS. See liquefied hydrocarbon gas.

BOTULINUM TOXIN A.
Acute tox data: Ip LD_{50} (mouse) = 0.00016 mg/kg. [*3*]
THR = VERY, VERY HIGH via ip route. Probably the most toxic material known. Not absorbed via the skin of man. Extremely HIGH via oral route.

BOVOLIDE. See 3,4-dimethyl-5-pentylidene-2(5H) furanone.

BOX TOE GUM. Liquid.
THR = U.
Fire Hazard: Dangerous, when exposed to heat or flame; can react with oxidizing materials.

BOXWOOD. See sawdust.

BRACKEN FERN.
THR = HIGH oral to cattle. An exper neo. [*23, 3*] See also plant and fungal products.

BRADYKIN.
THR = An exper teratogen. [*3*]

BRASS(MASSIVE). Pale gold to red metal alloy. Comp: Zn + Cu (mainly), Pb (small amounts), As, Sb, P, Sn, CN^- (traces). mp: 940°–1030°, d: 8.40–8.75.
THR = See brass scrapings.

BRASS POWDER. See brass scrapings.

BRASS SCRAPINGS. Syn: *brass powder*. Finely divided particles of metal. Comp: Zn, Cu_x (mainly), Pb (minor constituent), As, P, Sn, Sb and CN^- (traces).
THR = See specific element, particularly Zn, Cu, Pb. Where quantities of brass scrapings can come in contact with acid or acid fumes, significant quantities of toxic fumes can be evolved. See As, Sb, P, Sn and CN^-.
Fire Hazard: Mod, when exposed to heat or flame. See also powdered metals. Forms an explosive with

C_2H_2. Reacts vigorously with Cl_2, H_2O_2, $Pb(N_3)_2$. [*19*]
Explosion Hazard: Slight, when exposed to flame. See also powdered metals.
Disaster Hazard: Mod dangerous; can react with oxidizing materials, and on contact with acid or acid fumes, can emit toxic fumes.

BRAZIL NUT.
THR = These nuts have been found to contain a concentration of about 1000 times as much ^{226}Ra and ^{222}Th and their associated daughter products as usual foods. These nuts are also relatively HIGH in Ba content. A MILD allergen.
Fire Hazard: Slight, when exposed to heat or flame.

BRAZIL WAX. See carnauba wax.

BRAZIL WOOD. Syns: *redwood*, *pernambuco*, *fernambuco*. See sawdust.

BRAZING AND SOLDERING VENTILATION is discussed in Section 2.

BREITHAUPTITE. See nickel antimonide.

BRESTAN. Syn: *triphenyltin acetate*. Practically insol, crystalline solid. $C_{20}H_{18}O_2Sn$, mw: 409.0; mp: 120°.
Acute tox data: Dermal LD_{50} (rat) = 450 mg/kg; oral LD_{50} (rat) = 125 mg/kg; ip LD_{50} (rat) = 8.5 mg/kg; iv LD_{50} (rat) = 18 mg/kg; dermal LD_{LO} (guinea pig) = 180 mg/kg. [*3*]
THR = HIGH via dermal, oral, ip and iv routes. An exper carc via oral route. [*3*] A fungicide and algicide.

BRILLIANT BLUE, FCF. See FD and C blue #1.

BRILLIANT CRESYL BLUE BB. A dye. $C_{15}H_{16}ON_3Cl$, mw: 289.8.
Acute tox data: Iv LD_{50} (mice) = 235 mg/kg. [*3*]
THR = HIGH via iv route. An allergen.
Disaster Hazard: Mod dangerous; when heated to decomp, emits toxic fumes.

BRILLIANT FAST YELLOW. See N,N-dimethyl-*p*-phenylazo aniline.

BRIMSTONE. See sulfur.

BRITISH ANTI-LEWISITE. See 2,3-dimercapto-1-propanol.

BROMACETIC ACID. Syn: *bromoacetic acid*. Hygroscopic crystals, very sol in water or alcohol. $CH_2BrCOOH$, mw: 158.96, d: 1.93, mp: 50°, bp: 208°.
Acute tox data: Oral LD_{50} (mouse) = 100 mg/kg. [*3*]
THR = HIGH via oral and inhal routes. A strong irr and corrosive.
Disaster Hazard: Dangerous; see bromides.

BROMACETIC ACID, SOLUTION. See bromacetic acid.

BROMACETYLENE. See bromoacetylene.

BROMACIL. Syn: *5-bromo-3-sec-butyl-6-methyluracil*. Slightly water sol crystal. $C_9H_{13}BrN_2O_2$, mw: 261.
Acute tox data: Oral LD_{50} (rat) = 5200 mg/kg.
THR = MOD via oral route. An herbicide.
Disaster Hazard: Dangerous. When heated to decomp, emits highly toxic fumes.

BROMATES.
THR = Generally considered to be more toxic than chlorates, causing CNS paralysis. They may form methemoglobin, but less actively than chlorates. See also specific compounds as listed.
Fire Hazard: Mod in the form of gas, vapor or dust by chemical reaction with (powdered metals + acids), Al, As, CaH_2, C, Cu, powdered metals, metal sulfides, organic matter, PH_4I, P, SrH, S, (H_2SO_4 + metals). [*19*]
Disaster Hazard: Dangerous; when heated to decomp, they emit toxic fumes of bromine; they can react with reducing material.

***p*-BROMOBENZYL CHLORIDE.** Syns: *p-bromo-α-chlorotoluene*, *α-chloro-4-bromotoluene*. Needles, freely sol in hot alcohol. $C_6H_4BrCH_2Cl$, mw: 206, mp: 40°–41°, bp: 105°–115° @ 12 mm.
THR = No data.

BROMELLITE. See beryllium oxide.

BROMIC ACID, SODIUM SALT. Colorless or slightly yellow liquid, turns yellow on exposure. $NaBrO_3$, mw: 150.9.
Acute tox data: Ip LD_{50} = 140 mg/kg; oral LD_{LO} (rabbit) = 250 mg/kg. [*3*]
THR = HIGH via ip and oral routes.
Fire Hazard: See bromates. Reacts violently with phosphonium iodide.
Disaster Hazard: See bromates.

BROMIC ETHER. See ethyl bromide.

BROMIDES.
THR = The most common inorganic bromides are Na, K, NH_4, Ca and Mg bromides. Methyl and ethyl bromides are among the most common organic bromides. The inorganic bromides produce depression, emaciation and, in severe cases, psychoses and mental deterioration. Bromide rashes (bromoderma), especially of the face, and resembling acne and furunculosis, often occur when bromide inhalation or administration is prolonged. Organic bromides such as methyl bromide and ethyl bromide are volatile liquids of relatively high toxicity. See also specific compounds.
Disaster Hazard: When strongly heated, they emit highly toxic fumes.

BROMINE. Rhombic crystals or dark red liquid. Br_2, mw: 159.83, fp: −7.3°, bp: 58.73°, d: 2.928 @ 59°, 3.12 @ 20°, vap. press: 175 mm @ 21°, 1 atm @ 58.2°, vap. d: 5.5.

Acute tox data: Oral LD_{LO} (human) = 14 mg/kg; inhal LC_{LO} (human) = 1000 ppm; inhal LC_{LO} (rabbit) = 180 ppm for 7 hrs. [*3*]

THR = HIGH via oral and inhal routes. The action of bromine is essentially the same as that of chlorine, being an irr to the mu mem of the eyes and upper respiratory tract. Severe exposures may result in pulmonary edema. Usually, however, the irr qualities of the chemical force the workman to leave the exposure before serious poisoning can result. Chronic exposure is similar to therapeutic ingestion of excessive bromides. See also bromides. Regular physical examinations should be made upon people who work with bromine or bromides.

Radiation Hazard: For permissible levels see Table 5A.5.

Fire Hazard: Mod, in the form of liquid or vapor by spont chemical reaction with reducing materials. May react violently with acetaldehyde, C_2H_2, acrylonitrile, Al, NH_3, Sb, B, Ca_3N_2, Cs_2O, Cs_2C_2, CsC_2H, ClF_3C_2, CuH_2, Cu_2C_2, dimethyl formamide, ethyl phosphine, F_2, Ge, H_2, Fe_2C, isobutyrophenone, Li, Li_2C_2, Li_2Si_2, Mg_3P_2, CH_3OH, $Ni(CO)_4$, NI_3, olefins, OF_2, O_3, PH_3, P, PO_x, K, Rb_2C_2, RbC_2H, AgN_3, Na, Na_2C_2, NaC_2H, Sr_3P, Sn, UC_2, ZrC_2. [*19*] A very powerful oxidizer.

Disaster Hazard: Highly dangerous; when heated, emits highly toxic fumes; will react with water or steam to produce toxic and corrosive fumes; can react vigorously with reducing materials.

BROMINE ANALOG OF DDT. See 1-trichloro-2,2-bis(*p*-bromophenyl)ethane.

BROMINE AZIDE. Syn: *bromoazide.* Crystals or red liquid. BrN_3, mw: 121.94, mp: 45°, bp: explodes.

THR = HIGH. See bromine and azides.

Fire Hazard: Mod, in the form of vapor by chemical reaction. A powerful oxidant. See also bromine. Can explode spont. Reacts vigorously with Sb, As, ethyl ether, P, Ag, Na. [*19*]

Explosion Hazard: Mod, when exposed to heat.

Disaster Hazard: Dangerous; when heated to decomp, emits highly toxic fumes of bromine and explodes; will react with water or steam to produce toxic and corrosive fumes; can react on contact with reducing materials.

BROMINE CHLORIDE. Reddish-yellow liquid or gas. BrCl, mw: 115.37, bp: decomp @ 10°.

THR = See bromine, chlorine.

Fire Hazard: Mod, in the form of vapor by spont chemical reaction. A powerful oxidant. See also bromine.

Disaster Hazard: Dangerous; when heated to decomp, emits highly toxic fumes of bromine and chlorine; will react with water or steam to produce toxic and corrosive fumes; on contact with reducing material, it can react vigorously.

BROMINE CYANIDE. See cyanogen bromide.

BROMINE DIOXIDE. Light yellow crystals. BrO_2, mw: 111.92, mp: 0° (decomp).

THR = See bromine.

Fire Hazard: Mod, in the form of vapor by chemical reaction with reducing agents. A strong oxidant.

Disaster Hazard: Dangerous; when heated to decomp, emits highly toxic fumes of bromine; will react with water or steam to produce toxic and corrosive fumes; can react vigorously with reducing materials.

BROMINE HYDRATE. Red crystals. $Br_2 \cdot 10H_2O$, mw: 340.0, mp: 6.8° (decomp).

THR = See bromine.

Fire Hazard: Mod, in the form of vapor by chemical reaction with reducing materials. A strong oxidant.

Disaster Hazard: Dangerous; see bromine, on contact with water or steam, will react to produce toxic and corrosive fumes; can react vigorously with reducing materials.

BROMINE MONOFLUORIDE. BrF, mw: 98.9.

THR = HIGH. Powerful irr. See bromine, fluorine. Reacts violently with organic matter, water. [*19*]

BROMINE MONOXIDE. Dark brown crystals. Br_2O, mw: 175.83, mp: −18° (decomp).

THR = See bromine.

Fire Hazard: Mod, by chemical reaction with reducing agents. A strong oxidant.

Disaster Hazard: Dangerous; see bromine; will react with water or steam to produce toxic and corrosive fumes; can react vigorously with reducing materials.

BROMINE OCTOXIDE. White crystals. Br_3O_8, mw: 367.75, mp: stable @ −40°.

THR = See bromine.

Fire Hazard: Mod, upon contact with reducing agents. A strong oxidant.

Disaster Hazard: Dangerous; see bromine; will react with water or steam to produce toxic and corrosive fumes; can react vigorously with reducing materials.

BROMINE PENTA FLUORIDE. Colorless fuming liquid. BrF_5, mw: 174.92, mp: −61.3, bp: 40.5, d: 2.466 @ 25°, vap. d: 6.05.

THR = HIGH and corrosive. See bromine and fluorides. Reacts violently with acetic acid, NH_3, As, C_6H_6, H_2S, cellulose, charcoal, C_2H_5OH, I, alkaline halides, metallic halides, metal oxides, metals, CH_4, HNO_3, organic matter, Se, S, H_2SO_4, H_2O. [*19*]

Disaster Hazard: Dangerous; see bromine and fluorine; will react with water or steam to produce toxic and corrosive fumes. Liquefied gas reacts violently with many organic compounds and some inorganic compounds. It is a powerful oxidizer.

BROMINE TRIFLUORIDE. Colorless, fuming liquid. BrF_3, mw: 136.92, mp: 8.8°, bp: 127°, d: 2.84.

THR = HIGH and corrosive. See bromine pentafluoride.

Disaster Hazard: Very dangerous; see bromides and fluorides. Very reactive; a powerful oxidizer which reacts violently with NH_4Br, NH_4Cl, NH_4I, Sb, SbOCl, Sb_2O_3, As, $BaCl_2$, Bi_2O_5, B, Br, $CdCl_2$, $CaCl_2$, Co, CCl_4, CI_4, CsCl, I, LiCl, $MnIO_3$, metals, Mo, Nb, Nb_2O_5, organic matter, P, $PtBr_4$, $PtCl_4$, (Pt + KFO), KBr, KCl, KI, $RhBr_4$, RbCl, AgCl, NaBr, NaCl, NaI, $SnCl_2$, S, Ta, Ta_2O_5, Sn, Ti, W, UO_x, V, H_2O. [*19*]

N-BROMOACETAMIDE. Syn: *NBA*. White powder, odor similar to bromine. $CH_3CONHBr$, mw: 138.0, mp: 105°–108°.

THR = U.

Disaster Hazard: Dangerous; see bromine.

BROMOACETIC ACID. See bromacetic acid.

BROMOACETONE. Colorless liquid when pure, pungent odor, slightly sol in water, sol in alcohol and acetone. $CH_2BrCOCH_3$, mw: 136.99, mp: −54°, bp: 136°, d: 1.631 @ 0°.

THR = Intensely irr. Powerful lacrimator. A chemical warfare agent.

Disaster Hazard: Dangerous; see bromides.

BROMOACETOPHENONE. Syn: *phenacyl bromide*. White rhombic prisms changing to greenish color under the influence of light. $BrCH_2COC_6H_5$, mw: 199.05, mp: 50°, bp: 140° @ 12 mm, d: 1.647 @ 20°/4°.

THR = U. See bromides and phenyl methyl ketone.

Disaster Hazard: Dangerous; see bromides.

BROMOACETYLENE. Syns: *bromoethyne, bromacetylene*. Gas. CHCBr, mw: 104.9, bp: −2°, vap. d: 4.684.

THR = U. Probably similar to dibromoacetylene.

Fire Hazard: Dangerous by spont chemical reaction. A spont flam gas.

Explosion Hazard: High on contact with air even at very low temperatures, or upon distillation.

Disaster Hazard: Dangerous; when heated to decomp, burns and emits toxic fumes; can react with oxidizing materials.

BROMOANILINE. Syn: *o-bromoaniline*. Crystals. $BrC_6H_4NH_2$, mw: 172.0, mp: 32°, bp: 229°.

THR = U. See also aniline.

Disaster Hazard: Dangerous; when heated to decomp or on contact with acid or acid fumes, emits toxic fumes.

BROMOAURIC ACID. Red-brown crystals. $HAuBr_4 \cdot 5H_2O$, mw: 607.95, mp: 27°.

THR = See bromides and gold compounds.

Disaster Hazard: Dangerous; see bromine.

BROMOAZIDE. See bromine azide.

***p*-BROMOAZOBENZENE.** $BrC_6H_4NNC_6H_5$, mw: 261.1.

THR = Details U. An insecticide. See also bromobenzene.

Disaster Hazard: See bromides.

7-BROMOBENZ(a)ANTHRACENE. $C_{18}H_{11}Br$, mw: 307.2.

THR = An exper carc. [*3, 23*]

BROMOBENZENE. Syn: *phenyl bromide*. Colorless, clear, mobile liquid. C_6H_5Br, mw: 157.02, mp: −30.7, bp: 156.2°, flash p: 124°F, d: 1.497, vap. press: 10 mm @ 40°, vap. d: 5.41, autoign. temp.: 1051°F.

THR = MOD via oral and inhal routes. An irr to eyes and mu mem.

Fire Hazard: Mod, when exposed to heat or flame.

Disaster Hazard: Dangerous; see bromides; can react with oxidizing materials.

To Fight Fire: Water to blanket fire, foam, CO_2, water spray or mist, dry chemical.

***p*-BROMOBENZENE TETRATHIO-*o*-STANNATE.** Solid. $Sn(SC_6H_4Br)_4$, mw: 871.0, mp: 217°.

THR = Details U. See tin compounds.

Disaster Hazard: Dangerous. SO_x and bromine fumes.

***p*-BROMOBENZONITRILE.** Crystals. BrC_6H_4CN, mw: 182, mp: 113°.

THR = An insecticide. See also nitriles.

Disaster Hazard: See nitriles.

6-BROMOBENZO(a)PYRENE. $C_{20}H_{11}Br$, mw: 331.2.

THR = An exper neo. [*3*]

***p*-BROMOBENZOYL ACETANILIDE.** $C_{15}H_{12}O_2NBr$, mw: 318.2.

THR = No data. In (dimethyl sulfoxide + 100° heat) on steam bath, it explodes. [*19*]

***p*-BROMOBENZYL BROMIDE.** Syn: *p, α-dibromotoluene*. Crystals, aromatic odor, sol in water and cold alcohol, more sol in hot alcohol, ether, carbon disulfide, benzene and glacial acetic acid. $C_7H_6Br_2$, mw: 249.5, mp: 61°, bp: 115°–124° @ 12 mm.

THR = See benzyl bromide.

BROMOBENZYL CYANIDE. Syns: *o-bromo-2-phenylacetonitrile, BBC, bromobenzyl nitride*. Pure:

yellowish-white crystals; tech: brown, oily liquid, pungent odor of soured fruit. $C_6H_4CH_2CNBr$, mw: 196.1, mp: 29°, bp: 242°, fp: 25.5°, d: 1.5160 @ 20°, flash p: none, vap. d: 6.8, vap. press: 0.011 mm @ 20°.
Acute tox data: Inhal LC_{50} (human) = 3500 mg/m^3, oral LD_{LO} (rat) = 100 mg/kg. [*3*]
THR = HIGH via inhal and oral routes. A strong lachrymator. See cyanides.
Disaster Hazard: See cyanides.

***o*-BROMOBENZYL CYANIDE.** See bromobenzyl cyanide.

BROMOBENZYL NITRILE. See bromobenzyl cyanide.

BROMO BENZYL TRIFLUORIDE. $BrC_6H_4CF_3$, mw: 225.
THR = Probably HIGH. Reacts violently with Mg. [*19*] See also fluorides.

α-BROMO-β,β-BIS(*p*-ETHOXY PHENYL)STYRENE. $C_{24}H_{23}O_2Br$, mw: 423.4.
THR = An exper carc. [*3*]

4-BROMO-7-BROMOETHYL BENZ(a)ANTHRACENE.
THR = An exper carc. [*23*]

4-BROMO-7-BROMOMETHYL BENZ(a)ANTHRACENE. $C_{19}H_{12}Br_2$, mw: 400.1.
THR = An exper neo. [*3, 23*]

BROMOBUTANE. See butyl bromide.

α-BROMOBUTYRIC ACID. Colorless, oily liquid, sol in alcohol and ether, sparingly sol in water. $CH_3CH_2CHBrCOOH$, mw: 167, d: 1.54, bp: 181° @ 250 mm, mp: −4°.
Acute tox data: Oral LD_{50} (mouse) = 310 mg/kg. [*3*]
THR = HIGH via oral route.
Disaster Hazard: Dangerous. See bromides.

5-BROMO-3-*sec*-BUTYL-6-METHYL URACIL. See bromocil.

2-BROMO-4-*tert*-BUTYL PHENOL. Straw colored liquid. $(CH_3)_3CC_6H_3BrOH$, mw: 229.1, bp: 104°–105° @ 5 mm, flash p: 240°F, fp: <−20°, d: 1.319 @ 25°/25°.
THR = U. See also phenol.
Fire Hazard: Mod, when exposed to heat or flame.
Disaster Hazard: Mod dangerous; when heated to decomp, emits toxic fumes; can react with oxidizing materials.
To Fight Fire: Water, foam, CO_2, dry chemical.

α-BROMO-*n*-CAPROIC ACID. Syn: *2-bromohexanoic acid.* Liquid, sol in alcohol and ether. $(CH_2)_3CHBrCOOH$, mw: 195.06, bp: 240°.
Acute tox data: Oral LD_{50} (mouse) = 590 mg/kg. [*3*]
THR = HIGH to MOD via oral route.
Disaster Hazard: Dangerous; see bromides.

1-BROMO-4-CHLOROBENZENE. White crystals. C_6H_4BrCl, mw: 191.5, mp: 67.4°, bp: 196.3°, fp: 64.5°.
THR = U.
Disaster Hazard: Dangerous; see bromine.

1-BROMO-2-CHLOROETHANE. See ethylene chlorobromide.

BROMOCHLOROMETHANE. See methylene chlorobromide.

BROMOCHLORO DIFLUOROMETHANE. Colorless gas. $BrClCF_2$, mw: 146.4.
THR = U. Must be considered toxic.

2-BROMO-2-CHLORO-1,1,1-TRIFLUOROETHANE. $C_2HBrClF_3$, mw: 197.4.
THR = An exper teratogen. [*3*]

BROMOCUMENE. See bromoisopropyl benzene.

BROMOCYCLOHEXANE. Yellowish liquid. $C_6H_{11}Br$, mw: 163.1, bp: 165.8–167.3°, flash p: 145°F, d: 1.337.
THR = U. Probably narcotic in high conc. See also cyclohexane.
Fire Hazard: Mod, when exposed to heat or flame.
Disaster Hazard: Dangerous; see bromides; can react with oxidizing materials.
To Fight Fire: Water, foam, CO_2, dry chemical.

5-BROMO-2′-DEOXYCYTIDINE. $C_9H_{12}O_4N_3Br$, mw: 306.2.
THR = An exper teratogen. [*3*]

5-BROMO-1-(2-DEOXY-β-D-RIBOFURANOSYL)-5-FLUORO-6-METHOXY HYDROURACIL. $C_9H_{14}O_6N_2FBr$, mw: 345.2.
THR = An exper teratogen. [*3*]

5-BROMO-2-DEOXY URIDINE. $C_9H_{11}O_5N_2Br$, mw: 307.1.
THR = An exper teratogen. [*3*]

BROMO DIBORANE. $BrBH(H_2)BH_2$, mw: 106.6.
THR = No data. See diborane. Burns in air.

BROMO DICHLOROMETHANE. Colorless liquid. $CHBrCl_2$, mw: 163.8, bp: 89.2°–90.6°, d: 1.971 @ 25°/25°.
THR = U. Probably narcotic in HIGH conc. See also methylene chloride.
Disaster Hazard: Dangerous; see bromides and chlorides.

BROMO DIETHYL ALUMINUM. $(C_2H_5)_2AlBr$, mw: 165.4.
THR = No data. Ignites spont in air, alcohol, water. [*19*]

2-BROMO-3,5-DIMETHYOXY ANILINE. $NH_2C_6H_2Br(OCH_3)_2$, mw: 232.

THR = No data. Probably HIGH. A very unstable compound.

8-BROMO-7,12-DIMETHYL BENZ(α)ANTHRACENE. $C_{20}H_{15}Br$, mw: 335.3.
THR = An exper carc. [*3*, *23*]

3-BROMO-5,5-DIMETHYL HYDANTOIN. White to buff powder, mild odor but like bromine. $C_5H_7N_2O_2Br$, mw: 207.03.
THR = U. See also dimethyl hydantoin.
Disaster Hazard: Dangerous; see bromine and bromides.

4-BROMO DIPHENYL. White crystals. $C_6H_5C_6H_4Br$, mw: 233.1, mp: 89.8°–90.3°, bp: 310.8°, flash p: 290°F.
THR = U.
Fire Hazard: Low, when exposed to heat or flame.
Disaster Hazard: Dangerous; when heated to decomp, emits toxic fumes; can react with oxidizing materials.
To Fight Fire: CO_2, water spray or mist, dry chemical.

BROMOETHANE. See ethyl bromide.

BROMOETHENE. See vinyl bromide.

2-BROMOETHYL ACETATE. $BrC_2H_4O_2C_2H_3$, mw: 165.
THR = See esters.
Disaster Hazard: Dangerous; see bromine and bromides.

BROMOETHYL BENZENE. Colorless to pale yellow liquid. $BrC_6H_4CH_2CH_3$, mw: 185.1, mp: −65°, bp: 201°, d: 1.35–1.4 @ 25°/25°, flash p: 205°F, vap. d: 6.4.
THR = U. See also bromobenzene.
Fire Hazard: Mod, when exposed to heat or flame.
Disaster Hazard: Dangerous; see bromides; can react with oxidizing materials.
To Fight Fire: Water, foam, CO_2, water spray or mist, dry chemical.

(2-BROMOETHYL)BUTYRYL UREA. $C_7H_{13}O_2N_2Br$, mw: 237.1.
THR = HIGH acute. An exper neo. [*3*]

BROMOETHYL CHLOROSULFONATE. Liquid. $BrCH_2CH_2OSO_2Cl$, mw: 223.5, bp: 100°–105° @ 18 mm.
THR = A very powerful tissue irr to skin and mu mem.
Disaster Hazard: Dangerous; when heated to decomp, emits highly toxic fumes.
Treatment and Antidotes: Remove victim to fresh air and if breathing has stopped, administer artificial respiration. Give oxygen if necessary. Call a physician.

BROMOETHYLENE. See vinyl bromide.

2-BROMOETHYLETHYL ETHER. Liquid. $BrCH_2CH_2OCH_2CH_3$, mw: 153, vap. d: 5.25.
THR = An insecticide. See also ethers.
Fire Hazard: See ethers.
Explosion Hazard: See ethers.
Disaster Hazard: Dangerous; see bromides.

BROMOETHYNE. See bromoacetylene.

BROMOFORM. Syn: *tribromoethane*. Colorless liquid or hexagonal crystals. $CHBr_3$, mw: 252.77, mp: 6°–7°, bp: 149.5°, flash p: none, d: 2.890 @ 20°/4°.
Acute tox data: Sc LD_{50} (mouse) = 1820 mg/kg. [*3*]
THR = HIGH via sc route. Violent reaction with Li, NaK alloy. [*19*] This material causes lachrymation. It can damage the liver to a serious degree and cause death. It has been said that its medicinal application has resulted in numerous poisonings. It has anesthetic properties similar to those of chloroform, but it is not sufficiently volatile for inhalation purposes and is far too toxic to be recommended. In addition to its narcotic effects, it is a metabolic poison. Petroleum geologists working in closed rooms with a large number of funnels or open separatory flasks, as required for routine procedure in the separation of minerals, can be subjected to appreciable conc of bromoform as an atmospheric contaminant. Inhal of small amounts of this material causes irr, provoking the flow of tears and saliva and reddening of the face. In dogs, 29,000 ppm caused a deep narcosis after 8 minutes, and death after 1 hr.; in 30 minutes there was deep narcosis and recovery on the next day.
Disaster Hazard: Dangerous, when heated to decomp, emits highly toxic fumes.
Treatment and Antidotes: Remove patient to fresh air and if breathing has stopped, administer artificial respiration. Give oxygen if necessary. Call a physician.

BROMOGERMANE. Colorless liquid. GeH_3Br, mw: 155.54, mp: −32.0°, bp: 52.0°, d: 2.34 @ 29.5°.
THR = See germanium compounds and bromides.
Fire Hazard: Mod, when exposed to heat or flame.
Disaster Hazard: Dangerous; see bromine; can react vigorously with oxidizing materials.

***p*-BROMOHYDROAZO BENZENE.**
THR = U. An insecticide.
Disaster Hazard: Dangerous; see bromides.

2-BROMOISO BUTYRIC ACID. Crystals, sparingly sol in cold water, decomp by hot water into the hydroxy acid, sol in alcohol and ether. $(CH_3)_2CBrCOOH$, mw: 167.01, d: 1.52, mp: 48°–49°, bp: 198°–200°.
THR = U. Limited animal exper suggest LOW.
Disaster Hazard: Dangerous; see bromides.

BROMOISOPROPYL BENZENE. Syn: *ar-bromo cumene*. Colorless liquid. $BrC_6H_4CH(CH_3)_2$, mw: 199.1, mp: −20°, bp: 212°–216°, flash p: 207°F, d: 1.27–1.3 @ 25°/25°, vap. d: 6.90.

THR = U. See also bromobenzene.

Fire Hazard: Mod, when exposed to heat or flame; can react with oxidizing materials.

To Fight Fire: Water, foam, CO_2, water spray or mist, dry chemical.

5-BROMO-3-ISOPROPYL-6-METHYL URACIL. Syns: *isocil, hyvar*. Crystals, sol in absolute alcohol. $OCCBrC(CH_3)NHC(O)NCH(CH)_2$, mw: 247, mp: 158°.

Acute tox data: Oral LD_{50} (rat) = 3400 mg/kg. [*3*]

THR = MOD via oral route.

BROMOISOVALERIC ACID. Syn: *2-bromo-3-methyl butyric acid*. Exists in two isomeric forms (α and β). (α) lustrous crystals, sparingly sol in water, sol in alcohol and ether; (β) needles, slightly sol in water, sol in alcohol, benzene and ether.

(α) $(CH_3)_2CHCHBrCOOH$;

(β) $(CH_3)_2CBrCH_2COOH$. Mw: 181.04 (α and β), mp: (α) 44°, (β) 73.74°, bp: (α) about 230° (slight decomp).

THR = U. Limited animal exper suggest LOW.

Disaster Hazard: Dangerous, see bromides.

BROMOL. See tribromo phenol.

2-BROMO-D-LYSERGIC ACID DIETHYLAMIDE.

THR = An exper teratogen. [*3*]

BROMOMETHANE. See methyl bromide.

9-BROMOMETHYL ANTHRACENE. Syn: *ICR 506*. $C_{15}H_{11}Br$, mw: 271.2.

THR = An exper carc to mice via iv route. [*103*]

7-BROMO METHYL BENZ(a)ACRIDINE.

THR = An exper carc. [*23*]

7-BROMO METHYL BENZ(a)ANTHRACENE. $C_{19}H_{13}Br$, mw: 321.2.

THR = An exper neo. [*3, 23*]

6-BROMO METHYL BENZO(a)PYRENE. $C_{21}H_{13}Br$, mw: 345.3.

THR = An exper carc. [*3*]

2-BROMO-3-METHYL BUTYRIC ACID. See α-bromoisovaleric acid.

7-BROMO METHYL-4-CHLOROBENZ(a)ANTHRACENE. $C_{19}H_{12}ClBr$, mw: 355.7.

THR = An exper neo. [*3*]

BROMO METHYL ETHYL KETONE. Colorless to pale yellowish liquid. $BrCH_2COC_2H_5$, mw: 135.01, bp: 145°–146° decomp, d: 1.43.

THR = HIGH irr to skin, eyes and mu mem. See also ketones.

Disaster Hazard: Dangerous; when heated to decomp, emits toxic fumes of bromides; can react with oxidizing materials.

7-BROMO METHYL-6-FLUOROBENZ(a)ANTHRACENE. $C_{19}H_{12}BrF$, mw: 339.2.

THR = An exper neo. [*3*]

7-BROMO METHYL-1-METHYL BENZ(a)ANTHRACENE. $C_{20}H_{15}Br$, mw: 335.2.

THR = An exper neo. [*3*]

7-BROMO METHYL-12-METHYL BENZ(a)ANTHRACENE. $C_{20}H_{15}Br$, mw: 335.2.

THR = An exper neo and carc. [*3, 23*]

1-BROMO-3-NITROBENZENE. Crystals. $BrC_6H_4NO_2$, mw: 202.0, mp: 56°, bp: 256.5°, d: 1.70 @ 20°/4°.

THR = Details U. See also nitrobenzene.

Disaster Hazard: Dangerous; when heated to decomp, emits highly toxic fumes of NO_x and bromides; it can react vigorously with oxidizing materials.

3-BROMO-4-NITROQUINOLINE-1-OXIDE. $C_9H_5O_3N_2Br$, mw: 269.1.

THR = An exper neo. [*3*]

1-BROMOPENTANE. See amyl bromide.

BROMO PHENOLS(*m,p,o*). (*m*) crystals, insol in water, sol in alcohol, ether and alkalis. (*p*) crystals, slightly sol in water, sol in alcohol, ether, chloroform and glacial acetic acid. (*o*) yellow to oily red liquid, unpleasant odor, insol in water, sol in alcohol, ether and chloroform. $HO(C_6H_4)Br$, mw: 173 (*m,p,o*), d: (*p*) 1.840 (15°), 1.5875 (80°), d: (*o*) 1.5, mp: (*m*) 33°, mp: (*p*) 64°, mp: (*o*) 6°, bp: (*m*) 236°, bp: (*p*) 238°, bp: (*o*) 194°.

Acute tox data: Sc LD_{50} (rat) = 1500 mg/kg for *o*-form. [*3*]

THR = MOD via sc route.

Disaster Hazard: Dangerous; see bromides.

***o*-BROMO-2-PHENYL ACETONITRILE.** See bromo benzyl cyanide.

2-BROMO-4-PHENYL PHENOL. Crystals. $C_{12}H_9BrO$, mw: 249.1, mp: 95°, bp: 195° decomp, flash p: 405°F, vap. press: 1 mm @ 100.0°.

THR = Details U. See also phenol.

Disaster Hazard: Dangerous; see bromides.

BROMOPICRIN. Syns: *nitrobromoform, tribromonitromethane*. Prismatic crystals. CBr_3NO_2, mw: 297.77, mp: 103°, bp: 127°, d: 2.79 @ 18°.

THR = A powerful irr.

Fire Hazard: U.

Explosion Hazard: Severe, when heated rapidly.

Disaster Hazard: Dangerous; see NO_x and bromides.

BROMOPLATINIC ACID. Monoclinic deliquescent red crystals. $H_2PtBr \cdot 9H_2O$, mw: 438.9, mp: <100° decomp.

THR = See bromides and platinum compounds.
Disaster Hazard: Dangerous, see bromides.

1-BROMOPROPANE. See propyl bromide.

5-(3-BROMO-1-PROPENYL)-1,3-BENZO DIOXOLE. $C_{10}H_9O_2Br$, mw: 241.1.
THR = An exper neo. [*3*]

3-BROMO PROPIONIC ACID. $C_3H_5O_2Br$, mw: 153.
THR = An exper carc. [*3, 23*]

3-BROMO-1-PROPYNE. See propargyl bromide.

BROMO SILANE. $BrSiH_3$, mw: 111.
THR = No data. See silanes. Spont ignites in air. [*19*]

BROMO SUCCINIMIDE. White to pale buff, fine crystalline powder with faint odor of bromine. $C_4H_4O_2NBr$, mw: 178.0, mp: 173°–175°, d: 2.098.
Acute tox data: Ip LD_{50} (mouse) = 256 mg/kg. [*3*]
THR = HIGH irr to skin, eyes and mu mem. Also via oral and inhal routes.
Disaster Hazard: Dangerous; see bromides.

α-BROMO TOLUENE. See benzyl bromide.

***m*-BROMO TOLUENE.** Syns: *3-bromo toluene, 3-bromo-1-methyl benzene, m-tolyl bromide.* Liquid, sol in alcohol ether and benzene. $BrC_6H_4CH_3$, mw: 171.04, d: 1.4099 @ 20°/4°, mp: 39.8°, bp: 183.7°.
THR = MOD via inhal route. An irr and narcotic.
Disaster Hazard: Dangerous; see bromides.

***o*-BROMO TOLUENE.** Colorless liquid. $CH_3C_6H_4Br$, mw: 171.0, bp: 180°–182°, fp: −27°, flash p: 175°F, d: 1.422 @ 25°/25°, vap. d: 5.9.
THR = U. Limited animal exper suggest MOD. See *m*-bromotoluene.
Fire Hazard: Mod, when exposed to heat, flame or powerful oxidizers.
Disaster Hazard: Dangerous; see bromides, can react with oxidizing materials.
To Fight Fire: Water to blanket fire, foam, CO_2, dry chemical.

***p*-BROMO TOLUENE.** Syn: *p-tolyl bromide.* White crystals. $CH_3C_6H_4Br$, mw: 171.0, bp: 183.3°–184.1°, fp: 28.5°, flash p: 185°F, d: 1.400 @ 27°/25°, vap. d: 5.9.
THR = U. See *m*-bromotoluene.
Fire Hazard: Mod, when exposed to heat, flame or powerful oxidizers.
Disaster Hazard: Dangerous; see bromides; can react with oxidizing materials.
To Fight Fire: Water to blanket fire, foam, CO_2, dry chemical.

BROMO TRICHLOROMETHANE. Colorless liquid. $CBrCl_3$, mw: 198.3, bp: 103.8°–105.1°, d: 1.997 @ 25°/25°.
THR = U. Probably narcotic in HIGH conc. See also chloroform.
Disaster Hazard: Dangerous; when heated to decomp, emits toxic fumes. It has exploded when added to ethylene. [*19*]

3-BROMO TRICYCLO QUINAZOLINE. $C_{21}H_{11}N_4Br$, mw: 399.3.
THR = An exper carc. [*3*]

BROMO TRIFLUOROETHYLENE. Syn: *BFE.* Colorless, flam, highly reactive (in liquefied state) gas. $BrFC{:}CF_2$, mw: 160.94.
THR = HIGH.
Fire Hazard: Flam gas or liquid.
Disaster Hazard: Dangerous. Reacts with powerful oxidizer. When heated to decomp, emits highly toxic fumes of bromine, fluorine and $COCF_2$.

BROMO TRIFLUOROMETHANE. $CBrF_3$, mw: 148.9.
Acute tox data: Inhal LC_{LO} (rat) = 4000 ppm for $\frac{1}{2}$ hr. [*3*]
THR = HIGH via inhal route. An irr. See also bromotrichloromethane. Strong reaction with aluminum. [*19*]

BROMOXYNIL. Syn: *3,5-dibromo-4-hydroxybenzonitrile.* $C_7H_3ONBr_2$, mw: 276.9.
Acute tox data: Oral LD_{50} (mouse) = 110 mg/kg. [*3*]
THR = HIGH via oral route. An herbicide.
Disaster Hazard: When heated to decomp, emits highly toxic fumes.

BRONZE. See copper, tin and zinc.

BRONZE LIQUID PAINT. Comp: liquid vehicle and bronze. Flash p: below 100°F.
THR = See specific vehicle employed.
Fire Hazard: Dangerous, when exposed to heat or flame; can react with oxidizing materials.
Explosion Hazard: U.
Explosive Range: Depends upon components used.
To Fight Fire: Foam, CO_2, dry chemical.

BRUCINE. Monoclinic prisms. $C_{23}H_{26}N_2O_4$, mw: 394.45, mp: 178°.
Acute tox data: Oral LD_{50} (rat) = 1 mg/kg; ip LD_{50} (rat) = 91 mg/kg. [*3*]
THR = A deadly poison. HIGH via oral and ip routes. An alkaloid like strychnine but $\frac{1}{6}$ as toxic.
Disaster Hazard: Dangerous; when heated, emits toxic fumes.

BRUCINE COMPOUNDS. See brucine.

BRUCINE NITRATE. White powder. $C_{23}H_{26}N_2O_4 \cdot HNO_3 \cdot 2H_2O$, mw: 493.51, mp: (anhy) 230° decomp.
THR = See brucine.
Fire Hazard: Mod by chemical reaction with reducing agents. A powerful oxidant.
Disaster Hazard: See nitrates.

BRUNSWICK GREEN. See copper oxychloride.

BTC. See alkyl dimethyl benzyl ammonium chloride.

BUCLODIN. $C_{28}H_{33}N_2Cl \cdot 2HCl$, mw: 506.
Acute tox data: Oral LD_{50} (rat) = 1000 mg/kg. [*3*]
THR = MOD via oral route. An exper teratogen. [*3*]

BULAN. See 1,1-bis(*p*-chlorophenyl)-2-nitrobutane.

BUFFERING FOOD ADDITIVES. [*109*]

BULAZOLIDINE.
THR = A recog carc. [*14*] See benzene.

BUNKER C FUEL OILS.
THR = A recog carc. [*14*] See mineral oils.

BUNSENITE. See nickel monoxide.

BUNQUINOLATE. Syn: *ethyl-4-hydroxy-6,7-diiso-butoxy-3-quinoline carboxylate.*
THR = U. Used as a food additive; permitted in the feed and drinking water of animals and/or for the treatment of food-producing animals. Also permitted in food for human consumption. [*109*]

BURNT COTTON (NOT RE-PICKED).
THR = U.
Fire Hazard: Mod, when exposed to heat or flame.

BURNT LIME. See calcium oxide.

BURSTERS, EXPLOSIVES. See explosives, high.

BURWEED MARSH ELDER.
THR = A mild allergen.
Fire Hazard: Slight, when exposed to heat or flame.

BUSULFAN. See 1,4-butanediol dimethyl sulfonate.

BUTACIDE. See piperonyl butoxide.

L-BUTADIENE DIEPOXIDE. $C_4H_6O_2$, mw: 86.1.
THR = An exper neo and carc. [*3, 13*]

***meso*-BUTADIENE DIEPOXIDE.**
THR = An exper carc. [*3, 13*]

BUTADIENE-1,3-UNINHIBITED. Syn: *erythrene.* Colorless gas, mild aromatic odor. Very reactive. $CH_2CHCHCH_2$, mw: 54.09, bp: −4.5°, mp: −113°, fp: −108.9°, flash p: −105°F, lel = 2.0%, ucl = 11.5%, d: 0.621 @ 20°/4°, autoign. temp.: 788°F, vap. d: 1.87, vap. press: 1840 mm @ 21°.
THR = The vapors are irr to eyes and mu mem. Inhalation of high conc can cause unconsciousness and death. If spilled on skin or clothing, can cause burns or frost bite (due to rapid vaporization). Chronic systemic poisoning in humans has not been reported.
Fire Hazard: Dangerous, when exposed to heat, flame or powerful oxidizers.
Spont Heating: No.
Explosion Hazard: High if heated under pressure, in air, mixed with phenol, ClO_2, crotonaldehyde. [*19*] May form explosive peroxides upon exposure to air.
Disaster Hazard: Mod dangerous; when heated, emits acrid fumes; can react with oxidizing materials.
To Fight Fire: Stop flow of gas.

BUTADIENE DIOXIDE. Syn: *1,2,3,4-diepoxy butane.* Colorless liquid. $C_4H_6O_2$, mw: 86.1, bp: 142°, mp: 19°, d: 1.113 @ 18/4°.
Acute tox data: Oral LD_{50} (rat) = 78 mg/kg; inhal LC_{50} (rat) = 90 ppm for 4 hrs; dermal LD_{50} (rabbit) = 89 mg/kg. [*3*]
THR = HIGH via oral, inhal and dermal routes. An exper neo. [*3, 23*]

BUTADIENE DIOXIDE(*dl*). Colorless liquid. $C_4H_6O_2$, mw: 86, bp: 138°, mp: 4°, d: 1.112 @ 18°/4°.
THR = An exper mutagen, carc and neo. [*3, 13*] HIGH acute dermal.

BUTADIENE, INHIBITED. See butadiene.

BUTADIENE MONOXIDE Syn: *vinylethylene oxide.* Liquid. $CH_2CHCHCH_2O$, mw: 70.09, mp: −135°, bp: 67°, flash p: <−58°F (CC), d: 0.869, autoign. temp.: 806°F, vap. d: 2.41.
THR = See butadiene.
Fire Hazard: Dangerous, when exposed to heat or flame.
Spont Heating: No.
Disaster Hazard: Mod dangerous; when heated to decomp, emits acrid fumes; can react with oxidizing materials.
To Fight Fire: CO_2, dry chemical, water spray.

1,3-BUTADIEN-1-OL ACETATE. CH_3COOHC: $CHHC{:}CH_2$, mw: 112.
Acute tox data: Dermal LD_{50} (rabbit) = 420 mg/kg; inhal LD_{LO} (rat) = 63 ppm for 4 hrs. [*3*]
THR = HIGH via dermal and inhal routes.

BUTADIONE. Syn: *butalidon.* $C_{19}H_{20}O_2N_2$, mw: 308.4.
Acute tox data: Oral LD_{50} (rat) = 375 mg/kg; ip LD_{50} (rat) = 250 mg/kg; sc LD_{50} (rat) = 270 mg/kg; iv LD_{50} (rat) = 150 mg/kg; iv LD_{50} (dog) = 121 mg/kg. [*103*]
THR = HIGH via oral, ip, sc, iv routes. An exper carc to women via oral route. [*103*]

BUTADIYNE. See diacetylene.

BUTANAL. See butyraldehyde.

2-BUTANAL. See croton aldehyde.

BUTANAL OXIDE. See butyraldoxime.

BUTANE. Syns: *n-butane, methylethyl methane, butyl hydride.* Colorless gas, faint disagreeable odor. C_4H_{10}, mw: 58.1, bp: −0.5°, fp: −138°, lel = 1.9%, uel = 8.5%, flash p: −76°F (CC), d: 0.599, autoign. temp.: 761°F, vap. press: 2 atm @ 18.8°, vap. d: 2.046.
Acute tox data: Inhal LC_{50} (rat) = 659 mg/kg. [*3*]

THR = MOD via inhal. Causes drowsiness. An asphyxiant. A general purpose food additive. [*109*]
Fire Hazard: Very dangerous, when exposed to heat, flame or oxidizers.
Spont Heating: No.
Explosion Hazard: High when exposed to flame; also when mixed with [$Ni(CO)_4 + O_2$].
Disaster Hazard: Mod dangerous; when heated, emits acrid fumes; can react with oxidizing materials.
To Fight Fire: Stop flow of gas.

1,4-BUTANEDICARBOXYLIC ACID. See adipic acid.

1,2,4-BUTANEDICARBOXYLIC ACID TRI(2-ETHYL HEXYL)ESTER.
THR = U. Limited animal exper suggest LOW. See also esters.

BUTANEDINITRILE. See succinonitrile.

BUTANEDIOIC ACID. See succinic acid.

BUTANEDIOIC ANHYDRIDE. See succinic anhydride.

BUTANEDIOIC PEROXIDE. See succinic acid peroxide.

BUTANEDIOL. See 1,4-butanediol.

1,2-BUTANEDIOL. Syns: *dihydroxybutane-1,2, 1,2-butylene glycol.* $CH_3CH_2CHOHCH_2OH$, mw: 90.1, d: 1.0, vap. d: 3.1, bp: 194°, flash p: 194°F.
Fire Hazard: Mod, when exposed to heat or flame.
To Fight Fire: Alcohol foam.

1,3-BUTANEDIOL. See 1,3-butylene glycol.

1,4-BUTANEDIOL. Syns: *butanediol, 1,4-butylene glycol, tetramethylene glycol.* Nearly odorless, colorless, viscid liquid. $HOCH_2CH_2CH_2CH_2OH$, mw: 90.1, bp: 228°, fp: 20.9°, flash p: 250°F (OC), d: 1.0154 @ 25°/4°, vap. d: 3.1.
Acute tox data: Oral LD_{50} (rat) = 1525 mg/kg, oral LD_{50} (guinea pig) = 1200 mg/kg. [*3*] No chronic effects data.
THR = MOD via oral route.
Fire Hazard: Low, when exposed to heat or flame; can react with oxidizing materials.
To Fight Fire: Alcohol foam, mist, foam, CO_2, dry chemical.

2,3-BUTANEDIOL. See 2,3-butylene glycol.

BUTANEDIOL DICAPRYLATE. A clear liquid. $(CH_2)_2(CH_2OCOC_7H_{15})_2$, mw: 342, bp: 220° @ 5 mm, fp: 10.5°, flash p: 390°F, d: 0.929 @ 20°/20°, vap. d: 11.
THR = U.
Fire Hazard: Slight, when exposed to heat or flame; can react with oxidizing materials.

1,4-BUTANEDIOL DIMETHYL SULFONATE. Syns: *myleran, busulphan.* White crystals. $C_6H_{14}O_6S_2$, mw: 246.3, mp: 114°–118°.
Acute tox data: Ip LD_{50} (rat) = 18 mg/kg. [*3*]
THR = An exper teratogen, neo and (S) carc. [*3, 10, 23*] HIGH via ip route.

BUTANEDIONE. See diacetyl.

BUTANENITROLE. See N-butyronitrile.

1,4-BUTANE SULTONE. $C_4H_8O_3S$, mw: 136.2.
THR = An exper neo. [*3, 23*]

1-BUTANETHIOL. See *n*-butyl mercaptan.

2-BUTANETHIOL. See *sec*-butyl mercaptan.

BUTANETRIOL. See 1,2,4-butanetriol.

1,2,4-BUTANETRIOL. Syn: *butanetriol.* Colorless, nearly odorless, thick liquid. $HOCH_2CHOHCH_2CH_2OH$, mw: 106.1, bp: 312°, fp: supercools, flash p: 332°F (COC), d: 1.184 @ 4°.
THR = Practically none. Resembles glycerine.

BUTANOIC ANHYDRIDE. See butyric anhydride.

BUTANOL. See butyl alcohol.

2-BUTANOL. See *sec*-butyl alcohol.

2-BUTANOL ACETATE. See *sec*-butyl acetate.

BUTANOLAL. See aldol.

2-BUTANONE. Syn: *methylethyl ketone.* Colorless liquid, acetone-like odor. $CH_3COCH_2CH_3$, mw: 72.10, bp: 79.57°, fp: −85.9°, lel = 1.8%, uel = 10%, flash p: 22°F (TOC), d: 0.80615 @ 20°/20°, vap. press: 71.2 mm @ 20°, autoign. temp.: 960°F, vap. d: 2.42, ULC: 85–90.
Acute tox data: Oral LD_{50} (rat) = 3400 mg/kg; dermal LD_{50} (rabbit) = 13,000 mg/kg. [*3*]
THR = MOD via oral; LOW via dermal routes. A strong irr. Affects peripheral nervous system, CNS. is an exper teratogen. [*3*] See ketones.
Fire Hazard: Dangerous when exposed to heat or flame; reacts violently with chlorosulfonic acid, oleum, potassium-*tert*-butoxide. [*19*]
Spont Heating: No.
Disaster Hazard: Highly dangerous upon exposure to heat or flame.
Explosion Hazard: Mod, when exposed to flame.
To Fight Fire: Alcohol foam, CO_2, dry chemical.

2-BUTANONE PEROXIDE. See methylethyl ketone peroxide.

BUTANOYL CHLORIDE. See butyryl chloride.

BUTAZOLIDIN. See phenylbutazone.

2-BUTENAL. See crotonaldehyde.

1-BUTENE. See α-butylene.

cis-**BUTENE-2.** Syns: *dimethylethylene, pseudo-butylene.* Colorless flam gas, slightly aromatic odor.

$CH_3CHCHCH_3$, mw: 56.1, bp: 1°, fp: −139°, flash p: −100°F, d: 0.627 @ 15.5°/15.5°, vap. press: 1410 mm @ 21°, autoign. temp.: 615°F, lel = 1.7%, uel = 9.0%, vap. d: 1.9.

THR = U. May act as a simple asphyxiant.

Fire Hazard: Very dangerous, when exposed to heat or flame; can react with oxidizing materials.

Explosion Hazard: U.

To Fight Fire: Stop flow of gas.

***trans*-BUTENE-2.** A colorless, flam gas, slightly aromatic odor. C_4H_8, mw: 56.1, bp: 2.5°, fp: −105.6°, flash p: −100°F, d: 0.613 @ 15.5°/15.5°, vap. d: 1.95, vap. press: 1592 mm @ 21°, autoign. temp.: 615°F, lel = 1.8%, uel = 9.7%, vap. d: 1.9.

THR = U. May act as a simple asphyxiant.

Fire Hazard: Very dangerous, when exposed to heat or flame; can react with oxidizing materials.

Explosion Hazard: U.

To Fight Fire: Stop flow of gas.

***trans*-BUTENEDIOIC ACID.** See fumaric acid.

***cis*-BUTENEDIOIC ANHYDRIDE.** See maleic anhydride.

2-BUTENE-1,4-DIOL. Colorless, odorless liquid. $HOCH_2CHCHCH_2OH$, mw: 88.1, bp: 234°, fp: 12.5°, flash p: 263°F (COC), d: 1.07 @ 25°/15°, vap. d: 3.04.

THR = U. This material is an irr to the skin.

Fire Hazard: Low when exposed to heat or flame; can react with oxidizing materials.

To Fight Fire: Alcohol foam, water mist or spray, foam, CO_2, dry chemical.

***trans*-BUTENEDIOYL CHLORIDE.** See fumaryl chloride.

2-BUTENE NITRILE. See crotonitrile.

3-BUTENE-2-ONE. See methyl vinyl ketone.

2-BUTEN-1-OL. See α-methallyl alcohol.

1-BUTEN-3-YNE. Syn: *vinyl acetylene.* C_4H_6, mw: 54, lel = 2%, uel = 100%, d: 0.68 @ 1.7 atm, vap. d: 1.8, bp: 11°.

THR = No data.

Fire Hazard: Very dangerous via spont combustion, heat, flame or oxidizers.

3-(2-BUTENYL)-4-METHYL-2-OXO-3-CYCLOPENTEN-1-YL ESTER OF CHRYSANTHEMUM DICARBOXYLIC ACID MONOMETHYL ESTER. See cinerin II.

BUTESIN. Syn: *butyl-p-aminobenzoate.* Yellow, amorphous powder. $C_6H_4NH_2COOC_4H_9$, mw: 193.3, mp: 55°.

Acute tox data: Oral LD_{50} (mouse) = 67 mg/kg. [*3*]

THR = HIGH via oral route. An allergen.

Fire Hazard: Slight.

BUTONATE. Syn: *o,o-dimethyl (2,2,2-trichloro-1,n-butyryl oxyethyl) phosphonate.* $C_7H_8O_5Cl_3P$, mw: 309.4.

Acute tox data: Oral LD_{50} (rat) = 1100 mg/kg; dermal LD_{50} (rat) = 7000 mg/kg.

THR = MOD via oral and LOW via dermal routes.

Disaster Hazard: When heated to decomp, emits highly toxic fumes.

BUTOPYRONOXYL. Syn: *butyl mesityl oxide.* Yellow to reddish liquid. $C_{12}H_{18}O_4$, mw: 226.3, bp: 113°, d: 1.052–1.060 @ 25°/25°, flash p: 315°F.

Acute tox data: Oral LD_{50} (rabbit) = 5400 mg/kg; oral LD_{50} (guinea pig) = 3200 mg/kg. [*3*]

THR = MOD via oral, LOW irr. Mildly irr to skin. In exper it has produced liver necrosis in animals.

Fire Hazard: Slightly dangerous.

1-BUTOXYBUTANE. See butyl ether.

2-BUTOXY-M-(2-DIETHYL AMINOETHYL)CINCHONINAMIDE HYDROCHLORIDE. See dibucaine hydrochloride.

BUTOXYETHANOL. See butyl "cellosolve."

2,β-BUTOXY ETHOXY ETHYL CHLORIDE. $C_4H_9C_2H_4OC_2H_4Cl$, mw: 165, d: 1.0, vap. d: 6.1, bp: 200°–225°, flash p: 190°F.

Fire Hazard: Mod, when exposed to heat, flame or oxidizers.

To Fight Fire: Water, foam, fog, mist, spray, dry chemical.

1-(BUTOXY ETHOXY)-2-PROPANOL. Sol in water. $CH_3CH(OH)CH_2OC_2H_4OC_2H_4C_2H_5$, mw: 176, d: 0.9310 @ 20°/20°, bp: 230.3°, fp: −90°, flash p: 250°F (OC).

Acute tox data: Oral LD_{50} (rat) = 4000 mg/kg; dermal LD_{50} (rabbit) = 2830 mg/kg. [*3*]

THR = MOD via oral and dermal routes.

Fire Hazard: Slight, when exposed to heat or flame.

To Fight Fire: Alcohol foam, dry chemical, spray or mist.

BUTOXY ETHYL ACETATE. Syn: *ethylene glycol monobutyl ether acetate.* Colorless liquid, fruity odor, sol in hydrocarbons and organic solvents, insol in water. $C_4H_9OCH_2CH_2OOCCH_3$, mw: 160, bp: 192.3°, d: 0.9424 @ 20°/20°, fp: −63.5°, flash p: 190°F.

Acute tox data: Oral LD_{50} (mouse) = 3200 mg/kg; dermal LD_{50} (rat) = 1580 mg/kg. [*3*]

THR = MOD via oral and dermal routes.

Fire Hazard: Mod, when exposed to heat, flame or oxidizers.

To Fight Fire: Alcohol foam.

BUTOXYETHYL ACRYLATE. Liquid. $CH_2CHCO_2CH_2OC_4H_9$, mw: 172.2, bp: 62° @ 2 mm, flash p: 185°F (OC), d: 0.948 @ 20°, vap. d: 5.93.

Acute tox data: Oral LD_{50} (rat) = 6500 mg/kg; dermal LD_{50} (rabbit) = 640 mg/kg. [3]
THR = MOD via oral and MOD to HIGH via dermal routes.
Fire Hazard: Mod, when exposed to heat or flame; can react with oxidizing materials.
To Fight Fire: Foam, CO_2, dry chemical.

BUTOXY ETHYL-2,4-DICHLORO PHENOXY ACETATE. $C_{16}H_{16}O_4Cl_2$, mw: 343.2.
THR = HIGH oral acute to rats. An exper teratogen. [3]

BUTOXY ETHYL DIGLYCOL CARBONATE. See diethylene glycol bis (2-butoxy ethyl carbonate).

β-BUTOXY ETHYL SALICYLATE. $C_8H_{23}O_4$, mw: 183.2, flash p: 315°F, bp: 188°.
THR = U. See also esters.
Fire Hazard: Low.
To Fight Fire: Dry chemical, mist, fog.

BUTOXY ETHYL STEARATE. Light-colored liquid with mild odor. $CH_3(CH_2)_{16}COOC_2H_4OC_4H_9$, mw: 384, mp: 10°–16°, bp: 215°–245° @ 4 mm, flash p: 420°F, d: 0.882 @ 20°/20°, vap. press: 0.03 mm @ 150°, vap. d: 13.2.
THR = See esters.
Fire Hazard: Slight, when exposed to heat or flame; can react with oxidizing materials.
To Fight Fire: Foam, CO_2, dry chemical.

BUTOXYL. See methoxybutyl acetate.

***p*-BUTOXY PHENYL.** $C_8H_{14}O_3$, flash p: 180°F (OC), d: 0.9, vap. d: 5.2, bp: 210°, mw: 158.2.
Acute tox data: Oral LD_{50} (rat) = 1680 mg/kg; dermal LD_{50} (rabbit) = 1250 mg/kg. [3]
THR = MOD via oral and dermal routes.
Fire Hazard: Mod via exposure to heat, flame, oxidizers.
To Fight Fire: Water, alcohol foam, mist, fog, dry chemical.

BUTOXY POLYPROPYLENE GLYCOL. Syn: *crag fly repellant*. Colorless liquid.
Acute tox data: Oral LD_{50} (rat) = 11.2 g/kg; dermal LD_{50} (rabbit) = 21 g/kg. [3]
THR = LOW via oral and dermal routes.

β-BUTOXY-β-THIOCYANODIETHYL ETHER. Syns: *lethane 384, β,β-tubatoxythio cyanodiethyl ether*.
Acute tox data: Oral LD_{50} (rat) = 90 mg/kg; dermal LD_{50} (rat) = 250 mg/kg; oral LD_{50} (dog) = 30 mg/kg. [3]
THR = HIGH via oral and dermal routes. Irr to skin and mu mem. HIGH conc can cause CNS depression. An insecticide. See also thiocyanates.
Fire Hazard: U.
Disaster Hazard: Dangerous; see cyanides.

BUTTER OF ANTIMONY. See antimony trichloride.

BUTTER YELLOW. See dimethyl amino azo benzene.

BUTTER OF ZINC. See zinc chloride.

BUTYL ACETAMIDE. Liquid. $CH_3CONHC_4H_9$, mw: 117.2, bp: 234°, flash p: 240°F, d: 0.89.
THR = U. Probably LOW. See amides.
Fire Hazard: Low, when exposed to heat or flame; can react with oxidizing materials.
To Fight Fire: Water spray or mist, dry chemical.

BUTYL ACETANILIDE. Slightly yellow liquid. $CH_3(CH_2)_3N(C_6H_5)C(O)CH_3$, mw: 191.3, mp: 20.8°, bp: 277°–281°, flash p: 286°F, d: 0.992 @ 25°/25°, vap. d: 6.6.
Acute tox data: Oral LD_{50} (mouse) = 800 mg/kg; acute oral LD_{50} (guinea pig) = 300 mg/kg. [3]
THR = MOD to HIGH via oral route.
Fire Hazard: Low, when exposed to heat or flame; can react with oxidizing materials.
To Fight Fire: CO_2, dry chemical.

BUTYL ACETATE. Syn: *butyl ethanoate*. Colorless liquid. $CH_3COOC_4H_9$, mw: 116.16, bp: 126°, fp: −73.5°, ulc: 50–60, lel = 1.7%, uel = 7.6%, flash p: 72°F, d: 0.88 @ 20°/20°, autoign. temp.: 797°F, vap. press: 15 mm @ 25°.
Acute tox data: Inhal LC_{LO} (human) = 200 ppm ⟶ irr effects; oral LD_{50} (rat) = 14000 mg/kg. [3]
THR = MOD irr to lungs and mu mem; LOW via oral route and a MILD allergen. High conc are irr to eyes and respiratory tract and cause narcosis. Evidence of chronic systemic toxicity is inconclusive.
Fire Hazard: Dangerous, when exposed to heat or flame; can react with oxidizing materials. Violent reaction with potassium-*tert*-butoxide. [19]
Spont Heating: No.
Explosion Hazard: Mod when exposed to flame.
To Fight Fire: Alcohol foam, CO_2, dry chemical.

***sec*-BUTYL ACETATE.** Syns: *acetic acid secondary butyl ester, 2-butanol acetate*. Colorless liquid, mild odor. $CH_3COOC_4H_9$, mw: 116.1, bp: 112°, flash p: 88°F (OC), lel = 1.7%, d: 0.862–0.866 @ 20°/20°, vap. d: 4.00.
THR = A MILD irr and allergen. LOW oral and MOD inhal.
Fire Hazard: Dangerous; when exposed to heat or flame, can react with oxidizing materials.
Explosion Hazard: Mod, when exposed to flame.
To Fight Fire: Alcohol foam, CO_2, dry chemical.

***tert*-BUTYL ACETATE.** See *sec*-butyl acetate.

BUTYL ACETOACETATE. Liquid. $CH_3COCH_2COOCH_2CH_2CH_2CH_3$, mw: 160.2, bp: 214°, flash p: 185°F, d: 0.96, vap. d: 5.55.

THR = U. See esters, acetoacetic acid and butyl alcohol. Probably MOD tox.

Fire Hazard: Mod, when exposed to heat or flame; can react with oxidizing materials.

To Fight Fire: Alcohol foam, CO_2, dry chemical.

BUTYL ACETYL RICINOLEATE. Yellow, oil liquid. $C_{18}H_{33}O_3C_4H_8CH_3CO$, mw: 396.6, mp: −32°, bp: 220°, flash p: 230°F (OC), d: 0.940, autoign. temp.: 725°F, vap. d: 13.7.

THR = U. See esters.

Fire Hazard: Low, when exposed to heat or flame.

To Fight Fire: CO_2, dry chemical, mist or fog.

***n*-BUTYL ACID PHOSPHATE.** See acid butyl phosphate.

N-*tert*-BUTYL ACRYLAMIDE. White crystalline solid. $C_7H_{13}ON$, mw: 127.18, mp: 128°–130° with polymerization, d: 1.015 @ 30°.

THR = MOD irr to eyes and mu mem and via oral and inhal routes.

Fire Hazard: Slight.

BUTYL ACRYLATE. Water white, extremely reactive monomer. $C_7H_{12}O_2$, mw: 128.2, bp: 69° @ 50 mm, fp: −64.6°, flash p: 120°F (OC), d: 0.89 @ 25°/25°, vap. press: 10 mm @ 35.5°, vap. d: 4.42.

Acute tox data: Oral LD_{50} (rat) = 3730 mg/kg; dermal LD_{50} (rabbit) = 2000 mg/kg. [*3*]

THR = MOD via oral and dermal routes.

Fire Hazard: Mod, when exposed to heat or flame; can react with oxidizing materials.

To Fight Fire: Foam, CO_2, dry chemical.

BUTYL ALCOHOL. Syn: *n-butanol.* Colorless liquid. $CH_3(CH_2)_2CH_2OH$, mw: 74.12, bp: 117.5°, ulc: 40, lel = 1.4%, uel = 11.2%, fp: −88.9°, flash p: 95°–100°F, d: 0.80978 @ 20°/4°, autoign. temp.: 689°F, vap. press: 5.5 mm @ 20°, vap. d: 2.55.

Acute tox data: Oral LD_{50} (rat) = 790 mg/kg; dermal LD_{50} (rabbit) = 4200 mg/kg; 25 ppm ⟶ irr to humans.

THR = MOD to HIGH via oral and MOD via dermal routes. MOD irr via inhal route to humans. Though animal exper have shown the butyl alcohols to possess toxic properties, butyl alcohols have produced few cases of poisoning in industry because of their low volatility. The use of normal butyl alcohol is reported to have resulted in irr of the eyes, with corneal inflam, slight headache and dizziness, slight irr of the nose and throat, and dermatitis about the fingernails and along the side of the fingers. Keratitis has also been reported.

Fire Hazard: Dangerous, when exposed to heat, flame, or oxidizers.

Spont Heating: No.

Explosion Hazard: Mod, when exposed to flame.

Disaster Hazard: Mod dangerous; when heated to decomp, emits toxic fumes; can react with oxidizing materials.

To Fight Fire: Water spray, alcohol foam, CO_2, dry chemical.

***sec*-BUTYL ALCOHOL.** Syns: *2-butanol, ethylmethyl carbinol.* Colorless liquid. $CH_3CH_2CHOHCH_3$, mw: 74.12, mp: −89°, bp: 99.5°, flash p: 75°F (CC), d: 0.808 @ 20°/4°, autoign. temp.: 763°F, vap. press: 10 mm @ 20°, vap. d: 2.55, lel = 1.7% @ 212°F, uel = 9.8% @ 212°F.

Acute tox data: Oral LD_{50} (rat) = 6480 mg/kg; inhal LC_{LO} (rat) = 16000 ppm for 4 hrs. [*3*]

THR = MOD via oral and inhal routes. See also butyl alcohol.

Fire Hazard: Dangerous, when exposed to heat or flame; can react with oxidizing materials.

Explosion Hazard: U.

To Fight Fire: Water spray, alcohol foam, CO_2, dry chemical.

***tert*-BUTYL ALCOHOL.** Syn: *2-methyl-2-propanol.* Colorless liquid or rhombic prisms or planes. $(CH_3)_3COH$, mw: 74.12, mp: 25.3°, bp: 82.8°, flash p: 52°F (CC), d: 0.7887 @ 20°/4°, autoign. temp.: 896°F, vap. press: 40 mm @ 24.5°, vap. d: 2.55, lel = 2.4%, uel = 8.0%.

Acute tox data: Oral LD_{50} (rat) = 3500 mg/kg; ip LD_{50} (mouse) = 933 mg/kg. [*3*]

THR = MOD via oral and ip routes.

Fire Hazard: Dangerous, when exposed to heat or flame; can react with oxidizing materials.

Spont Heating: No.

Explosion Hazard: Mod, in the form of vapor when exposed to flame. Violent reaction with H_2O_2. [*19*]

To Fight Fire: Alcohol foam, CO_2, dry chemical.

BUTYL ALDEHYDE. See butyraldehyde.

BUTYLAMINE. Syn: *1-aminobutane, tutane.* Liquid, ammonia-like odor. $C_4H_9NH_2$, mw: 73.1, mp: −50°, bp: 77°, flash p: 10°F (OC), 10°F (CC), d: 0.74–0.76 @ 20°/20°, autoign. temp.: 594°F, vap. d: 2.52, lel = 1.7%, uel = 9.8%.

Acute tox data: Oral LD_{50} (rat) = 500 mg/kg; inhal LD_{50} (rat) = 4000 ppm for 4 hrs; dermal LD_{50} (guinea pig) = 500 mg/kg. [*3*]. No data on chronic effects.

THR = HIGH via oral and dermal routes. MOD via inhal route.

Fire Hazard: Dangerous, when exposed to heat or flame; can react with oxidizing materials.

To Fight Fire: Alcohol foam, CO_2, dry chemical.

***sec*-BUTYLAMINE.** Syn: *2-aminobutane.* Liquid. $CH_3CH_2CH(NH_2)CH_3$, mw: 73.1, mp: −104°, bp: 63°, flash p: 15°F, d: 0.724 @ 20°.

Acute tox data: Oral LD_{50} (rat) = 380 mg/kg. [*3*]
THR = HIGH via oral route. A powerful irr. See also butyl amine, and amines.
Fire Hazard: Dangerous; when exposed to heat or flame, can react with oxidizing materials.
Explosion Hazard: U.
To Fight Fire: Alcohol foam, water spray or mist, dry chemical.

***tert*-BUTYLAMINE.** Syn: *2-aminoisobutane, trimethyl aminomethane*. Colorless liquid. $(CH_3)_3CNH_2$, mw: 73.14, mp: −67.5°, bp: 44°–46°, d: 0.700 @ 15°, lel = 1.7% @ 212°F, uel = 8.9% @ 212°F. vap. d: 2.5, autoign. temp.: 716°F.
Acute tox data: Oral LD_{50} (rat) = 180 mg/kg; oral LD_{50} (mouse) = 900 mg/kg. [*3*] No chronic data.
THR = HIGH via oral route. See also butylamine and amines.
Fire Hazard: Very dangerous, when exposed to heat or flame.
Explosion Hazard: U.
To Fight Fire: Alcohol foam.

BUTYLAMINE OLEATE. Syn: *monobutylamine oleate*. Liquid. $C_{17}H_{33}COONH_3C_4H_9$, mw: 355.6, flash p: 150°F (OC), d: 0.891.
THR = Details U. See esters and butylamine.
Fire Hazard: Mod, when exposed to heat or flame; can react with oxidizing materials.
To Fight Fire: Alcohol foam, CO_2, dry chemical.

BUTYL AMINOBENZOATE. See butesin.

2-BUTYLAMINO ETHANOL. See butyl ethanolamine.

***tert*-BUTYLAMINO ETHYL METHACRYLATE.** Liquid. $C_{10}H_{19}NO_2$, mw: 185, bp: 100°–105°, d: 0.914, fp: 205°F (OC).
Fire Hazard: Low, when exposed to heat or flame.
To Fight Fire: Alcohol foam, water spray or mist, dry chemical.

4-BUTYLAMINO-N-NITROSO BUTYRIC ACID. $C_8H_{16}O_3N_2$, mw: 188.3.
THR = An exper carc. [*3*]

***N,n*-BUTYL-*p*-AMINOPHENOL.** Crystals. $HOC_6H_4HNCH_2CH_2CH_2CH_3$, mw: 165.2, mp: −33°, flash p: 61°F.
THR = U. May be an irr or skin sensitizer.
Fire Hazard: Dangerous, when exposed to heat, flame or oxidizers.
To Fight Fire: Water spray, fog, foam, dry chemical.
Disaster Hazard: Mod dangerous; when heated to decomp, emits toxic fumes; can react with oxidizing materials.

N-BUTYL ANILINE. Colorless liquid. $(C_4H_9)NHC_6H_5$, mw: 149.3, bp: 241°, flash p: 225°F (COC), fp: −15.1°, d: 0.9288 @ 20°/20°, vap. press: 0.02 mm @ 20°, vap. d: 5.15.
Acute tox data: Oral LD_{50} (rat) = 1620 mg/kg. [*3*]
THR = MOD via oral route.
Fire Hazard: Low, when exposed to heat or flame.
Disaster Hazard: Dangerous; when heated to decomp, emits highly toxic fumes of aniline; can react vigorously with oxidizing materials.
To Fight Fire: Alcohol foam, CO_2, dry chemical.

***p-tert*-BUTYL ANILINE.** Syn: *1-amino-4-tert-butyl benzene*. Oil. $(CH_3)_3CC_6H_4NH_2$, mw: 149.2, mp: 17°, bp: 241°, d: 0.9525 @ 15°/4°.
Acute tox data: Oral LD_{50} (mouse) = 81 mg/kg. [*3*]
THR = HIGH via oral route.
Disaster Hazard: Dangerous; when heated to decomp, emits highly toxic fumes of aniline.

BUTYLATED HYDROXYTOLUENE. See di-*tert*-butyl-*p*-cresol.

***tert*-BUTYL AZIDO FORMATE.** $(CH_3)_3COCONNN$, mw: 143.1.
THR = No data. In presence of excess $COCl_2$ during synthesis of this material, an explosive can be formed. [*19*]

8-BUTYL BENZ(a)ANTHRACENE. $C_{22}H_{20}$, mw: 284.4.
THR = An exper carc. [*3*]

N-BUTYL BENZENE. Syn: *1-phenyl butane*. Colorless liquid. $C_6H_5CH_2CH_2CH_2CH_3$, mw: 134.2, mp: −81.2°, bp: 182.1°, fp: −88.2°, flash p: 160°F (TOC), d: 0.8601 @ 20°/4°, vap. press: 1 mm @ 22.7°, autoign. temp.: 774°F, lel = 0.8%, uel = 5.8%, vap. d: 4.6.
Acute tox data: Oral LD_{LO} (rat) = 5000 mg/kg. [*3*]
THR = MOD via oral route.
Fire Hazard: Mod, when exposed to heat or flame; can react with oxidizing materials.
To Fight Fire: Alcohol foam, CO_2, dry chemical.

***sec*-BUTYL BENZENE.** Syn: *2-phenyl butane*. Colorless liquid. $C_6H_5CH(CH_3)C_2H_5$, mw: 134.2, mp: −82.7°, bp: 173.5°, fp: −75.8°, flash p: 126°F (TOC), d: 0.8621 @ 20°, vap. press: 1 mm @ 18.6°, vap. d: 4.62, autoign. temp.: 788°F, lel = 0.8%, uel = 6.9%.
Acute tox data: Oral LD_{50} (rat) = 2240 mg/kg. [*3*]
THR = MOD via oral route.
Fire Hazard: Mod, when exposed to heat or flame; can react with oxidizing materials.
Spont Heating: No.
To Fight Fire: Foam, CO_2, dry chemical, water spray or mist.

***tert*-BUTYL BENZENE.** Syn: *2-methyl-2-phenyl propane*. Colorless liquid. $C_6H_5C(CH_3)_3$, mw: 134.2, bp: 168.2°, fp: −58°, flash p: 140°F (TOC), d: 0.8665 @ 20°, vap. press: 1 mm @ 13.0°, vap. d: 4.62, autoign. temp.: 842°F, lel = 0.7% @ 212°F, uel = 5.7% @ 212°F.

Acute tox data: Oral LD_{LO} (rat) = 5000 mg/kg. [*3*]
THR = MOD via oral route.
Fire Hazard: Mod, when exposed to heat or flame; can react with oxidizing materials.
To Fight Fire: Foam, CO_2, dry chemical, water spray, fog, mist.

***p*-BUTYL BENZENE TETRATHIO-*o*-STANNATE.** Solid. $(SC_6H_4C_4H_9)_4Sn$, mw: 779.8, mp: 106°.
THR = Details U. See tin compounds.
Disaster Hazard: Dangerous; see sulfur compounds.

BUTYL BENZOATE. Liquid. $C_6H_5COOC_4H_9$, mw: 178.22, mp: −21.5°, bp: 250°, flash p: 225°F (OC), d: 1.0073 @ 20°/20°, vap. press: <0.01 mm @ 20°, vap. d: 6.15.
Acute tox data: Oral LD_{50} (rat) = 5140 mg/kg. [*3*]
THR = MOD via oral route. See also esters, benzoic acid and *n*-butyl alcohol.
Fire Hazard: Low, when exposed to heat or flame; can react with oxidizing materials.
To Fight Fire: CO_2, dry chemical, water mist, fog, spray.

***p-tert*-BUTYL BENZOIC ACID.** Syn: *p-tert-BBA*. Colorless, fine crystalline powder. $HOOCC_6H_4C(CH_3)_3$, mw: 178.11, mp: 166.3°, d: 1.142 @ 20°/4°.
Acute tox data: Oral LD_{50} (rat) = 735 mg/kg. [*3*]
THR = MOD via oral route. A MILD irr.
Fire Hazard: Slight, when exposed to heat or flame; can react with oxidizing materials.
To Fight Fire: Foam, CO_2, dry chemical.

1-BUTYL BENZ(a)PYRENE.
THR = An exper carc. [*23*]

BUTYL BENZYL PHTHALATE. Clear, oily liquid. $C_4H_9COOC_6H_4COOCH_2C_6H_5$, mw: 312.4, mp: <−35°, bp: 370°, flash p: 390°, d: 1.116 @ 25°/25°, vap. d: 10.8.
Acute tox data: Ip LD_{50} (mouse) = 3160 mg/kg. [*3*]
THR = MOD via ip route. See also esters.
Fire Hazard: Slight, when exposed to heat or flame; can react with oxidizing materials.
To Fight Fire: Spray or mist, CO_2, dry chemical.

BUTYL BENZYL SEBACATE. Liquid. $(CH_2)_8CO_2C_4H_9CO_2C_6H_5CH_2$, mw: 348.5, bp: 245°–285° @ 10 mm, flash p: 395°F, d: 1.004 @ 20°/20°, vap. d: 12.0.
THR = See esters.
Fire Hazard: Slight, when exposed to heat or flame; can react with oxidizing materials.
To Fight Fire: Foam, CO_2, dry chemical.

2-BUTYL BIPHENYL. $C_{16}H_{18}$, mw: 210.2, flash p: >212°F, autoign. temp.: 806°F, vap. d: 7.26, bp: 290°.
THR = No data. May be very toxic.
Fire Hazard: Low.
To Fight Fire: Water, spray, fog, alcohol foam.

BUTYL BORIC ACID. Colorless crystals. $C_4H_9B(OH)_2$, mw: 101.95, mp: 92°–94°, bp: decomp.
Acute tox data: Oral LD_{50} (mouse) = 1740 mg/kg. [*3*] See boron compounds and butyl alcohol.
THR = MOD via oral route.
Fire Hazard: Slight; when heated, emits toxic fumes; may have flam decomp products.

***tert*-BUTYL BORIC ACID.** White crystals. $C_4H_9B(OH)_2$, mw: 101.95, mp: 105° decomp, bp: decomp.
Acute tox data: Oral LD_{50} (mouse) = 2100 mg/kg. [*3*]
THR = MOD via oral route. See boron compounds and butyl alcohol.
Fire Hazard: Slight; when heated, emits toxic fumes; may emit flam products of decomp.

***tert*-BUTYL BORON OXIDE.** Colorless liquid hydrolyzed by water, sol in ether. $(C_4H_9)B_3O_3$, mw: 251.8, mp: 20°, bp: 67° @ 5 mm.
THR = Details U. See boron compounds.

***n*-BUTYL BORON OXIDE.** Colorless liquid hydrolyzed by water, sol in ether. $(C_4H_9)B_3O_3$, mw: 251.8, bp: 154° @ 30 mm.
THR = Details U. See boron compounds.

BUTYL BROMIDE. Syn: *1-bromobutane*. Colorless to pale straw-colored liquid. $CH_3(CH_2)_2CH_2Br$, mw: 137.03, mp: −112.4, bp: 101.4, flash p: 65°F (OC), d: 1.274 @ 25°/25°, autoign. temp.: 509°F, vap. d: 4.72, lel = 2.6% @ 212°F, uel = 6.6% @ 212°F.
THR = Details U. See also chlorinated HC, aliphatic.
Fire Hazard: Dangerous, when exposed to heat or flame or oxidizers.
Disaster Hazard: Dangerous; see bromides; can react with oxidizing materials.
To Fight Fire: CO_2, dry chemical, mist or spray.

***sec*-BUTYL BROMIDE.** Colorless liquid. $C_2H_5CHBrCH_3$, mw: 137.0, fp: <−50°, bp: 91.4°, flash p: 70°F, d: 1.257 @ 25°/25°.
THR = U. Narcotic in high conc. See also chlorinated HC aliphatic.
Fire Hazard: Dangerous, when exposed to heat or flame.
Explosion Hazard: U.
Disaster Hazard: Dangerous, when heated to decomp, emits toxic fumes; can react with oxidizing materials.
To Fight Fire: Water, spray or mist, foam, CO_2, dry chemical.

***tert*-BUTYL BROMIDE.** Colorless liquid. $(CH_3)_3CBr$, mw: 137.0, mp −20°, bp: 73.3°, fp: −18°, d: 1.215 @ 25°/25°.

Acute tox data: Ip LD_{50} (rat) = 1250 mg/kg. [*3*]
THR = MOD via ip route. See also chlorinated HC, aliphatic.
Disaster Hazard: Mod; when heated to decomp, emits toxic fumes.

BUTYL BUTANOATE. See *n*-butyl butyrate.

***n*-BUTYL BUTYRATE.** Syn: *butyl butanoate*. Liquid. $CH_3CH_2CH_2CO_2CH_2CH_2CH_2CH_3$, mw: 144.2, bp: 166°, flash p: 128°F (OC), d: 0.874, vap. d: 5.0.
THR = MOD irr to eyes and mu mem and via inhal route. Irr and narcotic in high conc.
Fire Hazard: Mod, when exposed to heat or flame; can react with oxidizing materials.
To Fight Fire: Alcohol foam, foam, CO_2, dry chemical.

BUTYL CARBAMATE. $C_5H_{11}O_2N$, mw: 117.2.
Acute tox data: Sc LD_{50} (mouse) = 540 mg/kg. [*3*]
THR = MOD via sc route. An exper neo. [*3*]

BUTYL CARBINOL. See amyl alcohol.

***n*-BUTYL CARBINOL.** See 1-pentanol.

***sec*-BUTYL CARBINOL.** See *sec*-amyl alcohol.

***tert*-BUTYL CARBINOL.** See *tert*-amyl alcohol.

BUTYL "CARBITOL." Syn: *diethylene glycol monobutyl ether*. Colorless liquid.
$C_4H_9OCH_2CH_2OCH_2CH_2OH$; mw: 162.2, fp: −68.1°, bp: 230.6°, flash p: 172°F, d: 0.9553 @ 20°/4°, autoign. temp.: 442°F, vap. press: 0.02 mm @ 20°, vap. d: 5.58.
Acute tox data: Dermal LD_{50} (rabbit) = 4120 mg/kg; oral LD_{50} (rat) = 6560 mg/kg. [*3*] No chronic effects data.
THR = MOD via dermal and oral routes.
Fire Hazard: Mod, when exposed to heat or flame; emits degradation products; can react with oxidizing materials.
Spont Heating: No.
To Fight Fire: Alcohol foam, CO_2, dry chemical.

BUTYL "CARBITOL" ACETATE. Syn: *diethylene glycol monobutyl ether acetate*. Colorless liquid. $C_4H_9O(CH_2)_2O(CH_2)_2OOCCH_3$, mw: 204.26, fp: −32.2°, bp: 247°, flash p: 240°F (OC), d: 0.981 @ 20°/20°, autoign. temp.: 570°F, vap. press: 0.01 mm @ 20°.
Acute tox data: Oral LD_{50} (rat) = 12 g/kg; dermal LD_{50} (rabbit) = 15 g/kg. [*3*]
THR = LOW via oral and dermal routes.
Fire Hazard: LOW; when exposed to heat or flame, emits degradation products; can react with oxidizing materials.
To Fight Fire: Foam, CO_2, dry chemical.

BUTYL "CARBITOL" THIOCYANATE. See β-butoxy-β-thiocyano diethyl ether.

BUTYL CARBITYL-6-PROPYL PIPERONYL ETHER. See piperonyl butoxide.

BUTYL CARBO BUTOXY METHYLPHTHALATE. $C_{18}H_{25}O_6$, mw: 337.4.
THR = An exper teratogen. [*3*]

***p-tert*-BUTYL CATECHOL.** White crystalline solid. $(CH_3)_3CC_6H_3(OH)_2$, mw: 166.2, fp: 52°, flash p: 265°F, bp: 285°, d: 1.049 @ 60°/25°.
Acute tox data: Oral LD_{50} (rat) = 2820 mg/kg; dermal LD_{50} (rabbit) = 630 mg/kg. [*3*]
THR = MOD via oral and dermal routes.
Fire Hazard: Low, when exposed to heat or flame.
Disaster Hazard: Mod dangerous; when heated to decomp, emits toxic fumes; can react with oxidizing materials.
To Fight Fire: CO_2, dry chemical, fog, mist.

BUTYL "CELLOSOLVE." See ethylene glycol-*n*-butyl ether.

BUTYL "CELLOSOLVE" ACETATE. See 2-butoxy ethanol acetate.

BUTYL CHLORIDE. Syn: *1-chlorobutane*. Colorless liquid. $CH_3(CH_2)_2CH_2Cl$, mw: 92.57, mp: −123.1°, bp: 78°, lel = 1.9%, uel = 10.1%, flash p: 15°F (OC), d: 0.884, autoign. temp.: 860°F, vap. d: 3.20.
Acute tox data: Oral LD_{50} (rat) = 2670 mg/kg. [*3*]
THR = MOD via oral route. See chlorinated HC, aliphatic.
Fire Hazard: Dangerous, when exposed to heat or flame.
Explosion Hazard: Mod, when exposed to flame.
Disaster Hazard: Dangerous; when heated to decomp, emits highly toxic fumes of phosgene; can react vigorously with oxidizing materials.
To Fight Fire: Foam, CO_2, dry chemical.

***sec*-BUTYL CHLORIDE.** Syn: *2-chlorobutane*.
C_4H_9Cl, mw: 92.6, flash p: 32°F, d: 0.87, vap. d: 3.2, bp: 68.50.
Acute tox data: Oral LD_{50} (rat) = 20,000 mg/kg; inhal LC_{LO} (rat) = 8000 ppm for 4 hrs. [*3*]
THR = NONE via oral and MOD via inhal routes.
Fire Hazard: Dangerous from heat, open flame (sparks) or oxidizers.
To Fight Fire: Water, water spray, fog, mist, dry chemical, alcohol foam.

***tert*-BUTYL CHLORIDE.** Syn: *2-chloro methyl propane*. C_4H_9Cl, mw: 92.6, flash p: 32°F, d: 0.87, vap. d: 3.2, bp: 51°.
THR = U. See also *n*- and *sec*-butyl chlorides.
Fire Hazard: Dangerous via heat, flame (sparks) and oxidizers.
To Fight Fire: Water, spray, fog, alcohol foam, dry chemical.

4-*tert*-BUTYL-2-CHLOROPHENOL. $C_{10}H_{13}OCl$, mw: 184.5, bp: 234°–251°, d: 1.1, flash p: 225°F.
Fire Hazard: Low, when exposed to heat or flame.
Disaster Hazard: Dangerous. See chlorides.
To Fight Fire: Spray or mist, CO_2, dry chemical, foam.

20-*tert*-BUTYL CHOLANTHRENE. $C_{24}H_{22}$, mw: 310.5.
THR = An exper neo. [*3*]

***tert*-BUTYL CHROMATE.** $C_8H_{18}O_4Cr$, mw: 230.3.
THR = HIGH. See chromium compounds.

BUTYL CITRATE. See tributyl citrate.

6-*n*-BUTYL-*m*-CRESOL. $C_{11}H_{16}O$, mw: 164, d: 1.0, bp: 142°, flash p: 244°F.
Acute tox data: Oral LD_{50} (rat) = 1100 mg/kg. [*3*]
THR = MOD via oral route.
Fire Hazard: Slight, when exposed to heat or flame.
To Fight Fire: Water, foam.

***tert*-BUTYL-*m*-CRESOL.** Syn: *2-tert-butyl-p-cresol.* Clear liquid, sol in organic solvents and aqueous potassium hydroxide. $C_{11}H_{16}O$, mw: 164.3, fp: 23.1°, bp: 244°, d: 0.922 80°, flash p: 116°F.
Acute tox data: Oral LD_{50} (rat) = 2390 mg/kg; oral LD_{50} (mouse) = 700 mg/kg; dermal LD_{50} (rabbit) = 2200 mg/kg. [*3*]
THR = MOD via oral and dermal routes.
Fire Hazard: Mod, when exposed to heat, flame or oxidizers.
To Fight Fire: Alcohol foam, foam, water spray, fog, dry chemical.

BUTYL CROTONATE. Water white liquid, pleasant persistent odor. $CH_3CHCHCOOC_4H_9$, mw: 142.2, bp: 180.5°, d: 0.9037 @ 20°/20°, vap. d: 4.9.
THR = U.
Fire Hazard: Dangerous, when exposed to heat or flame; can react with oxidizing materials.

***n*-BUTYL CYCLOHEXYL AMINE.** $C_4H_9C_6H_{10}NH$, mw: 154, flash p: 200°F (OC), d: 0.8, bp: 210°.
THR = Limited animal exper suggest at least MOD tox, irr. See also amines.
Fire Hazard: Low, when exposed to heat or flame.
To Fight Fire: Alcohol foam.

4-*tert*-BUTYL CYCLOHEXANOL. $C_{10}H_{20}O$, mw: 156.3.
Acute tox data: Oral LD_{50} (rat) = 4200 mg/kg. [*3*]
THR = MOD via oral route.

BUTYL DECALIN. Liquid. $C_4H_9C_{10}H_{17}$, mw: 194.35, flash p: 500°F (CC), vap. d: 6.7.
THR = U. See also decalin.
Fire Hazard: Slight, when exposed to heat or flame; can react with oxidizing materials.
To Fight Fire: Spray, mist, CO_2, dry chemical.

***tert*-BUTYL DECALIN.** Liquid. $C_4H_9C_{10}H_{17}$; mw: 194.4, flash p: 640°F (CC), vap. d. 6.7.
THR = U.
Fire Hazard: Slight, when exposed to heat or flame; can react with oxidizing materials.
To Fight Fire: Spray, mist, CO_2, dry chemical.

BUTYL DIAMYLAMINE. Liquid. $(C_4H_9)(C_5H_{11})_2N$, mw: 213.4, bp: 229°, flash p: 200°F, d: 0.78, vap. d: 7.3.
THR = U. See also amines.
Fire Hazard: Mod, when exposed to heat or flame; can react with oxidizing materials.
To Fight Fire: Foam, CO_2, dry chemical.

14-BUTYL DIBENZ(*a,h*)ACRIDINE. $C_{25}H_{21}N$, mw: 335.5.
THR = An exper neo. [*3*] to mice via sc route.

BUTYL DICHLOROARSINE. Oily liquid, somewhat agreeable odor. $C_4H_9AsCl_2$, mw: 202.945, bp: 194°.
THR = See arsenic compounds. HIGH, military poison gas.
Disaster Hazard: See arsenic compounds.

BUTYL DICHLOROBORANE. $C_4H_9Cl_2B$, mw: 138.7.
THR = No data. Spont flam in air. [*19*]

BUTYL DICHLORO PHENOXYACETATE.
Acute tox data: Oral LD_{50} (rat) = 1500 mg/kg. [*3*]
THR = MOD via oral route.

BUTYL-2,4-DICHLORO PHENOXY ACETATE. Syn: *2,4-d-butyl ester*. Light brown liquid. $C_6H_3(Cl)_2OCH_2COO(CH_2)_3CH_3$, mw: 277.15, d: 1.235–1.245 @ 25°/25°.
THR = An oral teratogen to rats. [*3*]
Disaster Hazard: Dangerous; when heated to decomp, emits toxic fumes.

1-*m*-BUTYL-3-(3,4-DICHLOROPHENYL)-1-METHYLUREA. See neburon.

BUTYL DIETHANOLAMINE. Liquid. $C_4H_9N(CH_2CH_2OH)_2$, mw: 161.24, bp: 262°, flash p: 245°F (OC), d: 0.97, vap. d: 5.55.
Acute tox data: Oral LD_{50} (rat) = 4250 mg/kg. [*3*]
THR = MOD via oral route. No chronic effects data.
Fire Hazard: Slight, when exposed to heat or flame; can react with oxidizing materials.
To Fight Fire: Alcohol foam, foam, CO_2, dry chemical.

***tert*-BUTYL DIETHANOLAMINE.** Syn: [*2,2-(tert-butylimino) diethanol*]. $C_8H_{10}NO_2$, mw: 152, mp: 47.5°, bp: 165°–170° @ 33 mm, d: 1.0, flash p: 285°F (OC).
Fire Hazard: Slight, when exposed to heat or flame.
To Fight Fire: Alcohol foam.

BUTYL DIETHYLHEPTYLATE. Liquid. $C_4H_8[COOCH_2CH(C_2H_5)C_4H_8]_2$, mw: 370.56, mp: −60°, bp: 214° @ 5 mm, flash p: 385°F (OC), d: 0.9268 @ 20°/20°, vap. d: 12.8.
THR = U. See also esters.
Fire Hazard: Slight, when exposed to heat or flame; can react with oxidizing materials.
To Fight Fire: Foam, dry chemical, CO_2.

1-BUTYL-3,3-DIMETHYL-1-NITROSOUREA. $C_7H_{15}O_2N_3$, mw: 173.3. An exper carc to rats via oral route. [*103*]

2-*sec*-BUTYL-4,6-DINITROPHENOL. See dinitro-*o*-*sec*-butyl phenol.

6-*sec*-BUTYL-2,4-DINITROPHENYL ACETATE. $C_{12}H_{14}O_6N_2$, mw: 282.3.
Acute tox data: Oral LD_{50} (rat) = 55 mg/kg. [*3*]
THR = HIGH via oral route.

***n*-BUTYL DISULFIDE.** Syn: *1-butyl dithiobutane.* $[CH_3(CH_2)_3]SS[CH_3(CH_2)_3]$, mw: 178.4, bp: 103° @ 15 mm.
THR = Probably toxic. See alkyl disulfides.
Fire Hazard: Probably mod.
Disaster Hazard: Dangerous; when heated to decomp, emits highly toxic fumes of sulfides. Probably reacts strongly with powerful oxidizers.
To Fight Fire: Water spray, foam, CO_2, dry chemical.

α-BUTYLENE. Syn: *1-butene.* A colorless flam gas, slightly aromatic odor. $CH_3CH_2CHCH_2$, mw: 56.10, bp: −6.3°, fp: −185.3°, lel = 1.6%, uel = 10%, flash p: −112°F, d: 0.668 @ 0°/1°, vap. d: 1.93, vap. press: 3480 mm @ 21°, autoign. temp.: 723°F.
THR = An asphyxiant.
Fire Hazard: Very dangerous, when exposed to heat, flame or oxidizers.
Spont Heating: No.
Explosion Hazard: Mod, when exposed to flame.
Disaster Hazard: Mod dangerous; can react with oxidizing materials.
To Fight Fire: Stop flow of gas.

β-BUTYLENE. See 2-butene.

γ-BUTYLENE. See isobutylene.

BUTYLENE CHLORIDE. Colorless liquid. C_4H_7Cl, mw: 90.6, lel = 2.3%, uel = 9.3%, bp: 72°, d: 0.926, vap. d: 3.13.
THR = U. See also chlorinated HC, aliphatic.
Fire Hazard: Dangerous, when exposed to heat or flame.
Explosion Hazard: Mod, when exposed to flame.
Disaster Hazard: Dangerous; when heated to decomp, emits highly toxic fumes of phosgene; can react vigorously with oxidizing materials.
To Fight Fire: CO_2, dry chemical.

α-BUTYLENE DIBROMIDE. Syn: *1,2-dibromo butane.* Yellowish liquid, insol in water, miscible with water. d: 1.820 @ 20°/4°, mp: −65°, bp: 166°.
THR = An irr and narcotic. See also bromides.
Disaster Hazard: Dangerous; see bromides.

α-BUTYLENE GLYCOL. See 1,2-butanediol.

β-BUTYLENE GLYCOL. See 1,3-butylene glycol.

1,2-BUTYLENE GLYCOL. See 1,2-butanediol.

1,3-BUTYLENE GLYCOL. Syns: *1,3-butanediol, β-butylene glycol.* Viscous liquid. $CH_3CH(OH)CH_2CH_2OH$, mw: 90.12, bp: 207.5°, fp: <−50°, flash p: 250°F, d: 1.006 @ 20°/20°, autoign. temp.: 741°F, vap. press: 0.06 mm @ 20°, vap. d: 3.2.
Acute tox data: Oral LD_{50} (rat) = 23 g/kg. [*3*]
THR = LOW via oral route. See also glycols. A food additive permitted in food for human consumption. [*109*]
Fire Hazard: Slight, when exposed to heat or flame; can react with oxidizing materials.
Spont Heating: No.
To Fight Fire: Foam, alcohol foam, CO_2, dry chemical.

1,4-BUTYLENE GLYCOL. See 1,4-butanediol.

2,3-BUTYLENE GLYCOL. Syns: *2,3-butanediol, pseudo butylene glycol.* Colorless liquid or solid. $CH_3CH(OH)CH(OH)CH_3$, mw: 90.12, bp: 180°, fp: 19°, flash p: 185°F (TOC), d: 1.0095 @ 20°/20°, autoign. temp.: 756°F, vap. press: 0.17 mm @ 20°, vap. d: 3.1.
THR = U. See also glycols.
Fire Hazard: Mod, when exposed to heat or flame; can react with oxidizing materials.
To Fight Fire: Alcohol foam, CO_2, dry chemical.

BUTYLENE OXIDE. See tetrahydrofuran.

1,2-BUTYLENE OXIDE. See 1,2-epoxy butane.

N-BUTYLENE PYRROLIDINE. Colorless to light yellow liquid, penetrating amine-like odor. $C_4H_8NC_4H_7$, mw: 125.2, bp: 154°, fp: −75°, flash p: 93°F, d: 0.837.
THR = A strongly alkaline material. See ammonia. Probably HIGH. An irr.
Fire Hazard: Dangerous, when exposed to sparks, heat, open flame or powerful oxidizers.
Disaster Hazard: Dangerous; when heated to decomp, emits highly toxic fumes.
To Fight Fire: Water, spray, foam, CO_2, dry chemical.

BUTYL-9,10-EPOXY STEARATE. $C_{22}H_{42}O_3$, mw: 354.6.
THR = An exper neo. [*3*]

BUTYL ESTER-2,4-D. See butyl-2,4-dichlorophenoxy acetate.

BUTYL ETHANEDIOATE. See butyl oxalate.

BUTYL ETHANOATE. See butyl acetate.

BUTYLETHANOLAMINE. Syn: *2-butylaminoethanol.* Liquid. $CH_3(CH_2)_3NHCH_2CH_2OH$, mw: 117.19, bp: 192°, flash p: 170°F (OC), d: 0.89, vap. d: 4.03.
Acute tox data: Oral LD_{50} (rat) = 1150 mg/kg. [*3*]
THR = MOD via oral route. See also amines.
Fire Hazard: Mod, when exposed to heat or flame; can react with oxidizing materials.
Spont Heating: No.
To Fight Fire: Alcohol foam, foam, CO_2, dry chemical.

BUTYL ETHER. Syns: *1-butoxybutane, n-dibutyl ether.* Colorless liquid. $CH_3(CH_2)_3O(CH_2)_3CH_3$, mw: 130.23, mp: −95°, bp: 142°, flash p: 77°F, d: 0.769 @ 20°/20°, autoign. temp.: 382°F, vap. d: 4.48, lel = 1.5%, uel = 7.6%.
Acute tox data: Oral LD_{50} (rat) = 11 g/kg; dermal LD_{50} (rabbit) = 10 g/kg. [*3*]
THR = LOW via oral and dermal routes; MOD via inhal route.
Fire Hazard: Dangerous; see ethers. Reacts violently with NCl_3. [*19*]
Spont Heating: No.
Explosion Hazard: Mod, see also ethers.
Disaster Hazard: Mod dangerous; when heated, emits acrid fumes; can react with oxidizing materials.
To Fight Fire: Alcohol foam, dry chemical.

BUTYL ETHYL ACETALDEHYDE. See 2-ethylhexaldehyde.

BUTYL ETHYL ACETIC ACID. See 2-ethylhexoic acid.

BUTYL ETHYL "CELLOSOLVE." Liquid. $C_4H_9OCH_2CH_2OC_2H_5$, mw: 146.22, mp: −90°, bp: 164.2°, d: 0.8389.
THR = See glycols and "cellosolves."
Fire Hazard: Slight; can react with oxidizing materials.

BUTYL ETHYLENE. See hexene-1.

BUTYL ETHYL ETHER. See ethyl-*n*-butylether.

BUTYL ETHYL KETENE. Syn: *BEK.* Yellow liquid with pungent odor. $C_8H_{14}O$, mw: 126.1, d: 0.8266 @ 20°/4°, bp: 36° @ 12 mm; mp: <−80°, flash p: 64°F (TOC).
THR = U. An irr.
Fire Hazard: Dangerous, via heat, flames or oxidizers.
To Fight Fire: Foam, fog, mist, dry chemical.

BUTYL FLUORIDE. C_4H_9F, mw: 76.
THR = No data. See fluorides. Reacts violently with $MgClO_4$. [*19*]

BUTYL FORMAL. See formaldehyde and butyl alcohol.

BUTYL FORMATE. Syn: *butyl methanoate.* Colorless liquid. $HCOOCH_2CH_2CH_2CH_3$, mw: 101.12, mp: −90°, bp: 106.0°, flash p: 64°F (CC), d: 0.911, autoign. temp.: 612°F, vap. press: 40 mm @ 31.6°, vap. d: 3.52, lel = 1.7%, uel = 8%.
Acute tox data: Oral LD_{50} (rabbit) = 2660 mg/kg. [*3*]
THR = MOD via oral route. An irr and narcotic in high conc. See esters, butyl alcohol and formic acid.
Fire Hazard: Dangerous, when exposed to heat or flame; can react with oxidizing materials.
Spont Heating: No.
To Fight Fire: Alcohol foam, foam, CO_2, dry chemical.

***n*-BUTYL GLYCIDYL ETHER.** Syn: *BGE.* $C_7H_{14}O_2$, mw: 130.1.
Acute tox data: Oral LD_{50} (mice) = 1520 mg/kg; oral LD_{50} (rat) = 2050 mg/kg; inhal LC (rat) = 670 ppm. [*3*]
THR = MOD via oral and inhal routes. No chronic data.

BUTYL GLYCOLATE. $C_6H_{12}O_3$, mw: 132.2, flash p: 142°, d: 1.01, vap. d: 4.45, bp: 184°.
THR = No data. Probably LOW.
Fire Hazard: Mod, via heat, open flame (sparks), oxidizers.
To Fight Fire: Water spray, mist, fog, alcohol foam, dry chemical.

BUTYL HYDRAZINE HYDROCHLORIDE. $C_4H_{12}N_2 \cdot HCl$, mw: 124.6.
THR = An exper carc. [*3*]

BUTYL HYDRIDE. See butane.

***tert*-BUTYL HYDROPEROXIDE.** Water white liquid, slightly sol in water, very sol in esters and alcohols. $(CH_3)_3COOH$, mw: 90.12, flash p: 80°F or above, fp: −35°, d: 0.860, vap. d: 2.07.
Acute tox data: Oral LD_{50} (rat) – 406 mg/kg; inhal LC_{50} (rat) = 500 ppm for 4 hrs. [*3*]
THR = HIGH via oral and inhal routes. Limited animal exper indicate MOD via oral administration. At highest dosage levels, symptoms noted were severe depression, incoordination and cyanosis. Death was due to respiratory arrest.
Fire Hazard: Very dangerous, when exposed to heat or flame, or by spont chemical reaction; can react with reducing materials.
Explosive Hazard: Mod.
To Fight Fire: Alcohol foam, CO_2, dry chemical.

6-BUTYL-4-HYDROXYAMINO QUINOLINE-1-OXIDE. $C_{13}H_{16}O_2N_2$, mw: 232.3.
THR = An exper neo. [*3*]

BUTYL HYDROXYANISOLE. Syn: *BHA*.
$C_{11}H_{16}O_2$, mw: 130.2.
Acute tox data: Oral LD_{LO} (rat) = 1000 mg/kg. [*3*]
THR = MOD via oral route. Used as an antioxidant in foods. [*109*]

BUTYL-12-HYDROXY-9-OCTADECENOATE. See butyl ricinoleate.

***tert*-BUTYL HYPOCHLORITE.** $CH_3C(CH_3)_2OCl$, mw: 106.6.
THR = HIGH. See hypochlorites. Spont explosive. [*19*]

***p,p-sec*-BUTYLIDENEDIPHENOL.** Tan, granular solid. $(HOC_6H_4)_2C(CH_3)(C_2H_5)$, mw: 242.3, mp: 118.9°–121.7°.
THR: Details U. See also phenols.
Fire Hazard: Slight; when heated, emits acrid fumes; can react with oxidizing materials.

***n*-BUTYL ISOCYANATE.** Colorless liquid. $CH_3CH_2CH_2CH_2NCO$, mw: 99.1, bp: 115°, d: 0.880 @ 20°/4°.
THR = HIGH. A powerful irr to eyes, skin and mu mem. Flam liquid.
Disaster Hazard: See cyanates.

***tert*-BUTYL ISOPROPYL BENZENE HYDROPEROXIDE.** Crystals. $(C_4H_9)(C_6H_4)(C_3H_6OOH)$, mw: 208.3.
THR = Powerful irr.
Fire Hazard: Mod, when exposed to heat or flame or by chemical reaction; can react with oxidizing or reducing materials.

BUTYL ISOTHIOCYANATE. Syn: *butyl mustard oil.* $C_2H_5CH_2CH_2CH_2NCS$, mw: 115, d: 1.0, vap. d: 4.0, flash p: 150°F, mp: 342°–347°F.
THR = Powerful local irr. No chronic effects data.
Fire Hazard: Mod, when exposed to heat or flame.
Disaster Hazard: Dangerous; see thiocyanates.

BUTYL ISOVALERATE. $C_9H_{18}O_2$, mw: 158.2, d: 0.87, vap. d: 5.45, bp: 150°.
THR = No data.
Fire Hazard: Mod, via heat, flame (sparks) and oxidizers.
To Fight Fire: Alcohol foam, dry chemical, spray, mist, fog.

BUTYL LACTATE. Liquid. $CH_3CH(OH)COOC_4H_9$, mw: 146.18, mp: −43°, bp: 188°, flash p: 160°F (OC), d: 0.968, autoign. temp.: 720°F, vap. d: 5.04, vap. press: 0.4 mm @ 20°.
THR = HIGH via inhal route. A strong irr. Toxic conc in air for humans = 4 ppm. [*3*] See esters, butyl alcohol and lactic acid.
Fire Hazard: Mod, when exposed to heat or flame; can react with oxidizing materials.
To Fight Fire: Alcohol foam, foam, CO_2, dry chemical.

BUTYL LITHIUM. $CH_3CH_2CH_2CH_2Li$, mw: 63.94.
THR = U. Probably toxic.
Fire Hazard: Very dangerous. Extremely flam. Ignites on contact with moist air.
Disaster Hazard: Very dangerous. Heat or moisture can cause it to ignite and burn rapidly.
To Fight Fire: Dry chemical; see special instructions of manufacturer.

***n*-BUTYL MAGNESIUM CHLORIDE.** Colorless liquid, sol in ether and tetrahydrofuran. C_4H_9MgCl, mw: 116.9, d: 0.88.
THR = U. Probably toxic.
Disaster Hazard: Dangerous. Heat, source of ignition or powerful oxidizer can cause fires. See chlorides.
To Fight Fire: Water spray, alcohol foam and dry chemical.

BUTYL MERCAPTAN. Syns: *1-butanethiol*, *n-butyl thioalcohol.* Colorless liquid, skunk-like odor. $C_4H_{10}S$, mw: 90.18, mp: −116°, bp: 98°, d: 0.8365 @ 25°/4°, flash p: 35°F, vap. d: 3.1.
Acute tox data: Inhal LC_{50} (rat) = 4020 ppm for 4 hrs; oral LD_{50} (rat) = 1500 mg/kg; ip LD_{50} (rat) = 399 mg/kg. [*3*] Inhal LC_{50} (dogs) = 700 ppm for 30 min; TC (human) = 10 mg/m^3 for 3 hrs. [*3*]
THR = MOD via inhal and oral routes; HIGH via ip route. Reacts violently with HNO_3 [*19*]
Fire Hazard: Dangerous by exposure to heat, flame, sparks or powerful oxidizers.
Disaster Hazard: Dangerous; when heated to decomp or on contact with acid or acid fumes, emits highly toxic fumes; can react vigorously with oxidizing materials.
To Fight Fire: Alcohol foam.

***sec*-BUTYL MERCAPTAN.** Syn: *2-butanethiol.* Mobile liquid, skunk-like odor. $C_4H_{10}S$, mw: 90.2, mp: −165°, bp: 85°, d: 0.83 @ 17°, flash p: −10°F.
THR = No specific data. Probably MOD via oral and inhal routes. An irr.
Fire Hazard: Dangerous, when exposed to heat, flame or oxidizers.
Disaster Hazard: Dangerous; when heated to decomp or on contact with acid or acid fumes, emits highly toxic fumes; can react vigorously with oxidizing materials.
To Fight Fire: Water spray, fog, foam, dry chemical.

***tert*-BUTYL MERCAPTAN.** Syn: *2-methyl-2-propanethiol.* Liquid, skunk-like odor. $C_4H_{10}S$, mw: 90.2, mp: −0.5°, d: 0.79–0.82 @ 15.5°/15.5°, flash p: <−20°F, bp: 62–67°, vap. d: 3.1.

Acute tox data: Oral LD_{50} (rat) = 4729 mg/kg; ip LD_{50} (rat) = 590 mg/kg. [*3*]
THR = MOD via oral and HIGH via ip routes.
Fire Hazard: Dangerous, when exposed to heat or flame.
Disaster Hazard: Dangerous; when heated to decomp or on contact with acid or acid fumes, emits highly toxic fumes; can react vigorously with oxidizing materials.
To Fight Fire: Alcohol foam, dry chemical, mist, fog.

9-BUTYL-6-MERCAPTOPURINE. $C_9H_{12}N_4S$, mw: 208.3.
THR = An exper teratogen. [*3*]

BUTYL MESITYL OXIDE. See butopyronoxyl.

***n*-BUTYL MESITYL OXIDE OXALATE.** See α,α-dimethyl-α-carbobutoxyhydro-γ-pyrone.

BUTYL METHACRYLATE, MONOMER. Colorless liquid, ester odor. $CH_2C(CH_3)COOC_4H_9$, mw: 142.19, bp: 163°, flash p: 126°F (TOC), lel = 2%, uel = 8%, autoign. temp.: 562°F, vap. press: 4.9 mm @ 20°, d: 0.895 @ 20°/4°, vap. d: 4.8.
Acute tox data: Ip LD_{50} (rat) = 2304 mg/kg; oral LD_{LO} (rabbit) = 6270 mg/kg. [*3*]
THR = MOD via ip and oral routes. MOD irr to skin, eyes and mu mem. An exper teratogen. [*3*]
Fire Hazard: Mod, when exposed to flame; can react with oxidizing materials.
Explosion Hazard: Some, when exposed to flame or sparks. Heat, moisture or oxidizers can cause violent polymerization.
To Fight Fire: Foam, dry chemical, CO_2.

BUTYL METHANE SULFONATE.
THR = An exper carc. [*23*]

BUTYL METHANOATE. See butyl formate.

***n*-BUTYL-α-METHYLBENZYLAMINE.** $C_{12}H_{19}N$, mw: 177.3.
Acute tox data: Oral LD_{50} (rat) = 360 mg/kg; dermal LD_{50} (rabbit) = 570 mg/kg. [*3*]
THR = HIGH via oral, MOD via dermal routes.

BUTYL METHYL HYDRAZINE. $C_5H_{14}N_2$, mw: 102.2.
THR = An exper carc. [*3*]

BUTYL METHYL KETONE. See methyl butyl ketone.

BUTYL MONO ETHANOLAMINE. See butyl ethanol amine.

BUTYL NAPHTHALENE. Liquid. $C_4H_9C_{10}H_7$, mw: 184.3, flash p: 680°F (CC), vap. d: 6.2.
THR = U. See also naphthalene.
Fire Hazard: Slight, when exposed to heat or flame; can react with oxidizing materials.
To Fight Fire: Spray or mist, CO_2, dry chemical.

BUTYL NICOTINATE. Colorless liquid. $C_{10}H_{13}NO_2$, mw: 179.3, bp: 122°–123° @ 8 mm, d: 1.0471 @ 25°/4°.
THR = Details U. See also esters.
Disaster Hazard: Dangerous; when heated to decomp, emits highly toxic fumes; can react vigorously with oxidizing materials.

BUTYL NITRATE. Liquid. $CH_3(CH_2)_3ONO_2$, mw: 119.1, bp: 136°, flash p: 97°F, d: 1.048 @ 0°/4°, vap. d: 4.0.
THR = Details U. See also nitrates and butyl alcohol.
Fire Hazard: Dangerous, when exposed to heat or flame or by spont chemical reaction. An oxidizer.
Explosion Hazard: See nitrates.
Disaster Hazard: Dangerous; see nitrates; can react vigorously with oxidizing or reducing materials.
To Fight Fire: Alcohol foam, spray, mist, fog, dry chemical.

***sec*-BUTYL NITRATE.** Syn: *α-methyl propyl nitrate.* Liquid. $C_2H_5CH(CH_3)ONO_2$, mw: 119.1, bp: 124°, d: 1.0382 @ 0°/4°, vap. d: 4.0.
THR = Details U. See nitrates and *sec*-butyl alcohol.
Fire Hazard: Mod, when exposed to heat or flame or by spont chemical reaction. An oxidizer.
Explosion Hazard: See nitrates.
Disaster Hazard: See nitrates.
To Fight Fire: Water, spray, foam, dry chemical.

N-BUTYL NITRITE. Oily liquid, characteristic odor, miscible in alcohol and ether. $CH_3(CH_2)_3ONO$, mw: 103.1, bp: 75°, d: 0.9114 @ 0°/4°, vap. d: 3.5.
THR = MOD via oral and inhal routes. An irr. Resembles amyl nitrite in causing fall in blood pressure, headache, throbbing and weakness. See also nitrites and *sec*-butyl alcohol.
Fire Hazard: Mod, when exposed to heat or flame or by spont chemical reaction.

***tert*-BUTYL NITRITE.** Yellow liquid, agreeable odor, bp: 63°, d: 0.8671 @ 20°/4°.
THR = See nitrites, amyl nitrite and *n*-butyl nitrite. Probably MOD via oral and inhal routes. An oxidizer.
Explosion Hazard: See nitrites.
Disaster Hazard: See nitrites.

***sec*-BUTYL NITRITE.** Syn: *α-methyl propyl nitrite.* Liquid. $C_2H_5CH(CH_3)ONO$, mw: 103.1, bp: 68°, d: 0.8981 @ 0°/4°, vap. d: 3.5.
THR = Details U. See butyl nitrite.
Fire Hazard: Mod, when exposed to heat or flame or by spont chemical reaction. An oxidizer.
Explosion Hazard: See nitrites.
Disaster Hazard: See nitrites.
To Fight Fire: Water, spray, foam, dry chemical.

***tert*-BUTYL NITRITE.** Syn: *α,α-dimethyl ethyl nitrite*. Yellowish liquid. $(CH_3)_3CONO$, mw: 103.1, bp: 63°, d: 0.8941 @ 0°/4°, vap. d: 3.5.
THR = Details U. See butyl nitrite.
Fire Hazard: Mod, when exposed to heat or flame or by spont chemical reaction. An oxidizer.
Explosion Hazard: See nitrites.
Disaster Hazard: See nitrites.
To Fight Fire: Water spray, foam, dry chemical, fog.

6-BUTYL-4-NITROQUINOLINE-1-OXIDE. $C_{13}H_{14}O_3N_2$, mw: 246.3.
THR = An exper neo via sc route. [*3, 23*]

4-(BUTYLNITROSAMINO)-1-BUTANOL. $C_8H_{18}O_2N_2$, mw: 174.3.
THR = An exper carc. [*3*]

2-(BUTYLNITROSAMINO)ETHANOL. $C_6H_{14}O_2N_2$, mw: 146.2.
THR = An exper carc via oral route. [*3, 23*]

1-(BUTYLNITROSAMINO)-2-PROPANONE. $C_7H_{14}O_2N_2$, mw: 158.2.
THR = An exper carc via oral route. [*3, 23*]

N-BUTYL-N-NITROSO-1-BUTANAMINE. A pale yellow liquid. $C_8H_{18}ON_2$, mw: 158.3; bp: 234°–237°.
THR = An exper carc. [*3, 10*]

N-BUTYL-N-NITROSOETHYL CARBAMATE. $C_7H_{14}O_3N_2$, mw: 174.2.
THR = An exper carc. [*3*]

N-BUTYL-N-NITROSO PENTYLAMINE. $C_9H_{20}ON_2$, mw: 172.3.
Acute tox data: Sc LD_{50} (rat) = 2500 mg/kg. [*103*]
THR = MOD via sc route.

1-BUTYL-1-NITROSOUREA. $C_5H_{11}O_2N_3$, mw: 145.2.
THR = HIGH acute oral. An exper carc and neo. [*3, 23*]

2-BUTYL OCTANOL. Liquid. $C_6H_{13}CH(C_4H_9)(CH_2OH)$, mw: 186.33, mp: −80°, flash p: 230°F (OC), bp: 253.3°, d: 0.8355 @ 20°/20°, vap. d: 6.42.
Fire Hazard: Slight, when exposed to heat or flame; can react with oxidizing materials.
To Fight Fire: CO_2, dry chemical.

BUTYL OLEATE. Liquid. $C_{17}H_{33}COOC_4H_9$, mw: 338.56, bp: 173°, flash p: 356°F (OC), d: 0.873, vap. d: 11.3.
THR = Details U. See also esters, butyl alcohol and oleic acid.
Fire Hazard: Slight, when exposed to heat or flame; can react with oxidizing materials.
To Fight Fire: CO_2, dry chemical.

BUTYL OXALATE. Syn: *butyl ethane dioate*. Liquid. $(COOC_4H_9)_2$, mw: 202.24, flash p: 265°F (OC), d: 0.989–0.993, vap. d: 7.0.
THR = Details U. See also esters, butyl alcohol and oxalic acid.
Fire Hazard: Slight, when exposed to heat or flame; can react with oxidizing materials.
To Fight Fire: CO_2, dry chemical, mist, fog, spray.

11-BUTYL-17-OXO-16,17-*d*-HYDRO-15H-CYCLO PENTA(*a*)PHENANTHRENE.
THR = An exper carc. [*23*]

BUTYLOXYETHYL SALICYLATE. $OCH_6H_4COOCH_2CH_2OC_4H_9$, mw: 183, d: 1.0+, bp: 367°–378F, flash p: 315°F.
Fire Hazard: Slight, when exposed to heat or flame.
To Fight Fire: Water, foam.

***tert*-BUTYL PERACETATE** (solution in benzene). Clear, colorless solution, insol in water, sol in organic solvents. $CH_3CO(O_2)C(CH_3)_3$ + benzene, mw: 132.2, d: 0.923, vap. press: 50 mm @ 26°, flash p: <80°F (COC).
THR = Sensitive to shock and heat; can explode in contact with organic matter. [*19*]
Fire Hazard: Dangerous via heat, flame, reducers.
To Fight Fire: Dry chemical, alcohol foam, spray and mist.

***tert*-BUTYL PERBENZOATE.** Colorless to slight yellow liquid, mild aromatic odor. Insol in water, sol in organic solvents. $C_6H_5COOOC(CH_3)_3$, mw: 194.2, bp: 112° (decomp), flash p: <190°F (OC), fp: 8°, vap. press: 0.33 mm @ 50°, d: 1.0+.
THR = Mod acute oral. An exper neo. [*3*] See peroxides, organic.
Fire Hazard: See peroxides, organic.
Explosion Hazard: Dangerous in contact with organic matter. [*19*]
Disaster Hazard: See peroxides, organic.
To Fight Fire: See peroxides, organic.

***tert*-BUTYL PEROXIDE.** See di-*tert* butyl peroxide.

***tert*-BUTYL PEROXYISOBUTYRATE.** Solution is colorless to yellow liquid, insol in water or glycerine, sol in alcohols, hydrocarbons, esters, ethers and ketones. $C_8H_{16}O_3$ + benzene, mw: (of solute): 160.2, d: 0.90 @ 25°, flash p: <80°F (micro OC).
THR = See benzene.
Fire Hazard: Highly flam; see benzene and peroxides, organic.
Explosion Hazard: See peroxides, organic.
Disaster Hazard: See peroxides, organic.
To Fight Fire: Foam, spray, dry chemicals.

***tert*-BUTYL PEROXYPIVALATE.** Colorless liquid, insol in water and ethylene glycol, sol in most organic solvents. $C_9H_{18}O_3$, mw: 174.3, d: 0.854 @ 25°/25°, fp: <19°, flash p: >155°F (OC), rapid decomp @ 21°.

Acute tox data: Oral LD_{50} (rat) = 4300 mg/kg. [*3*]
THR = MOD via oral route.
Fire Hazard: Moderate via heat, flame (sparks), oxidizers.
Explosion Hazard: Explodes on heating.
To Fight Fire: Water, fog, mist, alcohol foam, dry chemical.

***o-sec*-BUTYLPHENOL.** Colorless liquid. $(CH_3CHC_2H_5)C_6H_4OH$, mw: 150.2, bp: 226°–228° @ 25 mm, fp: 12°, flash p: 225°F, d: 0.981 @ 25°/25°.
Acute tox data: Oral LD_{50} (rat) = 2700 mg/kg. [*3*]
THR = Irr via oral route. See phenol.
Fire Hazard: Low, when exposed to heat or flame; can react with oxidizing materials.
To Fight Fire: Foam, CO_2, dry chemical.

***p-sec*-BUTYL PHENOL.** Nearly white flakes. $(CH_3CHC_2H_5)C_6H_4OH$, mw: 150.2, bp: 135.4°–136.5° @ 25 mm, fp: 51°, flash p: 240°F, d: 0.963 @ 60°/60°.
Acute tox data: Oral LD_{50} (rat) = 2700 mg/kg. [*3*]
THR = MOD via oral route.
Fire Hazard: Slight, when exposed to heat or flame.
Disaster Hazard: Mod; when heated to decomp, emits toxic fumes; can react with oxidizing materials.
To Fight Fire: Foam, CO_2, dry chemical.

***p-tert*-BUTYLPHENOL.** Crystals or practically white flakes. $C_4H_9C_6H_4OH$, mw: 150.2, bp: 238°, fp: 97°, d: 0.9081 @ 114°/4°, vap. press: 1 mm @ 70.0°, vap. d: 5.1.
Acute tox data: Oral LD_{50} (rat) = 3250 mg/kg; dermal LD_{50} (rabbit) 2520 mg/kg. [*3*]
THR = MOD via oral and dermal routes.
Disaster Hazard: Dangerous; when heated to decomp, emits toxic fumes.

2-(*p-sec*-BUTYL PHENOXY)ETHANOL. See ethylene glycol-*p-sec*-butyl phenyl ether.

2-(*p-tert*-BUTYL PHENOXY)ETHYL ACETATE. See ethylene gylcol-*p-tert*-butyl phenyl ether acetate.

1-(*o-sec*-BUTYL PHENOXY)-2-PROPANOL. See propylene glycol-*sec*-butyl phenyl ether.

1-(*p-sec*-BUTYL PHENOXY)-2-PROPANOL. See propylene glycol-*tert*-butyl phenyl ether.

***p-tert*-BUTYLPHENYL DIPHENYL PHOSPHATE.** See diphenyl mono-*p-tert*-butyl phenyl phosphate.

BUTYL PHENYL ETHER. See butoxy phenyl.

***p*-((*p*-BUTYLPHENYL)AZO)-N,N-DIMETHYL ANILINE.** $C_{18}H_{23}N_3$, mw: 281.4.
THR = An exper neo. [*3*]

***p*-((*p*-BUTYL-*tert*-PHENYL)AZO)-N,N-DIMETHYL ANILINE.**
THR = An exper neo. [*3*]

4-*tert*-BUTYL-2-PHENYL PHENOL. $C_6H_5C_6H_3OHC(CH_3)_3$, mw: 226, d: 1.0+, bp: 385°–388°F, flash p: 320°F.
Fire Hazard: Slight, when exposed to heat or flame.
To Fight Fire: Dry chemical, CO_2, mist, fog.

***n*-BUTYL PHTHALYL BUTYL GLYCOLATE.** Liquid. $C_{18}H_{24}O_6$, mw: 336.4, bp: 345°, flash p: 390°F (OC), d: 1.097, vap. d: 11.6.
Acute tox data: Oral LD_{50} (rat) = 15 g/kg. [*3*]
THR = LOW via oral route.
Fire Hazard: Slight, when exposed to heat or flame; can react with oxidizing materials.
Spont Heating: No.
To Fight Fire: Spray or mist, CO_2, dry chemical.

BUTYL PIPERONYLAMIDE. Crystals. $CH_2(O_2)C_6H_2(C_4H_9)CONH_2$, mw: 221.3, vap. d: 7.4.
THR = U. An insecticide.
Disaster Hazard: Mod dangerous; when heated to decomp, emits toxic fumes.

BUTYL PROPANOATE. See butyl propionate.

BUTYL PROPIONATE. Syn: *butyl propanoate*. Water white liquid, apple-like odor. $C_2H_5CO_2C_4H_9$, mw: 130.2, mp: −89.6°, bp: 145.4°, flash p: 90°F, d: 0.875 @ 20°, autoign. temp.: 800°F, vap. d: 4.49.
THR = MILD irr. See also esters, *n*-butyl alcohol and propionic acid.
Fire Hazard: Dangerous, when exposed to heat or flame; can react with oxidizing materials.
Spont Heating: No.
To Fight Fire: Foam, CO_2,dry chemical.

BUTYL RICINOLEATE. Syn: *butyl-12-hydroxy-9-octadecenoate*. Liquid. $C_{18}H_{33}O_3C_4H_9$, mw: 354.56, bp: 275° @ 13 mm, flash p: 230°F, d: 0.906, vap. d: 12.2.
THR = Details U. See also esters and butyl alcohol.
Fire Hazard: Slight, when exposed to heat or flame, can react with oxidizing materials.
To Fight Fire: CO_2, dry chemical, mist, fog.

BUTYL STEARAMIDE. Liquid. $C_4H_9C_{17}H_{34}CONH_2$, mw: 339.6, bp: 195°, flash p: 430°F, vap. d: 11.7.
THR = U. See also butyl stearate and amides.
Fire Hazard: Slight, when exposed to heat or flame; can react with oxidizing materials.
To Fight Fire: Foam, CO_2, dry chemical.

BUTYL STEARATE. Liquid. $C_{17}H_{35}COOC_4H_9$, mw: 340.57, mp: 19.5°, bp: 220°–225° @ 25 mm, flash p: 320°F (CC), d: 0.855 @ 25°/25°, vap. d: 11.4, autoign. temp.: 671°F.
THR = U. Limited animal exper suggest LOW. See also esters and *n*-butyl alcohol.
Fire Hazard: Slight, when exposed to heat or flame; can react with oxidizing materials.
To Fight Fire: CO_2, dry chemical, fog or mist.

1-BUTYL-3-SULFANILYL UREA. $C_{11}H_{17}O_3N_3S$, mw: 271.4.
THR = An exper teratogen. [*3*]

BUTYL SULFIDE. Syns: *dibutyl sulfide, butyl thiobutane*. Liquid. $(CH_3CH_2CH_2CH_2)_2S$, mw: 146.3, mp: −80°, bp: 182°, d: 0.839 @ 16°/0°, vap. d: 4.9.
THR = U. See also alkyl disulfides.
Fire Hazard: Mod, when exposed to heat or flame or on contact with acid or acid fumes.
Disaster Hazard: Dangerous; when heated to decomp, emits highly toxic fumes of SO_x; can react vigorously with oxidizing materials.

BUTYL TARTRATE. See dibutyl tartrate.

***tert*-BUTYL TETRALIN.** Liquid. $C_4H_9C_{10}H_{11}$, mw: 188.30, flash p: 680°F (CC), vap. d: 6.3.
THR = U. See also tetrahydronaphthalene.
Fire Hazard: Slight, when exposed to heat or flame; can react with oxidizing materials.
To Fight Fire: Water spray or mist, CO_2, dry chemical.

***n*-BUTYL TETRATHIO-*o*-STANNATE.** Solid. $(SC_4H_9)_4Sn$, mw: 475.4, bp: 136° @ 0.001 mm.
THR = Details U. See tin compounds.

***sec*-BUTYL TETRATHIO-*o*-STANNATE.** Solid. $(SC_4H_9)_4Sn$, mw: 475.4, bp: 136° @ 0.001 mm.
THR = Details U. See tin compounds.
Disaster Hazard: Dangerous. See sulfur compounds.

BUTYL TITANATE. Syn: *titanium butylate*. Colorless to light yellow liquid, odor of butanol. $Ti(OC_4H_9)_4$, mw: 340.4, mp: −55°, bp: 312°, flash p: 170°F, vap. d: 11.5.
THR = See butyl alcohol and titanium compounds.
Fire Hazard: Mod, when exposed to heat or flame; can react with oxidizing materials.
To Fight Fire: Water, spray, foam, dry chemical.

***sec*-BUTYL TITANATE.** A clear liquid. $Ti(OCH(CH_3)(C_2H_5))_4$, mw: 340, mp: −25° to −30°, bp: 138° @ 10 mm, d: 0.93.
THR = See esters and *sec*-butanol.
Fire Hazard: Mod, when exposed to heat or flame; can react with oxidizing materials.
To Fight Fire: Water, spray, foam, dry chemical.

***n*-BUTYL THIOALCOHOL.** See *n*-butyl mercaptan.

BUTYL THIOBUTANE. See butyl sulfide.

***m-tert*-BUTYL TOLUENE.**
Acute tox data: Inhal LC_{LO} (rat) = 4000 ppm for 1 hr. [*3*]
THR = MOD via inhal route.

***p-tert*-BUTYLTOLUENE.** Colorless liquid. $C_4H_9C_6H_4CH_2$, mw: 148.3.
Acute tox data: Oral LD_{50} (rat) = 1500 mg/kg; inhal LC_{50} (rat) = 15 mg/m^3 for 4 hrs; inhal TC_{LO} (human) = 10 ppm for 3 min ⟶ irr; and 20 ppm for 5 min ⟶ CNS symptoms. [*3*]
THR = MOD via oral; HIGH via inhal routes. An irr to eyes and mu mem. Inhal of vapors causes irr of lungs and depression of CNS. Prolonged exposure may result in damage to liver and kidneys.
Fire Hazard: Mod, when exposed to heat or flame.
Disaster Hazard: Dangerous; when heated, emits highly toxic fumes; can react with oxidizing materials.

***n*-BUTYL-*p*-TOLUENE SULFONATE.** $C_{11}H_{16}O_3S$, mw: 228.3.
Acute tox data: Sc LD_{50} (rat) = 5000 mg/kg. [*3*]
THR = MOD via sc route. An exper carc. [*23*]

BUTYL TRICHLORO SILANE. Liquid. $C_4H_9SiCl_3$, mw: 191.6, vap. d: 6.4, flash p: 130°F (OC), d: 1.2.
THR = MOD. A corrosive material. See also chlorosilanes.
Fire Hazard: Mod, via heat, flame (sparks), oxidizers.
To Fight Fire: Water to blanket fire, fog, mist, dry chemical, alcohol foam.
Disaster Hazard: Dangerous; when heated to decomp, emits highly toxic fumes of chlorides; will react with water or steam to produce heat and toxic and corrosive fumes.

3-*tert*-BUTYL TRICYCLO QUINAZOLINE. $C_{25}H_{21}N_4$, mw: 377.5.
THR = An exper carc. [*3*] An exper carc to mice via dermal route. [*103*]

BUTYL TRIPHENYL GERMANIUM. Colorless crystals, insol in water. $Ge(C_4H_9)(C_6H_5)_3$, mw: 361.
THR = Details U. See germanium compounds.

BUTYL URETHANE. Syns: *carbonic acid butyl ethyl ester, ethyl-n-butylcarbamate*.
$CH_3(CH_2)_3NHCOOC_2H_5$, mw: 145, d: 0.9, vap. d: 5.0, bp: 202°, flash p: 197°F.
Fire Hazard: Mod, when exposed to heat or flame.
To Fight Fire: Water spray, fog, foam, alcohol foam, dry chemical.

BUTYL VINYL ETHER. Liquid. $CHCH_2OC_4H_9$, mw: 100.2, mp: −92°, bp: 93.3°, flash p: 15°F (OC), d: 0.77, vap. d: 3.4.
THR = U. See also ethers.
Explosion Hazard: Mod, by spont chemical reaction. See also ethers.
Fire Hazard: Dangerous. See ethers.
To Fight Fire: Alcohol foam. See also ethers.

BUTYN. Syn: *p-aminobenzoyl-γ-di-n-butyl aminopropanol sulfate*. Colorless, odorless powder. $C_{18}H_{30}O_2N_2$, mw: 306.5, mp: 98°–100°.
Acute tox data: Sc LD (rat) = 150 mg/kg. [*3*]

THR = HIGH via sc route. A weak allergen.
Fire Hazard: Slight.

1-BUTYNE. See ethylacetylene.

2-BUTYNE. See crotonylene.

2-BUTYNE-1,4-DIOL. Straw to amber crystals. $C_4H_6O_2$, mw: 86.1, mp: 57.5°, bp: 194° @ 100 mm.
Acute tox data: Oral LD_{50} (rat) = 104 mg/kg. [*3*]
THR = HIGH via oral route. No chronic effects data.
Explosion Hazard: Mod, when exposed to heat or by spont chemical reaction in contract with certain materials; i.e., mercury salts, strong acids and alkali earth hydroxides and halides at high temp.
Disaster Hazard: Dangerous; when heated to decomp, emits acrid fumes and may explode.

BUTYRALDEHYDE. Syns: *butanal, butyric aldehyde, n-butyl aldehyde.* Colorless liquid. $CH_3(CH_2)_2CHO$, mw: 72.1, mp: −100°, bp: 74.7°, flash p: 20°F (CC), d: .902 @ 20°/4°, autoign. temp.: 446°F, vap. d: 2.5, lel = 2.5%, uel = 12.5%.
Acute tox data: Dermal LD_{50} (rabbit) = 3560 mg/kg. [*3*]
THR = MOD via dermal route.
Fire Hazard: Dangerous, when exposed to heat or flame; can react with oxidizing materials. Reacts vigorously with chlorosulfonic acid, HNO_3, oleum, H_2SO_4. [*19*]
To Fight Fire: Foam, CO_2, dry chemical.

BUTYRALDEHYDE ANILINE. Liquid. $CH_3(CH_2)_2CONHC_6H_5$, mw: 163.2, vap. d: 5.5.
THR = Details U. May resemble aniline. See also aniline.
Fire Hazard: Slight, when exposed to heat or flame.
Disaster Hazard: Dangerous; when heated to decomp, emits highly toxic fumes of aniline; can react vigorously with oxidizing materials.

BUTYRALDOL. Slightly sol in water. $C_8H_{16}O_2$, mw: 144, d: 0.9, bp: 280°F @ 50 mm, flash p: 165°F (OC).
Fire Hazard: Mod, when exposed to heat, flame or oxidizers.
To Fight Fire: Alcohol foam.

BUTYRALDOXIME. Syn: *butanal oxime.* Liquid. C_4H_8NOH, mw: 87.1, mp: −29.5°, bp: 152°, flash p: 136°F (CC), d: 0.923, vap. d: 3.01.
Acute tox data: Ip LD (mouse) = 200 mg/kg. [*3*]
THR = HIGH via ip route.
Fire Hazard: Mod, when exposed to heat or flame; can react with oxidizing materials.
To Fight Fire: Alcohol foam, dry chemical.

BUTYRIC ACID. Syns: *butanoic acid, n-butyric acid, ethyl acetic acid, propyl formic acid.* Liquid. $CH_3(CH_2)_2COOH$, mw: 88.10, mp: −7.9°, bp: 163.5°, flash p: 161°F, fp: −5.5°, d: 0.9590 @ 20°/20°, autoign. temp.: 846°F, vap. press: 0.43 mm @ 20°, vap. d: 3.04, lel = 2.0%, uel = 10.0%.
Acute tox data: Oral LD_{50} (rat) = 2940 mg/kg; dermal LD_{50} (rabbit) = 530 mg/kg; sc and ip LD_{50} (mouse) = 3180 mg/kg; iv LD_{50} (mouse) = 800 mg/kg. [*3*]
THR = MOD via oral, dermal, sc, ip and iv routes. A synthetic flavoring substance and adjuvant. [*109*]
Fire Hazard: Mod, when exposed to heat or flame; can react with oxidizing materials.
Spont Heating: No.
To Fight Fire: Alcohol foam, CO_2, dry chemical.

BUTYRIC ALDEHYDE. See butyraldehyde.

BUTYRIC ANHYDRIDE. Syn: *butanoic anhydride.* Liquid, decomp in water. $(CH_3(CH_2)_2CO)_2O$, mw: 158.19, mp: −73.3°, bp: 198°, flash p: 190°F (CC), d: 0.978, vap. d: 5.4.
THR = MOD irr to skin, eyes and mu mem.
Fire Hazard: Mod, when exposed to heat or flame; can react with oxidizing materials.
To Fight Fire: Alcohol foam, CO_2, dry chemical.

BUTYRIC ETHER. See ethyl butyrate.

β-BUTYROLACTONE. $C_4H_6O_2$, mw: 86.1.
THR = An exper carc. [*23, 3*]

γ-BUTYROLACTONE. Colorless liquid, mild odor. $C_4H_6O_2$, mw: 86.1, mp: −44°, bp: 206°, flash p: 209°F (OC), d: 1.124 @ 25°/4°, vap. d: 3.0.
THR = Less than β-propiolactone. An exper (±) carc. [*23, 3, 13*]
Fire Hazard: Low, when exposed to heat or flame; can react with oxidizing materials.
To Fight Fire: Foam, alcohol foam, CO_2, dry chemical.

BUTYRONE. Syns: *4-heptanone, dipropyl ketone.* Colorless, refractive liquid. $(C_3H_7)_2CO$, mw: 114.18, bp: 144°, mp: −32.6°, vap. press: 5.2 mm @ 20°, flash p: 120°F (CC), d: 0.815, vap. d: 3.93.
Acute tox data: Oral LD_{50} (rat) = 3730 mg/kg; inhal LC (rat) = 4000 ppm for 4 hrs. [*3*]
THR = MOD via oral and inhal routes. See also ketones.
Fire Hazard: Mod, when exposed to heat or flame; can react with oxidizing materials.
Spont Heating: No.
To Fight Fire: CO_2, dry chemical, alcohol foam, fog and mist.

***n*-BUTYRONITRILE.** Syns: *butanenitrile, n-propyl cyanide.* Colorless liquid, slightly sol in water, sol in alcohol and ether. C_3H_7CN, mw: 69, d: 0.796 @ 15°, mp: −112.6°, bp: 117°, flash p: 79°F (OC).
Acute tox data: Oral LD_{50} (rat) = 500 mg/kg; dermal LD_{50} (rabbit) = 500 mg/kg. [*3*]
THR = MOD via oral and dermal routes. See nitriles and cyanides. A poison.

Fire Hazard: Dangerous, when exposed to heat, flame or oxidizers.
Disaster Hazard: Dangerous; see nitriles.
To Fight Fire: Alcohol foam.

BUTYRYL AZIRIDINE. $C_6H_{11}ON$, mw: 113.2.
THR = An exper. neo. [*3*]

BUTYRYL CHLORIDE. Syn: *butanoyl chloride*. Clear, colorless liquid with sharp odor. C_3H_7COCl, mw: 106.6, mp: −89°, bp: 101°, d: 1.028 @ 20°/4°, vap. d: 3.67.
THR = HIGH irr to skin, eyes and mu mem.
Disaster Hazard: Dangerous; when heated to decomp, emits highly toxic fumes of chlorides; will react with water or steam to produce toxic and corrosive fumes; can react vigorously with oxidizing materials.

***n*-BUTYRYL ETHYLENE IMINE.**
THR = An exper carc. [*23*]

BUTYRYL TRIGLYCERIDE. See tributyrin.

BUX. Syn: *Mixture of m-(1-ethyl propyl)phenyl methyl carbamate + m-(1-methyl butyl) phenyl methyl carbamate.* A low-melting amber solid, very sol in xylene ethanol, nearly insol in water. $C_{13}H_{19}O_2N$, mw: 221, mp: 26.4°.
Acute tox data: Oral LD_{50} (rat) = 170 mg/kg; dermal LD_{50} (rabbit) = 400 mg/kg; dermal LD_{50} (dog) = 1400 mg/kg. [*3*]
THR = HIGH via oral and dermal routes.

BUYO QUID.
THR = A susp carc. [*14*] See also plant and fungal products.

C

CA. See cellulose acetate.

CACODYL. Syns: *dicacodyl, tetramethyl diarsyl.* Oily liquid, colorless to yellow, slightly sol in water. $(CH_3)_2As\text{-}As(CH_3)_2$, mw: 210.0, bp: 165°, fp: −6°, d: 1.15.
THR = HIGH via oral and inhal routes. An herbicide. See arsenic compounds.
Fire Hazard: Dangerous, by spont chemical reaction. Ignites spont in dry air.
Spont Heating: Yes.
Explosion Hazard: U.
Disaster Hazard: Dangerous; see arsenic; can react vigorously with oxidizing materials i.e., air, Cl_2. [*19*]

CACODYLATES. See arsenic compounds.

CACODYL BROMIDE. See dimethyl bromarsine.

CACODYL CHLORIDE. See dimethyl chlorarsine.

CACODYL DIOXIDE. A liquid.
THR = HIGH via oral and inhal routes. An herbicide. See arsenic compounds.
Fire Hazard: Dangerous, by spont chemical reaction. Ignites spont in air.
Spont Heating: Yes.
Explosion Hazard: U.
Disaster Hazard: Dangerous; see arsenic; can react vigorously with oxidizing materials.

CACODYL HYDRIDE. See dimethyl arsine.

CACODYLIC ACID. Syns: *hydroxy dimethyl arsine oxide, dimethyl arsinic acid.* Colorless crystals, odorless and sol in water. $(CH_3)_2AsOOH$, mw: 138.0, mp: 192°.
Acute tox data: Oral LD_{50} (rat) = 1350 mg/kg. [*3*]
THR = MOD via oral route. An exper neo. [*3*] Used as an herbicide, defoliant and silvicide. See arsenic compounds.
Disaster Hazard: High, when water solution is in contact with active metals; i.e., Fe, Al, and Zn, or when heated to decomp.

CACODYL OXIDE. Syn: *dicacodyl oxide.* Colorless liquid, slightly sol in water. $O((CH_3)_2AsI)_2$, mw: 226, d: 1.486 @ 15°, mp: −25°, bp: 150°.
THR = HIGH via oral and inhal routes. An herbicide. See arsenic compounds.
Disaster Hazard: See cacodyl dioxide.

CACODYL SULFIDE. Syn: *dicacodyl sulfide.* Oily liquid, slightly sol in water. $((CH_3)_2As)_2S$, mw: 242, bp: 211°.
THR = HIGH via oral and inhal routes. An herbicide. See arsenic compounds and sulfides.
Fire Hazard: Dangerous, when exposed to heat or by spont chemical reaction. Ignites spont in air.
Spont Heating: Yes.
Explosion Hazard: U.
Disaster Hazard: Dangerous; see arsenic and oxides of sulfur; can react vigorously with oxidizing materials.

CADAVERINE. See pentamethylene diamine.

CADE OIL. Syn: *juniper tar.* Dark brown, viscous volatile oil, d: 0.950–1.055 @ 25°/25°.
Acute tox data: Oral LD_{50} (rat) = 8014 mg/kg. [*3*]
THR = LOW via oral route. An allergen.
Fire Hazard: A combustible material; can react with oxidizing materials.

CADIA DEL PERRO. Aqueous extraction from dried leaves of the plant.
THR = An exper neo and carc. [*3*]

CADMIUM. Hexagonal crystals, silver-white malleable metal. Cd, atwt: 112.41, mp: 320.9°, bp: 767 ± 2°, d: 8.642, vap. press: 1 mm @ 394°.
Radiation Hazard: For permissible levels, see Section 5, Table 5A.5. Artificial isotope ^{109}Cd, $T\frac{1}{2}$ − 450d, decays to stable ^{109}Ag via ec. Emits γ's of 0.09 MeV. Artificial isotope ^{115}Cd, $T\frac{1}{2}$ = 43d, decays to ^{115}In via β's of 1.6 meV.
THR = See cadmium compounds. An exper (+) carc. [*3, 6, 22, 23, 95*]
Fire Hazard: Mod, in the form of dust when exposed to heat or flame or by chemical reaction with oxidizing agents, metals, HN_3, Te, Zn. [*19*]
Explosion Hazard: Mod, in the form of dust when exposed to flame.
Disaster Hazard: Dangerous; cadmium dust can react vigorously with oxidizing materials.

CADMIUM ACETATE. Monoclinic colorless crystals, odor of acetic acid. $Cd(C_2H_2O_2)_2$, mw: 230.50, mp: 256°, bp: decomp, d: 2.341.
THR = See cadmium compounds.

CADMIUM ALLOYS. See cadmium compunds.

CADMIUM AMIDE. White solid. $Cd(NH_2)_2$, mw: 144.46, mp: decomp @ 120°, d: 3.05 @ 25°.

THR = See cadmium compounds and ammonia. Can explode when moistened.

CADMIUM AMMONIUM BROMIDE. Syn: *ammonium cadmium bromide*. Colorless crystals. $CdBr_2 \cdot 4NH_4Br$, mw: 664.0.

THR = See cadmium compounds and bromides.

CADMIUM ARSENIDE. Dark gray cubes. Cd_3As_2, mw: 487.05, mp: 721°, d: 6.21 @ 15°/4°.

THR = HIGH via oral and inhal routes. See As and Cd compounds.

Fire Hazard: Mod, when exposed to heat or flame. May evolve arsine upon contact with moisture or acids.

Explosion Hazard: Mod, when exposed to flame. See also arsine.

Disaster Hazard: Dangerous, when heated to decomp or on contact with acids, it emits high tox fumes which will react violently with water, steam or oxidizing materials.

CADMIUM AZIDE. $Cd(N_3)_2$, mw: 196.5.

THR = Probably HIGH. See cadmium. Very explosive by touch. [*19*]

CADMIUM BENZOATE. White solid. $Cd(C_7H_5O_2)_2 \cdot 2H_2O$, mw: 390.66.

THR = See cadmium compounds.

CADMIUM BOROTUNGSTATE. Yellow triclinic crystals. $Cd_5(BW_{12}O_{40})_2 \cdot 18H_2O$, mw: 6602.06, mp: 75°.

THR = See cadmium compounds and boron compounds.

CADMIUM BROMATE. Rhombic white crystals. $Cd(BrO_3)_2 \cdot H_2O$, mw: 386.26, mp: decomp, d: 3.758.

THR = See cadmium compounds and bromates.

Fire Hazard: Mod, by chemical reaction with reducing agents. A powerful oxidizing agent.

Disaster Hazard: Dangerous, see cadmium, bromine and bromides, it can react vigorously with reducing materials.

CADMIUM BROMIDE. Yellow crystals. $CdBr_2$, mw: 272.24, mp: 567°, bp: 863°, d: 5.192 @ 25°.

THR = See cadmium compounds and bromides. Can explode if mixed with K. [*19*]

CADMIUM CARBONATE. Trigonal, white crystals. $CdCO_3$, mw: 172.42, mp: decomp $< 500°$, d: 4.258 @ 4°.

THR = See cadmium compounds.

CADMIUM CHLORATE. Colorless, deliquescent prisms. $Cd(ClO_3)_2 \cdot 2H_2O$, mw: 315.36, mp: 80°, d: 2.28 @ 18°.

THR = See cadmium compounds and chlorates.

Fire Hazard: Mod, by chemical reaction with reducing agents. A powerful oxidizing agent. Reacts violently with Sb_2S_3, As_2S_3, CuS, SnS_2, SnS. [*19*]

Explosion Hazard: Mod, when shocked or exposed to heat. See also chlorates.

Disaster Hazard: Dangerous; heat or shock will explode it; see cadmium, chlorine and chlorides; can react vigorously with reducing materials.

CADMIUM CHLORIDE. Hexagonal, colorless crystals. $CdCl_2$, mw: 183.32, mp: 568°, d: 4.047 @ 25°, vap. press: 10 mm @ 656°, bp: 960°.

Acute tox data: Oral LD_{50} (rat) = 88 mg/kg; im LD_{50} (rat) = 25 mg/kg; dermal LD_{LO} (guinea pig) = 233 mg/kg; inhal LC_{90} (dog) = 8 mg/m^3 for 30 min. [*3*]

THR = HIGH via dermal, oral and inhal routes. An exper (+) carc, neo and teratogen. [*3, 6*] Reacts violently with BrF_3, K. [*19*] See cadmium compounds and chlorides.

CADMIUM COBALTINITRITE. See cadmium nitrocobaltate (III).

CADMIUM COMPOUNDS

THR = The oral toxicity of Cd and its compounds is HIGH. However, when these materials are ingested, the irr and emetic action is so violent that little of the Cd is absorbed and fatal poisoning does not as a rule ensue. Cases of human Cd poisoning have been reported from ingestion of food or beverages prepared or stored in Cd-plated containers. The inhal of fumes or dusts of Cd primarily affects the respiratory tract; the kidneys may also be affected. Even brief exposure to high conc may result in pulmonary edema and death. Usually the edema is not massive, with little pleural effusion. In fatal cases, fatty degeneration of the liver and acute inflammatory changes in the kidneys have been noted. Ingestion of Cd results in a gastro-intestinal type of poisoning resembling food poisoning in its symptoms. Inhal of dust or fumes [*54*] may cause dryness of the throat, cough, headache, a sense of constriction in the chest, shortness of breath (dyspnea) and vomiting. More severe exposure results in marked lung changes, with persistent cough, pain in the chest, severe dyspnea and prostration which may terminate fatally. X-ray changes are usually similar to those seen in broncho-pneumonia. The urine is frequently dark. These symptoms are usually delayed for some hours after exposure, and fatal conc may be breathed without sufficient discomfort to warn the workman to leave the exposure. See also cadmium. Many Cd compounds are exper (+) carc and neo [*3, 6, 14*] of the connective tissue, lungs and liver. [*50, 51, 2, 17, 52*] There is some evidence of teratogenicity. [*53*] Ingestion of Cd results in sudden nausea, salivation, vomiting and diarrhea and abdominal pain and discomfort. Symptoms begin almost immediately after ingestion. A yellow discoloration of the teeth has been reported in workers exposed to Cd. Cadmium oxide fumes can cause metal fume fever resembling that caused by zinc oxide fumes.

CADMIUM CYANIDE. Crystals. $Cd(CN)_2$, mw: 164.45, mp: >200° decomp, d: 2.226.
THR = See cyanides and cadmium compounds. Reacts violently with Mg. [*19*]

CADMIUM DIETHYL DITHIO CARBAMATE. $C_{10}H_{20}N_2S_2$, mw: 296.6.
THR = An exper neo. [*3*]

CADMIUM DIHYDROGEN PHOSPHATE. Triclinic crystals. $Cd(H_2PO_4)_2 \cdot 2H_2O$, mw: 342.4, mp: decomp 100°, d: 2.74 @ 15°/4°.
THR = See cadmium compounds and phosphoric acid.

CADMIUM-9,10-EPOXYSTEARATE.
THR = See cadmium compounds.
Disaster Hazard: Dangerous. See cadmium compounds.

CADMIUM (II) EDTA COMPLEX.
Acute tox data: ip LD_{50} (mouse) = 7.8 mg/kg. [*3*]
THR = HIGH via ip and oral routes.

CADMIUM ETHYLENE BIS DITHIOCARBAMATE.
THR = See cadmium compounds. An exper neo. [*3*]
Disaster Hazard: Dangerous; see cadmium and oxides of sulfur.

CADMIUM FERROCYANIDE. Solid. $Cd_2Fe(CN)_6 \cdot xH_2O$.
THR = See cadmium compounds and ferrocyanides.

CADMIUM FLUOBORATE.
Acute tox data: Oral LD_{LO} (rat) = 250 mg/kg; inhal LC_{LO} (mouse) = 670 mg/m^3 for 10 min. [*3*]

CADMIUM FLUOGALLATE. Colorless crystals. $((Cd(H_2O)_6)(GaF_5H_2O)$, mw: 403.24, mp: $-5H_2O$ @ 110°, d: 2.79.
THR: See fluorides and cadmium compounds.

CADMIUM FLUORIDE. Cubic white crystals. CdF_2, mw: 150.41, mp: 1100°, bp: 1758°, d: 6.64, vap. press: 1 mm @ 1112°.
Acute tox data: Oral LD_{50} (guinea pig) = 150 mg/kg. [*3*]
THR = HIGH via oral route. Violent reaction with K. [*19*] See fluorides and cadmium compounds.

CADMIUM FLUOSILICATE. Hexagonal, colorless crystals. $CdSiF_6 \cdot 6H_2O$, mw: 362.6.
Acute tox data: Oral LD_{LO} (rat) = 100 mg/kg; inhal LC_{LO} (mouse) = 670 mg/m^3 for 10 min. [*3*]
THR = HIGH via oral and inhal routes. See cadmium compounds and fluosilicates.

CADMIUM FORMATE. Monoclinic crystals. $Cd(CHO_2)_2 \cdot 2H_2O$, mw: 238.48, mp: decomp, d: 2.44.
THR = See cadmium compounds and formic acid.

CADMIUM FUMARATE. Solid. $CdC_4H_2O_4$, mw: 226.47.
THR = See cadmium compounds.

CADMIUM HEXAMMINE CHLORATE. $Cd(NH_3)_6(ClO_3)_2$, mw: 381.4.
THR = No data. See cadmium compounds. Explodes when struck. [*19*]

CADMIUM HEXAMMINE PERCHLORATE. $Cd(NH_3)_6(ClO_4)_2$, mw: 403.
THR = No data. See cadmium compounds. Explodes when struck. [*19*]

CADMIUM HYDROGEN ARSENATE. Solid. $CdHAsO_4 \cdot H_2O$, mw: 270.34, d: 4.164 @ 15/4°.
THR = See arsenic and cadmium compounds.

CADMIUM HYDROXIDE. Trigonal white crystals or amorphous $Cd(OH)_2$, mw: 146.43, mp: decomp @ 300°, d: 4.79 @ 15°.
THR = See cadmium compounds.

CADMIUM IODATE. White crystals. $Cd(IO_3)_2$, mw: 462.25, mp: decomp, d: 6.43.
THR = See cadmium compounds and iodates.

CADMIUM IODIDE. Hexagonal, brownish crystals. CdI_2, mw: 366.25, mp: 385°, bp: 796°, d: 5.670 @ 30°, vap. press: 1 mm @ 416°.
THR = See cadmium compounds and iodides. Violent reaction with K. [*19*]

CADMIUM LACTATE. Needles. $Cd(C_3H_5O_3)_2$, mw: 290.56.
Acute tox data: sc LD_{50} (mouse) = 13.9 mg/kg. [*3*]
THR = HIGH via sc route. See cadmium compounds.

CADMIUM MALEATE. Solid. $CdC_4H_2O_4 \cdot 2H_2O$, mw: 262.50.
THR = See cadmium compounds.

CADMIUM MOLYBDATE. Solid. $CdMoO_4$, mw: 272.36, d: 5.347.
THR = See cadmium and molybdenum compounds.

CADMIUM NITRATE. White, prismatic needles, hygroscopic. $Cd(NO_3)_2$, mw: 236.43, mp: 350°.
Acute tox data: Oral LD_{50} (rat) = 300 mg/kg. [*3*]
THR = HIGH via oral route. See also cadmium and nitrates.

CADMIUM NITRIDE. Cd_3N_2, mw: 365.2.
THR = See cadmium compounds. Reacts explosively with acids, bases, water. [*19*]

CADMIUM NITROCOBALTATE (III). Syn: *cadmium cobaltinitrite*. Yellow crystals. $Cd_3((Co(NO_2)_6)_2$, mw: 1007.21, mp: decomp @ 175°.
THR = See cadmium, cobalt compounds and nitrates.

CADMIUM ORANGE. See cadmium selenide.

CADMIUM OXALATE. Colorless crystals. CdC_2O_4, mw: 200.43, mp: decomp @ 340°, d: 3.32 @ 18°/4°.
THR = See cadmium compounds and oxalates.

CADMIUM OXIDE. (1) amorphous, brown crystals; (2) cubic, brown crystals. CdO, mw: 128.41, mp(1): < 1426°, mp(2): decomp @ 950°, bp: 1559°, d(1): 6.95, d(2): 8.15, vap. press: 1 mm @ 1000°.
Acute tox data: Oral LD_{50} (rat) = 72 mg/kg. [*3*]
THR = HIGH via oral and inhal routes. See also cadmium. An exper (+) carc. [*3, 6*] Reacts violently with Mg. [*19*]

CADMIUM OXIDE FUME.
Acute tox data: Inhal TC_{LO} (man) = 8.63 mg/m^3 for 5 hrs ⟶ pulmonary damage, inhal LC_{50} (rat) = 500 mg/m^3; inhal LC_{50} (dog) = 4000 mg/m^3 for 10 min. [*3*]
THR = HIGH irr via inhal route.

CADMIUM PERCHLORATE. Crystals. $Cd(ClO_4)_2 \cdot 6H_2O$, mw: 419.9, mp: 129.4°.
THR = See cadmium compounds and perchlorates.

CADMIUM PERMANGANATE. Violet crystals. $Cd(MnO_4)_2 \cdot 6H_2O$, mw: 458.4, mp: decomp @ 95°, d: 2.81.
THR = See cadmium and manganese compounds.

CADMIUM PHOSPHATE. Amorphous or colorless crystals. $Cd_3(PO_4)_2$, mw: 527.19, mp: 1500°.
Acute tox data: Inhal LC_{LO} (mouse) = 650 mg/m^3. [*3*]
THR = HIGH via inhal route. See cadmium compounds and phosphates.

CADMIUM PICRATE. Yellow solid. $Cd(C_6H_2(NO_2)_3OH)_2$.
THR = See cadmium compounds and picric acid.

CADMIUM PIGMENTS.
THR = Some are (S) carc. [*14, 3*] See also cadmium compounds.

CADMIUM POTASSIUM CYANIDE. Cubic crystals, sol in 3 parts cold and 1 part hot water, insol in alcohol. $Cd(CN)_2 \cdot 2KCN$, mw: 294.67.
THR = See cyanides and cadmium compounds.

CADMIUM POTASSIUM IODIDE. See potassium cadmium iodide.

CADMIUM RED. See cadmium selenide.

CADMIUM RICINOLEATE. Fine white powder. $Cd(C_{18}H_{33}O_3)_2$, mw: 706.5, mp: 103°, d: 1.11 @ 25°/25°.
THR = See cadmium compounds.

CADMIUM SALICYLATE. White needles. $Cd(C_7H_5O_3)_2 \cdot H_2O$, mw: 404.65.
THR = See cadmium compounds.

CADMIUM SELENATE, DIHYDRATE. Rhombic crystals. $CdSeO_4 \cdot 2H_2O$, mw: 291.40, mp: $-H_2O$ @ 100°, d: 3.632.
THR = See cadmium and selenium compounds.

CADMIUM SELENIDE. Red powder, or it may be gray to brown. CdSe, mw: 191.37, mp: > 1350°, d: 5.81 @ 15°/4°.
THR = See cadmium and selenium compounds.
Fire Hazard: Mod, when exposed to heat or flame. See hydrogen selenide, which can be evolved on contact with acids or moisture.
Explosion Hazard: Mod, when exposed to flame.
Disaster Hazard: Dangerous; see cadmium; can emit hydrogen selenide and selenium compounds on contact with moisture, acid or acid fumes.

CADMIUM-*m*-SILICATE. Colorless, rhombic crystals. $CdSiO_3$, mw: 188.47, mp: 1242°, d: 4.93.
THR = See cadmium compounds and silicates.

CADMIUM SOAP.
THR = A (S) carc. [*14*] See cadmium compounds.

CADMIUM SOLDER.
THR = A (S) carc. [*14*] See cadmium compounds.

CADMIUM SUCCINATE. $C_4H_6O_4 \cdot Cd$, mw: 230.5.
Acute tox data: Oral LD_{50} (rat) = 660 mg/kg; oral LD_{50} (mouse) = 312 mg/kg; ip LD_{50} (mouse) = 270 mg/kg. [*3*]
THR = HIGH via oral and ip routes.

CADMIUM SULFATE. Rhombic white crystals. $CdSO_4$, mw: 208.5, mp: 1000°, d: 4.691.
Acute tox data: sc LD_{50} (dog) = 27 mg/kg; ip LD_{50} (mouse) = 69 mg/kg. [*3*]
THR = HIGH via sc and ip routes. See cadmium compounds and sulfates. An exper (+) neo, carc and teratogen. [*3, 6*]

CADMIUM SULFATE TETRAHYDRATE. $CdSO_4 \cdot 4H_2O$, mw: 280.5.
THR = An exper (+) carc. [*3, 6*] See cadmium sulfate.

CADMIUM SULFIDE. Syn: *greenockite*. Hexagonal, yellow-orange crystals. CdS, mw: 144.48, mp: 1750 @ 100 atm, bp: subl in N_2, d: 4.82.
THR = An exper (+) carc. [*3, 6*] Reacts violently with ICl. [*19*] See cadmium compounds and sulfides.

CADMIUM SULFITE. Crystals. $CdSO_3$, mw: 192.48, mp: decomp.
THR = See cadmium compounds and sulfites.

CADMIUM SULFOSELENIDE.
THR = See selenium and cadmium compounds.

CADMIUM TARTRATE. White, crystalline powder. $CdC_4H_4O_6$, mw: 260.48.
THR = See cadmium compounds.

CADMIUM TELLURIDE. Black, cubic crystals. CdTe, mw: 240.0, mp: 1041°, d: 6.20 @ 15°.
THR = See cadmium and tellurium compounds.
Fire Hazard: Mod, when exposed to heat or flame. See hydrogen telluride, which can be evolved on contact with acids or moisture.
Explosion Hazard: Mod, when exposed to flame.
Disaster Hazard: Dangerous; see cadmium; can emit hydrogen telluride and tellurium compounds on contact with acid, acid fumes, or moisture.

CADMIUM TRICHLORO ACETATE. Rhombic crystals. $Cd(C_2Cl_3O_2)_2 \cdot 1\frac{1}{2}H_2O$, mw: 464.22, d: 2.093 @ 25°.
THR = See cadmium compounds and trichloroacetic acid.

CADMIUM TRIHYDRAZINE CHLORATE. $Cd(N_2H_4)_3(ClO_3)_2$, mw: 375.5.
THR = See cadmium compounds. Explodes when struck. [*19*]

CADMIUM TRIHYDRAZINE PERCHLORATE. $Cd(N_2H_4)_3(ClO_4)_2$, mw: 407.5.
THR = See cadmium compounds. High explosive salt. [*19*]

CADMIUM TUNGSTATE. Yellow crystals. $CdWO_4$, mw: 360.3.
THR = See cadmium and tungsten compounds.

"CADMOLITH." See barium sulfate and cadmium.

CADOX. See *tert*-butyl peroxide.

CADWALADER, THOMAS. See Section 1.

CAFFEIC ACID. Syn: *3,4-dihydroxy cinnamic acid.* Yellow crystals, sol in hot water, alcohol. $C_9H_8O_4$, mw: 180.2, mp: 195°.
THR = MOD allergen. LOW via oral route.

CAFFEINE. Syn: *theine.* White, fleecy masses. $C_8H_{10}N_4O_2$, mw: 194.2, mp: 236.8°.
Acute tox data: Oral LD_{50} (rat) = 192 mg/kg; ip LD_{50} (rat) = 260 mg/kg; iv LD_{50} (rat) = 105 mg/kg; iv LD_{50} (mouse) = 68 mg/kg. [*3*]
THR = HIGH via oral, ip and iv routes. An exper teratogen, mutagen. [*3*] Implicated in increased fetal losses. An exper carc. [*23*] A general purpose food additive. Large doses (above 1.0 g) cause palpitation, excitement, insomnia, dizziness, headache and vomiting. Continued excessive use of caffeine in tea or coffee may lead to digestive disturbances, constipation, palpitations, shortness of breath and depressed mental states. It is also implicated in cardiac disorders under those conditions.
Treatment and Antidotes: Evacuate stomach with emetic or stomach tube. Call a physician.
Fire Hazard: Slight; when heated to decomp, emits toxic fumes.

CAKE ALUM. See aluminum sulfate, dry.

CALABAR BEAN. See physostigma.

CALABRINE. See physostigmine.

CALCIC LIVER OF SULFUR. See calcium sulfide.

CALCIFEROL. See vitamin D_2.

CALCIMINE. White powder. Comp: chalk and glue.
THR = LOW via inhal route.

CALCITE. See calcium carbonate.

CALCIUM. Silver-white, soft metal. Ca, atwt. 40.08, mp: 842°, bp: 1484°, d: 1.54 @ 20°, vap. press: 10 mm @ 983°.
Radiation Hazard: For permissible levels see Section 5, Table 5A.5.
Artificial isotope ^{45}Ca, $T\frac{1}{2}$ = 165d, decays to stable ^{45}Sc via β's of 0.25 MeV.
THR = See calcium compounds.
Fire Hazard: Mod, when heated or in intimate contact with moisture or acids, evolves hydrogen. See also hydrogen.
Explosion Hazard: Mod, in intimate contact with very powerful oxidizing agents; i.e., Cl_2, ClF_3, F_2, O_2, Si, S, V_2O_3. [*19*]
Disaster Hazard: Dangerous; reacts with moisture or acids to liberate large quantities of hydrogen; can develop explosive pressure in containers. See also hydrogen.
To Fight Fire: Special mixtures of dry chemical.

CALCIUM ACETATE. Syns: *vinegar salts, gray acetate, lime acetate, calcium diacetate.* Brown, gray or white powder, slight odor of acetic acid, sol in water, slightly sol in alcohol. $Ca(C_2H_3O_2)_2 \cdot H_2O$, mw: 176.
Acute tox data: iv LD_{50} (mouse) = 52 mg/kg. [*3*]
THR = HIGH via iv route. A sesquestrant food additive. [*109*]

CALCIUM ACRYLATE, MONOHYDRATE. White crystals. $Ca(CO_2CHCH_2)_2 \cdot H_2O$, mw: 200.2.
Acute tox data: Oral LD_{50} (rat) = 4920 mg/kg. [*3*]
THR = MOD via oral route. See calcium compounds and acrylic acid.
Fire Hazard: Slight; will react with water or steam to produce heat.

CALCIUM ACRYLATE DIHYDRATE. White powder. $(CH_2CHCOO)_2Ca \cdot 2H_2O$; mw: 218.2.
THR = See calcium acrylate, monohydrate.

CALCIUM ALGINATE. White or cream colored powder, or filaments, grains or granules, slight character-

istic odor, colloidal, insol in water and acids, sol in alkaline solutions, mw: 32,000–35,000.
THR = Probably LOW. A stabilizer food additive. [*109*]

CALCIUM ALLOYS.
THR = Violent reaction with water and acids. [*19*]

CALCIUM ALUMINUM ANHYDRIDE. See hydrides.

CALCIUM ARSENATE. Syns: *tricalcium-o-arsenate, calcium-o-arsenate.* White, amorphous powder. $Ca_3(AsO_4)_2$, mw: 398.06, d: 3.620.
Acute tox data: Oral LD_{50} (rat) = 20 mg/kg; oral LD_{50} (mammal) = 35 mg/kg. [*3*]
THR = HIGH via oral and inhal routes. A (S) human carc, exper neo. [*3, 6*] An insecticide and an herbicide. See arsenic compounds.

CALCIUM ARSENIDE. Red crystals. Ca_3As_2, mw: 270.06, mp: decomp, d: 3.031 @ 25°.
THR = HIGH via oral and inhal routes. See also arsine.
Fire Hazard: Mod; on contact with moisture, acid or acid fumes, it will evolve arsine. See also arsine.
Explosion Hazard: Mod; on contact with moisture, acid or acid fumes, it will evolve arsine. See also arsine.
Disaster Hazard: Dangerous; see arsenic; will react with water, steam, acid or acid fumes to produce toxic and flam vapors of arsine.

CALCIUM ARSENITE. White, granular powder. $Ca(As_2H_2O_4)$, mw: 256.
THR = HIGH. A poison. See arsenic compounds.
Disaster Hazard: See arsenic compounds.

CALCIUM ASCORBATE. A white to yellow crystalline powder, odorless, sol in water, slightly sol in alcohol, insol in ether. $Ca(C_6H_7O_6)_2 \cdot 2H_2O$, mw: 426.2.
THR = U. Probably LOW. A chemical preservative food additive. [*109*]

CALCIUM AZIDE. Rhombic, colorless crystals. $Ca(N_3)_2$, mw: 124.13, mp: explodes 144°–156°.
THR = See azides.

CALCIUM BIPHOSPHATE. See calcium phosphate monobasic.

CALCIUM BISULFITE. Syn: *calcium hydrogen sulfite.* Colorless or slightly yellowish liquid, strong sulfur dioxide odor. $Ca(HSO_3)_2$, mw: 202.21, d: 1.06.
THR = HIGH irr via contact, oral and inhal routes. See bisulfites and sulfurous acid.

CALCIUM-*m*-BORATE. Colorless, rhombic or long, flat plates. CaB_4O_7, mw: 195.3, mp: 1154°.
THR = See boron compounds.

CALCIUM BORIDE. Cubic, black crystals. CaB_6, mw: 105.00, d: 2.3 @ 20°, mp: 2235°.
THR = See calcium and boron compounds.
Fire Hazard: Mod; on contact with moisture, acid or acid fumes, will evolve boron hydrides, some of which are highly toxic and flam.

CALCIUM BROMATE. Monoclinic crystals. $Ca(BrO_3)_2 \cdot H_2O$, mw: 313.9, mp: $-H_2O$ @ 180°, d: 3.329.
THR = Reacts violently with Al, As, C, Cu, metal sulfides, organic matter, P, S. [*19*] See bromates.

CALCIUM BROMIDE. Deliquescent needles. $CaBr_2$, mw: 199.91, mp: 765°, d: 3.353 @ 25°, bp: 806°–812°.
THR = See bromides.

CALCIUM CARBAMATE. Crystals, sol in water, insol in alcohol. $Ca(NH_2CO_2)_2$, mw: 160.2.
THR = Probably toxic. See carbamates.

CALCIUM CARBIDE. Rhombic, gray crystals. CaC_2, mw: 64.10, mp: approx. 2300°, d: 2.222.
THR = Acetylene is evolved when calcium carbide is in contact with moisture. See also calcium hydroxide and acetylene.
Fire Hazard: Mod; on contact with moisture, acid or acid fumes, evolves heat or flam vapors. It incandesces with HCl gas, PbF_2, Mg. Reacts violently with Se, (KOH + Cl_2), $AgNO_3$, Na_2O_2, $SnCl_2$, S, water. [*19*]
Explosion Hazard: Mod; see also acetylene.

CALCIUM CARBIMIDE. See calcium cyanamide.

CALCIUM CARBONATE. Syns: *calcite, aragonite.* White powder. $CaCO_3$, mw: 100.09, mp: 825° (α); 1339° (β) @ 102.5 atm, d: 2.7–2.95.
THR = See calcium compounds. A nutrient and/or dietary supplement food additive. Also a general purpose food [*109*] additive. Calcium carbonate is a common air contaminant. Reacts violently with F_2, (Mg + H_2). [*19*]

CALCIUM CHLORATE. Monoclinic, white-yellowish, deliquescent crystals. $Ca(ClO_3)_2 \cdot 2H_2O$, mw: 243.03, mp: $-H_2O$ @ > 100°, d: 2.711.
Acute tox data: Oral LD_{LO} (rat) = 4500 mg/kg; ip LD_{LO} (rat) = 625 mg/kg. [*3*]
THR = MOD via oral and ip routes.
Fire and Explosion Hazard: See chlorates. Violent reaction with Al, As, C, Cu, charcoal, MnO_2, metal sulfides, S, dibasic organic acids, organic matter, P. [*19*]
Disaster Hazard: See chlorates.

CALCIUM CHLORIDE. Cubic, colorless, deliquescent crystals. $CaCl_2$, mw: 110.99, mp: 772°, bp: > 1600°, d: 2.512 @ 25°.
Acute tox data: Oral LD_{50} (rat) = 1 g/kg; ip LD_{50}

(mice) = 280 mg/kg; iv LD_{50} (mouse) = 42 mg/kg. [*3*]

THR = HIGH via ip and iv; MOD via oral routes. See calcium compounds. A sequestrant food additive, also a general purpose food additive and substance migrating [*109*] to food from packaging materials. Reacts violently with (B_2O_3 + CaO), BrF_3. [*19*]

CALCIUM CHLORIDE FLUORIDE-*o*-PHOSPHATE. Colorless crystals. $CaClF \cdot 3Ca_3(PO_4)_2$, mw: 1025.14, mp: 1270°, d: 3.14.

THR = See fluorides and chlorides and phosphates.

CALCIUM CHLORIDE HYPOCHLORITE. See bleaching powder.

CALCIUM CHLORITE. White solid. $CaClO_2$, mw: 107.6.

THR = See chlorites.

Fire Hazard: See chlorites. With $Cl_2 \longrightarrow$ explosive ClO_2. [*19*]

Disaster Hazard: See chlorites.

CALCIUM CHROMATE. Monoclinic prisms, yellow color. $CaCrO_4$, mw: 158.1.

THR = HIGH via oral route. An exper carc. [*3*] See chromium compounds.

CALCIUM CHROMATE, DIHYDRATE. $CaCrO_4 \cdot 2H_2O$, mw: 192.1.

THR = HIGH via all routes. An exper (+) carc. [*3, 6*]

CALCIUM CITRATE. Syns: *lime citrate*, *tricalcium citrate*. White powder, odorless, almost insol in water, insol in alcohol. $Ca_3(C_6H_5O_7)_2 \cdot 4H_2O$, mw: 570.5, mp: $-4H_2O$ @ 120°.

THR = Probably LOW. A nutrient and/or dietary supplement food additive. Also a sequestrant food additive and a general purpose food additive. [*109*]

CALCIUM COMPOUNDS.

THR = The fumes evolved by burning calcium in air are composed of calcium oxide (quick lime). This material is irr to the skin, eyes and mu mem. Many calcium compounds are used medicinally. Generally speaking, calcium compounds should be considered toxic only when they contain toxic components (such as arsenic, etc.) or as calcium oxide or hydroxide. Calcium compounds are common air contaminants.

Treatment and Antidotes: Any calcium residue left on the body or clothing should be brushed off immediately.

CALCIUM CYANAMIDE. Syn: *calcium carbimide*. Hexagonal, rhombohedral, colorless crystals. $CaCN_2$, mw: 80.11, mp: 1300°, subl > 1500°.

Acute tox data: Oral LD_{LO} (human) = 571 mg/kg; LD_{50} (rat) = 1000 mg/kg; oral LD_{50} (rabbit) = 1400 mg/kg. [*3*]

THR = MOD via oral route. An irr. An exper carc. [*3*] An herbicide. Calcium cyanamide acts locally on the skin as a primary irr, and the lesions produced vary from erythema to acute and subacute eczema. Usually the moist skin areas are attacked first, but the material is spread by scratching and parts of the body not ordinarily exposed may be affected. In severe cases ulceration may develop; the ulcers are usually covered by a black, necrotic crust. There is frequently irritations of the conjunctivae and of the mu mem of the nose and throat, with production of conjunctivitis, inflamed ulcers in the nose and throat, rhinitis and gingivitis. Systematically, headache, flushing of the skin of the head and neck, shortness of breath, vasodilation with lowered blood pressure, and rapid pulse have been described among exposed persons who have consumed alcohol. No fatalities have been reported. Calcium cyanamide is not believed to have a cumulative action. The fatal dose, by ingestion, is probably around 20 to 30 grams for an adult. It does not have a cyanide effect. See also amides.

Disaster Hazard: See cyanides.

CALCIUM CYANIDE. Syn: *cyanogas*. Rhombohedral crystals, white powder. $Ca(CN)_2$, mw: 92.12, mp: decomp > 350°.

Acute tox data: Oral LD_{50} (rat) = 39 mg/kg. [*3*]

THR = A deadly poison. HIGH via oral and inhal routes. A fumigant. See cyanides.

CALCIUM CYANOPLATINITE. Rhombic, yellow-green fluorescent crystals. $CaPt(CN)_4 \cdot 5H_2O$, mw: 429.46, mp: $-5H_2O$ @ 100°.

THR = See cyanides, and platinum compounds.

CALCIUM CYCLAMATE. Syn: *calcium cyclohexyl sulfamate*. White crystalline powder, almost odorless, freely sol in water, practically insol in alcohol, benzene, chloroform and ether. $(C_6H_{11}NHSO_3)_2Ca \cdot 2H_2O$, mw: 432.

THR = An exper carc and neo. [*3*] A non-nutritive sweetener food additive. [*109*]

CALCIUM CYCLOHEXYL SULFAMATE. See calcium cyclamate.

CALCIUM DIACETATE. See calcium acetate.

CALCIUM DICYANO GUANIDINE. Crystals. $Ca[NH_2CN(NCN)CN]_2 \cdot 4H_2O$, mw: 328.3, mp: > 300°.

THR = See cyanides.

CALCIUM DISODIUM EDETATE. See calcium disodium EDTA.

CALCIUM DISODIUM EDTA. Syns: *calcium disodium edetate, edathamil calcium disodium, calcium disodium ethylene diamine tetraacetate, versene.* White, odorless powder or flakes, sol in water, insol in organic solvents. $C_{10}H_{12}O_8N_2 \cdot Ca \cdot 2Na$, mw: 374.3.
Acute tox data: Oral LD_{50} (rabbit) = 7000 mg/kg; iv LD_{50} (rat) = 3000 mg/kg; ip LD_{50} (rat) = 3800 mg/kg. [*3*]
THR = MOD via oral, iv and ip routes. A food additive permitted in food for human consumption. [*109*]

CALCIUM DISODIUM ETHYLENE DIAMINE TETRA ACETATE. See calcium disodium EDTA.

CALCIUM ETHYLENE BIS DITHIOCARBAMATE. Solid. $Ca[(C_2H_3)_2NCS_2]_2$, mw: 330.5.
THR = HIGH via oral and inhal routes. A fungicide. See bis(dimethyl thiocarbamyl)disulfide and carbamates.
Disaster Hazard: Mod dangerous; when heated to decomp, emits toxic fumes.

CALCIUM FERRICYANIDE. Red, deliquescent needles. $Ca_3[Fe(CN)_6]_2 \cdot 12H_2O$, mw: 760.4.
THR = See ferricyanides and calcium compounds.

CALCIUM FERROCYANIDE. Yellow, triclinic crystals. $Ca_2Fe(CN)_6 \cdot 11H_2O$, mw: 490.3, mp: decomp, d: 1.68.
THR = See calcium compounds and ferrocyanides.

CALCIUM FLUORIDE. Cubic, colorless crystals, luminous with heat. CaF_2, mw: 78.1, mp: 1360°, d: 3.180.
THR = See fluorides.

CALCIUM FLUORO ACETATE. $C_{21}H_{21}O_{21}CaF_2$, mw: 687.5.
Acute tox data: Oral LD_{LO} (rat) = 2 mg/kg. [*3*]
THR = HIGH via oral and inhal routes. See also fluorides.

CALCIUM FLUOROPHOSPHATE. Syn: *calcium monofluorophosphate.* Colorless monoclinic crystals, slightly water sol, insol in organic solvents. $CaPO_3F$, mw: 138.1.
THR = See fluorides.
Disaster Hazard: Dangerous; see fluorides and phosphates.

CALCIUM FLUOSILICATE. White, crystalline powder. $CaSiF_6$, mw: 182.14, d: 2.662 @ 17.5°.
Acute tox data: sc LD_{LO} (guinea pig) = 450 mg/kg; oral LD_{LO} (guinea pig) = 250 mg/kg. [*3*]
THR = HIGH via sc and oral routes. See also fluosilicates.

CALCIUM GLUCONATE. White fluffy powder or granules, odorless, sol in hot water, less sol in cold water, insol in alcohol, acetic acid and other organic solvents. $C_{12}H_{22}O_{14} \cdot Ca$, mw: 430.2, mp: $-H_2O$ @ 120°.
Acute tox data: im LD_{LO} (human) = 20 mg/kg; affects CNS, skin and lungs. im TD_{LO} (child) = 50 mg/kg; ip LD_{LO} (dog) = 1100 mg/kg. [*3*]
THR = HIGH via im, MOD via ip routes. A sequestrant food additive; also a general purpose food additive. [*109*]

CALCIUM GLYCEROPHOSPHATE. Syn: *calcium glycerino phosphate.* White crystalline powder, odorless, slightly sol in water, insol in alcohol. $CaC_3H_7O_2PO_4$, mw: 210, mp: decomp @ 170°.
THR = No data. A nutrient and/or dietary supplement food additive. [*109*]
Disaster Hazard: Dangerous. See phosphates.

CALCIUM HEXAMETA PHOSPHITE.
THR = See phosphites. A sequestrant food additive. [*109*]
Disaster Hazard: Dangerous; see phosphites.

CALCIUM HEXAMMONIATE. $Ca(NH_3)_6$, mw: 142.3.
THR = No data. Self-ignites in air. [*19*]

CALCIUM HYDRIDE. Syn: *hydrolith.* Gray-white crystals, powder. CaH_2, mw: 42.10, mp: 816° in H, decomp @ about 600°, d: 1.8 approx.
THR = See calcium compounds and hydrides. Burns fiercely in air when heated. Violent reaction with bromates, chlorates, perchlorates, AgF. [*19*]

CALCIUM HYDROGEN SULFITE. See calcium bisulfite.

CALCIUM HYDROSULFIDE. See calcium sulfhydrate.

CALCIUM HYDROXIDE. Syns: *hydrate lime, slaked lime.* Rhombic, trigonal, colorless crystals. $Ca(OH)_2$, mw: 74.10, mp: $-H_2O$ @ 580°, bp: decomp, d: 2.343.
THR = A general purpose food additive, also a substance migrating to food from packaging materials. [*109*] See calcium compounds.
Acute tox data: Oral LD_{50} (rat) = 7340 mg/kg. [*3*]
THR = LOW via oral route. Calcium hydroxide has a caustic reaction and therefore is irr to the skin and respiratory system. In the form of dust it is considered to be an important industrial hazard. It can cause dermatitis, and irr of the eyes and mu mem. It is a common air contaminant. Violent reaction with maleic anhydride, nitroethane, nitromethane, nitroparaffins, nitropropane, P. [*19*]
Treatment and Antidotes: Irrigate any areas which have come in contact with this material. If the eyes are involved, they should be washed at once with copious amounts of warm water. If the skin is involved, a shower is recommended. See also calcium compounds.

CALCIUM HYPOCHLORITE. See bleaching powder.

CALCIUM HYPOCHLORITE COMPOUNDS, DRY. Compounds containing more than 8.80% available oxygen, 39% available chlorine. See hypochlorites.

CALCIUM HYPOPHOSPHITE. $Ca(H_2PO_2)_2$, mw: 170.1.
THR = No data. Decomp in air when heated, to yield PH_3, which self-ignites. Reacts violently with HNO_3, $KClO_3$ [*19*]

CALCIUM IODATE. Syn: *autarite*. Triclinic crystals. $Ca(IO_3)_2$, mw: 389.9, mp: decomp, d: 4.519 @ 15°.
THR = A trace mineral added to animal feed. See iodates and calcium compounds. [*109*] Reacts violently with Al, As, C, Cu, metal sulfides, P, S, organic matter. [*19*]

CALCIUM IODOBEHENATE. White or yellowish powder, odorless or slightly fatty odor, contains 24% iodine, sol in warm chloroform, slightly sol in alcohol and ether, insol in water. $Ca(OOCC_{21}H_{42}I)_2$, mw: 960.
THR = No data. Not highly toxic. Used in medicine. Can cause GI distrubances or Iodism. A trace mineral added to animal feeds. [*109*]
Disaster Hazard: Dangerous; see iodides.

CALCIUM LACTATE. White powder, almost odorless, sol in water, practically insol in alcohol.
$Ca(C_3H_5O_3)_2 \cdot 5H_2O$, mw: 308, mp: $-5H_2O$ @ 120°.
THR = No data. A general purpose food additive. [*109*]

CALCIUM LACTOBIONATE. The calcium salt of lactobionic acid, produced by the oxidation of lactone.
THR = U. A food additive permitted in food for human consumption. [*109*]

CALCIUM LIGNOSULFATE.
THR = U. A food additive permitted in food for human consumption. [*109*]

CALCIUM MAGNESIUM SILICON.
THR = An alloy which may evolve a self-igniting gas when treated with acid. [*19*]

CALCIUM METALLIC. See calcium.

CALCIUM METALLIC, CRYSTALLINE. See calcium.

CALCIUM METHANE ARSONATE. CH_3O_3AsCa, mw: 178.
Acute tox data: ip LD_{50} (mouse) = 500 mg/kg. [*3*]
THR = HIGH via ip and oral routes. See also arsenic compounds.

CALCIUM MONOFLUOROPHOSPHATE. See calcium fluorophosphate.

CALCIUM NITRATE. Cubic, colorless, hygroscopic crystals. $Ca(NO_3)_2 \cdot 4H_2O$, mw: 236.2, mp: 561.0°, d: 2.36.
Acute tox data: Oral LD_{50} (rat) = 3900 mg/kg. [*3*]
THR = MOD via oral route. See nitrates and calcium compounds.
Fire Hazard: See nitrates.
Disaster Hazard: See nitrates.

CALCIUM NITRIDE. Brown crystals. Ca_3N_2, mw: 148.3, mp: 1195°, d: 2.63 @ 17°.
THR = See calcium compounds and ammonia. Ammonia can be evolved upon contact with mositure. It self ignites in air. Reacts violently with halogens. [*19*]

CALCIUM NITRITE. Hexagonal, colorless to yellowish crystals. $Ca(NO_2)_2 \cdot H_2O$, mw: 150.11, mp: $-H_2O$ @ 100°, d: 2.23 @ 34° (anh); 2.53 @ 30°.
THR = See nitrites and calcium compounds.

CALCIUM OXALATE. Cubic, colorless crystals.
CaC_2O_4, mw: 128.10, mp: decomp, d: 2.2 @ 4°.
THR = See oxalates and calcium compounds.

CALCIUM OXIDE. Syns: *unslaked lime, quick lime, burnt lime, calx*. Cubic, colorless crystals. CaO, mw: 56.08, mp: 2580°, d: 3.37, bp: 2850°.
THR = A nutrient and/or dietary supplement food additive. [*109*] See calcium compounds. A common air contaminant. A powerful caustic to living tissue.
Disaster Hazard: Violent reaction with (B_2O_3 + $CaCl)_2$, BF_3, ClF_3, F_2, HF, P_2O_5, water. [*19*]

CALCIUM PANTOTHENATE. White, slightly hygroscopic powder, odorless, sol in water and glycerol, insol in alcohol, chloroform and ether.
$(C_9H_{16}NO_5)_2Ca$, mw: 476, mp: 170°–172°, decomp @ 195°–196°.
Acute tox data: ip LD_{50} (rat) = 820 mg/kg. [*3*]
THR = MOD via ip route. A nutrient and/or dietary supplement food additive. [*109*]

CALCIUM PERCHLORATE. Colorless crystals.
$Ca(ClO_4)_2$, mw: 239.0, d: 2.651, mp: 270° (decomp).
THR = See perchlorates and calcium compounds.

CALCIUM PERCHROMATE. Buff-colored powder.
$Ca_3(CrOO_3)_2$, mw: 352.2.
THR = See chromium compounds. Explodes at 100°. [*19*]

CALCIUM PERMANGANATE. Syn: *acerdol*. Violet, deliquescent crystals. $Ca(MnO_4)_2 \cdot 5H_2O$, mw: 368.02, mp: decomp, d:2.4.
THR = See calcium compounds and permanganates.

CALCIUM PEROXIDE. Syn: *calcium superoxide*. Yellow crystals or powder or white crystals. CaO_2, mw: 72.08, mp: decomp @ 275°.
THR = Although this material is used in dentifrices, the conc form would be an irr. It would react with moisture to form slaked lime. See also calcium

compounds, calcium hydroxide and peroxides, inorganic.

Fire Hazard: Mod, if hot and mixed with finely divided combustible material; decomp rapidly above 200°. Like calcium oxide, it is a strong alkali. An oxidizer. See also peroxides, inorganic.

Explosion Hazard: An explosion hazard when intimately mixed with finely divided reducing agents such as organic matter.

Disaster Hazard: See peroxides, inorganic.

CALCIUM PHENATE. See calcium phenoxide.

CALCIUM-1-PHENOL-4,*p*-SULFONATE. White to pinkish powder. $Ca(C_6H_4(OH)SO_3)_2 \cdot H_2O$, mw: 404.43.

THR = See phenol sulfonates.

CALCIUM PHENOXIDE. Syn: *calcium phenate.* Powder. $Ca(OC_6H_5)_2$, mw: 226.28.

THR = See phenol.

***sec*-CALCIUM PHOSPHATE.** See calcium phosphate, dibasic.

CALCIUM PHOSPHATE, DIBASIC. Syns: *dicalcium-o-phosphate, bicalcium phosphate, sec-calcium phosphate.* White crystalline powder, odorless, sol in dilute hydrochloric, nitric and acetic acids, insol in alcohol, slightly sol in water. (1) $CaHPO_4$, (2)$CaHPO_4 \cdot 2H_2O$, mw(1): 135, mw(2): 172.

THR = U. A nutrient and/or dietary supplement food additive. Also a general purpose food additive. [*109*]

Disaster Hazard: Dangerous. See phosphates.

CALCIUM PHOSPHATE, MONOBASIC. Syn: *calcium biphosphate, acid calcium phosphate, calcium phosphate, primary.* Colorless, pearly scales or powder, sol in water and acids.

$CaH_4(PO_4)_2 \cdot H_2O$, mw: 252, mp: loses H_2O @ 100°, decomp @ 200°, d: 2.20.

THR = U. A nutrient and/or dietary supplement food additive. Also a general purpose food additive and sequestrant food additive. [*109*]

Disaster Hazard: Dangerous; see phosphates.

CALCIUM PHOSPHATE PRIMARY. See calcium phosphate, monobasic.

CALCIUM PHOSPHATE, TRIBASIC. Syns: *bone ash, tricalcium phosphate.* Amorphous, odorless powder. $Ca_3(PO_4)_2$, mw: 310.2, mp: 1670°, d: 3.14.

THR = A nutrient and/or dietary supplement food additive. [*109*] also a general purpose food additive. See calcium compounds and phosphates. A caustic and irr material.

CALCIUM-*m*-PHOSPHATE.

THR = U. A sequestrant food additive. [*109*]

Disaster Hazard: Dangerous. See phosphates.

CALCIUM PHOSPHIDE. Red crystals. Ca_3P_2, mw: 182.2, mp: >1600°, d: 2.238 @ 25°.

THR = See phosphides. Violent reaction with Cl_2, ClO, HCl, O_2, S, water. [*19*]

CALCIUM PHYTATE. Syn: *hexacalcium phytate.* Free flowing white powder, slightly sol in water. $(CaC_2H_4O_8P_2)_6$, mw: 1548 (approx).

THR = U. A sequestrant food additive. [*109*]

Disaster Hazard: Dangerous; see phosphates.

CALCIUM-*o*-PLUMBATE. Reddish brown crystals. Ca_2PbO_4, mw: 351.37, mp: decomp, d: 5.71.

THR = See lead compounds.

CALCIUM POLYSULFIDES. See calcium sulfide.

CALCIUM PROPIONATE. White powder, sol in water, slightly sol in alcohol. $Ca(OOCCH_2CH_3)_2$ (occurs also with H_2O molecule), mw: 186.

Acute tox data: Oral LD_{50} (rat) = 5160 mg/kg. [*3*]

THR = MOD via oral route. Used as a chemical preservative food additive. [*109*]

CALCIUM PYROPHOSPHATE. White powder, sol in dilute hydrochloric acid and nitric acid, insol in water. $Ca_2P_2O_7$, mw: 254. d: 3.09, mp: 1353°.

THR = U. A nutrient and/or dietary supplement food [*109*] additive. A mild caustic and irr.

Disaster Hazard: Dangerous; see phosphates.

CALCIUM RESINATE. Yellowish white amorphous powder or lumps. $Ca(C_{44}H_{62}O_4)_2$, mw: 1349.50.

THR = U.

Fire Hazard: Mod, when heated; can react with oxidizing materials.

CALCIUM RESINATE, FUSED. See calcium resinate.

CALCIUM RICINOLEATE. Fine, white powder. $Ca[CO_2(CH_2)_7CHCHCH_2CHOH(CH_2)_5CH_3]_2$, mw: 630, mp: 84°, d: 109 @ 25°.

THR = See calcium compounds.

CALCIUM SACCHARIN. White crystalline powder, odorless or faint aromatic odor, sol in water. $Ca(C_6H_4COSO_2N)_2$, mw: 464.

THR = A non-nutritive sweetener food additive. [*109*] See saccharin.

Disaster Hazard: Dangerous; see sulfates.

CALCIUM SELENATE. Colorless crystals. $CaSeO_4$, mw: 184.04, d: 2.93.

THR = See selenium compounds.

CALCIUM SELENIDE. Simple cubic crystals. CaSe, mw: 119.04, d: 7.593.

THR = See selenium compounds.

Fire Hazard: Mod, when heated. See hydrogen selenide, which can be evolved on contact with acids or moisture.

Explosion Hazard: Mod, when exposed to flame.

Disaster Hazard: Dangerous; can emit highly toxic and flam fumes of hydrogen selenide, selenium compounds on contact with water, steam, acid or acid fumes.

CALCIUM SILICATE. White to cream colored, free flowing powder. Ca_2SiO_4, $Ca_3Si_2O_7$, $Ca_3(Si_3O_9)$, $Ca_4(H_2Si_4O_{13})$, d: 2.10 @ 25°/4°.
THR = MILD irr via inhal and oral routes. An anti-caking agent food additive. Also a food additive permitted in feed and drinking water of animals and/or for the treatment of food-producing animals. [*109*]

CALCIUM SILICIDE. Glassy solid. $CaSi_2$, mw: 96.2, d: 2.5.
THR: See calcium hydroxide and silanes. Can evolve self-igniting gas with acids, reacts violently with F_2. [*19*]

CALCIUM SILICOFLUORIDE. See calcium fluosilicate.

CALCIUM SILICON.
THR = Evolves self-igniting gas with acids. [*19*]

CALCIUM SORBATE. $Ca(OOC_5H_7)_2$, mw: 262.
THR = U. Used as a chemical preservative food additive. [*109*]

CALCIUM STEARYL-2-LACTYLATE. Syn: *verv-Ca.* Free flowing white powder, sparingly sol in water. $(C_{24}H_{43}O_6)Ca$, mw: 895.
THR = Practically non-toxic. A food additive. [*109*]

CALCIUM SULFATE. Pure anhydrous white powder or odorless crystals. $CaSO_4$, mw: 136, d: 2.964, mp: 1450°.
THR = U. A nutrient and/or dietary supplement food [*109*] additive. Violent reaction with Al.
Disaster Hazard: Dangerous; see sulfates.

CALCIUM SULFHYDRATE. Syn: *calcium hydrosulfide.* Colorless, transparent crystals, decomp in air. $Ca(HS)_2$, mw: 106.21.
THR = See sulfides.

CALCIUM SULFIDE. Syns: *oldhamite, hepar calcis. calcic liver of sulfur.* Cubic, colorless crystals. CaS, mw: 72.14, bp: decomp, d: 2.18 @ 15°.
THR = See sulfides. Reacts violently with PbO_2, $KClO_3$, KNO_3. [*19*]

CALCIUM SULFITE. Hexagonal, colorless crystals. $CaSO_3 \cdot 2H_2O$, mw: 156.18, mp: $-2H_2O$ @ 100°.
THR = See sulfites.

CALCIUM SUPEROXIDE. See calcium peroxide.

CALCIUM SUPERPHOSPHATE. Syn: *calcium tri-o-phosphate (fertilizer grade).* Amorphous, white powder. $Ca_3(PO_4)_2$, mw: 310.28, mp: 1670°, d: 3.14.
THR = See calcium compounds and phosphates.

CALCIUM TELLURIDE. Simple cubic crystals. CaTe, mw: 167.69, d: 7.593.
THR = See tellurium compounds.
Fire Hazard: Mod, when heated. See hydrogen telluride, which can be evolved on contact with acids or moisture.
Explosion Hazard: Slight, when exposed to flame.
Disaster Hazard: Dangerous; emits highly toxic and flam fumes of hydrogen telluride and tellurium compounds on contact with water, steam, acid or acid fumes.

CALCIUM TELLURITE. White flakes. $CaTeO_3$, mw: 215.69, mp: >960°.
THR = See tellurium compounds.

CALCIUM TETRABORATE. White solid. CaB_4O_7, mw: 195.36, mp: 986°.
THR = See boron compounds.

CALCIUM THIOCYANATE. White crystals, deliquescent. $Ca(SCN)_2 \cdot 3H_2O$, mw: 210.30.
Acute tox data: Oral LD_{LO} (mouse) = 120 mg/kg. [*3*]
THR = HIGH via oral route. See thiocyanates.

CALCIUM TITANATE. Powder. $CaTiO_3$, mw: 136, d: 3.98, mp: 1800°.
THR = U. Animal exper show LOW. See also titanium compounds.

CALCIUM TRI-*o*-PHOSPHATE. See calcium superphosphate.

CALCIUM ZINC ARSENATE MONOHYDRATE. Syn: *austinite.* A solid. $2CaO \cdot 2ZnO \cdot As_2O_5 \cdot H_2O$, mw: 523.
THR = HIGH. See arsenic compounds.
Disaster Hazard: Dangerous; see arsenic

CALCO URANITE. See autunite.

CALGON. See sodium hexa-*m*-phosphate.

CALIFORNIUM. Cf.
Radiation Hazard: For permissible levels see Section 5, Table 5A.5. Artificial isotope ^{249}Cf, $T\frac{1}{2}$ – 360 y, decays to ^{245}Cm via α's of 5.8 MeV. Artificial isotope ^{250}Cf, $T\frac{1}{2}$ = 13 y, decays to ^{246}Cm via α's of 6.0 MeV. Artificial isotope ^{251}Cf, $T\frac{1}{2}$ = 800 y, decays to ^{247}Cm via α's of 5.7–5.8 MeV. Artificial isotope ^{252}Cf, $T\frac{1}{2}$ = 2.6 y, decays to ^{248}Cm via α's of 6.1 MeV.

CALOMEL. See mercurous chloride.

CALOMEL + MAGNESIUM SULPHATE.
THR = An exper neo to mice via oral route. [*103*]

CALVACIN. Glycopeptide from giant puffball mushroom.
Acute tox data: ip LD_{50} (rat) = 65 mg/kg. [*3*]
THR = HIGH via ip and oral routes.

CALX. See calcium oxide.

CAMBOGIA. See gamboge.

α-CAMPHANOL. See *d*-borneol.

CAMPHENE. Syn: *3,3-dimethyl-2-methylene norcamphone.* Cubic crystals. $C_{10}H_{16}$, mw: 136, mp: 50°–51°, bp: 159°, d: 0.842 @ 54°/4°.
THR = U.
Fire Hazard: Slight; yields flam vapors when heated, can react with oxidizing materials.

2-CAMPHONONE. See camphor.

CAMPHOR. Syns: *2-camphonone, gum camphor, laurel camphor.* White, transparent, crystalline masses, penetrating odor, pungent, aromatic taste. $C_{10}H_{16}O$, mw: 152.23, mp: 180°, bp: 204°, lel = 0.6%, uel = 3.5%, flash p: 150°F(CC), d: 0.992 @ 25°/4°, autoign. temp.: 871°F, vap. d: 5.24.
Acute tox data: ip LD_{LO} (rat) = 900 mg/kg; sc LD_{LO} (mouse) = 2200 mg/kg. [*3*]
THR = MOD via ip and sc routes. Locally, camphor is an irr. When swallowed it causes nausea, vomiting, dizziness, excitation and convulsions.
Fire Hazard: Mod, when exposed to heat or flame; can react with oxidizing materials.
Spont Heating: No.
Explosion Hazard: In the form of vapor when exposed to heat or flame or CrO_3.
Treatment and Antidotes: Evacuate stomach with emetic or stomach tube. Give sedatives if necessary. Call a physician.
To Fight Fire: Foam, carbon dioxide, dry chemical.

CAMPHORIC ANHYDRIDE. Acicular, odorless, crystals, almost insol in water, freely sol in alcohol, benzene, ether, carbon disulfide, chloroform, ethyl acetate. $C_{10}H_{14}O_3$, mw: 182.21, d: 1.19, mp: 220°–221°, bp: 270°.
THR = See camphor.

CAMPHOR OIL (LIGHT). Syn: *liquid camphor.* An oily, fragrant liquid, bp: 175°–200°, flash p: 117°F (CC), d: 0.88.
THR = A (S) carc. [*3*] See safrol and camphor.
Fire Hazard: Mod, when exposed to heat or flame; can react with oxidizing materials.
Spont Heating: No.
To Fight Fire: Foam, CO_2, dry chemical, mist, fog.

CANADA BALSAM. Syns: *Canada turpentine, balm of Gilead.* Yellowish to greenish viscid, transparent, fluorescent, aromatic liquid. d: 0.987–0.994.
THR = MILD irr and allergen.

CANADA TURPENTINE. See canada balsam.

CANDLE PITCH.
THR = A (S) carc. [*14*]

CANNABIS. Syns: *cannabin resin, Indian hemp, Indian cannabis, hashish, guaza, marihuana.* A resinous, bitter substance from Cannabis Sativa, greenish black mass.
Acute tox data: Oral LD_{50} (rat) = 1380 mg/kg. [*3*]
THR = MOD via oral route. An allergen. An exper teratogen. [*3*] Cannabis is a narcotic analgesic and sedative, it causes the pupils to be widely dilated and to barely react to light. When ingested or inhaled as smoke can cause euphoria, delirium, hallucinations, drowsiness, weakness and hyporeflexia. An overdose can cause coma and death. Like the other narcotics, this material can be habit-forming. It can be grown in the temperate zones and acquired illegally for narcotic purposes.
Fire Hazard: Slight; because dried material can burn, it can react with oxidizing materials.
Treatment and Antidotes: Removal from exposure or cessation of ingestion of this material will generally cause the symptoms to disappear. In case of a large dose, or if overdose is suspected, a physician should be consulted at once.

CANNABIS RESIN. See cannabis.

CANOPY HOOD. See Section 2.

CANTHARIDES. Syns: *blistering flies, blistering beetles, Spanish fly.* Brown to black powder and scales. $C_{10}H_{12}O_4$, mw: 196.15, mp: 218°, bp: subl @ 90°.
THR = HIGH irr via dermal, oral, inhal and contact with skin, eyes and mu mem. An allergen. Extremely irr to the eyes. Can cause conjunctivitis, keratitis, blepharitis, slight swelling of cornea and inflammation of iris. A powerful irr. Used externally chiefly as a blistering agent. It is often and mistakenly used as an aphrodisiac, but it is much too dangerous and irr a material for this purpose. Symptoms of intoxication are extreme irr of intestines and kidneys.
Disaster Hazard: Slight; on decomp, emits toxic fumes.
Treatment and Antidotes: If it has been ingested, a stomach syphon is recommended, or emetics, demulcents but not oils. Administer morphine, stimulants or poultices to the abdomen. Call a physician.

CANTHARIDES CAMPHOR. $C_{10}H_{12}O_4$, mw: 196.2.
THR = HIGH via oral route. An exper (S) carc. [*3, 7*]

CAOUTCHOUC. See rubber, crude.

CAPRIC ETHER. See ethyl caprate.

CAPRINIC ETHER. See ethyl caprate.

CAPROIC ACID. Syn: *hexanoic acid.* Colorless liquid, odor of limburger cheese. $C_6H_{12}O_2$, mw: 116.2, bp: 205.0°, fp: −3.4°, flash p: 215°F (COC), d: 0.9295 @ 20°/20°, vap. press: 0.18 mm @ 20°, vap. d: 4.0, autoign. temp.: 716°F.

Acute tox data: Oral LD_{Lo} (rat) = 3000 mg/kg; ip LD_{50} (mouse) = 3180 mg/kg; dermal LD_{50} (rabbit) = 630 mg/kg. [*3*]
THR = MOD via oral, ip and dermal routes.
Fire Hazard: Low, when exposed to heat or flame; can react with oxidizing materials.
To Fight Fire: CO_2, dry chemical, fog, mist.

ε-CAPROLACTAM. Syn: *2-oxohexamethylenimine.* White crystals. $C_6H_{11}NO$, mw: 113.16, mp: 69°, vap. press: 6 mm @ 120°.
Acute tox data: Oral LD_{50} (rat) = 2140 mg/kg; dermal LD_{Lo} (rabbit) = 1410 mg/kg; 7 ppm ⟶ irr to humans via inhal. [*3*]
THR = MOD via oral and dermal; HIGH via inhal routes.
Disaster Hazard: Mod dangerous; when heated to decomp. emits toxic fumes.

CAPROYL ETHYLENE IMINE. $C_8H_{15}ON$, mw: 141.2.
THR = An exper neo. [*3, 23*]

CAPRYL ALCOHOL. Syn: *2-octanol.* Clear, colorless, oily liquid, pungent, aromatic odor.
$CH_3HCOH(CH_2)_5CH_3$, mw: 130.2, bp: 178.5°, fp: −38.6°, d: 0.8193 @ 20°/4°, vap. d: 4.48, flash p: 190°F.
Acute tox data: Oral LD_{50} (rat) = 1790 mg/kg. [*3*]
THR = MOD via oral route.
Fire Hazard: Mod, when exposed to heat or flame; can react with oxidizing materials.
To Fight Fire: Water foam, fog, alcohol foam, dry chemical, CO_2.

CAPRYL DINITRO PHENYL CROTONATE. $C_{18}H_{24}O_6N_2$, mw: 364.4.
THR = MOD via oral route. An exper neo. [*3*]

1-CAPRYLENE. See 1-octene.

CAPRYLIC ACID. Syn: *octanoic acid.* Colorless, oily liquid, unpleasant odor, burning rancid taste.
$CH_3(CH_2)_6COOH$, mw: 144.2, d: 0.91 @ 20°, bp: 240°, mp: 17°.
Acute tox data: iv LD_{50} (mouse) = 600 mg/kg. [*3*]
THR = MOD via iv route. Yields irr vapors which can cause coughing. Used as a chemical preservative food additive. [*109*]

CAPRYLYL CHLORIDE. See ethyl hexanoyl chloride.

CAPRYLYL PEROXIDE SOLUTION. $C_8H_{18}O_2$, mw: 146.3.
THR = See peroxides, organic.
Fire Hazard: See peroxides, organic.
Disaster Hazard: See peroxides, organic.

CAPSICUM. Syn: *tincture of tabasco pepper.* An alcoholic solution of capsicum.
THR = MOD irr via oral and inhal routes.
Fire Hazard: Mod, when exposed to heat. See also alcohol.
Disaster Hazard: See alcohol.

CAPTAFOL. $C_{10}H_9O_2NSCl_4$, mw: 349.1.
Acute tox data: ip LD_{Lo} (mouse) = 3 mg/kg; oral LD_{50} (rat) = 2500 mg/kg. [*3*]
THR = MOD via oral; HIGH via ip routes. An exper teratogen. [*3*]
Disaster Hazard: High; see chlorides and sulfur compounds.

CAPTAN. See *cis*-N-((trichloromethyl)thio)-4-cyclohexene-1,2-dicarboximide.

CARAMEL. Syn: *sugar coloring, burnt sugar.* Deliquescent, dark brown powder or thick liquid, burnt sugar odor, sol in water and dilute alcohol, immiscible with most organic solvents. d: approx 1.35.
THR = U. A general purpose food additive. [*109*]

CARBAMATES. Compounds based on carbamic acid. H_2NCOOH, only of use in the form of its numerous derivatives, and some of these materials are (S) carc of lung and hematopoietic organs. [*14, 23*]

CARBAMAZEPINE. $C_{15}H_{12}ON_2$, mw: 236.3.
Acute tox data: Oral LD_{50} (dog) = 5620 mg/kg; oral LD_{50} (rabbit) = 2680 mg/kg; oral LD_{50} (rat) = 4025 mg/kg. [*3*]
THR = MOD via oral route, also to women, affecting red blood cells and skin. Has been implicated in development of aplastic anemia.

CARBAMIC ACID BUTYL ETHYL ESTER. See butyl urethane.

CARBAMIDE. Syn: *urea.* White crystals. $(NH_2)_2CO$, mw: 60.1, mp: 132.7, bp: decomp, d: (solid) 1.335.
Acute tox data: sc LD_{Lo} (dog) = 3000 mg/kg; iv LD_{Lo} (dog) = 3000 mg/kg; oral LD_{Lo} (domestic animals) = 511 mg/kg. [*3*]
THR = MOD via sc, iv and oral routes. Reacts violently with gallium perchlorate. [*19*]

CARBAMIDE PHOSPHORIC ACID. White crystals. $CO(NH_2)_2 \cdot H_3PO_4$, mw: 158.
THR = See phosphoric acid, urea and amides.
Disaster Hazard: Dangerous; when heated to decomp, emits highly toxic fumes of cyanides or PO_x.

CARBAMOYL OXIME. $C_{16}H_{20}O_4N_2S_2 \cdot HCl \cdot H_2O$, mw: 459.4.
Acute tox data: sc LD_{50} (rat) = 3000 mg/kg. [*3*]
THR = MOD via oral route. An exper teratogen. [*3*]

CARBAMYL CHLORIDE. Syn: *chloroformamide.* Liquid, offensive odor. H_2NCOCl, mw: 79.5, mp: 50° (approx), bp: 61°–62° (decomp).
THR = See hydrochloric acid, which is evolved upon standing.
Disaster Hazard: Dangerous; see chlorides.

CARBAMYL-2-PHENYL HYDRAZINE. Syn: *phenicarbazide.* Crystals. $C_7H_9ON_3$, mw: 151.2, mp: 172°.
THR = An exper neo via oral route. [*3*] producing lung tumors, angiomas and/or angiosarcomas in the liver. [*12*]

CARBARYL. Syn: *1-naphthyl methyl carbamate, sevin.* White crystals. $C_{10}H_7OOCNHCH_3$, mw: 201, mp: 142°, d: 1.232 @ 20°/20°.
Acute tox data: Oral LD_{50} (guinea pig) = 280 mg/kg. [*3*]
THR = An exper (±) carc and teratogen. [*12, 3*] Absorbed via all routes, although skin absorption is slow. No mammalian tissue accumulation. Symptoms include blurred vision, headache, stomachache, vomiting. Usually recovery is rapid. HIGH via oral route. See carbamates. A reversible cholinesterase inhibitor. Symptoms similar to those due to parathion, but much less severe from equal doses. [*55, 56, 57*]

CARBASONE. Syn: *p-ureidobenzene arsonic acid.* White, nearly odorless powder, slightly acid taste, sol in water and alcohol. $NH_2CONHC_6H_4AsO(OH)_2$, mw: 260, mp: 174°.
Acute tox data: Oral LD_{50} (rat) = 510 mg/kg. [*3*]
THR = MOD via oral route. An exper carc. [*3*] See also arsenic compounds.

CARBAZOLE. Syn: *dibenzopyrrole.* White crystals. $(C_6H_4)_2NH$, mw: 167.20, mp: 244.8°, bp: 354.8°, d: 1.10 @ 18°/4°, vap. press: 400 mm @ 323.0°.
Acute tox data: ip LD_{50} (mouse) = 200 mg/kg. [*3*]
THR = HIGH via ip and oral routes. An allergen.
Fire Hazard: Slight.

2-CARBAZOLYL ACETAMIDE.
THR = An exper carc. [*23*]

CARBAZOTIC ACID. See picric acid.

CARBIDES. See acetylides.

"CARBITOL." Syn: *diethylene glycol monoethyl ether.* Colorless liquid, mild pleasant odor. $C_6H_{14}O_3$, mw: 134.17, bp: 201.9°, flash p: 201°F (OC), d: 0.9902 @ 20°/4°, vap. d: 4.62.
Acute tox data: Oral LD_{50} (rabbit) = 3620 mg/kg. [*3*]
THR = MOD via oral route. See also glycols.
Fire Hazard: Low, when exposed to heat; can react with oxidizing materials.
Spontaneous Heating: No.
To Fight Fire: Alcohol foam, CO_2, dry chemical.

"CARBITOL" ACETATE. Syn: *diethylene glycol monoethylether acetate.* Liquid. $C_3H_8O_4(CH_2)_4$, mw: 176.21, bp: 217.4°, fp: −25°, flash p: 230°F (OC), d: 1.0114 @ 20°/20°, vap. press: 0.05 mm @ 20°, vap. d: 6.07.
Acute tox data: Oral LD_{50} (rat) = 11 g/kg; dermal LD_{50} (rabbit) = 15 g/kg. [*3*]
THR = LOW via dermal and oral routes. See glycols.
Fire Hazard: Low, when exposed to heat; can react with oxidizing materials.
Spont Heating: No.
To Fight Fire: Alcohol foam, water, CO_2, dry chemical.

"CARBITOL" PHTHALATE. Syn: *dicarbitol phthalate.* Liquid. $C_6H_4(COOC_2H_4OC_2H_4OC_2H_5)_2$, mw: 398.44, bp: 200°–260°, flash p: 406°F (CC), d: 1.121.
THR = See glycols.
Fire Hazard: Slight, when exposed to heat; can react with oxidizing materials.
Spont Heating: No.
To Fight Fire: Water, foam, CO_2, dry chemical.

CARBODIMIDE. See cyanimide.

CARBOLIC ACID, LIQUID. Liquid tar acid containing over 50% benzophenol. See phenol.

CARBOLIC ACID, SOLID. See phenol.

CARBOLINEUM. Syn: *chlorinated anthracene oil.*
THR = MOD irr. Action resembles that of phenol. May contain carcinogenic impurities.
Disaster Hazard: See chlorinated hydrocarbons, aromatic.

CARBOMETHANE. See ketene.

4-CARBOMETHOXY-2,3'-DIMETHYL AZO BENZENE. $C_{16}H_{16}O_3N_2$, mw: 284.3.
THR = An exper carc. [*3*]

2-CARBOMETHOXY-1-METHYL VINYL DIMETHYL PHOSPHATE. See phosdrin.

CARBOMETHOXY PHENYL TRICHLOROSTANNANE. A solid. $(CH_3OCOC_6H_4)SnCl_3$, mw: 360, mp: 164°.
THR = U. See tin compounds.
Disaster Hazard: Dangerous; see chlorides.

CARBON. Black crystals, powder or diamond form. C, atwt: 12.01, mp: 3652°–3697° (subl), bp: approx 4200°, d(amorphous): 1.8–2.1, d(graphite): 2.25, d(diamond): 3.51, vap. press: 1 mm @ 3586°.
THR = Nuisance dust slightly irr. In the form of graphite (which is one of the common forms of carbon) it can cause a dust irritation, particularly to the eyes. Carbon in the form of soot can cause conjunctivitis. It can also cause epithelial hyperplasia of cornea, as well as eczematous inflammation of eyelids. Some forms of carbon dust can cause irr of eyes and mu mem.
Radiation Hazard: For permissible levels see Section 5, Table 5A.5. Artificial and natural isotope ^{14}C, $T\frac{1}{2}$ = 5730 y, decays to stable ^{14}N via β's of 0.155

MeV. ^{14}C occurs in nature as a result of cosmic ray bombardment of ^{14}N. The natural activity in living things or atmospheric CO_2 = 12.5 dpm/g of C.

Fire Hazard: Slight, when exposed to heat.

Explosion Hazard: In the form of dust when exposed to heat or flame or (NH_4NO_3 + heat), (NH_4ClO_4 @ 240°), bromates, $Ca(OCl)_2$, chlorates, Cl_2, (Cl_2 + $Cr(OCl)_2$), ClO, F_2, iodates, IO_5, ($Pb(NO_3)_2$, $HgNO_3$, HNO_3, (oils + air), (K + air), Na_2S, $Zn(NO_3)_2$. [*19*]

CARBON ACTIVATED. See carbon.

CARBON ARC RADIATION.
THR = A recog carc. [*14*]

CARBON BISULFIDE. See carbon disulfide.

CARBON BLACK. See soot.

CARBON DIBROMIDE. See dicarbon tetrabromide.

CARBON DICHLORIDE. See perchloro ethylene.

CARBON DIOXIDE. Syn: *carbonic acid, carbonic anhydride.* Colorless, odorless gas. CO_2, mw: 44.01, mp: subl @ −78.5°, (−56.6° @ 5.2 atm), vap. d: 1.53.

THR = Asphyxiant. An exper teratogen. [*3*] Symptoms resulting only when such high conc are reached that there is insufficient oxygen in the atmosphere to support life. The signs and symptoms are those which precede asphyxia, namely, headache, dizziness, shortness of breath, muscular weakness, drowsiness and ringing in the ears. Removal from exposure results in rapid recovery. Contact of carbon dioxide snow with the skin may cause a "burn." See also discussion of simple asphyxiants under Argon. Reacts vigorously with (Al + Na_2O_2), Cs_2O, $Mg(C_2H_5)_2$, Li, (Mg + Na_2O_2), K, KHC_2, Na, Na_2C_2, NaK, Ti. [*19*]

CARBON DIOXIDE-NITROUS OXIDE MIXTURE. Gas. Composition: CO_2 + N_2O.

THR = See also components as listed.

Fire Hazard: Slight. An oxidizing mixture.

Disaster Hazard: Mod dangerous; can react with reducing materials.

CARBON DIOXIDE, SOLID. Syn: *dry ice.* CO_2, White, snow-like solid. d: 1.35.

THR = See carbon dioxide. Normally it is very cold and can cause frostbite if handled.

CARBON DISELENIDE. Greenish-yellow liquid. CSe_2, mw: 169.9, bp: 90° (approx).

THR = Probably HIGH. See selenium compounds and carbon disulfide.

CARBON DISULFIDE. Syn: *carbon bisulfide.* Clear, colorless liquid, nearly odorless when pure. CS_2, mw: 76.1, mp: −110.8°, bp: 46.5°, lel = 1.3%, uel = 50%, flash p: −22°F (CC), d: 1.261 @ 20°/20°, autoign. temp.: 194°F, vap. press: 400 mm @ 28°, vap. d: 2.64.

Acute tox data: Oral LD_{LO} (human) = 14 mg/kg; inhal LC_{LO} (human) = 4000 ppm for 30 min; inhal TC_{LO} (human) = 50 mg/m^3 for 7 hrs ⟶ CNS effects. [*3*]

THR = HIGH via oral and inhal routes. [*58, 59, 60*] An insecticide. The chief toxic effect is on the CNS, acting as a narcotic and anesthetic in acute poisoning with death following from respiratory failure. The anesthetic action is much more powerful than that of chloroform. In chronic poisoning, the effect on the nervous system is one of central and peripheral damage, which may be permanent if the damage has been severe. Sensory symptoms usually precede motor involvement. A secondary anemia may be caused.

In acute poisoning, early excitation of the CNS resembling alcoholic intoxication occurs, followed by depression, with stupor, restlessness, unconsciousness, and possibly death. If recovery occurs, the patient usually passes through the after-stage of narcosis, with nausea, vomiting, headache, etc. In chronic poisoning, the picture is that of involvement of the nervous system, with neuritis and disturbance of vision being the commonest early changes. Sensory changes such as a crawling sensation in the skin, sensations of heaviness and coldness, and visually, "veiling" of objects so that they appear indistinct, are noticed first. Often there is pain in the affected parts, particularly the limbs. These symptoms are followed by gradually increasing loss of strength. Wasting of the muscles may occur. Mental symptoms vary from simple excitation or depression and irritability in the mild cases to mental deterioration, Parkinsonian paralysis, and even insanity. These changes are accompanied by insomnia, loss of memory, and personality changes. Chronic fatigue is a very common complaint. A fumigant.

Fire Hazard: Dangerous, when exposed to heat, flame, sparks or friction.

Spont Heating: No.

Explosion Hazard: Severe; when exposed to heat or flame, reacts violently with Al, Cl_2, azides, CsN_3, ClO, ethylamine diamine, ethylene imine, F_2, $Pb(N_3)_2$, LiN_3, NO, N_2O_4, (H_2SO_4 + permangates), K, KN_3, RbN_3, NaN_3 Zn. [*19*]

Disaster Hazard: Dangerous; when heated to decomp, emits highly toxic fumes of SO_x; can react vigorously with oxidizing materials.

To Fight Fire: Water, CO_2, dry chemical, fog, mist.

CARBON HEXACHLORIDE. See hexachloroethane.

CARBONIC ACID. See carbon dioxide.

CARBONIC ANHYDRIDE. See carbon dioxide.

CARBONIC ETHER. See diethyl carbonate.

CARBONITES. See explosives, high.

CARBON MONOSULFIDE. Red powder. CS or $(CS)_x$, mw: 44.08, mp: decomp @ 200°, d: 1.66.
THR = See sulfides.

CARBON MONOXIDE. Colorless, odorless gas. CO, mw: 28.01, mp: −207°, bp: −191.3°, lel = 12.5%, uel = 74.2%, d: (gas) 1.250 g/liter @ 0°, (liq) 0.793, autoign. temp.: 1128°F.
Acute tox data: Inhal LC_{LO} (man) = 4000 ppm for 30 min; inhal TC_{LO} (man) = 655 ppm for 45 min⟶ CNS effects; inhal LC_{50} (rat) = 1807 ppm for 4 hrs; inhal LC_{50} (mouse) = 5718 ppm for 4 hrs. [*3*]
THR = HIGH via inhal route. Carbon monoxide has an affinity for hemoglobin 210 times that of oxygen, and by combining with the hemoglobin, renders the latter incapable of carrying oxygen to the tissues. The effect on the body is therefore predominantly one of asphyxia. In addition to this action, the presence of CO-hemoglobin in the blood interferes with the dissociation of the remaining oxyhemoglobin, so that the tissues are further deprived of oxygen.

A conc of 400 to 500 ppm in air can be inhaled without appreciable effect for 1 hour. An hour's exposure to 600 to 700 ppm will cause barely appreciable effects, and a similar exposure to 1,000 to 1,200 ppm is dangerous; conc of 4,000 ppm and over are fatal in less than an hour.

Carbon monoxide is eliminated through the lungs when air free from CO is inhaled. Over half the CO is eliminated in the first hour, when the exposure has been moderate.

With conc up to 10% of CO-hemoglobin in the blood, there rarely are any symptoms. Conc of 20 to 30% cause shortness of breath on moderate exertion and slight headache. Conc from 30 to 50% cause severe headache, mental confusion and dizziness, impairment of vision and hearing and collapse and fainting on exertion. With conc of 50 to 60%, unconsciousness results, and death may follow if exposure is long. Conc of 80% result in almost immediate death.

Acute cases of poisoning resulting from brief exposures to high conc seldom results in any permanent disability if recovery takes place. Chronic effects as the result of repeated exposure to lower concentrations have been described, particularly in the Scandinavian literature. Auditory disturbances and contraction of the visual fields have been demonstrated. Glycosuria does occur, and heart irregularities have been reported. Other workers have found that where the poisoning has been relatively long and severe, cerebral congestion and edema may occur, resulting in long-lasting mental or nervous damage. Repeated exposure to low conc of the gas, up to 100 ppm in air, is generally believed to cause no signs of poisoning or permanent damage. [*46*] Industrially, sequelae are rare, as exposure, though often severe, is usually brief. It is a common air contaminant.
Fire Hazard: Very dangerous, when exposed to flame.
Spont Heating: No.
Explosion Hazard: Severe, when exposed to heat or flame. Severe reaction with BrF_3, Cs_2O, ClF_3, IF_7, $(Li + H_2O)$, NF_3, O_2, OF_2, $(K + O_2)$, Ag_2O, $(Na + NH_3)$. [*19*]
To Fight Fire: Stop flow of gas.

CARBON OXYBROMIDE. See carbonyl bromide.

CARBON OXYCHLORIDE. See phosgene.

CARBON OXYCYANIDE. See carbonyl cyanide.

CARBON OXYSULFIDE. Syn: *carbonyl sulfide.* Gas or liquid. COS, mw: 60.07, mp: −138°, bp: 49.9°, lel = 12%, uel = 28.5%, d: liq 1.24 @ −87°, vap. d: 2.1.
Acute tox data: Inhal LC_{LO} (mouse) = 2900 ppm. [*3*]
THR = HIGH via inhal route. An irr. Narcotic in high conc. May liberate highly toxic hydrogen sulfide upon decomp.
Fire Hazard: Very dangerous, when exposed to heat or flame.
Spont Heating: No.
Explosion Hazard: Mod, when exposed to heat or flame. See also sulfides.
Disaster Hazard: Dangerous; see sulfides; can react vigorously with oxidizing materials.
To Fight Fire: Stop flow of gas. CO_2, dry chemical or water spray. See also sulfides.

CARBON PAPER.
THR = A MILD allergen.

CARBON REMOVER, LIQUID. Flash p: <80°F.
THR = U.
Fire Hazard: Dangerous, when exposed to heat or flame; can react with oxidizing materials.
To Fight Fire: CO_2, dry chemical.

CARBON SELENOSULFIDE. Yellow, oily liquid. CSeS, mw: 123.04, mp: −75.2°, bp: 85.6°, d: 1.9874, vap. press: 100 mm @ 28.3°.
THR = See selenium compounds and sulfides.

CARBON SUBOXIDE. Colorless gas or liquid. C_3O_2, mw: 68.03, mp: −111.3°, bp: 6.8°, d: liq 1.114 @ 0°.
THR = HIGH via inhal route.

CARBON SUBSULFIDE. Red liquid. C_3S_2, mw: 100.16, mp: 0.4°, d: 1.274, vap. press: 1 mm @ 14.0°.
THR = See also carbon disulfide.

CARBON SULFIDE TELLURIDE. Yellow-red crystals. CSTe. mw: 171.69, mp: >−54°, bp: decomp −54.0°, d: 2.9 @ 50°.

THR = See carbon disulfide, tellurium compounds and sulfides.

CARBON TETRABROMIDE. Syn: *tetrabromomethane.* Colorless monoclinic tablets. CBr_4, mw: 331.67, mp: α 48.4°, β 90.1°, bp: 189.5°, d: 3.42, vap. press: 40 mm @ 96.3°.

Acute tox data: sc LD_{50} (mouse) = 298 mg/kg; oral LD_{LO} (rat) = 1 g/kg. [*3*]

THR = HIGH via sc and MOD via oral routes. Severe reaction with Li. [*19*] Narcotic in high conc.

Disaster Hazard: See chlorinated HC, aliphatic.

CARBON TETRACHLORIDE. Syn: *tetrachloromethane.* Colorless liquid, heavy, ethereal odor. CCl_4, mw: 153.84, mp: −22.6°, bp: 76.8°, fp: −22.9°, flash p: none, d: 1.597 @ 20°, vap. press: 100 mm @ 23.0°.

Acute tox data: Oral LD_{50}, (rat) = 1770 mg/kg; ip LD_{50} (rat) = 1500 mg/kg; inhal LC_{50} (mouse) = 9526 ppm for 8 hrs; oral LD_{LO} (humans) = 60 mg/kg; inhal TC_{LO} (humans) = 20 ppm → CNS effects. [*3*]

THR = MOD via oral, in and inhal; HIGH via human oral and inhal routes. Carbon tetrachloride has a narcotic action resembling that of chloroform, though not as strong. Following exposures to high conc, the victim may become unconscious, and if exposure is not terminated, death can follow from respiratory failure. In cases of narcosis that recover, the after-effects are more serious than those of delayed chloroform poisoning, usually taking the form of damage to the kidneys, liver and lungs. Exposure to lower conc, insufficient to produce unconsciousness, usually results in severe gastro-intestinal upset, and may progress to serious kidney and hepatic damage. The kidney lesion is an acute nephrosis; the liver involvement consists of an acute degeneration of the central portions of the lobules. Where recovery takes place, there may be no permanent disability. Marked variation in individual susceptibility to carbon tetrachloride exists, some persons appear to be unaffected by exposures which seriously poison their fellow-workers. Alcoholism and previous liver and kidney damage seem to render the individual more susceptible. Conc of the order of 1,000 to 1,500 ppm are sufficient to cause symptoms if exposure continues for several hours. Repeated daily exposure to such conc may result in poisoning.

Though the common form of poisoning following industrial exposure is usually one of gastrointestinal upset, which may be followed by renal damage, other cases have been reported in which the CNS has been affected with production of polyneuritis, narrowing of the visual fields, and other neurological changes. Prolonged exposure to small amounts of carbon terachloride has also been reported as causing cirrhosis of the liver.

Locally, a dermatitis may be produced following long or repeated contact with the liquid. The skin oils are removed, and the skin becomes red, cracked and dry (Section 9). The effect of carbon tetrachloride on the eyes either as a vapor or as a liquid, is one of irr with lacrimation and burning.

Industrial poisoning is usually acute, with malaise, headache, nausea, dizziness, and confusion, which may be followed by stupor and sometimes loss of consciousness. Symptoms of liver and kidney damage may follow later, with development of dark urine, sometimes jaundice and liver enlargement, followed by scanty urine, albumenuria and renal casts, uremia may develop and cause death. Where the exposure has been less acute, the picture is usually one of headache, dizziness, nausea, vomiting epigastric distress, loss of appetite, and fatigue. Visual disturbances (blind spots, spots before the eyes, a visual "haze" and restriction of the visual fields), secondary anemia, and occasionally a slight jaundice may occur. Dermatitis may be noticed on the exposed parts. It is an exper (+) carc and teratogen and an exper neo. [*3, 9, 23*] Caution: Severe reaction with allyl alcohol, Al, $Al(C_2H_5)_3$, Ba, (benzoyl peroxide + C_2H_4), Be, BrF_3, $Ca(OCl)_2$, diborane, C_2H_4, dimethyl formamide, disilane, F_2, Li, Mg, liquid O_2, Pu, K, ($AgClO_4$ + HCl), potassium-*tert*-butoxide, Na, NaK, tetrasilane, trisilane, U, Zr, burning wax. [*19*] See chlorinated HC. It has been banned from household use by FDA.

Disaster Hazard: Dangerous; when heated to decomp, emits highly toxic fumes of phosgene.

CARBON TETRAFLUORIDE. Syn: *tetrafluoromethane.* Colorless gas. CF_4, mw: 88.01, mp: −184°, bp: −127.7°, d: 1.96 @ −184°.

THR = LOW via inhal route. Appears to be less chronically toxic than carbon tetrachloride. See halogenated HC. Violent reaction with Al. [*19*]

Disaster Hazard: See halogenated HC, aliphatic.

CARBON TETRAIODIDE. Syn: *tetraiodomethane.* Octahedral, red crystals. CI_4, mw: 519.69, mp: 171° decomp, d: 4.32.

THR = U. See also iodoform. Vigorous reaction with BrF_3. [*19*]

CARBON TRICHLORIDE. See hexachloroethane.

8-CARBONYLAMIDE-7-METHYL BENZ(a)ANTHRACENE.
THR = An exper carc. [*23*]

CARBONYL BROMIDE. Syn: *carbon oxybromide.* Liquid. $COBr_2$, mw: 187.84, bp: 64.5°, d: 2.44.
THR = HIGH irr and via inhal route.
Fire Hazard: See carbon monoxide and bromine.
Explosion Hazard: See carbon monoxide.
Disaster Hazard: Dangerous; when heated to decomp, emits highly toxic fumes; will react with water or steam to produce toxic and flam vapors.

CARBONYL CHLORIDE. See phosgene.

CARBONYL CYANIDE. Syn: *carbon oxycyanide.* Colorless liquid. $CO(CN)_2$, mw: 80.05, bp: 65.5°, d: 1.124.
THR = See cyanides and carbon monoxide.

CARBONYL FLUORIDE. Syn: *fluoroformyl fluoride.* Colorless gas, pungent, hygroscopic. COF_2, mw: 66.01, mp: −114°, bp: −83°, d: 1.139 @ −114°.
Acute tox data: Inhal LC_{50} (rat) = 360 ppm for 1 hr. [*3*]
THR = HIGH irr and via inhal route. A powerful irr. See hydrofluoric acid and fluorine. Hydrolyzes instantly upon contact with moisture.
Fire Hazard: See carbon monoxide.
Explosion Hazard: See carbon monoxide.
Disaster Hazard: See fluorides and carbon monoxide.

CARBONYLS. The (CO) group with a metal.
THR = Most carbonyls are highly toxic. The toxicity of carbonyls depends in part, but not always entirely, on their ready decomp which releases carbon monoxide. Symptoms are due in part to carbon monoxide and in part to the direct irritating action of the carbonyl. See also specific carbonyl in question.
Fire Hazard: Mod, when exposed to heat or flame. More or less readily evolves carbon monoxide. See also carbon monoxide and powdered metals.
Explosion Hazard: Mod, when exposed to heat or flame.
Explosive Range: See carbon monoxide. Powerful oxidizers; can react violently. Heat can cause carbonyls to decompose.
Disaster Hazard: Dangerous; when heated to decomp, they emit highly toxic fumes of carbon monoxide; they react with water or steam to produce toxic and flammable vapors; they can react vigorously with oxidizing materials.

CARBONYL SULFIDE. See carbon oxysulfide.

CARBOPHENOTHION. Syns: *S-((p-chlorophenyl) thio)methyl-o,o-diethyl phosphorodithioate, o,o-diethyl-S-(p-chlorophenyl thio methyl)phosphorodithioate, S-(4-chlorophenyl thiomethyl)diethyl phosphorothiothionate.* Amber liquid, essentially insol in water, miscible in common solvents.
$(C_2H_5O)_2P(S)SCH_2S(C_6H_4)Cl$, mw: 343, bp: 82° @ 0.1 mm, d: 1.29 @ 20°.
Acute tox data: Oral LD_{50} (rat) = 10 mg/kg; dermal LD_{50} (rat) = 27 mg/kg; oral LD_{50} (wild birds) = 6 mg/kg [*3*] ip LD_{50} (rat) = 11 mg/kg.
THR = HIGH via oral, dermal and ip routes. A cholinesterase inhibitor. Used as an insecticide and acaricide. A food additive permitted in feed and drinking water of animals and/or for the treatment of food-producing animals. [*109*]
Disaster Hazard: Dangerous. See sulfur compounds, phosphates and chlorides.

CARBOPOLS. See carboxy polymethylene.

CARBOPROPOXIDE, STABILIZED. Liquid.
THR = HIGH irr via oral and inhal routes.
Fire Hazard: Mod, when exposed to heat or flame.
Disaster Hazard: Mod dangerous; when heated to decomp, emits toxic fumes; can react with oxidizing materials.

CARBOPROPOXIDE, UNSTABILIZED. Solid.
THR = HIGH irr via oral and inhal routes.
Fire Hazard: Dangerous, when exposed to heat.
Disaster Hazard: Mod dangerous; when heated to decomp, emits toxic fumes; can react with oxidizing materials.

"CARBORUNDUM." See silicon carbide.

"CARBOWAX" METHOXY POLYETHYLENE GLYCOL 350. Slightly viscous liquid.
$CH_3OCH_2(CH_2OCH_2)_xCH_2OH$, mw: 330–370, mp: −5° to −10°, flash p: 440°F (OC), d: 1.09 @ 20°/20°, vap. press: < 50 mm @ 37°.
THR = See glycols.
Fire Hazard: Slight, when exposed to heat or flame; can react with oxidizing materials.
To Fight Fire: Water, foam, CO_2, dry chemical.

"CARBOWAX" METHOXY POLYETHYLENE GLYCOL 550. Soft, wax-like solid.
$CH_3OCH_2(CH_2OCH_2)_xCH_2OH$. mw: 525–575, mp: 15°–25°, flash p: 460°F (OC), d: 1.07 @ 55°/20°, vap. press: <50 mm @ 37°.
THR = See glycols.
Fire Hazard: Slight, when exposed to heat or flame; can react with oxidizing materials.
To Fight Fire: Water, foam, CO_2, dry chemical.

CARBOXIDE. Liquid. 1 part ethylene oxide + 9 parts carbon dioxide.
THR = A powerful irr. May cause CNS damage, including hemorrhages into the brain, hemorrhagic nasal discharge, unsteadiness, dyspnea and death

following sufficient exposure. 30,000 ppm may be tolerated for perhaps 1 hour.

3′-CARBOXY-4-DIMETHYL AMINOAZO BENZENE. $C_{15}H_{15}O_2N_3$, mw: 269.3.
THR = An exper neo. [*3*]

3-[(2-CARBOXYETHYL)THIO]ALANINE. $C_6H_{11}O_4NS$, mw: 193.2.
THR = An exper neo. [*3*]

4,6-CARBOXYL-4-HYDROXYLAMINO QUINOLINE-1-OXIDE. $C_{10}H_8O_4N_2$, mw: 220.2.
THR = An exper neo. [*3*]

6-CARBOXYL-4-NITROQUINOLINE-1-OXIDE. $C_{10}H_6O_5N_2$, mw: 234.2.
THR = An exper neo via sc route. [*3*, *23*]

CARBOXY METHYL MERCAPTOSUCCINIC ACID. White powder. $C_6H_8O_6S$, mw: 208.2, mp: 128°–136°.
THR = U. Preliminary tests indicate toxicity is somewhat less than citric acid.
Disaster Hazard: See sulfur compounds.

α-CARBOXY-*p*-PHENYL-β-PROPIOLACTONE.
THR = An exper carc. [*23*]

CARBOXY POLYMETHYLENE. Syn: *carbopol.* White powder, a vinyl polymer with active carboxyl groups.
THR = U. Animal experiments show LOW toxicity.

CARBOYS, ACID, EMPTY.
THR = See specific material shipped.
Disaster Hazard: Slightly dangerous; when heated, they may emit toxic fumes.

CARBURIZING SALT. See cyanides.

CARBUTAMIDE. $C_{11}H_{17}O_3N_3S$, mw: 271.4.
THR = LOW to rats. An exper teratogen [*3*] and has been implicated in aplastic anemia.

CARCINOGENESIS. Carcinogenic air pollutants are discussed in Section 8.

CARE OF RESPIRATORS. See Section 2.

CARITOL. See caroten.

CARNAUBA WAX. Syn: *Brazil wax.* flash p: 540°F (CC), mp: 85°.
THR = A MILD allergen. A general purpose food additive. [*109*]
Fire Hazard: Slight, when exposed to heat; can react with oxidizing materials.
To Fight Fire: Water, CO_2, dry chemical, mist, fog.

CAROB BEAN GUM. Syns: *locust bean gum, carob seed gum.* A galactomannan polysaccharide; in powdered form, nearly pure white, insol in most organic solvents. mw: 310,000 (approx).
THR = U. A stabilizer food additive. Also a substance migrating to food from packaging materials. [*109*]

CAROB SEED GUM. See carob bean gum.

CARO'S ACID. See peroxysulfuric acid.

CAROTENE. Syn: *caritol, provitamin A.* Ruby red crystals, insol in water, slightly sol in alcohol and ether, sol in oils, chloroform, benzene. A precursor of vitamin A, occurring naturally in plants. It consists of 3 isomers (15%α, 85%β, 0.1γ) $C_{40}H_{56}$, mw: 536, mp(α): 188°, mp(β): 184°, mp(γ): 178°.
THR = U. A nutrient and/or dietary supplement food additive. [*109*]

CARRAGEENIN. See chondrus extract.

CARTAP. See padan.

CARVONE. Optically active ketone occurring in *d* and *l* forms. Pale yellowish or colorless liquid, strong characteristic odor, sol in alcohol, ether, chloroform, propylene glycol and mineral oils, insol in glycerin and water. $CH_3C{:}CHCH_2CH[C(CH_3){:}CH_2]CH_2CO$, mw: 150, d: 0.960 @ 20°, bp: 227°–230°.
Acute tox data: Oral LD_{50} (rat) = 1640 mg/kg; oral LD_{50} (guinea pig) = 766 mg/kg. [*3*]
THR = MOD via oral route. A synthetic flavoring substance and adjuvant. [*109*]

CARYOPHYLLIC ACID. See eugenol.

CASE-HARDENING COMPOUNDS. See cyanides.

CASE HISTORIES OF NOISE CONTROL. Treatment discussed in Section 3.

CASEIN. The principle protein in milk; a phospho protein. White, odorless, amorphous solid, sol in dilute alkalies and conc acids, almost insol in water. Composition: approx 85% phosphorous and 0.76% sulfur. Consists of approx 15 amino acids. mw: 75,000–375,000, d: 1.25–1.31.
THR = Probably LOW, A substance migrating to food from packaging materials. [*109*]

CASEIN-SODIUM. See sodium caseinate.

CASHEW NUT SHELL OIL. See anacardic acid.

CASSELL'S GREEN. See barium manganate.

CASSEL YELLOW LAURIONITE. See lead oxychlorides.

CASTOR-BEAN MEAL.
THR = HIGH. It contains the very toxic materials ricin and ricinine. Also a very potent allergen.

CASTOR OIL. Syn: *ricinus oil.* A colorless to pale yellow viscous liquid, characteristic odor. mp: −12°, bp: 313°, flash p: 445°F (CC), d: 0.96, autoign. temp.: 840°F.
THR = MOD via oral route. An allergen. A food

additive permitted in food for human consumption. [*109*]
Fire Hazard: Slight, when exposed to heat.
Spont Heating: Yes.
To Fight Fire: CO_2, dry chemical, fog, mist.

CASTOR OIL, HYDROGENATED. Insol in water. $(C_{18}H_{35}O_3)_3C_3H_5$, mw: 938, flash p: 401°F.
Fire Hazard: Slight, when exposed to heat or flame.
To Fight Fire: Dry chemical, CO_2, fog, mist.

CASTOR OIL PLANT DUST.
THR = LOW irr and allergen.
Fire Hazard: Mod, when exposed to heat or flame; can react vigorously with oxidizing materials.
Explosion Hazard: Slight, when exposed to flame.

CASTRIX. See 2-chloro-4-dimethylamino-6-methyl pyrimidine.

CATALYTICALLY CRACKED OILS.
THR = Recog carc. [*14*] See mineral oils.

CATECHOL. See pyrocatechol.

CATEGORIES OF LABEL-USERS. See Section 11.

CATHODE TUBE RADIATIONS.
THR = Recog carc. [*14*]

CATIONIC AMINE 200. Syn: *1-hydroxyethyl-1,2-heptadecenyl glyoxaldene.*
$C_{17}H_{33}CH(CH_2)_2NHC_2CH_2OH$, mw: 350.58, flash p: 465°F (OC).
THR = U. See also amines.
Fire Hazard: Slight, when exposed to heat.
To Fight Fire: Water, foam, CO_2, dry chemical.

CATIONIC SP. See steramidopropyl dimethyl-β-hydroxyethyl ammonium phosphate.

CATTLE FEED DUST.
THR = LOW irr an allergen.
Fire Hazard: Mod, when exposed to heat or flame; can react with oxidizing materials.
Spont Heating: Yes; dangerous when very dry or very moist.
Explosion Hazard: Slight, when exposed to flame.

CAUSTIC ALCOHOL. See sodium ethylate.

CAUSTIC ALCOHOL SOLUTION. See sodium hydroxide and ethyl alcohol.

CAUSTIC ALKALI SOLUTION. See alkaline caustic liquid, n.o.s.

CAUSTIC BARYTA. See barium hydroxide.

CAUSTIC BARLEY. See sabadilla seed.

CAUSTIC POTASH. See potassium hydroxide.

CAUSTIC POTASH, LIQUID. See potassium hydroxide.

CAUSTIC SODA. See sodium hydroxide.

CAUSTIC SODA, LIQUID. See sodium hydroxide.

CAVALITE BRILLIANT BLUE R. $C_{22}H_{18}O_{11}N_2S_3$, mw: 628.6.
THR = An exper neo. [*3*]

CBP. See chlorobromopropene.

CDAA. See N,N-diallyl-2-chloroacetamide.

CEDAR. A wood.
THR = LOW irr and allergen. An exper carc. [*23*] The wood dust contains chemical irr which can cause conjunctivitis, intense lachrymation, iridocyclitis and keratitis, as well as dermatitis.
Fire Hazard: Mod, when exposed to heat.
Explosion Hazard: Slight in the form of dust when exposed to flame. See also dust.
Treatment and Antidotes: Removal from exposure will cause the symptoms to disappear.

CEDAR WOOD OIL. Colorless or slightly yellow, viscid liquid. Composition: Cedrene and cedrol. d: 0.940–0.950 @ 20°/20°.
THR = LOW irr and allergen. An exper carc. [*23*]
Fire Hazard: Slight, when heated.

CEDAR LEAF OIL.
Acute tox data: Oral LD_{50} (rat) = 830 mg/kg. [*3*]
THR = MOD via oral route.

"CELANESE SOLVENT 203." Liquid, odor of isobutanol. bp: 110°–120°, flash p: 100°F (TOC), d: 0.805–0.815 @ 20°/4°.
THR = See isobutyl alcohol, *n*-butyl alcohol and amyl alcohol.
Fire Hazard: Mod, when exposed to heat or flame; can react with oxidizing materials.
To Fight Fire: Foam, CO_2, dry chemical.

"CELANESE SOLVENT 301." Liquid. bp: 97°–100°, flash p: 90°F (TOC), d: 0.804–0.808 @ 20°/4°.
THR = See *n*-propanol, *sec*-butanol.
Fire Hazard: Dangerous, when exposed to heat or flame, can react with oxidizing materials.
To Fight Fire: Foam, CO_2, dry chemical.

"CELANESE SOLVENT 601." Liquid, ether-like odor. bp: 74°–84°, flash p: 10°F (TOC), d: 0.840–0.850 @ 20°/4°.
THR = U.
Fire Hazard: Dangerous, when exposed to heat or flame; can react with oxidizing materials.
To Fight Fire: Foam, CO_2, dry chemical.

"CELANESE SOLVENT 901." Liquid. bp: 125°–155°, flash p: 115°F (TOC), d: 0.835–0.840 @ 20°/4°.
THR = U.
Fire Hazard: Mod, when exposed to heat or flame; can react with oxidizing materials.
To Fight Fire: Foam, CO_2, dry chemical.

CELERY SEED. Volatile and fixed oil, bitter extractive and resin.
THR = MILD allergen.
Fire Hazard: Slight, when exposed to heat.

CELLOIDIN. See nitrocellulose.

CELLOPHANE. $(C_6H_{10}O_5)_x$.
THR = An exper carc by imp. [*3*] See polymers.

"CELLOSOLVE" ACETATE. Syn: *ethylene glycol monoethyl ether acetate, ethoxy acetate.* Colorless liquid with a mild, pleasant ester-like odor. $C_6H_{13}O_3$, mw: 132.17, bp: 156.4°, flash p: 117°F (COC), lel = 1.7%, fp: −61.7°, d: 0.9748 @ 20°/20°, autoign. temp.: 715°F, vap. press: 1.2 mm @ 20°, vap. d: 4.72.
Acute tox data: Oral LD_{50} (rat) = 5.1 g/kg; dermal LD_{50} (rabbit) = 10.5 g/kg. [*3*]
THR = MOD via oral and dermal routes. See glycols.
Fire Hazard: Mod, when exposed to heat or flame; can react with oxidizing materials.
Spont Heating: No.
Explosion Hazard: Mod, in the form of vapor when heated.
To Fight Fire: Alcohol foam, CO_2, dry chemical.

"CELLOSOLVE" SOLVENT. Syns: *ethylene glycol monoethyl ether, 2-ethoxy ethanol.* Colorless liquid, practically odorless. $C_4H_{10}O_2$, mw: 90.12, bp: 135.1°, lel = 1.8%, uel − 14%, flash p: 202°F(CC), d: 0.9360 @15°/15°, autoign. temp.: 455°F, vap. press: 3.8 mm @20°, vap. d: 3.10.
Acute tox data: Oral LD_{50} (rat) = 3 g/kg; dermal LD_{50} (rabbit) = 3.5 g/kg. [*3*]
THR = MOD via dermal and oral routes. See glycols. Animal exper indicate that exposure to air saturated with "Cellosolve" vapor (0.6%) for periods of 18–24 hrs may produce congestion and edema of the lungs and congestion of the kidneys. Exposure of humans to the same conc for a few seconds resulted in irr of the eyes. No cases of poisoning have so far been reported in industry. Absorbed by skin.
Fire Hazard: Low, when exposed to heat or flame; can react with oxidizing materials.
Spont Heating: No.
Explosion Hazard: Mod, in the form of vapor when exposed to heat or flame.
To Fight Fire: Alcohol foam, dry chemical.

"CELLULOID." Clear or colored cellulose nitrate. d: 1.35–1.60.
THR = LOW unless burned or in solution.
Fire Hazard: Mod, when exposed to heat.
Disaster Hazard: Dangerous; when heated to decomp, emits tox fumes; can react with oxidizing materials.

CELLULOSE. $(C_6H_{12}O_5)_x$.
THR = Reacts violently with BrF_5, F_2, H_2O_2, NaOCl, $NaNO_2$. [*19*]

CELLULOSE ACETATE. Syn: *CA.* White flakes or powder, sol in acetone, ethyl acetate, cyclohexanol, nitropropane, ethylene dichloride. Melts approx 260°, d: 1.27–1.34.
THR = U. A substance migrating to food from packaging materials. [*109*]

CELLULOSE DINITRATE. Syn: *cellulose tetranitrate.* White amorphous solid. $C_{12}H_{16}(ONO_2)_4O_6$, mw: 504.3, d: 1.66, flash p: 55°F.
THR = See nitrocellulose.
Fire Hazard: Dangerous; see nitrocellulose.
Explosion Hazard: See nitrocellulose
Disaster Hazard: See nitrocellulose.
To Fight Fire: Alcohol foam.

CELLULOSE GUM. See sodium carboxymethyl cellulose.

CELLULOSE HEXANITRATE. See nitrocellulose.

CELLULOSE METHYL ETHER. See methyl cellulose.

CELLULOSE NITRATE. See nitrocellulose.

CELLULOSE PENTANITRATE. See nitrocellulose.

CELLULOSE TETRANITRATE. See cellulose dinitrate and nitrocellulose.

CELLULOSE TRINITRATE. See nitrocellulose.

CEMENT, LEATHER.
Fire Hazard: Dangerous, when exposed to heat or flame; can react with oxidizing materials.
Explosion Hazard: U.

CEMENT, LIQUID, N.O.S.
Fire Hazard: Dangerous, when exposed to heat or flame; can react with oxidizing materials.

CEMENT, PORTLAND. Syn: *cement hydraulic.* Fine gray powder composed of compounds of lime, aluminum, silica and iron oxide as $(4CaO \cdot Al_2O_3 \cdot Fe_2)_3$, $(3CaOAl_2O_3)$, $(3CaO \cdot SiO_2)$, and $(2CaOSiO_2)$. Small amounts of magnesia, sodium, potassium, chromium and sulfur are also present in combined form.
THR = MOD irr. An allergen. LOW via inhal route. Significant pulmonary fibrosis due to cement dust occurs rarely if at all. Eczema is by no means rare and is believed to be due to chromium in the cement. Cement dust is a common air contaminant.

CEMENT, PYROXYLIN.
Fire Hazard: Dangerous, when exposed to heat or flame; can react with oxidizing materials.

CEMENT, ROOFING, LIQUID.
Fire Hazard: Dangerous, when exposed to heat or flame; can react with oxidizing materials.

CEMENT, RUBBER. Flash p: 50°F or less.
THR = Often contains benzene or other toxic solvents. See specific constituent.

Fire Hazard: Dangerous, when exposed to heat or flame; can react with oxidizing materials.

CEPORIN. $C_{19}H_{17}O_4N_3S_2$, mw: 415.5.
Acute tox data: Oral LD_{50} (rat) = 2500 mg/kg. [*3*]
THR = MOD via oral route. An exper teratogen. [*3*]

CERA ALBA. See beeswax, white.

CERESIN WAX. Syn: *earth wax, mineral wax.* White or yellow waxy material. mp: 68°–72°, d: 0.92–0.94, flash p: 236°F.
THR = An allergen.
Fire Hazard: Slight, when exposed to heat or flame.
To Fight Fire: Water, foam.

CERIC AMMONIUM NITRATE. Syn: *cerium-ammonium nitrate.* Small prismatic, orange-red crystals. $Ce(NO_3)_4 \cdot 2NH_4NO_3 \cdot 1\frac{1}{2}H_2O$, mw: 575.40.
THR = See cerium compounds and nitrates.

CERIC FLUORIDE. Syn: *cerium trifluoride.* Tiny crystals, nearly insol in water. CeF_3, mw: 197.11, d: 4.77, mp: 650°.
Acute tox data: oral LD_{LO} (guinea pig) = 5000 mg/kg; sc LD_{LO} (guinea pig) = 5000 mg/kg. [*3*]
THR = MOD via oral and sc routes. Low sol reduces toxicity. See also fluorides.
Disaster Hazard: Dangerous; see fluorides.

CERIC IODATE. Colorless crystals. $Ce(IO_3)_4$, mw: 839.81.
THR = See cerium compounds and iodates.

CERIC NITRATE, BASIC. Long, red needles. $Ce(OH)(NO_3)_3 \cdot 3H_2O$, mw: 397.1.
THR = See cerium compounds and nitrates.

CERIUM. Cubic or hexagonal, steel gray crystals. Ce, atwt: 140.13, mp: 815°, bp: 3257°, d: (cubic form): 6.90; hexagonal form 6.75. Cerium resembles aluminum in its pharmacological action as well as in its chemical properties. The insol salts such as the oxalate are stated to be non-toxic even in large doses. It is used to prevent vomiting in pregnancy. The average dose is from 0.05 to 0.5 g. Cerium tartrate has been found to produce a direct injurous action on the hearts of small animals. The effect on the CNS of the rare-earth metals following inhal may preclude welding operations with these materials to any large extent. Cerium is stated to produce polycythemia but is useless in the treatment of anemia owing to its toxic effects. The salts of cerium increase the blood coagulation rate.
Radiation Hazard: For permissible levels, see Section 5, Table 5A.5. Artificial isotope ^{141}Ce, $T\frac{1}{2}$ = 33d, decays to stable ^{141}Pr via β's of 0.44 MeV (70%), 0.58 MeV (30%). Emits γ's of 0.14 MeV. Artificial isotope ^{144}Ce, $T\frac{1}{2}$ = 284d, decays to ^{144}Pr via β's of 0.17 MeV (25%), 0.25 MeV (5%), 0.31 MeV (70%). Emits γ's of 0.03–0.13 MeV. ^{144}Ce exists in radioactive equilibrium with its daughter, and permissible levels include the daughter activity.
Fire Hazard: Mod; ignites spont in air at 150°–180°. A strong reducing agent. See also powdered metals. Severe reaction with halogens, P. [*19*]
Explosion Hazard: Mod, in the form of dust when exposed to flame. See also powdered metals.

CERIUM AMMONIUM NITRATE. See ceric ammonium nitrate.

CERIUM CARBIDE. Hexagonal, red crystals. CeC_2, mw: 164.15, d: 5.23.
THR = See cerium and acetylides.

CERIUM CHLORIDE. Syn: *cerous chloride* Colorless, deliquescent crystals. $CeCl_3$, mw: 246.50, mp: 848°, bp: 1727°, d: 3.92.
Acute tox data: iv LD_{50} (rabbit) = 35 mg/kg. [*3*]
THR = HIGH via iv route. See cerium and compounds.

CERIUM COMPOUNDS.
The toxicity of cerium compounds may be taken to be that of cerium, except when the anion has a toxicity of its own. See cerium.

CERIUM HYDRIDE. CeH_3, mw: 143.1.
THR = See cerium and hydrides. Self-ignites in air. [*19*]

CERIUM NITRATE, HEXAHYDRATE.
$Ce(NO_3)_3 \cdot 6H_2O$, mw: 434.3.
Acute tox data: Oral LD_{50} (rat) = 4200 mg/kg; ip LD_{50} (rat) = 290 mg/kg; iv LD_{50} (rat) = 4 mg/kg. [*3*]
THR = HIGH via iv, ip; MOD via oral routes.

CERIUM NITRIDE. CeN, mw: 154.1. See cerium and nitrides. Reacts violently with acids, air, (air + moisture). [*19*]

CERIUM TARTRATE. White powder. $Ce_2(C_4H_4O_6)_3$, mw: 724.5.
THR = See cerium.

CERIUM TETRABORIDE. Tetragonal crystals. CeB_4, mw: 183.41, d: 5.74.
THR = See cerium and boron compounds.

CERIUM TETRAFLUORIDE. See ceric fluoride.

CERIUM TETRAHYDRO ALUMINATE. $Ca(AlH_4)_3$, mw: 233.2.
THR = No data. Self-ignites in air. [*19*]

CERIUM TRIFLUORIDE. See cerous fluoride.

CEROUS AMMONIUM NITRATE. Monoclinic, white, transparent crystals. $2NH_4NO_3 \cdot Ce(NO_3)_3 \cdot 4H_2O$, mw: 588.32, mp: 74°.
THR = See cerium nitrates.

CEROUS BROMATE. Hexagonal, reddish-white crystals. $Ce(BrO_3)_3 \cdot 9H_2O$, mw: 686.02, mp: 49°.
THR = See cerium and bromates.

CEROUS CHLORIDE. See cerium chloride.

CEROUS CYANOPLATINITE. Monoclinic, yellow-blue, lustrous crystals. $Ce(Pt(CN)_4)_3 \cdot 18H_2O$, mw: 1502.45, d: 2.657, $-13\frac{1}{2}H_2O$ @ 110°.
THR = See cyanides and cerium.

CEROUS FLUORIDE. Syn: *cerium trifluoride.* Colorless crystals or powder, nearly insol in water. CeF_3, mw: 197.13, mp: 1460°, d: 6.16, bp: 2300°.
Acute tox data: Oral LD_{LO} (guinea pig) = 5 g/kg; $scLD_{LO}$ (guinea pig) = 5000 mg/kg. [*3*]
THR = MOD via oral and sc routes. See also fluorides.
Disaster Hazard: See fluorides.

CEROUS HYDRIDE. Amorphous, dark blue powder. CeH_3, mw: 143.15, mp: ignites.
THR = See cerium and hydrides.

CEROUS IODIDE. Reddish-white crystals. $CeI_3 \cdot 9H_2O$, mw: 683.03, mp: 752°.
THR = See cerium and iodides.

CEROUS NITRATE. See cerium nitrate, hexahydrate.

CEROUS OXYCHLORIDE. Purple, leaf-like crystals. CeOCl, mw: 191.6.
THR = See cerium and hydrochloric acid.

CEROUS PHOSPHIDE. CeP, mw: 171.1.
THR = HIGH. See phosphides. Violent reaction with water. [*19*]

CEROUS SELENATE. Rhombic crystals. $Ce_2(SeO_4)_3$, mw: 709.14, d: 4.456.
THR = See selenium and cerium.

CEROUS SULFIDE. Red crystals, brown, dark purple powder. Ce_2S_3, mw: 376.46, mp: decomp, d: 5.020 @ 11°.
THR = See sulfides and cerium.

CERUSSITE. See lead carbonate.

CESIUM. Hexagonal crystals, silver-white, ductile metal or possibly a silvery liquid. Cs, atwt: 132.19, mp: 28.5°, bp: 705°, d: 1.873, vap. press: 1 mm @ 279°.
THR = Cesium is quite similar to potassium in its elemental state. It has been shown, however, to have pronounced physiological action in exper with animals. Hyperirritability, including marked spasms, has been shown to follow the administration of cesium in amounts equal to the potassium content of the diet. It has been found that replacing the potassium in the diet of rats with cesium, caused death after 10–17 days.
Radiation Hazard: For permissible levels, see Section 5, Table 5A.5. Artificial isotope ^{134}Cs, $T\frac{1}{2}$ = 2y, decays to stable ^{134}Ba via β's of 0.09 MeV (28%), 0.66 MeV (71%). Emits γ's of 0.48–1.36 MeV.
Artificial isotope ^{137}Cs, $T\frac{1}{2}$ = 30y, decays to stable ^{137}Ba via β's of 0.51 MeV (93%), 1.18 MeV (7%). Emits γ's of 0.66 MeV.
Fire Hazard: Dangerous, by chemical reaction; reacts with Cl_2, O_2, P. Can ignite spont in moist air, See also sodium.
Explosion Hazard: Mod, by chemical reaction; reacts with moisture to liberate hydrogen.
Disaster Hazard: Mod dangerous; will react with water or steam to produce heat and hydrogen; on contact with oxidizing materials, it can react vigorously.

CESIUM ACETYLIDE. CsC_2H, mw: 157.9.
THR = No data. Reacts violently with As, halogens, CuO, HCl, PbO_2, N_2O_4, P, SO_2, H_2SO_4. [*19*]

CESIUM AMIDE. $CsNH_2$, mw: 148.9.
THR = No data. Incandesces in water. [*19*]

CESIUM AZIDE. CsN_3, mw: 174.9.
THR = No data. Decomp @ 326° and reacts vigorously with CS_2. [*19*]

CESIUM BROMATE. Solid. $CsBrO_3$, mw: 260.83.
THR = See bromates and cesium.

CESIUM BROMOCHLOROIODIDE. Rhombic, yellow-red crystals. CsBrClI, mw: 375.20, mp: 235°, bp: decomp @ 290°.
THR = See bromides, chlorides, iodides and cesium.

CESIUM BROMODIIODIDE. CsI_2Br, mw: 466.67, mp: 195.5°.
THR = See bromides, iodides and cesium.

CESIUM CARBIDE. Cs_2C_2, mw: 289.8.
THR = No data. Reacts violently with B, Fe_2O_3, HCl, I_2, HNO_3, Si. [*19*]

CESIUM CHLORIDE. Cubic crystals. CsCl, mw: 168.4, d: 3.99, mp: 646°, bp: 1303°.
Acute tox data: Oral LD_{50} (mouse) = 2300 mg/kg. [*3*]
THR = MOD via oral route. Reacts violently with BF_3. [*19*]

CESIUM CHLOROAURATE. Monoclinic, yellow crystals. $CsAuCl_4$, mw: 471.94.
THR = See gold compounds and cesium.

CESIUM CHLORODIBROMIDE. Yellow crystals. $CsBr_2Cl$, mw: 328.20, mp: 191°.
THR = See bromides, chlorides and cesium.

CESIUM CHLOROPLATINATE. Cubic, yellow crystals. Cs_2PtCl_6, mw: 673.79, mp: decomp.
THR = See platinum compounds, cesium and chlorides.

CESIUM CHLOROSTANNATE. Cubic, white crystals. Cs_2SnCl_6, mw: 597.26, d: 3.33.
THR = See tin compounds and chlorides.

CESIUM CHROMATE. (α) yellow prisms, (β) yellow, rhombic crystals. Cs_2CrO_4, mw: 381.83, d: 4.237.
THR = See chromium compounds and cesium.

CESIUM COMPOUNDS.
THR = Probably have the toxicity of cesium unless they contain a more toxic radical. Toxicity is of minor importance in industry. See also cesium.

CESIUM CYANIDE. Colorless crystals. CsCN, mw: 158.93.
THR = See cyanides.

CESIUM DIBROMO IODIDE. Rhombic crystals. $CsIBr_2$, mw: 419.66, mp: 248°, bp: decomp @ 320°.
THR = See bromides, iodides and cesium.

CESIUM DICHLORO BROMIDE. Crystals. $CsBrCl_2$, mw: 283.74, mp: 205°.
THR = See bromides, chlorides and cesium.

CESIUM DICHLORO IODIDE. Rhombic, pale orange crystals. $CsICl_2$, mw: 330.74, mp: 230°, bp: decomp @ 290°, d: 3.86.
THR = See iodides, cesium and chlorides.

CESIUM DISULFIDE. Amorphous or dark red crystals. Cs_2S_2, mw: 329.95, bp: >800°, mp: 460°.
THR = See sulfides.

CESIUM FLUOGERMANATE. Isotropic crystals, regular, octahedral. Cs_2GeF_6 mw: 452.42, d: 4.10.
THR = See fluorides.

CESIUM FLUORIDE NITROSYL FLUORIDE COMPLEX. CsF · NOF, mw: 200.9.
THR = See fluorides. At room temperatures it can explode. [*19*]

CESIUM FLUOSILICATE. Cubic, white crystals. Cs_2SiF_6, mw: 407.88, d: 3.372 @ 17°.
THR = See fluosilicates.

CESIUM FLUOTELLURITE. Colorless needles. $CsTeF_5$, mw: 355.52.
THR = See tellurium compounds and fluorides.

CESIUM GALLIUM SELENATE. Colorless crystals. $CsGa(SeO_4)_2 \cdot 12H_2O$, mw: 704.74.
THR = See selenium and gallium compounds.

CESIUM GALLIUM SULFATE. Cubic, colorless crystals. $CsGa(SO_4)_2 \cdot 12H_2O$, mw: 610.95, d: 2.113.
THR = See cesium, gallium compounds and sulfates.

CESIUM GRAPHITE. Violet to black platelets, highly reactive with water, air and alcohols. CsC_8, mw: 228.9.
THR = See cesium. For all other properties and associated hazards see cesium.

CESIUM HEXAHYDRO ALUMINATE. Cs_3AlH_6, mw: 431.8.
THR = No data. Very unstable compound. [*19*]

CESIUM HEXASULFIDE. Bright, red crystals. Cs_2S_6, mw: 458.22, mp: 186°.
THR = See sulfides and cesium.

CESIUM HYDRIDE. White crystals. CsH, mw: 133.92, mp: decomp, d: 3.41.
THR = See cesium and hydrides. Reacts violently with (water + C_2H_2), O_2. [*19*]

CESIUM HYDROXIDE. Colorless, yellowish, very deliquescent crystals. CsOH, mw: 149.92, mp: 272.3°, d: 3.675.
Acute tox data: Oral LD_{50} (rat) = 1026 mg/kg; ip LD_{50} (rat) = 100 mg/kg; oral LD_{50} (mouse) = 800 mg/kg. [*3*]
THR = HIGH via ip, MOD via oral routes. See cesium compounds.

CESIUM IODATE. Monoclinic, white crystals. $CsIO_3$, mw: 307.83, d: 4.85.
THR = See iodates and cesium.

CESIUM MERCURIC BROMIDE. Rhombic crystals. $CsBr \cdot 2HgBr_2$, mw: 933.71.
THR = See mercury compounds, inorganic; also see bromides and cesium.

CESIUM MERCURIC CHLORIDE. Cubic or rhombic, colorless crystals. $CsCl \cdot HgCl_2$, mw: 439.89.
THR = See mercury compounds, inorganic and chlorides.

CESIUM METAL. See cesium.

CESIUM METAL IN CARTRIDGES. See cesium.

CESIUM MONOBROMIDE. Cubic, colorless crystals. CsBr, mw: 212.83, mp: 636°, bp: 1300°, d: 4.44 (liq), 3.04 @ 700°.
Acute tox data: ip LD_{50} (rat) = 1400 mg/kg. [*3*]
THR = MOD via ip route. See bromides and cesium.

CESIUM MONOIODIDE. Cubic, colorless crystals. CsI, mw: 259.83, mp: 621°, bp: 1280°, d: 4.510.
Acute tox data: Oral LD_{50} (rat) = 2386 mg/kg; ip LD_{50} (rat) = 1400 mg/kg. [*3*]
THR = MOD via oral and ip routes. See iodides and cesium.

CESIUM MONOXIDE. Cs_2O, mw: 281.8.
THR = A powerful caustic like sodium oxide. Violent reaction with halogens, CO_2, CO, SO_2, H_2O.[*19*]

CESIUM NITRATE. Colorless, hexagonal or cubic, glittering crystalline powder. $CsNO_3$, mw: 194.92, mp: 414°, bp: decomp, d: 3.685; 2.71 @ 500° (liq).
Acute tox data: ip LD_{50} (rat) = 1200 mg/kg. [*3*]
THR = MOD via ip route. See cesium and nitrates.

CESIUM NITRIDE. Cs_3N, mw: 412.7.
THR = No data. Reacts violently with air, Cl_2, P, S. [*19*]

CESIUM NITRITE. Yellow crystals. $CsNO_2$, mw: 178.92.
THR = See nitrites.

CESIUM OXALATE. White solid. $Cs_2C_2O_4$, mw: 353.84, d: 3.230 @ 15°.
THR = See oxalates.

CESIUM PENTASULFIDE. Solid. Cs_2S_5, mw: 426.15, mp: 210°, d: 2.806 @ 15°.
THR = See sulfides.

CESIUM PERCHLORATE. Rhombic colorless crystals. $CsClO_4$, mw: 232.37, mp: decomp, d: 3.327.
THR = See perchlorates and cesium.

CESIUM PERFLUORO PROPOXIDE. $CsOC_3F_7$, mw: 317.9.
THR = No data. Violent reaction with F_2. [*19*] See also fluorides.

CESIUM-*m*-PERIODATE. Rhombic, white plates. $CsIO_4$, mw: 323.83, d: 4.259.
THR = See iodates and cesium.

CESIUM PERMANGANATE. A violet solid. $CsMnO_4$, mw: 251.84, mp: 320° (decomp), d: 3.597.
THR = See permanganates.

CESIUM PHOSPHIDE. PCs_3, mw: 429.73.
THR = See phosphine which is produced by contact with water. Phosphine is self-flam [*19*] and highly toxic. [*2*]

CESIUM SILICIDE. Cs_2Si_2, mw: 322.
THR = No data. Explodes with water. [*19*]

CESIUM SULFIDE. White, deliquescent crystals. $Cs_2S \cdot 4H_2O$, mw: 369.95.
THR = See sulfides.

CESIUM TETRASULFIDE. Yellow crystals. Cs_2S_4, mw: 394.08, mp: 160° (decomp).
THR = See sulfides.

CESIUM TRIBROMIDE. Rhombic crystals. $CsBr_3$, mw: 372.66, mp: 180°.
THR = See bromides and cesium.

CESIUM TRISULFIDE. Yellow leaf. Cs_2S_3, mw: 362.02, mp: 217°, bp: 780°.
THR = See sulfides.

CETAB. See cetyl trimethyl ammonium bromide.

CETANE. See *n*-hexadecane.

CETYL ALCOHOL. Syn: *hexadecanol.* Solid or leaf-like crystals. $C_{16}H_{33}OH$, mw: 242.4, mp: 49.3°, bp: 344°, d: 0.8176 @ 50°/4°.
Acute tox data: Dermal LD_{50} (rabbit) = 2600 mg/kg. [*3*]
THR = MOD via dermal route.
Fire Hazard: Mod, when exposed to heat or flame; can react with oxidizing materials.
To Fight Fire: Foam, CO_2, dry chemical.

CETYL BROMIDE. Dark yellow liquid. $CH_3(CH_2)_{15}Br$, mw: 305.3, bp: 186°–197° @ 10 mm, fp: 15°, flash p: 350°F, d: 0.991 @ 25°/25°.
THR = U. See bromides.
Fire Hazard: Slight, when exposed to heat or flame.
Disaster Hazard: Dangerous; reacts violently with oxidizing materials; see bromides.

CETYL ISOQUINOLINIUM BROMIDE.
THR = U. A fungicide.
Fire Hazard: Slight.
Disaster Hazard: Dangerous. See bromides.

CETYL TRIMETHYL AMMONIUM BROMIDE. Syn: *cetab.* Creamy white powder, water sol. $C_{16}H_{35}(CH_3)_3NBr$ (at least 80%).
Acute tox data: Oral LD_{50} (rat) = 410 mg/kg; ip LD_{50} (mouse) = 106 mg/kg; ip LD_{50} (rabbit) = 125 mg/kg. [*3*]
THR = HIGH via oral, ip routes. An exper teratogen. [*3*] See also benzethonium chloride.
Disaster Hazard: Dangerous; when strongly heated, emits highly toxic fumes.

CETYL TRIMETHYL AMMONIUM PENTACHLOROPHENATE. See pentachlorophenol.

CEVADILLA. See sabadilla seed.

CEVADINE. See veratrine.

CG. See phosgene.

CHALCOCITE. See cuprous sulfide.

CHAMBER CRYSTALS. See nitrosyl sulfuric acid.

CHAMOMILE OIL. Blue liquid, turning brownish-yellow. Composition: Amyl and butyl esters of angelic and tiglic acids, butyric acid, etc. d: 0.905–0.915 @ 15°/15°.
THR = A MILD allergen.
Fire Hazard: Slight, when heated.

CHANNEL BLACK. See soot.

CHARCOAL. Black amorphous solid. Composition: C + impurities. mw: 12.0, mp: >3500°, bp: 4200°, d: 3.51.
THR = Carbon itself has no toxic action, but if it contains impurities, these may be toxic.
Fire Hazard: Reacts with liquid air, $Ba(ClO_3)_2$, BrF_5, ClO, $Ca(ClO_3)_2$, ClF_2, F_2, H_2O_2, $Mg(ClO_3)_2$, (O_2 + wood), perchlorates, peroxides, (P + air), K, $KClO_3$, KNO_3, RuO_4, $AgNO_3$, $NaClO_3$, (AgCl + NaO_2), S, (S + $NaNO_3$), $Zn(ClO_3)_2$. [*19*]
Spont Heating: Yes; particularly when wet, freshly calcined or tightly packed, it can ignite and burn.
Explosion Hazard: Slight, when exposed to heat or flame.
To Fight Fire: Water, mist, foam or dry chemical.

CHARCOAL, ACTIVATED. See activated carbon.

CHARCOAL BRIQUETTES. See charcoal.

CHARCOAL SHELL. See charcoal.

CHARCOAL, WET. See charcoal.

CHARCOAL, WOOD (GROUND, CRUSHED, GRANULATED OR PULVERIZED). See charcoal.

CHARCOAL, WOOD (LUMP). See charcoal.

CHARCOAL, WOOD SCREENINGS, MADE FROM "PINON." See charcoal.

CHARCOAL, WOOD SCREENINGS, OTHER THAN "PINON." See charcoal.

CHARCOAL WOOD SCREENINGS, WET. See charcoal.

CHELIDONINE. White crystalline powder. $C_{20}H_{19}O_5N \cdot H_2O$, mw: 371.38 mp: 135°–136°.
Acute tox data: sc LD_{LO} (rat) = 300 mg/kg; sc LD_{LO} (mouse) = 300 mg/kg. [*3*]
THR = HIGH via sc route. A CNS depressant, causing sleepiness, depression, slowing of the pulse, and in large doses, coma and circulatory failures.
Fire Hazard: Slight.
Treatment and Antidotes: Stimulants, epinephrine, pilocarpine. Call a physician.

CHELIDONINE HYDROCHLORIDE. See chelidonine.

CHELIDONINE PHOSPHATE. See chelidonine.

CHELIDONINE TANNATE. See chelidonine.

"CHEM-D." See DDT.

CHEM-HEX. See 1,2,3,4,5,6-hexachlorocyclohexane.

CHEMICAL STRUCTURE vs CARCINOGENICITY. See Section 8.

CHEMISTRY DISCIPLINE CONTRIBUTES TO INDUSTRIAL TOXICOLOGY. See Section 9.

CHEMOPROPHYLAXIS of CHEMICAL CARCINOGENESIS. See Section 8.

CHEWING GUM BASE. Made up of a number of substances both natural and synthetic. Used in the manufacture of chewing gum.
THR = U. A food additive permitted in food for human consumption. [*109*]

CHILI PEPPER.
THR = A (S) carc. [*14*] See plant and fungal products.

CHINA CLAY. See kaolin.

CHINA GREEN. See malachite green.

CHINA WOOD OIL. See tung oil.

CHINESE BEAN OIL. See soy bean oil.

CHINESE WAX. White to yellowish-white solid. Composition: Excretion of an insect, primarily ceryl cerotate. mp: 92°, d: 0.91.
THR = An allergen.
Fire Hazard: Slight, when exposed to heat.
To Fight Fire: Foam, CO_2, dry chemical.

CHINESE WHITE. See zinc oxide.

CHINOLINE. See quinoline.

CHINONE. See quinone.

CHINOSOL. See quinosol.

CHIOLITE. See sodium aluminum fluoride.

CHLORACETO PHENONE. See chloroaceto phenone.

CHLORACETYL CHLORIDE. See chloroacetyl chloride.

CHLORAL. Syn: *trichloroacetaldehyde.* Oily liquid, irr odor. CCl_3CHO, mw: 147.4, mp: −57.5°, bp: 97.8°. flash P: 167°F, d: 1.510 @ 20°/4°, vap. d: 5.1.
THR = HIGH via oral route. An irr.
Disaster Hazard: Dangerous; when heated to decomp, emits toxic fumes.

CHLORAL ALCOHOLATE. Syns: *chloral ethylalcoholate, trichloroacetaldehyde monoethyl acetal.* Crystals, less sol in water than chloral hydrate, sol in organic solvents. $C_4H_7Cl_3O_2$, mw: 193.47, d: 1.143, mp: 47.5°, bp: 116°.
Acute tox data: Oral LD_{50} (rat) = 880 mg/kg; oral LD_{50} (dog) = 1200 mg/kg. [*3*]
THR = MOD via oral route.

CHLORAL HYDRATE. Transparent, colorless crystals, aromatic, penetrating, slightly acrid odor and slightly bitter, caustic taste. $CCl_3CH(OH)_2$, mw: 165.41, mp: 52°, bp: 97.5°, d: 1.9.
Acute tox data: Oral LD_{50} (rat) = 480 mg/kg; sc LD_{50} (rat) = 620 mg/kg; oral LD_{50} (mouse) = 1190 mg/kg. [*3*]
THR = MOD via oral, sc, iv, rectal routes. An exper neo. [*3*, *23*] Chloral hydrate is a poisonous drug popularly called "knockout drops." The liquid and vapors are both very dangerous to the eyes. If this material is ingested, the pupils contract during sleep, but dilate upon waking. A number of incidents have been reported where it was given in alcoholic beverages and death was caused when only its hypnotic effect was desired. Recently there has been much addiction to chloral hydrate. These addicts are mostly between the ages of 21 and 40 years. In the majority of cases they are also addicted to alcohol or opium. Occasionally this material may produce excitement and delirium. Nausea and vomiting may be the result of local action of the drug upon the stomach. The respiration is irregular and shallow and the pulse is scarcely perceptible. The pupils are moderately contracted, but rarely dilated. The face is cyanotic or in the early stages, flushed. On account of the dilation of the vessels, the extremities are cold

and the blood pressure and temperature are slightly lower, but little more so than in natural sleep. If the dose exceeds 2 to 3 g, the patient passes into stupor and coma with complete muscular relaxation. The action is very similar to that of chloroform. Death is ordinarily caused by paralysis of the respiratory center. The average fatal dose is placed at about 10 grams. Use caution in giving over the therapeutic dose.

Fire Hazard: Slight; when heated.

Treatment and Antidotes: A physician should be called at once. Wash out the stomach. Administer strychnine hypodermically, also give caffeine, and caffeine with sodium benzoate. Maintain the temperature of the patient with the use of electric pads, hot water bottles and blankets if necessary.

CHLORAL HYDROCYANIDE. See chlorocyanohydrin.

α-CHLORALOSE. $C_8H_{11}O_6Cl_3$, mw: 309.5.

Acute tox data: ip LD_{50} (mice) = 190 mg/kg; oral LD_{50} (birds) = 42 mg/kg; oral LD_{50} (wild birds) = 32 mg/kg; oral LD_{50} (pigeon) = 178 mg/kg. [*103*]

THR = HIGH via ip and oral routes, probably sc route also. An exper neo to mice via sc route. [*3, 103*]

CHLORAMBUCIL. Syn: *N,N-di(2-chloroethyl)p-amino phenylbutyric acid.* $C_{14}H_{19}O_2NCl_2$, mw: 304.1.

THR = An exper (+) carc, neo and teratogen.[*3, 8*]

CHLORAMIPHENE. $C_{32}H_{36}O_8NCl$, mw: 598.1.

THR = MOD to LOW via oral route. An exper teratogen. [*3*]

CHLORAMINE. See chloramine-T.

CHLORAMINE-T. Syns: *chloramine, sodium p-toluene sulfon chloramide.* White or faintly yellow crystals. Slight chlorine odor, water sol. $C_7H_7ClNNaO_2S \cdot 3H_2O$, mw: 281.

THR = Inhal of vapors can cause vasomotor rhinitis and asthma. A mild irr and allergen.

Disaster Hazard: Dangerous; see sulfonates and chlorides.

CHLORAMPHENICOL. Syn: *chloromycetin,* Crystals, slightly sol in water. $C_{11}H_{12}Cl_2N_2O_5$, mw: 323.1, mp: 151°.

THR = Used medically as a broad spectrum antibiotic. Can cause GI disturbances and blood dyscrasias. An exper (±) carc. [*3, 7*]

Disaster Hazard: Dangerous; see chlorides.

CHLORANIL. See tetrachloroquinone.

CHLORAQUINE. $C_{18}H_{26}N_3Cl$, mw: 319.9.

THR = An exper teratogen. [*3*]

***cis*-CHLORAQUOTETRAMINE COBALT (III) CHLORIDE.** Rhombic violet crystals. $[Co(NH_3)_4(H_2O)Cl]Cl_2$, mw: 251.40, mp: decomp, d: 1.847.

THR = See cobalt compounds.

"CHLORASOL" FUMIGANT. See ethylene dichloride and carbon tetrachloride.

CHLORATE AND BORATE MIXTURES. See also chlorates, N.O.S. and boron compounds.

CHLORATE OF POTASH. See potassium chlorate.

CHLORATE OF SODA. See sodium chlorate.

CHLORATES, N.O.S. Chlorates are a combination of a metal or hydrogen and $^-ClO_3$ monovalent radical. They are crystalline and somewhat deliquescent. The principle toxic effects of chlorates are the production of methemoglobin in the blood and destruction of red blood corpuscles. The latter may lead to irr of the kidneys. Damage to heart muscle has been reported.

Fire Hazard: Dangerous, in contact with flam matter. When contaminated with oxidizable materials, they are particularly sensitive to friction, heat and shock; they are powerful oxidizing agents.

Explosion Hazard: Dangerous, when shocked, exposed to heat or rubbed, particularly when contaminated with sugar, charcoal, shellac, sulfur, starch, sawdust, sulfuric acid, ammonium compounds, cyanides, phosphorous or antimony sulfide, Al, (metals + acids), As_2S_3, CaH_2, MnO_2, metal sulfides, organic acids, powdered metals, Hg_3P_4, PHI_4, SCN, (S + Cu), Se, NaH_2PO_2, SrH, SO_2. [*19*] Chlorates when mixed with combustible materials may form explosive mixtures. For instance, potassium chlorate, when mixed with sulfur or with other combustible substances explodes on friction. Pure chlorates which have been spilled on the floor, or mixed with small amounts of impurities, become very sensitive to shock and friction. Water is considered the best agent for fighting fires involving chlorates. In the explosive industry, chlorates are used as oxidizing agents in the primer caps in combination with mercury fulminates, phosphorus, antimony sulfide and other combustible substances. They are used in pyrotechnic mixtures, as a component of airplane flares and aerial bombs. They are also used as a component of permissible explosives. Chlorates are used extensively in the manufacture of chlorate explosives. The chief constituent of such an explosive is from 60 to 80% chlorate. This can be the chlorate of ammonium, sodium or potassium. The other ingredients in such a mixture are combustible materials, such as metallic powders, powdered sulfur, powdered charcoal or possibly mixtures of organic matter. Nitro derivatives of benzene, toluene, and other aromatic compounds are also added. Paraffin may be added as a desensitizer. Recently, similar mixtures were used in Europe but with the addition of small amounts of nitroglycerin or collodion cotton. Chlorate explosives are more sensitive than modern permissible explosives, and therefore, not as safe as for instance the perchlorate explosives, or the permissibles. Plastic mixtures of chlorate explosives (containing nitroglycerin) are somewhat less sensitive to shock and friction, in spite of the nitroglycerin present, than the dryer explosives with no nitroglycerin. In this case the nitroglycerin or "explosive oil," as it is known, serves to wet the rest of the mixture. Barium chlorate is shipped and stored in wooden boxes, barrels, or kegs. It should have isolated storage in a cool, ventilated place, away from acute fire hazards and should not be stored in the same building with combustible materials, acids, sulfur, powdered magnesium or powdered aluminum. Examples of chlorates used in the explosive industry, would be potassium chlorate, sodium chlorate and barium chlorate.

Disaster Hazard: Dangerous; shock will explode them; when heated to decomposition, they can emit toxic fumes and explode; can react with reducing materials.

CHLORATES, N.O.S. WET. See also chlorates, N.O.S.

CHLORAURIC ACID. See gold compounds.

CHLORAZENE. See chloramine.

CHLORAZINE. See 2-chloro-4,6-bis(diethylamino)-s-triazine.

CHLORBENSIDE. See *p*-chlorobenzyl-*p*-chlorophenyl sulfide.

CHLORCYCLINE. $C_{18}H_{21}N_2Cl$, mw: 300.9.

THR = An exper teratogen. [*3*]

CHLORCYCLIZINE HYDROCHLORIDE. $C_{18}H_{21}N_2Cl \cdot HCl$, mw: 337.2.

THR = An exper teratogen. [*3*]

CHLORDAN. See chlordane.

CHLORDANE. Syns: *1,2,4,5,6,7,8,8-octachloro-4,7-methano-3a,4,7,7a-tetrahydroindane*, *chlordan*, *octachlorotetrahydro methano indane*, "*Octa Klor*," "*1068*," "*Velsicol 1068*," "*Dowklor*," "*Ortho-Klor*," and other trade names. Colorless to amber, odorless, viscous liquid. $C_{10}H_6Cl_8$, mw: 409.75, bp: 175°, d: 1.57–1.63 @ 15.5°/15.5°.

Acute tox data: Oral LD_{50} (rat) = 283 mg/kg; dermal LD_{50} (rat) = 700 mg/kg; ip LD_{50} (rat) = 343 mg/kg; oral LD_{50} (chicken) = 220 mg/kg. [*3*]

THR = HIGH via oral and ip; MOD via dermal routes. Has been implicated in develoment of aplastic anemia. An exper carc. [*93*] An insecticide. Must be considered quite toxic. Chlordane is readily absorbed through the skin as well as through other portals. It is a CNS stimulant whose exact mode of action is unknown, but it may involve microsomal enzyme stimulation. Animals poisoned by this and related compounds show an extremely marked loss of appetite and neurological symptoms. The fatal dose to man is unknown. It has been estimated to be between 6 to 60 g (1/5 to 2 oz.). One person receiving an accidental skin application of 25% solution (amounting to something over 30 g of technical chlordane) developed symptoms within about 40 min and died before medical attention was obtained. In two patients, death followed exposure to low oral doses of chlordane (2–4 g); on microscopic examination both patients showed severe chronic fatty degeneration of the liver, characteristic of chronic alcoholism. Although these two fatalities cannot be attributed exclusively to chlordane, they are entirely consistent with previous observations that the toxicity of other chlorinated hydrocarbons is much enhanced in the presence of chronic liver damage. The dangerous chronic dose in man is unknown.

One person poisoned by chlordane developed convulsions within 40 min of gross skin contamination and died, apparently of respiratory failure, before medical aid could be obtained.

Acutely poisoned exper animals show similar signs. Exper animals exposed to repeated small doses exhibit hyperexcitability, tremors, and convulsions, and those which survive long enough show marked anorexia and loss of weight. Symptoms in animals frequently occur within an hour of the administration of a large dose, but death often is delayed for several days depending on the dosage and route of administration. In any event, symptoms are of longer duration with chlordane than with DDT under similar conditions.

Laboratory findings are essentially normal, except that the insecticide may be demonstrated in tissues of poisoned animals by means of bioassay. A method for specific, quantitative chemical analysis for chlordane is now available using small amounts of subcutaneous fat. Chronically poisoned animals show degenerative changes in the liver and kidney tubules.

Disaster Hazard: Dangerous; see chlorides.

Treatment and Antidotes: Removal of the poison from the skin or the alimentary tract should be attempted. Oil laxatives should be avoided. The nervous symptoms may best be combatted with pentobarbital or phenobarbital.

CHLORDIAZEPOXIDE. Syn: *librium*. $C_{16}H_{14}ON_3Cl$, mw: 299.8.

THR = Mod to high via oral route to women ⟶ CNS effects. Has been implicated in development of aplastic anemia. [*3*]

CHLOROETHANE. See ethyl chloride.

CHLORETHYL BENZENE. Liquid. $C_6H_5ClC_2H_5$, mw:141.6.
THR = See chlorinated hydrocarbons, aromatic.
Fire Hazard: Mod, when exposed to heat or flame.
Explosion Hazard: U.
Disaster Hazard: Dangerous; when heated to decomp, emits toxic fumes; reacts with oxidizing materials.

1-(2-CHLORETHYL)-3-CYCLOHEXYL-1-NITROSO UREA. $C_9H_{16}O_2N_3Cl$, mw: 233.7.
THR = HIGH via oral route. An exper teratogen. [*3*]

"CHLOREX." See dichloroethyl ether.

CHLORFENVINFOS. Syn: *2-chloro-1-(2,4-dichlorophenyl)-vinyl diethyl phosphate*. $C_{12}H_{14}O_4PCl_3$, mw: 359.6.
Acute tox data: Oral LD_{50} (rat) = 10 mg/kg; dermal LD_{50} (rat) = 30 mg/kg; sc LD_{50} (rat) = 16 mg/kg; iv LD_{50} (rat) = 7 mg/kg; oral LD_{50} (chicken) = 29 mg/kg. [*3*]
THR = HIGH via all routes of exposure.

CHLORGUANIDE. Syn: *1-(p-chlorophenyl)-5-isopropyl biguanide hydrochloride*. White powder. $ClC_6H_4C_3H_7C_2H_5N_3HCl$, mw: 290.2, mp: 244°.
Acute tox data: Oral LD_{50} (mouse) = 50 mg/kg. [*3*]
THR = HIGH via oral route.
Disaster Hazard: Dangerous. See chlorides.

CHLORHYDROL ALUMINUM. Syn: *aluminum chlorohydroxide complex*. $Al(OH)_2Cl$, mw: 96.4.
THR = An allergen. Probably LOW.

CHLORIC ACID. Colorless solution. $HClO_3 \cdot 7H_2O$, mw: 210.58, mp: $<-20°$, bp: decomp @ 40°, d: 1.282 @ 14.2°.
THR = HIGH irr via oral and inhal routes. See also chlorates.
Fire Hazard: Dangerous; ignites organic matter upon contact; a very powerful oxidizing agent.
Explosion Hazard: $>40\%$ decomp, reacts violently with NH_3, Sb, Sb_2S_3, As_2S_3, Bi, CuS, PHI_4, SnS_2, SnS. [*19*]
Disaster Hazard: Dangerous; see chlorides; reacts vigorously with reducing material.

CHLORIC ETHER. A liquid solution of 60 cc chloroform and 940 cc alcohol.
THR = See also chloroform and ethanol.
Fire Hazard: Mod, when exposed to heat or flame.
Disaster Hazard: Dangerous; when heated to decomp, emits highly toxic fumes of phosgene; can react vigorously with oxidizing materials.

CHLORIDE OF LIME. See bleaching powder.

CHLORIDES.
THR = Varies widely. Sodium chloride (table salt) has very low toxicity, while carbonyl chloride (phosgene) is lethal in small doses. See specific entries.
Disaster Hazard: Dangerous; when heated to decomp or on contact with acids or acid fumes, they evolve highly toxic chloride fumes. Some organic chlorides decomp to yield phosgene.

CHLORIDINE. $C_{12}H_{13}N_4Cl$, mw: 248.7.
THR = An exper neo and teratogen. [*3*]

CHLORINATED ANTHRACENE OIL. See carbolineum.

CHLORINATED BIPHENOLS. See chlorinated diphenyls.

CHLORINATED CAMPHENE. See octachloro camphene.

CHLORINATED DIBENZO DIOXINS. Syns: *dibenzo-p-dioxin, 1-chlorodibenzo-p-dioxin, 2-chlorodibenzo-p-dioxin, 1,3-dichlorodibenzo-p-dioxin, 1,6-dichloro dibenzo-p-dioxin, 2,3-dichlorodibenzo-p-dioxin, 2,7-dichloro dibenzo-p-dioxin, 2,8-dichloro dibenzo-p-dioxin, 1,2,4-trichloro dibenzo-p-dioxin, 2,3,7-trichlorodibenzo-p-dioxin, 1,2,3,4-tetra chlorodibenzo-p-dioxin, 1,2,3,8-tetrachloro dibenzo-p-dioxin, 1,3,6,8-tetrachlorodibenzo-p-dioxin, 1,3,7,8-tetrachlorodibenzo-p-dioxin, 2,3,6,7-tetra chloro dibenzo-p-dioxin, 2,3,7,8-tetra chlorodibenzo-p-dioxin, 1,2,3,4,7-penta chlorodibenzo-p-dioxin, 1,2,3,7,8-penta chlorodibenzo-p-dioxin, 1,2,4,7,8-penta chlorodibenzo-p-dioxin, 1,2,3,4,7,8-hexachlorodibenzo-p-dioxin, 1,2,3,6,7,8-hexachlorodibenzo-p-dioxin, 1,2,3,6,7,9-hexachlorodibenzo-p-dioxin, 1,2,3,7,8,9-hexa chlorodibenzo-p-dioxin, 1,2,3,4,6,7,8-hepta chlorodibenzo-p-dioxin, 1,2,3,4,6,7,9-hepta chlorodibenzo-p-dioxin, 1,2,3,4,6,7,8,9-octachloro-dibenzo-p-dioxin.* For physical properties see individual entries. The chlorinated dibenzo dioxins are not manufactured on a commercial basis, but some are present as impurities in herbicide and fungicide formulations, such as 2,4,5-T, the penta chlorophenols, and hexachlorphene (from trichlorophenol). The chlorinated dibenzo dioxins include some with antibacterial action, flameproofing, insecticidal and fungicidal actions.
Acute tox data: MOD–HIGH; accumulate in organisms; some are carc, mutagens and teratogens. [*81*]

CHLORINATED DIPHENYL (AROCLOR 1221).
Acute tox data: Oral LD_{50} (rat) = 3980 mg/kg; dermal LD_{LO} (rabbit) = 3169 mg/kg. [*3*]
THR = MOD via oral and dermal routes. An exper (+) carc. [*1, 3*]

CHLORINATED DIPHENYL (AROCLOR 1232). Acute tox data: Oral LD_{50} (rat) = 4470 mg/kg; dermal LD_{LO} (rabbit) = 2000 mg/kg. [*3*]
THR = MOD via oral and dermal routes. An exper (+) carc. [*1, 3*]

CHLORINATED DIPHENYL (AROCLOR 1242). Acute tox data: Oral LD_{50} (rat) = 4250 mg/kg; inhal TC_{LO} (humans) = 10 mg/m^3 ⟶ irr; dermal LD_{LO} (rabbit) = 794 mg/kg. [*3*]
THR = MOD via oral, inhal and dermal routes. An exper (+) carc. [*1, 3*]

CHLORINATED DIPHENYL (AROCLOR 1248). Acute tox data: Oral LD_{50} (rat) = 11000 mg/kg; dermal LD_{LO} (rabbit) = 1269 mg/kg. [*3*]
THR = MOD via dermal and LOW via oral routes.

CHLORINATED DIPHENYL (AROCLOR 1254). Acute tox data: Oral LD_{50} (rat) = 1295 mg/kg; ip LD_{50} (mouse) = 2840 mg/kg; iv LD_{50} (rat) = 358 mg/kg. [*3*]
THR = HIGH via iv; MOD via ip, dermal and oral routes. An exper (+) neo via oral route. [*1, 3*]

CHLORINATED DIPHENYL (AROCLOR 1260). Acute tox data: Oral LD_{50} (rat) = 1315 mg/kg; dermal LD_{LO} (rabbit) = 2000 mg/kg. [*3*]
THR = MOD via oral and dermal routes. An exper (+) carc. [*3, 1*]

CHLORINATED DIPHENYL (AROCLOR 1262). Acute tox data: Oral LD_{50} (rat) = 11300 mg/kg; dermal LD_{LO} (rabbit) = 3160 mg/kg. [*3*]
THR = MOD via dermal and LOW via oral routes. An exper (+) carc via oral route. [*1, 3*]

CHLORINATED DIPHENYL (AROCLOR 1268). Acute tox data: Oral LD_{50} (rat) = 10900 mg/kg; dermal LD_{LO} (rabbit) = 2500 mg/kg. [*3*]
THR = MOD via dermal and LOW via oral routes. An exper (+) carc. [*1, 3*]

CHLORINATED DIPHENYL (AROCLOR 2565). Acute tox data: Oral LD_{50} (rat) = 6310 mg/kg; dermal LD_{LO} (rabbit) = 3160 mg/kg. [*3*]
THR = MOD via oral and dermal routes. An exper (+) carc. [*3, 1*]

CHLORINATED DIPHENYL (AROCLOR 4465). Acute tox data: Oral LD_{50} (rat) = 1600 mg/kg; dermal LD_{LO} (rabbit) = 3160 mg/kg. [*3*]
THR = MOD via dermal and LOW via oral routes. An exper (+) carc. [*3, 1*]

CHLORINATED DIPHENYL (KANECLOR 300). THR = An exper (S) carc via oral route. [*1, 3*]

CHLORINATED DIPHENYL (KANECLOR 400). THR = An exper (S) carc via oral route. [*1, 3*]

CHLORINATED DIPHENYL (KANECLOR 500). THR = An exper (+) carc via oral route. [*1, 3*]

CHLORINATED DIPHENYL OXIDE.
THR = HIGH via oral and inhal; MOD via dermal routes. A powerful irr.
Disaster Hazard: Dangerous; when heated to decomp, emits highly toxic fumes.

CHLORINATED DIPHENYLS. Syns: *aroclor, aroclor 1221, aroclor 1232, aroclor 1242, aroclor 1248, aroclor 1254, aroclor 1260, aroclor 1262, aroclor 1268, aroclor 2565, aroclor 4465, chlophen, clorinated biphenyl, chlorinated diphenyl, chlorinated diphenylene, chlorextol, chloro biphenyl, chloro-1,1-biphenyl, dykanol, fenclor, inerteen, kanechlor, kanechlor 300, kanechlor 400, kanechlor 500, montar, nonflamol, PCBs, phenochlor, phenoclor, polychlorobiphenyl, pyralene, pyranol, santotherm FR, sovol, therminol, therminol FR-1.* A series of technical mixtures, consisting of many isomers and compounds that vary from mobile oily liquids to white, crystalline solids and hard non-crystalline resins. They vary in composition and degree of chlorination and perhaps by batch. [*1, 3*]
bp: 340°–375°, flash p: 383°F (COC), d: 1.44 @ 30°.
THR = MOD via dermal and oral routes. A strong irr. Oral exposure can cause (+) neo and carc. [*3, 1*] Also causes a chloracne. Like the chlorinated naphthalenes, the chlorinated diphenyls have 2 distinct actions on the body, namely, a skin effect and a toxic action on the liver. The lesion produced in the liver is an acute yellow atrophy. This hepato toxic action of the chlorinated diphenyls appears to be increased if there is exposure to carbon tetrachloride at the same time. The higher the chlorine content of the diphenyl compound, the more toxic is it liable to be. Oxides of chlorinated diphenyls are more toxic than the unoxidized materials. The skin lesion is known as chloracne, and consists of small pimples and dark pigmentation of the exposed areas, initially. Later, comedones and pustules develop. In persons who have suffered systematic intoxication, the usual signs and symptoms are nausea, vomiting, loss of weight, jaundice, edema and abdominal pain. Where the liver damage has been severe the patient may pass into coma and die.
Fire Hazard: Slight, when exposed to heat or flame.
Disaster Hazard: Dangerous; when heated to decomp, they emit highly toxic fumes.

CHLORINATED HYDROCARBONS, ALIPHATIC. The substitution of a Cl (or other halogen) atom for a hydrogen greatly increases the anesthetic action of a member of the aliphatic hydrocarbons. In addition, the chlorine derivative is usually less specific in its action and may affect other tissues of the body in addition to those of the CNS; in many cases the chlorine derivative is quite toxic. Thus, chloroform, in addition to its narcotic qualities, may cause liver, heart, and kidney damage.

As a general rule, the unsaturated chlorine derivatives are highly narcotic but less toxic than the saturated derivatives, thus causing degenerative changes in the liver and kidneys less frequently. In the saturated group, the narcotic effect is enhanced with an increase in the number of chlorine atoms. However, there is less relationship between the number of chlorine atoms present and the toxicity of the compound.

In dealing with these chlorinated HC, it must be remembered that a toxic action may result from repeated exposure to conc which are too low to produce a narcotic effect, and which, consequently, are too low to give warning of danger. Individual susceptibility is also important when poisoning by this group of solvents is being considered. Certain workmen may be seriously affected by conc that seem to have no effect on fellow employees in the same exposure. A (S)carc of the liver, lung, skin and blood forming tissues. [*14*]

Disaster Hazard: Dangerous; when heated to decomp, they emit highly toxic fumes of phosgene; they react violently with Al, liquid O_2, K, Na. [*19*]

CHLORINATED HYDROCARBONS, AROMATIC. In most instances it is difficult to predict the toxicity of these compounds. However, in the case of most aromatic chlorine compounds, their toxicity is usually no greater, and frequently is less, than that of the corresponding aromatic hydrocarbons, with the notable exception of naphthalene. A (S) carc. [*14*]

Fire Hazard: U.

Explosion Hazard: React violently with Al, liquid O_2, K, Na. [*19*]

Disaster Hazard: Dangerous; when heated to decomp, they emit toxic fumes; they can react with oxidizing materials.

CHLORINATED HYDROCHLORIC ETHER. See ethylidene chloride.

CHLORINATED LIME. See bleaching powder.

CHLORINATED NAPHTHALENES.

THR = HIGH irr via oral, inhal and dermal routes. A (S) carc of liver. [*14*] Also see chlorinated diphenyls. The action of the chlorinated naphthalenes on the body is quite similar to that of the chlorinated diphenyls, the chief effects being the production of chloracne of the skin and, systematically, an acute yellow atrophy of the liver. [*14*]

Disaster Hazard: Dangerous; see chlorides.

CHLORINATED PHENOLS.

THR = HIGH irr via oral, inhal and dermal routes. An exper (+) carc. [*3, 1*]

Disaster Hazard: Dangerous; when heated to decomp, they emit highly toxic fumes.

CHLORINATED TRIPHENYLS.

THR = MOD to LOW oral; MOD via dermal routes. An exper (+) transplacental carc. [*3, 1*] See also chlorinated diphenyls.

Disaster Hazard: Dangerous; when heated to decomp, they emit highly toxic fumes of chlorides.

CHLORINE. Greenish-yellow gas, liquid, or rhombic crystals. Cl_2, mw: 70.914, mp: −101°, bp: −34.5°, d: (liq) 1.47 @ 0° (3.65 atm), vap. press: 4800 mm @ 20°, vap. d: 2.49.

Acute tox data: Inhal TC_{LO} (human) = 15 ppm ⟶ pulmonary problems; inhal LD_{LO} (humans) = 430 ppm for 30 min; inhal LC_{50} (rat) = 293 ppm for 1 hr. [*3*]

THR = HIGH irr via inhal route. Chlorine is extremely irr to the mu mem of the eyes and the respiratory tract. It combines with moisture to liberate nascent oxygen and form hydrochloric acid. Both these substances, if present in quantity, cause inflammation of the tissues with which they come in contact. If the lung tissues are attacked, pulmonary edema may result. A conc of 3.5 ppm produces a detectable odor; 15 ppm causes immediate irritation of the throat. Conc of 50 ppm are dangerous for even short exposures, 1,000 ppm may be fatal, even where the exposure is brief.

Because of its intensely irritating properties, severe industrial exposure seldom occurs, as the workman is forced to leave exposure before he can be seriously affected. In cases where this is impossible, the initial irr of the eyes and mu mem of the nose and throat is followed by cough, a feeling of suffocation, and later, pain and a feeling of constriction in the chest. If exposure has been severe, pulmonary edema may follow, with rales being heard over the chest. It is a common air contaminant.

Radiation Hazard: For permissible levels, see Section 5, Table 5A.5. Artificial isotope ^{30}Cl, $T_{\frac{1}{2}} = 3 \times 10^5 y$, decays to stable ^{36}A via β's of 0.71 MeV.

Fire Hazard: Can react to cause fires or explosions upon contact with turpentine, ether, ammonia gas, illuminating gas, hydrocarbons, hydrogen and powdered metals, polydimethyl siloxane, polypropylene, drawing wax, rubber, sulfamic acid, $As_2(CH_3)_4$, UC_2, acetaldehyde, C_2H_2, alcohols, alkyl isothiourea salts, alkyl phosphines, Al, Sb, As, AsS_2, AsH_3, Ba_3P_2, C_6H_6, Bi, B, BPI_2, B_2S_3, brass, BrF_5, Ca, (CaC_2 + KOH), $Ca(ClO_2)_2$, Ca_3N_2, Ca_3P_2, C, CS_2,

Cs, $CsHC_2$, Co_2O, Cs_3N, $(C + Cr(OCl)_2)$, Cu, CuH_2, CuC_2, dialkyl phosphines, diborane, dibutyl phthalate, $Zn(C_2H_5)_2$, C_2H_6, C_2H_4, ethylene imine, $C_2H_5PH_2$, F_2, Ge, glycerol, $(NH_2)_2$, $(H_2O + KOH)$, I_2, hydroxylamine, Fe, FeC_2, Li, Li_2C_2, Li_6C_2, Mg, Mg_2P_3, Mn, Mn_3P_2, HgO, HgS, Hg, Hg_3P_2, CH_4, Nb, NI_3, OF_2, H_2SiO, $(OF_2 + Cu)$, PH_3, P, $P(SNC)_3$, P_2O_3, PCB's, K, KHC_2, KH, Ru, $RuHC_2$, Si, SiH_2, Ag_2O, Na, $NaHC_2$, Na_2C_2, SnF_2, SbH_3, Sr_3P, Te, Th, Sn, WO_2, U, V, Zn, ZrC_2. [*19*]

Disaster Hazard: Dangerous; when heated, emits highly toxic fumes; will react with water or steam to produce toxic and corrosive fumes of hydrogen chloride.

CHLORINE AZIDE. Syn: *chlor(o)azide.* An explosive gas. ClN_3, mw: 77.48.

THR = HIGH irr via inhal route.

Explosion Hazard: Severe, when shocked, exposed to heat, flame or 1,3-butadiene, C_2H_6, C_2H_4, CH_4, C_3H_8. [*19*]

Disaster Hazard: Dangerous; shock can explode it; when heated to decomp, emits highly toxic fumes of chlorine and NO_x; will react with water or steam to produce toxic and corrosive fumes of hydrogen chloride.

CHLORINE CYANIDE. See cyanogen chloride.

CHLORINE DIOXIDE. Red-yellow gas or orange-red crystals. ClO_2, mw: 67.5, mp: −59°, bp: 9.9° @ 731 mm explodes, d: 3.09 g/liter @ 11°.

Acute tox data: Inhal TC_{LO} (human) = 19 ppm. [*3*]

THR = HIGH irr via inhal route.

Fire Hazard: Dangerous, a powerful oxidizer.

Disaster Hazard: Dangerous; reacts violently with P, KOH, S, conc @ from 0.1 to 1 atm of $>$10% in air explodes, also F_2Hg, organic matter, NHF_2. [*19*] When heated to decomp, emits highly toxic fumes of chlorine; will react with water or steam to produce toxic and corrosive fumes of hydrochloric acid.

CHLORINE DIOXIDE HYDRATE, FROZEN. See chlorine dioxide.

CHLORINE FLUOROXIDE. ClOF, mw: 70.5.

THR = HIGH irr via inhal route. Explosively unstable. [*19*]

CHLORINE HEPTAOXIDE. Colorless oil. Cl_2O_7, mw: 182.91, mp: −91.5°, bp: 82°, vap. press: 100 mm @ 29.1°.

THR = HIGH irr poison. Very unstable.

Fire Hazard: Dangerous; a very powerful oxidizing agent.

Explosion Hazard: Severe, when shocked or exposed to heat or flame.

Disaster Hazard: Dangerous; shock or heat will explode it; on decomp, emits highly toxic fumes of chlorine; will react with water or steam to produce toxic and corrosive fumes.

CHLORINE HYDRATE. Rhombic light yellow crystals. $Cl_2 \cdot 8H_2O$, mw: 215.04, mp: decomp @ 9.6°, d: 1.23.

THR = HIGH irr and tox. See also chlorine.

Disaster Hazard: Dangerous; see chlorine; will react with water or steam to produce toxic and corrosive fumes.

CHLORINE MONOFLUORIDE. Nearly colorless gas. ClF, mw: 54.46, mp: −154 ± 0.5°, bp: −100.8°, d: 1.62 @ −100°.

THR = HIGH irr via inhal route. A very irr poison. Very unstable. Reacts violently with water, Te, organic matter. [*19*] See fluorides and chlorine.

CHLORINE MONOXIDE. Yellow-red gas or red-brown liquid. Cl_2O, mw: 86.91, mp: −20°, bp: 2.2°, d: 3.89 g/liter @ 0°, lel = 23.5%, uel = 100%.

THR = See chlorine.

Explosion Hazard: Severe, when shocked or exposed to heat of 39°. Reacts violently with NH_3, Sb, Sb_2S_3, Se, BaS, Ca_3P_2, C, CS_2, charcoal, H_2S, HgS, metal sulfides, NO, organic matter, PH_3, P, S, SnS, SnS_2, turpentine. [*19*]

Fire Hazard: Very dangerous via heat, flame or reducing agents.

Disaster Hazard: Dangerous; see chlorine; will react with water or steam to produce toxic and corrosive fumes. When heated it explodes.

CHLORINE TETROXIDE. See chlorine dioxide.

CHLORINE TETROXYFLUORIDE. See fluorine perchlorate.

CHLORINE TRIFLUORIDE. Colorless gas to yellow liquid, sweet odor, ClF_3, mw: 92.46, mp: −83°, bp: 11.8°, d: 1.77 @ 13°.

THR = HIGH irr via inhal route. See also fluorides, chlorine and fluorine.

Fire Hazard: Dangerous. Spont flam.

Explosion Hazard: Reacts violently with organic matter, glass wool, acetic acid, Al, Sb, As, Cu, H_2, I_2, Ir, Fe, Pb, Mg, Mo, Os, P, K, Rh, Se, Si, Ag, Na, S, Te, Sn, W, Zn, oxides, NH_3, benzene, CO, CrO_3, ether, graphite, H_2S, HgI_2, HNO_3, K_2CO_3, KI, rubber, $AgNO_3$, NaOH, H_2SO_4, V_2P_5, water, WO_3. [*19*]

Disaster Hazard: Dangerous; when heated to decomp or on contact with acid or acid fumes, emits highly toxic fumes; will react with water or steam to produce much heat and toxic and corrosive fumes; reacts vigorously with reducing materials.

CHLORINE TRIOXIDE. ClO_3, mw: 83.5.
THR = HIGH via contact and inhal route; a corrosive material. Spont explosive. Reacts violently with C_2H_4OH, organic matter, P, PCl_5, water. [*19*]

CHLORITES. ClO_2.
THR = See individual chlorites. Reacts violently with NH_3, organic matter, metals. [*19*]

CHLORMADINONE ACETATE. $C_{23}H_{29}O_4Cl$, mw: 405.
THR = An exper (S) carc and teratogen. [*3, 15*]

CHLORNAPHTHAZINE. See naphthylamine mustard.

CHLOROACETALDEHYDE. Syn: *chloroaldehyde.* Clear, colorless liquid, pungent odor. C_2H_3OCl, mw: 78.5, bp: 90.0°–100.1° (40% sol), fp: −16.3° (40% sol), flash p: 190°F, d: 1.19 @ 25°/25° (40% sol), vap. press: 100 mm @ 45° (40% sol).
Acute tox data: Oral LD_{50} (rat) = 23 mg/kg; dermal LD_{50} (rabbit) = 67 mg/kg. [*3*]
THR = HIGH irr via oral, inhal and dermal routes.
Fire Hazard: Mod, when exposed to heat or flame.
To Fight Fire: Water, foam, CO_2, dry chemical.
Disaster Hazard: Dangerous; see chlorides; reacts with oxidizing materials.

CHLOROACETALDEHYDE OXIME.
$ClCH_2CHNOH$, mw: 93.5.
THR = No data. Probably very irr. Unstable, explosive. [*19*]

2-CHLOROACETAMIDE. Syn: *2-chloroethanamide.* Mod sol crystals. C_2H_4ONCl, mw: 93.5, mp: 120°, bp: 225° (with decomp).
Acute tox data: ip LD_{LO} (mouse) = 100 mg/kg. [*3*]
THR = HIGH irr via oral and ip routes.

CHLOROACETIC ACID. See monochloroacetic acid.

CHLOROACETONE. Syn: *chlorinated acetone.* Colorless liquid, pungent odor. CH_3COCH_2Cl, mw: 92.53, mp: −44.5°, bp: 119°, d: 1.162.
Acute tox data: Oral LD_{50} (rat) = 50 mg/kg. [*3*]
THR = HIGH via oral route. An exper neo. [*3, 23*] A lachrymator poison gas. See chlorinated hydrocarbons, aliphatic and acetone.
Fire Hazard: Mod, when exposed to heat or flame. Old material can explode. [*19*]
Disaster Hazard: Dangerous; when heated to decomp, emits highly toxic fumes of phosgene; can react vigorously with oxidizing materials.

CHLOROACETONITRILE. Crystals. CH_2ClCN, mw: 75.5.
Acute tox data: Oral LD_{50} (rat) = 220 mg/kg; dermal LD (rabbit) 71 mg/kg. [*3*]
THR = HIGH via oral and dermal routes.
Disaster Hazard: Dangerous; when heated to decomp, emits highly toxic fumes; will react with water, steam, acid or acid fumes to produce toxic and flam vapors.

CHLOROACETOPHENONE. Syns: *phenacylchloride, phenyl chloromethyl ketone.* Pale straw-colored liquid or white crystals, fragrant, non-persistent odor. C_8H_7ClO, mw: 154.6, mp: 56°, bp: 237°–247°, fp: 59°, d: 1.19 @ 25°/25°, vap. press: 0.012 mm @ 0°, vap. d: 5.2, flash p: 244°F.
Acute tox data: Oral LD_{50} (rat) = 52 mg/kg; inhal LD_{50} (rat) = 14 mg/kg. [*3*]
THR = HIGH via oral, inhal routes. A powerful irr and lachrymator and military poison.
Disaster Hazard: Dangerous; when heated to decomp, emits toxic fumes, will react with water or steam to produce toxic and corrosive fumes.
Fire Hazard: Low.
To Fight Fire: Water, foam, alcohol foam, dry chemical.

CHLOROACETYL CHLORIDE. Water white or slightly yellow liquid. $CH_2ClCOCl$, mw: 112.95, bp: 105°–106°, fp: −22.5°, flash p: none, d: 1.495 @ 0°.
Acute tox data: LC_{LO} (rat) = 1000 ppm for 4 hrs. [*3*]
THR = HIGH irr via inhal route. A lachrymator (tear gas).
Disaster Hazard: Dangerous; when heated to decomp, emits highly toxic fumes of chlorides; will react with water or steam to produce toxic and corrosive fumes.

CHLOROACETYLENE. HCCCl, mw: 60.5.
THR = No data. It is unstable and reacts violently in air. [*19*] Probably very irr via inhal route.

CHLOROACETYL ISOCYANATE.
THR = Irr. See also cyanates.
Disaster Hazard: Dangerous; see chlorides and isocyanates.

CHLOROACROLEIN. Colorless liquid. $CH_2CClCHO$, mw: 90.52, bp: 30°, d: 1.205 @ 15°.
THR = HIGH irr via oral and inhal routes. A poison gas.
Disaster Hazard: Dangerous; when heated to decomp, emits highly toxic fumes; will react with water or steam to produce toxic and corrosive fumes.

CHLOROALDEHYDE. See chloroacetaldehyde.

2-CHLOROALLYL DIETHYL DITHIOCARBAMATE. See vegadex.

2-CHLORO-4-*tert*-AMYL PHENOL. Liquid. $HOC_6H_3ClC_5H_{11}$, mw: 198.7, bp: 253°–265°, flash p: 225°F, d: 1.11.
THR = U. See chlorinated phenols.
Fire Hazard: Slight, when exposed to heat or flame.

Disaster Hazard: Dangerous; see chlorides; reacts with oxidizing materials.
To Fight Fire: Dry chemical, fog, mist, CO_2.

CHLORO-4-*tert*-AMYL PHENYL METHYL ETHER. $C_5H_{11}C_6H_3 \cdot ClOCH_3$, mw: 212.5, d: 1.1, vap. d: 7.3, bp: 518°–529°F, flash p: 230°F.
THR = No details. See also ethers.
Fire Hazard: Slight, when exposed to heat or flame.
Disaster Hazard: Dangerous; see chlorides and ethers.
To Fight Fire: Dry chemical, fog, mist, CO_2.

***m*-CHLOROANILINE.** Liquid. C_6H_6ClN, mw: 127.9, mp: −10°, bp: 230.5°, d: 1.223 @ 15°/15°, vap. press: 1 mm @ 63.5°.
Acute tox data: sc LD_{LO} (rat) = 310 mg/kg; oral LD_{50} (rat) = 880 mg/kg. [*3*]
THR = MOD via oral and HIGH via sc routes. An allergen. No dermal data. See also aniline.
Disaster Hazard: Dangerous, see chlorides and aniline.

***o*-CHLOROANILINE.** Liquid. C_6H_6ClN, mw: 127.6, mp: 1.94° (2 forms), bp: 209°, d: 1.213 @ 20°/4°, vap. press: 1 mm @ 46.3°.
Acute tox data: sc LD_{LO} (rat) = 310 mg/kg; oral LD_{50} (mouse) = 256 mg/kg. [*3*]
THR = HIGH via oral and sc routes.
Disaster Hazard: Dangerous; see aniline and chlorides.

***p*-CHLOROANILINE.** Crystalline solid. C_6H_6ClN. mw: 127.6, mp: 72.5°, bp: 230.5°, d: 1.427, vap. press: 1 mm @ 59.3°.
Acute tox data: An allergen. Inhal TC_{LO} (human) = 44 mg/m^3; oral LD_{50} (rat) = 300 mg/kg; dermal LD_{LO} (rabbit) = 36 mg/kg. Damages the blood. [*3*]
THR = HIGH via inhal, oral and dermal routes.
Disaster Hazard: Dangerous; see aniline and chlorides.

***m*-CHLOROANILINE DIAZONIUM CHLORIDE.** $NH_2C_6H_4NNCl$, mw: 155.6.
THR = No data. Probably toxic. Violent reaction with NaSH, Na_2S. [*19*]

β-(*o*-CHLOROANILINO)PROPIONITRILE. Colorless to red (on aging) liquid. $ClC_6H_4NHCH_2CH_2CN$, mw: 180.6, bp: 139°–141° @ 0.3 mm, d: 1.2103 @ 25°/25°, vap. d: 6.23.
THR = HIGH. See nitriles.
Fire Hazard: Mod, when exposed to heat or flame.
Disaster Hazard: Dangerous; see nitriles. Will react with water, steam, acid or acid fumes to produce toxic and flam vapors of cyanides.

CHLOROAURIC ACID. Bright yellow needles, deliquescent. $HAuCl_4 \cdot 4H_2O$, mw: 412.10, mp: decomp.
THR = See gold compounds and hydrochloric acid.

CHLOROAZIDE. See chlorine azide.

CHLOROAZODIN. See azochloramide.

7-CHLOROBENZ(a)ANTHRACENE. $C_{18}H_{11}Cl$, mw: 262.7.
THR = An exper neo. [*3*]

CHLOROBENZENE. Syns: *phenyl chloride*, *monochlorobenzene*, *chlorobenzol*. Clear, colorless liquid. C_6H_5Cl, mw: 112.56, bp: 131.7°, lel = 1.3%, uel = 7.1%, @ 150°, mp: −45°, flash p: 85°F (CC), d: 1.113 @ 15.5°/15.5°, autoign. temp.: 1180°F, vap. press: 10 mm @ 22.2°, vap. d: 3.88.
Acute tox data: Oral LD_{50} (rat) = 2910 mg/kg; oral LD_{50} (rabbit) = 2930 mg/kg. [*3*]
THR = MOD via oral, inhal and sc routes. Monochlorobenzol is a fairly strong narcotic and possesses only slight irr qualities. For cats, conc of 1,200 ppm are quite narcotic, and conc of 3,700 ppm are fatal after several hours. The dichlorobenzols are strongly narcotic, 1,000 ppm causing narcosis in guinea pigs followed by death after 20 hrs exposure. Knowledge of the effects on man of repeated exposure to subnarcotic concentrations is meager. In general, it appears that the chlorobenzols are not as toxic as benzol. Some of the symptoms described (methemoglobinemia) suggest that other substances, such as nitrobenzol, may have been partially responsible for the few cases of industrial illness reported. It is possible that prolonged exposure to chlorobenzol may cause kidney and liver damage.
Somnolence, loss of consciousness, twitchings of the extremities, cyanosis, deep, rapid respirations and a small, irregular pulse are the chief symptoms occurring in acute exposures. The urine may be burgundy red, and the red blood cells show degenerative and regenerative changes.
Fire Hazard: Dangerous, when exposed to heat or flame. Also violent reaction with $AgClO_4$, dimethyl sulfoxide. [*19*]
Spontaneous Heating: No.
Explosion Hazard: Mod, when exposed to heat or flame.
Disaster Hazard: Dangerous; see chlorine compounds; can react vigorously with oxidizing materials.
To Fight Fire: Foam, CO_2, dry chemical, water to blanket fire.

***p*-CHLOROBENZENE DISULFONAMIDE.** Crystals. $C_6H_7O_4N_2S_2Cl$, mw: 270.7.
Acute tox data: Oral LD_{50} (rat) = 340 mg/kg. [*3*]
THR = HIGH via oral route.
Disaster Hazard: Dangerous; see sulfonates and chlorides.

4-CHLOROBENZENE SULFONIC ACID CHLOROPHENYL ESTER. See trichlorofenson.

p-CHLOROBENZENE SULFONYL FLUORIDE. White solid. $ClC_6H_4SO_2F$, mw: 195, mp: 46°–51°, bp: 229°–230°, flash p: 340°F, d: 1.475, vap. press: 11 mm @ 100°, vap. d: 6.73.
THR = U. Does not appear to be highly toxic to test animals by inhal on an acute basis. May be irr to the skin.
Fire Hazard: Slight, when exposed to heat or flame.
Disaster Hazard: Dangerous; see fluorides, chlorides and sulfonates; can react vigorously with oxidizing materials.
To Fight Fire: Water, foam, CO_2, dry chemical.

p-CHLOROBENZENE TETRATHIO-o-STANNATE. Solid. $(SC_6H_4Cl)_4Sn$, mw: 693.1, mp: 189°.
THR = U. See tin compounds.
Disaster Hazard: Dangerous; see sulfates and chlorides.

CHLOROBENZILATE. Syn: *acaraben*. Yellow, viscous liquid. $C_{16}H_{14}O_3Cl_2$, mw: 325, bp: 148° @ 0.04 mm.
Acute tox data: Inhal LC_{50} (mammals) = 500 ppm, oral LD_{50} (rat) = 700 mg/kg. An exper. (+) carc. [*3*, *12*]
THR = MOD via inhal and oral routes.

2-CHLORO BENZO(a)(1)BENZOTHIO PYRANO-(4,3-6)INDOLE. $C_{19}H_{10}NSCl$, mw: 319.8.
THR = An exper carc. [*3*]

CHLOROBENZOL. See chlorobenzene.

CHLOROBENZONITRILES (o and p). Crystals. ClC_6H_4CN, mw: 137.6.
Acute tox data: (*o*)Oral LD_{50} (rat) = 435 mg/kg; ip LD_{50} (mouse) = 150 mg/kg. [*3*] (*p*)Oral LD_{50} (rat) = 887 mg/kg; ip LD_{50} (mouse) = 150 mg/kg. [*3*]
THR = HIGH via oral and ip routes.
Disaster Hazard: Dangerous; see cyanides and chlorides; they will react with water, steam, acid or acid fumes to produce toxic fumes.

5-CHLOROBENZO TRIAZOLE. $HN_3C_6H_3Cl$, mw: 153.5.
THR = Very unstable and self-igniting. [*19*]

m-CHLOROBENZOTRIFLUORIDE. Syn: *m-chlorotrifluoromethyl benzene*. Water-white aromatic liquid. $ClC_6H_4CF_3$, mw: 180.56, mp: −56°, bp: 138°, d: 1.351, vap. d: 6.24, flash p: 117°F.
THR = Only slightly toxic to exper animals by skin contact and oral route.
Disaster Hazard: Dangerous; see chlorides and fluorides.
Fire Hazard: Mod, via heat, flame, sparks or oxidizing agents.
To Fight Fire: Foam, alcohol foam, fog, mist, dry chemical.

o-CHLOROBENZOTRIFLUORIDE. Syns: *o-chlorotrifluoro methyl benzene*, *o-chloro-α,α-trifluorotoluene*. Colorless liquid, aromatic odor. $ClC_6H_4CF_3$, mw: 180.56, d: 1.379 @ 15.5°/15.5°, bp: 152°, fp: −7.4°, flash p: 138°F, vap. d: 6.2.
THR = See *m*-chlorobenzo trifluoride.
Fire Hazard: Mod, when exposed to heat, flame or powerful oxidizers.
Disaster Hazard: Dangerous; see chlorides and fluorides.
To Fight Fire: Water, fog, foam, spray, alcohol foam, dry chemical.

p-CHLOROBENZO TRIFLUORIDE. Syn: *p-chlorotrifluoromethyl benzene*. Clear water white liquid. $ClC_6H_4CF_3$, mw: 180.56, mp: −36°, bp: 139.3°, flash p: 116°F, d: 1.353 @ 15.5°/15.5°, vap. d: 6.24.
THR = See *m*-chlorobenzo trifluoride.
Fire Hazard: Mod, when exposed to heat, flame or powerful oxidizers.
Disaster Hazard: Dangerous; see chlorides and fluorides; can react vigorously with oxidizing materials.
To Fight Fire: Water, foam, CO_2, dry chemical.

p-CHLOROBENZOYL PEROXIDE. Syn: *luperco BDB*. A white granular material, insol in water, sol in organic solvents. $(ClC_6H_4CO)_2O_2$, mw: 311.1.
Acute tox data: ip LD_{LO} (mouse) = 500 mg/kg. [*3*]
THR = HIGH via ip route. Limited animal exper indicate MOD toxicity via oral route. Probably an irr to skin and mu mem. See peroxides, organic.
Fire Hazard: Dangerous; a powerful oxidizer. Store in a cool place away from fire hazards, sparks, open flames and out of the direct rays of the sun.
Explosion Hazard: Dangerous; this material may be caused to explode by heat (over 38°) or contamination. Any contaminant which acts as an accelerator to the polymerization or decomposition of this material can cause an explosion.
Disaster Hazard: Dangerous; heat or contact with certain fumes or mists can cause it to explode.
To Fight Fire: For small quantities, carbon dioxide, or foam extinguishers may be used. Water spray or mist may also be used. Dry chemical is effective.
Storage and Handling: Must be kept below 38° in even cool storage. This material is dangerous when involved in a fire. Consult manufacturer's literature on storage and handling.

5-CHLORO BENZ(a)PYRENE.
THR = An exper carc. [*23*]

o-CHLOROBENZYL CHLORIDE. A colorless liquid or crystals. $ClC_6H_4CH_2Cl$, mw: 161.08, bp: 216°–222°, fp: <−30°, d: 1.270–1.280 @ 25°/15°, vap. d: 5.55.

THR = U. See also benzyl chloride.
Disaster Hazard: Dangerous; see phosgene.

***p*-CHLOROBENZYL CHLORIDE.** Syn: *α-dichlorotoluene*. Needle-like crystals or colorless liquid. $ClC_6H_4CH_2Cl$, mw: 161.0, mp: 29°, bp: 222°, d: 1.250–1.260 @ 25°/15°, vap. d: 5.55.
THR = U. See also benzyl chloride.
Disaster Hazard: Dangerous; see phosgene.

***p*-CHLOROBENZYL-*p*-CHLOROPHENYL SULFIDE.** Syns: *chlorosulfacide*, *mitox*, *chlorobenside*. Crystals, almond-like odor, insol in water, sol in most organic solvents. $ClC_6H_4CH_2SC_6H_4Cl$, mw: 269, mp: 75°–76°, d: 1.4210 @ 25°/4°, vap. press: 1.21×10^{-5} mm @ 30°.
Acute tox data: Oral LD_{50} (rat) = 2 g/kg. [*3*]
THR = MOD via oral route. Has caused liver and kidney injury, also skin irr in exper animals.
Disaster Hazard: Dangerous; see chlorides and sulfur compounds.

S-(4-CHLOROBENZYL)-N,N-DIETHYLTHIOL CARBAMATE. See saturn.

***o*-CHLOROBENZYLIDENE MALONITRILE.** Syns: *OCBM*, *CS gas*. White crystals, solid. $C_{10}H_5N_2Cl$, mw: 187.6, mp: 95°, bp: 313°.
Acute tox data: Oral LD_{50} (rat) = 178 mg/kg; iv LD_{50} (rat) = 35 mg/kg. [*3*]
THR = HIGH via oral and iv routes. An exper carc. [*23*] Human exposure data suggest relatively LOW systematic toxicity, but intense irr of eyes, skin, mu mem. See nitriles.
Disaster Hazard: Dangerous; see chlorides and nitriles.

2-CHLOROBIPHENYL. Oily liquid. $C_{12}H_9Cl$, mw: 188.7, mp: 54°, bp: 268°.
THR = An exper carc. [*3*, *1*]

2-CHLORO-4,6-BIS (DIETHYLAMINE)-3-TRIAZINE. Syns: *princep*, *chlorazine*. Clear liquid, slight odor. $C_7H_{12}N_5Cl$, mw: 201.7, mp: 27°, d: 1.0956.
Acute tox data: Oral LD_{50} (rat) = 850 mg/kg. [*3*]
THR = MOD via oral route.
Disaster Hazard: Dangerous; see chlorides.

2-CHLORO-4,6-BIS(DIETHYLAMINO)-2-TRIAZINE. See 2-chloro-4,6-bis(diethylamine)-*s*-triazine.

2-CHLORO-4,6-BIS(ETHYLAMINO)-*s*-TRIAZINE. $C_7H_{12}N_5Cl$, mw: 201.7.
Acute tox data: Oral LD_{50} (rat) = 5000 mg/kg. [*3*]
THR = MOD via oral route. An exper neo via sc route. [*3*]

CHLOROBROMOMETHANE. See methylene chlorobromide.

4-CHLORO-7-BROMOMETHYL BENZ(*a*) ANTHRACENE.
THR = An exper carc. [*23*]

CHLOROBROMOPHOSGENE. Liquid. COClBr, mw: 143.39, bp: 25°, d: 1.82 @ 15°.
THR = HIGH. See carbon oxychloride.
Disaster Hazard: Dangerous; see phosgene.

CHLOROBROMOPROPENE. Syn: *CBP*. Liquid. $ClBrC_3H_4$, mw: 155.4.
THR = HIGH irr to skin, eyes and mu mem.
Disaster Hazard: Dangerous; see chlorides and bromides.

2-CHLORO-1,3-BUTADIENE. See chloroprene.

1-CHLOROBUTANE. See butyl chloride.

CHLOROBUTANOL. Syn: *acetone chloroform*. Crystals, camphor odor. $Cl_3CC(CH_3)_2OH$, mw: 177.5, mp: 97°, bp: 167°.
Acute tox data: Oral LD_{LO} (rabbit) = 213 mg/kg. [*3*]
THR = HIGH via oral route. Tox and narcotic. A food additive permitted in food for human consumption. [*109*] See chloral hydrate, which acts similarly.
Fire Hazard: Slight, when exposed to heat or flame.
Disaster Hazard: Dangerous; see phosgene; can react with oxidizing materials.

2-CHLOROBUTENE-2. See methallyl chloride.

1-CHLORO-2-(β-CHLOROETHOXY)ETHANE. See 2,2′-dichloroethyl ether.

***o*-CHLOROCINNAMIC ACID.** Light tan powder. $ClC_6H_4CHCHCOOH$, mw: 182.6, mp: 207°–212°.
Acute tox data: Oral LD_{50} (mouse) = 989 mg/kg. [*3*]
THR = MOD via oral route.
Disaster Hazard: Dangerous; see phosgene.

***p*-CHLORO-*m*-CRESOL.** Syns: *4-chloro-3-hydroxytoluene*, *PCMC*. Odorless crystals (when pure). Somewhat sol in water, very sol in organic solvents. C_7H_7ClO, mw: 142.6, mp: 66°, bp: 235°.
Acute tox data: Oral LD (rat) = 500 mg/kg; sc LD_{50} (rat) = 400 mg/kg. [*3*]
THR = HIGH via oral and sc routes. An allergen. See cresol.
Disaster Hazard: Dangerous; see phosgene.

CHLOROCYANOHYDRIN. Syns: *chloral hydrocyanide*, *trichloroacetonitrile*. Crystals, odor of chloral and hydrogen cyanide. $CCl_3CH(OH)CN$, mw: 174.4, mp: 61°, bp: 220°.
Acute tox data: sc LD (mouse) = 23 mg/kg. [*3*]
THR = HIGH via sc, oral and inhal routes.
Disaster Hazard: Dangerous; see cyanides; will react with water, steam, acid or acid fumes to produce toxic fumes.

CHLOROCYCLAMIDE-R.
THR = An exper teratogen. [*3*]

3-CHLOROCYCLOPENTENE. (C_5H_7Cl), mw: 102.5.
THR = Explodes spont. [*19*]

CHLORO DEOXYSCELERATINE.
THR = An exper carc. [*23*]

5-CHLORO-2-DEOXY URIDINE. $C_9H_{11}O_5N_2Cl$, mw: 246.2.
THR = An exper teratogen. [*3*]

α-CHLORODIALLYL ACETAMIDE. See *n,n*-diallyl chloroacetamide.

2-CHLORO-4,6-DI-*tert*-AMYL PHENOL.
$(C_5H_{11})_2C_6H_2ClOH$, mw: 268.5, bp: 320°–354°F @ 23 mm, flash p: 250°F.
THR = U.
Fire Hazard: Slight, when exposed to heat or flame.
Disaster Hazard: Dangerous; see chlorides.
To Fight Fire: Dry chemical, CO_2, mist, fog.

CHLORO DIBORANE. B_2H_5Cl, mw: 62.1.
THR = See diboranes. Spont flam in air. [*19*]

1-CHLORO-1,1-DIETHOXY ETHANE.
$(C_2H_5O)_2CHCH_2Cl$, mw: 152.5.
THR = Mixed with sodium amides and ammonia is pyroforic. [*19*]

CHLORODIETHYL ALUMINUM. See diethyl aluminum chloride.

CHLORODIETHYL BISMUTHINE. $(C_2H_5)_2BiCl$, mw: 302.5.
THR = Ignites in air. [*19*]

CHLORODIETHYL BORANE. $(C_2H_5)_2BCl$, mw: 104.3.
THR = Ignites in air. [*19*]

2-CHLORO-(2,4-DICHLOROPHENYL) VINYL DIETHYL PHOSPHATE. See chlorofenvinfos.

2-CHLORO-2-DIETHYL CARBAMYL-1-METHYL VINYL DIMETHYL-PHOSPHATE. Syns: *phosphamidon, dimercron*. Colorless liquid, sol in water and organic solvents. $(CH_3O)_2P(O)OC(CH_3)$: $C(Cl)C(O)N(C_2H_5)_2$, mw: 299.5, bp: 162° @ 1.5 mm.
Acute tox data: Organic phosphate insecticide. Oral LD_{50} (rat) = 17 mg/kg; dermal LD_{50} (rat) = 125 mg/kg; oral LD_{50} (wild birds) = 2 mg/kg. [*3*]
THR = HIGH via all routes. A poison.
Disaster Hazard: Dangerous; see chlorides and phosphates.

3-CHLORO-2,4-DIETHYLENE IMINO PYRAMIDINE.
THR = An exper mutagen. [*3*]

1-CHLORO-1,1-DIFLUOROETHANE. See difluoro monochloroethane.

CHLORODIFLUORO METHANE. Syn: *freon-22*. Gas. $ClHCF_2$, mw: 86.5, d: 3.87 air @ 0°, mp: −146°, bp: −40.8°, autoign. temp.: 1170°F.
Acute tox data: Inhal LC_{LO} (rat) = 25% for 4 hrs. [*3*]
THR = LOW via inhal route. Asphyxiant in high conc.
Fire Hazard: Low.
Disaster Hazard: Dangerous; see chlorides and fluorides.

CHLORODIFLUORO METHANE-CHLORO PENTAFLUORO ETHANE MIXTURE. See components as listed.

12-CHLORO-7-DIHYDRO-9,11-DIMETHYL BENZO(a)PHENARSAZINE. $C_{18}H_{15}NClAs$, mw: 355.7.
THR = An exper neo. [*3*]

CHLORODIISOBUTYL ALUMINUM. $AlCl(C_4H_9)_2$, mw: 176.7.
THR = Self-ignites in air. [*19*]

CHLORO(DIMETHYLAMINO)DIBORANE.
$(CH_3)_2NB_2H_4Cl$, mw: 105.2.
THR = Self ignites in air. [*19*]

2-CHLORO-4-DIMETHYLAMINO-6-METHYL PYRIMIDINE. Syn: *castrix*. Crystals, very slightly water-sol. $C_7H_{10}ClN_3$, mw: 171.6.
Acute tox data: Oral LD_{50} (rat) = 1.25 mg/kg; oral LD_{50} (mouse) = 1.2 mg/kg; oral LD_{50} (rabbit) = 5 mg/kg. [*3*]
THR = HIGH via oral route. Can cause CNS damage and convulsions. A rodenticide. It is intensely poisonous to mammals.
Disaster Hazard: Dangerous; see chlorides.

CHLORODIMETHYL ARSINE. See dimethyl chloroarsine.

9-CHLORO-8,12-DIMETHYL BENZ(a)ACRIDINE.
$C_{19}H_{14}NCl$, mw: 291.8.
THR = An exper neo. [*3*]

10-CHLORO-7,11-DIMETHYL BENZ(c)ACRIDINE.
THR = An exper neo. [*3*]

2-CHLORO-N,N-DIMETHYL-4-STILBENAMINE.
$C_{16}H_{16}NCl$, mw: 257.8.
THR = An exper carc. [*3*]

3-CHLORO-N,N-DIMETHYL-4-STILBENAMINE.
THR = An exper carc. [*3*]

4-CHLORO-N,N-DIMETHYL-4-STILBENAMINE.
THR = An exper neo. [*3*]

2,1,4-CHLORO DINITROBENZENE. See dinitrochloro benzene.

4-CHLORO-3,6-DINITROPHENYL DIAZONIUM-2-OXIDE. $C_6H_2(NO_2)_2ClN_2O$, mw: 245.6.
THR = Violent reaction with nitric acid. [*19*]

CHLORO-2,4-DINITROTOLUENE. $C_6H_3(NO_2)_2CH_2Cl$, mw: 216.6.
THR = Can detonate @ 150°. If confined, shock will explode it. [*19*]

CHLORODIPHENYL. See chlorinated diphenyls.

CHLORODIPHENYL ARSINE. See diphenyl chloroarsine.

CHLORODIPHENYL OXIDE. See chlorinated diphenyl oxide.

CHLORO DIPROPYL BORANE. $(C_3H_7)_2BCl$, mw: 132.5.
THR = Self ignites in air. [*19*]

1-CHLORO-2,3-EPOXYPROPANE. See epichlorohydrin.

2-CHLOROETHANAMIDE. See 2-chloroacetamide.

CHLOROETHANE. See ethyl chloride.

CHLOROETHANOIC ACID. See monochloroacetic acid.

2-CHLOROETHANOL. See ethylene chlorohydrin.

CHLOROETHENE. See vinyl chloride.

CHLOROETHYL ACETATE. See ethyl chloroacetate.

β-CHLOROETHYL ALCOHOL. See ethylene chlorohydrin.

N-(2-CHLOROETHYL)AMINO AZO BENZENE.
THR = An exper carc. [*23*]

9-[2-(2-CHLORO ETHYL)AMINO]ETHYLAMINO-6-CHLORO-2-METHOXY ACRIDINE.
THR = An exper carc. [*23*]

2-CHLORO-4-ETHYLAMINO-6-ISOPROPYL AMINO-*s*-TRIAZINE. Syn: *atrazine.* Colorless crystals, slightly sol in hot water, sol in methanol, chloroform. $C_8H_{14}N_5Cl$, mw: 215.6, mp: 175°.
Acute tox data: Oral LD_{50} (rat) = 1750 mg/kg; oral LD_{50} (mouse) = 1750 mg/kg. [*3*]
THR = MOD via oral route. A selective herbicide.

CHLOROETHYL AMINOPHENYL CARBOXYPHENYL CARBAMATE.
THR = An exper teratogen. [*3*]

2-(4-CHLORO-6-ETHYLAMINO-*s*-TRIAZIN-2-YL AMINO)-2-METHYL PROPIONITRILE. See bladex.

1-CHLORO-4-ETHYLBENZENE. Clear, colorless liquid. $ClC_6H_4CH_2CH_3$, mw: 140.6, mp: −62.6°, bp: 184.3°, flash p: 147°F, d: 1.05 @ 25°/25°, vap. press: 1 mm @ 19.2°, vap. d: 4.86.
THR = U. See chlorinated hydrocarbons, aromatic and chlorobenzene.
Fire Hazard: Mod, when exposed to heat or flame; reacts with oxidizing materials.
Disaster Hazard: Dangerous; see chlorides.
To Fight Fire: Foam, CO_2, dry chemical.

β-CHLOROETHYL CHLOROFORMATE. Colorless liquid. $COClOCH_2CH_2Cl$, mw: 142.99, bp: 152.5° @ 752 mm, d: 1.3825 @ 20°.
THR = MOD irr via oral and inhal routes.
Disaster Hazard: Dangerous; see chlorides.

β-CHLOROETHYL CHLOROSULFONATE. Chloropicrin-like odor. $ClCH_2CH_2OSO_2Cl$, mw: 179.04, bp: 101° @ 23 mm.
THR = HIGH irr via oral and inhal routes.
Disaster Hazard: Dangerous; see chlorides and sulfonates.

CHLOROETHYLENE. See vinyl chloride.

2-CHLOROETHYLENE IMINE. CHClCNH, mw: 75.6.
THR = An old sample exploded. [*19*]

N-CHLOROETHYLENE IMINE.
THR = Highly unstable. [*19*]

CHLOROETHYLENE POLYMER. Syns: *polyvinyl chloride, bakelite, PVC.* White powder. $(C_2H_3Cl)_n$, mw: 60,000, d: 1.406.
THR = An exper (±) carc and neo. [*3, 1, 14*] Is implicated as a possible teratogen. [*102*] Can cause allergic dermatitis. Reacts violently with F_2. [*19*]

4-CHLORO-6-ETHYLENE IMINO-2-PHENYL PYRIMIDINE. $C_{12}H_{16}N_3Cl$, mw: 231.7.
THR = An exper neo. [*3, 23*]

CHLOROETHYLENE POLYMER. See polyvinyl chloride.

2-CHLOROETHYL ETHER. See dichloroethyl ether.

7-[2-(2-CHLOROETHYL ETHYL AMINO)ETHYL-AMINO]BENZ(c)ACRIDINE DIHYDROCHLORIDE. Syn: *ICR 311.* $C_{23}H_{24}N_3Cl \cdot 2HCl$, mw: 450.9.
THR = An exper carc to mice via iv route. [*103*]

7-[3-(2-CHLOROETHYLETHYL)AMINO]PROPYL-AMINO BENZ(c)ACRIDINE DIHYDROCHLORIDE. Syn: *ICR 292.* $C_{24}H_{26}N_3Cl \cdot 2HCl$, mw: 464.9.
THR = An exper carc to mice via iv route. [*103*]

6-CHLOROETHYL HEXANOATE. $C_8H_{15}O_2Cl$, mw: 178.5.
THR = Limited animal exper suggest MOD toxicity and LOW irr. See also esters.
Disaster Hazard: Dangerous; see chlorides.

1-CHLORO-7-ETHYL-11-METHYL BENZ(c)ACRIDINE. $C_{20}H_{16}NCl$, mw: 305.8.
THR = An exper neo. [*3*]

2-CHLOROETHYL METHYL SULFIDE. See hemisulfur mustard.

N-(2-CHLOROETHYL)N-NITROSO URETHANE.
THR = An exper carc. [*23*]

(2-CHLOROETHYL)TRIMETHYL AMMONIUM CHLORIDE. $C_5H_{13}NCl_2$, mw: 158.1.
THR = MOD to HIGH via oral route to rats. An exper teratogen and carc. [*3, 23*]

2-CHLOROETHYL VINYL ETHER. Liquid. $CH_2ClCH_2OCHCH_2$, mw: 106.55, bp: 109° @ 740 mm, d: 1.0525, flash p: 80°F (OC), mp: −70.3°.
Acute tox data: Oral LD_{50} (rat) = 250 mg/kg; inhal LC (rat) = 250 ppm for 4 hrs. [*3*]
THR = HIGH via oral and inhal routes.
Fire Hazard: Dangerous, when exposed to heat, flame or oxidizers.
Explosion Hazard: See ethers.
Disaster Hazard: Dangerous; see chlorides; can react with oxidizing materials.
To Fight Fire: Alcohol foam, dry chemical.

2-CHLOROETHYL-2-XENYL ETHER. $C_6H_5C_6H_4OCH_2CH_2Cl$, mw: 232.5, flash p: 320°F, d: 1.1, bp: 323°.
THR = U. See also ethers.
Fire Hazard: Slight, when exposed to heat or flame.
Disaster Hazard: Dangerous; see chlorides and ethers.
To Fight Fire: Alcohol foam.

CHLOROETHYNE. See acetylene chloride.

2-CHLOROFLUORENE.
THR = U.
Disaster Hazard: Dangerous; see chlorides.

N-(7-CHLORO-2-FLUORENYL)ACETAMIDE. $C_{15}H_{12}ONCl$, mw: 257.7.
THR = LOW via oral route. An exper carc. [*3, 23*]

3-CHLORO-2-FLUORO-1-PROPENE. $CH_2ClCFCH_2$, mw: 94.5.
THR = U. Limited animal exper suggest HIGH via oral, dermal and inhal routes.
Disaster Hazard: Dangerous; see chlorides and fluorides.

CHLOROFORM. Syn: *trichloromethane*. Colorless liquid, heavy, ethereal odor. $CHCl_3$, mw: 119.39, mp: −63.5°, bp: 61.26°, fp: −63.5°, flash p: none, d: 1.49845 @ 15°, vap. press: 100 mm @ 10.4°, vap. d: 4.12.
Acute tox data: Oral LD_{50} (rat) = 800 mg/kg; inhal LC_{LO} (rat) = 8000 ppm for 4 hrs; inhal LC_{LO} (mouse) = 28 ppm. [*3*]
THR = MOD via oral and inhal routes. A exper (S) neo and carc. [*3, 9, 23*] Chloroform causes irr of the conjunctiva. Upon inhal, it causes dilation of the pupils with reduced reaction to light, as well as reduced intraocular pressure (exper). The material is well known as an anesthetic. In the initial stages there is a feeling of warmth of the face and body, then an irr of the mu mem and skin followed by nervous aberration. Prolonged inhal will bring on paralysis accompanied by cardiac respiratory failure and finally death.

It has been widely used as an anesthetic. However, due to its toxic effects, this use is being abandoned. 68,000–82,000 ppm kill most animals in a few min. 14,000 ppm is dangerous to life after an exposure of from 30 to 60 min. 5,000–6,000 ppm can be tolerated by animals for 1 hr without serious disturbances. The maximum conc tolerated for several hrs or for prolonged exposure with slight symptoms is 2,000–2,500 ppm. The harmful effects are narcosis, and damage to the liver and heart. Prolonged administration as an anesthetic may lead to such serious effects as profound toxemia and damage to the liver, heart and kidneys. Exper prolonged but light anesthesia in dogs produces a typical hepatitis. Inhal of the concentrated chloroform vapor results in irr of the mu surfaces exposed to it. The narcosis is ordinarily preceded by a stage of excitation which is followed by loss of reflexes, sensation, and consciousness. See chlorinated hydrocarbons. Reacts violently with (acetone + a base), Al, disilane, Li, Mg, nitrogen tetroxide, K, (perchloric acid + phosphorus pentoxide), (KOH + methanol), K-*tert*-butoxide, Na, (NaOH + methanol), sodium methylate, NaK. [*19*]
Fire Hazard: Slight, when exposed to high heat; otherwise practically non-flam.
Disaster Hazard: Dangerous; see phosgene.
Treatment and Antidotes: If it has been ingested, or there has been great overexposure, the following antidotes may be applied: emetics, stomach syphon, friction, cold douche, fresh air, strychnine (hypodermically—from 1/120 to 1/60 grain), rubefacients, artificial respiration, etc. If during exposure to unknown amounts of chloroform vapor, the patient should feel any of the symptoms noted above, he should immediately be moved to fresh air and kept under observation until the symptoms disappear.

CHLOROFORMAMIDE. See carbamyl chloride.

CHLOROFORMIC ACID DIMETHYLAMIDE. C_3H_6ONCl, mw: 107.6.
THR = MOD via oral route. An exper carc and neo. [*3*]

CHLOROFORMOXIME. Needles, odor resembling hydrocyanic acid. CClHNOH, mw: 79.5.
THR = HIGH irr via inhal route.
Disaster Hazard: Dangerous, see chlorides.

CHLOROGERMANE. Colorless liquid. GeH_3Cl, mw: 111.08, mp: −52.0°, bp: 28.0°, d: 1.75 @ −25°.
THR = See hydrochloric acid and germanium compounds.

1-CHLOROHEXANE. Syn: *n-hexyl chloride.* Mobile liquid, insol in water. $C_6H_{13}Cl$, mw: 120.62, d: 0.8789 @ 20°/4°, bp: 134°, vap. d: 4.2, flash p: 95°F.
THR = U.
Fire Hazard: Dangerous, when exposed to heat or flame.
Disaster Hazard: Dangerous; see chlorides.
To Fight Fire: Water may be ineffective. Use dry chemical, alcohol foam, mist, fog.

6-CHLOROHEXANOIC ACID. $CH_2Cl(CH_2)_4COOH$, mw: 150.5.
Acute tox data: Oral LD_{50} (rat) = 3080 mg/kg. [*3*]
THR = MOD via oral route.
Disaster Hazard: Dangerous; see chlorides.

CHLOROHYDRIC ACID. See hydrochloric acid.

α-CHLOROHYDRIN. See 1-chloropropane-2,3-diol.

5-CHLORO-4-(HYDROXYAMINO)QUINOLINE-1-OXIDE. $C_9H_7O_2N_2Cl$, mw: 210.6.
THR = An exper neo. [*3*]

6-CHLORO-4-(HYDROXYAMINO)QUINOLINE-1-OXIDE.
THR = An exper neo. [*3*]

7-CHLORO-4-(HYDROXYAMINO)QUINOLINE-1-OXIDE.
THR = An exper neo. [*3*]

2-CHLORO-5-HYDROXY-1,3-DIMETHYL BENZENE. See *p*-chloro-*m*-xylenol.

2-CHLORO-4-(HYDROXYMERCURI)PHENOL. See hydroxymercurichlorophenol.

4-CHLORO-3-HYDROXYTOLUENE. See *p*-chloro-*m*-cresol.

2-CHLORO-N-ISOPROPYL ACETANILIDE. See ramrod.

CHLOROISOPROPYL ALCOHOL. See propylene chlorohydrin.

6-CHLORO-4-ISOPROPYL-1-METHYL-3-PHENOL. See chlorothymol.

β-CHLORO-N-ISOPROPYLNAPHTHALENE ETHYLAMINE HYDROCHLORIDE.
THR = An exper neo. [*3*]

CHLOROMANGANO KALITE. See potassium manganous chloride.

CHLOROMERCURIPHENOL. See *o*-hydroxyphenol mercuric chloride.

CHLOROMETHANE. See methyl chloride.

7-CHLOROMETHYL BENZ(a)ANTHRACENE. Syn: *ICR 451.* $C_{19}H_{13}Cl$, mw: 276.8.
THR = An exper carc to mice via iv route. [*103*]

8-CHLORO-7-METHYL BENZ(a)ANTHRACENE.
THR = An exper carc. [*23*]

10-CHLORO-7-METHYL BENZ(a)ANTHRACENE.
THR = An exper carc. [*23*]

2-CHLORO-1-METHYLBENZENE. See *o*-chlorotoluene.

3-CHLORO-1-METHYLBENZENE. See *m*-chlorotoluene.

4-CHLORO-1-METHYLBENZENE. See *p*-chlorotoluene.

8-CHLORO-7-METHYL BENZ(a)ANTHRACENE. $C_{19}H_{13}Cl$, mw: 276.8.
THR = An exper carc. [*3*]

16-CHLORO-7-METHYL BENZ(a)ANTHRACENE.
THR = An exper carc. [*3*]

6-CHLOROMETHYL BENZO(a)PYRENE. $C_{21}H_{13}Cl$, mw: 300.8.
THR = An exper carc. [*3*]

10-CHLOROMETHYL-9-CHLOROANTHRACENE. Syn: *ICR 486.* $C_{15}H_{10}Cl_2$, mw: 261.2.
THR = An exper carc to mice via iv route. [*103*]

CHLOROMETHYL CHLOROFORMATE. Mobile, colorless liquid, penetrating irr odor. $Cl_2C_2O_2H_2$, mw: 129, bp: 106.5°–107°, d: 1.465 @ 15°.
THR = HIGH irr via oral and inhal routes.
Disaster Hazard: Dangerous; see chlorides.

CHLOROMETHYL-4-CHLOROPHENYL SULFONE. Crystals. $C_7H_6Cl_2SO_2$, mw: 225.1.
THR = An irr. No details. Probably MOD-HIGH toxicity. See also chloromethyl chlorosulfonate.
Disaster Hazard: Dangerous; see chlorides and sulfonates; will react with water or steam to produce toxic and corrosive fumes.

CHLOROMETHYL CHLOROSULFONATE. Colorless liquid. $Cl_2CH_2OSO_2$, mw: 165.01, bp: 49–50° @ 14 mm, d: 1.63.
THR = HIGH irr via oral and inhal routes.
Disaster Hazard: Dangerous; see chlorides and sulfonates; will react with water or steam to produce toxic and corrosive fumes.

3-CHLORO-2-METHYL FURAN. C_5H_5OCl, mw: 116.5.
THR = Very unstable. Violent reaction with (lithium Al hydride + ethyl acetate). [*19*] No toxicity data.

10-CHLOROMETHYL-9-METHYL ANTHRACENE. Syn: *ICR 433.* $C_{16}H_{13}Cl$, mw: 240.7.
THR = An exper carc to mice via iv route. [*103*]

7-(CHLOROMETHYL)-12-METHYL BENZ(a) ANTHRACENE. $C_{20}H_{15}Cl$, mw: 290.1.
THR = An exper neo. [*3, 23*]

CHLOROMETHYL METHYL ETHER. C_2H_5OCl, mw: 80.5.
Acute tox data: Oral LD_{50} (rat) = 817 mg/kg; inhal LC_{50} (rat) = 55 ppm for 7 hrs. [*3*]
THR = MOD via oral and HIGH via inhal routes. A (S) human and exper carc. [*3, 10, 23*]

2-CHLORO-N-METHYL-N-NITROSO ETHYLAMINE. $C_5H_7ON_2Cl$, mw: 146.6.
THR = An exper carc. [*3, 23*]

4-CHLORO METHYLPHENOXY ACETIC ACID. See methoxone.

CHLOROMETHYL SILICANE. Gas. CH_3SiH_2Cl, mw: 80.6, d: 0.935 @ −80°, mp: −134.1°, bp: 7°.
THR = See silanes and chlorides.

CHLOROMETHYL SULFONYL TRICHLORANILIDE.
THR = Probably toxic. Increased basal metabolism, sweating, weight loss reported in humans. See also aniline.
Disaster Hazard: Dangerous. See chlorides, sulfonates and aniline.

2-CHLOROMETHYL THIOPHENE. Crystals. $C_4H_3SCH_2Cl$, mw: 132.6.
THR = U. See also hydrochloric acid, which is liberated by this material upon storage. See also thiophene.
Fire Hazard: Mod, when exposed to heat or flame.
Explosion Hazard: Severe, when shocked, exposed to heat or by spont chemical reaction.
Disaster Hazard: Dangerous; shock will explode it; see chlorides and oxides of sulfur; can react vigorously with oxidizing materials.

3′-CHLORO-2-METHYL-*p*-VALEROTOLUIDIDE. Syns: *solan, N-(3-chloro-4-methyl phenyl)-2-methyl pentanamide*. Solid, insol in water, sol in pine oil, diisobutyl ketone, isophorone, xylene. $H_3CC_6H_3(Cl)NHCOCH(CH_3) \cdot CH_2CH_2CH_3$, mw: 239.5, mp: 86°.
Acute tox data: Oral LD_{50} (rat) => 10,000 mg/kg. [*94*]
THR = LOW via oral route. An herbicide.
Disaster Hazard: Dangerous; see chlorides.

CHLOROMYCETIN. See chloramphenicol.

CHLORONAB. Syn: *demosan, 1,4-dichloro-2,5-dimethoxy benzene*. White crystals, low sol in water. $C_8H_8O_2Cl_2$, mw: 207, mp: 134°.
Acute tox data: Oral LD_{50} (rat) = > 11,000 mg/kg. [*94*]
THR = LOW via oral route.
Disaster Hazard: Dangerous; see chlorides.

1-CHLORONAPHTHALENE. Syn: *α-chloronaphthalene*. Oily liquid, volatile with steam, sol in benzene, petroleum ether, alcohol, insol in water. $C_{10}H_7Cl$, mw: 162.61, flash p: 270°F (OC), autoign. temp.: 1036°F, d: 1.19382 @ 20°/4°, mp: −20°, bp: 259°.
Acute tox data: Oral LD_{50} (rat) = 1540 mg/kg; oral LD_{50} (mouse) = 1091 mg/kg. [*3*]
THR = MOD via oral route. See chlorinated naphthalenes.
Fire Hazard: Slight, when exposed to heat or flame.
Disaster Hazard: Dangerous; see chlorides.
To Fight Fire: Dry chemical, spray or mist, CO_2.

2-CHLORONAPHTHALENE.
Acute tox data: Oral LD_{50} (rat) = 2078 mg/kg. [*3*]
THR = MOD via oral route.

***o*-CHLORO-*p*-NITROANILINE.** Syn: *OCPN*. Yellow crystalline powder. $O_2NC_6H_3(Cl)NH_2$, mw: 172.6, mp: 108.4°.
Acute tox data: ip LD_{LO} (mouse) = 500 mg/kg; iv LD_{LO} (mouse) = 50 mg/kg. [*3*]
THR = HIGH via ip, iv, oral and inhal routes. An allergen. An exper carc. [*14*] See aromatic amines.
Fire Hazard: Slight, when exposed to heat or flame.
Disaster Hazard: Dangerous; when heated to decomp or on contact with acid or acid fumes, emits highly toxic fumes; can react with oxidizing materials.

***p*-CHLORO-*o*-NITROANILINE.** Orange crystalline powder. $ClC_6H_3(NO_2)NH_2$, mw: 172.6, mp: 116.3°.
Acute tox data: iv LD_{50} (mouse) = 63 mg/kg. [*3*]
THR = HIGH via iv, oral and inhal routes. An allergen. An exper carc. [*14*] See aromatic amines. Reacts with nitric acid to form 2 explosive products. [*19*]
Fire Hazard: Slight, when exposed to heat or flame.
Disaster Hazard: Dangerous; when heated to decomp or on contact with acid or acid fumes, emits highly toxic fumes; can react with oxidizing materials.

***p*-CHLORO-*m*-NITROANILINE.**
Acute tox data: Oral LD_{50} (wild birds) = 100 mg/kg. [*3*]
THR = HIGH via oral and inhal routes.

2-CHLORO-4-NITROBENZAMIDE. Syn: "*Aklomide*." $C_7H_5O_3 \cdot N_2Cl$, mw: 201, minimum bp: 170°.
THR = U. A food additive permitted in the feed and drinking water of animals and/or for the treatment of food-producing animals. Also a food additive permitted in food for human consumption. [*109*]
Disaster Hazard: Dangerous; see chlorides and nitrates.

1-CHLORO-3-NITROBENZENE. See *m*-chloronitrobenzene.

4-CHLORO-1-NITROBENZENE. See *p*-chloronitrobenzene.

m-CHLORONITROBENZENE. Syn: *1-chloro-3-nitrobenzene*. Yellowish crystals. $ClC_6H_4NO_2$, mw: 157.6, mp: 46°, bp: 236°, d: 1.534 @ 20°/4°.

Acute tox data: Inhal TC_{LO} (human) = 0.012 mg/m^3 ⟶ eye problems. [*3*]

THR = HIGH via inhal and oral routes. A poison. Intoxication from this material can be serious. When absorbed, it forms methemoglobin and gives rise to cyanosis and blood changes. Its effects are analogous to those of nitrobenzene. It can cause poisoning by the pulmonary route and its effects are cumulative. Chemically, it is probably reduced in the body to chloroaniline, which is also poisonous. The *p* compound is thought to be somewhat less toxic than the *o* compound. In industry it is the dust of this material that is most often the source of intoxication.

Fire Hazard: Slight, when exposed to heat or flame.

Disaster Hazard: Dangerous; see nitrates and phosgene; can react with oxidizing materials.

Treatment and Antidotes: Removal from exposure is important as soon as the symptoms appear. If cyanosis is evident, administer oxygen and call a physician as soon as possible. If breathing has stopped, give artificial respiration.

o-CHLORONITROBENZENE. Syn: *o-nitrochlorobenzene*. Yellow crystals. $C_6H_4ClNO_2$, mw: 157.6, mp 32°–33°, bp: 245°–246°, d: 1.305, flash p: 261°F.

Acute tox data: Oral LD_{50} (rat) = 288 mg/kg; oral LD_{50} (mouse) = 135 mg/kg. [*3*]

THR = HIGH via oral and inhal routes. See also *m*-chloronitro benzene.

To Fight Fire: Water, foam.

p-CHLORONITROBENZENE. See *p*-nitrochlorobenzene.

2-CHLORO-5-NITROBENZO TRIFLUORIDE. Liquid. $C_7H_3F_3ClNO_2$, mw: 225.6, bp: 231.9°, flash p: 275°F, d: 1.504 @ 30°/4°, vap. d: 7.8.

THR = See *m*-chloronitrobenzene and fluorides.

Fire Hazard: Low, when exposed to heat or flame.

Disaster Hazard: Dangerous; see chlorides, nitrates and fluorides; can react with oxidizing materials.

To Fight Fire: CO_2, dry chemical, fog, spray, mist.

4-CHLORO-3-NITROBENZO TRIFLUORIDE. Liquid. $C_7H_3F_3ClNO_2$, mw: 225.6, bp: 222.6°, mp: −2.5°.

THR = See *m*-chloronitro benzene and fluorides.

Disaster Hazard: Dangerous; see chlorides, nitrates, and fluorides.

1-CHLORO-1-NITROETHANE. Liquid. $CH_3CHCl(NO_2)$, mw: 109.6, bp: 129°, flash p: 133°F (OC), d: 1.258 @ 20°/20°, vap. d: 3.77.

Acute tox data: Oral LD_{50} (rat) = 50 mg/kg; sc LD_{50} (mouse) = 185 mg/kg. [*3*]

THR = HIGH via oral and sc routes. High conc can cause pulmonary edema and narcosis.

Fire Hazard: Dangerous, when shocked or exposed to heat or flame. Also when exposed to flames (sparks), and oxidizers.

Explosion Hazard: Dangerous; shock will explode it. See phosgene and oxides of nitrogen; can react vigorously with oxidizing materials.

To Fight Fire: Alcohol foam, water, CO_2, dry chemical.

1-CHLORO-1-NITROPROPANE. Liquid. $CH_3CH_2CH(NO_2)Cl$, mw: 123.5., bp: 139.5°, flash p: 144°F (OC), d: 1.209 @ 20°/20°, vap. d: 4.26.

Acute tox data: Oral LD_{50} (mouse) = 510 mg/kg; sc LD_{50} (mouse) = 165 mg/kg. [*3*]

THR = HIGH via oral, inhal and sc routes. In exper it has produced injury to kidneys, liver and cardiovascular system.

Fire Hazard: Mod, when exposed to heat, flame (sparks) and oxidizers.

Explosion Hazard: Mod, when exposed to heat.

Disaster Hazard: Dangerous; see chlorides; can react with oxidizing materials; explodes when heated.

To Fight Fire: Alcohol foam, water, CO_2, dry chemical.

1-CHLORO-2-NITROPROPANE.

Acute tox data: Oral LD_{50} (rat) = 197 mg/kg; inhal LC_{50} (rat) = 1070 mg/m^3; dermal LD_{50} (rabbit) = 362 mg/kg. [*3*]

THR = HIGH via oral, inhal and dermal routes.

2-CHLORO-2-NITROPROPANE. Liquid. $CH_3CCl(NO_2)CH_3$, mw: 123.5, bp: 134°, flash p: 135°F (OC), d: 1.197 @ 20°/20°, vap. d: 4.26.

Acute tox data: Oral LD_{50} (mouse) = 580 mg/kg; sc LD_{50} (mouse) = 195 mg/kg. [*3*]

THR = HIGH via oral, inhal and sc routes.

Fire Hazard: Mod, via heat, flame and oxidizers.

Explosion Hazard: Explodes on rapid heating.

Disaster Hazard: See nitrates.

3-CHLORO-4-NITROQUINOLINE-1-OXIDE. $C_8H_5O_3N_2Cl$, mw: 212.6.

THR = An exper neo via sc route. [*3, 23*]

5-CHLORO-4-NITRO QUINOLINE-1-OXIDE.

THR = An exper neo via sc route. [*3, 23*]

6-CHLORO-4-NITRO QUINOLINE-1-OXIDE.

THR = An exper carc and neo via dermal and sc routes. [*3, 23*]

7-CHLORO-4-NITRO QUINOLINE-1-OXIDE.

THR = An exper neo via sc route. [*3, 23*]

α-CHLORO-p-NITROSTYRENE. $C_6H_4ClCH{:}CHNO_2$, mw: 183.6.

Acute tox data: Oral LD_{LO} (rat) = 710 mg/kg; dermal LD_{50} (rabbit) = 390 mg/kg. [*3*]
THR = HIGH via dermal and MOD via oral and inhal routes.
Disaster Hazard: Dangerous; see chlorides and nitrates.

2-CHLORO-4-NITROTOLUENE. Liquid. $C_7H_6NO_2Cl$, mw: 171.6
Acute tox data: Oral LD_{50} (rat) = 3020 mg/kg. [*3*]
THR = MOD via oral and inhal routes. Reacts violently with NaOH. [*19*]
Fire Hazard: See nitrates.
Disaster Hazard: See nitrates and chlorides.

2-CHLORO-5-NITROTRIFLUORO METHYL BENZENE. See *m*-chloronitrobenzene and fluorides.

4-CHLORO-2-OXOBENZOTHIAZOLIN-3-YL ACETIC ACID. See benazolin.

CHLOROPENTA FLUOROETHANE. Syns: *monochloropenta fluoroethane*; *fluorocarbon 115*, *propellant 115*, *Refrigerant 115*. Colorless gas, insol in water, sol in alcohol and ether. $CClF_2CF_3$, mw: 155, bp: −39.3°, mp: −77°.
THR = Probably LOW. A food additive permitted in food for human consumption. [*109*] A non-flam gas.
Disaster Hazard: Dangerous; see chlorides and fluorides.

CHLOROPENTAMMINE CHROMIUM (III) CHLORIDE. Red crystals. $(Cr(NH_3)_5Cl)Cl_2$, mw: 243.54, d: 1.696.
THR = See chromium compounds.
Fire Hazard: Slight, when heated.
Disaster Hazard: Dangerous; see chlorides.

CHLOROPENTAMMINE COBALT (III) CHLORIDE. Syn: *purpureo*. Rhombic dark-red violet crystals. $[Co(NH_3)_5Cl]Cl_2$, mw: 250.47, mp: decomp, d: 1.819 @ 25°/25°.
THR = See cobalt compounds.

1-CHLOROPENTANE. See amyl chloride.

β-CHLOROPHENETOLE. Syn: *phenoxy ethyl chloride*. $C_6H_5OCH_2CH_2Cl$, mw: 133, flash p: 225°F, d: 1.1, bp: 155°.
THR = U.
Fire Hazard: Low, when exposed to heat or flame.
Disaster Hazard: Dangerous; see chlorides.
To Fight Fire: Alcohol foam.

***m*-CHLOROPHENOL.** Crystals. C_6H_4ClOH, mw: 128.6, mp: 32.5°, bp: 214°, d: 1.245, vap. press: 1 mm @ 44.2°.
Acute tox data: Oral LD_{50} (rat) = 570 mg/kg; ip LD_{50} (rat) = 355 mg/kg; sc LD_{50} (rat) = 1390 mg/kg. [*3*]
THR = HIGH via ip and MOD via sc, oral and inhal routes.
Disaster Hazard: Dangerous; when heated to decomp, emits highly toxic fumes.

***o*-CHLOROPHENOL.** Light amber liquid. C_6H_4OHCl, mw: 128.6, bp: 174.5°, fp: 7°, d: 1.256 @ 25°/25°, flash p: 147°F, vap. press: 1 mm @ 12.1°.
Acute tox data: Oral LD_{50} (rat) = 670 mg/kg; ip LD_{50} (rat) = 230 mg/kg; sc LD_{50} (rat) = 950 mg/kg. [*3*]
THR = HIGH via ip; MOD via oral, inhal and sc routes. An exper neo. [*3*]
Fire Hazard: Mod, when exposed to heat, flames or oxidizers.
Disaster Hazard: Dangerous; see phenol and chlorides. Can react with oxidizing materials.
To Fight Fire: Alcohol foam.

***p*-CHLOROPHENOL.** Needle-like, white to straw colored crystals, unpleasant odor. C_6H_4ClOH, mw: 128.6, fp: 42.8°, flash p: 250°F, d: 1.246 @ 60°/25°, vap. press: 1 mm @ 49.8°.
Acute tox data: Oral LD_{50} (rat) = 500 mg/kg; ip LD_{50} (rat) = 281 mg/kg; sc LD_{50} (rat) = 1030 mg/kg. [*3*]
THR = HIGH via oral and ip; MOD via sc routes.
Fire Hazard: Low, when exposed to heat or flame.
Disaster Hazard: Dangerous; see chlorides and phenol.
To Fight Fire: Water, spray, mist, fog, foam, dry chemical.

CHLOROPHENOTHANE. See DDT.

4-[3-(2-CHLOROPHENOTHIAZIN-10-YL)PROPYL-1-PIPERAZINE]ETHANOL MONOHYDROCHLORIDE. $C_{21}H_{26}ON_3SCl$, mw: 440.5.
THR = An exper teratogen. [3]

CHLOROPHENOXY ACETIC ACID. $C_8H_7O_3Cl$, mw: 186.6.
Acute tox data: Oral LD_{50} (rat) = 850 mg/kg. [*3*]
THR = MOD via oral and inhal routes.

2-(4-CHLOROPHENOXY)ETHANOL. $C_6H_4Cl(O)CH_2CH_2OH$, mw: 172.6.
Acute tox data: Oral LD_{LO} (rat) = 2830 mg/kg; dermal LD_{50} (rabbit) = 500 mg/kg. [*3*]
THR = HIGH via dermal and MOD via oral and inhal routes.
Disaster Hazard: Dangerous; see chlorides.

***p*-CHLOROPHENOXY ETHOXY ETHYL CHLORIDE.**
THR = See chlorinated hydrocarbons, aromatic.
Disaster Hazard: Dangerous; see chlorides.

1-(*o*-CHLOROPHENOXY)-2-PROPANOL. See propylene glycolchlorophenyl ether.

3-(α,*p*-CHLOROPHENYL-β-ACETYLETHYL)-4-HYDROXY COUMARIN. See tomorin.

2-CHLORO-4-PHENYL ANILINE.
THR = An exper carc. [*23*]

3-CHLOROPHENYL-N-CARBAMOYL AZIRIDINE. $C_9H_9ON_2Cl$, mw: 196.7.
THR = An exper neo. [*3*]

***p*-CHLOROPHENYL-*p*-CHLOROBENZENE SULFONATE.** See murvesco.

***p*-CHLOROPHENYL-*p*-CHLOROBENZYL SULFIDE.** $C_{13}H_{10}Cl_2S$, mw: 269.2.
Acute tox data: Oral LD_{50} (rat) = 2000 mg/kg. [*3*]
THR = MOD irr via oral and inhal routes. Exper caused liver and kidney damage.
Disaster Hazard: Dangerous; see chlorides and sulfates.

1-(4-CHLOROPHENYL)-3,3-DIMETHYL TRIAZENE.
THR = An exper carc. [*23*]

3-(*p*-CHLOROPHENYL)-1,1-DIMETHYL UREA. See monuron.

3-(*p*-CHLOROPHENYL)-1,1-DIMETHYL UREA TRICHLOROACETATE. See urox.

***o*-CHLOROPHENYL DIPHENYL PHOSPHATE.** Clear, pale straw-colored mobile liquid. $C_{18}H_{14}O_4ClP$, mw: 360.73, mp: <0°, bp: 240°–255° @ 5 mm, flash p: >419°F, d: 1.3 @ 25°/25°, vap. d: 12.5.
THR = U.
Fire Hazard: Slight, when exposed to heat or flame.
Disaster Hazard: Dangerous; see chlorides and phosphates; can react with oxidizing materials.
To Fight Fire: Water, foam, CO_2, dry chemical.

4-(2-CHLOROPHENYL HYDRAZONO)-3-METHYL-5-ISOXAZOLONE. See drazoxolon.

***o*-CHLOROPHENYL ISOCYANATE.** Colorless liquid, sol in organic solvents. C_7H_4ClNO, mw: 153.6, bp: 106° @ 30 mm.
Acute tox data: Oral LD_{50} (rat) = 4710 mg/kg; ocular TC_{LO} (human) = 1 ppm for 1 min. [*3*]
THR = HIGH irr via inhal, eyes and MOD via oral routes. A powerful irr to skin and mu mem. A very strong lachrymator.
Disaster Hazard: Dangerous; see chlorides and cyanates.

***m*-CHLOROPHENYL ISOCYANATE.** Water-white liquid, sol in organic solvents. C_7H_4ClNO, mw: 153.6, mp: −4°, bp: 101° @ 30 mm, flash p: 215°F (COC).
Acute tox data: Inhal LC_{LO} (rat) = 60 mg/m^3 for 4 hrs; inhal LC_{50} (mouse) = 69 mg/m^3 for 2 hrs. [*3*]
THR = HIGH via inhal and oral routes.
Disaster Hazard: Dangerous; see chlorides and cyanates.

***p*-CHLOROPHENYL ISOCYANATE.** White solid, sol in organic solvents. C_7H_4ClNO, mw: 153.6, mp: 28°, bp: 106.5° @ 30 mm, flash p: 230°F.
Acute tox data: Oral LD_{50} (rat) = 4710 mg/kg; oral LD_{50} (mouse) = 530 mg/kg; ocular TC_{LO} (human) = 1 ppm for 1 min. [*3*]
THR = HIGH via ocular and MOD via oral routes. Can explode on distillation. [*19*]
Disaster Hazard: Dangerous; see chlorides and cyanates.

1-(*p*-CHLOROPHENYL)-5-ISOPROPYLBIGUANIDE HYDROCHLORIDE. See chlorguanide.

CHLORO-2-PHENYL PHENOL. See 2-chloro-4-phenyl phenol.

2-CHLORO-4-PHENYL PHENOL. Syn: *dowicide 4*. White flakes. $C_{12}H_9OCl$, mw: 204.7, bp: 322° (decomp), fp: 74.2°, flash p: 345°F, d: <1, mp: 80°.
Acute tox data: Oral LD_{50} (rat) = 4220 mg/kg. [*3*]
THR = MOD via oral and inhal routes.
Fire Hazard: Slight, when exposed to heat or flame.
Disaster Hazard: Dangerous; see chlorides.
To Fight Fire: Alcohol foam, CO_2, dry chemical.

6-CHLORO-2-PHENYLPHENOL SODIUM SALT. Clear, colorless to straw-colored viscous liquid. $C_{12}H_8OCl \cdot Na$, mw: 226.6, bp: 162°–178° @ 10 mm, d: 1.234 @ 25°/25°, vap. d: 7.07.
Acute tox data: Oral LD_{50} (rat) = 3500 mg/kg. [*3*]
THR = MOD via oral route.
Fire Hazard: Mod, when exposed to heat or flame.
Disaster Hazard: Dangerous; see chlorides.
To Fight Fire: Water, foam, CO_2, dry chemical.

1-(4-CHLOROPHENYL)-1-PHENYL-2-PROPYNYL CARBAMATE. $C_{16}H_{12}O_2NCl$, mw: 285.7.
THR = An exper carc via oral route. [*3*, *12*]

***p*-CHLOROPHENYL PHENYL SULFONE.** See sulphenone.

CHLOROPHENYL TRICHLOROSILANE. Colorless to pale yellow liquid, readily hydrolyzed by moisture, with the liberation of hydrochloric acid. $ClC_6H_4SiCl_3$ (a mixture of 3 isomers), mw: 246, bp: 230°, d: 1.439 @ 25°/25°, flash p: 255°F (COC).
THR = HIGH irr via inhal and oral routes.
Fire Hazard: Slight, when exposed to heat or flame.
Disaster Hazard: Dangerous; see chlorides.

CHLOROPICRIN. Syns: *nitrotrichloro methane*, *trichloronitromethane*, *nitrochloroform*. Slightly oily, colorless liquid. CCl_3NO_2, mw: 164.39, mp: −64°, bp: 112°, d: 1.692, vap. press: 40 mm @ 33.8°, vap. d: 5.69.
Acute tox data: Oral LD_{50} (rat) = 250 mg/kg. An eye irr at 5 mg/m^3. [*3*]
THR = HIGH via oral, inhal and ocular routes. An insecticide. Reacts violently with propargyl bromide. [*19*] Above a critical volume it can be shock detonated. Chloropicrin is a powerful irr and affects all body surfaces. It causes lachrymation, vomiting,

bronchitis, and pulmonary edema. It is not only a lachrymator but also is irr to skin, gastro-intestinal, and respiratory tracts. An additional toxic effect is its reaction with SH-groups in hemoglobin thus interfering with oxygen transport. Photochemical transformation of chloropicrin into phosgene (carboxy chloride, $COCl_2$) has been reported. A conc of 1 ppm causes a smarting pain in the eyes and therefore in itself constitutes a good warning of exposure. It causes vomiting, probably due to swallowing saliva in which small amounts of chloropicrin have dissolved. It is called vomiting gas and has been extensively used by the military. Its primary lethal effect is to produce lung injury and it is a difficult gas to protect oneself against because it is chemically inert and does not react with the usual chemical used in gas masks. Four ppm is sufficient to render a man unfit for action and 20 ppm, when breathed from 1 to 2 minutes, causes definite bronchial or pulmonary lesions. Industrially it is used as a warning agent in commercial fumigants. It is more toxic than chlorine but less so than phosgene.

Disaster Hazard: Dangerous; when heated to decomposition, emits highly toxic fumes.

Treatment and Antidotes: Removal from exposure is an immediate necessity. If exposure has been severe, consult a physician.

6-CHLOROPIPERONYL CHRYSANTHEMUMATE. See barthrin.

CHLOROPLATINIC ACID. Red-brown, deliquescent prisms. $H_2PtCl_6 \cdot 6H_2O$, mw: 518.08, mp: 60°, d: 2.431.

THR = See platinum compounds and chlorides.

CHLOROPRENE. Syns: *2-chloro-1,3-butadiene, chlorobutadiene.* Colorless liquid. $CH_2CHCClCH_2$, mw: 88.54, bp: 59.4°, d: 0.9583, flash p: −4°F, lel = 4.0%, uel = 20.0%, vap. d: 3.0.

Acute tox data: Oral LD_{LO} (rat) = 1600 mg/kg; sc LD_{LO} (rat) = 500 mg/kg; inhal LD_{LO} (mouse) = 600 mg/m^3 for 8 hrs. [*3*]

THR = MOD via oral, sc and inhal routes. An exper mutagen. [*3*] Animal exper have shown that a conc of 250 ppm in air is toxic, and a conc of 75 ppm may be toxic with continued exposure. Exposure to the vapor first causes irritation of the respiratory tract, followed by depression of respiration and, if exposure is continued, asphyxia. The vapor is a CNS depressant; in animals it causes severe degenerative changes in the vital organs, particularly the liver and kidneys. Blood pressure is lowered. Lung changes accompany exposure to the higher conc. Humans exposed to chloroprene have been reported to develop dermatitis, conjunctivitis, corneal necrosis, anemia, temporary loss of hair, nervousness and irritability.

Disaster Hazard: Dangerous; see chlorides.

Fire Hazard: Dangerous, when exposed to heat or flame.

To Fight Fire: Alcohol foam.

1-CHLOROPROPANE. See *n*-propyl chloride.

2-CHLOROPROPANE. See isopropyl chloride.

1-CHLOROPROPANE-2,3-DIOL. Syn: *α-chlorohydrin.* Colorless liquid. $CH_2ClCHOHCH_2OH$, mw: 110.54, bp: 213° decomp, d: 1.326.

Acute tox data: Oral LD_{50} (rat) = 150 mg/kg; inhal LC_{LO} (rat) = 125 ppm for 4 hrs. [*3*]

THR = HIGH via oral and inhal routes.

Fire Hazard: Slight, when exposed to heat or flame.

1-CHLORO-2-PROPANOL. See *sec*-propylene chlorohydrin-primary.

1-CHLORO-2-PROPANONE. Syns: *chloroacetone, acetonyl chloride.* C_3H_5OCl, mw: 92.5.

Acute tox data: Inhal LC_{LO} (human) = 605 ppm for 1/6 hr; oral LD_{LO} (rat) = 50 mg/kg; dermal LD_{LO} (rat) = 100 mg/kg. [*103*]

THR = HIGH via inhal, oral and dermal routes. A powerful irr. An exper neo to mice via dermal route. [*103*]

1-CHLORO-1-PROPENE. Syn: *1-chloropropylene.* C_3H_5Cl, mw: 76.5, flash p: <21°F, lel = 4.5%, uel = 16%, d: 0.9.

Acute tox data: Oral LD_{50} (rat) = 1950 mg/kg; inhal LC_{LO} (mice) = 230,000 mg/m^3 for 10 min. [*3*]

THR = MOD via oral; LOW via inhal routes.

Fire Hazard: Very dangerous, via heat, flames (sparks) or oxidizers.

To Fight Fire: Alcohol foam, dry chemical, mist, spray, fog.

2-CHLOROPROPENE. Syn: *isopropenyl chloride.* Colorless liq. $CH_3CCl:CH_2$, mw: 76.53, bp: 22.65°, fp: −137.4°, d: 0.918 @ 9°, flash p: −4°.

Acute tox data: Oral LD_{LO} (mouse) = 63 mg/kg; inhal LC_{LO} (mouse) = 198 ppm for 6 hrs. [*3*]

THR = HIGH via oral and inhal routes. A (S) carc. [*102*]

Fire Hazard: Very dangerous, via heat, flame, sparks or powerful oxidizers.

Disaster Hazard: Dangerous; reacts with powerful oxidizers; see chlorides.

To Fight Fire: Water, spray, mist, fog, dry chemical, alcohol foam.

3-CHLOROPROPENE. See allyl chloride.

α-CHLOROPROPIONIC ACID. Syn: *2-chloropropionic acid.* Sol in water. $CH_3CHClCOOH$, mw:

108.5, d: 1.260–1.268 @ 20°, bp: 183–187°, flash p: 225°F.
Acute tox data: Oral LD_{50} (mouse) = 980 mg/kg. [*3*]
THR = MOD via oral route.
Fire Hazard: Mod, when exposed to heat or flame.
Disaster Hazard: Dangerous. See chlorides.
To Fight Fire: Water, foam, alcohol foam.

2-CHLOROPROPIONIC ACID. See α-chloropropionic acid.

4-CHLOROPROPIONIC ACID.
Acute tox data: Dermal LD_{LO} (mouse) = 1040 mg/kg. [*3*]
THR = MOD via dermal route. An exper carc. [*23*]

β-CHLOROPROPIONITRILE. Colorless liquid. $ClCH_2CH_2CN$, mw: 89.5, mp: −51°, bp: 176° decomp, flash p: 168°F (CC), d: 1.1363 @ 25°, vap. press: 6 mm @ 50°, vap. d: 3.09.
Acute tox data: Oral LD_{50} (rat) = 50 mg/kg. [*3*]
THR = HIGH via oral and inhal routes.
Fire Hazard: Mod, in its liquid form when exposed to heat or flame.
To Fight Fire: Alcohol foam, water, foam, CO_2, or dry chemical.

β-CHLOROPROPYL ALCOHOL. See propylene chlorohydrin-primary.

3-CHLOROPROPYLENE-1,2-OXIDE. See epichlorohydrin.

CHLOROPROPYLENE SULFIDE.
$S \cdot CH_2CHCH_2Cl$, mw: 108.5.
Acute tox data: Oral LD_{50} (rat) = 56 mg/kg; inhal LC_{50} (rat) = 750 ppm for 30 min. [*3*]
THR = HIGH via oral and inhal routes.
Disaster Hazard: Dangerous; see chlorides and sulfur compounds.

γ-CHLOROPROPYLTRICHLORO SILANE. Liquid. $ClC_3H_6SiCl_3$, mw: 212, bp: 180°, d: 1.336 @ 25°.
THR = See hydrochloric acid and chlorosilanes.
Disaster Hazard: Dangerous; see chlorosilanes.

3-CHLORO-1-PROPYNE. See propargyl chloride.

6-CHLORO PURINE. $C_5H_3N_4Cl$, mw: 154.6.
THR = An exper teratogen. [*3*]

2-CHLOROPYRIDINE. Colorless oily liquid.
C_5H_4NCl, mw: 113.6, bp: 170°, d: 1.205 @ 25°, vap. press: 1 mm @ 13.3°, vap. d: 3.93.
Acute tox data: ip LD_{50} (mouse) = 130 mg/kg; dermal LD_{50} (rabbit) = 64 mg/kg; inhal LC (rat) = 250 ppm for 7 hrs. [*3*]
THR = HIGH via ip, dermal and inhal routes.
Fire Hazard: Slight, when exposed to heat or flame.
Disaster Hazard: Dangerous; see phosgene; can react with oxidizing materials.

CHLOROQUINE MUSTARD. $C_{18}H_{24}N_3Cl_3 \cdot 2HCl$, mw: 461.7
THR = HIGH. An exper carc. [*3, 23*]

CHLOROSILANES. Compounds of Si, Cl and H where the total number of atoms of Cl and H add up to 4. SiH_xCl_{4-x}.
THR = HIGH irr to skin, eyes and mu mem, and via oral and inhal routes. Toxicity based on hydrochloric acid, which is formed upon hydrolysis of a chlorosilane. Self-ignites in air; with a little ammonia it forms a self-igniting product. [*19*]
Disaster Hazard: Dangerous; when heated to decomp, they emit highly toxic fumes of chlorides; they will react with water or steam to produce heat and toxic and corrosive fumes of hydrochloric acid.

CHLOROSOL. See ethylene dichloride, carbon tetrachloride.

CHLOROSTANNIC ACID. Colorless leaf-like crystals or liquid. $H_2SnCl_6 \cdot 6H_2O$, mw: 441.6, mp: 9°, d: 1.93.
THR = See tin compounds and hydrochloric acid.
Disaster Hazard: Dangerous; see chlorides; will react with water or steam to produce heat and toxic and corrosive fumes of hydrochloric acid.

2′-CHLORO-4-STILBENYL-N,N-DIMETHYL AMINE.
THR = An exper carc. [*23*]

3′-CHLORO-4-STILBENYL-N,N-DIMETHYL AMINE.
THR = An exper carc. [*23*]

4′-CHLORO-4-STILBENYL-N,N-DIMETHYL AMINE.
THR = An exper carc. [*23*]

n-CHLOROSUCCINIMIDE. Syn: *succinchlorimide.* White powder, mild odor of chlorine. $C_4H_4O_2NCl$, mw: 133.5, mp: 148°–149°.
Acute tox data: Oral LD_{LO} (rat) = 2700 mg/kg. [*3*]
THR = MOD via oral and inhal routes.
Disaster Hazard: See hypochlorites.

CHLOROSULFACIDE. See *p*-chlorobenzyl-*p*-chlorophenyl sulfide.

CHLOROSULFONIC ACID. Syn: *sulfuric chlorohydrin.* Clear to cloudy, colorless to pale yellow liquid, sharp odor. $ClSO_3H$. mw: 116.53, mp: −80°, bp: 151.0°, d: 1.766 @ 18°, vap. press: 1mm @ 32°, vap. d: 4.02.
THR = See sulfuric acid. HIGH via irr. Chlorosulfonic acid can cause severe acid burns and is very irr to the eyes, lungs and mu mem. It can cause acute toxic effects either in the liquid or vapor state. Inhal of conc vapor may cause loss of consciousness with serious damage to lung tissue. Contact of liquid with

the eyes can cause severe burns if not immediately and completely removed. It also causes severe skin burns due to its highly corrosive action. Upon ingestion, it will irr the mouth, esophagus and stomach to a serious degree and on contact with skin cause dermatitis. Even in the vapor form it may cause conjunctivitis.

Caution: Vigorous to violent reactions with acetic acid, acetic anhydride, acetonitrile, acrolein, acrylic acid, acrylonitrile, allyl alcohol, allyl chloride, 2-amino ethanol, ammonium hydroxide, aniline, *n*-butyraldehyde, creosote oil, cresol, cumene, dichloroethyl ether, diethylene glycol monomethyl ether, diisobutylene, diisopropylether, epichloro hydrin, ethyl acetate, ethyl acrylate, ethylene chlorohydrin, ethylene cyanohydrin, ethylene diamine, ethylene glycol, ethylene glycol monoethyl ether acetate, ethylene imine, glyoxal, HCl, HF, H_2O_2, isoprene, mesityl oxide, metal powders, methyl ethyl ketone, HNO_3, 2-nitropropane, P, β-propiolactone, propylene oxide, pyridine, NaOH, H_2SO_4, sulfolane, styrene monomer, vinyl acetate, vinylidene chloride, water, organic matter. [*19*]

Disaster Hazard: Dangerous. See sulfuric acid and hydrochloric acid and sulfonates. Decomposes explosively on contact with water.

CHLOROSULFONIC ACID-SULFUR TRIOXIDE MIXTURE. See chlorosulfonic acid and sulfur trioxide. Clear colorless mobile liquid.

CHLOROSULFONYL ISOCYANATE. Syn: *CSI.* CO_3NSCl, mw: 141.7, fp: −43°, bp: 107°.

THR = Reacts violently with water. A very strong irr.

CHLOROSULFURIC ACID. See sulfuryl chloride.

2-CHLORO-2-TETRACHLOROPHENOXY DIETHYL ETHER. Colorless, odorless liquid. $Cl(C_2H_4)_2O_2Cl_4$, mw: 338.4, mp: 31°, bp: 170°–176° @ 1 mm, flash p: 392°F, d: 1.506, vap. d: 11.6.

THR = U.

Fire Hazard: Slight, when exposed to heat or flame.

To Fight Fire: Water, foam, CO_2, dry chemical.

CHLOROTETRACYCLINE. See chlortetracycline.

CHLOROTETRAFLUOROETHANE. See monochloro tetrafluoroethane.

CHLOROTHIAZIDE. Syn: *diril.* $C_7H_6O_4N_3S_2Cl$, mw: 295.7.

Acute tox data: ip LD_{50} (rat) = 1386 mg/kg. [*3*]

THR = MOD via ip route. Has been implicated in aplastic anemia.

CHLOROTHION. $(CH_3O)_2PSOC_6H_3ClNO_2$, mw: 297.6.

THR = Very unstable to heating. [*19*]

CHLOROTHYMOL. Syn: *6-chloro-4-isopropyl-1-methyl-3-phenol.* Crystals. $C_{10}H_{13}ClO$, mw: 184.7, mp: 62°–64°.

THR = HIGH irr via oral and inhal routes.

Fire Hazard: Slight, when exposed to heat or flame.

Disaster Hazard: Dangerous; see chlorides.

(4-CHLORO-*o*-TOLOXY)ACETIC ACID. See 2,4-dichloromethyl phenoxy acetic acid.

α-CHLOROTOLUENE. See benzyl chloride.

***m*-CHLOROTOLUENE.** Syn: *3-chloro-1-methyl benzene.* Liquid. C_7H_7Cl. mw: 126.6, mp: −48°, bp: 162.3°, d: 1.0797 @ 13.90°/4°, vap. press: 10 mm @ 43.2°, flash p: 126°F (OC).

THR = U. Narcotic in high conc. See also chlorinated hydrocarbons, aromatic and α-chlorotoluene.

Fire Hazard: Mod, when exposed to heat, flame or oxidizers.

***o*-CHLOROTOLUENE.** Syn: *2-chloro-1-methyl benzene.* Liquid. C_7H_7Cl, mw: 126.6, mp: −35°, bp: 159.3°, d: 1.1018 @ 0°/4°, vap. press: 10 mm @ 43.2°, flash p: 126°F (OC).

THR = HIGH via inhal and probably oral routes. No data. See α-chlorotoluene.

Fire Hazard: Mod, when exposed to heat, flame or oxidizers. Violent reaction with dimethyl sulfoxide. [*19*]

***p*-CHLOROTOLUENE.** Syn: *4-chloro-1-methyl benzene.* Liquid. C_7H_7Cl, mw: 126.6, mp: 7.3°, bp: 162.3°, d: 1.0695 @ 24.4°/4°, vap. press: 10 mm @ 43.8°, flash p: 126°F (OC).

THR = See also benzyl chloride.

Fire Hazard: Mod, when exposed to heat, flame or oxidizers.

To Fight Fire: Water, mist, fog, alcohol foam, dry chemical.

2-CHLORO-*p*-TOLUIDINE. Grayish-white crystals. C_7H_8NCl, mw: 141.6, mp: 243°, bp: 257° (sublimes).

Acute tox data: Oral LD_{50} (rat) = 367 mg/kg. [*3*]

THR = HIGH via oral and inhal routes. May cause cyanosis, tachycardia, hematuria, albuminuria.

Disaster Hazard: Dangerous; see chlorides.

5-CHLORO-*o*-TOLUIDINE. Solid; bp: 241°, mp: 29°.

Acute tox data: Oral LD_{50} (rat) = 464 mg/kg. [*3*]

THR = HIGH via oral and inhal routes. An exper carc. [*23*]

Disaster Hazard: Dangerous; see chlorides.

6-CHLORO-*o*-TOLUIDINE.

Acute tox data: sc LD_{LO} (cat) = 200 mg/kg. [*3*]

THR = HIGH via sc and oral routes.

4-CHLORO-*o*-TOLUIDINE DIAZONIUM CHLORIDE. $H_2NC_6H_3ClCH_2N_2Cl$, mw: 204.1.

THR = Reacts violently with sodium sulfide, sodium polysulfide, sodium bisulfide. [*19*]

***p*-[(3-CHLORO-*m*-TOLYL)AZO]-N,N-DIMETHYL ANILINE.** $C_{14}H_{16}N_3Cl$, mw: 261.8.
THR = An exper carc. [*3*]

***p*-[(4-CHLORO-*m*-TOLYL)AZO]-N,N-DIMETHYL ANILINE.** $C_{14}H_{16}N_3Cl$, mw: 261.8.
THR = An exper carc. [*3*]

2-CHLORO-1,3,3-TRIETHOXY PROPANE. Liquid. $C_9H_{19}O_3Cl$, mw: 210.70.
Acute tox data: Oral LD_{50} (rat) = 1320 mg/kg; inhal LC (rat) = 2000 ppm for 4 hrs. [*3*]
THR = MOD via oral and inhal routes.
Fire Hazard: Slight, when exposed to heat or flame.

CHLOROTRIFLUORO ETHYLENE. A gas. $ClFCCF_2$, mw: 116.46, lel = 24%, uel = 40.3%, flash P: −18°F.
Acute tox data: Inhal LC_{50} (rat) = 1000 ppm for 4 hrs; inhal LC_{50} (mouse) = 1000 ppm for 24 hrs. [*3*]
THR = MOD via inhal route. Violent reaction when mixed with ($Br_2 + O_2$) or (ClF_3 + water). [*19*]
Fire Hazard: Very dangerous, via heat, flames (sparks) or oxidizers.
Disaster Hazard: Dangerous; see chlorides and fluorides.
To Fight Fire: Stop flow of gas.

CHLORO TRIFLUORO ETHYLENE PEROXIDE. $F_2CClCOF$, mw: 132.5.
THR = A heat-sensitive explosive. [*19*] No data. Probably toxic. See also peroxides.

CHLOROTRIFLUORO GERMANE. Colorless gas. GeF_3Cl, mw: 165.06, mp: −66.2°, bp: −20.3°.
THR = See fluorides, germanium compounds and chlorides.

CHLOROTRIFLUORO METHANE. See monochloro trifluoromethane.

***m*-CHLOROTRIFLUORO METHYL BENZENE.** See *m*-chlorobenzotrifluoride.

***o*-CHLOROTRIFLUORO METHYL BENZENE.** See *o*-chlorobenzotrifluoride.

***p*-CHLOROTRIFLUORO METHYL BENZENE.** See *p*-chlorobenzotrifluoride.

2-CHLORO-1,1,2-TRIFLUORO-3-METHYL-3-VINYL CYCLOBUTANE. Liquid. $C_7H_8F_3Cl$, mw: 184.5.
THR = U.
Disaster Hazard: Dangerous; when heated to decomp, emits highly toxic halide fumes.

***o*-CHLORO-2,α-TRIFLUORO TOLUENE.** See *o*-chlorobenzotrifluoride.

2-CHLORO-1,1,2-TRIFLUORO-3-VINYL CYCLOBUTANE. Liquid. $C_6H_6F_3Cl$, mw: 170.5.
THR = U.
Disaster Hazard: Dangerous; see chlorides and fluorides.

2-CHLORO-1,3,5-TRINITROBENZENE. See picryl chloride.

CHLORO TRIS(*p*-METHOXY PHENYL)ETHYLENE. $C_{23}H_{21}O_3Cl$, mw: 380.9.
THR = An exper neo. [*3*]

β-CHLOROVINYLDICHLOROARSINE. See dichloro-(2-chlorovinyl) arsine.

β-CHLOROVINYL METHYL CHLOROARSINE. Liquid. $CH_3AsClCHCHCl$, mw: 187.0, bp: 112°–115°.
THR = HIGH via oral, inhal and dermal routes.
Disaster Hazard: Dangerous; see arsenic compounds.

6-CHLORO-*o*-XENOL. See 2-phenyl-6-chlorophenol.

***p*-CHLORO-*m*-XYLENOL.** Syn: *2-chloro-5-hydroxy-1,3-dimethyl benzene*. Crystals. Phenolic odor, slightly water sol. C_8H_9OCl, mw: 156.6, mp: 115.5°, bp: 246°.
THR = MOD irr via oral and dermal routes. HIGH toxicity. See phenol.
Disaster Hazard: Dangerous; see chlorides.

CHLORPROMAZINE. Syn: *thorazine*. $C_{17}H_{19}N_2SCl$, mw: 318.9.
THR = HIGH via oral route to humans. It affects the blood picture (has been implicated in aplastic anemia) and affects the CNS. [*3*]

CHLORPROPAMIDE. Syn: *diabenese*. $C_{10}H_{13}O_3N_2SCl$, mw: 276.8.
Acute tox data: ipLD_{50} (rat) = 580 mg/kg. [*3*]
THR = MOD via ip route. Has been implicated in aplastic anemia.

CHLORTETRACYCLINE. Syns: *CTC, chlorotetracycline*. Golden yellow crystals, slightly sol in water, very sol in aqueous solution pH7.65, freely sol in the "cellosolves," dioxane, "Carbitol," sol in methanol, ethanol, butanol, acetone, ethyl acetate, and benzene, insol in ether and petroleum ether. $C_{22}H_{23}ClN_2O_8$, mw: 479, mp: 168°–169°.
Acute tox data: Oral LD_{50} (rat) = 807 mg/kg. [*3*]
THR = MOD via oral and inhal routes. A food additive permitted in the feed and drinking water of animals and/or for the treatment of food-producing animals. Also a food additive permitted in food for human consumption. [*109*]
Disaster Hazard: Dangerous; see chlorides.

CHLORTHION. Syn: *compound 22/190*. $C_8H_9O_5NPSCl$, mw: 297.7.
Acute tox data: Oral LD_{50} (rat) = 800 mg/kg; dermal LD_{50} (rat) = 1500 mg/kg. [*3*]

THR = MOD via oral and dermal routes. See parathion. This is a toxic organic phosphate cholinesterase inhibitor.
Disaster Hazard: Dangerous; when heated to decomp, emits highly toxic fumes.

CHOLANTHRENE. $C_{20}H_{14}$, mw: 254.3 An exper carc. [*3*, *23*] An exper carc to mice via dermal and sc routes. [*103*]

CHOLCHAMINE. $C_{21}H_{25}O_5N$, mw: 371.5.
An exper teratogen and neo. [*3*]

CHOLECALCIFEROL. See vitamin D_3.

CHOLEIC ACID. $C_{24}H_{40}O_4$, mw: 392.8.
THR = An exper neo. [*3*]

5,7-CHOLESTADUN-3-β-OL. See vitamin D_3.

5-α-CHOLEST-7-EN-3-β-OL. $C_{27}H_{47}O$, mw: 387.7.
THR = An exper neo. [*3*]

CHOLEST-5-EN-3-OL-14-METHYL HEXADECANOATE. $C_{44}H_{78}O_2$, mw: 638.7.
THR = An exper carc to mice via ip route. [*103*]

CHOLESTENONE. $C_{27}H_{44}O$, mw: 384.7.
THR = An exper carc. [*3*]

CHOLESTEROL. White or faint yellow, pearly leaflets. $C_{27}H_{46}O$, mw: 386.7, mp: 148.5°, bp: 360° decomp.
THR = An exper (±) neo and carc. [*3*, *7*]

CHOLESTEROL-5-HYDROPEROXIDE. $C_{27}H_{46}O_3$, mw: 418.7.
THR = An exper neo. [*3*]

CHOLESTERYL-*p*-BIS(2-CHLOROETHYL) AMINO PHENYL ACETATE.
THR = An exper neo. [*3*]

CHOLESTEROL ISOHEPTYLATE. $C_{34}H_{58}O_2$, mw: 498.9.
THR = An exper neo. [*3*]

CHOLESTEROL-α-OXIDE. $C_{27}H_{46}O_2$, mw: 402.7.
THR = An exper neo to mice via sc route. [*3*, *103*]

CHOLIC ACID, HYDRATE. The most abundant bile acid, the monohydrate crystallizes in plates from dilute acetic acid, sol in glacial acetic acid, acetone, and alcohol. Slightly sol in chloroform, practically insol in water and benzene. $C_{24}H_{40}O_5 \cdot H_2O$, mw: 426.7.
Acute tox data: sc LD_{LO} (frog) = 1600 mg/kg. [*3*]
THR = MOD via sc route. An emulsifying agent food additive. [*109*]

CHOLINE. $C_5H_{14}ON$, mw: 104.2.
Acute tox data: ip LD_{50} (rat) = 400 mg/kg. [*3*]
THR = HIGH via ip route. An exper carc. [*23*]

CHOLINE ACETATE. $C_7H_{16}O_2N$, mw: 146.2.
Acute tox data: Oral LD_{50} (mouse) = 3000 mg/kg; iv LD_{50} (rat) = 22 mg/kg; sc LD_{50} (rat) = 250 mg/kg. [*3*]
THR = HIGH via iv and sc; MOD via oral routes.

CHOLINE BITARTRATE. White crystalline powder, odorless or faint trimethylamine like odor, sol in water and alcohol, insol in ether, chloroform and benzene. ($C_5H_{14}NO \cdot C_4H_5O_8$), mw: 253.
THR = U. Probably LOW. A nutrient and/or dietary supplement food additive. [*109*]

CHOLINE XANTHATE. A food additive permitted in the feed and drinking water of animals and/or for the treatment of food-producing animals. [*109*]
Disaster Hazard: Dangerous; see sulfur compounds.

CHONDROITIN SULFATE.
THR = An exper teratogen. [*3*]

CHONDRUS EXTRACT. Syns: *carrageen, irish moss, chondrus, irish gum, pig wrack, rock salt moss.* A sulfated polysaccharide. Dried plant of seaweed chondrus crispus, yellow-white when powdered, insol in organic solvents.
Acute tox data: iv LD_{LO} (rabbit) = 50 mg/kg. [*3*]
THR = HIGH via ip route. An exper (S) neo and carc. [*3*, *7*] A stabilizer food additive. [*109*]

CHROMATED ZINC ARSENATE. Syn: *Boliden salt.* Powder. Arsenic acid + sodium arsenate + sodium dichromate + zinc sulfate.
THR = HIGH via oral and inhal routes. An exper carc. [*3*] See arsenic and chromium compounds.
Disaster Hazard: Dangerous; see arsenic compounds.

CHROMATES. See chromium compounds.

CHROM ALUM. Syn: *chromium-potassium sulfate.* Dark violet-red crystals. $CrK(SO_4)_2 \cdot 12H_20$, mw: 499.3, d: 1.813.
THR = See chromium compounds.

CHROME HYDROXIDE. Syn: *chromic oxide, hydrous.* Blue-green powder, nearly insol in water. $Cr(OH)_3$, mw: 103.0.
THR = HIGH via inhal, oral and imp routes. A recog human carc and neo. [*3*, *6*] See chromium compounds.

CHROME YELLOW. See lead chromate.

CHROMIC ACETATE (III). Gray-green powder or bluish green pasty mass. $Cr(C_2H_3O_2)_3 \cdot H_2O$, mw: 247.16.
THR = A recog human carc and neo. [*3*, *6*] See also chromium compounds.

CHROMIC ACID. Syns: *chromic anhydride, chromium trioxide.* Dark, purple-red crystals. CrO_3, mw: 100.01, mp: 196°, d: 2.70.
THR = VERY HIGH via inhal route. An exper (+) carc. [*3*, *6*] See also chromium compounds. It is a

common air contaminant. This material is usually caustic in its action on skin, mu mem or organic matter in general.

Fire Hazard: Dangerous; a very powerful oxidizing agent. In contact with organic matter or reducing agents, causes violent reactions with acetic acid, acetic anhydride, acetone, Al, NH_3, anthracene, As, C_6H_6, camphor, ClF_3, CrS, ethyl ether, dimethyl formamide, alcohol, glycerol, hydrocarbons, H_2S, CH_3OH, naphthalene, P, K, organic matter, potassium ferricyanide, pyridine, Se, Na, sodium amide, S, turpentine. [*19*]

Explosion Hazard: Upon intimate contact with powerful reducing agents, can cause violent explosions.

CHROMIC ACID SOLUTION. See also chromic acid.

CHROMIC ACID-TREATED FATS AND OILS.
THR = A (S) carc. [*14*]

CHROMIC ANHYDRIDE. See chromic acid.

CHROMIC BROMIDE. Hexagonal olive-green crystals. $CrBr_3$, mw: 291.76, mp: subl, d: 4.25 @ 0°.
THR = See chromium compounds and bromides.

CHROMIC CHLORIDE. See chromium chloride III.

CHROMIC CHROMATE. Cr_5O_{12}, mw: 452.
THR = An exper (+) neo and carc. [*3*, *6*]

CHROMIC FLUORIDE. Rhombic green crystals. CrF_3, mw: 109.01, mp: >1000°, subl 1100–1200°, d: 3.8.
THR = See chromium compounds and fluorides. Violent reaction with (K + NH_4Br). [*19*]

CHROMIC NITRATE. Monoclinic brown crystals. $Cr(NO_3)_3 \cdot 7\frac{1}{2}H_2O$, mw: 373.15, mp: 100°.
THR = See chromium compounds and nitrates.

CHROMIC-*o*-PHOSPHATE. Violet crystals. $Cr(PO_4) \cdot 2H_2O$, mw: 183.02, d: 2.42 @ 32.5°.
THR = See chromium compounds and phosphates.

CHROMIC SESQUESULFIDE III. Violet or red powder. $Cr_2(SO_4)_3$, mw: 392.22, d: 3.012.
THR = See chromium compounds and sulfates.

CHROMIC SULFITE. Greenish-white crystals. $Cr_2(SO_3)_3$, mw: 344.22, mp: decomp, d: 2.2.
THR = See chromium compounds and sulfites.

CHROMITE ORE. Syn: *ferric chromate.*
THR = An exper (+) neo and carc. [*3*, *6*] See chromium compounds.

CHROMITE ORE ROAST.
THR =An exper (+) carc. [*3*, *6*] See chromium compounds.

CHROMITE ROAST LEACHED BROMIDE.
THR = An exper (+) carc. [*3*, *6*, *2*] See chromium compounds.

CHROMIUM. Very hard metal, cubic steel, gray crystals. Cr, at wt: 52.01, mp: 1890°, bp: 2200°, d: 7.20, vap. press: 1 mm @ 1616°.
THR = HIGH pulmonary toxicity. An exper (+) neo and carc. [*3*, *6*, *23*, *95*] See also chromium compounds.

Radiation Hazard: For permissible levels, see Section 5, Table 5A.5. Artificial isotope ^{51}Cr, $T\frac{1}{2}$ = 28d, decays to stable ^{51}V via ec. Emits γ's of 0.32 MeV and x-rays.

Fire Hazard: Mod, in form of dust. Reacts violently with NH_4NO_3, H_2O_2, Li, NO, $KClO_3$, SO_2. [*19*]

CHROMIUM AMMINE NITRATES.
THR = May be heat- and impact-sensitive. [*19*]

CHROMIUM AMMINE PERCHLORATES.
THR = May be impact-sensitive. [*19*]

CHROMIUM BORIDE.
THR = See chromium compounds.

CHROMIUM CARBIDE.
THR = See chromium compounds.

CHROMIUM CARBONYL. Colorless crystals. $Cr(CO)_6$, mw: 220.07, mp: subl @ room temp., sinters @ 90°, decomp @ 130°, explodes @ 210°, bp: 151.0°, d: 1.77, vap. press: 1 mm @ 36.0°, vap. d: 7.6.
THR = VERY HIGH via iv route.

CHROMIUM CHLORIDE (II). Syn: *chromium dichloride.* $CrCl_2$, mw: 122.9.
Acute tox data: Oral LD_{50} (rat) = 1870 mg/kg. [*3*]
THR = MOD via oral route. See also chromium compounds.

CHROMIUM CHLORIDE (III). Syn: *chromic chloride.* $CrCl_3$, mw: 158.4, bp: 1300° (subl).
Acute tox data: Oral LD_{50} (rat) = 1870 mg/kg; dermal LD_{50} (guinea pig) = 202 mg/kg. [*3*]
THR = HIGH via dermal and MOD via oral routes. Violent reaction with Li. [*19*]

CHROMIUM CHLORIDE (IV). See chromium tetrachloride.

CHROMIUM COMPOUNDS. Chromic acid and its salts have a corrosive action on the skin and mu mem. The lesions are confined to the exposed parts, affecting chiefly the skin of the hands and forearms and the mu mem of the nasal septum. The characteristic lesion is a deep, penetrating ulcer, which, for the most part, does not tend to suppurate, and which is slow in healing.

Small ulcers, about the size of a matchhead or end of a lead pencil may be found, chiefly around the base of the nails, on the knuckles, dorsum of the hands and forearms. These ulcers tend to be clean, and progress slowly. They are frequently painless, even though quite deep. They heal slowly, and leave scars. On the mu mem of the nasal septum the ulcers are usually accompanied by purulent discharge and crusting. If exposure continues, perforation of the nasal septum may result, but produces no deformity of the nose. Chromate salts are recog carc of the lungs, nasal cavity and paranasal sinus, also exper carc of the stomach and larynx. [*14*, *23*, *95*, *62*] Hexavalent compounds are said to be more toxic than the trivalent. [*61*, *60*, *26*, *62*, *63*, *64*] Eczematous dermatitis due to trivalent chromium compounds has been reported.

CHROMIUM DIFLUORIDE. See chromous fluoride.

CHROMIUM FLUORIDE (III). CrF_3, mw: 109.
THR = HIGH via oral and sc routes. See also chromium compounds. Corrosive.

CHROMIUM FORMATE. Crystals. $Cr(CHO_2)_3$, mw: 187.1.
THR = See chromium compounds.

CHROMIUM METAL AND ALLOYS OF IRON, NICKEL AND COBALT.
THR = A recog carc. [*3*, *6*] See chromium and nickel compounds.

CHROMIUM MONOARSENIDE. Gray, hexagonal crystals. CrAs, mw: 126.92, d: 6.35, @ 16°.
THR = See arsenic and chromium compounds.
Fire Hazard: See arsine.
Explosion Hazard: Dangerous; when heated to decomp or on contact with water, steam, acid or acid fumes, will react to produce toxic and flam vapors of arsine.

CHROMIUM MONOBORIDE. Orthorhombic, silvery crystals. CrB, mw: 62.83, mp: 2760°, d: 6.17.
THR = See chromium and boron compounds.
Fire Hazard: See boron hydrides.
Explosion Hazard: See boron hydrides.
Disaster Hazard: Dangerous; on contact with water, steam, acid or acid fumes, will react to produce toxic and flam vapors of boron hydrides.

CHROMIUM MONOPHOSPHIDE. Gray-black crystals. CrP, mw: 82.99, d: 5.7 @ 15°.
THR = See chromium compounds and phosphides.
Fire Hazard: Dangerous; upon contact with moisture, acid or acid fumes, phosphine is evolved. See phosphine.
Explosion Hazard: See phosphides and phosphine.
Disaster Hazard: Dangerous; see phosphides.

CHROMIUM OXIDE III. Syn: *green chromium oxide*. Cr_2O_3, mw: 152.
THR = HIGH via inhal route. An exper (S) carc. [*3*, *6*] Reacts violently with ClF_3, glycerol, Li, OF_2. [*19*]

CHROMIUM OXIDE IV. See chromic acid.

CHROMIUM OXYCHLORIDE. Syn: *chromyl chloride*. Dark red liquid, musty burning odor. CrO_2Cl_2, mw: 154.92, mp: −96.5°, bp: 115.7°, d: 1.9145 @ 25°/4°, vap. press: 20 mm @ 20°.
Acute tox data: sc LD_{LO} (mice) = 545 mg/kg. [*3*]
THR = HIGH via sc and inhal routes. A strong irr. Hydrolyzes to form chromic and hydrochloric acids. See chromium compounds. Reacts violently with alcohol, ether, acetone, turpentine, NH_3, (Cl_2 + C), F_2, P, PCl_3, NaN_3, S, SCl. [*19*]
Disaster Hazard: Dangerous; see chlorides.

CHROMIUM-2,4-PENTANE DIONE DERIVATIVE. Syn: *acetylacetonate of chromium*. A solid. $Cr(C_5H_7O_2)_3$, mw: 349.33, mp: 216°, bp: 340°.
THR = See chromium compounds.

CHROMIUM PICRATE. Solid, $Cr[C_6H_2OH(NO_2)_3]_3$, mw: 739.4.
THR = See chromium compounds.
Fire Hazard: See nitrates.
Explosion Hazard: See explosives, high, and nitrates.
Disaster Hazard: See nitrates.

CHROMIUM POTASSIUM SULFATE. See chrome alum.

CHROMIUM SULFATE. See chromic sulfate.

CHROMIUM TETRACHLORIDE. $CrCl_4$, mw: 193.8.
THR = HIGH via inhal and oral routes. See chromium compounds and chlorides. Violent reaction with Na or K. [*19*]

CHROMIUM TETRAFLUORIDE. Brown, amorphous, hygroscopic mass, sol in water with hydrolysis. CrF_4, mw: 128.01, d: 2.89, mp: 200°, bp: approx 400° evolving intensely blue flame.
THR = HIGH irr via oral and inhal routes. See also chromium compounds.
Disaster Hazard: Dangerous; see fluorides.

CHROMIUM TRIAMMINO TETROXIDE. $Cr(NH_3)_3O_4$, mw: 167.
THR = See chromium compounds. Incandesces when heated. Detonates on impact. [*19*]

CHROMIUM TRIOXIDE. See chromic acid.

CHROMOMYCIN A3. Isolated from streptomyces griseus.
THR = HIGH via oral and inhal routes. An exper teratogen. [*3*]

CHROMOUS ACETATE. Red crystals. $Cr(C_2H_3O_2)_2$, mw: 170.10.
Acute tox data: Oral LD_{50} (rat) = 11,260 mg/kg. [*3*]
THR = LOW via oral and inhal routes. See chromium compounds.

CHROMOUS BROMIDE. White crystals. $CrBr_2$, mw: 211.84, mp: 842°, d: 4.356.
THR = See chromium compounds.

CHROMOUS CHLORIDE. See chromium chloride II.

CHROMOUS FLUORIDE. Syn: *chromium difluoride.* CrF_2, mw: 90.01, mp: 1100°, bp: >1300°, d: 4.11.
THR = See chromium compounds. A powerful irr. See fluorides.
Disaster Hazard: See fluorides.

CHROMOUS HYDROXIDE. Yellow-brown crystals. $Cr(OH)_2$, mw: 86.03.
THR = See chromium compounds.

CHROMOUS IODIDE. Grayish powder. CrI_2, mw: 305.85, d: 5.196.
THR = See chromium compounds.

CHROMOUS MONOSULFIDE. Black powder. CrS, mw: 84.08, d: 4.1.
THR = See chromium compounds and sulfides. Reacts violently with F_2, CrO_3. [*19*]

CHROMOUS MONOXIDE. Black crystals. CrO, mw: 68.01.
THR = Self ignites in air. [*19*] See chromium compounds.

CHROMOUS OXALATE. Yellow crystalline powder. $CrC_2O_4 \cdot H_2O$, mw: 158.05.
THR = See chromium compounds and oxalates.

CHROMOUS SULFATE. Blue crystals. $CrSO_4 \cdot 7H_2O$, mw: 274.19.
THR = See chromium compounds.

CHROMYL CHLORIDE. See chromium oxychloride.

CHROMYL FLUORIDE. Exists in 2 modifications; appears first as a reddish-black solid and polymerizes on exposure to light into a dirty white solid, forming reddish-brown vapors on melting. CrO_2F_2, mw: 122.01, mp: 200°.
THR = See chromium compounds.

CHRONIC TOXICITY. See Sections 9 and 1.

CHRYSAROBIN. Syn: *goa powder.* Brownish to orange-yellow crystals. $C_{15}H_{12}O_3$, mw: 240.3.
Acute tox data: ip LD_{LO} (mouse) = 4 mg/kg. [*3*]
THR = HIGH via ip route. An irr and allergen.
Fire Hazard: Slight; when heated, emits smoke.

CHRYSENAMINE. $C_{18}H_{14}N$, mw: 244.3.
THR = An exper carc. [*3*]

CHRYSENE. Syn: *1,2-benzphenathrene.* Crystals, slightly sol in ether, alcohol and glacial acetic acid, insol in water. $C_{18}H_{12}$, mw: 228.2, d: 1.274 @ 20°/4°, mp: 254°, bp: 448°.
THR = HIGH via sc and dermal and probably inhal routes. An exper (+) neo and carc. [*3, 11, 23*] A polycyclic hydrocarbon air pollutant.

CHRYSOIDINE B. $C_{12}H_{12}N_4 \cdot HCl$, mw: 248.7.
THR = An exper (+) carc, neo. [*3, 4*]

CHRYSOPHANIC ACID ANTHRANOL. See chrysarobin.

CHRYSOTILE. It composes 96% of all asbestos. See asbestos white and asbestos particles.
THR = An exper carc. [*23*]

CHYMOSIN. See rennet.

CI ACID BLUE 9(DISODIUM SALT). $C_{37}H_{36}O_9N_2S_3$, mw: 795.
THR = An exper neo. [*3*]

CI ACID GREEN 5. $C_{37}H_{36}C_9N_2S_3$, mw: 795.
THR = An exper carc. [*3*]

CICUTA. See coniine.

CICUTINE. See coniine.

CIGAR SMOKE.
THR = A carc. [*14*]

CIGARETTE SMOKE.
THR = A carc. [*14*]

CIMENE. See dipentene.

CINERIN I. Syn: *3-(2-butenyl)-4-methyl-2-oxo-3-cyclopenten-l-yl ester of chrysanthemum monocarboxylic acid.* Viscous liquid. $C_{20}H_{31}O_3$, mw: 319.5, bp: 200° @ 0.1 mm with decomp.
Acute tox data: LD_{50} (rat) = 1050 mg/kg. [*3*]
THR = MOD via inhal and oral routes. Large doses can cause diarrhea, convulsions and damage to kidneys and liver; prostration and death from respiratory paralysis. See also pyrethrin I.

CINERIN II. Syn: *3-(2-butenyl)-4-methyl-2-oxo-3-cyclopenten-l-yl ester of chrysanthemum dicarboxylic acid monomethyl ester.* A viscous liquid. $C_{21}H_{28}O_5$, mw: 360.4, bp: 200° @ 0.1 mm.
THR = U. An insecticide. See cinerin I.
Fire Hazard: Slight, when heated.

CINNABAR. See mercuric sulfide.

CINNAMALDEHYDE. Syns: *cinnamic aldehyde, 3-phenyl propenal, cinnamyl aldehyde.* Yellowish oil, cinnamic odor, sol in 5 volumes of 60% alcohol, very slightly sol in water. $C_6H_5CH{:}CHO$, mw: 115, d: 1.048–1.052 @ 25°/25°, mp: −8°, bp: 246°.
Acute tox data: Oral LD_{50} (rat) = 2220 mg/kg; ip LD_{50} (mouse) = 200 mg/kg. [*3*]
THR = HIGH via ip and MOD via oral and inhal routes. Synthetic flavoring substance and adjuvant. [*109*]

CINNAMAMIDE. Solid. $C_6H_5CHCHCONH_2$, mw: 147.2, mp: 147°.
THR = U. An insecticide.
Fire Hazard: Slight.

CINNAMEIN. See benzyl cinnamate.

CINNAMENE. See phenyl ethylene.

CINNAMIC ACID, SODIUM SALT. White crystalline scales. $C_9H_8O_2 \cdot Na$, mw: 171.2.

Acute tox data: MILD irr and allergen, ip LD_{50} (mouse) = 2000 mg/kg. [*3*]
THR = MOD via ip, oral and inhal routes.
Fire Hazard: Slight.

CINNAMON LEAF OIL.
Acute tox data: Oral LD_{50} (rat) = 2650 mg/kg. [*3*]
THR = MOD via oral and inhal routes. A (S) carc. [*14*]

CINNAMIC ALDEHYDE. See cinnamaldehyde.

CINNAMOYL CHLORIDE. Yellow crystals. $C_6H_5CH:CHCOCl$, mw: 166.6, mp: 35°–36°, bp: 170–171° @ 58 mm, d: 1.162 @ 45.3°/4°.
THR = MOD irr via oral route. May cause dermatitis. A mild irr.
Disaster Hazard: Dangerous; see chlorides.

CINNAMYL ALDEHYDE. See cinnamaldehyde.

CINNAMYL ANTHRANILATE. $C_{16}H_{15}O_2N$, mw: 253.3, bp: 332°, mp: >60°.
Acute tox data: Oral LD_{50} (rat) = 5 g/kg; dermal LD_{50} (rat) = 5 g/kg. [*3*]
THR = MOD via oral and dermal routes. An exper neo. [*3*, *108*] in mice via ip route.

CIPC. See isopropyl-*n*-(3-chlorophenyl)carbamate.

C.I. SOLVENT BROWN 1. $C_{16}H_{14}N_4$, mw: 262.3.
THR = An exper (±) carc. [*3*, *4*]

C.I. SOLVENT RED 19. $C_{24}H_{21}N_5$, mw: 379.5.
THR = An exper (±) carc and neo. [*3*, *4*]

CITRAL. Syns: *geranial*, *neral*, *geranialdehyde*, *2,6-dimethyl-2,6-octadienal.* Mobile, pale yellow liquid, strong lemon odor, sol in 5 volumes of 60% alcohol, sol in all proportions of benzyl benzoate, diethyl phthalate, glycerin, propylene glycol, mineral oil, fixed oils and 95% alcohol, insol in water. $(CH_3)_2C:CHC_2H_4C(CH_3):CHCHO$, mw: 152, d: 0.891–0.897 @ 15°.
Acute tox data: Oral LD_{50} (rat) = 4960 mg/kg. [*3*]
THR = MOD via oral and inhal routes. A synthetic flavoring substance and adjuvant. [*109*]

CITRAZINIC ACID. Syns: *CZA*, *2,6-dihydroxy-4-carboxy pyridine*. Buff to gray powder. $C_6H_5NO_4$, mw: 155.12.
Acute tox data: ip LD_{LO} (rat) = 800 mg/kg. [*3*]
THR = MOD via ip and inhal routes. A mod irr organic acid and with some allergenic properties.

CITRIC ACID. Syn: *β-hydroxytricarballylic acid.* Colorless, odorless crystals. $C_6H_8O_7 \cdot H_2O$, mw: 210.2, mp: 153°, bp: decomp, d: 1.542.
Acute tox data: Oral LD_{50} (rat) = 11,700 mg/kg; oral LD_{50} (mouse) = 5040 mg/kg. [*3*]
THR = MOD via oral and inhal routes. A MOD irr organic acid, some allergenic properties. A sequestrant food additive. Also a general purpose food additive. [*109*]
Fire Hazard: Slight, when heated.

CITRONELLA. Syn: *oil of citronella* (*Ceylon*). Colorless to pale yellow liquid, turns red on standing, pleasant odor. Composition: 60% geraniol, 15% citronellol, 15% camphene and dipentene, also linalool and borneol. d: 0.897–0.912.
THR = LOW. A MILD allergen.
Fire Hazard: Slight, when heated.

CITRUS OILS.
THR = A (S) carc. [*14*] See plant and fungal products.

CITRUS RED #2. Syn: *1-[(2,5-dimethoxy phenyl)azo]-2-naphthalenol.* $C_{18}H_{16}N_2O_3$, mw: 308.3, mp: 156°. Slightly water sol, moderately sol in alcohol.
THR = A stable, intense red dye used in foods, drugs and cosmetics. Presently under FDA regulation. An exper (+) carc, neo in mice and rats via oral and sc routes and bladder implantation. [*3*, *4*]

CK. See cyanogen chloride.

CLASSIFICATION OF MAJOR LABELING SYSTEMS IN TERMS OF USERS AND AREAS OF COVERAGE. See Section 11.

CLAUSTHALITE. See lead selenide.

CLEANING FLUID. Insol in water, flash p: <80°F.
Fire Hazard: Dangerous, when exposed to heat, flame or oxidizers.
To Fight Fire: Alcohol foam, dry chemical, fog, mist, spray.

CLEANING CONTAMINATED AIR. See Section 2.

CLEANING SOLVENTS (KEROSENE CLASS). See stoddard solvent.

CLEANING SOLVENTS, 140°F CLASS. Insol in water. flash p: 138.2°F or higher, autoign. temp.: 453.2°F or higher, lel = 0.8% @ 302°F, bp: (initial) 181° or higher.
Fire Hazard: Mod, when exposed to heat, flame or oxidizers.
To Fight Fire: Alcohol foam, dry chemical, fog, spray, mist.

CLINICAL MEDICINE AFFECTED BY INDUSTRIAL TOXICOLOGY. See Section 9.

"CLOROX." See hypochlorites.

CLOTH RED B. $(2Na)C_{24}H_{20}O_7N_4S_2$, mw: 586.6.
THR = An exper (−) carc. [*3*,*4*]

CLYSOLATE. See 4-chloro benzenesulfonic acid, chlorophenyl ester.

CMC. See sodium carboxymethyl cellulose.

CM CELLULOSE. See sodium carboxymethyl cellulose.

CMU. See monuron.

COAL BRIQUETTES, HOT. See carbon.

COAL CREOSOTE.
Acute tox data: Oral LD_{50} (rat) = 725 mg/kg. [*3*]
THR = MOD via oral and inhal routes. A recog carc. [*14*] See creosote.

COAL DUST. See anthracite particles.

COAL GAS. Contains hydrogen, methane, carbon monoxide, etc. lel = 5.3%, uel = 31%, autoign. temp.: 1200°F.
THR = HIGH. See carbon monoxide.
Fire Hazard: Very dangerous; see hydrogen.
Explosion Hazard: Mod, when exposed to heat or flame.
Disaster Hazard: Dangerous; see hydrogen and methane.
To Fight Fire: Stop flow of gas; CO_2, dry chemical or water spray.

COAL, GROUND BITUMINOUS, SEA COAL, COAL FACINGS, ETC. Black powder or chunks.
THR = Depends upon content of SiO_2. See also silica.
Fire Hazard: Mod, when exposed to heat; can react with oxidizing materials.
Spont Heating: Mod.
Explosion Hazard: Slight, when exposed to flame.

COALITES. See explosives, high.

COAL NAPHTHA. See benzene.

COAL SPECIALS. See explosives, high.

COAL TAR. Black, viscous liquid. Composition: benzene, toluene, naphthalene, anthracene, xylene, phenol, cresol, ammonia, pyridine, thiophene, etc.
THR = MOD via inhal and contact. A recog carc. [*14*, *23*] See also coal tar and pitch.
Fire Hazard: Mod, when exposed to heat.
Explosion Hazard: Mod, when vapor is exposed to heat or flame.
Disaster Hazard: Dangerous; when heated to decomp, emits highly toxic fumes; can react with oxidizing materials.

COAL TAR ACID OIL.
THR = A recog carc. [*14*] See coal tar and pitch.

COAL TAR AEROSOL. Benzene-sol. Fraction contains about 10% polycyclic hydrocarbons.
THR = An exper neo. [*3*]

COAL TAR AND PITCH.
THR = A recog carc of the skin, scrotum, lip, larynx and lungs. Also an exper carc of the bladder. [*14*]

COAL TAR AND PRODUCTS.
THR = A recog carc. [*14*] See aromatic amines.

COAL TAR CREOSOTE (PITCH). flash p: 405°F, d: >1.
Acute tox data: Oral LD_{50} (rat) = 725 mg/kg; oral LD_{LO} (dog) = 600 mg/kg. [*3*]
THR = MOD via oral route.
Fire Hazard: Low.
To Fight Fire: Dry chemical, CO_2, mist.

COAL TAR DISTILLATE. Flash p: <80°F, d: <1.
THR = An allergen. MOD via oral and inhal routes. A recog carc. [*14*] See also coal tar and pitch.
Fire Hazard: Dangerous, when exposed to heat or flame; can react vigorously with oxidizing materials.
To Fight Fire: Dry chemical, foam, alcohol foam, fog or mist.
Explosion Hazard: U.

COAL TAR DYES.
THR = Many of the coal tar dyes are quite harmless and are permitted for foods, drugs, and cosmetics. Some of them may be allergens or carc. [*14*]

COAL TAR GREASE.
THR = A recog carc. [*14*] See coal tar and pitch.

COAL TAR LIGHT OIL. lel = 1.3%, uel = 8%, d: <1, flash p: 60°–77°F.
THR = LOW via oral and inhal routes. An allergen. A recog carc. [*14*]
Fire Hazard: Dangerous, when exposed to heat or flame.
Explosion Hazard: Mod, when exposed to heat or flame.
Disaster Hazard: Dangerous; when heated to decomp, emits highly toxic fumes; can react vigorously with oxidizing materials.
To Fight Fire: Foam, CO_2, dry chemical.

COAL TAR NAPHTHA. See naphtha, coal tar.

COAL TAR NAPHTHALENES.
THR = A recog carc. [*14*] See coal tar and pitch.

COAL TAR OIL. See coal tar light oil.

COAL TAR PITCH. A black to brown tarry mass. Flash p: 405°F (CC), d: >1.
THR = LOW irr via inhal route. An allergen. A recog carc. [*14*] See coal tar and pitch.
Fire Hazard: Light, when exposed to heat.
Disaster Hazard: Dangerous; when heated to decomp, emits toxic fumes.
To Fight Fire: CO_2, dry chemical.

COAL TAR RESINS. See coumarone-indene resins.

COBALAMIN. See vitamin B_{12}.

COBALT. Silver-gray metal. Co, atwt: 58.9, mp: 1495°, bp: 2900°, d: 8.9.
THR = See cobalt compounds. An exper carc. [*3*, *23*, *95*]
Radiation Hazard: For permissible levels, see Section 5, Table 5A.5.

Artificial isotope ^{57}Co, T½ = 270d, decays to stable ^{57}Fe via ec, emits γ's of 0.01, 0.14 MeV and x-rays. Artificial isotope ^{58}Co, T½ = 71d, decays to stable ^{58}Fe via positrons (15%) of 0.47 MeV. Also decays via ec and emits γ's of 0.81 MeV. Artificial isotope ^{60}Co, T½ = 5.2y, decays to stable ^{60}Ni via β's of 0.32 MeV. Emits γ's of 1.17, 1.33 MeV.

Fire Hazard: Mod, when exposed to heat or flame or by spont chemical reaction; see also powdered metals. Pyroforic Co reacts violently with acetylene, air, NH_4NO_3. [*19*]

COBALT 60. See cobalt.

COBALT ACETATE. Syn: *cobaltous acetate.* Monoclinic, red-violet, deliquescent crystals. $Co(C_2H_3O_2)_2 \cdot 4H_2O$, mw: 249.1, mp: $-4H_2O$ @ 140°, d: 1.705.

Acute tox data: iv LD_{50} (mouse) = 31 mg/kg; iv LD_{50} (rabbit) = 25 mg/kg. [*3*]

THR = HIGH via iv route. A trace mineral added to animal feeds. [*109*] See cobalt compounds.

COBALT ALLOYS.

THR = Reacts violently with molten Li. [*19*] A (S) carc. [*14*] See also cobalt compounds.

COBALT ALUMINATE. Cubic blue crystals. $CoAl_2O_4$, mw: 176.88.

THR = See cobalt compounds and aluminum compounds.

COBALT ARSENIC SULFIDE. Syn: *cobaltite.* Gray-reddish crystals. CoAsS, mw: 165.92, mp: decomp, d: 6.2–6.3.

THR = HIGH. A recog carc. [*14*] See arsenic compounds and sulfides.

Fire Hazard: See sulfides and arsine.

Explosion Hazard: See sulfides and arsine.

Disaster Hazard: Dangerous; when heated to decomp, emits highly toxic fumes of arsine and oxides of sulfur; will react with water, steam, acid or acid fumes to produce toxic and flam vapors of arsine and hydrogen sulfide.

COBALT BROMOPLATINATE. Crystals. $CoPtBr_6 \cdot 12H_2O$, mw: 949.86, d: 2.762.

THR = See cobalt compounds, bromides and platinum compounds.

COBALT CARBONYL. Syns: *cobalt tricarbonyl; tetracobalt dodecacarbonyl.* Black crystals. $[Co(CO)_3]_4$ or $Co_4(CO)_{12}$, mw: 571.88.

THR = HIGH. See carbon monoxide, carbonyls and cobalt compounds.

Fire Hazard: See carbonyls and carbon monoxide.

Explosion Hazard: See carbonyls and carbon monoxide.

Disaster Hazard: Dangerous; see carbonyls.

COBALT CHLORIDE. Blue powder. $CoCl_2$, mw: 129.86, mp: 724°, bp: 1049°, d: 3.348.

THR = A trace mineral added to animal feeds. [*109*] See cobalt compounds and chlorides.

COBALT CHLOROPLATINATE. Crystals. $CoPtCl_6 \cdot 6H_2O$, mw: 575.01, mp: decomp, d: 2.699.

THR = See cobalt compounds, platinum compounds and chlorides.

COBALT CHLOROSTANNATE. Rhombic or trigonal crystals. $CoSnCl_6 \cdot 6H_2O$, mw: 498.48, mp: decomp @ 100°.

THR = See cobalt compounds, tin compounds and chlorides.

COBALT COMPOUNDS.

THR = An exper neo. [*3, 14, 22*] Exper show that the toxicity of Co by mouth is LOW. In animals, administration of cobalt salts produces polycythemia. In humans, a single case of poisoning, liver and kidney damage has been attributed to cobalt. Locally, Co has been shown to produce dermatitis and certain investigators have been able to demonstrate a hypersensitivity of the skin to Co. There have also been reports of hematologic, digestive, and pulmonary changes in humans. It is a (S) carc of the connective tissue and lungs.

COBALT FLUOBORATE II. $CoBF_5$, mw: 164.7.

Acute tox data: Oral LD_{50} (rat) = 100 mg/kg. [*3*]

THR = HIGH via oral and inhal routes.

COBALT HEXAMETHYLENE TETRAMINE. Ultramarine blue crystals, sol in water. $CoC_6H_{12}N_4$, mw: 199.

THR = See chlorides and cobalt compounds and hexamethylene tetramine.

COBALT HEXAMMINE CHLORATE. $Co(NH_3)_6(ClO_3)_2$, mw: 328.

THR = Detonates when impacted. [*19*] See also cobalt compounds.

COBALT HEXAMMINE CHLORITE. $Co(NH_3)_6ClO_2 \cdot 3H_2O$, mw: 282.6.

THR =Very unstable and explosive. [*19*] See also cobalt compounds.

COBALT HEXAMMINE PERCHLORATE. $Co(NH_3)_6(ClO_4)_2$, mw: 360.

THR = It is impact-sensitive. [*19*] See also cobalt compounds.

COBALT HYDROXY QUINONE. Ruby red crystals. $Co(C_{10}H_5O_3)_2$, mw: 405.2, mp: 210°–215° (decomp).

THR = See cobalt compounds.

COBALTIC ACETATE. Green crystals. $Co(C_2H_3O_2)_3$, mw: 237.07, mp: decomp @ 100°.

THR = See cobalt compounds.

COBALTIC CHLORIDE. Red crystals. $CoCl_3$, mw: 165.31, mp: subl, d: 2.94.
THR = See cobalt compounds and chlorides.

COBALTIC DIHYDRO PHOSPHIDE. $Co(PH_2)_3$, mw: 158.
THR = See phosphides. Self-ignites in air. [*19*] See also cobalt compounds.

COBALTIC HEXAMMINE IODATE.
$[Co(NH_3)_6](IO_3)_3$, mw: 685.8.
THR = Explodes at 355° and is impact-sensitive. [*19*] See also cobalt compounds and iodates.

COBALTIC HEXAMMINE PERCHLORATE.
$[Co(NH_3)_6](ClO_4)_3$, mw: 459.5.
THR = Heat (360°) and impact-sensitive. [*19*] See also cobalt compounds and perchlorates.

COBALTIC FLUORIDE. Hexagonal brown crystals. CoF_3, mw: 115.94, d: 3.88.
THR = See fluorides and cobalt compounds. Incandesces with Si. [*19*]

COBALTIC HYDROXIDE. Black-brown powder. $Co(OH)_3$, mw: 109.96, mp: $-1\frac{1}{2}H_2O$ @ 100°.
THR = See cobalt compounds.

COBALTIC OXIDE. Black-gray powder. Co_2O_3, mw: 165.88, mp: decomp @ 895°, d: 5.18.
THR = See cobalt compounds.

COBALTIC PENTAMMINE AZIDE.
$CoN_3(NH_3)_5(N_3)_2$, mw: 270.1.
THR = Explodes on impact. [*19*] Unstable. See also azides and cobalt compounds.

COBALTIC PENTAMMINE CHLORITE.
$Co(NH_3)_5Cl(ClO_2)_2$, mw: 314.4.
THR = Explosively unstable. [*19*] See also chlorites and cobalt compounds.

COBALTIC SESQUISULFIDE. Black crystals. Co_2S_3, mw: 214.08, d: 4.8.
THR = See sulfides and cobalt compounds.

COBALTIC SULFATE. Blue needles.
$Co_2(SO_4)_3 \cdot 18H_2O$, mw: 730.37.
THR = See cobalt compounds and sulfates.

COBALTIC TRIAMMINE AZIDE. $Co(N_3)_3(NH_3)_3$, mw: 236.1.
THR = Unstable. [*19*] No toxicity data. See also azides and cobalt compounds.

COBALTIC CYANIC ACID. See cyanocobaltic III acid.

COBALT IODOPLATINATE. Crystals.
$CoPtI_6 \cdot 9H_2O$, mw: 1177.83, d: 3.618.
THR = See cobalt compounds, iodides and platinum compounds.

COBALTITE. See cobalt arsenic sulfide.

COBALT MONOBORIDE. Prisms. CoB, mw: 69.76, d: 7.25 @ 18°.
THR = See cobalt compounds and boron compounds.
Fire Hazard: See boron hydrides.
Explosion Hazard: See boron hydrides.
Disaster Hazard: Dangerous; will react with water, steam, acid or acid fumes to produce toxic and flam vapors of boron hydrides.

COBALT MONOSELENIDE. Yellow crystals. CoSe, mw: 137.90, mp: red heat, d: 7.65.
THR = See selenium and cobalt compounds.
Fire Hazard: See hydrogen selenide.
Explosion Hazard: See hydrogen selenide.
Disaster Hazard: Dangerous; when heated or on contact with water, steam, acid or acid fumes, will react to produce highly toxic and flam vapors of hydrogen selenide.

COBALT MONOSULFIDE. Reddish, silver-white crystals. CoS, mw: 91.01, mp: >1116°, d: 5.45.
THR = See sulfides and cobalt compounds.

COBALT NAPHTHA. Syns: *cobalt naphthenate*, *cobaltous naphthenate*. Brown amorphous powder or bluish-red solid, insol in water, sol in alcohol, ether and oils.
Composition: indefinite. Flash p: 121°F, autoign. temp.: 529°F, d: 0.9.
THR = A (S) carc. [*14*] See cobalt compounds.
Fire Hazard: Mod, when exposed to heat or flame.
To Fight Fire: Alcohol foam.

COBALT NITRO PRUSSIDE.
Acute tox data: Oral LD_{50} (rat) = 147 mg/kg; ip LD_{50} (rat) = 15 mg/kg. [*3*]
THR = HIGH via oral and ip routes.

COBALTOCENE. Syn: *di-pi-cyclopenta dienyl cobalt*. $CoC_{10}H_{10}$, mw: 189.1.
Acute tox data: ip LD_{50} (rat) = 125 mg/kg; ip LD_{50} (mouse) = 80 mg/kg. [*3*]
THR = HIGH via ip and oral routes. See also cobalt compounds.

COBALT ORE.
THR = A recog carc. [*14, 3*] See arsenic and cobalt compounds.

COBALTOUS ACETATE. See cobalt acetate.

COBALTOUS-*o*-ARSENATE. Syn: *erythrite*. Monoclinic violet-red crystals. $Co_3(AsO_4)_2 \cdot 8H_2O$, mw: 598.77, mp: decomp, d: 3.178 @ 15°.
THR = See arsenic compounds and cobalt compounds.

COBALTOUS BENZOATE. Gray-red leaf.
$Co(C_7H_5O_2)_2 \cdot 4H_2O$, mw: 373.22, mp: $-4H_2O$ @ 115°.
THR = See cobalt compounds.

COBALTOUS BROMATE. Red crystals. $Co(BrO_3)_2 \cdot 6H_2O$, mw: 422.87.
THR = See cobalt compounds and bromates.

COBALTOUS BROMIDE. Green deliquescent crystals. $CoBr_2$, mw: 218.77, mp: decomp, d: 4.909 @ 25°.
THR = See cobalt compounds and bromides. Reacts violently with K, Na. [*19*]

COBALTOUS CARBONATE. Syn: *spherocobaltite.* Red crystals. $CoCO_3$, mw: 118.95, mp: decomp, d: 4.13.
THR = A trace mineral added to animal feeds. [*109*] See cobalt compounds.

COBALTOUS CHLORATE. Cubic, red, deliquescent crystals. $Co(ClO_3)_2 \cdot 6H_2O$, mw: 333.95, mp: 61°, bp: decomp at 100°, d: 1.92.
THR = See cobalt compounds and chlorates.
Fire Hazard: See chlorates. Can explode easily when contaminated.
Explosion Hazard: See chlorates.
Disaster Hazard: See chlorates.

COBALTOUS CHLORIDE. Black crystals. $CoCl_2$, mw: 129.85, mp: 735°, bp: 1050°, d: 3.356, vap. press: 40 mm @ 770°.
Acute tox data: Oral LD_{50} (rat) = 80 mg/kg. [*3*]
THR = HIGH via oral and inhal routes. See cobalt compounds and chlorides.

COBALTOUS CHROMATE. Gray-black crystals. $CoCrO_4$, mw: 174.95, mp: decomp.
THR = A recog carc. [*14*] See chromium compounds and cobalt compounds.

COBALTOUS CITRATE. Rose-red crystals. $Co_3(C_6H_5O_7)_2 \cdot 2H_2O$, mw: 591.05, mp: $-2H_2O$, @ 150°.
THR = See cobalt compounds.

COBALTOUS CYANIDE. Buff color, anhydrous; blue-violet powder. $Co(CN)_2 \cdot 2H_2O$, mw: 147.01, mp: $-2H_2O$ @ 280°, d: 1.872 @ 25°.
THR = See cobalt compounds and cyanides.
Fire Hazard: See cyanides.
Explosion Hazard: See cyanides. Reacts violently with Mg. [*19*]
Disaster Hazard: Dangerous; see cyanides.

COBALTOUS FERRICYANIDE. Red needles. $Co_3[Fe(CN)_6]_2$, mw: 600.74.
THR = See ferricyanides and cobalt compounds.

COBALTOUS FERROCYANIDE. Gray-green crystals. $Co_2[Fe(CN)_6] \cdot xH_2O$.
THR = See ferrocyanides and cobalt compounds.

COBALTOUS FLUOBORATE. See cobalt fluoborate II.

COBALTOUS FLUOGALLATE. Pink crystals. $[Co(H_2O)_6](GaF_5H_2O)$, mw: 349.8, mp: $-5H_2O$ @ 110°, d: 2.35.
THR = See fluogallates and cobalt compounds.

COBALTOUS FLUORIDE. Monoclinic rose-red crystals. $CoF_2 \cdot 2H_2O$, mw: 132.97, d: 4.46.
THR = See fluorides and cobalt compounds.

COBALTOUS FLUOSILICATE. Pink crystals. $CoSiF_6 \cdot 6H_2O$, mw: 309.10 d: 2.113 @ 19°.
THR = See fluosilicates and cobalt compounds.

COBALTOUS FORMATE. Red crystals. $Co(CHO_2)_2 \cdot 2H_2O$, mw: 185.01, mp: $-2H_2O$ @ 140°, bp: anhydrous decomp @ 175°, d: 2.129 @ 22°.
THR = See cobalt compounds and formic acid.

COBALTOUS HYDROXIDE. Rhombic rose-red crystals. $Co(OH)_2$, mw: 92.96, mp: decomp, d: 3.597 @ 15°.
THR = See cobalt compounds.

COBALTOUS HYPOPHOSPHITE. $Co(PH_2O_2)_2$, mw: 189.
THR = Above 150° it liberates PH_3, which is highly toxic and self-igniting. [*19*] See also cobalt compounds.

COBALTOUS IODATE. Black-violet needles. $Co(IO_3)_2$, mw: 408.78, d: 5.008 @ 18°.
THR = See cobalt compounds and iodates.

α-COBALTOUS IODIDE. Hexagonal black crystals. CoI_2, mw: 312.78, d: 5.68.
THR = See cobalt compounds and iodides.

COBALTOUS LINOLEATE. Brown, amorphous mass. $Co(C_{18}H_{31}O_2)_2$, mw: 617.80.
THR = See cobalt compounds.

COBALTOUS NAPHTHENATE.
Acute tox data: Oral LD_{50} (rat) = 3900 mg/kg. [*3*]
THR = MOD via oral route.

COBALTOUS NITRATE. Red crystals. $Co(NO_3)_2 \cdot 6H_2O$, mw: 291.05, mp: $<100°$; $-3H_2O$ @55°, d: 1.87.
THR = See nitrates and cobalt compounds. An exper neo. [*3*] HIGH via oral route.

COBALTOUS OLEATE. Brown amorphous powder. $Co(C_{18}H_{33}O_2)_2$, mw: 621.83.
THR = See cobalt compounds.

COBALTOUS OXALATE. Reddish-white crystals. CoC_2O_4, mw: 146.96, d: 3.021 @ 25°. See oxalates and cobalt compounds.

COBALTOUS OXIDE. Cubic green-brown crystals. Syn: *cobalt oxide.* CoO, mw: 74.94, mp: 1800° decomp, d: 5.7–6.7.
Acute tox data: Oral LD_{50} (rat) = 1700 mg/kg. [*3*]
THR = MOD via oral and inhal routes. An exper carc. [*3*] See cobalt compounds. A trace mineral added to animal feeds. [*109*]

COBALTOUS PENTAMMINE CHLOROPERCHLORATE. $Co(NH_3)_5Cl \cdot ClO_4$, mw: 279.
THR = It is heat- (310°) and impact-sensitive. [*19*] See also cobalt compounds and perchlorates.

COBALTOUS PERCHLORATE. Red needles. $Co(ClO_4)_2$, mw: 257.85, d: 3.327.
THR = See perchlorates and cobalt compounds.

COBALTOUS PERRHENATE. Dark pink crystals. $Co(ReO_4)_2 \cdot 5H_2O$, mw: 649.64, mp: decomp.
THR = See cobalt compounds and rhenium compounds.
Fire Hazard: Dangerous; an oxidizing agent.
Disaster Hazard: Dangerous; keep away from combustible materials.

COBALTOUS-*o*-PHOSPHATE. Reddish crystals. $Co_3(PO_4)_2$, mw: 366.78.
THR = See cobalt compounds and phosphates.

COBALTOUS PROPIONATE. Dark red crystals. $Co(C_3H_5O_2)_2 \cdot 3H_2O$, mw: 259.13, mp: ca. 250°.
THR = See cobalt compounds.

COBALTOUS RESINATE. Brown-red powder. $Co(C_{44}H_{62}O_4)_2$, mw: 1368.81.
THR = See cobalt compounds.
Fire Hazard: Dangerous; spont flam in air; reacts vigorously with oxidizing materials.

COBALTOUS SELENATE. Ruby-red crystals. $CoSeO_4 \cdot 5H_2O$, mw: 291.98, mp: decomp, d: 2.512.
THR = See selenium compounds and cobalt compounds.

COBALTOUS-*o*-SILICATE. Violet crystals. Co_2SiO_4, mw: 209.94, d: 4.63.
THR = See cobalt compounds and silicates.

COBALTOUS SULFATE. Red water-sol powder. $CoSO_4$, mw: 155, mp: 989°, d: 3.71 @ 25°/25°.
Acute tox data: ip LD_{50} (mouse) = 54 mg/kg. [*3*]
THR = HIGH via ip route. See cobalt compounds. A trace mineral added to animal feeds. [*109*]

COBALTOUS SULFIDE. CoS, mw: 91.
THR = An exper carc. [*3*]

COBALTOUS SULFITE. Red crystals. $CoSO_3 \cdot 5H_2O$, mw: 229.08.
THR = See cobalt compounds and sulfites.

COBALTOUS TARTRATE. Monoclinic reddish crystals. $CoC_4H_4O_6$, mw: 207.01.
THR = See cobalt compounds.

COBALTOUS THIOCYANATE. Rhombic violet crystals. $Co(SCN)_2 \cdot 3H_2O$, mw: 229.16, mp: $-3H_2O$ @ 105°.
THR = See thiocyanates and cobalt compounds.

COBALTOUS TRIHYDRAZINE CHLORATE. $Co(N_2H_4)_3(ClO_3)_2$, mw: 302.
THR = Detonates on impact. [*19*] See also cobalt compounds and chlorates.

COBALTOUS TUNGSTATE. Monoclinic blue-green crystals. $CoWO_4$, mw: 306.86, d: 8.42.
THR = See cobalt compounds and tungsten compounds.

COBALT OXIDE. See cobaltous oxide.

COBALT PALMITATE. A solid. $Co(C_{16}H_{31}O_2)_2$, mw: 569.76, mp: 70.5°.
THR = See cobalt compounds.

COBALT PHOSPHIDE. Small gray needles. Co_2P, mw: 148.9, mp: 138.6°
THR = See phosphides and cobalt compounds.
Fire Hazard: Dangerous; decomp to evolve phosphine.
Explosion Hazard: See phosphides.
Disaster Hazard: Dangerous; see phosphides.

COBALT RESINATE, PRECIPITATE.
THR = See cobalt compounds.
Fire Hazard: Dangerous; spont flam in air; reacts with oxidizing materials.

COBALT TETRACARBONYL. See dicobalt octacarbonyl.

COBALT-*o*-TITANATE. Cubic greenish-black crystals. Co_2TiO_4, mw: 229.78, d: 5.07–5.12.
THR = See cobalt compounds and titanium compounds.

COBALT TRICARBONYL. See cobalt carbonyl.

COCA CHEWING.
THR = A (S) carc. [*14*] See plant and fungal products.

COCAINE. Syn: *methyl benzoyl ecgonine.* Colorless to white crystals. $C_{17}H_{21}NO_4$, mw: 303.35, mp: 98°.
Acute tox data: ip LD_{50} (rat) = 70 mg/kg; sc LD_{50} (rat) = 250 mg/kg; iv LD_{50} (rat) = 18 mg/kg. [*3*]
THR = HIGH via oral, inhal, ip, sc and iv routes. A well known drug of abuse. A poisonous alkaloid. Can cause dilation and immobility of the pupils, blepharitis and blindness. Fatal cases run a very rapid course with anxiety, sudden fainting, extreme pallor, dyspnea, sometimes brief convulsions, arrest of respiration and death, generally in a few min. If death does not occur in a few min to a half hour, the patient almost always recovers.
Disaster Hazard: Dangerous; when heated to decomp, emits highly toxic fumes.
Treatment and Antidotes: Wash stomach out immediately with water and sodium bicarbonate (60 g to 1 pint). If a solution of permanganate is handy (1 crystal to 8 oz. of water), use it to wash out stomach. Give inhal of ammonia. Allay convulsions with chloral or chloroform. If breathing is disturbed, give artificial respiration and cardiac massage after respi-

ration fails. Inhal of oxygen plus 5% carbon dioxide should be given.

COCARCINOGENESIS. See Section 8.

COCCULUS SOLID. Syns: *fish berry, picrotoxin.* Dried ripe fruit of a woody climbing plant.
Acute tox data: Oral LD_{50} (mouse) = 15 mg/kg; ip LD_{50} (mouse) = 7 mg/kg. [*3*]
THR = HIGH via ip, oral and inhal routes. A convulsant poison.
Fire Hazard: Slight, when exposed to heat.

COCOA. Brown powder.
THR = An allergen. LOW via oral and inhal routes.
Fire Hazard: Slight, when heated.

COCO BEAN SHELL TANKAGE. A hygroscopic material.
THR = See cocoa.
Fire Hazard: Mod, when exposed to heat.
Spont Heating: Mod; extreme caution must be observed to keep the moisture content low; on contact with oxidizing materials, can react vigorously.

COCOLODE. Syn: *heliotropium angio spermum.* Aqueous extract of the plant without the root.
THR = An exper neo to rats via sc route. [*103*]

COCONUT OIL. Syns: *coconut palm oil, coconut butter.* White, semi-solid, lard-like fat, characteristic odor, sol in alcohol, ether, chloroform and carbon disulfide. d: 0.92, mp: 20°–25°, flash p: 420°F (crude), 548°F (refined).
Fire Hazard: Slight, when exposed to heat or flame.
To Fight Fire: Dry chemical, mist, fog.

COCO BOLO. See sawdust.

COCONUT BUTTER. See coconut oil.

COCONUT PALM OIL. See coconut oil.

CODEINE. Syn: *methyl morphine.* $C_{18}H_{21}NO_3 \cdot H_2O$, mw: 317.37, mp: 154.9°, d: 0.9, vap. d: 3.5, bp: 115°, flash p: 75°.
Acute tox data: sc LD_{50} (rat) = 500 mg/kg; oral LD_{50} (mouse) = 250 mg/kg; ip LD_{50} (mouse) = 200 mg/kg; iv LD_{50} (rabbit) = 34 mg/kg. [*3*]
THR = HIGH via sc, oral, ip and iv routes. An addictive drug.
Fire Hazard: Mod, when exposed to heat or flame.
To Fight Fire: Alcohol foam.

COD LIVER OIL. Syn: *morrhua oil.* Pale yellow liquid, fixed, non-drying oil, characteristic odor, sol in ether, chloroform, ethyl acetate, petroleum ether and carbon disulfide, slightly sol in alcohol. d: 0.9, flash p: 412°F.
Fire Hazard: Slight, when exposed to heat or flame.
To Fight Fire: Dry chemical, CO_2, mist.

COFFEE BEVERAGE.
THR = An allergen. Much used as a CNS stimulant. Professional coffee tasters can have amblyopia and coffee has been known to cause temporary blindness. Ingesting large quantities of coffee beverage over an extended period of time is being linked to circulatory disorders and hypertension and bladder tumors.
Disaster Hazard: Slightly dangerous in dry state; burns when heated.

COKE, HOT. A black amorphous mass. Composition: carbon and impurities.
THR = See carbon and charcoal.
Fire Hazard: See carbon. While still hot it may ignite easily in air or by contact with spark or flame.
Disaster Hazard: See carbon and charcoal.

COKE OVEN TAR AND COKE OVEN FUMES. lel = 4.4%, uel = 34%.
THR = A recog carc. [*14, 23*] See coal tar and pitch.
Fire Hazard: Very dangerous via heat, flames (sparks), or oxidizers.
To Fight Fire: Stop flow of gas.

COLCHICINE. Yellow, crystalline alkaloid, amorphous powder. $C_{22}H_{25}NO_6$, mw: 399.5, mp: 143.7° anhyd.
Acute tox data: iv LD_{50} (mouse) = 3.55 mg/kg; dermal TD_{LO} (mouse) = 144 mg/kg caused neo in 14 weeks; ip LD_{50} (rat) = 6.1 mg/kg. [*3*]
THR = HIGH via iv, dermal and ip routes. An exper neo, teratogen and mutagen. [*3*] An exper carc. [*23*] A poisonous alkaloid. Large doses cause diarrhea with griping in susceptible persons. Symptoms take several hrs and arise from the alimentary tract. Pain in the gastric region is followed by salivation, nausea, vomiting and diarrhea. Depression, apathy and collapse follow. In exper poisoning, paralysis starts with the posterior limbs and increases upward until the forelimbs and respiratory muscles are involved and death occurs from asphyxia. Externally, it is a local irr which causes redness and smarting when applied to the skin. Upon inhal, the dust causes sneezing and conjunctival hyperemia and a burning sensation of mouth and throat. 0.02 g is very likely to cause death in about 24 hrs from a single dose. Repeated doses may be cumulative.
Fire Hazard: Slight.
Treatment and Antidotes: Give a cathartic and an emetic at once; also give large quantities of water for the kidneys. Tannic acid should be given in large amounts. Opium and stimulants should also be given to counteract depression.

COLCHICINE TANNATE. See colchicine.

COLLIERS. See explosives, high.

COLLODION. Solution of nitrated cellulose in ether-

alcohol. $C_{12}H_{16}O_6(NO_3)_4C_{13}H_{17}O_7(NO_3)_3$, mw: 975, flash p: <0°F.
Fire Hazard: Very dangerous, when exposed to heat or flame.
To Fight Fire: Alcohol foam.

COLLODION COTTON. See nitrocellulose.

COLLODION WOOL. See nitrocellulose.

COLLOIDAL CLAY. See bentonite.

COLOCYNTH. Syns: *dried pulp of bitter apple, bitter cucumber.*
THR = A powerful cathartic, purgative. [*2*]

COLOGNE SPIRITS (ALCOHOL). See ethyl alcohol.

"COLONIAL SPIRITS." See methyl alcohol.

COLORING FOOD ADDITIVES. [*109*]

COLOPHONY POWDER. See rosin.

COLTSFOOT. A senecione herb.
THR = An exper carc to rats via oral route. [*103*]

COLUMBIAN SPIRITS. See methyl alcohol.

COLUMBIUM. See niobium.

COLUMBIUM CHLORIDE. Syn: *columbium pentachloride.* Yellow-white, deliquescent powder. $CbCl_5$, mw: 270.20, mp: 194°, bp: 240.5°.
Acute tox data: Oral LD_{50} (rat) = 1400 mg/kg; oral LD_{50} (mouse) = 829 mg/kg. [*3*]
THR = MOD via oral and inhal routes. See niobium and chlorides.

COLUMBIUM PENTACHLORIDE. See columbium chloride.

COLZA OIL. See rapeseed oil.

COMBINATION FUZES. See explosives, high.

COMBINATION PRIMERS. See explosives, high.

COMBUSTION PRODUCT GAS.
THR = A food additive manufactured by the controlled combustion in air of butane, propane or natural gas.
Acute tox data: U. A food additive permitted in food for human consumption. [*109*]

COMMON MALIC ACID. See malic acid.

COMPACTOR DANGERS IN HANDLING SOLID WASTE. See Section 6.

COMPAZINE. See prochlorperazine.

COMPLETE ENCLOSURE TYPE HOOD. See Section 2.

CONDURANGIN. Yellowish, amorphous, bitter powder from Condurango Bark. $C_{35}H_{54}O_{14}(OCH_3)_2$, mw: 760.9, mp: 187°.
THR = A MOD irr. An astringent and aromatic bitter. May cause convulsions and then paralysis depending upon dose.
Fire Hazard: Slight, when heated.

CONHYDRINE. Colorless, crystalline alkaloid. $C_8H_{17}NO$, mw: 143.23, mp: 121°, bp: 226°.
Acute tox data: sc LD_{LO} (guinea pig) = 400 mg/kg. [*3*]
THR = HIGH via sc route. A poisonous alkaloid. Toxic doses cause weakness and drowsiness but not actual sleep. Movements are weak and unsteady and gait is staggering. There are usually nausea and vomiting with profuse salivation. Intelligence usually remains clear. The pupils are ordinarily somewhat dilated and ptosis also occurs, indicating oculomotor paralysis. Speech is thick and hearing is imperfect. Tremors and occasionally convulsions occur. Breathing becomes weaker and slower and finally stops, causing death. It is said to have a depressant effect on bruised surfaces of the skin. One-half to one grain is considered a poisonous dose.
Fire Hazard: Slight, when heated.
Treatment and Antidotes: Administer emetics and wash stomach out, then give tannic acid freely and wash stomach out again. Strychnine and other stimulants are given hyperdermically. Keep patient warm and administer artificial respiration. A physician should be summoned.

CONICINE. See coniine.

CONIINE. Syns: *cicutine, cicuta, conicine.* Colorless, oily liquid with mousy odor. $C_8H_{17}N$, mw: 127.23, bp: 166.5°, fp: −2.5°, d: 0.844–0.848 @ 20°/4°.
THR = Tox principle of poison hemlock. Ingestion causes weakness, drowsiness, nausea, vomiting, labored respiration, paralysis, asphyxia, death from paralysis of the nervous system. [*2*] In small doses it is a sedative. Poisoning is treated by evacuating the stomach and administering tannic acid.
Fire Hazard: Slight, when heated.

CONIINE HYDROCHLORIDE. See coniine.

CONIUM. See coniine.

CONIUM MACULATIUM. See coniine.

CONSUMER PRODUCT SAFETY COMMISSION REGULATIONS. See Section 10A.

CONTAINERS, EMPTY. See acid carboys, empty; bottles, empty; drums, empty; cylinders, empty.

CONTAMINANTS OF HEATED FATS AND OILS.
THR = A (S) carc. [*14*]

CONTAMINANT GASES STRESS AIR ENVIRONMENT. See Section 4.

CONTAMINATED AIR CLEANING. See Section 2.

CONTAMINANT PARTICULATES STRESS UPON THE AIR ENVIRONMENT. See Section 4.

CONTAMINANTS AROUND BUILDINGS DUE TO SPECIFIC METEOROLOGICAL CONDITIONS AND CROSS-CONTAMINATION BETWEEN AIR EXHAUSTS AND INLETS. See Section 4.

CONTAMINANTS IN THE WORKING PLACE AND BUILDING VENTILATION. See Sections 2 and 4.

CONTROL OF INHALATION HAZARDS BY VENTILATION. See Section 2.

COOLING AND CUTTING OILS.
THR = A recog carc. [*14*] See mineral oils.

COPAIBA. Syn: *balsam copaiba.* Transparent, viscid yellow liquid. Peculiar odor. Bitter, acrid nauseating taste. Constituents: volatile oil, resin. d: 0.93–1.00.
THR = Large doses cause vomiting and diarrhea. Can also cause dermatitis and kidney damage. Incompatible with water, magnesia, mineral acids. [*2*]

COPAL. Syns: *resin copal, gum copal.* Yellowish-brown pieces.
THR = LOW. A MILD allergen.
Fire Hazard: Slight, when heated.

COPPER. A metal with a distinct reddish color. Cu atwt: 63.5, mp: 1083°, bp: 2324°, d: 8.92, vap. press: 1 mm @ 1628°.
Acutc tox data: ipLD_{50} (mouse) = 3.5 mg/kg. [*3*]
THR = HIGH via ip route. See copper compounds. A (S) mutagen. [*22*]
Radiation Hazard: For permissible levels, see Section 5, Table 5A.5.
Fire and Explosion Hazard: Reacts violently with C_2H_2, NH_4NO_3, bromates, chlorates, iodates, Cl_2, ClF_3, (Cl_2 + OF_2), ethylene oxide, F_2, H_2O_2, hydrazine mononitrate, hydrazoic acid, H_2S, $Pb(N_3)_2$, K_2O_2, NaN_3, Na_2O_2. [*19*]

COPPER ABIETINATE. Syn: *cupric abietinate.* Green scales. $Cu(C_{19}H_{27}O_2)_2$, mw: 637.69.
THR = See copper compounds.
Fire Hazard: Slight, when heated.

COPPER ACETATE. Syns: *cupric acetate, neutral verdigris.* Greenish-blue, fine powder or small crystals. $Cu(C_2H_3O_2)_2 \cdot H_2O$, mw: 199.64, mp: 115°, bp: 240°, d: 1.882, (anhy): 1.93.
Acute tox data: Oral LD_{50} (rat) = 710 mg/kg. [*3*]
THR = MOD via oral and inhal routes. See copper compounds.

COPPER ACETATE, BASIC. Syn: *verdigris.* Greenish-blue powder. $Cu(C_2H_3O_2)_2 \cdot CuO \cdot 6H_2O$, mw: 369.33.
THR = See copper compounds.

COPPER ACETOARSENITE. Syns: *emerald green, imperial, king's green, moss green, vienna green.* Emerald green powder. $C_4H_6O_{16}Cu_4As_6$, mw: 1013.8.
Acute tox data: Oral LD_{50} (rat) = 22 mg/kg; LD_{50} oral (mammal) = 18 mg/kg. [*3*]
THR = HIGH via oral and inhal routes. See arsenic compounds.

COPPER ACETONATE.
THR = See copper compounds.

COPPER ACETYLIDE. See cuprous acetylide.

COPPER AMMONIUM SULFATE. Crystals. $CuSO_4 \cdot 4NH_3 \cdot H_2O$, mw: 245.8.
THR = See copper compounds and sulfates.

COPPER ARSENATE, BASIC. Syn: *cuprous arsenate, basic.* A green solid. $Cu(CuOH)AsO_4$, mw: 283.0.
THR = See arsenic compounds and copper compounds.

COPPER ARSENIDE. Black crystals. Cu_5As_2, mw: 467.52, mp: decomp, d: 7.56.
THR = See arsenic compounds and copper compounds.
Fire Hazard: See arsine.
Explosion Hazard: See arsine.
Disaster Hazard: Dangerous; see arsenides.

COPPER ARSENITE. Syns: *cupric arsenite, Sheele's mineral.* Yellowish-green powder. $CuHAsO_3$, mw: 187.5, mp: decomp.
THR = HIGH. See also arsenic compounds.
Disaster Hazard: See arsenic compounds and copper compounds.

COPPERAS. See ferrous sulfate.

COPPER BORIDE. Syn: *cupric boride.* Yellow crystals. Cu_3B_2, mw: 212.26, d: 8.116.
THR = See copper and boron compounds.
Fire Hazard: See boron hydrides.
Disaster Hazard: Mod dangerous; on contact with acid, acid fumes, water or steam, it will react to produce toxic and flam vapors of boron hydrides.

COPPER CARBONATE HYDROXIDE. *Syn: cupric carbonate.* Green powder. $CuCO_3 \cdot Cu(OH)_2$, mw: 221.17, mp: decomp @ 200°, d: 4.0.
Acute tox data: Oral LD_{50} (rat) = 159 mg/kg; oral LD_{50} (birds) = 810 mg/kg. [*3*]
THR = HIGH to MOD via oral route. A fungicide. Also a trace mineral added to animal feeds. [*109*] See copper compounds.

COPPER CHELATE COMPLEX OF 8-QUINOLINOL.
THR = A (S) carc [*14*]. See also 8-hydroxy quinoline.

COPPER CHLORATE. See cupric chlorate.

COPPER CHLORIDE. Syn: *cupric chloride.* Yellowish-brown hygroscopic powder. $CuCl_2$, mw: 134.48, mp: 498°, d: 3.054.

Acute tox data: Oral LD_{50} (rat) = 140 mg/kg; oral LD_{50} (human) = 200 mg/kg. [*3*]
THR = HIGH via oral and inhal routes. Used as a fungicide. Also a trace mineral added to animal feed. [*109*] See copper compounds and chlorides. Can react violently with K, Na. [*19*]

COPPER-γ-CHLOROACETO ACETANILIDE. Solid. $Cu(C_8H_7 \cdot ClNO)_2$, mw: 400.7.
THR = See copper compounds, acetanilide and chlorides.

COPPER CHROMATE, BASIC. See cupric chromate, basic.

COPPER COMPOUNDS. As the sublimed oxide, copper may be responsible for one form of metal fume fever. Inhal of copper dust has caused, in animals, hemolysis of the red blood cells, deposition of hemofuscin in the liver and pancreas, and injury to the lung cells; injection of the dust has caused cirrhosis of the liver and pancreas, and a condition closely resembling hemochromatosis, or bronzed diabetes. However, considerable trial exposure to copper compounds has not resulted in such disease.

As regards local effect, copper chloride and sulfate have been reported as causing irr of the skin and conjunctivae which may be on an allergic basis (Section 9). Cuprous oxide is irr to the eyes and upper respiratory tract. Discoloration of the skin is often seen in persons handling copper, but this does not indicate any actual injury from copper. There is an excess of cancer cases in the Cu smelting industry. [*102*]

In man the ingestion of a large quantity of copper sulfate has caused vomiting, gastric pain, dizziness, exhaustion, anemia, cramps, convulsions, shock, coma and death. Symptoms attributed to damage to the nervous system and kidney have been recorded, jaundice has been observed and, in some cases, the liver has been enlarged. Deaths have been reported to have occurred following the ingestion of as little as 27 g of the salt, while other victims have recovered after having taken much larger amounts, up to 120 g. Many copper-containing compounds are used as fungicides. Many Cu salts form highly unstable acetylides. Those formed in basic solutions from (Cu^+ salts + C_2H_2) are less stable than those formed from Cu^{++} salts. (Cu salts + hydrazine) react strongly, and with nitro-methane are explosive. [*19*]

COPPER-8-CUNILATE.
THR = See copper compounds.

COPPER CYANIDE. Syn: *cupric cyanide.* Yellowish-green powder. $Cu(CN)_2$, mw: 115.61, mp: decomp before melting.
Acute tox data: ip LD_{LO} (rat) = 50 mg/kg. [*3*]
THR = HIGH via ip route. See cyanides and copper compounds.

COPPER DIAZO AMINO BENZENE. Orange crystals, insol in water, sol in benzene. $CuN_3(C_6H_5)_2$, mw: 259.8, mp: 270° (decomp).
THR = See copper compounds.

COPPER DICHLOROBENZOATE. Syn: *CDCB.*
THR = A toxic material. See copper compounds. Used as a fungicide.
Disaster Hazard: Dangerous; see chlorides.

COPPER DIMETHYL DITHIOCARBAMATE.
Acute tox data: ip LD_{LO} (rat) = 25 mg/kg. [*3*]
THR = HIGH via ip route. See carbamates and copper compounds.
Disaster Hazard: Dangerous; when heated to decomp, emits toxic fumes.

COPPER ETHYL XANTHOGENATE. See copper xanthate.

COPPER FLUORIDE. Syn: *cupric fluoride.* Monoclinic blue crystals. $CuF_2 \cdot 2H_2O$, mw: 137.60, d: 2.93.
THR = See fluorides and copper compounds.

COPPER FLUOROACETIC ACID. $CuFOOCCH_3$, mw: 141.7.
Acute tox data: Oral LD_{LO} (rat) = 10 mg/kg. [*3*]
THR = HIGH via oral and inhal routes. See fluorides.

COPPER GLUCONATE. Syn: *cupric gluconate.* Light blue, fine crystalline powder, sol in water, insol in acetone, alcohol and ether.
$[CH_2OH(CHOH)_4COO]_2Cu$, mw: 453.5.
THR = U. See copper compounds. A nutrient and/or dietary supplement food additive. Also a trace mineral added to animal feed. [*109*]

COPPER HYDRIDE. Red-brown crystals. CuH, mw: 64.55, mp: decomp @ 60°, d: 6.38.
THR = See copper compounds and hydrides.

COPPER HYDROSELENITE. Bluish-green, tiny prisms. $Cu(HSeO_3)_2$, mw: 319.5.
THR = See selenium and copper compounds.

COPPER HYDROXIDE. Syn: *cupric hydroxide.* Blue gelatinous or amorphous powder. $Cu(OH)_2$, mw: 97.59, d: 3.368.
Acute tox data: Oral LD_{LO} (human) = 200 mg/kg. [*3*]
THR = HIGH via oral and inhal routes. A trace mineral added to animal feeds. Used as a fungicide. [*109*] See copper compounds.

COPPER-8-HYDROXYQUINOLINE. See copper-8-quinolinolate.

COPPER MATTE NICKEL ORE.
THR = A recog carc. [*14*] See nickel compounds.

COPPER MERCURY IODIDE. See mercuric cuprous iodide.

COPPER NAPHTHENATE. A solid. $(C_6H_5COO)_2Cu$, mw: 221.9, flash p: 100°F, d: 1.055.
Acute tox data: Oral LD_{LO} (mouse) = 110 mg/kg. [*3*]
THR = HIGH via oral and inhal routes. See copper compounds.
Fire Hazard: Mod, when exposed to heat or flame; can react with oxidizing materials.
To Fight Fire: Foam, CO_2, dry chemical.

COPPER NITRATE. Syn: *cupric nitrate*. Blue, deliquescent crystals. $Cu(NO_3)_2 \cdot 3H_2O$, mw: 241.63, mp: 114.5°, d: 2.047.
Acute tox data: Oral LD_{50} (rat) = 940 mg/kg. [*3*]
THR = MOD via oral route. Can ignite on prolonged contact with paper. Can explode when finely mixed with potassium ferrocyanide and with (water + Sn); produces sparking. [*19*] See copper compounds and nitrates.

COPPER NITRIDE. Dark-green powder. Cu_3N, mw: 204.63, mp: decomp @ 300°, d: 5.84 @ 25°/4°.
THR = See copper compounds and nitrides. When mixed with HNO_3, (conc), explodes violently. [*19*]

COPPER NITRODITHIOACETATE. Solid.
THR = See copper compounds.
Disaster Hazard: Dangerous; when heated to decomp, emits toxic fumes.

COPPER OLEATE. Syn: *cupric oleate*. Brown powder or greenish-blue mass. $Cu(C_{18}H_{33}O_2)_2$, mw: 626.46.
THR = Used as a fungicide. See copper compounds and oleic acid.

COPPER ORE.
THR = A recog carc. [*14, 3*] See arsenic and selenium compounds.

COPPER OXALATE. Syn: *cupric oxalate*. Solid, light bluish-green powder. $CuC_2H_4 \cdot \frac{1}{2}H_2O$, mw: 160.57.
THR = See oxalates and copper compounds.

COPPER OXIDE, BLACK. Syns: *cupric oxide, paramelaconite*. Fine black powder. CuO, mw: 79.5, bp: decomp @ 1026°, d: 6.4.
THR = Used as fungicide. Also a trace mineral added [*109*] to animal feeds. See copper compounds. Reacts violently with Al, B, $CsHC_2$, hydrazine, Mg, K, $RbHC_2$, PN_2H, Na, Ti, Zr. [*19*]

COPPER OXIDE, RED. See cuprous oxide.

COPPER OXYCHLORIDE. Syns: *brunswick green, cupric oxychloride*. Emerald green to greenish-black powder. $CuCl_2 \cdot 2CuO \cdot 4H_2O$, mw: 365.60, mp: $-3H_2O$ @ 140°.
Acute tox data: Oral LD_{50} (rat) = 700 mg/kg; oral LD_{LO} (human) = 200 mg/kg. [*3*]
THR = HIGH to MOD via oral and inhal routes. Reacts violently with K. [*19*] See copper compounds and chlorides.

COPPER-2,4-PENTANEDIONE DERIVATIVE. Syn: *acetylacetonate of copper*. Blue crystals. $Cu(C_6H_7O_2)_2$, mw: 261.75, mp: >230°, bp: subl.
THR = See copper compounds.

COPPER PERCHLORATE. Crystalline. $Cu(ClO_4)_2 \cdot 6H_2O$, mw: 370.6, mp: 60°.
Acute tox data: ip LD_{50} (rat) = 29 mg/kg. [*3*]
THR = HIGH via ip route. See perchlorates and copper compounds.

COPPER PEROXIDE. Brown or brownish-black crystals. $CuO_2 \cdot H_2O$, mw: 113.56.
THR = See copper compounds and peroxides.

COPPER-3-PHENYL SALICYLATE. An odorless, non-volatile crystalline material. $C_{26}H_{18}CuO_6$, mw: 490.0, mp: 145°.
Acute tox data: Oral LD_{50} (rat) = 520 mg/kg. [*3*]
THR = MOD via oral and inhal routes. See copper compounds.
Disaster Hazard: Mod dangerous; when heated to decomp, emits toxic fumes.

COPPER PHOSPHATE. Syn: *cupric phosphate*. Solid, bluish-green powder. $Cu_3(PO_4)_2$, mw: 380.6.
THR = A trace mineral added to animal feeds. [*109*] See copper compounds and phosphates.

COPPER PHOSPHIDE. Syn: *cupric phosphide*. Solid. Cu_3P_2, mw: 252.6, mp: decomp, d: 6.67.
THR = See phosphides and copper compounds. Mixed with KNO_3 or $KClO_3$, can explode. With water, liberates highly toxic and self-igniting PH_3. [*19*]

COPPER PICRATE. A solid. $Cu(C_6H_2N_3O_7)_2$, mw: 519.7.
THR = See picrates and copper compounds.

COPPER POLYSULFIDE. See cuprous sulfide.

COPPER PROPARGYLATE. A solid.
THR = See copper compounds.
Explosion Hazard: Severe, when shocked or exposed to heat. See also explosives, high.
Disaster Hazard: Dangerous; shock and heat will explode it.

COPPER PROPIONYL ACETATE. Crystals. CuC_5H_9O, mw: 148.7.
THR = See copper compounds.

COPPER PYROPHOSPHATE.
THR = U. A trace mineral added to animal feeds. [*109*] See also copper compounds and phosphates.
Disaster Hazard: Dangerous; see phosphates.

COPPER-8-QUINOLINOLATE. Syn: *copper-8-hydroxyquinoline*. Yellow-green powder.

$C_{18}H_{12}N_2O_2Cu$, mw: 351.83, mp: decomp @ 210°.
Acute tox data: ip LD_{50} (mouse) = 67 mg/kg. [*3*]
THR = HIGH via ip route. An exper (±) neo via oral and sc routes. [*3, 81*] A fungicide. See copper compounds.

COPPER RESINATE. Green powder. $Cu(C_{20}H_{29}O_2)_2$, mw: 666.43.
THR = See copper compounds.

COPPER RICINOLEATE. A green plastic solid. $Cu_2(CO_2(CH_2)_7CHCHCH_2CHOH(CH_2)_5)$, mw: 654, softening p: 64°.
THR = See copper compounds.

COPPER SEBACATE. A solid. $Cu(CH_2)_8C_2O_4$, mw: 263.8.
THR = See copper compounds.

COPPER SILICATE. Greenish crystals. $CuSiO_3$, mw: 139.6.
THR = See copper compounds and silicates.

COPPER SILICIDE. White metallic crystals. Cu_4Si, mw: 282.22, mp: 850°, d: 7.53.
THR = See copper compounds and silanes.

COPPER SILICOFLUORIDE. See cupric fluosilicate.

COPPER STEARATE. Syn: *cupric stearate.* Light blue amorphous powder. $Cu(C_{18}H_{35}O_2)_2$, mw: 630.50, mp: 125°.
THR = See copper compounds.

COPPER SUBOXIDE. Olive green crystals. Cu_4O, mw: 270.16, mp: decomp.
THR = See copper compounds.

COPPER SUBSULFATE. Syn: *cupric sulfate, basic.* Light blue powder. $4CuO \cdot SO_3$
Acute tox data: Oral LD_{LO} (human) = 200 mg/kg. [*3*]
THR = HIGH via oral route. See copper compounds and sulfates.

COPPER SULFATE. Syns: *blue vitriol, blue stone, roman vitriol.* Blue crystals or blue, crystalline granules or powder. $CuSO_4 \cdot 5H_2O$, mw: 249.71, mp: $-4H_2O$ @ 110°, d: 2.284.
Acute tox data: Oral LD_{50} (rat) = 960 mg/kg; ip LD_{50} (mouse) = 33 mg./kg. [*3*]
THR = HIGH via ip; MOD via oral and inhal routes. Reacts violently with hydroxylamine, Mg. [*19*] Used as a trace mineral added to animal feeds. It is a substance which migrates to food from packaging materials. [*109*] It is also used as an herbicide. See copper compounds and sulfuric acid.

COPPER SULFATE, AMMONIATED. Syn: *cupric sulfate, ammoniated.* Dark blue crystals. $CuSO_4 \cdot 4NH_3 \cdot H_2O$, mw: 245.8.
THR = See copper compounds, ammonia and sulfates.

COPPER SULFATE, BASIC. See copper subsulfate.

COPPER SULFIDE. Syn: *cupric sulfide.* Black powder or crystals. CuS, mw: 95.6, mp: transition @ 103°, bp: decomp @ 220°, d: 4.6.
THR = See copper compounds and sulfides. Can react violently with $Cd(ClO_3)_2$, $HClO_3$, H_2O_2, (NH_4MgNO_3 + water), $Mg(ClO_3)_2$, $Zn(ClO_3)_2$. [*19*]

COPPER TELLURITE. Green solid. $CuTeO_3$, mw: 239.15.
THR = See tellurium compounds and copper compounds.

COPPER TETRAZOL.
THR = See copper compounds.
Explosion Hazard: Severe, when exposed to heat or shock.
Disaster Hazard: Very dangerous; shock and heat will explode it; when heated to decomp, emits toxic fumes.

COPPER THIOCYANATE. Syn: *cuprous thiocyanate.* White to yellowish powder. CuCNS, mw: 121.6, mp: 1084°, d: 2.85.
THR = See copper compounds and thiocyanates.

COPPER TRICHLOROPHENATE. A crystalline solid. $Cu(Cl_3C_6H_2O)_2$, mw: 456.5.
THR = See chlorinated phenols and copper compounds.

COPPER XANTHATE. Syn: *copper ethyl xanthogenate.* Yellow precipitate. $Cu(C_3H_5OS_2)_2$, mw: 305.94, mp: decomp.
THR = See copper compounds.
Disaster Hazard: Dangerous; see sulfates.

COPPER ZINC CHROMATE. Variable in composition.
THR = HIGH. A recog carc. [*14*] See chromium compounds. Used as a fungicide.

COPPER ZINC SULFATE.
THR = Toxic. See copper and zinc compounds. Used as a fungicide.
Disaster Hazard: Dangerous. See sulfates.

COPRA.
THR = U.
Fire Hazard: Mod, when exposed to heat; dangerous if stored wet and hot.
Spontaneous Heating: Slight.

CO-RAL. See coumaphos.

CORDEAU DETONANT FUSE. See explosives, high.

CORDITE. See smokeless powder.

CORN MEAL FEEDS.
THR = Weak allergens. Local contact can cause contact dermatitis. Inhal or ingestion may provoke

bronchial asthma, eczema, hives, angioedema, conjunctivitis, rhinorrhea, etc., in already sensitized persons.
Fire Hazard: Slight, when exposed to heat.
Spont Heating: High. A safe moisture content must be retained. Presence of oil may be dangerous during storage.

CORN OIL. Light yellow, clear, oily liquid, faint characteristic odor. mp: −10°, flash p: 490° F (CC), d: 0.92, autoign. temp.: 740° F.
THR = May be allergen.
Fire Hazard: Slight, when exposed to heat or flame. Dangerous when stored if leakage impregnates rags, waste, etc.
Spont Heating: Moderate.
To Fight Fire: CO_2, dry chemical.

CORN SUGAR. See dextrose.

CORONENE.
THR = A polycyclic hydrocarbon air pollutant. An exper carc. [*23*]

COROXON. Syn: *o,o-diethyl-o-(3-chloro-4-methylcoumarin-7-yl)phosphate.*
Acute tox data: Oral LD_{50} (rat) = 12 mg/kg; oral LD_{50} (chicken) = 2.2 mg/kg. [*3*]
THR = HIGH via oral route. A cholinesterase inhibitor. See parathion.
Dangerous Hazard: Dangerous; see phosphates.

CORROSIVE LIQUID, N.O.S. See sulfuric acid or sodium hydroxide.

CORROSIVE MATERIALS STORAGE AND HANDLING. See Section 7.

CORROSIVE SUBLIMATE. See mercuric chloride.

CORTISOL. $C_{21}H_{30}O_5$, mw: 362.5.
THR = An exper teratogen. [*3*]

CORUNDUM. See emery.

CORYNINE. See yohimbine.

COSMIC RADIATION.
THR = A recog carc. [*14*] See radiation, ionizing.

COSMETICS. See drugs, medicine, or cosmetics, N.O.S.

COTARNINE CHLORIDE. Syn: *stypticin.* Light yellow powder. $C_{12}H_{13}NO_3HCl$, mw: 255.64.
THR = Large doses cause paralysis of the CNS and death due to depression of the respiratory center.
Disaster Hazard: Dangerous; see chlorides and nitrates.

COTTON DUST.
THR = MOD via inhal route. Can cause a mild febrile condition of the lungs known as byssinosis or Monday fever. This resembles metal fume fever and is prevalent in plants where the dusts of such fibers are found. Immunity can be acquired after a few days of exposure. It does not ordinarily cause any fibrosis. Coarser grades of cotton contain more dust than the finer varieties, and therefore constitute a greater hazard. It is considered an inert dust and indeed it is, within the meaning of the term. However, it can cause some illness, due to the allergens or fungi in the cotton or on the dust. Workers in processing rooms may develop conjunctivitis or blepharitis from the burned products of the gassing of the double yarn. It is a mild allergen. Inhal may produce bronchial asthma, sneezing and eczema in sensitized persons. [*65, 66, 67*]
Fire Hazard: Mod, when exposed to heat or flame; can react with oxidizing materials.
Explosion Hazard: Mod, when exposed to heat or flame.
Treatment and Antidotes: Removal from exposure will usually cause symptoms to disappear. Medical attention should be given to ascertain that there is no fibrosis of the lungs.

COTTONSEED.
THR = A very powerful allergen. Inhal or ingestion may produce bronchial asthma, sneezing, rhinorrhea, conjunctivitis, eczema and hives in persons already sensitized to this material.
Fire Hazard: Slight, when exposed to heat.
Spont Heating: Low. If piled or stored wet and hot, it can generate dangerous amounts of heat.

COTTONSEED MEAL CONTAINING AFLATOXIN.
THR = A (S) carc. [*14*] See plant and fungal products.

COTTONSEED OIL (REFINED). Oily, pale yellow, nearly odorless liquid from seeds of species of gossypium. Flash p: 486° F (CC), fp: 0° to 5°, d: 0.915–0.921 @ 25°/25°, autoign. temp.: 650° F.
THR = An allergen.
Fire Hazard: Slight, when exposed to heat or flame. However, if allowed to impregnate rags or oily waste, it can be a dangerous hazard.
Spont Heating: Mod.
To Fight Fire: CO_2, dry chemical.

COTTON WASTE, OILY, with more than 5% animal or vegetable oil.
THR = No rating.
Fire Hazard: Mod, when exposed to heat or flame; can react with oxidizing materials.

COTUNNITE. See lead chloride.

COUMADIN SODIUM. $C_{19}H_{15}O_4Na$, mw: 330.3.
THR = HIGH via oral route. An exper teratogen. [*3*]

COUMAPHOS. Syn: *co-ral.* Tan crystalline solid, insol

in water. $C_{14}H_{13}O_5PSCl$, mw: 362.6, mp: 92°.
Acute tox data: Dermal LD_{50} (rat) = 860 mg/kg; oral LD_{50} (rabbit) = 80 mg/kg; oral LD_{50} (rat) = 16 mg/kg. [*3*]
THR = MOD via dermal; HIGH via oral routes. An insecticide.

COUMARIN. Syn: *1,2-benzopyrone.* Crystals, fragrant, pleasant odor, burning taste. $C_9H_6O_2$, mw: 146.1, mp: 70°, bp: 291.0°, vap. press: 1 mm @ 106.0°.
Acute tox data: Oral LD_{50} (mouse) = 196 mg/kg; dermal TD_{LO} (mouse) = 1800 mg/kg. [*3*]
THR = HIGH via oral and MOD via dermal routes. An exper (+) neo and carc. [*3, 7, 23*] The parent substance of dicoumarol, which causes disturbances in the clotting mechanism of the blood and hence can lead to spont bleeding.
Fire Hazard: Slight, when exposed to heat or flame.

COUMARONE-INDENE RESINS. Syns: *coal tar resins, indene resins, polycoumarone resins, polyvindene resins.* Vary from fairly viscous liquids to hard resins, color ranges from pale yellow to nearly black, sol in hydrocarbon solvents, pyridine, acetone, carbon disulfide and carbon tetrachloride, insol in water and alcohol.
THR = U. A food additive permitted in food for human consumption. [*109*]

COUMESTROL. Crystals. $C_{15}H_8O_5$, mw: 268.2, mp: 385°.
THR = An exper carc. [*23*]

CRAB-E-RAD. See disodium monomethyl arsonate.

CRAG HERBICIDE 974. See 3,5-dimethyl tetrahydro-1,3,5-2H-thiodiazine-2-thione.

CRAG I. See dichlorophenoxy ethyl sulfate.

CRAG FRUIT FUNGICIDE 341. See glyodin.

CRAG FUNGICIDE 974. See mylone.

CREAM OF TARTAR. See potassium acid tartrate.

CRESOL. Syns: *2-methoxy-4-methyl phenol, 4-methyl guaiacol, 2-methoxy-p-creosol.* Colorless to yellow liquid, slightly sol in water, sol in alcohol, benzene, chloroform, ether and acetic acid.
$CH_3O(CH_3)C_6H_3OH$, mw: 138.16, d: 1.092 @ 25°/4°, mp: 5.5°, bp: 220°.
THR = HIGH. See phenol.

CREOSOTE, COAL TAR. Syn: *creosote oil.* Colorless or yellow clear, oily liquid. Composition: a mixture of phenols from coal tar. bp: 200°–250°, flash p: 165°F (CC), d: 1.07, autoign. temp.: 637°F.
THR = MOD irr via oral and inhal routes. A recog carc of skin, forearm, scrotum, face, neck and penis. An exper carc of the lungs. [*14*]
Fire Hazard: Mod, when exposed to heat or flame. Reacts violently with chlorosulfonic acid. [*19*]
Disaster Hazard: Dangerous; when heated to decomp, emits tox fumes.
To Fight Fire: Water may be used to blanket the fire; dry chemical, mist, fog.

CREOSOTE (CRESOL AND PHENOL) MIXTURE. See cresol and phenol.

CREOSOTE OIL. See creosote, coal tar.

CRESOL. Syns: *cresylic acid, cresylol, tricresol.* Description (U.S.P. XVI): mixture of isomeric cresols obtained from coal tar, colorless or yellowish to brown-yellow or pinkish liquid, phenolic odor. $C_6H_4OHCH_3$, mw: 108.10, mp: 10.9°–35.5°, bp: 191°-203°, flash p: 178°F, d: 1.030–1.038 @ 25°/25°, vap. press: 1 mm @ 38–53°, vap. d: 3.72.
Acute tox data: Oral LD_{50} (rat) = 1454 mg/kg. [*3*]
THR = MOD via oral and inhal routes. Cresol is similar to phenol in its action on the body, but it is less severe in its effects. It has corrosive action on the skin and mu mem. Systemic poisoning has rarely been reported, but it is possible that absorption may result in damage to the kidneys, liver and nervous system. The main hazard accompanying its use in industry lies in its action on the skin and mu mem, with production of severe chemical burns and dermatitis.
Fire Hazard: Mod, when exposed to heat or flame.
Explosion Hazard: Slight, in the form of vapor when exposed to heat or flame. Reacts violently with HNO_3, oleum, chlorosulfonic acid. [*19*]
Explosive Range: 1.35% @ 300°F.
Disaster Hazard: Dangerous; when heated to decomp, emits highly toxic fumes; can react vigorously with oxidizing materials.
To Fight Fire: Foam, CO_2, dry chemical.

***m*-CRESOL.** Syn: *m-methyl phenol.* Colorless to yellowish liquid, phenolic odor. C_7H_8O, mw: 108.1, mp: 10.9° bp: 202.8°, lel: 1.1% @ 302°F, flash p: 202°F, d: 1.034 @ 20°/4°, autoign. temp.: 1038°F, vap. press: 1 mm @ 52.0°, vap. d: 3.72.
Acute tox data: Oral LD_{50} (rat) = 242 mg/kg; dermal LD_{50} (rat) = 620 mg/kg. [*3*]
THR = HIGH via oral; MOD via dermal routes. With 7,12-dimethyl benz(a)anthracene, it is an exper neo. [*3*] See cresol.
Fire Hazard: See cresol.
Explosion Hazard: Mod, in the form of vapor when exposed to heat or flame.
Disaster Hazard: See cresol.

***o*-CRESOL.** Syns: *o-cresylic acid; o-hydroxytoluene.* Crystals or liquid darkening with exposure to air and light. C_7H_8O, mw: 108.1, mp: 30.8°, bp: 190.8°, flash p: 178°F, d: 1.047 @ 20°/4°, autoign. temp.: 1110°F, vap. press: 1 mm @ 38.2°, vap. d. 3.72, lel = 1.4% @ 300°F.

Acute tox data: Oral LD_{50} (rat) = 121 mg/kg; dermal LD_{50} (rat) = 1100 mg/kg; dermal LD_{50} (rabbit) = 1380 mg/kg. [*3*]

THR = HIGH via oral and MOD via dermal routes. With 7,12-dimethyl benz(a)anthracene it is an exper neo. [*3*] See cresol.

Fire Hazard: Mod via heat, flame, oxidants.

Explosion Hazard: See cresol.

Disaster Hazard: See cresol.

To Fight Fire: Water may be used to blanket fire; foam, fog, mist, dry chemical.

***p*-CRESOL.** Syn: *4-cresol.* Crystals, phenolic odor. C_7H_8O, mw: 108.1, mp: 35.5°, bp: 201.8°, lel = 1.1% @ 302°F, flash p: 202°F, d: 1.0341 @ 20°/4°, autoign. temp.: 1038°F, vap. press 1 mm @ 53.0°, vap. d: 3.72.

Acute tox data: Oral LD_{50} (rat) = 207 mg/kg; dermal LD_{50} (rat) = 750 mg/kg. [*3*]

THR = HIGH via oral and MOD via dermal routes. With 7,12-dimethyl benz(a)anthracene it is an exper neo. [*3*] See cresol.

Fire Hazard: Low, when exposed to heat or flame.

Spont Heating: No.

Explosion Hazard: Mod, in the form of vapor when exposed to heat or flame.

Disaster Hazard: See cresol.

To Fight Fire: CO_2, dry chemical, alcohol foam.

4-CRESOL. See *p*-cresol.

CRESOLITE. See 2,4,6-trinitro-*m*-cresol.

CRESOTIC ACID. See *o*-cresotinic acid.

***o*-CRESOTINIC ACID.** Syns: *cresotic acid, hydroxytoluic acid, homosalicylic acid.* White to yellowish needle-like crystals. $C_6H_3(CH_3)(OH)COOH$, mw: 152.1, mp: 166° (subl).

THR = U. See salicylic acid.

Fire Hazard: Slight.

CRESYL DIPHENYL PHOSPHATE. Liquid. $(CH_3C_6H_4)(C_6H_5)_2 \cdot PO_4$, mw: 403.3, bp: 368°, flash p: 450°F, d: 1.208, vap. d: 11.7.

THR = See tri-*o*-cresyl phosphate.

Fire Hazard: Slight, when exposed to heat or flame.

Disaster Hazard: Dangerous; see phosphates.

To Fight Fire: CO_2, dry chemical.

CRESYLIC ACID. See cresol.

***o*-CRESYLIC ACID.** See o-cresol.

CRISTOBALITE. See quartz.

CROCIDOLITE. Syn: *blue asbestos.*

THR = Considered to be the most carc of all asbestos. [*23*] See asbestos particles.

CROCOITE. See lead chromate.

CROTONALDEHYDE. Syns: *2-butenal, crotonic aldehyde, β-methyl acrolein.* Water white mobile liquid, pungent suffocating odor. $CH_3CHCHCHO$, mw: 70.09, bp: 104°, fp: −76.0°, lel = 2.1%, uel = 15.5%, flash p: 55°F, d: 0.853 @ 20°/20°, vap. d: 2.41, autoign. temp.: 450°F.

Acute tox data: Inhal TC_{LO} (humans) = 4.1 ppm; oral LD_{50} (rat) = 300 mg/kg; dermal LD_{50} (rabbit) = 380 mg/kg. [*3*]

THR = HIGH via oral, inhal and dermal routes. Reacts violently with 1,3-butadiene. [*19*] A lachrymating material which is very dangerous to the eyes. Can cause corneal burns and is irr to the skin. Very irr.

Fire Hazard: Dangerous, when exposed to heat or flame; can react with oxidizing materials.

Spont Heating: No.

Disaster Hazard: Dangerous; keep away from heat and open flame.

Treatment and Antidotes: In case of contact, immediately flush the skin or eyes with water for at least 15 min. Get medical attention.

To Fight Fire: Alcohol foam, CO_2, dry chemical.

CROTONALIC ACID. See tiglic acid.

CROTONIC ACID. Syn: *β-methacrylic acid.* Colorless needle-like crystals. $CH_3CHCHCOOH$, mw: 86.09, bp: 185°, mp: 72°, flash p: 190°F (COC), d: 1.018 @ 15°/4°, vap. press: 0.19 mm @ 20°, vap. d: 2.97.

Acute tox data: Oral LD_{50} (rat) = 1000 mg/kg; dermal LD_{50} (guinea pig) = 600 mg/kg. [*3*]

THR = MOD via oral and dermal routes. Powerful irr and corrosive.

Fire Hazard: Mod, when exposed to heat or flame; can react with oxidizing materials.

To Fight Fire: Alcohol Foam, CO_2, dry chemical.

CROTONIC ACID-2,4-DINITRO-6-(1-METHYL HEPTYL)PHENYL ESTER. See dinocap.

CROTONIC ALDEHYDE. See crotonaldehyde.

CROTONIC ANHYDRIDE. $C_8H_{10}O_3$, mw: 154.2.

Acute tox data: Oral LD_{50} (rat) = 2830 mg/kg. [*3*]

THR = MOD via oral and inhal routes.

CROTONITRILE. Syn: *2-butene nitrile.* C_4H_5N, mw: 67, flash p: <212°F, d: 0.8, vap. d: 2.3, bp: 230°–240.8°F.

THR = See nitriles. Probably very toxic.

Fire Hazard: Low, when exposed to heat or flame.

Disaster Hazard: Dangerous. See nitriles.

To Fight Fire: Alcohol foam, dry chemical, fog or mist.

CROTON OIL. Syns: *tiglium oil, croton resin.* Brownish-yellow, viscid oil, slight offensive odor. Composition: croton resin, glycerides of fatty acids and crotin. d: 0.935 @ 25°/25°.
THR = HIGH irr via oral route. An allergen. An exper carc. [*3, 23*]
Fire Hazard: Slight.

CROTONYL ALCOHOL. Syn: *2-buten-1-ol.* C_4H_8O, mw: 72.1, flash p: 91°F, d: 0.85, vap. d: 2.49, bp: 114.5°.
THR = No data. See also alcohols.
Fire Hazard: Dangerous via heat, flames (sparks), oxidizers.

CROTONYLENE. Syn: *2-butyne.* Liquid. CH_3CCCH_3, mw: 54.09, bp: 27°, flash p: <-4°F, lel = 1.4%, d: 0.688 @ 25°, vap. d: 1.91.
THR = A simple asphyxiant. See also argon.
Fire Hazard: Very dangerous, when exposed to heat or flame; can react with oxidizing materials.
Explosion Hazard: Mod, in the form of vapor when exposed to heat or flame.
To Fight Fire: Foam, CO_2, dry chemicals.

CROTOXIN. Syn: *rattlesnake venom.*
THR = HIGH toxicity. When injected, as in a snake bite, local pain results, as well as inflammation, hemorrhage, necrosis and clouding of sensorium. Effects are vertigo, impairment of motor activity and collapse with signs of shock. May be fatal.

1-CROTYL BROMIDE. Syn: *1-bromo-2-butene.* C_4H_7Br, mw: 135, lel = 4.6%, uel = 12.0%, vap. d: 4.66.
THR = MOD. See also bromides.
Fire Hazard: Dangerous, via heat, flames, oxidizers.
To Fight Fire: Water, spray, fog, alcohol foam, dry chemical.

1-CROTYL CHLORIDE. Syn: *1-chloro-2-butene.* C_4H_7Cl, mw: 90.5, lel = 4.2%, uel = 19.0%, vap. d: 3.13.
THR = MOD.
Fire Hazard: Dangerous, via heat, flame, oxidizers.
To Fight Fire: Water, spray, fog, alcohol foam, dry chemical.

CROTYLIDENE DICROTONATE. $C_{12}H_{18}O_4$, mw: 226.3.
Acute tox data: oral LD_{50} (rat) = 2590 mg/kg. [*3*]
THR = MOD via oral route.

CRUDE ISOPROPYL ALCOHOL LIQUOR. See isopropyl alcohol.

CRUDE NITROGEN FERTILIZER SOLUTION. See compressed gases, N.O.S.

CRUDE OIL. Syns: *earth oil, seneca oil, petroleum.* Thick, flam liquid, dark yellow to brown or green-black. Composition: a mixture of hydrocarbons. d: 0.780–0.970, flash p: 20°–90°F.
THR = An exper neo. [*3*] A recog carc. [*14*] See mineral oils.
Fire Hazard: Dangerous, when exposed to heat, flame or oxidizers.
Disaster Hazard: Mod dangerous; when heated to decomp, emits toxic fumes; can react with oxidizing materials.
To Fight Fire: Foam, CO_2, dry chemical.

CRUDE STEAM CRACKED OILS.
THR = A recog carc. [*14*] See mineral oils.

CRUDE THERMO-CRACKED OILS.
THR = A recog carc. [*14*] See mineral oils.

CRYOLITE. See sodium aluminum fluoride.

CRYPTOHALITE. See ammonium fluosilicate.

CRYPTOPINE. White, crystalline powder. $C_{21}H_{23}NO_5$, mw: 369.4, mp: 221°, d: 1.35.
THR = HIGH via oral and inhal routes. Related to opium in derivation and action.
Disaster Hazard: Dangerous; when heated to decomp, emits highly toxic fumes.

CRYPTOPINE HYDROCHLORIDE. See cryptopine.

CRYSPATINE.
THR = An exper carc. [*23*]

CRYSTAL AMMONIA. See ammonium carbonate.

CRYSTAL CARBONATE. See sodium carbonate.

C-STUFF. Liquid. 50% hydrazine hydrate + 50% methanol.
THR = See hydrazine and methanol.
Fire Hazard: This material is a high energy rocket propellant. Special instructions on storage and handling and fire-fighting must be obtained before having anything to do with this material.
Explosion Hazard: Very dangerous. See fire hazard.
Disaster Hazard: Very dangerous. See fire hazard.

CUBE. See rotenone.

CUMENE. Syns: *isopropyl benzene, 2-phenyl propane, cumol.* Colorless liquid. $C_6H_5CH(CH_3)_2$, mw: 120.19, mp: −96.0°, bp: 152°, flash p: 111°F, d: 0.864 @ 20°/4°, vap. press: 10 mm @ 38.3°, autoign. temp.: 795°F, lel = 0.9%, uel = 6.5%, vap. d: 4.1.
Acute tox data: Oral LD_{50} (rat) = 1400 mg/kg; inhal LC_{50} (rat) = 8000 ppm for 4 hrs. [*3*]
THR = MOD via oral and inhal routes. Cumene has a potent narcotic action characterized by a slow induction period, although the effects are of long duration. It is a depressant to the CNS. The long duration of its action indicates a possible slow rate of elimination, meaning that possible cumulative effects must be considered. It is thought to have a

greater acute toxicity than benzene or toluene; there is no apparent difference between the toxicity of pure cumene or that derived from petroleum. See also benzene and toluene.

Fire Hazard: Mod, when exposed to flame; can react with oxidizing materials. Violent reaction with HNO_3, oleum, chlorosulfonic acid. [*19*]

To Fight Fire: Foam, CO_2, dry chemical.

CUMENE HYDROPEROXIDE. $C_9H_{12}O_2$, mw: 152.2. Syn: *α-dimethyl benzyl hydroperoxide*. bp: 153°, flash p: 175°F, d: 1.05.

Acute tox data: Oral LD_{50} (rat) = 382 mg/kg; inhal LC_{50} (rat) = 220 ppm for 4 hrs; ip LD_{50} (rat) = 95 mg/kg; sc LD_{50} (rat) = 400 mg/kg. [*3*]

THR = HIGH via oral, inhal, ip and sc routes. An exper neo. [*3*]

Fire Hazard: Mod, when exposed to heat or flame; can react with reducing materials. Can decompose violently @ 150° in conc of 91%, 95%. [*19*] A strong oxidizing agent. See peroxides.

To Fight Fire: Foam, CO_2, dry chemical.

***o*-CUMENOL.** See iospropyl phenol.

***p*-(*p*-CUMENYLAZO)-N,N-DIMETHYL ANILINE.** $C_{15}H_{16}N_2$, mw: 224.3.

THR = An exper carc. [*3*]

CUMOL. See cumene.

***p*-*α*-CUMYL PHENOL.** White to light tan crystals. $C_6H_5C(CH_3)_2C_6H_4OH$, mw: 212.3, bp: 187° @ 10 mm, fp: 70°, flash p: 320°F.

THR = U.

Fire Hazard: Slight, when exposed to heat or flame; can react with oxidizing materials.

To Fight Fire: Foam, CO_2, dry chemical.

CUPRAMATE. See copper dimethyl dithiocarbamate.

CUPRIC ABIETINATE. See copper abietinate.

CUPRIC ACETATE. See copper acetate.

CUPRIC ACETATE-*m*-ARSENITE. See copper acetoarsenite.

CUPRIC-*o*-ARSENATE. Bluish-green crystals. $Cu_3(AsO_4)_2 \cdot 4H_2O$, mw: 540.50.

THR = See arsenic and copper compounds.

Disaster Hazard: Dangerous; see arsenic.

CUPRIC ARSENATE, BASIC. Green powder. $Cu_3(AsO_4)_2 \cdot Cu(OH)_2$, mw: 566.

THR = See arsenic and copper compounds.

Disaster Hazard: Dangerous; see arsenic.

CUPRIC ARSENITE. See copper arsenite.

CUPRIC-*m*-ARSENITE. Powder. $Cu(AsO_2)_2 \cdot H_2O$, mw: 295.4.

THR = See arsenic and copper compounds.

Disaster Hazard: Dangerous; see arsenic.

CUPRIC AZIDE. $Cu(N_3)_2$, mw: 147.6.

THR = Can self-explode. [*19*] See also azides and copper compounds.

CUPRIC BENZOATE. Light blue crystalline powder. $Cu(C_7H_5O_2)_2 \cdot 2H_2O$, mw: 341.8, mp: $-2H_2O$ @ 110°.

THR = See copper compounds.

CUPRIC-*m*-BORATE. Bluish-green crystalline powder. $Cu(BO_2)_2$, mw: 149.18, d: 3.859.

THR = See copper and boron compounds.

CUPRIC BORIDE. See copper boride.

CUPRIC BROMATE. Cubic blue-green crystals. $Cu(BrO_3)_2 \cdot 6H_2O$, mw: 427.47, mp: decomp @ 180°; $-6H_2O$ @ 200°, d: 2.583.

THR = See bromates and copper compounds.

CUPRIC BROMIDE. Monoclinic, black, deliquescent crystals. $CuBr_2$, mw: 223.37, mp: 498°.

THR = See bromides and copper compounds. Reacts violently with K. [*19*]

CUPRIC BUTYRATE. Dark green crystals, odor of butyric acid. $Cu(C_4H_7O_2)_2 \cdot 2H_2O$, mw: 273.8.

THR = See copper compounds.

CUPRIC CARBONATE. See copper carbonate.

CUPRIC CHLORATE. Cubic, green, deliquescent crystals. $Cu(ClO_3)_2 \cdot 6H_2O$, mw: 338.55, mp: 65°, bp: decomp @ 100°.

THR = See chlorates and copper compounds.

CUPRIC CHLORIDE. See copper chloride.

CUPRIC CHLORITE. $Cu(ClO_2)_2$, mw: 198.5.

THR = Violent explosive by percussion. [*19*]

CUPRIC CHROMATE, BASIC. Yellow-brown crystals. $CuCrO_4 \cdot 2CuO \cdot 2H_2O$, mw: 374.66, mp: $-2H_2O$ @ 260°.

THR = See chromium and copper compounds.

CUPRIC CITRATE. Bluish-green powder. $2Cu_2C_6H_4O_7 \cdot 5H_2O$, mw: 720.42.

THR = See copper compounds.

CUPRIC CYANIDE. See copper cyanide.

CUPRIC DICHROMATE. Black deliquescent crystals. $CuCr_2O_7 \cdot 2H_2O$, mw: 315.59, mp: $-2H_2O$ @ 100°, d: 2.283.

THR = See chromium compounds and copper compounds.

CUPRIC DIHYDRAZINE CHLORATE. $Cu(NH_2)_4ClO_3$, mw: 211.1.

THR = HIGH explosive, especially when dry. [*19*]

CUPRIC DIHYDROGEN-*o*-ARSENATE. Blue crystals. $Cu_5H_2(AsO_4)_4 \cdot 2H_2O$, mw: 911.39.

THR = See arsenic and copper compounds.

CUPRIC ETHYL ACETOACETATE. Green needles. $Cu(C_6H_9O_3)_2$, mw: 321.80, mp: 192°, bp: subl.
THR = See copper compounds.

CUPRIC FERRICYANIDE. Yellow-green crystals. $Cu_3[Fe(CN)_6]_2$, mw: 870.76.
THR = See ferricyanides and copper compounds.

CUPRIC FLUORIDE. See copper fluoride.

CUPRIC FLUOSILICATE. Monoclinic prisms. $CuSiF_6 \cdot 4H_2O$, mw: 277.66, d: 2.158.
THR = See fluosilicates and copper compounds.

CUPRIC FORMATE. Monoclinic, blue crystals. $Cu(CHO_2)_2$, mw: 153.58, d: 1.831.
THR = See copper compounds.

CUPRIC GLUCONATE. See copper gluconate.

CUPRIC GLYCINE DERIVATIVE. Blue needles. $Cu(C_2H_4NO_2)_2 \cdot 2H_2O$, mw: 229.68, mp: $-H_2O$ @ 130°.
THR = See copper compounds.

CUPRIC HYDROXIDE. See copper hydroxide.

CUPRIC HYPOPHOSPHITE. $Cu(PH_2O_2)_2$, mw: 193.6.
THR = Explodes suddenly @ about 90°. An impact-sensitive explosive. [*19*]

CUPRIC IODATE. Monoclinic, green crystals. $Cu(IO_3)_2$, mw: 413.38. mp: decomp, d: 5.241 @ 15°.
THR = See iodates and copper compounds.

CUPRIC LACTATE. Monoclinic, dark blue crystals. $Cu(C_3H_5O_3)_2 \cdot 2H_2O$. mw: 277.71.
THR = See copper compounds.

CUPRIC LAURATE. Light blue powder. $Cu(C_{12}H_{23}O_2)_2$, mw: 462.15, mp: 111°–113°.
THR = See copper compounds.

CUPRIC NITRATE. See copper nitrate.

CUPRIC NITROPRUSSIDE. Greenish-white powder. $CuFe(CN)_5NO \cdot 2H_2O$, mw: 315.52.
THR = See cyanides and copper compounds.

CUPRIC OLEATE. See copper oleate.

CUPRIC OXALATE. See copper oxalate.

CUPRIC OXIDE. See copper oxide.

CUPRIC OXYCHLORIDE. See copper oxychloride.

CUPRIC PALMITATE. Green-blue powder. $Cu(C_{16}H_{31}O_2)_2$, mw: 574.36, mp: 120°.
THR = See copper compounds.

CUPRIC-*p*-PERIODATE. Green powder. Cu_2HIO_6, mw: 351.01, mp: decomp @ 110°.
THR = See copper compounds and iodates.

CUPRIC PHENOSULFONATE. Bluish-green crystals. $Cu[C_6H_4(OH)SO_3]_2 \cdot 6H_2O$, mw: 518.0.
THR = See copper compounds and phenosulfonates.

CUPRIC-*o*-PHOSPHATE. See copper phosphate.

CUPRIC PHOSPHIDE. See copper phosphide.

CUPRIC SALICYLATE. Blue-green needles. $Cu(C_7H_5O_3)_2 \cdot 4H_2O$, mw: 409.8.
THR = See copper compounds.

CUPRIC SELENATE. Blue, triclinic crystals. $CuSeO_4 \cdot 5H_2O$, mw: 296.58, d: 2.559.
THR = See selenium and copper compounds.

CUPRIC STEARATE. See copper stearate.

CUPRIC SULFATE. See copper sulfate.

CUPRIC SULFATE, AMMONIATED. See copper sulfate, ammoniated.

CUPRIC SULFATE, BASIC. See copper subsulfate.

CUPRIC SULFIDE. See copper sulfide.

CUPRIC TARTRATE. Light blue powder. $CuC_4H_4O_6$, mw: 211.6.
THR = See copper compounds.

CUPRIC TETRAMMINE CHLORATE. $Cu(NH_3)_4(ClO_3)_2$, mw: 298.6.
THR = Detonates on impact. [*19*] See also copper compounds and chlorates.

CUPRIC TETRAMMINE PERCHLORATE. $Cu(NH_3)_4(ClO_4)_2$, mw: 330.6.
THR = Can detonate on impact. [*19*] See also copper compounds and perchlorates.

CUPRIC THIOCYANATE. Black crystals. $Cu(SCN)_2$, mw: 179.71, mp: decomp @ 100°.
THR = See thiocyanates and copper compounds.

CUPRIC TUNGSTATE. Light green crystals. $CuWO_4 \cdot 2H_2O$, mw: 347.49, mp: @ red heat.
THR = See copper and tungsten compounds.

CUPRIETHYLENE DIAMINE. $(C_2H_{10}N_2)Cu$.
THR = An irr and corrosive material.

CUPRITE. See cuprous oxide.

CUPROUS ACETYLIDE. Amorphous red powder. Cu_2C_2, mw: 150.10, mp: explodes.
THR = See copper compounds.
Explosion Hazard: Severe, when shocked or exposed to heat. Reacts vigorously with C_2H_2, Br, Cl_2, I, $AgNO_3$. [*19*]
Disaster Hazard: Highly dangerous; shock or heat will explode it.

CUPROUS ARSENATE, BASIC. See copper arsenate, basic.

CUPROUS AZIDE. CuN_3, mw: 107.5.
THR = Decomp @ 205°. Highly unstable. [*19*] See also azides.

CUPROUS BROMIDE. White crystals. CuBr, mw: 143.66, mp: 504°, bp: 1355°, d: 4.718, vap. press: 1 mm @572°.

THR = See bromides and copper compounds. Reacts violently with K. [*19*]

CUPROUS CARBONATE. Yellow crystals. Cu_2CO_3, mw: 187.1, mp: decomp, d: 4.40.
THR = See copper compounds.

CUPROUS CHLORIDE. Syn: *nantokite*. Cubic white crystals. CuCl, mw: 99.00, mp: 422°, bp: 1366°, d: 3.53, vap. press: 1 mm @ 546°.
THR = See chlorides and copper compounds. Reacts violently with K. [*19*]

CUPROUS CYANIDE. Monoclinic white prisms. CuCn, mw: 89.56, mp: 473° in N_2, bp: decomp, d: 2.92.
THR = See cyanides and copper compounds. A poison. Reacts violently with Mg. [*19*]

CUPROUS FERRICYANIDE. Brown-red crystals. $Cu_3Fe(CN)_6$, mw: 402.59.
THR = See ferricyanides and copper compounds.

CUPROUS FLUOGALLATE. Pale blue crystals. $[Cu(H_2O)_6](GaF_5 \cdot H_2O)$, mw: 354.37, mp: $-5H_2O$ @ 110°, d: 2.20.
THR = See fluogallates and copper compounds.

CUPROUS FLUORIDE. Red crystals. CuF, mw: 82.54, mp: 908°, bp: subl @ 1100°.
THR = See fluorides and copper compounds.

CUPROUS FLUOSILICATE. Syn: *cuprous silicofluoride*. Red powder. Cu_2SiF_6, mw: 269.14, mp: decomp.
THR = See fluosilicates and copper compounds.

CUPROUS HYDRIDE. Yellow-brown crystals, insol in cold water, reacts in hot water. Cu_2H_2, mw: 128.2, mp: decomp slowly @ 25°, d: 6.29.
THR = See copper compounds and hydrides. Can react violently with halogens. Self-igniting in air. [*19*]

CUPROUS HYDROXIDE. Yellow crystals. CuOH, mw: 80.55, mp: $-\frac{1}{2}H_2O$ @ 360°, d: 3.37.
THR = See copper compounds.

CUPROUS IODIDE. Syn: *marshite*. Cubic white crystals. CuI, mw: 190.46, mp: 605°, bp: 1290°, d: 5.62, vap. press: 10 mm @ 656°.
THR = A nutrient and/or dietary supplement food additive. Also a trace mineral added to animal feeds. [*109*] See iodides and copper compounds. Reacts violently with K. [*19*]

CUPROUS NITRIDE. Cu_3N, mw: 204.6.
THR = Reacts violently with HNO_3, H_2SO_4. [*19*]

CUPROUS OXIDE. Syn: *cuprite*. Octahedral, cubic red crystals. Cu_2O, mw: 143.14, mp: 1235°, bp: $-O_2$ @ 1800°, d: 6.0.
Acute tox data: Oral LD_{50} (rat) = 470 mg/kg. [*3*]
THR = HIGH via oral route. See copper compounds.

CUPROUS SELENIDE. Black cubes. Cu_2Se, mw: 206.04, mp: 1113°, d: 6.749 @ 30°/4°.
THR = See selenium and copper compounds.

CUPROUS SILICOFLUORIDE. See cuprous fluosilicate.

CUPROUS SULFATE. Gray powder. Cu_2SO_4, mw: 223.15, mp: −O @ 200°, d: 3.605.
THR = See copper compounds and sulfates.

CUPROUS SULFIDE. Syn: *chalcocite*. Rhombic, black crystals. Cu_2S, mw: 159.15, mp: 1100°, d: 5.6.
THR = See sulfides and copper compounds.

CUPROUS SULFITE. Red prisms. $Cu_2SO_3 \cdot H_2O$, mw: 225.16, d: 4.46 @ 15°.
THR = See sulfites and copper compounds.

CUPROUS TETRAHYDRO ALUMINATE. $CuAlH_4$, mw: 94.6.
THR = Self-ignites in air. [*19*]

CUPROUS THIOCYANATE. See copper thiocyanate.

CURARE. Brown, brittle, resinous mass.
Acute tox data: iv LD_{50} (dog) = 1.2 mg/kg; iv LD_{50} (rabbit) = 1.3 mg/kg. [*3*]
THR = HIGH via iv route. A deadly poison. Visual disturbances, choking sensation, then generalized paralysis ensues from exposure.
Disaster Hazard: Dangerous; when heated to decomp, emits highly toxic fumes.
Treatment and Antidotes: Artificial respiration, physostigmine. Call a physician immediately.

CURIUM. Cm, mp: 1340°, d: (calc) = 13.51. A silvery metal, chemically reactive, and bone seeking when absorbed into the body.
THR = HIGH radio-toxicity.
Radiation Hazard: For permissible levels, see Section 5A, Table 5A.5. Artificial isotope ^{242}Cm, $T\frac{1}{2}$ = 160d, decays to ^{238}Pu via α's of 6.1 MeV. Artificial isotope ^{243}Cm, $T\frac{1}{2}$ = 32y, decays to ^{239}Pu via α's of 5.7–5.8 MeV. Artificial isotope ^{244}Cm, $T\frac{1}{2}$ = 18y, decays to ^{240}Pu via α's of 5.8 MeV. Artificial isotope ^{245}Cm, $T\frac{1}{2}$ = 9300y, decays to ^{241}Pu via α's of 5.3–5.4 MeV.

CURLED MINT OIL. See spearmint oil.

CUTTING OILS.
THR = This oil is the cause of "cutting oil" dermatitis. Although it is generally caused by an insoluble oil, it can occasionally be caused by a soluble one. Many have looked for a causative factor other than the oil itself. Bacteria have frequently been blamed, although insol oils are usually sterile while the sol oils may contain bacteria. The metal slivers which occur in these oils after use have also been blamed as well as the sulfur, chlorine and inhibitors which they con-

tain. The oil itself can plug the pores, forming boils. They are often carc. [*14*, *23*] See mineral oils.
Fire Hazard: Slight, when exposed to heat or flame.
Treatment and Antidotes: Disinfectant agents have been added to the oils to cut down the infectious effects. Filters have been installed in the cutting oil lines to filter out the tiny metal slivers which may irr and cause infection of the skin. General housecleaning measures are also thought advisable to promptly remove any excess oil which gets onto the hands and skin. It has been found that occasionally changing from one oil to another may bring relief from the irr. This would also indicate that the specific irr is not present in all oils.

CYACETAZIDE. See cyanoacetic acid hydrazide.

CYAMELIDE. Syn: *insoluble cyanuric acid.* White powder. $C_3H_3N_3O_3$, mw: 129.1.
THR = HIGH via oral and inhal routes. See also cyanides.
Disaster Hazard: Dangerous; when heated to decomp or on contact with acid or acid fumes, emits highly toxic fumes.

CYANAMIDE. Syns: *carbodiimide*, *cyanogenamide.* Deliquescent crystals. HNCNH, mw: 42.05, mp: 45°, bp: 260°, flash p: 285°F, d: 1.282, vap. d: 1.45.
Acute tox data: Oral LD_{50} (rat) = 125 mg/kg; ip LD_{LO} (rat) = 200 mg/kg. [*3*]
THR = HIGH via oral, inhal and ip routes. Does not contain free cyanide. Causes increase in respiration and pulse rate, lowered blood pressure and dizziness. There may be a flushed appearance of the face.
Fire Hazard: Slight, when exposed to heat or flame.
Disaster Hazard: Mod dangerous; when heated to decomp or on contact with acid or acid fumes, emits toxic fumes.
To Fight Fire: CO_2, dry chemical.

CYANATES.
THR = Variable. See individual entry.
Disaster Hazard: Dangerous; when heated to decomp or on contact with acid or acid fumes, they emit toxic fumes.

CYANIC ACID. Syn: *isocyanic acid.* A liquid. Very acrid odor. HOCN, mw: 43.01, bp: 23.3°, mp: −86°, d: 1.140 @ 0°/0°.
THR = HIGH via oral and inhal routes. HIGH irr.
Explosion Hazard: Severe.
Disaster Hazard: Dangerous; can explode; when heated to decomp or on contact with acid or acid fumes, emits highly toxic fumes and flam vapors.

CYANIDE OF CALCIUM OR CYANIDE OF CALCIUM, MIXTURE, SOLID. See calcium cyanide.

CYANIDE OF POTASSIUM. See potassium cyanide.

CYANIDE OF SODIUM. Syn: *sodium cyanide.* See also sodium cyanide.
THR = The volatile cyanides resemble hydrocyanic acid physiologically, inhibiting tissue oxidation and causing death through asphyxia. Cyanogen is probably as toxic as hydrocyanic acid; the nitriles are generally considered somewhat less toxic, probably because of their lower volatility. The non-volatile cyanide salts appear to be relatively non-toxic systemically, so long as they are not ingested and care is taken to prevent the formation of hydrocyanic acid. Workers, such as electroplaters and picklers, who are daily exposed to cyanide solutions may develop a "cyanide" rash, characterized by itching, and by macular, papular, and vesicular eruptions. Frequently there is secondary infection. Exposure to small amounts of cyanide compounds over long periods of time is reported to cause loss of appetite, headache, weakness, nausea, dizziness, and symptoms of irr of the upper respiratory tract and eyes. See also specific compounds.
Fire Hazard: Mod, by chemical reaction with heat, moisture, acid. Many cyanides evolve hydrocyanic acid rather easily. This is a flam gas and is highly toxic. Carbon dioxide from the air is sufficiently acidic to liberate hydrocyanic acid from cyanide solutions. See also hydrocyanic acid.
Explosion Hazard: See hydrocyanic acid. Explodes if melted with nitrite or chlorate @ about 450°. Violent reaction with F_2, Mg, nitrates, HNO_3, nitrites. [*19*]
Disaster Hazard: Dangerous; on contact with acid, acid fumes, water or steam, they will produce toxic and flam vapors.

CYANIDES OF COPPER, ZINC, LEAD AND SILVER. See specific entries.

CYANOACETAMIDE. Syns: *nitrilomalonamide*, *propionamide nitrile.* White powder. $CNCH_2CONH_2$, mw: 84.08, mp: 119°, bp: decomp.
Acute tox data: Oral LD_{50} (rat) = 1680 mg/kg; ip LD_{50} (mouse) = 750 mg/kg. [*3*]
THR = MOD via oral, inhal and ip routes. See cyanides.

CYANOACETIC ACID. Syn: *malonic acid mononitrile.* White, deliquescent needles. $CNCH_2COOH$, mw: 85.06, mp: 66°, bp: 108° @ 15 mm.
Acute tox data: ip LD_{50} (mouse) = 200 mg/kg. [*3*]
THR = HIGH via ip route. Violent reaction with furfuryl alcohol. [*19*]

CYANOACETIC ACID HYDRAZINE. Syn: *cyacetazide.* $C_3H_5ON_3$, mw: 99.1, mp: 115°.
Acute tox data: Oral LD_{50} (mouse) = 250 mg/kg. [*3*]
THR = HIGH via oral and inhal routes.

CYANOACETONITRILE. Syns: *malonitrile, malonic dinitrile, methylene dicyanide.* White powder. $CH_2(CN)_2$, mw: 66.1, d: 1.049 @ 34°/4°, mp: 30.5°, bp: 220°, flash p: 266°F (TOC).
Acute tox data: ip LD_{50} (mouse) = 13 mg/kg; oral LD_{50} (rat) = 61 mg/kg. [*3*]
THR = HIGH via ip, oral and inhal routes.
Disaster Hazard: Dangerous; when heated to decomp, emits highly toxic fumes.

CYANOACETYL CHLORIDE. $CNCH_2COCl$, mw: 101.5.
THR= Explodes spont. [*19*]

N-CYANOACETYL ETHYL CARBAMATE. $C_6H_8O_3N_2$, mw: 156.2.
THR = An exper neo. [*3*]

CYANOAURIC ACID. Syn: *auric cyanide.* Tablets. $HAu(CN)_4 \cdot 3H_2O$, mw: 356.33, mp: 50°, bp: decomp.
THR = HIGH. See cyanides and gold compounds.

7-CYANOBENZ(a)ANTHRACENE. $C_{19}H_{11}N$, mw: 253.3.
THR = An exper neo and carc. [*3, 23*]

CYANOCOBALTIC III ACID. Syn: *cobalticyanic acid.* Colorless needles, deliquescent. $[H_3Co(CN)_6]_2H_2O$, mw: 454.16, mp: decomp @ 100°.
THR = See cyanides and cobalt compounds.

N-CYANODIALLYL AMINE. See diallyl cyanamide.

N-CYANODIETHYL AMINE. See diethyl cyanamide.

CYANODIMETHYL AMINO ETHOXY PHOSPHINE OXIDE. See tabun.

CYANODIMETHYL ARSINE. $(CH_3)_2AsCN$, mw: 131.
THR = Self igniting in air. [*19*]

8-CYANO-7,12-DIMETHYL BENZ(a)ANTHRACANE. $C_{21}H_{15}N$, mw: 281.4.
THR = An exper neo. [*3, 23*]

2-CYANOETHANOL. $HOCH_2CH_2CN$, mw: 71.1.
THR = Reacts violently with H_2SO_4. [*19*]

2-(2-CYANO ETHOXY)ETHYL ACETATE.
THR = HIGH. Limited animal exper suggest HIGH toxicity. See also nitriles.
Disaster Hazard: Dangerous; see nitriles.

4-CYANOETHOXY-2-METHYL-2-PENTANOL.
THR = HIGH. Limited animal exper suggest toxicity less than that of cyanides. See also nitriles.
Disaster Hazard: Dangerous; see nitriles.

CYANOETHYL ACRYLATE. Sol in water. $C_4H_6O_2CN$, mw: 80, d: 1.069, bp: polymerizes when heated, fp: −16.9°, flash p: 255°F (COC), vap. d: 4.3.
Acute tox data: Oral LD_{50} (rat) = 180 mg/kg. [*3*]
THR = HIGH irr via oral and inhal routes. See also nitriles.
Fire Hazard: Slight, when exposed to heat or flame.
Disaster Hazard: Dangerous; see nitriles.
To Fight Fire: Dry chemical, CO_2, fog or mist.

***n*-(2-CYANOETHYL)CYCLOHEXYL AMINE.** Insol in water. $C_6H_{11}NHC_2H_4CN$, mw: 152, flash p: 255°F (OC), d: 0.9, vap. d: 5.2.
Fire Hazard: Slight, when exposed to heat or flame.
Disaster Hazard: Dangerous; see nitriles.
To Fight Fire: Dry chemical, CO_2, fog, mist, spray.

3-CYANOETHYL PROPIONATE. $CH_3CH_2COOCH_2CH_2CN$, mw: 127.
THR = Limited animal exper suggest MOD. See also nitriles.
Disaster Hazard: Dangerous; see nitriles.

CYANOFORMIC CHLORIDE. Oily liquid. COCNCl, mw: 89.49, bp: 126°–128° @ 750 mm.
THR = HIGH. See cyanides and chlorides.

CYANOGEN (GAS). Syns: *ethane dinitrile, prussite.* Colorless gas, pungent odor. NCCN, mw: 52.04, mp: −34.4°, bp: −21.0°, d: 0.866 @ 17°/4°, lel = 6.6%, uel = 32%, vap. d: 1.8.
Acute tox data: Inhal LC_{50} (rat) = 350 ppm for 1 hr.
THR = HIGH irr via inhal route. See cyanides. Violent reaction with F_2, O_2, [*19*]
Fire Hazard: Very dangerous, via heat, flames (sparks), oxidizers.
Disaster Hazard: Dangerous, when heated to decomp or on contact with acid, acid fumes, water or steam, will react to produce highly toxic fumes.
To Fight Fire: Stop flow of gas.

CYANOGENAMIDE. See cyanamide.

CYANOGEN AZIDE. NCN_3, mw: 68.
THR = Explodes when heated or shocked. [*19*]

CYANOGEN BROMIDE. Syn: *bromine cyanide.* Colorless needles. BrCN, mw: 105.93, mp: 52°, bp: 61.6°, d: 2.015 @ 20°/4°, vap. press: 100 mm @ 22.6°.
Acute tox data: Inhal LC_{LO} (human) = 92 ppm for 10 hrs; inhal LC_{LO} (mouse) = 500 mg/m^3 for 10 min. [*3*]
THR = HIGH via inhal route.

CYANOGEN CHLORIDE. Syns: *chlorine cyanide, CK.* Colorless liquid or gas, lacrimatory and irr odor. CNCl, mw: 61.48, mp: −6.5°, bp: 13.1°, d: 1.218 @ 4°/4°, vap. press: 1010 mm @ 20°, vap. d: 1.98.
Acute tox data: Inhal TC_{LO} (human) = 10 mg/m^3; inhal LC_{50} (rat) = 118 ppm for 30 min; inhal LC_{50} (mouse) = 177 ppm for 30 min. [*3*]
THR = HIGH irr via inhal and ocular routes. An insecticide. See cyanides and hydrochloric acid.
Disaster Hazard: Highly dangerous; when heated to decomp or on contact with water or steam, will react to produce highly toxic and corrosive fumes.

CYANOGEN FLUORIDE. Colorless gas. CNF, mw: 45.02, mp: subl @ −72°.
THR = See cyanides and fluorides.

CYANOGEN IODIDE. Syn: *iodine cyanide*. Colorless needles. CNI, mw: 152.94, mp: 146.5°, vap. press: 1 mm @ 25.2°.
Acute tox data: sc LD_{LO} (rat) = 44 mg/kg. [*3*]
THR = HIGH via sc route. Violent reaction with P. [*19*] See cyanides and iodides.

CYANOGEN SULFIDE. Syn: *cyanogen thiocyanate*. Crystals. NCSCN, mw: 84.1, mp: 65°.
Acute tox data: Oral LD (rat) = 100 mg/kg.
THR = HIGH. See cyanides and thiocyanates.

CYANOGEN THIOCYANATE. See cyanogen sulfide.

CYANOGUANIDINE. See dicyandiamide.

6-CYANOHEXANOIC ACID. $CH_2CN(CH_2)_4COOH$, mw: 141.
THR = Probably HIGH. See also nitriles.
Disaster Hazard: Dangerous; see nitriles.

CYANOMETHYL ACETATE. Syn: *methyl cyanoethanoate*. Colorless liquid. $CNCH_2COOCH_3$, mw: 99.1, mp: −22.5°, bp: 200°, d: 1.123 @ 15°.
Acute tox data: Oral LD_{50} (rat) = 32 mg/kg; inhal LC_{LO} (rat) = 16 ppm for 4 hrs; dermal LD_{50} (rabbit) = 43 ppm. [*3*]
THR = HIGH via oral, inhal and dermal routes.

7-CYANO-12-METHYL BENZ(a)ANTHRACENE. $C_{20}H_{14}N$, mw: 268.4.
THR = An exper neo and carc. [*3, 23*]

8-CYANO-7-METHYL BENZ(a)ANTHRACENE.
THR = An exper neo. [*3, 23*]

10-CYANO-7-METHYL BENZ(a)ANTHRACENE.
THR = An exper carc. [*23*]

N-CYANOMETHYL MORPHOLINE. $C_6H_{10}ON_2$, mw: 126.2.
Acute tox data: Oral LD_{50} (rat) = 1230 mg/kg; dermal LD_{50} (rabbit) = 200 mg/kg. [*3*]
THR = MOD via oral and HIGH via dermal routes. See also nitriles and morpholine.
Disaster Hazard: Dangerous; see nitriles.

3-CYANOPYRIDINE. Gray crystals. $C_6H_4N_2$, mw: 104.1, mp: 47°–49°, bp: 83°–84° @ 10 mm.
THR = U. Probably HIGH.
Fire Hazard: Slight, when exposed to heat or flame.
Disaster Hazard: Dangerous; see cyanides; can react with oxidizing materials.

2-CYANO-4-STILBENAMINE.
THR = An exper carc. [*23*]

CYANOTRIAMIDE. See melamine.

CYANOVINYL ACETATE. $CH_3COO(CNCH{:}CH_2)$, mw: 112.
Acute tox data: Oral LD (rat) = 100 mg/kg.
THR = HIGH. See also nitriles.
Disaster Hazard: Dangerous. See nitriles.

CYANURIC ACID. Syns: *sym-triazinetriol, tricyanic acid*. Off-white, odorless crystals. $C_3H_3N_3O_3$, mw: 129.1, mp: >360°, d: 2.500 @ 20°/4°.
THR = MOD via oral route. Dry form irr eyes and abraided skin. An exper neo. [*3*]
Disaster Hazard: Dangerous; Violent reaction with ethanol. [*19*]

CYANURIC CHLORIDE. Syns: *trichloro-s-triazine, trichlorocyanidine, tricyanogen chloride*. Monoclinic, colorless crystals, pungent odor. $C_3Cl_3N_3$, mw: 184.43, mp: 145.8°, bp: 190°, d: 1.32 @ 20°/4°, vap. press: 2 mm @ 70°, vap. d: 6.36.
Acute tox data: Oral LD_{50} (rat) = 485 mg/kg. [*3*]
THR = HIGH via oral and inhal routes. An allergen. A exper carc. [*3*] Has been reported as causing irr of mu mem and disturbed rhythm of the heart in humans. Violent reaction with dimethyl sulfoxide, water. [*19*]
Disaster Hazard: Dangerous; see chlorides and nitriles.

CYANURIC TRIAZIDE.
THR = U. See also cyanides and azides.
Explosion Hazard: Severe, when shocked or exposed to heat.
Disaster Hazard: Dangerous; shock or heat will explode it; when heated to decomp or on contact with acid or acid fumes, emits highly toxic fumes.

CYCAD NUT MEAL CONTAINING CYCASIN. See cycasin and plant and fungal products.

CYCASIN. $C_8H_{16}O_7N_2$, mw: 252.3.
Acute tox data: Oral LD_{50} (rat) = 270 mg/kg. [*3*]
THR = HIGH via oral route. An exper (S) carc, [*3, 23*] mutagen and teratogen. [*14, 23*]

CYCLAMATE SODIUM. $C_6H_{13}O_3NSNa$, mw: 202.3.
THR = MOD via oral route. An exper neo. [*3, 23*]

CYCLETHRIN. See 3-(2-cyclopentenyl)-2-methyl-4-oxo-2-cyclopentenyl ester of chrysanthemum monocarboxylic acid.

"CYCLINE OIL." See mineral oil, vegetable oil.

CYCLOBUTANE. Syn: *tetramethylene*. A gas. C_4H_8, mw: 56.10, mp: −50°, bp: 12.9°, flash p: <50°F (CC), d: 0.708 @ 11°, vap. d: 1.93.
THR = U. May be a simple asphyxiant. See also cycloparaffins.
Fire Hazard: Very dangerous, when exposed to heat or flame; can react with oxidizing materials.
Spont Heating: No.

To Fight Fire: Stop flow of gas; CO_2, dry chemical or water spray.

CYCLOBUTENE. Syn: *cyclobutylene.* Gas. C_4H_6, mw: 54.1, bp: 2.4°, d: 0.733 @ 0°/4°.
THR = U. May be a simple asphyxiant.
Fire Hazard: Dangerous, when exposed to heat or flame; can react with oxidizing materials.

CYCLOBUTYLENE. See cyclobutene.

CYCLOCHLOROTINE. White needles, $C_{24}H_{30}O_7N_5Cl_2$, mp: 251°, decomp; mw: 571.1.
THR = HIGH via oral route. An exper neo. [*3, 7*]

CYCLOCYTIDINE. $C_9H_{11}O_4N_3 \cdot HCl$, mw: 261.7.
THR = An exper teratogen. [*3*]

***p*-N-CYCLOETHYLENE UREIDIAZO BENZENE.** $C_{15}H_{14}ON_4$, mw: 266.3.
THR = An exper neo. [*3, 23*]

CYCLOHEPTANE. Syn: *suberane.* An oil. C_7H_{14}. mw: 98.2, mp: −12°, bp: 117°, flash p: <70°, d: 0.8099 @ 20°/4°, vap. d: 3.3.
THR = U. See cycloparaffins.
Fire Hazard: Dangerous, when exposed to heat or flame; can react with oxidizing materials.
To Fight Fire: Foam, CO_2, dry chemical.

CYCLOHEPTANONE. Syn: *suberone.* Liquid, nearly insol in water. $C_7H_{12}O$, mw: 112.2, bp: 181°, d: 0.9490 @ 20°/4°.
Acute tox data: ip LD_{50} (mouse) = 750 mg/kg. [*3*]
THR = MOD via ip route mainly CNS depression.

CYCLOHEXANE. Syns: *hexahydrobenzene, hexamethylene.* Colorless mobile liquid, pungent odor. C_6H_{12}, mw: 84.16, mp: 6.5°, bp: 80.7°, fp: 4.6°, flash: p: −4°F, ulc: 90–95, lel = 1.3%, uel = 8.4%, d: 0.7791 @ 20°/4°, autoign. temp.: 473°F, vap. press: 100 mm @ 60.8°, vap. d: 2.90.
Acute tox data: Oral LD_{50} (mouse) = 1297 mg/kg. [*3*]
THR = MOD irr via inhal and oral routes. Irr to skin. See also cycloparaffins.
Fire Hazard: Dangerous, when exposed to heat or flame; can react with oxidizing materials.
Spont Heating: No.
Explosion Hazard: Mod, in the form of vapor when exposed to flame. When mixed hot with liquid N_2O_4 an explosion resulted. [*19*]
To Fight Fire: Foam, CO_2, dry chemical, spray, fog.

***cis*-1,2-CYCLOHEXANE DICARBOXYLIC ANHYDRIDE.** See *cis*-hexahydrophthalic anhydride.

CYCLOHEXANOL. Syn: *hexahydrophenol.* Colorless needles of viscous liquid, hygroscopic, camphor-like odor. $C_6H_{11}OH$, mw: 100.16, mp: 24°, bp: 161.5°, flash p: 154°F (CC), d: 0.9449 @ 25°/4°, vap. press 1 mm @ 21.0°, vap. d: 3.45, autoign. temp.: 572°F.
Acute tox data: Oral LD_{50} (rat) = 2060 mg/kg; inhal TC_{LO} (human) = 75 ppm. [*3*]
THR = MOD via oral and inhal routes. Narcotic in high conc. Has caused damage to kidneys, liver and blood vessels in exper animals.
Fire Hazard: Mod, when exposed to heat or flame; can react with oxidizing materials. Violent reaction with HNO_3. [*19*]
Spont Heating: No.
To Fight Fire: Alcohol foam, foam, CO_2, dry chemical.

CYCLOHEXANOL ACETATE. Pale yellow liquid. $C_6H_{11}OOCCH_3$, mw: 142.19, bp: 177°, d: 0.996, vap. d: 4.9, flash p: 136°F, autoign. temp.: 633°F, vap. d: 4.9.
Acute tox data: Inhal TC_{LO} (human) = 516 ppm for 8 hrs. [*3*]
THR = MOD via inhal route. See cycloparaffins.
Fire Hazard: Mod via heat, open flames, oxidizers.

CYCLOHEXANONE. Syns: *ketohexamethylene, pimelic ketone.* Colorless liquid, acetone-like odor. $CO(CH_2)_4CH_2$, mw: 98.14, mp: −45.0°, bp: 115.6°. ulc: 35–40, lel = 1.1% @ 100°, flash p: 111°F, d: 0.9478 @ 20°/4°, autoign. temp.: 788°F, vap. press: 10 mm @ 38.7°, vap. d: 3.4.
Acute tox data: Inhal TC_{LO} (human) = 75 ppm; oral LD_{50} (rat) = 1620 mg/kg; dermal LD_{50} (rabbit) = 1000 mg/kg. [*3*]
THR = MOD via inhal, dermal and oral routes. MILD narcotic properties have also been described. See also cycloparaffins.
Fire Hazard: Mod, when exposed to heat or flame; can react vigorously with oxidizing materials such as HNO_3.
Spont Heating: No.
Explosion Hazard: Slight, in its vapor form, when exposed to flame.
To Fight Fire: Alcohol foam, dry chemical or CO_2.

CYCLOHEXANONE-Δ. Liquid. C_6H_8O, mw: 96.12, bp: 155.5°, flash p: 93°F (CC), vap. d: 3.31, vap. press: 4 mm @ 20°.
THR = Skin contact can cause a dermatitis. Irr to eyes, skin and mu mem. Can damage the liver and kidneys. See also cycloparaffins.
Fire Hazard: Dangerous, when exposed to flame and heat; can react with oxidizing materials.
Spont Heating: No.
To Fight Fire: CO_2, dry chemical.

CYCLOHEXANONE PEROXIDE (paste with dibutyl phthalate). Off-white thick paste, insol in water, sol in common organic solvents. $C_{12}H_{22}O_5$ + dibutyl phthalate, mw: 246.3.
THR = See peroxides, organic.

CYCLOHEXENE. Syn: *1,2,3,4-tetrahydrobenzene*. Colorless liquid. $CH_2CH_2CH_2CH_2CHCH$, mw: 82.14, bp: 83°, fp: −103.7°, flash p: 20°F, d: 0.8102 @ 20°/4°, vap. press: 160 mm @ 38°, autoign. temp.: 590°F, vap. d: 2.8.
THR = MOD via inhal route.
Fire Hazard: Dangerous, when exposed to flame; can react with oxidizing materials.
Disaster Hazard: Dangerous; keep away from heat and open flame.
To Fight Fire: Foam, CO_2, dry chemical.

4-CYCLOHEXENE-1-CARBOXALDEHYDE. Syn: *1,2,5,6-tetrahydrobenzaldehyde*. Liquid, slightly sol in water. $C_7H_{10}O$, mw: 110.2, d: 0.9721, bp: 164.2°, fp: −108°, flash p: 135°F (OC).
Acute tox data: Oral LD_{50} (rat) = 2460 mg/kg; dermal LD_{LO} (rabbit) = 1770 mg/kg. [3]
THR = MOD via oral and dermal routes. See also aldehydes.
Fire Hazard: Mod, when exposed to heat, flame or oxidizers.
To Fight Fire: Alcohol foam, foam, dry chemical.

CYCLOHEXENE-1,2-DICARBOXIMIDE. $C_6H_9O_2N$, mw: 127.2.
THR = An exper teratogen. [3]

CYCLOHEXENE OXIDE. Clear liquid. $C_6H_{10}O$, mw: 98.14, bp: 129.5°, flash p: 81°F, d: 0.9678 @ 25°/4°, vap. d: 3.5.

CYCLOHEXENYL TRICHLOROSILANE. Colorless fuming liquid, HCl odor. $C_6H_9Cl_3Si$, mw: 215.6, bp: 202°, d: 1.263 @ 25°/25°, flash p: 200°F (COC).
Acute tox data: Oral LD_{50} = 2800 mg/kg. [3]
THR = MOD via oral route. An irr. See also hydrochloric acid.
Disaster Hazard: Exposed to moisture it readily hydrolyzes to HCl. See hydrochloric acid.

CYCLOHEXIMIDE. Syn: *actidione*. Crystals, mod sol in water, sol in chloroform, ether and acetone. $C_{15}H_{23}NO_4$, mw: 281.3, mp: 116°.
Acute tox data: Oral LD_{50} (rat) = 2 mg/kg; ip LD_{50} (rat) = 3.7 mg/kg; sc LD_{50} (rat) = 3 mg/kg; iv LD_{50} (rat) = 150 mg/kg. [3]
THR = HIGH via iv, ip, sc, oral and dermal routes. A poison.

CYCLOHEXYL ACETATE. See cyclohexanol acetate.

CYCLOHEXYL ALCOHOL. See cyclohexanol.

CYCLOHEXYLAMINE. Syns: *hexahydroaniline, aminocyclohexane*. Liquid, strong fishy odor. $C_6H_{13}N$, mw: 99.2, mp: −17.7°, bp: 134.5°, flash p: 90°F (OC), d: 0.865 @ 25°/25°, autoign. temp.: 560°F, vap. d. 3.42.
Acute tox data: Oral LD_{50} (rat) = 710 mg/kg; inhal LC_{LO} (rat) = 8000 ppm for 4 hrs; dermal LD_{50} (rabbit) = 320 mg/kg. [3]
THR = MOD via oral and inhal routes; HIGH via dermal route. May cause dermatitis, convulsions.
Fire Hazard: Dangerous, when exposed to heat or flame.
Disaster Hazard: Dangerous; when heated to decomp, emits highly toxic fumes; can react vigorously with oxidizing materials.
To Fight Fire: Alcohol foam, CO_2, dry chemical.

CYCLOHEXYL AMINE SULFATE. $C_6H_{12}N \cdot H_2SO_4$, mw: 196.3.
THR = An exper carc. [3]

CYCLOHEXYLBENZENE. See phenyl cyclohexane.

***n*-CYCLOHEXYL-2-BENZOTHIAZOLE SULFENAMIDE.** Light tan or buff powder. $C_6H_4NSCSNHC_6H_{11}$, mw: 264.4, mp: 94°, d: 1.27 @ 25°.
Acute tox data: Oral LD_{50} (rabbit) = 4000 mg/kg. [3]
THR = MOD via oral route.
Disaster Hazard: Dangerous; see sulfonates.

N-CYCLOHEXYL-N-CARBAMOYL AZIRIDINE. $C_9H_{16}ON_2$, mw: 168.3.
THR = An exper neo. [3]

CYCLOHEXYL CHLORIDE. Clear liquid. $C_6H_{11}Cl$, mw: 118.61, mp: −43°, bp: 142°, flash p: 89°F, d: 0.9923 @ 25°/4°, vap. d: 4.0.
THR = See cyclohexane.
Fire Hazard: Dangerous, when exposed to heat or flame.
Disaster Hazard: Mod dangerous; see chlorides; can react with oxidizing materials.
To Fight Fire: Alcohol foam, dry chemical, water spray, fog, mist.

2-CYCLOHEXYL CYCLOHEXANOL. Colorless liquid. $C_6H_{11}C_6H_{10}OH$, mw: 182.3, bp: 271°–277°, flash p: 270°F, d: 0.977 @ 25°/25°.
THR = See alcohols.
Fire Hazard: Slight, when exposed to heat or flame; can react with oxidizing materials.
To Fight Fire: CO_2, dry chemicals.

2-CYCLOHEXYL-4,6-DINITROPHENOL. Crystals. $C_6H_{11}C_6H_2OH(NO_2)_2$, mw: 266.23.
Acute tox data: Oral LD_{50} (rat) = 65 mg/kg; oral LD_{50} (mammal) = 50 mg/kg; ip LD_{50} (mouse) = 25 mg/kg. [3]
THR = HIGH via oral and ip routes.
Fire Hazard: See nitrates.
Disaster Hazard: Dangerous; see nitrates; can react vigorously with oxidizing materials.

CYCLOHEXYL FORMATE. $HCOOC_6H_{11}$, mw: 128.2, flash p: 124°F.

THR = Compound is said to resemble amyl acetate in type and intensity of toxicity. See amyl acetate.
Fire Hazard: Mod, via heat, open flames, oxidants.
To Fight Fire: Dry chemical, foam, spray, fog, mist.

CYCLOHEXYL ISOCYANATE. $C_6H_{11}NCO$, mw: 125.
THR = HIGH. See isocyanates.
Disaster Hazard: Dangerous; see isocyanates.

CYCLOHEXYLMETHANE. See methyl cyclohexane.

3-CYCLOHEXYLOXY-1,2-PROPANEDIOL. $C_9H_{19}O_3$, mw: 175.3.
THR = An exper teratogen. [*3*]

***o*-CYCLOHEXYL PHENOL.** Nearly white crystals. $C_6H_{11}C_6H_4OH$, mw: 176.2, mp: 50°, bp: 169° @ 25 mm, flash p: 273°F, d: 1.018 @ 60°/25°.
THR = U. See phenol.
Fire Hazard: Slight, when exposed to heat or flame; can react with oxidizing materials.
To Fight Fire: Alcohol foam, foam, CO_2, dry chemical.

N-CYCLOHEXYL SULFAMIC ACID. See cyclohexane sulfamic acid.

CYCLOHEXYL TETRATHIO-*o*-STANNATE. Solid. $(SC_6H_{11})_4Sn$, mw: 579.5, mp: 54°.
THR = U. See tin compounds.
Disaster Hazard: Dangerous; emits sulfur fumes.

N-CYCLOHEXYL-*p*-TOLUENE SULFONYL UREA. Fine white crystals. $C_{14}H_{20}O_3N_2S$, mw: 296.4.
Acute tox data: ip LD_{50} (rat) = 920 mg/kg. [*3*]
THR = MOD via ip route.
Disaster Hazard: Dangerous; see sulfonates.

CYCLOHEXYL TRICHLOROSILANE. $C_6H_{11}SiCl_3$, mw: 218.6, bp: 208°, flash p: 196°F (OC), d: 1.2, vap. d: 7.5.
THR = A corrosive, flam material. HIGH irr via oral and inhal routes.
Fire Hazard: Mod, when exposed to flame, heat or oxidizers.
Disaster Hazard: Dangerous; see chlorosilanes.
To Fight Fire: Foam, CO_2, dry chemical, water as a blanket over the fire.

CYCLONITE. See cyclotrimethylene trinitramine.

CYCLOOCTATETRAENE. Liquid. C_8H_8, mw: 104.14, mp: −7°, bp: 140.6°, fp: −4.7°, vap. press: 7.9 mm @ 25°.
THR = U. May be a simple asphyxiant.
Fire Hazard: Mod, when exposed to heat or flame; can react with oxidizing materials.
To Fight Fire: Spray, mist, fog, foam, dry chemical.

CYCLOPAMINE. Derived from *Veratrum Californicum.*
THR = An exper teratogen. [*3*]

CYCLOPARAFFINS. Both the saturated and unsaturated members of the cycloparaffin series are narcotic, and may cause death through respiratory paralysis. For most of the members there appears to be little range between the conc causing deep narcosis and those causing death. There is very little information in the literature regarding the chronic effects resulting from exposure of humans to the cycloparaffins. Exper with rabbits indicate that barely demonstrable changes in the liver and kidneys may result from exposure to 786 ppm of cyclohexane for 6 hrs daily, repeated for 50 days.

4-H-CYCLOPENTA(def)CHRYSENE. $C_{19}H_{12}$, mw: 240.3.
THR = An exper neo. [*3*]

CYCLOPENTADIENE-1,3. Colorless liquid. C_5H_6, mw: 66.1, mp: −85°, bp: 42.5°, d: 0.80475 @ 19°/4°.
Acute tox data: Inhal TC_{LO} (human) = 250 ppm. [*3*]
THR = MOD via inhal route. An insecticide and fungicide.
Fire Hazard: Mod, when exposed to heat or flame; can react with oxidizing materials.
Explosion Hazard: Mod, in the form of gas when exposed to heat or by chemical reaction. It decomp violently at high temp. and pressure.

CYCLOPENTADIENYL CHROMIUM DINITROSYL DIMER. $C_6H_5(NO)_2Cr_2$, mw: 350.2.
THR = Sample exploded by He-Ne laser beam. [*19*]

CYCLOPENTAMETHYLENE GERMANIUM DICHLORIDE. Colorless liquid. $(CH_2)_3GeCl_2$, mw: 213.7, bp: 55°–60° @ 12 mm.
THR = U. See germanium compounds.
Disaster Hazard: Dangerous; see chlorides.

CYCLOPENTANE. Syn: *pentamethylene.* Colorless liquid. $(CH_2)_5$, mw: 70.08, bp: 49.3°, fp: −93.7°, flash p: <20°F, autoign. temp.: 716°F, d: 0.745 @ 20°/4°, vap. press: 400 mm @ 31.0°, vap. d: 2.42.
THR = MOD via oral and inhal routes. High conc have narcotic action.
Fire Hazard: Dangerous, when exposed to flame; can react with oxidizing materials.
Explosion Hazard: U.
Disaster Hazard: Dangerous. Keep away from heat and flame (open).
To Fight Fire: Foam, CO_2, dry chemical.

CYCLOPENTANE, METHYL. See methyl cyclopentane.

CYCLOPENTANOL. Clear liquid. C_5H_9OH, mw: 86.13, bp: 139.5°, flash p: 124°F, d: 0.9422 @ 25°/4°.
THR = U. See also alcohols.
Fire Hazard: Mod, in the presence of heat or flame; can react with oxidizing materials.

To Fight Fire: Dry chemical, alcohol foam, water spray, mist.

CYCLOPENTANONE. Syns: *dumasin, ketocyclopentane.* Liquid. C_5H_8O, mw: 84.1, mp: −58.2°, bp: 130.6°, flash p: 79°F, d: 0.9509 @ 18°/4°, vap. d: 2.3.
Acute tox data: ip LD_{50} (mouse) = 1950 mg/kg. [*3*]
THR = MOD via ip and probably oral and inhal routes also.
Fire Hazard: Dangerous, when exposed to flame; can react with oxidizing materials.
To Fight Fire: Alcohol foam, foam, CO_2, dry chemical.

CYCLOPENTANONE OXIME. Solid. C_5H_8NOH, mw: 99.13, mp: 57.5°.
THR = U.
Fire Hazard: Slight.

4H-CYCLOPENTA(def)PHENANTHRENE. $C_{15}H_{10}$, mp: 190.3.
THR = An exper carc. [*3*]

CYCLOPENTA(cd)PYRENE. $C_{18}H_{10}$, mw: 226.3.
THR = An exper neo. [*3*]

CYCLOPENTENE. Liquid. C_5H_8, mw: 68.1, mp: −93.3°, bp: 44.242°, fp: −135.2°, flash p: −20°F, d: 0.77199 @ 20°.
Acute tox data: Oral LD_{50} (rat) = 2140 mg/kg; dermal LD_{50} (rabbit) = 1590 mg/kg. [*3*]
THR = MOD via oral and dermal routes. Probably via inhal route too.
Fire Hazard: Dangerous, when exposed to flame or heat; can react with oxidizing materials.
Disaster Hazard: Dangerous. Keep away from heat and open flame.
To Fight Fire: Foam, CO_2, dry chemical.

2-CYCLOPENTENE-1-OL. $OHCHCH{:}CHCH_2CH_2$, mw: 84.
Acute tox data: Oral LD_{50} (rat) = 470 mg/kg; inhal LC_{LO} (rat) = 1000 ppm for 4 hrs; dermal LD_{LO} (rabbit) = 180 mg/kg. [*3*]
THR = HIGH via dermal, oral and inhal routes.

1,2-CYCLOPENTENO-5,10-ACEANTHRENE. $C_{19}H_{16}$, mw: 244.4.
THR = An exper neo. [*3*]

3-(2-CYCLOPENTENYL)-2-METHYL-4-OXO CYCLOPENTENYL ESTER OF CHRYSANTHEMUM MONOCARBOXYLIC ACID. Syn: *cyclethrin.*
THR = See pyrethrin I.

CYCLOPENTYL BROMIDE. Liquid. C_5H_9Br, mw: 149.04, bp: 137.5°, flash p: 108°F, d: 1.3866 @ 25°/4°, vap. d: 5.
THR = See bromides.
Fire Hazard: Mod, when exposed to heat or flame.
Disaster Hazard: Dangerous; see bromides; can react with oxidizing materials.

CYCLOPENTYL CHLORIDE. Liquid. C_5H_9Cl, mw: 104.58, bp: 113.5°, flash p: 60°F, d: 1.0024 @ 25°/4°, vap. d: 3.5.
THR = See chlorinated hydrocarbons, aliphatic and aromatic.
Fire Hazard: Dangerous; when exposed to heat or flame.
Explosion Hazard: U.
Disaster Hazard: Dangerous; see chlorides; can react with oxidizing materials.

CYCLOPENTYL ETHER. $(C_5H_9)_2O$, mw: 154.
Acute tox data: Oral LD_{50} (rat) = 470 mg/kg; inhal LC_{LO} (rat) = 250 ppm for 4 hrs; dermal LD_{LO} (rabbit) = 1410 mg/kg. [*3*]
THR = HIGH via oral and inhal; MOD via dermal routes. See also ethers.
Disaster Hazard: U. See ethers.

CYCLOPHOSPHAMIDE. See endoxan.

CYCLOPROPANE. Syn: *trimethylene.* Colorless gas. $CH_2CH_2CH_2$, mw: 42.08, mp: −126.6°, bp: −33.5°, lel = 2.4%, uel = 10.4%, d: 1.879 g/l @ 0°, autoign. temp.: 932°F.
THR = MOD via inhal route. High conc have narcotic action. Used as a surgical anesthetic.
Fire Hazard: Very dangerous, when exposed to heat or flame; can react with oxidizing materials.
Spont Heating: No.
Explosion Hazard: Mod, in the form of vapor when exposed to heat or flame.
Disaster Hazard: Dangerous. Keep away from heat and open flame.
To Fight Fire: Stop flow of gas. CO_2, dry chemical or water spray.

CYCLOPROPYL ETHYL ETHER. Liquid. $C_3H_5OC_2H_5$, mw: 86.1.
THR = See ethers.

CYCLOPROPYL METHYL ETHER. Syn: *cypronic ether.* Liquid $C_3H_5OCH_3$, mw: 72.1, mp: −119°, bp: 44.7°, d: 0.786 @ 25°/4°.
THR = See ethers.

CYCLOPROPYL PROPYL ETHER. Liquid. $C_3H_5OC_3H_7$, mw: 100.2.
THR = See ethers.

CYCLOTETRAMETHYLENE OXIDE. See tetrahydrofuran.

CYCLOTRIMETHYLENE TRINITRAMINE. Syns: *RDX, cyclonite, hexogen.* White, crystalline powder. $C_3H_6N_6O_6$, mw: 222.15, mp: 202°.
Acute tox data: Oral LD_{50} (rat) = 200 mg/kg; iv LD_{50}

(mouse) = 19 mg/kg; dermal LD_{LO} (guinea pig) = 465 mg/kg. [*3*]

THR = HIGH via oral, dermal and iv routes. An exper carc. [*23*] Cases of epileptiform convulsions have been reported from exposure.

Fire Hazard: See nitrates.

Explosion Hazard: It is one of the most powerful high explosives in use today. See explosives, high. Has more shattering power than TNT and is often mixed with TNT as a bursting charge for aerial bombs, mines and torpedoes. Because it is easily initiated by mercury fulminate it may be used as a booster.

Disaster Hazard: See nitrates.

***p*-CYMENE.** Syn: *isopropyl toluene.* Liquid. $CH_3C_6H_4CH(CH_3)_2$, mw: 134.21, mp: −68.2°, bp: 176°, lel = 0.7%, @ 100°, ulc: 30–35, flash p: 117°F (CC), d: 0.86, autoign. temp.: 817°F, vap. d: 4.62, vap. press: 1 mm @ 17.3°, flash p: (technical) 127°F, uel (technical) = 5.6%.

Acute tox data: Oral TD_{LO} (humans) = 86 mg/kg (affects the CNS), oral LD_{50} (rat) = 4750 mg/kg. [*3*]

THR = MOD via oral route, although humans sustain CNS effects at low dose rates.

Fire Hazard: Mod, when exposed to heat, flame or oxidizers.

Spont Heating: No.

Explosion Hazard: Slight, in the form of vapor.

Disaster Hazard: Mod dangerous; can react with oxidizing materials.

To Fight Fire: Foam, CO_2, dry chemical.

CYMOGENE. See liquefied petroleum gas.

CYPREX. See *n*-dodecyl guanidine acetate.

CYPROMID.

Acute tox data: Oral LD_{50} (rat) = 215 mg/kg. [*3*]

THR = HIGH via oral and probably inhal routes. An herbicide.

CYPRONIC ETHER. See cyclopropyl methyl ether.

CYPROSTERONE ACETATE. $C_{24}H_{29}O_4Cl$, mw: 417.

THR = An exper teratogen to rats. [*3*]

CYSTEINE. Syns: *α-amino-β-thiolpropionic acid, β-mercaptoalanine.* An amino acid derived from cystine, occurring naturally in the *l*-form, which will be considered here. Colorless crystals, sol in water, ammonium hydroxide and acetic acid, insol in ether, acetone, benzene, carbon disulfide and carbon tetrachloride. $HSCH_2CH(NH_2)COOH$, mw: 121.

THR = U. Probably not toxic. A nutrient and/or dietary supplement food additive. [*109*]

CYSTINE. Syn: *β,β′-dithiobisalanine, di-α-amino-β-thiolpropionic acid*). The chief sulfur-containing amino acid of protein. White crystalline plates, sol in water, insol in alcohol. Occurs in *dl*, *l* and *d* form. We consider the *l* and *dl* forms here. $HOOCCH(NH_2)CH_2SSCH_2CH \cdot (NH_2)COOH$, mw: 240, mp(*dl*): 260°, mp(*l*): 258°–261°.

THR = U. Probably not toxic. A nutrient and/or dietary supplement food additive. [*109*]

CYTARABINE. $C_9H_{13}O_5N_2$, mw: 243.3.

THR = An exper teratogen. [*3*]

CYTISUS. A wood dust.

THR = MILD irr and allergen.

Fire Hazard: Mod, when exposed to heat or flame.

Explosion Hazard: Slight, when exposed to flame.

CYTOSTASAN.

THR = An exper carc. [*3*]

CZA. See citrazinic acid.

2,4-D. See 2,4-dichlorophenoxyacetic acid.

DACRON. See polyethylene terphthalate (film).

DACTIN. See 1,3-dichloro-5,5-dimethylhydantoin.

DALAPON-Na. See 2,2-dichloropropionic acid sodium.

DALF. See methyl parathion.

DALMATIAN INSECT POWDER. See pyrethrum flowers.

D & C RED 14. $C_{24}H_{20}ON_4$, mw: 380.5.
THR = An exper carc to rats via oral route. [*3*]

DANGER POINTS IN SOLID WASTE HANDLING. See Section 6.

DAP. See ammonium phosphate, dibasic.

DAPHNIN. See 7,8-dihydroxycoumarin-7-*β*-*d*-glucoside.

DASANIT. See *o,o*-diethyl-*o*-[*p*(methyl sulfinyl) phenyl] phosphorothioate.

DATURINE. See atropine.

DAUNOMYCIN. Thin red needles. $C_{27}H_{28}O_{10}N$, mw: 526.6, mp: 190° (decomp).
Acute tox data: Acute iv LD_{50} (rat) = 13 mg/kg. ip LD_{50} (mouse) = 2.5 mg/kg. [*3*]
THR = HIGH via iv and ip routes. An exper (+) carc. [*3, 7*]

DAVY, SIR HUMPHREY. See Section 1.

DBH. See 1,2,3,4,5,6-hexachlorocyclohexane.

DBMC. See di-*tert*-butyl-*m*-cresol.

DBPC. See di-*tert*-butyl-*p*-cresol.

"DC 200 FLUID." Syn: *hexamethyldisiloxane*. Viscous liquid. $(CH_3)_3Si\text{-}O\text{-}Si(CH_3)_3$, mw: 162.4, vap. d: 5.5.
THR = MOD irr.

DCMX. See 2,4-dichloro-3,5-xylenol.

2,4-DDD. Syns: *2,2-bis-(p-chlorophenyl)-1,1-dichloroethane, dichloro diphenyl dichloroethylene, TDE.* Crystalline solid. $(ClC_6H_4)_2HCCHCl_2$, mw: 320.1, mp: 110°, vap. d: 11.
Acute tox data: Oral LD_{50} (rat) = 113 mg/kg. Dermal LD_{50} (rabbit) = 1200 mg/kg. [*3*]
THR = HIGH via oral, MOD via dermal routes. Used as a food additive permitted in the feed and drinking water of animals, and/or for the treatment of food-producing animals. [*109*] See also DDT.
Disaster Hazard: Dangerous; when heated to decomp, emits highly toxic fumes of chlorides. Similar to DDT. An exper (+) carc and neo. [*3, 12*]

DDE. See 1,1-dichloro-2,2-bis(*p*-chlorophenyl) ethylene.

4,4-DDM. See methylene dianiline.

DDT. Syns: *dichloro diphenyl trichloroethane, chlorophenothane, dicophane, 1,1,1-trichloro-2,2-bis-(p-chlorophenyl) ethane.* Colorless crystals or white to slightly off-white powder. Odorless or with slight aromatic odor. $(ClC_6H_4)_2CHCCl_3$, mw: 354.5, mp: 108.5°–109°.
Acute tox data: Oral LD_{LO} (infant) = 150 mg/kg. Oral TD_{LO} (humans) = 16 mg/kg ⟶ CNS damage; oral LD_{50} (rat) = 113 mg/kg. Dermal LD_{50} (rabbit) = 300 mg/kg. [*3*]
THR = HIGH via oral and dermal routes. Used as a food additive permitted in the food and drinking water of animals and/or for the treatment of food-producing animals. Also a food additive permitted in food for human consumption. [*109*] Note: DDT is a common air contaminant.

DDT is readily absorbed from the intestinal tract and, if it occurs in the air in the form of an aerosol or dust, it may be taken into the lung and readily absorbed. DDT is not, however, absorbed from the skin unless it is in solution. Solutions are absorbed from the skin and, by the same token, emulsions are absorbed to some extent. Likewise, fats and oils from whatever source increase the absorption of DDT from the intestine. DDT acts on the CNS, but the exact mechanism of this action either in man or in animals has not been elucidated. DDT is an exper (S) mutagen and carc. [*3, 12, 23*] See chlorinated hydrocarbons. Large doses of DDT also induce nausea and/or diarrhea in man; however, whether this is a central or local action is not yet clear. Chronically, DDT produces microscopic changes in the liver and kidneys in some exper animals. This has not been demonstrated in man. DDT is secreted in the milk and, as an acid derivative is excreted in the urine of rabbits, dogs and man. DDT and certain of its degradation products, particularly DDE, are stored in fat. Such storage results either from a single large dose or from repeated small doses. DDT stored in the fat is at least largely inactive since a greater total dose may be stored in an exper animal than is sufficient as a lethal dose for that same animal if given at one time. A study based on 75 human cases reported an average of 5.3 ppm of DDT stored in the fat. A higher content of DDT and its derivatives (up to 434 ppm of DDE and 648 ppm of DDT) was found in workers who had very extensive exposure. Without exception, the samples were taken from persons who were either asymptomatic or suffering from some disease completely unrelated to DDT. Careful hospital examination of workers, who had been very extensively exposed and who had volunteered for examination revealed no abnormality which could be attributed to DDT. Much higher levels than have been found in man have been observed in the fat of exper animals which were apparently asymptomatic. DDT stored in the fat is eliminated only very gradually when further dosage is discontinued. After a single dose, the secretion of DDT in the milk and its excretion in the urine reach their height within a day or two and continue at a lower level thereafter.

Dangerous Acute Dose in Man: A dose of 20 g has proved highly dangerous though not fatal to man. This dose was taken by 5 persons who vomited an unknown portion of the material and even so recovered only incompletely after 5 weeks. Smaller doses produced less important symptoms with relatively rapid recovery. Exper ingestion of 1.5 g resulted in great discomfort and moderate neurological changes including paraesthesia, tremor, moderate ataxia, exaggeration of part of the reflexes, headache, and fatigue. Vomiting followed only after 11 hours. Recovery was complete on the following day. The fatal dose of DDT for man is not known. Judging from the literature, no one has ever been killed by DDT in the absence of other insecticides and/or a variety of toxic solvents. However, these common solvent formulations are highly fatal when taken in small doses, partly because of the toxicity of the solvent, and perhaps because of the increased absorbability of the DDT; several fatal cases in man have been reported. Acute oral toxicity for man = 250 mg/kg. Acute oral LD_{50} (rat) = 113 mg/kg (tech grade). Federal fruit and vegetable tolerance = 7 ppm.

Dangerous Chronic Dose in Man: Even less is known of the hazard of chronic DDT poisoning. It is known that certain exper animals fed diets containing one part of DDT per million store the compound in their fat. The storage of DDT in man has been mentioned above. The exact significance of these findings is not known and their further investigation is of the greatest importance. Human volunteers have ingested up to 35 mg/day for 21 months with no ill effects.

Signs and Symptoms of Poisoning in Man: In patients who ate substantial doses of DDT in flour, the symptoms observed were vomiting, numbness and partial paralysis of the extremities, mild convulsions, loss of proprioception and vibratory sensation of the extremities, and hyperactive knee jerk reflexes. Symptoms appeared in 30 to 60 min after eating the DDT. The paralysis and numbness were most evident in the most distal portions of the extremities, and their intensity was directly proportional to the amount of DDT ingested. All the patients were apprehensive and excited; respiration was moderately rapid; pulse remained slow to normal. The immediate protective mechanism in man, following substantial doses, is vomiting. With smaller doses, nausea and vomiting are less prominent, but diarrhea has been observed. Signs and symptoms of chronic poisoning in man are unknown, although, judging from the observed microscopic changes in exper animals, liver and kidney dysfunctions should be looked for. The primary irr of DDT is practically nil, and it has little or no tendency to produce allergy. Dermatitis induced by DDT has occasionally been reported, but these reports are unconfirmed; nevertheless the phenomenon should be expected to occur in rare instances.

Laboratory Findings: Laboratory findings are essentially negative except for the presence of DDT which may be quantitatively measured in stomach contents, urine, or tissues.

Treatment of Poisoning: Depending on the condition of the patient, attention should first be given to the sedation or to the removal of poison which may have been taken internally. Stomach lavage and saline laxatives may be used. Oil laxatives should be avoided; they promote absorption of DDT and of many organic solvents. The five drugs of choice, arranged roughly in order of their effectiveness, are phenobarbital, pentobarbital, paraldehyde, urethane, and calcium gluconate. Phenobarbital, which has been used in doses up to 0.7 g per day in epilepsy, and pentobarbital (0.25 to 0.5 g) are the barbiturates known to control convulsions of central origin. Paraldehyde (average dosage 15 cc orally, 1 cc undiluted intravenously, 35 cc rectally in normal saline) controls the convulsions of DDT-poisoned animals. Urethane (human dosage 1 to 4 g) has proved very effective in rats, but it should be remembered that the hypnotic and narcotic effects of urethane are not correspondingly high in man. Urethane has an added advantage, however, of being tolerated in the young and the aged. The object of sedation is not to induce sleep but to restore a relative calm; however, the proper dosage in the presence of poisoning may be so large that it would induce anesthesia if poisoning were not present.

Calcium gluconate has been used less than the other antidotes, but it is reported to control DDT-induced convulsions in several animals. Since its mechanism of action is entirely different, it may be used in addition to sedatives. Epinephrine is contraindicated.

DDVP. See dimethyldichlorovinyl phosphate.

DEAC. See diethyl aluminum chloride.

DEADLY NIGHTSHADE.
Source of the alkaloids, atropine and belladonine.

DEAPASIL. See *p*-aminosalicylic acid.

DECARBORANE. Syns: *boron hydride, decaboron tetradecahydride*. Colorless needles. $B_{10}H_{14}$, mw: 122.3, mp: 99.7°, d: 0.94. (solid), d: 0.78 (liquid @ 100°), vap. press: 19 mm @ 100°.
Acute tox data: Oral LD_{50} (rat) = 64 mg/kg. Inhal LD_{50} (rat) = 46 ppm for 4 hrs. Dermal LD_{50} (rat) = 740 mg/kg. [*3*]
THR = HIGH via oral, inhal; MOD via dermal routes. Self-ignites in O_2. [*19*]

DECABORON TETRADECAHYDRIDE. See decaborane.

DECACHLOROOCTAHYDO-1,3,4-METHENO-2H-CYCLOBUTA(cd) PENTALEN-2-ONE. See kepone.

DECAHYDRONAPHTHALENE. See Decalin.

DECAHYDRONAPHTHALENE, trans. flash p: 129°F; autoign. temp.: 491°F, lel = 0.7%, uel = 5.4%, d: 0.87, vap. d: 4.77, bp: 195°.
THR = See Decalin.
Fire Hazard: Mod, via heat, flame and oxidizers.
To Fight Fire: Water spray, mist, alcohol foam, dry chemical.

DECALIN. Syn: *decahydronaphthalene*. Water white liquid. $C_{10}H_{18}$, mw: 138.3, mp (*cis*): −43.3°, mp (*trans.*): −30.7°, bp: (*cis*): 194.6°, flash p: 136°F, (CC), autoign. temp.: 482°F, vap. press: (*cis*) 1 mm @ 22.5°, (*trans*) 10 mm @ 47.2°, d: 0.8963, vap. d: 4.76, lel = 0.7% @ 212°F, uel = 4.9% @ 212°F.
Acute tox data: Oral LD_{50} (rat) = 4170 mg/kg. Inhal LD_{50} (rat) = 500 ppm for 2 hrs. [*3*]
THR = MOD via oral and inhal routes. Irr to skin, eyes and mu mem. Has caused kidney damage in exper animals.
Fire Hazard: Mod, when exposed to heat or flame; can react with oxidizing materials.
Spont Heating: No.
To Fight Fire: Foam, CO_2, dry chemical.

1-DECANAL. Syns: *caprylaldehyde, capric aldehyde, n-decylaldehyde, aldehyde-C-10*. Colorless to light yellow liquid, floral fatty odor, sol in 80% alcohol, fixed oils, volatile oils and mineral oils, insol in water and glycerol. $CH_3(CH_2)_8CHO$, mw: 156, d: 0.831–0.838 @ 15°.
Acute tox data: Oral LD_{50} (rat) = 3730 mg/kg. [*3*]
THR = MOD via oral route. Used as a synthetic flavoring substance and adjuvant. [*109*]

DECANE. Syn: *decyl hydride*. Liquid. $CH_3(CH_2)_8CH_3$, mw: 142.3, mp: −29.7°, bp: 174.1°, lel = 0.8%, uel = 5.4%, flash p: 115°F (CC), d: 0.730 @ 20°/4°, autoign. temp.: 410°F, vap. press: 1 mm @ 16.5°, vap. d: 4.90.
THR = A simple asphyxiant. Narcotic in high conc. See argon.
Fire Hazard: Mod, when exposed to heat or flame; can react with oxidizing materials.
Spont Heating: No.
Explosion Hazard: Mod, in its vapor form.
To Fight Fire: Foam, CO_2, dry chemical.

DECANOIC ACID. Syns: *decoic acid, decylic acid*. White crystals, unpleasant odor, sol in most organic solvents and in dilute nitric acid, insol in water. $CH_3(CH_2)_8COOH$, mw: 172.3, d: 0.8858, bp: 270°, mp: 31.4°.
Acute tox data: iv LD_{50} (mouse) = 129 mg/kg. [*3*]
THR = HIGH via iv route.

1-DECANOL. See *n*-decyl alcohol.

1-DECENE. Syn: *n-decylene*. Colorless liquid. $H_2CCH(CH_2)_7CH_3$, mw: 140.26, mp: −66.3°, bp: 172°, d: 0.7396 @ 20°/4°, vap. press: 1 mm @ 95.7°, vap. d: 4.83, flash p: < 131°F, autoign. temp.: 455°F.
THR = U. Compounds in this group generally have irr and narcotic action. See also hexene-1.
Fire Hazard: Mod, when exposed to heat or flame; can react with oxidizing materials.
To Fight Fire: Foam, CO_2, dry chemical.

DECYL ACRYLATE. Very slightly sol in water. $C_{13}H_{24}O_2$, mw: 212.4, flash p: 441°F (OC), d: 0.9, bp: 157° @ 50 mm.
Acute tox data: Oral LD_{LO} (rat) = 4930 mg/kg; dermal LD_{LO} (rabbit) = 4660 mg/kg. [*3*] See also esters.
THR = MOD via oral and dermal routes.
Fire Hazard: Slight, when exposed to heat or flame.
To Fight Fire: Dry chemical, CO_2, mist, spray.

***n*-DECYL ALCOHOL.** Syn: *1-decanol, nonyl carbinol*. Viscous, refractive liquid. $CH_3(CH_2)_8CH_2OH$, mw: 158.3, mp: 7°, bp: 232.9°, flash p: 180°F (OC), d:

0.8297 @ 20°/4°, vap. press: 1 mm @ 69.5°, vap. d: 5.3.
Acute tox data: Oral LD_{50} (rat) = 4720 mg/kg; inhal LD_{50} (mouse) = 4000 mg/m^3. [*3*]
THR = MOD via oral and inhal routes. See also alcohols.
Fire Hazard: Mod, when exposed to heat or flame; can react with oxidizing materials.
To Fight Fire: Foam, CO_2, dry chemical.

DECYLAMINE. Liquid. $CH_3(CH_2)_9NH_2$, mw: 157.3, mp: 17°, bp: 95° @ 10 mm, flash p: 210°F, d: 0.79 @ 20°, vap. d: 5.5.
Acute tox data: Oral LD_{50} (rat) = 280 mg/kg; dermal LD_{50} (rabbit) = 350 mg/kg. [*3*]
THR = HIGH via oral and dermal routes. See also amines and fatty amines.
Fire Hazard: Mod, when exposed to heat or flame; can react with oxidizing materials.
To Fight Fire: Alcohol foam, foam, dry chemical.

DECYL BENZENE. Insol in water. $C_{16}H_{26}$, d: 0.9, bp: 255°–280°, flash p: 225°F, mw: 218.1.
THR = U.
Fire Hazard: Mod, when exposed to heat, flame or oxidizers.
To Fight Fire: Dry chemical, CO_2, mist.

DECYL CHLORIDE. $C_{10}H_{21}Cl$, mw: 176.5.
Acute tox data: Oral LD_{50} (rat) = 45,000 mg/kg. Dermal LD_{50} (rabbit) = 5660 mg/kg. [*3*]
THR = LOW via oral and dermal routes.
Disaster Hazard: Dangerous; see chlorides.

***n*-DECYLENE.** See 1-decene.

DECYL HYDRIDE. See decane.

***tert*-DECYL MERCAPTAN.** $C_{10}H_{21}SH$, mw: 174, d: 0.9, vap. d: 6.0, bp: 210°–218°, flash p: 190°F.
THR = U. See mercaptans.
Fire Hazard: Mod, when exposed to heat, flame or powerful oxidizers.
Disaster Hazard: Dangerous; see sulfur compounds.
To Fight Fire: Fog, dry chemical, CO_2, alcohol foam.

DECYL NAPHTHALENE. Liquid. $C_{20}H_{28}$, mw: 268.5, bp: 330°–440°, flash p: 350°F, d: 1.0, vap. d: 9.6.
THR = U. See also naphthalene.
Fire Hazard: Slight, when exposed to heat or flame; can react with oxidizing materials.
To Fight Fire: CO_2, dry chemical.

DECYL NITRATE. Insol in water. $CH_3(CH_2)_9ONO_2$, mw: 203, d: 1.0, bp: 128° @ 11 mm, flash p: 235°F (OC).
THR = U. See also nitrates.
Fire Hazard: See nitrates.
Disaster Hazard: Dangerous; see nitrates, organic.
To Fight Fire: Dry chemical, CO_2.

DEET. See *n,n*-diethyl-*m*-toluamide.

DEFINITIONS AND PROCEDURAL AND INTERPRETATIVE REGULATIONS OF FOOD ADDITIVES. [*109*]

DEFINITION OF TOXICITY AND HAZARD. See Section 1.

DEFINITION OF INDUSTRIAL TOXICOLOGY. See Section 9.

DEFINITION OF LABEL. See Section 11.

DEGUELIN. Pale green crystals. $C_{23}H_{22}O_8$, mw: 394.41, mp: 171°.
THR = A strong irr. Inhal can produce pulmonary edema.
Fire Hazard: Slight, when heated; emits acrid fumes.

de HAENS SALT. See antimony salt.

DEHYDRITE. See magnesium perchlorate.

DEHYDROABIETYLAMINE. Syn: "*Rosinamine-D.*" Viscous amber liquid. $C_{20}H_{25}N$, mw: 279.4, bp: 344° decomp, flash p: 375°F (OC), d: 1.001 at 25°/15.6°, autoign. temp.: 430°F, vap. d: 10.
THR = MOD irr via oral and inhal routes.
Fire Hazard: Slight, when exposed to heat or flame; can react with oxidizing materials.
To Fight Fire: Foam, CO_2, dry chemical.

DEHYDROABIETYLAMINE ACETATE. Amber paste or solution. $C_{20}H_{25}NCH_3COO$, mw: 338.5, d: 1.029 @ 25°/15.6°, vap. d: 11.5.
THR = See dehydroabietylamine.
Fire Hazard: Slight.

DEHYDROABIETYLAMINE NAPHTHENATE. Viscous liquid. $C_{20}H_{25}NOOCC_6H_5$, mw: 400.6.
THR = U. See also dehydroabietylamine.
Fire Hazard: Slight, when exposed to heat or flame; can react with oxidizing materials.

DEHYDROABIETYLAMINE PENTACHLOROPHENATE. Crystals. $C_{20}H_{25}NC_6Cl_5$, mw: 528.8, vap. d: 18.
THR = U. See also pentachlorophenol.
Disaster Hazard: Dangerous; when heated to decomp or on contact with acid or acid fumes, emits highly toxic fumes of chlorides.

DEHYDROABIETYL NITRILE. Light amber, partially crystallized solid. See Nitriles.

DEHYDROACETIC ACID. Syn: *3-acetyl-6-methyl-1,2-pyran-2,2,4(3H)-dione*. Crystals, mod sol in water and organic solvents. $C_8H_8O_4$, mw: 168.2, mp: 109°, bp: 269.0°, vap. press: 1 mm @ 91.7°, vap. d: 5.8.
Acute Tox Data: Oral LD_{50} (rat) = 500 mg/kg. An exper carc. [*3, 23*]
THR = HIGH via oral route. Large internal doses have produced vomiting, ataxia and convulsions in

exper animals. Can depress kidney function. A food additive permitted in food for human consumption. [*109*]

Fire Hazard: Slight.

DEHYDROCHOLESTEROL, ACTIVATED. See vitamin D_2.

DEHYDRO FOLLICULINIC ACID. $C_{19}H_{22}O_3$, mw: 298.4.

THR = An exper carc. [*3*]

DEHYDRO RETRONECINE.

THR = An exper carc. [*23*]

DEKANITROCELLULOSE. See nitrocellulose.

DELANEY CLAUSE is discussed in Section 10.

DELAY ELECTRIC IGNITERS. See explosives, high.

DELNAV. Syn: *dioxathion-2,3-p-dioxanedithiol-s,s-bis-(o,o-diethyl phosphorodithioate)*. A non-volatile, stable solid, non-flam, insol in water. $C_{12}H_{26}O_4S_4P_2$, mw: 362.4.

Acute tox data: Oral LD_{50} (rat) = 20 mg/kg. Inhal LD_{50} (rat) = 1398 mg/m^3 for 1 hr. Dermal LD_{50} (rat) = 63 mg/kg. [*3*]

THR = HIGH via oral, inhal and dermal routes. A toxic insecticide and miticide. See parathion. A cholinesterase inhibitor.

***m*-DELPHENE.** See *n,n*-dimethyl-*m*-toluamide.

DELPHININE. White crystalline, rhombic plates. $C_{34}H_{47}NO_9$, mw: 613.73, mp: 191° decomp.

THR = An alkaloid poison.

Disaster Hazard: Dangerous; when heated to decomp, emits highly toxic fumes of NO_x.

DELPHINIUM. Syns: *larkspur*, *stagger weed*. Dried ripe seeds.

THR = HIGH via oral, inhal routes. An allergen.

Fire Hazard: Slight, when exposed to heat or flame.

DEMETON. Syns: *o,o-diethyl-o-2-(ethylthio)ethyl thiophosphate*, "*E-1059*," *systox*. A light brown liquid, sulfur compound odor.

Acute tox data: Oral LD_{50} (rat) = 1.7 mg/kg.; dermal (rat) LD_{50} = 8.2 mg/kg.; oral LD_{50} (birds) = 7 mg/kg. [*3*]

THR = HIGH via oral and dermal routes. An exper teratogen. [*3*] and carc. [*23*] A highly toxic insecticide. It is believed that most of the physiological actions of demeton resemble, generally, those of parathion, TEPP, and other related organic phosphorus poisons. The actions of this compound and its metabolites are based principally upon the inhibition of the enzyme cholinesterase, thus allowing the accumulation of large amounts of acetylcholine.

Dangerous Acute and Chronic Doses: Little is known regarding the acute and chronic dosages which would be dangerous to man. However, the acute dose is believed to be approximately equal to that of parathion (estimated at from 12 to 20 mg or $\frac{1}{5}$–$\frac{1}{3}$ grain). The compound is only slightly less toxic by the dermal route. Doses of organic phosphorus insecticides tend to be cumulative in their effect. However, if illness occurs, it is acute in nature, whether caused by a single large dose or by repeated exposure.

Signs and Symptoms of Poisoning: Persons poisoned with demeton may be expected to show the following symptoms: headache, giddiness, blurred vision, weakness, nausea, diarrhea, and discomfort in the chest. In addition to these symptoms, sweating, miosis, muscular fasciculation, incoordination, tearing, salivation, pulmonary edema, cyanosis, papilledema, convulsions, coma, and loss of reflexes and sphincter control may occur. There is one result of demeton poisoning in rats which may or may not be peculiar to the species. Shortly after other signs of poisoning appeared, rats fed 50 or 100 ppm of demeton in their regular diet developed severe exophthalmos, and perhaps associated changes, leading to drying and ulceration of the cornea so severe that some animals lost both eyes through necrosis.

Treatment of Poisoning: Keep the patient fully atropinized, using 1 to 2 mg of atropin sulfate per dose and up to 10 to 20 mg per day. Remove any remaining poison. Administer oxygen and other supportive measures early and be prepared for the immediate use of mechanical artificial respiration. Watch the patient continuously. See also parathion.

DENATURANTS.

THR = These materials are almost always toxic. See also specific denaturant.

DENATURED ALCOHOL. Ethyl alcohol to which a toxic ingredient has been added (denaturant). Flash p: 60°F, autoign. temp.: 750°F, d: 0.8, vap. d: 1.6, bp: 175°F.

THR = Almost always toxic.

Fire Hazard: Dangerous, when exposed to heat, flame or oxidizers.

To Fight Fire: Alcohol foam.

DEOXY CHOLIC ACID. $C_{24}H_{40}O_4$, mw: 392.6, mp: 178°.

Acute tox data: iv LD_{50} (rabbit) = 14 mg/kg. [*3*]

THR = HIGH via iv route. An exper carc. [*23*]

2-DEOXY-5-FLUORO URIDINE. $C_9H_{11}O_5N_2F$, mw: 246.2.

THR = HIGH via oral route. An exper teratogen. [*3*]

2-DEOXY-5-IODO URIDINE. $C_9H_{11}O_5N_2I$, mw: 354.1.

THR = An exper teratogen. [*3*]

11-DEOXY CORTICO STERONE ACETATE. $C_{23}H_{32}O_4$, mw: 372.6.
THR = An exper carc. [3]

1-DEOXY-1-(METHYL NITROSAMINO)-D-GLUCITOL. $C_7H_{16}O_6N_2$, mw: 224.3.
THR = An exper carc. [3, 23] via oral route.

β-DEOXY THIOGUANOSINE. $C_{10}H_{14}O_3N_5S \cdot H_2O$, mw: 302.4.
THR = An exper neo. [3]

DEPARTMENT OF LABOR REGULATIONS. See Section 11.

DEPARTMENT OF TRANSPORTATION LABELING SYSTEM. See Section 11.

DENATURED SPIRITS. See alcohol, denatured.

"DEPENDIP." Water white liquid. Flash p: 52°F, d: 0.758 @ 15.5°.
THR = LOW via irr.
Fire Hazard: Dangerous, when exposed to heat or flame; can react vigorously with oxidizing materials.
Disaster Hazard: Dangerous; when heated, emits acrid fumes.
To Fight Fire: Foam, CO_2, dry chemical.

DERMATOL (COMMERCIAL). See bismuth gallate, basic.

DERRIS. See rotenone.

DESIGN OF RADIATION INSTALLATIONS. See Section 5.

DESMODUR. See 1,5-naphthyl diisocyanate.

DESOXY CHOLIC ACID. See deoxycholic acid.

DESOXY PHENO BARBITONE. $C_{12}H_{14}O_2N_2$, mw: 218.3.
THR = An exper teratogen. [3]

DETERGENTS. See surfactants.

DEUTERIUM. Syn: D_2. A gas. mw: 4, lel = 5%; uel = 75%.
THR = See hydrogen.
Fire Hazard: Very dangerous, via ignition sources, oxidizers.
To Fight Fire: Stop flow of gas.

DEXEDRINE SULFATE. See *d-α*-methyl phenethylamine sulfate.

DEXTRANS. Syn: *macroses*. Certain polymers of glucose which have chain-like structures and very high molecular weights (up to 200,000 or higher).

DEXTRAN 1. mw: 200000.
THR = An exper neo. [3]

DEXTRAN 2. mw: 100000.
THR = An exper neo. [3]

DEXTRAN 5.
THR = An exper neo. [3]

DEXTRAN 10. mw: 89,400.
THR = An exper carc. [3]

DEXTRAN 11. mw: 71,400.
THR = An exper neo. [3]

DEXTRAN-IRON COMPLEX. See iron-dextran complex.

DEXTRINS. Syns: *starch gum, artificial gum, vegetable gum, tapioca, dextran*. An intermediate product formed by the hydrolysis of starches yellow or white powder or granules, sol in water, insol in alcohol and ether, colloidal in properties and describes a class of substances (hence it has no formula).
Acute tox data: iv LD_{50} (mouse) = 350 mg/kg. [3]
THR = HIGH via iv route. A substance migrating to food from packaging material. [109]

DEXTROSE. Syns: *dextroglucose, grape sugar, corn sugar*. Colorless crystals or white crystalline or granular powder, odorless, sol in water, slightly sol in alcohol. $C_6H_{12}O_6 \cdot H_2O$, mw: 198, d: 1.544, mp: 146°.
THR = Very LOW. Substance migrating to food from packaging materials. [109] Reacts violently with (Na_2O_2 + KNO_3). [19]

DEXTROSE NITRATE. See nitrolevulose.

DFDD. Syn: *difluorodiphenyldichloroethane*. Crystals. $C_{14}H_{10}Cl_2F_2$, mw: 287.2, mp: 77.5°, vap. d: 10.
THR = HIGH, a contact insecticide. See DDT.
Disaster Hazard: Dangerous; when heated to decomp, emits highly toxic fumes of fluorides and chlorides.

DFDT. Syns: *2,2-bis(p-fluorophenyl)-1,1,1-trichloroethane, difluorodiphenyl trichloroethane*. Crystals. $(FC_6H_4)_2CHCCl_3$, mw: 321.6, mp: 45.5°, vap. d: 10.5.
Acute tox data: Oral LD_{50} (rat) = 1120 mg/kg; LD_{50} (mammal) = 480 mg/kg. [3]
THR = MOD via oral route. Highly toxic; a contact insecticide. See DDT.
Disaster Hazard: Dangerous; when heated to decomp, emits highly toxic fumes of fluorides and chlorides.

DFP. See diisopropyl fluorophosphate.

DGE. See diglycidyl ether.

DHS. See dihydrostreptomycin.

2,4-DIACETAMIDO-6-(5-NITRO-2-FURYL)-*s*-TRIAZONE. $C_{11}H_{10}O_5N_6$, mw: 306.3.
THR = An exper carc [3, 23] via oral route.

DIABENZENE. See chlorpropamide.

DIACETIC ETHER. See ethyl acetoacetate.

DIACETIN. See glyceryl diacetate.

DIACETONE ALCOHOL. Syn: *4-hydroxy-4-methylpentanone-2*. Liquid, faint pleasant odor. $(CH_3)_2C(OH)CH_2COCH_3$, mw: 116.2, mp: −47° to −54°, bp: 167.9°, flash p: 148°F, d: 0.9306 @ 25°/4°,

autoign. temp.: 1118°F, vap. d: 4.00, vap. press: 1.1 mm @ 20°, lel = 1.8%, uel = 6.9%, flash p: (acetone free): 136°F.
Acute tox data: Oral LD_{50} (rat) = 4000 mg/kg; ip LD_{50} (mouse) = 933 mg/kg. [*3*]
THR = MOD via oral and ip routes. Irr to eyes and mu mem. Narcotic in high conc. Exper has caused anemia and damage to kidneys and liver.
Fire Hazard: Mod, when exposed to heat or flame; can react with oxidizing materials.
To Fight Fire: Alcohol foam, foam, CO_2, dry chemical.

1,1-DIACETOXY-2,3-DICHLOROPROPANE. $C_7H_{10}O_4Cl_2$, mw: 229.1.
Acute tox data: Oral LD_{LO} (rat) = 32 mg/kg.; dermal LD_{LO} (rabbit) = 1000 mg/kg. [*3*]
THR = HIGH via oral, MOD via dermal routes.

7-DIACETOXY METHYL BENZ(a)ANTHRACENE. $C_{23}H_{18}O_4$, mw: 358.4.
THR = An exper carc. [*3*]

DIACETYL. Syn: *butanedione*. Greenish-yellow liquid, strong odor. $CH_3COCOCH_3$, mw: 86.09, bp: 88°, flash p: 80°F, d: 0.9904 @ 15°/15°, vap. d: 3.00.
Acute tox data: Oral LD_{50} (rat) = 1580 mg/kg.; ip LD_{50} (rat) = 400 mg/kg. [*3*]
THR = MOD via oral, HIGH via ip routes.
Note: Used as a synthetic flavoring substance and adjuvant. [*109*]
Fire Hazard: Dangerous, when exposed to heat or flame.
Disaster Hazard: Dangerous. Keep away from heat and open flame.
To Fight Fire: Alcohol foam, CO_2, dry chemical.

DIACETYL AMINOAZO TOLUENE. Syn: *Fast orange*. Red crystals or needles. $C_{18}H_{19}O_2N_3$, mw: 309.4, mp: 65° (crystals), mp: 75° (needles).
THR = Used in human and veterinary medicine. An exper (S) hepatocarc. [*3, 4*]

DIACETYL BENZIDINE. See 4′,4‴-biacetanilide.

DIACETYLENE. Syns: *butadiyne*, *butadine*. Gas. C_4H_2, mw: 50.1, mp: −36.°, bp: 10.3°, d: 2.233.
THR = In high conc it acts as a simple asphyxiant.
Fire Hazard: Mod, when exposed to heat or flame, or by chemical reaction with oxidizers. Can ignite spont in contact with moist silver salts.
Explosion Hazard: Can explode @ > −25° in air. [*19*]

DIACETYLENE CARBONIC ACID.
THR = U.
Fire Hazard: U.
Explosion Hazard: Mod, @ 186°.

1,2-DIACETYLETHANE. See acetonyl acetone.

0,0-DIACETYL-4-HYDROXY AMINO QUINOLINE-1-OXIDE. $C_{13}H_{13}O_4N_2$, mw: 261.3.
THR = An exper neo. [*3*]

DIACETYL MORPHINE. Syns: *diamorphine*, *heroin*. White, odorless, bitter crystals or crystalline powder. $C_{17}H_{17}(OOCCH_3)_2NO$, mw: 369.40, mp: 173°, bp: 273° @ 12 mm.
Acute tox data: sc LD_{LO} (rabbits) = 150 mg/kg.
THR = Fatal dose between $\frac{1}{6}$ and 2 grains. Resembles morphine in its general results, but acts more strongly on the respiration and is therefore more poisonous. Its depressant effects on the cerebrum appear to be greater than that of codeine. Large doses cause excitement and convulsions in animals and man. The more common symptoms are headache, disturbance of vision, slow, small, regular pulse, restlessness, cramps in the extremities, slight cyanosis, respiration slow and deep, and death from respiratory paralysis. A poisonous habit-forming drug.
Disaster Hazard: Dangerous; when heated to decomp, emits toxic fumes of NO_x.
Treatment and Antidotes: Use the same treatment as for morphine, with special attention to control of the respiratory paralysis.

DIACETYLMORPHINE HYDROCHLORIDE. See diacetyl morphine.

DIACETYL PEROXIDE. See acetyl peroxide.

DIACETYL PEROXIDE SOLUTION. See acetyl peroxide 25% solution in dimethylphthalate.

DIACETYL TARTARIC ACID ESTERS OF MONO- and DI-GLYCERIDES FROM THE GLYCEROLS OF EDIBLE FATS OR OILS. [*109*]

DIAGNOSTIC X-RAY DOSES. See Section 5A.

DIALIFOR. $C_{12}H_{17}O_4NPS_2Cl$, mw: 369.8.
THR = An exper teratogen [*3*]

DIALLATE. See 2,3-dichloroallyl diisopropyl thiocarbamate.

DIALLYLAMINE. Syn: *di-2-propenylamine*. Liquid, sol in water. $(CH_2{:}CHCH_2)_2NH$, mw: 97.2, d: 0.7889 @ 20°, bp: 112°, fp: −100°.
Acute tox data: Inhal TC_{LO} (man) = 5 ppm for 5 min; oral LD_{50} (rat) = 650 mg/kg; dermal LD_{50} (rabbit) = 356 mg/kg. [*3*]
THR = HIGH via inhal and dermal routes, MOD via oral route. See also allylamine.

4-DIALLYLAMINO-3,5-DIMETHYL PHENYL-N-METHYL CARBAMATE. See hydrol.

N,N-DIALLYL-2-CHLOROACETAMIDE. Syns: *CDAA*, *"Randox," α-chloro-n,n-diallyl acetamide*. Amber liquid, slightly sol in water, sol in alcohol,

hexane and xylene. $ClCH_2CON(CH_2CH:CH_2)_2$, mw: 173.7, bp: 74° at 0.3 mm.
Acute tox data: Oral LD_{50} (rats) = 700 mg/kg; dermal LD_{50} (rat) = 360 mg/kg. [*3*]
THR = MOD via oral, LOW via dermal route.
Disaster Hazard: Dangerous; see chlorides.

DIALLYLCYANAMIDE. Syn: *N-cyanodiallylamine.* Colorless, mobile liquid when pure.
$C_7H_{10}N_2$, mw: 122.17, mp: $<-70°$, bp: 222° (slight decomp), d: 0.9021, vap. d: 4.1.
THR = It is said to be more toxic than cyanamide. See cyanamide and amines.
Disaster Hazard: Dangerous; when heated to decomp or on contact with acid or acid fumes, emits highly toxic fumes of cyanides.

DIALLYL DISULFIDE. See allyl sulfide.

DIALLYL ETHER. See allyl ether.

DIALLYL ETHYLENE BIS-GLYCOLATE. Liquid. $(CH_2OCH_2CO_2CH_2CHCH_2)_2$, mw: 258.3, bp: 130°–133° @ 1 mm, flash p: 320°F (OC), d: 1.1227 @ 25°/25°, vap. d: 8.9.
THR = U.
Fire Hazard: Slight, when exposed to heat or flame; can react with oxidizing materials.
To Fight Fire: Water, foam, CO_2, dry chemical.

DI-2-ALLYL-4-HYDROXY-3-METHYL-2-CYCLOPENTEN-1-ONE ESTER OF CIS AND TRANS-DI-CHRYSANTHEMUM MONOCARBOXYLIC ACID. See allethrin.

DIALLYL MALEATE. Syn: *allyl maleate.* Liquid. $(CHCOOCH_2CHCH_2)_2$, mw: 196.2, vap. d: 6.6.
Acute tox data: Oral LD_{50} (rat) = 300 mg/kg; dermal LD_{50} (rabbit) = 1150 mg/kg. [*3*]
THR = HIGH via oral, MOD via dermal routes.
Fire Hazard: Slight.

DIALLYL MELAMINE. White, crystalline solid. $(CH_2CHCH_2)_2N(CN)_2(NH_2)_2$, mw: 226.3, mp: 142°, d: 1.242 @ 30°, vap. d: 7.6.
THR = MOD irr by ingestion (oral route).
Disaster Hazard: Dangerous; when heated to decomp or on contact with acid or acid fumes, emits highly toxic fumes of cyanides.

DIALLYL METHYL CARBINOL. $C_8H_{14}O$, mw: 126.2.
THR = Has exploded under conditions of vacuum conc after addition of (O_3 + acetic acid). [*19*]

DIALLYL PHOSPHITE. $(C_3H_5O)_2POH$, mw: 162.1.
THR = Can explode while being vacuum distilled. [*19*]

DIALLYL PHTHALATE. Nearly colorless, oily liquid. $C_6H_4(CO_2CH_2CHCH_2)_2$, mw: 246.3, bp: 157°, flash p: 330°F, d: 1.120 @ 20°/20°, vap. d: 8.3.
Acute tox data: Oral LD_{50} (rat) = 770 mg/kg; dermal LD_{LO} (rabbit) = 2800 mg/kg. [*3*]
THR = MOD via oral and dermal routes.
Fire Hazard: Slight, when exposed to heat or flame; can react with oxidizing materials.
To Fight Fire: CO_2, dry chemical.

DIALLYL SULFIDE. See allyl sulfide.

DIALLYL TRISULFIDE. See allyl trisulfide.

DIAMIDO PHOSPHORUS ACID. $POH(NH_2)_2$, mw: 80.
THR = Incandesces upon dissolving in water. [*19*]

DIAMINE. See hydrazine.

DIAMINE BLUE 3B. See Trypan Blue.

DIAMINE SKY BLUE FF. $4Na \cdot C_{34}H_{24}O_{14}N_6S_4$, mw: 960.8.
THR = An exper (+) carc and teratogen. [*3, 4*]

2,4-DIAMINO ANISOLE. $C_7H_{10}ON_2$, mw: 138.1.
THR = An exper. carc. [*105*] Insufficient data. [*108*]

***p*-DIAMINOAZOBENZENE.** Syns: *p-azoaniline, 4,4'-azodianiline, 4,4'-diaminoazobenzene.* Golden yellow needles, slightly sol in water, benzene and petroleum ether, freely sol in alcohol. $C_{12}H_{12}N_4$, mw: 212.25, mp: 240° (decomp).
Acute tox data: Oral LD_{50} (rat) = 1650 mg/kg. [*3*]
THR = MOD via oral route.

DI-*p*-AMINOAZOBENZENE FLUOSILICATE. Brown crystals, slightly sol in water.
$(NH_2C_6H_4N_2C_6H_5)_2 \cdot H_2SiF_6$, mw: 538.5, mp: 220° (decomp).
THR = See fluosilicates.

***p*-DIAMINOBENZENE.** See *p*-phenylenediamine.

***m*-DIAMINOBENZENE.** See *m*-phenylene diamine.

***o*-DIAMINOBENZENE.** See *o*-phenylene diamine.

3,5-DIAMINOBENZOIC ACID. Needle-like crystals, slightly water-sol, sol in alcohol and ether. $C_7H_8N_2O_2 \cdot H_2O$, mw: 170.2, mp: 228°, $-H_2O$ @ 110°.
THR = See *p*-aminobenzoic acid.

DI-*p*-AMINOBENZOIC ACID FLUOSILICATE. White crystals. $(C_6H_4NH_2COOH)_2 \cdot H_2SiF_6$, mp: 242°, mw: 418.4.
THR = See fluosilicates.

1,3-DIAMINOBUTANE. Liquid. $C_4H_8(NH_2)_2$, mw: 88.2, bp: 142°–150°, flash p: 125°F, d: 0.85, vap. d: 3.04.
Acute tox data: Oral LD_{50} (rat) = 1350 mg/kg; dermal LD_{50} (rabbit) = 430 mg/kg. [*3*]
THR = MOD via oral and HIGH via dermal routes.
Fire Hazard: Mod, when exposed to heat or flame; can react with oxidizing materials.
To Fight Fire: Alcohol foam, foam, CO_2, dry chemical.

α,ε-DIAMINO CAPROIC ACID. See lysine.

2,4-DIAMINO-5-(3,4-DICHLOROPHENYL)-6-METHYL PYRIMIDINE.
THR = An exper teratogen. [*3*]

3-DIAMINO-4-DIHYDROXY-1-ARSENOBENZENE HYDROCHLORIDE. See arsphenamine.

***p*-DIAMINODIPHENYL.** See benzidine.

4,4′-DIAMINODIPHENYLAMINE. Crystals. $C_{12}H_{13}N_3$, mw: 199.3, mp: 158°.
THR = See diphenylamine.

4,4′-DIAMINO DIPHENYL METHANE. See methylene dianiline.

4,4′-DIAMINODIPHENYL SULFONE. Syn: *p,p′-sulfonyl dianiline*. Crystals, nearly insol in water, sol in acetone and alcohol. $C_{12}H_{12}O_2N_2S$, mw: 248.3, mp: 176°, vap. d: 8.3.
Acute tox data: Oral LD_{LO} (rat) = 1000 mg/kg; oral LD_{50} (mouse) = 496 mg/kg. [*3*]
THR = MOD via oral route. Mod doses can cause neuritis, dermatitis and hepatitis. Used in treating leprosy and in veterinary medicine. An exper neo. [*3*]
Disaster Hazard: Dangerous; when heated to decomp, emits highly toxic fumes of SO_x.

1,2-DIAMINOETHANE. See ethylenediamine.

2,4-DIAMINO-4′-ETHOXYAZOBENZENE. See ethoxazene.

1,3-DIAMINOISOPROPANOL. Sol in water. $NH_2CH_2CHOHCH_2NH_2$, mw: 90, d: 1.1, bp: 130°, flash p: 27°F.
Fire Hazard: Dangerous, when exposed to heat or flame.
To Fight Fire: Water, foam, alcohol foam.

1,8-DIAMINO-*p*-METHANE. $C_{10}H_{22}O_2$, mw: 170.3.
Acute tox data: Oral LD_{50} (rat) = 770 mg/kg; dermal LD_{50} (rabbit) = 630 mg/kg. [*3*]
THR = MOD via oral and dermal routes. See also amines.

1,8-DIAMINO NAPHTHALENE. See 1,8-naphthalene diamine.

4,6-DIAMINO-2-(5-NITRO-2-FURYL)-*s*-TRIAZINE. $C_7H_6O_3N_6$, mw: 222.2.
THR = An exper carc. [*3, 23*]

DIAMINOPHENOL HYDROCHLORIDE. See amidol.

2,6-DIAMINO-3-(PHENYLAZO)PYRIDINE. $C_{11}H_{11}N_5$, mw: 213.3.
THR = An exper (S) carc. [*3, 4*]

2,6-DIAMINO-3-(PHENYLAZO)PYRIDINE HYDROCHLORIDE. $C_{11}H_{11}N_5 \cdot HCl$, mw: 249.7.
THR = An exper (S) carc. [*3, 4*]

1,3-DIAMINOPROPANE. Liquid. $C_3H_6(NH_2)_2$, mw: 74.1, bp: 133°, flash p: 75°F (TOC), d: 0.88, vap. d: 2.5.
Acute tox data: Oral LD_{50} (rat) = 350 mg/kg; dermal LD_{50} (rabbit) = 200 mg/kg. [*3*]
THR = HIGH via oral and dermal routes.
Fire Hazard: Dangerous, when exposed to heat or flame.
Explosion Hazard: U.
Disaster Hazard: Dangerous! Keep away from heat and open flame!
To Fight Fire: Alcohol foam, foam, CO_2, dry chemical.

1,3-DIAMINO-2-PROPANOL. $C_3H_{10}ON_2$, mw: 90.1, flash p: 270°F, d: 1.1, bp: 130°.
THR = No data. See alcohols.
Fire Hazard: Low.
To Fight Fire: Alcohol foam, dry chemical.

N-(*p*(2,4-DIAMINO-6-PTERIDYL) METHYL METHYL AMINO BENZOYL) GLUTAMIC ACID. See methotrexate.

2,6-DIAMINOPYRIDINE. Crystals. $NC_5H_3(NH_2)_2$, mw: 109.08, mp: 120.8°, bp: 285°.
Acute tox data: ip LD_{50} (mouse) = 100 mg/kg. [*3*]
THR = HIGH via ip route. See also pyridine.
Disaster Hazard: Dangerous; when heated to decomp, emits highly toxic fumes of nitrogen compounds.

4,4′-DIAMINO STILBENE.
THR = An exper carc. [*23*]

DI(α-AMINO-β-THIOLPROPIONIC ACID). See cystine.

2,5-DIAMINOTOLUENE. See 2,5-toluenediamine.

DIAMMINECOBALT (II) CHLORIDE. Rose crystals. $CoCl_2 \cdot 2NH_3$, mw: 163.92, mp: 273°, d: 2.097.
THR = See cobalt compounds, chlorides and ammonia.

DIAMMINECOPPER (II) ACETATE. Violet-blue crystals. $Cu(C_2H_3O_2)_2 \cdot 2NH_3$, mw: 215.69, mp: decomp @ approx 175°.
THR = See copper compounds and ammonia.

DIAMMONIUM HYDROGEN PHOSPHATE. See ammonium phosphate, dibasic.

DIAMMONIUM PHOSPHATE. See ammonium phosphate, dibasic.

DIAMORPHINE. See diacetylmorphine.

DIAMYL ACETAMIDE. Liquid. $(C_5H_{11})CHCONH_2$, mw: 199.3, flash p: 235°F, vap. d: 6.7.
THR = U.
Fire Hazard: Slight, when exposed to heat or flame; can react with oxidizing materials.
To Fight Fire: Foam, CO_2, dry chemical.

DIAMYLAMINE. Syn: *di-n-amylamine*. Water white liquid. $(C_5H_{11})_2NH$, mw: 157.26, bp: 202°, flash p: 124°F, d: 0.777 @ 20°/20°, vap. d: 5.42.
Acute tox data: Oral LD_{50} (rat) = 270 mg/kg; inhal LC_{LO} (rat) = 63 ppm for 4 hrs; dermal LD_{LO} (rabbit) = 35 mg/kg. [*3*]
THR = HIGH via oral, inhal and dermal routes. See also amines.
Fire Hazard: Mod, when exposed to heat or flame; can react with oxidizing materials.
To Fight Fire: Alcohol foam, foam, CO_2, dry chemical.

DI-*n*-AMYLAMINE. See diamylamine.

DIAMYL ANILINE. Liquid. $C_6H_5N(C_5H_{11})_2$, mw: 233.4, bp: 277°, flash p: 260°F, d: 0.89, vap. d: 7.8.
THR = U. See also aniline.
Fire Hazard: Slight, when exposed to heat or flame.
Disaster Hazard: Mod dangerous; when heated to decomp, emits toxic fumes; can react with oxidizing materials.
To Fight Fire: Foam, CO_2, dry chemical.

DIAMYL BENZENE. Liquid. $C_6H_4(C_5H_{11})_2$, mw: 218.4, bp: 265°, flash p: 225°F, d: 0.85, vap. d: 7.3.
THR = U. See also benzene.
Fire Hazard: Slight, when exposed to heat or flame; can react with oxidizing materials.
To Fight Fire: CO_2, dry chemical.

DIAMYL BIPHENYL. Liquid. $C_5H_{11}(C_6H_4)_2C_5H_{11}$, mw: 294.5, mp: −30°, flash p: 340°F, bp: 355°–385°, d: 0.938 @ 20°/20°, vap. d: 10.2.
THR = U. See also diphenyl and amyl alcohol.
Fire Hazard: Slight, when exposed to heat or flame; can react with oxidizing materials.
To Fight Fire: CO_2, dry chemical.

DIAMYL CHLORONAPHTHALENE. Liquid. $(C_5H_{11})_2C_{10}H_5Cl$, mw: 302.9, bp: 350°, flash p: 330°F, d: 1.06, vap. d: 10.4.
THR = U. See also chloronaphthalenes.
Fire Hazard: Slight, when exposed to heat or flame.
Disaster Hazard: Dangerous; see chlorides; can react vigorously with oxidizing materials.
To Fight Fire: Water, foam, CO_2, dry chemical.

DI-*tert*-AMYLCYCLOHEXANOL. Insol in water. $(C_5H_{11})_2C_6H_9OH$, mw: 240.4, flash p: 270°F, bp: 554°–572°F, d: 0.9.
Fire Hazard: Slight, when exposed to heat or flame.
To Fight Fire: Dry chemical, CO_2.

DIAMYLENE. Liquid. $C_{10}H_{20}$, mw: 140.26, mp: <−50°, bp: 150°, flash p: 118°F (OC), d: 0.77–0.78, vap. d: 4.7.
THR = U. Compounds of this type have irr and narcotic action.
Fire Hazard: Mod, when exposed to heat or flame; can react with oxidizing materials.
To Fight Fire: Foam, CO_2, dry chemical.

DIAMYL ETHER. See amyl ether.

DIAMYL MALEATE. Liquid. $(CHCO_2C_5H_{11})_2$, mw: 256.3, flash p: 270°F, d: 0.981, vap. d: 8.6, bp: 500°–572°F.
THR = U. See also esters and amyl alcohol.
Fire Hazard: Slight, when exposed to heat or flame; can react to oxidizing materials.
To Fight Fire: CO_2, dry chemical.

DI-*n*-AMYL MERCURY. Crystals. $Hg(C_5H_{11})_2$, mw: 342.9, d: 1.6369, bp: 133° @ 10 mm.
THR = HIGH toxicity. See mercury compounds, organic.

DIAMYL NAPHTHALENE. Liquid. $C_{10}H_6(C_5H_{11})_2$, mw: 268.43, bp: 326°, flash p: 315°F (OC), d: 0.93–0.94, vap. d: 9.3, vap. press: 0.00124 mm @ 20°.
THR = U. See also naphthalene.
Fire Hazard: Slight, when exposed to heat or flame; can react with oxidizing materials.
To Fight Fire: CO_2, dry chemical.

DIAMYL-β-NAPHTHOL. Liquid. $(C_5H_{11})_2C_{10}H_5OH$, mw: 284.4, bp: 205°, flash p: 345°F, d: 0.97, vap. d: 9.6.
THR = U. See also β-naphthol.
Fire Hazard: Slight, when exposed to heat or flame; can react with oxidizing materials.
To Fight Fire: Foam, CO_2, dry chemical.

DI-N-AMYL NITROSAMINE. See *n*-nitrosodiphenylamine.

DIAMYL OXALATE. Liquid. $(C_5H_{11}CO_2)_2$, mw: 230.3, flash p: 257°F, vap. d: 7.7, d: 1.0.
THR = HIGH toxicity. See oxalates and amyl alcohol.
Fire Hazard: Slight, when exposed to heat or flame; can react with oxidizing materials.
To Fight Fire: CO_2, dry chemical.

2,4-DIAMYL PHENOL. Liquid. $(C_5H_{11})_2C_6H_3OH$, mw: 234.37, bp: 278°, flash p: 260°F (OC), d: 0.93–0.94, vap. d: 8.1.
Acute tox data: ip LD_{50} (rat) = 620 mg/kg. [*3*]
THR = MOD via ip route.
Fire Hazard: Slight, when exposed to heat or flame.
Disaster Hazard: Mod dangerous; when heated to decomp, emits toxic fumes; can react with oxidizing materials.
To Fight Fire: CO_2, dry chemical.

DI-*sec*-AMYL PHENOL. Liquid. $(C_5H_{11})_2C_6H_3OH$, mw: 234.4, bp: 280°, flash p: 260°F (OC), d: 0.91, vap. d: 8.1.
THR = Probably toxic. See also phenol.

Fire Hazard: Slight, when exposed to heat or flame; can react with oxidizing materials.
To Fight Fire: Foam, CO_2, dry chemical.

DI-*tert*-AMYLPHENOXYETHANOL. Liquid. $C_6H_3(C_5H_{11})_2OCH_2CH_2OH$, mw: 278.4, mp: −35°, flash p: 300°F (OC), bp: 321°, d: 0.959, vap. d: 9.6, vap. press: <0.01 mm.
THR = U. See also alcohols.
Fire Hazard: Slight, when exposed to heat or flame; can react with oxidizing materials.
To Fight Fire: CO_2, dry chemical.

DIAMYL PHTHALATE. Liquid. $C_6H_4(COOC_5H_{11})_2$, mw: 306.4, bp: 333°, flash p: 357°F (CC), vap. d: 10.5.
THR = See esters, amyl alcohol and phthalic acid.
Fire Hazard: Slight, when exposed to heat or flame: can react with oxidizing materials.
To Fight Fire: Water, foam, CO_2, dry chemical.

DI-*n*-AMYL PHTHALATE. Syn: *amyl phthalate*. Liquid. $C_6H_4(CO_2C_5H_{11})_2$, mw: 306.4, bp: 243°–255° @ 50 mm, flash p: 245°F (CC), d: 1.023, vap. d: 10.5.
THR = See esters, amyl alcohol and phthalic acid.
Fire Hazard: Slight, when exposed to heat or flame; can react with oxidizing materials.
To Fight Fire: Water, foam, CO_2, dry chemical.

DIAMYL SODIUM SULFOSUCCINATE. White waxy solid. $C_5H_{11}CO_2CH_2CHCO_2C_5H_{11}SO_3Na$, mw: 360.
THR = LOW via oral and inhal routes. A mild irr.
Disaster Hazard: Dangerous. See sulfonates.

DIAMYL SULFIDE. See amyl sulfide.

DIANILINE CALCIUM. White crystals, decomp by water. $(NHC_6H_5)_2Ca$, mw: 224.3.
THR = See aniline.

DIANILINE FLUOSILICATE. White crystals, water-sol. $(C_6H_5NH_2)_2 \cdot H_2SiF_6$, mw: 330, mp: sublimes @ 230°.
THR = See aniline and fluosilicates.

***o,o*-DIANISIDINE.** Syn: *dimethoxybenzidine*. Colorless crystals. $(NH_2(OCH_3)C_6H_3)_2$, mw: 244.29, mp: 137°–138°, flash p: 403°F, vap. d: 8.5.
Acute tox data: Oral LD_{50} (rat) = 1920 mg/kg. [*3*]
THR = MOD via oral route. An exper (+) neo. [*3, 10, 23*]
Fire Hazard: Slight, when exposed to heat or flame; can react with oxidizing materials.
To Fight Fire: Water, CO_2, dry chemical.

DI-*o*-ANISYL DICHLOROSTANNANE. Solid. $(CH_3OC_6H_4)_2SnCl_2$, mw: 404, mp: 113°.
THR = See chlorides and tin compounds.

2,2-DI-*p*-ANISYL-1,1,1-TRICHLOROETHANE. See methoxychlor.

DIATOMACEOUS EARTH. Syns: *infusorial earth, kieselguhr*. A soft, earthy rock, a silicious material. SiO_2K_2O, Al_2O_3, Fe_2O_3, CaO, d: 0.24–0.34.
Acute tox data: Oral LD_{50} (rat) = 3160 mg/kg. [*3*]
THR = MOD via oral and inhal routes. Diatomaceous earth dust may cause disabling fibrosis of the lungs, but it is less likely to do so than SiO_2 in the crystalline form. It is a substance which migrates to food from packaging materials. [*109*] Less dangerous before roasting or calcining than after.

DIAZEPAM. See Valium.

DI-*o*-AZIDOBENZOYL PEROXIDE. $C_{14}H_8O_4N_6$, mw: 324.2.
THR = Unstable; can be easily detonated. [*19*]

1,3-DIAZIDOPROPENE. $C_3H_4N_6$, mw: 124.
THR = Very unstable and may explode. [*19*]

DIAZINON. Syns: *G-24,480, o,o-diethyl-o,2-isopropyl-4-methylpyrimidyl thiophosphate*. Liquid with faint ester-like odor, miscible in organic solvents. $C_{12}H_{21}N_2O_3PS$, mw: 304.4, bp: 84° @ 0.002 mm, d: 1.116 @ 20°/4°.
Acute tox data: Oral LD_{50} (male rat) = 76 mg/kg; dermal LD_{50} (rat) = 455 mg/kg. [*3*]
THR = HIGH via oral and dermal routes. Action less severe than parathion. An exper teratogen. [*3*]

N-(DIAZO ACETYL)GLYCINE HYDRAZINE. $C_4H_7O_2N_5$, mw: 157.2.
THR = An exper carc. [*3*]

α-DIAZOAMIDOBENZOL. Syns: *diazoaminobenzene, 1,3-diphenyltriazine*. Golden yellow crystals. $C_6H_5NNNHC_6H_5$, mw: 197.2, mp: 98°–99°, bp: explodes, vap. d: 6.8.
THR = U. An insecticide.
Explosion Hazard: Severe, when shocked or exposed to heat.
Disaster Hazard: Dangerous; shock or heat will explode it.

β-DIAZOAMIDOBENZOL. Syn: *benzene diazoanilide*. Yellow crystals. $C_{12}H_{11}N_3$, mw: 197.2, mp: 80°–81°, bp: explodes, vap. d: 6.8.
THR = U.
Explosion Hazard: Severe, when shocked or exposed to heat.
Disaster Hazard: Dangerous; explodes when heated or emits toxic fumes.

1,1-DIAZOAMIDONAPHTHALENE. Syn: *1,3-di-1-naphthyltriazine*. Yellow leaflets. $C_{10}H_7NNNHC_{10}H_7$, mw: 297.4, mp: explodes @ 100°, vap. d: 10.25.
THR = U. See also naphthalene.
Explosion Hazard: Severe, when shocked or exposed to heat.

Disaster Hazard: Dangerous; explodes when heated or shocked; emits toxic fumes.

DIBENZ (a,i)ANTHRACENE.
THR = An exper carc. [*23*]

DIAZOAMINOBENZENE. See α-diazoamidobenzol.

DIAZOBENZENE CHLORIDE. Syn: *benzene diazonium chloride*. Colorless needles. $C_6H_5N(\vdots N)Cl$, mw: 140.6, mp: decomp, bp: explodes, vap. d: 4.9.
Acute tox data: Oral LD_{50} (rat) = 400 mg/kg. [*3*]
THR = HIGH via oral route. An exper carc. [*3*]
Explosion Hazard: Severe, when shocked or exposed to heat or H_2SO_4 or upon distillation.
Disaster Hazard: Dangerous; shock and heat will explode it; emits toxic fumes.

DIAZOBENZENEIMIDE. See phenyl azoimide.

DIAZOBENZENE NITRATE. Syn: *benzene diazonium nitrate*. Colorless needles. $C_6H_5N(\vdots N)NO_3$, mw: 167.1, mp: explodes @ 90°, d: 1.37 @ 20°/4°, vap. d: 5.8.
THR = U. See also nitrates.
Fire Hazard. See nitrates.
Explosion Hazard: Severe; explodes when slightly shocked or exposed to heat. See also nitrates.
Disaster Hazard: See nitrates.

DIAZOBENZENE SULFATE. See diazobenzol sulfate.

***p*-DIAZOBENZENESULFONIC ACID.** Syn: *sulfanilic acid diazide*. White paste (white or slightly red crystals). $C_6H_4NSO_3N$, mw: 184.17, vap. d: 6.3.
THR = U. Probably toxic.
Explosion Hazard: Severe, when shocked, exposed to heat, friction, chemical reaction.
Disaster Hazard: Dangerous; shock and heat will explode it; see sulfonates.

DIAZOBENZOIC NITRATE. See diazobenzene nitrate.

DIAZOBENZOL ANILIDE. See diazamidobenzol.

DIAZOBENZOL CHLORIDE. See diazobenzene chloride.

DIAZOBENZOL CHROMATE. Syn: *benzene diazonium chromate*. Crystals. $[C_6H_5N(\vdots N)]_2CrO_4$, mw: 326.3.
THR = See chromium compounds.
Fire Hazard: See chromates.
Explosion Hazard: Severe, when shocked or exposed to heat.
Disaster Hazard: Dangerous; shock and heat will explode it; emits highly toxic fumes.

DIAZOBEN. Yellow-brown crystals. $C_8H_{10}O_3S \cdot Na$, mw: 251.3.
Acute tox data: Oral LD_{50} (rat) = 60 mg/kg; dermal LD_{50} (rat) = 100 mg/kg. [*3*]
THR = HIGH via oral and dermal routes. An exper (±) carc. [*3, 4*]

DIAZOBENZOLIMIDE. See phenylazoimide.

DIAZOBENZOL NITRATE. See diazobenzene nitrate.

DIAZOBENZOL SULFATE. Syn: *diazobenzene sulfate*. Crystals. $[C_6H_5N(\vdots N)]_2SO_4$, mw: 306.3, mp: explodes @ 100°, vap. d: 10.6.
THR = U.
Explosion Hazard: Severe, when shocked or exposed to heat.
Disaster Hazard: Dangerous; shock and heat will explode it; see sulfates.

1,12-DIAZOBENZO(rst) PENTAPHENE.
THR = An exper carc. [*23*]

DIAZOBENZOSULFONIC ACID. See *p*-diazobenzenesulfonic acid.

DIAZO CYCLOPENTADIENE. $C_5H_4N_2$, mw: 92.1.
THR = Explosively unstable. [*19*]

4,11-DIAZODIBENZO(b,def)CHRYSENE.
THR = An exper carc. [*23*]

DIAZODINITROPHENOL. Syn: *dinol*. Crystals. $HOC_6H_3(NO_2)_2N(\vdots N)$, mw: 212.1, mp: explodes @ 180°, d: 1.63, vap. d: 7.3.
THR = U. See also dinitrophenol.
Fire Hazard: See nitrates.
Explosion Hazard: Severe, when shocked or exposed to heat or friction. A less sensitive high explosive to impact than lead azide or mercury fulminate. A powerful initiator.

5-DIAZOIMIDAZOLE-4-CARBOXAMIDE.
$C_4H_3ON_5$, mw: 137.1.
THR = An exper neo. [*3*]

5-DIAZOIMIDAZOLE-4-CARBOXAMIDE HYDROCHLORIDE.
THR = An exper neo. [*3*]

DIAZOMALONIC ACID. $C_3H_2O_4N_2$, mw: 130.1.
THR = Can explode during low pressure distillation. [*19*]

DIAZOMETHANE. Syn: *azimethylene*. Yellow gas at ordinary temp. CH_2N_2, mw: 42.04, mp: −145°, bp: −23°, d: 1.45.
THR = HIGH via irr, inhal routes. A powerful allergen. It can cause pulmonary edema and frequently causes hypersensitivity leading to asthmatic symptoms. An exper (+) carc and neo. [*3, 1, 23*]
Explosion Hazard: Severe, when shocked, exposed to heat or by chemical reaction. Undiluted liquid or gas may explode on contact with alkali metals, rough surfaces, heat (200°) and shock.

Disaster Hazard: Highly dangerous; shock and heat will explode it; when heated to decomp or on contact with acid or acid fumes, emits highly toxic fumes.

1-DIAZO-2-NAPHTHOL-4-SULFONIC ACID. Yellow needles in paste or dry form. $C_{10}H_5N_2OSO_3H$, mw: 250.23, bp: decomp. by heat above 100°, vap. d: 8.6.
THR = U.
Explosion Hazard: Mod, when exposed to heat.
Explosive Range: Above 100°.
Disaster Hazard: Dangerous; when heated to decomp, emits highly toxic fumes of SO_x and explodes.

DIAZONITROPHENOL. Crystals. $HOC_6H_4(NO_2)N(\vdots N)$, mw: 167.1, mp: explodes @ 180°, vap. d: 5.8.
THR = U.
Fire Hazard: See nitrates.
Explosion Hazard: Severe, when shocked, exposed to heat or by chemical reaction.
Disaster Hazard: See nitrates.

DIAZONIUM CHLORIDE SALTS.
THR = Reacts violently with Na_2S. [*19*]

DIAZONIUM PERCHLORATES.
THR = Very unstable, particularly when dry. [*19*]

DIAZONIUM SALTS.
THR = Can explode with xanthates, sodium bisulfide and sodium sulfide. [*19*] Reacts violently with thiophenates.

5-DIAZOSALICYLIC ACID. Yellow crystals or powder. $C_7H_4N_2O_3$, mw: 164.12, mp: explodes @ 155°.
THR = U. See also salicylic acid.
Explosion Hazard: Mod, when exposed to heat.
Disaster Hazard: Dangerous; emits toxic fumes and explodes when heated.

DIAZOTIZING SALTS. See sodium nitrite.

DIBAL. See diisobutyl aluminum hydride.

DIBASIC SODIUM-*o*-ARSENATE. White powder. $Na_2HAsO_4 \cdot 7H_2O$, mw: 312.0, d: 1.87.
THR = See arsenic compounds.

DIBASIC SODIUM-*o*-ARSENITE. White powder. Na_2HAsO_3, mw: 169.9.
THR = See arsenic compounds.

DIBENAMINE. See dibenzyl chloroethylamine.

DIBENZ(a,i)ACEANTHRYLENE. $C_{24}H_{14}$, mw: 302.4.
THR = An exper neo and carc. [*3*]

13-H-DIBENZ(bcj)ACEANTHRYLENE.
THR = An exper neo. [*3*]

DIBENZ(a,h)ACRIDINE.
THR = An exper carc. [*23*]

DIBENZ(c,h)ACRIDINE.
An exper carc. [*23*]

DIBENZ(a,j)ACRIDINE.
THR = An exper carc. [*23*]

DIBENZ(a,c)ANTHRACENE. Clear plates or leaflets. $C_{22}H_{14}$, mw: 278.4, mp: 267°.
THR = An exper carc. [*3, 23*]

DIBENZ(a,f)ANTHRACENE.
THR = An exper carc. [*23*]

DIBENZ(a,h)ANTHRACENE.
THR = An exper neo and carc. [*3, 11, 23*]

DIBENZ(a,j)ANTHRACENE.
THR = An exper carc. [*3, 23*]

1,2,5,6-DIBENZANTHRACENE CHOLEIC ACID. $C_{118}H_{174}O_{16}$, mw: 1848.9.
THR = An exper carc. [*3*]

1,2,5,6-DIBENZ ANTHRACENE ENDO-α,β-SUCCINO GLYCINE ETHYL ESTER.
THR = An exper carc. [*3*]

N-(DIBENZ(a,h)ANTHRACEN-7-YL CARBAMOYL) GLYCINE. $C_{25}H_{18}O_3N_2$, mw: 394.5.
THR = An exper neo. [*3*]

1,2,5,6-DIBENZANTHRANYL-9-ISO CYANATE.
THR = An exper carc. [*3*]

DIBENZANTHRENE. See perylene.

DIBENZ(c,f)INDENO(1,2,3-ij)(2,7)NAPHTHYRIDINE. $C_{22}H_{12}N_2$, mw: 304.4.
THR = An exper neo. [*3*]

DIBENZ(c,f)(1)BENZO PYRANO(2,3,3-ij)-2,7-NAPHTHYDRINE. $C_{22}H_{12}ON_2$, mw: 320.4.
THR = An exper neo. [*3*]

DIBENZO(b,def)CHRYSENE-7-CARBOXALDEHYDE. $C_{25}H_{14}O$, mw: 330.4.
THR = An exper neo. [*3*]

7H-DIBENZO(a,g)CARBAZOLE. $C_{20}H_{13}N$, mw: 267.34.
THR = An exper carc. [*3, 23*]

7H-DIBENZO(s,i)CARBAZOLE.
THR = An exper neo. [*3*]

7H-DIBENZO(c,g)CARBAZOLE. Needles. $C_{20}H_{13}N$, mw: 267, mp: 158°.
THR = An exper neo and carc. [*3, 11, 23*]

13H-DIBENZO(a,i)CARBAZOLE.
THR = An exper carc. [*23*]

DIBENZO(b,def)CHRYSENE. Syn: *dibenzo(a,h) pyrene*.
THR = An exper carc. [*23*]

DIBENZO(def,p)CHRYSENE.
THR = An exper (+) carc, neo. [*3, 11, 23*]

DIBENZO(b,def)CHRYSENE. Yellow plates. $C_{24}H_{14}$, mw: 302.4, mp: 163°.
THR = An exper (+) carc. [*3, 11*]

DIBENZO(def,p)CHRYSENE-10-CARBOXALDEHYDE.
THR = An exper neo. [*3*]

DIBENZO(def,mno)CHRYSENE-12-CARBOXALDEHYDE.
THR = An exper neo. [*3, 23*]

DIBENZO-*p*-DIOXIN. Syn: *dibenzo(b,e)(1,4)dioxin.* Crystals. $C_{12}H_8O_2$, mw: 184.2, mp: 122°–123°.
Acute tox data: ip LD_{50} (mice) = > 56 mg/kg. [*81*]
THR = HIGH via ip route. Has caused an exper lymphoma.

13H-DIBENZO(a,g)FLUORENE. $C_{21}H_{14}$, mw: 266.4.
THR = An exper neo. [*3*]

13H-DIBENZO(a,i)FLUORENE.
THR = An exper carc. [*3*]

DIBENZOFURAN. See diphenylene oxide.

2-DIBENZO FURANAMINE. $C_{12}H_9N$, mw: 183.2.
THR = An exper oral carc [*3, 23*] to mice.

3-DIBENZO FURANAMINE.
THR = An exper oral carc [*3*] to rats.

3-DIBENZO FURANYL ACETAMIDE.
THR = An exper carc. [*23*]

DIBENZO(a,c)NAPHTHACENE.
THR = An exper carc. [*23*]

DIBENZO(h,rst)PENTAPHENE. Pale yellow needles. $C_{28}H_{16}$, mw: 352.4, mp: 321°.
THR = An exper (+) neo, carc. [*3, 11, 23*] Via sc route.

DIBENZO(cd,lm)PERYLENE. $C_{26}H_{14}$, mw: 326.4.
THR = An exper neo. [*3, 23*]

DIBENZO(a,c)PHENAZINE. $C_{20}H_{12}N_2$, mw: 280.3.
THR = An exper neo. [*3*]

DIBENZO(a,h)PHENAZINE.
THR = An exper neo. [*3*]

DIBENZOPYRAZINE. See phenazine.

DIBENZO(a,e)PYRENE. $C_{24}H_{14}$, mw: 302.4.
THR = An exper (+) carc and neo. [*3, 11, 23*]

DIBENZO(a,h)PYRENE.
THR = An exper carc. [*23*]

DIBENZO(a,i)PYRENE.
THR = An exper carc. [*23*]

DIBENZO(a,l)PYRENE.
An exper carc. [*23*]

DIBENZOPYRROLE. See carbazole.

2,2-DIBENZOTHIAZYL DISULFIDE. Syn: *2,2'-dithio bis-benzothiazole.* Light yellow crystals. $C_{14}H_8N_2S_4$, mw: 332.5, mp: 180°, bp: decomp, d: 1.50 @ 20°/4°.
Acute tox data: Oral LD_{50} (rat) = 7000 mg/kg; ip LD_{LO} (mouse) = 100 mg/kg. [*3*]
THR = MOD via oral, HIGH via ip routes.
Disaster Hazard: Dangerous; when heated or on contact with acid or acid fumes, it emits highly toxic fumes of SO_x.

DI-(2-BENZOTHIAZYL THIOMETHYL) UREA. Syn: *1,3-bis(2-benzothiazolyl mercaptomethyl)urea.* Cream colored powder. $(C_6H_4NSCSCH_2NH)_2CO$, mw: 418.6, mp: 220°, d: 1.29 @ 20°.
THR = U. See also thiourea.
Disaster Hazard: Dangerous; when heated to decomp, emits highly toxic fumes of SO_x.

N-2-DIBENZOTHIENYL ACETAMIDE. Syn: *2-Acetylamino dibenzothiophene.* $C_{14}H_{11}ONS$, mw: 241.3.
THR = Acute oral tox to rats is MOD. An exper carc. [*3*]

N-3-DIBENZOTHIENYL ACETAMIDE. Syn: *3-acetylamino dibenzo thiophene.* $C_{14}H_{11}ONS$, mw: 241.3.
THR = MOD acute oral toxicity to animals. An exper carc, neo. [*3*]

N-3-DIBENZOTHIENYL ACETAMIDE-5-OXIDE. $C_{14}H_{11}O_2NS$, mw: 257.3.
THR = LOW acute oral toxicity to rats. An exper neo. [*3*]

2-DIBENZO THIO PHENYL ACETAMIDE.
THR = An exper carc. [*23*]

3-DIBENZO THIO PHENYL ACETAMIDE.
THR = An exper carc. [*23*]

DIBENZOTHIOXIN. See phenothioxin.

DIBENZOXY HYDROQUINONE. See dibenzyl ether of hydroquinone.

DIBENZOYL. See benzil.

DIBENZOYL DIETHYLENE GLYCOL ESTER. $C_{18}H_{18}O_5$, mw: 314.4.
Acute tox data: Oral LD_{50} (rat) = 2830 mg/kg. [*3*]
THR = MOD via oral route.

DIBENZOYL DIPROPYLENE GLYCOL ESTER. $C_{20}H_{22}O_5$, mw: 342.4.
Acute tox data: Oral LD_{50} (rat) = 9800 mg/kg. [*3*]
THR = LOW via oral route.

DIBENZOYL PEROXIDE. See benzoyl peroxide.

1,2,5,6-DIBENZ PHENANTHRENE. See benzo(c) chrysene.

DIBENZYL AMINE. An oily liquid with ammonia-like odor, slightly sol in water, sol in alcohol and ether. $C_{14}H_{15}N$, mw: 197.3, bp: 300° (partial decomp).
THR = See diphenyl amine.

N,N-DIBENZYL CHLOROETHYL AMINE. $C_{16}H_{18}NCl$, mw: 259.8.
Acute tox data: Oral LD_{50} (rat) = 2400 mg/kg; sc LD_{50} (mouse) = 800 mg/kg; iv LD_{50} (mouse) = 88 mg/kg. [*3*]
THR = MOD via oral, sc; LOW via iv routes. Can cause leucopenia.
Disaster Hazard: Dangerous; See chlorides.

DIBENZYL CHLOROPHOSPHONATE. $C_{14}H_{14}O_3ClP$, mw: 300.7.
THR = Can decomp violently during vacuum distillation. [*19*]

7,14-DIBENZYL DIBENZ(a,h)ANTHRACENE. $C_{36}H_{26}$, mw: 458.6.
THR = An exper carc. [*3*]

DIBENZYL DIETHYL STANNANE. A liquid sol in organic solvents. $(C_6H_5CH_2)_2Sn(C_2H_5)_2$, mw: 359.1, d: > 1, mp: < 20°, bp: 224° @ 20 mm.
THR = See tin compounds.

DIBENZYL ETHER. Liquid. $(C_6H_5CH_2)O(C_6H_5CH_2)$, mw: 198.25, mp: 5°, bp: 298°, flash p: 275°F (CC), d: 1.036, vap. d: 6.84.
Acute tox data: Oral LD_{50} (rat) = 2740 mg/kg. [*3*]
THR = MOD via oral route. An irr. Vapors probably narcotic in high conc. See also ethers.
Fire Hazard: Slight, when exposed to heat or flame; can react with oxidizing materials.
Spont Heating: No.
Explosion Hazard: Mod, by spont chemical reaction. See also ethers.
To Fight Fire: CO_2, dry chemical.

DIBENZYL ETHER OF HYDROQUINONE. Syn: *dibenzoxyhydroquinone*. Powder. $C_6H_4(OC_6H_5CH_2)_2$, mw: 290.3, vap. d: 10.
THR = U. See also hydroquinones and ethers.
Fire Hazard: Mod, when exposed to heat or flame; can react with oxidizing materials.

DIBENZYL ETHYL PROPYL STANNANE. Liquid miscible in organic liquids. $(C_6H_5CH_2)_2Sn(C_2H_5)(C_3H_7)$, mw: 373.1, mp: > 0°, bp: 225° @ 15 mm.
THR = See tin compounds.

DIBENZYL MERCURY. Colorless crystals, sol in organic solvents. $(C_7H_7)_2Hg$, mw: 382.9.
THR = HIGH toxicity. See mercury compounds, organic.

N,N-DIBENZYL NITROSAMINE. $C_{14}H_{14}ON_2$, mw: 226.3.
Acute tox data: Oral LD_{50} (rat) = 900 mg/kg. [*3*]
THR = MOD via oral route.

DIBENZYL PHOSPHITE. $C_{14}H_{15}O_3P$, mw: 266.2.
THR = Can explode during vacuum distillation. [*19*]

DIBENZYL SEBACATE. Plastic material or light straw-colored liquid. $[(CH_2)_4CO_2C_6H_5CH_2]_2$, mw: 382.5, bp: 265° @ 4 mm, flash p: 457°F, d: 1.055 @ 30°/20°.
THR = U. See also esters, benzyl alcohol and sebacic acid.
Fire Hazard: Slight, when exposed to heat or flame; can react with oxidizing materials.
To Fight Fire: Water, foam, CO_2, dry chemical.

DIBENZYL TIN ACETATE. Colorless crystals, sol in acetone and benzene. $(C_6H_5CH_2)_2Sn(CH_3CO_2)_2$, mw: 419, mp: 137°.
THR = See tin compounds.

DIBENZYL TIN DIBROMIDE. Colorless crystals sol in organic solvents. $(C_6H_5CH_2)_2SnBr_2$, mw: 460.8, mp: 130°.
THR = See tin compounds and bromides.

DIBENZYL TIN DICHLORIDE. Colorless crystals sol in organic solvents. $(C_6H_5CH_2)_2SnCl_2$, mw: 371.9, mp: 164°.
THR = See tin compounds and chlorides.

DIBENZYL TIN DIIODIDE. Colorless crystals sol in organic solvents. $(C_6H_5CH_2)_2SnI_2$, mw: 555, mp: 87°.
THR = See tin compounds and iodides.

DIBORANE. Syns: *boroethane, boron hydride*. Colorless gas, sickly sweet odor. B_2H_6, mw: 27.7, mp: −165.5°, bp: −92.5°, d: 0.447 (liquid @ −112°); 0.577 (solid @ −183°), vap. press: 224 mm @ −112°, autoign. temp.: 100°–125°F, lel = 0.9%, uel = 88%, flash p: −130°F.
Acute tox data: Inhal LC_{50} (rat) = 50 ppm for 4 hrs; inhal LC_{50} (mouse) = 30 ppm for 4 hrs. [*3*]
THR = HIGH via inhal route. Inhal causes irr of lungs and can lead to pulmonary edema. Severe exposures produce symptoms resembling metal fume fever. See also boron hydrides.
Fire Hazard: Very dangerous, when exposed to flame and oxidizers; see also boron hydrides.
Explosion Hazard: By chemical reaction with Al, CCl_4, Cl_2, halogenated hydrocarbons, HNO_3, NF_3, O_2, PF_3, H_2O. [*19*]
Disaster Hazard: Dangerous; when heated to decomp, emits toxic fumes of boron oxides; will react with water or steam to produce hydrogen; can react explosively with oxidizing materials.
To Fight Fire: Stop the flow of gas. Caution: diborane reacts violently with halogenated extinguishing agents.

DIBORANIDE. See potassium diborane.

DIBORON TETRACHLORIDE. B_2Cl_4, mw: 164.

THR = May liberate high irr fumes. Self-ignites in air. [*19*]
Disaster Hazard: Dangerous; See chlorides.

DIBROMETHYNE. See dibromoacetylene.

DIBROM. See dimethyl-1,2-dibromo-2,2-di-chloroethyl phosphate.

DIBROMOACETYLENE. Syn: *dibromethyne.* Liquid. BrC⋮CBr, mw: 183.9, mp: 76° (approx), bp: explodes, d: 2 (approx), vap. d: 6.35.
THR = HIGH irr via oral and inhal routes.
Explosion Hazard: Severe, when exposed to heat or by spont chemical reaction. Explodes easily with a trace of oxygen.
Disaster Hazard: Highly dangerous; when heated, emits toxic fumes of bromides; can react explosively with oxidizing materials.

***p*-DIBROMOBENZENE.** Crystals, odor of xylene. BrC_6H_4Br, mw: 235.9, mp: 87°, bp: 219°, d: 2.261, vap. press: 1 mm @ 61.0°, vap. d: 8.1.
THR = U. A fumigant. Limited animal exper suggest mod toxicity.
Disaster Hazard: Mod dangerous; when heated to decomp, emits toxic fumes of bromides.

2,6-DIBROMO-*p*-BENZOQUINONE-4-CHLORIMINE. C_6H_2OClN, mw: 139.5.
THR = Can spont explode. [*19*]

DIBROMO BORYL PHOSPHINE. Br_2BPH_2, mw: 203.6.
THR = Spont flam in air. [*19*]

2,6-DIBROMO-N-CHLOROBENZO QUINIMIME. $C_6H_2ONClBr_2$, mw: 299.3.
THR = Can explode violently at ordinary laboratory temp. [*19*]

DIBROMOCHLOROMETHANE. Colorless to pale yellow, heavy liquid. $CHClBr_2$, mw: 208.3, bp: 118°–122°, fp: <−20°, d: 2.440 @ 25°/25°.
THR = U. Compounds of this type are generally irr and narcotic. See also bromoform and chloroform.
Disaster Hazard: Mod dangerous; when heated to decomp, emits toxic fumes.

1,2-DIBROMO-3-CHLOROPROPANE. Syns: *nemagon, fumazone, DBCP.* $C_3H_5Br_2Cl$, mw: 236.4, bp: 196°, flash p: 170° F (TOC).
Acute tox data: Oral LD_{50} (rat) = 173 mg/kg; inhal LC_{50} (rat) = 103 ppm for 8 hrs; dermal LD_{50} (rabbit) = 1400 mg/kg. [*3*]
THR = HIGH via oral, inhal; MOD via dermal routes. Irr to skin and mu mem. Narcotic in high conc. Has been implicated in causing human male sterility in factory workers. [*31*] An exper (+) carc. [*81, 3*] and mutagen. [*90*]
Disaster Hazard: Dangerous; see bromides and chlorides.

DIBROMODIETHYL SULFIDE. White crystals. $(CH_2CH_2Br)_2S$, mw: 232.00, mp: 31°–34°, bp: 240°, decomp, d: 2.05 @ 15°, vap. d: 8.0.
THR = HIGH irr via oral and inhal routes. A military poison.
Fire Hazard: Mod, when exposed to heat or flame.
Disaster Hazard: Dangerous; when heated to decomp or on contact with acid or acid fumes, emits highly toxic fumes of oxides of sulfur and bromides; will react with water or steam; reacts vigorously with oxidizing materials.
To Fight Fire: CO_2, dry chemical.

DIBROMODIFLUOROMETHANE. Syn: *difluorodibromomethane.* Colorless, heavy liquid. CF_2Br_2, mw: 209.8, bp: 23.2°, fp: −141°, d: 2.288 @ 15°/4°.
Acute tox data: Inhal LC_{LO} (rat) = 1870 ppm for 15 min. [*3*]
THR = MOD irr via inhal route.
Disaster Hazard: Dangerous; when heated to decomp, emits toxic fumes.

DIBROMO-3,5-DIMETHOXY ANILINE. $C_8H_9O_2NBr_2$, mw: 311.
THR = Very explosive spont. [*19*]

DIBROMOMALONONITRILE. $CBr_2(CN)_2$, mw: 223.9.
THR = Violent reaction with NaN_3. [*19*]

1,6-DIBROMODULCITOL. $C_6H_{12}O_4Br_2$, mw: 308.
THR = HIGH, acute via oral route. An exper carc. [*3*]

1,1-DIBROMOETHANE. Syn: *ethylidene bromide.* Liquid, insol in water, sol in organic solvents. CH_3CHBr_2, mw: 188, d: 1.089 @ 20.5°/4°, bp: 110°.
THR = MOD irr via inhal route. Irr and narcotic material.
Disaster Hazard: See bromides.

1,2-DIBROMOETHANE. See ethylene dibromide.

1,2-DIBROMOETHYL BENZENE. Syn: *styrene dibromide.* Colorless liquid. $C_6H_5(Br_2C_2H_3)$, mw: 264.2, bp: 254.0°, fp: −43°, flash p: none, d: 1.744 @ 25°/25°, vap. press: 1 mm @ 86.0°, vap. d: 9.1.

DIBROMOFORMOXIME. Crystals. CBr_2NOH, mw: 202.87, mp: 70°–71°, vap. d: 7.0.
THR = HIGH irr via oral and inhal routes. A military poison.
Disaster Hazard: Dangerous; when heated, emits highly toxic fumes.

DIBROMOGERMANE. Colorless liquid. GeH_2Br_2, mw: 234.45, mp: −15.0°, bp: 89.0°, d: 2.80 @ 0°, vap. d: 8.1.
THR = See germanium compounds and bromides.

3,5-DIBROMO-4-HYDROXY BENZONITRILE. See bromoxynil.

DIBROMO KETONE. $(Br)_2C=O$, mw: 188, d: 6.5.
THR = HIGH irr via inhal and oral routes.
Disaster Hazard: Dangerous; when heated, emits highly toxic fumes; will react with water or steam to produce toxic and corrosive fumes.

DIBROMO MANNITOL. See myelobromol.

DIBROMOMETHANE. See methylene bromide.

DIBROMOMETHYL ETHER. Colorless liquid. $(CH_2Br)_2O$, mw: 203.90, mp: −34°, bp: 154°–155°, d: 2.2, vap. d: 7.0.
THR = HIGH irr via oral and inhal routes. A lacrimator type of military poison.
Fire Hazard: U. See also ethers.
Disaster Hazard: Dangerous; when heated, emits highly toxic fumes; can react vigorously with oxidizing materials.

DIBROMOMETHYL SULFIDE. Br_2CS, mw: 204.
THR = U. A strong irr.
Disaster Hazard: Dangerous; see sulfides.

DIBROMO NITROETHYLBENZENE. $(C_6H_3)(Br_2)(C_2H_4NO_2)$, mw: 309.2, vap. d: 10.7.
THR = U. An insecticide.
Fire Hazard: See nitrates.
Disaster Hazard: See nitrates.

3,4-DIBROMO NITROSO PIPERIDINE. $C_5H_8ON_2Br_2$, mw: 271.9.
THR = An exper carc to rats via oral route. [*103*]

2,3-DIBROMO-1-PROPANOL. $C_3H_6OBr_2$, mw: 217.9.
Acute tox data: ip LD_{LO} (mouse) = 125 mg/kg. [*3*]
THR = HIGH via ip route. An exper mutagen. [*90*] A metabolite of tris(2,3-dibromopropyl)phosphate also known as Tris-BP. May cause sterility and may be a carc. [*110*] See TRIS(2,3-DIBROMO PROPYL)PHOSPHATE.

DIBROMO TRIMETHYL ALUMINUM. $(CH_3)_3AlBr_2$, mw: 231.9.
THR = Self-ignites in air. [*19*]

DIBUCAINE HYDROCHLORIDE. Syns: *2-butoxy-N-(2-diethylaminoethyl) cinchoninamide hydrochloride, nupercaine hydrochloride, percaine.* Crystals. $C_{20}H_{29}O_2 \cdot HCl$, mw: 379.9, mp: decomp. @ 90°–98°, vap. d: 13.1.
THR = More toxic than procaine. See procaine hydrochloride.
Disaster Hazard: Dangerous; when heated to decomp, emits highly toxic fumes of chlorides.

1,1-DIBUTOXY ETHANE. Syn: *dibutyl acetal.* $CH_3CH_2(OC_4H_9)_2$, mw: 175.
Acute tox data: Oral LD_{50} (rat) = 8790 mg/kg. [*3*]
THR = LOW via oral route on test animals.
Fire Hazard: Slight, when exposed to heat or flame; can react with oxidizing materials.
To Fight Fire: Water, foam, CO_2, dry chemical.

DIBUTOXYMETHANE. Liquid. $CH_2(OC_4H_9)_2$, mw: 160.25, mp: 60°, bp: 165°–188°, flash p: 140°F (CC), d: 0.838, vap. d: 5.5.
THR = U. See also dibutoxyethane.
Fire Hazard: Mod, when exposed to heat or flame; can react with oxidizing materials.
To Fight Fire: Foam, CO_2, dry chemical.

DIBUTOXYTETRAGLYCOL. Practically colorless liquid with characteristic odor. $(C_2H_4OC_2H_4OC_4H_9)_2O$, mw: 306.44, mp: −20°, bp: 237° @ 50 mm, flash p: 305°F (OC), d: 0.9436 @ 20°/20°, vap. d: 10.6.
THR = U. See also glycols.
Fire Hazard: Slight, when exposed to heat or flame; can react with oxidizing materials.
To Fight Fire: Alcohol foam, foam, CO_2, dry chemical.

DIBUTYL ACETAL. See 1,1-dibutoxy ethane.

DIBUTYL ACETAMIDE. Liquid. $(C_4H_9)_2CHCONH_2$, mw: 171.3, bp: 242°, flash p: 225°F, d: 0.89, vap. d: 5.9.
THR = U. See also amines.
Fire Hazard: Low when exposed to heat or flame; can react with oxidizing materials.
To Fight Fire: CO_2, dry chemical.

DIBUTYLAMINE. Liquid. $(C_4H_9)_2NH$, mw: 129.2, mp: −59°; bp: 159°, flash p: 125°F (OC), d: 0.76, vap. d: 4.46, vap. press: 2 mm @ 20°.
Acute tox data: Oral LD_{50} (rat) = 360 mg/kg; inhal LD_{LO} (rat) = 500 ppm for 4 hrs; dermal LD_{50} (rabbit) = 1010 mg/kg. [*3*]
THR = HIGH via oral and inhal; MOD via dermal routes.
Fire Hazard: Mod, when exposed to heat or flame; can react with oxidizing materials.
To Fight Fire: Alcohol foam, foam, CO_2, dry chemical.

DI-*sec*-BUTYLAMINE. Liquid. $(C_4H_9)_2NH$, mw: 129.2, bp: 134°, flash p: 75°F (OC), d: 0.75, vap. d: 4.5.
THR = See dibutylamine.
Fire Hazard: Dangerous, when exposed to heat or flame; can react with oxidizing materials.
To Fight Fire: Alcohol foam, foam, CO_2, dry chemical.

2-*n*-DIBUTYLAMINOETHANOL. Liquid. $(C_4H_9)_2NCH_2CH_2OH$, mw: 173.3, bp: 222°, flash p: 220°F (OC), d: 0.85, vap. d: 6.0.
Acute tox data: Oral LD_{50} (rat) = 1070 mg/kg; ip LD_{50} (rat) = 140 mg/kg; dermal LD_{50} (rabbit) = 1680 mg/kg. [*3*]

THR = MOD via oral and dermal routes. Can be absorbed by skin in toxic quantities.
Fire Hazard: Mod, when exposed to heat or flame; can react with oxidizing materials.
To Fight Fire: CO_2, dry chemical.

3-(DIBUTYLAMINO)PROPYLAMINE. $C_{11}H_{26}N_2$, mw: 186.
Acute tox data: Oral LD_{50} (rat) = 820 mg/kg; dermal LD_{50} (rabbit) = 270 mg/kg. [*3*]
THR = MOD via oral, HIGH via dermal routes.

DIBUTYLANILINE. Liquid. $C_6H_5N(C_4H_9)_2$, mw: 205.3, bp: 266°, flash p: 230°F, d: 0.94, vap. d: 7.1.
THR = Probably toxic.
Fire Hazard: Slight, when exposed to heat or flame.
Disaster Hazard: Dangerous; when heated to decomp, emits highly toxic fumes of nitrogen compounds; can react vigorously with oxidizing materials.
To Fight Fire: Dry chemical, CO_2, mist.

DIBUTYL-1,2-BENZENEDICARBOXYLATE. See dibutyl-*o*-phthalate.

2,5-DI-*tert*-BUTYL BENZOQUINONE. Crystals. $(O)_2 \cdot C_6H_2 \cdot [C(CH_3)_3]_2$, mw: 220.3, mp: 150°, vap. d: 7.6.
THR = U. See also quinone.
Fire Hazard: Slight.
To Fight Fire: Water, spray, mist, dry chemical.

DIBUTYL BERYLLIUM. Colorless liquid. $Be(C_4H_9)_2$, mw: 123.25, bp: 170° @ 25 mm, vap. d: 4.2.
THR = HIGH. See beryllium compounds.
Fire Hazard: Mod, when in contact with heat, flame or oxidizers.
Disaster Hazard: Dangerous; when heated to decomp, emits highly toxic fumes of beryllium; can react vigorously with oxidizing materials.
To Fight Fire: Water spray, mist, dry chemical.

DI-*tert*-BUTYL BERYLLIUM. A clear, colorless, mobile liquid. $Be(C_4H_9)_2$, mw: 123.25, mp: −16°, d: 0.65, vap. press: 35 mm @ 25°, vap. d: 4.2.
THR = HIGH. See beryllium compounds.
Fire Hazard: Mod, when exposed to heat or flame.
Disaster Hazard: Dangerous; when heated to decomp, emits highly toxic fumes; can react vigorously with oxidizing materials.
To Fight Fire: Water spray, mist, dry chemical, foam.

DIBUTYL BORON CHLORIDE. $(C_4H_9)_2BCl$, mw: 160.5.
THR = Self-ignites in air. [*19*]

DIBUTYL BUTYL PHOSPHONATE. Colorless liquid, mild odor. $C_4H_9P(O)(OC_4H_9)_2$, mw: 250.3, bp: 128° @ 2.5 mm, flash p: 311° (COC), d: 8.62.
THR = U.
Fire Hazard: Slight, when exposed to heat or flame.
Disaster Hazard: Dangerous; when heated to decomp, emits highly toxic fumes of oxides of phosphorus; it can react vigorously with oxidizing materials.
To Fight Fire: Foam, CO_2, dry chemical.

DIBUTYL CADMIUM. Oily liquid, decomp by water. $(C_4H_9)_2Cd$, mw: 227, d: 1.3056, mp: −48°, bp: 104° @ 13 mm.
THR = HIGH. See cadmium compounds.
Fire Hazard: U. Probably dangerous when exposed to heat or flame.
Disaster Hazard: Dangerous. See cadmium compounds.

DIBUTYL "CARBITOL." See diethylene glycol dibutyl ether.

DIBUTYL "CELLOSOLVE." See glycols.

DI(BUTYL "CELLOSOLVE") ETHYLENE BIS-GLYCOLATE. Liquid. $(CH_2OCH_2CO_2CH_2CH_2OC_4H_9)_2$, mw: 378.5, bp: 199°–207° @ 1.3 mm, flash p: 410°F (OC), d: 1.0667 @ 25°/25°, vap. d: 13.1.
THR = U. See also glycols.
Fire Hazard: Slight, when exposed to heat or flame; can react with oxidizing materials.
To Fight Fire: Foam, CO_2, dry chemical.

DIBUTYL "CELLOSOLVE" PHTHALATE. See dibutoxy ethyl phthalate.

DIBUTYL "CELLOSOLVE" SUCCINATE.
THR = U. See also glycols.
Fire Hazard: Slight; when heated, can react with oxidizing materials.

DIBUTYL CHLOROPHOSPHATE. Water white liquid. $(C_4H_9O)_2POCl$, mw: 228.5, bp: 103°–106° @ 1.5 mm, d: 1.0742 @ 20°/20°.
THR = U.
Disaster Hazard: Dangerous; when heated to decomp, emits highly toxic fumes of chlorides and oxides of phosphorus.

DI-*tert*-BUTYL CHROMATE.
THR = HIGH, acute. See chromium compounds.

DI-*tert*-BUTYL-*m*-CRESOL. Syn: *DBMC*. Yellow, crystalline solid. $C_{15}H_{24}O$, mw: 220.3, mp: 62.1°, bp: 282°, flash p: 262°F (OC), d: 0.912 @ 80°/4°.
Acute tox data: Oral LD_{50} (mouse) = 1420 mg/kg. [*3*]
THR = MOD via oral route.
Fire Hazard: Slight, when exposed to heat or flame.
Disaster Hazard: Mod dangerous; when heated to decomp, emits toxic fumes; can react with oxidizing materials.
To Fight Fire: Foam, CO_2, dry chemical.

DI-*tert*-BUTYL-*p*-CRESOL. Syns: *DBPC, 4-methyl-2,6-di-tert-butyl phenol, BHT, butylated hydroxy* *toluene, o,o'-di-tert-butyl-p-cresol.* White, crystalline solid. $C_6H_2(C_4H_9)_2(CH_3)OH$, mw: 220.3, bp: 265°, fp: 68°, flash p: 260°F (TOC), d: 1.048 @ 20°/4°, vap. d: 7.6.

Acute tox data: Oral LD_{50} (rat) = 3510 mg/kg; oral LD_{50} (mouse) = 1040 mg/kg. [*3*]

THR = MOD via oral route. An exper teratogen [*3*] via oral route. Used as a food additive. [*109*] It may be a tumorigen. [*30*]

Fire Hazard: Slight, when exposed to heat or flame.

Disaster Hazard: Mod dangerous; when heated to decomp, emits tox fumes; can react with oxidizing materials.

To Fight Fire: CO_2, dry chemical.

o,o'-DI-*tert*-BUTYL-*p*-CRESOL. See di-*tert*-butyl-*p*-cresol.

DIBUTYL-1,2-CYCLOHEXANEDICARBOXYLIC ACID. See dibutyl hexahydrophthalate.

DIBUTYL DIACETOXYSTANNANE. See dibutyltin diacetate.

7,14-DIBUTYL DIBENZ(a,h)ANTHRACENE. $C_{30}H_{30}$, mw: 390.6.

THR = An exper neo. [*3, 23*]

DIBUTYL DICHLOROSTANNANE. See dibutyl tin dichloride.

DIBUTYL-2,3-DIHYDROXYBUTANEDIOATE. See dibutyl tartrate.

DI-*tert*-BUTYL DIPERPHTHALATE (solution in dibutyl phthalate). Clear liquid, immiscible in water but miscible in common organic solvents. $C_{16}H_{22}O_6$, mw: 310.3, d: 1.056 @ 25°, flash p: 145°F (MOC).

THR = See peroxides, organic; also dibutyl phthalate.

N,N-DIBUTYL ETHANOLAMINE. See dibutylaminoethanol.

N-DIBUTYL ETHER. See *n*-butyl ether.

DIBUTYL ETHYLENE BIS-GLYCOLATE. Liquid. $(CH_2OCH_2CO_2C_4H_9)_2$, mw: 290.4, bp: 145°–150° @ 0.6 mm, flash p: 248°F (OC), d: 1.0391 @ 25°/25°, vap. d: 10.0.

THR = U.

Fire Hazard: Slight, when exposed to heat or flame; can react with oxidizing materials.

To Fight Fire: Foam, CO_2, dry chemical.

DIBUTYL FUMARATE. Colorless, clear, mobile, liquid, typical odor. $C_4H_9OCOCHCHOOC_4H_9$, mw: 228.28, bp: 285.1°, fp: −19°, flash p: 300°F (OC), d: 0.986 @ 20°/20°, vap. d: 7.88.

Acute tox data: Oral LD_{50} (rat) = 8350 mg/kg; ip LD_{50} (mouse) = 250 mg/kg; dermal LD_{50} (rabbit) = 16,000 mg/kg. [*3*]

THR = LOW via oral and dermal routes; HIGH via ip route. Can cause slight irr of eyes and mu mem.

Fire Hazard: Slight, when exposed to heat or flame; can react with oxidizing materials.

To Fight Fire: Foam, CO_2, dry chemical.

DIBUTYL HEXAHYDROPHTHALATE. Syn: *dibutyl-1,2-cyclohexanedicarboxylic acid.* Liquid. $C_6H_{10}(CO_2C_4H_9)_2$, mw: 284.4, bp: 185°, flash p: 305°F, d: 1.0, vap. d: 9.8.

THR = U. Probably low acute.

Fire Hazard: Slight, when exposed to heat or flame; can react with oxidizing materials.

To Fight Fire: Foam, CO_2, dry chemical.

N,N-DIBUTYL-(2-HYDROXY PROPYL)AMINE. $CH_3CHOHCH_2NH(C_4H_9)_2$, mw: 187.

Acute tox data: Oral LD_{50} (rat) = 1990 mg/kg. [*3*]

THR = MOD via oral route.

DIBUTYL ISOPHTHALATE. See dibutyl-*m*-phthalate.

DIBUTYL ISOPROPANOLAMINE. Syn: *2-propanol-1-N,N-dibutylamine*. Liquid. $CH_3CHOHCH_2N(C_4H_9)_2$, mw: 187.3, mp: −80°, bp: 229°, flash p: 205°F (OC), d: 0.8419 @ 20°/20°, vap. press: $<$ 0.1 mm, vap. d: 6.5.

THR = U. See also amines.

Fire Hazard: Low, when exposed to heat or flame; can react with oxidizing materials.

To Fight Fire: Alcohol foam, foam, CO_2, dry chemical.

DIBUTYL LAURAMIDE. Liquid. $(C_4H_9)_2CH(CH_2)_{10}CONH_2$, mw: 311.5, bp: 200°, flash p: 375°F, d: 0.86, vap. d: 10.7.

THR = U.

Fire Hazard: Slight, when exposed to heat or flame; can react with oxidizing materials.

To Fight Fire: Foam, CO_2, dry chemical.

DIBUTYL MAGNESIUM. $(C_4H_9)_2Mg$, mw: 138.5.

THR = Self-ignites in air. [*19*]

DIBUTYL MALEATE. Liquid. $C_4H_9OOCHCCHCOOC_4H_9$, mw: 228.3, mp: −85° (sets to a glass), bp: 281°, flash p: 285°F (OC), d: 0.9964 @ 20°/20°, vap. d: 7.9.

Acute tox data: Oral LD_{50} (rat) = 3730 mg/kg; ip LD_{50} (mouse) = 150 mg/kg. [*3*]

THR = MOD via oral route; HIGH via ip route. Upon skin it causes mod irr, causing slight swelling and redness. See also esters and butyl alcohol.

Fire Hazard: Slight, when exposed to heat or flame; can react with oxidizing materials.

To Fight Fire: Foam, CO_2, dry chemical, alcohol foam.

DIBUTYL MALYL DIOXYSTANNANE. See dibutyl tin maleate.

DIBUTYL MERCURY. Liquid. $Hg(C_4H_9)_2$, mw: 314.8, bp: 105° @ 10 mm, d: 1.779, vap. d: 10.8.
Acute tox data: ip LD_{LO} (mouse) = 31 mg/kg. [3]
THR = HIGH via ip route. See mercury compounds, organic.
Fire Hazard: Mod, when exposed to heat or flame.
Disaster Hazard: Dangerous; when heated to decomp or on contact with acid or acid fumes, emits highly toxic fumes of mercury; can react vigorously with oxidizing materials.

DIBUTYL METHIONATE. Liquid. $CH_2(SO_3C_4H_9)_2$, mw: 288.3, vap. d: 10.0.
THR = U.
Disaster Hazard: Dangerous; when heated to decomp, emits highly toxic fumes of oxides of sulfur; will react with water or steam to produce toxic and corrosive fumes.

N,N-DIBUTYLMETHYLAMINE. Colorless liquid, amine odor, insol in water, sol in alcohol and ether, miscible with hydrocarbons. C_9H_2N, mw: 143.3, d: 0.7613 @ 20°/20°, bp: 159.6°, fp: −62°, flash p: 125°F (OC).
Acute tox data: Oral LD_{50} (rat) = 540 mg/kg; inhal LC_{LO} (rabbit) = 250 ppm for 4 hrs; dermal LD_{LO} (rabbit) = 800 mg/kg. [3]
THR = MOD via oral and dermal routes; HIGH via inhal route.
Fire Hazard: Mod, when exposed to heat or flame.

2,6-DI-*tert*-BUTYL-4-METHYL PHENOL. See di-*tert*-butyl-*p*-cresol.

N-BUTYL NITROSAMINE. $C_8H_{18}ON_2$, mw: 158.3.
Acute tox data: Oral LD_{50} (rat) = 1200 mg/kg. [3]
THR = MOD via oral route; an exper (+) carc. [3, 23]

2,6-DI-*tert*-BUTYL-4-NITROPHENOL. $C_{14}H_{21}O_3N$, mw: 251.3.
THR = Can explode spont. [19]

DIBUTYLOLEAMIDE. Liquid.
$C_8H_{17}CHCH(CH_2)_7CON(C_4H_9)_2$, mw: 393.7, bp: 220°, d: 0.85, vap. d: 13.6.
THR = U.
Fire Hazard: Slight, when exposed to heat or flame; can react with oxidizing materials.

DIBUTYL OXALATE. Water white, high-boiling liquid, mild odor. $(COOC_4H_9)_2$, mw: 202.3, mp: −30°, bp: 246°, flash p: 220°F (CC), d: 0.989–0.993 @ 20°/20°, vap. d: 7.0.
THR = See oxalic acid.
Fire Hazard: Slight, when exposed to heat or flame; can react with oxidizing materials.
Spont Heating: No.
To Fight Fire: CO_2, dry chemical.

DIBUTYL OXOSTANNANE. See dibutyltin oxide.

DI-*tert*-BUTYL PEROXIDE. Clear, water white liquid. $(C_4H_9O)_2$, mw: 146.2, mp: −40°, bp: 80° @ 284 mm, flash p: 65°F (OC), d: 0.79, vap. press: 19.51 mm @ 20°, vap. d: 5.03.
Acute tox data: ip LD_{50} (rat) = 3210 mg/kg. [3]
THR = MOD via ip route. Powerful irr via oral and inhal routes. An exper neo. [3]
Fire Hazard: See peroxides, organic; dangerous.
Explosion Hazard: See peroxides, organic.
Disaster Hazard: See peroxides, organic.
To Fight Fire: Water may not work.

DI-*sec*-BUTYL PHENOL. Amber liquid.
$(CH_3CHC_2H_5)_2C_6H_3OH$, mw: 206.3, bp: 152°–165° @ 25 mm, fp: −50°, flash p: 280°F, d: 0.936 @ 25°/4°.
Acute tox data: Oral LD_{50} (rat) = 1320 mg/kg. [3]
THR = MOD via oral route.
Fire Hazard: Slight, when exposed to heat or flame; can react with oxidizing materials.
To Fight Fire: Foam, CO_2, dry chemical.

2,4-DI-*tert*-BUTYL PHENOL. Tan crystals.
$HOC_6H_3(C_4H_9)_2$, mw: 206.3, mp: 51°, bp: 260.8°, flash p: 265°F, d: 0.907 @ 60°/4°, vap. press: 1 mm @ 84.5°.
Acute tox data: ip LD_{50} (mouse) = 25 mg/kg. [3]
THR = HIGH via ip route.
Fire Hazard: Slight, when exposed to heat or flame. Violent reaction with HNO_3. [19]
Disaster Hazard: Mod dangerous; when heated to decomp, emits toxic fumes; can react with oxidizing materials.
To Fight Fire: Foam, CO_2, dry chemical.

N,N′-DI-*sec*-BUTYL-*p*-PHENYLENEDIAMINE. Liquid. $C_6H_4(CHNHCH_2CH_3)_2$, mw: 220.4, mp: 17.8°, flash p: 285°F (OC), d: 0.94–0.95 @ 24°/24°.
Acute tox data: Oral LD_{LO} (rat) = 200 mg/kg; inhal LC_{LO} (rat) = 600 mg/m^3 for 6 hrs; dermal LD_{50} (guinea pig) = 5000 mg/kg. [3]
THR = HIGH via oral, inhal, MOD via dermal routes. Can cause severe burns of skin. Systemic symptoms are sweating, flushing, shortness of breath and slow pulse. A mild allergen.
Fire Hazard: Slight, when exposed to heat or flame; can react with oxidizing materials.
To Fight Fire: Foam, CO_2, dry chemical.

DI-(*p-tert*-BUTYL PHENYL)MONOPHENOL PHOSPHATE. Insol in water.
$(C_4H_9C_6H_4O)_2POOC_6H_5$, mw: 438, d: 1.1, bp: 260°–275° @ 5 mm, flash p: 482°F.
Fire Hazard: Slight, when exposed to heat or flame.
Disaster Hazard: Dangerous; see phosphorus compounds.
To Fight Fire: Water, foam.

DIBUTYL PHOSPHATE. Pale amber liquid. $C_8H_{19}PO_4$, mw: 210.2, bp: decomp > 100°.
Acute tox data: Oral LD_{50} (rat) = 3200 mg/kg. [*3*]
THR = MOD via oral route. An irr.
Disaster Hazard: Dangerous; see phosphorus compounds.

DIBUTYL PHOSPHITE. Liquid. $(C_4H_9O)_2P(O)H$, mw: 194.2, bp: 115° @ 10 mm, flash p: 120°F, d: 0.971 @ 35°/4°, vap. press: < 1 mm @ 20°, vap. d: 6.7.
Acute tox data: Oral LD_{50} (rat) = 3200 mg/kg; dermal LD_{50} (rabbit) = 2000 mg/kg. [*3*]
THR = MOD via oral and dermal routes.
Fire Hazard: Mod, when exposed to heat or flame or by chemical reaction. Many phosphites decomp to evolve phosphine when heated.
Explosion Hazard: See phosphine.
Disaster Hazard: Dangerous; when heated to decomp or on contact with acid or acid fumes, emits highly toxic fumes of PO_x; can react vigorously with oxidizing materials.
To Fight Fire: Foam, CO_2, dry chemical.

DIBUTYL-*m*-PHTHALATE. Syns: *dibutyl isophthalate*, *dibutyl phthalate*. Liquid. $C_6H_4(CO_2C_4H_9)_2$, mw: 278.3, flash p: 322°F (CC), vap. d: 9.58.
THR = See esters, butyl alcohol and phthalic acid. Limited animal exper show low toxicity.
Fire Hazard: Slight, when exposed to heat or flame; can react with oxidizing materials; violent reaction with Cl_2. [*19*]
To Fight Fire: CO_2, dry chemical.

DIBUTYL-*o*-PHTHALATE. Syns: *dibutyl-1,2-benzenedicarboxylate*, *n-butyl phthalate*. Oily liquid, mild odor. $C_6H_4(COOC_4H_9)_2$, mw: 278.3, bp: 340°, fp: −35°, flash p: 315°F (CC), d: 1.047–1.049 @ 20°/20°, autoign. temp.: 757°F, vap. d: 9.58.
Acute tox data: ip LD_{50} (rat) = 3050 mg/kg; TD_{LO} (human) = 140 mg/kg (CNS effects). [*3*] An exper teratogen. [*3*]
THR = MOD via ip; HIGH via oral routes. See esters, phthalic acid and butyl alcohol.
Fire Hazard: Slight, when exposed to heat or flame; can react with oxidizing materials. Violent reaction with Cl_2. [*19*]
To Fight Fire: CO_2, dry chemical.

3,5-DIBUTYLPYRIDINE. Colorless liquid with mild pyridine-like odor. $C_5H_3N(C_4H_9)_2$, mw: 191.3, bp: 271°, fp: −75°, flash p: 252°F, d: 0.882.
THR = Probably somewhat toxic. A mild irr. See also pyridine.
Fire Hazard: Low, but can be ignited.
Disaster Hazard: Dangerous; when heated to decomp, emits highly toxic fumes.
To Fight Fire: Spray, foam, CO_2 or dry chemical.

DIBUTYL SEBACATE. Clear liquid. $[(CH_2)_4COOC_4H_9]_2$, mw: 314.45, bp: 180° @ 3 mm, fp: −11°, flash p: 353°F (COC), d: 0.936 @ 20°/20°, vap. d: 10.8.
THR = See esters and butyl alcohol.
Fire Hazard: Slight, when exposed to heat or flame; can react with oxidizing materials.
To Fight Fire: CO_2, dry chemical.

DIBUTYL STEARAMIDE. Liquid. $CH_3(CH_2)_{16}CON(C_4H_9)_2$, mw: 395.7, bp: 175°, flash p: 420°F, d: 0.86, vap. d: 13.7.
THR = U. Probably low.
Fire Hazard: Slight, when exposed to heat or flame; can react with oxidizing materials.
To Fight Fire: CO_2, dry chemical.

DIBUTYL SUCCINATE. Syn: *tabatrex*. Liquid. $C_{12}H_{22}O_4$, mw: 230.3, bp: 275°, mp: −29°, d: 0.9760 @ 20°/4°.
Acute tox data: Oral LD_{50} (rat) = 8000 mg/kg.
THR = LOW via oral route.

DIBUTYL-*n*-SULFATE. $C_8H_{18} \cdot SO_4$, mw: 210.3.
THR = An exper carc and neo. [*3, 23*]

DIBUTYL SULFIDE. See butyl sulfide.

DIBUTYL SULFOXIDE. $(C_4H_9)_2SO$, mw: 162.2.
THR = Violent reaction with $HClO_4$. [*19*]

DIBUTYL SULFIDE. See butyl sulfide.

DIBUTYL TARTRATE. Syn: *dibutyl-2,3-dihydroxybutanedioate*. Light tan liquid, mild odor. $(COOC_4H_9)_2(CHOH)_2$, mw: 262.30, mp: 21°, bp: 204° @ 26 mm, flash p: 195°F (CC), d: 1.087–1.093 @ 20°/20°, autoign. temp.: 544°F, vap. d: 9.03.
THR = See esters, butyl alcohol and tartaric acid.
Fire Hazard: Mod, when exposed to heat or flame; can react with oxidizing materials.
Spont Heating: No.
To Fight Fire: Water, foam, CO_2, dry chemical.

DI-*n*-BUTYL TELLURIDE. Yellow oil. $(C_4H_9)_2Te$, mw: 241.8, d: 1.334 @ 40°, bp: 132°–135°.
THR = High tox. See tellurium compounds.
Disaster Hazard: Dangerous. See tellurium compounds.

1,3-DIBUTYLTHIOUREA. White to light tan powder. $C_4H_9NHCSNHC_4H_9$, mw: 188.4, mp: 60°, vap. d: 6.5.
Acute tox data: Oral LD_{50} (rat) = 350 mg/kg. [*3*]
THR = HIGH via oral route.
Disaster Hazard: Dangerous; see sulfur compounds.

DIBUTYLTIN DIACETATE. Syn: *dibutyldiacetoxystannane*. Clear, colorless liquid with a slight acetic acid odor. $(C_4H_9)_2Sn(CH_3CO_2)_2$, mw: 351.0, bp: decomp, fp: 5°–10°, flash p: 290°F (OC), d: 1.31 @ 25°, vap. d: 12.1.

Acute tox data: LD_{50} (rat) = 110 mg/kg; LD_{50} (rat) = 32 mg/kg. [3]

THR = HIGH via oral route.

Fire Hazard: Slight, when exposed to heat or flame; can react with oxidizing materials.

To Fight Fire: Water, foam, CO_2, dry chemical.

DIBUTYLTIN DIBROMIDE. Small crystals. $(C_4H_9)_2SnBr_2$, mw: 392.8, mp: 20°.

THR = U. See tin compounds, bromides and butyl alcohol.

Disaster Hazard; Dangerous; see bromides.

DIBUTYLTIN DICHLORIDE. Syn: *dibutyldichlorostannane*. White, crystalline solid. $(C_4H_9)_2SnCl_2$, mw: 303.8, mp: 43°, bp: 135° @ 10 mm, flash p: 335°F (OC), d: 1.36 @ 50°, vap. press: 2 mm @ 100°, vap. d: 10.5.

Acute tox data: Oral LD_{50} (rat) = 150 mg/kg; ip LD_{LO} (rat) = 7.5 mg/kg. [3]

THR = HIGH via oral and ip routes. See tin compounds, butyl alcohol, hydrochloric acid.

Fire Hazard: Slight, when exposed to heat or flame.

Disaster Hazard: Dangerous; emits highly toxic fumes of hydrochloric acid; will react with water or steam to produce heat and toxic fumes; can react vigorously with oxidizing materials.

To Fight Fire: Water, foam, CO_2, dry chemical.

DIBUTYLTIN DILAURATE. Pale yellow liquid to colorless solid (when pure).

$(C_4H_9)_2Sn[O_2C(CH_2)_{10}CH_2]_2$, mw: 631.5, mp: 23°, bp: non-distillable @ 10 mm, flash p: 455°F (OC), d: 1.066 @ 20°/20°, vap. d: 21.8.

Acute tox data: Oral LD_{50} (rat) = 243 mg/kg. [3]

THR = HIGH via oral route. Used against tapeworm in chickens in doses of 100 mg. See tin compounds and butyl alcohol. The vapors evolved when this material is heated should be avoided. A skin irr.

Fire Hazard: Slight, when exposed to heat or flame; can react with oxidizing materials.

To Fight Fire: Water, foam, CO_2, dry chemical.

DIBUTYLTIN MALEATE. Syn: *dibutylmalyldioxstannane*. A white amorphous powder or glass-like resin. $(C_4H_9)_2Sn(CHCO_2)_2$, mw: 347.0, mp: 110°, flash p: 400°F (OC), bulk density: 0.36, vap. d: 12.0.

THR = See tin compounds. Avoid vapors emitted when this material is hot.

Fire Hazard: Slight, when exposed to heat or flame.

Disaster Hazard: Mod dangerous; when heated to decomp, emits toxic fumes; can react with oxidizing materials.

To Fight Fire: Foam, CO_2, dry chemical.

DIBUTYLTIN OXIDE. Syn: *dibutyloxostannane*. White amorphous powder. $(C_4H_9)_2SnO$, mw: 248.9, mp: decomp without melting, bulk density: 0.5, vap. d: 8.6.

Acute tox data: Oral LD_{50} (rat) = 45 mg/kg; ip LD_{50} (rat) = 40 mg/kg. [3]

THR = HIGH via oral and ip routes. See tin compounds and butyl alcohol.

Fire Hazard: Ignites when exposed to flame; can react with oxidizing materials.

***n,n*-DIBUTYLTOLUENE SULFONAMIDE.** $CH_3C_6H_4SO_3N(C_4H_9)_2$, mw: 299, d: 1.1, bp: 392° @ 10 mm, flash p: 330°F.

THR = U.

Fire Hazard: Low, when exposed to heat or flame.

Disaster Hazard: Dangerous; see sulfonates.

To Fight Fire: Dry chemical, CO_2.

DIBUTYL UREA. Liquid. $C_4H_9NHCONHC_4H_9$, mw: 172.3, mp: 25°, bp: 119°, flash p: 279°F, vap. d: 5.95.

THR = U.

Fire Hazard: Slight, when exposed to heat or flame; can react with oxidizing materials.

To Fight Fire: Foam, CO_2, dry chemical.

DI-*n*-BUTYL ZINC. Liquid, decomp by cold water. $(C_4H_9)_2Zn$, mw: 179.6, bp: 82° @ 9 mm.

THR = U. See zinc compounds.

DICACODYL. See cacodyl.

DICACODYL OXIDE. See cacodyl oxide.

DICACODYL SULFIDE. See cacodyl sulfide.

DICALCIUM-*o*-PHOSPHATE. See calcium phosphate dibasic.

DICAPROATE. See triethylene glycol.

DICAPRYL ADIPATE. Liquid. $(CH_2CH_2CO_2C_8H_{17})_2$, mw: 370.6, bp: 211°–217° @ 4 mm, flash p: 352°F, d: 0.916 @ 20°/20°, vap. d: 12.8.

THR = See esters.

Fire Hazard: Slight, when exposed to heat or flame; can react with oxidizing materials.

To Fight Fire: Foam, CO_2, dry chemical.

DICAPRYL PHTHALATE. Clear liquid. $C_6H_4(CO_2C_5H_{17})_2$, mw: 390.6, bp: 222°–230° at 4 mm, fp: −60°, flash p: 394°F (COC), d: 0.978 @ 20°/20°, vap. d: 9.8.

Acute tox data: ip LD_{50} (mouse) = 14,000 mg/kg. [3]

THR = LOW via ip route. See esters.

Fire Hazard: Slight, when exposed to heat or flame; can react with oxidizing materials.

To Fight Fire: CO_2, dry chemical.

DICAPRYL SEBACATE. Liquid. $[(CH_2)_4CO_2C_8H_{17}]_2$, mw: 426.7, bp: 230°–240° @ 4 mm, flash p: 445°F, d: 0.907 @ 20°/20°, vap. d: 14.7.
THR = See esters.
Fire Hazard: Slight, when exposed to heat or flame; can react with oxidizing materials.
Disaster Hazard: Slight, can react with oxidizing materials.
To Fight Fire: Foam, CO_2, dry chemical.

DICAPTAN. See dicapthon.

DICAPTHON. Syns: *o-(2-chloro-4-nitrophenyl)-o,o-dimethylphosphorothioate, dicaptan.* White solid, insol in water, sol in acetone, cyclohexane, ethyl acetate, toluene and xylene.
$(CH_3O)_2P(:S)OC_6H_3(Cl)NO_2$, mw: 297.5, mp: 51°–52°.
Acute tox data: LD_{50} (rat) = 330 mg/kg; dermal LD_{50} (rat) = 790 mg/kg. [*3*]
THR = HIGH via oral; MOD via dermal routes. An insecticide. High tox. Cholinesterase inhibitor. See parathion.

DICARBITOL PHTHALATE.
See "Carbitol" phthalate.

DICARBON HEXABROMIDE. Syn: *hexabromoethane.* Rhombic crystals. C_2Br_6, mw: 503.52, mp: 149° decomp, bp: 210°, d: 3.823, vap. d: 18.2.
THR = U. See also bromides.
Disaster Hazard: Mod dangerous; when heated to decomp, emits toxic fumes of bromides and will react with water or steam to produce toxic and corrosive fumes.

DICARBON HEXACHLORIDE.
See hexachloroethane.

DICARBON TETRABROMIDE. Syns: *carbon dibromide, tetrabromoethylene.* Solid. C_2Br_4, mw: 343.68, mp: 57.5°, bp: 227°, vap. d: 11.9.
THR = U. See also bromides.
Disaster Hazard: Dangerous; see bromides.

DICARBON TETRACHLORIDE.
See perchloroethylene.

DICHLOBENIL. See 2,6-dichlorobenzonitrile.

DICHLONE. See 2,3-dichloro-1,4-naphthoquinone.

DICHLORACETYLENE. See dichloroacetylene.

DICHLORETHANAL. See dichloroacetaldehyde.

DICHLORINATED METHYL OXIDE.
See dichloromethyl ether.

DICHLORINE HEPTOXIDE. Cl_2O_7, mw: 182.9.
THR = Very unstable, powerful oxidizer. Heat- and impact-sensitive. [*19*]

DICHLOROACETALDEHYDE. Syn: *2,2-dichlorethanal.* Colorless liquid, polymerizes slowly to white solid. $CHCl_2CHO$, mw: 112.95, bp: 88°, fp: −50°, flash p: 140°F (CC), d: 1.436 @ 25°/4°, vap. press: 50 mm @ 20°, vap. d: 3.9.
THR = HIGH irr via oral and inhal routes.
Fire Hazard: Mod, when exposed to heat or flame.
Disaster Hazard: Dangerous; see chlorides; can react vigorously with oxidizing materials.
To Fight Fire: Water, foam, CO_2, dry chemical.

DICHLOROACETIC ACID. Colorless corrosive liquid, pungent odor. $CHCl_2COOH$, mw: 129.0, mp (a): 10°, mp (b): −4°, bp: 194°, d: 1.5634 @ 20°/4°, vap. press: 1 mm @ 44.0°, vap. d: 4.45.
Acute tox data: Oral LD_{50} (rat) = 2820 mg/kg; dermal LD_{50} (rabbit) = 510 mg/kg. [*3*]
THR = MOD via oral and dermal routes. A strong irr. The LD_{50} for dosage by mouth is of the same order of magnitude as acetic acid or trichloroacetic acid. It is corrosive and irr to the skin and mu mem.
Disaster Hazard: Dangerous; when heated to decomp, emits highly toxic fumes of chlorides; will react with water or steam to produce toxic and corrosive fumes.

1,1-DICHLOROACETONE. Syn: *1,1-dichloro-2-propanone.* Oily liquid. $CH_3COCHCl_2$, mw: 127.0, bp: 120°, d: 1.305 @ 18°/15°, vap. d: 4.38.
THR = U. See 1,3-dichloroacetone and ketones.
Disaster Hazard: Dangerous; see chloride; can react vigorously with oxidizing materials.

1,3-DICHLOROACETONE. Syn: *1,3-dichloro-2-propanone.* Crystals. $CH_2ClCOClCH_2$, mw: 127.0, mp: 45°, bp: 173°, d: 1.3826 @ 46°/4°, vap. d: 4.38.
THR = MOD irr via oral and inhal routes.
Disaster Hazard: Dangerous; when heated to decomp, emits highly toxic fumes of chlorides.

DICHLOROACETONITRILE. $CHCl_2CN$, mw: 110.0, vap. d: 3.8.
THR = U. Very probably HIGH. See nitriles and chlorides.

α,α-DICHLOROACETOPHENONE. Crystals. $C_6H_5COCHCl_2$, mw: 189.04, mp: 21°, bp: 247° decomp, d: 1.34 @ 15°, vap. d: 6.5.
Acute tox data: Inhal LC_{LO} (mouse) = 940 mg/m^3 for 10 min. [*3*]
THR = HIGH irr via oral and inhal routes.
Disaster Hazard: Dangerous; when heated to decomp, emits highly toxic fumes of chlorides.

2,2-DICHLOROACETYL CHLORIDE. Syn: *dichloroethanoyl chloride.* Fuming liquid, acrid odor, miscible in ether. $CHCl_2COCl$, mw: 147.4, d: 1.5315 @ 16°/4°, bp: 108°, flash p: 151°F, vap. d: 5.1.

Acute tox data: Oral LD_{50} (rat) = 2460 mg/kg; inhal LC_{LO} (rat) = 2000 ppm for 4 hrs; dermal LD_{50} (rabbit) = 650 mg/kg. [*3*]
THR = MOD via oral, inhal and dermal routes.
Fire Hazard: Mod, when exposed to heat, flame or powerful oxidizers.
Disaster Hazard: Dangerous; see chlorides.
To Fight Fire: Alcohol foam, dry chemical.

DICHLOROACETYLENE. Syn: *dichloracetylene.* ClC⋮CCl, mw: 94.94.
Acute tox data: Inhal TC_{LO} (human) = 0.5 ppm; inhal LC_{50} (mouse) = 19 ppm for 6 hrs; inhal LC_{50} (guinea pig) = 20 ppm for 4 hrs. [*3*]
THR = HIGH via inhal route. HIGH via CNS route. Can be formed by thermal decomp (> 70°) from trichloroethylene, as used in medicine. Symptoms include a disabling nausea and intense jaw pain. See also chlorinated hydrocarbons, aliphatic.
Explosion Hazard: Severe, when shocked or exposed to heat or air. [*19*]
Disaster Hazard: Highly dangerous; shock and heat will explode it; when heated to decomp or on contact with acid or acid fumes, emits highly toxic fumes of chlorides; can react vigorously with oxidizing matcrials.

2,3-DICHLOROALLYL DIISOPROPYL THIOCARBAMATE. Syns: *diallate, avadex.* Brown liquid, slightly sol in water, sol in organic solvents. $[(CH_3)_2CH]_2NCOSCH_2C(Cl){:}CHCl$, mw: 270.2, bp: 150° @ 9 mm, mp: 25°–30°.
Acute tox data: Oral LD_{50} (rat) = 395 mg/kg; dermal LD_{50} (rabbit) = 2000 mg/kg.
THR = HIGH via oral; MOD via dermal routes. An exper carc. [*3, 12*]
Disaster Hazard: Dangerous; when strongly heated, gives off highly toxic fumes.

2,4-DICHLOROANILINE. Needle-like crystals. $Cl_2C_6H_3NH_2$, mw: 162.0, bp: 251°, mp: 50°, vap. d: 5.6.
Acute tox data: Oral LD_{LO} (cat) = 157 mg/kg. [*3*]
THR = HIGH via oral route.
Disaster Hazard: Dangerous; when heated to decomp, emits toxic fumes.

3,4-DICHLOROANILINE. Crystals, insol in water, sol in most organic solvents. $Cl_2C_6H_3NH_2$, mw: 162.0, mp: 72°, bp: 272°, flash p: 331°F (OC).
Acute tox data: Oral LD_{50} (rat) = 648 mg/kg; inhal TC_{LO} (human) = 0.025 mg/kg.
THR = MOD via oral and inhal routes. In this case it affects the eyes. [*3*]
Fire Hazard: Slight, when exposed to heat or flame.
Disaster Hazard: Dangerous; see chlorides and aniline.
To Fight Fire: Dry chemical, CO_2.

DICHLORO-*o*-ANISIC ACID. Syn: *banvel D.* $C_6H_2COOH(Cl)_2OCH_3$, mw: 221.
Acute tox data: Oral LD_{50} (rat) = 1040 mg/kg. [*3*]
THR = MOD via oral route. An herbicide.

7,12-DICHLOROBENZ(a)ANTHRACENE.
THR = An exper carc. [*23*]

1,3-DICHLORO BENZENE. See *m*-dichlorobenzene.

1,4-DICHLORO BENZENE. See *p*-dichlorobenzene.

***m*-DICHLOROBENZENE.** Syn: *1,3-dichlorobenzene.* Colorless liquid. $C_6H_4Cl_2$, mw: 147.0, mp: −24.8°, bp: 173°, d: 1.288 @ 20°/4°, vap. press: 1 mm @ 12.1°, vap. d: 5.08.
THR = An insecticide and fumigant. See also *o*-dichlorobenzene and chlorobenzene.
Fire Hazard: Mod; can react violently with Al. [*19*]
Disaster Hazard: Dangerous; see chlorides; can react vigorously with oxidizing materials.
To Fight Fire: Water, foam, CO_2, dry chemical.

***o*-DICHLOROBENZENE.** Clear liquid. $C_6H_4Cl_2$, mw: 147.0, mp: −17.5°, bp: 180°–183°, fp: −22°, flash p: 151°F, d: 1.307 @ 20°/20°, vap. d: 5.05, autoign. temp.: 1198°F, lel = 2.2%, uel = 9.2%.
Acute tox data: LC_{LO} (rat) = 707 ppm for 7 hrs; oral LD_{LO} (guinea pig) = 2000 mg/kg. [*3*]
THR = MOD via inhal and oral routes. See chlorobenzene. The *o*-isomer is probably more toxic than the *m*- or *p*-forms. It is irr to skin and mu mem. Exper produced liver and kidney injury. An exper (±) carc. [*3, 1*]
Fire Hazard: Mod, when exposed to heat or flame.
Disaster Hazard: Dangerous; scc chloridc; can rcact vigorously with oxidizing materials.
To Fight Fire: Water, foam, CO_2, dry chemical.

***p*-DICHLOROBENZENE.** Syn: *1,4-dichlorobenzene.* White crystals, penetrating odor. $C_6H_4Cl_2$, mw: 147.0, mp: 53°, bp: 173.4°, flash p: 150°F (CC), d: 1.4581 @ 20.5°/4°, vap. press: 10 mm @ 54.8°, vap. d: 5.08.
Acute tox data: 500 mg/kg; ip LD_{50} (rat) = 2500 mg/kg. [*3*]
THR = MOD via ip and inhal; HIGH via oral routes. An insecticide. Has been reported to cause liver injury in humans. An exper (±) carc. [*3, 1*]
Fire Hazard: Mod, when exposed to heat, flame or oxidizers.
Spont Heating: No.
Disaster Hazard: Dangerous; see chlorides; can react vigorously with oxidizing materials.
To Fight Fire: Water, foam, CO_2, dry chemical.

4,5-DICHLORO-*m*-BENZENE DISULFONAMIDE. $C_6H_6O_4N_2S_2Cl_2$, mw: 305.2.
THR = An exper teratogen. [*3*]

2,2′-DICHLOROBENZIDINE. Crystalline, needle-like, insol in water, sol in alcohol and ether. $C_6H_3ClNH_2C_6H_3ClNH_2$, mw: 253.14, mp: 165°, vap. d: 8.73.
THR = MOD irr via oral and dermal routes. An allergen.

3,3′-DICHLOROBENZIDINE. Crystals, insol in water, sol in alcohol, benzene and glacial acetic acid. $C_{12}H_{10}Cl_2N_2$, mw: 253.1, mp: 133°.
Acute tox data: Oral LD_{LO} (rat) = 4740 mg/kg. [*3*]
THR = See 2,2′-dichlorobenzidine. An exper carc. [*3, 10, 23*]
Disaster Hazard: See 2,2′-dichlorobenzidine and benzidine.

4,4′-DICHLOROBENZILIC ACID ETHYL ESTER. See ethyl-4,4′-dichlorobenzilate.

2,6-DICHLOROBENZONITRILE. Syn: *dichlobenil.* White solid, almost insol in water, sol in organic solvents. $(Cl)_2C_6H_3CN$, mw: 172, mp: 144°.
Acute tox data: Oral LD_{50} (rat) = 2710 mg/kg; dermal LD_{50} (rabbit) = 1350 mg/kg. [*3*]
THR = MOD via oral and dermal routes. Does not hydrolyze to HCN in body. Less toxic than most aliphatic nitriles. See also benzonitrile.
Disaster Hazard: Dangerous; see chlorides and nitriles.

2,4-DICHLOROBENZOYL PEROXIDE PASTE WITH DIBUTYL PHTHALATE. Syn: *luperco CCC.* Thick white paste insol in water, sol in organic nonpolar solvents. $(Cl_2C_6H_3CO)_2O_2$ + dibutyl phthalate.
THR = See peroxides, organic.

3,4-DICHLOROBENZYL ALCOHOL. Crystals. $Cl_2C_6H_3CH_2OH$, mw: 177.1, vap, d: 6.1.
Acute tox data: Oral LD_{50} (rat) = 810 mg/kg; dermal LD_{50} (rabbit) = 400 mg/kg. [*3*]
THR = MOD via oral; HIGH via dermal routes. An insecticide. See also alcohols.
Disaster Hazard: Dangerous; see chlorides.

2,4-DICHLOROBENZYL CHLORIDE. Liquid. $Cl_2C_6H_3CH_2Cl$, mw: 195.5, vap. d: 6.76.
THR = U. An insecticide. See also *p*-dichlorobenzene.
Disaster Hazard: Dangerous; see chlorides.

3,4-DICHLOROBENZYL CHLORIDE. Liquid. $Cl_2C_6H_3CH_2Cl$, mw: 195.5, bp: 255°, vap. d: 6.76.
THR = U. An insecticide. See also chlorobenzene.
Disaster Hazard: Dangerous; see chlorides.

1,1-DICHLORO-2,2-BIS-(*p*-CHLOROPHENYL) ETHYLENE. Syn: *DDD.* $C_{14}H_8Cl_4$, mw: 318.
THR = An exper neo. [*3*]

1,1-DICHLORO-2,2-BIS-(*p*-ETHYLPHENYL) ETHANE. See perthane.

O-(2,5-DICHLORO-4-BROMOPHENYL)-*o*-METHYL PHENYL THIOPHOSPHONATE. Syn: *phosvel.* Slightly sol in water, very sol acetone. $C_{13}H_{10}O_2PSBrCl_2$, mw: 412.1, mp: 70.4°, d: 1.53 @ 25°.
Acute tox data: Oral LD_{50} (rat) = 42 mg/kg; oral LD_{50} (rabbit) = 124 mg/kg; dermal LD_{50} (rabbit) = 800 mg/kg. [*3*]
THR = HIGH via oral and dermal routes. On humans the effects include partial paralysis, failure in muscular coordination, suffocation, hardening of the arteries, encephalitis, delayed neuropathy, myelin degeneration. [*21, 3*]

1,1-DICHLOROBUTANE. $Cl_2CHCH_2CH_2CH_3$, mw: 127.0, autoign. temp.: 527°F, vap. d: 4.38.
THR = U. Animal experiments suggest moderate toxicity. See also chlorinated hydrocarbons, aliphatic.
Fire Hazard: Mod, via heat, ignition sources, oxidizers.
To Fight Fire: Alcohol foam, dry chemical, fog, mist, water spray.

1,4-DICHLOROBUTANE. Syn: *sym-dichlorobutane.* Colorless, mobile liquid with mild, pleasant odor. $CH_2ClCH_2CH_2CH_2Cl$, mw: 127.02, bp: 155°, flash p: 126°F (TOC), d: 1.141, vap. d: 4.4, vap. press: 4 mm @ 20°.
THR = U. See 1,1-dichlorobutane.
Fire Hazard: Mod, when exposed to heat or flame or oxidizers.
Disaster Hazard: Dangerous; see chlorides; can react vigorously with oxidizing materials.
To Fight Fire: Foam, CO_2, dry chemical, water.

2,3-DICHLOROBUTANE. Liquid. $C_4H_7Cl_2$, mw: 127, mp: −80°, bp: 116.0°, flash p: 194°F (OC), vap. press: 40 mm @ 35.0°, vap. d: 4.4, d: 1.1.
THR = U. See also chlorinated hydrocarbons, aliphatic and 1,1-dichlorobutane.
Fire Hazard: Mod, when exposed to heat or flame.
Explosion Hazard: U.
Disaster Hazard: Dangerous; see chlorides; can react vigorously with oxidizing materials.
To Fight Fire: Foam, CO_2, dry chemical.

sym-DICHLOROBUTANE. See 1,4-dichlorobutane.

1,3-DICHLORO-2-BUTENE. $CH_2ClCHCClCH_3$, mw: 125.01, bp: 123°, flash p: 80°F (CC), vap. d: 4.31.
Acute tox data: Inhal LC_{50} (rat) = 546 ppm for 4 hrs. [*3*]
THR = HIGH via inhal route. See also chlorinated hydrocarbons, aliphatic.
Fire Hazard: Dangerous, when exposed to heat, flame or oxidizers.
Spont Heating: No.

Explosion Hazard: U.
Disaster Hazard: Dangerous; see chlorides; can react with oxidizing materials.
To Fight Fire: Foam, CO_2, dry chemical.

1,4-DICHLORO-2-BUTENE. Colorless liquid. $C_4H_6Cl_2$, mw: 125, mp: 1°–3°, bp: 156°, d: 1.183 @ 25°/4°.
Acute tox data: Oral LD_{50} (rat) = 89 mg/kg; inhal LC_{LO} (rat) = 62 ppm for 4 hrs; dermal LD_{50} (rabbit) = 620 mg/kg. [3]
THR = HIGH via oral and inhal; MOD via dermal routes. An exper carc via sc and cp routes. [*81*]

1,4-DICHLORO-2-BUTENE (*trans*).
Acute tox data: Inhal LC_{50} (rat) = 86 ppm for 4 hrs. [*3*]
THR = HIGH via inhal route. An exper neo. [*3*]

2,4-DICHLORO-6-*o*-CHLOROANILINE)-*s*-TRIAZINE. See dyrene.

DICHLORO-(2-CHLOROVINYL) ARSINE. Syns: *β-chlorovinyl dichlorarsine, lewisite.* Liquid, faint odor of geranium. $C_2H_2AsCl_3$, mw: 207.32, bp: 190° decomp, fp: −13°, d: 1.888 @ 20°/4°, vap. press: 0.4 mm @ 20°, vap. d: 7.15.
THR = A blistering type military poison. Has a delayed action similar to distilled mustard gas. This gas exhibits a systemic poisoning effect. To decontaminate, use 2,3-dimercapto-1-propanol (BAL). In case of poisoning give BAL intramuscularly. Lewisite is absorbed through skin. As little as 2 ml can cause death. A recognized carc. [*14*]
Disaster Hazard: Dangerous; see arsenic.

3,3′-DICHLORO-4,4′-DIAMINO DIPHENYL ETHER. $C_{12}H_{10}ON_2Cl_2$, mw: 269.1, mp: 129°.
THR = An exper carc [*14, 108*] in rats via sc route.

3,3′-DICHLORO-4,4′-DIAMINO DIPHENYL METHANE. Syn: *4,4′-methylene-bis-(2-chloroaniline).* $C_{13}H_{12}N_2Cl_2$, mw: 267.2.
THR = An exper (+) neo and carc. [*3, 10, 23*]

2,2′-DICHLORO-4,4′-DIAMINO STILBENE.
THR = An exper carc. [*23*]

3,3′-DICHLORO-4,4′-DIAMINO STILBENE.
THR = An exper carc. [*23*]

2,7-DICHLORODIBENZO-*p*-DIOXIN. Syn: *DCDD.* Colorless crystals. $C_{12}H_6O_2Cl_2$, mw: 253, mp: 209°–210°.
Acute tox data: 1% DCDD fed to 50 male and 50 female mice ⟶ 1 male death and 2 female deaths in 17 weeks. Using 35 male and 35 female rats ⟶ no deaths. Against 0.5% DCDD as above ⟶ 0 male and 1 female death in mice and no deaths in rats. [*81*] Oral single dose (female rat) = 100–2000 mg/kg ⟶ no deaths. 100 mg/kg day were not teratogenic or embryotoxic.
THR = HIGH via oral and ip routes. See also chlorinated dibenzo-*p*-dioxins.

DICHLORODIETHYLDIAMMINE COBALT PERCHLORATE. $C_4H_{14}O_4N_2Cl_3Co$, mw: 319.4.
THR = Is heat = (300°) and impact-sensitive. [*19*]

DICHLORODIETHYLSILANE. See diethyldichlorosilane.

β,β′DICHLORODIETHYLSULFIDE. Syns: *mustard gas, HD.* Colorless (if pure), to light yellow, oily liquid. $S(CH_2CH_2Cl)_2$, mw: 159.1, bp: 228°, fp: 14.4°, flash p: 221°F, d: 1.2741 @ 20°/4°, vap. d: 5.4, vap. press: 0.09 mm @ 30°.
Acute tox data: Dermal LD_{50} (guinea pig) = 20 mg/kg; iv LD_{50} (rabbit) = 1.1 mg/kg; dermal LD_{50} (dog) = 20 mg/kg; sc LD_{50} (rat) = 3.2 mg/kg. Inhal LC_{LO} (humans) = 23 ppm for 10 min; dermal LC_{LO} (humans) = 64 mg/kg. [*3*]
THR = HIGH irr to skin, eyes and mu mem and via oral, sc, inhal and dermal as well as iv routes. A "blistering" gas. High irr to eyes, skin and lungs. Pulmonary lesions are often fatal. However, effects upon eyes and skin are likewise serious. There can also be severe gastric disturbances. Only slight evidence for systemic damage. This material penetrates deeply and injures blood vessels which can result in a severe inflammatory reaction and tissue necrosis. Exposure, even to minute traces, can lead to intense inflammation after some hours. It produces marked leucopenia with involution of the lymph nodes; secondary infections are common.

Treatment of local lesions is mainly by cleanliness and emollients, similar to that of burns. Oils protect the skin only slightly. Immediately after exposure, the poison may be partly removed by scrubbing the victim with kerosene, but penetration is so rapid that this treatment is not successful if delayed for 15 to 30 min. An exper (S) human carc of the lungs and larynx. [*3, 8, 23*] An exper mutagen. [*3, 23*]

Fire Hazard: Slight, when exposed to heat or flame; can be ignited by a large explosive charge.
Disaster Hazard: Dangerous; when heated to decomp or on contact with acid or acid fumes, emits highly toxic fumes of oxides of sulfur and chlorides; will react with water or steam to produce toxic and corrosive fumes; can react vigorously with oxidizing materials.
To Fight Fire: Water, foam, CO_2, dry chemical.

DICHLORODIETHYLSULFONE. Colorless crystals. $(CH_2CH_2Cl)_2O_2S$, mw: 191.08, mp: 52°, bp: 179°–181° @ 14–15 mm, vap. d: 6.6.
THR = HIGH irr to skin, eyes and mu mem and via oral and inhal routes.

Disaster Hazard: Dangerous; see chloride and sulfur compounds.

1,1-DICHLORO-2,2-DIFLUOROETHYLENE. Liquid. Cl_2CCF_2, mw: 133, vap. d: 4.6.
Acute tox data: Inhal LC_{LO} (rat) = 1000 ppm for 4 hrs. [*3*]
THR = MOD via inhal route.
Disaster Hazard: Dangerous; when heated to decomp, emits highly toxic fumes of fluorides and chlorides; will react with water or steam to produce toxic and corrosive fumes.

DICHLORODIFLUOROGERMANE. Colorless gas. $GeCl_2F_2$, mw: 181.51, mp: −51.8°, bp: −2.8°.
THR = See fluorides and germanium compounds.

DICHLORODIFLUOROMETHANE. Syn: "*Freon-12.*" Colorless, almost odorless gas. CCl_2F_2, mw: 120.92, mp: −158°, bp: −29°, vap. press: 5 atm @ 16.1°.
THR = Narcotic in high conc.
Disaster Hazard: Dangerous; when heated to decomp, emits highly toxic fumes of phosgene and fluorides. Can react violently with Al. [*19*]

DICHLORODIFLUOROMETHANE AND DIFLUOROETHANE, CONSTANT BOILING MIXTURE. See components as listed.

DICHLORODIFLUOROMETHANE-DICHLOROTETRAFLUOROETHANE MIXTURE. See components as listed.

DICHLORODIFLUOROMETHANE-MONOCHLORODIFLUOROMETHANE MIXTURE. See components as listed.

DICHLORODIFLUOROMETHANE-MONOFLUOROTRICHLOROMETHANE MIXTURE. See components as listed.

DICHLORODIFLUOROMETHANE-TRICHLOROMONOFLUOROMETHANE-MONOCHLORODIFLUOROMETHANE MIXTURE. See components as listed.

DICHLORODIFLUOROMETHANE-TRICHLOROTRIFLUOROETHANE MIXTURE. See components as listed.

2,2-DICHLORO-1,1-DIFLUORO-3-VINYL CYCLOBUTANE. $C_6H_6F_2Cl_2$, mw: 187.
THR = U.
Disaster Hazard: Dangerous; when heated to decomp, emits highly toxic fumes.

2,3-DICHLORO-2,3-DIHYDRO AFLATOXIN B_1.
THR = An exper carc. [*23*]

1,3-DICHLORO-5,5-DIMETHYL HYDANTOIN. Syns: *dactin, halane*. Crystals, liberates chlorine on contact with hot water. $C_5H_6N_2Cl_2O_2$, mw: 197, mp: 132°, sublimes @ 100°, conflagrates @ 212°, d: 1.5 @ 20°, vap. d: 6.8.
Acute tox data: LD_{50} (rats) = 542 ± 84 mg/kg. [*3*]
THR = Irr. Toxicity to warm-blooded animals appears to be MOD. Avoid excessive contact because of effects of active chlorine to irr the skin. Some of the hydantoins are CNS depressants. Readily releases chlorine on decomp.
Disaster Hazard: Dangerous; see chlorides; will react with water or steam to produce toxic and corrosive fumes.

DICHLORODIMETHYLSILANE. Liquid. $C_2H_6Cl_2Si$, mw: 129.1, mp: below −86°, bp: 70.5°, d: 1.07 @ 25°/25°, vap. d: 4.45, d: 1.1; lel = 3.4%, uel = >9.5%, flash p: <70°F.
Acute tox data: Oral LD_{LO} (rat) = 1000 mg/kg; inhal LC_{LO} (rat) = 930 ppm for 4 hrs. [*3*]
THR = MOD via oral and inhal routes. A flam liquid.
Fire Hazard: Dangerous via heat, flames (sparks) or oxidants.
Disaster Hazard: Dangerous; see chlorides; reacts violently with water or steam to produce heat and toxic and corrosive fumes.
To Fight Fire: Dry chemical, CO_2, alcohol foam.

1,8-DICHLORO-3,6-DINITROCARBAZOLE. Crystals. $(NO_2ClC_6H_2NH)_2$, mw: 341.1.
THR = U. An insecticide. See also nitro compounds of aromatic hydrocarbons and chlorides.
Disaster Hazard: Dangerous; see chlorides and NO_x can react vigorously with oxidizing materials.

2,3-DICHLORO-*p*-DIOXANE. $C_4H_6O_2Cl_2$, mw: 157.
THR = An exper neo. [*3*]

DICHLORODIPHENYLDICHLOROETHANE. See DDD.

DICHLORODIPHENYLENE OXIDE. See dichlorodiphenyl oxide.

DICHLORODIPHENYL OXIDE. Syn: *dichlorodiphenylene oxide*. Liquid. $Cl_2C_{12}H_8O$, mw: 239.1, vap. d: 8.2.
THR = U.
Disaster Hazard: Dangerous; see also chlorides.

DICHLORODIPHENYLTRICHLOROETHANE. See DDT.

DICHLORODI-*m*-TOLYLSTANNANE. Solid. $(C_6H_4CH_3)_2SnCl_2$, mw: 371.9, mp: 40°.
THR = U. See tin compounds and chlorides.
Disaster Hazard: Dangerous; see chlorides.

1,4-DICHLORO-2,3-EPOXYBUTANE. Syn: *1,4-dichloro-2,3-epoxy butane*. $C_4H_6OCl_2$, mw: 141.0.
Acute tox data: Oral LD_{LO} (rat) = 710 mg/kg; dermal LD_{LO} (rabbit) = 2830 mg/kg. [*3*]

THR = MOD via oral, inhal and dermal routes.
Disaster Hazard: Dangerous; see chlorides.

1,2-DICHLOROETHANE. See ethylene dichloride.

1,1-DICHLOROETHANE. See ethylidene chloride.

DICHLOROETHANOYL CHLORIDE. See 2,2-dichloro acetyl chloride.

1,2-DICHLOROETHYL ACETATE. Water white liquid. $CH_3COOCHClCH_2Cl$, mw: 157, mp: $<-32°$, bp: 58°-65° @ 13 mm, flash p: 307°F, d: 1.296 @ 20°, vap. d: 5.42.
Acute tox data: Inhal LC_{LO} (rat) = 16 ppm for 4 hrs. [*3*]
THR = HIGH via inhal route. See also esters, alcohols and acetic acid.
Fire Hazard: Slight, when exposed to heat or flame.
Disaster Hazard: Dangerous; see chlorides; can react vigorously with oxidizing materials.
To Fight Fire: Water, foam, CO_2, dry chemical.

α,β-DICHLOROETHYL ALCOHOL. See 1,2-dichloroethyl alcohol.

1,2-DICHLOROETHYL ALCOHOL. Syn: *α,β-dichloroethyl alcohol.* Liquid. $CH_2ClCHClOH$, mw: 115.0, vap. d: 3.97.
THR = U. See also alcohols.
Disaster Hazard: Dangerous; see chlorides.

2,2′-DICHLOROETHYLAMINE. See nitrogen mustard.

5-(DI-2-CHLOROETHYL)AMINO URACIL. See uracil mustard.

DICHLOROETHYLARSINE. See ethyldichloroarsine.

ar-DICHLOROETHYLBENZENE. Colorless liquid. $Cl_2C_6H_3CH_2CH_3$, mw: 175.1, mp: <70°, bp: 220°–224°, flash p: 205°F, d: 1.208 @ 25°/25°, vap. d: 6.05.
Acute tox data: Oral LD_{50} (rat) = 5500 mg/kg; dermal LD_{50} (rabbit) = 14000 mg/kg. [*3*]
THR = MOD via oral; LOW via dermal routes. See also chlorinated hydrocarbons, aromatic.
Fire Hazard: Mod, when exposed to heat, flame or oxidizers.
Disaster Hazard: Dangerous; see chlorides; can react vigorously with oxidizing materials.
To Fight Fire: Water, foam, CO_2, dry chemical.

1,2-DICHLOROETHYLBENZENE. Clear, colorless liquid. $Cl_2C_6H_3CH_2CH_3$, mw: 175.1, mp: $<-70°$, bp: 224°-226°, flash p: 401°F, d: 1.21 @ 25°/25°, vap. d: 6.05.
THR = U. See also chlorinated hydrocarbons, aromatic.
Fire Hazard: Slight, when exposed to heat or flame.
Disaster Hazard: Dangerous; see chlorides; can react vigorously with oxidizing materials.
To Fight Fire: Water, foam, CO_2, dry chemical.

DICHLOROETHYLBORANE. $C_2H_5BCl_2$, mw: 110.7.
THR = Self-ignites in air. [*19*] See also boron compounds.

N,N-DI-(2-CHLOROETHYL)-2,3-DIMETHOXYANILINE.
THR = An exper carc. [*23*]

1,1-DICHLOROETHYLENE. See vinylidene chloride.

1,2-DICHLOROETHYLENE. See *cis-* and *trans*-dichloroethylene.

***cis*-DICHLOROETHYLENE.** Syns: *1,2-dichloroethylene, acetylene dichloride.* Colorless liquid, pleasant odor. ClCHCHCl, mw: 97.0, mp: −80.5°, bp: 59°, lel = 9.7%, uel = 12.8%, flash p: 39°F, d: 1.2743 @ 25°/4°, vap. press: 400 mm @ 41.0°, vap. d: 3.34.
Acute tox data: Oral LD_{50} (rat) = 770 mg/kg. [*3*]
THR = MOD via oral route. In high conc it is irr and narcotic. Has produced liver and kidney injury in exper animals.
Fire Hazard: Dangerous, when exposed to heat or flame. Reacts violently with N_2O_4, KOH, Na, NaOH. [*19*]
Spont Heating: No.
Explosion Hazard: Mod, in the form of vapor when exposed to flame.
Disaster Hazard: Dangerous; see chlorides; can react vigorously with oxidizing materials.
To Fight Fire: Water spray, foam, CO_2, dry chemical.

***trans*-DICHLOROETHYLENE.** Syn: *acetylene dichloride.* Colorless liquid, pleasant odor. ClCHCHCl, mw: 97.0, mp: −50°, bp: 48°, flash p: 36°F, autoign. temp.: 860°F, lel = 9.7%, uel = 12.8%, d: 1.2743 @ 25°/4°, vap. press: 400 mm @ 30.8°, vap. d: 3.34.
Acute tox data: Inhal LC_{LO} (rat) = 10,000 ppm for 24 hrs; oral LD_{LO} (dog) = 5750 mg/kg. [*3*]
THR = LOW via inhal; MOD via oral routes. Exposure to high conc of vapor can cause nausea, vomiting, weakness, tremor and cramps. Recovery is usually prompt following removal from exposure. Dermatitis may result from de-fatting action on skin.
Fire Hazard: Dangerous, when exposed to heat, flame or oxidizers.
Spont Heating: No.
Explosion Hazard: Mod, in the form of vapor when exposed to flame.
Disaster Hazard: Dangerous; see chlorides; can react vigorously with oxidizing materials.
To Fight Fire: Water, foam; CO_2, dry chemical.

2,2′-DICHLOROETHYL ETHER. Syns: *chlorex, 1-*

chloro-2-(β-chloroethoxy)ethane. Colorless, stable liquid. $ClCH_2CH_2OCH_2CH_2Cl$, mw: 143.0, bp: 178.5°, fp: −51.9°, flash p: 131°F (CC), d: 1.2220 @ 20°/20°, autoign. temp: 696°F, vap. press: 0.7 mm @ 20°, vap. d: 4.93.

Acute tox data: Oral LD_{50} (rat) = 75 mg/kg; inhal LC_{LO} (rat) = 1000 ppm for 45 min; dermal LD_{50} (rabbit) = 720 mg/kg. [*3*]

THR = HIGH via oral and inhal; MOD via dermal routes. The vapor is irr to the mu mem of the eyes and nose. It affects the kidneys and liver in varying degrees, and is a mild narcotic. Guinea pigs cannot be killed immediately by exposure to conc which can be attained at ordinary room temp, but exposure to 1,000 ppm for 30 to 60 min may produce death after several days. Autopsy shows congestion of the lungs and upper respiratory tract, pulmonary edema, and congestion of the liver, brain and kidneys. The pulmonary edema apparently develops after a latent period of several hours, similar to the action of "nitrous fumes." In humans, exposure to 500 to 1,000 ppm causes severe irr of the eyes and nose after brief exposure, and deep inhal is nauseating and intolerable. A conc of 100 ppm produces slight nausea and irr; conc of 35 ppm are practically free from irr, though the odor is easily detectable. An exper (+) carc. [*3, 8*]

Fire Hazard: Mod, when exposed to heat, flame or oxidants.

Explosion Hazard: See also ethers. Reacts vigorously with oleum, chlorosulfonic acid. [*19*]

Disaster Hazard: Dangerous; when heated to decomp, emits highly toxic fumes; reacts with water or steam to evolve toxic and corrosive fumes; can react vigorously with oxidizing materials.

To Fight Fire: Water, foam, mist, fog, spray, dry chemical.

DI-1-CHLOROETHYL ETHER.

THR = An exper neo. [*3*]

DI-2-CHLOROETHYL FORMAL. Syn: *di-[2-chloroethyl] formal.* Liquid. $CH_2(OCH_2CH_2CH_2Cl)_2$, mw: 173.05, bp: 217.5°, flash p: 230°F (OC), d: 1.23, vap. d: 5.9.

Acute tox data: Oral LD_{50} (rat) = 65 mg/kg; dermal LD_{50} (guinea pig) = 170 mg/kg; inhal LC_{LO} (rat) = 62 ppm for 4 hrs. [*3*]

THR = HIGH via oral, dermal and inhal routes.

Fire Hazard: Slight, when exposed to heat or flame.

Disaster Hazard: Dangerous; see chlorides; can react vigorously with oxidizing materials.

To Fight Fire: Alcohol foam, foam, CO_2, dry chemical.

DI-(2-CHLOROETHYL)METHYL AMINE.

THR = An exper carc. [*23*]

2-(1,2-DICHLOROETHYL)-4-METHYL-1,3-DIOXOLANE.

Acute tox data: Oral LD_{50} (rat) = 620 mg/kg; dermal LD_{50} (rabbit) = 1010 mg/kg. [*3*]

THR = MOD via oral and dermal routes.

DI-(2-CHLOROETHYL)METHYL AMINE-N-OXIDE. Syn: *nitrogen mustard-N-oxide.*

THR = An exper carc. [*23*]

(N,N-DI-(2-CHLOROETHYL)-1-NAPHTHYLAMINE). See naphthylamine mustard.

N,N-DI-(2-CHLOROETHYL)-2-NAPHTHYLAMINE. See 2-naphthylamine mustard.

2-(1,2-DICHLOROETHYL)PYRIDINE HYDROCHLORIDE. $C_7H_7NCl \cdot HCl$, mw: 212.5.

THR = An exper carc. [*3*]

DICHLOROETHYL SULFIDE. See dichlorodiethyl sulfide.

N,N-DI-(2-CHLOROETHYL)-*p*-TOLUIDINE. See *p*-toluidine mustard.

DICHLOROFLUORAMINE. Cl_2NF, mw: 103.9.

THR = Explosive in the liquid state. [*19*] Very unstable.

DICHLOROFLUOROMETHANE. Syn: "*Freon 21.*" Heavy colorless gas. $HCCl_2F$, mw: 103, mp: −135°, bp: 8.9°, d: 1.48, vap. press: 2 atm @ 28.4°, vap. d: 3.82.

Acute tox data: Inhal LC_{LO} (guinea pig) = 10,000 ppm for 1 hr. [*3*]

THR = LOW by inhal route.

Disaster Hazard: Dangerous; see chlorides and fluorides.

DICHLOROFORMOXIME. Colorless, prismatic crystals, disagreeable penetrating odor. CCl_2NOH, mw: 113.94, mp: 39°–40°, bp: 53°–54° @ 28 mm, vap. d: 4.8.

THR = HIGH irr to skin, eyes and mu mem and via oral and inhal routes. A military poison.

Disaster Hazard: Dangerous; see chlorides.

DICHLOROGERMANE. Colorless liquid. GeH_2Cl_2, mw: 145.53, mp: −68.0°, bp: 69.5°, d: 1.90 @ −68°, vap. d: 5.0.

THR = See hydrochloric acid and germanium compounds.

1,3-DICHLORO-2,4-HEXADIENE. Liquid. $CH_2ClHCCClHCCHCH_3$, mw: 151.04, flash p: 168°F (OC), vap. d: 5.2.

THR = See chlorinated hydrocarbons, aliphatic.

Fire Hazard: Mod, when exposed to heat, flame or oxidizers.

Disaster Hazard: Dangerous; see chlorides. Will react

with water or steam to produce toxic and corrosive fumes; can react vigorously with oxidizing materials.
To Fight Fire: Water, foam, CO_2, dry chemical.

1,6-DICHLORO-2,4-HEXADIENE. $C_6H_4Cl_2$, mw: 146.9.
THR = A shock-sensitive explosive. [*19*]

α-DICHLOROHYDRIN. See 1,3-dichloropropanol-2.

6,7-DICHLORO-4-(HYDROXYAMINO)QUINOLINE-1-OXIDE. $C_9H_6O_2N_2Cl_2$, mw: 245.1.
THR = An exper neo. [*3*]

DICHLOROISOCYANURIC ACID. White crystals, chlorine odor, mod sol in water. $Cl_2H(NCO)_3$, mw: 198, mp: 225°.
Acute tox data: Oral LD_{50} (rats) = 1.4–1.7 g/kg. [*3*]
THR = Symptoms include emaciation, weakness, lethargy, diarrhea and weight loss. Autopsy findings include GI tract irr, tissue edema, liver and kidney congestion. Not a primary irr or sensitizer. Very irr to eyes and abraded skin.
Disaster Hazard: Dangerous; when heated to decomp, emits chlorides and carbon monoxide.

DICHLOROISOPROPYL ALCOHOL. See 1,3-dichloropropanol-2.

DICHLOROISOPROPYL ETHER. Colorless liquid. $(CH_2ClCH_3CH)_2O$, mw: 171.09, bp: 187.8°, fp: $<-20°$, flash p: 185°F (OC), d: 1.11 @ 25°/25°, vap. d: 6.0, vap. press: 0.10 mm @ 20°.
Acute tox data: Oral LD_{50} (rat) = 240 mg/kg; dermal LD_{50} (rabbit) = 3000 mg/kg; inhal LC_{LO} (rat) = 700 ppm for 5 hrs. [*3*]
THR = HIGH via oral; MOD via dermal and inhal routes. See also ethers. Animal exper show variable toxicity, depending on species and route of administration.
Fire Hazard: Mod, when exposed to heat, flame or powerful oxidizers.
Explosion Hazard: U. See also ethers.
Disaster Hazard: Dangerous; when heated to decomp, emits highly toxic fumes of chlorides; can react vigorously with oxidizing materials.
To Fight Fire: Water to blanket fire; foam, CO_2, dry chemical.

DICHLOROMETHANE. See methylene chloride.

3,9-DICHLORO-7-METHOXYACRIDINE. Syn: *halocrin*. Needle-like crystals sol in organic solvents. $C_{14}H_9Cl_2NO$, mw: 278, mp: 161°.
THR = No data. Probably HIGH irr to skin, eyes and mu mem and via oral and inhal routes.
Disaster Hazard: Dangerous; when heated to decomp, emits highly toxic chloride and NO_x fumes.

3′,4′-DICHLORO-2-METHYL ACRYL ANILIDE. Syn: *dicryl*. Solid, insol in water, sol in acetone, alcohol, isophorone and dimethyl sulfoxide. $Cl_2C_6H_3NHCOC(CH_3){:}CH_2$, mw: 230, mp: 128°.
Acute tox data: Oral LD_{50} (rat) = 1800 mg/kg. [*3*]
THR = MOD via oral route. It is an herbicide.
Disaster Hazard: Dangerous; see chlorides.

N,N-DICHLORO METHYL AMINE. CH_3NCl_2, mw: 100.
THR = Violent reaction with Na_2S. [*19*]

DICHLOROMETHYL ARSINE. Syns: *methylarsenic dichloride, MD*. Colorless liquid. CH_3AsCl_2, mw: 160.86, bp: 134.5°, fp: −59°, flash p: $>221°F$, d: 1.838 @ 20°/4°, vap. press: 10 mm @ 24.3°, vap. d: 5.40.
THR = HIGH irr to skin, eyes and mu mem and via oral and inhal routes. This is a blistering type of military poison. It is rapidly detoxified in the body. A mod persistent gas. For effects and antidotes see dichloro-(2-chloro-vinyl) arsine. See also arsenic compounds.
Fire Hazard: Slight, when exposed to heat or flame.
Disaster Hazard: Dangerous; when heated to decomp or on contact with acid or acid fumes, emits highly toxic fumes of chlorides and arsenic; can react vigorously with oxidizing materials.
To Fight Fire: Water, foam, CO_2, dry chemical.

DICHLOROMETHYLCHLOROFORMATE. Colorless liquid. $ClCOOCHCl_2$, mw: 163.40, bp: 110°–111°, d: 1.56 @ 15°, vap. d: 5.63.
THR = HIGH irr to skin, eyes and mu mem and via oral and inhal routes.
Disaster Hazard: Dangerous; see chlorides; will react with water or steam to produce toxic and corrosive fumes.

2,2-DICHLORO-N-METHYL DIETHYLAMINE HYDROCHLORIDE. See mechlorethamine hydrochloride.

2,2′-DICHLORO-N-METHYL DIETHYLAMINE, BUNTE SALT.
THR = An exper carc. [*3*]

2,2′-DICHLORO-N-METHYL DIETHYLAMINE-N-OXIDE. $C_5H_{11}ONCl_2$, mw: 172.1.
THR = HIGH acute via dermal, iv and sc routes. An exper (±) carc, neo. [*3, 8, 23*]

2,2′-DICHLORO-N-METHYL DIETHYLAMINE-N-OXIDE HYDROCHLORIDE. $C_5H_{11}ONCl_2 \cdot HCl$, mw: 208.5.
THR = An exper (±) [*3, 8*] neo and mutagen via ip route.

α,α-DICHLOROMETHYL ETHER. Syn: *dichlorinated methyl oxide*. Volatile liquid. $O(CH_2Cl)_2$, mw: 115, bp: 105°, d: 1.315 @ 20°, vap. d: 4.0.

Acute tox data: Oral LD_{50} (rat) = 210 mg/kg; inhal LC_{50} (rat) = 7 ppm for 7 hrs; dermal LD_{50} (rabbit) = 280 mg/kg. [*3*]
THR = HIGH via oral, inhal, and dermal routes. A human neo and carc. [*3, 10, 23, 102*]
Fire Hazard: U. See also ethers.
Explosion Hazard: U. See also ethers.
Disaster Hazard: Dangerous; see chlorides.

α,β-DICHLOROMETHYLETHYL KETONE. Liquid. $ClCH_2COCH_2CH_2Cl$, mw: 141, bp: 65° @ 3 mm, vap. d: 4.9.
THR = HIGH irr to skin, eyes and mu mem and via oral and inhal routes. A lacrimator type of military poison.
Disaster Hazard: Dangerous; when heated to decomp, burns and emits highly toxic fumes of chlorides.

DICHLOROMETHYL METHYL ETHER. $C_2H_4OCl_2$, mw: 115.0.
THR = An exper neo to mice via dermal route. [*103*]

2,3-DICHLORO-2-METHYL PROPIONALDEHYDE. CH_2ClCCH_3ClCHO, mw: 129.
THR = Probably HIGH irr to skin, eyes and mu mem.
Disaster Hazard: Dangerous; see chlorides.

DICHLOROMETHYL SILICANE. See methyl dichlorosilane.

DICHLOROMONOFLUOROMETHANE. See dichlorofluoromethane.

2,3-DICHLORO-1,4-NAPHTHOQUINONE. Syns: *dichlone, phygon.* Golden yellow crystals, insol in water, mod sol in organic solvents. $C_{10}H_4O_2Cl_2$, mw: 227.1, mp: 193°, vap. d: 7.8.
Acute tox data: Oral LD_{50} (rat) = 1300 mg/kg. [*3*]
THR = MOD via oral route. A fungicide. Irr to skin, eyes and mu mem. Large doses can cause CNS depression. An exper neo. [*3*]
Disaster Hazard: Dangerous; see chlorides.

1,2,4-DICHLORONITROBENZENE. Liquid. $C_6H_3Cl_2NO_2$, mw: 192.0, vap. d: 6.6.
THR = U. See also chlorinated hydrocarbons, aromatic, and nitrobenzene.
Fire Hazard: See nitrates.
Disaster Hazard: Dangerous; when heated to decomp, emits highly toxic fumes of oxides of nitrogen and chlorides; can react vigorously with oxidizing materials.

1,4,2-DICHLORONITROBENZENE. Liquid. $C_6H_3Cl_2NO_2$, mw: 192.0, vap. d: 6.6.
THR = U. See also chlorinated hydrocarbons, aromatic and nitrobenzene.
Fire Hazard: See nitrates.
Disaster Hazard: Dangerous; when heated to decomp, emits highly toxic fumes of oxides of nitrogen and chlorides; can react vigorously with oxidizing materials.

1,1-DICHLORO-1-NITROETHANE. Syn: *ethide.* Liquid. $H_3CC(Cl)_2NO_2$, mw: 143.96, bp: 124°, flash p: 168°F (OC), d: 1.4153 @ 20°/20°, vap. d: 4.97.
Acute tox data: Oral LD_{50} (rat) = 410 mg/kg; oral LD_{LO} (rabbit) = 150 mg/kg. [*3*]
THR = HIGH via oral route. Strong irr. Inhal causes pulmonary edema.
Fire Hazard: Mod, when exposed to heat, flame or oxidizers.
Spont Heating: No.
Disaster Hazard: Dangerous; when heated to decomp, emits highly toxic fumes of chlorides and oxides of nitrogen; can react vigorously with oxidizing materials.
To Fight Fire: Water, CO_2, dry chemical.

1,2-DICHLORO-3-NITRONAPHTHALENE. $C_{10}H_5O_2NCl_2$, mw: 242.1.
THR = An exper neo. [*3*]

1,1-DICHLORO-1-NITROPROPANE. Liquid. $C_2H_5CCl_2NO_2$, mw: 158.00, bp: 141°, flash p: 151°F (OC), d: 1.314, vap. d: 5.45.
THR = U. See also chlorinated hydrocarbons, aliphatic.
Fire Hazard: Mod, when exposed to heat, flame or oxidizers.
Disaster Hazard: Dangerous; when heated to decomp, emits highly toxic fumes of chlorides and oxides of nitrogen; can react vigorously with oxidizing materials.
To Fight Fire: Alcohol foam, CO_2, dry chemical.

6,7-DICHLORO-4-NITROQUINOLINE-1-OXIDE. $C_9H_4O_3N_2Cl_2$, mw: 259.1.
THR = An exper neo [*3, 23*] via sc route.

3,4-DICHLORO-1-NITROSO PIPERDINE. $C_5H_8ON_2Cl_2$, mw: 183.1.
THR = An exper carc to rats via oral route. [*103*]

4,5-DICHLORO-2-OXO-1,3-DIOXOLANE. $C_3H_2O_3Cl_2$ mw: 157.
THR = An exper neo. [*3*]

1,5-DICHLOROPENTANE. Syns: *amylene chloride, pentamethylene dichloride.* Insol in water. $CH_2Cl(CH_2)_3CH_2Cl$, mw: 141, d: 1.1, vap. d: 4.9, bp: 352°–358°F, flash p: > 80°F (OC).
Acute tox data: ip LD_{LO} (mouse) = 64 mg/kg. [*3*]
THR = HIGH via ip route.
Fire Hazard: Dangerous, when exposed to heat or flame.
Disaster Hazard: Dangerous; see chlorides.

To Fight Fire: Water ineffective except as a blanket; use alcohol foam or spray.

DICHLOROPENTANES (MIXED). Clear, light yellow-colored liquid. $C_5H_{10}Cl_2$, mw: 141.04, bp: 130°, flash p: 106°F (OC) d: 1.06–1.08 @ 20°, vap. d: 4.86.
THR = See 1,5-dichloropentane. See also chlorinated hydrocarbons, aliphatic.
Fire Hazard: Mod, when exposed to heat or flame.
Disaster Hazard: Dangerous; when heated to decomp, emits highly toxic fumes of phosgene; can react vigorously with oxidizing materials.
To Fight Fire: Water, foam, CO_2, dry chemical.

DICHLOROPHENE. See 2,2′-methylene-bis-(4-chlorophenol).

2,4-DICHLOROPHENOL. Colorless crystals. $C_6H_3OHCl_2$, mw: 163.0, mp: 45°, bp: 210°, flash p: 237°F, d: 1.383 @ 60°/25°, vap. d: 5.62 vap. press: 1 mm @ 53.0°.
Acute tox data: Oral LD_{50} (rat) = 580 mg/kg; ip LD_{50} (rat) = 430 mg/kg. [*3*]
THR = MOD via oral; HIGH via ip routes. See also chlorinated phenols. An exper carc. [*3*]
Fire Hazard: Slight, when exposed to heat or flame.
Disaster Hazard: Dangerous; when heated to decomp, or on contact with acid or acid fumes, emits highly toxic fumes of chlorides; can react vigorously with oxidizing materials.
To Fight Fire: Alcohol foam, foam, CO_2, dry chemical.

DICHLORO(PHENOL DIPHENYLETHER). See dichloro(phenol xenyl ether).

DICHLORO(PHENOLXENYL ETHER). Syn: *dichloro(phenol diphenyl ether)*. Light, viscous, straw-colored liquid. $C_{18}H_{12}OCl_2$, mw: 315.2, mp: $<0°$, bp: 222° @ 10 mm, flash p: 399°F, d: 1.233 @ 25°/25°, vap. d: 10.9.
THR = U. See also chlorinated hydrocarbons, aromatic and chlorinated phenols.
Fire Hazard: Slight, when exposed to heat or flame.
Disaster Hazard: Dangerous; see chlorides; can react with oxidizing materials.
To Fight Fire: Water, foam, CO_2, dry chemical.

2,4-DICHLOROPHENOXYACETIC ACID. Syn: *2,4-D*. White powder. $C_8H_6Cl_2O_3$, mw: 221, mp: 141°, bp: 160° @ 0.4 mm, vap. d: 7.63.
Acute tox data: Oral LD_{LO} (human) = 80 mg/kg; dermal LD_{50} (rat) = 1500 mg/kg. [*3*] Oral LD_{50} (rat) = 320 mg/kg.
THR = HIGH via oral; MOD via dermal routes. An exper (±) carc. [*81*] Can cause nausea, vomiting, and CNS depression. Liver and kidney injury have been reported in experimental animals. It is used as an herbicide.
Disaster Hazard: Dangerous; see chlorides.

2,4-DICHLOROPHENOXYETHYL BENZOATE. Syn: *sesin*. Crystals. $C_{15}H_{12}O_3Cl_2$, mw: 311.2.
THR = LOW on limited animal exper.
Disaster Hazard: Dangerous; see chlorides.

DICHLOROPHENOXYETHYL SULFATE. Syn: *crag I*. $C_6H_3Cl_2OC_2H_5SO_4$, mw: 287.1.
Acute tox data: Oral LD_{50} (rats) = 730 mg/kg. [*3*]
THR = MOD via oral route. Has produced liver and kidney damage in exper animals.
Disaster Hazard: Dangerous; see sulfates.

DI-(4-CHLOROPHENOXY)METHANE. See bis(*p*-chlorophenoxymethane).

1-(2,4-DICHLOROPHENOXY)-2-PROPANOL. See propylene glycol-2,4-dichlorophenyl ether.

2-(2,5-DICHLOROPHENOXY)PROPIONIC ACID. $C_9H_8O_3Cl_2$, mw: 235.1.
THR = An exper neo. [*3*]

***p*-[(3,4-DICHLOROPHENYL)AZO]-N,N-DIMETHYL ANILINE.** $C_{14}H_{13}N_3Cl_2$, mw: 294.2.
THR = An exper carc. [*3*]

DICHLOROPHENYL BENZENE SULFONATE. $C_{12}H_8Cl_2SO_3$, mw: 303.2.
Acute tox data: Oral LD_{50} (rat) = 1400 mg/kg; oral LD_{50} (rabbit) = 700 mg/kg. [*3*]
THR = MOD via oral route. No specific human toxicity data is available. Probably irr to skin and mu mem. Animal exper have shown effects similar to DDT. It is used as an insecticide. An exper neo. [*3*]
Disaster Hazard: Dangerous; see dichlorophenoxyethyl sulfate.

α,α-DI-(4-CHLOROPHENYL)DIETHYL ETHER. Crystals. $C_{16}H_{16}OCl_2$, mw: 295, vap. d: 11.3.
THR = U. See also chlorinated hydrocarbons, aromatic.
Disaster Hazard: Dangerous; see chlorides.

2,4-DICHLOROPHENYL-*o,o*-DIETHYL PHOSPHOROTHIOATE. See V-C13 nemacide.

3-(3,4-DICHLOROPHENYL)-1,1-DIMETHYL UREA. See diuron.

2,5-DICHLOROPHENYL ISOCYANATE. White to light green solid, sol in organic solvents. C_7H_3ClNO, mw: 189.1, mp: 27°, bp: 126° @ 30 mm.
THR = A powerful irr to the skin and mu mem. Very strong lacrimator.
Disaster Hazard: Dangerous; see chlorides and cyanates.

3,4-DICHLOROPHENYL ISOCYANATE. White to light brown solid, sol in organic solvents. $C_7H_3Cl_2NO$, mw: 189.1, mp: 42°, bp: 133° @ 30 mm.
THR = See 2,5-dichlorophenyl isocyanate.

3-(3,4-DICHLOROPHENYL)-1-METHOXY-1-METHYL UREA. Syns: *"Horox"®*, *linuron*. Solid, slightly sol in water, partially sol in acetone and alcohol. $C_6H_3ClNHC(O)N(OCH_3)CH_3$, mw: 214, mp: 93°–94°.

Acute tox data: Oral LD_{50} (rat) = 3300 mg/kg; oral LD_{50} (dog) = 500 mg/kg. [3]

THR = MOD to HIGH via oral route. See also monuron.

DI-(*p*-CHLOROPHENYL)METHYL CARBINOL. Syns: *dimite*, *DMC*. $C_{14}H_{12}Cl_2O$, mw: 267.2.

Acute tox data: Oral LD_{50} (rat) = 500 mg/kg. [3]

THR = HIGH via oral route. An insecticide.

Disaster Hazard: Dangerous; see chlorides.

3,4-DICHLOROPHENYL-N-METHYL SULFONAMIDE. Crystals. $Cl_2C_7H_5O_3NS$, mw: 254.2, vap. d: 6.4.

THR = See chlorinated hydrocarbons, aromatic.

Disaster Hazard: Dangerous; see sulfonates and chlorides.

2,4-DICHLOROPHENYL-*p*-NITROPHENYL ETHER. $C_{12}H_7O_3NCl_3$, mw: 284.1.

THR = HIGH acute via oral route. An exper teratogen. [3]

DICHLOROPHENYLPHOSPHINE. See phosphenyl chloride.

DICHLOROPHENYLTRICHLOROSILANE. Straw colored liquid, sol in benzene and perchloroethylene. $Cl_2C_6H_3SiCl_3$ (mixture of isomers), mw: 280.5, d: 1.562, bp: 260°, flash p: 286°F.

THR = HIGH. See silanes.

Fire Hazard: Slight, when exposed to heat or flame.

Disaster Hazard: Dangerous; see chlorides and silanes.

DICHLOROPHTHALIC ANHYDRIDE. White flakes. $C_6H_2Cl_2(CO)_2O$, mw: 217, mp: 80°–140°, bp: 220° @ 40 mm.

THR = U; probably an irr.

Disaster Hazard: Dangerous; see chlorides; will react with water or steam to produce heat and toxic fumes.

1,1-DICHLOROPROPANE. Syn: *propylidene chloride*.

Acute tox data: Oral LD_{50} (rat) = 6500 mg/kg; inhal LC_{LO} (rat) = 4000 ppm for 4 hrs; dermal LD_{50} (rabbit) = 1400 mg/kg. [3]

THR = MOD via oral, inhal and dermal routes.

1,2-DICHLOROPROPANE. Syn: *propylene dichloride*. Colorless liquid. $C_3H_6Cl_2$, mw: 113.0, bp: 96.8°, flash p: 60°F, d: 1.1593 @ 20°/20°, vap. press: 40 mm @ 19.4°, vap. d: 3.9, autoign. temp.: 1035°F, lel = 3.4%, uel = 14.5%.

Acute tox data: Oral LD_{50} (rat) = 1900 mg/kg; inhal LC_{LO} (rat) = 2000 ppm for 4 hrs; dermal LD_{50} (rabbit) = 8750 mg/kg. [3]

THR = MOD via oral, inhal and dermal routes.

1,2-dichloropropane can cause dermatitis, and is regarded as one of the more toxic chlorinated hydrocarbons. A suggested order of increasing toxicity is dichloromethane, trichloroethylene, carbon tetrachloride, dichloropropane, dichloroethane. Animals exposed to high conc often showed marked visceral congestion, fatty degeneration of the liver, kidney, and, less frequently, of the heart. They also showed areas of coagulation and necrosis of the liver. There was found to be a heavy mortality among mice exposed to 400 ppm conc.

Fire Hazard: Dangerous, when exposed to heat or flame. Reacts violently with Al. [19]

Disaster Hazard: Dangerous; see chlorides; can react vigorously with oxidizing materials.

To Fight Fire: Water, foam, CO_2, dry chemical.

1,3-DICHLOROPROPANE. Syn: *trimethylene chloride*. Colorless liquid. $CH_2ClCH_2CH_2Cl$, mw: 113.0, bp: 125°, d: 1.201 @ 15°, vap. d: 3.90.

Acute tox data: Oral LD_{LO} (dog) = 3000 mg/kg. [3]

THR = MOD via oral route. See chlorinated hydrocarbons, aliphatic, and 1,2-dichloropropane.

Disaster Hazard: Dangerous; when heated to decomp, emits highly toxic fumes of phosgene.

2,2-DICHLOROPROPANE. Syn: *acetone dichloride*. Liquid. $CH_3CCl_2CH_3$, mw: 113.0, mp: −34.6°, bp: 69.7°, d: 1.093 @ 20°/20°, vap. d: 3.9.

THR = See chlorinated hydrocarbons, aliphatic, and 1,2-dichloropropane.

Disaster Hazard: Dangerous; when heated to decomp, emits highly toxic fumes of phosgene; can react vigorously with oxidizing materials.

1,3-DICHLOROPROPANOL-2. Syns: *dichloroisopropyl alcohol*, *α-dichlorohydrin*. Colorless liquid, ether-like odor. $CH_2ClCHOHCH_2Cl$, mw: 129.0, bp: 174°, d: 1.367 @ 20°/4°, vap. press: 1 mm @ 28.0°, vap. d: 4.45, flash p: 165°F (OC), mp: −4°.

Acute tox data: Oral LD_{50} (rat) = 490 mg/kg; inhal LC_{LO} (rat) = 125 ppm for 4 hrs; dermal LD_{50} (rabbit) = 800 mg/kg. [3]

THR = HIGH via oral and inhal; MOD via dermal routes. Action may be similar to carbon tetrachloride but more irr to mu mem.

Fire Hazard: Mod, when exposed to heat, flame or oxidizers.

Disaster Hazard: Dangerous; when heated to decomp, emits highly toxic fumes of phosgene.

To Fight Fire: Alcohol foam, dry chemical, fog, mist, spray.

1,1-DICHLORO-2-PROPANONE. See 1,1-dichloroacetone.

1,3-DICHLORO-2-PROPANONE. See 1,3-dichloroacetone.

1,2-DICHLOROPROPENE. See 1,2-dichloropropylene.

1,3-DICHLOROPROPENE. Syn: *1,3-dichloropropylene.* Liquid. $C_3H_4Cl_2$, mw: 111.0, bp: 103°–110°, flash p: 95°F, d: 1.22, vap. d: 3.8.
Acute tox data: Oral LD_{50} (rat) = 250 mg/kg; inhal LC_{50} (man) = 1000 ppm for 4 hrs; dermal LD_{LO} (rabbit) = 2100 mg/kg. [*3*]
THR = HIGH via oral and inhal; MOD via dermal routes. A strong irr. Has produced liver and kidney injury in exper animals.
Fire Hazard: Dangerous, when exposed to heat, flame or oxidizers.
Disaster Hazard: Dangerous; see chlorides; can react vigorously with oxidizing materials.
To Fight Fire: Water, foam, CO_2, dry chemical.

2,3-DICHLOROPROPENE.
Acute tox data: Oral LD_{50} (rat) = 320 mg/kg; inhal LC_{LO} (rat) = 500 ppm for 4 hrs; dermal LD_{50} (rabbit) = 1580 mg/kg. [*3*]
THR = HIGH via oral and inhal; MOD via dermal routes.

2,3-DICHLOROPROPIONALDEHYDE. Liquid. CH_2Cl_2CHCHO, mw: 127, vap. d: 4.4.
Acute tox data: Oral LD_{50} (rat) = 160 mg/kg; inhal LC_{50} (rat) = 16 ppm for 4 hrs; dermal LD_{50} (rabbit) = 78 mg/kg. [*3*]
THR = HIGH via oral, inhal and dermal routes.
Disaster Hazard: Dangerous; see chlorides.

3,4-DICHLOROPROPIONANILIDE. Syn: *propanil.* Light brown solid (pure); liquid (technical grade). $Cl_2C_6H_3NHCOCH_2CH_3$, mw: 218, mp (pure): 85°–89°, bp (technical grade): 91°–95°.
Acute tox data: Oral LD_{50} (rat) = 560 mg/kg; oral LD_{50} (dog) = 1217 mg/kg. [*3*]
THR = HIGH MOD via oral route.

2,2-DICHLOROPROPIONIC ACID. Syns: *dalapon, dalapon-na.* White to tan powder. CH_3CCl_2COOH, mw: 143.0, bp: 98° @ 20 mm.
Acute tox data: Oral LD_{50} (female rat) = 970 mg/kg. [*3*]
THR = MOD irr to skin, eyes and mu mem and via oral and inhal routes.
Disaster Hazard: Dangerous; see chlorides.

2,3-DICHLOROPROPIONIC ACID.
Acute tox data: Oral LD_{50} (rat) = 420 mg/kg; dermal LD_{50} (rabbit) = 400 mg/kg. [*3*]
THR = HIGH via oral and dermal routes.

1,2-DICHLOROPROPYLENE. Syn: 1,2-dichloropropene. Liquid. $CHClCClCH_3$, mw: 111.0, bp: 75°, vap. d: 3.83.
THR = See chlorinated hydrocarbons, aliphatic.
Disaster Hazard: Dangerous; see chlorides.

1,3-DICHLOROPROPYLENE. See 1,3-dichloropropene.

4,7-DICHLORO QUINOLINE. Crystals. $C_9H_5Cl_2N$, mw: 198.1, mp: 105°, vap. d: 6.83.
Acute tox data: sc LD_{LO} (mouse) = 80 mg/kg. [*3*]
THR = HIGH via sc route.
Disaster Hazard: Dangerous; see chlorides.

α,β-DICHLOROSTYRENE. Liquid. $C_6H_5CClCHCl$, mw: 173.0, flash p: 225°F (OC), vap. d: 6.0.
Acute tox data: Oral LD_{50} (rat) = 4000 mg/kg; dermal LD_{50} (rabbit) = 9000 mg/kg. [*3*]
THR = MOD via oral and dermal routes. See also phenylethylene (styrene).
Fire Hazard: Slight, when exposed to heat, flame or oxidizers.
Disaster Hazard: Dangerous; see chlorides; can react vigorously with oxidizing materials.
To Fight Fire: CO_2, dry chemical.

sym-DICHLOROTETRAFLUOROACETONE. A colorless liquid. Miscible with water and many organic solvents. $CClF_2COCClF_2$, mw: 199.0, bp: 45.2°, fp: <−100°.
Acute tox data: Oral LD_{50} (rats) = 61 mg/kg; dermal LD_{50} (rat) = 91 mg/kg; inhal LC_{LO} (rat) = 50 ppm for 6 hrs. [*3*]
THR = HIGH via oral, dermal and inhal routes. See also fluorides.

DICHLOROTETRAFLUOROETHANE. Syn: *F-114.* Colorless gas. $F_2ClCCClF_2$, mw: 171, bp: 3.5°.
THR = a MILD irr, narcotic in high conc. An asphyxiant. Also, reacts violently with Al. [*19*]
Disaster Hazard: Dangerous; see fluorides and chlorides.

2,2-DICHLOROTETRAHYDROFURAN. $C_4H_6OCl_2$, mw: 141.
THR = An exper neo. [*3*]

α,4-DICHLOROTOLUENE. See *p*-chlorobenzyl chloride.

2,2′-DICHLOROTRIETHYLAMINE. See nitrogen mustard.

β,β-DICHLOROVINYLCHLOROARSINE. Syn: *lewisite.* Yellow or yellowish-brown liquid when pure, darker when impure. $(ClCHCH)_2AsCl$, mw: 233.35, bp: 230° decomp. d: 1.702 @ 20°, vap. d: 8.05.
Acute tox data: Oral LD_{50} (rat) = 50 mg/kg; inhal LC_{LO} (human) = 6 ppm for $\frac{1}{2}$ hr; dermal LD_{LO} (human) = 0.04 mg/kg. [*3*]

THR = VERY HIGH irr to skin, eyes and mu mem and via inhal, oral and dermal routes.
Disaster Hazard: Dangerous; see arsenic and chlorides.

β,β-DICHLOROVINYLMETHYLARSINE. Liquid. $(CHCHCl)_2CH_3As$, mw: 212.9, bp: 140°–145° @ 10 mm, vap. d: 7.35.
THR = No specific data. Known to be HIGH irr to skin, eyes and mu mem. See also β,β-dichlorovinylchloroarsine.
Disaster Hazard: Dangerous; see arsenic and chlorides.

DICHLOROVOS. Syns: *DDVP, 2,2-dichlorovinyldimethyl phosphate*. Liquid, slightly sol in water and glycerine; miscible with aromatic and chlorinated hydrocarbon solvents and alcohol. $(CH_3O)_2P(O)OCH{:}CCl_2$, mw: 221, bp: 120° @ 14 mm, bp: 77° @ 1 mm.
Acute tox data: Oral LD_{50} (rat) = 56 mg/kg; dermal LD_{50} (rat) = 75 mg/kg; also, TC_{LO} (humans) = 1 mg/m^3 for 8 hrs, caused blood problems. [*3*]
THR = HIGH via oral, dermal and inhal routes. Used in so-called pest collars or flea collars for pets. It is a cholinesterase inhibitor. No carc noted at 500 ppm dietary level. No neurotox observed. Very rapidly metabolized and excreted. An exper teratogen. [*3*]
Disaster Hazard: Dangerous; see chlorides and phosphorus compounds.

DICHLOROXYLENOL. See 2,4-dichloro-3,5-xylenol.

2,4-DICHLORO-3,5-XYLENOL. Syns: *dichloroxylenol, DCMX*. $(CH_3)_2(Cl)_2OHC_6H$, mw: 191.
THR = See chlorinated phenols.

DICOBALT OCTACARBONYL. Syn: *cobalt tetracarbonyl*. Orange crystals. $Co_2(CO)_8$, mw: 342.0, mp: 51°, bp: decomp @ 52°, d: 1.87, vap. press: 0.07 mm @ 15°.
Acute tox data: Gavage LD_{50} (mice) = 378 mg/kg; gavage LD_{50} (rats) = 754 mg/kg. [*2*]
THR = MOD to HIGH via gavage route of exposure. See carbonyls and cobalt compounds.
Fire Hazard: Mod, when exposed to heat or flame. See also carbonyls.
Explosion Hazard: Mod, when exposed to heat or flame. See also carbon monoxide and carbonyls.
Disaster Hazard: Dangerous; when heated to decomp, emits highly toxic fumes; can react vigorously with oxidizing materials.

DICOPHANE. See DDT.

DICOUMAROL. See bis-hydroxy coumarin.

DI-*m*-CRESYLTRICHLOROETHANE. Crystals. $Cl_3C_2H(C_6H_4CH_3)_2$, mw: 313.7, vap. d: 10.8.
THR = U. A fungicide. See also chlorinated hydrocarbons, aromatic and cresol.
Disaster Hazard: Dangerous; see chlorides.

DICROTALINE.
THR = An exper carc. [*23*]

DICRYL. See 3′,4′-dichloro-2-methyl acryl anilide.

DICUMYL PEROXIDE, SOLID. $[C_6H_5C(CH_3)_2O]_2$, mw: 270.
Acute tox data: Oral LD_{50} (rat) = 4100 mg/kg. [*3*]
THR = MOD via oral route. See also peroxides, organic.

DICYANDIAMIDE. Syn: *cyanoguanidine*. Pure white crystals. $H_2NC(NH)NHCN$, mw: 84.1, mp: 208°, d: 1.400 @ 25°.
THR = U. See cyanamides.
Disaster Hazard: Dangerous; see cyanides.

DICYANDIAZIDE. $NCNC(N_3)_2$, mw: 136.1.
THR = A shock-sensitive material. [*19*]

α-DICYANOBENZENE. Syns: *1,2-benzene dicarbonitrile, phthalonitrile*. Colorless crystals, insol in water, sol in acetone and benzene. $C_6H_4(CN)_2$, mw: 128.1, mp: 138°, bp: sublimes, vap. d: 4.42.
Acute tox data: Oral LD_{50} (rat) = 1860 mg/kg; ip LD_{50} (rat) = 62 mg/kg. [*3*]
THR = MOD via oral, HIGH via ip routes. An insecticide. See also nitriles.
Disaster Hazard: Dangerous; see cyanides.

p-DICYANOBENZENE. Crystals. $C_6H_4(CN)_2$, mw: 128.1, vap. d: 4.42.
Acute tox data: ip LD_{50} (mouse) = 699 mg/kg. [*3*]
THR = MOD via ip route. A fumigant. See also nitriles.
Disaster Hazard: See cyanides.

1,4-DICYANOBUTANE. See adiponitrile.

DI-(2-CYANOETHYL)AMINE. Liquid. $C_6H_9N_3$, mw: 123.2, mp: −5.5°, bp: 173° @ 10 mm, d: 1.0165 @ 30°, vap. d: 3.3.
Acute tox data: Oral LD_{50} (rat) = 2700 mg/kg; dermal LD_{50} (rabbit) = 2520 mg/kg.
THR = MOD via oral and dermal routes. An exper teratogen. [*3*]
Disaster Hazard: Dangerous; see cyanides; can react vigorously with oxidizing materials.

DICYANOETHYLSULFIDE. Liquid. $C_4H_3N_2S$, mw: 111.2, vap. d: 3.82.
Acute tox data: Oral LD_{50} (rat) = 4210 mg/kg. [*3*]
THR = MOD via oral route.
Disaster Hazard: Dangerous; see cyanides and oxides of sulfur.

DICYCLOHEXYL. See bicyclohexyl.

DICYCLOHEXYLAMINE. Liquid, fishy odor.

$(C_6H_{11})_2NH$, mw: 181.32, mp: −1°, bp: 256°, flash p: >210°F (OC), d: 0.910, vap. d: 6.27.
Acute tox data: Oral LD_{50} (rat) = 373 mg/kg; sc LD_{50} (mouse) = 135 mg/kg. [*3*]
THR = HIGH via oral and sc route. An exper neo. [*3*] Animal exper show intense local irr and nervous excitation. See also cyclohexylamine.
Fire Hazard: Low, when exposed to heat or flame; can react with oxidizing materials.
To Fight Fire: Alcohol foam, CO_2, dry chemical.

DICYCLOHEXYLAMINE NITRITE. $C_{12}H_{23}N \cdot NO_2H$, mw: 228.4.
Acute tox data: Oral LD_{50} (rat) = 284 mg/kg; oral LD_{50} (mouse) = 205 mg/kg. [*3*]
THR = HIGH via oral route. An exper neo. [*3*]
Fire Hazard: Dangerous; see nitrates.
Disaster Hazard: Dangerous; see nitrates.

DICYCLOHEXYL PHTHALATE. White solid. $C_6H_4(COOC_6H_{11})_2$, mw: 330.5, mp: 58°, bp: 200°–235° @ 4 mm, flash p: 405°F, d: 1.148 @ 20°/20°, vap. press: 0.1 mm @ 150°.
THR = Probably MOD to LOW.
Fire Hazard: Slight, when exposed to heat or flame; can react with oxidizing materials.
To Fight Fire: Water, foam, CO_2, dry chemical.

DICYCLOPENTADIENE. Colorless crystals. $C_{10}H_{12}$, mw: 132.3, mp: 32.9°, bp: 166.6°, d: 0.976 @ 35°, vap. press: 10 mm @ 47.6°, vap. d: 4.55, flash p: 90°F (OC).
Acute tox data: Oral LD_{50} (rat) = 353 mg/kg; inhal LC_{LO} (rat) = 500 ppm for 4 hrs; dermal LD_{50} (rabbit) = 5080 mg/kg. [*3*]
THR = HIGH via oral and inhal; MOD via dermal routes.
Fire Hazard: Dangerous, when exposed to heat or flame; can react with oxidizing materials.
To Fight Fire: Alcohol foam.

DICYCLOPENTADIENYL IRON. See ferrocene.

DICYCLOPENTENYL ALCOHOL. Liquid. $C_{10}H_{17}O$, mw: 153.2, bp: 238°, flash p: 426°F, d: 1.07, vap. d: 5.3.
THR = U. See also alcohols.
Fire Hazard: Slight, when exposed to heat or flame; can react with oxidizing materials.
To Fight Fire: Water, foam, CO_2, dry chemical.

DI-(DECANOYL)TRIETHYLENE GLYCOL ESTER. $C_{26}H_{50}O_6$, mw: 458.8.
Acute tox data: Oral LD_{50} (rat) = 7460 mg/kg; dermal LD_{50} (rabbit) = 11,000 mg/kg. [*3*]
THR = LOW via oral and dermal route.

DIDECYL ADIPATE. A clear liquid. $(C_{10}H_{21}CO_2)_2C_4H_8$, mw: 426, bp: 240° @ 4 mm, fp: −72°, flash p: 425°F (COC), d: 0.916–0.922 @ 20°/20°.
THR = U. Limited animal exper suggest LOW toxicity. See also esters and adipic acid.
Fire Hazard: Slight, when exposed to heat or flame; can react with oxidizing materials.
To Fight Fire: Foam, CO_2, dry chemical.

DI-*n*-DECYLAMINE. Liquid. $(C_{10}H_{21})_2NH$, mw: 297.5, bp: 195° @ 12 mm, vap. d: 10.3.
THR = U. See also amines.
Fire Hazard: Mod, when exposed to heat or flame; can react with oxidizing materials.
To Fight Fire: Water spray, mist, foam, dry chemical.

DIDECYL ETHER. $C_{20}H_{42}O$, mw: 298.4, autoign. temp.: 419°F, vap. d: 10.3.
THR = No data. See also ethers.
Fire Hazard: Low. See ethers.
To Fight Fire: Alcohol foam, water, mist or spray.

DIDECYL PHTHALATE. A clear liquid. $(C_{10}H_{21}CO_2)_2C_6H_4$, mw: 446, bp: 252° @ 4 mm, fp: −53°, flash p: 450°F (COC), d: 0.964–0.968 @ 20°/20°.
Acute tox data: Dermal LD_{50} (rabbit) = 17,000 mg/kg. [*3*]
THR = Virtually non-toxic via dermal route. See also esters and phthalic acid.
Fire Hazard: Slight, when exposed to heat or flame; can react with oxidizing materials.
To Fight Fire: Foam, CO_2, dry chemical.

3,8-DIDEHYDRO RETRONECINE. $C_8H_{11}O_2N$, mw: 153.2.
THR = An exper neo to rats via sc route. [*103*]

DIDIPHENYLAMINE FLUOSILICATE. White crystals. $[(C_6H_5)_2NH]_2 \cdot H_2SiF_6$, mw: 482.5, mp: 169°.
THR = HIGH. See fluosilicates.
Disaster Hazard: Dangerous; see fluosilicates.

DIDODECYLAMINE. White solid, slight ammoniacal odor. $(C_{12}H_{25})_2NH$, mw: 253.7, mp: 50°, bp: 210° @ 1 mm.
THR = U. See also amines.

DIDYMIUM NITRATE. Violet-red, hygroscopic crystals.
Comp: a mixture of praseodynium and neodymium nitrates.
THR = See nitrates.
Radiation Hazard: Didymium salts are usually slightly radioactive from the presence of thorium as an impurity.
Fire Hazard: See nitrates.
Disaster Hazard: See nitrates.

DIELDRIN. Syns: *1,2,3,4,10,10-hexachloro-6,7-epoxy-1,4,4a,5,6,7,8,8a-octahydro-1,4,5,8-di-*

methanonaphthalene, compound 497, Octalox and other trade names. White crystals, odorless, insol in water, sol in common organic solvents. $C_{12}H_8Cl_6O$, mw: 380.9, mp: 150°, vap. d: 13.2.

THR = An exper (+) neo and carc. [*3, 102, 12*] Used as an insecticide. Dieldrin is absorbed readily from the skin as well as through other portals. It acts as a CNS stimulant, but the exact mechanism of this action is entirely unknown. It also greatly reduces or eliminates appetite, apparently by an action on the CNS. Either nervous symptoms or anorexia may appear first. However, appetite may occasionally return in animals which are extremely sick and which eventually die. See also chlorinated hydrocarbons.

Dangerous Acute Dose in Man: The effects of dieldrin and aldrin are similar both quantitatively and qualitatively in animals as well as in man. Persons exposed to oral dosages which exceed 10 mg/kg frequently become acutely ill. Symptoms may appear within 20 min and in no instance has a latent period of more than 12 hrs been confirmed. No death or permanent sequelae have been reported following known poisoning by aldrin or dieldrin in man. In an attempted suicide by ingestion of dieldrin the dosage was estimated at 25.6 mg/kg.

The oral LD_{50} of dieldrin for rats is 40–50 mg/kg, indicating a toxicity roughly five times that of DDT. The dermal LD_{50} of dieldrin in xylene for rats is only slightly less than the oral toxicity (60 mg/kg for the female and 90 mg/kg for the male) indicating an acute dermal toxicity roughly 40 times that of DDT. Tests with certain other solvents indicate a factor of only six times.

Dangerous Chronic Dose: Nothing is known with certainty about the chronic toxicity of dieldrin for man but poisoning has occurred from use of 0.5–2.5% suspensions. Exper animals show a wide species variation in their susceptibility to dieldrin. Repeated dermal applications of 10 mg or even 20 mg/kg are tolerated by rats, whereas rabbits are killed by both of these dosages. Animals have shown convulsions as much as 120 days following the last dose of dieldrin indicating that dieldrin or its derivatives and/or residual toxicant-induced injury may persist in the body for a long time once severe poisoning has occurred.

Signs and Symptoms of Poisoning in Man: Early symptoms include headache, nausea, vomiting, general malaise, and dizziness. With more severe poisoning, clonic and tonic convulsions ensue or they may appear without the premonitory symptoms just mentioned. Coma may or may not follow the convulsions. Hyperexcitability and hyperirritability are common findings.

Disaster Hazard: Dangerous; when heated to decomp, emits highly toxic fumes of chlorides.

Treatment of Poisoning: Every effort should be made to remove dieldrin from the skin by thorough washing with soap and water or from the alimentary tract by the use of lavage and/or saline laxatives. Oil laxatives should be avoided. Exper with dogs and monkeys indicate that phenobarbital is effective as an antidote. It has been necessary to give the drug in large doses over a period of 2 weeks or more. The dosage which is required to keep poisoned animals from showing hyperexcitability or convulsions and which enables them to eat and behave normally is often a dosage which would induce sleep or even anesthesia in a normal animal of the same species. In human beings the dosage should be adjusted to the symptoms.

DIENESTRAL DIACETATE. Syn: *3,4-bis-(p-acetoxy phenyl)-2,4-hexadiene.*

THR = U. A food additive permitted in the feed and drinking water of animals, and/or for the treatment of food-producing animals. Also a food additive permitted in food for human consumption. [*109*]

DIEPOXY BUTANE.

THR = An exper carc. [*23*]

1,2,3,4-DIEPOXY BUTANE. See butadiene dioxide.

1,2,6,7-DIEPOXY HEPTANE. $C_7H_{12}O_2$, mw: 128.2.

THR = An exper neo [*3*] and carc. [*23*]

1,2,5,6-DIEPOXY HEXANE. $C_6H_{10}O_2$, mw: 114.2.

THR = An exper carc and neo. [*3, 23*]

DIEPOXY HYDRO MYRCENE. $C_{10}H_{18}O_2$, mw: 170.3.

THR = An exper neo. [*3*]

1,2,8,9-DIEPOXY LIMONENE. See dipentene dioxide.

1,2,8,9-DIEPOXY NONANE. $C_9H_{16}O_2$, mw: 156.3.

THR = An exper neo. [*3*]

9,10,12,13-DIEPOXY OCTADECANOIC ACID. $C_{18}H_{31}O_4$, mw: 311.5.

THR = An exper neo [*3*] to mice via dermal route.

1,2,7,8-DIEPOXY OCTANE.

$H_2COCH(CH_2)_4HCOCH_2$, mw: 142.

Acute tox data: Oral LD_{50} (rat) = 1070 mg/kg; dermal LD_{50} (rabbit) = 320 mg/kg. [*3*]

THR = MOD via oral, HIGH via dermal routes.

1,2,4,5-DIEPOXY PENTANE. $C_5H_8O_2$, mw: 100.1.

THR = An exper carc. [*3, 23*]

DIESEL OIL. Syn: *fuel oil no. 2.* Brown, slightly viscous liquid. flash p: 100°F, d: < 1, autoign. temp.: 494°F.
THR = A recog carc. See also mineral oils and kerosene. [*14*]
Fire Hazard: Mod, when exposed to heat, flame or oxidizers.
Explosion Hazard: See kerosene.
Disaster Hazard: See kerosene.
To Fight Fire: Foam, CO_2, dry chemical.

DIETHANOLAMINE. Syn: *di(2-hydroxyethyl)amine.* A faintly colored, viscous liquid. $(HOCH_2CH_2)_2NH$, mw: 105.14, mp: 28°, bp: 269.1° (decomp), flash p: 305°F (OC), d: 1.0919 @ 30°/20°, autoign. temp.: 1224°F, vap. press: 5 mm @ 138°, vap. d: 3.65.
Acute tox data: Oral LD_{50} (rat) = 710 mg/kg; sc LD_{50} (mouse) = 3553 mg/kg. [*3*]
THR = MOD via oral and sc routes. See also amines.
Fire Hazard: Slight, when exposed to heat or flame; can react with oxidizing materials.
Spont Heating: No.
To Fight Fire: Alcohol foam, water, CO_2, dry chemical.

DIETHOXYBORON CHLORIDE. Colorless liquid, decomp upon contact with moisture. $(C_2H_5O)_2BCl$, mw: 136.4, bp: 112°.
THR = U. See boron compounds.
Disaster Hazard: Dangerous; see chlorides.

DIETHOXYCHLOROSILANE. Liquid.
$(C_2H_5O)_2SiHCl$, mw: 154.7, vap. d: 5.33.
Acute tox data: Inhal LC_{LO} (rat) = 1000 ppm for 4 hrs. [*3*]
THR = MOD via inhal route.
Disaster Hazard: Dangerous; see chlorides; will react with water or steam to produce heat and toxic and corrosive fumes.

2,2-DIETHOXY CYCLODISILOXANE. $C_4H_{12}O_4Si_2$, mw: 180.3.
THR = Self-ignites in air. [*19*]

DIETHOXYDIMETHYLSILANE. Liquid.
$(C_2H_5O)_2Si(CH_3)_2$, mw: 148.2, bp: 113.5°, d: 0.834, vap. press: 10 mm @ 13.3°, vap. d: 5.1.
Acute tox data: Oral LD_{LO} (rat) = 3000 mg/kg. [*3*]
THR = MOD via oral route.
Fire Hazard: Mod, when exposed to heat or flame; can react with oxidizing materials.

1,1-DIETHOXYETHANE. See acetal.

1,2-DIETHOXYETHYLENE. See diethyl "cellosolve."

DIETHOXY ETHYL PHTHALATE. Light-colored liquid with a mild odor.
$C_6H_4(COOC_2H_4OC_2H_5)_2$, mw: 310, mp: 31°, bp: 198°–211° @ 4 mm, flash p: 356°F, d: 1.12, vap. d: 10.7.
THR = U. See esters.
Fire Hazard: Slight, when exposed to heat or flame; can react with oxidizing materials.
To Fight Fire: Water, foam, CO_2, dry chemical.

DIETHOXY TERAHYDROFURAN. Liquid.
$C_2H_5OCH(CH_2)OCH(CH_2)OC_2H_5$, mw: 160.21, mp: −26.9°, bp: 173.1°, flash p: 160°F (OC), d: 0.9686 @ 20°/20°.
THR = U.
Fire Hazard: Mod, when exposed to heat or flame; can react with oxidizing materials.
To Fight Fire: Foam, CO_2, dry chemical.

DIETHYL ACETAL. See acetal.

DIETHYLACETALDEHYDE.
See 2-ethylbutyraldehyde.

DIETHYLACETAMIDE. Liquid. $(C_2H_5)_2CHCONH_2$, mw: 115.2, mp: < 65°, bp: 180°, flash p: 170°F, d: 0.92, vap. d: 4.0.
Acute tox data: Oral TD_{LO} (rat) = 910 mg/kg; ip LD_{50} (rat) = 1840 mg/kg. [*3*]
THR = MOD via oral and ip routes. Chronic toxicity: An exper neo, teratogen. [*3*]
Fire Hazard: Mod, when exposed to heat or flame; can react with oxidizing materials.

DIETHYLACETIC ACID. See 2-ethylbutyric acid.

N,N-DIETHYLACETOACETAMIDE. Liquid, sol in water. $CH_3COCH_2CON(C_2H_5)_2$, mw: 157, d: 0.995 @ 20°/20°, bp: decomp, fp: −70°, flash p: 250°F (COC).
THR = Limited animal exper suggest LOW toxicity.
Fire Hazard: Slight, when exposed to heat or flame.
To Fight Fire: Alcohol foam.

DIETHYL ACETOACETATE. Very slightly sol in water. $CH_3COC(C_2H_5)_2COOC_2H_5$, mw: 186, d: < 1, vap. d: 6.4, bp: 412°–424°F (decomp), flash p: 170°F.
THR = U. See esters.
Fire Hazard: Mod, when exposed to heat, flame or oxidizers.
To Fight Fire: Foam, mist, dry chemical.

DIETHYL ACETYL ETHYLENE IMINE. $C_8H_{15}ON$, mw: 141.2.
THR = An exper carc and neo. [*3, 23*]

DIETHYL ALUMINUM BROMIDE. See bromodiethyl aluminum.

DIETHYL ALUMINUM CHLORIDE. Syns: *aluminum diethyl monochloride, DEAC.* Colorless, pyrophoric liquid. $(C_2H_5)_2AlCl$, mw: 120.5, bp: 208°, fp: −50°.
THR = U. Probably toxic.
Fire Hazard: Dangerous; ignites spont in air. Reacts violently with (C_6H_6 + allyl chloride), (toluene + allyl chloride), water. [*19*]

Disaster Hazard: Dangerous; see chlorides.
To Fight Fire: Do not use water, foam or halogenated extinguishing agents.

DIETHYL ALUMINUM HYDRIDE. A pyrophoric mixture with triethyl aluminum $(C_2H_5)_2AlH$, mw: 86.
THR = Probably HIGH irr to skin, eyes, mu mem and via inhal route. See also hydrides.
Fire Hazard: Dangerous; ignites spont in air.
To Fight Fire: Do not use water, foam or halogenated extinguishing agents.

DIETHYL ALUMINUM MALONATE. White crystals, insol in water, sol in organic solvents. $Al(C_7H_{11}O_4)_3$, mw: 504.5, mp: 98°, d: 1.084 @ 100°.
THR = U.

DIETHYLAMINE. Colorless liquid, ammoniacal odor. $(C_2H_5)_2NH$, mw: 73.1, mp: −38.9°, bp: 55.5°, flash p: −15°F, d: 0.7108 @ 20°/20°, autoign. temp.: 594°F, vap. press: 400 mm @ 38.0°, vap. d: 2.53, lel = 1.8%, uel = 10.1%.
Acute tox data: Oral LD_{50} (rat) = 540 mg/kg; dermal LD_{50} (rabbit) = 820 mg/kg; inhal LC_{50} (rat) = 4000 ppm for 4 hrs. [*3*]
THR = MOD via oral, dermal and inhal routes. See also amines.
Fire Hazard: Dangerous, when exposed to heat or flame; can react with oxidizing materials.
Spont Heating: No.
Explosion Hazard: U.
To Fight Fire: Alcohol foam, CO_2, dry chemical.

DIETHYLAMINE HYDROCHLORIDE. White crystals, sol in water and alcohol, slightly sol in chloroform, insol in ether. $(C_2H_5)_2NHHCl$, mw: 109.60, d: 1.048, mp: 223°–224°, bp: 320°–330°.
THR = U. See diethylamines and amines.
Disaster Hazard: Dangerous; see chlorides.

DIETHYLAMINOETHANOL.
See diethylethanolamine.

2-(2-DIETHYLAMINO)ETHOXY BENZANILIDE. $C_{19}H_{24}O_2N_2$, mw: 312.5.
THR = An exper teratogen. [*3*]

2-[(2-DIETHYLAMINO)ETHOXY]-5-BROMO-BENZANILIDE. $C_{19}H_{23}O_2N_2Br$, mw: 391.4.
THR = An exper teratogen. [*3*]

2-[(2-DIETHYLAMINO)ETHOXY]-5-CHLORO-BENZANILIDE. $C_{19}H_{23}O_2N_2Cl$, mw: 346.9.
THR = An exper teratogen. [*3*]

2-[(2-DIETHYLAMINO)ETHOXY)-5-METHYL BENZANILIDE. $C_{20}H_{25}O_2N_2$, mw: 325.5.
THR = An exper teratogen. [*3*]

N,N-DIETHYLAMINO ETHYL ACRYLATE. $CH_2CHCOOH_2CH_2CN(H_5C_2)_2O$, mw: 200, d: 0.9, vap. d: 5.9, flash p: 195°F (OC), bp: decomp.
Acute tox data: Oral LD_{50} (rat) = 770 mg/kg; dermal LD_{50} (rat) = 200 mg/kg. [*3*]
THR = MOD via oral, HIGH via dermal routes.
Fire Hazard: Mod, when exposed to heat, flame or oxidizers.
To Fight Fire: Decomp in water; use alcohol foam or dry chemical.

6-DIETHYLAMINO ETHYL-4-HYDROXY AMINO QUINOLINE-1-OXIDE. $C_{14}H_{20}O_2N_3$, mw: 266.3. $C_{14}H_{20}O_2N_3$, mw: 266.3.
THR = An exper neo to rats via iv route. [*103*]

DIETHYLAMINO ETHYL POLYMER.
THR = An exper neo. [*3*]

3-DIETHYLAMINOPROPYLAMINE. Liquid. $(C_2H_5)_2NC_3H_6NH_2$, mw: 130.3, bp: 165°–170°, flash p: 138°F (OC), d: 0.82, vap. d: 4.48.
Acute tox data: Oral LD_{50} (rat) = 1410 mg/kg; dermal LD_{50} (rabbit) = 750 mg/kg. [*3*]
THR = MOD via oral and dermal routes. Animal exper show irr and sensitizing properties. See also amines.
Fire Hazard: Mod, when exposed to heat or flame; can react with oxidizing materials.
To Fight Fire: Foam, CO_2, dry chemical.

2,6-DIETHYLANILINE. Colorless to yellow liquid. $(C_2H_5)_2NC_6H_5$, mw: 149.23, mp: −38°, flash p: 185°F, bp: 215.5°, d: 0.9351, vap. press: 1 mm @ 49.7°, vap. d: 5.15, autoign. temp.: 1166°F.
Acute tox data: Oral LD_{50} (rat) = 2690 mg/kg. [*3*]
THR = MOD via oral route. See also aniline.
Fire Hazard: Mod, when exposed to heat, flame or oxidizers.
Disaster Hazard: Dangerous; see aniline.
To Fight Fire: Foam, CO_2, dry chemical.

DIETHYLANILINE FLUOSILICATE. White crystals. $(C_6H_5NHC_2H_5)_2 \cdot H_2SiF_6$, mw: 386.4, mp: 165°.
THR = HIGH. See aniline and fluosilicates.
Disaster Hazard: Dangerous; see fluorides.

DIETHYL ARSINE. $(C_2H_5)_2AsH$, mw: 134.
THR = HIGH. See arsine. Spont flam in air. [*19*]

DIETHYL AZO DICARBOXYLATE. $C_6H_{10}O_4N_2$, mw: 174.1.
THR = Decomp violently during distillation. [*19*]

6,8-DIETHYL BENZ(a)ANTHRACENE. $C_{22}H_{20}$, mw: 284.4.
THR = An exper carc. [*3, 23*]

7,12-DIETHYL BENZ(a)ANTHRACENE.
THR = An exper neo. [*3, 23*]

8,12-DIETHYL BENZ(a)ANTHRACENE.
THR = An exper carc. [*3, 23*]

DIETHYLBENZENE (mixture of isomers). Colorless, mobile liquid. $C_6H_4(C_2H_5)_2$, mw: 134.21, bp: 183.8°, flash p: 134°F, d: 0.868 @ 25°/25°, autoign. temp.: 743°–842°F, vap. press: 1 mm @ 20.7°, vap. d: 4.62.
Acute tox data: Oral LD_{LO} (rat) = 5000 mg/kg. [*3*]
THR = LOW to MOD via oral route. See also ethyl benzene.
Fire Hazard: Mod, when exposed to heat or flame; can react with oxidizing materials.
To Fight Fire: CO_2, dry chemical.

N,N-DIETHYL BENZENE SULFONAMIDE. $C_{10}H_{15}O_2NS$, mw: 213.3.
THR = An exper teratogen. [*3*]

DIETHYLBERYLLIUM. Colorless liquid. $Be(C_2H_5)_2$, mw: 67.14, mp: 12°, bp: 110° @ 15 mm, vap. d: 2.3.
THR = See beryllium compds. VERY toxic.
Fire Hazard: Dangerous, when exposed to heat or flame. Spont flam in air.
Disaster Hazard: Dangerous; when heated to decomp, emits highly toxic fumes of beryllium; can react vigorously with oxidizing materials.
To Fight Fire: Special extinguishing agents, dry chemical.

DI(2-ETHYLBUTYL)AZELATE. Liquid. $(CH_2)_7(CO_2C_6H_{13})_2$, mw: 356.5, flash p: 385°F, d: 0.93, vap. d: 12.3.
THR = U.
Fire Hazard: Slight, when exposed to heat or flame; can react with oxidizing materials.
To Fight Fire: Foam, CO_2, dry chemical.

DI(2-ETHYLBUTYL)PHTHALATE. Clear, oily, slightly aromatic liquid. $C_6H_4[COOCH_2CH(C_2H_5)_2]_2$, mw: 334, mp: −50°, bp: 350°, flash p: 381°F, d: 1.01–1.016 @ 20°/20°, vap. d: 11.5.
THR = U. See also esters.
Fire Hazard: Slight, when exposed to heat or flame; can react with oxidizing materials.
To Fight Fire: Water, foam, CO_2, dry chemical.

DIETHYLCADMIUM. An oil, decomp by moisture. $(C_2H_5)_2Cd$, mw: 170.5, d: 1.6562, mp: −21°, bp: 64°.
THR = HIGH. See cadmium compounds.
Fire Hazard: Fumes explode in air or when exposed to heat or flame. [*19*]
Disaster Hazard: When heated to decomp, emits highly toxic fumes of cadmium.

DIETHYL CARBAMYL CHLORIDE. Liquid. $(C_2H_5)_2NCOCl$, mw: 119.6, mp: −44°, bp: 190°–195°, vap. d: 4.1.
Acute tox data: ip LD_{50} (mouse) = 750 mg/kg. [*3*]
THR = MOD via ip route.
Disaster Hazard: Dangerous; when heated to decomp, emits highly toxic fumes of chlorides; will react with water or steam to produce toxic and corrosive fumes.

DIETHYL CARBAMAZINE. $C_{10}H_{21}ON_3$, mw: 199.3.
Acute tox data: Oral LD_{50} (rat) = 1380 mg/kg; ip LD_{50} (rat) = 465 mg/kg; iv LD_{50} (rat) = 150 mg/kg. [*3*]
THR = MOD via oral, HIGH via ip and iv routes. A food additive permitted in the food and drinking water of animals and/or for the treatment of food-producing animals. [*109*]

N,N′-DIETHYLCARBANILIDE. Syn: *N,N′-diethyl-N,N′*-diphenyl urea. Colorless crystals. $CO[N(C_2H_5)(C_6H_5)]_2$, mw: 268.4, mp: 73°, d: 1.12, bp: 326°, flash p: 302°F (CC), vap. d: 9.3.
Acute tox data: ip LD_{50} (mouse) = 200 mg/kg. [*3*]
THR = HIGH via ip route.
Explosion Hazard: Probably quite low although it is a component of smokeless explosive mixtures.
Disaster Hazard: Highly dangerous; when heated to decomposition it burns and emits very toxic fumes.
Fire Hazard: Low.
To Fight Fire: Dry chemical, CO_2, spray or mist.

DIETHYL CARBINOL. See *sec*-amyl alcohol.

DIETHYL "CARBITOL." See "Carbitol."

DIETHYL CARBONATE. Syns: *ethyl carbonate, carbonic ether.* Colorless liquid, mild odor. $(C_2H_5)_2CO$, mw: 118.13, mp: 43°, bp: 125.8°, flash p: 77°F (OC), d: 0.975 @ 20°/4°, vap. press: 10 mm @ 23.8°, vap. d: 4.07.
Acute tox data: sc LD_{50} (rat) = 8500 mg/kg. [*3*]
THR = LOW via sc route. An exper neo [*3*] and carc. [*23*]
Fire Hazard: Dangerous, when exposed to heat or flame; can react with oxidizing materials.
Spont Heating: No.
To Fight Fire: Foam, CO_2, dry chemical.

DIETHYL "CELLOSOLVE." Syn: *1,2-diethoxyethylene*. Colorless liquid, slight ethereal odor. $C_2H_5OCH_2CH_2OC_2H_5$, mw: 119.17, mp: −74°, bp: 121.4°, flash p: 95°F (OC), d: 0.8417 @ 20°/20°, autoign. temp.: 406°F, vap. d: 6.56, vap. press: 9.4 mm.
Acute tox data: Oral LD_{50} (rat) = 4390 mg/kg; inhal LC_{LO} (rat) = 8000 ppm for 4 hrs. [*3*]
THR = MOD via oral, LOW via inhal routes. See also glycols and "Cellosolves."
Fire Hazard: Dangerous, when exposed to heat or flame; can react with oxidizing materials.
Spont Heating: No.
To Fight Fire: CO_2, dry chemical.

DIETHYL-β-CHLOROETHYLAMINE. Liquid. $(C_2H_5)_2NC_2H_4Cl$, mw: 135.64, vap. d: 4.69.

Acute tox data: Oral LD_{50} (rat) = 17 mg/kg; dermal LD_{50} (rat) = 300 mg/kg. [3]
THR = HIGH via oral and dermal routes.
Disaster Hazard: Dangerous; see chlorides.

o,o-DIETHYL-3-p-CHLOROPHENYL THIOMETHYL PHOSPHORODITHIOATE. See trithion.

DIETHYL CHLOROPHOSPHATE. Water white liquid. $(C_2H_5O)_2POCl$, mw: 172, bp: 60° @ 2 mm, d: 1.1915 @ 25°/25°, vap. d: 5.94.
Acute tox data: Oral LD_{50} (rat) = 110 mg/kg; dermal LD_{50} (rabbit) = 0.8 mg/kg. [3]
THR = HIGH via oral and VERY HIGH via dermal routes. A cholinesterase inhibitor. See parathion.

DIETHYL-2-CHLOROVINYL PHOSPHATE. Syn: *compound 1836*. Powder. $C_6H_{12}PO_4Cl$, mw: 214.6.
Acute tox data: Oral LD_{50} (rat) = 10 mg/kg; dermal LD_{50} (rabbit) = 18 mg/kg. [3]
THR = HIGH via oral and dermal routes. An organic phosphate insecticide and cholinesterase inhibitor. See parathion.
Disaster Hazard: Dangerous; see phosphates.

DIETHYL CYANAMIDE. Syn: *N-cyanodiethylamine*. Liquid. $CNN(C_2H_5)_2$, mw: 98.15, mp: −80°, bp: 186°, flash p: 176°F, d: 0.8591 @ 30°, vap. d: 3.4.
Acute tox data: ip LD_{50} (mouse) = 100 mg/kg. [3]
THR = HIGH via ip route.
Fire Hazard: Mod, when exposed to heat or flame.
Disaster Hazard: Dangerous; when heated to decomp or on contact with acid or acid fumes, emits highly toxic fumes of cyanides; will react with water or steam to produce toxic and corrosive fumes; can react vigorously with oxidizing materials.
To Fight Fire: Foam, CO_2, dry chemical.

DIETHYL CYCLOHEXANE. Liquid, insol in water. $(C_2H_5)_2C_6H_{10}$, mw: 140, d: 0.8037 @ 20°/20°, bp: 174°, fp: −100°, flash p: 120°F (OC), autoign. temp.: 465°F, lel = 0.8% @ 140°F, uel = 6.0% @ 230°F.
THR = U.
Fire Hazard: Mod, when exposed to heat, flame or oxidizers.
To Fight Fire: Foam, mist, dry chemical.

N,N-DIETHYL CYCLOHEXYLAMINE. Clear, colorless liquid. $(C_2H_5)_2N(C_6H_{11})$, mw: 155.3, bp: 194.5°, vap. d: 5.36.
THR = U. See also amines.
Fire Hazard: Mod, when exposed to heat or flame; can react with oxidizing materials.
To Fight Fire: Foam, CO_2, dry chemical.

1,1-DIETHYL CYCLOPENTAMETHYLENE GERMANIUM. Colorless liquid. $(CH_2)_5Ge(C_2H_5)_2$, mw: 200.9, bp: 52° @ 13 mm.
THR = U. See germanium compounds.
Fire Hazard: Probably mod dangerous. Details U.

DIETHYL DIBROMODIPYRIDINE TIN. Solid. $(C_2H_5)_2 \cdot Br_2 \cdot (C_5H_5N)_2$, mw: 494.9, mp: 140°.
THR = U. See tin compounds.
Disaster Hazard: Dangerous; emits bromides and NO_x fumes when heated to decomp.

o,o-DIETHYL-s-2,5-DICHLOROPHENYL THIOMETHYL PHOSPHORO DITHIOATE. See phencapton.

DIETHYL DICHLOROSILANE. Syn: *dichlorodiethylsilane*. Liquid. $C_4H_{10}Cl_2Si$, mw: 157.05, mp: −96°, bp: 131.0°, flash p: 70°F, d: 1.05, vap. d: 5.41.
Acute tox data: Oral LD_{LO} (rat) = 1000 mg/kg; ip LD_{LO} (rat) = 100 mg/kg. [3]
THR = MOD via oral, HIGH via ip routes. Corrosive. See chlorosilanes.
Fire Hazard: Dangerous, when exposed to heat, flame or oxidizers.
Explosion Hazard: U.
Disaster Hazard: Dangerous; see chlorides; will react with water or steam to produce heat and toxic and corrosive fumes; can react vigorously with oxidizing materials.
To Fight Fire: Foam, CO_2, dry chemical.

o,o-DIETHYL-s-(β-DIETHYLAMINO) ETHYL PHOSPHOROTHIOLATE HYDROGEN OXALATE. See tetram.

DIETHYL DIETHYL AMINO-3-PROPYL ALUMINUM. $C_{11}H_{26}NAl$, mw: 199.3.
THR = Self-igniting in air. [19]

DIETHYL DIISOAMYL TIN. Solid. $(C_2H_5)_2Sn(C_5H_{11})_2$, mw: 319.1, d: 1.0725 @ 19°, bp: 131° @ 13.5 mm.
THR = U. See tin compounds.

DIETHYL DIISOBUTYL TIN. Solid. $(C_2H_5)_2Sn(C_4H_9)_2$, mw: 291.1, d: 1.1030, bp: 108° @ 13 mm.
THR = U. See tin compounds.

DIETHYL DIMETHYL LEAD MIXTURE. See compounds, motor fuel, anti-knock.

DIETHYLDIMETHYL METHANE. See 2,3-dimethylpentane.

DIETHYL DIPHENYL GERMANIUM. Colorless liquid, insol in water. $(C_2H_5)_2Ge(C_6H_5)_2$, mw: 284.9, bp: 316°.
THR = U. See germanium compounds.
Fire Hazard: Probably mod dangerous. Details U.

DIETHYL DIPHENYL TIN. Solid. $(C_2H_5)_2Sn(C_6H_5)_2$, mw: 331.1, bp: 156° @ 4 mm.
THR = U. See tin compounds.

DIETHYL DIPHENYL UREA. See N,N′-diethylcarbanilide.

DIETHYL DISULFIDE. See 3,4-dithiahexane.

DIETHYLENE DIAMINE. See piperazine.

DIETHYL DITHIO BIS (THIONOFORMATE). See herbisan.

DIETHYLENE DIOXIDE. See dioxan.

1,3-DIETHYLENE DISULFIDE-2,2-DIPHENYL. See 1,3-dithiane-2,2-diphenyl.

DIETHYLENE GLYCOL. Syn: *diglycol.* Clear, colorless, practically odorless, syrupy liquid.
$CH_2OHCH_2OCH_2CH_2OH$, mw: 106.12, bp: 245.8°, fp: −8°, flash p: 255°F, d: 1.1184 @ 20°/20°, autoign. temp.: 444°F, vap. press: 1 mm @ 91.8°, vap. d: 3.66.
Acute tox data: Oral LD_{50} (humans) = 1000 mg/kg; oral LD_{50} (dogs) = 9000 mg/kg. [*3*]
THR = MOD to LOW via oral route. See glycols. A (S) carc of the bladder. [*14*]
Fire Hazard: Slight, when exposed to heat or flame; can react with oxidizing materials.
Spont Heating: No.
To Fight Fire: Alcohol foam, water, CO_2, dry chemical.

DIETHYLENE GLYCOL BIS (2-BUTOXYETHYL CARBONATE). Syn: *butoxyethyl diglycol carbonate.* Slightly sol in water.
$[CH_3(CH_2)_3O(CH_2)_2OOCOC_2H_4]_3$, mw: 378, flash p: 379°F, d: 1.1, bp: 327° @ 2 mm.
THR = U.
Fire Hazard: Slight, when exposed to heat or flame.
To Fight Fire: Alcohol foam.

DIETHYLENE GLYCOL BIS (BUTYL CARBONATE). Syn: *butyl diglycol carbonate.* Colorless liquid, slightly sol in water, widely sol in organic solvents. $[CH_3(CH_2)_3OOCOCH_2CH_2]_2O$, mw: 306, d: 1.07 @ 20°/4°, boiling range: 164°–166° @ 2 mm, flash p: 372°F.
THR = U. See esters.
Fire Hazard: Slight, when exposed to heat or flame.
To Fight Fire: Foam, alcohol foam.

DIETHYLENE GLYCOL BIS PHENYL CARBONATE. Syn: *phenyl diglycol carbonate.* Colorless solid, insol in water, widely sol in organic solvents. $(C_6H_5OOCOCH_2CH_2)_2O$, mw: 346, d: 1.23 @ 20°/4°, mp: 40°, bp: 225°–229° @ 2 mm, flash p: 460°F.
THR = See glycols and esters.
Fire Hazard: Slight, when exposed to heat or flame.
To Fight Fire: Dry chemical, CO_2

DIETHYLENE GLYCOL-*n*-BUTYL ETHER. See butyl carbitol.

DIETHYLENE GLYCOL BUTYL ETHER ACETATE. $C_{10}H_{20}O_4$, mw: 204.2, flash p: 241°F (OC), autoign. temp.: 563°F, d: 0.98, vap. d: 7.05, bp: 246°.
THR = No data. See also glycols.
Fire Hazard: Low.
To Fight Fire: Water, spray, mist, dry chemical.

DIETHYLENE GLYCOL CHLOROHYDRIN. Water white liquid. $ClCH_2CH_2OCH_2CH_2OH$, mw: 124.57, mp: −90° (sets to a glass), bp: 198.9°, flash p: 437°F, d: 1.1753 @ 20°/20°, vap. d: 4.3.
THR = U. See also glycols.
Fire Hazard: Slight, when exposed to heat or flame.
Disaster Hazard: Dangerous; see chlorides; can react with oxidizing materials.
To Fight Fire: Water, foam, CO_2, dry chemical.

DIETHYLENE GLYCOL DIABIETATE (20% XYLENE). Syn: *diglycol diabietate.* Liquid mixture. $(C_{19}H_{29}COOCH_2CH_2)_2O$, mw: 622.6, flash p: 97°F.
THR = U. See also xylene and glycols.
Fire Hazard: High, when exposed to heat or flame; can react with oxidizing materials.
To Fight Fire: Foam, CO_2, dry chemical.

DIETHYLENE GLYCOL DIACETATE. Syn: *diglycol diacetate.* Colorless liquid. $(CH_3COOCH_2CH_2)_2O$, mw: 190.20, mp: 19.1°, bp: 250°, flash p: 255°F (OC), d: 1.1159 @ 20°/20°, vap. d: 6.56, vap. press 0.02 mm @ 20°.
THR = U. See also glycols.
Fire Hazard: Slight, when exposed to heat or flame; can react with oxidizing materials.
To Fight Fire: Alcohol foam, water, foam, CO_2, dry chemical.

DIETHYLENE GLYCOL DIBENZOATE. See ethylene glycol dibenzoate.

DIETHYLENE GLYCOL DIBUTYL ETHER. Practically colorless liquid, characteristic odor, slightly sol in water. $C_4H_9O(C_2H_4O)_2C_4H_9$, mw: 218, d: 0.8853 @ 20°/20°, bp: 256°, fp: −60.2°, flash p: 245°F (OC).
Acute tox data: Oral LD_{50} (rat) = 3900 mg/kg. [*3*]
THR = MOD via oral route.
Fire Hazard: Slight, when exposed to heat or flame.
To Fight Fire: Foam, alcohol foam.

DIETHYLENE GLYCOL DI-2-ETHYL BENZOATE. Liquid. $C_6H_5COOCH_2CH_2OCH_2CH_2OCOC_6H_5$, mw: 317, mp: 15.9°, bp: 236° @ 5 mm, flash p: 455°F (OC), d: 1.1765 @ 20°/20°, vap. d: 10.0.
THR = U. See also glycols.
Fire Hazard: Slight, when exposed to heat or flame; can react with oxidizing materials.
To Fight Fire: Water, foam, CO_2, dry chemical.

DIETHYLENE GLYCOL DIETHYL ETHER. Colorless liquid, sol in hydrocarbons and water.
$CH_3(CH_2OCH_2)_3CH$, mw: 162, d: 0.9082 @ 20°/20°, bp: 189°, fp: −44°, flash p: 180°F (OC), vap. d: 5.6.
Acute tox data: Oral LD_{50} (rat) = 4970 mg/kg; oral LD_{50} (guinea pig) = 1850 mg/kg. [*3*]

THR = MOD via oral route.
Fire Hazard: Mod, when exposed to heat, flame or oxidizers.
To Fight Fire: Alcohol foam.

DIETHYLENE GLYCOL DIGLYCOLATE. Yellow liquid, faint odor. $(CH_2CH_2OOCCH_2OH)_2O$, mw: 222.2, d: 1.30 @ 30°, vap. d: 7.65.
THR = See glycols.
Fire Hazard: Slight.

DIETHYLENE GLYCOL DILEVULINATE. Amber liquid, pleasant odor.
$[CH_2CH_2OOC(CH_2)_2COCH_3]_2O$, mw: 270.4, flash p: 340°F (CC), d: 1.145 @ 25°, vap. d: 9.33.
THR = U. See also glycols.
Fire Hazard: Slight, when exposed to heat or flame; can react with oxidizing materials.
To Fight Fire: Alcohol foam, foam, CO_2, dry chemical.

DIETHYLENE GLYCOL DINITRATE. Syn: *dinitroglycol.* Liquid. $C_4H_8O_3(NO_2)_2$, mw: 196.1, vap. d: 6.76.
Acute tox data: Oral LD_{50} (rat) = 777 mg/kg. [*3*]
THR = MOD via oral route. This compound can cause a drop in blood pressure and possibly various cardiac disturbances. See also glycols and nitrates.
Fire Hazard: See nitrates, and explosives, high.
Explosion Hazard: Severe, when shocked or exposed to heat. Used in low freezing dynamites and some permissible explosives. See also explosives, high.
Disaster Hazard: Dangerous; shock and heat will explode it; when heated, emits toxic fumes of NO_x; can react vigorously with oxidizing or reducing materials.

DIETHYLENE GLYCOL DIOLEATE. Pale yellow liquid. $(C_{17}H_{33}CO_2C_2H_4)_2O$, mw: 635.0, d: 0.9319 @ 20°/4°.
THR = U. See also glycols.
Fire Hazard: Slight, when exposed to heat or flame; can react with oxidizing materials.

DIETHYLENE GLYCOL DIPELARGONATE. Syn: *plastolein X-55.* Liquid. $(C_8H_{17}COOC_2H_4)_2O$, mw: 386.3, mp: −15°, bp: 229° @ 5 mm, flash p: 410°F, d: 0.966 @ 20°, vap. d: 12.5.
THR = Animal exper suggest LOW toxicity. See also glycols.
Fire Hazard: Slight, when exposed to heat or flame; can react with oxidizing materials.
To Fight Fire: Foam, CO_2, dry chemical.

DIETHYLENE GLYCOL DIPROPIONATE. Liquid. $CH_3O(CH_2CH_2O)_2CH_3$, mw: 134.17, bp: 162°, mp: $<75°$, flash p: 260°F, d: 0.9440 @ 20°, vap. d: 4.62.
THR = U. See glycols.
Fire Hazard: Low, when exposed to heat or flame; can react with oxidizing materials.
Explosion Hazard: U. See also ethers.
To Fight Fire: Alcohol foam, CO_2, dry chemical.

DIETHYLENE GLYCOL DIVINYL ETHER.
$C_8H_{14}O_3$, mw: 158.
Acute tox data: Oral LD_{50} (rat) = 3730 mg/kg. [*3*]
THR: = MOD via oral route.

DIETHYLENE GLYCOL ETHYL ETHER.
See carbitol.

DIETHYLENE GLYCOL ETHYL VINYL ETHER.
$C_8H_{16}O_3$, mw: 148.2.
Acute tox data: Oral LD_{50} (rat) = 11000 mg/kg; dermal LD_{50} (rabbit) = 8410 mg/kg. [*3*]
THR = MOD via oral and dermal routes.

DIETHYLENE GLYCOL LAURATE. Syn: *diglycol laurate.* Light straw-colored oily liquid.
$HOCH_2CH_2OCH_2CH_2OOC(CH_2)_{10}CH_3$, mw: 228.4, mp: 17°, bp: 315–325°, flash p: 290°F, d: 0.965–0.975, vap. d: 10.0.
THR = U. See also glycols.
Fire Hazard: Slight, when exposed to heat or flame; can react with oxidizing materials.
To Fight Fire: CO_2, dry chemical.

DIETHYLENE GLYCOL METHYL ETHER. See methyl "Carbitol."

DIETHYLENE GLYCOL METHYL ETHYL ETHER.
$CH_3O(C_2H_4O)_2C_2H_5$, mw: 148.
THR = U. Limited animal exper suggest LOW toxicity. See also glycols and ethers.

DIETHYLENE GLYCOL MONOBUTYL ETHER.
See butyl "Carbitol."

DIETHYLENE GLYCOL MONOBUTYL ETHER ACETATE. See butyl "Carbitol" acetate.

DIETHYLENE GLYCOL MONO-2-CYANOETHYL ETHER. $(CH_2CH_2CN)OCH_2CH_2OCH_2CH_2OH$, mw: 159.
Acute tox data: Oral LD_{50} (rat) = 13,000 mg/kg. [*3*]
THR = LOW via oral route. See also nitriles.

DIETHYLENE GLYCOL MONOETHYL ETHER. See "Carbitol."

DIETHYLENE GLYCOL MONOMETHYL ETHER.
See methyl "Carbitol."

DIETHYLENE GLYCOL MONOMETHYL ETHER ACETATE. See methyl "Carbitol" acetate.

DIETHYLENE GLYCOL MONO-2-METHYL PENTYL ETHER. $C_{10}H_{22}O_3$, mw: 190.
Acute tox data: Dermal LD_{50} (rabbit) = 1580 mg/kg. [*3*]
THR = MOD via dermal route. See also glycols and ethers.

DIETHYLENE GLYCOL MONOPHENYL ETHER. See phenyl "Carbitol."

DIETHYLENE GLYCOL MYRISTATE. Syn: *diglycol myristate.* Light-colored wax, faint odor. $HOCH_2CH_2OCH_2CH_2OOC(CH_2)_{12}CH_3$, mw: 316.5, mp: 37°, flash p: 290°F, d: 0.938 @ 38°, vap. d: 10.9.
THR = U. See also glycols.
Fire Hazard: Slight, when exposed to heat or flame; can react with oxidizing materials.

DIETHYLENE GLYCOL PHTHALATE. $C_{16}H_{22}O_6$, mw: 310.3, flash p: 343°F, d: 1.1.
THR = No data. Probably MOD. See also glycols.
Fire Hazard: Low.
To Fight Fire: Alcohol foam.

DIETHYLENE OXIDE. See tetrahydrofuran.

1,3-DI(ETHYLENE SULFAMOYL)PROPANE.
THR = An exper mutagen. [3]

DIETHYLENETRIAMINE. Yellow, viscous liquid, mild ammoniacal odor. $(NH_2C_2H_4)_2NH$, mw: 103.17, mp: −39°, bp: 207°, flash p: 215°F (OC), d: 0.9586 @ 20°/20°, autoign. temp.: 750°F, vap. press: 0.22 mm @ 20°, vap d: 3.48.
Acute tox data: Oral LD_{50} (rat) = 1080 mg/kg; dermal LD_{50} (rabbit) = 1090 mg/kg. [3]
THR = MOD via oral and dermal routes. High conc of vapors causes irr of respiratory tract, nausea and vomiting. Repeated exposures can cause asthma and sensitization of skin.
Fire Hazard: Low, when exposed to heat or flame; can react with oxidizing materials.
To Fight Fire: Alcohol foam.

DIETHYLENIMIDE OXIDE. See morpholine.

DIETHYL-β,α-EPOXY PROPYL PHOSPHIONATE. $C_7H_{15}O_4P$, mw: 194.2.
THR = An exper neo. [3]

DIETHYL EPOXY SUCCINATE.
An exper carc. [23]

DIETHYL ETHANEDIOATE. See ethyl oxalate.

DIETHYLETHANOLAMINE. Syn: *diethyl amino ethanol.* Colorless, hygroscopic liquid. $(C_2H_5)_2NCH_2CH_2OH$, mw: 117.2, bp: 162°, flash p: 140°F (OC), d: 0.8851 @ 20°/20°, vap. press: 1.4 mm @ 20°, vap. d: 4.03.
Acute tox data: Oral LD_{50} (rats) = 1.3 g/kg; dermal LD_{50} (rabbit) = 1260 mg/kg. [3]
THR = MOD via oral and dermal routes. Causes toxic effects upon the CNS. [3]
Fire Hazard: Mod, when exposed to heat or flame; can react with oxidizing materials.
To Fight Fire: Alcohol foam, CO_2, dry chemical.

DIETHYL ETHER. See ethyl ether.

DIETHYL ETHOXYMETHYLENE MALONATE. Liquid. $C_2H_5OCH{:}C(COOC_2H_5)_2$, mw: 217, vap. d: 7.48.
THR = U. See also esters.
Fire Hazard: Slight; when heated, emits acrid fumes; can react with oxidizing materials.

DIETHYL-*s*-(2-ETHYLTHIOETHYL) PHOSPHORODITHIOATE. Syn: *disyston.*
Acute tox data: Oral LD_{LO} (rat) = 2 mg/kg; dermal LD_{50} (rat) = 6 mg/kg. [3]
THR = HIGH via oral and dermal routes. An exper ca [23]

DIETHYL ETHYLENE-BIS-GLYCOLATE. Liquid. $(CH_2OCH_2CO_2C_2H_5)_2$, mw: 234.2, bp: 152°–155° @ 5 mm, flash p: 255°F (OC), d: 1.1199 @ 25°/25°, vap. d: 8.09.
THR = U. See also esters and ethyl alcohol.
Fire Hazard: Slight, when exposed to heat or flame; can react with oxidizing materials.
To Fight Fire: Water, foam, CO_2, dry chemical.

N,N-DIETHYL ETHYLENE DIAMINE. Liquid. $(C_2H_5)_2NC_2H_4NH_2$, mw: 116.2, bp: 60° @ 40 mm, flash p: 115°F (OC), d: 0.82 @ 20°/20°, vap. d: 4.00.
Acute tox data: Oral LD_{50} (rat) = 2830 mg/kg; dermal LD_{50} (rabbit) = 820 mg/kg. [3]
THR = MOD via oral and dermal routes.
Fire Hazard: Mod, when exposed to heat or flame; can react with oxidizing materials.
To Fight Fire: Alcohol foam, CO_2, dry chemical.

DIETHYL ETHYL PHOSPHONATE. Colorless liquid; sweet odor. $C_2H_5P(O)(OC_2H_5)_2$, mw: 166.2, bp: 83° @ 11 mm, flash p: 221°F (COC), d: 1.025 @ 20°/4°, vap. d: 5.73.
THR = U.
Fire Hazard: Low, when exposed to heat or flame.
Disaster Hazard: Dangerous; see phosphorus compounds; can react with oxidizing materials.
To Fight Fire: Foam, CO_2, dry chemical.

***o,o*-DIETHYL-*s*,2-(ETHYLTHIO)ETHYL PHOSPHODITHIOATE.** See *o,o*-diethyl-2-ethyl thioethyl phosphorodithioate.

***o,o*-DIETHYL-*o*,2-(ETHYLTHIO)ETHYL THIOPHOSPHATE.** See demeton.

N,N-DIETHYL-*p*-(4-FLUOROPHENYL AZO)ANILINE.
THR = An exper neo. [3]

DIETHYL FLUOROPHOSPHATE. A liquid with a sweet or fruity odor. $(C_2H_5O)_2POF$, mw: 156.10, mp: low, bp: 170°, d: 1.15 (approx), vap. d: 5.38.
Acute tox data: Dermal LD_{50} (mouse) = 35 mg/kg.
THR = HIGH via dermal and oral routes.

Disaster Hazard: Dangerous; when heated to decomp or on contact with acid or acid fumes, emits highly toxic fumes of fluorides and oxides of phosphorus.

DIETHYL FUMARATE. White crystals or liquid. $C_2H_5OCOCHCHCOOC_2H_5$, mw: 172.2, mp: 0.6°, bp: 218.5°, flash p: 220°F, d: 1.0529 @ 20°/20°, vap. press: 1 mm @ 53.2°, vap. d: 5.93.
Acute tox data: Oral LD_{50} (rat) = 1781 mg/kg. [*3*]
THR = MOD via oral route. See esters, fumaric acid and ethyl alcohol.
Fire Hazard: Low, when exposed to heat or flame; can react with oxidizing materials.
To Fight Fire: Alcohol foam, foam, CO_2, dry chemical.

DIETHYL GERMANIUM BROMIDE. Colorless liquid, decomp by water. $(C_2H_5)_2GeBr_2$, mw: 290.6, mp: <−33°, bp: 202°.
THR = See germanium compounds and bromides.
Disaster Hazard: Dangerous; see bromides.

DIETHYL GERMANIUM CHLORIDE. Colorless liquid decomp by water. $(C_2H_5)_2GeCl_2$, mw: 201.6, mp: −38°, bp: 175°.
THR = See germanium compounds and chlorides.
Disaster Hazard: Dangerous; see chlorides.

DIETHYL GERMANIUM IMINE. Colorless liquid. $(C_2H_5)_2GeNH$, mw: 145.7, bp: 100° @ 0.01 mm.
THR = U. See germanium compounds.

DIETHYL GERMANIUM IODIDE. Colorless liquid, decomp by water. $(C_2H_5)_2GeI_2$, mw: 384.6, mp: −2°, bp: 252°.
THR = U. See germanium compounds and iodides.
Disaster Hazard: Dangerous; see iodides.

DIETHYL GERMANIUM OXIDE. Colorless liquid, insol in water. $[(C_2H_5)_2GeO]_3$, mw: 440.2, mp: 18°.
THR = U. See germanium compounds.

DIETHYL GLYCOL. $C_6H_{14}O_2$, mw: 118.1, flash p: 95°F, autoign. temp.: 401°F, d: 0.84, vap. d: 4.07, bp: 123°.
THR = No data. See also glycols.
Fire Hazard: Dangerous, via heat, ignition sources, oxidizers.
To Fight Fire: Foam, dry chemical, spray, mist.

DIETHYL GLYCOCOLL-*p*-AMINO-*o*-OXYBENZOIC METHYL ESTER. Colorless, prismatic crystals.
$C_6H_3OHCOOCH_3NHCOCH_2N(C_2H_5)_2$, mw: 280.3, mp: 185°, vap. d: 9.65.
THR = U. See also esters.
Fire Hazard: Slight.

DIETHYL GLYCOCOLLGUAIACOL HYDROCHLORIDE. White prisms.
$CH_3OC_6H_4OCOCH_2N(C_2H_5)_2 \cdot HCl$, mw: 273.70, vap. d: 9.45.
THR = U.
Disaster Hazard: Dangerous: see chlorides.

DIETHYL GLYCOL DIPROPRIONATE. Slightly sol in H_2O. $(C_2H_5COOC_2H_4)_2O$, mw: 218, d: 1.1, bp: 491°–529°F, flash p: 260°F.
THR = U.
Fire Hazard: Slight, when exposed to heat or flame.
To Fight Fire: Water, foam, alcohol foam.

DIETHYL GLYCOL PHTHALATE. Liquid.
$C_6H_4[COO(CH_2)_2OC_2H_5]_2$, mw: 310.34, flash p: 343°F, d: 1.11.
THR = See diethyl-*o*-phthalate and esters.
Fire Hazard: Slight, when exposed to heat or flame; can react with oxidizing materials.
To Fight Fire: Alcohol foam, water, CO_2, dry chemical.

DI(2-ETHYLHEXYL)ADIPATE. Liquid.
$[C_4H_9(C_2H_5)CHCH_2OCOC_2H_4]_2$, mw: 370.56, mp: 60°, bp: 214° @ 5 mm, flash p: 385°F, d: 0.9268 @ 20°/20°, vap. press: 2.6 mm @ 200°, vap. d: 12.8.
Acute tox data: Oral LD_{50} (rat) = 9110 mg/kg; dermal LD_{50} (rabbit) = 16000 mg/kg. [*3*]
THR = LOW via oral and dermal routes.
Fire Hazard: Slight, when exposed to heat or flame; can react with oxidizing materials.
To Fight Fire: Foam, CO_2, dry chemical.

DI(2-ETHYLHEXYL)AMINE. Syn: *dioctyl amine.* Water white liquid with slightly ammoniacal odor. $[C_4H_9CH(C_2H_5)CH_2]_2NH$, mw: 241.45, bp: 281.1°, flash p: 270°F (OC), d: 0.8062 @ 20°/20°, vap. d: 8.35.
Acute tox data: Oral LD_{50} (rat) = 1640 mg/kg; dermal LD_{50} (rabbit) = 1190 mg/kg. [*3*]
THR = MOD oral and dermal routes. See also amines.
Fire Hazard: Slight, when exposed to heat or flame; can react with oxidizing materials.
To Fight Fire: Alcohol foam, foam, CO_2, dry chemical.

DI(2-ETHYLHEXYL)AZELATE. Liquid. $C_{25}H_{48}O_4$, mw: 412.63, mp: −67.8°, bp: 237° @ 5 mm, flash p: 415°F, d: 0.918 @ 20°, vap. d: 18.7.
THR = U. See also esters.
Fire Hazard: Slight, when exposed to heat or flame; can react with oxidizing materials.
To Fight Fire: Foam, CO_2, dry chemical.

DI(2-ETHYLHEXYL)4,5-EPOXYCYCLOHEXANE-1,2-DICARBOXYLATE.
$C_6H_8O[COOC_4H_9CH(C_2H_5)CH_2]_2$, mw: 327.
THR = LOW. Based on limited animal exper.

DI(2-ETHYLHEXYL)ETHANOLAMINE. Liquid.
$[C_4H_9CH(C_2H_5)CH_2]NC_2H_4OH$, mw: 285.5, mp: −60°, bp: 216° @ 50 mm, flash p: 280°F (OC), d: 0.8573 @ 20°/20°, vap. press: <0.01 mm @ 20°, vap. d: 9.87.

Acute tox data: Oral LD_{50} (rat) = 4920 mg/kg; dermal LD_{50} (rabbit) = 2520 mg/kg. [*3*]
THR = MOD via oral and dermal routes. See also amines.
Fire Hazard: Slight, when exposed to heat or flame; can react with oxidizing materials.
To Fight Fire: Alcohol foam, CO_2, dry chemical.

DI(2-ETHYLHEXYL)ETHYLENE BIS-GLYCOLATE. Liquid. $(CH_2OCH_2CO_2C_8H_{17})_2$, mw: 402.6, bp: 192°–202° @ 0.3 mm, flash p: 390°F (OC), d: 0.9760 @ 25°/25°, vap. d: 13.9.
THR = Details U. See also glycols and esters.
Fire Hazard: Slight, when exposed to heat or flame; can react with oxidizing materials.
To Fight Fire: Foam, CO_2, dry chemical.

DI(2-ETHYLHEXYL)HEXAHYDROPHTHALATE. Liquid. $C_6H_{10}[COOCH_2CH(C_2H_5)C_4H_9]_2$, mw: 396.59, mp: −53°, bp: 216° @ 5 mm, flash p: 425°F, d: 0.9586 @ 20°/20°, vap. press: 2.2 mm @ 200°, vap. d: 13.7.
THR = Details U. See also esters.
Fire Hazard: Slight, when exposed to heat or flame; can react with oxidizing materials.
To Fight Fire: Foam, CO_2, dry chemical.

DI(2-ETHYLHEXYL)FUMARATE. See dioctyl fumarate.

DI(2-ETHYLHEXYL)MALEATE. Liquid. $[CHCO_2CH_2CH(C_2H_5)C_4H_9]_2$, mw: 340.5, mp: −60°, bp: 164° @ 10 mm, flash p: 365°F, d: 0.9436 @ 20°/20°, vap. d: 11.7.
THR = Details U. See also esters.
Fire Hazard: Slight, when exposed to heat or flame; can react with oxidizing materials.
To Fight Fire: Alcohol foam, dry chemical, mist or spray.

DI(2-ETHYLHEXYL)PHOSPHITE. Liquid. $(C_8H_{17}O)_2PHO$, mw: 306, bp: 150°–155° @ 2–3 mm, d: 0.929 @ 25°/4°, vap. d: 10.6, flash p: 330°F.
THR = Details U.
Disaster Hazard: Dangerous; see phosphorus compounds.
Fire Hazard: Low, via heat, flame, oxidizers.
To Fight Fire: Water spray, mist, foam, dry chemical.

DI(2-ETHYLHEXYL)PHOSPHORIC ACID. Syn: *dioctyl phosphoric acid.* Liquid, insol in water, sol in organic solvents. $[C_4H_9CH(C_2H_5)CH_2]_2HPO_4$, mw: 322, d: 0.973 @ 25°/25°, fp: −60°, flash p: 385°F (OC).
Acute tox data: Oral LD_{50} (rat) = 4940 mg/kg; dermal LD_{50} (rabbit) = 1250 mg/kg. [*3*]
THR = MOD via oral and dermal routes.
Fire Hazard: Slight, when exposed to heat or flame.
Disaster Hazard: Dangerous; see phosphorous compounds.
To Fight Fire: Alcohol foam.

DI(2-ETHYLHEXYL)PHTHALATE. Syns: *dioctyl phthalate, di-sec-octyl phthalate.* Stable, light-colored liquid, mild odor. $C_6H_4[CO_2CH_2CH(C_2H_5)C_4H_9]_2$, mw: 390.6, bp: 230° @ 5 mm, fp: −55°, flash p: 425°F (OC), d: 0.9861 @ 20°/20°, vap. press: 1.2 mm @ 200°, vap. d: 16.0.
Acute tox data: Oral LD_{50} (rat) = 31000 mg/kg; dermal LD_{50} (guinea pig) = 10,000 mg/kg. [*3*]
THR = LOW to NONE via oral and dermal routes. On a chronic basis, an exper teratogen. Also TD_{LO} (man) = 143 mg/kg to produce GI symptoms. [*3*]
Fire Hazard: Slight, when exposed to heat or flame; can react with oxidizing materials.
To Fight Fire: CO_2, dry chemical.

DI(2-ETHYLHEXYL)SEBACATE. Light, clear liquid, mild odor. $(CH_2)_8(CO_2C_8H_{17})_2$, mw: 426, bp: 248° @ 9 mm, fp: −55°, flash p: 410°F, d: 0.913 @ 25°/25°, vap. d: 14.7.
Acute tox data: Oral LD_{50} (rat) = 1280 mg/kg. [*3*]
THR = MOD via oral route.
Fire Hazard: Slight, when exposed to heat or flame; can react with oxidizing materials.
To Fight Fire: Foam, CO_2, dry chemical.

DI(2-ETHYLHEXYL)TETRAHYDROPHTHALATE. Liquid. $C_6H_8[COOCH_2CH(C_2H_5)C_4H_9]_2$, mw: 394.58, mp: −50°, bp: 219° @ 5 mm, flash p: 350°F (OC), d: 0.9685 @ 20°/20°, vap. d: 13.6.
Acute tox data: Oral LD_{50} (rat) = 114,000 mg/kg. [*3*]
THR = NONE. See also esters.
Fire Hazard: Slight, when exposed to heat or flame; can react with oxidizing materials.
To Fight Fire: Foam, CO_2, dry chemical.

DI(2-ETHYLHEXYL) TIN DICHLORIDE. Syn: *di-n-octyltin dichloride.* Crystals. $(C_8H_{17})_2SnCl_2$, mw: 656.2.
Acute tox data: iv LD_{50} (rat) = 5 mg/kg. [*3*]
THR = HIGH via iv route. See tin compounds.
Disaster Hazard: Dangerous; when heated to decomp, emits highly toxic fumes of chlorides.

1,2-DIETHYL HYDRAZINE. Sol in alcohol and ether. $C_4H_{12}N_2$, mw: 88.2, bp: 86°, d: 0.797 @ 26°.
THR = A toxic material. See also hydrazines. An exper (+) transplacental carc. [*3, 10, 23*]

1,2-DIETHYL HYDRAZINE DIHYDROCHLORIDE. $C_4H_{12}N_2 \cdot 2HCl$, mw: 161.1.
THR = An exper carc and teratogen to rats via iv route. [*103*]

DIETHYL ISOAMYLTIN BROMIDE. Solid. $(C_2H_5)_2(C_5H_{11})SnBr$, mw: 327.9, bp: 138° @ 17 mm, d: 1.4881 @ 17°.
THR = Details U. See tin compounds and bromides.
Disaster Hazard: Dangerous; see bromides.

DIETHYL ISOAMYLTIN CHLORIDE. Solid. $(C_2H_5)_2(C_5H_{11})SnCl$, mw: 283.4, d: 1.2994 @ 20°, bp: 126° @ 13 mm.
THR = Details U. See tin compounds.
Disaster Hazard: Dangerous; see chlorides.

DIETHYL ISOBUTYLTIN BROMIDE. Solid. $(C_2H_5)_2(C_4H_9)SnBr$, mw: 313.9, d: 1.5108, bp: 122° @ 17 mm.
THR = Details U. See tin compounds.
Disaster Hazard: Dangerous; see bromides.

***o,o*-DIETHYL-*s*-ISOPROPYLMERCAPTO-METHYL PHOSPHORODITHIOATE.** See thimet.

***o,o*-DIETHYL-*o*-2-ISOPROPYL-4-METHYL-6-PYRIMIDYL THIOPHOSPHATE.** See diazinon.

DIETHYL KETONE. Syns: *3-pentanone, metacetone, propione, ethyl propionyl.* Colorless, mobile liquid, acetone-like odor. $C_2H_5COC_2H_5$, mw: 86.13, mp: −42°, bp: 101°, flash p: 55°F, d: 0.8159 @ 19°/4°, vap. d: 2.96, autoign. temp.: 842°F, lel = 1.6%.
Acute tox data: Oral LD_{50} (rat) = 2140 mg/kg; inhal LD_{50} (rat) = 8000 ppm for 4 hrs. [*3*]
THR = MOD via oral and inhal routes.
Fire Hazard: Dangerous, when exposed to heat or flame; can react vigorously with oxidizing materials.
Explosion Hazard: U.
To Fight Fire: Alcohol foam, foam, CO_2, dry chemical.

DIETHYL LAURAMIDE. Liquid. $(C_2H_5)_2NOC(CH_2)_{10}CH_3$, mw: 255, bp: 165°, flash p: >150°F (OC), d: 0.86, vap. d: 8.8.
THR = Details U.
Fire Hazard: Mod, when exposed to heat, flame or oxidizers.
Disaster Hazard: Dangerous; when heated to decomp, emits highly toxic fumes of oxides of nitrogen; can react with oxidizing materials.
To Fight Fire: Foam, CO_2, dry chemical.

DIETHYL MAGNESIUM. See magnesium diethyl.

DIETHYL MALEATE. Water white liquid. $(HCCOOC_2H_5)_2$, mw: 172.2, mp: −11.5°, bp: 225.0°, flash p: 250°F (OC), d: 1.0687 @ 20°, vap. press: 1 mm @ 57.3°, vap. d: 5.93.
Acute tox data: Oral LD_{50} (rat) = 3200 mg/kg; dermal LD_{50} (rabbit) = 140 mg/kg. [*3*]
THR = MOD via oral and inhal; HIGH via dermal routes.
Fire Hazard: Low, when exposed to heat or flame; can react with oxidizing materials.
To Fight Fire: CO_2, dry chemical.

DIETHYL MALONATE. Syn: *ethyl malonate.* Clear, colorless liquid. $CH_2(COOC_2H_5)_2$, mw: 160.17, bp: 198.9°, fp: −49.8°, flash p: 200°F (OC), d: 1.055 @ 25°/25°, vap. press: 1 mm @ 40.0°, vap. d: 5.52.
Acute tox data: Oral LD_{50} (rat) = 15000 mg/kg. [*3*]
THR = LOW via oral route.
Fire Hazard: Low, when exposed to heat or flame; can react with oxidizing materials.
Spont Heating: No.
To Fight Fire: Water to blanket fire; foam, CO_2, dry chemical.

DIETHYL MERCURY. Colorless liquid, hazel-like odor. $Hg(C_2H_5)_2$, mw: 258.73, bp: 159°, d: 2.4660 @ 20°.
Acute tox data: Inhal LC_{LO} (human) = 1 mg/m^3 for a period of 14 weeks. [*3*]
THR = HIGH via inhal route. See mercury compounds, organic.
Fire Hazard: Mod, when exposed to heat or flame.
Disaster Hazard: Dangerous; when heated to decomp or on contact with acid or acid fumes, emits highly toxic fumes of mercury; can react with oxidizing materials.

DIETHYLMETHYLMETHANE.
See 3-methyl pentane.

1,1-DIETHYL-3-METHYL-3-NITROSO UREA. $C_6H_{13}O_2N_3$, mw: 159.2.
THR = An exper neo. [*3*]

***o,o*-DIETHYL-*o*-(3-METHYL-5-PYRAZOLYL) PHOSPHATE.** $C_8H_{15}O_4N_2P$, mw: 234.2.
Acute tox data: Oral LD_{50} (mice) = 4 mg/kg; oral LD_{50} (wild birds) = 40 mg/kg. [*3*]
THR = HIGH via oral route. An exper insecticide. A cholinesterase inhibitor. See parathion.
Disaster Hazard: See phosphorous compds.

***o,o*-DIETHYL-*o*-(3-METHYL-5-PYRAZOLYL) PHOSPHOROTHIOATE.** $C_8H_{15}O_3N_2PS$, mw: 250.3.
Acute tox data: Oral LD_{50} (rat) = 12 mg/kg. [*3*]
THR = HIGH via oral route. An exper insecticide. A cholinesterase inhibitor. See parathion.
Disaster Hazard: Dangerous; see phosphorous compounds.

***o,o*-DIETHYL-*o*-[*p*-(METHYL SULFINYL) PHENYL] PHOSPHOROTHIOATE.** Syns: *dasant, fensulfothione.* A brown liquid, slightly sol in water and most organic solvents.
$(C_2H_5O)_2PSOC_6H_4SOCH_3$, mw:308.2, d: 1.202.
Acute tox data: Oral LD_{50} (rat) = 2 mg/kg; dermal LD_{50} (rat) = 3 mg/kg; oral LD_{50} (chicken) = 1 mg/kg. [*3*]
THR = HIGH via oral and dermal routes. A cholinesterase inhibitor.
Disaster Hazard: See phosphates and sulfates.

DIETHYL-*p*-NITROPHENYLPHOSPHATE. Syn: *para-oxon.* Oily liquid, slight odor, slightly water-sol, freely sol in organic solvents. $C_{10}H_{14}NO_6P$, mw: 275.2, bp: 170° @ 1 mm, d: 1.2736 @ 20°/4°.

Acute tox data: Oral LD_{50} (rat) = 1.8 mg/kg; ipLD_{50} (rat) = 0.93 mg/kg. [3]
THR = HIGH via oral, ip, iv and sc routes. A cholinesterase inhibitor. See parathion.
Disaster Hazard: Dangerous; see phosphorous compounds.

o,o-DIETHYL-o,p-NITROPHENYL THIOPHOSPHATE. See parathion.

DIETHYL OXALATE. See ethyloxalate.

3,3-DIETHYLPENTANE. Liquid. $CH_3CH_2C(C_2H_5)_2CH_2CH_3$, mw: 128.3, autoign. temp.: 554°F, vap. d: 4.42, lel = 0.7%, uel = 5.7%.
THR = Details U. Probably has irr and narcotic effects.
Fire Hazard: Dangerous, when exposed to heat or flame; can react vigorously with oxidizing materials.
To Fight Fire: Foam, CO_2, dry chemical.

DIETHYL PEROXIDE. $C_2H_5OOC_2H_5$, mw: 90, lel = 2.3%, d: 0.8, vap. d: 7.7, bp: 65°.
THR = Very unstable, reacts violently with O_2. [19] See peroxides, organic.
Fire Hazard: Very dangerous. Explodes on heating.

DIETHYL PEROXY DICARBONATE. $C_6H_{10}O_6$, mw: 178.1.
THR = Decomp rapidly >10°. Do not allow it to crystallize. [19]

p-[(3,4-DIETHYLPHENYL)AZO]-N,N-DIMETHYL ANILINE. $C_{18}H_{23}N_3$, mw: 281.4.
THR = An exper carc. [3]

3,3-DIETHYL-1-PHENYL TRIAZENE. $C_{10}H_{15}N_3$, mw: 177.3.
THR = An exper transplacental carc. [3, 23] via sc route.

DIETHYL PHOSPHATE. Liquid. $(C_2H_5O)_2PHO$, mw: 138, bp: 138°, d: 1.071 @ 20°/20°, vap d: 4.76, flash p: 195°F (COC).
Acute tox data: Oral LD_{50} (rat) = 5190 mg/kg; dermal LD_{50} (rabbit) = 2020 mg/kg. [3]
THR = LOW to MOD via oral and dermal routes.
Fire Hazard: Mod, when exposed to heat or flame.
Disaster Hazard: Dangerous; see phosphorous compounds, can react with oxidizing materials.

DIETHYLPHOSPHINE. Colorless liquid. $(C_2H_5)_2PH$, mw: 90.1, bp: 85°, d: <1, vap. d: 3.11.
THR = HIGH via oral and inhal routes.
Fire Hazard: Dangerous, when exposed to heat or flame; spont flam in air.
Explosion Hazard: U.
Disaster Hazard: Dangerous; see phosphorus; can react vigorously with oxidizing materials.
To Fight Fire: Foam, CO_2, dry chemical.

DIETHYL-o-PHTHALATE. Syn: ethyl phthalate. Clear, colorless liquid. $C_6H_4(CO_2C_2H_5)_2$, mw: 222.2, mp: −40.5°, bp: 302°, flash p: 325°F (OC), d: 1.110, vap. d: 7.66.
Acute tox data: ipLD_{50} (rat) = 5058 mg/kg. [3]
THR = MOD via ip route. An irr to mu mem. Narcotic in high conc. An exper teratogen. [3]
Fire Hazard: Slight, when exposed to heat or flame; can react with oxidizing materials.
To Fight Fire: Water, spray, mist, foam, dry chemical.

DIETHYL-p-PHTHALATE. Liquid. $C_6H_4(COOC_2H_5)_2$, mw: 222.2, mp: −5°, bp: 296°, flash p: 243°F (CC), d: 1.117–1.121 @ 20°/20°, vap. d: 7.66.
THR = See diethyl-o-phthalate.
Fire Hazard: Low.
To Fight Fire: Dry chemical, CO_2.

N,N-DIETHYL PIPERONYLAMIDE. Crystals. $CH_2O_2:C_6H_3CON(C_2H_5)_2$, mw: 237.25, vap. d: 9.44.
THR = Details U, but piperonyl compounds are generally of a LOW order of toxicity; an insecticide.
Fire Hazard: Slight.

2,2-DIETHYL-1,3-PROPANEDIOL. Syn: *prenderol.* Crystals. $HOCH_2C(C_2H_5)_2CH_2OH$, mw: 132.2, bp: 125° @ 10 mm, fp: 61.3°, flash p: 215°F (OC), d: 1.052 @20°.
Acute tox data: Oral LD_{50} (rat) = 850 mg/kg; dermal LD_{50} (rabbit) = 4240 mg/kg; oral TD_{LO} (human) = 7.1 mg/kg causing psychic disturbances. [3]
THR = MOD via oral and dermal routes. In humans the oral is HIGH. Large doses can cause drowsiness, vertigo, nausea and vomiting.
Fire Hazard: Low, when exposed to heat or flame; can react with oxidizing materials.
To Fight Fire: Alcohol foam, foam, CO_2, dry chemical.

DIETHYL-n-PROPYLTIN BROMIDE. A powder. $(C_2H_5)_2(C_3H_7)SnBr$, mw: 300, d: 1.5910 @ 21°, bp: 112° @ 16 mm.
THR = See tin compounds and bromides.

DIETHYL-n-PROPYLTIN CHLORIDE. A powder. $(C_2H_5)_2(C_3H_7)SnCl$, mw: 255, d: 1.3848 @ 16°, bp: 108° @ 17 mm.
THR = See tin compounds and chlorides.

DIETHYL-n-PROPYLTIN FLUORIDE. Powder, sol in alcohol. $(C_2H_5)_2(C_3H_7)SnF$, mw: 239, mp: 271°.
THR = See tin compounds and fluorides.

o,o-DIETHYL-o-(2-PYRAZINYL)PHOSPHOROTHIOATE. See zinophos.

N,N′-DIETHYL-4-[4′-(PYRIDYL-1′-OXIDE)AZO] ANILINE. $C_{15}H_{12}ON_4$, mw: 264.3.
THR = An exper carc. [3]

DIETHYL-*m*-PYRIDYL TRIAZENE. $C_9H_{14}N_4$, mw: 178.3.
THR = An exper carc. [*3*]

3,3-DIETHYL-1-(*m*-PYRIDYL)TRIAZENE. $C_9H_{15}N_4$, mw: 179.3.
THR = An exper transplacental carc and neo. [*3, 23*] via oral, sc and iv routes.

DIETHYL PYROCARBONATE. Clear, colorless liquid, sweet ester-like odor, miscible with ethanol and methanol. $C_6H_{10}O_5$, mw: 162.
Acute tox data: Oral LD_{50} (rat) = 850 mg/kg; ipLD_{50} (rat) = 100 mg/kg. [*3*]
THR = MOD via oral; HIGH via ip routes. A food additive permitted in food for human consumption. [*109*]

DIETHYL SELENIDE. Liquid. $(C_2H_5)_2Se$, mw: 137.08, bp: 108.0°, d: 1.23, lel = 2.5%, vap. press: 40 mm @ 31.2°, vap. d: 4.73.
THR = HIGH. See selenium and its compounds.
Fire Hazard: Dangerous, when exposed to heat or flame.
Spont Heating: No.
Explosion Hazard: Mod, when exposed to flame.
Disaster Hazard: See selenium compounds.
To Fight Fire: Water, foam, CO_2, dry chemical.

DIETHYLSILANE. Liquid. Stable when pure. $(C_2H_5)_2SiH_2$, mw: 88.2, d: 0.6843 @ 20°/4°, bp: 56° @ 741 mm.
THR = A powerful irr. HIGH. See silanes.

DIETHYL STEARAMIDE. Liquid. $(C_2H_5)_2NOC(CH_2)_{16}CH_3$, mw: 339.6, bp: 140°, flash p: 375°F, d: 0.86, vap. d: 11.7.
THR = U. See also amides.
Fire Hazard: Slight, when exposed to heat or flame; can react with oxidizing materials.
To Fight Fire: CO_2, dry chemical.

N,N-DIETHYL-4-STILBENAMINE. $C_{18}H_{21}N$, mw: 251.4.
THR = An exper neo. [*3, 23*] via sc route.

DIETHYLSTILBESTROL. Syn: *stilbestrol.* Small crystals. $C_{18}H_{20}O_2$, mw: 268.3, mp: 169°–172°.
Acute tox data: Dermal TD_{LO} (human) = 0.06 mg/kg for 14 days caused glandular problems; TD_{LO} (women) = 240 mg/kg over 35 weeks was a teratogen; ip LD_{50} (rat) = 34 mg/kg. [*3*]
THR = HIGH via dermal and ip routes. Implicated in male impotence and enlargement of male breasts. [*41*] A human (+) carc [*3, 41, 15*] and an exper carc, neo and teratogen. [*3, 23*] A food additive permitted in the feed and drinking water of animals, and/or for the treatment of food-producing animals. Also a food additive permitted in food for human consumption. [*109*] It is a suspected transplacental carc. [*14*] See estrogens.
Fire Hazard: Slight.

DIETHYLSTILBESTROL DIPALMITATE. $C_{50}H_{80}O_4$, mw: 745.3.
THR = An exper carc. [*3*]

DIETHYLSTILBESTROL DIPROPIONATE. Syn: *stilbestrol diethyl dipropionate.* Crystals. $C_{24}H_{28}O_4$, mw: 380.4, mp: 104°.
THR = An exper neo. [*3*]
Fire Hazard: Slight.

DIETHYLSTILBESTROL TETRAIODATE. $C_{18}H_{16}O_2I_4$, mw: 771.9.
THR = An exper carc. [*3*]

DIETHYL SUCCINATE. Syn: *ethyl succinate.* Essentially colorless, clear mobile liquid, slight odor. $C_8H_{14}O_4$, mw: 174.2, mp: −20.8°, bp: 216.5°, flash p: 230°F (OC), d: 1.037 @ 25°/25°, vap. press: 1 mm @ 54.6°, vap. d: 6.0.
Acute tox data: Oral LD_{50} (rats) = 8.5 g/kg.
THR = Low via oral route.
Fire Hazard: Slight, when exposed to heat or flame; can react with oxidizing materials.
To Fight Fire: Alcohol foam, foam, CO_2, dry chemical.

DIETHYL SULFATE. Syn: *ethyl sulfate.* Colorless oily liquid, faint ethereal odor. $(C_2H_5)_2SO_4$, mw: 154.18, mp: −25°, bp: decomp to ethyl ether, flash p: 220°F (CC), d: 1.172 @ 25°/4°, autogin. temp.: 817°F, vap. press: 1 mm @ 47.0°, vap. d: 5.31.
Acute tox data: Oral LD_{50} (mouse) = 647 mg/kg; dermal LD_{50} (rabbit) = 600 mg/kg; inhal LD_{LO} (rat) = 250 ppm for 4 hrs. [*3*]
THR = MOD to HIGH via oral and dermal routes; HIGH via inhal route. An exper (+) carc. [*3, 23, 10*] and neo. [*3*]
Fire Hazard: Low, when exposed to heat or flame.
Spont Heating: No.
Disaster Hazard: Dangerous; see sulfates; can react with oxidizing materials; moisture causes liberation of H_2SO_4; violent reaction with potassium *tert*-butoxide. [*19*]
To Fight Fire: Alcohol foam, water, foam, CO_2, dry chemical.

DIETHYL SULFIDE. See ethyl sulfide.

DIETHYL TARTRATE. Colorless, thick, oily liquid, sol in water and alcohol. $C_8H_{14}O_6$, mw: 206, bp: 280°, mp: 17°, d: 1.204 @ 20°/4°, flash p: 200°F.
THR = U.
Fire Hazard: Low, when exposed to heat or flame.
To Fight Fire: Alcohol foam.

DIETHYL TELLURIDE. Liquid $(C_2H_5)_2Te$, mw: 186, bp: 138°.

THR = Self-ignites in air. [19] See tellurium compounds.

1,2-DIETHYL TETRAIODO DIALUMINUM. $C_4H_{10}I_4Al_2$, mw: 619.7.
THR = Spont flam in air. [19]

DIETHYLTHIOCARBAMYL CHLORIDE. Crystals. $(C_2H_5)_2NC(S)Cl$, mw: 151.7, mp: 46.2°, bp: 97°–103° @ 5 mm.
THR = HIGH.
Disaster Hazard: Dangerous; see chlorides and sulfates.

DIETHYLTIN. Yellowish oily liquid, not miscible with water. $(C_2H_5)_2Sn$, mw: 177, d: 1.654, mp: <−12°, bp: 150° (decomp).
THR = Probably toxic. Details U. See tin compounds.
Fire Hazard: Dangerous. A powerful reducing agent.

DIETHYLTIN DIBROMIDE. Colorless water-sol crystals. $(C_2H_5)_2SnBr_2$, mw: 337, d: 2.068 @ 74°, mp: 63°, bp: 233°.
THR = See tin compounds and bromides.

DIETHYLTIN DICHLORIDE. White water-sol crystals. $(C_2H_5)_2SnCl_2$, mw: 248, mp: 85°, bp: 220°.
Acute tox data: Oral LD_{LO} (rat) = 160 mg/kg; ivLD_{50} (rat) = 25 mg/kg. [3]
THR = HIGH via oral and iv routes. See tin compounds and chlorides.

DIETHYLTIN DIFLUORIDE. Crystals slightly sol in alcohol. $(C_2H_5)_2SnF_2$, mw: 215, mp: 229°.
THR = See tin compounds and fluorides.

DIETHYLTIN DIIODIDE. Very slightly sol white crystals. $(C_2H_5)_2SnI_2$, mw: 431, mp: 45°, bp: 240–245° (decomp).
Acute tox data: Oral LD_{LO} (rat) = 100 mg/kg; ip LD_{LO} (rat) = 26 mg/kg. [3]
THR = HIGH via oral and ip routes. See tin compounds and iodides.

DIETHYLTIN OXIDE. White, insol powder. $(C_2H_5)_2SnO$, mw: 193, mp: infusible.
THR = See tin compounds.

N,N-DIETHYL-*m*-TOLUAMIDE. Syn: *deet.* A liquid, sol in water, alcohol and ether. $C_{12}H_{17}NO$, mw: 191.3, bp: 160° @ 19 mm, d: 0.996 @ 20°/4°.
Acute tox data: Oral LD_{50} (rat) = 200 mg/kg; dermal LD_{50} (rabbit) = 3180 mg/kg. [3]
THR = HIGH via oral and MOD via dermal routes. An insect-repellent. An irr to eyes and mu mem. Can cause CNS disturbances.

DIETHYL TRIAZENE. $C_4H_{11}N_3$, mw: 101.2.
THR = An exper carc. [3]

***o,o*-DIETHYL-*o*-(3,5,6-TRICHLORO-2-PYRIDYL) PHOSPHOROTHIOATE.** See dursban.

DIETHYL VALERAMIDE. See valyl.

DIETHYL XANTHOGEN SULFIDE. See ethyl xanthogen disulfide.

2,4-DIETHYNYL-5-METHOXYTOLUENE. $C_{12}H_{10}O$, mw: 170.2.
THR = Forms an explosive polymer. [19]

2,4-DIETHYNYL-5-METHYL PHENOL. $C_{11}H_8O$, mw: 156.1.
THR = Unstable to light and to air. [19]

DIFFERENCES OF RAT RESPONSE TO CARCINOGENS. See Section 8.

DIFLUORAMINE. HNF_2, mw: 53.
THR = Exploded upon shock. Tends to explode in solid state or upon freezing. Violent reaction with ClO_2. [19]

DIFLUOROBENZENE. Liquid, pungent odor. $C_6H_4F_2$, mw: 114.09, mp: −23.7°, d: 1.17006 @ 20°, bp: 88.82°.
Acute tox data: Inhal LD_{LO} (mouse) = 55 mg/m^3 for 3 hrs. [3]
THR = HIGH via inhal route. Vapors may be irr and narcotic.
Disaster Hazard: Dangerous; see fluorides.

1,1,1-DIFLUOROCHLOROETHANE. See difluoro monochloroethane.

DIFLUOROCHLOROMETHYL METHANE. See difluorochloroethane.

DIFLUORODIAZENE. $(NF)_2$, mw: 66.
THR = Very unstable. [19]

DIFLUORODIBROMOMETHANE. See dibromodifluoromethane.

DIFLUORODIPHENLYDICHLOROETHANE. Scc DFDD.

DIFLUORODIPHENYL DISULFIDE. Crystals. $C_{12}H_6S_2F_2$, mw: 252.3, vap. d: 8.72.
THR = An insecticide. See also fluorides.
Disaster Hazard: Dangerous; see fluorides and oxides of sulfur; will react with water or steam to produce toxic fumes.

DIFLUORODIPHENYL TRICHLOROETHANE. See DFDT.

DIFLUOROETHANE. See 1,1-difluoroethane.

1,1-DIFLUOROETHANE. Syns: *difluoroethane, ethylene fluoride, ethylidene difluoride.* Colorless gas. $F_2C_2H_4$, mw: 66.1, mp: −117.0°, bp: −26.5°, d: 1.004 @ 25°, vap. d: 2.28.
Acute tox data: Inhal LC_{LO} (rat) = 500,000 ppm for 10 min. [3]
THR = LOW inhal; LOW to MOD irr. Narcotic in high conc.

Fire Hazard: Dangerous, when exposed to heat or flame.
Explosion Hazard: U.
Disaster Hazard: Dangerous; see fluorides; can react vigorously with oxidizing materials.

1,1-DIFLUOROETHYLENE. Syn: *vinylidene fluoride.* Colorless gas. CH_2CF_2, mw: 64.04, bp: <−70°, lel = 5.5%, uel = 21.3%
Acute tox data: Inhal LC_{LO} (rat) = 128,000 ppm for 4 hrs. [*3*]
THR = LOW via inhal route.
Fire Hazard: Very dangerous via heat, flames or oxidizers.
Disaster Hazard: Dangerous; when heated to decomp, emits toxic fumes of fluorides.
To Fight Fire: Stop flow of gas.

DIFLUOROMONOCHLOROETHANE. Syn: 1,1,1-difluorochloroethane. Gas. $C_2H_3F_2Cl$, mw: 100.5, mp: −131°, bp: −9.5°, d: 1.19, lel = 6.2%, uel = 17.9%.
Acute tox data: Inhal LC_{LO} (rat) = 500,000 ppm for 30 min.[*3*]
THR = VERY LOW via inhal route.
Fire Hazard: Very dangerous, when exposed to heat or flame.
Explosion Hazard: U.
Disaster Hazard: Dangerous; when heated to decomp, emits toxic fumes of fluorides and chlorides; can react vigorously with oxidizing materials.
To Fight Fire: Stop flow of gas.

DIFLUOROMONOCHLOROMETHANE. See chlorodifluoromethane.

***p*-[(2,4-DIFLUOROPHENYL)AZO]-N,N-DIMETHYL ANILINE.** $C_{14}H_{13}N_3F_2$, mw: 261.3.
THR = An exper carc. [*3*]

DIFLUOROPHOSPHORIC ACID, ANHYDROUS. Mobile, strongly fuming, colorless liquid. HPO_2F_2, mw: 102, mp: −75°, bp: 116°, d: 1.583 @ 25°/4°, vap. d: 3.52.
THR = HIGH via contact, inhal, oral routes. An irr and corrosive material. See fluorides and phosphoric acid.
Disaster Hazard: Dangerous; see fluorides and oxides of phosphorus; will react with water or steam to produce toxic and corrosive fumes.

3,8-DIFLUOROTRICYCLOQUINAZOLINE. $C_{21}H_{10}N_4F_2$, mw: 356.4
THR = An exper neo. [*3*]

DIFOLATAN. See captafol.

N,N′-DIFURFURYLIDENE-2-FURANMETHANE DIAMINE. See hydrofuramide.

DIGALLANE. See gallium hydride.

DIGERMANE. Syn: *Germanium hydride.* Liquid or gas. Ge_2H_6, mw: 151.25, bp: 29°, mp: −109°, d (gas): 6.74 g/l @ 20°, d (liquid): 1.98 @ −109°.
THR = See hydrides and germanium compounds. Spont flam in air. [*19*]

DIGITALIN. See digitalis.

DIGITALIS. Syns: *foxglove, purple foxglove, fairy gloves.* Dried leaves of digitalis purpurea. Composition: digitoxin (0.2–0.4%), etc.
THR = An overdose can be fatal. 2.5 g or 30 cc of the tincture is considered a toxic dose. It contains digitalin, digitalein, digitonin and digitoxin, the most toxic component. An overdose causes the eyes to become prominent, the pupils dilated and the sclera blue. Furthermore, it seems to affect the digestive system, causing nausea, vomiting and cardiac irregularities, and excessive doses cause heart failure. In addition to the nausea, great malaise and headache are often noted. If this material is taken for medical purposes and the above symptoms are noted, the dosage should be discontinued until medical advice can be obtained. Has been implicated in aplastic anemia.
Treatment and Antidotes: Wash stomach with tannic acid or strong tea. Keep patient lying down. Administer stimulants such as caffeine or ammonia, atropine if pulse falls below 50 per min, morphine for irr. Call a physician.

DIGITOXIN. See digitalis.

DIGLYCEROL TETRANITRATE. Liquid.
THR = Details U. See also TNT.
Fire Hazard: See nitrates.
Explosion Hazard: Severe, when shocked or exposed to heat. See explosives, high.
Disaster Hazard: Dangerous; shock will explode it; when heated to decomp, emits highly toxic fumes of NO_x; can react vigorously with oxidizing and reducing materials.

DIGLYCIDYL ETHER. Syn: *DGE.* Liquid. $C_6H_{10}O_3$, mw: 130.1.
Acute tox data: Oral LD_{50} (rat) = 450 mg/kg; dermal LD_{50} (rabbit) = 1500 mg/kg; inhal LC_{50} (rat) = 200 ppm for 8 hrs. [*3*]
THR = HIGH via oral, inhal routes. MOD via dermal route. Animal exper show severe irr of skin. Repeated exposures caused depression of bone marrow in rats, rabbits, and dogs. An exper neo. [*3*]

N,N-DIGLYCIDYL-*p*-TOLUENE SULFONAMIDE. $C_{13}H_{17}O_4NS$, mw: 283.4.
THR = An exper neo. [*3*]

DIGLYCOL. See diethylene glycol.

DIGLYCOL CHLOROFORMATE. Liquid. $ClCOOCH_2CH_2OCH_2CH_2OH$, mw: 191.0, bp: 125° @ 5 mm, flash p: 295°F (OC), vap. d: 5.83.
THR = Details U. See also glycols.
Fire Hazard: Slight, when exposed to heat or flame.
Disaster Hazard: Dangerous; see chlorides; will react with water or steam to produce toxic and corrosive fumes; can react with oxidizing materials.
To Fight Fire: CO_2, dry chemical.

DIGLYCOLCHLOROHYDRIN. Colorless liquid. $ClCH_2CH_2OCH_2CH_2OH$, mw: 124.57, bp: 196.8°, flash p: 225°F (OC), d: 1.1698, vap. press: 0.17 mm @ 20°.
Acute tox data: Oral LD_{50} (rats) = 6300 mg/kg; dermal LD_{50} (rabbit) = 3000 mg/kg. [*3*]
THR = MOD via oral and dermal routes.
Fire Hazard: Slight, when exposed to heat or flame.
Disaster Hazard: Dangerous; see chlorides; will react with water or steam to produce toxic and corrosive fumes; can react with oxidizing materials.
To Fight Fire: Alcohol foam, CO_2, dry chemical.

DIGLYCOL DIABIETATE. See diethylene glycol diabietate (20% xylene).

DIGLYCOL DIACETATE. See diethylene glycol diacetate.

DIGLYCOL DILEVULINATE. See diethylene glycol dilevulinate.

DIGLYCOLIC ACID. White crystalline solid. $O(CH_2COOH)_2$, mw: 134.1, mp: 148°, bp: decomp.
THR = Details U. See also glycols and acetic acid.
Fire Hazard: Slight.

DIGLYCOL LAURATE. See diethylene glycol laurate.

DIGLYCOL MYRISTATE. See diethylene glycol myristate.

DIGOXIN. Syn: *lanoxin.* White crystalline powder. $C_{41}H_{64}O_{14}$, mw: 780.9, mp: 265°.
Acute tox data: Oral LD_{50} (cat) = 0.2 mg/kg; ipLD_{50} (rat) = 11 mg/kg. [*3*]
THR = HIGH via oral and ip routes. See also digitalis.
Fire Hazard: Slight.

DI-*n*-HEPTYLAMINE. Liquid. $(C_7H_{15})_2NH$, mw: 213.4, vap. d: 7.35.
THR = Details U. See also amines.
Fire Hazard: Mod, when exposed to heat or flame; can react with oxidizing materials.

DIHEXYL. See dodecane.

DIHEXYL ADIPATE. Syn: *hexyl adipate.* Liquid. $(CH_2CH_2CO_2C_6H_{12})_2$, mw: 314.5, bp: 168° −170° @ 4 mm, flash p: 325°F, d: 0.926 @ 20°/20°, vap. d: 10.85.
THR = Details U. See also esters.
Fire Hazard: Slight, when exposed to heat or flame; can react with oxidizing materials.
To Fight Fire: Foam, CO_2, dry chemical.

DI-*n*-HEXYLAMINE. Liquid. $(C_6H_{13})_2NH$, mw: 185.4, bp: 233–243°, flash p: 220°F (OC), d: 0.78, vap. d: 6.38.
Acute tox data: Oral LD_{50} (rat) = 380 mg/kg; dermal LD_{50} (rabbit) = 170 mg/kg. [*3*]
THR = HIGH via oral and dermal routes.
Fire Hazard: Low, when exposed to heat or flame; can react with oxidizing materials.
To Fight Fire: CO_2, dry chemical.

DIHEXYL ETHER. See hexyl ether.

DIHEXYL MALEATE. Syn: *di-n-hexyl maleate.* Liquid, sol in water less than 0.01% by weight @ 20°. $C_6H_{13}OOCCH{:}CHCOOC_6H_{13}$, mw: 274, d: 0.9602 @ 20°/20°, bp: 179° @ 10 mm, vap. press: <0.01 mm @ 20°, fp: 70°, flash p: 290°F (OC).
Acute tox data: Oral LD_{50} (rat) = 7340 mg/kg; dermal LD_{50} (rabbit) = 12,000 mg/kg. [*3*]
THR = MOD via oral and LOW via dermal routes. A combustible material. See also esters and maleic acid.
To Fight Fire: Dry chemical, CO_2, water, fog, mist.

DI-*n*-HEXYLMERCURY. Liquid. $Hg(C_6H_{13})_2$, mw: 370.9, bp: 158° @ 10 mm, d: 1.5361, vap. d: 12.8.
THR = HIGH. See mercury compounds, organic.
Fire Hazard: Mod, when exposed to heat or flame.
Disaster Hazard: Dangerous; see mercury; can react with oxidizing materials.

DIHEXYL PHTHALATE. Liquid. $C_6H_4(CO_2C_6H_{13})_2$, mw: 334.5, mp: −58°, bp: 210° @ 5 mm, flash p: 350°F, d: 0.995 @ 20°/20°, vap. d: 11.5.
Acute tox data: Oral LD_{50} (rat) = 30,000 mg/kg; dermal LD_{50} (rabbit) = 20,000 mg/kg. [*3*]
THR = VERY LOW via oral and dermal routes. See phthalic acid and esters.
Fire Hazard: Slight, when exposed to heat or flame; can react with oxidizing materials.
To Fight Fire: Foam, CO_2, dry chemical.

DIHEXYL SEBACATE. Liquid $[(CH_2)_4CO_2C_6H_{13}]_2$, mw: 370.6, bp: 184° @ 1 mm, flash p: 415°F, d: 0.911 @ 20°/20°, vap. d: 12.8.
THR = Details U. See also esters.
Fire Hazard: Slight, when exposed to heat or flame; can react with oxidizing materials.
To Fight Fire: Foam, CO_2, dry chemical.

DIHEXYL SODIUM SULFOSUCCINATE. Clear viscous liquid. $C_6H_{13}CO_2CH_2CH(SO_3Na)CO_2C_6H_{13}$, mw: 388.
THR = LOW irr and via oral route.

Fire Hazard: Mod, via heat, flame and oxidizers.
To Fight Fire: Water, spray, foam, mist, dry chemical.

DIHYDRAZINE SULFATE. White crystalline flakes. $(N_2H_4)_2 \cdot H_2SO_4$, mw: 162.2, mp: 104°, bp: 180° decomp.
THR = MOD acute via oral and inhal routes. Irr of eyes, skin and mu mem. See hydrazine.
Disaster Hazard: Dangerous; see sulfates.

DIHYDRAZINO ZINC CHLORATE. $Zn(NH_2)_4(ClO_3)_2$, mw: 296.4.
THR = Shock-sensitive explosive. [*19*]

1,2-DIHYDRO-5-ACENAPHTHYLENAMINE. $C_{12}H_{11}N$, mw: 169.2; sol. in ethanol; colorless needles, mp: 108°.
THR = An exper carc in rats via imp and ip routes. [*108*]

9,10-DIHYDROANTHRACENE. Crystals. $C_6H_4CH_2C_6H_4CH_2$, mw: 180.2, vap. d: 6.21.
THR = Details U; an insecticide.
Fire Hazard: Slight, when exposed to heat; can react with oxidizing materials.

5,6-DIHYDROBENZ(e)ACEANTHRYLENE. $C_{20}H_{14}$, mw: 254.3.
THR = An exper neo. [*3*]

1,2-DIHYDROBENZ(i)ACEANTHRYLENE.
THR = An exper carc. [*3*]

4,5-DIHYDROBENZ(k)ACEPHENANTHRYLENE. $C_{20}H_{14}$, mw: 254.3.
THR = An exper carc. [*3*]

6,13-DIHYDROBENZO(e,i)BENZO THIOPYRANO (4,3-b) INDOLE. $C_{19}H_{13}NS$, mw: 287.4.
THR = An exper carc. [*3*]

2,3-DIHYDRO-1H-BENZO(a)CYCLOPENT(h)ANTHRACENE. $C_{20}H_{12}$, mw: 252.3.
THR = An exper neo. [*3*]

6,12b-DIHYDRO CHOLANTHRENE. $C_{20}H_{17}$, mw: 257.4
THR = An exper carc. [*3, 23*]

7,14-DIHYDRO DIBENZ(a,b)ANTHRACENE. $C_{22}H_{16}$, mw: 280.4.
THR = An exper carc. [*3*]

5,6-DIHYDRO DIBENZ(a,h)ANTHRACENE.
THR = An exper carc. [*3*]

5,6-DIHYDRO DIBENZ(a,i)ANTHRACENE.
THR = An exper neo. [*3*]

12,13-DIHYDRO-7H-DIBENZO(a,g)CARBAZOLE. $C_{20}H_{15}N$, mw: 269.4.
THR = An exper neo. [*3*]

7,*p*-DIHYDRO-7,8-DIHYDROXY BENZO(a) PYRENE. $C_{20}H_{14}O_2$, mw: 286.2.
THR = An exper carc. to mice via dermal route. [*103*]

5,6-DIHYDRO-7,12-DIMETHYL BENZ(a) ANTHRACENE. $C_{20}H_{18}$, mw: 258.4.
THR = An exper carc. [*3*]

2,3-DIHYDRO-2,2-DIMETHYL-7-BENZOFURANYL METHYL CARBAMATE. See furadan.

3,4-DIHYDRO-1,11-DIMETHYL CHRYSENE. $C_{20}H_{18}$, mw: 258.4
THR = An exper neo. [*3*]

7,12-DIHYDRO-7,12-DIMETHYL-7,12-EPOXY BENZ(a)ANTHRACENE. $C_{20}H_{16}O$, mw: 272.4.
THR = An exper carc. [*3*]

7,14-DIHYDRO-7,11-DIPROPYL DIBENZ(a,h) ANTHRACENE-7,14-DIOL. $C_{28}H_{28}O_2$, mw: 296.6.
THR = An exper neo. [*3*]

11,12-DIHYDRO-11,12-EPOXY-3-METHYL CHOLANTHRENE. $C_{21}H_{18}O$, mw: 286.4.
THR = An exper carc. [*3*]

7,14-DIHYDRO-7,14-ETHANO DIBENZ(a,h) ANTHRACENE-15,16-DICARBOXYLIC ACID, SODIUM SALT. $C_{26}H_{16} \cdot 2NaO_4$, mw: 438.4.
THR = An exper carc. [*3*] via sc to mice.

1,2-DIHYDRO-6-ETHOXY-2,2,4-TRIMETHYL QUINOLINE. See 6-ethoxy-1,2-dihydro-2,2,4-trimethyl quinoline.

15,16-DIHYDRO-11-ETHYL CYCLOPENTA(a) PHENANTHREN-17-ONE. $C_{19}H_{16}O$, mw: 260.4.
THR = An exper neo. [*3*]

2,3-DIHYDRO-3-ETHYL-6-METHYL-1H-CYCLOPENTA(a)ANTHRACENE. $C_{20}H_{20}$, mw: 260.4.
THR = An exper carc. [*3*]

2,3-DIHYDRO-2-FORMYL-1,4-PYRAN. Liquid. $OHCCH(CH_2)_2CHCHO$, mw: 112.12, mp: −90°, bp: 150.6°, d: 1.0776 @ 20°/20°, vap. press: 2.4 mm @ 20°, vap. d: 3.87.
THR = Details U. See also dihydropyran.
Fire Hazard: Slight.

DIHYDROGEN POTASSIUM ARSENATE. White solid. KH_2AsO, mw: 180, mp: 288°.
THR = An exper carc. See arsenic. [*3*]

15,16-DIHYDRO-16-HYDROXY-11-METHYL-17H-CYCLOPENTA(a)PHENANTHREN-17-ONE. $C_{18}H_{14}O_2$, mw: 262.3.
THR = An exper carc. [*3*]

15,16-DIHYDRO-6-METHOXY-17H-CYCLOPENTA(a)PHENANTHREN-17-ONE. $C_{19}H_{16}O_2$, mw: 276.4.
THR = An exper neo. [*3*]

15,16-DIHYDRO-11-METHOXY-7-METHYL-17H-CYCLOPENTA(a)PHENANTHREN-17-ONE.

$C_{19}H_{16}O_2$, mw: 276.4.
THR = An exper carc. [3]

1,3-DIHYDRO-4-METHOXY-1-METHYL-2-OXO-NICOTINONITRILE. See ricinine.

9,10-DIHYDRO-7-METHYL BENZ(a)PYRENE. $C_{21}H_{16}$, mw: 268.4.
THR = An exper carc. [3]

6,12b-DIHYDRO-3-METHYL CHOLANTHRENE. $C_{21}H_4$, mw: 270.4.
THR = An exper carc. [3, 23]

11,12-DIHYDRO-3-METHYL CHOLANTHRENE.
THR = An exper carc. [3]

16,17-DIHYDRO-7-METHYL-15H-CYCLOPENTA (a)PHENANTHRENE. $C_{18}H_{16}$, mw: 232.3.
THR = An exper carc. [3]

16,17-DIHYDRO-11-METHYL-15H-CYCLOPENTA (a)PHENANTHRENE.
THR = An exper carc. [3]

15,16-DIHYDRO-11-METHYL-17H-CYCLOPENTA (a)PHENANTHRENE. $C_{18}H_{14}$, mw: 230.3.
THR = An exper neo. [3]

15,16-DIHYDRO-7-METHYL-17H-CYCLOPENTA (a)PHENANTHREN-17-ONE. $C_{18}H_{14}O$, mw: 246.3.
THR = An exper carc. [3]

15,16-DIHYDRO-11-METHYL-17H-CYCLOPENTA (a)PHENANTHREN-17-ONE.
THR = An exper carc. [3]

16,17-DIHYDRO-11-METHYL CYCLOPENTA(a) PHENANTHREN-5-ONE.
THR = An exper neo. [3]

16,17-DIHYDRO-17-METHYLENE-15H-CYCLO-PENTA(a)PHENANTHRENE. $C_{18}H_{14}$, mw: 230.3.
THR = An exper neo. [3]

4,5-DIHYDRONAPHTHA(1,2-k)ACE PHENAN-THRYLENE. $C_{24}H_{16}$, mw: 304.4.
THR = An exper neo. [3]

3,6-DIHYDRO-2-NITROSO-2H-1,2-OXAZINE. $C_4H_6O_2N_2$, mw: 114.1.
THR = An exper neo [3, 23] via oral route.

1,2-DIHYDRO-2-(5-NITRO-2-THIENYL)QUIN-AZOLIN-4-(3H)-ONE. $C_{12}H_9O_3N_3S$, mw: 275.3.
THR = An exper oral carc to rats. [103]

5,13-DIHYDRO-5-OXOBENZO(e,2)BENZO PY-RANO-(4,3-b)INDOLE. $C_{19}H_{11}O_2N$, mw: 285.3.
THR = An exper neo. [3]

DIHYDRO PENTABORANE. See pentaborane, unstable.

9,10-DIHYDRO-2-PHENANTHRAMINE.
THR = An exper carc. [23]

DIHYDROPYRAN. Colorless, mobile liquid, ethereal odor. C_5H_8O, mw: 84.1, bp: 85.6°, flash p: 0°F, d: 0.923 @ 20°/4°, vap. d: 2.90.
Acute tox data: ipLDLo (mouse) = 256 mg/kg. [3]
THR = MOD via oral and inhal routes.
Fire Hazard: Dangerous, when exposed to heat or flame; can react vigorously with oxidizing materials.
Explosion Hazard: U.
Disaster Hazard: Dangerous! Keep away from heat and open flame!
To Fight Fire: Alcohol foam, CO_2, dry chemical.

1,2-DIHYDRO-3,6-PYRIDIAZINEDIONE. See maleic hydrazide.

2,5-DIHYDROPYRROLE. See pyrroline.

DIHYDROROTENONE. $C_{23}H_{24}O_6$, mw: 396.4.
Acute tox data: Oral LD_{50} (rat) = 2500 mg/kg. [3]
THR = MOD via oral route. May be more toxic than rotenone. See also rotenone.

DIHYDRO SAFROLE. See 1,2-methylene dioxy-4-propyl benzene.

DIHYDRO STILBESTROL. $C_{18}H_{22}O_2$, mw: 270.4.
THR = An exper carc, neo. [3]

DIHYDRO STREPTOMYCIN. Syn: *DHS*.
$C_{21}H_{41}N_7O_{12}$, mw: 583.7.
Acute tox data: ivLD$_{50}$ (rat) = 200 mg/kg; imLD$_{50}$ (mouse) = 350 mg/kg. [3]
THR = HIGH via im and iv routes. A food additive permitted in food for human consumption. [109] A derivative of streptomycin; has anesthetic properties.

DIHYDRO STREPTOMYCIN SULFATE. White or practically white powder, odorless or slight odor, freely sol in water, very slightly sol in alcohol, practically insol in chloroform. $(C_{21}H_{41}N_7O_{12})_2 \cdot 3H_2SO_4$, mw: 1386.
Acute tox data: ipLD$_{50}$ (mouse) = 1380 mg/kg; ivLD$_{50}$ (mouse) = 164 mg/kg. [3]
THR = MOD via ip; HIGH via iv routes. A food additive permitted in the feed and drinking water of animals, and/or for the treatment of food-producing animals. [109]
Disaster Hazard: Dangerous; see sulfates.

DIHYDROTETRABORANE. Syns: *borobutane, boron hydride, tetraborondecahydride, tetraborane.* Colorless gas, repulsive odor. B_4H_{10}, mw: 53.4, mp: −120°, bp: 18°, d: 0.59 (liquid @ −70°), 0.56 (liquid @ −35°), vap. press: 580 mm @ 6°, vap. d: 1.8.
THR = See boron hydrides.

DIHYDRO-5-TETRADECYL-2-(3H)-FURANONE.
$C_{18}H_{34}O_2$, mw: 282.5.
THR = An exper neo. [3]

5,6-DIHYDRO-2,3,6-TRIMETHYL-1,3,5-DITHIAZINE. See thialdine.

1,3-DIHYDRO-1,4,6-TRIMETHYL-2-OXO-NICOTINONITRILE.
THR = Probably HIGH toxicity. See also nitriles and nicotine.
Disaster Hazard: See cyanides.
Fire Hazard: Slight.

DIHYDROXYACETONE. Syn: *man-tan.* Crystalline powder, hygroscopic, characteristic odor, sweet cooling taste. Very water-sol, sol in alcohol, ether, and acetone. $HO \cdot CH_2COCH_2OH$, mw: 90.1, mp: 75°–80°.
Acute tox data: $ipLD_{LO}$ (mouse) = 1000 mg/kg. [*3*]
THR = MOD via ip route. Has been reported as causing dermatitis in man.

1,8-DIHYDROXYANTHRANOL. See anthralin.

3,4-DIHYDROXYANTHRANOL. See anthrarobin.

1,2-DIHYDROXYANTHRAQUINONE. See alizarin.

1,4-DIHYDROXYANTHRAQUINONE. Crystals. $C_{14}H_8O_4$, mw: 240.2, mp: 194°, bp: 450.0°, vap. press: 1 mm @ 196.7°, vap. d: 8.3.
Acute tox data: $ipLD_{LO}$ (mouse) = 500 mg/kg. [*3*]
THR = MOD via ip route. A weak allergen.
Fire Hazard: Slight.

1,5-DIHYDROXYANTHRAQUINONE. Green to yellow crystals. $C_{14}H_8O_4$, mw: 240.2, mp: 280°, bp: sublimes, vap. d: 8.3.
THR = U. A weak allergen.
Fire Hazard: Slight.

1,8-DIHYDROXYANTHRAQUINONE. Crystals. $C_{14}H_8O_4$, mw: 240.2, mp: 193°, vap. d: 8.3.
THR = U. A weak allergen.
Fire Hazard: Slight.

1,8-DIHYDROXY-9-ANTHRONE.
THR = An exper carc. [*23*]

***o*-DIHYDROXY BENZENE.** See pyrocatechol.

***p*-DIHYDROXY BENZENE.** See hydroquinone.

3,3′-DIHYDROXY BENZIDINE.
THR = An exper carc. [*23*]

DIHYDROXYBENZYL DIMETHYL METHANE. Flake material. $HOC_6H_4CH_2C(CH_3)_2CH_2C_6H_4OH$, mw: 256.3, mp: 152.5°, bp: 220° @ 4 mm.
THR = Details U.
Fire Hazard: Slight.

2,6-DIHYDROXY-4-CARBOXYPYRIDINE. See citrazinic acid.

3,4-DIHYDROXYCINNAMIC ACID. See caffeic acid.

7,8-DIHYDROXYCOUMARIN-7-β-D-GLUCOSIDE. Syn: *daphnin.* Crystals. $C_{15}H_{16}O_9$, mw: 340.3.
THR = Details U. See also warfarin.

DIHYDROXYDIAMINOMERCUROBENZENE. $OHNH_2C_6H_3HgC_6H_3OHNH_2$, mw: 416.9, vap. d: 14.4.
THR = See mercury compounds, organic.

2,2′-DIHYDROXY-5,5′-DICHLORODIPHENYLMETHANE. See 2,2′-methylene bis (4-chlorophenol).

2,4-DIHYDROXY-3,3-DIMETHYL BUTYRONITRILE. $NCCHOHC(CH_3)_2CH_2OH$, mw: 129.
THR = U. Limited animal exper suggest mod toxicity. See also nitriles.

DI(2-HYDROXY ETHYL)AMINE.
See diethanolamine.

N,N′-DIHYDROXYETHYLETHYLENE DIAMINE. Crystals. $HOC_2H_4NHCH_2CH_2NHC_2H_4OH$, mw: 148.2, mp: 98.1, bp: 196° @ 10 mm, flash p: 355°F, vap. d: 5.1.
THR = Details U. See also amines.
Fire Hazard: Slight, when exposed to heat or flame; can react with oxidizing materials.
To Fight Fire: Foam, CO_2, dry chemical.

DIHYDROXYETHYL NITRAMINE DINITRATE.
THR = Details U. See also tetryl.
Fire Hazard: See nitrates.
Explosion Hazard: Severe, when shocked or exposed to heat.
Disaster Hazard: Dangerous; shock will explode it; when heated to decomp, emits highly toxic fumes of oxides of nitrogen; can react vigorously with reducing materials.

1,4-DI(2-HYDROXYETHYL)PIPERAZONE. $N(C_2H_4OH)CH_2CH_2N(C_2H_4OH)CH_2CH_2$, mw: 174.
Acute tox data: Oral LD_{50} (rat) = 3730 mg/kg. [*3*]
THR = MOD via oral route. See also piperazine.

DI(HYDROXYETHYL)-*o*-TOLYLAMINE. Syn: *Bis-(hydroxyethyl)-o-tolylamine.*
Acute tox data: Oral LD_{LO} (rat) = 2200 mg/kg; dermal LD_{LO} (rabbit) = 1000 mg/kg. [*3*]
THR = MOD via oral and dermal routes. See also amines.

2,2′-DIHYDROXY-3,5,6,3′,5′,6′-HEXACHLORODIPHENYLMETHANE. See hexachlorophene.

1,6-DIHYDROXYHEXANE. See 1,6-hexanediol.

4′,7′-DIHYDROXY ISO FLAVONE.
THR = An exper carc. [*23*]

2,2′-DIHYDROXYISOPROPYL ETHER. See dipropylene glycol.

1,8-DIHYDROXY-3-METHYLANTHRAQUINONE. See chrysophanic acid.

7,12-DIHYDROXY METHYL BENZ(a)ANTHRACENE.
THR = An exper carc. [*23*]

cis-1,2-DIHYDROXY-3-METHYL CHOLANTHREN. $C_{21}H_{16}O_2$, mw: 300.4.
THR = An exper carc. [*3*]

1,3-DIHYDROXY NAPHTHALENE. See naphtha resorcinol.

2,2′-DIHYDROXY-N-NITROSO DIPROPYL AMINE. $C_6H_{14}O_3N_2$, mw: 186.2.
THR = An exper carc [*3*, *23*] via sc route.

DIHYDROXY OCTACHLORODIPHENYL. $C_{12}H_2O_2Cl_8$, mw: 461.8, vap. d: 15.9.
Acute tox data: Oral LD_{50} (rat) = 360 mg/kg. [*3*]
THR = HIGH via oral and inhal routes.
Disaster Hazard: Dangerous; when heated to decomp, emits toxic fumes of chlorides.

1,2-DIHYDROXYPROPANE. See propylene glycol.

4-(2,3-DIHYDROXY PROPYLAMINO)-2-(5-NITRO-2-THIENYL)QUINAZOLINE. $C_{15}H_{14}O_4N_4S$, mw: 346.4.
THR = An exper oral carc to rats. [*103*]

DIIMINO ISOINDOLINE. $C_8H_7N_3$, mw: 145.2.
THR = An exper neo and carc. [*3*]

DIIODOACETYLENE. Syn: *diiodoethyne.* Colorless, rhombic crystals, unpleasant odor. IC:CI, mw: 277.86, mp: 78°–82°, bp: 80°–100° decomp, vap. d: 9.6.
THR = HIGH irr to skin, eyes and mu mem and via oral, inhal routes. A military poison.
Explosion Hazard: Details U; many acetylene compounds are unstable.
Disaster Hazard: Dangerous; when heated to decomp, emits toxic fumes of iodine.

DIIODODIETHYL SULFIDE. Bright yellow prisms. $(CH_2CH_2I)_2S$, mw: 215.08, mp: 62°, vap. d: 7.41.
THR = HIGH irr and via oral and inhal routes.
Disaster Hazard: Dangerous; when heated to decomp or on contact with acid or acid fumes, emits highly toxic fumes of iodides, oxides of sulfur.

1,1-DIIODOETHANE. See ethylidine diiodide.

DIIODOETHYNE. See diiodoacetylene.

DIIODOFORMOXIME. Crystals. $CNHOI_2$, mw: 296.9, mp: 69°, vap. d: 10.2.
THR = HIGH irr to skin, eyes and mu mem and via oral and inhal routes.
Disaster Hazard: Dangerous; see iodides and NO_x.

1,6-DIIODO-2,4-HEXADIYNE. C_6H_4I, mw: 203.
THR = Shock-sensitive explosive. [*19*]

DIIODOMETHANE. See methylene iodide.

DIIODO-8-QUINOLINOL. THR = A (S) carc. [*14*] See also 8-hydroxy quinoline.

3,5-DIIODO SALICYLIC ACID. White to pale pink crystalline powder, slightly sol in water. $I_2C_6H_2(OH)COOH$, mw: 390.
THR = U. A trace mineral added to animal feeds. [*109*]

DIIRON PHOSPHIDE. Blue-gray crystals or powder. Fe_2P, mw: 142.68, mp: 1290°, d: 6.56.
THR = See phosphides.

DIISOAMYLAMINE. Syn: *isodiamylamine.* Liquid, sol in water, alcohol, ether, chloroform. $HNCH_2CH_2CH(CH_3)_2$, mw: 193.76, d: 0.767 @ 20°/4°, bp: 188°, mp: −44°.
THR = Probably MOD irr via oral and inhal routes. May cause flushing and rise in blood pressure.

DI-*sec*-ISOAMYLBORANE. See bis (1,2-dimethylpropyl) borine.

DIISOAMYLCADMIUM. An oil. $(C_5H_{11})_2Cd$, mw: 254.7, d: 1.2209, mp: −115°, bp: 121.5° @ 15 mm.
THR = Probably HIGH toxicity. Details U. See cadmium compounds.
Fire Hazard: Details U. Probably dangerous.
Disaster Hazard: Dangerous; see cadmium.

DIISOAMYLMERCURY. Liquid. $Hg(C_5H_{11})_2$, mw: 342.9, bp: 125° @ 10 mm, d: 1.6397, vap. d: 11.8.
THR = See mercury compounds, organic. HIGH toxicity.

DIISOAMYLOXYBORON CHLORIDE. Colorless liquid which decomp upon contact with moisture. $(C_5H_{11}O)_2BC1$, mw: 188.6, bp: 110° −115° @ 14 mm.
THR = Details U. See boron compounds.
Disaster Hazard: Dangerous; see chlorides.

DIISOAMYLTIN DIBROMIDE. Liquid. $(C_5H_{11})_2SnBr_2$, mw: 420.8, mp: −24°.
THR = Details U. See tin compounds and bromides.
Disaster Hazard: See bromides.

DIISOAMYLTIN DICHLORIDE. Solid. $(C_5H_{11})_2SnCl_2$, mw: 331.9, mp: 28°.
THR = Details U. See tin compounds.
Disaster Hazard: See chlorides.

DIISOAMYLTIN DIIODIDE. Oily liquid. $(C_5H_{11})_2SnI_2$, mw: 514.8, bp: 202°–205° @ 8 mm.
THR = Details U. See tin and iodides.
Disaster Hazard: Dangerous; see iodides.

DIISOAMYL ZINC. $C_{10}H_{22}Zn$, mw: 207.6.
THR = Self-ignites in air. [*19*]

DIISOBUTYL ALUMINUM CHLORIDE. Syn: *DIBAC.* Colorless liquid. $[(CH_3)_2CHCH_2]_2A1C1$, mw: 176.5, d: 0.905, fp: 39.5°.
THR = HIGH irr to skin, eyes and mu mem and via oral and inhal routes. A strong irr and corrosive to skin and lungs.

DIISOBUTYL ALUMINUM HYDRIDE. Syn: *DIBAL-H.* Colorless, pyrophoric liquid, miscible in hydrocarbon solvents. $[(CH_3)_2CHCH_2]_2AlH$, mw: 142, fp: −80°, d: 0.798, bp: 105° @ 2 mm.

THR = U. See also hydrides.

Fire Hazard: Dangerous; ignites spont in air.

To Fight Fire: Do *not* use water, foam, or halogenated extinguishing agents.

DIISOBUTYLAMINE. Water white liquid, amine odor. $[(CH_3)_2CHCH_2]_2NH$, mw: 129.24, mp: −70°, bp: 139°, flash p: 85°F, d: 0.745 @ 20°/4°, vap. press: 10 mm @ 30.6°, vap. d: 4.46.

Acute tox data: Oral LD_{50} (rat) = 258 mg/kg. [*3*]

THR = HIGH via oral route. See amines.

Fire Hazard: Dangerous, when exposed to heat or flame; can react vigorously with oxidizing materials.

Explosion Hazard: U.

To Fight Fire: Alcohol foam, CO_2, dry chemical.

DIISOBUTYLCADMIUM. An oil, decomp by water. $(C_4H_9)_2Cd$, mw: 226.6, d: 1.2690, mp: −37°, bp: 90.5° @ 20 mm.

THR = Probably HIGH toxicity. Details U. See cadmium compounds.

Fire Hazard: Details U. Probably dangerous.

Disaster Hazard: Dangerous; see cadmium.

DIISOBUTYL CARBINOL. Syns: *nonyl alcohol, 2,6-dimethyl heptanol-4.* Colorless liquid. $(CH_3)_2CHCH_2CHOHCH_2CH(CH_3)_2$, mw: 144.3, bp: 173.3°, fp: −65°, flash p: 165°F, d: 0.8121 @ 20°/20°, vap. press: 0.3 mm @ 20°, vap. d: 4.98, lel = 0.8% @ 212°F, uel = 6.1% @ 212°F.

Acute tox data: Oral LD_{50} (rat) = 3560 mg/kg; dermal LD_{50} (rabbit) = 5660 mg/kg. [*3*]

THR = MOD via oral and dermal routes. Limited animal exper suggest MOD toxicity. Has caused CNS and liver injury in animals. See also alcohols.

Fire Hazard: Mod, when exposed to heat or flame; can react with oxidizing materials.

To Fight Fire: Alcohol foam, foam, CO_2, dry chemical.

DIISOBUTYLENE. A colorless, liquid, hydrocarbon, C_8H_{16}, mw: 112.2, mp: −101°, bp: 102°, flash p: $<$20°F (CC), d: 0.7227 @ 15.6°, vap. d: 3.97, autoign. temp.: 779°F.

THR = MOD irr via oral and inhal routes. Irr and narcotic in high conc. Has caused liver and kidney damage in exper animals.

Fire Hazard: Dangerous, when exposed to heat or flame; can react vigorously with oxidizing materials such as oleum, chlorosulfonic acid, H_2SO_4. [*19*]

Disaster Hazard: Dangerous! Keep away from heat and open flame!

Explosion Hazard: U.

To Fight Fire: Foam, CO_2, dry chemical.

DIISOBUTYLENE OXIDE. Syn: *(2,4,4-trimethyl pentene-1)oxide.* $C_8H_{16}O$, mw: 128.2.

Acute tox data: Oral LD_{50} (rat) = 4920 mg/kg; inhal LD_{50} (rat) = 4000 ppm for 4 hrs. [*3*]

THR = MOD via oral and inhal routes.

DIISOBUTYL FUMARATE. $C_{12}H_{20}O_4$, mw: 228.3.

Acute tox data: Oral LD_{50} (rat) = 8120 mg/kg; dermal LD_{50} (rabbit) = 7490 mg/kg. [*3*]

THR = LOW via oral and dermal routes. See also esters and fumaric acid.

DIISOBUTYL KETONE. Syns: *isovalerone, 2,6-dimethyl-4-heptanone.* Liquid. $[(CH_3)_2CHCH_2]_2CO$, mw: 142.2, bp: 166°, flash p: 140°F, d: 0.81, vap. d: 4.9, lel = 0.8% @ 212°F, uel = 6.2% @ 212°F.

Acute tox data: Oral LD_{50} (mice) = 1416 mg/kg; inhal LC_{LO} (rat) = 2000 ppm for 4 hrs; dermal LD_{50} (rabbit) = 20,000 mg/kg. [*3*]

THR = MOD via oral and inhal routes; LOW via dermal route. A mild irr. Narcotic in high conc.

Fire Hazard: Mod, when exposed to heat or flame; can react with oxidizing materials.

To Fight Fire: CO_2, dry chemical, water spray, mist or fog.

DIISOBUTYLMERCURY. Colorless liquid. $Hg(C_4H_9)_2$, mw: 314.8, mp: volatile @ 100°, bp: 207°, d: 1.7678, vap. d: 10.8.

THR = HIGH. See mercury compounds, organic.

DIISOBUTYL PHENOL. See octyl phenol.

DIISOBUTYL PHTHALATE. Liquid. $C_6H_4[COOCH_2CH(CH_3)_2]_2$, mw: 278, mp: −64°, flash p: 385°F, d: 1.039–1.043, vap. d: 9.59.

Acute tox data: Oral LD_{50} (mouse) = 13000 mg/kg; dermal LD_{50} (guinea pig) = 10,000 mg/kg. [*3*]

THR = LOW via oral and dermal routes. See phthalic acid and esters. An exper teratogen. [*3*]

Fire Hazard: Slight.

To Fight Fire: Foam, CO_2, dry chemical.

DIISOBUTYL SODIUM SULFOSUCCINATE. Syn: *aerosol IB, alphasol IB, dibutyl sodium sulfosuccinate.* Available as 3 esters of sulfosuccinic acid. The mixture is white, powder-like, sol in water, glycerol, pine oil and oleic acid, insol in acetone, kerosene, liquid petrolatum, carbon tetrachloride, ethanol, benzene and olive oil. $C_{12}H_{21}NaO_7S$, mw: 332.35.

THR = MOD via oral and inhal routes. An irr.

Disaster Hazard: Dangerous; see sulfates.

DIISOBUTYLTIN DIIODIDE. Solid. $(C_4H_9)_2SnI_2$, mw: 486.8, bp: 290°–295°.

THR = Details U. See tin compounds.

Disaster Hazard: Dangerous; see iodides.

DIISOBUTYL ZINC. $C_8H_{18}Zn$, mw: 179.6.

THR = Self-ignites in air. [*19*]

***p,p'*-DIISOCYANATODIPHENYL METHANE.** Crystals. $OCNC_6H_4CH_2C_6H_4NCO$, mw: 250.3, bp:

194°–199° @ 5 mm, fp: 37.2°, d: 1.19 @ 50°, vap. d: 8.63.
THR = MOD via oral route; an irr.
Disaster Hazard: Dangerous; see cyanides.

2,4-DIISOCYANOTOLUENE.
See 2,4-tolylene diisocyanate.

DIISODECYL ADIPATE. Clear, mobile, oily liquid. $(CH_2CH_2CO_2C_{10}H_{21})_2$, mw: 426, bp: 240° @ 4 mm, flash p: 225°F (OC), vap. d: 14.7.
THR = Details U. See also esters.
Fire Hazard: Slight, when exposed to heat or flame; can react with oxidizing materials.
To Fight Fire: CO_2, dry chemical.

DI-(ISODECYL)-4,5-EPOXYCYCLOHEXANE-1,2-DICARBOXYLATE.
THR = LOW via all routes.

DIISODECYL PHTHALATE. Clear, oily liquid, mild characteristic odor. $(C_{10}H_{21}CO_2)_2C_6H_4$, mw: 446.7, flash p: 450°F, d: 0.966 @ 25°/25°, vap. d: 15.4.
THR = MOD via oral route; an irr.
Fire Hazard: Slight, can react with oxidizing materials.

DIISOOCTYL ACID PHOSPHATE. A corrosive liquid. $(C_8H_{17})_2HPO_4$, mw: 322.
THR = MOD irr to skin, eyes and mu mem.
Disaster Hazard: Dangerous; see phosphorous compounds.

DIISOOCTYL ADIPATE. Liquid.
$[(CH_2)_2CO_2C_8H_{17}]_2$, mw: 370.6, bp: 207°–213° @ 4 mm, flash p: 370°F, d: 0.930 @ 20°/20°, vap. d: 12.8.
THR = See esters.
Fire Hazard: Slight, when exposed to heat or flame; can react with oxidizing materials.
To Fight Fire: Foam, CO_2, dry chemical.

DIISOOCTYL AZELATE. Liquid.
$(CH_2)_7[CO_2(CH_2)_5CH(CH_3)_2]_2$, mw: 412, bp: 235° @ 5 mm, mp: <−59.5°, flash p: 415°F, d: 0.92 @ 20°, vap. d: 18.7.
THR = Details U. See also esters.
Fire Hazard: Slight, when exposed to heat or flame; can react with oxidizing materials.
To Fight Fire: Foam, CO_2, dry chemical.

DIISOOCTYL PHENYL PHOSPHONATE. Liquid. $C_6H_5PO(OC_8H_{17})_2$, mw: 382, mp: <0°, bp: 180°–190° @ 1 mm, d: 0.973 @ 25°/25°, vap. d: 13.2.
THR = See esters, phosphates.

DIISOOCTYL PHTHALATE. Liquid, mild odor. $C_6H_4(CO_2C_8H_{17})_2$, mw: 390.5, bp: 230–240° @ 4 mm, fp: −45°, flash p: 450°F, d: 0.986 @ 20°/20°, vap. d: 13.5.
THR = Details U. See phthalic acid and esters.
Fire Hazard: Slight, when exposed to heat or flame; can react with oxidizing materials.
To Fight Fire: CO_2, dry chemical.

DIISOOCTYL SEBACATE. Liquid.
$[(CH_2)_4CO_2C_8H_{17}]_2$, mw: 426.7, bp: 248°–255° @ 4 mm, flash p: 470°F, d: 0.917 @ 20°/20°, vap. d: 14.7.
THR = Details U. See esters.
Fire Hazard: Slight, when exposed to heat or flame; can react with oxidizing materials.
To Fight Fire: Foam, CO_2, dry chemical.

DIISOOCTYL STYRYLPHOSPHONATE. Liquid. $C_6H_5C_2H_2PO(OC_8H_{17})_2$, mw: 408, mp: <0°, bp: 185–190° @ 1 mm, d: 0.977 @ 25°/25°.
THR = See esters and phosphates.

DIISOPROPANOLAMINE. $[CH_3CH(OH)CH_2]_2NH$, mw: 133.19, mp: 42°, bp: 249°, flash p: 260°F (OC), d: 0.9890 @ 45°/20°, vap. d: 4.59.
Acute tox data: Oral LD_{50} (rat) = 6720 mg/kg. [*3*]
THR = LOW to MOD via oral route.
Fire Hazard: Slight, when exposed to heat or flame; can react with oxidizing materials.
To Fight Fire: Alcohol foam, CO_2, dry chemical.

DIISOPROPYL. See 2,3-dimethylbutane.

DIISOPROPYLAMINE. Colorless liquid.
$[(CH_3)_2CH]_2NH$, mw: 101.19, bp: 83–84°, flash p: 30°F (OC), d: 0.722 @ 220.0°, vap. d: 3.5.
Acute tox data: Oral LD_{50} (rat) = 700 mg/kg; inhal LC_{LO} (rat) = 1000 ppm for 4 hrs. [*3*]
THR = MOD to HIGH via oral and inhal routes. Inhal of fumes can cause pulmonary edema. See also amines.
Fire Hazard: Dangerous, when exposed to heat or flame; can react vigorously with oxidizing materials.
Explosion Hazard: U.
Disaster Hazard: Dangerous! Keep away from heat and open flame!
To Fight Fire: Alcohol foam, foam, CO_2, dry chemical.

DIISOPROPYLBENZENE. Clear, colorless liquid. $[(CH_3)_2CH]_2C_6H_4$, mw: 162.26, mp: <−55°, bp: 205°, flash p: 170°F (OC), d: 0.863–0.867 @ 25°/25°, autoign, temp.: 840°F, vap. d: 5.6.
Acute tox data: Oral LD_{LO} (rat) = 5000 mg/kg. [*3*]
THR = LOW to MOD via oral route. See also cumene.
Fire Hazard: Mod, when exposed to heat or flame; can react with oxidizing materials.
To Fight Fire: Foam, CO_2, dry chemical, water spray or mist.

DIISOPROPYL BENZENE HYDROPEROXIDE.
Colorless to pale yellow liquid.
THR = Probably toxic. See also peroxides, organic.
Fire Hazard: A powerful oxidizer.

DIISOPROPYL BERYLLIUM. See beryllium diisopropyl.

DIISOPROPYL CARBINOL. Syn: *2,4-dimethyl-3-pentanol.* Colorless liquid. $(C_3H_7)_2CHOH$, mw: 116.2, mp: −70°, bp: 140°, flash p: 120°F, d: 0.8288 @ 20°/4°, vap. d: 4.0.
THR = See alcohols.
Fire Hazard: Mod, when exposed to heat or flame; can react with oxidizing materials.
To Fight Fire: Foam, CO_2, dry chemical.

DIISOPROPYL CYANAMIDE. Colorless, mobile liquid, characteristic odor. $(C_3H_7)_2NCN$, mw: 126.20, mp: −27.3°, bp: 207°, flash p: 179.5°F, d: 0.8451 @ 30°, vap. press: 9 mm @ 80°, vap. d: 4.34.
THR = Probably HIGH irr to skin, eyes and mu mem. See also cyanides.
Fire Hazard: Mod, when exposed to heat, flame or oxidizers.
Disaster Hazard: Dangerous; when heated to decomp or on contact with acid or acid fumes, emits highly toxic fumes of cyanides; will react with water or steam to produce toxic and corrosive fumes; can react with oxidizing materials.
To Fight Fire: Foam, CO_2, dry chemical.

DIISOPROPYL DIXANTHOGEN. Yellow to greenish pellets. $(C_4H_7OS_2)_2$, mw: 270.5, mp: 52°, d: 1.28, vap. d: 9.35.
THR = No data. Probably HIGH irr to skin, eyes and mu mem.
Fire Hazard: Mod, when exposed to heat or flame.
To Fight Fire: Water spray, mist, foam, dry chemical.
Disaster Hazard: Dangerous; when heated to decomp or on contact with acid or acid fumes, emits highly toxic fumes of oxides of sulfur; will react with water or steam to produce toxic fumes; can react with oxidizing materials.

N,N-DIISOPROPYL ETHANOLAMINE. Syn: *n,n-diisopropylamino ethanol.* Colorless liquid, slightly sol in water. $[(CH_3)_2CH]_2NC_2H_4OH$, mw: 145, d: 0.8742 @ 20°; vap. press: 0.08 mm @ 20°, fp: −39.3°, bp: 191°, flash p: 175°F (OC).
Acute tox data: Oral LD_{50} (rat) = 1070 mg/kg; dermal LD_{50} (rabbit) = 450 mg/kg. [*3*]
THR = HIGH via dermal and MOD via oral routes. See also amines.
Fire Hazard: Mod, when exposed to heat, flame or oxidizers.
To Fight Fire: Dry chemical, CO_2.

DIISOPROPYL ETHER. See isopropyl ether.

DIISOPROPYL FLUOROPHOSPHATE. Oily liquid. $C_6H_{14}FPO_2$, mw: 208.17, mp: −82°, bp: 46° @ 5 mm, d: 1.07 (approx), vap. d: 5.24.
Acute tox data: Oral LD_{50} (rat) = 6 mg/kg; inhal LD_{50} (rat) = 360 mg/m^3 for 10 min; dermal LD_{50} (mouse) = 72 mg/kg. [*3*]
THR = HIGH via oral, inhal and dermal routes. Used in Germany as a basis for "nerve gases." An insecticide. Ingestion can cause damage to eyes, nausea, vomiting, diarrhea and CNS disturbances. See parathion. An exper carc. [*23*]
Disaster Hazard: Dangerous; see fluorides and phosphorus compounds.

DIISOPROPYL MALEATE. Insol in water. $(CH_3)_2CHOCOCH:COOCH(CH_3)_2$, mw: 187, d: 1.0 +, bp: 444°F, flash p: 220°F (OC).
THR = U. See esters.
Fire Hazard: Low, when exposed to heat or flame.
To Fight Fire: Alcohol foam.

DIISOPROPYLMERCURY. Liquid. $Hg(C_3H_7)_2$, mw: 286.8, bp: 63° @ 10 mm, d: 2.0024, vap. d: 9.9.
THR = HIGH tox. See mercury compounds, organic.

DIISOPROPYL PEROXYDICARBONATE. Syns: *isopropyl percarbonate, isopropyl peroxydicarbonate, IPP.* Colorless, crystalline solid, almost insol in water, miscible with aliphatic and aromatic hydrocarbons, esters, ethers, chlorinated hydrocarbons. $(CH_3)_2CHOCOOCOOCH(CH_3)_2$, mw: 120, rapid decomp @ 63°F, mp: 8°–10°, d: 1.080 @ 15.5°/4°.
Acute tox data: Oral LD_{50} (rat) = 2140 mg/kg; dermal LD_{50} (rabbit) = 2025 mg/kg. [*3*]
THR = MOD via oral and dermal routes. See peroxides, organic.
Explosion Hazard: Dangerous; explodes on heating. Is unstable >10°. DO NOT crystallize. [*19*] Explodes with organic matter.
Fire Hazard: Very dangerous:

2,6-DIISOPROPYL PHENOL. A colorless liquid or solid. $HC(CH_3)_2C_6H_3(OH)(CH_3)_2CH$, mw: 178.3, bp: 242.4°, fp: 17.9°, flash p: 235°F (CC), d: 0.955 @ 20°/4°.
THR = U. See also phenol.
Fire Hazard: Slight, when exposed to heat or flame; can react with oxidizing materials.
To Fight Fire: Foam, CO_2, dry chemical.

DIISOPROPYL PHOSPHOFLUORIDATE. See diisopropyl fluorophosphate.

DIISOPROPYL SULFATE. $C_6H_{14}O_4S$, mw: 182.3.
Acute tox data: Oral LD_{50} (rat) = 1090 mg/kg; dermal LD_{50} (rabbit) = 1410 mg/kg. [*3*]
THR = MOD via oral and dermal routes. An exper carc. [*3*]

DIISOPROPYLTIN DIBROMIDE. Pale yellow crystals, decomp by water. $(C_3H_7)_2SnBr_2$, mw: 364.7, mp: 54°.

THR = Details U. See tin compounds.
Disaster Hazard: Dangerous; see bromides.

DIISOPROPYLTIN DICHLORIDE. Colorless crystals, sol in water. $(C_3H_7)_2SnCl_2$, 275.8, mp: 84°.
Acute tox data: ivLD$_{50}$ (rat) = 15 mg/kg. [*3*]
THR = HIGH via iv route. See tin compounds.
Disaster Hazard: Dangerous; see chlorides.

DIISOPROPYLTIN OXIDE. Solid, insol in water. $(C_3H_7)_2SnO$, mw: 220.9.
THR = Details U. See tin compounds.

DIKERYL BENZENE-12. Liquid. flash p: 320°F.
THR = Details U. Probably irr and narcotic.
Fire Hazard: Slight, when exposed to heat or flame; can react with oxidizing materials.
To Fight Fire: Foam, CO_2, dry chemical.

DIKETENE. Syn: *acetyl ketene*. Colorless, nonhygroscopic liquid, pungent odor, decomp in water. $C_4H_4O_2$, mw: 84.07, mp: −6.5°, bp: 127.4°, d: 1.0897, vap. d: 2.9, flash p: 93°F (TOC).
Acute tox data: Oral LD$_{50}$ (rat) = 560 mg/kg. [*3*]
THR = MOD to HIGH via oral route. An irr.
Fire Hazard: Mod, when exposed to heat or flame; can react with oxidizing materials.
To Fight Fire: Alcohol foam.

DILAN. Syns: *Prolan, Bulan.* Dilan is a mixture of 2-nitro-1,1-bis(*p*-chlorophenyl)propane (1 part) and 2-nitro-1,1-bis(*p*-chlorophenyl)butane (2 parts). The first of the two components is known as Prolan, the second as Bulan. The former is not to be confused with the pituitary hormone of the same name.
Acute tox data: The acute dose in man is U. The LD$_{50}$ to rats by oral administration about 475 mg/kg for Dilan as compared to 135 mg/kg for DDT and about 3000 mg/kg for methoxychlor. In rats it has been found that neither Prolan nor Bulan is irr to the skin. The LD$_{50}$ by the dermal route for each is 74000 mg/kg. The dermal LD$_{50}$ for Dilan in white rats is 5900 mg/kg for females, and 6900 mg/kg for males.
THR = HIGH via oral and MOD via dermal routes. Exact physiologic action of this compound is U. Toxicity tests in lower animals indicate that it can be absorbed from the digestive tract.
Signs and Symptoms: No signs and symptoms are as yet known in man. In animals they simulate those caused by other chlorinated hydrocarbons.
Disaster Hazard: Dangerous; see chlorides and nitrates.
Treatment of Poisoning: The treatment of exposed individuals should be directed along similar lines to those recommended for chlorinated hydrocarbons. [*94*]

DILANTIN. See diphenyl hydantoin.

DILANTIN SODIUM. See diphenyl hydantoin sodium.

DILAURYL THIODIPROPIONATE. Syns: *didodecyl-3,3'-thiodipropionate, thiodipropionic acid, dilauryl ester.* White flakes, sweet ester odor, sol in most organic solvents. $(C_{12}H_{25}OOCCH_2CH_2)_2S$, mw: 504, d (solid, 25°): 0.975, mp: 40°.
THR = U. Used as a chemical preservative food additive. [*109*]
Disaster Hazard: Dangerous; see sulfur compounds.

DIMAGNESIUM PHOSPHATE. See magnesium phosphate, dibasic.

DIMAGNESIUM-*o*-PHOSPHATE. See magnesium phosphate, dibasic.

DIMANGANESE ARSENIDE. Solid. Mn_2As, 184.77, mp: 1400°.
THR = HIGH Toxicity. See arsenic compounds and manganese compounds.
Fire Hazard: Dangerous; when exposed to heat or by chemical reaction.
Explosion Hazard: See arsine.
Disaster Hazard: Dangerous; see arsenic; will react with water or acid to produce arsine; can react vigorously with oxidizing materials.

DIMAZINE. See syn-dimethyl hydrazine.

DIMECRON. See 2-chloro-2-diethyl carbamoyl-1-methylvinyl dimethyl phosphate.

DIMEFOX. See bis(dimethylamino) fluorophosphine oxide.

DIMELONE. See dimethyl carbate.

1,3-DIMERCAPTOPROPANE.
See 1,3-propanedithiol.

2,3-DIMERCAPTO-1-PROPANOL. Syns: *BAL, british anti-lewisite.* Viscous, oily liquid, pungent odor. $C_3H_8OS_2$, mw: 124.2, bp: 140° @ 40 mm, vap. d: 4.3, d: 1.2385 @ 25°/4°.
THR = Applied locally to the skin it causes redness and swelling, but does not produce blisters or ulcers. It is intensely irr to the eyes and mu mem. Systemic symptoms are caused by injection of the drug and usually occur within a few min, reaching their max in 15 to 30 min. At first there is a feeling of warmth, with tingling sensations in the nose, mouth and skin. There may be nausea, vomiting, restlessness, weakness, rapid pulse and rise in blood pressure. In severe cases there may be tremors and convulsions. An exper teratogen. [*3*] Treatment is symptomatic. Epinephrine and antihistamines may be helpful. It may be necessary to give large doses of barbiturates or even anesthetics to control convulsions.
Disaster Hazard: Dangerous; see sulfur compounds.

DIMERCURIC AMMONIUM OXIDE.
THR = See mercury compounds, inorganic.
Explosion Hazard: Severe, when shocked or exposed to heat.
Disaster Hazard: Dangerous; shock or heat will explode it; when heated to decomp, emits highly toxic fumes of mercury.

DIMETAN. Syn: *5,5-dimethyl-3-oxo-1-cyclohexen-1-yl dimethyl carbamate.* Crystals, sol in water and organic solvents. $C_{11}H_{17}NO_3$, mw: 211.3, mp: 46°, bp: 170°–180° @ 11 mm.
Acute tox data: Oral LD_{50} (rat) = 120 mg/kg. [*3*]
THR = HIGH via oral and inhal routes. Causes symptoms similar to parathion, but less severe. A recog carc. [*14*] See carbamates.
Disaster Hazard: Dangerous; when heated to decomp, emits highly toxic fumes.

DIMETHANE SULFONATE METHANE DIOL.
$C_3H_8O_6S_2$, mw: 204.2.
THR = An exper teratogen. [*3*]

***cis*-1,4-DIMETHANE SULFONOXY-2-BUTENE.**
$C_6H_{12}O_6S_2$, mw: 212.3.
THR = An exper neo. [*3*]

***trans*-1,4-DIMETHANE SULFONOXY-2-BUTENE.**
THR = An exper neo. [*3*]

1,4-DIMETHANE SULFONOXY-2-BUTANE.
THR = An exper neo. [*3*]

1,4-DIMETHANE SULFONOXY-2-BUTYNE.
An exper carc. [*23*]

1,6-DIMETHANE SULFONOXY HEXANE.
THR = An exper carc. [*23*]

1,6-DIMETHANE SULFONOXY-*d*-MANNITOL.
Syn: *mannitol myleran.* $C_8H_{18}O_{10}S_2$, mw: 338.3.
THR = An exper carc. [*23*]

DIMETHICONE. Water white, viscous, oil-like liquid, not miscible with water. An anti-foaming agent, silicone oil. A mixture of dimethyl siloxane polymers. D: 0.965–0.970.
THR = Details U. Limited animal exper suggest LOW toxicity.

DIMETHOATE. See *o,o*-dimethyl-S-(N-methyl carbamoyl methyl) phosphorodithioate.

DIMETHOXANE. See 2,6-dimethyl-*m*-dioxan-4-yl acetate.

3,4-DIMETHOXY-4-AMINOAZO BENZENE.
$C_{14}H_{15}O_2N_3$, mw: 257.3.
THR = An exper carc. [*3*]

7,12-DIMETHOXY BENZ(a)ANTHRACENE.
$C_{20}H_{16}O_2$, mw: 288.4.
THR = An exper carc. [*3, 23*]

1,4-DIMETHOXY BENZENE. See hydroquinone dimethyl ether.

3,3′-DIMETHOXY BENZIDINE. See dianisidine.

3,3′-DIMETHOXY BENZIDINE HYDROCHLORIDE. $C_{14}H_{16}O_2N_2 \cdot 2HCl$, mw: 317.2.
THR = An exper carc. [*3*]

1,6-DIMETHOXY BENZ(a)PYRENE.
THR = An exper carc. [*23*]

3,6-DIMETHOXY BENZ(a)PYRENE.
THR = An exper carc. [*23*]

3,4-DIMETHOXYBENZYL ALCOHOL. Colorless liquid. $(CH_3O)_2C_6H_3CH_2OH$, mw: 168.2, mp: 22°, bp: 174° @ 15 mm, d: 1.17 @ 25°/25°, vap. d: 5.8.
THR = Details U. See also alcohols.
Fire Hazard: Mod; can react with oxidizing materials, flame and heat.
To Fight Fire: Water, foam, CO_2, dry chemical.

3,4-DIMETHOXY BENZYL CHLORIDE. Syn: *veratryl chloride.* $C_{19}H_{11}O_2Cl$, mw: 186.7.
THR = An exper neo. [*3, 23*]

DIMETHOXYBORINE. Colorless, unstable liquid, decomp upon contact with moisture. $(CH_3O)_2BH$, mw: 73.9, mp: −131°, bp: 26°.
THR = Details U. See boron compounds.

DIMETHOXYBORON CHLORIDE. Colorless liquid, decomp upon contact with moisture. $(CH_3O)_2BCl$, mw: 108.4, mp: −87.5°, bp: 74.7°.
THR = Details U. See boron compounds.
Disaster Hazard: Dangerous; when heated to decomp, emits highly toxic chloride fumes.

1,3-DIMETHOXYBUTANE.
$CH_3OCH_2CH_2CH(OCH_3)CH_3$, mw: 118.
THR = U. Limited animal exper suggest mod toxicity.

DIMETHOXYCHLOROBENZENE-2,5. Slightly sol in water. $C_8H_9ClO_2$, mw: 172.5, vap. d: 5.9, bp: 460°–467°F, flash p: 243°F.
THR = U.
Fire Hazard: Slight, when exposed to heat or flame.
To Fight Fire: Water, foam, alcohol foam.

5,7-DIMETHOXY COUMARIN. Syn: *limettin.*
$C_{11}H_{10}O_4$, mw: 206.2.
THR = An exper carc. [*23*]

5,6-DIMETHOXY DIBENZ(a,h)ANTHRACENE.
$C_{24}H_{18}O_2$, mw: 338.4.
THR = An exper neo and carc. [*3, 23*]

DIMETHOXYDIPHENYL TRICHLOROETHANE. See methoxychlor.

1,2-DIMETHOXYETHANE. See ethylene glycol dimethyl ether.

DIMETHOXYETHYL PHTHALATE. Light colored, clear liquid, mild aromatic odor.
$C_6H_4(COOC_2H_4OCH_3)_2$, mw: 282, mp: −40° (forms gel), bp: 190°–210° @ 4 mm, flash p: 360°F, d: 1.171 @ 20°/20°, vap. press: 0.3 mm @ 150°, vap. d: 9.75.
THR = An exper teratogen. [*3*] See also esters.
Fire Hazard: Slight, when exposed to heat or flame; can react with oxidizing materials.
To Fight Fire: Water, foam, CO_2, dry chemical.

DIMETHOXY METHANE. See methylal.

7,14-DIMETHOXY DIBENZ(a,h)ANTHRACENE.
THR = An exper carc. [*23*]

β-3,4-DIMETHOXYPHENYLETHYLAMINE. Colorless to pale yellow liquid. $(CH_3O)_2C_6H_3CH_2CH_2NH_2$, mw: 181.2, mp: 15°, bp: 156° @ 10 mm, d: 1.08 @ 28°/4°, vap. d: 6.25.
Acute tox data: ip LD_{50} (mouse) = 160 mg/kg. [*3*]
THR = HIGH via ip route.
Fire Hazard: Slight.

1,2-DIMETHOXYPROPANE. Liquid.
$CH_3CHOCH_3CH_2OCH_3$, mw: 104.15, bp: 95°, d: 0.8461, vap. d: 3.59.
THR = Details U. Probably irr and narcotic in high conc.
Fire Hazard: Mod, when exposed to heat or flame; can react with oxidizing materials such as $[Mn(ClO_4)_2 + C_2H_5OH]$, $Ni(ClO_4)_2$. [*19*]
Explosion Hazard: U.
To Fight Fire: Foam, CO_2, dry chemical.

2,5-DIMETHOXY-4-STILBENAMINE.
THR = An exper carc. [*23*]

DIMETHOXY STRYCHININE. See brucine.

DIMETHOXYTETRAETHYLENE GLYCOL. See dimethoxytetraglycol.

DIMETHOXYTETRAGLYCOL. Syns: *tetraethylene glycol dimethyl ether, dimethoxytetraethylene glycol.* Water white, practically odorless liquid.
$(CH_3OC_2H_4OC_2H_4)_2O$, mw: 222.28, bp: 275.3°, fp: −29.7°, flash p: 285°F (OC), d: 1.0132 @ 20°/20°, vap. press: 0.01 mm @ 20°, vap. d: 7.7.
Acute tox data: Oral LD_{50} (rat) = 5140 mg/kg. [*3*]
THR = MOD via oral route. See glycols.
Fire Hazard: Slight, when exposed to heat or flame; can react with oxidizing materials.
To Fight Fire: Alcohol foam, CO_2, dry chemical.

2,5-DIMETHOXYTETRAHYDROFURAN. Colorless liquid. $(CH_3O)_2C_4H_6O$, mw: 132.16, bp: 35° @ 10 mm, vap. d: 4.56.
THR = Details U. See also tetrahydrofuran.
Fire Hazard: Slight.

DIMETHRIN. Syn: *2,4-dimethyl benzyl-2,2-dimethyl-3-(2-methyl propenyl)cyclopropane carboxylate.*
THR = HIGH Toxicity. Details U.

DIMETHYL. See ethane.

DIMETHYLACETAL. Syn: *ethylidene dimethyl ether.* Colorless liquid, strong aromatic odor.
$CH_3(OCH_3)_2CH$, mw: 90.12, bp: 61.8°, flash p: <80°F, d: 0.848 @ 25°, vap d: 3.1.
Acute tox data: Inhal LC_{50} (rat) = 3000 ppm for 4 hrs; oral LD_{50} (rabbit) = 4507 mg/kg. [*3*]
THR = MOD via oral and inhal routes.
Fire Hazard: Dangerous, when exposed to heat or flame; can react vigorously with oxidizing materials.
Explosion Hazard: U.
To Fight Fire: Foam, CO_2, dry chemical.

N,N-DIMETHYL ACETAMIDE. Liquid.
$CH_3CON(CH_3)_2$, mw: 87.12, mp: −20°, bp: 165°, d: 0.9448 @ 15.5°, vap. d: 3.01, vap. press: 1.3 mm @ 25°, flash p: 171°F (TOC), lel = 2.0%, uel = 11.5% @ 740 mm and 160°.
Acute tox data: Dermal LD_{50} (rabbit) = 2240 mg/kg; inhal TC_{LO} (human) = 20 ppm. [*3*]
THR = HIGH via inhal and MOD via oral routes. Somewhat less acutely toxic than dimethyl formamide, based upon animal tests. An exper teratogen. [*3*]
Fire Hazard: Mod, via heat, flame or oxidizers.

3,4-DIMETHYL ACETANILIDE. $C_{10}H_3ON$, mw: 163.2.
THR = An exper neo. [*3*]

N,N-DIMETHYLACETOACETAMIDE. Liquid, miscible in water and organic solvents.
$CH_3COCH_2CON(CH_3)_2$, mw: 129, bp: 220°, d: 1.049–1.052 @ 20°/20°, flash p: 252°F (COC).
Acute tox data: Oral LD_{50} (rat) = 2300 mg/kg; dermal LD_{50} (rabbit) = 14000 mg/kg. [*3*]
THR = LOW via oral and dermal routes.
Fire Hazard: Slight, when exposed to heat or flame.

7,12-DIMETHYL ACETO BENZ(a)ANTHRACENE.
THR = An exper carc. [*23*]

S,S′-[(DIMETHYL AMINO)TRIMETHYLENE] BIS THIOCARBAMATE. See padan.

DIMETHYL ACROLEIN. See senecidaldehyde.

DIMETHYL ALLYL ARSINE. See allyl dimethyl arsine.

DIMETHYLAMINE, ANHYDROUS. Colorless gas. $(CH_3)_2NH$, mw: 45.08, bp: 6.88°, flash p: 0°F, fp: −92.19°, d: 0.6804 @ 0°/4°, autoign. temp.: 752°F, vap. d: 1.55, lel = 2.8%, uel = 14.4%.
Acute tox data: Oral LD_{50} (rat) = 698 mg/kg; oral LD_{50} (rabbit) = 240 mg/kg. [*3*]

THR = MOD to HIGH oral route. A MOD irr.
Fire Hazard: Very dangerous, when exposed to heat or flame; can react vigorously with oxidizing materials.
Explosion Hazard: Mod, when exposed to flame.
Disaster Hazard: Dangerous! Keep away from heat and open flame!
To Fight Fire: Stop flow of gas, foam, CO_2, dry chemical.

DIMETHYLAMINE-(2,4-DICHLOROPHENOXY) ACETATE. $C_{10}H_{11}O_3NCl_2$, mw: 264.1.
THR = HIGH via oral route. An exper teratogen. [*3*]

DIMETHYLAMINOACETONITRILE.
$(CH_3)_2NCH_2CN$, mw: 84.
Acute tox data: Oral LD_{50} (rat) = 50 mg/kg; inhal LC_{50} (rat) = 250 ppm for 4 hrs; dermal LD_{50} (rabbit) = 170 mg/kg. [*3*]
THR = HIGH via oral, inhal and dermal routes. See also nitriles.

DIMETHYL AMINO ANTIPYRINE.
See aminopyrine.

4-DIMETHYL AMINO AZO BENZENE. Syns: *butter yellow, methyl yellow, n,n-dimethyl-p-phenyl azoaniline.* Yellow crystalline tablets, insol in water, sol in strong mineral acids and oils. $C_{14}H_{15}N_2$, mw: 229.1.
Acute tox data: Oral LD_{50} (rat) = 200 mg/kg; dermal TD_{LO} (rat) = 155 mg/kg. [*3*]
THR = HIGH via oral and dermal routes. An exper (+) carc, neo. [*3, 4, 14, 23*]

DIMETHYLAMINOBENZALDEHYDE. Syns: *d-dimethylaminobenzenecarbinol, Ehrlich's reagent.*
Small, granular lemon-colored crystals (may turn pink upon exposure to light), slightly sol in water, sol in alcohol, ether, chloroform, acetic acid and many other organic solvents. $C_6H_4[N(CH_3)_2]CHO$, mw: 149.19, mp: 73°, bp: 176°–177° @ 17 mm.
Acute tox data: Oral LD_{LO} (rat) = 500 mg/kg; ipLD_{50} (rat) = 620 mg/kg. [*3*]
THR = HIGH via oral and ip routes.

DIMETHYLAMINO BENZENE. See xylidene.

1-(4-DIMETHYLAMINO BENZYLIDENE)INDENE.
$C_{18}H_{17}N$, mw: 247.4.
THR = An exper neo. [*3*]

***p*-DIMETHYL AMINO BENZYLIDENE-3,4,5,6-DIBENZ-9-METHYL ACRIDINE.** $C_{31}H_{24}N_2$, mw: 424.6.
THR = An exper neo. [*3*]

4-DIMETHYLAMINO BIPHENYL. $C_{14}H_{15}N$, mw: 197.3.
THR = All exper neo. [*3*]

DIMETHYLAMINOCHLOROPROPANE. Liquid.
$C_5H_{12}NCl$, mw: 121.6, vap. d: 4.20.
THR = MOD irr to skin, eyes and mu mem.
Disaster Hazard: Dangerous; see chlorides; can react with oxidizing materials.

2-DIMETHYLAMINOETHANOL. Syn: *dimethylethanolamine.* Liquid. $(CH_3)_2NCH_2CH_2OH$, mw: 89.14, bp: 131°, flash p: 105°F (OC), d: 0.8866 @ 20°/4°, vap. d: 3.03.
Acute tox data: Oral LD_{50} (rat) = 2340 mg/kg; dermal LD_{50} (rabbit) = 1370 mg/kg. [*3*]
THR = MOD via oral and dermal routes. Used medically as a CNS stimulant.
Fire Hazard: Mod, when exposed to heat or flame; can react vigorously with oxidizing materials.
Explosion Hazard: U.
To Fight Fire: Alcohol foam, foam, CO_2, dry chemical.

DIMETHYLAMINOETHYL METHACRYLATE.
Liquid, sol in water and organic solvents.
$CH_2(CH_3)CCO_2(CH_2)_2N(CH_3)_2$, mw: 157, d: 0.933 @ 25°, bp: 182°–190°, flash p: 165°F (TOC), vap. d: 5.4.
Acute tox data: ipLD_{50} (mouse) = 104 mg/kg. [*3*]
THR = HIGH via ip route. It is an irr to skin, eyes and mu mem. A powerful lachrymator.
Fire Hazard: Mod, when exposed to sparks, heat, open flame or oxidizers.
To Fight Fire: Alcohol foam, dry chemical, spray.

1-[2-(N,N-DIMETHYL AMINO)ETHYL]-4-METHYL PIPERAZINE.
THR = U. Limited animal exper suggest HIGH toxicity and irr. See also piperazine.

2-[(DIMETHYLAMINO)METHYL IMINO]-5-[2-(5-NITRO-2-FURYL)VINYL]-ε-1,3,4,-OXIDIAZOLE. $C_{11}H_{11}O_4N_5$, mw: 277.3.
THR = An exper (+) carc. [*3, 1*]

1,3-DIMETHYL-4-AMINO-5-NITROSOURACIL.
$C_5H_8O_3N_4$, mw: 172.2.
THR = An exper neo. [*3*]

***p*-((*p*-(DIMETHYLAMINO)PHENYL)AZO)BENZENE SULFONAMIDE.** $C_{14}H_{16}O_2N_4S$, mw: 304.4.
THR = An exper carc. [*3*]

4-((*p*-(DIMETHYLAMIN)PHENYL)AZO)BENZIMIDAZOLE. $C_{15}H_{15}N_5$, mw: 265.4.
THR = An exper carc. [*3*]

6-((*p*-(DIMETHYLAMINO)PHENYL)AZO)BENZOTHIAZOLE. $C_{15}H_{14}N_4S$, mw: 282.4.
THR = An exper carc. [*3*]

7-((*p*-(DIMETHYLAMINO)PHENYL)AZO)BENZOTHIAZOLE.
THR = An exper carc. [*3*]

4-((*p*-(DIMETHYLAMINO)PHENYL)AZO)ISOQUINOLINE. $C_{16}H_{16}N_4$, mw: 264.4.
THR = An exper carc. [*3*]

5-((*p*-(DIMETHYLAMINO)PHENYL)AZO)ISOQUINOLINE.
THR = An exper carc. [*3*, *23*]

7-((*p*-(DIMETHYLAMINO)PHENYL)AZO)ISOQUINOLINE.
THR = An exper carc. [*3*]

5-((*p*-(DIMETHYLAMINO)PHENYL)AZO)ISOQUINOLINE-2-OXIDE. $C_{16}H_{16}ON_4$, mw: 280.4.
THR = An exper carc. [*3*, *23*]

4-((*p*-(DIMETHYLAMINO)PHENYL)AZO)-N-METHYL ACETANILIDE. $C_{17}H_{20}ON_4$, mw: 296.4.
THR = LOW via oral route. An exper neo. [*3*]

5-[(*p*-(DIMETHYLAMINO)PHENYL]AZO-3-METHYL QUINOLINE. $C_{18}H_{18}N_4$, mw: 290.4.
THR = An exper carc. [*3*]

5-[(*p*-(DIMETHYLAMINO)PHENYL]AZO-6-METHYL QUINOLINE.
THR = An exper neo. [*3*]

5-[(*p*-(DIMETHYLAMINO)PHENYL]AZO-7-METHYL QUINOLINE.
THR = An exper carc. [*3*]

5-((*p*-(DIMETHYLAMINO)PHENYL)AZO-8-METHYL QUINOLINE.
THR = An exper carc. [*3*]

4-((*p*-(DIMETHYLAMINO)PHENYL)AZO)-2-PICOLINE. $C_{14}H_{16}N_4$, mw: 240.3.
THR = An exper carc. [*3*]

4-[(4-DIMETHYLAMINO)PHENYL]AZO-2-PICOLINE-1-OXIDE. $C_{14}H_{16}ON_4$, mw: 256.3.
THR = An exper carc. [*3*]

4-[(*o*-DIMETHYLAMINO)PHENYL]AZO-2-PICOLINE-1-OXIDE.
THR = An exper carc. [*3*]

3-[(*p*-(DIMETHYLAMINO)PHENYL]AZOPYRIDINE. $C_{13}H_{14}N_4$, mw: 226.3.
THR = An exper carc. [*3*]

4-[(*p*-(DIMETHYLAMINO)PHENYL]AZOPYRIDINE.
THR = An exper carc. [*3*]

5-((*p*-(DIMETHYLAMINO)PHENYL)AZO) QUINOLDINE. $C_{18}H_{18}N_4$, mw: 290.4.
THR = An exper carc. [*3*]

4-((*p*-(DIMETHYLAMINO)PHENYL)AZO)QUINOLINE. $C_{17}H_{16}N_4$, mw: 276.4.
THR = An exper carc. [*3*]

5-((*p*-(DIMETHYLAMINO)PHENYL)AZO)QUINOLINE.
THR = An exper carc. [*3*]

6-((*p*-(DIMETHYLAMINO)PHENYL)AZO) QUINOLINE.
THR = An exper carc. [*3*]

4-[(*p*-DIMETHYLAMINO)PHENYL]AZO QUINOLINE-1-OXIDE. $C_{17}H_{16}ON_4$, mw: 292.4.
THR = An exper carc. [*3*]

5-[(*p*-DIMETHYLAMINO)PHENYL]AZO QUINOLINE-1-OXIDE.
THR = An exper carc. [*3*]

6-[(*p*-DIMETHYLAMINO)PHENYL]AZO QUINOLINE-1-OXIDE.
THR = An exper carc. [*3*]

5-(*p*-DIMETHYLAMINO)PHENYL]AZO QUINOXALINE. $C_{16}H_{15}N_5$, mw: 277.4.
THR = An exper carc. [*3*]

6-[(*p*-DIMETHYLAMINO)PHENYL]AZO QUINOXALINE.
THR = An exper carc. [*3*]

DIMETHYLAMINOPHENYL MERCURIC ACETATE. Colorless crystals, insol in water. $C_6H_4N(CH_3)_2HgO_2C_2H_3$, mw: 379.8, mp: 165°.
THR = HIGH toxicity. See mercury compounds, organic.

1-DIMETHYLAMINO-2-PROPANOL. Clear, amber-colored, volatile liquid. $(CH_3)_2NCH_2CHOHCH_3$, mw: 103.2, bp: 122.5°–126.2°, flash p: 90°F, fp: <−20°, d: 0.850 @ 25°/25°, vap. d: 3.52.
Acute tox data: Oral LD_{50} (rat) = 1890 mg/kg. [*3*]
THR = MOD via oral route. See also amines.
Fire Hazard: Dangerous, when exposed to heat or flame; can react vigorously with oxidizing materials.
Explosion Hazard: U.
To Fight Fire: Foam, CO_2, dry chemical.

DIMETHYLAMINOPROPIONITRILE. Liquid. $(CH_3)_2NHC_2CH_2CN$, mw: 98.2, mp: −43°, bp: 170°, flash p: 149°F (OC), d: 0.8617, vap. d: 3.35.
Acute tox data: Oral LD_{50} (rat) = 2600 mg/kg; dermal LD_{50} (rabbit) = 1410 mg/kg. [*3*]
THR = MOD via oral and dermal routes. See nitriles.
Fire Hazard: Mod, when exposed to heat, flame or oxidizers.
Disaster Hazard: Dangerous; when heated to decomp, emits highly toxic fumes; can react with oxidizing materials.
To Fight Fire: Foam, CO_2, dry chemical.

3-DIMETHYLAMINOPROPYLAMINE. Colorless liquid. $(CH_3)_2NHC_2CH_2CH_2NH_2$, mw: 102.8, mp: <−70°, bp: 123°, flash p: 100°F (OC), d: 0.8100 @ 30°, vap. press: 10 mm @ 30°, vap. d: 3.52.
Acute tox data: Oral LD_{LO} (rat) = 1870 mg/kg. [*3*]
THR = MOD via oral route.
Fire Hazard: Mod, when exposed to heat, flame or oxidizers.
Explosion Hazard: U.

Disaster Hazard: Dangerous; reacts vigorously with oxidizing materials; emits toxic fumes when heated.
To Fight Fire: Alcohol foam, CO_2, dry chemical.

6-[*p*-(DIMETHYLAMINO)STYRYL]BENZ(c)ACRIDINE. $C_{27}H_{22}N_2$, mw: 374.5.
THR = An exper neo. [*3*]

12-[*p*-(DIMETHYLAMINO)STYRYL]BENZ(c)ACRIDINE.
THR = An exper neo. [*3*]

2-[*p*-[DIMETHYLAMINO)STYRYL]BENZOTHIAZOLE. $C_{17}H_{16}N_2S$, mw: 280.4.
THR = An exper carc. [*3*]

4-(*p*-DIMETHYLAMINO)STYRYL-6,8-DIMETHYL QUINOLINE. $C_{21}H_{22}N_2$, mw: 302.5.
THR = An exper neo. [*3*]

4-[(*p*-DIMETHYLAMINO)STYRYL]QUINOLINE. $C_{19}H_{18}N_2$, mw: 274.4.
THR = An exper neo. [*3, 23*]

4-[(*p*-DIMETHYLAMINO)STYRYL]QUINOLINE MONOHYDROCHLORIDE. $C_{19}H_{18}N_2 \cdot HCl$, mw: 310.9.
THR = An exper carc. [*3*]

4-((4-(DIMETHYLAMINO)-*m*-TOLYL)AZO)-2-PICOLINE-1-OXIDE. $C_{15}H_{18}ON_4$, mw: 270.4.
THR = An exper carc. [*3*]

4-((4-(DIMETHYLAMINO)-*o*-TOLYL)AZO)-2-PICOLINE-1-OXIDE.
THR = An exper carc. [*3*]

4-((4-DIMETHYLAMINO)- *m*-TOLYL)AZO)-3-PICOLINE-1-OXIDE.
THR = An exper neo. [*3*]

4-((4-(DIMETHYLAMINO)-*o*-TOLYL)AZO-3-PICOLINE-1-OXIDE.
THR = An exper neo. [*3*]

5-((*p*-DIMETHYLAMINO)-*m*-TOLYL)AZO) QUINOLINE. $C_{18}H_{18}N_4$, mw: 290.4.
THR = An exper neo. [*3*]

4-(DIMETHYLAMINO)-3,5-XYLENOL. $C_{10}H_{15}ON$, mw: 165.3.
THR = An exper neo. [*3*]

4-((4-(DIMETHYLAMINO)-3,5-XYLYL)AZO)PYRIDINE-1-OXIDE.
THR = An exper neo. [*3*]

4-((4-(DIMETHYLAMINO)-2,3-XYLYL)AZO)PYRIDINE-1-OXIDE. $C_{15}H_{18}ON_4$, mw: 270.4.
THR = An exper neo. [*3*]

4-((4-(DIMETHYLAMINO)-2,5-XYLYL)AZO)PYRIDINE-1-OXIDE.
THR = An exper neo. [*3*]

5-((4-(DIMETHYLAMINO)-*o*-TOLYL)AZO)QUINOLINE.
THR = An exper neo. [*3*]

DIMETHYL AMMONIUM DIMETHYL CARBAMATE. Liquid or crystals. $(CH_3)_2NCOONH_2(CH_3)_2$, mw: 134.17, mp: −38°, bp: 60.2°, d: 1.026 @ 25°/4°, vap. d: 4.62.
THR = See also carbamates.
Disaster Hazard: Slight; on decomp, emits toxic fumes.

DI(METHYL AMYL)MALEATE.
See dihexyl maleate.

N,N-DIMETHYLANILINE. Liquid. $C_6H_5N(CH_3)_2$, mw: 121.18, mp: 2.5°, bp: 193.1°, flash p: 145°F (CC), d: 0.9557 @ 20°/4°, ulc: 20–25, autoign. temp.: 700°F, vap. press: 1 mm @ 29.5°, vap. d: 4.17.
Acute tox data: Dermal LD_{50} (rabbit) = 1770 mg/kg; oral LD_{LO} (human) = 50 mg/kg. [*3*]
THR = HIGH via oral; MOD via dermal routes. Its physiological action is similar to that of aniline, although it is believed to be less toxic. It acts as a depressant on the CNS. Oral or sc administration of 2 g/kg of body weight in guinea pigs has caused weakness tremors, tonic and clonic convulsions, slowing of the respiration and finally death due to respiratory paralysis. The hazard associated with this material is that untrained personnel will disregard small splashes of it upon the shoes, clothing, or body and will use improperly ventilated equipment. Industrial accidents involving dimethyl aniline are dangerous in that they can release sudden massive quantities of the oil or its vapor from breaks in the pipes of a closed system. See also aniline.
Fire Hazard: Mod, when exposed to heat, flame or oxidizers.
Spont Heating: No.
Disaster Hazard: Dangerous; when heated to decomp, emits highly toxic fumes of aniline; can react with oxidizing materials such as benzoyl peroxide. [*19*]
To Fight Fire: Foam, CO_2, dry chemical.

DIMETHYL ANALOG OF PARATHION. See *o,o*-dimethyl-*o,p*-nitrophenyl thiophosphate.

3,4-DIMETHYL ANILINE. See 3,4-xylidene.

DIMETHYLANILINE FLUOSILICATE. White crystals, sol in hot alcohol. $(C_6H_5NHCH_3)_2 \cdot H_2SiF_6$, mw: 358.4.
THR = See fluosilicates.

DIMETHYLANILINE MERCURY. Colorless crystals, sol in chloroform. $Hg[C_6H_4N(CH_3)_2]_2$.
THR = HIGH toxicity. See mercury compounds, organic.

6,12-DIMETHYL ANTHANTHRENE.
THR = An exper carc. [*23*]

9,10-DIMETHYL ANTHRACENE.
An exper carc. [*23*]

DIMETHYLARSINE. Syn: *cacodyl hydride*. Colorless liquid. $(CH_3)_2AsH$, mw: 106.0, bp: 36°, d: 1.213 @ 29°/4°, vap. d: 3.65.
THR = HIGH toxicity. See arsine and arsenic compounds.
Fire Hazard: Dangerous; ignites spont in air.
Disaster Hazard: Dangerous; extremely toxic. More hazardous than its oxidation products; reacts vigorously with oxidizing agents.
To Fight Fire: Exclude O_2 or allow fire to burn or apply water, foam, dry chemical, water spray, CO_2.
Spont Heating: No.

DIMETHYLARSINIC ACID. See cacodylic acid.

7,9-DIMETHYLBENZ(c)ACRIDINE. $C_{19}H_{15}N$, mw: 257.4.
THR = An exper neo. [*3, 23*]

7,10-DIMETHYLBENZ(c)ACRIDINE.
THR = An exper neo. [*3, 23*]

7,11-DIMETHYLBENZ(c)ACRIDINE.
THR = An exper neo. [*3, 23*]

1,12-DIMETHYLBENZ(a)ANTHRACENE. $C_{20}H_{16}$, mw: 256.4.
THR = An exper neo. [*3*]

4,5-DIMETHYLBENZ(a)ANTHRACENE.
THR = An exper carc. [*3, 23*]

6,7-DIMETHYLBENZ(a)ANTHRACENE.
THR = VERY HIGH acute exper toxicity. An exper carc. [*3, 23*]

6,8-DIMETHYLBENZ(a)ANTHRACENE.
THR = An exper carc. [*3, 23*]

6,12-DIMETHYLBENZ(a)ANTHRACENE.
THR = An exper neo and carc. [*3, 23*]

7,8-DIMETHYLBENZ(a)ANTHRACENE.
THR = An exper neo and carc. [*3, 23*]

7,11-DIMETHYLBENZ(a)ANTHRACENE.
THR = An exper carc. [*3, 23*]

7,12-DIMETHYLBENZ(a)ANTHRACENE.
THR = An exper teratogen, neo and carc. [*3, 23*]

7,12-DIMETHYLBENZ(a)ANTHRACENE
THR = An exper carc. [*3*]

8,9-DIMETHYLBENZ(a)ANTHRACENE.
THR = An exper carc. [*3, 23*]

8,12-DIMETHYLBENZ(a)ANTHRACENE.
THR = An exper carc and neo. [*3, 23*]

9,10-DIMETHYLBENZ(a)ANTHRACENE.
THR = An exper neo. [*3, 23*]

9,11-DIMETHYLBENZ(a)ANTHRACENE.
THR = An exper carc. [*23*]

9,12-DIMETHYLBENZ(a)ANTHRACENE.
THR = An exper carc. [*23*]

1,2-DIMETHYLBENZENE. See *o*-xylene.

1,3-DIMETHYLBENZENE. See *m*-xylene.

1,4-DIMETHYLBENZENE. See *p*-xylene.

3,3′-DIMETHYLBENZIDINE. See *o*-tolidine.

N,N-DIMETHYL-4-(4′-BENZIMIDAZOLYL AZO) ANILINE.
THR = An exper carc. [*23*]

6,12-DIMETHYLBENZO(1,2-b:5,4:b)BIS(1)BENZOTHIOPHENE. $C_{20}H_{14}S_2$, mw: 318.4.
THR = An exper neo to mice via sc route [*3*]

6,12-DIMETHYLBENZO(1,2-b:4,5-b)DITHIONAPHTHENE. $C_{20}H_{14}S_2$, mw: 318.5.
THR = An exper neo. [*3*]

1,2-DIMETHYLBENZO(a)PYRENE. $C_{22}H_{16}$ mw: 280.4.
THR = An exper carc. [*3, 23*]

1,3-DIMETHYLBENZO(a)PYRENE.
THR = An exper carc. [*23, 3*]

1,4-DIMETHYLBENZO(a)PYRENE.
THR = An exper carc. [*3, 23*]

1,6-DIMETHYLBENZO(a)PYRENE.
THR = An exper carc. [*3, 23*]

2,3-DIMETHYLBENZO(a)PYRENE.
THR = An exper carc. [*3, 23*]

3,6-DIMETHYLBENZO(a)PYRENE.
THR = An exper carc. [*3*]

3,2-DIMETHYLBENZ(a)PYRENE.
THR = An exper carc. [*3*]

3,6-DIMETHYLBENZ(a)PYRENE.
THR = An exper carc. [*23*]

3,12-DIMETHYLBENZ(a)PYRENE.
An exper carc. [*23*]

4,5-DIMETHYLBENZO(a)PYRENE.
An exper carc. [*3, 23*]

N,N-DIMETHYL-4-(6′-BENZTHIAZOLYL AZO) ANILINE.
THR = An exper carc. [*23*]

N,N-DIMETHYL-4-(7′-BENZTHIAZOLYL AZO) ANILINE.
THR = An exper carc. [*23*]

6,12-DIMETHYLBENZO(b)THIONAPHTHENO (3,2-f)THIONAPHTHEN.
THR = An exper carc. [*23*]

6,12-DIMETHYLBENZO(b)THIONAPHTHENO (2,3-1)THIONAPHTHEN.
THR= An exper carc. [*23*]

4,9-DIMETHYL-2,3-BENZTHIOPHENANTHRENE. $C_{18}H_{14}S$, mw: 262.4.
THR = An exper neo. [*3*]

***p,α*-DIMETHYLBENZYL ALCOHOL.** Syns: *p-tolylmethyl carbinol, methyl-p-tolylcarbinol, 4-(α-hydroxyethyl)toluene, 4-methyl-α-phenethyl alcohol, l-p-tolyl-l-ethanol.* Viscous liquid, menthol-like odor, very sparingly sol in water, miscible with absolute alcohol and ether, sol in isopropanol and liquid petrolatum. $H_3C(C_6H_4)CH(CH_3)(OH)$, mw: 136.19, d: 0.9668 @ 15.5°/4°, bp: 219° @ 756 mm.
Acute tox data: Oral LD_{50} (rat) = 1280 mg/kg. [*3*]
THR = MOD via oral route. Probably irr and narcotic. See also toluene.

DIMETHYLBENZYLCETYL AMMONIUM CHLORIDE. $(CH_3)_2(C_6H_5CH_2)C_{16}H_{33}NCl$, mw: 436.1, vap. d:1.5.
THR = MOD irr and via oral route.
Fire Hazard: Slight.

2,4-DIMETHYLBENZYL-2,2-DIMETHYL-3-(2-METHYL PROPENYL)CYCLOPROPANE CARBOXYLATE. See dimethrin.

α-DIMETHYLBENZYL HYDROPEROXIDE. See cumene hydroperoxide.

DIMETHYLBERYLLIUM. White needles. $Be(CH_3)_2$, mw: 39.09, bp: sublimes @ 200°.
THR = HIGH toxicity. See beryllium compounds.
Fire Hazard: Mod, when exposed to heat or flame.
Disaster Hazard: Dangerous; when heated to decomp, emits highly toxic fumes of beryllium oxides; can react with oxidizing materials.

3′,3‴-DIMETHYL-4′,4‴-BIACETANILIDE. $C_{18}H_{18}O_2N_2$, mw: 294.4.
THR = An exper neo. [*3*]

1,1-DIMETHYL BIGUANIDINE. $C_4H_{11}N_5$, mw: 129.2.
THR = An exper teratogen. [*3*]

3,2-DIMETHYL-4-BIPHENYLAMINE. $C_{14}H_{15}N$, mw: 197.3.
THR = An exper carc. [*3, 23*]

DIMETHYLBORIC ACID. Syn: *dimethyl hydroxyborine.* Colorless liquid miscible with water. $(CH_3)_2BOH$, mw: 57.9, bp: 0° @ 36 mm.
THR = Details U. See boron compounds.

DIMETHYLBORIC ANHYDRIDE. Colorless crystals, hydrolyze in water. $(CH_3)_2BOB(CH_3)_2$, mw: 97.8, mp: −37.3°, bp: 43°.
THR = Details U. See boron compounds.

DIMETHYLBORINE TRIMETHYL AMMINE. Colorless liquid, decomp in water. $(CH_3)_2HB \cdot N(CH_3)_3$, mw: 101.0, mp: −18°, bp: 172° (decomp).
THR = Details U. See boron compounds.

DIMETHYLBORON BROMIDE. Colorless liquid or gas decomp by moisture. $(CH_3)_2BBr$, mw: 120.8, mp: −123.4°, bp: 22°.
THR = Details U. See boron compounds and bromides.
Disaster Hazard: Dangerous; see bromides.

DIMETHYLBORON IODIDE. Colorless liquid decomp by water. $(CH_3)_2BI$, mw: 167.8, mp: −111°, bp: 65°.
THR = Details U. See boron compounds.
Disaster Hazard: Dangerous; see iodides.

DIMETHYLBROMARSINE. Syn: *cacodyl bromide.* Yellow oily liquid. $(CH_3)_2AsBr$, mw: 184.9, bp: 130°.
THR = HIGH toxicity. See arsenic compounds.
Disaster Hazard: Dangerous; see arsenic, bromides.

2,2-DIMETHYLBUTANE. Syn: *neohexane.* Liquid. $(CH_3)_3CCH_2CH_3$, mw: 86.2, bp: 49.7°, mp: −98.2°, flash p: −54°F, fp: −101.9°, d: 0.649, autoign. temp.: 797°F, vap. press: 400 mm @ 31.0°, vap. d: 3.00, lel = 1.2%, uel = 7.0%.
THR = Details U. Probably is irr and narcotic in HIGH conc.
Fire Hazard: Dangerous, when exposed to heat or flame; can react vigorously with oxidizing materials.
Explosion Hazard: U.
Disaster Hazard: Dangerous! Keep away from heat or open flame.
To Fight Fire: Foam, CO_2, dry chemical.

2,3-DIMETHYLBUTANE. Syn: *diisopropyl.* Liquid. $(CH_3)_2CHCH(CH_3)_2$, mw: 86.17, mp: −135°, bp: 58.0°, flash p: −20°F, d: 0.662 @ 20°/4°, autoign. temp.: 788°F, vap. press: 400 mm @39.0°, vap. d: 3.0, lel = 1.2%, uel = 7.0%.
THR = U. Probably irr and narcotic in HIGH conc.
Fire Hazard: Dangerous, when exposed to heat or flame; can react vigorously with oxidizing materials.
Explosion Hazard: U.
Disaster Hazard: Dangerous; keep away from heat and open flame!
To Fight Fire: Foam, CO_2, dry chemical.

2,2-DIMETHYL-1,3-BUTANEDIOL. Liquid, miscible in water. $CH_3CH(OH)C(CH_3)_2CH_2OH$, mw: 119, d: 0.9700, bp: 202.4°, fp: −12.8°.
Acute tox data: Oral LD_{50} (rat) = 10,000 mg/kg. [*3*]
THR = LOW via oral route.

2,3-DIMETHYLBUTANOL. $CH_3CH_2C(CH_3)_2CH_2OH$, mw: 102.

Acute tox data: Oral LD_{50} (rat) = 2330 mg/kg; dermal LD_{50} (rabbit) = 1770 mg/kg. [*3*]
THR = MOD via oral and dermal routes.

2,3-DIMETHYL-1-BUTENE. C_6H_{12}, mw: 84.1, flash p: $<-4°F$, autoign. temp.: 680°F, d: 0.68, vap. d: 2.91, bp: 55°.
THR = No data.
Fire Hazard: Dangerous, via heat, open flames or oxidizers.
To Fight Fire: Dry chemical, spray, mist, CO_2.

2,3-DIMETHYL-2-BUTENE. flash p: $<4°F$, autoign. temp.: 753°F, d: 0.71, vap. d: 2.91, bp: 73.3°.
THR = No data.
Fire Hazard: Dangerous, via heat, flames (sparks) or oxidizers.
To Fight Fire: Dry chemical, spray, mist, CO_2.

1,3-DIMETHYLBUTYL ACETATE. Slightly sol in water. $CH_3COOCH(CH_3)CH_2CH(CH_3)_2$, mw: 144, d: 0.9, vap. d: 5.0, bp: 284–297°F, flash p: 113°F.
Acute tox data: Inhal TC_{LO} (human) = 100 ppm (affects eyes); inhal LD_{LO} (rat) = 4000 ppm. [*3*]
THR = HIGH to MOD via inhal route.
Fire Hazard: Mod, when exposed to heat, flame or oxidizers.
To Fight Fire: Alcohol foam.

1,3-DIMETHYLBUTYLAMINE. A liquid. $CH_3CH(NH_2)CH_2CH(CH_3)_2$, mw: 101.2, bp: 106°–109°, flash p: 55°F (OC), d: 0.750 @ 20°/20°.
Acute tox data: Oral LD_{LO} (rat) = 600 mg/kg; oral LD_{50} (mouse) = 470 mg/kg; inhal LC_{LO} (mouse) = 1278 ppm for 15 min; dermal LD_{LO} (rabbit) = 600 mg/kg. [*3*]
THR = HIGH via oral, inhal and dermal routes. See amines.
Fire Hazard: Dangerous, when exposed to heat or flame; can react vigorously with oxidizing materials.
Explosion Hazard: U.
To Fight Fire: Foam, CO_2, dry chemical.

3,3-DIMETHYLBUTYRIC ACID. See *tert*-butyl acetic acid.

DIMETHYLCADMIUM. Oil decomp by water, foul odor. $(CH_3)_2Cd$, mw: 142.5, d: 1.984, mp: −4.5°, bp: 106°.
THR = See cadmium compounds. Expl. >150°. [*19*]

DIMETHYL-α-CAPROLACTONE.
Acute tox data: Oral LD_{50} (rat) = 11000 mg/kg; dermal LD_{50} (rabbit) = 3540 mg/kg. [*3*]
THR = LOW via oral and MOD via dermal routes.

DIMETHYLCARBAMOYL CHLORIDE. See dimethyl carbamyl chloride.

2-DIMETHYLCARBAMOYL-3-METHYL-5-PYRAZOLYL DIMETHYL CARBAMATE. See dimetilan.

DIMETHYLCARBAMYL CHLORIDE. Syn: *dimethyl carbamoyl chloride.* Liquid. $(CH_3)_2NCOCl$, mw: 107.6, mp: −33°, bp: 165°–167°, d: 1.678 @ 20°/4°, vap. d: 3.73.
THR = Lachrymator; strong local irr action. It is an exper carc. [*12, 3*] Causes skin and papillary tumors in mice due to skin exposure. Squamous cell carcinoma in rats due to inhal. [*3, 12*]
Disaster Hazard: Dangerous; see chlorides; will react with water or steam to produce toxic and corrosive fumes.

DIMETHYLCARBATE. Syn: *dimelone.* $C_{11}H_{14}O_4$, mw: 210.3.
Acute tox data: Oral LD_{50} (rat) = 1000 mg/kg; oral LD_{50} (mouse) = 1400 mg/kg. [*3*]
THR = MOD via oral route and LOW irr. An insect-repellent.
Fire Hazard: Slight.

DIMETHYLCARBINOL. See isopropyl alcohol.

3-DIMETHYLCARBINOLPROPYLAMINE. See 3-isopropoxypropylamine.

2,2-DIMETHYL-6-CARBOBUTOXY-2,3-DIHYDRO-4-PYRONE. See butopyronoxyl.

DIMETHYL-1-CARBOMETHOXY-1-PROPENYL-2-PHOSPHATE. Syns: *mevinphos, phosdrin.* $C_7H_{13}O_6P$, mw: 224.2.
Acute tox data: Oral LD_{50} (rat) = 4 mg/kg; inhal LC_{50} (rat) = 14 ppm for 1 hr; dermal LD_{50} (rat) = 4 mg/kg. [*3*]
THR = HIGH via oral, inhal and dermal routes. An organo P pesticide of VERY HIGH mamalian toxicity. Absorbed via all routes. Early symptoms of poisoning include excessive sweating, headache, weakness, nausea, giddiness, vomiting, stomach ache, blurred vision, slurred speech and muscle twitching, coma and loss of sphincter control A cholinesterase inhibitor. See parathion.
Disaster Hazard: Dangerous; see phosphorus compounds.

DIMETHYLCARBONATE. Syn: *methyl carbonate.* Colorless liquid, pleasant odor, miscible with acids and alkalies, sol in most organic solvents, insol in water. Flam liquid. $CO(OCH_3)_2$, mw: 90, mp: 0.5°, bp: 90°, flash p: 66°F (OC).
THR = An irr. MOD via oral and inhal routes. Violent reaction with potassium-*tert*-butoxide. [*19*]
Fire Hazard: Dangerous, via heat, open flames (sparks) or oxidizers.
To Fight Fire: Alcohol foam.

DI(METHYL "CELLOSOLVE") MALEATE. Liquid. $CH_3O(CH_2)_2OCOCH{:}CHCOO(CH_2)_2OCH_3$, mw: 232.23, mp: −50°, bp: 152° @ 5 mm, d: 1.1413 @ 20°/20°, vap. d: 8.01.
THR = Details U. See also glycols.
Fire Hazard: Slight, when heated; can react with oxidizing materials.

DIMETHYL "CELLOSOLVE" PHTHALATE. See dimethylglycol phthalate.

DIMETHYLCHLORACETAL. Colorless liquid. $ClCH_2CH(OCH_3)_2$, mw: 124.6, bp: 126°–132°, flash p: 110°F, d: 1.082–1.092 @ 25°/4°, autoign. temp.: 450°F, vap. d: 4.3.
THR = Details U. As an aldehyde it may have irr and narcotic action.
Fire Hazard: Mod, when exposed to heat or flame or oxidizers.
Disaster Hazard: Dangerous; see chlorides; can react with oxidizing materials.
To Fight Fire: Foam, CO_2, dry chemical.

DIMETHYLCHLOROARSINE. Syns: *cacodyl chloride, chlorodimethylarsine.* Colorless liquid, insol in water. $(CH_3)_2AsCl$, mw: 140.44, mp: <40°, bp: 106.5°, d: >1, vap. d: 4.84.
THR = HIGH toxicity. See arsenic compounds. See also dichloro(2-chlorvinyl) arsine.
Fire Hazard: Flam liquid. Keep away from powerful oxidizers, heat and flame.
Spont Heating: Yes.
Disaster Hazard: See arsenic compounds.
To Fight Fire: Water, water spray, foam, dry chemical, CO_2.

8,12-DIMETHYL-9-CHLOROBENZ(a)ACRIDINE.
THR = An exper carc. [*23*]

DIMETHYL-β-CHLOROETHYLAMINE. Liquid. $(CH_3)_2NC_2H_4Cl$, mw: 107.6, vap. d: 3.72.
THR = Probably very irr. No data.
Disaster Hazard: Dangerous; when heated to decomp, emits highly toxic fumes of chlorides.

***o,o*-DIMETHYL-*o*-(3-CHLORO-4-NITROPHENYL) THIOPHOSPHATE.** See chlorthion.

N,N-DIMETHYL-*p*-[(*m*-CHLOROPHENYL)AZO] ANILINE. $C_{19}H_{14}N_3Cl$, mw: 199.7.
THR = An exper neo. [*3*]

N,N-DIMETHYL-*p*-[(*o*-CHLOROPHENYL)AZO] ANILINE.
THR = An exper neo. [*3*]

N,N-DIMETHYL-*p*-[(*p*-CHLOROPHENYL)AZO] ANILINE.
THR = An exper neo. [*3*]

3,3-DIMETHYL-1-*p*-CHLOROPHENYL TRIAZINE. $C_8H_{10}N_3Cl$, mw: 183.7.
THR = An exper carc. [*3*]

1,3-DIMETHYLCHOLANTHRENE. $C_{22}H_{18}$, mw: 282.4.
THR = An exper neo. [*3, 23*]

2,3-DIMETHYLCHOLANTHRENE.
THR = An exper neo. [*3, 23*]

4,5-DIMETHYLCHOLANTHRENE.
THR = An exper carc. [*23*]

1,2-DIMETHYLCHRYSENE. $C_{20}H_{16}$, mw: 256.4.
THR = An exper (±) carc. [*3, 6*]

1,1-DIMETHYLCHRYSENE.
THR = An exper neo. [*3*]

4,5-DIMETHYLCHRYSENE.
THR = An exper carc. [*23*]

5,6-DIMETHYLCHRYSENE.
THR = An exper neo [*3*] and carc. [*23*]

DIMETHYLCYANAMIDE. Colorless, mobile liquid. $(CH_3)_2NCN$, mw: 74.09, mp: −41.0°, bp: 160°, flash p: 160°F (TCC), d: 0.8767 @ 30°, vap. press: 40 mm @ 80°, vap. d: 2.55.
Acute tox data: Oral LD_{50} (rat) = 146 mg/kg; inhal LC_{50} (rat) = 250 mg/m^3. [*3*]
THR = HIGH via oral and inhal routes.
Fire Hazard: Mod, when exposed to heat, flame or oxidizers.
Disaster Hazard: Dangerous; see cyanides; will react with water or steam to produce toxic and flam vapors; can react with oxidizing materials.
To Fight Fire: Foam, CO_2, dry chemical.

***p*-DIMETHYLCYCLOHEXANE.** Syn: *hexahydroxylene.* Liquid. $(CH_3)_2C_6H_{10}$, mw: 112.21, mp: −86°, bp: 119.5°, flash p: 52°F (CC), d: 0.77, vap. press: 10 mm @ 10.2°, vap. d: 3.86.
THR = See cycloparaffins.
Fire Hazard: Dangerous, when exposed to heat or flame; can react vigorously with oxidizing materials.
Spont Heating: No.
Explosion Hazard: U.
Disaster Hazard: Dangerous. Keep away from heat and open flame.
To Fight Fire: Foam, CO_2, dry chemical.

DIMETHYL-1-CYCLOHEXENE-1,2-DICARBOXYLIC ACID. See dimethyl tetrahydrophthalate.

11,17-DIMETHYL-15H-CYCLOPENTA(a)PHENANTHRENE.
THR = An exper carc. [*23*]

12,17-DIMETHYL-15H-CYCLOPENTA(a)PHENANTHRENE.
THR = An exper carc. [*23*]

3,4-DIMETHYL-1,2-CYCLO PENTENOPHENANTHRENE.
THR = An exper neo. [3]

DIMETHYLDECALIN. $C_{12}H_{18}$, mw: 162, bp: 455°, d: 1.0, flash p: 184°F, autoign. temp.: 455°F, lel = 0.7% @ 200°F, uel = 5.3% @ 300°.
THR = U.
Fire Hazard: Mod, when exposed to heat, flame or oxidizers.
To Fight Fire: Dry chemical, water spray or mist, carbon dioxide.

7,14-DIMETHYLDIBENZ(a,h)ANTHRACENE. $C_{24}H_{18}$, mw: 306.4.
THR = An exper carc. [3, 23]

7,14-DIMETHYLDIBENZ(a,j)ANTHRACENE.
THR = An exper carc. [23]

4,9-DIMETHYL-2,3,5,6-DIBENZOTHIOPHENTHRENE. $C_{22}H_{16}S$, mw: 312.4.
THR = An exper neo. [3]

1,1-DIMETHYLDIBORANE. Colorless gas, decomp by water. $B_2H_4(CH_3)_2$, mw: 55.7, mp: −150°, bp: −2.6°.
THR = Details U. See boron compounds and boron hydrides.
Fire Hazard: Details U. See also boron hydrides.

1,2-DIMETHYLDIBORANE. Colorless gas, decomp by water. $B_2H_4(CH_3)_2$, mw: 55.7, mp: −125°, bp: −49°.
THR = Details U. See boron hydrides.

DIMETHYL-1,2-DIBROMO-2,2-DICHLOROETHYL PHOSPHATE. Syns: *"DiBrom," naled.* Slightly sol in aliphatic hydrocarbons, very sol in aromatic hydrocarbons. $C_4H_7O_4PBr_2Cl_2$, mw: 380.7, mp: 27.0°.
Acute tox data: Oral LD_{50} (rat) = 250 mg/kg; dermal LD_{50} (rat) = 800 mg/kg. [3]
THR = HIGH via oral and dermal routes. A cholinesterase inhibitor. It is an insecticide and nonsystemic acaricide. See parathion.
Disaster Hazard: Dangerous; see phosphorus compounds and chlorides.

DIMETHYLDIBROMODIPYRIDINE TIN. Solid. $(CH_3)_2SnBr(C_5H_5N)_2$, mw: 466.8, mp: 172°.
THR = Details U. See tin compounds.
Disaster Hazard: Dangerous; see bromides.

2,5-DIMETHYL-2,5-DI-(*tert*-BUTYLPEROXY) HEXANE. Colorless to light yellow liquid, insol in water, sol in many organic solvents. $C_{16}H_{34}O_4$, mw: 290.5, d: 0.85, fp: 8°, flash p: >180°F (MOC), bp: 250°.
Acute tox data: $ipLD_{LO}$ (mouse) = 1700 mg/kg. [3]
THR = MOD via ip route. See peroxides, organic.
Fire Hazard: Mod, via heat, flames, reducing agents.
To Fight Fire: Water spray, foam, dry chemical.

2,5-DIMETHYL-2,5-DI-(*tert*-BUTYL PEROXY) HEXYNE-3. $C_{16}H_{30}O_4$, mw: 286.4, fp: 9.7, bp: 240°.
Acute tox data: $ipLD_{50}$ (mouse) = 1850 mg/kg. [3]
THR = MOD via ip route. See peroxides, organic.

DIMETHYL-2,2-DICHLOROETHENYL PHOSPHATE. See dichlorvos.

***o,o*-DIMETHYL-*o*-(2,5-DICHLORO-4-IODOPHENYL) THIOPHOSPHATE.** See iodofenphos.

DIMETHYLDICHLOROSILANE. See dichlorodimethyl silane.

DIMETHYLDICHLOROVINYL PHOSPHATE. See dimethyl-2,2-dichloroethenylphosphate.

DIMETHYL DICHLORODIPYRIDINE TIN. Solid. $(CH_3)_2(C_5H_5N)_2SnCl_2$, mw: 377.9, mp: 163°.
THR = Details U. See tin compounds.
Disaster Hazard: Dangerous; see chlorides.

DIMETHYLDIDODECYLAMMONIUM CHLORIDE. Solid. $(CH_3)_2(C_{12}H_{25})_2NCl$, mw: 418.2, vap. d: 14.4.
THR = MOD irr to skin, eyes and mu mem.
Disaster Hazard: Dangerous; see chlorides.

2,5-DIMETHYL-1,2,5,6-DIEPOXYHEX-3-YNE. $C_8H_{10}O_2$, mw: 138.2.
THR = An exper neo. [3]

DIMETHYLDIETHYLTIN. Colorless liquid, insol in water. $(CH_3)_2Sn(C_2H_5)_2$, mw: 206.9, d: 1.2319 @ 19°, mp: < −13°, bp: 146°.
THR = See tin compounds.

N,N-DIMETHYL-2,5-DIFLUORO-*p*-(2,5-DIFLUOROPHENYL AZO)ANILINE. $C_{14}H_{11}N_3F_4$, mw: 297.3.
THR = An exper carc. [3]

N,N-DIMETHYL-*p*-(3,4-DIFLUOROPHENYL AZO) ANILINE. $C_{14}H_{13}N_3F_2$, mw: 261.3.
THR = An exper carc. [3]

N,N′-DIMETHYL-*p*-(2,5-DIFLUOROPHENYLAZO) ANILINE.
THR = An exper carc. [3]

11,12-DIMETHYL-16,17-DIHYDRO-15H-CYCLOPENTA(a)PHENANTHRENE.
THR = An exper carc. [23]

11,17-DIMETHYL-16,17-DIHYDRO-15H-CYCLOPENTA(a)PHENANTHRENE.
THR = An exper carc. [23]

DIMETHYLDIISOBUTYLTIN. Solid. $(CH_3)_2Sn(C_4H_9)_2$, mw: 263.0, d: 1.1179 @ 20°, bp: 85° @ 16.5 mm.
THR = See tin compounds.

8,9-DIMETHYL-7,12-DIMETHYL BENZ(a)ANTHRACENE.
THR = An exper carc. [*23*]

9,10-DIMETHYL-7,12-DIMETHYL BENZ(a) ANTHRACENE.
THR = An exper carc. [*23*]

N,N-DIMETHYL-4-[4′-(2′,5′-DIMETHYL PYRIDYL-1′-OXIDE)AZO]ANILINE.
THR = An exper neo. [*3, 23*]

N,N-DIMETHYL-4-(4′-(3′,5′-DIMETHYL PYRIDYL-1′-OXIDE)AZO)ANILINE.
THR = An exper neo. [*3*]

***o,o*-DIMETHYL-*o,p*-(DIMETHYLSULFAMOYL) PHENYLPHOSPHOROTHIOATE.** Syn: *famphur.* Crystalline powder, very sol in chloroform and carbon tetrachloride, slightly sol in water. $(CH_3O)_2P(S)OC_6H_4SO_2N(CH_3)_2$, mw: 325, mp: 55°.
Acute tox data: Oral LD_{50} (rat) = 35 mg/kg; dermal LD_{50} (rabbit) = 1460 mg/kg; oral LD_{50} (wild birds) = 2 mg/kg. [*3*]
THR = HIGH via oral and MOD via dermal routes. A cholinesterase inhibitor. See parathion.
Disaster Hazard: Dangerous; see phosphates.

N,N′-DIMETHYL-N,N′-DINITROSO ETHYLENE DIAMINE. $C_4H_{10}O_2N_4$, mw: 146.2.
THR = An exper carc. [*3*]

DIMETHYLDINITROSO OXAMIDE. $C_4H_6O_4N_4$, mw: 174.1.
Acute tox data: Oral LD_{50} (rat) = 96 mg/kg. [*3*]
THR = HIGH via oral route.
Fire Hazard: See nitrates.
Explosion Hazard: Severe, when shocked or exposed to heat. See also explosives, high.
Disaster Hazard: Dangerous; shock will explode it; when heated, emits highly toxic fumes; can react vigorously with reducing materials.

2,5-DIMETHYL-1,4-DINITROSO PIPERAZINE. $C_6H_{14}O_2N_4$, mw: 174.2.
THR = An exper carc. [*3*]

2,6-DIMETHYL-1,4-DINITROSO PIPERAZINE.
THR = An exper carc. [*3*]

N,N-DIMETHYL-N,N-DINITROSO PHTHALAMIDE. $C_{10}H_{10}O_4N_4$, mw: 250.2.
THR = An exper neo. [*3*]

N,N′-DIMETHYL-N,N′-DINITROSO-1,3-PROPANE DIAMINE. $C_5H_{12}O_2N_4$, mw: 160.2.
THR = An exper carc. [*3*]

DIMETHYLDIOXANE. Water white liquid. $OCH(CH_3)CH_2OCH_2CH(CH_3)$, mw: 116.16, bp: 117.5°, flash p: 75°F, d: 0.9268, vap. press: 15.4 mm @ 20°, vap. d: 4.0.
Acute tox data: Oral LD_{50} (rat) = 3000 mg/kg; inhal LC_{LO} (rat) = 8000 ppm for 4 hrs. [*3*]
THR = MOD via oral and LOW via inhal routes.
Fire Hazard: Dangerous, when exposed to heat or flame; can react vigorously with oxidizing materials.
To Fight Fire: Foam, CO_2, dry chemical.

3,6-DIMETHYL-2,5-*p*-DIOXANEDIONE. Syn: *lactide.* Pale yellow, crystalline solid. $C_6H_8O_4$, mw: 144.1, mp: 96°–104°.
THR = U.
Disaster Hazard: Slight.

2,6-DIMETHYL-*m*-DIOXAN-4-YL-ACETATE. Syns: *dimethoxane, dioxin.* Clear yellow to light amber liquid, sol in or miscible with water and organic solvents. $C_8H_{14}O_4$, mw: 174.2, d: 1.068–1.075 @ 25°/25°, bp: 66°–68° @ 3 mm, fp: $<-25°$.
Acute tox data: Inhal LC_{50} (guinea pigs) = 1 ppb. [*3*]
THR = HIGH. This is one of the most toxic materials known. Damaging to guinea pigs at 0.6 ppb. An exper (+) carc, neo and teratogen. [*3, 48, 81*]

N,N-DIMETHYL-2,2-DIPHENYL ACETAMIDE. Syns: *diphenamid, dymid.* White solid, very slightly sol in water, mod sol in acetone, dimethyl formamide, and phenyl cellosolve. $(C_6H_5)_2CHCON(CH_3)_2$, mw: 303, mp: 134.5°–135.5°.
Acute tox data: LD_{50} (mammal) = 1190 mg/kg; oral LD_{50} (rat) = 975 mg/kg. [*3*]
THR = MOD via oral route.

1,3-DIMETHYL-1,3-DIPHENYL CYCLOBUTANE. Insol in water. $C_{18}H_{20}$, mw: 236, mp: 120°F, d: 1.0 @ 122°F, bp: 585°–588°F, flash p: 289°F.
THR = U.
Fire Hazard: Slight, when exposed to heat or flame.
To Fight Fire: Dry chemical, CO_2.

N,N-DIMETHYL-4-(DIPHENYLMETHYL)ANILINE. $C_{21}H_{21}N$, mw: 287.4.
THR = An exper neo. [*3*]

1,1′-DIMETHYL-4,4′-DIPYRIDINIUM DICHLORIDE. See paraquat.

DIMETHYLDISULFIDE. See 2,3-dithiabutane.

DIMETHYLDITHIOCARBAMIC ACID + DIMETHYL AMINE (1:1). $C_3H_7NS_2 + NC_2H_7$.
THR = An exper neo. [*3*]

DIMETHYLDITHIOPHOSPHATE.
THR = HIGH toxicity. An insecticide. Similar to parathion. See parathion.

DIMETHYLDITHIOPHOSPHATE OF DIETHYL MERCAPTOSUCCINATE. Syn: *malathion.* Brown to yellow liquid, characteristic odor, miscible in organic solvents, slightly water-sol. $C_{10}H_{19}O_6PS_2$, mw: 330, d: 1.23 @ 25°/4°, mp: 2.9°, bp: 156° @ 0.7 mm.

Acute tox data: Dermal LD_{50} (rat) = 4444 mg/kg; oral LD_{50} (rat) = 1401 mg/kg; LD_{50} (mammal) = 500 mg/kg; oral TD_{LO} (man) = 857 mg/kg. [3]
THR = MOD via oral and dermal routes. Affects the CNS. Has caused allergic sensitization of the skin. An organic phosphate cholinesterase inhibitor. Less toxic than parathion. Used as a food additive permitted in the feed and drinking water of animals and/or for the treatment of food-producing animals. [*109*] An exper carc. [*23*]
Disaster Hazard: Dangerous; see phosphates and parathion.

DIMETHYLENEIMINE. See ethyleneimine.

DIMETHYLENE METHANE. See allene.

DIMETHYLETHANOLAMINE. See dimethylaminoethanol.

DIMETHYLETHANOL OCTADECYL AMMONIUM CHLORIDE. Solid.
$(CH_3)_2(C_2H_5O)(C_{18}H_{37})NCl$, mw: 378.1, vap. d: 13.0
THR = Probably a MOD irr to skin, eyes and mu mem.
Disaster Hazard: Dangerous; see chlorides.

DIMETHYL ETHER. Syns: *methyl ether, methyl oxide.* Colorless gas, ether odor. CH_3OCH_3, mw: 46.07, mp: −138.5°, bp: −23.7°, lel = 3.4%, uel = 27%, flash p: −42°F (CC), d: 0.661, autoign. temp.: 662°F, vap. d: 1.617.
Acute tox data: A MILD irr and MOD via oral and inhal routes. See also ethyl ether.
Fire Hazard: Very dangerous, when exposed to heat, flame or oxidizers.
Spont Heating: No.
Explosion Hazard: Dangerous, when exposed to flame, sparks, etc. Violent reaction with AlH_3, $LiAlH_2$. [*19*] See also ethers.
Disaster Hazard: Highly dangerous. Keep in closed container away from heat and open flame.
To Fight Fire: Stop flow of gas.

N,N-DIMETHYL-*p*-[(3-ETHOXY PHENYL)AZO] ANILINE. $C_{16}H_{19}ON_3$, mw: 269.4.
THR = An exper neo. [*3*]

DI-N-METHYLETHYL CARBAMATE. $C_5H_{11}O_2N$, mw: 117.2.
Acute tox data: $scLD_{50}$ (mouse) = 1050 mg/kg. [*3*]
THR = MOD via sc route. An exper neo. [*3*]

DIMETHYLETHYL CARBINOL. See *tert*-amyl alcohol.

DIMETHYLETHYLENE. See *cis*-butene-2.

DIMETHYLETHYLENE BISGLYCOLATE. Liquid. $(CH_2OCH_2CO_2CH_3)_2$, mw: 206.2, bp: 145°–146° @ 4 mm, flash p: 290°F (OC), d: 1.1994 @25°/25°, vap. d: 7.11.
THR = Details U. See esters.
Fire Hazard: Slight, when exposed to heat or flame; can react with oxidizing materials.
To Fight Fire: Water, foam, CO_2, dry chemical.

N,N'-DIMETHYLETHYLENE UREA.
THR = An exper mutagen. [*3*]

α,α-DIMETHYLETHYL NITRITE.
See *tert*-butyl nitrite.

1,1-DIMETHYL-3-ETHYL-3-NITROSO UREA.
$C_5H_{11}O_2N_3$, mw: 145.2.
THR = An exper neo. [*3*]

2,3-DIMETHYL-3-ETHYLPENTANE. Liquid. C_9H_{20}, mw: 128.25, bp: 142°, flash p: 47°F, d: 0.754 @ 20°, vap. d: 4.43, autoign. temp.: 734°F.
THR = Details U. Probably irr and narcotic in high conc.
Fire Hazard: Dangerous, when exposed to heat or flame; can react vigorously with oxidizing materials.
Explosion Hazard: U.
Disaster Hazard: Dangerous! Keep away from heat and open flame!
To Fight Fire: Foam, CO_2, dry chemical.

DIMETHYLETHYLPROPYL TIN. A powder.
$(CH_3)_2(C_2H_5)(C_3H_7)Sn$, mw: 221, d: 1.2014 @ 20°/20°, bp: 150°.
THR = See tin compounds.

***o,o*-DIMETHYL-*s*-(2-ETHYLTHIOETHYL)PHOSPHORODITHIOATE.** See ekatin.

DIMETHYLETHYL TIN IODIDE. Crystals.
$(CH_3)_2(C_2H_5)SnI$, mw: 305, d: 2.026 @ 20°/20°, bp: 78° @ 11 mm.
THR = See tin compounds and iodides.

7,8-DIMETHYLFLUORANTHENE.
THR = An exper neo. [*3*]

2,2-DIMETHYLFLUORANTHENE. $C_{18}H_{14}$, mw: 230.3.
THR = An exper neo. [*3*]

N,N-DIMETHYL-*p*-(2-FLUOROPHENYLAZO) ANILINE. $C_{14}H_{14}N_3F$, mw: 243.3.
THR = An exper carc. [*3*]

N,N-DIMETHYL-2-FLUORO-4-PHENYLAZO ANILINE.
THR = An exper carc. [*3*]

N,N-DIMETHYL-*p*-(3-FLUOROPHENYLAZO) ANILINE.
THR = An exper carc. [*3*]

N,N-DIMETHYL-*p*-(*p*-FLUOROPHENYLAZO) ANILINE.
THR = An exper carc. [*3*]

N,N-DIMETHYL FLUOREN-2-AMINE. $C_{15}H_{15}N$, mw: 209.3.
THR = An exper carc. [*3*]

DIMETHYL FLUOROPHOSPHATE. Liquid. $(CH_3O)_2POF$, mw: 128.1, mp: low, bp: 149°, d: 1.28, vap. d: 4.42.
Acute tox data: Inhal LC_{LO} (rat) = 280 mg/m^3 for 10 min; ivLD_{50} (mouse) = 0.45 mg/kg. [*3*]
THR = HIGH via inhal and iv routes.
Disaster Hazard: Dangerous; see fluorides and phosphates.

DIMETHYL FORMAMIDE. Colorless, mobile liquid. $(CH_3)_2NCHO$, mw: 73.1, bp: 152.8°, lel = 2.2% @ 100°, uel = 15.2% @ 100°, flash p: 136°, fp: −61°, d: 0.9445 @ 25°/4°, autoign. temp.: 833°F, vap. press: 3.7 mm @ 25°, vap. d: 2.51.
Acute tox data: Inhal TC_{LO} (human) = 20 ppm causing CNS symptoms; oral LD_{50} (rat) = 2800 mg/kg; dermal LD_{50} (rabbit) = 5000 mg/kg; ipLD_{50} (cat) = 400 mg/kg. [*3*]
THR = HIGH via inhal and ip; and MOD via oral and dermal routes.
Fire Hazard: Mod, when exposed to heat or flame; can react with oxidizing materials.
Explosion Hazard: When exposed to flame.
Avoid contact with halogenated hydrocarbons, inorganic nitrates, Br_2, CCl_4, CrO_3, (2,5-dimethyl pyrrole + $P(OCl)_3$), C_6Cl_6, organic nitrates, methylene diisocyanates, P_2O_3, $Al(C_2H_5)_3$. [*19*]
To Fight Fire: Foam, CO_2, dry chemical.

DIMETHYL FORMOCARBOTHIALDINE.
See mylone.

2,5-DIMETHYLFURAN. Colorless liquid. $C_4H_2O(CH_3)_2$, mw: 96.12, bp: 94°, flash p: 45°F (OC), d: 0.9026 @ 17.7°/4°, vap. d: 3.31.
Acute tox data: Inhal LC_{LO} (rat) = 500 ppm for 4 hrs. [*3*]
THR = MILD irr, MOD via inhal route.
Fire Hazard: Dangerous, when exposed to heat or flame; can react vigorously with oxidizing materials.
Explosion Hazard: U.
Disaster Hazard: Dangerous! Keep away from heat and open flame!
To Fight Fire: Alcohol foam, foam, CO_2, dry chemical.

DIMETHYLGALLIUM AMIDE. White crystals. $Ga(CH_3)_2NH_2$, mw: 115.8.
THR = See gallium compounds and amides.

DIMETHYLGALLIUM CHLORIDE DIAMMINE. White crystals, decomp by water. $Ga(CH_3)_2Cl \cdot 2NH_3$, mw: 169.3, mp: 112°.
THR = See gallium compounds and chlorides.

DIMETHYLGALLIUM CHLORIDE MONAMMINE. White crystals, decomp by water. $Ga(CH_3)_2Cl \cdot NH_3$, mw: 152.3, mp: 54°.
THR = See gallium compounds and chlorides.

DIMETHYLGLYCOL PHTHALATE. Syn: *dimethyl "Cellosolve" phthalate*. Liquid. $C_6H_4[COO(CH_2)_2OCH_3]_2$, mw: 282.3, bp: 230°, flash p: 369°F (CC), d: 1.8, vap. d: 9.72.
THR = Details U. See also glycols.
Fire Hazard: Slight, when exposed to heat or flame; can react with oxidizing materials.
Spont Heating: No.
To Fight Fire: CO_2, dry chemical.

2,6-DIMETHYL-4-HEPTANONE.
See diisobutyl ketone.

2,6-DIMETHYL-2,5-HEPTADIEN-4-ONE.
See phorone.

2,5-DIMETHYLHEPTANE. Liquid. C_9H_{20}, mw: 128.25, bp: 136°, flash p: 73.5°F, d: 0.715 @ 20°, vap. d: 4.42.
THR = No details. Probably narcotic in high conc. MILD irr.
Fire Hazard: Dangerous, when exposed to heat or flame; can react vigorously with oxidizing materials.
Explosion Hazard: U.
To Fight Fire: Foam, CO_2, dry chemical.

3,5-DIMETHYLHEPTANE. Liquid. C_9H_{20}, mw: 128.25, bp: 136°, flash p: 73.5°F, d: 0.723 @ 20°, vap. press: 9.5 mm @ 25°, vap. d: 4.42.
THR = No details. Probably irr and narcotic in high conc.
Fire Hazard: Dangerous, when exposed to heat or flame; can react vigorously with oxidizing materials.
Explosion Hazard: U.
To Fight Fire: CO_2, dry chemical.

4,4-DIMETHYLHEPTANE. Liquid. C_9H_{20}, mw: 128.25, bp: 135.2°, flash p: 73.5°F, d: 0.72 @ 25°/4°, vap. press: 10.4 mm @ 25°, vap. d: 4.42.
THR = No details. Probably irr and narcotic in high conc.
Fire Hazard: Dangerous, when exposed to heat or flame; can react vigorously with oxidizing materials.
Explosion Hazard: U.
To Fight Fire: CO_2, dry chemical.

2,6-DIMETHYLHEPTANOL-4.
See diisobutylcarbinol.

2,6-DIMETHYLHEPTENE-3 (*cis* and *trans* ISOMERS). A clear liquid. C_9H_{18}, mw: 126.23, bp: 128.5°–129°, flash p: 70°F (TOC), d: 0.722 @ 15.5°/15.5°, vap. press: 28.4 mm @ 38°, vap. d: 4.38.
THR = Details U. Probably irr and narcotic in high conc.

Fire Hazard: Dangerous, when exposed to heat or flame; can react vigorously with oxidizing materials.
To Fight Fire: Foam, CO_2, dry chemical.

2,5-DIMETHYLHEXADIENE-2,4. Liquid. $(CH_3)_2CCHCHC(CH_3)_2$, mw: 110.12, mp: −91.3°, bp: 102.5°, d: 0.762 @ 20°/20°, vap. d: 3.8.
THR = Details U. Probably irr and narcotic in high conc.
Fire Hazard: Mod; can react with oxidizing materials.
To Fight Fire: Foam, CO_2, dry chemical.

2,3-DIMETHYLHEXANE. A clear liquid. C_8H_{18}, mw: 114.23, bp: 116°, flash p: 45°F (OC), d: 0.716 @ 15.5°/15.5°, vap. d: 4.1, autoign. temp.: 820°F.
THR = Details U. Probably irr and narcotic in high conc.
Fire Hazard: Dangerous, when exposed to heat or flame; can react vigorously with oxidizing materials.
Explosion Hazard: U.
To Fight Fire: Foam, CO_2, dry chemical.

2,4-DIMETHYLHEXANE. A liquid. C_8H_{18}, mw: 114.23, bp: 109°, flash p: 50°F (OC), d: 0.705 @ 15.5°/15.5°, vap. d: 3.9.
THR = U. Probably irr and narcotic in high conc.
Fire Hazard: Dangerous, when exposed to heat or flame; can react vigorously with oxidizing materials.
To Fight Fire: Foam, CO_2, dry chemical.

2,5-DIMETHYLHEXANE-2,5-DIHYDROPEROXIDE. Fine white crystals, insol in hydrocarbons, slightly sol in water, esters and glycerine, sol in other organic solvents. $C_8H_{18}O_4$, mw: 178.2, mp: 104°.
THR = See peroxides, organic.

2,5-DIMETHYLHEXANE-2,5-DIOL. Crystals. $(CH_3)_2(OH)C(CH_2)_2C(OH)(CH_3)_2$, mw: 146.14, mp: 88°–89°, vap. d: 5.03.
THR = Details U.

2,5-DIMETHYL-1,2,6-HEXANETRIOL. $CH_2(OH)CH(CH_3)CH_2CH_2C(CH_3)(OH) \cdot CH_2(OH)$, mw: 162.2.
THR = Limited animal exper suggest LOW toxicity. See also glycols.

DIMETHYLHEXYNEDIOL. White crystals. $C_8H_{14}O_2$, mw: 142.2, mp: 94°–95°, bp: 205°–206°, vap. d: 4.9.
THR = Details U.
Fire Hazard: Slight.

DIMETHYLHEXYNOL. Syn: *3,5-dimethyl-1-hexyn-3-ol.* Colorless liquid, camphor-like odor. $CH_3CH(CH_3)CH_2C(CH_3)(OH)C \vdots CH$, mw: 126.5, bp: 150°–151°, fp: −68°, flash p: 134°F (TOC), d: 0.8545 @ 20°/20°.
THR = U.
Fire Hazard: Mod, when exposed to heat or flame; can react with oxidizing materials.
To Fight Fire: Foam, CO_2, dry chemical.

DIMETHYLHYDANTOIN. Syn: *α-ureidoisobutyric acid lactam.* White crystalline solid. $C_5H_8O_2N_2$, mw: 128.13, mp: 178°, vap. d: 4.4.
THR = Details U. Hydantoins are related to barbiturates and have a depressant action on the CNS.
Disaster Hazard: Slight.

DIMETHYLHYDANTOIN-FORMALDEHYDE RESIN. Colorless to light yellow brittle lumps, odorless or faint caramel-like odor. mw: 240–300 (average), mp: 60° (min), d: 1.30.
THR = MOD via oral and inhal routes. An irr and allergen. This resin contains a small percentage of formaldehyde (about 0.3% max) which may cause dermatitis in individuals sensitive to formaldehyde. When crushed or ground it gives a fine dust which may pass through ordinary respirators, irr mu mem and lead to sore throat, coughing and occasionally vomiting. In case of contact with the body, individuals are advised to flush skin or eyes with water. In grinding or crushing operations, adequate ventilation should be provided.
Disaster Hazard: Slight.

1,1-DIMETHYLHYDRAZINE.
See asym-dimethyl hydrazine.

asym-DIMETHYLHYDRAZINE. Syns: *1,1-dimethylhydrazine, dimazine.* Colorless liquid, ammonia-like odor. Hygroscopic, water-miscible. $(CH_3)_2NNH_2$, mw: 60.1, bp: 63.3°, fp: −58°, flash p: 5°F, d: 0.782 @ 25°/4°, vap. press: 157 mm @ 25°, vap. d: 1.94, autoign. temp.: 480°F, lel = 2%, uel = 95%.
Acute tox data: Oral LD_{50} (rat) – 122 mg/kg; inhal LC_{50} (rat) = 252 ppm for 4 hrs; dermal LD_{50} (guinea pig) = 1329 mg/kg; ip LD_{50} (rat) = 102 mg/kg. [*3*]
THR = HIGH via oral, inhal, and ip routes. MOD via dermal route. An exper (+) carc. [*3, 10*]
Fire Hazard: Dangerous, when exposed to heat, flame or oxidizers.
Explosion Hazard: U.
Disaster Hazard: Highly dangerous; when heated to decomp, emits highly toxic fumes; can react vigorously with oxidizing materials such as air, H_2O_2, HNO_3 fuming HNO_3, $(HNO_3 + N_2O_4)$, NO. [*19*]
To Fight Fire: Alcohol foam, CO_2, dry chemical.

1,2-DIMETHYLHYDRAZINE. Clear, colorless, flam, hygroscopic liquid, fishy ammonia odor. $C_2H_8N_2$, mw: 60.1, bp: 81°, mp: −9°, d: 0.8274 @ 20°/4°.
Acute tox data: Oral LD_{50} (rat) = 100 mg/kg; sc LD_{50} (rat) = 220 mg/kg; im LD_{50} (hamster) = 95 mg/kg; inhal LC_{LO} (rat) = 280 ppm for 4 hrs; imp of 2 mg/kg (rabbit) ⟶ neo. [*3*]

THR = HIGH via oral, sc, im, inhal routes. An exper (+) carc. [*3, 10, 23*]

1,2-DIMETHYLHYDRAZINE HYDROCHLORIDE. $C_2H_8N_2 \cdot HCl$, mw: 96.6.
THR = HIGH via oral acute route. An exper carc. [*3, 23*]

1,1-DIMETHYLHYDRAZINE HYDROCHLORIDE. $C_2H_8N_2 \cdot HCl$, mw: 96.6.
THR = HIGH acute via oral route. An exper neo. [*3*]

1,2-DIMETHYLHYDRAZINE DIHYDROCHLORIDE. $C_2H_8N_2 \cdot 2HCl$, mw: 133.
THR = An exper carc. [*3*]

2-(2,2-DIMETHYLHYDRAZINO)-4-(5-NITRO-2-FURYL)THIAZOLE. $C_9H_{10}O_3N_4S$, mw: 254.3.
THR = An exper carc. [*3, 23*]

N,N-DIMETHYLHYDROXYACETAMIDE. Crystals. $HOCH_2CON(CH_3)_2$, mw: 103.12, mp: 45°, bp: 213°, d: 1.076 @ 50°/4°.
THR = Details U. See also amides.
Disaster Hazard: Slight; when heated, emits toxic fumes.

2,5-DIMETHYL-2-HYDROXYADIPALDEHYDE. $OHCCH(CH_3)(CH_2)_2C(CH_3)OHCHO$, mw: 158.
THR = Limited animal exper suggest LOW toxicity. See also aldehydes.

DIMETHYLHYDROXYBORINE. See dimethylboric acid.

***o,o*-DIMETHYL-1-HYDROXY-2,2,2-TRICHLOROETHYL PHOSPHONATE.** See dipterex.

N,N-DIMETHYL-*p*-(6-INDAZYLAZO)ANILINE. $C_{15}H_{15}N_5$, mw: 265.4.
THR = An exper carc. [*3, 23*]

DIMETHYLISOPROPANOLAMINE. Liquid. $C_3H_7ON(CH_3)_2$, mw: 103.2, mp: −85°, bp: 125.8°, flash p: 95°F (OC), d: 0.86, vap. press: 9.0 mm @ 20°, vap. d: 3.55.
THR = Details U. See also amines.
Fire Hazard: Dangerous, when exposed to heat or flame; can react with oxidizing materials.
To Fight Fire: Alcohol foam, CO_2, dry chemical.

N,N-DIMETHYL-*m*-ISOPROPYLPHENYL CARBAMATE. $C_6H_3(CH_3)_2NHCOOHC(CH_3)_2$, mw: 207.
THR = A recognized carc. [*14*] See carbamates. Limited animal exper suggest HIGH acute toxicity.
Disaster Hazard: Dangerous; see carbamates.

DIMETHYL KETOL. See acetoin.

DIMETHYL KETONE. See acetone.

N,N-DIMETHYL-4-[4′-(2′,5′-LUTIDYL)AZO] ANILINE.
THR = An exper carc. [*23*]

N,N-DIMETHYL-4-[4′-(2′,6′-LUTIDYL-1′-OXIDE) AZO]ANILINE.
THR = An exper carc. [*23*]

N,N-DIMETHYL-4-(3′,5′-LUTIDYL-1′-OXIDE)AZO ANILINE.
THR = An exper carc. [*23*]

DIMETHYLMAGNESIUM. $(CH_3)_2Mg$, mw: 54.4.
THR = Self-ignites in air. [*19*]

DIMETHYLMALEATE. Liquid. $(:CHCOOCH_3)_2$, mw: 144.12, mp: −17.5°, bp: 205.0°, flash p: 235°F (OC), d: 1.153, vap. press: 1 mm @ 45.7°, vap. d: 4.97.
THR = Details U. Limited animal exper suggest MOD toxicity. See also esters and maleic acid.
Fire Hazard: Slight, when exposed to heat or flame; can react with oxidizing materials.
To Fight Fire: CO_2, dry chemical.

***α,β*-DIMETHYL MALEIC ANHYDRIDE.** $C_6H_6O_3$, mw: 126.1.
THR = An exper carc. [*3, 23*]

DIMETHYLMALONATE. $C_5H_8O_4$, mw: 132.1.
THR = Violent reaction with CH_3N_3 occurred with $NaOCH_3$ present. [*19*]

DIMETHYL MANGANESE. $(CH_3)_2Mn$, mw: 85.
THR = Self-ignites in air. [*19*] See also manganese compounds.

DIMETHYL MERCURY. Colorless liquid, sweet odor, sol in alcohol. $Hg(CH_3)_2$, mw: 230.7, d: 3.069, bp: 96°.
THR = HIGH tox. See mercury compounds, organic.

DIMETHYL METHANE. See propane.

N,N-DIMETHYL-*p*-(2-METHOXYPHENYL AZO) ANILINE. $C_{15}H_{17}ON_3$, mw: 255.4.
THR = An exper carc. [*3*]

N,N-DIMETHYL-*p*-(3-METHOXY PHENYL AZO) ANILINE.
THR = An exper carc. [*3*]

N,N-DIMETHYL-*p*-(4-METHOXY PHENYL AZO) ANILINE.
THR = An exper carc. [*3*]

3,3-DIMETHYL-1-*p*-METHOXY PHENYL TRIAZENE. $C_9H_{13}ON_3$, mw: 179.3.
THR = An exper carc. [*3*]

***o,o*-DIMETHYL-S-(N-METHYL CARBAMOYL METHYL)PHOSPHORODITHIOATE.** Syn: *dimethoate.* White solid, mod sol in water, sol in most organic solvents except hydrocarbons. $(CH_3O)_2PSSCH_2CONHCH_3$, mw: 197, mp: 50°.
Acute tox data: Oral LD_{50} (human) = 30 mg/kg; oral LD_{50} (rat) = 152 mg/kg; dermal LD_{50} (rat) = 353 mg/kg; oral LD_{50} (wild birds) = 7 mg/kg. [*3*]

THR = VERY HIGH via oral and dermal routes. An organic phosphate ester insecticide. A cholinesterase inhibitor. See parathion. An exper carc. [*3*]
Disaster Hazard: Dangerous; see phosphates.

4,12-DIMETHYL-7-METHYLBENZ(a)ANTHRACENE.
THR = An exper carc. [*23*]

5,12-DIMETHYL-7-METHYLBENZ(a)ANTHRACENE.
THR = An exper carc. [*23*]

6,8-DIMETHYL-7-METHYLBENZ(a)ANTHRACENE.
THR = An exper carc. [*23*]

6,12-DIMETHYL-7-METHYLBENZ(a)ANTHRACENE.
THR = An exper carc. [*23*]

8,12-DIMETHYL-7-METHYLBENZ(a)ANTHRACENE.
THR = An exper carc. [*23*]

9,12-DIMETHYL-7-METHYL BENZ(a)ANTHRACENE.
THR = An exper carc. [*23*]

10,12-DIMETHYL-7-METHYLBENZ(a)ANTHRACENE.
THR = An exper carc. [*23*]

3,3-DIMETHYL-1-0-METHYLPHENYL TRIAZENE.
Acute tox data: Oral LD_{50} (rat) = 350 mg/kg; sc LD_{50} (rat) = 500 mg/kg. [*3*]
THR = HIGH via oral and sc routes. An exper carc [*3, 23*] via sc route.

3,3-DIMETHYL-1-*o*-METHYLPHENYL TRIAZENE.
Acute tox data: Oral LD_{50} (rat) = 350 mg/kg; sc LD_{50} (rat) = 500 mg/kg. [*3*]
THR = HIGH via oral and sc routes. An exper carc. [*3, 23*]

N,N-DIMETHYL-4-[4′-(2′-METHYL PYRIDYL) AZO]ANILINE.
THR = An exper carc. [*23*]

N,N-DIMETHYL-4-[4′-(2′-METHYL PYRIDYL-1′-OXIDE)AZO]ANILINE. $C_{14}H_{16}ON_4$, mw: 256.4.
THR = An exper carc. [*3, 23*]

N,N-DIMETHYL-4-[4′-(3′-METHYL PYRIDYL-1′-OXIDE)AZO]ANILINE.
THR = An exper carc. [*3, 23*]

N,N-DIMETHYL-4-[4′-(2′-METHYL PYRIDYL-1-OXIDE)AZO]-*o*-TOLUIDINE.
THR = An exper carc. [*23*]

N,N-DIMETHYL-4-[5′-(3′-METHYL QUINOLYL) AZO]ANILINE.
THR = An exper carc. [*23*]

N,N-DIMETHYL-4-[5′-(6′-METHYL QUINOLYL) AZO]ANILINE.
THR = An exper carc. [*23*]

N,N-DIMETHYL-4-[5′-(7′-METHYL QUINOLYL) AZO]ANILINE.
THR = An exper carc. [*23*]

N,N-DIMETHYL-4-[5′-(8′-METHYL QUINOLYL) AZO]ANILINE.
THR = An exper carc. [*23*]

N,N-DIMETHYL-2-METHYL STILBENAMINE.
$C_{17}H_{19}N$, mw: 237.4.
THR = An exper neo. [*3*]

***o,o*-DIMETHYL-*o*-[4-(METHYL THIO)-*m*-TOLYL] PHOSPHOROTIOATE.** Syns: *fenthion, baytex*. Yellow to tan oily liquid, insol in water, sol in most organic solvents. $(CH_3O)_2P(S)OC_6H_3(CH_3)SCH_3$, mw: 278, bp: 105° @ 0.01 mm, d: 1.245 @ 20°/20°.
Acute tox data: Oral LD_{50} (rat) = 215 mg/kg; dermal LD_{50} (rat) = 330 mg/kg; oral LD_{50} (wild birds) = 2 mg/kg. [*3*]
THR = HIGH via oral and dermal routes, particularly to wild birds. A cholinesterase inhibitor. See also parathion. An exper teratogen. [*3*]
Disaster Hazard: Dangerous; see phosphates and sulfur compounds.

2,6-DIMETHYL MORPHOLINE. Liquid, very sol in water. $C_6H_{13}ON$, mw: 115, d: 0.9346, bp: 146.6, fp: −85°, flash p: 112°F (OC), vap. d: 4.0.
Acute tox data: Oral LD_{50} (rat) = 2830 mg/kg; dermal LD_{50} (rabbit) = 710 mg/kg. [*3*]
THR = MOD via oral and dermal routes.
Fire Hazard: Mod, when exposed to heat, flame or oxidizers.
To Fight Fire: Alcohol foam.

N,N-DIMETHYL-*p*-(1-NAPHTHYLAZO)ANILINE.
$C_{18}H_{17}N_3$, mw: 275.4.
THR = An exper carc. [*3*]

N,N-DIMETHYL-*p*-(2′-NAPHTHYLAZO)ANILINE.
THR = An exper carc. [*3, 23*]

N,N-DIMETHYL-*p*-[2-(1-NAPHTHYL)VINYL]-ANILINE. $C_{20}H_{19}N$, mw: 273.4.
THR = An exper neo. [*3*]

4,6-DIMETHYL-2-(NITRO-2-FURYL)PYRIMIDINE. $C_{10}H_9O_3N_3$, mw: 219.2.
THR = An exper oral carc to rats. [*103*]

1,2-DIMETHYL-5-NITROIMIDAZOLE.
See dimetridiazole.

N,N-DIMETHYL-*p*-[(*m*-NITROPHENYL)AZO] ANILINE. $C_{14}H_{14}O_2N_4$, mw: 270.3.
THR = An exper neo. [*3*]

N,N-DIMETHYL-*p*-[(*o*-NITROPHENYL)AZO] ANILINE.
THR = An exper neo. [*3*]

3,3-DIMETHYL-1-*p*-NITRO PHENYL TRIAZENE. $C_8H_{10}O_2N_4$, mw: 194.2.
Acute tox data: Oral LD_{50} (rat) = 1660 mg/kg; sc LD_{50} (rat) = 350 mg/kg. [*3*]
THR = HIGH via sc and MOD via oral routes. An exper neo [*3, 23*] via sc route.

***o,o*-DIMETHYL-*o,p*-NITROPHENYL THIOPHOSPHATE.** Syn: *dimethyl analog of parathion.* Crystals. $(CH_3O)_2SPOC_6H_4NO_2$, mw: 263.2, vap. d: 9.1, mp: 38°, d: 1.235 @ 20°/4°.
Acute tox data: Oral LD_{50} (rat) = 9 mg/kg; inhal LD_{50} (rat) = 120 mg/m^3 for 4 hrs; dermal LD_{50} (rat) = 67 mg/kg; ip LD_{50} (rat) = 3.5 mg/kg. [*3*] Adult LD = < 1.8 g [*101*]
THR = An organo phosphorus insecticide of HIGH mammalian toxicity via all routes. A cholinesterase inhibitor. HIGH via oral, inhal, ip and dermal routes. [*101*] See also parathion. An exper teratogen. [*3*]
Disaster Hazard: Dangerous; when heated to decomp or on contact with acid or acid fumes, emits highly toxic fumes. See also parathion.

DIMETHYLNITROSAMINE. See N-nitrosodimethyl amine.

N,N-DIMETHYL-*p*-NITROSO ANILINE. $C_8H_{10}ON_2$, mw: 150.2.
THR = An acute oral, deadly poison to rats. An exper neo. [*3*]

2,2′-DIMETHYL-N-NITROSO DIPROPYL AMINE. $C_8H_{18}ON_2$, mw: 158.3.
THR = An exper neo. [*3*]

2,6-DIMETHYL-N-NITROSO MORPHOLINE. $C_6H_{12}O_2N_2$, mw: 144.2.
THR = An exper carc. [*3*]

2,6-DIMETHYL-1-NITROSO PIPERAZINE. $C_7H_{14}ON_2$, mw: 142.2.
THR = An exper neo. [*3*]

1,3-DIMETHYL-1-NITROSO UREA. $C_3H_7O_2N_3$, mw: 117.1.
THR = HIGH acute via oral route. An exper carc [*23*] and teratogen. [*3*]

2,6-DIMETHYL-2,6-OCTADIENOL. See citral.

3,7-DIMETHYL-1,6-OCTADIEN-3-OL. See linalool.

3,7-DIMETHYL-2,6-OCTADIEN-1-OL. See geraniol.

3,7-DIMETHYL-3,6-OCTADIEN-1-OL. See geraniol.

DIMETHYLOCTANE. $C_{10}H_{22}$, mw: 142.2, flash p: < 131°F, autoign. temp.: 437°F, d: 0.74, vap. d: 4.91, bp: 164°.
THR = No data.
Fire Hazard: Mod, via heat, flames or oxidizers.
To Fight Fire: Dry chemical, CO_2, water spray, foam, fog, mist.

3,4-DIMETHYLOCTANE. flash p: < 131°F.
Fire Hazard: Mod, via heat, flames or oxidizers.
To Fight Fire: Dry chemical, CO_2, water spray, fog, mist, foam.

DIMETHYLOCTYNEDIOL. Syn: *3,6-dimethyl-4-octyne-3,6-diol.* White crystals. $C_{10}H_{18}O_2$, mw: 170.3, mp: 55°–56°, bp: 135° @ 20 mm, vap. d: 5.9.
THR = Details U. Probably toxic.
Fire Hazard: Slight, when heated.

2,2-DIMETHYLOLPROPANOL-1. See trimethylolethane.

***o,o*-DIMETHYL-S-(4-OXOBENZOTRIAZINO-3-METHYL)PHOSPHORODITHIOATE.** See guthion.

5,5-DIMETHYL-3-OXO-1-CYCLOHEXEN-1-YL DIMETHYL CARBAMATE. See dimetan.

2,3-DIMETHYLPENTALDEHYDE. Liquid, slightly sol in water. $CH_3CH_2CH(CH_3)CH(CH_3)CHO$, mw: 114, d: 0.8293, bp: 140.5°, fp: −110°, flash p: 94°F (OC).
THR = U. See also aldehydes.
Fire Hazard: Dangerous, when exposed to heat, flame or oxidizers.
To Fight Fire: Foam, mist, dry chemical.

2,3-DIMETHYL-4-PENTANAL. $CH_3CH(CH_3)CH(CH_3)OCHOH$, mw: 115.
Acute tox data: Inhal LC_{LO} (rat) = 10,000 ppm for 4 hrs. [*3*]
THR = LOW via inhal route. Limited animal exper suggest LOW tox and irr. See also aldehydes.

2,3-DIMETHYLPENTANE. Syn: *diethyldimethylmethane.* Liquid. C_7H_{16}, mw: 100.20, mp: −135°, bp: 89.8°, d: 0.69 @ 15.5°/15.5°, autoign. temp.: 635°F, flash p: < 20°F, vap. press: 40 mm @ 13.9°, vap. d: 3.45, lel = 1.1%, uel = 6.7%.
THR = U. Probably irr and narcotic in high conc.
Fire Hazard: Dangerous, when exposed to heat, flame or oxidizers.
Explosion Hazard: U.
Disaster Hazard: Dangerous; keep away from heat and open flame; can react vigorously with oxidizing materials.
To Fight Fire: Foam, CO_2, dry chemical.

2,4-DIMETHYLPENTANE. A clear liquid. C_7H_{16}, mw: 100.2, mp: −123.4°, bp: 80.3°, fp: −119.4°, flash p: 10°F, d: 0.6728 @ 20°/4°, vap. press: 8.2 mm @ 21°, vap. d: 3.48.
THR = U. Probably irr and narcotic in high conc.

Fire Hazard: Dangerous, when exposed to heat, flame or oxidizers.
Explosion Hazard: U.
Disaster Hazard: Dangerous; keep away from heat and open flame; can react vigorously with oxidizing materials.
To Fight Fire: Foam, CO_2, dry chemical.

2,3-DIMETHYLPENTANOL.
$CH_3CH(CH_3)CH(CH_3)CH_2CH_2OH$, mw: 116.
Acute tox data: Oral LD_{50} (rat) = 2850 mg/kg; dermal LD_{50} (rabbit) = 2500 mg/kg. [*3*]
THR = MOD via oral and dermal routes.

2,4-DIMETHYL-3-PENTANOL.
See diisopropyl carbinol.

3,4-DIMETHYL-5-PENTYLIDENE-2-(5H)FURANONE. Syn: *Bovolide*. $C_{11}H_{16}O_2$, mw: 180.3.
THR = An exper carc [*3*] which occurs in cow's butter.

3,5-DIMETHYLPHENOL. See 3,5-xylenol.

2,4-DIMETHYLPHENOL.
THR = MOD via oral route. An exper carc. [*3*]

2,5-DIMETHYLPHENOL.
THR = HIGH via oral route. An exper carc. [*3*]

2,6-DIMETHYLPHENOL.
THR = HIGH via oral route. An exper neo. [*3*]

3,4-DIMETHYLPHENOL.
THR = HIGH via oral route. An exper carc. [*3*]

N,N-DIMETHYL-*p*-PHENYL AZO ANILINE. See dimethylamino azo benzene.

2,3-DIMETHYL-4-PHENYL AZO ANILINE.
$C_{14}H_{15}N_3$, mw: 225.2.
THR = An exper carc to mice via oral route. [*103*]

N,N-DIMETHYL-*p*-PHENYLAZO ANILINE-N-OXIDE. $C_{14}H_{15}ON_3$, mw: 241.3.
THR = An exper carc and neo. [*3*]

N,N-DIMETHYL-4-PHENYLAZO-*o*-ANISIDINE.
$C_{15}H_{17}ON_3$, mw: 255.4.
THR = An exper carc. [*3*]

N,N-DIMETHYL-4-(PHENYLAZO)-*m*-TOLUIDINE.
$C_{15}H_{17}N_3$, mw: 239.4.
THR = An exper neo. [*3*]

N,N-DIMETHYL-4-(PHENYLAZO)-*o*-TOLUIDINE.
THR = An exper carc. [*3*]

DIMETHYL-*p*-PHENYLENE DIAMINE. Syn: *p-aminodimethyl aniline*. Crystalline mass.
$C_6H_4NH_2N(CH_3)_2$, mw: 136.20, mp: 53°, bp: 150° @ 17 mm, d: 1.036 @ 20°/4°, vap. d: 4.69.
Acute tox data: Oral LD_{50} (rat) = 935 mg/kg; ip LD_{50} (rat) = 21 mg/kg. [*3*]
THR = HIGH via ip and MOD via oral routes.
Disaster Hazard: Mod; when heated to decomp, emits tox fumes.

2,4-DIMETHYL PHENYL MALEIMIDE.
Acute tox data: Oral LD_{50} (rat) = 710 mg/kg. [*3*]
THR = MOD via oral route.

3,3-DIMETHYL-1-*m*-PHENYL METHYL TRIAZENE. $C_9H_{13}N_3$, mw: 163.3.
THR = An exper carc. [*3*]

3,3-DIMETHYL-1-PHENYL TRIAZENE. $C_8H_{11}N_3$, mw: 149.2.
THR = HIGH via oral route. An exper transplacental carc. [*3, 23*] via oral, ip, sc and rec routes.

1,1-DIMETHYLPHENYL UREA. Syns: *fenuron, 3-phenyl-1,1-dimethyl urea.* White crystalline solid, almost insol in water, sparingly sol in hydrocarbon solvents. $C_6H_5NHCON(CH_3)_2$, mw: 164, mp: 127°–129°.
Acute tox data: Oral LD_{50} (rat) = 6400 mg/kg. [*3*]
THR = MOD via oral route.

DIMETHYLPHOSPHINE. Colorless liquid.
$(CH_3)_2PH$, mw: 62.1, bp: 25°, d: < 1, vap. d: 2.14.
THR = HIGH via oral and inhal routes. See also phosphine.
Fire Hazard: Dangerous, when exposed to heat or flame; spont flam in air.
Explosion Hazard: U.
Disaster Hazard: Dangerous; see phosphates; can react vigorously with oxidizing materials.
To Fight Fire: Foam, CO_2, dry chemical.

DIMETHYLPHOSPHOROAMIDOTHIOATE.
See tamaron.

***o*-DIMETHYLPHTHALATE.** Syn: *DMP*. Colorless, odorless liquid. $C_6H_4(COOCH_3)_2$, mw: 194.18, bp: 283.7°, flash p: 295°F (CC), d: 1.189 @ 25°/25°, autoign. temp.: 1032°F, vap. d: 6.69, vap. press: 1 mm @ 100.3°.
Acute tox data: Oral LD_{50} (rabbit) = 4400 mg/kg; ip LD_{50} (rat) = 3375 mg/kg; inhal LC_{LO} (cat) = 10,000 ppm. [*3*]
THR = MOD via oral and ip routes; LOW via inhal route. An exper teratogen. [*3*]
Fire Hazard: Slight, when exposed to heat or flame; can react with oxidizing materials.
Spont Heating: No.
To Fight Fire: CO_2, dry chemical.

N,N-DIMETHYL-4-(3′-PICOLYL-1′-OXIDE)AZO-*o*-TOLUIDINE. An exper carc. [*23*]

N,N-DIMETHYL-4-(3′-PICOLYL-1-OXIDE)AZO-*m*-TOLUIDINE.
THR = An exper carc. [*23*]

1,4-DIMETHYLPIPERAZINE. Colorless, mobile liquid. $CH_3N(CH_2CH_2)_2NCH_3$, mw: 114.2, d: 0.8565 @ 20°/4°, flash p: 85°F (TOC), bp: approx. 130°.

THR = Details U. See also *trans*-2,5-dimethylpiperazine.
Fire Hazard: Dangerous, when exposed to heat, sparks, powerful oxidizers.
To Fight Fire: Foam, spray, CO_2, dry chemical.

***cis*-2,5-DIMETHYLPIPERAZINE.** A liquid, typical amine odor. $C_6H_{14}N_2$, mw: 114.2, mp: 17.5°, bp: 164.5° @ 746 mm, flash p: 154.5°F (COC), d: 0.9195 @ 25°/25°.
THR = Details U. See also *trans*-2,5-dimethylpiperazine.
Fire Hazard: Mod, when exposed to heat or flame; can react with oxidizing materials.
To Fight Fire: Foam, CO_2, dry chemical.

***trans*-2,5-DIMETHYLPIPERAZINE.** Crystals, typical amine odor. $HNCH_2CH(CH_3)NHCH_2CH(CH_3)$, mw: 114.19, mp: 117.5°, bp: 161.9° @ 746 mm, flash p: 210°F (OC).
Acute tox data: Oral LD_{50} (rat) = 3160 mg/kg; dermal LD_{50} (rabbit) = 800 mg/kg. [*3*]
THR = MOD via oral and dermal routes.
Fire Hazard: Mod, when exposed to heat or flame; can react with oxidizing materials.
To Fight Fire: CO_2, dry chemical.

2,2-DIMETHYLPROPANE. Syn: *neopentane*. Gas. $(CH_3)_4C$, mw: 72.2, bp: 9.5°, fp: −18.2°, flash p: < 20°F, d: 0.590 @ 20°/4°, autoign. temp.: 842°F, vap. press: 1100 mm @ 21°, vap. d: 2.48, lel = 1.4%, uel = 7.5%.
THR = U. Probably irr and narcotic in high conc.
Fire Hazard: Very dangerous, when exposed to heat or flame; can react vigorously with oxidizing materials.
Explosion Hazard: U.
To Fight Fire: Stop flow of gas.

2,2-DIMETHYL-1,3-PROPANEDIOL. See neopentyl glycol.

7,12-DIMETHYLPROPYL BENZ(a)ANTHRACENE. $C_{23}H_{21}$, mw: 297.4.
THR = An exper carc. [*3*]

DIMETHYLPROPYL METHANE.
See 2-methylpentane.

***p*-(α,α-DIMETHYLPROPYL)PHENOL.** See *p*-tert-amyl phenol.

N,N-DIMETHYL-*p*-[(*p*-PROPYLPHENYL)AZO] ANILINE. $C_{17}H_{21}N_3$, mw: 267.5.
THR = An exper carc. [*3*]

2,5-DIMETHYL PYRAZINE. $C_6H_8N_2$, mw: 108.2, flash p: 147°F (OC), d: 0.99, vap. d: 3.72, bp: 182.2°, water-sol.
Acute tox data: ip LD_{50} (mice) = 1350 mg/kg. [*3*]
THR = MOD via ip route.
Fire Hazard: Mod, via heat, open flames, oxidizers.
To Fight Fire: Water spray, mist, dry chemical, CO_2, foam.

2,6-DIMETHYL PYRIDINE. See 2,6-lutidine.

N,N-DIMETHYL-4-(3′-PYRIDYL AZO)ANILINE.
THR = An exper carc. [*23*]

N,N-DIMETHYL-4-(4′-PYRIDYL AZO-1′-OXIDE) AZO ANILINE.
THR = An exper carc. [*23*]

N,N-DIMETHYL-4-(4′-PYRIDYL-1′-OXIDE)AZO-2,3-XYLIDENE.
THR = An exper carc. [*23*]

N,N-DIMETHYL-4-(4′-PYRIDYL-1′-OXIDE)AZO-2,5-XYLIDENE.
THR = An exper carc. [*23*]

N,N-DIMETHYL-4-(4′-PYRIDYL-1′-OXIDE)AZO-3,5-XYLIDENE.
THR = An exper carc. [*23*]

3,3-DIMETHYL-10-(*m*-PYRIDYL-N-OXIDE)TRIAZENE. $C_7H_{11}ON_4$, mw: 167.2.
THR = An exper carc. [*3*]

3,3-DIMETHYL-1-(*m*-PYRIDYL)TRIAZENE. $C_7H_{10}N_4$, mw: 150.2.
THR = An exper teratogen and neo. [*3, 23*]

N-(4,6-DIMETHYL-2-PYRIMIDYL)SULFANILAMINE. See sulfamethazine.

2,5-DIMETHYL PYRROLE. C_6H_9N, mw: 95.1.
THR = Reacted violently with [$P(OCl)_3$ + dimethyl formamide]. [*19*]

N,N-DIMETHYL-4-(5′-QUINALDYL AZO)ANILINE.
THR = An exper carc. [*23*]

N,N-DIMETHYL-4-(3′-QUINOLYL AZO)ANILINE.
THR = An exper carc. [*23*]

N,N-DIMETHYL-4-(4′-QUINOLYL AZO)ANILINE.
THR = An exper carc. [*23*]

N,N-DIMETHYL-4-(4′-QUINOLYL AZO)ANILINE.
THR = An exper carc. [*23*]

N,N-DIMETHYL-4-(5′-QUINOLYL AZO)ANILINE.
THR = An exper carc. [*23*]

N,N-DIMETHYL-4-(6′-QUINOLYL AZO)ANILINE.
THR = An exper carc. [*23*]

DIMETHYL-4-(5′-QUINOLYL AZO)-*m*-TOLUIDINE.
THR = An exper carc. [*23*]

DIMETHYL-4-(4′-QUINOLYL-1′-OXIDE AZO) ANILINE.
THR = An exper carc. [*23*]

DIMETHYL-4-(5′-QUINOLYL-1′-OXIDE AZO) ANILINE.
THR = An exper carc. [*23*]

DIMETHYL-4-(6′-QUINOLYL-1′-OXIDE AZO) ANILINE.
THR = An exper carc. [*23*]

DIMETHYL-4-(5′-QUINOXALYL AZO)ANILINE.
THR = An exper carc. [*23*]

DIMETHYL-4-(6′-QUINOXALYL AZO)ANILINE.
THR = An exper carc. [*23*]

DIMETHYL-4-(2′-QUINOXALYL AZO)ANILINE.
THR = An exper carc. [*23*]

7,8-DIMETHYL-10-(1-*d*-RIBITYL)ISOALLOXAZINE. See riboflavin.

DIMETHYLSEBACATE. Syn: *methyl sebacate*. Liquid. $[(CH_2)_4CO_2CH_3]_2$, mw: 230.3, mp: 38°, bp: 293.5°, flash p: 293°F, d: 0.986 @ 30°/25°, vap. press: 1 mm @ 104.0°, vap. d: 7.95.
THR = See esters and methyl alcohol.
Fire Hazard: Slight, when exposed to heat or flame; can react with oxidizing materials.
To Fight Fire: Foam, CO_2, dry chemical.

DIMETHYLSELENIDE. Syn: *methylselenide*. Liquid. $(CH_3)_2Se$, mw: 109.0, bp: 58°, d: 1.4077 @ 14.6°/4°, vap. d: 3.75.
Acute tox data: ip LD_{LO} (rat) = 552 mg/kg. [*3*]
THR = MOD via ip route. See selenium and its compounds.
Fire Hazard: Dangerous, when exposed to heat or flame.
Explosion Hazard: U.
Disaster Hazard: Dangerous; see selenium; will react with water, steam, acid or acid fumes to produce toxic fumes; can react vigorously with oxidizing materials.
To Fight Fire: Foam, CO_2, dry chemical.

DIMETHYLSILICANE. Gas. $H_2Si(CH_3)_2$, mw: 60.1, d: 0.68 @ −80°, mp: −150°, bp: −20°.
THR = See silanes.

N,N-DIMETHYLSTILBENAMINE. $C_{16}H_{17}N$, mw: 223.3.
THR = An exper carc. [*3*]

N,N-DIMETHYLSTILBENAMINE (*trans*).
THR = An exper carc. [*3, 23*] via oral route.

N,N-DIMETHYLSTILBENAMINE (*cis*).
THR = An exper neo. [*3*]

DIMETHYL SULFATE. Syn: *methyl sulfate*. Colorless, odorless liquid. $(CH_3)_2SO_4$, mw: 126.13, mp: −31.8°, bp: 188°, flash p: 182°F (OC), d: 1.3322 @ 20°/4°, vap. d: 4.35, autoign. temp.: 370°F.
Acute tox data: Oral LD_{50} (rat) = 440 mg/kg; sc LD_{50} (rat) = 100 mg/kg; inhal LC_{LO} (human) = 97 ppm for 10 min. [*3*]
THR = HIGH via oral, sc and inhal routes. Contact of the skin and mu mem with the liquid or vapor, even for short periods, results in intense irr of these tissues several hours later. There is no odor or initial irr to give warning of exposure. On brief, mild exposures, conjunctivitis, catarrhal inflammation of the mu mem of the nose, throat, larynx and trachea and possibly some reddening of the skin develop after the latent period. With longer, heavier exposures, the cornea shows clouding, the irr changes to the nasopharynx are more marked and after 6 to 8 hrs pulmonary edema may develop. Death may occur in 3 or 4 days. The liver and kidneys are frequently damaged. Spilling of the liquid on the skin can cause ulceration and local necrosis. The fatal conc for cats and monkeys is in the range of 25 to 200 ppm of the vapor in air. After a latent period of several hrs, there is severe lacrimation, conjunctivitis, photophobia, coughing and hoarseness, followed, in the case of more severe exposures, by chest pain, dyspnea, cyanosis and possibly death. In patients surviving severe exposures, there may be serious injury of the liver and kidneys, with suppression of urine, jaundice, albuminuria and hematuria appearing. Death, resulting from the kidney or liver damage, may be delayed for several weeks. An exper (S) human carc. [*10, 14, 23*] Also an exper carc. [*3*]
Fire Hazard: Mod, when exposed to heat, flame or oxidizers.
Spont Heating: No.
Disaster Hazard: Dangerous; see sulfates. Can react with oxidizing materials. Violent reaction with NH_4OH, NaN_3. [*19*]
To Fight Fire: Water, foam, CO_2, dry chemical.

DIMETHYL SULFIDE. See methyl sulfide.

DIMETHYL SULFITE. See methyl sulfite.

2,4-DIMETHYL SULFOLANE. Solid. $C_6H_{12}SO_2$, mw: 148.22, bp: 280°, flash p: 290°F (OC), d: 1.1362 @ 20°/4°, vap. press: 0.006 mm @ 20°.
Acute tox data: Oral LD_{50} (mouse) = 140 mg/kg; oral LD_{50} (rabbit) = 115 mg/kg. [*3*]
THR = HIGH via oral route.
Fire Hazard: Slight, when exposed to heat or flame.
Disaster Hazard: Dangerous; see sulfates; can react with oxidizing materials.
To Fight Fire: Water, foam, CO_2, dry chemical.

DIMETHYL SULFOXIDE. Syn: *DMSO*. Clear, water white, hygroscopic liquid. $(CH_3)_2SO$, mw: 78.1, mp: 18.5°, bp: 189°, flash p: 203°F (OC), d: 1.100 @ 20°,

vap. press: 0.37 mm @ 20°, lel = 2.6%, uel = 28.5%, autoign. temp.: 419°F.
Acute tox data: iv LD_{50} (dog) = 2500 mg/kg; oral LD_{50} (mouse) = 21,000 mg/kg. [3]
THR = MOD via iv, VERY LOW via oral routes. An allergen. An exper teratogen. [3] Freely penetrates skin. Acts as a primary irr on skin, causing redness, burning, itching, and scaling; also causes urticaria; systemic symptoms are nausea, vomiting, chills, cramps, and lethargy. A case of anaphylactic reaction has been reported. Has caused corneal opacity in exper animals.
Fire Hazard: Low, when exposed to heat or flame. Reacts violently with many acyl and aryl halides, bromobenzoyl acetanilide, cyanuric chloride, IF_5, $Mg(ClO_4)_2$, CH_3Br, NIO_4, P_2O_3, AgF, NaH. [19]
Disaster Hazard: Mod dangerous; when heated to decomp, emits toxic fumes; can react with oxidizing materials.
To Fight Fire: Foam, alcohol foam, CO_2, dry chemical.

DIMETHYL TELLURIDE. Syn: *methyl telluride*. Yellowish oil with odor of garlic. $(CH_3)_2Te$, mw: 157.7, bp: 82°, vap. d: 5.45.
THR = See tellurium compounds.

α-DIMETHYL TELLURONIUM DIBROMIDE. Orange crystals, sol in alcohol. $C_2H_6Br_2Te$, mw: 317.5, mp: 142° (decomp).
THR = See tellurium compounds and bromides.

α-DIMETHYL TELLURONIUM DICHLORIDE. Crystals, sol in water. $C_2H_6Cl_2Te$, mw: 228.6, mp: 92°.
THR = See tellurium compounds and chlorides.

β-DIMETHYL TELLURONIUM DICHLORIDE. Crystals, soluble in alcohol and ether. $C_2H_5Cl_2Te$, mw: 228.6, mp: 134°.
THR = See tellurium compounds and chlorides.

α-DIMETHYL TELLURONIUM DIIODIDE. Red crystals; insol in cold water. $C_2H_6I_2Te$, mw: 411.5, mp: 125° (decomp).
THR = See tellurium compounds and iodides.

DIMETHYL-2,3,5,6-TETRACHLOROTEREPHTHALATE. Crystals, insol in water, slightly sol in acetone and benzene. $C_6Cl_4(COOCH_3)_2$, mw: 324, mp: 156°.
Acute tox data: Oral LD_{50} (rat) = 3000 mg/kg; dermal LD_{50} (rabbit) = 10,000 mg/kg. [3]
THR = MOD via oral and LOW via dermal routes.
Disaster Hazard: Dangerous; see chlorides.

1,11-DIMETHYL-1,2,3,4-TETRAHYDROCHRYSENE. $C_{20}H_{19}$, mw: 259.4.
THR = An exper neo. [3]

DIMETHYLTETRAHYDROPHTHALATE. Syn: *dimethyl-1-cyclohexene-1,2-dicarboxylic acid*. Crystals. $C_6H_8(COOCH_3)_2$, mw: 198.2, vap. d: 6.83.
Acute tox data: Oral LD_{50} (rat) = 700 mg/kg. [3]
THR = MOD via oral route.
Fire Hazard: Slight, when exposed to heat or flame; can react with oxidizing materials.

2,6-DIMETHYLTETRAHYDRO-1,4-PYRONE. $C_7H_{14}O_2$, mw: 130.2, vap. d: 4.48.
Acute tox data: Inhal LC_{LO} (rat) = 8000 ppm for 4 hrs. [3]
THR = MOD via inhal route.
Fire Hazard: Slight.

3,5-DIMETHYLTETRAHYDRO-1,3,5,2-H-THIODIAZINE-2-THIONE. Syn: *crag herbicide 974*. White crystalline powder, slightly sol in water or alcohol, sol in acetone. $C_5H_{10}N_2S_2$, mw: 162.2, mp: 156°, vap. d: 5.59, d: 1.3.
Acute tox data: Oral LD_{50} (rat) = 500 mg/kg; ip LD_{50} (rat) = 87 mg/kg; oral LD_{50} (guinea pig) = 160 mg/kg. [3]
THR = HIGH via oral and ip routes. A fungicide, an herbicide, and a food additive resulting from contact with containers or equipment. [109]
Disaster Hazard: Dangerous; when heated to decomp, emits highly toxic fumes of oxides of sulfur and nitrogen.

2,4-DIMETHYLTHIOPHENE. See *p*-thioxene.

DIMETHYLTIN. Yellow solid, insol in water. A polymer of $(CH_3)_2Sn$, mw: a multiple of 148.8.
THR = See tin compounds.

DIMETHYLTIN DIBROMIDE. Colorless crystals, sol in water and organic solvents. $(CH_3)_2SnBr_2$, mw: 308.6, mp: 76°, bp: 208°–213°.
THR = See tin compounds and bromides.

DIMETHYLTIN DICHLORIDE. Solid, water-sol. $(CH_3)_2SnCl_2$, mw: 219.7, mp: 90°, bp: 190°.
Acute tox data: iv LD_{50} (rat) = 40 mg/kg; oral LD_{LO} (rat) = 160 mg/kg. [3]
THR = HIGH via iv and oral routes. See tin compounds and chlorides.

DIMETHYLTIN DIFLUORIDE. White crystals, water-sol. $(CH_3)_2SnF_2$, mw: 186.8, bp: decomp $< 360°$.
THR = See fluorides and tin compounds.

DIMETHYLTIN DIIODIDE. White crystals, sol in hot water. $(CH_3)_2SnI_2$, mw: 402.6, d: 2.872, mp: 43°, bp: 228°.
THR = See tin compounds and iodides.

DIMETHYLTIN OXIDE. White powder, insol in water. $(CH_3)_2SnO$, mw: 164.8.
THR = Details U. See tin compounds.

DIMETHYLTIN SULFIDE. Solid. $(CH_3)_2SnS$, mw: 180.8, mp: 148°.
THR = See sulfides and tin compounds.

N,N-DIMETHYL-*m*-TOLUAMIDE. $C_6H_2(CH_3)_3CONH_2$, mw: 163.
THR = MILD irr to eyes and mu mem. Ingestion can cause CNS depression.

N,N-DIMETHYL-4-(*o*-TOLYLAZO)ANILINE. $C_{15}H_{17}N_3$, mw: 239.4.
THR = An exper neo. [*3*]

N,N-DIMETHYL-4-(*p*-TOLYLAZO)ANILINE.
THR = An exper neo. [*3*]

N,N-DIMETHYL-4-(*p*-TOLYLAZO)-*m*-TOLUIDINE. $C_{16}H_{19}N_3$, mw: 253.4.
THR = An exper neo. [*3*]

5-(3,3-DIMETHYL-1-TRIAZENO)IMIDAZOLE-4-CARBOXAMIDE. $C_6H_{10}ON_6$, mw: 182.2.
THR = An exper carc. [*3*]

5-(3,3-DIMETHYL-1-TRIAZENO)IMIDAZOLE-4-CARBOXAMIDE CITRATE. $C_{12}H_{10}O_8N_6$, mw: 274.4.
THR = An exper carc and teratogen. [*3*]

DIMETHYLTRIBORINE TRIAMINE (B). Colorless liquid, hydrolyzed by water. $(CH_3)_2B_3N_3H_4$, mw: 108.6, mp: −48°, bp: 107°.
THR = Details U. See boron compounds and amines.

DIMETHYLTRIBORINE TRIAMINE (N). Colorless liquid, hydrolyzed by water. $(CH_3)_2B_3N_3H_4$, mw: 108.6, bp: 108°.
THR = Details U. See also boron compounds and boron hydrides and amines.

DIMETHYL (2,2,2-TRICHLORO-1-HYDROXY-ETHYL)PHOSPHONATE. See dylox.

***o,o*-DIMETHYL-*o*-(2,4,5-TRICHLOROPHENYL) PHOSPHOROTHIOATE.** See ronnel.

DIMETHYLTRIMETHYLENE OXIDE.
THR = An exper carc. [*23*]

DIMETHYLTRITHIOCARBONATE. Crystals. $CS_3(CH_3)_2$, mw: 139.26, vap. d: 4.8.
THR = HIGH irr to skin, eyes and mu mem and via oral route. [*3*]
Fire Hazard: Mod, when exposed to heat.
Disaster Hazard: Dangerous; see sulfates; can react with oxidizing materials.

N,N-DIMETHYL-*p*-(2,4,6-TRIFLUOROPHENYL AZO)ANILINE. $C_{14}H_{12}N_3F_3$, mw: 279.3.
THR = An exper carc. [*3*]

***n,n*-DIMETHYL UREA.** Colorless crystals, water and alcohol sol. $(CH_3NH)_2CO$, mw: 88, d: 1.14, mp: 106°, bp: 270°.
Acute tox data: Oral LD_{50} (rat) = 6400 mg/kg. [*3*]
THR = Probably MOD via oral route.

2,3-DIMETHYLVALERALDEHYDE. Syn: *2,3-dimethyl pentaldehyde*. Liquid, slightly sol in water. $CH_3CH_2CH(CH_3)CH(CH_3)CHO$, mw: 111, d: 0.8293, bp: 140.5°, fp: 110°, flash p: 94°F.
Acute tox data: Oral LD_{50} (rat) = 3540 mg/kg. [*3*]
THR = MOD via oral route.
Fire Hazard: Mod, when exposed to heat or flame.

β,β-DIMETHYLVINYL CHLORIDE. Syn: *isocrotyl chloride*. Liquid. C_4H_7Cl, mw: 90.6, d: 0.919 @ 20°/4°, bp: 68°.
Acute tox data: Inhal LC_{LO} (mouse) = 181,000 mg/m^3 for 10 min. [*3*]
THR = LOW via inhal route. Causes local irr and is narcotic in high conc.
Disaster Hazard: Dangerous; when heated to decomp, emits highly toxic chloride fumes.

1,3-DIMETHYLXANTHINE. See theophylline.

N,N-DIMETHYL-*p*-(2,3-XYLYLAZO)ANILINE. $C_{16}H_{19}N_3$, mw: 253.4.
THR = An exper carc. [*3*]

N,N-DIMETHYL-*p*-(3,4-XYLYL AZO)ANILINE.
THR = An exper carc. [*3*]

DIMETILAN. Syn: *2-dimethylcarbamoyl-3-methyl-5-pyrazolyl dimethylcarbamate*. $C_{10}H_{16}O_3N_4$, mw: 240.1.
Acute tox data: Oral LD_{50} (rat) = 25 mg/kg; dermal LD_{50} (rat) = 600 mg/kg; dermal LD_{50} (rabbit) = 2000 mg/kg. [*3*]
THR = HIGH via oral, MOD via dermal routes.

3,6-DIMETHYL XANTHINE. See theobromine.

DIMETRIDIAZOLE. Syn: *1,2-dimethyl-5-nitroimidazole*. $C_5H_7O_2N_3$, mw: 141.2.
Acute tox data: Oral LD_{50} (turkey) = 500 mg/kg. [*3*]
THR = HIGH via oral route. A food additive permitted in the feed and drinking water of animals and/or for the treatment of food-producing animals. Also a food additive permitted in food for human consumption. [*109*] An exper neo. [*3, 23*]

"DIMITE." See di(*p*-chlorophenyl)methylcarbinol.

DIMORPHOLINETHIURAM DISULFIDE.
THR = Details U. A fungicide; see disulfiram.
Disaster Hazard: Dangerous; see sulfates.

1,2,5,6-DINAPHTHACRIDINE. Yellow crystals. $C_{21}H_{13}N$, mw: 279.4, mp: 228°.
THR = An exper (+) neo and carc. [*3, 11*]

3,4,6,7-DINAPHTHACRIDINE.
THR = An exper (+) neo and carc. [*3, 11*]

DINAPHTHAZINE. $C_{20}H_{12}N_2$, mw: 280.3.
THR = An exper carc. [*3*]

DI-α-NAPHTHYLAMINE FLUOSILICATE. White crystals, slightly sol in alcohol. $(C_{10}H_7NH_2)_2 \cdot H_2SiF_6$, mw: 430.4, mp: 218°.
THR = See fluosilicates.

DI-β-NAPHTHYLAMINE FLUOSILICATE. White crystals. $(C_{10}H_7NH_2)_2 \cdot H_2SiF_6$, mw: 430.4, mp: 236°.
THR = See fluosilicates.

DINAPHTHYLMERCURY. White crystals, insol in cold water. $(C_{10}H_7)Hg(C_{10}H_7)$, mw: 454.9, d: 1.929, mp: 188°, bp: 249°.
THR = See mercury compounds, organic. HIGH toxicity.

DI-β-NAPHTHYL-*p*-PHENYLENE DIAMINE. Solid. $(C_{10}H_7)_2NC_6H_4NH_2$, mw: 360.42, vap. d: 12.4.
Acute tox data: Oral LD_{50} (mouse) = 1800 mg/kg; ip LD_{50} (rat) = 4500 mg/kg. [*3*]
THR = MOD via oral and ip routes.
Disaster Hazard: Mod dangerous; when heated to decomp, emits toxic fumes.

DI-α-NAPHTHYLTIN. A powder. $(C_{10}H_7)_2Sn$, mw: 373, mp: 200°, bp: decomp @ 255°.
THR = See tin compounds.

1,3-DI-1-NAPHTHYL TRIAZINE. See 1,1′-diazoamidonaphthalene.

DINICKEL PHOSPHIDE. Gray crystals. Ni_2P, mw: 148.36, mp: 1112°, d: 6.31 @ 15°.
THR = See phosphine, phosphides and nickel compounds.

DI-*m*-NITRANILINE FLUOSILICATE. White crystals, slightly sol in alcohol. $(C_6H_4NH_2NO_2)_2 \cdot H_2SiF_6$, mw: 420.3, mp: 200°.
THR = See fluosilicates and nitrates, organic and *m*-nitroaniline.

DINITROAMINOPHENOL. See picramic acid.

DINITRO-*o*-*sec*-AMYLPHENOL. $(NO_2)_2C_6H_2OHC_5H_{11}$, mw: 253.2.
Acute tox data: ip LD_{LO} (mouse) = 4 mg/kg. [*3*]
THR = HIGH via ip route. An herbicide. See dinitrophenol and nitrates.

2,4-DINITROANILINE. Syn: *2,4-dinitrophenylamine*. Yellow, needle-like crystals, insol in water. $(NO_2)_2C_6H_3NH_2$, mw: 183.13, mp: 180°, bp: 56.7°, flash p: 435°F (CC), d: 1.615, vap. d: 6.31.
Acute tox data: Oral LD_{50} (rat) = 418 mg/kg; ip LD_{LO} (rat) = 250 mg/kg. [*3*]
THR = HIGH via oral and ip routes. A powerful poison. See also nitroanilines.
Fire Hazard: Slight, when exposed to heat or flame.
Disaster Hazard: Dangerous; when heated to decomp, emits highly toxic fumes; can react with oxidizing materials.
To Fight Fire: CO_2, dry chemical.

DINITROANILINE HYDROCHLORIDE. $C_6H_6O_4N_3Cl$, mw: 219.6.
THR = Reacts violently with nitrosyl sulfuric acid. [*19*]

2,4-DINITROANISOLE. Syn: *2,4-dinitrophenylmethyl ether*. Colorless to yellow crystals. $CH_3OC_6H_3(NO_2)_2$, mw: 198.1, mp: 89°, bp: sublimes, d: 1.341 @ 20°/4°, vap. d: 6.83.
Acute tox data: Oral LD_{LO} (rat) = 100 mg/kg. [*3*]
THR = HIGH via oral route. See nitro compounds of aromatic hydrocarbons and nitrates.

3,5-DINITROBENZAMIDE. $C_7H_5N_3O_5$, mw: 211.13, mp range: 180°–184°.
THR = U. A food additive permitted in the feed and drinking water of animals and/or for the treatment of food-producing animals. [*109*]
Disaster Hazard: Dangerous; see nitrates.

1,2-DINITROBENZENE. Syn: *o-dinitrobenzol*. Colorless needles or plates. $C_6H_4(NO_2)_2$, mw: 168.11, mp: 118°, bp: 319°, flash p: 302°F (CC), d: 1.571 @ 0°/4°, vap. d: 5.79.
Acute tox data: Oral LD_{50} (cat) = 29 mg/kg. [*3*]
THR = HIGH via oral and inhal routes; MOD via dermal route. Produces a wide variety of pathological changes, including anemia, jaundice, enlarged liver or yellow atrophy, degeneration of kidneys and injury to CNS.
Fire Hazard: Slight, when exposed to heat or flame.
Explosion Hazard: Severe, when shocked or exposed to heat or flame. This compound is highly explosive and is used in bursting charges and to fill artillery shells. It is a useful industrial explosive when mixed with more powerful explosives or with oxygen carriers, such as inorganic nitrates or chlorates, or with ammonium nitrate.
Disaster Hazard: Dangerous; when heated to decomp, emits highly toxic fumes of oxides of nitrogen and explodes; can react vigorously with oxidizing materials.
To Fight Fire: Water, CO_2, dry chemical.

***m*-DINITROBENZENE.** Yellowish crystals, mp: 89°, bp: 301°.
Acute tox data: Oral LD_{50} (wild birds) = 42 mg/kg; oral LD_{LO} (cat) = 27 mg/kg. [*3, 2*]
THR = HIGH via oral route.

***p*-DINITROBENZENE.** White crystals. mp: 173°, bp: 299°. Volatile with steam.
Acute tox data: Oral LD (cats) = 29.4 mg/kg. [*3*]
THR = HIGH via oral route. See also 2,4-dinitrobenzene.

2,4-DINITROBENZENE SULFENYL CHLORIDE. Bright yellow, crystalline solid. $C_6H_3SCl(NO_2)_2$, mw: 234, mp: 95°–96°, vap. d: 8.08.
THR = Details U. See also nitro compounds of aromatic hydrocarbons.
Fire Hazard: See nitrates.

Explosion Hazard: When shocked; also during solvent-removal. [*19*]

Disaster Hazard: Dangerous; shock will explode it; when heated to decomp, emits highly toxic fumes of oxides of sulfur and nitrogen; will react with water or steam to produce toxic and corrosive fumes; can react with reducing materials.

***o*-DINITROBENZOL.** See 1,2-dinitrobenzene.

3,5-DINITROBENZOYL CHLORIDE. Yellow crystals. $(NO_2)_2C_6H_3COCl$, mw: 230.6, mp: 69° @ 12 mm, vap. d: 7.96.

THR = HIGH via contact, oral and inhal routes. A powerful irr.

Disaster Hazard: Dangerous; see chlorides.

4,4′-DINITRO BIPHENYL. $C_{12}H_{18}O_4N_2$, mw: 244.2.

THR = An exper neo. [*3*]

4,6-DINITRO-*o*-*sec*-BUTYL PHENOL. Crystals. $C_{10}H_{12}O_2N_2$, mw: 192.2, vap. d: 7.73.

Acute tox data: Oral LD_{50} (rat) = 25 mg/kg; dermal LD_{50} (rat) = 80 mg/kg; oral LD_{50} (wild birds) = 7 mg/kg. [*3*]

THR = HIGH via oral and dermal routes. An exper neo and teratogen to mice via oral and ip routes.

Fire Hazard: See nitrates.

Disaster Hazard: See nitrates.

DINITROCAPRYL CROTONATE. Crystals. $(NO_2)_2CH(CH_2)_6CH_2O_2CCH{:}CHCH_3$, mw: 288.3, vap. d: 9.95.

THR = HIGH via oral and inhal routes, and as irr to skin, eyes and mu mem. A fungicide.

Fire Hazard: See nitrates.

Disaster Hazard: See nitrates.

4,6-DINITRO(2-CAPRYL)PHENYL CROTONATE. Liquid. $C_{18}H_{24}N_2O_6$, mw: 364.4.

Acute tox data: Oral LD_{50} (rat) = 980 mg/kg; iv LD_{50} (rat) = 23 mg/kg; oral LD_{50} (mammal) = 400 mg/kg. [*3*]

THR = MOD to HIGH via oral and HIGH via iv routes. See nitrates.

2,4-DINITROCHLOROBENZENE. Syn: *1-chloro-2,4-dinitrobenzene*. Yellow rhombic crystals, insol in water. $(NO_2)_2C_6H_3Cl$, mw: 202.56, mp(α): 53.4°, mp(β): 43°, mp(γ): 27°, bp: 315°, lel = 2.0%, uel = 22%, flash p: 382°F (CC), d(α): 1.687 @ 22°, d(β): 1.680 @ 20°/4°, vap. d: 6.98.

Acute tox data: Oral LD_{50} (rat) = 1070 mg/kg; dermal LD_{50} (rabbit) = 130 mg/kg; ip LD_{50} (mammal) = 280 mg/kg. [*3*]

THR = HIGH via dermal and ip; and MOD via oral routes. A poison. Acts as a primary irr as well as a sensitizer of skin. An allergen.

Fire Hazard: Slight, when exposed to heat or flame.

Explosion Hazard: When exposed to flame or sparks or heated to 150° or if confined, by shock. Reacts violently with hydrazine hydrate. [*19*]

Disaster Hazard: See nitrates.

To Fight Fire: CO_2, dry chemical.

DINITROCHLOROHYDRIN. $CH_2ClCHNO_2CH_2NO_2$, mw: 168.6, bp: 180°, d: 1.5, vap. d: 5.83.

THR = HIGH irr to skin, eyes and mu mem.

Fire Hazard: See nitrates.

Explosion Hazard: Dangerous; shock will explode it; when heated to decomp, burns and emits highly toxic fumes; can react vigorously with reducing materials.

4,6-DINITRO-*o*-CRESOL. Syns: *2-methyl-4,6-dinitrophenol, DNOC*. Yellow prismatic crystals. $(NO_2)_2C_6H_2(CH_3)OH$, mw: 198.1, mp: 85.8°, vap. d: 6.82.

Acute tox data: Inhal TC_{LO} (human) = 1 mg/m^3 $\longrightarrow$ CNS effects; oral LD_{50} (rat) = 25 mg/kg; dermal LD_{50} (rat) = 200 mg/kg; ip LD_{50} (mice) = 19 mg/kg. [*3*]

THR = HIGH via inhal, oral, dermal and ip routes. An insecticide and an herbicide. Appears to be somewhat less toxic than the *p*-form, but is still highly toxic.

Fire Hazard: See nitrates.

Disaster Hazard: See nitrates.

2,6-DINITRO-*p*-CRESOL.

Acute tox data: ip LD_{50} (mouse) = 24.8 mg/kg. [*3*]

THR = HIGH via ip route.

3,5-DINITRO-*o*-CRESOL.

Acute tox data: sc LD_{LO} (rat) = 40 mg/kg. [*3*]

THR = HIGH via sc route.

3,5-DINITRO-*p*-CRESOL. Crystals. $(NO_2)_2C_6H_2CH_3OH$, mw: 198.1.

Acute tox data: LD_{LO} (dog) = 15 mg/kg; LD_{LO} (rabbit) = 250 mg/kg. [*3*]

THR = HIGH irr to skin, eyes and mu mem. Can cause brain damage, as well as damage to liver and kidneys. See 4,6-dinitro-*o*-cresol.

3,5-DINITRO-*o*-CRESOL ACETATE. $CH_3COOC_6H_2CH_3(NO_2)_2$, mw: 240.2, vap. d: 8.28.

Acute tox data: ip LD_{LO} (mouse) = 63 mg/kg. [*3*]

THR = HIGH via ip route.

2,4-DINITRO-6-CYCLOHEXYL PHENOL. See 2-cyclohexyl-4,6-dinitrophenol.

2,4-DINITRO-1-FLUOROBENZENE. See 1-fluoro-2,4-dinitrobenzene.

2,5-DINITROFLUORENE.

THR = An exper carc. [*23*]

2,7-DINITROFLUORENE. $C_{13}H_8O_4N_2$, mw: 256.2.
THR = An exper carc. [*3*, *23*]

DINITROGEN PENTOXIDE. N_2O_5, mw: 108.
THR = Reacts violently with O_3. [*19*]

DINITROGEN TRIOXIDE. Syn: *nitrous anhydride.* Red-brown gas, blue solid or liquid. N_2O_3, mw: 76.02, mp: −102°, bp: 3.5° decomp, d: 1.447 @ 20°.
Acute tox data: HIGH irr to skin, eyes and mu mem. See also nitrogen oxides.
Disaster Hazard: Dangerous; see nitric oxide.

DINITROGLYCOL. See diethylene glycol dinitrate.

1,5-DINITRONAPHTHALENE.
See dinitronaphthalene.

1,8-DINITRONAPHTHALENE.
See dinitronaphthalene.

DINITRONAPHTHALENE. Syns: (*a*) *1,5-dinitronaphthalene*, (*b*) *1,8-dinitronaphthalene*. (*a*) yellowish needles, (*b*) yellow rhombic crystals. $C_{10}H_6(NO_2)_2$, mw: 218.16, mp(*a*): 217.5°, mp(*b*): 173°–173.5°, bp(*a*): sublimes, bp(*b*): decomp, vap. d: 7.51.
THR = Details U. See also nitro compounds of aromatic hydrocarbons.
Fire Hazard: Mod, when exposed to heat or flame.
Explosion Hazard: Mod, when shocked or exposed to heat. It is used mixed with chlorates and perchlorates and in combination with picric acid. Dinitronaphthalene is an ingredient of permissible explosives and is also used in combination with ammonium nitrate.
Disaster Hazard: See explosives, high, and nitrates.

2,4-DINITRONAPHTHOL-1. Yellow needles or leaflets. $(NO_2)_2C_{10}H_5OH$, mw: 234.16, mp: 138°, vap. d: 8.08.
Acute tox data: Dermal TD_{LO} (human) = 50 mg/kg; im LD_{50} (pigeon) = 1.85 mg/kg; iv LD_{LO} (dog) = 30 mg/kg. [*3*]
THR = HIGH via dermal, im, and iv routes.
Fire Hazard: See nitrates.
Explosion Hazard: See nitrates.
Disaster Hazard: See nitrates.

2,4-DINITROPHENETOLE. Crystals.
$(NO_2)_2C_6H_3OC_2H_5$, mw: 212.2, vap. d: 7.32.
Acute tox data: Oral LD_{LO} (rat) = 250 mg/kg. [*3*]
THR = HIGH via oral route.

2,3-DINITROPHENOL. Syn: *1-hydroxy-2,3-dinitrobenzene.* Yellow needles. $(NO_2)_2C_6H_3OH$, mw: 184.11, mp: 144°, d: 1.681 @ 20°, vap. d: 6.35.
Acute tox data: LD_{50} (rat) = 190 mg/kg; LD_{50} (mouse) = 200 mg/kg. [*3*]
THR = HIGH toxicity. The harmful effects, which can be fatal, are damage to the liver and induced fever. Fatal cases have been reported in the literature from the inhal of the dust in a conc estimated at approx 40 mg/cubic meter. It is a powerful stimulant of metabolism. It is the excessive oxidation effect of this material upon the metabolism and nutrition that damages the liver and kidney cells. Like other nitrated phenols, it is an irr to the skin and has been known to cause dermatitis. Ingestion will cause dilation of the pupils or posterior subcapsular opacities or cataracts. It also exhibits some allergic manifestations. Wood preservative. See also nitro compounds of aromatic hydrocarbons.
Fire Hazard: See nitrates.
Explosion Hazard: Severe, when exposed to heat. A high explosive used as a component of some shell and bomb charges. See also explosives, high.
Disaster Hazard: See explosives, high and nitrates.

2,4-DINITROPHENOL. Yellow crystals.
$(NO_2)_2C_6H_3OH$, mw: 184.11, mp: 112°, d: 1.683 @ 24°, vap. d: 6.35.
Acute tox data: Oral LD_{LO} (human) = 36 mg/kg; oral LD_{50} (rat) = 30 mg/kg; ip LD_{50} (rat) = 20 mg/kg; sc LD_{50} (rat) = 25 mg/kg; dermal LD_{LO} (guinea pig) = 700 mg/kg; oral LD_{50} (wild birds) = 13 mg/kg. [*3*]
THR = HIGH via oral, ip and sc; MOD via dermal routes. It is phytotoxic. See 2,3-dinitrophenol.
Fire Hazard: See nitrates.
Explosion Hazard: Mod, when exposed to heat.
Disaster Hazard: See nitrates.

2,5-DINITROPHENOL.
Acute tox data: LD_{50} (rat) = 150 mg/kg; LD_{50} (mouse) = 273 mg/kg. [*3*]
THR = HIGH mammalian toxicity.

2,6-DINITROPHENOL. Yellow crystals.
$(NO_2)_2C_6H_3OH$, mw: 184.11, mp: 63°, vap. d: 6.35.
Acute tox data: im LD_{LO} (pigeon) = 40 mg/kg. [*3*]
THR = HIGH via im route.
Fire Hazard: See nitrates.
Explosion Hazard: Mod, when exposed to heat.
Disaster Hazard: See nitrates.

3,4-DINITROPHENOL.
Acute tox data: LD_{50} (rat) = 98 mg/kg; LD_{50} (mouse) = 112 mg/kg. [*3*]
THR = HIGH mammalian toxicity.

3,5-DINITROPHENOL.
Acute tox data: LD_{50} (rat) = 45 mg/kg; LD_{50} (mouse) = 50 mg/kg. [*3*]
THR = HIGH mammalian toxicity.

2,4-DINITROPHENYLACETATE. Crystals.
$CH_3CO_2C_6H_3(NO_2)_2$, mw: 228.2, vap. d: 7.87.
THR = See nitro compounds of aromatic hydrocarbons and nitrates.

DINITROPHENYLAMINE. See 2,4-dinitroaniline.

2,4-DINITROPHENYL DISULFIDE. $C_{12}H_6O_8N_4S_2$, mw: 398.1.
THR = Spont explosive.

DINITROPHENYLHYDRAZINE. Red crystalline powder, slightly sol in water and alcohol. $(NO_2)_2C_6H_3NHNH_2$, mw: 198.2, mp: 200° (approx).
THR = U.
Explosion Hazard: Dangerous explosive.

2,4-DINITROPHENYL METHYL ETHER. See 2,4-dinitroanisole.

4,6-DINITROQUINOLINE-1-OXIDE. $C_9H_5O_5N_3$, mw: 235.2.
THR = An exper neo. [*3*]

2,4-DINITRORESORCINOL. Syn: *2,4-dinitro-1,3-benzenediol.* Yellow crystals.
$(NO_2)_2C_6H_2(OH)_2$, mw: 200.11, mp: 148°, bp: sublimes and explodes, vap. d: 6.79.
THR = Probably toxic. See nitro compounds of aromatic hydrocarbons.
Fire Hazard: See nitrates.
Explosion Hazard: Severe, when shocked or exposed to heat. Dinitroresorcinol and its lead salt are used in commercial priming compositions and blasting caps. It is also used to facilitate the ignition of lead azide. See also explosives, high.
Disaster Hazard: See explosives, high, and nitrates.

N,N′-DINITROSO-N,N′-DIETHYLETHYLENEDIAMINE.
THR = An exper carc. [*23*]

N,N′-DINITROSO-N,N′-DIMETHYLETHYLENE-DIAMINE.
THR = An exper carc. [*23*]

N,N′-DINITROSO DIMETHYL OXAMIDE.
THR = An exper carc. [*23*]

N,N′-DINITROSO-N,N′-DIMETHYL PHTHALAMIDE.
THR = An exper carc. [*23*]

N,N′-DINITROSO-N,N′-DIMETHYL-1,3-PROPANE DIAMINE.
THR = An exper carc. [*23*]

DINITROSODIPHENYLAMINE FLUOSILICATE. Indigo colored crystals, slightly sol in alcohol. $[(C_6H_5)_2N{:}NO]_2 \cdot H_2SiF_6$, mw: 540.5, mp: 124.5°.
THR = See fluosilicates and nitro compounds of aromatic hydrocarbons.

1,4-DINITROSOHOMOPIPERIDINE. $C_5H_{10}O_2N_4$, mw: 158.2.
THR = An exper carc. [*3*]

N,4-DINITROSO-N-METHYL ANILINE. Syn: *elastopax.* Green crystals. $C_7H_7O_2N_3$, mw: 165.2, mp: 101°.
THR = An exper (±) carc and neo. [*3, 1*]

1,4-DINITROSO-2-METHYL PIPERAZINE. $C_5H_{11}O_2N_4$, mw: 159.2.
THR = An exper carc. [*3*]

N,N′-DINITROSOPENTAMETHYLENE TETRAMINE. Syn: *3,7-dinitroso-1,3,5,7-tetra azo bicyclo-3,3,1-nonane.* $C_5H_{10}O_2N_6$, mw: 186.2.
THR = An exper (−) carc. [*13, 3, 23*] Can explode @ > 203°. [*19*]

1,4-DINITROSOPIPERAZINE. White crystals. $C_4H_8O_2N_4$, mw: 144.1, mp: 158°, vap. d: 4.97.
THR = Details U; a stomach insecticide. It is a recog carc. [*14*] An exper carc and neo. [*3, 23*]
Fire Hazard: See nitrates.
Disaster Hazard: See nitrates.

4,4′-DINITRO-2,2′-STILBENE DISULFONIC ACID. Yellow paste or brownish crystals, mw: 430.
THR = U. Probably toxic.
Fire Hazard: See nitrates.
Disaster Hazard: Dangerous; when heated to decomp, emits highly toxic fumes of oxides of nitrogen and sulfur; can react vigorously with reducing materials.

3,5-DINITRO-*o*-TOLUAMIDE. See zoalene.

2,3-DINITROTOLUENE. $C_7H_6O_4N_2$, mw: 182.2.
Acute tox data: Oral LD_{50} (rat) = 1122 mg/kg; oral LD_{50} (mouse) = 1072 mg/kg. [*3*]
THR = MOD via oral route.

2,4-DINITROTOLUENE. Syn: *dinitrotoluol.* Yellow needles. $(NO_2)_2C_6H_3CH_3$, mw: 182.13, mp: 69.5°, bp: 300°, d: 1.521 @ 15°, vap. d: 6.27, flash p: 404°F.
Acute tox data: Oral LD_{50} (rat) = 268 mg/kg; oral LD_{50} (mouse) = 1625 mg/kg. [*3*]
THR = HIGH via oral, inhal, dermal routes. An irr and an allergen. Can cause anemia, methemoglobinemia, cyanosis and liver damage. See also trinitrotoluene.
Fire Hazard: Low.
To Fight Fire: Water spray or mist, dry chemical.

2,5-DINITROTOLUENE.
Acute tox data: Oral LD_{50} (rat) = 707 mg/kg; oral LD_{50} (mouse) = 1231 mg/kg. [*3*]
THR = MOD via oral route.

2,6-DINITROTOLUENE.
Acute tox data: Oral LD_{50} (rat) = 177 mg/kg. [*3*]
THR = HIGH via oral route.

3,4-DINITROTOLUENE.
Acute tox data: Oral LD_{50} (rat) = 177 mg/kg. [*3*]
THR = HIGH via oral route.
Fire Hazard: Mod, when exposed to heat or flame; an oxidizer.

Explosion Hazard: Mod, when exposed to heat.
Disaster Hazard: See explosives, high, and nitrates.
To Fight Fire: Water, CO_2, dry chemical.

DINITROTOLUOL. See 2,4-dinitrotoluene.

4,6-DINITRO-1,2,3-TRICHLORO BENZENE. Syn: *vancide P.B.* $C_6HO_4N_2Cl_3$, mw: 271.4.
THR = HIGH acute tox. An exper neo. [*3*]

DINOCAP. See dinitrocapryl crotonate.

DINOL. See diazodinitrophenol.

DINONYL NAPHTHALENE. Dark straw-colored viscous liquid. $(C_9H_{19})_2C_{10}H_6$, mw: 380.6, bp: 200°–270° @ 20 mm, d: 0.92–0.95 @ 30°/20°, vap. d: 13.1.
THR = U. See also naphthalene.
Disaster Hazard: Slight; when heated, emits acrid fumes; can react with oxidizing materials.

DINONYL PHENOL. Clear liquid. $C_6H_3OH(C_9H_{19})_2$, mw: 346.6, bp: 180°–220° @ 10 mm, d: 0.914 @ 25°, vap. d: 12.0.
THR = Details U. See also phenols.
Fire Hazard: Mod, when exposed to heat or flame; can react with oxidizing materials.
To Fight Fire: Foam, CO_2, dry chemical.

DINOSEB. See 2-*sec*-butyl-6,4-dinitrophenol.

DIOCTYL ADIPATE. A clear liquid. $(C_8H_{17}CO_2)_2C_4H_8$, mw: 370, bp: 214° @ 5 mm, fp: −79°, flash p: 400°F (COC), d: 0.924–0.930 @ 20°/20°.
THR = See esters and adipic acid.
Fire Hazard: Slight, when heated; emits acrid fumes; can react with oxidizing materials.
To Fight Fire: Foam, CO_2, dry chemical.

DIOCTYLAMINE. See di-2-ethylhexylamine.

DIOCTYLCHLOROPHOSPHATE. Liquid. $(C_8H_{17}O)_2POCl$, mw: 341, bp: 125° @ 0.2 mm, d: 0.9838 @ 25°/25°.
THR = Details U. An insecticide.
Disaster Hazard: Dangerous; when heated to decomp, emits highly toxic fumes of PO_x and chlorides.

DI(*n*-OCTYL-*n*-DECYL)PHTHALATE. Clear oily liquid. $C_{26}H_{42}O_2$, mw: 418.3, bp: 232°–267° @ 5 mm, fp: −30°, flash p: 426°F (OC), d: 0.968–0.977 @ 25°/25°, vap. d: 14.4.
THR = See esters.
Fire Hazard: Slight, when exposed to heat or flame; can react with oxidizing materials.
To Fight Fire: Foam, CO_2, dry chemical.

DIOCTYL ETHER. Syn: *octyl ether.* $C_{16}H_{34}O_2$, mw: 258.4, flash p: >212°F, autoign. temp.: 401°F, d: 0.82, vap. d: 8.36, bp: 300°.
THR = No data. See ethers.
Fire Hazard: Low. See also ethers.
To Fight Fire: Dry chemical, CO_2, water spray or mist, foam.

DIOCTYL FUMARATE. Syn: *di-2-ethylhexyl fumarate*. Clear mobile liquid, mild odor. $(C_8H_{17}CO_2)_2C_2H_2$, mw: 340, bp: 211°–220°, flash p: 365°F (COC), d: 0.942 @ 20°/20°.
THR = See esters and fumaric acid.
Fire Hazard: Slight; can react with oxidizing materials, heat, flame.
To Fight Fire: Foam, CO_2, dry chemical.

DIOCTYL PHOSPHITE. $(C_8H_{17}O)_2PHO$, mw: 306.3, bp: 128° @ 0.5 mm, d: 0.9291 @ 25°/25°.
THR = U.
Disaster Hazard: Dangerous; when heated to decomp, emits highly toxic fumes of oxides of phosphorus.

DIOCTYL PHTHALATE.
See di-2-ethylhexyl phthalate.

DI-*sec*-OCTYL PHTHALATE.
See di-2-ethylhexyl phthalate.

DIOCTYL SEBACATE. See di-(2-ethylhexyl) sebacate.

DIOCTYL SODIUM SULFOSUCCINATE. White solid or clear viscous liquid.
$C_8H_{17}CO_2CH_2CH(CO_2C_8H_{17})SO_3Na$, mw: 444.
Acute tox data: Oral LD_{50} (rat) = 1900 mg/kg; oral LD_{50} (mouse) = 4800 mg/kg; iv LD_{50} (mouse) = 60 mg/kg. [*3*]
THR = MOD via oral and HIGH via iv routes. Used as a food additive permitted in food for human consumption. [*109*]
Disaster Hazard: Mod dangerous; when heated to decomp, emits toxic fumes.

DIOCTYL TETRAHYDROPHTHALATE. Crystals. $C_6H_8(COOC_8H_{17})_2$, mw: 394.6, vap. d: 13.6.
THR = A MILD irr.
Fire Hazard: Slight, when exposed to heat or flame; can react with oxidizing materials.
To Fight Fire: Spray, foam, dry chemical, mist.

DI-*n*-OCTYLTIN DICHLORIDE. See di-2-ethylhexyltin dichloride.

DIONIN. Syn: *ethylmorphine hydrochloride*. White, microscopic, crystalline powder. $C_{19}H_{23}NO_3 \cdot HCl \cdot 2H_2O$, mw: 385.9, mp: 125° decomp, vap. d: 13.3.
Acute tox data: sc LD_{50} (mouse) = 200 mg/kg. [*3*]
THR = HIGH via sc route. May cause habituation. See also codeine.
Disaster Hazard: Dangerous; see chlorides.

1,4-DIOXANE. Syns: p-*dioxane*, *diethylene oxide*, *diethylene dioxide*. Colorless liquid, pleasant odor. $OCH_2CH_2OCH_2CH_2$, mw: 88.10, mp: 12°, bp: 101.1°, lel = 2.0%, uel = 22.2%, flash p: 54°F (CC), d: 1.0353 @ 20°/4°, autoign. temp.: 356°F, vap. press: 40 mm @ 25.2°, vap. d: 3.03.

Acute tox data: Oral LD_{50} (rat) = 7.1 g/kg; oral LD_{50} (cat) = 2.0 g/kg; dermal LD_{50} (rabbit) = 7.6 g/kg. [*3*]

THR = MOD via oral and dermal routes. Eye effects from 5500 ppm for 1 min. Exposure of animals to conc of 0.1 to 3% of dioxâne vapor causes irr of the eyes and nose, followed by narcosis and/or pulmonary edema and death. The irr effects probably provide sufficient warning, in acute exposures, to enable the workman to leave exposure before he is seriously affected. On the other hand, repeated exposure to low conc has resulted in human fatalities, the organs chiefly affected being the liver and kidneys. Death resulted from acute hemorrhagic nephritis. The hepatic lesion consists of an acute central necrosis of the lobules. The brain and lungs may show acute edema.

In acute exposures, the signs and symptoms consist of irr of the eyes and naso-pharynx, which may later subside, to be followed by headache, drowsiness, dizziness, and occasionally nausea and vomiting. In chronic exposures, there may be loss of appetite, nausea and vomiting, pain and tenderness in the abdomen and lumbar region, malaise, and enlargement of the liver without jaundice. There may be changes in the blood picture. Further exposure may result in suppression of urine, followed by uremia and death. An exper (+) carc and neo. [*3, 13*]

Fire Hazard: Dangerous, when exposed to heat or flame; can react vigorously with oxidizing materials.

Explosion Hazard: When exposed to flame or by chemical reaction with oxidizers. Violent reaction with (H_2 + Raney Ni), $AgClO_4$. [*19*]

To Fight Fire: Alcohol foam, CO_2, dry chemical.

2,3-*p*-DIOXANEDITHION-S,S-BIS-(*o,o*-DIETHYL) PHOSPHORODITHIOATE. See delnav.

1,3,2-DIOXATHIOLANE-2,2-DIOXIDE. $C_2H_4O_4S$, mw: 124.1.

THR = An exper neo. [*3*]

DIOXATHION. See delnav.

DIOXIN. See (2,6-dimethyl-dioxan-4-yl)acetate.

1,1′,1″-(3,6-DIOXO-1,4-CYCLOHEXADIENE-1,2,4-TRIYL)-TRIS AZIRIDINE. Purple needles. $C_{12}H_{13}O_2N_3$, mw: 231.3, mp: 163°.

THR = An exper (+) mutagen and carc. [*3, 8*]

2-(1,3-DIOXOLAN-2-YL)PHENYL-N-METHYL CARBAMATE. See elocron.

DIOXOLANE. Water white liquid. $C_3H_6O_2$, mw: 74.08, mp: −26.4°, bp: 75°, flash p: 35°F (OC), d: 1.065, vap. press: 70 mm @ 20°, vap. d: 2.6.

Acute tox data: Oral LD_{50} (rat) = 7460 mg/kg; inhal LC_{LO} (rat) = 32000 ppm for 4 hrs; dermal LD_{50} (rabbit) = 8480 mg/kg. [*3*]

THR = MOD via oral; LOW via dermal and inhal routes.

Fire Hazard: Dangerous, when exposed to heat or flame; can react vigorously with oxidizing materials.

Explosion Hazard: U.

Disaster Hazard: Dangerous! Keep away from heat and open flame!

To Fight Fire: Alcohol foam, CO_2, dry chemical.

DIOXOLONE-2. See ethylene carbonate.

2,2′-DIOXO-N-NITROSODIPROPYLAMINE. $C_6H_{10}O_3N_2$, mw: 158.2.

THR =An exper neo. [*3*]

2-(2,6-DIOXOPIPERIDEN-3-YL)PHTHALIMIDINE. $C_{13}H_9O_3N_2$, mw: 241.2.

THR = An exper tetratogen. [*3*]

DIOXYGEN DIFLUORIDE. See fluorine dioxide.

DIPACINONE. See 2-diphenylacetyl-1,3-indandione.

DIPAXIN. See 2-diphenylacetyl-1,3-indandione.

DIPENTAERYTHRITOL. White powder, odorless. $C_{10}H_{22}O_7$, mw: 254.3, mp: 212°–220°, d: 1.33 @ 25°/4°, vap. d: 8.77.

THR = Dctails U.

Fire Hazard: Slight.

DIPENTENE. Syn: *cinene*. Colorless liquid, pleasant lemon-like odor. The racemic mixture of *d* and *l* limonene is called dipentene. $C_{10}H_{16}$, mw: 136.23, bp: 174.6°, flash p: 113°F, d: 0.865 @ 18°, 0.845 @ 20°, vap. press: 1 mm @ 14.0°, vap. d: 4.66, autoign. temp.: 458°F, lel = 0.7% @ 302°F, uel = 6.1% @ 302°F.

Acute tox data: Oral LD_{50} (rat) = 5000 mg/kg. [*3*]

THR = MOD via oral route. An allergen and irr. A synthetic flavoring substance and adjuvant. [*109*] Can react violently with a mixture of (IF_5 + tetrafluoroethylene). [*19*]

Fire Hazard: Mod, when exposed to heat or flame; can react with oxidizing materials.

To Fight Fire: Foam, CO_2, dry chemical.

DIPENTENE DIOXIDE. $C_{10}H_{16}O_2$, mw: 168.3.

Acute tox data: Oral LD_{50} (rat) = 5630 mg/kg; dermal LD_{LO} (rabbit) = 1770 mg/kg. [*3*]

THR = MOD via oral and dermal routes. An exper neo. [*3*]

DIPENTYL MALEATE. Syn: *diamyl maleate*. Liquid, water white, faintly alcoholic odor. $(CHCOOC_5H_{11})_2$, mw: 256, d: 0.981 @ 20°, boiling range: 263°–300°, flash p: 270°F.

THR = Limited animal exper suggest LOW toxicity. See also esters and maleic acid.

Fire Hazard: Slight, when exposed to heat or flame.

DIPHACIN. See 2-diphenylacetyl-1,3-indandione.

DIPEROXY TERAPHTHALIC ACID. $C_8H_6O_6$, mw: 198.1.
THR = Shock- and heat-sensitive explosive. [*19*]

DIPHENADIONE.
See 2-diphenylacetyl-1,3-indandione.

DIPHENAMIDE.
See N,N-dimethyl-2,2-diphenylacetamide.

DI-*p*-PHENETYL DITELLURIDE. Orange-brown crystals. $(C_2H_5OC_6H_4)_2Te_2$, mw: 497.5, d: 1.666, mp: 108°.
THR = See tellurium compounds.

DIPHENOLIC ACID. Syn: *DPA*. Crystals, sol in water, acetone, alcohol and acetic acid. $C_{17}H_{18}O_4$, mw: 286.3, mp: 172°.
THR = Details U. See phenol.

DIPHENYL. Syn: *biphenyl*. White scales, pleasant odor. $C_6H_5C_6H_5$, mw: 154.2, mp: 70°, bp: 255°, flash p: 235°F (CC), d: 0.991 @ 75°/4°, autoign. temp.: 1004°F, vap. d: 5.31, lel = 0.6% @ 232°, uel = 5.8% @ 331°F.
Acute tox data: Oral LD_{50} (rat) = 3280 mg/kg; oral LD_{50} (rabbit) = 2400 mg/kg; inhal TD_{LO} (human) = 4.4 mg/m^3. [*3*]
THR = MOD via oral; HIGH via inhal routes. Causes convulsions; and paralysis. An exper neo and irr. [*3*]
Fire Hazard: Slight, when exposed to heat or flame; can react with oxidizing materials.
Spont Heating: No.
To Fight Fire: CO_2, dry chemical, water spray, mist, fog.

DIPHENYLACETIC ACID. Syn: *diphenylmethane-α-carboxylic acid*. White crystals. $(C_6H_5)_2CHCOOH$, mw: 212.2, mp: 147°–148.2°, bp: sublimes, vap. d: 7.3.
THR = Details U. See also diphenyl.
Fire Hazard: Slight.

DIPHENYLACETONITRILE. $C_{14}H_{11}N$, mw: 193.3.
THR = An exper neo. [*3*]

DIPHENYLACETYLENE. Syns: *tolan, diphenyl ethyne*. Monoclinic pseudorhombic crystals, insol in water, freely sol in ether and hot alcohol. $(C_6H_5)C\vdots C(C_6H_5)$, mw: 178.22, mp: 60°–61°, bp: 300°, d: 0.966 @100°/4°.
THR = See diphenylethylene.

2-DIPHENYLACETYL-1,3-INDANDIONE. Syns: *diphenadione, dipaxin*. Pale yellow crystals, sol in acetone and acetic acid. $C_{23}H_{16}O_3$, mw: 340.4, mp: 147°.
Acute tox data: Oral LD_{50} (rat) = 1.4 mg/kg; oral LD_{50} (cat) = 15 mg/kg. [*3*]
THR = HIGH via oral route. Inhibits blood clotting, leading to hemorrhages. Action similar to that of warfarin.

DIPHENYLAMINE. Syns: *phenylaniline, anilinobenzene*. Crystals, floral odor. Sol in benzene, ether and carbon disulfide. $(C_6H_5)_2NH$, mw: 169.24, mp: 52.9°, bp: 302.0° flash p: 307°F (CC), d: 1.16, autoign. temp.: 1173°F, vap. press: 1 mm @ 108.3°, vap. d: 5.82.
Acute tox data: Oral LD_{LO} (rat) = 3000 mg/kg. [*3*]
THR = MOD via oral route. Action similar to aniline but less severe. See also aniline and amines. See also aromatic amines. An exper teratogen and carc. [*3*] A possible carc. [*14*, *23*] which may be due to traces of 4-aminobiphenyl. See 4-aminobiphenyl.
Fire Hazard: Slight, when exposed to heat or flame. Can react violently with hexachloromelamine, trichloromelamine. [*19*]
Spont Heating: No.
Disaster Hazard: Dangerous; when heated to decomp, emits highly toxic fumes; can react with oxidizing materials.
To Fight Fire: CO_2, dry chemical.

DIPHENYLAMINE ARSENIOUS OXIDE.
THR = See arsenic compounds.

DIPHENYLAMINECHLOROARSINE. Syn: *adamsite*. Light yellow to green granules, irr odor. $NH(C_6H_4)_2AsCl$, mw: 277.6, mp: 195°, bp: 410° (decomp), d: 1.65, vap. press: very low @ 20°, vap. d: 9.6.
THR = HIGH irr to skin, eyes and mu mem. See also arsenic. A vomiting type of military poison and a non-persistent military gas. See also arsenic compounds.
Disaster Hazard: Dangerous; when heated, emits highly toxic fumes of arsenic.

DIPHENYLAMINE HYDROCHLORIDE. White crystals become blue in air, freely sol in water and alcohol. $(C_6H_5)_2NHHCl$, mw: 205.68.
THR = See diphenylamine.
Disaster Hazard: Dangerous; see chlorides.

DIPHENYLAMINE SULFATE. White to yellowish powder, insol in water, sol in alcohol and sulfuric acid. $(C_6H_5)_2NHH_2SO_4$, mw: 267.30 mp: 123°–125°.
Acute tox data: ip LD_{LO} (mouse) = 200 mg/kg. [*3*]
THR = HIGH via ip route.
Disaster Hazard: Dangerous; see sulfates.

1,2-DIPHENYLBENZENE. See *o*-terphenyl.

DIPHENYLBORIC ACID. Syn: *diphenylhydroxyborine*. Colorless crystals, insol in water. $(C_6H_5)_2BOH$, mw: 182.0, mp: 226°, bp: 215°–235° @ 17 mm.
THR = Details U. See also boron compounds.

DIPHENYLBORON BROMIDE. Colorless thick liq-

uid or crystals, decomp by water. $(C_6H_5)_2BBr$, mw: 244.9, mp: 25°, bp: 150°–160° @ 8 mm.
THR = Details U. See also boron compounds.
Disaster Hazard: Dangerous; see bromides.

DIPHENYLBORON CHLORIDE. Colorless liquid, decomp by water. $(C_6H_5)_2BCl$, mw: 200.5, bp: 271°.
THR = Details U. See boron compounds.
Disaster Hazard: Dangerous; see chlorides.

DIPHENYLBROMOARSINE. Crystals. $(C_6H_5)_2AsBr$, mw: 309.1, mp: 54°, vap. d: 10.7.
THR = A powerful irr. An allergen. HIGH tox via oral and inhal routes. A poison. See also arsenic compounds.
Fire Hazard: Mod, when exposed to heat, flame or oxidizers.
Disaster Hazard: Dangerous; see arsenic.
To Fight Fire: Spray, foam, dry chemical.

DIPHENYLBUTANE. $C_{16}H_{18}$, mw: 210.2, flash p: >212°F, autoign. temp.: 851°F, d: 0.98, vap. d: 7.26, bp: 294°.
THR = No data.
Fire Hazard: Low.
To Fight Fire: Water spray, mist, foam, dry chemical.

DIPHENYLCARBINOL. See benzhydrol.

DIPHENYLCARBONATE. $C_{13}H_{10}O_2$, mw: 198.2.
THR = An exper carc. [3]

DIPHENYLCHLOROARSINE. Colorless crystals when pure, technical product is dark brown liquid. $(C_6H_5)_2AsCl$, mw: 264.57, bp: 333° (decomp), fp: 44°, d: 1.363 @ 40° (solid): 1.358 @ 45° (liquid), vap. press: 0.00049 mm @ 20°, vap. d: 9.15.
Acute tox data: Inhal LC_{LO} (human) = 55 ppm for 30 min; dermal TD_{LO} (mouse) = 33 mg/kg ⟶ neo; iv LD_{LO} (rat) = 0.5 mg/kg. [3]
THR = HIGH via inhal, dermal and iv routes. An exper neo. [3] A powerful irr military poison. Can be decontaminated by use of chlorine or caustic soda in confined spaces. A non-persistent gas. Exposure yields cold-like symptoms, plus headache, vomiting and nausea.
Disaster Hazard: Dangerous; see arsenic.

DIPHENYLCYANOARSINE. Colorless prisms, characteristic odor resembling that of a mixture of bitter almonds and garlic. $(C_6H_5)_2AsCN$, mw: 255.0, bp: 213° @ 21 mm, fp: 31°, d: 1.33 @ 20°, vap. press: 0.0002 mm @ 20°, vap. d: 8.75.
THR = HIGH irr to eyes and mu mem. A military poison. Acts as an irr only to the eyes. Exposure yields cold-like symptoms, plus headache, vomiting and nausea. A vomiting gas of the non-persistent variety. Decontamination needed only in confined space by alkali solution. See also cyanides and arsenic compounds.
Disaster Hazard: Dangerous; see cyanides and arsenic.

DIPHENYLDODECYL PHOSPHITE. Nearly water white liquid, insol in water. $C_{22}H_{31}O_3P$, mw: 374.4, d: 1.023 @ 25°/15.5°, mp: 18°, flash p: 425°F (OC).
THR = U.
Fire Hazard: Slight when exposed to heat or flame.
Disaster Hazard: Dangerous; see phosphates.
To Fight Fire: Dry chemical, CO_2, mist.

DIPHENYLDIAZO SULFIDE. $C_{12}H_{10}N_4S$, mw: 242.2.
THR = Exploded during air drying. [*19*]

DIPHENYLDICHLOROSILANE. Colorless liquid. $(C_6H_5)_2SiCl_2$, mw: 245.22, mp: −22°, bp: 303°, d: 1.19 @ 20°, vap. d: 8.45.
THR = HIGH irr to skin, eyes and mu mem. See also chlorosilanes.
Disaster Hazard: Mod dangerous; when heated to decomp or on contact with acid or acid fumes, emits toxic fumes; will react with water or steam to produce heat, toxic and corrosive fumes; can react vigorously with oxidizing materials.

DIPHENYL DISULFIDE. White powder. C_6H_5-S-S-C_6H_5, mw: 218.3, mp: 59.5°–60.0°, bp: 310°.
Acute tox data: ip LD_{LO} (mouse) = 100 mg/kg. [3]
THR = HIGH via ip route.
Fire Hazard: Mod, when exposed to heat or flame.
Disaster Hazard: Dangerous; see sulfur compounds. Can react with oxidizing materials.

DIPHENYLENE OXIDE. Syn: *dibenzofuran*. Colorless crystals. $C_6H_4OC_6H_4$, mw: 168.2, mp: 87°, bp: 288° = vap. d: 5.8.
THR = Details U; an insecticide.
Fire Hazard: Slight.

1,1-DIPHENYLETHANE. Pale, yellowish liquid, aromatic odor. $(C_6H_5)_2CHCH_3$, mw: 182.3, mp: −20°, bp: 272°, flash p: 264°F, autoign. temp.: 824°F, d: 0.987 @ 25°/25°, vap. d: 6.28.
Acute tox data: ip LD_{LO} (mouse) = 1000 mg/kg. [3]
THR = MOD via ip route. Probably irr and narcotic in high conc.
Fire Hazard: Slight, when exposed to heat or flame.
Disaster Hazard: Mod dangerous; can react with oxidizing materials.
To Fight Fire: Foam, CO_2, dry chemical.

1,2-DIPHENYLETHANE. flash p: 264°F, autoign. temp.: 896°F, d: 1.0, vap. d: 6.29, bp: 285°.
THR = No data. See also 1,1-diphenylethane.
Fire Hazard: Low.
To Fight Fire: Water, spray, mist, alcohol foam, dry chemical.

DIPHENYL ETHER. See diphenyl oxide.

DI-β-PHENYLETHYLAMINE. Colorless to slightly yellow liquid. $(C_6H_5CH_2CH_2)_2NH$, mw: 225.3, bp: 204° @ 22 mm, d: 1.0 @ 25°/25°, vap. d: 7.76.
THR = Details U. See also amines.
Fire Hazard: Slight.

DIPHENYLETHYLENE. Syns: *stilbene, tolulyne, trans form of α,β-diphenyl ethylene.* Colorless or slightly yellow crystals, insol in water, sol in 90 parts cold alcohol and 13 parts boiling alcohol, freely sol in benzene and ether. $C_6H_5CH{:}CHC_6H_5$, mw: 180.24, mp: 124°-125°, bp: 306°-307°, d: 0.9707.
THR = Animal exper show low oral and inhal toxicity but greater effects from skin absorption. Reacts violently with O_2. [*19*]

DIPHENYLGERMANIUM. White crystals, insol in water. $[(C_6H_5)_2Ge]_4$, mw: 907.2, mp: 295°.
THR = See germanium compounds.

DIPHENYLGERMANIUM DIBROMIDE. Colorless liquid, hydrolyzed by water. $(C_6H_5)_2GeBr_2$, mw: 386.6, bp: 207° @ 512 mm.
THR = See germanium compounds and bromides.

DIPHENYLGERMANIUM DICHLORIDE. Colorless liquid, hydrolyzed by water. $(C_6H_5)_2GeCl_2$, mw: 297.7, d: 0.71, mp: 9°, bp: 223° @ 12 mm.
THR = See germanium compounds and chlorides.

DIPHENYLGERMANIUM DIFLUORIDE. Colorless liquid, hydrolyzed by water. $(C_6H_5)_2GeF_2$, mw: 264.8, bp: 100° @ 0.007 mm.
THR = See fluorides and germanium compounds.

DIPHENYLGLYCOLIC ACID. See benzilic acid.

DIPHENYLGUANIDINE. White powder. $NH{:}C(NHC_6H_5)_2$, mw: 211.1, mp: 145°, d: 1.115 @ 25°.
Acute tox data: Oral LD_{50} (rat) = 375 mg/kg; oral LD_{50} (mouse) = 290 mg/kg. [*3*]
THR = HIGH via oral route.
Disaster Hazard: Mod dangerous; when heated to decomp, emits toxic fumes.

DIPHENYLGUANIDINE PHTHALATE. White to light gray powder. $C_6H_4(COOH)_2 \cdot C_{26}H_{26}N_6 \cdot \frac{1}{2}H_2O \cdot$ mw: 597.5, mp: 178°, d: 1.2 @ 25°.
THR = Details U. See also esters and diphenyl guanidine. Possibly a skin sensitizer.
Fire Hazard: Slight.

DIPHENYL HYDANTOIN. Syn: *Dilantin.* $C_{15}H_{12}O_2N_2$, mw: 252.3.
THR = A HIGH toxicity exper material via oral, ip, sc, and iv routes. An exper carc, neo, teratogen. Affects the CNS. [*3*]

DIPHENYLHYDANTOIN SODIUM.
THR = An exper teratogen. [*3*]

DIPHENYLHYDROXYBORINE. See diphenylboric acid.

DIPHENYLINE. See 2,4′-biphenyl diamine.

DIPHENYL KETENE. Syn: *diphenylethenone.* Reddish-yellow liquid. $(C_6H_5)_2C{:}CO$, mw: 194.22, d: 1.1107 @ 13.7°/4°, bp: 265°-276° (decomp).
THR = U. Vapors may have a narcotic effect.

DIPHENYL KETONE. See benzophenone.

DIPHENYLMERCURY. White crystals, insol in water. $(C_6H_5)_2Hg$, mw: 354.8, d: 2.318, mp: 122° (sublimes), bp: 204° @ 10.5 mm.
THR = HIGH toxicity. See mercury compounds, organic.

DIPHENYLMETHANE. Syn: *benzylbenzene.* Liquid. $(C_6H_5)_2CH_2$, mw: 168.23, mp: 26.5°, bp: 264.5°, flash p: 266°F (CC), d: 1.006, vap. press: 1 mm @ 76.0°, vap. d: 5.79, autoign. temp.: 905°F.
Acute tox data: Oral LD_{LO} (rat) = 5000 mg/kg. [*3*]
THR = MOD via oral route.
Fire Hazard: Slight, when exposed to heat or flame; can react with oxidizing materials.
Spont Heating: No.
To Fight Fire: CO_2, dry chemical.

DIPHENYLMETHANE-α-CARBOXYLIC ACID. See diphenylacetic acid.

DIPHENYLMETHANE DIISOCYANATE. See methylene bis(4-phenyl isocyanate).

DIPHENYLMETHYL BROMIDE. Syn: *benzhydryl bromide.* Solid, decomp in hot water, sol in alcohol, very sol in benzene. $BrCH(C_6H_5)_2$, mw: 247, mp: 45°, bp: 193° @ 26 mm.
THR = HIGH. A corrosive poison.
Disaster Hazard: Dangerous; see bromides.

DIPHENYL MONO(*p-tert*-BUTYLPHENYL)PHOSPHATE. Syn: *p-tert-butyl-phenyldiphenyl phosphate.* Colorless pale straw-colored liquid. $(C_6H_5)_2[(CH_3)_3CC_6H_4]PO_4$, mw: 382.4, bp: 245°-260° @ 5 mm, fp: 0°, flash p: 435°F, d: 1.16 @ 25°/25°.
THR = Details U.
Fire Hazard: Slight, when exposed to heat or flame.
Disaster Hazard: Mod dangerous; when heated to decomp, emits toxic fumes; can react with oxidizing materials.
To Fight Fire: Water, foam, CO_2, dry chemical.

DIPHENYL MONO(*o*-XENYL)PHOSPHATE. $(C_6H_5O)_2PO(OC_6H_4C_6H_5)$, mw: 402, d: 1.2, bp: 482°-545°F @ 5 mm, flash p: 473°F.
THR = U.
Fire Hazard: Slight, when exposed to heat or flame.
Disaster Hazard: Dangerous; see phosphates.
To Fight Fire: Water, foam.

DIPHENYL NITROSAMINE. See *n*-nitrosodiphenylamine.

3,3-DIPHENYL-2-OXETANONE. $C_{15}H_{12}O_2$, mw: 224.3.
THR = An exper. carc. [*3*]

DIPHENYL OXIDE. Colorless crystals, geranium odor. $(C_6H_5)_2O$, mw: 170.20, mp: 28°, bp: 257°, flash p: 239°F, d: 1.0728 @ 20°, vap. d: 5.86, autoign. temp.: 1148°F, lel = 0.8%, uel = 1.5%.
Acute tox data: Oral LD_{50} (rat) = 3370 mg/kg. [*3*]
THR = MOD via oral and inhal routes. Prolonged exposure damages liver, spleen, kidneys and thyroids and upsets GI tract. A MILD irr.
Fire Hazard: Mod, when exposed to heat or flame; can react with oxidizing materials.
Spont Heating: No.
Explosion Hazard: See ethers.
To Fight Fire: Water, foam, CO_2, dry chemical.

1,1-DIPHENYLPENTANE. $C_{17}H_{20}$, mw: 224.2, flash p: >212°F, autoign. temp.: 824°F, d: 0.97, vap. d: 7.74, bp: 308°.
THR = No data.
Fire Hazard: Low.
To Fight Fire: Dry chemical, water spray or mist, foam.

N,N-DIPHENYL-*p*-PHENYLENEDIAMINE. Solid. $(C_6H_5)_2NC_6H_4NH_2$, mw: 260.3, d: 1.20, vap. d: 9.0.
Acute tox data: Oral LD_{50} (rat) = 2370 mg/kg. [*3*]
THR = MOD via oral route. A weak allergen. An exper neo. [*3*]
Fire Hazard: Slight, when exposed to heat or flame.
Disaster Hazard: Mod dangerous; when heated to decomp, emits toxic fumes; can react with oxidizing materials.

DIPHENYL PHTHALATE. Solid. $C_6H_4(COOC_6H_5)_2$, mw: 318.3, mp: 70.7°, bp: 405°, flash p: 435°F (CC), d: 1.28.
THR = MOD. See phenol and phthalic acid.
Fire Hazard: Slight, when exposed to heat or flame; can react with oxidizing materials.
To Fight Fire: CO_2, dry chemical.

N,N′-DIPHENYL PIPERAZINE. Crystals. $C_6H_5N(CH_2)_4NC_6H_5$, mw: 238.3, mp: 167°, vap. d: 8.22.
THR = Details U.

1,1-DIPHENYL PROPANE. $C_{12}H_{16}$, mw: 160.1, flash p: >212°F, autoign. temp.: 860°F, d: 0.97, vap. d: 6.77, bp: 283°.
THR = No data.
Fire Hazard: Low.
To Fight Fire: Dry chemical, CO_2, water spray, fog, foam.

α,α-DIPHENYL-β-PROPIOLACTONE.
THR = An exper carc. [*23*]

DIPHENYL-*sec*-PROPYLGERMANIUM BROMIDE. Colorless liquid. $(C_6H_5)_2(C_3H_7)GeBr$, mw: 349.8, bp: 215°–250° @ 13 mm.
THR = See bromides and germanium compounds.

1,1-DIPHENYL-2-PROPYNYL-N-CYCLOHEXYL CARBAMATE.
THR = An exper carc. [*3*, *12*] via oral route.

3,5-DIPHENYLPYRIDINE. Pale yellow to green solid, insol in water. $(C_6H_5)_2C_5H_3N$, mw: 231.3, mp: 137°.
THR = Based upon animal exper, it appears to be practically non-toxic.
Disaster Hazard: Dangerous; when heated to decomp, emits toxic fumes.

1,4-DIPHENYLSEMICARBAZIDE. Crystals. $C_6H_5NHNHCONHC_6H_5$, mw: 227.3, vap. d: 7.8.
THR = Details U; an insecticide.
Disaster Hazard: Dangerous; when heated to decomp, emits toxic fumes.

DIPHENYLSULFOXIDE. Crystals. $(C_6H_5)_2SO$, mw: 202.3, vap. d: 7.0.
Acute tox data: ip LD_{50} (mouse) = 650 mg/kg. [*3*]
THR = MOD via ip route. A fungicide. See also diphenyl.
Disaster Hazard: Dangerous; see sulfates.

DIPHENYLTETRAACETYLENE. $C_{20}H_{10}$, mw: 250.2.
THR = Spont exploded after standing 13 months. [*19*]

DIPHENYLTHIOCARBAZONE. Syn: *octhiozone.* Bluish-black crystalline powder, insol in water, sparingly sol in alcohol; freely sol in carbon tetrachloride and chloroform. $C_6H_5N{:}NCSNHNHC_6H_5$, mw: 256.32.
Acute tox data: ip LD_{LO} (mouse) = 200 mg/kg. [*3*]
THR = HIGH via ip route. Has caused eye injury and glycosuria in animals.
Disaster Hazard: Dangerous; when heated to decomp, emits highly toxic fumes.

5,5-DIPHENYL-2-THIOHYDANTOIN. $C_{15}H_{12}ON_2S$, mw: 268.4.
THR = An exper neo. [*3*]

DIPHENYLTHIOL DIPHENYL STANNANE. Solid. $(C_6H_5)_2Sn(C_6H_5S)_2$, mw: 491.2, mp: 65°.
THR = See tin compounds and sulfides.

N,N′-DIPHENYLTHIOUREA. See thiocarbanilide.

1,3-DIPHENYL-2-THIOUREA. See thiocarbanilide.

1,3-DIPHENYLTRIAZINE. See α-diazoamidoben.

DIPHENYLTIN. Yellow powder, insol in water. $(C_6H_5)_2Sn$, mw: 272.9, mp: 226°.
Acute tox data: ip LD_{LO} (rat) = 15 mg/kg. [*3*]
THR = HIGH via ip route. See tin compounds.

DIPHENYLTIN BROMIDE. Colorless crystals, sol in alcohol and ether. $(C_6H_5)_2SnBr_2$, mw: 432.7, mp: 38°, bp: 230° @ 42 mm.
THR = See tin compounds and bromides.

DIPHENYLTIN DICHLORIDE. Colorless crystals, decomp by water. $(C_6H_5)_2SnCl_2$, mw: 343.8, mp: 42°, bp: 333°–337° (decomp).
Acute tox data: Oral LD_{LO} (rat) = 410 mg/kg. [*3*]
THR = HIGH via oral route.

DIPHENYLTIN DIFLUORIDE. Solid. $(C_6H_5)_2SnF_2$, mw: 310.9, mp: 360°.
THR = See fluorides and tin compounds.

DIPHENYLTIN DIIODIDE. Colorless crystals, insol in water. $(C_6H_5)_2SnI_2$, mw: 526.7, mp: 72°, bp: 176°–182° @ 12 mm.
THR = See tin compounds and iodides.

DIPHENYLTIN HYDROXYCHLORIDE. White powder, insol in water. $(C_6H_5)_2Sn(OH)(Cl)$, mw: 325.4, mp: 187°.
THR = See tin compounds and chlorides.

DIPHENYLTIN OXIDE. Colorless powder, insol in water. $(C_6H_5)_2SnO$, mw: 288.9, mp: infusible.
THR = Details U. See tin compounds.

1,3-DIPHENYLTRIAZENE. $C_{12}H_{11}N_3$, mw: 197.3.
THR = An exper carc and neo. [*3*] HIGH acute via oral route.

DIPHENYLZINC. White crystals, decomp by cold water. $(C_6H_5)_2Zn$, mw: 219.6, mp: 107°.
THR = Details U. See zinc compounds and phenol.

DIPHOSGENE. Syn: *trichloromethyl chloroformate*. Colorless liquid, odor of phosgene. $ClCO_2CCl_3$, mw: 197.9, mp: −57°, bp: 128°, d: 1.653 @ 14°/4°, vap. p: 10.3 mm @ 20°.
Acute tox data: Inhal LD_{LO} (mouse) = 344 mg/m^3. [*3*]
THR = HIGH irr to skin, eyes and mu mem. A military poison. A lung irr. It is but slightly lacrimatory. Its physiological action, like phosgene's, is delayed. To decontaminate in enclosed spaces, use ammonia or steam. It is moderately persistent. See also phosgene.
Disaster Hazard: Dangerous; when heated, emits highly toxic fumes; will react with water or steam to produce toxic and corrosive fumes.

DIPHOSPHINE. Liquid. P_2H_4, mw: 66.
THR = Spont flam in air. Decomp by moisture. [*19*] See also phosphine.

DIPHOSPHORUS PENTASELENIDE. See phosphorus pentaselenide.

DIPHOSPHORYL CHLORIDE. $P_2O_3Cl_2$, mw: 180.5.
THR = HIGH via oral and inhal routes. A powerful irr to skin, eyes and mu mem. Hydrolyzes violently. [*19*]

DIPICRYLAMINE.
See 2,4,6,2′,4′,6′-hexanitrodiphenylamine.

DIPICRYL SULFIDE. See hexanitrodiphenyl sulfide.

1,10-DIPIPERIDINEDECANE. Solid. $(CH_2)_5NCH_2(CH_2)_8CH_2N(CH_2)_5$, mw: 308.5, vap. d: 10.7.
THR = Details U. Piperidines usually have LOW toxicity.

DIPOTASSIUM CHROMIC ACID. K_2CrO_4, mw: 194.2.
THR = An exper (+) neo and carc. [*3, 6*]

DIPOTASSIUM NITROACETATE. Solid. $K_2(NO_2)C_2O_2H$, mw: 181.2.
THR = U.
Explosion Hazard: Severe, when shocked or exposed to heat.
Disaster Hazard: Highly dangerous; shock and heat will explode it; when heated to decomp, emits highly toxic fumes.

DIPOTASSIUM-*o*-PHOSPHATE. See potassium phosphate, dibasic.

DIPPING ACID. See sulfuric acid.

DIPROPARGYL. Syn: *1,5-hexadiyne*. Colorless liquid. $CH\vdots CCH_2CH_2C\vdots CH$, mw: 78.1, bp: 86°, fp: −6°, d: 0.8049 @ 20°/4°, vap. d: 2.69.
THR = Details U.
Fire Hazard: Mod, when exposed to heat, flame or oxidizers.
Explosion Hazard: Mod, when exposed to heat.
Disaster Hazard: Mod dangerous; when heated to decomp, explodes; can react vigorously with oxidizing materials.

DIPROPARGYL ETHER. C_6H_6O, mw: 94.
THR = Has exploded during distillation. [*19*]

DIPROPYL ALUMINUM HYDRIDE. $(C_3H_7)_2AlH$, mw: 114.
THR = See also hydrides.
Fire Hazard: Dangerous; ignites spont in air and reacts violently with water. [*19*]
To Fight Fire: DO NOT use water, foam, or halogenated extinguishing agents.

DI-*n*-PROPYLAMINE. Syn: *dipropylamine*. Water white liq., amine odor. $(C_3H_7)_2NH$, mw: 101.19, mp: −40°, bp: 105°, flash p: 63°F (OC), d: 0.741 @ 20°, vap. d: 3.5.
Acute tox data: Oral LD_{50} (rat) = 930 mg/kg; dermal LD_{50} (rabbit) = 1250 mg/kg; inhal LC_{LO} (rat) = 100 ppm for 4 hrs. [*3*]
THR = MOD via oral, dermal and inhal routes. An irr.
Fire Hazard: Dangerous, when exposed to heat or flame.

Explosion Hazard: U.
Disaster Hazard: Dangerous! Keep away from heat and open flame!
To Fight Fire: Foam, CO_2, dry chemical.

DIPROPYLBERYLLIUM. Liquid. $Be(C_3H_7)_2$, mw: 95.19, mp: <−17°, bp: 245°, vap. d: 3.29.
THR = HIGH toxicity. See beryllium compounds.
Fire Hazard: Mod, when exposed to heat, flame or oxidizers.
Disaster Hazard: Dangerous; when heated to decomp, emits highly toxic fumes of beryllium oxide; can react vigorously with oxidizing materials.
To Fight Fire: Dry chemical.

DIPROPYLCADMIUM. An oil, decomp by water. $(C_3H_7)_2Cd$, mw: 198.6, d: 1.420 mp: −83°, bp: 84° @ 21.5 mm.
THR = HIGH. See cadmium compounds.

DI-*n*-PROPYL DIBROMODIPYRIDINE TIN. Solid. $(C_3H_7)_2SnBr_2(C_5H_5N)_2$, mw: 522.9, mp: 128°.
THR = See tin compounds and bromides.

DIPROPYLENE GLYCOL. Syn: *2,2′-dihydroxyisopropyl ether*. Colorless, slightly viscous liquid. Practically no odor. $(CH_3CHOHCH_2)_2O$, mw: 134.17, bp: 231.8°, flash p: 280°F (OC), d: 1.0252 @ 20°/20°, vap. press: 1 mm @ 73.8°, vap. d: 4.63.
Acute tox data: Oral LD_{50} (rat) = 1500 mg/kg. [*3*]
THR = LOW via oral route. See also glycols.
Fire Hazard: Slight, when exposed to heat or flame; can react with oxidizing materials.
Spont Heating: No.
To Fight Fire: Alcohol foam, CO_2, dry chemical.

DIPROPYLENE GLYCOL DIBENZOATE. Liquid. $CH_3CHC_6H_5CO_2CH_2OCH_2CHC_6H_5CO_2CH_3$, mw: 342.4, mp: <−30° (sets to a glass), bp: 250° @ 10 mm, d: 1.1260 @ 20°/20°, vap. press: 1.2 mm @ 200°, vap. d: 11.8.
Acute tox data: Oral LD_{50} (rat) = 9800 mg/kg. [*3*]
THR = LOW via oral route.
Fire Hazard: Slight, when exposed to heat or flame; can react with oxidizing materials.
To Fight Fire: Spray, mist, foam.

DI(PROPYLENE GLYCOL)-4,4′-ISOPROPYLIDENE BIS-PHENYL. Very viscous liquid. $(CH_3CHOHCH_2OC_6H_4)_2C_3H_6$, mw: 344.4, mp: 54°, bp: decomp, flash p: 450°F, d: 1.058 @ 80°/4°.
THR = U.
Fire Hazard: Slight, when exposed to heat or flame; can react with oxidizing materials.
To Fight Fire: Water, foam, CO_2, dry chemical.

DIPROPYLENE GLYCOL METHYL ETHER. Syn: *dowanol 50B*. Liquid. $CH_3OC_3H_6OC_3H_6OCH_3$, mw: 148.2, bp: 190°, d: 0.951, vap. d: 5.11, flash p: 185°F.
Acute tox data: Oral LD_{50} (dog) = 7500 mg/kg. [*3*]
THR = MOD via oral route. See also glycols. A mild allergen. Absorbed via skin.
Fire Hazard: Mod, when exposed to heat or flame; can react with oxidizing materials.
To Fight Fire: Dry chemical, CO_2, mist, foam.

DIPROPYLENE GLYCOL MONOMETHYL ETHER. See dipropylene glycol, methyl ether.

DIPROPYLENE GLYCOL PHENYL ETHER. Slightly yellow liquid. $C_6H_5OC_3H_6OC_3H_6OH$, mw: 210.2, mp: <−25°, bp: 285.7°, flash p: 315°F, d: 1.044 @ 25°/25°, vap. d: 7.25.
THR = See glycols.
Fire Hazard: Slight, when exposed to heat or flame; can react with oxidizing materials.
Explosion Hazard: Details U. See also ether.
To Fight Fire: CO_2, dry chemical.

DIPROPYLENE TRIAMINE. See bis-(3-amino propyl)amine.

DIPROPYL ETHER. See propyl ether.

DIPROPYL KETONE. See butyrone.

DIPROPYL MERCURY. Colorless liquid, insol in water. $(C_3H_7)_2Hg$, mw: 286.8, d: 2.0208, bp: 190°.
Acute tox data: ip LD_{LO} (mouse) = 2 mg/kg. [*3*]
THR = HIGH via ip route. See mercury compounds, organic.

DIPROPYL METHANE. See heptane.

DI-*m*-PROPYL NITROSAMINE. $C_6H_{14}ON_2$, mw: 130.2.
THR = HIGH acute via oral route. An exper carc and neo. [*3, 23*]

DI-*n*-PROPYL PHOSPHOTHIONIC CHLORIDE. $(n\text{-}C_3H_7O)_2PSCl$, mw: 216.67, bp: 70°–75° @ 1 mm, d: 1.123 @ 25°/4°.
THR = Very likely a strong irr. Details U.
Disaster Hazard: Dangerous; when heated to decomp or on contact with acid or acid fumes, emits highly toxic fumes of oxides of phosphorus and sulfur.

DIPROPYL PHTHALATE. Liquid. $C_6H_4(COOC_3H_7)_2$, mw: 250.28, bp: 129°–132° @ 1 mm, d: 1.071 @ 25°.
THR = See esters, phthalic acid and propyl alcohol.
Fire Hazard: Slight, when exposed to heat or flame; can react with oxidizing materials.
To Fight Fire: Water, foam, CO_2, dry chemical.

DI-*n*-PROPYL SULFATE. See propyl sulfate.

DIPROPYLTIN DIBROMIDE. Colorless crystals, sol in organic solvents. $(C_3H_7)_2SnBr_2$, mw: 364.7, mp: 49°.
THR = See tin compounds and bromides.

DIPROPYLTIN DICHLORIDE. Colorless crystals,

sol in organic solvents. $(C_3H_7)_2SnCl_2$, mw: 275.8, mp: 81°.
Acute tox data: Oral LD_{LO} (rat) = 160 mg/kg; iv LD_{50} (rat) = 7 mg/kg. [*3*]
THR = HIGH via iv and oral routes. See tin compounds and chlorides.

DIPROPYLTIN DIFLUORIDE. Crystals, nearly insol in water. $(C_3H_7)_2SnF_2$, mw: 242.9, mp: 205°.
THR = See fluorides and tin compounds.

DIPROPYLTIN DIIODIDE. Colorless oily liquid, insol in water, sol in organic solvents. $(C_3H_7)_2SnI_2$, mw: 458.7, mp: < −15°, bp: 273°.
THR = See tin compounds and iodides.

DI-*n*-PROPYLZINC. Liquid, decomp in cold water. $(C_3H_7)_2Zn$, mw: 151.6, bp: 146°.
THR = Details U. See zinc compounds and propyl alcohol. Spont flam in air. [*19*]

DIPTEREX. Syn: *o,o-dimethyl-l-hydroxy-2,2,2-trichloroethyl phosphonate*. Crystals, sol in water, chloroform and ether. $C_4H_8Cl_3O_4P$, mw: 257.5, d: 1.73, mp: 84°.
Acute tox data: Oral LD_{50} (rat) = 400 mg/kg; ip LD_{50} (rat) = 190 mg/kg; oral LD_{50} (wild birds) = 40 mg/kg. [*3*]
THR = HIGH via oral and ip routes. An exper carc. [*3, 23*] A cholinesterase inhibitor. Is believed to be less toxic than parathion. See related compound: parathion.
Disaster Hazard: Dangerous; when heated to decomp, emits highly toxic fumes of chlorides and oxides of phosphorus.

α,α-DIPYRIDYL. Syns: *2,2′-bipyridine, 2,2′-dipyridyl*. White crystals, sol in 2200 parts water, very sol in alcohol, ether, benzene, chloroform and petroleum ether. $C_{10}H_8N_2$, mw: 156.18, mp: 69.7°, bp: 272°–273°.
Acute tox data: ip LD_{50} (mouse) = 200 mg/kg; oral LD_{LO} (rat) = 250 mg/kg; sc TD_{LO} (mouse) = 8000 mg/kg ⟶ neo. [*3*]
THR = HIGH via ip and oral routes.

2,2′-DIPYRIDYLAMINE. Crystals. $C_{10}H_9N_3$, mw: 171.2, mp: 95.3°, bp: 222°.
THR = Details U. See also amines.
Disaster Hazard: Mod dangerous; when heated to decomp, emits toxic fumes.

DIPYRIDYLETHYL SULFIDE. Liquid. $H_4NC_5(CH_2)_2S(CH_2)_2C_5NH_4$, mw: 241.4, mp: 1.5°, d: 1.113 @ 25°, vap. d: 8.31.
THR = Details U; an insecticide.
Disaster Hazard: Dangerous; when heated to decomp or on contact with acid or acid fumes, emits highly toxic fumes of SO_x and NO_x.

2,2-DIPYRIDYL PERRHENATE. Colorless crystals, water-sol. $(C_5H_4N)_2HReO_4$, mw: 407.5.
THR = Details U.

2,2′-DIPYRIDYL RHENICHLORIDE. Yellow crystals, slightly water-sol. $(C_5H_4N)_2ReCl_6$, mw: 557.3.
THR = See chlorides.

DIQUAT DIBROMIDE. Syns: *6,7-dihydrodipyrido (1,2-a:2′,1′-c) pyrazidinium dibromide, 1,1′-ethylene-2,2′-dipyridinium dibromide*. Yellow crystals, sol in water. $(C_5H_4NCH_2)_2Br_2$, mw: 344, mp: 355°.
Acute tox data: Oral LD_{50} (rat) = 231 mg/kg; sc LD_{50} (rat) = 20 mg/kg; oral LD_{50} (rabbit) = 188 mg/kg. [*3*] A tolerance of 0.05 ppm has been established on sugar cane.
THR = HIGH on contact, oral and sc routes. An exper teratogen. [*3*] See also paraquat.

DIRECT BLACK 30. A dye.
THR = An exper carc. [*107*]

DIRECT BLUE 6. A dye.
THR = An exper carc. [*107*]

DIRECT BROWN 95. A dye.
THR = An exper carc. [*107*]

DIRECT FOOD ADDITIVES. *See* [*109*]

DISALICYLAL PROPYLENE DIIMINE. Syn: *Schiff's base*. $C_{17}H_{18}O_2N_2$, mw: 282.3, mp: −23.3°.
Acute tox data: Oral LD_{50} (rat) =4560 mg/kg. [*3*]
THR = MOD irr and allergen.
Fire Hazard: Slight.

DISASTER CONTROL PROGRAM MODEL. See Section 7.

DISASTER EMERGENCY HEADQUARTERS. See Section 7.

DISASTER EXECUTIVE AUTHORITY AND DUTIES. See Section 7.

DISASTER FIRE DEPT. OPERATION. See Section 7.

DISASTER MAINTENANCE STAFF OPERATION. See Section 7.

DISASTER MEDICAL DEPT. OPERATION. See Section 7.

DISASTER MONITORING GROUP FOR TOXIC INHALANTS AND RADIOACTIVITY. See Section 7.

DISASTER POLICE FORCE OPERATION. See Section 7.

DISASTER PUBLIC INFORMATION AND INDUSTRIAL RELATIONS GROUP-OPERATION. See Section 7.

DISILANE. Syn: *silicoethane*. Gas, repulsive odor.

Si_2H_6, mw: 62.2, mp: −132.5°, bp: −14.5°, d: 0.686 @ −25°/4°.
THR = HIGH via inhal route. See also hydrides and silanes.
Fire Hazard: Dangerous, when exposed to heat or flame or by chemical reaction; can react with oxidizing materials. Ignites spont in air.
Explosion Hazard: Reacts violently with CCl_4, $CHCl_3$, O_2, SF_6. [*19*]

DISILICON HEXABROMIDE. Syn: *hexabromodisilane*. Rhombic white crystals. Si_2Br_6, mw: 535.62, mp: 95°, bp: 240°, vap. d: 18.5.
THR = See bromides and silanes.
Disaster Hazard: Dangerous; see bromides; will react with water or steam or produce heat and toxic and corrosive fumes.

DISILICON HEXACHLORIDE. Syn: *hexachlorodisilane*. Colorless liquid. Si_2Cl_6, mw: 268.86, mp: −1°, bp: 139°, d(liquid): 1.58 @ 0°, vap. d: 9.29.
THR = Details U. See also silanes.
Disaster Hazard: Dangerous; see chlorides; will react with water or steam to produce toxic and corrosive fumes.

DISILICON HEXAFLUORIDE. Syn: *hexafluorodisilane*. Gas. Si_2F_6, mw: 170.12, mp: −18.7°, bp: −18.1° sublimes, vap. d: 5.87.
THR = See fluorides and silanes.

DISILICON HEXAIODIDE. Syn: *hexaiododisilane*. Hexagonal, colorless crystals. Si_2I_6, mw: 817.64, mp: 250°, bp: decomp, vap. d: 28.2.
THR = See iodides and silanes.

DISILICON TETRAIODIDE. Syn: *tetraiododisilane*. Orange-red crystals. Si_2I_4, mw: 563.80, vap. d: 19.4.
THR = Details U. See iodides and silanes.
Disaster Hazard: Dangerous; see iodides; will react with water or steam to produce toxic and corrosive fumes.

DISILOXANE. See silicyl oxide.

(DISILYLAMINO)DIBORANE. $B_2H_{11}Si_2N$, mw: 102.8.
THR = Self-ignites in air. [*19*]

DISILYLAMINODICHLOROBORINE. $(SiH_3)_2NBCl_2$, mw: 157.9.
THR = Spont flam in air. [*19*]

DISODIUM ACETOARSENATE. Syn: *aricyl*. White crystalline powder. $COONaCH_2AsOOHONa$, mw: 227.95.
THR = HIGH toxicity. See arsenic compounds.

DISODIUM CHROMATE. Na_2CrO_4, mw: 162.
THR = HIGH acute toxicity. See also chromates. An exper carc. [*3, 6*]

DISODIUM DIHYDROGEN PYROPHOSPHATE. See sodium acid phosphate.

DISODIUM DIPHOSPHATE. See sodium acid pyrophosphate.

DISODIUM ETHYLENE BIS(DITHIOCARBAMATE). See nabam.

DISODIUM ETHYLENEDIAMINE TETRAACETATE. Syns: *sequestrene*, *ethylenediamine tetraacetic acid*. White crystals. $C_{10}H_{14}O_8Na_2 \cdot 2H_2O$, mw: 372, vap. d: 12.8.
Acute tox data: Oral LD_{50} (rat) = 2000 mg/kg; oral LD_{50} (mouse) = 21 mg/kg; oral LD_{50} (rabbit) = 2300 mg/kg; ip LD_{50} (mouse) = 2.6 mg/kg; iv LD_{50} (rabbit) = 47 mg/kg. [*3*]
THR = HIGH and MOD depending on species of test animal and route of exposure. There have been reports of kidney injury in man following use of this type of compound in treating lead poisoning. Used as a food additive permitted in the feed and drinking water of animals and/or for the treatament of food-producing animals. Also used as a food additive permitted in food for human consumption. [*109*]

DISODIUM GUANYLATE. See sodium guanylate.

DISODIUM HYDROGEN PHOSPHATE. See sodium phosphate, dibasic.

DISODIUM HYDROXYMERCURY SALICYLOXY ACETATE. See "Mercurosal."

DISODIUM INOSINATE. See sodium inosinate.

DISODIUM METHANE ARSONATE. Syns: *Crab-e-rad*, *DSMA*. Crystals, water-sol hydrate. $CH_3AsO(ONa)_2 \cdot 6H_2O$, mw: 292, mp: 132°–139°.
Acute tox data: Oral LD_{50} (rat) = 1800 mg/kg; oral LD_{50} (mammal) = 1000 mg/kg. [*3*]
THR = MOD via oral route. See also arsenic compounds.

DISODIUM NITRITE. Na_2NO_2, mw: 92.
THR = Reacts violently with ($K + NH_3$). [*19*]

DISODIUM PHOSPHATE. See sodium phosphate, dibasic.

DISODIUM-*o*-PHOSPHATE. See sodium phosphate, dibasic.

DISODIUM PYROPHOSPHATE. See sodium acid pyrophosphate.

DISODIUM TARTRATE. See sodium tartrate.

DISPERSANT GAS, N.O.S. Gas.
THR = U.

DISTILLED AND CRACKED NATIVE ASPHALTS.
THR = A recog carc. [*3, 14*] See mineral oils.

DISTILLED MUSTARD GAS. See dichlorodiethyl sulfide.

DISULFIRAM. See bis(diethylthiocarbamyl) disulfide.

DISULFOTON. See *o,o*-diethyl-*s*-2-(ethylthio)ethyl-phosphorodithioate.

DISULFUR DECAFLUORIDE.
See sulfur pentafluoride.

DISULFURIC ACID. See pyrosulfuric acid.

DISULFURYL AZIDE. $N_6S_2O_5$, mw: 228.2.
THR = HIGH irr to skin, eyes and mu mem. Explosive decomp at $< 80°$.

DISULFURYL CHLORIDE.
See pyrosulfuryl chloride.

DITAINE. Syn: *echitamine*. Thick, white, glistening, crystalline alkaloid. $C_{22}H_{29}N_2O_4$, mw: 385.5, mp: 206° decomp.
THR = HIGH via oral route. Acts much like curare. [*2*]
Disaster Hazard: Dangerous; when heated to decomp, emits toxic fumes.

DITCH POWDER. See explosives, high.

DITELLUROMETHANE. Dark red solid, insol in water. $TcCH_2Te$, mw: 269.3, mp: 214° (decomp).
THR = See tellurium compounds.

2,3-DITHIABUTANE. Syn: *dimethyl disulfide*. Liquid. CH_3-S-S-CH_3, mw: 94.2, bp: 109.7°, d: 1.0569 @ 25°, vap. press: 28.6 mm @ 25°, vap. d: 3.24.
THR = Details U. See also sulfides and alkyl disulfides.
Fire Hazard: Mod.
Disaster Hazard: Dangerous; can react vigorously with oxidizing materials. See sulfates.

3,4-DITHIAHEXANE. Syn: *diethyl disulfide*. Liquid. C_2H_5-S-S-C_2H_5, mw: 122.3, bp: 154°, fp: −101.5°, d: 0.99267 @ 20°/4°, vap. d: 4.22, vap. press: 4.28 mm @ 25°.
THR = Details U. See also sulfides and alkyl disulfides.
Fire Hazard: Mod.
Disaster Hazard: Dangerous; can react vigorously with oxidizing materials. See sulfates.
To Fight Fire: Foam, dry chemical.

1,3-DITHIANE-3,2-DIPHENYL. Syn: *1,3-diethylene disulfide-2,2-diphenyl*. Crystals.
$S[CH_2C(C_6H_5)_2CH_2CH_2]S$, mw: 272.4, vap. d: 9.4.
THR = Details U. See also sulfides and alkyl disulfides.
Disaster Hazard: Dangerous; see sulfates.

β,β-DITHIO BIS ALANINE. See cystine.

2,2′-DITHIO BIS(BENZOTHIAZOLE). See 2,2′-dibenzothiazyl disulfide.

2,4-DITHIOBIURET. Crystals. $H_2NC(S)NHC(S)NH_2$, mw: 135.20, mp: 181°, bp: decomp, d: 1.522 @ 30°.
Acute tox data: Oral LD_{50} (rat) = 5 mg/kg; ip LD_{LO} (rat) = 20 mg/kg. [*3*]
THR = HIGH via oral and inhal routes ⟶ respiratory paralysis.
Disaster Hazard: Dangerous; when heated to decomp, emits highly toxic fumes.

DITHIOCARBAMIC ACID, SODIUM SALT. Syn: *sodium dithiocarbamate*. Colorless needles. $CH_2NS_2 \cdot Na$, mw: 115.2.
Acute tox data: Oral LD_{LO} (rat) = 500 mg/kg. [*3*]
THR = MOD via oral route.
Disaster Hazard: Dangerous; when heated to decomp, emits toxic fumes.

β-DITHIOCYANOETHYL ETHER.
$O[C(CNS)_2CH_3]_2$, mw: 302.5, vap. d: 10.4.
THR = Details U; an insecticide.
Fire Hazard: Mod.
Disaster Hazard: Dangerous; when heated to decomp or on contact with acid or acid fumes, emits toxic fumes; can react with oxidizing materials.

4,4-DITHIOMORPHOLINE. Syns: *sulfasan*; *4,4-dithiodimorpholine*. Tan to gray powder.
$C_4H_8ONSSNOC_4H_8$, mp: 122° min, d: 1.36 @ 25°.
Acute tox data: ip LD_{LO} (mouse) = 50 mg/kg. [*3*]
THR = HIGH via ip route. See also morpholine.
Disaster Hazard: Dangerous; when heated to decomp, emits highly toxic fumes.

DITHIONE. Syns: *7-hydroxy-2,4-tetramethylene coumarin*, *o,o-diethylthiophosphate*, *bladafume*, *sulfotepp*. Crystals nearly insol in water. $C_{17}H_{21}O_5PS$, mw: 368.4, mp: 88°.
Acute tox data: Oral LD_{50} (rat) = 5 mg/kg; im LD_{50} (rat) = 0.055 mg/kg; ip LD_{50} (mouse) = 0.94 mg/kg; dermal LD_{50} (rabbit) = 20 mg/kg; oral LD_{50} (wild birds) = 100 mg/kg. [*3*]
THR = VERY HIGH via oral, im, ip and dermal routes. See also parathion.
Disaster Hazard: Dangerous; when heated to decomp, emits highly toxic fumes of oxides of sulfur and phosphorus.

DITHIOZONE. See diphenylthiocarbazone.

DITHRANOL. See anthralin.

DI-*o*-TOLUIDINE FLUOSILICATE. White crystals, sol in hot alcohol. $(C_6H_4NH_2CH_3)_2 \cdot H_2SiF_6$, mw: 358.4.
THR = See fluosilicates.

DI-*m*-TOLUIDINE FLUOSILICATE. White crystals, sol in hot alcohol. $(C_6H_4NH_2CH_3)_2 \cdot H_2SiF_6$, mw: 358.4.
THR = See fluosilicates.

DI-*p*-TOLUIDINE FLUOSILICATE. White crystals. $(C_6H_4NH_2CH_2)_2 \cdot H_2SiF_6$, mw: 358.4.
THR = See fluosilicates.

DI-*p*-TOLYLBORIC ANHYDRIDE. White powder, insol in water, sol in ether. $(C_7H_7)_2BOB(C_7H_7)_2$, mw: 402.2, mp: 78°.
THR = Details U. See also boron compounds.

DI-*p*-TOLYLGERMANIUM DIBROMIDE. Yellowish liquid hydrolyzed by water. $(CH_3C_6H_4)_2GeBr_2$, mw: 414.7, bp: 230°–233° @ 13 mm.
THR = See germanium compounds and bromides.

1,3-DI-*o*-TOLYLGUANIDINE. White crystals. $C_{15}H_{17}N_2$, mw: 239.3, mp: 179°, d: 1.10 @ 20°/4°, vap. d: 8.24.
Acute tox data: Oral LD_{50} (rat) = 500 mg/kg; ip LD_{LO} (mouse) = 25 mg/kg. [3]
THR = HIGH oral and ip routes. An allergen.
Disaster Hazard: Dangerous; when heated to decomp, emits highly toxic fumes of oxides of nitrogen.

***o*-DITOLYLMERCURY.** White crystals, insol in water. $Hg(C_7H_7)_2$, mw: 382.9, mp: 107°, bp: 219° @ 14 mm.
THR = HIGH. See mercury compounds, organic.

***m*-DITOLYLMERCURY.** Colorless to light yellow crystals, insol in water. $(C_7H_7)_2Hg$, mw: 382.9, mp: 102°.
THR = HIGH. See mercury compounds, organic.

***p*-DITOLYLMERCURY.** Crystals, insol in water. $(C_7H_7)_2Hg$, mw: 382.9, mp: 238°.
THR = HIGH. See mercury compounds, organic.

DI-*p*-TOLYLPHENYL GERMANIUM BROMIDE. Colorless crystals. $(CH_2C_6H_4)_2(C_6H_5)GeBr$, mw: 411.9, mp: 119°.
THR = See bromides and germanium compounds.

DI-*o*-TOLYLTHIOUREA. Syn: *sym-di-o-tolylthiourea.* Crystals. $CS(NHC_6H_4CH_3)_2$, mw: 256.4, mp: 178°, vap. d: 8.85.
Acute tox data: Oral LD_{50} (rat) = 2000 mg/kg. [3]
THR = MOD via oral route. An allergen.
Disaster Hazard: Dangerous; when heated to decomp, emits highly toxic fumes.

DI-*m*-TOLYLTHIOUREA.
Acute tox data: Oral LD_{50} (rat) = 100 mg/kg. [3]
THR = HIGH via oral route.

DI-*p*-TOLYLTIN. Orange-yellow powder, sol in benzene. $(CH_3C_6H_4)_2Sn$, mw: 301.0, mp: 111.5°.
THR = Details U. See tin compounds.

DI-*o*-TOLYLTIN DICHLORIDE. Solid. $(CH_3C_6H_4)_2SnCl_2$, mw: 371.9, mp: 50°.
THR = See tin compounds and chlorides.

DI-*p*-TOLYLTIN DICHLORIDE. Solid. $(CH_3C_6H_4)_2SnCl_2$, mw: 371.9, mp: 50°.
THR = See tin compounds and chlorides.

DI-*m*-TOLYLTIN OXIDE. White powder, insol in water. $(CH_3C_6H_4)_2SnO$, mw: 317.0.
THR = Details U. See tin compounds.

DI-*m*-TOLYLTIN SULFIDE. Solid, sol in chloroform and benzene. $(CH_3C_6H_4)_2SnS$, mw: 333.0, mp: 122°.
THR = See tin compounds and sulfides.

2,2-DI-*p*-TOLYL-1,1,1-TRICHLOROETHANE. Crystals. $(C_6H_4CH_3)_2C_2HCl_3$, mw: 313.7, vap. d: 10.8.
THR = Probably toxic. An insecticide. See also chlorinated hydrocarbon insecticides and DDT.
Disaster Hazard: Mod dangerous; when heated to decomp, emits toxic fumes.

DI-*o*-TOLYLZINC. White crystals, sol in xylene and petroleum ether. $(C_7H_7)_2Zn$, mw: 247.6, mp: 210°.
THR = Details U. See zinc compounds.

DI-(TRIDECYL)AMINE. $C_{26}H_{55}N$, mw: 381.8.
Acute tox data: Oral LD_{50} (rat) = 9850 mg/kg; dermal LD_{50} (rabbit) = 3540 mg/kg. [3]
THR = LOW via oral; MOD via dermal routes.

DI-(TRIDECYL)PHTHALATE. Syn: *DTDP.* $C_6H_4(COOC_{13}H_{27})_2$, mw: 538, d: 0.951 @ 20°/20°, bp: > 285° @ 5 mm, flash p: 470°F (OC).
THR = U. Limited animal exper suggest LOW toxicity.
Fire Hazard: Slight, when exposed to heat or flame.
To Fight Fire: Dry chemical, CO_2.

DITRIPHENYL GERMANYL METHANE. Colorless crystals, insol in water. $[(C_6H_5)_3Ge]_2CH_2$, mw: 621.8, mp: 133°.
THR = Details U. See germanium compounds.

DITRIPHENYL STANNYL METHANE. White crystals, sol in benzene, ether and chloroform. $[(C_6H_5)_3Sn]_2CH_2$, mw: 714.0, mp: 104.5°.
THR = Details U. See tin compounds.

DIURIL. See chloroethiazide.

DIURON. Syn: *3-(3,4-dichlorophenyl)-1,1-dimethyl urea.* Crystals, low sol in water and hydrocarbon solvents. $C_9H_{10}Cl_2N_2O$, mw: 233.1, mp: 159°.
Acute tox data: Oral LD_{50} (mouse) = 640 mg/kg; dermal LD_{LO} (rabbit) = 140 mg/kg. [3]
THR = MOD via oral and HIGH via dermal routes. Used as a food additive permitted in the feed and drinking water of animals and/or for the treatment of food-producing animals. [109]
Disaster Hazard: Dangerous; when heated to decomp, emits highly toxic fumes.

1,2-DIVALERATE DIETHYL SULFIDE. Liquid. $[CH_2(COOC_4H_9)CH(COOC_4H_9)]_2S$, mw: 490.64, mp: −45°, bp: 246° @ 5 mm, flash p: 430°F (OC), d: 1.0543 @ 20°/20°, vap. d: 16.95.

THR = Details U. See also sulfides and alkyl disulfides.
Fire Hazard: Slight, when exposed to heat or flame.
Disaster Hazard: Dangerous; see sulfates. Can react with oxidizing materials.
To Fight Fire: Water, foam, CO_2, dry chemical.

DIVANADIUM DODECACARBONYL. $V_2(CO)_{12}$, mw: 438.
THR = HIGH toxicity. See also nickel carbonyl. Self-ignites in air. [*19*]

DIVINYL-*b*. See 1,3-butadiene.

***m*-DIVINYLBENZENE.** Syn: *vinylstyrene*. Water white liquid. $C_6H_4(CHCH_2)_2$, mw: 130.1, bp: 199.5°, fp: −66.9°, lel = 0.3%, flash p: 165°F, d: 0.9289 @ 20°, vap. press: 1 mm @ 32.7°, vap. d: 4.48.
Acute tox data: Oral LD_{50} (rat) = 4040 mg/kg. [*3*]
THR = MOD via oral and inhal routes.
Fire Hazard: Mod, when exposed to heat or flame or by spont chemical reaction; can react with oxidizing materials.
Spont Heating: Yes.
Explosion Hazard: Mod, in the form of vapor when exposed to flame.
To Fight Fire: Foam, CO_2, dry chemical. Commercial DVB without inhibitor may start to polymerize and gel. Heat can then build up rapidly to possibly 300° and the vap. press may attain dangerous levels. In storage this material should be kept below 90°F, possibly by water spray.

DIVINYL DISULFIDE. Mobile liquid, characteristic odor. $(CHCH_2)_2S$, mw: 86.2, bp: 86°, d: 0.9174 @ 15°, vap. d: 2.97.
Acute tox data: Inhal LC_{LO} (rat) = 2 ppm. [*3*]
THR = HIGH via inhal or oral routes. See also sulfides and alkyl disulfides.
Fire Hazard: Mod, when exposed to heat or flame.
Disaster Hazard: Dangerous; when heated to decomp, emits highly toxic fumes of oxides of sulfur; on contact with oxidizing materials, can react vigorously.
To Fight Fire: Spray, mist, foam, dry chemical.

DIVINYL ETHER. See vinyl ether.

DI(VINYL OCTYLATE)CAPRYLAMIDE. Liquid. $(C_7H_{15}COOC_2H_3)_2NCOC_7H_{15}$, mw: 483.71, mp: −33°, bp: 256° @ 5 mm, flash p: 420°F, d: 0.9564 @ 20°/20°, vap. d: 16.7.
THR = U. See also amides.
Fire Hazard: Slight, when exposed to heat or flame; can react with oxidizing materials.
To Fight Fire: Foam, CO_2, dry chemical.

DIVINYL OXIDE. See vinyl ether.

DIVINYL ZINC. $Zn(C_4H_6)$ mw: 119.4.
THR = HIGH. Self-ignites in air. [*19*]

DI(*o*-XENYL)MONOPHENYL PHOSPHATE. Insol in water. $(C_6H_5C_6H_4)_2PO(OC_6H_5)$, mw: 446, d: 1.20 @ 60°, boiling range: 285°–330° @ 5 mm, flash p: 482°F.
THR = U.
Fire Hazard: Slight, when exposed to heat or flame.
Disaster Hazard: Dangerous; see phosphates.
To Fight Fire: Dry chemical, CO_2.

DI-*p*-XYLYLTIN. Solid. $[(CH_3)_2C_6H_3]_2Sn$, mw: 329.0. mp: 157°, bp: 240° (decomp).
THR = Details U. See tin compounds.

DKP. See potassium phosphate, dibasic.

"DM." See diphenylamine chloroarsine.

DMC. See di(*p*-chlorophenyl)methyl carbinol.

DMDT. See methoxychlor.

DMF. See dimethylformamide.

DMP. See dimethyl phthalate.

DN-111. See 2,4-dinitro-6-cyclohexyl phenol.

DNBP. See 2-*sec*-butyl-6,4-dinitrophenol.

DNOC. See dinitro-*o*-cresol.

DNTP. See parathion.

DODECACHLOROOCTAHYDRO-1,3,3-METHENO-2H-CYCLABUTA(c,d)PENTALENE. See mirex.

DODECANE. Syn: *dihexyl*. Liquid. $C_{12}H_{26}$, mw: 170.3, mp: −12°, bp: 216.2°, lel = 0.6%, flash p: 165°F, d: 0.750, vap. press: 1 mm @ 47.8° vap. d: 5.96, autoign. temp.: 399°F.
THR = Irr and narcotic in high conc. An exper carc. [*23*]
Fire Hazard: Mod, when exposed to heat, flame or oxidizers.
Spont Heating: No.
Explosion Hazard: Mod, in the form of vapor when exposed to flame.
Disaster Hazard: Dangerous; keep away from heat and open flame.
To Fight Fire: Foam, CO_2, dry chemical.

DODECANETHIOL. See lauryl mercaptan.

DODECANOIC ACID. See lauric acid.

1-DODECANOL. Syns: *dodecyl alcohol, lauryl alcohol, 1°-n-lauryl alcohol*. Colorless solid, floral odor, sol in 2 parts of 70% alcohol, insol in water. $CH_3(CH_2)_{11}OH$, mw: 186.33, d: 0.8201 @ 24°/4°, mp: 24°, bp: 259°, flash p: 260°F, autoign. temp.: 527°F.
Acute tox data: ip LD_{50} (rat) = 800 mg/kg. [*3*]
THR = MOD via ip route.

Fire Hazard: Slight, when exposed to heat or flame.
To Fight Fire: Dry chemical, CO_2.

DODECANOYL PEROXIDE. See lauryl peroxide.

DODECENE. Syn: *dodecylene*. Colorless liquid. $C_{12}H_{24}$, mw: 168.31, mp: −31.5°, bp: 213°, d: 0.76 @ 20°/4°, vap. press: 1 mm @ 47.2°, vap. d: 5.81, flash p: < 212°F, autoign. temp.: 491°F.
THR = Details U. Probably irr and narcotic in high conc.
Fire Hazard: Low, when exposed to heat or flame; can react with oxidizing materials.
To Fight Fire: Foam, CO_2, dry chemical.

DODECENE EPOXIDE. $C_{12}H_{24}O$, mw: 184.4.
THR = An exper neo. [*3*]

DODECENYL SUCCINIC ANHYDRIDE. Light yellow, clear viscous oil. $C_{16}H_{27}NO_2$, mw: 265.4, bp: 180°–182° @ 5 mm, flash p: 352°F (COC), d: 1.002 @ 25°/4°.
Acute tox data: ip LD_{50} (mouse) = 320 mg/kg. [*3*]
THR = HIGH via ip route. Animal exper show irr and sensitizing properties as well.
Fire Hazard: Slight, when exposed to heat or flame; can react with oxidizing materials.
To Fight Fire: Foam, CO_2, dry chemical.

DODECYL ALCOHOL. See 1-dodecanol.

DODECYLAMINE. Oil, amine odor. $C_{12}H_{25}NH_2$, mw: 185, fp: 28.3°, vap. d: 64 mm @ 170°.
Acute tox data: Oral LD_{50} (rat) = 1960 mg/kg; ip LD_{50} (mouse) = 50 mg/kg. [*3*]
THR = MOD via oral and HIGH via ip routes.

DODECYLBENZENE, CRUDE. Liquid. $C_{12}H_{25}C_6H_5$, mw: 246.4, bp: 290°–410°, flash p: 285°F, d: 0.9, vap. d: 8.47.
THR = Details U. Probably MOD.
Fire Hazard: Slight, when exposed to heat or flame; can react with oxidizing materials.
To Fight Fire: Foam, CO_2, dry chemical.

DODECYL BENZOL DERIVATIVES.
THR = See surfactants. A (S) carc. [*14*]

DODECYL BROMIDE. See lauryl bromide.

DODECYL CHLORIDE. See lauryl chloride.

DODECYLENE. See dodecene.

***n*-DODECYL GUANIDINE ACETATE.** Syns: *dodine*, *cyprex*. Crystals, sol in hot water and alcohol. $C_{12}H_{25}NHC(:NH)NH_2CH_3COOH$, mw: 287, mp: 136°.
Acute tox data: Oral LD_{50} (rat) = 566 mg/kg. [*3*] Dermal LD_{50} (rabbit) = > 1500 mg/kg via oral single 24 hr contact. [*94*]
THR = MOD via oral and dermal routes. A fungicide.

DODECYL MERCAPTAN. See lauryl mercaptan.

***tert*-DODECYL MERCAPTAN.** White to light yellow liquid. $C_{12}H_{25}SH$, mw: 202.4, bp: 200°–235°, flash p: 205°F (OC), d: 0.85 @ 25°/25°, vap. d: 6.98.
THR = Probably toxic. See also mercaptans.
Disaster Hazard: Dangerous; see sulfates. Can react vigorously with oxidizing materials.
Fire Hazard: LOW, when exposed to heat or flame.
To Fight Fire: Foam, CO_2, dry chemical.

DODECYL PHENOL. Straw-colored liquid, phenolic odor. $C_{12}H_{25}C_6H_4OH$, mw: 262.4, bp: 154°–168°, flash p: 325°F (OC), d: 0.93 @ 20°/20°, vap. d: 9.04.
Acute tox data: Oral LD_{50} (rat) = 2140 mg/kg; dermal LD_{50} (rabbit) = 5000 mg/kg. [*3*]
THR = MOD via oral and dermal routes.
Fire Hazard: Slight, when exposed to heat or flame.
Disaster Hazard: Mod dangerous; when heated to decomp, emits toxic fumes; can react with oxidizing materials.
To Fight Fire: CO_2, dry chemical.

DODECYL PYRIDINIUM CHLORIDE. See lauryl pyridinium chloride.

DODECYL TRICHLOROSILANE. Colorless to yellow liquid, readily hydrolyzed by moisture with the production of hydrochloric acid. $C_{12}H_{25}SiCl_3$, mw: 303.5, bp: 288°, d: 1.026 @ 25°/25°.
THR = An irr, HIGH toxicity, corrosive. See silanes.
Disaster Hazard: Dangerous; see chlorides.

DODINE. See *n*-dodecyl guanidine acetate.

DOISYNOLIC ACID. $C_{18}H_{24}O_3$, mw: 288.4.
THR = An exper neo. [*3*]

DOMEYKITE. See tricopper arsenide.

DORMISON. See methylpentynol.

DOSE-RESPONSE RELATIONSHIP IN INDUSTRIAL TOXICOLOGY. See Section 9.

DOT-OFFICE OF HAZARDOUS MATERIALS. See Section 11.

DOWANOL-50B. See dipropylene glycol monomethyl ether.

DOWICIDE 1. See *o*-phenylphenol.

DOWICIDE 4. See 2-chloro-4-phenylphenol.

DOWKLOR. See chlordane.

DOWLAP. See 3,4,6-trichloro-2-nitrophenol.

DPA. See diphenolic acid.

DRAZOXOLON. Syn: *4-(2-chlorophenyl hydrazono)-3-methyl-5-isoxazolone*. A solid, almost insol in water and acids, sol in alkalis. $C_{10}H_8O_2N_3$, mw: 202.1.
Acute tox data: Oral LD_{50} (rat) = 126 mg/kg; ip LD_{50} (rat) = 20 mg/kg; oral LD_{50} (mouse) = 129 mg/kg. [*3*]

THR = HIGH via oral and ip routes. A highly toxic protectant fungicide.
Disaster Hazard: Dangerous; see chlorides.

DRIED GRASS. See hay.

DRY ICE. See carbon dioxide, solid.

"DRYOLENE." See naphtha, VM & P.

DRYORTH. See sodium-*o*-silicate.

DSP. See sodium phosphate, dibasic.

DTMC.
See 4,4′-dichloro-α-(trichloromethyl)benzhydrol.

DUBOISINE. See hyoscyamine.

DUAZOMYCIN.
THR = An exper carc. [*3*]

DULCIN. See (*p*-ethoxy phenyl)urea.

DULCIN-*p*-PHENETYL UREA. A (S) carc. [*14*] See sweetners.

DULCITOL HEXANITRATE. Syn: *1,2,3,4,5,6-hexane hexanitrate*. Crystals. $C_6H_8(NO_2)_6$, mw: 356.2; explodes @ 205°.
THR = Details U. See also nitrates.
Fire Hazard: Dangerous, when exposed to heat or flame or by chemical reaction. A powerful oxidizing agent.
Explosion Hazard: Severe, when shocked or exposed to heat.
Disaster Hazard: Highly dangerous; shock or heat will explode it; when heated to decomp, emits highly toxic fumes of oxides of nitrogen; can react vigorously with oxidizing materials.

DUMASIN. See cyclopentanone.

DUPHAR. See tedion.

DUPONOL. See sodium lauryl sulfate.

DURASET. Syn: *N-m-tolylphthalamic acid*.
Acute tox data: Oral LD_{50} (rat) = 5320 mg/kg. [*3*]
THR = MOD via oral route.

DURATION OF EXPOSURE EFFECTS. See Section 9.

DURSBAN. Syn: *o,o-diethyl-o-(3,5,6-trichloro-2-pyridyl) phosphorothioate*. $C_9H_{10}O_3PSNCl_3$, mw: 349.5, mp: 42.5°. White crystals, mild mercaptan odor.
Acute tox data: Dermal LD_{50} (rabbit) = 202 mg/kg; oral LD_{50} (wild birds) = 8 mg/kg; oral LD_{50} (rat) = 145 mg/kg; LD_{50} (mouse) = 74 mg/kg. [*3*]
THR = HIGH via oral and dermal routes. Absorbed via ingestion, skin, inhal. No mutagen or teratogen effects noted. A HIGH toxicity cholinesterase inhibitor, insecticide. See parathion.
Disaster Hazard: Dangerous; see chlorides and parathion.

DUST EXPLOSIONS. See Section 7.

DUSTS, METALLIC. See specific metal and powdered metals.

DUST, TOTAL (below 5% FREE SiO_2).
THR = The effects of dust vary according to the composition. If it is organic dust, it is primarily a nuisance, but may produce asthma by being an allergen. If the dust contains toxic compounds, either organic or inorganic, the hazard is that of the toxic component. A common air contaminant.
Fire and Explosion Hazard: Dangerous; both organic and inorganic dusts can explode or burn violently if exposed to sparks or open flame.

DVPy. See divinyl pyrethrin I.

DYE ORANGE #1. $C_{16}H_{11}O_4N_2SCl \cdot Na$, Red-brown crystals. mw: 350. [*3*]
THR = Acute systemic = cathartic. An exper (±) carc. [*3, 4*]

DYFONATE. Syn: *o-ethyl-s-phenyl ethyl phosphonodithioate*. $C_6H_5SPS(C_2H_5O)_2$, mw: 238.1.
Acute tox data: Oral LD_{50} (rat) = 3 mg/kg; dermal LD_{50} (rat) = 147 mg/kg; dermal LD_{50} (rabbit) = 25 mg/kg. [*3*]
THR = HIGH via oral and dermal routes. Very toxic cholinesterase inhibitor, soil insecticide.
Disaster Hazard: Dangerous; see phosphates.

DYLOX. See dipterex.

DYMID. See N,N-dimethyl-2,2-diphenyl acetamide.

DYNALTONE. See aralkonium chloride.

DYNAMAGNITE. See dynamite.

DYNAMITE. A high explosive used industrially in construction and mining. The name generally refers to a mixture containing as its principal explosive ingredient either glyceryl trinitrate (nitroglycerin) or ammonium nitrate, suitably sensitized. It does not apply to black blasting powders, chlorate powders, and other deflagrating mixtures.

An ordinary blasting cap or an electric blasting cap is used for detonating a charge of dynamite. The various classes and grades of dynamites are made from mixtures composed of an explosive compound or a mixture of explosive compounds, a dope, and an antacid. If any of the explosive ingredients are in a liquid state they are referred to as the "explosive oil," which is usually composed of glyceryl trinitrate (nitroglycerin) and about 25–30% of ethylene glycol dinitrate. The latter compound depresses the freezing point of the nitroglycerin and renders the dynamite low-freezing. Other compounds may also be used as freezing point depressants. The explosive oil is absorbed by carbonaceous materials that have entirely replaced kieselguhr (diatomaceous earth), formerly used exclusively as the absorbent or dope in dynamites. This type of dope does not enter into the explosive reaction. Wood pulp is now most commonly used as the absorber, either alone or mixed in suitable proportions with flour, starch, etc.

The absorbents may be mixed with an oxidizer such as sodium nitrate, in which case an active dope is formed. For neutralizing any acid that may be present, about 1% of an antacid (calcium carbonate or zinc oxide) is added to the mixture. The explosive oil is mixed into the dope. The strength of a kieselguhr dynamite, when detonated, is derived only from the explosive oil, since kieselguhr is inert. A mixture of this kind is known as a straight dynamite. See dynamite, straight. On the other hand, an active dope, (an admixture of carbonaceous absorbents with an oxidizer), furnishes explosive strength in addition to that derived from the explosive ingredients.

By replacing a part of the explosive oil of a straight dynamite with ammonium nitrate, so that the latter becomes the principal explosive ingredient, a mixture known as an ammonia dynamite is obtained. See dynamite, ammonia.

When the explosive oil is gelatinized the explosive is known as a gelatin or an ammonia gelatin dynamite. See also dynamite, ammonia gelatin and dynamite, gelatin.

Blasting gelatin is a gelatinized mass of an elastic nature obtained by incorporating nitrocotton with an explosive oil into which is mixed about 1% of antacid. See also dynamite, blasting gelatin.

Dynamites may be in bulk form (bag powder) or put up in cartridge form, the most common size being $1\frac{1}{4}$ in. in diameter and 8 in. long, although for holes of small diameter, cartridges as small as $\frac{7}{8}$ in. in diameter are also used. In large diameter well-drill holes for quarry blasting, cartridge diameters up to 10 in. and lengths up to 30 in. may be used. These upper limits or 50 lbs in weight of each cartridge are imposed by the DOT Regulations, and the maximum length of 30 in. applies to all cartridge diameters between 4 and 10 in.

An integral part of a stick of dynamite is the paraffined paper wrapper that not only holds the ingredients together but enters into the explosive reaction.

The wrapper also affords some measure of protection from moderate exposure to dampness. For blasting in wet operations, a gelatinized dynamite which resists the absorption of water should be used.

The strength of straight dynamite is graded by its explosive oil content (% by weight), while for any other class of dynamite, the strength is determined exper in comparison with the various grades of the straight dynamites. For example, a 40% straight dynamite is one which contains 40% of explosive oil; a 40% strength ammonia dynamite, as determined by tests, equals a 40% straight dynamite in strength. In other words a 40% strength ammonia dynamite will release the same energy as an equivalent weight of a 40% straight dynamite.

THR = See nitrates.

Fire Hazard: See explosives, high, and nitrates.

Explosion Hazard: While this material is a powerful explosive when detonated by shock or heat, it is only mod hazardous. See also explosives, high and nitrates.

Disaster Hazard: Dangerous; shock and heat will explode it; when heated to decomp emits highly toxic fumes of NO_x and CO, etc. It can react vigorously with oxidizing materials.

DYNAMITE, AMMONIA.

This class of dynamite contains ammonium nitrate as the principal explosive ingredient, which may be considered as replacing some of the sodium nitrate and at least 60% of the explosive oil as found in the straight dynamites.

A suitable sensitizer must be added because the firing of an ordinary No. 6 blasting cap embedded in pure ammonium nitrate is not capable of bringing about its detonation. If the sensitizer in the dynamite is a liquid ingredient, carbonaceous absorbent materials are added in amounts adequate to avoid leakage of the liquid ingredients and to obtain a suitable oxygen balance. Only a few of the various substances which have been proposed as sensitizers for ammonium nitrate have really been successful. The most successful one takes the form of an explosive oil, like that used in the straight dynamites. Other sensitizers are in solid form and may be either of an explosive nature or a nonexplosive nature. Certain of these solid sensitizers are explosive and may be dissolved by the nitroglycerin. Among the solid sensitizers are the organic compounds such as diphenyl and diphenylamine, both of which are non-explosive; or nitrated organic compounds, such as nitrostarch (as used in nitrostarch dynamites), nitrotoluenes, nitronaphthalenes, etc., which are explosive. Inorganic solid substances like calcium silicide, ferrosilicon, aluminum, and sulfur have also been used as sensitizers, although none of them are explosive except when mixed with other substances.

The following is a typical formula for a 40% ammonia dynamite: 1% moisture: 15% explosive oil; 31% ammonium nitrate; 38% sodium nitrate; 1% antacid; 4% sulfur; 9% carbonaceous material. For a 1¼ by 8 in. cartridge, the weight of the wrapper per 100 g of explosive ingredient averages 7 g.

The ingredients of ammonia dynamites are of a granular nature that have very little cohesion and do not readily resist the absorption of water. Owing to the hygroscopic nature of ammonium nitrate, ammonia dynamites are not usually used in wet holes. Under this condition the gelatinized types of dynamites (ammonia gelatin or gelatin dynamites) should be used and for very wet work it is advisable to use only the gelatin dynamites.

The regular grades of ammonia dynamites range from 15 to 60% strength. The ammonia gelatin, or gelatin dynamites of 40 to 60% grades are used to greatest extent in blasting operations.

High count powders are modifications of ammonia dynamites. The cartridge count per 50 lb case for the regular grades of ammonia dynamite is about 100 in comparison with 115 to 172 for the high count ammonia dynamites. Those having a cartridge count of 115 have a bulk strength of 60% and those of the 172 count have a strength of 20% on the same basis. In weight strength all these powders are approximately equivalent to a 60% straight dynamite.

THR = See nitrates.

Fire Hazard: Dangerous; when exposed to heat or flame (Section 7). See also explosives, high.

Explosion Hazard: Moderate when shocked or exposed to heat or flame (Section 7). See also explosives, high.

Disaster Hazard: Highly dangerous; shock and heat will explode it; when heated to decomp, emits highly toxic fumes of NO_x.

DYNAMITE, AMMONIA GELATIN. An ammonia gelatin dynamite is a gelatinized explosive mixture that contains ammonium nitrate. At least 60% of the ammonium nitrate and all of the explosive oil entering into the composition of an ammonia dynamite can be considered as being replaced by a plastic material which then becomes the main explosive ingredient. This material is obtained by gelatinizing about 0.7% of a low-nitrated cellulose with an explosive oil which is warmed slightly to produce the jelly-like mass. The other ingredients are then mixed with the gelatinized material. The final product is dense, plastic and more water-resistant than an ammonia dynamite. The regular grades of ammonia gelatin dynamites range from 30 to 90% strength.

The following formula is typical of a 40% ammonia gelatin: 1% moisture; 26% gelatinized ingredient; 8% ammonium nitrate; 50% sodium nitrate; 1% antacid; 6% sulfur; 8% carbonaceous combustible material.

For a 1¼ by 8 in. cartridge the wrapper per 100 g of explosive ingredient averages 4.5 g. The cartridge count per 50 lb case averages 100.

When detonated, this class of dynamite emits a low volume of poisonous gases and is therefore recommended along with the gelatin dynamites for underground blastings. On the other hand, quarry gelatin, which is a midified type of ammonia gelatin dynamite should be used only in open work and should never be used underground because it emits large volumes of poisonous gases on detonation.

THR = See nitrates.

DYNAMITE, BLASTING GELATIN. Blasting gelatin, which is the most powerful explosive used industrially, is a translucent material of an elastic texture. It is obtained by incorporating about 7% of a low-nitrated cellulose with an explosive oil which is slightly warmed during the process of gelatinization. This explosive is employed only where its high density and water-resisting properties can be advantageously used, such as in submarine work and other operations that are very wet. There is only one grade and it is rated as 100%.

THR = See nitrates.

DYNAMITE, GELATIN. In the gelatin dynamites all the explosive oil contained in a straight dynamite is replaced by a gelatinized material prepared from 0.7% of soluble cellulose nitrate. The latter substance is dissolved in the explosive oil to form a jelly-like mass. Wood pulp or other carbonaceous materials, sodium nitrate, and an antacid are mixed together to form the dope.

The grades of gelatin dynamites range from 20 to 90%, depending on the relative proportions of the gelatinized ingredient and the sodium nitrate.

Gelatin dynamites have a higher density than the ammonia dynamites and are more suitable for blasting in hard ground. They have the further advantage of being plastic and can therefore be packed solidly in the hole so that the explosive is concentrated to best advantage with the least loss of force. Gelatin dynamites are the most suitable dynamites to use in wet holes due to their superior water-resisting properties. On detonation they emit a very low volume of poisonous gases which makes them suitable for use in underground operations where trouble might otherwise be experienced from fumes.

DYNAMITE, SEMI-GELATIN. The principal explosive ingredient of a semi-gelatin dynamite is ammonium nitrate, sensitized by a small quantity of a gelatinized explosive oil. For semigelatin dynamite only about 0.3% of a solution of nitrocellulose is used for the gelatinization process. Wood pulp and other carbonaceous materials, sodium nitrate, and an antacid are also added to the dope.

The two most commonly used grades correspond to a bulk strength of 45 and 60% and have cartridge count per 50 lb case of 105 and 120, respectively, for the $1\frac{1}{4}$ by 8 in. size.

These dynamites are cohesive and are superior in water-resisting properties to the ammonia dynamites. They also emit a smaller volume of poisonous gases. Semi-gelatin dynamites are usually used for blasting rocks and ores of moderate hardness, such as limestones and gypsum.

DYNAMITE, STRAIGHT. A straight dynamite contains no ammonium nitrate, nitroglycerin being the explosive ingredient. Straight dynamites have good water-resisting properties and their mixed ingredients are definitely cohesive. Their present use is confined mainly to propagated ditch blasting and submarine blasting. However, for virtually all other purposes, this class of dynamite has been replaced by the ammonia dynamites and the gelatin types of dynamites. See also dynamite, ammonia and dynamite, gelatin. On account of the large volume of poisonous gases emitted on detonation the straight dynamites are not recommended for blasting operations in underground workings. For blasting ditches, 50% straight dynamite (containing 50% explosive oil) has been found to be generally suitable and for other purposes, straight dynamites ranging from 15 to 60% can be procured. A typical formula for a 40% straight dynamite is as follows: 1% moisture; 40% explosive oil; 44% sodium nitrate; 1% antacid; 14% carbonaceous material. The wrapper of straight dynamite constitutes from 5 to 6% of the weight of a $1\frac{1}{4}$ by 8 in. cartridge which weighs nearly 0.5 lb.

DYPNONE. Syn: *β-methylchalcone*. Liquid.
$C_6H_5COCH{:}C(CH_3)C_6H_5$, mw: 222.27, mp: −30°, bp: 246° @ 50 mm, flash p: 350°F (OC), d: 1.093 @ 20°/20°, vap. press: < 0.01 @ 20°, vap. d: 7.67.
Acute tox data: Oral LD_{50} (rat) = 3600 mg/kg. [*3*]
THR = MOD via oral route.
Fire Hazard: Slight, when exposed to heat or flame; can react with oxidizing materials.
To Fight Fire: Alcohol foam, CO_2, dry chemical.

DYRENE. Syn: *2,4-dichloro-6-(o-chloroanilino)-s-triazine*. White to tan crystals, insol in water.
$C_9H_5N_4Cl_3$, mw: 275.5, mp: 160°.
Acute tox data: Oral LD_{50} (rat) = 2.71 mg/kg; ip LD_{50} (rat) = 25 mg/kg; ip LD_{50} (mouse) = 50 mg/kg. [*3*]
THR = HIGH via oral and ip routes.
Disaster Hazard: See chlorides.

DYSPROSIUM. Dy, atwt: 162.5, mp: 1409°, bp: 2335°, d: 8.540 @ 25°. Bright, lustrous, soft, silvery metal.
THR = U. It may exhibit an anti-coagulant effect.
Fire Hazard: Mod. An active reducing agent. Reacts violently in air and to halogens. [*19*]
Radiation Hazard: For permissible levels, see Section 5, Table 5A.5.

DYSPROSIUM BROMATE. Yellow, hexagonal needles. $Dy(BrO_3)_3 \cdot 9H_2O$, mw: 708.35, mp: 78°, $-6H_2O$ @ 110°.
THR = Details U. See bromates.
Fire Hazard: Mod, by spont chemical reaction. A powerful oxidizing agent.
Disaster Hazard: Dangerous; see bromides; can react vigorously with reducing materials.

DYSPROSIUM BROMIDE. Colorless crystals. $DyBr_3$, mw: 402.2, mp: 881°, bp: 1480°.
THR = See bromides.
Disaster Hazard: Dangerous, see bromides.

DYSPROSIUM CHLORIDE. Shiny yellow crystals. $DyCl_3$, mw: 268.9, d: 3.67 @ 0°/4°, mp: 718°, bp: 1500°. A sol salt.
THR = U. See chlorides and dysprosium.
Disaster Hazard: Dangerous, see chlorides.

DYSPROSIUM CHROMATE. Yellow crystals.
$Dy_2(CrO_4)_3 \cdot 10H_2O$, mw: 853.11, mp: $-3\frac{1}{2}H_2O$ @ 150°, bp: decomp.
THR = HIGH.
Fire Hazard: Slight, by chemical reaction; an oxidizing agent; can react with reducing materials.

DYSPROSIUM FLUORIDE. Colorless crystals. DyF_3, mw: 219.5, mp: 1360°, bp: > 2200°. Insol in water.
THR = U. As a fluoride it could be very toxic. See fluorides.
Disaster Hazard: Dangerous, see fluorides.

DYSPROSIUM IODIDE. Green, yellow crystals. DyI_3, mw: 543.2, mp: 955°, bp: 1320°. Water sol.
THR = U. See iodides.
Disaster Hazard: Dangerous, see iodides.

DYSPROSIUM NITRATE. Yellow crystals. $Dy(NO_3)_3 \cdot 5H_2O$, mw: 438.6, mp: 88.6°.
Acute tox data: Oral LD_{50} (rat) = 3100 mg/kg; ip LD_{50} (rat) = 295 mg/kg; ip LD_{50} (mouse) = 310 mg/kg. [*3*]
THR = MOD via oral and HIGH via ip routes.
Fire Hazard: Mod, by chemical reaction. A powerful oxidizing agent.

Disaster Hazard: Dangerous; when heated to decomp, emits highly toxic fumes of NO_x; can react vigorously with reducing materials.

DYSPROSIUM OXALATE. Prisms. $Dy_2(C_2O_4)_3 \cdot 10H_2O$, mw: 769.2, mp: $-H_2O$ @ 40°.
THR = See oxalates.

DYSPROSIUM SELENATE. Yellow needles. $Dy_2(SeO_4)_3 \cdot 8H_2O$, mw: 898, mp: $-8H_2O$ @ 200°.
THR = See selenium and its compounds.

E-60. See ethylene glycol dibenzoate.

E-605. See parathion.

E-1059. See demeton.

EADC. See ethyl aluminum dichloride.

EASC. See ethyl aluminum sesquichloride.

EARTHNUT OIL. See peanut oil.

EARTH OIL. See crude oil (petroleum).

EARTH WAX. See ceresin wax.

ECGONINE. Syn: *3β-hydroxy-1αH-5αH-tropane-2β-carboxylic acid.* White crystalline alkaloid, slightly bitter taste. $C_9H_{15}NO_3 \cdot H_2O$, mw: 203.24, mp: 198°, d: 1.370 @ 12°/4°, vap. d: 7.0.
THR = HIGH via oral and inhal routes. An allergen.
Disaster Hazard: Dangerous; when heated to decomp, emits highly toxic fumes of NO_x.

ECHITAMINE. See ditaine.

EDATHMIL CALCIUM DISODIUM. See calcium disodium EDTA.

EDTA. See ethylene diamine tetraacetic acid.

EDTN. See ethylene diamine tetraacetonitrile.

EFFECTS OF ACUTE WHOLE BODY EXTERNAL RADIATION EXPOSURE. See Section 5A.

EFFECTS OF AGE ON CARCINOGENIC RESPONSE. See Section 8.

EICOSANE. $C_{20}H_{42}$, mw: 282.4, flash p: >212°F, d: 0.79, vap. d: 9.75, bp: 344°.
THR = No data.
Fire Hazard: Low.
To Fight Fire: Dry chemical, CO_2, spray, mist.

EINSTEINIUM. Es.
Radiation Hazard: For permissible levels, see Section 5, Table 5A.5

EKABORON. See scandium.

EKATIN. Syn: *o,o-dimethyl-s-(2-ethylthio)ethyl phosphorodithioate.* A blue oily liquid. $(CH_3O)_2PS_2C_2H_4$, SC_2H_5, mw: 246.3, d: 1.208 @ 20°.
Acute tox data: Oral LD_{50} (rat) = 25 mg/kg; dermal LD_{50} (rat) = 680 mg/kg. [*3*]
THR = HIGH via oral and MOD via dermal route. A highly toxic cholinesterase inhibitor, systemic, and contact insecticide and acaricide. See parathion.
Disaster Hazard: Dangerous; see parathion.

ELAIDIC ACID. See oleic acid.

ELAIOMYCIN. See 4-methoxy-3-(1-octenyl azoxy)-*d*-threo-2-butanol.

ELASTOSPAR. See N,*p*-dinitroso-N-methyl aniline.

ELAYL. See ethylene.

ELECTRIC BLASTING CAPS (1,000 OR LESS). See explosives, high.

ELECTRIC BLASTING CAPS (MORE THAN 1,000). See explosives, high.

ELECTROLYTE (ACID) BATTERY FLUID. See sulfuric acid.

ELECTROLYTE (ALKALINE) CORROSIVE BATTERY FLUID. See Section 11.

ELECTROLYTE (ACID OR ALKALINE) CORROSIVE BATTERY FLUID PACKED WITH STORAGE BATTERIES. See Section 11.

ELEMENTS OF GOOD PRACTICE IN OCCUPATIONAL SAFETY. See sections 1,2,4,7.

ELIXIR OF VITRIOL. See sulfuric acid, aromatic.

ELOCRON. Syn: *2-(1,3-dioxolan-2-yl)phenyl-n-methyl carbamate.* $C_6H_4OCONHCH_3(O_2)$, mw: 182.1
Acute tox data: Oral LD_{50} (rat) = 90 mg/kg; oral LD_{50} (mouse) = 85 mg/kg; dermal LD_{50} (mouse) = 1660 mg/kg; dermal LD_{50} (rabbit) = 1950 mg/kg. [*3*]
THR = HIGH via oral and MOD via dermal route. A toxic contact and systemic insecticide.

EMERALD, SYNTHETIC. See also beryllium and chromium.

EMERALD GREEN. See copper acetoarsenite.

EMERY. Syn: *corundum.* A varicolored mineral. Al_2O_3, d: 3.95–4.10.
THR = LOW. It is mainly a nuisance dust.

EMETINE. White powder or lumps, bitter taste, darkens on exposure. $C_{29}H_{40}O_4N_2$, mw: 480.63. mp: 74°.
Acute tox data: ip LD_{50} (mouse) = 12 mg/kg; sc LD_{LO} (mouse) = 25 mg/kg; oral LD_{LO} (guinea pig) = 20 mg/kg. [*2, 3*]
THR = HIGH via ip, oral and sc routes. Probably inhal as well. This is one of the two potent alkaloids obtained from the Brazilian plant ipecac. The therapeutic use of various ipecac preparations has given rise to many cases of poisoning, in some instances with fatal results. The toxic effects are particularly prominent if the drug is given intravenously. Special care should therefore be exercised

when administering it in this manner. The symptoms of intoxication are gastrointestinal irr and salivation, as well as general edema, which follows renal insufficiency, hemoptysis, flaccid paralysis, peripheral neuritis, aphonia, difficulties in swallowing, delirium, coma and failure of the heart. The fatal dose is considered to be approximately 2 g, whether administered over a short or relatively long period. The drug seems to have a cumulative effect.

Disaster Hazard: Dangerous; when heated to decomp, emits highly toxic fumes of NO_x.

Treatment and Antidotes: Since poisoning occurs generally only after repeated dosage, discontinuing use of the drug usually stops symptoms and recovery follows. When acute intoxication occurs, the remedial measures are purely symptomatic. Heart depression is the most serious symptom and is to be most guarded against.

EMILENE. $C_7H_{10}O_2N_2S \cdot HCl$, mw: 222.7.
Acute tox data: iv LD_{50} (rat) = 1170 mg/kg. [*3*]
THR = MOD via iv route.

EMILINE. $C_{10}H_{16}O_3$, mw: 184.3.
THR = An exper carc. [*23*]

EMMATOS. See *o,o-dimethyldithiophosphate of diethyl mercapto succinate.*

EMPTY CARTRIDGE BAGS, BLACK POWDER IGNITERS. See explosives, low.

EMPTY CARTRIDGE CASES, PRIMED. See explosives, high.

EMULSIFYING FOOD ADDITIVES. [*109*]

ENAVID.
THR = An exper (±) carc and neo. [*3, 15*]

ENCLOSURES FOR SOUND CONTROL. See Section 3.

ENDO-*cis*-BICYCLO(2,2,1)-5-HEPTENE-2,3-DICARBOXYLIC ANHYDRIDE. White crystals. $C_9H_8O_3$, mw: 164.2, mp: 165°.
THR = U.
Fire Hazard: Slight; will react with water or steam to produce heat.

ENDO-DICYCLOPENTADIENE DIOXIDE. White crystalline powder, slightly sol in water, sol in acetone and benzene. $C_{10}H_{12}O$, mw: 164, mp: 180°–184°, d: 1.331 @ 25°.
THR = Probably HIGH via oral route, and MOD via inhal and dermal routes.

2,5-ENDOMETHYLENE CYCLOHEXENE CARBOXYLIC ACID, ETHYL ESTER.
Acute tox data: Oral LD_{LO} (rat) = 4290 mg/kg. [*3*]
THR = MOD via oral route.

(2,5-ENDOMETHYLENE CYCLOHEXYL METHYL)AMINE.
Acute tox data: Oral LD_{LO} (rat) = 1410 mg/kg; dermal LD_{50} (rabbit) = 520 mg/kg. [*3*]
THR = MOD irr via oral and dermal route.

ENDOSULFAN. Syn: *thiodan.* A mixture of 2 isomers, brown crystals, nearly insol in water, sol in most organic solvents. $C_9H_6Cl_6O_3S$, mw: 407.0, mp(α): 106°, mp(β): 212°, d: 1.745 @ 20°/20°.
Acute tox data: Oral LD_{50} (rat) = 18 mg/kg; dermal LD_{50} (rabbit) = 74 mg/kg; oral LD_{50} (wild birds) = 35 mg/kg. [*3*] oral LD_{50} (rat) = 100 mg/kg. [*2*]
THR = VERY, VERY HIGH via oral; VERY HIGH via dermal routes. An exper neo. [*3*] A CNS stimulant producing convulsions. A highly toxic organochlorine pesticide which does not accumulate significantly in human tissue. Absorption is normally slow but is increased by alcohols, oil, emulsifiers.
Disaster Hazard: Dangerous; see chlorides and S compounds.

ENDOTHAL. Syn: *3,6-endoxo hexanhydrophthalic acid, disodium salt.* A water-sol solid. $C_8H_8O_5Na_2$, mw: 230.1, mp: 144°.
Acute tox data: Oral LD_{50} (rat) = 51 mg/kg; dermal LD_{50} (rat) = 750 mg/kg; dermal LD_{50} (rabbit) = 100 mg/kg. [*3*] oral LD_{50} (rat) = ca 35 mg/kg. [*2*]
THR = HIGH via oral and dermal routes. Very irr to eyes, skin and mu mem. A defoliant and an herbicide.

ENDOXAN. Syn: *cyclophosphamide.* Crystals. Water-sol, slightly sol in organic solvents. $C_7H_{15}Cl_2N_2O_2P$, mw: 261.1, mp: 41°–45°.
Acute tox data: 45 mg/kg → glandular symptoms in women. Oral LD_{50} (rat) = 94 mg/kg; dermal LD_{50} (rat) = 60 mg/kg; ip LD_{50} (rat) = 40 mg/kg; iv LD_{50} (rat) = 160 mg/kg. [*3*]
THR = HIGH via oral, dermal, iv routes. An exper (S) mutagen, teratogen, neo and carc. [*23, 3, 8*] Can cause GI disturbances and leukopenia, nausea, alopecia and hepatic dysfunction.
Disaster Hazard: When heated to decomp, emits highly toxic fumes of PO_x and NO_x.

3,6-ENDOXO HEXAHYDROPHTHALIC ACID DISODIUM SALT. See endothal.

ENDRIN. Syn: *1,2,3,4,10,10-hexachloro-6,7-epoxy-1,4,4a,5,6,7,8,8a-octahydro-1,4,5,8-endo-endo-dimethanonaphthalene.* White crystals. $C_{12}H_8Cl_6O$, mw: 380.9, mp: decomp @ 200°.
Acute tox data: Oral LD_{50} (rat) = 3 mg/kg; dermal LD_{50} (rat) = 15 mg/kg; oral LD_{50} (mouse) = 1.37 mg/kg; oral LD_{50} (wild birds) = 2 mg/kg; iv LD_{50} (mouse) = 2 mg/kg. [*3*]

THR = EXTREMELY HIGH via oral, iv; VERY VERY HIGH via dermal routes. A CNS stimulant. HIGH toxicity to birds, fish and man. Many cases of fatal poisoning attributed to it. Does not accumulate in human tissue. An exper (±) carc. [*12, 3*] In man, ingestion of 1 mg/kg caused symptoms.

Disaster Hazard: Dangerous. See aldrin. Has reacted violently with parathion. [*19*]

ENERGY-DEMAND EFFECTS UPON THE AIR ENVIRONMENT. See Section 4.

ENGINE DISTILLATE GREEN OIL.

THR = A recog carc. [*14*] See mineral oils.

EOSINE. Syn: *Red 11731.* $C_{20}H_8O_5Br_4$, mw: 647.9, Yellow-red crystals. mp: 296°. Insol in water and ether, sol in ethanol.

THR = An exper (±) carc. [*81*]

ENVIRONMENTAL PROTECTION AGENCY (EPA). See Section 10A.

EOSINE DWC 73. $C_{20}H_8O_5Br_4 \cdot 2Na$, mw: 693.9.

Acute tox data: ip LD_{LO} (rat) = 500 mg/kg; sc LD_{LO} (rat) = 1500 mg/kg. [*81, 3*]

THR = MOD via ip and sc routes. An exper neo via sc route. [*81, 3*]

EPHEDRINE. Syn: *1-phenyl-2-methyl aminopropanol.* White granules. $C_{10}H_{15}NO$, mw: 165.23, mp: 79° (dl), bp: 255° (decomp).

Acute tox data: sc LD_{LO} (rat) = 300 mg/kg; oral LD_{50} (mouse) = 400 mg/kg; iv LD_{50} (mouse) = 74 mg/kg. [*3*]

THR = VERY HIGH via iv and HIGH via sc and oral routes. Causes rapid pulse, rise in blood pressure, and other actions similar to epinephrine. Has been known to cause allergic sensitization.

Disaster Hazard: Slight; when heated, emits toxic fumes.

EPICHLOROHYDRIN. Syn: *1-chlor-2,3-epoxypropane.* Colorless, mobile liquid, irr chloroform-like odor. C_3H_5OCl, mw: 92.52, bp: 117.9°, fp: −57.1°, flash p: 105°F (OC), d: 1.1761 @ 20°/20°, vap. press: 10 mm @ 16.6°, vap. d: 3.29.

Acute tox data: Oral LD_{50} (rat) = 90 mg/kg; oral LD_{50} (mouse) = 238 mg/kg; dermal LD_{50} (rabbit) = 1300 mg/kg. [*78, 3*]

THR = HIGH via oral and MOD via dermal routes. Can produce at least temporary sterility. [*78*] In acute poisoning, death may be caused by respiratory paralysis. In chronic poisoning there is kidney damage. Inflammatory changes in the eyes and lungs have been observed. Primary irr and sensitization of the skin has been described. An exper (±) carc, neo. [*3, 13, 23, 78*]

Fire Hazard: Mod, when exposed to heat or flame. Can react violently with HNO_3, 2-aminoethanol, chlorosulfonic acid, ethylene diamine, ethylene imine, oleum, H_2SO_4, K-*tert*-butoxide. [*19*]

Disaster Hazard: Dangerous; when heated to decomp, emits highly toxic fumes of phosgene; can react with oxidizing materials.

To Fight Fire: Foam, alcohol foam, CO_2, dry chemical.

EPINEPHRINE. See "adrenaline."

EPN. Syn: *ethyl-p-nitrophenyl thionobenzene phosphate.* $C_{14}H_{14}O_4NPS$, mw: 323.3. Liquid or pale yellow crystals with an aromatic odor, nearly insol in water, sol in organic solvents. d: 1.268 @ 25°, mp: 36°.

Acute tox data: Oral LD_{50} (rat) = 8 mg/kg; dermal LD_{50} (rat) = 25 mg/kg; ip LD_{50} (rat) = 8 mg/kg; oral LD_{50} (wild birds) = 13 mg/kg. [*3*]

THR = VERY HIGH via oral, dermal, and ip routes. A highly toxic insecticide. A cholinesterase inhibitor. See related compound, parathion. This material is extremely hazardous on contact with skin, inhal or ingestion.

Disaster Hazard: Dangerous; when heated to decomp, emits highly toxic fumes of SO_x and PO_x.

EPON RESINS, CURED. See epoxy resins, cured.

EPON RESINS, UNCURED. See epoxy resins, uncured.

1,2-EPOXYBUTANE. Syn: *1,2-butylene oxide.* Colorless liquid, sol in water, miscible with most organic solvents. $H_2COCHCH_2CH_3$, mw: 72, d: 0.8312 @ 20°/20°, bp: 63°, flash p: 5°F, lel = 1.5%, uel = 18.3%.

Acute tox data: Dermal LD_{50} (rabbit) = 2100 mg/kg; inhal LC_{LO} (rat) = 4000 ppm; oral LD_{LO} (rat) = 1410 mg/kg. [*3*]

THR = MOD irr via dermal, inhal and oral routes.

Fire Hazard: Dangerous; when exposed to heat, flame or powerful oxidizers.

To Fight Fire: Dry chemical, water spray, mist or fog, alcohol foam.

2,3-EPOXYBUTYRIC ACID, BUTYL ESTER.

THR = U. Limited animal exper suggest MOD toxicity and LOW irr. See also epoxy resins and esters.

1,2-EPOXYBUTYRONITRILE. C_4H_5ON, mw: 83.1.

THR = An exper neo. [*3*]

4,5-EPOXYCYCLOHEXANE-1,2-DICARBOXYLIC ACID, DI(DECYL) ESTER.

Acute tox data: Oral LD_{LO} (rat) = 640 mg/kg. [*3*]

THR = MOD via oral route. Limited animal exper suggest LOW toxicity and irr. See also epoxy resins and esters.

3,4-EPOXYCYCLOHEXANE CARBONITRILE. Liq-

uid, sol in water. $O(C_6H_9)CN$, mw: 123, d: 1.0929 @ 19°/4°, bp: 244.5°, fp: −33°.
Acute tox data: Oral LD_{50} (rat) = 1230 mg/kg; dermal LD_{50} (rabbit) = 990 mg/kg. [*3*]
THR = MOD irr via oral and dermal routes.

3,4-EPOXYCYCLOHEXYLMETHYL-3,4-EPOXY-CYCLOHEXANE CARBOXYLATE.
Acute tox data: Oral LD_{50} (rat) = 4490 mg/kg. [*3*]
THR = MOD via oral routes.

5,6-EPOXY-5,6-DIHYDROBENZ(a)ANTHRACENE. $C_{18}H_{12}O$, mw: 244.3.
THR = An exper carc. [*3*]

5,6-EPOXY-5,6-DIHYDROBENZ(a,h)ANTHRA-CENE. $C_{22}H_{14}O$, mw: 294.4.
THR = An exper carc. [*3*]

5,6-EPOXY-5,6-DIHYDROCHRYSENE. $C_{18}H_{12}O$, mw: 244.3.
THR = An exper carc. [*3*]

5,6-EPOXY-5,6-DIHYDRO-7-METHYL BENZ(a) ANTHRACENE. $C_{19}H_{14}O$, mw: 258.3.
THR = An exper neo and carc. [*3*]

1,3-EPOXY-2,2-DIMETHYL PROPANE. $C_5H_{10}O$, mw: 86.2.
THR = An exper carc. [*3*]

1,2-EPOXYETHANE. See ethylene oxide.

1,2-EPOXYETHYL BENZENE. See styrene oxide.

2-(α,β-EPOXYETHYL)-5,6-EPOXYBENZENE.
Acute tox data: Oral LD_{LO} (rat) = 2830 mg/kg; dermal LD_{50} (rabbit) = 620 mg/kg. [*3*]
THR = MOD via oral and dermal routes.

1,2-EPOXYHEXADECANE. $C_{16}H_{32}O$, mw: 240.5.
THR = An exper carc. [*3, 23*]

2,3-EPOXY-2-ETHYL HEXANOL.
$C_3H_7CHOC(C_2H_5)CH_2OH$, mw: 144.
Acute tox data: Dermal LD_{50} (rabbit) = 3150 mg/kg. [*3*]
THR = MOD via dermal route. Probably also via oral route.

2,3-EPOXY-2-ETHYLHEXYL-9,10-EPOXY STEA-RATE.
THR = Probably LOW.

4,5-EPOXY-3-HYDROXY VALERIC ACID-β-LAC-TONE. $C_5H_6O_3$, mw: 114.1.
THR = An exper neo. [*3*]

3,4-EPOXY-6-METHYLCYCLOHEXANE CAR-BOXYLIC ACID, ALLYL ESTER. See allyl-3,4-epoxy-6-methylcyclohexane carboxylate.

3,4-EPOXY-6-CYCLOHEXYLMETHYL-3,4-ACRY-LATE.
THR = Limited animal exper suggest LOW toxicity and irr. See also epoxy resins.

3,4-EPOXY-6-METHYLCYCLOHEXYLMETHYL-3, 4-EPOXY-6-METHYLCYCLOHEXANE.
Acute tox data: Oral LD_{50} (rat) = 4920 mg/kg. [*3*]
THE = MOD via oral and probably via dermal as well. An exper (±) carc. [*3, 23*]

1,2-EPOXYOCTANE. $C_8H_{16}O$, mw: 128.2.
THR = An exper neo. [*3*]

1,2-EPOXYPROPANE. See propylene oxide.

1,3-EPOXYPROPANE. See trimethylene oxide.

2,3-EPOXY-1-PROPANOL. See glycidol.

***p*-(2,3-EPOXYPROPOXY)-N-PHENYL BENZYL AMINE.** $C_{15}H_{15}O_2N$, mw: 241.3.
THR = An exper neo. [*3*]

2,3-EPOXYPROPYL ACRYLATE. Insol in water. $H_2C\underset{O}{\bigtriangledown}CHCH_2OOCHC{:}CH_2$, mw: 123, bp: 57.2° @ 2 mm, flash p: 141°F (OC), d: 1.1, vap. d: 4.4.
Acute tox data: Oral LD_{50} (rat) = 210 mg/kg; inhal LC_{LO} (rat) = 125 ppm, 4 hrs; dermal LD_{50} (rabbit) = 400 mg/kg. [*3*]
THR = HIGH irr via oral, inhal and dermal routes.
Fire Hazard: Mod, via heat, flames, oxidizers.
To Fight Fire: Foam, dry chemical, CO_2.

2,3-EPOXYPROPYL ESTER TRIDECANOIC ACID. $C_{16}H_{30}O_3$, mw: 270.5.
THR = An exper neo. [*3*]

2,3-EPOXYPROPYL ETHER. $H_2\underbrace{COC}HOCH_3$, mw: 74.
THR = Limited animal exper suggest MOD. See also epoxy resins and ethers.

2,3-EPOXYPROPYLHEXANOATE. $C_9H_{16}O_3$, mw: 172.3.
THR = An exper neo. [*3*]

EPOXY RESINS, CURED.
THR = Most cured resins have little or no toxic effects. If curing is incomplete there may be residues of high toxicity curing agents such as the organic amines: *m*-phenylene diamine, diethylene triamine, tetraethylene pentamine, and hexamethylene tetramine, as well as phthalic anhydride and related compounds. An (S) carc. [*14*] See also polymers.
Disaster Hazard: Dangerous; when heated to decomp, they emit highly toxic fumes.

EPOXY RESINS, UNCURED. Syn: *polymers of epichlorohydrin and 2,2-bis(4-hydroxy phenyl) piperazine.*
THR = Animal exper have shown disturbed blood formation. The degree of toxicity of uncured epoxy resins varies, and is partly dependent on the extent of

unreacted curing agents. See also epoxy resins, cured, and specific agents. A (S) carc. [*14*] See also polymers.
Disaster Hazard: Dangerous; When heated to decomp, they emit highly toxic fumes.

9,10-EPOXYSTEARIC ACID. $C_{18}H_{34}O_3$, mw: 298.5.
THR = An exper (±) carc, neo. [*3*, *13*]

9,10-EPOXY STEARIC ACID, ALLYL ESTER.
Acute tox data: Oral LD_{50} (rat) = 1410 mg/kg. [*3*]
THR = MOD via oral route.

9,10-EPOXY STEARIC ACID-2-ETHYLHEXYL ESTER.
Acute tox data: Oral LD_{50} (rat) = 31000 mg/kg. [*3*]
THR = VERY LOW via oral route.

3,4-EPOXY SULFOLANE. $C_4H_3O_3S$, mw: 134.2.
THR = An exper neo. [*3*]

1,2-EPOXY-4-VINYL CYCLOHEXANE. See 4-vinyl cyclohexene monoxide.

EPSOM SALTS. See magnesium sulfate.

EQUILENIN. $C_{18}H_{18}O_2$, mw: 266.4.
THR = An exper neo. [*3*]

EQUILENIN BENZOATE. $C_{25}H_{22}O_3$, mw: 370.3.
THR = An exper carc. [*3*]

EQUILIN. $C_{18}H_{20}O_2$, mw: 268.4.
THR = An exper carc. [*3*]

***d*-EQUILIN.** See equilin.

EQUILIN BENZOATE. $C_{25}H_{24}O_3$, mw: 372.5.
THR = An exper carc. [*3*]

ERADICATOR (PAINT OR GREASE), LIQUID.
Fire Hazard: Dangerous, when exposed to heat or flame; can react vigorously with oxidizing materials.
To Fight Fire: Water spray, mist, foam, dry chemical.

ERASOL. See mechlorethamine hydrochloride.

ERBIUM. Soft, malleable, bright silvery metal, imparts a pink color to salts, glass, etc. Er, atwt: 167.26, mp: 1522°, bp: 2510°, d: 9.16 @ 25°, vap. press: 2 mm @ 1530°.
Radiation Hazard: For permissible levels, see Section 5, Table 5A.5.
Fire Hazard: Mod, when exposed to heat or flame by chemical reaction with oxidizers. Like other rare earth metals it will ignite in air @ 150–180°. Is pyroforic on cutting. Burns vigorously in halogen vapors at >200°.
Explosion Hazard: Mod, when as dust it is exposed to flame or sparks.

ERBIUM CHLORIDE. $ErCl_3$, mw: 273.6.
Acute tox data: ip LD_{50} (mouse) = 226 mg/kg; ip LD_{50} (guinea pig) = 128 mg/kg. [*3*]
THR = HIGH via ip route.

ERBIUM CITRATE.
Acute tox data: ip LD_{50} (mouse) = 122 mg/kg; ip LD_{50} (guinea pig) = 63 mg/kg. [*3*]
THR = HIGH via ip route.

ERBIUM EDETATE.
Acute tox data: ip LD_{50} (guinea pig) = 134 mg/kg. [*3*]
THR = HIGH via ip route.

ERBIUM NITRATE. Reddish crystals. $Er(NO_3)_3 \cdot 5H_2O$, mw: 443.30, $-4H_2O$ @ 130°.
Acute tox data: iv LD_{LO} (rat) = 63 mg/kg. [*3*]
THR = HIGH via iv route. See nitrates.

ERBIUM OXALATE. Reddish powder. $Er_2(C_2O_4)_3 \cdot 10H_2O$, mw: 778.8, mp: decomp @ 575°.
THR = See oxalates.

ERBIUM OXIDE.
Acute tox data: ip LD_{LO} (rat) = 600 mg/kg. [*3*]
THR = MOD via ip route.

ERBIUM PERCHLORATE. $Er(ClO_4)_3$, mw: 465.6.
THR = Violent reaction with acetonitrile. [*19*] See also perchlorates.

ERBON. Syn: *2-(2,4,5-trichlorophenoxy)ethyl-2,2-dichloropropionate.* Crystals, insol in water, sol in acetone, alcohol, kerosene and xylene. $C_{11}H_9Cl_5O_3$, mw: 366.5, mp: 50°, bp: 161°–164° @ 0.5 mm.
Acute tox data: Oral LD_{50} (rat) = 1120 mg/kg; oral LD_{50} (mouse) = 912 mg/kg; oral LD_{50} (rabbit) = 2193 mg/kg. [*3*]
THR = MOD via oral route. It may be toxic via dermal route as well.
Disaster Hazard: Dangerous; see chlorides.

ERGAMINE. See histamine.

ERGOCALCIFEROL. See vitamin D.

ERGOT. Syn: *spurred rye.* Composition: ergot amine, ergosine, ergocristine, ergocryptine, ergocornine, ergosinine, ergocristinine, ergocryptinine, ergotaminine, etc.
THR = Vomiting, diarrhea, thirst, tachycardia, confusion, coma, CNS symptoms, GI disturbances, gangrene, circulatory changes. [*2*] Can follow ingestion.

ERGOTAMINE TARTRATE. $C_{66}H_{70}O_{10}N_{10}$, mw: 1313.6.
THR = VERY HIGH via oral route. An exper teratogen. [*3*] See also ergot.

ERGOTININE. Syn: *ergotoxine.* White crystalline alkaloid. $C_{35}H_{39}N_5O_5$, mw: 609.53, mp: 205°.
Acute poisoning is very rare and is caused accidentally, except in some cases where it has been employed to produce abortion. The symptoms are vomiting, burning pains in the abdomen, tingling of the extremities, great thirst, diarrhea, collapse with

weak, rapid pulse, cold skin, hemorrhage from the uterus and abortion. There is ecchymosis in many organs and in the subcutaneous tissues, suppression of urine, prostration, coma and death from respiratory and cardiac failure. Death may occur in a few hours or be delayed a few days. In cases of recovery, abnormal symptoms persist for a few days. Sometimes a cataract forms in the eyes. Chronic poisoning is very rarely caused by the medicinal use of ergot. Some epidemics have occurred from eating bread made with grain containing ergot. A severe epidemic causing 11,319 cases of poisoning occurred in Russia. Of those, 1,618 were admitted to the hospital and 93 died. There are two types of symptoms from ergot poisoning, one which causes a nervous disorder and one which causes gangrene. In some cases both types have been found, but as a rule one prevails. The gangrenous type generally develops in the fingers and toes; sometimes the entire leg or arm becomes cold and numb, dark, dry, hard and shrunken and falls off with little or no pain and no hemorrhage. Gangrene also occurs in the internal organs. Cataracts are common. The symptoms of spasmodic ergotism are drowsiness, depression, weakness, giddiness, headache, and cramps in the limbs, with itching. In severe cases, convulsions generally clonic and often epileptiform occur. There is mental weakness and sometimes complete dementia. In animals, restlessness, salivation, vomiting and purging, depression, ataxia, clonic convulsions and death by paralysis of the respiratory center occur. The fatal dose cannot be stated accurately. However, 30 grains has caused severe poisoning, but recovery has followed 150 grains (9.72 g). Gangrene and death have been said to follow 10 grains. Death is more likely to follow prolonged use of small medicinal doses than one large dose.

Disaster Hazard: Dangerous; when heated to decomp, emits highly toxic fumes of NO_x.

Treatment and Antidotes: Wash out the stomach with water containing 80 to 100 grains of sodium bicarbonate. Call a physician, give 0.5 to 1.0 g of sodium tetrathionate dissolved in 10 cc of sterile distilled water intravenously daily. Empty the bowels by soap-suds enemas and purgatives, such as calomel and castor oil. Treat symptomatically. Give stimulants, such as strychnine and coffee. If gangrene sets in, the parts should be bathed with warm water and wrapped in soothing ointment.

ERGOTOXINE. See ergotinine.

ERYTHORBIC ACID. Syn: *d-erythro ascorbic acid.* Shiny granular crystals, sol in water, alcohol and pyridine, moderately sol in acetone; slightly sol in glycerol. $C_6H_8O_6$, mw: 176.

THR = U. Used as a chemical preservative food additive. [*109*]

ERYTHRENE. See butadiene-1,3.

ERYTHRITE. See cobaltous-*o*-arsenate.

ERYTHROMYCIN. An antibiotic, white or slightly yellow crystalline powder, odorless, freely sol in alcohol, chloroform and ether, very slightly sol in water. $C_{37}H_{67}NO_{13}$. mw: 733, mp: 133°–138°.

Acute tox data: Oral LD_{50} (mouse) = 4600 mg/kg; ip LD_{50} (mouse) = 710 mg/kg; sc LD_{50} (mouse) = 1800 mg/kg; im LD_{50} (mouse) = 394 mg/kg. [*3*]

THR = MOD via oral, ip, sc and im routes. A food additive permitted in food for human consumption. [*109*]

ERYTHROMYCIN THIOCYANATE. The thiocyanate salt of the antibiotic substance produced by the growth of streptomyces erythreus.

THR = U. A food additive permitted in the feed and drinking water of animals and/or for the treatment of food producing animals. [*109*]

ESERINE. See physostigimine.

ESSENTIAL OILS, OLEORESINS (SOLVENT FREE) AND NATURAL EXTRACTIVES (INCLUDING DISTILLATES). See food additives.

ESSENTIAL SALT OF LEMON. See potassium binoxalate.

ESTER GUM. Syn: *rosin ester.* Hard resin. mp: 141°, bp: 257°, flash p: 375°F (CC).

THR = U.

Fire Hazard: Slight, when exposed to heat or flame; can react with oxidizing materials.

Spont Heating: No.

To Fight Fire: Water, CO_2, dry chemical.

ESTERS. A large group of organic compounds which correspond structurally to salts in inorganic chemistry. They are considered as being derived from acids by the replacement of hydrogen by an organic alkyl radical.

THR = No general statement can be made as to the tox of esters. Many of them are high violatile and hence can act as asphyxiants or narcotics. Skin absorption, as well as inhal, may be an important route of absorption for those esters which are volatile and have a high solvent action. The degree of tox of the esters covers a wide range, from low to high. Esters generally hydrolyze upon contact with moisture; hence a rough guide to the tox of a given ester may be the sum of the toxicities of the products of hydrolysis. May explode on mixing with nitrates. [*19*]

ESTRADIOL. Syns: *femestral, progynon.* $C_{18}H_{24}O_2$, mw: 272.4.
THR = An exper neo, teratogen and carc. [*3, 15, 23*].

17-α-ESTRADIOL. See estradiol.

ESTRADIOL BENZOATE. White or slightly yellow to brownish crystalline powder; odorless. Almost insol in water; sol in alcohol, acetone, and dioxane; sparingly sol in vegetable oils; slightly sol in ether. $C_{18}H_{23}O \cdot C_7H_5O_2$, mw: 376, mp: 191°–196°.
THR = An exper neo and carc. [*3*] A food additive permitted in the feed and drinking water of animals and/or for the treatment of food-producing animals. [*109*] See also estradiol.

ESTRADIOL-17-BENZOATE-3-*m*-BUTYRATE. $C_{29}H_{34}O_4$, mw: 446.6.
THR = An exper neo. [*3*]

ESTRADIOL-3-BENZOATE + PROGESTERONE (1:14).
THR = An exper neo. [*3*]

ESTRADIOL-17-CAPRYLATE. $C_{26}H_{38}O_3$, mw: 398.6.
THR = An exper neo. [*3*]

ESTRADIOL DIPROPIONATE. $C_{24}H_{32}O_4$, mw: 384.6.
THR = An exper neo and carc. [*3*]

ESTRADIOL MONOPALMITATE. Syn: *1,3,5-(10)-estratriene-3,17β-diol-17-palmitate.* $C_{34}H_{54}O_3$, mw: 510, mp: 81°–85°.
THR = U. A food additive permitted in the feed and drinking water of animals and/or for the treatment of food-producing animals. Also a food additive permitted in food for human consumption. [*109*] See also estradiol.

ESTRADIOL MUSTARD. $C_{42}H_{50}O_4N_2Cl_4$, mw: 788.7, mp: 40–65° (freeze dried).
THR = An exper carc. [*3, 8*]

ESTRAGON OIL. Syn: *tarragon oil.*
Acute tox data: Oral LD_{50} (rat) = 1900 mg/kg. [*3*]
THR = MOD via oral route.

α-ESTRA-1,3,5,7,9-PENTANE-3,17-DIOL. $C_{18}H_{19}O_2$, mw: 267.4.
THR = An exper neo. [*3*]

β-ESTRA-1,3,5,7,9-PENTANE-3,17-DIOL.
THR = An exper neo. [*3*]

ESTRIOL. Syn: *estratriol.* Small white crystals. $C_{18}H_{24}O_3$, mw: 288.4.
THR = An exper (±) carc. [*3, 15*]

ESTROGENS.
THR = A (S) carc of uterus, breast, prostate, blood-forming tissues, bladder and kidney. [*14, 3, 15*]

ESTRONE. White crystals. $C_{18}H_{22}O_2$, mw: 270.4, mp: 254°.
THR = An exper (+) neo, teratogen and carc. [*3, 15, 23*]

ESTRONE BENZOATE. $C_{25}H_{26}O_3$, mw: 374.5.
THR = An exper neo and carc. [*3*]

ETHANE. Syns: *bimethyl, methyl methane, dimethyl, ethyl hydride.* Colorless, odorless gas. C_2H_6, mw: 30.07, mp: −172°, bp: −88.6°. lel = 3.0%, uel = 12.5%, fp: −183.2°, d: 0.446 @ 0° (liquid), autoign. temp.: 959°F, vap. d: 1.04.
THR = A simple asphyxiant. See argon for properties of simple asphyxiants.
Fire Hazard: Very dangerous, when exposed to heat or flame; can react vigorously with oxidizing materials.
Spont Heating: No.
Explosion Hazard: Mod, when exposed to flame.
Disaster Hazard: Dangerous, upon exposure to heat or flame.
To Fight Fire: Stop flow of gas.

ETHANE ARSONIC ACID. Syn: *ethylarsonic acid.* Crystals, water-sol. $C_2H_5AsO\ (OH)_2$, mw: 154.0, mp: 99.5°.
Acute tox data: ip LD_{50} (mouse) = 500 mg/kg. [*3*]
THR = HIGH - MOD via ip route. See arsenic compounds.

1,2-ETHANE DIAMINE. See ethylene diamine.

ETHANEDIOIC ACID. See oxalic acid.

ETHANEDIAL. See glyoxal.

1,2-ETHANEDIOL. See ethylene glycol.

ETHANEDIOYL BROMIDE. See oxalyl bromide.

ETHANEDIOYL CHLORIDE. See oxalyl chloride.

ETHANEDINITRILE. See cyanogen.

ETHANE HEXAMERCARBIDE. Yellowish-white powder, insol in water. $C_2Hg_6O_2(OH)_2$, mw: 1293.7, mp: explodes @ 230°, vap. d: 44.6.
THR = See mercury compounds, organic. HIGH toxicity.
Explosion Hazard: Dangerous. When shocked or heated to 230°, explodes.
Disaster Hazard: Dangerous; when heated to decomp or on contact with acid or acid fumes, below 230°, emits highly toxic fumes of mercury; can react vigorously with oxidizing materials.

ETHANE NITRILE. See methyl cyanide.

ETHANE TETRACARBONITRILE. See tetracyanoethylene.

ETHANETHIOL. See ethyl mercaptan.

ETHANETHIOLIC ACID. See thioacetic acid.

ETHANOIC ACID. See acetic acid.

ETHANOIC ANHYDRIDE. See acetic anhydride.

ETHANOL. See ethyl alcohol.

ETHANOLAMINE. See monoethanolamine.

ETHANOL FORMAMIDE. Liquid. $HOCH_2CH_2NHCHO$, mw: 89.09, mp: −72°, bp: 143° @ 2.5 mm, flash p: 347°F, d: 1.17 @ 25°/4°, vap. press: 0.1 mm @ 20°, vap. d: 3.07.
THR = U. Probably LOW. See also amides.
Fire Hazard: Low, when exposed to heat or flame; can react with oxidizing materials.
To Fight Fire: Water, foam, CO_2, dry chemical.

ETHANOL MERCURIC CHLORIDE. Syns: *ceresam, grandsan.* C_2H_5HgCl, mw: 265.1.
Acute tox data: Oral LD_{50} (rat) = 30 mg/kg; dermal LD_{50} (rat) = 200 mg/kg; ip LD_{50} (mouse) = 16 mg/kg. [*3*]
THR = HIGH via dermal, oral and ip routes.

ETHANOXYTRIPHETOL. $C_{27}H_{33}O_3N$, mw: 419.6.
THR = An exper teratogen. [*3*]

ETHANOYL BROMIDE. See acetyl bromide.

ETHANOYL CHLORIDE. See acetyl chloride.

ETHANOYL IODIDE. See acetyl iodide.

ETHANOYL PEROXIDE. See acetyl peroxide.

ETHENE. See ethylene.

ETHENOL. See vinyl alcohol.

ETHENONE. See ketone.

ETHENOXYETHANE. See ethyl vinyl ether.

ETHER. See ethyl ether.

ETHERS. Organic compounds in which an oxygen atom is interposed between two carbon atoms in the structure of the molecule.
THR = The simpler ethers such as ethyl ether, isopropyl ether, etc. are powerful narcotics which in large doses can cause death. The danger from ethers is usually acute and seldom chronic.
There are seldom after-effects to ether intoxication although continued exposure to small concentrations (not enough to cause an overt symptom) has been known to cause loss of appetite, excessive thirst and fatigue.
Fire Hazard: The most common ethers such as ethyl and methyl are particularly dangerous fire hazards. The common ones are easily ignited and have low flash points. It is necessary to control smoking, open flames or even the use of hot plates in areas where low molecular weight ethers are apt to reach 1% conc or more in air. Only electrical equipment of explosion-proof type (Group C classification) is permitted to be operated in ether areas. Ethers should not be stored near powerful oxidizers or in areas of high fire hazard. They should be kept cool and the containers electrically grounded to avoid sparks.
Explosion Hazard: Dangerous; when heated or exposed to flame or sparks. Besides the risk of explosion from air mixtures of ether vapors, ethers tend to form peroxides upon standing. When ethers containing peroxides are heated they can detonate. See also ethyl ether. Violent reaction with BI_3. [*19*]
Disaster Hazard: Dangerous; shock or heat can cause gaseous ethers to escape from their containers; they can react vigorously with oxidizing materials.

ETHERIN. See ethylene.

ETHIDE. See 1,1-dichloro-1-nitroethane.

ETHIDIUM CHLORIDE. $C_{21}H_{20}N_3Cl$, mw: 349.9.
Acute tox data: iv LD_{50} (rat) = 21 mg/kg. [*3*]
THR = VERY HIGH via iv route. An exper carc. [*23*]

ETHINE. See acetylene.

17-α-ETHINYL-3,17-DIHYDROXY-4-ESTRENE DIACETATE. White crystals. $C_{24}H_{32}O_4$, mw: 384.5, mp: 127°.
THR = An exper (+) carc. [*3, 15*]

ETHINYLESTRADIOL. $C_{20}H_{24}O_2$, mw: 296.4.
THR = An exper (+) carc, neo. [*3, 15*]

17-α-ETHINYLESTRADIOL-3-METHYL ETHER. $C_{21}H_{26}O_2$, mw: 310.5.
THR = An exper (+) carc, neo. [*3, 15*]

17-α-ETHINYL-5,19-ESTRENOLONE. White crystals. $C_{20}H_{26}O_2$, mw: 298.5.
THR = An exper (+) carc. [*3, 15, 23*]

17-α-ETHINYL-19-NORTESTOSTERONE. White crystals. $C_{20}H_{26}O_2$, mw: 298.4, mp: 204°.
THR = An exper (+) carc. [*3, 15*]

17-α-ETHINYL-19-NORTESTOSTERONE-17-β-ACETATE. White crystals. $C_{22}H_{26}O_3$, mw: 340.5, mp: 161°.
THR = An exper (±) carc. [*3, 15*]

ETHINYL TRICHLORIDE. See trichloroethylene.

ETHION. Syn: *O,O,O',O'-tetraethyl-S,S-methylene diphosphorodithioate.* Liquid, slightly sol in water, sol in xylene, chloroform, acetone. $C_9H_{22}O_4P_2S_4$, mw: 384.5, mp: −13°, d: 1.220 @ 20°/4°.
Acute tox data: Oral LD_{50} (rat) = 13 mg/kg; dermal LD_{50} (rat) = 62 mg/kg; ip LD_{50} (rat) = 26 mg/kg; oral LD_{50} (wild birds) = 45 mg/kg. [*3*]
THR = VERY HIGH via oral, ip; HIGH via dermal routes. An insecticide. Used also as a food additive permitted in the feed and drinking water of animals and/or for the treatment of food-producing animals. Also permitted in food for human consumption. [*109*]
Disaster Hazard: Dangerous; when heated to decomp, emits highly toxic fumes of oxides of sulfur and phosphorus. See parathion.

ETHIONINE. Syn: *2-amino-4-(ethylthio)butyric acid.* A crystalline material. $C_6H_{13}O_2NS$, mw: 162.3, decomp @ 273°.

THR = An exper carc and teratogen. [*14, 23, 3*] Use with extreme caution.

Disaster Hazard: Dangerous, see SO_x and PO_x.

ETHOPABATE. Syn: *methyl-4-acetamide-2-ethoxybenzoate.* Oily liquid. $C_{12}H_{15}NO_4$, mw: 273.3.

THR = U. Used as a food additive permitted in food for human consumption. [*109*]

ETHOXAZENE. Syn: *2,4-diamino-4',4'-ethoxyazobenzene.* Reddish powder or crystals, insol in water. $C_{14}H_{16}N_4O$, mw: 256.3.

THR = U. Large doses are said to cause irr of urinary tract.

ETHOXYACETYLENE. Insol in water. C_4H_6O, mw: 70, flash p: <20°, d: 0.8, vap. d: 2.4, bp: 51°.

Fire Hazard: Dangerous, when exposed to heat, flame or oxidizers.

To Fight Fire: Foam, dry chemical, CO_2.

6-ETHOXY-2-BENZOTHIAZOLE SULFONAMIDE. $C_9H_{10}O_3N_2S_2$, mw: 258.3.

THR = An exper teratogen. [*3*]

ETHOXYBORON DICHLORIDE. Colorless liquid, decomp by water. $C_2H_5OBCl_2$, mw: 126.8, bp: 78°.

THR = See chlorides and boron compounds. Can react violently with $(C_2H_5)MgI$. [*19*]

(4-ETHOXYBUTYL)DIETHYL ALUMINUM. $C_{10}H_{23}OAl$, mw: 186.2.

THR = Self-ignites in air. [*19*]

ETHOXYCARBONYL ETHYLENE IMINE. $C_5H_9O_2N$, mw: 115.2.

THR = An exper mutagen. [*3*]

2-ETHOXY-3,4-DIHYDRO-2,2-PYRAN. Syn: *2-ethoxy-3,4-dihydro-2H-pyran.* Liquid, very slightly sol in water. $OCH{:}CHCH_2CH_2CHOC_2H_5$, mw: 128, d: 1.0, bp: 289°F, flash p: 111°F (OC).

Acute tox data: Inhal LD_{LO} (rat) = 8000 ppm for 4 hrs; oral LD_{50} (rat) = 6160 mg/kg; dermal LD_{50} (rabbit) = 3560 mg/kg. [*3*]

THR = MOD via inhal, oral and dermal routes.

Fire Hazard: Mod, when exposed to heat, flame or oxidizers.

To Fight Fire: Dry chemical, CO_2, foam, fog.

6-ETHOXY-1,2-DIHYDRO-2,2,4-TRIMETHYLQUINOLINE. Clear light yellow liquid. $C_{14}H_{19}NO$. mw: 217.3, mp: <0°, bp: 125° @ 2 mm, vap. d: 7.48, d: 1.030 @ 25°.

Acute tox data: Oral LD_{50} (rat) = 800 mg/kg; ip LD_{LO} (mouse) = 200 mg/kg. [*3*]

THR = HIGH via ip; MOD via oral routes. Used as a food additive permitted in the feed and drinking water of animals and/or for the treatment of food-producing animals. Also permitted in food for human consumption. [*109*]

Fire Hazard: Slight; when heated to decomp, emits toxic fumes; can react with oxidizing materials.

2-ETHOXYETHANOL. See "Cellosolve" solvents.

ETHOXYETHOXYETHYL ACETATE. $C_8H_{16}O_4$, mw: 176.2.

THR = Reacts violently with chlorosulfonic acid, oleum. [*19*]

2-ETHOXYETHYL ACETATE. See "Cellosolve" acetate.

ETHOXYETHYNYL CARBINOL. $C_5H_8O_2$, mw: 100.1.

THR = Can explode violently. [*19*]

7-(ETHOXYMETHYL)BENZ(a)ANTHRACENE. $C_{21}H_{18}O$, mw: 286.4.

THR = An exper carc. [*3, 23*]

7-(ETHOXY)-12-METHYL BENZ(a)ANTHRACENE.

THR = HIGH. An exper carc via im and sc routes. [*3, 23*]

2-ETHOXY-4-METHYL-3,4-DIHYDROPYRAN.

Acute tox data: Oral LD_{50} (rat) = 3400 mg/kg; dermal LD_{50} (rabbit) = 1060 mg/kg. [*3*]

THR = MOD via oral and dermal routes.

2-ETHOXYMETHYL-2,4-DIMETHYLPENTANE DIOL-1,5. $HOCH_2CH(CH_3)CH_2C(CH_3)(CH_2CH_2OCH_3)$, mw: 190.3, mp: −40°, bp: 208° @ 100 mm, flash p: 295°F (OC), d: 1.0165, vap. d: 6.08.

THR = U.

Fire Hazard: Slight, when exposed to heat or flame; can react with oxidizing materials.

To Fight Fire: Foam, CO_2, dry chemical.

N-ETHOXY-N-MORPHOLINO DIAZENIUM FLUOROBORATE. $C_6H_{13}O_2N_2 \cdot BF_4$, mw: 232.

THR = An exper carc. [*3*]

(*p*-ETHOXYPHENYL)UREA. $C_9H_{12}O_2N_2$, mw: 180.2, mp: 174°. Needles.

Acute tox data: Oral LD_{50} (rat) = 1900 mg/kg; In man up to 600 mg/day → no effects. In children 8000–10,000 mg caused death. In adults 20g–40g → dizziness, nausea, methemoglobinemia, cyanosis, hypotension.

THR = HIGH-MOD via oral route. An exper carc via oral route. [*3, 12*]

2-ETHOXY-1-PROPANOL. $C_5H_{12}O_2$, mw: 104.2

Acute tox data: Oral LD_{50} (rat) = 4900 mg/kg. [*3*]

THR = MOD via oral route.

1-ETHOXY-2-PROPANOL

Acute tox data: Oral LD_{50} (rat) = 4900 mg/kg. [*3*]

THR = MOD via oral route.

3-ETHOXY-1-PROPANOL.
Acute tox data: Dermal LD_{50} (rabbit) = 2830 mg/kg. [*3*]
THR = MOD via dermal route.

3-ETHOXYPROPIONALDEHYDE. Liquid.
$C_2H_5OCH_2CH_2CHO$, mw: 102.13, mp: −69.4°, bp: 135.2°, flash p: 100°F (OC), d: 0.918 @ 20°/20°, vap. d: 3.63, vap. press: 5.5 mm @ 20°.
Acute tox data: Oral LD_{50} (rat) = 900 mg/kg; inhal LC_{LO} (rat) = 500 ppm for 4 hrs; dermal LD_{50} (rabbit) = 1000 mg/kg. [*3*]
THR = MOD via oral, inhal and dermal routes. See also aldehydes.
Fire Hazard: Dangerous, when exposed to heat or flame; can react with oxidizing materials.
To Fight Fire: Alcohol foam, CO_2, dry chemical.

3-ETHOXYPROPIONIC ACID. Liquid.
$C_2H_5OCH_2CH_2COOH$, mw: 118.13, mp: −10.7°, bp: 219°, flash p: 225°F (OC), d: 1.0474, vap. d: 4.08.
Acute tox data: Oral LD_{50} (rat) = 4800 mg/kg; dermal LD_{50} (rabbit) = 750 mg/kg. [*3*]
THR = MOD via oral and dermal routes.
Fire Hazard: Slight when exposed to heat or flame; can react with oxidizing materials.
To Fight Fire: Alcohol foam, CO_2, dry chemical.

3-(ETHOXYPROPYL)MERCURY BROMIDE. Syn: *aagrano.*
THR = HIGH. A fungicide. See mercury compounds, organic.

N′-6-(ETHOXY-3-PYRIDAZINYL)SULFANILAMIDE. See sulfaethoxy pyridazine.

ETHOXYTRIETHYL DIPHOSPHINYL OXIDE.
$C_8H_{20}O_2P_2$, mw: 210.2.
THR = Self-ignites in air. [*19*]

ETHOXYTRIGLYCOL. $C_2H_5O(C_2H_4O)_3H$. mw: 178.21, bp: 255.4°, flash p: 275°F (OC), d: 1.0208 @ 20°/20°, vap. press: 0.01 mm @ 20°.
Acute tox data: Oral LD_{50} (rat) = 11,000 mg/kg; dermal LD_{50} (rabbit) = 8000 mg/kg. [*3*]
THR = LOW via oral and dermal routes.
Fire Hazard: Slight, when exposed to heat or flame; can react with oxidizing materials.
To Fight Fire: Foam, alcohol foam, CO_2, dry chemical.

ETHOXYTRIMETHYLSILANE. $C_5H_{14}OSi$, mw: 118.3, bp: 75.7°, vap. press: 100 mm @ 22.1°, vap. d: 4.1.
Acute tox data: Oral LD_{LO} (rat) = 1400 mg/kg; inhal LD_{LO} (rat) = 4000 ppm for 8 hrs. [*3*]
THR = MOD via oral and inhal routes. See also silanes.
Fire Hazard: Mod, when heated, can ignite; can react with oxidizing materials.

ETHYL ABIETATE. Syn: *abietic acid ethyl ester.* Amber colored, viscous liquid. $C_{19}H_{29}CO_2C_2H_5$, mw: 330.4, mp: 45°, bp: 350°, flash p: 352.4°F (OC), d: 1.02, vap. d: 11.4.
THR = See abietic acid and ethyl alcohol.
Fire Hazard: Slight, when exposed to heat or flame; can react with oxidizing materials.
To Fight Fire: CO_2, dry chemical.

N-ETHYL ACETAMIDE. Water white liquid. $CH_3CONHC_2H_5$, mw: 87.1, mp: −32°, bp: 206°–208°, flash p: 230°F, d: 0.920 @ 20°/20°. vap. d: 3.0.
THR = U. See also amides.
Fire Hazard: Slight, when exposed to heat or flame; can react with oxidizing materials.
To Fight Fire: Alcohol foam.

N-ETHYL ACETANILIDE. Syn: *ethyl phenyl acetamide.* White crystals, faint odor.
$CH_3CON(C_2H_5)(C_6H_5)$, mw: 163.21, mp: 54°, bp: 258°, flash p: 126°F, d: 0.994, vap. d: 5.62.
Acute tox data: Oral LD_{50} (mouse) = 409 mg/kg. [*3*]
THR = HIGH via oral route.
Fire Hazard: Mod, when exposed to heat or flame; can react with oxidizing materials.
To Fight Fire: Foam, CO_2, dry chemical.

ETHYL ACETATE. Syns: *acetic ether, ethyl ester, ethyl ethanoate.* Colorless liquid, fragrant odor.
$CH_3COOC_2H_5$, mw: 88.10, mp: −83.6°, bp: 77.15°, ulc: 85–90, lel = 2.2%, uel = 11%, flash p: 24°F, d: 0.8946 @ 25°, autoign. temp.: 800°F, vap. press: 100 mm @ 27.0°, vap. d: 3.04.
Acute tox data: Oral LD_{50} (rat) = 11,000 mg/kg; inhal LC_{50} (rat) = 1600 ppm; ip LD_{50} (mouse) = 709 mg/kg; inhal LC_{LO} (mouse) = 31 mg/m^3 for 1 hr. [*3*]
THR = LOW via oral; MOD via inhal and ip routes. Ethyl acetate is irr to mu surfaces, particularly the eyes, gums and respiratory passages, and is also mildly narcotic. On repeated or prolonged exposures, it causes conjunctival irr and corneal clouding. It can cause dermatitis. High conc have a narcotic effect and can cause congestion of the liver and kidneys. Chronic poisoning has been described as producing secondary anemia, leucocytosis and cloudy swelling, and fatty degeneration of the viscera. Used as a synthetic flavoring substance and adjuvant. [*109*]
Fire Hazard: Dangerous, when exposed to heat or flame; can react vigorously with chlorosulfonic acid, ($LiAlH_2$ + 2-chloromethyl furan), oleum, K-*tert*-butoxide. [*19*]
Spont Heating: No.
Explosion Hazard: Mod, when exposed to flame.

Disaster Hazard: Dangerous, upon exposure to heat or flame.
To Fight Fire: CO_2, dry chemical or alcohol foam.

n-ETHYL ACETOACETAMIDE. $CH_3CONH_2CO_2C_2H_5$, mw: 132.14.
Acute tox data: Oral LD_{50} (rat) = 14,000 mg/kg. [*3*]
THR = LOW via oral route.

ETHYL ACETOACETATE. Syns: *diacetic ether, acetoacetic ester.* Colorless liquid, fruity odor. $CH_3COCH_2COOC_2H_5$, mw: 130.14, bp: 180.8°, fp: −45°, flash p: 185°F (COC), autoign. temp.: 563°F, d: 1.0261 @ 20°/20°, vap. press: 1 mm @ 28.5°, vap. d: 4.48.
Acute tox data: Oral LD_{50} (rat) = 3980 mg/kg. [*3*]
THR = MOD via oral and probably inhal routes.
Fire Hazard: Mod, when exposed to heat or flame; can react with oxidizing materials. Violent reaction with (Zn + tribromoneopentyl alcohol). [*19*]
Spont Heating: No.
To Fight Fire: Alcohol foam, CO_2, dry chemical.

ETHYL ACETYLENE. Syn: *1-butyne.* A colorless, highly flam gas. $C_2H_5C : CH$, mw: 54, bp: 8.3°, d: 0.669 @ 0°/0°, mp: −130°, flash p: <20°F (TOC).
THR = U. Probably an asphyxiant; see acetylene.
Fire Hazard: Dangerous, when exposed to heat, open flame or powerful oxidizers.
Explosion Hazard: Dangerous; details not known.
To Fight Fire: Stop flow of gas.

ETHYL ACETYL GLYCOLATE. $C_6H_{10}O_4$, mw: 146.2, flash p: 180°F, d: 1.09, vap. d: 5.04, bp: 185°.
THR = No data. See glycols.
Fire Hazard: Mod, via heat, flames, oxidizers.
To Fight Fire: Dry chemical, CO_2, foam.

ETHYL ACRYLATE. Syn: *ethyl propenoate.* Colorless liquid. Acrid, penetrating odor. $CH_2CHCOOC_2H_5$, mw: 100.11, bp: 99.8°, fp: <−72°, lel = 1.8%, flash p: 60°F (OC), d: 0.941 @ 20°/4°, vap. press: 29.3 mm @ 20°, vap. d: 3.45.
Acute tox data: Oral LD_{50} (rat) = 1000 mg/kg; inhal LC_{LO} (rat) = 1000 ppm for 4 hrs; dermal LD_{50} (rabbit) = 1950 mg/kg. [*3*]
THR = MOD via oral, inhal and dermal routes. Oral administration of 0.42 g or more per kg of body weight in the case of rabbits resulted in fatal poisoning. This was characterized in its terminal stages by dyspnea, cyanosis, and convulsive movements. It caused severe local irr of the gastro-enteric tract; and toxic degenerative changes of cardiac, hepatic, renal, and splenic tissues were observed. It gave no evidence of cumulative effects in rabbits. When applied to the intact skin of rabbits, the ethyl ester caused marked local irritation, erythema, edema, thickening, and vascular damage. Animals subjected to a fairly high conc of these esters suffered irr of the mu mem of the eyes, nose, and mouth as well as lethargy, dyspnea, and convulsive movements. A substance migrating to food from packaging materials. [*109*]
Fire Hazard: Dangerous, when exposed to heat or flame; can react vigorously with oxidizing materials. Violent reaction with chlorosulfonic acid. [*19*]
To Fight Fire: Alcohol foam.

ETHYL ALCOHOL. Syns: *ethanol, methyl carbinol, spirit of wine.* Clear, colorless, fragrant liquid, burning taste. CH_3CH_2OH, mw: 46.07, bp: 78.32°, ulc: 70, lel = 3.3%, uel = 19%, fp: <−130°, flash p: 55°F, d: 0.7893 @ 20°/4°, autoign. temp.: 793°F, vap. press: 40 mm @ 19°, vap. d: 1.59.
Acute tox data: Oral LD_{50} (rat) = 21,000 mg/kg; iv LD_{50} (rat) = 1440 mg/kg; oral LD_{50} (rabbit) = 6300 mg/kg; dermal LD_{50} (rabbit) = 20,000 mg/kg; oral LD_{LO} (human) = 6000 mg/kg. [*3*]
THR = MOD - LOW via oral, iv and dermal routes. Probably also via inhal route. It is an exper carc. [*3*] The systemic effect of ethyl alcohol differs from that of methyl alcohol. Ethyl alcohol is rapidly oxidized in the body to carbon dioxide and water, and in contrast to methyl alcohol, no cumulative effect occurs. Though ethyl alcohol possesses narcotic properties, conc sufficient to produce this effect are not reached in industry. Exposure to conc of 5,000–10,000 ppm results in irr of the eyes and mu mem of the upper respiratory tract. If continued for an hour, stupor and drowsiness may result. Conc below 1,000 ppm usually produce no signs of intoxication. There is no concrete evidence that repeated exposure to ethyl alcohol vapor results in cirrhosis of the liver. Large doses can cause alcohol poisoning. Repeated ingestions can lead to alcoholism. It is a CNS depressant.

Exposure to conc of over 1,000 ppm may cause headache, irr of the eyes, nose and throat, and, if long continued, drowsiness and lassitude, loss of appetite and inability to concentrate.
Fire Hazard: Dangerous, when exposed to heat or flame; can react vigorously with acetyl chloride, $(Ag_2O + NH_4OH)$, BrF_5, $Ca(OCl)_2$, ClO_3, CrO_3, $Cr(OCl)_2$, (cyanuric acid + H_2O), H_2O_2, HNO_3, $(H_2O_2 + H_2SO_4)$, $(I + CH_3OH + HgO)$, [$Mn(ClO_4)_2$ + 2,2-dimethoxy propane], $Hg(NO_3)_2$, $HClO_4$, perchlorates, (H_2SO_4 + permanganates), $HMnO_4$, KO_2, $KOC(CH_3)_3$, $(Ag + HNO_3)$, $AgNO_3$, $AgClO_4$, NaH_3N_2, $UO_2(ClO_4)_2$. [*19*]
Disaster Hazard: Dangerous, when exposed to heat or flame.
Spont Heating: No.

Explosion Hazard: Mod, when exposed to flame.
To Fight Fire: Alcohol foam, CO_2, dry chemical.

ETHYL ALDEHYDE. See acetaldehyde.

ETHYL-α-ALLYL ACETOACETATE. Liquid. $CH_3COCH(CH_2CHCH_2) \cdot CO_2C_2H_5$. mw: 170.23, bp: 208°–218°, d: 0.989 @ 25°/20°, vap. d: 5.88.
THR = U. See also esters.
Fire Hazard: Slight; can react with oxidizing materials, heat and flame.
To Fight Fire: Water spray, mist, foam.

ETHYL ALUMINUM DICHLORIDE. Syn: *EADC.* Clear, yellow liquid. $(C_2H_5)_3Al_2Cl_3$, mw: 247.5.
THR = HIGH irr via skin, eyes and mu mem.
Fire Hazard: Dangerous, ignites spont in air. See also ethyl aluminum dichloride. Reacts violently with (benzene + allylchloride), CCl_4, (toluene + allyl chloride), water. [*19*]
To Fight Fire: Do not use water, foam or halogenated extinguishing agents. Use dry chemical preparations.

ETHYL AMINE. Syns: *aminoethane, monoethylamine.* Colorless liquid, strong ammoniacal odor. $C_2H_5NH_2$, mw: 45.08, mp: −80.6°, bp: 16.6°, flash p: <0°F, d: 0.7059 @ 0°/4°, autoign. temp.: 725°F, vap. press: 400 mm @ 20°, vap. d: 1.56, lel = 3.5%, uel = 14.0%.
Acute tox data: Dermal LD_{50} (rabbit) = 390 mg/kg; oral LD_{LO} (rat) = 400 mg/kg; inhal LC_{LO} (rat) = 3000 ppm for 4 hrs. [*3*]
THR = HIGH via dermal, oral, skin, eyes and mu mem routes.
Fire Hazard: Very dangerous, when exposed to heat or flame; can react vigorously with oxidizing materials.
Disaster Hazard: Dangerous; keep away from heat and open flame.
To Fight Fire: Alcohol foam, dry chemical.

ETHYL-*p*-AMINOBENZOATE. See anesthesin.

ETHYL AMINO ETHANOL. $C_4H_{11}ON$, mw: 89.1, flash p: 160°F (OC), d: 0.92, vap. d: 3.06, bp: 161°.
Acute tox data: Oral LD_{50} (rat) = 1000 mg/kg; ip LD_{50} (rat) = 1170 mg/kg; dermal LD_{50} (rabbit) = 360 mg/kg. [*3*]
THR = HIGH via dermal; MOD via oral and ip routes.
Fire Hazard: Mod, via heat, flames and oxidizers.
To Fight Fire: Alcohol foam, dry chemical, CO_2.

2-ETHYL AMINO-1,3,4-THIADIAZOLE.
THR = An exper teratogen. [*3*]

ETHYL AMYL KETONE. Syn: *5-methyl-3-heptanone.* Liquid with mild fruity odor, sol in many organic solvents. $C_8H_{16}O$, mw: 128.2, bp: 157°–162°, d: 0.822 @ 20°/20°, flash p: 138°F.
Acute tox data: Oral LD_{50} (rat) = 3500 mg/kg; inhal LC_{LO} (rat) = 3484 ppm for 8 hrs; oral LD_{50} (mouse) = 3800 mg/kg.
THR = MOD irr to skin, eyes, mu mem via oral and inhal routes. Lower conc, about 100 ppm, can cause headache and nausea. High conc are narcotic. It is mod irr to eyes and mu mem.

N-ETHYLANILINE. Syn: *ethyl phenyl amine.* Clear, yellow-brown oil. $C_2H_5NH(C_6H_5)$, mw: 121.2, mp: −63.5°, bp: 204°, d: 0.958 @ 25°/25°, vap. press: 1 mm @ 38.5°, vap. d: 4.18, flash p: 185°F (OC).
Acute tox data: Oral LD_{LO} (rat) = 300 mg/kg. [*3*]
THR = HIGH irr to skin, eyes, mu mem via oral, inhal and probably dermal routes. An allergen.
Fire Hazard: Mod, via heat, flame, oxidizers, Violent reaction with HNO_3. [*19*]
Disaster Hazard: Highly dangerous; on decomp or on contact with acid or acid fumes, emits highly toxic fumes of aniline and oxides of nitrogen; can react with oxidizing materials.
To Fight Fire: Dry chemical, CO_2, foam.

***o*-ETHYLANILINE.** Yellow liquid, darkens upon standing. $C_6H_4NH_2C_2H_5$, mw: 121.2, mp: −63.5°, bp: 215°, flash p: 185°F (OC), d: 0.98 @ 25°/25°, vap. d: 4.17.
Acute tox data: Oral LD_{50} (rat) = 1260 mg/kg. [*3*]
THR = MOD via oral, probably other routes as well. See also *n*-ethyl amine.
Fire Hazard: Mod, when exposed to heat or flame.
Disaster Hazard: Dangerous; when heated to decomp, emits highly toxic fumes of aniline and oxides of nitrogen; can react with oxidizing materials.
To Fight Fire: Foam, CO_2, dry chemical.

4-ETHYLANILINE.
Acute tox data: LD_{50} (mouse) = 133 mg/kg. [*3*]
THR = HIGH via oral and inhal routes. See also *n,o*-ethylaniline.

ETHYL ARSENIOUS OXIDE. Colorless oil, garlic-like odor. C_2H_5AsO, mw: 120.0, bp: 158° @ 10 mm, d: 1.802 @ 11°., vap. d: 4.14.
THR = An exper carc. [*3*] See arsenic.

ETHYL AZIDES.
THR = Very unstable. [*19*]

7-ETHYL BENZ(c)ACRIDINE. $C_{19}H_{15}N$, mw: 257.4.
THR = An exper neo. [*3*]

7-ETHYL BENZ(a)ANTHRACENE. $C_{20}H_{16}$, mw: 256.4.
THR = An exper neo. [*3, 23*]

8-ETHYL BENZ(a)ANTHRACENE.
THR = An exper carc. [*3, 23*]

12-ETHYL BENZ(a)ANTHRACENE.
THR = An exper neo. [*23*]

ETHYL BENZENE. Syns: *ethyl benzol, phenylethane.* Colorless liquid, aromatic odor. $C_6H_5C_2H_5$, mw: 106.16, bp: 136.2°, fp: −94.9°, flash p: 59°F, d: 0.8669 @ 20°/4°, autoign. temp.: 810°F, vap. press: 10 mm @ 25.9°, vap. d: 3.66, lel = 1.0%, uel = 6.7%.

Acute tox data: Oral LD_{50} (rat) = 3500 mg/kg; dermal LD_{50} (rabbit) = 5000 mg/kg; inhal TC_{LO} (human) = 100 ppm for 4 hrs. [*3*]

THR = MOD via irr to skin, eyes, mu mem and via oral and inhal routes. The liquid is an irr to the skin and mu mem. A conc of 0.1% of the vapor in air is an irr to the eyes of humans, and a conc of 0.2% is extremely irr at first, then causes dizziness, irr of the nose and throat and a sense of constriction of the chest. Exposure of guinea pigs to 1% conc has been reported as causing ataxia, loss of consciousness, tremor of the extremities and finally death through respiratory failure. The pathological findings were congestion of the brain and lungs, with edema. No data are available regarding the effect of chronic exposure.

Erythema and inflammation of the skin may result from contact of the skin with the liquid (Section 9). Exposure to the vapor causes lachrymation and irr of the nose and throat, dizziness, and a sense of constriction of the chest. The irr properties are sufficient to cause workers to leave an atmosphere containing 0.5% of the vapor.

Fire Hazard: Dangerous, when exposed to heat or flame; can react vigorously with oxidizing materials.

Spont Heating: No.

Disaster Hazard: Dangerous; keep away from heat and open flame.

To Fight Fire: Foam, CO_2, dry chemical.

ETHYL BENZOATE. Syn: *benzoic ether.* Colorless, aromatic liquid. $C_6H_5CO_2C_2H_5$, mw: 150.2, mp: −34.6, bp: 213.4°, flash p: >204°F, d: 1.048 @ 20°/20°, vap. press: 1 mm @ 44.0°, vap. d: 5.17, autoign. temp.: 914°F.

Acute tox data: Oral LD_{50} (rat) = 6400 mg/kg. [*2*] Dermal LD_{50} (rabbit) = 1940 mg/kg. [*3*]

THR = MOD via dermal and oral routes. See also esters.

Fire Hazard: Low, when exposed to heat or flame; can react with oxidizing materials.

To Fight Fire: Foam, CO_2, dry chemical.

ETHYL BENZOL. See ethyl benzene.

5-ETHYL BENZO(c)PHENANTHRENE. $C_{20}H_{17}$, mw: 257.4.

THR = An exper carc. [*3, 23*]

ETHYL BENZOYL ACETATE. Clear liquid. $C_6H_5COCH_2COOC_2H_5$, mw: 192.21, bp: 265.0°, decomp; flash p: 285°F (OC), d: 1.11, vap. press: 1 mm @ 107.6°, vap. d: 6.63.

THR = U. See also esters.

Fire Hazard: Slight, when exposed to heat or flame; can react with oxidizing materials.

To Fight Fire: CO_2, dry chemical.

α-ETHYLBENZYL ALCOHOL. $C_9H_{12}O$, mw: 136.2.

Acute tox data: Oral LD_{50} (rat) = 1500 mg/kg. [*3*]

THR = MOD via oral route.

ETHYL BENZYL ANILINE. Clear, colorless oil. $C_6H_5N(C_2H_5)CH_2C_6H_5$, mw: 211.3, bp: 286°, d: 1.034, vap. d: 7.28.

THR = See aniline.

ETHYL BORATE. Syns: *triethyl borate, triethoxyborine.* Colorless liquid, mild odor, decomp in water. $B(OC_2H_5)_3$, mw: 146.00, bp: 120°, flash p: 52°F (CC), d: 0.864 @ 26.5°, vap. d: 5.04.

THR = U. See also boron compounds.

Fire Hazard: Dangerous, when exposed to heat or flame; will react with water or steam to produce flam vapors; can react vigorously with oxidizing materials.

Spont Heating: No.

Disaster Hazard: Dangerous; keep away from heat and open flame.

To Fight Fire: CO_2, dry chemical.

ETHYL BORIC ACID. White crystals, water-sol. $(C_2H_5)B(OH)_2$, mw: 73.9, mp: subl @ 40°.

THR = U. See also boron compounds.

ETHYL BROMIDE. Syns: *bromoethane, hydrobromic ether, bromic ether.* Colorless, volatile liquid. CH_3CH_2Br, mw: 108.98, mp: −119°, bp: 38.4°, lel = 6.7%, uel = 11.3%, flash p: <−4°F, d: 1.451 @ 20°/4°, autoign. temp: 952°F, vap. press: 400 mm @ 21°, vap. d. 3.76.

Acute tox data: Inhal LC_{LO} (rat) = 16,700 ppm for 15 min; inhal LC_{LO} (guinea pig) = 2200 ppm. [*3*]

THR = MOD irr via inhal; irr to eyes, mu mem. It readily decomp into volatile toxic products, such as hydrobromic acid and bromine, particularly in the presence of hot surfaces or open flame. Physiologically, it is an anesthetic and narcotic. Its vapors are markedly irr to the lungs on inhal for even short periods. It can produce acute congestion and edema. Liver and kidney damage in humans has been reported. It is much less toxic than methyl bromide, but more toxic than ethyl chloride.

Spont Heating: No.

Fire Hazard: Dangerous, via heat, open flame (sparks), oxidizers.

Explosion Hazard: Mod, when exposed to flame.

Disaster Hazard: Dangerous; when heated to decomp, emits highly toxic fumes of bromine; will react with water or steam to produce toxic and corrosive fumes; can react vigorously with oxidizing materials.

To Fight Fire: CO_2, dry chemical.

ETHYL BROMOACETATE. Syn: *ethyl bromoethanoate.* Colorless to straw-colored liquid.

$CH_2BrCOOC_2H_5$, mw: 167.01 bp: 158.8°, fp: <−20°, flash p: 118°F, d: 1.514 @ 13°/4°, vap. d: 5.8.

THR = A strong irr to skin, eyes, mu mem. An exper neo. [*3*]

Fire Hazard: Mod, via heat, flame, oxidizers.

Disaster Hazard: Dangerous; when heated to decomp or on contact with acid or acid fumes, emits highly toxic fumes of bromides; will react with water or steam to produce toxic and corrosive fumes.

To Fight Fire: Water as a fire blanket.

ETHYL BROMOETHANOATE. See ethyl bromoacetate.

ETHYL BUTANOATE. See ethyl-*n*-butyrate.

2-ETHYL BUTANOIC ACID. See 2-ethyl butyric acid.

2-ETHYL BUTANOL. Syn: *2-ethyl butyl alcohol.* Clear liquid. $(C_2H_5)_2CHCH_2OH$, mw: 102.17, bp: 148.9°, flash p: 135°F (COC), d: 0.8328, vap. press: 0.9 mm @ 20°, vap. d: 3.4.

Acute tox data: Oral LD_{50} (rat) = 1850 mg/kg; oral LD_{50} (rabbit) = 1200 mg/kg; dermal LD_{50} (rabbit) = 1260 mg/kg. [*3*]

THR = MOD via oral, dermal and probably inhal routes.

Fire Hazard: Mod, when exposed to heat or flame; can react with oxidizing materials.

To Fight Fire: Dry chemical, CO_2, foam, fog.

2-ETHYL-1-BUTENE. C_6H_{12}, mw: 84.1, flash p: <−4°, autoign. temp.: 599°F, d: 0.69, vap. d: 2.9, bp: 62°.

THR = No data.

Fire Hazard: Dangerous via heat, flames, oxidizers.

To Fight Fire: Dry chemical, CO_2, foam, spray.

2,2-ETHYL BUTOXY ETHANOL. See 2-ethylbutyl "Cellosolve."

3-(2-ETHYL BUTOXY)PROPIONIC ACID. Water-white liquid, insol in water.

$CH_3CH_2CH(C_2H_5)CH_2OCH_2CH_2COOH$, mw: 174, d: 0.96 @ 20°/20°, bp: 200° @ 100 mm, vap. press: <0.1 mm @ 20°, flash p: 280°F.

Acute tox data: Oral LD_{50} (rat) = 3730 mg/kg; dermal LD_{50} (rabbit) = 530 mg/kg. [*3*]

THR = MOD via oral and HIGH via dermal routes.

Fire Hazard: Slight, when exposed to heat or flame.

To Fight Fire: Dry chemical, CO_2.

2-ETHYL BUTYL ACETATE. Colorless liquid, mild odor. $(C_2H_5)_2CHCH_2OOCCH_3$, mw: 144.21, bp: 163°, flash p: 130°F (OC), d: 0.875–0.881 @ 20°/20°, vap. d: 5.0.

THR = U. See esters.

Fire Hazard: Mod, when exposed to heat or flame; can react with oxidizing materials.

To Fight Fire: CO_2, foam, dry chemical.

2-ETHYL BUTYL ACRYLATE. Clear, colorless liquid. $H_2C = CHCOOCH_2C(C_2H_5)HC_2H_5$, mw: 156.22, bp: 82° @ 10 mm, fp: −70°, flash p: 125°F (OC), d: 0.8964 @ 20°/20°, vap. press: 1.7 mm @ 20°.

Acute tox data: Oral LD_{LO} (rat) = 4950 mg/kg; dermal LD_{LO} (rabbit) = 2980 mg/kg. [*3*]

THR = MOD via oral and dermal routes.

Fire Hazard: Mod, when exposed to heat or flame; can react with oxidizing materials.

To Fight Fire: Foam, CO_2, dry chemical.

ETHYL BUTYL ALCOHOL. See amyl carbinol.

2-ETHYL BUTYL ALCOHOL. See 2-ethyl butanol.

ETHYL-*n*-BUTYL AMINE. Water-white liquid. $C_2H_5NHCH_2 \cdot CH_2CH_2CH_3$, mw: 101.2, bp: 110°–113°, flash p: 64°F (OC), d: 0.739 @ 20°/20°, vap. d: 3.5.

Acute tox data: Oral LD_{50} (rat) = 390 mg/kg; dermal LD_{50} (rabbit) = 2000 mg/kg. [*3*]

THR = HIGH via oral; MOD via dermal routes.

Fire Hazard: Dangerous; when exposed to heat or flame; can react vigorously with oxidizing materials.

Explosion Hazard: U.

Disaster Hazard: Dangerous; keep away from heat and open flame.

To Fight Fire: Dry chemical, CO_2, foam.

ETHYL-*n*-BUTYL CARBAMATE. See butyl urethane.

ETHYL BUTYL CARBONATE. Liquid.

$(C_2H_5)(C_4H_9)CO_3$, mw: 146.48, bp: 135°, flash p: 122°F (CC), d: 0.92, vap. d: 5.03.

THR = U.

Fire Hazard: Mod, when exposed to heat or flame; can react with oxidizing materials.

Spont Heating: No.

To Fight Fire: Foam, CO_2, dry chemical.

2-ETHYL BUTYL "CELLOSOLVE." Syn: *2,2-ethyl butoxy ethanol.* Liquid. $(C_2H_5)_2CHCH_2OC_2H_4OH$. mw: 146.2, bp: 197°, fp: −90°, flash p: 180°F (OC), d: 0.8954 @ 20°/20°, vap. press: 0.17 mm @ 20°, vap. d: 5.04.

Acute tox data: Oral LD_{50} (rat) = 1910 mg/kg; dermal LD_{50} (rabbit) = 320 mg/kg. [*3*]

THR = HIGH via dermal; MOD via oral routes.

Fire Hazard: Mod, when exposed to heat or flame; can react with oxidizing materials.

To Fight Fire: Foam, CO_2, dry chemical.

ETHYL-*n*-BUTYL ETHER. Syn: *butyl ethyl ether.* Liquid. $C_2H_5OC_4H_9$, mw: 102.2, bp: 92°, mp: −124°, flash p: 40°F, d: 0.7528 @ 20°/20°, vap. d: 3.52.
Acute tox data: Oral LD_{50} (rat) = 1870 mg/kg. [*3*]
THR = MOD via oral route. See also ethers.
Fire Hazard: Dangerous, when exposed to heat or flame; can react vigorously with oxidizing materials. See also ethers.
Explosion Hazard: U. See also ethers.
Disaster Hazard: Dangerous; keep away from heat and open flame.
To Fight Fire: Alcohol foam, CO_2, dry chemical.

ETHYL BUTYL KETONE. Clear liquid. $(C_2H_5)(C_4H_9)CO$, mw: 114.2, mp: −36.7°, bp: 148°, flash p: 115°F (OC), d: 0.8198 @ 20°/20°, vap. d: 3.93.
Acute tox data: Oral LD_{50} (rat) = 2760 mg/kg; inhal LC_{LO} (rat) = 2000 ppm for 4 hrs. [*3*]
THR = MOD via oral and inhal routes.
Fire Hazard: Mod, when exposed to heat or flame; can react with oxidizing materials.
To Fight Fire: Foam, CO_2, dry chemical.

2-ETHYL-2-BUTYL-1,3-PROPANEDIOL. Crystals or liquid. $HOCCH_2C(C_2H_5)(C_4H_9)CH_2OH$, mw: 160.25, bp: 178° @ 50 mm, fp: 40.1°, flash p: 280°F (OC), d: 0.931 @ 50°/20°, vap. press: <0.01 mm @ 20°, vap. d: 5.53.
THR = U.
Fire Hazard: Slight, when exposed to heat or flame; can react with oxidizing materials.
To Fight Fire: Alcohol foam, dry chemical.

α-ETHYL-α-*s*-BUTYL STILBENE. $C_{20}H_{24}$, mw: 264.4
THR = An exper carc. [*3*]

2-ETHYL BUTYL TITANATE. Syn: *hexyltitanate.* A liquid. $Ti[OCH_2CH(C_2H_5)_2]_4$, mw: 453, mp: −55°, bp: 208° @ 10 mm, d: 0.96.
THR = See titanium compounds and organometals.
Fire Hazard: Mod, when exposed to heat or flame; can react with oxidizing materials.

2-ETHYL BUTYRALDEHYDE. Syns: *2-ethylbutyric aldehyde, diethyl acetaldehyde.* Colorless liquid. $(C_2H_5)_2CHCHO$. mw: 100.16, bp: 116.8°, flash p: 70°F (OC), fp: −89°, d: 0.8164 @ 20°/20°, vap. press: 13.7 mm @ 20°, vap. d: 3.45, lel = 1.2%, uel = 7.7%,
Acute tox data: Oral LD_{50} (rat) = 3980 mg/kg; inhal LC_{LO} (rat) = 8000 ppm for 4 hrs. [*3*]
THR = MOD via oral and inhal routes.
Fire Hazard: Dangerous, when exposed to heat or flame; can react vigorously with oxidizing materials.
Disaster Hazard: Dangerous; keep away from heat and open flame.
To Fight Fire: Alcohol foam, CO_2, dry chemical.

ETHYL-*n*-BUTYRATE. Syns: *butyric ether, ethyl butanoate.* Colorless, volatile liquid, pineapple-like odor. $C_3H_7CO_2C_2H_5$, mw: 116.16, mp: −93°, bp: 121.0°, flash p: 78°F (CC), d: 0.8788, vap. press: 10 mm @ 15.3°, vap. d: 4.0, autoign. temp.: 865°F.
Acute tox data: Oral LD_{50} (rat) = 2200 mg/kg; dermal LD_{50} (rabbit) = 520 mg/kg. [*3*]
THR = MOD via oral and dermal routes. See esters. Irr to mu mem and narcotic in high conc. Used as a synthetic flavoring substance and adjuvant. [*109*]
Fire Hazard: Dangerous, when exposed to heat or flame; can react vigorously with oxidizing materials.
Spont Heating: No.
Explosion Hazard: U.
Disaster Hazard: Dangerous; keep away from heat and open flame.
To Fight Fire: CO_2, dry chemical, alcohol foam.

2-ETHYL BUTYRIC ACID. See ethyl-*n*-butyrate.

2-ETHYL BUTYRIC ALDEHYDE. See 2-ethyl butyraldehyde.

ETHYL CACODYL. See arsenic diethyl.

ETHYL CALCIUM IODIDE. A powder, decomp by water. C_2H_5CaI, mw: 196.1.
THR = See iodides.

ETHYL CAPRATE. Syns: *capric ether, caprinic ether.* Colorless, fragrant liquid. $C_9H_{19}CO_2C_2H_5$, mw: 200.3, bp: 243°, d: 0.862, vap. d: 6.9.
THR = See esters and ethers.
Fire Hazard: Slight, can react with oxidizing materials.

ETHYL CARBAMATE. Syn: *urethane, ethyl urethane.* Colorless, odorless crystals. $CO(NH_2)OC_2H_5$, mw: 89.1, mp: 49°, bp: 184°, d: 0.9862, vap. press: 10 mm @ 77.8°, vap. d: 3.07.
Acute tox data: ip LD_{50} (rat) = 1500 mg/kg; oral LD_{50} (mouse) = 2700 mg/kg; sc LD_{50} (mouse) = 2230 mg/kg. [*3*]
THR = MOD via ip, oral and sc routes. An exper (+) carc, and teratogen [*3, 1*] and mutagen. [*4, 23*] Causes depression of bone marrow and occasionally focal degeneration in the brain. Can also produce CNS depression, nausea and vomiting. See carbamates.
Disaster Hazard: Mod; when heated, emits toxic fumes.

ETHYL CARBANILATE. See phenyl urethane.

ETHYL CARBINOL. See *n*-propyl alcohol.

ETHYL CARBONATE. See diethyl carbonate.

ETHYL CELLULOSE. White, granular solid. Formula is variable. mp: 240°+.
THR = U. Used as a food additive permitted in the feed and drinking water of animals and/or for the treatment of food-producing animals. It migrates to food from packaging materials. [*109*]

Fire Hazard: Mod, when exposed to heat or flame or by chemical reaction with oxidizing agents. Flam varies with degree of replacement of OH^- radicals of cellulose by ethyl radicals.

ETHYL CHLORIDE. Syns: *chloroethane, hydrochloric ether, muriatic ether.* Colorless liquid or gas; ether-like odor, burning taste. CH_3CH_2Cl, mw: 64.52, bp: 12.3°, lel = 3.8%, uel = 15.4%, fp: −139°, flash p: −58° F (CC), d: 0.9214 @ 0°/4°, autoign. temp.: 966°F, vap. press: 1000 mm @ 20° vap. d: 2.22

Acute tox data: Inhal TC_{LO} (human) = 13,000 ppm → CNS symptoms. [*3*]

THR = MOD via oral and inhal routes. An irr of skin, eyes and mu mem. The liquid is harmful to the eyes and can cause some irr. In the case of guinea pigs, the symptoms attending exposure are similar to those caused by methyl chloride, except that the signs of lung irr are not as pronounced. It gives some warning of its presence because it is irr, but it is possible to tolerate exposure to it until one becomes unconscious. It is the least toxic of all the chlorinated hydrocarbons. It can cause narcosis, although the effects are usually transient. Animal exper show some evidence of kidney irr and accumulation of fat due to this material in the kidneys, cardiac muscles and liver.

Fire Hazard: Highly dangerous, when exposed to heat, flame or oxidizers.

Spont Heating: No.

Explosion Hazard: Severe, when exposed to flame.

Disaster Hazard: Highly dangerous! Keep away from heat and open flame; forms phosgene on combustion; reacts with water or steam to produce toxic and corrosive fumes; can react vigorously with oxidizing materials.

To Fight Fire: Carbon dioxide.

ETHYL CHLOROACETAL. Water-white liquid, pleasant odor. $ClCH_2CH(OC_2H_5)_2$, mw: 156.6, mp: −32°, bp: 151°, flash p: 117° F, d: 1.022 @ 20°, vap. d: 5.41.

THR = U. See aldehydes.

Fire Hazard: Mod, when exposed to heat or flame.

Disaster Hazard: Dangerous; when heated to decomp, emits highly toxic fumes of chlorides; will react with water or steam to produce toxic and corrosive fumes; can react with oxidizing materials.

To Fight Fire: Foam, CO_2, dry chemical.

ETHYL CHLOROACETATE. Syn: *ethyl chloroethanoate.* Colorless liquid, fruity pungent odor. $CH_2ClCOOC_2H_5$, mw: 122.55, bp: 143.6°, fp: −26.6° flash p: 100° F, d: 1.159 @ 20°/4° vap. press: 10 mm @ 37.5° vap. d: 4.3.

THR = MOD irr via oral and inhal routes.

Fire Hazard: Dangerous, when exposed to heat or flame.

Disaster Hazard: Dangerous; when heated to decomp, emits highly toxic fumes of chlorides; will react with water or steam to produce toxic and corrosive fumes; can react with oxidizing materials.

To Fight Fire: Water, foam, CO_2, dry chemical.

ETHYL-2-CHLOROACETATE. Syn: *2-chloroethyl acetate.* Insol in water. d: 1.2, vap. d: 4.2, bp: 144°, flash p: 151° F.

THR = See also ethyl chloroacetate.

Fire Hazard: Mod, via heat, flame, powerful oxidizers.

To Fight Fire: Water, foam, fog, mist, dry chemical.

ETHYL CHLOROCARBONATE. Syns: *ethyl chloromethanoate, ethyl chloroformate.* Colorless liquid, decomp in water. $ClCOOC_2H_5$, mw: 108.53, mp: −80.6°, bp: 94°, flash p: 61° F (CC), d: 1.138 @ 20°/4°, vap. d: 3.74, autoign. temp.: 932° F.

Acute tox data: ip LD_{LO} (mouse) = 15 mg/kg. [*3*]

THR = HIGH via ip route. A powerful irr of skin, eyes and mu mem.

Fire Hazard: Dangerous; when exposed to heat, flame, oxidizers.

Explosion Hazard: U.

Disaster Hazard: Dangerous; when heated to decomp, emits highly toxic fumes of chlorides; will react with water or steam to produce toxic and corrosive fumes; can react vigorously with oxidizing materials.

To Fight Fire: CO_2, dry chemical.

ETHYL-N-β-CHLOROETHYL-N-NITROSOCARBAMATE. $C_5H_9O_3N_2Cl$, mw: 180.6.

Acute tox data: Inhal LD_{50} (mice) = 75 mg/m^3 for 10 min; inhal LC_{LO} (cat) = 330 mg/m^3 for 10 min; ip LD_{LO} (rat) = 6.5 mg/kg; oral LD_{LO} (rat) = 10 mg/kg. [*103*]

THR = HIGH via inhal, ip and oral routes. An exper carc to rats via oral route. [*103*]

ETHYL CHLOROETHANOATE. See ethyl chloroacetate.

ETHYL CHLOROFORMATE. See ethyl chlorocarbonate.

ETHYL CHLOROMETHANOATE. See ethyl chlorocarbonate.

7-ETHYL-10-CHLORO-11-METHYL BENZ(c)ACRIDINE.

THR = An exper carc. [*23*]

ETHYL CHLOROSILICONE.

THR = A (S) carc. See polymers. [*14*]

ETHYL CHLOROSTANNIC ACID. Colorless deli-

quescent crystals, decomp by water. $H_2SnC_2H_5Cl_5$, mw: 327.1.
THR = See tin compounds and chlorides.

ETHYL CHLOROSULFONATE. Colorless oily liquid, pungent odor. $C_2H_5ClSO_3$, mw: 144.57, bp: 152°–153°, d: 1.379 @ 0°, vap. d: 5.0.
THR = HIGH irr of skin, eyes, mu mem via oral and inhal routes.
Disaster Hazard: Dangerous; when heated, emits highly toxic fumes of phosgene; will react with water or steam to produce toxic and corrosive fumes.

ETHYL CHLOROTHIOFORMATE. C_3H_5OSCl, mw: 124.6.
THR = HIGH irr to skin, eyes and mu mem and via oral and inhal routes.

3-ETHYL CHOLANTHRENE. $C_{22}H_{18}$, mw: 282.4.
THR = An exper neo. [*3, 23*]

ETHYL CROTONATE. Colorless, monoclinic prisms or water white liquid, pungent odor.
$CH_3CHCHCOOC_2H_5$, mw: 114.14, mp: 45° (solid). bp: 209° (solid), 139° (liq), flash p: 36.0°F, d: 0.9207 @ 20°/20°, vap. d: 3.93.
Acute tox data: Oral LD_{50} (rat) = 3000 mg/kg. [*3*]
THR = MOD via oral, inhal routes. Also irr to skin, eyes, mu mem.
Fire Hazard: Dangerous, when exposed to heat or flame; can react vigorously with oxidizing materials.
Spont Heating: No.
Disaster Hazard: Dangerous, upon exposure to heat or flame.
To Fight Fire: Foam, CO_2, dry chemical.

ETHYL CYANIDE. Syns: *propionitrile, propanenitrile.* Colorless liquid, ethereal odor. CH_3CH_2CN, mw: 55.08, mp: −103.5°, bp: 97.1°, d: 0.783 @ 21°/4°, vap. d: 1.9, flash p: 36°F, lel = 3.1%.
Acute tox data: Oral LD_{50} (rat) = 39 mg/kg; inhal LC_{LO} (rat) = 500 ppm, for 4 hrs; ip LD_{50} (mouse) = 50 mg/kg; dermal LD_{50} (rabbit) = 210 mg/kg. [*3*]
THR = HIGH via oral, inhal and dermal routes. See also nitrites.
Fire Hazard: Dangerous, via heat, flame (sparks), oxidizers.
To Fight Fire: Water spray, foam, mist, CO_2, dry chemical.

ETHYL CYANOACETATE. Syn: *malonic ethyl ester nitrile.* Colorless to pale straw-colored liquid.
$CNCH_2CO_2C_2H_5$, mw: 113.1, mp: −22.5°, bp: 206°, flash p: 230°F, d: 1.06 @ 25°/25°, vap. press: 1 mm @ 67.8°, vap. d: 3.9.
Acute tox data: ip LD_{50} (mouse) = 750 mg/kg. [*3*]
THR = MOD via ip route. Probably MOD–HIGH via oral route. See also nitriles.
Fire Hazard: Slight, when exposed to heat or flame.
Disaster Hazard: Dangerous; when heated to decomp or on contact with acid or acid fumes, emits highly toxic fumes of cyanides; will react with water or steam to produce toxic and flam vapors; can react with oxidizing materials.
To Fight Fire: CO_2, dry chemical.

ETHYL CYCLOBUTANE. Insol in water. $C_2H_5C_4H_7$, mw: 84.1, d: 0.7284 @ 20°4°, autoign. temp.: 410°F, lel = 1.2%, uel = 7.7%, mp: −142.4°, bp: 70.7°, flash p: <4°F.
THR = Probably an asphyxiant; see argon. No data.
Fire Hazard: Dangerous, if exposed to heat, open flame or oxidizers.
Explosion Hazard: Dangerous; easily forms explosive mixtures in air.
To Fight Fire: Foam, spray, mist, dry chemical.

ETHYL CYCLOHEXANE. Colorless liquid, insol in water. $C_2H_5C_6H_{11}$, mw: 112, d: 0.787, bp: 131.8°, flash p: 95°F, autoign. temp.: 504°F, lel = 0.9%, uel = 6.6%, vap. d: 3.9.
THR = Probably an asphyxiant. See argon.
Fire Hazard: Dangerous, when exposed to heat or flame.
Explosion Hazard: Dangerous; easily forms explosive mixtures in air.
To Fight Fire: Foam, dry chemical, mist.

***n*-ETHYL(CYCLOHEXYL)AMINE.** Slightly sol in water. $C_2H_5(C_6H_{10})NH_2$, mw: 126, flash p: 86°F (OC), d: 0.8, vap. d: 4.4.
Acute tox data: Oral LD_{50} (rat) = 590 mg/kg; inhal LC_{LO} (rat) = 500 ppm for 4 hrs; dermal LD_{50} (rabbit) = 750 mg/kg. [*3*]
THR = MOD-HIGH via oral, inhal and dermal routes.
Fire Hazard: Dangerous, when exposed to heat, flame or oxidizers.
To Fight Fire: Alcohol foam, mist, spray, dry chemical.

ETHYL CYCLOPENTANE. $C_2H_5C_5H_9$, mw: 98, autoign. temp.: 504°F, lel = 1.1%, uel = 6.7%, d: 0.8, vap. d: 3.4, bp: 104°.
THR = Probably an asphyxiant; see argon.
Fire Hazard: Dangerous; easily forms explosive mixtures in air.
To Fight Fire: Foam, dry chemical, mist.

ETHYL DIAZOACETATE. See diazoacetic ester.

ETHYL DIAZO ETHANOATE. See diazoacetic ester.

1-ETHYL DIBENZ(a,h)ACRIDINE. $C_{23}H_{16}N$, mw: 306.4.
THR = An exper neo. [*3*]

1-ETHYL DIBENZ(a, j)ACRIDINE.
THR = An exper neo. [3]

ETHYL DICHLOROARSINE. Syn: *dichloroethyl arsine.* Colorless liquid, fruity, biting, irr odor. $C_2H_5AsCl_2$, mw: 174.89, mp: −65°, bp: 156° decomp, d: 1.742 @ 14°, vap. press: 2.29 mm @ 21.5°, vap. d: 6.03.
Acute tox data: Inhal LC_{LO} (human) = 14 ppm for ½ hr; inhal LC_{LO} (mouse) = 160 mg/m^3; inhal LC_{LO} (dog) = 94 ppm for ⅓ hr; iv LD_{50} (cat) = 1 mg/kg. [3]
THR = VERY HIGH via inhal and iv; probably HIGH via oral routes as well. Very irr. See arsenic compounds. Used as a military poison gas.
Disaster Hazard: Dangerous; on contact with acid or acid fumes, emits highly toxic fumes of arsenic and phosgene; will react with water or steam to produce toxic and corrosive fumes. Can react with oxidizing materials.

ETHYL-4,4′-DICHLOROBENZILATE. Syns: *4,4′-dichlorobenzilic acid ethyl ester, compound G23,922.* Viscous liquid, sometimes yellow, slightly sol in water. $(C_6H_4Cl)_2 \cdot C(OH)COOC_2H_5$, mw: 325.2, bp: 156°–158°, vap. press: 2.2×10^{-6} mm @ 20°.
THR = HIGH via oral route. An exper carc. [3]

ETHYL DICHLOROSILANE. Liquid. $C_2H_5SiHCl_2$, mw: 129.1, vap. d: 4.45, flash p: 30°F.
THR = HIGH irr to skin, eyes, mu mem and via oral and inhal routes.
Fire Hazard: Dangerous, if exposed to heat, open flames or powerful oxidizers.
Disaster Hazard: Dangerous; when heated, emits highly toxic fumes of phosgene; will react with water or steam to produce heat and toxic and corrosive fumes.
To Fight Fire: Foam, dry chemical, mist, spray.

ETHYL DIETHANOLAMINE. Clear white liquid. $C_2H_5N \cdot (C_2H_4OH)_2$, mw: 133.19, bp: 240°, flash p: 280°F (OC). d: 1.015 @ 20°, vap. d: 4.59.
THR = U. See also amines.
Fire Hazard: Slight, when exposed to heat or flame; can react with oxidizing materials.
Disaster Hazard: Mod dangerous; when heated to decomp, emits toxic fumes of NO_x.
To Fight Fire: Alcohol foam, CO_2, dry chemical.

ETHYL-N,N-DIETHYL CARBAMATE. $C_7H_{15}O_2N$, mw: 145.2.
THR = An exper neo. [3]

4-ETHYL-N,N-DIETHYL-*p*-(PHENYLAZO)ANILINE. $C_{18}H_{23}N_3$, mw: 281.4.
THR = An exper carc. [3]

ETHYL DIISOAMYL TIN BROMIDE. Solid. $(C_2H_5)(C_2H_{11})_2 \cdot SnBr$, mw: 370.0, d: 1.3650, bp: 155° @ 16 mm.
THR = See tin compounds and bromides.

ETHYL DIISOBUTYL TIN BROMIDE.
$(C_2H_5)(C_4H_9)_2SiCl$, mw: 122.7, bp: 89.2°, vap. d: 4.24.
THR = HIGH irr of skin, eyes and mu mem.
Disaster Hazard: Dangerous; when heated, emits highly toxic fumes of phosgene; will react with water or steam to produce heat and toxic and corrosive fumes.

ETHYL DIMETHYL-7-OCTADECENYL AMMONIUM BROMIDE. Syn: *onyxide.* $C_{22}H_{46}NBr$, mw: 404.6.
Acute tox data: Oral LD_{50} (rat) = 0.5 mg/kg; oral LD_{50} (guinea pig) = 158 mg/kg. [3]
THR = VERY HIGH via oral route.

4-ETHYL-N,N-DIMETHYL-4-(PHENYLAZO)-*m*-TOLUIDINE. $C_{17}H_{21}N_3$, mw: 267.4.
THR = An exper carc. [3]

***o*-ETHYL-*s,s*-DIPHENYLDITHIOPHOSPHATE.**
see hinosan.

ETHYL DIPHENYL ETHER. Light green to yellow liquid, aromatic odor. $C_6H_5OC_6H_4C_2H_5$, mw: 198.3, mp: <−20°, flash p: 295°F, d: 1.032 @ 25°/25°, vap. d: 6.84.
THR = U. See also ethers.
Fire Hazard: Low, when exposed to heat or flame; can react with oxidizing materials.
Explosion Hazard: Mod, by chemical reaction with oxidizing agents.
To Fight Fire: Foam, CO_2, dry chemical.

ETHYL DIPHENYLPHOSPHINE. $(C_6H_5)_2PC_2H_5$, mw: 214.2, bp: 293°, vap. d: 7.39.
THR = HIGH via inhal and oral routes.
Fire Hazard: U.
Explosion Hazard: U.
Disaster Hazard: Dangerous; when heated, emits highly toxic fumes of PH_3 and PO_x; can react with oxidizing materials.

ETHYL DISULFIDE. Syns: *diethyl disulfide, ethyl dithioethane.* Oily liquid, very slightly sol in water, sol in alcohol and ether. $C_2H_5SSC_2H_5$, mw: 122.2, bp: 154°, d: 0.99267 @ 20°/4°.
Acute tox data: Oral LD_{50} (rat) = 2030 mg/kg. [3]
THR = MOD via oral route.
Fire Hazard: Probably mod.
Disaster Hazard: Dangerous; when heated to decomp, emits highly toxic fumes of sulfides. Probably reacts strongly with powerful oxidizers.
To Fight Fire: Water, water spray, CO_2, foam, dry chemical.

ETHYL DODECANOATE. See ethyl laurate.

ETHYLENE. Syns: *ethene, elayl, etherin.* Colorless gas, sweet odor and taste. CH_2CH_2, mw: 28.05, bp: −103.9°, mp: −169.4°, lel = 2.7%, uel = 36%, d: 0.610 @ 0°, autoign. temp.: 914°F, vap. d: 0.98, fp: −181°.

THR = MOD via inhal route. High conc cause anesthesia. A simple asphyxiant. A common air contaminant. It is phytotoxic.

Fire Hazard: Very dangerous, when exposed to heat or flame; can react vigorously with $AlCl_3$, (CCl_4 + benzoyl peroxide), (bromtrichloro methane + $AlCl_3$), O_3. [*19*]

Explosion Hazard: Mod, when exposed to flame.

Disaster Hazard: Dangerous! Flam gas!

To Fight Fire: Stop flow of gas, CO_2, dry chemical or fine water spray.

ETHYLENE ACETATE. See ethylene glycol diacetate.

ETHYLENE ALCOHOL. See ethylene glycol.

N,N-ETHYLENE BIS(1-AZIRIDINYL)-N-METHYL PHOSPHINIC AMIDE. See aphamide.

ETHYLENE BIS(GLYCOLIC ACID). Light colored, viscous liquid. $(CH_2OCH_2COOH)_2$, mw: 178.14, mp: 30°–35°, d: 1.3 @ 75°/25°, vap. d: 6.14.

THR = U.

Fire Hazard: Slight.

ETHYLENE BROMIDE. See ethylene dibromide.

ETHYLENE BUTYRATE. See ethylene dibutyrate.

ETHYLENE CARBONATE. Syns: *glycol carbonate, dioxolone-2.* Colorless liquid or crystalline solid. $C_3H_4O_3$, mw: 88.06, bp: 244° @ 740 mm, fp: 35.7°, flash p: 290°F (OC), d: 1.322 @ 40°/20°, vap. press: 0.01 mm @ 20°, vap. d: 3.04.

THR = U. Probably LOW as ordinarily used.

Fire Hazard: Slight, when exposed to heat or flame; can react with oxidizing materials.

To Fight Fire: Alcohol foam, CO_2, dry chemical.

ETHYLENE CHLORIDE. See ethylene dichloride.

ETHYLENE CHLOROBROMIDE. Syn: *1-bromo-2-chloroethane.* Colorless, volatile liquid, sweet chloroform-like odor. $BrCH_2CH_2Cl$, mw: 143.4, bp: 106.1°, fp: −18.4°, flash p: none, d: 1.7272 @ 25°/4°, vap. press 40 mm @ 29.7°, vap. d: 4.94.

Acute tox data: Oral LD_{50} (rat) = 64 mg/kg. [*3*]

THR = HIGH via oral and probably inhal routes. An irr to skin, eyes and mu mem. May cause injury to liver and kidneys.

Disaster Hazard: Dangerous; when heated to decomp, emits highly toxic fumes of chlorides and bromides.

ETHYLENE CHLOROHYDRIN. Syns: *β-chloroethyl alcohol, glycol chlorohydrin, 2-chloroethanol.* Colorless liquid, faint ethereal odor. CH_2ClCH_2OH, mw: 80.52, mp: −69°, bp: 128.8°, flash p: 140°F (OC), d: 1.197 @ 20°/4°, autoign. temp.: 797°F, vap. press: 10 mm @ 30.3°, vap d: 2.78, lel = 4.9%, uel = 15.9%.

Acute tox data: Dermal LD_{LO} (human) = 5000 mg/kg; oral LD_{LO} (rat) = 0.58 mg/kg; sc LD_{50} (rat) = 84 mg/kg; oral LD_{50} (mouse) = 81 mg/kg; inhal LD_{50} (mouse) = 385 mg/m^3; ip LD_{50} (mouse) = 81 mg/kg; dermal LD_{50} (rabbit) = 56 mg/kg. [*3*]

THR = HIGH via dermal, oral, ip, inhal and sc routes. An exper carc. [*23*] Ethylene chlorohydrin is a narcotic poison affecting the nervous system and the liver, spleen and lungs. Exposure to the vapor may result in irr of the mu mem, followed by sleepiness, drowsiness, giddiness, nausea and vomiting. The initial symptoms may be slight. After a latent period of several hours, dyspnea, severe headache, stupor, cyanosis, and pain over the heart may develop. Autopsy shows pulmonary edema, ulceration of the mu mem of the larger bronchi, and acute liver and kidney lesions. Fatal amounts of ethylene chlorohydrin may be absorbed through the skin. Violent reaction with chlorosulfonic acid, ethylene diamine, sodium hydroxide. [*19*]

Fire Hazard: Mod, when exposed to heat, flame or oxidizers.

Spont Heating: No.

Disaster Hazard: Dangerous; when heated to decomp, emits highly toxic fumes of phosgene; will react with water or steam to produce toxic and corrosive fumes; can react with oxidizing materials.

To Fight Fire: Alcohol foam, CO_2, dry chemical.

ETHYLENE CYANIDE. See succinonitrile.

ETHYLENE CYANOHYDRIN. Syn: *β-hydroxypropionitrile, hydracrylonitrile, glycol cyanohydrin.* Colorless to straw-colored liquid. $HOCH_2CH_2CN$, mw: 71.08, bp: 228° decomp, fp: −46°, flash p: 265°F (OC), d: 1.0404 @ 25°, vap. press: 0.08 mm @ 25°, vap. d: 2.45.

Acute tox data: Oral LD_{50} (rat) = 6600 mg/kg; dermal LD_{50} (rabbit) = 5000 mg/kg. [*3*]

THR = MOD via oral and dermal routes. See also nitriles.

Fire Hazard: Low, when exposed to heat or flame.

Disaster Hazard: Dangerous; when heated or on contact with acid or acid fumes, emits highly tox fumes of cyanides; will react with water or steam to produce toxic and flam vapors; can react vigorously with chlorosulfonic acid, oleum, NaOH, H_2SO_4. [*19*]

To Fight Fire: CO_2, dry chemical, alcohol foam.

ETHYLENE DIACETATE. See ethylene glycol diacetate.

ETHYLENE DIAMINE. Syns: *1,2-ethane diamine, 1,2-diamino ethane.* Volatile, colorless, hygroscopic liquid, ammonia-like odor. $NH_2CH_2CH_2NH_2$, mw: 60.10, mp: 8.5°, bp: 117.2°, flash p: 110°F (CC), d: 0.8994 @ 20°/4°, vap. press: 10.7 mm @ 20°, vap. d: 2.07, autoign. temp.: 725°F.

Acute tox data: Oral LD_{50} (rat) = 0.76 mg/kg; dermal LD_{50} (rabbit) = 730 mg/kg; sc LD_{50} (mouse) = 424 mg/kg; oral LD_{50} (guinea pig) = 470 mg/kg. [*3*]

THR = HIGH via oral, sc; MOD via dermal routes. An irr and allergen. In addition to being an irr it can cause sensitization leading to allergic dermatitis, asthma. Used as a food additive permitted in food for human consumption. [*109*]

Fire Hazard: Mod, when exposed to heat, flame or oxidizers. Can react violently with acetic acid, acetic anhydride, acrolein, acrylic acid, acrylonitrile, allyl chloride, CS_2, chlorosulfonic acid, epichlorohydrin, ethylene chlorohydrin, HCl, mesityl oxide, HNO_3, oleum, $AgClO_4$, H_2SO_4, β-propiolactone, vinyl acetate. [*19*]

Spont Heating: No.

To Fight Fire: CO_2, dry chemical, alcohol foam.

ETHYLENE DIAMINE DIHYDROIODIDE. $(-CH_2NH_2)_2 \cdot 2HI$, mw: 316.

THR = U. A trace material added to animal feeds for iodine content. [*109*]

ETHYLENE DIAMINE DIPERCHLORATE. $N_2H_8C_2Cl_2O_8$, mw: 259.

THR = A very powerful explosive. [*19*]

ETHYLENE DIAMINE TETRAACETIC ACID. Syn: *EDTA.* Colorless crystals, slightly water-sol, insol in common organic solvents. $(HOOCCH_2)_2NCH_2CH_2N(CH_2COOH)_2$, mw: 292.3 (decomp @ 240°).

Acute tox data: Oral LD_{50} (rat) = 2000 mg/kg; ip LD_{50} (rat) = 397 mg/kg. [*3*]

THR = HIGH via ip; MOD via oral routes. An exper teratogen. [*3*] See disodium ethylene diamine tetraacetate. The calcium disodium salt of EDTA is used as a chelating agent in treating lead poisoning.

ETHYLENE DIAMINE TETRAACETONITRILE. Syn: *EDTN.* White to cream-colored powder. $(NCCH_2)_2NCH_2CH_2N(CH_2CN)_2$, mw: 216, mp: 132°, d: 0.5 @ 25°.

THR = HIGH. See nitriles.

ETHYLENE DIAMINO DIBUTYL GOLD BROMIDE. Colorless crystals, water-sol. $(CH_2NH_2)_2Au(C_4H_9)_2Br$, mw: 451.4, mp: 190° decomp.

THR = Dangerous. See bromides and gold compounds.

ETHYLENE DIAMINO DIPROPYL GOLD BROMIDE. Colorless crystals $(CH_2NH_2)_2Au(C_3H_7)_2Br$, mw: 423.4, mp: volatile @ 130°, decomp @ 190°.

THR = Dangerous. See bromides and gold compounds.

ETHYLENE DIBENZOATE. Crystals. $(C_6H_5CO_2)_2C_2H_4$, mw: 270.3, mp: 73°–74°, bp: 360° (decomp), vap. d: 9.35.

THR = Probably mod. See ethylene glycol and benzoic acid.

Fire Hazard: Slight; can react with oxidizing materials.

ETHYLENE DIBROMIDE. Syns: *1,2-dibromoethane,* glycol dibromide. Colorless, heavy liquid, sweet odor. CH_2BrCH_2Br, mw: 187.88, bp: 131.4°, fp: 9.3°, flash p: none, d: 2.172 @ 25°/25°, 2.1707 @ 25°/4°, vap. press: 17.4 mm @ 30°, vap. d: 6.48.

Acute tox data: Oral LD_{50} (rat) = 140 mg/kg; dermal LD_{50} (rabbit) = 300 mg/kg; oral LD_{50} (chicken) = 79 mg/kg; inhal LC_{LO} (rat) = 400 ppm for 2 hrs. [*3*]

THR = HIGH irr via oral, dermal, inhal routes. An exper (+) carc. [*3, 23, 81*] An insecticide. Action resembles that of ethylene dichloride. Has been implicated in cases of sterility among factory workers. [*31*]

Disaster Hazard: Dangerous; when heated to decomp, emits highly toxic fumes of bromides.

ETHYLENE DIBUTYRATE. Syn: *ethylene butyrate.* Liquid. $(CH_2O_2CCH_2CH_2CH_3)_2$, mw: 202.3, bp: 240°, d: 1.024 @ 0°/4°. vap. d: 6.96.

THR = U. See esters.

Fire Hazard: Slight; can react with oxidizing materials.

ETHYLENE DICARBOXYLIC ACID. See succinic acid.

ETHYLENE DICHLORIDE. Syns: *ethylene chloride, 1,2-dichlorethane.* Colorless liquid, pleasant odor, sweet taste. CH_2ClCH_2Cl, mw: 99.0, bp: 83.5°, ulc: 60–70, lel = 6.2%, uel = 15.9%, fp: −35.7°, flash p: 56°F, d: 1.257 @ 20°/4°, autoign. temp.: 775°F, vap. press: 100 mm @ 29.4°, vap. d: 3.35.

Acute tox data: Oral LD_{50} (rat) = 770 mg/kg. [*3*]

THR = HIGH - MOD via oral and dermal routes. Causes a pulmonary edema upon inhal. [*79*] An exper transplacental carc. [*23, 3*], mutagen [*111*] and teratogen. Ethylene dichloride has a distinctive odor and strong local irr effects, which give warning of its presence in relatively safe conc. There Ethylene dichloride has a specific effect on the cornea. Exposure to the vapor, or, in animals, injection under the skin, produces a clouding which may progress to endothelial necrosis and infiltration of the cornea by lymphocytes and connective tissue cells. The narcotic action of the compound is strong, probably of the same order as chloroform. Its toxic effects upon the liver and kidneys are less than that of carbon tetrachloride, but animal exper indicate that these organs may show congestion and fatty degeneration. Edema of the lungs has also been reported in animals. Dermatitis in man has been observed. In short exposures to high conc the picture is one of irr of the eyes, nose and throat, followed by dizziness, nausea, vomiting, increasing stupor, cyanosis, rapid pulse, and loss of consciousness. [*79*]

Chronic poisoning, where exposure has occurred over a period of several months, may cause loss of appetite, nausea and vomiting, epigastric distress, tremors, nystagmus, leucocytosis, low blood sugar levels, and possibly dermatitis if there has been skin contact. A soil fumigant. Used as a food additive permitted in food for human consumption. [*109*] It is to be treated as a human carc. [*111*]

Fire Hazard: Dangerous, if exposed to heat, flame or oxidizers. Violent reaction with Al, N_2O_4, NH_3, dimethylaminopropylamine. [*19*]

Spont Heating: No.

Explosion Hazard: Mod, in the form of vapor when exposed to flame.

Disaster Hazard: Dangerous; when heated to decomp, emits highly toxic fumes of phosgene; can react vigorously with oxidizing materials and emit vinyl chloride and hydrochloric acid.

To Fight Fire: Water, foam, CO_2, dry chemicals.

ETHYLENE DIFLUORIDE. See 1,1-difluoroethane.

ETHYLENE DIFORMATE. See ethylene glycol diformate.

ETHYLENE DIIODIDE. Syn: *1,2-diodoethane*. Yellow crystals. CH_2ICH_2I, mw: 281.9, mp: 82°, bp: decomp, d: 2.132 @ 10°/4°, vap. d: 9.72.

THR = See iodides.

ETHYLENE DILAURATE. Syn: *ethylene laurate*. Solid. $(C_{11}H_{23}CO_2CH_2)_2$, mw: 426.7, mp: 52°, bp: 188° @ 20 mm, vap. d: 14.8.

THR = U.

Disaster Hazard: Slight; when heated, emits acrid fumes. Can react with oxidizing materials.

ETHYLENE DINITRATE. Syn: *ethylene nitrate*. Yellow liquid. $(ONO_2)_2C_2H_4$, mw: 152.07, mp: −20°, bp: explodes @ 114°, d: 1.483 @ 8°, vap. d: 5.25.

Acute tox data: Oral LD_{50} (rat) = 616 mg/kg. [*3*]

THR = MOD via oral route. Probably MOD via inhal and dermal routes. Can cause lowered blood pressure, leading to headache, dizziness and weakness.

Fire Hazard: See nitrates.

Disaster Hazard: See nitrates.

Explosion Hazard: See nitrates.

(ETHYLENE DINITRILO)TETRAACETIC ACID. See ethylene diamine tetraacetic acid.

ETHYLENE DIOXYAMINE PERCHLORATE. $C_2H_5O_2N \cdot HClO_4$, mw: 175.6.

THR = Has exploded during purification. [*19*]

ETHYLENE DITHIOCYANATE. Colorless crystals. $(CH_2SCN)_2$, mw: 144.2, mp: 90° (decomp), vap. d: 4.98.

THR = No data.

Disaster Hazard: Dangerous! Will decomp, when heated, to give off highly toxic fumes.

ETHYLENE FLUORIDE. See 1,1-difluorethane.

ETHYLENE FORMATE. See ethylene glycol diformate.

ETHYLENE GLYCOL. Syns: *1,2-ethanediol, glycol, ethylene alcohol, glycol alcohol.* Colorless, sweet-tasting liquid. Hygroscopic. CH_2OHCH_2OH, mw: 62.1, bp: 197.5°, lel = 3.2%, fp: −13°, flash p: 232°F (CC), d: 1.113 @ 25°/25°, autoign. temp.: 752°F, vap. d: 2.14, vap. press: 0.05 mm @ 20°.

Acute tox data: Oral LD_{50} (rat) = 5840 mg/kg; iv LD_{50} (mouse) = 3000 mg/kg; ip LD_{LO} (rabbit) = 1000 mg/kg. [*3*]

THR = MOD irr via skin, eyes and mu mem, and via oral, iv and ip routes. (Lethal dose for man reported to be 100 ml). If ingested it causes initial CNS stimulation followed by depression. Later, it causes kidney damage which can terminate fatally. Very toxic in particulate form upon inhal.

Fire Hazard: Slight, when exposed to heat or flame; can react violently with chlorosulfonic acid, oleum, H_2SO_4. [*19*]

Spont Heating: No.

Explosion Hazard: Mod, when exposed to flame.

To Fight Fire: Alcohol foam, water, foam, CO_2, dry chemical.

ETHYLENE GLYCOL ACETATE. See ethylene glycol diacetate.

ETHYLENE GLYCOL BIS(2,3-EPOXY-2-METHYL PROPYL)ETHER. $C_{10}H_{18}O_4$, mw: 202.3.

Acute tox data: Oral LD_{50} (rat) = 7460 mg/kg; dermal LD_{50} (rabbit) = 3150 mg/kg. [*3*]

THR = MOD via dermal and LOW via oral routes.

ETHYLENE GLYCOL-*n*-BUTYL ETHER. Clear, mobile liquid, pleasant odor. $C_4H_9OCH_2CH_2OH$, mw: 118.2, bp: 168.4°–170.2°, fp: −74.8°, flash p: 160°F (COC), d: 0.9012 @ 20°/20°, vap. press: 300 mm @ 140°.

Acute tox data: Oral LD_{50} (rat) = 1480 mg/kg; inhal LC_{LO} (rat) = 500 ppm for 4 hrs; dermal LD_{50} (guinea pig) = 230 mg/kg; inhal LC_{50} (mouse) = 700 ppm. [*3*]

THR = HIGH via dermal: MOD via inhal and oral routes.

Fire Hazard: Mod, when exposed to heat or flame; can react with oxidizing materials.

To Fight Fire: Foam, CO_2, dry chemical.

ETHYLENE GLYCOL-*p-sec*-BUTYLPHENYL ETHER. Syn: *2-(p-sec-butylphenoxy)ethanol.* $C_2H_5CH(CH_3)C_6H_4OCH_2CH_2OH$, mw: 194.3, mp: <−20°, bp: 151°–161° @ 10 mm; d: 1.008 @ 25°/25°, flash p: 248°F (OC), vap. d: 6.7.

THR = See glycols and ethers.

Fire Hazard: Slight, when exposed to heat or flame; can react with oxidizing materials.

To Fight Fire: Foam, CO_2, dry chemical.

ETHYLENE GLYCOL-*p-tert*-BUTYL PHENYL ETHER. Yellow liquid. $(CH_3)_3CC_6H_4OCH_2CH_2OH$, mw: 194.3, mp: 12.2°, bp: 289°–328° @ 10 mm, flash p: 248°F (OC), d: 1.017 @ 25°/25°, vap. d: 6.7.

THR = U. See glycols and ethers.

Fire Hazard: Slight, when exposed to heat or flame; can react with oxidizing materials. See ethers.

To Fight Fire: CO_2, dry chemical.

ETHYLENE GLYCOL-*p-tert*-BUTYL PHENYL ETHER ACETATE. Syn: *2-(p-tert-butylphenoxy) ethyl acetate*. Slightly yellow liquid. $(CH_3)_3CC_6H_4CH_2OOCCH_3$, mw: 236.3, mp: <−20°, bp: 208°, flash p: 324°F (OC), d: 1.029 @ 25°/25°, vap. d: 8.15.

THR = U. See also glycols.

Fire Hazard: Slight, when exposed to heat or flame; can react with oxidizing materials.

To Fight Fire: CO_2, dry chemical, fog, mist.

ETHYLENE GLYCOL DIABIETATE. Syn: *glycol diabietate.* Liquid. $(CH_2CO_2C_{19}H_{29})_2$, mw: 630.9, flash p: 520°F vap. d: 21.8.

THR = U. See also glycols.

Fire Hazard: Slight, when exposed to heat or flame; can react with oxidizing materials.

To Fight Fire: Foam, CO_2, dry chemical.

ETHYLENE GLYCOL DIACETATE. Syns: *ethylene diacetate, glycol diacetate.* Colorless liquid or crystals. $(CH_2OOCCH_3)_2$, mw: 146.14, mp: −31°, bp: 191°, flash p: 205°F (OC), d: 1.128 @ 0°/4°, vap. press: 1 mm @ 38.3°, vap. d: 5.04.

Acute tox data: Oral LD_{50} (rat) = 6850 mg/kg; dermal LD_{50} (rabbit) = 8480 mg/kg. [*3*]

THR = MOD via oral; LOW via dermal routes.

Fire Hazard: Low, when exposed to heat or flame; can react with oxidizing materials.

To Fight Fire: Alcohol foam, CO_2, dry chemical.

ETHYLENE GLYCOL DIBENZOATE. Syn: *benzoflex,* Crystalline material. $C_{18}H_{18}O_5$, mw: 314.4 bp: 208°–211°, fp: 69°–71°, flash p: 365°F, vap. d: 9.38.

Acute tox data: Oral LD_{50} (rat) = 2830 mg/kg. [*3*]

THR = Mod via oral route. See also glycols.

Fire Hazard: Slight, when exposed to heat or flame; can react with oxidizing materials.

To Fight Fire: Water, foam, CO_2, dry chemical.

ETHYLENE GLYCOL DIFORMATE. Syns: *ethylene diformate, glycol diformate.* Liquid. $HCOOCH_2CH_2OOCH$, mw: 118.09, mp: −10°, bp: 177°, flash p: 200°F (OC), d: 1.2277 @ 20°/20°, vap d: 4.07.

Acute tox data: Oral LD_{50} (rat) = 1510 mg/kg; oral LD_{50} (guinea pig) = 390 mg/kg. [*3*]

THR = HIGH–MOD via oral route.

Fire Hazard: Mod, when exposed to heat or flame; can react with oxidizing materials.

To Fight Fire: CO_2, dry chemical.

ETHYLENE GLYCOL DIMETHYL ETHER. Syn: *1,2-dimethoxy ethane.* Water white liquid, ethereal odor. $CH_3OCH_2CH_2OCH_3$, mw: 90.1, bp: 83°, flash p: 104°F, fp: −60°, d: 0.8672 @ 20°/20°, vap. press: 48 mm @ 20°, vap. d: 3.11. [*2*]

THR = U. See also glycols.

Fire Hazard: Mod, when exposed to heat or flame; can react vigorously with oxidizing materials and $LiAlH_2$.

Explosion Hazard: See ethers.

Disaster Hazard: Dangerous; upon exposure to heat or flame.

To Fight Fire: Alcohol foam, CO_2, dry chemical.

ETHYLENE GLYCOL DINITRATE. See ethylene dinitrate.

ETHYLENE GLYCOL ETHYL ETHER. See "cellosolve" solvent.

ETHYLENE GLYCOL METHYL BUTYL ETHER. Liquid. $C_4H_9OCH_2 \cdot CH_2OCH_3$, mw: 132.2, bp: 147°, d: 0.8487, vap. d: 4.57

THR = See also glycols and ethers.

ETHYLENE GLYCOL METHYL ETHER. See ethylene glycol monomethyl ether.

ETHYLENE GLYCOL METHYL-*n*-HEXYL ETHER. Liquid. $C_9H_{20}O_2$, mw: 160.25, bp: 190°, d: 0.855, vap. d: 5.53.

THR = U. See also glycols and ethers.

Fire Hazard: Mod; can react with oxidizing materials. See ethers.

To Fight Fire: Foam, CO_2, dry chemical.

ETHYLENE GLYCOL METHYL PHENYL ETHER. Liquid. $C_9H_{12}O_2$, mw: 152.19, bp: 215°, d: 1.03.

THR = U. See also glycols and ethers.

ETHYLENE GLYCOL MONOACETATE. Syn: *glycol monoacetate.* Colorless, almost odorless liquid.

$CH_3COOCH_2CH_2OH$, mw: 104.10, bp: 182°, flash p: 215°F (OC), d: 1.108, @ 15°, vap. d: 3.59.

Acute tox data: ip LD_{50} (mouse) = 1310 mg/kg; oral LD_{50} (guinea pig) = 3800 mg/kg. [*3*]

THR = MOD via ip and oral routes. MILD irr.

Fire Hazard: Low, when exposed to heat or flame; can react with oxidizing materials.

Spont Heating: No.

To Fight Fire: Alcohol foam, foam, water, CO_2, dry chemical.

ETHYLENE GLYCOL MONOBENZYL ETHER. See benzyl "Cellosolve."

ETHYLENE GLYCOL MONOBUTYL ETHER. See butyl "Cellosolve."

ETHYLENE GLYCOL MONOETHYL ETHER. See "Cellosolve" solvent.

ETHYLENE GLYCOL MONOETHYL ETHER ACETATE. See "Cellosolve" acetate.

ETHYLENE GLYCOL MONOISOPROPYL ETHER. Colorless liquid, mild, agreeable odor. $(CH_3)_2CHOCH_2CH_2OH$, mw: 104.14, bp: 139°, d: 0.906, vap. d: 3.6.

Acute tox data: Oral LD_{50} (rat) = 5660 mg/kg; inhal LC_{50} (mouse) = 1930 ppm; dermal LD_{50} (rabbit) = 1600 mg/kg. [*3*]

THR = MOD via oral, inhal and dermal routes.

Fire Hazard: Mod; can react with oxidizing materials. See ethers.

To Fight Fire: Spray, foam, dry chemical.

ETHYLENE GLYCOL MONOMETHYL ETHER. Syns: *2-methoxyethanol, methyl "Cellosolve."* Colorless liquid, mild, agreeable odor. $CH_3OCH_2CH_2OH$, mw: 76.09, bp: 124.5°, fp: −86.5°, flash p: 115°F (OC), lel = 2.5%, uel = 14%, d: 0.9660 @ 20°/4°, autoign. temp.: 545°F, vap. press: 6.2 mm @ 20°,vap. d: 2.62

Acute tox data = Oral LD_{50} (rat) = 2460 mg/kg; iv LD_{50} (rat) = 2140 mg/kg; inhal LC_{50} (mouse) = 1480 ppm; Oral LD_{LO} (human) = 3570 mg/kg; inhal TC_{LO} (human) = 25 ppm → CNS effects. [*3*]

THR = MOD via oral, iv and inhal routes. When used under conditions which do not require the application of heat, this material probably presents little hazard to health. However, in the manufacture of fused collars, which require pressing with a hot iron, cases have been reported showing disturbance of the hemopoietic system with or without neurological signs and symptoms. The blood picture may resemble that produced by exposure to benzene. Two cases reported had severe aplastic anemia with tremors and marked mental dullness. One case had multiple neuritis and four others had normal reflexes. The commonest change in the blood picture was the finding of immature neutrophils (shift to the left); in other cases there were a reduction in number of the blood platelets and a macrocytic anemia. The persons affected had been exposed to vapors of methyl "Cellosolve" (76 ppm), ethyl and methyl alcohol, ethyl acetate and petroleum naphtha.

The first signs of poisoning are probably abnormalities found in the blood picture, as mentioned above. Reflexes may be exaggerated or abnormal in character, and may be accompanied by complaints of drowsiness and fatigue. Tremors may be present. Severe damage probably takes the form of aplastic anemia.

Fire Hazard: Mod, when exposed to heat or flame; can react with oxidizing materials to form explosive peroxides.

Spont Heating: No.

To Fight Fire: Alcohol foam, CO_2, dry chemical.

ETHYLENE GLYCOL MONOMETHYL ETHER ACETAL. Liquid. $CH_3CH(OCH_2CH_2OCH_2)_2$, mw: 178.2, mp: −85°, (glass), bp: 207.2°, flash p: 205°F, d: 0.9773 @ 20°/20°, vap. press: 0.1 mm @ 20°, vap. d: 6.16.

THR = U. See also glycols.

Fire Hazard: Mod, when exposed to heat or flame; can react with oxidizing materials.

To Fight Fire: Foam, CO_2, dry chemical.

ETHYLENE GLYCOL MONOMETHYL ETHER ACETATE. See methyl "Cellosolve" acetate.

ETHYLENE GLYCOL MONO-2-METHYL PENTYL ETHER. $HOCH_2CH_2OCH_3C_5H_{10}$, mw: 146.

Acute tox data: Oral LD_{50} (rat) = 3730 mg/kg; dermal LD_{LO} (rabbit) = 440 mg/kg. [*3*]

THR = HIGH via dermal; MOD via oral routes.

ETHYLENE GLYCOL MONOOCTYL ETHER. Colorless, odorless liquid. $C_4H_9CHC_2H_5CH_2OCH_2CH_2OH$, mw: 174.28, bp: 228.3°, flash p: 230°F, d: 0.8859 @ 20°, vap. press: 0.02 mm @ 20°.

THR = U. See also glycols and ethers.

Fire Hazard: Slight, when exposed to heat or flame; can react with oxidizing materials.

Explosion Hazard: See ethers.

To Fight Fire: Foam, CO_2, dry chemical.

ETHYLENE GLYCOL MONO-2,6,8-TRIMETHYL-4-NONYL ETHER. $C_{14}H_{30}O_2$, mw: 230.5.

Acute tox data: Dermal LD_{50} (rabbit) = 3150 mg/kg. [*3*]

THR = MOD via dermal route.

ETHYLENE GLYCOL PHENYL ETHER ACETATE. Syn: *2-phenoxyethyl acetate*. Colorless, mobile liquid. $C_6H_5OCH_2CH_2OOCCH_3$, mw: 180.2, mp: $<-20°$, bp: 231°, d: 1.103 @ 25°/25°, flash p: 275°F, vap. d: 6.2.

Acute tox data: Oral LD_{50} (rat) = 4900 mg/kg; oral LD_{50} (mouse) = 3700 mg/kg. [*3*]

THR = MOD via oral, probably via inhal routes as well.

Fire Hazard: Slight, when exposed to heat or flame; can react with oxidizing materials.

To Fight Fire: Water, foam, CO_2, dry chemical.

ETHYLENE IMINE. Syns: *ethylenimine, dimethylenimine.* Oil, pungent ammoniacal odor, water-white liquid. $NHCH_2CH_2$, mw: 43.07, bp: 55°–56°, fp: −71.5°, flash p: 12°F, d: 0.832 @ 20°/4°, autoign. temp.: 608°F, vap. press: 160 mm @ 20°, vap. d: 1.48, lel = 3.6%, uel = 46%.

Acute tox data: Oral LD_{50} (rat) = 15 mg/kg; dermal LD_{50} (guinea pig) = 14 mg/kg; inhal LC_{LO} (rat) = 25 ppm for 8 hrs. ip LD_{50} (rat) = 3.8 mg/kg. [*3*]

THR = HIGH irr of skin, eyes, mu mem and via oral, dermal, ip and inhal routes. An exper (+) carc and neo. [*3, 8, 23*] An allergic sensitizer of skin. Causes opaque cornea, keratoconus and necrosis of cornea (exper). Has been known to cause severe human eye injury. Drinking of carbonated beverages is recommended as antidote to this material in stomach.

Fire Hazard: Dangerous, when exposed to heat, flame or oxidizers. Reacts violently with acetic acid, acetic anhydride, acrolein, acrylic acid, allyl chloride, CS_2, Cl_2, chlorosulfonic acid, epichlorohydrin, glyoxal, HCl, HF, HNO_3, oleum, β-propiolactone, Ag, NaOCl, H_2SO_4, vinyl acetate. [*19*]

Disaster Hazard: Dangerous. Heat and/or the presence of catalytically active metals or chloride ions can cause a violent exothermic reaction. This material should be handled as per instructions of the manufacturer.

To Fight Fire: Alcohol foam, CO_2, dry chemical.

ETHYLENE IMINOSULFONYL HEPTANE.

THR = An exper carc. [*23*]

ETHYLENE IMINOSULFONYL PENTANE.

THR = An exper carc. [*23*]

ETHYLENE IMINOSULFONYL PROPANE.

THR = An exper carc. [*23*]

ETHYLENE LAURATE. See ethylene dilaurate.

ETHYLENE METHANE SULFONATE.

THR = An exper teratogen. [*3*]

ETHYLENE MONOCHLORIDE. See vinyl chloride.

ETHYLENE NITRATE. See ethylene dinitrate.

ETHYLENE NITRITE. See ethylene dinitrite.

ETHYLENE OXIDE. Syns: *1,2-epoxyethane, oxirane.* Colorless gas at room temperature. $(CH_2)_2O$, mw: 44.05, mp: −111.3°, bp: 10.7°, ulc: 100, lel = 3.0%, uel = 100%, flash p: $<$0°F, d: 0.8711 @ 20°/20°, autoign. temp.: 804°F, vap. press: 1095 mm @ 20°, vap. d: 1.52.

Acute tox data: Oral LD_{50} (rat) = 330 mg/kg; inhal LC_{50} (rat) = 1462 ppm for 4 hrs; inhal LC_{50} (mouse) = 836 ppm for 4 hrs. [*3*]

THR = HIGH via oral and MOD via inhal routes. An exper (+) carc and mutagen. [*3, 13, 23*] Irr to eyes and mu mem of respiratory tract. High conc can cause pulmonary edema. See aliphatic and aromatic epoxies.

Fire Hazard: Very dangerous, when exposed to heat or flame; can react with acids and bases, alcohols, aluminum chloride, aluminum oxide, ammonia, copper, iron chlorides, iron oxides, magnesium perchlorate, mercaptans, potassium, tin chlorides. [*19*]

Spont Heating: No.

Explosion Hazard: Severe, when exposed to flame.

Disaster Hazard: Highly dangerous, upon exposure to heat or flame.

To Fight Fire: Alcohol foam, CO_2, dry chemical.

ETHYLENE OXIDE POLYMER. The polymer of ethylene oxide.

THR = No data. A food additive permitted in foods for human consumption. [*109*]

ETHYLENE SUCCINIC ACID. See succinic acid.

ETHYLENE SULFIDE. Syn: *thiirane.* Colorless liquid. $(CH_2)_2S$, mw: 60.11, bp: 55°–56° decomp, d: 1.0368 @ 0°/4°, vap. d: 2.07.

Acute tox data: Oral LD_{50} (rat) = 178 mg/kg; inhal LC_{50} (rat) = 690 ppm; ip LD_{50} (rat) = 42 mg/kg; sc LD_{50} (rat) = 90 mg/kg. [*3*]

THR = HIGH irr of skin, eyes and mu mem, and via oral, inhal, ip and sc routes. An exper carc and neo. [*3, 13, 23*]

Fire Hazard: U.

Disaster Hazard: Dangerous; when heated to decomp, or on contact with acid or acid fumes; emits highly toxic fumes of SO_x; can react with oxidizing materials.

ETHYLENE TETRACHLORIDE. See perchloroethylene.

ETHYLENE THIOUREA. White crystals, water-sol = 9 g/100 ml @ 30°. Often occurs as a main degradation product of the metal salts of ethylene bis-dithiocarbamic acid. $C_3H_6N_2S$, mw: 102.2.

THR = MOD systemic toxicity. An exper (+) carc and teratogen. [*1, 2, 3*]

ETHYLENE TRICHLORIDE. See trichloroethylene.

ETHYLENE TRIFLUORIDE. CH_3CF_3, mw: 84.
THR = Animal exper show LOW irr and narcotic effects.

ETHYLENE TRITHIOCARBONATE. Bright yellow to brown crystalline solid. $C_3H_4S_3$, mw: 136.3, mp: 34.5° bp: 117.5° @ 0.6 mm, flash p: 330°F (TOC), d: 1.48, vap. d: 4.70.
THR = U. Avoid excessive skin contact, inhal or ingestion of material.
Fire Hazard: Slight, when exposed to heat or flame.
Disaster Hazard: Dangerous; when heated to decomp, emits highly toxic fumes of oxides of sulfur; can react with oxidizing materials.
To Fight Fire: CO_2, water, dry chemical.

ETHYLENIMINE. See ethylene imine.

ETHYL-2,3-EPOXYBUTYRATE. $C_5H_{10}O_3$, mw: 118.2.
Acute tox data: Oral LD_{50} (rat) = 500 mg/kg. [*3*]
THR = HIGH via oral route. An exper neo. [*3*]

ETHYL ESTER. See ethyl acetate.

ETHYL ETHANEDIOATE. See ethyl oxalate.

ETHYL ETHANEOATE. See ethyl acetate.

ETHYL ETHANOLAMINE-1. Liquid. $C_2H_5NHC_2H_4OH$, mw: 89.14, mp: −7.8°, bp: 167°, flash p: 160°F (OC), d: 0.9182 @ 20°/20°, vap. press: 0.4 mm @ 20°, vap. d: 3.00.
THR = U. See also amines.
Fire Hazard: Mod, when exposed to heat or flame; can react with oxidizing materials.
To Fight Fire: Foam, CO_2, dry chemical.

ETHYL ETHER. Syns: *sulfuric ether*, *anesthesia ether*, *ethyl oxide*. A clear, volatile liquid, sweet, pungent odor. $C_2H_5OC_2H_5$, mw: 74.12, mp: −116.2°, bp: 34.6°, ulc: 100, lel = 1.85%, uel = 36%, flash p: −49°F, d: 0.7135 @ 20°/4°, autoign. temp.: 320°F, vap. press: 442 mm @ 20°, vap. d: 2.56.
Acute tox data: Oral LD_{LO} (human) = 420 mg/kg; oral LD_{50} (rat) = 1700 mg/kg; inhal LC_{50} (mouse) = 110,000 ppm for 10 min. [*3*]
THR = An irr. MOD via oral and LOW via inhal routes. Ether is not corrosive or dangerously reactive. However, it must not be considered safe for individuals to inhale or ingest. It is not toxic in the sense of being a poison. It is, however, a depressant of the CNS and is capable of producing intoxication, drowsiness, stupor, and unconsciousness. Death due to respiratory failure may result from severe and continued exposure.
Fire Hazard: Very dangerous when exposed to heat or flame; can react vigorously with acetyl peroxide, air, liquid air, bromoazide, Cl_2, ClF_3, CrO_3, $Cr(OCl)_2$, $LiAlH_2$, $NOClO_4$, O_2, $NClO_2$, O_3, $HClO_4$, (H_2SO_4 + permanganates), K_2O_2, Na_2O_2, [$(C_2H_5)_3Al$ + air], [$(CH_3)_3Al$ + air]. [*19*]
Explosion Hazard: Severe, when exposed to heat or flame. See ethers.
Disaster Hazard: Highly dangerous, in the presence of heat or flame. See ethers.
To Fight Fire: Alcohol foam, CO_2, dry chemical.
Treatment and Antidotes: Removal from exposure almost always produces rapid and complete recovery.

ETHYL ETHER OF PROPYLENE GLYCOL-α. Liquid. $C_5H_{12}O_2$, mw: 104.42, vap. d: 3.59
Acute tox data: Oral LD_{50} (rat) = 7110 mg/kg. [*3*]
THR = LOW via oral route.
Fire Hazard: Slight, can react with oxidizing materials.

ETHYL ETHER OF PROPYLENE GLYCOL-β.
Acute tox data: Dermal LD_{50} (rabbit) = 2830 mg/kg. [*3*]
THR = MOD via dermal route.

ETHYL-3-ETHOXYPROPIONATE. Liquid. $C_7H_{14}O_3$, mw: 146.18, mp: −100°, bp: 170.1°, flash p: 180°F (OC), d: 0.9496 @ 20°/20°, vap. d: 5.03.
Acute tox data: Oral LD_{50} (rat) = 5000 mg/kg. [*3*]
Fire Hazard: Mod, when exposed to heat or flame; can react with oxidizing materials.
To Fight Fire: Foam, CO_2, dry chemical.

ETHYL-N-ETHYL CARBAMATE. $C_5H_{11}O_2N$, mw: 117.2.
THR = An exper neo. [*3*]

ETHYLETHYLENE ACETATE. $CH_2(CO_2CH_3)CH_2CHCH_2$, mw: 114.1, vap. d: 3.93.
THR = A very MILD irr.
Disaster Hazard: Slight.

ETHYL FLUORIDE. Syn: *fluoroethane*. Colorless gas. CH_3CH_2F, mw: 48.06, mp: −143.2°, bp: −37.7°, d: 0.8158 @ −37.7°, vap. d: 1.66.
THR = LOW via inhal route.
Fire Hazard: Very dangerous, via heat, flames, oxidizers.
To Fight Fire: Stop flow of gas.
Disaster Hazard: Dangerous; when heated to decomp, emits toxic fumes of fluorides.

ETHYL FLUOROFORMATE. Liquid. $FCOOC_2H_5$, mw: 92.07, bp: 57°, d: 1.11 @ 33°, vap. d: 3.18.
THR = HIGH irr to skin, eyes and mu mem.
Disaster Hazard: Dangerous; when heated to decomp, or on contact with acid fumes, emits toxic fumes of fluorides; will react with water or steam to produce toxic and corrosive fumes.

ETHYL-*p*-FLUOROPHENYL SULFONE. Syn: *fluoresone*. Crystals. $C_8H_9FO_2S$. mw: 188.2, mp: 41°.
Acute tox data: Oral LD_{50} (mouse) = 542 mg/kg; ip

LD_{50} (mouse) = 343 mg/kg; iv LD_{50} (mouse) = 320 mg/kg. [*3*]
THR = HIGH via oral, ip and iv routes.

ETHYL FLUOROSULFONATE. Liquid, ethereal odor. $C_2H_5SO_3F$, mw: 128.1, vap. d: 4.42.
THR = HIGH irr to skin, eyes and mu mem. Narcotic in high conc. A general purpose food additive. [*109*]
Fire Hazard: Dangerous, when exposed to heat or flame; can react vigorously with oxidizing materials.
Spont Heating: No.
Explosion Hazard: Severe, when exposed to flame.
Disaster Hazard: Highly dangerous, upon exposure to heat or flame.
To Fight Fire: Alcohol foam, foam, CO_2, dry chemical.

ETHYL FORMATE. Syns: *formic ether, ethyl methanoate.* Water white liquid, pleasant aromatic odor. $HCOOC_2H_5$, mw: 74.08, mp: −79°, bp: 54.3°, lel = 2.7%, uel = 16%, flash p: −4°F (CC), d: 0.9236 @ 20°/20°, autoign. temp.: 851°F, vap. press: 100 mm @ 5.4°, vap. d: 2.55.
THR = Probably LOW. An exper neo to mice via dermal route. [*103*] See also esters and formic acid.
Fire Hazard: Dangerous, when exposed to heat, flame or oxidizers.
To Fight Fire: Alcohol foam, spray, mist, dry chemical.

ETHYL FORMATE, ANHYDROUS. See ethyl formate.

ETHYL FORMYL PROPIONATE. Liquid. Somewhat sol in water. $C_2H_5OOCC_2H_4C(O)H$, mw: 130, d: 1.0625 @ 20°/20°, flash p: 200°F.
Acute tox data: Oral LD_{50} (rat) = 1850 mg/kg; inhal LC_{LO} (rat) = 8000 ppm for 4 hrs. [*3*]
THR = MOD via oral and inhal routes.
Disaster Hazard: Dangerous; when heated to decomp or on contact with acid or acid fumes, emits highly toxic fumes of fluorides and SO_x; will react with water or steam to produce toxic and corrosive fumes.

5-ETHYL-2-(5H)-FURANONE. $C_6H_8O_2$, mw: 112.1.
THR = An exper carc. [*3*]

ETHYL GERMANIUM OXIDE. Colorless crystals, water-sol. $(C_2H_5GeO)_2O$, mw: 251.3, mp: >300°.
THR = U. See germanium compounds.

ETHYL GERMANIUM TRIBROMIDE. Colorless liquid, decomp by water. $(C_2H_5)GeBr_3$, mw: 341.4, mp: <−33°, bp: 200° @ 763 mm.
THR = Dangerous. See bromides and germanium compounds.

ETHYL GERMANIUM TRICHLORIDE. Colorless liquid, decomp by water. $(C_2H_5)GeCl_3$, mw: 208.0°, mp: <33°, bp: 144° @ 762 mm.
THR = Dangerous. See chlorides and germanium compounds.

ETHYL GERMANIUM TRIFLUORIDE. Colorless liquid, decomp by water. $(C_2H_5)GeF_3$, mw: 158.7, mp: −16°, bp: 112° @ 750 mm.
THR = HIGH. See fluorides and germanium compounds.

ETHYL GERMANIUM TRIIODIDE. Yellow liquid, decomp by water. $(C_2H_5)GeI_3$, mw: 482.2 mp: −2°, bp: 281° @ 755 mm
THR = Dangerous. See iodides and germanium compounds.

ETHYL GLYCOL. See "Cellosolve."

ETHYL GLYCOL ACETATE. See "Cellosolve" acetate.

ETHYL GUTHION. Syns: *ethyl homolog of guthion, azinphosethyl.* $C_{12}H_{16}O_3N_3PS_2$, mw: 345.4.
Acute tox data: Oral LD_{50} (rat) = 9 mg/kg; dermal LD_{50} (rat) = 250 mg/kg; ip LD_{50} (rat) = 7.5 mg/kg; oral LD_{50} (chicken) = 34 mg/kg. [*3*]
THR = VERY HIGH via oral and ip; HIGH via dermal routes. A cholinesterase inhibitor. See parathion.

ETHYL HEPTANOIC ACID. Syns: *pelagonic acid, nonanoic acid.* Oily colorless liquid, very slightly sol in water. $CH_3(CH_2)_7COOH$, mw: 158.2, bp: 254°, mp: 12°, d: 0.9055 @ 20°/4°.
THR = No data: Probably LOW.

2-ETHYL HEXALDEHYDE. Syns: *butyl ethyl acetaldehyde, 2-ethylhexanal.* $C_4H_9CH(C_2H_5)CHO$, mw: 128.21, bp: 163.4°, flash p: 125°F (OC), autoign. temp.: 387°F, d: 0.8205, vap. press: 1.8 mm @ 20°, vap. d: 4.42.
Acute tox data: Oral LD_{50} (rat) = 3200 mg/kg; inhal LC_{LO} (rat) = 4000 ppm for 4 hrs. [*3*]
THR = MOD via oral and inhal routes.
Fire Hazard: Dangerous; spont flam in air. [*19*] Can react with oxidizing materials.
To Fight Fire: Foam, CO_2, dry chemical, water spray, mist, fog.

2-ETHYLHEXANAL. See 2-ethyl hexaldehyde.

2,2-(2-ETHYL HEXANAMIDO)DIETHYL DI(2-ETHYLHEXANOATE). Light brown liquid. $(C_7H_{15}OCOC_2H_4)_2NCOC_7H_{15}$, mw: 483.71, mp: −33°, bp: 255° @ 5 mm, flash p: 420°F (OC), d: 0.9564 @ 20°/20°, vap. press: 0.31 mm @ 200°, vap. d: 16.7.
THR = No data.
Fire Hazard: Slight; can react with oxidizing materials.
To Fight Fire: Foam, CO_2, dry chemical.

2-ETHYLHEXANEDIOL-1,3. Syn: *octylene glycol.* Practically colorless, somewhat viscous, odorless

liquid. $(C_2H_5)C_6H_{12}O(OH)$, mw: 146.22, bp: 243.1°, flash p: 260°F (OC), fp: −40°, d: 0.9422 @ 20°/20°, vap. press: <0.01 mm @ 20°, vap. d: 5.03.
Acute tox data: Oral LD_{50} (rat) = 2400 mg/kg; oral

Sax: DANGEROUS PROPERTIES OF INDUSTRIAL MATERIALS
5th edition

Errata sheet for page 665

The entry below should be inserted following 2-ETHYL HEXAL ACRYLATE:

2-ETHYL HEXYL ALCOHOL. Syn: 2-ethyl hexanol. Clear liquid. $C_4H_9CH(C_2H_5)CH_2OH$, mw: 130.26, bp: 179°-185°, mp: <-76°, flash p: 178°F, d: 0.834 @ 20°/20°, vap. press: 0.2 mm @ 20°, vap. d: 4.49
Acute tox data: Oral LD_{50}(rat) = 3200 mg/kg; dermal LD_{50}(rabbit) = 2380 mg/kg; oral LD_{LO} (mouse) = 3200 mg/kg (3)
THR = MOD via ingestion and skin absorption
Fire Hazard: Mod, when exposed to heat or flame; reacts with oxidizers.
To Fight Fire: Foam, CO_2, dry chemical

2-ETHYL HEXYL ACRYLATE. Syn: *octyl acrylate.* Liquid. $CH_2CHCO_2CH_2CH(C_2H_5)C_4H_9$, mw: 184.3, bp: 130° @ 50 mm, fp: −90°, flash p: 180°F (OC), d: 0.8869 @ 20°/20°, vap. press: 1 mm @ 50.0°, vap. d:

(rat) = 6500 mg/kg; dermal
mg/kg; ip LD_{50} (mouse) =

l ip; VERY LOW via dermal

xposed to heat or flame; can
terials.
am, CO_2, dry chemical.
. A clear, miscible liquid.
mw: 129.24, bp: 169.2°, flash
@ 20°/20°, vap. press: 1.2

(rat) = 450 mg/kg; dermal
/kg; inhal LC_{LO} (rat) = 250

nd inhal; MOD via dermal

xposed to heat or flame; can
terials.
am, CO_2, dry chemical.
OPROPYL ETHER. Liq-
$C_3H_6NH_2$, mw: 187.32, bp:
0°F (OC), d: 0.8483, vap. d:

.
xposed to heat or flame; can
terials.
2, dry chemical.
LINE. Liquid, mild odor.
H_9), mw: 205.3, bp: 194°
p: 325°F (COC), d: 0.9119
0.01 @ 20°, vap. d: 6.9.
(rat) = 2410 mg/kg. [3]
te.
exposed to heat or flame.
us; when heated to decomp,
; it will react with water or
c fumes; can react with

al, CO_2, mist or fog.
SOLVE". Water-white liq-
CH_2CH_2OH, mw: 175.28,
d: 0.8859 vap. press: 0.02

te.
exposed to heat or flame;
materials.
, dry chemical.

ection.

2-ETHYL HEXYL-1-CHLORIDE. Colorless liquid. $C_4H_9CH \cdot (C_2H_5)CH_2Cl$, mw: 148.67, bp: 172.9°, fp: −135°, flash p: 140°F (OC), d: 0.8833, @ 20°, vap. d: 5.14.
Acute tox data: Inhal LC_{LO} (rat) = 4000 ppm for 4 hrs; Oral LD_{LO} (rat) = 4640 mg/kg. [3]
THR = MOD via inhal and oral routes.
Fire Hazard: Mod, when exposed to heat, flame or oxidizers.
Disaster Hazard: Dangerous; when heated to decomp, emits highly toxic fumes of phosgene; can react with oxidizing materials.
To Fight Fire: Foam, CO_2, dry chemical.

***n*-(2-ETHYL HEXYL)CYCLOHEXYLAMINE.** Insol in water. $C_{14}H_{29}N$, mw: 211.3, d: 0.8, bp: 172° @ 20 mm, flash p: 265°F (OC).
Acute tox data: Dermal LD_{50} (rabbit) = 110 mg/kg; Oral LD_{LO} (rat) = 710 mg/kg. [3]
THR = HIGH via dermal and MOD via oral routes.
Fire Hazard: Slight, when exposed to heat or flame.
To Fight Fire: Dry chemical.

ETHYL HEXYL DIPHENYLPHOSPHATE. $C_{20}H_{27}O_4P$, mw: 362.5.
Acute tox data: iv LD_{LO} (rabbit) = 272 mg/kg. [3]
THR = HIGH via iv route. An exper carc. [23]

2-ETHYL HEXYL-9,10-EPOXYSTEARATE. $C_{26}H_{50}O_3$, mw: 410.8.
Acute tox data: Oral LD_{50} (rat) = 31,000 mg/kg. [3]
THR = None.

2-ETHYL HEXYL ETHER. Liquid. $[C_4H_9CH(C_2H_5)CH_2]_2CO$, mw: 242.43, mp: −95°, bp: 269.4°, flash p: 235°F (OC), d: 0.8121 @ 20°/20°, vap. d: 8.36.
THR = U. See also ethers.
Fire Hazard: Slight, when exposed to heat or flame; can react with oxidizing materials.
Explosion Hazard: Slight, when exposed to heat, spark or flame.
To Fight Fire: CO_2, dry chemical.

2-ETHYL HEXYL-2-ETHYL HEXANOATE. Liquid. $C_4H_9CH(C_2H_5) \cdot CH_2CO_2(C_2H_5)CHC_4H_9$, mw: 256.4, vap. d: 8.85.
THR = A MILD irr.
Fire Hazard: Slight; can react with oxidizing materials.

2-ETHYL HEXYL MERCAPTAN. Colorless liquid. $CH_3CH(C_2H_5) \cdot (CH_2)_4SH$, mw: 146.3, bp: 90° @ 35 mm, d: 0.8543 @ 20°/20°, vap. d: 5.04.
THR = U. See also mercaptans.
Disaster Hazard: Dangerous; when heated to decomp, emits highly toxic fumes of SO_x; can react with oxidizing materials.

2-ETHYL HEXYL PHTHALATE. See di-(2-ethyl hexyl)phthalate.

2-ETHYL HEXYL TITANATE. Syn: *octyl titanate*. Liquid. $Ti[OCH_2CH(C_2H_5)(C_4H_9)]_4$, mw: 565, mp: −55°, bp: 254°, d: 0.93.
THR = See titanium compounds.
Fire Hazard: Mod, when exposed to heat or flame; can react with oxidizing materials.
To Fight Fire: Dry chemical.

ETHYL HYDRAZINE HYDROCHLORIDE. $C_2H_8N_2 \cdot HCl$, mw: 96.6.
THR = An exper carc. [3]

ETHYL HYDRIDE. See ethane.

ETHYL HYDROGEN SULFATE. See ethyl sulfuric acid.

ETHYL HYDROSULFIDE. See ethyl mercaptan.

ETHYL-*p*-HYDROXYBENZOATE SODIUM. $C_9H_9O_3Na$, mw: 188.2.
Acute tox data: Oral LD_{LO} (mouse) = 2500 mg/kg; ip LD_{50} (mouse) = 520 mg/kg. [3]
THR = MOD via oral and ip routes.

N-ETHYL-N-(4-HYDROXYBUTYL)NITROSO AMINE. $C_6H_{14}O_2N_2$, mw: 146.2.
THR = An exper carc. [3]

ETHYL-N-HYDROXY CARBAMATE. $C_3H_7O_3N$, mw: 105.1.
Acute tox data: ip LD_{50} (rat) = 800 mg/kg. [3]
THR = MOD via ip route. An exper teratogen, carc and neo. [3]

ETHYL-2-HYDROXY PROPANOATE. See ethyl lactate.

ETHYL HYPOCHLORITE. C_2H_5OCl, mw: 80.5.
THR = Very unstable. Explodes when exposed to light. [19]

ETHYLIDENE ACETONE. See 1-pentene-2-one.

ETHYLIDENE CHLORIDE. Syns: *ethylidene dichloride, chlorinated hydrochloric ether, 1,1-dichlorethane*. Colorless liquid, aromatic, ethereal odor, hot saccharine taste. CH_3CHCl_2, mw: 99.0, mp: −97.7°, lel = 5.6%, bp: 57.3°, flash p: 22°F (TOC), d: 1.174 @ 20°4°, vap. press: 230 mm @ 25°, vap. d: 3.44, autoign. temp.: 856°F.
Acute tox data: Oral LD_{50} (rat) = 725 mg/kg. [3]
THR = MOD via oral route. An exper teratogen. [3] Liver injury has been reported in exper animals.
Fire Hazard: Dangerous, when exposed to heat or flame.
Explosion Hazard: Mod, when exposed to heat or flame.
Disaster Hazard: Dangerous; when heated to decomp,

emits highly toxic fumes of phosgene; can react vigorously with oxidizing materials.
To Fight Fire: Alcohol foam, water, foam, CO_2, dry chemical.

ETHYLIDENE CYCLOHEXANE. $C_6H_{11}CH_3CH_2$, mw: 112.2, vap. d: 4.
THR = HIGH narcotic action on acute inhal of vapors and injury to livers and kidneys on chronic exposure.

1,1-ETHYLIDENE DICHLORIDE. See ethylidene chloride.

1,2-ETHYLIDENE DICHLORIDE. $C_2H_4Cl_2$, mw: 99, flash p: 55°F, autoign. temp.: 824°F, lel = 6.2%, uel = 16%.
THR = See 1,1-ethylidene chloride.
Fire Hazard: Dangerous, via heat, flames or oxidizers.
To Fight Fire: Water as a blanket; alcohol foam.

ETHYLIDENE DIETHYL ETHER. See acetal.

ETHYLIDENE DIFLUORIDE. See 1,1-difluoroethane.

ETHYLIDENE DIIODIDE. Syn: *1,1-diiodoethane.* Liquid. CH_3CHI_2, mw: 281.9, bp: 179°, d: 2.84 @ 0°/4°, vap. d: 9.74.
THR = U. Probably irr and narcotic in high conc. See also iodides.
Disaster Hazard: Dangerous; when heated to decomp, or on contact with acid or acid fumes; emits highly toxic fumes of iodides; will react with water or steam to produce toxic and corrosive fumes.

ETHYLIDENE DIMETHYL ETHER. See dimethyl acetal.

ETHYLIDENE DIURETHANE. $C_8H_{16}O_4N_2$, mw: 204.3.
THR = An exper neo. [*3*]

5-ETHYLIDENE-2-(5H)-FURANONE. $C_6H_6O_2$, mw: 110.1.
THR = An exper carc via sc route. [*3, 23*]

5-ETHYLIDENE DIHYDRO-2-(3H)-FURANONE. $C_6H_8O_2$, mw: 112.1.
THR = An exper carc. [*3*]

ETHYL IODO METHYL ARSINE. C_3H_8AsI, mw: 245.9.
THR = Sometimes self-ignites in air. [*19*] Highly toxic. See arsenic and iodides.

ETHYLIDENE HYDROXYLAMINE. See acetaldehyde oxime.

ETHYLIDENE LACTIC ACID. See lactic acid.

ETHYL IODIDE. Syns: *hydriodic ether, iodoethane.* Clear, colorless liquid, turns brown on exposure to light, refractive, heavy. C_2H_5I, mw: 156.0, mp: −108°, bp: 72.4°, d: 1.90-1.93 @ 25°/25°, vap. press: 100 mm @ 18.0°, vap. d: 5.38.
THR = MOD irr to skin, eyes, mu mem. Narcotic in high conc.
Fire Hazard: Mod, when exposed to heat or flame.
Disaster Hazard: Dangerous; when heated to decomp, emits highly toxic fumes of iodides; will react with water or steam to produce toxic and corrosive fumes; can react vigorously with oxidizing materials.
To Fight Fire: Water, CO_2, dry chemical.

ETHYL IODOACETATE. Dense, colorless liquid. $CH_2ICOOC_2H_5$, mw: 214.01, bp: 179°, d: 1.80, vap. press: 0.54 mm @ 20°, vap. d: 7.4.
Acute tox data: ip LD_{50} (mouse) = 45 mg/kg. [*3*]
THR = HIGH via ip route. An irr to skin, eyes, mu mem.
Disaster Hazard: Dangerous; when heated to decomp or on contact with acid or acid fumes, emits highly toxic fumes of iodides; will react with water or steam to produce toxic and corrosive fumes.

5-ETHYL-5-ISOAMYL BARBITURIC ACID. See amobarbital.

ETHYL ISOBUTYRATE. Syns: *isobutyric ether, ethyl-2-methyl propanoate.* Colorless, volatile liquid, fruity, aromatic odor. $(CH_3)_2CHCOOC_2H_5$, mw: 116.16, mp: −88°, bp: 110°–111°, d: 0.870, vap. press: 40 mm @ 33.8°, vap. d: 4.01, flash p: <70°F.
THR = Probably MOD via oral route. See also esters.
Fire Hazard: Dangerous, when exposed to heat or flame; can react vigorously with oxidizing materials.
To Fight Fire: Foam, CO_2, dry chemical.

ETHYL ISOCYANATE. C_2H_5NCO, mw: 71.1, bp: 60°, d: 0.90 @ 20°/4°, vap. d: 2.45.
THR = No data. See also isocyanates.

ETHYL ISOCYANIDE. Syn: *ethyl isonitrile.* Colorless liquid, sol in water and organic solvents. C_2H_5NC, mw: 55.1, d: 0.7402 @ 20°/4°, mp: <−66°, bp: 79°.
THR = No data. Probably very toxic. See also nitriles. Has exploded upon heating. [*19*]

ETHYL ISONITRILE. See ethyl isocyanide.

ETHYL ISOTHIOCYANATE. Syn: *ethyl thiocarbimide.* Colorless liquid, pungent odor. C_2H_5NCS, mw: 87.2, mp: −5.9°, bp: 131°, d: 1.004 @ 15°/4°, vap. press: 10 mm @ 22.8°, vap. d: 3.01.
THR = VERY HIGH irr to skin, eyes and mu mem; also via oral and inhal routes. A military poison.
Disaster Hazard: Dangerous; when heated to decomp or on contact with acid or acid fumes, emits highly toxic fumes of oxides of sulfur and cyanides; can react with oxidizing materials.

ETHYL ISOTHIOUREA SULFATE. $N_2H_{10}C_3O_4S_2$, mw: 202.2.
THR = No data. Violent reaction with Cl_2. [*19*]

ETHYL LACTATE. Syn: *ethyl-2-hydroxypropionate.* Colorless liquid, mild odor. $CH_3CHOHCOOC_2H_5$, mw: 118.13, bp: 154°, ulc: 30-35, lel = 1.55% @ 212°F, flash p: 115°F (CC), flash p(technical): 131°F, d: 1.020–1.036 @ 20°/20°, autoign. temp.: 752°F, vap. d: 4.07.
Acute tox data: Oral LD_{50} (mouse) = 2500 mg/kg; sc LD_{50} (mouse) = 2500 mg/kg; iv LD_{50} (mouse) = 600 mg/kg. [*3*]
THR = MOD via oral, sc and iv routes.
Fire Hazard: Mod, when exposed to heat or flame; can react with oxidizing materials.
Spont Heating: No.
Explosion Hazard: Slight, when exposed to flame.
To Fight Fire: Foam, CO_2, dry chemical.

ETHYL LAURATE. Syn: *ethyl dodecanoate.* Oily liquid, insol in water. $CH_3(CH_2)_{10}COOC_2H_5$, mw: 228.4, d: 0.8615 @ 20°/4°, mp: −10.7°, bp: 269°.
THR = No data. Probably MOD to LOW. See also esters.

ETHYL LEVULINATE. Colorless liquid. $CH_3CO(CH_2)_2CO_2C_2H_5$, mw: 144.2, bp: 206.2°, d: 1.012, vap. press: 1 mm @ 47.3°, vap. d: 4.97.
THR = No data. See also esters.
Fire Hazard: Slight; can react with oxidizing materials.

ETHYL LITHIUM. See lithium ethyl.

ETHYL MAGNESIUM BROMIDE. Solid. C_2H_5MgBr, mw: 133.3.
THR = No data. See also bromides.
Fire Hazard: Mod; can evolve flam ethane.
Disaster Hazard: Mod dangerous; when heated to decomp, emits toxic fumes of bromides; will react with water or steam to produce flam vapors.

ETHYL MAGNESIUM CHLORIDE. Solid. C_2H_5MgCl, mw: 88.8.
THR = No data. See also chlorides.
Fire Hazard: Mod; can react with moisture to evolve ethane.
Disaster Hazard: Dangerous; when heated to decomp, emits highly toxic fumes of chlorides; will react with water or steam to produce flam vapors.

ETHYL MAGNESIUM IODIDE. C_2H_5MgI, mw: 180.2.
THR = No data. Has exploded with ethoxyacetylene. [*19*]

***n*-ETHYL MALEIMIDE.** Crystals, irr odor. $C_6H_7NO_2$, mw: 125.1, mp: 45°.
Acute tox data: ip LD_{LO} (mouse) = 25 mg/kg; ip LD_{LO} (rat) = 17 mg/kg. [*3*]
THR = HIGH via ip route. Probably also via oral and inhal routes.
Disaster Hazard: Dangerous; on heating, liberates high irr vapors.

ETHYL MALONATE. See diethyl malonate.

ETHYL MANDELATE. Crystals. $C_6H_5CH(OH)CO_2CH_2CH_3$, mw: 180.2.
THR = U. See esters.
Fire Hazard: Slight; can react with oxidizing materials.

ETHYL MERCAPTAN. Syns: *ethanethiol, ethyl hydrosulfide, ethyl thioalcohol, ethyl sulfhydrate.* Colorless liquid, penetrating garlic-like odor. C_2H_5SH, mw: 62.13, mp: −147°, bp: 36.2°, lel = 2.8%, uel = 18.2%, flash p: <80°F (CC), d: 0.83907 @ 20°/4°, autoign. temp.: 570°F, vap. d: 2.14.
Acute tox data: Oral LD_{50} (rat) = 682 mg/kg; inhal LC_{50} (rat) = 4420 ppm for 4 hrs; ip LD_{50} (rat) = 450 mg/kg. [*3*]
THR = MOD via oral and inhal; HIGH via ip routes.
Fire Hazard: Very dangerous, when exposed to heat, flame or oxidizers.
Explosion Hazard: Mod, when exposed to spark or flame. Violent reaction with $Ca(OCl)_2$. [*19*]
Disaster Hazard: Dangerous; when heated to decomp or on contact with acid or acid fumes, emits highly toxic fumes of oxides of sulfur; will react with water or steam to produce toxic and flam vapors; can react vigorously with oxidizing materials.
To Fight Fire: CO_2, dry chemical.

9-ETHYL-6-MERCAPTOPURINE. $C_7H_8N_4S$, mw: 180.3.
THR = An exper teratogen. [*3*]

ETHYL MERCURIC CHLORIDE. See ethyl mercury chloride.

ETHYL MERCURY CHLORIDE. Syns: *ethyl mercuric chloride, lignasan.* Silvery, irridescent leaflets. C_2H_5HgCl, mw: 265.13, mp: 192.5°.
Acute tox data: Oral LD_{50} (rat) = 30 mg/kg; dermal LD_{50} (rat) = 200 mg/kg; sc LD_{50} (rat) = 66 mg/kg. [*3*]
THR = HIGH via oral, dermal, sc and ip routes. A fungicide. See also mercury compounds, organic.

ETHYL MERCURY CRESOL. $C_2H_5HgC_6H_4CH_3$, mw: 320.8, vap. d: 11.05.
THR = HIGH. See mercury compounds, organic.

ETHYL MERCURY HYDROXIDE. Silvery crystals, insol in water. C_2H_5HgOH, mw: 246.7, mp: 37°.
THR = HIGH. See mercury compounds, organic.

ETHYL MERCURY IODIDE. Crystals, sol in alcohol. C_2H_5HgI, mw: 356.6, mp: 186°.
THR = HIGH. See mercury compounds, organic.

ETHYL MERCURY ISOTHIOCARBAMIDE. Crystals. $C_2H_5HgNHCSNH_2$, mw: 304.8, vap. d: 10.5.
THR = HIGH. See mercury compounds, organic.

ETHYL MERCURY NITROPHENOL. Crystals. $C_2H_5HgC_6H_4NO_2$, mw: 351.8, vap. d: 12.1.
THR = HIGH. See mercury compounds, organic.

ETHYL MERCURY PHOSPHATE. Solid. $C_2H_5HgH_2PO_4$, mw: 326.7.
Acute tox data: Oral LD_{50} (rat) = 30 mg/kg; oral LD_{50} (mouse) = 61 mg/kg; ip LD_{50} (mouse) = 63 mg/kg. [*3*]
THR = HIGH via oral and ip routes. A fungicide. See also mercury compounds, organic.

ETHYL MERCURY SULFATE. Crystals. $(C_2H_5Hg)_2SO_4$, mw: 555.4.
THR = HIGH. See mercury compounds, organic, and sulfates.

ETHYL MERCURY-*p*-TOLUENE SULFONANILIDE. Syn: *ceresan M.* Crystals, pungent, garlic-like odor, water-insol. $C_{15}H_{17}HgNO_2S$, mw: 476.0.
Acute tox data: Oral LD_{50} (rat) = 100 mg/kg; inhal LD_{50} (rat) = 50 mg/kg. [*3*]
THR = HIGH via oral route. See also mercury compounds, organic.

ETHYL METHACRYLATE. A liquid. $H_2CCCH_3COOC_2H_5$, mw: 114.07, mp: $<-75°$, bp: 119°, lel = 1.8%, uel = saturation, flash p: 68°F (OC), d: 0.911 @ 25°/25°, vap. d: 3.94.
Acute tox data: ip LD_{50} (rat) = 1223 mg/kg; oral LD_{LO} (rabbit) = 3630 mg/kg. [*3*]
THR = MOD via oral and ip routes. An exper teratogen. [*3*]
Fire Hazard: Dangerous, when exposed to heat or flame; can react with oxidizing materials.
Explosion Hazard: Dangerous, when exposed to heat, sparks or flame.
To Fight Fire: CO_2, dry chemical.

ETHYL METHANE SULFONATE. $C_3H_8O_3S$, mw: 124.2.
THR = An exper (+) carc, neo, teratogen and mutagen. [*3, 1, 23*]

ETHYL METHANOATE. See ethyl formate.

7-ETHYL-5-METHOXYBENZ(a)ANTHRACENE. $C_{21}H_8O$, mw: 276.3.
THR = An exper neo. [*3*]

4-[((4-ETHYL-4-METHYL AMINO)PHENYL)AZO] PYRIDINE-1-OXIDE. $C_{14}H_{16}ON_4$, mw: 256.3.
THR = An exper carc. [*3*]

ETHYL METHYL ARSINE. C_3H_9As, mw: 120.
THR = Self-ignites in air. [*19*] See arsenic.

9-ETHYL-7-METHYL BENZ(c)ACRIDINE. $C_{20}H_{16}N$, mw: 270.4.
THR = An exper neo. [*3, 23*]

7-ETHYL-12-METHYL BENZ(a)ANTHRACENE. $C_{21}H_{18}$.
THR = HIGH. An exper carc. [*3, 23*]

12-ETHYL-7-METHYL BENZ(a)ANTHRACENE.
THR = An exper carc. [*3, 23*]

ETHYL-2-METHYL-4-CHLOROPHENOXY ACETATE. $C_{11}H_{13}O_3Cl$, mw: 228.7.
THR = HIGH via oral route to rats. An exper teratogen. [*3*]

***n*-ETHYL METHYL (α-METHYL BENZYL)AMINE.** $C_2H_5NH_2C_6H_4CH_3$, mw: 136.
THR = No data. Limited animal exper suggest HIGH and an irr. See also amines.

ETHYL METHYL CARBINOL. See *sec*-butyl alcohol.

2-ETHYL-2-METHYL-1,3-DIOXOLANE. $C_7H_{12}O_2$, mw: 128.
Acute tox data: Oral LD_{LO} (rat) = 2830 mg/kg; inhal LC_{LO} (rat) = 4000 ppm for 4 hrs. [*3*]
THR = MOD via oral and inhal routes.

ETHYL METHYL ETHER. Syn: *methoxy ethane.* Colorless liquid or gas. $CH_3OC_2H_5$, mw: 60.09, bp: 11°, lel = 2.0%, uel = 10.1%, flash p: −35°F (CC), d: 0.7260 @ 0°/4°, autoign. temp.: 374°F, vap. d: 2.07.
THR = No data. Has anesthetic properties. See also ethers.
Fire Hazard: Highly dangerous, when exposed to heat or flame; can react vigorously with air, O_2. [*19*] See also ethers.
Explosion Hazard: Mod, when exposed to flame. See also ethers.
Disaster Hazard: Highly dangerous, upon exposure to heat or flame.
To Fight Fire: Alcohol foam, CO_2, dry chemical.

7-ETHYL METHYL ETHER BENZ(a)ANTHRACENE.
THR = An exper carc. [*23*]

7-ETHYL-2-METHYL-4-HENDECANOL. Slightly water-sol. $C_{14}H_{30}O$, mw: 214, d: 0.8, bp: 265°F, flash p: 285°F.
Fire Hazard: Low.
To Fight Fire: Dry chemical, CO_2, mist.

7-ETHYL-2-METHYL-4-HENDECANOL SODIUM SULFATE. Syn: *tergitol.* Very slightly water-sol. $C_{14}H_{29} \cdot NaSO_4$, mw: 316.5.
Acute tox data: Oral LD_{50} (rat) = 1250 mg/kg; oral LD_{50} (guinea pig) = 650 mg/kg; dermal LD_{50} (guinea pig) = 650 mg/kg. [*3*]
THR = MOD via oral and dermal routes.

ETHYL METHYL KETONE. See butanone.

ETHYL METHYL KETONE PEROXIDE (in dimethyl phthalate). Water-white liquid, insol in water, sol in

benzene, alcohols, esters and ethers. d: 1.091, flash p: 122°F (COC), decomp mildly @ 130°.
THR = See peroxides, organic.

2-ETHYL-4-METHYL PENTANOL. $C_8H_{18}O$, mw: 130.3.
Acute tox data: Oral LD_{50} (rat) = 5190 mg/kg; dermal LD_{50} (rabbit) = 5660 mg/kg. [*3*]
THR = MOD via oral and dermal routes.

N-ETHYL-N-METHYL-*p*-(PHENYL AZO)ANILINE. $C_{15}H_{17}N_3$, mw: 239.4.
THR = An exper carc. [*3*]

ETHYL METHYL PHENYL GLYCIDATE. Syns: *3-methyl-3-phenylglycidic acid ethyl ester*, *"strawberry aldehyde," aldehyde C-16.* Colorless to yellowish liquid, strawberry-like odor, sol in 3 volumes of 60% alcohol. $CH_3(C_6H_5)\underline{COC}HCOOC_2H_5$, mw: 206, d: 1.104–1.123.
Acute tox data: Oral LD_{50} (guinea pig) = 4050 mg/kg. [*3*]
THR = MOD via oral route. A synthetic flavoring substance and adjuvant. [*109*]

5-ETHYL-3-METHYL-5-PHENYL HYDANTOIN. See phenantoin.

ETHYL-2-METHYL PROPANOATE. See ethyl isobutyrate.

ETHYL METHYL PROPYL TIN IODIDE. Crystals. $(C_2H_5)(CH_3)(C_3H_7)SnI$, mw: 332.8, d: 1.8182 @ 20°/20°, bp: 110° @ 11 mm.
THR = See tin compounds and iodides.

5-ETHYL-2-METHYL PYRIDINE. Syn: *aldehydine.* Liquid. $(C_2H_5)(CH_3)C_3H_3N$, mw: 121.2, bp: 174°, d: 0.9184 @ 23°/4°, flash p: 165° (OC).
Acute tox data: Oral LD_{50} (rat) = 1540 mg/kg; dermal LD_{50} (rat) = 1000 mg/kg. [*3*]
THR = MOD via oral and dermal routes.
Disaster Hazard: Dangerous; when heated to decomp, emits toxic fumes; violent reaction with HNO_3 [*19*] and other oxidizers.
Fire Hazard: Mod, via heat, open flames, oxidizers.
To Fight Fire: Alcohol foam.

5-ETHYL-2-METHYL PYRIDINE-1-OXIDE. $C_8H_{11}ON$, mw: 137.2.
Acute tox data: Oral LD_{50} (rat) = 2000 mg/kg; dermal LD_{50} (rabbit) = 1770 mg/kg. [*3*]
THR = MOD via oral and dermal routes.

3-ETHYL-3-METHYL-1-PYRIDYL TRIAZENE. $C_8H_{11}N_4$, mw: 164.2.
THR = An exper teratogen. [*3*]

ETHYL METHYL TELLUROPHETONE. Dark yellow oil. $C_2H_5CTeCH_3$, mw: 183.7, d: 1.8711, bp: 63°–66°.
THR = HIGH. See tellurium compounds.

ETHYL MONOBROMOACETATE. See ethyl bromoacetate.

ETHYL MONOCHLOROACETATE. See ethyl chloroacetate.

ETHYL MORPHINE HYDROCHLORIDE. See dionin.

ETHYL MORPHOLINE. Colorless liquid. $\underline{CH_2CH_2OCH_2CH_2}NCH_2CH_3$, mw: 115.2, bp: 138°, flash p: 89.6°F (OC), d: 0.916 @ 20°/20°, vap. d: 4.00.
Acute tox data: Oral LD_{50} (rat) = 1780 mg/kg; inhal LC_{LO} (rat) = 2000 ppm for 4 hrs; inhal TC_{LO} (human) = 100 ppm. [*3*]
THR = MOD irr to skin, eyes, mu mem, and via oral and inhal routes. Possibly via dermal route too.
Fire Hazard: Dangerous, when exposed to heat or flame; can react vigorously with oxidizing materials.
To Fight Fire: Alcohol foam, foam, CO_2, dry chemical.

ETHYL MUSTARD OIL. See dichlorodiethyl sulfide.

1-ETHYL NAPHTHALENE. $C_{12}H_{12}$, mw: 156.2, autoign. temp.: 896°F, d: 1.02, vap. d: 5.39, bp: 257.8°.
Acute tox data: Oral LD_{LO} (rat) = 5000 mg/kg. [*3*]
Fire Hazard: Low, via heat, flame or oxidizers.
To Fight Fire: Water, spray, mist, foam, dry chemical.

ETHYL NITRATE. Syn: *nitric ether.* Colorless liquid, pleasant odor, sweet taste. $C_2H_5ONO_2$, mw: 91.07, mp: −112°, bp: 88.7°, explodes @ 185°F, lel = 3.8%, flash p: 50°F (CC), d: 1.105 @ 20°/4°, vap. d: 3.14.
THR = No data. Probably at least MOD via oral and inhal routes. See also nitrates, organic.
Fire Hazard: Dangerous, when exposed to heat or flame.
Spont Heating: No.
Explosion Hazard: Moderate, when exposed to heat.
Disaster Hazard: See nitrates.
To Fight Fire: Foam, CO_2, dry chemical, water to blanket fire.

ETHYL NITRITE. Syns: *nitrous ether, hyponitrous ether.* Colorless or yellowish liquid, highly aromatic, ethereal odor. Decomp on standing. C_2H_5ONO, mw: 75.07, bp: 16.4°, lel = 4.1%, −3.0%, uel = 50%, explodes at 194°F, flash p: −31°F (CC), d: 0.900 @ 15.5°, autoign. temp.: 194°F, vap. d: 2.59.
Acute tox data: LD_{LO} (child) = 200 mg/kg. [*3*]
THR = HIGH to children, probably via inhal route. Narcotic in high conc and lowers blood pressure. Methemoglobinemia has been reported.
Fire Hazard: Highly dangerous when exposed to heat or flame. A powerful oxidizer.
Spont Heating: No.
Explosion Hazard: Severe, especially @ >90°. [*19*]
Disaster Hazard: Highly dangerous; when heated to decomp or on contact with acid or acid fumes, emits

highly toxic fumes of NO_x; reacts vigorously with oxidizers.

To Fight Fire: Foam, CO_2, dry chemical or water spray.

ETHYL NITRITE SPIRIT. See spirits of niter (sweet).

1-ETHYL-2-NITROBENZENE. See *o*-ethyl nitrobenzene.

***o*-ETHYL NITROBENZENE.** Syn: *1-ethyl-2-nitrobenzene.* Clear yellow to green liquid. $C_6H_4NO_2C_2H_5$, mw: 151.16, mp: 23°, bp: about 228°, fp: −13°, d: 1.1 @ 25°/25°, vap. d: 5.2.

THR = U. May resemble nitrobenzene. See nitrobenzene.

Fire Hazard: See nitrates.

Disaster Hazard: See nitrates.

To Fight Fire: Water, foam, CO_2, dry chemical.

ETHYL-*p*-NITROBENZOATE. See nitrobenzoic acid ethyl ester.

ETHYL-*p*-NITROPHENYL THIONOBENZENE PHOSPHATE. See EPN.

2-(ETHYL NITROSAMINO)ETHANOL. $C_4H_{10}O_2N_2$, mw: 118.2.

THR = An exper neo. [*3, 23*]

4-[(ETHYL NITROSAMINO)METHYL]PYRIDINE. $C_8H_{11}ON_3$, mw: 165.2.

THR = An exper carc. [*3, 23*]

4-(ETHYL NITROSOAMINO)BUTYRIC ACID. $C_6H_{12}O_3N_2$, mw: 160.2.

THR = An exper carc. [*3*]

N-ETHYL-N-NITROSO ANILINE. $C_8H_{10}ON_2$, mw: 150.2.

THR = An exper teratogen. [*3*]

1-ETHYL-1-NITROSO BIURET. $C_4H_8O_3N_4$, mw: 160.2.

THR = An exper transplacental carc. [*3, 23*]

N-ETHYL-N-NITROSO BUTYLAMINE. $C_6H_{14}ON_2$, mw: 130.2.

THR = An exper carc via oral and iv routes. [*3, 23*]

N-ETHYL-N-NITROSO-*tert*-BUTYLAMINE.

THR = An exper neo via oral route. [*3, 23*]

ETHYL NITROSO CYANAMIDE. $C_3H_5N_3$, mw: 99.1.

THR = An exper carc. [*3*]

1-ETHYL-3-NITRO-3-NITROSO GUANIDINE. $C_3H_7O_3N_5$, mw: 161.2.

THR = An exper neo and carc. [*3, 23*]

ETHYL-4-NITROSO-PIPERAZINE CARBOXYLATE. $C_7H_{13}O_3N_3$, mw: 187.2.

THR = An exper carc. [*3*]

ETHYL NITROSO UREA. Syn: *ENU.* $C_3H_7O_2N_3$. Pale crystals, yellow. mw: 117.1, mp: 103° (decomp), water-sol (1.3% @ 20°).

THR = An exper teratogen and transplacental carc. [*3, 9, 23*]

N-ETHYL-N-NITROSO URETHANE. $C_5H_{10}O_3N_2$, mw: 146.2.

THR = An exper carc, neo and teratogen. [*3, 23*]

N-ETHYL-N-NITROSO VINYL AMINE. $C_4H_8ON_2$, mw: 100.1.

THR = HIGH. An exper carc via oral and iv routes. [*3, 23*]

3-ETHYL OCTANE. $C_{10}H_{22}$, mw: 142.2, autoign. temp.: 446°F, d: 0.74, vap. d: 4.91, bp: 167.2°.

THR = No data.

Fire Hazard: Mod, via heat, flames (sparks), oxidizers.

To Fight Fire: Foam, dry chemical, CO_2, water mist.

4-ETHYL OCTANE. Autoign. temp.: 445°F, d: 0.74, vap. d: 4.91, bp: 165°.

THR = No data.

Fire Hazard: See 3-ethyl octane.

To Fight Fire: See 3-ethyl octane.

ETHYL OLEATE. Light-colored oleaginous liquid. $C_{20}H_{38}O_2$, mw: 310.5, mp: −32° (approx), bp: 205°, flash p: 347.5°F, d: 0.867.

THR = No data; probably LOW.

Fire Hazard: Slight; can react with oxidizing materials.

ETHYL OXALATE. Syns: *diethyl oxalate, diethyl ethanedioate, oxalic ether.* Colorless, oily aromatic liquid, decomp in water. $(COOC_2H_5)_2$, mw: 146, mp: −40.6°, bp: 185.4°, flash p: 168°F (OC), d: 1.08426 @ 15°, 1.0785 @ 20°/4°, vap. d: 5.04.

Acute tox data: Oral LD_{50} (rat) = 400 mg/kg. [*3*]

THR = HIGH via oral route.

Fire Hazard: Mod, when exposed to heat or flame; can react with oxidizing materials.

Spont Heating: No.

Disaster Hazard: On combustion, will give off toxic fumes.

To Fight Fire: Foam, CO_2, dry chemical.

ETHYL OXIDE. See ethyl ether.

11-ETHYL-17-OXO-16,17-DIHYDRO-15H-CYCLOPENTA(a)PHENANTHRENE.

THR = An exper carc. [*23*]

ETHYL PARABEN. See ethyl-*p*-hydroxy benzoate.

ETHYL PARASEPT. See ethyl-*p*-hydroxy benzoate.

ETHYL PERCHLORATE. $C_2H_5ClO_4$, mw: 128.5.

THR = VERY HIGH explosive. [*19*] No toxicity data. See also perchlorates.

***o*-ETHYL PHENOL.** Syn: *phloral.* Colorless liquid, phenol odor, insol in water, freely sol in alcohol, benzene and glacial acetic acid. $C_8H_{10}O$, mw: 122.16, d: 1.037 @ 12°, bp: 207°–208°, solidifies <18°.
THR = See phenol.

***p*-ETHYL PHENOL.** Colorless needles, sol in alcohol or ether, slightly sol in water. $HOC_6H_4C_2H_5$, mw: 122.6, mp: 46°, bp: 219°, d: 1.0 @ 140°F, flash p: 219°F.
THR = No data.
Fire Hazard: Low, when exposed to powerful oxidizers, heat or source of ignition.
To Fight Fire: Alcohol foam.

ETHYL PHENYL ACETAMIDE. See ethyl acetanilide.

ETHYL PHENYL ACETATE. Colorless liquid. $C_6H_5CH_2COOC_2H_5$, mw: 164.2, bp: 276°, d: 1.033 @ 20°, vap. d: 5.67.
Acute tox data: Oral LD_{50} (rat) = 3300 mg/kg. [*3*]
THR = MOD via oral route. A weed-killer.
Disaster Hazard: Mod dangerous; when heated to decomp, may emit toxic fumes.

ETHYL PHENYLAMINE. See *n*-ethylaniline.

***p*-[(*m*-ETHYL PHENYL)AZO]-N,N-DIMETHYL ANILINE.** $C_{16}H_{19}N_3$, mw: 253.4.
THR = an exper carc. [*3*]

***p*-[(*p*-ETHYLPHENYL)AZO]-N,N-DIMETHYL ANILINE.**
THR = An exper carc. [*3*]

***p*-[(4-ETHYL PHENYL)AZO]-N-METHYL ANILINE.** $C_{15}H_{17}N_3$, mw: 239.4.
THR = An exper neo and carc. [*3*]

***p*-((*p*-ETHYL PHENYL)AZO)PHENOL.** $C_{14}H_{14}ON_2$, mw: 226.3.
THR = An exper neo. [*3*]

5-ETHYL-5-PHENYL BARBITURIC ACID. $C_{12}H_{12}O_3N_2$, mw: 232.3.
THR = HIGH via dermal route to CNS. An exper teratogen. [*3*]

5-ETHYL-5-PHENYL BARBITURIC ACID SODIUM. See sodium phenobarbital.

ETHYL PHENYL DICHLOROSILANE. Liquid. $C_8H_{10}SiCl_2$, mw: 158.3.
THR = HIGH irr to skin, eyes and mu mem, and via oral and inhal routes.
Disaster Hazard: Dangerous; when heated to decomp, emits highly toxic fumes of chlorides and phenol; will react with water or steam to produce toxic and corrosive fumes; can react with oxidizing materials.

ETHYL PHENYL DI-*p*-TOLYL GERMANIUM. White crystals. $(C_2H_5)(C_6H_5)Ge(C_6H_4CH_3)_2$, mw: 361.0 mp: 55°.
THR = See germanium compounds.

ETHYL PHENYL ETHER. See phenyl ethyl ether.

ETHYL PHENYL KETONE. See propiophenone.

ETHYL PHOSPHATE. See triethyl phosphate.

ETHYL PHOSPHINE. Syn: *phosphinoethane.* Colorless liquid. $C_2H_5PH_2$, mw: 62.1, bp: 25°, d: 1, vap. d: 2.14.
THR = No data. Probably HIGH. See also phosphine.
Fire Hazard: Dangerous, when exposed to heat or flame.
Explosion Hazard: Reacts violently with Br_2, Cl_2, HNO_3. [*19*]
Disaster Hazard: Dangerous; when heated to decomp, emits highly toxic fumes of oxides of phosphorus; can react vigorously with oxidizing materials.
To Fight Fire: Foam, CO_2, dry chemical.

ETHYL PHOSPHONOTHIOIC DICHLORIDE. $C_2H_5OCl_2PS$, mw: 179.0.
Acute tox data: Oral LD_{50} (rat) = 900 mg/kg; inhal LC_{LO} (rat) = 32 ppm for 4 hrs. [*3*]
THR = HIGH via inhal and MOD via oral routes. A strong irr to skin, eyes and mu mem.
Disaster Hazard: Dangerous, see SO_x and PO_x and chlorides.

ETHYL PHOSPHORUS DICHLORIDE.
THR = U, but it is a very corrosive liquid and must be considered as having HIGH toxicity.
Disaster Hazard: Dangerous; see PO_x and chlorides.

ETHYL PHOSPHORODICHLORIDITE. Syn: *diethyl chloro phosphinate*, A fuming (in moist air) liquid, odor of chlorides, reacts violently with water. $(C_2H_5O)_2PCl$, mw: 156.6, d: 1.0816 @ 20°/4°, bp: 154°.
THR = HIGH irr to skin, eyes and mu mem, and via oral and inhal routes. A corrosive.
Disaster Hazard: Dangerous, see PO_x and chlorides.

ETHYL PHTHALATE. See diethyl-*o*-phthalate.

ETHYL PHTHALYL ETHYL GLYCOLLATE. Liquid. $C_{14}H_{16}O_6$, mw: 280.3, bp: 320°, flash p: 365°F (CC), d: 1.189, vap. d: 9.6.
THR = U. See also esters.
Fire Hazard: Slight, when exposed to heat or flame; can react with oxidizing materials.
To Fight Fire: Foam, alcohol foam, CO_2, dry chemical.

ETHYL PICOLINE. 3-ethyl-6-methyl pyridine.

ETHYL PROPENOATE. See ethyl acrylate.

ETHYL-1-PROPENYL ETHER. $CH_3CH{:}CHOCH_2CH_3$, mw: 86, d: 0.8, bp: 158°F, flash p: >19°F (OC).
Acute tox data: Oral LD_{50} (rat) = 4660 mg/kg; inhal LC_{LO} (rat) = 8000 ppm for 4 hrs. [*3*]
THR = MOD via oral and inhal routes.
Fire Hazard: Dangerous, when exposed to heat, flame or oxidizers.
To Fight Fire: Water spray or mist, dry chemical, foam, CO_2.

ETHYL PROPIONATE. Syn: *propionic ether*. Water-white liquid, pineapple-like odor. $C_2H_5COOC_2H_5$, mw: 102.11, bp: 99°, mp: −72.6°, flash p: 54°F (CC), d: 0.895 @ 15.5°, autoign. temp.: 824°F, vap. press: 40 mm @ 27.2°, vap. d: 3.52, lel = 1.9%, uel = 11%.
Acute tox data: Oral LD_{50} (rat) = 3500 mg/kg; ip LD_{50} (mouse) = 1300 mg/kg. [*3*]
THR = MOD via oral and ip routes.
Fire Hazard: Dangerous, when exposed to heat or flame; can react vigorously with oxidizing materials.
Spont Heating: No.
Explosion Hazard: U.
Disaster Hazard: Dangerous, upon exposure to heat or flame.
To Fight Fire: Foam, CO_2, dry chemical.

ETHYL PROPIONYL. See diethyl ketone.

2-ETHYL-3-PROPYL ACROLEIN. Syn: *2-ethyl-hexenal.* Colorless liquid, powerful odor.
$CH_3(CH_2)_2CHC(C_2H_5)CHO$, mw: 126.19, bp: 175°, flash p: 155°F (OC), d: 0.848 @ 20°/4°, vap. d: 4.35, vap. press: 1.0 mm @ 20°.
Acute tox data: Oral LD_{50} (rat) = 3000 mg/kg. [*3*]
THR = MOD via oral route. A strong irr to skin, eyes and mu mem.
Fire Hazard: Mod, when exposed to heat or flame; can react with oxidizing materials.
To Fight Fire: Alcohol foam, CO_2, dry chemical.

2-ETHYL-3-PROPYL ACRYLIC ACID. Liquid. $C_3H_7CHC(C_2H_5) \cdot COOH$, mw: 142.19, mp: −7.8°, bp: 232.1°, d: 0.948 @ 20°/20°, vap. d: 4.91, flash p: 330°F (OC).
THR = U. See also acrylic acid.
Fire Hazard: Slight.
To Fight Fire: Alcohol foam, dry chemical, CO_2.

ETHYL-*n*-PROPYL DIISOAMYL TIN. $(C_2H_5)(C_3H_7)(C_5H_{11})_2Sn$, mw: 333.1, d: 1.0654 @ 22°, bp: 141° @ 17 mm.
THR = U. See tin compounds.

ETHYL-*sec*-PROPYL DIPHENYL GERMANIUM. Liquid. $(C_2H_5)(C_3H_7)Ge(C_6H_5)_2$, mw: 299.0, mp: 175°–190°.
THR = U. See germanium compounds.

ETHYL PROPYL ETHER. Syn: *ethoxy propane-1.* Sol in water. $C_2H_5OC_3H_7$, mw: 88, lel = 1.7%, uel = 9%, d: 0.8, bp: 147°F, flash p: <−4°F.
THR = U. See ethers.
Fire Hazard: Dangerous when exposed to heat or open flame. See ethers.
To Fight Fire: Alcohol foam.

1-ETHYL-2-PROPYL ETHYLENE. See 1,3-heptene (mixture of *cis* and *trans* isomers).

ETHYL PROPYL KETONE. See 3-hexanone.

ETHYL PROPYL TIN DICHLORIDE. Crystals, water-sol. $(C_2H_5)(C_3H_7)SnCl_2$, mw: 261.8, mp: 58°.
THR = See tin compounds and chlorides.

4-ETHYL PYRIDINE. Liquid. $CH_3CH_2C_5H_4N$, mw: 107.1, bp: 168.3°, d: 0.9460 @ 20°/20°, vap. d: 3.70.
THR = U. See also pyridine.

2-(5′-ETHYL PYRID-2′-YL)ETHYL ACRYLATE. Liquid, very slightly sol in water. $C_{12}H_{15}O_2N$, mw: 205.3, d: 1.0458 @ 20°, bp: 181° @ 50 mm, fp: −75°.
Acute tox data: Oral LD_{50} (rat) = 4920 mg/kg; dermal LD_{50} (rabbit) = 2230 mg/kg. [*3*]
THR = MOD via oral and dermal routes.
Fire Hazard: Mod.
Explosion Hazard: U.
Disaster Hazard: Dangerous; when heated to decomp, emits highly toxic fumes of oxides of nitrogen; can react with oxidizing materials.
To Fight Fire: Foam, CO_2, dry chemical.

ETHYL SELENIDE. See diethyl selenide.

ETHYL SILICATE. Syn: *tetraethyl-o-silicate.* Colorless liquid, faint odor, decomp in water. $(C_2H_5)_4SiO_4$, mw: 208.30, bp: 165.5°, flash p: 125°F (OC), d: 0.933 @ 20°/4°, vap. press: 1.0 mm @ 20°, vap. d: 7.22.
Acute tox data: Inhal LC_{LO} (rat) = 1000 ppm for 4 hrs; oral LD_{LO} (rat) = 1000 mg/kg. [*3*]
THR = MOD via inhal and oral routes. A strong irr to skin, eyes, mu mem and upper respiratory tract. Conc of 2,500 ppm produce narcosis in guinea pigs after 1½ hrs exposure. This is the maximum conc available at ordinary room temp by evaporation. The maximum conc which these animals can inhale for 60 min without the production of serious disturbance is 2,000 ppm. For an exposure of several hours, conc of 500 ppm or more are required to cause serious injury. Animals dying after acute exposures show pulmonary edema, sometimes with secondary pneumonia, acute nephritis and evidence of injury to the liver. For humans, conc of 85 ppm produce a detectable odor, and 700 ppm causes stinging of the eyes and nose. No cases of human poisoning have been reported nor have exper dealing with the effects of prolonged exposure been reported.

Irr of the eyes and nose, tremors, respiratory difficulty or irregularity, anemia, leucocytosis, and narcosis have been reported in animal experiments. Brief exposures of humans to 3,000 ppm cause only extreme irr of the eyes and nose.
Fire Hazard: Mod, when exposed to heat, flame or oxidizers.
Disaster Hazard: Dangerous; when heated to decomp, emits toxic fumes; can react with oxidizing materials.
To Fight Fire: CO_2, dry chemical.

ETHYL SODIO-ACETO ACETATE. $C_6H_9O_3Na$, mw: 152.1.
THR = It can explode upon mixing with 2-iodo-3,5-dinitrobiphenyl. [*19*]

ETHYL SODIUM. C_2H_5Na, mw: 52.
THR = Self-ignites in air. [*19*]

ETHYL STANNIC ACID. White amorphous powder, insol in water. C_2H_5SnOOH, mw: 180.8.
THR = U. See tin compounds.

ETHYL STYRENE. See *m*-ethyl vinyl benzene.

ETHYL SUCCINATE. See diethyl succinate.

ETHYL SULFATE. See diethyl sulfate.

ETHYL SULFHYDRATE. See ethyl mercaptan.

ETHYL SULFIDE. Syns: *diethyl sulfide, thioethyl ether.* Liquid, garlic-like odor. $(C_2H_5)_2S$, mw: 90.2, mp: −102°, bp: 92-93°, d: 0.837 @ 20°/4°, vap. d: 3.11.
THR = No details. Probably HIGH. See also sulfides.
Fire Hazard: Probably mod, via heat, flame (sparks) or oxidizers.
Disaster Hazard: When heated to decomp, yields highly toxic fumes of SO_x; also from the action of acids or acid fumes. Can react with water or steam to produce toxic and flam vapors; can react with oxidizing materials.
To Fight Fire: Water spray or mist, dry chemical, CO_2, foam.

2-(ETHYL SULFONYL)ETHANOL. Syn: *ethyl sulfonyl ethyl alcohol.* Hygroscopic crystals, water-sol. $CH_3CH_2SO_2CH_2CH_2OH$, mw: 138.2, d: 1.24 @ 20°/4°, mp: 43°, bp: 153° @ 2.5 mm, flash p: 371°F.
Acute tox data: Oral LD_{50} (rat) = 18,000 mg/kg. [*3*]
THR = VERY LOW via oral and ip routes.
Disaster Hazard: Dangerous; when heated to decomp, emits highly toxic fumes of oxides of sulfur.
Fire Hazard: Slight, when exposed to heat or flame.
To Fight Fire: Water, foam, CO_2, dry chemical.

ETHYL SULFONYL ETHYL ALCOHOL. See 2-(ethyl sulfonyl) ethanol.

4-ETHYL SULFONYL NAPTHALENE-1-SULFONAMIDE. $C_{12}H_{13}O_4NS_2$, mw: 299.4.
THR = An exper carc and neo. [*3*] An exper carc to mice via oral route; and an exper neo to mice via imp route. [*103*]

ETHYL SULFURIC ACID. Syns: *ethyl hydrogen sulfate, acid ethyl sulfate.* Colorless oily liquid. $C_2H_5OSO_3H$, mw: 126.13, bp: 280° decomp, d: 1.316 @ 17°/4°.
THR = HIGH irr to skin, eyes and mu mem and via oral and inhal routes.
Disaster Hazard: Dangerous; when heated to decomp, emits highly toxic fumes of SO_x; will react with water or steam to produce heat.

ETHYL TETRAPHOSPHATE. See hexaethyl tetraphosphate.

ETHYL TETRATHIO-*o*-STANNATE. Solid $(SC_2H_5)_4Sn$, mw: 363.2, bp: 105° @ 0.001 mm.
THR = See tin and sulfur compounds.

ETHYL THIOALCOHOL. See ethyl mercaptan.

ETHYL THIOCARBIMIDE. See ethyl isothiocyanate.

***o*-[2-(ETHYL THIO)ETHYL]-*o,o*-DIMETHYL PHOSPHOROTHIOATE.** See demeton.

2-ETHYL THIOISONICOTINAMIDE. $C_8H_{10}N_2S$, mw: 166.3.
THR = An exper neo. [*3*]

ETHYL THIOL CARBAMATE. C_3H_7ONS, mw: 105.2.
THR = An exper neo. [*3*]

ETHYL TIN TRIBROMIDE. Colorless crystals, water-sol. $(C_2H_5)SnBr_3$, mw: 387.5, mp: 310°.
THR = See tin compounds and bromides.

ETHYL TIN TRIIODIDE. Crystals. $(C_2H_5)SnI_3$, mw: 528.5, bp: 181°–184° @ 19 mm.
THR = See tin compounds and iodides.

ETHYL TOLUENE. C_9H_{12}, mw: 120.2, autoign. temp.: 824°F, d: 0.88, vap. d: 4.15, bp: 164.1°.
Acute tox data: Oral LD_{LO} (rat) = 5000 mg/kg. [*3*]
THR = MOD via oral route.
Fire Hazard: Mod, via heat, flame or oxidizers.
To Fight Fire: Water spray or mist, dry chemical, CO_2.

***p*-ETHYL TOLUENE.** Autoign. temp: 887°F, bp: 162.2°.
Acute tox data: See *o*-ethyl toluene.
THR = See *o*-ethyl toluene.
Fire Hazard: See *o*-ethyl toluene.
To Fight Fire: See *o*-ethyl toluene.

***m*-ETHYL TOLUENE.** Autoign. temp: 896°F.
Fire Hazard: See *o*-ethyl toluene.
To Fight Fire: See *o*-ethyl toluene.

ETHYL-*p*-TOLUENE SULFONAMIDE. Liquid. $C_7H_7SO_2NHC_2H_5$, mw: 199.3, bp: 98° @ 745 mm, flash p: 260°F (CC), d: 1.253, vap. d: 5.6.

THR = U. See also amides.
Fire Hazard: Slight, when exposed to heat or flame.
Disaster Hazard: Dangerous; when heated to decomp, emits highly toxic fumes of oxides of sulfur; can react with oxidizing materials.
To Fight Fire: CO_2, dry chemical.

ETHYL-*p*-TOLUENE SULFONATE. Liquid. $C_7H_7SO_3C_2H_5$, mw: 200.25, mp: 33°, bp: 221.3°, flash p: 316°F (CC), d: 1.17, vap. d: 6.98.
Acute tox data: scLD_{50} (rat) = 500 mg/kg. [*3*]
THR = HIGH via sc route. An exper carc [*23*] and neo. [*3*]
Fire Hazard: Slight, when exposed to heat or flame.
Disaster Hazard: Dangerous; when heated to decomp, emits highly toxic fumes of SO_x; can react with oxidizing materials.
To Fight Fire: CO_2, dry chemical.

***p*-[(3-ETHYL-*p*-TOLYL)AZO]-N,N-DIMETHYL ANILINE.** $C_{17}H_{21}N_3$, mw: 267.4.
THR = An exper carc. [*3*]

***p*-[(4-ETHYL-*p*-TOLYL)AZO]-N,N-DIMETHYL ANILINE.**
THR = An exper carc. [*3*]

1-ETHYL-3-*p*-TOLYL TRIAZENE. $C_9H_{13}N_3$, mw: 163.3.
THR = An exper neo. [*3*]

ETHYL-*p*-TOSYLATE. See ethyl-*p*-toluene sulfonate.

ETHYL TRIBENZYL GERMANIUM. Colorless crystals, sol in methyl alcohol. $(C_2H_5)Ge(CH_2C_6H_5)_3$, mw: 375.0, mp: 57°.
THR = U. See germanium compounds.

ETHYL TRI-*n*-BUTYL TIN. Crystals. $(C_2H_5)(C_4H_9)_3Sn$, mw: 319.1, d: 1.0783, bp: 129° @ 10 mm.
THR = See tin compounds.

ETHYL TRICHLOROSILANE. Liquid. $C_2H_5Cl_3Si$, mw: 163.47, mp: −105.6°, bp: 99.5°, flash p: 72°F (OC), d: 1.24 @ 25°/25°, vap. d: 5.6.
Acute tox data: Inhal LC_{LO} (rat) = 500 ppm for 4 hrs; oral LD_{50} (rat) = 1330 mg/kg; ip LD_{LO} (rat) = 30 mg/kg. [*3*]
THR = MOD via inhal and oral and HIGH via ip routes.
Fire Hazard: Dangerous, when exposed to heat, flame or oxidizers.
Explosion Hazard: U.
Disaster Hazard: Dangerous; when heated to decomp, emits highly toxic fumes of phosgene; will react with water or steam to produce heat and toxic and corrosive fumes; can react vigorously with oxidizing materials.
To Fight Fire: CO_2, dry chemical.

3-ETHYL TRICYCLO QUINAZOLE. $C_{23}H_{16}N_4$, mw: 348.4.
THR = An exper carc. [*3*]

ETHYL TRIPHENYL GERMANIUM. Colorless powder, insol in water. $(C_2H_5)Ge(C_6H_5)_3$, mw: 330.0, mp: 78.5°.
THR = U. See germanium compounds.

ETHYL TRIPHENYL SILANE. Crystals, insol in water. $(C_6H_5)_3Si(C_2H_5)$, mw: 288.4, mp: 76°.
THR = U. See silanes.

ETHYL TRI-*n*-PROPYL TIN. $(C_2H_5)Sn(C_3H_7)_3$, mw: 277.0, mp: 101° @ 10 mm.
THR = U. See tin compounds.

ETHYL TRIS-*p*-BIPHENYL GERMANIUM. Colorless crystals. $(C_6H_5C_6H_4)_3(C_2H_5)Ge$, mw: 561.2, mp: 155°.
THR = U. See germanium compounds.

ETHYL UREA. Crystals. $NH_2CONHC_2H_5$, mw: 88.1, mp: 92°, flash p: >200°F, d: 1.213 @ 18°.
THR = No data. When mixed with $NaNO_2$ (2:1) it is an exper transplacental carc and teratogen. [*3, 23*]
Fire Hazard: Mod, when exposed to heat, flame or oxidizers.
Disaster Hazard: Dangerous; when heated to decomp, emits highly toxic fumes of cyanides; can react with oxidizing materials.
To Fight Fire: Foam, CO_2, dry chemical.

ETHYL URETHANE. See ethyl carbamate.

ETHYL VANILLIN. Fine, crystalline needles. $C_6H_3(OH)(CHO)(OC_2H_5)$, mw: 166.17, mp: 76.5°.
Acute tox data: Oral LD_{50} (rat) = 1590 mg/kg; ip LD_{50} (mouse) = 750 mg/kg; iv LD_{LO} (dog) = 760 mg/kg. [*3*]
THR = MOD via oral, ip and iv routes. Used as a synthetic flavoring substance and adjuvant. It is a substance which migrates to food from packaging materials. [*109*]

***m*-ETHYL VINYL BENZENE.** Syn: *ethylstyrene*. Water white liquid. $C_2H_5C_6H_4CHCH_2$, mw: 132.1, bp: 191.5°, fp: −127°, d: 0.8955 @ 20°, vap. press: 2.17 mm @ 40°, vap. d: 4.56.
Acute tox data: Oral LD_{50} (rat) = 4360 mg/kg. [*3*]
THR = MOD via oral route.
Fire Hazard: Mod; can react with oxidizing materials.
To Fight Fire: Foam, CO_2, dry chemical.

ETHYL VINYL ETHER. Syn: *ethenoxy ethane*. Liquid. $CH_2CHOC_2H_5$, mw: 72.1, mp: −115°, bp: 35.5°, flash p: <−50°F, d: 0.763 @ 15°/18°, vap. d: 2.49, autoign. temp.: 395°F, lel = 1.7%, uel = 28%.
Acute tox data: Oral LD_{50} (rat) = 8160 mg/kg. [*3*]

THR = LOW via oral route. May act as an anesthetic. See also ethers.
Fire Hazard: Very dangerous, when exposed to heat or flame; can react vigorously with oxidizing materials.
Explosion Hazard: U. See also ethers.
Disaster Hazard: Highly dangerous, upon exposure to heat or flame.
To Fight Fire: Alcohol foam, CO_2, dry chemical.

ETHYL XANTHOGEN DISULFIDE. Syn: *diethyl xanthogen sulfide.*
Acute tox data: Oral LD_{50} (rat) = 480 mg/kg; dermal LD_{LO} (rat) = 2100 mg/kg; ip LD_{50} (mouse) = 60 mg/kg. [*3*]
THR = HIGH via oral and ip; MOD via dermal routes.
Disaster Hazard: Dangerous; when heated to decomp, emits highly toxic fumes of SO_x.

ETHYL ZINC. See zinc diethyl.

ETHYNE. See acetylene.

2-ETHYNYL-2-BUTANOL. See methyl pentynol.

1-ETHYNYL CYCLOHEXAN-1-OL. See ethynyl cyclohexanol.

ETHYNYL CYCLOHEXANOL. Syn: *1-ethynyl cyclohexan-1-ol.* $C_8H_{12}O$, mw: 124.18, mp: 32°, bp: 180°, vap. d: 3.73.
Acute tox data: Oral LD_{50} (rat) = 600 mg/kg; dermal LD_{50} (rabbit) = 1000 mg/kg; ip LD_{LO} (mouse) = 500 mg/kg. [*3*]
THR = MOD via oral, dermal and ip routes.
Fire Hazard: Mod; can react with oxidizing materials.
To Fight Fire: Foam, CO_2, dry chemical.

17-α-ETHYNYL-4-ESTRENE-3-β,17-DIOL DIACETATE. $C_{20}H_{28}O_2$, mw: 300.3.
THR = An exper teratogen to mice via oral route. [*103*]

17-ETHYNYL-18-METHYL-19-NORTESTOSTERONE. Syn: *norgestrel.* White crystals; $C_{21}H_{28}O_2$, mw: 312.5, mp: 206°.
THR = An exper (±) neo. [*3, 15*]

ETIOLOGIC AGENTS. Cultures and infectious materials causative of serious communicable diseases.

EUCALYPTUS OIL. See oil of eucalyptus.

EUCLASE. See beryllium aluminum silicate.

EUGENIC ACID. See eugenol.

EUGENOL. Syns: *4-allyl-2-methoxy phenol, coryophyllic acid, eugenic acid.* Colorless or yellowish liquid, oily, spicy odor, sol in alcohol, chloroform, ether and volatile oils, very slightly sol in water. $C_3H_5C_6H_3(OH)OCH_3$, mw: 164, d: 1.064–1.070, bp: 253.5°.
Acute tox data: Oral LD_{50} (rat) = 2680 mg/kg; ip LD_{LO} (rat) = 800 mg/kg; sc LD_{LO} (rat) = 5000 mg/kg. [*3*]
THR = MOD via oral, ip and sc routes. A synthetic flavoring and adjuvant. [*109*]

EUGENOL ACETATE. $C_{12}H_{14}O_3$, mw: 206.3.
Acute tox data: Oral LD_{50} (rat) = 1670 mg/kg. [*3*]
THR = MOD via oral route.

EUROPIUM. A silvery-white metal, mod soft and malleable, about as reactive as Ca. Eu, atwt: 151.96, mp: 822°, bp: 1597°, d: 5.253 @ 25°.
THR = U.
Radiation Hazard: For permissible levels, see Section 5, Table 5A.5.
Artificial isotope ^{152}Eu, $T_{\frac{1}{2}}$ = 12y, decays to stable ^{152}Sm via ec. Also decays to ^{152}Gd via β's of 0.22–1.48 MeV. Emits γ's of 0.12–1.4 MeV and x-rays. Can react violently with air, halogens. [*19*]

EUROPIUM CHLORIDE. $EuCl_2$, mw: 258.3.
Acute tox data: Oral LD_{50} (mouse) = 500 mg/kg; ip LD_{50} (mouse) = 387 mg/kg. [*3, 20*]
THR = HIGH via ip: MOD via oral routes.

EUROPIUM (III) NITRATE, HEXAHYDRATE. $Eu(NO_3)_3 \cdot 6H_2O$, mw: 446.1.
Acute tox data: ip LD_{50} (rat) = 210 mg/kg. [*3*]
THR = HIGH via ip route.

EXCESS CANCER IN CHEMISTS. See Section 8

EXEMPT QUANTITIES OF RADIOACTIVITY. See Section 5A.

EXPANSIN. Syns: *clairformin, patulin.* Colorless crystals, $C_7H_6O_4$, mw: 154, mp: 111°.
THR = An exper carc. [*3, 7*]

EXPERIMENTAL INDUCTION OF TUMORS vs DIET. See Section 8.

EXPERIMENTAL INDUSTRIAL CARCINOGENS. See Section 8.

EXPLOSIONS IN SOLID WASTE. See Section 6.

EXPLOSIVE RANGE. See Section 7.

EXPLOSIVES STORAGE AND HANDLING. See Section 7.

EXPLOSIVES, HIGH. High explosives (HE) are those which decompose by detonation. This is a very rapid (nearly instantaneous) process; hence the action is fast and violent. An explosion may be initiated by sudden shock, by high temp. or by a combination of the two. For many explosives the conditions under which they will explode are well known, for example:

An explosion may be initiated by elevated temp. alone:

(1) In the case of mercury fulminate, 15 sec exposure to 200° or 1 sec exposure to 340° will set it off.

(2) Trinitrotoluene will be set off by exposure to 500° for 1 sec.

(3) Tetryl will detonate in 1000 sec at 160° or in 0.1 sec at 500°.

(4) Picric acid will detonate in 9 sec at 300° or 1 sec at 355°.

An explosion of HE may also be initiated by severe shock. Sensitivity of explosives to shock may be measured in several ways, such as the impact pendulum method and the drop test. The impact pendulum test operates by allowing a heavy pendulum to swing down over a sample of explosive in a dished, inclined container so arranged that there is very little clearance between the pendulum and the sample. Thus the effect of contact between the sample and the pendulum bob is one of a combination of shock and rubbing. The height from which the pendulum is allowed to swing to explode the sample is a measure of the sensitivity of the sample to this test. The drop test consists of placing a sample upon an anvil and allowing a 5 lb. weight to drop on it. The height from which the weight must drop to explode the sample is a measure of the sample's sensitivity to shock.

Below is a table of the results of a drop test upon several samples. These results must be considered as relative and not by any means absolute. Solid explosive in a tightly fitting container is much more sensitive to shock.

(1) mercury fulminate = 2 in. at 5 lbs.
(2) nitroglycerin = 4 in. at 5 lbs.
(3) tetryl = 8 in. at 5 lbs.
(4) picric acid = 14 in. at 5 lbs.
(5) trinitrotoluene = 20 in. at 5 lbs.
(6) black powder (a low explosive) = 30 in. at 5 lbs.*

*From Explosions, Their Anatomy and Destructiveness," by C. S. Robinson (McGraw-Hill).

Another test for explosives is the speed at which a detonation travels. This speed is usually in the range of thousands of m/sec. Speed of detonation is found to be a function of kind of explosive and state of compaction. There is an optimum state of compaction beyond which the explosive tends to become "deadpressed," in which state it is difficult to make the whole sample explode. Below the point of optimum compaction the rate of detonation is found to be directly proportional to the density of the sample. Below are listed some maximum detonation rates, in meters/second for some common explosives:

(1) nitroglycerin	8500
(2) PETN	8100
(3) tetryl	7700
(4) picric acid	7400
(5) trinitrotoluene	7400
(6) lead azide	4900
(7) mercury fulminate	4800
(8) ammonium nitrate	1100
(9) low explosives	1000

It has been found that upon detonation an explosive can cause a nearby sample of explosive to detonate "sympathetically." The distance over which one charge can detonate another is a function of the amount of energy produced by the first explosion and the medium through which the shock wave is propagated to the second charge of explosive. For instance the relationship for air (very approximately) would be expected to be: Weight of explosive in lbs/(distance in ft)^3 = 4. Thus to calculate the maximum distance for a possible sympathetic detonation of 40,000 lbs of explosive, the calculation is:

$$D^3 = (40{,}000)/4$$
$$D^3 = 10{,}000$$
$$D = 22 \text{ ft (approximately).}$$

According to C. S. Robinson the formula is more nearly:

$$\text{weight of explosive} = 4 \times (\text{distance})^{2.25}$$

The power of the shock wave is much more rapidly attenuated in water, wood, etc., than in air, which means that if a shield of water or wood is interposed between piles of explosive the distance between them may be lessened.

Liquid Oxygen. Though not by itself explosive, liquid oxygen can be dangerous when blended with highly flam or carbonaceous materials. In this combination it is used in coal mining, quarrying, strip mining, and open-cut ore mining, and in rocket fuels. Its use underground or in confined places is not recommended by the U.S. Bureau of Mines, because it evolves a lot of carbon monoxide. (See carbon monoxide). This type of explosive has many safety advantages. For instance, it is not itself an explosive until mixed with a flam absorbent which can be done at the last moment before firing. However, once the explosive has been made up, it is very flam and when it catches fire it will usually detonate. Liquid oxygen explosives are not stored, as they deteriorate rapidly and lose a great deal of their explosive power in a short time.

General Fire Hazard: Severe, when exposed to heat or by chemical reaction with powerful oxidizing or reducing agents.

General Explosion Hazard: Mod to dangerous when severely shocked or heated, depending upon kind of explosive, state of compaction, degree of confinement, etc. Practically all high explosives used commercially require a detonator or cap to set them off, as compared to an igniter needed to set off black blasting powder (See also Explosives, Permissible;

Dynamite; Nitroglycerin; Ammonium Nitrate; Nitrates; Trinitrotoluene).

Detonating Devices: To develop the desired disruptive effect of an explosive, some means must be adopted to "set off," "fire," or "detonate" it without killing or maiming the persons doing the blasting. Several devices or methods are being utilized, all with a view to having this work done as safely and efficiently as possible. There are two general types of devices or methods of getting explosives into action, namely, igniters and detonators. The former merely conveys a flame to the explosive mass and ignites it, while the latter transmits (originally through ignition of a small quantity of highly explosive substance by an arc, a flame, or spark) a sharp blow that causes the explosive to disassociate, or detonate, or burn with very great rapidity. Igniters are squibs (plain and electric), fuse, and delay igniters; detonators are blasting caps (plain and electric), delay electric blasting caps, delay electric igniters with caps, and Cordeau-Bickford detonating fuse.

The squib is a small-diameter tube of straw or paper filled with quick-burning powder and having a relatively slow burning "match head" attached to one end; the latter is ignited or lighted by an ordinary match or other flame, and its relatively slow burning allows the person handling the ignition to retire before the fire is communicated to the quick-burning material in the tube. Squibs are by no means either safe or efficient, even though still used to a considerable extent, especially in coal mining. Electric squibs are somewhat similar to ordinary squibs, except that the ignition is accomplished by means of an electric arc; electric squibs are much more satisfactory from a safety viewpoint than ordinary squibs.

A fuse (or, as it is sometimes called, "safety fuse") consists of a fine-grained black powder core covered with cotton hemp or jute to form a ropelike material about 3/16 in. in diameter; one end of the fuse is brought in contact with the powder charge or with a detonating "cap," and the other end (usually several feet away from the explosive) is lighted by a flame from a match or open light. The fine-grained black powder burns gradually and somewhat slowly (about 30 to 40 sec. to the linear foot of fuse) until it reaches the explosive (black powder) or the detonating cap (if some form of dynamite is used), giving the blaster time to get in the clear before the main explosion takes place. Fuses are much safer than squibs, but have their own hazards and must be used with care.

Delay electric igniters usually are a combination of electric igniters and fuses, the latter being ignited by the igniters within the blasting hole, the fuse transmitting the ignition to the explosive. Delay igniters usually are much safer than fuse, particularly for coal-mine use; but they, too, have their hazards. Delay blasting is by no means a safe procedure in coal mining, though it is a standard and relatively safe practice in metal mining and tunneling, if sensible precautions are taken.

Blasting caps or detonators are metallic cylinders (usually copper) closed at one end, about 3/16 in. in diameter, and usually less than 2 in. in length, partly filled with a small amount of relatively easily fired or "detonated" compound, the resultant shock or blow when fired being sufficient, when embedded in dynamite, to fire or detonate the dynamite mass. Ordinary blasting caps usually are fired or detonated by the flame of the fuse, the end of the latter being inserted into the open end of the detonator or cap and placed in contact with the highly explosive material in the interior of the metallic capsule or cap. Caps are extremely hazardous to handle, as they are likely to be detonated by heat, friction, or a relatively moderate blow; however, they are relatively safe if handled carefully. Partial proof of this is the fact that they are manufactured and shipped by the thousands daily and accidents are decidedly rare, primarily because the caps are at all times handled with utmost care.

Electric blasting caps are somewhat similar to ordinary caps or detonators, but the cap is fired by electricity. The electric wires are so placed in the capsule or cap that when attached to an electric current an arc is formed within the cap, which detonates the sensitive explosive material in the cap. A hazard in the use of electric blasting caps is unexpected explosions due to radio or radar induced electric currents which may activate the cap.

Delay electric blasting caps or detonators are somewhat similar to ordinary electric blasting caps, except that several time intervals in blasting are obtained by having the electric arc ignite a short piece of fuse or some slow-burning substance before it reaches the highly sensitive detonating material in the capsule or cap. Numerous time-interval delays are obtained; in general delay electric blasting caps are relatively safe and effective even in wet holes, though they ought not be used in coal mining if explosions of gas or dust are to be avoided. Delay electric igniters with caps or detonators are a combination of electric igniter and blasting cap, usually with suitable lengths of fuse between to give the desired delay; they have some advantages but are relatively unsafe and should not be used in coal mining.

The Cordeau-Bickford denotating fuse is a combined fuse and detonator in the form of a lead tube about 1/4 in. in diameter filled throughout its length with a high explosive, trinitrotoluene (TNT). It is fired by fuse and an ordinary detonator or cap or by an electric cap; when fired, it detonates throughout its length (which may be up to or over 100 ft) almost instantaneously, the explosion wave traveling at a rate of about 17,500 ft/sec. Although somewhat expensive, it is relatively safe to handle and is particularly effective in deep-well drill holes in quarry and similar work, as it detonates simultaneously throughout its length, adding effectiveness to the main body of explosive which it detonates. It fires black powder as well as high explosives (dynamite, etc.) and is obtainable in lengths of approximately 500 ft, wound on spools.

Transportation of Explosives. In the transportation of explosives, the manufacturers and the railroads have set a most excellent example in safety, as is proved by the fact that accidents in rail transportation of explosives were so rare as to have been almost nonexistent in the past decade. The problem of transporting explosives from the railroad car to the users' magazines and from the magazines to the places of use (the mine, quarry, tunnel) is decidedly serious and should be given much more consideration than it usually receives.

Users of explosives who transport them by truck would profit by familiarizing themselves with the regulations established by the Interstate Commerce Commission. Railroad companies formulate and issue their own regulations for the transportation of explosives and other dangerous articles, and the restrictions of some of the carriers are drastic in the extreme. The Bureau of Explosives of the American Railway Association has issued much interesting information relating to the transportation of dangerous substances, including explosives of various kinds. The Bureau of Mines and other organizations have also issued information on the transportation of explosives, but much careless explosives-transportation practice occurs, not only in mining work but to an even greater extent in other activities in which explosives are used. State laws on transportation of explosives usually are confusing and inadequate; in fact, state laws on this subject usually are almost nonexistent or are phrased so broadly as to be almost meaningless.

EXPLOSIVES, LOW.

Syns: *Black blasting powder*, *gunpowder*, *"A" blasting powder*, *"B" blasting powder*. Composition: Black powder is composed of saltpeter, charcoal and sulfur in the approximate proportions of 6:1:1. ("A" blasting powder uses KNO_3 and "B" blasting powder uses $NaNO_3$). Low explosives are explosives which deflagrate; this differentiates them both in composition and properties from high explosives, which detonate. A deflagrating explosive is one that burns progressively over a relatively sustained period of time in comparison with a detonating explosive, which decomposes almost instantaneously.

THR = Variable.

Fire Hazard: Dangerous, when exposed to heat or flame or by chemical reaction.

Although most safety men now look upon black blasting powder with disfavor, it is one of the oldest and most generally used explosives in commercial work. It burns with extreme rapidity instead of detonating as high explosives do; it is highly sensitive to flame or sparks or friction and gives off much flame, which is hot and of great length of duration. These properties make it extremely hazardous for use in mines (especially coal mines) and quarries. The gases given off in detonation are not only very hot but frequently contain harmful constituents. Notwithstanding its numerous deficiencies, from a safety standpoint, it has action characteristics that make it valuable in both coal mining and quarrying, though it has relatively little utility in metal mining. It is difficult to use effectively in wet places and this is its main disadvantage from an efficiency standpoint. Pellet powder is black blasting powder in consolidated (pellet or stick) form rather than in grains or granules, and it has few if any real advantages over black blasting powder, notwithstanding the fairly prevalent idea that it is a "safe" explosive.

Smokeless powders have a somewhat different composition from that of black blasting powder and are used chiefly for sportsmen's ammunition and, more widely, for military purposes. They are decidedly sensitive to flame and impact but ordinarily are so packaged that if reasonable judgment is used they are relatively harmless.

Most black powder fires start from sparks. Ignition results in an explosion so quickly that no attempt can be made to fight the fire. Every effort should be made to prevent fires from reaching stores of black powder; but if this fails, fire-fighting forces should be withdrawn at least 800 ft from the fire and should protect themselves against an explosion by seeking any cover available or by lying flat on the ground. If an explosion does occur, every effort should be made to prevent flames from spreading to neighboring magazines. Fire-fighting forces should be cautioned against approaching a fire which may involve black powder to avoid being trapped or injured by an explosion. Most explosions from black powder originate from sparks and the following safety rules should be strictly enforced and obeyed.

(1) Open no containers in a magazine in which explosives or ammunition are stored. This should be done only in a building free from all other explosives or ammunitions; or in suitable weather in the open, at least 100 ft from the nearest magazine. The quantity at or near such an operation should be limited to 100 lbs. Safety tools only should be used in opening or closing containers or in other operations involving black powder. Safety shoes (non-insulating) should be worn in all rooms where black powder is handled and by all persons engaged in handling black powder. The wearing of all non-conductive shoes, such as rubber is prohibited. If the handling of black powder is carried on over a concrete floor; the floor should be covered with a tarpaulin or other suitable material.

Loose black powder is extremely dangerous. Whenever it is necessary to handle loose black powder, not over 50 lbs should be permitted at or near such operations. If black powder is spilled on benches or floors, all work should be stopped until it has been removed and the explosive hazard of any remaining dust or particles has been neutralized with water. Rooms or buildings in which black powder is handled should be inspected frequently for dust, and all such dust should be immediately removed with water. (See also Section 7).

Explosive Hazard: Severe, when exposed to heat or flame or by chemical reaction. Black powder is the slowest acting of all explosives. It has a shearing and heaving action, tending to blast materials in large, firm fragments. The action derives from a relatively slow development of gas pressure so that it must be carefully loaded and closely confined. The application is limited because it disintegrates in water and therefore cannot be used in wet work, except with special precautions. It also produces more smoke and fumes than most dynamites.

Precautions in Handling: Black powder is the most treacherous explosive material used today and it is regarded as one of the worst known explosive hazards. When ignited unconfined it burns with explosive violence and will explode if ignited under even slight confinement. It can be ignited easily by very small sparks, heat and friction. It is subject to rapid deterioration in the presence of moisture, but if kept dry it retains its explosive properties for many years. It is used to ignite smokeless powder, propelling charges, airplane flares and bursting charges of hand grenades, as a bursting charge in shrapnel, practice bombs, practice trench-mortar shells, in saluting charges, smoke-puff charges, time and percussion fuses, pellets, primers and primer detonators, and in expelling charges of pyrotechnic signals.

When black powder is shipped or received, each container should be inspected for holes and weak spots. It should be examined particularly for small holes, such as those made by nails which are visible only upon close examination. Damaged containers should not be repaired. The contents should instead be transferred to new or serviceable containers. Metal containers for export shipment should be crated. Usually two containers are packed in each crate. Repainting of containers and repacking of black powder, contained in damaged or unserviceable containers, constitutes the principal maintenance activity for such stocks. Black powder containers are also subjected to sweating, which rusts metal drums or kegs, so that repainting is necessary to keep containers serviceable. Repainting should not be done in a magazine in which explosives or ammunition are stored. It may be done in a nearly empty magazine, or in clear weather in the open, at least 100 ft from the nearest magazine. The quantity of black powder, at or near such operations should be limited to 100 lbs.

The marking on repainted containers should be checked carefully to see that it is a facsimile of the old. Furthermore, the metal caps on certain types of black powder containers deteriorate in storage. Replacement of these caps may be allowed, but the same safety precautions as outlined above for repainting containers should be followed. Operations, such as removal of black powder from containers and its transfer from unserviceable to serviceable drums should be conducted in strict compliance with applicable portions of the above outlined safety regulations. Black powder operations of any kind should be conducted in special buildings which are not used for other purposes at the same time. The floors of such buildings should be covered with suitable materials. Intraplant-quantity-distance requirements for high explosives as given in Section 7 should be followed. Absolute cleanliness should be maintained at all times in and around each operation. Non-insulating safety shoes should be worn by personnel in all assembly operations. All equipment should be electrically grounded and this should be determined by tests. Empty metal containers which have held black powder should be thoroughly washed inside with water before they are disposed of. Serious explosions have occurred from supposedly empty cans. Wooden containers should be destroyed by burning. Safety tools only should be used in opening or closing containers or in handling black powder. Processes should be so laid out as to bring about frequent grounding of all operators handling this material.

Destruction: Black powder is best destroyed by dumping it into water, preferably a stream or large body of water, but some states have anti-stream pollution laws that may be violated in this manner. It is well to be informed on this subject and to be certain that no damage will be done to water supplies.

If it is inconvenient or inadvisable to destroy black powder thus, it may be burned. If that is done the contents of only one container holding 25 lbs or less should be burned at a time. The powder should be poured on the ground in a trail not wider than 2 in., and no part of such a trail should parallel another part at a distance of less than 10 ft. A train of combustible material should be laid to the explosive for igniting it, as with dynamite.

"Pellet powder" (black powder in compressed stick form) may be destroyed either by throwing it into water after removing the wrapper, or as described for dynamite. If dry and in good condition, black powder burns rapidly, especially in small grain size, with a yellow or pinkish-blue flame and dense smoke.

The empty powder containers should be washed out, as explosions are said to have occurred from "empty" containers.

EXPLOSIVES, PERMISSIBLE. "Permissible" explosives are essentially high explosives (dynamite) modified by the introduction of "dopes." The function of the dopes in general is to decrease flame temperature and to a smaller extent, the length and duration of flame, when the explosive is converted from a solid into a gas; in other words, when it is fired or detonated. The designation "permissible" is given to an explosive of modified dynamite type after it has passed certain tests made by the Federal Bureau of Mines. The permissible character of such explosives depends not only upon the ingredients in the explosive, but also on certain well-defined specifications as to handling and use. As with the dynamites, there are several different types and grades; "permissibles," hydrated "permissibles," organic nitrate "permissibles," nitroglycerin "permissibles," ammonium perchlorate "permissibles," and gelatin "permissibles." Essentially all those now used to any extent are in either the ammonium nitrate or the gelatin classes. (See also dynamite.)

The ammonium nitrate "permissible" explosives contain relatively little nitroglycerin and relatively large proportions of ammonium nitrate. The latter is an explosive but one less sensitive to impact, sparks, and flames than nitroglycerin. This type of permissible" explosive is now used extensively, as it has a rather wide range of strength, rate of detonation, density, size of cartridge, etc., and can be utilized not only in dry but also to some extent in fairly wet holes if charged carefully and fired promptly. Gelatin "permissibles," explosives are more suitable than ammonium nitrate "permissibles," for wet holes, and in general are stronger and more violent than the ammonium nitrate types.

All "permissible" explosives are strong, must be used in relatively small quantities (less than 1½ lbs) per hole to retain their permissibility, give off considerable quantities of toxic gases on detonation, and, while much safer than black blasting powder or dynamite, must be stored, handled, and used with care.

Classification Upon Basis of Toxic Gases: All "permissible" explosives, when detonated, emit some toxic gases and a much larger volume of non-toxic gases. In order that the toxic products may not become a menace to the life or health of miners, no explosive is now or can become "permissible" if upon detonation it evolves more than 158 liters (5½ cubic ft) of toxic gases per 1½ lb charge as determined by tests in the Bichel pressure gage.

Classification upon the basis of the volume of toxic gases produced by 680 g (1½ lbs) of explosive is as follows: Class A, not more than 53 liters; Class B, between 53 and 106 liters; and Class C, between 106 and 158 liters. (These classifications are not to be confused with the I.C.C. Classification of explosives).

Field tests were made with a 1½ lb charge of a "permissible explosive that produced, in the Bichel gage, the maximum allowable quantity of poisonous gases (158 liters per 1½ lbs); these tests indicated that in a narrow entry, without artificial ventilation, 1800 ppm of carbon monoxide (the only poisonous gas present) was produced, as shown by analysis of an air sample taken 2 min. after the shot. Another sample of the air taken 2 min. later contained 800 ppm of carbon monoxide. Under no conditions should miners or shot firers return to the face until the poisonous gases have been removed by adequate ventilation.

It is provided further that, in accordance with the provisions and conditions, explosives enumerated on the "permissible" lists of the Bureau of Mines are "permissible" in use only when they satisfy the following requirements:

1. That the explosive be in all respects similar to the sample submitted by the manufacturer for test, and that the diameters of the cartridges used must be those that have been approved.

2. That electric detonators (not fuze and detonators)

be used of not less efficiency than No. 6, the detonation charge of which shall consist of a 1 gram mixture of 80 parts of mercury fulminate and 20 parts of potassium chlorate (or their equivalents), and that the required electric firing must be done by means of a "permissible" type blasting unit.

3. That the explosive be stored in surface magazines under proper conditions, so that it will not undergo change in character, and that after taking underground it be used in less than 36 hrs.

4. That the coal to be blasted be undercut or equivalently relieved; that, to prevent blow-through, all portions of the borehole must be at least 18 in. from relief in any direction; that, to prevent blowouts, the charge be properly confined with not less than 2 ft of clay (if the length of the hole will not permit the charge desired and 2 ft of stemming, at least half the length of the hole shall be filled with stemming) or other incombustible stemming and not be on the solid; that, to prevent the hole being on the solid, it shall be at least 6 in shorter than the depth of the undercut or equivalent relief, and, when placed adjacent to the roofs, ribs, or floor, all but 12 in at the rear of the hole must be at least 6 in from the adjacent surface as projected into the coal to be blasted, and all parts of the hole shall be free from the adjacent surface as projected into the coal to be blasted; that the shot be not a dependent shot; and that the shot hole be cleaned before charging.

5. That the quantity used for a shot (1) be not in excess of 680 g ($1\frac{1}{2}$ lbs) when fired in accordance with these requirements and (2) when used under certain additional requirements or restrictions be not in excess of 1,361 g (3 lbs). The use of charges over $1\frac{1}{2}$ lbs and not exceeding 3 lbs is approved tentatively pending further investigation. For charges of over $1\frac{1}{2}$ lbs, the following additional requirements must be observed; (a) Shot holes must be 6 ft or more in length. (b) Explosive must be charged in a continuous train, with no cartridges deliberately deformed or crushed, with all cartridges in contact with each other, and with the end cartridges touching the rear of the hole and the stemming, respectively. (c) Examination for gas must be made in the blasting area before and after a shot is fired. (d) The "permissible" explosive must be one showing toxic gas emission that will place it either in Class A or Class B.

6. That the region in which the blasting is done be kept well-protected by rock dust or otherwise in accordance with Bureau of Mines inspection standards.

7. That the shot not be fired in the presence of a dangerous percentage of firedamp. Examination for firedamp to be made at the blasting area before shooting in a gassy mine.

See also articles on dynamite, nitroglycerin, trinitrotoluene, pentaerythritol tetranitrate, nitrates, ammonium nitrates, picrates, azides, fulminates.

EXTRACT D AND C RED #14. $C_{18}H_{16}ON_2$, mw: 276.4.
THR = An exper (+) carc and neo. [*3*, *4*]

EXTRACTS, LIQUID FLAVORING. Flash p: 80° F.
THR = Variable; an allergen.
Fire Hazard: Dangerous, when exposed to heat or flame; can react vigorously with oxidizing materials.
Explosion Hazard: U.

f

F-114. See dichlorotetrafluoroethane.

FACTORS OF TOXICITY IN HUMANS. See Sections 1, 8, 9.

FAIRY GLOVES. See digitalis.

FAMPHUR. see *o,o*-dimethyl-*o,p*-(dimethylsulfanoyl) phenyl phosphorothioate.

FANFT. $C_8H_5O_4N_3S$, mw: 239.2.
THR = An exper carc via oral route. [*3, 23*]

FANS FOR EXHAUST SYSTEMS. See Section 2.

FARNESOL. Syn: *3,7,11-trimethyl-2,6,10-dodecatrien-1-ol.* An insect hormone.
THR = U.

FAST GREEN FCF. $C_{37}H_{34}N_2O_{10}S_3 \cdot 2Na$; mw: 808.9; a dark green powder.
THR = An exper (+) carc of subcutaneous tissue [*14*] in rats. [*108*]

FAST LIGHT ORANGE. $2Na \cdot C_{16}H_{10}O_7N_2S_2$, mw: 452.4.
THR = An exper (±) carc. [*3, 4*]

FAST ORANGE. See diacetyl aminoazo toluene.

FAST OIL RED B. $C_{24}H_{20}ON_4$, mw: 380.5.
THR = An exper (+) carc. [*3, 4*]

FAST RED R. $C_{22}H_{16}ON_4$, mw: 352.4.
THR = An exper (±) carc. [*3, 4*]

FATTY ACIDS. Monobasic organic acids derived from natural fats and oils. The term is also applied to all the monobasic acids of the general formula: $C_nH_{2n+1}COOH$; e.g., stearic acid, oleic acid, linoleic acid, etc.
THR = Variable. Food additives permitted in foods for human consumption. [*109*]

FATTY ACID ESTERS OF PHORBOL.
THR = A cocarc. [*23*]

FATTY AMINES. Syn: *aliphatic amines.* Look like ordinary oils or fats. $R\text{-}NH_2$ where R = 8–22 C atoms. Can be primary, secondary or tertiary.
THR = As organic bases they are MILD irr to skin and mu mem. Regard these compounds as mild alkalies. See also amines.

FD and C BLUE #1. $C_{37}H_{36}O_9N_2$, mw: 782.8.
THR = An exper carc and neo. [*3*] in rats via sc route. [*108*]

FD and C GREEN #3. $C_{37}H_{36}O_{10}N_2S_3$, mw: 810.9
THR = An exper carc. [*3*]

FD and C ORANGE #2. $C_{17}H_{14}ON_2$, mw: 262.3.
THR = An exper (+) carc. [*3, 4*]

FD and C RED #1. $C_{19}H_{16}O_7N_2S_2 \cdot 2Na$, mw: 494.5.
THR = An exper (+) carc and neo. [*3, 4*]

FD and C RED #2. $C_{20}H_{11}O_{10}N_2S_3$, mw: 604.5.
THR = An exper (±) carc and teratogen. [*3, 4*]

FD and C RED #4. $2Na \cdot C_{18}H_{14}O_7N_2S_2$, mw: 480.4.
THR = An exper (−) carc. [*3, 4*]

FD and C RED #32. $C_{18}H_{16}ON_2$, mw: 276.4.
THR = An exper neo, carc. [*3*]

FD and C YELLOW #6. $2Na \cdot C_{16}H_{10}O_7N_2S_2$, mw: 452.4.
THR = An exper (±) carc and neo. [*3, 4*]

FEDERAL COAL MINE HEALTH AND SAFETY ACT OF 1969. See Section 1.

FEDERAL RADIATION COUNCIL. See FRC, Section 5A.

FEDERAL REGULATORY AGENCIES. See Sections 11, 10A.

FELT WASTE, WET. See waste wool, wet.

FENAC. See 2,3,6-trichlorophenylacetic acid.

FENAZO RED C. $2Na \cdot C_{20}H_{12}O_7N_2S_2$, mw: 502.4.
THR = An exper (±) carc. [*3, 4*]

FENAZO SCARLET 2R. $2Na \cdot C_{18}H_{14}O_7N_2S_2$, mw: 480.4.
THR = An exper (+) carc, neo. [*3, 4*]

FENCHLORPHOS. See ronnel.

FENITROTHION. Syns: *nitrophos, folithion.*
$C_9H_{12}O_5NPS$, mw: 277.3.
Acute tox data: LD_{50} (rat) = 250 mg/kg; dermal LD_{LO} (rat) = 300 mg/kg; oral LD_{50} (wild birds) = 25 mg/kg; oral LD_{50} (chicken) = 280 mg/kg. [*3*]
THR = HIGH via oral and dermal routes. Has recently been implicated in Reyes Syndrome when sprayed over forest areas, although the connection may be due to petrochemical used as a vehicle. [*39*]

FENOPROP. See 2-(2,4,5-trichlorophenoxy)propionic acid.

FENSON. See murvesco.

FENTHION. See *o,o*-dimethyl-*o*-[(4-methylthio)-*m*-tolyl] phosphorothioate.

FENURON. See 1,1-dimethyl-3-phenyl urea.

FERACONITINE. See pseudoaconitine.

FERBAM. Syn: *ferric dimethyl dithiocarbamate.* Black solid, slightly sol in water. $Fe[(CH_3)_2NCS_2]_3$, mw: 416.5, mp: decomp 180°.

Acute tox data: Oral LD_{50} (rat) = 4000 mg/kg; ip LD_{50} (rat) = 2700 mg/kg. [*3*]

THR = MOD via oral and ip routes. An exper neo via oral route. [*3, 12*] A fungicide. See also bis(diethyl thiocarbamyl)disulfide.

Disaster Hazard: Dangerous; when heated to decomp, emits highly toxic fumes.

FERMATE. See ferbam.

FERMIUM. A synthetic, highly radioactive element. Fm.

Radiation Hazard: For permissible levels, see Section 5, Table 5A.5.

FERNAMBUCO. See brazil wood.

FERRIC AMMONIUM CITRATE. See iron ammonium citrate.

FERRIC ARSENATE. Syn: *scorodite.* Rhombic green crystals. $FeAsO_4 \cdot 2H_2O$, mw: 230.78, mp: decomp, d: 3.18.

THR = HIGH via oral and inhal routes. See also arsenic compounds.

Disaster Hazard: See arsenic compounds.

FERRIC ARSENIDE. Syn: *iron arsenide.* White powder. FeAs, mw: 130.8, mp: 1020°, d: 7.83.

THR = No data but probably HIGH via oral and inhal routes. See arsenic compounds.

FERRIC ARSENITE. Brown-yellow powder. $2FeAsO_3 \cdot Fe_2O_3 \cdot 5H_2O$.

THR = HIGH via oral and inhal routes. A poison.

Disaster Hazard: See arsenic compounds.

FERRIC BROMIDE. Dark red-brown, deliquescent crystals. $FeBr_3$, mw: 295.60, mp: sublimes with decomp.

THR = A powerful irr to skin, eyes and mu mem. See also bromides. Can explode with K, Na. [*19*]

Fire Hazard: Slight.

FERRIC CACODYLATE. Yellowish-brown odorless powder. $Fe[(CH_3)_2AsO_2]_3$, mw: 466.8.

THR = HIGH via oral and inhal routes. See also arsenic compounds.

FERRIC CHLORIDE. Black-brown solid. $FeCl_3$, mw: 162.2, mp: 292°, bp: 319.0°, d: 2.90 @ 25°, vap. press: 1 mm @ 194.0°.

Acute tox data: Oral LD_{50} (rat) = 900 mg/kg; ip LD_{50} (mouse) = 68 mg/kg; oral LD_{50} (mouse) = 440 mg/kg. [*3*]

THR = HIGH–MOD via oral route; HIGH via ip route. Used as a trace mineral added to animal feeds. [*109*] Violent reaction with allyl chloride, Na, K. [*19*]

Disaster Hazard: Dangerous; when heated to decomp, emits highly toxic fumes of hydrochloric acid; will react with water to produce toxic and corrosive fumes.

FERRIC DIARSENIDE. Syn: *iron diarsenide.* Cubic, silver-gray crystals. $FeAs_2$, mw: 205.7, mp: 990°, d: 7.4.

THR = HIGH. See also arsenic and arsine.

FERRIC DICHROMATE. Red-brown granules. $Fe_2(Cr_2O_7)_3$, mw: 759.8.

THR = HIGH via oral and inhal routes. See also chromium compounds.

Fire Hazard: Mod, by chemical reaction with reducing agents.

FERRIC DIMETHYL DITHIOCARBAMATE. See ferbam.

FERRIC ETHYLENE BIS-DITHIOCARBAMATE.

THR = A fungicide. HIGH via oral and inhal routes. See ferbam and bis(diethylthiocarbamyl)disulfide.

Disaster Hazard: Dangerous; when heated to decomp, emits highly toxic fumes of SO_x.

FERRIC FERRICYANIDE. Syn: *berlin green.* Cubic crystals. $Fe[Fe(CN)_6]$, mw: 267.8.

THR = See ferricyanides.

FERRIC FERROCYANIDE. Dark blue crystals. $Fe_4[Fe(CN)_6]_3$, mw: 859.3, mp: decomp.

THR = See ferrocyanides. Upon being ground with $PbCrO_3$, it sparked and burned violently. [*19*]

FERRIC FLUORIDE. Syn: *iron fluoride.* Green crystals. FeF_3, mw: 112.9 d: 3.87.

THR = See fluorides.

FERRIC FLUOSILICATE. Gelatinous, flesh-colored solid. $Fe_2(SiF_6)_3$, mw: 537.9.

THR = See fluosilicates.

FERRIC FORMATE. Syn: *iron formate.* Red crystals. $Fe(CHO_2)_3$, mw: 190.9.

THR = See formic acid.

FERRIC HYDROXIDE NITRILOPROPIONIC ACID COMPLEX.

THR = An exper neo. [*3*]

FERRIC HYPOPHOSPHITE. $Fe(PH_2O_2)_3$, mw: 250.9.

THR = Impact-sensitive. [*19*] See also hypophosphites.

FERRIC METHANE ARSENATE. Reddish-brown lustrous scales. $Fe_2(CH_3AsO_3)_3$, mw: 525.5.

THR = HIGH. See arsenic compounds.

FERRIC NITRATE NONAHYDRATE. Crystals. $Fe(NO_3)_3 \cdot 9H_2O$, mw: 404.1.

Acute tox data: Oral LD_{50} (rat) = 3250 mg/kg. [*3*]

THR = MOD via oral and inhal routes.

Fire Hazard: See nitrates.

Disaster Hazard: See nitrates.

FERRIC OXALATE. Yellow powder. $Fe_2(C_2O_4)_2 \cdot 5H_2O$, mw: 465.8, mp: decomp @ 100°.
THR = HIGH via oral and inhal routes. A MILD irr to skin, eyes and mu mem.
Fire Hazard: Slight.

FERRIC OXIDE. Syn: *iron oxide pigment.* Fe_2O_3, mw: 159.70.
THR = Reacts violently with Al, $Ca(OCl)_2$, Cs_2C_2, N_2H_4, ethylene oxide. [*19*] An exper (±) carc. [*3, 9*]

FERRIC PHOSPHATE. Syn: *iron phosphate.* Yellowish to white powder, insol in water, sol in acids. $FePO_4 \cdot 2H_2O$, mw: 187, d: 2.87.
THR = No data. A nutrient and/or dietary supplement food additive. Also used as a trace mineral added to animal feeds. [*109*]

FERRIC PYROPHOSPHATE. Syn: *iron pyrophosphate.* Yellowish-white powder, insol in water, sol in dilute acids. $Fe_4(P_2O_7)_3$, mw: 745.3.
THR = No data. A nutrient and/or dietary supplement food additive. Also used as a trace mineral added to animal feeds. [*109*]

FERRIC SODIUM PYROPHOSPHATE.
THR = No data. Used as a nutrient and/or dietary supplement food additive. [*109*]

FERRIC SULFATE. Rhombic gray crystals, slightly water-sol, hygroscopic. $Fe_2(SO_4)_3$, mw: 399.9, d: 3.097 @ 18°, mp: 480° (decomp).
THR = LOW via oral route. A MILD irr. May cause local irr. Practically non-toxic systemically. See iron compounds. It is a substance which migrates to food from packaging materials. [*109*]

FERRIC SULFIDE. Yellow-green crystals. Fe_2S_3, mw: 207.90, mp: decomp, d: 4.3.
THR = Easily evolves H_2S, which is very toxic. See also hydrogen sulfide.
Fire Hazard: See sulfides.
Explosion Hazard: See sulfides.
Disaster Hazard: See sulfides.

FERRIC THIOCYANATE. Cubic, black-red deliquescent crystals. $Fe(SCN)_3$, mw: 230.10, mp: decomp.
THR = Probably LOW via oral and inhal routes. A MILD irr.
Disaster Hazard: Dangerous; when heated to decomp or on contact with acid or acid fumes, emits highly toxic fumes of cyanides.

FERRIC-*m*-VANADATE. Syn: *iron m-vanadate.* Grayish-brown powder. $Fe(VO_3)_3$, mw: 352.70.
THR = See vanadium compounds.

FERRICYANIC ACID. Green-brown, deliquescent needles. $H_3Fe(CN)_6$, mw: 215.0, mp: decomp.
THR = No data. Probably LOW.
Disaster Hazard: Dangerous; when heated to decomp or on contact with acid fumes, emits hightly toxic fumes of cyanides.

FERRICYANIDES.
THR = No data: Ferricyanides as such are of low toxicity since the CN is bound. It has been stated but not conclusively proven that HCN can be liberated in the stomach as a result of contact with gastric acidity.
Disaster Hazard: Dangerous; when heated to decomp or on contact with acid or acid fumes, they emit highly toxic fumes of cyanides.

FERRITIN.
THR = An exper neo. [*3*]

FERROCENE. Syn: *dicyclopentadienyl iron.* Orange crystals, camphor odor, insol in water, sol in alcohol and ether. $C_{10}H_{10}Fe$, mw: 186.0, mp: 174°, sublimes >100°, volatile in steam.
Acute tox data: Oral LD_{50} (rat) = 1320 mg/kg; ip LD_{50} (rat) = 500 mg/kg; ip LD_{50} (mouse) = 335 mg/kg. [*3*]
THR = HIGH via ip and MOD via oral routes.
Fire Hazard: Mod. Reacts violently with NH_4ClO_4. [*19*]
Disaster Hazard: Dangerous; when heated to decomp, emits toxic fumes.

FERROCERIUM. An alloy of iron and misch metal.

FERROCYANIC ACID. White needles turning blue in moist air. $H_4Fe(CN)_6$, mw: 216.0, mp: decomp.
THR = No data. Probably LOW. See also iron.
Disaster Hazard: Dangerous; when heated to decomp or on contact with acid or acid fumes, emits highly toxic fumes of cyanides.

FERROCYANIDES.
THR = Ferrocyanides as such are of a LOW order of toxicity. But highly toxic decomp products can form upon mixing them with hot conc acids. Acid, basic or neutral solutions of ferrocyanides liberate hydrocyanic acid upon strong irradiation.
Disaster Hazard: Dangerous; when heated to decomp or on contact with acid or acid fumes, they emit highly toxic fumes of cyanides.

FERROSILICON. Crystalline, metallic solid. Fe + Si, d: 5.4.
THR = U. This material is decomp by moisture, under which conditions impurities may liberate such poisonous gases as phosphine and arsine.
Fire Hazard: Mod, by chemical reaction with moisture.
Explosion Hazard: Mod, by chemical reaction.
Disaster Hazard: Dangerous; will react with water or steam to produce hydrogen and other flam vapors;

can react with oxidizing materials; on contact with acid or acid fumes, can emit toxic fumes.

FERROUS-*o*-ARSENATE. Green, amorphous powder. $Fe_3(AsO_4)_2 \cdot 6H_2O$, mw: 553.5, mp: decomp.
THR = HIGH via oral and inhal routes. See also arsenic compounds.
Disaster Hazard: See arsenic compounds.

FERROUS BROMIDE. Hexagonal, green-yellow crystals. $FeBr_2$, mw: 215.7, mp: 684°, d: 4.636 @ 25°.
THR = No data. See also bromides and iron. Can react violently with Na, K. [*19*] See also bromides.
Disaster Hazard: Dangerous; when heated to decomp, emits highly toxic fumes of hydrobromic acid.

FERROUS CHLORIDE. Syn: *lawrencite.* Green to yellow, deliquescent crystals. $FeCl_2$, mw: 126.8, mp: 614°–670°, bp: 1026°, d: 3.16, vap. press: 10 mm @ 700°.
Acute tox data: ip LD_{50} (mouse) = 59 mg/kg. [*3*]
THR = HIGH via ip route. See also chlorides and iron. Can react violently with ethylene oxide, K, Na. [*19*]
Disaster Hazard: Dangerous; see chlorides.

FERROUS CHLORIDE TETRAHYDRATE. See iron (II) chloride tetrahydrate.

FERROUS CHLOROPLANTINATE. Yellow hexagonal crystals. $FePtCl_6 \cdot 6H_2O$, mw: 571.9, mp: decomp, d: 2.714.
THR = No data. Chloroplatinates generally tend to be high irr to the skin, but systemic poisoning is unknown. See also platinum and iron.
Disaster Hazard: Dangerous; when heated to decomp, emits toxic fumes of chlorides.

FERROUS FERRICYANIDE. Deep blue crystals. $Fe[Fe(CN)_6]_2$, mw: 591.5, mp: decomp.
THR = No data. See also cyanides and iron.

FERROUS FERROCYANIDE. Amorphous, blue-white crystals. $Fe_2[Fe(CN)_6]$, mw: 323.7, d: 1.201 @ 25°/4°.
THR = No data. See also ferrocyanides and iron.

FERROUS FLUORIDE. Crystals. FeF_2, mw: 93.84, mp: >1100°, d: 3.95–4.33.
THR = No data. See also iron and fluorides.

FERROUS FLUOSILICATE. Trigonal colorless crystals, $FeSiF_6 \cdot 6H_2O$, mw: 306.0, d: 1.961.
THR = No data. See also iron and fluosilicates.

FERROUS FUMARATE. Reddish-brown, anhydrous powder, odorless, insol in alcohol, very slightly sol in water. $FeC_4H_2O_4$, mw: 170, mp: >280°.
THR = U. A food additive permitted in the feed and drinking water of animals and/or for the treatment of food-producing animals. [*109*]

FERROUS GLUCONATE. Syn: *iron gluconate.* Yellowish-gray or pale greenish-yellow, fine powder or granules with slight odor, sol in water and glycerin, insol in alcohol. $C_{12}H_{22}O_{14}Fe$, mw: 446.2.
Acute tox data: Oral LD_{50} (rat) = 4600 mg/kg. [*3*]
THR = MOD via oral route. An exper neo. [*3*] A nutrient food additive. Also used as a trace mineral added to animal feeds. [*109*]

FERROUS GLUTAMATE. $C_5H_9O_4N \cdot Fe$, mw: 203.
THR = An exper neo. [*3*]

FERROUS IODIDE. Red-violet crystals. Hygroscopic. FeI_2, mw: 309.7, d: 5.315.
THR = No data. See also iron and iodides. Can react violently with Na, K. [*19*]

FERROUS LACTATE. Syn: *iron lactate.* Greenish-white crystals, slight peculiar odor, mod sol in water, slightly sol in alcohol. $Fe(C_3H_5O_3)_2 \cdot 3H_2O$, mw: 288.
Acute tox data: ip LD_{LO} (rabbit) = 287 mg/kg. [*3*]
THR = HIGH via iv route. An exper neo. [*3*] A nutrient and/or dietary supplement food additive. [*109*]

FERROUS NITRATE HEXAHYDRATE. Rhombic green crystals. $Fe(NO_3)_2 \cdot 6H_2O$, mw: 288.0, mp: decomp @ 190°, d: 2.28.
THR = No data. See nitrates.

FERROUS OXALATE. Pale yellow powder. $FeC_2O_4 \cdot 2H_2O$; mw: 179.9; mp: decomp @ 190°, d: 2.28.
THR = No data, probably high. See oxalates.

FERROUS OXIDE. Black crystals. FeO, mw: 71.84, mp: 1420°, d: 5.7.
THR = No data. Probably LOW. See also iron. A trace mineral added to animal feeds. [*109*] Self-ignites in air. Reacts violently with ethylene oxide, F_2, HNO_3, SO_2 @ 300°. [*19*]

FERROUS PERCHLORATE. Green crystals. $Fe(ClO_4)_2 \cdot 6H_2O$, mw: 362.86, mp: decomp >100°.
THR = No data: A powerful oxidizer. MOD irr to skin, eyes and mu mem. See also iron and perchlorates.
Fire Hazard: Mod.
Explosion Hazard: Mod, when shocked, exposed to heat or by chemical reaction.
Disaster Hazard: Dangerous; shock or heat will explode it; when heated to decomp, emits highly toxic fumes of oxides of chlorides; can react with reducing materials.

FERROUS PYROARSENITE. Green-white crystals. $Fe_2As_2O_5$, mw: 341.52.
THR = No data. Probably HIGH. See also arsenic.

FERROUS REDUCTUM. See iron, reduced.

FERROUS SULFATE. Syns: *iron sulfate, copperas,*

green vitriol. Monoclinic crystals. $FeSO_4$, mw: 151.93, d: 2.99–3.08.

Acute tox data: Oral LD_{50} (rat) = 1480 mg/kg; ip LD_{50} (mouse) = 100 mg/kg; oral LD_{LO} (child) = 50 mg/kg. [*3*]

THR = MOD via oral; HIGH via ip routes. An exper neo. [*3*] Used as a nutrient and/or dietary supplement food as well as a trace mineral added to animal feed; it is a substance migrating to foods from packaging materials. [*109*]

FERROUS SULFIDE. Syn: *troilite.* Black-brown crystals. FeS, mw: 87.9, mp: 1193°, bp: decomp, d: 4.84.

THR = No data. Probably toxic. See also iron and sulfides. When moist, readily heats to incandescence. Reacts violently with H_2O_2, Li. [*19*]

FERROUS THIOCYANATE. Rhombic green crystals. $Fe(SCN)_2 \cdot 3H_2O$, mw: 226.07, mp: decomp.

THR = No data. Probably toxic. See also thiocyanates.

FERROVANADIUM DUST. A gray to black dust. FeV, mw: 106.8.

THR = See also vanadium and iron. Can cause pulmonary damage.

Fire Hazard: Mod, when exposed to heat or flame.

Explosion Hazard: Slight, when exposed to flame.

FERRUM. See iron.

FERTILIZER AMMONIATING SOLUTION CONTAINING FREE AMMONIA. See ammonia and compressed gases, N.O.S.

FIBER, BURNT.

THR = A MILD irr via inhal route.

Fire Hazard: Mod, when exposed to heat or flame; can react with oxidizing materials.

Explosion Hazard: Slight, when exposed to heat or flame.

"FIBERGLAS." A brand of fibrous glass; also proprietary glass flakes.

THR = this proprietary product is glass in the form of fine fibers. When handled by workers for a period of time, it can cause considerable skin irr, particularly when in intimate contact with the skin. The possibility of lung injury due to inhal of fine particles of this material has been raised repeatedly and it now appears that durable particles of fibrous glass, especially of a particular size range, can produce an effect upon the lungs like that of asbestos. [*23*] See asbestos. It has been reported that the dermatitis due to this material is caused by the mechanical impingement of tiny pieces of glass upon the skin during handling. Occasionally, it is thought to be due to the tricresyl phosphate which is added to the glass wool as a dust adhesive, or it is sometimes blamed upon the tar or phenolformaldehyde-resin binder which this material may contain. An examination of the minute lesions resulting from the handling of glass wool shows that these are never of an allergic eczematous character. With continued use, most workers become hardened to it and the temporary itching, swelling and redness subside. Reassurance and treatment with calamine lotion and phenol plus other protective measures take care of about 95% of the cases.

FIBERS OR FABRICS, WITH ANIMAL OR VEGETABLE OILS.

THR = Variable; possible allergen.

Fire Hazard: Mod, when exposed to heat or flame; can react vigorously with oxidizing materials.

Explosion Hazard: Mod, when finely divided and exposed to flame.

FIRES, CLASSIFICATION OF. See Section 7.

FIRES, DEFINITION OF. See Section 7.

FIRE DISCOVERY. See Section 7.

FIRE AND EXPLOSION HAZARDS OF ACID AND ACID FUME-SENSITIVE MATERIALS. See Section 7.

FIRE EXTINGUISHMENT. See Section 7.

FIRE HOSES. See Section 7.

FIREMASTER T 23P. See tris(2,3-dibromopropyl) phosphate.

FIRE POINT DEFINED. See Section 7.

FIRE PREVENTION. See Section 7.

FIRE PROTECTION. See Section 7.

FIRE PROTECTION TERMINOLOGY. See Section 7.

FIRES AND EXPLOSIONS DUE TO SOLID WASTES. See Section 6.

FISCHER'S SALT. See potassium cobaltinitrate.

FISCHER-TROPSCH HYDROGENATED COAL OILS.

THR = A recog carc. [*14*] See also mineral oils.

FISCHER-TROPSCH SYNTHETIC HYDROGENATED COAL WAXES.

THR = A recog carc. [*14*] See paraffin and petroleum waxes.

FISH BERRY. See cocculus solid.

FISH MEAL. See fish scrap.

FISH OILS. flash p: about 420°F.

THR = May act as irr to the skin or as an allergen; as an insecticide when combined with sodium or potassium base as a soap. A substance migrating to food from packaging materials. [*109*] A (S) carc. [*14*]

Fire Hazard: Slight, when exposed to heat or flame.
Spont Heating: Yes.
To Fight Fire: CO_2, dry chemical.

FISH SCRAP.
THR = No data. A MILD allergen. A nuisance dust.
Fire Hazard: Mod, when exposed to heat or flame; can react with oxidizing materials.

FLAMMABLE MATERIALS STORAGE AND HANDLING. See Section 7.

FLAMMABILITY LIMITS. See Section 7. See also explosive range.

FLASH POINT DEFINED. See Section 7.

FLAVIN MONONUCLEOTIDE. See riboflavin-5′-phosphate.

FLAVOMYCELIN. Anthraquinoid hepatotoxic of Penicillium Islandicum. Yellow rectangular crystals. $C_{30}H_{22}O_{12}$, mw: 574.5, mp: 278° (decomp).
THR = An exper (+) carc. [*3, 7*]

FLAVORING FOOD ADDITIVES. [*109*]

FLAX.
THR = A (S) carc. [*14*] See plant and fungal products.

FLECTOL H. POLYMER. $(C_{11}H_{15}N)_n$.
THR = An exper neo. [*3*]

"FLEXOL" PLASTICIZER CC-55. See hexyldiethyl heptylate.

FLOUR.
THR = A MILD allergen and irr. Some systemic effects.
Fire Hazard: Mod in the form of dust.
Explosion Hazard: Mod in the form of dust, when exposed to heat or flame. Severe flour dust explosions have occurred in grain elevators and processing plants.

FLOWERS OF SULFUR. See sulfur.

FLOWERS OF ZINC. See zinc oxide.

FLUE DUST, POISONOUS.
THR = Variable, depends on composition. A common air contaminant.
Fire Hazard: Mod.
Explosion Hazard: Mod, when exposed to flame or spark.

FLUOACETATES.
THR = HIGH via oral and inhal routes. These are irr to skin, eyes and mu mem. See also sodium fluoacetate.
Disaster Hazard: Dangerous; when heated to decomp or on contact with acid or acid fumes, they emit highly toxic fumes of fluorides.

FLUOACETIC ACID. See fluoroacetic acid.

FLUOALUMINATES.
THR = U. See also fluorides.
Disaster Hazard: Dangerous; when heated to decomp or on contact with acid or acid fumes, emits highly toxic fumes of fluorides.

FLUOANTIMONATES.
THR = See fluorides and antimony compounds.
Disaster Hazard: See fluorides and antimony compounds.

FLUOBERYLLATES.
THR = VERY HIGH irr to skin, eyes and mu mem and via oral and inhal routes. See also fluorides and beryllium compounds.
Disaster Hazard: See fluorides and beryllium compounds.

FLUOBORATES.
THR = HIGH irr to skin, eyes and mu mem and via oral and inhal routes. See also fluorides and boron compounds.
Disaster Hazard: See fluorides.

FLUOBORIC ACID. Colorless liquid. HBF_4, mw: 87.8, bp: decomp @ 130°.
THR = HIGH irr. See also fluoborates.
Disaster Hazard: See fluorides.

FLUOGALLATES.
THR = See fluorides.

FLUOGERMANATES.
THR = See fluorides.

FLUOOROTIC ACID. $C_5H_3O_4N_2F$, mw: 174.1.
THR = An exper teratogen. [*3*]

FLUOPHOSPHATES.
THR = See fluorides.

FLUOPHOSPHORIC ACIDS. A term used to designate several acids containing fluorine and phosphorous (from H_2PO_3F to HPF_6).
THR = See fluorides and phosphates.

FLUORACETANILIDE.
THR = An exper carc. [*23*]

FLUORANTHENE. Colorless solid. $C_{16}H_{19}$, mw: 202.24, mp: 120°, bp: 367°, vap. press: 0.01 mm @ 20°.
THR = A polycyclic hydrocarbon found in air pollution studies.
Fire Hazard: Slight, when exposed to heat or flame.

N-3-FLUORANTHENYL ACETAMIDE. See 3-acetylamino fluoranthene.

FLUOREN-2-AMINE. $C_{13}H_{11}N$, mw: 181.3.
THR = An exper carc and neo. [*3, 23*]

FLUORENE. Syn: *o-biphenylene methane*. White shining flakes. $C_{13}H_{10}$, mw: 166.2, mp: 116°, bp: 295°, d: 1.202, vap: press: 10 mm @ 146.0°.

THR = No data.
Fire Hazard: Slight.

FLUORENE-2,7-DIAMINE. $C_{13}H_{11}N_2$, mw: 195.3.
THR = An exper carc. [*3*]

9H-FLUOREN-9-ONE. $C_{13}H_8O$, mw: 180.2.
THR = An exper neo. [*3*]

FLUORENO(9a,1-gh)QUINOLINE. $C_{19}H_{11}N$.
THR = An exper neo. [*3, 23*]

N-FLUOREN-2-YL ACETAMIDE. See acetamidofluorene.

N-FLUOREN-1-YL ACETAMIDE. $C_{15}H_{13}ON$, mw: 223.3.
THR = LOW via oral to rats. An exper carc. neo. [*3, 23*]

3-FLUORENYL ACETAMIDE.
THR = An exper carc. [*23*] An exper carc via im route. [*103*]

N-FLUOREN-YL-ACETAMIDE. See 4-acetamidofluorene.

N-FLUOREN-1-YL-ACETOHYDROXYAMIC ACID. $C_{15}H_{12}O_2N$, mw: 238.3.
THR = An exper carc. [*3, 23*]

N-FLUOREN-2-YL-ACETOHYDROXAMIC ACID. $C_{15}H_{13}O_2N$, mw: 239.3.
THR = MOD to HIGH for animal studies. An exper carc and neo. [*3*]

N-FLUOREN-3-YL-ACETOHYDROXAMIC ACID. $C_{15}H_{13}O_2N$, mw: 239.3.
THR = An exper carc and neo. [*3, 23*]

N-FLUOREN-2-YL-ACETOHYDROXAMIC ACID. (cobalt^{++} complex). $C_{30}H_{42}O_4N_2 \cdot Co$, mw: 535.5.
THR = An exper carc. [*3*]

N-FLUOREN-2-YL-ACETOHYDROXAMIC ACID. (Copper^{++} complex). $C_{20}H_{24}O_4N_2 \cdot Cu$., mw: 540.1.
THR = An exper carc and neo. [*3*]

FLUOREN-2-YL-ACETOHYDROXAMIC ACID-*o*-GLUCURONIDE. $C_{21}H_{21}O_9N$, mw: 431.4.
THR = An exper carc. [*3*]

N-FLUOREN-2-YL-ACETOHYDROXAMIC ACID IRON^{+++} complex). $C_{30}H_{24}O_4N_2 \cdot Fe$, mw: 532.4.
THR = An exper carc. [*3*]

N-FLUOREN-2-YL-ACETOHYDROXAMIC ACID. (Manganese^{++} complex). $C_{30}H_{24}O_4N_2 \cdot Mn$, mw: 531.5.
THR = An exper carc. [*3*]

N-FLUOREN-2-YL-ACETOHYDROXAMIC ACID (Nickel^{++} complex). $C_{30}H_{24}O_4N_2 \cdot Ni$, mw: 535.3.
THR = An exper carc. [*3*]

N-FLUOREN-2-YL-ACETOHYDROXAMIC ACID (Potassium salt). $C_{15}H_{12}O_2N \cdot K$, mw: 278.4.
THR = An exper carc. [*3*]

N-FLUOREN-2-YL-ACETOHYDROXAMIC ACID (Zinc complex). $C_{30}H_{24}O_4N_2 \cdot Zn$. mw: 541.9.
THR = An exper carc. [*3*]

5-(FLUOREN-2-YL-AZO)TOLUENE-2,4-DIAMINE. $C_{20}H_{18}N_4$, mw: 314.4.
THR = An exper neo. [*3*]

4-FLUOREN-2-YL-BENZAMIDE. $C_{20}H_{15}ON$, mw: 285.4.
THR = An exper neo. [*3*]

N-FLUOREN-1-YL-BENZOHYDROXAMIC ACID. $C_{20}H_{15}O_2N$, mw: 301.4.
THR = An exper carc. [*3*]

N-FLUOREN-2-YL-BENZOHYDROXAMIC ACID.
THR = An exper carc. [*3, 23*]

N-FLUOREN-2-YL-BENZO HYDROXAMIC ACID ACETATE. $C_{22}H_{17}O_3N$, mw: 343.4.
THR = An exper neo. [*3*]

N-FLUOREN-2-YL-DIACETAMIDE. $C_{17}H_{14}O_2N$, mw: 264.3.
THR = Mod to rats via oral route. An exper (S) carc. [*3, 23*]

N,N′-FLUOREN-2,5-YLENE BIS ACETAMIDE. $C_{17}H_{16}O_2N_2$, mw: 280.4.
THR = LOW to rats via oral route. An exper (S) neo. [*3, 23*]

N,N′—FLUOREN-2,7-YLENE BIS ACETAMIDE. $C_{17}H_{16}O_2N_2$, mw: 280.4.
THR = MOD to rats via oral route. An exper (S) carc. [*3, 23*]

2′,7-FLUORENYL DIAMINE.
THR = An exper carc. [*23*]

2-FLUORENYL DIETHYLAMINE.
THR = An exper carc. [*23*]

2-FLUORENYL DIMETHYL AMINE.
THR = An exper carc. [*23*]

N,N′-FLUOREN-2,7-YLENE BIS(TRIFLUOROACETAMIDE). $C_{15}H_{10}O_2N_2F_6$, mw: 364.3.
THR = LOW to rats via oral route. An exper carc. [*3*]

N,2-FLUORENYL(2′-CARBOXYBENZ(AMIDE).
THR = An exper carc. [*23*]

N,2-FLUORENYL FORMAMIDE.
THR = An exper carc. [*23*]

2-FLUOREN-2-YL-HYDROXYLAMINE-*o*-GLUCURONIDE. $C_{19}H_{15}O_3N$, mw: 389.4.
THR = An exper neo. [*3*]

2-FLUORENYL MONOMETHYLAMINE.
THR = An exper carc. [*23*]

N-FLUOREN-2-YL-PHTHALAMIC ACID. $C_{21}H_{15}O_3N$, mw: 329.4.
THR = An exper carc. [*3, 23*]

N,2-FLUORENYL-SUCCINAMIC ACID. $C_{17}H_{15}O_3N$, mw: 281.3.
THR = An exper carc. [*3, 23*]

N-FLUOREN-2-YL-2,2,2-TRIFLUORO ACETAMIDE. $C_{15}H_{10}ONF_3$, mw: 277.3.
THR = An exper carc. [*3, 23*]

FLUORESCEIN. Syn: *Resorcinol phthalein.* Orange-red crystalline powder. $C_{20}H_{12}O_5$, mw: 332.3, mp: 314°–316° with decomp.
Acute tox data: ip LD_{LO} (rat) = 600 mg/kg; oral LD_{LO} (rabbit) = 2500 mg/kg; iv LD_{LO} (rat) = 300 mg/kg. [*3*]
THR = HIGH via iv and MOD via oral and ip routes. An allergen.
Fire Hazard: Slight.

FLUORESCEIN SODIUM. $C_{20}H_{10}O_5 \cdot 2Na$, mw: 376.3.
THR = An exper neo. [*3*]

FLUORESCENT LAMP RADIATION.
THR = A recog carc. [*14*] See Section 8. See also radiation, U V.

FLUORESONE. See ethyl-*p*-fluorophenyl sulfone.

2-FLUORETHYL(4-BIPHENYLIL)ACETATE. See Lambrol.

FLUORIDES.
THR = Inorganic fluorides are generally highly irr and toxic. Acute effects resulting from exposure to fluorine compounds are due to hydrogen fluoride. Chronic fluorine poisoning, or "fluorosis," occurs among miners of cryolite, and consists of a sclerosis of the bones, caused by fixation of the calcium by the fluorine. There may also be some calcification of the ligaments. The teeth are mottled, and there is osteosclerosis and ostemalacia. The bony and ligamentous changes are demonstrable by x-ray. Estimated LD (man) = 2.5 to 5.9 g of F. Large doses can cause very severe nausea, vomiting, diarrhea, abdominal burning and cramp-like pains. It is not taken up by the thyroid and does not interfere with iodine uptake. Can cause or aggravate attacks of asthma. Can cause severe bone changes, making normal movements painful. Some signs of pulmonary fibrosis are noted. Some enzyme systems effects are reported. An irr to the eyes, skin and mu mem. [*36*] Also loss of weight, anorexia, anemia, wasting and cachexia, and dental defects are among the common findings in chronic fluorine poisoning. There may be an eosinophilia, and impairment of growth in young workers. Symptoms of intoxication include gastric, intestinal, circulatory, respiratory and nervous complaints and skin rashes. Organic fluorides are generally less toxic than other halogenated hydrocarbons. Common air contaminants.
Disaster Hazard: Dangerous; when heated to decomp, or on contact with acid or acid fumes, they emit highly toxic fumes.

FLUORINE. Pale yellow gas. F_2, mw: 38.0, mp: −218°, bp: −187°, d: 1.14 @ −200°, 1.108 @ −188°, vap. d: 1.695.
Acute tox data: Inhal 25 ppm for 5 min yields serious eye symptoms; inhal LC_{50} (rat) = 185 ppm for 1 hr; inhal LC_{50} (mouse) = 150 ppm for 1 hr. [*3*]
THR = HIGH irr to skin, eyes, mu mem, via oral and inhal routes. See also fluorides. A most powerful caustic irr.
Fire and Explosion Hazard: Dangerous; reacts violently with many materials such as most organic matter, H-containing molecules, oxides of S,N,P, alkali metals and alkaline earths. It reacts violently with metals and their sulfides, phosphides, oxides, hydrides, acetylides and carbides. It also reacts violently with halogens and halogen acids, P, S, hydrazine, ClO_2, C, coke, charcoal, cyanamide, cyanides, KNO_3, (PbO + glycerol), CCl_4, silicides, silicates, alkenes, alkyl benzenes, CS_2, $Cr(OCl)_2$, B, Al, Tl, Sn, Sb, Te, Se, S, P, As, natural gas, H_2O, liquid air, perfluoropropionyl fluoride, phenol formaldehyde resin, polyamides, polychloroprene, polyethylene, polyvinyl chloride acetate, polyurethane. Many reactions go on even at $<-160°$. [*19*]
Disaster Hazard: Highly dangerous; when heated, emits highly toxic fumes; will react with water or steam to produce heat and toxic and corrosive fumes.
Radiation Hazard: For permissible levels, see Section 5A, Table 5A.5.

FLUORINE ANALOG OF DDT. See 1-trichloro-2,2-bis (p-fluorophenyl)ethane.

FLUORINE AZIDE. Syn: *azido fluorine.* N_3F, mw: 61.
THR = HIGH irr and toxicity. See also fluorine. Extremely unstable, light-sensitive, heat-sensitive. [*19*]

FLUORINE DIOXIDE. Syn: *dioxygen difluoride.* Brown gas, cherry red liquid, orange solid. F_2O_2, mw: 70.0, mp: −163.5°, d(solid): 1.912 @ −165°, d(liquid): 1.45 @ −57°, bp: −57°, decomp @ −100°.
THR = See fluorine.

FLUORINE FLUOROSULFATE. FSO_2OF, mw: 118.1. A vapor.
THR = Can explode at room temp. [*19*] See also fluorine.

FLUORINE MONOXIDE. Syn: *oxygen fluoride.* Colorless gas, yellowish-brown liquid. Reacts slowly with water, OF_2, mw: 54.0, d: (liq) 1.90 @ −224°, mp: −223.8°, bp: −144.8°.
Acute tox data: Inhal LC_{50} (rat) = 3 ppm for 1 hr; inhal LC_{50} (mouse) = 1.5 ppm for 1 hr; inhal LC_{50} (dog) = 26 ppm for 1 hr. [*3*]
THR = VERY HIGH irr to skin, eyes and mu mem. HIGH corrosive to tissue. Attacks lungs with delayed appearance of symptoms. See also fluorides.
Fire Hazard: A very powerful oxidizer. Must be kept away from contact with reducing agents. Reacts violently with $AlCl_3$, NH_3, As_2O_3, Br_2, CO, Cl_2, (Cl_2 + Cu), CrO_3, H_2, H_2S, I, Ir, CH_4, O_3, (O_2 + H_2O), Pd, P_2O_5, Pt, Rh, Ru, SiO_2. [*19*]
Disaster Hazard: Dangerous; when heated to decomp, emits highly toxic fumes of fluorine.

FLUORINE NITRATE. Syn: *nitrogen trioxyfluoride.* Colorless gas, acrid odor. Hydrolyzes upon contact with water. NO_3F, mw: 81.0, d: (liquid) 1.507 @ −45.9°, (solid) 1.951 @ −193.2°, mp: −175°, bp: −45.9°.
THR = VERY HIGH irr to skin, eyes and mu mem and via oral and inhal routes. See also fluorine.
Fire Hazard: Dangerous; very powerful oxidizer. Conflagrates upon contact with reducers such as alcohol, ether, aniline.
Explosion Hazard: Very dangerous. Liquid explodes upon slight shock.
Disaster Hazard: Dangerous. Upon warming, emits highly toxic fumes.

FLUORINE PERCHLORATE. Syn: *chlorine tetroxyfluoride.* Colorless gas, pungent, acrid odor. Very unstable. $FClO_4$, mw: 118.5, mp: −167.3°, bp: −15.9°.
THR = No data. Known to be very corrosive and a powerful oxidizer. Probably HIGH irr to skin, eyes and mu mem and via inhal and oral routes. See also fluorine.
Fire Hazard: Very powerful oxidizer. See perchlorates.
Explosion Hazard: Very dangerous. Explodes on slightest provocation, such as contact with rough surfaces, dirt and grease, heating, melting, KI, rubber tubing. [*19*]
Disaster Hazard: Very dangerous. When warm, emits highly toxic fumes.

FLUOROACETAMIDE. Syn: *1081.* $H_2FCCONH_2$, mw: 77.
Acute tox data: Oral LD_{50} (rat) = 5.7 mg/kg; dermal LD_{50} (rat) = 80 mg/kg; iv LD_{50} (rabbit) = 0.25 mg/kg; ip LD_{50} (mouse) = 85 mg/kg. [*3*]
THR = HIGH via oral, dermal, iv and ip routes. See also fluorides.
Disaster Hazard: See fluorides.

FLUOROACETANILIDE. C_8H_8ONF, mw: 153.2.
Acute tox data: Oral LD_{50} (rat) = 10 mg/kg. [*2*] Oral LD_{LO} (rat) = 3 mg/kg. [*3*]
THR = HIGH via oral and inhal routes.

FLUOROACETIC ACID. Syns: *fluoroethanoic acid, fluoacetic acid.* Colorless solid, water-sol. CH_2FCOOH, mw: 78.9, mp: 33°, bp: 165°.
Acute tox data: iv LD_{50} (rabbit) = 0.25 mg/kg; sc LD_{50} (rat) = 0.28 mg/kg; ip LD_{50} (mouse) = 6.6 mg/kg. [*3*]
THR = VERY HIGH via iv, sc, ip and oral routes. See also sodium fluoroacetate. It is said to cause convulsions and ventricular fibrillation.
Disaster Hazard: Dangerous. When heated to decomp, emits highly toxic fumes of fluorides.

FLUOROACETIC ACID PHENYL HYDRAZINE. $C_8H_9FO_2N_2$, mw: 184.1.
Acute tox data: Oral LD (rat) = 9.1 mg/kg. [*94*]
THR = HIGH via oral and inhal routes.
Disaster Hazard: See fluorides and phenol.

FLUOROACETOPHENONE. Syn: *phenacyl fluoride.* Brown liquid, pungent odor. $C_6H_5COCH_2F$, mw: 138.2, bp: 98° @ 8 mm.
Acute tox data: ip LD_{LO} (rat) = 225 mg./kg. [*3*]
THR = HIGH via ip route. Also a MOD irr to skin, eyes and mu mem.
Disaster Hazard: See fluorides.

7-FLUORO-2-ACETYL AMINO FLUORINE. See *n*-(7-fluorofluoren-2-yl-acetamide).

FLUOROACETYLENE. HC:CF, mw: 44, fp: −196°, bp: −80°.
THR = A liquid which can explode violently. The Ag salt is heat-sensitive; the Hg salt is also heat-sensitive. [*19*] No toxicity data.

FLUOROANILINE. Liquid. $FC_6H_4NH_2$, mw: 111 d: 1.1724, bp: 187.4°, mp: −1.9°.
THR = No data. Probably HIGH. See also fluorine and aniline.

4-FLUOROBENZ(a)ANTHRACENE. $C_{18}H_{11}F$, mw: 246.3.
THR = An exper [*3, 23*] carc.

FLUOROBENZENE. C_6H_5F, mw: 96.1, d: 1.08, vap. d: 3.31, bp: 82.8°.
Acute tox data: inhal LD_{LO} (mice) = 45 mg/m^3 for 2 hrs. [*3*]
THR = HIGH via inhal route.
Fire Hazard: Dangerous, via heat, flame or oxidizers.

To Fight Fire: Water spray, mist, foam, dry chemical, CO_2.

4′-FLUOROBIPHENYL AMINE. See 4-amino-4-fluorodiphenyl.

6-FLUORO-7-BROMOMETHYL BENZ(a)ANTHRACENE.
THR = An exper carc. [*23*]

FLUOROCARBON 115. See chloropentafluorethane.

FLUOROCHLORO LUBRICANTS.
THR = See aluminum for details of violent interaction. [*19*]

5-FLUOROCYTOSINE. $C_4H_4O_3NF$, mw: 129.1.
THR = An exper teratogen. [*3*]

5-FLUORODEOXYCYTIDINE. $C_9H_{13}O_4N_3F$, mw: 246.3.
THR = An exper teratogen. [*3*]

6-FLUORODIBENZ(a,h)ANTHRACENE. $C_{22}H_{13}F$, mw: 296.4.
THR = An exper carc. [*3*]

FLUORODIMETHYL ARSINE. $FAs(CH_3)_2$, mw: 124.
THR = Self-ignites in air. [*19*] Must be very toxic. See also arsenic and fluorides.

3-FLUORO-2,10-DIMETHYL-5,6-BENZACRIDINE. $C_{18}H_{12}NF$, mw: 261.3.
THR = An exper neo. [*3, 23*]

5-FLUORO-7,12-DIMETHYL BENZ(a)ANTHRACENE.
THR = An exper carc. [*23*]

4-FLUORO-N,N-DIMETHYL-4-(PHENYLAZO)-*m*-TOLUIDINE. $C_{15}H_{13}N_3F$, mw: 254.3.
THR = An exper neo. [*3*]

2-FLUORO-N,N-DIMETHYL-4-STILBENAMINE. $C_{16}H_{16}NF$, mw: 241.3.
THR = An exper carc. [*3*]

4-FLUORO-N,N-DIMETHYL-4-STILBENAMINE.
THR = An exper carc. [*3*]

1-FLUORO-2,4-DINITROBENZENE. Syns: *2,4-dinitro-1-fluorobenzene, sanger's reagent*. Crystals, sol in ether, benzene, propylene glycol. $C_6H_3F(NO_2)_2$, mw: 186.1, mp: 26°, bp: 137° @ 20 mm.
Acute tox data: Oral LD_{LO} (rat) = 50 mg/kg; sc LD_{LO} (mouse) = 100 mg/kg. [*3*]
THR = HIGH via oral and sc routes. An exper carc and mutagen. [*33*] A powerful irr and vesicant. [*2*] See also fluorides.
Disaster Hazard: Dangerous. When heated to decomp, emits highly toxic fumes of NO_x and fluorides.

1-FLUORO-2,2-DINITROETHANE. $CH_2FCH(NO_3)_2$, mw: 170.2
THR = Can explode on contact with air. [*19*] No toxicity data. Probably HIGH.

FLUOROETHANE. See ethyl fluoride.

FLUOROETHANOIC ACID. See fluoroacetic acid.

N-(1-FLUOROFLUOREN-2-YL)ACETAMIDE. $C_{15}H_{12}ONF$, mw: 241.3.
THR = MOD via oral route to rats. An exper carc. [*3, 23*]

N-(3-FLUOROFLUOREN-2-YL)ACETAMIDE.
THR = An exper carc. [*3, 23*]

N-(4-FLUOROFLUOREN-2-YL)ACETAMIDE.
THR = An exper carc. [*3, 23*]

N-(5-FLUOROFLUOREN-2-YL)ACETAMIDE.
THR = An exper carc. [*3, 23*]

N-(6-FLUOROFLUOREN-2-YL)ACETAMIDE.
THR = An exper carc. [*3, 23*]

N-(7-FLUOROFLUOREN-2-YL)ACETAMIDE.
THR = An exper carc. [*3, 23*]

N-(8-FLUOROFLUOREN-2-YL)ACETAMIDE.
THR = An exper carc. [*3, 23*]

N-(7-FLUOROFLUOREN-2-YL)ACETOHYDROXAMIC ACID. $C_{15}H_{12}O_2NF$, mw: 257.3.
THR = An exper oral carc to rats. [*3, 23*]

FLUOROFORM. Syn: *trifluoromethane*. Colorless, odorless gas. CHF_3, mw: 70.02, mp: −163°, bp: −82.2°, d: 1.52 (liq) @ −100°.
THR = A MILD irr to respiratory tract; narcotic in high conc.
Disaster Hazard: Dangerous; when heated to decomp, emits highly toxic fumes of fluorides.

FLUOROFORMYL FLUORIDE. See carbonylfluoride.

FLUOROISOPROPOXY METHYL PHOSPHINE OXIDE. See sarin.

FLUOROISOPROPYL ALCOHOL. See propylene fluorohydrin.

FLUOROMETHANE. See methyl fluoride.

9-FLUORO-7-METHYL BENZ(c)ACRIDINE. $C_{18}H_{12}NF$, mw: 261.3.
THR = An exper neo. [*3, 23*]

10-FLUORO-12-METHYL BENZ(c)ACRIDINE.
THR = An exper neo. [*3, 23*]

11-FLUORO-7-METHYL BENZ(c)ACRIDINE.
THR = An exper neo. [*3, 23*]

3-FLUORO-7-METHYL BENZ(a)ANTHRACENE. $C_{19}H_{13}F$, mw: 260.3.
THR = An exper carc. [*3, 23*]

4-FLUORO-7-METHYL BENZ(a)ANTHRACENE.
THR = An exper carc. [*23*]

5-FLUORO-12-METHYL BENZ(a)ANTHRACENE.
THR = An exper carc. [*23*]

6-FLUORO-7-METHYL BENZ(a)ANTHRACENE.
THR = An exper neo and carc. [*3, 23*]

9-FLUORO-7-METHYL BENZ(a)ANTHRACENE.
THR = An exper carc. [*3, 23*]

10-FLUORO-7-METHYL BENZ(a)ANTHRACENE.
THR = An exper carc. [*3, 23*]

12-FLUORO-5-METHYL CHRYSENE. $C_{19}H_{13}F$, mw: 260.3.
THR = An exper neo. [*3*]

5-FLUORO-4-METHYL-2′-DEOXYCYTIDINE. $C_9H_{14}O_4N_3F$, mw: 247.3.
THR = An exper teratogen. [*3*]

FLUOROMETHYL PINACOLYLOXY PHOSPHINE OXIDE. See soman.

3-FLUORO-4-NITROQUINOLINE-1-OXIDE. $C_9H_5O_3N_2F$, mw: 208.2.
THR = An exper neo via sc route. [*3, 23*]

8-FLUORO-4-NITROQUINOLINE-1-OXIDE.
THR = An exper neo via sc route. [*3, 23*]

2′-FLUORO-4′-PHENYL ACETANILIDE. $C_{14}H_{12}ONF$, mw: 229.3.
THR = MOD via oral route to rats. An exper carc. [*3*]

4′-(*m*-FLUOROPHENYL) ACETANILIDE. $C_{14}H_{12}ONF$, mw: 229.3.
THR = MOD via oral route to rats. An exper neo. [*3*]

4′-(*p*-FLUOROPHENYL)ACETANILIDE. $C_{14}H_{12}ONF$, mw: 229.3.
THR = An exper carc. [*3*]

2-FLUOROPHENYL ANILINE.
THR = An exper carc. [*23*]

4′-FLUORO-*p*-PHENYL ANILINE.
THR = An exper carc. [*23*]

3′-FLUORO-4-PHENYL ANILINE.
THR = An exper carc. [*23*]

FLUOROPHOSPHORIC ACID. Colorless, viscous liquid, miscible with water. H_2PO_3F, mw: 100, d: 1.1818 @ 25°.
THR = HIGH irr to skin, eyes and mu mem. HIGH toxicity. See also fluorides.

3-FLUOROPROPANE. See allyl fluoride.

2-FLUORO-2-PROPEN-1-OL. $CH_2OHCFCH_2$, mw: 76.
Acute tox data: Oral LD_{LO} (mouse) = 63 mg/kg; inhal LC_{LO} (mouse) = 198 ppm for 6 hrs; dermal LD_{LO} (rabbit) = 25 mg/kg. [*3*]
THR = HIGH via oral, dermal and inhal routes.

4′-FLUORO-4-STILBENAMINE.
THR = An exper carc. [*23*]

2′-FLUORO-4-STILBENYL-N,N-DIMETHYLAMINE.
THR = An exper carc. [*23*]

4′-FLUORO-4-STILBENYL-N,N-DIMETHYL AMINE.
THR = An exper carc. [*23*]

FLUOROTRICHLOROMETHANE. Syn: *trichlorofluoromethane*. Colorless liquid. $FCCl_3$, mw: 137.38, mp: −111°, bp: 24.1°, d: 1.484 @ 17.2°.
Acute tox data: Inhal LC_{LO} (rat) = 100,000 ppm for 20 min. [*3*]
THR = MILD irr; LOW via inhal routes. Reacts violently with Al, Li. [*19*] High conc cause narcosis and anesthesia.
Disaster Hazard: Dangerous; when heated to decomp, emits highly toxic fumes of fluorides and chlorides.

2-FLUOROTRICYCLOQUINAZOLINE. $C_{21}H_{11}N_4F$, mw: 338.4.
THR = An exper neo. [*3*]

3-FLUORO TRICYCLO QUINAZOLINE.
THR = An exper carc. [*3*]

2-FLUOROURACIL. $C_4H_3O_2N_2F$, mw: 130.1
THR = An exper teratogen and neo. [*3*]

5-FLUOROURIDINE. $C_9H_{11}O_6N_2F$, mw: 262.2.
THR = An exper teratogen. [*3*]

FLUOROXY TRIFLUOROMETHANE. See trifluoromethyl hypofluoride.

2-FLUORYLAMINE. $C_{13}H_{11}NF$, mw: 200.2.
THR = U. A stomach insecticide. See also amines.
Disaster Hazard: Slight; when heated, emits acrid fumes.

FLUOSILICATES. See silicofluorides.

FLUOSILICIC ACID. See hydrofluosilicic acid.

FLUOSULFONATES.
THR = No data. These are salts of a strong acid. Probably HIGH. See also fluorides.
Disaster Hazard: Dangerous; when heated to decomp, they emit highly toxic fumes of fluorides and oxides of sulfur; will react with water or steam to produce toxic and corrosive fumes.

FLUOSULFONIC ACID. Syn: *fluosulfuric acid*. Colorless, fuming, highly corrosive liquid. HSO_3F, mw: 100.07, mp: −87.3°, bp: 163.5°, d: 1.743 @ 15°.
THR = HIGH irr to skin, eyes, mu mem and via oral and inhal routes. See also fluorides and sulfuric acid.
Disaster Hazard: See fluosulfonates.

FLUOSULFURIC ACID. See fluosulfonic acid.

FLUOTELLURITES.
THR = HIGH–MOD. See also fluorides and tellurium compounds.
Disaster Hazard: See fluorides.

FLUOXY PREDNISOLONE. $C_{21}H_{27}O_5F$, mw: 394.5.
THR = An exper teratogen. [3]

"FLURAL." Dry, free-flowing powder. $AlFSO_4 \cdot xH_2O$ to $Al_2F_7SO_4 \cdot xH_2O$.
THR = HIGH. See fluorides.

FLUX, BLACK.
THR = Dangerous! A low explosive. For details, see explosives, low.

FLUX, WHITE. Composition: sodium nitrate and sodium nitrite.
THR = Variable. See also nitrites and nitrates.
Fire Hazard: Dangerous. This is a strong oxidizing agent.
Disaster Hazard: Keep away from heat and open flame; do not store near reducing agents or easily oxidized materials.

FMN. See riboflavin-5′-phosphate.

FOAM. See Section 7.

FOLACIN. See pteroylglutamic acid.

FOLIC ACID. See pteroyglutamic acid.

FOLIMAT. Colorless to slightly yellow oily liquid, water-sol. $C_5H_{12}O_4PSN$, mw: 213.1, d: 1.32.
Acute tox data: Oral LD_{LO} (rat) = 50 mg/kg; dermal LD_{50} (rat) = 700 mg/kg. [3]
THR = HIGH via oral and MOD via dermal routes. An insecticide. Toxic to bees.
Disaster Hazard: See phosphorous compounds and sulfur compounds.

FOLPET. See phaltan.

FOOD AND DRUG ADMINISTRATION. See Section 10A.

FOOD COLOR RED #102. $3Na \cdot C_{20}H_{14}N_2S_3$, mw: 607.5.
THR = An exper carc. [3]

FOOD, DRUG AND COSMETIC ACT. See Section 10A.

FOOTS OIL.
THR = A recog carc. [14] See paraffin and petroleum waxes.

FORENSIC TOXICOLOGY AFFECTED BY INDUSTRIAL TOXICOLOGY. See Section 9.

FORMAL. See methylal.

FORMALDEHYDE (COMMERCIAL SOLUTIONS). Syns: *methanal, methyl aldehyde, formalin.* HCHO. Clear, water white, very slightly acid, gas or liquid, pungent odor. Pure formaldehyde is not available commercially because of its tendency to polymerize. It is sold as aqueous solutions containing from 37% to 50% formaldehyde by weight and varying amounts of methanol. Some alcoholic solutions are used industrially and the physical properties and hazards may be greatly influenced by the solvent. mw: 30.03, lel = 7.0%, uel = 73.0%, autoign. temp.: 806°F, d: 1.0, bp: −3°F, flash p: (37% methanol-free): 185°F, flash p: (15% methanol-free): 122°F.
Acute tox data: Oral LD_{50} (rat) = 800 mg/kg; inhal LD_{LO} (rat) = 250 ppm for 4 hrs; sc LD_{50} (rat) = 420 mg/kg; dermal LD_{50} (rabbit) = 270 mg/kg; oral LD_{LO} (women) = 36 mg/kg. [3, 20]
THR = HIGH irr to skin, eyes, mu mem. If swallowed it causes violent vomiting and diarrhea which can lead to collapse. A fungicide. A common air contaminant. An exper carc of the lung. [3] Frequent or prolonged exposure can cause hypersensitivity leading to contact dermatitis, possibly of an eczematoid nature.
Fire Hazard: Very dangerous for gas, moderate for vapors. Will burn above flash point if exposed to flame, sparks, etc. Should formaldehyde be involved in a fire, irr gaseous formaldehyde may be evolved.
Spont Heating: No.
Explosion Hazard: When aqueous formaldehyde solutions are heated above their flash points, a potential explosion hazard exists. High formaldehyde conc of methanol content lowers flash point. Reacts with NO_2 @ about 180°; the reaction becomes explosive. Also reacts violently with ($HClO_4$ + aniline) and performic acid. [19]
Disaster Hazard: Mod dangerous; because of irr vapor which may be in toxic conc locally if storage tank is ruptured.
To Fight Fire: Stop flow of gas for pure form; alcohol foam for 37% methanol-free form.

FORMALDEHYDE ACETAMIDE. See formicin.

FORMALDEHYDE CYANOHYDRIN. See glycolonitrile.

FORMALDEHYDE DIMETHYL ACETAL. See methylal.

FORMALDEHYDE GAS. Syns: *methanal, methyl aldehyde.* Colorless gas, pungent suffocating odor. Very sol in water. HCHO, mw: 30.3, d(air = 1.000): 1.067, d(water = 1.000): 0.815 @ −20°/4°, mp: −92°, bp: −19.5°, autoign. temp.: 572.°F.
THR = See formaldehyde (commercial solutions).

FORMALIN. See formaldehyde.
Formalin was originally used as a European trade name for 35% formaldehyde solution. It is commonly employed as a synonym for commercial formaldehyde.

FORMAMIDE. Syn: *methanamide.* Colorless, hygroscopic and oily liquid. $HCONH_2$, mw: 45.04, fp: 2.6°, vap. press: 29.7 mm @ 129.4°, flash p: 310°F (COC), bp: 210° decomp, d: 1.134 @ 20°/40°; 1.1292 @ 25°/4°.

Acute tox data: im LD_{50} (guinea pig) = 2539 mg/kg; oral LD_{50} (rat) = 7500 mg/kg. [*3*, *20*]

THR = MOD to LOW via im and oral routes. An irr skin, eyes and mu mem. An exper teratogen. [*3*]

Fire Hazard: Mod. Vapors will burn in air at temp. above 310°F.

Disaster Hazard: Mod. When heated to decomp, emits toxic fumes.

FORMAMIDINE THIOLACETIC ACID HYDROCHLORIDE. $C_3H_6O_2N_2S \cdot HCl$, mw: 170.6.

THR = Violent reaction with Cl_2. [*19*]

FORMIC ACID. Syns: *methanoic acid, hydrogen carboxylic acid.* Colorless, fuming liquid, pungent penetrating odor. HCOOH, mw: 46.03, bp: 100.8°, fp: 8.2°, flash p: 156°F (OC), d: 1.2267 @ 15°/4°, 1.220 @ 20°/4°, autoign. temp.: 1114°F, vap. press: 40 mm @ 24.0°, vap. d: 1.59, flash p:(90% solution): 122°F, autoign. temp. (90% solution): 813°F, lel (90% solution) = 18%, uel (90% solution) = 57%.

Acute tox data: Oral LD_{50} (rat) = 1210 mg/kg; ip LD_{50} (mouse) = 940 mg/kg; iv LD_{50} (mouse) = 145 mg/kg. [*3*]

THR = HIGH via iv and MOD via oral and ip routes. An irr to skin, eyes and mu mem. [*20*] A substance migrating to food from packaging materials. [*109*]

Fire Hazard: Mod, when exposed to heat or flame; can explode with furfuryl alcohol, H_2O_2, $Tl(NO_3)_3 \cdot 3H_2O$. [*19*]

To Fight Fire: CO_2, dry chemical, alcohol foam.

FORMIC ETHER. See ethyl formate.

FORMICIN. Syn: *formaldehyde acetamide.* Colorless, very hygroscopic mass, very sol in water and organic solvents. $CH_3CONHCH_2OH$, mw: 89.1, d: 1.14–1.18.

THR = See formaldehyde gas and formaldehyde.

FORMONONETIN. See 7-hydroxy-4′-methoxyisoflavone.

FORMOTHION. See 5-(*n*-formyl-N-methcarbamoylmethyl)-*o,o*-dimethyl phosphorodithioate.

2-FORMYL AMINO FLUORENE. $C_{14}H_{11}ON$, mw: 209.3.

THR = An exper oral carc to rats. [*3*]

5-FORMYL BENZO(r,st)PENTAPHENE.

THR = An exper carc. [*23*]

5-FORMYL BENZO(c)PHENANTHRENE.

THR = An exper carc. [*23*]

7-FORMYL DIBENZO(b,def)CHRYSENE.

THR = An exper carc. [*23*]

10-FORMYL DIBENZO(def,p)CHRYSENE.

THR = An exper carc. [*23*]

FORMYL FLUORIDE. Colorless gas. HCOF, mw: 48.02, bp: −29°, mp: −142°.

THR = HIGH irr to skin, eyes and mu mem and via oral and inhal routes.

Fire Hazard: Mod, by chemical reaction.

Disaster Hazard: Dangerous; when heated to decomp or on contact with acid or acid fumes, emits highly toxic fumes of fluorides; will react with water or steam to produce toxic fumes, such as carbon monoxide.

FORMYL FORMIC ACID. See glyoxylic acid.

FORMYL HYDROPEROXIDE. See performic acid.

N-FORMYL-N-HYDROXYL GLYCINE. $C_3H_5O_4N$, mw: 119.1.

THR = An exper teratogen. [*3*]

7-FORMYL-11-METHYL BENZ(c)ACRIDINE.

THR = An exper neo. [*3*, *23*]

7-FORMYL-9-METHYL BENZ(c)ACRIDINE. $C_{19}H_{13}ON$, mw: 271.3.

THR = An exper neo. [*3*, *23*]

5-FORMYL-8-METHYL BENZO(rst)PENTAPHENE.

THR = An exper carc. [*23*]

7-FORMYL-14-METHYL DIBENZO(b,def)CHRYSENE.

THR = An exper carc. [*23*]

5-(*n*-METHYL CARBAMOYL METHYL)-*o,o*-DIMETHYL PHOSPHORODITHIOATE. $C_6H_{12}O_4NPS$, mw: 257.3.

Acute tox data: Oral LD_{50} (rat) = 330 mg/kg; dermal LD_{50} (rat) = 100 mg/kg; oral LD_{50} (dog) = 15

THR = HIGH via oral and dermal routes. A cholinesterase inhibitor. See also parathion.

N-FORMYL-N-METHYL-*p*-(PHENYLAZO)ANILINE. $C_{13}H_{11}ON_3$, mw: 225.3.

THR = An exper carc and neo. [*3*]

FORMYL TRICHLORIDE. See chloroform.

1-FORMYL-3-THIO SEMICARBAZIDE. $C_2H_5ON_3S$, mw: 119.2.

THR = An exper neo. [*3*]

FORSTERITE. Mg_2SiO_3, mw: 124.7.

THR = An exper neo via imp route. [*3*]

FOSTION. Syn: *o,o-diethyl-s-(n-isopropyl carbamoyl methyl)phosphorodiethioate.* White crystals, sol in most organic solvents. $(C_2H_5O)_2SPSCH_2SCH(CH_3)_2$, mw: 243.2, mp: 24°.

Acute tox data: Oral LD_{50} (rat) = 8 mg/kg; dermal LD_{50} (rat) = 100 mg/kg; oral LD_{50} (dog) = 15 mg/kg. [*3*]

THR = HIGH via oral and dermal routes. A systemic insecticide; a cholinesterase inhibitor. See also parathion. A powerful ovicide.

Disaster Hazard: See parathion.

FOSVEX. See tetraethyl pyrophosphate.

FOX GLOVE. See digitalis.

FP SALT #4. See potassium hexafluorophosphate.

FRC. See Section 5A.

FREE HANGING HOOD. See Section 2.

FREMY'S SALT. See potassium bifluoride.

FRENCH CHALK. See soapstone dust.

FRENCH POLISH. Shellac dissolved in alcohol.
THR = See alcohol and shellac.
Fire Hazard: Mod, when exposed to heat or flame; can react with oxidizing materials.

"FREON-11." See fluorotrichloromethane.

"FREON-12." See dichlorodifluoromethane and dichlorotetrafluoroethane.

"FREON 21." See dichlorofluoromethane.

"FREON 22." See dichlorodifluoromethane.

"FREON 112." See tetrachlorodifluoroethane.

"FREON 113." See 1,1,2-trichloro-1,2,2-trifluoroethane.

"FREON 114." See 1,2-dichloro-1,1,2,2-tetrafluoroethane.

β-FRUCTOPYRANOSE. Syns: *fruit sugar, levulose.* $C_6H_{12}O_6$, mw: 180.2.
THR = An exper neo via sc routes. [*3*]

FUEL OIL NO. 1. See kerosene.

FUEL OIL NO. 2. See diesel oil.

FUEL OIL NO. 3. Somewhat viscous, brown, odoriferous liquid. flash p: 110°–230°F (CC), d: <1, autoign. temp.: 498°F.
THR = U.
Fire Hazard: Mod, when exposed to heat or flame.
Explosion Hazard: U.

FUEL OIL NO. 4. Moderately viscous, dark, odoriferous liquid. flash p: 130°F, d: <1, autoign. temp.: 505°F.
THR = U.
Fire Hazard: Mod, when exposed to heat, flame or oxidizers.
Explosion Hazard: U.
To Fight Fire: Foam, CO_2, dry chemical.

FUEL OIL NO. 5. Syn: *navy special.* Viscous, oily liquid. flash p: 130 + °F, d: <1.
THR = U.
Fire Hazard: Mod, when exposed to heat, flame or oxidizers.
To Fight Fire: Foam, CO_2, dry chemical.

FUEL OIL NO. 6. Very viscous, dark-colored, odoriferous liquid. flash p: 150 + °F, d: <1, autoign. temp.: 765°F.
THR = U.
Fire Hazard: Mod, when exposed to heat, flame or oxidizers.
To Fight Fire: Foam, CO_2, dry chemical.

FULMINATE OF MERCURY. White solid. $HgC_2N_2O_2$, mw: 284.7, mp: explodes, d: 4.42.
THR = See mercury compounds, organic.
Fire Hazard: Dangerous! Materials should be kept moist until used. See fulminates.
Explosion Hazard: See fulminates.
Disaster Hazard: Highly dangerous. See fulminates.

FULMINATES.
THR = Variable.
Fire Hazard: Dangerous. Keep away from heat and open flame!
Explosion Hazard: Severe, when shocked or exposed to heat or flame (see explosives, high, and Section 7).
Storage and Handling: The fulminates are a group of explosives which are very sensitive to heat, impact, and friction when dry. They should be kept moist till ready for use. If compressed beyond 25,000 psi they become what is known as "dead-pressed," i.e., not capable of being exploded by flame. Fulminates are subject to deterioration when stored in hot climates. They decompose completely when they detonate and do so with great violence. They can be ignited with a flame or "spit," with a fuse, or with an electrically heated wire. They are widely used as initiators or primers for bringing about the detonation of high explosives or the ignition of powder. They are commonly used in combination with substances which provide a more prolonged blow and a bigger flame than fulminates alone. In the reinforced type of detonator, fulminates are made more effective by the addition of a more sensitive and powerful high explosive such as tetryl. This material is generally used in the manufacture of caps and detonators for initiating explosions for military, industrial, and sporting purposes. All precautions required for protection of magazines apply to storage of these materials. They should not be handled when frozen. Wet fulminate of mercury or wet floor coverings containing small quantities of fulminates may be burned on windrows of flammable material. Nonexplosive products are formed by neutralizing fulminates with cold sodium thiosulfate. All floors, tables and walls where the dry fulminates have been used should be washed with this solution.

In the manufacture of mercury fulminate, the fumes given off are toxic and flam. Care is required to prevent fulminate dust from being carried off in the exhaust system; deposits thus made have caused explosions. Much attention should be given to cleanliness as foreign or gritty materials in the product may cause an unexpected explosion. The floors on which fulminates are used, should be covered with 1/16 in. cloth inserted rubber packing or its equal. All cracks and crevices should be covered. The walls of these rooms should be covered with glazed, water-proof material. Frequent washing with neutralizing solution is necessary. In manufacture, the fulminate is dried on muslin squares on a drying table. Drying tables may be heated with hot water or the dry house may be heated with an air blower system to between 50 and 60°. Primer caps and detonators loaded with fulminate of mercury are less sensitive than the dry bulk material but must be handled with great care. Fires involving these assemblies should be treated the same as for the bulk material. They will explode as soon as fire reaches them. Stocks in an assembly or loading room should be kept as small as possible. Examples of fulminates commonly used in the explosive industry are mercury fulminate, copper fulminate and silver fulminate.

FULVINE. $C_{16}H_{23}O_5N$, mw: 309.4.
Acute tox data: LD_{50} (rat) = 40 mg/kg. [*3*]
THR = HIGH. An exper mutagen, teratogen. [*23*]

FUMARIC ACID. Syn: *trans-butenedioic acid.* Colorless, odorless crystals. HOOCCH:CHCOOH. mw: 116.1, mp: 287°, d: 1.635 @ 20°/4°.
Acute tox data: ip LD_{50} (mouse) = 200 mg/kg. [*3*]
THR = HIGH via ip route. Probably LOW via oral route. A food additive permitted in food for human consumption. [*109*]
Fire Hazard: Slight.

FUMARIN. See 3-(α-acetonyl furfuryl)hydroxy coumarin.

FUMARYL CHLORIDE. Syn: *trans-butenedioyl chloride.* Clear, straw-colored liquid. $(ClOCCH)_2$, mw: 153, mp: 160°, d: 1.408 @ 20°/4°.
THR = HIGH irr to skin, eyes and mu mem. See also hydrochloric acid.
Disaster Hazard: Dangerous; when heated to decomp, emits highly toxic fumes of phosgene and hydrochloric acid; will react with water or steam to produce toxic and corrosive fumes.

FUMING SULFURIC ACID. See oleum.

FURACYCLINE. Odorless, lemon-yellow crystals. Slightly sol in water. $C_6H_6O_4N_4$, mw: 198.2.
Acute tox data: Oral LD_{50} (rat) = 590 mg/kg; sc LD_{50} (rat) = 3000 mg/kg. [*3*]
THR = MOD via oral and sc routes. An exper (±) carc, neo via oral routes. [*3*, *23*] A sensitizer and food additive.

FURADAN. Syn: *2,3-dihydro-2,2-dimethyl-7-benzofuranylmethyl carbamate.* An odorless, white crystalline solid, slightly water-sol. $C_{12}H_{15}O_3N$, mw: 221.3, mp: 105°–152°, d: 1.180 @ 20°/20°, vap. press: 2×10^{-5} mm @ 33°.
Acute tox data: Oral LD_{50} (human) = 11 mg/kg; dermal LD_{50} (rat) = 120 mg/kg. [*3*]
THR = HIGH via oral and dermal routes. A carbamate insecticide. See also carbamates.

FURAL. See furfural.

2-FURALDEHYDE. See furfural.

FURALTADONE. Yellow crystals. $C_{13}H_{16}N_4O_6$, mw: 324.3, decomp @ 206°.
THR = A food additive permitted in food for human consumption. [*109*]

FURAN. Syns: *furfurane, oxole.* Water white liquid. C_4H_7O, mw: 68.07, mp: −85.65°, bp: 31.36°, lel = 2.3%, uel = 14.3%, flash p: <32°F, d: 0.937 @ 20°/4°, vap. d: 2.35.
Acute tox data: Inhal LC (rat) = 30,400 ppm. [*2*]
THR = LOW via inhal route. A narcotic. Absorbed via dermal route. Probably at least MOD via oral and dermal routes. The exposure conc limit of 10 ppm together with its low boiling point requires that adequate ventilation be provided in areas handling this chemical. Contact with liquid must be avoided since this chemical can be absorbed through the skin. Thorough washing with soap and water followed by prolonged rinsing should be done immediately after accidental contact.
Fire Hazard: Highly dangerous when exposed to heat or flame; can react with oxidizing materials. Unstabilized, it may form unstable peroxides on exposure to air and should always be tested before distillation. Washing with an aqueous solution of ferrous sulfate slightly acidified with sodium bisulfate will remove these peroxides. Confirm by test. Contact with acids can initiate a violent exothermic reaction.
Explosion Hazard: Moderate when exposed to flame. The low boiling point of this material makes it easy to obtain explosive concentrations of the vapor in inadequately ventilated areas.
Disaster Hazard: Highly dangerous, upon exposure to heat or flame; can react vigorously with oxidizing materials.
To Fight Fire: CO_2, dry chemical.

2-FURANCARBONAL. See furfural.

2-FURAN METHYLAMINE. See furfurylamine.

FURAN-2-PEROXY CARBOXYLIC ACID. $C_5H_4O_4$, mw: 128.1.

THR = Explodes when heated to 30° or on contact with organic materials, C, $CaCl_2$, $BaCl_2$, $SrCl_2$, $MgCl_2$. [*19*]

FURAZOLIDONE. Syn: *N-5-nitro-2-furfurylidene-3-amino-2-oxazolidinone.* Yellow powder, odorless, slightly sol in polyethylene glycol, insol in water, alcohol and peanut oil. $C_8H_7N_3O_5$, mw: 225, mp: 255°.

Acute tox data: Oral LD_{50} (rat) = 2336 mg/kg; ip LD_{50} (mouse) = 300 mg/kg. Causes pulmonary symptoms in man @ 11 mg/kg. [*3*]

THR = HIGH via ip and MOD via oral route.

FURANYL BORIC ACID. White crystals, water-sol. $(C_4H_3O)B(OH)_2$, mw: 111.9, mp: 110° (decomp).

THR = U. See also boron compounds. Probably LOW. A food additive permitted in the feed and drinking water of animals and/or for the treatment of food-producing animals. Also permitted in food for human consumption. [*109*]

FURCELLARON. Vegetable gum available as an odorless white powder, sol in warm water.

THR = A food additive permitted in food for human consumption. [*109*]

FURFURAL. Syns: *2-furancarbonal, 2-furaldehyde, fural, furale.* Colorless-yellowish liquid, almond-like odor. C_4H_3OCHO, mw: 96.1, mp: −36.5°, bp: 161.7° @ 764 mm, lel = 2.1%, uel = 19.3%, flash p: 140°F (CC), d: 1.161 @ 20°/20°, autoign. temp.: 600°F, vap. d: 3.31.

Acute tox data: Oral LD_{50} (rat) = 127 mg/kg; inhal LC_{LO} (rat) = 153 ppm for 4 hrs; oral LD_{50} (mouse) = 425 mg/kg. In humans 0.6 mg/m^3 causes eye symptoms. [*3*]

THR = MOD via dermal route and MOD irr to skin, eyes and mu mem. The liquid is dangerous to the eyes. The vapor is irr to mu mem and is a CNS poison. However, its low volatility reduces its toxicity effect. Furfural which has been ingested has produced cirrhosis of the liver in rats. In industry there is a tendency to minimize the danger of acute effects resulting from exposure to it. This is true, particularly, because of its low volatility. Little is known concerning the possibility of chronic effects, such as nervous disturbances, following prolonged or severe exposure to this material.

Fire Hazard: Mod, when exposed to heat or flame; can react with oxidizing materials.

Spont Heating: No.

Explosion Hazard: Mod, when exposed to heat or flame or by chemical reaction. An exothermic resinification of almost explosive violence can occur upon contact with strong mineral acids or alkalies.

Disaster Hazard: Mod dangerous. Keep away from heat and open flames.

To Fight Fire: Alcohol foam, CO_2, dry chemical.

FURFURAMIDE. See hydrofuramide.

FURFURANE. See furan.

FURFURYL ACETATE. Colorless liquid turning brown on exposure to light and air, pungent odor, insol in water, sol in alcohol and ether. $C_4H_3OCH_2OOCCH_3$, mw: 136, d: 1.1175 @ 20°/4°, bp: 175°–177°, vap. d: 4.8, flash p: 185°F.

THR = U.

Fire Hazard: Mod, when exposed to heat, flame or oxidizers. Reacts violently with cyanoacetic acid, formic acid, mineral acids, HNO_3,($HNO_3 + N_2O_4 + H_2SO_4$). [*19*]

To Fight Fire: Water may be used to blanket fire.

FURFURYL ALCOHOL. Syn: *2-furyl carbinol.* Clear, colorless, mobile liquid. $C_4H_3OCH_3OH$, mw: 98.1, mp: −31°, lel = 1.8%, uel = 16.3%, both between 72°–122°, bp: 171° @ 750 mm, flash p: 167°F (OC), d: 1.129 @ 20°/4°, autoign. temp.: 915°F, vap. press: 1 mm @ 31.8°, vap. d: 3.37.

Acute tox data: Oral LD_{50} (rat) = 275 mg/kg; inhal LC_{50} (rat) = 233 ppm for 4 hrs; oral LD_{50} (mouse) = 40 mg/kg. [*3*] dermal LD (rabbit) = 400 mg/kg. [*20*] sc LD_{50} (rat) = 85 mg/kg. [*20*]

THR = HIGH via oral, inhal, sc and dermal routes. MOD irr to skin, eyes and mu mem.

Fire Hazard: Mod, when exposed to heat; can react with oxidizing materials.

Explosion Hazard: Mod, when exposed to heat or flame.

To Fight Fire: Alcohol foam, CO_2, dry chemical.

FURFURYLAMINE. Syn: *2-furanmethylamine.* Light straw-colored liquid. $C_4H_3OCH_2NH_2$, mw: 97.1, bp: 146°, flash p: 99°F (OC), fp: −70°, d: 1.0502 @ 25°, vap. d: 3.35.

Acute tox data: ip LD_{LO} (mouse) = 200 mg/kg. [*3*]

THR = HIGH via ip route. MOD irr to skin, eyes and mu mem.

Fire Hazard: Dangerous; when exposed to heat or flame; can react with oxidizing materials.

Explosion Hazard: U.

To Fight Fire: Foam, CO_2, dry chemical.

FURNACE BLACK. See soot.

FURNITURE POLISH.

THR = Variable; allergen.

FUROIC ACID. Syn: *pyromucic acid.* White solid. $C_4H_3OCO_2H$. mw: 112.1, mp: 133°, bp: 230°–232°.

THR = U. A bactericide.
Fire Hazard: Slight.

FUROYL CHLORIDE. Colorless liquid. C_4H_3OCOCl, mw: 130.5, mp: −2°, bp: 176°.
THR = HIGH irr to skin, eyes and mu mem.
Disaster Hazard: Dangerous; when heated to decomp, emits highly toxic fumes of phosgene and chlorides; will react with water or steam to produce toxic and corrosive fumes.

FURS.
THR = An allergen.
Fire Hazard: Mod, when exposed to heat or flame.

2-FURYL CARBINOL. See furfuryl alcohol.

FURYL FURAMIDE. $C_{11}H_8O_5N_2$, mw: 248.2.
Acute tox data: Oral LD_{50} (rat) = 1554 mg/kg; oral LD_{50} (mouse) = 475 mg/kg. [*3*]
THR = MOD via oral route. Affects animal pregnancy. An exper neo. [*3*]

FUSARENON. $C_{17}H_{22}O_8$, mw: 354.4.
Acute tox data: Oral LD_{50} (rat) = 4 mg/kg; ip LD_{50} (mouse) = 3.3 mg/kg, sc LD_{50} (mouse) = 4.2 mg/kg. [*3*]
THR = HIGH via oral, sc and ip routes. An exper (±) carc. [*3*]

FUSEL OIL.
THR = An exper mutagen. [*3*]

FUSES. See explosives, high (detonating devices).

FUZES. Official U.S. Army spelling of "fuses." See explosives, high (detonating devices).

g

G-4. See 2,2′-methylene bis(4-chlorophenol).

G-11. See 2,2′-methylene bis(3,4,6-trichlorophenol).

G-19258. See dimetan.

G-22008. See pyrolan.

G-24480. See diazinon.

GABIAN OIL.
THR = U.
Fire Hazard: Mod, when exposed to heat or flame; can react with oxidizing materials.
Explosion Hazard: U.

GADOLINIUM. A yellow-white malleable and ductile metal, stable in dry air, reacts slowly with H_2O. atwt: 157.25, mp: 1311°, bp: 3233°, d: 7.898 @ 25°.
THR = An exper neo. [*3*] It may act as an anti-coagulant. It can react violently with air, halogens. [*19*]
Radiation Hazard: For permissible levels, see Section 5, Table 5A.5.

GADOLINIUM BROMIDE. Rhombic plates. $GdBr_3 \cdot 6H_2O$, mw: 505.1, d: 2.844 @ 15°.
THR = See bromides and gadolinium.

GADOLINIUM CHLORIDE. $GdCl_3$, mw: 263.6.
Acute tox data: ip LD_{50} (mouse) = 378 mg/kg. [*3*]
THR = HIGH via ip route. See also chlorides and gadolinium.

GADOLINIUM CITRATE.
Acute tox data: ip LD_{50} (mouse) = 153 mg/kg; ip LD_{50} (guinea pig) = 60 mg/kg. [*3*]
THR = HIGH via ip route. No oral or inhal data, but probably HIGH.

GADOLINIUM FLUORIDE. White, gelatinous mass. GdF_3, mw: 214.3.
THR = See fluorides and gadolinium.

GADOLINIUM NITRATE. Triclinic crystals. $Gd(NO_3)_3 \cdot 6H_2O$, mw: 451.36, mp: 92°, d: 2.406 @ 15°.
Acute tox data: ip LD_{50} (rat) = 230 mg/kg. [*3*]
THR = HIGH via ip route. Probably HIGH via oral and inhal routes. See also nitrates and gadolinium.

GADOLINIUM OXALATE. Monoclinic crystals. $Gd_2(C_2O_4)_3 \cdot 10H_2O$, mw: 758.72, mp: $-6H_2O$ @ 110°.
THR = HIGH via oral route. See also oxalates and gadolinium.

GADOLINIUM OXIDE. White- to cream-colored hygroscopic power, insol in water, sol in acids. Gd_2O_3, mw: 362.5, d: 7.407 @ 15°, mp: 2330°.
Acute tox data: ip LD_{50} (rat) = 1000 mg/kg. [*3*]
THR = MOD via ip route. Probably also via oral and inhal routes as well.

GADOLINIUM SELENATE. Monoclinic pearly crystals. $Gd_2(SeO_4)_3 \cdot 8H_2O$, mw: 887.5, mp: $-8H_2O$ @ 130°, d: 3.309.
THR = See selenium and its compounds and gadolinium.

GADOLINIUM SULFIDE. Yellow hygroscopic mass. Gd_2S_3, mw: 410.69, d: 3.8.
THR = See sulfides and gadolinium.

D-GALACTOSAMINE HYDROCHLORIDE. $C_5H_{13}O_5N \cdot HCl$, mw: 215.7.
THR = An exper neo. [*3*]

GALENA. See lead compounds.

GALLIC ACID. Syn: *3,4,5-trihydroxybenzoic acid.* White- to pale fawn-colored odorless crystals, somewhat water-sol. $C_7H_6O_5 \cdot H_2O$, mw: 188.1, d: 1.694, mp: 225°–250° (decomp), $-H_2O$ @ 100°–120°.
Acute tox data: Oral LD_{50} (rat) = 5000 mg/kg; [*3*] sc LD (rat) = 4000 mg/kg. [*2*]
THR = MOD via sc and oral routes. A MILD irr. Has produced weakness and neurological disturbances in exper animals. Reacts violently with $KClO_3$. [*19*]

GALLIUM. A beautiful, lustrous, silvery liquid or metal or a gray solid. Ga, atwt: 69.72, mp: 29.78°, bp: 2403°, d(solid): 5.904 @ 29.6°, d(liquid): 6.905 @ 29.8°.
Acute tox data: sc LD_{LO} (rat) = 110 mg/kg. [*3*]
THR = HIGH via sc route. It has a metallic taste, causes dermatitis and depression of bone marrow function. See Ga compounds.
Radiation Hazard: For permissible levels, see Section 5A. Table 5A.5

GALLIUM ACETATE, BASIC. White crystals, $4Ga(C_2H_3O_2)_3 \cdot 2Ga_2O_3 \cdot 5H_2O$, mw: 1452.4, mp: decomp @ 160°.
THR = No data. Probably LOW.

GALLIUM ARSENIDE. Cubic crystals with dark gray metallic sheen. GaAs, mw: 144.6, mp: 1238°, d: 5.31.
Acute tox data: Oral LD_{50} (rat) = 4700 mg/kg. [*3*]
THR = MOD via oral route. See also arsenic compounds.

GALLIUM CHLORIDE. $GaCl_3$, mw: 176.1.
Acute tox data: Inhal LD_{LO} (rat) = 191 mg/kg for 3 hrs; ip LD_{50} (mouse) = 37 mg/kg. [3]
THR = HIGH via inhal, oral and ip routes.

GALLIUM CITRATE. $C_6H_7O_7Ga$, mw: 260.9.
Acute tox data: sc LD_{50} (rat) = 220 mg/kg. [3]
THR = HIGH via sc and probably oral and inhal routes. Preliminary investigations were done with the oxide, tartrate, benzoate, and anthranilate, which were used by some investigators in the experimental treatment of syphilis. Amounts up to 15 mg/kg of body weight were injected intravenously and were tolerated without harm by laboratory animals. Larger doses produce hemorrhagic nephritis. In the case of gallium lactate, work done at the Naval Medical Research Institute showed that intravenous injections of about 40 mg of gallium per kg of body weight in rats or rabbits was lethal. Metallic gallium as well as the nitrate produced no skin injury and subcutaneous injection of relatively large amounts could be tolerated both by rabbits and rats without evidence of injury. It has, however, been demonstrated that gallium remains in the tissues for long periods of time following intramuscular injection of soluble gallium salts. Tissue distribution experiments indicate that it behaves like bismuth and mercury in that one respect.

GALLIUM HYDRIDE. Colorless liquid. Ga_2H_5, mw: 145.59, mp: −21.4°, bp: 139°, decomp >130°, vap. press: 2.5 mm @ 0°.
THR = See hydrides and gallium. Spont flam in air. [19]

GALLIUM MONOSELENIDE. Syn: *gallium selenide.* Dark red-brown, greasy leaf. GaSe, mw: 148.68, mp: 960°, d: 5.03.
THR = Probably HIGH via oral and inhal routes. See selenium and its compounds and gallium.
Fire Hazard: Mod, by chemical reaction which liberated hydrogen selenide.
Explosion Hazard: Slight, by chemical reaction.
Disaster Hazard: Mod dangerous; when heated to decomp, emits highly toxic fumes of selenium.

GALLIUM MONOSULFIDE. Sublimate, light yellow. GaS, mw: 107.79, mp: 956 ± 10°, d: 3.86 @ 25°.
THR = See sulfides and gallium.

GALLIUM MONOTELLURIDE. Soft, black, greasy leaves. GaTe, mw: 197.33, mp: 824°, d: 5.44.
THR = See tellurium compounds and gallium.

GALLIUM NITRATE. White, deliquescent crystals. $Ga(NO_3)_3 \cdot xH_2O$, mp: decomp @ 110°, bp: ⟶ Ga_2O_3 @ 200°.
Acute tox data: sc LD_{LO} (rat) = 72 mg/kg. [3]
THR = HIGH via sc route. See nitrates and gallium.

GALLIUM NITRITE.
Acute tox data: Oral LD_{50} (guinea pig) = 3280 mg/kg. [3]
THR = MOD via oral route.

GALLIUM OXALATE. White powder. $Ga_2(C_2O_4)_3$, mw: 403.50, bp: decomp @ 195°.
THR = HIGH. See oxalates and gallium.

GALLIUM OXIDE. The sesquioxide Ga_2O_3 and suboxide Ga_2O are known. Both are stable at room temp. Ga_2O_3 is found in α and β forms; both are white crystals, insol in water, sol in alcohol, very slightly sol in hot acid. Ga_2O_3 (sesquioxide), mw: 188. d(α): 6.44, d(β): 5.88, mp: 1900°, changes to β @ 600°.
THR = Animal exper suggest LOW toxicity. See also gallium compounds and gallium.

GALLIUM OXYCHLORIDE. Crystals. $6GaOCl \cdot 14H_2O$, mw: 979.3.
THR = May evolve hydrochloric acid. See also hydrochloric acid and gallium compounds.
Fire Hazard: Slight. An oxidizer.
Disaster Hazard: Dangerous, see chlorides.

GALLIUM PERCHLORATE. White deliquescent crystals. $Ga(ClO_4)_3 \cdot 6H_2O$, mw: 476.19, mp: decomp @ 175°.
THR = See perchlorates and gallium. Forms a violently explosive product with urea. [19]

GALLIUM PHOSPHIDE. Pale orange transparent crystals or whiskers up to 2 cm long. GaP, mw: 100.7, mp: 1400°.
THR = Probably HIGH. See gallium compounds and phosphides.

GALLIUM SELENATE. Colorless crystals. $Ga_2(SeO_4)_3 \cdot 16H_2O$, mw: 856.58.
THR = See selenium compounds.

GALLIUM SELENIDE. See gallium monoselenide.

GALLIUM SESQUISELENIDE. Reddish-black crystals. Ga_2Se_3, mw: 376.32, mp: >1020°, d: 4.92.
THR = Probably HIGH. See selenium compounds and gallium.
Fire Hazard: Mod; can liberate a flam gas upon contact with moisture or acids.
Explosion Hazard: Slight; can liberate a flam gas upon contact with moisture or acids.
Disaster Hazard: Dangerous; when heated to decomp or on contact with acid or acid fumes, emits highly toxic fumes.

GALLIUM SESQUISULFIDE. Yellow crystals, white amorphous mass. Ga_2S_3, mw: 235.64, mp: 1255° ± 10°, d: 3.65 @ 25°.
THR = See sulfides and gallium.

GALLIUM SESQUITELLURIDE. Hard, black, brit-

tle crystals. Ga_2Te_3, mw: 522.3, mp: 790°, d: 5.57.
THR = See tellurium compounds and gallium.

GALLIUM SUBOXIDE. Brown-black powder. Ga_2O, mw: 155.44, mp: >660°, sublimes >500°, d: 4.77.
THR = See gallium compounds.

GALLIUM SUBSELENIDE. Crystals. Ga_2Se, mw: 218.40, d: 5.02.
THR = See selenium compounds and gallium.

GALLIUM SUBSULFIDE. Green crystals or black powder. Ga_2S, mw: 171.51, d: 4.18 @ 25°, decomp: >800°.
THR = See sulfides and gallium.

GALLIUM SULFATE. White powder. $Ga_2(SO_4)_3$, mw: 427.6. mp: decomp <700°.
THR = See gallium compounds.

GALLIUM TRIBROMIDE. Colorless deliquescent crystals. $GaBr_3$, mw: 309.47, mp: 121.5 ± 0.6°, bp: 278.8°, d: 3.69 @ 25°.
THR = See bromides and gallium.

GALLIUM TRIFLUORIDE. White powder. GaF_3, mw: 126.7, mp: >1000°, bp: sublimes approx 950°, d: 4.47.
THR = See fluorides and gallium.

GALLIUM TRIFLUORIDE TRIAMMINE. White powder. $GaF_3 \cdot 3NH_3$, mw: 177.82, mp: $-NH_3$ @ 100°.
THR = See fluorides and gallium.

GALLIUM TRIIODIDE. Colorless to lemon yellow, hygroscopic needles. GeI_3, mw: 450.48, mp: 212 ± 1°, bp: 345°, (sublimes), d: 4.15 @ 25°.
THR = See iodides and gallium.

GALLOETHANE. See gallium hydride.

GALLOTANNIC ACID. See tannic acid.

GAMBALA. See sawdust.

GAMBOGE. Syn: *gambogia.*
THR = HIGH via oral route. A violent purgative.
Fire Hazard: Mod, when exposed to heat or flame.

GAMBOGIA. See gamboge.

GAMMA HEXANE. See 1,2,3,4,5,6-hexachlorocyclohexane.

GAMMA ISOMER OF BHC. See 1,2,3,4,5,6-hexachlorocyclohexane.

GAMMA RADIATION.
THR = A recog carc. [*14*] See radiation, ionizing; Sections 8 and 5.

GAMMA RAY CONCRETE SHIELDING. See Section 5A.

GAMMA RAY RADIOGRAPHY. See Section 5A.

GAMMA RAYS. Electromagnetic radiation emitted by certain radioactive isotopes. Gamma rays are similar to x-rays, but are more energetic and penetrating than the x-rays ordinarily used in medical diagnosis and therapy.
Radiation Hazard: See Section 5, for complete discussion.

GAMMEXANE. See 1,2,3,4,5,6-hexachlorocyclohexane.

GAMOXO. See 1,2,3,4,5,6-hexachlorocyclohexane.

GAMTOX. See 1,2,3,4,5,6-hexachlorocyclohexane.

GARBAGE TANKANGE CONTAINING LESS THAN 8% MOISTURE.
Fire Hazard: Dangerous, when exposed to heat or flame; can react vigorously with oxidizing materials.
Explosion Hazard: U.

GARDENIA OXIDE. See amyl benzyl ether.

GARDINOL. See sodium lauryl sulfate.

GARLIC OIL. See allyl sulfide.

GARLON. See slivex.

GAS DRIPS, HYDROCARBON.
THR = U.
Fire Hazard: Dangerous, when exposed to heat or flame; can react vigorously with oxidizing materials.
Explosion Hazard: U.

GAS FIRES EXTINGUISHMENT. See Section 7.

GASHOUSE TANKAGE. See spent oxide.

GAS HOUSE TAR.
THR = A recog carc. [*14*] See coal tar and pitch.

GAS OIL. Yellow liquid. flash p: 150°F, d: <1, lel = 6.0%, uel = 13.5%, autoign. temp.: 640°F, bp: 230°–250°.
THR = U. See also kerosene.
Fire Hazard: Mod, when exposed to heat or flame; can react with oxidizing materials.
Explosion Hazard: Mod, when exposed to heat or flame.
To Fight Fire: Foam, CO_2, dry chemical.

GASOLINE (from 50–100 octane). Syn: *petrol.* Clear, aromatic, volatile liquid, a mixture of aliphatic hydrocarbons. flash p: −45°F, d: <1.0, vap. d: 3.0–4.0, ulc: 95-100, lel = 1.4%, uel = 7.6%, autoign. temp.: 536°–853°F.
Acute tox data: Inhal TC_{LO} (man) = 900 ppm for 1 hr; caused CNS symptoms. [*3*]
THR = HIGH to MOD via inhal route. Repeated or prolonged dermal exposure causes dermatitis. Can cause blistering of skin. [*20*] Inhal and via oral routes causes CNS depression. Pulmonary aspiration can cause severe pneumonitis. Some addiction

has been reported to inhal of fumes. Even brief inhal of high conc can cause a fatal pulmonary edema. [20] It can cause hyperemia of the conjunctiva and other disturbances of the eyes. The vapors are considered to be mod poisonous. If its conc in air is sufficiently high to reduce the oxygen content below that needed to maintain life, it acts as a simple asphyxiant. Gasoline is a common air contaminant. See mineral oils.

Fire Hazard: Dangerous, when exposed to heat or flame; can react vigorously with oxidizing materials.

Explosion Hazard: Mod, when exposed to heat or flame.

Disaster Hazard: Dangerous, in the presence of heat or flame.

To Fight Fire: Foam, CO_2, dry chemical.

GASOLINE (100–130 octane). flash p: −50° F, autoign. temp.: 824° F, lel = 1.3%, uel = 7.1%.

Fire Hazard: Dangerous, via heat, flame or oxidizers.

To Fight Fire: Water spray or mist, CO_2, dry chemical.

GASOLINE (115–145 octane). flash p: −50° F, autoign. temp.: 880° F, lel = 1.2%, uel = 7.1%.

Fire Hazard: See Gasoline (100–130 octane).

To Fight Fire: See Gasoline (100–130 octane).

GASOLINE (CASINGHEAD). Aromatic, volatile liquid. flash p: 0° F or less.

THR = See gasoline.

Fire Hazard: Very dangerous, when exposed to heat or flame; can react with oxidizing materials.

Explosion Hazard: U.

Disaster Hazard: Dangerous, in the presence of heat or flame.

To Fight Fire: Dry chemical, CO_2, foam.

GASOLINE ENGINE EXHAUST TAR.

THR = An exper carc. [*3*]

GASOLINE VAPORS. See gasoline.

GAS PRODUCTION FROM SOLID WASTES. See Section 6.

GBH. See benzene hexachloride.

GEDUNIN. $C_{28}H_{36}O_7$, mw: 484.6.

THR = An exper neo. [*3, 23*]

GELATINE DYNAMITE. See explosives, high.

GEL COALITES. See explosives, high.

GELSEMINE. An alkaloid. $C_{20}H_{22}N_2O_2$, mw: 322.4, mp: 178°.

Acute tox data: sc LD_{LO} (rabbit) = 0.5 mg/kg; iv LD_{LO} (rabbit) = 0.8 mg/kg. [*3*]

THR = HIGH via sc and iv routes. A poisonous alkaloid. CNS stimulant, can cause muscular weakness, respiratory arrest.

GELSEMIUM. A mixture of alkaloids. Composition: chief constituents are gelsemine, gelseminine and gelsemic acid.

THR = A poison. See gelsemine.

GELSEMIUM SEMPERVIRONS. Syn: *yellow jasmine.*

THR = See gelsemine. Can seriously affect the eyes.

GERMERINE. Crystals. $C_{37}H_{59}NO_{11}$, mw: 694, decomp @ 202°–205°.

Acute tox data: Oral LD_{50} (rat) = 30 mg/kg; sc LD_{50} (rat) = 3.7 mg/kg; iv LD_{LO} (rabbit) = 0.3 mg/kg. [*3*]

THR = HIGH via oral, sc and iv routes.

Disaster Hazard: Dangerous; when heated to decomp, emits highly toxic fumes.

GENERAL GUIDELINES TO USE OF LABELS. See Section 11.

GENERAL PRINCIPLES OF INDUSTRIAL HYGIENE. See Section I.

GENERALS. See explosives, high.

GENIPHENE. See toxaphene.

GENISTEIN. $C_{15}H_{10}O_5$, mw: 270.3, mp: 297°.

THR = An exper carc. [*23*]

GENITHION. See parathion.

GEORGIUS AGRICOLA VOLUME OF METALLURGY. See Section 1.

GERANIAL. See citral.

GERANIOL. Syn: *3,7-dimethyl-trans-2,6-octadien-1-ol.* Colorless to pale yellow, liquid oil, pleasant geranium odor. $C_{10}H_{18}O$, mw: 154.3, d: 0.870–0.890 @ 15°, mp: 15°, bp: 230°.

Acute tox data: Oral LD_{50} (rat) = 3600 mg/kg; im LD_{50} (mouse) = 4000 mg/kg; iv LD_{50} (rabbit) = 50 mg/kg. [*3*]

THR = MOD via oral and im and HIGH via iv routes. A synthetic flavoring substance and adjuvant. [*109*]

GERANIOL ACETATE. See geranyl acetate.

GERANYL ACETATE. Syn: *geraniol acetate.* Colorless, sweet, clear liquid, odor of lavender, sol in alcohol and ether, insol in water and glycerol. $CH_3COOC_{10}H_{17}$, mw: 196, d: 0.907–0.918 @ 15°, bp: 128°–129° @ 16 mm.

Acute tox data: Oral LD_{50} (rat) = 6330 mg/kg. [*3*]

THR = MOD via oral route. A synthetic flavoring substance and adjuvant. [*109*]

GERMANIUM. A gray-white metalloid; crystalline and brittle, stable at room temp. Ge, atwt: 72.59, mp: 937.4°, bp: 2830°, d: 5.323 @ 25°.

Acute tox data: sc LD_{LO} (rabbit) = 586 mg/kg. [*3*]

THR = MOD via sc route.

Radiation Hazard: For permissible levels, see Section 5, Table 5A.5.
Fire Hazard: Mod, in the form of dust, when exposed to heat or flame. See also powdered metals. Reacts violently with Br_2, O_2, Cl_2, HNO_3, $KClO_3$, KNO_3. [*19*]

GERMANIUM CHLOROFORM. See trichlorogermane.

GERMANIUM COMPOUNDS.
THR = Little is known about the toxicity of organic germanium compounds; but they may resemble other organometals in having higher toxicity than inorganic forms.

Germanium compounds are considered to be of a low order of toxicity, but rare instances of poisoning have been reported in the literature. Interest is high in this material because of its close chemical relationship to arsenic. It has been found that the dioxide stimulates the generation of red blood cells, but it is believed to be relatively nontoxic. When germanium is given in sublethal amounts, it causes a pronounced tolerance to be exhibited. Germanium compounds are considered much less toxic than the corresponding lead and tin compounds. Buffered germanium dioxides in solution have been found to be nonirritating to the skin. Germanium hydride is a hemolytic gas and has been shown to have toxic properties at concentration levels of 100 ppm. It can cause death in concentration of 150 ppm.

GERMANIUM DIBROMIDE. Colorless needles or plates. $GeBr_2$, mw: 232.4, mp: 122.0°, bp: decomp.
THR = See bromides and germanium.
Disaster Hazard: See bromides.

GERMANIUM DICHLORIDE. White powder. $GeCl_2$, mw: 143.5, mp: decomp to Ge + $GeCl_4$.
THR = See germanium compounds.
Disaster Hazard: Slight; when heated, emits toxic fumes.

GERMANIUM DIFLUORIDE. White, deliquescent crystals. GeF_2, mw: 110.6, mp: decomp >350°, bp: sublimes.
THR = See fluorides and germanium.
Disaster Hazard: See fluorides.

GERMANIUM DIIODIDE. Yellow hexagonal crystals. GeI_2, mw: 326.44, mp: sublimes and decomp, d: 5.37.
THR = See iodides and germanium
Disaster Hazard: See iodides.

GERMANIUM DIOXIDE (INSOLUBLE). Tetragonal crystals. GeO_2, mw: 104.60, mp: 1086 ± 5°, d: 6.239.
THR = No data.

GERMANIUM DIOXIDE (SOLUBLE). Hexagonal colorless crystals. GeO_2, mw: 104.60, mp: 1115.0°, d: 4.703 @ 18°.
Acute tox data: ip LD_{50} (rat) = 750 mg/kg. [*3*]
THR = MOD via ip route.

GERMANIUM DISULFIDE. White powder, orthorhombic white crystals. GeS_2, mw: 136.73, mp: ca. 800°, bp: sublimes > 600°, d: 2.094 @ 14°.
THR = See sulfides and germanium.

GERMANIUM HYDRIDE. See digermane.

GERMANIUM HYDRIDES. GeH_4, Ge_2H_6, Ge_3H_8.
THR = See hydrides and germanium compounds and germanium. Self-ignites in air. [*19*]

GERMANIUM MONOSULFIDE. Yellow-red amorphous or black crystals. GeS, mw: 104.67, mp: 530°, bp: sublimes >430°, d(amorphous): 3.31, d(rhombic): 4.0. @ 14°/14°.
THR = See sulfides and germanium.

GERMANIUM MONOXIDE. Black crystalline powder. GeO, mw: 88.60, mp: sublimes @ 710°.
Acute tox data: ip LD_{LO} (guinea pig) = 300 mg/kg. [*3*]
THR = HIGH via ip route. See also germanium compounds and germanium.

GERMANIUM OXYCHLORIDE. Colorless liquid. $GeOCl_2$, mw: 159.51, mp: −56.0°, bp: decomp >20°.
THR = See hydrochloric acid and germanium.
Disaster Hazard: Dangerous; when heated, emits toxic fumes of chlorides.

GERMANIUM SULFIDE. GeS_2, mw: 136.7.
THR = Probably toxic. See also sulfides and germanium. Can explode when mixed with KNO_3. [*19*]

GERMANIUM TETRABROMIDE. Gray-white crystals. $GeBr_4$, mw: 392.26, mp: 26.1°, bp: 186.5°, d: 3.232 @ 29°/29°.
THR = See bromides and germanium.
Disaster Hazard: Dangerous; when heated, emits toxic fumes of bromides.

GERMANIUM TETRACHLORIDE. Colorless liquid. $GeCl_4$, mw: 214.43, mp: −49.5°, bp: 83.1°, d: 1.879.
THR = MOD irr to skin, eyes and mu mem. LOW systemic. See also germanium compounds, hydrochloric acid and germanium.
Disaster Hazard: Dangerous; when heated to decomp, emits toxic fumes of chlorides; will react violently with water or steam to produce toxic and corrosive fumes.

GERMANIUM TETRAETHYL. See tetraethyl germanium.

GERMANIUM TETRAFLUORIDE. Colorless gas or solid, not liquid at atmospheric pressure. GeF_4, mw: 148.60, mp: sublimes, d: 6.65 g/liter.

THR = HIGH irr to skin, eyes and mu mem. See fluorides, germanium compounds, and germanium.
Disaster Hazard: Dangerous; emits highly toxic fumes of fluorides.

GERMANIUM TETRAHYDRIDE. See monogermane.

GERMANIUM TETRAIODIDE. Cubic yellow crystals. GeI_4, mw: 580.28, mp: 144.0°, bp: decomp, d: 4.322 @ 26°/26°.
THR = See iodides and germanium. Vapors may self-ignite in air. [*19*]
Disaster Hazard: Mod dangerous; when heated to decomp, emits toxic fumes of iodides.

GERMANIUM TETRAPHENYL. $Ge(C_6H_5)_4$, mw: 380.6.
THR = No data. There is inhibition of blood formation. See also germanium compounds and phenol.

GESAROL. See DDT.

GHATTI GUM. The gummy exudation from the stem of *anogeissus latifolia.* Colorless to pale yellow tears, almost odorless, partially sol in water.
THR = No data. Probably LOW. Used as a stabilizer food additive. [*109*]

GIBBERELLIC ACID. A plant growth-promoting hormone, crystals, slightly sol in water, sol in methanol, ethanol, acetone, aqueous solutions of sodium bicarbonate and sodium acetate. $C_{19}H_{22}O_6$, mw: 346, mp: 233°–235°.
THR = No data. Probably LOW. A food additive permitted in food for human consumption [*109*]

GIBBERELLIC ACID, POTASSIUM SALT. Syn: *"Gibrel."*
THR = No data. Probably LOW. A food additive permitted in food for human consumption. [*109*]

GILSONITE. Syn: *uintaite.* A black solid hydrocarbon mineral formed from petroleum millions of years ago by geologic processes.
THR = MOD irr to skin, eyes and mu mem. An allergen. Has been known to cause photosensitization of skin.
Fire Hazard: Mod, when exposed to heat or open flame.
Disaster Hazard: Mod; when heated to decomp, emits toxic fumes.
To Fight Fire: Water, foam, dry chemical and CO_2.

GIN. Composition: 40–50% ethyl alcohol; beverage made from juniper and coriander berries, etc. bp: 82°, flash p: 89°F, ulc: 70, lel = 3.3%, uel = 19%, autoign. temp.: 799°F.
THR = MOD via oral route. Acts as a kidney irr. Has had some medicinal use as a diuretic.
Fire Hazard: Mod, when exposed to heat or flame; can react vigorously with oxidizing materials.
Explosion Hazard: Mod, when heated or exposed to flame.
To Fight Fire: CO_2, dry chemical.

GINGER. Syn: *zingiber.* Irregularly bunched pieces; aromatic odor.
THR = An allergen and MILD irr. A flavoring. Has been used in medicine.
Fire Hazard: Slight.

GLASS FIBER, DUST. See "Fiberglas."

GLASS WOOL. See "Fiberglas."

GLUCAGON. $C_{151}H_{224}O_{50}N_{42}S$, mw: 3460.2.
THR = Effects the white blood cells. An exper teratogen. [*3*]

GLUCINUM. See beryllium.

GLUCINUM POTASSIUM FLUORIDE. See beryllium potassium fluoride.

GLUCINUM SODIUM FLUORIDE. See beryllium sodium fluoride.

GLUCOSE PENTAPROPIONATE. Syns: *pentapropionyl glucose, tetrapropionyl glucosyl propionate.* Insol in water. $C_6H_7O_6(COC_2H_5)_5$, mw: 460, d: 1.2, bp: 401°F @ 2 mm, flash p: 509°F.
THR = No data.
Fire Hazard: Slight when exposed to heat or flame.
To Fight Fire: Spray, dry chemical, CO_2.

GLUE. A colloidal suspension of proteins in water, from white to brown slabs, flakes or chips.
THR = An allergen.
Disaster Hazard: Slight; when heated, emits smoke.

GLUSIDE. See sodium saccharin.

GLUTAMIC ACID. Syn: *α-aminoglutaric acid.* A non-essential amino acid present in all complete proteins. Crystals. $COOH(CH_2)_2CH(NH_2)COOH$, mw: 147, mp(*dl* form): 194°, d(*dl* form): 1.4601 @ 20°/4°, mp(l form): 224°–225°, d(l form): 1.538 @ 20°/4°.
THR = U. A general purpose food additive. [*109*]

GLUTAMIC ACID HYDROCHLORIDE. White crystal powder, sol in water (liberating hydrochloric acid), almost insol in alcohol and ether.
$COOH(CH_2)_2CH(NH_2)COOH$, mw: 185, d: 1.525 mp: 202°–213° (decomp).
THR = An irr. A general purpose food additive. [*109*] See also hydrochloric acid.

GLUTARALDEHYDE. $C_5H_8O_2$, mw: 100.1.
Acute tox data: Oral LD_{50} (rat) = 600 mg/kg; inhal LC_{LO} (rat) = 5000 ppm for 4 hrs; dermal LD_{50} (rabbit) = 2560 mg/kg. [*3*]
THR = MOD via oral, inhal and dermal routes.

GLUTARIC ANHYDRIDE. Syns: *pentanedioic acid anhydride, 1,3-propane dicarboxylic acid anhydride.* Sol in benzene and toluene, highly sol in water on complete hydrolysis. $C_5H_6O_3$, mw: 114.1, bp: 144°–146° @ 13 mm, d: 0.989.

Acute tox data: Oral LD_{LO} (rat) = 4460 mg/kg; dermal LD_{LO} (rabbit) = 1780 mg/kg. [*3*]

THR = MOD via oral and dermal routes.

GLUTARONITRILE. Syn: *pentanedinitrile.* Colorless liquid, sol in water, insol in ether. $CN(CH_2)_3CN$, mw: 94.1, d: 0.989 @ 15°/4°, mp: −29°, bp: 286.4°.

THR = No data. See also nitriles.

GLYCERIN DICHLOROHYDRIN. Insol in water. $CH_2ClCHClCH_2OH$, mw: 129, d: 1.4, bp: 360°F, flash p: 200°F.

Acute tox data: Oral LD_{50} (rat) = 150 mg/kg; inhal LC_{LO} (rat) = 125 ppm for 4 hrs; ip LD_{50} (mouse) = 73 mg/kg; sc LD_{50} (mouse) = 5 mg/kg. [*3*]

THR = HIGH via oral, inhal, ip and sc routes.

Fire Hazard: Low, when exposed to heat or flame.

To Fight Fire: Alcohol foam, mist, spray, dry chemical.

GLYCERINE. Syns: *1,2,3-propanetriol, glycerol.* Colorless or pale yellow liquid, odorless, syrupy, sweet and warm taste. $CH_2OHCHOHCH_2OH$, mw: 92.09, mp: 17.9 (solidifies @ a much lower temp.), bp: 290°, ulc: 10–20, flash p: 320°F, d: 1.260 @ 20°/4°, autoign. temp.: 698°F, vap. press: 0.0025 mm @ 50°, vap. d: 3.17.

Acute tox data: Oral LD_{50} (mouse) = 470 mg/kg; sc LD_{50} (mouse) = 10,000 mg/kg; iv LD_{50} (mouse) = 4250 mg/kg. [*3*]

THR = MOD via oral, sc and iv routes. In the form of mist it is an inhal-irr. A general purpose food additive. [*109*] It migrates to food from packaging materials.

Fire Hazard: Low, when exposed to heat, flame or powerful oxidizers. Can react violently with acetic anhydride, (aniline + nitrobenzene), $Ca(OCl)_2$, CrO_3, Cr_2O_3, $(F_2 + PbO)$, $(HClO_4 + PbO)$, $KMnO_4$, K_2O_2, $AgClO_4$, Na_2O_2. [*19*]

To Fight Fire: Alcohol foam, CO_2, dry chemical.

GLYCERIN PITCH. A (S) carc. [*14*]

GLYCEROL. See glycerine.

GLYCEROL ESTER OF WOOD ROSIN.

THR = U. Used as a food additive permitted in food for human consumption. [*109*]

GLYCEROL MONOSTEARATE. See glyceryl monostearate.

GLYCOL PERCHLORATES.

THR = HIGH and violently unstable. [*19*]

GLYCEROL TRICHLOROHYDRIN. See 1,2,3-trichloropropane.

GLYCEROL TRINITRATE. See nitroglycerine.

GLYCERYL DIACETATE. Syn: *diacetin.* Colorless liquid, very sol in water. $C_7H_{12}O_5$, mw: 176.2, d: 1.178 @ 15°/15°, bp: 280°, mp: 40°.

Acute tox data: sc LD_{50} (rat) = 4000 mg/kg; sc LD_{50} (mouse) = 2500 mg/kg. [*3*]

THR = MOD via sc route. Probably MOD via other routes as well.

GLYCERYL MONOACETATE. Syns: *monoacetin, acetin.* Colorless, very hygroscopic liquid, characteristic odor. $C_5H_{10}O_4$, mw: 134.2, d: 1.206 @ 20°/4°, bp: 158° @ 17 mm.

Acute tox data = sc LD_{50} (rat) = 5500 mg/kg; sc LD_{50} (mouse) = 3500 mg/kg. [*3*]

THR = MOD via sc routes. An irr to skin, eyes and mu mem.

GLYCERYL MONOFLUOROACETATE. $C_5H_9O_4F$, mw: 152.1.

Acute tox data: sc LD_{50} (mouse) = 12 mg/kg. [*3*]

THR = HIGH via sc route. Probably other routes also.

GLYCERYL MONOSTEARATE. Syns: *g.m.s., glycerol monostearate, monostearin.* Pure white- or cream-colored wax-like solid, faint odor, sol in (hot) alcohol, oils and hydrocarbons. $(C_{17}H_{35})COOCH_2CHOHCH_2OH$, mw: 358, mp: 58°–59°, d: 0.97.

Acute tox data: ip LD_{LO} (mouse) = 200 mg/kg. [*3*]

THR = HIGH via ip route. Probably HIGH via other routes. A general purpose food additive. [*109*]

GLYCERYL TRIACETATE. Syn: *triacetin.* $(C_3H_5)(OOCCH_3)_3$, mw: 218.2, mp: −78°, bp: 258°, flash p: 280°F (COC), d: 1.161, autoign. temp.: 812°F, vap. d: 7.52.

Acute tox data: Oral LD_{50} (rat) = 3000 mg/kg; sc LD_{50} (rat) = 2800 mg/kg; ip LD_{50} (mouse) = 1400 mg/kg; iv LD_{50} (mouse) = 1600 mg/kg. [*3*]

THR = MOD via oral, sc, ip and iv routes. Used as a general food additive. [*109*]

Fire Hazard: Slight, when exposed to heat, flame or powerful oxidizers.

To Fight Fire: Alcohol foam, water, CO_2, dry chemical.

GLYCERYL TRIBUTYRATE. See tributyrin.

GLYCERYL TRINITRATE. See nitroglycerin.

GLYCERYL TRISTEARATE. $C_{57}H_{110}O_6$, mw: 891.7.

THR = An exper neo via sc route. [*3*]

GLYCIDALDEHYDE. Syn: *2,3-epoxypropanal.* $C_3H_4O_2$, mw: 72.1, bp: 113°, d: 1.1403 @ 20°/4°. Colorless liquid.

THR = Powerful skin irr and sensitizer. Irr to eyes and mu mem. An exper (+) neo and carc. [*3*, *19*]

GLYCIDOL. Syn: *2,3-epoxy-1-propanol.* Colorless liquid, entirely sol in water, alcohol and ether. $C_3H_6O_2$, mw: 74.1, d: 1.165 @ 0°/4°, bp: 167° (decomp).

Acute tox data: Oral LD_{50} (rat) = 850 mg/kg; inhal LC_{50} (rat) = 580 ppm for 8 hrs; dermal LD_{50} (rabbit) = 1980 mg/kg. [*3*]

THR = MOD via oral, inhal and dermal routes. See also diglycidyl ether. Animal exper suggest somewhat lower toxicity than related epoxy compounds. Readily absorbed through the skin. Causes nervous excitation followed by depression.

GLYCIDYL ACRYLATE. See 2,3-epoxypropyl acrylate.

GLYCIDYL ALDEHYDE. $C_4H_6O_2$, mw: 86.2.

THR = HIGH irr to skin, eyes and mu mem. An exper carc. [*23*] Large doses caused convulsions in exper animals. Repeated injections caused bone marrow depression.

GLYCIDYL BENZOATE. $C_{10}H_{10}O_3$, mw: 178.2.

Acute tox data: Oral LD_{50} (rat) = 1410 mg/kg. [*3*]

THR = MOD via oral route.

***n*-GLYCIDYL DIETHYLAMINE.** $C_7H_{15}ON$, mw: 129.2.

Acute tox data: Oral LD_{50} (rat) = 420 mg/kg; inhal LC_{LO} (rat) = 2000 ppm for 4 hrs; dermal LD_{50} (rabbit) = 790 mg/kg. [*3*]

THR = MOD via oral, inhal and dermal routes.

GLYCIDYL PHENYL ETHER. Liquid. $C_6H_4CH_2CHOCH_2$, mw: 150.17, bp: 245°, flash p: 175°F (OC), d: 1.1092 @ 20°/4°, vap. d: 5.18.

Acute tox data: Oral LD_{50} (rat) = 3850 mg/kg; dermal LD_{50} (rabbit) = 1500 mg/kg. [*3*]

THR = MOD via oral and dermal routes.

Fire Hazard: Mod, when exposed to heat or flame; can react with oxidizing materials.

Disaster Hazard: Dangerous; when heated to decomp, emits tox fumes.

To Fight Fire: Water, foam, CO_2, dry chemical.

GLYCIDYL SORBATE (DIMER). $C_{18}H_{26}O_6$, mw: 338.4.

Acute tox data: Oral LD_{50} (rat) = 5750 mg/kg. [*3*]

THR = MOD via oral routes.

GLYCIDYL STEARATE. Syn: *glycidyl octadecanoate.* $C_{21}H_{40}O_3$, mw: 340.6.

THR = An exper (±) carc and neo. [*3*, *13*, *23*]

GLYCINE. Syns: *amino acetic acid*, *glycocoll.* The principal amino acid in sugar cane. White crystals, odorless, sol in water, insol in alcohol and ether. NH_2CH_2COOH, mw: 75, mp: 232°–236° (decomp), d: 1.1607.

THR = U. A nutrient and/or dietary supplement food additive. [*109*]

GLYCINONITRILE. $C_2H_4N_2$, mw: 56.1.

THR = An exper teratogen. [*3*]

GLYCINONITRILE SULFATE. $C_2H_4N_2 \cdot \frac{1}{2}H_2SO_4$, mw: 105.1.

THR = HIGH via oral route. An exper teratogen. [*3*]

GLYCOCHOLIC ACID. Syn: *cholyglycine.* Occurs as sodium salt in bile, practically insol in water; the sodium salt is sol in water and alcohol. $C_{26}H_{43}NO_6$, mw: 465.

THR = U. An emulsifying agent food additive. [*109*]

GLYCOCOLL. See glycine.

GLYCOCOLL CALCIUM. Slightly water-sol crystals. $Ca(CH_2NHCO_2)$, mw: 113.1.

THR = U. See calcium compounds.

GLYCOL. See ethylene glycol.

GLYCOL ALCOHOL. See ethylene glycol.

GLYCOL BENZYL ETHER. Water-insol. $C_9H_{12}O_2$, mw: 152.1, flash p: 264°F (OC), autoign. temp.: 662°F, d: 1.07, vap. d: 5.20, bp: 256.1°.

THR = No data. See also ethers.

Fire Hazard: Low.

To Fight Fire: Water spray or mist, dry chemical, CO_2, foam.

GLYCOL BIS(HYDROXYETHYL)ETHER. See triethylene glycol.

GLYCOL CARBONATE. See ethylene carbonate.

GLYCOL CHLOROHYDRIN. See ethylene chlorohydrin.

GLYCOL DIABIETATE. See ethylene glycol diabietate.

GLYCOL DIBROMIDE. See ethylene dibromide.

GLYCOL DIACETATE. See ethylene glycol diacetate.

GLYCOL DIFORMATE. See ethylene glycol diformate.

GLYCOLIC ACID. Syn: *hydroxyethanoic acid.* Rhombic leaflets from ether. Odorless. $HOCH_2COOH$, mw: 76.05, bp: decomp, mp(α): 63°, mp(β): 79°.

Acute tox data: iv LD_{50} (rat) = 1000 mg/kg; [*2*] oral LD_{50} (rat) = 1950 mg/kg. [*3*]

THR = MOD via oral and iv routes.

Disaster Hazard: Slight, when heated, emits acrid fumes.

GLYCOL MONOACETATE. See ethylene glycol monoacetate.

GLYCOL MONOBUTYL ETHER. See butyl "Cellosolve."

GLYCOL MONOBUTYL ETHER ACETATE. See butyl "Cellosolve" acetate.

GLYCOLONITRILE. Syn: *formaldehyde cyanohydrin.* Mobile, colorless, odorless oil. $HOCH_2CN$, mw: 57.1, bp: 183° (with slight decomp), fp: −67°, d: 1.104, vap. d: 1.97.

Acute tox data: Oral LD_{50} (rat) = 16 mg/kg; dermal LD_{50} (rabbit) = 5 mg/kg; inhal LC_{LO} (rat) = 250 ppm. [*3*]

THR = HIGH via oral, dermal and inhal routes.

Explosion Hazard: Mod, when exposed to heat or by spont chemical reaction in the presence of alkalies if uninhibited (polymerizes).

Disaster Hazard: Highly dangerous; emits highly toxic fumes on decomp.

GLYCOLS.

THR = Dihydric alcohols are physically and chemically related to glycerol. Most of the glycols have low volatility and consequently present little danger from inhal of vapors. Severe and even fatal poisoning has occurred following ingestion of ethylene glycol and severe occupational poisoning has been caused by ethylene glycol monomethyl ether (methyl "Cellosolve"). Under ordinary conditions of industrial use, most of the commonly used glycols are not considered to be toxic.

There is some evidence that the glycol ethers are more toxic than other members of the group. Violent reaction with $HClO_4$. [*19*]

GLYCOL URIL. See acetylene diureine.

GLYODIN. Syns: *2-heptadecyl glyoxalidine acetate, "Crag," fruit fungicide 341, 2-heptadecyl-2-imidoazoline acetate.* $C_{22}H_{44}N_2O_2$, mw: 368.6, mp: 94°.

Acute tox data: Oral LD_{50} (rat) = 3170 mg/kg; mammalian LD_{50} = 1000 mg/kg. [*3*]

THR = MOD via oral route. An irr to skin, eyes and mu mem. High conc irr to skin and damaging to cornea. A fungicide.

GLYOXAL. Syns: *ethanedial, oxalaldehyde, biformyl.* Yellow crystals or light yellow liquid. OCHCHO, mw: 58.0, mp: 15°, bp: 50.4°, d: 1.14 @ 20°/4°.

Acute tox data: Dermal LD_{50} (guinea pig) = 6600 mg/kg; [*20*] oral LD_{50} (rat) = 1100 mg/kg; [*20*] oral LD_{50} (guinea pig) = 760 mg/kg; ip LD_{50} (mouse) = 200 mg/kg; oral LD_{LO} (rat) = 100 mg/kg. [*3*]

THR = MOD via dermal, oral; HIGH via ip routes.

Fire Hazard: Violent reactions with water, air, chlorosulfonic acid, ethylene imine, HNO_3, oleum, NaOH. [*19*]

GLYOXALIDINE DERIVATIVES. See 1-hydroxethyl-2-heptadecyl glyoxalidine.

GLYOXYLIC ACID. Hemihydrate. $C_2H_2O_3 \cdot \frac{1}{2}H_2O$. Also obtained in anhydrous form as monoclinic crystals from water. Deliquesces quickly and forms a syrup on short exposure to air. Freely sol in water, insol in ether and hydrocarbons. $C_2H_2O_3$, mw: 74.04, d: 1.42 @ 20°/4°, mp: 73°.

Acute tox data: im LD_{LO} (rat) = 25 mg/kg. [*3*]

THR = HIGH to skin, eyes and mu mem.

GMP. See sodium guanylate.

GMS. See glyceryl monostearate.

GOALS OF INDUSTRIAL HYGIENE. See Section 1.

GOA POWDER. See chrysarobin.

GOAT HAIR. See mohair.

GOLD. Cubic yellow, ductile, metallic crystals. Au, atwt: 197.20, mp: 1063°, bp: 2966°, d: 19.3 (liquid); 17.0 @ 1063°, vap. press: 1 mm @ 1869°.

Radiation Hazard: For permissible levels, see Section 5, Table 5A.5. Can form an explosive compound with NH_3, (NH_4OH + aqua regia), H_2O_2. [*19*]

GOLD CHLORIDE. See auric chloride.

GOLD COMPOUNDS.

THR = Salts of gold, particularly gold sodium thiosulfate, have been used extensively in the treatment of some forms of arthritis and for other diseases. Toxic reactions are not infrequent. Various allergic manifestations have occurred, such as urticaria (hives), itching, purpura and other types of skin rashes, some of which may be quite severe. Damage to the blood forming organs resulting in aplastic anemia has been described. The liver, kidneys and nervous system may be affected.

If gold therapy is being used, careful watch must be kept for the first signs of toxic reaction and the drug be promptly discontinued. BAL has been reported to be of value in treating gold poisoning. Can react violently with nitromethane. [*19*]

GOLD CYANIDE. See aurous cyanide.

GOLD DITELLURIDE. Crystal forms: rhombic, monoclinic, triclinic. $AuTe_2$, mw: 452.2, mp: 472° decomp; d: 8.2–9.3.

THR = See tellurium compounds and gold compounds.

Fire Hazard: See hydrides.

Explosion Hazard: See hydrides.

Disaster Hazard: Dangerous; when heated to decomp, emits toxic fumes of tellurium; will react with water, steam or acid to produce toxic and flam vapors of tellurium hydride.

GOLDEN RAGWORT. See senecio.

GOLD, EXPLOSIVE. Syn: *fulminating gold.* A dark brown powder. Explodes on heating or rubbing. Very unstable. Comp: Au, N, Cl. [*2*]

GOLD IODIDE. Syn: *aurous iodide*. Greenish-yellow powder. AuI, mw: 324, mp: 120° (decomp), d: 8.25.
THR = See iodides.
Disaster Hazard: See iodides.

GOLD NITRATE. See auric hydrogen nitrate.

GOLD ORE. A recog carc. [*14*] See arsenic compounds.

GOLD OXIDE. See auric oxide.

GOLD PAINT. See paint, enamel, lacquer, stain, shellac, varnish, etc.

GOLD PHOSPHIDE. Gray crystals. Au_2P_3, mw: 486.9, mp: decomp, d: 6.67.
THR = See phosphides.

GOLD POTASSIUM BROMIDE. Syn: *potassium auribromide*. Violet crystals. $AuBr_3 \cdot KBr \cdot 2H_2O$, mw: 592.01.
THR = See bromides.

GOLD POTASSIUM CHLORIDE. Syn: *potassium aurichloride*. Yellow crystals. $AuCl_3 \cdot KCl \cdot 2H_2O$, mw: 414.17.
THR = See gold compounds.
Disaster Hazard: See chlorides.

GOLD POTASSIUM CYANIDE. Syns: *potassium cyanaurite, potassium aurocyanide*. White crystalline powder. $KAu(CN)_2 \cdot 2H_2O$, mw: 324.36.
THR = See cyanides.

GOLD SELENIDE. Solid. Au_2Se_3, mw: 630.8, d: 4.65 @ 22°.
THR = See selenium compounds.
Fire Hazard: Mod, by chemical reaction with moisture or acids.
Explosion Hazard: Slight; see hydrogen selenide.
Disaster Hazard: Dangerous; when heated, emits highly toxic fumes of selenium; will react with water, steam or acid to produce toxic and flam vapors of hydrogen selenide.

GOLD SODIUM BROMIDE. Syn: *sodium auribromide*. Black-brown crystals. $AuBr_3 \cdot NaBr \cdot 2H_2O$, mw: 575.91.
THR = See bromides.

GOLD SODIUM CHLORIDE. Syns: *sodium aurichloride, sodium chloraurate*. Yellow crystals. $NaAuCl_4 \cdot 2H_2O$, mw: 398.05.
THR = See gold compounds and chlorides.

GOLD SODIUM THIOSULFATE. Syn: *aurous sodium thiosulfate*. White crystals, odorless. $Na_3Au(S_2O_3)_2 \cdot 2H_2O$, mw: 527.47.
Acute tox data: sc LD_{LO} (rat) = 30 mg/kg; iv LD_{LO} (rat) = 80 mg/kg; im LD_{LO} (rat) = 35 mg/kg. [*3*]
THR = HIGH via sc, iv and im route.
Disaster Hazard: Mod dangerous; when heated to decomp, emits toxic fumes.

GOLD SULFIDE. See auric sulfide.

GOLD TELLURIDE. Triclinic, rhombic, monoclinic crystals. $AuTe_2$, mw: 452.2, d: 9.04.
THR = See tellurium compounds.
Fire Hazard: Mod, on contact with moisture.
Explosion Hazard: Slight, upon contact with moisture.
Disaster Hazard: Mod dangerous; when heated to decomp, emits tox fumes of tellurium; on contact with water, steam or acid, can emit toxic and flam fumes of hydrogen telluride.

GOLD TRIBROMIDE. Gray powder, crystalline. $AuBr_3$, mw: 436.7, mp: $-Br_2$ @ 160°.
THR = See bromides.

GRAIN SMUTS. These are minute, threadlike, parasitic plants, the mycelium of which enters the growing portion of the seedling and grows up with the host. They remain invisible until the heads of grain appear, at which time they partially or wholly destroy the heads, appearing at that stage as masses of dark substance. These masses contain spores with further propagate by getting on or into the seeds where they reenact their life after the seed is planted. Due to these smut spores, contact dermatitis, an atopic form of skin allergy, perennial rhinitis, and asthma may occur in those exposed. These conditions may be fairly abundant wherever favorable climate for infestation exists, which can be in the Middle West, the Pacific Northwest, the South and the Southwest.

GRANITE DUST.
THR = Depends upon silica content. See also nuisance dusts.

GRAPE SUGAR. See dextrose.

GRAPHITE. Syns: *black lead, plumbago*. A crystalline form of carbon, soft, greasy feel, steel gray to black color. Usually contains up to 10% free silica. C, atwt: 12.00, d: 2.0–2.25.
THR = A nuisance dust. Possibly allergenic. Cases of pulmonary fibrosis, emphysema and cor pulmonale have resulted from prolonged inhal of graphite dust; also lung modulation. Reacts violently with ClF_3, F_2, K, KO_2. [*19*]

GRAS EXEMPTIONS REVIEWS. [*109*]

GRAS LISTINGS. [*109*]

GRASSELLIS. See explosives, high.

GRAY ACETATE. See calcium acetate.

GREEK FIRE. Essentially, this is a mixture of nitrates and pitch, which has been in use since the 8th century. This material is a deflagrating mixture and therefore properly belongs in the class of low explosives.
THR = See explosives, low.

GREENOCKITE. See cadmium sulfide.

GREEN OIL. See anthracene.

GREEN VITRIOL. See ferrous sulfate.

GRENADES (ALL TYPES). See explosives, high.

GRIGNARD REAGENTS.
Since all Grignard reagents react rapidly with both water and oxygen, contact with these substances must be avoided. So far as is known the ordinary materials of construction are satisfactory for use with Grignard reagents. Adequate ventilation should, of course, be employed.

Because the heat of decomposition of Grignard reagents with water is great and the ether they are dissolved in highly volatile and flam, they must be handled with extreme care. Some of these reagents, especially the solution of MeMgBr in ethyl ether, may ignite spont on contact with water or even damp floors; nearly all of these compounds will ignite on a wet rag or similar material.
Disaster Hazard: Dangerous. Avoid exposure to air or moisture.

GRISEOFULVIN. See spirofulvin.

GUAIACOL. Syns: *methyl catechol, 2-methoxyphenol.* Clear, pale yellow liquid or solid. $OHC_6H_4OCH_3$, mw: 124.1, mp: 28°, bp: 202°–209°, flash p: 180°F (OC), d: 1.097 @ 25°/25°.
Acute tox data: Oral LD_{50} (rat) = 725 mg/kg. [*3*] LD (adult human) = 3–10 g. [*20*]
THR = HIGH via oral route. Ingestion produces burning in mouth and throat, GI distress, tremors and collapse. See also phenol.
Fire Hazard: Mod, when exposed to heat or flame; can react with oxidizing materials.
To Fight Fire: Foam, CO_2, dry chemical.

GUAIACOL VALERATE. Oily, pale yellow liquid. $C_6H_4OCH_3OCOC_4H_9$, mw: 208.3, bp: 265°, d: 1.05.
THR = U. See also guaiacol and valerian.
Disaster Hazard: Slight; when heated, emits acrid fumes.

GUANAJUATITE. See bismuth triselenide sulfide.

GUANIDINE AMINO VALERIC ACID. See arginine.

GUANIDINE CARBONATE. Columnar crystals. $[H_2NC(NH)NH_2]_2H_2CO_3$, mw: 180.2, mp: 333°, d: 1.24.
Acute tox data: LD_{LO} (mammal) = 500 mg/kg. [*3*]
THR = At least MOD.
Disaster Hazard: Mod; when heated to decomp, emits toxic fumes.

GUANIDINE HYDROCHLORIDE. White powder. $H_2NC(NH)NH_2 \cdot HCl$, mw: 95.6, mp: 183°.
Acute tox data: sc LD_{LO} (rat) = 404 mg/kg; ip LD_{50} (mouse) = 500 mg/kg; oral LD_{LO} (mammal) = 300 mg/kg. [*3*]
THR = HIGH via oral and sc; MOD via ip routes. Can cause nausea, diarrhea and neurological disturbances.
Disaster Hazard: Dangerous; when heated to decomp, emits highly toxic fumes of HCl.

GUANIDINE NITRATE. White granules.
$H_2NC(NH)NH_2 \cdot HNO_3$, mw: 122.1, mp: 214°.
THR = See nitrates and guanidine hydrochloride.
Fire Hazard: See nitrates.
Explosion Hazard: Mod, when shocked or exposed to heat or flame; a stable, flashless, non-hygroscopic high explosive used as a blasting explosive in combination with charcoal and inorganic nitrates.
Disaster Hazard: Dangerous; keep away from heat and open flame.

GUANIDINE PERCHLORATE. $C(NH_2)_3 \cdot ClO_4$, mw: 159.5.
THR = A very powerful, sensitive, unstable explosive. [*19*]

GUANINE OXIDE. $C_5H_5O_2N_5$, mw: 167.2.
THR = An exper carc. [*3*] HIGH via ip route.

GUANYL NITROSAMINO GUANYLIDENE HYDRAZINE. See explosives, initiating.

GUANYL NITROSOAMINO GUANYL TETRAZENE. See explosives, initiating.

GUANYL UREA NITRATE. Crystals.
$NH_2C(NH)NHCONH_2 \cdot HNO_3$, mw: 165.1, mp: 230°, d: 1.579 @ 30°.
THR = See nitrates.

GUANYL UREA PHOSPHATE. Crystals.
$NH_2C(NH)NHCONH_2 \cdot H_3PO_4$, mw: 200.1, mp: 184° (decomp), d: 1.614.
THR = See urea and phosphates.

GUANYL UREA SULFATE. White powder.
$NH_2C(NH)NHCONH_2 \cdot H_2SO_4$, mw: 302.28, mp: 190° (decomp), d: 1.609.
THR = U.
Disaster Hazard: Dangerous; when heated to decomp, emits highly toxic fumes of SO_x.

GUAR GUM. Syn: *guar flour.* Yellowish-white powder, dispersible in hot or cold water, obtained from the ground endosperms of *cyanopsis tetragonoloan.*
THR = U. A stabilizer food additive. It migrates to food from packaging materials. [*109*]

GUAR FLOUR. See guar gum.

GUAZA. See cannabis.

GUM ARABIC. See acacia gum.

GUM CAMPHOR. See camphor.

GUM COPAL. See copal.

GUM GUAIAC. Syns: *guaiac gum, guaiac resin.* A resin from certain Mexican and West Indian trees, sol in alcohol, ether, acetone, chloroform and caustic soda.

THR = U. A chemical preservative food additive. [*109*]

GUM OPIUM. See opium.

GUM ROSIN. See rosin.

GUM THUS. See turpentine gum.

GUM TRAGACANTH. See tragacanth.

GUN COTTON. See nitrocellulose. See also explosives, high.

GUNPOWDER. See explosives, high.

GUTHION. Syns: *o,o-dimethyl-S-(4-oxobenzotriazino-3-methyl)-phosphorodithioate, azinphos methyl.* Crystals, slightly water-sol but sol in organic solvents, brown, waxy solid. $C_{10}H_{12}N_3O_3PS_2$, mw: 317.3, d: 1.44, mp: 74°.

Acute tox data: Oral LD_{50} (rat) = 16 mg/kg; dermal LD_{50} (rat) = 300 mg/kg; ip LD_{50} (mouse) = 4 mg/kg; oral LD_{50} (wild birds) = 8 mg/kg. [*3*]

THR = HIGH via dermal, oral and ip routes. A cholinesterase inhibitor. See parathion. A food additive permitted in the food and drinking water of animals and/or for the treatment of food-producing animals. [*109*]

GYPLURE. Syn: *cis-9-octadecene-1,2-iol-12-acetate.*

THR = U. See gypsy moth attractant.

GYPSOM. Syn: *calcium sulfate dihydrate.* Colorless crystals. $CaSO_4 \cdot 2H_2O$, mw: 172.2, d: 2.32, mp: 128° ($-1\frac{1}{2}H_2O$), bp: 163° ($-2H_2O$).

THR = Depends upon silica content. See nuisance dusts.

h

HAFNIA. See hafnium oxide.

HAFNIUM. A silvery ductile, lustrous metal. Hf, atwt: 178.49, mp: 2227°, bp: 4602°, d: 13.31 @ 20°.
THR = See hafnium compounds.
Radiation Hazard: For permissible levels, see Section 5A, Table 5A.5.

HAFNIUM CARBIDE. Solid. HfC, mw: 190.61, mp: 3887°.
THR = See hafnium compounds.
Fire Hazard: Slight, will react with water or steam to produce flam vapors.

HAFNIUM CHLORIDE. $HfCl_4$, mw: 320.3. [*20*]
Acute tox data: ip LD_{50} (rabbit) = 112 mg/kg. CTD: liver changes @ 1% diet level after 90 days. [*20*] Ocular irr is transient; dermal application produced transient irr. Application to abraded skin → non-healing ulcers. [*20*]
THR = HIGH via ip.

HAFNIUM CHLORIDE OXIDE. $HfOCl_2$, mw: 265.4.
Acute tox data: ip LD_{50} (mouse) = 112 mg/kg; ip LD_{50} (rabbit) = 112 mg/kg. [*3*]
THR = HIGH via ip and probably via oral as well.

HAFNIUM COMPOUNDS.
Acute tox data: Oral LD_{50} (mouse) = 76 mg/kg; iv LD_{50} (rat) = 100 mg/kg.
THR = HIGH via oral and iv routes. Liver damage was reported.

HAFNIUM DIOXIDE. See hafnium oxide.

HAFNIUM HYDRIDE. Metallic crystals, insol in water. HfH_2, mw: 180.6, d: 11.48, mp: decomp >40°.
THR = See hafnium compounds and hydrides.

HAFNIUM METAL, IN VARIOUS STATES AND FORMS. See hafnium.

HAFNIUM OXIDE. Syns: *hafnium dioxide, hafnia.* White crystals. HfO_2, mw: 210.60, mp: 2810°, d: 9.68.
THR = See hafnium compounds.

HAFNIUM OXYCHLORIDE OCTAHYDRATE. Colorless crystals. $HfOCl_2 \cdot 8H_2O$, mw: 409.6.
THR = An exper neo. [*3*] See also hafnium chloride oxide.
Disaster Hazard: Dangerous; see chlorides.

HAFNIUM TETRAHYDROBORATE. $Hf(BH_4)_4$, mw: 237.8.
THR = Burns violently in air. [*19*] No data. See also hafnium and boron compounds.

HAIR.
THR = No data. A known allergen.
Fire Hazard: Slight, when exposed to heat or flame.

HALANE. See 1,3-dichloro-5,5-dimethyl hydantoin.

HALITE. See sodium chloride.

HALOCARBONS. See polychlorotrifluoroethylene.

HALOCRIN. See 3,9-dichloro-7-methoxyacridine.

HALOWAX. See hexachloronaphthalene.

HAMAMELIA. See witch hazel.

HAMBERGITE. See beryllium-*o*-borate, basic.

HAMILTON, DR. ALICE. See Section 1.

HANANE. Syns: *dimefox, pestox IV.* $C_4H_{12}ON_2PF$, mw: 154.2.
Acute tox data: Oral LD_{50} (rat) = 1 mg/kg; dermal LD_{50} (rat) = 2 mg/kg; ip LD_{50} (rat) = 5 mg/kg; sc LD_{50} (mouse) = 1 mg/kg. [*3*]
THR = VERY HIGH via oral, dermal, ip and sc. A cholinesterase inhibitor. See also parathion.
Disaster Hazard: Dangerous. See parathion.

HAND GRENADES. See Section 11.

HANDLING THE DISASTER ALARM. See Section 7.

HAND SIGNAL DEVICES. See Section 11.

HANDY REFERENCE CHART OF DOT HAZARDOUS MATERIALS REGULATIONS. See Section 11.

HASHISH. See cannabis.

HAT GLAZING LACQUER.
THR = U. See also lacquers.
Fire Hazard: Dangerous, when exposed to heat or flame; can react with oxidizing materials.
Explosion Hazard: U.

HAY. Syn: *dried grass.* Yellow, fibrous material.
THR = An allergenic nuisance dust.
Fire Hazard: Mod, when exposed to heat or flame; can react with oxidizing materials.
Spont Heating: Occurs when wet or improperly packed.

HAZARD CONTROL BY ATMOSPHERIC DISPERSION AND DILUTION. See Section 2.

HAZARD CONTROL BY METHODS OTHER THAN VENTILATION. See Section 2.

HAZARD CONTROL DESIGNED INTO PLANT OR PROCESS. See Section 2.

HAZARD CONTROL FOR SPECIFIC INDUSTRIAL PROCESSES. See Section 2.

HAZARD CONTROL VIA HOUSEKEEPING. See Section 2.

HAZARD CONTROL VIA LOCAL SUPPRESSION OF CONTAMINANTS. See Section 2.

HAZARD CONTROL VIA PROCESS CHANGE. See Section 2.

HAZARD CONTROL VIA PROCESS ISOLATION IN TIME OR SPACE. See Section 2.

HAZARD CONTROL VIA RESPIRATORY PROTECTION. See Section 2.

HAZARD CONTROL VIA SEGREGATION OF PERSONNEL. See Section 2.

HAZARD CONTROL VIA SUBSTITUTION. See Sections 1 and 2.

HAZARD OF SOLID WASTE TREATMENT BIBLIOGRAPHY. See Section 6.

HAZARDOUS MATERIALS SHIPPING DOCUMENT REQUIREMENTS. See Section 11.

HD. See dichlorodiethyl sulfide.

HEARING DAMAGE. See Section 3.

HEAVY OIL.
THR = A recog carc. [*14*] See creosote.

HEAZLEWOODITE. See nickel subsulfide.

HELIANTHINE B. See methyl orange.

HELIOTRIDINE.
THR = An exper carc. [*23*]

HELIOTRINE. $C_{16}H_{27}O_5N$: mw: 313.4.
THR = HIGH. An exper teratogen and neo. [*3*, *23*]

HELIOTROPIN. See piperonal.

HELIOTROPIUM SUPINUM L.
THR = An exper neo to rats via oral route. [*103*]

HELIUM. Colorless, odorless, tasteless, inert gas. He, atwt: 4.003, mp: −272.2° @ 26 atm, bp: −268.9°, d(gas): 0.1785 g/liter @ 0°, d(liquid): 0.147 @ −270.8°
THR = A simple asphyxiant. For properties see argon. Used as a general purpose food additive. [*109*]

HELIUM, LIQUID. See helium.

HELIUM-OXYGEN MIXTURE. See helium and oxygen.

HELLEBORE, BLACK. See helleborein.

HELLEBOREIN. Glucoside crystallizable in yellow prisms. $C_{37}H_{56}O_{18}$, mw: 788.63, mp: 270°.
Acute tox data: Iv LD_{LO} (cat) = 1.9 mg/kg. [*3*]
THR = HIGH via iv route. An irr to skin, eyes and mu mem.
Fire Hazard: Slight.

HELLEBORIN. Colorless, lustrous crystals. $C_{28}H_{36}O_6$, mw: 468.6, mp: >250°.
THR = Details U. See also helleborein.

HELVELLIC ACID.
THR = This is a poisonous material and a powerful hemolytic agent. It can cause superficial punctate hepatitis. The symptoms of poisoning from this material usually occur within the first few hours after ingestion and consist of nausea, vomiting, gastric pains and diarrhea. Further, there is stupor, dilation of the pupils, enlargement of the liver and spleen, and tubular damage to the kidneys with albuminuria. In some cases death may follow from an acute yellow atrophy of the liver. In the absence of this last symptom, the prognosis is favorable.
Treatment and Antidotes: Injections of glucose to protect the liver can be given.

HEMATITE. Syns: *bloodstone*, *iron ore*. Mainly Fe_2O_3.
THR = An exper (±) carc. [*3*, *9*]

HEMATOXYLIN. $C_{16}H_{14}O_6$, mw: 302.3.
THR = An exper carc. [*3*]

HEMEL. See hexamethyl melamine.

HEMICELLULOSE EXTRACT. A natural substance occurring in woody tissue.
THR = U. A food additive permitted in the feed and drinking water of animals and/or for the treatment of food-producing animals. [*109*]

HEMISULFUR MUSTARD. Liquid, odor of mustard gas. C_4H_9OSCl, mw: 140.6.
THR = HIGH irr. An exper carc. [*23*] See also dichlorodiethyl sulfide.
Disaster Hazard: Dangerous; see chlorides and sulfur compounds.

HEMP.
THR = An allergen and MILD irr.
Fire Hazard: Slight, when exposed to heat or flame; can react with oxidizing materials.

HEMPA. See hexamethyl phosphoric triamide.

HENDECANE. Syn: *undecane-n*. Colorless liquid, insol in water. $CH_3(CH_2)_9CH_3$, mw: 156, d: 0.7402 @ 20°/4°, fp: −25.75°, bp: 195.6°, flash p: 149°F (OC), vap. d: 5.4.
THR = See aliphatic hydrocarbons.
Fire Hazard: Mod, when exposed to heat, flame or oxidizers.
To Fight Fire: Foam, mist, dry chemical.

HENDECANOL-2. See undecanol-2.

HENNA. Dried powdery leaves of *Lawsonia alba, Lawsonia inermis* and *Lawsonia spinosa.*
THR = An allergen. Not a harmful dye. Makes the hair brittle and can irr the skin. Absorbed by the skin.

HEPAR CALCIS. See calcium sulfide.

HEPAR SULFURIS. See potassium sulfide.

HEPTACHLOR. Syn: *1,4,5,6,7,8,8-heptachloro-3a,4,7,7a-tetrahydro-4,7-methanoindene.* Crystals, nearly insol in water, sol in organic solvents. $C_{10}H_7Cl_7$, mw: 375.3, mp: 96°.
Acute tox data: Oral LD_{50} (rat) = 40 mg/kg; dermal LD_{50} (rat) = 195 mg/kg; ip LD_{50} (rat) = 27 mg/kg; oral LD_{50} (chicken) = 62 mg/kg. [*3*]
THR = High via oral, dermal and ip routes. An exper (±) carc. [*3, 5*] Acute exposure causes liver damage. Chronic doses have caused liver damage in exper animals. See closely related chlordane. In man, a dose of 1–3 g can cause serious symptoms, especially where liver impairment is the case. Acute symptoms include tremors, convulsions, kidney damage, respiratory collapse and death.
Disaster Hazard: Dangerous; when heated to decomp, emits highly toxic fumes.

HEPTACHLOROPROPANE. C_3HCl_7, mw: 285.3.
THR = Details U. See also chlorinated hydrocarbons, aliphatic.
Disaster Hazard: Dangerous; when heated to decomp, emits highly toxic fumes of phosgene.

1,4,5,6,7,8,8-HEPTACHLORO-3a,4,7,7a-TETRAHYDRO-4,7-METHANOINDENE. See heptachlor.

HEPTADECANOL. Syn: *heptadecyl alcohol.*
$C_{17}H_{35}OH$, mw: 256.46, mp: 54°, bp: 309°, flash p: 310°F (COC), d: 0.8475 @ 20°/20°, vap. press: 0.01 mm @ 20°, vap. d: 8.84.
Acute tox data: Oral LD_{50} (rat) = 52000 mg/kg; dermal LD_{50} (rabbit) = 17000 mg/kg. [*3*]
THR = LOW to NONE via oral and dermal.
Fire Hazard: Slight; when exposed to heat or flame, can react with oxidizing materials.
To Fight Fire: CO_2, dry chemical.

HEPTADECYL ALCOHOL. See heptadecanol.

2-HEPTADECYLIMIDOAZOLINE. See glyodin.

HEPTADONE. $C_{21}H_{27}ON$, mw: 309.5.
THR = An exper teratogen. [*3*] HIGH via oral, sc, id, ip and iv routes.

HEPTAFLUOROBUTYRIC ACID. Colorless liquid, sharp odor similar to butyric acid. $CF_3CF_2CF_2COOH$, mw: 214.00, bp: 210.0° @ 735 mm.
THR = HIGH irr to skin, eyes and mu mem.
Disaster Hazard: Dangerous; when heated to decomp, emits highly toxic fumes of fluorides; will react with water or steam to produce corrosive fumes.

HEPTAMETHYLENE HYDROCHLORIDE.
THR = Mixed (1:1) with $NaNO_2$, an exper carc. [*3*]

HEPTANE. Syns: *heptyl hydride, dipropyl methane.* Colorless liquid. $CH_3(CH_2)_5CH_3$, mw: 100.20, bp: 98.52, lel = 1.05%, uel = 6.7%, fp: −90.5°, flash p: 25°F (CC), d: 0.684 @ 20°/4°, autoign. temp.: 419°F, vap. press: 40 mm @ 22.3°, vap. d: 3.45.
Acute tox data: Inhal TC_{LO} (human) = 1000 ppm, for 6 min ⟶ CNS symptoms. [*3*] Human inhal of 5000 ppm for 1/4 hr ⟶ marked vertigo, incoordination and hilarity. 5000 ppm for 7 min ⟶ marked vertigo, incoordination, hilarity. 5000 ppm for 4 min ⟶ marked vertigo, inability to walk straight, hilarity. 3500 ppm for 4 min ⟶ mod vertigo. 2000 ppm for 4 min ⟶ slight vertigo.
THR = MOD via inhal. Irr to the respiratory tract. Narcotic in HIGH conc. CNS irr.
Fire Hazard: Dangerous, when exposed to heat or flame.
Spont Heating: No.
Explosion Hazard: Mod, when exposed to heat or flame. Violent reaction with (P + Cl). [*19*]
Disaster Hazard: Dangerous, upon exposure to heat or flame; can react vigorously with oxidizing materials.
To Fight Fire: Foam, CO_2, dry chemical.

3-HEPTANECARBOXYLIC ACID. See 2-ethylhexoic acid.

HEPTANOIC ACID. Syns: *heptoic acid, oenanthylic acid.* Oily liquid, disagreeable rancid odor, less odor when very pure. $CH_3(CH_2)_5COOH$, mw: 130.2, d: 0.9345 @ 0°/4°, mp: −7.5°, bp: 223.0°.
Acute tox data: Oral LD_{50} (mouse) = 160 mg/kg; iv LD_{50} (mouse) = 1200 mg/kg. [*3*]
THR = HIGH via oral and MOD via iv routes.

1-HEPTANOL. Syn: *heptyl alcohol.* Liquid. $C_7H_{15}OH$, mw: 116.2, mp: −34.6°, bp: 175.8°, d: 0.824 @ 20°/4°.
Acute tox data: Oral LD_{50} (rat) = 3.25 mg/kg; oral LD_{50} (mouse) = 1500 mg/kg; oral LD_{50} (rabbit) = 750 mg/kg. [*3*]
THR = MOD via oral route.
Fire Hazard: Mod, when exposed to heat or flame; can react with oxidizing materials.

2-HEPTANOL. Syn: *amyl methyl carbinol.* Liquid. $CH_3(OH)CH(CH_2)_4CH_3$, mw: 116.2, bp: 160.4°, flash p: 160°F (OC), d: 0.8344 @ 0°, vap. press: 1 mm @ 14.6°, vap. d: 4.01.
Acute tox data: Oral LD_{50} (rat) = 2580 mg/kg; dermal LD_{50} (rabbit) = 1780 mg/kg. [*3*]

THR = MOD via oral and dermal routes.
Fire Hazard: Mod, when exposed to heat or flame; can react with oxidizing materials.
Explosion Hazard: U.
To Fight Fire: Foam, CO_2, dry chemical.

3-HEPTANOL. Liquid. $CH_3CH_2CH(OH)C_4H_9$, mw: 116.2, bp: 156.2°, flash p: 140°F (COC), fp: −70°, d: 0.8224 @ 20°/20°, vap. press: 0.5 mm @ 20°, vap. d: 4.01.
Acute tox data: Oral LD_{50} (rat) = 1870 mg/kg; dermal LD_{50} (rabbit) = 4360 mg/kg.
THR = MOD via oral and dermal routes.
Fire Hazard: Mod, when exposed to heat or flame; can react with oxidizing materials.
To Fight Fire: Foam, CO_2, dry chemical.

2-HEPTANONE. See methyl amyl ketone.

4-HEPTANONE. See butyrone.

1-HEPTENE. See α-heptylene.

2-HEPTENE. Clear liquid. C_7H_{14}, mw: 98.18, bp: 98.2°, flash p: 32°F, d: 0.709 @ 20°/4°, vap. d: 3.4.
THR = No specific data. Probably LOW via oral and inhal routes.
Fire Hazard: Dangerous, when exposed to heat, flame or oxidizers.
Disaster Hazard: Dangerous, upon exposure to heat or flame; can react vigorously with oxidizing materials.
To Fight Fire: Dry chemical, CO_2, foam.

3-HEPTENE (mixture of *cis* and *trans* isomers). Syn: *1-ethyl-2-propyl ethylene.* Liquid.
$CH_3CH_2CHCHCH_2CH_2CH_3$, mw: 98.18, bp: 96°, flash p: 20°F, d: 0.705 @ 15.5°/25.5°, vap. d: 3.38.
THR = Probably irr and narcotic in HIGH conc. See also 2-heptene.
Fire Hazard: Dangerous, when exposed to heat or flame; can react vigorously with oxidizing materials.
Disaster Hazard: Dangerous, upon exposure to heat or flame.
To Fight Fire: Foam, CO_2, dry chemical.

HEPTOIC ACID. See heptanoic acid.

HEPTYL ACETATE. $CH_3COOCH_3(CH_2)_6$, mw: 159.0.
THR = MOD irr to skin, eyes and mu mem.

HEPTYL ALCOHOL. See 1-heptanol.

HEPTYL AMINE. Syn: *1-amino heptane.* Colorless liquid, sol in water, sol in alcohol or ether.
$CH_3(CH_2)_6NH_2$, mw: 115, d: 0.727 @ 20°/4°, mp: −23°, bp: 155°, vap. d: 4.0, flash p: 130°F (OC).
THR = U. See also amines.
Fire Hazard: Mod, when exposed to heat, flame or oxidizers.
To Fight Fire: Alcohol foam.

2-HEPTYLAMINE.
Acute tox data: Sc LD_{50} (rat) = 130 mg/kg; ip LD_{LO} (mouse) = 3.9 mg/kg. [*3*]
THR = HIGH via sc and ip routes. Probably also via oral route.

3-HEPTYLAMINE.
Acute tox data: Ip LD_{50} (mouse) = 90 mg/kg. [*3*]
THR = HIGH via ip and probably other routes.

8-HEPTYL BENZ(a)ANTHRACENE. $C_{25}H_{26}$, mw: 326.5.
THR = An exper neo. [*3*]

α-HEPTYLENE. Syn: *1-heptene.* Colorless liquid; insol in water; sol in ether. $CH_2CH(CH_2)_4CH_3$, mw: 98.2, d: 0.6969 @ 20°, mp: −10°, bp: 93.6, flash p: 32°F, autoign. temp.: 707°F.
THR = A simple asphyxiant. For effects of exposure, see argon.
Fire Hazard: Dangerous, via heat, flame or oxidizers.
Explosion Hazard: Details U.
To Fight Fire: Foam, dry chemical, CO_2.

HEPTYL HYDRIDE. See heptane.

***n*-HEPTYL-*p*-HYDROXYBENZOATE.**
THR = U. Used as a food additive permitted in food for human consumption. [*109*]

HERBAN. Syn: *norea.* A crystalline material.
$C_{13}H_{22}N_2O$, mw: 222.3, mp: 178°.
Acute tox data: Oral LD_{50} (rat) = 2000 mg/kg; oral LD_{50} (dog) = 3700 mg/kg. [*3*]
THR = MOD via oral route. See also monuron. A very selective herbicide.

HERBISAN. See ethylxanthogen disulfide.

HEROIN. See diacetyl morphine.

HESAMITE. See tetraethyl pyrophosphate.

HETP. See hexaethyl tetraphosphate.

HEXAANTIPYRINE CERIUM III PERCHLORATE. Colorless, hexagonal crystals.
$[Ce(C_{11}H_{12}N_2O)_6] \cdot (ClO_4)_3$, mw: 1567.83, mp: 295°–300° (decomp).
THR = See perchlorates.

HEXAANTIPYRINE LANTHANUM IODIDE. Yellow crystals, water-sol. $La(C_{66}H_{72}O_6N_{12})I_3$, mw: 1649.0, mp: 269° (decomp).
THR = See iodides and lanthanum compounds.

HEXABENZYL DIGERMANE. Colorless crystals. $Ge_2(C_6H_5CH_2)_6$, mw: 692.0, mw: 184°.
THR = Details U. See germanium compounds.

HEXABORANE. Syns: *boron hydride, hexaboron decahydride.* Colorless liquid, turns yellow on standing. B_6H_{10}, bp: 0° @ 7.2 mm, mw: 75.0, mp: −65.1°, d: 0.69 @ 0°, vap. press: 7.2 mm @ 0°, vap. d: 2.6.

THR = See boron hydride. Self-ignites in air. [*19*] The $C_{16}H_{12}$ compound is also spont flam in air. [*19*]

HEXABORON DECAHYDRIDE. See hexaborane.

HEXABROMODISILANE. See disilicon hexabromide.

HEXABROMOETHANE. See dicarbon hexabromide.

HEXACALCIUM PHYTATE. See calcium phytate.

γ-HEXACHLORIDE. See benzene hexachloride.

HEXACHLOROACETONE. Colorless liquid. CCl_3COCCl_3, mw: 264.8, bp: 202°–204°, fp: −2°, vap. d: 9.2.

Acute tox data: Oral LD_{50} (rat) = 1290 mg/kg; inhal LC_{50} (rat) = 360 ppm for 6 hrs; dermal LD_{50} (rat) = 2980 mg/kg. [*3*]

THR = MOD via and dermal routes; HIGH via inhal route.

Disaster Hazard: Dangerous; when heated to decomp, emits highly toxic fumes of phosgene.

HEXACHLOROBENZENE. Syn: *perchlorobenzene.* Monoclinic prisms. C_6Cl_6, mw: 284.80, mp: 230°, bp: 326°, flash p: 468°F, d: 1.57, vap. press: 1 mm @ 114.4°, vap. d: 9.8.

Acute tox data: Oral LD_{50} (rat) = 3500 mg/kg; oral LD_{50} (mouse) = 4000 mg/kg.

THR = MOD via oral route. An exper teratogen. [*3*]

Fire Hazard: Slight, when exposed to heat or flame. Violent reaction with dimethylformamide. [*19*]

Disaster Hazard: Dangerous; when heated to decomp, emits highly toxic fumes of chlorides.

To Fight Fire: CO_2, dry chemical.

HEXACHLOROBUTADIENE. C_4Cl_6, mw: 260.7, autoign. temp.: 1130°F, vap. d: 8.99.

Acute tox data: Oral LD_{50} (rat) = 90 mg/kg; oral LD_{50} (mice) = 110 mg/kg; inhal LC_{LO} (mice) = 235 ppm for 4 hrs. [*3*]

THR = HIGH via oral and inhal routes.

Fire Hazard: Low.

To Fight Fire: Dry chemical, CO_2, alcohol foam, water spray, fog, mist.

1,2,3,4,7,7-HEXACHLORO-5,6-BIS(CHLOROMETHYL)-2-NORBORNENE.

THR = U. Limited data suggest similar to but less toxicity than DDT. See also DDT.

Disaster Hazard: Dangerous; see chlorides.

1,2,3,4,5,6-HEXACHLOROCYCLOHEXANE-γ. Syns: *benzene hexachloride, gammahexane lindane, streunex, BHC, DBH, HCCH, HCH, "666," "Gammexane," "Benzahex," "Cemhex," "Gamoxol," "Hexadon," other trade names.* White crystalline powder. $C_6H_6Cl_6$, mw: 290.9, mp: 113°, vap. press: 0.0317 mm @ 20°.

Acute tox data: Oral LD_{50} (cattle) = 5-25 mg/kg; oral LD_{50} (rat) = 88 mg/kg; dermal LD_{50} (rat) = 500 mg/kg; dermal LD_{50} (rabbit) = 50 mg/kg; oral LD_{LO} (wild birds) = 100 mg/kg; LD for a child was 188 mg/kg via oral route. [*3*]

THR = Hexachlorocyclohexane, a toxic organochlorine pesticide which is persistent in the environment and accumulates in mammalian tissue. For cattle the oral LD_{50} = 5–25 mg/kg. The several isomers of hexachlorocyclohexane have different actions. The γ and α isomers are CNS stimulants, the principal symptom being convulsions. The β and δ isomers are depressants of the CNS.

The dangerous acute dose of the technical mixture has been estimated at about 30 g and the dangerous dose of lindane at about 7 to 15 g. These estimates may be too high, for a young man suffered a serious illness, including convulsions, following a single carefully measured dose of 45 mg intended as a vermifuge. Thus 40 mg of a purified commercial preparation was tolerated daily for 10 days. This mixture of isomers burned the tongue, and unpurified mixtures were found to be very irr. Doses of 70 mg of the preparation caused attacks of dizziness, slight nausea, headache and a sensation of pressure in the temples. No change was indicated in the blood or urine. Preparations with more than 25% γ isomer were slightly less toxic. Forty mg of pure γ isomer were given for 14 days with no bodily disturbance. A daily dose of 180 mg caused diarrhea and a feeling of dizziness. A dose of 30 mg three times a day for a week caused no injurious signs in the patient, and no change in the urine or blood. However, as already mentioned, a single dose of 45 mg (or approximately 0.65 mg/kg) of lindane caused convulsions.

It is interesting to note that lindane shows a marked difference in toxicity to different species. Its toxic effect on laboratory animals compares favorably with that of DDT, but for several domestic animals, notably calves, lindane is more toxic than DDT or dieldrin.

Four children drank undetermined amounts of a home-made soft drink containing lindane, which had been mixed with sugar. Within less than 6 hrs, three vomited and had convulsions. The fourth did likewise within about 12 hrs. All recovered without treatment. It was not possible to determine the dosage of lindane.

The fatal poisoning of a 5-year-old girl weighing 55 lbs was caused by the accidental ingestion of 4.5 g of hexachlorocyclohexane as a 30% solution in an unspecified organic solvent. This represents a dosage of 180 mg/kg. Shortly afterward, she developed dyspnea, cyanosis, and clonic-tonic convulsions. In spite of evacuation of her stomach and therapy to restore failing circulation, she died. Autopsy showed pulmonary edema, dilation of the heart, fatty infiltration of the liver, and extensive necrosis of blood vessels in the lungs, kidney and liver.

The use of thermal vaporizers with lindane has caused some clear-cut instances of acute poisoning. For example, two refreshment-stand operators suffered severe headache, nausea, and irritation of eyes, nose, and throat shortly after exposure to lindane vapors from a dispenser in which the insecticide apparently became overheated. The symptoms abated 2 hrs after the device was removed. Overheated lindane is more apt to cause respiratory distress than lindane which is vaporized at lower temperatures. This is true because heat releases more of the compound and also causes some splitting of the molecule into highly irr decomp products. On a chronic systemic basis the α, β and γ isomers are exper carc. Human carc not definite. Has been implicated in aplastic anemia. [*3, 5*]

Dermatitis, and perhaps other manifestations based on sensitivity represent a sort of chronic, though probably not systemic intoxication, which has been observed in human beings. Dermatitis has been reported in workers who came in contact with benzene hexachloride and its precursors during manufacture and without proper hygienic precautions. Shortly after a lindane vaporizer was installed in her place of employment, a 35-year-old woman developed urticaria. The dermatitis improved during weekends, but recurred when she returned to work. Patch tests were positive. Complete elimination of exposure resulted in permanent recovery.

The signs and symptoms of confirmed acute poisoning in man have paralleled those of experimental animals. These signs and symptoms are: excitation, hyperirritability, loss of equilibrium, clonic-tonic convulsions, and later depression. There is some evidence that the pulmonary edema and vascular collapse may be of neurogenic origin also. The symptoms in animals systemically poisoned by the γ isomer alone are essentially similar to those caused by mixtures, although the onset may be earlier. Men acutely exposed to high air conc of lindane and its decomp products show headache, nausea, and irr of eyes, nose and throat.

Urticaria has followed exposure to lindane vapor in rare instances. Unlike the signs and symptoms already mentioned, this allergic manifestation occurs only in susceptible individuals, and usually only after a period of sensitization.

Disaster Hazard: Dangerous; when heated to decomp, emits highly toxic fumes of phosgene.

HEXACHLOROCYCLOHEXANE. Syns: *BHC, DBH, benzenehexachloride.*
Acute tox data: Oral LD_{50} (rat) = 100 mg/kg; oral LD_{50} (wild birds) = 75 mg/kg.
THR = HIGH via oral route.

1,2,3,4,5,6-HEXACHLOROCYCLOHEXANE (β).
THR = An exper (±) carc. [*3, 5*] See also γ isomer.

1,2,3,4,5,6-HEXACHLOROCYCLOHEXANE (α).
THR = An exper (+) carc. [*3, 5*] See also γ isomer.

1,2,3,4,5,6-HEXACHLOROCYCLOHEXANE (MIXED). Composed of 64% α, 10% β, 13% γ, 9% Δ and 1% ϵ isomers. Technical grade BHC.
THR = An exper (+) carc. [*3, 5*]

1,2,3,4,5,6-HEXACHLORO CYCLOHEXANE (Δ).
Acute tox data: Oral LD_{50} (rat) = 1000 mg/kg. [*3*]
THR = MOD via oral and ip routes.

HEXACHLOROCYCLOPENTADIENE. Yellow- to amber-colored liquid, pungent odor. C_5Cl_6, mw: 272.79, mp: 9.9°, bp: 239°, fp: −2°, flash p: none (OC), d: 1.715 @ 15.5°/15.5°, vap. d: 9.42.
Acute tox data: Oral LD_{50} (rat) = 113 mg/kg; dermal LD_{50} (rabbit) = 430 mg/kg. [*3*]
THR = HIGH via oral and dermal routes.

HEXACHLORODIBENZO-*p*-DIOXIN. Syn: *HCDD.* Colorless solid. $C_{12}H_2O_2Cl_6$, mw: 390.9, mp: 239°.
Acute tox data: Oral LD_{LO} (rat) = 100 mg/kg. [*3*] 0.01 mg/kg in benzine was active in a rabbit ear bioassay. Daily doses of 0.01 and 0.1 mg/kg→+ assay in chick oedema bioassay. A teratogen. [*81*] See also chlorinated dibenzo-*p*-dioxins.

HEXACHLORODIPHENYL OXIDE. Light yellow, very viscous liquid. $C_{12}H_4Cl_6O$, mw: 376.9, bp: 230°–260° @ 8 mm, d: 1.60 @ 20°/60°, autoign. temp.: 1148°F, vap. d: 13.0.
Acute tox data: Oral LD_{LO} (guinea pig) = 50 mg/kg. [*3*]
THR = HIGH via oral route. See closely related compound, aldrin.
Fire Hazard: Low.
To Fight Fire: Water spray, fog, foam, dry chemical, CO_2.
Disaster Hazard: Dangerous; see chlorides.

1,2,3,4,10,10-HEXACHLORO-1,4,4a,5,8,8a-HEXAHYDRO-1,4,5,8-ENDO-EXO-DIMETHANO NAPHTHALENE.
THR = Animal exper show that all compounds of this class are HIGH, causing liver injury and acne-

like skin rashes. See also chlorinated phenols and aldrin.
Disaster Hazard: Dangerous, see chlorides.

HEXACHLORODISILANE. See disilicon hexachloride.

1,2,3,4,10,10-HEXACHLORO-6,7-EPOXY-1,4,4a,5,6,7,8,8a-OCTAHYDRO-1,4,5,8-ENDO-ENDO-DIMETHANONAPHTHALENE. See endrin.

1,2,3,4,10,10-HEXACHLORO-6,7-EPOXY-1,4,4a,5,6,7,8,8a-OCTAHYDRO-1,4,5,8-ENDO-EXO-DIMETHANONAPHTHALENE. See dieldrin.

HEXACHLOROETHANE. Syns: *carbon trichloride, carbon hexachloride.* Rhombic triclinic or cubic crystals, colorless, camphor-like odor. CCl_3CCl_3, mw: 236.76, mp: 186.6° (sublimes), d: 2.091, vap. press: 1 mm @ 32.7°.
Acute tox data: Iv MLD (dogs) = 325 mg/kg; sc MLD (rabbit) = 4000 mg/kg. [*3*]
THR = HIGH via iv and MOD via sc; probably MOD to HIGH via oral and dermal routes. Liver injury has been described from exposure to this material. See also chlorinated hydrocarbons.
Explosion Hazard: Slight, by spont chemical reaction. Dehalogenation of this material by reaction with alkalies, metals, etc., will produce spont explosive chloroacetylenes.
Disaster Hazard: Dangerous; when heated to decomp, emits highly toxic fumes of phosgene.

1,2,3,4,10,10-HEXACHLORO-1,4,4a,8,8a-HEXAHYDRO-1,4,5,8-ENDO-DIMETHANO-NAPHTHALENE. Syn: *isodrin.* Crystals. $C_{12}H_8Cl_6$, mw: 364.9, mp: 241°.
Acute tox data: Oral LD_{50} (rat) = 7 mg/kg; dermal LD_{50} (rat) = 23 mg/kg.
THR = HIGH via oral and dermal routes. An insecticide.

HEXACHLOROMELAMINE. $C_3N_6Cl_6$, mw: 332.7. Reacts violently with acetone, NH_3, aniline, diphenylamine, organic contaminants, turpentine. [*19*]

HEXACHLOROMETHYL CARBONATE. Syn: *triphosgene.* White crystals. $(OCCl_3)_2CO_3$, mw: 329.8, mp: 78°–79°, bp: 205°–206° (partial decomp), d: 2 (approx).
THR = HIGH irr to skin, eyes and mu mem.
Disaster Hazard: Dangerous; when heated to decomp, emits highly toxic fumes of chlorides.

HEXACHLOROMETHYL ETHER. Liquid. $O(CCl_3)_2$, mw: 252.76, bp: 98° (partial decomp), d: 1.538 @ 18°.
THR = MOD to HIGH irr to skin, eyes and mu mem.
Disaster Hazard: Dangerous; when heated to decomp, emits highly toxic fumes of chlorides.

HEXACHLORONAPHTHALENE. White solid. $C_{10}H_2Cl_6$, mw: 334.9.
Acute tox data: Oral LD_{LO} (cattle) = 11 mg/kg.
THR = HIGH via oral route. Probably HIGH via dermal and inhal routes as well. Causes severe acne form eruptions and toxic narcosis of liver. Absorbed by skin.

HEXACHLOROPHENE. See 2,2′-methylene bis (3,4,6-trichlorophenol).

HEXACHLOROPROPENE. C_3Cl_6, mw: 248.8.
Acute tox data: Inhal LC_{50} (rat) = 425 ppm for 1/2 hour; ip LD_{50} (rat) = 40 mg/kg.
THR = HIGH via inhal and ip routes. Probaby HIGH via oral route as well. A powerful irr.
Disaster Hazard: Dangerous, see chlorides.

***n*-HEXADECANE.** Syn: *cetane.* Colorless liquid, sol in alcohol, acetone and ether, insol in water. $C_{16}H_{34}$, mw: 226, d: 0.77335 @ 20°/4°, bp: 286.5°, mp: 18.14°, vap. d: 7.8, autoign. temp.: 401°F, flash p: >212°F.
THR = U.
Fire Hazard: Low, when exposed to heat or flame.
To Fight Fire: Water spray, mist, CO_2, dry chemical, foam.

***tert*-HEXADECANETHIOL.** Syn: *tert-hexadecylmercaptan.* Colorless liquid, insol in water. $C_{16}H_{33}SH$, mw: 258, boiling range: 121°–149° @ 5 mm, d: 0.874 @ 15°/15°, flash p: 265°F (OC).
THR = Details U. See mercaptans.
Fire Hazard: Slight, when exposed to heat or flame.
Disaster Hazard: Dangerous; when heated to decomp, emits highly toxic fumes.
To Fight Fire: Dry chemical, CO_2.

HEXADECANOL. See cetyl alcohol.

1-HEXADECENE. Syn: *1-hexadecyne.* Liquid or crystals. $CH_3(CH_2)_{13}C:CH$, mw: 222.40, mp: 15°, bp: 274.0°, d: 0.797 @ 20°, vap. press: 1 mm @ 101.6°, vap. d: 7.68, flash p: >212°F, autoign. temp.: 464°F.
THR = U.
Fire Hazard: Low, when exposed to heat or flame; can react with oxidizing materials.
To Fight Fire: Foam, CO_2, dry chemical.

HEXADECYL ACETYLENE. See 1-octadecene.

***tert*-HEXADECYL MERCAPTAN.** See *tert-hexadecanethiol.*

HEXADECYL TETRATHIO-*o*-STANNATE. Solid. $Sn(SC_{16}H_{33})_4$, mw: 1148.7, mp: 54°.
THR = See tin and sulfur compounds.

HEXADECYL TRICHLOROSILANE. Colorless to yellow liquid. $C_{16}H_{33}SiCl_3$, mw: 259.5, d: 0.996 @ 25°/25°, bp: 269°, flash p: 295°F (COC).
THR = MOD irr to skin, eyes and mu mem.

Fire Hazard: Low, when exposed to heat or flame.
Disaster Hazard: Dangerous; see chlorides.

1-HEXADECYNE. See 1-hexadecine.

2,4-HEXADIENAL. Very slightly sol in water. $CH_3CH{:}CHCH{:}CHC(O)H$, mw: 96, d: 0.9, bp: 170°, lel = 1.3%, uel = 8.1%, flash p: 154°F (OC).
Acute tox data: Oral LD_{50} (rat) = 730 mg/kg; dermal LD_{50} (rabbit) = 270 mg/kg; inhal LC_{LO} (rat) = 2000 ppm for 4 hrs. [*3*]
THR = HIGH via dermal and MOD via oral and inhal routes.
Fire Hazard: Mod, when exposed to heat, flame or oxidizers.
To Fight Fire: Spray, foam, CO_2, dry chemical.

1,4-HEXADIENE. Syn: *allyl propenyl.* Colorless liquid, insol in water. $CH_3CH{:}CHCH_2CH{:}CH_2$, mw: 82, d: 0.6996 @ 20°/4°, bp: 64° @ 745 mm, flash p: −6°F, vap. d: 2.8, lel = 2.0%, uel = 6.1%.
THR = U.
Fire Hazard: Dangerous, when exposed to heat, flame or oxidizers.
To Fight Fire: Foam, fog.

2,4-HEXADIENOIC ACID. See sorbic acid.

1,5-HEXADIEN-3-YNE. C_6H_6, flash p: <−4°F, vap. d: 2.69, bp: 84°; mw: 78.1.
THR = Reacts with O_2 to form an explosive polymer. [*19*]
Fire Hazard: Dangerous, via heat, open flame, oxidizers.
To Fight Fire: Spray, CO_2, foam.

1,5-HEXADIYNE. See dipropargyl.

2,4-HEXADIYN-1,6-BISCHLOROSULFITE. $C_6H_4O_4S_2Cl_2$, mw: 275.1.
THR = Shock-sensitive. Explodes on distillation. [*19*]

2,4-HEXADIYN-1,6-DIOL. $C_6H_6O_2$, mw: 110.1.
THR = Reacts violently with $COCl_2$, thionyl chloride. [*19*]

"HEXADOW." See 1,2,3,4,5,6-hexachlorocyclohexane.

HEXAEPOXY SQUALENE. $C_{30}H_{50}O_6$, mw: 506.8.
THR = An exper neo. [*3*]

HEXAETHYL BENZENE. Colorless crystals. $C_6(C_2H_5)_6$, mw: 246.42, mp: 130°, bp: 298°, d: 0.831 @ 130°, vap. press: 10 mm @ 150.3°.
THR = U.
Fire Hazard: Mod, when exposed to heat or flame; can react with oxidizing materials.

HEXAETHYL DIGERMANE. Colorless liquid, insol in water. $Ge_2(C_2H_5)_6$, mw: 319.6, mp: <−60°, bp: 266°.
THR = Details U. See germanium compounds.

HEXAETHYL DILEAD. See triethyl lead.

HEXAETHYL DISTANNANE. Liquid $[Sn(C_2H_5)_3]_2$, mw: 411.8, d: 1.412 @ 20°, bp: 160° @ 23 mm.
THR = Details U. See tin compounds.

HEXAETHYLENE GLYCOL. Liquid. $(OH)_2(CH_2)_{12}O_5$, mw: 282.3.
THR = Details U. See also glycols.
Fire Hazard: Mod, when exposed to heat or flame; can react with oxidizing materials.

HEXAETHYL TETRAPHOSPHATE. Syns: *ethyl tetraphosphate, HETP.* Liquid. $(C_2H_5O)_6P_4O_7$, mw: 506.4, mp: −40°, bp: decomp above 150°.
Acute tox data: Oral LD_{50} (rat) = 7 mg/kg; sc LD_{50} (rat) = 0.64 mg/kg; ip LD_{50} (mouse) = 6.1 mg/kg; dermal LD_{LO} (rat) = 15 mg/kg. [*3*]
THR = HIGH via oral, dermal, sc and ip routes. See parathion.
Disaster Hazard: See tetraethyl pyrophosphate.

HEXAETHYL TETRAPHOSPHATE AND COMPRESSED GAS MIXTURES.

HEXAETHYL TETRAPHOSPHATE MIXTURE, DRY.

HEXAETHYL TETRAPHOSPHATE MIXTURE, LIQUID.

HEXAFLUOROACETONE. A colorless, non-flam solvent liquid. C_3F_6O, mw: 166.0, d: 1.65 @ 25°.
Acute tox data: Oral LD_{50} (rat) = 191 mg/kg; inhal LC_{50} (rat) = 275 ppm for 3 hrs. [*3*] Inhal LC_{50} (albino rat) = 900 ppm for 1/2 hr. [*20*]
THR = HIGH irr to skin, eyes and mu mem and via oral and inhal and possibly dermal routes. [*20*] See also fluorine and fluorides.

HEXAFLUORODICHLOROBUTENE.
Acute tox data: Inhal LC_{50} (rat) = 16 ppm for 4 hrs; inhal LC_{50} (mouse) = 26 ppm for 4 hrs; inhal LC_{50} (mouse) = 54 ppm for 3 hrs. [*3*]
THR = HIGH irr to skin, eyes and mu mem and via inhal route.

HEXAFLUORODIETHYL ETHER. $C_4H_4OF_6$, mw: 182.1.
Acute tox data: Ip LD_{50} (rat) = 1260 mg/kg; iv LD_{50} (mouse) = 46 mg/kg. [*2*]
THR = HIGH via iv and MOD via ip routes. Probably MOD via oral and inhal routes.

***tert*-HEXADECANETHIOL.** Syn: *tert-hexadecylmercaptan.* Colorless liquid, insol in water. $C_{16}H_{33}SH$, mw: 258, boiling range: 121°-149° @ 5 mm, d: 0.874 @ 15°/15°, flash p: 265°F (OC).
THR = U. See mercaptans.
Fire Hazard: Slight, when exposed to heat or flame.
Disaster Hazard: Dangerous; when heated to decomp, emits highly toxic fumes.
To Fight Fire: Water or foam.

HEXADECANOL. See cetyl alcohol.

HEXAFLUORO BENZENE. C_6F_6, mw: 186.1.
Acute tox data: Inhal LD_{LO} (mouse) = 95 mg/m^3 for 2 hrs. [*3*]
THR = HIGH via inhal route.

HEXAFLUOROISOPROPANOL. $C_3H_2OF_6$, mw: 168.1.
Acute tox data: Inhal LC_{LO} (rat) = 3200 ppm for 4 hrs. [*3*]
THR = HIGH irr to skin, eyes and mu mem. Also via inhal route. [*20*]

HEXAFLUOROPHOSPHORIC ACID. Corrosive, colorless, clear liquid. HPF_6, mw: 145.99, mp: 31°, d: 1.65.
THR = HIGH irr to skin, eyes and mu mem and via all routes. See also hydrofluoric and phosphoric acids.
Disaster Hazard: Dangerous; when heated to decomp, emits highly toxic fumes of fluorides and oxides of phosphorus.

HEXAFLUOROPROPYLENE. Syn: *perfluoropropene.* Gas. $CF_3CF{:}CF_2$, mw: 150.0, mp: −156°, bp: −29°, d: 1.583 @ −40°/4°.
Acute tox data: Inhal LC_{50} (rat) = 2800 ppm for 4 hrs; inhal LC_{50} (mouse) = 750 ppm for 4 hrs. [*3*]
THR = MOD via inhal route.

***cis*-HEXAHYDROPHTHALIC ANHYDRIDE.** Syn: *cis-1,2-cyclohexane dicarboxylic anhydride.* Clear, colorless, viscous liquid or glassy solid. $C_8H_{10}O_3$, mw: 154.1, bp: 158° @ 17 mm, mp: 35°–36°, vap. d: 5.31.
THR = A primary skin irr and possibly dangerous skin-sensitizer. Vapors are irr to skin and mu mem. Inhal of high conc as vapor may result in serious injury to respiratory apparatus. Particularly dangerous to eyes. It appears to have a low systemic toxicity by ingestion.
Fire Hazard: Slight, when exposed to heat or flame; can react with oxidizing materials.

HEXAHYDROPYRAZINE. See piperazine.

HEXAHYDROPYRIDINE. See piperidine.

HEXAHYDROTHYMOL. See menthol.

HEXAHYDROTOLUENE. See methyl cyclohexane.

HEXAHYDRO-1,3,5-TRIETHYL-SYM-TRIAZINE. See vancide.

HEXAHYDRO-1,3,5-TRIPHENYL-SYM-TRIAZINE. See methylene aniline.

HEXAHYDROXY CYCLOHEXANE. See inositol.

HEXAHYDROXYLOL. See p-dimethyl cyclohexane.

HEXAIODOSILANE. See disilicon hexaiodide.

***n*-HEXALDEHYDE.** Liquid. $C_5H_{11}CHO$, mw: 100.16, mp: −56.3°, bp: 128.7°, flash p: 90°F (OC), d: 0.8156 @ 20°/20°, vap. press: 8.6 mm @ 20°, vap. d: 3.45.
Acute tox data: Oral LD_{50} (rat) = 4890 mg/kg; inhal LC_{LO} (rat) = 2000 ppm for 4 hrs. [*3*]
THR = MOD via oral and inhal routes. See also aldehydes.
Disaster Hazard: Dangerous. See fluorides.

HEXAHYDRIC ALCOHOL. See sorbitol.

HEXAHYDROANILINE. See cyclohexylamine.

1,2,4,5,6,7-HEXAHYDROBENZ(e)ACEANTHRALENE. $C_{20}H_{18}$, mw: 258.4.
THR = An exper neo. [*3*]

1,2,3,4,12,13-HEXAHYDRO DIBENZ(a,h)ANTHRACENE. $C_{22}H_{20}$, mw: 284.4.
THR = An exper carc. [*3*]

HEXAHYDROBENZENE. See cyclohexane.

HEXAHYDROBENZOIC ACID. See naphthenic acid.

HEXAHYDROCRESOL. See methyl cyclohexanol.

HEXAHYDRO-1,4-DIAZENE. See piperazine.

6,7,8,9,10,12b-HEXAHYDRO-3-METHYL CHOLANTHRENE. $C_{20}H_{21}$, mw: 261.4.
THR = An exper carc. [*3*]

HEXAHYDROMETHYL PHENOL. See methylcyclohexanol.

HEXAHYDRO-1-NITROSO-1H-AZEPINE. $C_6H_{12}ON_2$, mw: 128.2.
THR = HIGH acute oral to rats. An exper carc via oral and sc routes. [*3*, *23*]

HEXAHYDROPHENOL. See cyclohexanol.
Fire Hazard: Dangerous, when exposed to heat or flame; can react with oxidizing materials.
To Fight Fire: Foam, CO_2, dry chemical.

HEXALIN. See cyclohexanal.

HEXALIN ACETATE. See cyclohexyl acetate.

HEXAMETHYL BENZENE. $C_6(CH_3)_6$, mw: 162.3.
THR = Can explode with nitromethane. [*19*]

HEXAMETHYL BICYCLO(2,2,0)HEXA-2,5-DIENE. $C_{12}H_{18}$, mw: 162.3.
THR = An exper neo. [*3*]

HEXAMETHYL DISILANE. Liquid. $(CH_3)_3$-Si-Si-$(CH_3)_3$, mw: 146.3, mp: 13°, bp: 112.5°.
THR = Details U. See silanes.

HEXAMETHYL DISILOXANE. See "DC 200 fluid."

HEXAMETHYLENE. See cyclohexane.

HEXAMETHYLENEDIAMINE. Syn: *1,6-hexanediamine.* Colorless silk leaf. $NH_2(CH_2)_6NH_2$, mw: 116.21, mp: 42°, bp: 205°.

THR = MOD irr to skin, eyes and mu mem via oral and inhal routes.

Fire Hazard: Slight, when exposed to heat or flame; can react with oxidizing materials.

HEXAMETHYLENE DIISOCYANATE. $(CH_2)_6(NCO)_2$, mw: 168.2.

Acute tox data: Oral LD_{50} (rat) = 710 mg/kg; dermal LD_{50} (rabbit) = 570 mg/kg; inhal LC_{50} (mouse) = 1570 mg/m^3. [*3*]

THR = MOD via oral, inhal and dermal routes. Violent reaction with alcohols. [*19*]

HEXAMETHYLENE GLYCOL. See hexanediol-1,6.

HEXAMETHYLENETETRAMINE. Syns: *methenamine, formamine, hexamine, urotropin, metramine.* Odorless rhombic crystals from alcohol. $(CH_2)_6N_4$, mw: 140.19, mp: 280° sublimes, flash p: 482°F, d: 1.33 @ −5°.

Acute tox data: Iv LD_{50} (rat) = 9200 mg/kg; sc LD_{50} (cat) = 200 mg/kg. [*3*]

THR = HIGH via sc and LOW via iv routes. Probably a MOD irr to skin, eyes and mu mem. An exper neo. [*3*] Some persons suffer a skin rash if they come in contact with this material or the fumes evolved when it is heated. Pure hexamethylenetetramine may be taken internally in small amounts and is used in medicine as a urinary antiseptic. Its major industrial use is in the manufacture of phenolic resins. It is combustible and can be readily ignited when a flame is applied directly to its surface. It liberates formaldehyde on decomp.

Fire Hazard: Mod, when exposed to heat or flame; can react with oxidizing materials. Reacts violently with Na_2O_2. [*19*]

HEXAMETHYLENIMINE. Syn: *HMI.* Colorless liquid, water-sol, ammonia-like odor. $C_6H_{12}N$, mw: 99.2, bp: 138°, mp: −37°.

Acute tox data: Oral LD (rat) = 450 mg/kg. [*20*] Oral LD_{50} (rat) = 33 mg/kg; inhal LC_{LO} (rat) = 4800 ppm for 4 hrs. [*3*]

THR = HIGH via oral and MOD via inhal routes. A powerful irr to skin, eyes and mu mem.

HEXAMETHYL MELAMINE. Syn: *hemel.* A solid material, insol in water, sol in acetone. $C_3N_3[N(CH_3)_2]_3$, mw: 210.

Acute tox data: Ip LD_{50} (rat) = 265 mg/kg; oral LD_{50} (mouse) = 437 mg/kg; oral LD_{50} (chicken) = 341 mg/kg. [*3*] A chemosterilant for insects.

THR = HIGH via ip and oral routes. An exper neo. [*3, 23*]

HEXAMETHYL PHOSPHORAMIDE. Clear, colorless, mobile liquid, spicy odor. $[(CH_2)_2N]_3PO$, mw: 179.2, bp: 233°, fp: 6°, d: 1.024 @ 25°/25°, vap. d: 6.18.

Acute tox data: Oral LD_{50} (M rat) = 2650 mg/kg; = 3360 mg/kg for (F rat); iv LD_{50} (mice) = 800 mg/kg; dermal LD_{50} (rabbit) = 2600 mg/kg. [*81*]

THR = MOD via oral, iv and dermal routes. An exper (+) carc [*3*] via inhal route. [*81*]

Disaster Hazard: Dangerous; see phosphorus compounds.

HEXAMETHYL PHOSPHORIC TRIAMIDE. See hexamethyl phosphoramide.

HEXAMINE. See hexamethylenetetramine.

HEXAMMINE CHROMIUM III CHLORIDE. Yellow crystals. $[Cr(NH_3)_6]Cl_3 \cdot H_2O$, mw: 278.59, d: 1.585.

THR = See chromium compounds and chlorides.

HEXAMMINE CHROMIUM NITRATE. $Cr(NH_3)_6(NO_3)_3$, mw: 340.2.

THR = Heat- (265°) and impact-sensitive explosive. [*19*]

HEXAMMINE COBALT TRICHLORIDE. See cobalt and chlorides.

HEXAMMINE NICKEL (II) NITRATE. Cubic blue crystals. $[Ni(NH_3)_6](NO_3)_2$, mw: 284.90.

THR = See nitrates and nickel compounds.

HEXANAL. See *n*-hexaldehyde.

***n*-HEXANE.** Syn: *hexyl hydride.* Colorless liquid, faint odor. $CH_3(CH_2)_4CH_3$, mw: 86.17, bp: 68.7, ulc: 90–95, lel = 1.2%, uel = 7.5%, fp: −95.6°, flash p: −7°F, d: 0.6603 @ 20°/4°, autoign. temp.: 437°F, vap. press: 100 mm @ 15.8°, vap. d: 2.97.

Acute tox data: Inhal LC_{LO} (mouse) = 120,000 mg/kg. [*3*]

THR = LOW via oral route. Used as a food additive permitted in food for human consumption. [*109*] Can cause a motor neuropathy in exposed workers. [*9*] Inhal of 5000 ppm for 1/6 hr ⟶ marked vertigo. 2500–1000 ppm for 12 hrs/day ⟶ drowsiness, fatigue, loss of appetite, paresthesia in distal extremities. 2500–500 ppm ⟶ muscle weakness, cold pulsation in extremities, blurred vision, headache, anorexia and onset of polyneuropathy. 2000 ppm for 1/6 hr ⟶ no symptoms. 1000–500 ppm for 3–6 months ⟶ fatigue, loss of appetite, distal paresthesia. Dermal exposure ⟶ no anesthesia, blister formation, irr, itching, erythema, pigmentation, pain. [*98*] Dangerous if abused.

Fire Hazard: Dangerous, when exposed to heat or flame.

Spont Heating: No.

Explosion Hazard: Mod, when exposed to heat or flame.

Disaster Hazard: Dangerous, when heated or exposed

to flame, can react vigorously with oxidizing materials.
To Fight Fire: CO_2, dry chemical.

1,6-HEXANEDIAMINE. See hexamethylenediamine.

HEXANEDIOIC ACID. See adipic acid.

HEXANEDIOL-1,6. Syns: *1,6-dihydroxyhexane, hexamethylene glycol.* Clear liquid.
$CH_2OH(CH_2)_4CH_2OH$, mw: 118.2, mp: 42°, bp: 250°, flash p: 266°F, d: 0.967 @ 0°/4°, vap. d: 4.07.
Acute tox data: Oral LD_{50} (rat) = 3730 mg/kg. [*3*]
THR = MOD via oral route.
Fire Hazard: Mod, when exposed to heat or flame; can react with oxidizing materials.
Spont Heating: No.
To Fight Fire: Foam, CO_2, dry chemical.

HEXANEDIOL-2,5. Liquid.
$CH_3CH(OH)CH_2CH_2CH(OH)CH_3$, mw: 118.17, bp: 220.8, flash p: 230°F, d: 0.9617 @ 20°/20°, vap. d: 4.07.
Acute tox data: Oral LD_{50} (rat) = 5000 mg/kg; dermal LD_{50} (rabbit) = 16,000 mg/kg. [*3*]
THR = MOD via oral and VERY LOW via dermal route.
Fire Hazard: Low, when exposed to heat or flame; can react with oxidizing materials.
To Fight Fire: Alcohol foam, CO_2, dry chemical.

HEXANEDIONE-2,5. See acetonyl acetone.

1,2,3,4,5,6-HEXANE HEXANITRATE. See dulcitol hexanitrate.

HEXANETHIOL. $C_6H_{14}S$, mw: 118.3.
Acute tox data: Oral LD_{50} (cat) = 1254 mg/kg; inhal LC_{50} (rat) = 1080 ppm for 4 hrs. [*3*]
THR = MOD via oral and inhal routes.

HEXANETRIOL-1,2,6. Colorless liquid.
$HOCH_2CH(OH)(CH_2)_3CH_2OH$, mw: 134.17, bp: 178° @ 5 mm, fp: −20°, flash p: 375°F (COC), d: 1.1063 @ 20°/20°, vap. press: <0.01 mm @ 20°, vap. d: 4.63.
Acute tox data: Oral LD_{50} (rat) = 16,000 mg/kg. [*3*]
THR = VERY LOW via oral route.
Fire Hazard: Slight, when exposed to heat or flame; can react with oxidizing materials.
To Fight Fire: Alcohol foam, spray, mist, dry chemical.

2,4,6,2′,6′-HEXANITRODIPHENYLAMINE. Syns: *hexil, hexite.* $C_{12}H_5O_{12}N_7$, mw: 439.2.
THR = An exper neo. [*3*] See also nitro compounds of aromatic hydrocarbons.
Fire Hazard: See nitrates.
Explosion Hazard: See nitrates. A powerful and violent explosive used as a booster explosive, in which use it is superior to TNT. It is not as good for this purpose as tetryl, but is extremely stable and much safer to handle.
Disaster Hazard: See nitrates.

HEXANITRODIPHENYLAMINO ETHYL NITRATE.
THR = Details U. See also nitro compounds of aromatic hydrocarbons.
Fire Hazard: See nitrates.
Explosion Hazard: See nitrates. This high explosive is stable and not very sensitive. It is used in detonating compositions and is considered more powerful than picric acid. See also explosives, high.
Disaster Hazard: See nitrates.

HEXANITRODIPHENYL SULFIDE. Syn: *picryl sulfide.* $C_{12}H_4O_{12}N_6S$, mw: 456.3.
Acute tox data: Oral LD_{50} (rat) = 1200 mg/kg. [*3*] Oral LD_{50} (mouse) = 470 mg/kg. [*3*]
THR = MOD via oral route.
Fire Hazard: See nitrates.
Explosion Hazard: See nitrates. This material is a powerful explosive and has an added military advantage in that its explosion gases contain irr and very toxic sulfur dioxide. See also explosives, high.
Disaster Hazard: See nitrates.

HEXANITRODIPHENYL SULFONE.
THR = Details U. See also nitro compounds of aromatic hydrocarbons.
Fire Hazard: See nitrates.
Explosion Hazard: See nitrates. This material is a stable very powerful high explosive, used in detonating compositions. See also explosives, high.
Disaster Hazard: See nitrates.

HEXANITROMANNITOL. See mannitol hexanitrate.

HEXANITRO OXANILITE.
THR = Details U. See also nitro compounds of aromatic hydrocarbons.
Fire Hazard: See nitrates.
Explosion Hazard: See nitrates. This material is about as powerful an explosive as TNT. See also explosives, high.

HEXANOIC ACID. See caproic acid.

1-HEXANOL. Syns: *n-hexyl alcohol, amyl carbinol.* Colorless liquid. $CH_3(CH_2)_4CH_2OH$, mw: 102.2, bp: 157.2°, fp: −44.6°, flash p: 145°F, d: 0.8186 @ 20°/4°, vap. press: 1 mm @ 24.4°, vap. d: 3.52.
Acute tox data: Oral LD_{50} (rat) = 720 mg/kg; dermal LD_{50} (rabbit) = 3100 mg/kg. [*3*]
THR = MOD via oral and dermal routes.
Fire Hazard: Mod, when exposed to heat or flame; can react with oxidizing materials.
Spont Heating: No.
To Fight Fire: Alcohol foam, CO_2, dry chemical.

2-HEXANOL. See 2-ethylbutyl alcohol.

2-HEXANONE. See methyl-*n*-butyl ketone.

3-HEXANONE. Syn: *ethyl propyl ketone.* Colorless liquid. $C_2H_5CO(CH_2)_2CH_3$, mw: 100.16, bp: 124°, d: 0.813 @ 21.8°/4°, flash p: 95°F (OC).
Acute tox data: Oral LD_{50} (rat) = 3360 mg/kg; inhal LC_{LO} (rat) = 4000 ppm for 4 hrs. [*3*]
THR = MOD via oral and inhal routes. See also ketones.
Fire Hazard: Dangerous, when exposed to heat or flame; can react with oxidizing materials.
To Fight Fire: Foam, CO_2, dry chemical.

HEXAPHENYL DIGERMANE. White crystals, insol in water. $[(C_6H_5)_3Ge]_2$, mw: 607.8, mp: 340°.
THR = Details U. See germanium compounds.

HEXAPHENYL DIGERMANE(TRIBENZENE). Colorless crystals, insol in water, sol in benzene. $[(C_6H_5)_3Ge]_2 \cdot 3C_6H_6$, mw: 842.1, mp: loses benzene upon warming.
THR = See benzene and germanium compounds.

HEXAPHENYL DITIN. Crystals. $[(C_6H_5)_6Sn]_2$, mw: 700, mp: 232.5°.
THR = See tin compounds.

HEXA-*p*-TOLYL DIGERMANE. Colorless crystals. $[(CH_3C_6H_4)_3Ge]_2$, mw: 692, mp: 227°.
THR = See germanium compounds.

HEXA-*p*-TOLYL DITIN. Crystals, slightly sol in some organic solvents. $[(C_6H_4CH_3)_3Sn]_2$, mw: 784.2, mp: 143.5°.
THR = See tin compounds.

HEXA-*p*-XYLYL DITIN. Crystals, sol in benzene. $[(CH_3)_2(C_6H_3)_3Sn]_2$, mw: 868.4, mp: 192.5°.
THR = See tin compounds.

HEXENE-1. Syns: *hexene, butyl ethylene, hexylene.* Colorless liquid. $CH_2CH(CH_2)_3CH_3$, mw: 84.16, bp: 64.5°, mp: −139.9°, flash p: <20°F, d: 0.6732 @ 20°/4°, vap. press: 310 mm @ 38°, vap. d: 3.0.
THR = MOD irr to skin, eyes and mu mem.
Fire Hazard: Dangerous, when exposed to heat, flame or oxidizers.
Disaster Hazard: Dangerous, upon exposure to heat or flame; can react vigorously with oxidizing materials.
To Fight Fire: Dry chemical, CO_2, foam.

2-HEXENE-*cis*.
Flash p: −4°F.
Fire Hazard: Dangerous, via heat, flame or oxidizers.
To Fight Fire: Dry chemical, CO_2, foam.

HEXENE-2 (mixed isomers). A liquid. C_6H_{12}, mw: 84.16, bp: 68.5°, flash p: <20°F, d: 0.686 @ 15.5°, vap. d: 2.92, mp: 141.35°.
THR = MOD to LOW irr to skin, eyes and mu mem.
Fire Hazard: Dangerous, when exposed to heat, flame or oxidizers.
Explosion Hazard: U.
Disaster Hazard: Dangerous, upon exposure to heat or flame; can react vigorously with oxidizing materials.
To Fight Fire: Foam, CO_2, dry chemical.

4-HEXENE-1-YNE-3-OL. HC:CH_2(OH)C:$CHCH_2$, mw: 95.
Acute tox data: Oral LD_{50} (rat) = 34 mg/kg; inhal LC_{LO} (rat) = 63 mg/kg; dermal LD_{LO} (rabbit) = 71 mg/kg. [*3*]
THR = HIGH via oral, inhal and dermal routes.

5-HEXEN-2-ONE. See allyl acetone.

HEXESTROL. See dihydrostilbesterol.

HEXIL. See 2,4,6,2′,4′,6′-hexanitro diphenylamine.

HEXITE. See 2,4,6,2′,4′,6′-hexanitrodiphenylamine.

HEXOGEN. See cyclotrimethylene trinitramine.

HEXONE. See isobutyl methyl ketone.

***sec*-HEXYL ACETATE.** See methyl amyl acetate.

HEXYL ADIPATE. See dihexyl adipate.

HEXYL ALCOHOL. See hexanol.

***n*-HEXYL AMINE.** Liquid. $C_6H_{13}HN_2$, mw: 101.19, mp: −22.9°, bp: 131.4°, flash p: 85°F (OC), d: 0.7675 @ 20°/20°, vap. d: 3.49.
Acute tox data: Oral LD_{50} (rat) = 670 mg/kg; inhal LC_{LO} (rat) = 500 ppm for 4 hrs; dermal LD_{50} (rabbit) = 420 mg/kg. [*3*]
THR = MOD via oral and HIGH via dermal and inhal routes.
Fire Hazard: Dangerous, when exposed to heat or flame; can react with oxidizing materials.
Explosion Hazard: U.
To Fight Fire: Alcohol foam, CO_2, dry chemical.

***n,n*-HEXYL-2-AMINOHEPTANE.** See *n,n*-hexyl-*n*-heptyl amine.

8-HEXYL BENZ(a)ANTHRACENE. $C_{24}H_{24}$, mw: 312.5.
THR = An exper carc. [*3*]

HEXYLBORIC ACID. White crystals, slightly sol in water. $C_6H_{13}B(OH)_2$, mw: 130.0, mp: 90° (decomp).
THR = Details U. See also boron compounds.

HEXYLBORIC OXIDE. Colorless liquid, hydrolyzed by water but sol in organic solvents. $(C_6H_{13})_3B_3O_3$, mw: 336.0, d: 0.8876, bp: 180° @ 24 mm.
THR = Details U. See also boron compounds.

***n*-HEXYL "CARBITOL."** Liquid. $C_6H_{13}O(C_2H_4O)_2H$, mw: 190.03, bp: 258.2°, fp: −33.3°, flash p: 285°F (OC), d: 0.9346 @ 20°/20°, vap. press: 0.01 mm @ 20°.

Acute tox data: Oral LD_{50} (rat) = 4920 mg/kg; dermal LD_{50} (rabbit) = 1500 mg/kg. [*3*]
THR = MOD via oral and dermal routes.
Fire Hazard: Slight, when exposed to heat or flame; can react with oxidizing materials.
To Fight Fire: Foam, CO_2, dry chemical.

***n*-HEXYL "CELLOSOLVE."** Liquid. $C_6H_{13}OCH_2CH_2OH$, mw: 146.22, bp: 208.3°, fp: −45.1°, flash p: 195°F (OC), d: 0.8894 @ 20°/20°, vap. press: 0.1 mm @ 20°, vap. d: 5.04.
Acute tox data: Oral LD_{50} (rat) = 1480 mg/kg; dermal LD_{50} (rabbit) = 890 mg/kg. [*3*]
THR = MOD via oral and dermal routes.
Fire Hazard: Mod, when exposed to heat or flame, can react with oxidizing materials.
To Fight Fire: Foam, CO_2, dry chemical.

HEXYL DIETHYL HEPTYLATE. Syn: *flexol plasticizer CC-55*. Liquid. $C_6H_{10}[COOCH_2CH(C_2H_5)C_4H_9]_2$; mw: 376.59, mp: −53°, bp: 216° @ 5 mm, flash p: 425°F (OC), d: 0.9586 @ 20°/20°, vap. d: 13.7.
THR = Details U. See also esters.
Fire Hazard: Slight, when exposed to heat or flame; can react with oxidizing materials.
To Fight Fire: Foam, CO_2, dry chemical.

HEXYLENE. See hexene-1.

HEXYLENE GLYCOL. Syn: *2-methylpentanediol-2,4*. Mild odor, colorless liquid, water-sol. $CH_3COH(CH_3)CH_2CHOHCH_3$, mw: 118.17, bp: 197.1°, fp: −50°, flash p: 205°F (OC), d: 0.9234 @ 20°/20°, vap. press: 0.05 mm @ 20°, d: 4.
Acute tox data: Inhal TC_{LO} (human) = 50 ppm for 15 min → eye symptoms; oral LD_{50} (rat) = 3696 mg/kg; oral LD_{50} (mouse) = 3860 mg/kg; ip LD_{50} (mouse) = 1299 mg/kg. [*3, 20*]
THR = MOD via oral and ip routes. Irr to skin, eyes and mu mem. Large oral doses produce narcosis. See also glycols.
Fire Hazard: Low, when exposed to heat or flame; can react with oxidizing materials.
To Fight Fire: Foam, CO_2, dry chemicals.

HEXYL ETHER. Syn: *dihexyl ether*. $C_6H_{13}OC_6H_{13}$, mw: 186.33, mp: −43.0°, bp: 227°, flash p: 170°F (OC), d: 0.794, autoign. temp.: 365°F, vap. d: 6.4.
Acute tox data: Oral LD_{50} (rat) = 30,900 mg/kg; dermal LD_{50} (rabbit) = 6900 mg/kg. [*3*]
THR = NIL oral and LOW via dermal routes.
Fire Hazard: Mod, when exposed to heat or flame; can react with oxidizing materials.
Explosion Hazard: Details U. See also ethers.
To Fight Fire: Foam, CO_2, dry chemical.

***n,n*-HEXYL-*n*-HEPTYL AMINE.** Syn: *n,n-hexyl-2-amino heptane*. $(C_6H_{13})_2(C_7H_{15})N$, mw: 283.5.
THR = Details U. See also amines.
Fire Hazard: Slight, when exposed to heat or flame.

HEXYL HYDRIDE. See *n*-hexane.

HEXYL MANDELATE. $C_6H_5CHOHCOOC_6H_{13}$, mw: 220.
THR = Limited animal exper suggest LOW. See also esters and mandelic acid.

HEXYL METHACRYLATE. Liquid. $C_6H_{13}OOCC(CH_3):CH_2$, mw: 170, d: 0.88, vap. d: 5.9, bp: 67°–85° @ 8 mm, flash p: 180°F (OC).
THR = U. See also esters.
Fire Hazard: Mod, when exposed to heat, flame or oxidizers.
To Fight Fire: Dry chemical, CO_2, foam, spray.

HEXYL METHYL KETONE. See 2-octanone.

***n*-HEXYL PYRROLIDINE.** Colorless liquid, sol in water. $C_4H_8NC_6H_{13}$, mw: 155.3, bp: 201°, d: 0.835 @ 20°, fp: −75°, flash p: 154°F.
THR = MOD irr to skin, eyes and mu mem. HIGH via oral route.
Fire Hazard: Mod dangerous, when exposed to heat, flame, sparks or powerful oxidizers.
To Fight Fire: Spray, foam, CO_2, dry chemical.

4-HEXYL RESORCINOL. $C_{12}H_{18}O_2$, mw: 194.3.
THR = HIGH acute oral. An exper neo. [*3*]

HEXYL TITANATE. See 2-ethylbutyl titanate.

HEXYL TRICHLOROSILANE. $C_6H_{13}SiCl_3$, mw: 219.7.
THR = HIGH irr to skin, eyes and mu mem and via oral and inhal routes.
Disaster Hazard: Dangerous; see chlorides; will react with water or steam to produce toxic and corrosive fumes.

HIERATITE. See potassium fluosilicate.

HI-FLASH NAPHTHA. See naphtha (coal tar).

HIGH EXPLOSIVES. See explosives, high.

HIGH TEST HYPOCHLORITES. See hypochlorites.

HIGH WINES. See wines, high, and alcohols, N.O.S.

HINOSAN. Syn: *o-ethyl-s,s-diphenyl dithiophosphate*. A clear yellow to light brown liquid. $C_{14}H_{15}O_2PS_2$, mw: 310.2, d: 1.23.
Acute tox data: Oral LD_{50} (rat) = 150 mg/kg; ip LD_{50} (rat) = 26 mg/kg. [*3*]
THR = HIGH via oral and ip routes. A cholinesterase inhibitor. See parathion.
Disaster Hazard: See parathion.

HIPPOCRATES AND PUBLIC HEALTH. See Section 1.

HISTAMINE. Syns: *4-imidazole ethylamine, ergamine*.

Deliquescent needles. $C_5H_9N_3$, mw: 111.2, mp: 84°, bp: 210° @ 18 mm.

Acute tox data: Iv LD_{50} (monkey) = 50 mg/kg; sc LD_{50} (guinea pig) = 0.5 mg/kg.

THR = HIGH via iv and sc routes. The most potent capillary dilator known. [*20*] Ingestion or inhal of this material produces the following effects: Flushing followed by pallor, dizziness, fainting, fall in blood pressure, headache, rapid, weak pulse. Allergic effects on skin (hives) may occur.

Disaster Hazard: Dangerous; when heated to decomp, emits highly toxic fumes.

Treatment and Antidotes: If swallowed, wash out stomach. Administer stimulants (epinephrine or ephedrine). Call a physician.

HISTAMINE DIHYDROCHLORIDE.
$C_5H_9N_3 \cdot 2HCl$, mw: 184.1.
THR = An exper neo. [*3*] See also histamine.

HISTAMINE SALTS. See histamine.

HISTIDINE. Syn: *α-amino-β-imidazolepropionic acid.* An amino acid essential for rats. Occurs in *dl*, *l* and *d* forms. Both the *dl* and *l* forms are colorless crystals, sol in water, insol in alcohol and ether.

$HOOCCH(NH_2)CH_2C_3H_3N_2$, mw: 155, mp(*dl*): 285°–286°, mp (*l*): 277°.

THR = U. Used as a nutrient and/or dietary supplement food additive. [*109*]

HISTORICAL PERSPECTIVE OF INDUSTRIAL HYGIENE. See Section 1.

HMI. See hexamethylenimine.

HN-1 See nitrogen mustard.

HOERNESITE. See magnesium-*o*-arsenate.

HOLMIUM. Bright metallic luster, soft, malleable metal, stable in dry air, oxidizes rapidly in moist air.

Ho, atwt: 164.93, mp: 1470°, bp: 2720°, d: 8.78 @ 25°, vap. press: 2 mm @ 1630°.

THR = U. It may be an anti-coagulant like the lanthanides. Can react violently with air, halogens. [*19*]

Radiation Hazard: For permissible levels, see Section 5A, Table 5A.5.

HOLMIUM CHLORIDE. $HoCl_3$, mw: 271.3.
Acute tox data: Ip LD_{50} (mouse) = 312 mg/kg. [*3*]
THR = HIGH via ip route.
Disaster Hazard: Dangerous; when heated to decomp, emits highly toxic fumes.

HOLMIUM CITRATE. $C_6H_7O_7$; mw: 356.1.
Acute tox data: ip LD_{50} (rat) = 117 mg/kg; ip LD_{50} (guinea pig) = 63 mg/kg. [*3*]
THR = HIGH via ip route.

HOLMIUM NITRATE HEXAHYDRATE.
$Ho(NO_3)_3 \cdot 6H_2O$, mw: 459.1.
Acute tox data: Oral LD_{50} (rat) = 3000 mg/kg; ip LD_{50} (rat) = 270 mg/kg; ip LD_{50} (mouse) = 320 mg/kg. [*3*]
THR = HIGH via ip and MOD via oral routes.

HOMATROPINE COMPOUNDS. Syns: *mandelytropeine*, *homoatropine*. Deliquescent prisms from ether, glistening prisms from alcohol. $C_{16}H_{21}NO_3$, mw: 275.34, mp: 95.5°–98.5°.

Acute tox data: ip LD_{LO} (rat) = 180 mg/kg. [*3*]

THR = HIGH via ip route. Poisoning can occur from drops instilled into children's eyes. [*20*] A poisonous alkaloid. Resembles atropine in its action. See also atropine.

HOME FUEL OILS.
THR = A recog carc. [*14*] See mineral oils.

HORMODIN. Syn: *indolebutyric acid.* White crystals or powder, insol in water and chloroform. $C_{12}H_{13}NO_2$, mw: 203.2, mp: 124°.

Acute tox data: Ip LD_{LO} (mouse) = 100 mg/kg. [*3*]

THR = HIGH via ip route. MOD irr to skin, eyes and mu mem.

Disaster Hazard: Dangerous; when heated to decomp, emits toxic fumes.

HORN DUST.
THR = A nuisance dust and allergen. See also nuisance dust.

HORNSTONE. See magnesium fluosilicate and zinc fluosilicate.

HORSEHAIR. See hair.

HOW NOISE AFFECTS PEOPLE. See Section 3.

HTH. See calcium hypochlorite.

HTH-15. A less conc formulation of calcium hypochlorite than HTH. See calcium hypochlorite.

HUMAN AUDITORY SYSTEM. See Section 3.

HUMAN CARCINOGENS. See Section 8.

HUMIDITY AS A STRESS ON THE AIR ENVIRONMENT. See Section 4.

HVL'S FOR X-RAYS. See Section 5.A.

HYAMINE 1622. See benzethonium chloride.

HYCOL. Syn: *saponified cresol.*
THR = See phenol.

HYDRACRYLIC ACID-*p*-LACTONE. See oxetanone.

HYDRACRYLONITRILE. See ethylene cyanohydrin.

HYDRASTINE. Colorless rhombic prisms, white pulverulent alkaloid. $C_{21}H_{21}NO_6$, mw: 383.39, mp: 132°.

Acute tox data: ip LD_{50} (rat) = 104 mg/kg. [*3*]

THR = High via ip and probably oral and inhal routes.

Disaster Hazard: Dangerous; when heated to decomp, emits highly toxic fumes.

HYDRASTININE. White-yellowish needles. $C_{11}H_{13}NO_3$, mw: 207.2, mp: 116°–117°.
THR = HIGH via oral route. Can cause paralysis of vasomotor nerves and vagus endings.
Disaster Hazard: Dangerous; when heated to decomp, emits highly toxic fumes.

HYDRATED ALUMINA. See aluminum hydroxide.

HYDRATED ALUMINUM HYDROXIDE. See aluminum hydroxide.

HYDRATED ALUMINUM SILICATE. See kaolin.

HYDRATED LEAD OXIDE. See lead hydroxide.

HYDRATED LIME. See calcium hydroxide.

HYDRAZIDE-*o*-ANISIC ACID. $C_8H_{10}N_2O_2$, mw: 166.2.
THR = An exper carc. [*3*]

HYDRAZIDE-*p*-ANISIC ACID.
THR = An exper carc. [*3*]

HYDRAZINE. Syns: *hydrazine base, diamine, hydrazine anhydrous.* Colorless, fuming liquid, white crystals. NH_2NH_2, mw: 32.05, mp: 1.4°, bp: 113.5°, flash p: 100°F (OC), d: 1.1011 @ 15° (liq), autoign. temp.: can vary from 74°F in contact with iron rust to 270°F in contact with black iron to 313°F in contact with stainless steel to 518°F in contact with glass, vap. d: 1.1; lel = 4.7%, uel = 100%.
Acute tox data: Oral LD_{50} (rat) = 60 mg/kg; inhal LC_{50} (rat) = 570 ppm for 4 hrs; dermal LD_{50} (rabbit) = 91 mg/kg; iv (rabbit) = 20 mg/kg. [*3*]
THR = HIGH via oral, iv and dermal routes. May cause skin sensitization as well as systemic poisoning. Hydrazine and some of its derivatives may cause damage to the liver and destruction of red blood cells. See phenyl hydrazine. An exper (+) carc. [*3, 10, 14*] of lung, nervous system, liver, kidney, hematopoietic organs, breast, subcutaneous tissue.
Fire Hazard: Dangerous, when exposed to heat, flame or oxidizing agents; a powerful reducing agent.
Explosion Hazard: Severe, when exposed to heat or flame or by chemical reaction with alkali metals, NH_3, Cl_2, chromates, CuO, Cu ++ salts, F_2, H_2O_2, iron rust, metallic oxides, Ni, $Ni(ClO_4)_2$, HNO_3, N_2O, O_2, liquid O_2, $K_2Cr_2O_7$, $Na_2Cr_2O_7$, tetryl, zinc diamide, $Zn(C_2H_5)_2$. Can self-ignite when absorbed on earth, asbestos, cloth, wood. [*19*] It is a powerful explosive, much used in rocket fuels. It is very sensitive and must not be used without full and complete instructions from the manufacturer as to handling, storage and disposal.
Disaster Hazard: Dangerous; when heated to decomp, emits highly toxic fumes of nitrogen compounds; may explode by heat or chemical reaction.
To Fight Fire: Foam, CO_2, dry chemical.

HYDRAZINE AZIDE. $H_2NNH_3N_3$, mw: 75.1, mp: 75.4.
THR = An explosive salt; even when moist. A hot wire decomp it violently. [*19*]

HYDRAZINE BASE. See hydrazine.

HYDRAZINE DIBORANE. White crystalline powder, hygroscopic. $B_2N_2H_{10}$, mw: 59.7, d: 0.91.
THR = See boron hydrides, boron compounds, hydrazine.

HYDRAZINE CHLORITE. $H_2NNH_3ClO_2$, mw: 100.5.
THR = A very flam, explosive salt. [*19*]

HYDRAZINE HYDRATE. H_6ON_2, mw: 50.1, mp: −40°, bp: 118.5° @ 740 mm. Colorless, fuming liquid.
THR = HIGH via oral route. An exper carc. [*3*] Violent reaction with HgO, Na, $SnCl_2$, 2,4-dintrochlorobenzene. [*19*]

HYDRAZINE NITRATE. $H_2NNH_3NO_3$, mw: 95.1.
THR = Violent reactions with Cu, metal carbides, metal nitrides, metal oxides, metal sulfides, Zn. [*19*] See hydrazine.

HYDRAZINE PERCHLORATE. Solid, decomp in water, sol in alcohol, insol in ether, benzene, chloroform and carbon disulfide. $N_2H_4 \cdot HClO_4 \cdot \frac{1}{2}H_2O$, mw: 141.5, d: 1.939, mp: 137°, bp: 145° decomp.
THR = HIGH. See hydrazine, perchlorates.
Disaster Hazard: An explosive; can be detonated by shock or friction. [*19*]

HYDRAZINE SELENATE. $N_2H_4 \cdot H_2SeO_4$, mw: 177.
THR = An explosive salt. [*19*] HIGH irr to skin, eyes and mu mem. HIGH via oral and inhal routes also. See selenium compounds and hydrazine.

HYDRAZINE SULFATE. Colorless crystals, water-sol, insol in alcohol. $2N_2H_4 \cdot H_2SO_4$; mw: 162.2, mp: 85°.
THR = MOD to HIGH acute oral. An exper carc. [*3*] See hydrazine.

HYDRAZINE TARTRATE. Syn: *hydrazine acid tartrate.* Crystals, water-sol. $H_2NNH_2 \cdot C_4H_6O_6$, mw: 182.1, mp: 183°.
THR = HIGH. See hydrazine.

2-HYDRAZINO ETHANOL. See 2-hydroxyethyl hydrazine.

2-HYDRAZINO-4-(5-NITRO-2-FURYL)THIAZOLE. $C_7H_6O_3N_4S$, mw: 226.2.
THR = An exper carc and neo. [*3, 23*]

2-HYDRAZINO-4-(*p*-NITROPHENYL)THIAZOLE. $C_9H_8O_2N_4S$, mw: 236.3.
THR = An exper carc and neo. [*3, 23*]

HYDRAZO BENZENE. $C_{12}H_{12}N_2$, mw: 184.3.
THR = HIGH via oral. An exper neo. [*3*]

HYDRAZO DICARBOXY BIS (METHYL NITROSAMIDE).
THR = An exper carc. [*23*]

HYDRAZOIC ACID. Syns: *azoimide, hydrogen azide.* Colorless liquid, very sol in water, intolerable pungent odor. HN_3, mw: 43.02, mp: −80°, bp: 37°, d: 1.09 @ 25°/4°.
Acute tox data: Inhal LC_{LO} (rat) = 1100 ppm for 1 hr; ip LD_{50} (mouse) = 22 mg/kg; inhal TC_{LO}(human) = 0.3 ppm → CNS problems.
THR = HIGH irr to skin, eyes and mu mem and via oral and inhal routes. Exposures to vapors causes irr of the eyes and mu mem. Continued inhal causes cough, chills and fever. High conc can cause fatal convulsions. Chronic exposure has been reported as causing injury to kidneys and spleen.
Explosion Hazard: Dangerous, when shocked or exposed to heat. Reacts violently with Cd, Cu, Ni, HNO_3, F_2. [*19*]
Disaster Hazard: Dangerous; shock or heat will explode it.

HYDRIDES.
THR = Variable. The hydrides of phosphorus, arsenic, sulfur, selenium, tellurium and boron which are of HIGH toxicity produce local irr and destroy red blood cells. They are particularly dangerous because of their volatility and ease of entry into the body. The hydrides of the alkali metals, alkaline earths, aluminum, zirconium and titanium react with moisture to evolve hydrogen and leave behind the hydroxide of the metallic element. This hydroxide is usually caustic. See also sodium hydroxide. Hydrides, metallic, primary type.

This group includes the hydrides of calcium, lithium, magnesium, potassium, sodium and strontium. In the presence of moisture they are readily converted to hydroxides which are highly irr to the skin by caustic and thermal action. Similar effects can occur on contact with eyes and respiratory mu mem.
Fire Hazard: The volatile hydrides are flam, some spont so in air. All hydrides react violently on contact with powerful oxidizing agents. When heated or on contact with moisture or acids an exothermic reaction evolving hydrogen occurs. Often enough heat is evolved to cause ignition. Hydrides require special handling instructions which should be obtained from the manufacturers (Section 7).
Explosion Hazard: The volatile hydrides (such as hydrides of boron, arsenic, phosphorus, selenium, tellurium) form explosive mixtures with air. The nonvolatile hydrides (such as sodium, lithium, calcium) readily liberate hydrogen when heated or on contact with mositure or acids. Furthermore, hydrides form dust clouds which can explode due to contact with flames, sparks, heat or oxidizers.
Disaster Hazard: Highly dangerous; when heated, they can ignite at once or liberate hydrogen: they react with moisture or acids to evolve heat and hydrogen; on contact with powerful oxidizers violent reactions can occur.

HYDRIODIC ACID. Colorless gas or pale yellow liquid. HI, mw: 127.93, mp: −50.8°, bp: −35.38° @ 5 atm, d: 5.66g/liter @ 0°.
THR = HIGH irr to skin, eyes and mu mem and via oral and inhal routes. Violent reaction with F_2, ($HClO_4$ + Mg), HNO_3, O_3, K, $KClO_3$. [*19*]
Disaster Hazard: Dangerous; when heated to decomp, emits highly toxic fumes of iodides; will react with water or steam to produce toxic and corrosive fumes.

HYDRIODIC ETHER. See ethyl iodide.

HYDROABIETYL ALCOHOL. Syn: *abitol.* Colorless, tacky, viscous liquid. Mixture of tetra-, di- and dehydroabietyl alcohols. Flash p: 380°F (OC), d: 1.007–1.008 @ 20°/20°, softening p: 33°.
THR = Limited animal exper indicate LOW order.
Fire Hazard: Slight, when exposed to heat or flame; can react with oxidizing materials.
To Fight Fire: Foam, CO_2, dry chemical.

HYDROBROMIC ACID. Syn: *hydrogen bromide.* Colorless gas or pale yellow liquid. HBr, mw: 80.92, mp: −87°, bp: −66.5°, d: 3.50 g/liter @ 0°.
Acute tox data: Inhal LC_{50} (mouse) = 814 ppm for 1 hr; inhal LC_{50} (rat) = 2858 ppm for 1 hr; inhal TC_{LO} (human) = 5 ppm ⟶ irr. [*3*]
THR = HIGH irr and via oral and inhal routes. Reacts violently with F_2, NH_3, O_3. [*19*]
Disaster Hazard: Dangerous; see bromides; will react with water or steam to produce toxic and corrosive fumes.

HYDROBROMIC ETHER. See ethyl bromide.

HYDROCARBON GAS, LIQUEFIED. See specific gas.

HYDROCARBON GAS, NONLIQUEFIED. See specific material.

HYDROCHLORIC ACID. Syns: *muriatic acid, chlorohydric acid, hydrogen chloride.* Colorless gas or colorless fuming liquid, strongly corrosive. HCl, mw: 36.47, mp: −114.3°, bp: −84.8°, d: 1.639 g/liter (gas) @ 0°, 1.194 @ −26° (liquid), vap. press: 4.0 atm @ 17.8°.
Acute tox data: Oral LD_{50} (rabbit) = 900 mg/kg; inhal

LC_{50} (rat) = 3124 ppm for 1 hr; inhal LC_{LO} (human) = 1300 ppm for ½ hr. [*3*]

THR = MOD irr to skin, eyes and mu mem and via oral and inhal routes. Hydrochloric acid is an irr to the mu mem of the eyes and respiratory tract, and a conc of 35 ppm causes irr of the throat after short exposure. Conc of 50–100 ppm are tolerable for 1 hr. More severe exposures result in pulmonary edema, and often laryngeal spasm. Conc of 1,000–2,000 ppm are dangerous, even for brief exposures. Mists of hydrochloric acid are considered less harmful than the anhydrous hydrogen chloride, since the droplets have no dehydrating action. In general, hydrochloric acid causes little trouble in industry, other than from accidental splashes and burns. It is used as a general purpose food additive. [*109*] It is a common air contaminant. Violent reactions with acetic anhydride, 2-amino ethanol, NH_4OH, Ca_3P_2, chlorosulfonic acid, ethylene diamine, ethylene imine, oleum, $HClO_4$, β-propiolactone, propylene oxide, ($AgClO_4$ + CCl_4), NaOH, H_2SO_4, U_3P_4, vinyl acetate. [*19*] Also CaC_2, CsC_2H, Cs_2C_2, Li_6Si, Mg_3B_2, $HgSO_4$, RbC_2H, Rb_2C_2, Na. [*19*]

Disaster Hazard: Dangerous; see chlorides; will react with water or steam to produce toxic and corrosive fumes.

HYDROCHLORIC ACID MIXTURES. See hydrochloric acid.

HYDROCHLORIC ETHER. See ethyl chloride.

HYDROCERUSSITE. See lead carbonate, basic.

HYDROCORTISONE ACETATE. White, odorless, crystalline powder, very slightly sol in ether, practically insol in water, slightly sol in alcohol and chloroform. $C_{23}H_{32}O_6$, mw: 404, mp: 216°–233°.

Acute tox data: Ip LD_{50} (mouse) = 2300 mg/kg. [*3*]

THR = MOD via ip route. Used as a food additive permitted in food for human consumption. [*109*]

HYDROCORTISONE SODIUM SUCCINATE. White, odorless, hygroscopic, amorphous solid, very sol in water and alcohol, insol in chloroform, very slightly sol in acetone. $C_{25}H_{33}NaO_8$, mw: 484.5, mp: 169°–171°.

THR = Details U. Used as a food additive permitted in food for human consumption. [*109*]

HYDROCOTARNINE. Monoclinic prisms. $C_{12}H_{15}NO_3 \cdot \frac{1}{2}H_2O$, mw: 230.3, mp: 56°.

THR = An alkaloid obtained by reducing cotarnine. Made from the poppy.

Disaster Hazard: Dangerous; when heated to decomp, can emit highly toxic fumes of nitrogen compounds.

HYDROCYANIC ACID (96%). Syns: *hydrogen cyanide, prussic acid.* Colorless liquid, faint odor of bitter almonds. HCN, mw: 27.03, mp: −13.2°, bp: 25.7°, lel = 5.6%, uel = 40%, flash p: 0°F (CC), d: 0.6876 @ 20°/4°, autoign. temp.: 1000°F, vap. press: 400 mm @ 9.8°, vap. d: 0.932.

Acute tox data: Oral LD_{LO} (human) = 0.57 mg/kg; inhal LC_{LO} (human) = 120 mg/m^3 for 1 hr; inhal LC_{LO} (human) = 200 mg/m^3 for 10 min; iv LD_{50} (human) = 1 mg/kg; inhal LC_{50} (rat) = 544 ppm for 5 min; oral LD_{50} (mouse) = 3.7 mg/kg; ip LD_{50} (mouse) = 3 mg/kg; im LD_{50} (rabbit) = 1.1 mg/kg. [*3*]

THR = VERY HIGH via oral, dermal, inhal, iv, ip and im routes. Hydrocyanic acid and the cyanides are true protoplasmic poisons, combining in the tissues with the enzymes associated with cellular oxidation. They thereby render the oxygen unavailable to the tissues, and cause death through asphyxia. The suspension of tissue oxidation lasts only while the cyanide is present; upon its removal, normal function is restored provided death has not already occurred. Hydrocyanic acid does not combine easily with hemoglobin, but it does combine readily with methemoglobin to form cyanmethemoglobin. This fact is utilized in the treatment of cyanide poisoning, when an attempt is made to induce methemoglobin formation. The presence of cherry-red venous blood in cases of cyanide poisoning is due to the inability of the tissues to remove the oxygen from the blood. Exposure to conc of 100–200 ppm for periods of 30–60 min can cause death.

In cases of acute cyanide poisoning, death is extremely rapid; though sometimes breathing may continue for a few minutes. In less acute cases, there is headache, dizziness, unsteadiness of gait, a feeling of suffocation, and nausea. Where the patient recovers, there is rarely any disability. An insecticide.

Fire Hazard: Very dangerous, when exposed to heat, flame or oxidizers. Can polymerize @ 50°–60°, or catalyze with traces of alkali. Reacts violently with acetaldehyde.

Explosion Hazard: Severe, when exposed to heat or flame or by chemical reaction with oxidizers. Under certain conditions, particularly contact with alkaline materials, hydrogen cyanides can polymerize or decomp explosively. The liquid is commonly stabilized by addition of acids.

Disaster Hazard: Highly dangerous; the gas forms explosive mixtures with air; will react with water, steam, acid or acid fumes to produce highly toxic fumes of cyanides.

To Fight Fire: CO_2, non-alkaline dry chemical, foam.

HYDROCYANIC ACID SOLUTION. See hydrocyanic acid.

HYDROCYANIC ACID, UNSTABILIZED. See hydrocyanic acid.

HYDROFERROCYANIC ACID. White crystalline material. $H_4Fe(CN)_6$, mw: 216.
THR = See ferrocyanides.

HYDROFLUORIC ACID. Syns: *hydrogen fluoride, fluorohydric acid.* Clear, colorless, fuming corrosive liquid or gas. HF, mw: 20.01, mp: −83.1°, bp: 19.54°, d: 0.901 g/liter (gas); 0.699 @ 22° (liquid), vap. press: 400 mm @ 2.5°.
Acute tox data: Inhal TC_{LO} (human) = 32 ppm for 1 min; inhal TC_{LO} (human) = 110 ppm for 1 min; inhal LC_{50} (rat) = 1276 ppm for 1 hr. [*3*]
THR = HIGH irr to skin, eyes and mu mem and via oral route. [*84*] It is extremely irr and corrosive to the skin and mu mem. Inhal of the vapor may cause ulcers of the upper respiratory tract. Conc of 50–250 ppm are dangerous, even for brief exposures. Hydrofluoric acid produces severe skin burns which are slow in healing. The subcutaneous tissues may be affected, becoming blanched and bloodless. Gangrene of the affected areas may follow. See also fluorides. It is a common air contaminant. [*84, 57, 85, 86*] Violent reaction with As_2O_3, P_2O_5, acetic anhydride, 2-amino ethanol, NH_4OH, $HBiO_3$, CaO, chlorosulfonic acid, ethylene diamine, ethylene imine, F_2, (HNO_3 + lactic acid), oleum, β-propiolactone, propylene oxide, Na, NaOH, H_2SO_4, vinyl acetate. [*19*]
Disaster Hazard: Dangerous; when heated, emits highly corrosive fumes of fluorides; will react with water or steam to produce toxic and corrosive fumes.

HYDROFLUORIC ACID, ANHYDROUS. See hydrofluoric acid.

HYDROFLUORIC AND SULFURIC ACIDS, MIXTURES. See components.

HYDROFLUOSILICIC ACID. Syns: *fluosilicic acid, silicofluoric acid.* Transparent, colorless, fuming liquid. H_2SiF_6, mw: 144.08, bp: decomp.
Acute tox data: Oral LD_{LO} (guinea pig) = 200 mg/kg; sc LD_{LO} (guinea pig) = 250 mg/kg. [*3*]
THR = HIGH irr to skin, eyes and mu mem and via inhal route.
Disaster Hazard: Dangerous; when heated to decomp, emits highly toxic and corrosive fumes of fluorides; will react with water or steam to produce toxic and corrosive fumes.

HYDROFURAMIDE. Syns: *n,n′-difurfurylidene-2-furan methane diamine, furfuramide.* Light brown crystals. $(C_4H_3OCH)_3N_2$, mw: 268.26, mp: 117°, bp: 250° decomp.
Acute tox data: Oral LD_{50} (rat) = 40 mg/kg; oral LD_{50} (mouse) = 950 mg/kg. [*3*]
THR = HIGH irr to skin, eyes and mu mem and via oral route. A component of fungicides. Causes intense pulmonary irr and reported to cause liver and kidney damage. See also amines and amides.
Disaster Hazard: Slight; when heated to decomp, emits toxic fumes of NO_x.

HYDROGEN. Colorless gas. H_2, mw: 2.0162, mp: −259.18°, bp: −252.8°, lel = 4.1%, uel = 74.2%, d: 0.0899 g/liter, autoign. temp.: 752°F, vap. d: 0.069.
THR = Practically NONE.
Radiation Hazard: For permissible levels, see Section 5A, Table 5A.5. Artificial and natural isotope ^3H (tritium), $T\frac{1}{2}$ = 12.3y, decays to stable ^3He via β's of 0.019 MeV. Tritium occurs naturally as a result of cosmic ray bombardment of ^2H.
Fire Hazard: Highly dangerous, when exposed to heat, flame or oxidizers.
To Fight Fire: Stop flow of gas.
Explosion Hazard: Severe, when exposed to heat or flame. Violent reaction with (air + Pt), Br_2, Cl_2, ClF_3, (dioxane + Ni), F_2, Li, (Mg + $CaCO_3$), nitroanisole, NF_3, OF_2 (Pd + isopropyl alcohol), 3-methyl-2-penten-4-yn-1-ol. [*19*]
Disaster Hazard: Dangerous; can react vigorously with oxidizing materials.

HYDROGEN ANTIMONIDE. See antimony hydride.

HYDROGEN ARSENIDE (GAS). See arsine.

HYDROGEN AZIDE. See hydrazoic acid.

HYDROGEN BISMUTHIDE. Syn: *bismuthine.* Liquid. H_3BI, mw: 212.02, bp: 22°.
THR = See bismuth compounds and hydrides.

HYDROGEN BROMIDE. See hydrobromic acid.

HYDROGEN BROMIDE (CONSTANT BOILING MIXTURE). Colorless liquid. HBr(47%) + H_2O; mp: −11°, bp: 126°, d: 1.49.
THR = See hydrobromic acid.

HYDROGEN BROMIDE, HYDRATED. Colorless liquid. $HBr \cdot H_2O$, mw: 98.94, bp: −3.3° to −15°, d: 1.78.
THR = See hydrobromic acid.

HYDROGEN CARBOXYLIC ACID. See formic acid.

HYDROGEN CHLORIDE. See hydrochloric acid.

HYDROGEN CYANIDE. See hydrocyanic acid.

HYDROGEN DIOXIDE. See hydrogen peroxide.

HYDROGEN DIPHOSPHIDE. HP_2, mw: 63.
THR = Ignites @ 100° or by impact. [*19*] HIGH via oral route. See also phosphides.

HYDROGEN DISULFIDE. Yellow oil. H_2S_2, mw: 66.15, mp: −89.7°, bp: 74.5°, d: 1.376, vap. press: 100 mm @ 22.0°.
THR = HIGH. See hydrogen sulfide and sulfides.

HYDROGEN FLUORIDE. See hydrofluoric acid.

HYDROGEN IODIDE. See hydroiodic acid.

HYDROGEN NITRATE. See nitric acid.

HYDROGEN PENTASULFIDE. Clear yellow oil. H_2S_5, mw: 162.35, mp: decomp, d: 1.67 @ 16°.
THR = HIGH. See hydrogen sulfide and sulfides.

HYDROGEN PEROXIDE. Syns: *hydrogen dioxide, t-stuff.* Colorless heavy liquid, or, at low temp., a crystalline solid. H_2O_2, mw: 34.016, bp: 158°, d: 1.71 @ −20°, 1.46 @ 0°, vap. press: 1mm @ 15.3°, mp: −2°.
THR = HIGH irr to skin, eyes and mu mem and via oral and inhal routes. A very powerful oxidizer. Pure H_2O_2, its solutions, vapors and mists are irr to body tissue. This irr can vary from mild to severe depending upon the conc of H_2O_2. For instance solutions of H_2O_2 of 35 wt% and over can easily cause blistering of the skin. Irr caused by H_2O_2 which does not subside upon flushing of the affected part with water should be treated by a physician. The eyes are particularly sensitive to irr by this material. It is used as a general purpose food additive; it is a substance which migrates to food from packaging materials. [*109*] It is a common air contaminant.
Fire Hazard: Dangerous by chemical reaction with flammable materials. H_2O_2 is a powerful oxidizer, particularly in the concentrated state. It is important to keep containers of this material covered because (1) uncovered containers are much more prone to react with flam vapors, gases, etc.; (2) because if uncovered, the water from an H_2O_2 solution can evaporate, concentrating the material and thus increasing the fire hazard of the remainder.

For instance, solutions of H_2O_2 of conc in excess of 65 wt% heat up spont when decomp to $H_2O + \frac{1}{2}O_2$. Thus 90 wt% solutions, when caused to decompose rapidly due to the introduction of a catalytic decomposition agent, can get quite hot and perhaps start fires.
Explosion Hazard: Severe, when highly conc or pure H_2O_2 is exposed to heat, mechanical impact, detonation of a blasting cap, or caused to decomp catalytically by metals, or on contact with acetic acid, acetic anhydride, acetone, (alcohols + H_2SO_4), Sb_2S_3, As_2S_3, *tert*-butyl alcohol, cellulose, charcoal, (Cl_2 + KOH), chlorosulfonic acid, CuS, ethanol, FeS, (formic acid + organic matter), H_2Se, hydrazine, (ketones + HNO_3), PbO_2, PbO, PbS, MnO_2, HgO, Hg_2O, MoS_2, HNO_3, organic matter, $KMnO_4$, $NaIO_3$, thiodiglycol, uns-dimethyl hydrazine. [*19*]

Although many mixtures of H_2O_2 and organic materials do not explode upon contact, the result-and combination is detonatable either upon catching fire or by impact.

The detonation velocity of aqueous solutions of H_2O_2 has been found to be about 6500 m/sec. for solutions of between 96 wt% and 100 wt% H_2O_2.

Another source of H_2O_2 explosions is from sealing the material in strong containers. Under such conditions even gradual decomposition of H_2O_2 to $H_2O + \frac{1}{2}O_2$ can cause large pressures to build up in the containers which may then burst explosively.
Disaster Hazard: Highly dangerous because when heated, or shocked or contaminated, the concentrated material can explode or start fires.

HYDROGEN PHOSPHIDE (DI). Colorless liquid. H_4P_2, mw: 65.99, mp: −10°, bp: 57.5° @ 735 mm, d: 1.012.
THR = HIGH. See phosphine and phosphides.

HYDROGEN PHOSPHIDE. See phosphine.

HYDROGEN PHOSPHIDE (POLYMER) (DI). Yellow solid. $(H_2P_4)_3$, mw: 377.81, mp: ignites @ 160°, bp: decomp, d: 1.83 @ 19°
THR = See phosphine and phosphides.

HYDROGEN SELENIDE. Colorless gas. H_2Se, mw: 80.98, mp: −64°, bp: −41.4°, d: 3.614 g/liter (gas); 2.12 @ −42° (liquid), vap. press: 10 atm @ 23.4°.
Acute tox data: Inhal TD_{LO} (human) = 0.2 ppm ⟶ CNS symptoms; inhal LC_{50} (guinea pig) = 1 mg/m^3 for 8 hrs. [*3*]
THR = VERY HIGH irr to skin, eyes and mu mem and via inhal route. An allergen. This material is a hazardous compound of selenium which can cause damage to the lungs and liver as well as conjunctivitis. It has been found that repeated 8 hr exposures to conc of 0.3 ppm prove fatal to guinea pigs by causing a pneumonitis, as well as injury to the liver and spleen. Conc of 0.3 ppm are readily detected by odor, but there is no noticeable irr effect at that level. Conc of 1.5 ppm or higher are strongly irr to the eyes and nasal passages.

As in the case of hydrogen sulfide, the odor of hydrogen selenide in concentrations below 1 ppm disappears rapidly because of olfactory fatigue. Although the odor and irr effects are both useful to an experienced investigator for estimating the conc, they do not offer a dependable warning to workmen who may be exposed to gradually increasing amounts and therefore become used to it. Due to its extreme toxicity and irr effects, it seldom is allowed to reach a conc in which it is flam in air.

Very few data are available on possible chronic effects of this material, but it is logical to assume that when the conc of this gas is low enough to avoid the irr effects, only the systemic effects will be noticeable.

Fire Hazard: Dangerous; will react vigorously with powerful oxidizing agents, such as H_2O_2, HNO_3.

Explosion Hazard: Dangerous; forms explosive mixtures with air. See also hydrides.

Disaster Hazard: Highly dangerous; keep away from heat and open flame.

HYDROGEN SULFIDE. Syn: *sulfuretted hydrogen.* Colorless, flam gas, offensive odor. H_2S, mw: 34.08, mp: −85.5°, bp: −60.4°, lel = 4%, uel = 44%, autoign. temp.: 500°F, d: 1.539 g/liter @ 0°, vap. press: 20 atm @ 25.5°, vap. d: 1.189.

Acute tox data: Inhal LC_{LO} (human) = 600 ppm for ½ hr; inhal LC_{50} (rat) = 713 ppm for 1 hr; inhal LC_{50} (mouse) = 673 ppm for 1 hr. [*3*]

THR = HIGH irr to eyes and mu mem and via inhal route. Hydrogen sulfide is both an irr and an asphyxiant. Low conc of 20–150 ppm cause irr of the eyes; slightly higher conc may cause irr of the upper respiratory tract, and if exposure is prolonged, pulmonary edema may result. The irr action has been explained on the basis that H_2S combines with the alkali present in moist surface tissues to form sodium sulfide, a caustic.

With higher conc the action of the gas on the nervous system becomes more prominent, and a 30-min exposure to 500 ppm results in headache, dizziness, excitement, staggering gait, diarrhea and dysuria, followed sometimes by bronchitis or bronchopneumonia. The action on the nervous system is, with small amounts, one of depression; in larger amounts, it stimulates, and with very high amounts the respiratory center is paralyzed. Exposures of 800–1000 ppm may be fatal in 30 min, and high conc are instantly fatal. Fatal hydrogen sulfide poisoning may occur even more rapidly than that following exposure to a similar concentration of hydrogen cyanide. H_2S does not combine with the hemoglobin of the blood; its asphyxiant action is due to paralysis of the respiratory center.

With repeated exposures to low concentrations, conjunctivitis, photophobia, corneal bullae, tearing, pain and blurred vision are the commonest findings. High conc may cause rhinitis, bronchitis, and occasionally pulmonary edema. Exposure to very high conc results in immediate death. Chronic poisoning results in headache, inflammation of the conjunctivae and eyelids, digestive disturbances, loss of weight and general debility. It is a common air contaminant. [*83*, *59*]

Fire Hazard: Very dangerous, when exposed to heat, flame or oxidizers.

Explosion Hazard: Mod, when exposed to heat or flame. Reacts violently with Na_2O_2, NI_3, NCl_3, NF_3, OF_2, HNO_3, PbO_2, F_2, Cu, CrO_3, ClF_3, ClO, BrF_5, acetaldehyde, ($BaO + Hg_2O$ + air), (BaO + NiO + air), hydrated iron oxide, phenyl diazonium chloride, (NaOH + CaO + air), Na. [*19*]

Disaster Hazard: Highly dangerous; when heated to decomp, emits highly toxic fumes of oxides of sulfur; can react vigorously with oxidizing materials.

To Fight Fire: Stop flow of gas.

HYDROGEN TELLURIDE. Colorless gas or yellow needles. H_2Te, mw: 129.63, mp: −49.0°, bp: −2.0°, d: 5.81 g/liter; (liquid): 2.57 @ −20°.

THR = See tellurium compounds and hydrides. Reacts violently with HNO_3. [*19*]

HYDROGEN TRISULFIDE. Bright yellow liquid. H_2S_3, mw: 98.21, mp: −52°, bp: decomp @ 90°, d: 1.496 @ 15°.

THR = HIGH. See hydrogen sulfide. Violent reaction with amyl alcohol, Fe_3O_4, PbO, $KMnO_4$, SnO_2. [*19*]

HYDROHYDRASTININE. White crystalline alkaloid. $C_{11}H_{13}NO_2$, mw: 191.22, mp: 66°, bp: 303° @ 752 mm.

THR = HIGH via oral and inhal routes.

Disaster Hazard: Dangerous; when heated to decomp, emits highly toxic fumes of NO_x.

HYDROL. Syn: *4-diallylamino-3,5-dimethyl phenyl-n-methyl carbamate.* A powder, insol in water, sol in alcohol and benzene. $C_{16}H_{23}O_2N_2$, mw: 275.1.

Acute tox data: Oral LD_{50} (rat) = 89 mg/kg; oral LD_{50} (wild birds) = 13 mg/kg. [*3*]

THR = HIGH via oral route. An insecticide. See also carbamates.

HYDROLITH. See calcium hydride.

HYDRONIUM PERCHLORATE. See perchloric acid, monohydrate.

6-HYDROPEROXY-4-CHOLESTEN-3-ONE. $C_{27}H_{46}O_2$, mw: 402.7.

THR = An exper neo. [*3*]

1-HYDROPEROXY-3-CYCLOHEXENE. $C_6H_9O_2$, mw: 113.2.

THR = An exper neo. [*3*]

4-HYDROPEROXY-4-VINYL-1-CYCLOHEXENE. $C_8H_{12}O_2$, mw: 132.2.

THR = An exper carc. [*3*]

HYDROQUINOL. See *p*-hydroquinone.

***p*-HYDROQUINONE.** Syns: *1,4-benzenediol, quinol, hydroquinol.* Colorless hexagonal prisms. $C_6H_4(OH)_2$, mw: 110.1, mp: 170.5°, bp: 286.2°, flash p: 329°F (CC), d: 1.358 @ 20°/4°, autoign. temp.: 960°F (CC), vap. press: 1 mm @ 132.4°, vap. d: 3.81.

Acute tox data: Oral LD_{50} (rat) = 320 mg/kg; ip LD_{50} (rat) = 170 mg/kg; iv LD_{50} (rat) = 115 mg/kg; oral LD_{50} (pigeon) = 300 mg/kg; sc LD_{50} (mouse) = 190 mg/kg. [*3*]

THR = HIGH via oral, ip, iv and sc routes. An active allergen and a strong irr. An exper neo [*3*] via imp route. [*81*] Absorption of this material by tissues can cause symptoms of illness, which resemble those induced by its *o* or *m* isomers. For instance, the ingestion of 1 g by an adult or a smaller quantity by a child may induce tinnitis, nausea, dizziness, a sensation of suffocation, an increased rate of respiration, vomiting, pallor, muscular twitchings, headache, dyspnea, cyanosis, delirium, and collapse. The literature contains reports of fatal cases which have been caused by the ingestion of 5–12 g. Cases of dermatitis have resulted from skin contact with this material, and have also followed the application of an antiseptic oil which apparently contained traces of hydroquinone added as an antioxidant. The report also contains cases of keratitis and discoloration of the conjunctiva among personnel exposed to this material in conc ranging from 10 to 30 mg of the vapor or dust per cubic meter of air. It is considered to be more toxic than phenol. The inhal of vapors of this material, particularly when liberated at high temp., must be avoided.

Fire Hazard: Slight, when exposed to heat or flame; can react with oxidizing materials. Violent reaction with NaOH. [*19*]

Explosion Hazard: Slight, when exposed to heat.

Treatment and Antidotes: When personnel working with this material exhibit some of the symptoms listed above, they should immediately be removed to fresh air. If the symptoms do not subside quickly, consult a physician. In cases of dermatitis due to this material, removal from exposure will quickly clear up the symptoms. If this material accidentally comes into contact with the skin, it should be removed at once and the affected area washed with plenty of soap and water.

To Fight Fire: Water, CO_2, dry chemical.

HYDROQUINONE (IN OIL). See antiseptic oil.

HYDROQUINONE DIMETHYL ETHER. Syn: *1,4-dimethoxybenzene*. Colorless leaflets, odor of sweet clover. $C_6H_4(OCH_3)_2$, mw: 138.16, mp: 56°, bp: 212.6, d: 1.053 @ 55°/55°.

Acute tox data: Ip LD_{50} (rat) = 1100 mg/kg. [*3*]

THR = MOD via ip route. A mild allergen and irr.

Fire Hazard: Mod, when exposed to heat or flame; can react with oxidizing materials.

Explosion Hazard: Details U. See also ethers.

HYDROQUINONE MONOBENZYL ETHER. See benzyl hydroquinone.

HYDROQUINONE MONOMETHYL ETHER. Syns: *4-methoxy phenol, hydroxy anisole*. White, waxy solid. $CH_3OC_6H_4OH$, mw: 124.13, mp: 52.5°, bp: 246°, d: 1.55 @ 20°/20°.

Acute tox data: Oral LD_{50} (rat) = 1600 mg/kg; ip LD_{50} (rat) = 725 mg/kg. [*3*]

THR = MOD via oral and ip routes.

Fire Hazard: Slight, when exposed to heat or flame; can react with oxidizing materials.

Explosion Hazard: Details U. See also ethers.

HYDROQUINONE MUSTARD. Syn: *2,5-bis[bis-(2-chloroethyl)] amino methyl hydroquinone*. $C_{16}H_{24}O_2N_2Cl_4$, mw: 418.2.

THR = HIGH. An exper carc. [*3, 23*] A powerful irr.

HYDROUSMAGNESIUM SILICATE. See talc.

***o*-HYDROXYACETANILIDE.** $C_8H_9O_2N$, mw: 151.2.

Acute tox data: Oral LD_{50} (rat) = 4960 mg/kg. [*3*]

THR = MOD via oral route. An exper carc. [*23*]

HYDROXYACETIC ACID. See glycolic acid.

***o*-HYDROXYACETOPHENONE.** Greenish-yellow liquid, highly refractive, minty odor. $C_6H_4(OH)COCH_3$, mw: 136.14, mp: 95°, vap. d: 4.69.

Acute tox data: Ip LD_{LO} (mouse) = 100 mg/kg. [*3*]

THR = HIGH via ip route.

***n*-HYDROXY-2-ACETYL AMINO-FLUORENE.** See *n*-fluoren-2-yl acetylhydroxamic acid.

N-HYDROXYADENINE. $C_5H_4ON_5$, mw: 150.1.

THR = An exper teratogen. [*3*]

2-HYDROXY ADIPALDEHYDE(25% AQUEOUS SOLUTION). Liquid. $OHC(CH_2)_3CH(OH)CHO$, mw: 130.1, fp: −3.5°, d: 1.066 @ 20°/20°, vap. press: 17 mm @ 20°.

Acute tox data: Inhal LC_{LO} (rat) = 3000 ppm for 4 hrs; inhal LC_{LO} (mouse) = 1500 ppm for 14 hrs. [*3*]

THR = MOD via inhal route.

Disaster Hazard: Slight; when heated, emits acrid fumes.

***β*-HYDROXYALANINE.** See serine.

N-HYDROXY-3-AMINO FLUORENE. $C_{13}H_{11}ON$, mw: 197.3.

THR = An exper neo. [*3*]

N-HYDROXY-2-AMINO FLUORENE.

THR = An exper neo. [*3*]

4-(HYDROXYAMINO)-5-METHYL QUINOLINE-1-OXIDE. $C_{10}H_{10}O_2N_2$, mw: 190.2.

THR = An exper neo. [*3*]

4-(HYDROXYAMINO)-6-METHYL QUINOLINE-1-OXIDE.

THR = An exper neo. [*3*]

4-(HYDROXYAMINO)-7-METHYL QUINOLINE-1-OXIDE.
THR = An exper neo. [*3*]

4-(HYDROXYAMINO)-8-METHYL QUINOLINE-1-OXIDE.
THR = An exper neo. [*3*]

4-(HYDROXYAMINO)-6-NITROQUINOLINE OXIDE. $C_9H_7O_4N_3$, mw: 221.2.
THR = An exper neo. [*3*]

4-(HYDROXYAMINO)-7-NITROQUINOLINE-1-OXIDE.
THR = An exper neo. [*3*]

4-(HYDROXYAMINO)QUINOLINE-1-OXIDE. $C_9H_8O_2N_2$, mw: 176.2.
THR = An exper neo and carc. [*3*]

4-(HYDROXYAMINO)QUINOLINE-1-OXIDE HYDROCHLORIDE. $C_9H_8O_2N_2 \cdot HCl$, mw: 212.7.
THR = An exper neo and carc. [*3*]

N-HYDROXY-4-AMINOSTILBENE. $C_{14}H_{13}ON$, mw: 211.3.
THR = An exper carc. [*3*, *23*] via sc route.

***p*-HYDROXYANILINE.** See *p*-amino phenol.

HYDROXYANISOLE. See hydroquinone monomethyl ether.

β-HYDROXYANTHRAQUINONE.
THR = A MILD allergen.
Disaster Hazard: Slight; when heated, emits acrid fumes.

3-HYDROXY-β-AZAXANTHINE. $C_4H_3O_3N_5$, mw: 169.1.
THR = An exper neo. [*3*]

***o*-HYDROXYBENZAMIDE.** See salicylamide.

3-HYDROXY-1,2-BENZANTHRACENE. $C_{18}H_{12}O$, mw: 244.3.
THR = An exper neo. [*3*]

5-HYDROXYBENZ(a)ANTHRACENE.
THR = An exper carc. [*23*]

N-HYDROXYBENZENE SULFONALIDE. $C_{12}H_{11}O_3NS$, mw: 249.3.
THR = An exper neo. [*3*]

3-HYDROXYBENZ(a)PYRENE.
THR = An exper carc. [*23*]

5-HYDROXYBENZ(a)PYRENE.
THR = An exper carc. [*23*]

6-HYDROXYBENZ(a)PYRENE.
THR = An exper carc. [*23*]

***p*-HYDROXYBENZOATE, BUTYL ESTER.** $C_{11}H_{14}O_3$, mw: 194.3.
Acute tox data: Oral LD_{50} (mouse) = 5000 mg/kg; ip LD_{50} (mouse) = 230 mg/kg. [*3*]
THR = HIGH via ip and MOD via oral routes. An allergen.

***p*-HYDROXYBENZOATE, ETHYL ESTER.** $C_9H_{10}O_3$, mw: 166.2.
Acute tox data: Oral LD_{50} (dog) = 5000 mg/kg; oral LD_{50} (rabbit) = 5000 mg/kg. [*3*]
THR = MOD via oral route. An allergen.

***p*-HYDROXYBENZOATE, METHYL ESTER.** $C_8H_8O_3$, mw: 152.3.
Acute tox data: Oral LD_{50} (dog) = 3000 mg/kg; ip LD_{50} (mouse) = 960 mg/kg. [*3*]
THR = MOD via oral and ip routes. An allergen.

***p*-HYDROXYBENZOATE, PROPYL ESTER.** $C_{10}H_{12}O_3$, mw: 180.2.
Acute tox data: Oral LD_{50} (dog) = 6000 mg/kg; ip LD_{50} (mouse) = 200 mg/kg. [*3*]
THR = HIGH via ip and MOD to LOW via oral routes. An allergen.

***o*-HYDROXYBENZOIC ACID.** See salicylic acid.

β-HYDROXY-3,4-BENZ PYRENE. $C_{20}H_{12}O$, mw: 268.3.
THR = An exper neo and carc. [*3*]

3-HYDROXY BUTANAL. See aldol.

3-HYDROXY-2-BUTANONE. See acetoin.

β-HYDROXYBUTYRALDEHYDE. See aldol.

6-HYDROXYCHOLEST-4-EN-3-ONE + CHOLEST-5-EN-3-α-OL.
THR = An exper neo. [*3*]

3-HYDROXYCHRYSENE.
THR = An exper carc. [*23*]

4-HYDROXY-3,5-DI-*tert*-BUTYL TOLUENE. See di-*tert*-butyl-*p*-cresol.

4-HYDROXY-3,5-DIIODOBENZONITRILE. See ioxynil.

HYDROXYDIMETHYLARSINE OXIDE. See cacodylic acid.

1-HYDROXY-3,5-DIMETHYL BENZENE. See 3,5-xylenol.

5-HYDROXY-1,3-DIMETHYL BENZENE. See 3,5-xylenol.

3-HYDROXY-N,N-DIMETHYL-*cis*-CROTONAMIDEDIMETHYL PHOSPHATE. Syn: *bidrin*. Brown liquid, ester odor, miscible with water and xylene, very slightly sol in kerosene and diesel fuel. $(CH_3O)_2P(O)OC(CH_3)CHC(O)N(CH_3)_2$, mw: 237, bp: 400°.
Acute tox data: Oral LD_{50} (rat) = 16 mg/kg; dermal LD_{50} (rat) = 42 mg/kg; sc LD_{50} (rat) = 8.1 mg/kg; oral LD_{50} (wild birds) = 2 mg/kg; oral LD_{50} (duck) = 4 mg/kg. [*3*]

THR = HIGH via oral, dermal, sc and probably inhal routes. Resembles other organic phosphate pesticides. See parathion.
Disaster Hazard: Dangerous. See PO_x.

1-HYDROXY-2,3-DINITROBENZENE. See 2,3-dinitrophenol.

4-HYDROXY-3,5-DINITROBENZENE ARSONIC ACID. 1,3,4,5-$C_6H_2(NO_2)_2(OH)H_2AsO_3$, mw: 308.
THR = A HIGH explosive. [*19*]

***p*-HYDROXYDIPHENYLAMINE.** Syn: *anilinophenol.* Gray solid leaflets. $C_6H_5NHC_6H_4OH$, mw: 185.2, mp: 70°, bp: 330°.
THR = Details U. See also amines.
Disaster Hazard: Dangerous; when heated to decomp, emits highly toxic fumes of aniline.

***p*-HYDROXYDIPHENYLAMINE ISOPROPYL ETHER.** $C_{15}H_{17}O_2N$, mw: 243.3.
THR = An exper neo. [*3*]

(2-HYDROXYDIPHENYL)METHANE. See *o*-benzylphenol.

(4-HYDROXYDIPHENYL)METHANE. See *p*-benzylphenol.

HYDROXYETHANOIC ACID. See glycolic acid.

N-β-HYDROXYETHYL ACETAMIDE. Syn: *n-acetyl ethanolamine.* Brown viscous liquid.
$CH_3CONHC_2H_4OH$, mw: 103.12, bp: 151°, flash p: 355°F (OC), fp: 15.8°, d: 1.122 @ 20°/20°, autoign. temp.: 860°F.
Acute tox data: Oral LD_{50} (rat) = 28,000 mg/kg. [*3*]
THR = Practically NONE via oral route.
Fire Hazard: Slight, when exposed to heat or flame.
To Fight Fire: Alcohol foam, CO_2, dry chemical.

4-(2-HYDROXYETHYL AMINO)-2-(5-NITRO-2-THIENYL)QUINAZOLINE. $C_{14}H_{12}O_3N_4S$, mw: 316.4.
THR = An exper carc to rats via oral route. [*103*]

β-HYDROXYETHYL ANILINE. See 2-anilinoethanol.

7-HYDROXYETHYL BENZ(a)ANTHRACENE.
THR = An exper carc. [*23*]

β-HYDROXYETHYL CARBAMATE. $C_3H_7O_3N$, mw: 105.1.
THR = An exper teratogen. [*3*]

N-(2-HYDROXYETHYL)CYCLOHEXYLAMINE. Sol in water. $C_6H_{11}NHC_2H_4OH$, mw: 143, mp: 97–102°F, flash p: 249°F (OC).
Acute tox data: Oral LD_{50} (rat) = 38,000 mg/kg. [*3*]
THR = Practically NONE via oral route.
Fire Hazard: Slight, when exposed to heat or flame.
To Fight Fire: Alcohol foam.

HYDROXYETHYLENE DIAMINE. See aminoethyl ethanolamine.

(2-HYDROXYETHYL)ETHYLENE DIAMINE. See aminoethyl ethanolamine.

2-HYDROXYETHYL-5-ETHYL PYRIDINE. $C_9H_{13}ON$, mw: 151.2.
THR = U. Limited animal exper suggest LOW. See also pyridine.

1-HYDROXYETHYL-1,2-HEPTADECENYL GLYOXALIDINE. See amine 220.

1-HYDROXYETHYL-2-HEPTADECENYLIMIDAZOLINE. See amine 220.

2-HYDROXY-3-ETHYLHEPTANOIC ACID. $CH_3CH(OH)CH(C_2H_5)(CH_2)_3COOH$.
Acute tox data: Oral LD_{50} (rat) = 3400 mg/kg; dermal LD_{LO} (rabbit) = 1780 mg/kg. [*3*]
THR = MOD via oral and dermal routes.

β-HYDROXYETHYL HYDRAZINE. Colorless, slightly viscous liquid. $C_2H_8N_2$, mw: 76.1, mp: 70°, bp: 145°–153° @ 25 mm, flash p: 224°F, vap. d: 2.63.
THR = HIGH. See hydrazine. An exper carc. [*3*]
Fire Hazard: Mod, when exposed to heat or flame; can react with oxidizing materials.
To Fight Fire: Foam, CO_2, dry chemical.

2-HYDROXYETHYL MERCAPTAN. See mercaptoethanol.

3-(2-HYDROXYETHYL)-3-METHYL-1-PHENYL TRIAZENE. $C_9H_{13}ON_3$, mw: 179.3.
THR = An exper carc. [*3*]

7-HYDROXYETHYL-12-METHYL BENZ(a)ANTHRACENE.
THR = An exper carc. [*23*]

N-HYDROXYETHYL MORPHOLINE. See morpholine ethanol.

1-(2-HYDROXYETHYL)-1-NITROSOUREA. $C_3H_7O_3N_3$, mw: 133.1.
THR = An exper carc. [*3*]

1-(2-HYDROXYETHYL)-3-(5-NITROFURYLIDENE)AMINO-2-IMIDAZOLINE. $C_{10}H_{12}O_5N_2$, mw: 240.2.
Acute tox data: Ip LD_{LO} (mouse) = 100 mg/kg. [*3*]
THR = HIGH via ip route. An exper carc. [*23*]

1-(2-HYDROXYETHYL)-2-METHYL-5-NITROIMIDAZOLE. See metronidazole.

N-(2-HYDROXYETHYL)PHENYLAMINE. See 2-anilinoethanol.

3-(2-HYDROXYETHYL)-3-METHYL-1-PHENYL TRIAZENE. $C_9H_{13}ON_3$, mw: 179.3.
THR = An exper carc. [*3*]

N,β-HYDROXYETHYL PIPERAZINE. Light colored liquid. $HOCH_2CH_2N(CH_2)_4NH$, mw: 130.2, d: 1.10610, bp: 240°, flash p: 255°F (OC), vap. d: 4.5.

Acute tox data: Oral LD_{50} (rat) = 4920 mg/kg. [*3*]
THR = MOD via oral route.
Fire Hazard: Slight, when exposed to heat or flame.
To Fight Fire: Foam, alcohol foam.

***n*-HYDROXYETHYL PROPYLENEDIAMINE.** Liquid. $CH_3CH(NHC_2H_4OH)CH_2NH_2$, mw: 118.18, mp: −65°, bp: 240°, flash p: 260° F. (OC), d: 0.9938, vap. d: 4.07.
Acute tox data: Oral LD_{50} (rat) = 4920 mg/kg. [*3*]
THR = MOD via oral route.
Fire Hazard: Slight, when exposed to heat or flame; can react with oxidizing materials.
To Fight Fire: Alcohol foam, CO_2, dry chemical.

N-HYDROXYETHYL PYRROLIDINE. Colorless liquid. $C_4H_8NCH_2CH_2OH$, mw: 115.2, bp: 187°, fp: −75°, flash p: 160° F, d: 0.9800 @ 20°.
THR = MOD irr to skin, eyes and mu mem and via oral route.
Fire Hazard: Mod, when exposed to heat sources, sparks or open flame.
Disaster Hazard: Dangerous; when heated to decomp, emits highly toxic fumes.
To Fight Fire: Foam, spray, dry chemical or CO_2.

HYDROXYETHYL SULFONIC ACID. Syn: *isethonic acid.* $C_2H_6O_4S$, mw: 126.1.
Acute tox data: Ip LD_{LO} (mouse) = 250 mg/kg. [*3*]
THR = HIGH via ip route and irr to skin, eyes and mu mem.
Disaster Hazard: Dangerous, see sulfonates.

2-HYDROXYETHYL TRIMETHYLAMMONIUM CHLORIDE CARBAMATE. $C_6H_{15}O_2N_2Cl$, mw: 182.7.
Acute tox data: Oral LD_{50} (rat) = 40 mg/kg; sc LD_{50} (rat) = 4 mg/kg; iv LD_{50} (rat) = 0.1 mg/kg. [*3*]
THR = HIGH via oral, sc and iv routes.

N-(9-HYDROXYFLUOREN-2-YL)ACETAMIDE. $C_{15}H_{13}O_2N$, mw: 239.3.
THR = LOW to MOD to rats on an acute oral basis. An exper neo. [*3, 23*]

2-(7-HYDROXYFLUOREN-2-YL)ACETAMIDE. $C_{15}H_{13}O_2N$, mw: 239.3.
THR = An exper neo via oral route. [*103*]

N-(7-HYDROXYFLUOREN-2-YL)ACETOHYDROXAMIC ACID. $C_{15}H_{13}O_3N$, mw: 255.3.
THR = An exper carc. [*3*]

3-HYDROXYGUANINE.
THR = An exper carc. [*23*]

HYDROXYHEPTYL PEROXIDE. Syn: *"Luperco HDB."* Paste with dibutyl phthalate, thick, white, slightly water-sol, sol in common organic solvents.
THR = An organic peroxide. See also dibutyl phthalate. See peroxides, organic.

α-HYDROXYISOBUTYRONITRILE. See acetone cyanhydrin.

3-HYDROXYKYNURENINE. $C_{10}H_{12}O_4N_2$, mw: 224.2.
THR = An exper neo. [*3*]

3-HYDROXY-L-KYNURENINE. $C_{10}H_{12}O_4N_2$, mw: 224.2.
THR = An exper neo. [*3*]

HYDROXYLAMINE. Syn: *oxammonium.* Colorless liquid or white needles. NH_2OH, mw: 33.02, mp: 34.0°, bp: 110.0°, flash p: explodes at 265° F, d: 1.227, vap. press: 10 mm @ 47.2°.
Acute tox data: Ip LD_{50} (rat) = 59 mg/kg; sc LD_{50} (rat) = 29 mg/kg. [*3*]
THR = HIGH via ip and sc routes. A MOD irr to skin, eyes and mu mem. Locally it is irr, and systemically, it can cause methemoglobinemia. An exper teratogen. [*3*]
Explosion Hazard: Dangerous, when exposed to heat or open flame. Can react violently with BaO_2, BaO, Cl_2, $CuSO_4$, PbO_2, PCl_5, PCl_3, $K_2Cr_2O_7$, $KMnO_4$, Na, Zn. [*19*]
Fire Hazard: Dangerous, via heat, flame and oxidizers.
Disaster Hazard: Dangerous upon exposure to heat or flame.

HYDROXYLAMINE FLUOGERMANATE. Monoclinic prisms. $2NH_2OHH_2GeF_6 \cdot 2H_2O$, mw: 290.71, d: 2.229 @ 25°/25°.
THR = See fluogermanates and hydroxylamine.

HYDROXYLAMINE FLUOSILICATE. Scales. $2NH_2OH \cdot H_2SiF_6 \cdot 2H_2O$, mw: 246.17.
THR = See fluosilicates, hydroxylamine, fluorides.

HYDROXYLAMINE HYDROCHLORIDE. Syn: *hydroxyl ammonium chloride.* Colorless hygroscopic crystals. $NH_2OH \cdot HCl$, mw: 69.50, mp: 151°, bp: decomp, d: 1.67 @ 17°.
Acute tox data: Oral LD_{50} (mouse) = 400 mg/kg; ip LD_{50} (mouse) = 100 mg/kg. [*3*]
THR = HIGH via oral and ip routes. Very irr to skin, eyes and mu mem.

HYDROXYLAMINE HYPOPHOSPHITE. $NH_2OHPH_2O_2$, mw: 98.
THR = Can detonate @ 100°. [*19*] See also hypophosphites.

HYDROXYLAMINE NITRATE. White crystals. $NH_2OH \cdot HNO_3$, mw: 96.05, mp: 48°, bp: 100° (decomp).
THR = See hydroxylamine and nitrates.

HYDROXYLAMINE SULFATE. Syn: *hydroxyl ammonium sulfate.* A crystalline material. $(NH_2OH)_2 \cdot H_2SO_4$, mw: 164.14, mp: 177°.
Acute tox data: Ip LD_{LO} (mouse) = 102 mg/kg. [*3*]

THR = HIGH via ip route and irr to skin, eyes and mu mem.

Explosion Hazard: Mod, when exposed to heat or by chemical reaction. In the presence of alkalies at elevated temp., free hydroxylamine is liberated and may decomp explosively.

Disaster Hazard: Dangerous; see sulfur compounds.

4-HYDROXYLAMINO BIPHENYL.
THR = An exper carc. [*3*]

4-HYDROXYLAMINO-6-NITROQUINOLINE-1-OXIDE.
THR = An exper carc. [*23*]

4-HYDROXYLAMINO-7-NITROQUINOLINE-OXIDE.
THR = An exper carc. [*23*]

4-HYDROXYLAMINO-6-*n*-BUTYL QUINOLINE-1-OXIDE.
THR = An exper carc. [*23*]

4-HYDROXYLAMINO-6-CARBOXYLIC ACID QUINOLINE-1-OXIDE.
THR = An exper carc. [*23*]

4-HYDROXYLAMINO-5-CHLOROQUINOLINE-1-OXIDE.
THR = An exper carc. [*23*]

4-HYDROXYLAMINO-6-CHLOROQUINOLINE-1-OXIDE.
THR = An exper carc. [*23*]

4-HYDROXYLAMINO-7-CHLOROQUINOLINE-1-OXIDE.
THR = An exper carc. [*23*]

4-HYDROXYLAMINO-6,7-DICHLOROQUINOLINE-1-OXIDE.
THR = An exper carc. [*23*]

4-HYDROXYLAMINO-2-METHYLQUINOLINE-1-OXIDE.
THR = An exper carc. [*23*]

4-HYDROXYLAMINO-3-METHYLQUINOLINE-1-OXIDE.
THR = An exper carc. [*23*]

4-HYDROXYLAMINO-5-METHYL QUINOLINE-1-OXIDE.
THR = An exper carc. [*23*]

4-HYDROXYLAMINO-6-METHYL QUINOLINE-1-OXIDE.
THR = An exper carc. [*23*]

4-HYDROXYLAMINO-7-METHYL QUINOLINE-1-OXIDE.
THR = An exper carc. [*23*]

4-HYDROXYLAMINO-8-METHYL QUINOLINE-1-OXIDE.
THR = An exper carc. [*23*]

HYDROXYL AMMONIUM CHLORIDE. See hydroxyl amine hydrochloride.

HYDROXYL AMMONIUM SULFATE. See hydroxyl amine sulfate.

HYDROXYMERCURICHLOROPHENOL. Syns: *2-chloro-4-(hydroxymercuri)phenol, semesan.* Solid, insol in water and common organic solvents. $C_6H_5ClHgO_2$, mw: 345.2.
THR = HIGH. A fungicide. See mercury compounds, organic.

HYDROXYMERCURICRESOL.
$HOHgC_6H_4OHCH_3$, mw: 325.8.
THR = HIGH. See mercury compounds, organic.

HYDROXYMERCURINITROPHENOL.
$HOHgC_6H_3OHNO_2$, mw: 356.7.
THR = HIGH. A fungicide. See mercury compounds, organic.

7-HYDROXY-4-METHOXYISOFLAVONE. Syn: *formononetin.*
THR = An exper carc. [*23*]

4-HYDROXY-8-METHOXYQUINALDIC ACID.
$C_{11}H_9O_4N$, mw: 219.2.
THR = An exper carc. [*3*]

HYDROXYMETHYL ACRYLAMIDE. $C_4H_7O_2N$, mw: 101.1.
THR = Heat and/or contamination can cause exothermic polymerization. [*19*]

7-HYDROXYMETHYLBENZ(a)ANTHRACENE.
THR = An exper carc. [*23*]

10-HYDROXYMETHYL-1,2-BENZANTHRACENE.
$C_{19}H_{14}O$, mw: 258.3.
THR = An exper carc. [*3*]

1-HYDROXY-3-METHYL CHOLANTHRENE.
THR = An exper carc. [*23*]

1-HYDROXY-4-METHYL-2,6-DI-*tert*-BUTYL BENZENE. See di-*tert*-butyl-*p*-cresol.

7-HYDROXYMETHYL-4,12-DIMETHYL BENZ(a) ANTHRACENE.
THR = An exper carc. [*23*]

1-(HYDROXYMETHYL)-5,5-DIMETHYL HYDANTOIN. Syn: *MDMH.*
THR = Liberates formaldehyde on decomp. See formaldehyde.

4-HYDROXY-5-METHYL-2-(SH)-FURANONE.
$C_5H_6O_3$, mw: 114.1.
THR = An exper carc [*3, 23*] via sc route.

3-HYDROXY-1-METHYL GUANINE. $C_6H_7O_2N_5$, mw: 181.2.
THR = An exper neo. [*3, 23*]

3-HYDROXY-9-METHYL GUANINE.
THR = An exper neo. [*3*]

3-HYDROXY METHYL-1-[(3-(5-NITRO-2-FURYL) ALLYLIDENE)AMINO]HYDANTOIN.
$C_{11}H_{10}O_6N_4$, mw: 294.3.
THR = An exper carc. [*3*]

2-HYDROXYMETHYL NORCAMPHANE.
Acute tox data: Oral LD_{50} (rat) = 1620 mg/kg; dermal LD_{LO} (rabbit) = 800 mg/kg. [*3*]
THR = MOD via oral and dermal routes.

13-HYDROXYMETHYL-9,11-OCTADECADIENO-ATE. $C_{19}H_{34}O_3$, mw: 310.4.
THR = An exper neo to mice via dermal route. [*103*]

4-HYDROXY-4-METHYL-2-PENTANONE. See diacetone alcohol.

3-HYDROXY-3-METHYL-1-PHENYL TRIAZENE.
$C_7H_9ON_3$, mw: 151.2.
THR = An exper carc. [*3*]

2-HYDROXYMETHYL TETRAHYDROPYRAN.
Syn: *tetrahydropyran-2-methanol.*
Acute tox data: Oral LD_{50} (rat) = 1620 mg/kg; dermal LD_{LO} (rabbit) = 800 mg/kg. [*3*]
THR = MOD via oral and dermal routes.

2-HYDROXY-4-METHYLTHIOBUTYRIC ACID, CALCIUM SALT. See methionine hydroxy analog and its calcium salts.

3-HYDROXY-1-METHYL XANTHINE. $C_6H_5O_3N_4$, mw: 181.2.
THR = An exper neo. [*3*, *23*] via sc route.

3-HYDROXY-8-METHYL XANTHINE. $C_6H_6O_3N_4$, mw: 182.2.
THR = An exper neo [*3*] via sc route.

1-HYDROXYNAPHTHALENE. See α-naphthol.

β-HYDROXYNAPHTHALENE. See β-naphthol.

***n*-HYDROXYNAPHTHALIMIDE DIETHYL PHOSPHATE.** See rawetin.

3-HYDROXY-2-NAPHTHOIC ACID. Syns: *β-oxynaphthoic acid, 2-naphthol-3-carboxylic acid.* Yellow needle-like crystals. $HOC_{10}H_6COOH$, mw: 188.2, mp: 216°.
THR = U. Animal exper show MOD and irr.
Fire Hazard: Slight; when heated, emits acrid fumes.

4-HYDROXYPENTAMETHYL FLAVAN. Syn: *2-hydroxy-2,4,4,4,7-pentamethyl flavan.*
THR = U. An insecticide.
Disaster Hazard: Slight; when heated, emits acrid fumes.

4-HYDROXY-2-PENTENOIC ACID-γ-LACTONE.
$C_5H_6O_2$, mw: 98.1.
THR = An exper neo. [*3*]

β-*p*-HYDROXYPHENYL LACTIC ACID. $C_9H_{10}N_4$, mw: 174.2.
THR = An exper carc. [*3*]

***o*-HYDROXYPHENYL MERCURIC CHLORIDE.** Syn: *o-chloro mercuriphenol.* White to faint pink crystals. $C_6H_4OHHgCl$, mw: 329.2, mp: 143.5°.
THR = HIGH. See mercury compounds, organic.

2-HYDROXYPROPANE NITRILE. See lactonitrile.

β-HYDROXYPROPIONITRILE. See ethylene cyanohydrin.

HYDROXYPROPYL ALGINATE. See propylene glycol alginate.

2-HYDROXYPROPYL AMINE. See 1-amino-2-propanol.

HYDROXYPROPYL CELLULOSE. White powder, sol in water and organic solvents.
Acute tox data: Oral LD_{50} (rat) = 10,200 mg/kg. [*3*]
THR = LOW via oral route. Used as a food additive permitted in food for human consumption. [*109*]

***n*-HYDROXYPROPYL DIETHYLENE TRIAMINE.**
Acute tox data: Oral LD_{50} (rat) = 3730 mg/kg; dermal LD_{50} (rabbit) = 1500 mg/kg. [*3*]
THR = MOD via oral and dermal routes.

***n*-(2-HYDROXYPROPYL)ETHYLENE DIAMINE.**
Colorless to faint yellow liquid.
$H_2NCH_2CH_2NH(CH_2CHOHCH_3)$, mw: 118, bp: 180° @ 114 mm, flash p: 254°F (COC), d: 0.9858 @ 25°/25°.
THR = Details U. See also amines.
Fire Hazard: Slight, when exposed to heat or flame; can react with oxidizing materials.

HYDROXYPROPYL METHYL CELLULOSE. Syn: *methyl cellulose propylene glycol ether.* White fibrous or granular powder, insol in anhydrous alcohol, ether and chloroform.
THR = Details U. Used as a food additive permitted in food for human consumption. [*109*]

N-(2-HYDROXYPROPYL)-N-NITROSO-PROPYLAMINE. $C_6H_{14}O_2N_2$, mw: 146.2.
THR = An exper carc [*3*, *23*] via sc route.

***n*-(3-HYDROXYPROPYL)-1,2-PROPANE DIAMINE.** $CH_3CH(CHNH_2CHNH_2CH_3)CH_2CH$, mw: 132.
THR = Limited animal exper suggest LOW and MOD irr.

3-HYDROXYPYRIDINE. Tan-colored lumps.
C_5H_5ON, mw: 95.1, mp: 126°–129°, bp: 151°–153° @ 3 to 4 mm.
THR = Details U. See also pyridine.
Fire Hazard: Slight, when exposed to heat or flame.
Disaster Hazard: Dangerous; when heated to decomp,

emits highly toxic fumes of nitrogen oxides; can react vigorously with oxidizing material.

8-HYDROXYQUINALDIC ACID. $C_{10}H_7O_3N$, mw: 189.2.
THR = An exper neo. [*3*]

8-HYDROXYQUINOLINE. See β-quinolinol.

HYDROXYSAFROLE.
THR = A (S) carc of the esophagus. [*23*]

12-HYDROXYSTEARIC ACID. $C_{18}H_{36}O_3$, mw: 300.5.
THR = An exper neo. [*3*]

HYDROXYSENKIRKINE. From the plant *crotalaria laburnifolis*. $C_{19}H_{27}O_7N$, mw: 381.5.
THR = An exper (S) carc. [*3, 7, 23*]

4′-(*p*-HYDROXYSTYRYL)ACETANILIDE. $C_{16}H_{15}O_2N$, mw: 253.3.
THR = Acute oral to rats is MOD. An exper carc. [*3*]

4-HYDROXY-4,N-STILBENYL ACETAMIDE.
THR = An exper carc. [*23*]

HYDROXYSUCCINIC ACID. See malic acid.

6-HYDROXYTETRAHYDROPYRAN-2-CARBOXYLIC ACID LACTONE. $C_6H_8O_3$, mw: 128.1.
Acute tox data: Dermal LD_{50} (rabbit) = 3970 mg/kg. [*3*]
THR = MOD via dermal route.

7-HYDROXY-3,4-TETRAMETHYLENE COUMARIN. See dithione.

7-HYDROXYTHEOPHYLLINE. $C_7H_8O_3N_4$, mw: 196.2.
THR = An exper neo. [*3*]

HYDROXY TOLUENE. See benzyl alcohol.

α-HYDROXY-*o*-TOLUIC ACID LACTONE. See phthalide.

β-HYDROXY TRICARBALLYLIC ACID. See citric acid.

HYDROXY UREA. $CH_4O_2N_2$, mw: 76.1.
THR = An exper teratogen. [*3*]

3-HYDROXYXANTHINE. $C_5H_4O_3N_4$, mw: 168.1.
THR = An exper neo and carc. [*3, 23*]

1-HYDROXYXANTHINE DIHYDRATE. $C_5H_4O_3N_4 \cdot 2H_2O$, mw: 204.2.
THR = An exper neo. [*3*]

3-HYDROXYXANTHINE HYDRATE. $C_5H_4O_3N_4 \cdot H_2O$, mw: 186.2.
THR = An exper carc. [*3*]

HYDROXYZINE PAMOATE. $C_{44}H_{41}O_7N_2Cl$, mw: 745.3.
THR = An exper teratogen. [*3*]

HYGROMYCIN B.
THR = U. Used as a food additive permitted in the feed and drinking water of animals and/or for the treatment of food-producing animals. Also permitted in food for human consumption. [*109*]

HYOSCINE. Syn: *scopolamine*. Thick, colorless, syrupy liquid alkaloid. $C_{17}H_{21}NO_4$, mw: 303.35, mp: 55°.
Acute tox data: Iv LD_{50} (mouse) = 163 mg/kg; sc LD_{50} (mouse) = 1700 mg/kg; sc TD_{LO} (human) = 0.002 mg/kg; oral TD_{LO} (human) = 0.014 mg/kg. [*3*]
THR = This material is a poisonous alkaloid. It can produce profound depression of the central nervous system and occasionally it causes excitement. It can cause hallucinations and the individual affected loses a certain amount of his normal inhibitory control. It is for that reason that it has been called "truth serum." In many cases of poisoning from this material, and even to a certain extent following its medical application, there is retention of the urine caused by paralysis of the bladder, and catheterization is necessary. The fatal dose is variable. Death has occurred from as little as 0.6 mg, while recovery has occurred from doses of 7–15 mg. There is even a reported case of recovery following a 500-mg ingestion of the hydrobromide by mouth.
Disaster Hazard: Dangerous; when heated to decomp, emits highly toxic fumes of NO_x.
Treatment and Antidotes: A physician should be called at once. Wash the stomach with a large volume of warm water containing tannin, then give 15% magnesium sulfate (approximately 8 oz.). A solution of iodine may be used instead of tannin. Charcoal may also be used. If iodine is used, the amount should be roughly 4 times the weight of the hyoscine to be precipitated. The iodine solution should not be allowed to remain in the stomach. In moderate cases of poisoning, morphine is of advantage if given hypodermically. If the respiration is markedly depressed, caffeine and artificial respiration should be employed. Coramine may also be of value. If there is extreme excitement or violent delirium, chloroform or ether may be employed.

HYOSCINE HYDROBROMIDE. See hyoscine.

HYOSCINE SULFATE. See hyoscine.

HYOSCYAMINE. Syns: *l-hyoscyamine*, *daturine*, *duboisine*. White crystalline alkaloid. $C_{17}H_{23}NO_3$, mw: 289.36, mp: 106°–108°.
Acute tox data: Iv LD_{50} (mouse) = 95 mg/kg. [*3*]
THR = HIGH via iv route. This is one of the atropine alkloids and is very toxic, acting very much like atropine. It has the same effect on the CNS but twice the effect on the peripheral nerves. The symptoms of

poisoning from this material are those of dryness of the throat and mouth, marked difficulty in swallowing, and a sensation of burning and thirst. The vision becomes impaired through dilation and loss of accommodation, and the eyes present a rather prominent, brilliant, staring appearance. The voice is husky and the tongue is red.

Disaster Hazard: Dangerous; when heated to decomp, emits highly toxic fumes of NO_x.

HYOSCYAMINE HYDROBROMIDE. See hyoscyamine.

HYOSCYAMINE HYDROCHLORIDE. See hyoscyamine.

HYOSCYAMINE SULFATE. See hyoscyamine.

HYPERSENSITIVITY. See Section 9.

HY-PHOS. See sodium hexa-*m*-phosphate.

HYPNONE. See phenyl methyl ketone.

HYPO. See sodium thiosulfate.

HYPOBROMOUS ACID. Colorless to yellow solution. HBrO, mw: 96.92, mp: 40° (in vacuum).

THR = HIGH irr to skin, eyes and mu mem and via oral and inhal routes. See also bromides.

Disaster Hazard: Dangerous; see bromides; will react with water or steam to produce toxic and corrosive fumes.

HYPOCHLORITES. Salts of hypochlorous acid.

THR = MOD irr to skin, eyes and mu mem and via oral and inhal routes.

Fire Hazard: Mod, by chemical reaction with reducing agents. These are powerful oxidizers particularly at higher temp., even when chlorine and then oxygen are evolved or in the presence of moisture or carbon dioxide. With urea, it forms the highly explosive NCl_3. [*19*]

Disaster Hazard: Dangerous; when heated or on contact with acid or acid fumes, emits highly toxic fumes of chlorine and chlorides; will react with water or steam to produce toxic and corrosive fumes; can react vigorously with oxidizing materials.

HYPOCHLORITE SOLUTIONS containing more than 7% available chlorine by weight. See hypochlorites.

HYPOCHLOROUS ACID. Aqueous solutions are greenish yellow in color. HClO, mw: 52.5.

THR = HIGH irr to skin, eyes and mu mem and via inhal route. See also hypochlorites. Violent reaction with alcohols, NH_3, As. [*19*] See hypochlorites and hydrochloric acid.

HYPOCHLOROUS ANHYDRIDE. See chlorine tetroxide.

HYPOIODOUS ACID. Syn: *iodine hydroxide*. Yellow to greenish solution. HOI, mw: 143.93.

THR = HIGH irr to skin, eyes and mu mem and via inhal and oral route. See also iodides.

Disaster Hazard: Dangerous; see iodides; will react with water or steam to produce toxic and corrosive fumes.

HYPONITROUS ACID. White solid. $H_2N_2O_2$, mw: 62.03, mp: explodes.

THR = Details U. See also nitric oxide.

Fire Hazard: U.

Explosion Hazard: Severe, when exposed to heat or flame.

Disaster Hazard: Highly dangerous; when heated to decomp, emits highly toxic fumes of oxides of nitrogen and may explode.

HYPONITROUS ETHER. See ethyl nitrite.

HYPOPHOSPHITES. Salts of hypophosphorus acid. H_2PO_2.

THR = Details U. Generally have little or none.

Fire Hazard: Mod, when exposed to heat.

Explosion Hazard: Mod, when exposed to heat. See also phosphine. Forms a violent explosive with $HClO_4$. [*19*]

Disaster Hazard: Dangerous; when heated to decomp, emits highly toxic fumes of phosphine and PO_x.

HYPOPHOSPHORIC ACID. Crystals. $H_4P_2O_6$, mw: 161.99, mp: 55°, bp: decomp @ 100°.

THR = Details U. See also phosphoric acid.

Explosion Hazard: Mod, when exposed to heat. Reacts violently with $Hg(NO_3)_2$. [*19*]

Disaster Hazard: Dangerous; see phosphorus compounds.

HYPOPHOSPHORUS ACID. Colorless, oily liquid or deliquescent crystals. $H(H_2PO_2)$, mw: 66.0, mp: 26.5°, bp: decomp, d: 1.493 @ 19°.

THR = Details U. See also phosphoric acid.

Fire Hazard: Mod; will react violently with oxidizing agents. Reacts violently with HgO.[*19*]

Disaster Hazard: Dangerous; on decomp, emits highly toxic fumes of phosphine and may explode. Keep away from oxidizing agents.

HYPOVANADIC HYDROCHLORIDE. See vanadium oxydichloride.

HYPOXANTHINE-3-OXIDE.

THR = An exper carc. [*23*]

HYVAR. See 5-bromo-3-isopropyl-6-methyl uracil.

ICR 340. $C_{20}H_{24}ON_4Cl_2 \cdot 2HCl$, mw: 480.3.
THR = An exper carc to mice via iv route. [*103*]

ICR 394. $C_{24}H_{27}N_4Cl$, mw: 516.4.
THR = An exper carc to mice via iv route. [*103*]

ICRP. See Section 5A.

ICTHYOL. Thick brown liquid, bituminous odor.
THR = LOW. A mild allergen.
Fire Hazard: Slight, when exposed to heat or flame; can react with oxidizing materials.

IGE. See isopropyl glycidyl ether.

IGNITER CORD. See Section 11.

IGNITER FUSE-METAL CLAD. See Section 11.

IGNITERS. See Section 11.

IGNITERS, AIRCRAFT ROCKET ENGINE (COMMERCIAL). See Section 11.

IGNITERS, JET THRUST (JATO). See Section 11.

IGNITERS, ROCKET MOTOR. See Section 11.

IGNITION SOURCES. See Section 7.

ILLUMINATING GAS. See carbon monoxide and methane.

4-IMIDAZOLE ETHYLAMINE. See histamine.

IMIDAZOLE MUSTARD. $C_8H_6N_2S$, mw: 279.2.
THR = An exper neo. [*3*]

IMIDAZOLIDINE THIONE. See ethylene thiourea.

β-IMIDAZOLYL-(4)-β-METHYLETHYL AMINE.
THR = An exper teratogen. [*3*]

4,4′-(IMIDOCARBONYL)BIS(N,N-DIMETHYL)-ANILINE. Syn: *auramine*. Yellow needles. $C_{17}H_{21}N_3$, mw: 267.4, mp: 136°.
THR = An exper and human (+) carc. [*3, 9, 23*]

4,4′-(IMIDOCARBONYL)BIS(N,N-DIMETHYL)-ANILINE HYDROCHLORIDE. Syn: *ADC auramine D.* $C_{17}H_{21}N_3 \cdot HCl \cdot H_2O$, mw: 321.9.
THR = An exper carc. [*3*]

3,3′-IMINO-BIS-PROPYLAMINE. Colorless liquid. $C_6H_{17}N_3$, mw: 131.23, mp: −14.5°, bp: 151° @ 5 mm, flash p: > 175°F (CC), d: 0.9237 @ 25°, vap. d: 4.52.
THR = See also amines.
Fire Hazard: Mod, when exposed to heat or flame; can react with oxidizing materials.
To Fight Fire: Foam, CO_2, dry chemical.

IMINODIACETIC ACID. Syn: *aminodiacetic acid*. Crystalline material, decomp @ 247.5°. $HOOCCH_2NHCH_2COOH$, mw: 133.1.
Acute tox data: ip LD_{LO} (mice) = 250 mg/kg; inhal TC_{LO} (human) = 200 ppm ⟶ irr. [*3*]
THR = It is a chelating agent which might disturb trace element metabolism if taken into the system.

3,3′-IMINODI-1-PROPANOL DIMETHANE SULFONATE HYDROCHLORIDE. $C_8H_{19}O_6NS \cdot HCl$, mw: 293.8.
THR = An exper neo. [*3*]

IMINODIPROPIONITRILE. See di-(2-cyandiethyl)-amine.

IMIPRAMINE-N-OXIDE HYDROCHLORIDE. $C_{19}H_{24}ON_2 \cdot HCl$; mw: 332.9.
THR = An exper teratogen. [*3*]

IMPERIAL GREEN. See copper acetoarsenite.

IMPLEMENTATION OF FOOD ADDITIVES LAW. [*109*]

INCOMPLETE COMBUSTION PRODUCTS OF ALL KINDS OF CARBONACEOUS MATTER. See soot.

INDALONE. See butopyronoxyl.

5-INDANOL. $C_9H_{10}O$, mw: 134.2.
Acute tox data: Oral LD_{50} (rat) = 3250 mg/kg; dermal LD_{50} (rabbit) = 450 mg/kg. [*3*]
THR = HIGH via dermal and MOD via oral routes.

1,2,3-INDANTRIONE HYDRATE. See ninhydrin.

INDENE. Syn: *indonaphthene*. Liquid from coal tars. Water-insol,but miscible in organic solvents. C_9H_8, mw: 116.2, d: 0.9968 @ 20°/4°, mp: −1.8°, bp: 182°.
Acute tox data: Sc LD_{LO} (rat) = 1000 mg/kg. [*3*]
THR = MOD via sc route and irr to skin, eyes and mu mem, oral and inhal routes also. During nitration with (H_2SO_4 + HNO_3) it exploded. [*19*]

INDENE RESINS. See coumorone-indene resin.

13H-[INDENO(1,2-1)]PHENANTHRENE. $C_{21}H_{14}$, mw: 266.4.
THR = An exper neo. [*3*]

INDIAN ACONITE. See aconitum ferox.

INDIAN CANNABIS. See cannabis.

INDIAN HEMP. See cannabis.

INDIAN TRAGACANTH. See karaya gum.

INDIAN RUBBER. See rubber, crude.

INDIGO CARMINE. $C_{16}H_{10}O_8N_2S_2$, mw: 468.4.
THR = An exper carc. [*3*]

INDIRECT FOOD ADDITIVES. [*109*]

INDIUM. Soft, silvery-white metal. In, atwt: 114.76, mp: 156.61°, bp: 2080°, d: 7.31 @ 20°.
Acute tox data: Oral LD_{50} (rat) = 4200 mg/kg; sc LD_{LO} (mouse) = 10 mg/kg. [*3*]
THR = HIGH via sc and MOD via oral routes. When this material is injected into animals via ip or iv, it is found to be highly toxic. It affects the liver, heart, kidneys and the blood. However, available data are scanty. Toxicity is based on animal exper.
Radiation Hazard: For permissible levels, see Section 5, Table 5A.5. Natural isotope (96%) ^{115}In, $T_{\frac{1}{2}} = 6 \times 10^{14}$ y; decays to stable ^{115}Sn via β's of 0.48 MeV.
Fire Hazard: Mod, in the form of dust when exposed to heat or flame. Incandesces. See powdered metals.

INDIUM ANTIMONIDE. Crystals. InSb, mw: 236.58, mp: 535°.
Acute tox data: Ip LD_{50} (mouse) = 3760 mg/kg. [*3*]
THR = MOD via ip route. See also antimony and indium compounds.

INDIUM ARSENIDE. Crystals. InAs, mw: 189, mp: 943°.
THR = U. See indium and arsenic compounds.

INDIUM CYANIDE. White precipitate. $In(CN)_3$, mw: 192.81.
THR = See cyanides.

INDIUM DIBROMIDE. Pale yellow solid. $InBr_2$, mw: 235°, bp: 632° sublimes, d: 4.22 @ 25°.
THR = See indium and bromides.

INDIUM DICHLORIDE. White powder. $InCl_2$, mw: 185.7, mp: 235°, bp: 550°–570°, d: 3.655 @ 25°.
THR = See indium and chlorides.

INDIUM DIHYDROGEN SULFATE. Powder. $In_2(SO_4)_3 \cdot H_2SO_4 \cdot 7H_2O$, mw: 741.9, mp: $-H_2SO_4 \cdot 7H_2O$ @ 250°.
Acute tox data: Oral LD_{LO} (rat) = 1200 mg/kg; sc LD_{LO} (rat) = 13 mg/kg; oral LD_{LO} (rat) = 1300 mg/kg. [*3*]
THR = HIGH via sc and MOD via oral routes. See indium sulfates.

INDIUM DIIODIDE. Solid InI_2, mw: 368.6, mp: 212°, d: 4.71 @ 25°.
THR = See indium and iodides.

INDIUM FLUORIDE. Solid. InF_3, mw: 171.8, mp: 1170 ± 10°, bp: > 1200°, d: 4.39 ± 0.01 @ 25°.
THR = See indium and fluorides.

INDIUM HYDROXIDE. White precipitate. $In(OH)_3$, mw: 165.78, mp: $-H_2O$ @ 150°.
THR = See indium.

INDIUM IODATE. White crystals. $In(IO_3)_3$, mw: 639.52, bp: decomp.
THR = See indium and iodates.
Fire Hazard: Mod, by chemical reaction with reducing agents. A powerful oxidizing agent.
Explosion Hazard: U.
Disaster Hazard: Mod dangerous; keep away from reducing agents.

INDIUM MONOBROMIDE. Red-brown crystals. InBr, mw: 194.7, mp: 220°, bp: 662° (sublimes), d: 4.96.
THR = See indium and bromides.

INDIUM MONOCHLORIDE. White powder. InCl, mw: 150.2, mp: 225°, bp: 550°, d: 4.18 (red); 4.19 (yellow) @ 25°.
THR = See indium and chlorides.

INDIUM MONOIODIDE. Brown-red solid. InI, mw: 241.7, mp: 351°, bp: 714°, d: 5.31.
THR = See indium and iodides.

INDIUM MONOXIDE. Gray-white crystals. InO, mw: 130.8.
THR = See indium. Self-ignites in air. [*19*]

INDIUM MONOSULFIDE. Dark crystals. InS, mw: 146.8, mp: 692°, bp: 850°, d: 5.18 @ 25°.
THR = See indium and sulfides.

INDIUM NITRATE.
THR = Acute oral is MOD. An exper teratogen. [*3*]

INDIUM PERCHLORATE. Colorless, deliquescent crystals. $In(ClO_4)_3 \cdot 8H_2O$, mw: 557.26, mp: ca 80°, bp: 200°.
THR = See indium and perchlorates.

INDIUM SELENATE. Deliquescent crystals. $In_2(SeO_4)_3 \cdot 10H_2O$, mw: 838.7.
THR = See selenium and indium.

INDIUM SESQUIOXIDE. Red-brown crystals. In_2O_3, mw: 277.5, bp: 850°, d: 7.179.
THR = See indium.

INDIUM SESQUISULFIDE. Yellow crystals. In_2S_3, mw: 325.8, mp: 1050°, bp: sublimes @ 850° in high vacuum, d: 4.90 @ 25°.
THR = See sulfides and indium.

INDIUM SUBOXIDE. Black crystals. In_2O, mw: 245.5, d: 6.99 @ 25°, mp: sublimes in vacuum @ 656°.
THR = See indium.

INDIUM SUBSULFIDE. Yellow to black needles. In_2S, mw: 261.7, mp: 653°, d: 5.87 @ 25°.
THR = See sulfides and indium.

INDIUM SULFATE. Grayish, deliquescent powder. $In_2(SO_4)_3$, mw: 517.7, d: 3.438.
THR = See indium dihydrogen sulfate.
Disaster Hazard: Dangerous; see sulfates.

INDIUM SULFATE, BASIC. Crystals. $2In_2O_3 \cdot 3SO_2 \cdot 8H_2O$, mw: 891.59, mp: $-3H_2O$ @ 100°, bp: $-8H_2O$ @ 260°.
THR = See indium and sulfites.

INDIUM TRIBROMIDE. White to yellow needle-like crystals. $InBr_3$, mw: 354.5, mp: 436°, bp: sublimes, d: 4.74 @ 25°.
THR = See indium and bromides.

INDIUM TRICHLORIDE. White powder. $InCl_3$, mw: 221.1, mp: 586°, bp: volatile @ 600°, d: 3.46.
Acute tox data: Sc LD_{LO} (rat) = 10 mg/kg; ip LD_{LO} (mouse) = 5 mg/kg; iv LD_{LO} (rabbit) = 0.64 mg/kg. [*3*]
THR = HIGH via sc, ip and iv routes. No data on oral or inhal routes. See indium and chlorides.

INDIUM TRIIODIDE. Yellow, deliquescent crystals. InI_3, mw: 495.5, mp: 210°, d: 4.69.
THR = See indium and iodides.

INDIUM TRIMETHYL. Colorless crystals, decomp by water. $In(CH_3)_3$, mw: 160.0, d: 1.568 @ 19°/19°, mp: 90°.
THR = U. See also organometals.

INDOCENE. bp: 171°, flash p: 170°F (CC), d: 1.064.
THR = U.
Fire Hazard: Mod, when exposed to heat or flame.

INDOLE. Syn: *1-benzo-β-pyrrole*. Colorless to yellowish scales, fecal odor, sol in hot water, alcohol, ether and petroleum ether, insol in mineral oil and glycerol. CHCHCHCHCCCHCHNH, mw: 117.14, mp: 52°, bp: 254°.
Acute tox data: Oral LD_{50} (rat) = 1000 mg/kg; dermal LD_{50} (rabbit) = 790 mg/kg. [*3*]
THR = MOD via oral and dermal routes. An exper neo. [*3*]

INDOLE-α-AMINOPROPIONIC ACID.
See tryptophane.

INDOLEBUTYRIC ACID. See hormodin.

INDOL-3-OL POTASSIUM SULFATE.
$K\text{-}C_8H_7NSO_4$, mw: 252.3.
THR = An exper carc. [*3*]

2(3)-INDOLONE. Syn: *oxindole*. Colorless needle-like crystals. $C_6H_4NHCOCH_2$, mw: 133.1, mp: 120°, bp: decomp, flash p: 205°–275°F, d: 1.058 @ 20°/20°.
THR = Details U, but probably toxic; an insecticide.
Fire Hazard: Slight, when exposed to heat or flame; can react with oxidizing materials.
To Fight Fire: Water, foam, CO_2, dry chemical.

INDONAPHTHENE. See indene.

INDUSTRIAL AIR CLEANING. See Section 2.

INDUSTRIAL AIR CONTAMINANT CONTROL.
See Section 2.

INDUSTRIAL CANCER RISKS. See Section 8.

INDUSTRIAL DISASTER CONTROL. See Section 7.

INDUSTRIAL FUEL OILS.
THR = A recog carc. [*14*] See mineral oils.

INDUSTRIAL WASTES DANGERS. See Section 6.

INDUSTRIAL WASTES HAZARDS PREVENTIVE MEASURES. See Section 6.

INFECTIOUS MATERIALS. See etiologic agents, N.O.S.

INFUSORIAL EARTH. See diatomaceous earth.

INITIALLY HAZARDOUS SOLID WASTES. See Section 6.

INITIATING EXPLOSIVE. See explosive, initiating.

INK, PRINTING. Liquid or paste. Carbon, black dye, glue, gum arabic, solvent.
THR = SLIGHT due to solvent or dye used.
Fire Hazard: Dangerous, when exposed to heat or flame; can react with oxidizing materials.
Explosion Hazard: Variable.

INOSITOL. Syn: *hexahydroxy cyclohexane*. A constituent of body tissue. There are nine isomeric forms, of which the optically inactive 1-inositol is the one having vitamin activity. White crystals, odorless, sol in water, insol in absolute alcohol and ether. $C_6H_6(OH)_6 \cdot 2H_2O$, mw: 216, mp: 218°, d: 1.524.
Acute tox data: Ip LD_{LO} (mouse) = 500 mg/kg. [*3*]
THR = HIGH-MOD via ip route. Used as a nutrient and/or dietary supplement food additive. [*109*]

INSECTICIDE, LIQUEFIED GAS.
See specific components.

INSECTICIDES, DRY. See specific components listed on container.

INSECTICIDES, LIQUID. See specific components.

INSOLUBLE CYANURIC ACID. See cyamelide.

INSTANTANEOUS FUSE. See explosives, high.

INSULIN ZINC SUSPENSION.
THR = An exper carc. [*3*]

INTANGIBLE FIRE LOSSES. See Section 7.

INTERNATIONAL AIR TRANSPORT ASSOCIATION(IATA)REGULATIONS. See Section 11.

INTERNATIONAL COMMISSION ON RADIOLOGICAL PROTECTION. See Section 5A.

INVERT SUGAR. A mixture of 50% glucose and 50% fructose obtained by the hydrolysis of sucrose.
THR = Details U. It is a substance which migrates to food from packaging materials. [*109*]

IODATES.
THR = Variable; similar to bromates and chlorates. See also specific compound.

Fire Hazard: These materials are a dangerous fire hazard because they are powerful oxidizers. In contact with flam or even combustible materials (i.e., Al, As, C, Cu, metal sulfides, organic matter, P, S, PIH_4) they can start fires. [*19*]
Disaster Hazard: Dangerous; see iodine compounds; they can react vigorously with reducing materials.

IODIACETYLENE. HC⋮CC⋮CI, mw: 175.9.
THR = Detonates in light if scratched. [*19*]

IODIC ACID. Colorless to pale yellow crystals. HIO_3, mw: 175.9, mp 110°, d: 4.629.
THR = HIGH irr to skin, eyes and mu mem and via oral route.
Fire Hazard: See iodates.
Explosion Hazard: Reacts violently with B, PIH_4. [*19*]
Disaster Hazard: See iodides.

IODIC-*m*-PERACID. Rhombic, colorless to pale yellow crystals. HIO_4, mw: 191.9, bp: sublimes @ 110°.
THR = HIGH irr to skin, eyes and mu mem.
Fire Hazard: See periodates.
Explosion Hazard: See periodates.
Disaster Hazard: See periodates.

IODIC-*o,p*-PERACID. Powder. H_5IO_6, mw: 228.0, mp: decomp @ 140°.
THR = HIGH irr to skin, eyes and mu mem.
Fire Hazard: Mod, by chemical reaction. A powerful oxidizer.
Explosion Hazard: U.
Disaster Hazard: Dangerous; see iodine compounds; can react vigorously with reducing agents.

IODIDES.
THR = Similar in toxicity to bromides. Prolonged absorption of iodides may produce "iodism" which is manifested by skin rash, running nose, headache and irr of mu mem. In severe cases the skin may show pimples, boils, redness, black and blue spots, hives and blisters. Weakness, anemia, loss of weight and general depression may occur.
Disaster Hazard: When heated to decomp, they can emit highly toxic fumes of iodine and iodine compounds.

IODINE. Rhombic, violet-black crystals, metallic luster. I_2, mw: 253.8, mp: 113.5°, bp: 184°, d: 4.93, vap. press: 1 mm @ 38.7°.
THR = HIGH irr to skin, eyes and mu mem. The effect of iodine vapor upon the body is similar to that of chlorine and bromine, but it is more irr to the lungs. Serious exposures are seldom encountered in industry, due to the low volatility of the solid at ordinary room temp. Signs and symptoms are irr and burning of the eyes, lachrymation, coughing and irr of the nose and throat. See iodides.
Radiation Hazard: For permissible levels, see Section 5, Table 5A.5. Artificial isotope ^{125}I, $T\frac{1}{2}$ = 60d, decays to stable ^{125}Te via ec and emits γ's of 0.04 MeV. Artificial isotope ^{129}I, $T\frac{1}{2} = 1.7 \times 10^7$y, decays to stable ^{129}Xe via β's of 0.15 MeV; emits γ's of 0.04 MeV. Artificial isotope ^{131}I, $T\frac{1}{2}$ = 8.05d, decays to stable ^{131}Xe via β's of 0.33 MeV (7%), 0.61 MeV (90%) emits γ's of 0.08–0.72 MeV. It reacts violently with acetaldehyde, C_2H_2, Al, NH_3, NH_4OH, Sb, BrF_5, $CsHC_2$, Cs_2C_2, Cs_2O, Cl_2, ClF_3, Cu_2C_2, (ethanol + methanol + HgO), F_2, Li, Li_2C_2, Li_6C, Mg, OF_2, P, K, $RbHC_2$, Rb_2C_2, AgN_3, NaH, ZrC. [*19*]
Disaster Hazard: Dangerous; when heated, emits highly toxic fumes of iodine and iodine compounds; can react vigorously with reducing materials.

IODINE ACIDS.
THR = HIGH irr to skin, eyes and mu mem.
Explosion Hazard: U.
Disaster Hazard: Dangerous; by decomp, they can emit highly toxic iodine compounds; they react with water or steam to produce corrosive fumes.

IODINE AZIDE. Yellow crystals. Explodes easily. IN_3, mw: 168.94.
THR = HIGH irr to skin, eyes and mu mem.
Explosion Hazard: Severe, when shocked or exposed to heat. [*19*]
Disaster Hazard: Highly dangerous; shock and heat will explode it; when heated to decomp, emits highly toxic fumes of iodine and iodine compounds.

IODINE CYANIDE. See cyanogen iodide.

IODINE DIOXIDE. See iodine oxide.

IODINE HEPTAFLUORIDE. Colorless crystals or gas. IF_7, mw: 259.92, mp: 5.5°, bp: sublimes, d: 2.8 @ 6° (as liquid).
THR = HIGH irr to skin, eyes and mu mem. See fluorides and iodine.
Explosion Hazard: Mod, when exposed to heat. Reacts violently with NH_4Br, NH_4Cl, NH_4I, CO, organic matter, H_2SO_4, H_2O. [*19*]
Disaster Hazard: Dangerous; when heated to decomp or on contact with acid or acid fumes, emits highly toxic fumes of iodine and fluorides, and may explode; will react with water or steam to produce toxic and corrosive fumes.

IODINE HYDROXIDE. See iodous hypoacid.

IODINE MONOBROMIDE. Dark gray crystals. IBr, mw: 206.84, mp: 42°, sublimes @ 50°, bp: 116° decomp.
THR = HIGH irr to skin, eyes and mu mem. See also bromides and iodine.
Explosion Hazard: Mod, when exposed to heat. Reacts violently with P, K, Na. [*19*]

Disaster Hazard: Dangerous; when heated to decomp, emits highly toxic fumes of bromine, iodine and their compounds; will react with water or steam to produce toxic and corrosive fumes.

IODINE MONOCHLORIDE. Exists in α, β forms; 1Cl; mw: 162.4; red-brown cryst or oily liquid. mp(α): 27°, (β): 14°, bp: 97.4 decomp @ 100°; d(α): 3.1822 @ 0°, (β): 3.24 @ 34°.

Acute tox data: Oral LD_{LO} (rat) = 50 mg/kg; dermal LD_{LO} (rat) = 500 mg/kg. [*3*]

THR = HIGH irr to skin, eyes and mu mem, also oral and dermal routes. See iodine and chlorine.

Explosion Hazard: Mod, when exposed to heat. Reacts violently with Al foil, CdS, PbS, organic matter, P, PCl_3, K, rubber, Ag_2S, Na, ZnS. [*19*]

Disaster Hazard: Dangerous; when heated to decomp, emits highly toxic fumes of chlorine and iodine, and may explode; will react with water or steam to produce toxic and corrosive fumes.

IODINE NONOXIDE. Yellow, hygroscopic powder. I_4O_9, mw: 651.6, bp: 75° decomp.

THR = HIGH. See iodine.

Fire Hazard: Mod, by spont chemical reaction. A powerful oxidizing agent.

Explosion Hazard: Mod, when exposed to heat.

Disaster Hazard: Dangerous; when heated to decomp, emits highly toxic fumes of iodine and iodine compounds, and may explode; can react vigorously with reducing materials.

IODINE OXIDE. Syns: *iodine dioxide, iodine tetroxide*. Lemon-yellow crystals. IO_2, mw: 158.9, mp: decomp slowly @ 75°, rapidly @ 130°, d: 4.2 @ 10°/10°.

THR = HIGH. See iodine.

Fire Hazard: Mod, by spont chemical reaction. An oxidizing agent.

Explosion Hazard: Slight, when exposed to heat.

Disaster Hazard: Dangerous; when heated to decomp, emits highly toxic fumes of iodine and iodine compounds; can react vigorously with reducing materials.

IODINE PENTABROMIDE. Crystals. IBr_5, mw: 526.5.

THR = HIGH. See iodine and bromine.

Fire Hazard: Dangerous, by spont chemical reaction.

Explosion Hazard: Slight, when exposed to heat or flame.

Disaster Hazard: Dangerous; when heated to decomp, emits highly toxic fumes of bromine, iodine and their compounds; will react with water or steam to produce toxic and corrosive fumes; can react vigorously with reducing materials.

IODINE PENTACHLORIDE. ICl_5, mw: 304.2.

THR = A poison. HIGH irr to skin, eyes and mu mem.

IODINE PENTAFLUORIDE. Colorless fuming liquid. IF_5, mw: 221.9, mp: 9.43°, bp: 100.5°, d: 3.19 @ 25°, vap. press: 10 mm @ 8.5°.

THR = HIGH. See iodine and fluorides. Very toxic and corrosive; attacks glass.

Explosion Hazard: Reacts violently with As, Bi, dimethyl sulfoxide, organic matter, P, K, Si, Na, S, (tetra fluoroethylene + limonene), tetraiodoethylene, W, H_2O. [*19*]

Disaster Hazard: Dangerous; when heated to decomp or on contact with acid or acid fumes, emits highly toxic fumes of iodine, fluorine and their compounds; will react with water or steam to produce toxic and corrosive fumes.

IODINE PENTOXIDE. White crystals. I_2O_5, mw: 333.8, mp: 300°–350° (decomp), d: 4.799 @ 25°.

THR = HIGH. See iodine.

Fire Hazard: Dangerous, by spont chemical reaction. A powerful oxidizing agent.

Explosion Hazard: When exposed to heat or flame, reacts violently with C, S, organic matter. [*19*]

Disaster Hazard: Dangerous; see iodine compounds; will react with water or steam to produce toxic fumes; can react vigorously with reducing materials.

IODINE PHOSPHIDE. H_4PI, mw: 161.9.

Acute tox data: Oral LD_{LO} (rabbit) = 5 mg/kg. [*3*]

THR = HIGH via oral route. See also phosphides and iodine.

IODINE POLYVINYL PYRROLIDONE.

THR = A susp carc. [*19*] See polymers, water-sol.

IODINE TETRAOXIDE. See iodine oxide.

IODINE TRIBROMIDE. Dark brown liquid. IBr_3, mw: 366.63.

THR = HIGH. See iodine and bromine.

Fire Hazard: Mod, by spont chemical reaction.

Explosion Hazard: U.

Disaster Hazard: Dangerous; when heated to decomp or on contact with acid or acid fumes, emits highly toxic fumes and explodes; will react with water or steam to produce toxic and corrosive fumes; can react vigorously with reducing materials.

IODINE TRICHLORIDE. Orange-yellow, deliquescent, crystalline powder, pungent, irr odor. ICl_3, mw: 233.39, mp: 101° @ 16 atm, bp: decomp @ 77°, d: 3.203 @ −4°.

THR = HIGH. See iodine and chlorine. Corrosive to human skin. [*2*] Very irr.

Explosion Hazard: Slight, when exposed to heat or flame.

Disaster Hazard: Dangerous; when heated to decomp or on contact with acid or acid fumes, emits highly toxic fumes; will react with water or steam to produce toxic and corrosive fumes; can react with reducing materials.

IODOACETAMIDE. C_2H_4ONI, mw: 185.
Acute tox data: Ip LD_{LO} (mouse) = 50 mg/kg. [*3*]
THR = HIGH via ip route and probably oral route as well.

IODOACETIC ACID. Colorless or white crystals, sol in water and alcohol. CH_2ICOOH, mw: 186, mp: 82°.
Acute tox data: Iv LD_{50} (dog) = 45 mg/kg; sc LD_{50} (rat) = 60 mg/kg; oral LD_{50} (mouse) = 83 mg/kg. [*3*]
THR = HIGH irr to skin, eyes and mu mem. HIGH via oral, iv and sc routes. An exper teratogen and neo. [*3, 23*] See iodine.

IODOACETONE. Syn: *1-iodo-2-propanone.* CH_2ICOCH_3, mw: 183.9.
THR = HIGH. A strong irr.
Disaster Hazard: Dangerous; when heated to decomp, emits highly toxic fumes.

3-IODO-2-ACETYL AMINO FLUORENE.
THR = A carc. [*14*]

***p*-IODOANISOLE.** See isoform.

IODOCHLORO-8-QUINOLINOL.
THR = A susp carc. [*14*] See 8-hydroxyquinoline.

IODODIMETHYL ARSINE. $(CH_3)_2AsI$, mw: 231.9.
THR = Self-ignites in air. [*19*]. HIGH via inhal and oral routes. See also arsine.

2-IODO-3,5-DINITROBIPHENYL. $C_6H_5C_6H_2(NO_2)_2I$, mw: 370.1.
THR = Can explode with ethyl sodioacetate. [*19*]

IODOETHANE. See ethyl iodide.

IODOETHENE. See vinyl iodide.

IODOFENPHOS. Syn: *o,o-dimethyl-o-(2,5-dichloro-4-iodo phenyl) thiophosphate.* A crystalline powder, sol in kerosene. $C_8H_8O_3PSCl_2I$, mw: 412.9.
Acute tox data: Oral LD_{50} (rat) = 2000 mg/kg. [*3*]
THR = MOD via oral route. An insecticide. See also parathion.
Disaster Hazard: See parathion.

N-(IODO-2-FLUORENYL) ACETAMIDE.
THR = MOD via oral to rats. An exper carc. [*3, 23*]

IODOFORM. Syn: *triiodo methane.* Yellow powder or crystals. Characteristic disagreeable odor. CHI_3, mw: 393.8, d: 4.1, mp: 120° (approx).
Acute tox data: Oral LD_{LO} (rabbit) = 450 mg/kg; sc LD_{50} (mouse) = 629 mg/kg; oral TD_{LO} (human) = 114 mg/kg ⟶ GI symptoms.
THR = HIGH via oral and MOD via sc routes.
Disaster Hazard: Dangerous; see iodides. Reacts violently with Li. [*19*]

IODOMETHANE. See methyl iodide.

7-(IODOMETHYL)-12-METHYL BENZ(a)ANTHRACENE. $C_{20}H_{15}I$, mw: 382.3.
THR = An exper neo. [*3, 23*]

IODONITROTETRAZOLIUM. Syn: *2-(p-iodophenyl)-5-phenyl tetrazolium chloride.* Lemon yellow crystals, slightly sol in water. $C_{19}H_{13}N_5O_2ICI$, mw: 505.7, mp: 239° (decomp).
THR = U. Probably toxic.
Disaster Hazard: Dangerous; when heated to decomp, emits highly toxic fumes of chlorides, iodides and NO_x.

3-IODO-1-PHENYL-1-PROPYNE. $C_6H_5C{:}CCH_2I$, mw: 242.1.
THR = Has detonated during distillation @ about 180°. [*19*]

IODOPLATINIC ACID. Monoclinic, black, deliquescent crystals. $H_2PtI_6 \cdot 9H_2O$; mw: 1120.91, mp: 100°.
THR = HIGH. See iodine and platinum compounds.
Disaster Hazard: Dangerous; see iodine.

1-IODOPROPANE. See propyl iodide.

1-IODO-2-PROPANONE. See iodoacetone.

IODOPROPENE. See allyl iodide.

3-IODOPROPIONIC ACID. $C_3H_5O_2I$, mw: 200.
THR = An exper neo. [*3, 23*]

N-IODOSUCCINIMIDE. Syn: *NIS.* Off-white, nearly odorless powder. $C_4H_4O_2NI$, mw: 225.0, mp: 201°.
THR = HIGH irr to skin, eyes and mu mem.

α-IODOTOLUENE. See benzyl iodide.

IODOUS HYPOACID. Syn: *iodine hydroxide.* Yellowish to greenish solution. HOI, mw: 143.9.
THR = Details U. See also iodine and iodides.

IONOL. See 2,6-di-*tert*-butyl-4-methylphenol.

IOXYNIL. Syn: *4-hydroxy-3,5-diiodobenzonitrile.* A colorless solid. $C_6H_2OHI_2CN$, mw: 370.9, mp: 213°, slightly water-sol.
Acute tox data: Oral LD_{50} (rat) = 110 mg/kg; dermal LD_{LO} (rat) = 0.2 mg/kg. [*3*]
THR = VERY HIGH via dermal and HIGH via oral routes. An herbicide. See also nitriles.
Disaster Hazard: See nitriles and iodides.

IPC. Syns: *isopropyl carbanilate, propham.* A white crystalline solid, sol in acetone and benzene. $C_6H_5NHCOOCH(CH_3)_2$, mw: 179, mp: 90°.
Acute tox data: Oral LD_{50} (rat) = 1000 mg/kg. [*3*]
THR = MOD via oral route. An exper neo. [*3*] A pre-emergence and post-emergence herbicide. An initiator in 2-stage-carc via oral route on mice. [*12*]

IPE. See isopropyl glycidyl ether.

IPECAC. Syn: *ipecacuanha.* Dried rhizome and roots of Rio or Brazilian ipecac. Emetine, cephaline, emeta-

mine, ipecacuanic acid, psychotrine, methyl psychotaine, resin.
THR = A centrally acting emetic. Has caused fatalities. Symptoms include retention of urine, fever, diarrhea, violent abdominal pain, dehydration and cardiac irregularities. [*20*] HIGH via oral route. Can cause conjunctivitis with opacity of the cornea. See also emetine.

IPECACUANHA. See ipecac.

IPP. See diisopropyl peroxydicarbanate.

IRIDIUM AMMINE NITRATE. $Ir(NH_3)_5OH(NO_3)_3$, mw: 490.4.
THR = May be impact-sensitive; also may detonate @ red heat. [*19*]

IRIDIUM AMMINE PERCHLORATES.
THR = May be impact-sensitive. [*19*]

IRIDIUM. Slightly yellowish-white, hard brittle metal. Ir, atwt: 192.22, mp: 2410°, bp: 4130°, d: 22.65.
THR = NO data. Probably MOD via oral and inhal routes. Soluble iridium compounds are said to be toxic. However, there are no industrial data available upon which to base a maximum allowable conc in air.
Radiation Hazard: For permissible levels, see Section 5A, Table 5A.5. Artificial isotope ^{192}Ir, $T_{\frac{1}{2}}$ = 74d, decays to stable ^{192}Pt via β's of 0.24 MeV (8%), 0.54 MeV (41%), 0.67 MeV (46%) emits γ's of 0.30–0.61.
Fire Hazard: Mod, in the form of dust when exposed to heat or flame. See also powdered metals. Incandesces with OF_2 or ClF_3. Reacts violently with F_2 @ 260°. [*19*]

IRIDIUM CHLORIDE. $IrCl_3$, mw: 298.6.
Acute tox data: Iv LD_{LO} (dog) = 778 mg/kg. [*3*]
THR = MOD via oral route.

IRISH MOSS. See chondrus extract.

IROKO. See sawdust.

IRON, DUST. Syn: *ferrum*. Silvery-white, tenacious, lustrous, ductile metal. Fe, atwt: 55.8, mp: 1535°, bp: 3000°, d: 7.86, vap. press: 1 mm @ 1787°.
Acute tox data: Ip LD_{50} (mouse) = 26 mg/kg. [*3*]
THR = HIGH via ip route. Iron dust can cause conjunctivitis, choroiditis, retinitis and siderosis of tissues if iron remains in these tissues. Iron ore dust can cause palpebral conjunctivitis, massive pulmonary fibrosis and an increased incidence of lung cancer. An iron oxide fume is generated in welding operations and continued exposure to conc above 30 mg/m^3 of air can cause chronic bronchitis. Fresh iron oxide fume can cause metal fume fever. Iron compounds are susp carc of the lung, liver, connective tissue and reticuloendothelial tissue. [*14, 3*]
Radiation Hazard: For permissible levels, see Section, 5, Table 5A.5. Artificial isotope ^{55}Fe, $T_{\frac{1}{2}}$ = 2.6y, decays to stable ^{55}Mn via ec and emits x-rays. Artificial isotope ^{59}Fe, $T_{\frac{1}{2}}$ = 45d, decays to stable ^{59}Co via β's of 0.27 MeV (48%), 0.48 MeV (51%) and γ's. Emits γ's of 1.10 and 1.29 MeV.
Fire Hazard: Mod, in the form of dust when exposed to heat or flame. See also powdered metals. Reacts violently with Cl_2, ClF_3, F_2, H_2O_2, NO_2, P, Na_2C_2, H_2SO_4. [*19*]
Explosion Hazard: Mod in the form of dust when exposed to heat or flame. See also powdered metals.
To Fight Fire: Special mixtures of dry chemical.

IRON AMMONIUM CITRATE. Syn: *ferric ammonium citrate*. Thin, transparent, garnet red scales or granules or brownish-yellow powder, odorless or slight ammonia odor, sol in water, insol in alcohol.
THR = U. Used as a trace mineral added to animal feeds. [*109*]

IRON ARSENIDE. See ferric arsenide.

IRON BORIDE. Gray crystals. FeB, mw: 66.67.
THR = Details U. See boron hydrides and borides.
Fire Hazard: Mod; borides can react with moisture and acids to evolve toxic boron hydrides.
Explosion Hazard: A possible explosion hazard.
Disaster Hazard: Dangerous; can react with water, steam or acids to evolve toxic and flam fumes.

IRON CARBIDE. FeC_2, mw: 79.9.
THR = Violent reaction with Br_2, Cl_2. [*19*]

IRON CARBONATE.
THR = U. Used as a trace mineral added to animal feeds. See iron. [*109*]

IRON CARBONYL. See iron pentacarbonyl.

IRON (II) CHLORIDE. See ferrous chloride.

IRON (II) CHLORIDETETRAHYDRATE. $FeCl_2 \cdot 4H_2O$, mw: 198.8.
Acute tox data: Oral LD_{50} (rat) = 984 mg/kg; ip LD_{50} (mouse) = 93 mg/kg. [*3*]
THR = HIGH via ip and MOD via oral routes.

IRON COMPOUNDS. See corresponding ferric and/or ferrous compound.

IRON CONTAINING ASBESTOS.
THR = A susp carc for iron, a recog carc for asbestos. See iron and compounds; also asbestos. [*3, 14*]

IRON-DEXTRAN GLYCEROL GLYCOSIDE.
THR = An exper carc. [*3*]

IRON DEXTRAN COMPLEX. For human use, it is a sterile dark brown colloidal solvent, water-sol. mw: 180,000.
THR = An exper carc. A susp human carc. [*3, 6*]

IRON DEXTRIN COMPLEX. For human use, it is a clear, brown colloidal solvent. mw: 230,000.
THR = An exper (+) carc. [*3*, *6*]

IRON DIARSENIDE. See ferric diarsenide.

IRON DIMETHYL DITHIOCARBAMATE. See ferbam.

IRON FLUORIDE. See ferric fluoride.

IRON FORMATE. See ferric formate.

IRON GLUCONATE. See ferrous gluconate.

IRON HYDROGENATED DEXTRAN.
THR = An exper neo. [*3*]

IRON LACTATE. See ferrous lactate.

IRON MASS, SPENT. See iron sponge, spent.

IRON MONOPHOSPHIDE. Rhombic crystals. FeP, mw: 86.8, d: 5.2 @ 20°.
THR = See phosphides and iron.

IRON NONACARBONYL. Orange, hexagonal crystals. $Fe_2(CO)_9$, mw: 363.8, mp: decomp @ 100°, d: 2.805 @ 18°.
THR = See carbon monoxide and carbonyls and iron.

IRON ORE.
THR = A recog carc. [*14*] See arsenic and ferric oxide. [*3*]

IRON OXIDE. See also ferrous oxide.

IRON OXIDE (III). See ferric oxide.

IRON OXIDE (MAGNETITE). Fe_3O_4, mw: 231.6.
THR = Violent reaction with H_2S_3. [*19*]

IRON OXIDE, SACCHARATED.
THR = An exper (+) carc. [*3*, *6*]

IRON PENTACARBONYL. Syn: *iron carbonyl.* Yellow to dark red viscous liquid. $Fe(CO)_5$, mw: 195.9, mp: −25°, bp: 103.0°, flash p: 5°F, d: 1.453 @ 25°/4°, vap. press: 40 mm @ 30.3°.
Acute tox data: Oral LD_{50} (guinea pig) = 22 mg/kg; oral LD_{50} (rabbit) = 12 mg/kg. [*20*] Inhal LC_{50} (mouse) = 7 mg/kg. [*3*] Iv LD_{50} (rabbit) = 240 mg/kg; oral LD_{LO} (rabbit) = 18 mg/kg. [*3*] Iv LD_{LO} (rabbit) = 17 mg/kg. [*3*]
THR = HIGH via inhal, dermal, oral and iv routes. Inhal of this material causes dizziness, nausea and vomiting. If continued, unconsciousness follows. Often there is a delayed reaction of chest pain, cough and difficult breathing. There may be cyanosis and circulatory collapse. In fatal cases death occurs from the fourth to eleventh day with pneumonitis and injury to kidneys, liver and brain. Iron carbonyl is less toxic than nickel carbonyl.
Fire Hazard: Dangerous. See carbonyls. Pyrophoric in air!
Explosion Hazard: Moderate. See carbonyls.
Disaster Hazard: See carbonyls.
To Fight Fire: Water, foam, CO_2, dry chemical.

IRON PHOSPHATE. See ferric phosphate.

IRON-POLYMALTOSE.
THR = A susp carc. [*14*] See iron and compounds.

IRON POLYSACCHARIDE COMPLEX. mw: 20,000.
THR = An exper neo. [*3*]

IRON PYROPHOSPHATE. See ferric pyrophosphate.

IRON REDUCED. Syn: *ferrum reductum.* Grayish-black, amorphous, fine granular powder.
THR = U. Used as a trace mineral added to animal feeds and a nutrient and/or dietary supplement food additive. It migrates to food from packaging materials. See also iron. [*109*]

IRON-SORBITOL-CITRIC ACID COMPLEX.
THR = A susp carc. [*3*, *14*] See iron compounds.

IRON SPONGE (NOT PROPERLY OXIDIZED).
THR = Details U. See also iron.
Fire Hazard: Dangerous, when exposed to heat or flame; can react vigorously with oxidizing materials.

IRON SPONGE, SPENT. Iron + other materials.
THR = Details U. See also iron.
Fire Hazard: Dangerous, when exposed to heat or flame; can react vigorously with oxidizing materials.

IRON SULFATE. See ferrous sulfate.

IRON TETRACARBONYL. Dark green, lustrous crystals. $Fe(CO)_4$, mw: 167.88, mp: decomp @ 140°–150°, d: 1.996 @ 18°.
THR = See carbonyls and iron pentacarbonyl.

IRON-*m*-VANADATE. See ferric-*m*-vanadate.

IRRITATING AGENT, N.O.S. See Section 11.

ISANO OIL. Fatty oil from African tree of same name. d: 1.0.
THR = U.
Explosion Hazard: Dangerous; exothermic reaction above 502°F may become explosive.

ISATININE.
THR = An exper carc. [*23*]

ISETHONIC ACID. See hydroxyethyl sulfonic acid.

ISLANDI TOXIN.
THR = An hepato toxin and exper carc. [*23*]

ISOAMYL ACETATE. Syns: *amyl acetate, iso; banana oil.* Liquid, banana-like odor. $CH_3COO(C_5H_{11})$, mw: 130.2, bp: 142.0°, ulc: 55–60, lel = 1% @ 212°F, uel = 7.5%, flash p: 77°F, d: 0.876, autoign. temp.: 680°F, vap. d: 4.49.
Acute tox data: Inhal TC_{LO} (human) = 200 ppm ⟶ irr of lungs.
THR = MOD irr via oral and inhal routes. Exposure to conc of about 1,000 ppm for 1 hr can cause head-

ache, fatigue, pulmonary irr and serious toxicity effects.

Fire Hazard: Dangerous, when exposed to heat or flame; can react vigorously with reducing materials.

Explosion Hazard: Mod, when exposed to heat or flame.

Disaster Hazard: Dangerous; keep away from heat and open flame.

To Fight Fire: Alcohol foam, CO_2, dry chemical.

ISOAMYL ALCOHOL. Syn: *isopentyl alcohol.* Clear liquid, pungent, repulsive taste. $C_5H_{12}O$, mw: 88.15, bp: 132°, ulc: 35–40, lel = 1.2%, uel = 9.0% @ 212°F, flash p: 109°F (CC), d: 0.813, autoign. temp.: 662°F, vap. d: 3.04, mp: −117.2°.

Acute tox data: Oral LD_{50} (rat) = 1300 mg/kg; dermal LD_{50} (rabbit) = 3970 mg/kg; inhal TC_{LO} (rabbit) = 150 ppm ⟶ irr. [*3*] Oral LD_{50} (rats) = 7070 mg/kg. [*2*]

THR = MOD via oral, dermal routes. An exper carc. [*3*] Vapors are poisonous. [*2*]

Fire Hazard: Mod, when exposed to heat or flame; can react vigorously with reducing materials.

Explosion Hazard: Slight, when exposed to flame.

To Fight Fire: Alcohol foam, CO_2, dry chemical.

***tert*-ISOAMYL ALCOHOL.** See 2-methyl-2-butanol.

ISOAMYL AMINE. Colorless liquid, strong odor. $CH_3C_4H_8NH_2$, mw: 87, d: 0.751, bp: 95°.

THR = HIGH irr to skin, eyes and mu mem. [*3*] Intravenous injection in man has caused flushing and excitability.

8-ISOAMYL BENZ(a)ANTHRACENE.

THR = An exper carc. [*23*]

ISOAMYL BROMIDE. Colorless liquid. $(CH_3)_2CHC_2H_4Br$, mw: 151.1, bp: 119.7°–121.8°, flash p: 90°F, mp: −112°, d: 1.208 @ 25°/25°, vap. d: 5.25.

THR = U. Probably an irr and narcotic in HIGH conc. See also bromine compounds.

Fire Hazard: Mod, when exposed to heat or flame.

Disaster Hazard: Dangerous; when heated to decomp, emits toxic fumes of bromides; can react with oxidizing materials.

To Fight Fire: Water, foam, CO_2, dry chemical.

ISOAMYL BUTYRATE. Water-white liquid. $C_5H_{11}COOC_3H_7$, mw: 158.23, bp: 178.6°, d: 0.866 @ 15.5°, vap. press: 1 mm @ 21.2°, flash p: 138°F.

THR = LOW to MOD via oral route. See also esters.

Fire Hazard: Mod, when exposed to heat, flame or oxidizers.

To Fight Fire: Dry chemical, CO_2, mist, foam.

ISOAMYL CHLORIDE. $C_5H_{11}Cl$, mw: 106.6, flash p: < 70°F, lel = 1.5%, uel = 7.4%, d: 0.89, vap. d: 3.67, bp: 100°.

THR = No data.

Fire Hazard: Dangerous, via heat, open flame, oxidizers.

To Fight Fire: Dry chemical, CO_2, foam, mist.

ISOAMYL DICHLOROARSINE. Oily liquid, somewhat agreeable odor, decomp by water. $C_5H_{11}AsCl_2$, mw: 217.0, bp: 88°–91.5°.

THR = See also arsenic compounds.

α-ISOAMYLENE. See 3-methyl butene-1.

β-ISOAMYLENE. See 2-methyl butene-2.

ISOAMYL FORMATE. Clear liquid. Fruity odor. $C_5H_{11}OOCH$, mw: 116.16, bp: 123.3°, d: 0.877 @ 20°, vap. press: 10 mm @ 17.1°.

Acute tox data: Oral LD_{50} (rabbit) = 3020 mg/kg. [*3*]

THR = MOD via oral route. This material is very irr and can cause narcosis. The symptoms are usually transient in nature, but it is possible upon severe or prolonged exposure to have serious consequences.

Fire Hazard: Mod, when exposed to heat or flame; can react with oxidizing materials.

ISOAMYL HYDRIDE. See isopentane.

3-ISOAMYL-5-(METHYLENE DIOXYPHENYL)-2-CYCLOHEXENONE. See piperonyl cyclonene.

ISOAMYL NITRITE. See amyl nitrite.

ISOBENZAN. See telodrin.

ISOBORNEOL. A geometrical isomer of borneol. White solid, camphor odor, more sol in most solvents than borneol. $C_{10}H_{17}OH$, mw: 154.24, mp: 216° (sublimes).

THR = See borneol.

ISOBORNYL THIOCYANOACETATE. Syn: *thanite.* Yellow, oily liquid. $C_{13}H_{19}NO_2S$, mw: 253.4, d: 1.1465 @ 25°/4°.

Acute tox data: Oral LD_{50} (rat) = 0.2 mg/kg; oral LD_{50} (rabbit) = 630 mg/kg; oral LD_{50} (guinea pig) = 551 mg/kg. [*3*]

THR = VERY HIGH to MOD via oral route depending upon test species. HIGH irr to eyes, mu mem and skin. See also thiocyanates. An insecticide and fly spray.

Disaster Hazard: See thiocyanates.

ISOBUTANE. Syns: *2-methyl propane, trimethyl methane.* C_4H_{10}, mw: 58.12, bp: −11.7°, lel = 1.9%, uel = 8.5%, fp: −160°, d: 0.5572 @ 20°, autoign. temp.: 864°F, vap. d: 2.01.

THR = Practically NONE. An asphyxiant. A common air contaminant.

Fire Hazard: Very dangerous, when exposed to heat, flame or oxidizers.

Explosion Hazard: Severe, when exposed to heat or flame.
Disaster Hazard: Dangerous; on contact with oxidizing materials, can react vigorously.
To Fight Fire: Stop flow of gas.

ISOBUTANE THIOL. $(CH_3)_2CHCH_2SH$, mw: 90.2.
THR = Violent reaction with $Ca(OCl)_2$. [*19*]

ISOBUTANOL. See isobutyl alcohol.

ISOBUTANOLAMINE. See 2-amino-2-methyl-1-*n*-propanol.

ISOBUTENE. See isobutylene.

ISOBUTYL ACETATE. Colorless, neutral liquid, fruit-like odor. $C_4H_9OOCCH_3$, mw: 116.2, mp: −98.9°, bp: 118°, flash p: 64°F (CC), d: 0.8685 @ 15°, vap. press: 10 mm @ 12.8°, autoign. temp.: 793°F, vap. d: 4.0, lel = 2.4%, uel = 10.5%.
Acute tox data: Oral LD_{50} (rat) = 15,000 mg/kg; inhal LC_{LO} (rat) = 8000 ppm for 4 hrs. [*3*]
THR = LOW via oral and inhal routes. Upon absorption it can hydrolyze to acetic acid and isobutanol. See *n*-butyl acetate.
Fire Hazard: Dangerous, when exposed to heat, flame or oxidizers.
Spont Heating: No.
Disaster Hazard: Dangerous; keep away from heat or open flame.
To Fight Fire: Alcohol foam, CO_2, dry chemical.

ISOBUTYL ALCOHOL. Syns: *isopropylcarbinol, 2-methyl propanol-1, isobutanol.* Clear liquid, sweet odor. $(CH_3)_2CHCH_2OH$, mw: 74.12, bp: 107.90°, flash p: 82°F, ulc: 40–45, lel = 1.2%, uel = 10.9% @ 212°F, fp: −108°, d: 0.805 @ 20°/4°, autoign. temp.: 800°F, vap. press: 10 mm @ 21.7°, vap. d: 2.55.
Acute tox data: Oral LD_{50} (rat) = 2460 mg/kg; inhal LC_{LO} (rat) = 8000 ppm for 4 hrs; dermal LD_{50} (rabbit) = 4240 mg/kg. [*3*]
THR = MOD via oral, inhal and dermal routes. An exper carc. [*3*]
Fire Hazard: Dangerous, when exposed to heat or flame.
Explosion Hazard: Mod, in the form of vapor when exposed to heat, flame or oxidizers.
Disaster Hazard: Mod dangerous; emits toxic fumes when heated. Keep away from heat and open flame.
To Fight Fire: Alcohol foam, CO_2, dry chemical.

ISOBUTYL ALDEHYDE. Syns: *isobutyraldehyde, 2-methylpropanol.* Transparent, colorless, highly refractive liquid, pungent odor. $(CH_3)_2CHCHO$; mw: 72.10, mp: −65°, bp: 64°, flash p: −40°F (CC), d: 0.7938 @ 20°/4°, autoign. temp.: 490°F, lel = 1.6%, uel = 10.6%, vap. d: 2.5.
Acute tox data: Oral LD_{50} (rat) = 2810 mg/kg; inhal LC_{LO} (rat) = 8000 ppm for 4 hrs. [*3*]
THR = MOD via oral and inhal routes. See also aldehydes.
Fire Hazard: Dangerous, when exposed to heat, flame or oxidizers.
Explosion Hazard: Mod, in the form of vapor when exposed to heat or flame.
Disaster Hazard: Dangerous; keep away from heat and open flame; can react vigorously with reducing materials.
To Fight Fire: Dry chemical, CO_2, mist, foam.

ISOBUTYLAMINE. Syn: *1-amino-2-methylpropane.* Colorless liquid. $(CH_3)_2CHCH_2NH_2$, mw: 73.14, mp: −85.5°, bp: 68.6°, flash p: 15°F, d: 0.731 @ 20°/20°, vap. press: 100 mm @ 18.8°, autoign. temp.: 712°F, vap. d: 2.5.
THR = A powerful irr to skin, eyes and mu mem. Skin contact can cause blistering. Inhal can cause headache and dryness of nose and throat. See also amines. [*2*]
Fire Hazard: Dangerous, when exposed to heat or flame.
Spont Heating: No.
Disaster Hazard: Dangerous; keep away from heat and open flame; can react vigorously with oxidizing materials.
To Fight Fire: Dry chemical, foam, CO_2, alcohol foam.

8-ISOBUTYL BENZ(a)ANTHRACENE.
THR = An exper carc. [*23*]

ISOBUTYL BENZENE. Liquid. $C_4H_9C_6H_5$, mw: 134.23, mp: −51.5°, bp: 172.8°, flash p: 131°F (CC), d: 0.867, autoign. temp.: 806°F, vap. press: 1 mm @ 14.1°, vap. d: 4.62, lel = 0.8%, uel = 6.0%.
Acute tox data: Oral LD_{LO} (rat) = 5000 mg/kg. [*3*]
THR = MOD via oral route. An irr and possibly narcotic.
Fire Hazard: Mod, when exposed to heat or flame; can react with oxidizing materials.
To Fight Fire: Foam, CO_2, dry chemical.

ISOBUTYL BORIC ACID. Colorless crystals, sol in water. $C_4H_9B(OH)_2$, mw: 102.0, mp: 106°–112° (decomp).
THR = Details U. See also boron compounds.

ISOBUTYL BROMIDE. Colorless liquid. $(CH_3)_2CHCH_2Br$, mw: 137.0, mp: −118.5°, bp: 91°–93°, d: 1.258 @ 25°/25°, vap. d: 4.76.
THR = Details U. See bromides.
Fire Hazard: Mod, when exposed to heat or flame.
Disaster Hazard: Mod dangerous; when heated to decomp, emits highly toxic fumes of bromide; can react vigorously with oxidizing materials.
To Fight Fire: Water, foam, CO_2, dry chemical.

ISOBUTYL BUTYRATE. $C_8H_{16}O_2$, mw: 144.1, flash p: 122°F, d: 0.87, vap. d: 5.0, bp: 157.20.
THR = No data. See also esters.
Fire Hazard: Mod via heat, flame and oxidizers.
To Fight Fire: Dry chemical, CO_2, mist, foam.

ISOBUTYL CARBINOL. See isoamyl alcohol.

ISOBUTYL CHLORIDE. A liquid. $(CH_3)_2CHCH_2Cl$, mw: 92.5, lel = 2.0%, uel = 8.8%, d: 0.9, vap. d: 3.2, bp: 69°, mp: −131°.
THR = NO details. See esters.
Fire Hazard: Dangerous, when exposed to heat or flame.
To Fight Fire: Alcohol foam, water spray, fog, mist, dry chemical.
Disaster Hazard: Dangerous; see chlorides.

3-ISOBUTYL CHOLANTHRENE.
THR = An exper carc. [*23*]

***n*-ISOBUTYL-2,6,8-DECATRIENEAMIDE.**
THR = An insecticide. Limited data suggest LOW.

ISOBUTYLENE. Syns: *isobutene*, *2-methylpropene*. Volatile liquid or easily liquefied gas. $(CH_3)_2CCH_2$, mw: 56.1, bp: −6.9°, fp: −140.3°, flash p: −105°F, d: 0.600, autoign. temp.: 869°F, lel = 1.8%, uel = 9.6%.
THR = A simple asphyxiant; may have narcotizing action.
Fire Hazard: Very dangerous, when exposed to heat or flame.
Disaster Hazard: Dangerous; keep away from heat and open flame; can react vigorously with oxidizing materials.
To Fight Fire: Stop flow of gas.

ISOBUTYL FORMATE. $C_5H_{10}O_2$, mw: 102.2, flash p: < 70°F, autoign. temp.: 608°F, lel = 1.7%, uel = 8%.
Acute tox data: Oral LD_{50} (rabbit) = 3060 mg/kg. [*3*]
THR = MOD via oral route.
Fire Hazard: Dangerous, via heat, open flame, oxidizers.
To Fight Fire: Water spray, foam, CO_2, dry chemical.

ISOBUTYL ISOBUTYRATE. Liquid with fruity odor. $(CH_3)_2CHCOOCH_2CH(CH_3)_2$, mw: 144.2, mp: −80.7°, bp: 147.5°, d: 0.850–0.860 @ 20°/20°, vap. press: 10 mm @ 39.9°.
Acute tox data: Inhal LC_{LO} (rat) = 5000 ppm for 6 hrs. [*3*]
THR = MOD via inhal route. An insect-repellent. See also isobutyl alcohol.
Fire Hazard: Slight, when exposed to heat or flame. Can react with oxidizing materials.

ISOBUTYL METHACRYLATE. $C_8H_{14}O_2$, mw: 142.2.
THR = An exper teratogen. [*3*]

ISOBUTYLMETHYL CARBINOL. See methylamyl alcohol.

ISOBUTYLMETHYL KETONE. Syns: *methyl isobutyl ketone*, *hexone*, *4-methyl-2-pentanone*, *isopropylacetone*. Clear liquid. $(CH_3)_2CHCH_2COCH_3$, mw: 100.2, bp: 118°, lel = 1.4%, uel = 7.5%, flash p: 73°F, d: 0.803, fp: −80.2°, autoign. temp.: 858°F, vap. press: 16 mm @ 20°, d: 3.45.
Acute tox data: Oral LD_{50} (rat) = 2080 mg/kg; inhal LC_{LO} (rat) = 4000 ppm for 1/4 hr; ip LD_{50} (mouse) = 268 mg/kg. [*3*]
THR = MOD via oral and inhal routes and HIGH via ip route. Irr to eyes and mu mem. Narcotic in HIGH conc. See also ketones.
Fire Hazard: Dangerous, when exposed to heat, flame or oxidizers. Violent reaction with potassium *tert*-butoxide. [*19*]
Explosion Hazard: Mod, in the form of vapor when exposed to heat or flame.
Disaster Hazard: Dangerous; keep away from heat and open flame; can react vigorously with reducing materials.
To Fight Fire: Alcohol foam, CO_2, dry chemical.

ISOBUTYL PHOSPHATE. $C_{12}H_{27}O_4P$, mw: 266.2, flash p: 275°F (OC), d: 0.98, vap. d: 9.12, bp: 150° @ 20 mm.
THR = NO data.
Fire Hazard: Low.
To Fight Fire: Water spray, mist, dry chemical, CO_2, foam.

ISOBUTYL VINYL ETHER. Liquid. $C_4H_9OCHCH_2$, mw: 100.2, mp: −112°, bp: 82.9°–83.2°, flash p: 15°F, d: 0.76 @ 25°/4°, vap. d: 3.45.
Acute tox data: Oral LD_{50} (rat) = 17,000 mg/kg; dermal LD_{50} (rabbit) = 20,000 mg/kg. [*3*]
THR = Practically NONE via oral and dermal routes.
Fire Hazard: Dangerous, via heat, flame, oxidizers.
Explosion Hazard: Severe, when exposed to sparks or open flame. See ethers.
Disaster Hazard: Dangerous. Keep away from heat and open flame; can react vigorously with oxidizing materials.
To Fight Fire: Alcohol foam, CO_2, dry chemical.

ISOBUTYRALDEHYDE. See isobutyl aldehyde.

ISOBUTYRIC ACID. Colorless liquid, pungent odor. $(CH_3)_2CHCOOH$: mw: 88.1, mp: −47°, bp: 154.5°, flash p: 132°F (TOC), d: 0.949 @ 20°/4°, vap. press: 1 mm @ 14.7°, vap. d: 3.04, autoign. temp.: 935°F.
Acute tox data: Oral LD_{50} (rat) = 280 mg/kg; dermal LD_{50} (rabbit) = 500 mg/kg. [*3*]
THR = HIGH via oral and dermal routes.
Fire Hazard: Mod, when exposed to heat or flame; can react with oxidizing materials.
To Fight Fire: Alcohol foam, CO_2, dry chemical.

ISOBUTYRIC ANHYDRIDE. Liquid, decomp in water. $[(CH_3)_2CHCO]_2O$; mw: 158.19, bp: 360°F, d: 0.951–0.956 @ 20°/20°, vap. d: 5.5, flash p: 139°F, autoign. temp.: 665°F.
THR = HIGH irr to skin, eyes and mu mem.
Fire Hazard: Mod, when exposed to heat, flame, oxidizers.
To Fight Fire: Alcohol foam, fog, dry chemical, CO_2.

ISOBUTYRIC ETHER. See ethyl isobutyrate.

ISOBUTYRONITRILE. Syns: *2-methylpropanenitrile, isopropyl cyanide*. Colorless liquid, slightly sol in water, very sol in alcohol and ether. $(CH_3)_2CHCN$, mw: 69, d: 0.773 @ 20°/20°, bp: 107°, mp: −75°.
Acute tox data: Oral LD_{50} (rat) = 102 mg/kg; inhal LC_{Lo} (rat) = 1000 ppm for 4 hrs; dermal LD_{50} (rabbit) = 310 mg/kg. [*3*]
THR = HIGH via oral, inhal and dermal routes.
Disaster Hazard: Dangerous; see nitriles.

ISO BUTYROPHENONE. $C_6H_5CHCHO(CH_3)_2$, mw: 149.2.
THR = Violent reaction with Br_2. [*19*]

ISOCIL. See 5-bromo-3-isopropyl-6-methyl uracil.

ISOCINCHOMERONIC ACID. Off-white powder. $C_5H_3N(COOH)_2$, mw: 167.12, mp: 254°.
THR = U.
Disaster Hazard: Slight; when heated, emits acrid fumes.

ISOCROTONIC ACID. Syns: *2-butenoic acid, β-methacrylic acid*. The *cis*-isomeric form of crotonic acid. Liquid. $CH_3CH{:}CHCOOH$: mw: 86, mp: 15°, bp: 169°.
THR = HIGH irr to skin, eyes and mu mem.

ISOCROTYL CHLORIDE. See *β,β*-dimethyl vinyl chloride.

ISOCYANATES.
THR = Violent reaction with alcohols. [*19*]

ISOCYANIC ACID. See cyanic acid.

ISODECALDEHYDE. Insol in water. $C_9H_{19}CHO$, mw: 156, d: 0.829, bp: 197.0°, vap. d: 5.4, flash p: 185°F (OC).
THR = Probably LOW. See also aldehydes.
Fire Hazard: Mod, when exposed to heat, flame, oxidizers.
To Fight Fire: Dry chemical, CO_2, foam, fog.

ISODECANE. $C_{10}H_{22}$, mw: 142.2, autoign. temp.: 410°F, d: 0.73, vap. d: 4.91, bp: 167.2°.
THR = NO data. See also decane.
Fire Hazard: Mod, via heat, flame, oxidizers.
To Fight Fire: Dry chemical, CO_2, foam, mist.

ISODECANOIC ACID. Liquid; very slightly sol in water. $C_9H_{19}COOH$, mw: 172, bp: 254°, d: 0.9019 @ 20°/20°, fp: glass below −60°, vap. d: 5.9, flash p: 300°F (OC).
THR = U.
Fire Hazard: Slight, when exposed to heat or flame.
To Fight Fire: Dry chemical, CO_2, foam, mist.

ISODECANOL (MIXED ISOMERS). Insol in water. $C_{10}H_{21}OH$, mw: 158.3, d: 0.8395, bp: 220°, flash p: 220°F.
Acute tox data: Oral LD_{50} (rat) = 6400 mg/kg; dermal LD_{50} (rabbit) = 3150 mg/kg. [*3*]
THR = MOD via dermal and oral routes.
Fire Hazard: Low, when exposed to heat or flame.
To Fight Fire: Dry chemical, CO_2, foam, mist.

ISODECYL ADIPATE. See dodecyl adipate.

ISODECYL METHACRYLATE. $C_{14}H_{26}O_2$, mw: 226.4.
Acute tox data: Ip LD_{50} (rat) = 2467 mg/kg. [*3*]
THR = MOD via ip route. An exper teratogen. [*3*]

ISODECYL OCTYL ADIPATE. Liquid, low viscosity, water-white. $(C_{10}H_{21})(C_8H_{17})(CO_2)_2(CH_2)_4$, mw: 398.6, mp: −60°, bp: 215°–240° @ 4 mm, flash p: 400°F, d: 0.924 @ 20°/20°, vap. d: 13.8.
THR = Details U. See esters.
Fire Hazard: Slight, when exposed to heat or flame; can react with oxidizing materials.
To Fight Fire: Foam, CO_2, dry chemical.

ISODRIN. See 1,2,3,4,10,10-hexachloro-1,4,4a,5,8,8a-hexahydro-1,4,5,8-endo-endo-dimethanonaphthalene.

ISOFLURANE. Syn: *forane*. $C_3H_2OClF_4$, mw: 165.5.
THR = A susp carc. [*102*]

ISOFORM. Syn: *p-iodoanisole*. White powder. $C_6H_4(OCH_3)IO_2$, mw: 265.98.
THR = U. See iodides.
Disaster Hazard: Mod; see iodides; explodes at 225°.

ISOHEPTANE. See 2-methylhexane.

8-ISOHEPTYL BENZ(a)ANTHRACENE.
THR = An exper carc. [*23*]

ISOHEXANE. Syn: *mixture of hexane isomers*. Liquid. C_6H_{14}, mw: 86.17, lel = 1.0%, uel = 7.0%, bp: 54°–60°, flash p: 20°F (CC), d: 0.669, vap. d: 3.00, autoign. temp.: 583°F.
THR = See hexane.
Fire Hazard: Dangerous, when exposed to heat, flame or oxidizers.
Explosion Hazard: Severe, when exposed to heat or flame.
Disaster Hazard: Dangerous; keep away from sparks, heat or open flame; can react vigorously with oxidizing materials.
To Fight Fire: Foam, CO_2, dry chemical.

ISOHEXYL ALCOHOL. See amyl methyl alcohol.

***tert*-ISOHEXYL ALCOHOL.** $C_6H_{14}O$; mw: 102.1, flash p: 115°F, d: 0.77, vap. d: 3.53, bp: 116.7.

THR = NO data. See also alcohols.
Fire Hazard: Mod, via heat, flame, oxidizers.
To Fight Fire: Water spray, foam, dry chemical, CO_2.

8-ISOHEXYL BENZ(a)ANTHRACENE.
THR = An exper carc. [*23*]

1,3-ISOINDOLEDIONE. See phthalimide.

ISOLAN. Syn: *1-isopropyl-3-methyl-5-pyrazolyl dimethyl carbamate*. A liquid, miscible in water.
$C_{10}H_{17}N_3O_2$, mw: 211.3, bp: 103° @ 0.7 mm, d: 1.07 @ 20°.
Acute tox data: Oral LD_{50} (rat) = 13 mg/kg; dermal LD_{50} (rat) = 5.6 mg/kg; oral LD_{50} (wild birds) = 9 mg/kg; oral LD_{50} (chicken) = 3.3 mg/kg. [*3*]
THR = HIGH via oral and dermal routes. A cholinesterase inhibitor. Although this is not an organic phosphate, it resembles that group in action. See parathion for effects. See carbamates.
Disaster Hazard: Dangerous; when heated to decomp, emits highly toxic fumes.

ISOLEUCINE. Syns: *α-amino-β-methylvaleric acid, 2-amino-3-methyl pentanoic acid*. An essential amino acid; many isomeric forms; we consider the *dl* and *l* forms. Crystalline, slightly sol in water, nearly insol in alcohol, insol in ether.
$CH_3CH_2CH(CH_3)CH(NH_2)COOH$, mw: 131; mp: (*dl*): 292° (decomp), mp (*l*): 283°–284° (decomp).
THR = U. A nutrient and/or dietary supplement food additive. [*109*]

ISOMER OF YOHIMBINE. See yohimbene.

ISOMETHEPTENE. Syns: *octin, 1,5-trimethyl-4-hexenylamine*. Colorless oily liquid, characteristic amine odor, water-insol.
$(CH_3)_2CCHCH_2CH_2CH_2CH(NHCH_3)CH_3$, mw: 141.3, d: 0.795, bp: 177°.
Acute tox data: Oral LD_{50} (rat) = 150 mg/kg. [*3*]
THR = HIGH via oral route. Can cause headache, nausea, dizziness.

ISONAPHTHOL. See β-naphthol.

ISONIAZID SODIUM METHANE SULFONATE. $Na \cdot C_7H_9O_4N_3S$, mw: 254.2.
THR = HIGH via oral route. An exper neo. [*3*]

ISONICAZID. See isonicotinic acid hydrazide.

ISONICOTINIC ACID HYDRAZIDE. Syn: *isonicazide*. $C_6H_7ON_3$, mw: 137.2.
THR = An exper (+) carc and neo. [*3, 10*]

ISONICOTINIC ACID SODIUM SALT. $Na \cdot C_6H_5O_2N$. mw: 146.1.
THR = An exper neo. [*3*]

ISOOCTANE. Syns: *2,2,4-trimethyl pentane, 2-methyl heptane*. Clear liquid, odor of gasoline. $(CH_3)_2CH(CH_2)_4CH_3$, mw: 114.2, bp: 99.2°, fp: −116°, flash p: 10°F, d: 0.692 @ 20°/4°, autoign. temp.: 779°F, vap. press: 40.6 mm @ 21°, vap. d: 3.93, lel = 1.1%, uel = 6.0%.
THR = MOD via oral and inhal routes. High conc can cause narcosis.
Fire Hazard: Dangerous, when exposed to heat, flame, oxidizers.
Spont Heating: No.
Disaster Hazard: Dangerous; keep away from heat and open flame; can react vigorously with reducing materials.
To Fight Fire: CO_2, dry chemical.

ISOOCTANOIC ACID (MIXED ISOMERS). Liquid, water-sol. $C_6H_{15}COOH$, mw: 132, d: 0.9, vap. d: 5.0, bp: 428°F (decomp), flash p: 270°F (OC).
THR = U.
Fire Hazard: Slight, when exposed to heat or flame.
To Fight Fire: Dry chemical, CO_2, fog.

ISOOCTENES. C_8H_{16}, mw: 112.1, bp: 102°–107°, flash p: $<$20°F, d: 0.723 @ 15.5°/15.5°, vap. press: 102 mm @ 21°, d: 3.9.
THR = U. May have narcotic properties.
Fire Hazard: Dangerous, when exposed to heat or flame.
Disaster Hazard: Dangerous; keep away from heat and open flame; can react with oxidizing materials.
To Fight Fire: Dry chemical, CO_2, fog, foam.

ISOOCTYL DECYLPHTHALATE. Water white, low viscosity liquid. $C_{10}H_{21}O_2CC_6H_4CO_2C_8H_{17}$, mw: 418, bp: 232°–243° @ 4 mm, pour p: −43°, flash p: 445°F (COC), d: 0.973 @ 20°, vap. d: 14.4.
THR = See esters.
Fire Hazard: Slight, when exposed to heat or flame; can react with oxidizing materials.
To Fight Fire: Foam, CO_2, dry chemical.

ISOOCTYL-2,4-DICHLOROPHENOXYACETATE. $C_{16}H_{22}O_3Cl_2$, mw: 383.3.
THR = HIGH via oral to rats. An exper teratogen. [*3*]

ISOOCTYL NITRATE. Insol in water. $C_8H_{17}NO_3$, bp: 106°–109° @ 1 mm, d: 1.0, flash p: 205°F (OC), mw: 175.
THR = U. See nitrates, organic.
Fire Hazard: Low, when exposed to heat or flame.
Disaster Hazard: Dangerous. See nitrates.
To Fight Fire: Water spray, dry chemical, CO_2.

ISOOCTYL THIOGLYCOLATE. A clear, water white liquid, fruity odor. $HSCH_2COOC_8H_{17}$, mw: 204.33, bp: 52° @ 17 mm, d: 0.9736 @ 25°.
THR = See thioglycolic acid. Probably toxic.
Disaster Hazard: Dangerous; see SO_x; can react vigorously with oxidizing materials.

ISOPENTANE. Syns: *2-methyl butane, isoamyl hydride*. Colorless liquid with pleasant odor.

$CH_3CHCH_3CH_2CH_3$, mw: 72.15, bp: 27.8°, fp: −160.5°, flash p: < −60°F (CC), vap. press: 595 mm @ 21.1°, vap. d: 2.48, lel = 1.4%, uel = 7.6%.
THR = See pentane.
Fire Hazard: Highly dangerous, when exposed to heat, flame or oxidizers.
Explosion Hazard: U.
Disaster Hazard: Dangerous; keep away from sparks, heat or open flame; can react with oxidizing materials.
To Fight Fire: Foam, CO_2, dry chemical.

ISOPENTYL ALCOHOL. See isoamyl alcohol.

ISOPESTEX.
See bis(isopropylamido)fluorophosphate.

ISOPHORONE. Practically water-white liquid. $COCHC(CH_3)CH_2C(CH_3)_2CH_2$, mw: 138.1, bp: 215.2°, flash p: 184°F (OC), d: 0.9229, autoign. temp.: 864°F, vap. press: 1 mm @ 38.0°, vap. d: 4.77, lel = 0.8%, uel = 3.8%.
Acute tox data: Oral LD_{50} (rat) = 2330 mg/kg; dermal LD_{50} (rabbit) = 1500 mg/kg; inhal LD_{LO} (rat) = 1840 ppm for 4 hrs. [*3*]
THR = MOD via oral, dermal and inhal routes. Considered to be more toxic than mesityl oxide. However, due to its low volatility, it is not a dangerous industrial hazard. The response of guinea pigs and rats to repeated inhal of the vapors indicates that it is one of the most toxic of the ketones. It is chiefly a kidney poison. It can cause irr, lachrymation, possible opacity of the cornea and necrosis of the cornea (exper). It is irr at the level of 25 ppm to humans. In animal exper death during exposure was usually due to narcosis, but occasionally to irr of the lungs.
Fire Hazard: Mod, when exposed to heat or flame; can react with oxidizing materials.
Explosion Hazard: U.
To Fight Fire: Foam, CO_2, dry chemical.

ISOPHOSPHAMIDE. $C_7H_{14}O_2N_2PCl_2$, mw: 260.1.
Acute tox data: oral TD_{LO} (human) = 150 mg/kg ⟶ CNS effects. Oral TD_{LO} (humans) = 100 mg/kg ⟶ blood dyscrasias; ip LD_{LO} (mice) = 646 mg/kg. [*3*]
THR = An exper neo [*3*] to mice via ip route. [*113*]

ISOPHTHALIC ACID. Syns: *m-phthalic acid, IPA*. Colorless crystals, slightly sol in water, sol in alcohol and acetic acid, insol in benzene and petroleum ether. $C_6H_4(COOH)_2$, mw: 166, mp: 345°–348°.
Acute tox data: Ip LD_{50} (mouse) = 4200 mg/kg. [*3*]
THR = MOD via ip route and probably oral route. Mild irr.

ISOPHTHALOYL CHLORIDE. Syn: *m-phthalyl chloride*. White crystalline solid, sol in benzene and carbon tetrachloride. $C_8H_4O_2Cl_2$, mw: 203.0, bp: 276°, fp: 43.3°, d: 1.387 @ 46.9°, flash p: 356°F (COC), vap. d: 6.9.
Acute tox data: Oral LD_{50} (rat) = 4290 mg/kg; dermal LD_{50} (rabbit) = 1410 mg/kg. [*3*]
THR = MOD via oral and dermal routes.
Disaster Hazard: Dangerous; see chlorides.
Fire Hazard: Low.
To Fight Fire: Dry chemical, CO_2, spray, mist.

ISOPRAL. See 1,1,1-trichloroisopropyl alcohol.

ISOPRENE. Syn: *2-methyl-1,3-butadiene*. Colorless, volatile liquid. $CH_2C(CH_3)CHCH_2$, mw: 68.11, mp: −146.7°, bp: 34°, flash p: −65°F, d: 0.6806 @ 20°/4°, autoign. temp.: 428°F, vap. press: 400 mm @ 15.4°, vap. d: 2.35.
THR = MOD irr to skin, eyes and mu mem. Isoprene is irr to mu mem of the eyes, nose and upper respiratory passages. Conc of 2% in air not narcotic to mice but produces bronchial irr. Conc of 5% fatal to mice. No data on human exposures.
Fire Hazard: Highly dangerous, when exposed to heat, flame or oxidizers; i.e., violent reaction with chlorosulfonic acid, HNO_3, oleum, H_2SO_4. [*19*]
Disaster Hazard: Dangerous; keep away from sparks, heat or open flame; can react vigorously with reducing materials.
To Fight Fire: CO_2, dry chemical.

ISOPROPANOL. See isopropyl alcohol.

ISOPROPANOLAMINE. See 1-amino-2-propanol.

ISOPROPANOLAMINES, MIXED. Clear liquid. $CH_3CH(OH)CH_2NH_2$, mw: 75.1, mp: 29.5°, bp: 159°, flash p: 160°F (OC), d: 0.962, vap. d: 2.58.
Acute tox data: Oral LD_{50} (rat) = 5240 mg/kg; dermal LD_{50} (rabbit) = 8900 mg/kg; oral LD_{50} (guinea pig) = 1520 mg/kg. [*3*]
THR = MOD via oral and LOW via dermal routes.
Fire Hazard: Mod, when exposed to heat or flame; can react with oxidizing materials.
To Fight Fire: Foam, CO_2, dry chemical.

ISOPROPENYL ACETATE. Water white liquid. $C_5H_2O_2$, mw: 94.1, mp: −92.9°, bp: 96.6° @ 746 mm, flash p: 60°F (OC), d: 0.9226 @ 20°/20°, vap. d: 3.45.
Acute tox data: Oral LD_{50} (rat) = 3000 mg/kg. [*3*]
THR = MOD via oral route. An irr to eyes and mu mem.
Fire Hazard: Dangerous, when exposed to heat, flame, oxidizers.
Disaster Hazard: Dangerous; keep away from heat and open flame; can react vigorously with oxidizing materials.
To Fight Fire: Alcohol foam, CO_2, dry chemical.

ISOPROPENYL ACETYLENE. Syn: *2-methyl-1-buten-3-yne*. Colorless liquid, very slightly sol in water, miscible with acetone, alcohol, benzene, carbon tetrachloride and kerosene. $CH_2:C(CH_3)C:CH$, mw: 66, bp: 33°–34°, fp: −113°, d: 0.695 @ 20°/20°, flash p: < 19°F (OC), vap. d: 2.3.

THR = U.

Fire Hazard: Very dangerous, when exposed to heat or flame.

To Fight Fire: Alcohol foam.

ISOPROPENYL CHLORIDE. See 2-chloropropene.

ISOPROPENYL CHLOROFORMATE. Liquid. $ClCOOC(CH_3)CH_2$, mw: 120.5, bp: 93° @ 746 mm, d: 1.103 @ 20°.

THR = HIGH irr to skin, eyes and mu mem.

Disaster Hazard: Dangerous; when heated to decomp, emits toxic fumes of chlorides.

***o*-ISOPROPOXYPHENYL METHYL CARBAMATE.** See baygon.

2-ISOPROPOXYPROPANE. See isopropyl ether.

β-ISOPROPOXY PROPIONITRILE. Liquid. $C_3H_7OCH_2CH_2CN$, mw: 113.16, mp: −67°, bp: 82°–86°, flash p: 155°F, d: 0.9058, vap. d: 3.91.

THR = HIGH. See nitriles.

Fire Hazard: Mod, when exposed to heat, flame, oxidizers.

Disaster Hazard: Dangerous; see cyanides.

To Fight Fire: Foam, CO_2, dry chemical.

3-ISOPROPOXY PROPYLAMINE. Syn: *3-dimethyl carbinol propylamine.* Colorless liquid. $(CH_3)_2CHOCH_2CH_2CH_2NH_2$, mw: 118.2, bp: 147°, d: 0.845 @ 20°/20°, vap. d: 4.07.

THR = HIGH irr to skin, eyes and mu mem. See also amines.

Disaster Hazard: Dangerous; when heated, emits toxic fumes; can react with oxidizing materials.

ISOPROPYL ACETATE. Colorless aromatic liquid. $CH_3COOCH(CH_3)_2$, mw: 102.13, mp: −73°, bp: 88.4°, lel = 1.8%, uel = 7.8%, fp: −69.3°, flash p: 40°F, d: 0.874 @ 20°20°, autoign. temp.: 860°F, vap. press: 40 mm @ 17.0°, d: 3.52.

Acute tox data: Oral LD_{50} (rat) = 3000 mg/kg; inhal LC_{LO} (rat) = 32,000 ppm for 4 hrs; 200 ppm ⟶ TD_{LO} (human) Affects the eyes. [*3*]

THR = MOD via oral and LOW via inhal routes. Narcotic in high conc. Chronic exposure can cause liver damage.

Fire Hazard: Dangerous, when exposed to heat, flame or oxidizers.

Spont Heating: No.

Explosion Hazard: Mod, when exposed to heat or flame.

Disaster Hazard: Dangerous; keep away from heat and open flame; can react vigorously with oxidizing materials.

To Fight Fire: Foam, CO_2, dry chemical.

ISOPROPYL ACETONE. See isobutyl methyl ketone.

ISOPROPYL ALCOHOL. Syns: *dimethyl carbinol, sec-propyl alcohol, isopropanol.* Clear colorless liquid, slight odor. $CH_3CHOHCH_3$, mw: 60.09, mp: −88.5°–−89.5°, bp: 80.3°, lel = 2.0%, uel = 12%, flash p: 53°F, d: 0.7854 @ 20°/4°, vap. d: 2.07, ulc: 70.

Acute tox data: Dermal LD_{50} (rabbit) = 16,000 mg/kg; oral LD_{50} (rat) = 5840 mg/kg; ip LD_{50} (mouse) = 933 mg/kg; oral LD_{50} (dog) = 6000 mg/kg. [*3*]

THR = LOW via dermal and MOD via oral and ip routes. The single LD for a human adult = about 250 ml. [*20*] An irr to the eyes. [*87*] Acts as a local irr and in high conc as a narcotic. It can cause corneal burns and often eye damage. It has good warning properties because it causes a mild irr of the eyes, nose and throat, at conc levels of 400 ppm. It may induce a mild narcosis, the effects of which are usually transient, and it is somewhat less toxic than the normal isomer, but twice as volatile. It is not considered an important toxic hazard. There is some evidence that personnel can acquire a slight tolerance to this material, and single or repeated applications of it on the skin of rats, rabbits, dogs or human beings induced no untoward effects. It acts very much like ethanol in regard to absorption, metabolism and elimination but with a stronger narcotic action. Chronic injuries due to it have been detected in animals. Workers producing isopropyl alcohol show an excess of sinus cancers and laryngeal cancers. This may all or in part be due to the by-product, isopropyl oil. [*81, 87*] Humans have ingested up to 20 ml diluted with water and noticed only a sensation of heat and slight lowering of the blood pressure. There are, however, reports of serious illness from as little as 10 ml taken internally. A food additive permitted in food for human consumption. [*109*] A common air contaminant. Absorbed by skin.

Fire Hazard: Dangerous, when exposed to heat, flame or oxidizers.

Spont Heating: No.

Explosion Hazard: Mod, when exposed to heat or flame. Reacts violently with (H_2 + Pd), nitroform, oleum, $COCl_2$, potassium-*tert*-butoxide. [*19*]

Disaster Hazard: Dangerous; keep away from heat and open flame; can react vigorously with oxidizing materials.

To Fight Fire: CO_2, dry chemical, alcohol foam.

ISOPROPYLAMINE. Syn: *2-aminopropane.* Colorless liquid, amino odor. $(CH_3)_2CHNH_2$, mw: 59.11,

mp: −101.2°, bp: 31.7°, flash p: −35°F (OC), d: 0.694 @ 15°/4°, autoign. temp.: 756°F, d: 2.03.
Acute tox data: Oral LD_{50} (rat) = 820 mg/kg; inhal LD_{LO} (rat) = 800 ppm for 8 hrs; dermal LD_{50} (rabbit) = 550 mg/kg. [*3*]
THR = MOD via oral, inhal and dermal routes. A strong irr. Occasionally causes sensitization. Narcotic in HIGH conc.
Fire Hazard: Very dangerous, when exposed to heat, flame or oxidizers.
Disaster Hazard: Dangerous; keep away from sparks, heat or open flame; can react vigorously with oxidizing materials.
To Fight Fire: Alcohol foam, foam, CO_2, dry chemical.

β-ISOPROPYLAMINO PROPIONITRILE. Liquid. $C_3H_7NHCH_2CH_2CN$, mw: 112.18, mp: −20° bp: 86° @ 17 mm, flash p: 100°F, d: 0.864, vap. d: 3.87.
THR = HIGH. See nitriles.
Fire Hazard: Mod, when exposed to heat or flame.
Disaster Hazard: Dangerous; see cyanides; will react with water or steam to produce toxic and flam vapors; reacts vigorously with oxidizing materials.
To Fight Fire: Foam, CO_2, dry chemical.

3-ISOPROPYLAMINO PROPYLAMINE. Colorless liquid. $(CH_3)_2CHNHCH_2CH_2CH_2NH_2$, mw: 116.2, bp: 159°, vap. d: 4.00, d: 0.827 @ 25°/4°.
THR = HIGH irr to skin, eyes and mu mem. See also amines.
Fire Hazard: Slight, when exposed to heat or flame.
Disaster Hazard: Slight, can react with oxidizing materials.

ISOPROPYL ANTIMONITE. $(C_3H_7O)_3Sb$, mw: 299.1, bp: 82° @ 7 mm.
THR = See antimony compounds and organometals.

8-ISOPROPYL BENZ(a)ANTHRACENE. $C_{21}H_{18}$, mw: 270.4.
THR = An exper carc and neo. [*3, 23*]

9-ISOPROPYL BENZ(a)ANTHRACENE.
THR = An exper carc. [*3, 23*]

ISOPROPYL BENZENE. See cumene.

ISOPROPYL BENZOATE. Liquid. $C_6H_5COOCH(CH_3)_2$, mw: 164.2, mp: −26.4°, bp: 219°, flash p: 210°F (OC), d: 1.0112, vap. press: 0.12 mm @ 20°, vap. d: 5.67.
Acute tox data: Oral LD_{50} (rat) = 3730 mg/kg. [*3*]
THR = MOD via oral route.
Fire Hazard: Low, when exposed to heat or flame, can react with oxidizing materials.
To Fight Fire: Water, spray, CO_2, dry chemical.

5-ISOPROPYL BENZO(c)PHENANTHRENE. $C_{21}H_{18}$, mw: 270.4.
THR = An exper carc and neo. [*3, 23*]

2-ISOPROPYL BICYCLOHEXYL. $C_{15}H_{28}$, mw: 208.2, flash p: 255°F, autoign. temp.: 446°F, lel = 0.5% @ 302°F, uel = 41% @ 400°F, d: 0.9.
THR = NO data. Probably LOW.
Fire Hazard: Low.
To Fight Fire: Dry chemical, CO_2, foam, mist.

2-ISOPROPYL BIPHENYL. $C_{15}H_{16}$, mw: 196.2, flash p: 285°F, autoign. temp.: 815°F, lel = 0.5% @ 347°F, uel = 3.2% @ 392°F, d: 1.0, bp: 270°.
THR = NO data. Probably LOW.
Fire Hazard: Low.
To Fight Fire: Dry chemical, CO_2, mist.

ISOPROPYL BROMIDE. Colorless liquid. $CH_3CHBrCH_3$, mw: 123.0, bp: 58.5°–60.5°, fp: −90°, d: 1.304 @ 25°/25°, vap. d: 4.27.
THR = Details U. Animal exper suggest LOW systemic. Narcotic in HIGH conc.
Disaster Hazard: Dangerous; see bromides.

ISOPROPYL CARBAMATE. $C_4H_9O_2N$, mw: 103.1.
THR = An exper carc. [*3*]

ISOPROPYL CARBANILATE. See IPC.

ISOPROPYL CARBINOL. See isobutyl alcohol.

ISOPROPYL CHLORIDE. Syn: *2-chloropropane*. Clear liquid. $CH_3CHClCH_3$, mw: 78.55, bp: 35.3°, fp: −117.6°, lel = 2.8%, uel = 10.7%, flash p: −26°F (CC), d: 0.858 @ 25°/25°, autoign. temp.: 1100°F, vap. d: 2.71.
THR = Details U. Animal exper suggest MOD systemic. Has been used as a surgical anesthetic.
Fire Hazard: Highly dangerous; when exposed to heat, flame or oxidizers.
Explosion Hazard: Mod, when exposed to heat or flame.
Disaster Hazard: Dangerous; keep away from sparks, heat or open flame; emits highly toxic fumes of chlorides, such as phosgene; can react vigorously with oxidizing materials.
To Fight Fire: Dry chemical, CO_2, mist.

ISOPROPYL-*m*-CHLOROCARBANILATE. See isopropyl-N-(3-chlorophenyl)carbamate.

ISOPROPYL CHLOROFORMATE. Clear colorless liquid; a phosgene derivative. $(CH_3)_2CHOOCCl$, mw: 122.5.
Acute tox data: Oral LD_{50} (rat) = 1070 mg/kg; inhal LC_{LO} (rat) = 200 ppm for 5 hrs.
THR = HIGH irr via inhal and MOD via oral routes. Can cause pulmonary edema.
Disaster Hazard: Dangerous; see chlorides. Has exploded while stored in a refrigerator. Reacts violently with phosgene. [*19*]

ISOPROPYL-N-(3-CHLOROPHENYL)CARBAMATE. Syn: *isopropyl-m-chlorocarbanilate*. Light

brown crystalline solid, faint characteristic odor. $C_{10}H_{12}ClNO_2$, mw: 213.66, mp: 41°, bp: 247° (decomp).
THR = See carbamates. An insecticide and an herbicide. MOD via oral route. An exper neo. [*3*] via oral route. Has initiated skin carc via oral route. [*12*]
Disaster Hazard: Dangerous; when heated to decomp, emits highly toxic fumes of phosgene.

3-ISOPROPYL CHOLANTHRENE. $C_{23}H_{20}$, mw: 296.4.
THR = An exper neo. [*3, 23*]

ISOPROPYL CITRATE.
THR = U. Used as a sequestrant food additive. [*109*]

ISOPROPYL CYANIDE. See isobutyronitrile.

ISOPROPYL CYCLOHEXYL AMINE. Insol in water. $C_6H_{11}NHCHC_2H_6$, mw: 141, d: 0.8, vap. d: 4.9, flash p: 93°F (OC).
THR = U. See also amines.
Fire Hazard: Dangerous, when exposed to heat, flame or oxidizers.
To Fight Fire: Dry chemical, CO_2, foam.

14-ISOPROPYL DIBENZ(a,j)ACRIDINE. $C_{24}H_{19}N$, mw: 321.4.
THR = An exper neo. [*3*]

ISOPROPYL-2,4-DICHLOROPHENOXY ACETATE. Nearly colorless crystals.
$C_6H_3Cl_2OCH_2COOCH(CH_3)_2$, mw: 263.13, d: 1.255–1.270 @ 25°/25°.
Acute tox data: Oral LD_{50} (rat) = 700 mg/kg; oral LD_{50} (mouse) = 541 mg/kg; oral LD_{50} (chicken) = 1420 mg/kg. [*3*]
THR = MOD via oral route.
Disaster Hazard: Mod dangerous; when heated to decomp, emits toxic fumes of chlorides.

8-ISOPROPYL-7,12-DIMETHYL BENZ(a)ANTHRACENE.
THR = An exper carc. [*23*]

ISOPROPYL ETHER. Syns: *2-isopropoxy propane, diisopropyl ether*. Colorless liquid, ethereal odor, miscible in water. $(CH_3)_2CHOCH(CH_3)_2$, mw: 102.17, mp: −60°, bp: 68.5°, lel = 1.4%, uel = 7.9%, flash p: −18°F (CC), d: 0.719 @ 25°, autoign. temp.: 830°F, vap. press: 150 mm @ 25°, vap. d: 3.52.
Acute tox data: Oral LD_{50} (rat) = 8470 mg/kg; ip LD_{50} (mouse) = 812 mg/kg; dermal LD_{50} (rabbit) = 20,000 mg/kg. [*3*]
THR = MOD via ip and LOW via oral and dermal routes. 800 ppm ⟶ irr to humans; 16,000 ppm is narcotic to rats. An irr.
Fire Hazard: Dangerous, when exposed to heat, flame or oxidizers.
Spont Heating: No.
Explosion Hazard: Severe, when exposed to heat or flame. Can form peroxides and explode upon shaking unless treated with sodium sulfite. Violent reaction with chlorosulfonic acid, HNO_3. [*19*]
Disaster Hazard: Dangerous; keep away from heat, sparks or open flame; under some conditions shock will explode it; emits highly toxic fumes; reacts vigorously with oxidizing materials.
To Fight Fire: Alcohol foam, CO_2, foam, dry chemical.

ISOPROPYL ETHYL URETHANE. $C_5H_{13}O_2N$, mw: 119.2.
THR = An exper neo. [*3*]

ISOPROPYL FORMATE. Syn: *isopropyl methanoate*. Clear liquid. $C_3H_7CHO_2$, mw: 88.1, bp: 68.3°, flash p: 22°F (CC), d: 0.873, autoign. temp.: 905°F, vap. press: 100 mm @ 17.8°, vap. d: 3.03.
Acute tox data: Oral LD_{50} (guinea pig) = 1.4 mg/kg. [*3*]
THR = HIGH via oral route. A toxic fumigant. See also esters.
Fire Hazard: Dangerous, when exposed to heat or flame.
Spont Heating: No.
Disaster Hazard: Dangerous; keep away from heat and open flame; can react vigorously with oxidizing materials.
To Fight Fire: Alcohol foam, foam, CO_2, dry chemical.

ISOPROPYL GLYCIDYL ETHER. Syn: *IGE.* Liquid. $C_6H_{12}O_2$, mw: 116.
Acute tox data: Oral LD_{50} (rat) = 4200 mg/kg; inhal LC_{50} (rat) = 1100 ppm for 8 hrs. [*3*]
THR = MOD via oral and inhal routes. Also skin irr and sensitizer. Can cause CNS depression.

4-ISOPROPYL HEPTANE. $C_{10}H_{22}$, mw: 142.1, autoign. temp.: 491°F, d: 0.87, vap. d: 3.04, bp: 68.3°.
THR = NO data. Probably very LOW.
Fire Hazard: Mod, via heat, flame, oxidizers.
To Fight Fire: Alcohol foam.

ISOPROPYL HYPOCHLORITE. $(CH_3)_2CHOCl$, mw: 94.5.
THR = MOD unstable. In light, it explodes. Reacts violently with alcohols. [*19*]

4-ISOPROPYLIDENE-3,3-DIMETHYL-2-OXETANONE. $C_8H_{12}O_2$, mw: 140.2.
THR = An exper neo. [*3*]

ISOPROPYL LACTATE. Syn: *isopropyl-2-hydroxypropanoate*. Sol in water. $CH_3CHOHCOOCH(CH_3)_2$, mw: 132, d: 1.0, vap. d: 4.2, bp: 331°–334°F, flash p: 130°F (OC).
THR = U. See esters.
Fire Hazard: Mod, when exposed to heat, flame or oxidizers.
To Fight Fire: Alcohol foam.

ISOPROPYL MANDELATE. Crystals. $C_6H_5CH(OH)COOC_3H_7$, mw: 194.2.
THR = An irr. See also esters.
Fire Hazard: Mod.

ISOPROPYL MERCAPTAN. Liquid, extremely powerful unpleasant odor. $(CH_3)_2CH(HS)$, mw: 76, d: 0.814 @ 60°/60°F, boiling range: 51°–55°, flash p: −30°F.
THR = Probably MOD via inhal route.
Fire Hazard: Dangerous, when exposed to heat, flame or oxidizers.
Disaster Hazard: Dangerous; when heated to decomp, emits highly toxic fumes. See mercaptans.

ISOPROPYL METHANOATE. See isopropyl formate.

ISOPROPYL METHANE SULFONATE. $C_4H_{10}O_3S$, mw: 138.2.
THR = An exper mutagen and teratogen. [*3*, *23*]

N-ISOPROPYL-α-(2-METHYLAZO)-*p*-TOLUAMIDE. $C_{12}H_{17}ON_3$, mw: 219.3.
THR = An exper carc. [*3*]

7-ISOPROPYL-12-METHYL BENZ(a)ANTHRACENE.
THR = An exper carc. [*23*]

8-ISOPROPYL-7-METHYL BENZ(a)ANTHRACENE.
THR = An exper carc. [*23*]

1-ISOPROPYL-2-METHYLETHYLENE. See 4-methyl-2-pentene.

N-ISOPROPYL-α-(2-METHYL HYDRAZINO)-*p*-TOLUAMIDE. $C_{12}H_{19}ON_3$, mw: 221.3.
THR = An exper transplacental carc, teratogen and neo. [*3*, *23*]

N-ISOPROPYL-α-(2-METHYL HYDRAZINO)-*p*-TOLUAMIDE HYDROCHLORIDE. $C_{12}H_{19}ON_3 \cdot HCl$; mw: 257.8.
THR = An exper carc. [*3*]

ISOPROPYL NITRATE. Syn: *2-propanol nitrate.* Liquid. $(CH_3)_2CHNO_3$, mw: 105, bp: 102°.
THR = U. See nitrates, organic.
Fire Hazard: See nitrates.
Disaster Hazard: Dangerous. See nitrates.

ISOPROPYL NITRITE. Syn: *2-propanol nitrite.* Pale yellow oil. $C_3H_7NO_2$, mw: 89.1, d: 0.856 @ 0°/4°, bp: 39° @ 752 mm.
THR = See nitrites and also amyl nitrite.

ISOPROPYL OIL.
THR = A by-product of isopropyl alcohol manufacture composed of trimeric and tetrameric polypropylene + small amounts of benzene, toluene, alkyl benzenes, polyaromatic ring compounds, hexane, heptane, acetone, ethanol, isopropyl ether and isopropyl alcohol. A recog carc of para nasal sinuses, larynx and lung. [*14*] An exper (±) carc of para nasal sinuses. [*81*]

ISOPROPYL PERCARBONATE.
See diisopropyl peroxydicarbonate.

ISOPROPYL PEROXYDICARBONATE. See diisopropyl peroxydicarbonate.

***o*-ISOPROPYL PHENOL.** Syn: *o-cumenol.* Colorless to amber liquid. $(CH_3)_2CHC_6H_4OH$, mw: 136.2, mp: 15.5°, bp: 209°, flash p: 220°F, d: 0.995 @ 25°/25°, vap. press: 1 mm @ 56.6°.
THR = U. See also phenol and esters.
Fire Hazard: Mod, when exposed to heat, flame or oxidizers.
Disaster Hazard: Dangerous; when heated to decomp, emits highly toxic fumes of phenol; can react with oxidizing materials.

***o*-ISOPROPYL-N-PHENYL CARBAMATE.** See IPC.
THR = An exper neo. [*3*] MOD acute oral. See also carbamates.

3-ISOPROPYL PHENYL-N-METHYL CARBAMATE. See UC 10854.

4-ISOPROPYL PYRIDINE. Liquid. $(CH_3)_2CHC_5H_4N$, mw: 121.3, bp: 182.2°, d: 0.9282 @ 20°/20°, vap. d: 4.18.
THR = See pyridine.

ISOPROPYL STANNIC ACID. White powder, insol in water. C_3H_7SnOOH, mw: 194.8.
THR = Details U. See tin compounds.

N-ISOPROPYL TEREPHTHALAMIC ACID.
$C_{11}H_{13}O_3N$, mw: 207.3.
THR = An exper neo. [*3*]

N-ISOPROPYL TEREPHTHALALDEHYDAMIDE-4-METHYL HYDRAZONE. $C_{12}H_{17}ON_3$, mw: 219.3.
THR = An exper carc. [*3*]

ISOPROPYL TETRATHIO-*o*-STANNATE. Solid. $(C_3H_7)_4Sn$, mw: 419.3, bp: 92° @ 0.001 mm.
THR = See tin and sulfur compounds.

ISOPROPYLTIN TRIBROMIDE. Pale yellow crystals, water-sol. $(C_3H_7)SnBr_3$, mw: 401.5, mp: 112°.
THR = See tin compounds and bromides.

ISOPROPYLTIN TRICHLORIDE. Solid, insol in water. $(C_3H_7)SnCl_3$, mw: 286.2, bp: 75° @ 16 mm.
THR = See tin compounds and chlorides.

ISOPROPYL TITANATE. Syn: *titanium isopropylate.* Colorless liquid. $Ti(OC_3H_7)_4$, mw: 284.2, mp: 14.8°, bp: 104° @ 10 mm, d: 0.954, vap. d: 9.8.
THR = Details U. See organometals.
Fire Hazard: Mod, when exposed to heat or flame.
Disaster Hazard: Will react with water or steam to

produce flam vapors; can react with oxidizing materials.

ISOPROPYL TOLUENE. See *p*-cymene.

ISOPROPYL TRICHLOROSILANE. Liquid.
$C_3H_7SiCl_3$, mw: 177.5, mp: −87.7°, bp: 119.4°, d: 1.196.
THR = HIGH irr to skin, eyes and mu mem. See also silanes.
Disaster Hazard: Dangerous; when heated to decomp, emits highly toxic fumes of chlorides; will react with water or steam to produce toxic and corrosive fumes.

ISOQUINOLINE. Syns: *2-benzazine, benzo(c)pyridine.* Liquid, pungent odor, almost insol in water, miscible with many organic solvents, sol in dilute acids.
$C_6H_4C_3H_3N$, mw: 129.15, d: 1.09101 @ 30°/4°, mp: 27°, bp: 243°.
Acute tox data: Oral LD_{50} (rat) = 350 mg/kg; dermal LD_{50} (rabbit) = 590 mg/kg. [*3*]
THR = HIGH via oral and MOD via dermal routes.

***trans*-ISOSAFROLE.** Liquid, odor of anise.
$C_{10}H_{10}O_2$, mw: 162.2, bp: 253°, mp: 8.2°.

ISOSAFROLE-*n*-OCTYL SULFOXIDE.
See sulfoxcide.

ISOTHAN. See lauryl isoquinolinium bromide.

ISOTHIOCYANATES. See thiocyanates.

ISOTOC. See 1,2,3,4,5,6-hexachlorocyclohexane.

ISOTRICYCLOQUINAZOLINE NITRATE.
$C_{21}H_{12}N_4HNO_3$.
THR = An exper neo. [*3*]

ISOVALERALDEHYDE. See isovaleric aldehyde.

ISOVALERIC ALDEHYDE. Syns: *isovaleraldehyde, 3-methylbutanal.* Colorless liquid, apple-like odor.
$(CH_3)_2CHCH_2CHO$, mw: 86.13, mp: −51°, bp: 92.5°, d: 0.803 @ 17°/4°, vap. d: 2.96.
THR = Details U. See also aldehydes.
Disaster Hazard: Slight; when heated, emits acrid fumes.

ISOVALERONE. See diisobutyl ketone.

2-ISOVALERYL-1,3-INDANDIONE. See valone.

ITACONIC ACID. Syn: *methylene succinic acid.* White crystalline powder, hygroscopic.
$CH_2C(COOH)CH_2COOH$, mw: 130.10, mp: 167.5° (decomp), d: 1.63.
THR = LOW via oral and inhal routes.
Fire Hazard: Slight.

j

JACOBINE. An alkaloid. $C_{18}H_{25}O_6N$, mw: 351.4.
THR = An exper (±) carc. [*3, 7, 23*]

JAPAN LACQUER.
THR = MOD irr to skin, eyes and mu mem. An allergen. Dermatitis is frequently caused by natural Japan lacquer due to a highly irr chemical, urushiol. Synthetic Japan lacquer contains linseed oil, lead oxide and pigments, and solvents such as kerosene or turpentine.
Fire Hazard: Mod, when exposed to heat or flame; can react with oxidizing materials.

JAVELLE WATER. A solution of chlorinated potash.
THR = See hypochlorites.

JET FUELS. Petroleum products similar to kerosene; a number of different types are used. Jet A and Jet A-1: flash p: 110°-150°F, bp: 400°-550°F.
Fire Hazard: Mod.
Jet B: flash p: −16°-−30°F.
Fire Hazard: Dangerous.
JP-1: flash p: 95°-145°F, autoign. temp.: 442°F.
JP-4: 65% gasoline, 35% light petroleum distillate. flash p: −10° to 30°F, autoign. temp.: 468°F, lel = 1.3%, uel = 8%.
Fire Hazard: Dangerous.
JP-5: Specially refined kerosene. flash p: 95°-145°F, autoign. temp.: 475°F (approx).
Fire Hazard: Mod.
JP-6: A higher kerosene cut than JP-4, with fewer impurities. flash p: 100°F (OC), autoign. temp.: 446°F, lel = 0.6%, uel = 3.7%, d: 0.8, vap. d: < 1, bp: 250°F.
THR = Violent reaction with F_2. [*19*]

J-O. A mixture containing white phosphorus.
THR = See phosphorus, white.

JUNIPER BERRIES. Dried ripe fruit of *Juniperus communis L. pinaceae.*
THR = MOD irr to skin, eyes and mu mem. An allergen. A systemic irr. The oil of juniper berries can be irr to the skin. If taken internally, a severe kidney irr similar to that caused by turpentine may result.
Fire Hazard: U.
Disaster Hazard: Slightly dangerous; when heated, emits toxic fumes.

JUNIPER TAR. See cade oil.

JUTE. A combustible fiber. In the form of dust, it may be highly flam. See cotton dust.

K-101. See *p*-chlorophenyl-*p*-chlorobenzene sulfonate.

KAOLIN. Syns: *china clay, hydrated aluminum silicate*. White or yellowish-white earthy mass of powder, insol in water. $H_2Al_2Si_2O_8 \cdot H_2O$, mw: 258.2.

THR = LOW. A nuisance dust. Toxicity depends upon SiO_2 content. See also nuisance dusts. An allergen. Prolonged inhal of HIGH conc may cause deposits to form on lungs.

KAPOK. A fibrous material.

THR = A source of nuisance dust. See also nuisance dusts. An allergen. Gives off a nuisance dust. Inhal can cause coughing and sneezing.

Fire Hazard: Slight, when exposed to heat or flame; can react with oxidizing materials.

Explosion Hazard: Mod, in the form of dust when exposed to flame.

KARATHANE. See dinocap.

KARAYA GUM. Syn: *Indian tragacanth*. Fine white powder, slight odor of acetic acid. Dried exudate of the tree, *Sterculia ureus*.

THR = A MILD allergen. A stabilizer food additive. [*109*]

KARMEX. See diuron.

KATCHUNG OIL. See peanut oil.

KEL-F. $(C_2ClF_3)_n$.

THR = An exper carc, [*3*] by implantation.

KELP. A large, coarse seaweed. Dried kelp contains 2–4% ammonia, 1–2% phosphoric acid, 15–20% potash, and traces of iodine.

THR= U. A food additive permitted in food for human consumption. [*109*]

KELP DUST. See sawdust.

KELTHANE.

See 1,1-bis (*p*-chlorophenyl)-2,2,2-trichloroethanol.

KEPONE. Syn: *decachlorooctahydro-1,3,4-metheno-2H-cyclobuta(cd)pentalen-2-one*. A chlorinated polycyclic ketone, a crystalline material, slightly water-sol, but sol in alcohols, ketones and acetic acid. $C_{10}Cl_{10}O$, mw: 490.7, mp: decomp @ 350°.

Acute tox data: Oral LD_{50} (rat) = 96 mg/kg (in corn oil); dermal LD_{50} (rat) = 250 mg/kg; an exper carc. [*3*]

THR = In man, inhal, absorption or ingestion can lead to CNS, liver and kidney damage, including bizarre symptoms caused by damage to the nervous system. Usually, the symptoms are tremors, ataxia, skin changes, hyperexcitability, hyperactivity, muscle spasms, testicular atrophy, low sperm count, estrogenic effects, sterility, breast enlargement, liver lesions and cancer. An insecticide and a fungicide.

Disaster Hazard: See chlorides.

KERMES MINERAL. See antimony, sulfurated.

KEROSENE. Syn: *fuel oil # 1*. A pale yellow to water-white oily liquid. A mixture of petroleum hydrocarbons, chiefly of the methane series, having 10–16 atoms of C. bp: 175°–325°, ulc: 40, flash p: 100°–165°F (CC), d: 0.80 to <1.0, lel = 0.7%, uel = 5.0%, autoign. temp.: 410°F, vap. d: 4.5.

Acute tox data: Oral LD_{50} (rat or rabbit) = 28 g/kg; [*2*] iv LD_{50} (rabbit) = 180 mg/kg. [*3*]

THR = HIGH via iv and VERY LOW via oral routes. Inhal of HIGH conc of vapor can cause headache and stupor. Ingestion causes irr of the stomach and intestines with nausea and vomiting. Aspiration of vomitus can cause serious pneumonitis, particularly in young children. It is a susp carc. [*14*] See mineral oils.

Fire Hazard: Mod, when exposed to heat or flame; can react with oxidizing materials.

Explosion Hazard: Mod, when exposed to heat or flame.

To Fight Fire: Foam, CO_2, dry chemical.

KEROSENE, DEODORIZED. See ultrasene.

KERYL BENZENE-12. Syn: *monodecyl benzene*. Very light straw color, practically odorless. $C_{12}H_{25}C_6H_5$, mw: 246.4, flash p: 280°F, d: 0.894 @ 20°, boiling range: 291°–354°.

THR = Details U. Probably irr and narcotic in HIGH conc. See benzene.

Fire Hazard: Slight, when exposed to heat or flame; can react with oxidizing materials.

To Fight Fire: Foam, CO_2, dry chemical.

KERYL NAPHTHALENE 10. Syn: *monodecyl naphthalene*. Color of straw, odor of naphthalene. $C_{10}H_7C_{10}H_{21}$, mw: 268.4, flash p: 325°F, d: 0.968 @ 20°, boiling range: 341°–420°.

THR = Details U. See also naphthalene.

Fire Hazard: Slight, when exposed to heat or flame; can react with oxidizing materials.

To Fight Fire: Foam, CO_2, dry chemical.

KETEN. See ketene.

KETENE. Syns: *ethenone, carbomethane, keten*. Colorless gas of disagreeable taste. Decomp in water. CH_2CO, mw: 42.04, mp −151°, bp: −56°, vap. d: 1.45.
Acute tox data: Oral LD_{50} (rat) = 1300 mg/kg; inhal LC_{LO} (rat) = 53 ppm for 100 min; inhal LC_{LO} (guinea pig) = 53 ppm for 4 hrs. [*3*]
THR = MOD via oral and HIGH via inhal routes. Can cause pulmonary edema.

β-KETOBUTYRANILIDE. See acetoacetanilide.

KETOCYCLOPENTANE. See cyclopentanone.

β-KETOGLUTARIC ACID. See acetone dicarboxylic acid.

KETOHEXAMETHYLENE. See cyclohexanone.

2′-KETO-3,4-IMIDOAZOLIDO-2-TETRAHYDROTHIOPHENE-*n*-VALERIC ACID. See biotin.

KETONE OILS. See acetone oils.

KETONE PROPANE. See acetone.

KETONES. Organic compounds containing the chemical group =CO derived from secondary alcohols by oxidation. Acetone, which is dimethyl ketone, is the most familiar of this group of compounds. See acetone. No general statement can be made as to the toxicity of ketones. Some are highly volatile and hence may have narcotic or anesthetic effects. Skin absorption as well as inhal may be an important route of entry into the body. None of the ketones has been shown to have a high degree of chronic toxicity. Some of them are dangerous fire hazards. See diethyl ketone. They react violently with aldehydes, HNO_3, (HNO_3 + H_2O_2), $HClO_4$. [*19*] Common air contaminants.

KHAINI.
THR = A susp carc. [*14*] See plant and fungal products.

KIESELGUHR. See diatomaceous earth.

KILL-TONE POWDER. See arsenic oxide and copper compounds.

KING'S GREEN. See copper acetoarsenite.

KOLKER ACARICIDE. See *p*-chlorophenyl-*p*-chlorobenzene sulfonate.

KOLLIDON. See polyvinyl pyrrolidone.

KRYOFIN. Syn: *methoxy acetyl-p-phenetidine*. White needles, tasteless but becoming bitter on chewing. $CH_3OCH_2CONHC_6H_4OC_2H_5$, mw: 209.24, mp: 98°.
THR = MOD via oral route.
Fire Hazard: Slight.

KRYPTON. Colorless inert gas. Kr, atwt: 83.8, mp: −156.7°, bp: −152.0°, d: 3.708 g/liter @ 0°, (liq) 2.155 @ 152.9°.
THR = A simple asphyxiant in large conc. See also argon.
Radiation Hazard: For permissible levels, see Section 5, Table 5A.5. Artificial isotope ^{85}Kr, $T_{\frac{1}{2}}$ = 10.8 y. Decays to stable ^{85}Rb via β's of 0.67 MeV. ^{85}Kr, like ^{41}A is produced by activation of natural krypton in air-cooled reactors. Also, a direct fission product.

K-STROPHANTHIN. See strophanthin.

KURON. See 2-(2,4,5-trichlorophenoxy) propionic acid.

L

LABELING SYSTEMS. See Section 11.

LABORATORY HOODS. See Section 2.

LACQUER BASE OR CHIPS, DRY. See lacquers, nitrocellulose.

LACQUER BASE OR CHIPS, PLASTIC (wet with alcohol or solvents). See specific components.

LACQUER DILUENT. Insol in water. d: 0.7, bp: 190°–225°F, flash p: 12°F, autoign. temp.: 450°–550°F, lel = 1.2%, uel = 6.0%.
THR = U.
Fire Hazard: Dangerous, when exposed to heat or flame. See lacquers.

LACQUERS. Solutions of resins, gums or plastics in an organic solvent. flash p: 0°–80°F.
THR = Variable; they may have allergic effects. They are common air contaminants.
Fire Hazard: Dangerous, when exposed to heat or flame. A large part of the great fire hazard of lacquers is due to the solvents commonly used. Even when solvent free, however, nitrocellulose lacquers are highly flam. See lacquers, nitrocellulose.
Explosion Hazard: Severe, in the form of vapor when exposed to flame. See also individual solvents as well as individual solid components.
Disaster Hazard: Dangerous. Keep away from heat and open flame; they can react vigorously with oxidizing materials.
Explosive Range: Variable.
To Fight Fire: CO_2, dry chemical, water spray.

LACQUERS, NITROCELLULOSE. flash p: 40°F.
THR = Variable. They may have allergic effects.
Fire Hazard: Highly dangerous, when exposed to heat or flame.
Explosion Hazard: Mod, when exposed to heat or flame.
Disaster Hazard: Dangerous; on decomp, they emit highly toxic fumes.
To Fight Fire: CO_2, dry chemical.

LACTIC ACID. Syn: *ethylidene lactic acid.* Yellow to colorless, thick liquid. $CH_3CHOHCOOH$, mw: 90.08, mp: 18°, bp: 122° @ 15 mm. d: 1.249 @ 15°.
Acute tox data: Oral LD_{50} (rat) = 3730 mg/kg; oral LD_{50} (mouse) = 4875 mg/kg; sc LD_{50} (mouse) 4500 mg/kg; oral LD_{50} (guinea pig) = 1810 mg/kg. [*3*]
THR = MOD via oral and sc routes. Used as a general purpose food additive. [*109*] Reacts violently with (NHO_3 + HF). [*19*]
Disaster Hazard: Slight; emits acrid fumes when heated.

LACTIDE. See 3,6-dimethyl-2,5-p-dioxanedione.

LACTOL SPIRITS. See naphtha, petroleum.

LACTONES. Lactones are esters of hydroxy acids. Toxicity is variable but in general seems to be relatively LOW. An exception is β-propiolactone.

LACTONITRILE. Syns: *2-hydroxypropane nitrile, acetaldehydecyanohydrin.* Straw-colored liquid. $CH_3CH(OH)CN$, mw: 71.08, mp: −40°, bp: 103° @ 50 mm, fp: −34°, flash p: 170°F (TCC), d: 0.9834 @ 25°, vap. d: 2.45.
Acute tox data: Oral LD_{50} (rat) = 87 mg/kg; inhal LC_{LO} (rat) = 125 ppm for 4 hrs; dermal LD_{LO} (rabbit) = 140 mg/kg. [*3*]
THR = HIGH via oral, dermal and inhal routes. See also nitriles. In the presence of alkali, evolves hydrocyanic acid.
Fire Hazard: Mod, when exposed to heat, flame or oxidizers.
Explosion Hazard: See cyanides.
Disaster Hazard: Dangerous; see cyanides; can react vigorously with oxidizing material.
To Fight Fire: Foam, CO_2, dry chemical.

LACTOSE. Syn: *milk sugar.* $C_{12}H_{22}O_{11}$, mw: 342.3.
Acute tox data: iv LD_{LO} (rabbit) = 1500 mg/kg. [*3*]
THR = MOD via iv route. An exper neo via sc route. [*3*]

LAGGING. See Section 3.

LAMP BLACK. See soot.

LAMBROL. Syn: *2-fluoroethyl(4-biphenylyl)acetate.* A brown solid, sol in organic solvents. An insecticide. $C_{16}H_{15}O_2F$, mw: 258, mp: 60.6°.
Acute tox data: Oral LD_{50} (rat) = 5 mg/kg; dermal LD_{50} (rat) = 4 mg/kg; dermal LD_{50} (rabbit) = 7 mg/kg; oral LD_{50} (dog) = 2 mg/kg. [*3*]
THR = HIGH via oral and dermal routes.

LANDFILLS AS DANGER POINTS IN SOLID WASTE HANDLING. See Section 6.

LANDRIN. Syn: *trimethyl phenyl methyl carbamate (mixture of isomers).* $C_{11}H_{15}O_2N$, mw: 193.1.
Acute tox data: Oral LD_{50} (rat) = 208 mg/kg. [*3*]
THR = HIGH via oral routes.

LANNATE. Syn: *s-methyl-n[(methylcarbamoyl)oxy]-thioacetimidate*. White crystal solid, slightly sulfurous odor, mod water-sol. $C_6H_{10}O_2N_2S$, mw: 162.1, mp: 79°.

Acute tox data: Oral LD_{50} (rat) = 17 mg/kg; oral LD_{50} (wild birds) = 10 mg/kg; oral LD_{50} (duck) = 15 mg/kg; [*3*] dermal LD_{50} (rabbit) = 1500 mg/kg. [*95*]

THR = HIGH via oral and MOD via dermal routes. An insecticide and nematocide.

LANOLIN. Syn: *wool grease.* mp: 37.0°, flash p: 460°F (CC), d: < 1, autoign. temp.: 833°F.

THR = A MILD allergen.

Fire Hazard: Light, when exposed to heat or flame; can react with oxidizing materials.

Spont Heating: Yes.

To Fight Fire: CO_2, dry chemical.

LANOXIN. See digoxin.

LANTHANA. See lanthanum oxide.

LANTHANUM. Silvery-white, malleable and ductile metal, soft enough to cut with a knife. Very reactive rare earth metal; reacts with H_2O, C, N, B, Se, Si, P, S and halogens. La, atwt: 138.906, mp: 920°, bp: 3454°, d: 6.166 @ 25°.

Acute tox data: iv LD_{50} (rat) = 3.5 mg/kg. [*3*]

THR = HIGH via iv route. Lanthanum and other lanthanons can cause delayed blood clotting, leading to hemorrhages. Has caused liver injury in exper animals.

Radiation Hazard: For permissible levels, see Section 5, Table 5A.5. Artificial isotope ^{140}La, $T_{\frac{1}{2}}$ = 40 h decays to stable ^{140}Ce by β's of 0.83–2.2 MeV. Emits γ's of 0.32–2.5 MeV. ^{140}La usually exists in equilibrium with its parent, ^{140}Ba. Permissible levels are given for the equilibrium mixture.

Fire Hazard: Dangerous, in the form of dust when exposed to flame; can react vigorously with oxidizing materials. See also powdered metals. Reacts violently with air, P, halogens. [*19*]

Explosion Hazard: Mod, in the form of dust when exposed to flame or by chemical reaction. See also powdered metals.

LANTHANUM ACETATE.

Acute tox data: ip LD_{50} (rat) = 475 mg/kg; sc LD_{50} (mouse) = 3500 mg/kg. [*3*]

THR = HIGH via ip and MOD via sc routes. See also lanthanum.

LANTHANUM BORIDE. Syn: *lanthanum hexaboride.* Red-purple metallic crystals. LaB_6, mw: 203.8, mp: 2210°, bp: decomp: d: 2.61.

THR = See boron hydrides and lanthanum.

LANTHANUM BROMATE. Crystals. $La(BrO_3)_3 \cdot 9H_2O$, mw: 684.81, mp: 37.5°.

THR = Probably MOD via oral route. See also lanthanum.

Fire Hazard: See bromates.

Explosion Hazard: See bromates.

Disaster Hazard: See bromates.

LANTHANUM BROMIDE. Colorless crystals. $LaBr_3 \cdot 7H_2O$, mw: 504.78, mp: anh. 783 ± 3°, d: 5.057 @ 25°.

THR = See bromides and lanthanum.

LANTHANUM CARBIDE. Yellow crystals. LaC_2, mw: 162.94, d: 5.02.

THR = See lanthanum and carbides.

LANTHANUM AMMONIUM NITRATE.

Acute tox data: Oral LD_{50} (rat) = 3400 mg/kg; ip LD_{50} (rat) = 625 mg/kg. [*3*]

THR = MOD via oral and ip routes.

LANTHANUM HEXAANTIPYRINE PERCHLORATE. Colorless, hexagonal crystals. $(La[(C_{11}H_{12}N_2O)_6](ClO_4)_3$, mw: 1566.62, mp: 290°–305° decomp.

THR = See perchlorates and lanthanum.

LANTHANUM HEXABORIDE.

See lanthanum boride.

LANTHANUM IODATE. Colorless crystals. $La(IO_3)_3$, mw: 663.68.

THR = See iodates and lanthanum.

LANTHANUM IODIDE. Crystals. LaI_3, mw: 519.68, mp: 761 ± 2°, d: 5.057 @ 25°.

THR = See iodides and lanthanum. Violent reaction with ClF_3. [*19*]

LANTHANUM NITRATE. $La(NO_3)_3$, mw: 324.9.

Acute tox data: Oral LD_{50} (rat) = 4500 mg/kg; ip LD_{50} (rat) = 450 mg/kg; ip LD_{50} (mouse) = 410 mg/kg. [*3*]

THR = HIGH via ip and MOD via oral routes.

LANTHANUM OXIDE. Syns: *lanthana, lanthanum trioxide.* White, amorphous oxide, hisses in moist air like quicklime. La_2O_3, mw: 325.84, mp: 2315°, bp: 4200°, d: 6.51.

THR = See lanthanum. An irr and caustic.

LANTHANUM PHOSPHIDE. LaP, mw: 169.9.

THR = Violent reaction with water. [*19*] See also phosphides. Probably HIGH via oral and inhal routes.

LANTHANUM SULFATE. White, hygroscopic powder. $La_2(SO_4)_3$, mw: 566, mp: 1150° (decomp), d: 3.60 @ 15°.

Acute tox data: ip LD_{50} (rat) = 275 mg/kg. [*3*]

THR = HIGH via ip and probably via oral route. See also lanthanum. Sulfuric acid is formed upon hydrolysis of this material.

LANTHANUM SULFIDE. Red-yellow crystals. La_2S_3, mw: 374.04, mp: 2110°–2150° (in vacuo), d: 5.0 @ 0°/0°.
THR = See sulfides and lanthanum.

LANTHANUM TRIOXIDE. See lanthanum oxide.

LARAHA. Syn: *Citrus Aurantium.*
THR = An exper neo to rats via im route. [*3*]

LARD OIL (COMMERCIAL OR ANIMAL). Colorless or pale yellow liquid.
Composition: olein, stearin, mp: −2°, ulc: 10–20, flash p: 395°F (CC), d: 0.905–0.915, autoign. temp.: 838°F.
THR = A very weak allergen. It is an exper carc. [*14*] See thermic and oxidation products, etc.
Fire Hazard: Slight, when exposed to heat or flame; can react with oxidizing materials.
Spont Heating: Yes.
To Fight Fire: CO_2, dry chemical.

LARD OIL (PURE). Syn: *No 2 mineral oil.* Colorless or pale yellow liquid. mp: −2°, flash p: 500°F, d: 0.9, ulc: 10–20, flash p: (#1) = 440°F, (#2) = 419°F.
THR = A substance migrating to food from packaging materials. [*109*] It is a (S) carc. [*14*] See thermic and oxidation products, etc.
Fire Hazard: Slight, when exposed to heat or flame; can react with oxidizing materials.
Spont Heating: Yes.
To Fight Fire: CO_2, dry chemical.

LARKSPUR. See delphinium.

LASIOCARPINE. Alkaloid. $C_{21}H_{33}O_7N$, mw: 411.6.
THR = An exper (+) carc. [*3, 7, 23*]

LATERAL HOOD. See Section 2.

LATEX RUBBER. Syn: *polydimethyl siloxane.*
THR = MOD irr to skin, eyes and mu mem. An allergen. An exper neo via sc route. [*3*]

LAUGHING GAS. See nitrous oxide.

LAUREL CAMPHOR. See camphor.

LAURIC ACID. Syn: *dodecanoic acid.* Colorless, needle-like crystals. $CH_3(CH_2)_{10}COOH$, mw: 200.3, mp: 48°, bp: 299.2°, d: 0.883, vap. press: 1 mm @ 121.0°.
Acute tox data: iv LD_{50} (mouse) = 131 mg/kg. [*3*]
THR = HIGH via iv route. Probably LOW via oral route. An exper tumor promoting agent. [*23*]
Fire Hazard: Slight, when exposed to heat or flame; can react with oxidizing materials.

LAURIC ACID-2-THIOCYANOETHYL ESTER. Syn: *lethane 60.* $C_{15}H_{27}O_2NS$, mw: 285.5.
Acute tox data: Oral LD_{50} (rat) = 500 mg/kg. [*3*]
THR = HIGH via oral route.

LAURITE. See ruthenium sulfide.

LAUROYL ETHYLENE IMINE. $C_{14}H_{27}ON$, mw: 225.4.
THR = An exper neo. [*3*]

LAUROYL PEROXIDE. Syns: *dodecanoyl peroxide, Alperox C.* White, tasteless, coarse powder, faint odor. $C_{24}H_{46}O_4$, mw: 398.7, mp: 53°–55°.
THR = HIGH irr to skin, eyes, mu mem. See peroxides, organic. An exper neo. [*3*] It can irr and cause burns upon the skin and mu mem; it is a powerful oxidizing agent. See also peroxides, organic.
Fire Hazard: See peroxides, organic.
Disaster Hazard: See peroxides, organic.

LAURYL ALCOHOL. See 1-dodecanol.

LAURYL BROMIDE. Syn: *dodecyl bromide.* An amber liquid. $CH_3(CH_2)_{10}CH_2Br$, mw: 249.2, mp: $< -40°$, bp: 136.6°–193.5° @ 10 mm, flash p: 291°F, d: 1.021 @ 25°/25°, vap. d: 8.6.
THR = U. Probably MOD. See also bromides.
Fire Hazard: Low, via heat, flames, oxidizers.
To Fight Fire: Dry chemical, CO_2.

LAURYL CHLORIDE. Syn: *dodecyl chloride.* Clear, water-white, oily liquid. $CH_3(CH_2)_{10}CH_2Cl$, mw: 204.8, mp: −19°, bp: 112°–160°, flash p: 235°F, d: 0.87, vap. d: 7.07.
THR = U. Probably LOW.
Fire Hazard: Slight, when exposed to heat or flame.
Disaster Hazard: Mod dangerous; when heated to decomp, emits toxic fumes; can react with oxidizing materials.
To Fight Fire: Foam, CO_2, dry chemical.

N-LAURYL ETHYLENE IMINE.
THR = An exper carc. [*23*]

LAURYL ISOQUINOLINIUM BROMIDE. $C_{21}H_{32}NBr$, mw: 378.5. Syn: *isothan.* Deep amber, water-sol liquid, pleasant characteristic odor.
Acute tox data: Oral LD_{50} (rat) = 230 mg/kg; oral LD_{50} (guinea pig) = 200 mg/kg. [*3*]
THR = HIGH via oral route.
Fire Hazard: Slight, when exposed to heat or flame.
Disaster Hazard: Dangerous; when heated to decomp, emits toxic fumes, can react with oxidizing materials.

LAURYL MERCAPTAN. Syn: *dodecyl mercaptan.* Water-white to pale yellow liquid. $C_{12}H_{25}SH$, mw: 202.5, mp: −7°, bp: 115°–177°, flash p: 262°F (OC), d: 0.849 @ 15.5°/15.5°.
THR = See mercaptans.
Fire Hazard: Low.
To Fight Fire: Alcohol foam.

LAURYL PYRIDINIUM CHLORIDE. Syn: *dodecyl pyridinium chloride.* Mottled tan semi-solid. $C_5H_5NClC_{12}H_{25}$, mw: 283.7, flash p: 347°F.
THR = U.

Fire Hazard: Slight, when exposed to heat or flame.
Disaster Hazard: Mod dangerous; when heated to decomp, emits toxic fumes; can react with oxidizing materials.
To Fight Fire: Foam, CO_2, dry chemical.

LAURYL QUINALDINIUM BROMIDE.
THR = U. See also bromides.
Fire Hazard: U.
Disaster Hazard: Dangerous. See bromides.

LAURYL QUINOLINIUM CHLORIDE. U. A fungicide.
Fire Hazard: U.
Disaster Hazard: Dangerous. See chlorides.

LAURYL THIOCYANATE. $CH_3(CH_2)_{10}CH_2SCN$, mw: 227.3.
Acute tox data: oral LD_{50} (rat) = 1250 mg/kg. [*3*]
THR = MOD via oral route. An insecticide.

LAWRENCITE. See ferrous chloride.

LAWRENCIUM. A synthetic transuranium element of atomic number 103 and atomic mass 257. Lw.
THR = Radioactive.
Radiation Hazard: Intensely radioactive and therefore highly radiotoxic.

LD-813. A mixture of aromatic amines. (approx 40% MOCA).
THR = An exper carc to rats via oral route. [*3*]

LEACHATE PRODUCTION FROM SOLID WASTE. See Section 6.

LEAD. Syn: *plumbum*. Bluish-gray, soft metal. Pb, atwt: 207.21, mp: 327.43°, bp: 1620°, d: 11.288 @ 20°/20°. vap. press: 1 mm @ 973°.
THR = See lead compounds. A common air contaminant. It is a (S) carc of the lungs and kidney and an exper teratogen. [*3, 23*]
Radiation Hazard: For permissible levels, see Section 5, Table 5A.5. Natural isotope ^{210}Pb (radium-D, uranium series), $T\frac{1}{2}$ = 21y. Decays to radioactive ^{210}Pb via β's of 0.0015 (19%) MeV. Emits γ's of 0.046 MeV. ^{210}Pb usually exists in equilibrium with its daughters, ^{210}Bi and ^{210}Po. Natural isotope ^{212}Pb (Thorium-B, thorium Series), $T\frac{1}{2}$ = 10.6 h. Decays to radioactive ^{212}Bi via β's of 0.16 (5%). 0.34 (81%), 0.58 (14%) MeV. Emits γ's of 0.24, 0.34 MeV and x-rays.
Fire Hazard: Mod, in the form of dust when exposed to heat or flame. See also powdered metals.
Explosion Hazard: Mod, in the form of dust when exposed to heat or flame. Violent reactions with NH_4NO_3, ClF_3, H_2O_2, NaN_3, Na_2C_2, Zr. [*19*]
Disaster Hazard: Dangerous; when heated, emits highly toxic fumes; can react vigorously with oxidizing materials.

LEAD ACETATE. Syn: *sugar of lead*. White crystals, sol in water. Commercial grades are frequently brown or gray lumps. $Pb(C_2H_3O_2)_2 \cdot 3H_2O$, mw: 379.35, mp: 75°, anhydrous mp: 280°. d: 2.55.
Acute tox data: ip LD_{LO} (rat) = 204 mg/kg; iv LD_{50} (rat) = 120 mg/kg. [*3*]
THR = HIGH via ip and iv routes. See also lead compounds. A poison. An exper (+) carc and teratogen. [*3, 9*] Violent reaction with $KBrO_3$. [*19*] An insecticide.

LEAD ACETATE, BASIC. White powder. $Pb_2OH(C_2H_3O_2)_3$, mw: 608.6.
THR = An exper (+) carc. [*3, 9*] See also lead acetate. A poison.

LEAD ACETATE (III) TRIHYDRATE.
THR = An exper (+) carc. [*3, 9*] See also lead acetate.

LEAD ANTIMONATE. Syns: *naples yellow*, *antimony yellow*. Orange yellow powder. $Pb_3(SbO_4)_2$, mw: 993.2.
THR = See lead and antimony compounds.

LEAD ARSENATES. Syn: *lead-o-arsenate*. White crystals. $PbHAsO_4$, mw: 327.1.
Acute tox data: Oral LD_{LO} (human) = 1.4 mg/kg; oral LD_{50} (rat) = 100 mg/kg. [*3*]
THR = HIGH via oral route. See also lead and arsenic compounds. A poison. An exper carc. [*3, 9*]
Disaster Hazard: Dangerous; on heating, emits highly toxic fumes.

LEAD-*m*-ARSENATE. $AsH_3O_4 \cdot (Pb)x$.
Acute tox data: Oral LD_{50} (rat) = 100 mg/kg; oral LD_{50} (mouse) = 1000 mg/kg; oral LD_{50} (rabbit) = 125 mg/kg. [*3*]
THR = HIGH via oral to MOD via oral routes depending upon species. See also lead arsenate. A poison.

LEAD-*o*-ARSENATE. See lead arsenates.

LEAD ARSENITE. Syns: *lead-o-arsenite*, *lead-m-arsenite*. White powder; $PbAs_2O_4$, mw: 421.
THR = HIGH. See lead compounds and arsenic compounds.
Disaster Hazard: Dangerous; on heating, emits highly toxic fumes.

LEAD-*m*-ARSENITE. See lead arsenite.

LEAD-*o*-ARSENITE. See lead arsenite.

LEAD AZIDE. Colorless needles. $Pb(N_3)_2$, mw: 291.26.
THR = See lead compounds and azides.
Fire Hazard: U.
Explosion Hazard: Severe, when shocked or exposed to heat or flame. Explodes at 250°. Violent reaction with brass, calcium stearate. CS_2, Cu, Zn. [*19*]
Disaster Hazard: Highly dangerous; shock and heat

will explode it; when heated, emits highly toxic fumes of lead.

LEAD BENZOATE. White crystals. $Pb(C_7H_5O_2)_2 \cdot H_2O$, mw: 467.5, mp: $-H_2O$ @ 100°.
THR = See lead compounds and lead.

LEAD-*m*-BORATE. White powder. $Pb(BO_2)_2 \cdot H_2O$, mw: 310.87, d: 5.598 (anhydrous).
THR = See lead and boron compounds. A poison.

LEAD BROMATE. Monoclinic crystals. $Pb(BrO_3)_2 \cdot H_2O$, mw: 481.06, mp: 180° (decomp), d: 5.53.
THR = See lead compounds and bromates. A poison.

LEAD BUTYRATE. $C_8H_{14}O_4Pb$, mw: 381.4.
THR = A poison. See also lead compounds.

LEAD CAPRATE. $Pb(C_{10}H_{19}O_2)_2$, mw: 549.71, mp: 103°–104°.
THR = See lead compounds and lead.

LEAD CAPROATE. Crystals. $Pb(C_6H_{11}O_2)_2$, mw: 437.51, mp: 73°–74°.
THR = See lead compounds and lead.

LEAD CAPRYLATE. White leaf. $Pb(C_8H_{15}O_2)_2$, mw: 493.61, mp: 83.5°–84.5°.
THR = See lead compounds and lead.

LEAD CARBONATE. Syn: *cerussite*. White powdery crystals. $PbCO_3$, mw: 267.22, mp: decomp @ 315°, d: 6.61.
THR = An exper (±) carc. [*3, 9*] A poison. Violent reaction with F_2. [*19*] See lead compounds and lead.

LEAD CARBONATE, BASIC. Syns: *white lead, hydrocerussite*. White powder, amorphous.
$2PbCO_3Pb(OH)_2$, mw: 775.67, mp: decomp @ 400°, d: 6.14.
THR = See lead compounds and lead. A poison. Violent reaction with F_2. [*19*]

LEAD CEROTATE. White crystals. $Pb(C_{26}H_{51}O_2)_2$, mw: 998.55, mp: 113.5°.
THR = See lead compounds and lead.

LEAD CHLORATE. Monoclinic white crystals.
$Pb(ClO_3)_2$, mw: 374.12, mp: decomp. d: 3.89.
THR = See lead compounds, chlorates and lead. A poison. Reacts violently with S. [*19*]

LEAD CHLORIDE. Syn: *cotunnite*. White crystals, $PbCl_2$, mw: 278.1, mp: 501°, bp: 954°, d: 5.85, vap. press: 1 mm @ 547°.
THR = See lead compounds. A poison. An exper teratogen. [*3*]

LEAD CHLORITE. Monoclinic yellow crystals.
$Pb(ClO_2)_2$, mw: 342.12, mp: explodes @ 126°.
THR = See lead compounds and chlorites. Reacts violently with S. [*19*]

LEAD CHROMATE. Syns: *crocoite, chrome yellow*. Yellow crystals. $PbCrO_4$, mw: 323.22, mp: 844°, bp: decomp, d: 6.3.
Acute tox data: ip LD_{50} (guinea pig) = 400 mg/kg. [*3*]
THR = HIGH via ip route. An exper (±) neo and carc. [*3, 6*] Reacts violently with ferric ferrocyanide. [*19*]

LEAD CHROMATE, BASIC. Red, amorphous or crystals. $Pb_2(OH)_2 \cdot CrO_4$, mw: 564.45, mp: 920°.
THR = See lead and chromium compounds. An exper neo. [*3*]

LEAD CITRATE. White crystalline powder.
$Pb_3(C_6H_5O_7)_2 \cdot 3H_2O$, mw: 1053.88.
THR = See lead compounds.

LEAD COMPOUNDS.
THR = Poisons. Lead poisoning is one of the commonest of occupational diseases. The presence of lead-bearing materials or lead compounds in an industrial plant does not necessarily result in exposure on the part of the workman. The lead must be in such form, and so distributed, as to gain entrance into the body or tissues of the workman in measurable quantity, otherwise no exposure can be said to exist. Some are exper (+) carc of the lungs and kidneys. [*14, 23, 9, 95*]
Mode of entry into body:
1. By inhal of the dusts, fumes, mists or vapors. (Common air contaminants.)
2. By ingestion of lead compounds trapped in the upper respiratory tract or introduced into the mouth on food, tobacco, fingers or other objects.
3. Through the skin; this route is of special importance in the case of organic compounds of lead, as lead tetraethyl. In the case of the inorganic forms of lead, this route is of no practical importance.

When lead is ingested, much of it passes through the body unabsorbed, and is eliminated in the feces. The greater portion of the lead that is absorbed is caught by the liver and excreted, in part, in the bile. For this reason, larger amounts of lead are necessary to cause poisoning if absorption is by this route, and a longer period of exposure is usually necessary to produce symptoms. On the other hand, upon inhal, absorption takes place easily from the respiratory tract and symptoms tend to develop more quickly. From the point of view of industrial poisoning, inhal of lead is much more important than is ingestion.

Lead is a cumulative poison. Increasing amounts build up in the body and eventually a point is reached where symptoms and disability occur. Lead produces a brittleness of the red blood cells so that they hemolyze with but slight trauma; the hemoglobin is not affected. Due to their increased fragility, the red cells are destroyed more rapidly in the body than normally, producing an anemia which is rarely severe. The loss of circulating red cells stimulates the production of new young cells which, on entering the blood stream, are acted upon by the circulating lead, with resultant coagulation of their basophilic material. These cells after suitable staining, are recognized as "stippled cells." As regards the effect of lead on the white blood cells, there is no uniformity of opinion. In addition to its effect on the red cells of the blood, lead produces a damaging effect on the organs or tissues with which it comes in contact. No specific or characteristic lesion is produced. Autopsies of deaths attributed to lead poisoning and experimental work on animals, have shown pathological lesions of the kidneys, liver, male gonads, nervous system, blood vessels and other tissues. None of these changes, however, have been found consistently.

In cases of lead poisoning, the amount of lead found in the blood is frequently in excess of 0.07 mg per 100 cc of whole blood. The urinary lead excretion generally exceeds 0.1 mg per liter of urine.

The toxicity of the various lead compounds appears to depend upon several factors: (1) the sol of the compound in the body fluids; (2) the fineness of the particles of the compound; sol is greater, of course, in proportion to the fineness of the particles; (3) conditions under which the compound is being used; where a lead compound is used as a powder; contamination of the atmosphere will be much less where the powder is kept damp. Of the various lead compounds, the carbonate, the monoxide and sulfate are considered to be more toxic than metallic lead or other lead compounds. Lead arsenate is very toxic, due to the presence of the arsenic radical.

Signs and Symptoms: Industrial lead poisoning commonly occurs following prolonged exposure to lead or its compounds. The common clinical types of lead poisoning may be classified according to their clinical picture as (a) alimentary; (b) neuromotor; and (c) encephalic. Some cases may show a combination of clinical types. The alimentary type occurs most frequently, and is characterized by abdominal discomfort or pain. Severe cases may present actual colic. Other complaints are constipation and/or diarrhea, loss of appetite, metallic taste, nausea and vomiting, lassitude, insomnia, weakness, joint and muscle pains, irritability, headache and dizziness. Pallor, lead line on the gums, pyorrhea, loss of weight, abdominal tenderness, basophilic stippling, anemia, slight albuminuria, increased urinary excretion, and an increase in the lead content of the whole blood, are signs which may accompany the above symptoms.

In the neuromuscular type, the chief complaint is weakness, frequently of the extensor muscles of the wrist and hand, unilateral or bilateral. Other muscle groups which are subject to constant use may be affected. Gastroenteric symptoms are usually present, but are not as severe as in the alimentary type of poisoning. Joint and muscle pains are likely to be more severe. Headache, dizziness and insomnia are frequently prominent. True paralysis is uncommon, and usually is the result of prolonged exposure.

Lead encephalopathy is the most severe but the rarest manifestation of lead poisoning. In the industrial worker it follows rapid and heavy lead absorption. Organic lead compounds, such as tetraethyl lead, are absorbed rapidly through the skin as well as through the lungs, and are selectively absorbed by the CNS. The clinical picture in these cases is usually an encephalopathy. With inorganic lead compounds, comparable conc in the CNS are reached only when the workplace is heavily contaminated with vapor, fume and dust. Encephalopathy begins abruptly, and is characterized by signs of cerebral and meningeal involvement. There is usually stupor, progressing to coma, with or without convulsion, and often terminating in death. Excitation, confusion and mania are less common. In milder cases of short duration, there may be symptoms of headache, dizziness, somnolence and insomnia. The cerebrospinal pressure may be increased. See also specific compound.

Diagnosis: A diagnosis of lead poisoning should not be made on the basis of any single clinical or laboratory finding. There must be a history of significant exposure, signs, and symptoms (as described above) compatible with the diagnosis, and confirmatory laboratory tests. Increase of stippled red blood cells, mild anemia, and elevated lead in blood and urine, i.e., more than 0.07 mg/100 ml blood and similar values per liter of urine. An increase of coproporphyrins and certain amino acids in urine may be present. Diagnostic mobilization of lead with calcium EDTA may be useful in questionable cases.

Treatment of Lead Poisoning: It has been found that the chelating agent, calcium ethylenediaminetetracetate, and related compounds are highly efficacious in removing absorbed lead from the tissues of the body. (The therapeutic agents of this group are also known as versene, versenate, edathamil and Ca EDTA. Ca EDTA is effective only when administered intravenously. Various dosage schedules have been proposed. An effective regime is 3–6 g of Na Ca EDTA in 300 cc–500 cc of 5% glucose by intravenous drip over a period of 3–8 hrs. Treatment may

be given daily for 5–10 days with an interval of one week between courses. Another plan is to give treatment at intervals of 3–5 days until deleading has been accomplished.
Disaster Hazard: See lead.

LEAD CYANATE. White needles. $Pb(OCN)_2$, mw: 291.25, mp: decomp.
THR = A poison. See lead compounds and cyanates.

LEAD CYANIDE. White powder. $Pb(CN)_2$, mw: 259.25.
Acute tox data: ip LD_{LO} (rat) = 100 mg/kg. [*3*]
THR = HIGH via ip route. A poison. See lead compounds and cyanides. Violent reaction with Mg. [*19*]
Fire Hazard: Mod, by chemical reaction with reducing agents.
Disaster Hazard: Dangerous; see lead; can react vigorously with reducing materials.

LEAD DI-*o*-ARSENATE. See lead arsenates.

LEAD DICHROMATE. Red crystals. $PbCr_2O_7$, mw: 423.23.
THR = See lead and chromium compounds.
Fire Hazard: Mod, by chemical reaction with reducing agents.
Disaster Hazard: Dangerous; see lead; can react vigorously with reducing materials.

LEAD DICYANOGUANIDINE. Crystals. $Pb[NH_2CN(NCN)CN]_2$, mw: 423.4, mp: $>300°$.
THR = This material may be quite toxic by skin absorption. See also lead compounds.
Disaster Hazard: Dangerous; see lead and cyanides.

LEAD DIIODIDE. Golden-yellow crystals or powder. PbI_2, mw: 461.05, mp: 402°, bp: 954°, d: 6.16.
THR = See lead compounds and iodides.

LEAD DIMETHYL DITHIOCARBAMATE. Syn: *ledate*. $PbS_2[(CH_3)_2CN]_2$, mw: 447.6, mp: 258°, d: 2.5.
THR = See lead compounds and disulfiram. A poison. An exper neo via oral route. [*3, 14*]
Fire Hazard: Slight, when exposed to heat or flame.
Disaster Hazard: Dangerous; see lead and SO_x.

LEAD DIOXIDE. Syns: *plattnerite*, *lead peroxide*. Brown hexagonal crystals. PbO_2, mw: 239.21, mp: decomp @ 290°, d: 9.375.
Acute tox data: ip LD_{50} (guinea pig) = 200 mg/kg. [*3*]
THR = HIGH via ip route. See also lead compounds.
Fire Hazard: Violent reactions with Al_4C_3, BaS, B, CaS, $CsHC_2$, ClF_3, H_2O_2, H_2S, hydroxylamine, Mo, performic acid, phenyl hydrazine, P, PCl_3, S, $S(OCl)_2$, W, Zr. [*19*]
Disaster Hazard: Dangerous; see lead; can react with reducing material.

LEAD DI-*o*-PHOSPHATE. Colorless crystals. $PbHPO_4$, mw: 303.20, mp: decomp, d: 5.661 @ 15°.
THR = See lead compounds.

LEAD DITHIONATE. Crystals. $PbS_2O_6 \cdot 4H_2O$, mw: 439.39, mp: decomp, d: 3.22.
THR = See lead compounds.
Disaster Hazard: Dangerous; see lead and SO_x.

LEAD DUST. See lead.

LEAD ENANTHATE. White leaf-like crystals. $Pb(C_7H_{13}O_2)_2$, mw: 465.56, mp: 79°–80°.
THR = See lead compounds.

LEAD FERRICYANIDE. Black-brown to red crystals. $Pb_3[Fe(CN)_6]_2$, mw: 1135.65.
THR = See lead compounds and ferricyanides.

LEAD FERRITE. Hexagonal crystals. $PbFe_2O_4$, mw: 382.89, mp: 1530° (decomp).
THR = See lead compounds.

LEAD FERROCYANIDE. Yellowish-white powder. $Pb_2Fe(CN)_6 \cdot 3H_2O$, mw: 680.43, mp: $-H_2O$ @ 100°.
THR = See lead compounds and ferrocyanides.

LEAD FLUOBORATE. $Pb(BF_4)_2$, mw: 380.8.
Acute tox data: Oral LD_{LO} (rat) = 50 mg/kg. [*3*]
THR = HIGH via oral route. A poison. See lead and fluorides.

LEAD FLUORIDE. Colorless solid. PbF_2, mw: 245.2, mp: 855°, bp: 1293°, vap. press: 10 mm @ 904°, d: 8.24.
Acute tox data: Oral LD_{LO} (guinea pig) = 4000 mg/kg; sc LD_{LO} (guinea pig) = 2800 mg/kg. [*3*]
THR = MOD via oral and sc routes. See also lead compounds and fluorides. Violent reaction with CaC_2, F_2. [*19*]

LEAD FLUOSILICATE. Monoclinic, colorless powder. $PbSiF_6 \cdot 2H_2O$, mw: 385.30, mp: decomp.
Acute tox data: Oral LD_{LO} (guinea pig) = 200 mg/kg; sc LD_{LO} (guinea pig) = 250 mg/kg; sc LD_{LO} (frog) = 140 mg/kg. [*3*]
THR = HIGH via oral and sc routes. A poison. See lead compounds and fluosilicates.

LEAD FORMATE. Rhombic, white, lustrous crystals. $Pb(CHO_2)_2$, mw: 297.25, mp: decomp @ 190°, d: 4.63.
THR = See lead compounds.

LEAD HYDRATE. See lead hydroxide.

LEAD HYDROXIDE. Syns: *lead hydrate*, *hydrated lead oxide*. White, bulky powder. $Pb(OH)_2$, mw: 241.23, mp: decomp @ 145°, d: 7.592.
THR = See lead compounds.

LEAD HYPOPHOSPHITE. $Pb(PH_2O_2)_2$, mw: 337.2.
THR = An impact-sensitive explosive. Forms a powerful explosive with $Pb(NO_3)_2$. [*19*] A poison. See also lead compounds.

LEAD HYPOSULFITE. See lead thiosulfate.

LEAD IMIDE. PbNH, mw: 222.2.
THR = Explodes when heated with water or dilute acid. [*19*] A poison. See also lead compounds.

LEAD IODATE. White solid. $Pb(IO_3)_2$, mw: 557.05, mp: decomp @ 300°.
THR = See lead compounds and iodates.

LEAD ISOBUTYRATE. White prisms. $Pb(C_4H_7O_2)_2$, mw: 381.40, mp: < 100°.
THR = See lead compounds.

LEAD LACTATE. $C_3H_4O_6Pb$, mw: 343.3.
Acute tox data: Oral LD_{LO} (guinea pig) = 1000 mg/kg. [*3*]
THR = MOD via oral route. See also lead compounds.

LEAD LAURATE. Chalky white powder. $Pb(C_{12}H_{23}O_2)_2$, mw: 605.8, mp: 117°.
THR = See lead compounds.

LEAD LIGNOCERATE. White powder. $Pb(C_{24}H_{47}O_2)_2$, mw: 942.22, mp: 117°.
THR = See lead compounds.

LEAD LINOLEATE. Yellowish-white paste. $Pb(C_{18}H_{31}O_2)_2$, mw: 766.07.
THR = See lead compounds.

LEAD MELISSATE. White powder. $Pb(C_{31}H_{61}O_2)_2$, mw: 1138.81, mp: 115°–116°.
THR = See lead compounds.

LEAD MERCAPTIDE.
THR = See lead compounds and mercaptans.

LEAD METAL. See lead.

LEAD MOLYBDATE. Syn: *wulfenite*. Yellow powder. $PbMoO_4$, mw: 367.16, mp: 1070°.
THR = See lead compounds.

LEAD MONO-*o*-ARSENATE. See lead arsenates.

LEAD MONOIODIDE. Pale yellow. PbI, mw: 334.13, mp: decomp @ 300°.
THR = See lead compounds and iodides.

LEAD MONONITRORESORCINATE. See explosives, high.

LEAD MONO-*o*-PHOSPHATE. Needles. $Pb(H_2PO_4)_2$, mw: 401.28.
THR = See lead compounds and phosphates.

LEAD MONOXIDE. See litharge.

LEAD MYRISTATE. White powder. $Pb(C_{14}H_{27}O_2)_2$, mw: 661.92, mp: 107°.
THR = See lead compounds.

LEAD-β-NAPHTHALENE SULFONATE. White crystalline powder. $Pb(C_{10}H_7SO_3)_2$, mw: 621.64.
THR = See lead compounds.
Disaster Hazard: Dangerous; see lead and SO_x.

LEAD NAPHTHENATE. 24% lead.
Acute tox data: Oral LD_{50} (rat) = 5100 mg/kg; dermal LD_{50} (rat) = 520 mg/kg. [*3*]
THR = MOD via oral and dermal routes. See also lead compounds.

LEAD NITRATE. White crystals. $Pb(NO_3)_2$, mw: 331.23, mp: decomp @ 470°, d: 4.53 @ 20°.
Acute tox data: ip LD_{LO} (rat) = 432 mg/kg; oral LD_{LO} (guinea pig) = 1330 mg/kg. [*3*]
THR = HIGH via ip and MOD via oral routes. An exper teratogen. [*3*] Reacts violently with ammonium thiocyanate, carbon, lead hypophosphite. [*19*] See lead compounds and nitrates.

LEAD OLEATE. White, ointment-like granules or mass. $Pb(C_{18}H_{33}O_2)_2$, mw: 770.10.
Acute tox data: Oral LD_{LO} (guinea pig) = 8000 mg/kg. [*3*]
THR = LOW via oral route. See lead compounds.

LEAD ORE.
THR = A recog carc. [*14*] See arsenic compounds.

LEAD OXALATE. Heavy white powder. PbC_2O_4, mw: 295.23, mp: decomp @ 300°, d: 5.28.
THR = See lead compounds and oxalates.

LEAD OXIDE. Syns: *red lead, minium, lead tetroxide*. Bright red powder. Pb_3O_4, mw: 685.6, mp: 890° (decomp), bp: 1472°, d: 8.32–9.16, vap. press: 1 mm @ 943°.
Acute tox data: ip LD_{50} (guinea pig) = 220 mg/kg. [*3*]
THR = HIGH via ip route. See also lead compounds.
Fire Hazard: Slight, by chemical reaction with reducing agents. An oxidizing agent. Reacts violently with Al, $CsHC_2$, (F_2 + glycerol), H_2S_3, (glycerine + $HClO_4$), $RbHC_2$, (Si + Al), Na, SO_3, Ti, Zr. [*19*]
Disaster Hazard: See lead.

LEAD OXYCHLORIDES. Syns: *mendipite, cassel yellow, laurionite, matlockite*. Yellow or white solid. Composition: $PbCl_2$ + varying amounts of PbO and possibly some H_2O.
THR = See lead compounds and chlorides. Violent reaction with K. [*19*]

LEAD PALMITATE. Chalky white powder. $Pb(C_{16}H_{31}O_2)_2$, mw: 718.03, mp: 112.3°.
THR = See lead compounds.

LEAD PERCHLORATE. Crystals. $Pb(ClO_4)_2 \cdot 3H_2O$, mw: 460.17, mp: decomp @ 100°, d: 2.6.
Acute tox data: ip LD_{LO} (mouse) = 275 mg/kg. [*3*]
THR = HIGH via ip route. See lead compounds and perchlorates. Can explode with methanol. [*19*] See lead compounds.

LEAD-*p*-PERIODATE. Crystals. $PbHIO_5$, mw: 415.14, mp: decomp @ 130°.
THR = HIGH. See lead compounds and iodates.

LEAD PEROXIDE. See lead dioxide.

LEAD PHENATE. Yellowish to grayish-white powder. $Pb(OH)OC_6H_5$, mw: 317.28.
THR = See lead compounds and phenol.

LEAD-*o*-PHOSPHATE. Hexagonal, colorless or white powder. $Pb_3(PO_4)_2$, mw: 811.59, mp: 1014, d: 6.9–7.3.
THR = A poison. See also lead compounds.

LEAD-*m*-PHOSPHATE. Colorless crystals. $Pb(PO_3)_2$, mw: 365.17, mp: 800°.
THR = See lead compounds. An exper (±) carc. [*3, 9*]

LEAD-*o*-PHOSPHITE. White powder. $PbHPO_3$, mw: 287.20, mp: decomp.
THR = See lead compounds and phosphites. A fiber drum of it self-ignited. [*19*]

LEAD PICRATE. Yellow crystals. $Pb(C_6H_2O_7)_2 \cdot H_2O$, mw: 681.43, mp: $-H_2O$ @ 130°, bp: explodes, d: 2.831 @ 20°.
THR = See lead compounds and picric acid.

LEAD PROTOXIDE. See litharge.

LEAD PYROARSENATE. See lead arsenates.

LEAD PYROPHOSPHATE. White crystals. $Pb_2P_2O_7$, mw: 588.38, mp: 824°, d: 5.8.
THR = See lead compounds and phosphates.

LEAD RED. See lead oxide.

LEAD RESINATE. Yellowish-white paste. $Pb(C_{20}H_{29}O_2)_2$, mw: 810.07.
THR = See lead compounds.
Fire Hazard: Mod, when exposed to heat or flame.
Disaster Hazard: See lead.

LEAD SELENATE. White crystals. $PbSeO_4$, mw: 350.17, mp: decomp. d: 6.37.
THR = See lead compounds and selenium compounds.

LEAD SELENIDE. Syn: *clausthalite*. Cubic crystals. PbSe, mw: 286.17, mp: 1065°, d: 8.10 @ 15°.
THR = See lead and selenium compounds.
Fire Hazard: Mod, in the form of dust when exposed to flame or by chemical reaction with moisture to evolve the hydrides. See also hydrogen selenide.
Explosion Hazard: Slight, by chemical reaction with moisture. See also hydrogen selenide.
Disaster Hazard: See lead and selenium.

LEAD SILICATE. See lead-*m*-silicate.

LEAD-*m*-SILICATE. Syn: *alamosite*. White crystalline powder. $PbSiO_3$, mw: 283.27, mp: 766°, d: 6.49.
THR = See lead compounds.

LEAD STEARATE. White powder. $Pb(C_{18}H_{35}O_2)_2$, mw: 774.1, mp: 115.7°.
THR = See lead compounds.

LEAD STYPHNATE. See lead trinitroresorcinate.

LEAD SULFATE. Syn: *anglisite*. White rhombic crystals. $PbSO_4$, mw: 303.27, mp: decomp @ 1000°, d: 6.2.
Acute tox data: ip LD_{50} (guinea pig) = 300 mg/kg. [*3*]
THR = HIGH via ip route. A strong irr to skin, eyes and mu mem. See lead compounds. Violent reaction with K. [*19*]

LEAD SULFIDE. Syns: *galena, plumbous sulfide*. Silvery, metallic crystals or black powder. PbS, mw: 239.27, mp: 1114°, bp: 1281° (sublimes), d: 7.5, vap. press: 1 mm @ 852°.
Acute tox data: ip LD_{LO} (rat) = 1847 mg/kg. [*3*]
THR = MOD via ip route. See also sulfides and lead compounds. Violent reaction with ICl, H_2O_2. [*19*]

LEAD SULFITE. White powder. $PbSO_3$, mw: 287.28.
THR = See lead compounds and sulfites.

LEAD SULFOCYANATE. Syn: *lead thiocyanate*. Monoclinic white crystals. $Pb(SCN)_2$, mw: 323.37, d: 3.82.
THR = See lead compounds and thiocyanates.

LEAD TARTRATE. White crystalline powder. $PbC_4H_4O_6$, mw: 355.28, d: 2.54 @ 19°.
Acute tox data: ip LD_{LO} (rat) = 1200 mg/kg. [*3*]
THR = MOD via ip route. See lead compounds.

LEAD TELLURIDE. Syn: *altaite*. White cubic crystals. PbTe, mw: 334.82, mp: 917°, d: 8.16.
THR = See lead compounds and tellurium compounds.

LEAD TETRAACETATE. Colorless to faintly pink monoclinic crystals. $Pb(CH_3COO)_4$, mw: 443.39, mp: 175°, d: 2.228 @ 17°.
THR = See lead compounds.

LEAD TETRAAZIDE. $Pb(N_3)_4$, mw: 355.3.
THR = Very unstable. [*19*] Probably HIGH toxicity. See also lead compounds and azides.

LEAD TETRACHLORIDE. Yellow, oily liquid. $PbCl_4$, mw: 349.04, mp: −15°, bp: explodes @ 105°, d: 3.18 @ 0°.
THR = See lead and hydrochloric acid.

LEAD TETRAETHYL. Syn: *TEL*. Colorless, oily liquid, pleasant characteristic odor. $Pb(C_2H_5)_4$, mw: 323.5, mp: 125°–150°, bp: 198°–202° with decomp, d: 1.659 @ 18°, vap. press: 1 mm @ 38.4°, flash p: 200°F.
Acute tox data: Oral LD_{LO} (rat) = 17 mg/kg; inhal LC_{50} (rat) = 6 ppm; pa LD_{50} (rat) = 15 mg/kg; dermal LD_{LO} (dog) = 500 mg/kg; dermal LD_{LO} (guinea pig) = 990 mg/kg. [*3*]
THR = HIGH via oral, inhal, pa and dermal routes. This material is a powerful poison and a solvent for fatty materials. It has some solvent action on rubber as well. The fact that it is a lipoid sol-

vent makes it an industrial hazard, because it can cause intoxication not only by inhal but also by absorption through the skin. Decomp when exposed to sunlight or allowed to evaporate; forms triethyl lead, which is also a poisonous compound, as one of its decomp products. This liquid lead compound, when handled in undiluted form or concentrated solution as when it is manufactured or in the plants where it is mixed with gasoline, may cause lead exposure intoxication by coming in contact with the skin. Therefore, any open receptacle which contains these liquids in high conc or any container, article of clothing, or any other object which is not kept clean, particularly in contact with this material, may subject personnel to serious lead exposure. An exper (±) carc. [*3*, *6*] A common air contaminant.
Fire Hazard: Mod, when exposed to heat, flame or oxidizers.
Disaster Hazard: Dangerous; see lead; can react vigorously with oxidizing materials.
To Fight Fire: Dry chemical, CO_2, mist, foam.

LEAD TETRAFLUORIDE. Syn: *plumbic fluoride*. White crystals, reacts with moisture. PbF_4, mw: 283.2, d: 6.7, mp: 600° (approx).
THR = See lead compounds and fluorides.

LEAD TETRAMETHYL. Syn: *tetramethyl lead*. Colorless liquid. $Pb(CH_3)_4$, mw: 267.33, mp: −18°F, lel = 1.8%, bp: 110°, d: 1.99, vap. d: 9.2, flash p: 100°F.
Acute tox data: Oral LD_{50} (rat) = 109 mg/kg; pa LD_{50} (rat) = 105 mg/kg; ip LD_{LO} (rat) = 73 mg/kg; iv LD_{LO} (rabbit) = 90 mg/kg. [*3*]
THR = HIGH via oral, pa, ip and iv routes. See also lead and lead tetraethyl. An exper (±) carc. [*3*, *6*]
Fire Hazard: Dangerous, when exposed to heat, flame or oxidizers.
Explosion Hazard: Mod, in the form of vapor when exposed to flame.
Disaster Hazard: Dangerous; see lead; can react vigorously with oxidizing materials.
To Fight Fire: Water, foam, CO_2, dry chemical.

LEAD THIOCYANATE. See lead sulfocyanate.

LEAD THIOSULFATE. Syn: *lead hyposulfite*. White crystals. PbS_2O_3, mw: 319.33, mp: decomp, d: 5.18.
THR = See lead compounds.
Disaster Hazard: Dangerous; See lead and SO_x.

LEAD-*m*-TITANATE. Pale yellow solid. $PbTiO_3H$, mw: 304.1, d: 7.52.
Acute tox data: ip LD_{50} (rat) = 2000 mg/kg. [*3*]
THR = MOD via ip route. See also lead and titanium compounds.

LEAD TRINITRORESORCINATE. Syn: *lead styphnate*. Orange-yellow, monoclinic crystals.
$C_6H(NO_2)_3(O_2Pb)$, mw: 450.30, mp: explodes @ 311°, d: 3.1–2.9.
THR = See lead compounds and nitrates. Very sensitive explosive. It is shock-sensitive and has detonated spont when dry. [*19*]
Fire Hazard: See nitrates and explosives, high.
Explosion Hazard: Severe, when heated.
Disaster Hazard: Dangerous; explodes at 311°.

LEAD TUNGSTATE. Syn: *lead wolframate*. Yellowish powder. $PbWO_4$, mw: 455.13, d: 8.235.
THR = See lead compounds and tungsten compounds.

LEAD-*m*-VANADATE. Yellow powder. $Pb(VO_3)_2$, mw: 405.11.
THR = See lead and vanadium compounds.

LEAD WOLFRAMATE. See lead tungstate.

LEATHER.
THR = A MILD allergen. Handling "green" hides from certain parts of the world can bring about contact with anthrax spores. Furthermore, tanning can introduce materials such as chromium, formaldehyde, etc.
Fire Hazard: Slight, when exposed to heat or flame; can react with oxidizing materials.

LEATHER BLEACH.
THR = U.
Fire Hazard: Dangerous, when exposed to heat or flame.
Explosion Hazard: U.
Disaster Hazard: Mod dangerous; when heated to decomp, emits toxic fumes; can react with oxidizing materials.
To Fight Fire: Foam, CO_2, dry chemical.

LEATHER DRESSING. Flash p: < 80°F.
THR = U. Some dressings may act as irr or allergens.
Fire Hazard: Dangerous, when exposed to heat or flame; can react with oxidizing materials.
Explosion Hazard: U.
To Fight Fire: Foam, CO_2, dry chemical.

LECITHIN. The lecithins are mixtures of diglycerides of fatty acids linked to the choline ester of phosphoric acid. They are classified as phosphoglycerides or phosphatides.
$CH_2(R)CH(R')CH_2OPO(OH)O(CH_2)_2N(OH)(CH_3)_3$
where R and R′ are fatty acid groups.
THR = U. Probably LOW. Used as a general purpose food additive. It is a substance which migrates to food from packaging materials. [*109*]

"LEDATE." See lead dimethyl dithiocarbamate.

LEGAL MEDICINE. See forensic toxicology.

LENACIL. See venzar.

"LETHANE 60." See lauric acid-2-thiocyanatoethyl ester.

"LETHANE 384." See β-butoxy-β-thiocyanodiethyl ether.

"LETHANES." Insecticides with *n*-butyl carbitolthiocyanate, etc., in a light petroleum base. Accidental and suicidal poisonings have occurred. Symptoms include drowsiness, followed by coma, the limbs becoming flaccid and the appearance of twitching and convulsions. The pupils may dilate and respiration may become labored. Cyanosis and vomiting occur.
Acute tox data: Oral LD_{50} (rat) = 500 mg/kg for "Lethane 60" and 90 mg/kg for "Lethane 384." The lethanes are mild irr and in higher doses narcotic.

"LETHANE." $C_9H_{17}O_2NS$, mw: 203.3.
Acute tox data: Oral LD_{50} (rat) = 90 mg/kg; dermal LD_{50} (rat) = 250 mg/kg; dermal LD_{50} (rabbit) = 150 mg/kg. [*3*]
THR = HIGH via oral and dermal routes.
Disaster Hazard: See thiocyanates.

L-LEUCINE. Syns: *α-amino-γ-methylvaleric acid, α-aminoisocaproic acid.* An essential amino acid; occurs in isomeric forms. Below we consider the *l* and *dl* forms. White crystals, sol in water, slightly sol in alcohol, insol in ether. $(CH_3)_2CHCH_2CH(NH_2)COOH$, mw: 131, mp(*dl*): 332° with decomp, mp(*l*): 295°, d: 1.239 @ 18°/4°.
THR = An exper teratogen. [*3*] A nutrient and/or dietary supplement food additive. [*109*]

LEUKOL. See quinoline.

LEUROCRISTINE. $C_{46}H_{56}O_{10}N_4$, mw: 825.1.
Acute tox data: ip and iv LD_{50} (rat) = 1.3 mg/kg; ip LD_{50} (mouse) = 5.2 mg/kg. [*3*]
THR = LOW via ip and iv routes. An exper teratogen. [*3*] in monkeys.

LEWISITE. See dichloro-(2-chlorovinyl)arsine.

LEXONE. See 1,2,3,4,5,6-hexachlorocyclohexane.

"LEYTOSAN." Syn: *phenyl mercury urea.*
Acute tox data: oral LD_{50} (rat) = 50 mg/kg. [*3*]
THR = Probably HIGH via oral and inhal routes. See mercury compounds, organic. A fungicide.

LIBRIUM. See chlordiazepoxide.

LIGHT GREEN SF.
THR = A (S) carc of subcutaneous tissue [*14*] and ip, oral routes. [*108*]

LIGHT LIGROIN. See petroleum spirits.

LIGHT OILS. bp: 110°–160°.
THR = See kerosene.
Fire Hazard: Mod, when exposed to heat or flame; can react with oxidizing materials.
Explosion Hazard: U.

LIGNASAN. See ethyl mercury chloride.

LIGNITE DUST. autoign. temp.: 302°F.
THR = A nuisance dust. See also nuisance dusts.
Fire Hazard: Mod, when exposed to heat or flame; can react with oxidizing materials.
Explosion Hazard: U.

LIGNITE AND OIL SHALE RETORTING PRODUCTS.
THR = A recog carc. [*14*] See petroleum asphalt.

LIGNITE OILS.
THR = A recog carc. [*14*] See mineral oils.

LIGNITE TAR.
THR = A recog carc. [*14*] See petroleum asphalt.

LIME. See calcium oxide.

LIME ACETATE. See calcium acetate.

LIME CITRATE. See calcium citrate.

LIME OIL.
THR = An exper carc. [*23*]

LIMESTONE. See calcium carbonate.

LIME URANITE. See autunite.

LIMONENE.
THR = Violent reaction with (IF_3 + tetrafluoroethylene). [*19*] See dipentene.

LIMONENE DIOXIDE. See dipentene dioxide.

LIMONITE. Mainly hydrated sesquioxide of iron.
THR = An exper (±) carc. [*3, 9*]

LINALOL. See linalool.

LINALOOL. Syns: *linalol, 3,7-dimethyl-1,6-octadien-3-ol.* Colorless liquid, odor similar to that of bergamot oil and French lavender, sol in alcohol and ether. $(CH_3)_2C{:}CHCH_2CH_2C(CH_3)OHCH{:}CH_2$, mw: 154, d: 0.858–0.868 @ 25°, bp: 195°–199°.
Acute tox data: Oral LD_{50} (rat) = 2790 mg/kg; ip LD_{50} (rat) = 307 mg/kg; im LD_{50} (mouse) = 8000 mg/kg; dermal LD_{50} (rabbit) = 5610 mg/kg. [*3*]
THR = HIGH via ip, MOD via oral and dermal, LOW via im routes. A synthetic flavoring substance and adjuvant. [*109*]

LINALYL ACETATE. Clear, colorless, oily liquid, odor of bergamot, sol in alcohol, ether, diethyl phthalate, benzyl benzoate, mineral oil, fixed oils and alcohol, slightly sol in propylene glycol, insol in water and glycerin. $C_{10}H_{17}C_2H_3O_2$, mw: 196, bp: 108°–110°, d: 0.908–0.920.
Acute tox data: Oral LD_{50} (rat) = 14550 mg/kg. [*3*]
THR = LOW via oral route. An exper carc. [*23*] Used

as a synthetic flavoring substance and adjuvant. [*109*]

LINALYL OLEATE.
THR = An exper carc. [*23*]

LINDANE. See 1,2,3,4,5,6-hexachlorocyclohexane.

LINOLEIC ACID. Syn: *linolic acid.* Colorless oil, easily oxidized by air, sol in ether and ethanol. $CH_3(CH_2)_4CH{=}CHCH_2CH{=}CH(CH_2)_7COOH$, mw: 280.4, d: 0.9038 @ 18°/4°, mp: −12°, bp: 230° @ 16 mm.
THR = U. Ingestion can cause nausea and vomiting. A nutrient and/or dietary supplement food additive. [*109*]

LINOLIC ACID. See linoleic acid.

LINSEED OIL. bp: 343°, mp: −19°, d: 0.93, flash p: (raw oil): 432°F (CC), flash p (boiled): 403°F (CC), autoign. temp.: 650°F.
THR = An allergen and a MILD irr.
Fire Hazard: Slight, when exposed to heat or flame; can react with oxidizing materials.
Spont Heating: Yes.
Explosion Hazard: Violent reaction with Cl_2. [*19*]
To Fight Fire: CO_2, dry chemical.

LINTOX. See 1,2,3,4,5,6-hexachlorocyclohexane.

LIQUEFIED CARBON DIOXIDE. Syn: *liquid carbonic gas.* Heavy gas or liquid under pressure. CO_2, mw: 44.0, mp: −56.6° @ 3952 mm, bp: −78.5° (sublimes), d: 1.977 g/liter @ 0°, d (liquid): 1.101 @ −37°.
THR = This material is very cold and can damage tissue exposed to it. Solid CO_2 goes directly to gaseous CO_2 (sublimes) which is mainly an asphyxiant. See also carbon dioxide.
Disaster Hazard: Mod dangerous.

LIQUEFIED HYDROCARBON GAS. Syns: *LP gas, bottle gas, LPG.*
THR = Olefinic impurities may lend a narcotic effect or it may act as a simple asphyxiant.
Fire Hazard: Dangerous, when exposed to heat or flame.
Explosion Hazard: U.
Disaster Hazard: Mod dangerous; can react with oxidizing materials.
To Fight Fire: CO_2, dry chemical, water spray.

LIQUEFIED NONFLAMMABLE GASES (CHARGED WITH NITROGEN, CARBON DIOXIDE OR AIR). See specific component.

LIQUID CAMPHOR. See camphor oil (light.)

LIQUID CARBONIC GAS.
See liquefied carbon dioxide.

LIQUID ROSIN. See tall oil.

LITHARGE. Syns: *lead oxide, yellow plumbous oxide, lead protoxide, lead monoxide.* Tetragonal yellow crystals. PbO, mw: 223.21, mp: 888°, d: 9.53.
Acute tox data: ip LD_{LO} (rat) = 430 mg/kg. [*3*]
THR = HIGH via ip route. See lead compounds. Reacts with H_2O_2, Li_2C_2. [*19*]

LITHIC ACID. See uric acid.

LITHIUM. Silver-colored light metal, mixture of isotopes Li^6 and Li^7. Li, atwt: 6.94, mp: 179°, bp: 1317°, d: 0.534 @ 25°, vap. press: 1 mm @ 723°.
THR = See lithium compounds for a discussion of the toxicity of the lithium ion. See sodium for a discussion which applies to the toxicity of metallic lithium. An exper teratogen. [*3*]
Fire Hazard: Dangerous, when exposed to heat or flame. It also can react violently with air, As, Be, Br_2, $CHBr_3$, maleic anhydride, carbides, CO_2, ($CO + H_2O$), CBr_4, CCl_4, Cl_2, $CHCl_3$, CrO_3, Cr, $CrCl_3$, cobalt alloys, FeS, diborane, Mn alloys, CH_2Cl_2, CH_2I_2, Mo_2O_3, Ni alloys, Nb_2O_5, $CFCl_3$, HNO_3, N_2, organic matter, O_2, P, Pt, rubber, silicates, NaCl, $NaNO_2$, S, Ta_2O_5, TiO_2, trichloroethylene, tetrachloroethylene, trichlorotrifluoroethane, WO_3, V, V_2P_5, H_2O, $ZrCl_4$, Fe alloys, CHI_3, I, H_2. [*19*]
Disaster Hazard: Dangerous; when burned, emits toxic fumes of lithium oxide and hydroxide, will react with water or steam to produce heat and hydrogen; can react vigorously with oxidizing materials. Reacts with nitrogen at high temp.
To Fight Fire: Special mixtures of dry chemical, soda ash, graphite.

LITHIUM ACETYLENE CARBIDE DIAMMINO. $LiC{\equiv}CH(NH_3)_2$, mw: 66.1.
THR = It will burn on contact with CO_2, Cl_2, SO_2, H_2O. [*19*]

LITHIUM ACETYL SALICYLATE. Slightly hygroscopic powder, decomp in moist air. $LiC_9H_7O_4$, mw: 186.09.
THR = An irr material. See also lithium and acetol.
Fire Hazard: Slight.

LITHIUM ACID OXALATE. Colorless crystals. $LiHC_2O_4 \cdot H_2O$, mw: 113.98, mp: decomp.
THR = See oxalates and lithium compounds.

LITHIUM ALUMINATE. Syn: *lithium-m-aluminate.* White powder. $LiAlO_2$, mw: 65.91, mp: > 1625°, d: 2.55 @ 25°.
THR = See lithium and aluminum compounds.

LITHIUM ALUMINUM DEUTERIDE. White to gray-white microcrystalline lumps or powder. $LiAlD_4$, mw: 42.0, mp: 150° with decomp (ignites in air), d: 1.02 @ 25°.

THR = See lithium and aluminum compounds and hydrides.

LITHIUM ALUMINUM HYDRIDE. White microcrystalline lumps, solid, stable in dry air at room temp., above 125° it decomp forming Al, H_2 and lithium hydride. $LiAlH_4$, mw: 37.94.

THR = Reacts violently with air, acids, alcohols, benzoyl peroxide, boron trifluoride etherate, (2-chloromethyl furan + ethyl acetate), diethylene glycol dimethyl ether, diethyl ether, 1,2-dimethoxyethane, dimethyl ether, methyl ethyl ether, (nitriles + H_2O), perfluorosuccinamide, (perfluorosuccinamide + H_2O), tetrahydrofuran, water. [*19*] See aluminum and lithium compounds and hydrides.

To Fight Fire: Use dry chemical, including special formulations of dry chemicals as recommended by the supplier of the lithium aluminum hydride. *Do not* use water, fog, spray or mist.

LITHIUM ALUMINUM HYDRIDE, ETHEREAL. Liquid solution of $LiAlH_4$ in ether. Composition: $LiAlH_4$ + an ether.

THR = See lithium aluminum hydride and ethyl ether.

LITHIUM AMIDE. White crystalline solid or powder. $LiNH_2$, mw: 23.0, mp: 380°–400°. [*2*] d: 1.178 @ 17.5°.

THR = Ammonia is liberated and lithium hydroxide is formed when this compound is exposed to moisture. See also ammonia and lithium hydroxide. A powerful irr to skin, eyes and mu mem.

Fire Hazard: See ammonia.

Explosion Hazard: See ammonia. Reacts violently with H_2O. [*19*]

Disaster Hazard: Mod dangerous; will react with water or steam to produce toxic and flam vapors; on contact with oxidizing materials, can react vigorously; on contact with acid or acid fumes, can evolve much heat.

LITHIUM ANTIMONIDE. Solid. Li_3Sb, mw: 142.58, mp: > 950°, d: 3.2 @ 17°.

THR = See antimony compounds and lithium compounds.

Fire Hazard: In contact with moisture, stibine is liberated. See also antimony hydride.

Explosion Hazard: See antimony hydride.

Disaster Hazard: Dangerous; when heated to decomp or on contact with acid or acid fumes, emits highly toxic fumes of antimony; will react with water or steam to produce toxic and flam vapors; can react vigorously with oxidizing materials.

LITHIUM ARSENATE. Syn: *lithium-o-arsenate*. White powder. Li_3AsO_4, mw: 159.73, d: 3.07 @ 17°.

THR = See arsenic and lithium compounds.

LITHIUM AZIDE. LiN_3, mw: 49.

THR = Violent reaction with CS_2. [*19*]

LITHIUM BENZOATE. White crystals, water-sol. $C_7H_5LiO_2$, mw: 128.1.

THR = See lithium compounds.

LITHIUM BICHROMATE. See lithium dichromate. See also azides and lithium compounds.

LITHIUM BORATE. See lithium-*m*-borate.

LITHIUM-*m*-BORATE. Syn: *lithium borate*. White powder. $LiBO_2$, mw: 49.8, mp: 845°, d: 1.397 @ 41.7°.

THR = See boron and lithium compounds.

LITHIUM-*m*-BORATE OCTAHYDRATE. See lithium-*m*-borate.

LITHIUM BOROHYDRATE. White to grayish microcrystalline powder and lumps. $LiBH_4$, mw: 21.8, mp: 279° (decomp), d: 0.66.

THR = See boron and lithium compounds and hydrides. Can ignite on contact with cellulose, water. [*19*] A hazardous material.

LITHIUM BROMIDE. White, hygroscopic, granular powder. LiBr, mw: 86.86, mp: 549°, bp: 1265°, d: 3.46 @ 25°, vap. press: 1 mm @ 748°.

THR = See bromides and lithium compounds.

LITHIUM CACODYLATE. White powder. $Li(CH_3)_2AsO_2$, mw: 143.96.

THR = HIGH via oral route. See also arsenic and lithium compounds.

LITHIUM CARBIDE. Crystals or white powder. Li_2C_2, mw: 37.90, d: 1.65 @ 18°.

THR = See lithium compounds and carbides. Reacts violently with Br_2, Cl_2, F_2, I, PbO, P, Se, S. [*19*]

LITHIUM CARBOLATE. See lithium phenate.

LITHIUM CARBONATE. Li_2CO_3, mw: 73.9, mp: 723°, d: 2.11 @ 17.5°, white cryst.

Acute tox data: Oral LD_{LO} (rat) = 710 mg/kg; oral TD_{LO} (human) = 7 mg/kg $\longrightarrow$ psychological symptoms. [*3*]

THR = MOD via oral route. An exper teratogen. [*3*] Reacts violently with F_2. [*19*]

LITHIUM CARBONYL. LiCOCOLi, mw: 69.9.

THR = Very unstable. [*19*] Probably HIGH. See also carbonyls and lithium compounds.

LITHIUM CARMINE.

THR = An exper teratogen. [*3*]

LITHIUM CHLORATE. White crystals, $LiClO_3$, mw: 90.4, mp: 127.6°, bp: decomp @ 300°.

THR = See chlorates and lithium compounds.

LITHIUM CHLORIDE. Cubic, white deliquescent crystals. LiCl, mw: 42.4, mp: 605°, bp: 1350°, d: 2.068 @ 25°, vap. press: 1 mm @ 547°.

Acute tox data: Oral LD_{50} (rat) = 751 mg/kg; ip LD_{50} (mouse) = 604 mg/kg; oral LD_{50} (rabbit) = 850 mg/kg. [*3*]

THR = MOD via oral and ip routes. An exper teratogen. [*3*] This material has been recommended and used as a substitute for sodium chloride in "salt-free" diets, but cases have been reported in which the ingestion of lithium chloride has produced dizziness, ringing in the ears, visual disturbances, tremors and mental confusion. In most cases, the symptoms disappeared when use was discontinued. Reaction is violent with BrF_3. [*19*]

LITHIUM CHLOROPLATINATE. Red deliquescent crystals. $Li_2PtCl_6 \cdot 6H_2O$, mw: 529.8, bp: $-6H_2O$ @ 180°.

THR = See platinum and lithium compounds.

Disaster Hazard: Dangerous. See chlorine compounds.

LITHIUM CHROMATE. Yellow, crystalline, deliquescent powder. $Li_2CrO_4 \cdot 2H_2O$, mw: 165.92, mp: $-2H_2O$ @ 150°.

THR = A toxic material. See also lithium and chromium compounds.

Fire Hazard: Slight, an oxidizer. It can react with reducing materials.

LITHIUM CITRATE. White crystals, feebly alkaline taste, deliquescent, water-sol. $C_6H_5Li_3O_7 \cdot 4H_2O$, mw: 282.0, mp: decomp; $-4H_2O$ @ 105°.

THR = See lithium compounds.

LITHIUM COBALTITE. Blue-black powder. $LiCoO_2$, mw: 97.9.

THR = See cobalt and lithium compounds.

LITHIUM COMPOUNDS.

THR = Has been implicated in development of aplastic anemia. Lithium oxide, hydroxide, carbonate, etc., are strong bases and these solutions in water are very caustic. See also potassium compounds. Lithium ion has CNS toxicity. Large doses of lithium compounds have caused dizziness and prostration. Can cause kidney damage, anorexia, nausea, apathy, coma and death.

LITHIUM DEUTERIDE. Bluish-gray, fine powder. LiD, mw: 9.0, mp: $>600°$, d: 0.906 @ 25°.

THR = See also lithium hydride.

LITHIUM DICHROMATE. Syn: *lithium bichromate*. Yellowish-red, crystalline, deliquescent powder. $Li_2Cr_2O_7 \cdot 2H_2O$, mw: 265.93, mp: $-2H_2O$ @ 110°, bp: decomp, d: 2.34 @ 30°.

THR = See also lithium and chromium compounds.

LITHIUM ETHYL. Syn: *ethyl lithium*. Transparent crystals, decomp by water. LiC_2H_5, mw: 36.00, mp: 95°.

THR = See lithium compounds and organometallics.

Fire Hazard: A spont flam material; can react with oxidizing materials.

Explosion Hazard: U.

LITHIUM FERROSILICON. Dark, crystalline, brittle, metallic lumps or powder, evolves a flam gas when in contact with moisture.

Fire & Disaster Hazard: Dangerous! Evolves a flam gas when exposed to moisture, steam or acid fumes.

LITHIUM FLUORIDE. Fine white powder. LiF, mw: 25.94, mp: 845°, bp: 1676°, d: 2.635 @ 20°, vap. press: 1 mm @ 1047°.

Acute tox data: Oral LD_{LO} (guinea pig) = 200 mg/kg; sc LD_{LO} (frog) = 280 mg/kg; sc LD_{LO} (guinea pig) = 2000 mg/kg. [*3*]

THR = HIGH via oral and sc routes. Possibly MOD via sc route. See also fluorides.

LITHIUM FLUOSILICATE. Syn: *lithium silicofluoride*. Monoclinic, white crystals. $Li_2SiF_6 \cdot 2H_2O$, mw: 191.97, mp: $-2H_2O$ @ 100°, bp: decomp, d: 2.33 @ 12°.

THR = See fluosilicates and lithium compounds.

LITHIUM GALLIUM HYDRIDE. White crystals. $LiGaH_4$, mw: 80.69.

THR = See gallium and lithium compounds and hydrides.

LITHIUM GALLIUM NITRIDE. Light gray powder. Li_3GaN_2, mw: 118.56, mp: decomp @ 800°, d: 3.35.

THR = See gallium and lithium compounds and nitrides.

LITHIUM-*m*-GERMANATE. Crystals. Li_2GeO_3, mw: 134.48, mp: 1239°, d: 3.53 @ 21°.

THR = See germanium and lithium compounds.

LITHIUM HEXAPHENYL TUNGSTATE ETHERATE. $Li_3W(C_6H_5)_6 \cdot 3(C_2H_5)_2O$, mw: 857.4.

THR = Spont flam in air. [*19*]

LITHIUM HYDRIDE. White, translucent, crystalline mass. LiH, mw: 7.95, mp: 680°, d: 0.82.

THR = In contact with moisture, lithium hydroxide is formed. Can ignite spont in moist air. See also lithium compounds. The LiOH formed is very caustic and therefore highly toxic, particularly to lungs and respiratory tract, skin and mu mem.

Fire Hazard: See hydrides. Reacts violently with N_2O, liquid O_2 and air. [*19*]

Disaster Hazard: See hydrides.

Explosion Hazard: See hydrides.

To Fight Fire: Special mixtures of dry chemical.

LITHIUM HYDROSULFIDE. White hygroscopic powder. LiHS, mw: 40.01.

THR = See sulfides and lithium compounds.

LITHIUM HYDROXIDE. Syn: *lithium hydroxide*

monohydrate. Colorless crystals. $LiOH \cdot H_2O$, mw: 41.96, mp: 680°, bp: decomp, d: 1.46.
THR = Very caustic and toxic. See lithium compounds and sodium hydroxide.

LITHIUM HYDROXIDE MONOHYDRATE. See lithium hydroxide.

LITHIUM IODATE. White crystals. $LiIO_3$, mw: 181.86, d: 4.502 @ 27°/4°.
THR = See iodates and lithium compounds.

LITHIUM MANGANITE. Reddish-brown powder. Li_2MnO_3, mw: 116.8.
THR = See manganese and lithium compounds.

LITHIUM METHYLIDE. CH_2Li_2, mw: 27.9.
THR = Spont flam in air. [*19*]

LITHIUM MOLYBDATE. White, crystalline powder. Li_2MoO_4, mw: 173.83, mp: 705°, d: 2.66.
THR = See molybdenum and lithium compounds.

LITHIUM NITRATE. Colorless powder, deliquescent granules. $LiNO_3$, mw: 69.0, mp: 264°, d: 2.38.
THR = See nitrates and lithium compounds.

LITHIUM NITRIDE. Brownish-red hexagonal crystals, slowly decomp on contact with moisture. Li_3N, mw: 34.8, d: 1.3, mp: 845°.
THR = Upon contact with moisture, it decomp into lithium hydroxide, lithium compounds and ammonia.
Fire Hazard: Mod; at elevated temp., ignites and burns intensely in air.
To Fight Fire: Dry chemicals, sand, graphite. *Avoid use* of water or carbon tetrachloride.

LITHIUM NITRITE. Colorless crystals. $LiNO_2 \cdot H_2O$, mw: 71.0, mp: > 100°, bp: decomp, d: 1.615.
THR = See nitrites and lithium compounds.

LITHIUM OXALATE. Colorless crystals. $Li_2C_2O_4$, mw: 101.90, mp: decomp: d. 2.121 @ 17.5°.
THR = See oxalates and lithium compounds.

LITHIUM OXIDE. White cubic crystals. Li_2O, mw: 29.9, mp: >1700°, d: 2.013 @ 25.2°, bp: 1200° @ 600 mm.
THR = See lithium compounds.

LITHIUM PERCHLORATE. Colorless deliquescent crystals. $LiClO_4$, mw: 106.40, mp: 236°, bp: decomp @ 430°, d: 2.429.
THR = Irr to skin, eyes and mu mem. See perchlorates and lithium compounds.

LITHIUM PERMANGANATE. Purple crystals. $LiMnO_4 \cdot 3H_2O$, mw: 179.92, mp: decomp @ 190°, d: 2.06.
THR = See manganese and lithium compounds.
Fire Hazard: Mod; a powerful oxidizing agent; can react with reducing materials.

LITHIUM PEROXIDE. Fine white powder or sandy yellow, granular material. Li_2O_2, mw: 45.88, mp: decomp, d: 2.14 @ 20°.
THR = A powerful oxidizer and irr to skin, eyes and mu mem. See also lithium compounds and peroxides.
Fire Hazard: Dangerous, because it is an extremely powerful oxidizing agent. See also peroxides, inorganic.
Disaster Hazard: Dangerous; will react with water or steam to produce heat; on contact with reducing materials, can react vigorously.

LITHIUM PHENATE. Syns: *lithium carbolate, phenol lithium*. White or reddish powder, water-sol. C_6H_5OLi, mw: 100.0.
THR = HIGH irr to skin, eyes and mu mem. See also lithium compounds.

LITHIUM PHENYL. C_6H_5Li, mw: 84.
THR = Spont flam in air. With N_2O forms an explosively unstable product. [*19*] No toxicity data. Probably very irr to skin, eyes and mu mem.

LITHIUM PHENYL AZOXIDE. $LiON{:}NC_6H_5$, mw: 128.
THR = No toxicity data. Probably HIGH. Forms a very unstable product with N_2O. [*19*]

LITHIUM PHOSPHIDE. Li_3P, mw: 51.8.
THR = HIGH, See phosphides. It reacts with water to form highly toxic and spont flam PH_3. [*19*]

LITHIUM PROPYL. C_3H_7LI, mw: 50.
THR = Spont flam in air. [*19*]

LITHIUM SALICYLATE. White or gray-white, odorless, sweet tasting powder, water-sol. $C_7H_5LiO_3$, mw: 144.1, mp: decomp.
THR = See lithium compounds.

LITHIUM SELENIDE. Colorless, deliquescent crystals. $Li_2Se \cdot 9H_2O$, mw: 255.0.
THR = See selenium compounds, lithium compounds and hydrogen selenide.

LITHIUM SILICATE. White powder. Li_2SiO_3, mw: 90.0, mp: 1204°, d: 2.52 @ 25°.
THR = See silicates and lithium compounds.

LITHIUM SILICIDE. Black, hygroscopic crystals. Li_6Si_2, mw: 97.8.
THR = Reacts violently with Br_2, F_2, HCl, HNO_3, P, Se, H_2SO_4, Te, water (yields a spont flam gas). [*19*] See also lithium silicon.

LITHIUM SILICOFLUORIDE. See lithium fluosilicate.

LITHIUM SILICON. Solid. Composition: Li + Si.
THR = See lithium and silicon.

Fire Hazard: Dangerous, in the form of dust when exposed to heat or flame or by chemical reaction with moisture or acids. In contact with water, silane and hydrogen are evolved.

Explosion Hazard: Slight, in the form of dust when exposed to flame. See also powdered metals.

Disaster Hazard: Mod dangerous; will react with water or steam to produce flam vapors; on contact with oxidizing materials, can react vigorously; on contact with acid or acid fumes, can emit toxic and flam fumes.

To Fight Fire: CO_2, dry chemical.

LITHIUM SODIUM FLUOALUMINATE. Cubic crystals. $Li_3Na_3(AlF_6)_2$, mw: 371.75, mp: 710°, d: 2.774.

THR = See fluoaluminates and lithium compounds.

LITHIUM SULFIDE. Cubic, white-yellow crystals. Li_2S, mw: 45.9, d: 1.66, mp: 900°–975°.

THR = See sulfides and lithium compounds.

LITHIUM SULFITE. Needles. $Li_2SO_3 \cdot H_2O$, mw: 112.0, mp: 455° (slight decomp).

THR = See sulfides and lithium compounds.

LITHIUM TETRAAZIDIBORATE. $LiB(N_3)_4$, mw: 185.8.

THR = No toxicity data. A sensitive explosive. [*19*] See components.

LITHIUM TETRABORATE. White crystalline powder. $Li_2B_4O_7 \cdot 5H_2O$, mw: 259.24, mp: $-2H_2O$ @ 200°.

THR = See boron and lithium compounds.

LITHIUM TETRAMETHYL BORATE. $LiB(CH_3)_4$, mw: 77.9.

THR = No toxicity data. May self-ignite in moist air. [*19*] See also lithium and boron compounds.

LITHIUM THALLIUM-*dl*-TARTRATE. Crystals. $LiTlC_4H_4O_6 \cdot 2H_2O$, mw: 395.43, d: 3.144.

THR = See thallium compounds.

LITHIUM THIOCYANATE. Deliquescent, white crystals. LiSCN, mw: 65.0.

Acute tox data: Oral LD_{LO} (mouse) = 210 mg/kg. [*3*]

THR = HIGH via oral route. See thiocyanates.

LITHIUM TRI-*tert*-BUTOXYALUMINATE. $LiAl[OC(CH_3)_3]_3H$, mw: 254.2.

THR = Reacts with water $\longrightarrow H_2$, which can ignite [*19*] and possibly explode. See also hydrogen. No toxicity data. See also lithium compounds.

LITHIUM-*m*-VANADATE. Yellowish powder. $LiVO_3 \cdot 2H_2O$, mw: 141.92.

THR = See vanadium and lithium compounds.

LITHIUM ZIRCONATE. Creamy white powder. Li_2ZrO_3, mw: 153.10.

THR = See zirconium and lithium compounds.

LITHIUM ZIRCONIUM SILICATE. White powder. $2Li_2O \cdot ZrO_2SiO_2$, mw: 243.0.

THR = See lithium compounds, zirconium compounds and silicates.

LITHOPONE. White powder. Composition: zinc sulfide, barium sulfate, zinc oxide.

THR = HIGH because it can liberate hydrogen sulfide upon decomp by heat, moisture and acids. See hydrogen sulfide.

Disaster Hazard: Dangerous; when heated to decomp, emits highly toxic fumes.

LOBELIA.

THR = See lobeline.

Disaster Hazard: Dangerous; when heated, emits highly toxic fumes.

LOBELINE. Yellow, syrupy liquid. $C_{22}H_{27}NO_2$, mw: 285.28, mp: 131°.

Acute tox data: ip LD_{50} (mouse) = 40 mg/kg. [*3*]

THR = HIGH via ip route. Probably HIGH via oral route as well. Causes stimulation which leads to convulsions in severe cases. Nausea and vomiting are frequent.

Disaster Hazard: Dangerous; when heated, emits highly toxic fumes.

Treatment and Antidotes: Wash out stomach if vomiting has not occurred.

LOBELINE SULFATE. See lobeline.

LOCUST BEAN GUM. See carob bean gum.

LOCAL EXHAUST VENTILATION. See Section 2.

LOCATION AND CONSTRUCTION OF EXPLOSIVE MAGAZINES. See Section 7.

LOGWOOD. See sawdust.

LOMITE #1. See explosives, high.

LONDON PURPLE, SOLID.

THR = See arsenic compounds and aniline.

LOROX. See 3-(3,4-dichlorophenyl)-1-methoxy-1-methyl urea.

LOSS LIMITATION FROM FIRE AND EXPLOSION. See Section 7.

LOST. See mustard gas.

LOW EXPLOSIVES. See explosives, low.

LPG. See liquefied hydrocarbon gas.

LP GAS. See liquefied hydrocarbon gas.

LSD. Syns: *pearly gates*, *royal blue*, *wedding bells*, *heavenly blue*. $C_{20}H_{25}ON_3$, mw: 323.5.

THR = VERY HIGH via iv route. An exper mutagen, teratogen. [*3*] An hallucinogen.

LUBRICATING OIL. Syn: *straw oil*. flash p: 315°–366°F, d: < 1.00, autoign. temp.: 783°F.

THR = An exper carc. [*23*] Can cause dermatitis. See also petroleum hydrocarbons.
Fire Hazard: Slight, when exposed to heat or flame; can react with oxidizing materials.
Spont Heating: No.
To Fight Fire: Spray, foam, CO_2, dry chemical.

LUBRICATING OIL, CYLINDER. d: < 1.
THR = See lubricating oils. Can cause dermatitis. See also petroleum hydrocarbons.
Fire Hazard: Slight, when exposed to heat or flame; can react with oxidizing materials.
Spont Heating: No.
To Fight Fire: Water spray, foam, CO_2, dry chemical.

LUBRICATING OIL, LIGHT. See lubricating oil.

LUBRICATING OIL (MAINLY MINERAL). d: < 1, autoign. temp.: 700° F, flash p: 300°–450° F.
THR = See also petroleum hydrocarbons and mineral oils.
Fire Hazard: Slight, when exposed to heat or flame; can react with oxidizing materials.
Spont Heating: No.
To Fight Fire: Water spray, foam, CO_2, dry chemical.

LUBRICATING OIL, MINERAL. d: < 1, flash p: 300°–450° F, autoign. temp.: 500°–700° F.
THR = See mineral oil. Can cause dermatitis. See also petroleum hydrocarbons.
Fire Hazard: Slight, when exposed to heat or flame; can react with oxidizing materials.
Spont Heating: No.
To Fight Fire: Spray, CO_2, dry chemical.

LUBRICATING OIL, MOTORS. d: < 1, autoign. temp.: 500°–700° F, flash p: 300°–450° F.
THR = See mineral oil. Can cause dermatitis. See also petroleum hydrocarbons.
Fire Hazard: Slight, when exposed to heat or flame; can react with oxidizing materials.
Spont Heating: No.
To Fight Fire: Water spray, foam, CO_2, dry chemical.

LUBRICATING OILS AND GREASES.
THR = An exper carc. [*23*, *14*] See oils, mineral.

LUBRICATING OILS, AUTO.
THR = Can cause dermatitis. See also petroleum hydrocarbons.
Fire Hazard: Slight, when exposed to heat or flame.

LUBRICATING OIL, SPINDLE. d: < 1, autoign. temp.: 478° F, flash p: 169° F.
THR = U. Can cause dermatitis. See also petroleum hydrocarbons.
Fire Hazard: Mod, when exposed to heat or flame; can react with oxidizing materials.
Spont Heating: No.
To Fight Fire: Water spray, foam, CO_2, dry chemical.

LUBRICATING OIL, TURBINE. d: < 1, autoign. temp.: 700° F, flash p: 400° F (OC).
THR = U. Can cause dermatitis. See also petroleum hydrocarbons.
Fire Hazard: Slight, when exposed to heat or flame; can react with oxidizing materials.
Spont Heating: No.
To Fight Fire: Spray, CO_2, dry chemical.

LUCIDOL. See benzoyl peroxide, dry.

"LUCITE." See methyl methacrylate.

"LUMINAL." See ethyl phenyl barbituric acid.

"LUPERCO." Paste or solution in organic or inorganic solvents.
THR = HIGH irr to skin, eyes and mu mem. See also peroxides, organic.
Fire Hazard: Mod; a powerful oxidizing agent; can react with reducing materials.
Explosion Hazard: U.

LUPERCO BDB. See chlorobenzoyl peroxide.

LUPERCO CCC. See 2,4-dichlorobenzoyl peroxide.

LUPERCO HDB. See hydroxyheptyl peroxide (paste with dibutyl phthalate).

"LUPERSOL" SOLUTIONS.
THR = HIGH irr to skin, eyes and mu mem. See also peroxides, organic.
Fire Hazard: Mod; a powerful oxidizing agent; can react with reducing materials.
Explosion Hazard: U.

LUPININE. White crystalline alkaloid. $C_{10}H_{19}NO$, mw: 169.29, mp: 69°–71°, bp: 269°.
THR = A toxic alkaloid via oral route.
Disaster Hazard: Dangerous; when heated, emits toxic fumes.

LUTETIUM. A silvery-white metal, stable in air. Lu, atwt: 174.97, mp: 1656°, bp: 3315°, d: 9.835 @ 25°.
THR = U, but very likely nearly non-toxic.
Radiation Hazard: For permissible levels, see Section 5, Table 5A.5. Natural (2.6%) isotope ^{176}Lu, $T_{\frac{1}{2}}$ = 3×10^{10}y. Decays to stable ^{176}Hf via β's of 0.43 MeV. Emits γ's of 0.09, 0.31 MeV.
Fire Hazard: Mod, in the form of dust when exposed to heat or flame or by chemical reaction with oxidizers. Reacts violently with air, halogens. [*19*]
Explosion Hazard: See powdered metals.

LUTETIUM CHLORIDE. $LuCl_3$, mw: 281.3.
Acute tox data: ip LD_{50} (mouse) = 315 mg/kg; ip LD_{50} (guinea pig) = 161 mg/kg. [*3*]
THR = HIGH via ip route.

LUTETIUM (III) NITRATE. $Lu(NO_3)_3$, mw: 361.
Acute tox data: ip LD_{50} (mouse) = 290 mg/kg. [*3*]
THR = HIGH via oral route.

Disaster Hazard: Mod dangerous; when heated to decomp., emits toxic fumes.

LYNESTRENOL + MESTRANOL.
THR = An exper neo to women via oral route. [*103*]

LYE. See sodium hydroxide.

LYSINE. Syn: *α,ε-diamino caproic acid.* An essential amino acid, occurs in isomeric forms; the *l* and *dl* forms are considered here. Colorless crystals, sol in water, slightly sol in alcohol, insol in ether.
$NH_2(CH_2)_4CH(NH_2)COOH$, mw: 146.
THR = U. A nutrient and dietary supplement food additive. [*109*]

MACASSAR GUM. See agar.

MACHETE. Syn: *2-chloro-2',6'-diethyl-N-(butoxymethyl) acetanilide.* Amber-colored oily liquid, very slightly sol in water. $C_{17}H_{26}O_2NCl$, mw: 311.6.
Acute tox data: Oral LD_{50} (rat) = 3120 mg/kg; dermal LD_{50} (rabbit) = 2510 mg/kg. [*3*]
THR = MOD via oral and dermal routes.
Disaster Hazard: Dangerous; see chlorides.

MACQUER'S SALT. See potassium arsenate.

***p*-MAGENTA.** $C_{19}H_{16}N_3 \cdot HCl$, mw: 323.8.
THR = An exper (±) carc. [*3*, *10*]

MAGENTA I. $C_{20}H_{18}N_3 \cdot HCl$, mw: 337.
THR = An exper (±) carc. [*3*, *10*]

MAGENTA II. $C_{21}H_{20}N_3 \cdot HCl$, mw: 351.9.
THR = An exper (±) carc. [*3*, *10*]

MAGISTERY OF BISMUTH. See bismuth subnitrate.

MAGNESIA ALBA. See magnesium carbonate.

MAGNESIA MAGNA. See magnesium hydroxide.

MAGNESITE. See magnesium carbonate hydroxide.

MAGNESIUM. Hexagonal, silvery-white crystals of metal. Mg, atwt: 24.32, mp: 651°, bp: 1107°, d: 1.74 @ 5°, vap. press: 1 mm @ 621°.
Acute tox data: Oral LD_{LO} (dog) = 230 mg/kg. [*3*]
THR = MOD via oral route. See also magnesium compounds.
Fire Hazard: Dangerous, in the form of dust or flakes, when exposed to flame, or by violent chemical reaction with oxidizing agents. In solid form, magnesium is difficult to ignite because heat is conducted rapidly away from the source of ignition; it must be heated above its melting point before it will burn. However, in finely divided form it may be ignited by a spark or the flame of a match. Magnesium fires do not flare up violently unless there is moisture present. Therefore, it must be kept away from water, moisture, etc. It may be ignited by a spark, match flame, or even spont when the material is finely divided and damp, particularly with water-oil emulsion. Also, magnesium reacts with moisture, acids, etc., to evolve hydrogen, which is a highly dangerous fire and explosion hazard.
Explosion Hazard: Mod, in the form of dust, when exposed to flame or by violent chemical reaction with metals, air, (Al + $KClO_4$), NH_4NO_3, BeO, [$Ba(NO_3)_2$ + BaO_2 + Zn], BPI_2, bromobenzyl trifluoride, $Cd(CN)_2$, CdO, CaC, carbonates, CCl_4, Cl_2, ClF_3, $CHCl_3$, $Co(CN)_2$, $Cu(CN)_2$, CuO, [$CuSO_4$(anhydrous) + NH_4NO_3 + $KClO_3$ + H_2O], $CuSO_4$, F_2, AuCN, (H_2 + $CaCO_3$), HI, H_2O_2, I, $Pb(CN)_2$, HgO, $Hg(CN)_2$, CH_3Cl, MoO_3, $Ni(CN)_2$, HNO_3, NO_2, O_2 liquid, performic acid, phosphates, $KClO_3$, $KClO_4$, $AgNO_3$, Ag_2O, $NaClO_4$, (Na_2O_2 + CO_2), SnO_2, sulfates, trichloroethylene, $Zn(CN)_2$, ZnO, Na_2O_2. [*19*]
Disaster Hazard: Dangerous; when heated, burns violently in air and emits fumes; will react with water or steam to produce hydrogen; on contact with oxidizing materials, it can react vigorously.
To Fight Fire: Operators and fire fighters can approach a magnesium fire to within a few ft if no moisture is present. Water and ordinary extinguishers, such as CO_2, carbon tetrachloride, etc., should not be used on magnesium fires. G-1 powder or powdered talc should be used on open fires.

Magnesium turnings, borings, etc., should be collected frequently during the working hours and always at the end of the shift from machines and surrounding area and placed in clean, dry metal covered containers plainly labeled, "For Magnesium Only." Good housekeeping is essential. If more than 50 cubic ft (6 fifty-gal drums) of turnings, borings, etc., are allowed to accumulate, it is necessary that they be left in drums and removed to a separate fire resistant storage room or separated from other occupancies by a space of at least 50 ft. Magnesium grinding dust should always be collected in a dust collector approved specifically for this service. The collected material should be removed frequently and burned. This may be accomplished by spreading it three or four in. thick on a layer of firebrick or hard burnt paving brick and igniting some combustible refuse placed on top. The burning refuse ignites the top surface of the magnesium and the layer dries as it burns. Danger of a sudden flare up exists when the sludge has been allowed to become partially dry.

During machining, magnesium fires result from the use of dull cutting tools, machining with light cuts at high speeds, or rubbing of the tool on the work after the cutting operation is finished. Chip fires on machine tools can best be controlled by the use of a newly developed liquid extinguishant, by flooding with coolant, by using G-1 powder or powdered talc, or, if necessary, by dry sand. These dry materials should be shoveled onto a fire until it is completely blanketed. If the fire is on a combustible floor, best results can be obtained by spreading a layer of extinguishing material on the floor and on the fire, and then shoveling the fire on the layer of extinguishing material. Oven fires can easily be controlled by the use of boron trifluoride or boron trichloride gas.

MAGNESIUM ACETATE. $C_4H_6O_4Mg$, mw: 142.4.
Acute tox data: iv LD_{50} (mouse) = 16 mg/kg. [*3*]
THR = HIGH via iv route.

MAGNESIUM ALUMINUM PHOSPHIDE.
Mg_3AlP_3, mw: 192.8.
THR = In contact with water, evolves spont flam PH_3. [*19*] See also phosphides and phosphine.

MAGNESIUM-*o*-ARSENATE. Syn: *hoernesite*. Monoclinic, white crystals. $Mg_3(AsO_4)_2 \cdot 8H_2O$, mw: 494.91, d: 2.60–2.61.
Acute tox data: Oral LD_{LO} (rat) = 280 mg/kg; oral LD_{50} (mouse) = 315 mg/kg. [*3*]
THR = HIGH via oral route. See also arsenic compounds.

MAGNESIUM-*o*-ARSENITE. Solid. $Mg_3(AsO_3)_2$, mw: 318.78.
THR = HIGH via oral route. See also arsenic compounds.

MAGNESIUM-*o*-BORATE. Rhombic, colorless crystals. $Mg(BO_3)_2$, mw: 190.60, d: 2.99 @ 21°.
THR = See boron compounds and magnesium.

MAGNESIUM BORIDE. Mg_3B_2, mw: 94.6.
THR = HIGH toxicity. See boranes. Reacts with water to form a spont flam gas. [*19*]

MAGNESIUM BROMATE. Colorless crystals. $Mg(BrO_3)_2 \cdot 6H_2O$, mw: 388.25, mp: $-6H_2O$ @ 200°, bp: decomp, d: 2.29.
THR = See bromates. Violent reactions with Al, As, C, Cu, metal sulfides, organic matter, P, S. [*19*]

MAGNESIUM BROMIDE. Large lustrous, white, deliquescent crystals. $MgBr_2$, mw: 184.15, mp: 700°, d: 3.72.
THR = See bromides.

MAGNESIUM BROMOPLATINATE. Trigonal crystals. $MgPtBr_6 \cdot 12H_2O$, mw: 915.24, d: 2.802.
THR = See platinum compounds and bromides.

MAGNESIUM CARBONATE. Syns: *magnesium carbonate, precipitated; magnesia alba*. Very light, odorless, white powder, sol in acids, insol in water and alcohol. $MgCO_3$, mw: 84.4, d: 3.04, decomp @ 350°.
THR = U. Probably LOW. A general purpose food additive; it migrates to food from packaging materials. [*109*]

MAGNESIUM CARBONATE, PRECIPITATED. See magnesium carbonate.

MAGNESIUM CARBONATE HYDROXIDE. Syn: *magnesite*. White bulky powder. $(MgCO_3)_4 \cdot Mg(OH)_2 \cdot 5H_2O$ (approx).
THR = A nuisance dust.

MAGNESIUM CHLORATE. White deliquescent crystals or powder. $Mg(ClO_3)_2 \cdot 6H_2O$, mw: 299.33, mp: 35°, bp: decomp @ 120°, d: 1.80 @ 25°.
Acute tox data: Oral LD_{LO} (rat) = 5250 mg/kg; ip LD_{LO} (rat) = 1100 mg/kg; it LD_{LO} (rat) = 700 mg/kg. [*3*]
THR = MOD via oral, ip and it routes. A defoliant. See magnesium compounds and chlorates. Reacts violently with Al, Sb_2S_3, As, As_2S_3, C, charcoal, Cu, CuS, MnO_2, metal sulfides, dibasic organic acids, organic matter, P, SnS_2, SnS, S. [*19*]

MAGNESIUM CHLORIDE. Thin white to opaque gray granules and/or flakes. $MgCl_2$, mw: 95.2, mp: 708°, bp: 1412°, d: 2.325.
Acute tox data: Oral LD_{50} (rat) = 2800 mg/kg; ip LD_{50} (mouse) = 342 mg/kg; iv LD_{50} (mouse) = 14 mg/kg. [*3*]
THR = HIGH via ip and iv and MOD via oral routes. A substance which migrates to food from packaging materials. [*109*] See magnesium compounds and chlorides.

MAGNESIUM CHLORIDE, HYDRATED. See magnesium chloride.

MAGNESIUM CHLOROPALLADATE. Hexagonal crystals. $MgPdCl_6 \cdot 6H_2O$, mw: 451.86, mp: decomp, d: 2.12.
THR = See palladium compounds, magnesium compounds and chlorides.

MAGNESIUM CHLOROPLATINATE. Crystals. $MgPtCl_6 \cdot 6H_2O$, mw: 540.39 (decomp), d: 2.437.
THR = See platinum, magnesium compounds and chlorides.

MAGNESIUM CHLOROSTANNATE. Crystals. $MgSnCl_6 \cdot 6H_2O$, mw: 463.86, mp: decomp @ 100°, d: 2.08.
THR = See tin, magnesium compounds and chlorides.

MAGNESIUM CHROMATE. Rhombic, yellow crystals. $MgCrO_4 \cdot 7H_2O$, mw: 266.44, d: 1.695.
THR = HIGH via oral route. See chromium and magnesium compounds.
Fire Hazard: Slight, by chemical reactions; can react with reducing materials.

MAGNESIUM COMPOUNDS.
THR = The inhal of fumes of freshly sublimed magnesium oxide may cause metal fume fever. There is no evidence that magnesium produces true sys-

temic poisoning. Occupational health hazards may exist in magnesium foundries, probably from the presence of atmospheric contaminants such as fluorides, sulfur dioxide, carbon tetrachloride and chromium compounds.

Particles of metallic magnesium or magnesium alloy which perforate the skin or gain entry through cuts and scratches may produce a severe local lesion characterized by the evolution of gas and acute inflammatory reaction, frequently with necrosis. The condition has been called a "chemical gas gangrene." Gaseous blebs may develop within 24 hrs of the injury. The inflammatory response is marked at the site of injury and there may be signs of lymphangitis. The lesion is very slow to heal.

The most serious hazard presented by magnesium is the danger from burns. Protection necessary for personnel handling and processing magnesium is usually no different from that which is necessary for other metals. It is recommended that smooth clothing and leather or fire resistant, easily removable aprons be worn in grinding operations on magnesium. The toxicity of magnesium compounds is usually that of the anion. Refer to magnesium and anion. See also specific compounds.

MAGNESIUM CYANIDE. Solid. $Mg(CN)_2$, mw: 76.4.
THR = See cyanides.

MAGNESIUM CYCLAMATE. Syn: *magnesium cyclohexyl sulfamate.*
THR = A non-nutritive sweetner food additive. [*109*] See also sodium cyclamate. See also cyclamates. An exper carc. [*3*]
Disaster Hazard: Dangerous; see sulfates.

MAGNESIUM DIAMIDE. White powder. $Mg(NH_2)_2$, mw: 56.4, mp: decomp.
THR = HIGH irr due to ammonia, which it evolves in contact with moist living tissue. See also ammonia, which is evolved upon contact with water and amides.
Fire Hazard: Dangerous, by chemical reaction to evolve ammonia; spont flam in air.
Explosion Hazard: Mod, by spont chemical reaction; reacts violently with water.
Disaster Hazard: Dangerous; will react with water or steam to produce toxic and flam vapors; can react vigorously with oxidizing materials.

MAGNESIUM DIETHYL. Crystals or liquid. $Mg(C_2H_5)_2$, mw: 82.4, mp: 0°.
THR = Details U. See also magnesium and organometals.
Fire Hazard: Dangerous; spont flam in air or in carbon dioxide.
Explosion Hazard: Mod, by spont chemical reaction or upon contact with water.
Disaster Hazard: Dangerous; will react violently with water or steam; can react vigorously with oxidizing materials.

MAGNESIUM DIOXIDE. See magnesium peroxide.

MAGNESIUM DIPHENYL. Feathery crystals. $Mg(C_6H_5)_2$, mw: 178.5.
THR = U. See magnesium and organometals.
Fire Hazard: Mod; spont flam in moist (but not dry) air.
Explosion Hazard: Mod, when exposed to flame, or upon contact with moisture.
Disaster Hazard: Mod dangerous; when heated to decomp or on contact with acid or acid fumes, emits toxic fumes; decomp violently in water; can react with oxidizing materials.

MAGNESIUM DROSS. See magnesium.

MAGNESIUM DUST. See magnesium.

MAGNESIUM-9,10-EPOXYSTEARATE.
Acute tox data: Oral LD_{50} (rat) = 39000 mg/kg. [*3*]
THR = None via oral route.

MAGNESIUM ETHYLENE BIS DITHIOCARBAMATE.
THR = Details U. A skin irr. See also carbamates.
Disaster Hazard: Dangerous; when heated to decomp, emits highly toxic fumes of SO_x.

MAGNESIUM FERROCYANIDE. Pale yellow crystals. $Mg_2Fe(CN)_6 \cdot 12H_2O$, mw: 476.79, mp: decomp @ approx 200°.
THR = See ferrocyanides.

MAGNESIUM FLUORIDE. Syns: *seelaite, afluon.* Faint violet, luminous crystals. MgF_2, mw: 62.32, mp: 1396°, bp: 2239°, d: 2.9–3.2.
Acute tox data: Oral LD_{LO} (guinea pig) = 1000 mg/kg; sc LD_{LO} (guinea pig) = 3000 mg/kg. [*3*]
THR = MOD via oral and sc routes. See also fluorides.

MAGNESIUM FLUOROACETIC ACID.
$Mg(OOCH_2F)$, mw: 178.4.
Acute tox data: Oral LD_{LO} (rat) = 2 mg/kg. [*3*]
THR = HIGH via oral route. See also fluorides.
Disaster Hazard: See fluorides.

MAGNESIUM FLUOSILICATE. White crystals or powder. $MgSiF_6$, mw: 166.38.
Acute tox data: Oral LD_{50} (guinea pig) = 200 mg/kg; sc LD_{LO} (guinea pig) = 400 mg/kg. [*3*]
THR = HIGH via oral and sc routes. See fluosilicates.

MAGNESIUM FORMATE. Rhombic, colorless crystals. $Mg(CHO_2)_2 \cdot 2H_2O$, mw: 150.4.
THR = See formic acid.
Fire Hazard: Slight.

MAGNESIUM-*o*-GERMANATE. White precipitate. Mg_2GeO_4, mw: 185.24.
THR = See germanium compounds.

MAGNESIUM HYDRATE.
See magnesium hydroxide.

MAGNESIUM HYDRIDE. White crystals, reacts with water, sol in isopropylamine. MgH_2, mw: 26.3, d: 1.419, mp: decomp $> 200°$.
THR = See magnesium compounds and hydrides. Reacts violently with air, water. [*2, 19*]

MAGNESIUM HYDROGEN PHOSPHATE. See magnesium phosphate, dibasic.

MAGNESIUM HYDROXIDE. Syns: *magnesium hydrate, magnesia magna.* White powder, odorless, sol in solutions of ammonium salts and dilute acids; almost insol in water and alcohol. $Mg(OH)_2$, mw: 58, d: 2.36, mp: decomp @ 350°.
THR = A general purpose food additive; it is a substance which migrates to food from packaging materials. [*109*] See magnesium compounds. Violent reaction with maleic anhydride, P. [*19*]

MAGNESIUM HYPOPHOSPHITE. White crystals. $Mg(H_2PO_2)_2 \cdot 6H_2O$, mw: 262.36, d: 1.59.
THR = See hypophosphites.
Fire Hazard: Mod, when exposed to heat. It liberates a spont flam gas when heated. [*19*]
Explosion Hazard: Mod, by chemical reaction.
Disaster Hazard: Dangerous; when heated to 100° it evolves flam gas and highly toxic fumes of phosphine; can react with oxidizing materials.

MAGNESIUM IODATE. Colorless crystals. $Mg(IO_3)_2 \cdot 4H_2O$, mw: 446.22, mp: $-4H_2O$ @ 210°, bp: decomp, d: 3.3 @ 13.5°.
THR = See iodates and magnesium compounds. Reacts violently with Al, As, C, Cu, metal sulfides, organic matter, P, S. [*19*]

MAGNESIUM IODIDE. White deliquescent crystals. MgI_2, mw: 278.16, mp: $> 700°$ (decomp), d: 4.244.
THR = See iodides and magnesium compounds.

MAGNESIUM LACTATE. $C_6H_6O_{10}Mg$, mw: 202.5.
Acute tox data: iv LD_{LO} (mouse) = 45 mg/kg. [*3*]
THR = HIGH via iv route.

MAGNESIUM, METALLIC, POWDERED, PELLETS, TURNINGS OR RIBBON. See magnesium.

MAGNESIUM METHYLIDE. $Mg(CH_3)_2$, mw: 54.4.
THR = Self-ignites in air. [*19*]

MAGNESIUM NITRATE. (1) white crystals (prisms); (2) monoclinic, colorless, deliquescent crystals. (1) $Mg(NO_3)_2 \cdot 2H_2O$, (2) $Mg(NO_3)_2 \cdot 6H_2O$; mw: (1) 184.37, (2) 256.43; d: (1) 2.0256 @ 25°, (2) 1.464; mp: (1) 129.0°, (2) 95°; bp: (2) $-5H_2O$ @ 330°.
THR = See nitrates. Violent reaction with dimethyl formamide. [*19*]

MAGNESIUM OXALATE. White powder. $MgC_2O_4 \cdot 2H_2O$, mw: 148.37, mp: decomp, d: 2.45.
THR = See oxalates.

MAGNESIUM OXIDE. White powder. MgO, mw: 40.32, mp: 2500°–2800°, d: 3.65–3.75.
Acute tox data: Inhal TC_{LO} (human) = 400 mg/m^3. [*3*]
THR = MOD via inhal route. A nutrient and/or dietary [*109*] supplement food additive. See magnesium compounds. Inhal of the fumes can produce in man a febrile reaction and a leukocytosis. Reacts violently with ClF_3, PCl_5. [*19*]

MAGNESIUM PEMOLINE. $C_9H_{10}O_4N_2Mg$, mw: 234.5.
Acute tox data: Oral LD_{50} (rat) = 550 mg/kg; oral LD_{50} (monkey) = 400 mg/kg; ip LD_{50} (mouse) = 487 mg/kg. [*3*]
THR = HIGH via oral and ip routes.

MAGNESIUM PERBORATE. White powder. $Mg(BO_3)_2 \cdot 7H_2O$, mw: 268.02.
THR = See magnesium compounds and boron compounds.
Fire Hazard: Slight, by chemical reaction; an oxidant.

MAGNESIUM PERCHLORATE. Syn: *dehydrite.* White, hygroscopic crystals. $Mg(ClO_4)_2$, mw: 223.33, mp: decomp @ 251°, d: 2.60 @ 25°.
Acute tox data: ip LD_{50} (mouse) = 1500 mg/kg. [*3*]
THR = MOD via oral route. See also magnesium compounds and perchlorates.
Fire Hazard: See perchlorates.
Explosion Hazard: See perchlorates. Avoid contact with mineral acids, ammonia, butyl fluorides, P, dimethyl sulfoxide, ethylene oxide, hydrocarbons, organic matter, trimethyl phosphite. [*19*]
Disaster Hazard: See perchlorates.

MAGNESIUM PERMANGANATE. Dark purple, deliquescent needles. $Mg(MnO_4)_2 \cdot 6H_2O$, mw: 370.28, mp: decomp, d: 2.18.
THR = See manganese compounds.
Fire Hazard: Mod, by chemical reaction; a powerful oxidizer.

MAGNESIUM PEROXIDE. Syn: *magnesium dioxide.* White powder. MgO_2 (theoretical), mw: 56.32.
THR = See peroxides, inorganic, and magnesium compounds.
Fire Hazard: Mod, by chemical reaction with acidic materials and moisture; an oxidizing agent.
Disaster Hazard: Dangerous; can react with reducing agents; will decomp violently in or near a fire.

MAGNESIUM PHOSPHATE, DIBASIC. Syns: *dimagnesium-o-phosphate, dimagnesium phosphate,*

magnesium phosphate (sec), magnesium hydrogen phosphate. White crystalline powder, sol in dilute acids, slightly sol in water. $MgHPO_4 \cdot 3H_2O$, mw: 173, d: 2.13, decomp @ 550°–560°.
THR = Details U. Used as a nutrient and/or dietary supplement food additive. [*109*]
Disaster Hazard: Dangerous; see phosphates.

MAGNESIUM PHOSPHATE, NEUTRAL. See magnesium phosphate, tribasic.

MAGNESIUM PHOSPHATE, SEC. See magnesium phosphate, dibasic.

MAGNESIUM PHOSPHATE, TRIBASIC. Syns: *magnesium phosphate, neutral; trimagnesium phosphate.* Fine, soft, bulky, white powder, odorless, sol in acids, insol in water. $Mg_3(PO_4)_2 \cdot 8H_2O$, mw: 406, loses all water @ 400°.
THR = U. Used as a nutrient and/or dietary supplement food additive. [*109*]
Disaster Hazard: Dangerous; see phosphates.

MAGNESIUM PHOSPHIDE. Mg_3P_2, mw: 134.9.
Acute tox data: Inhal LC_{LO} (rat) = 580 ppm for 1 hr; inhal LC_{LO} (cat) = 173 ppm for 2 hrs. [*3*]
THR = HIGH via inhal and oral route. See phosphides. Reacts violently with Br_2, Cl_2, HNO_3, H_2O. [*19*]

MAGNESIUM RICINOLEATE. Coarse white granules.
$Mg[CO_2(CH_2)_7CH=CHCH_2CHOH(CH_2)_5CH_3]_2$, mw: 614, mp: 95°, d: 1.06 @ 25°/25°.
THR = See magnesium compounds.

MAGNESIUM SCRAP (SHAVINGS, BORINGS, OR TURNINGS). See magnesium.

MAGNESIUM SELENATE. Colorless crystals.
$MgSeO_4 \cdot 6H_2O$, mw: 275.38, d: 1.928.
THR = HIGH. See selenium compounds.

MAGNESIUM SILICATE, HYDRATED.
See asbestos.

MAGNESIUM SILICATE, HYDROUS. Solid.
$Mg_2Si_3O_8 \cdot 5H_2O$.
THR = No data. An anti-caking agent food additive. [*109*] See silicates and Mg compounds.

MAGNESIUM STEARATE. Soft, white, light powder, odorless, insol in water and alcohol.
$Mg(C_{18}H_{35}O_2)_2$, mw: 590, d: 1.028, mp: 88.5° (pure), 132° (technical).
THR = U. A general purpose food additive. [*109*]

MAGNESIUM SULFATE. Syn: *epsom salts.* Opaque needles. $MgSO_4 \cdot 7H_2O$, mw: 246.5, mp: $-7H_2O$ @ 200°, d: 1.68.
Acute tox data: sc LD_{LO} (rabbit) = 1750 mg/kg; ip LD_{LO} (dog) = 1600 mg/kg; iv LD_{LO} (dog) = 750 mg/kg. [*3*]
THR = MOD via sc, ip and iv routes. A nutrient and/or dietary supplement food additive, it migrates to food from packaging materials. [*109*] See magnesium compounds. Anti convulsant, purgative. [*2*]

MAGNESIUM SULFIDE. Pale red to reddish-brown. MgS, mw: 56.39, mp: decomp, d: 2.79–2.85.
THR = See magnesium compounds and sulfides.

MAGNESIUM SULFITE. White crystalline powder. $MgSO_3 \cdot 6H_2O$, mw: 212.48, mp: $-6H_2O$ @ 200°, bp: decomp, d: 1.725.
THR = See magnesium compounds and sulfites.

MAGNESIUM THIOTELLURITE. Pale yellow, crystalline mass. Mg_3TeS_5, mw: 360.9.
THR = See sulfides and tellurium compounds.
Fire Hazard: Slight, when exposed to heat or flame.
Disaster Hazard: Dangerous; when heated to decomp, or on contact with acid or acid fumes, emits highly toxic fumes of oxides of sulfur and tellurium; can react with oxidizing materials.

MAGNESIUM-THORIUM ALLOYS IN FORMED SHAPES (not powdered and which shall contain not more than 4% nominal thorium 232). See thorium and magnesium.

MAGNESIUM TUNGSTATE. Colorless, monoclinic crystals. $MgWO_4$, mw: 272.24, d: 5.66.
THR = See magnesium and tungsten compounds.

MAGNETITE. See black iron oxide.

MAGNUS' SALT.
See tetrammine platinum II chloroplatinate.

MALACHITE GREEN. Syns: *aniline green, china green.* Green crystals. $C_{23}H_{25}N_2Cl$, mw: 364.9.
Acute tox data: Oral LD_{LO} (rabbit) = 75 mg/kg. [*3*]
THR = HIGH via oral route.
Disaster Hazard: Dangerous; when heated to decomp, emits highly toxic fumes of aniline and NO_x.

MALATHION. See *o,o*-dimethyl dithiophosphate of diethyl mercaptosuccinate.

MALEANILIC ACID. See *n*-phenylmaleamic acid.

MALE FERN. See aspidium.

MALEIC ACID. Syns: *maleinic acid, toxilic acid.* White crystals, faint acidulous odor.
HOOCCHCHCOOH, mw: 116.1, mp: 130.5°, bp: 135° decomp, d: 1.590 @ 20°/4°, vap. d: 4.0.
Acute tox data: Oral LD_{50} (rat) = 708 mg/kg; dermal LD_{50} (rabbit) = 1560 mg/kg. [*3*]
THR = MOD via oral and dermal routes. Believed to be more toxic than its isomer, fumaric acid.
Fire Hazard: Slight.

MALEIC ANHYDRIDE. Syns: *toxilic anhydride, cis-butenedioic anhydride.* Fused block or white crystals. OCOCHCHCO, mw: 98.1, mp: 53°, bp: 202°, flash p: 215°F (CC), d: 1.48 @ 20°/4°, autoign. temp.: 890°F, vap. press: 1 mm @ 44.0°, vap. d: 3.4, lel = 1.4%, uel = 7.1%.

Acute tox data: Oral LD_{LO} (rat) = 850 mg/kg; oral LD_{50} (mouse) = 60 mg/kg. [*3*]

THR = HIGH irr to skin, eyes and mu mem. Inhal of vapor can cause pulmonary edema. Causes burns to skin and eyes. An exper carc. [*3, 23*]

Fire Hazard: Low, when exposed to heat or flame; will react with water or steam to produce heat; emits toxic fumes when heated; can react on contact with oxidizing materials. Violent reaction with alkali metals, amines, $Ca(OH)_2$, Li, K, KOH, Na, NaOH, pyridine. [*19*]

To Fight Fire: Alcohol foam.

MALEIC HYDRAZIDE. Syn: *1,2-dihydro-3,6-pyridazinedione.* Crystals, somewhat sol in water and alcohol. $C_4H_4N_2O_2$, mw: 112.1, mp: >300°.

Acute tox data: Oral LD_{50} (rat) = 3800 mg/kg. [*3*]

THR = MOD via oral route. Animal studies show acute CNS and chronic liver damage. A food additive permitted in foods for human consumption. [*109*] See hydrazine. An exper (±) carc. [*3, 10, 23*]

Disaster Hazard: Dangerous; when heated to decomp, emits highly toxic fumes.

MALEIC HYDRAZIDE DIETHANOL AMINE SALT. $C_4H_{11}O_2N \cdot N_2O_2 \cdot C_4H_4$, mw: 217.3.

THR = MOD via oral route. An exper carc. [*3*]

MALEINIC ACID. See maleic acid.

MALIC ACID. Syns: *common malic acid, hydroxysuccinic acid, apple acid.* Colorless crystals, very sol in water and alcohol, slightly sol in ether. Exhibits isomeric forms (*dl*, *l* and *d*). $COOHCH_2CH(OH)COOH$, mw: 134, d(*dl*): 1.601, d(*d* or *l*): 1.595 @ 20°/40°, mp(*dl*): 128°, mp(*d* or *l*): 100°, bp(*dl*): 150°, bp(*d* or *l*): 140° (decomp).

Acute tox data: Oral LD_{LO} (rabbit) = 5000 mg/kg. [*3*]

THR = MOD via oral route. A general purpose food additive. Also a synthetic flavoring substance and adjuvant. [*109*]

MALONALDEHYDE. See propandial.

MALONIC ACID. Syns: *methanedicarbonic acid, methanedicarboxylic acid, propanedioic acid.* Small crystals, sol in water, alcohol and ether. $CH_2(COOH)_2$, mw: 104.06, mp: 132°–134°, bp: decomp, d: 1.67.

Acute tox data: Oral LD_{50} (rat) = 1310 mg/kg; ip LD_{50} (mouse) = 300 mg/kg. [*3*]

THR = HIGH via ip and MOD via oral routes, powerful irr. [*2*]

MALONIC ACID MONONITRILE. See cyanoacetic acid.

MALONIC DINITRILE. See cyanoacetonitrile.

MALONIC ETHYL ESTERNITRILE. See ethyl cyanoacetate.

MALONONITRILE. See cyanoacetonitrile.

MALONYLUREA. See barbituric acid.

MALT EXTRACT. A powder. Contains diastase, dextrin, dextrose, protein bodies and barley salts.

THR = An allergen. Malt has been associated with grain or malt fever, an illness existing among grain workers. This is described in the literature as follows. It occurs several hours after the patient leaves work, possibly during sleep. It causes headache, weakness, fever, chills and cold sweats, as well as nausea, coughing and often vomiting. By morning these symptoms have subsided. The victim feels normal and generally reports back for work. This strange illness occurs most frequently among men never before exposed to grain dust, or even among experienced men at the beginning of a new season or upon returning to work after a short absence. It resembles metal fume fever because only those constantly exposed appear to develop an immunity. No really satisfactory explanation can be offered for this reaction to malt dust, except that it is probably due to the presence of a foreign protein.

Fire Hazard: Slight, when exposed to heat or flame; can react with oxidizing materials.

Explosion Hazard: Mod, in the form of dust, when exposed to flame.

MALTOSE. Syn: *malt sugar.*

THR = An exper neo. [*3*] via sc route.

MANDELIC ACID. Syns: *amygdalic acid, phenylhydroxy acetic acid, phenylglycolic acid, dl-mandelic acid, α-hydroxy-α-toluic acid, racemic mandelic acid, uromaline.* Large white crystals or powder, faint odor, sol in ether, slightly sol in water and alcohol. $C_6H_5CH(OH)COOH$, mw: 142, d: 1.30, mp: 117°–119°.

Acute tox data: im LD_{50} (rat) = 300 mg/kg; oral LD_{LO} (rat) = 3000 mg/kg. [*3*]

THR = HIGH via im and MOD via oral routes. Continued absorption can cause kidney irr. Used medicinally. Ingestion of large doses causes nausea, diarrhea and possibly kidney damage.

MANDELONITRILE. See benzaldehyde cyanhydrin.

MANDELYLTROPEINE. See homatropine and compounds.

MANEB. Syns: *manganous ethylene bis-dithio carbamate, manzate.* Yellow powder or crystals, water-sol. $C_4H_6MnN_2S_4$, mw: 265.3.

Acute tox data: Oral LD_{50} (rat) = 6750 mg/kg. [3] LD_{50} (rat) = 4500 mg/kg. [12]
THR = MOD via oral route. See also manganese compounds and carbamates. An exper teratogen and carc. [3, 12] via oral route.
Disaster Hazard: Dangerous; when heated to decomp, emits highly toxic fumes of NO_x and SO_x.

MANGANESE. Reddish-grey or silvery, brittle, metallic element. Mn, atwt: 54.93, mp: 1260°, bp: 1900°, d: 7.20, vap. press: 1 mm @ 1292°.
Acute tox data: ip LD_{50} (mouse) = 53 mg/kg; inhal TC_{LO} (human) = 11 mg/m^3 ⟶ CNS symptoms. [3]
THR = HIGH via ip and inhal routes. A known mutagen and (S) carc. [22, 23] See manganese compounds.
Radiation Hazard: For permissible levels, see Section 5, Table 5A.5. Artificial isotope ^{54}Mn, $T\frac{1}{2}$ = 300d. Decays to stable ^{54}Cr by ec. Emits γ's of 0.84 MeV and x-rays.
Fire Hazard: Mod, in the form of dust or powder, when exposed to flame.
Spont Heating: No.
Explosion Hazard: Mod, in the form of dust, when exposed to flame. See also powdered metals. Violent reaction with (Al + air), Cl_2, F_2, H_2O_2, HNO_3, NO_2, P, SO_2. [19]
Disaster Hazard: Mod dangerous; will react with water or steam to produce hydrogen; can react with oxidizing materials.
To Fight Fire: Special dry chemical.

MANGANESE ACETATE. Pale red crystals, very sol in water and alcohol. $Mn(C_2H_3O_2)_2 \cdot 4H_2O$, mw: 245, d: 1.54, mp: 80°.
THR = See manganese compounds. Used as a trace mineral added to animal feeds. [109]

MANGANESE ARSENATE. Reddish-white, crystalline solid. $MnHAsO_4$, mw: 194.9.
THR = HIGH tox. See arsenic and manganese compounds.

MANGANESE BACITRACIN.
THR = U. Used as a food additive permitted in food for human consumption. [109] See also manganese compounds.

MANGANESE BENZOATE.
See manganous benzoate.

MANGANESE BROMIDE. See manganese dibromide.

MANGANESE CACODYLATE. Reddish-white crystals. $Mn[(CH_3)_2AsO_2]_2$, mw: 328.9.
THR = HIGH. See arsenic and manganese compounds.

MANGANESE CHLORIDE.
See manganese dichloride.

MANGANESE COMPOUNDS.
THR = Chronic manganese poisoning is a clearly characterized disease which results from the inhal of fumes or dusts of manganese. Exposure to heavy conc of dusts or fumes for as little as three months may produce the condition, but usually cases develop after 1–3 yrs of exposure. The CNS is the chief site of damage. If cases are removed from exposure shortly after the appearance of symptoms, some improvement in the patient's condition frequently occurs, though there may be some residual disturbances in gait and speech. When well established, however, the disease results in permanent disability.

Individuals exposed to dusts and fumes of manganese have been reported by several investigators to suffer from a much higher incidence of upper respiratory infections and pneumonia than does the general population. It has not yet been possible to prove that a definite pneumonitis results in humans from exposure to manganese dusts or fumes under industrial conditions. However, experiments with mice have produced definite and striking lung pathology which varied in intensity with the length of exposure to the dust.

Chronic manganese poisoning begins usually with complaints of languor and sleepiness. This is followed by weakness in the legs and the development of a stolid, mask-like facies, and the patient speaks with a slow monotonous voice. Then muscular twitchings appear, varying from a fine tremor of the hands to coarse, rhythmical movements of the arms, legs and trunk. Nocturnal cramps of the legs appear about the same time. There is a slight increase in tendon reflexes, ankle and patellar clonus, and a typical Parkinsonian slapping gait. The handwriting may be quite minute. There are no sensory disturbances, and no eye, gastrointestinal or genitourinary complaints. The urine and spinal fluid are normal, and the blood shows no abnormality or only a slight leucopenia. The symptoms may simulate progressive bulbar paralysis, postencephalitic Parkinsonism, multiple sclerosis, amyotrophic lateral sclerosis and progressive lenticular degeneration (Wilson's Disease). An exper (+) carc. [12, 14, 23, 117] Often a history of exposure is the only aid in establishing the diagnosis. The blood may show increased erythrocyte formation and increased osmotic fragility. Early administration of EDTA can hasten recovery, but it is of little value in cases of long standing.

Manganese compounds are common air contaminants.

MANGANESE CYCLOPENTADIENYL TRICARBONYL. $C_8H_5O_3Mn$, mw: 204.1.

Acute tox data: Oral LD_{50} (mouse) = 150 mg/kg; inhal LC_{LO} (rat) = 120 mg/m^3 for 2 hrs. [3]

THR = HIGH via oral and inhal routes. Animal exper show MILD narcotic effect and impaired kidney function. See also manganese compounds.

MANGANESE DIBROMIDE. Syn: *manganese bromide*. Rose-red crystals. $MnBr_2 \cdot 4H_2O$, mw: 286.8, mp: 64° (decomp), d: 4.385 @ 25°.

THR = Reacts violently with K. [19] See also manganese compounds.

MANGANESE DICHLORIDE. Syns: *scacchite, manganese chloride*. Cubic, deliquescent, pink crystals. $MnCl_2$, mw: 125.84, mp: 650°, bp: 1190°, d: 2.977 @ 25°.

Acute tox data: ip LD_{50} (mouse) = 121 mg/kg. [3, 103] pa LD_{LO} (dog) = 56 mg/kg; [3] iv LD_{50} (dog) = 202 mg/kg; pa LD_{LO} (rabbit) = 18 mg/kg; sc LD_{LO} (mice) = 210 mg/kg. [103]

THR = HIGH via ip, iv, pa and sc routes. An exper carc to mice via ip and sc routes. [103] A trace mineral added to animal feeds, also a nutrient and/or dietary supplement food additive. [109] See manganese compounds and chlorides. Reacts violently with K, Na, Zn. [19]

MANGANESE DIFLUORIDE. Syn: *manganese fluoride*. Red tetragonal crystals or reddish powder. MnF_2, mw: 92.93, mp: 856°, d: 3.98.

Acute tox data: Oral LD_{LO} (guinea pig) = 200 mg/kg; sc LD_{LO} (guinea pig) = 700 mg/kg. [3]

THR = HIGH via oral and MOD via sc routes. See fluorides and manganese compounds.

MANGANESE DIIODIDE. Syn: *manganese iodide*. Yellowish-brown or pink, deliquescent crystalline mass or white powder. $MnI_2 \cdot 4H_2O$, mw: 380.8, mp: decomp @ approx 80°, d: 5.01.

THR = See manganese compounds and iodides. Reacts violently with K. [19]

MANGANESE DIOXIDE. Syn: *manganese oxide*. Black powder. MnO_2, mw: 86.93, mp: $-O_2$ @ 535°, d: 5.0.

Acute tox data: iv LD_{LO} (rabbit) = 45 mg/kg. [3]

THR = HIGH via iv route. See manganese compounds.

Fire Hazard: Mod, by chemical reaction; a powerful oxidizer. It must not be heated or rubbed in contact with easily oxidizable matter. Avoid contact with chlorates, ClF_3, H_2O_2, H_2SO_5, KN_3, $RbHC_2$, Na_2O_2. [19]

Disaster Hazard: Mod dangerous; keep away from heat and flam materials.

MANGANESE FLUORIDE.
See manganese difluoride.

MANGANESE GLUCONATE. Light pinkish powder or coarse granules, sol in water, insol in alcohol and benzene. $Mn(C_6H_{11}O_7)_2 \cdot 2H_2O$, mw: 481.

THR = Used as a nutrient and/or dietary supplement food additive. A trace mineral added to animal feeds. [109] See also manganese compounds.

MANGANESE GLYCEROPHOSPHATE. Yellow-white or pinkish powder, odorless, sol in citric acid solution; slightly sol in water, insol in alcohol. $CH_2OHCHOHCH_2OP(O)O_2Mn$, mw: 195.

THR = A nutrient and/or dietary supplement food additive. [109] See also manganese compounds.

MANGANESE GREEN. See barium manganate.

MANGANESE HEPTOXIDE. A dark red oil. Mn_2O_7, mw: 221.9, mp: 5.9°, bp: explodes @ 95°, d: 2.596 @ 20°/4°.

THR = See manganese compounds.

Fire Hazard: Dangerous; by chemical reaction, a powerful oxidizer.

Explosion Hazard: Severe, when shocked or exposed to heat. Also can self-explode. Avoid contact with organic matter. [19]

Disaster Hazard: Highly dangerous; shock or heat will explode it; can react vigorously with reducing materials.

MANGANESE HYPOPHOSPHITE. Pink odorless crystals. $Mn(H_2PO_2)_2 \cdot H_2O$, mw: 202.9, mp: $-H_2O >$ 150°.

THR = A nutrient and/or dietary supplement food additive. [109] See manganese compounds and hypophosphites. Can detonate @ > 200°.

MANGANESE IODIDE. See manganese diiodide.

MANGANESE IODOPLATINATE. Crystals. $MnPtI_6 \cdot 9H_2O$, mw: 1173.82, mp: decomp, d: 3.60.

THR = See platinum and manganese compounds and iodides.

MANGANESE LEAD RESINATE. Dark brown to black mass.

THR = See lead and manganese compounds.

Fire Hazard: Mod, when exposed to heat or flame.

Disaster Hazard: Dangerous; when heated, emits highly toxic fumes of lead; can react with oxidizing materials.

To Fight Fire: Water, foam, CO_2, dry chemical.

MANGANESE MONOARSENIDE. Black hexagonal crystals. MnAs, mw: 129.84, mp: decomp @ 400°, d: 6.2.

THR = HIGH. See arsenic and manganese compounds.
Fire Hazard: Mod, when exposed to flame. It will react with water and acids to form hydrogen arsenide.
Disaster Hazard: Dangerous; when heated to decomp, emits highly toxic fumes of arsenic; will react with water, steam or acids to produce toxic and flam vapors; can react with oxidizing materials.

MANGANESE MONOBORIDE. Crystalline powder. MnB, mw: 65.75, d: 6.2 @ 15°.
THR = See manganese and boron compounds.
Fire Hazard: Mod; will react with acids or steam to form boron hydride. It will ignite in presence of powerful oxidizers.
Explosion Hazard: Mod, by chemical reaction. See boron hydride.
Disaster Hazard: Dangerous; will react with water, steam or acids to produce toxic and flam vapors; reacts vigorously with oxidizing materials.

MANGANESE MONOPHOSPHIDE. Dark gray crystals. MnP, mw: 85.91, mp: 1190°, d: 5.39 @ 21°.
THR = See phosphides and manganese compounds.
Fire Hazard: Dangerous, when exposed to flame or by chemical reaction.
Explosion Hazard: Mod, by chemical reaction with moisture or acids.
Disaster Hazard: Dangerous; when heated to decomp, emits highly toxic fumes of oxides of phosphorus; will react with water, steam or acids to produce toxic and flam vapors; can react vigorously with oxidizing materials.

MANGANESE MONOSULFIDE.
See manganese sulfide.

MANGANESE MONOXIDE. See manganous oxide.

MANGANESE OXALATE. See manganous oxalate.

MANGANESE OXIDE. See manganese dioxide.

MANGANESE PHOSPHATE, DIBASIC. Syns: *manganous phosphate, acid*; *manganese hydrogen phosphate*, *manganous phosphate (sec)*, *manganese phosphate*. Pink powder, sol in acids, slightly sol in water. $MnHPO_4 \cdot 3H_2O$, mw: 205.
THR = See manganese compounds. A trace mineral added to animal feeds. [*109*]
Disaster Hazard: Dangerous; see phosphates.

MANGANESE PROTOXIDE. See manganous oxide.

MANGANESE PYROPHOSPHATE.
See manganous pyrophosphate.

MANGANESE RESINATE. Dark, brownish-black mass. $Mn(C_{20}H_{29}O_2)_2$, mw: 657.8.
THR = See manganese compounds.
Fire Hazard: Slight, when exposed to heat or flame or by reaction with powerful oxidizers.

MANGANESE SELENATE. Rhombic crystals. $MnSeO_4 \cdot 2H_2O$, mw: 233.92, d: 2.95–3.01.
THR = See manganese and selenium compounds.

MANGANESE SELENIDE. Gray cubic crystals. MnSe, mw: 133.89, d: 5.59 @ 15°.
THR = See manganese and selenium compounds.
Fire Hazard: Mod, by chemical reaction either with moisture or acids to liberate selenium hydride, or on contact with powerful oxidizers.
Explosion Hazard: Mod, by chemical reaction.
Disaster Hazard: Dangerous; when heated to decomp, emits highly toxic fumes of selenium; will react with water or acids to produce toxic and flam vapors; can react with oxidizing materials.

MANGANESE SELENITE. See manganous selenite.

MANGANESE SULFATE. See manganous sulfate.

MANGANESE SULFATE, HYDRATE. $MnSO_4 \cdot H_2O$, mw: 169.0.
THR = An exper carc to mice via ip route. [*103*]

MANGANESE TARTRATE. See manganous tartrate.

MANGANIC FLUORIDE. Red crystals. MnF_3, mw: 112, mp: decomp, d: 3.54.
THR = HIGH toxicity. When heated, can violently attack glass. [*19*] See manganese compounds and fluorides.

MANGANIC-*m*-PHOSPHATE. Pink crystals. $Mn_2(PO_3)_6 \cdot 2H_2O$, mw: 619.8.
THR = See manganese compounds.

MANGANIC-*o*-PHOSPHATE. Greenish-gray crystalline powder. $MnPO_4 \cdot H_2O$, mw: 167.93.
THR = See manganese compounds and phosphates.

MANGANIC SULFATE. Green deliquescent crystals. $Mn_2(SO_4)_3$, mw: 398.06, mp: decomp @ 160°.
THR = See manganese compounds and sulfates.

MANGANOCENE. $Mn(C_5H_5)_2$, mw: 185.1.
THR = No data. See also manganese compounds. Self-ignites in air. [*19*]

MANGANOCYANIC ACID. $H_4Mn(CN)_6$, mw: 215.07, mp: decomp.
THR = See cyanides and manganese compounds.

MANGANOLANGBEINITE. See potassium manganous sulfate.

MANGANOUS ACETATE TETRAHYDRATE. $C_4H_6O_4Mn \cdot 8H_2O$, mw: 317.2.
Acute tox data: Oral LD_{50} (rat) = 3730 mg/kg. [*3*]
THR = MOD via oral route.

MANGANOUS ARSENATE. See manganese arsenate.

MANGANOUS BENZOATE. Syn: *manganese benzoate*. Flat prisms. $Mn(C_7H_5O_2)_2 \cdot 3H_2O$, mw: 351.20.
THR = See manganese compounds.

Disaster Hazard: Slight; when heated, emits acrid fumes.

MANGANOUS CHLOROPLATINATE. Crystals. $MnPtCl_6 \cdot 6H_2O$, mw: 571.00, mp: decomp, d: 2.692.
THR = See platinum compounds, manganese compounds and chlorides.

MANGANOUS CITRATE. White-reddish powder. $Mn_3(C_6H_5O_7)_2$, mw: 542.99.
THR = See manganese compounds. Used as a nutrient and/or dietary supplement food additive. Also as a trace mineral added to animal feeds. [*109*]
Disaster Hazard: Slight; when heated, emits acrid fumes.

MANGANOUS ETHYLENE BIS-DITHIOCARBAMATE. See maneb.

MANGANOUS FERROCYANIDE. Greenish-white powder. $Mn_2Fe(CN)_6 \cdot 7H_2O$, mw: 447.93.
THR = See manganese compounds and ferrocyanides.

MANGANOUS FLUOGALLATE. Pink crystals. $[Mn(H_2O)_6][GaF_5 \cdot H_2O]$, mw: 345.76, mp: decomp @ 230°, d: 2.22.
THR = See fluogallates and manganese compounds.

MANGANOUS FLUOSILICATE. Syn: *manganous silicofluoride*. Hexagonal, rose-red prisms. $MnSiF_6 \cdot 6H_2O$, mw: 305.09, mp: decomp, d: 1.903.
THR = See fluosilicates and manganese compounds.

MANGANOUS FORMATE. Rhombic crystals. $Mn(CHO_2)_2 \cdot 2H_2O$, mw: 181.00, mp: decomp, d: 1.953.
THR = See manganese compounds.
Disaster Hazard: Slight; when heated, emits acrid fumes.

MANGANOUS HYDROXIDE. Syn: *pyrochroite*. White-pink crystals. $Mn(OH)_2$, mw: 88.95, mp: decomp, d: 3.258 @ 13°.
THR = See manganese compounds.

MANGANOUS IODATE. $Mn(IO_3)_2$, mw: 404.8.
THR = See manganese compounds and iodates. Reacts violently with K. [*19*]

MANGANOUS NITRATE. Monoclinic rose-white or colorless crystals. $Mn(NO_3)_2 \cdot 4H_2O$, mw: 251.0, mp: 25.8°, bp: 129.4°, d: 1.82.
THR = See manganese compounds and nitrates.

MANGANOUS OXALATE. Syn: *manganese oxalate*. White crystals or powder. $MnC_2O_4 \cdot 2H_2O$, mw: 179.0, mp: $-2H_2O$ @ 100°, d: 2.43 @ 21.7°.
THR = See oxalates and manganese compounds.

MANGANOUS OXIDE. Syns: *manganese protoxide, manganese monoxide*. Grass green powder, sol in acids, insol in water. MnO, mw: 71, d: 5.45, mp: 1650°, converted to Mn_3O_4 if heated in air.
THR = Not fully known. Used as a nutrient and/or dietary supplement food additive as well as a trace mineral added to animal feeds. [*109*] See also manganese compounds. Reacts violently with $Ca(OCl)_2$, F_2. [*19*]

MANGANOUS PERCHLORATE. $Mn(ClO_4)_2 \cdot 6H_2O$, mw: 361.9.
THR = See manganese compounds and perchlorates. Reacts violently with (ethanol + 2,2-dimethoxy propane). [*19*]

MANGANOUS-*o*-PHOSPHATE. Rhombic crystals or pale rose-pink or yellowish-white granules. $Mn_3(PO_4)_2 \cdot 3H_2O$, mw: 408.80, d: 3.102.
THR = See manganese compounds and phosphates.

MANGANOUS PHOSPHIDE. Mn_3P_2, mw: 226.8.
THR = See also manganese compounds and phosphides. Violent reaction with Cl_2. [*19*]

MANGANOUS PYROPHOSPHATE. Syn: *manganese pyrophosphate*. Monoclinic, brown-pink crystals. $Mn_2P_2O_7$, mw: 283.82, mp: 1196°, d: 3.707 @ 25°.
THR = See manganese compounds and phosphates.

MANGANOUS SELENITE. Syn: *manganese selenite*. Crystals. $MnSeO_3 \cdot 2H_2O$, mw: 217.92.
THR = See manganese and selenium compounds.

MANGANOUS SILICOFLUORIDE. See manganous fluosilicate.

MANGANOUS SULFATE. Syn: *manganese sulfate*. Reddish crystals. $MnSO_4$, mw: 151.00, mp: 700°, bp: decomp @ 850°, d: 3.25.
Acute tox data: ip LD_{50} (mouse) = 120 mg/kg. [*3*]
THR = HIGH via ip route. Used as a nutrient and/or dietary supplement food additive and as a trace mineral added to animal feeds. [*109*] See manganese compounds and sulfates.

MANGANOUS SULFIDE. Syns: *alabandite, manganese sulfide*. Cubic, green, amorphous crystals. MnS, mw: 87.00, mp: decomp, d: 3.99.
THR = See manganese compounds and sulfides. The dried material becomes red-hot in air. [*19*]

MANGANOUS TARTRATE. Syn: *manganese tartrate*. White powder. $MnC_4H_4O_6$, mw: 203.00.
THR = See manganese compounds.
Disaster Hazard: Slight; when heated, emits acrid fumes.

MANGANOUS TETRAHYDROALUMINATE. $Mn(AlH_4)_2$, mw: 117.
THR = See manganese compounds. Spont flam in air. [*19*]

MANGANOUS THIOCYANATE. Deliquescent crystals. $Mn(SCN)_2 \cdot 3H_2O$, mw: 225.15, mp: $-3H_2O$ @ 160° to 170°.
THR = See manganese compounds and thiocyanates.

MANNA SUGAR. See mannitol.

MANNITE. See mannitol.

MANNITOL. Syns: *manna, mannite.* White crystalline powder, odorless, sol in water, slightly sol in lower alcohols and amines, almost insol in organic solvents. $C_6H_8(OH)_6$ (straight chain hexahydric alcohol), mw: 182, d: 1.52, mp: 165°–167°, bp: 290°–295° @ 3–5 mm.

Acute tox data: Oral LD_{50} (rat) = 17000 mg/kg; ip LD_{50} (mouse) = 14000 mg/kg; iv LD_{50} (mouse) = 17000 mg/kg. [*3*]

THR = LOW via oral, ip and iv routes. Not fully known, but LOW. Used as a nutrient and/or dietary supplement food additive. [*109*]

MANNITOL BUSULFAN. $C_8H_{18}O_{10}S_2$, mw: 338.4.

THR = An exper neo. [*3*]

MANNITOL HEXANITRATE. Syns: *mannitol nitrate, nitromannite, nitromannitol.* Colorless crystals. $C_6H_8(NO_3)_6$, mw: 452.17, mp: 112°, bp: explodes @ 120°, d: 1.603 @ 0°.

THR = MOD via oral and inhal routes ⟶ fall in blood pressure, which may result in weakness, headache and dizziness. Chronic exposure may produce methemoglobinemia with cyanosis.

Fire Hazard: See nitrates.

Explosion Hazard: Severe, when shocked or exposed to heat. A very sensitive high explosive used as a substitute for fulminates and in combination with tetracine. See also explosives, high, and nitrates.

Disaster Hazard: Highly dangerous! See nitrates and explosives, high.

MANNITOL MONOLAURATE. Colorless solid. $C_6H_8(OH)_5CO_2C_{11}H_{23}$, mw: 364.5.

THR = Details U. A fly spray.

Disaster Hazard: Slight; when heated, emits acrid fumes; can react with oxidizing materials.

MANNITOL MUSTARD. Syns: *mannomustine, degranol.* $C_{10}H_{22}O_4N_2Cl_2$, mw: 305.2.

THR = HIGH acute toxicity. An exper (±) carc and neo. [*3, 8, 23*]

MANNITOL MYLERAN. See 1,6-dimethane sulfonoxy-*d*-mannitol.

MANNITOL NITRATE. See mannitol hexanitrate.

MANNITOL NITROGEN MUSTARD.

$C_{10}H_{22}O_4N_2Cl_2 \cdot 2HCl$, mw: 378.2.

THR = HIGH acute toxicity. An exper (+) carc. [*3, 8*]

MAN-TAN. See dihydroxyacetone.

"MANZATE." See maneb.

MAPP. See methyl acetylene-propadiene mixture.

MAPO.

See tris[1-(2-methyl)aziredinyl]-phosphine oxide.

MARBLE. See calcium carbonate.

MARGARINE EMULSIFIER.

THR = See thermic and oxidation products. A (S) carc. [*14*]

MARIHUANA. See cannabis.

MARSH GAS. See methane.

MARSH ROSEMARY.

THR = Contains tannin. An exper sc neo to rats. [*103*]

MARSHITE. See cuprous iodide.

MATCHES (SAFETY). See matches (strike anywhere).

MATCHES (STRIKE ANYWHERE). Usually contain phosphorus, antimony and sulfur.

THR = See individual components.

Fire Hazard: Dangerous, when exposed to heat or flame or by spont chemical reaction. Keep away from rodents!

Disaster Hazard: Dangerous; will tend to propagate fires and emit toxic fumes of phosphorus, sulfur and antimony.

MATCHES, STRIKE ANYWHERE, SHIP'S LIFEBOAT TYPE AND SIMILAR MATCHES. See matches (strike anywhere).

MATCHES (TRICK). See matches (strike anywhere).

MATERIAL SAFETY DATA SHEETS. See Section 11.

MATLOCHITE. See lead oxychlorides.

MATURING FOOD ADDITIVES. [*109*]

MATTING ACID. A special mixture of acids having properties similar to white acid.

MAUVINE HYDROBROMIDE. Yellowish powder.

THR = HIGH via oral and inhal routes.

Disaster Hazard: Dangerous; when heated, emits highly toxic fumes.

MBA. See nitrogen mustard.

MBDSA. See *m*-benzenedisulfonic acid.

McABEES. See explosives, high.

MCPA. See methyl chlorophenoxyacetic acid.

MD. See dichloromethyl arsine.

MDMH.

See 1-(hydroxymethyl)-5,5-dimethyl hydantoin.

MEAT INSPECTION ACT. [*109*]

MECHLORETHAMINE. See nitrogen mustard.

MECHLORETHAMINE ACETATE.

THR = An exper teratogen. [*3*]

MECHLORETHAMINE FLUOROACETATE.

THR = An exper teratogen. [*3*]

MECHLORETHAMINE HYDROCHLORIDE. Syns: *erasol, 2,2-dichloro-N-methyl diethylamine hydro-*

chloride, *nitol*. Hygroscopic leaf-life crystals, sol in water and alcohol. $CH_3N(CH_2CH_2Cl)_2 \cdot HCl$, mw: 192.5.

Acute tox data: Oral LD_{50} (rat) = 10 mg/kg; ip LD_{50} (rat) = 1.5 mg/kg; sc LD_{50} (rat) = 1.9 mg/kg; iv LD_{50} (dog) = 1 mg/kg; dermal TD_{LO} (mouse) = 60 mg/kg ⟶ neo. [*3*]

THR = VERY HIGH via oral, ip, sc and iv routes. An exper (+) carc and teratogen. [*3*, *8*]

Disaster Hazard: Dangerous; when heated to decomp, emits highly toxic fumes of chlorides.

MEDICINES, N.O.S. See drugs, chemicals.

MEDROXY PROGESTERON. $C_{22}H_{32}O_3$, mw: 344.5.

THR = An exper teratogen. [*3*]

MEDROXYPROGESTERONE ACETATE. Syn: *17-hydroxy-6,α-methyl-preg-4-ene-3,20-dione-17-acetate*. White to off-white, odorless crystalline powder, insol in water, freely sol in chloroform, sparingly sol in alcohol. $C_{24}H_{34}O_4$, mw: 386, melting range: 200°–205°.

THR = Used as a food additive permitted in the feed and drinking water of animals and/or for the treatment of food-producing animals; also permitted in food for human consumption. [*109*] An exper (S) carc. [*3*, *15*]

MELAMINE. Syns: *2,4,6-triamino-s-triazine*, *cyanurotriamide*. Monoclinic, colorless prisms. $NC(NH_2)NC(NH_2)NC(NH_2)$, mw: 126.13, mp: 250°, bp: sublimes, d: 1.573 @ 20°/4°, vap. press: 50 mm @ 315°, vap. d: 4.34.

Acute tox data: Oral LD_{LO} (mouse) = 1600 mg/kg. [*3*]

THR = MOD via oral route. An irr to skin, eyes and mu mem. Causes dermatitis in humans.

Disaster Hazard: Dangerous; when heated to decomp, emits highly toxic fumes of cyanides.

MELIPRAMINE HYDROCHLORIDE. $C_{19}H_{24}N_2 \cdot HCl$, mw: 316.9.

THR = An exper teratogen. [*3*]

MEMTETRAHYDROPHTHALIC ANHYDRIDE. See tetrahydrophthalic anhydride.

MENHADEN OIL. Syns: *pogy oil*, *moss bunker oil*. Thick liquid derived from fish. Autoign. temp.: 828°F, fp: −5°, flash p: 435°F (CC), d: 0.923–0.933.

THR = U.

Fire Hazard: Slight, when exposed to heat or flame; can react with oxidizing materials.

Spont Heating: Yes.

To Fight Fire: CO_2, dry chemical.

1(7)-2-*p*-MENTHADIENE. Syn: *β-phellandrene*. $C_{15}H_{20}$, mw: 200.2, flash p: 120°F, d: ~0.9, vap. d: 4.68, bp: 171.1°.

THR = No data.

Fire Hazard: Mod, via heat, flames, oxidizers.

To Fight Fire: Water spray or fog, dry chemical, CO_2, foam.

MENTHOL. Syn: *hexahydrothymol*. Crystals, peppermint taste and odor. $C_{10}H_{19}OH$, mw: 156.26, mp: 42.5°, bp: 212.0°, d: 0.890 @ 15°/15°, vap. press: 1 mm @ 56.0°, vap. d: 5.38.

Acute tox data: Oral LD_{50} (rat) = 3180 mg/kg; sc LD_{LO} (rat) = 2000 mg/kg; ip LD_{LO} (mouse) = 1800 mg/kg; iv LD_{LO} (cat) = 37 mg/kg. [*3*]

THR = MOD via oral, sc and ip; HIGH via iv routes. An allergen and an irr.

Fire Hazard: Slight.

MEPAZINE. $C_{19}H_{22}N_2S$, mw: 310.5.

THR = HIGH toxicity to animals by iv route. Has been implicated in aplastic anemia. [*3*]

MEPROBAMATE. $C_9H_{18}O_4N_2$, mw: 218.3. Syn: *Miltown*.

THR = MOD to HIGH toxicity to test animals via various routes. Causes CNS effects, is an exper teratogen and has been implicated in aplastic anemia. [*3*]

MERBROMIN. See "Mercurochrome."

MERCAPTAN MIXTURES, ALIPHATIC. See mercaptans.

MERCAPTANS.

THR = Generally have a very offensive odor, which may cause nausea and headache. High conc can produce unconsciousness with cyanosis, cold extremities and rapid pulse. A common air contaminant.

Disaster Hazard: Dangerous; when heated to decomp, they almost always emit highly toxic fumes of SO_x; they will react with water, steam or acids to produce toxic and flam vapors; they can react violently with powerful oxidizers such as $Ca(OCl)_2$. [*19*]

MERCAPTOACETATES. See thioglycolates.

MERCAPTOACETIC ACID. See thioglycolic acid.

β-MERCAPTOALANINE. See cysteine.

2-MERCAPTOBENZOTHIAZOLE. Light yellow powder. C_6H_4SCNSH, mw: 167.3, mp: 170°, d: 1.42 @ 25°.

THR = An exper neo. [*3*]

Disaster Hazard: Dangerous; when heated to decomp, or on contact with acid or acid fumes, emits highly toxic fumes of sulfur compounds; can react with oxidizing materials.

2-MERCAPTOBENZOTHIAZOLE, ZINC SALT. $C_{14}H_{10}N_2S_4Zn$, mw: 399.9.

THR = An exper neo. [*3*] See also zinc compounds and mercaptans.

2-MERCAPTOETHANOIC ACID.
See thioglycolic acid.

2-MERCAPTOETHANOL. Syn: *2-hydroxyethylmercaptan*. Water white, mobile liquid. $HSCH_2CH_2OH$, mw: 78.13, bp: 157.1°, flash p: 165°F (COC), d: 1.1168 @ 20°/20°, vap. press: 1.0 mm @ 20°, vap. d: 2.69.
Acute tox data: Oral LD_{50} (rat) = 300 mg/kg; dermal LD_{50} (rabbit) = 150 mg/kg; dermal LD_{50} (guinea pig) = 300 mg/kg; iv LD_{50} (mouse) = 480 mg/kg. [*3*]
THR = HIGH via oral, dermal and iv routes.
Fire Hazard: Mod, when exposed to heat, flame or oxidizers.
Disaster Hazard: Dangerous; when heated to decomp, emits highly toxic fumes of oxides of sulfur; can react with oxidizing materials.
To Fight Fire: Alcohol foam, CO_2, dry chemical.

2-MERCAPTOETHYLAMINE HYDROCHLORIDE. White, slightly hygroscopic crystals. $HSCH_2CH_2NH_2 \cdot HCl$, mw: 113.6, mp: 70.2°–70.7°.
Acute tox data: ip LD_{LO} (mouse) = 425 mg/kg. [*3*]
THR = HIGH via ip and probably oral routes.
Disaster Hazard: Dangerous; when heated to decomp, emits highly toxic fumes of SO_x; will react with water or steam to produce toxic fumes.

2-MERCAPTO-4-HYDROXYPYRAMIDINE.
See thiouracil.

MERCAPTOPURINE-3-N-OXIDE. $C_6H_3ON_4$, mw: 179.2.
THR = An exper teratogen. [*3*]

MERCAPTOPURINE RIBONUCLEOSIDE.
$C_{10}H_{12}O_4N_4S$, mw: 284.3.
THR = An exper teratogen. [*3*]

MERCURIC ACETATE. White crystalline powder. $Hg(C_2H_3O_2)_2$, mw: 318.70, mp: decomp, d: 3.270.
Acute tox data: Oral LD_{50} (rat) = 76 mg/kg; oral LD_{50} (mouse) = 62 mg/kg; iv LD_{50} (mouse) = 4.4 mg/kg. [*3*]
THR = HIGH via oral and iv routes. See also mercury compounds, organic.

MERCURIC ACETYLIDE. White powder. $3HgC_2 \cdot H_2O$, mw: 691.85, mp: explodes, d: 5.3.
THR = HIGH. See mercury compounds, organic.
Fire Hazard: Details U. See also acetylides.
Explosion Hazard: Severe, when shocked or exposed to heat.
Disaster Hazard: Dangerous! Heat or shock will cause detonation; when heated, emits highly toxic fumes of mercury.

MERCURIC AMINOACETATE. Syn: *mercuric glycocollate*. Crystals, sol in water. $Hg[CH_2(NH_2)COO]_2$, mw: 348.7.
THR = HIGH. See mercury compounds, organic.

MERCURIC AMINOPROPIONATE. Syn: *mercury alanine*. $Hg[CH_3CH(NH_2)COO]_2$, mw: 374.77.
THR = HIGH. See mercury compounds, organic.

MERCURIC AMMONIUM CHLORIDE. White, pulverulent lumps or powder. $HgNH_2Cl$, mw: 252.1.
THR = HIGH. See mercury compounds, inorganic.

MERCURIC ARSENATE. Yellow powder. $HgHAsO_4$, mw: 340.5, mp: decomp.
THR = HIGH. See mercury compounds, inorganic, and arsenic compounds.
Disaster Hazard: Dangerous; when heated to decomp, or on contact with acid or acid fumes, emits highly toxic fumes of mercury and arsenic.

MERCURIC AZIDE. $Hg(N_3)_2$, mw: 284.6.
THR = HIGH. See mercury compounds and azides. It is explosively unstable. Also, decomp violently @ 190°. [*19*]

MERCURIC BARIUM BROMIDE. Syn: *barium mercury bromide*. Colorless, very hygroscopic crystals. $HgBr_2 \cdot BaBr_2$, mw: 657.6.
THR = HIGH. See mercury compounds, inorganic, barium compounds and bromides.

MERCURIC BARIUM IODIDE. Reddish or yellow, unstable, deliquescent, crystalline mass. $HgI_2 \cdot BaI_2 \cdot 5H_2O$, mw: 935.73.
THR = HIGH. See mercury compounds, inorganic, iodides and barium compounds.

MERCURIC BENZOATE. White crystalline powder. $Hg(C_7H_5O_2)_2$, mw: 442.83, mp: 165°.
THR = HIGH. See mercury compounds, organic.

MERCURIC BROMATE. Crystals. $Hg(BrO_3)_2 \cdot 2H_2O$, mw: 492.47, mp: decomp @ 130°–140°.
THR = HIGH. See mercury compounds, inorganic, and bromates.

MERCURIC BROMIDE. Rhombic, colorless crystals. $HgBr_2$, mw: 360.44, mp: 237°, bp: 322° (sublimes), d: 6.109 @ 25°, vap. press: 1 mm @ 136.5°.
THR = HIGH. See mercury compounds, inorganic, and bromides. Reacts violently with Na, K. [*19*]

MERCURIC BROMIDE, AMMONOBASIC. White powder. $Hg(NH_2)Br$, mw: 296.55, mp: decomp.
THR = HIGH. See mercury compounds, inorganic, and bromides.

MERCURIC BROMIDE DIAMMINE. White powder. $Hg(NH_3)_2Br_2$, mw: 394.51, mp: 180°.
THR = HIGH. See mercury compounds, inorganic, and bromides.

MERCURIC BROMIDE IODIDE. Rhombic yellow crystals. HgBrI, mw: 407.45, mp: 229°, bp: 360°.
THR = HIGH. See mercury compounds, inorganic, and bromides.

MERCURIC CACODYLATE. Hygroscopic, somewhat unstable crystalline powder, sol in water and alcohol, insol in ether. $Hg[(CH_3)_2AsO_2]_2$, mw: 474.52.

THR = HIGH. A recognized carc. [*14*] See mercury compounds, organic and arsenic compounds.

MERCURIC CHLORATE. Colorless, needle-like crystals. $Hg(ClO_3)_2$, mw: 367.52, mp: decomp, d: 4.998.

THR = HIGH. See mercury compounds, inorganic and chlorates.

MERCURIC CHLORIDE. Syn: *corrosive sublimate.* White crystals or powder. $HgCl_2$, mw: 271.52, mp: 276°, bp: 302°, d: 5.440 @ 25°, vap. press: 1 mm @ 136.2°.

Acute tox data: Oral LD_{50} (rat) = 37 mg/kg; $scLD_{50}$ (rat) = 14 mg/kg; $ipLD_{50}$ (mouse) = 5 mg/kg; dermal LD_{LO} (guinea pig) = 345 mg/kg; oral LD_{LO} (human) = 29 mg/kg. [*3*]

THR = HIGH via all routes. A deadly poison. See mercury compounds, inorganic, and chlorides. Reacts violently with K, Na. [*19*]

MERCURIC CHLORIDE DIAMMINE. Fusible white precipitate. $Hg(NH_3)_2Cl_2$, mw: 305.59, mp: 300°.

THR = HIGH. See mercury compounds, inorganic, and chlorides.

MERCURIC CHLORIDE IODIDE. Rhombic red crystals. HgClI, mw: 362.99, mp: 153°, bp: 315°.

THR = HIGH. See mercury compounds, inorganic, chlorides and iodides.

MERCURIC CHLORITE. $Hg(ClO_2)_2$, mw: 335.5.

THR = HIGH. See mercury compounds, inorganic. This salt is explosive. [*19*]

MERCURIC CHLOROIODIDE. White crystals. $2HgCl_2 \cdot HgI_2$, mw: 997.48.

THR = HIGH. See mercury compounds, inorganic.

MERCURIC CHROMATE. Rhombic red crystals. $HgCrO_4$, mw: 316.62, mp: decomp.

THR = HIGH. See mercury compounds, inorganic, and chromates.

Fire Hazard: Mod, by chemical reaction; an oxidizer.

Disaster Hazard: Dangerous; when heated to decomp, emits highly toxic fumes of mercury; can react with reducing materials.

MERCURIC CUPROUS IODIDE. Syn: *copper-mercury iodide.* Dark red, crystalline powder. $HgI_2 \cdot 2CuI$, mw: 835.42.

THR = See mercury compounds, inorganic.

MERCURIC CYANIDE. Colorless, transparent prisms, darkened by light. $Hg(CN)_2$, mw: 252.65, mp: decomp, d: 3.996.

Acute tox data: Oral LD_{50} (rat) = 25 mg/kg; $ipLD_{LO}$ (rat) = 7.5 mg/kg; $ivLD_{LO}$ (rabbit) = 2 mg/kg; TD_{LO} = 10 mg/kg (human) → GI symptoms. [*3*]

THR = HIGH via oral, ip and iv routes. See cyanides and mercury compounds, organic.

Fire Hazard: See cyanides.

Explosion Hazard: See cyanides. Reacts violently with Mg, F_2. [*19*]

Disaster Hazard: Dangerous; when heated to decomp, or on contact with acid or acid fumes, emits highly toxic fumes of cyanides and mercury; will react with water or steam to produce toxic fumes.

MERCURIC CYANIDE OXIDE. $Hg_2(CN)_2O$, mw: 469.2.

Acute tox data: $ivLD_{LO}$ (rabbit) = 2.5 mg/kg. [*3*]

THR = HIGH via iv route. See mercury compounds and cyanides. Often explodes, possibly if rubbed. [*19*]

MERCURIC DINAPHTHYLAMINE DISULFONATE. $C_{21}H_{14}O_6S_2 \cdot (Hg)_x$.

Acute tox data: Oral LD_{50} (mouse) = 30 mg/kg. [*3*]

THR = HIGH via oral route. See also mercury compounds, organic.

MERCURIC ETHYL MERCAPTIDE. Crystals, insol in water. $(SC_2H_5)_2Hg$, mw: 322.9, mp: 77°.

THR = HIGH. See mercury compounds, organic, and sulfides.

MERCURIC FLUORIDE. Transparent crystals. HgF_2, mw: 238.61, mp: 645°, bp: >650°, d: 8.95 @ 15°.

THR = HIGH. See fluorides and mercury compounds, inorganic. Reacts violently with K, Na. [*19*]

MERCURIC FLUOSILICATE. Colorless, deliquescent crystals. $HgSiF_6 \cdot 6H_2O$, mw: 450.77.

THR = HIGH. See mercury compounds, inorganic, and fluosilicates.

MERCURIC FORMAMIDE. A mixture of mercuric oxide + formamide.

THR = HIGH. See mercury compounds, organic.

MERCURIC GALLATE. Gray-green powder, insol in water. $Hg[C_6H_2(OH)_3COO]_2$, mw: 538.8.

THR = HIGH. See mercury compounds, organic.

MERCURIC GLYCOCOLLATE. See mercuric aminoacetate.

MERCURIC GUAIACOLSULFONATE. Brown crystals, sol in water. $Hg[C_6H_3(OH)(OCH_3)SO_3]_2$, mw: 606.99.

THR = HIGH. See mercury compounds, organic, and sulfonates.

MERCURIC IODATE. White crystals, slightly water-sol. $Hg(IO_3)_2$, mw: 550.45.

THR = HIGH. See mercury compounds, inorganic, and iodates.

MERCURIC IODIDE. (1) Rhombic yellow crystals; (2) tetragonal red crystals or powder. HgI_2, mw: 454.9, d: (1) 6.271, (2) 6.283; mp: (1) 259°, (2) 126°–127°; bp: 354° (sublimes), vap. press: (1) 1 mm @ 157.5°.
Acute tox data: Oral LD_{LO} (human) = 357 mg/kg; oral LD_{50} (rat) = 40 mg/kg; ipLD_{50} (mouse) = 60 mg/kg. [*3*]
THR = HIGH via oral and ip routes. See mercury compounds, inorganic, and iodides. Violent reaction with ClF_3, K, Na. [*19*]

MERCURIC IODIDE AMMONOBASIC. Dirty white crystals. $Hg(NH_2)I$, mw: 343.55.
THR = HIGH. See mercury compounds, inorganic, and iodides.

MERCURIC IODIDE, AQUOBASIC-AMMONOBASIC. Yellow to brown crystals. OHg_2NH_2I, mw: 560.16, mp: >128°, bp: explodes.
THR = HIGH. See mercury compounds, inorganic, and iodides.

MERCURIC IODIDE DIAMMINE. Colorless or pale yellow powder or needles. $Hg(NH_3)_2I_2$, mw: 488.51.
THR = HIGH. See mercury compounds, inorganic, and iodides.

MERCURIC IODIDE SOLUTION. See mercuric iodide.

MERCURIC LACTATE. White crystalline powder. $Hg(C_3H_5O_3)_2$, mw: 378.71.
Acute tox data: Oral LD_{50} (rat) = 200 mg/kg. [*3*]
THR = HIGH via oral route. See mercury compounds, organic.

MERCURIC NAPHTHOLATE. Syn: *mercury-β-naphthol.* Brown powder, insol in water. $Hg(C_{10}H_7O)_2$, mw: 486.9.
THR = HIGH. See mercury compounds, organic and β-naphthol.

MERCURIC NITRATE. White-yellowish, deliquescent powder. $Hg(NO_3)_2$, mw: 324.6, mp: 79°, bp: decomp, d: 4.39.
Acute tox data: ipLD_{50} (mouse) = 8 mg/kg. [*3*]
THR = HIGH via ip route. See mercury compounds, inorganic, and nitrates. Forms a sensitive explosive product with acetylene, ethanol, PH_3, S; reacts violently with hypophosphoric acid, unsaturates, aromatics. [*19*]

MERCURIC NITRIDE. Hg_3N_2, mw: 629.8.
THR = HIGH. See mercury compounds. Very explosive. Explodes on contact with H_2SO_4. [*19*]

MERCURIC OLEATE. Syn: *mercury oleate.* Yellowish to red liquid, semisolid or solid mass. $(C_{17}H_{33}CO_2)_2Hg$, mw: 763.6.
THR = HIGH. See mercury compounds, organic.

MERCURIC OXALATE. White powder. HgC_2O_4, mw: 288.63, mp: decomp.
THR = HIGH. See mercury compounds, organic, and oxalates.

MERCURIC OXIDE. Heavy, bright orange-red powder or orange-yellow powder. HgO, mw: 216.61, mp: decomp @ 500°, d: 11.14.
Acute tox data: Oral LD_{50} (rat) = 18 mg/kg; imLD_{LO} (rat) = 2.5 mg/kg. [*3*]
THR = HIGH via oral and im routes. See mercury compounds, inorganic.
Fire Hazard: By chemical reactions; an oxidizer. Reacts violently with Cl_2, hydrazine hydrate, H_2O_2, hypophosphorus acid, ($I + CH_3OH + C_2H_5OH$), Mg, P, phospham, NaK, S. [*19*]
Disaster Hazard: Dangerous; when heated to decomp, emits highly toxic fumes of mercury; can react with reducing materials.

MERCURIC OXYBROMIDE. Yellow crystals. $HgBr_2 \cdot 3HgO$, mw: 1010.27.
THR = HIGH. See mercury compounds, inorganic, and bromides.

MERCURIC OXYCHLORIDE. Hexagonal yellow crystals. $HgCl_2 \cdot 3HgO$, mw: 921.35, mp: decomp @ 260°, d: 7.93.
THR = HIGH. See mercury compounds, inorganic, and chlorides.

MERCURIC OXYCYANIDE. See mercuric cyanide oxide.

MERCURIC OXYFLUORIDE. Yellow crystals. $HgF_2 \cdot HgO \cdot H_2O$, mw: 473.24, mp: decomp @ 100°.
THR = HIGH. See fluorides and mercury compounds, inorganic.

MERCURIC OXYIODIDE. Yellow-brown crystals. $HgI_2 \cdot 3HgO$, mw: 1104.28.
THR = HIGH. See mercury compounds, inorganic, and iodides.

MERCURIC PERCHLORATE. Colorless deliquescent crystals. $Hg(ClO_4)_2 \cdot 6H_2O$, mw: 507.62.
THR = HIGH. See mercury compounds, inorganic, and perchlorates.

MERCURIC PHENATE. Syn: *mercury carbolate.* Gray or reddish-gray powder, nearly insol in water. $Hg(C_6H_5O)_2$, mw: 386.8.
THR = HIGH. See mercury compounds, organic.

MERCURIC PHENYL CYANAMIDE. Powder. $C_6H_5HgNCNH$, mw: 318.8.
THR = HIGH. See mercury compounds, organic, and cyanides.

MERCURIC PHENYL MERCAPTIDE. Light yellow crystals, insol in water. $(C_6H_5S)_2Hg$, mw: 418.9, mp: 153° (decomp).
THR = HIGH. See mercury compounds, organic.

MERCURIC PHOSPHATE. Syn: *mercuri-o-phosphate*. Heavy white or yellowish powder. $Hg_3(PO_4)_2$, mw: 791.87.
THR = HIGH. See mercury compounds, inorganic.

MERCURIC-*o*-PHOSPHATE. See mercuric phosphate.

MERCURIC POTASSIUM CYANIDE. Colorless crystals. $Hg(CN)_2 \cdot 2KCN$, mw: 382.85.
THR = HIGH. See cyanides and mercury compounds, organic.
Fire Hazard: See cyanides.
Disaster Hazard: Dangerous; when heated to decomp, or on contact with acid or acid fumes, emits highly toxic fumes of cyanides and mercury; will react with water or steam to produce toxic and flam vapors.
Explosion Hazard: See cyanides.

MERCURIC POTASSIUM IODIDE. Heavy, bright orange-red to yellow powder. $HgI_2 \cdot 2KI$, mw: 786.4.
Acute tox data: Dermal LD_{LO} (guinea pig) = 1000 mg/kg; ipLD_{LO} (guinea pig) = 1000 mg/kg. [*3*]
THR = MOD via ip and dermal routes. See mercury compounds, inorganic, and iodides. A poison.

MERCURIC POTASSIUM THIOSULFATE. White powder. $HgS_2O_3K_2S_2O_3$, mw: 503.1.
THR = HIGH. See mercury compounds, inorganic, and sulfates.

MERCURIC RESORCINOL ACETATE. Syn: *resorcinol mercury acetate*. Contains approx 69% mercury, yellow crystalline powder, insol in water, sol in hot glacial acetic acid and solutions of fixed alkali hydroxides.
THR = HIGH. See mercury compounds, organic.

MERCURIC SALICYLATE. Syn: *mercury subsalicylate*. White to yellowish powder. $C_7H_4O_3Hg$, mw: 336.7.
THR = HIGH. See mercury compounds, organic.

MERCURIC SELENIDE. Syn: *tiemannite*. Gray plates. HgSe, mw: 279.57, mp: sublimes, d: 8.21.
THR = HIGH toxicity. See mercury compounds, inorganic. See also selenium.
Fire Hazard: This material readily liberates hydrogen selenide upon contact with acids or moisture.
Disaster Hazard: Dangerous; when heated to decomp, emits highly toxic fumes of mercury and selenium; will react with water, steam or acid to produce toxic fumes; can react with oxidizing materials.

MERCURIC SESQUIIODIDE. See mercury compounds, inorganic, and iodides.

MERCURIC SILVER IODIDE. Deep yellow powder. $HgI_2 \cdot 2AgI$, mw: 924.1.
THR = HIGH. See mercury compounds, inorganic, silver and iodides.

MERCURIC STEARATE. Yellowish granular powder. $(C_{18}H_{35}O_2)_2Hg$, mw: 767.5.
THR = HIGH toxicity. See mercury compounds, organic.

MERCURIC SUBSULFATE. Lemon yellow powder. $HgSO_4 \cdot 2HgO$, mw: 729.90, bp: volatilizes, d: 6.44, vap. d: 25.2.
THR = HIGH. See mercury compounds, inorganic, and sulfates.

MERCURIC SUCCINATE. White to yellowish crystalline powder, slightly sol in water, sol in aqueous sodium chloride. $Hg(CH_2COO)_2$, mw: 316.68.
THR = HIGH. See mercury compounds, organic.

MERCURIC SULFATE. White crystalline powder. $HgSO_4$, mw: 296.67, mp: decomp, d: 6.47.
Acute tox data: Oral LD_{50} (rat) = 57 mg/kg; oral LD_{50} (mouse) = 40 mg/kg. [*3*]
THR = HIGH via oral route. See mercury compounds, inorganic, and sulfates. Absorption of HCl can become violent @ 125°. [*19*]

MERCURIC SULFIDE, BLACK. Syn: *metacinnabarite*. Black powder. HgS, mw: 232.67, mp: 583.5°, d: 7.73.
THR = HIGH. See mercury compounds, inorganic, and sulfides. Can react violently with Cl_2, ClO, Ag_2O. [*19*]

MERCURIC SULFIDE, RED. Syns: *cinnabar*, *vermillion*. Hexagonal red crystals or powder. HgS, mw: 232.68, vap. d: 8.0, mp: sublimes @ 583.5°, d: 8.10.
THR = HIGH. See mercury compounds, inorganic, and sulfides. See also mercuric sulfide, black.

MERCURIC SULFOCYANATE. Syns: *mercury rhodanide*, *mercuric thiocyanate*. White powder. $Hg(SCN)_2$, mw: 316.77, mp: decomp.
THR = HIGH. See mercury compounds, organic and cyanates.

MERCURIC TELLURATE. Amber cubic crystals. Hg_3TeO_6, mw: 825.44.
THR = HIGH. See mercury compounds, inorganic and tellurium.

MERCURIC THALLIUM IODIDE. Red crystalline lumps. $HgI_2 \cdot TlI$, mw: 708.9.
THR = HIGH. See mercury compounds, inorganic, and thallium.

MERCURIC THIOCYANATE. See mercuric sulfocyanate.

MERCURIC THYMOLATE. Yellowish gray powder, variable composition.
THR = HIGH. See mercury compounds, organic.

MERCURIC THYMOL NITRATE. Amorphous, white to reddish-white powder.
THR = HIGH. See mercury compounds, organic and nitrates.

MERCURIC THYMOLSALICYLATE. White to reddish powder.
THR = HIGH. See mercury compounds, organic.

MERCURIC THYMOLSULFATE. White to reddish-white powder.
THR = HIGH. See mercury compounds, organic.

MERCURIC TUNGSTATE. Yellow crystals. $HgWO_4$, mw: 448.53, mp: decomp.
THR = HIGH. See mercury compounds, inorganic.

MERCURIC ZINC ACETATE. White crystals. $Hg(C_2H_3O_2)_2 + Zn(C_2H_3O_2)_2$.
THR = HIGH. See mercury compounds, organic.

MERCURIC ZINC CYANIDE. White powder, a mixture of zinc cyanide with varying quantities of mercuric cyanide. Water dissolves the mercuric cyanide.
THR = HIGH. See mercury compounds and cyanides.

MERCURIC PHENYL ACETATE. See phenyl mercuric acetate.

"MERCUROCHROME." Syn: *merbromin.* $C_{20}H_8Br_2HgNa_2O_6 \cdot 3H_2O$, mw: 804.4.
Acute tox data: $ivLD_{LO}$ (rabbit) = 15 mg/kg. [*3*]
THR = HIGH via iv and oral routes. Relatively nonirr and nontoxic to damaged skin or tissue. An antiseptic. See mercury compounds, organic.

MERCUROL, SOLID. Colorless to brownish powder. Contains 20% mercury.
THR = HIGH. See mercury compounds, organic.

"MERCUROSAL." Syn: *disodium hydroxymercury salicyloxyacetate.* White amorphous powder. $(HOHg)NaOOCC_6H_3OCH_2COONa$, mw: 456.69.
THR = HIGH. See mercury compounds, organic.

MERCUROUS ACETATE. Colorless scales or plates. $Hg(C_2H_3O_2)$, mw: 259.7, mp: decomp.
THR = See mercury compounds, organic.

MERCUROUS ARSENITE. Unstable brown powder.
THR = HIGH. See mercury compounds, inorganic, and arsenic.

MERCUROUS AZIDE. White crystals. HgN_3, 242.6, mp: explodes @ 210°.
THR = HIGH. See mercury compounds, inorganic, and azides. A sensitive explosive.

MERCUROUS BROMATE. Crystals. $Hg_2(BrO_3)_2$, mw: 657.06, mp: decomp.
THR = See mercury compounds, inorganic, and bromates.

MERCUROUS BROMIDE. Tetragonal, white-yellow crystals. Hg_2Br_2, mw: 561.07, mp: sublimes @ 345°, d: 7.307, vap. d: 19.3.
THR = See mercury compounds, inorganic, and bromides.

MERCUROUS CARBONATE. Yellow-brown crystals. Hg_2CO_3, mw: 461.23, mp: decomp @ 130°, d: 5.07 @ 218°.
THR = See mercury compounds, inorganic.

MERCUROUS CHLORATE. White crystals. $HgClO_3$, mw: 284.1, mp: decomp @ 250°, d: 6.409.
THR = See mercury compounds, inorganic, and chlorates.

MERCUROUS CHLORIDE. Syn: *calomel.* White rhombic crystals or crystalline powder. Hg_2Cl_2, mw: 472.14, mp: sublimes @ 400°, d: 7.150.
Acute tox data: Oral LD_{50} (rat) = 210 mg/kg. [*3*]
THR = HIGH via oral route. See mercury compounds, inorganic, and chlorides. Violent reaction with K, Na. [*19*]

MERCUROUS CHROMATE. Red needles or powder. Hg_2CrO_4, mw: 517.23, mp: decomp.
THR = HIGH. See mercury compounds, inorganic, and chromates.
Fire Hazard: Mod, by chemical reaction; an oxidizer.
Disaster Hazard: Dangerous; when heated to decomp, emits highly toxic fumes of mercury; can react with reducing materials.

MERCUROUS FLUORACETYLIDE. FCCHg, mw: 243.6.
THR = HIGH. See mercury compounds and fluorides. It is unstable and may explode.

MERCUROUS FLUORIDE. Cubic yellow crystals. Hg_2F_2, mw: 439.22, mp: 570°, d: 8.73.
THR = See fluorides and mercury compounds, inorganic.

MERCUROUS FLUOSILICATE. Colorless prisms. $Hg_2SiF_6 \cdot 2H_2O$, mw: 579.31.
THR = See mercury compounds, inorganic, and fluosilicates.

MERCUROUS FORMATE. Glistening scales. $Hg_2(CHO_2)_2$, mw: 491.26, mp: decomp.
THR = See mercury compounds, organic.

MERCUROUS GLUCONATE. White solid. $HgCO_2C_5H_6(OH)_5$, mw: 395.8.
THR = See mercury compounds, organic.

MERCUROUS HYPOPHOSPHATE. $Hg_4P_2O_6$, mw: 960.3.
THR = HIGH. See mercury compounds. Unstable. Decomp explosively. [*19*]

MERCUROUS IODATE. Yellowish crystals. $Hg_2(IO_3)_2$, mw: 751.06, mp: decomp.
THR = See mercury compounds, inorganic, and iodates.

MERCUROUS IODIDE. Yellow tetragonal crystals or amorphous powder. HgI, mw: 327.50, mp: sublimes @ 140°, bp: decomp @ 290°, d: 7.70.
Acute tox data: Oral LD_{50} (mouse) = 110 mg/kg; ipLD_{50} (mouse) = 50 mg/kg. [*3*]
THR = HIGH via oral and ip routes. See mercury compounds, inorganic, and iodides.

MERCUROUS MONOHYDROGEN-*o*-ARSENATE. Yellow-red crystals. Hg_2HAsO_4, mw: 541.14.
THR = HIGH. See arsenic compounds and mercury compounds, inorganic.

MERCUROUS NITRATE. Short, colorless, efflorescent crystals. $Hg_2(NO_3)_2 \cdot 2H_2O$, mw: 561.26, mp: 70°, d: 4.79 @ 4°.
Acute tox data: Oral LD_{50} (rat) = 297 mg/kg; oral LD_{50} (mouse) = 388 mg/kg; ipLD_{50} (mouse) = 5 mg/kg. [*3*]
THR = HIGH via oral and ip routes. See mercury compounds, inorganic, and nitrates. Violent reaction with C, P. [*19*]

MERCUROUS NITRATE, AMMONIATED. Syn: *black precipitate*. Black powder, $Hg_2ONH_2 \cdot Hg_2(NO_3)_2$, mw: 958.4.
THR = See mercury compounds, inorganic, and nitrates.

MERCUROUS NITRITE. Yellow crystals. $Hg_2(NO_2)_2$, mw: 493.24, mp: decomp @ 100°, d: 7.33.
THR = HIGH. See mercury compounds, inorganic, and nitrites.

MERCUROUS OXALATE. White crystals. $Hg_2C_2O_4$, mw: 489.24.
THR = See oxalates and mercury compounds, organic.

MERCUROUS OXIDE, BLACK. Black to grayish-black powder. Hg_2O, mw: 417.22, mp: decomp @ 100°, d: 9.8.
THR = HIGH. See mercury compounds, inorganic.
Fire Hazard: Mod, by chemical reaction; an oxidizer. Reacts violently with H_2O_2, K, Na, S, (H_2S + BaO + air). [*19*]
Disaster Hazard: Dangerous; when heated to decomp, emits highly toxic fumes of mercury; can react with reducing materials.

MERCUROUS PHOSPHATE. Heavy white powder. Hg_3PO_4, mw: 696.85.
THR = See mercury compounds, inorganic.

MERCUROUS SULFATE. White crystalline powder. Hg_2SO_4, mw: 497.28, mp: decomp, d: 7.56.
THR = See mercury compounds, inorganic, and sulfates.

MERCUROUS SULFIDE. Black crystals. Hg_2S, mw: 433.24, mp: decomp.
THR = See mercury compounds, inorganic, and sulfides.

MERCUROUS TARTRATE. Yellowish-white crystalline powder. $Hg_2C_4H_4O_6$, mw: 549.29.
THR = See mercury compounds, organic.

MERCURY. Silvery liquid, metallic element. Hg, atwt: 200.7, mp: −38.89°, bp: 356.9°, d: 13.546, vap. press: 1 mm @ 126.2°.
Acute tox data: Oral LD_{LO} (human) = 1429 mg/kg; inhal TC_{LO} (human) = 0.17 mg/m^3 for 40 yrs ⟶ CNS problems; ivTD_{LO} (human) = 29 mg/kg; ⟶ GI symptoms. [*3*]
THR = HIGH to CNS, GI tract. See mercury compounds. An exper neo. [*3*] Reacts violently with acetylene, NH_3, BPI_2, Cl_2, ClO_2, CH_3N_3, Na_2C_2, nitromethane (butyne diol + acid). [*19*]
Radiation Hazard: For permissible levels, see Section 5, Table 5A.5. Artificial isotope ^{203}Hg, $T\frac{1}{2}$ = 47d. Decays to stable ^{203}Tl by emitting β's of 0.21 MeV. Emits γ's of 0.28 MeV.
Disaster Hazard: Dangerous; when heated emits highly toxic fumes.

MERCURY ACETAMIDE. White powder. CH_3CONHg, mw: 257.7.
THR = HIGH. See mercury compounds, organic.

MERCURY ACETATE. See mercurous acetate or mercuric acetate.

MERCURY ALANINE. See mercury-α-aminopropionate.

MERCURY-*p*-AMINOPHENOL ARSENATE. See mercury atoxylate.

MERCURY-α-AMINOPROPIONATE. Syn: *mercury alanine*. White crystals, water-sol. $Hg[CH_2CH(NH_2)COO]_2$, mw: 374.8.
THR = HIGH. See mercury compounds, organic.

MERCURY, AMMONIATED. See mercuric ammonium chloride.

MERCURY ANTIMONY SULFIDE. Gray-black powder. Mixture of equal parts of black mercury sulfide and gray antimony sulfide.
THR = See mercury compounds, antimony and sulfides.

MERCURY ARC RADIATION. A recog carc. [*14*] See radiation, ultra-violet.

MERCURY ATOXYLATE. Syn: *mercury-p-aminophenol arsenate.* White powder. $C_{12}H_{14}O_6N_2As_2Hg$, mw: 632.71.

THR = HIGH. See arsenic compounds and mercury compounds, organic.

MERCURY BENZAMIDE. White powder. C_6H_5CONHg, mw: 319.7.

THR = HIGH. See mercury compounds, organic.

MERCURY BISULFATE. See mercuric sulfate.

MERCURY CARBOLATE. See mercuric phenate.

MERCURY COLLOIDAL. See mercury.

MERCURY COMPOUNDS, INORGANIC.

THR = Mercury is a general protoplasmic poison; after absorption it circulates in the blood and is stored in the liver, kidneys, spleen and bone. It is eliminated in the urine, feces, sweat, saliva and milk. In industrial poisoning, the chief effect is upon the CNS and upon the mouth and gums. Colitis has been reported frequently; a nephritis or nephrosis is rarely reported. Fulminate of mercury rarely produces symptoms of systemic poisoning, but frequently causes a dermatitis. The cardinal symptoms of industrial mercury poisoning are stomatitis, tremors, and psychic disturbances. Usually the first complaints are of excessive salivation and pain on chewing; in severe cases there may be gingivitis, with loosening of the teeth, and a dark line on the gum margins, resembling the "lead line." In slow poisoning the salivation may be absent, and the only complaint dryness of the throat and mouth. Tremor and psychic disturbances are commonly seen in the slow chronic form of the poisoning; the tremor is of the intention type, and may be seen when the patient spreads the outstretched fingers or protrudes the tongue, or attempts to perform specified movements. Muscles of the face, hands and arms are chiefly affected. In more severe cases there may also be convulsive or shaking movements; writing is frequently illegible. Hyperactive kneejerks and scanning speech may be present in advanced cases. The psychic disturbance (so called "erethism") includes such changes as loss of memory, insomnia, lack of confidence, irritability, vague fears and depression.

The dermatitis produced by fulminate of mercury takes the form of small, discrete ulcers on the exposed parts, and is usually accompanied by conjunctivitis and inflammation of the mu mem of the nose and throat.

Elemental mercury is probably not absorbed through the gastrointestinal tract, but many mercury compounds are. A number of mercury compounds, in addition to the fulminate, can cause skin irr and can be absorbed through the skin; they are strong allergens (Section 9).

These are common air contaminants.

Disaster Hazard: Dangerous; when heated to decomp, they emit highly toxic fumes of mercury.

MERCURY COMPOUNDS, N.O.S. LIQUID. See mercury compounds, inorganic and organic.

MERCURY COMPOUNDS, N.O.S. SOLID. See mercury compounds, inorganic and organic.

MERCURY COMPOUNDS, ORGANIC.

THR = The customary grouping of all organic mercurials in a single category is not fully justified by the toxicity of the compounds. Alkyl mercurials have very high toxicity, aryl compounds, particularly the phenyls, are much less toxic, and the organomercurials used in therapeutics are less toxic. The alkyls and aryls commonly cause skin burns and other forms of irr, and both can be absorbed through the skin. Fatal poisoning has occurred due to exposure to alkyl mercurials and permanent damage to the brain has been reported. Extensive human observation on exposure to the phenyl mercurials have shown no greater toxicity than is caused by metallic mercury. In fact, up to the present time there has not been an authenticated case of occupational poisoning due to the phenyl mercurials reported in the literature. Organic mercury compounds, like organic lead compounds, seem to have an affinity for lipoid-containing organs, resulting in CNS disturbances such as from tetraethyl lead. These are common air contaminants.

Disaster Hazard: Dangerous; when heated to decomp, they emit highly toxic fumes of mercury.

MERCURY CYANIDE. See mercuric cyanide.

MERCURY ETHYLENE DIAMINE SULFATE. See sublamin.

MERCURY FULMINATE, MERCURIC FULMINATE. See fulminate of mercury, dry.

MERCURY MORPHINE OLEATE. An oleaginous mass.

THR = See mercury compounds, organic, and morphine.

MERCURY NAPHTHENATE. White crystals; syn: *dinaphthyl mercury*, mw: 454.9, d: 1.929, bp: 249°, mp: 188°.

THR = See mercury compounds, organic.

MERCURY-β-NAPHTHOL. See mercuric naphtholate.

MERCURY NITRIDE. Brown powder. Hg_3N_2, mw: 629.85, mp: explodes.
THR = HIGH. See mercury compounds, inorganic.
Explosion Hazard: Severe, when exposed to heat.
Disaster Hazard: Dangerous; when heated to decomp, emits highly toxic fumes of mercury and may explode.

MERCURY NUCLEATE. See mercurol, solid.

MERCURY OLEATE. See mercuric oleate.

MERCURY RHODANIDE. See mercuric sulfocyanate.

MERCURY SUBSALICYLATE. See mercuric salicylate.

MERCURY TETRAPHOSPHIDE. Hg_3P_4, mw: 725.8.
THR = HIGH toxicity. See mercury compounds and phosphides. Ignites if warmed in air. Burns in Cl_2. With $KClO_3$ it explodes by percussion. [*19*]

MERCURY SUBSALICYLATE. See mercuric salicylate.

MESCALINE. $C_{11}H_{17}O_3N$, mw: 211.3.
Acute tox data: Oral TD_{LO} (human) = 5 mg/kg ⟶ CNS effects; ivTD_{LO} (human) = 7 mg/kg ⟶ CNS effects; imTD_{LO} (human) = 214 mg/kg ⟶ CNS effects; oral LD_{50} (mouse) = 880 mg/kg. [*3*]
THR = HIGH via oral, iv, im routes. An exper teratogen. [*3*]

MESIDINE. Syn: *2,4,6-trimethyl aniline, mesityl amine.* $C_9H_{13}N$, mw: 135.2.
Acute tox data: oral LD_{50} (rat) = 743 mg/kg; LD_{50} (mouse) = 372 mg/kg; LD_{50} (rabbit) = 338 mg/kg; LD_{50} (guinea pig) = 338 mg/kg. [*3*]
THR = HIGH to most species. An exper neo to rats via oral route. [*103*]

MESITYLENE. Syn: *1,3,5-trimethylbenzene.* Liquid, peculiar odor, insol in water, sol in alcohol and benzene. C_9H_{12}, mw: 120.2, mp: −44.8°, d: 0.8637 @ 20°/4°, bp: 164.7°, autoign. temp.: 1022°F.
Acute tox data: Inhal LC_{LO} (rat) = 2400 ppm for 24 hrs; ipLD_{LO} (rat) = 1500 mg/kg; inhal TC_{LO} (human) = 10 ppm ⟶ CNS effects. [*3*]
THR = MOD via ip and inhal routes. Causes CNS disturbances. Data based on animal exper show narcotic effects and some disturbances of blood. Leucopenia and thrombocytopenia have been reported in exper animals. Reacts violently with NHO_3. [*19*]
Fire Hazard: Mod, via heat, flames, oxidizers.
To Fight Fire: Water spray, fog, foam, CO_2.

MESITYL OXIDE. Syn: *4-methyl-3-penten-2-one.* Oily colorless liquid, strong odor. $(CH_3)_2CCHCOCH_3$, mw: 98.14, mp: −59°, bp: 130.0°, flash p: 87°F (CC), d: 0.8539 @ 20°/4°, autoign. temp.: 652°F, vap. press: 10 mm @ 26.0°, vap. d: 3.38.
Acute tox data: Oral LD_{50} (rat) = 1120 mg/kg; inhal LC_{LO} (rat) = 1000 ppm for 4 hrs; ipLD_{50} (mouse) = 354 mg/kg; dermal LD_{50} (rabbit) = 5990 mg/kg. [*3*]
THR = MOD via oral, inhal and dermal routes; HIGH via ip route. It can cause opaque cornea, keratoconus, and extensive necrosis of cornea. Guinea pigs and rats were subject to repeated inhal of the vapors of this material. No effect upon them was found after the conc had been reduced to 50 ppm even with 30 8-hr exposures. Single exposures tend to indicate that this ketone has greater acute and narcotic action than isophorone. It can have harmful effects upon the kidneys and liver, and may damage the eyes and lungs to a serious degree. Animals exposed to 100 ppm for 30 8-hr exposures showed internal damage as noted above. Exposure to 500 ppm caused their death. This compound is high irr to all tissues on contact and its vapors also are irr. High conc are narcotic. Prolonged exposure can injure liver, kidneys and lungs. It is readily absorbed through intact skin.
Fire Hazard: Dangerous, when exposed to heat or flame; can react with oxidizing materials. Reacts violently with 2-amino ethanol, chlorosulfonic acid, ethylene diamine, HNO_3, oleum, H_2SO_4. [*19*]
To Fight Fire: Alcohol foam, CO_2, dry chemical.

MESTRANOL + ETHYNODIOL DIACETATE.
THR = An exper neo to women via oral route. [*103*]

METABOLISM OF TOXIC MATERIALS. See Section 9.

METABORATE PEROXYHYDRATE. See sodium perborate.

METACETONE. See diethyl ketone.

METACIDE. See parathion.

METACINNABARITE. See mercuric sulfide, black.

METALDEHYDE. Syns: *m-acetaldehyde, polymerized acetaldehyde.* Colorless crystals. $(C_2H_4O)_4$, mw: 176.21, mp: 246° sealed in glass, sublimes @ 122°, flash p: 97°F (CC), vap. d: 6.06.
Acute tox data: Oral LD_{50} (rat) = 630 mg/kg; oral LD_{50} (dog) = 600 mg/kg; oral LD_{LO} (human) = 60 mg/kg. [*2, 3*]
THR = HIGH via oral to humans; MOD via oral to test animals. An irr to skin, eyes and mu mem. Can cause kidney and liver damage. See also acetaldehyde. Used as a food additive permitted in food for human consumption. [*109*]
Fire Hazard: Dangerous, when exposed to heat or flame; can react with oxidizing materials.

Spont Heating: No.
To Fight Fire: CO_2, dry chemical.

METAL OXIDES.
THR = React violently with BrF_5, performic acid. [*19*]

METAL POLISH LIQUID. flash p: <0°F.
THR = Details U. Some metal polishes can act as irr and allergens.
Fire Hazard: Dangerous, when exposed to heat or flame; can react vigorously with oxidizing materials.
Explosion Hazard: U.

METALS.
THR = Violent reactions with NH_4NO_3, BrF_5, BrF_3, chlorates, performic acid. [*19*]

METAL SULFIDES.
THR = Reacts violently with bromates, chlorates, $HClO_3$, hydrazine mononitrate, iodates, Ag_2O. [*19*] See individual sulfides.

METANILIC ACID, SODIUM SALT. See *m*-aniline sodium sulfonate.

METASQUALENE. $C_{27}H_{32}O_2NCl$, mw: 438.1.
THR = An exper teratogen. [*3*]

META-SYSTOX. Syn: *o-[2-(ethylthio) ethyl]-o,o-dimethyl phosphorothioate.* An oily liquid, slightly sol in water. $C_6H_{15}O_3PS_2$, mw: 230.1, d: 1.20.
Acute tox data: Oral LD_{50} (rat) = 180 mg/kg. [*3*]
THR = HIGH via oral route. A cholinesterase inhibitor. See parathion. An insecticide and acaricide.

α-METHACROLEIN. Syn: *methylpropenal.* Colorless liquid. $CH_2C(CH_3)CHO$, mw: 70.1, bp: 73.5°, flash p: 35°F (OC), d: 0.830 @ 20°/4°, vap. press: 120 mm @ 20°, vap. d: 2.42.
Acute tox data: Oral LD_{50} (rat) = 111 mg/kg; inhal LC_{LO} (rat) = 125 ppm for 4 hrs; dermal LD_{50} (rabbit) = 430 mg/kg. [*3*]
THR = HIGH irr to skin, eyes and mu mem and via oral, inhal and dermal routes.
Fire Hazard: Dangerous; when exposed to heat, flame or oxidizers.
Explosion Hazard: U.
Disaster Hazard: Dangerous; on decomp, emits toxic fumes; can react vigorously with oxidizing materials.
To Fight Fire: CO_2, alcohol foam, foam, dry chemical.

METHACRYLIC ACID. Syn: *α-methylacrylic acid.* Corrosive liquid or colorless crystals. $(CH_3)(CH_2)CCOOH$, mw: 86.1, mp: 15°, bp: 161°, flash p: 171°F (COC), d: 1.014 @ 25° (glacial), vap. press: 1 mm @ 25.5°.
Acute tox data: ipLD_{50} (mouse) = 48 mg/kg. [*3*]
THR = HIGH irr to skin, eyes and mu mem and via ip route.
Fire Hazard: Mod, when exposed to heat, flame or oxidizers.
Disaster Hazard: Dangerous; when heated to decomp, emits toxic fumes.
To Fight Fire: Alcohol foam, spray, mist, dry chemical.

β-METHACRYLIC ACID. See crotonic acid.

METHACRYLONITRILE. Syn: *2-methylpropenitrile.* $H_2C = C(CH_3)C \equiv N$, mw: 67.09, mp: −36°, bp: 90.3°, d: 0.805, vap. press: 40 mm @ 12.8°, flash p: 55°F.
Acute tox data: Oral LD_{50} (rat) = 250 mg/kg; inhal LC_{50} (rat) = 328 ppm for 4 hrs; dermal LD_{50} (rabbit) = 320 mg/kg. [*3*]
THR = HIGH via oral, inhal and dermal routes. A poison, see nitriles.

1,4(8)-*p*-METHADIENE. See terpinolene.

METHAHEXAMIDE. $C_{14}H_{21}O_3N_9S$, mw: 311.4.
THR = An exper teratogen. [*3*]

1-METHALLYL ALCOHOL. Syn: *2-buten-1-ol.* Colorless liquid. $CH_3CHCHCH_2OH$, mw: 72.1, mp: <30°, bp: 118°, flash p: 92°F, d: 0.8726 @ 0°/4°, vap. d: 2.49.
Acute tox data: Oral LD_{LO} (mouse) = 500 mg/kg; inhal LC_{LO} (mouse) = 2924 ppm for 2 hrs; dermal LD_{LO} (rabbit) = 2000 mg/kg. [*3*]
THR = MOD via oral, inhal and dermal routes.
Fire Hazard: Dangerous, when exposed to heat or flame; can react with oxidizing materials.
To Fight Fire: Alcohol foam, CO_2, dry chemical.

METHALLYL CHLORIDE. Syn: *2-chlorobutene-2.* Colorless, volatile liquid, disagreeable odor. C_4H_7Cl, mw: 90.55, bp: 72.17°, lel = 3.2%, uel = 8.1%, fp: <−80°, flash p: 11°F, d: 0.9257 @ 20°/4°, vap. press: 101.7 mm @ 20°, vap. d: 3.12.
Acute tox data: Inhal LC_{50} (rat) = 2000 mg/m^3 for 24 hrs; inhal LC_{LO} (mouse) = 9100 mg/m^3 for 10 min. [*3*]
THR = MOD via inhal route. An irr.
Fire Hazard: Dangerous, when exposed to heat, flame or oxidizers.
Explosion Hazard: Mod, when exposed to heat or flame.
Disaster Hazard: Dangerous; on decomp, emits highly toxic fumes of chlorides; can react vigorously with oxidizing materials.
To Fight Fire: Alcohol foam, CO_2, dry chemical.

METHAMPHETAMINE HYDROCHLORIDE. $C_{10}H_{15}N \cdot HCl$, mw: 185.7.
THR = HIGH acute via oral route. An exper teratogen. [*3*]

METHAM SODIUM. See methyldithiocarbamic acid sodium salt.

METHANAL. See formaldehyde.

METHANAMIDE. See formamide.

METHANDIENONE. Syn: *Danabol.*
THR = An exper neo. [*3*]

METHANE. Syns: *marsh gas, methyl hydride.* Colorless, odorless, tasteless gas. CH_4, mw: 16.05, bp: −161.5°, lel = 5.3%, uel = 15%, fp: −183.2°, d: 0.415 @ −164°, 0.7168 g/liter, autoign. temp.: 1004°F, vap. d: 0.6, flash p: −306°F.
THR = A simple asphyxiant. See argon.
Fire Hazard: Very dangerous, when exposed to heat or flame. Reacts violently with BrF_5, Cl_2, ClO_2, NF_3, liquid O_2, OF_2. [*19*]
Spont Heating: No.
Explosion Hazard: Dangerous, when exposed to heat or flame.
Disaster Hazard: Dangerous.
To Fight Fire: Stop flow of gas, CO_2 or dry chemical.

METHANE ARSONIC ACID. Syn: *methyl arsonic acid.* Crystals, water- and alcohol-sol. $CH_3AsO(OH)_2$, mw: 140.0, mp: 155°.
Acute tox data: $scLD_{50}$ (mouse) = 3350 mg/kg. [*3*]
THR = MOD via sc route. A MILD irr and an allergen. See also arsenic.
Disaster Hazard: Hazardous when water solution is in contact with active metals, i.e., iron, aluminum or zinc, or when heated to decomp.

METHANE CARBOTHIOLIC ACID. See thioacetic acid.

METHANE CARBOXYLIC ACID. See acetic acid.

p-**METHANE HYDROPEROXIDE.** Clear, pale yellow liquid. $C_{10}H_{20}O_2$, mw: 172.3, d: 0.910–0.925 @ 15.5°/4°.
THR = An exper neo. [*3*] An irr.
Fire Hazard: Dangerous. A powerful oxidizer.
Disaster Hazard: Dangerous. A powerful oxidizer.

METHANE SULFONIC ACID. Syn: *methyl sulfonic acid.* Solid, sol in water, alcohol and ether. CH_3SO_2OH, mw: 96.10, d: 1.485 @ 20°/20°, mp: 20°, bp: 167° @ 10 mm.
Acute tox data: Oral LD_{LO} (rat) = 200 mg/kg; $ipLD_{LO}$ (rat) = 50 mg/kg. [*3*]
THR = HIGH irr to skin, eyes and mu mem. HIGH via oral and ip routes.
Disaster Hazard: Dangerous; see sulfonates.

METHANE SULFONIC ACID-2-CHLOROETHYL ESTER.
THR = An exper carc. [*23*]

METHANETHIOL. See methyl mercaptan.

METHANOIC ACID. See formic acid.

METHANO INDANE. See chlordane.

METHANOL. See methyl alcohol.

METHAPYRILENE. Syn: *diethylandiamine,* $C_{14}H_{19}SN_3$, mw: 261.4.
Acute tox data: oral LD_{50} (rat) = 525 mg/kg; ip LD_{50} (mice) = 77 mg/kg. [*3*]
THR = HIGH oral and ip routes. May cause liver tumors in rats. It is an antihistamine.

METHENAMINE. See hexamethylene tetramine.

***n,n'*-METHENYL-*o*-PHENYLENE DIAMINE.** See benzimidazole.

METHIMAZOLE. Syn: *1-methyl imidazole-2-thiol.* $C_4H_6N_2S$, mw: 114.2.
THR = MOD toxic to mice. Has been implicated in aplastic anemia.

METHIONINE. Syn: *2-amino-4-(methyl thio) butyric acid.* An essential sulfur-containing amino acid; white crystalline powder or platelets, faint odor, sol in water, dilute acids and alkalis, very slightly sol in alcohol, practically insol in ether.
$CH_3SCH_2CH_2CH(NH_2)COOH$, mw: 149.
THR = U. A nutrient and/or dietary supplement food additive. [*109*]
Disaster Hazard: Dangerous; see sulfides.

METHIONINE HYDROXY ANALOG CALCIUM. Syns: *di-α-hydroxy-γ-methyl mercaptobutyric acid, calcium salt; 2-hydroxy-4-methyl thiobutyric acid, calcium salt.* Light tan powder, sol in water, insol in common organic solvents.
$(CH_3SCH_2CH_2CHOHCOO)_2Ca$, mw: 338.
THR = U. A nutrient and/or dietary supplement food additive. [*109*]

METHIOTEPA.
THR = An exper insect chemosterilant. No toxicity details.

METHIOTRIAZAMINE. Syn: *4,6-diamino-1-(4-methylmercaptophenyl)-1,2-dihydro-2,2-dimethyl-1,3,5-triazine hydrochloride.*
THR = U. Used as a food additive permitted in the feed and drinking water of animals and/or for the treatment of food-producing animals. Also permitted in food for human consumption. [*109*]
Disaster Hazard: Dangerous; see chlorides and sulfides.

METHOTREXATE. Syn: *n-[p-(2,4-diamino-6-pteridyl) methyl methylamino benzoyl] glutamic acid.*
Acute tox data: $ipLD_{50}$ (rat) = 17 mg/kg. [*3*]
THR = HIGH via ip route. An insect chemosterilant. An exper teratogen and neo. [*3*]

METHOXONE. Syn: *4-chloro-2-methylphenoxyacetic acid.* Crystals. $C_9H_9O_3Cl$, mw: 200.6.
Acute tox data: Oral LD_{50} (rat) = 700 mg/kg; $scLD_{LO}$ (mouse) = 28 mg/kg. [3]
THR = HIGH via sc and MOD via oral routes.
Disaster Hazard: Dangerous; when heated to decomp, emits highly toxic fumes of chlorides.

METHOXYACETYL-*p*-PHENETIDINE. See kryofin.

METHOXYAMINE. Syn: *methoxylamine.* Mobile liquid, fishy amine odor, miscible with water. CH_3ONH_2, mw: 47.1, bp: 50°.
THR = A powerful irr to skin, eyes and mu mem. Details U. See related compound, hydroxylamine.

3-METHOXY-4-AMINOAZO BENZENE. $C_{13}H_{13}ON_3$, mw: 227.3.
THR = An exper carc. [3, 23] via oral and dermal routes.

METHOXY ANALOG OF DDT. See 1-trichloro-2,2-bis-(*p*-methoxyphenyl) ethane.

***o*-METHOXY ANILINE.** See *o*-anisidine.

***p*-METHOXY ANILINE.** See *p*-anisidine.

METHOXY BENZALDEHYDE-*o*. Syns: *o-anisaldehyde, o-anisic aldehyde.* White to light tan solid, slight phenolic odor, slightly sol in water. $CH_3OC_6H_4CHO$, mw: 136, bp: 238°, mp: 38°–39° and 30° (2 crystalline forms), d(liquid): 1.1274 @ 25°/25°, d(solid): 1.258 @ 25°/25°, flash p: 244°F (OC).
Acute tox data: Oral LD_{50} (rat) = 1510 mg/kg; oral LD_{50} (guinea pig) = 1260 mg/kg. [3]
THR = MOD via oral route.
Fire Hazard: Slight, when exposed to heat or flame.
To Fight Fire: Alcohol foam, spray, mist, dry chemical.

5-METHOXY BENZ(a)ANTHRACENE. $C_{19}H_{14}O$, mw: 258.3.
THR = An exper neo. [3, 23]

7-METHOXY BENZ(a)ANTHRACENE.
THR = An exper neo. [3, 23]

8-METHOXY BENZ(a)ANTHRACENE.
THR = An exper carc. [3, 23]

2-METHOXY-3-BENZO FURANYL AMINE.
THR = An exper carc. [23]

***p*-METHOXY BENZOIC ACID.** See anisic acid.

***p*-METHOXY BENZOIC ACID PIPERIDINO-ESTER.** $C_{13}H_{17}O_3N$, mw: 235.3.
THR = An exper neo. [3]

3-METHOXY BENZ(a)PYRENE. $C_{21}H_{14}O$, mw: 282.4.
THR = An exper carc. [3, 23]

4-METHOXY BENZ(a)PYRENE.
THR = An exper carc. [3, 23]

5-METHOXY BENZ(a)PYRENE.
THR = An exper carc. [3]

4-METHOXY BIPHENYL. $C_{13}H_{12}O$, mw: 184.3.
THR = An exper carc. [3]

3-METHOXY-4-BIPHENYL AMINE.
THR = An exper carc. [23]

METHOXY BORON DICHLORIDE. Colorless liquid, decomp by water. CH_3OBCl_2, mw: 112.8, mp: −15°.
THR = See boron compounds and chlorides.

METHOXY BORON DIFLUORIDE. Colorless liquid, decomp by water. CH_3OBF_2, mw: 79.9, d: 1.417 @ 36°, bp: 86°.
THR = See fluorides and boron compounds.

1-METHOXY-1,3-BUTADIENE.
$CH_2 : CHCH : CH(OCH_3)$, mw: 84.
THR = Details U. Limited animal exper suggest MOD toxicity. See also butadiene.

3-METHOXY BUTANOL. Liquid.
$CH_3CH(OCH_3)CH_2CH_2OH$, mw: 104.15, bp: 161.1°, fp: −85°, flash p: 165°F (COC), d: 0.9229 @ 20°/20°, vap. press: 0.9 mm @ 20°, vap. d: 3.59.
Acute tox data: Oral LD_{LO} (rat) = 4000 mg/kg. [3]
THR = MOD via oral route.
Fire Hazard: Mod, when exposed to heat or flame; can react with oxidizing materials.
To Fight Fire: Alcohol foam, CO_2, dry chemical.

METHOXY BUTYL ACETATE. Syn: *butoxyl.* Liquid, bitter taste, acrid odor. $CH_3CO_2C_4H_8OCH_3$, mw: 146.2, bp: 135°, flash p: 170°F, d: 0.952–0.958 @ 20°/20°, vap. d: 5.05.
Acute tox data: Oral LD_{50} (rat) = 4210 mg/kg. [3]
THR = MOD via oral route.
Fire Hazard: Mod, when exposed to heat or flame; can react with oxidizing materials.
To Fight Fire: Alcohol foam, CO_2, dry chemical.

3-METHOXY BUTYL ALCOHOL. See 3-methoxy butanol.

3-METHOXYBUTYRALDEHYDE.
$CH_3CH(OCH_3)CH_2CHO$, mw: 102, flash p: 140°F.
THR = Details U. Limited animal exper suggest HIGH toxicity and irr. See also aldehydes.
Fire Hazard: Mod via heat, open flame, oxidizers.
To Fight Fire: Foam, water spray or mist, dry chemical, CO_2.

3-METHOXY BUTYRIC ACID. Liquid.
$CH_3OC_3H_6COOH$, mw: 118.3, mp: 12°, bp: 139° @ 50 mm, d: 1.053 @ 20°/20°.
Acute tox data: Oral LD_{50} (rat) = 3030 mg/kg. [3]
THR = MOD via oral route.
Disaster Hazard: Slight, when heated, emits acrid fumes; can react with oxidizing materials.

METHOXYCHLOR. Syns: *DMDT; 2,2-bis-(p-methoxyphenyl)-1,1,1-trichloroethane.* Crystals.
$C_{15}H_{15}O_3Cl_3$, mw: 345.7, mp: 78°, vap. d: 12.
Acute tox data: Oral LD_{50} (rat) = 5000 mg/kg; dermal TD_{LO} (human) = 2414 mg/kg; $ipLD_{LO}$ (rat) = 500 mg/kg. [*3*]
THR = MOD via oral, ip and dermal routes. An irr and an allergen. An insecticide. Prolonged exposure may cause kidney injury. See also DDT. An exper (±) carc. [*3, 12*]
Disaster Hazard: Dangerous; when heated to decomp, emits highly toxic fumes of chlorides.

2-METHOXY-6-CHLORO-9-[4-BIS(2-CHLOROETHYL)AMINO-1-METHYL BUTYLAMINO] ACRIDINE DIHYDROCHLORIDE.
$C_{23}H_{28}ON_3Cl_3 \cdot 2HCl$, mw: 541.8.
THR = An exper carc via iv route. [*103*]

2-METHOXY-6-CHLORO-9-[3-(2-CHLOROETHYL) MERCAPTO PROPYLAMINO]ACRIDINE HYDROCHLORIDE. $C_{19}H_{20}ON_2SCl_2 \cdot HCl$, mw: 431.8.
THR = An exper carc via iv route. [*103*]

2-METHOXY-6-CHLORO-9-[3-(ETHYL-2-CHLOROETHYL)AMINO PROPYLAMINO]-2-METHOXY ACRIDINE HYDROCHLORIDE.
$C_{21}H_{25}ON_3Cl_2 \cdot 2HCl$, mw: 479.3.
Acute tox data: $ivLD_{LO}$ (mice) = 5 mg/kg. [*103*]
THR = HIGH via iv route. An exper carc via iv route. [*103*]

5-METHOXY CHRYSENE. $C_{19}H_{14}O$, mw: 258.3.
THR = An exper neo. [*3*]

5-METHOXY DIBENZ(a,h)ANTHRACENE.
$C_{23}H_{16}O$, mw: 308.4.
THR = An exper neo. [*3, 23*]

7-METHOXY DIBENZ(a,h)ANTHRACENE.
THR = An exper carc. [*3, 23*]

3-METHOXY-2-DIBENZOFURANAMINE.
$C_{13}H_{11}O_2N$, mw: 213.3.
THR = An exper neo. [*3*]

2-METHOXY-3-DIBENZOFURANAMINE.
THR = An exper carc. [*3*]

11-METHOXY-16,17-DIHYDRO-15H-CYCLOPENTA(a)PHENANTHRENE.
THR = An exper carc. [*23*]

METHOXYETHANE. See ethyl methyl ether.

2-METHOXYETHANOL. See ethylene glycol monomethyl ether.

2-(2-METHOXYETHOXY)ETHANOL. See methyl carbitol.

METHOXYETHYL ACETYL RICINOLEATE. Light straw-colored liquid.
$C_{17}H_{32}(OOCCH_3)COOC_2H_4OCH_3$, mw: 398, mp: −60° (gels), bp: 200°–260° @ 4 mm, flash p: 446°F, d: 0.966 @ 20°/20°, vap. press: <0.01 mm @ 150°, vap. d: 13.8.
Acute tox data: Oral LD_{50} (rat) = 20,000 mg/kg; oral LD_{50} (guinea pig) = 12,000 mg/kg. [*3*]
THR = VERY LOW via oral route.
Fire Hazard: Slight, when exposed to heat or flame.
To Fight Fire: Foam, CO_2, dry chemical.

METHOXYETHYL ACRYLATE. Liquid.
$CH_2CHCO_2CH_2CH_2OCH_3$, bp: 61° @ 17 mm, flash p: 180°F (OC), d: 1.0134 @ 20°, vap. d: 4.49, mw: 130.1.
THR = Details U. See ethyl acrylate.
Fire Hazard: Mod, when exposed to heat or flame; can react with oxidizing materials.
To Fight Fire: Foam, CO_2, dry chemical.

5-METHOXY-7-ETHYL BENZ(a)ANTHRACENE.
THR = An exper carc. [*23*]

METHOXYETHYL MERCURY ACETATE. Crystals, water-sol. $CH_3OCH_2CH_2HgOOCCH_3$, mw: 318.7.
Acute tox data: Oral LD_{50} (rat) = 16 mg/kg. [*3*]
THR = HIGH via oral route. See mercury compounds, organic.

METHOXYETHYL MERCURY CHLORIDE. Crystals. $OCH_3CH_2CH_2HgCl$, mw: 295.2
Acute tox data: Oral LD_{50} (rat) = 30 mg/kg; oral LD_{50} (mouse) = 47 mg/kg; $scLD_{50}$ (mouse) = 60 mg/kg. [*3*]
THR = HIGH via oral and sc routes. See mercury compounds, organic, and chlorides.

2-METHOXYETHYL MERCURY SILICATE. Syn: *verisan.* $C_9H_{22}O_7SiHg_3$, mw: 872.2.
Acute tox data: LD_{50} (rat) = 50 mg/kg. [*3*]
THR = HIGH. See mercury compounds, organic.

METHOXYETHYL OLEATE. Light-colored liquid.
$C_8H_{17}CHCH(CH_2)_7COOC_2H_4OCH_3$, mw: 340, mp: −20°, bp: 188°–225° @ 4 mm, flash p: 386°F, d: 0.902 @ 20°/20°, vap. press: 0.04 mm @ 150°, vap. d: 11.8.
Acute tox data: Oral LD_{LO} (rat) = 16,000 mg/kg. [*3*]
THR = VERY LOW via oral route. See also esters.
Fire Hazard: Slight, when exposed to heat or flame; can react with oxidizing materials.
To Fight Fire: Foam, CO_2, dry chemical.

METHOXYETHYL PHTHALATE. Liquid.
$(OCH_3CH_2CH_2CO_2)_2C_6H_4$, mw: 282.3, bp: 190°–220°, flash p: 275°F (CC), d: 1.17, vap. d: 9.75.
Acute tox data: Oral LD_{50} (guinea pig) = 1600 mg/kg; $ipLD_{LO}$ (mouse) = 300 mg/kg. [*3*]
THR = HIGH via ip and MOD via oral route.
Fire Hazard: Slight, when exposed to heat or flame; can react with oxidizing materials.

Spont Heating: No.
To Fight Fire: CO_2, dry chemical.

1-METHOXYFLUOREN-2-AMINE HYDROCHLORIDE. $C_{14}H_{14}ONCl$, mw: 247.7.
THR = An exper carc. [*3*]

N-(7-METHOXYFLUOREN-2-YL)ACETAMIDE. $C_{16}H_{15}O_2N$, mw: 253.3.
THR = MOD toxic to rats on an acute oral basis. An exper carc. [*3, 23*]

1-METHOXY-2-FLUORENYL AMIDE.
THR = An exper carc. [*23*]

3-METHOXY-4-HYDROXYBENZALDEHYDE. See vanillin.

METHOXYLAMINE. See methoxyamine.

5-METHOXY-7-METHYLBENZ(a)ANTHRACENE. $C_{20}H_{16}O$, mw: 272.4.
THR = An exper neo. [*3, 23*]

7-METHOXY-12-METHYL BENZ(a)ANTHRACENE.
THR = An exper carc and neo. [*3, 23*]

8-METHOXY-7-METHYL BENZ(a)ANTHRACENE.
THR = An exper carc. [*23, 3*]

11-METHOXY-17-METHYL-15H-CYCLOPENTA(a) PHENANTHRENE.
THR = An exper carc. [*23*]

7-(METHOXYMETHYL)-12-METHYL BENZ(a)ANTHRACENE. $C_{21}H_{18}O$, mw: 266.4.
THR = An exper neo. [*3*]

2-METHOXY-4-METHYL PHENOL. See creosol.

1-METHOXY-2-NAPHTHYLAMINE.
THR = An exper carc. [*23*]

2-METHOXY-4-NITROANILINE. See 1-amino-2-methoxy-4-nitrobenzene.

4-METHOXY-3-(1-OCTENYL-ONN-AZOXY)-D-THREO-2-BUTANOL. $C_{13}H_{26}O_3N_2$, mw: 258.4.
THR = An exper carc. [*3, 23*] HIGH acute toxicity.

2-METHOXYPHENOL. See guaiacol.

4-METHOXYPHENOL. See hydroquinone monomethyl ether.

***p*-METHOXYPHENYLACETIC ACID.** $C_9H_{10}O_3$, mw: 166.2.
THR = An exper neo. [*3*]

N-(*p*-METHOXYPHENYL)-1-AZIRIDINE CARBAMIDE. $C_9H_6ON_2Cl_2$, mw: 231.1.
THR = An exper neo. [*3*]

4-[(*p*-METHOXYPHENYL)AZO]-*o*-ANISIDINE.
THR = An exper carc. [*23*]

***m*-METHOXYPHENYL BORIC ACID.** See *m*-anisyl boric acid.

***o*-METHOXYPHENYL BORIC ACID.** See *o*-anisyl boric acid.

***p*-METHOXYPHENYL BORIC ACID.** See *p*-anisyl boric acid.

4-(*p*-METHOXYPHENYL)-2-BUTANONE. See anisyl acetone.

***p*-METHOXYPHENYL-N-CARBAMOYL AZIRIDINE.** $C_{10}H_{12}O_2N_2$, mw: 192.2.
THR = An exper neo. [*3*]

2-(*p*-METHOXYPHENYL)-3,3-DIPHENYL ACRYLONITRILE. $C_{22}H_{17}ON$, mw: 311.4.
THR = An exper carc. [*3*]

4-METHOXY-*m*-PHENYLENE DIAMINE SULFATE. $C_7H_{10}ON_2 \cdot (H_2SO_4)_n$.
Acute tox data: Oral LD_{50} (rat) = 515 mg/kg. [*3*]
THR = HIGH via oral route. An exper carc to rats via oral route. [*114*] Absorbed via human skin.

1-METHOXY-2-PROPANOL. See monopropylene glycol methyl ether.

***p*-METHOXYPROPENYL BENZENE.** See anethole.

METHOXYPROPIONITRILE. Liquid.
$OCH_3CH_3CH_2CN$, mw: 85.1, mp: −63°, bp: 160°, flash p: 149°F (OC), d: 0.92, vap. d: 2.94.
THR = HIGH. See cyanides.
Fire Hazard: Mod, when exposed to heat, flame or oxidizers.
Disaster Hazard: Dangerous; when heated to decomp, emits highly toxic fumes of cyanides; will react with water, steam, or acids to produce toxic and flam vapors; can react with oxidizing materials. See also nitriles.
To Fight Fire: CO_2, dry chemical.

3-METHOXYPROPYLAMINE. Colorless liquid.
$CH_3OCH_2CH_2CH_2NH_2$, mw: 89.14, mp: −75.7°, bp: 116°, flash p: 90°F (TOC), d: 0.8615 @ 30°, vap. press: 20 mm @ 30°, vap. d: 3.07.
THR = MOD irr to skin, eyes and mu mem.
Fire Hazard: Dangerous, when exposed to heat or flame; can react with oxidizing materials.
Explosion Hazard: U.
To Fight Fire: CO_2, dry chemical.

5-METHOXYPSORALEN. Syn: *bergapten.*
THR = Lime pickers dermatitis. An exper carc. [*23*]

2-METHOXYTRICYCLOQUINAZOLINE. $C_{22}H_{14}ON_4$, mw: 350.4.
THR = An exper neo. [*3*]

3-METHOXYTRICYCLOQUINAZOLINE.
THR = An exper neo. [*3*]

METHOXY TRIGLYCOL. Miscible with water. $C_7H_{16}O_4$, mw: 122, d: 1.0494, bp: 249°, fp: −44°, flash p: 245°F (OC).
Acute tox data: Oral LD_{50} (rat) = 11,000 mg/kg; dermal LD_{50} (rabbit) = 7100 mg/kg. [*3*]
THR = LOW via oral and MOD via dermal routes.
Fire Hazard: Slight, when exposed to heat or flame.
To Fight Fire: Alcohol foam, dry chemical.

METHOXYTRIGLYCOLACETATE. Syn: *triethylene glycol methyl ether acetate*. Liquid. $C_9H_{18}O_5$, mw: 206.23, bp: 130°, flash p: 260°F (OC), d: 1.094, vap. d: 7.11.
THR = U. See also glycols.
Fire Hazard: Slight, when exposed to heat or flame; can react with oxidizing materials.
To Fight Fire: Alcohol foam, CO_2, dry chemical.

***p*-METHOXY-α-VINYL BENZYL ALCOHOL.** $C_{10}H_{12}O_2$, mw: 164.1.
THR = An exper carc to mice via sc route. [*103*]

METHYL ABIETATE. Liquid. $CH_3C_{20}H_{29}O_2$, mw: 316.5, bp: 361°, flash p: 356°F (OC), d: 1.02, vap. d: 10.9.
THR = Probably LOW. See also abietic acid.
Fire Hazard: Slight, when exposed to heat or flame; can react with oxidizing materials.
To Fight Fire: CO_2, dry chemical.

METHYL-4-ACETAMIDO-2-ETHOXYBENZOATE. See ethopabate.

METHYL ACETATE. Colorless, volatile liquid. $CH_3CO_2CH_3$, mw: 74.08, mp: −98.7°, lel = 3.1%, uel = 16%, bp: 57.8°, ulc: 85-90, flash p: 14°F, d: 0.92438, autoign. temp.: 935°F, vap. press: 100 mm @ 9.4°, vap. d: 2.55.
Acute tox data: Oral LD_{50} (rabbit) = 3700 mg/kg; inhal LC_{LO} (cat) = 67 mg/m^3 for 56 min; $scLD_{LO}$ (cat) = 3000 mg/kg. [*3*]
THR = MOD via oral, sc and inhal routes. An irr. Methyl acetate is narcotic, but is less so than the higher members of the acetate series. It has an irr effect upon the mu mem of the eyes and upper respiratory tract, and in this respect its action is stronger than that of the higher members of the series. The irr conc is about 10,000 ppm. Signs and symptoms are irr and burning of the eyes, lachrymation, dyspnea, palpitation of the heart, and complaints of depression or dizziness.
Fire Hazard: Dangerous, when exposed to heat, flame or oxidizers.
Spont Heating: No.
Explosion Hazard: Mod, when exposed to heat or flame.
Disaster Hazard: Dangerous, upon exposure to heat or flame; can react vigorously with oxidizing materials.
To Fight Fire: Alcohol foam, CO_2, dry chemical.

METHYL ACETIC ACID. See propionic acid.

METHYL ACETOACETATE. Colorless liquid. $CH_3COCH_2COOCH_3$, mw: 116.11, mp: 27.5°, bp: 170°, flash p: 170°F, autoign. temp.: 536°F, d: 1.077, vap. d: 4.00.
Acute tox data: Oral LD_{50} (rat) = 3000 mg/kg. [*3*]
THR = MOD via oral route.
Fire Hazard: Mod, when exposed to heat, flame or oxidizers.
Spont Heating: No.
To Fight Fire: Foam, CO_2, dry chemical.

7-METHYL ACETOBENZ(a)ANTHRACENE.
THR = An exper carc. [*23*]

7-METHYL ACETO-12-METHYL BENZ(a) ANTHRACENE.
THR = An exper carc. [*23*]

METHYL ACETONE. See 2-butanone.

METHYL ACETYLENE. See allylene.

METHYL ACETYLENE-PROPADIENE MIXTURE. Syns: *methyl acetylene propadiene, stabilized*; *MAPP*. Clear, colorless, liquefied gas. d(liquid): 0.576 @ 15°/15°, boiling range: −39° to −20°.
THR = No data. Probably LOW.
Fire Hazard: Flam when exposed to heat or flame.

METHYL ACETYLENE-PROPADIENE, STABILIZED. See methyl acetylene-propadiene mixture.

METHYL ACETYL RICINOLEATE. Crystals. $CH_3CH_2CH_2CO_2(CH_2)_7CHCHCH_2CHOH(CH_2)_5CH_3$, mw: 340.5, vap. d: 11.9.
THR = A MILD irr. Probably MOD via oral route.
Fire Hazard: Slight, when exposed to heat or flame; can react with oxidizing materials.

α-METHYL ACROLEIN. See methacrolein.

β-METHYL ACROLEIN. See crotonaldehyde.

METHYL ACRYLATE. Syn: *acrylic acid methyl ester*. Colorless liquid. $CH_2CHCO_2CH_3$, mw: 86.1, bp: 80°, lel = 2.8%, uel = 25%, fp: −75°, flash p: 27°F (OC), d: 0.949 @ 25°, vap. press: 100 mm @ 28°, vap. d: 2.97
Acute tox data: Oral LD_{50} (rat) = 300 mg/kg; inhal TC_{LO} (human) = 75 ppm → an irr; dermal LD_{50} (rabbit) = 1300 mg/kg; $ipLD_{50}$ (mouse) = 265 mg/kg. [*3*]
THR = HIGH via oral, inhal and ip; MOD via dermal routes. Chronic exposure has produced injury to lungs, liver and kidneys in exper animals. It is a substance which migrates to food from packaging material. [*109*]

Fire Hazard: Dangerous, when exposed to heat, flame or oxidizers.

Explosion Hazard: Dangerous, when exposed to heat, sparks or flame.

Disaster Hazard: Dangerous; when heated to decomp, emits toxic fumes; can react vigorously with oxidizing materials.

To Fight Fire: Foam, CO_2, dry chemical.

METHYL ACRYLATE, INHIBITED. See methyl acrylate.

METHYL ACRYLATE, UNINHIBITED. See methyl acrylate.

METHYL ACRYLIC ACID. See methacrylic acid.

METHYLAL. Syns: *formal, methylene dimethyl ether, formaldehyde dimethylacetal.* Colorless liquid, pungent odor. $CH_2(OCH_3)_2$, mw: 76.1, mp: −104.8°, bp: 42.3°, flash p: 0°F (OC), d: 0.864 @ 20°/4°, vap. press: 330 mm @ 20°, vap. d: 2.63, autoign. temp.: 459°F.

Acute tox data: Oral LD_{50} (rabbit) = 5708 mg/kg; $scLD_{LO}$ (guinea pig) = 3013 mg/kg. [*3*]

THR = MOD via sc and oral routes. Narcotic in high conc. Has produced injury to lungs, liver, kidneys and heart in exper animals. High conc of vapors act as an anesthetic.

Fire Hazard: Dangerous, when exposed to heat, flame or oxidizers.

Explosion Hazard: Mod, when exposed to heat or flame.

Disaster Hazard: Dangerous, upon exposure to heat or flame; can react vigorously with oxidizing materials.

To Fight Fire: Foam, CO_2, dry chemical.

METHYL ALCOHOL. Syn: *methanol.* Clear, colorless, very mobile liquid. CH_3OH, mw: 32.04, bp: 64.8°, lel = 6.7%, uel = 36%, ulc: 70, fp: −97.8°, flash p: 52°F, d: 0.7913 @ 20°/4°, autoign. temp.: 725°F, vap. press: 100 mm @ 21.2°, vap. d: 1.11.

Acute tox data: Oral LD_{50} (rat) = 13,000 mg/kg; $scLD_{50}$ (mouse) = 9800 mg/kg; inhal LC_{50} (monkey) = 1000 ppm; dermal LD_{50} (rabbit) = 20,000 mg/kg. [*3*]

THR = LOW via oral, sc and dermal; MOD via inhal routes. Methyl alcohol possesses distinct narcotic properties. It is also a slight irr to the mu mem. Its main toxic effect is exerted upon the nervous system, particularly the optic nerves and possibly the retinae. The effect upon the eyes has been attributed to optic neuritis, which subsides but is followed by atrophy of the optic nerve. Once absorbed, methyl alcohol is only very slowly eliminated. Coma resulting from massive exposures may last as long as 2–4 days. In the body, the products formed by its oxidation are formaldehyde and formic acid, both of which are toxic. Because of the slowness with which it is eliminated, methyl alcohol should be regarded as a cumulative poison. Though single exposures to fumes may cause no harmful effect, daily exposure may result in the accumulation of sufficient methyl alcohol in the body to cause illness. [*100*]

Severe exposures may cause dizziness, unconsciousness, sighing respiration, cardiac depression, and eventually death. Where the exposure is less severe, the first symptoms may be blurring of vision, photophobia and conjunctivitis, followed by the development of definite eye lesions. There may be headache, gastrointestinal disturbances, dizziness and a feeling of intoxication. The visual symptoms may clear temporarily, only to recur later and progress to actual blindness. Irr of the mu mem of the throat and respiratory tract, peripheral neuritis, and occasionally, symptoms referable to other lesions of the nervous system have been reported. The skin may become dry and cracked due to the solvent action of methyl alcohol.

Methyl alcohol is a common air contaminant. It is used as a food additive permitted in foods for human consumption. [*109*]

Fire Hazard: Dangerous, when exposed to heat, flame or oxidizers.

Spont Heating: No.

Explosion Hazard: Mod, when exposed to flame. Violent reaction with CrO_3, (I + ethanol + HgO), $Pb(ClO_4)_2$, $HClO_4$, P_2O_3, (KOH + $CHCl_3$), (NaOH + $CHCl_3$). [*19*]

Disaster Hazard: Dangerous, upon exposure to heat or flame; can react vigorously with oxidizing materials.

To Fight Fire: Alcohol foam.

METHYL ALDEHYDE. See formaldehyde.

METHYL ALLENE. See butadiene-1,3.

METHYL ALLYL CHLORIDE. See methallyl chloride.

METHYL ALUMINUM SESQUIBROMIDE. Cloudy yellow liquid @ 25°. $(CH_3)_2Al_2Br_3$, mw: 204, fp: −40°, bp: 166°, d: 1.514.

THR = Toxic. See bromides, organometals.

Fire Hazard: Dangerous; ignites spont in air.

Disaster Hazard: Dangerous; spont flam. See also bromides.

To Fight Fire: Do not use water, foam or halogenated extinguishing agents.

METHYL ALUMINUM SESQUICHLORIDE. Clear, colorless liquid @ 25°. $(CH_3)_3Al_2Cl_3$, mw: 203.5, fp: 22.80°, bp: 143.7°, d: 1.1629 @ 25°.

THR = Very unstable. Probably toxic. See also organometals.
Fire Hazard: Dangerous; ignites spont in air.
Disaster Hazard: Dangerous; ignites spont. See also chlorides.
To Fight Fire: Do not use water, foam or halogenated extinguishing agents.

METHYLAMINE. See monomethylamine.

N-METHYLAMINO ACETIC ACID. See sarcosine.

METHYL-2-AMINOBENZOATE. See methyl anthranilate.

2-METHYLAMINOETHANOL. Viscous liquid, fishy odor, miscible with water, alcohol and ether. $CH_3NHCH_2CH_2OH$, mw: 75, d: 0.9, vap. d: 2.9, bp: 156°, flash p: 165°F (OC).
Acute tox data: Oral LD_{50} (rat) = 2340 mg/kg; ipLD_{50} (rat) = 1330 mg/kg; scLD_{50} (mouse) = 1802 mg/kg. [*3*]
THR = MOD via ip, sc and oral routes.
Fire Hazard: Mod, when exposed to heat, flame or oxidizers.
To Fight Fire: Alcohol foam.

***o*-METHYL-2-AMINO-1-NAPHTHOL HYDROCHLORIDE.** $C_{11}H_{11}ON \cdot HCl$, mw: 209.7.
THR = An exper neo. [*3*]

METHYL-*p*-AMINOPHENOL. Colorless needles. $CH_3NHC_6H_4OH$, mw: 123.17, mp: 87°.
Acute tox data: LD_{50} (mouse) = 40 mg/kg. [*3*]
THR = HIGH irr to skin, eyes and mu mem.
Disaster Hazard: Mod; when heated, emits toxic fumes.

METHYL-*p*-AMINOPHENOL SULFATE. Colorless needles. $(CH_3NHC_6H_4OH)_2 \cdot H_2SO_4$, mw: 344.38, mp: 127° (decomp).
Acute tox data: Oral LD_{LO} (rat) = 200 mg/kg; ipLD_{LO} (rat) = 50 mg/kg. [*3*]
THR = HIGH via oral and ip routes.
Disaster Hazard: Dangerous; when heated to decomp, emits highly toxic fumes of SO_x.

4-(*p*-METHYLAMINO PHENYL AZO)-N-METHYL ACETANILIDE. $C_{16}H_{18}N_4O$, mw: 282.4.
THR = An exper neo. [*3*]

METHYL AMYLACETATE. Syns: *4-methyl pentyl-2-acetate, sec-hexyl acetate.* Clear liquid, pleasant odor. $(CH_3)_2CH(CH_2)_3OOCCH_3$, mw: 144.21, bp: 146.3°, fp: −63.8°, flash p: 113°F (COC), d: 0.8598 @ 20°/20°, vap. press: 3.8 mm @ 20°, vap. d: 4.97.
Acute tox data: Oral LD_{50} (rat) = 6160 mg/kg; inhal LC_{LO} (rat) = 2000 ppm for 4 hrs; dermal LD_{50} (rabbit) = 20,000 mg/kg. [*3*]
THR = MOD via oral, inhal and VERY LOW via dermal routes.
Fire Hazard: Mod, when exposed to heat or flame; can react with oxidizing materials.
To Fight Fire: Alcohol foam, CO_2, dry chemical.

METHYL AMYL ALCOHOL. Syns: *methyl isobutyl carbinol, 4-methyl pentanol-2.* Clear liquid. $(CH_3)_2CHCH_2CHOHCH_3$, mw: 102.2, bp: 131.8°, fp: <−90° (sets to a glass), flash p: 106°F, d: 0.8079 @ 20°/20°, vap. press: 2.8 mm @ 20°, vap. d: 3.53, lel = 1.0%, uel = 5.5%.
Acute tox data: Oral LD_{50} (rat) = 2600 mg/kg; inhal LC_{LO} (rat) = 2000 ppm for 4 hrs; ipLD_{50} (mouse) = 812 mg/kg; dermal LD_{50} (rabbit) = 3560 mg/kg. [*3*]
THR = MOD via oral, inhal, ip and dermal routes. A strong irr. High conc cause anesthesia.
Fire Hazard: Mod, when exposed to heat or flame; can react with oxidizing materials.
To Fight Fire: Alcohol foam.

METHYL-*n*-AMYL KETONE. See *n*-amyl methyl ketone.

N-METHYLANILINE. See *m*-toluidine.

***o*-METHYLANILINE.** See *o*-toluidine.

***p*-METHYLANILINE.** See *p*-toluidine.

9-METHYL ANTHRACENE. $C_{15}H_{12}$, mw: 192.3.
THR = An exper neo. [*3*] to mice via ip route. [*103*]

METHYL ANTHRANILATE. Syns: *methyl-2-amino benzoate; neroli oil, artificial.* Colorless to pale yellow liquid with light bluish fluorescence, grape type odor, sol in 5 volumes or more of 60% alcohol, sol in fixed oils, volatile oils and propylene glycol, slightly sol in water and mineral oil, insol in glycerin. $H_2NC_6H_4CO_2CH_3$, mw: 151.
Acute tox data: Oral LD_{50} (rat) = 2910 mg/kg; oral LD_{50} (mouse) = 3900 mg/kg; oral LD_{50} (guinea pig) = 2780 mg/kg. [*3*]
THR = MOD via oral route. Used as a synthetic flavoring substance and adjuvant. [*109*]

METHYL ARSENIC DICHLORIDE. See dichloromethyl arsine.

METHYL ARSINE. Colorless liquid, insol in water, sol in ether and alcohol. CH_3AsH_2, mw: 92.0, mp: 2°.
THR = HIGH. See arsine.
Fire Hazard: Ignites spont in air.
Disaster Hazard: Dangerous. Extremely toxic. More hazardous than its oxidation products. Reacts vigorously with oxidizing agents. In case of fire, exclude O_2 or allow fire to burn out.

METHYL ARSONIC ACID. See methane arsonic acid.

METHYL AZIDE. H_3CN_3, mw: 57.
THR = Detonates upon rapid heating. Reacts vio-

lently with (dimethyl malonate + sodium methylate), Hg. [*19*]

METHYL-2-AZIDO BENZOATE. $C_8H_7O_2N_3$, mw: 177.1.
THR = Has exploded during distillation. [*19*]

2-METHYL AZIRIDINE. See propylene imine.

METHYL AZIRIDINYL PHOSPHINE OXIDE.
THR = An exper teratogen. [*3*]

METHYL AZOXY BUTANE. $C_5H_{12}ON_2$, mw: 116.13.
THR = An exper carc. [*3, 23*]

METHYL AZOXY METHANOL. $C_2H_6O_2N_2$, mw: 90.1.
THR = An exper carc and teratogen; [*3*] an exper mutagen. [*23*]

METHYL AZOXY METHANOL ACETATE. $C_4H_8O_3N_2$, mw: 132.1.
THR = An exper (+) carc, mutagen and teratogen. [*3, 9, 23*]

METHYL AZULENO (5,6,7-cd)PHENALENE. $C_{21}H_{14}$, mw: 266.4.
THR = An exper carc. [*3*]

6-METHYL-11H-BENZ(bc)ACEANTHRYLENE. $C_{20}H_{14}$, mw: 254.3.
THR = An exper neo. [*3*]

3-METHYL BENZ(j)ACEANTHRYLENE. $C_{21}H_{14}$, mw: 266.4.
THR = An exper carc. [*3*]

7-METHYL BENZ(c)ACRIDINE. $C_{18}H_{13}N$, mw: 243.3.
THR = An exper neo. [*3, 23*]

12-METHYL BENZ(c)ACRIDINE.
THR = An exper neo. [*3*]

1-METHYL BENZ(a)ANTHRACENE. $C_{19}H_{14}$, mw: 242.3.
THR = HIGH acute toxic. An exper neo. [*3, 23*]

2-METHYL BENZ(a)ANTHRACENE.
THR = An exper neo. [*3, 23*]

4-METHYL BENZ(a)ANTHRACENE.
THR = An exper neo. [*3, 23*]

5-METHYL BENZ(a)ANTHRACENE.
THR = HIGH acute toxicity. An exper neo and carc. [*3, 23*]

6-METHYL BENZ(a)ANTHRACENE.
THR = HIGH acute toxicity. An exper carc and neo. [*3*]

7-METHYL BENZ(a)ANTHRACENE.
THR = HIGH acute toxicity. An exper carc and neo. [*3, 23*]

8-METHYL BENZ(a)ANTHRACENE.
THR = HIGH acute toxicity. An exper carc and neo. [*3, 23*]

9-METHYL BENZ(a)ANTHRACENE.
THR = An exper neo and carc. [*3, 23*]

10-METHYL BENZ(a)ANTHRACENE.
THR = An exper neo. [*3, 23*]

11-METHYL BENZ(a)ANTHRACENE.
THR = An exper neo. [*3, 23*]

12-METHYL BENZ(a)ANTHRACENE.
THR = An exper neo and carc. [*3, 23*]

12-METHYL BENZ(a)ANTHRACENE-12-ALDEHYDE.
THR = An exper carc. [*23*]

7-METHYL BENZ(a)ANTHRACENE-8-CARBONITRILE. $C_{20}H_{13}N$, mw: 267.3.
THR = An exper neo. [*3*]

7-METHYL BENZ(a)ANTHRACENE-10-CARBONITRILE. $C_{20}H_{13}N$, mw: 267.3.
THR = An exper neo. [*3*]

12-METHYL BENZ(a)ANTHRACENE-7-CARBOXALDEHYDE. $C_{20}H_{14}O$, mw: 270.3.
THR = An exper carc. [*3*]

12-METHYL BENZ(a)ANTHRACENE-7-METHANOL. $C_{20}H_{16}O$, mw: 272.4.
THR = An exper carc and neo. [*3, 23*]

7-METHYL BENZ(a)ANTHRACENE-12-METHANOL.
THR = An exper carc. [*3, 23*]

12-METHYL BENZ(a)ANTHRACENE-7-METHANOL ACETATE. $C_{22}H_{18}O_2$, mw: 314.4.
THR = An exper neo. [*3, 23*]

12-METHYL BENZ(a)ANTHRACENE-7-METHANOL BENZOATE. $C_{26}H_{20}O_2$, mw: 364.5.
THR = An exper neo. [*3*]

7-METHYL BENZ(a)ANTHRACENE-8-YL CARBAMIDE. $C_{20}H_{15}ON$, mw: 285.4.
THR = An exper neo. [*3*]

METHYL BENZENE. See toluene.

METHYL BENZENE SULFONIC ACID. See *o*-toluene sulfonic acid.

METHYL BENZOATE. Syn: *niobe oil.* Colorless, liquid, fragrant odor. $C_6H_5COOCH_3$, mw: 136.14, mp: −12.5°, bp: 199.6°, flash p: 181°F, d: 1.0937, vap. press: 1 mm @ 39.0°, vap. d: 4.69.
Acute tox data: Oral LD_{50} (rat) = 1350 mg/kg; oral LD_{50} (mouse) = 3330 mg/kg; oral LD_{50} (guinea pig) = 4100 mg/kg. [*3*]
THR = MOD via oral route.
Fire Hazard: Mod, when exposed to heat or flame; can react with oxidizing materials.
To Fight Fire: Foam, CO_2, dry chemical, water to blanket fire.

7-METHYL BENZOATE-12-METHYL BENZ(a)ANTHRACENE.
THR = An exper carc. [*23*]

10-METHYL-7H-BENZOCARBAZOLE. $C_{17}H_{13}N$, mw: 231.3.
THR = An exper neo. [*3*]

9-METHYL-1,2-BENZOCARBAZOLE.
THR = An exper carc. [*3*]

6-METHYL-4,5-BENZO-2,3,1′,2′-INDENOINDOLE. $C_{20}H_{13}N$, mw: 267.3.
THR = An exper neo. [*3*]

5-METHYL BENZO(rat)PENTAPHENE. $C_{25}H_{16}$, mw: 316.4.
THR = An exper neo. [*3*]

5-METHYL BENZO(rst)PENTAPHENE.
THR = An exper carc. [*23*]

8-METHYL BENZO(rat)PENTAPHENE-5-CARBOXALDEHYDE. $C_{26}H_{16}O$, mw: 344.4.
THR = An exper neo [*3*] via sc to mice.

METHYL BENZO(ghi)PERYLENE. $C_{23}H_{14}$, mw: 290.4.
THR = An exper neo. [*3*]

7-METHYL BENZO(a)PHENALENO(1,9-hi)ACRIDINE. $C_{28}H_{17}N$, mw: 367.5.
THR = An exper neo. [*3*]

7-METHYL BENZO(h)PHENALENO(1,9-bc)ACRIDINE.
THR = An exper neo. [*3*]

2-METHYL BENZO(c)PHENANTHRENE. $C_{19}H_{14}$, mw: 242.3.
THR = An exper neo. [*3, 23*]

3-METHYL BENZO(c)PHENANTHRENE.
THR = An exper carc. [*3, 23*]

4-METHYL BENZO(c)PHENANTHRENE.
THR = An exper neo. [*3, 23*]

5-METHYL BENZO(c)PHENANTHRENE.
THR = An exper neo and carc. [*3, 23*]

6-METHYL BENZO(c)PHENANTHRENE.
THR = An exper neo and carc. [*3, 23*]

7-METHYL BENZO(pqr)PICENE.
THR = An exper carc. [*23*]

2-METHYL BENZO(a)PYRENE. $C_{21}H_{14}$, mw: 266.4.
THR = An exper carc. [*3, 23*]

3-METHYL BENZO(a)PYRENE.
THR = An exper carc. [*3, 23*]

4-METHYL BENZO(a)PYRENE.
THR = An exper carc. [*3, 23*]

5-METHYL BENZO(a)PYRENE.
THR = An exper carc. [*23*]

6-METHYL BENZO(a)PYRENE.
THR = An exper neo. [*3, 23*]

7-METHYL BENZO(a)PYRENE.
THR = An exper neo. [*3, 23*]

11-METHYL BENZO(a)PYRENE.
THR = An exper carc. [*3, 23*]

12-METHYL BENZO(a)PYRENE.
THR = An exper carc. [*3, 23*]

METHYLBENZOYL ECGONINE. See cocaine.

α-METHYLBENZYL ALCOHOL. See phenyl methyl carbinol.

α-METHYLBENZYLAMINE. Liquid.
$C_6H_8CH(CH_3)NH_2$, mw: 121.18, mp: −65°, bp: 188.5°, flash p: 175°F (OC), d: 0.9535, vap. press: 0.5 mm @ 20°, vap. d: 4.18.
Acute tox data: Oral LD_{50} (rat) = 940 mg/kg; dermal LD_{50} (rat) = 780 mg/kg. [*3*] Mixed with $NaNO_2$ (1:1) or (2:2) is an exper neo and carc. [*3*]
THR = MOD via oral and dermal routes.
Fire Hazard: Mod, when exposed to heat or flame; can react with oxidizing materials.
To Fight Fire: Alcohol foam, CO_2, dry chemical.

α-METHYLBENZYL BENZOIC ACID. Syn: *benzoic acid-α-methyl benzyl ester*. Crystals.
$CH_3C_6H_5CHC_6H_4COOH$, mw: 226.3.
THR = Details U. An insecticide. See also esters and benzoic acid.
Fire Hazard: Slight.

***p*-METHYLBENZYLBROMIDE.** See xylyl bromide.

***o*-METHYLBENZYL "CELLOSOLVE."** Liquid.
$CH_3C_6H_4CH_2OCH_2CH_2OH$, mw: 166.21, bp: 253.5°, fp: −50°, d: 1.0395 @ 20°/20°, vap. press: 0.02 mm @ 20°, vap. d: 5.73.
THR = Details U. See also glycols.
Fire Hazard: Slight, when exposed to heat or flame; can react with oxidizing materials.
To Fight Fire: Foam, CO_2, dry chemical.

***o*-METHYLBENZYL CHLORIDE.** See *o*-xylyl chloride.

α-METHYLBENZYL DIETHANOLAMINE. Liquid.
$C_6H_5CH(CH_3)N(C_2H_4OH)_2$, mw: 209.28, mp: −7°, bp: 244° @ 50 mm, flash p: 370°F (OC), d: 1.0812 @ 20°/20°, vap. press: <0.01 mm @ 20°, vap. d: 7.2.
THR = Details U. See also amines.
Fire Hazard: Slight, when exposed to heat or flame; can react with oxidizing materials.
To Fight Fire: Foam, CO_2, dry chemical.

α-METHYLBENZYL DIMETHYLAMINE. Liquid.
$C_6H_5CH(CH_3)N(CH_3)_2$, mw: 149.23, mp: −75°, bp: 195°, flash p: 175°F (OC), d: 0.9044 @ 20°/20°, vap. press: 0.6 mm @ 20°, vap. d: 5.15.

THR = Details U. See also amines.
Fire Hazard: Mod, when exposed to heat or flame; can react with oxidizing materials.
To Fight Fire: Alcohol foam, CO_2, dry chemical.

METHYLBENZYL ETHER. See α-methyl benzyl ether.

α-METHYLBENZYL ETHER. Syn: *methyl benzyl ether.* Liquid. $C_6H_5CH(CH_3)OCH(CH_3)C_6H_5$, mw: 226.3, mp: −30°, bp: 286.3°, flash p: 275°F (OC), d: 1.0017 @ 20°/20°, vap. press: <0.01 mm @ 20°, vap. d: 7.82.
Acute tox data: Oral LD_{50} (rat) = 9300 mg/kg. [*3*]
THR = LOW via oral route.
Fire Hazard: Slight, when exposed to heat or flame; can react with oxidizing materials. See ethers.
To Fight Fire: Alcohol foam, CO_2, dry chemical.

1-METHYL-2-BENZYL HYDRAZINE. $C_8H_{12}N_2$, mw: 136.2.
Acute tox data: Oral LD_{50} (mouse) = 254 mg/kg; ipLD_{50} (mouse) = 89 mg/kg; ivLD_{50} (mouse) = 85 mg/kg; oral TD_{LO} = 0.14 mg/kg ⟶ mental effects. [*3*]
THR = HIGH oral, ip and iv routes. An exper carc. [*23*]

α-METHYLBENZYL MONOETHANOLAMINE. Liquid. $C_6H_5CH(CH_3)NHC_2H_4OH$, mw: 165.23, mp: −15°, bp: 182° @ 50 mm, flash p: 280°F (OC), d: 1.0327 @ 20°/20°, vap. press: <0.01 mm @ 20°, vap. d: 5.7.
THR = Details U. See also amines.
Fire Hazard: Slight, when exposed to heat or flame; can react with oxidizing materials.
To Fight Fire: CO_2, dry chemical.

METHYLBENZYL NITROSAMINE. $C_8H_{10}ON_2$, mw: 150.2.
THR = An exper carc. [*3, 23*]

2′-METHYL-4′,4‴-BIACETANILIDE. $C_{17}H_{18}O_2N_2$, mw: 282.4.
THR = An exper carc. [*3, 23*]

2-METHYL-4-BIPHENYL AMINE. $C_{13}H_{13}N$, mw: 183.3.
THR = An exper carc. [*3*]

3-METHYL-4-BIPHENYL AMINE.
THR = An exper carc. [*3, 23*]

METHYL-BIS(DIETHYLBORYL)AMINE. $C_9H_{23}NB_2$, mw: 156.1.
THR = Self-ignites in air. [*19*] No toxicity data.

METHYL BISMUTHINE. Liquid. $CH_3 \cdot BiH_2$, mw: 226.05, bp: 110°, d: 2.30 @ 18°, vap. d: 7.8.
THR = See bismuth compounds and metal organics.
Fire Hazard: Mod via heat, flame or oxidizers.
Disaster Hazard: Dangerous; when heated to decomp, or on contact with acid or acid fumes, emits toxic fumes of bismuth; can react with oxidizing materials.
To Fight Fire: Dry chemical, foam, CO_2.

METHYL BORATE. Syns: *trimethyl borate, trimethoxyborine.* Colorless liquid, decomp in water. $B(OCH_3)_3$, mw: 103.92, mp: −29°, bp: 68°, flash p: <80°F, d: 0.92 @ 20°, vap. d: 3.59.
Acute tox data: Oral LD_{50} (mouse) = 1290 mg/kg; dermal LD_{50} (rabbit) = 1980 mg/kg. [*3*]
THR = MOD via oral and dermal routes. See also esters and boron compounds.
Fire Hazard: Dangerous, when exposed to heat, flame or oxidizers.
Explosion Hazard: Mod, when exposed to flame.
Disaster Hazard: Dangerous; will react with water or steam to produce toxic and flam vapors; can react vigorously with oxidizing materials.
To Fight Fire: Dry chemical, CO_2, spray, foam.

METHYLBORIC ACID. White crystals, slightly water-sol. $CH_3B(OH)_2$, mw: 59.9, mp: decomp before melting.
THR = Details U. See also boron compounds.

METHYLBORIC ANHYDRIDE. Colorless liquid, hydrolyzed by water, sol in ether. $(CH_3)_3B_3O_3$, mw: 125.6.
THR = Details U. See boron compounds.

METHYLBORINE TRIMETHYL AMMINE. Colorless liquid, decomp by water. $(CH_3)_3NBH_2CH_3$, mw: 87.0, mp: 0.8°, bp: 177°.
THR = Details U. See also boron compounds.

METHYL BROMIDE. Syn: *bromomethane.* Colorless, transparent, volatile liquid or gas, burning taste, chloroform-like odor. CH_3Br, mw: 94.95, bp: 3.56°, lel = 10%, uel = 16%, fp: −93°, flash p: none, d: 1.732 @ 0°/0°, autoign. temp.: 998°F, vap. d: 3.27, vap. press: 1824 mm @ 25°.
Acute tox data: Inhal LC_{LO} (rat) = 3120 ppm for $\frac{1}{4}$ hr; inhal LC_{LO} (rabbit) = 6425 ppm for 1 hr; inhal LC_{LO} (guinea pig) = 3000 ppm for 9 hrs; dermal TD_{LO} (human) = 8000 ppm → skin irr. [*3*] Acute oral LD_{50} (rat) = <100 mg/kg.
THR = A powerful fumigant gas which is one of the most toxic of the common organic halides. It is nemotoxic and narcotic with delayed action; it is cumulative and damaging to nervous system, kidneys, lung. CNS effects include blurred vision, mental confusion, numbness, tremors, speech defects. Methyl bromide is reported to be 8 times more toxic on inhal than ethyl bromide. Moreover, because of its greater volatility, methyl bromide is a much more frequent cause of poisoning. Death fol-

lowing acute poisoning is usually caused by its irr effect on the lungs. In chronic poisoning, death is due to injury to the CNS. Fatal poisoning has always resulted from exposures to relatively high conc of methyl bromide vapors (from 8,600 to 60,000 ppm). Nonfatal poisoning has resulted from exposure to conc as low as 100–500 ppm. In addition to the lung and CNS injury mentioned, the kidneys may be damaged with development of albuminuria and, in fatal cases, cloudy swelling and/or tubular degeneration. The liver may be enlarged. There are no characteristic blood changes.

The onset of symptoms following the inhal of methyl bromide vapor is usually delayed for 4–6 hrs though the latent period may vary from 2–48 hrs. In fatal poisoning, the early symptoms are headache, visual disturbances, nausea and vomiting, smarting of the eyes, irr of the skin, listlessness, vertigo and tremor. Progress is nearly always rapid, with the development of convulsions, fever, pulmonary, edema, cyanosis, unconsciousness and death. Signs of involvement of the nervous system may be present before death. The clinical picture in non-fatal poisoning is extremely variable. Fatigue, blurred or double vision, nausea and vomiting are frequent; incoordination, tremors, convulsions, exaggeration of the patellar reflexes and a positive Babinski's sign may develop. Nearly every type of nervous disturbance has been reported. The pulmonary symptoms are comparatively slight. Recovery is frequently prolonged and there may be permanent injury, commonly characterized by sensory disturbances, weakness, disturbances, of gait, irritability, and blurred vision. Locally, methyl bromide is an extreme irr to the skin and may produce severe burns (Section 9).

Fire Hazard: Mod, when exposed to heat or flame.

Spont Heating: No.

Explosion Hazard: Mod, when exposed to sparks or flame. Violent reaction with Al, dimethyl sulfoxide. [*19*]

Disaster Hazard: Dangerous; when heated to decomp, emits highly toxic fumes of bromides.

To Fight Fire: Water, foam, CO_2, dry chemical.

METHYL BROMIDE AND CHLOROPICRIN MIXTURE, LIQUID. See components.

METHYL BROMIDE AND ETHYLENE DIBROMIDE MIXTURE, LIQUID. See components.

METHYL BROMIDE AND NONFLAMMABLE NONLIQUEFIED COMPRESSED GAS MIXTURES, LIQUID. See components.

1-METHYL-7-BROMOMETHYL BENZ(a)ANTHRACENE.

THR = An exper carc. [*23*]

2-METHYL BUTADIENE-1,3. See isoprene.

3-METHYLBUTANAL. See isovaleric aldehyde.

2-METHYLBUTANE. See isopentane.

2-METHYL-1-BUTANOL. Syns: *amyl alcohol, primary, active*; *sec-butyl carbinol.* Colorless liquid, slightly sol in water, miscible with alcohol and ether. $CH_3CH_2CH(CH_3)CH_2OH$, mw: 88: d: 0.81–0.82 @ 20°, fp: <−70°, bp: 128°, flash p: 122°F (OC), vap. d: 3.0.

Acute tox data: Oral LD_{50} (rat) = 4920 mg/kg; dermal LD_{LO} (rabbit) = 3540 mg/kg. [*3*]

THR = MOD irr to skin, eyes and mu mem; MOD via oral and dermal routes. Irr to eyes and respiratory passages. Can cause giddiness, headache, cough, nausea and even vomiting. Has caused deafness, and delerium. Said also to cause methemoglobinuria, glycosuria.

Fire Hazard: Mod, when exposed to heat, flame or oxidizers.

To Fight Fire: Alcohol foam, spray, mist, dry chemical.

2-METHYL-2-BUTANOL. See *tert*-amyl alcohol, refined.

3-METHYL-1-BUTANOL. See isoamyl alcohol.

METHYL BUTANONE. See butanone.

2-METHYL BUTENE-1 (TECHNICAL GRADE). Colorless, extremely volatile liquid or gas, insol in water. $CH_2C(CH_3)CH_2CH_3$, mw: 70, d: 0.7, vap. d: 2.4, bp: 38°, flash p: <20°F.

THR = No details. A simple asphyxiant.

Fire Hazard: Very dangerous, when exposed to heat, flame or oxidizers.

To Fight Fire: Dry chemical, CO_2, foam.

3-METHYL BUTENE-1. Syns: *isopropyl ethylene, isopentene.* Colorless, very volatile flam liquid, disagreeable odor, insol in water, sol in alcohol. $(CH_3)_2CHCH{:}CH_2$, mw: 70, bp: 31.11°, d: 0.65 @ 20°/20°, fp: −137.5°, flash p: <20°F, vap. d: 2.4, lel = 1.6%, uel = 9.1%.

THR = Not known. See also 2-methyl butene-1.

Fire Hazard: Very dangerous, when exposed to heat, flame or oxidizers.

To Fight Fire: Alcohol foam, mist, spray, dry chemical, CO_2.

2-METHYL BUTENE-2. See α,*n*-amylene.

3-METHYL BUTENONITRILE.

THR = Details unknown. Limited animal exper suggest HIGH toxicity. See also nitriles.

***n*-METHYL BUTYLAMINE.** Liquid, sol in water. $CH_3(CH_2)_3NHCH_3$, mw: 87, d: 0.7335, bp: 91.1°, vap. d: 3.0, flash p: 55°F (OC).

Acute tox data: Oral LD_{50} (rat) = 420 mg/kg; inhal LC_{LO} (rat) = 2000 ppm for 4 hrs; dermal LD_{50} (rabbit) = 1260 mg/kg. [*3*]
THR = MOD via oral, inhal and dermal routes.
Fire Hazard: Dangerous, when exposed to heat, flame or oxidizers.
To Fight Fire: Alcohol foam.

1-METHYL-2-*n*-BUTYL HYDRAZINE.
THR = An exper carc. [*23*]

METHYL-*n*-BUTYL KETONE. Syns: *2-hexanone, n-butyl methyl ketone.* Clear liquid. $CH_3OC(CH_2)_3CH_3$, mw: 100.16, mp: −56.9°, bp: 127.2°, lel = 1.22%, uel = 8.0%, flash p: 95°F (OC), d: 0.830 @ 0°/4°, vap. press: 10 mm @ 38.8°, vap. d: 3.45, autoign. temp.: 991°F.
Acute tox data: Oral LD_{50} (rat) = 2590 mg/kg; inhal LC_{LO} (guinea pig) = 6000 ppm for 7 hrs; ipLD_{LO} (rat) = 914 mg/kg. [*3*]
THR = MOD via oral, inhal and ip routes.
Fire Hazard: Dangerous, when exposed to heat or flame; can react with oxidizing materials.
Spont Heating: No.
Explosion Hazard: Mod, when exposed to heat or flame.
Disaster Hazard: Dangerous fire and explosion hazard.
To Fight Fire: Alcohol foam, CO_2, dry chemical.

METHYL-N-BUTYL NITROSAMINE. $C_5H_{12}ON_2$, mw:116.2.
Acute tox data: Oral LD_{50} (rat) = 420 mg/kg; dermal LD_{50} (rabbit) = 1260 mg/kg. [*3*]
THR = HIGH via oral and MOD via dermal routes. An exper carc. [*23, 3*]

3-METHYL BUTYNOL. Syn: *2-methyl-3-butyn-2-ol.* Colorless liquid. $(CH_3)_2(OH)C\text{-}C{\equiv}CH$, mw: 84.1, mp: 2.6°, bp: 104°–105°, vap. d: 2.49, d: 0.9, flash p: 77°F (OC), d: 0.8672 @ 20°/20°.
Acute tox data: ipLD_{50} (mouse) = 3600 mg/kg. [*2*]
THR = MOD via ip route.
Fire Hazard: Dangerous; can react with oxidizing materials, heat, flames.
To Fight Fire: Alcohol foam, mist, spray, CO_2.

METHYL BUTYRATE. Colorless liquid.
$CH_3COOC_3H_7$, mw: 102.13, mp: <−97°, bp: 102.3°, flash p: 57°F (CC), d: 0.898, vap. press: 40 mm @ 29.6°, vap. d: 3.53.
Acute tox data: Oral LD_{50} (rabbit) = 3370 mg/kg. [*3*]
THR = MOD via oral route.
Fire Hazard: Dangerous, when exposed to heat, flame or oxidizers.
Spont Heating: No.
Explosion Hazard: U.
Disaster Hazard: Dangerous, upon exposure to heat or flame; can react vigorously with oxidizing materials.
To Fight Fire: Alcohol foam, CO_2, dry chemical.

METHYL-ϵ-CAPROLACTONE. $C_7H_{12}O_2$, mw: 128.2.
Acute tox data: Oral LD_{50} (rat) = 11,000 mg/kg; dermal LD_{50} (rabbit) = 7100 mg/kg. [*3*]
THR = LOW via oral and dermal routes. See also lactones.

METHYL CARBAMATE. See methyl urethan.

METHYL CARBINOL. See ethyl alcohol.

METHYL "CARBITOL." Syns: *diethylene glycol methyl ether, 2-(2-methoxy ethoxy)ethanol.* Hygroscopic, water-white liquid.
$CH_3OCH_2CH_2OCH_2CH_2OH$, mw: 120.15, bp: 194.2°, flash p: 200°F (OC), d: 1.0354 @ 20°/4°, vap. press: 0.2 mm @ 20°, vap. d: 4.14, mp: <−84°.
Acute tox data: Oral LD_{50} (rat) = 9210 mg/kg; dermal LD_{50} (rabbit) = 650 mg/kg; oral LD_{50} (guinea pig) = 4160 mg/kg. [*3*]
THR = MOD via dermal; MOD–LOW via oral routes.
Fire Hazard: Mod, when exposed to heat or flame; can react with oxidizing materials. Reacts violently with $Ca(OCl)_2$, chlorosulfonic acid, oleum. [*19*]
To Fight Fire: Dry chemical, alcohol foam, water spray or mist, CO_2.

METHYL "CARBITOL" ACETATE. Syn: *diethylene glycol monomethyl ether acetate.* Colorless liquid. $CH_2COOC_2H_4OC_2H_4C_2H_4OCH_3$, mw: 162.19, bp: 209.1°, flash p: 180°F (OC), d: 1.0396 @ 20°/20°, vap. press: 0.12 mm @ 20°.
Acute tox data: Oral LD_{50} (guinea pig) = 3460 mg/kg. [*3*]
THR = MOD via oral route.
Fire Hazard: Mod, when exposed to heat or flame; can react with oxidizing materials.
Spont Heating: No.
To Fight Fire: Foam, CO_2, dry chemical, mist, spray.

METHYL "CARBITOL" FORMAL. Liquid.
$CH_2(CH_3OCH_2CH_2OCH_2CH_2O)_2$, mw: 252.3, mp: −37.4°, bp: 305°, flash p: 310°F, d: 1.040, vap. d: 8.7.
THR = Details U. See also glycols.
Fire Hazard: Slight, when exposed to heat or flame; can react with oxidizing materials.
To Fight Fire: Foam, CO_2, dry chemical.

METHYL CARBONATE. See dimethyl carbonate.

METHYL CATECHOL. See guaiacol.

METHYL "CELLOSOLVE." See ethylene glycol monomethyl ether.

METHYL "CELLOSOLVE" ACETAL. See ethylene glycol monomethyl ether acetal.

METHYL "CELLOSOLVE" ACETATE. Syn: *ethylene glycol monomethyl ether acetate.* Colorless liquid. $CH_3CO_2CH_2CH_2OCH_3$, mw: 118.2, bp: 143°, fp: −70°, flash p: 111°F (CC), d: 1.005 @ 20°/20°, vap. d: 4.07, lel = 1.7%, uel = 8.2%.

Acute tox data: Oral LD_{50} (rat) = 3390 mg/kg; dermal LD_{50} (rabbit) = 5250 mg/kg; inhal LC_{LO} (rat) = 7000 ppm for 4 hrs. [*3*]

THR = MOD via oral, dermal and inhal routes.

Fire Hazard: Mod, when exposed to heat or flame; can react with oxidizing materials.

Spont Heating: No.

To Fight Fire: CO_2, dry chemical.

METHYL "CELLOSOLVE" ACETYLRICINOLEATE. See methoxyethyl acetylricinoleate.

METHYL CELLULOSE. Syn: *cellulose methyl ether.* Grayish-white, fibrous powder, insol in alcohol, ether and chloroform, sol in glacial acetic acid. Molecular weights vary from 40,000–180,000. Here we consider U.S.P. methyl cellulose, except that the methoxy content shall not be less than 27.5% and not greater than 31.5% on a dry weight basis.

THR = U. Used as a general purpose food additive. [*109*]

METHYL CELLULOSE, PROPYLENE GLYCOL ETHER. See hydroxypropyl methylcellulose.

β-METHYL CHALCONE. See dypnone.

METHYL CHLORIDE. Syn: *chloromethane.* Colorless gas. CH_3Cl, mw: 50.49, bp: −23.7°, lel = 10.7%, uel = 17.2%, fp: −97.7°, flash p: <32°F (OC), d: 0.918 @ 20°/4°, autoign. temp.: 1170°F, vap. d: 1.78.

Acute tox data: Inhal LC_{LO} (mouse) = 3146 ppm for 7 hrs; inhal LC_{LO} (guinea pig) = 20,000 ppm for 2 hrs. [*3*]

THR = Methyl chloride has very slight irr properties and may be inhaled without noticeable discomfort. It has some narcotic action, but this effect is weaker than that of chloroform. Acute poisoning, characterized by the narcotic effect, is rare in industry. Repeated exposure to low conc causes damage to the CNS, and, less frequently, to the liver, kidneys, bone marrow and cardiovascular system. Hemorrhages into the lungs, intestinal tract and dura have been reported. Sprayed on the skin, methyl chloride produced anesthesia through freezing of the tissues as it evaporates. In exposures to high conc, dizziness, drowsiness, incoordination, confusion, nausea and vomiting, abdominal pains, hiccoughs, diplopia and dimness of vision are followed by delirium, convulsions and coma. Death may be immediate, but if the exposure is not fatal, recovery is usually slow, and degenerative changes in the CNS are not uncommon. The liver, kidneys and bone marrow may be affected, with resulting acute nephritis and anemia. Death may occur several days after exposure, resulting from degenerative changes in the heart, liver and especially the kidneys. In repeated exposures to lower conc there is usually fatigue, loss of appetite, muscular weakness, drowsiness and dimness of vision. After-effects are commonly the result of damage to the CNS, with visual changes and attacks of depression and other psychic disturbances being reported. Used as a food additive permitted in food for human consumption. [*109*]

Fire Hazard: Very dangerous when exposed to heat, flame or powerful oxidizers.

Spont Heating: No.

Explosion Hazard: Mod, when exposed to heat or flame. Violent reactions with Al, Mg, K, Na, NaK. [*19*]

Disaster Hazard: Dangerous; when heated to decomp, emits highly toxic fumes of chlorides; can react vigorously with oxidizing materials.

To Fight Fire: Stop flow of gas; CO_2, dry chemical or water spray.

METHYL CHLORIDE-METHYLENE CHLORIDE MIXTURE. See components.

METHYL CHLOROACETATE. Syn: *methyl chloroethanoate.* Colorless liquid, sweet pungent odor; slightly sol in water, miscible with alcohol and ether. $C_2H_5O_2Cl$, mw: 108.5, mp: −33°, bp: 131°, d: 1.24 @ 20°/20°, vap. d: 3.8, flash p: 122°F (OC).

THR = A solvent material. Details of toxicity U. However, it must be considered of HIGH toxicity. An irr.

Fire Hazard: Mod, when exposed to heat, flame or oxidizers.

Disaster Hazard: Dangerous; see chlorides.

To Fight Fire: Water spray, mist, foam, CO_2.

METHYL CHLOROBENZENE. See monochlorotoluene.

METHYL CHLOROCARBONATE. See methyl chloroformate.

METHYL CHLOROETHANOATE. See methyl chloroacetate.

METHYL CHLOROFORM. See α-trichloroethane.

METHYL CHLOROFORMATE. Syns: *methyl chloromethanoate, methyl chlorocarbonate.* Colorless liquid. $ClCOOCH_3$, mw: 94.50, bp: 71.4°, d: 1.223 @ 20°/4°, vap. d: 3.26, flash p: 54°F, autoign. temp.: 940°F.

Acute tox data: Inhal LC_{LO} (mouse) = 1000 mg/m^3; ipLD_{50} (mouse) = 40 mg/kg. [*3*]

THR = HIGH irr via ip and to skin, eyes and mu mem. MOD via inhal; HIGH via ip routes.
Disaster Hazard: Dangerous; when heated to decomp, emits highly toxic fumes of methyl chloroformate and phosgene; will react with water or steam to produce toxic and corrosive fumes.
Fire Hazard: Very dangerous when exposed to heat sources, sparks, flame or oxidizers.

METHYL CHLOROMETHANOATE. See methyl chloroformate.

METHYL CHLOROMETHOXYSTEARATE. Straw-yellow liquid. $C_{20}H_{39}O_2Cl$, mw: 354, fp: 8.0°, flash p: 266°F (OC), d: 0.980 @ 20°/40°, vap. d: 12.2.
THR = Details U. See also esters.
Fire Hazard: Slight, when exposed to heat or flame.
Disaster Hazard: Dangerous; when heated to decomp, emits highly toxic fumes of chlorides; can react with oxidizing materials.
To Fight Fire: Foam, CO_2, dry chemical.

METHYL CHLOROMETHYL ETHER, ANHYDROUS. Clear, colorless liquid, decomp in water, sol in alcohol and ether. C_2H_5OCl, mw: 80.5, d: 1.0625 @ 10°/4°, mp: −103.5°, bp: 59.5°.
Acute tox data: Oral LD_{50} (rat) = 817 mg/kg; inhal LC_{50} (rat) = 55 ppm for 7 hrs. [*3*]
THR = MOD via oral and HIGH via inhal routes. A human (S) carc. [*3, 10*] Has been implicated in oat-cell carcinoma of the lung. [*3, 14*] A poison.
Explosion Hazard: Dangerous; see also ethers.

4-METHYL-6-[((2-CHLORO-4-NITRO)PHENYL) AZO]-*m*-ANISIDINE. $C_{14}H_{13}O_3N_4Cl$, mw: 320.8.
THR = An exper neo. [*3*]

METHYL CHLOROPHENOXYACETIC ACID. See methoxone.

METHYL CHLOROSULFONATE. Colorless liquid, pungent odor. CH_3OSO_2Cl, mw: 130.55, mp: −70°, bp: 135° (decomp), d: 1.492 @ 10°, vap. d: 4.51.
THR = HIGH irr to skin, eyes and mu mem. Used as a military poison gas.
Disaster Hazard: Dangerous; when heated to decomp, emits highly toxic fumes of chlorides; will react with water, steam or acids to produce toxic and corrosive fumes.

3-METHYLCHOLANTHRENE. Yellow crystals. $C_{21}H_{16}$, mw: 268.3, bp: 280° @ 80 mm, mp: 180°, d: 1.28 @ 20°.
THR = A powerful irr. Carc via oral, dermal, sc, iv, pa, it, imp, ir and ipl routes. An exper neo and carc. [*3, 23*]
Fire Hazard: Slight; can react with oxidizing materials.

4-METHYL CHOLANTHRENE. $C_{21}H_{16}$, mw: 268.4.
THR = An exper carc. [*3*]

5-METHYL CHOLANTHRENE.
THR = An exper carc. [*3*]

20-METHYL CHOLANTHRENE CHOLEIC ACID. $C_{96}H_{160}O_{16} \cdot C_{21}H_{16}$, mw: 1838.9.
THR = An exper neo. [*3*]

3-METHYL-11,12-CHOLANTHRENE-OXIDE. $C_{21}H_{14}O$, mw: 283.4.
THR = An exper carc to mice via dermal route. [*103*]

3-METHYL CHOLANTHRENE + 1,3,5-TRINITROBENZENE (1:1).
THR = An exper neo. [*3*]

3-METHYL-1-CHOLANTHRENOL. $C_{21}H_{15}O$, mw: 283.4.
THR = An exper carc. [*3*]

3-METHYL CHOLANTHREN-2-OL. $C_{21}H_{16}O$, mw: 284.4.
THR = An exper carc. [*3*]

3-METHYL CHOLANTHREN-1-ONE. $C_{21}H_{14}O$, mw: 282.4.
THR = An exper carc. [*3*]

3-METHYL-2-CHOLANTHRENONE.
THR = An exper carc. [*3*]

1-METHYL CHRYSENE.
THR = An exper carc. [*23*]

2-METHYL CHRYSENE. $C_{19}H_{14}$, mw: 242.3.
THR = An exper neo. [*3, 23*]

3-METHYL CHRYSENE.
THR = An exper neo. [*3, 23*]

4-METHYL CHRYSENE.
THR = An exper neo. [*3, 23*]

5-METHYL CHRYSENE.
THR = An exper neo. [*3, 23*]

6-METHYL CHRYSENE.
THR = An exper neo. [*3, 23*]

METHYL COPPER. $CuCH_3$, mw: 78.6.
THR = Explodes violently in air when dry. [*19*] See also copper compounds and metal organics.

METHYL CYANIDE. Syns: *ethanenitrile*, *acetonitrile*. Colorless liquid, aromatic odor. CH_3CN, mw: 41.05, mp: −45°, bp: 81.1°, flash p: 42°F (COC), d: 0.7868 @ 20°/20°, vap. d: 1.42, vap. press: 100 mm @ 27°, lel = 4.4%, uel = 16%, autoign. temp.: 975°F.
Acute tox data: Oral LD_{50} (rat) = 3800 mg/kg; inhal LC_{LO} (rat) = 8000 ppm for 4 hrs; ipLD_{50} (mouse) = 500 mg/kg; dermal LD_{50} (rabbit) = 1250 mg/kg. [*3*]
THR = MOD via oral, inhal, ip and dermal routes.
Fire Hazard: Dangerous, when exposed to heat, flame or oxidizers.
Explosion Hazard: See cyanides; can react violently

with H_2SO_4, oleum, chlorosulfonic acid and perchlorates. [*19*]

Disaster Hazard: Dangerous; when heated to decomp, emits highly toxic fumes of cyanides; will react with water, steam or acids to produce toxic and flam vapors.

To Fight Fire: Foam, CO_2, dry chemical.

METHYL CYANOACETATE. Liquid. $CNCH_2COOCH_3$, mw: 99.09, mp: −22.5°, bp: 203°, vap. d: 3.41.

Acute tox data: ipLD_{50} (mouse) = 750 mg/kg. [*3*]

THR = MOD via ip route.

METHYL CYANOACRYLATE. Liquid. $C_5H_5NO_2$, mw: 111.1, bp: 48° @ 1.8 mm.

Acute tox data: ipLD_{LO} (rat) = 100 mg/kg. [*3*]

THR = HIGH via ip and inhal routes.

METHYL CYANOETHANOATE. See cyanomethyl acetate.

METHYL CYANOFORMATE. Colorless liquid, ethereal odor. $COOCH_3CN$, mw: 77.06, bp: 100°, d: approx 1.00 @ 20°, vap. d: 2.66.

THR = No data. See also nitriles.

METHYL CYCLOHEXANE. Syns: *hexahydrotoluene, cyclohexylmethane.* Colorless liquid. $CH_3C_6H_{11}$, mw: 98.18, mp: −126.4°, lel = 1.2%, uel = 6.7%, bp: 100.3°, flash p: 25°F (CC), d: 0.7864 @ 0°/4°, 0.769 @ 20°/4°, vap. press: 40 mm @ 22.0°, vap. d: 3.39, autoign. temp.: 482°F.

Acute tox data: Oral LD_{LO} (rabbit) = 4000 mg/kg. [*3*]

THR = MOD via oral route. The minimum lethal dose for rabbits via the oral route is from 4.0–4.5 g/kg of body weight; by inhal, 15,000 ppm caused the death of rabbits in about 70 min. As to chronic effects, these are also present, since all rabbits exposed to 10,000 ppm conc for 6 hrs/day, 5 days/week, for 2 weeks died. This material does not cause irr to the eyes and nose, and even at the level of 500 ppm, exhibits only a very faint odor. Therefore, it cannot be said to have any warning properties. It is believed to be about 3 times as toxic as hexane and can cause death by tetanic spasm as has been noted in animals. In sublethal conc it causes narcosis and anesthesia.

Fire Hazard: Dangerous, when exposed to heat, flame or oxidizers.

Spont Heating: No.

Explosion Hazard: Mod, when exposed to heat or flame.

Disaster Hazard: Dangerous; upon exposure to heat or flame, can react vigorously with oxidizing materials.

To Fight Fire: Foam, CO_2, dry chemical.

METHYL CYCLOHEXANOL. Syns: *hexahydromethyl phenol, hexahydrocresol.* Colorless, viscous liquid, aromatic menthol-like odor. $CH_3C_6H_{10}OH$, mw: 114.1, bp: 155°–180°, flash p: 154°F (CC), autoign. temp.: 565°F, d: 0.924 @ 15.5°/15.5°, vap. d: 3.93.

Acute tox data: Oral LD_{LO} (rabbit) = 1750 mg/kg. [*3*]

THR = MOD via oral route.

Fire Hazard: Mod, when exposed to heat, flame or oxidizers.

Spont Heating: No.

Disaster Hazard: On heating, emits acrid fumes; can react with oxidizing materials.

To Fight Fire: Alcohol foam, CO_2, dry chemical.

METHYL CYCLOHEXANOL ACETATE. See methyl cyclohexyl acetate.

METHYL CYCLOHEXANONE. Water-white to pale yellow liquid, acetone-like odor. $COCH(CH_3)CH_2CH_2CH_2CH_2$, mw: 112.17, bp: 160°–170°, flash p: 118°F (CC), d: 0.925 @ 15°/5°, vap. d: 3.86.

Acute tox data: Oral LD_{50} (rat) = 2140 mg/kg; ipLD_{50} (mouse) = 200 mg/kg; dermal LD_{LO} (rabbit) = 1770 mg/kg. [*3*]

THR = HIGH via ip; MOD via oral and dermal routes. This is a toxic compound which can damage the kidneys and the liver. It is similar to cyclohexanol in its toxic action, although it is somewhat less active. Harmful exposures in industry are rare. Exper on rabbits have found that they could withstand prolonged exposures to conc of from 0.02–0.05% by volume in air.

Fire Hazard: Mod, when exposed to heat or flame. Can react violently with HNO_3 and other oxidizers. [*19*]

Spont Heating: No.

To Fight Fire: Foam, CO_2, dry chemical.

4-METHYL CYCLOHEXENE-1. A clear liquid. C_7H_{12}, mw: 96.17, bp: 102.5°, flash p: 30°F (TOC), d: 0.804 @ 15.5°/15.5°, vap. press: 10.3 mm @ 38°, vap. d: 3.34.

THR = U. Probably an irr and narcotic in HIGH conc.

Fire Hazard: Dangerous, when exposed to heat or flame; can react vigorously with oxidizing materials.

To Fight Fire: Foam, CO_2, dry chemical.

2-METHYL-4-CYCLOHEXENE-1-CARBOXALDEHYDE. $CHCHOCH_3CHCH_2CH{:}CHCH_2$, mw: 124.

THR = Details U. Limited animal exper suggest LOW toxicity and MOD irr. See also aldehydes.

METHYL CYCLOHEXYL ACETATE. Syn: *methyl cyclohexanol acetate.* Liquid. $C_9H_{16}O_2$, mw: 156.22, flash p: 147°F (CC), vap. d: 5.37, d: 0.9.
THR = See esters.
Fire Hazard: Mod, when exposed to heat or flame; can react with oxidizing materials.
Spont Heating: No.
To Fight Fire: Foam, CO_2, dry chemical.

N-METHYL CYCLOHEXYL AMINE. $CH_3(C_6H_{11})NH$, mw: 113.2.
Acute tox data: Oral LD_{50} (male rats) = 520 mg/kg; (female rats) = 400 mg/kg; inhal LC_{50} (guinea pigs) = 7 mg/liter for 1 hr (aerosol). [*20*]
THR = HIGH irr to skin, eyes and mu mem and via dermal route. [*20*] Caused ptosis, lacrimation, gasping, irregular respiration, nose-bleeding, prostration, convulsions. Contact can cause severe eye damage. @ dose levels of 2000 mg/kg (dermally) rabbits died in 24 hrs, and skin was badly burned.

METHYCYCLOHEXYL FORMATE. Liquid. $H_3CC_6H_{10}COOH$, mw: 142.2.
THR = Details U. Is said to resemble amyl acetate in type and severity of effects. See esters.

METHYL CYCLOHEXYL LACTATE. Liquid. $C_{10}H_{18}O_3$, mw: 186.3, flash p: 208°F, d: 1.02, vap. d: 6.43.
THR = See esters.
Fire Hazard: Low, when exposed to heat or flame; can react with oxidizing materials.
To Fight Fire: Foam, CO_2, dry chemical.

1-(2-METHYL CYCLOHEXYL)-3-PHENYLUREA. See tupersan.

METHYL CYCLOHEXYL STEARATE. Clear, oily liquid, straw yellow. $C_{17}H_{25}COOC_6H_{10}CH_3$, mw: 380, mp: 10°, bp: 105°–116° @ 4 mm, flash p: 338°F, d: 0.890 @ 15°/15°, vap. d: 13.1.
THR = See esters.
Fire Hazard: Slight, when exposed to heat or flame; reacts with oxidizing materials.
To Fight Fire: Foam, CO_2, dry chemical.

METHYL CYCLOPENTADIENE. C_6H_8, mw: 80, d: 0.9, bp: 163°F, flash p: 120°F, autoign. temp.: 834°F, lel = 1.3% @ 212°F, uel = 7.6% @ 212°F.
THR = U.
Fire Hazard: Mod, when exposed to heat, flame or oxidizers.
To Fight Fire: Dry chemical, CO_2, foam, spray or mist.

2-METHYL CYCLOPENTADIENYL MANGANESE TRICARBONYL. $MnO_3C_9H_7$, mw: 218.1.
Acute tox data: Oral LD_{50} (rat) = 23 mg/kg; dermal LD_{50} (rabbit) = 1692 mg/kg; inhal LC_{50} (rat) = 350 mg/kg. [*3*]
THR = HIGH via oral and inhal routes and MOD via dermal route. See also manganese compounds and carbonyls.

METHYL CYCLOPENTADIENYL MANGANESE TRICARBONYL. Syn: *MMT.*
Acute tox data: Oral LD_{50} (rat) = 58 mg/kg; inhal LC_{50} (rat) = 220 mg/m^3; dermal LD_{50} (rat) = 665 mg/kg; oral LD_{50} (guinea pig) = 905 mg/kg; ivLD_{50} (rabbit) = 6.6 mg/kg. [*3*]
THR = HIGH via oral, inhal and iv routes; MOD via dermal route. A fuel additive.

METHYL CYCLOPENTANE. Colorless liquid or solid. C_6H_{12}, mw: 84.16, mp: −142.5°, bp: 71.8°, flash p: <20°F, d: 0.750 @ 20°/4°, vap. press: 100 mm @ 17.9°, vap. d: 2.9.
THR = U. Probably irr and narcotic in high conc.
Fire Hazard: Dangerous, when exposed to heat, flame or oxidizers.
Disaster Hazard: Dangerous; upon exposure to heat or flame; can react vigorously with oxidizing materials.
To Fight Fire: Foam, CO_2, dry chemical.

17-METHYL-15H-CYCLOPENTA(a)PHENANTHRENE. $C_{18}H_{14}$, mw: 230.3.
THR = An exper neo. [*3, 23*]

2-METHYL DECANE. $C_{11}H_{24}$, mw: 156.2, autoign. temp.: 437°F, d: 0.74, vap. d: 5.39, bp: 190°.
THR = No data. Probably LOW.
Fire Hazard: Mod, via heat, open flames or oxidizers.
To Fight Fire: Dry chemical, CO_2, foam, water spray or mist.

METHYL DEMETON. See *m*-systox.

METHYL DIACETOACETATE. Colorless liquid. $(CH_3CO)_2CHCO_2CH_3$, mw: 158.2, vap. d: 5.45.
Acute tox data: Oral LD_{50} (rat) = 1700 mg/kg. [*3*]
THR = MOD via oral route.
Fire Hazard: Slight.

7-METHYL DIBENZ(c,h)ACRIDINE. $C_{22}H_{15}N$, mw: 293.4.
THR = An exper neo. [*3*]

10-METHYL DIBENZ(a,c)ANTHRACENE. $C_{23}H_{16}$, mw: 292.4.
THR = An exper neo. [*3, 23*]

2-METHYL DIBENZ(a,h)ANTHRACENE.
THR = An exper carc. [*3, 23*]

3-METHYL DIBENZ(a,h)ANTHRACENE.
THR = An exper neo. [*3, 23*]

6-METHYL DIBENZ(a,h)ANTHRACENE.
THR = An exper carc. [*3, 23*]

7-METHYL DIBENZ(a,h)ANTHRACENE.
THR = An exper carc. [*23*]

N-METHYL-7H-DIBENZO(c,g)CARBAZOLE. $C_{21}H_{15}N$, mw: 281.4.
THR = An exper carc. [*3*]

8-METHYL DIBENZO(def)*p*-CHRYSENE. $C_{25}H_{16}$, mw: 316.4.
THE = An exper carc. [*3*]

10-METHYL DIBENZO(def,p)CHRYSENE.
THR = An exper neo. [*3, 23*]

12-METHYL DIBENZO(def,mno)CHRYSENE.
THR = An exper neo. [*3, 23*]

7-METHYL DIBENZO(b,def)CHRYSENE.
THR = An exper carc. [*23*]

14-METHYL DIBENZO(b,def)CHRYSENE-7-CARBOXALDEHYDE. $C_{26}H_{16}O$, mw: 344.4.
THR = An exper neo. [*3*]

6-METHYL DIBENZO(def,mno)CHRYSENE-12-CARBOXALDEHYDE.
THR = An exper neo. [*3, 23*]

7-METHYL DIBENZO(h,rst)PENTAPHENE. $C_{29}H_{18}$, mw: 366.5.
THR = An exper neo. [*3*]

2-METHYL-1,2,4,5-DIBENZOPYRENE. $C_{27}H_{18}$, mw: 342.5.
THR = An exper neo. [*3, 23*]

3-METHYL-1,2,4,5-DIBENZOPYRENE.
THR = An exper neo. [*3, 23*]

METHYL DIBORANE. Colorless gas, very unstable, decomp by water. $B_2H_5CH_3$, mw: 41.7, bp: −80° @ 50 mm (decomp).
THR = See boron hydrides and boron compounds.

4-METHYL-2,6-DI-*tert*-BUTYLPHENOL. See di-*tert*-butyl-*p*-cresol.

METHYL DICHLOROACETATE. Syn: *methyl dichloroethanoate.* Colorless liquid, ethereal odor. $Cl_2CHCOOCH_3$, mw: 143.0, bp: 143.0°, d: 1.3809 @ 19.2°/19.2°, vap, d: 4.93.
THR = HIGH irr to skin, eyes and mu mem. This material hydrolyzes upon contact with moisture to form a product corrosive to tissue. See dichloroacetic acid. See also esters.
Disaster Hazard: Dangerous; when heated to decomp, emits highly toxic fumes of phosgene.

METHYL DICHLOROARSINE. Syn: *methylarsenic dichloride.* Colorless, mobile liquid. CH_3AsCl_2, mw: 160.86, mp: −59°, bp: 136°, d: 1.838 @ 20°/4°, vap. press: 8.5 mm @ 20°.
THR = HIGH irr to skin, eyes and mu mem. A poison. See also arsenic compounds.
Disaster Hazard: Dangerous; when heated to decomp, emits highly toxic fumes of arsenic and chlorine; will react with water, steam or acids to produce toxic and corrosive fumes.

METHYL DICHLOROETHANOATE. See methyl dichloroacetate.

METHYL DICHLOROSILANE. Colorless liquid, sol in benzene, ether and heptane. CH_3SiHCl_2, mw: 115, bp: 41°, d: 1.10 @ 27°, flash p: −26°F.
Acute tox data: Inhal LC_{LO} (rat) = 300 ppm for 4 hrs. [*3*]
THR = HIGH irr via inhal and to skin, eyes and mu mem.
Fire Hazard: Dangerous, when exposed to heat or flame.

METHYL DICHLORSTEARATE. Light yellow, oily liquid. $C_{17}H_{33}Cl_2CO_2CH_2$, mw: 367.40, mp: −5° to +7°, bp: 250° (decomp), flash p: 358°F, d: 0.997 @ 15.5°/15.5°, vap. d: 12.7.
THR = Probably LOW. See esters.
Fire Hazard: Slight, when exposed to heat or flame.
Disaster Hazard: Dangerous; when heated to decomp, emits highly toxic fumes of phosgene; can react with oxidizing materials.
To Fight Fire: Foam, CO_2, dry chemical.

N-METHYL DICYCLOHEXYL AMINE. $C_{13}H_{25}N$, mw: 195.4.
Acute tox data: Oral LD_{50} (rat) = 446 mg/kg; [*3*] oral LD_{50} (male rats) = 521 mg/kg; oral LD_{50} (female rats) = 446 mg/kg. [*20*] All exposed guinea pigs survived exposure to 2 mg/liter for 1 hr. [*20*]
THR = HIGH to MOD irr to skin, eyes and mu mem. Symptoms of intoxication include depressed activity and sporadic lung congestion unrelated to dose level, @ ⩾215 mg/kg ⟶ muscular tremors, ataxia, convulsions. During exposure to aerosol, there was excessive preening and depression. [*20*]

METHYL DIEPOXYDIALLYLACETATE. $C_9H_{14}O_4$, mw: 186.2.
THR = An exper neo. [*3*]

METHYL DIETHANOLAMINE. Clear liquid. $CH_3N(CH_2CH_2OH)_2$, mw: 119.16, bp: 240°, flash p: 260°F (OC), d: 1.043, vap. press: 0.01 mm @ 20°.
THR = Details U. See amines.
Fire Hazard: Slight, when exposed to heat or flame; can react with oxidizing materials.
To Fight Fire: Alcohol foam, CO_2, dry chemical.

1-METHYL-3,5-DIETHYL BENZENE. $C_{11}H_{16}$, mw: 148.2, autoign. temp.: 851°F, d: 0.86, vap. d: 5.12, bp: 201.1°.
THR = No data.

Fire Hazard: Mod, via heat, open flame or oxidizers.
To Fight Fire: Dry chemical, CO_2, foam, spray or mist.

METHYL DIHYDROABIETATE. Liquid.
$CH_3CO_2C_{19}H_{32}$, mw: 319.5, bp: 365°, flash p: 361°F (OC), d: 1.02, vap. d: 11.0.
THR = U. See esters.
Fire Hazard: Slight, when exposed to heat or flame; can react with oxidizing materials.
To Fight Fire: CO_2, dry chemical.

7-METHYL-16,7-DIHYDRO-15H-CYCLOPENTA(a) PHENANTHRENE.
THR = An exper carc. [*23*]

11-METHYL-16,17-DIHYDRO-15H-CYCLOPENTA (a)PHENANTHRENE.
THR = An exper carc. [*23*]

17-METHYL-16,17-DIHYDRO-15H-CYCLOPENTA (a)PHENANTHRENE.
THR = An exper carc. [*23*]

2-METHYL-4-DIMETHYLAMINO AZOBENZENE.
$C_{15}H_{17}N_3$, mw: 239.4.
THR = An exper carc and neo. [*3*]

2-METHYL-4-DIMETHYLAMINO STILBENE.
$C_{17}H_{19}N$, mw: 237.4.
THR = An exper neo. [*3, 23*] via oral route.

3-METHYL-4-DIMETHYLAMINO STILBENE.
THR = An exper neo [*3, 23*] via oral route.

4′-METHYL-4-DIMETHYLAMINO STILBENE.
$C_{17}H_{19}N$, mw: 237.4.
THR = An exper neo [*3, 23*] via oral route.

METHYL-4-DIMETHYLAMINO-3,5-XYLYL CARBAMATE. Syns: *zectran, mexacarbate*. Crystals. $C_{12}H_{18}O_2N_2$, mw: 222.3, mp: 85°, vap. press: <0.1 mm @ 139°.
Acute tox data: Oral LD_{50} (rat) = 14 mg/kg; dermal LD_{50} (rat) = 1500 mg/kg; ipLD_{50} (rat) = 15 mg/kg; oral LD_{50} (wild birds) = 1 mg/kg and 10 mg/kg; oral LD_{50} (duck) = 3 mg/kg. [*3*]
THR = HIGH via oral and ip; MOD via dermal routes. An exper carc [*3*] via oral route. [*12*]

4-METHYL-7,12-DIMETHYL BENZ(a)ANTHRACENE.
THR = An exper carc. [*23*]

5-METHYL-7,12-DIMETHYL BENZ(a)ANTHRACENE.
THR = An exper carc. [*23*]

6-METHYL-7,12-DIMETHYL BENZ(a)ANTHRACENE.
THR = An exper carc. [*23*]

8-METHYL-7,12-DIMETHYL BENZ(a)ANTHRACENE.
THR = An exper carc. [*23*]

9-METHYL-7,12-DIMETHYL BENZ(a)ANTHRACENE.
THR = An exper carc. [*3*]

10-METHYL-7,12-DIMETHYL BENZ(a)ANTHRACENE.
THR = An exper carc. [*23*]

3-METHYL-4,6-DINITROPHENOL. See 4,6-dinitro-*o*-cresol.

***n*-METHYL-*n*,4-DINITROSOANILINE.** See N,4-dinitroso-N-methyl aniline.

3-METHYL-1,3-DIOXOLANE. Water-white liquid.
$OCH(CH_3)OCH_2CH_2$, mw: 88.10, bp: 82.5°, d: 1.002 @ 0°/4°, vap. d: 3.03.
Acute tox data: Oral LD_{50} (rat) = 3000 mg/kg. [*3*]
THR = MOD via oral route.
Fire Hazard: Slight.

METHYL DISULFIDE. Syns: *dimethyl disulfide, methyl dithiomethane*. Liquid, insol in water, sol in alcohol and ether. CH_3SSCH_3, mw: 94.2, bp: 117°, d: 1.057 @ 16°/4°.
THR = See alkyl disulfides.
Fire Hazard: Mod, via heat, flame or oxidizers.
Disaster Hazard: Dangerous; see sulfides.
To Fight Fire: Water, spray, CO_2, foam, dry chemical.

METHYL DITHIOCARBAMIC ACID, SODIUM SALT. Syns: *vapam, carbation*. $C_2H_5NS_2 \cdot Na$, mw: 130.2.
Acute tox data: Oral LD_{50} (rat) = 820 mg/kg; oral LD_{50} (mouse) = 50 mg/kg; dermal LD_{50} (rabbit) = 800 mg/kg; oral LD_{50} (rabbit) = 320 mg/kg. [*3*]
THR = MOD via oral and dermal routes. Irr to skin and mu mem. Accompanied by alcohol intake, causes violent vomiting and shock. See also bis(diethyl thiocarbamyl) disulfide.

3-O-METHYL DOPA. Syn: *3-methoxy-L-tyrosine*.
$C_{10}H_{13}O_4N$, mw: 211.2.
THR = MOD via oral route (rats). Has been implicated in aplastic anemia. [*3*]

***N,N′*-METHYLENE BIS-ACRYLAMIDE.** Colorless, crystalline, stable, white powder.
$H_2C(CH_2CHCONH)_2$, mw: 154.17, mp: 185° (with decomp), d: 1.235 @ 30°, vap. d: 5.31.
THR = No data. Probably MOD via oral route.
Disaster Hazard: Slight; when heated, emits toxic fumes.

4,4′-METHYLENE-BIS-(2-CHLOROANILINE). See 3,3′-dichloro-4,4′-diaminodiphenyl methane.

2,2′-METHYLENE-BIS(4-CHLOROPHENOL). See bis (5-chloro-2-hydroxyphenyl)methane.

4,4′-METHYLENE BIS(2-METHYL ANILINE).
$C_{15}H_{24}N_2$, mw: 226.4, mp: 149°.
THR = An exper carc. [*3, 23*]

2,2′-METHYLENE BIS(4-METHYL-6-*tert*-BUTYL PHENOL). Pale cream to white crystals. $C_{25}H_{36}O_2$, mw: 368.6.

Acute tox data: ipLD$_{LO}$ (mouse) = 50 mg/kg. [*3*]

THR = HIGH via ip route.

Disaster Hazard: Slight; when heated, may exit toxic fumes.

METHYLENE BIS(4-PHENYL ISOCYANATE). Syns: *diphenyl methane diisocyanate*, *MDI*. Crystals or yellow fused solid. $OCNC_8H_4CH_2C_6H_4NCO$, mw: 250.25, mp: 37.2°, bp: 194°–199° @ 5 mm, d: 1.19 @ 50°, vap. press: 0.001 mm @ 40°.

THR = An irr and allergic sensitizer. MOD toxic. See isocyanates.

Disaster Hazard: See isocyanates.

2,2′-METHYLENE BIS(3,4,6-TRICHLORO-PHENOL). Syns: *G-11*, *hexachlorophene*. Crystals, water insol. $C_{13}H_8Cl_6O_2$, mw: 406.9, mp: 165°.

Acute tox data: Oral LD$_{50}$ (rat) = 60 mg/kg; dermal LD$_{LO}$ (rat) = 600 mg/kg; ipLD$_{50}$ (rat) = 22 mg/kg; ivLD$_{50}$ (rat) = 7.5 mg/kg; oral TD$_{LO}$ (infant) = 259 mg/kg for 7 days (intermittant) ⟶ CNS problems; dermal TD$_{LO}$ (child) = 300 mg/kg ⟶ CNS problems; oral TD$_{LO}$ (humans) = 0.043 mg/kg⟶ CNS problems. [*3*]

THR = HIGH via oral, ip and iv routes. MOD via dermal route. Strong conc may be irr but ordinary use of 1–2% is not irr. Used as a food additive permitted in the feed and drinking water of animals and/or for the treatment of food-producing animals; also permitted in food for human consumption. [*109*]

For many years, the toxicologic hazard of hexachlorophene was unrecognized and the compound had a wide and virtually unrestricted use. However, recent studies by FDA scientists have shown that brain lesions occur in both rats and monkeys treated with hexachlorophene at levels only slightly higher than those of persons using soaps, tooth paste, shampoos, and a variety of other household products and cosmetics containing hexachlorophene. The FDA has now restricted sale of hexachlorophene and most preparations containing higher levels of the compound are available only through prescription.

In the recent FDA studies, it was found that 2 weeks after onset of exposure, rats fed 500 ppm (25 mg/kg/day) hexachlorophene in their diet showed weakness in their hindquarters, which progressed to paralysis. Microscopic examination of the brain and spinal cord of these rats revealed a particular edema of the white matter resembling spongy degeneration of the white matter in infants. When the animals were removed from the poisoned diet, they recovered gradually over a period of weeks. A similar symptom picture was noted in the monkey. In correlated studies in many, hexachlorophene was detected in the blood of volunteers in the range of 5–89 ppb. The levels found corresponded generally to the individual's reported use of hexachlorophene.

A number of cases of human poisoning, some of them fatal, have been reported for hexachlorophene. Most of the fatal poisonings have resulted from ingestion of hexachlorophene through accident or with suicidal intent. The use of hexachlorophene solution to wash patients with extensive burns has also been reported to cause poisoning.

Following ingestion of hexachlorophene, early symptoms are primarily gastrointestinal in nature, and include anorexia, nausea, vomiting, abdominal cramps, and diarrhea. Dehydration is sometimes severe and may be associated with shock. Later symptoms in patients who survive acute ingestion or in persons with severe chronic exposure are primarily associated with the CNS. These effects include diplopia, irritability, weakness of the lower extremities, and convulsions.

Since no data is presently available on specific antidotes for hexachlorophene poisoning, treatment would have to be generalized and supportive in nature, and directed toward removal of the poison and relief of the signs and symptoms which occur.

Disaster Hazard: Dangerous; when heated to decomp, emits highly toxic fumes of chlorides.

METHYLENE BROMIDE. Syn: *dibromomethane*. Colorless, heavy liquid. CH_2Br_2, mw: 173.9, bp: 95.6°–97.4°, fp: <50°, d: 2.485 @ 25°/25°, vap. d: 6.05.

Acute tox data: scLD$_{50}$ (mouse) = 3738 mg/kg. [*3*]

THR = MOD via sc route.

Disaster Hazard: Mod dangerous; when heated to decomp, emits toxic fumes of bromides.

METHYLENE CHLORIDE. Syn: *dichloromethane*. Colorless volatile liquid. CH_2Cl_2, mw: 84.94, bp: 39.8°, lel = 15.5% in O_2, uel = 66.4% in O_2, fp: −96.7°, d: 1.326 @ 20°/4°, autoign. temp.: 1139°F, vap. press: 380 mm @ 22°, vap. d: 2.93.

Acute tox data: Oral LD$_{50}$ (rat) = 2136 mg/kg; ipLD$_{50}$ (mouse) = 1500 mg/kg; scLD$_{50}$ (mouse) = 6440 mg/kg; inhal LC$_{LO}$ (guinea pig) = 5000 ppm for 2 hrs; inhal TC$_{LO}$ (humans) = 500 ppm for 8 hrs ⟶ blood problems. Prolonged inhal of 500 ppm ⟶ CNS problems.

THR = MOD via oral, ip, sc and inhal routes. An exper carc. [*23*] This material is very dangerous to the eyes. Except for its property of inducing narcosis, it has very few other acute toxicity effects. Its narcotic powers are quite strong, and in view of its great volatility, care should be taken in its use. It will not form explosive mixtures with air at ordinary temp. However, it can be decomp by contact with hot surfaces and open flame, and it can then yield toxic fumes, which are irr and will thus give warning of their presence. It has been used as an anesthetic in Europe and is still used there for local anesthesia. Exper have shown that 25,000 ppm conc for 2 hr exposures were not lethal. Conc of 7,200 ppm after 8 min caused paresthesia of the extremities; after 16 min, acceleration of the pulse to 100; during the first 20 min, congestion in the head, a sense of heat and slight irr of the eyes. At a level of 2,300 ppm, there was no feeling of dizziness during 1-hr exposures, but nausea did occur after 30 min of exposure. The limit of perception by smell is set at 25–50 ppm conc. Can cause a dermatitis upon prolonged skin contact. A respirator for organic vapors and fumes should be worn to avoid excessive inhal. Used as a food additive permitted in food for human consumption. [*109*]

Fire Hazard: Reacts violently with Li, NaK, potassium-*tert*-butoxide, (KOH + *n*-methyl-*n*-nitrosourea). [*19*]

Explosion Hazard: None under ordinary conditions, but will form explosive mixtures in atmosphere having high oxygen content, in liquid O_2, N_2O_4, K, Na, NaK. [*19*]

Disaster Hazard: Dangerous; when heated to decomp, emits highly toxic fumes of phosgene.

METHYLENE CHLOROBROMIDE. Syns: *bromochloromethane, chlorobromomethane.* Clear, colorless liquid, sweet odor. $BrCH_2Cl$, mw: 129.4, bp: 67.8°, fp: −88°, flash p: none: d: 1.930 @ 25°/25°, vap. d: 4.46.

Acute tox data: Oral LD_{50} (rat) = 5000 mg/kg; inhal LC_{LO} (rat) = 65170 ppm for ¼ hr; inhal LC_{LO} (guinea pig) = 8000 ppm for 2 hrs. [*3*]

THR = This material has a narcotic action of mod intensity, although of prolonged duration. Animals exposed for several weeks to 1,000 ppm of this substance had blood bromide levels as high as 350 mg/100 g of blood Therefore, until further data are available concerning it, it should be considered at least as toxic as carbon tetrachloride and more than minimal exposure to its vapors should be avoided.

Disaster Hazard: Dangerous; when heated to decomp, emits highly toxic fumes of halides.

2-METHYLENE CYCLOPROPANE ALANINE. $C_7H_{11}O_2N$, mw: 141.2.

THR = An exper teratogen. [*3*]

***p,p'*-METHYLENE DIANILINE.** Tan flakes or lumps, faint amine-like odor. $CH_2(C_6H_4NH_2)_2$, mw: 198.3, mp: 90°, flash p: 440°F.

Acute tox data: Oral LD_{50} (rat) = 347 mg/kg; scLD_{50} (rat) = 200 mg/kg. [*3*]

THR = HIGH via oral and sc routes. An exper (±) neo and carc. [*3, 10, 23*] It does not seem to be rapidly absorbed through skin in dangerous quantities or to irr the eyes. No information regarding allergy.

Disaster Hazard: Dangerous; when heated to decomp, emits highly toxic fumes of aniline.

METHYLENE DIISOCYANATE. $CH_2(NCO)_2$, mw: 98, flash p: 185° (OC).

THR = Probably MOD. See isocyanates.

Fire Hazard: Mod, when exposed to heat, flame or oxidizers. Reacts violently with dimethyl formamide. [*19*]

To Fight Fire: Foam, CO_2, dry chemical.

METHYLENE DIMETHYL ETHER. See methylal.

4,4'-METHYLENE DIMORPHOLINE. $C_9H_{18}O_2N_2$, mw: 186.3.

THR = An exper carc. [*3*]

3,4-METHYLENE DIOXYBENZALDEHYDE. See piperonal.

1,2-METHYLENE DIOXY-4-(3-OCTYL SULFINYL)-PROPENYL BENZENE. See sulfoxide.

3,4-(METHYLENE DIOXY)PHENOL. $C_7H_6O_3$, mw: 138.1.

THR = An exper neo. [*3*]

2-(3,4-METHYLENE DIOXYPHENOXY)-3,6,9-TRIOXAUNDECANE. See sesoxane.

1,2-(METHYLENE DIOXY)-4-PROPENYL BENZENE. See safrol.

4,4-METHYLENE DIPHENOL. $C_6H_5CH_2C_6H_5$, mw: 174.

Acute tox data: Oral LD_{50} (rat) = 4950 mg/kg. [*3*]

THR = MOD via oral route.

1,2-(METHYLENE DIOXY)-4-PROPYL BENZENE. Syn: *dihydrosafrol.* An oily liquid. $C_{10}H_{12}O_2$, mw: 164.2, bp: 228°, d: 1.0695 @ 20°.

THR = An exper (+) carc. [*3, 9*]

METHYLENE DISALICYLIC ACID. Light tan, coarse powder. $C_{15}H_{12}O_6$, mw: 288.25, mp: 225°–238°, d: 1.430 @ 25°/4°.

THR = Probably LOW. A MILD irr.

Fire Hazard: Slight.

METHYLENE DIURETHAN. $C_7H_{14}O_4N_2$, mw: 190.2.

THR = An exper neo. [*3*]

METHYLENE IODIDE. Syn: *diiodomethane.* Light straw-colored to clear heavy refractive liquid. CH_2I_2, mw: 267.9, mp: 5°–6°, bp: 181°, d: 3.325 @ 20°/4°, vap. d: 9.25.

THR = U. Probably irr and narcotic in high conc.
Disaster Hazard: Dangerous; when heated to decomp, emits toxic fumes of iodides. Reacts violently with NaK and Li. [*19*]

METHYLENE SUCCINIC ACID. See itaconic acid.

N-METHYLENE UREA.
THR = An exper mutagen. [*3*]

N-METHYL ETHANOLAMINE. Liquid. $CH_3NHC_2H_4OH$, mw: 75.11, mp: −4.5°, bp: 159.5°, flash p: 165° F (OC), d: 0.9414 @ 20°/20°, vap. press: 0.7 mm @ 20°, vap. d: 2.59.
THR = Details U. See amines.
Fire Hazard: Mod, when exposed to heat or flame; can react with oxidizing materials.
To Fight Fire: Foam, CO_2, dry chemical.

METHYL ETHER. See dimethyl ether.

METHYL ETHER OF PROPYLENE GLYCOL. See propylcne glycol methyl ether.

7-METHYL-9-ETHYL BENZ(c)ACRIDINE.
THR = An exper carc. [*23*]

7-METHYL-12-ETHYL BENZ(a)ANTHRACENE.
THR = An exper carc. [*23*]

METHYL ETHYL CELLULOSE. A white to pale cream-colored fibrous solid or powder, practically odorless.
THR = Details U. Used as a food additive permitted in food for human consumption. [*109*]

2-METHYL-2-ETHYL-1,3-DIOXOLANE. Insol in water. $C_6H_{12}O_2$, mw: 116, d: 0.9392, bp: 117.6°, fp: −81.96°, vap. d: 4.0, flash p: 74° (OC).
THR = U.
Fire Hazard: Dangerous, when exposed to heat, flame or oxidizers.
To Fight Fire: Dry chemical, CO_2, foam.

sym-METHYL ETHYL ETHYLENE. See 2-pentene.

METHYL ETHYL ETHER. See ethyl methyl ether.

2-METHYL-4-ETHYL HEXANE. C_9H_{20}, mw: 128.1, flash p: <70° F, autoign. temp.: 536° F, lel = 0.7%, d: 0.72; vap. d: 4.43, bp: 133.9°.
THR = No data. Probably LOW.
Fire Hazard: Dangerous via heat, flame, oxidizers.
To Fight Fire: Dry chemical, CO_2, foam, spray.

3-METHYL-4-ETHYLHEXANE. Clear, colorless liquid. C_9H_{20}, mw: 128.25, bp: 140°, flash p: 75°, d: 0.738 @ 25°/4°, vap. press: 8.1 mm @ 25°, vap. d: 4.43.
THR = U. Probably irr and narcotic in high conc.
Fire Hazard: Dangerous, when exposed to heat, flame or oxidizers.
Explosion Hazard: U.
Disaster Hazard: Dangerous; keep away from heat and open flame; can react with oxidizing materials.
To Fight Fire: Foam, CO_2, dry chemical.

METHYL ETHYL KETONE. See butanone.

METHYL ETHYL KETONE PEROXIDE. $C_4H_8O_2$, mw: 88.1.
Acute tox data: Oral LD_{50} (rat) = 484 mg/kg; inhal LC_{50} (rat) = 200 ppm for 4 hrs; ipLD_{50} (rat) = 65 mg/kg. [*3*]
THR = HIGH via oral, inhal and ip routes. An irr. An exper neo. [*3*]
Fire Hazard: See peroxides, organic.

METHYL ETHYL METHANE. See butane.

2-METHYL ETHYL PENTANE. C_8H_{18}, mw: 114.1, flash p: <70° F, autoign. temp.: 860° F, d: 0.72, vap. d: 3.94, bp: 116.1°.
THR = No data. Probably LOW.
Fire Hazard: Dangerous, via heat, flame or oxidizers.
To Fight Fire: Dry chemical, CO_2, foam, mist.

2-METHYL-5-ETHYL PIPERIDINE. Slightly sol in water. $\overline{NHCH(CH_3)CH_2CH_2CH(C_2H_5)CH_2}$, mw: 127, flash p: 126° F (OC), d: 0.8, vap. d: 4.4, bp: 326° F.
THR = U.
Fire Hazard: Mod, when exposed to heat or flame or oxidizers.
To Fight Fire: Alcohol foam.

METHYL ETHYL PYRIDINE. See 5-ethyl-2-methyl pyridine.

METHYL EUGENOL. Syn: *4-allylveratrole.*
Acute tox data: Oral LD_{50} (rat) = 1179 mg/kg; ipLD_{50} (mouse) = 540 mg/kg; ivLD_{50} (mouse) = 112 mg/kg. [*3*]
THR = HIGH via iv and MOD via oral and ip routes. An attractant for the oriental fruit fly.

2-METHYL FLUORANTHENE. $C_{17}H_{12}$, mw: 216.3.
THR = An exper carc. [*3, 23*]

3-METHYL FLUORANTHENE.
THR = An exper neo. [*3, 23*]

N-METHYL FLUORENAMINE. $C_{14}H_{13}N$, mw: 195.2.
THR = An exper carc. [*3*]

METHL FLUORIDE. Syn: *fluoromethane.* Colorless gas. CH_3F, mw: 34.0, mp: −142°, bp: −78.2°, d: 0.8774 @ 78.6°/4°.
THR = Details U. May act as a simple asphyxiant.
Disaster Hazard: Dangerous; when heated to decomp, emits highly toxic fumes of fluorides.

METHYL FLUOROFORMATE. Liquid. CH_3COOF, mw: 78.04, bp: 40°, d: 1.06 @ 33°, vap. d: 2.69.
THR = HIGH irr to skin, eyes and mu mem.
Disaster Hazard: Dangerous; when heated to decomp, emits highly toxic fumes of fluorides; will react with

water, steam or acids to produce toxic and corrosive fumes.

METHYL FLUOROSULFONATE. Liquid, ethereal odor. CH_3OSO_2F, mw: 114.09, bp: 92°, d: 1.427 @ 16°, vap. d: 3.94.
THR = HIGH irr to skin, eyes and mu mem.
Disaster Hazard: Dangerous; when heated to decomp, emits highly toxic fumes of fluorides and oxides of sulfur; will react with water, steam or acids to produce toxic and corrosive fumes.

N-METHYL FORMAMIDE. C_2H_5ON, mw: 59.1.
THR = An exper teratogen. [*3*] Violent reaction with benzene sulfonyl chloride. [*19*]

METHYL FORMATE. Syn: *methyl methanoate.* Colorless liquid, agreeable odor. $HCOOCH_3$, mw: 60.05, mp: −99.8°, bp: 32.0°, lel = 5.5%, uel = 23%, flash p: −2°F, d: 0.98149 @ 15°/4°, 0.975 @ 20°/4°, autoign. temp.: 869°F, vap. press: 400 mm @ 16°/0°, vap. d: 2.07.
Acute tox data: Oral LD_{50} (rabbit) = 1620 mg/kg; inhal LC_{LO} (guinea pig) = 10,000 ppm. [*3*]
THR = MOD via oral and inhal routes. It can cause irr to the conjunctiva and optic neuritis. Industrial exposures resulting in fatalities from this material are extremely rare, having occurred only in instances where high conc are encountered, as in painting the inside of a tank or working in a tank containing residue of this material.

Exposure of guinea pigs to 5% conc of methyl formate vapor in air proved lethal in 20–30 min, whereas 1.5–2.5% was dangerous in 30–60 min. The max conc tolerated without serious disturbance was 0.5% and the max conc tolerated for several hours without serious disturbances was 0.15–0.20% by volume in air. Used as a food additive permitted in food for human consumption. [*109*]
Fire Hazard: Very dangerous, when exposed to heat, flame or oxidizers.
Spont Heating: No.
Explosion Hazard: Mod, when exposed to heat or flame.
Disaster Hazard: Dangerous; upon exposure to heat or flame, emits toxic fumes and can react vigorously with oxidizing materials.
To Fight Fire: Alcohol foam, CO_2, dry chemical.

2-METHYL FURAN. Colorless, mobile liquid, ether-like odor. C_5H_6O, mw: 82.1, bp: 63.7°, fp: −88.7°, flash p: −22°F, d: 0.914 @ 20°/4°, vap. press: 139 mm @ 20°, vap. d: 2.8.
Acute tox data: Oral LD_{50} (rat) = 167 mg/kg; inhal LC_{LO} (rat) = 377 ppm for 4 hrs. [*3*]
THR = HIGH via oral and inhal routes.
Fire Hazard: Dangerous, when exposed to heat, flame or oxidizers.
Explosion Hazard: See ethers.
Disaster Hazard: Dangerous; upon exposure to heat or flame, will emit toxic fumes and can react vigorously with oxidizing materials.
To Fight Fire: CO_2, dry chemical.

METHYL GAG. $C_3H_{12}N_{10} \cdot 2HCl$, mw: 297.2.
THR = An exper neo. [*3*]

METHYL GALLIUM DICHLORIDE. White crystals, decomp by water. $Ga(CH_3)Cl_2$, mw: 155.7, mp: 75°.
THR = Details U. See gallium compounds.
Disaster Hazard: See chlorides.

METHYL GALLIUM DICHLORIDE MONAMMINE. White crystals decomp by water. $Ga(CH_3)Cl_2 \cdot NH_3$, mw: 172.7.
THR = See gallium compounds and chlorides.

METHYL GALLIUM DICHLORIDE PENTAMMINE. White crystals, decomp by water. $5NH_3 \cdot Cl_2Ga(CH_3)$, mw: 240.8, mp: >80° (decomp).
THR = See gallium compounds, chlorides and ammonia.

β-METHYGLYCIDOL. Liquid. $\overline{OCHCH_3CH}CH_2OH$, mw: 88.1, vap. d: 3.04.
THR = U.
Disaster Hazard: Slight; when heated, emits acrid fumes; can react with oxidizing materials.

***n*-METHYLGLYCINE.** Syn: *sarcosine.* Crystals. CH_3NHCH_2COOH, mw: 89.1, mp: 210° (decomp), bp: decomp.
THR = Details U.
Disaster Hazard: Slight; when heated, emits acrid fumes; can react with oxidizing materials.

METHYL GLYCOL ACETATE. See methyl "cellosolve" acetate.

METHYL GLYCOCOLL. See *n*-methylglycine.

METHYL GUTHION. See guthion.

METHYL HEPTADECYL KETONE. White solid. $(CH_3)(C_{17}H_{35})CO$, mw: 282, bp: 165° @ 3 mm, mp: 54°, flash p: 255°F (CC), vap. d: 9.72.
THR = U. See ketones.
Fire Hazard: Slight, when exposed to heat or flame; can react with oxidizing materials.
To Fight Fire: CO_2, dry chemical.

2-METHYLHEPTANE. See isooctane.

METHYL HEPTANOL.
$CH_3CH_2CH(CH_3)CH_2CH_2CH_2OH$, mw: 130.
THR = Details U. Limited animal exper suggest MOD toxicity. See alcohols.

5-METHYL-3-HEPTANONE. See ethyl amyl ketone.

METHYL HEPTYL KETONE. Colorless liquid. $(CH_3)(C_7H_{15})CO$, mw: 142, mp: −9°, bp: 194°, flash p: 160° F (CC), vap. d: 4.9, d: 0.832 @ 30°.
Acute tox data: Oral LD_{50} (rat) = 3200 mg/kg. [*3*]
THR = MOD via oral route.
Fire Hazard: Mod, when exposed to heat or flame; can react with oxidizing materials.
To Fight Fire: Foam, CO_2, dry chemical.

2-METHYL HEXANE. Syns: *ethylisobutylmethane, isoheptane*. Colorless liquid. $(CH_3)_2CH(CH_2)_2CH_2CH_3$, mw: 100.20, fp: −118.2, bp: 90.0°, flash p: <0° F, d: 0.6789 @ 20°/4°, vap. press: 40 mm @ 14.9°, vap. d: 3.45, lel = 1.0%, uel = 6.0%, autoign. temp.: 536° F.
THR = Probably LOW via inhal and oral routes.
Fire Hazard: Dangerous. Keep from sparks and flame.
To Fight Fire: Foam, CO_2, dry chemical.

3-METHYL HEXANE.
Fire Hazard: See 2-methyl hexane.
To Fight Fire: See 2-methyl hexane.

5-METHYL-2-HEXANONE. Syns: *MIAK, methyl isoamyl ketone*. Colorless, stable liquid, pleasant odor, slightly sol in water, miscible with most organic solvents. $(CH_3COC_2H_4CH)(CH_3)_2$, mw: 114, bp: 144°, d: 0.8132 @ 20°/20°, fp: −73.9°, flash p: 110° F (OC).
Acute tox data: Oral LD_{50} (rat) = 4760 mg/kg; inhal LC_{LO} (rat) = 2000 ppm for 4 hrs; dermal LD_{50} (rabbit) = 10,000 mg/kg. [*3*]
THR = MOD via oral and inhal; LOW via dermal routes.
Fire Hazard: Mod, when exposed to heat, flame or oxidizers.
To Fight Fire: Dry chemical, CO_2, foam, fog.

METHYL HEXYL KETONE. See 2-octanone.

METHYL HYDRAZINE. Syns: *monomethyl hydrazine, MMH*. Colorless, hygroscopic liquid, ammonia-like odor, slightly sol in water, sol in alcohol and ether. CH_3NHNH_2, mw: 46, d: 0.874 @ 20°/4°, mp: −20.9°, bp: 87.8°, lel = 4%, vap. d: 1.6, flash p: <80° F.
Acute tox data: Inhal LC_{50} (rat) = 74 ppm for 4 hrs; dermal LD_{50} (rat) = 183 mg/kg; oral LD_{50} (rat) = 33 mg/kg; ipLD_{50} (rat) = 21 mg/kg; ivLD_{50} (rat) = 17 mg/kg; scLD_{50} (rat) = 35 mg/kg. [*3*]
THR = HIGH via inhal, dermal, oral, ip, sc and iv routes. An exper teratogen, neo and carc. [*3*] May self-ignite in air. [*19*] See also hydrazine.
Fire Hazard: Dangerous, when exposed to heat or flame.
To Fight Fire: Alcohol foam, CO_2, dry chemical.

METHYL HYDRAZINE SULFATE. $CH_6N_2H_2SO_4$, mw: 144.2.
THR = An exper carc. [*3*]

α-(2-METHYL HYDRAZINO)-*p*-TOLUOYL UREA HYDROBROMIDE. $C_{10}H_{14}O_2N_4 \cdot HBr$, mw: 303.2.
THR = An exper carc. [*3*]

METHYL HYDRIDE. See methane.

METHYL HYDROGEN SULFATE. See methyl sulfuric acid.

METHYL-3-HYDROXYBUTYRATE. Colorless liquid. $C_5H_{10}O_3$, mw: 118.2, bp: 174.9°, flash p: 180° F (OC), d: 1.0559, vap. press: 0.85 mm @ 20°, vap. d: 4.1.
Acute tox data: Oral LD_{50} (rat) = 3030 mg/kg. [*3*]
THR = MOD via oral route. An irr. See also esters.
Fire Hazard: Mod, when exposed to heat or flame; can react with oxidizing materials.
To Fight Fire: Alcohol foam, CO_2, dry chemical.

4-METHYL-7-HYDROXYCOUMARIN DIETHOXYTHIOPHOSPHATE. See potasan.

11-METHYL-16-HYDROXY-16,17-DIHYDRO-15H-CYCLOPENTA(a) PHENANTHRENE.
THR = An exper carc. [*23*]

METHYL-(β-HYDROXYETHYL)AMINE. See methyl aminoethanol.

2-METHYL HYDROXYLAMINO QUINOLINE-1-OXIDE. $C_{10}H_{10}O_2N_2$, mw: 190.2.
THR = An exper neo. [*3*]

11-METHYL-16-HYDROXY-17-OXO-16,17-DIHYDRO-15H-CYLOPENTA(a)PHENANTHRENE.
THR = An exper carc. [*23*]

METHYL HYDROXYSTEARATE. $C_{19}H_{34}O_3$, mw: 310.5.
THR = An exper neo. [*3, 23*]

METHYL HYPOCHLORITE. CH_3OCl, mw: 66.6.
THR = Very unstable in light. Somewhat unstable in absence of light. [*19*] See also hypochlorites.

1-METHYL IMIDAZOLE-2-THIOL. See methimazole.

METHYL IODIDE. Syn: *iodomethane*. Colorless liquid, turns brown on exposure to light. CH_3I, mw: 141.95, mp: −66.4°, bp: 42.5°, d: 2.279 @ 20°/4°, vap. press: 400 mm @ 25.3°, vap. d: 4.89.
Acute tox data: Oral LD_{50} (rat) = 220 mg/kg; inhal LC_{LO} (rat) = 3790 ppm for ¼ hr; dermal LD_{LO} (rat) = 800 mg/kg; scLD_{50} (rat) = 110 mg/kg. [*3*]
THR = HIGH via oral, inhal, sc; MOD via dermal routes. A strong narcotic and anesthetic. An exper (+) neo. [*3, 81, 23*]
Fire Hazard: U.
Disaster Hazard: Dangerous; when heated to decomp, emits highly toxic fumes of iodides.

METHYL ISOAMYL KETONE. See 5-methyl-2-hexanone.

METHYL ISOBUTYL CARBINOL. See methyl amyl alcohol.

METHYL ISOBUTYL CARBINOLACETATE. $C_8H_{16}O_2$, mw: 144.2, flash p: 110°F (OC), d: 0.9, vap. d: 5.0, bp: 146.1°.
THR = No data. Probably LOW.
Fire Hazard: Mod, via heat, flames, oxidizers.
To Fight Fire: Dry chemical, CO_2, water, fog, spray, mist, foam.

METHYL ISOBUTYL KETONE. See isobutyl methyl ketone.

METHYL ISOCYANATE. Liquid, reacts with water. CH_3NCO, mw: 57, d: 0.9599 @ 20°/20°, bp: 39.1°, flash p: <20°F.
Acute tox data: Oral LD_{50} (rat) = 305 mg/kg. [*3*] Inhal LC_{50} (rat) = 5 ppm for 4 hrs. [*3*]
THR = HIGH irr to skin, eyes and mu mem and can cause pulmonary edema. Absorbed via skin.
Fire Hazard: Dangerous, when exposed to heat, flame or oxidizers.
To Fight Fire: Spray, foam, CO_2, dry chemical.

METHYL ISOCYANIDE. Syn: *methyl isonitrile.* Colorless liquid. CH_3NC, mw: 41.1, mp: −45°, bp: 59.6°, d: 0.7464 @ 20°/4°.
THR = HIGH. See cyanides.
Explosion Hazard: Severe, when shocked or exposed to heat. Has exploded upon distillation.
Disaster Hazard: Highly dangerous; shock will explode it; when heated to decomp, emits highly toxic fumes of cyanides.

METHYL ISONITRILE. See methyl isocyanide.

METHYL ISOPROPENYL KETONE. $CH_3COC:CH_2(CH_3)$, mw: 84, lel = 1.8%, uel = 9.0%, vap. d: 2.9, bp: 98°.
THR = Probably MOD. See also ketones.

METHYL ISOPROPYL CARBINOL. $CH_3C_3H_7CHOH$, mw: 88.2, flash p: 103°F (CC), vap. d: 3.04, d: 0.8.
THR = Details U. See also alcohols.
Fire Hazard: Mod, when exposed to heat or flame; can react with oxidizing materials.
To Fight Fire: Alcohol foam, CO_2, dry chemical.

1-METHYL-3-ISOPROPYL PHENANTHRENE.
THR = An exper carc. [*23*]

METHYL ISOTHIOCYANATE. Syn: *methyl mustard oil.* Crystalline, slightly water-sol. CH_3NCS, mw: 73.1, 36°, bp: 119°.
Acute tox data: Oral LD_{50} (rat) = 305 mg/kg; oral LD_{50} (mouse) = 97 mg/kg. [*3*]
THR = HIGH irr to skin, eyes and mu mem. A powerful irr. A military poison.
Disaster Hazard: Very dangerous. See cyanides and sulfur compounds for effects when heated to decomp.

METHYL LACTATE. Colorless liquid, sol in water, decomp. $C_4H_8O_3$, mw: 104.1, bp: 144°, lel = 2.2% @ 212°F, flash p: 121°F (CC), d: 1.09, autoign. temp.: 725°F, vap. d: 4.02, ulc: 30–35.
THR = MOD irr to skin eyes and mu mem.
Fire Hazard: Mod, when exposed to heat, flame or oxidizers.
Explosion Hazard: Mod, when exposed to heat or flame.
Disaster Hazard: Mod dangerous; vapor-air mixtures may explode; can react with oxidizing materials.
To Fight Fire: CO_2, dry chemical.

METHYL LITHIUM. H_3CLi, mw: 22.
THR = Self-ignites in air. [*19*]

METHYL MAGNESIUM BROMIDE IN ETHYL ETHER. CH_3MgBr + ETHYL ETHER.
THR = A dangerously flam material. See also ethers, bromides.
Fire Hazard: See ether.
Disaster Hazard: See ether and bromides.

METHYL MERCAPTAN. Syn: *methanethiol.* Liquid or gas. Odor of rotten cabbage. CH_3SH, mw: 48.10, bp: 7.6°, mp: −123.1°, flash p: 0°F, d: 0.868 @ 20°/4°, vap. d: 1.66, lel = 3.9%, uel = 21.8%.
Acute tox data: Inhal LC_{LO} (rat) = 10,000 ppm; scLD_{50} (mouse) = 2.4 mg/kg. [*3*]
THR = HIGH via sc and MOD via inhal. Methyl mercaptan is a common air contaminant.
Fire Hazard: Very dangerous, when exposed to heat, flame or oxidizers.
Explosion Hazard: U.
Disaster Hazard: Dangerous; on decomp, emits highly toxic fumes of SO_x; will react with water, steam or acids to produce toxic and flam vapors; can react vigorously with oxidizing materials.
To Fight Fire: Alcohol foam, CO_2, dry chemical.

3-METHYL MERCAPTO-N-METHYL-4-AMINOAZOBENZENE.
THR = An exper carc. [*23*]

β-METHYL MERCAPTOPROPIONALDEHYDE. C_4H_8OS, mw: 104.3, flash p: 142°F, autoign. temp.: 491°F, d: 1.03, vap. d: 3.6, bp: 165° (approx).
THR = No data.
Fire Hazard: Mod, via heat, flames, oxidizers.
To Fight Fire: Water spray or mist, dry chemical, CO_2, foam.

METHYL MERCURIC CHLORIDE. White crystals, characteristic odor. $CH_3 \cdot Hg \cdot Cl$, mw: 241, d: 4.063, mp: 170°.

Acute tox data: ipLD$_{50}$ (rat) = 11 mg/kg; oral LD$_{50}$ (guinea pig) = 21 mg/kg. [*3*]
THR = HIGH via oral and ip routes. See also mercury compounds, organic. An exper teratogen. [*3*]

METHYL MERCURIC CYANOGUANIDINE. $C_3H_6N_4Hg$, mw: 298.7.
Acute tox data: Oral LD$_{50}$ (rat) = 32 mg/kg; ipLD$_{50}$ (rat) = 13 mg/kg. [*3*]
THR = HIGH via oral and ip routes. An exper teratogen. [*3*] See also mercury compounds, organic.

METHYL MERCURY DICYANDIAMINE. See methyl mercuric cyanoguanidine.

METHYL MERCURY IODIDE. Crystals. CH_3HgI, mw: 342.6, vap. d: 11.8.
THR = HIGH. See mercury compounds, organic, and iodides.

METHYL METHACRYLATE (MONOMER). Colorless liquid, very slightly sol in water. $CH_2C(CH_3)COOCH_3$, mw: 100.11, mp: −50°, bp: 101.0°, lel = 1.7%, uel = 8.2%, flash p: 50°F (OC), d: 0.936 @ 20°/4°, vap. press: 40 mm @ 25.5°, vap. d: 3.45.
Acute tox data: In humans 125 ppm is an irr and 150 mg/m^3 causes CNS symptoms. [*3*] Inhal LC$_{50}$ (rat) = 3750 ppm; ipLD$_{50}$ (rat) = 1328 mg/kg; scLD$_{50}$ (rat) = 7500 mg/kg; oral LD$_{50}$ (guinea pig) = 6300 mg/kg. [*3*]
THR = MOD via inhal, ip, sc and oral routes. An exper carc via ip. [*3*] Methyl methacrylate is a common air contaminant.
Fire Hazard: Dangerous, when exposed to heat or flame; can react with oxidizing materials.
Explosion Hazard: Mod, when exposed to heat, sparks or flame. Reacts violently with benzoyl peroxide. [*19*]
To Fight Fire: Foam, CO_2, dry chemical.

METHYL METHACRYLATE, POLYMER.
THR = An exper neo [*3*] via imp route. [*103*] Violent reaction with F_2. [*19*]

METHYL METHANE. See ethane.

METHYL METHANE SULFONATE. $C_2H_6O_3S$, mw: 110.1.
THR = An exper (+) carc, mutagen, neo [*3, 1, 23*] via sc, iv, ip and oral routes.

METHYL METHANOATE. See methyl formate.

1-METHYL-6-(METHYLALLYL)-2,5-DITHIOBIUREA. $C_7H_{14}N_4S_2$, mw: 218.4.
THR = An exper teratogen [*3*] via oral route.

S-METHYL-*n*-([METHYLCARBAMOYL]OXY) THIOACETAMIDE. See lannate.

2-METHYL-2-(METHYLTHIO)PROPIONALDEHYDE-*o*-(METHYLCARBAMOYL) OXIME. See temik.

METHYL MONOCHLOROACETATE. Clear liquid. $CH_2ClCOOCH_3$, mw: 108.53, bp: 132°, flash p: 116°F (CC), d: 1.227, vap. d: 3.74.
THR = MOD irr to skin, eyes and mu mem.
Fire Hazard: Mod, when exposed to heat, flame or oxidizers.
Disaster Hazard: Dangerous; when heated to decomp, emits highly toxic fumes of phosgene; can react with oxidizing materials.
To Fight Fire: CO_2, dry chemical.

***n*-METHYLMORPHINE.** See codeine.

4-METHYL MORPHOLINE. $C_5H_{11}ON$, mw: 101.2, flash p: 75°F, d: 0.9, vap. d: 3.5, bp: 115°.
Acute tox data: Oral LD$_{50}$ (rat) = 2720 mg/kg; inhal LC$_{LO}$ (rat) = 2000 ppm for 4 hrs; dermal LD$_{50}$ (rabbit) = 1350 mg/kg. [*3*]
THR = MOD via oral, inhal and dermal routes.
Fire Hazard: Dangerous, via heat, flames, oxidizers.
To Fight Fire: Alcohol foam.

METHYL MUSTARD OIL. See methyl isothiocyanate.

1-METHYL NAPHTHALENE. Colorless liquid, insol in water, sol in alcohol and ether. $C_{10}H_7CH_3$, mw: 142, d: 1.0202 @ 20°/4°, mp: −22°, bp: 244.6°, autoign. temp.: 984°F.
THR = No details. See also 2-methyl naphthalene.
Fire Hazard: Mod, via heat, flames, oxidizers.
To Fight Fire: Dry chemical, CO_2, water spray or mist, foam.

2-METHYL NAPHTHALENE. Solid, insol in water, sol in alcohol and ether. $C_{10}H_7CH_3$, mw: 142, d: 1.0058 @ 20°/4°, bp: 241.1°, mp: 34.58°.
THR = Details U. Limited animal exper suggest HIGH toxicity. See also naphthylamine.

3-METHYL-2-NAPHTHYLAMINE. $C_{11}H_{11}N$, mw: 157.2.
THR = An exper carc [*3, 23*] via sc route.

3-METHYL-2-NAPHTHYLAMINE HYDROCHLORIDE. $C_{11}H_{11}N \cdot HCl$, mw: 193.7.
THR = An exper carc [*3*] via oral route.

METHYL-1-NAPHTHYL CARBAMATE. See carbaryl.

***n*-METHYL-*n*-(1-NAPHTHYL)MONOFLUOROACETAMIDE.** See MNFA.

METHYL NITRATE. Colorless liquid. CH_3NO_3, mw: 77.04, bp: 65° (explodes), d: 1.208 @ 20°/4°, vap. d: 2.66, mp: −83°.
Acute tox data: Oral LD$_{50}$ (rat) = 115 mg/kg; dermal LD$_{50}$ (rat) = 213 mg/kg; scLD$_{50}$ (rat) = 41 mg/kg;

$ipLD_{50}$ (mouse) = 164 mg/kg; $scLD_{50}$ (guinea pig) = 1 mg/kg. [*3*]
THR = HIGH irr to skin, eyes and mu mem and via oral, dermal, sc and ip routes.
Fire Hazard: See nitrates.
Explosion Hazard: Severe, when shocked or exposed to heat. A rocket fuel.
Disaster Hazard: Dangerous! See nitrates.

METHYL NITRITE. Gas. CH_3ONO, mw: 61.04, mp: −17°, bp: −12°, d: 0.991 @ 15°.
Acute tox data: Inhal LC_{LO} (rat) = 250 ppm for 4 hrs. [*3*]
THR = HIGH via inhal route. See nitrites. Narcotic in high conc. See also amyl nitrite.
Fire Hazard: Dangerous, when exposed to heat or flame.
Explosion Hazard: Severe, when exposed to heat or flame.
Disaster Hazard: Highly dangerous; when heated, emits highly toxic fumes of NO_x; can react vigorously with oxidizing materials.

METHYL NITROBENZENE. See *p*-nitrotoluene.

1-METHYL-3-NITRO-1-NITROSOGUANIDINE. Crystals. $C_2H_5N_5O_3$, mw: 147.1.
Acute tox data: Oral LD_{50} (rat) = 400 mg/kg; $scLD_{50}$ (rat) = 420 mg/kg. [*3*]
THR = HIGH via oral and sc routes. See nitrates. An exper (+) carc [*3, 10*] via oral and ip, sc and rectal routes.

4-METHYL-1-[(5-NITROFURFURYLIDENE) AMINE]-2-IMIDAZOLIDINONE. $C_8H_{10}O_4N_4$, mw: 226.2.
THR = An exper neo [*3, 23*] via oral route.

METHYL NITROIMIDAZOLYL MERCAPTOPURINE. $C_9H_7O_2N_7S$, mw: 277.3.
THR = An exper carc [*3, 23*] via im route.

3-METHYL-4-NITROPYRIDINE-1-OXIDE. $C_6H_6O_3N_2$, mw: 154.1.
THR = An exper neo [*3*] via sc route.

5-METHYL-4-NITROQUINOLINE-1-OXIDE. $C_{10}H_8O_3N_2$, mw: 204.2.
THR = An exper neo [*3*] via sc route.

6-METHYL-4-NITROQUINOLINE-1-OXIDE.
THR = An exper neo [*3*] via sc route.

7-METHYL-4-NITROQUINOLINE-1-OXIDE.
THR = An exper neo [*3*] via sc route.

8-METHYL-4-NITROQUINOLINE-1-OXIDE.
THR = An exper neo [*3*] via sc route.

N-METHYL-N-NITROSOACETAMIDE. $C_3H_6O_2N_2$, mw: 102.1.
THR = HIGH acute via oral route. An exper carc [*3, 23*] via oral route to rats.

N-METHYL-N-NITROSOALLYL AMINE. $C_4H_8ON_2$, mw: 100.1.
THR = HIGH via oral route. An exper carc and neo [*3, 23*] via oral and iv routes to rats.

2-(N-METHYL-N-NITROSO)AMINO ACETONITRILE. $C_3H_5ON_3$, mw: 99.1.
THR = An exper carc [*3*] via oral route. [*23*]

2-(N-METHYL-N-NITROSOAMINO)-PHENYL-1-PROPANOL. $C_{10}H_{14}O_2N_2$, mw: 194.3.
THR = An exper carc [*3*] via ip route.

4-(4-N-METHYL-N-NITROSOAMINOSTYRYL) QUINOLINE. $C_{18}H_{15}ON_3$, mw: 289.4.
THR = An exper neo [*3, 23*] via oral route.

N-METHYL-N-NITROSOANILINE. $C_7H_8ON_2$, mw: 136.2.
THR = An exper neo, teratogen and carc [*3, 23*] via oral, sc and ip routes. HIGH via oral and ip routes.

N-METHYL-N-NITROSO-β,D-GALACTOSAMINE.
THR = An exper carc. [*23*]

1-METHYL-1-NITROSOBIURET. $C_3H_6O_3N_4$, mw: 146.1.
THR = HIGH acute via oral route. An exper carc [*3, 23*] via oral route.

METHYL NITROSOCYANAMIDE. $C_2H_3ON_3$, mw: 85.1.
Acute tox data: Oral LD_{50} (mice) = 69 mg/kg. [*103*]
THR = HIGH via oral route. A poison. An exper carc to rats via oral route. [*103*]

METHYL-N-NITROSOCYCLOHEXYLAMINE. $C_7H_{14}ON_2$, mw: 142.2.
Acute tox data: Oral LD_{50} (rat) = 30 mg/kg; $ipLD_{50}$ (rat) = 28 mg/kg. [*3*]
THR = HIGH via oral and ip routes. An exper carc [*3*] via oral route. [*23*]

1-METHYL-N-NITROSODIETHYLAMINE. $C_5H_{12}ON_2$, mw: 116.2.
Acute tox data: Oral LD_{50} (rat) = 1100 mg/kg. [*3*]
THR = MOD via oral route. An exper carc to rats via oral route. [*3, 23*]

N-METHYL-N-NITROSOETHYLAMINE. $C_3H_8ON_2$, mw: 88.1.
Acute tox data: Oral LD_{50} (rat) = 90 mg/kg. [*3*]
THR = HIGH via oral route. An exper carc via oral route. [*3, 23*]

N-METHYL-N-NITROSO-β,D-GLUCOSYL AMINE. $C_7H_{14}O_6N_2$, mw: 222.2.
THR = An exper carc [*3, 23*] via oral route.

N-METHYL-N′-NITROSOGUANIDINE. An exper mutagen and exper carc. [23]

N-METHYL-N-NITROSOHEPTYL AMINE. $C_8H_{18}ON_2$, mw: 158.3.
THR = HIGH via sc route. An exper neo [3] via sc route. [23]

4-METHYL-4-(N-NITROSOMETHYL AMINO)-2-PENTANONE. $C_7H_{14}O_2N_2$, mw: 158.2.
THR = MOD via oral route. An exper neo [3] via oral route.

N-METHYL-N-NITROSOPENTYL AMINE. $C_6H_{14}ON_2$, mw: 130.2.
Acute tox data: Oral LD_{50} (rat) = 120 mg/kg; scLD_{50} (rat) = 120 mg/kg. [3]
THR = HIGH via oral and sc routes. An exper carc [3] via sc and oral routes. [23]

N-METHYL-N-NITROSOPHENETHYL AMINE. $C_9H_{12}ON_2$, mw: 164.2.
Acute tox data: Oral LD_{50} (rat) = 48 mg/kg. [3]
THR = HIGH via oral route. An exper carc [3] via oral route. [23]

N-METHYL-N-NITROSO-4-PHENYLAZO ANILINE. $C_{13}H_{12}ON_4$, mw: 240.3.
THR = MOD via acute oral route to rats. An exper neo [3] via oral route. [23]

1-METHYL-4-NITROSOPIPERAZINE. $C_5H_{11}ON_3$, mw: 129.2.
THR = MOD via acute oral route. An exper neo [3] via oral route. [23]

2-METHYL-N-NITROSOPIPERIDINE. $C_6H_{12}ON_2$, mw: 128.2.
THR = An exper carc [3] via oral route. [23]

2-METHYL-N-NITROSOPIPERIDINE (−).
Acute tox data: Oral LD_{50} (rat) = 600 mg/kg. [3]
THR = MOD via oral route. An exper carc. [3, 23] via oral route.

2-METHYL-N-NITROSOPIPERIDINE (+).
Acute tox data: Oral LD_{50} (rat) = 600 mg/kg. [3]
THR = MOD via oral route. An exper carc [3, 23] via oral route.

3-METHYL-N-NITROSOPIPERIDINE.
THR = An exper carc [3] via oral route.

4-METHYL-N-NITROSOPIPERIDINE.
THR = An exper carc [3] via oral route.

N-METHYL-N-NITROSOPROPIONAMIDE. $C_4H_8O_2N_2$, mw: 116.1.
THR = An exper carc [3] via oral route.

N-METHYL-N-NITROSOPROPYLAMINE. $C_4H_{10}ON_2$, mw: 102.2.
Acute tox data: scLD_{50} (hamster) = 493 mg/kg. [3]
THR = MOD–HIGH via sc route. An exper carc [3] via sc route. [23]

N-METHYL-*n*-NITROSOUREA. $NNOCH_3CONH_2$, mw: 103.1.
THR = Can detonate with ($KOH + CH_2Cl_2$). [19] An exper teratogen, neo, and exper (+) transplacental carc [23] via oral, dermal, ip, sc, iv, pa and imp routes. [3]

N-METHYL-N-NITROSOURETHANE. See nitrosomethyl urethane.

N-METHYL-N-NITROSOVINYL AMINE. $C_3H_6ON_2$, mw: 86.1.
Acute tox data: Oral LD_{50} (rat) = 24 mg/kg; inhal LC_{50} (rat) = 22 mg/kg. [3]
THR = HIGH via oral and inhal routes. An exper carc [3] via inhal and oral route. [23]

METHYL NONYL KETONE. Colorless liquid, insol in water. $(CH_3)(C_9H_{19})CO$, mw: 170, mp: 12°, bp: 223°, flash p: 192°F (CC), d: 0.829 @ 30°, vap. d: 5.9.
Acute tox data: Oral LD_{50} (rat) = 5000 mg/kg; oral LD_{50} (mouse) = 3880 mg/kg. [3]
THR = MOD via oral route. See ketones. A dog- and cat-repellant. Used in training of pets.
Fire Hazard: Mod, when exposed to heat or flame; can react with oxidizing materials.
To Fight Fire: CO_2, dry chemical.

METHYLNORBORNENE DICARBOXYLIC ANHYDRIDE.
THR = HIGH toxicity material. An irr.

2-METHYL OCTANE. C_9H_{20}, mw: 138.2, autoign. temp.: 428°F, d: 0.71, vap. d: 4.43, bp: 143.3.°.
Fire Hazard: Dangerous, via heat, flames, oxidizers.
To Fight Fire: Water spray, foam, fog, dry chemical, CO_2.

3-METHYL OCTANE. autoign. temp.: 428°F, d: 0.72, vap. d: 4.43, bp: 143.8°.
Fire Hazard: See 2-methyl octane.
To Fight Fire: See 2-methyl octane.

4-METHYL OCTANE. autoign. temp.: 428°F, bp: 142.4°.
Fire Hazard: See 2-methyl octane.
To Fight Fire: See 2-methyl octane.

METHYL OCTYL DIAZINE-1-OXIDE. $C_9H_{20}ON_2$, mw: 172.3.
THR = An exper carc [3] via ip route.

METHYL OLEATE. $C_{19}H_{36}O_2$, mw: 296.6.
THR = An exper carc [3] via dermal route.

9-METHYLOL CARBAZOLE. Crystals. $(C_6H_4)_2NCH_2OH$, mw: 197.2.
THR = Details U; a fumigant. Probably toxic.
Disaster Hazard: Dangerous; when heated, emits

highly toxic fumes of NO_x; can react on contact with oxidizing materials.

METHYLOL PINENE. Water white to light straw-colored liquid. $C_{11}H_{17}OH$, mw: 146.3, bp: 111° @ 10 mm, d: 0.963 @ 25°, vap. d: 5.05.
THR = U.
Fire Hazard: Slight, when exposed to heat or flame; can react with oxidizing materials.

METHYL ORANGE. See diazoben.

METHYL OXIDE. See dimethyl ether.

METHYL OXIRANE. See propylene oxide.

4-METHYL-2-OXETANONE. $C_4H_5O_2$, mw: 86.1.
THR = An exper (+) carc, neo [*3, 13*] via oral, sc and dermal routes. LOW via oral route.

7-METHYL-17-OXO-16,17-DIHYDRO-15H-CYCLOPENTA(a)PHENANTHRENE.
THR = An exper carc. [*23*]

11-METHYL-15-OXO-16,17-DIHYDRO-15H-CYCLOPENTA(a)PHENANTHRENE.
THR = An exper carc. [*23*]

METHYL-12-OXO-*trans*-1-OCTADECENOATE. $C_{19}H_{34}O_3$, mw: 310.5.
THR = An exper carc via dermal route. [*3, 23*]

11-METHYL-1-OXO-1,2,3,4-TETRAHYDROCHRYSENE. $C_{19}H_{16}O$, mw: 260.4.
THR = An exper carc [*3*] via dermal route.

7-METHYL-11-OXYMETHYL-16,17-DIHYDRO-15H-CYCLOPENTA(a)PHENANTHRENE.
THR = An exper carc. [*23*]

1-(4-METHYL OXYPHENYL)-3,3-DIMETHYL TRIAZENE.
THR = An exper carc. [*23*]

METHYL PARABEN. See *p*-hydroxybenzoic acid, methyl ester.

METHYL PARATHION. See *o,o*-dimethyl-*o,p*-nitrophenyl phosphorothioate.

METHYL PARATHION MIXTURE. See methyl parathion.

METHYL PENTADECYL KETONE. Insol in water. $C_{15}H_{31}COCH_3$, mw: 254, bp: 313°F @ 3 mm, flash p: 248°F.
THR = U. See also ketones.
Fire Hazard: Slight, when exposed to heat or flame.
To Fight Fire: Dry chemical, CO_2, water spray or fog.

α-METHYL PENTADIENE-1,3. Liquid. C_6H_{10}, mw: 82.14, bp: 75-77°, flash p: <−4°F, d: 0.7184 @ 20°/4°, vap. d: 2.83.
THR = Probably irr and narcotic in high conc.
Fire Hazard: Dangerous, when exposed to heat, flame or oxidizers.
Disaster Hazard: Dangerous! Keep away from heat and open flame.
To Fight Fire: Foam, CO_2, dry chemical.

4-METHYL-1,3-PENTADIENE. Flash p: −30°F.
Fire Hazard: See 2-methyl-1,3-pentadiene.
To Fight Fire: See 2-methyl-1,3-pentadiene.

METHYL PENTALDEHYDE. Syn: *methyl pentanal.* Very slightly sol in water.
$CH_3CH_2CH_2C(CH_3)HCHO$, mw: 100, d: 0.8092, bp: 118.3°, fp: −100°, flash p: 68°F (OC).
THR = U. See also aldehydes.
Fire Hazard: Dangerous, when exposed to heat, flame or oxidizers.
To Fight Fire: Dry chemical, CO_2.

METHYL PENTANAL. See methyl pentaldehyde.

2-METHYL PENTANE. See isohexane.

3-METHYLPENTANE. Syn: *diethylmethylmethane.* Colorless liquid. C_6H_{14}, mw: 86.17, bp: 63.3°, fp: −118° (sets to a glass), flash p: <20°F, d: 0.664 @ 20°/4°, vap. press: 100 mm @ 10.5°, vap. d: 2.97.
THR = Details U; may have narcotic or anesthetic properties.
Fire Hazard: Dangerous, when exposed to heat or flame. Can react vigorously with oxidizing agents.
Explosion Hazard: U.
Disaster Hazard: Dangerous; keep away from heat and open flame; can react vigorously with oxidizing materials.
To Fight Fire: Foam, CO_2, dry chemical.

2-METHYL-1,3-PENTANEDIOL. Sol in water.
$CH_3CH_2CH(OH)CH(CH_3)CH_2OH$, mw: 118.17, d: 0.9745, bp: 220.30°, fp: −30°, flash p: 230°F.
THR = U. See also alcohols.
Fire Hazard: Slight, when exposed to heat or flame.
To Fight Fire: Dry chemical, CO_2, water mist.

2-METHYL-2,4-PENTANEDIOL. See hexylene glycol.

3-METHYL PENTANEDIOL-1,5. Clear liquid.
$HOCH_2CH_2CH(CH_3)CH_2CH_2OH$, mw: 118.17, bp: 248.4°, fp: −60°, d: 0.9755 @ 20°/20°, vap. press: <0.01 mm @ 20°, vap. d: 4.
THR = Details U. See also alcohols.
Fire Hazard: Mod, when exposed to heat or flame; can react with oxidizing materials.
To Fight Fire: Foam, dry chemical, CO_2.

2-METHYL PENTANOIC ACID. Water white liquid, insol in water. $C_3H_7CH(CH_3)COOH$, mw: 116, d: 0.8947 @ 20°/4°, bp: 126.5° @ 750 mm, vap. press: 0.02 mm @ 20°, fp: sets to glass <−85°, flash p: 225°F (OC).
Acute tox data: Oral LD_{50} (rat) = 2040 mg/kg. [*3*]

Fire Hazard: Low, when exposed to heat, flame or oxidizers.
To Fight Fire: Dry chemical, CO_2, foam.

2-METHYL PENTANOL-1. See amyl methyl alcohol.

4-METHYL PENTANOL-2. See methyl amyl alcohol.

4-METHYL-2-PENTANONE. See methyl isobutyl ketone.

2-METHYL-1-PENTENE. Syn: *1-methyl-1-propyl ethylene*. Liquid. $CH_2C(CH_3)CH_2CH_2CH_3$, mw: 84.2, mp: −135.8°, bp: 62°, flash p: <20°F, d: 0.684 @ 15.5°/15.5°, vap. press: 326 mm @ 37.3°, vap. d: 2.9, autoign. temp.: 572°F.
THR = U. Probably irr and narcotic in high conc.
Fire Hazard: Dangerous, when exposed to heat, flame or oxidizers.
Explosion Hazard: U.
Disaster Hazard: Dangerous; keep away from heat and open flame; can react vigorously with oxidizing materials.
To Fight Fire: CO_2, dry chemical.

2-METHYL PENTENE-2. A liquid. C_6H_{12}, mw: 84.2, bp: 66.9°, flash p: <20°F, d: 0.690 @ 15.5°/15.5°, vap. press: 326 mm @ 38°, vap. d: 2.9.
THR = Irr and narcotic in high conc.
Fire Hazard: Dangerous, when exposed to heat, flame or oxidizers.
Explosion Hazard: U.
Disaster Hazard: Dangerous, upon exposure to heat or flame; can react vigorously with oxidizing materials.
To Fight Fire: Foam, CO_2, dry chemical.

4-METHYL PENTENE. A liquid. C_6H_{12}, mw: 84.2, mp: −153.6°, bp: 54°, flash p: <20°F, d: 0.668 @ −15.5°/15.5°, vap. press: 424 mm @ 38°, vap. d: 2.9, autoign. temp.: 572°F.
THR = Irr and narcotic in high conc.
Fire Hazard: Dangerous, when exposed to heat, flame or oxidizers.
Explosion Hazard: U.
Disaster Hazard: Dangerous, upon exposure to heat or flame; can react vigorously with oxidizing materials.
To Fight Fire: Foam, CO_2, dry chemical.

4-METHYL-2-PENTENE. Syn: *1-isopropyl-2-methyl ethylene*. Liquid. $CH_3CHCHCH(CH_3)_2$, mw: 84.16, mp: −134.4°, bp: 58°, d: 0.670 @ 20°/4°, vap. d: 2.90, flash p: <20°F.
THR = Irr and narcotic in high conc.
Fire Hazard: Dangerous via heat, flames, oxidizers.
Disaster Hazard: Mod; when heated, emits acrid fumes; can react with oxidizing materials.
To Fight Fire: Dry chemical, CO_2, foam.

4-METHYL-3-PENTEN-2-ONE. See mesityl oxide.

4-METHYL PENTYL-2-ACETATE. See methyl amyl-acetate.

3-METHYL PENTYN-3-OL. Syns: *dormison, 2-ethynyl-2-butanol*. Colorless, mobile liquid, acrid odor, burning taste, sol in water, ether, etc. $HC{:}CC(OH)CH_3C_2H_5$, mw: 98.14, mp: −30.6°, bp: 122°, flash p: 101°F (OC), d: 0.8688 @ 20°, vap. d: 3.38.
Acute tox data: Oral LD_{50} (rat) = 300 mg/kg; oral LD_{50} (mouse) = 698 mg/kg; scLD_{50} (mouse) = 750 mg/kg. [*3*]
THR = HIGH–MOD via oral and sc routes. Used as a soporific. Average doses may produce dermatitis, eructations, psychoses and CNS abnormalities. Overdoses can produce coma and death.
Fire Hazard: Mod; can react with oxidizing materials, heat or flame.
To Fight Fire: Alcohol foam, mist, fog.

3-METHYL-1-PENTYN-3-OL CARBAMATE. Syn: *oblivon*. Crystals, slightly water-sol. $C_7H_{11}NO_2$, mw: 141.2, mp: 57°, bp: 121° @ 16 mm.
Acute tox data: Oral LD_{50} (mouse) = 445 mg/kg. [*3*] Oral TD_{LO} (human) = 2.9 mg/kg → psychological disturbance.
THR = HIGH via oral route. Used as a tranquilizer. Overdoses may cause CNS depression and death. Potentiates action of alcohol and barbiturates. See carbamates.
Disaster Hazard: Dangerous; when heated to decomp, forms highly toxic fumes of NO_x.

METHYL PERCHLORATE. CH_3ClO_4, mw: 114.5.
THR = High explosive. Very sensitive to heat, shock and friction. [*19*] See also perchlorates.

2-METHYL PHENANTHRO(2,1-d)THIAZOLE. $C_{16}H_{11}NS$, mw: 249.3.
THR = An exper carc [*3*] via dermal route.

***d*,2-METHYL PHENETHYL AMINE SULFATE.** Syn: *dexedrine sulfate*. $C_{18}H_{26}N_2 \cdot H_2SO_4$, mw: 368.5.
Acute tox data: Oral LD_{50} (rat) = 38 mg/kg; scLD_{50} (rat) = 200 mg/kg; oral LD_{50} (dog) = 3 mg/kg; ivLD_{50} (mouse) = 30 mg/kg. [*3*]
THR = HIGH via oral, sc and iv routes. An exper teratogen. [*3*] Habit forming.

***o*-METHYL PHENOL.** See *o*-cresol.

***m*-METHYL PHENOL.** See *m*-cresol.

METHYL PHENYLACETATE. Colorless liquid. $C_6H_5CH_2COOCH_3$, mw: 150.17, d: 1.062, vap. d: 5.18.
Acute tox data: Oral LD_{50} (rat) = 2550 mg/kg; dermal LD_{50} (rabbit) = 2400 mg/kg. [*3*]
THR = MOD via oral and dermal routes.
Disaster Hazard: Mod dangerous; when heated to decomp, emits toxic fumes.

2-METHYL-4-PHENYL ANILINE.
THR = An exper carc. [*23*]

2′-METHYL-4-PHENYL ANILINE.
THR = An exper carc. [*23*]

3-METHYL-4-PHENYL ANILINE.
THR = An exper carc. [*23*]

N-METHYL-*p*-(PHENYLAZO)ANILINE. $C_{13}H_{13}N_3$, mw: 211.3.
THR = An exper carc [*3*] via oral and sc routes.

N-METHYL-4-(PHENYLAZO)ANILINE.
THR = An exper neo. [*3*]

N-METHYL-4-(PHENYLAZO)ANISIDINE.
$C_{14}H_{15}ON_3$, mw: 241.3.
THR = An exper carc [*3*] via oral route.

N-METHYL-4-(PHENYLAZO)-*o*-TOLUIDINE.
$C_{14}H_{14}N_3$, mw: 224.3.
THR = An exper neo [*3*] via oral route.

7-METHYL-9-PHENYL BENZ(c)ACRIDINE.
$C_{24}H_{17}N$, mw: 319.4.
THR = An exper neo [*3, 23*] via dermal route.

METHYL PHENYL CARBINOL. See phenyl methyl carbinol.

METHYL PHENYL CYCLOSILOXANE.
THR = An exper teratogen [*3*] via im route.

1-METHYL-1-PHENYL ETHYLENE. See α-methyl styrene.

METHYL PHENYLETHYL NITROSAMINE.
$C_9H_{12}ON_2$, mw: 164.2.
THR = HIGH via oral route. An exper carc [*3*] via oral route.

3-METHYL-3-PHENYL GLYCIDIC ACID, ETHYL ESTER. See ethyl methyl phenylglycidate.

METHYL PHENYL KETONE. See phenyl methyl ketone.

2-METHYL-2-PHENYLPROPANE. See *tert*-butylbenzene.

3-METHYL-1-PHENYL-5-PYRAZOLONE. White powder. $C_6H_5N_2COCH_2CCH_3$, mw: 174.2, bp: 191° @ 7 mm, mp: 128.9°.
Acute tox data: Oral LD_{50} (rat) = 1420 mg/kg; ivLD_{50} (mouse) = 220 mg/kg. [*3*]
THR = HIGH via iv and MOD via oral routes.
Disaster Hazard: Dangerous; when heated to decomp, emits highly toxic fumes of NO_x; can react with oxidizing materials.

2-METHYL-3-PHENYL-1-PROPANOL. $C_{10}H_{14}O$, mw: 150.2.
THR = Violent reaction with (H_2O_2 + H_2SO_4). [*19*]

3-METHYL-1-PHENYL-3-(2-SULFOETHYL) TRIAZENE, SODIUM SALT. $C_9H_{12}O_3N_3S \cdot Na$, mw: 265.3.
THR = An exper neo [*3, 23*] via sc route.

3-METHYL-1-PHENYL TRIAZENE. $C_7H_9N_3$, mw: 135.2.
THR = An exper neo and carc. [*3*]

METHYL PHOSPHINE. Colorless gas. CH_3PH_2, mw: 48.0, bp: −14°.
THR = HIGH via inhal route. See also phosphine.
Fire Hazard: Dangerous, when exposed to heat or flame. Self-ignites in air. [*19*]
Disaster Hazard: Dangerous; when heated, emits highly toxic fumes of PO_x; can react vigorously with oxidizing materials.

METHYL PHOSPHONIC DICHLORIDE.
Acute tox data: Inhal LC_{50} (rat) = 26 ppm for 4 hrs. [*3*]
THR = HIGH irr via skin, eyes and mu mem and via inhal route. Animal exper show strong irr. See also hydrochloric acid.
Disaster Hazard: Dangerous; see phosphorous compounds and hydrochloric acid.

METHYL PHOSPHONOTHIONIC DICHLORIDE. ANHYDROUS. CH_3PSCl_2, mw: 149.0.
THR = HIGH irr to skin, eyes and mu mem. A corrosive material.

METHYL PHOSPHONOUS DICHLORIDE.
THR = HIGH irr to skin, eyes and mu mem. A corrosive material.

3-METHYLPHTHALANILIC ACID. See duraset.

METHYL PHTHALATE. See dimethyl phthalate.

METHYL PHTHALYL ETHYL GLYCOLATE. Liquid. $CH_3CO_2C_6H_4CO_2CH_2CO_2C_2H_5$, mw: 266.2, bp: 310°, flash p: 380°F (CC), d: 1.220, vap. d: 9.16.
Acute tox data: Oral LD_{50} (rat) = 3200 mg/kg. [*3*]
THR = MOD via oral route.
Fire Hazard: Slight, when exposed to heat or flame; can react with oxidizing materials.
To Fight Fire: CO_2, dry chemical.

METHYL PICRATE. See trinitroanisole.

***n*-METHYL PIPERAZINE.** A hygroscopic solid, typical amine-like odor. $C_5H_{12}N_2$, mw: 100.1, d: 0.9031 20°/20°, mp: 65.5°, bp: 139°, flash p: 108°F (OC), vap. d: 3.5.
Acute tox data: Oral LD_{50} (rat) = 2830 mg/kg; dermal LD_{50} (rabbit) = 1490 mg/kg. [*3*]
THR = MOD via oral and dermal routes.
Fire Hazard: Mod, when exposed to heat or flame; can react with oxidizing materials.
To Fight Fire: Alcohol foam, CO_2, dry chemical.

2-METHYL PROPANE. See isobutane.

2-METHYL-2-PROPANETHIOL. See *tert*-butyl mercaptan.

2-METHYL PROPANAL. See isobutyl aldehyde.

2-METHYL-1-PROPANOL. See isobutyl alcohol.

2-METHYL-2-PROPANOL. See *tert*-butyl alcohol.

2-METHYL PROPANOIC ACID. See isobutyric acid.

2-METHYL PROPENE. See isobutylene.

2-METHYL PROPENITRILE. See methacrylonitrile.

METHYL PROPIONATE. Colorless liquid. $CH_3CH_2COOCH_3$, mw: 88.15, mp: −87.5°, bp: 79.8°, flash p: 28°F (CC), d: 0.937 @ 4°, autoign. temp.: 876°F, vap. press: 40 mm @ 11.0°, vap. d: 3.03, lel = 2.50%, uel = 13%.
Acute tox data: Oral LD_{LO} (rabbit) = 2550 mg/kg. [*3*]
THR = MOD via oral route.
Fire Hazard: Dangerous, when exposed to heat, flame or oxidizers.
Spont Heating: No.
Explosion Hazard: U.
Disaster Hazard: Dangerous; keep away from heat and open flame; can react vigorously with oxidizing materials.
To Fight Fire: Foam, CO_2, dry chemical.

METHYL PROPYL ACETYLENE. Syn: *2-hexyne*. C_6H_{10}, mw: 82.1, d: 0.73, flash p: <14°F, vap. d: 2.83, bp: 85°.
THR = No data.
Fire Hazard: Dangerous, via heat, flames, oxidizers.
To Fight Fire: Dry chemical, CO_2, foam.

12-METHYL-7-PROPYLBENZ(a)ANTHRACENE. $C_{22}H_{20}$, mw: 284.4.
THR = An exper carc [*3*] via im route.

METHYL PROPYL CARBINOL. See 2-pentanol.

METHYL PROPYL CARBINYLAMINE. See *sec*-amylamine.

METHYL PROPYL ETHER. Syn: *neothyl*. Mobile liquid, slightly water-sol, miscible in alcohol and ether. $C_4H_{10}O$, mw: 74.1, d: 0.7494 @ 0°/4°, bp: 38.5°, flash p: <−4°F.
THR = See ethers.
Fire Hazard: Dangerous, via heat, flames, oxidizers.
To Fight Fire: See ethers.

1-METHYL-1-PROPYL ETHYLENE. See 2-methyl-1-pentene.

METHYL-*n*-PROPYL KETONE. See pentanone-2.

α-METHYL PROPYL NITRITE. See *sec*-butyl nitrite.

2-METHYL PYRAZINE. Liquid, pyridine-like odor. $C_5H_6N_2$, mw: 94.12, mp: −29°, bp: 133° @ 737 mm, flash p: 122°F (COC), d: 1.0224 @ 25°/25°, vap. d: 3.2.
Acute tox data: ipLD_{50} (mouse) = 1820 mg/kg. [*3*]
THR = MOD via ip route.
Fire Hazard: Mod, when exposed to heat, flame or oxidizers.
Disaster Hazard: Dangerous; when heated to decomp, emits highly toxic fumes of NO_x; can react with oxidizing materials.
To Fight Fire: Water spray, foam, dry chemical, CO_2.

α-METHYL PYRIDINE. See α-picoline.

3-METHYL PYRIDINE. See β-picoline.

4-METHYL PYRIDINE. See γ-picoline.

1-METHYL-2,β-PYRIDYLPYRROLIDINE. Pale straw-colored liquid. $C_{10}H_{10}N_2$, mw: 158.2, bp: 247°, d: 1.01 @ 20°/20°, vap. d: 5.46.
THR = HIGH. See closely related compound, nicotine.
Disaster Hazard: Dangerous; when heated to decomp, emits highly toxic fumes of NO_x; can react with oxidizing materials.

METHYL PYRROLE-*n*. Liquid, insol in water. N(CH_3)CH:CHCH:CH, mw: 81, d: 0.9, vap. d: 2.8, bp: 112°, fp: −57°, flash p: 61°F.
THR = U. See pyrrole.
Fire Hazard: Dangerous, when exposed to heat, flame or oxidizers.
To Fight Fire: Dry chemical, CO_2, foam.

***n*-METHYL PYRROLIDINE.** Colorless to yellow liquid with penetrating amine-like odor. $C_4H_8NCH_3$, mw: 85.2, bp: 80.5°, fp: −90°, d: 0.8054 @ 20°/20°, flash p: 7°F, vap. d: 2.9.
THR = This material is strongly alkaline. Contact of liquid or vapors of this material with skin, eyes or mu mem should be avoided. See also ammonia.
Fire Hazard: Dangerous; keep away from sparks, heat sources and powerful oxidizers. Keep in closed containers.
Disaster Hazard: Dangerous; when heated to decomp, emits highly toxic fumes.
To Fight Fire: Alcohol foam.

***n*-METHYL-2-PYRROLIDONE.** Syn: *1-methyl-2-pyrrolidone*. Colorless liquid, mild odor. $H_3CNCH_2CH_2CH_2CO$, mw: 99.13, bp: 202°, fp: −24°, flash p: 204°F (OC), d: 1.027 @ 25°/4°, vap. d: 3.4.
Acute tox data: Oral LD_{50} (rat) = 7000 mg/kg. [*3*]
THR = MOD–LOW via oral route.
Fire Hazard: Low, when exposed to heat or flame; can react with oxidizing materials.
To Fight Fire: Foam, CO_2, dry chemical.

2-METHYL QUINOLINE. Syn: *quinaldine*. Colorless, oily liquid, quinoline odor, insol in water, sol in chloroform and ether. $C_6H_4(CH_3)C_3H_2N$, mw: 143.18, d: 1.06, bp: 246°–247°.

Acute tox data: Oral LD_{50} (rat) = 1230 mg/kg; dermal LD_{50} (rabbit) = 1870 mg/kg. [*3*]
THR = MOD via oral and dermal routes.

METHYL RED. Shiny violet crystals. $C_{15}H_{15}O_2N_3$, mw: 269.3.
THR = An exper (±) neo [*3*] via oral route.

METHYL SALICYLATE. Syn: *oil of wintergreen.* Colorless, yellowish or reddish oily liquid. $C_8H_8O_3$, mw: 152.14, bp: 223.3°, ulc: 20–25, flash p: 214°F (CC), fp: −1.2°, d: 1.1840 @ 20.2°/4°, autoign. temp.: 850°F, vap. press: 1 mm @ 54.0°, vap. d: 5.24.
Acute tox data: Oral LD_{50} (rat) = 887 mg/kg; oral LD_{LO} (human) = 170 mg/kg; oral LD_{50} (dog) = 2100 mg/kg. [*3*]
THR = MOD via oral route, possibly HIGH to humans. An exper teratogen [*3*] via pa route. Accidental acute poisoning is not uncommon. Kidney irr, vomiting and convulsions occur.
Fire Hazard: Slight, when exposed to heat or flame; can react with oxidizing materials.
Spont Heating: No.
To Fight Fire: CO_2, dry chemical.

METHYL SEBACATE. See dimethyl sebacate.

METHYL SELENIDE. See dimethyl selenide.

METHYL SILICANE. Liquid. $H_3Si(CH_3)$, mw: 46.1, d: 0.62 @ −57°, mp: −156.4°, bp: 31°.
THR = Details U. See silanes.

***o*-METHYL SILICATE.** Clear liquid. $Si(OCH_3)_4$, mw: 152.2, vap. d: 5.25.
Acute tox data: ipLD_{50} (mouse) = 250 mg/kg; inhal LC_{LO} (rat) = 250 ppm for 4 hrs.
THR = HIGH via inhal and ip routes. This material can cause extensive necrosis (exper), keratoconus and opaque cornea; it also causes severe human eye injuries, as well as necrosis of corneal cells, which progresses long after the exposure has ceased. It is destructive and its effects resist treatment. Permanent blindness is possible from exposure to it. Both the liquid and vapor of this material are dangerous to the eyes. From animal exper, the minimum lethal dose by oral administration has been found to be 0.07 ml/100 g of body weight for the rat and 0.01 ml/100 g of body weight intravenously for the rabbit. Administration of this material to animals resulted in death from within a few hrs to a few days.

The kidney seems to be the organ which is usually injured, regardless of the mode of administration of the toxic material. In less severe cases, degeneration of the convoluted tubules was found, with complete degeneration of the organ in the most severe cases. Pulmonary edema also occurred in those animals who had received intravenous injections. It has been reported that exposure to methyl silicate vapor under certain conditions of humidity or exposure to the liquid may cause the eye damages noted above. This material was found to be more toxic to animals than either ethyl silicate or silicic acid, although it has been thought that the injury caused is largely due to the action of the silicic acid.

METHYL SILYL AMINO DIBORANE. $B_2H_{11}NSiC$, mw: 86.8.
THR = Self-ignites in air. [*19*]

METHYL SODIUM. CH_3Na, mw: 38.
THR = Spont flam in air. [*19*] Highly flam, corrosive and irr material.

METHYL STANNIC ACID. White powder, insol in water. $(CH_3)SnOOH$, mw: 166.7.
THR = Details U. See tin compounds.

METHYL STEARATE. Liquid to semi-solid. $C_{17}H_{35}CO_2CH_3$, mw: 298.5, mp: 38°, bp: 215° @ 15 mm, flash p: 307°F (CC), d: 0.860.
THR = An exper neo via sc route. [*3*]
Fire Hazard: Slight, when exposed to heat or flame; can react with oxidizing materials.
To Fight Fire: CO_2, dry chemical.

2-METHYL-4-STILBENAMINE.
THR = An exper carc. [*23*]

3-METHYL-4-STILBENAMINE.
THR = An exper carc. [*23*]

METHYL STYRENE. See vinyl toluene.

α-METHYL STYRENE. Syn: *1-methyl-1-phenyl ethylene.* Colorless liquid, insol in water. $C_6H_5C(CH_3)CH_2$, mw: 118.17, d: 0.9062 @ 25°/25°, bp: 165.4°, flash p: 129°F, autoign. temp.: 1066°F, lel = 1.9%, uel = 6.1%.
Acute tox data: Oral LD_{50} (rat) = 4900 mg/kg; inhal LC_{LO} (rat) = 3000 ppm. An irr to humans at 600 ppm. [*3*]
THR = MOD via oral and inhal routes.
Fire Hazard: Mod, when exposed to heat, flame or oxidizers.
To Fight Fire: Water, foam, mist, dry chemical.

METHYLSTYRYL PHENYL KETONE. See dypnone.

METHYL SULFATE. See dimethyl sulfate.

METHYL SULFIDE. Syns: *methyl thiomethane, dimethyl sulfide.* Colorless liquid, disagreeable odor, sol in alcohol and ether. $(CH_3)_2S$, mw: 62.13, mp: −83.2° lel = 2.2%, uel = 19.7%, flash p: <0°F, bp: 37.5°–38°, d: 0.8458 @ 21°/4°, vap. d: 2.14, autoign. temp.: 403°F.
Acute tox data: Oral LD_{50} (rat) = 535 mg/kg; oral LD_{50} (mouse) = 3700 mg/kg. [*3*]
THR = MOD via oral route.

Fire Hazard: Very dangerous, when exposed to heat or flame.
Explosion Hazard: Mod.
Disaster Hazard: Dangerous; when heated to decomp, emits highly toxic fumes of SO_x and may explode; can react vigorously with oxidizing materials.
To Fight Fire: CO_2, dry chemical.

METHYL SULFITE. Syn: *dimethyl sulfite.* Colorless liquid. $(CH_3)_2SO_3$, mw: 110.1, bp: 126°, d: 1.242 @ 0°/0°, vap. d: 3.8.
THR = Details U; an irr and probably toxic material.
Disaster Hazard: Dangerous; when heated to decomp, emits highly toxic fumes of SO_x; will react with water, steam or acids to produce toxic and corrosive fumes; can react with oxidizing materials. Reacts with P_2O_3 violently. [*19*]

METHYL SULFOXIDE. See dimethyl sulfoxide.

METHYL SULFURIC ACID. Syns: *methyl hydrogen sulfate, acid methyl sulfate.* Oily liquid, freely sol in water, less sol in alcohol. CH_3OSO_2OH, mw: 112.10.
THR = See dimethyl sulfate. Highly toxic.

N-METHYL TAURINE. See 2-(N-methylamino) ethane sulfonic acid.

METHYL TELLURIDE. See dimethyl telluride.

17-METHYL TESTOSTERONE. $C_{20}H_{28}O_2$, mw: 300.5.
THR = An exper neo [*3*] via oral route.

6-METHYL-1,2,3,4-TETRAHYDROBENZ(a)ANTHRACENE. $C_{19}H_{18}$, mw: 246.4.
THR = An exper carc [*3*] via imp route.

2-METHYL TETRAHYDROFURAN. Syns: *tetrahydromethylfuran, tetrahydrosylvan.* Colorless, mobile liquid, ether-like odor. $OCH_2CH_2CH_2CHCH_3$, mw: 86.13, bp: 80°, flash p: 12°F, d: 0.853 @ 20°/4°, vap. d: 2.97.
Acute tox data: Inhal LC_{50} (rat) = 600 ppm for 4 hrs; dermal LD_{50} (rabbit) = 4500 mg/kg; oral LD_{LO} (rat) = 4860 mg/kg. [*3*]
THR = MOD via inhal, dermal and oral routes.
Fire Hazard: Dangerous, when exposed to heat, flame or oxidizers.
Explosion Hazard: U.
Disaster Hazard: Dangerous; keep away from heat and open flame; can react vigorously with oxidizing materials.
To Fight Fire: Alcohol foam, CO_2, dry chemical.

METHYL-1,2,5,6-TETRAHYDRO-1-METHYL NICOTINATE. See arecoline base.

1-*n*-METHYL TETRAHYDROPAPAVERINE. See laudanosine.

METHYL TETRATHIO-*o*-STANNATE. Solid. $(CH_3S)_4$,Sn, mw: 307.1, mp: 31°, bp: 81° @ 0.001 mm.
THR = See tin compounds and sulfur compounds.

2-(4-METHYL-2-THIAZOLYL)HYDRAZIDE FORMIC ACID. $C_5H_7ON_3S$, mw: 157.2.
THR = An exper neo. [*3*]

2-METHYL THIOETHYL ACRYLATE. $C_6H_{10}O_2S$, mw: 146.2.
Acute tox data: Oral LD_{50} (rat) = 1230 mg/kg; dermal LD_{50} (rabbit) = 1490 mg/kg. [*3*]
THR = MOD via oral and dermal routes.
Disaster Hazard: Dangerous. See sulfur compounds.

6-METHYL-2-THIOURACIL. $C_5H_6ON_2S$, mw: 142.2.
THR = An exper (+) carc; [*1*] a teratogen [*3*] via oral route.

METHYLTIN TRIBROMIDE. White crystals, sol in water, ether, alcohol and benzene. $(CH_3)SnBr_3$, mw: 373.5, mp: 55°, bp: 211° @ 746 mm.
THR = See tin compounds, hydrobromic acid and bromides.

METHYLTIN TRICHLORIDE. Colorless crystals, sol in water and organic solvents. $(CH_3)SnCl_3$, mw: 240.1, mp: 43°.
THR = See tin compounds, hydrochloric acid and chlorides.

METHYLTIN TRIIODIDE. Light yellow crystals, sol in water, ether, alcohol and benzene. $(CH_3)SnI_3$, mw: 514.5, mp: 86.5°.
THR = See tin compounds, hydriodic acid and iodides.

METHYL TOLUENE SULFONATE. Light brown crystals. $CH_3C_6H_4SO_3CH_3$, mw: 186.22, d: 1.230–1.238 @ 25°/25°, vap. d: 6.45, mp: 28°.
Acute tox data: Oral LD_{LO} (rat) = 1000 mg/kg; scLD_{50} (rat) = 250 mg/kg. [*3*]
THR = HIGH via sc and MOD via oral routes. A vesicant and skin-sensitizer. An exper carc [*23, 3*] via sc route.
Disaster Hazard: Dangerous; see sulfonates.

N-METHYL-*p*-(*m*-TOLYLAZO)ANILINE. $C_{14}H_{15}N_3$, mw: 225.3.
THR = An exper carc [*3*] via oral route.

N-METHYL-*p*-(*o*-TOLYLAZO)ANILINE.
THR = An exper carc [*3*] via oral route.

N-METHYL-*p*-(*p*-TOLYLAZO)ANILINE.
THR = An exper neo [*3*] via oral route.

METHYL TOSYLATE. See methyl toluene sulfonate.

METHYL TRIBORINE TRIAMINE (β). Colorless liquid, hydrolyzed by water. $CH_3B_3N_3H_5$, mw: 94.6, mp: −59°, bp: 87°.
THR = Details U. See also boron compounds and amines.

METHYL TRIBORINE TRIAMINE (N). Colorless liquid, hydrolyzed by water. $CH_3B_3N_3H_5$, mw: 94.6, bp: 84°

THR = Details U. See also boron compounds and amines.

METHYL TRIBROMODIPYRIDINE TIN. Solid. $CH_3SnBr_3(C_5H_5N)_2$, mw: 531.7, mp: 203°.
THR = See tin compounds and bromides.

METHYL TRI-*n*-BUTYL TIN. Liquid. $(CH_3)Sn(C_4H_9)_3$, mw: 305.1, d: 1.0898 @ 20°/4°, bp: 121° @ 10 mm.
THR = Details U. See tin compounds.

METHYL TRICHLOROSILANE. A liquid. CH_3Cl_3Si, mw: 149.46, mp: −90°, bp: 66.5°, d: 1.28 @ 25°/25°, flash p: <70°F, lel = 7.6%, vap. d: 5.17.
Acute tox data: Oral LD_{LO} (rat) = 1000 mg/kg; inhal LC_{LO} (rat) = 450 ppm for 4 hrs. [*3*]
THR = MOD via oral and inhal routes. Fumes in moist air, reacts violently with H_2O. [*19*] See chlorosilanes.
Fire Hazard: Dangerous, via heat, flame, oxidizers.
To Fight Fire: Dry chemical, CO_2, foam.

1-METHYL TRICYCLOQUINAZOLINE. $C_{22}H_{14}N_4$, mw: 334.4.
THR = An exper carc [*3*] via dermal route.

3-METHYL TRICYCLOQUINAZOLINE.
THR = An exper carc [*3*] via dermal route.

4-METHYL TRICYCLOQUINAZOLINE.
THR = An exper carc [*3*] via dermal route.

METHYL TRIETHOXYSILANE. Liquid. $C_7H_{18}O_3Si$, mw: 178.3, bp: 141°, d: 0.890, vap. d: 6.14.
Acute tox data: Oral LD_{LO} (rat) = 5000 mg/kg; inhal LD_{LO} (rat) = 4000 ppm for 8 hrs. [*3*]
THR = MOD via oral and inhal routes.
Fire Hazard: Slight; can react with oxidizing materials.

METHYL TRIPHENYL GERMANIUM. Colorless crystals, insol in water. Sol in organic solvents. $(CH_3)Ge(C_6H_5)_3$, mw: 318.9, mp: 71.0°.
THR = Details U. See germanium compounds and organometals.

METHYL TRIPHENYL SILANE. See methyl triphenyl silicane.

METHYL(TRIFLUOROPROPYL)POLYSILOXANE.
THR = Violent reaction with F_2. [*19*]

METHYL TRIPHENYL SILICANE. Syn: *methyltriphenylsilane.* Crystals. $CH_3(C_6H_5)_3Si$, mw: 274.4, mp: 67.3°.
THR = LOW to MOD irr.
Disaster Hazard: Mod dangerous; when heated to decomp, emits toxic fumes; can react vigorously with oxidizing materials.

METHYL TRI-*n*-PROPYL TIN. Liquid. $(CH_3)Sn(C_3H_7)_3$, mw: 263.0, bp: 93° @ 10 mm.
THR = U. See tin compounds.

***n*-METHYL TROPOLINE.** See tropine.

METHYL TRITHION. Syn: *S-[((p-chlorophenyl)thio) methyl]-o,o-dimethyl phosphorodithioate.* $C_9H_{12}O_2PS_3Cl$, mw: 314.7.
Acute tox data: Oral LD_{50} (rat) = 98 mg/kg; dermal LD_{50} (rat) = 190 mg/kg; oral LD_{50} (wild birds) = 18 mg/kg; dermal LD_{50} (rabbit) = 2420 mg/kg; dermal LD_{50} (mouse) = 2420 mg/kg. [*3*]
THR = HIGH oral (rats, wild birds), dermal (rat), and MOD dermal (rabbit, mice). A cholinesterase inhibitor.
Disaster Hazard: Dangerous; see chlorides and phosphates. See also parathion.

METHYL UNDECYL KETONE. Colorless liquid or white solid. $(CH_3)(C_{11}H_{23})CO$, mw: 198, mp: 28°, bp: 120° @ 5 mm, flash p: 225°F (CC), vap. d: 6.84, d: 0.825 @ 30°.
THR = Details U. See ketones.
Fire Hazard: Slight, when exposed to heat or flame; can react with oxidizing materials.
To Fight Fire: CO_2, dry chemical.

METHYL UREA. Crystals. $NH_2CONHCH_3$, mw: 74.1, mp: 101°, bp: decomp, d: 1.205 @ 20°/20°.
Acute tox data: Oral LD_{LO} (rat) = 500 mg/kg; paLD_{LO} (mouse) = 10189 mg/kg. [*3*]
THR = MOD via oral and LOW via pa routes.
Fire Hazard: Slight; can react with oxidizing materials.

6-METHYL URACIL. $C_5H_6O_2N_2$, mw: 126.1.
THR = An exper neo [*3*] via oral route.

METHYL URETHAN. Needles. $C_4H_9O_2N$, bp: 177°, mp: 54°, mw: 103.1.
Acute tox data: Oral LD_{50} (mice) = 6200 mg/kg. [*12*]
THR = MOD via oral route. An exper neo and teratogen [*12, 3*] via dermal and ip routes.

4-METHYL VALERALDEHYDE. $CH_3CH(CH_3)CH_2CH_2CHO$, mw: 100.
Acute tox data: Dermal LD_{50} (rabbit) = 4460 mg/kg. [*3*]
THR = MOD via dermal routes.

N-METHYL-*n*-VINYLACETAMIDE. $CH_3CH{:}CHCH_2CONH_2$, mw: 99.2.
Acute tox data: Oral LD_{LO} (rat) = 2830 mg/kg; dermal LD_{LO} (rabbit) = 1410 mg/kg. [*3*]
THR = MOD via oral and dermal routes.

1-METHYL VINYL ACETATE. See isopropenyl acetate.

4-METHYL-2-VINYL-1,3-DIOXOLANE. $\underline{CH_2OCH(CH{:}CH_2)CH(CH_3)O}$, mw: 114.

Acute tox data: Oral LD_{LO} (rat) = 890 mg/kg; inhal LC_{LO} (rat) = 1000 ppm for 4 hrs; dermal LD_{50} (rabbit) = 790 mg/kg. [*3*]
THR = MOD via oral, inhal and dermal routes.

METHYL VINYL ETHER. See vinyl methyl ether.

METHYL VINYL KETONE. Syn: *3-butene-2-one.* Colorless liquid, powerfully irr odor. $CH_3COCHCH_2$, mw: 70.09, bp: 81.4°, flash p: 20° F (CC), d: 0.8393 @ 25°/4°, vap. d: 2.41.
Acute tox data: ipLD_{LO} (mouse) = 16 mg/kg; ipLD_{50} (mouse) = 80 mg/kg. [*3*]
THR = HIGH irr to skin, eyes and mu mem and via ip route. A lachrymator. See also ketones.
Fire Hazard: Dangerous, when exposed to heat, flame or oxidizers.
Disaster Hazard: Dangerous, upon exposure to heat or flame; emits toxic and irr fumes; can react with oxidizing materials.
To Fight Fire: CO_2, dry chemical.

METHYL VINYL SULFONE. $C_3H_6O_2S$, mw: 106.2.
Acute tox data: Oral LD_{50} (rat) = 570 mg/kg; dermal LD_{50} (rabbit) = 32 mg/kg. [*3*]
THR = HIGH via dermal and MOD via oral routes.
Disaster Hazard: Dangerous; see sulfur compounds.

METHYL VIOLOGEN. See paraquat.

METRAMINE. See hexamine.

METRONIDAZOLE. $C_6H_9O_3N_3$, mw: 171.2.
THR = An exper carc via oral route. [*3*, *23*]

MEVINPHOS. See phosdrin.

MGK. See *n*-octyl bicycloheptene dicarboximide.

MICA DUST (<5% FREE SILICA). Light gray to dark flakes or particles.
THR = See also silica. Can cause a chronic fibrosis of the lungs.

MICROCRYSTALLINE WAXES.
THR = A recognized carc. [*14*] See paraffin and petroleum waxes.

MIDDLE OIL.
THR = A recognized carc. [*14*] See creosote.

MILD DETONATING FUSE, METAL CLAD. See explosives, high.

MILD DUST.
THR = MOD irr to lungs. A MILD allergen. See also dust, mineral dusts, nuisance dusts.

MILD SILVER PROTEIN. See argyrol.

MILITARY GUNCOTTON. See cellulose nitrate.

MILLERITE. See nickel monosulfide.

MILLON'S BASE. Crystals. $(HO)_2Hg_2NH_2OH$, mw: 468.27, d: 4.083 @ 18°.
THR = HIGH toxicity. See mercury compounds, inorganic.

MILTOWN. See meprobamate.

MINERAL DUSTS.
THR = Variable. From the economic and toxicity standpoints, the most important are those containing free silica, which can cause silicosis upon inhal of sufficient quantity. These include sand, sandstone, quartz and flint. They consist mainly of silica in the form of quartz; diatomaceous earth, which is essentially amorphous silica; and granite, which contains 20–40% quartz. Minerals that contain combined silica in the form of silicates but no free silica are generally less capable of causing silicosis. Asbestos, however, can cause a fibrotic lung condition of its own, known as asbestosis, also implicated in lung cancer. (See asbestos.) Mica and talc dust are also considered somewhat hazardous. (See mica dust.) Non-siliceous minerals, like limestone, marble, dolomite, etc., which do not contain toxic elements, do not ordinarily present any significant dust hazard, although minerals containing toxic elements, such as cryolite, which contains fluorine, and pyrolusite, which contains manganese, may cause systemic poisoning upon inhal or ingestion in sufficient quantity. In any event, the minerals are usually less reactive than synthetic compounds of the same elements, and in fact, may be relatively inert by comparison. These are common air contaminants.

MINERAL OILS. Syns: *petrolatum liquid, mineral oil, paraffin oil.* Colorless, oily liquid, practically tasteless and odorless, insol in water and alcohol, sol in benzene, chloroform and ether. A mixture of liquid hydrocarbons from petroleum. d: 0.83–0.86 (light), 0.875–0.905 (heavy), flash p: 444° F (OC), ulc: 10–20.
THR = A laxative. Inhal of vapor or particulates can cause aspiration pneumonia. Mineral oils are recog carc of the skin and scrotum. Also exper carc of the larynx, lung and alimentary tracts. [*14*, *23*]
Fire Hazard: Low.
To Fight Fire: Dry chemical, CO_2, foam.

#2 MINERAL OIL. See lard oil (pure).

MINERAL SEAL OIL. Syn: *signal oil.* A viscous liquid. Flash p: 275° F (OC), d: 0.8.
THR = Details U. See also kerosene.
Fire Hazard: Mod, when exposed to heat or flame; can react with oxidizing materials.
To Fight Fire: Foam, CO_2, dry chemical.

MINERAL SPIRITS NO. 10. See naphtha V.M.&P. and petroleum spirits.

MINERAL WAX. See ceresin wax.

MINIUM. See lead oxide.

MIPA. See 1-amino-2-propanol.

MIPAFOX. See bis(isopropylamido)fluorophosphate.

MIRBANE OIL. See nitrobenzene.

MIRESTROL.
THR = An exper carc. [*23*]

MIREX. Syn: *dodecachlorooctahydro-1,3,3-metheno-2H-cyclabuta(cd)pentalene.* Very white, odorless crystals. Water-insol, sol in dioxane and benzene. $C_{10}Cl_{12}$, mw: 545.6, decomp @ 485°.
Acute tox data: Oral LD_{LO} (rat) = 306 mg/kg; dermal LD_{50} (rabbit) = 800 mg/kg. [*3*]
THR = HIGH via oral and MOD via dermal routes. An exper (+) carc and teratogen [*3, 5*] via oral route. HIGH toxicity stomach insecticide. A persistent chlorinated HC which is toxic to non-target species. It can bioaccumulate.
Disaster Hazard: Dangerous; see chlorides.

MISCHMETAL. A commercial form of mixed rare earth metals, such as cerium, etc., used for making lighter flints and as an addition to alloys for improving their characteristics.

MITOMYCIN C.
THR = An exper carc. [*23*]

MITOX. See *p*-chlorobenzyl and *p*-chlorophenyl sulfide.

MIXED ACID. See sulfuric acid and nitric acid.

MNFA. See *n*-methyl-*n*-(1-naphthyl)monofluoroacetamide.

MOCA. See 3,3-dichloro-4,4′-diamino diphenyl methane.

MOCAP. Syn: *o-ethyl-s,s-dipropylphosphodithioate.* A powder, insol in water, sol in most organic solvents. $C_8H_{19}O_2PS_2$, mw: 242.1.
Acute tox data: Oral LD_{50} (rat) = 34 mg/kg; oral LD_{50} (chicken) = 6 mg/kg; dermal LD_{50} (rat) = 60 mg/kg. [*3*]
THR = HIGH via oral and dermal routes. A nematocide, soil insecticide material. It is a cholinesterase inhibitor; see parathion.
Disaster Hazard: Dangerous; see parathion.

MODIFIED HOP EXTRACT. Manufactured from a hexane extract of hops.
THR = U. A flavoring agent in the brewing of beer.

MOHAIR. Syn: *goat hair.* A fiber.
THR = A nuisance dust. A MILD allergen.
Fire Hazard: Slight, when exposed to heat or flame; can react with oxidizing materials.
Explosion Hazard: Slight, in the form of dust, when exposed to flame. See also dust explosions.

MOHICAN RED A-8008. Orange-red crystals. $C_{17}H_{13}O_4N_2SCl + \frac{1}{2}Ba$, mw: 445.5.
THR = An exper (±) carc. [*3, 4*]

MOLYBDENITE. See molybdenum sulfide.

MOLYBDENUM. Cubic, silver-white metallic crystals or gray-black powder. Mo, atwt: 95.95, mp: 2622°, bp: 5560°, d: 10.2, vap. press: 1 mm @ 3102°.
THR = See molybdenum compounds.
Radiation Hazard: For permissible levels, see Section 5, Table 5A.5.
Fire Hazard: Mod, in the form of dust, when exposed to heat or flame; violent reactions with BrF_3, ClF_3, F_2, PbO_2. [*19*] See also powdered metals.
Explosion Hazard: Slight, in the form of dust, when exposed to flame. See also powdered metals.

MOLYBDENUM CARBONYL. Diamagnetic crystals. $Mo(CO)_6$, mw: 264.01, mp: decomp @ 150°, d: 1.96, vap. press: 2.3 mm @ 55°, vap. d: 9.1.
THR = HIGH. See carbonyls. This compound may liberate carbon monoxide because it decomp easily.
Fire Hazard: See carbon monoxide.
Explosion Hazard: See carbonyls and carbon monoxide.
Disaster Hazard: See carbonyls.

MOLYBDENUM CHLORIDE. Syn: *molybdenum dichloride.* Amorphous or yellow crystals. $MoCl_2$, mw: 166.86, mp: decomp, d: 3.714 @ 25°.
THR = See molybdenum compounds.
Disaster Hazard: Dangerous; see chlorides.

MOLYBDENUM COMPOUNDS.
Acute tox data: $ipLD_{50}$ (mouse) = 160 mg/kg; $scLD_{50}$ (mouse) = 266 mg/kg. [*3*]
THR = HIGH via ip and sc routes. Molybdenum and its compounds are said to be somewhat toxic, but in spite of their considerable use in industry, industrial poisoning by molybdenum has yet to be reported. Some studies have been made of its effects, and it is suggested that suitable precautions should be taken against the inhal of considerable amounts of the more sol molybdenum compounds. From animal exper, it was found that no fatalities occurred to those subject to molybdic oxide fumes for 25 1-hr exposures at an average conc of 1.5 mg/cu. ft of air, and only 1 fatality in 24 1-hr exposures to molybdenite dust at an average conc of 8.1 mg/cu. ft of air. Molybdenum is not stored in the body to any extent because it is rapidly excreted. Exper with the sodium salts of hexavalent chromium, tungsten and molybdenum have shown that sodium molybdate is the least toxic of the three, and is less toxic following intraperitoneal injection than either sodium chromate or sodium tungstate in equivalent conc. All molybdenum compounds can be referred to molyb-

denum or to more toxic anions if present. Recent studies have shown that molybdenum has importance as a trace element in the normal growth and development of certain forms of plant life. It is found also in animal tissue, although its precise function is U. It is a common air contaminant.

MOLYBDENUM DICHLORIDE. See molybdenum chloride.

MOLYBDENUM DISULFIDE. See molybdenum sulfide.

MOLYBDENUM HEXAFLUORIDE. Colorless crystals or liquid, very hygroscopic. MoF_6, mw: 209.95, mp: 17° @ 406 mm, bp: 35°, d: 2.55.
THR = See fluorides.

MOLYBDENUM MONOBORIDE. Crystals. MoB, mw: 106.77, d: 8.65.
THR = See boron compounds and molybdenum compounds.
Fire Hazard: See borides.
Explosion Hazard: Details U. See also borides.
Disaster Hazard: See borides.

MOLYBDENUM OXIDE. Syn: *molybdenum sesquioxide.* Gray-black powder. Mo_2O_3, mw: 239.90.
THR = See molybdenum compounds.

MOLYBDENUM OXYTETRAFLUORIDE. Colorless to white deliquescent crystals. $MoOF_4$, mw: 187.95, mp: 98°, bp: 180°, d: 3.0.
THR − See fluorides.

MOLYBDENUM PENTACHLORIDE. Green-black solid, dark red as liquid or vapor, hygroscopic, reacting with water and air, sol in dry ether, dry alcohol and other anhydrous organic solvents. $MoCl_5$, mw: 273.2, mp: 194°, bp: 268°, d: 2.9.
THR = Toxic. See also hydrochloric acid.
Disaster Hazard: Dangerous; see hydrochloric acid.

MOLYBDENUM PENTASULFIDE. Dark brown powder. Mo_2S_5, mw: 352.23.
THR = See sulfides.

MOLYBDENUM PHOSPHIDE. Gray-green crystalline powder. MoP, mw: 126.93, d: 6.167.
THR = See phosphides.

MOLYBDENUM SESQUIOXIDE. See molybdenum oxide.

MOLYBDENUM SESQUISULFIDE. Steel-gray needles. Mo_2S_3, mw: 288.10, mp: decomp @ 1100°, bp: volatile @ 1200°, d: 5.91 @ 15°.
THR = See sulfides.

MOLYBDENUM SULFIDE. Syns: *molybdic sulfide, molybdenite.* Black, lustrous powder. MoS_2, mw: 160.07, mp: 1185°, bp: decomp in air, d: 4.80 @ 14°.
THR = See sulfides. Violent reaction with H_2O_2. [*19*]

MOLYBDENUM TETRABROMIDE. Black deliquescent needles. $MoBr_4$, mw: 416.61, mp: decomp, bp: volatilizes.
THR = See bromides and molybdenum compounds.

MOLYBDENUM TETRASULFIDE. Brown powder. MoS_4, mw: 224.19, mp: decomp.
THR = See sulfides.

MOLYBDENUM TRIOXIDE. Syn: *molybdenum anhydride.* White or yellow to slightly bluish powder or granules, sol in 1000 parts water, sol in conc mineral acids and solutions of alkali hydroxides, ammonia or potassium bitartrate. MoO_3, mw: 143.95, d: 4.7, mp: 795°, bp: 1155°.
Acute tox data: Oral LD_{50} (rat) = 125 mg/kg; $ipLD_{LO}$ (guinea pig) = 400 mg/kg; inhal TC_{LO} (man) = 6 mg/m^3. [*3*]
THR = HIGH via oral, ip and inhal routes. An irr. See molybdenum compounds. Violent reaction with ClF_3, Li, Mg, K, Na. [*19*] An exper carc to mice via ip route. [*103*]

MOLYBDENUM TRISULFIDE. Red dark-brown crystals. MoS_3, mw: 192.13, mp: decomp.
THR = See sulfides.

MOLYBDIC ACID. Hexagonal, white or slightly yellowish crystals. H_2MoO_4, mw: 162.0, mp: decomp @ 115°, d: 3.112.
THR = See molybdenum compounds.

MOLYBDIC ANHYDRIDE. See molybdenum trioxide.

MOLYBDIC SULFIDE. See molybdenum sulfide.

MONACETIN. See glyceryl monoacetate.

MOND GAS. See carbon monoxide and hydrogen.

MONEL. An alloy.
THR = Violent reaction with F_2. [*19*]

MONKSHOOD. See aconite.

MONOAMMONIUM GLUTAMATE. Syn: *ammonium glutamate.*
Acute tox data: $ipLD_{50}$ (rat) = 1000 mg/kg. [*3*]
THR = MOD via ip route. See also sodium glutamate. Used as a general purpose food additive. [*109*]

MONO-*sec*-AMYLAMINE. Colorless liquid, amine-like odor. $CH_3(CH_2)_2CH(CH_3)NH_2$, mw: 87.16, bp: 92°, flash p: 20° F, d: 0.73839 @ 20°/0°, vap. d: 3.01.
THR = Details U. See amines.
Fire Hazard: Dangerous, when exposed to heat, flame or oxidizers.
Explosion Hazard: U.
Disaster Hazard: Dangerous, upon exposure to heat or flame; can react vigorously with oxidizing materials.
To Fight Fire: Foam, CO_2, dry chemical.

***n*-MONOAMYL ANILINE.** Liquid. $C_5H_{11}NHC_6H_5$, mw: 163.3, bp: 245°, flash p: 225° F, d: 0.9, vap. d: 5.63.
THR = Probably toxic. See aniline.
Fire Hazard: Slight, when exposed to heat or flame.
Disaster Hazard: Dangerous; when heated to decomp, emits highly toxic fumes of aniline; can react with oxidizing materials.
To Fight Fire: Foam, CO_2, dry chemical.

MONOBROMOTRIFLUOROMETHANE. Syn: *bromotrifluoromethane*. Colorless, non-corrosive gas. $CBrF_3$, mw: 149, fp: −168°, bp: −58°, gas d:@ bp: 8.71 g/liter.
Acute tox data: Inhal LC_{LO} (rat) = 4000 ppm for ¼ hr. [*3*]
THR = MOD via inhal route.
Disaster Hazard: Dangerous; see bromides and fluorides.

MONO-*n*-BUTYLAMINE. Colorless liquid. $CH_3(CH_2)_2CH_2NH_2$, mw: 73.2, mp: −50.5°, flash p: 10° F (OC), d: 0.7401 @ 20.4°.
Acute tox data: Oral LD_{50} (rat) = 500 mg/kg; inhal LC_{LO} (rat) = 4000 ppm for 4 hrs; dermal LD_{50} (rabbit) = 850 mg/kg; dermal LD_{50} (guinea pig) = 500 mg/kg. [*3*]
THR = MOD via oral, inhal and dermal routes.
Fire Hazard: Dangerous, when exposed to heat or flame.
Disaster Hazard: Dangerous; keep away from heat and open flame; can react vigorously with oxidizing materials.
To Fight Fire: Foam, CO_2, dry chemical.

MONO-*sec*-BUTYLAMINE. Colorless liquid. $CH_3CH(NH_2)CH_2CH_3$, mw: 73.2, mp: −104.5°, bp: 63°, flash p: 15° F, d: 0.724 @ 20°/4°.
Acute tox data: Oral LD_{50} (rat) = 380 mg/kg. [*3*]
THR = HIGH via oral route. See also amines.
Fire Hazard: Dangerous, when exposed to heat or flame.
Explosion Hazard: U.
Disaster Hazard: Dangerous; keep away from heat and open flame; can react vigorously with oxidizing materials.
To Fight Fire: Foam, CO_2, dry chemical.

MONO-*tert*-BUTYL AMINE.
Acute tox data: Oral LD_{50} (rat) = 180 mg/kg; oral LD_{50} (mouse) = 900 mg/kg. [*3*]
THR = HIGH to MOD via oral route depending on test species. See also amines.

MONOBUTYLAMINE OLEATE. See butylamine oleate.

MONO-*tert*-BUTYL-*m*-CRESOL. Clear liquid. $(CH_3)_3CC_6H_3(CH_3)OH$, mw: 164.3, mp: 27.0°, bp: 137° @ 25 mm, d: 0.969 @ 30°/25°, vap. d: 5.77.
Acute tox data: Oral LD_{50} (rat) = 2390 mg/kg; dermal LD_{50} (rabbit) = 2200 mg/kg; oral LD_{50} (mouse) = 700 mg/kg. [*3*]
THR = MOD via oral and dermal routes.
Disaster Hazard: Mod dangerous; when heated, emits toxic fumes; can react with oxidizing materials.

MONOBUTYL DIPHENYL SODIUM MONOSULFONATE. Syn: *aresket*. Powder. $C_{16}H_{17}O_4SNa$, mw: 328.3.
Acute tox data: Oral LD_{50} (mouse) = 3500 mg/kg; ivLD_{50} (mouse) = 250 mg/kg. [*3*]
THR = HIGH via iv and MOD via oral routes. Can cause dermatitis by defatting of skin. Toxic by ingestion.
Disaster Hazard: Dangerous; see sulfonates.

MONOBUTYL PHENYL PHENOL SODIUM SULFATE. Syn: *areskap*. Liquid. $C_{16}H_{17}SO_4Na$, mw: 328.3.
Acute tox data: Oral LD_{50} (mouse) = 3800 mg/kg; ivLD_{50} (mouse) = 180 mg/kg. [*3*]
THR = HIGH via iv and MOD via oral routes.

MONOCALCIUM PHOSPHATE. See calcium phosphate, monobasic.

MONOCHLORATED ACETONE. See monochloroacetone, stabilized.

MONOCHLOROACETIC ACID. Syns: *chloroethanoic acid*, *chloroacetic acid*. Colorless cyrstals. $CH_2ClCOOH$, mw: 94.5, mp: $\alpha = 63°$, $\beta = 56°$, $\gamma = 50°$, bp: 189°, flash p: 259° F, d: 1.58 @ 20°/20°, vap. d: 3.26.
Acute tox data: Oral LD_{50} (rat) = 76 mg/kg; scLD_{50} (rat) = 5 mg/kg; oral LD_{50} (mouse) = 165 mg/kg. [*3*]
THR = HIGH irr to skin, eyes and mu mem. HIGH via oral and sc routes. An exper neo via sc route. [*3*]
Fire Hazard: Low.
Disaster Hazard: Dangerous; when heated to decomp, emits highly toxic fumes of phosgene and chlorides.
To Fight Fire: Water, spray, fog, mist, dry chemical, foam.

MONOCHLOROACETONE. See chloroacetone.

MONOCHLOROACETALDEHYDE DIETHYL ACETAL.
THR = An exper carc. [*23*]

MONOCHLOROBENZENE. See chlorobenzene.

MONOCHLOROBROMOMETHANE. See monochloromonobromomethane.

MONOCHLORODIFLUOROMETHANE. See chlorodifluoromethane.

MONOCHLOROETHYLENE CHLORIDE. See ethylene monochlorochloride.

MONOCHLOROMALEIC ANHYDRIDE. Nearly white liquid or solid. $ClCCH(CO)_2O$, mw: 132.4, mp: 33°, bp: 196°–197.5°, d: 1.54 @ 25°/25°.

THR = U. Probably HIGH irr and toxic.

Disaster Hazard: Dangerous; when heated to decomp, emits highly toxic fumes of phosgene.

MONOCHLOROMETHANE. See methyl chloride.

MONOCHLOROMONOBROMOMETHANE. See methylene chlorobromide.

MONOCHLORONAPHTHALENE. See chlorinated naphthalenes.

MONOCHLOROPENTAFLUOROACETONE. Syn: *chloropentafluoroacetone.* A colorless, non-flam, hygroscopic gas. C_3OClF_5, mw: 182.5, bp: 7.8°, fp: −133°, d(liquid): 1.43 @ 25°.

Acute tox data: Inhal LC_{50} (rat) = 125 ppm for 3 hrs. [*3*]

THR = HIGH irr to skin, eyes and mu mem and via inhal route.

Disaster Hazard: Reacts with water exothermally.

MONOCHLOROPENTAFLUOROETHANE. See chloropentafluoroethane.

MONOCHLOROTETRAFLUOROETHANE. Colorless gas. C_2HF_4Cl, mw: 136.49.

THR = Details U; probably acts as a simple asphyxiant. See also chlorinated hydrocarbons.

Disaster Hazard: Dangerous; when heated to decomp, emits highly toxic fumes of chlorides and fluorides.

MONOCHLOROTOLUENE. Syn: *methyl chlorobenzene.* Clear, colorless to straw-colored liquid. $CH_3C_6H_4Cl$, mw: 126.5, bp: 158°–165°, fp: <−45°, flash p: 125°F, d: 1.080 @ 15.5°/15.5°, vap. d: 4.38.

THR = Details U. See chlorinated hydrocarbons, aromatic.

Fire Hazard: Mod, when exposed to heat or flame.

Explosion Hazard: U.

Disaster Hazard: Dangerous; when heated to decomp, emits highly toxic fumes of chlorides; can react with oxidizing materials.

To Fight Fire: Foam, CO_2, dry chemical.

MONOCHLOROTRIFLUOROETHYLENE. See trifluorochlorethylene.

MONOCHLOROTRIFLUOROMETHANE. Syn: *chlorotrifluoromethane.* Colorless gas, ethereal odor. $CClF_3$, mw: 104.47, mp: −181°, bp: −80°.

THR = A MILD irr, narcotic in high conc. Reacts violently with Al.[*19*]

Disaster Hazard: Dangerous; when heated to decomp, emits highly toxic fumes of chlorides and fluorides.

MONOCRESYL DIPHENYL PHOSPHATE. Liquid. $CH_3C_6H_4O(C_6H_5O)_2PO_4$, mw: 388.3, flash p: 450°F (CC), d: >1, vap. d: 13.4.

THR = Details U; probably tox and an irr.

Fire Hazard: Slight, when exposed to heat or flame.

Spont Heating. No.

Disaster Hazard: Dangerous; when heated to decomp, emits highly toxic fumes of PO_x; can react with oxidizing materials.

To Fight Fire: Foam, CO_2, dry chemical.

MONOCROTALINE. $C_{16}H_{23}O_6N$, mw: 325.4.

THR = HIGH via oral and iv routes. An exper (+) carc and neo via ip route. [*3*, *7*]

MONOCROTOPHOS. See azodrin.

MONODECYL NAPHTHALENE. See keryl naphthalene.

MONO AND DIGLYCERIDES FROM THE GLYCEROLYSIS OF EDIBLE FATS AND OILS.

THR = U. Used as an emulsifying agent food additive. [*109*]

MONODODECYL BENZENE. See keryl benzene-12.

MONOETHANOL AMINE. Syns: *2-aminoethanol, ethanol amine.* Colorless liquid, ammoniacal odor. Hygroscopic. $NH_2CH_2CH_2OH$, mw: 61.08, bp: 170.5°, fp: 10.5°, flash p: 200°F (OC), d: 1.0180 @ 20°/4°, vap. press: 6 mm @ 60°, vap. d: 2.11.

Acute tox data: Oral LD_{50} (rat) = 2100 mg/kg; ipLD_{50} (rat) = 981 mg/kg; scLD_{50} (mouse) = 2537 mg/kg; dermal LD_{50} (rabbit) = 1000 mg/kg. [*3*]

THR = MOD via oral, dermal, ip and sc routes.

Fire Hazard: Mod, when exposed to heat or flame; can react violently with acetic acid, acetic anhydride, acrolein, acrylic acid, acrylonitrile, chlorosulfonic acid, epichlorohydrin, HCl, HF, mesityl oxide, HNO_3, oleum, H_2SO_4, β-propiolactone, vinyl acetate. [*19*]

To Fight Fire: Foam, alcohol foam, dry chemical.

MONOETHYLAMINE. See ethylamine.

MONOFLUOPHOSPHORIC ACID. See fluorophosphoric acid.

MONOFLUOROPHOSPHORIC ACID, ANHYDROUS. See monofluophosphoric acid.

MONOFLUOROTRICHLOROMETHANE. See fluorotrichloromethane.

MONOGERMANE. Syn: *germanium tetrahydride.* Gas. GeH_4, mw: 76.63, mp: −165°, bp: −88.5°, d(liquid): 1.523 @ −142°, d(gas): 3.43 g/liter.

THR = A poisonous gas. See also hydrides and germanium compounds.

Fire Hazard: Dangerous; see hydrides.

Explosion Hazard: Dangerous; see hydrides.

Disaster Hazard: See hydrides.

MONO-*sec*-HEXYLAMINE. See 1,3-dimethyl butylamine.

MONOISOPROPANOLAMINE. See 1-amino-2- propanol.

MONOISOPROPYL BICYCLOHEXYL. $C_{15}H_{28}$, mw: 208, d: 0.9, bp: 275°–280°, flash p: 255°F, autoign. temp.: 442°F, lel = 0.5% @ 347°F, uel = 4.1% @ 400°F.
THR = U.
Fire Hazard: Slight, when exposed to heat or flame.
To Fight Fire: Alcohol foam, CO_2, fog, mist, dry chemical.

2-MONOISOPROPYL BIPHENYL. $C_{15}H_{16}$, mw: 196, bp: 270°, flash p: 285°F, autoign. temp.: 513°F, d: 1.0, lel = 0.5% @ 347°F, uel = 3.2% @ 392°F.
THR = U.
Fire Hazard: Slight, when exposed to heat or flame.
To Fight Fire: Foam, spray, mist, dry chemical.

MONOISOPROPYL CITRATE.
THR = Details U. A sequestrant food additive. [*109*]

MONOMETHYLAMINE. Syn: *aminomethane.* Colorless gas or liquid, strong ammoniacal odor. CH_3NH_2, mw: 31.1, bp: −6.79°, lel = 4.95%, uel = 20.75%, fp: −93.5°, flash p: 32°F (CC), d: 0.662 @ 20°/4°, autoign. temp.: 806°F, vap. d: 1.07.
Acute tox data: $scLD_{LO}$ (mouse) = 2500 mg/kg; $scLD_{LO}$ (frog) = 2000 mg/kg. [*3*]
THR = MOD irr to skin, eyes and mu mem and via sc route of exposure.
Fire Hazard: Very dangerous, when exposed to heat or flame.
Explosion Hazard: Mod, when exposed to spark or flame.
Disaster Hazard: Dangerous; keep away from heat and open flame; can react vigorously with oxidizing materials.
To Fight Fire: Stop flow of gas.

MONOMETHYLAMINE, AQUEOUS SOLUTION. See monomethyl amine.

MONOMETHYLANILINE. See *n*-methylaniline.

MONOMETHYLOL DIMETHYL HYDANTOIN. Odorless, white crystalline solid. $C_6H_{10}N_2O_3$, mw: 158.16, mp: 99°–103°, d: 0.78.
THR = HIGH irr to skin, eyes and mu mem.
Disaster Hazard: Slight; when heated, emits toxic fumes.

MONONITROTOLUENE. See *p*-nitrotoluene.

MONONONYL NAPHTHALENE. Light straw-colored viscous liquid. $C_9H_{19}C_{10}H_7$, mw: 254.4, bp: 300°–350°, d: 0.93–0.94 @ 20°, vap. d: 8.8.
THR = Details U. See also naphthalene.
Fire Hazard: Slight, when exposed to heat or flame; can react with oxidizing materials.

MONOPOTASSIUM GLUTAMATE. Syn: *potassium glutamate.* White, free flowing, crystalline powder, practically odorless, freely sol in water, slightly sol in alcohol. $KOOC(CH_2)_2CH(NH_2)COOH \cdot H_2O$, mw: 203.3.
THR = U. Used as a general-purpose food additive. [*109*]

MONO-*n*-PROPYLAMINE. See propylamine-*n*.

MONOPROPYLENE GLYCOL METHYL ETHER. Syn: *1-methoxy-2-propanol.* $CH_3OCH_2CH(OH)CH_3$, mw: 90.1, bp: 120°, flash p: 100°F (OC), d: 0.92 @ 25°/20°, vap. d: 3.1.
Acute tox data: Oral LD_{50} (rat) = 7510 mg/kg; inhal LC_{LO} (rat) = 7000 ppm for 4 hrs; oral LD_{LO} (dog) = 10,000 mg/kg. [*3*]
THR = LOW via oral and inhal routes.
Fire Hazard: Dangerous, when exposed to heat or flame; can react with oxidizing materials.
Explosion Hazard: Details U. See also ethers.
To Fight Fire: Foam, CO_2, dry chemical.

MONOSODIUM ACID METHANE ARSONATE. Syn: *monosodium methyl arsonate.* Crystalline hydrate, water-sol. $CH_3AsO(OH)(ONa) \cdot 1\frac{1}{2}H_2O$, mw: 189, mp: 113°–116°.
Acute tox data: Oral LD_{50} (rat) = 700 mg/kg; LD_{50} (mammal) = 50 mg/kg. [*3*]
THR = HIGH via oral route. See arsenic.

MONOSODIUM GLUTAMATE. $C_5H_9O_4N \cdot Na$, mw: 170.1.
Acute tox data: $ipLD_{50}$ (rat) = 4253 mg/kg; $scLD_{50}$ (rat) = 8000 mg/kg. [*3*]
THR = MOD via ip and sc routes. An exper teratogen via pa route. [*3*]

MONOSODIUM METHYL ARSONATE. See monosodium acid methane arsonate.

MONOSTEARIN. See glyceryl monostearate.

MONOTHIOGLYCEROL. Liquid. $C_3H_8O_2S$, mw: 108.2, bp: 118° @ 5 mm, d: 1.248 @ 20°/20°.
Acute tox data: $ipLD_{LO}$ (mouse) = 400 mg/kg. [*3*]
THR = HIGH via ip route.
Fire Hazard: Mod.
Disaster Hazard: Dangerous; when heated to decomp, emits highly toxic fumes of SO_x; can react with oxidizing materials.

MONOURANYL-*o*-PHOSPHATE. See uranium phosphate.

MONTANITE. See bismuth tellurate.

MONURON. Syns: *CMU, 3-(p-chlorophenyl)-1,1-dimethyl urea.* Crystals, nearly water-insol, slight odor. $C_9H_{11}ClN_2O$, mw: 198.6, mp: 171°, vap. press: 2×10^{-3} mm @ 100°.

Acute tox data: Oral LD_{50} (rat) = 1480 mg/kg; ipLD_{50} (mouse) = 1000 mg/kg. [*3*] Oral LD_{50} (rat) = 3600 mg/kg.

THR = MOD via oral and ip routes. An exper (+) carc via oral route. [*3*, *12*] Has produced anemia and methemoglobinemia in exper animals.

Disaster Hazard: See chlorides.

MOROCIDE. See binapacryl.

MORPHINE. White crystalline alkaloid. $C_{17}H_{19}NO_2 \cdot H_2O$, mw: 303.35, mp(anhydrous): 254° (decomp), bp: 191°–192° (vacuum), d: 1.317.

Acute tox data: Oral LD_{50} (mouse) = 745 mg/kg; ipLD_{50} (mouse) = 480 mg/kg; scLD_{50} (mouse) = 360 mg/kg; ivLD_{50} (mouse) = 199 mg/kg. [*3*]

THR = HIGH via ip, sc and iv routes; MOD via oral route. Morphine is the constituent of opium most responsible for its poisonous effects. When taken orally, the effects of morphine poisoning begin to appear in 20–40 min; if taken hypodermically, the symptoms appear much earlier and narcotism is more likely to follow the early symptoms. Individual susceptibility varies greatly and children are more susceptible than adults. The usual symptoms due to an overdose of morphine are a sense of mental exhilaration and physical ease with a quickening and strengthening of the pulse and then a depression of the brain with special reference to its higher functions. Smaller doses which are not sufficient to cause this depression do cause a diminished sensibility to lasting impressions, such as pain, cold, hunger and discomfort. Following this there may be dizziness and heaviness of the head, nausea, languor and drowsiness, the pulse being reduced in force. Individuals with great susceptibility may complain of itching of the skin and even erythema and as the action continues, it produces sleep. There is a gradual loss of muscular power and a diminished sense of feeling. The pupils become contracted and fail to respond to light. The respirations become less and less frequent, finally being reduced in some cases to four or five a minute with stertorous breathing. If the patient who has taken an overdose of morphine survives for 48 hrs, the prognosis is favorable.

Treatment and Antidotes: A physician should be called at once. The stomach should be washed repeatedly and at short intervals, whether the poison was taken by mouth or injection, since it is excreted into the stomach anyway. Emetics should be given. Recommended is apomorphine hydrochloride, hypodermically. If narcosis has already set in, the effect of the emetic may be interfered with. A dilute solution of potassium permanganate in the form of a lavage which should be 0.5 g/liter has been used successfully. Tincture of belladonna by mouth, 2 cc by volume, may be repeated every 15 min. but this antidote must be used cautiously. Solubility of the morphine in the stomach may be decreased by giving tannic acid, coffee, tea and finally, powdered charcoal in water; also an iodine solution may be given. Every effort should be made to arouse the patient and keep him awake, especially by walking him around, pinching the skin, slapping him, and by ammonia inhal and artificial respiration if necessary. Keep the patient warm under any conditions. The fatal dose may be from 1–2 to 4–6 grains, depending very much upon the individual.

MORPHINE ACETATE. See morphine.

MORPHINE BENZYL ETHER HYDROCHLORIDE. See peronine.

MORPHINE HYDROCHLORIDE. See morphine.

MORPHINE MECONATE. See morphine.

MORPHINE NITRATE. White powder, darkens when exposed to light. $C_{17}H_{19}NO_3 \cdot HNO_3$, mw: 348.26.

THR = See morphine and nitrates.

MORPHINE SULFATE. See morphine.

MORPHOLINE. Syns: *tetrahydro-1,4-oxazine*, *diethylenimide oxide*. Colorless, hygroscopic oil. Amine odor. $\overline{OCH_2CH_2NHCH_2CH_2}$, mw: 87.12, bp: 128.9°, fp: −7.5°, flash p: 100°F (OC), d: 0.998 @ 25°/25°, autoign. temp.: 590°F, vap. press: 10 mm @ 23°, vap. d: 3.00.

Acute tox data: Oral LD_{50} (rat) = 1050 mg/kg; dermal LD_{50} (rabbit) = 500 mg/kg. [*3*]

THR = HIGH via dermal and MOD via oral routes. Irr to skin, eyes and mu mem. Has produced kidney damage in exper animals. Used as a food additive permitted in food for human consumption. [*109*]

Fire Hazard: Dangerous, when exposed to flame, heat or oxidizers.

Disaster Hazard: Dangerous; when heated to decomp, emits highly toxic fumes of NO_x; can react with oxidizing materials.

To Fight Fire: Alcohol foam, CO_2, dry chemical.

MORPHOLINE ETHANOL. Syn: *n-hydroxy ethyl morpholine*. Colorless liquid. $O(CH_2CH_2)_2NCH_2CH_2OH$, mw: 131.17, bp: 225.5°, mp: 1.6°, vap. press: 0.1 mm @ 20°, flash p: 210°F (OC), vap. d: 4.54, d: 1.071.

Acute tox data: Oral LD_{50} (rat) = 12000 mg/kg; scLD_{50} (mouse) = 2650 mg/kg; dermal LD_{50} (guinea pig) = 2500 mg/kg. [*3*]

THR = MOD via dermal and sc; LOW via oral routes.

Fire Hazard: Low, when exposed to heat or flame; can react with oxidizing materials.

To Fight Fire: Alcohol foam, CO_2, dry chemical.

MORPHOLINE + SODIUM NITRITE.
THR = An exper carc. [*3*]
Disaster Hazard: Dangerous; see NO_x.

1,5-MORPHOLINOMETHYL-3-[(5-NITRO-2-FURYLIDINE)AMINO]-2-OXAZOLIDINE.
THR = An exper carc. [*23*]

5-(MORPHOLINOMETHYL)-3-[(5-NITROFURFURYLIDINE)AMINO]-1,2-OXAZOLIDINONE HYDROCHLORIDE. $C_{13}H_{16}O_6N_4 \cdot HCl$, mw: 360.8.
THR = An exper (+) carc. [*3, 1*]

4-MORPHOLINO-2-(5-NITRO-2-THIENYL)QUINAZOLINE. $C_{16}H_{14}O_3N_4S$, mw: 342.4.
THR = An exper carc to rats via oral route. [*103*]

MORPHOTHION. Syn: *o,o-dimethyl-s-(morpholinocarbonyl methyl)phosphorodithioate.* A colorless solid, sol in acetone, dioxane, acetonitrile. $C_8H_{16}O_4NPS_2$, mw: 285.3, mp: 65°.
Acute tox data: Oral LD_{50} (rat) = 200 mg/kg; dermal LD_{50} (rat) = 283 mg/kg; oral LD_{50} (rabbit) = 190 mg/kg; oral LD_{50} (mouse) = 130 mg/kg. [*3*]
THR = HIGH via oral and dermal routes. A cholinesterase inhibitor. See parathion.

MORRHUA OIL. See cod liver oil.

MORTAR STAIN, LIQUID.
THR = Variable.
Fire Hazard: Mod, when exposed to heat or flame; can react with oxidizing materials.
Explosion Hazard: U.

MOSS BUNKER OIL. See menhaden oil.

MOSS GREEN. See copper acetoarsenite.

MOTH BALLS. See naphthalene.

MOTH FLAKES. See naphthalene.

MOTION PICTURE FILM, NITROCELLULOSE BASE.
Fire Hazard: Dangerous, when exposed to heat or flame; See also nitrocellulose.
Explosion Hazard: Mod, when exposed to heat or flame. See also nitrocellulose.
Disaster Hazard: Dangerous; when heated to decomp, emits highly toxic fumes of NO_x; can react vigorously with oxidizing materials.

MOTION PICTURE FILM, SLOW BURNING.
Fire Hazard: Mod, when exposed to heat or flame.
Disaster Hazard: Dangerous; when heated to decomp, emits highly toxic fumes of NO_x; can react with oxidizing materials.

MOTOR FUEL ANTI-KNOCK COMPOUND. See lead tetraethyl.

MOTOR FUEL, N.O.S. See gasoline.

MOUNTAIN TOBACCO. See arnica.

MSP. See sodium phosphate, monobasic.

MTIC. Syn: *NSC-407347*, $C_5H_8ON_6$, mw: 168.2.
THR = An exper carc and neo via oral and ip routes. [*3*]

MUCOCHLORIC ANHYDRIDE. Monoclinic prisms, slightly sol in cold water, sol in hot water, hot benzene and alcohol. $C_8H_2O_5Cl_4$, mw: 319.9.
Acute tox data: Oral LD_{50} (rat) = 2000 mg/kg. [*3*]
THR = MOD via oral route.
Disaster Hazard: Dangerous; see chlorides.

MUFFLERS. See Section 3.

MUIRA PUAMA. A wood. Aromatic resin, muira puamine, fat.
THR = An irr wood (dust) to eyes and mu mem and via oral route.
Fire Hazard: Mod, when exposed to heat or flame; can react with oxidizing materials.
Explosion Hazard: Slight, in the form of dust, when exposed to heat or flame. See also dust explosions.

MURIATIC ACID. See hydrochloric acid.

MURIATIC ETHER. See ethyl chloride.

MURVESCO. Syns: *fenson, p-chlorophenyl benzenesulfonate.* Colorless crystals, insol in water, sol in organic solvents. $C_{12}H_9O_3SCl$, mw: 267.7, mp: 62°.
Acute tox data: Oral LD_{50} (rat) = 1350 mg/kg. [*3*]
THR = MOD via oral route. An acaricide.
Disaster Hazard: Dangerous, see chlorides and sulfonates.

MUSTARD GAS. See β,β'-dichloroethyl sulfide.

MUTAGENESIS. See Sections 8 and 9.

MYELO BROMOL. $C_6H_{12}O_4Br_2$, mw: 308.
THR = MOD via oral route. An exper neo via ip route. [*3, 23*]

MYLAR. See polyethylene terphthalate (film).

MYLERAN. See 1,4-butane-diol dimethyl sulfonate.

MYLONE. Syns: *Crag fungicide 974, tetrahydro-3,5-dimethyl-2H-thiadiazine-2-thione.* Crystals, sol in alcohol. $C_5H_{10}N_2S_2$, mw: 162.3, mp: 107°.
Acute tox data: Oral LD_{50} (rat) = 500 mg/kg; ipLD_{50} (rat) = 87 mg/kg; oral LD_{50} (mouse) = 180 mg/kg; oral LD_{50} (cat) = 79 mg/kg; oral LD_{50} (rabbit) = 120 mg/kg. [*3*]
THR = HIGH via oral and ip routes. A mild primary skin irr and sensitizer. A nematocide and slimicide.
Disaster Hazard: Dangerous; see sulfates.

NABAM. Syn: *disodium ethylene bis(dithiocarbamate).* Crystals, sol in water. $C_4H_6N_2Na_2S_4$, mw: 256.4.
Acute tox data: Oral LD_{50} (rat) = 395 mg/kg; oral LD_{50} (mouse) = 580 mg/kg. [*3*]
THR = HIGH-MOD via oral route. See also carbamates.

NAFENOPIN. $C_{20}H_{22}O_3$, mw: 310.4.
THR = An exper carc. [*3*] via oral route.

NaK. Low-melting alloy of sodium and potassium metals.
THR = NaK, as its name implies, is a low-melting alloy of Na and K. Its toxicity is that of either Na or K alone. In contact with moisture, this material reacts very violently to evolve hydrogen and much heat, leaving behind a high caustic residue of NaOH or KOH. Of course, the heat would damage tissue as would the caustic residue. Even fine particles of NaK damage the eyes, irr the lungs, etc. Finally, when NaK burns, it evolves fumes of Na_2O and K_2O, which are powerful caustics.
Fire Hazard: Dangerous. In the presence of O_2, moisture, halogens, oxidizers, acids or acid fumes, etc., it will react violently, giving off much heat, often spattering either red hot particles or actually flaming particles.
Explosion Hazard: Severe; will react explosively under many conditions, such as contact with moisture, CO_2, CCl_4, $CHCl_3$, $CHBr_3$, (ammonium sulfate + NH_4NO_3), CH_2Cl_2, CH_2I_2, HgO, CH_3Cl, oxalyl bromide, oxalyl chloride, pentachloroethane, K oxides, KO_2, Si, Ag halides, $NaHCO_3$, polytetrafluoroethylene, tetrachloroethane, 1,1,1-trichloroethane, trichlorotrifluoroethane. [*19*]
Disaster Hazard: Dangerous; when heated, emits highly toxic fumes of sodium and potassium oxides; will react explosively with water, steam, acid, acid fumes or mists to produce heat, hydrogen and toxic and corrosive fumes; can react vigorously with oxidizing materials. See also sodium and potassium.
To Fight Fire: G-1 powder, dry sodium chloride or dry soda ash. Never use water or foam.

NALED. See dimethyl-1,2-dibromo-2,2-dichloroethyl phosphate.

NANTOKITE. See cuprous chloride.

NAPHTHACENE. See tetracene.

NAPHTHA (COAL-TAR). Syn: *naphtha solvent.* Dark straw-colored to colorless liquid. Sol in benzene, toluene, xylene, etc. bp: 149°–216°, flash p: 107°F (CC), d: 0.862–0.892, autoign. temp.: 531°F.
Acute tox data: Inhal LC_{LO} (rat) = 1600 ppm. [*3*]
THR = MOD via inhal route. Can cause unconsciousness which may go to coma, stentorious breathing and bluish tint to the skin. Recovery follows removal from exposure. In mild form, intoxication resembles drunkeness. On a chronic basis no true poisoning; sometimes headache, lack of appetite, dizziness, sleeplessness, indigestion and nausea. [*27*] A common air contaminant. A recog carc. [*14*] See oils, mineral.
Fire Hazard: Mod, when exposed to heat or flame; can react with oxidizing materials. Keep containers tightly closed.
Explosion Hazard: Slight.
To Fight Fire: Foam, CO_2, dry chemical.

NAPHTHA DISTILLATE. See naphtha.

NAPHTHALENE. Syns: *moth flakes, white tar, tar camphor.* Aromatic odor, white, crystalline, volatile flakes. $C_{10}H_8$, mw: 128.16, mp: 80.1°, bp: 217.9°, flash p: 174°F (CC), d: 1.162, lel = 0.9%, uel = 5.9%, autoign. temp.: 979°F, vap. press: 1 mm @ 52.6°, vap. d: 4.42.
Acute tox data: Oral LD_{LO} (child) = 100 mg/kg; oral LD_{50} (rat) = 1780 mg/kg; ip LD_{LO} (mouse) = 150 mg/kg. [*3*]
THR = MOD via oral and HIGH via ip routes. An exper neo [*3*] via sc route. May be used as an insecticide. Systemic reactions include nausea, headache, diaphoresis, hematuria, fever, anemia, liver damage, convulsions and coma.
Fire Hazard: Mod, when exposed to heat or flamc; reacts with oxidizing materials. Reacts violently with CrO_3. [*19*]
Spont Heating: No.
Explosion Hazard: Mod, in the form of dust, when exposed to heat or flame.
To Fight Fire: Water, CO_2, dry chemical.

***p*-NAPHTHALENE.** See anthracene.

α-NAPHTHALENE ACETIC ACID. Syns: *naphthyl acetic acid, planofix.* White odorless crystals, only slightly water-sol. $C_{10}H_7CH_2COOH$, mw: 186.2, mp: 134°.

Acute tox data: Oral LD_{50} (rat) = 1000 mg/kg. [*3*]
THR = MOD irr to skin, eyes and mu mem, and via oral route. Can cause CNS depression.

β-NAPHTHALENE AZO-β-NAPHTHOL.
THR = A recog carc. [*14*] See aromatic amines.

2,7-NAPHTHALENEDIAMINE. Syn: *2,7-diamine-naphthalene*. Crystals. Turns brown upon standing. Slightly water-sol. $C_{12}H_{12}N_4$, mw: 212.3, mp: 66.5°, bp: 205° @ 12 mm.
Acute tox data: LD_{LO} (mouse) = 50 mg/kg. [*3*]
THR = HIGH via unknown route. .

1,5-NAPHTHALENE DIISOCYANATE. Syn: *desmodur*. $C_{10}H_6(NCO)_2$, mw: 210.14, white to light yellow crystals.
THR = A powerful allergen. An irr. See also 2,4-tolylene diisocyanate.

1,4-NAPHTHALENEDIOL.
Acute tox data: Oral LD_{LO} (rat) = 100 mg/kg. [*3*]
THR = HIGH via oral route.

1,3-NAPHTHALENE-1,5-DISULFONIC ACID.
See naphthoresorcinol.

NAPHTHALENE-1,5-DISULFONIC ACID. Crystals. $C_{10}H_6(SO_3H)_2$, mw: 288.3.
THR = Details U; see also naphthalene.
Disaster Hazard: Dangerous; when heated to decomp, emits highly toxic fumes of SO_x.

NAPHTHALENE ETHYLENE. See acenapthene.

NAPHTHA (PETROLEUM). See petroleum spirits.

NAPHTHA, SAFETY SOLVENT.
See stoddard solvent.

NAPHTHA, SOLVENT. See naphtha (coal-tar).

NAPHTHA, V.M.&P. Syns: *benzine*, *76° naphtha*. Volatile liquid. bp: 100°–140°, flash p: 20°F (CC), d: 0.67–0.80, lel = 0.9% @ 212°F, uel = 6.0% @ 212°F, autoign. temp.: 450°F.
THR = See petroleum spirits.
Fire Hazard: Dangerous, when exposed to heat or flame.
Explosion Hazard: Mod, when exposed to flame.
Disaster Hazard: Dangerous, upon exposure to heat or flame; can react vigorously with oxidizing materials.
To Fight Fire: Foam, CO_2, dry chemical.

NAPHTHA, V.M.&P., 50° FLASH. Insol in water. flash p: 50°F, autoign. temp.: 450°F, lel = 0.9%, uel = 6.7%, d: < 1, vap. d: 4.1, bp: 115°–143°. (flash p and autoign. temp. will vary depending on the manufacturer.)
THR = See also petroleum spirits.

NAPHTHA, V.M.&P., HIGH FLASH. Insol in water. flash p: 85°F, autoign. temp.: 450°F, lel = 1.0%, uel = 6.0%, d: < 1, vap. d: 4.3, bp: 138°–165°. (flash p and autoign. temp. will vary depending on the manufacturer.)
THR = See also petroleum spirits.

NAPHTHENATES. See driers, paint, varnish, enamel, etc.

NAPHTHENE. See cyclohexane.

NAPHTHENIC ACID. Syn: *hexahydrobenzoic acid*. Odorless crystals, slightly water-sol. $C_7H_{12}O_2$, mw: 128.2, d: 1.034, mp: 31°, bp: 233°.
Acute tox data: Oral LD_{50} (rat) = 3000 mg/kg; ip LD_{50} (rat) = 640 mg/kg. [*3*]
THR = MOD via oral and ip routes.

NAPHTHO (1,8-gh:4,5-g'h')DIQUINOLINE.
$C_{22}H_{12}N_2$, mw: 304.4.
THR = An exper neo. [*3*] via sc route.

NAPHTHOIC ACID. Syn: *naphthylene-carboxylic acid*. Two forms: α and β. α: needles, slightly sol in hot water, sol in hot alcohol and ether; β: plates or needles, slightly sol in hot water, sol in alcohol and ether. $C_{11}H_8O_2$, mw: 172.17, mp(α): 160°–161°, mp(β): 184°–185°, bp(α): 300°, bp(β): > 300°.
Acute tox data: Oral LD_{50} (mouse) = 2370 mg/kg. [*3*]
THR = MOD via oral route.

NAPHTHITE. See trinitronaphthalene.

α-NAPHTHOL. Syns: *1-naphthol*, *1-hydroxynaphthalene*. Colorless crystals, disagreeable taste. $C_{10}H_7OH$, mw: 144.2, mp: 96°, bp: 282.5°, d: 1.0954 @ 98.7°/4°, vap. press: 1 mm @ 94.0°.
Acute tox data: Oral LD_{50} (rat) = 2590 mg/kg; dermal LD_{50} (rabbit) = 880 mg/kg. [*3*]
THR = MOD via oral and dermal routes. In humans, large ingestions cause nephritis, vomiting, diarrhea, circulatory collapse, anemia, convulsions and death. It is reported that sufficient naphthol may be absorbed through the skin to cause irr of the kidneys and injury to the cornea and lens of the eye.
Fire Hazard: Slight.

β-NAPHTHOL. Syns: *β-hydroxynaphthalene*, *isonaphthol*. White to yellowish-white crystals, slight phenolic odor. $C_{10}H_7OH$, mw: 144.2, mp: 122.5°, bp: 288.0°, flash p: 307°F, d: 1.22, vap. press: 10 mm @ 145.5°, vap. d: 4.97.
Acute tox data: Oral LD_{50} (rat) = 2420 mg/kg; sc LD_{LO} (rat) = 2940 mg/kg. [*3*]
THR = MOD via oral and sc routes. See also α-naphthol.
Fire Hazard: Slight, when exposed to heat or flame.
To Fight Fire; CO_2, dry chemical.

2-NAPHTHOL-3-CARBOXYLIC ACID. See 3-hydroxy-β-naphthoic acid.

2-NAPHTHOL-3,6-DISULFONIC ACID. See sodium-β-naphthol disulfonate.

NAPHTHOL PITCH.
THR = A recog carc. [*14*] See aromatic amines.

α-NAPHTHOL-1,2,3,4-TETRAHYDRIDE.
See tetrahydronaphthol.

NAPHTHO (8,1,2-cde)NAPHTHACENE.
THR = An exper carc. [*23*]

NAPHTHO(2,3-f)QUINOLINE. $C_{17}H_{11}N$, mw: 229.3.
THR = An exper carc to rats via sc route. [*103*]

1,2-NAPHTHOQUINONE.
Acute tox data: Oral LD_{LO} (rat) = 250 mg/kg. [*3*]
THR = HIGH via oral route.

1,4-NAPHTHOQUINONE. $C_{10}H_6O_2$, mw: 158.2.
Acute tox data: Oral LD_{50} (rat) = 190 mg/kg; ip LD_{LO} (mouse) = 10 mg/kg. [*3*]
THR = HIGH via ip and oral routes. An exper neo [*3*] via dermal route.
Disaster Hazard: Slight; when heated, emits toxic fumes.

NAPHTHORESORCINOL. Syns: *1,3-dihydroxynaphthalene, 1,3-naphthalenediol.* White crystalline leaves or plates. $C_{10}H_8O_2$, mw: 160.2, mp: 124°–125°.
Acute tox data: Ip LD_{50} (mouse) = 200 mg/kg. [*3*]
THR = HIGH via ip route.
Disaster Hazard: Slight; when heated, emits toxic fumes.

1-(2-NAPHTHOYL)AZIRIDINE. $C_{13}H_{11}ON$, mw: 197.3.
THR = An exper neo [*3*] via sc route.

NAPHTHYL ACETIC ACID. See α-naphthalene acetic acid.

α-NAPHTHYLAMINE. White crystals, reddening on exposure. $C_{10}H_7NH_2$, mw: 143.18, mp: 50°, bp: 300.8°, flash p: 315°F, d: 1.131, vap. press: 1 mm @ 104.3°, vap. d: 4.93.
Acute tox data: Oral LD_{50} (rat) = 779 mg/kg; sc LD_{LO} (rabbit) = 300 mg/kg; oral LD_{LO} (mammal) = 4000 mg/kg. [*3*]
THR = HIGH via sc route. MOD via oral route. Along with β-naphthylamine and benzidene, it has been incriminated as a cause of urinary bladder cancer. See also β-naphthylamine. See also aromatic amines. A (S) human carc and neo via sc route. [*3, 10, 23*]
Fire Hazard: Low.
To Fight Fire: Dry chemical, CO_2, mist, spray.

β-NAPHTHYLAMINE. White to faint pink, lustrous leaflets; faint aromatic odor. $C_{10}H_7NH_2$, mw: 143.18, mp: 111.5°, bp: 306.0°, d: 1.061 @ 98°/4°, vap. press: 1 mm @ 108.0°.
Acute tox data: Oral LD_{50} (rat) = 727 mg/kg; ip LD_{50} (mouse) = 200 mg/kg. [*3*]
THR = HIGH via ip and MOD via oral routes. It is not corrosive or dangerously reactive, but it is a very toxic chemical in any of its physical forms, such as flake, lump, dust, liquid or vapor. It can be absorbed into the body through the lungs, the gastrointestinal tract or the skin. Long and continued exposure to even small amounts may produce tumors and cancers of the bladder. A human (+) carc via sc, oral routes. [*3, 10, 23*] It is combustible, and at elevated temp. evolves a vapor which is flam and explosive. The explosive limit of these vapors has not yet been determined. See aromatic amines. See also α-naphthylamine.
Fire Hazard: Mod, when exposed to heat or flame; when heated, emits toxic fumes.

β-NAPHTHYL AMINE STILL COKE.
THR = A recog carc. [*14*] See aromatic amines.

1-NAPHTHYLAMINE-8-SULFONIC ACID. Syns: *peri acid, S-acid, Schoelkopf's acid, 8-amino-1-naphthalenesulfonic acid.* $NH_2C_{10}H_6SO_3H$ (bicyclic), mw: 223.25.
THR = Details U. Animal exper show irr properties. No definite evidence of carc.
Disaster Hazard: Dangerous. See sulfonic acid.

α-NAPHTHYLAMINE THIOUREA.
THR = A recog carc. [*14*] See aromatic amines.
Fire Hazard: Slight, when exposed to heat or flame; can react with oxidizing material; when heated, emits toxic fumes.
Spont Heating: No.
To Fight Fire: Water, CO_2, dry chemical.

NAPHTHYL AMINE ALDEHYDE CONDENSATE.
THR = A recog carc. [*14*] See also aromatic amines.

2-NAPHTHYLAMINE-1-D-GLUCOSIDURONIC ACID. $C_{16}H_{17}O_7N$, mw: 335.3.
THR = An exper neo [*3*] via implantation route.

NAPHTHYLAMINE MUSTARD. Syns: *chlornaphthazine*, [*N,N-di(2-chloroethyl)-1-naphthylamine*]. $C_{14}H_{15}NCl_2$, mw: 268.2.
THR = A human (+) carc via ip route. [*3, 10, 23*]

4-(1-NAPHTHYLAZO)-2-NAPHTHYL AMINE. $C_{10}H_{15}N_3$, mw: 177.3.
THR = An exper neo via oral route. [*3*]

5-(β-NAPHTHYLAZO)-2,4,6-TRIAMINOPYRIMIDINE. $C_{14}H_{13}N_7$, mw: 279.3.
THR = An exper carc via ip route. [*3*]

N-1-NAPHTHYL HYDROXYL AMINE. $C_{10}H_9ON$, mw: 159.2.
THR = An exper neo via pa and imp routes. [*3, 23*]

N-2-NAPHTHYL HYDROXYLAMINE.
THR = An exper neo and carc. [*3, 23*]

α-NAPHTHYL ISOCYANATE. $C_{11}H_7NO$, mw: 169.2, mp: 4°, bp: 269°–270°, d: 1.179 @ 25°.
THR = See cyanates.

α-NAPHTHYL ISOTHIOCYANATE. White, odorless, tasteless needles. $C_{10}H_7NCS$, mw: 185.2, mp: 58°, d: 1.81.
Acute tox data: Oral LD_{50} (mouse) = 245 mg/kg. [*3*]
THR = HIGH via oral route. Is said to cause dermatitis, chills, fever, and kidney injury. Can be absorbed via the intact skin when in solution.
Disaster Hazard: See thiocyanates.

2-NAPHTHYL MERCAPTAN.
Acute tox data: Ip LD_{LO} (mouse) = 200 mg/kg. [*3*]
THR = HIGH via ip route. A mosquito larvicide. See also mercaptans.
Disaster Hazard: Dangerous; when heated to decomp, emits highly toxic fumes of SO_x.

NAPHTHYL MERCURIC ACETATE. Crystals; insol in water. $C_{10}H_7HgO_2C_2H_3$, mw: 386.8, mp: 154°.
THR = HIGH tox. See mercury compounds, organic.

NAPHTHYL MERCURIC CHLORIDE. Silky crystals, insol in water. $(C_{10}H_7)HgCl$, mw: 363.2, mp: 189°.
THR = HIGH toxicity. See mercury compounds, organic.

1-NAPHTHYL-N-METHYL CARBAMATE. See carbaryl.

N-1-NAPHTHYLPHTHALAMIC ACID. Syn: *α-naphthylphthalamic acid*. Colorless crystals from alcohol. $C_{18}H_{13}NO_3$, mw: 291.3, mp: 203°, d: 1.40 @ 20°/4°.
Acute tox data: Oral LD_{50} (rat) = 8200 mg/kg. [*3*]
THR = LOW via oral route.
Fire Hazard: Mod, when exposed to heat or flame.
To Fight Fire: Water, foam, CO_2, dry chemical, water spray.

α-NAPHTHYLTHIOUREA. Syns: *antu*, *1-(1-naphthyl)-2-thiourea*. Crystals, bitter taste. $C_{11}H_{10}N_2S$, mw: 202.3, mp: 198°.
Acute tox data: Oral LD_{50} (rat) = 6 mg/kg; ip LD_{50} (rat) 2.5 mg/kg; oral LD_{50} (dog) = 0.4 mg/kg. [*3*]
THR = HIGH via oral and ip routes. An exper neo via sc route. [*3*] A rodenticide used extensively. Death is caused by pulmonary edema. Chronic toxicity has been known to cause drug rashes and a decrease in the white blood cells.
Disaster Hazard: Dangerous; on decomp, emits toxic fumes.

1-NAPHTHYL-2-THIOUREA. See α-naphthyl thiourea.

α-NAPHTHYLUREA.
THR = An exper carc. [*23*]

NAPLES YELLOW. See lead antimonate.

NATIONAL INSTITUTE FOR OCCUPATIONAL SAFETY AND HEALTH (NIOSH). See Section 10A.

NATIONAL FIRE PROTECTION ASSOCIATION (704M) LABELING SYSTEM. See Section 11.

NATRIUM. See sodium.

NATURAL EXTRACTIVES (SOLVENT-FREE) USED IN CONJUNCTION WITH SPICES, SEASONINGS, AND FLAVORINGS. [*109*]

NATURAL GAS. 85% methane, 10% ethane + propane, butane, nitrogen. autoign. temp.: 900°–1170°F, lel = 3.6–6.5%, uel = 13–17%.
THR = A simple asphyxiant. Upon incomplete combustion yields carbon monoxide. See also specific components.
Fire Hazard: Dangerous; see methane.
Explosion Hazard: See methane.
Explosive Range: See methane.
To Fight Fire: Stop flow of gas.

NATURAL GAS FURNACE BLACK. See soot.

NATURAL RADIATION DOSE. See Section 5A.

NATURAL RADIOACTIVE SUBSTANCES RADIATION.
THR = A recog carc. [*8*]

NATURE OF RADIATION. See Section 5A.

NAVY SPECIAL. See fuel oil #5.

NBA. See *n*-bromoacetamide.

NCRP. See Section 5A.

NEATSFOOT OIL. Pale yellow liquid. mp: 29°–41°, flash p: 470°F (CC), d: 0.92, autoign. temp.: 828°F.
THR = Details U. May cause dermatitis in sensitive individuals.
Fire Hazard: Slight, when exposed to heat or flame.
Spont Heating: Yes.
To Fight Fire: CO_2, dry chemical.

NEBURON. Syn: *1-n-butyl-3-(3,4-dichlorophenyl)-1-methylurea*. White or colorless crystals, low sol in water or hydrocarbon solvents. $C_{12}H_{16}ON_2Cl_2$, mw: 275, mp: 103°.
Acute tox data: Oral LD_{50} (rat) = 11,000 mg/kg.
THR = LOW via oral route. [*94*]
Disaster Hazard: Dangerous; see chlorides.

NEMAGON. See 1,2-dibromo-3-chloropropane.

NEMALITE. Syn: *brucite*. $Mg(OH)_2$, mw: 58.3.
THR = An exper neo to rats and mice via ip and ipl routes. [*103*]

NEOARSPHENAMINE. Syn: *neosalvarsan*. Yellow, almost odorless powder. $C_{13}H_{14}O_4N_2As_2$, mw: 467.2.
Acute tox data: Iv LD_{50} (rat) = 300 mg/kg. [*3*]
THR = HIGH via iv route. See also arsenic compounds.

NEOCOID. See DDT.

NEODYMIUM. It is a bright, silvery, lustrous, very reactive metal. Nd, atwt: 144.24, mp: 1010°, bp: 3127°, d: 6.80 and 7.00.
Acute tox data: iv LD_{LO} (guinea pig) = 70 mg/kg; ic TD_{LO} (human) = 0.02 mg/kg ⟶ blood effects. [*3*]
THR = HIGH via iv and ic routes. It may be an anticoagulant lanthanon. Care in handling is advised.
Radiation Hazard: For permissible levels (Section 5A, Table 5A.5). Isotope ^{144}Nd, $T_{\frac{1}{2}} = 24 \times 10^{15}$y. Decays to stable ^{140}Ce by α's of 1.8 MeV. Artificial isotope ^{147}Nd, $T_{\frac{1}{2}}$ = 11d. Decays to radioactive ^{147}Pm by emitting β's of 0.36 (20%), 0.80 (76%) MeV. Emits γ's of 0.09–0.53 MeV.
Fire Hazard: Mod, in the form of dust, when exposed to heat or flame. See also powdered metals. Can react violently with air, halogens, N_2, P. [*19*]
Explosion Hazard: Slight, in thc form of dust, when exposed to flame. See also powdered metals.

NEODYMIUM BROMATE. Hexagonal, red crystals. $Nd(BrO_3)_3 \cdot 9H_2O$, mw: 690.16, mp: 66.7°, bp: $-9H_2O$ @ 150°.
THR = See bromates and neodymium.

NEODYMIUM BROMIDE. Green crystals. $NdBr_3$, mw: 384.02.
THR = See bromides and neodymium.

NEODYMIUM CARBIDE. Hexagonal yellow leaflets. NdC_2, mw: 168.29, mp: decomp, d: 5.15.
THR = U. See neodymium.
Fire Hazard: A mod fire hazard by chemical reaction. See carbides.
Explosion Hazard: Slight, by chemical reaction with moisture and acids.
Disaster Hazard: Slight; will react with water or steam to produce flam vapors.

NEODYMIUM CHLORIDE. $NdCl_3$, mw: 250.6.
Acute tox data: iv LD_{50} (rat) = 3.5 mg/kg; ip LD_{50} (mouse) = 347 mg/kg.
THR = HIGH via iv, sc and ip routes.

NEODYMIUM CITRATE CHLORIDE. $(C_6H_8O_7)$ x(Cl)x(Nd).
Acute tox data: ip LD_{50} (mouse) = 138 mg/kg; ip LD_{50} (guinea pig) = 41 mg/kg. [*3*]
THR = HIGH via ip route.

NEODYMIUM HEXAANTIPYRINE IODIDE. Rose-colored crystals, sol in water. $[Nd(COC_{10}H_{12}N_2)_6I_3]$, mw: 1654.4, mp: 271°.
THR = See iodides and neodymium.

NEODYMIUM IODIDE Black crystalline powder. NdI_3, mw: 525.03, mp: 775 ± 3°.
THR = See iodides and neodymium.

NEODYMIUM LACTATE.
Acute tox data: iv LD_{LO} (cat) = 64 mg/kg. [*3*]
THR = HIGH via iv route.

NEODYMIUM MALEATE.
Acute tox data: sc LD_{50} (mouse) = 700 mg/kg; iv LD_{LO} (cat) = 56 mg/kg. [*3*]
THR = HIGH via iv and MOD via sc routes.

NEODYMIUM NITRATE. Triclinic crystals. $Nd(NO_3)_3 \cdot 6H_2O$, mw: 438.39.
Acute tox data: Oral LD_{50} (rat) = 2750 mg/kg; ip LD_{50} (rat) = 270 mg/kg; iv LD_{50} (rat) = 6 mg/kg. [*3*]
THR = HIGH via ip and iv routes, MOD via oral route. See nitrates and neodymium.

NEODYMIUM NITRIDE. Black powder. NdN, mw: 158.28.
THR = See ammonia, anhydrous, and neodymium.

NEODYMIUM OXALATE. Rose crystals. $Nd_2(C_2O_4)_3 \cdot 10H_2O$, mw: 732.76.
THR = See oxalates and neodymium.

NEODYMIUM PHOSPHIDE. NdP, mw: 175.2.
THR = HIGH toxicity. See phosphides. Can react violently with HNO_3 and also with water, in which case PH_3 is evolved. See phosphine. [*19*]

NEODYMIUM SULFIDE. Olive-green powder. Nd_2S_3, mw: 384.72, mp: decomp, d: 5.179 @ 11°.
THR = See sulfides and neodymium.

NEOHEXANE. See 2,2-dimethylbutane.

NEOMYCIN. An antibiotic.
Acute tox data: Oral LD_{50} (rat) = 2750 mg/kg; sc LD_{50} (mouse) = 275 mg/kg. [*3*]
THR = HIGH via sc and MOD via oral route. Used as a food additive permitted in foods for human consumption. [*109*]

NEON. Colorless, wholly inert gas. Ne, atwt: 20.18, mp: −248.67°, bp: −245.9°, d: 0.9002 g/liter @ 0°, d (liquid): 1.204 @ −245.9°.
THR = An inert asphyxiant gas. See also argon.

NEONICOTINE. See anabasine.

NEONICOTINE SULFATE. See anabasine sulfate.

NEON, LIQUID. See neon.

NEOPENTANE. See 2,2-dimethylpropane.

NEOPENTYL ALCOHOL. See amyl alcohol.

NEOPENTYL GLYCOL. Syn: *2,2-dimethyl-1,3-propanediol*. White crystalline solid. $C_5H_{12}O_2$, mw: 104.2, mp: 121°–126°, bp: 207°–212°.
Acute tox data: Oral LD_{LO} (rat) = 3200 mg/kg. [*3*]
THR = MOD via oral route. An insect-repellent. See also glycols.
Fire Hazard: Slight; can react with oxidizing materials.

NEOPRENE. Syn: *duprene*. An oil-resistant synthetic rubber made by the polymerization of chloroprene. $[CH_3ClC{:}CHCH_3]_x$.
THR = See chloroprene. Reacts violently with F_2, liquid F. [*19*]

NEOSALVARSAN. See neoarsphenamine.

NEOTHYL. See methyl propyl ether.

NEOTRAN. See bis(*p*-chlorophenoxy)methane.

NEPALINE. See pseudoaconitine.

NEPTUNIUM. Exists in α, β, γ forms. Np, mp: 640°, bp: 3902°, d: 20.45 @ 20°. The first synthetic transuranium element discovered. It is a silvery, radioactive, chemically active metal.
THR = HIGH radiotoxicity.
Radiation Hazard: For permissible levels, see Section 5A, Table 5A.5. Artificial isotope ^{237}Np (Neptunium Series), $T\frac{1}{2}$ = 2.1 × 10^6y, decays to radioactive ^{233}Pa by emitting α's of 4.8 MeV. Artificial isotope ^{239}Np, $T\frac{1}{2}$ = 2.3d, decays to radioactive ^{239}Pu by emitting β's of 0.33–0.71 MeV. Emits γ's of 0.04–0.3 MeV.

NERAL. See citral.

NEROLI OIL, ARTIFICIAL. See methyl anthranilate.

NERVE GAS. See tabun and sarin.

NEUROTOXICITY. See Section 9.

NEUTRALIZING FOOD ADDITIVES. [*109*]

NEUTRAL VERDIGRIS. See copper acetate.

NFN. $K \cdot C_{16}H_{10}O_6N_3$, mw: 379.4.
THR = An exper carc [*3*] via oral route.

NIACIN. Syns: *nicotinic acid*, *pyridene-3-carboxylic acid*. The anti-pellagra vitamin, colorless needles, odorless, sol in water and alcohol, insol in most lipid solvents. C_5H_4NCOOH, mw: 123, mp: 236°, sublimes above mp, d: 1.473.
Acute tox data: Oral LD_{50} (mouse) = 5000 mg/kg; sc LD_{LO} (mouse) = 4000 mg/kg. [*3*]
THR = MOD via oral and sc routes. A nutrient and/or dietary supplement food additive. [*109*]

NIACINIMIDE. Syns: *nicotinamide*, *nicotinic acid amide*. Colorless needles, very sol in water, ethyl alcohol and glycerol. $C_5H_4NCONH_2$, mw: 122, mp: 129°, d: 1.40.
Acute tox data: Oral LD_{50} (rat) = 3500 mg/kg; sc LD_{50} (rat) = 1680 mg/kg; iv LD_{50} (rat) = 2200 mg/kg. [*3*]
THR = MOD via oral, sc and iv routes. Used as a nutrient and/or dietary supplement food additive. [*109*]

NICCOLITE. See nickel arsenide.

NICKEL. A silvery-white, hard, malleable and ductile metal. Ni, atwt: 58.71, mp: 1453°, bp: 2732°, d: 8.90 @ 25°, vap. press: 1 mm @ 1810°.
THR = See nickel compounds. An exper (+) carc and neo via inhal, sc, im, ip and pa routes. [*3, 23, 13, 6, 95, 14*]
Radiation Hazard: For permissible levels, see Section 5A, Table 5A.5. Artificial isotope ^{59}Ni, $T\frac{1}{2}$ = 8 × 10^4y, decays to stable ^{59}Co by ec. Emits x-rays. Artificial isotope ^{63}Ni, $T\frac{1}{2}$ = 92y, decays to stable ^{63}Cu by emitting β's of 0.07 MeV. Reacts violently with F_2, NH_4NO_3, hydrazine, HN_3, (H_2 + dioxane), performic acid, P, Se, S, (Ti + $KClO_3$). [*19*]

NICKEL ACETATE. Green prisms. $Ni(C_2H_3O_2)_2$, mw: 176.8, mp: decomp, d: 1.798.
Acute tox data: Oral LD_{50} (rat) = 350 mg/kg; ip LD_{50} (rat) = 23 mg/kg; im LD_{50} (rat) = 35 mg/kg. [*3*]
THR = HIGH via oral, ip and im routes. An exper carc via im, ip routes. [*103, 118*]

NICKEL ALLOYS OF COPPER.
THR = See nickel compounds.

NICKEL ALLOYS OF CHROMIUM.
THR = See chromium and nickel compounds.

NICKEL ALLOYS OF IRON.
THR = See nickel and iron compounds.

NICKEL AMMONIUM NITRATE. Syn: *nickel nitrate tetrammine*. Green crystals. $Ni(NO_3)_2 \cdot 4NH_3 \cdot 2H_2O$, mw: 286.87.
THR = See nickel compounds and nitrates.

NICKEL AMMONIUM SULFATE. Syns: *nickel salts (double)*, *ammonium nickel sulfate*. Green crystals, sol in water, less sol in ammonium sulfate solution, insol in alcohol. $NiSO_4 \cdot (NH_4)_2SO_4 \cdot 6H_2O$, mw: 395, d: 1.923.
THR = See nickel compounds and sulfates.

NICKEL ANTIMONIDE. Syn: *breithauptite*. Hexagonal, light copper-red crystals. NiSb, mw: 180.45, mp: 1158°, bp: decomp @ 1400°, d: 7.54.
THR = HIGH toxicity. See antimony compounds and nickel compounds.
Fire Hazard: Mod, by chemical reaction with water to form stibine. See also hydrides.
Explosion Hazard: Slight, when exposed to flame; can evolve stibine. See also hydrides.

Disaster Hazard: Dangerous; when heated to decomp, emits highly toxic fumes of antimony; will react with water, steam or acids to produce toxic and flam vapors; can react with oxidizing materials.

NICKEL ARSENATE. Syn: *nickel-o-arsenate*. Yellowish-green powder. $Ni_3(AsO_4)_2$, mw: 435.89, d: 4.98.
THR = HIGH. See arsenic and nickel compounds.

NICKEL ARSENIDE. Syn: *niccolite*. Hexagonal crystals. NiAs, mw: 133.60, mp: 968°, d: 7.57 @ 0°.
THR = HIGH toxicity. See arsenic and nickel compounds.
Fire Hazard: Mod, by chemical reaction to form arsine. See also hydrides.
Explosion Hazard: Mod, by chemical reaction to evolve arsine. See also hydrides.
Disaster Hazard: Dangerous; when heated to decomp, emits highly toxic fumes of arsenic; will react with water, steam or acids to produce toxic and flam vapors; can react with oxidizing materials.

NICKEL-*o*-ARSENITE, ACID. Green-white crystals. $Ni_3H_6(AsO_3)_4 \cdot H_2O$, mw: 691.77, mp: decomp.
THR = See arsenic and nickel compounds.

NICKEL BENZENE SULFONATE. Green monoclinic crystals. $Ni(C_6H_5SO_3)_2 \cdot 6H_2O$, mw: 481.11, bp: decomp, d: 1.628 @ 25°.
THR = See nickel compounds and sulfonates.

NICKEL BIS(DIBUTYL DITHIOCARBAMATE). $Ni \cdot C_{18}H_{36}N_2S$, mw: 467.5, mp: 90°.
THR = An exper neo via oral route. [*103*, *118*]

NICKEL BORIDE. Prisms. NiB, mw: 69.51, d: 7.39 @ 18°.
THR = See nickel compounds, boron compounds and boron hydride.

NICKEL BROMATE. Monoclinic crystals. $Ni(BrO_3)_2 \cdot 6H_2O$, mw: 422.62, mp: decomp, d: 2.575.
THR = See nickel compounds and bromates.

NICKEL BROMIDE. Syn: *nickelous bromide*. Yellowish-brown deliquescent crystals. $NiBr_2$, mw: 218.52, mp: 963°, d: 4.64 @ 28°.
THR = See bromides and nickel compounds. Reacts violently with K. [*19*]

NICKEL BROMIDE AMMINE. Violet powder. $NiBr_2 \cdot 6NH_3$, mw: 320.72, d: 1.837.
THR = See bromides and nickel compounds.

NICKEL BROMOPLATINATE. Trigonal crystals. $NiPtBr_6 \cdot 6H_2O$, mw: 841.51, d: 3.715.
THR = See nickel compounds, platinum compounds and bromides.

NICKEL CARBIDE. Dark gray powder. Ni_3C, mw: 188.08, d: 7.957 @ 25°.
THR = See nickel compounds and carbides.

NICKEL CARBONATE. Rhombic, light green crystals. $NiCO_3$, mw: 118.70, mp: decomp.
Acute tox data: sc LD_{LO} (guinea pig) = 32 mg/kg. [*3*]
THR = HIGH via sc route. See also nickel compounds. An exper neo to rats via imp route. [*103*, *13*]

NICKEL CARBONYL. Colorless, volatile liquid or needles. $Ni(CO)_4$, mw: 170.8, mp: −25°, bp: 43°, lel = 2% @ 20°, d: 1.3185 @ 17°, vap. press: 400 mm @ 25.8°, flash p: $<-4°$.
Acute tox data: inhal LC_{50} (rat) = 240 mg/m^3 for ½ hr; ip LD_{50} (rat) = 39 mg/kg; sc LD_{50} (rat) = 63 mg/kg; iv LD_{50} (rat) = 66 mg/kg; inhal LC_{50} (cat) = 1900 mg/m^3 for ½ hr. [*3*]
THR = HIGH via inhal, ip and iv routes. See also nickel compounds and carbonyls. Toxicity symptoms from inhal are believed to be caused both by the nickel and the carbon monoxide when liberated in the lungs. In severe acute cases there is headache, dizziness, nausea, vomiting, fever and difficult breathing. Later there is coughing and cyanosis. Chronic exposure is associated with a high incidence of cancer of the respiratory tract and nasal sinuses. It is an exper (+) carc. [*3*, *6*, *13*, *23*] Sensitization dermatitis is fairly common. This is a common air contaminant.
Fire Hazard: Dangerous, when exposed to heat, flame or oxidizers.
Explosion Hazard: Mod, when exposed to heat or flame. Can react violently with air, O_2, Br_2 (*n*-butane + O_2). [*19*]
Disaster Hazard: Dangerous; when heated or on contact with acid or acid fumes, emits highly toxic fumes; can react with oxidizing materials.
To Fight Fire: Water, foam, CO_2, dry chemical.

NICKEL CATALYST, FINELY DIVIDED, ACTIVATED OR SPENT. See nickel.

NICKEL CHLORATE. Dark red crystals. $Ni(ClO_3)_2 \cdot 6H_2O$, mw: 333.70, mp: decomp @ 80°, d: 2.07.
THR = See nickel compounds and chlorates.

NICKEL CHLORIDE. A: Yellow deliquescent scales; B: monoclinic green crystals. A: $NiCl_2$, B: $NiCl_2 \cdot 6H_2O$; mw(A): 129.60, mw(b): 237.70, mp(A): sublimes, bp(A): 987°, d(A): 3.55, vap. press: 1 mm @ 671°.
Acute tox data: ip LD_{50} (mouse) = 26 mg/kg; im LD_{50} (rabbit) = 27 mg/kg. [*3*]
THR = HIGH via ip and im routes. See nickel compounds and chlorides. Violent reaction with K. [*19*]

NICKEL CHLORITE. $Ni(ClO_2)_2$, mw: 193.6.
THR = See nickel compounds. Dihydrate explodes @ 100°. [*19*]

NICKEL CHLOROPALLADATE. Hexagonal crystals. $NiPdCl_6 \cdot 6H_2O$, mw: 486.23, d: 2.353.

THR = See nickel compounds, palladium compounds and chlorides.

NICKEL CHLOROPLATINATE. Trigonal crystals. $NiPtCl_6 \cdot 6H_2O$, mw: 574.76, d: 2.798.

THR = See platinum compounds, nickel compounds and chlorides.

NICKEL COBALT SULFATE. Reddish-brown crystalline mass. $NiSO_4 \cdot CoSO_4 \cdot 14H_2O$, mw: 561.99.

THR = See nickel compounds, cobalt compounds and sulfates.

NICKEL COMPOUNDS.

THR = Nickel and most salts of nickel are generally considered not to cause systemic poisoning. Ingestion of large doses of nickel as nickel compounds (1–3 mg/kg) has been shown to cause intestinal disorders, convulsions and asphyxia in dogs. Nickel has been found in the hair of persons exposed to nickel oxide dust. Many Ni compounds are exper carc and some are human carc via inhal route. [*14*, *2*, *23*, *95*, *71*] All airborn Ni contaminating dusts are regarded as carc via inhal. [*71*] Use personal hygiene to minimize worker contact with Ni; eliminate wound contamination by Ni. The most common effect resulting from exposure to nickel compounds is the development of the "nickel itch." This form of dermatitis occurs chiefly in persons doing nickel-plating. There is marked variation in individual susceptability to the dermatitis. It occurs more frequently under conditions of high temp. and humidity, when the skin is moist, and chiefly affects the hands and arms. Nickel carbonyl is a (+) carc and a high irr to the lungs and also can produce asphyxia by decomp with the formation of carbon monoxide. These compounds are common air contaminants.

NICKEL CYANIDE. Apple-green plates or powder. $Ni(CN)_2$, mw: 110.8.

THR = A poison. See cyanides and nickel compounds. Reacts violently with Mg. [*19*]

NICKEL DI-pi-CYCLOPENTADIENYL. Syn: *nickelocene*. $Ni \cdot C_{10}H_{10}$, mw: 188.9.

Acute tox data: Oral LD_{50} (rat) = 490 mg/kg; ip LD_{50} (rat) = 50 mg/kg; im LD_{50} (mouse) = 150 mg/kg. [*3*]

THR = HIGH via oral, ip and im routes. See also nickel compounds. An exper (+) carc [*23*, *3*, *6*] via im route. [*103*, *118*]

NICKEL DIFLUORIDE. See nickel fluoride.

NICKEL DITHIONATE. Green triclinic crystals. $NiS_2O_6 \cdot 6H_2O$, mw: 326.92, mp: decomp, d: 1.908.

THR = See nickel compounds.

Disaster Hazard: Dangerous; see sulfates.

NICKEL FERRITE. See nickel compounds.

NICKEL FERROCYANIDE. Green-white crystals. $Ni_2Fe(CN)_6 \cdot xH_2O$, d: 1.892.

THR = See ferrocyanides and nickel compounds.

NICKEL FLUOBORATE.

Acute tox data: Oral LD_{LO} (rat) = 500 mg/kg; inhal LC_{LO} (mouse) = 530 mg/m^3 for 10 min. [*3*]

THR = HIGH via oral and inhal routes. See also fluorides, boron and nickel compounds.

NICKEL FLUOGALLATE. Pale green crystals. $[Ni(H_2O)_6][GaF_5H_2O]$, mw: 349.52, mp: $-5H_2O$ @ 110°, d: 2.45.

THR = See fluogallates and nickel compounds.

NICKEL FLUORIDE. Syn: *nickel difluoride*. Green crystals, slightly water-sol, decomp by boiling water. NiF_2, mw: 96.69, d: 4.63.

Acute tox data: iv LD_{50} (mouse) = 130 mg/kg. [*3*, *103*]

THR = HIGH via iv route. See also fluorides and nickel compounds. Reacts violently with K. [*19*]

NICKEL FLUOSILICATE. $NiSiF_6$, mw: 200.8.

Acute tox data: Oral LD_{LO} (rat) = 100 mg/kg. [*3*]

THR = HIGH via oral route. See also fluosilicates and nickel compounds.

NICKEL FORMATE. Green crystals. $Ni(CHO_2)_2 \cdot 2H_2O$, mw: 184.76, mp: decomp, d: 2.154.

THR = See nickel compounds.

Disaster Hazard: Slight; when heated, emits acrid fumes.

NICKEL GALLIUM ALLOY.

THR = An exper carc to rats via imp route. [*103*]

NICKEL HEXAMMINE CHLORATE.

$Ni(NH_3)_6(ClO_3)_2$, mw: 327.8.

THR = See nickel compounds. Detonates on impact. [*19*]

NICKEL HEXAMMINE PERCHLORATE.

$Ni(NH_3)_6(ClO_4)_2$, mw: 359.8.

THR = See nickel compounds. Detonates when struck. [*19*]

NICKEL HYDROXIDE. See nickelous hydroxide.

NICKEL HYPOPHOSPHITE. Green crystals. $Ni(H_2PO_2)_2 \cdot 6H_2O$, mw: 296.78, mp: explodes @ 100°, d: 1.82 @ 19.8°.

THR = See nickel compounds. See phosphine. Liberates PH_3 @ 100°. [*19*]

Explosion Hazard: Mod, when exposed to heat.

Disaster Hazard: Dangerous; may explode @ 100°. Keep away from heat and open flame.

NICKELIC HYDROXIDE. See nickelous hydroxide.

NICKELIC OXIDE. See nickel peroxide.

NICKEL IODATE. Yellow needles. $Ni(IO_3)_2$, mw: 408.53, d: 5.07.

THR = See nickel compounds and iodates.

NICKEL IODIDE. Black deliquescent crystals. NiI_2, mw: 312.53, mp: 797°, d: 5.834.
THR = See nickel compounds. Reacts violently with K. [*19*]

NICKEL IRON SULFIDE. $FeNi_4S_4$, mw: 418.8.
THR = An exper carc to rats via im route. [*103*]

NICKEL MONOSULFIDE. Syn: *millerite*. Trigonal or amorphous black crystals. NiS, mw: 90.76, mp: 797°, d: 5.3–5.65.
THR = See sulfides and nickel compounds.

NICKEL MONOXIDE. Syns: *bunsenite, nickel oxide*. Cubic green-black crystals. NiO, mw: 74.69, mp: 1900°, d: 7.45.
Acute tox data: iv LD_{LO} (cat) = 10 mg/kg; iv LD_{LO} (dog) = 7 mg/kg. [*3*]
THR = HIGH via iv route. See also nickel compounds. An exper (+) carc. [*22, 23, 3, 6, 13, 95*] Reacts violently with I_2, H_2S, (BaO + air). [*19*]

NICKEL NITRATE. Green deliquescent crystals. $Ni(NO_3)_2 \cdot 6H_2O$, mw: 290.8, mp: 56.7°, bp: 136.7°, d: 2.05.
Acute tox data: Oral LD_{50} (rat) = 1620 mg/kg. [*3*]
THR = MOD via oral route. See nickel compounds and nitrates.

NICKEL NITRATE TETRAMMINE. See nickel ammonium nitrate.

NICKEL NITRIDE. Ni_3N, mw: 190.1.
THR = See nickel compounds. Reacts violently with acids. [*19*]

NICKEL OLEATE. Green oil. $Ni(C_{18}H_{33}O_2)_2$, mw: 621.58, mp: 18°–20°.
THR = See nickel compounds.

NICKEL ORE.
THR = See arsenic compounds and nickel compounds.

NICKELOUS BROMIDE. See nickel bromide.

NICKELOUS HYDROXIDE. Light green crystals or amorphous. $NI(OH)_2$, mw: 92.7, d: 4.1.
THR = See nickel compounds. An exper (+) carc. [*103, 13*]

NICKELOUS NICKELICSULFIDE. Syn: *polydymite*. Cubic gray-black crystals. Ni_3S_4, mw: 304.33, d: 4.7.
THR = See nickel compounds and sulfides.

NICKELOUS PHOSPHATE. See nickel phosphate.

NICKEL OXALATE. Light green powder. $NiC_2O_4 \cdot 2H_2O$, mw: 182.74.
THR = HIGH toxicity. See oxalates and nickel compounds.
Disaster Hazard: Dangerous; when heated, emits toxic fumes of carbon monoxide.

NICKEL OXIDE. See nickel monoxide.

NICKEL PERCHLORATE. Hexagonal green needles. $Ni(ClO_4)_2 \cdot 6H_2O$, mw: 365.70, mp: 140°.
Acute tox data: ip LD_{LO} (mouse) = 100 mg/kg. [*3*]
THR = HIGH via ip route. See nickel compounds and perchlorates. Can react violently with 2,2-dimethoxypropane and hydrazine. [*19*]

NICKEL PEROXIDE. Syn: *nickelic oxide*. Gray-black powder. Ni_2O_3, mw: 165.38, mp: $-O_2$ @ 600°, d: 4.83.
THR = See nickel compounds and peroxides.

NICKEL PHOSPHATE. Syn: *nickelous phosphate*. Apple-green plates or emerald crystals or granules. $Ni_3(PO_4)_2 \cdot 8H_2O$, mw: 510.24, mp: decomp.
THR = See nickel compounds and phosphates.

NICKEL POTASSIUM SULFATE. Blue-green crystals. $NiSO_4 \cdot K_2SO_4 \cdot 6H_2O$, mw: 437.10, mp: decomp @ < 100°, d: 2.124.
THR = See nickel compounds and sulfates.

NICKEL REFINERY DUST. Composition: CuO 3.4%; $NiSO_4$ 20%; NiS 57%; NiO 6.3%; CoO 1%; Fe_2O_3 1.8%.
THR = A human (+) carc [*6, 3*] via im route.

NICKEL SELENATE. Green, tetragonal, monoclinic crystals. $NiSeO_4 \cdot 6H_2O$, mw: 341.75, d: 2.314.
THR = See selenium compounds and nickel compounds.

NICKEL SELENIDE. Cubic white or gray crystals. NiSe, mw: 137.65, d: 8.46.
THR = See selenium compounds and nickel compounds. HIGH toxicity.
Fire Hazard: Dangerous; can easily evolve hydrogen selenide, which is high flam.
Explosion Hazard: Slight; can evolve hydrogen selenide. See also hydrides.
Disaster Hazard: Dangerous; when heated to decomp, emits highly toxic fumes of selenium; will react with water, steam or acids to produce toxic and flam vapors.

NICKEL STEARATE. See nickel compounds.

NICKEL SUBSULFIDE. Syn: *heazlewoodite*. Pale yellowish-bronze, metallic, lustrous crystals. Ni_3S_2, mw: 240.19, d: 5.82.
THR = HIGH. See sulfides and nickel compounds. An exper (+) carc. [*3, 13, 23, 6*] Considered by some as one of the deadliest carc known. [*22*]

NICKEL SULFATE. Cubic yellow crystals. $NiSO_4$, mw: 154.76, mp: $-SO_3$ @ 840°, d: 3.68.
Acute tox data: ip LD_{50} (mice) = 21 mg/kg; sc LD_{LO} (dog) = 38 mg/kg; iv LD_{LO} (dog) = 38 mg/kg. [*3*]
THR = HIGH via ip, sc and iv routes. An exper neo to rats via imp route. [*103*] Used as a food additive permitted in the feed and drinking water of animals

and/or for the treatment of food-producing animals; [*109*] also permitted in food for human consumption. See nickel compounds and sulfates.

NICKEL SULFATE, ANHYDROUS. See nickel sulfate.

NICKEL SULFITE. Tetragonal green crystals. $NiSO_3 \cdot 6H_2O$, mw: 246.85.
THR = See nickel compounds and sulfites.

NICKEL TRIHYDRAZINE CHLORATE. $Ni(NH_2)_6(ClO_3)_2$, mw: 321.7.
THR = See nickel compounds. Detonates when struck. [*19*]

NICKEL TRIHYDRAZINE NITRATE. $Ni(NH_2)_6NO_3$, mw: 216.8.
THR = See nickel compounds. This material can detonate. [*19*]

NICOTINAMIDE. See niacinamide.

NICOTINE. Syn: *β-pyridyl-α-methyl pyrrolidine*. In its pure state, a colorless and almost odorless oil, sharp burning taste, alkaloid from tobacco. $C_{10}H_{14}N_2$, mw: 162.23, mp: $< -80°$, bp: 247.3°, lel = 0.75%, uel = 4.0%, d: 1.0092 @ 20°, autoign. temp.: 471°F, vap. press: 1 mm @ 61.8°, vap. d: 5.61.
Acute tox data: Oral LD_{50} (rat) = 53 mg/kg; dermal LD_{50} (rat) = 140 mg/kg; ip LD_{50} (mice) = 5.9 mg/kg; iv LD_{50} (mice) = 0.81 mg/kg; oral LD_{50} (dog) = 9.2 mg/kg; dermal LD_{50} (rabbit) = 50 mg/kg; oral LD_{50} (pigeon) = 75 mg/kg. [*3*]
THR = HIGH via all routes. Causes nausea, vomiting, diarrhea, mental disturbances and convulsions. May be absorbed via the intact skin. HIGH toxicity. An exper teratogen [*3*] via pa route.
Fire Hazard: Low, when exposed to heat or flame.
Explosion Hazard: Mod, when exposed to heat or flame.
Disaster Hazard: Dangerous; when heated, emits highly toxic fumes; can react with oxidizing materials.
To Fight Fire: Alcohol foam, dry chemical, CO_2.

NICOTINE COMPOUNDS, SOLID, AND PREPARATIONS THEREOF. See nicotine.

NICOTINE CUPROCYANIDE.
THR = HIGH. See nicotine, cyanides and copper compounds.

NICOTINE HYDROCHLORIDE. See nicotine.

NICOTINE ISOMER. See anabasine.

NICOTINE SALICYLATE. See nicotine.

NICOTINE SHEEP DIPS. See nicotine.

NICOTINE SULFATE. See nicotine.

NICOTINE TARTRATE. See nicotine.

NICOTINIC ACID. See niacin.

NICOTINIC ACID AMIDE. See niacinamide.

NIFOS. See tetraethyl pyrophosphate.

NIFURADINE. $C_8H_8O_4N_4$, mw: 224.2.
THR = MOD acute toxicity. An exper (+) carc [*3, 1*] via oral route.

NIFURTHIAZOLE. Bright yellow plates. $C_8H_6O_4N_4S$, mw: 254.2, mp: 215.5°.
THR = An exper (+) carc [*3, 1, 23*] via oral route.

NIFURDAZIL. $C_{10}H_{12}O_5N_4$, mw: 268.3.
THR = An exper neo [*3*] via oral route.

NIGRIN. $C_{25}H_{22}O_8N_4$, mw: 506.5.
THR = An exper teratogen [*3*] via ip route.

NIHYDRAZONE.
THR = U. Used as a food additive permitted in the food and drinking water of animals and/or for the treatment of food-producing animals; also permitted in food for human consumption. [*109*]

NINHYDRIN. Syn: *1,2,3-indantrione hydrate*. Crystals. $C_9H_4O_3 \cdot H_2O$, mw: 178.1, turn reddish @ 125°, swells @ 139°, decomp @ 240°.
Acute tox data: Oral LD_{LO} (rat) = 250 mg/kg. [*3*]
THR = HIGH via oral route.

NIOBE OIL. See methyl benzoate.

NIOBIUM. Syn: *columbium*. A shiny, white, soft, ductile metal. Nb, atwt: 92.9064, mp: 2468°, bp: 3300°, d: 8.57 @ 20°.
THR = U. Limited animal exper show HIGH toxicity for some salts of niobium. Animal exper have also shown toxicity effects related to disturbed enzyme action.
Radiation Hazard: See Section 5, Table 5A.5, for permissible levels. Artificial isotope ^{95}Nb, $T_{\frac{1}{2}}$ = 35d. Decays to stable ^{95}Mo by emitting β's of 0.16 MeV, emits γ's of 0.77 MeV.
Fire Hazard: Mod in the form of dust, when exposed to flame or by chemical reaction; can react violently with BrF_3, Cl_2, F_2. [*19*] See also powdered metals.
Explosion Hazard: Mod, in the form of dust, when exposed to flame.

NIOBIUM CARBIDE. Cubic, black or lavender-gray powder. NbC, mw: 104.91, mp: approx 3900°, d: 7.82.
THR = See carbides and niobium.

NIOBIUM CHLORIDE. Syn: *niobium pentachloride*. Yellow-white deliquescent powder. $NbCl_5$, mw: 270.20, mp: 194°, bp: 241°.
Acute tox data: Oral LD_{50} (rat) = 1400 mg/kg; oral LD_{50} (mice) = 829 mg/kg. [*3*]
THR = MOD via oral route. Animal exper show

kidney injury from small doses. See hydrochloric acid and niobium.

NIOBIUM HYDRIDE. Gray powder. NbH, mw: 93.92, mp: infusible, d: 6.6.
THR = See hydrides and niobium.

NIOBIUM HYDROGEN OXALATE. Monoclinic, colorless crystals. $Nb(HC_2O_4)_5$, mw: 538.05.
THR = HIGH. See oxalates and niobium.
Disaster Hazard: Dangerous; when heated, emits toxic fumes.

NIOBIUM NITRIDE. Cubic black crystals. NbN, mw: 106.92, mp: 2573°, d: 8.4.
THR = See nitrides and niobium.

NIOBIUM OXYBROMIDE. Yellow crystals. $NbOBr_3$, mw: 348.66, mp: sublimes.
THR = See bromides and niobium.

NIOBIUM PENTACHLORIDE. See niobium chloride.

NIOBIUM PENTAFLUORIDE. Colorless, monoclinic prisms. NbF_5, mw: 187.91, mp: 76°, bp: 236°, d: 3.293.
THR = See fluorides and niobium.

NIOBIUM PENTOXIDE. Syn: *niobium oxide*. White *o*-rhombic crystals, insol in water, sol in hydrofluoric acid or hot sulfuric acid. Nb_2O_5, mw: 265.8, d: 4.6, mp: 1520°.
THR = See niobium. Reacts violently with (Al + S), Li BrF_3. [*19*]

NIRAN. See parathion.

NIRIDAZOLE. $C_6H_6O_3N_4S$, mw: 214.2.
THR = An exper mutagen and carc [*3*] via oral route.

NIRVANOL. Syn: *phenylethylhydantion*. Colorless, odorless, crystalline powder.
$(NHCO)(CONH)C(C_2H_5)(C_6H_5)$, mw: 204.17, mp: 199°–200°.
THR = MOD irr to skin, eyes and mu mem.
Fire Hazard: Slight.

NITER. See potassium nitrate.

NITOL. See mechlorethamine hydrochloride.

NITON. See radon.

NITRAMINE. See tetryl.

NITRAMON. See explosives, high.

NITRANILIC ACID. Syn: *2,5-dehydroxy-3,6-dinitroquinone*. Flat yellow crystals, sol in water and alcohol, insol in ether. $C_6O_2(NO_2)_2(OH)_2$, mw: 230.09.
THR = Forms hydrogen cyanide on decomp. See cyanides.

***m*-NITRANILINE.** See *m*-nitroaniline.

***o*-NITRANILINE.** See *o*-nitroaniline.

NITRATE OF ALUMINUM. See aluminum nitrate.

NITRATE OF SODA AND POTASH. See nitrates, N.O.S.

NITRATES, N.O.S. Organic nitrates are usually termed nitro compounds. These compounds are a combination of the nitro ($-NO_2$) group and an organic radical. However, this term is often used to denote nitric acid esters of an organic material. Inorganic nitrates are compounds of metals which are combined with the mono-valent $-NO_3$ radical.
THR = Large amounts taken by mouth may have serious or even fatal effects. The symptoms are dizziness, abdominal cramps, vomiting, bloody diarrhea, weakness, convulsions and collapse. Small, repeated doses may lead to weakness, general depression, headache and mental impairment. Also there is some implication of increased cancer incidence among those exposed.
Fire Hazard: Mod, by spont chemical reaction; practically all nitrates are powerful oxidizing agents (Section 7).
Explosion Hazard: Nitrates may explode when shocked, exposed to heat or flame or by spont chemical reaction (See also explosives, high). All the inorganic nitrates act as oxygen carriers; under proper conditions these can give up their oxygen to other materials, which may in turn detonate. For example, potassium or barium nitrate are added to doublebase powders for the purpose of reducing flash and rendering the powder more ignitable. A further use for these materials is to mix them with a smokeless powder which is not completely colloided, for the purpose of granulation. An example of such a powder is "E. C. Powder," used for loading blank cartridges and hand grenades. Sodium and potassium nitrate are also used in black powder as the oxygen carrier to support the combustion of the sulfur and the charcoal.

Ammonium nitrate has all the properties of the other nitrates, but is also able to detonate by itself under certain conditions. It is therefore a high explosive, although very insensitive to impact and difficult to detonate. In the pure state, it requires a combination of an initiator and a high explosive. This combination is known as a reinforced detonator. Ammonium nitrate in combination with nitro compounds (such, perhaps, as trinitrotoluene) forms one of the major high explosives for military use. Ammonium nitrate is widely used also as the chief component of "ammonia permissibles," and of "ammonia dynamites"; as a component of many pyrotechnic mixtures; and in combination with smokeless powder, as a granular blasting explosive. It is a relatively safe high explosive which, however, must be stored in a cool, ventilated place, away from acute fire hazards and easily oxidized materials. Ammonium nitrate must not be confined, because if a fire should start, confinement can cause detonation with extremely violent results. Also reacts violently with Al, BP, cyanides, esters, PN_2H, P, NaCN, $SnCl_2$, sodium hypophosphite, thiocyanates. [*19*]

Disaster Hazard: Dangerous, due to fire and explosion hazard. On decomp, they emit toxic fumes. They are powerful oxidizing agents which may cause violent reaction with reducing materials. Nitrates should be protected carefully, as discussed in detail in Section 7.

NITRATINE. See sodium nitrate.

NITRATING (MIXED) ACID. See nitric acid and sulfuric acid.

NITRE. See potassium nitrate.

NITRIC ACID. Syns: *aqua fortis*, *hydrogen nitrate*, *azotic acid*. Transparent colorless or yellowish, fuming, suffocating, caustic and corrosive liquid. HNO_3, mw: 63.02, mp: −42°, bp: 86°, d: 1.502.

THR = VERY HIGH irr to skin, eyes and mu mem. Can affect the teeth. It destroys tissue, causes burns, stains skin, destroys eyes. Causes upper respiratory irr which may seem to clear up only to return in a few hours and more severely. [*88*] The exact composition of the "fumes" or vapor produced by nitric acid depends upon such factors as temp., humidity and whether or not the acid comes in contact with other materials, such as heavy metals or organic compounds. Depending upon these factors, the vapor will consist of a mixture of the various oxides of nitrogen and of nitric acid vapor. Nitric acid vapor is high irr to the mu mem of the eyes and respiratory tract and to the skin. It is corrosive to the teeth. Because of its irr properties, chronic exposure to dangerous conc of the acid vapor seldom occur.

Fire Hazard: Mod, by chemical reaction with reducing agents. It is a powerful oxidizing agent.

Explosion Hazard: Reacts violently with acetic acid, acetic anhydride, (acetone + acetic acid), (acetone + H_2SO_4), acetylene, acrolein, acrylonitrile, allyl alcohol, allyl chloride, 2-amino ethanol, NH_3, NH_4OH, aniline, anion exchange resins, (dichromate + anion exchange resins), Sb, AsH_3, Bi, B, boron decahydride, BP, BrF_5, *n*-butyraldehyde, Ca hypophosphite, C, Cs_2C_2, 4-chloro-2-nitroaniline, ClF_3, chlorosulfonic acid, cresol, cumene, Cu_3N_2, CuN_3, cyanides, cyclic ketones, cyclohexanol, cyclohexanone, diborane, 2,6-di-*tert*-butyl phenol, diisopropyl ether, epichlorohydrin, ethanol, *m*-ethylaniline, ethylene diamine, ethylene imine, 5-ethyl-2-methyl pyridine, 5-ethyl-2-picoline, $C_2H_5PH_2$, FeO, F_2, furfuryl alcohol, Ge, glyoxal, hydrazine, HN_3, HI, H_2O_2, H_2Se, H_2S, H_2Te, (indane + H_2SO_4), isoprene, (ketones + H_2O_2), (lactic acid + HF), Li, Li_6Si_2, Mg, Mg_3P_2, Mg-Ti alloy, Mn, mesitylene, mesityl oxide, 2-methyl-5-ethyl pyridine, 4-methylcyclohexanone, NdP, nitrobenzene, oleum, organic matter, PH_3, PH_4I, P, P_4I_3, PCl_3, phthalic acid, phthalic anhydride, KH_2PO_2, β-propiolactone, propylene oxide, pyridine, Rb_2C_2, Se, selenium iodophosphide, (Ag + ethanol), Na, NaN_3, NaOH, SbH_3, sulfamic acid, (H_2SO_4 + glycerides), terpenes, B_4H_{10}, thiocyanates, thiophene, Ti, Ti alloy, Ti-Mg alloy, (H_2SO_4 + $C_6H_5CH_3$), toluidine, triazine, uns-dimethyl hydrazine, U, U-Nd alloy, U-Nd-Zr alloy, vinylacetate, vinylidene chloride, Zn, Zr-U alloys. [*19*]

Disaster Hazard: Dangerous; when heated to decomp, emits highly toxic fumes of NO_x and hydrogen nitrate; will react with water or steam to produce heat and toxic and corrosive fumes.

To Fight Fire: Water.

NITRIC ACID, ANHYDROUS. See nitric acid, fuming.

NITRIC ACID, FUMING RED. Syn: *nitric acid, anhydrous*. Colorless to yellow to red corrosive liquid. $NHO_3 + N_2O_5$, d: > 1.480.

Acute tox data: Inhal LC_{50} (rat) = 65 ppm of NO_2 for 4 hrs. [*3*]

THR = VERY HIGH irr to skin, eyes and mu mem. A corrosive poison.

Fire Hazard: Dangerous; very powerful oxidizing agent.

Explosion Hazard: Mod; can react explosively with many reducing agents.

Disaster Hazard: Dangerous; when heated to decomp, emits highly toxic fumes of NO_x; will react with water or steam to produce heat and toxic, corrosive and flam vapors.

NITRIC ACID, FUMING WHITE.

Acute tox data: Inhal LC_{50} (rat) = 244 ppm of NO_2 for 30 min. [*3*]

THR = VERY HIGH irr to skin, eyes and mu mem. A corrosive poison.

NITRIC ANHYDRIDE. See nitrogen pentoxide.

NITRIC ETHER. See ethyl nitrate.

NITRIC OXIDE. Syn: *NO_x*. Colorless gas, blue liquid and solid. NO, mw: 30.01, mp: −161°, bp: −151.18, d: 1.3402 g/liter, liquid: 1.269 @ −150°.

Acute tox data: Inhal LD_{LO} (mouse) = 320 ppm; inhal LC_{50} (rabbit) = 315 ppm for $\frac{1}{4}$ hr. [*3*]

THR = HIGH irr via inhal route and to skin, eyes and mu mem. A poison gas. Exposure to such fumes may occur whenever nitric acid acts upon organic material, such as wood, sawdust and refuse; it occurs when nitric acid is heated, and when organic nitro compounds are burned, as for example, celluloid, cellulose nitrate (guncotton) and dynamite. The action of nitric acid upon metals, as in metal etching and pickling, also liberates the fumes. In high-temp welding, as with the oxyacetylene or electric torch, the nitrogen and oxygen of the air unite to form oxides of nitrogen. Exposure may also occur in many manufacturing processes when nitric acid is made or used.

The oxides of nitrogen are somewhat sol in water, reacting with it in the presence of oxygen to form nitric and nitrous acids. This is the action that takes place deep in the respiratory system. The acids formed are irr, causing congestion of the throat and bronchi, and edema of the lungs. The acids are neutralized by the alkalis present in the tissues, with the formation of nitrates and nitrites. The latter may cause some arterial dilation, fall in blood pressure, headache and dizziness, and there may be some formation of methemoglobin. However, the nitrite effect is of secondary importance.

Because of their relatively low sol in water, the nitrogen oxides are only slightly irr to the mu mem of the upper respiratory tract. Their warning power is therefore low, and dangerous amounts of the fumes may be breathed before the workman notices any real discomfort. Higher conc (60 to 150 ppm) cause immediate irr of the nose and throat, with coughing and burning in the throat and chest. These symptoms often clear up on breathing fresh air, and the workman may feel well for several hours. Some 6–24 hrs after exposure, he develops a sensation of tightness and burning in the chest, shortness of breath, sleeplessness and restlessness. Dyspnea and air hunger may increase rapidly, with development of cyanosis and loss of consciousness, followed by death. In cases which recover from the pulmonary edema, there is usually no permanent disability, but pneumonia may develop later. Conc of 100–150 ppm are dangerous for short exposures of 30–60 min. Conc of 200–700 ppm may be fatal after even very short exposures.

Continued exposure to low conc of the fumes, insufficient to cause pulmonary edema, is said to result in chronic irr of the respiratory tract, with cough, headache, loss of appetite, dyspepsia, corrosion of the teeth and gradual loss of strength.

Exposure to nitrous fumes is always potentially serious, and persons so exposed should be kept under close observation for at least 48 hrs.

Can react violently with Al, B, CS_2, ClO, Cr, F_2, fuels, hydrocarbons, NCl_3, O_3, PH_3, P, Rb_2C_2, Na_2O, uns-dimethyl hydrazine, U. Also, acetic anhydride, NH_3, BaO, BCl_3, $CsHC_2$, $CHCl_3$, 1,2-dichloroethane, dichloroethylene, ethylene, Fe, Mg, Mn, CH_2Cl_2, olefins, PNH_2, K, propylene, Na, S, WC, trichloroethylene, 1,1,1-trichloroethane, uns-tetrachloroethane. [*19*]

Disaster Hazard: Dangerous; when heated to decomp, emits highly toxic fumes of NO_x; will react with water or steam to produce heat and corrosive fumes; can react vigorously with reducing materials.

NITRIDES. Compounds of N≡ as the anion, such as Li_3N, Ca_3N_2, etc.

THR = The details of toxicity of nitrides as a group are U. However, many nitrides react with moisture to evolve ammonia. This gas is an irr to mu mem. See ammonia.

Fire Hazard: To the extent that many nitrides evolve flam ammonia gas upon contact with moisture, nitrides can be fire hazards. See ammonia.

Explosion Hazard: Mod. See ammonia.

Disaster Hazard: Mod dangerous due to fire and explosion hazard. On decomp, they emit toxic fumes of ammonia.

NITRILES. Nitriles are organic compounds containing the (–CN) grouping; e.g., acrylonitrile (CH_2:CHCN).

THR = Nitriles are organic cyanides; acrylonitrile, propionitrile and some others resemble cyanides in toxicity. Other nitriles, such as cyanamides and cyanates, have no cyanide effect. When nitriles are heated to decomp, they emit highly toxic cyanide fumes. See specific compounds; also cyanides. The nitriles may be used as insecticides. Can react violently with ($LiAlH_4$ + H_2O). [*19*]

NITRILOMALONAMIDE. See cyanoacetamide.

NITRILOTRIACETIC ACID. Syn: *NTA*. $C_6H_9O_6N$, mw: 191.2.

Acute tox data: Oral LD_{50} (rat) = 1470 mg/kg; oral LD_{50} (mouse) = 3160 mg/kg.

THR = Chronic toxicity; a possible teratogen, mutagen and carc. This question is presently under study. [*3*]

NITRILOTRIACETIC ACID, SODIUM SALT.

Acute tox data: ip LD_{50} (mouse) = 460 mg/kg. [*3*] See also nitrilotriacetic acid.

NITRILOTRIACETIC ACID, DISODIUM SALT.

Acute tox data: Oral LD_{50} (rat) = 1460 mg/kg. [*3*] See also nitrilotriacetic acid.

NITRILOTRIACETIC ACID, TRISODIUM SALT.

Acute tox data: ip LD_{50} (mouse) = 500 mg/kg. [*3*] See also nitrilotriacetic acid.

NITRILOTRISILANE. See trisilicylamine.

3,3′,3-NITRILOTRISPROPIONAMIDE. $C_9H_{19}O_3N_4$, mw: 230.3.
THR = An exper neo [*3*] via oral route.

NITRITES.
THR = Large amounts taken by mouth may produce nausea, vomiting, cyanosis (due to methemoglobin formation) collapse and coma. Repeated small doses cause a fall in blood pressure, rapid pulse, headache and visual disturbances. There appears to be some implication of increased cancer incidence associated with ingestion of nitrites.
Fire Hazard: Details U. They are generally powerful oxidizers. In contact with readily oxidized materials, a violent reaction such as a fire or explosion may ensue.
Explosion Hazard: Details U. Organic nitrites may decomp violently, in contact with NH_4^+ salts, cyanides, KCN. [*19*]
Disaster Hazard: Dangerous; shock may explode them; when heated to decomp, they emit highly toxic fumes of NO_x; can react vigorously with reducing materials.

5-NITROACENAPHTHENE. $C_{12}H_9NO_2$, mw: 199.2, mp: 103°.
THR = An exper carc in rats, hamsters via oral, ip routes. [*108*]

***p*-NITROACETANILIDE.** White crystals. $CH_3CONHC_6H_4NO_2$, mw: 180.16, mp: 215°.
Acute tox data: ip LD_{LO} (rat) = 500 mg/kg. [*3*]
THR = MOD via ip route. See also acetanilide and nitro compounds of aromatic hydrocarbons.
Disaster Hazard: Dangerous; when heated to decomp, emits highly toxic fumes of NO_x; can react with oxidizing materials.

9-NITROACRIDINE-9-OXIDE.
THR = An exper carc. [*23*]

5-NITRO-2-AMINOANISOLE. See 1-amino-2-methoxy-4-nitrobenzene.

***m*-NITROANILINE.** Syn: *m-nitraniline*. Yellow crystals. $C_6H_6N_2O_2$, mw: 138.1, mp: 114°, bp: 305.7°, d: 1.42, vap. press: 1 mm @ 119.3°.
Acute tox data: Oral LD_{50} (rat) = 535 mg/kg; oral LD_{50} (mice) = 308 mg/kg; ip LD_{LO} (dog) = 70 mg/kg. [*3*]
THR = HIGH via oral and ip routes.
Fire Hazard: Mod, when exposed to heat or flame or by chemical reaction with oxidizers. Causes nitration of materials which can then ignite spont.
Explosion Hazard: See nitrates.
Disaster Hazard: See nitrates.
To Fight Fire: Water, CO_2, dry chemical.

***o*-NITROANILINE.** Syn: *o-nitraniline*. Orange-yellow crystals. $C_6H_6N_2O_2$, mw: 138.1, mp: 71°–72°, bp: 284.5°, d: 1.442, vap. press: 1 mm @ 104.0°.
Acute tox data: Oral LD_{50} (rat) = 535 mg/kg; oral LD_{50} (mice) = 308 mg/kg. [*3*]
THR = HIGH via oral route.
Fire Hazard: See nitrates.
Explosion Hazard: See nitrates.
To Fight Fire: Water, CO_2, dry chemical.

***p*-NITROANILINE.** Syn: *1-amino-4-nitrobenzene*. $C_6H_6N_2O_2$, mw: 138.1, mp: 148.5°, bp: 332.0°, flash p: 390°F (CC), d: 1.424, vap. press: 1 mm @ 142.4°.
Acute tox data: Oral LD_{50} (rat) = 3249 mg/kg; oral LD_{50} (mice) = 812 mg/kg; ip LD_{50} (mice) = 250 mg/kg; oral LD_{50} (wild birds) = 75 mg/kg. [*3*]
THR = HIGH to wild birds via oral, and mice via ip routes. MOD via oral route to rats and mice. Acute symptoms are headache, nausea, vomiting, weakness and stupor, cyanosis and methemoglobinemia. Chronic exposure can cause liver damage. See also aniline.
Fire Hazard: Slight, when exposed to heat or flame.
Explosion Hazard: See nitrates.
Disaster Hazard: See nitrates.
To Fight Fire: Water spray or mist, foam, dry chemical, CO_2.

***o*-NITROANILINE DIAZONIUM SALT.**
THR = Reacts violently with NaHS, Na polysulfide, Na_2S. [*19*]

4-NITRO-*o*-ANISIDINE. See 1-amino-2-methoxy-4-nitrobenzene.

NITROANISOLE. $C_6H_4OCH_3ONO$, mw: 153.1.
Acute tox data: Oral LD_{50} (mammal) = 4700 mg/kg; ip LD_{50} (mammal) = 1400 mg/kg. [*3*]
THR = MOD via oral and ip routes. Has exploded with Ni catalyst. [*19*]

NITROBARITE. See barium nitrate.

***p*-NITROBENZALDEHYDE.** White to yellow crystals, slightly sol in water or ether, sol in alcohol, benzene, glacial acetic acid. $NO_2C_6H_4CHO$, mw: 151.12, mp: 106°–107°.
Acute tox data: ip LD_{50} (rat) = 545 mg/kg. [*3*]
THR = MOD via ip route. Animal exper show injury to eyes, skin and liver.
Disaster Hazard: Dangerous; see nitrates, organic.

***p*-NITROBENZALDOXIME.** $C_7H_6O_3N_2$, mw: 166.2.
Acute tox data: Oral LD_{50} (mammal) = 180 mg/kg; ip LD_{50} (mammal) = 120 mg/kg. [*3*]
THR = HIGH via oral and ip routes. An irr. Animal exper show injury to eyes, skin and liver.
Disaster Hazard: Dangerous; see nitrates, organic.

NITROBENZENE. Syns: *oil of mirbane, nitrobenzol.* Bright yellow crystals or yellow, oily liquid. $C_6H_5NO_2$, mw: 123.11, mp: 5.7°, bp: 210.9°, ulc: 20–30, lel = 1.8% @ 200°F, flash p: 190°F (CC), d: 1.205 @ 25°/4°, autoign. temp.: 900°F, vap. press: 1 mm @ 44.4°, vap. d: 4.25.

Acute tox data: Oral TD_{LO} (women) = 200 mg/kg ⟶ blood problems. Oral LD_{LO} (dog) = 750 mg/kg; dermal LD_{LO} (rabbit) = 600 mg/kg; dermal LD_{LO} (mouse) = 400 mg/kg; oral LD_{LO} (cat) = 2000 mg/kg. [*3*]

THR = MOD via oral, dermal, sc and iv routes. Causes cyanosis due to formation of methemoglobin. A common air contaminant.

Fire Hazard: Mod, when exposed to heat, flame or oxidizers.

Explosion Hazard: Mod, when exposed to heat or flame. Reacts violently with HNO_3, ($AlCl_3$ + C_6H_5OH), (aniline + glycerine), N_2O_4, $AgClO_4$. [*19*]

Disaster Hazard: See nitrates.

To Fight Fire: Water, foam, CO_2, dry chemical.

NITROBENZENE SULFONIC ACID. $ONOC_6H_4SO_3H$, mw: 203.2.

THR = No data. Probably HIGH irr. Decomp violently at about 200°. [*19*]

***m*-NITROBENZENE SULFONYL FLUORIDE.** Yellow solid. $O_2NC_6H_4SO_2F$, mw: 205, mp: 45°–47°, bp: 143° @ 7 mm, flash p: 335°F, d: 1.582, vap. press: 5 mm @ 130°, vap. d: 7.07.

THR = Apparently LOW acute inhal toxicity to test animals. Possibly slightly irr to skin.

Fire Hazard: Slight, when exposed to heat or flame.

Disaster Hazard: Dangerous; when heated to decomp, emits highly toxic fumes of fluorides and SO_x; can react with oxidizing materials.

To Fight Fire: Water, foam, CO_2, dry chemical.

***m*-NITROBENZOIC ACID.**

Acute tox data: ip LD_{50} (rat) = 670 mg/kg; iv LD_{50} (rat) = 680 mg/kg; pa LD_{50} (rat) = 1820 mg/kg. [*3*]

THR = MOD via ip, iv and pa routes.

***p*-NITROBENZOIC ACID.** Crystals. $NO_2C_6H_4COOH$, mw: 167.1, mp: 242.4°, bp: sublimes, d: 1.550 @ 32°/4°.

Acute tox data: Oral LD_{50} (rat) = 1960 mg/kg; ip LD_{50} (rat) = 1210 mg/kg; pa LD_{50} (rat) = 1960 mg/kg; iv LD_{50} (mice) = 1470 mg/kg. [*3*]

THR = MOD via oral, ip, pa and iv routes. See also nitro compounds of aromatic hydrocarbons.

Disaster Hazard: See nitrates.

***p*-NITROBENZOIC ACID ETHYL ESTER.** Syn: *ethyl-p-nitrobenzoate.*

Acute tox data: ip LD_{50} (mice) = 250 mg/kg. [*3*]

THR = HIGH via ip route. Larvicide for European corn borer. See also nitro compounds of aromatic hydrocarbons.

Fire Hazard: See nitrates.

Disaster Hazard: See nitrates.

NITROBENZOL. See nitrobenzene.

***m*-NITROBENZOTRIFLUORIDE.** Syn: *m-nitrotrifluoromethyl benzene.* Thin, pale straw-colored oily liquid with aromatic odor. $C_7H_4F_3NO_2$, mw: 191.1, mp: −5°, bp: 202.8°, flash p: 217°F (OC), d: 1.437 @ 15.5°/15.5°, vap. press: 0.3 mm @ 25°.

THR = U. See also nitro compounds of aromatic hydrocarbons and fluorides.

Fire Hazard: Low.

Disaster Hazard: Dangerous; when heated to decomp, emits highly toxic fumes of NO_x and fluorides.

To Fight Fire: Water spray, fog, foam, CO_2.

***o*-NITROBENZOYL ACETIC ACID.** $ONOC_6H_4COCH_2COOH$, mw: 209.1.

THR = Upon reaction with thionyl chloride, a violently explosive product is formed. [*19*]

***m*-NITROBENZOYL CHLORIDE.** Yellow to brown liquid, partially crystallized at room temp. $NO_2C_6H_4COCl$, mw: 185.5, mp: 28°–31°, bp: 278°, vap. d: 6.43.

Acute tox data: Oral LD_{50} (rat) = 2460 mg/kg; dermal LD_{50} (rabbit) = 790 mg/kg. [*3*]

THR = MOD via oral and dermal routes. See also nitro compounds. Very unstable. Has spont exploded. [*19*] See also nitro compounds of aromatic hydrocarbons.

Disaster Hazard: Dangerous; when heated to decomp, emits highly toxic fumes of chlorides and NO_x; will react with water or steam to produce toxic and corrosive fumes.

***p*-NITROBENZOYL CHLORIDE.** Yellow crystalline solid. $NO_2C_6H_4COCl$, mw: 185.5, mp: 70°, bp: 154° @ 15 mm.

THR = Details U. See also nitro compounds of aromatic hydrocarbons.

Disaster Hazard: Dangerous; when heated to decomp, emits highly toxic fumes of chlorides and NO_x; will react with water or steam to produce toxic and corrosive fumes.

N-(4-NITRO)BENZOYL OXYPIPERIDINE. $C_{12}H_{14}O_4N_2$, mw: 250.3.

THR = An exper neo [*3*] via dermal route.

***o*-NITROBENZYL CHLORIDE.** Crystals. $NO_2C_6H_4CH_2Cl$, mw: 171.59, mp: 48°–49°.

THR = HIGH irr to skin, eyes and mu mem. See also nitro compounds of aromatic hydrocarbons.

Disaster Hazard: Dangerous; see chlorides and nitrates.

***o*-NITROBIPHENYL.** Syns: *ONB, o-nitrodiphenyl.* Light yellow- to reddish-colored liquid or crystalline solid. $C_6H_5C_6H_4NO_2$, mw: 199.2, mp: 35°, bp: 330°, flash p: 290°F, d: 1.189 @ 40°/15.5°, autoign. temp.: 356°F, vap. press: 2 mm @ 140°, vap. d: 5.9.
Acute tox data: Oral LD_{50} (rat) = 1230 mg/kg; oral LD_{50} (rabbit) = 1580 mg/kg. [*3*]
THR = MOD via oral route. An irr. See aromatic amines.
Fire Hazard: Slight, when exposed to heat or flame.
Disaster Hazard: See nitrates.
To Fight Fire: CO_2, dry chemical.

4-NITROBIPHENYL.
Acute tox data: Oral LD_{50} (rat) = 2230 mg/kg; oral LD_{50} (rabbit) = 1970 mg/kg. [*3*]
THR = MOD via oral route. An exper (+) carc via oral route. [*3, 10, 23*]

NITROBROMOFORM. See bromopicrin.

4-NITRO-5-BROMOQUINOLINE-1-OXIDE.
THR = An exper carc. [*23*]

4-NITRO-6-BROMOQUINOLINE-1-OXIDE.
THR = An exper carc. [*23*]

4-NITRO-7-BROMOQUINOLINE-1-OXIDE.
THR = An exper carc. [*23*]

1-NITROBUTANE. Liquid. $C_4H_9NO_2$, mw: 103.12, bp: 151°.
Acute tox data: Oral LD_{LO} (rabbit) = 500 mg/kg. [*3*]
THR = MOD via oral route.
Fire Hazard: Mod, when exposed to heat or flame.
Explosion Hazard: See nitrates.
Disaster Hazard: See nitrates.
To Fight Fire: Foam, CO_2, dry chemical.

2-NITROBUTANE. Liquid. $C_4H_9NO_2$, mw: 103.1, bp: 139°.
Acute tox data: Oral LD_{LO} (rabbit) = 500 mg/kg. [*3*]
THR = MOD via oral route. See 1-nitrobutane.
Fire Hazard: Mod, when exposed to heat or flame.
Explosion Hazard: See nitrates.
Disaster Hazard: See nitrates.
To Fight Fire: Foam, CO_2, dry chemical.

2-NITRO-1-BUTANOL. Colorless liquid. $CH_3CH_2CH(NO_2)CH_2OH$, mw: 119.12, mp: −47°, bp: 105°, d: 1.133 @ 20°/20°.
THR = Details U. See also alcohols and nitrates, organic.
Fire Hazard: Mod, when exposed to heat or flame.
Disaster Hazard: See nitrates.

2-NITRO-2-BUTENE. $CH_3(NO_2)C{:}CHCH_3$, mw: 101.
Acute tox data: Oral LD_{LO} (rat) = 280 mg/kg; inhal LC_{LO} (rat) = 1400 ppm; dermal LD_{LO} (rabbit) = 620 mg/kg; ip LD_{LO} (rat) = 80 mg/kg. [*3*]
THR = HIGH irr via oral, ip and inhal routes. MOD via dermal route. Animal exper show marked pulmonary irr and some CNS irr.
Disaster Hazard: Dangerous; see nitrates, organic.

2-NITRO-1-BUTENE.
Acute tox data: Inhal LC_{LO} (mice) = 500 mg/m^3 for 10 min. [*3*]
THR = HIGH irr via inhal route and to skin, eyes and mu mem.

NITROCARBONITRATE. A solid explosive. Composition: a mixture of ammonium nitrate, dinitrocotton, etc.
THR = U. See nitrates.
Fire Hazard: Mod; an oxidizing agent. See nitrates.
Explosion Hazard: See explosives, high; nitrates.
Disaster Hazard: See explosives, high; nitrates.

NITROCELLULOSE. Syns: *cellulose nitrate, cellulose hexanitrate, cellulose pentanitrate, cellulose tetranitrate, cellulose trinitrate, guncotton, nitrocotton.* Dry: white amorphous powder or cotton-like solid. Wet: water white liquid. $C_{12}H_{14}(ONO_2)_6O_4$ to $C_{12}H_{17}(ONO_2)_3O_7$, mw: 459.28°–594.28°, mp: ignites @ 160°–170°, d: 1.66, flash p: 55°F.
Fire Hazard: Highly dangerous in the dry state, when exposed to heat, flame or powerful oxidizers. When wet with 35% of denatured ethyl alcohol it is about as hazardous as ethyl alcohol alone or gasoline. Dry nitrocellulose burns rapidly with intense heat and ignites easily.
Explosion Hazard: Mod dangerous. See explosives, high.
Disaster Hazard: The dry material is highly dangerous; the wet, somewhat less so. Keep away from heat and open flame!
To Fight Fire: Use copious volumes of water; CO_2 is effective in extinguishing fires of nitrocellulose solvents.

NITROCELLULOSE, COLLOIDED, GRANULAR OR FLAKE, DRY OR WET WITH LESS THAN 20% SOLVENT OR WATER. See nitrocellulose.

NITROCELLULOSE, DRY OR WET WITH LESS THAN 30% SOLVENT OR 20% WATER.
See nitrocellulose.

NITROCELLULOSE FLAKES, DRY OR WET WITH LESS THAN 20% ALCOHOL OR SOLVENT.
See nitrocellulose.

***m*-NITROCHLOROBENZENE.**
See *m*-chloronitrobenzene.

***o*-NITROCHLOROBENZENE.**
See *o*-chloronitrobenzene.

***p*-NITROCHLOROBENZENE.** Syn: *1-chloro-4-nitrobenzene.* Liquid. $ClC_6H_4NO_2$, mw: 157.6, mp: 83°, bp: 242°, flash p: 261°F (CC), d: 1.520 @ 22°/4°, vap. d: 5.43.

Acute tox data: Oral LD_{50} (mouse) = 1414 mg/kg; ip LD_{50} (mammal) = 420 mg/kg. [*3*]

THR = MOD via oral route; HIGH via ip and dermal routes.

Fire Hazard: Slight, when exposed to heat or flame.

Spont Heating: No.

Disaster Hazard: Dangerous; when heated to decomp, emits highly toxic fumes of NO_x and chlorides.

To Fight Fire: CO_2, dry chemical.

3-NITRO-4-CHLOROBENZOTRIFLUORIDE. Thin, yellow, oily liquid. $C_6H_3CF_3(NO_2)(Cl)$, mw: 225.6, mp: −7.5°, bp: 222°, d: 1.542 @ 15.5°/15.5°.

THR = Details U. See also nitro compounds of aromatic hydrocarbons.

Disaster Hazard: Dangerous; see chlorides; fluorides and nitrates.

NITROCHLOROFORM. See chloropicrin.

NITROCHLOROMETHANE. See chloropicrin.

***p*-NITROCHLOROPHENYL DIMETHYL THIONOPHOSPHATE.** See dicapthon.

NITRO COMPOUNDS OF AROMATIC HYDROCARBONS.

THR = The di- and trinitrobenzenes, like mononitrobenzene, are absorbed chiefly through the skin and through inhal of the dust or vapor when these materials are heated. The dinitrobenzenes are believed to be somewhat more toxic than the mononitrobenzene, and more toxic than aniline. The effect of di- and trinitrobenzene on the body is similar to that of aniline and mononitrobenzene, with reduction of the oxygen-carrying power of the blood and depression of the nervous system being responsible for most of the symptoms following acute exposure. Poisoning with the solid nitro compounds is usually slower and less severe than is the case with the liquid nitro and amino benzenes since absorption is less rapid. Thus, chronic poisoning occurs more frequently than acute, the picture in the chronic form being one of anemia, moderate cyanosis, fatigue, slight dizziness, headache, insomnia and loss of weight. The urine is frequently dark in color; the skin on the exposed parts is often yellowish-brown, and the hair yellowish-red. There may be irr of the nose and throat, nausea and vomiting, sclerotic icterus, and complaints related to the nervous system. Prolonged chronic exposure may result in damage to the liver and kidneys, with production of acute yellow atrophy, toxic hepatitis, and fatty degeneration of the kidneys.

The introduction of one or more chlorine atoms into the nitrobenzene ring results in the formation of chloronitrobenzene compounds of nitrochlors. The chloro-mono-nitrobenzenes have essentially the same toxic effect as nitrobenzene. The chlorine derivatives of dinitrobenzene, on the other hand, while resembling dinitrobenzene in their systemic effects are much more irr to the skin. They act as direct irr, and in addition may cause sensitization. For further information see specific nitro compounds.

Disaster Hazard: Dangerous; many of these compounds are highly flam and some are explosive. When heated to decomp, they evolve highly toxic fumes of NO_x.

NITROCOTTON. See cellulose nitrate.

NITROCRESOL METHYL ETHER. Pale yellow crystals. $C_6H_3CH_3NO_2OCH_3$, mw: 167.2, mp: 8.5°, bp: 274°.

THR = Details U; probably toxic. See also ethers.

Fire Hazard: Details U. See ethers and nitrates.

Disaster Hazard: See nitrates.

NITROCYCLOHEXANE. Straw-colored to water white mobile liquid, mild odor. $O_2NCH(CH_2)_5$, mw: 129.2, bp: 205.5° (decomp), fp: −35.7°, flash p: 190.4°F (TOC), d: 1.06656 @ 25°, vap. press: 0.5 mm @ 25°, vap. d: 4.46.

THR = Details U.

Fire Hazard: Mod, when exposed to heat, flame or oxidizers.

Explosion Hazard: See nitrates.

Disaster Hazard: See nitrates.

To Fight Fire: Foam, CO_2, dry chemical.

N-NITRODIMETHYL AMINE. $C_2H_6O_2N_2$, mw: 90.1.

THR = An exper neo. [*3*]

4-NITRO-6,7-DIMETHYLQUINOLINE-1-OXIDE.

THR = An exper carc. [*23*]

NITRO-2,4-DINITROPHENYL UREA.

$(ONO)_3C_6H_3CO(NH)_2$, mw: 271.1.

THR = A sensitive explosive compound. [*19*]

***o*-NITRODIPHENYL.** See *o*-nitrobiphenyl.

2-NITRODIPHENYLAMINE. Crystals.

$O_2NC_6H_4NHC_6H_5$, mw: 214.22, mp: 74.7°, bp: 223° @ 20 mm, d: 1.366.

THR = See nitro compounds of aromatic hydrocarbons and nitrates.

4-NITRODIPHENYLAMINE.

See 4-amino-4-nitrobiphenyl.

NITROETHANE. Colorless liquid. Disagreeable odor. $C_2H_5NO_2$, mw: 75.07, bp: 114.0°, fp: −50°, flash p: 82°F, d: 1.052 @ 20°/20°, autoign. temp.: 778°F, vap. press: 15.6 mm @ 20°, vap. d: 2.58, lel = 3.4%.

Acute tox data: Oral LD_{50} (rat) = 1100 mg/kg; oral LD_{50} (mice) = 860 mg/kg. [*3*]

THR = MOD via oral route. Has caused injury to liver and kidneys in exper animals.
Fire Hazard: Dangerous, when exposed to heat, flame or oxidizers. Explodes when heated.
Explosion Hazard: See nitrates. Reacts violently with $Ca(OH)_2$, hydrocarbons, hydroxides, inorganic bases, KOH, NaOH. [*19*] Explodes when heated.
Disaster Hazard: See nitrates.
To Fight Fire: Alcohol foam, CO_2, dry chemical; water can blanket fire.

2-NITRO-2-ETHYL-1,3-PROPANEDIOL. Crystalline material. $CH_2OHC(C_2H_5)NO_2CH_2OH$, mw: 149.15, mp: 56°–57°, bp: decomp.
THR = See nitrates, organic.

2-NITRO-9H-FLUORENE. $C_{13}H_9O_2N$, mw: 211.3.
THR = An exper neo and carc [*3*, *23*] via oral and dermal routes.

NITROFORM. See trinitromethane.

5-NITRO-2-FURALDEHYDESEMICARBAZONE. See furacycline.

NITROFURAZONE. See furacycline.

1-[(5-NITROFURYLIDENE)AMINO]-2-IMIDAZOLIDINONE.
THR = An exper carc. [*23*]

N-[(3-(5-NITRO-2-FURYL)-1,2,4-OXADIAZOLE-5-YL)METHYL]ACETAMIDE. $C_9H_8O_5N_4$, mw: 252.2.
THR = Low acute via oral route to rats. An exper carc [*3*] via oral route.

N-[5-(5-NITRO-2-FURYL)-1,3,4-THIADIAZOL-2-YL]-ACETAMIDE. $C_7H_6O_4N_4S$, mw: 242.2.
THR = LOW acute via oral route to rats. An exper carc and neo [*3*] via oral route.

N-[4-(5-NITRO-2-FURYL)-2-THIAZOLYL]ACETAMIDE. $C_9H_7O_4N_3$, mw: 253.3.
THR = LOW acute via oral route to rats. An exper carc and neo [*3*, *23*] via oral route.

[4-(5-NITRO-2-FUROYL)THIAZOL-2-YL]HYDRAZONE ACETONE. $C_{10}H_{10}O_3N_4S$, mw: 266.3.
THR = An exper neo. [*3*]

N-[4-(5-NITRO-2-FURYL)-2-THIAZOLYL]-2,2,2-TRIFLUOROACETAMIDE. $C_9H_4O_4N_3SF_3$, mw: 307.2.
THR = Nearly NO acute oral toxicity to rats. An exper carc. [*3*]

NITROGEN. Colorless gas, colorless liquid or cubic crystals at low temp. N_2, mw: 28.02, mp: −210.0°, −195.8°, d: 1.2506 g/liter @ 0°, d(liquid): 0.808 @ −195.8°.
THR = NONE. In high conc it is a simple asphyxiant. The release of nitrogen from solution in the blood, with formation of small bubbles, is the cause of most of the symptoms and changes found in compressed air illness (caisson disease). See also argon. It is used as a general purpose food additive. [*109*] Can react violently with Li, Nd, Ti [*19*] under the proper conditions.

NITROGEN BROMIDE. Syn: *nitrogen tribromide*. Crystalline solid. NBr_3, mw: 253.8.
THR = Details U; probably HIGH toxicity. See also bromides.
Explosion Hazard: Severe, when shocked, exposed to heat or flame or by spont chemical reaction. Violent reaction with P, As. [*19*]
Disaster Hazard: Dangerous; shock will explode it; on decomp, emits highly toxic fumes of bromine, can react vigorously with reducing materials.

NITROGEN CHLORIDE. Syn: *nitrogen trichloride*. Volatile yellowish oil or rhombic crystals, pungent odor. NCl_3, mw: 120.38, mp: $< -40°$, explodes@93°, bp: $< 71°$, d: 1.653, vap. press: 150 mm @ 20°.
THR = MOD irr to skin, eyes and mu mem. An irr to CNS.
Explosion Hazard: Severe, when shocked or exposed to heat or flame or by spont chemical reaction. Certain common materials catalyze its decomp. It is particularly sensitive when it contains impurities. Intense light can explode it. Reacts violently with NH_3, As, H_2S, NO_2, organic matter, O_3, PH_3, P, KCN, KOH, Se, dibutyl ether, grease. [*19*]
Disaster Hazard: Dangerous; even slight shock will explode it; on decomp, emits highly toxic fumes of chlorine; can react vigorously with reducing materials.

NITROGEN DIOXIDE. Syns: *nitrogen tetroxide*, *NO_x*, *nitrogen peroxide*. Colorless solid to yellow liquid. NO_2, mw: 46, mp: −9.3° (yellow liquid), bp: 21° (red-brown gas with decomp), d: 1.491 @ 0°, vap. press: 400 mm @ 80°.
Acute tox data: Inhal LC_{50} (rabbit) = 315 ppm for $\frac{1}{4}$ hr; inhal LC_{LO} (monkey) = 44 ppm for 6 hrs; inhal TC_{LO} (human) = 64 ppm ⟶ pulmonary problems. [*3*]
THR = HIGH via inhal route. See also nitric oxide. Violent reaction with cyclohexane, F_2, formaldehyde and alcohols, nitrobenzene, petroleum, toluene. [*19*]
Disaster Hazard: See NO_x.

NITROGEN FLUORIDE. N_2F_4, mw: 104.0.
Acute tox data: Inhal LC_{50} (rat) = 50 ppm for 4 hrs; inhal LC_{50} (guinea pig) = 900 ppm for 1 hr. [*3*]
THR = VERY HIGH via inhal route and HIGH irr to skin, eyes and mu mem.

NITROGEN FLUORIDE OXIDE. NOF_3, mw: 87.0.
Acute tox data: Inhal LC_{50} (rat) = 24 ppm for 4 hrs; ip LD_{50} (rat) = 38 mg/kg; inhal LC_{50} (mouse) = 18 ppm for 4 hrs. [*3*]
THR = HIGH irr via inhal and ip routes and irr to skin, eyes and mu mem. See also fluorides.

NITROGEN IODIDE. See nitrogen triiodide.

NITROGEN, LIQUID. See nitrogen.

NITROGEN MUSTARD. Syn: *HN-1*. Dark liquid. $C_5H_{11}NCl_2$, mw: 156.1, mp: 1° @ 10 mm, d: 1.09 @ 25°, vap. press: 0.17 mm @ 25°, vap. d: 5.9.
Acute tox data: Dermal LD_{50} (rat) = 22 mg/kg; iv LD_{50} (rat) = 1.1 mg/kg; inhal LC_{LO} (rat) = 300 mg/m^3. [*3*]
THR = HIGH irr via dermal, iv, inhal, and sc routes. An exper carc and teratogen [*23, 3, 8*] via sc and iv routes.
Disaster Hazard: Highly dangerous! A military poison!

NITROGEN MUSTARD-N-OXIDE. See 2,2′-dichloro-*n*-methyl diethylamine-*n*-oxide.

NITROGEN OXIDES (except nitrous). See nitric oxide and nitrogen dioxide.

NITROGEN OXYCHLORIDE. See nitrosyl chloride.

NITROGEN OXYFLUORIDE. See nitrosyl fluoride.

NITROGEN PENTOXIDE. Syns: *nitric anhydride*, *NO_x*. Hexagonal (rhombic) white crystals. N_2O_5, mw: 108.02, mp: 30°, bp: 47° (decomp), d: 1.642 @ 18°, vap. press: 400 mm @ 24.4°.
THR = See nitric oxide.
Fire Hazard: Mod, when heated, liberates oxygen.
Explosion Hazard: Dangerous; @ 122°F, decomp violently, liberating oxygen. Reacts violently with Na_2C_2. [*19*]
Disaster Hazard: Dangerous; when heated to decomp, emits highly toxic fumes of NO_x; will react with water or steam to produce toxic and corrosive fumes; can react with reducing materials.

NITROGEN PEROXIDE. See nitric oxide.

NITROGEN SELENIDE. Orange-red powder. $(NSe)_4$, mw: 371.9, bp: explodes.
THR = HIGH. See selenium compounds.
Explosion Hazard: Dangerous, when shocked or exposed to a temp. of 230°.
Disaster Hazard: Dangerous; shock will explode it; on decomp or on contact with acid or acid fumes, emits highly toxic fumes of selenium and NO_x.

NITROGEN SULFIDE (TETRA). Yellow crystals. N_4S_4, mw: 184.3, d: 2.24, mp: 178° (decomp).
THR = See sulfides.
Fire Hazard: See sulfides.
Explosion Hazard: Severe, when shocked or exposed to heat or flame.
Disaster Hazard: Dangerous; shock or heat (205°) will explode it; on decomp or on contact with acid or acid fumes, emits highly toxic fumes of NO_x and SO_x.

NITROGEN TETROXIDE. See nitrogen dioxide.

NITROGEN TETROXIDE AND NITRIC OXIDE, MIXTURE. See components.

NITROGEN TRIBROMIDE. See nitrogen bromide.

NITROGEN TRICHLORIDE. See nitrogen chloride.

NITROGEN TRIFLUORIDE. Colorless gas. Odor of mold. NF_3, mw: 71.01, mp: −208.5°, bp: −129°, d (liquid): 1.537 @ −129°.
Acute tox data: Inhal LC_{50} (rat) = 6700 ppm for 1 hr; inhal LC_{LO} (dog) = 9600 ppm for 1 hr. [*3*]
THR = MOD via inhal route. See also fluorides.
Explosion Hazard: Severe by chemical reaction, explosively with reducing agents, particularly when under pressure.
Fire Hazard: Dangerous; a very powerful oxidizer; otherwise inert at normal temp. and pressures. Reacts violently when ignited with H_2. When pure (dry) it does not attack glass or mercury at normal temp. Can react violently with NH_3, CO, diborane, H_2, H_2S, CH_4, tetrafluorohydrazine. [*19*]
Disaster Hazard: Dangerous; on decomp, emits highly toxic fumes of fluorides; can react vigorously with reducing materials. Particularly hazardous under pressure.

NITROGEN TRIIODIDE. Black crystals. NI_3, mw: 394.7, mp: explodes, bp: sublimes in vacuo.
THR = Details U. See also iodides.
Explosion Hazard: Severe; when shocked, exposed to heat or flame or by spont chemical reaction. It has no known uses as an explosive, because it is far too sensitive in the dry state to store or handle safely. If this material must be worked with, it should be kept wet. A convenient way of keeping it wet is with ether; when it is needed in the dry state, it simply has to be taken out into the open and the ether will evaporate, leaving it perfectly dry. When dry, it will explode when given the slightest touch, vibration or rise in temp. Even a puff of air directed into it can cause it to detonate. It is a high explosive and is very violent. It can be destroyed by throwing a quantity of it into large bodies of water, flowing streams or rivers. Reacts violently with O_3, H_2S, Cl_2, Br_2, acids. [*19*]

NITROGEN TRIIODIDE MONOAMMINE. Dark red rhombic crystals. NI_3NH_3, mw: 411.80, mp: decomp $< 20°$, bp: explodes, d: 3.5.

THR = Details U. See iodine, ammonia.
Explosion Hazard: Severe. This material is extremely unstable when dry. The slightest shock or heat will cause it to decomp explosively. It should be kept moist.
Disaster Hazard: Dangerous; on decomp, emits highly toxic fumes of iodine and ammonia.

NITROGEN TRIOXIDE. Syn: NO_x. Bluish gas. NO_3, mw: 62.01, mp: decomp slightly at ordinary temp.
THR = See nitric oxide. Even a trace can cause PH_3 to self-ignite. [*19*]

NITROGEN TRIOXYFLUORIDE.
See fluorine nitrate.

NITROGLYCERIN. Syns: *glycerol trinitrate*, *blasting oil*, *soup*. Colorless to yellow liquid, sweet taste. $C_3H_5(ONO_2)_3$, mw: 227.09, mp: 13°, bp: explodes @ 218°, d: 1.601, vap. press: 1 mm @ 127°, vap. d: 7.84, autoign. temp.: 518°F.
Acute tox data: Oral LD_{LO} (rat) = 80 mg/kg; iv LD_{LO} (rabbit) = 40 mg/kg; im LD_{LO} (rabbit) = 450 mg/kg; sc LD_{LO} (cat) = 150 mg/kg. [*3*]
THR = HIGH via oral, iv, im and sc routes. The symptoms of nitroglycerin poisoning are headaches and reduced blood pressure, excitement, vertigo, fainting, respiratory rales and cyanosis. If this material is taken internally, it causes respiratory difficulties and death due to respiratory paralysis. Severe poisoning often manifests itself at first by confusion, pugnaciousness, hallucinations, and maniacal manifestations. The most common complaint is headache which is noted upon commencing work but soon passes off. A break in the work interrupts this acclimatization and workers sometimes resort to the device of moistening their hat bands with nitroglycerin when they are off the job so as to maintain this effect during absence from their occupation. Furthermore it can be absorbed through uninjured skin and may produce eruptions on the palms and intradigital spaces of the hands. In normal manufacture and use of dynamite, the physiological effects of nitroglycerine cause only temporary discomfort and are not injurious to health.
Fire Hazard: Dangerous, when exposed to heat or flame or by spont chemical reaction.
Spont Heating: No.
Explosion Hazard: Severe, when shocked or exposed to O_3, heat or flame. Nitroglycerine is a powerful explosive, very sensitive to mechanical shock. Small quantities of it can readily be detonated by a hammer blow on a hard surface, particularly when it has been absorbed in filter paper. Frozen nitroglycerine is somewhat less sensitive than the liquid. However, a half or partially thawed out mixture is more sensitive than either one. See also explosives, high and dynamites.
Disaster Hazard: Highly dangerous; shock, heat and flame will explode it, and toxic fumes evolved on decomp.

NITROGLYCERIN, LIQUID, DESENSITIZED.
See nitroglycerin.

NITROGLYCERINE, SPIRITS OF.
See nitroglycerine.

NITROGUANIDINE. Yellow solid, high explosive. $H_2NC(NH)NHNO_2$, mw: 104.1, mp: 246°.
Acute tox data: Oral LD_{50} (rat) = 500 mg/kg. [*3*]
THR = HIGH via oral route. See also nitrates, organic.
Fire Hazard: Dangerous, when exposed to heat, flame or by chemical reaction with oxidizers.
Explosion Hazard: Severe, when shocked or exposed to heat or flame. Nitroguanidine is known as a flashless or cool explosive. It is about as powerful as TNT and is normally used mixed with colloided nitrocellulose in which form it yields a propellant powder which gives no flash from the muzzle of the gun, thus serving as a great advantage to the military. It has also been used mixed with ammonium nitrate and paraffin wax as a trench mortar ammunition.
Disaster Hazard: Dangerous; shock will explode it; when heated to decomp, emits highly toxic fumes; can react vigorously with oxidizing materials.

3-NITRO-3-HEXENE. $CH_3CH_2NO_2C{:}CHCH_2CH_3$, mw: 129.2.
Acute tox data: Oral LD_{LO} (rat) = 420 mg/kg; ip LD_{LO} (rat) = 80 mg/kg; dermal LD_{LO} (rabbit) = 940 mg/kg. [*103*, *3*]
THR = HIGH via ip and MOD via oral and dermal routes. An exper neo to mice via inhal route. [*103*]
Disaster Hazard: Dangerous; see nitrates, organic.

2-NITRO-2-HEXANE.
Acute tox data: Oral LD_{LO} (rat) = 420 mg/kg; ip LD_{LO} (rat) = 120 mg/kg; dermal LD_{LO} (rabbit) = 1400 mg/kg. [*3*]
THR = HIGH via ip and MOD via oral and dermal routes.

4-NITRO-6-HEXYLQUINOLINE-1-OXIDE.
THR = An exper carc. [*23*]

NITROHYDRENE. An oil. Composition: nitroglycerin + nitrosucrose.
THR = U. See nitroglycerin.
Fire Hazard: Dangerous, when exposed to heat or flame or by chemical reaction.
Explosion Hazard: Severe, when shocked or exposed

to heat. It is a powerful explosive, approx as powerful as nitroglycerin, and is used to stretch glycerin supplies. It is made up by dissolving up to 25% of sucrose in glycerin and nitrating the resulting mixture to give an explosive oil. This procedure saves considerable quantities of glycerin. The material is almost exactly like nitroglycerin. See also explosives, high.

Disaster Hazard: Dangerous; shock and heat will explode it; on decomp, emits highly toxic fumes of NO_x; can react vigorously with oxidizing materials.

NITROHYDROCHLORIC ACID. See aqua regia.

NITROHYDROCHLORIC ACID, DILUTED. See aqua regia.

3-NITRO-4-HYDROXYPHENYLARSONIC ACID. Syn: *4-hydroxy-3-nitrobenzene arsonic acid.* Pale yellow crystals. $HOC_6H_3(NO_2)AsO(OH)_2$, mw: 263.

Acute tox data: Oral LD_{50} (rat) = 155 mg/kg; ip LD_{50} (rat) = 66 mg/kg; oral LD_{50} (turkey) = 6 mg/kg. [*3*]

THR = HIGH via oral and ip routes. Used as a food additive permitted in the feed and drinking water of animals and/or for the treatment of food-producing animals. [*109*]

Disaster Hazard: Dangerous; see nitrates, organic.

NITRO JUTE.

THR = U. See nitrates, organic.

Fire Hazard: Dangerous, when exposed to heat or flame.

Explosion Hazard: Severe, when shocked or exposed to heat. See also explosives, high.

Disaster Hazard: Dangerous; shock or heat will explode it; on decomp, emits highly toxic fumes of NO_x; can react vigorously with oxidizing materials.

NITROLEVULOSE. Syn: *dextrose nitrate.*

THR = U. See nitrates, organic.

Fire Hazard: Dangerous, when exposed to heat or flame.

Explosion Hazard: Severe, when shocked or exposed to heat. See also explosives, high, and nitrates.

Disaster Hazard: Dangerous; shock or heat will explode it; on decomp, emits highly toxic fumes of NO_x; can react vigorously with oxidizing materials.

NITROMANNITE. See mannitolhexanitrate.

NITROMANNITOL. See mannitol hexanitrate.

NITROMETHANE. An oily liquid. CH_3NO_2, mw: 61.04, bp: 101°, lel = 7.3%, fp: −29°, flash p: 95°F (CC), d: 1.130 @ 20°/4°, autoign. temp.: 785°F, vap. press: 27.8 mm @ 20°, vap. d: 2.11.

Acute tox data: Oral LD_{50} (rat) = 940 mg/kg; oral LD_{50} (mouse) = 950 mg/kg; sc LD_{LO} (dog) = 565 mg/kg; inhal LC_{LO} (monkey) = 2446 mg/m^3 [*3*]

THR = MOD via oral, sc and inhal routes. In humans it may cause anorexia, nausea, vomiting and diarrhea. Also kidney injury and liver damage.

Fire Hazard: Dangerous, when exposed to heat, oxidizers or flame. See also nitroparaffins.

Explosion Hazard: Mod, when shocked or exposed to heat or flame. High temp. can detonate it. Can react violently with ($AlCl_3$ + organic matter), $Ca(OH)_2$, $Ca(OCl)_2$, hexamethylbenzene, hydrocarbons, inorganic bases, hydroxides, organic amines, KOH, NaOH. [*19*]

To Fight Fire: Alcohol foam.

4-NITRO-2-METHOXYQUINOLINE-1-OXIDE.

THR = An exper carc. [*23*]

4-NITRO-5-METHOXYQUINOLINE-1-OXIDE.

THR = An exper carc. [*23*]

4-NITRO-6-METHOXYQUINOLINE-1-OXIDE.

THR = An exper carc. [*23*]

Disaster Hazard: Dangerous; shock or heat will explode it, on decomp, emits highly toxic fumes of NO_x; can react with oxidizing materials.

To Fight Fire: Alcohol foam, CO_2, dry chemical.

2-NITRO-2-METHYL-1,3-PROPANEDIOL. Crystalline material. $CH_2OHC(CH_3)NO_2CH_2OH$, mw: 135.12, mp: 147°–149°, bp: decomp.

Acute tox data: Oral LD_{LO} (rat) = 2000 mg/kg; oral LD_{LO} (rabbit) = 1000 mg/kg. [*3*]

THR = MOD via oral route.

Fire Hazard: Mod, when exposed to heat or flame.

Disaster Hazard: Dangerous; on decomp, emits highly toxic fumes of NO_x; can react with oxidizing materials.

2-NITRO-2-METHYL-1-PROPANOL. Crystalline material. $CH_3C(CH_3)NO_2CH_2OH$, mw: 119.12, mp: 90°–91°, bp: 95° @ 10 mm.

Acute tox data: Oral LD_{LO} (rabbit) = 1000 mg/kg. [*3*]

THR = MOD via oral route.

Fire Hazard: See nitrates.

Disaster Hazard: See nitrates.

NITROMOLASSES.

THR = U. See nitrates, organic.

Fire Hazard: Dangerous, when exposed to heat or flame.

Explosion Hazard: Mod, when shocked or exposed to heat or flame. See also explosives, high, and nitrates.

Disaster Hazard: Dangerous; shock or heat will explode it; on decomp, emits highly toxic fumes of NO_x; can react vigorously with oxidizing materials.

To Fight Fire: Mist, foam, CO_2.

NITRONAPHTHALENE. Yellow crystals. $C_{10}H_7NO_2$, mw: 173.16, mp: 58.8°, bp: 304°, flash p: 327°F (CC), d: 1.331 @ 4°/4°, vap. d: 5.96.

THR = MOD irr to skin, eyes and mu mem.
Fire Hazard: Low, when exposed to heat or flame.
Explosion Hazard: See nitrates.
Disaster Hazard: See nitrates.
To Fight Fire: CO_2, dry chemical or water spray.

β-NITRONAPHTHALENE. $C_{10}H_7NO_2$, mw: 173.16.
THR = HIGH via oral route. An exper carc. [*3*] Animal exper show high oral toxicity and low irr of skin and lungs.
Fire Hazard: See nitronaphthalene.
Explosion Hazard: See nitrates.
Disaster Hazard: See nitrates.

3-NITRO-2-NAPHTHYLAMINE. $C_{10}H_8O_2N_2$, mw: 188.2.
THR = An exper carc. [*3, 23*]

4-NITRO-6-NITROQUINOLINE-1-OXIDE.
THR = An exper carc. [*23*]

2-NITRO-2-NONENE. $CH_3NO_2C{:}CH(CH_2)_5CH_3$, mw: 171.3.
Acute tox data: Oral LD_{LO} (rat) = 2100 mg/kg; dermal LD_{LO} (rabbit) = 620 mg/kg; inhal LC_{LO} (chicken) = 43 ppm. [*3*]
THR = HIGH via inhal; MOD via oral and dermal routes.
Disaster Hazard: Dangerous; see nitrates, organic.

3-NITRO-2-NONENE.
THR = MOD via oral and dermal routes.

3-NITRO-3-NONENE.
THR = MOD via oral and dermal routes.

4-NITRO-4-NONENE.
THR = MOD via oral and dermal routes.

5-NITRO-4-NONENE.
THR = MOD via oral and dermal routes.

NITROPENTAMMINE COBALT NITRATE. $[Co(NH_3)_5NO_2]NO_3$, mw: 252.1.
THR = Heat- (310°) and impact-sensitive. [*19*]

NITROPEROXYBENZOIC ACID. $C_7H_5O_5N$, mw: 183.1.
THR = An exper neo. [*3*] via sc route.

NITROPARAFFINS. Syns: *2-nitropropane, nitroethane, nitromethane, 1-nitropropane*. flash p: 103°–120°F.
THR = For toxicity data, see individual nitroparaffins.
Fire Hazard: Mod; see nitrates.
Explosion Hazard: Dry salts of nitromethane and inorganic bases are explosive. Otherwise the mononitroparaffins are relatively stable. Impact under confined conditions can cause explosions of mononitromethane. Avoid heating it to above 450°F. Can react violently with hydrocarbons, ($AlCl_3$ + organic matter). [*19*]
Disaster Hazard: See nitrates.
To Fight Fire: Foam, CO_2, dry chemical.

NITRONIUM PERCHLORATE. White crystals, odor of NO_x and chlorine. Sol (by reaction) in water. NO_3ClO_4, mw: 145.5, mp: decomp @ 120°–140°, d: 2.25 g/cc, vap. press: $<$ 0.05 mm @ 20°.
THR = A powerful irr to skin and mu mem. See nitric and perchloric acids, which are this material's products of hydrolysis due to the moisture of skin or mu mem.
Disaster Hazard: Dangerous; decomp by heat or steam to highly toxic fumes.

4-NITRO-2-METHYL QUINOLINE-1-OXIDE.
THR = An exper carc. [*23*]

4-NITRO-5-METHYL QUINOLINE-1-OXIDE.
THR = An exper carc. [*23*]

4-NITRO-6-METHYL QUINOLINE-1-OXIDE.
THR = An exper carc. [*23*]

4-NITRO-7-METHYL QUINOLINE-1-OXIDE.
THR = An exper carc. [*23*]

4-NITRO-8-METHYL QUINOLINE-1-OXIDE.
THR = An exper carc. [*23*]

***m*-NITROPHENOL.** Monoclinic prisms. $C_6H_5NO_3$, mw: 139.1, mp: 97°, bp: 194° @ 70 mm; d: 1.485 @ 20°/4°.
Acute tox data: Oral LD_{50} (rat) = 447 mg/kg; oral LD_{50} (mice) = 1414 mg/kg; iv LD_{LO} (dog) = 83 mg/kg. [*3*]
THR = HIGH via oral and iv routes.
Disaster Hazard: See nitrates.

***o*-NITROPHENOL.** Light yellow crystals, aromatic odor. $C_6H_5NO_3$, mw: 139.1, mp: 45°, bp: 214.5°, d: 1.495 @ 20°, vap. press: 1 mm @ 49.3°.
Acute tox data: Oral LD_{50} (rat) = 2828 mg/kg; oral LD_{50} (rat) = 1297 mg/kg; iv LD_{LO} (dog) = 100 mg/kg. [*3*]
THR = MOD via oral route. Has produced kidney and liver injury in exper animals. Violent reaction with KOH. [*19*]
Disaster Hazard: See nitrates.

***p*-NITROPHENOL.** Colorless to slightly yellow crystals. mp: 113°–114°, bp: 279° (decomp), d: 1.270 @ 120°/4°.
Acute tox data: Oral LD_{50} (rat) = 350 mg/kg; oral LD_{50} (mice) = 467 mg/kg; im LD_{LO} (pigeon) = 65 mg/kg; iv LD_{LO} (dog) = 10 mg/kg. [*3*]
THR = HIGH via oral, im and iv routes. Exper produced hyperthermia, methemoglobinemia and CNS depression.
Disaster Hazard: See nitrates.

***m*-NITROPHENOL DIAZONIUM PERCHLORATE.** A solid.

THR = See perchlorates and nitro compounds of aromatic hydrocarbons.
Fire Hazard: Dangerous, when exposed to heat or flame.
Explosion Hazard: Severe, when shocked or exposed to heat. See also explosives, high.
Disaster Hazard: Dangerous; shock will explode it; when heated to decomp, emits highly toxic fumes of chlorides and NO_x; can react vigorously with oxidizing materials.

***o*-NITROPHENYL ACETIC ACID.**
$ONOC_6H_4CH_2COOH$, mw: 181.1.
THR = With thionyl chloride it forms an explosive product. [*19*]

***o*-NITROPHENYL ACETYL CHLORIDE.**
$NO_2C_6H_4CH_2COCl$, mw: 201.6.
THR = Solvent-free material decomp violently. [*19*]

4-*p*-[(NITROPHENYL)AZO]DIPHENYL AMINE.
$C_{18}H_{14}O_2N_4$, mw: 318.4.
THR = An exper neo. [*3*] via oral route.

7-(*p*-NITROPHENYL AZO)METHYLBENZ(c)-ACRIDINE. $C_{24}H_{16}O_2N_4$, mw: 392.4.
THR = An exper neo. [*3*] via sc route.

NITROPHENYL BORIC ACID. Yellow crystals, slightly water-sol. $NO_2C_6H_4B(OH)_2$, mw: 166.9, mp: decomp before melting.
THR = See boron compounds, nitro compounds of aromatic hydrocarbons, and nitrates.

***m*-NITROPHENYL DIAZONIUM PERCHLORATE.**
$NO_2C_6H_4N{\equiv}NClO_4$, mw: 249.5.
THR = An explosive. Very sensitive to heat and shock. [*19*]

1-(*p*-NITROPHENYL)ETHYL NITRATE.
$(NO_2)C_6H_3(OH)C_2H_4NO_3$, mw: 214.
THR = Details U. Limited animal exper suggest mod toxicity. See also nitro compounds of aromatic hydrocarbons.
Disaster Hazard: Dangerous; see nitrates, organic.

5-NITRO-2-*n*-PROPOXYANILINE. Syns: *P-4000, 1-propoxy-2-amino-4-nitrobenzene*. Orange crystals, slightly water-sol, very sweet. $C_9H_{12}N_2O_2$, mw: 196.2, mp: 48°.
THR = Details U. A synthetic sweetener; suspected of having carc activity.
Disaster Hazard: Dangerous; when heated to decomp, emits toxic fumes.

NITROPRUSSIDES.
THR = HIGH via oral, iv, inhal routes. See also sodium nitroprusside.
According to some observers, sodium nitroprusside decomp in the body liberating cyanide; hence the tox is that of cyanides. There may be a nitrite effect in nonfatal cases. See also cyanides and nitrites.
Disaster Hazard: Dangerous; when heated to decomp, or on contact with acid or acid fumes; they emit high tox fumes of cyanides and NO_x.

1-NITROPROPANE. Colorless liquid.
$CH_3CH_2CH_2NO_2$, mw: 89.09, bp: 132°, fp: −108°, flash p: 120°F (TOC), d: 1.003 @ 20°/20°, autoign. temp.: 789°F, vap. press: 7.5 mm @ 20°, vap. d: 3.06, lel = 2.2%.
Acute tox data: Oral LD_{50} (mice) = 800 mg/kg.
THR = MOD via oral route.
Fire Hazard: Mod, when exposed to heat, open flame or oxidizers.
Explosion Hazard: See nitrates. Reacts violently with $Ca(OH)_2$, hydrocarbons, hydroxides, inorganic bases. [*19*] May explode on heating.
Disaster Hazard: See nitrates.
To Fight Fire: Alcohol foam, CO_2, dry chemical.

2-NITROPROPANE. Colorless liquid. $(CH_3)_2CHNO_2$, mw: 89.09, bp: 120°, fp: −93°, flash p: 103°F (TOC), d: 0.992 @ 20°/20°, autoign. temp.: 802°F, vap. press: 10 mm @ 15.8°, vap. d: 3.06, lel = 2.6%.
Acute tox data: Oral LD_{LO} (rat) = 500 mg/kg; inhal LC_{LO} (rat) = 1513 ppm for 3 hrs; inhal LC_{LO} (cat) = 714 ppm for 5 hrs; oral LD_{LO} (rabbit) = 500 mg/kg. [*3*]
THR = MOD via oral and inhal routes. An exper carc. [*3*] Causes hepatocellular carcinoma via inhal route to 207 ppm for 6 months. [*34, 43*] Can cause gastrointestinal disturbances and injury to liver and kidneys. Large doses produce methemoglobinemia and cyanosis. See also nitrates.
Fire Hazard: Mod, when exposed to heat, open flame or oxidizers.
Explosion Hazard: May explodc on heating; also violent reactions with chlorosulfonic acid, oleum. [*19*]
Disaster Hazard: See nitrates.
To Fight Fire: Alcohol foam, CO_2, dry chemical.

NITROPYRIDINE. Liquid. $C_5H_4N_2O_2$, mw: 124.1, autoign. temp.: 725°F.
THR = Details U. Probably highly toxic. See also pyridine.
Fire Hazard: Mod, when exposed to heat or flame.
Explosion Hazard: U. See nitrates.
Disaster Hazard: Dangerous; when heated to decomp, emits highly toxic fumes of NO_x; can react with oxidizing materials.
To Fight Fire: Water, CO_2, dry chemical.

4-NITROQUINALDINE-1-OXIDE. $C_{10}H_8O_3N_2$, mw: 204.2.
THR = An exper neo [*3*] via sc route.

2-NITROQUINOLINE. $C_9H_6O_2N_2$, mw: 174.2.
THR = An exper carc. [*3*] via sc route.

4-NITROQUINOLINE-1-OXIDE. $C_9H_6O_3N_2$, mw: 190.2.
THR = An exper carc and neo [*3, 23*] via sc, iv, imp, im and dermal routes.

5-NITRO-8-QUINOLINOL. $C_9H_6O_3N_2$, mw: 190.2.
THR = An exper neo [*3*] via oral route.

NITROSACCHAROSE. Syn: *nitrosugar*. White solid.
THR = U. See nitrates.
Fire Hazard: Dangerous, when exposed to heat or flame.
Explosion Hazard: Mod, when shocked or exposed to heat. See also explosives, high, and nitrates.
Disaster Hazard: See nitrates.

NITROSAMINES. These organic compounds are suspected of causing cancers of the lung, nasal sinuses, brain, esophagus, stomach, liver, bladder and kidney. [*14*]

N-NITROSOARABASIN. Syn: *n-nitroso-2-(e'-pyridyl) piperdine*. $C_{10}H_{13}ON$, mw: 191.3.
THR = An exper carc [*3, 23*] via oral route.

1-NITROSOAZACYCLOTRIDECANE. $C_{12}H_{24}ON_2$, mw: 212.4.
THR = An exper carc. [*3*]

1-NITROSOAZETIDINE. $C_3H_3ON_2$, mw: 83.1.
THR = An exper carc [*3, 23*] via oral route.

8-NITROSOBENZO(rst)PENTAPHENE.
THR = An exper carc. [*23*]

N-NITROSOBENZTHIAZURON. $C_9H_8ON_4S$, mw: 220.3.
THR = An exper carc [*3, 23*] via oral route.

N-NITROSOBIS(ACETOXYETHYL)AMINE. $C_8H_{14}O_5N_2$.
THR = An exper carc. [*23*]

N-NITROSOBIS(ACETOXYPROPYL)AMINE. See BAP.

N-NITROSO-*n*-BUTYL-(3-CARBOXYPROPYL) AMINE. $C_7H_{16}O_3N_2$,
THR = An exper carc. [*23*]

N-NITROSO-*n*-BUTYL-(4-HYDROXYBUTYL) AMINE. $C_8H_{18}O_2N_2$.
THR = An exper carc. [*23*]

N-NITROSO-*n*-BUTYL-(3-HYDROXYPROPYL) AMINE. $C_7H_{16}O_2N_2$.
THR = An exper carc. [*23*]

N-NITROSO-N-BUTYL-N-PENTYL AMINE. Syn: *n-butyl-n-pentyl nitrosamine*. $C_9H_{20}ON_2$, mw: 172.3.
THR = An exper carc. [*23*]

N-NITROSOCARBARYL. $C_{12}H_{10}O_3N_2$, mw: 230.2.
THR = An exper neo. [*3*] via sc route.

N-NITROSO-N-CARBETHOXY PIPERAZINE.
THR = An exper carc. [*23*]

NITROSO COMPOUNDS.
THR = Many of these are exper carc. [*14*]

N-NITROSODECAMETHYLONE IMINE. $C_{10}H_{20}ON_2$.
THR = An exper carc. [*23*]

N-NITROSODIACETONITRILE. $C_4H_4ON_4$.
THR = An exper carc. [*23*]

N-NITROSODICYCLOHEXYL NITROSAMINE. $C_{12}H_{22}ON$, mw: 210.4.
Acute tox data: Oral LD_{50} (rat) = 5000 mg/kg. [*3*]
THR = MOD via oral route. An exper carc. [*23*]

N-NITROSODIETHANOLAMINE. $C_8H_8O_2N_2$, mw: 164.2.
THR = An exper carc. [*23*]

N-NITROSODIETHYL AMINE. $C_4H_{10}ON_2$, mw: 102.2.
THR = An exper transplacental (+) teratogen, neo and carc. [*3, 23, 9*]
THR = HIGH acute oral toxicity.

N-NITROSODIISOPROPYL AMINE. $C_6H_{14}ON_2$, mw: 130.2.
THR = An exper carc [*3, 23*] via oral route.

N-NITROSODIMETHYLAMINE. Syn: *dimethyl nitrosamine*. Yellow liquid, sol in water, alcohol and ether. $C_2H_6N_2O$, mw: 74.1, bp: 152°, d: 1.005 @ 20°/4°.
Acute tox data: Oral LD_{50} (rat) = 26 mg/kg; inhal LC_{50} (rat) = 78 ppm for 4 hrs; ip LD_{50} (rat) = 36 mg/kg; sc LD_{50} (rat) = 45 mg/kg. [*3*]
THR = HIGH via oral, inhal, ip and sc routes. Has caused fatal liver disease in humans. An exper transplacental teratogen, carc and neo. [*3, 9, 23*]
Disaster Hazard: Dangerous; when heated to decomp, yields highly toxic fumes. May decomp violently under some disaster conditions.

***p*-NITROSODIMETHYLANILINE.** Syns: *accellerene, dimethyl-p-nitrosoaniline*. Green leaflets. $NOC_6H_4N(CH_3)_2$, mw: 150.14, mp: 92.5°.
Acute tox data: Oral LD_{50} (rat) = 65 mg/kg. [*3*]
THR = HIGH via oral route. An active allergen. An exper neo [*3*] via oral route.
Fire Hazard: Mod, when exposed to heat or flame.
Explosion Hazard: See nitrates.
Disaster Hazard: See nitrates.

N-NITROSO-N-(1,1-DIMETHYL-3-OXOBUTYL METHYLAMINE). Syn: *diphenyl nitrosamine*. $C_7H_{14}O_2N_2$.
THR = An exper carc. [*23*]

N-NITROSODIPHENYLAMINE. Green crystals. $C_6H_5NHC_6H_4NO$, mw: 198.2, mp: 144°.
Acute tox data: Oral LD_{50} (rat) = 1650 mg/kg. [*3*]
THR = MOD via oral route. An exper carc and neo. [*3, 14, 23*] See nitrosamines.
Fire Hazard: Dangerous, when exposed to heat or flame or by chemical reaction.
Disaster Hazard: Dangerous; when heated to decomp, emits highly toxic fumes of NO_x; can react vigorously with oxidizing materials.

N-NITROSOFLUORENE. $C_{13}H_9ON$, mw: 195.2.
THR = An exper carc and neo [*3, 23*] via ip and sc routes.

N-NITROSOFOLIC ACID. $C_{19}H_{18}O_7N_8$, mw: 470.5.
THR = An exper neo [*3*] via ip route. HIGH acute toxicity via ip route.

NITROSOGUANIDINE. A solid. CH_4ON_4, mw: 88.1.
Acute tox data: ip LD_{LO} (mice) = 21 mg/kg. [*3*]
THR = HIGH via ip route. See also nitrates.
Fire Hazard: Dangerous, when exposed to heat or flame.
Explosion Hazard: Severe, when shocked or exposed to heat or flame. See also explosives, high, and nitrates.
Disaster Hazard: Dangerous; shock will explode it; on decomp, emits highly toxic fumes of NO_x; can react with oxidizers.

N-NITROSOHEPTAMETHYLENE IMINE. $C_7H_{14}ON_2$, mw: 142.2.
THR = An exper carc. [*3, 23*] HIGH acute toxicity via oral route to rats.

1-NITROSOHYDANTOIN. $C_3H_3O_3N_3$, mw: 129.1.
THR = An exper carc [*3*] via oral route.

1-NITROSOHYDROURACIL. $C_4H_5O_3N_3$, mw: 143.1.
THR = An exper carc. [*3*]

N-NITROSOIMIDAZOLIDINONE. $C_3H_7O_2N_3$, mw: 117.1.
THR = An exper neo. [*3, 23*]

2,2-N-(NITROSOIMINO)DIACETONITRILE. $C_4H_4ON_4$, mw: 124.1.
Acute tox data: Oral LD_{50} (rat) = 163 mg/kg. [*3*]
THR = HIGH via oral route. An exper carc. [*3*] via oral route.

N-NITROSOIMINODIETHANOL. $C_4H_{10}O_3N_2$, mw: 134.2.
THR = An exper carc. [*3*]

N-NITROSOIMINODIETHANOL DIACETATE. $C_8H_{14}O_5N_2$, mw: 218.2.
THR = An exper carc. [*3*]

N-NITROSOINDOLE. $C_8H_8ON_2$.
THR = An exper carc. [*23*]

4-NITROSOMETHYL AMINO BENZALDEHYDE.
Acute tox data: Oral LD_{50} (rat) = 2 g/kg. [*3*]
THR = MOD via oral route. An exper carc. [*23*]

N-NITROSOMETHYLAMINO SULFOLANE. $C_5H_{10}O_3N_2S$, mw: 178.2.
THR = An exper carc. [*3, 23*]

NITROSOMETHYL-*n*-DODECYL AMINE. $C_{13}H_{28}ON_2$, mw: 228.4.
THR = An exper carc. [*3*]

N-NITROSO-N-METHYL-O-METHYL HYDROXYAMINE. $C_2H_6O_2N_2$, mw: 90.1.
Acute tox data: iv LD_{50} (rat) = 130 mg/kg. [*3*]
THR = HIGH via iv route.

N-NITROSO-N-METHYL-4-TOLYL SULFONAMIDE. $C_8H_{10}O_3N_2S$, mw: 214.3.
THR = An exper neo. [*3*]

NITROSOMETHYLUREA. Powder. $C_2H_5O_2N_3$, mw: 103.1.
Acute tox data: Oral LD_{50} (rat) = 180 mg/kg; ip LD_{50} (rat) = 110 mg/kg; iv LD_{50} (rat) = 110 mg/kg.
THR = HIGH via oral, ip and iv routes. Has been known to cause a contact dermatitis of the poison ivy type. An exper (+) teratogen, neo and carc. [*3, 9*] See nitrosamines.
Fire Hazard: Mod, when exposed to heat or flame.
Disaster Hazard: See nitrates.

NITROSOMETHYLURETHANE. Syn: *N-methyl-n-nitrosourethane*. Solid. $CH_3NNOCOOC_2H_5$, mw: 132.2.
Acute tox data: Oral LD_{50} (rat) = 180 mg/kg; iv LD_{50} (rat) = 4 mg/kg; ip LD_{50} (mice) = 37 mg/kg. [*3*]
THR = HIGH via oral, iv and ip routes. See also nitrosamines. An exper (+) teratogen, neo and transplacental carc. [*3, 10, 23*]
Fire Hazard: Slight when exposed to heat or flame.
Disaster Hazard: Dangerous; when heated to decomp, emits highly toxic fumes of NO_x.

N-NITROSOMORPHOLINE. $C_4H_8O_2N_2$, mw: 116.1.
THR = HIGH acute via oral route. An exper carc and neo. [*3, 23*]

1-NITROSONAPHTHALENE. $C_{10}H_7ON$, mw: 157.2.
THR = An exper neo via ip route. [*3, 23*]

2-NITROSONAPHTHALENE.
THR = An exper neo via ip route. [*3, 23*]

2-NITROSO-1-NAPHTHOL. $C_{10}H_7O_2N$, mw: 173.2.
THR = An exper carc. [*3, 23*]

N-NITROSONORNICOTINE. $C_9H_{11}ON_3$, mw: 177.2.
THR = An exper neo. [*3, 23*]

N-NITROSOOCTAMETHYLENE IMINE. Syn: *octahydro-1-nitroso-1,H-azonine*. $C_8H_{16}ON_2$, mw: 156.3.

THR = HIGH acute toxicity via oral route to rats. An exper carc. [*3*]

N-NITROSO-N-PENTYL UREA. $C_6H_{13}O_2N_3$, mw: 159.2.
THR = HIGH via oral route. An exper carc. [*3*]

***p*-NITROSOPHENOL.** Syn: *quinone monoxime*. Yellow, rhombic needles. NOC_6H_4OH, mw: 123.11, mp: 144° (decomp).
Acute tox data: ip LD_{LO} (mice) = 250 mg/kg; pa LD_{LO} (mice) = 200 mg/kg. [*3*]
THR = HIGH via ip and pa routes. An irr and sensitizer.
Fire Hazard: Dangerous, when exposed to heat or flame; burns explosively.
Explosion Hazard: Dangerous, when exposed to heat, conc acid, alkali. See also nitrates.
Disaster Hazard: See nitrates.

NITROSOPHENOL DIMETHYLPYRAZOLE. A solid.
THR = A (S) carc. [*14*] See nitroso compounds.
Fire Hazard: See nitrates.
Explosion Hazard: Severe, when exposed to heat. See also explosives, high, and nitrates.
Disaster Hazard: Dangerous; when heated to decomp, emits highly toxic fumes of NO_x; can react vigorously with oxidizing materials. A high explosive. See explosives, high.

N-NITROSO-N-PHENYL HYDROXYLAMINE AMMONIUM SALT. $C_6H_6O_2N_2 \cdot NH_4$, mw: 156.2.
THR = An exper neo. [*3, 23*]

N-NITROSOPHENYL UREA.
THR = An exper carc. [*23*]

1-NITROSOPIPERAZINE. $C_4H_9ON_3$, mw: 115.2.
THR = An exper neo. [*3, 23*]

1-NITROSOPIPERIDINE. $C_5H_{10}ON_2$, mw: 114.2.
THR = An exper carc. [*3, 23*]

N-NITROSO-3-PIPERIDINOL. $C_5H_{10}O_2N_2$, mw: 130.2.
THR = An exper carc. [*3*]

N-NITROSO-4-PIPERIDINOL.
THR = An exper carc. [*3*]

N-NITROSO-L-PROLINE. $C_6H_4O_3N$.
THR = An exper carc. [*23*]

N-NITROSO-L-PROLINE ETHYL ESTER. $C_7H_{12}O_3N_2$, mw: 172.2.
Acute tox data: Oral LD_{50} (rat) = 5000 mg/kg. [*3*]
THR = MOD via oral route. An exper carc. [*23*]

N-NITROSO-N-PROPYL ETHYL CARBAMATE. $C_6H_{12}O_3N_2$, mw: 160.2.
THR = An exper carc to rats via oral route. [*103*]

N-NITROSO-N-PROPYL-2-OXOPROPYLAMINE. $C_6H_{12}O_2N_2$, mw: 144.2.
THR = An exper carc. [*23*]

N-NITROSO-*n*-PROPYLUREA. $C_6H_{11}O_3N_3$, mw: 173.2.
THR = An exper transplacental teratogen and carc. [*3, 23*]

NITROSOPYRAZOLE.
THR = MOD irr to skin, eyes and mu mem. See nitrosamines, nitroso compounds.
Fire Hazard: Slight, when exposed to heat or flame.
Disaster Hazard: Dangerous; when heated to decomp, emits highly toxic fumes of NO_x; can react with oxidizing materials.

1-NITROSOPYRROLIDINE. $C_4H_8ON_2$, mw: 100.1.
THR = An exper carc. [*3, 23*]

N-NITROSO-3-PYRROLINE. $C_4H_6ON_2$, mw: 98.1.
THR = An exper carc. [*3*]

N-NITROSOSARCOSINE. $C_3H_5O_3N_2$, mw: 117.1.
THR = An exper carc and neo. [*3, 23*]

N-NITROSOSARCOSINE ETHYLATE. $C_5H_{10}O_3N_2$, mw: 146.2.
THR = An exper carc. [*3, 23*]

1-NITROSO-1,2,3,6-TETRAHYDROPYRIDINE. $C_5H_9ON_2$, mw: 113.1.
THR = An exper carc to rats via oral route. [*103*]

1-NITROSO-2,2,6,6-TETRAMETHYL PIPERIDINE. $C_9H_{18}ON_2$, mw: 170.3.
THR = An exper neo. [*3*]

4-NITROSOTHIOMORPHOLINE. $C_4H_8ON_2S$, mw: 132.2.
THR = An exper neo. [*3, 23*]

3-NITROSO-1,1,3-TRIETHYL UREA. $C_7H_{15}O_2N_3$, mw: 173.3.
THR = An exper neo. [*3*]

***p*-NITROSO-2,2,4-TRIMETHYL-1,2-DIHYDROQUINOLINE.** Syn: *curetard*. $(C_{12}H_{14}ON_2)_n$.
THR = An exper carc. [*3, 23*]

1-NITROSO-1,2,2-TRIMETHYL HYDRAZINE. $C_3H_9ON_3$, mw: 103.2.
THR = An exper carc via oral route. [*3*]

NITROSO URETHANE. $C_3H_6O_3N_2$, mw: 151.2.
THR = An exper neo and carc. [*3*]

N-NITROSO-N-VINYL ETHYL AMINE. $C_4H_8ON_2$, mw: 100.13.
Acute tox data: sc LD_{50} (hamster) = 109 mg/kg. [*103*]
THR = HIGH via sc route. An exper carc to hamsters via sc route. [*103*]

NITROSTARCH. Syn: *starch nitrate*. Solid. $C_{12}H_{12}(NO_2)_8O_{10}$, mw: 684.30.
THR = U. See nitrates.

Fire Hazard: Dangerous, when exposed to heat, flame or oxidizers.

Explosion Hazard: Severe, when shocked or exposed to heat or flame. It is a powerful high explosive. Nitrostarch is not a definite compound, but a mixture of various nitric acid esters of starch of different degrees of nitration. It is used as an ingredient of blasting explosives, for quarrying, and as a demolition explosive. See also nitrates.

4-NITROSTILBENE. $C_{14}H_{11}O_2N$, mw: 225.3.

THR = An exper neo. [*3*]

4′-NITRO-4-STILBENYL-N,N-DIMETHYLAMINE.

THR = An exper carc. [*23*]

NITROSTYRENE.

THR = See nitrates, organic.

7-(*m*-NITROSTYRYL)BENZ(c)ACRIDINE. $C_{25}H_{16}O_2N_2$, mw: 376.4.

THR = An exper neo. [*3*]

7-(*o*-NITROSTYRYL)BENZ(c)ACRIDINE.

THR = An exper neo. [*3*]

7-(*p*-NITROSTYRYL)BENZ(c)ACRIDINE.

THR = An exper neo.[*3*]

12-(*m*-NITROSTYRYL)BENZ(c)ACRIDINE.

THR = An exper neo. [*3*]

12-(*o*-NITROSTYRYL)BENZ(c)ACRIDINE.

THR = An exper neo. [*3*]

12-(*p*-NITROSTYRYL)BENZ(c)ACRIDINE.

THR = An exper neo. [*3*]

NITROSUGAR. See nitrosaccarose.

NITROSYL AZIDE. O=N—N_3, mw: 72. Yellow compound.

THR = Spont decomp even @ −50°. [*19*]

NITROSYL BROMIDE. Brown gas or dark brown liquid. NOBr, mw: 109.92, mp: −55.5°, bp: −2°, d: > 1.0.

THR = HIGH irr to skin, eyes and mu mem. See also bromides.

NITROSYL CHLORIDE. Syn: *nitrogen oxychloride*. Yellow gas or yellow liquid or crystals, irr odor. NOCl, mw: 65.47, mp: −64.5°, bp: −5.8°, d (liquid): 1.250 @ 30°, vap. press: 76 mm @ 50°, vap. d: 2.3.

THR = HIGH irr to skin, eyes and mu mem. Intensely irr. Can cause fatal pulmonary edema. Can react violently with acetone, Al. [*19*]

Disaster Hazard: See chlorides and NO_x.

NITROSYL FLUORIDE. Syn: *nitrogen oxyfluoride*. Colorless, highly toxic, irr gas. NOF, mw: 49.01, mp: −134°, bp: −60°, d: 2.176 g/l.

THR = See fluorides and hydrofluoric acid. Violent reaction with B, P, Si, Na, halogenated olefins. [*19*]

NITROSYL PERCHLORATE. O=N—ClO_4, mw: 129.5.

THR = Below 100° it slowly decomp; @ 115°–120°, it speeds up and explodes. Reacts violently with acetone and also amines, diethyl ether, metal salts, (cobalt pentammine triazoperchlorate + phenyl isocyanate). [*19*]

NITROSYL SULFURIC ACID. Syn: *chamber crystals*. Crystalline, colorless solid. $ONOSO_3H$, mw: 127.08, mp: 73° (for crystalline material).

THR = HIGH irr to skin, eyes and mu mem.

Fire Hazard: See nitrates. Violent reaction with dinitroaniline hydrochloride. [*19*]

Disaster Hazard: See nitrates and sulfates.

NITROSYLSULFURIC ANHYDRIDE. Tetragonal crystals. $(NOSO_3)_2O$, mw: 236.15, mp: 217°, bp: 360°.

THR = HIGH irr to skin, eyes and mu mem. Will react with water or steam to produce toxic and corrosive fumes.

***m*-NITROTOLUENE.** Liquid. $CH_3C_6H_4NO_2$, mw: 137.1, mp: 15.1°, bp: 231.9°, flash p: 233°F (CC), d: 1.1630 @ 15°/4°, vap. press: 1 mm @ 50.2°, vap. d: 4.72.

Acute tox data: Oral LD_{50} (rat) = 1072 mg/kg; oral LD_{50} (mice) = 330 mg/kg; oral LD_{50} (rabbit) = 2400 mg/kg; oral LD_{50} (guinea pig) = 3600 mg/kg. [*3*]

THR = MOD via oral and dermal routes.

Fire Hazard: Low, when exposed to heat, flame or oxidizers.

Disaster Hazard: See nitrates.

To Fight Fire: Water, CO_2, dry chemical.

***o*-NITROTOLUENE.** Yellowish liquid. $C_7H_7NO_2$, mw: 137.1, mp: −4.1°, bp: 222.3°, flash p: 223°F (CC), d: 1.1622 @ 19°/15°, vap. press: 1 mm @ 50°, vap. d: 4.72.

Acute tox data: Oral LD_{50} (rat) = 891 mg/kg; oral LD_{50} (mice) = 2462 mg/kg. [*3*]

THR = MOD via oral and dermal routes.

Fire Hazard: Low when exposed to heat or open flame.

Disaster Hazard: See nitrates.

To Fight Fire: Water spray, fog, foam, CO_2.

***p*-NITROTOLUENE.** Syn: *methyl nitrobenzene*. Yellowish crystals. $C_7H_7NO_2$, mw: 137.1, mp: 51.9°, bp: 238.3°, flash p: 223°F (CC), d: 1.286, vap. press: 1 mm @ 53.7°, vap. d: 4.72.

Acute tox data: Oral LD_{50} (rat) = 2144 mg/kg; oral LD_{50} (mice) = 1231 mg/kg. [*3*]

THR = MOD via oral and dermal routes.

Fire Hazard: Low, when exposed to heat or open flame. Can explode with H_2SO_4. [*19*]

Disaster Hazard: Dangerous; see nitrates.

To Fight Fire: CO_2, dry chemical.

NITROTOLUENE, LIQUID. See *m*- and *o*-nitrotoluene.

***m*-NITRO-*p*-TOLUIDINE.** Clear liquid. $C_7H_6NH_2NO_2$, mw: 168.15, mp: 116°, flash p: 315°F (CC), d: 1.312, vap. d: 5.80.
Acute tox data: Oral LD_{50} (wild birds) = 3 mg/kg. [*3*]
THR = HIGH via oral, particularly to wild birds. See also nitro compounds of aromatic hydrocarbons.
Fire Hazard: Slight, when exposed to heat or flame.
Disaster Hazard: See nitrates.
To Fight Fire: CO_2, dry chemical.

5-NITRO-*o*-TOLUIDINE.
Acute tox data: Oral LD_{50} (rat) = 574 mg/kg. [*3*]
THR = MOD–HIGH via oral route. See also nitro compounds of aromatic hydrocarbons.

NITROTRICHLOROMETHANE. See chloropicrin.

6-NITRO-2,4-(TRICHLOROMETHYL)-1,3-BENZODIOXANE.
THR = Details U. See also nitro compounds of aromatic hydrocarbons.
Disaster Hazard: Dangerous; when heated to decomp, emits highly toxic fumes of chlorides and NO_x.

NITROUREA. Syn: *m-nitrocarbamide*. Crystals. $CH_3O_3N_3$, mw: 105.1.
THR = U. See nitrates.
Fire Hazard: Dangerous; when exposed to heat or flame.
Explosion Hazard: Severe, when shocked or exposed to heat. See also explosives, high, and nitrates.
Disaster Hazard: Dangerous; shock will explode it; when heated to decomp, emits highly toxic fumes of NO_x; can react vigorously with oxidizing materials. It is a high explosive. See explosives, high.

NITROUS ACID. Pale blue solution. HNO_2, mw: 47.02.
THR = See nitrates.
Fire Hazard: Mod, by chemical reaction; a powerful oxidizer. Reacts violently with PH_3, PCl_3. [*19*]
Disaster Hazard: Dangerous; when heated to decomp, emits highly toxic fumes of NO_x.

NITROUS ANHYDRIDE. See dinitrogen trioxide.

NITROUS ETHER. See ethyl nitrite.

NITROUS FUMES. See nitric oxide.

NITROUS OXIDE. Syn: *laughing gas*. Colorless gas or liquid or cubic crystals. N_2O, mw: 44.02, mp: −90.8°, bp: −88.49°, d: 1.977 g/l, d liq: 1.226 @ −89°.
Acute tox data: Inhal LC_{LO} (mice) = 1500 ppm. [*3*]
THR = MOD via inhal route. An exper teratogen [*3*] via inhal route. May be used as a general purpose food additive. [*109*] An asphyxiant. Used as an anesthetic.
Fire Hazard: Mod, by chemical reaction; it supports combustion.
Explosion Hazard: Mod; it can form an explosive mixture with air. Violent reaction with Al, B, hydrazine, LiH, LiC_6H_5, PH_3, Na, WC. [*19*] Also self-explodes at high temp.
Disaster Hazard: Slightly dangerous; can form explosive mixtures with air.

***m*-NITROXYLENE.** See *m*-nitroxylol.

***o*-NITROXYLENE.** See *o*-nitroxylol.

***p*-NITROXYLENE.** See *p*-nitroxylol.

***m*-NITROXYLOL.** Syn: *n-nitroxylene*. Light yellow liquid. $C_6H_3(CH_3)_2NO_2$, mw: 151.2, mp: 2°, bp: 244°, d: 1.135 @ 15°/4°.
THR = Details U. A poison. See also nitro compounds of aromatic hydrocarbons. Said to be less toxic than nitrobenzene.
Fire Hazard: Mod, when exposed to heat or flame.
Explosion Hazard: See nitrates.
Disaster Hazard: See nitrates.

***o*-NITROXYLOL.** Syn: *o-nitroxylene*. Yellowish liquid or solid. $C_6H_3(CH_3)_2NO_2$, mw: 151.2, mp: 29°, bp: 258°, d: 1.139.
THR = See nitro compounds of aromatic hydrocarbons.

***p*-NITROXYLOL.** Syn: *p-nitroxylene*. Liquid. $C_6H_3(CH_3)_2NO_2$, mw: 151.2, bp: 240°, d: 1.132.
THR = See nitro compounds of aromatic hydrocarbons.

NITRYL CHLORIDE. Pale yellow or brown gas. NO_2Cl, mw: 81.47, mp: <−31°, bp: 5°, d: 2.57 g/l, d (liquid): 1.32 @ 14°.
THR = HIGH irr to skin, eyes and mu mem and via inhal route. See also chlorine.
Disaster Hazard: Dangerous; when heated to decomp, emits highly toxic fumes of chlorides and nitrogen dioxide; will react with water or steam to produce corrosive fumes. Violent reaction with NH_3, $SnBr_4$, SnI_4, SO_3. [*19*]

NITRYL FLUORIDE. Colorless gas, pungent odor. NO_2F, mw: 65.01, d: 2.90 g/l, mp: −139°, bp: −56°.
THR = HIGH irr to skin, eyes and mu mem, and via inhal route. See also fluorine and fluorides.
Disaster Hazard: Dangerous; see NO_x and fluorine. This gas is intensely reactive. It will spont ignite I, Se, P, As, B, Sb, Th, Mo; upon warming, will ignite Pb, Bi, Cr, Mn, Fe, Ni, W, S, C. [*19*]

NITRYL PERCHLORATE. NO_2ClO_4, mw: 145.5.
THR = Reacts violently with acetone, benzene, diethyl ether, organic matter. [*19*]

NONADECANE. $C_{19}H_{40}$, mw: 268.3, flash p: >212°F, autoign. temp.: 446°F, d: 0.79, vap. d: 9.27, bp: 331.1°.

THR = No data. Probably LOW.
Fire Hazard: Low.
To Fight Fire: Dry chemical, CO_2, foam, fog.

NONAETHYLENE GLYCOL.
Acute tox data: Oral LD_{50} (rat) = 37000 mg/kg. [*3*]
THR = Practically NONE via oral route. See also glycols.
Fire Hazard: Slight; when heated, emits acrid fumes.

N-NONANE. Colorless liquid. C_9H_{20}, mw: 128.25, mp: −53.7°, bp: 150.7°, lel = 0.8%, uel = 2.9%, flash p: 88°F (CC), d: 0.718 @ 20°/4°, autoign. temp.: 403°F, vap. press: 10 mm @ 38.0°, vap. d: 4.41.
THR = Irr to respiratory tract. Narcotic in high conc.
Fire Hazard: Dangerous, when exposed to heat or flame; can react with oxidizing materials.
Spont Heating: No.
Explosion Hazard: Mod, in the form of gas, when exposed to flame.
To Fight Fire: CO_2, dry chemical.

NONANEDIOIC ACID. See azelaic acid.

NONANOIC ACID. See ethyl heptanoic acid.

1-NONANOYL AZIRIDINE. $C_{11}H_{21}ON$, mw: 183.3.
THR = An exper neo. [*3, 23*]

NONETHYLENE GLYCOL HEXARICNOLEATE. flash p: 530°F (OC).
THR = Details U. See also glycols.
Fire Hazard: Slight, when exposed to heat or flame.
To Fight Fire: Foam, CO_2, dry chemical.

NONETHYLENE GLYCOL MONOSTEARATE. flash p: 505°F (OC).
THR = Details U. See also glycols.
Fire Hazard: Slight, when exposed to heat or flame.
To Fight Fire: Foam, CO_2, dry chemical.

NONYL ACETATE. Liquid. $CH_3COOCH[CH_2CH(CH_3)_2]_2$, mw: 186.29, mp: −48.1°, bp: 192.4°, flash p: 155°F (OC), d: 0.8530 @ 20°/20°, vap. d: 6.42.
THR = Details U. See also esters and acetic acid.
Fire Hazard: Mod, when exposed to heat or flame; can react with oxidizing materials.
To Fight Fire: CO_2, dry chemical.

NONYL ALCOHOL. See diisobutyl carbinol.

NONYL ALDEHYDE CYANOHYDRIN. Liquid. $(CH_3)_3CCH_2CH(CH_3)CH_2CH(OH)CN$, mw: 169, bp: decomp to HCN, d: 0.8976, vap. d: 5.83.
THR = See cyanides.

N-NONYLAMINE. Syn: *1-amino nonane*. Colorless liquid. $CH_3(CH_2)_7CH_2NH_2$, mw: 143.3, bp: 202.2°, d: 0.785.
THR = Details U. See also amines.
Fire Hazard: Slight, when exposed to heat or flame; can react with oxidizing materials.

NONYL BENZENE. See nonyl benzol.

NONYL BENZOL. Syn: *nonyl benzene*. Liquid. $C_6H_5C_9H_{19}$, mw: 204.3, bp: 245°, flash p: 210°F (CC), d: 0.86.
THR = Details U. Probably irr and narcotic in high conc.
Fire Hazard: Low, when exposed to heat or flame; can react with oxidizing materials.
To Fight Fire: Foam, CO_2, dry chemical.

NONYL CARBINOL. See *n*-decyl alcohol.

NONYL NAPHTHALENE. Liquid. $C_{10}H_7C_9H_{19}$, mw: 254.4, bp: 330°, flash p: $<$200°F, d: 0.94, vap. d: 8.8.
THR = Details U. See also naphthalene.
Fire Hazard: Mod, when exposed to heat or flame; can react with oxidizing materials.
To Fight Fire: Foam, CO_2, dry chemical.

NONYL PHENOL. Clear, viscous, straw-colored liquid. $C_6H_4OHC_9H_{19}$, mw: 220.34, bp: 290°−302°, pour p: 2°, flash p: 285°F, d: 0.949 @ 20°/4°, vap. d: 7.59.
Acute tox data: Oral LD_{50} (rat) = 1620 mg/kg; dermal LD_{50} (rabbit) = 2140 mg/kg. [*3*]
THR = MOD via oral and dermal routes.
Fire Hazard: Slight, when exposed to heat or flame; can react with oxidizing materials.
To Fight Fire: Alcohol foam, CO_2, dry chemical.

NONYLTRICHLOROSILANE. $C_9H_{19}SiCl_3$, mw: 261.7, d: 1.072 @ 25°.
THR = HIGH irr to skin, eyes and mu mem. See also chlorosilanes.

NORBORMIDE. Syns: *raticate, 5-(a-hydroxy-a-2-pyridylbenzyl)-7-(a-2-pyridylbenzylidene)-5-norbornene-2,3-dicarboximide*. $C_{33}H_{25}N_3O_3$, mw: 511.6, mp: 194°.
Acute tox data: Oral LD_{50} (rat) = 5.3 mg/kg; iv LD_{50} (rat) = 0.65 mg/kg. [*3*]
THR = VERY HIGH via oral and iv routes. A rodenticide.

NORDHAUSEN ACID. See oleum.

NORDIHYDROGUAIARETIC ACID. Syns: *NDGA, 4,4′-(2,3-dimethyl tetramethylene)dipyrocatechol*. Crystals from acetic acid, sol in methanol, ethanol and ether, slightly sol in hot water and chloroform, nearly insol in benzene and petroleum ether. $[C_6H_3(OH)_2CH_2CH(CH_3)]_2$, mw: 302.4, mp: 184°−185°.
Acute tox data: Oral LD_{50} (guinea pig) = 830 mg/kg; ip LD_{50} (mice) = 550 mg/kg. [*3*]
THR = MOD via ip and oral routes. Used as an antioxidant food additive. [*109*]

NOREA. See herban.

NORETHINDRONE + ETHYNYL ESTRADIOL.
THR = An exper neo. [*3*]

NORETHINDRONE + MESTRANOL.
THR = An exper neo. [*3*]

NORGESTREL + ETHINYL ESTRADIOL.
THR = An exper neo. [*3*]

NORLUTIN ENANTHATE. $C_{27}H_{38}O_3$, mw: 410.7.
THR = An exper teratogen. [*3*]

NORNICOTINE. Syn: *β-pyridyl-α-pyrrolidine*. Hygroscopic, viscous liquid. $C_9H_{12}N_2$, mw: 148.2, bp: 270°, d: 1.0737 @ 20°/4°.
Acute tox data: ip LD (rabbit) = 23.5 mg/kg. [*3*] iv LD (rabbit) = 3 mg/kg. [*2*]; ip LD_{LO} (rat) = 11 mg/kg. [*3*]
THR = HIGH via ip route. About ⅓ the toxicity of nicotine. Causes faintness, prostration, muscular weakness, severe nausea, vomiting, diarrhea and collapse without convulsions. [*2*] An insecticide.
Disaster Hazard: Dangerous; when heated to decomp, emits highly toxic fumes of NO_x.

19-NOR-17-α-PREGN-4-EN-20-YNE-3,β,17-DIOL + 3-METHOXY-17-α-19-NORPREGNA-1,3,5(10)-TRIEN-20-YN-OL.
THR = An exper carc to women via oral route. [*103*]

19-NORPROGESTERONE.
THR = An exper carc. [*23*]

F-NORSTERANTHRENE. $C_{22}H_{16}$, mw: 280.4.
THR = An exper carc. [*3*]

19-NORTESTOSTERONE.
THR = An exper neo. [*3, 23*]

NOT ENOUGH ANIMAL EXPERIMENTATION. See Section 8.

NOVEX. See 2,2′-thiobis(4-chlorophenol).

NOVOBIOCIN. Light yellow to white antibiotic crystals. $C_{31}H_{36}N_2O_{11}$, mw: 612.7, d: 1.345, mp: decomp @ 152°.
THR = An antibiotic with serious side effects which include liver and blood disease. It may also cause the development of resistant strains of staphylococcus. Used as a food additive permitted in the feed and drinking water of animals and/or for the treatment of food-producing animals; also permitted in food for human consumption. [*109*]

NSC-101983. $C_{17}H_{14}ON_2$, mw: 262.3.
THR = An exper carc [*3, 23*] via oral route.

NSC-34372. See quinacrine-ethyl mustard.

NUISANCE DUSTS AND AEROSOLS.
THR = Variable, depending upon composition. Cause local irr of eyes, nose, throat and lungs. Some may lead to chronic bronchitis, emphysema and bronchial asthma. Dermatitis may result from short contact. Asthma, angioneurotic edema, hives, etc. may result from short periods of inhal. A topic eczema, angioneurotic edema, hives, etc., may also result from prolonged contact. See also Section 9. A common air contaminant. Nuisance aerosols do evoke some tissue response in the lung upon inhal of sufficient amounts. However, this reaction is potentially reversible and leaves no scar tissue.

NUPERCAINE HYDROCHLORIDE. See dibucaine hydrochloride.

NYLON. See polyamides.

NYSTATIN. Syn: *fungiciden*. Yellow to light tan powder, odor suggestive of cereals, sparingly sol in methanol and ethanol, very slightly sol in water, insol in chloroform, ether and benzene. $C_{46}H_{77}NO_{19}$, mw: 947, mp: decomp > 160°.
Acute tox data: ip LD_{50} (rat) = 0.043 mg/kg; ip LD_{50} (mouse) = 0.015 mg/kg; sc LD_{50} (mouse) = 2000 mg/kg. [*3*]
THR = HIGH via ip route and MOD via sc route. Used as a food additive permitted in the feed and drinking water of animals and/or for the treatment of food-producing animals; also permitted in food for human consumption. [*109*]

OAT DUST.
THR = A mild irr and allergen. See also nuisance dusts.
Fire Hazard: Mod, when exposed to heat or flame.
Explosion Hazard: Slight, when exposed to flame.

OBLIVON. See 3-methyl-1-pentyn-3-ol-carbamate.

OCBM. See *o*-chlorobenzylidene malonitrile.

OCHRATOXIN A. $C_{20}H_{18}O_6NCl$, mw: 403.8.
Acute tox data: Oral LD_{50} (rat) = 20 mg/kg; oral LD_{50} (chicken) = 3.3 mg/kg. [*3*]
THR = HIGH via oral and sc routes. A rat poison. An exper (±) carc. [*3, 7*]

OCHRATOXIN B.
Acute tox data: Oral LD_{50} (chicken) = 54 mg/kg. [*3*]
THR = HIGH via oral route.

OCPN. See *o*-chloro-*p*-nitroaniline.

OCTABROMOBIPHENYL. $C_{12}H_2Br_8$, mw: 785.4.
Acute tox data: Oral LD_{LO} (rat) = 2000 mg/kg. [*3*] Oral ALD (male rat) = > 12500 mg/kg; dermal ALD (rat) = > 10,000 mg/kg.
THR = A MILD irr to skin, eyes and mu mem. MOD-LOW via oral, dermal routes. Can cause kidney and liver enlargement and thyroid hyperplasia; it is stored in fatty tissue. See also bromides.

OCTACHLOROCAMPHENE. Syns: *chlorinated camphene, toxadust, others.* Yellow, waxy solid, pleasant piney odor. $C_{10}H_{10}Cl_8$, mw: 413.84; mp: 65°–90°.
Acute tox data: Oral LD_{LO} (human) = 40 mg/kg; oral LD_{50} (rat) = 60 mg/kg; dermal LD_{50} (rat) = 780 mg/kg; oral LD_{50} (duck) = 71 mg/kg; inhal LD_{LO} (mouse) = 2000 mg/m^3 for 2 hrs. [*3*]
THR = HIGH via oral; MOD via dermal and inhal routes. A pesticide. An exper carc. [*32*] The Food and Drug Administration (1955) proposes a tolerance of 7 ppm for residues on fruits and vegetables.
THR = Toxic and, in some instances, lethal amounts of toxaphene can enter the body through the mouth, lungs and skin. Systemic absorption of the insecticide is increased by the presence of digestible oils, and liquid preparations of the insecticide which penetrate the skin more readily than do dusts and wettable powders. A toxic mixture of organochlorine pesticides stored to some extent in body fat, it is an exper carc. [*93*] It resembles chlordane and, to some extent, camphor in its physiological action. It causes diffuse stimulation of the brain and spinal cord resulting in generalized convulsions of a tonic or clonic character. Death usually results from respiratory failure. Detoxification appears to occur in the liver.

Dangerous Acute and Chronic Dose: The lethal oral dose for man is estimated to be 2–7 g, a toxicity of about four times that of DDT. Toxaphene causes moderate irr on the skin but little or no sensitization. At least seven human deaths have been reported due to toxaphene, all in children. Two families have been made ill by eating vegetables containing a large residue of toxaphene.

Signs and Symptoms of Poisoning: The symptoms are excitement followed by epileptiform convulsions associated with depression; these symptoms are aggravated by external stimuli. Death usually results from respiratory failure and may occur as early as 4 hrs and as late as 24 hrs after poisoning. In calves, sheep, and goats, internal hemorrhage and high temps have been reported, the latter also being one of the symptoms of camphor poisoning in man.

Pathology: Degenerative changes in the liver parenchyma and renal tubules were reported in laboratory animals chronically receiving toxaphene.

Treatment of poisoning: The stomach and intestinal tract should be evacuated. Oily laxatives should be avoided. Contaminated clothing should be removed and contaminated areas should be scrubbed with soap and water. Anticonvulsant drugs (bromides, barbiturates, etc.) should be administered. Sodium phenobarbital is the drug of choice to prevent convulsions, and sodium pentobarbitol is best to control those already in evidence. Dosages may have to be quite high in the presence of poisoning in order to control symptoms but should not be high enough to interfere with continuous sleep.

Disaster Hazard: Dangerous; when heated to decomp, emits highly toxic fumes of chlorides.

1,2,3,4,6,7,8,9-OCTACHLORODIBENZO-*p*-DIOXIN. Colorless crystals. $C_{18}H_{12}O_2$, mw: 459.7, mp: 239°.
Acute tox data: Single oral dose of 2000 mg/kg not lethal. 1% diet ⟶ 100% lethal to male mice ⟶

90% lethal to female mice ⟶ 100% lethal to male, female rats. 0.5% diet ⟶ 100% lethal to male mice, 10% lethal to female mice, 100% lethal to male, female rats. 0.25% diet ⟶ 98% lethal to male mice, 100% lethal to female mice ⟶ 20% lethal to male rats, 56% lethal to female rats. 500 mg/kg/day is embryo toxic. [*81*]

THR = Appears to be non-mutagenic. See also chlorinated dibenzo-*p*-dioxins. An exper neo [*3, 81*] via dermal route.

OCTACHLORODIPROPYL ETHER. $C_6H_6OCl_8$, mw: 377.7.

Acute tox data: Oral LD_{50} (rat) = 3630 mg/kg; oral LD_{50} (rabbit) = 2500 mg/kg. [*3*]

THR = MOD via oral route.

1,2,4,5,6,7,8,8-OCTOCHLORO-4,7-METHANO-3a,4,7,7a-TETRAHYDROINDANE. See chlordane.

OCTACHLORONAPHTHALENE. $C_{10}Cl_8$, mw: 403.8.

THR = HIGH via inhal, ingestion and skin absorption.

Disaster Hazard: Dangerous. When heated to decomp, emits highly toxic fumes.

OCTACHLOROTETRAHYDRO INDANE. See chlordane.

OCTACHLOROTRISILANE. Si_3Cl_8, mw: 368.3, mp: −67°, bp: 211.4°, vap. press: 1 mm @ 46.3°.

THR = HIGH irr to skin, eyes and mu mem. See also chlorosilanes.

Disaster Hazard: Dangerous; when heated to decomp, emits highly toxic fumes of chlorides; will react with water or steam to produce toxic and corrosive fumes. See silanes.

OCTADECANE. $C_{18}H_{38}$, mw: 254.5, flash p: > 212°F, autoign. temp.: 455°F, mp: 50°, d: 0.78, vap. d: 8.73, bp: 317.2°.

THR = No data. Probably LOW.

Fire Hazard: Low.

To Fight Fire: Foam, water spray or mist, CO_2, dry chemical.

OCTADECANOIC ACID. See stearic acid.

1-OCTADECANOL. See stearyl alcohol.

1-OCTADECENE. Syns: *1-octadecyne, hexadecyl acetylene.* HC⋮C$(CH_2)_{15}CH_3$, mw: 250.5, mp: 26°, bp: 180° @ 15 mm, d: 0.7884 @ 20°/4°, flash p: > 212°F, autoign. temp.: 482°F.

THR = U.

Fire Hazard: Slight, when exposed to heat or flame; can react with oxidizing materials.

To Fight Fire: Foam, water spray, fog, CO_2.

OCTADECENYL ALDEHYDE. Liquid. $C_{15}H_{36}O$, mw: 266.5, bp: 167°, d: 0.847, vap. d: 9.18.

THR = Details U. See also aldehydes.

Fire Hazard: Mod, when exposed to heat or flame; can react with oxidizing materials.

OCTADECYL ALCOHOL. Crystals. $CH_3(CH_2)_{16}CH_2OH$, mw: 270.5, mp: 58°, bp: 349.5°, d: 0.8124 @ 59°/4°, vap. press: 1 mm @ 150.3°.

THR = Probably LOW. See also alcohols.

N-OCTADECYL DISODIUM SULFOSUCCINAMATE. Cream-colored soft paste. $C_{22}H_{41}O_6NSNa_2$, mw: 493.

THR = U.

Disaster Hazard: Dangerous; when heated to decomp, emits highly toxic fumes of SO_x and NO_x.

OCTADECYLISOCYANATE. $C_{18}H_{35}NCO$, mw: 295.5, mp: 15°–18°, bp: 150°–180° @ 0.75 mm, d: 0.86 @ 25°.

THR = See cyanates.

OCTADECYL TRICHLOROSILANE. $C_{18}H_{37}SiCl_3$, mw: 268.01, bp: 223° @ 10 mm, d: 0.984 @ 25°.

THR = HIGH irr to skin, eyes and mu mem. See also chlorosilanes.

Disaster Hazard: Dangerous; when heated to decomp, emits highly toxic fumes of chlorides; will react with water or steam to produce toxic and corrosive fumes. See silanes.

1-OCTADECYNE. See 1-octadecene.

OCTAFLUOROCYCLOBUTANE. Colorless, nonflam gas, odorless. C_4F_5(cyclic), mw: 200, bp: −6.04°, mp: −41.4°, d(liquid): 1.513 @ −70°F.

Acute tox data: Inhal of 100,000 ppm for 6 hrs/day and 5 days/week ⟶ no untoward effects.

THR = LOW via oral and inhal routes. Can cause slight transient effects at high conc. No anesthesia or CNS effects. [*20*] A food additive permitted in food for human consumption. [*109*]

OCTAFLUOROPROPANE. See perfluoropropane.

OCTAHYDRODIBENZ(a,h)ANTHRACENE. $C_{22}H_{22}$, mw: 286.4.

THR = An exper neo. [*3*]

"OCTA-KLOR." See chlordane.

"OCTALENE." See aldrin.

"OCTALOX." See dieldrin.

OCTAMETHYL PYROPHOSPHORAMIDE. Syns: *OMPA, schradan, pestox, others.* Water white liquid. $C_8H_{24}N_4P_2O_3$, mw: 286.34, mp: 20°–21°, bp: 137°–142° @ 2 mm, d: 1.137 @ 25°/4°.

Acute tox data: Oral LD_{50} (rat) = 5 mg/kg; dermal LD_{50} (rat) = 15 mg/kg; ip LD_{50} (rat) = 8 mg/kg; oral LD_{50} (wild birds) = 11 mg/kg; iv LD_{50} (rabbit) = 6 mg/kg. [*3*]

THR = HIGH via oral, dermal, ip and iv routes. A

cholinesterase inhibitor. See parathion. An insecticide.
Disaster Hazard: Dangerous; see phosphates.

OCTAMETHYL TRISILOXANE. Liquid. $C_8H_{24}O_2Si_3$, mw: 236.5, bp: 153.2°, fp: −80°, d: 0.82, vap. press: 1 mm @ 7.4°, 10 mm @ 43.1°.
THR = Details U. See also silicones.
Disaster Hazard: Slightly dangerous; when heated to decomp, emits acrid fumes.

OCTANAL. See octylaldehyde.

OCTANE. Clear liquid. $CH_3(CH_2)_6CH_3$, mw: 114.23, bp: 125.8°, lel = 1.0%, uel = 6.5%, fp: −56.5°, flash p: 56°F, d: 0.7036 @ 20°/4°, autoign. temp.: 428°F, vap. press: 10 mm @ 19.2°, vap. d: 3.86.
THR = May act as a simple asphyxiant. See also argon. A narcotic in high conc. In humans via dermal exposure to undiluted octane, 5 hrs ⟶ blister formation but no anesthesia. 1 hr ⟶ diffuse burning sensation. In animals inhal of 3.2% for 5 min ⟶ resp arrest. 1.6% for 5 min ⟶ resp arrest in 1 out of 4 mice. 1.28% for 185 min ⟶ decrease resp followed by death. 1% for 2 hrs ⟶ loss of reflexes. 8560 ppm for 55 min ⟶ narcosis; 6634 ppm for 1 hr ⟶ lying down. [*98*] See also alkanes.
Fire Hazard: Dangerous; when exposed to heat, flame or oxidizers.
Explosion Hazard: Severe, when exposed to heat or flame.
Disaster Hazard: Dangerous, upon exposure to heat or flame; can react vigorously with oxidizing materials.

4,5-OCTANEDIOL. See octylene glycol.

OCTANOIC ACID. See caprylic acid.

2-OCTANOL. See capryl alcohol.

2-OCTANONE. Syns: *methyl hexyl ketone*, *hexyl methyl ketone*. Colorless liquid, pleasant odor, slightly sol in water, sol in alcohol, hydrocarbons, ether, esters, etc. $CH_3CO(CH_2)_5CH_3$, mw: 128, d: 0.82 @ 20°/4°, mp: −20.9°, bp: 173.5°, vap. d: 4.4, flash p: 160°F.
Acute tox data: ip LD_{50} (mice) = 406 mg/kg. [*3*]
THR = HIGH via ip route. See also ketones.
Fire Hazard: Mod, when exposed to heat, flame or oxidizers.
To Fight Fire: Foam, alcohol foam.

OCTANOYL CHLORIDE.
See ethyl hexanoyl chloride.

OCTAPHENYL CYCLOTETRASILOXANE. Crystals. $[(C_6H_5)_2SiO]_4$, mw: 793.12, mp: 202°, bp: 335°.
THR = Details U. See also silicones.
Disaster Hazard: Mod dangerous; when heated to decomp, emits toxic fumes.

OCTAPHENYL TRIGERMANE. White crystals, insol in water. $(C_6H_5)_8Ge_3$, mw: 834.6, mp: 248°.
THR = Details U. See germanium compounds.

1-OCTENE. Syns: *1-octylene*, *1-caprylene*. Colorless liquid. C_8H_{16}, mw: 112.21, mp: −101.9°, bp: 121.27°, flash p: 70°F (TOC), d: 0.716 @ 20°/4°, vap. d: 3.87, vap. press: 36.2 mm @ 38°.
THR = U.
Fire Hazard: Dangerous, when exposed to heat, flame or oxidizers.
Disaster Hazard: Dangerous, upon exposure to heat or flame; can react vigorously with oxidizing materials.
To Fight Fire: Foam, CO_2, dry chemical.

2-OCTENE. Colorless liquid. C_5H_{16}, mw: 112.21, mp: −94.04°, bp: 125.2°, flash p: 70°F (TOC), d: 0.7192 @ 20°/4°, vap. d: 3.87.
THR = No data. Probably LOW.
Fire Hazard: Dangerous, when exposed to heat, flame or oxidizers.
Disaster Hazard: Dangerous, upon exposure to heat or flame; can react vigorously with oxidizing materials.
To Fight Fire: Foam, CO_2, dry chemical.

OCTIN. See isometheptene.

OCTOIC ANHYDRIDE.
See 2-ethyl hexanoic anhydride.

N-OCTYL ACETATE. See 2-ethyl hexyl acetate.

OCTYL ACRYLATE. See 2-ethyl hexyl acrylate.

OCTYL ALCOHOL. See 2-ethyl hexyl alcohol.

OCTYL ALDEHYDE. Syns: *2-ethylhexanal*, *caprylaldehyde*, *octanal*. Liquid. $C_7H_{15}CHO$, mw: 128.2, bp: 163.4, flash p: 125°F (CC), d: 0.821 @ 20°/4°, vap. d: 4.41.
Acute tox data: Oral LD_{50} (rat) = 5630 mg/kg; dermal LD_{50} (rabbit) = 6350 mg/kg. [*3*]
THR = MOD via oral and dermal routes. See also aldehydes.
Fire Hazard: Mod, when exposed to heat or flame; can react with oxidizing materials.
To Fight Fire: Foam, CO_2, dry chemical.

N-OCTYLAMINE. Water white liquid, amine-like odor. $CH_3(CH_2)_7NH_2$, mw: 129.3, bp: 170°–179°, flash p: 140°F, d: 0.779 @ 20°/20°, vap. d: 4.46.
Acute tox data: ip LD_{50} (mouse) = 100 mg/kg. [*3*]
THR = HIGH via ip route. Administration to humans has caused headache, fall in blood pressure, rapid pulse, urticaria, and itching of skin due to release of histamine. See also amines and fatty amines.
Fire Hazard: Mod, when exposed to heat or flame; can react with oxidizing materials.
To Fight Fire: Alcohol foam, dry chemical.

***tert*-OCTYLAMINE.** Clear liquid, amine-like odor. $(CH_3)_3CCH_2C(CH_3)_2NH_2$, mw: 129.3, bp: 140°, flash p: 91°F (OC), d: 1.4213, vap. d: 4.46.
THR = Details U. See also octylamine.
Fire Hazard: Dangerous, when exposed to heat or flame; can react with oxidizing materials.
To Fight Fire: Water, foam, dry chemical.

N-OCTYL BICYCLOHEPTENE DICARBOXIMIDE. Syn: *MGK.*
Acute tox data: Oral LD_{50} (rat) = 2800 mg/kg; dermal LD_{50} (rabbit) = 470 mg/kg. [*3*]
THR = HIGH via dermal and MOD via oral routes. Large doses can cause CNS stimulation followed by depression.

OCTYL CARBAMATE. $C_9H_{19}O_2N$, mw: 173.3.
Acute tox data: ip LD_{LO} (mouse) = 100 mg/kg. [*3*]
THR = HIGH via ip route.

N-OCTYL CHLORIDE. Colorless liquid, sol in most organic solvents, insol in water. $CH_3(CH_2)_7Cl$, mw: 148.5, d: 0.8697 @ 25°/25°, fp: −62°, vap. d: 5.1, bp: 181.6°, flash p: 158°F.
THR = U. See esters.
Fire Hazard: Mod, when exposed to heat or flame.
Disaster Hazard: Dangerous; see chlorides.
To Fight Fire: Foam, alcohol foam.

OCTYL DECYL PHTHALATE. A clear liquid. $C_6H_4(COOC_8H_{17})(COOC_{10}H_{21})$, mw: 418.3, bp: 239° @ 4 mm, fp: −50°, flash p: 455°F (COC), d: 0.980 @ 20°/20°.
THR = U. See phthalic acid and esters.
Fire Hazard: Slight, when exposed to heat or powerful oxidizing agents.

1-OCTYLENE. See 1-octene.

OCTYLENE GLYCOL. See 2-ethyl hexanediol-1,3.

OCTYLENE GLYCOL BIBORATE. A white solid. $C_{24}H_{46}O_6B_2$, mp: 69°–104°, mw: 454.3, bp: 68°–78° @ 0.4 mm.
THR = See esters and boron compounds.

***tert*-OCTYL MERCAPTAN.** Liquid, insol in water. $C_8H_{17}SH$, mw: 146, boiling range: 159°–166°, d: 0.848 @ 60°/60°F, vap. d: 5.0, flash p: 115°F (OC).
THR = See mercaptans.
Fire Hazard: Mod, when exposed to heat, flame or oxidizers.
Disaster Hazard: Dangerous; see mercaptans.
To Fight Fire: Foam, alcohol foam.

OCTYL PHENOL. Syn: *diisobutyl phenol.* White or light pink flakes. $C_6H_4(C_8H_{17})OH$, mw: 206.3, bp: 280°–283°, fp: 72°–74°, d: 0.941 @ 24°/4°.
THR = U. See also phenol.
Fire Hazard: Slight, when exposed to heat or flame.
Disaster Hazard: Dangerous; when heated to decomp, emits highly toxic fumes of phenol; can react with oxidizing materials.

OCTYL SILICATE. $Si(OC_8H_{17})_4$, mw: 544.3, bp: 204° @ 3 mm, d: 0.8208 @ 20°.
THR = Details U; an insecticide.
Fire Hazard: Slight.

OCTYL TITANATE. See 2-ethyl butyl titanate.

OCTYL THIOCYANATE. Syn: *octyl rhodanate.* $C_9H_{17}NS$, mw: 171.3.
Acute tox data: Oral LD_{LO} (rat) = 1500 mg/kg; sc LD_{LO} (rat) = 5500 mg/kg. [*3*]
THR = MOD via oral and sc routes.

OCTYL TRICHLOROSILANE. Fuming liquid. $C_8H_{17}SiCl_3$, mw: 247.7.
THR = HIGH irr to skin, eyes and mu mem.
Disaster Hazard: Dangerous; when heated to decomp, emits highly toxic fumes of chlorides; will react with water or steam to produce toxic and corrosive fumes. See silanes.

OENANTHYLIC ACID. See heptanoic acid.

OIL BLACK. See soot.

OILED CLOTHING, FABRICS, RAGS, ETC.
Fire Hazard: Mod, when exposed to heat or flame, by spont chemical reaction or on contact with oxidizing materials.
Spont Heating: Yes, particularly when wet.

OILED MATERIALS, NOT PROPERLY DRIED. See Article No. 1315 in IATA list of restricted articles.

OILED MATERIALS, PROPERLY DRIED. See Article No. 1316 in IATA list of restricted articles.

OIL FURNACE BLACK. See soot.

OIL GAS. A gas derived from petroleum. Composition; illuminants 4.2%, carbon monoxide 10.4%, hydrogen 47.6%, methane 27.0%, carbon dioxide 4.6%, nitrogen 5.8%, oxygen 0.4%. lel = 4.8%, uel = 32.5%, autoign. temp.: 637°F.
THR = HIGH. See carbon monoxide.
Fire Hazard: Dangerous, when exposed to heat or flame.
Spont Heating: No.
Explosion Hazard: Mod, when exposed to heat or flame.
Disaster Hazard: Dangerous; can react vigorously with oxidizing materials.
To Fight Fire: CO_2, dry chemical, water spray.

OIL, N.O.S. See crude oil.

OIL OF AMBER. See amber oil.

OIL OF AMERICAN WORMSEED. See oil of chenopodium.

OIL OF BITTER ALMOND. Colorless to yellow liquid. Composition: 95% benzaldehyde, 2–4% HCN. d: 1.038–1.060 @ 25°/25°.
THR = See benzaldehyde and cyanides.

OIL OF CAJEPUT.
Acute tox data: Oral LD_{50} (rat) = 387 mg/kg. [*3*]
THR = HIGH via oral route to mu mem.

OIL OF CALAMUS. Syns: *calamus oil, oil of sweet flag.*
Acute tox data: Oral LD_{50} (rat) = 777 mg/kg. [*3*]
THR = MOD via oral route. An exper carc to rats via oral route. [*103*]

OIL OF CAMPHOR.
Acute tox data: Oral TD_{LO} (human) = 29 mg/kg ⟶ CNS effects. Oral LD_{LO} (child) = 50 mg/kg; oral LD_{LO} (rabbit) = 200 mg/kg. [*3*]
THR = HIGH via oral to humans and MOD via oral to rabbits.

OIL OF CAMPHOR SASSAFRASS.
Acute tox data: Oral LD_{50} (rat) = 3730 mg/kg. [*3*]
THR = MOD via oral route.

OIL OF CARAWAY.
Acute tox data: Oral LD_{50} (rat) = 3500 mg/kg; dermal LD_{50} (rabbit) = 1780 mg/kg. [*3*]
THR = MOD via oral and dermal routes.

OIL OF CHENOPODIUM. Syn: *oil of American wormseed.* Colorless or pale yellow liquid, characteristic disagreeable odor and taste, not water-sol. Composition: 60–70% ascaridol. d: 0.950–0.980 @ 25°/25°.
THR = HIGH irr via oral route. See also ascaridol. An exper carc. [*14*]
Fire Hazard: Slight.

OIL OF CHERRY LAUREL. Pale yellow liquid. Composition: hydrogen cyanide, benzaldehyde, benzaldehyde cyanhydrin, benzyl alcohol. d: 1.054–1.066 @ 20°/20°.
THR = HIGH toxicity. See cyanides.
Disaster Hazard: Dangerous; when heated, emits highly toxic fumes.

OIL OF CITRONELLA (CEYLON). See citronella.

OIL OF CLOVE.
Acute tox data: Oral LD_{50} (rat) = 3720 mg/kg. [*3*]
THR = MOD via oral route. An irr.

OIL OF CLOVE STEM.
Acute tox data: Oral LD_{50} (rat) = 2020 mg/kg. [*3*]
THR = MOD via oral route.

OIL OF CRISP MINT. See spearmint oil.

OIL OF EUCALYPTUS. Syn: *dinkum oil.* Colorless to pale yellow liquid. Spicy odor. Composition: eucalyptol, aldehydes, d-pinene. mp: −15.4° (approx), d: 0.905–0.925 @ 25°/25°.
Acute tox data: Oral LD_{50} (rat) = 4440 mg/kg. [*3*]
THR = MOD via oral route.

OIL OF FENNEL.
Acute tox data: Oral LD_{50} (rat) = 3120 mg/kg. [*3*]
THR = MOD via oral route.
Disaster Hazard: Slight; when heated to decomp, emits toxic fumes.

OIL OF GARLIC. Composition: allyl propyl disulfide, diallyl disulfide. d: 1.046–1.057 @ 15°/15°.
THR = Details U. See individual components.
Disaster Hazard: Dangerous; when heated to decomp, emits highly toxic fumes of SO_x.

OIL OF GERANIUM REUNION.
Acute tox data: Dermal LD_{50} (rabbit) = 2500 mg/kg. [*3*]
THR = MOD via dermal route.

OIL OF HARTSHORN.
Acute tox data: Oral LD_{LO} (rat) = 800 mg/kg. [*3*]
THR = MOD via oral route.

OIL OF LEMON.
Acute tox data: Oral LD_{50} (rat) = 2840 mg/kg. [*3*]
THR = MOD via oral route.

OIL OF MACE.
Acute tox data: Oral LD_{50} (rat) = 3640 mg/kg. [*3*]
THR = MOD via oral route.

OIL OF MUSTARD + OIL OF ARGEMONE.
THR = An exper neo [*3*] via dermal route.

OIL OF MIRBANE. See nitrobenzene.

OIL OF PEPPERMINT. Colorless to pale yellow liquid. d: 0.896–0.908 @ 25°/25°.
THR = LOW irr and allergen.

OIL OF TANSY. See tansy oil.

OIL OF VITRIOL. See sulfuric acid.

OIL ORANGE E. See 1-(phenylazo)-2-naphthol.

OIL ORANGE TX. See FD & C orange 2.

OIL VIOLET. $C_{24}H_{21}N_5$, mw: 379.5.
THR = An exper (±) carc and neo. [*3, 4*]

OIL OF WINTERGREEN. See methyl salicylate.

OIL YELLOW H A. A (S) carc of the liver. [*14*]

OLDHAMITE. See calcium sulfide.

OLEANDOMYCIN. Syns: *amimycin, matromycin, romicil.* White amorphous powder, mod sol in water, sol in dilute acids, freely sol in methanol, ethanol, butanol and acetone, practically insol in hexane, carbon tetrachloride and dibutyl ether. $C_{35}H_{61}NO_{12}$, mw: 687.8, mp: 134°–135°.
THR = Details U. A food additive permitted in food for human consumption. [*109*]

OLEFINS.
THR = Unsaturated aliphatic hydrocarbons do not differ greatly from paraffins, particularly insofar as their toxic effect on working personnel is concerned. Ethylene and some of its homologs occur in manufactured and natural gases. Ethylene can be used as an anesthetic, and on inhal in sufficient quantity it can be an asphyxiant. However, the greatest hazard from its use is the danger of fire and explosion. Prolonged or repeated exposures to high conc of various olefins have caused certain toxic effects in animals, such as liver damage and hyperplasia of the bone marrow (due to butene-2), but no corresponding effects have been discoverable in human beings due to industrial exposures. The diolefins, butadiene and isoprene, are more irr than paraffins or mono-olefins of the same volatility. In general, it may be stated that the olefins are comparatively innocuous materials.

OLEIC ACID. Syn: *red oil* (*commercial grade*). Colorless, odorless liquid when pure. $C_{17}H_{33}COOH$, mw: 282.45, mp: 6°, bp: 360.0°, flash p: 372°F (CC), d: 0.895 @ 25°/25°, autoign. temp.: 685°F, vap. press: 1 mm @ 176.5°.
Acute tox data: Oral LD_{50} (rat) = 74000 mg/kg; iv LD_{50} (mouse) = 230 mg/kg; iv LD_{LO} (rat) = 50 mg/kg. [*3*]
THR = Practically NONE via oral route; HIGH via iv route. An exper neo [*3*] via sc route. A substance which migrates to food from packaging materials.
Fire Hazard: Slight, when exposed to heat or flame.
To Fight Fire: CO_2, dry chemical.

OLEO OIL. See tallow oil.

OLEOYL ETHYLENE IMINE. $C_{20}H_{37}ON$, mw: 307.6.
THR = An exper neo. [*3, 23*]

OLEUM. Syn: *fuming sulfuric acid*. Heavy, fuming yellow liquid. H_2SO_4 + up to 80% SO_3.
THR = HIGH irr to skin, eyes and mu mem, and via oral and inhal routes. See also sulfonic acid.
Fire Hazard: Dangerous, by chemical reaction with reducing agents and carbohydrates.
Explosion Hazard: Severe, by chemical reaction with acetic acid, acetic anhydride, acetonitrile, acrolein, acrylic acid, acrylonitrile, allyl alcohol, allyl chloride, 2-amino ethanol, NH_4OH, aniline, cresol, *n*-butyraldehyde, cumene, dichloroethyl ether, diethylene glycol monomethyl ether, diisobutylene, epichlorohydrin, ethyl acetate, ethylene cyanohydrin, ethylene diamine, ethylene glycol, ethylene glycol monoethyl ether acetate, ethylene imine, glyoxal, HCl, HF, isoprene, isopropyl alcohol, mesityl oxide, methyl ethyl ketone, HNO_3, 2-nitropropane, β-propiolacetone, propylene oxide, pyridine, NaOH, styrene monomer, vinylidene chloride, sulfolane, vinyl acetate. [*19*]
Disaster Hazard: Dangerous; when heated to decomp, emits highly toxic fumes of oxides of sulfur; will react with water or steam to produce heat and toxic and corrosive fumes; can react vigorously with reducing materials.

OLIVE OIL. Syn: *sweet oil*. Yellow oil. mp: −6°, flash p: 437°F (CC), d: 0.910, autoign. temp.: 650°F.
Fire Hazard: Slight, when exposed to heat or flame; can react with oxidizing materials.
Spont Heating: Some.
To Fight Fire: CO_2, dry chemical.

OLIVINE. See chrysolite.

OMAFLORA. See β-hydroxyethylhydrazine.

"OMILITE." Syn: *polyethylene polysulfide*.
THR = Details U; a fungicide. See also sulfides.
Fire Hazard: Details U. See also sulfides.
Disaster Hazard: Dangerous; when heated to decomp, emits highly toxic fumes of SO_x.

OMPA. See octamethylpyrophosphoramide.

ONB. See *o*-nitrobiphenyl.

ONION OIL. See allyl propyl disulfide.

ONYXIDE. See ethyl dimethyl-9-octadecenyl ammonium bromide.

OPIUM. Syn: *gum opium*. A habit forming drug.
THR = HIGH via oral route. A narcotic, sedative, analgesic, hypnotic. Source of morphine, codeine, papaverine, etc. An habituating drug. Can cause nausea, vomiting, constipation and respiratory problems.
Fire Hazard: Slight, when exposed to heat or flame.

ORANGE I.
THR = A (S) carc of sc tissue. [*14*]

ORANGE III. See methyl orange.

ORDEAL BEAN. See physostigma.

ORGANIC PERCHLORATES. $RClO_4$ where R = An organic molecule.
THR = Almost all decomp more or less violently with heat; ammoniacal and aminiacal seem more sensitive. [*19*]

ORGANIC SULFIDES. RSR where R is an organic molecule.
THR = They react violently with $Ca(OCl)_2$. [*19*]

ORGANOMETALS. Compounds containing carbon and a metal. Ordinarily metallic carbonates (calcium carbonate, etc.) are excluded and also metallic salts of common organic acids. Examples of organic metal compounds are Grignard compounds such as methyl magnesium iodide (CH_3MgI) and metallic alkyls such as butyllithium (C_4H_9Li). Also we have many organotin compounds such as monoalkyltins, monoaryltins, dialkyltins, diaryltins, trialkyltins, triaryltins, tetraalkyltins and tetraaryltins. Trialkyltins are most toxic as a group. Next are the dialkyltins and the monoalkyltins. In each major organotin group the ethyltin derivative is the most toxic, followed by the methyltins.

THR = This group of compounds is constantly growing in importance but there is relatively little toxicity information on most of them. Alkyl compounds of lead, tin, mercury and aluminum are known to be highly toxic. Less is known about other organometals, but for the most part they are highly reactive chemically and therefore dangerous, if only on direct contact. Until specific toxicity data become available, it is prudent to exercise great caution in handling organometals, particularly the alkyl forms.

Thus iv LD (rats) = 10–40 mg/kg for dialkyltin dichlorides; oral LD (rats) = > 160 mg/kg for dialkyltin dichlorides. [*68*]

THR = HIGH via iv and oral routes. They are irr to skin, eyes and mu mem. Can damage lung tissue and the liver. See also individual compounds.

ORPHENADRINE HYDROCHLORIDE. $C_{18}H_{23}ON \cdot HCl$, mw: 305.9.

THR = HIGH acute via oral route. An exper teratogen. [*3*]

ORINASE. See tolbutamide.

ORPIMENT. See arsenic trisulfide.

ORRIS. Syn: *white flag*. Composition: iridin, irone, ionone, resin, starch, volatile oil.

THR = An active allergen.

ORTHODICHLOROBENZENE. See *o*-dichlorobenzene.

ORTHOFORMIC ESTER. See triethyl-*o*-formate.

"ORTHOKLOR." See chlordane.

ORTHONITROANILINE. See *o*-nitroaniline.

ORTHOXENOL. See *o*-phenyl phenol.

ORTOL. $C_{10}H_{14}O_3$, mw: 182.2.

Acute tox data: Oral LD_{50} (rat) = 945 mg/kg; ip LD_{50} (rat) = 283 mg/kg; iv LD_{50} (rat) = 195 mg/kg; sc LD_{50} (mice) = 285 mg/kg. [*3*]

THR = HIGH via ip, iv and sc routes and MOD via oral route.

OSMIC ACID. See osmium tetraoxide.

OSMIC ACID ANHYDRIDE. See osmium tetraoxide.

OSMIUM AMMINE NITRATE. $Os(NH_3)_4O_2(NO_3)_2$, mw: 414.3.

THR = See osmium compounds and nitrates. May be impact-sensitive. [*19*]

OSMIUM AMMINE PERCHLORATE. $Os(NH_3)_4O_2(ClO_4)_2$, mw: 489.2.

THR = See also osmium compounds and perchlorates. May be impact-sensitive. [*19*]

OSMIUM (COMPOUNDS). A lustrous, bluish-white, extremely hard and dense brittle metal. Os, atwt: 190.2, mp: 3045°, bp: 5027°, d: 22.57.

Acute tox data: iv LD_{LO} (dog) = 17 mg/kg. [*3*]

THR = HIGH via iv route. An irr to eyes and mu mem. The principal effects of exposure are ocular disturbances and an asthmatic condition caused by inhal. Furthermore, it causes dermatitis and ulceration of the skin upon contact. When osmium is heated, it gives off a pungent, poisonous fume of osmium tetraoxide. One case of osmium poisoning reported in the literature resulted from the inhal of osmium tetraoxide, which gave rise to a capillary bronchitis and dermatitis. The vapor has a pronounced and nauseating odor which should be taken as a warning of a toxic conc in the atmosphere, and personnel should immediately remove to an area of fresh air. Osmium compounds, other than the tetraoxide, are probably safe, particularly as ordinarily handled in industry. The metal itself is not highly toxic.

Radiation Hazard: For permissible levels, see Section 5, Table 5A.5. Artificial isotope ^{185}Os, $T\frac{1}{2}$ = 94d, decays to stable ^{185}Re by ec capture. Emits γ's of 0.65, 0.88 MeV and x-rays. Artificial isotope ^{191}Os, $T\frac{1}{2}$ = 15d, decays to stable ^{191}Ir via β's of 0.14 MeV. Emits γ's of 0.04–0.13 MeV.

Fire Hazard: Mod, in the form of dust, when exposed to heat or flame.

Explosion Hazard: Slight, in the form of dust when exposed to heat or flame. Os dust reacts violently with ClF_3, OF_2. [*19*]

OSMIUM DICHLORIDE. Dark brown, deliquescent crystals. $OsCl_2$, mw: 261.1, mp: decomp.

THR = See osmium and chlorides.

OSMIUM DISULFIDE. Cubic black crystals. OsS_2, mw: 254.3, bp: decomp.

THR = See osmium and sulfides.

OSMIUM HEXAFLUORIDE. Green crystals. OsF_6, mw: 304.20, mp: > 50°, bp: 205°.

THR = See fluorides and osmium.

OSMIUM OCTAFLUORIDE. Yellow crystals. OsF_8, mw: 342.20, mp: 34.4°, bp: 47.3°, d: 3.87.

THR = See fluorides and osmium. HIGH toxicity.

OSMIUM SULFITE. Blue-black crystals. $OsSO_3$, mw: 270.3, mp: decomp.

THR = See osmium and sulfites.

OSMIUM TETRACHLORIDE. Red-brown crystals. $OsCl_4$, mw: 332.03, mp: sublimes.
THR = See hydrochloric acid and osmium.

OSMIUM TETRAFLUORIDE. Brown powder. OsF_4, mw: 266.2.
THR = See fluorides and osmium.

OSMIUM TETRAOXIDE. Syn: *osmic acid.* (A) monoclinic, colorless crystals; (B) yellow mass, pungent, chlorine-like odor. OsO_4, mw: 254.20, mp (A): 39.5°, mp (B): 41°, bp: 130° (sublimes), d: 4.906 @ 22°, vap. press (A): 10 mm @ 26.0°, vap. press (B): 10 mm @ 31.3°.
Acute tox data: inhal TC_{LO} (man) = 0.133 mg/m^3 ⟶ eye effects; inhal TC_{LO} (man) = 0.1 mg/m^3 ⟶ irr effects; oral LD_{50} (rat) = 14 mg/kg; oral LD_{50} (mouse) = 162 mg/kg; ip LD_{50} (mouse) = 14 mg/kg. [*3*]
THR = HIGH irr via inhal, oral and ip routes and to skin, eyes and mu mem.
Disaster Hazard: Dangerous; when heated to decomp, emits toxic fumes.

OSMIUM TETRASULFIDE. Brown or black crystals. OsS_4, mw: 318.4, mp: decomp.
THR = See sulfides and osmium.

OSMIUM TRICHLORIDE. Cubic brown crystals. $OsCl_3$, mw: 296.57, mp: decomp @ 560°–600°.
THR = See osmium and chlorides.

OSMOFERRIN. A ferric sodium gluconate composition.
THR = An exper neo [*3*] via sc route.

OVERHANG TYPE OF HOOD. See Section 2.

OVEX. See murvesco.

OVOTRAN. See murvesco.

OXALDEHYDE. See glyoxal.

OXALATES. Salts of oxalic acid.
THR = HIGH via oral and inhal routes. Powerful irr. See also oxalic acid. Oxalates are corrosive to tissue and produce local irr. When taken by mouth they have a caustic effect on the mouth, esophagus and stomach. The soluble oxalates are readily absorbed from the gastro-intestinal tract and can cause severe damage to the kidneys.
Disaster Hazard: Dangerous; when heated to decomp, they emit toxic fumes.

OXALIC ACID. Syn: *ethanedioic acid.* $COOHCOOH \cdot 2H_2O$, mw: 126.1, mp: 101°, 189° (anhydrous), bp: sublimes @ 150°, d: 1.653.
Acute tox data: Oral LD_{LO} (human) = 100 mg/kg; oral LD_{LO} (dog) = 1000 mg/kg.
THR = HIGH irr to humans via oral route but MOD to dogs via oral route. Acute oxalic poisoning results from ingestion of a solution of the acid. There is marked corrosion of the mouth, esophagus and stomach, with symptoms of vomiting, burning and abdominal pain, collapse and sometimes convulsions. Death may follow quickly. The systemic effects are attributed to the removal by the oxalic acid of the calcium in the blood. The renal tubules become obstructed by the insoluble calcium oxalate, and there is profound kidney disturbance. The inhal of the dust or vapor may cause chronic symptoms of irr of the upper respiratory tract, gastro-intestinal disturbances, albuminuria, gradual loss of weight, increasing weakness and nervous system complaints. Oxalic acid has a caustic action on the skin and may cause dermatitis; a case of early gangrene of the fingers resembling that caused by phenol has been described.

The chief effects of inhal of the dusts or vapor are irr of the eyes and upper respiratory tract, ulceration of the mu mem of the nose and throat, epistaxis, headache, irr and nervousness. More severe cases may show albuminuria, chronic cough, vomiting, pain in the back and gradual emaciation and weakness. The skin lesions are characterized by cracking and fissuring of the skin and the development of slow-healing ulcers. The skin may be bluish in color, and the nails brittle and yellow. Violent reaction with furfuryl alcohol, Ag, $NaClO_3$, NaOCl. [*19*]

OXALIC ETHER. See ethyl oxalate.

OXALYL-*o*-AMINOAZO TOLUENE. $C_{16}H_8O_3N_3$, mw: 290.3.
THR = An exper neo. [*3*]

OXALYL BROMIDE. Syn: *ethanedioyl bromide.* A liquid. $(COBr)_2$, mw: 215.85.
THR = HIGH See oxalic acid. Reacts violently with NaK alloy. [*19*]
Disaster Hazard: Dangerous; when heated to decomp, emits highly toxic fumes of bromides; will react with water or steam to produce toxic and corrosive fumes.

OXALYL CHLORIDE. Syn: *ethanedioyl chloride.* Colorless, fuming liquid. COClCOCl, mw: 126.93, mp: −12°, bp: 64°, d: 1.488 @ 13°.
THR = HIGH toxicity. See oxalic acid. Violent reaction with K-Na alloy. [*19*]
Disaster Hazard: Dangerous; when heated to decomp, emits highly toxic fumes of chlorides; will react with water or steam to produce toxic and corrosive fumes.

OXAMMONIUM. See hydroxylamine.

OXAZOLIDINONE. $C_3H_5O_2N$, mw: 87.1.
THR = An exper neo [*3*] via ip route.

1,2-OXATHIALONE-2,2-DIOXIDE. See 1,3-propane sultone.

OX BILE EXTRACT.
THR = U. Used as an emulsifying agent food additive. [*109*]

OXETANE. Syn: *trimethylene oxide*. C_3H_6O, mw: 58.1.
THR = An exper neo [*3, 23*] via sc route.

2-OXETANONE. $C_3H_4O_2$, mw: 72.1.
THR = An exper (+) carc [*3, 10*] via oral, sc, it, dermal and iv routes.

OXIDIZING MATERIALS, N.O.S.
THR = Reacts violently with carbides. [*19*]

OXINDOLE. See 2(3)-indolone.

OXINE. See 8-quinolinol.

OXIRANE. See ethylene oxide.

7-OXIRANYL BENZ(a)ANTHRACENE. $C_{20}H_{14}O$, mw: 270.3.
THR = An exper neo to rats via sc route. [*103*]

6-OXIRANYL BENZO(a)PYRENE. $C_{22}H_{14}O$, mw: 294.4.
THR = An exper neo to rats via sc route. [*103*]

OXOATE. See amidooxyl benzoate.

8-OXO-8,13-DIHYDROBENZO(g)BENZO PYRANO-(4,3-b)INDOLE. $C_{19}H_{11}O_2N$, mw: 285.
THR = An exper neo via sc route. [*3, 23*]

5-OXO-5,13-DIHYDROBENZO(e)(2)BENZOPYRANO(4,3-b)INDOLE.
THR = An exper neo via sc route. [*3, 23*]

5-OXO-5H-BENZO(g)ISOCHROMENO(4,3-6)-INDOLE.
THR = An exper neo via sc route. [*23*]

OXODISILANE. H_4Si_2O, mw: 76.2.
THR = Spont flam in air. [*19*]

N-(9-OXO-2-FLUORENYL)ACETAMIDE.
$C_{15}H_{11}O_2N$, mw: 237.3.
THR = LOW acute via oral route to rats. An exper carc. [*3, 23*]

4-OXO-HEPTANEDIOIC ACID ISONICOTINOYL HYDRAZONE. $C_{13}H_{15}O_5N_3$, mw: 293.3.
THR = An exper carc. [*3*]

2-OXOHEXAMETHYLENIMINE.
See α-caprolactane.

OXOLE. See furan.

1-OXO-3-METHYLCHOLANTHRENE.
THR = An exper carc. [*23*]

2-OXOPROPYL METHANE SULFONATE.
THR = An exper carc. [*23*]

N-(2-OXOPROPYL)-N-NITROSOPROPYLAMINE.
$C_6H_{12}O_2N_2$, mw: 144.2.
THR = An exper carc. [*3*]

OXOSILANE. H_2SiO, mw: 46.1.
THR = Spont flam in air or Cl_2. [*19*]

OXYBUTYRIC ALDEHYDE. See aldol.

OXYDEMETHON-METHYL. See meta-systox R.

4,4′-OXYDIANILINE. $C_{12}H_{12}ON_2$, mw: 200.2, bp: > 300°, mp: 187°, colorless cryst.
THR = HIGH acute toxicity via oral route to rats. An exper neo. [*3, 108*] in rats and mice via oral and sc routes.

5-OXYDIBENZOTHIOPHENYL-2-ACETAMIDE.
THR = An exper carc. [*23*]

β,β'-OXYDIPROPIONITRILE. Colorless liquid. $O(CH_2CH_2CN)_2$, mw: 124.1, mp: −26.3°, bp: 172° @ 10 mm, d: 1.041 @ 30°.
Acute tox data: Oral LD_{50} (rat) = 2830 mg/kg. [*3*]
THR = MOD via oral routc. See also nitriles.

"OXYFUME" FUMIGANT. See ethylene oxide and carbon dioxide or ethylene oxide and dichloro-difluoromethane.

OXYGEN. Colorless, odorless, tasteless gas or liquid or hexagonal crystals. O_2, mw: 32.00, mp: −218°, bp: −183.0°, d (liquid): 1.14 @ −183.0°, d (solid): 1.426 @ −252.5°, vap. d: 1.429 @ 0°.
THR = NONE, as gas. In liquid form it can cause severe "burns" and tissue damage on contact with the skin, due to extreme cold.
Fire Hazard: Though itself nonflam, it is essential to combustion. Exclusion of O_2 from the neighborhood of a fire is one of the principal methods of extinguishment.
Explosion Hazard: Liquid O_2 can explode on contact with readily oxidizable material, especially at high temp., under the proper conditions of temp., pressure and reagent conc it can react violently with Al, $Al(BH_4)_3$, AlH_3, $Be(BH_4)_2$, BAs_2Br_3, B_2H_{10}, BCl_3, CsH, [butane + $Ni(CO)_4$], Ca, Cs, Ca_3P_2, (chlorotrifluoroethylene + Br_2), $C_{10}H_{14}$, B_2H_6, Ge, ethyl ether, diphenyl ethylene, disilane, ethers, hydrazine, H_2, Li, ($Ni(CO)_4$ + butane), organic matter, (OF_2 + H_2O), PH_3, P, PF_3, P_2O_3, polyurethane, polyvinyl chloride, (O_2 + CO), K_2O_2, Rb, Se, NaH, teflon, tetrafluorohydrazine, tetrasilane, Ti alloy, 1,1,1-trichloroethane, trichloroethylene, trisilane, asphalt, benzene, CO, CCl_4, chlorinated hydrocarbons, cyanogen, fuels, hydrocarbons, LiH, Mg, CH_4, CH_2Cl_2, oil, paraformaldehyde, wood, charcoal. [*19*]
Disaster Hazard: Compressed O_2 is shipped in steel cylinders under high pressure. If these containers

are broken due to shock or exposed to high temp., an explosion and fire may result.

OXYGEN DIFLUORIDE. See fluorine monoxide.

OXYGEN FLUORIDE. See fluorine monoxide.

OXYGEN, LIQUID. See explosives, high.

OXYISOBUTYRIC NITRILE. See acetone cyanhydrin.

6-OXYMETHYL-11-METHYL-16,17-DIHYDRO-15H-CYCLOPENTA(a)PHENANTHRENE.
THR = An exper carc. [*23*]

β-OXYMETHYL-β-PYRIDIL PROPIONIC ACID. See ecgonine.

β-OXYNAPHTHOIC ACID. See 3-hydroxy-2-naphthoic acid.

***p*-OXYPHENYL-β-NAPHTHYLAMINE.**
THR = Details U. Animal exper suggest LOW acute toxicity.

8-OXYQUINOLINE. See 8-quinolinol.

8-OXYQUINOLINE BENZOATE. See 8-quinolinol benzoate.

OXYQUINOLINE POTASSIUM SULFATE. See quinosol.

OXYSTEARIN. A mixture of the glycerides of partially oxidized stearic and other fatty acids.
THR = U. A food additive permitted in food for human consumption. [*109*]

OXYTETRACYCLINE. An antibiotic. Dull, yellow, odorless crystalline powder, sol in acids and alkalies, very slightly sol in acetone, alcohol, chloroform and water, practically insol in ether. $C_{22}H_{24}N_2O_9 \cdot 2H_2O$, mw: 496, mp: 179°–182° (decomp).
Acute tox data: ip LD_{50} (rat) = 660 mg/kg. [*3*]
THR = MOD via ip route. Used as a food additive permitted in the feed and drinking water of animals and/or for the treatment of food-producing animals; also permitted in food for human consumption. [*109*] Side effects include nausea, stomatitis, diarrhea.

OXYTETRACYCLINE HYDROCHLORIDE + SODIUM NITRATE.
THR = An exper neo. [*3*]

OZOKERITE. See ceresin wax.

OZONE. Colorless gas or dark blue liquid. O_3, mw: 48.00, mp: −193°, bp: −111.9°, d (gas): 2.144 g/1, 1.71 @ −183°.
Acute tox data: inhal TC_{LO} (humans) = 100 ppm for 1 min ⟶ skin effects; inhal TC_{LO} (man) = 1.86 ppm for 75 min ⟶ CNS effects; inhal LC_{50} (rat) = 4.8 ppm for 4 hrs; inhal LC_{50} (hamster) = 10.5 ppm for 4 hrs; inhal TC_{LO} (human) = 1 ppm ⟶ pulmonary effects. [*3*]
THR = HIGH irr via inhal and to skin, eyes and mu mem. Can be a safe water disinfectant in low conc. An exper neo via inhal. [*3*] Ozone is a strong irr to the eyes and upper respiratory system. Conc of 0.015 ppm of ozone in air produce a barely detectable odor. Conc of 1 ppm produce a disagreeable sulfur-like odor and may cause headache and irr of the upper respiratory tract; symptoms disappear after leaving the exposure. Exposure of guinea pigs to high conc may cause death from lung congestion and edema. No systemic effects have been reported following industrial exposures. Ozone is a common air contaminant.
Fire Hazard: Dangerous, by chemical reaction with aniline, C_6H_6, Br_2, (diallyl methyl carbinol + acetic acid), diethyl ether, N_2O_5, ethylene, HBr, HI, NO_2, NO, NCl_3, NI_3, nitroglycerin, organic liquids, organic matter, Sb. [*19*]
Explosion Hazard: Severe, in liquid form, when shocked, exposed to heat or flame or in conc form by chemical reaction with powerful reducing agents.

P-4000. See 5-nitro-2, N-propoxyaniline.

P.A.B.A. See *p*-aminobenzoic acid.

PACKAGING MATERIALS COMPONENTS WHICH MIGRATE INTO FOOD AND BECOME ADDITIVES. [*109*]

PADAN. Syns: *cartap; s,s'-[2-(dimethylamino)trimethylene]bis(thiocarbamate)*. $C_7H_{15}O_2N_2S_2Cl$, mw: 258.7.

Acute tox data: Oral LD_{50} (rat) = 250 mg/kg; oral LD_{50} (mouse) = 165 mg/kg. [*3*]

THR = HIGH via oral route. An insecticide. See also carbamates.

PAINT. A fluid mixture of a pigment and a vehicle. flash p: 0°–80°F.

THR = For lead-base paints, see lead; usually the toxicity is that of the solvent used. Common air contaminant.

Fire Hazard: Dangerous, when exposed to heat or flame.

Spont Heating: Yes.

Explosion Hazard: U.

Disaster Hazard: Dangerous; when heated, the solvent emits acrid fumes; can react vigorously with oxidizing materials.

To Fight Fire: Foam, CO, dry chemical.

PAINT, ENAMEL, LACQUER, STAIN, SHELLAC, VARNISH, ALUMINUM, BRONZE, GOLD, WOOD FILLER LIQUID, LACQUER BASE LIQUID. See under components.

PAINT AND GREASE ERADICATORS. flash p: 0°–80°F.

THR = Variable.

Fire Hazard: Dangerous, when exposed to heat or flame.

Explosion Hazard: U.

Disaster Hazard: Dangerous, upon exposure to heat or flame; can react vigorously with oxidizing materials.

To Fight Fire: Foam, CO_2, dry chemical.

PAINT, LACQUER AND VARNISH REMOVING, REDUCING OR THINNING COMPOUNDS. See compounds: lacquer, paint or varnish.

PAINT SCRAPINGS.

THR = Variable.

Fire Hazard: Mod, when exposed to heat or flame; avoid large unventilated piles. Danger of spont heating depends upon state of dryness.

Spont Heating: Mod.

PALLADIUM AMMINE NITRATES.

$Pd(NH_3)_2(NO_3)_2$, mw: 264.5. $Pd(NH_3)_4(NO_3)_2$, mw: 298.5.

THR = May be impact-sensitive and are heat-detonated. [*19*] See also nitrates and palladium.

PALLADIUM AMMINE PERCHLORATES.

THR = May be impact-sensitive. [*19*] See also perchlorates and palladium.

PALLADIUM. A steely white, stable metal, can be annealed to be soft and ductile. Pd, atwt: 106.4, mp: 1552°, bp: 3140°, d: 12.02 @ 20°.

THR = This metal in the form of palladium chloride has been administered orally in dosage of about 1 grain daily in the treatment of tuberculosis without apparent ill effects. Applied locally to the skin, palladium chloride shows little or no irr. In exper animals, palladium chloride has been given via iv producing damage to bone marrow, liver and kidneys when the dosage was of the order of 0.5–1.0 mg/kg. In the laboratory, Pd appears to bind to many cell components; blocks the action of a number of enzymes and interferes with the use of energy by nerves and muscles; induces lung malfunction and produces abnormal fetuses. [*70*]

Radiation Hazard: For permissible levels, see Section 5A, Table 5A.5. Artificial isotope ^{103}Pd, $T\frac{1}{2}$ = 17d, decays to stable ^{103}Rh by ec. Emits low energy γ's and x-rays.

Fire Hazard: Slight, in the form of dust, when exposed to heat or flame. Violent reaction with Al, (H_2 + isopropyl alcohol), OF_2, S. [*19*]

PALLADIUM BROMIDE. Red-brown crystals. $PdBr_2$, mw: 266.23, mp: decomp, d: 5.173.

THR = See palladium compounds and bromides.

PALLADIUM CHLORIDE. Syns: *palladous chloride, palladium dichloride*. Dark brown deliquescent crystals, sol in water, alcohol, acetone and hydrochloric acid. $PdCl_2 \cdot 2H_2O$ or $PdCl_2$ (anhydrous salt), mw: 213.35, mw (anhydrous salt): 177.35, d: 4.0 @ 18°, mp: 501° (decomp).

Acute tox data: ip LD_{50} (mice) = 104 mg/kg; iv LD_{LO} (rabbit) = 19 mg/kg. [*3*]

THR = HIGH via ip and iv routes. An exper carc via oral route. [*3*]

Disaster Hazard: Dangerous; when heated to decomp, emits highly toxic fumes.

PALLADIUM CYANIDE. Yellowish-white crystals. $Pd(CN)_2$, mw: 158.44, mp: decomp.
THR = See cyanides.

PALLADIUM DIFLUORIDE. Brown crystals. PdF_2, mw: 144.40, mp: volatil, d: 5.80.
THR = See fluorides and palladium.

PALLADIUM HYDRIDE. Silver, metallic crystals. $Pd\ H_2$, mw: 213.81, mp: decomp, d: 10.76.
THR = See hydrides and palladium.

PALLADIUM IODIDE. Black powder. PdI_2, mw: 360.21, mp: decomp @ 350°, d: 6.0.
THR = See palladium compounds and iodides.

PALLADIUM MONOSULFIDE. Brown-black crystals. PdS, mw: 138.5, mp: decomp @ 950°.
THR = See sulfides and palladium.

PALLADIUM MONOXIDE. Syns: *palladous oxide, palladium oxide.* Black powder, insol in water and acids, slightly sol in aqua regia, sol in 48% hydrobromic acid. PdO, mw: 122.40, d: 8.3, mp: 750° (decomp).
THR = See palladium.

PALLADIUM NITRATE. Rhombic, brown-yellow deliquescent crystals. $Pd(NO_3)_2$, mw: 230.41, mp: decomp.
THR = See nitrates and palladium.

PALLADIUM SELENIDE. Dark gray crystals. PdSe, mw: 185.36, mp: $<$960°.
THR = See selenium compounds.

PALLADIUM SUBSULFIDE. Green-gray crystals. Pd_2S, mw: 244.86, mp: decomp @ 800°, d: 7.303 @ 15°.
THR = See sulfides and palladium.

PALM BUTTER. See palm oil.

PALM KERNEL OIL. Syn: *palm nut oil.* mp: 25.5°–30°, flash p: 398°F (CC), d: 0.95.
THR = LOW irr and allergen.
Fire Hazard: Slight; when exposed to heat or flame; can react with oxidizing materials.
Spont Heating: Yes.
To Fight Fire: CO_2, dry chemical.

PALM NUT OIL. See palm kernel oil.

PALM OIL. Syn: *palm butter.* Reddish-yellow to dirty-red, fatty mass, faint odor of violet. Composition: palmitin, stearin, linolein. mp: 26.6°–43.3°, flash p: 323°F, d: 0.92, autoign. temp.: 600°F.
THR = LOW irr and allergen.
Fire Hazard: Slight, when exposed to heat or flame; can react with oxidizing materials.
Spont Heating: Low.
To Fight Fire: CO_2, dry chemical.

2-PAM. Syn: *2-pyridine aldoxime methiodide.* Crystals, water-sol. $C_7H_9IN_2O$, mw: 264.1, mp: 214°.
Acute tox data: Oral LD_{50} (mouse) = 1500 mg/kg; ip LD_{50} (rat) = 305 mg/kg; sc LD_{50} (mouse) = 140 mg/kg; iv LD_{50} (mouse) = 145 mg/kg. [*3*]
THR = MOD via oral route; HIGH via ip, iv and sc routes. Has been proven highly effective as an antidote vs cholinesterase inhibitors of the parathion group.
Disaster Hazard: Dangerous; when heated to decomp, emits highly toxic fumes.

***p*-PANTOTHENAMIDE.**
THR = U. Used as a food additive permitted in food for human consumption. [*109*]

PANTHENOL. See *d*-pantothenyl alcohol.

PANTOTHENOL. See *d*-pantothenyl alcohol.

***d*-PANTOTHENYL ALCOHOL.** Syns: *panthenol, pantothenol, 2,4-dihydroxy-n-(3-hydroxypropyl)-3,3-dimethyl butyramide.* A viscous liquid, sol in water, methanol and ethanol. $C_9H_{19}O_4N$, mw: 205.2.
THR = U. Used as a nutrient and/or dietary supplement food additive. [*109*]

PAPAIN. Syn: *papayotin.* White to gray, slightly hygroscopic powder, sol in water and glycerin, insol in other common organic solvents; the most thermostatic enzyme known; digests protein.
THR = U. Used as a general-purpose food additive. [*109*] Possible allergen.

PAPAVERINE. Colorless, rhombic needles. $C_{20}H_{21}NO_4$, mw: 339.38, mp: 147°, bp: decomp, d: 1.337.
Acute tox data: Oral LD_{50} (mice) = 350 mg/kg; ip LD_{50} (rat) = 64 mg/kg; iv LD_{50} (mice) = 32 mg/kg. [*3*]
THR = HIGH via oral, ip and iv routes. This material is considered to be a comparatively mild poison. Its CNS action is about midway between morphine and codeine and even large doses do not produce the amount of excitement caused by codeine or the soporific action of morphine. Small doses are followed by sleep and slow respiration. The heart beat is slowed, but the blood pressure is scarcely affected. It is thought that doses up to 15 grains are nonfatal. See also morphine.
Fire Hazard: Slight.

PAPAYOTIN. See papain.

PAPER STOCK, WET. See waste paper, wet.

PAPP. See *p*-aminopropiophenone.

PARABEN. See *p*-hydroxybenzoic acid esters.

PARACELSUS AND INDUSTRIAL HYGIENE. See Section 1.

PARADICHLOROBENZENE. See *p*-dichlorobenzene.

PARADUST. See parathion.

PARAFLOW. See parathion.

PARAFFIN OIL. See oil (mineral).

PARAFFINS.

THR = The effects of the paraffin hydrocarbons vary with the volatility. The gaseous hydrocarbons, such as methane, ethane, etc., have but slight anesthetic effects and are hazardous only when present in sufficient conc to dilute the oxygen to a point below that which is necessary to sustain life. With the volatile liquid hydrocarbons, or with the next higher fraction, the anesthetic action predominates, and with the higher molecular weights or with the less volatile compounds, the anesthetic increases, but at the same time an irr action becomes more pronounced. For information concerning toxic and hazardous properties of these materials, see the individual compounds. The semi-refined, fully-refined and the crude paraffins are exper carc. [*14*, *3*] They are also exper neo to mice via imp route. [*103*] See also paraffin waxes and petroleum waxes. Paraffins are common air contaminants.

PARAFFIN WAX (FUME). Colorless or white, somewhat transluscent, odorless mass, greasy feel. mp: 43.3°–60°, bp: > 370°, flash p: 390°F (CC), d: 0.9, autoign. temp.: 473°.

THR = See also paraffins. These materials are exper carc of the lung and stomach [*14*] and skin. [*3*]

Fire Hazard: Slight, when exposed to heat or flame; can react with oxidizing materials.

Spont Heating: No.

To Fight Fire: CO_2, dry chemical.

PARAFORM. See paraformaldehyde.

PARAFORMALDEHYDE. Syn: *paraform.* $(CH_2O)_n$, white crystals, odor of formaldehyde, slightly sol in cold water, mod sol in hot water ⟶ formaldehyde. flash p: 158°F, autoign. temp.: 572°F.

Acute tox data: Oral LD_{50} (rat) = 800 mg/kg; dermal LD_{LO} (rabbit) = 10,000 mg/kg. [*3*]

THR = MOD via oral route; LOW via dermal route.

Fire Hazard: Mod, when exposed to heat or flame. Reacts violently with O_2 liquid. [*19*]

Spont Heating: No.

Disaster Hazard: Mod dangerous; when heated, it forms formaldehyde gas and oxides of carbon; can react with oxidizing materials.

To Fight Fire: Alcohol foam, CO_2, dry chemical.

PARALDEHYDE. Syns: *p-acetaldehyde; 2,4,6-trimethyl-1,3,5-trioxane*. Colorless liquid. $C_6H_{12}O_3$, mw: 132.16, mp: 12.6°, lel = 1.3%, bp: 124.4° @ 752 mm, flash p: 96°F (OC), d: 0.9943 @ 20°/4°, autoign. temp.: 460°F, vap. d: 4.55.

Acute tox data: Dermal LD_{50} (rabbit) = 1400 mg/kg; oral LD_{50} (rat) = 3304 mg/kg; oral LD_{50} (rat) = 1530 mg/kg; Oral TD_{LO} (human) = 14 mg/kg ⟶ psychological effects; iv TD_{LO} (human) = 14 mg/kg ⟶ psychological effects; rec TD_{LO} (human) = 14 mg/kg. [*3*]

THR = MOD via oral and LOW via dermal routes. Psychotropic effects on humans. An exper neo via dermal route. [*3*] Paraldehyde has hypnotic and analgesic properties. However, poisoning due to it is very rare. There is a wide range between the hypnotic dose and the toxic dose; 2–5 grains produces a soporific effect; recovery has been observed following the ingestion of 50 g. Continued use of the drug has been known to result in habituation. There have been no cases reported of industrial poisoning.

Signs and symptoms are incoordination and drowsiness, followed by sleep. With larger doses, the pupils dilate, reflexes are lost, and patient lapses into coma. Weak pulse and shallow respiration are followed by cyanosis. Death results from respiratory paralysis. The symptoms of chronic intoxication from this material are disturbances of the digestion, continual thirst, general emaciation, muscular weakness and mental fatigue, followed by tremors of the hands and tongue. It can cause skin eruptions similar to those caused by chloral hydrate.

Fire Hazard: Dangerous, when exposed to heat, flame or oxidizers.

Explosion Hazard: Slight, when exposed to heat or flame.

Disaster Hazard: Dangerous; keep away from heat and open flame; emits toxic fumes on heating; can react vigorously with oxidizing materials.

To Fight Fire: Alcohol foam, CO_2, dry chemical.

PARAMAGENTA. $C_{19}H_{16}N_3 \cdot HCl$, mw: 322.8.

THR = An exper (S) carc [*10*] and neo [*3*] via sc route.

PARAMENTHANE HYDROPEROXIDE. See *p*-menthane hydroperoxide.

PARAMINOL. See *p*-aminobenzoic acid.

PARAMORPHINE. See thebaine.

PARANITRANILINE. See *p*-nitroaniline.

PARA-OXON. See diethyl-*p*-nitrophenyl phosphate.

PARAPERIODIC ACID. See periodic acid.

PARAQUAT. Syns: *1,1'-dimethyl-r,r'-dipyridium dichloride, 1,1'-dimethyl-4,4'-bipyridium salt, methyl viologen*. Yellow solid, sol in water. $[CH_3(C_5H_4N)_2CH_3] \cdot 2CH_3SO_4$, mw: 408.

Acute tox data: Oral LD_{50} (rat) = 57 mg/kg; dermal

LD_{50} (rat) = 80 mg/kg; ip LD_{50} (mice) = 30 mg/kg; oral LD_{50} (monkey) = 50 mg/kg. [*3*]

THR = HIGH via oral, ip and dermal routes. It is a highly toxic bipyridyl herbicide which is absorbed by the skin, by inhal or ingestion. Has a delayed damaging effect on the lung alveoli. Has caused fatal poisoning in humans with severe injury to lungs. Has been implicated in aplastic anemia. An exper teratogen. [*3*]

Disaster Hazard: Dangerous; see chlorides.

PARASCORBIC ACID. Syn: *parasorbic acid*. $C_6H_8O_2$, mw: 112.1.

THR = An exper (S) carc, neo, via oral and sc routes. [*3, 7, 23*]

PARASEPT. See *p*-hydroxybenzoic acid esters.

PARASPRAY. See parathion.

PARATHION. Syns: *O,O-diethyl-O, p-nitrophenyl thiophosphate, "Alkron," compound 3422, DNTP, DPP, E-605, genithion, niran, paradust, paraflow, paraspray, parawet, penphos, phos-kil, thiophos, vapophor, many others*. Yellowish liquid. $C_{10}H_{14}NO_5PS$, mw: 291.3, mp: 375°, d: 1.26 @25°/4°.

Acute tox data: Oral LD_{50} (wild birds) = 2 mg/kg; oral LD_{50} (duck) = 2.34 mg/kg; oral LD_{50} (quail) = 6 mg/kg; dermal LD_{50} (guinea pig) = 600 mg/kg; dermal LD_{50} (rabbit) = 40 mg/kg; ip LD_{50} (rat) = 1.5 mg/kg; dermal LD_{50} (rat) = 7 mg/kg; oral LD_{LO} (human) = 0.24 mg/kg; inhal LC_{LO} (rat) = 10 mg/m^3 for 2 hrs. [*3*] Oral LD_{50} (rat) = 2 mg/kg.

THR = HIGH via oral, dermal, ip and inhal routes. An exper teratogen via ip route and an exper carc. [*23*] Parathion, like the other organic phosphorus poisons, acts as an irreversible inhibitor of the molecules of the enzyme cholinesterase and thus allows the accumulation of large amounts of acetylcholine. When a critical level of cholinesterase depletion is reached, grave symptoms appear. Whether death is actually caused entirely by cholinesterase depletion or by the disturbance of a number of enzymes is not yet known. Recovery is apparently complete if a poisoned animal or man has time to reform his critical quota of cholinesterase. However, if a second small dose is administered before recovery from the first is complete, the effect is partially additive.

Dangerous chronic dose in man: Exposure to parathion reduces the cholinesterase level and the organism exposed remains susceptible to relatively low dosages of parathion until the cholinesterase level has regenerated. Small doses at frequent intervals are, therefore, more or less additive. There is not, however, at the present time, any indication that, when recovery from a given exposure is entirely complete, the exposed organism is prejudiced in any way. Laboratory findings are essentially normal except that by special techniques the cholinesterase level of the blood or serum may be shown to be greatly reduced. At autopsy, the same may be demonstrated for the cholinesterase level of the brain or other tissues provided fresh, unfixed tissue is employed.

The mean cholinesterase values of normal persons living without exposure to organic phosphorus insecticides have been found by various workers, using the Michel method, to range as follows:

Red blood cell	0.67–0.86	pH units/hr
Plasma	0.70–0.97	pH units/hr

It is believed that cholinesterase values of 0.5 or less for either cells or plasma represent abnormal depressions for most individuals. Nevertheless, people may experience far greater depressions (to 0.2 or less) without the onset of clinical signs or symptoms; this is especially true of workers who are exposed daily over a period of weeks but whose exposure at any one time is kept at a minimum.

Treatment for Poisoning: Keep the patient fully atropinized. Give 2–4 mg and repeat until signs of atropinization appear. The intravenous route is most rapid. The dosage of atropine is greater than that conventionally employed for other purposes but is within safe limits. Atropine relieves many of the distressing symptoms, reduces heart block, and dries secretions of the respiratory tract. Never give morphine, theophylline, or theophylline-ethylene-diamine (*Aminophylline*). 2-PAM, alone or with atropine, is also an effective antidote.

If the patient has not yet shown symptoms or they have been allayed by the first dose of atropine, he must be completely and quickly decontaminated. Wearing rubber gloves, remove the patient's clothing and, with due regard for his condition at the moment, bathe him thoroughly with soap and water. If washing soda is available use it, for parathion is hydrolyzed more rapidly in the presence of alkali. Any relatively mild alkali may be used.

If there is any suspicion that parathion has been ingested, induce vomiting, give some neutral material such as milk or water and induce vomiting again. Nausea may, of course, be anticipated on the basis of the systemic action of parathion but if vomiting is not profuse, gastric lavage may be used. If the pulmonary secretions have accumulated before atropine has become effective, the patient must be turned upside down or in some other positions of postural drainage in order to drain out mucus. Use suction and a catheter if necessary. If the stomach is distended, empty it with a Levine tube. Atropine does not protect against muscular weakness. The mechanism of death appears to be respiratory failure. The use of an oxygen tent or even the use of oxygen under slight positive pressure is advisable and should be started early. Watch the patient *constantly, for the need of artificial respiration may appear suddenly*. Equipment for artificial respiration should be placed by the patient's bed in readiness as soon as he is hospitalized. Cyanosis (anoxia) should be prevented by the most suitable means, since it aggravates the other signs of poisoning. Complete recovery may occur even after many hours of artificial respiration have been necessary. The acute emergency lasts 24–48 hrs, and the patient must be *watched continuously* during that time. Favorable response to one or more doses of atropine does not guarantee against sudden and fatal relapse. Medication must be continued during the entire emergency. Following exposure heavy enough to produce symptoms, further organic phosphorus insecticide exposure of any sort should be avoided. The patient remains susceptible to relatively small exposures of parathion until regeneration of cholinesterase is complete or nearly so. Persons exposed to other organic phosphorus insecticides before complete recovery from a previous exposure are made more susceptible and vice-versa.

Decomp by heat. Violent reaction with endrin. [*19*]

Disaster Hazard: Highly dangerous; shock can shatter the container, releasing the contents. When heated to decomp, emits highly toxic fumes of NO_x, PO_x, SO_x.

PARATHION AND COMPRESSED GAS MIXTURES. See parathion.

PARATHION MIXTURE. See parathion.

PARATHION MIXTURE, LIQUID. See parathion.

PARAWET. See parathion.

PARAZOL. See dinitrodichlorobenzene.

PARIS GREEN. See copper acetate-*m*-arsenate.

PARMELACONITE. See copper oxide.

PARTIALLY ENCLOSED HOODS are discussed in Section 2.

PAS. See *p*-aminosalicylic acid.

PATENT BLUE V. $C_{27}H_{32}O_6N_2S_2$, mw: 566.7.

Acute tox data: ip LD_{50} (mice) = 3000 mg/kg; iv LD_{50} (mice) = 1200 mg/kg. [*3*]

THR = MOD via ip and iv routes. An exper neo via sc and im routes. [*3*]

PATHOLOGY AS A DISCIPLINE CONTRIBUTING TO INDUSTRIAL TOXICOLOGY. See Section 9.

PATULIN. See expansin.

22-PB. See 2,2-bis(*tert*-butylperoxy)butane.

PCB's. See chlorinated diphenyls.

PCMC. See *p*-chloro-*m*-cresol.

PCP. See pentachlorophenol.

PCP. See phencyclidine.

PDU. See 3-phenyl-1,1-dimethyl urea.

PE. See polyethylene.

PEANUT MEAL. A (S) carc. [*14*] See plant and fungal products.

PEANUT OIL. Syns: *katchung oil, earthnut oil.* Straw-yellow to greenish-yellow or nearly colorless oil, nutty odor and bland taste. mp: 2.7°, flash p: 540°F, d: 0.92, autoign. temp.: 833°F.

THR = LOW via irr and a mild allergen. A substance which migrates to food from packaging materials. [*109*]

Fire Hazard: Slight, when exposed to heat or flame; can react with oxidizing materials.

Spont Heating: Slight.

To Fight Fire: CO_2, dry chemical.

PEANUT RED SKIN. A red skin-like layer between nut and shell.

THR = LOW irr and a MILD allergen. Improperly stored peanuts can be a source of aflatoxin via the red skin of peanuts. See also plant and fungal products.

Fire Hazard: Mod, by chemical reaction; can react with oxidizing materials.

Spont Heating: High.

PEARL ASH. See potassium carbonate.

PEAR OIL. See amyl acetate.

PELARGONIC ACID. See ethyl heptanoic acid.

PELARGONIC MORPHOLIDE. $C_{13}H_{25}O_3N$, mw: 243.4.

Acute tox data: Inhal LD_{50} (guinea pig) = 6 mg/kg; oral LD_{50} (rat) = 23 mg/kg; inhal LD_{50} (rat) = 23 mg/kg; iv LD_{50} (rabbit) = 21 mg/kg. [*3*]

THR = HIGH irr to skin, eyes and mu mem and via oral, inhal, and iv routes. Observations show intense irr to mu mem.

PELIGOT's SALT. See potassium chlorochromate.

PENETED. See pentaerythritol technical.

PENICILLAMINE HYDROCHLORIDE. $C_5H_{11}O_2NS \cdot HCl$, mw: 185.7.

THR = An exper teratogen. [*3*]

PENICILLIN. A group of isomeric and closely related antibiotic compounds with outstanding bacterial activity. $(CH_3)_2C_5H_3NSO(COOH)NHCOOR$ (bicyclic).

Acute tox data: sc LD_{LO} (mice) = 24 mg/kg; iv LD_{LO} (mice) = 8 mg/kg. [*3*]
THR = HIGH via sc and iv routes. Has been implicated in aplastic anemia. Used as a food additive permitted in food for human consumption. [*109*]

PENICILLIC ACID. $C_8H_{10}O_4$, mw: 170.2.
THR = An exper (+) carc. [*3, 7, 23*]

PENICILLIUM ISLANDICUM. A (S) carc. [*14*] See plant and fungal products.

PENICILLIUM VIRIDICATUM.
THR = An exper neo. [*3*]

PENPHOS. See parathion.

PENTABORANE, STABLE. Syn: *pentaboron enneahydride*. Colorless gas or liquid, bad odor. B_5H_9, mw: 63.2, mp: −46.6°, bp: 0° @ 66 mm, d: 0.61 @ 0°, vap. d: 2.2, vap. press: 66 mm @ 0°, lel = 0.42%.
Acute tox data: Inhal LC_{50} (rat) = 7 ppm for 4 hrs; ip LD_{50} (rat) = 11.1 mg/kg; inhal LC_{50} (mouse) = 37 mg/m^3 for 2 hrs. [*3*]
THR = HIGH via inhal and ip routes.
Fire Hazard: Dangerous, by chemical reaction; spont flam in air.
Explosion Hazard: Dangerous. Details U.
Disaster Hazard: Dangerous; on decomp, emits toxic fumes and can react vigorously with oxidizing materials.
To Fight Fire: Water is not effective; reacts violently with halogenated extinguishing agents. Get instructions from supplier.

PENTABORANE, UNSTABLE. Syn: *dihydropentaborane*. Colorless liquid, turns yellow on standing. B_5H_{11}, mw: 65.2, bp: 63°.
THR = See boron hydrides.
Fire Hazard: Dangerous, by chemical reaction; ignites spont in air.
Explosion Hazard: Dangerous. Details U.
Disaster Hazard: Dangerous; on decomp, emits highly toxic fumes; can react vigorously with oxidizing materials.
To Fight Fire: Get instructions from supplier.

PENTABORANIDE. See potassium pentaborane.

PENTABORON ENNEAHYDRIDE. See pentaborane, stable.

"PENT-ACETATE." A mixture of amyl acetate isomers and amyl alcohol isomers. flash p: 98°F, bp: 126.7°.
THR = See specific materials.
Fire Hazard: Dangerous fire hazard via heat, flames or oxidizers.
To Fight Fire: Use alcohol foam.

PENTACHLOROBENZENE. C_6HCl_5, mw: 250.3.
THR = An exper teratogen [*3*] via oral route.

PENTACHLOROETHANE. Syn: *pentalin*. Colorless liquid. $CHCl_2CCl_3$, mw: 202.3, mp: −29°, bp: 162°, d: 1.6728 @ 25°/4°.
Acute tox data: Inhal LC_{LO} (rat) = 4238 ppm for 2 hrs; inhal LC_{LO} (mice) = 35 mg/m^3 for 2 hrs; oral LD_{LO} (dog) = 500 mg/kg; iv LD_{LO} (dog) = 100 mg/kg; sc LD_{LO} (rabbit) = 700 mg/kg. [*3*]
THR = MOD via oral and sc routes; HIGH via inhal and iv routes.
Fire Hazard: Mod, when exposed to heat or flame.
Explosion Hazard: Mod, by spont chemical reaction. Dehalogenation by reaction with alkalies, metals, etc., will produce spont explosive chloroacetylenes. Violent reaction with (NaK alloy + bromoform). [*19*]
Disaster Hazard: Dangerous; when heated to decomp, emits highly toxic fumes of chlorides.
To Fight Fire: Water, CO_2, dry chemical.

PENTACHLOROETHYL BENZENE. White crystalline solid. $Cl_5C_6CH_2CH_3$, mw: 278.4, mp: 53.4°, bp: 305°, d: 1.552 @ 60°/25°, vap. press: 1 mm @ 96.2°, vap. d: 9.6.
THR = Details U. See chlorinated hydrocarbons, aromatic.
Disaster Hazard: Dangerous; when heated to decomp, emits highly toxic fumes of chlorides.

PENTACHLOROMETHYL ETHER. Liquid. CCl_3OCHCl_2, mw: 218.32, bp: 158.5–159.5°, d: 1.64 @ 20°.
THR = Probably HIGH via oral and inhal routes. An irr to skin, eyes and mu mem. See also ethers.
Fire Hazard: Mod, when exposed to heat or flame.
Explosion Hazard: U. See ethers.
Disaster Hazard: Dangerous; when heated to decomp, emits highly toxic fumes of chlorides; can react with oxidizing materials.
To Fight Fire: Water, foam, CO_2, dry chemical.

PENTACHLORONAPHTHALENE. White solid. $C_{10}H_3Cl_5$, mw: 300.41.
THR = HIGH via oral, inhal and dermal routes. An irr. Action similar to chlorinated naphthalenes and chlorinated diphenyls. See also chlorinated diphenyl.
Disaster Hazard: Dangerous; when heated to decomp, emits highly toxic fumes of chlorides.

PENTACHLORONITROBENZENE. Colorless crystals. $C_6Cl_5NO_2$, mw: 295.4, mp: 146°, bp: 328°, vap. press: 0.013 mm @ 25°.
Acute tox data: Oral LD_{50} (rat) = 1650 mg/kg. [*3*]
THR = MOD via oral route. An exper teratogen, carc and neo via oral and dermal routes. [*3, 12*] An insecticide. See nitro compounds of aromatic hydrocarbons.

Disaster Hazard: Dangerous; when heated to decomp, emits highly toxic fumes of chlorides and NO_x.

PENTACHLOROPHENOL. Dark-colored flakes and sublimed needle crystals with a characteristic odor. Cl_5C_6OH, mw: 266.4, mp: 191°, bp: 310° (decomp), d: 1.978, vap. press: 40 mm @ 211.2°.

Acute tox data: Oral LD_{50} (rat) = 50 mg/kg; dermal LD_{50} (rat) = 105 mg/kg; ip LD_{50} (rat) = 56 mg/kg; sc LD_{50} (rat) = 100 mg/kg. [*3*]

THR = HIGH via oral, dermal, ip and sc routes. An exper neo via sc route. [*3*] Acute poisoning is marked by weakness and respiratory, blood pressure and urinary output changes. Also causes dermatitis, convulsions and collapse. Chronic exposure can cause liver and kidney injury. See also phenols. A fungicide. A contact herbicide, wood preservative, molluscicide.

Disaster Hazard: Dangerous; when heated to decomp, emits highly toxic fumes of chlorides.

***o*-PENTADECADIENYL SALICYLIC ACID.** See anacardic acid.

3-PENTADECYL CATECHOL. Syn: *tetrahydrourushiol*. Crystals, sol in alcohol, ether and benzene. $C_{21}H_{36}O_2$, mw: 320.5, mp: 60°.

Acute tox data: ip LD_{LO} (mice) = 50 mg/kg. [*3*]

THR = HIGH via ip route. Causes a dermatitis similar to that of poison ivy.

3-PENTADECYL PHENOL. Flaked pink to white waxy solid. $C_{15}H_{31}C_6H_4OH$, mw: 304.5, mp: 49°–51°, d: < 1, bp: 190°–195° @ 1 mm.

THR = Details U. See also phenol.

Fire Hazard: Slight, when exposed to heat or flame.

Disaster Hazard: Dangerous; when heated to decomp, emits acrid fumes; can react with oxidizing materials.

5-PENTADECYL RESORCINOL. Pale amber to white crystalline solid. $C_{15}H_{31}C_6H_3(OH)_2$, mw: 320.5, mp: 91°–93°, bp: 220°–225° @ 1 mm, d: > 1.

THR = Details U; a weak allergen.

Fire Hazard: Slight, when exposed to heat or flame; can react with oxidizing materials.

PENTAERYTHRITOL. Syns: *pentek*, *penetek*. Crystalline material. $C(CH_2OH)_4$, mw: 136.1, mp: 262°, d: 1.38 @ 25°/4°.

THR = LOW. A nuisance dust.

Fire Hazard: Mod, when exposed to heat or flame; can react with oxidizing materials.

PENTAERYTHRITOL DIACETAL. White flakes. $C(CH_2O)_4(CHCH_3)_2$, mw: 188.22, mp: 42°, bp: 216°.

THR = See esters.

Fire Hazard: Mod, when exposed to heat or flame; can react with oxidizing materials.

PENTAERYTHRITOL DIBUTYRAL. White flakes. $C(CH_2O)_4(CHC_3H_7)_2$, mw: 244.3, mp: 59°, bp: 290°.

THR = See esters and aldehydes.

Fire Hazard: Slight, when exposed to heat or flame; can react with oxidizing materials.

PENTAERYTHRITOL DIFORMAL. White flakes. $C(CH_2O)_4(CH_2)_2$, mw; 160.17, mp: 50°, bp: 80°–83° @ 1 mm.

THR = Probably toxic; see formaldehyde.

Fire Hazard: Slight, when exposed to heat or flame; can react with oxidizing materials.

PENTAERYTHRITOL DIPROPIONAL. Slightly viscous, colorless liquid. $C(CH_2O)_4(CHC_2H_5)_2$, mw: 216.3, mp: 19°, bp: 259°, vap. d: 7.47.

THR = No data. See also aldehydes.

Fire Hazard: Slight, when exposed to heat or flame; can react with oxidizing materials.

PENTAERYTHRITOL TETRANITRATE. Syn: *PETN*. Crystals. $C(CH_2NO_3)_4$, mw: 316.2, mp: 138°–140°, bp: explodes @ 205°–215°, d: 1.773 @ 20°/4°.

THR = See also nitrates. Effects are similar to nitroglycerine; i.e., headache, weakness, and fall in blood pressure.

Fire Hazard: See nitrates.

Explosion Hazard: Severe, when shocked or exposed to heat. One of the most powerful high explosives, it is particularly sensitive to shock. It is used in detonating and priming compositions, as a base-charge in anti-aircraft shells and mixed with TNT (70-30) in mines, explosive bombs and torpedoes. It is a very effective demolition explosive. It is also used in blasting caps combined with lead azide and diazodinitrophenol. It explodes @ 215°. (See explosives, high.)

Disaster Hazard: Highly dangerous; shock or heat will explode it; on decomp, emits highly toxic fumes of NO_x; can react vigorously with oxidizing materials.

PENTAFLUOROPROPINOIC ACID. $F_5C_3HO_2$, mw: 164.

Acute tox data: ip LD_{50} (rabbit) = 68 mg/kg. [*3*]

THR = HIGH via ip route. See also fluorides.

Disaster Hazard: Dangerous; when heated to decomp, emits highly toxic fumes of fluorides.

"PENTALARM." A liquid. Composition: mixture of isomeric amyl mercaptans. bp: 100°, flash p: 63°F (OC), d: 0.83–0.85.

THR = MOD via inhal route. See also amyl mercaptan.

Fire Hazard: Dangerous, when exposed to heat or flame.

Explosion Hazard: U.

Disaster Hazard: Dangerous, keep away from heat

and open flame; can react vigorously with oxidizing materials.

To Fight Fire: CO_2, dry chemical.

PENTALIN. See pentachloroethane.

PENTAMETHYL DIALUMINUM HYDRIDE. $(CH_3)_5Al_2H$, mw: 129.1.

THR = See hydrides. Spont flam in air. [*19*]

PENTAMETHYLENE. See cyclopentane.

PENTAMETHYLENE DIAMINE. Syn: *cadaverine*. Colorless, thick liquid, characteristic odor. $C_5H_{14}N_2$, mw: 102.2, mp: 9°, bp: 178°–180°, d: 0.873 @ 25°/4°.

THR = HIGH via oral; MOD via dermal routes. An irr, an allergen. A toxic ptomaine. Can be absorbed via dermal route. and may cause irr and sensitization.

Disaster Hazard: Dangerous; when heated to decomp, emits highly toxic fumes of nitrogen compounds and cyanides.

PENTAMETHYLENE GLYCOL. See 1,5-pentanediol.

PENTAMETHYLENE OXIDE. See tetrahydropyran.

***n*-PENTANAL.** Syn: *valeraldehyde*. Liquid. $C_5H_{10}O$, mw: 86.2, flash p: 54°F.

Acute tox data: Oral LD_{50} (rat) = 3200 mg/kg; dermal LD_{50} (rat) = 6000 mg/kg. [*3*]

THR = MOD via oral and dermal routes. See also aldehydes.

Fire Hazard: Dangerous, via heat, flames, oxidizers.

To Fight Fire: Dry chemical, CO_2.

***n*-PENTANE.** Syn: *amyl hydride*. Colorless liquid. $CH_3(CH_2)_3CH_3$, mw: 72.15, bp: 36.1°, flash p: <−40°F, fp: −129.8°, d: 0.626 @ 20°/4°, autoign. temp.: 500°F, vap. press: 400 mm @ 18.5°, vap. d: 2.48, lel = 1.5%, uel = 7.8%.

Acute tox data: Inhal TC_{LO} (humans) = 130,000 ppm ⟶ CNS effects. [*3*]

THR = LOW via inhal route. Narcotic in high conc. In humans, inhal of 5000 ppm for 10 min ⟶ no symptoms; via dermal route, undiluted for 5 hrs ⟶ blisters, no anesthesia for 1 hr ⟶ irr, itching, erythema, pigmentation, swelling, burning, pain. [*98*]

Fire Hazard: Highly dangerous, when exposed to heat, flame or oxidizers.

Spont Heating: No.

Explosion Hazard: Severe, when exposed to heat or flame.

Disaster Hazard: Highly dangerous; keep away from heat, sparks or open flame; shock can shatter metal containers and release contents.

To Fight Fire: Foam, CO_2, dry chemical.

1,5-PENTANEDIAMINE.
See pentamethylene diamine.

PENTANEDINITRILE. See glutaronitrile.

1,5-PENTANEDIOL. Syn: *pentamethylene glycol*. Colorless, viscous, odorless liquid. $HOCH_2CH_2CH_2CH_2CH_2OH$, mw: 104.15, bp: 239°, flash p: 265°F (OC), fp: −15.6°, d: 0.994 @ 20°/4°, vap. press: < 0.01 mm @ 20°, vap. d: 3.59, autoign. temp.: 635°F.

Acute tox data: Oral LD_{50} (rat) = 5890 mg/kg. [*3*]

THR = MOD–LOW via inhal route. See also glycols.

Fire Hazard: Slight, when exposed to heat or flame; can react with oxidizing materials.

To Fight Fire: Foam, CO_2, dry chemical.

PENTANE DIOL-2,4. Syn: *2,4-amylene glycol*.

Acute tox data: Oral LD_{50} (rat) = 6860 mg/kg; dermal LD_{50} (rabbit) = 14100 mg/kg. [*3*]

THR = MOD via oral and LOW via dermal routes.

PENTANEDIONE-2,4. Syn: *acetylacetone*. Colorless liquid. $CH_3COCH_2COCH_3$, mw: 100.11, mp: −23.2°, bp: 139° @ 746 mm, flash p: 105°F (OC), d: 0.976, vap. d: 3.45, autoign. temp.: 644°F.

Acute tox data: Oral LD_{50} (rat) = 1000 mg/kg; inhal LC_{LO} (rat) = 1000 ppm for 4 hrs; dermal LD_{50} (mice) = 5000 mg/kg; ip LD_{50} (mouse) = 750 mg/kg. [*3*]

THR = MOD via oral, inhal and ip routes.

Fire Hazard: Mod, when exposed to heat or flame; can react with oxidizing materials.

Explosion Hazard: U.

To Fight Fire: Alcohol foam, CO_2, dry chemical.

1-PENTANETHIOL. See amyl mercaptan.

PENTANICKEL DIPHOSPHIDE. Needles, tablets or crystals. Ni_5P_2, mw: 355.41, mp: 1185°.

THR = See nickel compounds and phosphides.

PENTANOIC ACID. See valeric acid.

PENTANOL-2. See *sec*-amyl alcohol.

PENTANOL-3. See *tert-n*-amyl alcohol, refined.

PENTANONE-2. Syns: *methyl-n-propyl ketone, ethyl acetone*. Slightly sol in water, water white liquid. $CH_3COC_3H_7$, mw: 86, d: 0.8, vap. d: 3.0, bp: 216°F, flash p: 45°F, autoign. temp.: 941°F, lel = 1.5%, uel = 8.2%.

Acute tox data: Oral LC_{50} (rat) = 3730 mg/kg; inhal LC_{50} (rat) = 2000 ppm for 4 hrs. [*3*]

THR = MOD via oral and inhal routes. Inhal causes narcosis and irr of the respiratory passages. Also irr to eyes.

Fire Hazard: Dangerous, when exposed to heat or flame.

To Fight Fire: Alcohol foam.

PENTANONE-3. See diethyl ketone.

PENTAPROPIONYL GLUCOSE. See glucose pentapropionate.

PENTASILVER TRIHYDROXY DIAMIDOPHOSPHATE. $Ag_5O_3PN_3H_3$, mw: 663.4.
THR = Explodes with heat, friction, H_2SO_4. [*19*]

"PENTASOL." See amyl alcohol.

PENTATRIACONTENE. A white solid. $C_{17}H_{35}CHCHC_{16}H_{33}$, mw: 491.0, bp: 260°–265° @ 0.5 mm, flash p: 239°F (OC), mp: 75°, d: 0.8.
THR = U.
Fire Hazard: Slight, when exposed to heat or flame; can react with oxidizing materials.
To Fight Fire: Foam, CO_2, dry chemical.

PENTEK. See pentaerythritol, technical.

1-PENTENE. See α, *n*-amylene.

2-PENTENE. Syn: *sym-methylethylethylene*. Liquid. C_5H_{10}, mw: 70.1, mp(*cis*): −179°, mp(*trans*): −135°, bp(*cis*): 37°, bp(*trans*): 35.85°, flash p: <-4°F, d(*cis*): 0.6503 @ 20°/4°, d(*trans*): 0.6482 @ 20°/4°, vap. d: 2.41.
THR = U. Probably narcotic in high conc.
Fire Hazard: Very dangerous, when exposed to heat, flame or oxidizers.
Explosion Hazard: U.
Disaster Hazard: Dangerous; keep away from heat and open flame; can react vigorously with oxidizing materials.
To Fight Fire: Alcohol foam, spray or mist, dry chemical.

2-PENTENE-3-CARBOXYLIC ACID. See ethyl crotonate.

3-PENTENE-2-ONE. Syn: *ethylidene acetone*. Colorless liquid. $CH_3CHCHCOCH_3$, mw: 84.11, bp: 122°–124°, d: 0.856, vap. d: 2.89.
Acute tox data: Oral LD_{50} (rat) = 3730 mg/kg; inhal LC_{LO} (rat) = 250 ppm for 4 hrs; dermal LD_{50} (rabbit) = 500 mg/kg. [*3*]
THR = MOD via oral, dermal and inhal routes.
Fire Hazard: Mod, when exposed to heat or flame; can react with oxidizing materials.
To Fight Fire: Foam, CO_2, dry chemical.

PENTOL. $HC{\equiv}CC(CH_3){=}CHCH_2OH$, mw: 96.1.
THR = Violent reaction with H_2, NaOH. [*19*]

PENTRYL. See picric acid.

PENTYL ALCOHOL. See amyl alcohol.

PENTYL BENZENE. See amyl benzene.

***n*-PENTYL BENZENE.** See amyl benzene.

PENTYL HYDRAZINE HYDROCHLORIDE. $C_5H_{14}N_2 \cdot HCl$, mw: 138.7.
THR = An exper neo. [*3*]

1-PENTYNE. See *n*-propyl acetylene.

PENTYL PROPIONATE. See amyl propionate.

PEPPERMINT OIL. See oil of peppermint.

PERACETIC ACID (40% SOLUTION). Syns: *peroxyacetic acid, acetyl hydroperoxide*. Colorless liquid, strong odor, water-sol. CH_3COOOH, mw: 76.05, bp: 105°, explodes @ 110°, flash p: 105°F (OC), d: 1.15 @ 20°.
Acute tox data: Oral LD_{50} (rat) = 1540 mg/kg; dermal LD_{50} (rabbit) = 1410 mg/kg; oral LD_{50} (guinea pig) = 10 mg/kg. [*3*]
THR = HIGH via oral for guinea pig; MOD via oral and dermal routes for rats and rabbits. An exper neo via dermal route. [*3*]
Fire Hazard: Mod via heat, flames. See peroxides, organic.
Explosion Hazard: Severe; when exposed to heat or by spont chemical reaction. A powerful oxidizing agent. Violent reaction with acetic anhydride, olefins, organic matter. [*19*]
Disaster Hazard: Dangerous; keep away from combustible materials.
To Fight Fire: Water, foam, CO_2.

PERBENZOIC ACID. Syn: *benzoyl hydroperoxide*. Leaflets. C_6H_5COOOH, mw: 138.1, mp: 42°, bp: explodes @ 80°–100°.
THR = MOD irr to skin, eyes and mu mem and via oral and inhal routes. See also peroxides, organic.
Fire Hazard: A powerful oxidizing agent. See peroxides, organic.
Explosion Hazard: Severe, when exposed to heat or flame. Violent reaction with olefins. [*19*]
Disaster Hazard: Dangerous; heat will cause it to explode; can react vigorously with reducing materials.

PERCAINE. See dibucane hydrochloride.

PERCHLORATES, N.O.S. Composition: combinations with the monovalent $^-ClO_4$ radical.
THR = Perchlorates are unstable materials, and are irr to the skin and mu mem of the body wherever they come in contact with it. Avoid skin contact with these materials.
Fire Hazard: Mod, by chemical reaction; powerful oxidizers. See also explosives, high.
Explosion Hazard: Mod, when shocked or exposed to heat or by chemical reaction. Perchlorates, when mixed with carbonaceous material, form explosive mixtures. They are considered a fire and explosive hazard when associated with carbonaceous materials or finely divided metals. This is also true of the presence of sulfur, powdered magnesium and aluminum. See explosives, high. React violently with benzene, CaH_2, charcoal, olefins, ethanol, SrH_2, S, H_2SO_4. [*19*]

Disaster Hazard: Dangerous; shock will explode them; when heated, they emit highly toxic fumes of chlorides; they can react with reducing materials.
To Fight Fire: Water or foam.

PERCHLORIC ACID. Colorless, fuming, unstable liquid. $HClO_4$, mw: 100.47, mp: −112°, bp: 19° @ 11 mm, d: 1.768 @ 22°.
THR = VERY irr to skin, eyes and mu mem. HIGH via oral and inhal routes.
Fire Hazard: See perchlorates.
Explosion Hazard: See perchlorates. React violently with acetic acid, (acetic acid + acetic anhydride), acetic anhydride, alcohols, (aniline + HCHO), Sb compounds, Bi, cellulose, charcoal, dibutyl sulfoxide, ethyl ether, dimethyl sulfoxide, F_2, (PbO + glycerine), glycolethers, glycols, HI, HCl, H_2, P_2O_5, hypophosphites, ketones, CH_3OH, NI_3, nitrosophenol, paper, (P_2O_5 + $CHCl_3$), NaI, steel, sulfoxides, H_2SO_4, SO_3, wood. [*19*] The anhydrous form can explode spont.
Disaster Hazard: See perchlorates.

PERCHLORIC ACID DIHYDRATE. See perchloric acid (over 72%).

PERCHLORIC ACID, MONOHYDRATE. Syn: *hydronium perchlorate*. Fairly stable needles. $HClO_4 \cdot H_2O$, mw: 118.48, mp: 50°, bp: explodes @ 110°, d: 1.88, d. liq: 1.776 @ 50°.
THR = See perchloric acid, sulfuric acid, and perchlorates.

PERCHLORIC ACID (NOT OVER 72%). Clear liquid. $HClO_4 \cdot 3H_2O$, mw: 154.5, mp: −18°, bp: 200°, d: 1.5967 @ 25°/4°.
THR = See perchloric acid.

PERCHLORIC ACID (OVER 72%). Syn: *perchloric acid dihydrate*. Stable liquid. $HClO_4 \cdot 2H_2O$, mw: 136.5, mp: −17.8°, bp: 200°, d: 1.729 @ 25°/4°.
THR = See perchloric acid.

PERCHLOROBENZENE. See hexachlorobenzene.

PERCHLOROETHYLENE. Syns: *tetrachloroethylene, ethylene tetrachloride, carbon dichloride*. Colorless liquid, chloroform-like odor. CCl_2CCl_2, mw: 165.82, mp: −23.35°, bp: 121.20°, flash p: none, d: 1.6311 @ 15°/4°+, vap. press: 15.8 mm @ 22°, vap. d: 5.83.
Acute tox data: Inhal TC_{LO} (human) = 230 ppm ⟶ systemic effects; inhal TC_{LO} (man) = 280 ppm for 2 hrs ⟶ eye effects; inhal TC_{LO} (man) = 600 ppm for $\frac{1}{6}$ hr ⟶ CNS effects; oral LD_{LO} (dog) = 4000 mg/kg; sc LD_{LO} (rabbit) = 2000 mg/kg; iv LD_{LO} (dog) = 85 mg/kg. [*3*]
THR = MOD via inhal, oral, sc, ip and dermal routes. HIGH via iv route. Not corrosive or dangerously reactive, but toxic by inhal, by prolonged or repeated contact with the skin or mu mem, or when ingested by mouth. The liquid can cause injuries to the eyes; however, with proper precautions it can be handled safely. The symptoms of acute intoxication from this material are the result of its effects upon the nervous system.

Exposures to higher conc than 200 ppm cause irr, lachrymation and burning of the eyes and irr of the nose and throat. There may be vomiting, nausea, drowsiness, an attitude of irresponsibility, and even an appearance resembling alcoholic intoxication. This material also acts as an anesthetic, through the inhalation of excessive amounts within a short time. The symptoms of fatal intoxication are irritation of the eyes, nose and throat, then fullness in the head, mental confusion; there may be headache stupefaction, nausea and vomiting, personnel suffering from subacute poisoning may suffer from such symptoms as headache, fatigue, nausea, vomiting, mental confusion and temporary blurring of the vision. This can occur when inadequate ventilation results in concentrations higher than 200 ppm, or where the vapor conc are intermittently high due to faulty handling of the material, or when an individual fails to take adequate precautionary measures.

This material can cause dermatitis, particularly after repeated or prolonged contact with the skin. The dermatitis is preceded by a reddening and burning and more rarely, a blistering of the skin. In any event, the skin becomes rough and dry, due largely to the removal of skin oils by material. The skin then cracks easily and is readily susceptible to infection (Section 9). Upon ingestion it causes irr of the gastrointestinal tract, which, in turn, causes nausea, vomiting, diarrhea and bloody stools. However, such effects are usually less severe than the effects of swallowing similar amounts of other chlorinated hydrocarbons.

It may be handled in the presence or absence of air, water, and light with any of the common construction materials at temp. up to 140°C. This material is extremely stable and resists hydrolysis. A common air contaminant. Reacts violently with Ba, Be, Li. [*19*]
Disaster Hazard: Dangerous; when heated to decomp, it emits high tox fumes of chlorides.

PERCHLOROMETHYL MERCAPTAN. Syn: *thiocarbonyl tetrachloride, trichloromethane sulfonyl chloride*. Yellow oily liquid. $CHSCl_3$, mw: 151.4, bp: slight decomp @ 149°, d: 1.700 @ 20°, vap. d: 6.414.
Acute tox data: Inhal TC_{LO} (human) = 45 ppm ⟶ eye effects; inhal LC_{LO} (human) = 483 ppm for $\frac{1}{6}$

hr; inhal LC_{LO} (cat) = 58 ppm for $\frac{1}{4}$ hr; inhal LC_{LO} (mice) = 58 ppm for $\frac{1}{6}$ hr. [3]

THR = HIGH via irr to skin, eyes and mu mem and via inhal routes. An exper carc via inhal route. [3]

Disaster Hazard: Dangerous; when heated to decomp, emits highly toxic fumes.

PERCHLORYL FLUORIDE. Syn: *PF.* Colorless, non-corrosive gas with a characteristic sweet odor. ClO_3F, mw: 102.5, mp: −146°, bp: −46.8°, d(liquid): 1.434.

Acute tox data: Inhal LC_{LO} (rat) = 2000 ppm for $\frac{2}{3}$ hr; inhal LC_{50} (mice) = 630 ppm for 4 hrs. [3]

THR = HIGH via inhal route. Toxic via dermal route also. Forms methemoglobin in the body and destroys red cells causing anemia, anorexia and cyanosis. Recovery is said to be rapid, leaving no permanent physiological damage. Can be absorbed through the skin. Its odor can be detected as low as 10 ppm although this cannot be relied upon as an indication of toxic conc in air. See also fluorine.

Fire Hazard: Mod; while nonflam, it supports combustion. It is a powerful oxidizer.

Explosion Hazard: Mod; contact with readily oxidized substances, such as hydrogen sulfide, charcoal, sawdust, lampblack, etc., yield explosive products down to −78° in some cases. Can explode in contact with (C_6H_6 + $AlCl_3$), (sodium methylate + CH_3OH). [19]

Disaster Hazard: Dangerous; when heated to decomp, emits highly toxic fumes of chlorides and fluorides.

PERCHROMATES.

THR = HIGH toxicity. See chromium compounds. Violent reaction with aniline, olefins, pyridine, quinoline. [19]

PERCUSSION CAPS AND FUSES. See explosives, high.

PERFLUORO-2-BUTENE. Syn: *octafluoro-2-butene.* Colorless gas. $CF_3CF{:}CFCF_3$, mw: 200, bp: 1.2°, fp: −135°.

THR = U. See also fluorides.

Disaster Hazard: Dangerous. When heated to decomp, emits toxic fumes.

PERFLUORETHYLENE.

See tetrafluorethylene, inhibited.

PERFLUOROISOBUTYLENE. C_4F_8, mw: 200.

Acute tox data: Inhal LC_{LO} (rat) = 0.5 ppm for 6 hrs; inhal LC_{LO} (mice) = 1.2 ppm for 2 hrs. [3]

THR = HIGH via irr to skin, eyes and mu mem and via inhal route. Acute exposure in man has produced marked irr of conjunctiva, thoat and lungs.

PERFLUOROPROPANE. Syn: *octafluoropropane.* Colorless gas. C_3F_8, mw: 188, bp: −36.7°, fp: −160°.

THR = U. Probably a simple asphyxiant.

PERFLUOROPROPIONYL FLUORIDE. C_3OF_6, mw: 166.

THR = Addition of (F_2 + N_2) @ −50° caused an explosion. [19]

PERFLUOROSUCCINAMIDE. $C_4O_2N_2F_8$, mw: 260.

THR = Reacts violently with ($LiAlH_4$ + H_2O). [19]

PERFORMIC ACID. Syn: *formyl hydroperoxide.* Liquid. HCOOOH, mw: 62.0.

THR = HIGH irr to skin, eyes and mu mem.

Fire Hazard: See peroxides, organic.

Explosion Hazard: See peroxides, organic. Violent reaction with Al, Mg, aniline, benzaldehyde, HCHO, PbO_2, metal oxides, metals, Ni, olefins, P, Na_3N, Zn, and is unstable and may explode spont. [19]

Disaster Hazard: See peroxides, organic.

PERFUMES. Composition: essential oils and alcohol.

THR = Usually MOD via oral route. Sometimes allergenic.

Fire Hazard: Mod. See alcohol.

Disaster Hazard: Mod dangerous due to flam of alcohol.

PERI ACID. See 1-naphthylamine-8-sulfonic acid.

PERILLA KETONE. $C_{10}H_{13}O$, mw: 149.2.

Acute tox data: ip LD_{50} (mice) = 6 mg/kg (males); ip LD_{50} (mice) = 2.5 mg/kg (females); ip LD (male rat) = 10 mg/kg; iv LD (heifer) = $<$ 30 mg/kg. [47]

THR = HIGH via ip, iv and oral routes. Affects the lungs. Is a potent pulmonary edemagenic agent for test animals and livestock. Possible hazard to humans. [47]

PERILLALDEHYDE. Liquid with a pungent odor. $C_{10}H_{14}O$, mw: 150.2, bp: 237° @ 745 mm, d: 0.953 @ 20°/4°.

THR = U. Probably LOW. See also aldehydes. An irr.

PERILLARTINE. Syn: *perilla sugar.* $C_{10}H_{15}NO$, mw: 165.3, mp: 102°; about 2000 times as sweet as sucrose.

Acute tox data: Oral LD_{50} (rats) = 2500 mg/kg. [2, 3]

THR = MOD via oral route.

PERILLA OIL. Essential oil of perilla frutescens. Contains perillaldehyde, perillartine and possibly perilla ketone. mp: −5°, flash p: 522°F (CC), d: 0.93–0.94.

THR = U. See under constituents listed.

Fire Hazard: Slight, when exposed to heat or flame; can react with oxidizing materials.

Spont Heating: Yes.

To Fight Fire: CO_2, dry chemicals.

PERIODATES.

THR = Many are heat-sensitive, or friction sensitive explosives. [19]

PERIODIC ACID. Syn: *paraperiodic acid.* White deliquescent crystals. $HIO_4 \cdot 2H_2O$, mw: 228.0, mp: 122°, bp: decomp @ 140°.

THR = HIGH irr to skin, eyes and mu mem. See also iodides.

Fire Hazard: Mod, by chemical reaction; an oxidizing agent. Violent reaction with dimethyl sulfoxide. [*19*]

Disaster Hazard: Dangerous; when heated to decomp, emits highly toxic fumes of iodides; reacts vigorously with reducing materials.

PERISTON. See polyvinyl pyrrolidine.

PERMANGANATE OF POTASH.
See potassium permanganate.

PERMANGANATE OF SODA.
See sodium permanganate.

PERMANGANATES. Composition: compounds containing an MnO_4 radical.

THR = HIGH. Many are strong oxidizing agents, hence, irr. See manganese compounds.

Fire Hazard: Mod, by chemical reaction with reducing agents.

Explosion Hazard: Mod, when shocked or exposed to heat. Silver permanganate and other metallic permanganates may detonate when exposed to high temp. or when they are involved in fires or severely shocked. They should be stored in a cool, ventilated area, away from acute fire hazards and easily oxidized materials. They may be disposed of by dissolving in water. Practically all permanganates are sol in water. React violently with acetic acid, acetic anhydride, ($H_2SO_4 + C_6H_6$). [*19*]

Disaster Hazard: Dangerous; shock or heat may explode them; they can react vigorously on contact with reducing materials.

PERMANGANIC ACID. In solution only. $HMnO_4$, mw: 119.9.

THR = See manganese compounds and permanganates. Unstable $> 3°$. Explodes in contact with alcohols, alkanes, aryl hydrocarbons, greases, cycloalkanes, alkyl amines, amides, ethers, CCl_4, $CHCl_3$, and dichloromethane; do not burn. [*19*]

PERMANGANYL CHLORIDE.

THR = HIGH. See manganese compounds.

Explosion Hazard: Mod, when exposed to heat.

Disaster Hazard: Dangerous; when heated to decomp, emits highly toxic fumes of chlorides; will react with water or steam to produce toxic and corrosive fumes.

PERMISSIBLE LEVELS OF RADIATION. See Section 5A.

PERMISSIBLE RADIONUCLIDE CONCENTRATIONS IN AIR AND WATER ABOVE NATURAL BACKGROUND. See Section 5A.

PERMONO SULFURIC ACID. Syn: *Caro's acid.* H_2SO_5, mw: 114.1.

THR = HIGH corrosive, toxic material. See sulfuric acid. It reacts violently with acetone, alcohols, cotton, MnO_2, organic matter, Pt, Ag. [*19*]

PERNAM BUCO. See brazil wood.

PERONINE. Syn: *morphine benzyl ether hydrochloride.* White, prismatic, crystalline, odorless powder, bitter taste. $C_{17}H_{17}NO(OH)OC_7H_7 \cdot HCl$, mw: 411.9.

THR = HIGH via oral route. A habit-forming drug. Narcotic and analgesic.

Disaster Hazard: Dangerous; when heated to decomp, emits highly toxic fumes.

PEROXIDES, INORGANIC.

THR = Variable. They may cause injury on contact with skin or mu mem. See also hydrogen peroxide.

Fire Hazard: Mod to dangerous, by chemical reaction with reducing agents and contaminants; strong oxidizing agents; contact with moisture may produce much heat. See also hydrogen peroxide and sodium peroxide.

Explosion Hazard: Mod; heat, shock or catalysts can cause violent decomp. Contact with reducing agents may give rise to explosively violent reactions.

Disaster Hazard: Dangerous; shock, heat or moisture may cause explosion; reacts with reducing agents.

PEROXIDES, ORGANIC.

THR = These materials are irr to the skin, eyes and mu mem.

Fire Hazard: Dangerous, by chemical reaction with reducing agents or exposure to heat. They are powerful oxidizers.

Explosion Hazard: Severe, when shocked, exposed to heat or by spont chemical reaction. Many peroxides are very unstable. Upon contact with reducing materials, such as organic matter, thiocyanates, [*19*] an explosion can occur.

Disaster Hazard: Dangerous; shock or heat may explode them; react with reducing materials.

PEROXYACETIC ACID.
See peracetic acid (40% solution).

PEROXYCAMPHORIC ACID. $C_5H_{16}O_6$, mw: 172.2.

THR = Explodes on rapid heating to 80°–100°. [*19*]

PEROXYDISULFURIC ACID. Syn: *persulfuric acid.* Hygroscopic crystals. $H_2S_2O_8$, mw: 194.15, mp: decomp @ 65°, bp: decomp.

THR = HIGH irr to skin, eyes and mu mem.

Fire Hazard: Mod, by chemical reaction; a powerful oxidizer.

Disaster Hazard: Dangerous; on decomp, emits highly toxic fumes of SO_x.

PEROXYSULFURIC ACID. See permonosulfuric acid.

PEROXYTRICHLOROACETIC ACID. $C_2HO_3Cl_3$, mw: 179.4.

THR = Very unstable. Decomp into HIGH toxicity PCl_3, Cl_2, HCl, CO. [*19*]

PERSIMMON. Tannin-containing fraction.

THR = An exper neo to rats via sc route. [*103*]

"PERSISTO SPRAY." See DDT.

PERSULFURIC ACID. See peroxy disulfuric acid.

PERSULFUR HEPTOXIDE. See sulfur heptoxide.

PERTHANE. See 2,2-bis (*p*-ethylphenyl)-1,1-dichloroethane.

PERUVIAN BALSAM. See balsam of Peru.

PERYLENE. See dibenz(a,h)anthracene.

PESTICIDE FOOD ADDITIVES. [*109*]

PESTOX. See octamethyl pyrophosphoramide.

PETASITES JAPONICUS MAXIM. Syn: *Fuki-No-Toh.*

THR = An exper neo [*3*] to rats via oral route.

PETN. See pentaerythritol tetranitrate.

"PETROHOL" 91%. See isopropyl alcohol.

"PETROHOL" 96%. See isopropyl alcohol.

PETROL. See gasoline.

PETROLATUM. Syns: *mineral fat*, *petroleum jelly*, *mineral jelly*. Almost colorless to amber-colored gelatinous, oily, translucent, semisolid, amorphous mass, sol in chloroform, ether, benzene, carbon disulfide, very slightly sol in alcohol, insol in water. d: 0.815–0.880 @ 60°, mp: 38°–60°.

THR = Details U. Used as a food additive permitted in the feed and drinking water of animals and/or for the treatment of food-producing animals; also permitted in food for human consumption. [*109*] An exper carc. [*14*] See paraffin and petroleum waxes. This refers to the jelly, liquid and solid material.

PETROLATUM LIQUID. See oil (mineral).

PETROLEUM. See crude oil.

PETROLEUM ASPHALT.

THR = An exper neo [*3*] of skin and scrotum and an exper carc of lung and larynx. [*14*]

PETROLEUM BENZINE. See petroleum spirits.

PETROLEUM CARBON. A recog carc. [*14*] See petroleum asphalt.

PETROLEUM COKE. A recognized carc. [*14*] See petroleum asphalt.

PETROLEUM DISTILLATION AND CRACKING PRODUCTS. A recog carc. [*14*] See petroleum asphalt.

PETROLEUM ETHER. See petroleum spirits.

PETROLEUM FUEL ADDITIVE (OXIDATION INHIBITORS). A recog carc. [*14*] See aromatic amines.

PETROLEUM NAPHTHA. See petroleum spirits.

PETROLEUM NAPHTHAS AND SOLVENTS. See petroleum spirits.

PETROLEUM OILS. A carc material. [*14*]

PETROLEUM PITCH. A recog carc. [*3*, *14*] See petroleum asphalt.

PETROLEUM SPIRITS. Syns: *petroleum benzine*, *petroleum naphtha*, *light ligroin*, *petroleum ether*. Volatile, clear, colorless and non-fluorescent liquid. mp: $<-73°$, bp: 40°–80°, ulc: 95–100, lel = 1.1%, uel = 5.9%, flash p: $<0°$F, d: 0.635–0.660, autoign. temp.: 550°F, vap. d: 2.50.

THR = MOD irr via oral and inhal and to skin, eyes and mu mem. Ingestion can cause a burning sensation, vomiting, diarrhea, drowsiness, and, in severe cases, pulmonary edema. Inhal of conc vapors causes intoxication resembling that from alcohol, headache, nausea, and coma. Hemorrhages into various vital organs have been reported.

Fire Hazard: Highly dangerous, when exposed to heat, flame sparks, etc.

Spont Heating: No.

Explosion Hazard: Mod, when exposed to heat or flame.

Disaster Hazard: Highly dangerous; keep away from heat or flame!

To Fight Fire: Foam, CO_2, dry chemical.

PETROLEUM SULFONATE. Water-insol. flash p: 400°F (OC).

THR = An exper carc. [*14*] See oils (mineral).

Fire Hazard: Slight, when exposed to heat or flame; can react with oxidizing materials.

To Fight Fire: CO_2, dry chemical.

PETROLEUM TAR. See petroleum asphalt.

PETROLEUM WAXES.

THR = An exper carc [*14*] of lung, skin and stomach.

PETRYL. See trinitrophenyl nitramine ethyl nitrate.

PF. See perchloryl fluoride.

PGA. See pterolglutamic acid.

PGE. See glycidyl phenyl ether.

PHAEOMELANIN. Red-brown pigment that colors the hair and skin of redheads. May turn carc when exposed to sunlight. [*105*]

PHALTAN. Syns: *n-(trichloromethylthio)phthalamide, folpet.* White crystals nearly water-insol.
$C_9H_4Cl_3NO_2S$, mw: 296.6, mp: 177°.
Acute tox data: sc TD_{LO} (mice) = 1000 mg/kg ⟶ neo. Oral TD_{LO} (hamster) = 500 mg/kg ⟶ exper teratogen. [3]
Disaster Hazard: Dangerous; when heated to decomp, emits highly toxic fumes of SO_x and NO_x.

PHARMACOLOGY, A CONTRIBUTING DISCIPLINE TO INDUSTRIAL TOXICOLOGY. See Section 9.

α-PHELLANDRENE. Syns: *4-isopropyl-1-methyl-1,5-cyclohexadiene, 1,5-p-menthadiene.* Colorless oil, insol in water, sol in ether, two isomeric forms (*d* and *l*).
$CH_3C{:}CHCH_2CH_2CH[CH(CH_3)_2]CH{:}CH$, mw: 136.23, d(*d*): 0.8463 @ 25°, bp(*d*): 66°–68° @ 16 mm, d(*l*): 0.8324 @ 20°, bp(*l*): 59° @ 16 mm.
THR = MOD irr via oral and dermal routes.

β-PHELLANDRENE. See 1(7)2-*p*-menthadiene.

PHENACAINE HYDROCHLORIDE. Small white crystals. $(C_2H_5OC_6H_4)_2N_2HCCH_3 \cdot HCl \cdot H_2O$, mw: 352.8, mp: 190°.
Acute tox data: Oral LD_{50} (rat) = 80 mg/kg. [3]
THR = HIGH via oral route. A local anesthetic.
Disaster Hazard: Dangerous; when heated to decomp, emits highly toxic fumes of chlorides.

PHENACETIN. See acetophenetidine.

PHENACYL AMINE. See *o*-aminoacetophenone.

PHENACYL BROMIDE. See bromoacetophenone.

PHENACYL CHLORIDE. See chloroacetophenone.

PHENACYL FLUORIDE. See fluoroacetophenone.

PHENANTHRENE. Solid or monoclinic crystals. $(C_6H_4CH)_2$, mw: 178.22, mp: 97°, bp: 339°, d: 1.179 @ 25°, vap. press: 1 mm @ 118.3°, vap. d: 6.14.
Acute tox data: Oral LD_{50} (mice) = 700 mg/kg. [3]
THR = MOD via oral route. A skin photo-sensitizer. An exper carc via dermal route. [3, 23]
Fire Hazard: Slight, when exposed to heat or flame; can react with oxidizing materials.
To Fight Fire: Water, foam, CO_2 or dry chemical.

2-PHENANTHRENYL ACETAMIDE.
THR = An exper carc. [23]

PHENANTHRO(2,1-d)THIAZOLE. $C_{15}H_{19}NS$, mw: 235.3.
THR = An exper neo. [3, 23]

N,2-PHENANTHRYL ACETAMIDE. $C_{16}H_{13}ON$, mw: 235.3.
THR = Mod acute via oral route to rats. An exper carc. [3, 23]

N,3-PHENANTHRYL ACETAMIDE. See N,2-phenanthryl acetamide.
THR = An exper carc and neo. [3]

N,9-PHENANTHRYL ACETAMIDE.
THR = See N,2-phenanthryl acetamide. HIGH toxicity to mice via im route. An exper carc and neo. [3, 23]

N-(2-PHENANTHRYL)ACETOHYDROXAMIC ACETATE. $C_{18}H_{15}O_3N$, mw: 293.3.
THR = HIGH acute toxicity to animals. An exper neo. [3]

N,2-PHENANTHRYL ACETOHYDROXAMIC ACID. $C_{16}H_{13}O_2N$, mw: 251.3.
THR = An exper carc. [3, 23]

1-PHENANTHRYL AMINE. $C_{14}H_{11}N$, mw: 193.3.
THR = An exper neo. [3, 23]

2-PHENANTHRYL AMINE.
THR = An exper neo. [3, 23]

3-PHENANTHRYL AMINE.
THR = An exper neo. [3, 23]

9-PHENANTHRYL AMINE.
THR = An exper neo. [3, 23]

PHENANTOIN.
Syn: *5-ethyl-3-methyl-5-phenylhydantoin.*
$C_{12}H_{14}O_2N_2$, mw: 218.3.
THR = HIGH via oral route to mice. Toxic effect on blood. Has been implicated in aplastic anemia. [3]

PHENATOX. See octachlorocamphene.

PHENEANTHRENEQUINONE. Yellow-orange, needle-like crystals, sol in sulfuric acid, benzene, glacial acetic acid and hot alcohol, slightly sol in ether, insol in water. $C_{14}H_8O_2$, mw: 208, d: 1.4045, mp: 206°–207°, bp: >360° (sublimes).
THR = An exper, neo [3] via dermal route.

PHENARSAZINE CHLORIDE. See diphenylamine chlorarsine. See also phenanthrene.

PHENAZINE. Syn: *dibenzopyrazine.* Pale yellow crystals. $C_{12}H_8N_2$, mw: 180.2, mp: 171°, bp: >360° (sublimes).
THR = An exper neo via imp route. [3] A larvicide.
Disaster Hazard: Mod dangerous; when heated to decomp, emits toxic fumes.

PHENAZIN-5-OXIDE. $C_{12}H_8ON_2$, mw: 196.2.
Acute tox data: Oral LD_{50} (mice) = 12 mg/kg. [3]
THR = HIGH via oral route. A fungicide.
Fire Hazard: Slight; when heated, emits toxic fumes.

PHENCAPTON. Syn: *o,o-diethyl-s-(2,5-dichlorophenyl thiomethyl)phosphorodithioate.* $C_{11}H_{15}O_2PS_3$, mw: 306.3.
Acute tox data: Oral LD_{50} (rat) = 61 mg/kg; oral

LD_{50} (mice) = 175 mg/kg; oral LD_{50} (chicken) = 886 mg/kg. [*3*]
THR = HIGH via oral route. A cholinesterase inhibitor. See parathion.

PHENCYCLIDINE. Syns: *elysion*, *PCP*. Crystals. $C_{17}H_{25}N$, mw: 243.4, mp: 46°–46.5°, bp: 135°–137°.
Acute tox data: iv TD_{LO} (human) = 0.01 mg/kg ⟶ CNS effects. Oral LD_{50} (wild birds) = 42 mg/kg; oral LD_{50} (ducks) = 237 mg/kg; oral LD_{50} (pigeons) = 75 mg/kg; oral LD_{50} (birds) = 133 mg/kg. [*3*]
THR = HIGH via oral route. Used as an anesthetic and analgesic in medicine. Even in low doses it causes serious psychologic disturbances. [*2*] Effects are characterized by loss of motor nerve coordination, amnesia and sudden violence following a fleeting exhilaration. A drug often abused in conjunction with other drugs of abuse, yielding totally unpredictable, often tragic results, such as confusion, paranoia, psychoses. [*106*]

PHENETHYL ALCOHOL. Syns: *2-phenyl ethanol*, *benzyl carbinol*. Colorless liquid, floral odor of roses. $C_6H_5CH_2CH_2OH$, mw: 122.14, mp: −27°, bp: 220°, flash p: 216°F, d: 1.0245 @ 15°, vap. d: 4.21.
Acute tox data: Oral LD_{50} (rat) = 1790 mg/kg; ip LD_{50} (mice) = 800 mg/kg; dermal LD_{50} (rabbit) = 790 mg/kg; dermal LD_{50} (guinea pig) = 5000 mg/kg. [*3*]
THR = MOD via oral, dermal and ip routes. Being studied for additional oncological information. [*3*] Reported as causing severe CNS injury to exper animals. A local anesthetic.
Fire Hazard: Low, when exposed to heat or flame; can react with oxidizing materials.
To Fight Fire: CO_2, dry chemical.

PHENETHYL HYDRAZINE SULFATE. $C_8H_{11}N_2 \cdot SO_4$, mw: 231.3.
THR = HIGH and an exper carc. [*3*]

N-(*p*-PHENETHYL)PHENYL ACETOHYDROXAMIC ACID.
THR = An exper neo. [*3*]

***o*-PHENETIDINE.** Syn: *2-aminophenetole*. Oily liquid. $C_8H_{11}NO$, mw: 137.2, mp: $<-20°$, bp: 229°, vap. press: 1 mm @ 67.0°.
THR = HIGH irr via oral, inhal and dermal routes and to skin, eyes and mu mem. See also aromatic amines.
Disaster Hazard: Dangerous; when heated to decomp, emits toxic fumes of NO_x.

***p*-PHENETIDINE.** Syn: *4-aminophenetole*. Colorless liquid. $C_8H_{11}NO$, mw: 137.2, mp: 3°, bp: 254°, flash p: 241°F, d: 1.0652 @ 16°/4°, vap. d: 4.73.
THR = HIGH irr via oral, inhal and dermal routes and to skin, eyes and mu mem. See also aromatic amines.
Fire Hazard: Slight, when exposed to heat or flame.
Disaster Hazard: Dangerous; when heated to decomp, emits highly toxic fumes of NO_x; reacts vigorously with powerful oxidizers.
To Fight Fire: Dry chemical, spray, mist.

PHENETOLE. See phenyl ethyl ether.

PHENIC ACID. See phenol.

PHENIDONE. Syn: *1-phenyl-3-pyrazolidone*. Crystals, water-soluble. $C_9H_{10}N_2O$, mw: 162.2, mp: 121°.
Acute tox data: Oral LD_{50} (rat) = 200 mg/kg; ip LD_{50} (rat) = 200 mg/kg. [*3*]
THR = HIGH via oral and ip routes.
Disaster Hazard: Dangerous; when heated to decomp, emits highly toxic fumes.

PHENOBARBITAL. White, shining, crystalline, odorless powder, bitter taste. $[CO(NHCO)_2C(C_2H_5)C_6H_5]$, mw: 232.2, mp: 174°–178°.
Acute tox data: Oral TD_{LO} (child) = 10 mg/kg ⟶ CNS effects; oral TD_{LO} (human) = 18 mg/kg ⟶ skin effects; oral TD_{LO} (human) = 0.214 mg/kg ⟶ psychotropic effects; oral LD_{50} (rat) = 162 mg/kg; sc LD_{50} (rat) = 200 mg/kg; rec LD_{50} (rat) = 284 mg/kg; ip LD_{50} (mice) = 340 mg/kg. [*3*]
THR = HIGH via oral, sc, rec and ip routes. An exper teratogen via oral route. [*3*] Repeated ingestion may lead to habituation.
Fire Hazard: Slight; when heated to decomp, emits toxic fumes.

PHENOL. Syns: *carbolic acid*, *phenic acid*, *phenylic acid*. White, crystalline mass which turns pink or red if not perfectly pure, burning taste, distinctive odor. C_6H_5OH, mw: 94.11, mp: 40.6°, bp: 181.9°, flash p: 175°F (CC), d: 1.072, autoign. temp.: 1319°F, vap. press: 1 mm @ 40.1°, vap. d: 3.24.
Acute tox data: Oral LD_{50} (rat) = 414 mg/kg; dermal LD_{50} (rat) = 669 mg/kg; ip LD_{50} (rat) = 250 mg/kg; sc LD_{50} (mice) = 344 mg/kg.[*3*]
THR = HIGH via oral, ip, sc and dermal routes. A co-carc [*23*] and an exper carc [*3*, *23*] via dermal route. In acute phenol poisoning, the main effect is on the CNS. Absorption from spilling phenolic solutions on the skin may be very rapid, and death results from collapse within 30 min to several hrs. Death has resulted from absorption of phenol through a skin area of 64 in.^{2} Where death is delayed, damage to the kidneys, liver, pancreas and spleen and edema of the lungs may result. Absorbed phenol is partly excreted by the kidneys, partly oxidized. Part of the excreted portion is combined

with sulfuric and glycuronic acids; the remainder is excreted unchanged. The symptoms develop rapidly, frequently within 15–20 min following spilling of phenol on the skin. Headache, dizziness, muscular weakness, dimness of vision, ringing in the ears, irregular and rapid breathing, weak pulse, and dyspnea may all develop, and may be followed by loss of consciousness, collapse and death. When taken internally, there is also nausea, with or without vomiting, severe abdominal pain, and corrosion of the lips, mouth, throat, esophagus and stomach. There may be perforation. On the skin, the affected area is white, wrinkled and softened, and there is usually no immediate complaint of pain; later, intense burning is felt, followed by local anesthesia and still later, by gangrene. [*75*, *76*, *77*, *20*] Chronic poisoning, following prolonged exposures to low concs of the vapor or mist, results in digestive disturbances (vomiting, difficulty in swallowing, excessive salivation, diarrhea, loss of appetite), nervous disorders (headache, fainting, dizziness, mental disturbances) and skin eruptions. Chronic poisoning may terminate fatally in cases where there has been extensive damage to the kidneys or liver. Dermatitis resulting from contact with phenol or phenol-containing products is fairly common in industry. A common air contaminant. As little as 1.5 g (oral) has killed.

Fire Hazard: Mod, when exposed to heat, flame or oxidizers and reacts violently with ($AlCl_3$ + nitrobenzene), butadiene. [*19*]

Spont Heating: No.

Disaster Hazard: Dangerous; when heated, emits toxic fumes; can react with oxidizing materials.

To Fight Fire: Alcohol foam, CO_2, dry chemical.

PHENOL LITHIUM. See lithium phenate.

PHENOLPHTHALEIN. Syn: *3,3-bis(p-hydroxyphenyl)phthalide*. Small crystals. $C_{20}H_{14}O_4$, mw: 318.3, mp: 260°, d: 1.277.

Acute tox data: ip LD_{LO} (rat) = 500 mg/kg. [*3*]

THR = MOD via ip and oral routes. A laxative.

PHENOLS, HIGH-BOILING. Liquid. Composition: alkyl-substituted phenols. mw: 150 (average), bp: 238°–288°, fp: $<-40°$, flash p: 250°F (OC), d: 1.033 @ 20°/20°, vap. press: 0.01 mm @ 20°, vap. d: approx 5.2.

THR = See phenol.

Fire Hazard: Slight, when exposed to heat or flame; can react with oxidizing materials.

Disaster Hazard: Dangerous; when heated, emits toxic fumes.

To Fight Fire: Foam, CO_2, dry chemical.

PHENOLSULFONIC ACID. Syn: *sulfocarbolic acid*. Yellowish liquid, a mixture of *o*- and *p*-phenolsulfonic acids. $C_6H_5SO_4H$, mw: 174.17, mp: 50°, d: 1.155.

THR = Details U; less irr and less toxic than phenol. See phenol and sulfuric acid.

Disaster Hazard: Dangerous; see sulfonates; will react with water or steam to produce heat.

PHENOTHIAZINE. Syn: *thiodiphenylamine*. Yellow crystals. $C_{12}H_9NS$, mw: 199.3, mp: 185.1°, bp: 371° decomp.

Acute tox data: Oral LD_{50} (rat) = 5000 mg/kg; oral TD_{LO} (child) = 425 mg/kg. ⟶ CNS effects after 5 days. [*3*]

THR = MOD via oral route. An insecticide. Large doses, i.e., heavy exposure may cause hemolytic anemia and toxic degeneration of the liver. Can cause skin irr and photo-sensitization. Used as a food additive permitted in the feed and drinking water of animals and/or for the treatment of food-producing animals. Also permitted in food for human consumption. [*109*]

Disaster Hazard: Dangerous; when heated to decomp or on contact with acid or acid fumes, emits highly toxic fumes of SO_x and NO_x.

PHENOTHIOXIN. Syn: *dibenzothioxin*. A powder. $C_{12}H_8OS$, mw: 200.3.

Acute tox data: ip LD_{50} (mice) = 200 mg/kg. [*3*]

THR = HIGH irr to skin, eyes and mu mem and via ip route. Animal exper have shown evidence of liver damage and skin irr.

Disaster Hazard: Dangerous; when heated to decomp, emits highly toxic fumes of sulfur compounds.

2-PHENOXYETHANOL. See phenyl cellosolve.

2-PHENOXYETHYL ACETATE. See ethylene glycol phenyl ether acetate.

N-(β-PHENOXYETHYL)ANILINE. Insol in water. $C_6H_5OCH_2CH_2C_6H_4NH_2$, mw: 213, d: 1.1, bp: 202°, flash p: 338°F.

THR = No details. Probably toxic. See also aniline.

Fire Hazard: Slight, when exposed to heat or flame.

To Fight Fire: Dry chemical, CO_2, mist.

PHENOXYETHYL CHLORIDE. See β-chlorophenetole.

PHENOZONE. See antipyrine.

N-PHENYLACETAMIDE. See acetanilide.

2′-PHENYL ACETANILIDE. $C_{14}H_{12}ON$, mw: 210.3.

THR = An exper neo. [*3*]

3′-PHENYL ACETANILIDE.

THR = An exper neo. [*3*]

4′-PHENYL ACETANILIDE.

THR = An exper carc. [*3*]

PHENYL ACETATE. Syn: *acetyl phenol.* Water white liquid, infinitely sol in alcohol and ether, slightly sol in water, highly refractive. $CH_3COOC_6H_5$, mw: 136, d: 1.073 @ 25°/25°, bp: 195°–196°, vap. d: 4.7, flash p: 176°F.
Acute tox data: oral LD_{50} (rat) = 1630 mg/kg; dermal LD_{50} (rabbit) = 8000 mg/kg. [*3*]
THR = MOD via oral and LOW via dermal routes.
Fire Hazard: Mod, when exposed to heat, flame or oxidizers.
To Fight Fire: Alcohol foam.

PHENYLACETIC ACID. Syn: *α-toluic acid.* White crystals, sweet, floral, honey-like odor.
$C_6H_5CH_2COOH$, mw: 136.14, mp: 76.7°, bp: 265.5°, d: 1.228 @ 20°/4°, vap. press: 1 mm @ 97.0°.
Acute tox data: Oral LD_{50} (rat) = 400 mg/kg. [*3*]
THR = HIGH via oral route.
Disaster Hazard: Dangerous; when heated to decomp, emits toxic fumes.

PHENYLACETONITRILE. Syn: *benzyl cyanide.* Oily liquid, aromatic odor. $C_6H_5CH_2CN$, mw: 118.2, mp: −23.8°, bp: 233.5°, d: 1.0214 @ 15°/15°, vap. press: 1 mm @ 60.0°.
Acute tox data: Oral LD_{50} (rat) = 270 mg/kg; inhal LC_{50} (rat) = 430 mg/m^3 for 2 hrs; oral LD_{50} (mice) = 78 mg/kg; ip LD_{50} (mice) = 40 mg/kg. [*3*]
THR = HIGH via oral, inhal and ip routes. See also nitriles.

4′-PHENYL-*o*-ACETOTOLUIDIDE. $C_{15}H_{15}ON$, mw: 225.
THR = An exper neo. [*3*]

2-PHENYL ACRYLIC ACID. Syn: *atropic acid.* Tabular or acicular crystals, sol in 790 parts water, alcohol, benzene, chloroform, ether and carbon disulfide. $C_6H_5:CH_2COOH$, mw: 148.15, mp: 106°–107°, bp: 267° (decomp).
THR = See acrylic acid and phenols.

PHENYLALANINE. Syn: *α-amino-β-phenylpropionic acid.* An essential amino acid, occurs in isomeric forms, leaflets or prisms, sol in water.
$C_6H_5CH_2CH(NH_2)COOH$, mw: 165, *l*-form decomp @ 283°, *dl* form decomp @ 318°–320°.
THR = Details U. A nutrient and/or dietary supplement food additive. [*109*]

***d*-PHENYLALANINE MUSTARD.** White crystals. $C_{13}H_{18}O_2N_2Cl_2$, mw: 305.2, mp: 182°.
THR = An exper carc. Human leukemogen. [*3, 8, 23*] A chemotherapeutic agent.

***dl*-PHENYL ALANINE MUSTARD.** See the *d* form.

***l*-PHENYL ALANINE MUSTARD.** See the *d* form.

***l*-PHENYL ALANINE MUSTARD HYDROCHLORIDE.** $C_{13}H_{18}O_2N_2Cl_2 \cdot HCl$, mw: 341.7.
THR = Deadly oral human poison. An exper carc. [*3*]

PHENYLAMINE. See aniline.

PHENYLANILINE. See diphenylamine.

PHENYLARSONIC ACID. Syn: *benzene arsonic acid.* Colorless crystals, water-sol. $C_6H_5AsO(OH)_2$, mw: 202, d: 1.760, mp: 160° decomp.
Acute tox data: Oral LD_{50} (mouse) = 0.27 mg/kg; iv LD_{50} (rabbit) = 16 mg/kg. [*3*]
THR = HIGH via oral and iv routes. A deadly poison. [*3*] See also arsenic compounds.

4′-PHENYLAZO ACETANILIDE. $C_{14}H_{13}ON_3$, mw: 239.3.
THR = An exper neo. [*3*]

***p*-PHENYLAZO ANILINE.** See *p*-aminoazo benzene.

1-(PHENYLAZO)-2-ANTHROL. $C_{20}H_{14}ON_2$, mw: 298.4.
THR = An exper neo and carc. [*3*]

1-(PHENYLAZO)-2-NAPHTHOL. $C_6H_{12}ON_2$, mw: 248.3.
THR = An exper (+) carc, neo. [*3, 4*]

PHENYL AZOIMIDE. Syns: *diazobenzeneimide, triazobenzene.* Yellow oil. $C_6H_5N_3$, mw: 119.1, mp: 59° (explodes), d: 1.078 @ 22.5°.
THR = Details U. Probably toxic.
Fire Hazard: Mod, when exposed to heat or flame.
Explosion Hazard: Severe, when shocked, exposed to heat or flame or by chemical reaction.
Disaster Hazard: Dangerous; shock or heat will explode it; can react with oxidizing materials; on contact with acid or acid fumes, can emit toxic fumes.

1-(PHENYLAZO)-2-NAPHTHYLAMINE. See yellow A B.

***p*-(PHENYLAZO) PHENOL.** $C_{12}H_{10}ON_2$, mw: 198.2.
THR = An exper (±) carc. [*3, 4*]

4-PHENYL BENZANILIDE. $C_{19}H_{15}ON$, mw: 273.4.
THR = An exper neo. [*3*]

8-PHENYL BENZ(a)ANTHRACENE. $C_{24}H_{16}$, mw: 304.4.
THR = An exper neo. [*3, 23*]

PHENYL BENZENE. See biphenyl.

PHENYL BENZOATE. Colorless crystals, geranium odor. $C_6H_5COOC_6H_5$, mw: 198.2, mp: 70°, bp: 314°, d: 1.235, vap. press: 1 mm @ 106.8°.
THR = See phenol and benzoic acid.
Disaster Hazard: Slight; when heated, emits acrid fumes.

2-PHENYL BENZOTHIAZOLE. $C_{13}H_9NS$, mw: 211.3.
Acute tox data: ip LD_{50} (mice) = 1800 mg/kg. [*3*]
THR = MOD via ip route. A larvicide.

2-PHENYL-3,1-BENZOXAZINONE-4. See bentranil.

PHENYL BENZOYL CARBINOL. See benzoin.

PHENYL BENZYL TIN DICHLORIDE. Colorless crystals. $(C_6H_5)(C_6H_5CH_2)SnCl_2$, mw: 357.8, mp: 84°.
THR = See tin compounds and chlorides.

PHENYLBIGUANIDE. Colorless crystals. $C_5H_{11}N_5$, mw: 177.2, mp: 144°.
Acute tox data: ip LD_{LO} (mice) = 250 mg/kg. [*3*]
THR = HIGH via ip route.
Disaster Hazard: Dangerous; when heated to decomp, emits highly toxic fumes of nitrogen compounds.

PHENYLBIGUANIDE HYDROCHLORIDE. See phenylbiguanide.

PHENYLBIGUANIDE MERCAPTOBENZOTHIAZOLE SALT. See phenylbiguanide.

***m*-PHENYLBIPHENYL.** See *m*-terphenyl.

PHENYLBORON DIBROMIDE. Colorless crystals, decomp by water. $C_6H_5BBr_2$, mw: 247.8, mp: 34°, bp: 100° @ 20 mm.
THR = See boron compounds and bromides.

PHENYLBORON DICHLORIDE. Colorless liquid, decomp by water. $C_6H_5BCl_2$, mw: 158.8, mp: 0°, bp: 175°.
THR = See boron compounds and chlorides.

PHENYL BROMIDE. See bromobenzene.

1-PHENYLBUTANE. See butylbenzene.

2-PHENYLBUTANE. See *sec*-butylbenzene.

PHENYL BUTAZONE. Syn: *butazolidin*. $C_{19}H_{20}O_2N_2$, mw: 308.4.
Acute tox data: Oral LD_{50} (rat) = 375 mg/kg; ip LD_{50} (rat) = 215 mg/kg; iv LD_{50} (rat) = 150 mg/kg; sc LD_{50} (rabbit) = 270 mg/kg. [*3*]
THR = HIGH via oral, ip, iv and sc routes. Has been implicated in aplastic anemia.

1-PHENYLBUTENE-2. A liquid. $C_{10}H_{12}$, mw: 132.2, bp: 175°, flash p: 160°F (TOC), d: 0.888 @ 15.5°/15.5°.
Acute tox data: Oral LD_{LO} (rat) = 5000 mg/kg. [*3*]
THR = MOD via oral route. In high conc, probably narcotic.
Fire Hazard: Mod, in the presence of heat or flame; can react with oxidizing materials.
To Fight Fire: CO_2, dry chemical.

PHENYL-N-CARBAMOYL AZIRIDINE. $C_9H_{10}ON_2$, mw: 162.2.
THR = An exper neo. [*3*]

PHENYL CARBINOL. See benzyl alcohol.

PHENYL "CARBITOL." Syn: *diethylene glycol monophenyl ether*. Liquid. $C_6H_5OC_2H_4OC_2H_4OH$, mw: 182.21, bp: 207° @ 55 mm, fp: −50°, d: 1.1158 @ 20°/20°, vap. press: < 0.01 mm @ 20°, vap. d: 6.28.
Acute tox data: Oral LD_{50} (rat) = 2140 mg/kg; dermal LD_{50} (rabbit) = 2120 mg/kg. [*3*]
THR = MOD via oral and dermal routes. See also glycols.
Disaster Hazard: Mod; when heated, emits acrid fumes. See also ethers.

PHENYLCARBYLAMINE. Syn: *phenyl isocyanide*. Colorless to greenish liquid, sol in ether. $C_6H_5C{\equiv}N$, mw: 103.1, d: 0.9775 @ 15°, bp: 166° (decomp).
THR = Probably HIGH irr to skin, eyes and mu mem.

PHENYL CARBYLAMINE CHLORIDE. Syn: *phenyliminophosgene*. Pale yellow, oily liquid. $C_7H_7N \cdot HCl$, mw: 141.6, bp: 208°–210°, d: 1.30 @ 15°, vap. d: 6.03.
Acute tox data: Inhal TC_{LO} (human) = 7 ppm for 10 min ⟶ irr effects. [*3*]
THR = HIGH via irr to skin, eyes and mu mem and via inhal route.
Disaster Hazard: Dangerous; when heated to decomp, emits highly toxic fumes of chlorides. A tear gas.

PHENYL "CELLOSOLVE." Syn: *ethylene glycol phenyl ether*. Clear liquid. $C_6H_5O(CH_2)_2OH$, mw: 138.2, mp: 14°, bp: 242°, flash p: 250°F.
Acute tox data: Oral LD_{50} (rat) = 1260 mg/kg; dermal LD_{50} (rabbit) = 5000 mg/kg. [*3*]
THR = MOD via oral and dermal routes.
Fire Hazard: Slight, when exposed to heat or flame.
Spont Heating: No.
Disaster Hazard: Mod, when heated emits acrid fumes.
To Fight Fire: CO_2, dry chemical.

PHENYL CHLORIDE. See chlorobenzene.

PHENYLCHLOROFORM. See benzotrichloride.

2-PHENYL-6-CHLOROPHENOL. Syn: *6-chlor-o-xenol*. Pale yellow, viscous liquid. $C_{12}H_9ClO$, mw: 204.7, mp: 6°, bp: 318° (decomp), d: 1.24 @ 25°/4°.
Acute tox data: Oral LD_{50} (rat) = 4220 mg/kg. [*3*]
THR = MOD via oral route. A germicide and fungicide. See also chlorophenol.
Disaster Hazard: Dangerous; see chlorides.

PHENYLCHLOROPHENYL TRICHLOROETHANE. Crystals. $C_6H_5ClC_6H_4Cl_3C_2H_3$, mw: 320.1.
THR = Details U; an insecticide. See also DDT, which is a related compound.
Disaster Hazard: Dangerous; see chlorides.

PHENYL CYANIDE. See benzonitrile.

PHENYL CYCLOHEXANE. Syn: *cyclohexylbenzene*. Colorless, oily liquid. $C_{12}H_{16}$, mw: 150.3, mp: 7.5°, bp:

240.0°, flash p: 210°F (OC), d: 0.938 ± 0.02 @ 25°/15.6°, vap. press: 1 mm @ 67.5°.
Acute tox data: Oral LD_{LO} (rat) = 5000 mg/kg. [3]
THR = MOD via oral route.
Fire Hazard: Low, when exposed to heat or flame; can react with oxidizing materials.
To Fight Fire: Alcohol foam, fog, mist, dry chemical.

2-PHENYLCYCLOHEXANOL. Colorless to pale straw-colored liquid. $C_6H_{10}C_6H_5OH$, mw: 176.2, bp: 276°–281°, flash p: 280°F, d: 1.033 @ 25°/25°, vap. d: 6.13.
Acute tox data: Oral LD_{50} (rat) = 3500 mg/kg; oral LD_{50} (rabbit) = 2700 mg/kg; oral LD_{50} (guinea pig) = 1600 mg/kg. [3]
THR = MOD via oral route. Can cause dermatitis.
Fire Hazard: Slight; when exposed to heat or flame; can react with oxidizing materials.
To Fight Fire: Foam, CO_2, dry chemical.

PHENYL CYCLOHEXENE. See cyclohexylbenzene.

PHENYL CYCLOHEXYL HYDROPEROXIDE. Available as solution. $C_{12}H_{16}O_2$, mw: 192, flash p: >212°F.
THR = Details U. See also peroxides, organic.
Fire Hazard: Slight, when exposed to heat or flame or by spont chemical reaction.
Disaster Hazard: Dangerous; when heated, burns and emits acrid fumes; can react with oxidizing or reducing materials.
To Fight Fire: Foam, CO_2, dry chemical.

PHENYL CYCLOTETRAMETHYLENE BORINE. $C_6H_5B(CH_2)_4$, mw: 144.
THR = Spont flam in air. [19]

PHENYL DIAZOSULFIDE. Red crystals. $(C_6H_5N_2)_2S$, mw: 242.3.
THR = U. See also sulfides.
Fire Hazard: Slight, when exposed to heat.
Explosion Hazard: Severe, when shocked or exposed to heat; when dry, can explode at room temp.
Disaster Hazard: Dangerous; shock will explode it; on decomp, emits highly toxic fumes of SO_x and NO_x; can react with oxidizing materials.

PHENYL DICHLOROARSINE. Colorless gas or liquid, changes to yellow. $C_6H_5AsCl_2$, mw: 222.92, bp: 255°–275°, fp: −15.6°, d: 1.654 @ 20°, vap. press: 0.021 mm @ 20°, vap. d: 7.7.
Acute tox data: Inhal LC_{LO} (mice) = 340 mg/m^3; dermal LD_{LO} (rabbit) = 8 mg/kg; inhal LC_{LO} (guinea pig) = 100 mg/m^3 for 10 min. [3]
THR = HIGH via inhal, dermal and oral routes. See also arsenic. A lacrymator type of military poison gas.
Disaster Hazard: Dangerous; when heated to decomp, emits highly toxic fumes of arsenic; will react with water or steam to produce corrosive fumes.

PHENYL DICHLOROPHOSPHINE. See benzene phosphorus dichloride.

PHENYL DICHLOROPHOSPHINE OXIDE. See benzene phosphorus oxydichloride.

PHENYL DICHLOROPHOSPHINE SULFIDE. See benzene phosphorus thiodichloride.

PHENYL DICYCLOPENTADIENYL VANADIUM. $(C_6H_5)_2V(C_{10}H_{10})$, mw: 335.2.
THR = Spont flam in air. [19]

PHENYL DIDECYL PHOSPHITE. Nearly water white liquid, alcohol odor, insol in water. $C_6H_5OP(OC_{10}H_{21})_2$, mw: 433, d: 0.940 @ 25°/15.5°, mp: < 0°, flash p: 425°F (OC).
THR = U. See phosphites.
Fire Hazard: Slight, when exposed to heat or flame.
Disaster Hazard: Dangerous; see phosphites.
To Fight Fire: Dry chemical, CO_2, fog.

PHENYL DIETHANOLAMINE. Liquid. $C_6H_5N(CH_2CH_2OH)_2$, mw: 181.23, mp: 57.8°, bp: 192°, flash p: 375°F (OC), d: 1.1203, vap. press: < 0.01 mm @ 20°.
Acute tox data: Oral LD_{50} (rat) = 980 mg/kg. [3]
THR = MOD via oral route. See also amines.
Fire Hazard: Slight, when exposed to heat or flame; can react with oxidizing materials.
To Fight Fire: CO_2, dry chemical.

PHENYL DIGLYCOL CARBONATE. See diethylene glycol bis(phenyl carbonate).

O-PHENYL-N,N-DIMETHYL PHOSPHORODIAMIDATE. Syn: *DOWCO 169*. $C_8H_{13}O_2N_2P$, mw: 200.2.
Acute tox data: Oral LD_{50} (rat) = 250 mg/kg; oral LD_{50} (wild bird) = 13 mg/kg. [3]
THR = HIGH via oral route.

3-PHENYL-1,1-DIMETHYL UREA. See dimethyl phenyl urea.

PHENYL DI-*o*-XENYL PHOSPHATE. Insol in water. $(C_{12}H_9O)_2POOC_6H_5$, mw: 484, d: 1.2, bp: 545°–626°F, flash p: 482°F.
THR = U. See also phosphates.
Fire Hazard: Slight, when exposed to heat or flame.
Disaster Hazard: Dangerous. See phosphates.
To Fight Fire: Dry chemical, CO_2, fog.

***m*-PHENYLENDIAMINE.** Syn: *m-diaminobenzene*. White crystals. $C_6H_4(NH_2)_2$, mw: 108.2, mp: 63°, bp: 286°, d: 1.139, vap. press: 1 mm @ 99.8°.
Acute tox data: Oral LD_{LO} (rat) = 80 mg/kg; sc LD_{LO} (rat) = 80 mg/kg; iv LD_{LO} (dog) = 17 mg/kg; oral LD_{LO} (cat) = 300 mg/kg. [3]

THR = HIGH via oral, sc, iv routes. Being tested for carc, teratogen. [*108*]
Fire Hazard: Slight; when heated to decomp, emits toxic fumes.

***o*-PHENYLENEDIAMINE.** Syn: *o-diaminobenzene*. Brownish crystals. mp: 104°, bp: 257°.
Acute tox data: sc LD_{LO} (rat) = 600 mg/kg; sc LD_{LO} (mouse) = 600 mg/kg. [*3*]
THR = MOD via sc route.
Fire Hazard: Slight.

***p*-PHENYLENEDIAMINE.** Syns: *diaminobenzene, ursol D*. Colorless crystals. $C_6H_4(NH_2)_2$, mw: 108.1, mp: 146°, flash p: 312°F, vap. d: 3.72, bp: 267°.
Acute tox data: Oral LD_{LO} (rat) = 100 mg/kg; ip LD_{LO} (rat) = 50 mg/kg; sc LD_{LO} (dog) = 170 mg/kg; iv LD_{LO} (dog) = 17 mg/kg. [*3*]
THR = HIGH via oral, ip, sc and iv routes. Has been implicated in aplastic anemia. Of the three phenylene diamines, the *p*- form has proved to be an especially powerful skin irr. This material is also responsible for asthmatic symptoms and other respiratory symptoms of workers in the fur dye industry, where it is commonly used. The *p*-form will cause keratoconjunctivitis, swollen conjunctiva, and eczema of the eyelids. At least one fatal case of liver involvement due to the *p*-form has been reported in the literature; *m*- and *o*-phenylenediamine are somewhat less toxic than the *p*-isomer. For this reason, and possibly because they are much less used in industry, they are not considered to be industrial hazards. [*108*]
Fire Hazard: Slight, when exposed to heat or flame.
Spont Heating: No.
Disaster Hazard: Dangerous; when heated, burns and emits highly toxic fumes of nitrogen compounds; can react with oxidizing materials.
Treatment and Antidotes: If removal from exposure is prompt, the symptoms will be relieved. If the exposure has been severe, a physician should be called at once. If the material has come in contact with the eyes, mu mem, or the skin, the contaminated area or the involved skin areas should be washed promptly with plenty of lukewarm water and soap.
To Fight Fire: Water, CO_2, dry chemical.

***m*-PHENYLENEDIAMINE FLUOSILICATE.** Brown crystals, slightly sol in alcohol. $C_6H_4(NH_2)_2 \cdot H_2SiF_6$, mw: 252.2, mp: 274°.
THR = See fluosilicates.

***p*-PHENYLENEDIAMINE FLUOSILICATE.** Pink irregular crystals. Slightly sol in alcohol. $C_6H_4(NH_2)_2 \cdot H_2SiF_6$, mw: 252.2.
THR = See fluosilicates.

***p*-PHENYLENEDIAMINE PERCHLORATE.** $C_6H_4(NH_2)_2 \cdot ClO_4$, mw: 207.6.
THR = One of the most powerful explosives known. [*19*] See also perchlorates.

2,3-PHENYLENE PYRENE. $C_{22}H_{12}$, mw: 276.3.
THR = An exper (+) carc. [*3, 11, 23*]

PHENYLETHANE. See ethyl benzene.

2-PHENYLETHANOL. See phenethyl alcohol.

PHENYLETHANOLAMINE. See 2-anilino ethanol.

PHENYL ETHER. See diphenyl oxide.

PHENYL ETHER-DIPHENYL MIXTURE.
THR = MOD irr to skin, eyes and mu mem. Vapors cause nausea and vomiting and destroy appetites for food. See components as listed.

PHENYL ETHYL ACETATE. Colorless liquid. $C_6H_5C_2H_4OOCCH_3$, mw: 164.2, bp: 223.6°, fp: $<-20°$, flash p: 230°F, d: 1.032 @ 25°/25°.
THR = U. See esters.
Fire Hazard: Low, when exposed to heat or flame.
Disaster Hazard: Mod; when heated, emits acrid fumes; can react with oxidizing materials.
To Fight Fire: Alcohol foam, CO_2, dry chemical.

β-PHENYLETHYLAMINE. Colorless to slightly yellow liquid. $C_6H_5(CH_2)_2NH_2$, mw: 121.2, bp: 200°, d: 0.96 @ 15.5°/15.5°, vap. d: 4.18.
Acute tox data: Oral LD_{LO} (rat) = 800 mg/kg; oral LD_{50} (mice) = 400 mg/kg; ip LD_{50} (mice) = 366 mg/kg. [*3*]
THR = HIGH via oral and ip routes. A skin irr and possible sensitizer. See also amines.
Fire Hazard: Slight, when exposed to heat or flame; can react with oxidizing materials.

PHENYLETHYLBARBITURIC ACID.
See "Luminal."

PHENYLETHYL BENZOATE. Crystals. $C_2H_5C_6H_4C_6H_4COOH$, mw: 226.3.
Acute tox data: Oral LD_{50} (rat) = 5000 mg/kg. [*3*]
THR = MOD via oral route. An insecticide. See benzoic acid.
Fire Hazard: Slight, when exposed to flame.
Disaster Hazard: Slightly dangerous; when heated to decomp, emits acrid fumes; can react with oxidizing materials.

PHENYL ETHYL CARBAMATE.
See phenylurethane.

PHENYL ETHYLENE. Syns: *vinyl benzene, styrene (monomer), cinnamene*. Colorless, refractive, oily liquid. $C_6H_5CHCH_2$, mw: 104.14, mp: −31°, bp: 146°, lel = 1.1%, uel = 6.1%, flash p: 88°F, d: 0.9074 @ 20°/4°, autoign. temp.: 914°F, vap. d: 3.6, fp: −33°, ulc: 40–50.

Acute tox data: Inhal LC_{LO} (human) = 10,000 ppm for ½ hr; inhal TC_{LO} (humans) = 600 ppm ⟶ irr effects; inhal TC_{LO} (humans) = 376 ppm ⟶ CNS effects; inhal TC_{LO} (women) = 20 mg/m^3 ⟶ glandular effects; oral LD_{50} (rat) = 5000 mg/kg; oral LD_{50} (mice) = 316 mg/kg; inhal LC_{LO} (mice) = 10,000 ppm. [*3*]

THR = MOD via inhal and oral routes. It can cause irr, violent itching of the eyes, lachrymation, and severe human eye injuries. Its toxic effects are usually transient and result in irr and possible narcosis. It is not considered a very toxic material, because under ordinary conditions it does not vaporize sufficiently to reach a conc that can kill animals, such as rats and guinea pigs, in a few min. Exper have found that 10,000 ppm was dangerous to animal life in from 30–60 min, 2,500 ppm was dangerous to life in 8 hrs, while 1,300 ppm was the high amount which was found to cause no serious systemic disturbances in 8 hrs. However, all animals exposed to these amounts did evidence eye and nasal irr, while those exposed to 2,500 ppm or more showed varying degrees of weakness and stupor, followed by incoordination, tremors and unconsciousness. To produce this unconsciousness required 10 hrs at a conc of 2,500 ppm. From a study to determine the chronic effects of this material, it was discovered that rats exposed to 1,300 ppm for from 7–8 hrs/day, 5 days/week, for 26 weeks, showed evidence and definite signs of eye and nasal irr and appeared unkempt, though they made a normal gain in weight and presented no significant microscopic tissue changes or changes in the blood picture. Twelve rabbits exposed to 1,300 ppm for the same period of time showed similar results with one unexplained exception.

Fire Hazard: Dangerous, when exposed to flame, heat or oxidants.

Explosion Hazard: Reacts violently with chlorosulfonic acid, oleum, H_2SO_4. [*19*]

Disaster Hazard: Dangerous, upon exposure to heat or flame; on decomp, emits acrid fumes; can react vigorously with oxidizing materials.

Treatment and Antidotes: Personnel who show symptoms of irr or beginning narcosis due to exposure to this material should be removed from exposure and the symptoms will disappear. If the symptoms persist, consult a physician.

To Fight Fire: Foam, CO_2, dry chemical.

PHENYL ETHYLENE OXIDE. See styrene oxide.

PHENYL ETHYL ETHANOLAMINE. Crystals. $C_6H_5N(C_2H_5C_2H_4OH)$, mw: 165.23, mp: 37.2°, bp: 268° @ 740 mm, flash p: 270°F (OC), d: 1.04 @ 20°/20°, vap. press: < 0.01 mm @ 20°.

THR = Details U. See also amines.

Fire Hazard: Slight, when exposed to heat or flame; can react with oxidizing materials.

To Fight Fire: Alcohol foam, foam, CO_2, dry chemical.

PHENYL ETHYL ETHER. Syns: *ethyl phenyl ether, phenetole.* Colorless liquid, insol in water, sol in alcohol and ether. $C_6H_5OC_2H_5$, mw: 122.16, d: 0.967 @ 20°/4°, mp: −30°, bp: 172°, flash p: 145°F.

Acute tox data: sc LD_{LO} (rat) = 3500 mg/kg; oral LD_{50} (mice) = 2200 mg/kg. [*3*]

THR = MOD via oral and sc routes.

Fire Hazard: Mod via heat, flames, oxidizers.

To Fight Fire: Dry chemical, CO_2, foam, spray or mist. See ethers.

PHENYL ETHYL HYDANTOIN. See nervanol.

PHENYL ETHYL MALONYLUREA. See "Luminal."

PHENYL ETHYL *sec*-PROPYL GERMANIUM BROMIDE. Colorless oil. $(C_6H_5)(C_2H_5)[CH(CH_3)_2]GeBr$, mw: 301.8, bp: 130°–135° @ 13 mm.

THR = See germanium compounds and bromides.

PHENYLFORMIC ACID. See benzoic acid.

PHENYLGERMANIUM TRIBROMIDE. Colorless liquid, hydrolyzed by water. $(C_6H_5)GeBr_3$, mw: 389.5, bp: 121° @ 13 mm.

THR = See phenol, germanium compounds and bromides.

PHENYLGERMANIUM TRICHLORIDE. Colorless liquid, hydrolyzed by water. $(C_6H_5)GeCl_3$, mw: 256.1, bp: 106° @ 12 mm.

THR = See phenol, germanium compounds, and chlorides.

PHENYLGERMANIUM TRIIODIDE. White solid, decomp by actinic rays, hydrolyzed by water. $(C_6H_5)GeI_3$, mw: 530.5, mp: 56°.

THR = See phenol, germanium compounds and iodides.

PHENYL GLYCIDYL ETHER. Syn: *PGE.*

Acute tox data: Oral LD_{50} (rat) = 3850 mg/kg; oral LD_{50} (mice) = 1400 mg/kg; dermal LD_{50} (rabbit) = 1500 mg/kg. [*3*]

THR = MOD via oral and dermal routes.

PHENYL GLYCOLIC ACID. See mandelic acid.

PHENYLHYDRAZINE. Yellow, monoclinic crystals or oil. $C_6H_5NHNH_2$, mw: 108.14, mp: 19.6°, bp: 243.5° (decomp), flash p: 192°F (CC), d: 1.0978 @ 20°/4°, vap. press: 1 mm @ 71.8°, vap. d: 3.7.

Acute tox data: Oral LD_{50} (rat) = 188 mg/kg; sc LD_{LO} (rat) = 40 mg/kg; iv LD_{LO} (dog) = 120 mg/kg. [*3*]

THR = HIGH via oral, sc and iv routes. An irr; dermal toxicity probably MOD. The ingestion or subcutaneous injection of phenyl hydrazine has been shown to cause hemolysis of the red blood cells, an effect which has been utilized in the treatment of polycythemia. The erythrocytes frequently contain Heinz bodies. Part of the hemoglobin is converted to methemoglobin. Pathological changes seen in animals include congestion of the spleen with hyperplasia of the reticuloendothelial system, degeneration and necrosis of the liver cells with extensive pigmentation, early damage to the tubules of the kidneys with fatty changes in the cortical portion, and hyperplasia of the bone marrow. The most common effect of occupational exposure is the development of dermatitis which, in sensitized persons, may be quite severe. Systemic effects include anemia and general weakness, gastrointestinal disturbances and injury to the kidneys.

Fire Hazard: Mod, when exposed to heat, flame or oxidizers; reacts violently with PbO_2. [*19*]

Disaster Hazard: Dangerous; when heated to decomp, emits highly toxic fumes of nitrogen compounds; can react with oxidizing materials.

To Fight Fire: Alcohol foam.

PHENYLHYDRAZINE HYDROCHLORIDE. Leaflets. $C_6H_5NHNH_2 \cdot HCl$, mw: 144.6, mp: 245°.

THR = An exper neo. [*3*] See also phenyl hydrazine.

Disaster Hazard: Dangerous; when heated to decomp, emits toxic fumes of nitrogen compounds and chlorides.

PHENYL HYDRIDE. See benzene.

PHENYLHYDROXYACETIC ACID. See mandelic acid.

***m*-PHENYL HYDROXYLAMINE HYDROCHLORIDE.** $C_6H_5NHOH \cdot HCl$, mw: 145.6.

THR = Can explode spont. [*19*]

PHENYL-α-HYDROXYBENZYL KETONE. See benzoin.

PHENYLIC ACID. See phenol.

PHENYLIMINOPHOSGENE. See phenyl carbylamine chloride.

PHENYL ISOCYANATE. Liquid, acrid odor. C_6H_5NCO, mw: 119.1, mp: −30° approx, bp: 166°, d: 1.1 @ 20°, vap. press: 1 mm @ 10.6°, flash p: 132°.

Acute tox data: Oral LD_{50} (rat) = 940 mg/kg. [*3*]

THR = MOD via oral route. An irr. It exploded when stirred with (cobalt pentammine triazoperchlorate + nitrosyl perchlorate). [*19*]

PHENYL ISOCYANIDE. See phenyl carbylamine.

PHENYL ISOTHIOCYANATE. See phenyl mustard oil.

PHENYL KETONE. See benzophenone.

PHENYLMAGNESIUM BROMIDE. A solid. C_6H_5MgBr, mw: 181.3.

THR = Probably HIGH. See also bromides and phenol.

Fire Hazard: Dangerous, by chemical reaction.

Explosion Hazard: Mod, by chemical reaction.

Disaster Hazard: Dangerous; will react with water, steam or acids to produce heat and toxic and flam vapors; can react vigorously with oxidizing materials; on decomp, emits toxic fumes of bromides.

To Fight Fire: CO_2, dry chemical.

PHENYLMAGNESIUM CHLORIDE. Crystals, sol in ether. C_6H_5MgCl, mw: 136.9.

THR = See grignard reagents.

N-PHENYLMALEAMIC ACID. Syn: *maleanilic acid*. Yellow crystalline solid. $C_{10}H_9O_3N$, mw: 191.18, mp: 190°, d: 1.418 @ 30°.

THR = Probably MOD irr and via inhal and oral routes.

Fire Hazard: Slight.

PHENYL MERCAPTAN. Syns: *thiophenol*, *benzenethiol*. Liquid, repulsive odor. C_6H_5SH, mw: 110.2, bp: 168.3°, d: 1.0728 @ 25°/4°.

THR = Can cause severe dermatitis and exposure is said to be capable of causing headache and dizziness; mosquito larvicide. See also mercaptans.

Fire Hazard: U.

Disaster Hazard: Dangerous; when heated to decomp, or on contact with acids, emits toxic fumes of sulfur compounds.

PHENYL MERCAPTOACETIC ACID. White powder. $C_6H_5SCH_2COOH$, mw: 168.2, mp: 63°.

THR = Details U; a fungicide and bactericide; probably HIGH toxicity. See also mercaptans.

Disaster Hazard: Dangerous; when heated to decomp, or on contact with acids, emits highly toxic fumes of SO_x.

PHENYLMERCURIC ACETATE. Lustrous crystals, slightly sol in water. $(C_6H_5)HgC_2H_3O_2$, mw: 336.8, mp: 149°.

Acute tox data: Oral LD_{50} (rat) = 30 mg/kg; ip LD_{50} (mouse) = 8 mg/kg; sc LD_{50} (mice) = 37 mg/kg. [*3*]

THR = HIGH via oral, ip and sc routes. A fungicide and herbicide. See mercury compounds, organic. An exper teratogen and neo via iv route. [*3*]

PHENYLMERCURIC ACETOXYDECANOIC ACID.

THR = A fungicide. See mercury compounds, organic.

PHENYLMERCURIC AMMONIUM ACETATE.

THR = A fungicide. See mercury compounds, organic.

PHENYLMERCURIC BROMIDE. Lustrous crystals, insol in water. $(C_6H_5)HgBr$, mw: 357.6, mp: 276°.
Acute tox data: LD_{50} (rat) = 55 mg/kg. [*3*]
THR = HIGH. See mercury compounds, organic.

PHENYLMERCURIC CHLORIDE. Satiny crystals, sol in some organic solvents. $(C_6H_5)HgCl$, mw: 313.2, mp: 251°.
Acute tox data: Oral LD_{50} (rat) = 60 mg/kg; sc LD_{50} (rat) = 47 mg/kg. [*3*]
THR = HIGH via oral and sc routes. A fungicide. See mercury compounds, organic.

PHENYLMERCURIC CYANIDE. Crystals, slightly sol in hot water. $(C_6H_5)HgCN$, mw: 303.7, mp: 204°.
THR = See mercury compounds, organic and cyanides.

PHENYLMERCURIC FORMAMIDE.
THR = A fungicide. See mercury compounds, organic.

PHENYLMERCURIC HYDROXIDE. Fine white to cream crystals, slightly sol in water, sol in acetic acid and alcohol. C_6H_5HgOH, mw: 294.6, mp: 197°–205°.
THR = Can cause skin burns. See mercury compounds, organic.

PHENYLMERCURIC IODIDE. Satiny crystals, insol in water. $(C_6H_5)HgI$, mw: 404.6, mp: 266°.
THR = HIGH. See mercury compounds, organic.

PHENYLMERCURICMONOETHANOL AMMONIUM ACETATE. White crystalline solid, sol in water. $[(HOC_2H_4)NH_2(C_6H_5Hg)]OOCCH_3$, mw: 398.
THR = A fungicide. See mercury compounds, organic.

PHENYLMERCURIC NAPHTHENATE. Prepared by interaction of phenylmercuric acetate and naphthenic acid, producing colored solutions.
Acute tox data: Oral LD_{50} (rat) = 390 mg/kg; ip LD_{50} (rat) = 30 mg/kg. [*3*]
THR = HIGH via oral and ip routes. See mercury compounds, organic. A fungicide.

PHENYLMERCURIC NITRATE. Crystals, insol in cold water. $(C_6H_5)HgNO_3$, mw: 339.7, mp: 176–186°.
Acute tox data: sc LD_{50} (rat) = 63 mg/kg; iv LD_{50} (mice) = 27 mg/kg. [*3*]
THR = HIGH via sc and iv and oral routes. See mercury compounds, organic, and nitrates.

PHENYLMERCURIC OLEATE. White crystalline powder, insol in water, sol in organic solvents and some oils. $C_6H_5HgOOC(CH_2)_7CH{:}CHC_8H_{17}$, mw: 558.
THR = HIGH. A fungicide. See mercury compounds, organic, and phenols.

PHENYLMERCURIC PHTHALATE.
THR = HIGH. A fungicide. See mercury compounds, organic.

PHENYLMERCURIC SALICYLATE.
$C_6H_4(OH)(COOHgC_6H_5)$, mw: 415.
THR = HIGH. A fungicide. See mercury compounds, organic.

PHENYLMERCURIC STEARATE.
$CH_3(CH_2)_{16}COOHgC_6H_5$, mw: 561.
THR = HIGH. A fungicide. See mercury compounds, organic.

PHENYLMERCURIC TRIETHANOLAMINE.
$(HOCH_2CH_2)_3NHgC_6H_5$, mw: 427.
THR = HIGH. A fungicide. See mercury compounds, organic.

PHENYLMERCURIC TRIETHANOL AMMONIUM LACTATE. Syn: [*tris(2-hydroxyethyl)(phenylmercuri)ammonium lactate*]. White crystalline solid, sol in water. $[(HOC_2H_4)_3NHgC_6H_5]OOCCHOHCH_3$, mw: 516.
Acute tox data: Oral LD_{50} (rat) = 30 mg/kg. [*3*]
THR = HIGH via oral route. A fungicide. See mercury compounds, organic.

PHENYLMERCURY UREA. See "Leytosan."

PHENYLMETHANE. See toluene.

PHENYLMETHANE THIOL. See benzyl mercaptan.

1-PHENYL-2-METHYLAMINOPROPANOL. See ephedrine.

PHENYLMETHYL CARBINOL. Syns: *methylphenylcarbinol*, *1-phenylethanol*, *styralyl alcohol*. Colorless liquid. $C_6H_5CH(CH_3)OH$, mw: 122.16, bp: 204°, fp: 21.4°, d: 1.015 @ 20°/20°, vap. press: 0.1 mm @ 20°, vap. d: 4.21, flash p: 205°F (OC).
Acute tox data: Oral LD_{50} (rat) = 400 mg/kg; dermal LD_{50} (rabbit) = 2500 mg/kg. [*3*]
THR = HIGH via oral and MOD via dermal routes.
Fire Hazard: Low, when exposed to heat or flame; can react with oxidizing materials.
To Fight Fire: Alcohol foam, foam, CO_2, dry chemical.

PHENYLMETHYL ETHANOLAMINE. Liquid. $C_6H_5N(CH_3)C_2H_4OH$, mw: 151.2, bp: 192° @ 100 mm, mp: −30°, flash p: 280°F (OC), d: 1.0661 @ 20°/20°, vap. press: < 0.01 mm @ 20° vap. d: 5.21.
THR = See amines.
Fire Hazard: Slight, when exposed to heat or flame; can react with oxidizing materials.
Disaster Hazard: Dangerous; on heating to decomp, emits highly toxic fumes of nitrogen compounds.
To Fight Fire: Foam, CO_2, dry chemical.

PHENYLMETHYL ETHER. See anisole.

PHENYLMETHYL KETONE. Syns: *acetophenone*, *methylphenyl ketone*, *hypnone*, *acetylbenzene*. Colorless liquid or plates. $CH_3COC_6H_5$, mw: 120.14, mp: 19.7°, bp: 202.3°, flash p: 180°F (OC), d: 1.026 @

20°/4°, vap. d: 4.14, vap. press: 1 mm @ 15°, autoign. temp.: 1060°F.
Acute tox data: Oral LD_{50} (rat) = 900 mg/kg. [*3*]
THR = MOD via oral route. Narcotic in high conc. A hypnotic. See ketones.
Fire Hazard: Mod, when exposed to heat or flame; can react with oxidizing materials.
Spont Heating: No.
To Fight Fire: Foam, CO_2, dry chemical.

1-PHENYL-3-METHYL-5-PYRAZOLYL DIMETHYL CARBAMATE. See pyrolan.

4-PHENYLMORPHOLINE. Syn: *N-phenylmorpholine*. Crystals. $C_6H_5\overline{NC_2H_4OCH_2CH_2}$, mw: 163.2, mp: 57°, bp: 270°, flash p: 220°F (OC), d: 1.0599 @ 57°/20°, vap. press: < 0.1 mm @ 20°, vap. d: 5.63.
Acute tox data: Oral LD_{50} (rat) = 930 mg/kg; dermal LD_{50} (rabbit) = 360 mg/kg. [*3*]
THR = HIGH via dermal and MOD via oral routes. An insecticide.
Fire Hazard: Slight, when exposed to heat or flame; can react with oxidizing materials.
Disaster Hazard: Dangerous; on decomp, emits toxic fumes of nitrogen compounds.
To Fight Fire: Alcohol foam, foam, CO_2, dry chemical.

PHENYL MUSTARD OIL. Syns: *thiocarbanil, phenylisothiocyanate, phenylthiocarbonimide*. Pale yellow liquid. C_6H_5NCS, mw: 135.18, mp: −21°, bp: 221°, d: 1.1282.
Acute tox data: Oral LD_{50} (rat) = 400 mg/kg; ip LD_{50} (mice) = 100 mg/kg. [*3*]
THR = HIGH via oral and ip routes.
Disaster Hazard: Dangerous; when heated to decomp, or on contact with acid or acid fumes, emits highly toxic fumes of cyanides and SO_x.

PHENYL-α-NAPHTHYLAMINE. Crystals. $C_{10}H_7NHC_6H_5$, mw: 219.3.
Acute tox data: Oral LD_{50} (rat) = 1625 mg/kg; oral LD_{50} (mice) = 1231 mg/kg. [*3*]
THR = MOD via oral route. See also aromatic amines. Reported capable of producing dermatitis, nephritis, anemia and cyanosis. Cases of bladder tumor have been reported. Listed as a bladder tumorigen. [*14*]

PHENYL-β-NAPHTHYLAMINE. Syns: *n-phenyl-2-naphthylamine, agerite powder*. Needle-like crystals. $C_{10}H_7NHC_6H_5$, mw: 219.3, mp: 108°, bp: 395.0°, d: 1.20.
Acute tox data: Oral LD_{50} (mice) = 1450 mg/kg. [*3*]
THR = MOD via oral route. An exper carc and neo via oral and sc routes. [*3, 108*] See aromatic amines, phenyl-α-naphthylamine.
Fire Hazard: Mod, when exposed to heat or flame.
Disaster Hazard: Dangerous; when heated to decomp, emits highly toxic fumes of nitrogen compounds; can react with oxidizing materials.

PHENYL PENTANE. See amyl benzene.

***o*-PHENYLPHENOL.** Syn: *orthoxenol*. White, flaky crystals, mild odor. $C_6H_5C_6H_4OH$, mw: 170.2, bp: 286°, fp: 57.2°, flash p: 255°F, d: 1.217 @ 25°/25°, vap. press: 1 mm @ 100.0°.
Acute tox data: Oral LD_{50} (rat) = 2700 mg/kg. [*3*]
THR = MOD via oral route.
Fire Hazard: Slight, when exposed to heat or flame; can react with oxidizing materials.
To Fight Fire: Alcohol foam.

***p*-PHENYLPHENOL.** Flaky material. $C_{12}H_9OH$, mw: 170.2, mp: 164.5°, bp: 308°, flash p: 330°F, vap. press: 10 mm @ 176.2°.
Acute tox data: ip LD_{50} (mice) = 150 mg/kg. [*3*]
THR = HIGH via ip route. MOD via oral route. An exper neo [*3*] via sc route.
Fire Hazard: Slight; when exposed to heat or flame; can react with oxidizing materials.

PHENYL PHOSPHINE. Syn: *phosphaniline*. Liquid. $C_6H_5PH_2$, mw: 110.1, bp: 160°, d: 1.001 @ 15°, vap. d: 3.79.
Acute tox data: Inhal LC_{50} (rat) = 38 ppm for 4 hrs. [*3*]
THR = HIGH via inhal route. See also phosphine.
Fire Hazard: Mod, when exposed to heat or flame.
Disaster Hazard: Dangerous; when heated to decomp, emits highly toxic fumes of phosphorus compounds; can react with oxidizing materials.

1-PHENYLPIPERAZINE. Pale yellow oil, insol in water, sol in alcohol and ether. $C_{10}H_{14}N_2$, mw: 162.2, d: 1.0621 @ 20°/4°, bp: 286.5°, mp: 18.8°, flash p: 285°F.
Acute tox data: Oral LD_{50} (rat) = 210 mg/kg; dermal LD_{50} (rabbit) = 140 mg/kg. [*3*]
THR = HIGH via oral and dermal routes. A powerful irr. See also piperazine.
Fire Hazard: Slight, when exposed to heat or flame.
Disaster Hazard: Mod. It supports combustion and decomp to yield toxic fumes.
To Fight Fire: Water, foam, dry chemical.

1-PHENYLPROPANE. See *n*-propylbenzene.

2-PHENYLPROPANE. See cumene.

3-PHENYLPROPENAL. See cinnamaldehyde.

1-PHENYL-3-PYRAZOLIDONE. See phenidone.

PHENYL SALICYLATE. Syn: *salol*. Small white crystals, aromatic odor. $C_{13}H_{10}O_3$, mw: 214.2, mp: 42°, bp: 173° @ 12 mm, d: 1.25.
Acute tox data: LD_{LO} (human) = 50 mg/kg. [*3*]

THR = HIGH. See also salicylic acid and phenol.
Fire Hazard: Slight.

PHENYL SELENIDE. $C_{12}H_{10}Se$, mw: 186.3.
Acute tox data: Oral LD_{50} (rat) = 360 mg/kg. [*3*]
THR = HIGH via oral route. See also selenium compounds.

PHENYL SILICATE. Solid. $Si(OC_6H_5)_4$, mw: 401.5, mp: 48°.
THR = Details U. See also phenol and silicates.
Fire Hazard: Slight.

PHENYL SILVER. C_6H_5Ag, mw: 185.
THR = It is explosive @ room temp. [*19*]

PHENYL SULFIDE. Syn: *diphenyl sulfide*. Colorless liquid, almost no odor, insol in water, sol in hot alcohol, miscible with benzene, ether and carbon disulfide. $(C_6H_5)_2S$, mw: 186.27, d: 1.118 @ 15°/15°, mp: 40°, bp: 295°–297°.
Acute tox data: Oral LD_{50} (rat) = 2140 mg/kg. [*3*]
THR = MOD via oral route.
Disaster Hazard: Dangerous; when heated to decomp, emits highly toxic fumes. Reacts violently with diazonium salts. [*19*]

PHENYLTETRATHIO-*o*-STANNATE. Solid. $(SC_6H_5)_4Sn$, mw: 555.3, mp: 67°.
THR = See tin compounds and sulfur compounds.

PHENYLTHIOARSENATE.
Acute tox data: Inhal LC_{LO} (mice) = 1000 mg/m^3. [*3*]
THR = MOD via inhal route. See also arsenic compounds.

PHENYLTHIOCARBAMIDE. See phenylthiourea.

PHENYLTHIOCARBONIMIDE. See phenyl mustard oil.

PHENYLTHIOUREA. Syn: *phenylthiocarbamide*. Needle-like crystals, bitter taste. $C_6H_5NHCSNH_2$, mw: 152.2, mp: 154°, d: 1.3.
Acute tox data: Oral LD_{50} (rat) = 3 mg/kg; ip LD_{50} (rat) = 5 mg/kg; oral LD_{50} (mice) = 10 mg/kg; oral LD_{50} (rabbit) = 40 mg/kg. [*3*]
THR = HIGH via oral and ip routes. A rodenticide.
Disaster Hazard: Dangerous; when heated to decomp, or on contact with acid or acid fumes, emits highly toxic fumes of SO_x and NO_x.

PHENYLTIN TRIBROMIDE. Solid. $(C_6H_5)SnBr_3$, mw: 435.6, bp: 183° @ 29 mm.
THR = See tin compounds and bromides.

PHENYLTIN TRICHLORIDE. Solid, sol in water. $(C_6H_5)SnCl_3$, mw: 302.2, bp: 143° @ 25 mm.
THR = See tin compounds, hydrochloric acid and chlorides.

***o*-PHENYL TOLUENE.** $C_{13}H_{12}$, mw: 168.2, flash p: > 212°F, autoign. temp.: 923°F, d: 1.01, vap. d: 5.82, bp: 260°.
THR = No data.
Fire Hazard: Low.
To Fight Fire: Water spray, mist, CO_2, dry chemical.

PHENYLTRIBENZYLTIN. A liquid, sol in organic solvents (except alcohol). $(C_6H_5)Sn(C_6H_5CH_2)_3$, mw: 469.2, bp: 290 @ 5 mm.
THR = HIGH. See tin compounds and phenol.

PHENYLTRICHLOROSILANE. Liquid. $C_6H_5SiCl_3$, mw: 211.6, bp: 201°, flash p: 196°F (OC), d: 1.321 @ 25°.
Acute tox data: Oral LD_{50} (rat) = 2390 mg/kg; dermal LD_{50} (rabbit) = 890 mg/kg. [*3*]
THR = MOD via oral and dermal routes. See also chlorosilanes.
Fire Hazard: Mod, when exposed to heat, flame or oxidizers.
Disaster Hazard: Dangerous; when heated to decomp, emits highly toxic fumes of chlorides; will react with water or steam to produce toxic and corrosive fumes; can react with oxidizing materials.
To Fight Fire: Water spray, mist, dry chemical, CO_2.

PHENYLTRIFLUOROSILANE. $C_6H_5SiF_3$, mw: 162.2.
Acute tox data: Oral LD_{50} (rat) = 310 mg/kg; inhal LC_{LO} (rat) = 1000 ppm for 4 hrs; dermal LD_{50} (rat) = 640 mg/kg. [*3*]
THR = HIGH via oral route and MOD via inhal and dermal routes. See also fluorides.

PHENYLTRI-*p*-TOLYL GERMANIUM. White crystals, insol in water. $(C_6H_5)Ge(C_6H_4CH_3)_2$, mw: 423.1, mp: 191°.
THR = Details U. See germanium compounds.

PHENYLURETHANE. Syn: *phenyl ethyl carbamate*. Crystals. $C_6H_5NHCOOC_2H_5$, mw: 165.2, mp: 53°, bp: 238° (slight decomp), d: 1.106.
THR = MOD for mice. An exper neo [*4*] via dermal route. A citrus fruit fungicide. See carbamates.
Disaster Hazard: Dangerous; when heated to decomp, emits highly toxic fumes of nitrogen compounds.

PHENYL VINYL KETONE. C_9H_8O, mw: 132.2.
THR = An exper carc [*3*] via sc route.

PHISOHEX. See 2,2′-methylene bis(3,4,6-trichlorophenol).

PHLOROGLUCINOL. Syn: *1,3,5-benzene triol*. White crystalline material, sweet taste, slightly water-sol. $C_6H_6O_3$, mw: 126.1, mp: 218°.
Acute tox data: sc LD_{LO} (rat) = 1550 mg/kg. [*3*]
THR = MOD via sc route.

PHORATE. Syns: *o,o-diethyl-s-(ethyl thiomethyl)-phosphorodithioate*, *thimet*. Liquid, insol in water, miscible with carbon tetrachloride, dioxane and xylene. $(C_2H_5O)_2P(S)SCH_2SC_2H_5$, mw: 260, bp: 118°–120° @ 0.8 mm, d: 1.156.

Acute tox data: Oral LD_{50} (rat) = 1 mg/kg; dermal LD_{50} (rat) = 3 mg/kg; oral LD_{50} (wild birds) = 1 mg/kg. [*3*]

THR = VERY HIGH via oral and dermal routes. A cholinesterase inhibitor, see parathion. An insecticide. A food additive permitted in the feed and drinking water of animals and/or for the treatment of food-producing animals. [*109*]

Disaster Hazard: See parathion.

PHORBOL. $C_{20}H_{27}O_6$, mw: 363.5.
THR = An exper carc. [*3, 23*]

PHORBOL ACETATE CAPRATE. $C_{32}H_{48}O_8$, mw: 560.8.
THR = An exper neo. [*3*]

PHORBOL ACETATE LAURATE. $C_{33}H_{52}O_8$, mw: 576.9.
THR = An exper neo.[*3*]

PHORBOL ACETATE MYRISTATE. $C_{35}H_{56}O_8$, mw: 604.9.
THR = An exper neo. [*3*]

PHORBOL CAPRATE + 2-METHYL BUTYRATE. $C_{34}H_{56}O_8$, mw: 592.9.
THR = An exper neo. [*3*]

PHORBOL CAPRATE TIGLATE. $C_{34}H_{52}O_8$, mw: 588.9.
THR = An exper neo. [*3*]

PHORBOL DIACETATE.
THR = An exper carc. [*23*]

PHORBOL DIBENZOATE.
THR = An exper carc. [*23*]

PHORBOL DIDECANOATE.
THR = An exper carc. [*23*]

PHORBOL LAURATE + METHYL BUTYRATE. $C_{36}H_{58}O_8$, mw: 618.9.
THR = An exper neo. [*3*]

PHORBOL MYRISTATE ACETATE.
See phorbol acetate myristate.

PHORONE. Syn: *2,6-dimethyl-2,5-heptadien-4-one.* Solid or greenish liquid. $(CH_3)_2CCHCOCHC(CH_3)_2$, mw: 138.20, mp: 28°, bp: 197.2°, flash p: 185°F (OC), d: 0.879, vap. press: 1 mm @ 42.0°, vap. d: 4.8.

THR = Tox. See also isophorone.

Fire Hazard: Mod, when exposed to heat or flame; can react with oxidizing materials.

To Fight Fire: Foam, CO_2, dry chemical.

"PHOSDRIN." See dimethyl-1-carbomethoxy-1-propen-2-yl phosphate.

PHOSGENE. Syns: *carbon oxychloride, carbonyl chloride, CG.* Colorless gas or volatile liquid, odor of new mown hay or green corn. $COCl_2$, mw: 98.92, mp: −118°, bp: 8.3°, d: 1.37 @ 20°, vap. press: 1180 mm @ 20°, vap. d: 3.4.

Acute tox data: Inhal LC_{50} (human) = 3200 mg/m^3; inhal TC_{LO} (human) = 25 ppm for ½ hr ⟶ irr effects; inhal LC_{50} (mice) = 110 ppm for ½ hr; inhal LC_{50} (monkey) = 1087 ppm for 1 min; inhal LC_{50} (rat) = 1482 ppm for 1 min; inhal LC_{50} (rabbit) = 3211 ppm for 1 min; inhal LC_{50} (guinea pig) = 141 ppm for ½ hr. [*3*]

THR = HIGH via inhal route. HIGH irr to eyes and mu mem. [*72, 73, 74*] In the presence of moisture, phosgene decomp to form hydrochloric acid and carbon monoxide. This action takes place within the body, when the gas reaches the bronchioles and the alveoli of the lungs. There is little irr effect upon the respiratory tract, and the warning properties of the gas are therefore very slight. The liberation of hydrochloric acid in the lung tissues results in the development of pulmonary edema, which may be followed by bronchopneumonia, and occasionally lung abscess. Degenerative changes in the nerves have been reported as later sequelae. Conc of 3–5 ppm of phosgene in air cause irr of the eyes and throat, with coughing; 25 ppm is dangerous for exposure lasting 30–60 min, and 50 ppm is rapidly fatal after even short exposure. There may be no immediate warning that dangerous conc of the gas are being breathed. After a latent period of 2–24 hrs, the patient complains of burning in the throat and chest, shortness of breath and increasing dyspnea. There may be moist rales in the chest. Where the exposure has been severe, the development of pulmonary edema may be so rapid that the patient dies within 36 hrs after exposure. In cases where the exposure had been less, pneumonia may develop several days after the occurrence of the accident. In patients who recover, no permanent residual disability is thought to occur. A common air contaminant. Reacts violently with Al, *tert*-butyl azido formate, 2,4-hexadiyn-1,6-diol, isopropyl alcohol, K, Na. [*19*]

Disaster Hazard: Highly dangerous; when heated to decomp, or on contact with water or steam, will react to produce toxic and corrosive fumes.

PHOS-KIL. See parathion.

PHOSPHAM. White amorphous crystals. PN_2H, mw: 60.00.

THR = U. Reacts violently with air, CuO, HgO, NO_2, nitrates. [*19*]

Disaster Hazard: Dangerous; on decomp, emits highly toxic fumes of phosphorus and nitrogen compounds.

PHOSPHAMIDON. See 2-chloro-2-diethylcarbamoyl-1-methyl vinyl dimethyl phosphate.

PHOSPHATES.
THR = See individual compounds. Violent reaction with Mg. [*19*]
Disaster Hazard: Dangerous; when heated to decomp, can emit highly toxic fumes of PO_x.

PHOSPHATES ORGANIC. See parathion.

PHOSPHENILINE. See phenylphosphine.

PHOSPHENYL CHLORIDE. Syn: *dichlorophenyl phosphine*. Fuming liquid. $C_6H_5PCl_2$, mw: 179.0, bp: 225°, d: 1.319, vap. d: 6.17.
THR = HIGH irr to skin, eyes and mu mem and via oral and inhal routes.
Disaster Hazard: Dangerous; when heated to decomp, emits highly toxic fumes of PO_x and chlorides; will react with water or steam to produce heat and toxic and corrosive fumes.

PHOSPHIDES. Composition: combination of a cation + elemental phosphorus.
THR = Phosphides are particularly dangerous because they tend to decomp to phosphine upon contact with moisture or acids. See also phosphine.
Fire Hazard: Dangerous by chemical reaction; they react with water and acids to liberate phosphine.
Explosion Hazard: Mod; see phosphine.
Disaster Hazard: Dangerous; when heated, these may emit highly toxic fumes of PO_x; they react with water or steam to produce toxic and flam vapors; on contact with oxidizing materials, they can react vigorously; on contact with acid or acid fumes, they can emit toxic fumes.

PHOSPHINE. Syns: *hydrogen phosphide, phosphoretted hydrogen*. Colorless gas. PH_3, mw: 34.04, mp: −132.5°, bp: −87.5°, d: 1.529 g/l @ 0°, autoign. temp.: 212°F.
Acute tox data: Inhal LC_{50} (rat) = 11 ppm for 4 hrs; inhal TC_{LO} (humans) = 8 ppm for 1 hr ⟶ pulmonary effects; inhal LC_{LO} (rabbit) = 2500 ppm for $\frac{1}{3}$ hr. [*3*]
THR = HIGH via inhal route. Phosphine is a very toxic gas, but its action on the body has not been fully worked out. It appears to cause, chiefly, a depression of the CNS and irr of the lungs; autopsy findings in human cases may be entirely negative, or there may be pulmonary edema, dilation of the heart and hyperemia of the visceral organs. Inhal of phosphine causes restlessness, followed by tremors, fatigue, slight drowsiness, nausea, vomiting and, frequently, severe gastric pain and diarrhea. There is often headache, thirst, dizziness, oppression in the chest and burning substernal pain; later the patient may become dyspneic and develop cough and sputum. Coma or convulsions may precede death. Most cases recover without after-effects. Chronic poisoning, characterized by anemia, bronchitis, gastro-intestinal disturbances and visual, speech and motor disturbances, may result from continued exposure to very low conc.
Fire Hazard: Very dangerous, by spont chemical reaction.
Explosion Hazard: Mod, when exposed to flame. Reacts violently with air, BCl_3, Br_2, Cl_2, ClO, $Hg(NO_3)_2$, HNO_3, NO, NCl_3, $^-NO_3$, N_2O, HNO_2, O_2, ($K + NH_3$), $AgNO_3$. [*19*]
Explosive Range: Not known.
Disaster Hazard: Dangerous; when heated to decomp, emits highly toxic fumes of PO_x; can react vigorously with oxidizing materials.
To Fight Fire: CO_2, dry chemical or water spray.

PHOSPHINOETHANE. See ethyl phosphine.

PHOSPHOLEUM. See polyphosphoric acid.

PHOSPHON. Syn: *tributyl-2,4-(dichlorobenzyl)phosphoniumchloride*. Crystals, readily sol in water, acetone and ethanol. $C_{19}H_{32}Cl_3P$, mw: 397.8.
Acute tox data: Oral LD_{50} (rat) = 178 mg/kg. [*3*]
THR = HIGH via oral route. Plant growth regulant.
Disaster Hazard: Dangerous; see chlorides and phosphorus compounds.

PHOSPHONITRILE AZIDE TRIMER.
THR = HIGH explosive readily detonated by friction. [*19*]

PHOSPHONIUM BROMIDE. Cubic, colorless crystals. PH_4Br, mw: 114.93, bp: 38.8° @ 794 mm (sublimes approx 30°), d: 2.464 g/l, vap. press: 400 mm @ 28.0°.
THR = HIGH via oral route. A powerful irr. See also bromides.

PHOSPHONIUM CHLORIDE. Cubic, colorless crystals, decomp in water. PH_4Cl, mw: 70.47, mp: 28° @ 46 atm, bp: sublimes.
THR = HIGH via oral route. A powerful irr. See chlorides and phosphine.

PHOSPHONIUM IODIDE. Tetragonal, colorless, deliquescent crystals. PH_4I, mw: 161.93, mp: sublimes @ 18.5°, d: 2.86, vap. press: 40 mm @ 16.1°, 760 mm @ 62.5°.
Acute tox data: Oral LD_{LO} (rabbit) = 5 mg/kg. [*3*]
THR = HIGH via oral route. Rapid heating causes detonation. Violent reaction with bromates, $HClO_3$, chlorates, iodates, HIO_2, HNO_3, $AgNO_3$. [*19*]

PHOSPHONIUM PERCHLORATE. $(PH_4)_2 \cdot (ClO_4)_3$, mw: 368.5.
THR = A very explosive salt and cannot be dried. [*19*] HIGH via oral route. A powerful irr.

PHOSPHONIUM SULFATE. Colorless, deliquescent crystals. $(PH_4)_2SO_4$, mw: 166.09.

THR = HIGH via oral route. A powerful irr.
Disaster Hazard: Dangerous; when heated to decomp, emits highly toxic fumes of sulfuric acid and phosphine.

PHOSPHORAMIDE. Syn: *phosphoryl amide*. Amorphous white crystals. $PO(NH_2)_3$, mw: 95.05, mp: decomp.
THR = U. See amides.
Disaster Hazard: Dangerous; when heated to decomp, emits highly toxic fumes of NO_x and PO_x.

PHOSPHORAMIDE MUSTARD.
THR = An exper carc. [*23*]

PHOSPHORETTED HYDROGEN. See phosphine.

PHOSPHORIC ACID. Colorless liquid or rhombic crystals. H_3PO_4, mw: 98.04, mp: 42.35°, $-\frac{1}{2}H_2O$ @ 213°, fp: 42.4°, d: 1.864 @ 25°, vap. press: 0.0285 mm @ 20°.
Acute tox data: Inhal TC_{LO} (human) = 100 mg/m^3 $\longrightarrow$ irr effects. Oral LD_{50} (rat) = 1530 mg/kg; dermal LD_{50} (rabbit) = 2740 mg/kg. [*3*]
THR = MOD via oral and dermal routes. Used as a general purpose food additive. [*109*] It is a common air contaminant.
Disaster Hazard: Dangerous; when heated to decomp, emits toxic fumes of PO_x.

PHOSPHORIC ACID ANHYDRIDE. See phosphorus pentoxide.

PHOSPHORIC ANHYDRIDE.
See phosphorus pentoxide.

PHOSPHORODIAMIDIC ACID, CYCLOHEXYL AMINE SALT. $C_6H_{13}N \cdot PH_5O_2N_2$, mw: 195.2.
THR = An exper neo to mice via ip route. [*103*]

***m*-PHOSPHORUS ACID.** Feather-like crystals. HPO_2, mw: 63.99.
THR = See phosphorus compounds, inorganic.

***o*-PHOSPHORUS ACID.** Colorless to yellow deliquescent crystals, sol in water and alcohol, taste of garlic. $H_2(HPO_3)$, mw: 82.0, d: 1.651 @ 21°, mp: 73.6°, bp: decomp @ 200°.
THR = See phosphorus compounds, inorganic.

PHOSPHORUS (AMORPHOUS, RED). Syn: *red phosphorus*. Reddish-brown powder. P_4, mw: 124.08, bp: 280° (with ignition), mp: 590° @ 43 atm, d: 2.34, autoign. temp.: 500°F in air, vap. d: 4.77.
THR = Relatively harmless unless it contains white phosphorus as an impurity. Details U.
Fire Hazard: Dangerous; when exposed to heat or by chemical reaction with oxidizers.
Spont Heating: No.
Explosion Hazard: Mod, by chemical reaction or on contact with organic materials.
Disaster Hazard: Dangerous; when heated, emits highly toxic fumes of PO_x; can react with reducing materials.
To Fight Fire: Water.

PHOSPHORUS (WHITE). Cubic crystals, colorless to yellow, wax-like solid. P_4, mw: 124.08, mp: 44.1°, bp: 280°, flash p: spont in air, d: 1.82, autoign. temp.: 86°F, vap. press: 1 mm @ 76.6°, vap. d: 4.42.
Acute tox data: Dermal LD_{50} (rat) = 100 mg/kg; oral LD_{LO} (human) = 1.4 mg/kg; oral LD_{LO} (dog) = 50 mg/kg; sc LD_{LO} (dog) = 2 mg/kg. [*3*]
THR = HIGH via dermal, oral, sc and inhal routes. This material is dangerously reactive in air and turns red in sunlight. If combustion occurs in a confined space, it will remove the oxygen and render the air unfit to support life. High conc of the vapors evolved by burning it are irr to the nose, throat and lungs as well as the skin, eyes and mu mem. If phosphorus is ingested, it can be absorbed from the gastrointestinal tract or through the lungs. The absorption of toxic quantities of phosphorus has an acute effect on the liver and is accompanied by vomiting and marked weakness. The long-continued absorption of small amounts of phosphorus can result in necrosis of the mandible or jaw bone, and is known as "phossy-jaw." Long-continued absorption, particularly through the lungs and through the gastrointestinal tract, can cause a chronic poisoning. This gives rise to a generalized form of weakness attended by anemia, loss of appetite, gastrointestinal weakness and pallor. The most common symptom, however, of chronic phosphorus poisoning is necrosis of the jaw. It can also cause changes in the long bones and seriously affected bones may become brittle, leading to spont fractures. It is especially hazardous to the eyes and can damage them severely. It also has adverse effects on the teeth, and workers should have periodic dental examinations. The yellow form of phosphorus, when it comes into external contact with the eyes, can cause conjunctivitis with a yellow tint. If the material is inhaled, it can cause photophobia with myosis, dilation of pupils, retinal hemorrhage, congestion of the blood vessels and rarely an optic neuritis.
Radiation Hazard: For permissible levels, see Section 5A, Table 5A.5. Artificial isotope ^{32}P, $T\frac{1}{2}$ = 14.3d, decays to stable ^{32}S emitting β's of 1.71 MeV.
Fire Hazard: Dangerous, when exposed to heat or by chemical reaction with oxidizers. Ignites spont in air. Very reactive.
Spont Heating: No.
Explosion Hazard: Dangerous, by chemical reac-

tion with alkaline hydroxides, NH_4NO_3, SbF_5, $Ba(BrO_3)_2$, Be, BI_3, $Ca(BrO_3)_2$, $Mg(BrO_3)_2$, $K(BrO_3)$, $NaBrO_3$, $Zn(BrO_3)_2$, Br_2, halogens, BrF_3, BrN_3, (chlorates of Ba, Ca, Mg, K, Na, Zn), (iodates of Ba, Ca, Mg, K, Na, Zn), Ce, Cs, $CsHC_2$, Cs_3N, (charcoal + air), ClO_2, (Cl_2 + heptane), ClO, ClF_3, ClO_3, chlorosulfonic acid, CrO_3, $Cr(OCl)_2$, Cu, NCl, IBr, ICl, IF_5, Fe, La, PbO_2, Li, Li_2C_2, Li_6C_2, $Mg(ClO_4)_2$, Mn, HgO, $HgNO_3$, Nd, Ni, nitrates, NBr, NO_2, NBr_3, NCl_3, NOF, FNO_2, O_2, performic acid, Pt, K, KOH, K_3N, $KMnO_4$, K_2O_2, Rb, $RbHC_2$, Se_2Cl_2, $SeOCl_2$, $SeOF_2$, SeF_4, $AgNO_3$, Ag_2O, Na, Na_2C_2, $NaClO_2$, NaOH, Na_2O_2, S, SO_3, H_2SO_4, Th, $VOCl_2$, Zr. [*19*]

Disaster Hazard: Dangerous; emits highly toxic fumes of PO_x; can react vigorously with oxidizing materials.

To Fight Fire: Water.

PHOSPHORUS CHLORIDE.
See phosphorus trichloride.

PHOSPHORUS CHLORIDE NITRIDE. Prisms. $P_6N_7Cl_9$, mw: 603.05, mp: 237.5°, bp: 251°–261° @ 13 mm.

THR = HIGH. See hydrochloric acid and phosphorus.

Disaster Hazard: Dangerous; when heated to decomp, emits highly toxic fumes of PO_x and chlorides; will react with water or steam to produce toxic and corrosive fumes.

PHOSPHORUS COMPOUNDS, INORGANIC.

THR = Variable. Most inorganic phosphates, except phosphine, have low toxicity but in large doses they may cause serious disturbances, particularly in calcium metabolism. Metaphosphates may be highly toxic, causing irr and hemorrhages in the stomach, as well as liver and kidney damage. Common air contaminants.

PHOSPHORUS CYANIDE. White needles. $P(CN)_3$, mw: 110, mp: sublimes @ 130°.

THR = See cyanides. Ignites in air and reacts violently with H_2O. [*19*]

PHOSPHORUS DIBROMIDE TRICHLORIDE. Orange crystals. PBr_2Cl_3, mw: 297.18, mp: 35° (decomp).

THR = See hydrobromic acid and hydrochloric acid.

Disaster Hazard: Dangerous; when heated to decomp or on contact with acid or acid fumes, emits highly toxic fumes of PO_x, bromides and chlorides.

PHOSPHORUS DIBROMIDE TRIFLUORIDE. Pale yellow crystals. PBr_2F_3, mw: 247.81, mp: −20°, bp: decomp @ 15°.

THR = See fluorides and hydrobromic acid.

PHOSPHORUS DICHLORIDE. Colorless crystals. PCl_2, mw: 101.89, bp: 180°, mp: −28°.

THR = Probably toxic. See hydrochloric acid.

Disaster Hazard: Dangerous; when heated to decomp, emits highly toxic fumes of PO_x and chlorides; will react with water, steam or acids to produce toxic and corrosive fumes.

PHOSPHORUS DICHLORIDE TRIFLUORIDE. PCl_2F_3, mw: 158.89, bp: −8°, decomp @ 200°.

THR = See fluorides and hydrochloric acid.

PHOSPHORUS DIIODIDE. Crystals. P_2I_4, mw: 569.64, mp: 124.5°, bp: decomp.

THR = See phosphorus and iodides.

PHOSPHORUS HEMITRISELENIDE. Dark red mass. P_2Se_3, mw: 298.8.

THR = See Selenium compounds.

Fire Hazard: Mod, by chemical reaction.

Explosion Hazard: Mod, by chemical reaction.

Disaster Hazard: Dangerous; when heated to decomp, or on contact with acid or acid fumes, emits highly toxic fumes of selenium and PO_x.

PHOSPHORUS HEPTABROMIDE DICHLORIDE. Prisms. PBr_7Cl_2, mw: 661.31.

THR = HIGH. See bromides and chlorides.

Disaster Hazard: Dangerous; when heated to decomp, emits highly toxic fumes of bromides and chlorides; will react with water, steam or acids to produce toxic and corrosive fumes.

PHOSPHORUS HEPTASULFIDE. Solid, light yellow crystals, light gray powder or fused solid. P_4S_7, mw: 348.34, mp: 310°, bp: 523°, d: 2.19 @ 17°.

THR = See sulfides and phosphorus compounds.

PHOSPHORUS ISOCYANATE. $P(OCN)_3$, mw: 157.1.

THR = Reacts violently with acetaldehyde, acetic acid, Cl_2, $AgNO_3$, H_2SO_4, H_2O. [*19*]

PHOSPHORUS MONOBROMIDE TETRACHLORIDE. Yellow crystals. $PBrCl_4$, mw: 252.72.

THR = HIGH. See hydrochloric acid and bromides.

Disaster Hazard: Dangerous; when heated to decomp, emits highly toxic chlorides, PO_x and bromides; will react with water, steam or acids to produce toxic and corrosive fumes.

PHOSPHORUS NITRIDE. Amorphous white crystals. P_3N_5, mw: 162.98, bp: decomp @ 800°, d: 2.51 @ 18°.

THR = Probably toxic. See phosphorus compounds, inorganic, and nitrides.

Fire Hazard: Mod, by chemical reaction. See ammonia.

Explosion Hazard: Mod, by chemical reaction.

Disaster Hazard: Dangerous; when heated to decomp, emits highly toxic fumes of NO_x and PO_x; will react with water or steam to produce toxic and corrosive fumes.

PHOSPHORUS OXYBROMIDE. Colorless plates. $POBr_3$, mw: 286.73, mp: 56°, bp: 190°, d: 2.882.
THR = HIGH irr to skin, eyes and mu mem.
Disaster Hazard: Dangerous; when heated to decomp, emits highly toxic fumes of bromides and PO_x; will react with water or steam to produce toxic and corrosive fumes.

PHOSPHORUS OXYCHLORIDE. Syn: *phosphoryl chloride*. Colorless to slightly yellow fuming liquid. $POCl_3$, mw: 153.39, mp: 1.2°, bp: 105.1°, d: 1.685 @ 15.5°, vap. press: 40 mm @ 27.3°, vap. d: 5.3.
Acute tox data: Oral LD_{50} (rat) = 380 mg/kg; inhal LC_{50} (rat) = 48 ppm for 4 hrs. [*3*]
THR = HIGH irr to skin, eyes and mu mem and via oral and inhal routes. Reacts violently with BI_3, (2,5-dimethyl pyrrole + dimethyl formamide), dimethyl sulfoxide, organic matter, Na, water. [*19*]
Disaster Hazard: Dangerous; when heated to decomp, emits highly toxic fumes of chlorides and PO_x; will react with water or steam to produce heat and toxic and corrosive fumes.

PHOSPHORUS OXYFLUORIDE. Syn: *phosphoryl fluoride*. Colorless gas. POF_3, mw: 103.98, mp: −68°, bp: −39.8°, d: 4.69 g/l.
THR = Hydrolysis of this material and halofluorides, i.e., (phosphoryl chlorodifluoride, phosphoryl dichlorofluoride and the bromo analogs) is vigorous to violent. [*19*]

PHOSPHORUS OXYNITRIDE. Amorphous white crystals. PON, mw: 60.99, mp: red heat.
THR = See nitrides and phosphates.
Fire Hazard: Mod, by chemical reaction.
Explosion Hazard: Mod, by chemical reaction.
Disaster Hazard: Dangerous; on decomp, emits highly toxic fumes of PO_x and NO_x; on contact with acid or acid fumes, can emit toxic fumes.

PHOSPHORUS OXYSULFIDE. Tetragonal deliquescent crystals. $P_4O_6S_4$, mw: 348.18, mp: 102°, bp: 295°.
THR = See sulfides and phosphorus. Decomp rapidly in moist air. [*19*]

PHOSPHORUS PENTABROMIDE. Rhombic yellow crystals. PBr_5, mw: 430.56, mp: decomp above 100°, bp: decomp @ 106°.
THR = HIGH caustic. See bromides. VERY toxic.
Fire Hazard: Mod, by chemical reaction. Contact with moisture can cause a violent reaction and evolution of heat.
Disaster Hazard: Dangerous; when heated to decomp, emits highly toxic fumes of bromides; will react with water or steam to produce heat and toxic and corrosive fumes.

PHOSPHORUS PENTACHLORIDE. Yellowish-white, fuming, crystalline mass. PCl_5, mw: 208.2, mp: 166.8° decomp, bp: subl @ 162°, d: 4.65 g/liter @ 296°, vap. press: 1 mm @ 55.5°.
Acute tox data: Oral LD_{50} (rat) = 660 mg/kg; inhal LC_{50} (rat) = 205 mg/m^3; inhal LC_{Lo} (mice) = 120 ppm. [*3*]
THR = HIGH irr to skin, eyes and mu mem and via inhal route. MOD via oral route.
Fire Hazard: Mod, by chemical reaction. Reacts violently with moisture, ClO_3, F_2, hydroxylamine, MgO, P_2O_3, K, Na. [*19*]
Disaster Hazard: Dangerous; when heated to decomp, emits highly toxic fumes of chlorides; will react with water or steam to produce heat and toxic and corrosive fumes.
To Fight Fire: CO_2, dry chemical.

PHOSPHORUS PENTAFLUORIDE. Colorless gas, fumes strongly in air. PF_5, mw: 125.98, mp: −93.7°, bp: −84.5°, vap. d: 5.81 g/l.
THR = HIGH. Violently irr to skin, eyes and mu mem. See fluorides.
Disaster Hazard: Dangerous; when heated to decomp, emits highly toxic fumes of fluorides and phosphorus compounds; will react with water or steam to produce toxic and corrosive fumes.

PHOSPHORUS PENTASELENIDE. Syn: *diphosphorus pentaselenide*. Dark red-black needles. P_2Se_5, mw: 456.76, mp: decomp.
THR = See selenium compounds and phosphorus.

PHOSPHORUS PENTASULFIDE. Gray to yellow-green, crystalline, deliquescent mass. P_2S_5, mw: 222.3, mp: 276°, bp: 514°, d: 2.03, autoign. temp.: 287° F.
THR = HIGH irr to skin, eyes and mu mem. See also hydrogen sulfide. Readily liberates hydrogen sulfide and phosphorus pentoxide on contact with moisture.
Fire Hazard: Dangerous, in the form of dust when exposed to heat or flame. Evolves heat in contact with moisture.
Spont Heating: Yes, in the presence of moisture.
Explosion Hazard: Mod, in solid form by spont chemical reaction.
Disaster Hazard: Dangerous; when heated to decomp, emits highly toxic fumes of SO_x and PO_x; will react with water, steam or acids to produce toxic and flam vapors; can react vigorously with oxidizing materials.
To Fight Fire: CO_2 snow, dry chemical or sand.

PHOSPHORUS PENTOXIDE. Syns: *phosphoric anhydride*, *PO_x*. A fluffy, crystalline, deliquescent powder. P_2O_5, mw: 143.0, mp: 563°, d: 0.77–1.39, vap. press: 1 mm @ 384°.

THR = HIGH caustic. See phosphorus compounds, inorganic.

Fire Hazard: Dangerous, by chemical reaction. Reacts violently with water, NH_3, CaO, ClF_3, HF, OF_2, $HClO_4$, ($HClO_4$ + $CHCl_3$), K, propargyl alcohol, Na_2CO_3, Na, NaOH, (organic matter + moisture). [*19*]

Disaster Hazard: Dangerous; reacts with water or steam to produce heat; on contact with reducing materials, can react vigorously.

PHOSPHORUS SEQUISULFIDE. Syn: *tetraphosphorus trisulfide*. Yellow crystalline mass. P_4S_3, mw: 220.26, mp: 172.5°, bp: 407°, d: 2.03, autoign. temp.: 212°F.

Acute tox data: Oral LD_{LO} (dog) = 1000 mg/kg; oral LD_{LO} (rabbit) = 100 mg/kg. [*3*]

THR = HIGH-MOD via oral route. See also sulfides.

Fire Hazard: Dangerous, when exposed to heat or flame; ignites by friction.

Spont Heating: No.

Explosion Hazard: Mod, by chemical reaction.

Disaster Hazard: Dangerous; when heated to decomp, emits highly toxic fumes of PO_x and SO_x; will react with water or steam to produce toxic fumes; can react vigorously with oxidizing materials.

To Fight Fire: Water.

PHOSPHORUS SULFOCHLORIDE. See thiophosphoryl chloride.

PHOSPHORUS SULFOFLUORIDE. Solid. PSF_3, mw: 120.

THR = HIGH. Details U. See also phosphorus pentafluoride.

Disaster Hazard: Dangerous; when heated to decomp, emits highly toxic fumes of SO_x and PO_x as well as fluorides.

PHOSPHORUS TETROXIDE. Syn: *PO_x*. Rhombic, colorless, deliquescent crystals. P_2O_4, mw: 125.96, mp: > 100°, bp: sublimes @ 180°, d: 2.54 @ 23°.

THR = See phosphorus compounds, inorganic, and see also phosphorus pentoxide.

PHOSPHORUS THIOCYANATE. Liquid. $P(SCN)_3$, mw: 205.23, mp: approx −4°, bp: 265°, d: 1.625 @ 18°.

THR = Probably toxic. See thiocyanates.

Fire Hazard: Slight, when exposed to heat or flame.

Disaster Hazard: Dangerous; when heated to decomp, or on contact with acids, emits highly toxic fumes of cyanides; can react with oxidizing materials.

PHOSPHORUS TRIBROMIDE. Colorless, fuming liquid. PBr_3, mw: 270.7, mp: −40°, bp: 175.3°, d: 2.852 @ 15°, vap. press: 10 mm @ 47.8°.

THR = See hydrobromic acid. HIGH corrosive and toxic.

Violent reaction with K, RuO_4, Na. [*19*]

Disaster Hazard: Dangerous; when heated to decomp, emits highly toxic fumes of bromides and PO_x; will react with water, steam or acids to produce heat and toxic and corrosive fumes.

PHOSPHORUS TRICHLORIDE. Syn: *phosphorus chloride*. Clear, colorless, fuming liquid. PCl_3, mw: 137.39, mp: −111.8°, bp: 74.2°, d: 1.574 @ 21°, vap. press: 100 mm @ 21°, vap. d: 4.75.

Acute tox data: Oral LD_{50} (rat) = 550 mg/kg; inhal LC_{50} (rat) = 140 ppm for 4 hrs; inhal LC_{50} (guinea pig) = 50 ppm for 4 hrs. [*3*]

THR = HIGH irr to skin, eyes and mu mem and via oral and inhal routes.

Fire and Explosion Hazard: Violent reactions with acetic acid, Al, $Cr(OCl)_2$, (diallyl phosphite + allyl alcohol), F_2, dimethyl sulfoxide, hydroxylamine, ICl, PbO_2, HNO_3, HNO_2, organic matter, K, Na, water. [*19*]

Disaster Hazard: Dangerous; when heated to decomp, emits highly toxic fumes of chlorides and PO_x; will react with water, steam or acids to produce heat and toxic and corrosive fumes; can react with oxidizing materials.

To Fight Fire: CO_2, dry chemical.

PHOSPHORUS TRIFLUORIDE. Colorless gas. PF_3, mw: 87.98, mp: −152°, bp: −102°, d: 3.907 g/l.

Acute tox data: Inhal LC_{LO} (mice) = 1900 mg/m^3 for $\frac{1}{6}$ hr. [*3*]

THR = HIGH irr to skin, eyes and mu mem and via inhal. See hydrofluoric acid, fluorides and phosphorus pentafluoride. Violent reaction with diborane, F_2, O_2. [*19*]

Disaster Hazard: Dangerous; when heated to decomp, emits highly toxic fumes of fluorides and PO_x; will react with water or steam to produce toxic and corrosive fumes.

PHOSPHORUS TRIIODIDE. Hexagonal, red, deliquescent crystals. PI_3, mw: 411.74, mp: 61°, bp: decomp.

THR = Irr to skin, eyes and mu mem. See iodides and phosphorus compounds.

PHOSPHORUS TRIOXIDE. Syn: *PO_x*. Monoclinic colorless crystals or white deliquescent powder. P_2O_3, mw: 110.0, mp: 22.5°, bp: 173°, d: 2.135 @ 21°, vap. press: 10 mm @ 53.0°.

THR = HIGH irr to skin, eyes and mu mem. See phosphorus compounds. Can decomp to phosphine. Violent reaction with O_2, air, NH_3, AsF_3, Br_2, Cl_2, dimethyl formamide, dimethyl sulfoxide, dimethyl sulfite, CH_3OH, PCl_5, S, SCl, H_2O. [*19*]

Disaster Hazard: Very dangerous; will react with water or steam to produce heat and phosphine.

PHOSPHORUS TRISULFIDE. Syn: *tetraphosphorus hexasulfide*. Gray-yellow crystals. P_4S_6, mw: 316.44, mp: 290°, bp: 490°.
THR = HIGH. See sulfides.
Fire Hazard: Dangerous, when exposed to heat or by chemical reaction.
Explosion Hazard: Dangerous, by chemical reaction.
Disaster Hazard: Dangerous; when heated to decomp, emits highly toxic fumes of PO_x and sulfides; will react with water or steam to produce toxic fumes; can react with oxidizing materials.

PHOSPHORYL AMIDE. See phosphoramide.

PHOSPHORYL CHLORIDE.
See phosphorus oxychloride.

PHOSPHOTUNGSTIC ACID. Triclinic, yellow-green crystals. $H_3PW_{12}O_{40} \cdot 14H_2O$, mw: 3133.27.
THR = HIGH irr to skin, eyes and mu mem. A strong acid with caustic properties.
Disaster Hazard: Dangerous; when heated to decomp, emits highly toxic fumes of PO_x.

PHOSVEL. See *o*-(2,5-dichloro-4-bromophenyl)-*o*-methyl phenyl thiophosphonate.

PHOTOGRAPHIC AND X-RAY FILMS, SCRAP.
Fire Hazard: Dangerous, when exposed to heat or flame.
Disaster Hazard: Dangerous; when heated to decomp, they emit highly toxic fumes; they can react vigorously with oxidizing materials.

PHTHALANDIONE. See phthalic anhydride.

PHTHALIC ACID. Syn: *benzene dicarboxylic acid*. Crystals. $C_8H_6O_4$, mw: 166.1, mp: 206°–208°, d: 1.59.
Acute tox data: Oral LD_{LO} (rat) = 4600 mg/kg; ip LD_{50} (mice) = 1670 mg/kg. [*3*]
THR = MOD via oral and ip routes.
Fire Hazard: Slight; when heated in the form of dust (anhydride) it can explode.
Explosion Hazard: Mod; violent reaction with HNO_3. [*19*]

PHTHALIC ANHYDRIDE. Syn: *phthalandione*. White crystalline needles. $C_8H_4(CO)_2O$, mw: 148.11, mp: 131.2°, lel = 1.7%, uel = 10.4%, bp: 295° (sublimes), flash p: 305°F (CC), d: 1.527 @ 4°, autoign. temp.: 1058°F, vap. press: 1 mm @ 96.5°, vap. d: 5.10.
Acute tox data: Oral LD_{50} (rat) = 4020 mg/kg. [*3*]
THR = MOD via oral route. A common air contaminant.
Fire Hazard: Slight, when exposed to heat or flame; can react with oxidizing materials.
Explosion Hazard: Mod, in the form of dust, when exposed to flame. Violent reaction with HNO_3. [*19*]
To Fight Fire: CO_2, dry chemical.

PHTHALIMIDE. Syn: *1,3-isoindoledione*. Light tan to white powder. $C_6H_4(CO)_2NH$, mw: 147.1, mp: 238°, bp: sublimes.
THR = U.
Disaster Hazard: Slight; when heated to decomp, emits toxic fumes.

PHTHALODINITRILE. See *o*-dicyanobenzene.

PHTHALONITRILE. See *o*-dicyanobenzene.

***m*-PHTHALYL DICHLORIDE.**
See isophthaloyl chloride.

***p*-PHTHALYL DICHLORIDE.**
See terephthaloyl chloride.

PHYGON. See 2,3-dichloro-1,4-naphthoquinone.

PHYSIOLOGY CONTRIBUTING TO INDUSTRIAL TOXICOLOGY. See Section 9.

PHYSOSTIGMA. Syns: *calabar bean, ordeal bean*. Dried ripe seeds.
THR = HIGH via oral and inhal routes. An allergen.
Fire Hazard: Slight, when exposed to heat or flame.

PHYSOSTIGMINE. Syns: *eserine, calabrine*. Colorless, hygroscopic crystals. $C_{15}H_{21}N_3O_2$, mw: 275.3, mp: 105°–106°.
Acute tox data: Oral LD_{50} (mice) = 4.5 mg/kg; sc LD_{LO} (rat) = 2.2 mg/kg; iv LD_{LO} (mice) = 0.5 mg/kg; oral TD_{LO} (human) = 20 mg/kg ⟶ CNS effects. [*3*]
THR = HIGH via oral, sc and iv routes. Poisoning can occur as the result of a mistake in dosage or due to hypersensitivity of the patient. Symptoms of intoxication from this material, which usually manifest themselves in from 5–25 min after injection, are a marked cutaneous hyperaesthesia, vomiting, convulsions and diarrhea, paralysis of the diaphragm, spasm of the glottis with transient dyspnea, increased salivation, sweating, slowing of the respiration and lowering of the temperature. Death usually occurs from respiratory paralysis. It has a central and peripheral nervous system action.
Fire Hazard: Slight.
Disaster Hazard: When heated to decomp, emits toxic fumes of NO_x.
Treatment and Antidotes: Wash out stomach, keep body warm. Atropine by injection. Artificial respiration and oxygen if needed.

PICENE. Syns: *3,4-benzchrysene; 1,2-7,8-dibenzphenanthrene; dibenzo(a,i)-phenanthrene; β,β-binaphthylene ethene*. Leaflets. $C_{22}H_{14}$ (polycyclic), mw: 278.33, bp: 518°–520°, mp: 364°.
THR = An exper neo via dermal route. [*3, 23*]

PICLORAM. See "Tordon."

α-PICOLINE. Syn: *methyl pyridine*. Colorless liquid, strong, unpleasant odor. $C_5H_4NCH_3$, mw: 93.13, mp: −70°, bp: 129°, flash p: 102°F (OC), d: 0.95 @ 15°/4°, autoign. temp.: 1000°F, vap. press: 10 mm @ 24.4° vap. d: 3.2.

Acute tox data: Oral LD_{50} (rat) = 790 mg/kg; dermal LD_{50} (rabbit) = 410 mg/kg; oral LD_{50} (guinea pig) = 900 mg/kg; inhal LC_{LO} (rat) = 4000 ppm for 4 hrs. [*3*]

THR = HIGH via dermal route. MOD via oral and inhal routes. A respiratory irr.

Fire Hazard: Mod, when exposed to heat or flame; can react with oxidizing materials.

Explosion Hazard: U.

Disaster Hazard: Dangerous; when heated to decomp, emits toxic fumes of NO_x.

To Fight Fire: CO_2, dry chemical.

β-PICOLINE. Syns: *3-picoline, 3-methyl pyridine*. Colorless liquid. $CH_3C_5H_4N$, mw: 93.12, bp: 143.5°, d: 0.9613 @ 15°/4°, vap. d: 3.21.

THR = Details U. See pyridine and α-picoline.

Fire Hazard: Details U. Probably mod to dangerous.

Disaster Hazard: Dangerous; when heated to decomp, emits toxic fumes of NO_x.

γ-PICOLINE. Syns: *4-picoline; 4-methylpyridine*. Colorless liquid, disagreeable odor. $CH_3C_5H_4N$, mw: 93.06, bp: 145°, fp: 3.7°, d: 0.9571 @ 15°/4°, vap. d: 3.21, flash p: 134°F (OC).

Acute tox data: Oral LD_{50} (rat) = 1290 mg/kg; dermal LD_{50} (rabbit) = 270 mg/kg; inhal LC_{LO} (rat) = 1000 ppm for 4 hrs. [*3*]

THR = HIGH via dermal, MOD via oral and inhal routes. See also α-picoline.

Fire Hazard: Mod via heat, flames, oxidizers.

Disaster Hazard: See α-picoline.

To Fight Fire: Alcohol foam.

3-PICOLINE. See β-picoline.

4-PICOLINE. See γ-picoline.

4-PICOLINE-*n*-OXIDE. Crystals. C_6H_7NO, mw: 109.1, bp: 191° @ 738 mm.

THR = Details U. See also pyridine.

Fire Hazard: Slight, when exposed to heat or flame.

Disaster Hazard: Dangerous; when heated to decomp, emits highly toxic fumes of NO_x; can react with oxidizing materials.

PICRAMIC ACID. Syns: *picraminic acid; dinitroaminophenol*. Red monoclinic crystals. $NH_2(NO_2)_2C_6H_2OH$, mw: 199.12, mp: 168°, flash p: 410°F.

THR = See nitrates and 2,3-dinitrophenol.

PICRAMIDE. See trinitroaniline.

PICRAMINIC ACID. See picramic acid.

PICRATES. See nitrates.

PICRIC ACID. Syns: *picronitric acid, trinitrophenol, carbazotic acid*. Yellow crystals or yellow liquid, very bitter. $(NO_2)_3C_6H_2OH$, mw: 229.11, mp: 121.8°, bp: explodes > 300°, flash p: 302°F, d: 1.763, autoign. temp.: 572°F, vap. d: 7.90.

Acute tox data: Oral LD_{LO} (rabbit) = 120 mg/kg; sc LD_{LO} (frog) = 200 mg/kg. [*3*]

THR = HIGH via oral, dermal and sc routes. An irr and an allergen. See also nitrates. Dermal exposure ⟶ local and systemic allergic reactions. Can cause allergic as well as irr dermatitis. Symptoms of systemic poisoning are nausea, vomiting, diarrhea, suppressed urine, yellow discoloration of skin and convulsions, as well as stupor, skin eruptions, pruritis, anuria, abdominal pain and oliguria. [*2*]

Spont Heating: No.

Explosion Hazard: Dangerous, when shocked or exposed to heat (see explosives, high). Picric acid forms salts readily. Many of its salts, known as picrates, are more sensitive explosives than picric acid. This is particularly true of copper, lead and zinc. Therefore, this material must be kept out of contact with metals. It forms unstable salts with concrete, NH_3 and bases. [*19*] Picric acid is a more powerful explosive than TNT. Its ability to form the sensitive picrates in contact with metal has somewhat limited its usefulness as an explosive. In America, its primary use is in the converted form of ammonium picrate in base-fused shells for seacoast cannon and in all armor-piercing shells. Picric acid has also found use as a booster explosive and as a substitute for a part of the mercury fulminate charge in detonators. It has been used extensively in the form of mixtures with other nitro compounds. Such mixtures, having a lower melting point than picric acid, can be melted and cast at temps below 100°. These mixtures are more generally practical for use because of the hazard involved in melting straight picric acid, which has a relatively high melting temp. Guanidine picrate is an example of a picrate which is used in shells instead of pure picric acid. See also explosives, high.

Disaster Hazard: Highly dangerous; shock will explode it; on decomp, emits highly toxic fumes and explodes; can react vigorously with reducing materials.

PICRONITRIC ACID. See picric acid.

PICROTOXIN. See cocculus solid.

PICRYL CHLORIDE. Syn: *2-chloro-1,3,5-trinitrobenzene*. White needles. $C_6H_2ClN_3O_6$, mw: 247.6, mp: 83°, d: 1.797.

THR = See picric acid, nitrates and chlorides.

PICRYL SULFIDE. See hexanitrodiphenyl sulfide.

PIECE RESIN. See rosin.

PILOCARPINE. Colorless or yellow, hygroscopic, needle-like crystals. $C_{11}H_{16}N_2O_2$, mw: 208.26, mp: 34°, bp: 260° @ 5 mm.

Acute tox data: Oral LD_{50} (rat) = 91 mg/kg; ip LD_{50} (mice) = 500 mg/kg; im TD_{LO} (human) = 140 mg/kg ⟶CNS effects. [*3*]

THR = MOD via oral and ip routes. A very poisonous alkaloid. It can cause contraction of the pupils and lachrymation. The ingestion of toxic quantities of this material causes a marked secretion of saliva, excessive perspiration and tears, nausea, retching and vomiting, pain in the abdomen and violent movement of the intestines with profuse watery evacuation. The pulse is sometimes quickened, sometimes slow and irregular. The respiration is quick and labored with rales over the bronchi. Giddiness and confusion with tremors and feeble convulsions occur. Eventually slower respiration and weakness occur, but consciousness remains until breathing ceases. The action of pilocarpine on the sweat glands renders it the most powerful sudorific in the pharmacopeia. It is used to remove excess fluid accumulations in the body. It very rarely causes death, but when it does, it is by paralysis of the heart or edema of the lungs.

Disaster Hazard: Dangerous; on heating to decomp, emits toxic fumes of NO_x.

Treatment and Antidotes: Give general treatment for alkaloidal poisoning; in addition, atropine may be administered.

PIMELIC KETONE. See cyclohexanone.

PIMENTA. See allspice.

PINANE. $C_{10}H_{18}$, mw: 138, lel = 0.7% @ 320°F, uel = 7.2% @ 320°F, d: 0.8, bp: 168.9°.

"PINAP." flash p: 80°F (CC).

THR = U.

Fire Hazard: Dangerous, when exposed to heat or flame.

Explosion Hazard: U.

Disaster Hazard: Dangerous, upon exposure to heat or flame; can react vigorously with oxidizing materials.

PINDONE. See 2-pivalyl-1,3-indandione.

α-PINENE. Syn: *2,6,6-trimethylbicyclo-(3,1,1)-2-heptene*. Liquid, odor of turpentine. $C_{10}H_{16}$, mw: 136.2, mp: −55°, bp: 155°, flash p: 91°F, d: 0.8592 @ 20°/4°, vap. press: 10 mm @ 37.3°, vap. d: 4.7, autoign. temp.: 491°F.

THR = MOD irr to skin, eyes and mu mem and via oral, inhal and dermal routes. Irr to skin and mu mem. Can cause dizziness, palpitation, bronchitis and nephritis, GI upset, delerium, ataxia, coma. Fatal dose said to be 180 g (oral).

Fire Hazard: Dangerous, when exposed to heat or flame; can react with oxidizing materials.

To Fight Fire: Foam, CO_2, dry chemical.

PINE OIL. Pale yellow liquid, penetrating odor. Bp: 200°–220°, flash p: 172°F (CC), d: 0.86, flash p (steam distilled): 138°F.

THR = A weak allergen and an irr. See also turpentine.

Fire Hazard: Mod, when exposed to heat or flame; can react with oxidizing materials.

Spont Heating: Mod.

To Fight Fire: Foam, CO_2, dry chemical.

PINE PITCH. mp: 64.4°, bp: 255°, flash p: 285°F (CC), d: 1.1.

THR = A weak allergen.

Fire Hazard: Slight, when exposed to heat or flame; can react with oxidizing materials.

Spont Heating: Yes.

To Fight Fire: CO_2, dry chemical.

PINE RESIN. See rosin.

PINE TAR. Black-brown, viscous liquid, piny odor. bp: 240°–400°, flash p: 130°F (CC), d: > 1, autoign. temp.: 671°F.

THR = MILD irr and allergen.

Fire Hazard: Mod, when exposed to heat or flame; can react with oxidizing materials.

Spont Heating: Yes.

To Fight Fire: Foam, CO_2, dry chemical.

PINE TAR OIL. Syn: *tar oil rectified*. Dark, reddish-brown liquid. flash p: 144°F (CC), d: 0.96–0.99 @ 25°/25°.

THR = MILD irr and allergen.

Fire Hazard: Mod, when exposed to heat or flame; can react with oxidizing materials.

To Fight Fire: CO_2, dry chemical.

PINTSCH GAS. Typical composition: 30.0% illuminants, 0.1% carbon monoxide, 13.2% hydrogen, 45.0% methane, 9.0% ethane, 0.2% carbon dioxide, 1.6% nitrogen.

THR = See carbon monoxide.

Fire Hazard: Mod, when exposed to flame.

Explosion Hazard: U.

Disaster Hazard: Mod dangerous; when heated, emits toxic fumes; can react with oxidizing materials.

PIPERAZINE. Syn: *hexahydropyrazine, diethylenediamine*. Colorless rhombic crystals.

$\overline{NHCH_2CH_2NHCH_2CH_2}$, mw: 86.14, mp: 104°, bp: 145°, flash p: 190°F (OC), d: 1.1, vap. d: 3.0.

Acute tox data: Oral LD_{50} (rat) = 3800 mg/kg; sc

LD_{50} (mice) = 1100 mg/kg; dermal LD_{50} (rabbit) = 4000 mg/kg. [*3*]
THR = MOD via oral, sc and dermal routes. Excessive absorption can cause urticaria, vomiting, diarrhea, blurred vision and weakness.
Disaster Hazard: Dangerous; when heated to decomp, emits highly toxic fumes of NO_x.
Fire Hazard: Mod, via heat, flames, oxidizers.
To Fight Fire: Alcohol foam, mist, dry chemical, spray.

PIPERIDINE. Syn: *hexahydropyridine*. Clear, colorless liquid, amine-like odor. $C_5H_{11}N$, mw: 85.2, mp: −7°, bp: 106°, flash p: 61°F, d: 0.8622 @ 20°/4°, vap. press: 40 mm @ 29.2°, vap. d: 3.0.
Acute tox data: Oral LD_{50} (rat) = 400 mg/kg; dermal LD_{50} (rabbit) = 320 mg/kg; inhal LC_{LO} (rat) = 4000 ppm for 4 hrs. [*3*]
THR = HIGH via oral and dermal, MOD via inhal routes.
Fire Hazard: Dangerous, when exposed to heat, flame or oxidizers.
Disaster Hazard: Dangerous; when heated to decomp, emits highly toxic fumes of NO_x; can react vigorously with oxidizing materials.
To Fight Fire: Alcohol foam, CO_2, dry chemical.

PIPERIDINIUM PENTAMETHYLENE DITHIOCARBAMATE. White to cream crystals.
$C_{11}H_{22}N_2S_2$, mw: 246.5, mp: 170°, d: 1.13 @ 25°.
THR = HIGH via oral route. See also carbamates.
Disaster Hazard: Dangerous; when heated to decomp, emits highly toxic fumes of SO_x and NO_x.

PIPERINE. Syn: *1-piperoyl piperidine*. Crystals.
$C_{17}H_{19}NO_3$, mw: 285.3, mp: 130°, d: 1.193.
THR = Details U; an insecticide. Probably not of HIGH toxicity.
Disaster Hazard: Dangerous; when heated to decomp, emits highly toxic fumes of NO_x.

PIPERONAL. Syns: *3,4-methylene dioxybenzaldehyde, heliotropin*. Colorless, lustrous crystals. $C_8H_6O_3$, mw: 150.1, mp: 37°, bp: 263°, vap. press: 1 mm @ 87.0°.
Acute tox data: Oral LD_{50} (rat) = 2700 mg/kg. [*3*]
THR = MOD via oral route. A mosquito repellent. Used as a synthetic flavoring substance and adjuvant. [*109*] See also aldehydes. Can cause CNS depression.
Fire Hazard: Slight, when exposed to heat or flame; can react with oxidizing materials.

PIPERONYL. $C_{21}H_{30}O_2N_3F$, mw: 375.5.
Acute tox data: Oral LD_{50} (rat) = 160 mg/kg; sc LD_{50} (rat) = 160 mg/kg; iv LD_{50} (rat) = 48 mg/kg. [*3*]
THR = HIGH via oral, sc and iv routes.

PIPERONYL BUTOXIDE. Syn: *butylcarbityl (6-propyl piperonyl) ether butacide*. Light brown liquid, mild odor. $C_{19}H_{30}O_5$, mw: 338.4, bp: 180° @ 1 mm, flash p: 340°F, d: 1.04–1.07 @ 20°/20°.
Acute tox data: Oral LD_{50} (mice) = 3800 mg/kg; oral LD_{50} (rabbit) = 7500 mg/kg. [*3*]
THR = MOD oral route. An exper neo and carc via oral and sc routes. [*3*] Limited acute animal exper show LOW toxicity and little irr. Excessive ingestion can cause GI disturbances. Used as a food additive permitted in the feed and drinking water of animals and/or for the treatment of food-producing animals; also permitted in food for human consumption. [*109*]
Fire Hazard: Slight, when exposed to heat or flame; can react with oxidizing materials.
To Fight Fire: Foam, CO_2, dry chemical.

PIPERONYL CYCLONENE. Syn: *3-isoamyl-5-(methylene dioxyphenyl)-2-cyclohexanone*.
$C_5H_{11}CH_2(OC_6H_5)_2C_6H_8O$, mw: 367.5, flash p: 290°–300°F, d: 1.109–1.20 @ 20°/20°.
THR = Details U. See piperonyl butoxide, which is similar in action.
Fire Hazard: Slight, when exposed to heat or flame; can react with oxidizing materials.
To Fight Fire: Foam, CO_2, dry chemical.

PIPERONYL ETHER BUTOXIDE. $C_{19}H_{30}O_5$, mw: 338.5.
THR = An exper neo via oral route. [*3*]

1-PIPEROYL PIPERIDINE. See piperine.

PIPE SMOKE. A recog carc. [*14*]

PITCH. A recog carc. See coal tar and pitch. [*14*]

PITCH BLENDE. An ore of uranium, radium, etc.
THR = Continuous exposure by inhal may cause carcinoma of the lungs. [*14*] See uranium.

PITCH COKE. A recog carc. [*14*] See coal tar and pitch.

PIVAL. See 2-pivalyl-1,3-indandione.

PIVALIC ACID. See trimethylacetic acid.

2-PIVALYL-1,3-INDANDIONE. Syns: *pival, pindone*. Yellow crystals. $C_{14}H_{14}O_3$, mw: 230.3, mp: 108°.
Acute tox data: Oral LD_{50} (rat) = 280 mg/kg; pa LD_{50} (rat) = 50 mg/kg. [*3*]
THR = HIGH via pa and oral routes. Reduces blood-clotting leading to hemorrhages. For symptoms of exposure, see warfarin.

PLANOFIX. See α-naphthaleneacetic acid.

PLANT AND FUNGAL PRODUCTS. A (S) carc of the liver, oral cavity and stomach. [*14*]

PLASMOSAN. See polyvinyl pyrrolidone.

PLASTER OF PARIS. See calcium sulfate.

PLASTICIZER SC. See triethylene glycol caprylate.

PLASTIC POWDERS.
THR = Variable; a weak allergen.
Fire Hazard: Mod, when exposed to heat or flame.
Disaster Hazard: Mod dangerous; when heated, they burn and may emit acrid fumes; they may react with oxidizing materials.

PLASTOLEIN X-55.
See diethylene glycol dipelargonate.

PLATINIC AMMONIUM CHLORIDE. See ammonium chloroplatinate.

PLATINIC CHLORIDE. Syns: *acid platinic chloride, chloroplatinic acid.* Brownish-yellow, very deliquescent, crystalline mass, easily sol in water and alcohol. $H_2PtCl_6 \cdot 6H_2O$, mw: 517.94, d: 2.431, mp: 60°.
THR = See platinum compounds and chlorides. Violent reaction with BrF_3. [*19*]

PLATINIC DISULFIDE. Black-brown powder. PtS_2, mw: 259.36, mp: 225°–250° (decomp), d: 7.22.
THR = See platinum compounds and sulfides.

PLATINIC TETRABROMIDE. Dark brown crystals. $PtBr_4$, mw: 514.89, mp: decomp @ 180°, d: 5.69.
THR = See platinum compounds and bromides. Violent reaction with BrF_3. [*19*]

PLATINIC TETRACHLORIDE. Brown-red crystals. $PtCl_4$, mw: 337.06, mp: decomp @ 370°.
THR = See platinum compounds and chlorides.

PLATINIC TETRAFLUORIDE. Deep red fused mass or yellowish-light brown, deliquescent crystals. PtF_4, mw: 271.23, mp: decomp.
THR = See fluorides and platinum compounds.

PLATINIC TETRAIODIDE. Amorphous or brown to black crystals. PtI_4, mw: 702.91, mp: decomp @ 370°, d: 6.064 @ 25°.
THR = See platinum compounds and iodides.

PLATINOUS-AMMONIUM CHLORIDE. See ammonium chloroplatinate.

PLATINOUS CYANIDE. Yellow-brown crystals. $Pt(CN)_2$, mw: 247.13.
THR = See cyanides.

PLATINOUS DIBROMIDE. Brown crystals. $PtBr_2$, mw: 355.05, mp: decomp @ 250°, d: 6.65.
THR = See platinum compounds and bromides.

PLATINOUS DICHLORIDE. Olive green crystals. $PtCl_2$, mw: 266.14, mp: decomp @ 581°, d: 5.87 @ 11°.
THR = See platinum compounds and chlorides.

PLATINOUS DIFLUORIDE. Yellowish-green crystals. PtF_2, mw: 233.23.
THR = See fluorides and platinum compounds.

PLATINOUS DIIODIDE. Black crystals. PtI_2, mw: 449.07, mp: 300°–350° (decomp), d: 6.4.
THR = See platinum compounds and iodides.

PLATINOUS HYPOPHOSPITE. $Pt(PH_2O_2)_2$, mw: 325.1.
THR = HIGH toxicity. See phosphine. Liberates PH_3 spont @ > 130°. [*19*]

PLATINOUS MONOSULFIDE. Black crystals. PtS, mw: 227.29, mp: decomp, d: 8.847.
THR = See platinum compounds and sulfides.

PLATINUM. Silvery-white, malleable, ductile metal, stable in air. Pt, atwt: 195.09, mp: 1772°, bp: 3827°, d: 21.45 @ 20°.
THR = See platinum compounds.
Radiation Hazard: For permissible levels, see Section 5A, Table 5A.5.

PLATINUM ARSENIDE. Syn: *sperrylite.* Cubic white crystals. $PtAs_2$, mw: 345.05, mp: > 800°, d: 10.602.
THR = HIGH toxicity. See arsenic compounds and platinum compounds.
Fire Hazard: Mod, by chemical reaction. Reacts violently with (BrF_3 + KF), ($H_2 + O_2$), H_2O_2, Li, OF_2, H_2SO_5, P. [*19*]
Disaster Hazard: Dangerous; see arsenic compounds.

PLATINUM COMPOUNDS.
THR = Exposure to *complex* platinum salts has been shown to cause symptoms of intoxication such as wheezing, coughing, running of the nose, tightness of the chest, shortness of breath and cyanosis; exposure to dust of *pure* metallic platinum causes no intoxication. Furthermore, many people working with platinum salts are troubled with dermatitis. This seems only to be true of complex platinum salts. It does not include the complex salts of the other precious metals. Platinum ammine nitrates and perchlorates either detonate when heated or are impact-sensitive. [*19*]
Treatment and Antidotes: Removal from exposure effectively causes the symptoms to disappear.

PLATINUM FULMINATE. Solid. $Pt(C_2N_2O_2)_2$, mw: 363.3.
THR = See platinum compounds.
Fire Hazard: U.
Explosion Hazard: Severe, when shocked or exposed to heat. See also explosives, high, and fulminates.
Disaster Hazard: Dangerous; shock will explode it; when heated to decomp, emits highly toxic fumes of NO_x.

PLATINUM PYROPHOSPHATE. Green-yellow crystals. PtP_2O_7, mw: 369.27, mp: decomp @ 600°, d: 4.85.
THR = See platinum.

PLATINUM SALTS. See platinum compounds.

PLATINUM SESQUISULFIDE. Gray crystals (exist quest.). Pt_2S_3, mw: 486.66, mp: decomp, d: 5.52.
THR = See platinum compounds and sulfides.

PLINY THE ELDER WRITES OF OCCUPATIONAL MEDICINE. See Section 1.

PLIOFILM. $(C_3H_5Cl)_n$.
THR = An exper carc [*3*] via im route.

PLUMBAGO. See graphite.

PLUMBIC FLUORIDE. See lead tetrafluoride.

PLUMBOUS SULFIDE. See lead sulfide.

PLUMBUM. See lead.

PLUTONIUM. A silvery radioactive metal, chemically reactive. Pu, d: 19.84 @ 25°, mp: 641°, bp: 3232°.
Acute tox data: Permissible body burden = 0.6 μg; permissible lung burden = 0.25 μg. Chemical toxicity is trivial compared to radiation effects. LD (human) is calculated at 50 μg. HIGH carc and leukemogen. [*28*]
THR = The permissible levels for plutonium are the lowest for any of the radioactive elements. This is occasioned by the conc of plutonium directly on bone surfaces, rather than the more uniform bone distribution shown by other heavy elements. This increases the possibility of damage from equivalent activities of plutonium and has led to the adoption of the extremely low permissible levels given.
Radiation Hazard: For permissible levels, see Section 5A, Table 5A.5. Artificial isotope ^{238}Pu, $T\frac{1}{2}$ = 86y decays to radioactive ^{234}U via α's of 5.5 MeV. Artificial isotope ^{239}Pu, $T\frac{1}{2}$ = 24,000y decays to radioactive ^{235}U via α's of 5.1 MeV. Artificial isotope ^{240}Pu, $T\frac{1}{2}$ = 6600y decays to radioactive ^{236}U via α's of 5.1 MeV. Artificial isotope ^{241}Pu (Neptunium Series), $T\frac{1}{2}$ = 13y decays to radioactive ^{241}Am via β's of 0.02 MeV. Artificial isotope ^{242}Pu, $T\frac{1}{2}$ = 3.8×10^5y decays to radioactive ^{238}U via α's of 4.9 MeV. Reacts violently with air, CCl_4, H_2O. [*19*]

PLUTONIUM COMPOUNDS.
THR = The toxicity of Pu compounds is based first upon the very high radiotoxicity of the Pu atom and secondly upon whatever atoms or combinations of atoms they might contain. See also plutonium.
Disaster Hazard: Very dangerous! Any disaster which could cause quantities of Pu or Pu compounds to be scattered about the environment can cause great ecological stress and render areas of the land unfit for public occupancy.

PLUTONIUM HYDRIDE. PuH_2, mw: depends on isotope used.
THR = Very HIGH radiotoxicity. See also plutonium. May be spont flam in air. [*19*]

PMA. See pyromellitic acid.

PODOPHYLLIN. Syn: *podophyllum resin*. Light yellow powder or small, yellow, bulky, fragile lumps, bitter, acrid taste.
Acute tox data: Oral LD_{50} (mice) = 68 mg/kg; sc LD_{50} (mice) = 58 mg/kg. [*3*]
THR = HIGH irr to skin, eyes and mu mem, via oral and sc routes.
Fire Hazard: Slight, when exposed to heat or flame; can react with oxidizing materials.

PODOPHYLLOTOXIN. $C_{22}H_{22}O_8$, mw: 414.4.
Acute tox data: Oral LD_{50} (mice) = 90 mg/kg; sc LD_{50} (rat) = 8 mg/kg. [*3*]
THR = HIGH via oral and sc routes. An exper carc. [*23*]

PODOPHYLLUM RESIN. See podophyllin.

POGY OIL. See menhaden oil.

POISON IVY. Syns: *poison oak*, *poison sumac*. Leafy plant, vine or bush.
THR = Can cause a dermatitis in susceptible individuals, from bodily contact with the plant, exposure to the smoke from burning plants and contact with material which has been in contact with the plant. A common air contaminant.

POISON OAK. See poison ivy.

POISONOUS ARTICLES (IATA)

POISONOUS ARTICLES, CLASS A cover extremely poisonous gases and liquids, and their carriage is not acceptable either by passenger or cargo aircraft.

POISONOUS ARTICLES, CLASS B are substances, liquids or solids (including pastes and semisolids) other than Classes A or C, which are known to be so toxic to man as to afford a hazard to health during transportation or which, in the absence of adequate data on human toxicity, are presumed to be toxic to man.

The criteria used to determine whether an article warrants treatment as a Poison B are generally based on the standards used by the Public Health Authorities and Transport regulatory organisations in North America, such as the Department of Transportation (DOT) of USA and the Board of Transport Commissioners in Canada and are generally referred to as the LD_{50} (Lethal Dose for 50% of animals tested) values. These criteria are:

(i) Oral Toxicity—Those which produce death within 48 hrs in half or more than half of a group of 10 or more white laboratory rats weighing

200–300 g at a single dose of 50 mg or less per kg of body weight, when administered orally.

(ii) Toxicity on inhal—Those which produce death within 48 hrs in half or more than half of a group of 10 or more white laboratory rats weighing 200–300 g, when inhaled continuously for a period of 1 hr or less tested at a dosage of 2 mg or less per litre of vapor, mist or dust, provided such conc is likely to be encountered by man when the chemical product is used in any reasonable foreseeable manner.

(iii) Toxicity by skin absorption—Those which produce death within 48 hrs in half or more than half of a group of 10 or more rabbits tested at a dosage of 200 mg or less per kg body weight, when administered by continuous contact with the bare skin for 24 hrs or less.

The foregoing categories shall not apply if the physical characteristics or the probable hazards to humans, as shown by experience, indicate that the substance will not cause serious sickness or death.

In the case of vaporizing liquids or gases having toxic properties the Threshold Limit Values (TLV's) are also taken into account. However, vap. press is recog as an important aspect of this assessment and due regard is paid to this.

POISONOUS ARTICLES, CLASS C generally cover dangerous lachrymatory gases or solids, very few of which are permitted carriage on passenger aircraft, but some of which are permitted on cargo aircraft.

POISONOUS LIQUID OR GAS, N.O.S. See Section 11.

POISONOUS LIQUIDS, N.O.S. See Section 11.

POISONOUS SOLIDS, N.O.S. See Section 11.

POISON SUMAC. See poison ivy.

POLICE GRENADES. See Section 11.

POLISHES, LIQUID (for metal, stove or furniture).
THR = Variable.
Fire Hazard: Dangerous, when exposed to heat or flame; they can react with oxidizing materials.
Explosion Hazard: U.

POLONIUM. Syn: *radium F.* A low melting, volatile, natural metallic element. Po, at mass: (210), mp: 254°, bp: 962°, d: 9.4.
THR = HIGH radiotoxicity. Very dangerous to handle. An exper carc. [*23*] See also plutonium compounds.
Radiation Hazard: For permissible levels, see Section 5A, Table 5A.5. Natural isotope ^{210}Po (radium-F, Uranium Series), $T_{\frac{1}{2}}$ = 138d. Decays to stable ^{206}Pb via α's of 5.3 MeV.

POLONIUM CARBONYL. Radioactive material. PoCO, mw: 238.
THR = HIGH. See carbonyls and polonium.
Radiation Hazard: See polonium.
Fire Hazard: Dangerous, when exposed to heat or by chemical reaction with oxidizing agents. See carbonyls.
Explosion Hazard: See carbonyls.
Disaster Hazard: Dangerous; when heated to decomp, emits highly toxic fumes of carbon monoxide; can react vigorously with oxidizing materials.

POLONIUM HYDRIDE. Volatile, unstable material. PoH_2, mw: 212.02.
THR = See hydrides and polonium.

POLONIUM HYDROXIDE. Radioactive solid. $Po(OH)_4$, mw: 278.
THR = See polonium.

POLYACRYLAMIDE. White solid, water-sol high polymer. $(CH_2CHCONH_2)_x$.
THR = U. A food additive permitted in food for human consumption. [*109*]

POLYAMIDES. Syns: *nylon*, *nylonfilm*, *amylan*, *caprolon.* $(NOC_6N_{11})_n$.
THR = An exper neo [*103*, *3*] by implantation. Reacts violently with F_2. [*19*]

POLYAMYL NAPHTHALENE (mixed polymers). flash p: 360° F (OC), d: 0.9.
Fire Hazard: Low.
To Fight Fire: Dry chemical, CO_2.

POLYAMYL NAPHTHALENE. A mixture of polymers. flash p: 360° F (OC), d: 0.92–0.93.
THR = Details U. See also naphthalene.
Fire Hazard: Slight, when exposed to heat or flame; can react with oxidizing materials.
To Fight Fire: Foam, CO_2, dry chemical.

POLY-*sec*-AMYL PHENOL. bp; 305°–355°, flash p: 520° F, d: 0.90–0.92 @ 30°/20°.
THR = Details U. See also phenol.
Fire Hazard: Slight, when exposed to heat or flame.
Disaster Hazard: Dangerous; when heated to decomp, emits toxic fumes; can react with oxidizing materials.
To Fight Fire: Foam, CO_2, dry chemical.

POLYCHLORINATED BIPHENYLS. See chlorinated diphenyls.

POLYCHLORINATED TRIPHENYL (AROCHLOR 5442).
Acute tox data: Oral LD_{50} (rat) = 10600 mg/kg; dermal LD_{LO} (rabbit) = 3160 mg/kg. [*3*]
THR = MOD via dermal and LOW via oral routes. An exper (+) carc. [*3*, *1*]

POLYCHLORINATED TRIPHENYL (AROCHLOR 5460).
Acute tox data: Oral LD_{50} (rat) = 19200 mg/kg. [*3*]
THR = Practically NONE acute via oral route. An exper (+) carc. [*3, 1*]

POLYCHLOROPENTANES. bp: 174°, flash p: 175°F (OC), d: 1.33.
THR = See chlorinated hydrocarbons (aliphatic).
Fire Hazard: Mod, when exposed to heat or flame; can react with oxidizing materials.

POLYCHLORO BIPHENYLS.
See chlorinated diphenyls.

POLYCHLOROTRIFLUOROETHYLENE. Syn: *halocarbon*. Oils, greases and waxes which are generally chemically inert except to molten sodium, liquid fluorine, and liquid chlorine trifluoride. $(CF_2—CFCl)_n$.
THR = Little or no detail.
Disaster Hazard: Dangerous. Decomp to highly toxic volatile fumes occurs rapidly at temp. above 300°.

POLYCOUMARONE RESIN. See coumorone-indene resin.

POLYDIMETHYL SILOXANE. See latex.

POLYDYMITE. See nickelous, nickelic sulfide.

POLYETHYL CHLOROSILICONE. A (S) carc. [*14*] See polymers.

POLYETHYLENE. Syns: *PE, polythene*. Odorless. The high molecular weight materials are tough, white, leathery, resinous materials. $(C_2H_4)_x$.
THR = An exper neo. [*3, 14*] A food additive resulting from contact with packaging materials. [*109*] Violent reaction with F_2. [*19*] An exper neo to rats and mice via imp route. [*103*]

POLYETHYLENE GLYCOL 400. Syn: *carbowax*. Liquid. $HOCH_2(CH_2OCH_2)_nCH_2OH$ (*n* varies from 8–10), mw: 285–315, d: 1.110–1.140 @ 20°, mp: 4°–10°, flash p: 471°F.
THR = An exper carc. [*3*] A food additive permitted in food for human consumption. [*109*]
Fire Hazard: Mod, when exposed to heat or flame.
To Fight Fire: Water, foam, dry chemical.

POLYETHYLENE GLYCOL 6000. White, waxy solid, sol in water. mp: 58°–62°, flash p: >475°F.
Acute tox data: Oral LD_{LO} (rat) = 50000 mg/kg. [*3*]
THR = Practically NONE via oral route. See also other polyethylene glycols. A food additive permitted in food for human consumption. [*109*]
Fire Hazard: Slight, when exposed to heat or flame.

POLYETHYLENE GLYCOL CHLORIDE 210. Liquid. $ClC_2H_4OC_2H_4OH$, mw: 124.57, mp: −90°, bp: 198.9°, flash p: 225°F (OC), d: 1.1753 @ 20°/20°, vap. d: 4.31.
Acute tox data: Oral LD_{50} (rat) = 1070 mg/kg; dermal LD_{50} (rabbit) = 3180 mg/kg. [*3*]
THR = MOD via oral and dermal routes.
Fire Hazard: Slight, when exposed to heat or flame.
Disaster Hazard: Dangerous; when heated to decomp, emits highly toxic fumes of chlorides; can react with oxidizing materials.
To Fight Fire: Water, foam, CO_2, dry chemical.

POLYETHYLENE GLYCOL MONOSTEARATE.
THR = An exper carc and neo. [*3*]

POLYETHYLENE POLYSULFIDE. See "omilite."

POLYETHYLENE TERPHTHALATE (film). Syns: *Mylar, Dacron*. $(C_{10}H_8O_4)_n$.
THR = An exper neo [*3*] via imp route.

POLYETHYLENE Y-141-A.
THR = An exper neo. [*3*]

POLYISOBUTYLENE. $[CH_2C(CH_3)_2]_x$.
THR = Violent reaction with Ag_2O_2. [*19*]

POLYMERIC DIALDEHYDE.
THR = An exper neo. [*3*]

POLYINDENE RESINS. See coumorone-indene resin.

POLYMERIZABLE MATERIALS. Any liquid, solid, or gaseous material, which under conditions incident to transportation may polymerize (combine or react with itself) so as to cause dangerous evolution of gas or heat.

POLYMERIZED ACETALDEHYDE.
See metaldehyde.

POLYMERIZED AND OXIDIZED ALIPHATIC AND AROMATIC HYDROCARBONS AND EPOXIDES FROM VOLATILIZED GASOLINE EXPOSED TO OZONIDES.
THR = A recog carc. [*14*] See oils (mineral).

POLYMERIZED AND OXIDIZED FATTY ACID EMULSIFIERS.
THR = A (S) carc. [*14*] See thermic and oxidation products.

POLYMERS, WATER-INSOLUBLE.
THR = These are (S) carc [*14*] at sight of bodily implantation.

POLYMERS, WATER-SOLUBLE.
THR = A (S) carc of soft tissues around implant, lung, mucosal contact areas, organs and tissues of retention and deposition. [*14*]

POLYMETHACRYLATES.
THR = A (S) carc. [*14*] See polymers.

POLYMETHYLCHLOROSILICONE.
THR = A (S) carc. [*14*] See polymers.

POLYMIXIN. A series of antibiotic substances, polypeptide (basic), sol in water.

Acute tox data: ip LD_{LO} (mice) = 13 mg/kg; sc LD_{LO} (mice) = 68 mg/kg; iv LD_{LO} (mice) = 6 mg/kg; ic LD_{LO} (rabbit) = 0.6 mg/kg. [*3*]

THR = HIGH via ip, sc, iv and ic routes. A food additive permitted in food for human consumption. [*109*]

POLYOXYALKYLENE SORBITAN ALKYLATE. See Tween 80.

POLYOXYETHYLENE.

THR = An exper carc. [*14*] See surfactants.

POLYOXYETHYLENE GLYCOL (400) MONO- AND DI-OLEATES.

THR = U. Used as food additives permitted in the feed and drinking water of animals and/or for the treatment of food-producing animals. [*109*]

POLYOXYETHYLENE(20)SORBITAN TRISTEARATE. Syn: *polysorbate 65*. Tan colored, waxy solid @ 25°, faint odor, sol in mineral oil, petroleum ether, acetone, ether and ethanol. $C_{100}H_{194}O_{28}$ (approx), mw: 1842 (approx), d: 1.05 @ 25°.

THR = U. A food additive permitted in food for human consumption. [*109*]

POLYOXYETHYLENE SORBITAN MONOOLEATE. See Tween 80.

POLYOXYETHYLENE (8) STEARATE.

THR = An exper carc [*3*] of the bladder via oral, ip and im routes. [*14*]

POLYOXYMETHYLENE GLYCOL.

See paraformaldehyde.

POLYPHENOL FORMALDEHYDE PLASTICS. A (S) carc. [*14*] See polymers.

POLYPHENOL CHLOROSILICONE. A (S) carc. [*14*] See polymers.

POLYPHOSPHORIC ACID. Syn: *phospholeum*. Viscous liquid, water-sol with evolution of heat. $H_3PO_4 + P_2O_5$.

THR = MOD irr to skin, eyes and mu mem. See also phosphorus compounds, inorganic.

Disaster Hazard: Dangerous. When heated to decomp, emits highly toxic fumes of PO_x.

POLYPROPYLENE. The lightest plastic produced by stereoselective catalysts, white to yellow in color. $(C_3H_6)_n$, mw(commercial material): 40,000 or more, d: 0.90, mp: 165°–171° (isostatic).

THR = U. A food additive permitted in food for human consumption. [*109*] Reacts violently with Cl_2, $KMnO_4$. [*19*]

POLYPROPYLENE GLYCOL. Liquid.

$HO(C_3H_6O)_nH$, mw: 400–2000, mp: does not crystallize, flash p: 390°–> 440°F, d: 1.002–1.007.

Acute tox data: Oral LD_{50} (rat) = 419 mg/kg. [*3*]

THR = HIGH via oral route.

Fire Hazard: Slight, when exposed to heat or flame; can react with oxidizing materials.

To Fight Fire: Foam, CO_2, dry chemical.

POLYSILICONE. Syns: *silastic, silicone rubber*.

THR = An exper carc. [*3, 14*]

POLYSORBATE 60. See polyoxyethylene (20) sorbitan monostearate.

POLYSORBATE 65. See polyoxyethylene (20) sorbitan tristearate.

POLYSORBATE 80. See Tween 80.

POLYSTYRENE. Syns: *polystyrol, styrene polymer*.

THR = An exper carc [*3, 14*] via imp route. See also polymers.

POLYTETRAFLUOROETHYLENE. Syn: *Teflon*. $(C_2F_4)_n$, grayish-white tough plastic. Chemically very inert.

THR = An exper carc [*3, 14*] via imp route. The finished polymerized compound is inert under ordinary conditions. There have been reports of "polymer fume fever" in humans exposed to the unfinished product dust or to pyrolysis products which also are irr. Smoking should be prohibited in areas where this material is being fabricated or, in general, where there may be dust from it. It appears the chief health-related problem from "Teflon" is exposure to its pyrolysis or decomp products. It is an exper neo. [*3*]

Disaster Hazard: Dangerous; when heated to above 750°F, decomp to yield highly toxic fumes of fluorides.

POLYTETRAFLUOROMETHYLENE. A (S) carc. [*14*] See polymers.

POLYTHENE. See polyethylene.

POLYURETHANE FOAM. Syn: *M Polyurethane ether*.

THR = An exper neo. [*3, 14*]

POLYURETHANE PLASTIC. Syns: *polyfoam sponge, polyurethane sponge*. Sheet plastic.

THR = An exper neo [*3*] via imp route. Reacts violently with $(F_2 + O_2)$. [*19*]

POLYURETHANE Y-195.

THR = An exper neo [*3*] via imp route.

POLYURETHANE Y-217.

THR = An exper neo [*3*] via imp route.

POLYURETHANE Y-218.

THR = An exper neo [*3*] via imp route.

POLYURETHANE Y-221.

THR = An exper neo [*3*] via imp route.

POLYURETHANE Y-222.

THR = An exper neo [*3*] via imp route.

POLYURETHANE Y-223.
THR = An exper neo [*3*] via imp route.

POLYURETHANE Y-224.
THR = An exper neo [*3*] via imp route.

POLYURETHANE Y-225.
THR = An exper neo [*3*] via imp route.

POLYURETHANE Y-226.
THR = An exper neo [*3*] via imp route.

POLYURETHANE Y-229.
THR = An exper neo [*3*] via imp route.

POLYURETHANE Y-238.
THR = An exper neo [*3*] via imp route.

POLYURETHANE Y-290.
THR = An exper neo [*3*] via imp route.

POLYURETHANE Y-302.
THR = An exper neo [*3*] via imp route.

POLYURETHANE Y-304.
THR = An exper neo [*3*] via imp route.

POLYVINYL ALCOHOL (mixture of isomers). Syn: *PVA*. Colorless, amorphous powder. mp: decomp over 200°, flash p: 175°F (OC), d: 1.329.
THR = Not acutely toxic. MOD via oral route. An exper carc [*3*, *14*] via sc route. However, technical grades should not be used in foods or drugs without careful testing to assure conformity with the Federal Food, Drug and Cosmetic Act and similar state acts. See polymers, water-sol.
Fire Hazard: Mod, when exposed to heat or flame; can react with oxidizing materials.
Explosion Hazard: Slight, in the form of dust, when exposed to flame.
To Fight Fire: Alcohol foam, CO_2, dry chemical.

POLYVINYL CHLORIDE.
See chloroethylene polymer.

POLYVINYL METHYL ETHER. Light yellow to amber-colored, balsam-like liquid. $(C_3H_6O)_x$, mw: $(58.1)_x$, d: 1.05 @ 25°/4°.
THR = Details U. See also ethers.
Fire Hazard: Mod, when exposed to heat or flame; can react with oxidizing materials.
Explosion Hazard: Details U. See also ethers.

POLYVINYL PYRROLIDONE. Syn: *PVP*. A free flowing, white, amorphous powder, sol in water, chlorinated hydrocarbons, alcohols, amines, nitroparaffins and lower molecular weight fatty acids. $(C_6H_9ON)_n$, d: 1.23–1.29.
Acute tox data: Oral LD_{50} (rat) = 10,000 mg/kg; oral LD_{50} (guinea pig) = 10,000 mg/kg. [*3*]
THR = LOW via oral route. A food additive permitted in food for human consumption. [*109*] See also PVP's 1-7.

POPPY OIL. See poppy seed oil.

POPPY SEED OIL. Syn: *poppy oil*. Very pale golden yellow liquid, pleasant odor, sol in ether, chloroform, petroleum ether and carbon disulfide, insol in water. d: 0.924–0.928, flash p: 491°F, fp: −18°.
THR = U.
Fire Hazard: Slight, when exposed to heat or flame.
To Fight Fire: Mist, dry chemical, CO_2.

PORTLAND CEMENT. See cement, portland.

POTABA. See potassium-*p*-amino benzoate.

POTASAN. Syn: *4-methyl-7-hydroxycoumarin diethoxythiophosphate*. Crystals, weak aromatic odor. $C_{14}H_{17}O_5PS$, mw: 328.3, mp: 38°, bp: 210° @ 1 mm, d: 1.260 @ 38°/4°.
Acute tox data: Oral LD_{50} (rat) = 19 mg/kg; ip LD_{50} (rat) = 15 mg/kg; oral LD_{50} (mice) = 99 mg/kg; sc LD_{50} (mice) = 25 mg/kg; dermal LD_{50} (rabbit) = 300 mg/kg. [*3*]
THR = HIGH via oral, ip, sc and dermal routes. See also parathion.
Disaster Hazard: Dangerous; see parathion.

POTASH. See potassium carbonate.

POTASH MURIATE. See potassium chloride.

POTASSAMIDE. See potassium amide.

POTASSIUM. Soft ductile, silvery-white, very reactive metal. K, atwt: 39.10, mp: 63.65°, bp: 774°, d: 0.862 @ 20°.
THR = The toxicity of potassium compounds is almost always that of the anion.
Radiation Hazard: For permissible levels, see Section 5A, Table 5A.5. Natural (0.012%) isotope ^{40}K, $T\frac{1}{2} = 1.27 \times 10^9$y. Decays to stable ^{40}A by ec (11%). Also decays to stable ^{40}Ca via β's of 1.32 MeV. Emits γ's of 1.46 MeV and x-rays.
Fire Hazard: Dangerous. Metallic potassium reacts with moisture to form potassium hydroxide and hydrogen. The reaction evolves much heat, causing the potassium to melt and spatter. It also ignites the hydrogen, which burns, or—if there is any confinement—an explosion can occur. Burning potassium is difficult to extinguish; dry powdered soda ash or graphite or special mixtures of dry chemical are recommended. It can ignite spont in moist air.
Explosion Hazard: It reacts violently with the following materials under required conditions of temp., pressure, state of division: C_2H_2, air, (moist air), $AlBr_3$, metallic halides, ammonium chlorocuprate, NH_4Br, NH_4I, $[(NH_4)_2SO_4 + NH_4NO_3]$, Sb and As halides, $(AsH_3 + NH_3)$, Bi_2O_3, boric acid, BBr_3, Br_2, C, CO_2, CS_2, $(CO + O_2)$, CCl_4, charcoal, chlorinated hydrocarbons, Cl_2, ClO, ClF_3, $CHCl_3$, $CrCl_4$, CrO_3, Cu_2OCl_2, CuO, dichloromethane,

ethylene oxide, F_2, graphite, (graphite + air), I, (graphite + K_2O_2), HI, H_2O_2, IBr, ICl, IF_5, Pb_2OCl_2, PbO_2, $PbSO_4$, maleic anhydride, Hg_2O, CH_3Cl, MoO_3, NO_2, P_2NF, peroxides $COCl_2$, (PH_3 + NH_3), P, PCl_5, P_2O_5, PBr_3, PCl_3, potassium chlorocuprate, K oxides, KO_3, K_2O_2, KO_2, Se, $SeOCl_2$, $SiCl_4$, $AgIO_3$, $NaIO_3$, (NH_3 + $NaNO_2$), Na_2O_2, (SnI_4 + S), SnO_2, S, SBr_2, SCl_2, Te, tetrachloroethane, thiophosphoryl fluoride, $VOCl_2$, H_2O. [*19*] Potassium metal will form the peroxide (K_2O_2) and the superoxide (KO_3 or K_2O_4) at room temp. even when stored under mineral oil. Metal which has oxidized on storage under oil may explode violently when handled or cut. Oxide-coated potassium should be destroyed by burning.

Disaster Hazard: Dangerous; a highly reactive alkali metal. See sodium and lithium. In the presence of moist air it can spont catch fire and burn with great intensity. It may even explode. Reacts violently with moisture, acid fumes and oxidizers.

POTASSIUM ACETATE. White powder. CH_3COOK, mw: 98.14, mp: 292°, d: 1.8 @ 20°/20°.
Acute tox data: Oral LD_{50} (rat) = 3250 mg/kg. [*3*]
THR = MOD via oral route.
Disaster Hazard: Slight; when heated, emits acrid fumes.

POTASSIUM ACETYLIDE. KCCH, mw: 64.1.
THR = In CO_2 it can incandesce; reacts violently with Cl_2, SO_2. [*19*]

POTASSIUM ACID CARBONATE. See potassium bicarbonate.

POTASSIUM ACID FLUORIDE. See potassium bifluoride.

POTASSIUM ACID OXALATE.
See potassium binoxalate.

POTASSIUM ACID TARTRATE. Syns: *cream of tartar, potassium bitartrate.* White crystals or powder, sol in water, insol in alcohol. $KHC_4H_4O_6$, mw: 188, d: 1.984 @ 18°.
THR = U. Used as a general purpose food additive. [*109*]

POTASSIUM ALGINATE. Syn: *potassium polymannuronate.* A hydrophilic colloidal substance, occurs in filamentous, grainy, granular or powder forms, colorless or slightly yellow, slowly sol in water, insol in alcohol. $(C_6H_7P_6K)_n$, mw range: 32,000–35,000.
THR = U. A stabilizer food additive. [*109*]

POTASSIUM ALUMINUM BORATE, BASIC. Cubic white crystals. $K(AlO)_2(BO_2)_3$, mw: 253.50, mp: < 1800°, d: 3.415.
THR = See boron compounds and aluminum compounds.

POTASSIUM ALUMINUM FLUORIDE. White powder. K_3AlF_6, mw: 258.3.
THR = See fluorides.

POTASSIUM ALUMINUM SULFATE. See alum.

POTASSIUM AMALGAM. Silvery liquid or solid. K + Hg.
THR = HIGH. See potassium and mercury.
Fire Hazard: Mod, by spont chemical reaction; on contact with moisture, hydrogen is liberated. See also potassium.
Explosion Hazard: Mod; liberates hydrogen upon contact with moisture, acids, etc. See also potassium.
Disaster Hazard: Dangerous; when heated to decomp, emits highly toxic fumes of mercury and oxides of potassium; will react with water, steam or acids to produce hydrogen; can react with oxidizing materials.

POTASSIUM AMIDE. Syn: *potassamide.* Colorless-white or yellow-green crystals. KNH_2, mw: 55.12, mp: 335°, bp: sublimes @ 400°.
THR = See ammonia.
Fire Hazard: Mod, by chemical reaction; can react with moisture to liberate ammonia.
Explosion Hazard: Mod, due to the ammonia liberated. Violent hydrolysis. [*19*]
Disaster Hazard: Dangerous; will react with water or steam to produce toxic, corrosive and flam vapors; can react with oxidizing materials.

POTASSIUM-*p*-AMINOBENZOATE. Syn: *potaba.* Crystals, saline taste, water-sol. $C_7H_6KNO_2$, mw: 175.2.
THR = Details U. See *p*-aminobenzoic acid.

POTASSIUM AMMONIUM NITRATE. Syn: *BasF.* Gray to light or dark brown crystals. KNO_2 + NH_4NO_3.
THR = See nitrates.

POTASSIUM AMMONIUM SELENOSULFIDE. A solid. $(KNH_3S)_5Se$, mw: 525.01.
THR = See selenium compounds and sulfides.

POTASSIUM ANTIMONIDE. Yellow-green crystals. K_3Sb, mw: 239.05, mp: 812°.
THR = HIGH. See antimony compounds.
Fire Hazard: Mod, when exposed to heat or flame.
Explosion Hazard: Mod, when exposed to flame.
Disaster Hazard: Dangerous; when heated to decomp, or on contact with moisture or acids, emits highly toxic fumes of antimony; can react with oxidizing materials.

POTASSIUM ANTIMONY OXALATE. See antimony potassium oxalate.

POTASSIUM ANTIMONY TARTRATE. See antimony potassium tartrate.

POTASSIUM ARSENITE. Syn: *potassium-m-arsenite. gen arsenate; Macquer's salt.* Colorless crystals. KH_3AsO_4, mw: 181.1, mp: 288°, d: 2.867.
THR = A poison B. See also arsenic compounds.

POTASSIUM ARSENITE. Syn: *potassium-m-arsenite.* White powder. $KAsO_2 \cdot HAsO_2$, mw: 253.8.
Acute tox data: Oral LD_{50} (rat) = 14 mg/kg; dermal LD_{50} (rat) = 150 mg/kg. [*3*]
THR = HIGH via oral and dermal routes. See arsenic compounds. A human (S) carc. [*6, 3*] An exper neo and a deadly poison via oral and dermal routes. A human carc via oral route. [*103*] An exper carc to mice via dermal route. [*103*]

POTASSIUM AURATE. Light yellow needles. $KAuO_2 \cdot 3H_2O$, mw: 322.34, mp: decomp.
THR = See gold compounds.

POTASSIUM AURIBROMIDE. See gold potassium bromide.

POTASSIUM AURICHLORIDE. See gold potassium chloride.

POTASSIUM AUROCYANIDE. See gold potassium cyanide.

POTASSIUM AZIDE. Colorless crystals. KN_3, mw: 81.12, mp: 350°, d: 2.04.
THR = See azides. Violent reaction with CS_2, MnO_2. [*19*]

POTASSIUM AZIDOSULFATE. $KSO_4SO_2N_3$, mw: 241.3.
THR = Moisture can explode it. [*19*]

POTASSIUM BERYLLIUM FLUORIDE. White crystalline masses. $BeF_2 \cdot 2KF$, mw: 163.21.
THR = HIGH. See beryllium and fluorides.

POTASSIUM BERYLLIUM SULFATE. White powder. $K_2Be(SO_4)_2$, mw: 279.4.
THR = HIGH. See beryllium and sulfates.

POTASSIUM BICARBONATE. Syn: *potassium acid carbonate.* Colorless, transparent crystals or white powder, odorless, sol in water, insol in alcohol. $KHCO_3$, mw: 100, d: 2.17, mp: decomp between 100° and 120°.
THR = Probably LOW. Used as a general purpose food additive. [*109*] A very vigorous reaction with $NH_4H_2PO_4$. [*19*]

POTASSIUM BICHROMATE. Syns: *potassium dichromate, red potassium chromate.* Bright, yellowish-red, transparent crystals, bitter, metallic taste. $K_2Cr_2O_7$, mw: 294.21, mp: 398°, bp: decomp @ 500°, d: 2.69.
Acute tox data: Oral LD_{LO} (child) = 50 mg/kg; sc LD_{LO} (monkey) = 40 mg/kg; sc LD_{LO} (rabbit) = 10 mg/kg. [*3*]
THR = HIGH via oral, sc and inhal routes.
Fire Hazard: Mod, by chemical reaction. A powerful oxidizer. Reacts violently with (H_2SO_4 + acetone), hydrazine, hydroxylamine. [*19*]

POTASSIUM BIFLUORIDE. Syns: *potassium acid fluoride, Fremy's salt.* Colorless crystals. KHF_2, mw: 78.10, mp: decomp.
Acute tox data: Oral LD_{LO} (guinea pig) = 150 mg/kg; sc LD_{LO} (guinea pig) = 250 mg/kg. [*3*]
THR = HIGH via oral and sc routes. See also fluorides. In solution it is highly reactive and very irr to skin, eyes and mu mem.

POTASSIUM BINOXALATE. Syns: *potassium acid oxalate, sal acetosella, salt of sorrel, essential salt of lemon.* White, somewhat hygroscopic crystals, bitter, sharp taste. KHC_2O_4, mw: 128.12, mp: decomp, d: 2.0.
THR = See oxalates.

POTASSIUM BISULFIDE.
See potassium hydrosulfide.

POTASSIUM BISULFITE. See potassium hydrogen sulfite.

POTASSIUM BITARTRATE. See potassium acid tartrate.

POTASSIUM-*m*-BORATE. Colorless crystals. KBO_2, mw: 81.92, mp: 947°–950°.
THR = See boron compounds.

POTASSIUM BOROHYDRIDE. White crystals, sol in water by reaction. KBH_3, mw: 52.9, d: 1.177, mp: decomp $>$ 400°.
Acute tox data: Oral LD_{LO} (rat) = 160 mg/kg. [*3*]
THR = HIGH via oral route. Burns quietly in air. [*19*] See boron compounds and hydrides.

POTASSIUM BROMATE. White crystals or crystalline powder. $KBrO_3$, mw: 167.01, mp: 434°, decomp @ 370°, d: 3.27 @ 17.5°.
THR = A powerful oxidizer. An irr to skin, eyes and mu mem. A food additive permitted in food for human consumption. [*109*] Violent reaction with Al, As, C, Cu, $Pb(C_2H_3O_2)_2$, metal sulfides, organic matter, P, Se, S. [*109*]

POTASSIUM BROMIDE. Cubic, colorless, slightly hygroscopic crystals. KBr, mw: 119.01, mp: 730°, bp: 1380°, d: 2.75 @ 25°, vap. press: 1 mm @ 795°.
THR = See bromides. Violent reaction with BrF_3. [*19*]

POTASSIUM BROMOAURATE. Rhombic red-brown crystals. $KAuBr_4$, mw: 555.96, mp: decomp.
THR = See gold compounds and bromides.

POTASSIUM BROMOPLATINATE. Cubic, dark red-brown crystals. K_2PtBr_6, mw: 752.92, mp: decomp $> 400°$, d: 4.66 @ 24°.
THR = See platinum compounds and bromides.

POTASSIUM BROMOPLATINITE. Rhombic brown crystals. K_2PtBr_4, mw: 593.09.
THR = See platinum compounds and bromides.

POTASSIUM BROMOSTANNATE. Crystals. K_2SnBr_6, mw: 676.39, d: 3.783.
THR = See tin compounds and bromides.

POTASSIUM-*tert*-BUTOXIDE. $KO(CH_3)_3C$, mw: 112.2.
THR = Ignites on contact with acetone, ethyl methyl ketone, methyl isobutyl ketone, CH_3OH, C_2H_5OH, C_3H_7OH, isopropanol, ethyl acetate, *n*-butyl acetate, *n*-propyl formate, CH_3COOH, H_2SO_4, CH_2Cl_2, $CHCl_3$, CCl_4, epichlorohydrin, dimethyl carbonate, diethyl sulfate. [*19*]

POTASSIUM CACODYLATE. White crystals. $K[(CH_3)_2AsO_2] \cdot H_2O$, mw: 194.09.
THR = HIGH toxicity. See arsenic.

POTASSIUM CADMIUM IODIDE. See cadmium potassium iodide.

POTASSIUM CARBIDE. K_2C_2, mw: 102.2.
THR = Possible explosive reaction with water. [*19*]

POTASSIUM CARBONATE. Syns: *potash*, *pearl ash*. White, deliquescent, granular, translucent powder, sol in water, insol in alcohol. (*a*) K_2CO_3, mw: 138; (*b*) $K_2CO_3 \cdot \frac{1}{2}H_2O$, mw: 147, d: 2.428 @ 19°, mp: 891°, bp: decomp.
Acute tox data: Oral LD_{50} (rat) = 1870 mg/kg. [*3*]
THR = MOD via oral route. A strong caustic. Used as a general purpose food additive. [*109*] Violent reaction with ClF_3. [*19*]

POTASSIUM CARBONYL. Gray-red solid. $(KCO)_6$, mw: 402.64, mp: explodes.
THR = HIGH. See carbonyls.
Fire Hazard: Dangerous, when exposed to heat or flame.
Explosion Hazard: Mod, when shocked, exposed to heat, on contact with water, or by chemical reaction with air and O_2.
Disaster Hazard: Dangerous; shock will explode it; will react with water or steam to produce heat; can react vigorously with oxidizing materials.

POTASSIUM CHLORATE. Syn: *potassium oxymuriate*. Transparent colorless crystals or white powder, cooling, saline taste. $KClO_3$, mw: 122.55, mp: 368.4°, bp: decomp @ 400°, d: 2.32.
Acute tox data: Oral LD_{LO} (rat) = 7000 mg/kg; ip LD_{LO} (rat) = 1500 mg/kg; oral LD_{LO} (rabbit) = 2000 mg/kg. [*3*]
THR = MOD via oral and ip routes. See chlorates. Reacts violently with Al, NH_3, NH_4Cl, NH_4^+ salts, $(NH_4)_2SO_4$, Sb_2S_3, As, barium hypophosphite, BaS, B, calcium hypophosphite, CaS, C, charcoal, Cr, Cu, Cu_3P_2, gallic acid, Ge, HI, Mg, (Mg + $CuSO_4$ (anhydrous) + NH_4NO_3 + H_2O), MnO_2, Hg_3P_4, metal sulfides, dibasic organic acids, organic matter, P, Ag_2S, $NaNH_2$, S, SO_2, H_2SO_4, Zn, Zr, Ti, thiocyanates. [*19*]

POTASSIUM CHLORIDE. Syns: *potassium muriate*, *salt substitute*. Colorless or white crystals or powder, sol in water, slightly sol in alcohol, insol in absolute alcohol. KCl, mw: 74.5, d: 1.987, mp: 773° (sublimes @ 1500°).
Acute tox data: Oral LD_{50} (guinea pig) = 2500 mg/kg; ip LD_{50} (mice) = 552 mg/kg; iv LD_{LO} (rat) = 100 mg/kg. [*3*] A nutrient and/or dietary supplement food additive. [*109*] Large oral doses cause GI irr, purging, weakness and circulatory problems. Also affects blood picture. [*3*] Violent reaction with BrF_3, (H_2SO_4 + $KMnO_4$). [*19*]

POTASSIUM CHLOROAURATE. Monoclinic yellow crystals. $KAuCl_4$, mw: 378.12, mp: decomp @ 357°.
THR = See gold compounds and chlorides.

POTASSIUM CHLOROCHROMATE. Syn: *Peligot's salt*. Monoclinic red crystals. $KCrO_3Cl$, mw: 174.56, mp: decomp, d: 2.497.
THR = HIGH via oral and inhal routes. See chromium compounds.
Fire Hazard: Mod, by chemical reaction; a powerful oxidizer.
Disaster Hazard: Dangerous. See chlorides, chromates.

POTASSIUM CHLOROIODATE (III). Rhombic yellow crystals. $KICl_4$, mw: 307.84, mp: decomp, d: 1.76 @ 45°.
THR = Details U.
Disaster Hazard: Dangerous; when heated to decomp, or on contact with acid or acid fumes, emits highly toxic fumes of chlorides and iodides.

POTASSIUM CHLOROIRIDATE. Cubic black crystals. K_2IrCl_6, mw: 484.03, mp: decomp, d: 3.546.
THR = See iridium compounds and chlorides.

POTASSIUM CHLOROOSMATE (IV). Cubic red crystals. K_2OsCl_6, mw: 481.1, mp: decomp.
THR = See osmium compounds and chlorides.

POTASSIUM CHLOROPALLADATE. Cubic red crystals. K_2PdCl_6, mw: 397.63, mp: decomp, d: 2.738.
THR = See palladium compounds and chlorides.

POTASSIUM CHLOROPLATINATE. Cubic yellow crystals. K_2PtCl_6, mw: 486, mp: decomp @ 250°, d: 3.499 @ 24°

Acute tox data: id TD_{LO} (human) = 40 mg/kg ⟶ skin effects. [*3*]
THR = HIGH via id, oral and inhal routes. See platinum compounds and chlorides.

POTASSIUM CHLOROPLATINITE. K_2PtCl_4, mw: 415.1.
Acute tox data: Oral TD_{LO} (child) = 400 mg/kg; id TD_{LO} (human) = 40 mg/kg ⟶ skin effects. [*3*]

POTASSIUM CHLOROPLUMBATE. Cubes. K_2PbCl_6, mw: 498.14, mp: decomp @ 190°.
THR = See lead compounds and chlorides.

POTASSIUM CHLORORHENATE (IV). Yellow-green crystals. K_2ReCl_6, mw: 477.24, d: 3.34.
THR = See also rhenium compounds.

POTASSIUM CHLORORHENATE (IV). Cubic black crystals. K_2ReCl_6, mw: 392.63, mp: decomp.
THR = See rhenium compounds and chlorides.

POTASSIUM CHLOROSTANNATE. Cubic colorless crystals. K_2SnCl_6.
THR = See tin compounds and chlorides.

POTASSIUM CHLOROTELLURATE. Pale yellow crystals. K_2TeCl_6, mw: 418.54.
THR = See tellurium compounds and chlorides.

POTASSIUM CHROMATE. Syn: *tarapacaite.* Rhombic yellow crystals. K_2CrO_4, mw: 194.20, mp: 971°, d: 2.732 @ 18°.
Acute tox data: im LD_{50} (rabbit) = 11 mg/kg; sc LD_{LO} (dog) = 19 mg/kg. [*3*]
THR = HIGH via im, sc, oral and inhal routes. An exper (S) carc and neo. [*3, 6*] See chromium compounds and chromates.

POTASSIUM CHROME ALUM.
See chromium compounds.

POTASSIUM CHROMIC SULFATE. Red or green cubic crystals. $KCr(SO_4)_2 \cdot 12H_2O$, mw: 499.4, mp: 89°, bp: $-12H_2O$ @ 400°, d: 1.826 @ 25°.
THR = See chromium compounds and sulfates.

POTASSIUM CHROMIUM CHROMATE, BASIC. Violet-brown, amorphous powder. $K_2CrO_4 \cdot 2Cr(OH)CrO_4$, mw: 564.26, mp: 300°, d: 2.28 @ 14°.
THR = See chromium compounds.
Fire Hazard: Mod, by chemical reaction; a powerful oxidizer.

POTASSIUM CITRATE. Colorless or white crystals or powder, odorless, deliquescent, sol in water and glycerol, almost insol in alcohol. $K_3C_6H_2O_7 \cdot H_2O$, mw: 321.4, d: 1.98, decomp when heated to 230°.
Acute tox data: ip LD_{LO} (dog) = 167 mg/kg. [*3*]
THR = HIGH via ip; LOW via oral routes. A sequestrant food additive, also a general purpose food additive. [*109*]

POTASSIUM COBALTINITRITE. Syn: *Fischer's salt.* Yellow prisims. $K_3Co(NO_2)_6$, mw: 452.3.
THR = See cobalt compounds and nitrites.

POTASSIUM COBALTOUS SULFATE. Red monoclinic prisms. $K_2SO_4 \cdot CoSO_4 \cdot 6H_2O$, mw: 437.36, d: 2.218.
THR = See cobalt compounds and sulfates.

POTASSIUM COPPER CHLORIDE. Red needles. $KCl \cdot CuCl_2$, mw: 209, d: 2.86
THR = See copper compounds and chlorides.

POTASSIUM CUPROCYANIDE. Syns: *potassium copper cyanide, copper potassium cyanide.* White, crystalline, double salt of copper cyanide and potassium cyanide. Composition: copper (Cu) content (minimum): 25.8%; free KCN: 1.25–3.0%.
THR = HIGH. See cyanides.

POTASSIUM CYANATE. Colorless crystals. KOCN, mw: 81.11, mp: 700°–900° (decomp), d: 2.056 @ 20°.
Acute tox data: Oral LD_{50} (mice) = 841 mg/kg. [*3*]
THR = MOD via oral route. An exper neo [*3*] via ip route. An herbicide. Ingestion can cause irr of the GI tract. It is said to be slowly metabolized in body to cyanide but does not have high toxicity of cyanides.

POTASSIUM CYANIDE. White, deliquescent, crystalline solid, faint odor of bitter almonds. KCN, mw: 65.11, mp: 634.5°, d: 1.52 @ 16°.
Acute tox data: Oral LD_{50} (rat) = 10 mg/kg; ipLD_{50} (rat) = 4 mg/kg; scLD_{50} (rat) = 9 mg/kg. [*3*]
THR = HIGH via oral, ip and sc routes. A deadly poison. Toxic via dermal route. Violent reaction with chlorates, nitrites, NCl_3, $NaClO_3$. [*19*] See cyanides.

POTASSIUM CYANO-COMPOUNDS. With the exception of the ferri- and ferro-cyanides, and compound of potassium containing the cyanogen radical (CN) is highly toxic and should be handled with adequate ventilation and protective equipment. See cyanides.

POTASSIUM CYCLAMATE. Syn: *potassium cyclohexyl sulfamate.* $C_6H_{12}NO_3SK$, mw: 213.3.
THR = An exper carc. [*14*] See cyclamates. Used as a non-nutritive sweetener. [*109*] See also sodium cyclamate.

POTASSIUM DIBORANE. Syn: *diboranide.* White crystals. $K_2B_2H_6$, mw: 105.9, bp: decomp @ 300° @ 1 mm, d: 1.18.
THR = Details U. Probably highly toxic. See also boron compounds.

POTASSIUM DIBROMOIODIDE. Rhombic crystals. $KIBr_2$, mw: 325.85, mp: 60°, bp: decomp @ 180°.
THR = See bromides and iodides.

POTASSIUM DICHLOROIODIDE. Monoclinic crystals. $KICl_2$, mw: 236.93, mp: 60°, bp: decomp @ 215°.
THR = See chlorides and iodides.

POTASSIUM DICHLOROISOCYANURATE. White, slightly hygroscopic crystalline powder or granules, chlorine odor. $KC_3HO_3N_3Cl_2$, mw: 237.1, mp: 250° (decomp).

THR = The dry form can cause severe irr to eyes and abraided skin. Causes emaciation, weakness, lethargy, diarrhea, weight loss. Autopsy indicates GI tract irr, tissue edema, liver and kidney congestion. A powerful oxidizer.

POTASSIUM DICHROMATE.

See potassium bichromate.

POTASSIUM DICYANOGUANIDINE. Crystals. NCHNC(NH)NCNK, mw: 147.2, mp: decomp @ 265°.

THR = Details U. Possibly most toxic by skin absorption.

Disaster Hazard: Dangerous; when heated to decomp, or on contact with acid or acid fumes, it emits highly toxic fumes of cyanides.

POTASSIUM DIFLUOTELLURATE. Crystals. $K_2TeO_3F_2 \cdot 3H_2O$, mw: 345.85, mp: decomp.

THR = See fluorides and tellurium compounds.

POTASSIUM DIGERMANATE. White crystals. $K_2Ge_2O_5$, mw: 303.4, mp: >83°, d: 4.31 @ 21.5°.

THR = See germanium compounds.

POTASSIUM DIHYDROGEN-*o*-ARSENATE. See potassium arsenate.

POTASSIUM DIHYDROPHOSPHIDE. KPH_2, mw: 72.1.

THR = A spont flam solid. HIGH via oral and inhal routes due to evolution of phosphine.

POTASSIUM DIHYDROXYDIBORANE. Colorless cubic crystals. $K_2B_2H_6O_2$, mw: 137.88, mp: decomp to K, d: 1.39.

THR = See boron compounds.

POTASSIUM DIISOPROPYL DITHIOPHOSPHATE. Colorless, crystalline, nearly odorless solid. $[(CH_3)_2CHO]_2PSKS$, mw: 252.4, mp: decomp near 200°.

THR = Details U. Many derivatives of this material are highly toxic.

Disaster Hazard: Dangerous; when heated to decomp, emits highly toxic fumes of SO_x and PO_x.

POTASSIUM DISULFIDE. Red-yellow crystals. K_2S_2, mw: 142.3, mp: 470°.

THR = See sulfides.

POTASSIUM ETHYLXANTHATE.

See potassium xanthate.

POTASSIUM FERRIC SULFIDE. Purple hexagonal crystals. $KFeS_2$, mw: 159.1, d: 2.563.

THR = See sulfides and iron compounds.

POTASSIUM FERRICYANIDE. Syns: *red prussiate of potash, red potassium prussiate*. Bright red, lustrous crystals or powder. $K_2Fe(CN)_6$, mw: 329.24. mp: decomp, d: 1.894 @ 17°.

Acute tox data: Oral LD_{LO} (rat) = 1600 mg/kg. [*3*]

THR = MOD via oral route. Violent reaction with NH_3, CrO_3. [*19*] This is not a powerful poison as are the simple cyanides.

Disaster Hazard: Dangerous; when heated to decomp, or on contact with acid or acid fumes, emits highly toxic fumes of cyanides.

POTASSIUM FERROCYANIDE. Syn: *yellow prussiate of potash*. Lemon yellow crystals. $K_4Fe(CN)_6 \cdot 3H_2O$, mw: 422.39, mp: $-3H_2O$ @ 70°, bp: decomp, d: 1.85 @ 17°.

THR = Not as toxic as the simple cyanides. See ferrocyanides. Violent reaction with $Cu(NO_3)_2$. [*19*]

Disaster Hazard: Dangerous; when heated to decomp, or on contact with acid or acid fumes, emits highly toxic fumes of cyanides.

POTASSIUM FLUOBERYLLATE. Colorless rhombic crystals. K_2BeF_4, mw: 163.2, mp: red heat.

THR = See beryllium and fluorides.

POTASSIUM FLUOBORATE. Syn: *avogadrite*. Rhombic or cubic, colorless crystals. KBF_4, mw: 125.9, mp: 530°, d: 2.498.

Acute tox data: $ipLD_{50}$ (rat) = 240 mg/kg. [*3*]

THR = HIGH via ip and oral routes. See fluorides and boron compounds.

POTASSIUM FLUOGERMANATE. Hexagonal, white crystals. K_2GeF_6, mw: 264.8, mp: 730°, bp: approx 835°.

THR = See fluorides.

POTASSIUM FLUORIDE. White, crystalline, deliquescent powder, sharp, saline taste. KF, mw: 58.1, mp: 880°, bp: 1500°, d: 2.48, vap. press: 1 mm @ 885°.

Acute tox data: Oral LD_{50} (rat) = 245 mg/kg; $ipLD_{50}$ (rat) = 64 mg/kg. [*3*]

THR = HIGH via oral, inhal and ip routes. Violent reaction with (Pt + BrF_3). [*19*] See fluorides.

POTASSIUM FLUOROACETATE.

Acute tox data: $scLD_{50}$ (mice) = 16 mg/kg; oral LD_{LO} (rabbit) = 0.5 mg/kg; $ivLD_{LO}$ (rabbit) = 0.5 mg/kg. [*3*]

THR = HIGH via sc, oral and iv routes.

POTASSIUM FLUOSILICATE. Syn: *Hieratite*. Hexagonal or cubic, colorless crystals. K_2SiF_6, mw: 220.25, mp: decomp, d(hex): 3.08, d(cubic): 2.665 @ 17°.

Acute tox data: Oral LD_{50} (guinea pig) = 500 mg/kg; $scLD_{LO}$ (frog) = 500 mg/kg. [*3*]

THR = HIGH via oral and sc routes. See fluosilicate.

POTASSIUM FLUOSTANNATE. Monoclinic prisms. $K_2SnF_6 \cdot H_2O$, mw: 328.9, d: 3.058.
THR = See fluorides.

POTASSIUM FLUOSULFONATE. Short, thick prisms. $KFSO_2$, mw: 138.16, mp: 311°.
THR = See fluorides and sulfonates.

POTASSIUM FLUOTANTALATE. Rhombic, colorless crystals. K_2TaF_7, mw: 392.1, d: 4.56; 5.24.
Acute tox data: Oral LD_{50} (rat) = 2500 mg/kg; ip LD_{50} (rat) = 375 mg/kg; oral LD_{50} (mice) = 110 mg/kg. [*3*]
THR = HIGH via oral and ip routes. See fluorides.

POTASSIUM FLUOTHORATE. Colorless crystals. $K_2ThF_6 \cdot 4H_2O$, mw: 496.38.
THR = See fluorides and thorium.

POTASSIUM FLUOTITANATE HYDRATE. Colorless, small, lustrous leaflets. $K_2TiF_6 \cdot H_2O$, mw: 258.1, mp: 780°, bp: decomp.
Acute tox data: Oral LD_{LO} (guinea pig) = 200 mg/kg; sc LD_{LO} (guinea pig) = 450 mg/kg; sc LD_{LO} (frog) = 360 mg/kg. [*3*]
THR = HIGH–MOD via oral and sc routes. See fluorides.

POTASSIUM FLUOZIRCONATE. Monoclinic, colorless crystals. K_2ZrF_6, mw: 283.4, d: 3.48.
Acute tox data: Oral LD_{50} (mice) = 98 mg/kg. [*3*]
THR = HIGH via oral route. See fluorides.

POTASSIUM FORMATE. Colorless, deliquescent crystals. HCOOK, mw: 84.1, mp: 168°, bp: decomp, d: 1.91.
Acute tox data: Oral LD_{50} (mice) = 5000 mg/kg; iv LD_{50} (mice) = 95 mg/kg. [*3*]
THR = HIGH via iv and MOD via oral routes. See formic acid and potassium compounds.
Fire Hazard: Slight; when heated, emits acrid fumes.

POTASSIUM GALLIUM SULFATE. Colorless crystals. $KGa(SO_4)_2 \cdot 12H_2O$, mw: 517.14, d: 1.895.
THR = See gallium compounds and sulfates.

POTASSIUM-*m*-GERMANATE. White crystals. K_2GeO_3, mw: 198.78, mp: 823°, d: 3.40 @ 21.5°.
THR = See germanium compounds.

POTASSIUM GLUTAMATE.
See monopotassium glutamate.

POTASSIUM GLYCERINOPHOSPHATE. See potassium glycerophosphate.

POTASSIUM GLYCEROPHOSPHATE. Syn: *potassium glycerinophosphate*. Pale yellow, syrupy liquid, sol in alcohol, miscible with water. $K_2C_2H_5O_2 \cdot H_2PO_4 \cdot 3H_2O$, mw: 302.
THR = Details U. A nutrient and/or dietary supplement food additive. [*109*] See also phosphorus compounds.

POTASSIUM GRAPHITE. Brass colored platelets, highly reactive with water, air, alcohol. KC_8, mw: 135.1.
THR = See potassium. Spont flam in air. [*19*]

POTASSIUM HEXAFLUOROPHOSPHATE. Syn: *FP salt No. 4*. KPF_6, mw: 184.1, mp: about 575°, bp: decomp, d: 2.59.
THR = HIGH. See fluorides.
Disaster Hazard: Dangerous; when heated to decomp, or on contact with acid or acid fumes, emits highly toxic fumes of fluorides and PO_x.

POTASSIUM HEXAFLUOROTITANATE. See potassium fluotitanate.

POTASSIUM HEXAHYDROALUMINATE. K_3AlH_6, mw: 150.3.
THR = On standing, samples become very friction-sensitive. [*19*]

POTASSIUM HYDRATE. See potassium hydroxide.

POTASSIUM HYDRIDE. White needles. KH, mw: 40.1, mp: decomp, d: 1.43–1.47.
THR = See potassium and hydrides.
Fire Hazard: Dangerous, by chemical reaction. See potassium.
Explosion Hazard: Mod, when exposed to heat or by chemical reaction. Violent with air, Cl_2, F_2, acetic acid, acrolein, acrylonitrile, (CaC + Cl_2), ClO_2, (H_2O_2 + Cl_2), ($CHCl_3$ + CH_3OH), 1,2-dichloroethylene, maleic anhydride, (*n*-methyl-*n*-nitrosourea + CH_2Cl_2), nitroethane, NCl_3, nitromethane, nitroparaffins, *o*-nitrophenol, nitropropane, *n*-nitrosomethylurea, (nitrosomethylurea + CH_2Cl_2), P, ($K_2S_2O_3$ + H_2O), (NaN_3 + benzyl chloride), H_2O, trichloroethylene, tetrahydrofuran, tetrachlorethane. [*19*]
Disaster Hazard: Dangerous; when heated to decomp, emits highly toxic fumes of potassium oxide; will react with water, steam or acids to produce H_2; can react vigorously with oxidizing materials.
To Fight Fire: CO_2, dry chemical.

POTASSIUM HYDROGEN OXALATE. Monoclinic colorless crystals. KHC_2O_4, mw: 128.1, mp: decomp, d: 2.0.
THR = HIGH. See oxalates.

POTASSIUM HYDROGEN PHOSPHATE. See potassium phosphate, dibasic.

POTASSIUM HYDROGEN SULFATE. Syns: *potassium bisulfate, acid potassium sulfate*. Colorless crystals, sol in water, decomp in alcohol. $KHSO_4$, mw: 136, d: 2.245, mp: 214°, bp: decomp.
THR = See sulfates and sulfuric acid.

POTASSIUM HYDROGEN SULFITE. Syn: *potassium bisulfite*. Colorless crystals. $KHSO_3$, mw: 120.17, mp: decomp @ 190°.

THR = A chemical preservative food additive. [*109*] See sulfites and sulfurous acid.

POTASSIUM HYDROSULFIDE. Syns: *potassium bisulfide, potassium sulfhydrate.* Rhombic, yellow, deliquescent crystals (commercial). KHS, mw: 72.17, mp: 455°, d: 2.0.
THR = See sulfides.

POTASSIUM HYDROXIDE. Syn: *potassium hydrate.* White, deliquescent pieces, lumps or sticks having crystalline fracture. KOH, mw: 56.11, mp: 360° ±7°, bp: 1320°, d: 2.044.
Acute tox data: Oral LD_{50} (rat) = 365 mg/kg. [*3*]
THR = HIGH via oral route. HIGH irr to skin, eyes and mu mem. A very powerful caustic. Used as a general purpose food additive. [*109*] See sodium hydroxide.
Fire Hazard: Mod.
Disaster Hazard: Dangerous; will react with water or steam to produce caustic solution and heat.

POTASSIUM HYDROXOANTIMONATE. Syn: *pyroantimonate.* Granular, white, crystalline powder. $KSb(OH)_6 \cdot \frac{1}{2}H_2O$, mw: 271.9.
THR = See antimony compounds.

POTASSIUM HYDROXOPLATINATE. Rhombic yellow crystals. $K_2Pt(OH)_6$, mw: 375.47, mp: decomp.
THR = See platinum compounds.

POTASSIUM HYDROXOPLUMBATE. Colorless crystals. $K_2Pb(OH)_6$, mw: 387.45.
THR = See lead compounds.

POTASSIUM HYDROXOSTANNATE. Colorless crystals. $K_2Sn(OH)_6$, mw: 298.94, d: 3.197.
THR = See tin compounds.

POTASSIUM HYPERCHLORATE. See potassium perchlorate.

POTASSIUM HYPOBORATE. $K_4B_2O_4$, mw: 242.
THR = Very violent decomp upon heating. A very powerful reducing agent. [*19*]

POTASSIUM HYPOCHLORITE. Solution only. KClO, mw: 90.55, mp: decomp.
THR = HIGH irr and corrosive to skin, eyes and mu mem. See hypochlorites.

POTASSIUM HYPONITRITE. See nitrites.

POTASSIUM HYPOPHOSPHITE. White, opaque, very deliquescent crystals or powder, pungent saline taste. KH_2PO_2, mw: 104.09, mp: decomp.
THR = See hypophosphites. Decomp to PH_3 on heating, explodes on evaporation with HNO_3. [*19*]

POTASSIUM IODATE. Colorless crystals. KIO_3, mw: 214, mp: 560°, d: 3.89.
Acute tox data: ip LD_{50} (mice) = 130 mg/kg; oral LD_{LO} (mice) = 531 mg/kg. [*3*]
THR = HIGH via ip; MOD via oral routes. A trace mineral added to animal feeds. [*109*] Violent reaction with Al, As, C, Cu, metal sulfides, organic matter, P, S. [*19*]

POTASSIUM IODIDE. Colorless or white granules. KI, mw: 166.02, mp: 723°, bp: 1420°, d: 3.13, vap. press: 1 mm @ 745°.
Acute tox data: iv LD_{LO} (rat) = 120 mg/kg; oral LD_{LO} (mice) = 1862 mg/kg; ip LD_{LO} (mice) = 1117 mg/kg. [*3*]
THR = HIGH via iv; MOD via oral and ip routes. Violent reaction with BrF_3, ClF_3, $FClO_4$. [*19*] See iodides. A trace mineral added to animal feeds; a nutrient and/or dietary supplement food additive. [*109*]

POTASSIUM IODOAURATE. Lustrous black crystals. $KAuI_4$, mw: 743.98.
THR = See gold compounds and iodides.

POTASSIUM IODOCADMATE. See cadmium compounds and iodides.

POTASSIUM IODOIRIDITE. Green crystals. K_3IrI_6, mw: 1071.9, mp: decomp.
THR = See iridium compounds and iodides.

POTASSIUM IODOMERCURATE (II). See mercury (II) potassium iodide.

POTASSIUM IODOPLATINATE. Black crystals. K_2PtI_6, mw: 1034.94, d: 5.176.
THR = See platinum compounds and iodides.

POTASSIUM LEAD CHLORIDE. Syn: *pseudocotunnite.* Yellow crystals. $2KCl \cdot PbCl_2$, mw: 427.23, mp: 490°.
THR = See lead compounds and chlorides.

POTASSIUM MAGNESIUM CHROMATE. Triclinic crystals. $K_2CrO_4 \cdot MgCrO_4 \cdot 2H_2O$, mw: 370.56, d: 2.59.
THR = See chromium compounds.
Fire Hazard: Mod, by chemical reaction; a powerful oxidizer.
Disaster Hazard: Dangerous. Keep away from combustible materials.

POTASSIUM MANGANATE. Rhombic green crystals. K_2MnO_4, mw: 197.12, mp: decomp @ 190°.
THR = See manganese compounds and potassium permanganate.
Fire Hazard: Mod, by chemical reaction; a powerful oxidizer.
Disaster Hazard: Dangerous. Keep away from combustible materials.

POTASSIUM MANGANIC SULFATE. Violet crystals. $KMn(SO_4)_2 \cdot 12H_2O$, mw: 502.35.
THR = See manganese compounds and sulfates.

POTASSIUM MANGANICYANIDE. Rhombic red crystals. $K_2Mn(CN)_6$, mw: 328.33.
THR = See cyanides and manganese compounds.

POTASSIUM MANGANOCYANIDE. Deep blue crystals. $K_4Mn(CN)_6 \cdot 3H_2O$, mw: 421.47.
THR = See cyanides and manganese compounds.

POTASSIUM MANGANOUS CHLORIDE. Syn: *chloromanganokalite.* Crystals. $4KCl \cdot MnCl_2$, mw: 424.06, d: 2.31.
THR = See manganese compounds and chlorides.

POTASSIUM MANGANOUS SULFATE. Syn: *manganolangbeinite.* Rose-red crystals. $K_2SO_4 \cdot 2MnSO_4$, mw: 476.23, mp: 850°, d: 3.02.
THR = See manganese compounds and sulfates.

POTASSIUM MERCURICYANIDE. Colorless crystals. $K_2Hg(CN)_4$, mw: 382.87.
THR = HIGH. See cyanides and mercury compounds, organic. May explode with NH_3. [*19*]

POTASSIUM MERCUROUS TARTRATE. White crystalline powder. $KHgC_4H_4O_6$, mw: 387.78
THR = HIGH. See mercury compounds, organic.

POTASSIUM MERCURY IODIDE. See potassium iodomercurate (II).

POTASSIUM METABISULFITE.
See potassium pyrosulfite.

POTASSIUM, METALLIC, LIQUID ALLOY.
See potassium.

POTASSIUM METHAZONATE.
THR = U.
Fire Hazard: U.
Explosion Hazard: Severe, when shocked or exposed to heat.
Disaster Hazard: Dangerous; heat or shock will explode it.

POTASSIUM METHYL SULFATE. White crystals. $2KCH_3SO_4 \cdot H_2O$, mw: 318.41.
THR = Details U. Probably highly toxic. See also dimethyl sulfate.
Fire Hazard: Mod, when exposed to heat or flame; decomp readily to flam products.
Explosion Hazard: U.
Disaster Hazard: Dangerous; when heated to decomp, emits highly toxic fumes; can react with oxidizing materials.

POTASSIUM MONOHYDROGEN-*o*-ARSENATE. Colorless crystals. K_2HAsO_4, mw: 218.11.
THR = See arsenic compounds.

POTASSIUM MONOPHOSPHATE. See potassium phosphate, dibasic.

POTASSIUM MONOSULFIDE. Yellow-brown, deliquescent crystals. K_2S, mw: 110.26, mp: 840°, d: 1.805 @ 14°.
THR = See sulfides.

POTASSIUM MONOXIDE. See potassium oxide.

POTASSIUM MURIATE. See potassium chloride.

POTASSIUM NICKEL SULFATE. Blue crystals. $K_2SO_4 \cdot NiSO_4 \cdot 6H_2O$, mw: 437.11, mp: decomp $<100°$, d: 2.124.
THR = See nickel compounds and sulfates.

POTASSIUM NIOBATE. Syn: *potassium columbate.* Crystalline form of niobic acid solutions when treated with conc potassium hydroxide, sol in water. $4K_2O \cdot 3Nb_2O_5 \cdot 16H_2O$, mw: 1462.
Acute tox data: Oral LD_{50} (rat) = 3000 mg/kg; ip LD_{50} (rat) = 225 mg/kg. [*3*]
THR = HIGH via ip; MOD via oral routes. Can damage the kidneys.

POTASSIUM NITRATE. Syns: *niter, nitre, saltpeter.* Transparent, colorless or white crystalline powder or crystals, cooling, pungent, saline taste. KNO_3, mw: 101.10, mp: 334°, bp: decomp @ 400°, d: 2.109 @ 16°.
THR = See nitrates. A food additive permitted in food for human consumption. [*109*] Chronic exposure can cause anemia, nephritis and methemoglobinemia. See also nitrates. Can react violently with Sb, Sb_2S_3, As, AsS_2, BaS, B, BP, CaS, F_2, charcoal, Cu_3P_2, Ge, GeS, Na acetate, Na hypophosphite, (Na_2O_2 + dextrose), (S + As_2S_3), Ti, TiS_2, Zn, trichloroethylene, Zr. [*19*]

POTASSIUM NITRATE MIXED (FUSED) WITH SODIUM NITRITE. See potassium nitrate and sodium nitrite.

POTASSIUM NITRIDE. Greenish-black crystals. K_3N, mw: 131.30, mp: decomp.
THR = See ammonia, which is readily evolved on contact with moisture. Ignites spont in air and forms a very flam mixture with P. [*19*]

POTASSIUM NITRITE. White or slightly yellowish, deliquescent prisms or sticks. KNO_2, mw: 85.10, mp: 387°, bp: decomp, d: 1.915.
THR = See nitrites.
Fire Hazard: Mod; an oxidizing material. See nitrites.
Explosion Hazard: Slight, when exposed to heat. It will explode @ 1000° F or when mixed with cyanide salts and heated. Also reacts violently with B, $(NH_4)_2SO_4$. [*19*]
Disaster Hazard: Dangerous. See nitrites.

POTASSIUM NITROACETATE. $C_2H_2O_3NK_2$, mw: 166.3.
THR = Moisture can explode the dry salt. [*19*]

POTASSIUM-*p*-NITROBENZENE DIAZOSULFONATE. $C_6H_4O_5N_3SK$, mw: 269.3.
THR = The labile form is spont explosive. [*19*]

POTASSIUM NITROCYANIDE. A solid. mp: explodes @ 400°.
THR = HIGH. See cyanides.
Explosion Hazard: Mod; will explode @ 752°F.
Disaster Hazard: Dangerous; on decomp or on contact with acid or acid fumes, emits highly toxic fumes of cyanides. An explosive.

POTASSIUM NITROMETHANE. KCH_2NO_2, mw: 99.2.
THR = See nitroparaffins.
Fire Hazard: Dangerous, when exposed to heat or flame.
Explosion Hazard: Severe, when shocked or exposed to heat. Moisture can cause dry salt to explode. [*19*]
Disaster Hazard: Dangerous; shock will explode it; when heated to decomp, emits highly toxic fumes of NO_x; can react vigorously with oxidizing materials.

POTASSIUM-*m*-NITROPHENOXIDE. Flat orange needles. $KOC_6H_4NO_2 \cdot 2H_2O$, mw: 213.33, mp: $-2H_2O$ @ 130°, bp: decomp, d: 1.691 @ 20°.
THR = HIGH toxicity. See phenol.
Disaster Hazard: Dangerous; when heated to decomp, or on contact with acids or acid fumes, emits highly toxic fumes of NO_x.

POTASSIUM NITROPLATINATE. Monoclinic, colorless crystals. $K_2Pt(NO_2)_4$, mw: 457.45, mp: decomp.
THR = See platinum compounds and nitrates.

POTASSIUM NITROPRUSSIDE. Red hygroscopic crystals. $K_2[Fe(NO)(CN)_5] \cdot 2H_2O$, mw: 330.17.
THR = HIGH. See cyanides.
Disaster Hazard: Dangerous; when heated to decomp, or on contact with acids or acid fumes, emits highly toxic fumes of cyanides and NO_x.

POTASSIUM OSMATE. Syn: *potassium perosmate.* Violet hygroscopic crystals. $K_2OsO_4 \cdot 2H_2O$, mw: 368.42, mp: $-H_2O > 100°$.
THR = See osmium compounds.
Fire Hazard: Mod, by chemical reaction; a powerful oxidizer.

POTASSIUM OXALATE. Colorless, transparent crystals. $K_2C_2O_4 \cdot H_2O$, mw: 184.23 (decomp), d: 2.127 @ 39°.
THR = See oxalates.

POTASSIUM OXALATOURANATE (IV). Yellow crystals. $K_4[U(C_2O_4)_4] \cdot 5H_2O$, mw: 836.61, d: 2.563.
THR = See oxalates and uranium.

POTASSIUM OXIDE. K_2O, mw: 94.2. White crystals or powder.
THR = Violent reaction with water. HIGH toxicity. Corrosive to tissue. See also caustic soda.

POTASSIUM OXIDES. K_xO_y.
THR = They are all HIGH toxicity, corrosive materials. See also potassium hydroxide, peroxides. They react violently with K, Na, NaK alloy, organic matter. [*19*]

POTASSIUM OXYBORATE. White crystals. $KBO_3 \cdot \frac{1}{2}H_2O$, mw: 106.92.
THR = See boron compounds.

POTASSIUM OXYMURIATE. See potassium chlorate.

POTASSIUM HYDROXYFLUORO NIOBATE. See potassium pentafluoniobate.

POTASSIUM OZONIDE. KO_3, mw: 87.1.
THR = Violent reaction with K, Na, H_2O. Very unstable. [*19*]

POTASSIUM PENTABORANE. Syn: *pentaboranide.* White powder. $K_2B_5H_9$, mw: 141.36, mp: decomp <180°.
THR = See boron hydrides.

POTASSIUM PENTABORATE. Colorless crystals, fine granular structure. $KB_5O_8 \cdot 4H_2O$, mw: 449.66, mp: 780°.
THR = See boron compounds.

POTASSIUM PENTACHLOROAQUORUTHENATE (II). Rose prisms. $K_2Ru(H_2O)Cl_5$, mw: 375.19, mp: $-H_2O$ @ 200°.
THR = See ruthenium compounds and chlorides.

POTASSIUM PENTACHLOROHYDROXORUTHENATE (IV). Brown-red crystals. $K_2Ru(OH)Cl_5$, mw: 374.19, mp: decomp.
THR = See ruthenium compounds and chlorides.

POTASSIUM PENTACHLORONITROSYL RUTHENATE (III). Dark red rhombic crystals. $K_2Ru(NO)Cl_5$, mw: 387.19, mp: decomp.
THR = See ruthenium compounds and chlorides.

POTASSIUM PENTACHLORORHODITE. Red rhombic crystals. K_2RhCl_5, mw: 358.39, mp: decomp.
THR = See rhodium and chlorides.

POTASSIUM PENTAFLUONIOBATE. Syn: *potassium hydroxy fluoroniobate.* Monoclinic colorless leaflets. $K_2NbOF_5 \cdot H_2O$, mw: 300.12.
Acute tox data: Oral LD_{50} (mice) = 130 mg/kg. [*3*]
THR = HIGH via oral route. See fluorides and niobium.

POTASSIUM PENTASULFIDE. Orange crystals. K_2S_5, mw: 238.52, mp: 206°.
THR = See sulfides.

POTASSIUM PERCARBONATE. White, granular mass, sol in water with the evolution of oxygen. $K_2C_2O_6 \cdot H_2O$, mw: 216.23, mp: 200°–300°.

THR = HIGH irr to skin, eyes and mu mem. A strong caustic, hence caustic to skin and mu mem.

POTASSIUM PERCHLORATE. Syn: *potassium hyperchlorate*. Colorless crystals or white crystalline powder. $KClO_4$, mw: 138.55, mp: 610° ± 10°, d: 2.52 @ 10°.

THR = Powerful oxidizer. HIGH irr to skin, eyes and mu mem. Has been implicated in aplastic anemia. Violent reaction with (Al + Mg), charcoal, F_2, Mg, (Ni + Ti), reducing agents, S. [*19*] Irr to skin and mu mem. Absorption can cause methemoglobinemia and kidney injury.

POTASSIUM-*m*-PERIODATE. Tetragonal, colorless crystals. KIO_4, mw: 230.0, mp: 582°, bp: $-O_2$ @ 300°, d: 3.618 @ 15°.

THR = HIGH irr to skin, eyes and mu mem. A strong irr. See also iodates.

Fire Hazard: An oxidizing agent and mod fire hazard.

Disaster Hazard: Dangerous, when exposed to heat or flame; on decomp, emits toxic fumes of iodine compounds.

POTASSIUM PERMANGANATE. Dark purple crystals with a blue metallic sheen, sweetish astringent taste. $KMnO_4$, mw: 158.03, mp: decomp @ < 240°, d: 2.703.

Acute tox data: Oral LD_{50} (rat) = 1090 mg/kg; sc LD_{50} (mice) = 500 mg/kg. [*3*]

THR = HIGH via sc; MOD via oral routes. A strong irr because of oxidizing properties. See also manganese compounds.

Fire Hazard: Mod, by chemical reaction. A powerful oxidizing agent. Spont flam on contact with glycerine, ethylene glycol, Al_4C_3, Sb, As, dimethyl sulfoxide, H_2O_2, H_2S_3, NH_2OH, organic matter, P, polypropylene, S, H_2SO_4, (H_2SO_4 + organic matter), (H_2SO_4 + KCl), Ti. [*19*] See also permanganates.

Disaster Hazard: Dangerous; keep away from combustible materials.

POTASSIUM PEROSMATE. See potassium osmate.

POTASSIUM PEROXIDE. Yellow, amorphous mass (white crystals). K_2O_2, mw: 110.19, mp: 490°.

THR = See peroxides, inorganic.

Fire Hazard: Dangerous, by spont chemical reaction. It is a very powerful oxidizer. Fires of this material should be handled like sodium peroxide fires.

Explosion Hazard: Mod, by spont chemical reaction. Also violent reactions with air, Sb, As, O_2, K, water. [*19*]

Disaster Hazard: Dangerous; will react with water or steam to produce heat; on contact with reducing material, can react vigorously; on contact with acid or acid fumes, can emit toxic fumes.

POTASSIUM PEROXYCHROMATE. Brown-red crystals. K_3CrO_8, mw: 297.3, mp: decomp @ 170°.

THR = HIGH tox. See chromium compounds.

Fire Hazard: Mod, by chemical reaction; a powerful oxidizer.

Disaster Hazard: Dangerous; keep away from combustible materials.

POTASSIUM PEROXYDISULFATE. See potassium persulfate.

POTASSIUM PERRHENATE. White crystals. $KReO_4$, mw: 289.41, mp: 350°, d: 4.887.

Acute tox data: ip LD_{50} (mice) = 692 mg/kg. [*3*]

THR = MOD via ip route. See rhenium compounds.

Fire Hazard: Mod, by chemical reaction; a powerful oxidizer.

Disaster Hazard: Dangerous. Keep away from combustible materials.

POTASSIUM PERRUTHENATE. Black crystals. $KRuO_4$, mw: 204.8, mp: decomp @ 440°.

THR = See ruthenium compounds.

Fire Hazard: Mod, by chemical reaction; a powerful oxidizer.

Disaster Hazard: Dangerous; keep away from combustible materials.

POTASSIUM PERSELENATE. Crystals. $KSeO_4$, mw: 182.1.

THR = HIGH. See selenium compounds.

Fire Hazard: Mod, by chemical reaction; a powerful oxidizer.

Disaster Hazard: Dangerous; when heated to decomp, or on contact with acid or acid fumes, emits highly toxic fumes of selenium; keep away from combustible materials.

POTASSIUM PERSULFATE. Syns: *anthion, potassium peroxydisulfate*. White, odorless crystals. $K_2S_2O_8$, mw: 270.3, mp: decomp @ 100°, d: 2.477.

THR = MOD irr and an allergen.

Fire Hazard: Mod, when exposed to heat or by chemical reaction. It liberates oxygen above 100° when dry or @ about 50° when in solution.

Disaster Hazard: Dangerous; when heated to decomp, emits highly toxic fumes of SO_x; can react with reducing materials.

POTASSIUM PHENOL SULFONATE. See potassium phenyl sulfate.

POTASSIUM PHENYLACETATE. Dry powder. $C_6H_5CH_2COOK$, mw: 174.2.

THR = U.

Disaster Hazard: Mod dangerous; when heated to decomp, emits tox fumes.

POTASSIUM PHENYL SULFATE. Rhombic leaflets. $KC_6H_5SO_4$, mw: 212.26, mp: 150°–160° (decomp), bp: decomp.
Acute tox data: sc LD_{50} (mice) = 1890 mg/kg. [*3*]
THR = MOD via sc route. An irr material.
Disaster Hazard: Dangerous; see sulfates.

POTASSIUM PHOSPHATE, DIBASIC. Syns: *DKP*, *potassium hydrogen phosphate*, *potassium monophosphate*, *dipotassium-o-phosphate*. Deliquescent white crystals or powder, very sol in water and alcohol. K_2PO_4, mw: 173.
THR = U. A sequestrant food additive. [*109*]

POTASSIUM PHOSPHIDE. K_3P, mw: 148.3.
Acute tox data: Inhal LC_{LO} (rat) = 580 ppm for 1 hr; inhal LC_{LO} (cat) = 173 ppm for 2 hrs; inhal LC_{LO} (guinea pig) = 288 ppm for 2 hrs. [*3*]
THR = HIGH via inhal route. In contact with water ⟶ PH_3. See phosphine.

POTASSIUM PHTHALIMIDE. White powder. $C_6H_4(CO)_2NK$, mw: 185.23.
THR = Details U; an insecticide.
Disaster Hazard: Mod; when heated to decomp, emits toxic fumes.

POTASSIUM PICRATE. Yellow-reddish or greenish crystals. $KC_6H_2N_3O_7$, mw: 267.20, bp: explodes @ 310°, d: 1.852.
THR = HIGH. See picric acid.
Fire Hazard: See nitrates.
Explosion Hazard: Mod, when shocked or exposed to heat.
Disaster Hazard: Dangerous; shock will explode it; on decomp, emits highly toxic fumes of NO_x; can react vigorously with reducing materials.

POTASSIUM POLYMANNURATE. See potassium alginate.

POTASSIUM POLYSULFIDES. See sulfides.

POTASSIUM PYROPHOSPHATES.
Acute tox data: Oral LD_{50} (mice) = 1600 mg/kg. [*3*]
THR = MOD via oral route.

POTASSIUM PYROSULFITE. Syn: *potassium metabisulfite*. Monoclinic plates. SO_x odor. $K_2S_2O_5$, mw: 222.32, mp: decomp, d: 2.3.
THR = A chemical preservative food additive. [*109*] See sulfites.

POTASSIUM RHODANIDE.
See potassium sulfocyanate.

POTASSIUM RHODIUM SULFATE. Yellow cubes. $KRh(SO_4)_3 \cdot 12H_2O$, mw: 550.33, d: 2.23.
THR = See rhodium and sulfates.

POTASSIUM SELENATE. Rhombic, colorless crystals. K_2SeO_4, mw: 221.15, d: 3.066.
Acute tox data: iv LD_{LO} (rat) = 4.3 mg/kg; oral LD_{LO} (rabbit) = 1.8 mg/kg. [*3*]
THR = HIGH via oral and iv routes. See selenium compounds.

POTASSIUM SELENIDE. White crystals, reddens on exposure to air. K_2Se, mw: 157.15, d: 2.851 @ 15°.
THR = See selenium compounds.

POTASSIUM SELENITE. White deliquescent crystals. K_2SeO_3, mw: 205.15.
THR = See selenium compounds.

POTASSIUM SELENOCYANATE. Deliquescent needles. KSeCN, mw: 144.1, mp: decomp @ 100°, d: 2.347.
Acute tox data: ip LD_{LO} (rat) = 200 mg/kg. [*3*]
THR = HIGH via ip route. See selenium compounds and cyanates.

POTASSIUM SELENOCYANOACETATE.
Acute tox data: Oral LD_{LO} (rat) = 100 mg/kg. [*3*]
THR = HIGH via oral route.

POTASSIUM SELENOCYANOPLATINATE. Rhombic crystals. $K_2Pt(SeCN)_6$, mw: 903.29, mp: decomp @ 80°, d: 3.378 @ 12.5°.
THR = HIGH. For other properties, see potassium selenocyanate and platinum compounds.

POTASSIUM SESQUIOXIDE.
See potassium trioxide.

POTASSIUM SILICIDE. K_2Si_2, mw: 134.4.
THR = Violent reaction with acids, air, water. [*19*]

POTASSIUM SILICOFLUORIDE.
See potassium fluosilicate.

POTASSIUM SILVER CARBONATE. Rectangular plates. $KAgCO_3$, mw: 206.99, mp: decomp, d: 3.769.
THR = See silver compounds.

POTASSIUM SILVER CYANIDE. $KAg(CN)_2$, mw: 199.0.
Acute tox data: Oral LD_{50} (rat) = 21 mg/kg. [*3*]
THR = HIGH via oral and inhal routes.

POTASSIUM SILVER NITRATE. Monoclinic crystals. $KNO_3 \cdot AgNO_3$, mw: 270.99, mp: 125°, d: 3.219.
THR = See nitrates and silver compounds.

POTASSIUM-SODIUM ALLOY. See NaK.

POTASSIUM SODIUM ANTIMONY TARTRATE. White scales or powder. $KNaSbC_4H_3O_7$, mw: 346.9.
THR = See antimony compounds.

POTASSIUM SODIUM TARTRATE. See sodium potassium tartrate.

POTASSIUM SORBATE. Syn: *potassium-2,4-hexadienoate*. White powder, sol in water @ 25°. $C_6H_7O_2K$, mw: 150.2, mp: 270° (decomp), d: 1.36 @ 25°/20°.
Acute tox data: Oral LD_{50} (rat) = 4920 mg/kg. [*3*]

THR = MOD via oral route. A chemical preservative food additive; it is a substance which migrates to food from packaging materials. [*109*] A MILD irr.

POTASSIUM STRONTIUM CHROMIC OXALATE. Greenish-black crystals. $KSrCr(C_2O_4)_3 \cdot 6H_2O$, mw: 550.89, d: 2.155 @ 13°.
THR = See oxalates and chromium compounds.

POTASSIUM STYPHNATE. Syn: *potassium trinitroresorcinate*. Yellow monoclinic prisms. $KC_6H_2N_3O_8 \cdot H_2O$, mw: 301.21, mp: $-H_2O$ @ 120°, bp: explodes.
THR = See nitrates and explosives, high.

POTASSIUM SULFHYDRATE.
See potassium hydrosulfide.

POTASSIUM SULFATE. Colorless or white hard crystals or powder, sol in water, insol in alcohol. K_2SO_4, mw: 174, d: 2.66, mp: 1072°.
Acute tox data: sc LD_{LO} (guinea pig) = 3000 mg/kg. [*3*]
THR = MOD via sc and oral routes. Reacts violently with Al. [*19*] A general purpose food additive. [*109*]
Disaster Hazard: Dangerous. See sulfates.

POTASSIUM SULFIDE. Syns: *potassium sulfuret, hepar sulfuris*. Red crystalline mass, deliquescent in air. K_2S, mw: 110.25, mp: 840°, d: 1.805 @ 14°.
THR = See sulfides.

POTASSIUM SULFITE. White-yellowish crystals. $K_2SO_3 \cdot 2H_2O$, mw: 194.29, mp: decomp.
THR = See sulfites.

POTASSIUM SULFOCARBONATE. Syn: *potassium thiocarbonate*. Yellowish-red crystals. K_2CS_3, mw: 186.4, mp: decomp.
THR = HIGH irr and via oral route. A soil fumigant. A caustic.
Disaster Hazard: Dangerous; when heated to decomp, or on contact with acid or acid fumes, emits highly toxic fumes of SO_x.

POTASSIUM SULFOCYANATE. Syns: *potassium rhodanide, potassium thiocyanate*. Colorless crystals. KSCN, mw: 97.17, mp: 173.2°, d: 1.886, bp: decomp @ 500°.
Acute tox data: Oral LD_{50} (rat) = 850 mg/kg; oral LD_{LO} (human) = 80 mg/kg. [*3*]
THR = HIGH via oral route. Has been used medically to reduce blood pressure. Large doses can cause skin eruption, psychoses and collapse.
Disaster Hazard: Dangerous; on decomp, emits highly toxic fumes of cyanides. See sulfates.

POTASSIUM SUPEROXIDE. Yellow leaflets. KO_2, mw: 71.10, mp: approx 400°, bp: decomp.
THR = See peroxides, inorganic.
Fire Hazard: Dangerous; a powerful oxidizing agent.
Explosion Hazard: Can explode with ethanol, graphite, oil, NaK alloy, water, Na, K, Se, Cl. [*19*]
Disaster Hazard: Dangerous; reacts with water or steam to evolve heat; keep away from combustible materials.

POTASSIUM TELLURATE. Soft, glutinous mass. K_2TeO_4, mw: 269.80, mp: decomp @ 200°.
THR = See tellurium compounds.

POTASSIUM TELLURIDE. Colorless crystals. K_2Te, mw: 205.80, d: 2.51.
THR = See tellurium compounds.

POTASSIUM TELLURITE. White deliquescent crystals. K_2TeO_3, mw: 253.80, mp: 460°–470° (decomp).
THR = See tellurium compounds.

POTASSIUM TETRABORATE. Colorless crystals. $K_2B_4O_7 \cdot 8H_2O$, mw: 377.60, mp: decomp, d (anhydrous): 1.74.
THR = See boron compounds.

POTASSIUM TETRACHLOROCUPRATE.
K_2CuCl_4, mw: 283.5.
THR = Violent reaction with K. [*19*]

POTASSIUM TETRAETHYNYL NICKELATE.
$K_4[Ni(CCH_4)]$, mw: 327.3.
THR = Spont flammable in air. [*19*] See also nickel compounds.

POTASSIUM TETRAGERMANATE. White crystals. $K_2Ge_4O_9$, mw: 512.29, mp: 1033°, d: 4.12 @ 21.5°.
THR = See germanium compounds.

POTASSIUM TETRAIODOMERCURATE (II). See mercuric potassium iodide.

POTASSIUM TETRAPEROXYCHROMATE.
K_3CrO_8, mw: 297.3.
THR = Explodes @ 178°. [*19*] See also chromium compounds.

POTASSIUM TETRASULFIDE. Red-brown crystals. K_2S_4, mw: 206.46, mp: 145°, bp: decomp @ 850°.
THR = See sulfides.

POTASSIUM THIOANTIMONATE. Yellow crystals. $2K_3SbS_4 \cdot 9H_2O$, mw: 896.8.
THR = See antimony compounds and sulfides.

POTASSIUM THIOARSENATE. Deliquescent crystals. K_3AsS_4, mw: 320.46, mp: decomp.
THR = See arsenic compounds and sulfides.

POTASSIUM THIOARSENITE. A solid. K_2AsS_3, mw: 288.4, mp: decomp.
THR = See arsenic compounds and sulfides.

POTASSIUM THIOCARBONATE. See potassium sulfocarbonate.

POTASSIUM THIOCYANATE.
See potassium sulfocyanate.

POTASSIUM THIOPLATINATE. Blue-gray crystals.

$K_2Pt_4S_6$, mw: 1051.51, mp: decomp upon ignition, d: 6.44 @ 15°.
THR = See platinum compounds and sulfides.

POTASSIUM TRIIODIDE. Monoclinic, dark brown, deliquescent crystals. $KI_3 \cdot H_2O$, mw: 437.9, mp: 38°, bp: decomp @ 225°, d: 3.498.
THR = See iodides.

POTASSIUM TRINITRORESORCINATE. See potassium styphnate.

POTASSIUM TRIOXIDE. Syn: *potassium sesquioxide*. Red crystals (probably a mixture). K_2O_3, mw: 126.19, mp: 430°.
THR = HIGH. See potassium hydroxide.

POTASSIUM TRISULFIDE. Yellow crystals. K_2S_3, mw: 174.39, mp: 252°.
THR = See sulfides.

POTASSIUM-*m*-URANATE. Orange-yellow crystals. K_2UO_4, mw: 380.26.
THR = See uranium.

POTASSIUM URANIUM NITRATE. See uranium potassium nitrate.

POTASSIUM URANYL ACETATE. Radioactive crystals. $KUO_2(C_2H_3O_2)_2 \cdot H_2O$, mw: 504.31, mp: $-H_2O$ @ 275°, d: 2.396 @ 15°.
THR = See uranium.

POTASSIUM URANYL CARBONATE. Yellow, radioactive crystals. $2K_2CO_3 \cdot UO_2CO_3$, mw: 606.48, mp: $-CO_2$ @ 300°.
THR = See uranium.

POTASSIUM URANYL SULFATE. See uranium potassium sulfate.

POTASSIUM-*m*-VANADATE. Colorless crystals. KVO_3, mw: 138.05.
THR = See vanadium compounds.

POTASSIUM XANTHATE. Liquid. Syn: *potassium ethyl xanthate*. $C_2H_5OCS_2K$, mw: 161.3, mp: 200° (decomp), uel = 9.5%, flash p: 205°F (CC), d: 1.558.
Acute tox data: sc LD_{LO} (mouse) = 400 mg/kg. [*3*]
THR = HIGH-MOD via sc and oral routes.
Fire Hazard: Low, when exposed to heat or flame.
Spont Heating: No.
Explosion Hazard: Mod, when exposed to flame.
Disaster Hazard: Dangerous; when heated to decomp, emits highly toxic fumes of SO_x; can react with oxidizing materials.
To Fight Fire: Water, CO_2, dry chemical.

POTASSIUM ZINC CHROMATE. $K_2Zn_4Cr_4O_{17} \cdot 3H_2O$, mw: 873.8.
THR = An exper neo to mice via ir route. [*103*]

POWDERED METALS.
THR = See specific metals.
Fire Hazard: Dangerous, in dispersed form when exposed to flame or sparks or by chemical reaction with oxidizers. Many powdered metals can ignite spont and explode when suspended in air.
To Fight Fire: Use no water; use powdered graphite, dolomite, sodium chloride, etc. Get instructions from the supplier of the powdered metal.

POWDER OF ALGAROTH.
See antimony oxychloride.

PO_x. An abbreviation for oxides of phosphorus such as P_2O_3, P_2O_5.

PRASEODYMIUM. Soft, silvery, malleable metal. Pr, atwt: 140.9077, mp: 931°, bp: 3212°, d(α): 6.772, d(β): 6.64.
THR = As a lanthanon it may depress coagulation of the blood. See also lanthanum. Limited animal exper suggest low toxicity.
Radiation Hazard: For permissible levels, see Section 5A, Table 5A.5. Artificial isotope ^{142}Pr, $T\frac{1}{2}$ = 19h, decays to stable ^{142}Nd via β's of 0.64 (4%), 2.15 (96%) MeV, also via γ's of 1.57 MeV. Artificial isotope ^{143}Pr, $T\frac{1}{2}$ = 13.6d, decays to stable ^{143}Nd via β's of 0.93 MeV. Artificial isotope ^{144}Pr, $T\frac{1}{2}$ = 17m, decays to radioactive ^{144}Nd via β's of 3.0 MeV. ^{144}Pr exists in radioactive equilibrium with its parent, ^{144}Ce. Permissible levels are given for the equilibrium mixture.
Fire Hazard: Mod, in the form of dust, when exposed to heat or flame or by chemical reaction. Fine dust ignites readily. See also powdered metals. Violent reaction with air, halogens. [*19*]

PRASEODYMIUM BROMATE. Hexagonal green crystals. $Pr(BrO_3)_3 \cdot 9H_2O$, mw: 686.81, mp: 56.5°, bp: $-7H_2O$ @ 100°.
THR = See bromates. As a lanthanon it may depress coagulation of the blood. See also lanthanum.

PRASEODYMIUM CHLORIDE. $PrCl_3$, mw: 247.3.
Acute tox data: iv LD_{50} (rat) = 3.5 mg/kg; oral LD_{50} (mice) = 4500 mg/kg; ip LD_{50} (mice) = 359 mg/kg; ip LD_{50} (guinea pig) = 125 mg/kg. [*3*]
THR = HIGH via ip, iv and MOD via oral routes.

PRASEODYMIUM CHLORIDE CITRATE.
Acute tox data: ip LD_{50} (mice) = 141 mg/kg; ip LD_{50} (guinea pig) = 53 mg/kg. [*3*]
THR = HIGH via ip route.

PRASEODYMIUM MALEATE.
Acute tox data: sc LD_{50} (mice) = 1000 mg/kg. [*3*]
THR = MOD via sc route.

PRASEODYMIUM (III) NITRATE. $Pr(NO_3)_3$, mw: 326.9.
Acute tox data: Oral LD_{50} (rat) = 3500 mg/kg; ip

LD_{50} (rat) = 245 mg/kg; iv LD_{LO} (rat) = 4 mg/kg. [3]
THR = HIGH via iv and ip; MOD via oral routes.

PRASEODYMIUM OXALATE. Light green crystals. $Pr_2(C_2O_4)_3 \cdot 10H_2O$, mw: 726.06.
THR = See oxalates. As a lanthanon it may depress coagulation of the blood. See also lanthanum.

PRASEODYMIUM SELENATE. $Pr_2(SeO_4)_3$, mw: 710.72, d: 4.30 @ 15°.
THR = See selenium compounds. As a lanthanon it may depress coagulation of the blood. See also lanthanum.

PRASEODYMIUM SULFIDE. Brown powder. Pr_2S_3, mw: 378.04, mp: decomp, d: 5.042 @ 11°.
THR = See sulfides. As a lanthanon it may depress coagulation of the blood. See also lanthanum.

PREDNISOLONE. Syn: Δ^1-*dehydrocortisol.* White crystalline powder, odorless, very slightly sol in water, sol in alcohol, chloroform, acetone, methanol, and dioxane. $C_{21}H_{28}O_5$, mw: 360, mp: 235° with some decomp.
THR = U. A food additive permitted in food for human consumption. [*109*]

PREDNISONE. Syns: $\Delta^{1,4}$-*pregnadiene-17α, 21-diol-3,11,20-trione*. White, odorless, crystalline powder, very slightly sol in water, slightly sol in alcohol, chloroform, methanol and dioxane. $C_{21}H_{26}O_5$, mw: 358, mp: 235° (with some decomp).
THR = U. Has been implicated in aplastic anemia. A food additive permitted in food for human consumption. [*109*]

PRENDEROL. See 2,2-diethyl-1,3-propanediol.

PRESBYCUSIS. See Section 3.

PRESERVATIVE FOOD ADDITIVES. [*109*]

PREVENTION OF EXPOSURE TO TOXIC MATERIALS. See Sections 1, 2, 3, 8, 9.

PRIMERS. See explosives, high.

PRINCEP. See 2-chloro-4,6-bis(ethylamine)-*s*-triazine.

PRINCIPAL DECAY SCHEME OF ACTINIUM SERIES. See Section 5A.

PRINCIPAL DECAY SCHEME OF THORIUM SERIES. See Section 5A.

PRINCIPAL DECAY SCHEME OF URANIUM SERIES. See Section 5A.

PRINCIPLES OF RESPIRATORY PROTECTION. See Section 2.

PROBLEMS ASSOCIATED WITH PLATING, METAL FINISHING, CLEANING AND ALLIED PROCESSES. See Section 2.

PROCAINE BORATE. See borocaine.

PROCAINE HYDROCHLORIDE. Crystals. $C_{13}H_{20}N_2O_2 \cdot HCl$, mw: 272.8, mp: 153°–156°, d: 0.707 @ 17°.
Acute tox data: ip LD_{50} (rat) = 184 mg/kg; iv LD_{50} (rat) = 38 mg/kg; ip LD_{50} (mice) = 180 mg/kg. [*3*]
THR = HIGH via ip and iv routes. An allergen.
Disaster Hazard: Dangerous; when heated to decomp, emits highly toxic fumes of hydrochloric acid.

PROCAINE NITRATE. Crystals. $C_{13}H_{20}N_2O_2 \cdot HNO_3$, mw: 299.3, mp: 101°.
THR = See procaine hydrochloride and nitrates.

PROCAINE PENICILLIN G. Fine white crystals or powder, odorless, sparingly sol in water, slightly sol in alcohol, fairly sol in chloroform, an antibiotic. $C_{18}H_{18}N_2O_4S \cdot C_{13}H_2N_2O_2 \cdot H_2O$, mw: 788.
Acute tox data: ip LD_{50} (mice) = 146 mg/kg; iv LD_{50} (mice) = 70 mg/kg; im LD_{LO} (mice) = 54 mg/kg. [*3*]
THR = HIGH via ip, iv and im routes. A food additive permitted in the feed and drinking water of animals and/or for the treatment of food-producing animals. [*109*]
Disaster Hazard: Dangerous; when heated to decomp, emits highly toxic fumes.

PROCESSING MACHINERY LUBRICANTS WHICH GET INTO FOODS. [*109*] See Section 10.

PROCHLOROPERAZINE. Syn: *Compazine*. $C_{20}H_{24}N_3SCl$, mw: 374.
THR = MOD acute toxicity via oral route to rats. Has been implicated in aplastic anemia; an exper teratogen. [*3*]

PROCYTOX. See endoxan.

PRODUCER GAS. lel = 20–30%, uel = 70–80%.
THR = HIGH toxicity. See carbon monoxide, methane and hydrogen.
Fire Hazard: Dangerous, when exposed to flame.
Explosive Range: 20.7–73.7%.
Disaster Hazard: Dangerous; can react vigorously with oxidizing materials.
To Fight Fire: CO_2, dry chemical, water spray.

PROGESTERONE. Syn: Δ-*4-pregnene-3,20-dione*. A female sex hormone. White crystalline powder, odorless, practically insol in water, sol in alcohol, acetone and dioxane, sparingly sol in vegetable oils. $C_{21}H_{30}O_2$, mw: 314.
THR = An exper (S) carc and neo [*3, 15, 23*] via sc and im routes. A food additive permitted in the feed and drinking water of animals and/or for the treatment of food-producing animals. Also permitted in food for human consumption. [*109*]

PROLACTIN.
THR = An exper carc. [*23*]

PROLAN. See dilan.

PROLINE. Syn: *2-pyrrolidinecarboxylic acid.* A non-essential amino acid, colorless crystals, sol in water and alcohol, insol in ether, occurs in isomeric forms; we consider the *l* and *dl* forms. C_4H_8NCOOH, mw: 115, mp(*l*): 215°–200° (with decomp), mp(*dl*): 220°–222° (with decomp).
THR = U. A nutrient and/or dietary supplement food additive. [*109*]

PROMAZINE HYDROCHLORIDE. Syn: *10-(3-dimethylaminopropyl)-phenothiazine hydrochloride.* White to slightly yellow, practically odorless crystalline powder. $C_{17}H_{20}N_2S \cdot HCl$, mw: 321.
Acute tox data: Oral LD_{50} (rat) = 400 mg/kg; sc LD_{50} (rat) = 300 mg/kg; iv LD_{50} (rat) = 29 mg/kg; ip LD_{50} (mice) = 115 mg/kg; im LD_{50} (dog) = 4.4 mg/kg. [*3*]
THR = HIGH via oral, sc, iv, ip and im routes. A tranquilizer. A food additive permitted in food for human consumption; also permitted in the feed and drinking water of animals and/or for the treatment of food-producing animals. [*109*]
Disaster Hazard: See chlorides.

PROMIZOLE. See 2-amino-5-sulfanilylthiazole.

PROMETHIUM. A radioactive metal. Pr, atwt: 147, mp: 1080° (approx), bp: 2460°.
THR = A HIGH radiotoxicity metal.
Radiation Hazard: For permissible levels, see Section 5A, Table 5A.5. Artificial isotope ^{147}Pm, $T_{\frac{1}{2}}$ = 2.6y decays to radioactive ^{147}Sm via β's of 0.22 MeV.

PROPADIENE. See allene.

PROPANE. Syn: *dimethylmethane.* Colorless gas. $CH_3CH_2CH_3$, mw: 44.09, bp: −42.1°, lel = 2.3%, uel = 9.5%, fp: −187.1°, flash p: −156°F, d: 0.5852 @ −44.5°/4°, autoign. temp.: 842°F, vap. d: 1.56.
THR = A general purpose food additive. [*109*] An asphyxiant. At high conc has a CNS effect.
Fire Hazard: Highly dangerous when exposed to heat, flame or oxidizers.
Spont Heating: No.
Explosion Hazard: Severe, when exposed to flame or ClO_2. [*19*]
Disaster Hazard: Dangerous; can react vigorously with oxidizing materials.
To Fight Fire: Stop flow of gas.

PROPANEDIAL. $C_3H_4O_2$, mw: 72.1. Syn: *malonaldehyde.*
Acute tox data: Oral LD_{50} (rat) = 632 mg/kg. [*3*]
THR = MOD via oral route. An exper neo via dermal route. [*3*]

1,2-PROPANE DIAMINE. $C_3H_{10}N_2$, mw: 64.1, flash p: 92°F (OC), d: 0.9, vap. d: 2.6, bp: 118.9°.
Acute tox data: Oral LD_{50} (rat) = 2230 mg/kg; dermal LD_{50} (rabbit) = 500 mg/kg. [*3*]
THR = HIGH via dermal and MOD via oral routes. See amines.
Fire Hazard: Dangerous, via heat, flames, oxidizers.
To Fight Fire: Alcohol foam.

1,3-PROPANEDIAMINE. Water white liquid, amine odor, completely sol in water, methanol and ether. $NH_2(CH_2)_3NH_2$, mw: 74, d: 0.8881 @ 20°/20°, bp: 139.7°, fp: −12°, flash p: 120°F (TOC).
Acute tox data: Oral LD_{50} (rat) = 350 mg/kg; dermal LD_{50} (rabbit) = 200 mg/kg. [*3*]
THR = HIGH via oral and dermal routes.
Fire Hazard: Mod, when exposed to heat or flame.

PROPANE DIAMINE PERCHLORATE.
$C_3H_{10}O_8N_2Cl_2$, mw: 273.
THR = A very powerful explosive. [*19*]

1,2-PROPANEDIOL. See propylene glycol.

1,3-PROPANEDIOL. See trimethylene glycol.

1,3-PROPANEDITHIOL. Syn: *1,3-dimercaptopropane.* Oil with disagreeable odor, slightly water-sol. $HSCH_2CH_2CH_2SH$, mw: 108.2, d: 1.0722 @ 20°/4°, bp: 170°.
THR = Details U. See related compound, 2,3-dimercapto-1-propanol.
Disaster Hazard: Dangerous; when heated to decomp, emits highly toxic fumes.

PROPANENITRILE. See ethyl cyanide.

1,3-PROPANE SULTONE. Syn: *1,2-oxathialone-2,2-dioxide.* $C_3H_6O_3S$, mw: 122.2.
THR = An exper (+) carc, neo. [*3, 23, 10*]

1-PROPANETHIOL. C_3H_7SH, mw: 76.2.
Acute tox data: Oral LD_{50} (rat) = 1790 mg/kg; inhal LC_{50} (rat) = 7300 ppm for 4 hrs; ip LD_{50} (rat) = 515 mg/kg. [*3*]
THR = MOD via oral, inhal and ip routes. Violent reaction with $Ca(OCl)_2$. [*19*]

1,2,3-PROPANETRIOL. See glycerine.

PROPANIL. See 3,4-dichloropropionanilide.

PROPANOIC ANHYDRIDE.
See propionic anhydride.

PROPANOL. See propyl alcohol.

2-PROPANOL. See isopropyl alcohol.

2-PROPANOL-1,*n*,*n*-DIBUTYLAMINE. See dibutyl isopropanolamine.

2-PROPANOL NITRITE. See isopropyl nitrite.

4-PROPANOL PYRIDINE. Crystals.
$(CH_2CH_2CH_2OH)C_5H_4N$, mw: 137.1, bp: 289°, fp: 36.7°, d: 1.053 @ 40°.
THR = See pyridine.

PROPANONE. See acetone.

7-(2-PROPANONE)BENZ(a)ANTHRACENE.
THR = An exper carc. [23]

PROPANOYL CHLORIDE. See propionyl chloride.

PROPARGYL ALCOHOL. Syn: *2-propyn-1-ol.* A mod volatile liquid, geranium-like odor. $HC{\equiv}CCH_2OH$, mw: 56.1, mp: −50°, bp: 115°, flash p: 97°F (OC), d: 0.9715 @ 20°/4°, vap. press: 11.6 mm @ 20°, vap. d: 1.93.
Acute tox data: Oral LD_{50} (rat) = 0.07 mg/kg; inhal LC_{LO} (mice) = 2000 mg/m^3. [3]
THR = VERY HIGH via oral route. MOD via inhal route. Toxic also via dermal route. A CNS depressant.
Fire Hazard: Dangerous, when exposed to heat or flame; can react with oxidizing materials. Violent reaction with P_2O_5. [19]
To Fight Fire: Foam, CO_2, dry chemical.

PROPARGYL BROMIDE. An almost colorless liquid, sharp odor. $HC{\equiv}CCH_2Br$, mw: 118.97, bp: 88°–90°, fp: −61.07°, flash p: 65°F (COC), d: 1.564–1.570, vap. d: 6.87.
Acute tox data: Oral LD_{50} (guinea pig) = 0.029 mg/kg. [3]
THR = VERY HIGH via oral route.
Fire Hazard: Dangerous, when exposed to heat, flame or oxidizers.
Explosion Hazard: Detonates @ 220° or more; ignites by impact; mixed with chloropicrin, becomes shock-sensitive. [19] Detonates with other chemicals.
Disaster Hazard: Dangerous; when heated to decomp, emits highly toxic fumes of bromides; can react vigorously with oxidizing materials.
To Fight Fire: Water, foam, CO_2, dry chemical.

PROPARGYL CHLORIDE. Syn: *3-chloro-1-propyne.* Liquid, insol in water, sol in organic solvents. $HC{\equiv}CCH_2Cl$, mw: 74.5, mp: −78°, bp: 57°, flash p: < 140°F (CC), d: 1.03 @ 25°/4°.
THR = Probably HIGH. See propargyl bromide.
Fire Hazard: Dangerous, when exposed to heat, sparks or open flame.
Disaster Hazard: See chlorides.

PROPARGYL METHANE SULFONATE.
THR = An exper carc. [23]

PROPELLANT 115. See chloropentafluoroethane.

PROPENAL. See acrolein.

PROPENE. See propylene.

PROPENE ACID. See acrylic acid.

1-PROPENE-2-CHLORO-1,3-DIOL DIACETATE.
Acute tox data: Inhal LC_{LO} (rat) = 8 ppm for 4 hrs. [3]
THR = VERY HIGH via inhal route.
Disaster Hazard: Dangerous; when heated to decomp, emits highly toxic fumes of chlorides.

1-PROPENE-1,3-DIOL DIACETATE.
Acute tox data: Oral LD_{50} (rat) = 150 mg/kg; inhal LC_{LO} (rat) = 16 ppm for 4 hrs; dermal LD_{50} (rabbit) = 670 mg/kg; inhal LC_{LO} (rat) = 8 ppm for 4 hrs. [3]
THR = HIGH via oral, inhal and dermal routes.
Fire Hazard: Slight; when heated, emits acrid fumes.

PROPENE NITRILE. See acrylonitrile.

PROPENE OXIDE. See propylene oxide.

1,2,3-PROPENETRICARBOXYLIC ACID. See aconitic acid.

2-PROPEN-1-OL. See allyl alcohol.

PROPENYL ACETATE. See isopropenyl acetate.

2-PROPENYLAMINE. See allylamine.

***p*-PROPENYL ANISOLE.** See anethole.

1-PROPENYL-2-BUTENE-1-YL ETHER.
$CH_3CH{:}CHCH_2OCH{:}CHCH_3$, mw: 112.
Acute tox data: Oral LD_{50} (rat) = 8000 mg/kg; inhal LC_{LO} (rat) = 5000 ppm for 4 hrs. [3]
THR = MOD via oral and inhal routes. See also ethers.

2-PROPENYL ETHANOATE. See allyl acetate.

PROPENYL ETHYL ETHER. $CH_3CH{:}CHOCH_2CH_3$, mw: 86, d: 0.8, vap. d: 1.3, bp: 158°F, flash p: < 20°F (OC).
THR = U. See also ethers.
Fire Hazard: Dangerous, when exposed to heat, flame or oxidizers.
To Fight Fire: Foam, alcohol foam, mist.

2-PROPENYL METHANOATE. See allyl formate.

β-PROPIOLACTONE. $C_3H_4O_2$. A liquid. mw: 72.1, mp: −33.4°, bp: 155° (with rapid decomp), flash p: 165°F, d: 1.1460 @ 20°/4°, lel = 2.9%.
Acute tox data: Oral LD_{LO} (rat) = 50 mg/kg; ip LD_{LO} (mice) = 3 mg/kg; iv LD_{50} (mice) = 345 mg/kg. [3]
THR = HIGH via oral, ip and iv routes. A strong irr. It is considered to be the most toxic of the lactones. An exper carc [23, 14, 3] via oral, sc, it and dermal routes of the skin and connective tissue.
Fire Hazard: Mod, when exposed to heat or flame; can react violently with 2-aminoethanol, NH_4OH, aniline, chlorosulfonic acid, ethylene diamine, ethylene imine, HCl, HF, HNO_3, oleum, pyridine, NaOH, H_2SO_4. [19]
To Fight Fire: Alcohol foam.

PROPIONALDEHYDE. See propyl aldehyde.

PROPIONAMIDE NITRILE. See cyanoacetamide.

PROPIONE. See diethyl ketone.

PROPIONIC ACID. Syn: *methylacetic acid.* Colorless liquid, oily, rancid odor. CH_3CH_2COOH, mw: 74.1, mp: −22°, bp: 141°, d: 0.992, vap. press: 10 mm @ 39.7°, vap. d: 2.56, flash p: 130° F, autoign. temp.: 955° F.

Acute tox data: Oral LD_{50} (rat) = 1510 mg/kg; iv LD_{50} (mice) = 625 mg/kg; dermal LD_{50} (rabbit) = 500 mg/kg; oral LD_{50} (rabbit) = 1900 mg/kg. [*3*]

THR = HIGH via dermal; MOD via oral and iv routes. A chemical preservative food additive. A substance which migrates to food from packaging materials. [*109*]

Disaster Hazard: Slight; when heated, emits acrid fumes.

Fire Hazard: Mod, when exposed to heat, flame or oxidizers.

To Fight Fire: Alcohol foam.

PROPIONIC ANHYDRIDE. Syn: *propanoic anhydride.* Liquid, very rancid odor. $(CH_3CH_2CO)_2O$, mw: 130.14, mp: −45°, bp: 167.0°, flash p: 165° F (OC), d: 1.012, vap. press: 1 mm @ 20.6°, vap. d: 4.49.

Acute tox data: Oral LD_{50} (rat) = 2360 mg/kg. [*3*]

THR = MOD via oral route.

Fire Hazard: Mod, when exposed to heat or flame; can react with oxidizing materials.

To Fight Fire: CO_2, dry chemical.

PROPIONIC ETHER. See ethyl propionate.

PROPIONITRILE. See ethyl cyanide.

PROPIONYL CHLORIDE. Syn: *propanoyl chloride.* Colorless liquid. CH_3CH_2COCl, mw: 92.5, mp: −94°, bp: 80°, flash p: 54° F, d: 1.065, vap. d: 3.2.

THR = HIGH irr to skin, eyes and mu mem. See hydrochloric acid.

Fire Hazard: Dangerous, when exposed to heat, flame or oxidizers.

Explosion Hazard: U.

Disaster Hazard: Dangerous; when heated to decomp, emits highly toxic fumes of chlorides; will react with water or steam to produce toxic and corrosive fumes; can react vigorously with oxidizing materials.

To Fight Fire: CO_2, dry chemical; use no water.

PROPIOPHENONE. Syn: *ethyl phenyl ketone.* Water white to light amber liquid or crystals. $C_9H_{10}O$, mw: 134.2, mp: 21°, bp: 218.0°, flash p: 210° F (OC), d: 1.012 @ 20°/20°, vap. press: 1 mm @ 50.0°.

Acute tox data: Oral LD_{50} (rat) = 4490 mg/kg; ip LD_{LO} (mice) = 100 mg/kg. [*3*]

THR = HIGH via ip and MOD via oral routes. See also ketones.

Fire Hazard: Slight, when exposed to heat or flame; can react with oxidizing materials.

To Fight Fire: Foam, CO_2, dry chemical.

1-PROPOXY-2-AMINO-4-NITROBENZENE. See 5-nitro-2, *n*-propoxyaniline.

***n*-PROPYL ACETATE.** Clear, colorless liquid, pleasant odor. $CH_3COOC_3H_7$, mw: 102.13, mp: −92.5°, bp: 101.6°, flash p: 58° F, lel = 2.0%, uel = 8.0%, d: 0.887, autoign. temp.: 842° F, vap. press: 40 mm @ 28.8°, vap. d: 3.52.

Acute tox data: Oral LD_{50} (rat) = 9800 mg/kg; oral LD_{50} (rabbit) = 6630 mg/kg; inhal LC_{LO} (rat) = 8000 ppm for 4 hrs. [*3*]

THR = MOD–LOW via oral and inhal routes. This material causes narcosis and is somewhat irr. However, it is not likely to cause chronic poisoning, since there is definite evidence of habituation to this material. The after-effects are slight and recovery is quick from even deep narcosis. The symptoms noted are sleepiness, fatigue, slight stupefaction and retarded respiration. Repeated or prolonged inhal of high conc have been shown to produce irr and narcosis and, in certain cases, death, although no industrial injury has been reported as occurring to workmen exposed to it. Isopropyl acetate has been shown to have slightly less narcotic potency than normal propyl acetate.

Fire Hazard: Dangerous, when exposed to heat, flame or oxidizers.

Spont Heating: No.

Explosion Hazard: Mod, when exposed to heat or flame.

Disaster Hazard: Dangerous, upon exposure to heat or flame; can react vigorously with oxidizing materials.

Treatment and Antidotes: Personnel who show the symptoms of irr or narcosis from exposure to this material should immediately be removed to fresh air. Recovery is quick and complete. If the exposure has been very severe, consult a physician.

To Fight Fire: Alcohol foam, CO_2, dry chemical.

***n*-PROPYL ACETYLENE.** C_5H_8, mw: 68.1, flash p: $< -4°$F, d: 0.69, vap. d: 2.35, bp: 40°.

THR = No data.

Fire Hazard: Dangerous, via heat, flames, oxidizers.

To Fight Fire: Dry chemical, CO_2, fog, mist.

***n*-PROPYL ALCOHOL.** Syns: *1-propanol, ethyl carbinol.* Clear, odorless liquid, alcohol-like odor. mw: 60.1, mp: −127°, bp: 97.19°, flash p: 77° F (CC), ulc: 55–60, d: 0.8044 @ 20°/4°, lel = 2.1%, uel = 13.5%, autoign. temp.: 824° F, vap. press: 10 mm @ 14.7°, vap. d: 2.07.

Acute tox data: Oral LD_{LO} (woman) = 5700 mg/kg;

oral LD_{50} (rat) = 1870 mg/kg; inhal LC_{LO} (rat) = 4000 ppm for 4 hrs; dermal LD_{50} (rabbit) = 5040 mg/kg. [*3*]

THR = MOD via oral, inhal and dermal routes. An exper carc [*3*] via oral and sc routes.

Fire Hazard: Dangerous, when exposed to heat, flame or oxidizers.

Spont Heating: No.

Explosion Hazard: Mod, when exposed to flame. Reacts violently with potassium-*tert*-butoxide. [*19*]

Disaster Hazard: Dangerous, upon exposure to heat or flame; can react vigorously with oxidizing materials.

To Fight Fire: Alcohol foam, CO_2, dry chemical.

***sec*-PROPYL ALCOHOL.** See isopropyl alcohol.

PROPYL ALDEHYDE. Syns: *propylic aldehyde, propionaldehyde.* CH_3CH_2CHO, mw: 58.1, mp: −81°, bp: 48°, flash p: 15°–19°F (OC), d: 0.807 @ 20°/4°, lel = 2.9%, uel = 17%, vap. d: 2.0, autoign. temp.: 405°F.

Acute tox data: sc LD_{50} (rat) = 820 mg/kg; oral LD_{LO} (rat) = 800 mg/kg; inhal LC_{LO} (rat) = 8000 ppm for 4 hrs; dermal LD_{LO} (rabbit) = 3400 mg/kg. [*3*]

THR = MOD via sc, oral, inhal and dermal routes. See also aldehydes.

Fire Hazard: Dangerous, when exposed to heat, flame or oxidizers.

Disaster Hazard: Dangerous fire hazard; reacts vigorously with oxidizers.

To Fight Fire: Alcohol foam, CO_2, dry chemical.

PROPYLAMINE. Syn: *1-amino propane.* Colorless, alkaline liquid, strong ammonia odor, miscible with water, alcohol and ether. $CH_3CH_2CH_2NH_2$, mw: 59.11, d: 0.7191 @ 20°/20°, mp: −83°, bp: 48°–49°, vap. press: 248 mm @ 20°, flash p: −35°F, autoign. temp.: 604°F, lel = 2.0%, uel = 10.4%.

Acute tox data: Inhal LC_{50} (rat) = 2310 ppm for 4 hrs; dermal LD_{50} (rabbit) = 560 mg/kg; oral LD_{LO} (rat) = 570 mg/kg. [*3*]

THR = MOD via inhal, dermal and oral routes. A strong irr and possibly a skin-sensitizer. See also amines.

Fire Hazard: Dangerous, when exposed to heat, flame or oxidizers.

To Fight Fire: Alcohol foam.

8-PROPYL BENZ(a)ANTHRACENE. $C_{21}H_{18}$, mw: 270.4.

THR = An exper carc. [*3*]

***n*-PROPYLBENZENE.** Syn: *1-phenylpropane.* Clear liquid. $C_3H_7C_6H_5$, mw: 120.2, mp: −99.5°, bp: 159.2°, flash p: 86°F (CC), d: 0.862, vap. press: 10 mm @ 43.4°, vap. d: 4.14, autoign. temp.: 842°F, lel = 0.8%, uel = 6%.

Acute tox data: Oral LD_{LO} (rat) = 4830 mg/kg; inhal LC_{LO} (mice) = 4100 ppm. [*3*]

THR = MOD via oral and inhal routes. Limited animal exper show mod acute vapor toxicity.

Fire Hazard: Dangerous; when exposed to heat, flame or oxidizers; can react with oxidizing materials.

Spont Heating: No.

Explosion Hazard: U.

To Fight Fire: Foam, CO_2, dry chemical.

5-PROPYL BENZO(c)PHENANTHRENE. $C_{21}H_{18}$, mw: 270.4.

THR = An exper carc. [*3, 23*]

2-PROPYL BIPHENYL. $C_{15}H_{16}$, mw: 196.2, flash p: >212°F, autoign. temp.: 833°F, vap. d: 6.77, bp: 280°.

THR = No data.

Fire Hazard: Low.

To Fight Fire: Dry chemical, CO_2, foam.

PROPYL BORATE. See tri-*n*-propyl borate.

PROPYL BROMIDE. Syn: *1-bromopropane.* Liquid. $CH_3CH_2CH_2Br$, mw: 123.00, mp: −110°, bp: 70.9°, d: 1.353 @ 20°/4°, autoign. temp.: 914°F.

THR = HIGH irr to skin, eyes and mu mem. Probably dermal tox too.

Fire Hazard: Dangerous, when heated or exposed to flame or oxidizers.

Disaster Hazard: Dangerous; on decomp, emits highly toxic fumes of bromides; can react with oxidizing materials.

To Fight Fire: Water, foam, CO_2, dry chemical.

***s*-PROPYL BUTYLETHYL THIOCARBAMATE.** Syn: *tillam.* Liquid. $C_{10}H_{21}NOS$, mw: 203.4, bp: 142° @ 20 mm.

Acute tox data: Oral LD_{50} (rat) = 1020 mg/kg; oral LD_{50} (rat) = 1120 mg/kg. [*3*]

THR = MOD via oral route. An exper neo [*3*] via sc route. Causes violent vomiting when accompanied by alcohol ingestion.

Disaster Hazard: Dangerous. When heated to decomp, emits highly toxic fumes.

PROPYL BUTYRATE. Colorless liquid, slightly sol in water. $CH_3CH_2CH_2CO_2C_3H_7$, mw: 130.2, d: 0.879 @ 15°/4°, mp: −95°, bp: 143°, flash p: 99°F.

THR = Details U. Irr to mu mem. Narcotic in high conc.

Fire Hazard: Dangerous, via heat, flames, oxidizers.

To Fight Fire: Dry chemical, CO_2, foam.

***n*-PROPYL CARBAMATE.** Crystals. $C_4H_9O_2N$, bp: 196°, mp: 60°, vap. press: 1 mm @ 52.4°, mw: 103.1.

Acute tox data: sc LD_{50} (mice) = 1300 mg/kg. [*12*]

THR = An exper neo, carc and teratogen [*3*] via oral, ip routes. [*12*]

PROPYL "CELLOSOLVE."
Acute tox data: Oral LD_{50} (rat) = 4890 mg/kg; inhal LC_{LO} (rat) = 2000 ppm for 4 hrs; dermal LD_{50} (rabbit) = 940 mg/kg. [*3*]
THR = MOD via oral, inhal and dermal routes.
Fire Hazard: Mod; can react with oxidizing materials.

***n*-PROPYL CHLORIDE.** Syn: *1-chloropropane*. Colorless liquid, chloroform-like odor. $CH_3CH_2CH_2Cl$, mw: 78.54, mp: −122.8°, bp: 47.2°, lel = 2.6%, uel = 11.1%, flash p: < 0°F, d: 0.890, vap. d: 2.71, autoign. temp.: 968°F.
THR = MOD irr to skin, eyes and mu mem. Narcotic in high conc. See also chlorinated hydrocarbons, aliphatic.
Fire Hazard: Dangerous, when exposed to heat, flame or oxidizers.
Explosion Hazard: Mod, when exposed to flame.
Disaster Hazard: Dangerous. Keep away from heat and open flame; can react vigorously with oxidizing materials.
To Fight Fire: CO_2, dry chemical.

PROPYL CHLOROSULFONATE. Liquid.
$CH_3CH_2CH_2OSO_2Cl$, mw: 158.61, bp: 70°–72° @ 20 mm.
THR = HIGH irr to skin, eyes and mu mem.
Disaster Hazard: Dangerous; when heated to decomp, emits highly toxic fumes of chlorides and SO_x. A military poison.

***n*-PROPYL CHLOROTHIOLFORMATE.** Insol in water. C_3H_7SCOCl, mw: 122.5, d: 1.1, vap. d: 4.8, bp: 155°, flash p: 145°F.
THR = U.
Fire Hazard: Mod, when exposed to heat, flame or oxidizers.
Disaster Hazard: Dangerous; when heated to decomp, emits toxic fumes.
To Fight Fire: Water, foam, CO_2 and dry chemical.

***n*-PROPYL CYANIDE.** See *n*-butyronitrile.

PROPYLENE. Syn: *propene*. A gas. C_3H_6, mw: 42.1, mp: −185°, bp: −47.7°, d(liquid): 0.581 @ 0°, autoign. temp.: 860°F, vap. press: 10 atm @ 19.8°, lel = 2.0%, uel = 11.1%, vap. d: 1.5, flash p: −162°F.
THR = A simple asphyxiant. No irr effects from high conc in gaseous form. When compressed to liquid form, can cause skin burns from refrigerating effects on tissue of rapid evaporation. For effects of simple asphyxiants, see argon.
Fire Hazard: Very dangerous, when exposed to heat or flame.
Spont Heating: No.
Explosion Hazard: Mod, when exposed to heat or flame. Under unusual conditions, i.e., 955 atmospheres pressure and 327°, it has been known to explode. Reacts violently with NO_2, N_2O_4, N_2O. [*19*]
Disaster Hazard: Dangerous; can react vigorously with oxidizing materials.
To Fight Fire: Stop flow of gas.

PROPYLENE CARBONATE. A clear liquid.
$CH_3CHCH_2CO_3$, mw: 102.09, bp: 242.1°, fp: −48.8°, flash p: 275°F (OC), d: 1.2069 @ 20°/20°, vap. press: 0.03 mm @ 20°.
Acute tox data: Oral LD_{50} (rat) = 29000 mg/kg. [*3*]
THR = NONE via oral route.
Fire Hazard: Slight; when exposed to heat or flame.
Disaster Hazard: Can react with oxidizing materials.
To Fight Fire: Alcohol foam.

PROPYLENE CHLOROBROMIDE. Colorless liquid.
$CH_3CHBrCH_2Cl$, mw: 157.5, bp: 117°–118.5°, fp: < −20°, d: 1.540 @ 25°/25°, vap. d: 5.47.
THR = Probably tox. See bromides.
Disaster Hazard: Dangerous; when heated to decomp, emits highly toxic fumes of halides.

PROPYLENE CHLOROHYDRIN. Syns: *chloroisopropyl alcohol*, *2-chloro-1-propanol*. Colorless liquid, mild non-residual odor. $CH_2ClCHOHCH_3$, mw: 94.54, bp: 133.5°, flash p: 125°F (CC), d: 1.103 @ 20°, vap. d: 3.26.
Acute tox data: Dermal LD_{50} (rabbit) = 480 mg/kg. [*3*]
THR = HIGH via dermal route.
Fire Hazard: Mod, when exposed to heat, flame or powerful oxidizers.
Spont Heating: No.
Disaster Hazard: Dangerous; when heated to decomp, emits highly toxic fumes of chlorides; can react with oxidizing materials.
To Fight Fire: Alcohol foam, CO_2, dry chemical.

***sec*-PROPYLENE CHLOROHYDRIN.** Syn: *1-chloroisopropyl alcohol*. $CH_3CHClCH_2OH$, clear liquid, mw: 94.5, flash p: 125°F (OC), d: 1.1, vap. d: 3.3, bp: 126°.
Acute tox data: Oral LD_{50} (rat) = 220 mg/kg; inhal LC_{LO} (rat) = 500 ppm for 4 hrs; dermal LD_{50} (rabbit) = 480 mg/kg. [*3*]
THR = HIGH via oral, inhal and dermal routes. See propylene chlorohydrin.
Fire Hazard: Mod, when exposed to heat, flame or oxidizers.
Disaster Hazard: Dangerous. When heated to decomp, emits highly toxic fumes.
To Fight Fire: Alcohol foam.

PROPYLENE DIAMINE. See 1,2-propane diamine.

PROPYLENE DIBROMIDE. Colorless liquid.
$CH_3CHBrCH_2Br$, mw: 201.9, mp: −55°, bp: 139.6°–

142.6°, fp: <−75°, d: 1.940 @ 25°/25°, vap. d: 7.0.
THR = U. Probably irr and narcotic in high conc. See also bromides.
Disaster Hazard: Mod dangerous. See bromides.

PROPYLENE DICHLORIDE. See 1,2-dichloropropane.

PROPYLENE FLUOROHYDRIN. Syn: *fluoroisopropyl alcohol.* $CH_2FCHOHCH_3$, mw: 78.1.
Acute tox data: Oral LD_{50} (rat) = 3260 mg/kg; inhal LC_{LO} (rat) = 2000 ppm for 4 hrs. [*3*]
THR = MOD via oral and inhal routes.
Disaster Hazard: Dangerous; when heated to decomp, emits highly toxic fumes of fluorides.

PROPYLENE GLYCOL. Syns: *1,2-propanediol, 1,2-dihydroxypropane.* Colorless liquid, practically odorless, hygroscopic. $CH_2OHCHOHCH_3$, mw: 76.1, bp: 188.2°, flash p: 210°F (OC), lel = 2.6%, uel = 12.6%, d: 1.0362 @ 25°/25°, autoign. temp.: 700°F, vap. press: 0.08 mm @ 20°, vap. d: 2.62, fp: −59°.
Acute tox data: Oral LD_{50} (rat) = 21000 mg/kg; ip LD_{50} (rat) = 13000 mg/kg; im LD_{50} (rat) = 20,000 mg/kg; sc LD_{50} (mice) = 18,500 mg/kg; iv LD_{50} (mice) = 8000 mg/kg. [*3*]
THR = NONE via sc, im and oral routes. LOW via iv and ip routes. Used as an emulsifying agent and a general purpose food additive. It is a substance which migrates to food from packaging materials. [*109*]
Fire Hazard: Low, when exposed to heat or flame; can react with oxidizing materials.
Spont Heating: No.
Explosion Hazard: Mod, when exposed to flame.
To Fight Fire: Alcohol foam.

1,3-PROPYLENE GLYCOL. See trimethylene glycol.

PROPYLENE GLYCOL ALGINATE. Syn: *hydroxypropyl alginate.* White powder, odorless, sol in water and dilute organic acids. $(C_9H_{14}O_7)_8$.
THR = U. A food additive permitted in food for human consumption. [*109*]

PROPYLENE GLYCOL-4-BIPHENYLYL ETHER. Syn: *1-(4-biphenylyloxy)-2-propanol.* White crystalline solid. $C_6H_5C_6H_4OCH_2CH(OH)CH_3$, mw: 228.3, mp: 121°–123°.
THR = Details U. See also glycols.
Fire Hazard: Slight, when exposed to heat or flame; can react with oxidizing materials.

PROPYLENE GLYCOL-*sec*-BUTYLPHENYL ETHER. Syn: *1-(o-sec-butylphenoxy)-2-propanol.* Slightly yellow liquid. $C_2H_5CH(CH_3)C_6H_4OCH_2CH(OH)CH_3$, mw: 208.3, mp: <−20°, bp: 276.8°, flash p: 270°F, d: 0.992 @ 25°/25°, vap. d: 7.2.
THR = Details U. See also glycols.
Fire Hazard: Slight, when exposed to heat or flame; can react with oxidizing materials.
Explosion Hazard: Details U. See also ethers.
To Fight Fire: CO_2, dry chemical.

PROPYLENE GLYCOL-*p-tert*-BUTYLPHENYL ETHER. Syn: *1-(p-tert-butylphenoxy)-2-propanol.* White solid. $(CH_3)_3CC_6H_4OCH_2CH(OH)CH_3$, mw: 208.3, mp: 33°, bp: 288.8°, flash p: 290°F, d: 0.979 @ 60°/60°, vap. d: 7.2.
THR = Details U. See also glycols.
Fire Hazard: Slight, when exposed to heat or flame; can react with oxidizing materials.
To Fight Fire: Foam, CO_2, dry chemical.

PROPYLENE GLYCOL CHLOROPHENYL ETHER. Syn: *1-(o-chlorophenoxy)-2-propanol.* Slightly yellow liquid. $ClC_6H_4OCH_2CH(OH)CH_3$, mw: 186.6, mp: <−20°, bp: 272.3°, flash p: 270°F, d: 1.201 @ 25°/25°, vap. d: 6.45.
THR = Details U. See also glycols.
Fire Hazard: Slight, when exposed to heat or flame.
Disaster Hazard: Dangerous; on heating to decomp, emits toxic fumes of chlorides; can react with oxidizing materials.
To Fight Fire: Water, foam, CO_2, dry chemical.

PROPYLENE GLYCOL DIACETATE.
Acute tox data: Oral LD_{50} (guinea pig) = 3420 mg/kg. [*3*]
THR = MOD via oral route. An exper teratogen [*3*] via ip route. See glycols.
Disaster Hazard: Slight; when heated, emits acrid fumes.

PROPYLENE GLYCOL-2,4-DICHLOROPHENYL ETHER. Syn: *1-(2,4-dichlorophenoxy)-2-propanol.* Slightly yellow liquid. $Cl_2C_6H_3OCH_2CH(OH)CH_3$, mw: 221.1, mp: 9.5°, bp: 297.7°, flash p: 335°F, d: 1.309 @ 25°/25°, vap. d: 7.6.
THR = Details U. See glycols.
Fire Hazard: Slight, when exposed to heat or flame; can react with oxidizing materials.
To Fight Fire: Water, foam, CO_2, dry chemical.

PROPYLENE GLYCOL METHYL ETHER, ALPHA. Syns: *propylene glycol monomethyl ether, methyl ether of propylene glycol, Dowanal, PM, UCAR, solvent LM.* Colorless liquid. $CH_3OCH_2CHOHCH_3$, mw: 90.1, mp: −96.7°, bp: 120°, flash p: 100°F, d: 0.919 @ 25°/25°.
Acute tox data: Oral LD_{50} (rat) = 7510 mg/kg; inhal LC_{LO} (rat) = 7000 ppm for 4 hrs; oral LD_{LO} (dog) = 10,000 mg/kg. [*3*]
THR = LOW via oral and inhal routes. Rating based on extensive animal tests. No cases of human tox-

icity known. See also ethylene glycol monomethyl ether and glycol. See ethers.

Fire Hazard: Mod, when exposed to heat or flame; can react with oxidizing materials.

To Fight Fire: Foam, CO_2, dry chemical.

PROPYLENE GLYCOL METHYL ETHER, BETA.

Acute tox data: Oral LD_{50} (rat) = 5710 mg/kg; dermal LD_{50} (rabbit) = 5660 mg/kg.[*3*]

THR = MOD via oral and dermal routes. See also propylene glycol methyl ether, alpha.

PROPYLENE GLYCOL MONOACETATE.

Acute tox data: Oral LD_{50} (rat) = 18,000 mg/kg. [*3*]

THR = NONE via oral route.

PROPYLENE GLYCOL MONOACRYLATE. Syn: *hydroxypropyl acrylate*. Sol in water.

CH_2:$CHCOO(C_3H_6)OH$, mw: 130, d: 1.0+, vap. d: 4.5, bp: 171°F @ 5 mm, flash p: 210°F.

Acute tox data: Oral LD_{50} (rat) = 1230 mg/kg; sc LD_{50} (rabbit) = 160 mg/kg. [*3*]

THR = HIGH via sc and MOD via oral route.

Fire Hazard: Low, when exposed to heat or flame.

To Fight Fire: Alcohol foam, mist, spray, dry chemical.

PROPYLENE GLYCOL MONOETHYL ETHER, ALPHA. See ethoxypropanol.

PROPYLENE GLYCOL MONOETHYL ETHER, BETA. See 3-ethoxy-1-propanol.

PROPYLENE GLYCOL MONOMETHYL ETHER. See propylene glycol methyl ether.

PROPYLENE GLYCOL PHENYL ETHER. Liquid.

$C_6H_5OCH_2CH(OH)CH_3$, mw: 152.2, bp: 240°, fp: 13°–18°, flash p: 275°F, d: 1.063 @ 25°/4°, vap. d: 5.25.

THR = Details U. See glycols.

Fire Hazard: Slight, when exposed to heat or flame; can react with oxidizing materials.

To Fight Fire: Foam, CO_2, dry chemical.

PROPYLENE IMINE. Liquid. $NHCH_2CHCH_3$, mw: 58.10, vap. d: 2.0.

Acute tox data: Dermal LD_{50} (guinea pig) = 43 mg/kg; inhal LC_{LO} (rat) = 500 ppm for 4 hrs; oral LD_{50} (rat) = 19 mg/kg. [*3*]

THR = HIGH via dermal, inhal and oral routes. An exper (+) carc [*23, 3, 8*] via dermal route.

Fire Hazard: Mod, when exposed to heat or flame.

Disaster Hazard: Dangerous; when heated to decomp, emits toxic fumes of NO_x; can react with oxidizing materials.

PROPYLENE OXIDE. Syns: *1,2-epoxypropane, propene oxide, methyl oxirane*. Colorless liquid, ethereal odor, sol in water, alcohol and ether. OCH_2CHCH_3, mw: 58.08, bp: 33.9°, lel = 2.8%, uel = 37%, fp: −104.4°, flash p: −35°F (TOC), d: 0.8304 @ 20°/20°, vap. press: 400 mm @ 17.8°, vap. d: 2.0.

Acute tox data: Oral LD_{50} (rat) = 930 mg/kg; inhal LC_{LO} (rat) = 4000 ppm for 4 hrs; inhal LC_{50} (mice) = 1740 ppm for 4 hrs; dermal LD_{50} (rabbit) = 1500 mg/kg. [*3*]

THR = MOD via oral, inhal and dermal routes. An exper (s) carc and neo via sc route. [*3, 13, 23*] A food additive permitted in food for human consumption. [*109*] See aliphatic and aromatic epoxides. An insecticidal fumigant.

Fire Hazard: Highly dangerous, when exposed to heat or flame.

Spont Heating: No.

Explosion Hazard: Severe, when exposed to flame. Violent reaction with NH_4OH, chlorosulfonic acid, HCl, HF, HNO_3, oleum, H_2SO_4. [*19*]

Disaster Hazard: Dangerous; can react vigorously with oxidizing materials. Keep away from heat and open flame!

To Fight Fire: Alcohol foam, CO_2, dry chemical.

PROPYLENE SULFIDE. C_3H_6S, mw: 74.2.

Acute tox data: Oral LD_{50} (rat) = 269 mg/kg; ip LD_{50} (rat) = 47 mg/kg; inhal LC_{50} (rat) = 660 ppm. [*3*]

THR = HIGH via oral, ip and inhal routes.

Disaster Hazard: Dangerous; when heated to decomp, emits highly toxic fumes.

***n*-PROPYL ETHER.** Syn: *dipropyl ether*. Colorless liquid. $(C_3H_7)_2O$, mw: 102.2, mp: −122°, bp: 90°, d: 0.736 @ 20°/4°, flash p: 70°F.

THR = See ethers. Possibly narcotic.

Fire Hazard: Dangerous, when exposed to heat, flame or oxidizers. See ethers.

Explosion Hazard: Forms explosive peroxides.

Disaster Hazard: Dangerous, upon exposure to heat or flame; can react vigorously with oxidizing materials.

PROPYLETHYL CARBAMATE. $C_6H_{13}O_2N$, mw: 131.2

THR = An exper neo. [*3*]

PROPYLETHYLENE. See amylene.

***n*-PROPYL FORMATE.** Syn: *propyl methanoate*. Colorless liquid, pleasant odor. $C_3H_7CHO_2$, mw: 88.1, mp: −93°, bp: 82°, flash p: 27°F (CC), d: 0.901 @ 20°, vap. press: 100 mm @ 29.5°, vap. d: 3.03, autoign. temp.: 851°F.

Acute tox data: Oral LD_{50} (rat) = 3980 mg/kg; oral LD_{50} (mice) = 3400 mg/kg. [*3*]

THR = MOD via oral route. A MOD irr to skin, eyes and mu mem. Narcotic in high conc. See esters.

Fire Hazard: Dangerous; when exposed to heat, flame or oxidizers.

Spont Heating: No.
Explosion Hazard: Violent reaction with potassium-*tert*-butoxide. [*19*]
Disaster Hazard: Dangerous, upon exposure to heat or flame; can react vigorously with oxidizing materials.
To Fight Fire: Alcohol foam.

PROPYL GALLATE. Syn: *propyl-3,4,5-trihydroxy benzoate.* Odorless, fine, ivory powder. $(HO)_3C_6H_2COOC_3H_7$, mw: 212.2, mp: 147°–149°.
Acute tox data: Oral LD_{50} (rat) = 400 mg/kg; oral LD_{LO} (rat) = 500 mg/kg. [*3*]
THR = HIGH–MOD via oral route. Used in food as an antioxidant. [*109*] See *n*-propyl alcohol.
Fire Hazard: Slight, when exposed to heat or flame; can react with oxidizing materials.

2-PROPYL HEPTANOL.
$CH_2(CH_2)_4CH(C_3H_7)CH_2OH$, mw: 158.
Acute tox data: Oral LD_{50} (rat) = 6730 mg/kg. [*3*]
THR = MOD–LOW via oral route. See also alcohols.

PROPYL HYDRAZINE HYDROCHLORIDE.
$C_3H_{10}N_2 \cdot HCl$, mw: 110.6.
THR = An exper carc. [*3*]

PROPYLIC ALDEHYDE. See propyl aldehyde.

PROPYL IODIDE. Syn: *1-iodopropane.* Colorless to yellow liquid. $CH_3CH_2CH_2I$, mw: 170.0, mp: −98°, bp: 103°, d: 1.747 @ 20°/4°.
THR = Details U. Probably toxic.
Fire Hazard: Mod, when exposed to heat or flame.
Disaster Hazard: Dangerous; see iodides; can react with oxidizing materials.

PROPYL ISOMER. $C_{20}H_{26}O_6$, mw: 362.5.
Acute tox data: Oral LD_{50} (rat) = 1500 mg/kg. [*3*]
THR = MOD via oral route. An exper carc [*3*] via sc route.

***n*-PROPYL MERCAPTAN.** See 1-propanethiol.

PROPYL METHANOATE. See *n*-propyl formate.

PROPYL NITRATE. Liquid. Pale yellow, sickly odor. $CH_3CH_2CH_2NO_3$, mw: 105.1, bp: 110.5°, d: 1.054 @ 20°/4°, flash p: 68°F, autoign. temp.: 347°F (in air), lel = 2%, uel = 100%.
Acute tox data: Inhal LC_{LO} (dog) = 100 mg/kg. [*3*]
THR = HIGH via iv and inhal routes. See nitrates. Inhal can cause a hypotension and methemoglobinemia.
Fire Hazard: Dangerous; when exposed to heat, flame or oxidizers.
Explosion Hazard: Dangerous; may explode on heating.

PROPYL NITRITE. Liquid. $CH_3CH_2CH_2ONO$, mw: 89.1, bp: 46°, d: 0.886.
THR = MOD via inhal and oral routes. Can cause fall in blood pressure similar to amyl nitrite. See also nitrates, organic.

4-(PROPYL NITROSAMINO)-1-BUTANOL.
$C_7H_{16}O_2N_2$, mw: 160.3.
THR = An exper carc. [*3, 23*]

PROPYL PERCHLORATE. $C_3H_7ClO_4$, mw: 142.5.
THR = Explodes spont. [*19*]

PROPYL PARABEN. See *p*-hydroxybenzoic acid propyl ester.

***n*-PROPYL PROPIONATE.** Clear liquid. $CH_3CH_2COOCH_2CH_2CH_3$, mw: 116.16, mp: −76°, bp: 122.4°, flash p: 175°F (OC), d: 0.885, vap. press: 10 mm @ 19.4°, vap. d: 4.0.
Acute tox data: Oral LD_{50} (rabbit) = 3950 mg/kg. [*3*]
THR = MOD via oral route. See esters.
Fire Hazard: Mod, when exposed to heat or flame; can react with oxidizing materials.
Spont Heating: No.
To Fight Fire: Foam, CO_2, dry chemical.

PROPYL QUINACRINE HALF-MUSTARD.
THR = An exper carc. [*23*]

PROPYL QUINACRINE MUSTARD.
THR = An exper carc. [*23*]

***n*-PROPYL SILICATE.** $(C_3H_7O)_4Si$, mw: 264.4, bp: 226°, d: 0.915.
THR = U. See *n*-propyl alcohol.
Fire Hazard: Mod, when exposed to heat or flame; can react with oxidizing materials.

PROPYL SILANE. $C_3H_7SiH_3$, mw: 74.1.
THR = Spont flam in air. [*19*]

PROPYL SULFATE. Syn: *di-n-propyl sulfate.* Colorless, oily liquid. $(CH_3CH_2CH_2)_2SO_4$, mw: 182.2, mp: 140°–170° (decomp), bp: 120° @ 20 mm, d: 1.11 @ 22.5°, vap. d: 6.28.
THR = U. See esters.
Fire Hazard: Mod, when exposed to heat or flame.
Disaster Hazard: Dangerous; when heated to decomp, emits highly toxic fumes of SO_x; can react with oxidizing materials.
To Fight Fire: Water, foam, CO_2, dry chemical.

PROPYL SULFIDE. Syn: *1-propyl thiopropane.* Liquid. $(C_3H_7)_2S$, mw: 118.2, bp: 142°, fp: −101.9°, d: 0.814 @ 17°, vap. d: 4.08.
THR = U. See sulfides.
Fire Hazard: Mod, when exposed to heat or flame.
Disaster Hazard: Dangerous; when heated to decomp, emits highly toxic fumes of SO_x; can react with oxidizing materials.
To Fight Fire: Foam, CO_2, dry chemical.

PROPYL TETRATHIO-*o*-STANNATE. Solid. $(SC_3H_7)_4Sn$, mw: 419.3, bp: 123° @ 0.001 mm.
THR = See tin compounds and sulfur compounds.

PROPYL THIOCYANATE. Clear liquid. $CH_3CH_2CH_2SCN$, mw: 101.13, vap. d: 3.49.
THR = Probably slight. See thiocyanates.
Fire Hazard: Mod, when exposed to heat or flame.
Disaster Hazard: Dangerous; when heated to decomp, emits highly toxic fumes of cyanides; can react with oxidizing materials.
To Fight Fire: Foam, CO_2, dry chemical.

1-PROPYL THIOPROPANE. See propyl sulfide.

6-PROPYL-2-THIOURACIL. $C_7H_{10}ON_2S$, mw: 170.3.
THR = See thiourea. An exper (+) carc [*1*] and neo. [*3*]

PROPYLTIN TRIIODIDE. Solid. $(C_3H_7)SnI_3$, mw: 542.6, bp: 200° @ 16 mm (decomp).
THR = See tin compounds and iodides.

***n*-PROPYL TRI-*n*-AMYLTIN.** Liquid. $(C_3H_7)Sn(C_5H_{11})_3$, mw: 375.2, d: 1.0368, bp: 163° @ 10 mm.
THR = U. See tin compounds.

PROPYL TRICHLOROSILANE. $C_3H_7SiCl_3$, mw: 177.6, vap. d: 6.15, flash p: 100° F.
THR = HIGH irr to skin, eyes and mu mem.
Fire Hazard: Slight, when exposed to heat or flame.
Disaster Hazard: Dangerous; when heated to decomp, emits highly toxic fumes of chlorides; will react with water or steam to produce toxic and corrosive fumes; can react with oxidizing materials.
To Fight Fire: Foam, CO_2, dry chemical.

PROPYL-3,4,5-TRIHYDROXYBENZOATE. See propyl gallate.

PROPYL TRIPHENYL GERMANIUM. Colorless crystals, insol in water, sol in organic solvents. $(C_3H_7)Ge(C_6H_5)_3$, mw: 347.0, mp: 865°.
THR = U. See germanium compounds and phenols.

PROPYNE. See allylene.

2-PROPYN-1-OL. See propargyl alcohol.

PROPYNYL ADIPATE.
THR = U. Limited animal exper suggest higher toxicity than that of most esters. See also esters.

PROQUANIL. $C_9H_{18}O_4N_2$, mw: 218.3.
THR = An exper teratogen. [*3*]

PROSTAGLANDIN E2.
THR = An exper teratogen. [*3*]

PROTACTINIUM. A bright, lustrous metal. Pa, mp: 1600°, d: 15.37, vap. press: 5×10^{-5} mm @ 1927°.
THR = A highly radiotoxic metallic element. A carc. [*14*]
Radiation Hazard: For permissible levels, see Section 5A, Table 5A.5. Natural isotope ^{231}Pa (Actinium Series), $T\frac{1}{2} = 3 \times 10^4$y, decays to radioactive ^{227}Ac via α's of 5.0 MeV. Artificial isotope ^{233}Pa (Neptunium Series), $T\frac{1}{2} = 27$d, decays to radioactive ^{233}U via β's of 0.15 (37%), 0.26 (58%), 0.57 (5%) MeV; emits γ's of 0.02–0.42 MeV. Natural isotope ^{234}Pa (Uranium Series), $T\frac{1}{2} = 6.7$h; decays to radioactive ^{234}U via β's of 0.23–1.36 MeV; emits γ's of 0.04–0.8 MeV.

PROUSTITE. See silver thioarsenite.

PROVITAMIN A. See carotene.

PRUSSIATE OF SODA. See sodium ferricyanide and sodium ferrocyanide.

PRUSSIC ACID. See hydrocyanic acid.

PRUSSITE. See cyanogen.

PSEUDOACONITINE. Syns: *feraconitine*, *nepaline*. White crystals or syrupy mass. $C_{36}H_{51}NO_{12}$, mw: 689.78, mp: 214° (decomp).
THR = HIGH via oral, inhal and dermal routes. A deadly poison. [*2*]
Disaster Hazard: Dangerous; when heated, emits highly toxic fumes.

PSEUDOBUTYLENE. See *cis*-butene-2.

PSEUDOBUTYLENE GLYCOL. See 2,3-butylene glycol.

PSEUDOCOTUNNITE. See potassium lead chloride.

PSEUDOCUMENE. Syns: *1,2,4-trimethyl benzene, pseudocumol.* Liquid, insol in water, sol in alcohol, benzene and ether. $C_5H_3(CH_3)_3$, mp: 120.19, d: 0.888 @ 4°/4°, fp: −61°, bp: 168.89°, flash p: 130°F, autoign. temp.: 959°F.
Acute tox data: Oral LD_{LO} (rat) = 5000 mg/kg; ip LD_{LO} (rat) = 2000 mg/kg. [*3*]
THR = MOD via oral and ip routes. MOD–LOW via dermal route. Can cause CNS depression, anemia, bronchitis.
Fire Hazard: Mod, when exposed to heat, flame or oxidizers.
To Fight Fire: Foam, alcohol foam, mist.

PSEUDOIONONE. Syns: *citrylidene acetone; 2,6-dimethylene-deca-2,6,8-trien-10-one; 6,10-dimethyl-3,5,9-undecatrien-2-one.* Pale yellow liquid, sol in alcohol and ether. $(CH_3)_2C{:}CH(CH_2)_2C(CH_3){:}CHCH{:}CHCOCH_3$, mw: 192.29, d: 0.8984 @ 20°, bp: 144° @ 12 mm.
THR = No data. See also orris.

PTEROYLGLUTAMIC ACID. Syns: *folic acid, folacin, PGA.* A member of the vitamin B complex. Orange yellow needles or platelets, odorless, slightly sol in water, insol in lipid solvents, sol in dilute alkali hydroxide and carbonate solutions. $C_{19}H_{19}N_7O_6$, mw: 441.4.
Acute tox data: ip LD_{50} (mice) = 100 mg/kg; iv LD_{50} (mice) = 239 mg/kg. [*3*]
THR = HIGH via ip and iv routes. A food additive permitted in food for human consumption. [*109*]

PTOMAINES.
THR = These are exceedingly toxic compounds commonly formed in putrefying proteins, dead bodies, decayed meat and fish. They have been prepared synthetically and are derivatives of ethers of the polyhydric alcohols. "Ptomaine poisoning" is usually a misnomer for other forms of food poisoning.

PUBLIC HEALTH HAZARDS FROM SOLID WASTES. See Section 6.

PULSATILLA CAMPHOR. $C_{10}H_8O_4$, mw: 192.2.
Acute tox data: ip LD_{50} (mouse) = 150 mg/kg. [*3*]
THR = HIGH via ip route.
Fire Hazard: Mod, when exposed to heat or flame; can react with oxidizing materials.
Explosion Hazard: Slight, when exposed to flame.

PURINE-3-OXIDE.
THR = An exper carc. [*23*]

PURINE-6-THIOL. $C_5H_4N_4S$, mw: 152.2.
THR = An exper teratogen. [*3*]

PURPLE FOXGLOVE. Scc digitalis.

PURPUREO. See chloropentammine cobalt (III) chloride.

PVC. See polyvinyl chloride, polymer.

PVP 1. mw: 20,000.
THR = An exper carc via ip, sc and iv routes. [*3*]

PVP 2. mw: 20,000.
THR = An exper neo via ip, sc and iv routes. [*3*]

PVP 3. mw: 50,000.
THR = An exper carc via ip, sc and iv routes. [*3*]

PVP 4. mw: 300,000.
THR = An exper carc via ip, sc and iv routes. [*3*]

PVP 5. mw: 10,000.
THR = An exper carc via ip and sc routes. [*3*]

PVP 6. mw: 50,000.
THR = An exper carc via sc route. [*3*]

PVP 7.
THR = An exper neo via sc route. [*3*]

PYRAZINE CARBOXAMIDE. $C_4H_5ON_3$, mw: 123.1.
THR = An exper neo. [*3*]

PYRAZOTHION. See *o,o*-diethyl-*o*-(3-methyl-5-pyrazolyl) phosphorothioate.

PYRAZOXON. See *o,o*-diethyl-*o*-(3-methyl-5-pyrazolyl)phosphate.

PYRENE. Syn: *benzo(def)phenanthrene*. Colorless solid, solutions have a slight blue color, insol in water, fairly sol in organic solvents. $C_{16}H_{10}$ (a condensed ring hydrocarbon), mw: 202.24, mp: 156°, d: 1.271 @ 23°, bp: 404°.
THR = An exper neo via dermal route. [*3, 23*]

PYRENOLINE. $C_{19}H_{11}N$, mw: 253.3.
THR = An exper neo. [*3*]

N-PYREN-2-YL ACETAMIDE. Syn: *2-acetyl amino pyrene*. $C_{18}H_{13}ON$, mw: 259.3.
THR = LOW via oral route. An exper (S) neo. [*3*]

PYRETHRIN I. Syn: *pyrethrolone ester of chrysanthemum monocarboxylic acid.* Viscous liquid. $C_{21}H_{28}O_3$, mw: 328.4, bp: 170° @ 0.1 mm (decomp).
Acute tox data: Oral LD_{50} (rat) = 1200 mg/kg. [*3*]
THR = MOD via oral route. An allergen. Has produced diarrhea, convulsions, collapse and respiratory failure, nausea, tinnitus, headache and CNS upset. A highly insecticidal extract of weak mammalian toxicity. Rapidly detoxified in GI tract. For the long term, slight but definite liver damage occurs at 1000 ppm and 5000 ppm diet levels. Usual early symptoms are a contact dermatitis, asthma, sneezing, etc. A dose of 15 g was fatal to a child.
Fire Hazard: Slight.

PYRETHRIN II. Syn: *pyrethrolone ester of chrysanthemum dicarboxylic acid monomethyl ester.* Viscous liquid. $C_{22}H_{28}O_5$, mw: 372.4, bp: 200° @ 0.1 mm (decomp).
Acute tox data: Oral LD_{50} (rat) = 1200 mg/kg. [*3*]
THR = MOD via oral route. See also pyrethrum I. An allergen.

PYRETHRINS. See pyrethrin I and II.

PYRETHROLONE ESTER OF CHRYSANTHEMUM DICARBOXYLIC ACID MONOMETHYL ESTER. Scc pyrethrin II.

PYRETHROLONE ESTER OF CHRYSANTHEMUM MONOCARBOXYLIC ACID. See pyrethrin I.

PYRETHROSIN. Crystals, insol in water, sol in hot alcohol and chloroform, slightly sol in ether or petroleum ether. $C_{17}H_{22}O_5$, mw: 306.4, mp: 199°.
THR = See pyrethrin I.

PYRETHRUM FLOWERS. Syn: *dalmatian insect powder*. Fine powder.
Acute tox data: Oral LD_{50} (rat) = 200 mg/kg. [*3*]
THR − HIGH via oral route. An allergen. Can cause dermatitis of both allergic and contact types. See also pyrethrin I. Large doses can cause hyperexcitability, incoordination, tremors and muscular paralysis.

PYRIBENZAMINE. See tripelennamine.

PYRIDINE. Colorless liquid, sharp, penetrating, empyreumatic odor, burning taste. NCHCHCHCHCH, mw: 79.10, bp: 115.3°, lel = 1.8%, uel = 12.4%, fp: −42°, flash p: 68°F (CC), d: 0.982, autoign. temp.: 900°F, vap. press: 10 mm @ 13.2°, vap. d: 2.73.

Acute tox data: Oral LD_{50} (rat) = 891 mg/kg; inhal LC_{50} (rat) = 4000 ppm for 4 hrs; sc LD_{50} (rat) = 1000 mg/kg; dermal LD_{50} (rabbit) = 1121 mg/kg; iv LD_{50} (dog) = 880 mg/kg. [*3*]
THR = MOD via oral, dermal, sc, iv and inhal routes. Is mildly irr to skin and can cause CNS depression. Kidney, liver damage and GI upset also.
Fire Hazard: Dangerous; when exposed to heat, flame or oxidizers.
Spont Heating: No.
Explosion Hazard: Severe, in the form of vapor, when exposed to flame or spark. Reacts violently with chlorosulfonic acid, CrO_3, maleic anhydride, HNO_3, oleum, perchromates, β-propiolactone, $AgClO_4$, H_2SO_4. [*19*]
Disaster Hazard: Dangerous; when heated to decomp, emits highly toxic fumes of cyanides; can react vigorously with oxidizing materials.
To Fight Fire: Alcohol foam.

2-PYRIDINE ALDOXIME METHIODIDE. See 2-PAM.

PYRIDINE-3-CARBOXYLIC ACID. See niacin.

3-PYRIDINE METHANOL. Syn: *β-pyridyl carbinol.* Very hygroscopic liquid, water-sol. C_6H_7NO, mw: 109.1, bp: 154° @ 28 mm.
THR = Can cause gastrointestinal distress, flushing of skin, dizziness and paresthesias.
Disaster Hazard: Dangerous; when heated to decomp, emits highly toxic fumes of NO_x.

PYRIDINE-*n*-OXIDE. Water-sol crystals. C_5H_5NO, mw: 95.1. fp: 67.0°, bp: 102° @ 1 mm.
THR = U. See pyridine.
Fire Hazard: Mod, when exposed to heat or flame.
Disaster Hazard: Dangerous; when heated to decomp, emits toxic fumes of NO_x; can react with oxidizing materials.

PYRIDINE-1-OXIDE-3-AZO-*p*-DIMETHYLANILINE. $C_{13}H_{14}ON_4$, mw: 242.3.
THR = An exper carc. [*3*]

PYRIDINE-1-OXIDE-4-AZO-*p*-DIMETHYLANILINE.
THR = An exper carc. [*3*]

12H-PYRIDO(2,3-a)THIENO(2,3-i)CARBAZOLE. $C_{17}H_{10}N_2S$, mw: 274.4.
THR = An exper neo via sc route. [*3, 23*]

PYRIDINIUM PERCHLORATE. $C_6H_5NClO_4$, mw: 190.5.
THR = May explode upon contact with metals. May explode @ >335°. [*19*]

PYRIDINO TRIBROMOGOLD. Red crystals, water-sol. $(C_6H_5N)AuBr_3$, mw: 516.1, mp: 150° (decomp).
THR = See gold compounds and bromides.

PYRIDOXINE HYDROCHLORIDE. Commercial form of pyridoxine (Vitamin B_6), colorless–white platelets, sol in water, alcohol and acetone, slightly sol in other organic solvents. $C_8H_{11}O_3NHCl$, mw: 205.5, mp: 204°–206°.
Acute tox data: sc LD_{50} (rat) = 3700 mg/kg. [*3*]
THR = MOD via sc route.

3-PYRIDYL DIAZONIUM FLUOROBORATE. $C_6H_4NN_2BF_4$, mw: 204.9.
THR = A sample exploded @ 47°; another decomp @ room temp while standing. [*19*]

PYRIDYL MERCURIC ACETATE.
THR = A fungicide. HIGH toxicity. See mercury compounds, organic.

PYRIDYL MERCURIC CHLORIDE. See mercury compounds, organic.

β-PYRIDYL-α-METHYL PYRROLIDINE. See nicotine.

PYRIDYL METHYL STEARATE.
THR = A fungicide. Details U.
Disaster Hazard: Dangerous; when heated to decomp, emits highly toxic fumes.

β-PYRIDYL-α-PYRROLIDINE. See nornicotine.

PYROANTIMONATE. See potassium hydroxoantimonate.

PYROARSENIC ACID. Colorless crystals. $H_4As_2O_7$, mw: 265.9, mp: decomp @ 206°.
THR = HIGH. See arsenic compounds.

PYROCATECHIN. See pyrocatechol.

PYROCATECHOL. Syns: *1,2-benzendiol, catechol, pyrocatechin, o-dihydroxybenzene.* Colorless crystals. $C_6H_4(OH)_2$, mw: 110.11, mp: 105°, bp: 246°, flash p: 261°F (CC), d: 1.341 @ 15°, vap. press: 10 mm @ 118.3°, vap. d: 3.79.
Acute tox data: ip LD_{LO} (guinea pig) = 200 mg/kg; oral LD_{50} (rat) = 3890 mg/kg; sc LD_{50} (mice) = 247 mg/kg; oral LD_{50} (guinea pig) = 550 mg/kg. [*3*]
THR = HIGH via sc, ip; MOD via oral and dermal routes. An allergen. An exper carc. [*81, 23*] Can cause convulsions and injury to blood. See also phenol. Can cause dermatitis on skin contact. [*81*]
Fire Hazard: Slight, when exposed to heat or flame.
Spont Heating: No.
Disaster Hazard: Dangerous; when heated, emits highly toxic fumes; can react with oxidizing materials.
To Fight Fire: Water, CO_2, dry chemical.

PYROCELLULOSE. See cellulose nitrate.

PYROCHROITE. See manganous hydroxide.

PYROGALLIC ACID. See pyrogallol.

PYROGALLOL. Syns: *pyrogallic acid, 1,2,3-benzenetriol, trihydroxy benzene*. White, lustrous crystals. $C_6H_3(OH)_3$, mw: 126.11, mp: 133°–134°, bp: 309°, d: 1.453 @ 4°/4°, vap. press: 10 mm @ 167.7°.
Acute tox data: Oral LD_{50} (rat) = 789 mg/kg; ip LD_{50} (mouse) = 400 mg/kg; oral LD_{50} (rabbit) = 1600 mg/kg. [*3*]
THR = MOD via oral and ip routes. An exper carc. [*23*] If swallowed can cause vomiting and diarrhea. Convulsions, circulatory collapse, hemolysis, methemoglobinemia, kidney injury, liver damage and death. Readily absorbed via skin.

PYROLAN. Syns: *G-22008, 1-phenyl-3-methyl-5-pyrazolyl dimethyl carbamate*. Crystals, water-sol. $C_{13}H_{15}N_3O_2$, mw: 245.3, mp: 50°, bp: 161° @ 0.2 mm.
Acute tox data: Oral LD_{50} (rat) = 62 mg/kg; oral LD_{50} (chicken) = 11 mg/kg; oral LD_{50} (wild birds) = 39 mg/kg. [*3*]
THR = HIGH via oral route. An insecticide. See carbamates. A cholinesterase inhibitor.
Disaster Hazard: Dangerous; when heated to decomp, emits highly toxic fumes.

PYROLIGNEOUS ACID. Syn: *wood vinegar*. Yellowish, acidic liquid. $HC_2H_3O_2$ (6%).
THR = MOD irr and an allergen.

PYROMELLITIC ACID. Syns: (*PMA*), *1,2,4,5-benzene tetracarboxylic acid, phorbol myristate acetate*. Off-white powder. $C_6H_2(COOH)_4$, mw: 254.15, mp: 276°.
Acute tox data: ip LD_{LO} (mouse) = 300 mg/kg. [*3*]
THR = HIGH irr to skin, eyes and mu mem. An exper carc. [*23*]
Fire Hazard: Slight.

PYROMUCIC ACID. See furoic acid.

PYROMUCIC ALDEHYDE. See furfural.

PYROPHOSPHORIC ACID. Colorless, hygroscopic needles or liquid. $H_4P_2O_7$, mw: 178.0, mp: 61°.
THR = See phosphoric acid.

PYROPHOSPHOROUS ACID. Needles. $H_4P_2O_5$, mw: 145.99, mp: 38°, bp: decomp @ 139°.
THR = See phosphoric acid.

PYROPHYLLITE. Syn: *agalmatolete*. White, green, gray, brown; found in metamorphic rocks, a diluent. $Al_2Si_4O_{10}(OH)$, mw: 293, d: 2.8–2.9.
THR = U. A food additive permitted in the feed and drinking water of animals and for the treatment of food-producing animals. [*109*]

PYROSULFURIC ACID. Syn: *disulfuric acid*. Colorless to yellowish hygroscopic crystals. $H_2S_2O_7$, mw: 178.2, mp: 35°, bp: decomp, d: 1.89.
THR = Very corrosive. See sulfuric acid.
Fire Hazard: Mod, by chemical reaction. It reacts violently with organic materials containing hydrogen and oxygen.
Disaster Hazard: Dangerous. See sulfuric acid.

PYROSULFURYL CHLORIDE. Syn: *disulfuryl chloride*. Colorless, mobile, fuming liquid. $S_2O_5Cl_2$, mw: 215.03, mp: −39°–−37°, bp: 140°, d(gas): 9.6 g/l, d(liquid): 1.818 @ 11°/4°.
THR = HIGH irr to skin, eyes and mu mem. See also hydrochloric acid.

PYROXYLIN PLASTIC. See nitrocellulose.

PYROXYLIN PLASTIC, SCRAP. See nitrocellulose.

PYROXYLIN SOLUTION. flash p: 80°F or less.
Fire Hazard: Dangerous, via heat, flame or oxidizers.
To Fight Fire: Dry chemical, CO_2, foam.

PYRROLE. Colorless liquid, darkens on standing, mild odor. C_4H_5N, mw: 67.09, bp: 129°, fp: −24°, flash p: 102°F (TCC), d: 0.968 @ 20°/4°, vap. d: 2.31.
Acute tox data: sc LD_{50} (mice) = 61 mg/kg; ip LD_{LO} (rabbit) = 150 mg/kg. [*3*]
THR = HIGH via sc and ip routes. Probably MOD via oral and inhal routes.
Fire Hazard: Mod, when exposed to heat, flame or oxidizers.
Disaster Hazard: Dangerous; when heated to decomp, emits highly toxic fumes of NO_x; can react with oxidizing materials.
To Fight Fire: Foam, CO_2, dry chemical.

PYRROLIDINE. Colorless, mobile liquid, penetrating amine-like odor. C_4H_9N, mw: 71.12, bp: 86°–87°, fp: −63°, flash p: 37°F (TCC), d: 0.8618 @ 20°/4°, vap. press: 128 mm @ 39°, vap. d: 2.45.
Acute tox data: Oral LD_{50} (rat) = 300 mg/kg; inhal LC_{50} (mice) = 1300 mg/m^3 for 2 hrs. [*3*]
THR = HIGH via oral and inhal routes.
Fire Hazard: Dangerous, when exposed to heat, flame or oxidizers.
Explosion Hazard: U.
Disaster Hazard: Dangerous; when heated to decomp, emits highly toxic fumes of NO_x; can react vigorously with oxidizing materials.
To Fight Fire: Alcohol foam, CO_2, dry chemical.

2-PYRROLIDONE. A colorless liquid.
$HNCH_2CH_2CH_2CO$ (ring: N–CO), mw: 85.11, mp: 25°, bp: 245°, flash p: 265°F (OC).
Acute tox data: Oral LD_{50} (rat) = 6500 mg/kg; oral LD_{50} (guinea pig) = 6500 mg/kg. [*3*]
THR = MOD–LOW via oral route.
Fire Hazard: Slight, when exposed to heat or flame.
Disaster Hazard: Dangerous; when heated to decomp, emits highly toxic fumes of NO_x; can react with oxidizing materials.
To Fight Fire: Alcohol foam, CO_2, dry chemical.

2-PYRROLIDINE CARBOXYLIC ACID. See proline.

PYRROLINE. Syn: *2,2-dihydropyrrole*. Nearly colorless liquid, fumes in air with an unpleasant ammonia-like odor, miscible with water. C_4H_7N, mw: 69.1, bp: 91° @ 748 mm, d: 0.9097.
Acute tox data: sc LD_{LO} (rat) = 300 mg/kg. [*3*]
THR = HIGH via sc route.
Disaster Hazard: Dangerous; when heated to decomp, emits highly toxic fumes of NO_x.

3-PYRROL-2-YL PYRIDINE. $C_9H_8N_2$, mw: 144.2.
THR = An exper neo. [*3*]

PYRUVALDEHYDE. See pyruvic aldehyde.

PYRUVIC ALDEHYDE. Syn: *pyruvaldehyde*. Mobile yellow liquid, pungent odor. CH_3COCHO, mw: 72.06, bp: 72°, d: 1.06 @ 20°/20°.
THR = LOW irr via oral and inhal routes. See also aldehydes.
Fire Hazard: Slight.

q

QUAALUDE. $C_{16}H_{14}ON_2$, mw: 250.3.
Acute tox data: Oral LD_{50} (rat) = 255 mg/kg; ip LD_{50} (mice) = 626 mg/kg; iv LD_{50} (rabbit) = 100 mg/kg. [*3*]
THR = HIGH via oral and iv routes; MOD via ip route. It affects CNS. An exper teratogen [*3*] via oral route.

QUARTZ. Syns: *cristobalite, silicon dioxide*. Cubic, colorless crystals. SiO_2, mw: 60.1, mp: 1710°, bp: 2230°, d: 2.32.
Acute tox data: Causes pulmonary inhal effects on humans @ 16 mppcf over a long period. [*3*] An exper carc to rats via ipl route. [*103*] it LD_{LO} (rat) = 200 mg/kg; imp TD_{LO} (rat) = 900 mg/kg→neo; iv LD_{LO} (dog) = 20 mg/kg. [*3*]
THR = HIGH via iv and it routes. An exper neo via imp route; pulmonary effects via inhal. See silica. A food additive permitted in the feed and drinking water of animals and/or for the treatment of food-producing animals. Also permitted in food for human consumption. [*109*]

QUASSIA. Syn: *bitter wood tree*. Wood or bark.
THR = LOW-MOD via oral route. Large doses by mouth may produce nausea and vomiting; an insecticide.
Fire Hazard: Mod, when exposed to heat or flame.
Explosion Hazard: Slight, in the form of dust, when exposed to heat or flame.

QUEBRACHINE. See yohimbine.

QUENCHING OIL. An oil. flash p: 365°F (CC), d: 0.9.
THR = U.
Fire Hazard: Slight, when exposed to heat or flame.
To Fight Fire: CO_2, dry chemical.

QUICK LIME. See calcium oxide.

QUICKSILVER. See mercury.

QUINACRINE. See "Atabrine."

QUINACRINE ETHYL M/2. Syn: *LCR-125*. $C_{18}H_{14}ON_3Cl_2$, mw: 450.2.
THR = An exper neo.

QUINACRINE ETHYL MUSTARD. Syn: *NSC-35372*, $C_{20}H_{22}ON_3Cl_3 \cdot 2HCl \cdot H_2O$, mw: 517.7.
THR = An exper carc. [*3, 23*]

QUINACRINE MUSTARD. $C_{23}H_{27}ON_3Cl_3$, mw: 467.9.
Acute tox data: iv LD_{LO} (dog) = 0.91 mg/kg; iv LD_{LO} (monkey) = 0.91 mg/kg. [*3*]
THR = HIGH via iv route. An exper carc. [*23*]

QUINACRINE PROPYL MUSTARD.
THR = An exper carc. [*23*]

QUINALDINE. See methyl quinoline.

QUINAZOLINE. $C_8H_6N_2$, mw: 130.2.
THR = An exper neo. [*3*]

QUINHYDRONE. Dark green crystals, slightly sol in water, sol in alcohol, ether, hot water and ammonia. $C_6H_4O_2C_6H_4(OH)_2$, mw: 218, d: 1.40, mp: 171°.
Acute tox data: Oral LD_{LO} (rat) = 225 mg/kg; iv LD_{50} (rat) = 35 mg/kg. [*3*]
THR = HIGH via iv and oral routes. Small doses caused lowered metabolism.

QUINICARDINE. See quinidine.

QUINIDINE. Syn: *quinicardine*. $C_{20}H_{24}O_2N_2$, mw: 324.5.
Acute tox data: Oral LD_{50} (rat) = 1000 mg/kg; iv LD_{50} (rat) = 23 mg/kg; ip LD_{50} (mice) = 190 mg/kg; im LD_{50} (mice) = 200 mg/kg. [*3*]
THR = HIGH via im, iv and ip routes; MOD via oral route. Implicated in aplastic anemia.

QUININE. Bulky, white amorphous powder or crystals, bitter taste. $C_{20}H_{24}N_2O_2$, mw: 324.4, mp: 174.9°.
Acute tox data: Oral LD_{LO} (rabbit) = 800 mg/kg; sc LD_{LO} (rat) = 200 mg/kg; im LD_{LO} (rat) = 300 mg/kg. [*3*]
THR = HIGH via sc and im routes; MOD via oral route. An exper teratogen via oral route. [*3*]

Upon contact with this material, the eyes become swollen, watery and exude a sticky, viscous liquid which forms yellowish crusts. Upon ingestion, it causes dilation of the pupils. The optic nerve becomes pale and atrophic and retina shows thready arteries; ptosis and clonic spasms of the lids result. It may cause atrophy of the optic nerve. Vision returns in from 24–28 hrs and gradually improves. Quinine dermatitis is an occupational hazard to barbers particularly and generally to people who work with quinine tonics, medicaments, or cosmetics. Quinine has no influence upon sound skin, but it is distinctly irr to mu mem and raw surfaces. Internally it can cause a sense of fullness in the head, tinnitis aureum, slight deafness, disorders of vision and sometimes blindness. Its physiological effects vary with the individual. Occasionally, it can cause cutaneous eruptions, such as erythema, urticaria, herpes, purpuria, and even gangrenous

affections. Quinine is used as a food additive permitted in food for human consumption. [*109*]
Treatment and Antidotes: Wash out stomach. There are no specific antidotes. Treatment is symptomatic.
Fire Hazard: Slight; when heated to decomp, emits toxic fumes of NO_x.

QUININE ACETATE. See quinine.

QUININE ARSENATE. Fine needles or crystals. $3C_{20}H_{24}N_2O_2 \cdot 2H_3AsO_4 \cdot 5H_2O$, mw: 1347.2.
THR = HIGH. See arsenic compounds.

QUININE BISULFATE. See quinine.

QUININE CACODYLATE. White powder. $C_{20}H_{24}N_2O_2 \cdot (CH_3)_2 \cdot AsO_2H$, mw: 462.4.
THR = See arsenic compounds

QUININE CHLORATE. Crystals. $C_{20}H_{24}N_2O_2 \cdot HClO_3 \cdot 2H_2O$, mw: 444.9, mp: explodes.
THR = HIGH. See quinine.
Fire Hazard: Dangerous; see chlorates.
Explosion Hazard: Mod, when shocked or exposed to heat; see chlorates.
Disaster Hazard: Dangerous; shock will explode it; when heated to decomp, emits highly toxic fumes of NO_x and chlorides; can react vigorously with reducing materials.

QUININE CHLORIDE. $C_{20}H_{24}O_2N_2 \cdot HCl$, mw: 360.9.
Acute tox data: ip LD_{50} (rat) = 170 mg/kg; iv LD_{LO} (human) = 0.23 mg/kg → CNS effects; iv LD_{LO} (rat) = 75 mg/kg; im LD_{LO} (rat) = 300 mg/kg. [*3*]
THR = HIGH via ip, iv and im routes.

QUINOL. See *p*-hydroquinone.

QUINOLINE. Syns: *chinoline, leukol.* Refractive, colorless liquid, peculiar odor. $C_6H_4NCHCHCH$, mw: 129.2, mp: −14.5°, bp: 237.7°, d: 1.0900 @ 25°/4°, autoign. temp.: 896°F, vap. press: 1 mm @ 59.7°, vap. d: 4.45.
Acute tox data: Oral LD_{50} (rat) = 460 mg/kg; dermal LD_{50} (rat) = 540 mg/kg. [*3*]
THR = HIGH–MOD via oral and dermal routes. May produce retinitis similar to that caused by naphthalene but without causing opacity of the lens.
Fire Hazard: Slight, when exposed to heat. Violent reaction with perchromates. [*19*]
Disaster Hazard: Dangerous; when heated to decomp, emits toxic fumes of NO_x.

8-QUINOLINOL. Syns: *8-hydroxyquinoline, oxine.* White crystals or powder. Nearly insol in water. C_9H_7NO, mw: 145.2, mp: 76°, bp: 267°.
Acute tox data: Oral LD_{50} (rat) = 1200 mg/kg; ip LD_{50} (mice) = 48 mg/kg. [*3*]
THR = HIGH via ip route and MOD via oral route. A CNS stimulant. A fungicide. An exper carc and neo. [*3, 23*] of the uterus, rectum, brain and bladder via oral, imp and iv routes.
Fire Hazard: Slight.
Disaster Hazard: Dangerous; when heated to decomp, emits highly toxic fumes of NO_x.

QUINOLINOTRI BROMOGOLD. Deep red crystals, sol in chloroform. $(C_9H_7N)AuBr_3$, mw: 566.1, mp: >200°.
THR = Details U. See gold compounds.
Disaster Hazard: Dangerous. See bromides.

N-(4-QUINOLYL)ACETOHYDROXAMIC ACID. $C_{11}H_{10}O_2N_2$, mw: 202.2.
THR = An exper neo. [*3*]

QUINONE. Syns: *benzo quinone, chinone.* Yellow crystals, characteristic irr odor. OC_6H_4O, mw: 108.09, mp: 115.7°, bp: sublimes, d: 1.318 @ 20°/4°.
Acute tox data: Oral LD_{50} (rat) = 120 mg/kg; ip LD_{50} (mouse) = 8.5 mg/kg; iv LD_{50} (rat) = 25 mg/kg; LC_{LO} (mouse) = 320 mg/m^3. [*3*]
THR = HIGH via oral, ip, iv and inhal routes. An exper neo via dermal route. [*3*] An exper (±) carc. [*81*]

Quinone has a characteristic, irr odor. It can cause severe local damage to the skin and mu mem by contact with it in the solid state, in solution, or in the form of condensed vapors. Locally it can cause discoloration, severe irr, erythema, swelling and the formation of papules and vesicles, whereas prolonged contact may lead to necrosis. When the eyes become involved, it can cause dangerous disturbances of vision. A case is reported where ulceration of the cornea resulted from brief exposure to a high conc of the vapor. An accepted criterion for regulating workroom conc of this material in the air has been the comfort of personnel involved, as judged by eye irr, conjunctivitis, photophobia, moderate lachrymation, and burning sensations. It has been found that personnel can develop corneal injury of two types due to this material. One type is a typical superficial greenish-brown stain or grayish-white opacity varying in size and involving all the layers of the cornea. In a few cases there has been an appreciable loss of vision. The eye stain is probably an end product of the oxidation of quinone to hydroquinone and the subsequent polymerization of this material. Its odor becomes perceptible at or just above 0.1 ppm.

RACEMIC ACID. See tartaric acid.

RADAR TUBE RADIATIONS. See radiations, ionizing.

RADIATION EXPOSURE CONTROL. See Section 5A.

RADIATION EXPOSURE REDUCING TECHNICS. See Sections 5A and 5B.

RADIATION EXPOSURE RESPONSIBILITY. See Section 5A.

RADIATION FROM ELECTRON GUNS. See radiation, ionizing; also Section 5A.

RADIATION FROM PARTICLE ACCELERATORS. See radiation, ionizing.

RADIATION FROM RADIO TUBES. See radiation, ionizing.

RADIATION HAZARDS AND CONTROL. See Sections 5A and 5B.

RADIATION IN FOOD PRODUCTION, PROCESSING AND HANDLING. See Sections 10, 6, 5B.

RADIATION, IONIZING.

THR = A recog carc of the skin, subcutaneous tissue, bone, hematopoietic tissues, lung, liver, larynx, thyroid and kidney. An exper carc of the breast, uterus. [*14*] See also Section 5A. Has been implicated in aplastic anemia.

RADIATION MEASUREMENT. See Section 5A.

RADIATION PROTECTION GUIDES (RPG). See Section 5A.

RADIATION, ULTRAVIOLET.

THR = A recog carc of skin (basal cell and squamous cell carc). Also exper carc (melanoma carc). [*14*]

RADIOACTIVE AEROSOLS.

THR = An exper carc. [*23*]

RADIOACTIVE CONTAMINATION OF SURFACES. See Section 5A.

RADIOACTIVE DEVICES. See radiation, ionizing.

RADIOACTIVE ISOTOPE RADIATIONS.

THR = A recog carc. See radiation, ionizing. [*14, 3, 8*]

RADIOACTIVE MATERIALS, N.O.S. See individual isotopes.

RADIOACTIVE MATERIALS, GROUP IV. See Section 5A.

RADIOACTIVE WASTES. See Sections 5A and 5B.

RADIOACTIVE WASTE DISPOSAL. See Sections 5A and 5B.

RADIOACTIVE WASTE MONITORING. See Section 5.

RADIOACTIVITY CONTAMINATION OF PERSONNEL. See Section 5.

RADIOACTIVITY DECONTAMINATION. See Sec*tion 5*

RADIOISOTOPE STORAGE, HANDLING, AND SHIPPING. See Sections 5A and 5B.

RADIOLOGICAL PROTECTION AND CONTROLS. See Section 5A and 5B.

RADIUM. A brilliant white metal, tarnishes in air, decomp in H_2O, mp: 700°, bp: 1140°, d: 6.

THR = Common air contaminant. A highly radiotoxic element. 1 g = 3.7×10^{10} dps. Inhal, ingestion or bodily exposure to Ra can lead to lung cancer, bone cancer, osteitis, skin damage and blood dyscrasias.

Ra replaces calcium in the bone structure and is a source of irradiation to the blood forming organs. The ingestion of luminous dial paint prepared from radium was the cause of death of many of the early dial painters before the hazard was fully understood. The data on these workers has been the source of many of the radiation precautions and the maximum permissible levels for internal emitters which are now accepted. ^{226}Ra is the parent of radon and the precautions described under ^{222}Rn should be followed.

^{228}Ra is a member of the thorium series. It was a common constituent of luminous paints, and while its low β energy was not a hazard, its daughters in the series may have been a causative agent in the deaths of radium dial painters following World War I. Its metabolism is the same as any other radium isotope and it is a source of thoron. The precautions recommended under ^{220}Rn should be followed. A recog carc (Section 8).

Radiation Hazard: For permissible levels, see Section 5A, Table 5A.5. Natural isotope ^{223}Ra (Actinium-X, Actinium Series), $T\frac{1}{2}$ = 11.4d, decays to radioactive ^{219}Rn via α's of 5.5–5.7 MeV. Natural isotope ^{224}Ra (Thorium-X, Thorium Series), $T\frac{1}{2}$ = 3.6d, decays to

For Countermeasure Information and Abbreviations see the Directory at the Beginning of this Section.

radioactive ^{220}Rn via α's of 5.7 Mev. Natural isotope ^{226}Ra (Uranium Series), $T\frac{1}{2}$ = 1600y, decays to radioactive ^{222}Rn via α's of 4.8 MeV. Natural isotope ^{228}Ra (Mesothorium = I, Thorium Series), $T\frac{1}{2}$ = 6.7y, decays to radioactive ^{228}Ac via β's of 0.05 MeV.
Disaster Hazard: Highly dangerous; must be kept heavily shielded and stored away from possible dissemination by explosion, flood, etc. Section 5A.

RADIUM BROMIDE. Colorless yellowish crystals. $RaBr_2$, mw: 385.88, mp: 728°, d: 5.79.
THR = See radium and bromides.

RADIUM CARBONATE. White or slightly brownish crystals. $RaCO_3$, mw: 286.06.
THR = See radium.

RADIUM CHLORIDE. Monoclinic, colorless–yellowish crystals. $RaCl_2$, mw: 296.96, mp: 1000°, d: 4.91.
THR = See radium and chlorides.

RADIUM EMANATION. See radon.

RADIUM IODATE. White powder. $Ra(IO_3)_2$, mw: 575.89.
THR = See radium and iodates.

RADIUM SULFATE. Colorless crystals. $RaSO_4$, mw: 322.12.
THR = See radium and sulfates.

RADON. Syns: *nitrol, radium emanation.* A very dense, chemically inert, radioactive gas. Rn, mp: −71°, bp: −61.8°, d(gas): 9.73 g/l, d(liquid): 4.4 @ −62°, d(solid): 4.
THR = A common air contaminant. See Radiation Hazards below.
Radiation Hazard: For permissible levels, see Section 5A. Table 5A.5. Natural isotope ^{220}Rn (Thoron, Thorium Series), $T\frac{1}{2}$ = 55s, decays to radioactive ^{216}Po via α's of 6.3 MeV. Natural isotope ^{222}Rn (Uranium Series), $T\frac{1}{2}$ = 3.8d, decays to radioactive ^{218}Po via α's of 5.5 MeV. The permissible levels are given for ^{222}Rn in equilibrium with its daughters. The chief hazard from this isotope is inhal of the gaseous element and its solid daughters, which are collected on the normal dust of the air. This material is deposited in the lungs and has been considered to be a major causative agent in the high incidence of lung cancer found in uranium miners. Radon and its daughters build up to an equilibrium value in about a month from radium compounds, while the build-up from uranium compounds is negligible. Good ventilation of areas where radium is handled or stored is recommended to prevent accumulation of hazardous conc of Rn and its daughters. Section 5A.

RAGS, OILY.
THR = A variable allergen.
Fire Hazard: Mod, when exposed to heat or flame; can react with oxidizing materials.

RAGS, WET.
THR = A weak allergen.
Fire Hazard: Slight, when exposed to heat or flame or by chemical reaction.

RAMROD. Syns: *2-chloro-N-isopropylacetanilide.* A powder. $C_{11}H_{14}ONCl$, mw: 211.5.
Acute tox data: Oral LD_{50} (rat) = 1200 mg/kg; dermal LD_{50} (rabbit) = 380 mg/kg. [*3*]
THR = HIGH via dermal and MOD via oral routes. A pre-emergence selective herbicide.
Disaster Hazard: Dangerous; see chlorides.

RANDOX. See N,N-diallyl-2-chloroacetamide.

RANEY'S NICKEL. See nickel catalyst.

RAPESEED OIL. Syn: *colza oil.* Flash p: 325°F (CC), solidifying p: −2° to 10°, d: 0.915, autoign. temp.: 836°F.
THR = LOW. A fungicide for plants.
Fire Hazard: Slight, when exposed to heat or flame; can react with oxidizing materials.
Spont Heating: Yes.
To Fight Fire: CO_2, dry chemical.

RARE EARTH HYDRIDES.
THR = The hydrides of Ce, Pr, Nd, Pm, Sm, Eu, Gd, Tb, Dy, Ho, La, Er, Tm, Yb and Lu are stable in dry air but ignite in moist air. [*19*]

RARE EARTH METALS.
THR = Ce, Pr, Nd, Sm, Eu, Gd, Tb, Dy, Ho, Er, Tm, Yb and Lu ignite in air @ 150°. La ignites @ 450°. May be pyroforic upon cutting or filing. They burn vigorously in halogen vapors >200°. [*19*]

RATHANI. An aqueous extract of a plant.
THR = An exper carc. [*3*]

RATIONITE. See dimethyl sulfate and chlorosulfonic acid.

RATON. An aqueous extract of dried leaves of the plant.
THR = An exper neo. [*3*]

RATTLE SNAKE VENOM. See crotoxin.

RAWETIN. Syn: *N-hydroxy naphthalimide diethyl phosphate.* Tan crystalline powder, sol in methylene chloride, difficult sol in most organic solvents. mp: 177.0°.
Acute tox data: Oral LD_{50} (rat) = 70 mg/kg; oral LD_{50} (mice) = 50 mg/kg; oral LD_{50} (chicken) = 43 mg/kg. [*3*]
THR = HIGH via oral route. A ruminant antihelminitic; a cholinesterase inhibitor; See also parathion.
Disaster Hazard: Dangerous; see parathion.

RDX. See cyclotrimethylene trinitramine.

REALGAR. See arsenic bisulfide.

RECIRCULATION OF AIR. See Section 2.

RECLAMATION PLANTS, A DANGER POINT IN SOLID WASTE HANDLING. See Section 6.

RED LEAD. See lead oxide.

RED OIL (COMMERCIAL GRADE). See oleic acid.

RED PEPPER.
THR = A susp carc. [*14*] See plant and fungal products.

RED PHOSPHORUS. See phosphorus (amorphous red).

RED POTASSIUM CHROMATE. See potassium bichromate.

RED POTASSIUM PRUSSIATE. See potassium ferricyanide.

RED PRUSSIATE OF POTASH. See potassium ferricyanide.

RED PRUSSIATE OF SODA. See sodium ferricyanide.

RED PRUSSIATE OF SODIUM. See sodium ferricyanide.

REDWOOD. See brazil wood.

REFINED TARS.
THR = A recog carc. [*14*] See coal tar and pitch.

REFRIGERANT 115. See chloropentafluoroethane.

REFRIGERANT 116. See hexafluoroethane.

REFUSE CHUTES AS A DANGER POINT IN SOLID WASTE HANDLING. See Section 6.

REGULATION OF RADIATION EXPOSURE. See Section 5A.

REMAZOL YELLOW G. $C_{19}H_{21}O_{11}N_4S_3Cl \cdot 2Na$; mw: 659.1.
THR = An exper neo. [*3*]

REMAZOL RED B. $2Na \cdot C_{18}H_{16}O_{10}N_2S_3$, mw: 562.5.
THR = An exper neo. [*3*]

RENNASE. See rennet.

RENNET. Syns: *rennin, rennase, chymosin.* The enzyme secreted by the glands of the stomach which causes curdling of milk. A yellowish white powder, peculiar but not unpleasant odor, partially sol in water and dilute alcohol.
THR = U. A general purpose food additive. [*109*]

RENNIN. See rennet.

RESERPINE. White or pale buff to slightly yellow powder, odorless, insol in water, very slightly sol in alcohol, sol in chloroform and acetic acid. $C_{33}H_{40}N_2O_9$, mw: 608.7, mp: 264°–265° (decomp).
Acute tox data: iv LD_{50} (rat) = 15 mg/kg; oral LD_{50} (mouse) = 390 mg/kg; ip LD_{50} (mouse) = 70 mg/kg; iv LD_{50} (dog) = 0.5 mg/kg; oral LD_{50} (wild birds) = 100 mg/kg. [*3*]
THR = HIGH via oral, iv, ip routes. In humans, 0.014 mg/kg⟶psychotropic effects. An exper (±) carc [*3, 10*] and teratogen. [*3*] A medicine with side effects. Used as a food additive permitted in the feed and drinking water of animals and/or for the treatment of food-producing animals. Also permitted in food for human consumption. [*109*] A sedative.

RESERPIDINE. $C_{32}H_{38}O_8N_2$, mw: 578.7.
Acute tox data: Oral LD_{50} (mouse) = 500 mg/kg; ip LD_{50} (mouse) = 60 mg/kg; iv LD_{50} (rat) = 15 mg/kg. [*3*]
THR = HIGH via ip and iv routes; MOD via oral route. An exper teratogen. [*3*]

RESIDUAL ASPHALT.
THR = A recog carc. [*14*] See petroleum asphalt.

RESIDUAL OIL.
THR = A recog carc. [*14*] See oils (mineral).

RESIN COPAL. See copal.

RESORCIN. See resorcinol.

RESORCINOL. Syns: *1,3-benzenediol, resorcin, m-dihydroxybenzene.* Very white crystals, become pink on exposure to light when not perfectly pure, unpleasant sweet taste. $C_6H_4(OH)_2$, mw: 110.11, mp: 110°, bp: 276.5°, flash p: 261°F (CC), d: 1.285 @ 15°, autoign. temp.: 1126°F, vap. press: 1 mm @ 108.4°, vap. d: 3.79.
Acute tox data: Oral LD_{50} (rat) = 301 mg/kg; sc LD_{LO} (mouse) = 340 mg/kg; oral LD_{LO} (human) = 29 mg/kg. [*3*]
THR = HIGH via oral and sc routes. Does not initiate skin cancer in mice, nor show any carc effect in mice. [*81*] It is primarily a skin irr. However, it can cause systemic poisoning by acting both as a blood and nerve poison. It may also cause injury to the eyes and dermatitis, particularly to those who are sensitive to it. Such individuals are affected by even very slight traces of it. In a suitable solvent, this material can readily be absorbed through human skin, and can cause local hyperemia, itching, dermatitis, edema and corrosion associated with enlargement of regional lymph glands as well as serious systemic disorders such as restlessness, methemoglobinemia, cyanosis, convulsions, tachycardia, dyspnea and death. These same symptoms can be induced by ingestion of the material. For poisoning, treat symptomatically. Get medical advice.
Fire Hazard: Slight, when exposed to heat or flame; can react with oxidizing materials.

Spont Heating: No.
To Fight Fire: Water, CO_2, dry chemical.

RESORCINOL DIGLYCIDYL ETHER. $C_{12}H_{14}O_4$, mw: 222.3.
THR = An exper (±) carc. [*3, 13*] MOD acute oral.

RESORCINOL PHTHALEIN. See fluoroescein.

RESPIRATORY TESTING AND APPROVAL. See Section 2.

RETINOIC ACID (all-*trans*). $C_{20}H_{28}O_2$, mw: 300.5.
THR = An exper teratogen. [*3*]

RETINOIC ACID, SODIUM SALT. $C_{20}H_{27}O_2 \cdot Na$, mw: 322.5.
THR = An exper teratogen. [*3*]

RETINOL. $C_{20}H_{30}O$, mw: 286.5.
Acute tox data: Oral LD_{50} (rat) = 2000 mg/kg; oral LD_{50} (mice) = 4000 mg/kg. [*3*]
THR = MOD via oral route. An exper teratogen [*3*] via oral and ip routes.

RETINOL ACETATE. $C_{22}H_{32}O_2$, mw: 328.5.
THR = An exper teratogen. [*3*]

RETINOL PALMITATE. $C_{36}H_{60}O_2$, mw: 525.
THR = An exper teratogen. [*3*]

RETRONECINE. $C_8H_{13}O_2N$, mw: 155.2.
Acute tox data: iv LD_{50} (rat) = 1311 mg/kg; iv LD_{50} (mice) = 634 mg/kg. [*3*]
THR = MOD via iv route. An exper carc. [*23*]

RETRONECINE HYDROCHLORIDE. $C_8H_{13}O_2N \cdot HCl$, mw: 191.7.
An exper carc. [*3*]

RETROSINE. $C_{18}H_{25}O_6N$, mw: 351.4.
THR = An exper (+) carc and neo [*3, 7, 23*]

RETROSINE-N-OXIDE. $C_{18}H_{25}O_7N$, mw: 367.4.
THR = An exper (+) carc and neo [*3, 7, 23*] via oral ip, iv and other routes.

RHATHANI. Syn: *krameria triandra.* Aqueous extract from the plant root.
THR = An exper carc to rats via sc route. [*103*]

RHENIUM. Silvery-white, very dense, metallic, lustrous metal. Re, atwt: 186.2, mp: 3180°, bp: 5627°, d: 20.53.
THR = Not yet established but handle with care.
Radiation Hazard: For permissible levels, see Section 5A, Table 5A.5 Natural (63%) isotope ^{187}Re, $T\frac{1}{2}$ = 4×10^{10}y, decays to stable ^{187}Os via β's of less than 0.01 MeV.
Fire Hazard: Mod, in the form of dust, when exposed to heat or flame. Violent reaction with F_2 @ 125°. [*19*]

RHENIUM DISULFIDE. Black, hexagonal leaf. ReS_2, mw: 250.44, bp: decomp, d: 7.5.
THR = See sulfides.

RHENIUM HEPTASULFIDE. Black crystals. Re_2S_7, mw: 597.0 (decomp), d: 4.87 @ 24.5°.
THR = Ignites on heating. See sulfides.

RHENIUM HEXAFLUORIDE. Pale yellow crystals. ReF_6, mw: 300.2, mp: 18.8°, bp: 47.6°, d(liquid): 6.1573, d(solid): 3.616.
THR = See fluorides.

RHENIUM OXYTETRAFLUORIDE. Colorless crystals. $ReOF_4$, mw: 278.31, mp: 39.7°, bp: 62.7°, d(liquid): 3.717, d(solid): 4.032.
THR = See fluorides.

RHENIUM TETRAFLUORIDE. Green crystals. ReF_4, mw: 262.31, mp: 124.5°, d: 5.383, bp: decomp @ 500°.
THR = See fluorides.

RHENIUM TRICHLORIDE. $ReCl_3$, mw: 292.6.
Acute tox data: ip LD_{50} (rat) = 280 mg/kg. [*3*]
THR = HIGH via ip route.

RHENIUM TRIOXYBROMIDE. White crystals. ReO_3Br, mw: 314.23, mp: 39.5°, bp: 163°.
THR = See bromides.

RHIGOLENE. See pentane and isopentane.

RHIZOCTOL. Syn: *methylarsinic sulfide.* A powder. H_3CAsS; mw: 122.
Acute tox data: Oral LD_{50} (rat) = 100 mg/kg; dermal LD_{50} (rat) = 1400 mg/kg; oral LD_{50} (mammal) = 100 mg/kg. [*3*]
THR = HIGH via oral route; MOD via dermal route. See also arsenic compounds and sulfides.
Disaster Hazard: See arsenic and sulfides.

RHODAMINE 6G. $C_{28}H_{30}N_2O_3 \cdot HCl$, mw: 479; An exper carc in rats via sc route. [*108*]

RHODANINE. Syn: *2-thio-4-ketothiazolidine.* Yellow crystals. $C_3H_3ONS_2$, mw: 133.2, mp: 168° (can explode), d: 0.868.
Acute tox data: Oral LD_{50} (rat) = 326 mg/kg; ip LD_{LO} (mouse) = 200 mg/kg; sc LD_{LO} (mouse) = 200 mg/kg. [*3*]
THR = HIGH via oral, ip and sc routes.
Disaster Hazard: Dangerous; when heated to decomp, emits highly toxic fumes of SO_x.

RHODIUM. A silvery white metallic element. Rh, atwt: 102.906, mp: 1966°, bp: 3727°, d: 2.41 @ 20°.
THR = Not yet established but handle carefully. OSHA standard for air (TWA) = 100 μg Rh/m^3.
Radiation Hazard: For permissible levels, see Section 5A, Table 5A.5 Artificial isotope ^{103m}Rh, $T\frac{1}{2}$ = 57m, decays to stable ^{103}Rh via γ's of 0.04 MeV and x-rays. ^{103m}Rh exists in radioactive equilibrium with its parent, ^{103}Ru. Permissible levels are given for the equilibrium mixture. Artificial isotope ^{106}Rh, $T\frac{1}{2}$ = 30s, decays to stable ^{106}Pd via β's of 2.4 (11%), 3.1

(8%), 3.54 (78%) MeV. Emits γ's of 0.51, 0.62, 1.05 MeV. ^{106}Rh exists in radioactive equilibrium with its parent, ^{106}Ru. Permissible levels are given for the equilibrium mixture.

Fire Hazard: Mod, when exposed to heat or flame. Violent reaction with ClF_3, OF_2. [*19*]

RHODIUM AMMINE NITRATES.
THR = May be impact-sensitive explosives. [*19*]

RHODIUM AMMINE PERCHLORATES.
THR = May be impact-sensitive explosives. [*19*]

RHODIUM (III) CHLORIDE. $RhCl_3$, mw: 209.3.
Acute tox data: iv LD_{50} (rat) = 198 mg/kg; iv LD_{50} (rabbit) = 215 mg/kg. [*3*]
THR = HIGH via iv route. An exper carc [*3*] via oral route.

RHODIUM HYDROSULFIDE. Black crystals. $Rh(HS)_3$, mw: 202.13, mp: decomp.
THR = See sulfides.

RHODIUM MONOSULFIDE. Gray-black crystals. RhS, mw: 134.98, mp: decomp.
THR = See sulfides.

RHODIUM NITRATE. Brown-yellow crystals. $Rh(NO_3)_3$, mw: 288.9, mp: decomp.
THR = See nitrates.

RHODIUM SESQUISULFIDE. Black crystals. Rh_2S_3, mw: 302.02, mp: decomp, d: 6.4.
THR = See sulfides.

RHODIUM SULFITE. Yellow crystals. $Rh_2(SO_3)_3 \cdot 6H_2O$, mw: 554.11, mp: decomp.
THR = See sulfites.

RHODIUM TETRABROMIDE. $RhBr_4$, mw: 422.6.
THR = Reacts violently with BrF_3. [*19*] See also bromides.

RHODIUM TRIFLUORIDE. Rhombic red crystals. RhF_3, mw: 159.91, bp: >600° (sublimes), d: 5.38.
THR = See fluorides.

RIBOFLAVIN. Syns: *vitamin B_2*; *7,8-dimethyl-10-(1-d-ribityl)isoalloxozine.* Orange to yellow crystals, slightly sol in water and alcohols, insol in lipid solvents. $C_{17}H_{20}N_4O_6$, mw: 376, mp: 282° (decomp).
Acute tox data: ip LD_{50} (rat) = 560 mg/kg; sc LD_{50} (rat) = 5000 mg/kg. [*3*]
THR = MOD via ip and sc routes. A nutrient and/or dietary supplement food additive. [*109*]

RIBOFLAVIN-5′-PHOSPHATE. Syns: *FMN, flavin mononucleotide.* Yellow crystals, sol in water (sodium salt).
THR = U. A nutrient and/or dietary supplement food additive. [*109*]

RICHE GAS. See carbon monoxide.

RICIN. Syn: *agglutinin.* White powder.
Acute tox data: Oral LD_{LO} (rat) = 100 mg/kg; oral TD_{LO} (human) = 2 mg/kg. [*3*]
THR = HIGH via oral route. Inhal or ingestion of minute amounts causes violent purging which may lead to collapse and death. Small particle in eyes, nose or any skin abrasion may prove fatal. May cause destruction of red blood cells. ip MLD (mice) = 1 μg/kg in 48 hrs (as ricin "D" nitrogen). A very active military poison. [*2*]
Disaster Hazard: Dangerous; when heated to decomp, emits highly toxic fumes.

RICININE. Syn: *1,2-dihydro-4-methoxy-1-methyl-2-oxonicotinonitrile.* Alkaloid from castor bean plant. $C_8H_8N_2O_2$, mw: 164.2, mp: 201.5° (sublimes @ 170°–180° @ 20 mm).
Acute tox data: Oral LD_{LO} (rat) = 75 mg/kg. [*3*]
THR = HIGH via oral route. A deady poison. HIGH via ingestion and inhal. Can cause vomiting, nausea, liver and kidney damage, convulsions, coma, hypotension, respiratory distress, death. [*2*]
Disaster Hazard: Dangerous; when heated to decomp, emits highly toxic fumes of cyanide.

RICINOLEIC ACID. $C_{18}H_{34}O_3$, mw: 298.5.
THR = An exper neo. [*3*] via sc route.

RICINUS OIL. See castor oil.

RIDDELLINE. An alkaloid from *S. Riddelli.* $C_{18}H_{23}O_6N$, mw: 349.4.
THR = An exper (+) carc. [*3, 7*]

RIFAMYCIN. $C_{39}H_{49}O_{14}N$, mw: 755.9.
THR = An exper teratogen. [*3*]

RIFLE GRENADES. See explosives, high.

ROCKET AMMUNITION. See explosives, high.

RODINOL. See *p*-aminophenol.

ROENTGEN TUBE RADIATIONS.
THR = A recog carc. [*14*] See radiation, ionizing.

ROHRBACK'S SOLUTION. See mercuric barium iodide.

ROMAN VITRIOL. See copper sulfate.

RONNEL. Syns: *o,o-dimethyl o-(2,4,5-trichlorophenyl) phosphorothioate, fenchlorphos.* White powder. $C_8H_8Cl_3PS$, mw: 321.6, mp: 41°, vap. press: 8 × 10^{-4}mm.
Acute tox data: Oral LD_{50} (rat) = 906 mg/kg; dermal LD_{50} (rat) = 2000 mg/kg; ip LD_{50} (rat) = 2823 mg/kg; dermal LD_{50} (rabbit) = 1000 mg/kg; oral LD_{50} (wild birds) = 80 mg/kg. [*3*]
THR = HIGH to wild birds via oral route; MOD via dermal, oral and ip routes. A cholinesterase inhibitor. See parathion. A food additive permitted in the feed and drinking water of animals and/or for

the treatment of food-producing animals. Also permitted in food for human consumption. [*109*]

ROOT BEER FLAVORING.
THR = A susp carc. [*14*] See safrol.

ROSANILINE DYES. See triphenyl methane dyes.

ROSEMARY OIL.
Acute tox data: Oral LD_{50} (rat) = 5000 mg/kg. [*3*]
THR = MOD via oral route.

ROSIN. Syns: *gum rosin, colophony, piece resin.* Pale yellow to amber, translucent fragments, turpentine odor and taste. mp: 100°–150°, flash p: 370°F (CC), d: 1.08.
THR = LOW. An allergen. Pyrolysis products are irr. See formaldehyde.
Fire Hazard: Slight, can react with oxidizing materials. May ignite spont in air.
To Fight Fire: CO_2, dry chemical.

"ROSINAMINE-D." See dehydroabietyl amine.

ROSIN OIL. An oil, bp: 80°, flash p: 266°F (CC), d: 0.98, autoign. temp.: 648°F.
THR = MILD irr. May cause dermatitis in sensitive individuals.
Fire Hazard: Slight, when exposed to heat or flame.
Spont Heating: Yes.
To Fight Fire: CO_2, dry chemical.

ROTENONE. Syns: *tubatoxin, derris.* White, odorless crystals derived from derris root. $C_{23}H_{22}O_6$, mw: 394.4, mp: 163°, d: 1.27 @ 20°.
Acute tox data: Oral LD_{50} (rat) = 132 mg/kg; ip LD_{50} (rat) = 2.8 mg/kg; ip LD_{50} (mouse) = 2.8 mg/kg. [*3*]
THR = HIGH via oral and ip routes. Estimated LD (oral, man) = 200 mg/kg. Acute poisoning causes numbness, nausea, vomiting and tremors. A skin irr. Chronic exposure injures liver and kidneys. An insecticide. It is toxic to animals and very toxic to fish but leaves no harmful residue on vegetable crops.

ROUGE. See iron oxide.

ROUGH AMMONIATE TANKAGES.
THR = U.
Fire Hazard: Dangerous, when exposed to heat or flame; can react vigorously with oxidizing materials.

R SALT. See sodium-β-naphtholdisulfonate.

RUBBER CEMENT. See cement, rubber.

RUBBER, CRUDE. Syns: *India rubber, caoutchouc.* Light cream to dark amber, amorphous, elastic, dry loaves, sheets or slabs. d: about 0.9.
THR = NONE.
Fire Hazard: Very slight, but will support combustion. Will burn in Cl_2 and react violently with ClF_3, ICl, Li, F_2. [*19*]

RUBBER, CURED.
THR = Cured rubber products may have slight toxicity or cause superficial dermatitis due to the active vulcanizing agents used; for example, sulfur, diphenyl guanidine, phenyl-β-naphthylamine, mercaptobenzothiazole and its derivatives. The rubber itself is not toxic.
Fire Hazard: Slight, when exposed to flame. In the form of dust or fine particles, hard rubber is a more serious hazard and should be kept away from sparks and open flame. Will burn in liquid Cl_2 and react violently with ClF_3, ICl, Li, F_2. [*19*]
Disaster Hazard: Dangerous, in the form of dust. In burning, will emit toxic fumes of SO_x.

RUBBER LATEX. See latex, rubber.

RUBBER PITCH.
THR = A susp carc. [*14*] See thermic and oxidation products.

RUBBER, RECLAIMED.
THR = SLIGHT. A weak allergen.
Fire Hazard: Dangerous, when exposed to flame; emits acrid and toxic fumes of SO_x when burning.

RUBBER SCRAP OR BUFFINGS.
THR = NONE.
Fire Hazard: Dangerous, when exposed to heat or flame. See also rubber, cured.
Explosion Hazard: Mod, in the form of dust when exposed to flame.
Disaster Hazard: Dangerous; when heated to decomp, emits toxic fumes of sulfides.

RUBBER, SHODDY: REGENERATED RUBBER OR RECLAIMED RUBBER. See rubber, reclaimed.

RUBBER SOLVENT. A petroleum distillate used in making rubber cements and in tire manufacturer. Insol in water. flash p: −40°F (varies with manufacturer), autoign. temp.: 450°F (varies with manufacture), lel = 1.0%, uel = 7.0%, d: <1, bp: 100°–280°F.
THR = See specific material.
Fire Hazard: Dangerous, when exposed to heat or flame.
To Fight Fire: Foam, alcohol foam.

RUBIDIUM. Soft, silvery-white metal. Rb, atwt: 85.46, mp: 38.89°, bp: 688°, d(solid): 1.532 @ 20°, d(liquid): 1.475 @ 39°.
THR = See potassium.
Radiation Hazard: For permissible levels, see Section 5, Table 5A.5. Natural isotope (28%) ^{87}Rb, $T_{\frac{1}{2}}$ = 5×10^{10}y, decays to stable ^{87}Sr via β's of 0.27 MeV.
Fire Hazard: Dangerous, when exposed to heat or

flame or by chemical reaction with oxidizers. See also sodium.

Explosion Hazard: Mod; reacts explosively with moisture, acids and oxidizers. See also sodium and NaK. Violent reaction with Cl_2,O_2,P, water. [*19*]

Disaster Hazard: Dangerous; when heated, emits toxic fumes of rubidium oxide; will react with water or steam to produce hydrogen and flam vapors; reacts vigorously with oxidizing materials.

RUBIDIUM ACETYLIDE. RbC_2H, mw: 110.5.

THR = Violent reaction with As, Br, Cl_2, CuO, F_2, HCl, I, PbO, MnO_2, P, S, H_2SO_4. [*19*]

RUBIDIUM AZIDE. RbN_3, mw: 127.5.

THR = Very unstable. Decomp @ 321°. Reacts violently with CS_2. [*19*]

RUBIDIUM BROMATE. $RbBrO_3$, mw: 213.40, mp: 430°, d: 3.68.

THR = See bromides.

RUBIDIUM BROMOCHLOROIODIDE. Rhombic crystals. RbIBrCl, mw: 327.8, mp: 205° (decomp).

THR = See bromides, iodides and chlorides.

RUBIDIUM CARBIDE. Rb_2C_2, mw: 195.

THR = Violent reaction with As, As_2O_3, B, Br, HCl, I, PbO, HNO_3, NO, Se, Si, SO_2. [*19*]

RUBIDIUM CARBONATE. White deliquescent powder. Rb_2CO_3, mw: 230.96, mp: 837°, bp: decomp.

THR = See also potassium carbonate.

RUBIDIUM CHLORATE. Crystals. $RbClO_3$, mw: 168.94, d: 3.19.

THR = See chlorates.

RUBIDIUM CHLORIDE. RbCl, mw: 120.9.

Acute tox data: ip LD_{50} (rat) = 1200 mg/kg; oral LD_{50} (mice) = 3800 mg/kg; ip LD_{50} (mice) = 1160 mg/kg. [*3*]

THR = MOD via ip and oral routes. Reacts violently with BrF_3. [*19*]

RUBIDIUM CHLOROPLATINATE. Cubic yellow crystals. Rb_2PtCl_6, mw: 578.75, mp: decomp, d: 3.94 @ 17.5°.

THR = See platinum compounds and chlorides.

RUBIDIUM CHROMATE. Yellow crystals. Rb_2CrO_4, mw: 286.97, d: 3.518.

THR = See chromium compounds.

Fire Hazard: Mod, by chemical reaction with reducing agents.

RUBIDIUM CHROMIUM SULFATE. Cubic violet crystals. $RbCr(SO_4)_2 \cdot 12H_2O$, mw: 545.81, mp: 107°, d: 1.946.

THR = See chromium compounds and sulfates.

RUBIDIUM COPPER SULFATE. Monoclinic crystals. $Rb_2SO_4CuSO_4 \cdot 6H_2O$, mw: 534.73, d: 2.57.

THR = See copper compounds and sulfates.

RUBIDIUM DICHROMATE. Triclinic or monoclinic crystals. $Rb_2Cr_2O_7$, mw: 386.93, d: 3.02–3.13.

THR = See chromium compounds.

Fire Hazard: Mod, by chemical reaction with reducing agents.

RUBIDIUM DISULFIDE. Dark red crystals. Rb_2S_2, mw: 235.09, bp: vol @ 850°, mp: 420°.

THR = See sulfides.

RUBIDIUM FLUORIDE. Colorless crystals. RbF, mw: 104.48, mp: 775°, bp: 1410°, d: 3.557, vap. press: 1 mm @ 921°.

THR = See fluorides.

RUBIDIUM FLUOSILICATE. Cubic crystals. Rb_2SiF_6, mw: 313.02, d: 3.332.

THR = See fluosilicates.

RUBIDIUM FLUOSULFONATE. Needles. $RbFSO_3$, mw: 184.55, mp: 304°.

THR = See fluosulfonates.

RUBIDIUM GALLIUM SULFATE. Colorless crystals. $RbGa(SO_4)_2 \cdot 12H_2O$, mw: 563.52, d: 1.962.

THR = See gallium compounds and sulfates.

RUBIDIUM GRAPHITE. Violet to black platelets. Highly reactive with water, air and alcohols. RbC_8, mw: 181.5.

THR = See rubidium.

RUBIDIUM HEXASULFIDE. Brown-red crystals. Rb_2S_6, mw: 363.36, mp: 201°.

THR = See sulfides.

RUBIDIUM HYDRATE. See rubidium hydroxide.

RUBIDIUM HYDRIDE. Colorless needles; reacts with water. RbH, mw: 86.49, mp: decomp @ 300°, d: 2.60.

THR = See hydrides. With moisture, it reacts violently with C_2H_2 to −60°. Ignites on contact with water. [*19*]

RUBIDIUM HYDROGEN NITRATE. Tetragonal crystals. $RbNO_3HNO_3$, mw: 210.50, mp: 62°.

THR = See nitrates.

RUBIDIUM HYDROXIDE. Syn: *rubidium hydrate.* Grayish-white deliquescent mass, strong base, RbOH, mw: 102.49, mp: 300°, d: 3.203 @ 11°.

Acute tox data: Oral LD_{50} (rat) = 586 mg/kg; oral LD_{50} (mouse) = 900 mg/kg. [*3*]

THR = MOD via oral route. A powerful irr to skin, eyes and mu mem. See also potassium hydroxide.

RUBIDIUM IODATE. Cubic crystals. $RbIO_3$, mw: 260.40, mp: decomp, d: 4.33 @ 19.5°.

THR = See iodates.

RUBIDIUM IODIDE. Colorless crystals. RbI, mw: 212.40, mp: 642°, bp: 1300°, d: 3.55, d(liquid): 2.87 @ 825°, vap. press: 1 mm @ 748°.

Acute tox data: Oral LD_{50} (rat) = 4708 mg/kg. [*3*]
THR = MOD via oral route. See iodides.

RUBIDIUM MONOSULFIDE. Colorless crystals. Rb_2S, mw: 203.03, mp: 530° (decomp), d: 2.912.
THR = See sulfides.

RUBIDIUM MONOXIDE. Cubic, colorless-yellow crystals. Rb_2O, mw: 186.96, mp: decomp @ 400°, d: 3.72.
THR = HIGH and caustic. See also potassium oxide.

RUBIDIUM NITRATE. Hexagonal, cubic, rhombic or triclinic crystals. $RbNO_3$, mw: 147.49, mp: 310°, d: 3.11, d(liquid): 2.395 @ 400°.
THR = See nitrates.

RUBIDIUM NITRIDE. Rb_3N, mw: 270.4.
THR = Burns in air. [*19*]

RUBIDIUM PENTASULFIDE. Rhombic, red, deliquescent crystals. Rb_2S_5, mw: 33.29, mp: 225°, d: 2.618 @ 15°.
THR = See sulfides.

RUBIDIUM PERCHLORATE. Rhombic crystals. $RbClO_4$, mw: 184.94, mp: fuses, bp: decomp, d: 2.8.
THR = See perchlorates.

RUBIDIUM-*m*-PERIODATE. Tetragonal crystals. $RbIO_4$, mw: 276.40, d: 3.918 @ 16°.
THR = See iodates.

RUBIDIUM PERMANGANATE. Crystals. $RbMnO_4$, mw: 204.41, d: 3.235 @ 10.4°.
THR = See manganese compounds.
Fire Hazard: Mod, by chemical reaction with reducing agents. An oxidizing agent; keep away from flam materials.

RUBIDIUM PEROXIDE. Yellow crystals. Rb_2O_2, mw: 202.96, mp: 600°, d: 3.65 @ 0°.
THR = See peroxides, inorganic.

RUBIDIUM PHOSPHIDE. Rb_3P, mw: 287.4.
THR = HIGH. See phosphides and phosphine, Reacts vigorously with water to yield phosphine. [*19*]

RUBIDIUM PRASEODYMIUM NITRATE. Greenish, hygroscopic needles.
$2RbNO_3 \cdot Pr(NO_3)_3 \cdot 4H_2O$, mw: 693.98, mp: 63.5. $-4H_2O$ @ 60°, d: 2.50.
THR = See nitrates.

RUBIDIUM SELENATE. Colorless rhombic crystals. Rb_2SeO_4, mw: 313.92, d: 3.90.
THR = See selenium compounds.

RUBIDIUM SILICIDE. Rb_2Si_2, mw: 227.1.
THR = Violent reaction with water. [*19*]

RUBIDIUM SUPEROXIDE. Yellow crystals. RbO_2, mw: 117.48, mp: 280°, d: 3.05 @ 0°.
THR = A powerful irr and oxidizer. See peroxides, inorganic.

RUBIDIUM TRIBROMIDE. Rhombic crystals. $RbBr_3$, mw: 325.23, mp: decomp @ 140°.
THR = See bromides.

RUBIDIUM TRIIODIDE. Rhombic black crystals. RbI_3, mw: 466.24, mp: 190°, d: 4.03.
THR = See iodides.

RUBIDIUM TRISULFIDE. Reddish-yellow crystals. Rb_2S_3, mw: 267.16, mp: 213°.
THR = See sulfides.

RUBIDOMYCIN. $C_{27}H_{29}O_{10}N$, mw: 527.6.
Acute tox data: iv LD_{50} (rat) = 20 mg/kg; ip LD_{50} (mouse) = 5 mg/kg; sc LD_{50} (mouse) = 16 mg/kg. [*3*]
THR = HIGH via iv, ip and sc routes. An exper teratogen [*3*] via ip route.

RUBRATOXIN B. Isolate of *penicillium rubrum* and *P. Purpurogenum.* $C_{26}H_{30}O_{11}$, mw: 518.6.
Acute tox data: Oral LD_{50} (rat) = 400 mg/kg; ip LD_{50} (rat) = 0.350 mp/kg. [*3*]
THR = HIGH via ip and oral routes. An exper teratogen [*3*] via ip route.

RUM, DENATURED. flash p: 77°F (CC).
THR = U. See ethyl alcohol.
Fire Hazard: Dangerous, when exposed to heat or flame.
Disaster Hazard: Dangerous, upon exposure to heat or flame; can react vigorously with oxidizing materials.
To Fight Fire: CO_2, dry chemical.

RUTHENIUM. A hard, white metal, stable in air. Ru, atwt: 101.07, mp: 2310°, bp: 3900°, d: 12.30.
THR = See ruthenium compounds.
Radiation Hazard: For permissible levels, see Section 5A, Table 5A.5. Artificial isotope ^{103}Ru, $T_{\frac{1}{2}}$ = 40d, decays to radioactive ^{103m}Rh via β's of 0.11 (7%), 0.21 (89%) MeV. Emits γ's of 0.50, 0.61 MeV. ^{103}Ru exists in radioactive equilibrium with its daughter. Permissible levels are given for the equilibrium mixture. Artificial isotope ^{106}Ru, $T_{\frac{1}{2}}$ = 1.0y decays to radioactive ^{106}Rh via β's of 0.04 MeV. ^{106}Ru exists in radioactive equilibrium with its daughter. Permissible levels are given for the equilibrium mixture.
Fire Hazard: Mod, in the form of dust, when exposed to heat or flame. Violent reaction with OF_2. [*19*]

RUTHENIUM COMPOUNDS.
THR = Details U. Probably toxic, but such small amounts are used industrially that it does not constitute a hazard. It resembles osmium in that when it is heated in air, it evolves fumes which are injurious to the eyes and lungs.
Disaster Hazard: Dangerous; when heated to decomp, emits toxic fumes of ruthenium oxide.

RUTHENIUM HYDROXIDE. Black powder. $Ru(OH)_3$, mw: 152.72.
THR = See ruthenium compounds.

RUTHENIUM PENTAFLUORIDE. Dark green crystals. RuF_5, mw: 196.70, mp: 101°, bp: 270°, d: 2.963 @ 16.5°.
THR = See ruthenium compounds and fluorides.

RUTHENIUM SULFIDE. Syn: *laurite*. Cubic, gray-black crystals. RuS_2, mw: 165.83, d: 6.99.
THR = See sulfides and ruthenium compounds.

RUTHENIUM TETROXIDE. Yellow, volatile crystals, odor of ozone. RuO_4, mw: 165.70, mp: 25.5°, bp: approx 100° (decomp), d: 3.29 @ 21°.
THR = MOD irr to eyes and mu mem.
Fire Hazard: Mod, by chemical reaction with reducing agents. A powerful oxidizing agent. Reacts violently with charcoal, PBr_3. [*19*]
Disaster Hazard: See ruthenium compounds.

RUTILE. See titanium dioxide.

RYANIA. Ground wood of *Ryania spociosa*.
Acute tox data: Oral LD_{50} (rat) = 750 mg/kg; oral LD_{50} (mouse) = 550 mg/kg; oral LD_{50} (dog) = 150 mg/kg. [*3*]
THR = HIGH via oral to dogs; MOD via oral to rats and mice. HIGH via oral ⟶ weakness, respiratory changes, diarrhea, gastrointestinal disturbances, tremors, convulsions, coma and death. An insecticide. No tolerance limits have been established or proposed by the government for residues on foods.
Fire Hazard: Mod, when exposed to heat or flame.

RYANODINE. Crystals, sol in water, alcohol, acetone and chloroform, practically insol in benzene and petroleum ether. $C_{25}H_{35}NO_9$, mw: 493.54.
THR = See ryania.

S

SABADILLA. Syns: *cevadilla, caustic barley*. From the plant *Schoenocaulon officinale*. A botanical insecticide. The active ingredients are a group of alkaloids known as veratrin; i.e., cevadine and veratridine. A powder.

Acute tox data: Oral LD_{50} (rat) = 4000 mg/kg; ip LD_{50} (mice) = 7.5 mg/kg; iv LD_{50} (mice) = 0.42 mg/kg. [*3*]

THR = HIGH via ip and iv; MOD via oral routes. Ingestion causes severe GI disturbances, burning in the mouth, vomiting, diarrhea and cramps. Also produces headache, dizziness, slow pulse and weakness. Large doses cause death by circulatory and respiratory failure. It is a powerful irr to skin and mu mem and is less toxic than rotenone.

Disaster Hazard: Dangerous; when heated, emits toxic fumes.

SACCHARIN. Syns: *saxin, benzosulfimide*. Crystals or powder. $C_7H_5NO_3S$, mw: 183.2, mp: 228°(decomp), bp: sublimes.

THR = An exper neo and carc via oral and imp routes. [*3, 14*] A non-nutritive sweetener food additive. [*109*] See sweeteners.

Disaster Hazard: Dangerous; when heated to decomp, emits highly toxic fumes of SO_x and NO_x.

SAFETY EVALUATIONS BASED UPON INDUSTRIAL TOXICOLOGICAL FINDINGS. See Section 9.

SAFETY LAMP GASOLINE. See gasoline.

SAFETY SOLVENT. See Stoddard solvent.

SAFETY SQUIBS. See explosives, high.

SAFFLOWER OIL. Oil from the seed of *Carthamus tinctorious*. d: 0.9211 @ 25°/25°.

THR = U. If swallowed in large amounts, it causes vomiting.

SAFROLE. Syns: *1-allyl-3,4-methylenedioxybenzene, shikimole, 1,2-(methylene dioxy)-4-propenyl benzene*. Colorless liquid or crystals. $C_3H_5C_6H_3O_2OH_2$, mw: 162.18, mp: 11° bp: 234.5° d: 1.0960 @ 20° vap. press: 1 mm @ 63.8°.

Acute tox data: Oral LD_{50} (rat) = 1950 mg/kg; oral LD_{50} (mouse) = 2350 mg/kg; ip LD_{LO} (mice) = 64 mg/kg; sc LD_{LO} (rabbit) = 1000 mg/kg; iv LD_{LO} (rabbit) = 1000 mg/kg. [*3*]

THR = HIGH via ip and iv routes; MOD via oral and sc routes. A powerful irr via oral route. An exper (+) carc and neo. [*3, 9, 23*] of the liver via oral route.

Fire Hazard: Slight.

SAL ACETOSELLA. See potassium binoxalate.

SAL AMMONIA. See ammonium chloride.

SALI. Aqueous extract from dried leaves.

THR = An exper neo via sc route. [*3*]

SALICYL ALDEHYDE. Syn: *salicylic aldehyde*. Clear, colorless, oily liquid, burning taste. C_6H_4OHCOH, mw: 122.1, bp: 197°, fp: −7°, d: 1.167 @ 20°/4° vap. press: 1 mm @ 33.0°, flash p: 172°F.

Acute tox data: sc LD_{LO} (rat) = 1000 mg/kg. [*3*]

THR = MOD via sc route. See also aldehydes. An auxiliary fumigant.

Fire Hazard: Mod, when exposed to heat or flame; can react with oxidizing materials.

To Fight Fire: Alcohol foam, spray, mist, dry chemical.

SALICYLAMIDE. Syn: *o-hydroxybenzamide*. White to slightly pink crystals or powder, somewhat bitter taste. $C_7H_7NO_2$, mw: 137.1, mp: 140°.

Acute tox data: Oral LD_{50} (rat) = 1890 mg/kg; ip LD_{50} (rat) = 600 mg/kg; iv LD_{50} (mouse) = 315 mg/kg. [*3*]

THR = HIGH via iv, MOD via oral and ip routes. An exper teratogen. [*3*] Can cause dizziness, drowsiness, nausea, vomiting, epigastric distress, allergic reactions and blood dyscrasias in average to large doses.

Disaster Hazard: Dangerous; when heated to decomp, emits highly toxic fumes.

SALICYLANILIDE. Syns: *anasadol, salinidol*. White, odorless crystals. $C_6H_5NHCOC_6H_4OH$, mw: 213.2, mp: 135°, bp: decomp.

THR = A MOD irr via oral route. See also salicylic acid, aniline, and amides.

Disaster Hazard: Dangerous; when heated to decomp, emits highly toxic fumes.

SALICYLIC ACID. Syn: *o-hydroxy benzoic acid*. White needle crystals or powder, sweetish taste, acrid after-taste. HOC_6H_4COOH, mw: 138.12, mp: 159°, bp: 211° @ 20 mm, flash p: 315.°F, d: 1.443 @ 20°/4°, autoign. temp.: 1013°F, vap. press: 1 mm @ 113.7°, vap. d: 4.8.

Acute tox data: Oral LD_{50} (rat) = 891 mg/kg; sc LD_{50} (mice) = 520 mg/kg. [*3*]

THR = MOD via oral and sc routes. Symptoms of

poisoning are nausea and vomiting, ringing in the ears, dizziness, headache, dullness, confusion, sweating, rapid pulse and breathing and sometimes skin eruptions. Symptoms disappear when exposure or administration of the drug is terminated. Used as a food additive permitted in food for human consumption. [*109*]

Fire Hazard: Slight, when exposed to heat or flame; can react with oxidizing materials. Dust is explosive in air.

To Fight Fire: CO_2, dry chemical.

SALICYLIC ALDEHYDE. See salicyl aldehyde.

SALINIDOL. See salicylamide.

SALITHION. A light yellow crystalline powder, sol in acetone and hydrocarbons. $C_8H_9O_2PS$, mw: 200.1, mp: 54°.

Acute tox data: Oral LD_{50} (mouse) = 91 mg/kg; sc LD_{50} (mouse) = 81.6 mg/kg. [*3*]

THR = HIGH via oral and sc routes. An insecticide. See also parathion.

Disaster Hazard: See parathion.

SALOL. See phenyl salicylate.

SALT. See sodium chloride.

SAL TARTAR. See sodium tartrate.

SALT BATH EXPLOSIONS. See Section 7.

SALT BATHS (NITRATE OR NITRITE).

THR = U. See also nitrates and nitrites.

Fire Hazard: Dangerous, by spont chemical reaction. These baths are oxidizing in nature.

Explosion Hazard: Mod, by chemical reaction, due to contamination by cyanides or easily oxidizable materials or when heated to over 1000°F.

Disaster Hazard: Highly dangerous; in molten form will react with water, steam or acids to produce heat, hydrogen and toxic and corrosive fumes; can react vigorously with reducing materials.

SALT OF SORREL. See potassium binoxalate.

SALTPETER. See potassium nitrate.

SALTPETER, CHILE. See sodium nitrate.

SALT SUBSTITUTE. See potassium chloride.

SALUTES. See fireworks, N.O.S., and fireworks, common.

SALVARSAN. Syns: *606*, *arsphenamine*.
$C_{12}H_{12}O_2N_2As_2 \cdot 2HCl$, mw: 439.0.

Acute tox data: iv LD_{LO} (rat) = 100 mg/kg; iv LD_{LO} (cat) = 36 mg/kg. [*103*]

THR = HIGH via iv route. See also arsenic.

SAMARIUM. A bright, silvery, lustrous, stable metal. Sm, atwt: 150.35, mp: 1072°, bp: 1778°, d:(α) 7.536, d(β): 7.40.

THR = U. As a lanthanon it may cause impairment of blood clotting. See also lanthanum.

Radiation Hazard: For permissible levels, see Section 5A, Table 5A.5. Natural (15%) isotope ^{147}Sm, $T\frac{1}{2}$ = 1.0x10^{11}y, decays to stable ^{143}Nd via α's of 2.2 MeV. Artificial isotope ^{151}Sm, $T\frac{1}{2}$ = about 90y, decays to stable ^{151}Eu via β's of 0.076 MeV.

Fire Hazard: Mod, in the form of dust, when exposed to flame or by spont chemical reaction with oxidizers. See also powdered metals. Ignites at 150° in air, also releases H_2 in contact with water. Can react violently with halogens, 1,1,2-trichlorotrifluoroethane. [*19*]

SAMARIUM ACETATE. $C_6H_9O_6 \cdot Sm$, mw: 327.5.

Acute tox data: sc LD_{50} (mice) = 10,000 mg/kg; iv LD_{50} (mice) = 1000 mg/kg; iv LD_{LO} (cat) = 50 mg/kg. [*3*]

THR = MOD via iv and LOW via sc routes. The cat appears to be more sensitive to the material. See also samarium.

SAMARIUM BROMATE. Hexagonal yellow crystals. $Sm(BrO_3)_3 \cdot 9H_2O$, mw: 696.32, mp: 75°, $-9H_2O$ @ 150°.

THR = See bromates and samarium.

SAMARIUM BROMIDE. Yellow deliquescent crystals. $SmBr_3 \cdot 6H_2O$, mw: 498.27, d: 2.971 @ 22°.

THR = See bromides and samarium.

SAMARIUM CARBIDE. Yellow crystalline mass. SmC_2, mw: 174.45, d: 5.86.

THR = See carbides and samarium.

SAMARIUM (III)CHLORIDE. $SmCl_3$, mw: 256.7.

Acute tox data: ip LD_{50} (mice) = 365 mg/kg; sc LD_{LO} (frog) = 150 mg/kg; sc LD_{LO} (rat) = 2000 mg/kg. [*3*]

THR = HIGH via ip route and sc route for the frog. MOD via sc route for the dog and guinea pig. An irr.

SAMARIUM CITRATE.

Acute tox data: ip LD_{50} (mice) = 164 mg/kg; ip LD_{50} (guinea pig) = 75 mg/kg. [*3*]

THR = HIGH via ip route.

SAMARIUM DICHLORIDE. Dark reddish-brown, crystalline mass. $SmCl_2$, mw: 221.34, mp: 740°, d: 3.687 @ 22°.

THR = See hydrochloric acid and samarium.

SAMARIUM EDETATE.

Acute tox data: ip LD_{50} (mice) = 311 mg/kg; ip LD_{50} (guinea pig) = 75 mg/kg. [*3*]

THR = HIGH via ip route.

SAMARIUM IODIDE. Orange-yellow crystals SmI_3, mw: 531.19, mp: 816°–824°, bp: decomp.

THR = See iodides and samarium.

SAMARIUM NITRATE. Triclinic, pale yellow crystals. $Sm(NO_3)_3 \cdot 6H_2O$, mw: 444.55, mp: 78°–79°, d: 2.375.
Acute tox data: Oral LD_{50} (rat) = 2900 mg/kg; ip LD_{50} (rat) = 285 mg/kg; iv LD_{50} (rat) = 9 mg/kg. [*3*]
THR = HIGH via ip and iv routes; MOD via oral route.

SAMARIUM OXALATE. Crystals.
$Sm_2(C_2O_4)_3 \cdot 10H_2O$, mw: 745.08.
THR = See oxalates and samarium.

SAMARIUM SULFIDE. Yellowish-pink crystals. Sm_2S_3, mw: 397.06, mp: 1900°, d: 5.729.
THR = See sulfides and samarium.

SANGUINARINE. $C_{20}H_{14}O_4N$, mw: 332.4.
Acute tox data: ip LD_{50} (rat) = 18 mg/kg. [*3*]
THR = HIGH via ip route. An exper carc. [*23*]

SANTOBANE. See DDT.

SANTONIC LACTONE. See santonin.

L,α-SANTONIN. Syn: *santonic lactone*. Glossy, colorless crystals or white powder, turning yellow on exposure to light, odorless, tasteless at first, then bitter. $C_{15}H_{18}O_3$, mw: 246.30, mp: 170°, bp: sublimes, d: 1.187.
Acute tox data: sc LD_{LO} (mice) = 250 mg/kg. [*3*]
THR = HIGH via sc route. It can cause disturbance of color vision. Objects first show bluish tinge, then yellow, which is most prominent. Complete blindness may occur, lasting perhaps for nearly a week. Dizziness, drowsiness and nausea may also occur. Recovery is spont.
Fire Hazard: Slight; when heated, emits acrid fumes.

SANTOPHEN. See *o*-benzyl-*p*-chlorophenol.

SAPOGLYCOSIDES. See saponins.

SAPONIFIED CRESOL. See cresol.

SAPONIN. Syn: *sapogylcosides*.
Acute tox data: Oral LD_{LO} (mice) = 3000 mg/kg; sc LD_{LO} (mice) = 900 mg/kg; iv LD_{LO} (mice) = 1000 mg/kg. [*3*]
THR = MOD via oral, sc and iv routes. When administered by injection, saponins cause rapid and severe destruction of red blood cells.
Fire Hazard: Slight.

SAPOTOXIN.
Acute tox data: Oral LD_{LO} (mice) = 1000 mg/kg; sc LD_{LO} (mice) = 80 mg/kg; iv LD_{LO} (mice) = 20 mg/kg; oral LD_{LO} (dog) = 20 mg/kg; oral LD_{LO} (rabbit) = 50 mg/kg. [*3*]
THR = HIGH via oral, sc and iv routes.

SAPP. See sodium acid pyrophosphate.

SARAN. $(C_4H_5Cl_3)_n$. A plastic.
THR = An exper carc by implantation route. [*3*]

SARCOSINE. See N-methylglycine.

SARIN. Syn: *fluoroisopropoxy methyl phosphine oxide*. Colorless liquid. $[(CH_3)_2CHO](CH_3)(F)(O)P$, mw: 140.1, bp: 147°, fp: −58°, d: 1.100 @ 20°, vap. press: 1.57 mm @ 20°, vap. d: 4.86.
Acute tox data: Dermal LD_{50} (human) = 28 mg/kg; dermal TD_{LO} (human) = 0.1 mg/kg ⟶ CNS effects; inhal LC_{50} (human) = 70 mg/m^3; oral LD_{50} (rat) = 0.55 mg/kg; sc LD_{50} (rat) = 0.11 mg/kg; iv LD_{50} (rat) = 0.05 mg/kg; im LD_{50} (rat) = 0.2 mg/kg; dermal LD_{50} (rabbit) = 0.925 mg/kg. [*3*]
THR = HIGH via all routes. Est. LD (man) = 0.01 mg/kg. [*95*] This is a nerve gas, acting much like tabun. High toxicity to eyes. A small drop on skin will kill a man within 15 min. Liquid does not injure skin but penetrates it rapidly. See parathion.
Fire Hazard: Slight, when exposed to heat or flame.
Disaster Hazard: Highly dangerous, when heated, emits highly toxic fumes; will react with water or steam to produce toxic and corrosive fumes; can react with oxidizing materials.
To Fight Fire: Foam, CO_2, dry chemical.

SARKOMYCIN.
THR = An exper neo and teratogen. [*3*] An exper carc. [*23*]

SASSAFRAS OIL. A yellowish-reddish oil, volatile, pungent, aromatic odor and taste, sol in alcohol, ether, chloroform, glacial acetic acid and carbon bisulfide. Composed of safrole, eugenol, camphor, pinene and phellandrene. d: 1.065–1.077 @ 25°/25°.
THR = HIGH by ingestion. A (S) carc. [*14*] See safrole.

SATURN. Syn: *2-(4-chlorobenzyl)-n,n-diethylthiol carbamate*. $C_{12}H_{16}ONSCl$, mw: 257.8, bp: 127° @ 0.008 mm.
Acute tox data: Oral LD_{50} (rat) = 1300 mg/kg; dermal LD_{50} (rat) = 2900 mg/kg; oral LD_{50} (mouse) = 560 mg/kg. [*3*]
THR = MOD via oral and dermal routes.

SAWDUST. Syn: *wood dust*. Yellowish particles of wood.
THR = A nuisance dust and an allergen. The dust and chips of cedar wood are exper carc. [*23*] A common air contaminant.
Fire Hazard: Slight, when exposed to heat or flame; can react with oxidizing materials.
Spont Heating: Possible. Avoid hot, humid storage or contact with drying oils. Particularly dangerous if charred or partially burned.
Explosion Hazard: Severe, as dust, when exposed to flame.

SAXIN. See saccharin.

SCACCHITE. See manganese dichloride.

SCALE WAX. A recog carc. [*14*] See paraffins and petroleum waxes.

SCANDIUM. Syn: *ekaboron.* A silvery-white, soft, light metal. Sc, atwt: 44.956, mp: 1539°, bp: 2832°, d: 2.989 @ 25°.

THR = Not yet established, should be handled carefully.

Radiation Hazard: For permissible levels, see Section 5A, Table 5A.5. Artificial isotope ^{40}Sc, $T_{\frac{1}{2}}$ = 84d, decays to stable ^{46}Ti via β's of 0.36 MeV; emits γ's of 0.89, 1.12 MeV; decays to stable ^{47}Ti via β's of 0.44 (73%), 0.60 (27%) MeV; emits γ's of 0.16 MeV.

Fire Hazard: Mod, in the form of dust, when exposed to heat or flame or by chemical reaction with oxidizers. See also powdered metals. Can react violently with halogens, air.

SCANDIUM BROMIDE. Crystals. $ScBr_3$, mw: 284.65, mp: sublimes @ 1000°, d: 3.914.

THR = See bromides.

SCANDIUM (III) CHLORIDE. $ScCl_3$, mw: 151.31.

Acute tox data: Oral LD_{50} (mice) = 4000 mg/kg; ip LD_{50} (mice) = 93 mg/kg. [*3*]

THR = HIGH via ip and MOD via oral routes.

SCANDIUM NITRATE. Colorless crystals. $Sc(NO_3)_3$, mw: 231, mp: 150°.

THR = See nitrates.

SCANDIUM OXALATE. Crystals. $Sc_2(C_2O_4)_3 \cdot 5H_2O$, mw: 444.1, mp: $-4H_2O$ @ 140°.

THR = See oxalates.

SCHEELE'S GREEN. See copper arsenite.

SCHIFF'S BASE. See disalicylal propylene diimine.

SCHRADAN. See octamethyl pyrophosphoramide.

SCHWEINFORT GREEN. See cupric acetate-*m*-arsenate.

SCLERATINE.

THR = An exper carc. [*23*]

SCOPE OF INDUSTRIAL TOXICOLOGY. See Section 9.

SCOPOLAMINE. See hyoscine.

SCORODITE. See ferric arsenate.

SEA ONION. See squill, red.

SEA SALT. See sodium chloride.

SEBACIC ACID. Syn: *decanedioic acid.* Thin colorless crystals. $COOH(CH_2)_8COOH$, mw: 202.3, mp: 133°, bp: 295° @ 100 mm, vap. press: 1 mm @ 183.0°, d: 1.207.

Acute tox data: ip LD_{LO} (mice) = 500 mg/kg. [*3*]

THR = MOD via ip route.

SECONDARY BARIUM PHOSPHATE. See barium phosphate, dibasic.

SECONDARY HAZARDS FROM SOLID WASTES. See Section 6.

SELENIC ACID. Colorless hexagonal prisms. H_2SeO_4, mw: 144.98, mp: 58°, bp: 260° (decomp), d(solid): 2.951 @ 15°, d(liquid); 2.609 @ 15°.

THR = A corrosive and irr material. See selenium compounds.

SELENIFEROUS PLANTS. A (S) carc. [*14*] See selenium compounds.

SELENINYL CHLORIDE. See selenium oxychloride.

SELENIOUS ACID. Transparent, colorless crystals. H_2SeO_3, mw: 128.98, mp: decomp, d: 3.004 @ 15°/4°, vap. press: 2 mm @ 15°.

Acute tox data: Oral LD_{LO} (rat) = 25 mg/kg; ip LD_{LO} (rat) = 10 mg/kg. [*3*]

THR = HIGH via oral and ip routes. See selenium compounds.

SELENIUM. Steel gray, non-metallic element. Se_8, mw: 631.68, mp: 170°–217°, bp: 690°, d: 4.81–4.26, vap. press: 1 mm @ 356°.

THR = An exper carc as the selenide, selenate, selenite and sulfate. [*23, 14*] See selenium compounds. An exper neo to mice via oral route. [*103*]

Radiation Hazard: For permissible levels, see Section 5A, Table 5A.5. Artificial isotope ^{75}Se, $T_{\frac{1}{2}}$ = 120d, decays to stable ^{75}As by ec; emits γ's of 0.1–0.4 MeV and x-rays. Under proper conditions, can react violently with BaC_2, BrF_5, CaC_2, chlorates, ClF_3, CrO_3, F_2, Li_2C_2, Li_6Si_2, Ni, HNO_3, Na, NCl_3, O_2, K, $KBrO_3$, Rb_2C_2, Zn, $AgBrO_3$, SrC_2, ThC, U. [*19*]

SELENIUM ALLOYS WITH CADMIUM. See selenium and cadmium compounds.

SELENIUM ALLOYS WITH COPPER. See selenium compounds.

SELENIUM CARBIDE. SeC_2, mw: 103.

THR = Violent reaction with S. [*19*]

SELENIUM CHLORIDE. See selenium tetrachloride.

SELENIUM CHLORIDE OXIDE. See selenium oxychloride.

SELENIUM COMPOUNDS.

Acute tox data: Inhal LC_{LO} (rat) = 33 mg/kg for 8 hrs; iv LD_{50} (rat) = 6 mg/kg. [*3*]

THR = HIGH via iv and inhal routes. An exper carc. [*14, 23*] Selenium in small amounts is essential for normal growth of some animals. Deficiency or excess is associated with serious disease in livestock. Long term exposure may be a cause of amyotrophic lateral sclerosis in humans, just as it may cause "blind staggers" in cattle. [*82*] Elemental selenium has low acute systemic toxicity, but dust or fumes can cause serious irr of the respiratory tract. It is a (S) carc of the liver and thyroid. Hydrogen selenide resembles other hydrides in being highly toxic, and selenium oxychloride is a vesicant. Some organoselenium compounds have the high toxicity of other organometals. Inorganic selenium compounds can cause dermatitis. Garlic odor of breath is a common symptom. Pallor, nervousness, depression and digestive disturbances have been reported in cases of chronic exposure. Selenium compounds are common air contaminants. [*26*]

SELENIUM DIBUTYL DITHIOCARBAMATE. Liquid. $Se[SC(S)N(C_4H_9)_2]_4$, mw: 896.4, mp: −25°, flash p: 225°F, d: 1.14 @ 20°/20°, vap. d: 30.9.
THR = HIGH. See selenium compounds and carbamates.
Fire Hazard: Low, when exposed to heat or flame.
Disaster Hazard: Dangerous. See selenium compounds.
To Fight Fire: Water, foam, CO_2, dry chemical.

SELENIUM DIETHYL DITHIO CARBAMATE. Syn: *ethyl selenac.* Orange-yellow color. $Se[SC(S)N(C_2H_5)_2]_4$, mw: 672.1, d: 1.32 @ 20°/20°.
THR = An exper (±) carc of the liver and thyroid via the oral route. [*3, 12, 8, 14*] See selenium compounds and carbamates.

SELENIUM DIMETHYL DITHIOCARBAMATE. Syn: *methyl selenac.* Yellow powder or crystals. $Se[SC(S)N(CH_3)_2]_4$, mw: 559.9, melting range: 140°–172°, d: 1.58.
THR = HIGH via oral route. An exper (±) carc via oral and sc routes. [*13*] See selenium compounds and carbamates.

SELENIUM DIOXIDE. White to slightly reddish, lustrous crystalline powder or needles. SeO_2, mw: 110.96, mp: 340°–350° (sublimes), d: 3.95 @ 15°/15°, vap. press: 1 mm @ 157.0°.
Acute tox data: sc LD_{50} (rabbit) = 4 mg/kg. [*3*]
THR = HIGH via sc route. See selenium compounds.

SELENIUM DISULFIDE. Red-yellow crystals. SeS_2, mw: 143.09, mp: <100°, bp: decomp.
Acute tox data: Oral LD_{50} (rat) = 138 mg/kg. [*3*]
THR = HIGH via oral route. See selenium compounds and sulfides.

SELENIUM HEXAFLUORIDE. Colorless gas. SeF_6, mw: 192.96, mp: −39° (sublimes @ −46.6°), bp: −34.5°, d: 3.25 g/l @ −25°.
Acute tox data: Inhal LC_{LO} (rat) = 10 ppm. [*3*]
THR = HIGH via inhal route. See selenium compounds and fluorides.

SELENIUM HYDRIDE. See hydrogen selenide.

SELENIUM IODO PHOSPHIDE. $Se_3P_4I_2$, mw: 614.7.
THR = Explodes with HNO_3. [*19*] See also phosphides and selenium compounds.

SELENIUM MONOBROMIDE. Dark red liquid. Se_2Br_2, mw: 317.75, bp: 227° (decomp), d: 3.604 @ 15°.
THR = See selenium compounds and bromides.

SELENIUM MONOBROMIDE TRICHLORIDE. Yellow-brown crystals. $SeBrCl_3$, mw: 265.25, mp: 190°.
THR = See selenium compounds, bromides and chlorides.

SELENIUM MONOCHLORIDE. Brown-red liquid. Se_2Cl_2, mw: 228.83, mp: −85°, bp: 130° (decomp), d: 2.91 @ 17°, 2.77 @ 23°.
THR = See selenium compounds and chlorides. Violent reaction with P, K, KO_2, Na_2O_2. [*19*]

SELENIUM MONOSULFIDE. Orange-yellow tablets or powder. SeS, mp: 111.03, mp: decomp @ 118°–119°, d: 3.056 @ 0°.
Acute tox data: Oral LD_{50} (mice) = 370 mg/kg; oral LD_{LO} (rat) = 180 mg/kg. [*3*]
THR = HIGH via oral route. See selenium compounds and sulfides. Violent reaction with Ag_2O. [*19*]

SELENIUM NITRIDE. Amorphous orange-yellow to brick-red hygroscopic crystals. Se_4N_4, mw: 371.87, mp: explodes @ 160°–200°, bp: decomp.
THR = See selenium compounds.
Explosion Hazard: Mod, when exposed to heat.
Disaster Hazard: Dangerous; when heated to decomp, or on contact with acid or acid fumes, emits highly toxic fumes of selenium.

SELENIUM OXIDE. See selenium dioxide.

SELENIUM OXYBROMIDE. Red-yellow crystals. $SeOBr_2$, mw: 254.79, mp: 41.6°, bp: 217° @ 740 mm (decomp); d (liquid): 3.38 @ 50°.
THR = See selenium compounds and bromides.

SELENIUM OXYCHLORIDE. Colorless–yellowish liquid. $SeOCl_2$, mw: 165.87, mp: 8.5°, bp: 176.4°, d: 2.42 @ 22°, vap. press: 1 mm @ 34.8°.
Acute tox data: Dermal TD_{LO} (man) = 0.7 mg/kg; dermal LD_{LO} (rabbit) = 10 mg/kg; sc LD_{50} (rabbit) = 7 mg/kg. [*3*]
THR = HIGH via sc and dermal routes. See selenium compounds and chlorides. Violent reaction with P, K. [*19*]

SELENIUM OXYFLUORIDE. Colorless liquid. $SeOF_2$, mw: 132.96, mp: 4.6°, bp: 124°, d: 2.67.
THR = See selenium compounds and fluorides. Violent reaction with P. [*19*]

SELENIUM SULFIDE. See selenium monosulfide.

SELENIUM TETRABROMIDE. Orange-red-brown crystals. $SeBr_4$, mw: 398.62, mp: decomp @ 75°.
THR = See selenium compounds and bromides.

SELENIUM TETRACHLORIDE. Cubic, white-yellow deliquescent crystals. $SeCl_4$, mw: 220.79, mp: 305° (sublimes @ 170°–196°), bp: decomp @ 288°, d: 3.78 vap. press: 1 mm @ 74.0°.
Acute tox data: sc LD_{50} (guinea pig) = 19 mg/kg. [*3*]
THR = HIGH via sc route. See selenium compounds and chlorides.

SELENIUM TETRAFLUORIDE. Colorless liquid or white crystals. SeF_4, mw: 154.96, mp: −13.8°, bp: > 100°, fp: −90°.
THR = See selenium compounds and fluorides. Violent reaction with P. [*19*]

SELENIUM TRIOXIDE. Amorphous, pale yellow, hygroscopic solid. SeO_3, mw: 126.96, mp: decomp @ 120°, d: 3.6.
THR = See selenium compounds.

SELF LIGHTING CIGARETTES. See cigarettes, self lighting.

SELLAITE. See magnesium fluoride.

SEMESAN. See hydroxy mercurichlorophenol.

SEMICARBAZIDE. CH_5ON_3, mw: 75.1.
Acute tox data: sc LD_{50} (rat) = 173 mg/kg; oral LD_{50} (mice) = 176 mg/kg; ip LD_{50} (mice) = 123 mg/kg. [*3*]
THR = HIGH via sc, ip and oral routes. Toxic to CNS. An exper neo via oral route. [*3*]

SEMICARBAZIDE HYDROCHLORIDE. Prisms. $CH_5ON_3 \cdot HCl$, mw: 111.5, mp: 176° (decomp).
Acute tox data: sc, ip, iv and oral LD_{50} (mice) = 123-176 mg/kg. [*12*]
THR = HIGH via sc, ip, iv and oral routes. A convulsant. An exper (+) carc and teratogen via oral route. [*3, 12*]

SENECA OIL. See crude oil (petroleum).

SENECIOALDEHYDE. Syns: *2,2-dimethylaxolein*, *2-methyl crotonaldehyde*, *2-methyl-2-buten-4-al*. Liquid, pungent odor. $(CH_3)_2C{:}CHCHO$, mw: 84.11, d: 0.8722 20°/4°, bp: 135°.
THR = See acrolein. Can be narcotic.

SENECIO ALKALOIDS. Composed of senecifoline, senecine, resins and seneciolongilobus, senecionine.
THR = An exper carc via oral route. [*14, 3*]

SENECIPHYLLINE. $C_{18}H_{23}O_5N$, mw: 333.4.
THR = An exper (±) carc. [*3, 7, 23*]

SENKIRKINE. $C_{19}H_{27}O_8N$, mw: 365.5.
THR = An exper (±) carc. [*3, 7, 23*]

SENSORY THRESHOLD FOR SOUND. See Section 3.

SEQUESTRENE TRISODIUM. $C_{10}H_{13}O_8N_2 \cdot 3Na$, mw: 358.2.
Acute tox data: Oral LD_{50} (rat) = 2150 mg/kg; oral LD_{50} (mouse) = 2150 mg/kg; ip LD_{50} (mouse) = 300 mg/kg. [*3*]
THR = HIGH via ip and MOD via oral routes.

SERINE. Syns: *β-hydroxyalanine*, *α-amino-β-hydroxypropionic acid.* A non-essential amino acid, colorless crystals, sol in water, insol in alcohol and ether, optically active. Here we consider *l* and *dl* forms. $HOCH_2CH(NH_2)COOH$, mw: 105, mp (*dl*): 246° (decomp), mp(*l*): 228° (decomp).
THR = U. A nutrient and/or dietary supplement food additive. [*109*]

SERINE DIAZOACETATE. $C_5H_7O_4N_3$, mw: 173.2.
Acute tox data: Oral LD_{50} (rat) = 170 mg/kg; ip LD_{50} (rat) = 100 mg/kg. [*3*]
THR = Powerful CNS effects. An exper (+) carc and teratogen [*3*] via ip route. [*7*]

SERPENTINE. A mineral.
THR = An exper neo via ipl route. [*3*]

SES. See sodium-2,4-dichlorophenoxyethyl sulfate.

SESAMIN. Crystals. $C_{20}H_{18}O_6$, mw: 354.3, mp: 129°.
THR = U. A pyrethrum synergist.
Fire Hazard: Slight, when exposed to heat or flame.

SESIN. See sodium-2,4-dichlorophenoxy ethyl benzoate.

SESAME OIL. flash p: 491°F, d: 0.9.
Fire Hazard: Low.
To Fight Fire: Dry chemical, CO_2.

SESOXANE. Syn: *2-(3,4-methylene dioxy phenoxy)-3,6,9-trioxaundecane*. Crystals, sol in kerosene $C_{15}H_{21}O_6$, mw: 297.3.
Acute tox data: Oral LD_{50} (rat) = 2000 mg/kg. [*95*]
THR = MOD via oral route.

SESQUI. See sodium sesquicarbonate.

"SEVIN." See carbaryl.

SHALE OIL.
THR = MILD irr. Chronic exposure causes cancer. [*14, 23*] See oils (mineral) and petroleum asphalt.
Disaster Hazard: Dangerous; when exposed to heat or flame; emits acrid fumes.

SHAPED ENCLOSURE HOOD. See Section 2.

SHELLAC, LIQUID. flash p: 40°–70°F.
THR = U. May act as an allergen.
Fire Hazard: Dangerous, when exposed to heat or flame. See also ethyl alcohol, the usual solvent.
Explosion Hazard: See ethyl alcohol.
Disaster Hazard: See ethyl alcohol.
To Fight Fire: Foam, CO_2, dry chemical.

SHIKIMIC ACID. $C_7H_{10}O_5$, mw: 174.2.
Acute tox data: ip LD_{50} (mice) = 1000 mg/kg. [*3*]
THR = MOD via ip route. An exper neo [*3*] and mutagen via oral and ip routes. [*23*]

SHIKIMOLE. See safrole.

SHORTENINGS. See thermic and oxidation products.

SHREDDERS AS A DANGER POINT IN SOLID WASTE HANDLING. See Section 6.

SIGNALS, SHIP DISTRESS. See fireworks, N.O.S.

SIGNAL OIL. See mineral seal oil.

SILANE. Syns: *silicon tetrahydride, silicane.* Gas with repulsive odor, slowly decomp by water. SiH_4, mw: 32.1, d: 0.68 @ −185°, mp: −185°, bp: 112°, fp: −200°.
THR = See silanes. Easily ignitable in air. Violent reaction with Cl_2. [*19*]

SILANES. Syns: *silicon hydrides, disilane.* Gas or liquid.
Acute tox data: Inhal LC_{LO} (mice) = 9600 ppm for 4 hrs. [*3*]
THR = MOD irr to skin, eyes and mu mem and via inhal route.
Fire Hazard: Dangerous, by chemical reaction with oxidizers; often ignites spont in air.
Explosion Hazard: Variable.
Disaster Hazard: Dangerous; when heated, they can burn or explode and emit highly toxic fumes.

SILANOLS. They are the alcohol derivatives of the silanes.
THR = Details U.

SILASTIC. See polysilicone.

SILICA. Syns: *agate, amethyst, chalcedony, cherts, flint, glass, onyx, pure quartz, rose quartz, sand, silicon dioxide, silicic anhydride, cristobalite.* Crystals. SiO_2, mw: 60.09, mp: 1710°, bp: 2230°, d(amorphous): 2.2, d(crystalline): 2.6, vap. press.: 10 mm @ 1732°.
Acute tox data: Inhal TD_{LO} (human) = 400 particles/cc for 4 yrs (intermittent) → pulmonary effects. Inhal TD_{LO} (human) = 16 mppcf for 8 hrs/day for 17.9 yrs (intermittent) → pulmonary effects; itr LD_{LO} (rat) = 200 mg/kg; iv LD_{LO} (dog) = 20 mg/kg; [*3*] imp TD_{LO} (rat) = 900 mg/kg → neo. [*3, 103*]
THR = MOD as an irr dust. From the point of view of numbers of men exposed and cases of disability produced, silica is the chief cause of pulmonary dust disease. The prolonged inhal of dusts containing free silica may result in the development of a disabling pulmonary fibrosis known as silicosis. The Committee on Pneumoconiosis of the American Public Health Association defines silicosis as "a disease due to the breathing of air containing silica (SiO_2), characterized by generalized fibrotic changes and the development of miliary nodules in both lungs, and clinically by shortness of breath, decreased chest expansion, lessened capacity for work, absence of fever, increased susceptibility to tuberculosis (some or all of which symptoms may be present), and characteristic x-ray findings."

Silica occurs in the pure state in nature as quartz. It is the main constituent of sand, sandstone, tripoli and diatomaceous earth, and is present in high amounts (up to 35%) in granite. Exposure to silica occurs in hard rock mining, in foundries, in manufacture of porcelain and pottery, in the spraying of vitreous enamels, in sandblasting, in granite-cutting and tombstone-making, in the manufacture of silica firebrick and other refractories, in grinding and polishing operations where natural abrasive wheels are used and other occupations.

The duration of exposure which is associated with the development of silicosis varies widely for different occupations. Thus, the average duration of exposure required for the development of silicosis in sand-blasters is 2–10 yrs, in moulders and granite cutters, about 30 yrs, and in hard rock miners 10–15 yrs. There is also much variation in individual susceptibility, certain workers showing radiological evidence of the disease years before their fellow workmen who are similarly exposed. Such susceptible individuals are fortunately rather rare.

The action of silica on the lungs results in the production of a diffuse, nodular fibrosis in which the parenchyma and the lymphatic system are involved. This fibrosis is, to a certain extent, progressive, and may continue to increase for several yrs after exposure is terminated. Where the pulmonary reserve is sufficiently reduced, the worker complains of shortness of breath on exertion. This is the first and most common symptom in cases of uncomplicated silicosis. If severe, it may incapacitate the worker for heavy, or even light, physical exertion, and in extreme cases there may be shortness of breath even while at rest. The most common physical sign of silicosis is a limitation of expansion of the chest. There may be a dry cough, sometimes very troublesome. The characteristic radiographic appearance is one of diffuse, discrete nodulation, scattered throughout both lung fields. Where the disease advances, the shortness of breath becomes worse, and the cough more productive and troublesome. There is no fever or other evidence of systemic reaction. Further progress of the disease results in marked fatigue, extreme dyspnea and cyanosis, loss of appetite, pleuritic pain and total incapacity to work. If tuberculosis does not supervene, the condition may eventually cause death either from cardiac failure or from destruction of lung tissue, with resultant anoxemia. In the later stages, the x-ray may show large conglomerate shadows, due to the coalescence of the silicotic nodules, with areas of emphysema between them.

Silica is used as a food additive permitted in the feed and drinking water of animals and/or for the treatment of food-producing animals. It is also permitted in food for human consumption. [*109*] It is a common air contaminant. Reacts violently with ClF_3, MnF_3, OF_2. [*19*]

SILICA AEROGEL. A finely powdered microcellular silica foam having a minimum silica content of 89.5%.
Acute tox data: iv LD_{50} (rat) = 15 mg/kg; itr LD_{LO} (rat) = 50 mg/kg. [*3*]
THR = HIGH via iv and itr routes. A general purpose food additive. [*109*] See also silica.

SILICA GEL. Syn: *silicic acid (precipitated).* White powder or lustrous granules. H_2SiO_3, mw: 78.1.
THR = See silica.

SILICANE. See silane.

SILICANES. See silanes.

SILICATES.
THR = Soluble alkaline silicates act locally like mild alkalies. The dust of certain silicates, such as asbestos (hydrated magnesium silicate) and talc, can produce fibrotic changes in the lungs and are implicated as exper carc. [*14, 3*] React violently with Li. [*19*]

SILICIC ACID. See silica gel.

SILICIC ANHYDRIDE. See silica.

SILICIDES OF LIGHT METALS. Metallic, crystalline materials.
THR = Variable.
Fire Hazard: Mod, by chemical reaction. See also hydrogen.
Explosion Hazard: Mod, by chemical reaction. See also hydrogen.
Disaster Hazard: Mod dangerous; they will react with water or steam to produce H_2; on contact with acid or acid fumes, they can emit toxic fumes.

SILICOBROMOFORM. See tribromosilane.

SILICOCHLOROFORM. See trichlorosilane.

SILICOETHANE. See disilane.

SILICOFLUORIC ACID. See hydrofluorsilicic acid.

SILICOFLUORIDES. See fluosilicates.

SILICON. Cubic, steel-gray crystals or dark brown powder. Si, atwt: 28.09, mp: 1420°, bp: 2600°, d: 2.42 or 2.3 @ 20°, vap. press: 1 mm @ 1724°.
THR = U. Does not occur free in nature, but is found as silicon dioxide (silica), and as various silicates. See also silica and silicates.
Fire Hazard: Mod, when exposed to flame or by chemical reaction with oxidizers. See also powdered metals. Violent reactions with alkali carbonates, (Al + PbO), Ca, Cs_2C_2, Cl_2, CoF_2, F_2, IF_5, MnF_3, Rb_2C_2, FNO, AgF, NaK alloy. [*19*]
Disaster Hazard: Dangerous; when heated, will react with water or steam to produce H_2; can react with oxidizing materials.
Radiation Hazard: For permissible levels, see Section 5A, Table 5A.5.

SILICON BROMIDE. Syns: *tetrabromosilicane, tetrabromosilane* Colorless, fuming liquid, disagreeable odor. $SeBr_4$, mw: 347.72, mp: 5°, bp: 153°, d: 2.814, vap. d: 2.82, fp: −12°.
THR = U. Probably very irr to skin, eyes and mu mem. See also bromides.
Disaster Hazard: Dangerous; when heated to decomp, emits highly toxic fumes of hydrobromic acid; will react with water or steam to produce heat and toxic and corrosive fumes.

SILICON CARBIDE. Bluish-black, irridescent crystals. SiC, mw: 40.10, mp: 2600°, bp: sublimes > 2000°; decomp @ 2210°, d: 3.17.
THR = MILD irr via inhal route.

SILICON CHLORIDE. Syn: *silicon tetrachloride.* Colorless, fuming liquid, suffocating odor. $SiCl_4$, mw: 169.89, mp: −70°, bp: 57.57°, d: 1.482.
Acute tox data: Inhal LC_{50} (rat) = 8000 ppm for 4 hrs. [*3*]
THR = MOD via inhal route. Violent reaction with K, Na. [*19*]
Disaster Hazard: Dangerous; when heated to decomp, emits highly toxic fumes of hydrochloric acid; will react with water or steam to produce heat and toxic and corrosive fumes.

SILICON DIBROMIDE SULFIDE. Colorless plates. $SeSBr_2$, mw: 219.95, mp: 93°, bp: 150° @ 81.3 mm.
THR = No data. Probably very irr. See also sulfides and bromides.

SILICON DICHLORIDE SULFIDE. Colorless prisms. $SiSCl_2$, mw: 1310.04, mp: 75°, bp: 92° @ 22.5 mm.
THR = No data. Probably very irr. See also sulfides and chlorides.

SILICON DIOXIDE. See silica.

SILICON DISULFIDE. White needles. SiS_2, mw: 92.19, mp: sublimes, bp: white heat.
THR = No data. Probably very toxic as a sulfide. See also sulfides. For other properties, see sulfides.

SILICONE RUBBER. See polysilicone.

SILICONES. Syn: *siloxanes*. Organosilicon oxide polymers such as $-R_2Si-O-$ where R is a monovalent organic radical.
THR = Generally LOW. Most of the silicones that have been studied should have low toxicity or none at all and little or no irr effects. May be spont flam in air. [*19*]

SILICON FLUORIDE. See silicon tetrafluoride.

SILICON HYDRIDES. See silanes.

SILICON MONOSULFIDE. Yellow needles or black solid. SiS, mw: 60.13, bp: 940° @ 20 mm (sublimes), d: 1.854 @ 15°.
THR = No data. See also sulfides.

SILICON MONOXIDE. Solid, insol in water, sol in alkalis and hydrofluoric acid. SiO, mw: 44.1, d: 2.2, mp: > 1700°, bp: 1880°.
THR = U. Self-ignites in air. [*19*]

SILICON NITRIDE. $(Si_2N_2)_x$.
THR = Self-ignites in air. [*19*]

SILICON OXYHYDRIDE. White powder, insol in water. $(Si_2H_2O_3)_x$, mw: $(104.2)_x$, mp: 1635°.
THR = U. See also silica.

SILICON SODIUM FLUORIDE. $Na_2 SiF_6$, mw: 188.1.
Acute tox data: Oral LD_{50} (rat) = 125 mg/kg; sc LD_{LO} (frog) = 448 mg/kg. [*3*]
THR = HIGH via oral; MOD via sc routes. See also fluosilicates and fluorides.

SILICON TETRAACETATE. Hygroscopic crystals. $Si(C_2H_3O_2)_4$, mw: 264.24, mp: 110°, bp: 148° @ 6 mm.
THR = Violent reaction with water. A powerful irr. No data.

SILICON TETRAAZIDE. $Si(N_3)_4$, mw: 197.2.
THR = Has exploded spont. [*19*] See also azides.

SILICON TETRABROMIDE. See silicon bromide.

SILICON TETRACHLORIDE. See silicon chloride.

SILICON TETRAFLUORIDE. Syn: *tetrafluorosilane*. Colorless gas, very pungent odor. SiF_4, mw: 104.1, mp: −77°, bp: −65° @ 181 mm, d: 4.67 g/liter.
THR = HIGH. See fluorides and hydrofluoric acid.

SILICON TETRAIODIDE. Syn: *tetraiodosilane*. Cubic, colorless crystals. SiI_4, mw: 535.74, mp: 120.5°, bp: 288°, d: 4.2.
THR = No data. Probably irr and toxic. See also iodides.
Disaster Hazard: Dangerous; when heated to decomp, emits toxic fumes of iodides; will react with water or steam to produce toxic and corrosive fumes.

SILICON TETRATHIOCYANATE. Small prisms. $Si(SCN)_4$, mw: 260.40, mp: 143.8°, bp: 314.2°.
THR = No data. See also thiocyanates.

SILICON TRICHLORIDE HYDROSULFIDE. Colorless liquid. $SiCl_3HS$, mw: 167.51, bp: 96°–100°, d: 1.45.
THR = See sulfides and chlorides.

SILICYL OXIDE. Syn: *disiloxane*. Colorless gas. $(SiH_3)_2O$, mw: 78.17, mp: −144°, bp: −15.2°, d: 0.881 @ −80°.
THR = No data. Probably an irr. See also sulfides and hydrochloric acid.
Fire Hazard: Easily self-ignites in air. [*19*]

SILK DUST.
Acute tox data: imp TD_{LO} = 30 mg/kg ⟶ carc. [*3*]
THR = An allergen and nuisance dust. An exper carc via imp route. [*3*]
Fire Hazard: Mod, when exposed to flame; can react with oxidizing materials.
Explosion Hazard: Mod, in the form of a dust cloud, when exposed to flame.

SILOXANE. See silicones.

SILVER. Syn: *argentum*. Soft, ductile, malleable, lustrous, white metal. Ag, atwt: 107.88, mp: 961.93°, bp: 2212°, d: 10.50 @ 20°.
THR = See silver compounds. An exper neo by implantation route. [*3*]
Radiation Hazard: For permissible levels, see Section 5A, Table 5A.5. Artificial isotope ^{110m}Ag, $T\frac{1}{2}$ = 253d, decays to stable ^{110}Cd via β's of 0.085 (61%), 0.53 (36%) MeV; emits γ's of 0.66 to 1.50 MeV.
Fire Hazard: Mod, in the form of dust, when exposed to flame or by chemical reaction with C_2H_2, NH_3, bromoazide, ClF_3, ethylene imine, H_2O_2, oxalic acid, H_2SO_4, tartaric acid. [*19*] See also powdered metals.

SILVER ACETATE. White plates. $AgC_2H_3O_2$, mw: 166.92, mp: decomp, d: 3.259 @ 15°.
THR = See silver compounds.

SILVER ACETYLIDE. White precipitate. Ag_2C_2, mw: 239.78, mp: explodes.
THR = See silver compounds.
Fire Hazard: U.
Explosion Hazard: Severe, when shocked or exposed to heat. See acetylides. Very sensitive explosive.
Disaster Hazard: Dangerous; shock or heat will explode it.

SILVER AMALGAMS. Silvery liquid or solid. Ag + Hg.
THR = See mercury and silver.

SILVER AMMONIUM COMPOUNDS.
THR = See silver compounds.
Explosion Hazard: Severe, when shocked, exposed to heat or by chemical reaction.

Disaster Hazard: Dangerous; shock or heat will explode them.

SILVER-*o*-ARSENATE. Cubic, dark red crystals. Ag_3AsO_3, mw: 446.55, mp: 150° (decomp).

THR = See arsenic compounds and silver compounds.

SILVER ARSPHENAMINE. Syns: *silver diamino dihydroxy arsenobenzene, silver salvarsan, silver diarsenal.* Brownish-black powder, contains approximately 20% arsenic, 15% silver.

THR = See arsenic compounds and silver compounds.

SILVER AZIDE. White prisms. AgN_3, mw: 149.90, mp: 252° bp: 297°.

THR = See silver compounds and azides.

Explosion Hazard: Severe, when shocked or exposed to heat. Explodes @ 250°. See azides. Shock-sensitive when dry. Violent reaction with Br, I. [*19*]

Disaster Hazard: Dangerous; shock or heat will explode it.

SILVER BENZOATE. White powder. $AgC_7H_5O_2$, mw: 228.99.

THR = See silver compounds.

SILVER BROMATE. White powder. $AgBrO_3$, mw: 235.8, mp: decomp, d: 5.206.

THR = See silver compounds and bromates. Violent reaction with S, Te. [*19*]

SILVER BROMIDE. Yellowish, odorless powder, darkened by light. AgBr, mw: 187.8, mp: 432°, d: 6.47.

THR = Reacts violently with K, Na. [*19*]

SILVER CARBONATE. Yellow crystalline powder. Ag_2CO_3, mw: 275.8, mp: 218° (decomp), d: 6.077.

THR = See silver compounds.

SILVER CHLORATE. Tetragonal white crystals. $AgClO_3$, mw: 191.34, mp: 230°, bp: decomp @ 270°, d: 4.430.

THR = See silver compounds and chlorates. Violent reaction with S. [*19*]

SILVER CHLORIDE. White granular powder. AgCl, mw: 143.34, mp: 455°, bp: 1550°, d: 5.561, vap. press: 1 mm @ 912°.

THR = Violent reaction with Al, NH_3, BrF_3, K, Na, (charcoal + Na_2O_2). [*19*]

SILVER CHROMATE. Red crystals. Ag_2CrO_4, mw: 331.77, d: 5.625.

THR = See chromium compounds and silver compounds.

SILVER COMPOUNDS.

Acute tox data: Inhal TC_{LO} (human) = 1 mg/m^3 of Ag dust yields skin effects. [*3*]

THR = An exper neo via imp route. [*3, 22*]

The absorption of silver compounds into the circulation and the subsequent deposition of the reduced silver in various tissues of the body may result in the production of a generalized greyish pigmentation of the skin and mucous membranes—a condition known as argyria. The introduction of fine particles of silver through breaks in the skin produces a local pigmentation at the site of the injury.

Generalized argyria, rarely seen at the present time, was not infrequent in the past. The condition developed slowly, usually after some 2–25 yrs of exposure. Pigmentation was noticeable first in conjunctivae, and later in the mu mem of the mouth and gums and in the skin. There were no constitutional symptoms, and no physical disability. Persons exhibiting the condition, and who subsequently died from unrelated disease, showed, on autopsy, a deposition of silver in the blood vessel walls, kidneys, testes, pituitary, choroid plexus, and mu mem of the nose, maxillary antra, trachea and bronchi. Once deposited, there is no known method by which the silver can be eliminated; the pigmentation is permanent. These compounds may be irr to the skin and mu mem.

SILVER, COLLOIDAL.

Acute tox data: Oral LD_{50} (mice) = 100 mg/kg; iv LD_{LO} (human) = 0.7 mg/kg. [*3*]

THR = HIGH via iv and oral routes. See also silver compounds.

SILVER CYANATE. Colorless crystals. AgOCN, mw: 149.90, mp: decomp, d: 4.00.

THR = See cyanates and silver compounds.

SILVER CYANIDE. White, odorless, tasteless powder which darkens on exposure to light. AgCN, mw: 133.90, mp: 320° (decomp), d: 3.95.

Acute tox data: Oral LD_{50} (rat) = 123 mg/kg. [*3*]

THR = HIGH via oral route. A deadly poison. See cyanides and silver compounds. Violent reaction with F_2. [*19*]

SILVER DIAMINODIHYDROXYARSENOBENZENE. See silver arsphenamine.

SILVER DIARSENAL. See silver arsphenamine.

SILVER DICHROMATE. Red crystals. $Ag_2Cr_2O_7$, mw: 431.78, mp: decomp, d: 4.770.

THR = See chromium and silver compounds.

SILVER DIFLUORIDE. Brown powder. AgF_2, mw: 145.88, mp: 690°, d: 4.57–4.78.

Acute tox data: Oral LD_{LO} (guinea pig) = 3000 mg/kg; sc LD_{LO} (guinea pig) = 800 mg/kg. [*3*]

THR = HIGH via oral and MOD via sc routes. See fluorides and silver compounds.

SILVER FLUOGALLATE. Colorless crystals. $Ag_3(GaF_6) \cdot 10H_2O$, mw: 687.52, d: 2.90.
THR = See fluogallates and silver compounds.

SILVER FLUORACETYLIDE. FCCAg, mw: 150.9.
THR = Highly unstable, explosive. [*19*]

SILVER FLUORIDE. Yellow crystalline masses. AgF, mw: 126.88, mp: 435°, d: 5.852 @ 15.5°, bp: 1150°.
THR = See fluorides and silver compounds. Violent reaction with B, CaH_2, dimethyl sulfoxide, K, Si, Na. [*19*]

SILVER FLUOSILICATE. Colorless crystals or white, deliquescent powder. $Ag_2SiF_6 \cdot 4H_2O$, mw: 429.88, mp: <100°, bp: decomp.
THR = See fluosilicates and silver compounds.

SILVER FULMINATE. Small needles. $Ag_2C_2N_2O_2$, mw: 299.80, mp: explodes.
THR = See fulminates and silver compounds.
Fire Hazard: U.
Explosion Hazard: Severe, when shocked or exposed to 175°. See also explosives, high.
Disaster Hazard: Dangerous; shock or heat will explode it; when heated to decomp, emits highly toxic fumes.

SILVER HYPONITRITE. Yellow crystals. $Ag_2N_2O_2$, mw: 275.77, mp: decomp @ 110°, d: 5.75 @ 30°.
THR = See nitrites and silver compounds.
Fire Hazard: U.
Explosion Hazard: Mod, when exposed to heat. Explodes @ 302°F.
Disaster Hazard: Dangerous; when heated to decomp, emits highly toxic fumes of NO_x; can react with reducing agents. Heat can explode it.

SILVER HYPOPHOSPHITE. White crystals. AgH_2PO_2, mw: 172.9.
THR = See silver compounds and phosphorus compounds.
Fire Hazard: U.
Explosion Hazard: Mod, when exposed to heat.
Disaster Hazard: Dangerous; when heated to decomp, emits highly toxic fumes of PO_x. Heat can explode it.

SILVER IODATE. Rhombic colorless crystals. $AgIO_3$, mw: 282.80, mp: >200°, bp: decomp, d: 5.525.
THR = See silver compounds and iodates. Violent reaction with K. [*19*]

SILVER IODIDE. Pale yellow powder. AgI, mw: 234.8, mp: 556°, bp: 1506°, d: 5.675, vap. press: 1 mm @ 820°.
THR = See iodides and silver compounds. Violent reaction with K, Na. [*19*]

SILVER LACTATE. White or slightly gray crystalline powder. $AgC_3H_5O_3 \cdot H_2O$, mw: 214.97.
THR = See silver compounds.

SILVER NITRATE. Colorless, transparent, tabular, rhombic, odorless crystals, becoming gray or grayish-black on exposure to light in presence of organic matter; bitter, caustic metallic taste. $AgNO_3$, mw: 169.89, mp: 212°, bp: 444° (decomp), d: 4.352 @ 19°.
Acute tox data: Oral LD_{50} (mice) = 40 mg/kg; ip LD_{50} (mice) = 0.129 mg/kg. [*3*]
THR = VERY HIGH via ip and HIGH via oral routes. A powerful caustic. See also silver compounds and nitrates. This very reactive material can react violently with H_2C_2, (H_2C_2 + NH_4OH), NH_4OH, As, Ca_2C_2, charcoal, ClF_3, $CuHC_2$, ethanol, Mg, PH_3, PH_4I, P, $P(OCN)_3$, plastics, S. [*19*]

SILVER NITRIDE. Colorless solid. Ag_3N, mw: 337.7.
THR = See silver compounds and nitrides.
Explosion Hazard: Severe, when shocked or exposed to >100°. See also explosives, high.
Disaster Hazard: Shock or heat will explode it.

SILVER NITRIDOOSAMITE. $OsAgNO_3$, mw: 360.1.
THR = Violently explodes @ 80° or from shock. [*19*]

SILVER NITRITE. Rhombic white crystals. $AgNO_2$, mw: 153.89, mp: decomp @ 140°, d: 4.453 @ 26°.
THR = See nitrites and silver compounds.

SILVER NITROPRUSSIDE. Light pink crystals. $Ag_2[FeNO(CN)_5]$, mw: 431.71.
THR = HIGH. See hydrocyanic acid.
Disaster Hazard: Dangerous; emits highly toxic fumes on heating.

SILVER ORE.
THR = A recog carc. [*14*] See arsenic compounds.

SILVER OXALATE. Colorless crystals. $Ag_2C_2O_4$, mw: 303.78, d: 5.029 @ 4°.
THR = HIGH. See oxalates.
Explosion Hazard: Mod, when exposed to heat. Violent reaction with Cl_2. [*19*] Explodes at @ 140°.
Disaster Hazard: Dangerous; emits highly toxic fumes on heating and may explode.

SILVER OXIDE. A dark brown, odorless powder, metallic taste. Ag_2O, mw: 231.76, mp: decomp @ 300°, d: 7.143 @ 16.6°.
Acute tox data: Oral LD_{LO} (rat) = 2820 mg/kg. [*3*]
THR = MOD via oral route. See silver compounds.
Fire Hazard: Mod, by chemical reaction; an oxidizing agent.
Explosion Hazard: Explodes in contact with ammonia. Reacts violently in contact with SbS_3, (NH_4OH + ethanol), CO, Mg, HgS, P, SeS, S. [*19*]

SILVER PERCHLORATE. White deliquescent crystals. $AgClO_4$, mw: 207.34, mp: decomp @ 486°, d: 2.806 @ 25°.
THR = See perchlorates and silver compounds. Very unstable. Can explode on grinding. Reacts violently

with acetic acid, aniline, benzene, chlorobenzene, glycerol, nitrobenzene, pyridine, toluene, ethylene diamine, ethanol, (CCl_4 + HCl). [*19*]

SILVER PERMANGANATE. Violet crystalline powder. $AgMnO_4$, mw: 226.81, mp: decomp violently, d: 4.49.
THR = See silver compounds and permanganates. Explodes with vapor of H_2SO_4 and forms a very shock-sensitive product with NH_4OH. [*19*]

SILVER PEROXIDE. Ag_2O_2, mw: 247.7.
THR = Has exploded with polyisobutylene. [*19*]

SILVER PHENOSULFONATE. Syn: *silver sulfocarbolate.* White to faintly reddish crystal.
$AgC_6H_4SO_3OH$, mw: 281.1.
THR = See silver compounds and phenol sulfonates.
Disaster Hazard: Dangerous; see sulfonates.

SILVER PHOSPHATE. Yellow powder. Ag_3PO_4, mw: 418.6, mp: 849°, d: 6.37.
THR = See silver compounds and phosphates.

SILVER PICRATE. See picratol.

SILVER-POTASSIUM CYANIDE. See potassium silver cyanide.

SILVER PROPARGYLATE. Crystals. $CHCCH_2Ag$, mw: 147.
THR = See silver compounds.
Explosion Hazard: Mod, when exposed to heat.

SILVER SALVARSAN. See silver arsphenamine.

SILVER SELENATE. Crystals. Ag_2SeO_4, mw: 358.7, d: 5.72.
THR = See selenium compounds and silver compounds.

SILVER SELENIDE. Thin, cubic gray plates. Ag_2Se, mw: 294.72, mp: 880°, bp: decomp, d: 8.0.
THR = See selenium compounds and silver compounds.

SILVER SELENITE. Needle-like crystals. Ag_2SeO_3, mw: 342.7, d: 5.9297.
THR = See selenium compounds and silver compounds.

SILVER SULFATE. Rhombic white crystals. Ag_2SO_4, mw: 311.82, mp: 652°, bp: decomp @ 1085°, d: 5.45 @ 29.2°.
THR = See silver compounds and sulfates.

SILVER SULFIDE. Syn: *acanthite.* Rhombic gray-black crystals. Ag_2S, mw: 247.83, mp: 845°, bp: decomp, d: 7.326.
THR = See sulfides and silver compounds. Violent reaction with ICl, $KClO_3$. [*19*]

SILVER SULFITE. White crystals. Ag_2SO_3, mw: 295.82, mp: decomp @ 100°.
THR = See silver compounds and sulfites.

SILVER SULFOCARBOLATE. See silver phenosulfonate.

SILVER TETRAZOL. Solid. $AgCHN_4$, mw: 176.9.
THR = See silver compounds.
Explosion Hazard: Severe, when exposed to heat.
Disaster Hazard: Dangerous; when heated to decomp, emits highly toxic fumes and may explode.

SILVER-THALLIUM NITRATE. White crystalline powder. $AgNO_3 \cdot TlNO_3$, mw: 435.90, mp: 75°.
THR = See thallium compounds, silver compounds and nitrates.

SILVER THIOARSENITE. Syn: *proustite.* Crystals. Ag_3AsS_3, mw: 494.73, mp: > 175°, d: 5.49.
THR = See arsenic compounds and sulfides.

SILVEX. See 2-(2,4,5-trichlorophenoxy)propionic acid.

3-SILYL-1-PROPANEDIOL TRIMETHOXY ESTER.
THR = An exper carc. [*3*]

SIMAZINE. See 2-chloro-4,6-bis(ethylamino)-*s*-triazinc.

SILVER SULFIDE. Syn: *acanthite.* Rhombic gray-black crystals. Ag_2S, mw: 247.8.
Acute tox data: LD_{50} (mammal) = 5000 mg/kg; oral LD_{LO} (rat) = 7500 mg/kg. [*3*]
THR = Mod via oral route.
Disaster Hazard: Dangerous; see chlorides.

SISAL.
THR = MILD allergen.
Fire Hazard: Mod, in the form of dust, when exposed to heat or flame; keep cool and dry; partially burned or charred material is dangerous.

SLACK WAX.
THR = A recog carc. [*14*] See paraffins and petroleum waxes.

SLAKED LIME. See calcium hydroxide.

SLATE (BELOW 5% FREE SILICA). A fine-grained green, black or red sedimentary rock.
THR = A nuisance dust.

SLUDGE ACID. See sulfuric acid.

SLUDGE ASPHALT.
THR = A recog carc. [*14*] See petroleum asphalt.

SMALL INCINERATORS A DANGER POINT IN SOLID WASTE HANDLING. See Section 6.

SMOG. An atmospheric combination of smoke, fog, and industrial gases. Composition: contents vary, but sulfur dioxide is a common component; other sulfides, fluorides, chlorides, carbon particles and various hydrocarbons may be found in smog.
THR = MOD irr to eyes and mu mem. Chronic effects are presently under study. A common air contaminant. Possibly carc.

SMOKELESS POWDER. Nitrocellulose containing about 13.1% nitrogen produced by blending material of somewhat lower (12.6%) and slightly higher (13.2%) nitrogen content, converting to a dough with alcohol-ether mixture, extruding, cutting and drying to a hard horny product. Small amounts of stabilizers (amines) and plasticizers are usually present, as well as various modifying agents (nitrotoluene, nitroglycerine salts). See also nitrocellulose and explosives, high.

SNUFF.
THR = A (S) carc. [*14*] See plant and fungal products.

SOAPS.
THR = An allergen. An irr to eyes and mu mem. Ingestion of large amounts causes GI distress, diarrhea and vomiting. Use of some soaps can cause a photosensitive dermatitis. Antibacterial soaps should not be used routinely due to possible resulting dermatitis. [*20*]
Fire Hazard: Slight, by chemical reaction.
Spontaneous Heating: Mod.

SOAPSTONE DUST. See talc.

SODA CHLORATE. See sodium chlorate.

SODA LIME. Sodium hydroxide with lime. A mixture of calcium oxide with 5–20% sodium hydroxide and containing 6–18% water. White or gray granules.
THR = HIGH irr to skin, eyes and mu mem. See also lime and sodium hydroxide.

SODAMIDE. See sodium amide.

SODA MONOHYDRATE. See sodium carbonate.

SODA NITER. See sodium nitrate.

SODIUM. Syn: *natrium*. Light, soft, ductile, malleable, silver-white metal. Na, atwt: 23.0, mp: 97.81°, bp: 892°, d: 0.9710 @ 20°, autoign. temp.: >115° in dry air, vap. press: 1.2 mm @ 400°.
THR = Sodium in elemental form is highly reactive, particularly with moisture, with which it reacts violently and therefore attacks living tissue. Also, Na + HOH yields NaOH. See also sodium hydroxide. Metallic sodium reacts exothermally with the moisture of body or tissue surfaces, causing thermal and chemical burns due to the reaction with sodium and the sodium hydroxide formed.
Radiation Hazard: For permissible levels, see Section 5A. Table 5A.5. Artificial isotope ^{22}Na, $T\frac{1}{2} = 2.6y$, decays to stable ^{22}Ne by ec and positron emission (90%) of 0.54 MeV; emits γ's of 1.27 MeV and x-rays.
Fire Hazard: Dangerous, when exposed to heat and moisture. In dry air it reacts very slowly up to 550° or by chemical reaction with moisture, air, $AlBr_3$, $AlCl_3$, AlF_3, NH_4 chlorocuprate, NH_4NO_3, $SbBr_3$, $SbCl_3$, SbI_3, $AsCl_3$, AsI_3, $BiBr_3$, $BiCl_3$, BiI_3, Bi_2O_3, BBr_3, bromoazide, CO_2, ($CO + NH_3$), CCl_4, Cl_2, ClF_3, $CrCl_4$, CrO_3, CoBr, CoCl, $CuCl_2$, CuO, $FeBr_3$, $FeCl_3$, $FeBr_2$, $FeCl_2$, FeI_2, hydrazine hydrate, H_2O_2, H_2S, HCl, HF, F_2, 1,2-dichloroethylene, dichloromethane, Br_2, hydroxylamine, iodine, iodine monochloride, iodine pentafluoride, lead oxide, maleic anhydride, manganous chloride, mercuric bromide, mercuric chloride, mercuric fluoride, mercuric iodide, mercurous chloride, mercurous oxide, methyl chloride, molybdenum trioxide, monoammonium phosphate, nitric acid, nitrogen peroxide, nitrosyl fluoride, nitrous oxide, phosgene, phosphorus, phosphorous pentafluoride, phosphorus pentoxide, phosphorus tribromide, phosphorus trichloride, phosphoryl chloride, potassium oxides, potassium ozonide, potassium superoxide, selenium, silicon tetrachloride, silver bromide, silver chloride, silver fluoride, silver iodide, sodium peroxide, stannic chloride, (stannic iodide + sulfur), stannic oxide, stannous chloride, sulfur, sulfur dibromide, sulfur dichloride, sulfur dioxide, sulfuric acid, tellurium, tetrachloroethane, thallous bromide, thiophosphoryl bromide, trichlorethylene, vanadium pentachloride, vanadyl chloride, zinc bromide. [*19*] or any oxidizing material, decomp moisture to evolve hydrogen and heat; reacts exothermally with the halogens, acids and halogenated hydrocarbons. Heated sodium is spont flam in air. Can be safely stored under liquid hydrocarbons.
Spont Heating: No.
Explosion Hazard: Dangerous, when exposed to moisture in any form! Keep dry at all times!
Disaster Hazard: Dangerous; when heated in air, emits toxic fumes of sodium oxide; will react with water or steam to produce heat, hydrogen, and flam vapors; can react vigorously to explosively with oxidizing materials. See hydrogen.
To Fight Fire: Soda ash, dry sodium chloride or graphite, in order of preference.
Storage and Handling: In the absence of moisture, oxygen or halides, sodium is safe to handle. As to indoor storage of drums, the important thing in storing sodium is that the storage area must be kept dry, since explosions may result from the contact of sodium with water. No automatic sprinkler system, or water or steam pipes containing water should be allowed in the room. Sufficient heat should be provided (without the use of open flames) to prevent condensation of moisture in the room due to changes in atmospheric conditions. Empty sodium drums should be stored in this same area.

Fire extinguishers (preferably color-coded) must be provided in the storage area, but only those containing sodium chloride, sodium carbonate, or graphite may be used. Pails are adequate for storing extinguishant if special care is taken to insure that the materials are dry. Water, carbon dioxide, carbon tetrachloride, soda-acid, or conventional dry chemical (bicarbonate) extinguishers must be avoided, and signs should be posted in the storage area warning against their use.

Only that amount of sodium immediately needed should be removed from the storage area. Sodium should not be withdrawn for intermediate storage in reaction areas. A special metal container with a tight fitting cover should be used for transporting sodium bricks to other plant areas, once they have been removed from the original container.

Large-scale outdoor storage tanks such as tank cars are unloaded after melting the sodium by circulating hot oil and withdrawing the molten sodium by vacuum to storage tanks similar in construction to sodium tank cars. Although steam may be used to heat the circulating oil, for use on both tank cars and storage tanks, steam must not be used directly as the heating agent for sodium tanks.

SODIUM ACETATE. White crystals, sol in water. $NaC_2H_3O_2$. mw: 82.0, mp: 324°, d: 1.528, autoign. temp.: 1125°F.

Acute tox data: Oral LD_{50} (rat) = 3530 mg/kg; sc LD_{50} (mice) = 8000 mg/kg; iv LD_{50} (mice) = 335 mg/kg. [*3*]

THR = HIGH via iv; MOD via oral, and LOW via sc routes. Used as a general purpose food additive. It is a substance which migrates to food from packaging materials. [*109*] Violent reaction with F_2, KNO_3. [*19*]

7-SODIUM ACETOBENZ(a)ANTHRACENE.
THR = An exper carc. [*23*]

SODIUM-*p*-ACETYL AMINOPHENYL ANTIMONATE. Syn: *stibenyl.* Light yellow powder, antimony content 35%.
$CH_3CONHC_6H_4SbO_3HNa \cdot H_2O$, mw: 345.93.

Acute tox data: ivLD_{50} (mice) = 283 mg/kg; ip LD_{LO} (mice) = 133 mg/kg. [*3*]

THR = HIGH via ip and iv routes. See also antimony compounds.

SODIUM ACETYL ARSANILATE. See arsacetin (sodium salt).

SODIUM ACETYLIDE. NaC_2H, mw: 48.
THR = Reacts violently with Br_2, Cl_2. [*19*]

SODIUM ACID ARSENATE. Na_2HAsO_4, mw: 185.9.
Acute tox data: ip LD_{LO} (rat) = 30 mg/kg. [*3*]
THR = HIGH via ip route. A poison. See also arsenic.

SODIUM ACID CARBONATE. See sodium bicarbonate.

SODIUM ACID CHROMATE. See sodium dichromate.

SODIUM ACID FLUORIDE. See sodium difluoride.

SODIUM ACID PHOSPHATE. See sodium phosphate monobasic.

SODIUM ACID PYROPHOSPHATE. Syns: *sodium pyrophosphate acid, disodium pyrophosphate, disodium diphosphate, disodium dihydrogen pyrophosphate, SAPP.* White crystalline powder, sol in water. $Na_2P_2O_7 \cdot 6H_2O$, mw: 331, d: 1.862, mp: 220° (decomp).

Acute tox data: sc LD_{50} (mice) = 480 mg/kg. [*3*]

THR = HIGH via sc route. A general purpose food additive. [*109*] An irr.

Disaster Hazard: Dangerous. See phosphates.

SODIUM ACID SULFATE. See sodium bisulfate.

SODIUM ALGINATE. Syn: *sodium polymannurate.* Colorless or slightly yellow filamentous or granular solid or a powder, forms a viscous colloidal solution with water, insol in alcohol, ether, and chloroform. $(C_6H_7O_6Na)_x$, mw: 32,000–250,000.

THR = No data; probably LOW. A stabilizer food additive. [*109*]

SODIUM-*m*-ALUMINATE. White hygroscopic powder. $NaAlO_2$, mw: 82.0, mp: 1650°.

THR = MOD irr to skin, eyes and mu mem. A corrosive substance which migrates to food from packaging materials. [*109*]

SODIUM ALUMINOSILICATE. Syn: *sodium silicoaluminate.* Fine white amorphous powder or beads, odorless and tasteless, insol in water and in alcohol and other organic solvents.

THR = An irr to skin, eyes and mu mem. An anticaking agent food additive. [*109*] See also aluminum.

SODIUM ALUMINUM FLUORIDE. Syns: *chiolite, cryolite.* Very white, vitreous masses, sol in conc sulfuric acid. $3NaF \cdot AlF_3$, mw: 210.0, mp: 1000°, d: 2.95.

Acute tox data: Oral LD_{50} (rat) = 200 mg/kg; ip LD_{50} (rat) = 59 mg/kg; oral LD_{LO} (rabbit) = 9 mg/kg. [*3*]

THR = HIGH via oral and ip routes. See also fluoaluminates. See fluorides. An insecticide.

SODIUM ALUMINUM HYDRIDE. White crystalline, material, stable in dry air but sensitive to moisture, sol in tetrahydrofuran. $NaAlH_4$, mw: 54.0, d: 1.24, mp: 183°.

THR = See hydrides.

SODIUM ALUMINUM PHOSPHATE. White powder. $Na_3PO_4 \cdot AlPO_4$, mw: 285.92.

THR = A MILD irr via oral route. See also phosphates. A general purpose food additive; it is a substance which migrates to food from packaging materials. [*109*]

SODIUM ALUMINUM SULFATE. See aluminum sodium sulfate.

SODIUM AMALGAM. Silver-white liquid or porous, crystalline mass, contains 2–20% metallic sodium, decomp in water. Na_xHg_y.

THR = See sodium and mercury.

SODIUM AMIDE. Syn: *sodamide*. White crystalline powder. $NaNH_2$, mw: 39.02, mp: 210°, bp: 400°.

THR = See sodium hydroxide and ammonia, both of which are liberated by this material in the presence of moisture. An intense irr to tissue.

Fire Hazard: Mod, by chemical reaction. See also ammonia.

Explosion Hazard: Mod, when exposed to moisture or by chemical reaction with CrO_3, $KClO_3$, 1,1-diethoxy-2-chloroethane. [*19*] Can become explosive in storage.

Disaster Hazard: Dangerous; when heated to decomp, emits highly toxic fumes of ammonia and sodium oxide; will react with water or steam to produce heat and toxic and corrosive fumes; can react with oxidizing materials.

SODIUM AMINOPHENYL ARSONATE. See sodium arsanilate.

SODIUM ANILINE ARSONATE. See sodium arsanilate.

SODIUM ARSANILATE. Syns: *atoxyl, sodium aniline arsonate, sodium aminophenyl arsonate*. White, crystalline, odorless powder, faint salty taste. $C_6H_4NH_2(HOAsOONa) \cdot 4H_2O$, mw: 311.12.

Acute tox data: sc LD_{50} (rat) = 75 mg/kg. [*3*]

THR = HIGH via sc route. Can cause blindness. See arsenic compounds. A food additive permitted in the feed and drinking water of animals and/or for the treatment of food-producing animals. [*109*]

SODIUM-*m*-ARSENATE. Rhombic, efflorescent crystals. $NaAsO_3$, mw: 145.91, d: 2.301.

Acute tox data: ip LD_{50} (rat) = 9 mg/kg; oral LD_{LO} (rabbit) = 12.5 mg/kg. [*3*]

THR = HIGH via oral and ip routes. An exper teratogen and carc. [*3*] A (S) human carc via ip and iv routes. [*3, 6*] See arsenic compounds.

SODIUM ARSENITE. Colorless or grayish-white powder. $NaAsO_2$, mw: 129.91, d: 1.87.

Acute tox data: ip LD_{50} (mice) = 1.17 mg/kg. [*3*]

THR = HIGH via ip and oral routes. A deadly poison. An exper teratogen and carc. [*3, 6*] A (S) carc for humans. [*6*] See arsenic compounds. An herbicide.

SODIUM ARSPHENAMINE. Syns: *sodium diarsenal, sodium arsphenolamine*. Bright yellow powder, contains not less than 19% arsenic. $NaONH_2C_6AsAsC_6H_3NH_2ONa$, mw: 410.8.

THR = See arsenic compounds.

SODIUM ARSPHENOLAMINE. See sodium arsphenamine.

SODIUM ASCORBATE. White crystals or powder, odorless, sol in water, insol in alcohol. $C_6H_7NaO_6$, mw: 198, mp: 218° (decomp).

THR = U. A chemical preservative food additive. [*109*]

SODIUM AURIBROMIDE. See gold sodium bromide.

SODIUM AURIDE. Cubic yellow crystals. $NaAu_2$, mw: 417.40, mp: decomp @ 700°.

THR = See gold compounds.

SODIUM AUROCYANIDE. See sodium cyanoaurite.

SODIUM AZIDE. Colorless hexagonal crystals. NaN_3, mw: 65.02, mp: decomp, d: 1.846.

Acute tox data: Oral LD_{50} (mice) = 27 mg/kg; ip LD_{50} (mice) = 18 mg/kg; iv LD_{50} (mice) = 19 mg/kg. [*3*]

THR = HIGH via oral, ip and iv routes. A poison. See azides and sodium hydroxide. Very unstable explosive. Violent reaction with (benzoyl chloride + KOH), Br_2, CS_2, $Cr(OCl)_2$, Cu, Pb, HNO_3, $(CH_3)_2SO_4$, dibromomalononitrile. [*19*]

SODIUM BARBITAL. White powder. $NaC_8H_{11}N_2O_3$, mw: 206.18.

Acute tox data: Oral LD_{50} (mouse) = 800 mg/kg; iv LD_{50} (rat) = 280 mg/kg; ip LD_{50} (mouse) = 763 mg/kg; sc LD_{50} (mouse) = 880 mg/kg; ip LD_{50} (rabbit) = 250 mg/kg. [*3*]

THR = HIGH–MOD via oral, iv and ip and sc routes. Used medically as a sedative. Abuse can cause habituation. See also barbiturates.

Disaster Hazard: Slight; when heated to decomp, emits toxic fumes of NO_x.

SODIUM BENZOATE. Syn: *benzoate of soda*. White, odorless crystalline solid. $NaC_7H_5O_2$, mw: 144.1.

Acute tox data: Oral LD_{50} (rat) = 4100 mg/kg; sc LD_{LO} (rabbit) = 2000 mg/kg; ip LD_{LO} (guinea pig) = 1400 mg/kg. [*3*]

THR = MOD via oral, sc and ip routes. Larger doses of 8–10 g by mouth may cause nausea and vomiting. It is possible to tolerate as muc as 50 g/day. Small doses have little or no effect. A fungicide. See also benzoic acid. A chemical preservative food additive. [*109*]

Disaster Hazard: Slight; when heated to decomp, emits acrid fumes.

SODIUM BENZOSULFIMIDE. See sodium saccharin.

SODIUM BERYLLIUM FLUORIDE. White crystalline masses, water-sol, insol in alcohol. $BeF_2 \cdot 2NaF$, mw: 131.1, mp: 350°.
THR = HIGH. See beryllium and fluorides.

SODIUM BERYLLIUM MALATE.
$Na_2Be_4C_8H_6O_{12} \cdot 7H_2O$, mw: 502.3.
Acute tox data: iv LD_{50} (mouse) = 0.036 mg/kg. [*3*]
THR = VERY HIGH via iv route. See also beryllium compounds.

SODIUM BICARBONATE. Syns: *baking soda, sodium acid carbonate.* White powder or crystalline lumps, sol in water, insol in alcohol. $NaHCO_3$, mw: 84, d: 2.159, mp: loses CO_2 @ 270°.
Acute tox data: Oral LD_{50} (rat) = 4220 mg/kg. [*3*]
THR = MOD via oral routes. Used as a general purpose food additive; it is a substance which migrates to food from packaging materials. [*109*] Violent reaction with NaK alloy, $NH_4H_2PO_4$. [*19*]

SODIUM BICHROMATE. See sodium dichromate.

SODIUM BIFLUORIDE. See sodium difluoride.

SODIUM BINOXIDE. See sodium peroxide.

SODIUM BIPHOSPHATE. See sodium phosphate, monobasic.

SODIUM BISULFATE. Syns: *sodium hydrogen sulfate, sodium acid sulfate.* Colorless crystals. $NaHSO_4$, mw: 120.1, mp: >315° (decomp), d: 2.435 @ 13°.
THR = An irr. See sulfuric acid, which is liberated on contact with moisture.
Disaster Hazard: Dangerous; see sulfates; will react with water or steam to produce heat and toxic fumes.

SODIUM BISULFIDE. NaSH, mw: 56.1.
THR = See also sulfides. Violent reactions with diazonium salts. [*19*]

SODIUM BISULFITE. Syn: *sodium hydrogen sulfite.* White crystals. $NaHSO_3$, mw: 104.1, d: 1.48.
Acute tox data: ip LD_{50} (rat) = 650 mg/kg; iv LD_{50} (rat) = 115 mg/kg; iv LD_{50} (mice) = 130 mg/kg. [*3*]
THR = HIGH via iv; MOD via ip routes. An irr.

SODIUM-*m*-BISULFITE. Syn: *sodium pyrosulfite.* Chief constituent of commercial dry sodium bisulfite, with which most of its properties are identical. $Na_2S_2O_5$, mw: 190. d: 1.4, mp: >150° (decomp).
THR = HIGH. A chemical preservative food additive. [*109*]

SODIUM BORATE. See sodium tetraborate.

SODIUM BORATE HEPTAHYDRATE.
Acute tox data: ip LD_{50} (mouse) = 450 mg/kg. [*3*]
THR = HIGH via ip route. See also sodium tetraborate.

SODIUM BORATE PERHYDRATE. Crystals. $NaBO_2 \cdot H_2O_2$, mw: 99.8, mp: decomp @ 40°.
THR = No specific data. See also sodium tetraborate and peroxides.

SODIUM BORINATE. $NaBH_2O$, mw: 51.8.
THR = Violently unstable. [*19*]

SODIUM BOROHYDRIDE. White to gray-white microcrystalline powder or lumps. Reacts with hot water, sol in liquid ammonia and "Cellosolve" ether. $NaBH_4$, mw: 37.85, mp: 36°, d: 1.07.
Acute tox data: oral LD_{LO} (rat) = 160 mg/kg. [*4*]
THR = HIGH via oral route. Ignites in air. [*19*] See boron compounds and hydrides.
Fire Hazard: Mod, when exposed to heat or flame or by chemical reaction with oxidizers.
Disaster Hazard: Dangerous; when heated to decomp, emits toxic fumes; will react with water or steam to produce hydrogen; on contact with acid fumes, it can emit flam vapors.

SODIUM BROMATE. White crystals or crystalline powder. $NaBrO_3$, mw: 150.91, mp: 381°, d: 3.339 @ 17.5°.
Acute tox data: iv LD_{LO} (rabbit) = 360 mg/kg; oral LD_{LO} (rabbit) = 250 mg/kg; sc LD_{LO} (dog) = 120 mg/kg; ip LD_{50} (mice) = 140 mg/kg. [*3*]
THR = HIGH via iv, oral, sc and ip routes. See bromates. Violent reactions with Al, As, C, Cu, oil, metal sulfides, organic matter, P, S. [*19*] See bromates.

SODIUM BROMIDE. Cubic, colorless crystals. NaBr, mw: 102.91, mp: 755°, bp: 1390°, d: 3.203 @ 25°, vap. press: 1 mm @ 806°.
Acute tox data: Oral LD_{50} (rat) = 3500 mg/kg; sc LD_{50} (mice) = 5020 mg/kg; oral LD_{LO} (rabbit) = 580 mg/kg. [*3*]
THR = MOD via sc and oral routes. See bromides. Reacts violently with BrF_3. [*19*]

SODIUM BROMOAURATE. Brown-black crystals. $NaAuBr_4 \cdot 2H_2O$, mw: 575.89.
THR = See gold compounds and bromides.

SODIUM-2-BROMO-4-PHENYLPHENOL.
$NaOC_6H_3Br(C_6H_5)$, mw: 271.
THR = See phenols and bromides.

SODIUM BROMOPLATINATE. Dark red crystals. $Na_2PtBr_6 \cdot 6H_2O$, mw: 838.8, mp: decomp @ 150°, d: 3.323.
THR = See platinum compounds and bromides.

SODIUM CACODYLATE. Syn: *sodium dimethyl arsinate.* White crystalline hydrate, water-sol.
$NaAsC_2H_6O_2 \cdot 3H_2O$, mw: 214.02, mp: 82°–86°.
Acute tox data: Oral LD_{50} (rat) = 2600 mg/kg; oral LD_{50} (mouse) = 4 mg/kg. [*3*]
THR = HIGH via oral to mice and MOD via oral to rats. See also cacodylic acid.

SODIUM CALCIUM ALUMINOSILICATE HYDRATED. Syn: *sodium calcium silicoaluminate.*
THR = U. An anticaking agent food additive. [*109*] See also silicates.

SODIUM CARBIDE. White powder. Na_2C_2, mw: 70.01, bp: 700°, d: 1.575 @ 15°.
THR = See sodium hydroxide and acetylene (liberated on contact with water).
Fire Hazard: Mod, by chemical reaction with oxidizers.
Explosion Hazard: In contact with Al, Br_2, CO_2, Cl_2, Fe, Pb, Hg, N_2O_5, P, SO_2, water. [*19*] Also on contact with bromine. See acetylene.
Disaster Hazard: See carbides.
To Fight Fire: CO_2, dry chemical.

SODIUM CARBONATE. Syns: *soda monohydrate, crystal carbonate.* White, odorless, small crystals or crystalline powder, alkaline taste. Na_2CO_3, mw: 106.0, mp: 851°, bp: decomp, d: 2.509 @ 0°.
Acute tox data: ip LD_{50} (mice) = 117 mg/kg; oral LD_{LO} (rat) = 4000 mg/kg. [*3*]
THR = HIGH via ip; MOD via oral routes. A general purpose food additive, it migrates to food from packaging materials. [*109*] Can react violently with Al, P_2O_5, H_2SO_4. [*19*]

SODIUM CARBONATE PEROXIDE. Fine white powder. $2Na_2CO_3 \cdot 3H_2O_2$, mw: 314.
THR = See sodium carbonate and hydrogen peroxide.

SODIUM CARBONYL. NaCO, mw: 51.
THR = HIGH. See carbonyls.
Fire Hazard: Mod, when exposed to heat or by chemical reaction with oxidizers. Heat causes evolution of carbon monoxide.
Explosion Hazard: Contact with air, water or possibly heat. [*19*]
Disaster Hazard: Dangerous; when heated to decomp, emits highly toxic fumes of sodium oxide and carbon monoxide; may explode on heating.

SODIUM CARBOXY METHYL CELLULOSE. Syns: *CMC, sodium cellulose glycolate, cellulose gum, CM cellulose.* A synthetic cellulose gum (the sodium salt of carboxy methyl cellulose not <99.5% on a dry weight basis, with maximum substitution of 0.95 carboxymethyl groups per anhydroglucose unit, and with a minimum viscosity of 25 centipoises for 2% weight aqueous solutions at 25°). Colorless, odorless, hygroscopic powder or granules, insol in most organic solvents.
THR = See polymers, soluble. A general purpose food additive, it is a substance which migrates to food from packaging materials. [*109*] An exper neo via sc route. [*3*]

SODIUM CASEINATE COMPLEX. Syn: *casein-sodium.* Coarse white powder, odorless, sol in water.
THR = An exper carc to mice via sc route. [*103*] A general purpose food additive. [*109*]

SODIUM CELLULOSE GLYCOLATE. See sodium carboxymethyl cellulose.

SODIUM CHLORATE. Syn: *soda chlorate.* Colorless, odorless crystals, cooling saline taste. $NaClO_3$, mw: 104.65, mp: 248°–261°, bp: decomp, d: 2.490 @ 15°.
Acute tox data: Oral LD_{50} (rat) = 1200 mg/kg; ip LD_{50} (mice) = 596 mg/kg; ukLD_{50} (child) = 185 mg/kg; oral TD_{LO} (women) = 800 mg/kg ⟶ effects on red blood corpuscles. [*3*]
THR = HIGH–MOD via oral, ip and uk routes. Can cause local irr to skin, eyes and mu mem. Ingestion of large quantities can be fatal. Symptoms are abdominal pain, nausea, vomiting, cyanosis and collapse. An herbicide. Can react violently with Al, $NH_4S_2O_3$, Sb_2S_3, As, As_2O_3, C, charcoal, MnO_2, metal sulfides, dibasic organic acids, organic matter, P, KCN, S, H_2SO_4, thiocyanates, Zn. [*19*]

SODIUM CHLORAURATE. See gold sodium chloride.

SODIUM CHLORIDE. Syns: *salt, halite, sea salt.* Colorless, transparent crystals or white crystalline powder. NaCl, mw: 58.45, mp: 801°, bp: 1413°, d: 2.165, vap. press: 1 mm @ 865°.
Acute tox data: Oral LD_{50} (rat) = 3000 mg/kg; ip LD_{50} (mice) = 2602 mg/kg; sc LD_{LO} (rat) = 3500 mg/kg. [*3*] In a human TD_{LO} = 8200 mg/kg for 23 days ⟶ blood pressure problems.
THR = MOD via oral, ip and sc routes. An exper teratogen via pa routes. [*3*] When bulk sodium chloride is heated to high temp., a vapor is emitted which is irr, particularly to the eyes. Ingestion of large amounts of sodium chloride can cause irr of the stomach. Improper use of salt tablets may produce this effect. A substance which migrates to food from packaging materials. [*109*] Violent reaction with BrF_3, Li. [*19*]

SODIUM CHLORITE. White crystals or crystalline powder. $NaClO_2$, mw: 90.45, bp: decomp @ 175° ⟶ 0_2.
THR = U. May act as an irr due to oxidizing power.
Fire Hazard: A powerful oxidizing agent; ignited via friction, heat or shock.
Explosion Hazard: Dangerous from exposure to percussion, acids, organic matter, oxalic acid, P, S. [*19*]
Disaster Hazard: Dangerous; shock will explode it; when heated, emits highly toxic fumes of chlorides and may explode; can react vigorously on contact with reducing materials.

SODIUM CHLOROACETATE. White, odorless, free-flowing powder. $ClCH_2COONa$, mw: 116.49, mp: decomp @ 200°, flash p: none.
Acute tox data: Oral LD_{50} (rat) = 76 mg/kg; oral LD_{50} (mice) = 165 mg/kg; iv LD_{50} (mice) = 109 mg/kg; oral LD_{50} (guinea pig) = 80 mg/kg. [*3*]
THR = HIGH via oral and iv routes.

SODIUM CHLOROAURATE. Rhombic yellow crystals. $NaAuCl_4 \cdot 2H_2O$, mw: 398.06, mp: decomp.
THR = See gold compounds and chlorides.

SODIUM CHLOROIRIDATE. Dull red-black crystals. $Na_2IrCl_6 \cdot 6H_2O$, mw: 559.93, mp: decomp @ 600°.
THR = See iridium compounds and chlorides.

SODIUM-4-CHLORO-2-METHYL PHENOXY ACETATE.
Acute tox data: Oral LD_{50} (rat) = 800 mg/kg; ip LD_{50} (rat) = 400 mg/kg; oral LD_{50} (mice) = 560 mg/kg. [*3*]
THR = HIGH via ip and MOD via oral routes. A fungicide.
Disaster Hazard: Dangerous; see chlorides.

SODIUM CHLORO-2-PHENYLPHENATE. Syns: *sodium-2-chloro-o-phenylphenate, dowicide 31.*
Acute tox data: Oral LD_{50} (rat) = 3500 mg/kg. [*3*]
THR = MOD via oral route. A fungicide.
Disaster Hazard: Dangerous; see chlorides.

SODIUM CHLOROPLATINATE. Orange-yellow powder. Na_2PtCl_6, mw: 453.97, mp: 150°–160°.
Acute tox data: Inhal TC_{LO} (human) = 0.9 mg/m^3 ⟶ pulmonary effects such as asthma.
THR = HIGH via inhal route. See platinum compounds and chlorides.

SODIUM CHROMATE. Yellow rhombic crystals. Na_2CrO_4, mw: 162.00, d: 2.723 @ 25°, mp: 792°.
Acute tox data: Ip LD_{50} (mouse) = 32 mg/kg; sc LD_{LO} (rabbit) = 243 mg/kg; dermal LD_{LO} (guinea pig) = 206 mg/kg. [*3*]
THR = HIGH via ip, sc and dermal routes. An exper (±) carc. [*3, 6*] See also chromium.

SODIUM CITRATE. Syn: *disodium citrate.* White crystals or granular powder, odorless, sol in water, insol in alcohol. $C_5H_6O_7 \cdot 2Na$, mw: 236.1, mp: loses water @ 150°, bp: decomp @ red heat.
Acute tox data: ip LD_{50} (rat) = 1724 mg/kg; iv LD_{50} (mice) = 71 mg/kg; iv LD_{50} (rabbit) = 521 mg/kg. [*3*]
THR = HIGH via iv and MOD via ip routes. Used as a sequestrant and general purpose food additive. [*109*]

SODIUM COMPOUNDS.
THR = Variable. Sodium ion is practically nontoxic. The toxicity of sodium compounds is frequently, though not always, due to the anion involved. The hydroxide is very corrosive, being strongly basic. Even here it is the conc of hydroxyl ion which is responsible for the caustic action of this material.

SODIUM COPPER POLYPHOSPHATE. See copper compounds.

SODIUM CYANATE. Colorless needles. NaOCN, mw: 65.92, d: 1.937 @ 20°.
Acute tox data: imp LD_{50} (rat) = 310 mg/kg. [*3*]
THR = HIGH via imp route. See cyanates.

SODIUM CYANIDE. White, deliquescent, crystalline powder. NaCN, mw: 49.02, mp: 563.7°, bp: 1496°, vap. press: 1 mm @ 817°.
Acute tox data: Oral LD_{50} (rat) = 6.44 mg/kg; ip LD_{50} (mice) = 5 mg/kg; sc LD_{50} (guinea pig) = 4.5 mg/kg; iv LD_{LO} (dog) = 1.3 mg/kg. [*3*]
THR = HIGH via oral, ip, sc and iv routes. Toxic via dermal route. See also cyanides. Violent reaction with nitrates, nitrites. [*19*] Very poisonous.

SODIUM CYANOAURITE. Syn: *sodium aurocyanide.* White crystalline powder. $NaAu(CN)_2$, mw: 272.23.
THR = See cyanides and gold.

SODIUM CYCLAMATE. Syns: *sodium cyclohexyl sulfamate, sodium sucaryl.* White crystalline powder, practically odorless, practically insol in alcohol, benzene, chloroform, and ether, sol in water. $C_6H_{11}NHSO_3Na$, mw: 169.
Acute tox data: ip LD_{50} (rat) = 1350 mg/kg; oral LD_{50} (mice) = 1525 mg/kg; ip LD_{50} (mice) = 1150 mg/kg. [*3*]
THR = MOD via oral and ip routes. An exper neo via imp route. [*3*] See cyclamates. A non-nutritive sweetener food additive. [*109*]

SODIUM-2,4-D. $NaC_8H_6O_3Cl_2$, mw: 244.0.
Acute tox data: Oral LD_{50} (rat) = 666 mg/kg; ip LD_{50} (rat) = 666 mg/kg; sc LD_{50} (mice) = 280 mg/kg; iv LD_{50} (rabbit) = 400 mg/kg. [*3*]
THR = HIGH via sc and iv; MOD via oral and ip routes. See also 2,4-D.

SODIUM DIACETATE. White crystals, acetic odor, sol in water, slightly sol in alcohol, insol in ether. $CH_3COONa \cdot x(CH_3COOH)$ (anhydrous) or $CH_3COONa \cdot x(CH_3COOH) \cdot yH_2O$ (technical), decomp > 150°.
THR = U. A sequestrant food additive. [*109*]

SODIUM DIARSENAL. See sodium arsphenamine.

SODIUM DICHLORO ISOCYANURATE. White crystals; water sol, chlorine odor. $Cl_2Na(NCO)_3$, mw: 220.0, mp: 230°–250°.
Acute tox data: oral LD_{50} (rat) = 950–1670 mg/kg; oral LD_{50} (rabbit) = 600–2500 mg/kg; oral LD_{50} (guinea pig) = 400–600 mg/kg. [*91*]

THR = The main toxic effects were associated with gastrointestinal irr, including salivation, lachrymation, dyspnoea, weakness, emaciation, lethargy, diarrhea, coma and (following very high dosage) deaths after 1–8 days, showing irr of stomach and gastrointestinal tract, liver dysfunction and lung congestion. The conc material may be a little more toxic, due to greater gastrointestinal irr. In the dry form, it is not appreciably irr to dry skin. However, when moist, the conc material is irr to skin, and also may cause severe eye irr. Aqueous dilutions of 333 ppm or less sodium dichloroisocyanurate showed no irr to skin or eyes, despite daily application for 3 months. Rats and dogs were unaffected by the intake of up to 333 ppm in the diet for 6 months. Rats were unaffected by its inclusion in their drinking water for 30 days (33 ppm + 4000 ppm cyanuric acid). Rabbits were unaffected by daily dermal application of 5 ml of 333 ppm aqueous sodium dichloroisocyanurate on 5 days/week for 3 months. [*9*]

Disaster Hazard: Dangerous; When heated to decomp, emits carbon monoxide and chloride fumes.

SODIUM-2,4-DICHLOROPHENOXY ETHYL SULFATE. Syn: *SES*. Crystals. $C_8H_8O_5SCl_2Na$, mw: 310.11.

Acute tox data: Oral LD_{50} (rat) = 730 mg/kg. [*3*]

THR = MOD via oral route. Limited animal data indicate a mod degree of toxicity with kidney and liver injury. Strong solutions irr the skin. See also 2,4-dichlorophenoxyacetic acid.

Disaster Hazard: Dangerous; see chlorides.

SODIUM DICHLOROPROPIONATE. CH_3HCl_2COONa, mw: 165.

Acute tox data: Oral LD_{50} (rat) = 3860 mg/kg. [*3*]

THR = MOD via oral route. A food additive permitted in the feed and drinking water of animals and/or for the treatment of food-producing animals. [*109*]

SODIUM DICHROMATE. Syn: *sodium acid chromate*. Red crystals. $Na_2Cr_2O_7 \cdot 2H_2O$, mw: 298.1, mp: $-2H_2O$ @ 100°, mp(anhydrous): 356.7°, bp: decomp @ 400°, d: 2.35 @ 13°.

Acute tox data: sc LD_{LO} (guinea pig) = 51 mg/kg; ip LD_{LO} (guinea pig) = 335 mg/kg; dermal LD_{LO} (guinea pig) = 335 mg/kg. [*3*]

THR = HIGH via ip, dermal and sc routes. An exper (S) carc. [*3*, *6*, *14*] See chromium compounds. A caustic and irr. Reacts violently with hydrazine. [*19*]

SODIUM DICYANAMIDE. Colorless crystals. $NaN(CN)_2$, mw: 89.04, mp: 315° (decomp), d: 1.701 @ 30°.

THR = See calcium cyanamide and cyanides.

SODIUM DIETHYL DITHIOCARBAMATE. Crystals. $NaSC(S)N(C_2H_5)_2$. mw: 171.3, mp: 95°, d: 1.1 @ 20°/20°, vap. d: 5.9.

THR = U. See also bis(diethylthiocarbamyl)disulfide.

Disaster Hazard: Dangerous; when heated to decomp, emits highly toxic fumes of SO_x and NO_x.

SODIUM DIFLUORIDE. White powder. $NaF \cdot HF$. mw: 62.

Acute tox data: Oral LD_{LO} (guinea pig) = 2000 mg/kg; sc LD_{LO} (guinea pig) = 250 mg/kg. [*3*]

THR = HIGH via oral and sc routes.

SODIUM DIHYDROGEN-*o*-ARSENATE. Rhombic or monoclinic colorless crystals. $NaH_2AsO_4 \cdot H_2O$, mw: 181.94, mp: $-H_2O$ @ 100°–130°, bp: decomp @ 200°–289°, d: 2.53.

THR = See arsenic compounds.

SODIUM DIHYDROGEN PHOSPHATE. See sodium phosphate, monobasic.

SODIUM-6,7-DIHYDROXY-2-NAPHTHALENE SULFONATE. Dry paste. $C_{10}H_6(OH)_2SO_2Na$, mw: 263.2.

THR = U. Probably toxic.

Disaster Hazard: Dangerous; see sulfonates.

SODIUM-*p*-DIMETHYL AMINOBENZENE DIAZOSULFONATE. See methyl orange.

SODIUM DIMETHYL ARSINATE. See sodium cacodylate.

SODIUM DIMETHYL DITHIOCARBAMATE. Crystals. $(CH_3)_2NCS_2Na$, mw: 143.2.

Acute tox data: Oral LD_{50} (rat) = 1000 mg/kg. [*3*]

THR = MOD via oral route. An exper neo via oral route. [*4*]

Disaster Hazard: Dangerous; see sulfonates and NO_x.

SODIUM DINITRO-*o*-BUTYL PHENATE.

THR = U. An herbicide. See also phenols.

Disaster Hazard: Dangerous; see nitrates.

SODIUM DINITRO-*o*-CRESYLATE. Brilliant orange-yellow dye. $C_6H_2(ONa)(NO_2)_2(CH_3)$, mw: 220.1.

THR = Probably HIGH. Details U. An insecticide and selective herbicide. See also dinitrocresol.

Fire Hazard: Mod; see nitrates.

Disaster Hazard: Dangerous; see nitrates.

SODIUM DINITROPHENOL. $C_6H_3(ONa)(NO_2)_2$, mw: 206.1.

THR = Probably toxic. See dinithrophenol.

Fire Hazard: Mod; see nitrates.

Explosion Hazard: Severe, when shocked or exposed to heat. See also nitrates. Explodes @ 698°.

Disaster Hazard: Dangerous; when heated to decomp, emits toxic fumes and may explode.

SODIUM DIOCTYL SULFOSUCCINATE. See dioctyl sodium succinate.

SODIUM DIOXIDE. See sodium peroxide.

SODIUM DISPERSIONS. Finely divided metallic sodium suspended in toluene, xylene, naphtha, kerosene, etc.

THR = HIGH. See sodium and individual dispersant.

Fire Hazard: Dangerous, when exposed to heat or flame or by chemical reaction. These are very reactive forms of sodium, which if carelessly handled may catch fire. To extinguish, see sodium. After sodium has been extinguished, the burning organic vapor can be dealt with by very cautious use of a carbon dioxide extinguisher. Do *not* use carbon tetrachloride.

Explosion Hazard: Mod, by chemical reaction. See also sodium.

Disaster Hazard: Dangerous; when heated, it loses the solvent and emits highly toxic fumes of sodium, sodium oxide, etc.; will react with water or steam to produce heat and hydrogen; on contact with oxidizing materials, can react vigorously, and on contact with acid or acid fumes, can emit toxic fumes.

SODIUM DITHIOCARBAMATE. $CH_2NS_2 \cdot Na$, mw: 115.2.

Acute tox data: Oral LD_{LO} (rat) = 500 mg/kg. [*3*]

THR = MOD via oral route.

SODIUM DITHIONITE. See sodium hydrosulfite.

SODIUM DODECANOATE. $C_{12}H_{24}O_2 \cdot Na$, mw: 223.4.

Acute tox data: uk LD_{LO} (mice) = 400 mg/kg. [*3*]

THR = HIGH via uk route.

SODIUM DODECYL BENZENE SULFONATE. White to light yellow flakes, granules or powder. $C_{12}H_{25}C_6H_4SO_3Na$, mw: 348.

Acute tox data: Oral LD_{50} (rat) = 1260 mg/kg; oral LD_{50} (mice) = 2000 mg/kg; iv LD_{50} (mice) = 105 mg/kg. [*3*]

THR = HIGH via iv and MOD via oral routes. A detergent.

Disaster Hazard: Dangerous; see sulfonates.

SODIUM EDTA. $C_{10}H_{12}O_8N_2 \cdot 4Na$, mw: 380.2.

Acute tox data: ip LD_{50} (mice) = 330 mg/kg. [*3*]

THR = HIGH via ip route.

SODIUM ETHOXYACETYLIDE. NaC_4H_5O, mw: 92.1.

THR = Very unstable. [*19*]

SODIUM ETHYLATE. Syn: *caustic alcohol.* White powder, sometimes having brownish tinge. C_2H_5ONa, mw: 68.05.

THR = See sodium hydroxide and ethyl alcohol, into which it readily hydrolyzes.

Fire Hazard: Dangerous; when exposed to heat or flame.

Disaster Hazard: Dangerous; when heated to decomp, emits highly toxic fumes; can react vigorously with oxidizing materials.

SODIUM-2-ETHYLHEXENYL SULFONATE.

THR = U. Limited animal exper suggest LOW toxicity and MOD irr.

Disaster Hazard: Dangerous. See sulfonates.

SODIUM-2-ETHYLHEXYL SULFATE.

$C_8H_{18}O_4S \cdot Na$, mw: 233.3.

Acute tox data: Oral LD_{50} (rat) = 4125 mg/kg; ip LD_{50} (rat) = 320 mg/kg; sc LD_{50} (rat) = 4730 mg/kg; dermal LD_{LO} (guinea pig) = 1520 mg/kg. [*3*]

THR = HIGH via ip and MOD via oral, sc and dermal routes.

SODIUM ETHYL MERCURITHIOSALICYLATE. See mercury compounds, organic.

SODIUM ETHYL SULFATE. Syn: *sodium sulfovinate.* White hygroscopic crystalline material.

$NaC_2H_5SO_4 \cdot H_2O$, mw: 166.14.

THR = See sulfuric acid and ethyl alcohol.

SODIUM ETHYL XANTHATE. Syn: *sodium xanthogenate.* Yellowish powder, sol in water and alcohol. $C_2H_5OCSSNa$, mw: 144.2.

THR = MOD irr via oral, inhal and dermal routes.

Fire Hazard: Mod, when exposed to heat or flame. See sulfides.

Disaster Hazard: Dangerous; when heated to decomp, or on contact with acid or acid fumes, emits highly toxic fumes of SO_x.

SODIUM FERRICYANIDE. Syns: *red prussiate of sodium, red prussiate of soda.* Ruby red, deliquescent crystals. $Na_3Fe(CN)_6 \cdot H_2O$, mw: 298.96.

THR = See ferricyandides.

SODIUM FERROCYANIDE. Syn: *yellow prussiate of soda.* Yellow crystals. $Na_4Fe(CN)_6 \cdot 10H_2O$, mw: 484.1, d: 1.458.

THR = See ferrocyanides. Used as a food additive permitted in the feed and drinking water of animals and/or for the treatment of food producing animals, also permitted in food for human consumption. [*109*]

SODIUM FLUOACETATE. See sodium fluoroacetate.

SODIUM FLUOALUMINATE. See sodium aluminum fluoride.

SODIUM FLUOANTIMONATE. $NaSbF_6$, mw: 258.76, d: 3.375.

THR = See antimony compounds and fluorides.

SODIUM FLUOBERYLLATE. Rhombic or monoclinic white crystals Na_2BeF_4, mw: 131.01, mp: decomp.
THR = See beryllium compounds and fluorides.

SODIUM FLUOBORATE. White rhombic crystals. $NaBF_4$, mw: 109.82, mp: 384° (slight decomp), bp: decomp, d: 2.47 @ 20°.
THR = See fluorides.

SODIUM FLUORIDE. Syn: *villiaumite*. Clear, lustrous crystals or white powder or balls. NaF, mw: 42.00, mp: 993°, bp: 1700°, d: 2 @ 41°, vap. press: 1 mm @ 1077°.
Acute tox data: Oral LD_{50} (rat) = 180 mg/kg; ip LD_{50} (rat) = 22 mg/kg; dermal LD_{LO} (mice) = 300 mg/kg; sc LD_{LO} (dog) = 155 mg/kg; iv LD_{LO} (dog) = 80 mg/kg. [*3*]
THR = HIGH via oral, ip, sc, iv and dermal routes. See fluorides. Doses of 25–50 mg can cause severe vomiting, diarrhea and CNS manifestations. An insecticide. Lethal dose to man is taken as 75–150 mg/kg. It also is very phytotoxic.

SODIUM FLUOROACETATE. Also known as 1080. Fine, white, odorless powder. FCH_2COONa, mw: 100.
Acute tox data: Oral LD_{50} (rat) = 0.22 mg/kg; inhal LC_{50} (rat) = 300 mg/m^3 for 1/6 hr; ip LD_{50} (rat) = 0.8 mg/kg; sc LD_{50} (rat) = 5 mg/kg; im LD_{50} (rat) = 2500 mg/kg; oral LD_{50} (dog) = 0.066 mg/kg; oral LD_{50} (birds) = 2 mg/kg; oral LD_{50} (duck) = 4.81 mg/kg; dermal LD_{50} (rabbit) = 20 mg/kg. [*3*]
THR = HIGH via oral, dermal, inhal, ip, sc, im routes. A very highly toxic water-sol salt used mainly as a rodenticide. It is rapidly absorbed by GI tract but slowly via skin, unless the skin is abraided or cut. It operates by blocking the Krebs cycle via formation of fluorocitric acid, which inhibits aconitase. It has an effect on either or both the cardiovascular and nervous systems in all species and, in some species, the skeletal muscles. Man gives a mixed response with the cardiac feature predominating. By a direct action on the heart, notably in the rabbit, contractile power is lost which leads to declining blood pressure. Ventricular premature contractions and arrhythmias are seen in all species including man. The CNS, notably that of the dog, is directly attacked by sodium fluoroacetate. In man, the action on the CNS produces epileptiform convulsive seizures followed by severe depression.

The dangerous dose for man is 0.5–2 mg/kg. Other species vary considerably in their response to sodium fluoroacetate with primates and birds being the most resistant and carnivora and rodents being the most susceptible. Most domestic animals show a susceptibility falling between the two extremes indicated above.

The first indication of poisoning is nausea and mental apprehension followed by epileptiform convulsions. After a period of several hours, pulsus alternans may exist followed by ventricular fibrillation and death. Children appear to be more subject to cardiac arrest than to ventricular fibrillation.
Treatment and Antidotes: The treatment for sodium fluoroacetate poisoning is mainly symptomatic. Immediate emesis and stomach lavage followed by oral doses of magnesium sulfate are useful. Administration of certain compounds capable of supplying acetate ions has shown antidotal effects in animals, including monkeys; the choice drugs being monoacetin (glycerol monoacetate) (2–4 g/kg) and a combination of sodium acetate and theonol (2 g/kg of each). A single dose of magnesium sulfate (800 mg/kg) given intramuscularly as a 50% solution has saved the life of rats dosed with lethal amounts of sodium fluoracetate. Complete quiet and rest are indicated, but barbiturate anesthesia has proved disappointing when used as an antidote.

SODIUM FLUOSILICATE. See silicon sodium fluoride.

SODIUM FLUOSULFONATE. Shiny, hygroscopic leaflets. $NaSO_3F$, mw: 122.07, mp: decomp @ red heat.
THR = See fluosulfonates.

SODIUM FORMATE. White deliquescent crystals. HCOONa, mw: 68.0, mp: 253°, d: 1.92 @ 20°.
Acute tox data: iv LD_{50} (mice) = 807 mg/kg; oral LD_{LO} (dog) = 4000 mg/kg. [*3*]
THR = MOD via iv and oral routes. See formic acid.
Fire Hazard: Slight.

SODIUM-*m*-GERMANATE. Monoclinic, white, deliquescent crystals. Na_2GeO_3, mw: 166.59, mp: 1083°, d: 3.31 @ 22°.
THR = See germanium compounds.

SODIUM GLUCONATE. White to yellowish crystalline powder, readily sol in water, sparingly sol in alcohol. $NaC_6H_{11}O_7$, mw: 218.
THR = No data. Probably LOW. A sequestrant food additive. [*109*]

SODIUM GUANYLATE. Syns: *disodium guanylate*, *GMP*. A 5′-nucleotide, crystals, sol in cold water, very sol in hot water. $Na_2C_{10}H_{12}N_5O_8P \cdot 2H_2O$, mw: 443.
THR = U. A food additive permitted in food for human consumption. [*109*]
Disaster Hazard: Dangerous. See sulfates and phosphates.

SODIUM HEXAFLUORO PHOSPHATE. White solid. $NaPF_6 \cdot H_2O$ mw: 185.99, d: 2.369 @ 19°.
THR = See fluorides and phosphates.

SODIUM HEXA-*m*-PHOSPHATE. Syns: *hy-phos, calgon.* White powder or flakes, water-sol. $H_6(NaPO_3)_6$, mw: 617.8.

Acute tox data: ip LD_{50} (mice) = 870 mg/kg; iv LD_{LO} (rabbit) = 140 mg/kg. [*3*]

THR = HIGH via iv and MOD via ip routes. See also phosphates.

SODIUM HEXASELENIDE. Gray crystals forms a red solution, Na_2Se_6, mw: 519.8, mp: 258° (decomp).

THR = HIGH. See selenium and selenium compounds. Upon ingestion causes nausea, vomiting and diarrhea. Systemic effects are slight. A sequestrant food additive, it migrates to food from packaging materials. [*109*]

SODIUM HYDRATE. See sodium hydroxide.

SODIUM HYDRAZIDE. NaN_2H_3, mw: 54.

THR = Can explode on contact with air, alcohol, water. [*19*]

SODIUM HYDRIDE. Microcrystalline, white to brownish-gray powder, reacts with water. NaH, mw: 24.00, mp: 800° (decomp), d: 0.9.

THR = HIGH. See sodium hydroxide and hydrides.

Fire Hazard: Mod, when exposed to heat or flame or by chemical reaction with oxidizers.

Explosion Hazard: Mod with (C_2H_2 + moisture), air, Cl_2, F_2, dimethyl sulfoxide, O_2S, H_2O. [*19*]

Disaster Hazard: Dangerous; when heated to decomp, emits highly toxic fumes of oxides of sodium; will react with water or steam to produce heat, sodium hydroxide and H_2; can react with oxidizing materials.

To Fight Fire: Special mixtures of dry chemical.

SODIUM HYDROGEN FLUORIDE. See sodium difluoride.

SODIUM HYDROGEN OXALATE. Monoclinic white crystals. $NaHC_2O_4 \cdot H_2O$, mw: 130.04.

THR = HIGH. See oxalates.

SODIUM HYDROGEN SULFATE. See sodium bisulfate.

SODIUM HYDROSULFITE. See sodium bisulfite.

SODIUM HYDROXIDE. Syns: *caustic soda, sodium hydrate, lye, white caustic.* White deliquescent pieces, lumps or sticks. NaOH, mw: 40.01, mp: 318.4°, bp: 1390°, d: 2.120 @ 20°/4°, vap. press: 1 mm @ 739°.

Acute tox data: ip LD_{50} (mice) = 40 mg/kg; oral LD_{LO} (rabbit) = 500 mg/kg. [*3*]

THR = HIGH via ip and oral routes. This material, both solid and in solution, has a markedly corrosive action upon all body tissue. The symptoms of irr are frequently evident immediately. Its corrosive action on tissue causes burns and frequently deep ulceration, with ultimate scarring. Prolonged contact with dilute solutions has a destructive effect upon tissue. Mists, vapors, and dusts of this compound cause small burns, and contact with the eyes, either in the solid or solution form, rapidly causes severe damage to the delicate tissue. Ingestion either in the solid or solution form causes very serious damage to the mu mem or other tissues with which contact is made. It can cause perforation and scarring. Inhal of the dust or conc mist can cause damage to the upper respiratory tract and to lung tissue, depending upon the severity of the exposure. Thus, effects of inhal may vary from mild irr of the mu mem to a severe pneumonitis. It can cause an irr dermatitis (Section 9). It is a general purpose food additive; it migrates to food from packaging materials. [*109*] Caution: Under the proper conditions of temp., pressure and state of division, it can react violently with acetic acid, acetaldehyde, acetic anhydride, acrolein, acrylonitrile, allyl alcohol, allyl chloride, Al, ClF_3, ($CHCl_3$ + CH_3OH), chlorohydrin, chloronitrotoluenes, chlorosulfonic acid, 1,2-dichloroethylene, ethylene cyanhydrin, glyoxal, HCl, HF, hydroquinone, maleic anhydride, HNO_3, nitroethane, nitromethane, nitroparaffins, nitropropane, pentol, oleum, P, P_2O_5, β-propiolactone, H_2SO_4, (CH_3OH + tetrachlorobenzene), tetrahydrofuran, trichloroethylene, water. [*19*]

Disaster Hazard: Dangerous; will react with water or steam to produce heat and will attack living tissue.

Treatment and Antidotes: Speed in removing this caustic from contact with the skin of one who has come in contact with it is important to avoid injury. Remove all contaminated clothing at once and if possible give patient a shower under deluge type of shower using plenty of water. If the eyes are involved, they should be irrigated at once with plenty of warm water for 15 min. Persons so injured should be referred to a physician.

SODIUM HYDROXIDE, SOLUTION. See sodium hydroxide.

SODIUM HYDROXOPLUMBATE. Light yellow-white, fused, hygroscopic lumps. $Na_2Pb(OH)_6$, mw: 355.25.

THR = See lead compounds.

SODIUM HYDROXOSTANNATE. Hexagonal, colorless–white powder, crystals or lumps. $Na_2Sn(OH)_6$, mw: 166.74, mp: $-3H_2O$ @ 140°.

THR = See tin compounds.

SODIUM HYDROXYL AMINE. $NaONH_2$, mw: 55.

THR = Self-ignites in air. [*19*]

SODIUM-2-HYDROXY-3,6-NAPHTHALENE SULFONATE. See sodium β-naphtholdisulfonate.

SODIUM HYPOBROMITE. NaOBr, mw: 118.9.
THR = Violent reaction with acetone, cupric salts. [*19*]

SODIUM HYPOCHLORITE. NaClO, mw: 74.45, mp: decomp, bp: decomp.
THR = Corrosive and irr via ingestion and inhal. The anhydrous salt is highly explosive. Violent reaction with amines, ammonium acetate, $(NH_4)_2CO_3$, NH_4NO_3, ammonium oxalate, $(NH_4)_3PO_4$, cellulose, ethylene imine. [*19*]

SODIUM HYPONITRITE. Crystals. $Na_2N_2O_2$, mw: 106.01, mp: decomp @ 300°, d: 2.466 @ 4°.
THR = See nitrites.

SODIUM HYPOPHOSPHITE. Colorless, pearly crystalline plates or white granular powder, bittersweet, saline taste. $NaH_2PO_2 \cdot H_2O$, mw: 106.01.
Acute tox data: ip LD_{50} (mice) = 1584 mg/kg. [*3*]
THR = MOD via ip route. See hypophosphites.
Fire Hazard: Mod; see hypophosphites.
Explosion Hazard: Heat causes it to evolve PH_3. See phosphine. It can react violently with chlorates, nitrates. [*19*]
Disaster Hazard: Dangerous; see phosphates.

SODIUM HYPOSULFITE. See sodium bisulfate.

SODIUM INOSINATE. Syn: *disodium inosinate.* A 5'-nucleotide derived from sea tangle (a seaweed) or dried fish. $C_{10}H_{11}Na_2N_4O_8P$, mw: 392.
Acute tox data: iv LD_{LO} (rabbit) = 75 mg/kg. [*3*]
THR = HIGH via iv route. A food additive permitted in food for human consumption. [*109*]
Disaster Hazard: Dangerous. See phosphates.

SODIUM IODATE. Rhombic white crystals. $NaIO_3$, mw: 197.92, mp: decomp, d: 4.277 @ 17.5°.
THR = A trace mineral added to animal feeds. May react violently with Al, As, C, Cu, H_2O_2, metal sulfides, organic matter, P, K, S. [*19*]

SODIUM-5-IODO-2-THIOURACIL.
$C_4H_3ON_2Si \cdot Na$, mw: 277.
THR = An exper neo. [*3*]

SODIUM IODIDE. Cubic colorless crystals. NaI, mw: 149.92, mp: 651°, bp: 1300°, d: 3.667, vap. press: 1 mm @ 767°.
Acute tox data: Oral LD_{50} (rat) = 4340 mg/kg; ip LD_{50} (mice) = 869 mg/kg; iv LD_{LO} (rat) = 1300 mg/kg. [*3*]
THR = MOD via iv, ip and oral routes. See iodides. A trace mineral added to animal feeds. [*109*] Reacts violently with BrF_3, $HClO_4$. [*19*]

SODIUM ISOPROPYL XANTHATE. White deliquescent powder. $SC(OC_3H_7)SNa$, mw: 158.3.
THR = Causes local irr to skin and mu mem. Animal exper suggest LOW toxicity.
Fire Hazard: Mod, by chemical reaction with oxidizers. See also sulfides.
Disaster Hazard: Dangerous. See sulfides.

SODIUM LAURYL SULFATE. Syns: *gardinol, duponol.* White to cream-colored crystals, flakes or powder, water-sol. $C_{12}H_{25}NaO_4S$, mw: 288.4.
Acute tox data: Oral LD_{50} (rat) = 1288 mg/kg. [*3*]
THR = MOD via oral route. A MILD allergen and irr. A food additive permitted in food for human consumption; [*109*] a surfactant.
Disaster Hazard: Dangerous; when heated to decomp, emits highly toxic fumes. See sulfates.

SODIUM-LEAD ALLOY. Metallic material. $(Na)_x + (PB)_y$.
THR = HIGH. See lead compounds and sodium.
Fire Hazard: Mod; reacts with moisture and acids to evolve hydrogen and heat; can react with oxidizing materials.
Explosion Hazard: Mod, by chemical reaction to produce hydrogen.
Disaster Hazard: Dangerous. See lead compounds and sodium. See also hydrogen, which is liberated on contact with moisture.
To Fight Fire: CO_2, dry chemical.

SODIUM LEAD POLYPHOSPHATE. Dense white powder.
THR = See lead compounds and phosphates.

SODIUM MAGNESIUM HYDROXY AMPHIBOLE. Syn: *synthetic asbestos.*
THR = An exper carc via ipl rute. [*103*]

SODIUM MANGANATE. Monoclinic green crystals. $Na_2MnO_4 \cdot 1OH_2O$, mw: 345.08, mp: 17°.
THR = See manganese compounds.
Fire Hazard: Mod, by chemical reaction; an oxidizer.

SODIUM METALLIC, DISPERSION IN ORGANIC SOLVENT. See sodium dispersions.

SODIUM METAPHOSPHATE. Sodium metaphosphate is known as a number of different molecular species, some of which exhibit various crystalline forms. The vitreous sodium phosphates having a Na_2O/P_2O_3 mole ratio near unity are classified as sodium-*m*-phosphates. The term also extends to short-chain vitreous compositions, the compounds of which exhibit the polyphosphate formula $Na_{n+2}P_nO_{3n+1}$ with n as low as 4–5. In such as $(NaPO_3)$, n may be a small integer < 3 (cyclic molecules) or a large number (polymers).
Acute tox data: ip LD_{50} (rat) = 2630 mg/kg. [*3*]
THR = MOD via ip route. A sequestrant food additive. [*109*] See also phosphates.
Disaster Hazard: Dangerous. See phosphates.

SODIUM METHANE ARSONATE. See monosodium methane arsonate.

SODIUM-N-METHYL AMINE ACETATE. See sodium sarcosinate.

SODIUM-2-(N-METHYLAMINO)ETHANE SULFONATE. Syn: *sodium-N-methyl taurine.* Clear colorless liquid. $NH(CH_3)CH_2CH_2SO_3Na$, mw: 161, fp: −28°, d: 1.21 @ 25°/4°.
THR = U. Probably toxic.
Disaster Hazard: Dangerous; see sulfonates.

SODIUM METHYLARSONATE. See sodium methane arsonate.

SODIUM METHYLATE. Syn: *sodium methoxide.* White amorphous free-flowing powder, decomp by water, sol in methyl and ethyl alcohol, decomp in air above 260°F.
THR = A corrosive and irr material. It hydrolyzes into methanol and sodium hydroxide.
Fire Hazard: Violent reation with ($CHCl_3$ + CH_3OH), (methyl azide + dimethylmalonate), $FClO_3$; water causes ignition. [*19*]

SODIUM-N-METHYL DITHIOCARBAMATE DIHYDRATE. See "Vapam."

SODIUM METHYL SULFATE. White hygroscopic crystals, sol in water, alcohol or methanol. $NaCH_3SO_4 \cdot H_2O$, mw: 152.10.
THR = HIGH irr to skin, eyes and mu mem. A food additive permitted in food for human consumption. [*109*]

SODIUM-N-METHYL TAURINE. See sodium 2-(N-methylamino)ethane sulfonate.

SODIUM MOLYBDATE (VI). Crystals. $Na_2MoO_4 \cdot 2H_2O$, mw: 242.
Acute tox data: sc LD_{50} (mice) = 570 mg/kg; ip LD_{50} (mice) = 344 mg/kg. [*3*]
THR = HIGH via ip route and MOD via sc route.

SODIUM MONOFLUOROACETATE. See sodium fluoroacetate.

SODIUM MONOFLUOROPHOSPHATE. Colorless crystals. Na_2PO_3F, mw: 143.97, mp: approx 625°.
THR = See fluorides and phosphates.

SODIUM MONOHYDROGEN-*o*-ARSENATE. Monoclinic colorless crystals. $Na_2HAsO_4 \cdot 7H_2O$, mw: 312.02, mp: 120°–130° (−H_2O @ 180°), d: 1.88.
THR = See arsenic compounds.

SODIUM MONOIODOACETATE. $C_2H_3O_2I \cdot Na$, mw: 208.9.
Acute tox data: Oral LD_{50} (mouse) = 63 mg/kg; ip LD_{50} (mouse) = 75 mg/kg. [*3*]
THR = HIGH via oral and ip routes.

SODIUM MONOSULFIDE. See sodium sulfide.

SODIUM MONOXIDE. White-gray deliquescent crystals. Na_2O, mw: 61.99, bp: 1275° (sublimes), d. 2.27.
THR = HIGH irr to skin, eyes and mu mem. Can react violently with NO above 100°. [*19*] and also with water. See sodium hydroxide.

SODIUM-β-NAPHTHOLDISULFONATE. Syn: *R salt.* Gray paste. $(NaSO_3)_2C_{10}H_5OH$, mw: 348.26.
THR = U. See also β-naphthol.
Disaster Hazard: Dangerous; see sulfonates.

SODIUM NITRATE. Syns: *soda niter, nitratine.* Colorless, transparent, odorless crystals, saline, slightly bitter taste. $NaNO_3$, mw: 85.01, mp: 306.8°, bp: decomp @ 380°, d: 2.261.
Acute tox data: Oral LD_{LO} (rat) = 200 mg/kg. [*3*]
THR = HIGH via oral route.
Explosion Hazard: Explodes when heated to over 1000°F, or when mixed with cyanides, (S + charcoal), sodium hypophosphite, BP, Ba, rhodanide, Sb. [*19*]
Fire Hazard: Mod; when mixed with organic matter, it will ignite on friction. See nitrates.
Disaster Hazard: Dangerous; see nitrates.

SODIUM NITRIDE. Dark gray crystals. Na_3N, mw: 83.00, mp: decomp @ 300°.
THR = HIGH. See sodium hydroxide and ammonia.
Fire Hazard: Mod, by chemical reaction with water.
Explosion Hazard: Contact with air, performic acid. [*19*]
Disaster Hazard: Dangerous; when heated to decomp, emits highly toxic fumes of sodium; will react with water or steam to produce toxic, corrosive and flam vapors.

SODIUM NITRITE. Syn: *diazotizing salts.* Slightly yellowish or white crystals, sticks or powder. $NaNO_2$, mw: 69.01, mp: 271°, bp: decomp @ 320°, d: 2.168.
Acute tox data: oral LD_{50} (rat) = 85 mg/kg; ip LD_{50} (mice) = 158 mg/kg; iv LD_{LO} (dog) = 15 mg/kg; sc LD_{LO} (cat) = 35 mg/kg; oral LD_{LO} (human) = 3 mg/kg. [*3*]
THR = HIGH via oral, ip and sc routes. See nitrites. A food additive permitted in the feed and drinking water of animals and/or for the treatment of food-producing animals; also permitted in food for human consumption. [*109*]
Fire Hazard: Mod; a strong oxidizing agent. In contact with organic matter, will ignite by friction. See nitrites.
Explosion Hazard: Explodes when heated to over 1000°F or on contact with cyanides, NH_4^+ salts, cellulose, Li, (K + NH_3), $Na_2S_2O_3$. [*19*]
Disaster Hazard: Dangerous. See nitrites.

SODIUM NITRITE MIXED (FUSED) WITH POTASSIUM NITRITE. See nitrites.

SODIUM NITRITE MIXTURES (Sodium nitrate, sodium nitrite, and potassium nitrate.)

SODIUM-*m*-NITROBENZENE SULFONATE. Crystals. $Na_2C_6H_4OSO_2Na$, mw: 225.2.
THR = U. See also nitrobenzene.
Disaster Hazard: Dangerous; see sulfonates and nitrites.

SODIUM NITROMETHANE. $NaCH_2NO_2$, mw: 83.3.
THR = U. See also nitroparaffins.
Fire Hazard: Mod, when exposed to heat or flame.
Explosion Hazard: Severe, when shocked or exposed to heat. Moisture can make it explode. [*19*]
Disaster Hazard: Dangerous; shock will explode it; when heated to decomp, emits highly toxic fumes of NO_x; can react with reducing materials.

SODIUM NITROPHENATE. See sodium-*p*-nitrophenoxide.

SODIUM-*p*-NITROPHENOXIDE. Syn: *sodium nitrophenate*. Yellow prisms. $NaOC_6H_4NO_2 \cdot 4H_2O$, mw: 233.16, mp: $-2H_2O$ @ 36°, $-4H_2O$ @ 120°, bp: decomp.
THR = See *p*-nitrophenol, sodium hydroxide and nitrates.

SODIUM-*o*-NITROPHENYL SULFIDE.
$C_6H_4NaNO_2S$, mw: 177.1.
THR = Can explode during preparation. [*19*]

SODIUM NITROPRUSSIDE. Rhombic red crystals. $Na_2Fe(NO)(CN)_5 \cdot 2H_2O$, mw: 298.
Acute tox data: Oral LD_{LO} (rat) = 20 mg/kg; iv LD_{LO} (dog) = 1 mg/kg; iv LD_{LO} (cat) = 1 mg/kg; oral LD_{LO} (rabbit) = 40 mg/kg. [*3*]
THR = HIGH via oral and iv routes. The effects of this material are similar to that of nitrites, causing fall in blood pressure but no formation of methemoglobin. Large amounts, when taken internally, may form cyanide upon being metabolized. See also cyanides.
Disaster Hazard: Dangerous; see cyanides.

SODIUM OLEATE. White powder, slight tallow odor. $C_{17}H_{33}COONa$, mw: 304.5, mp: 232°–235°.
Acute tox data: iv LD_{50} (mice) = 152 mg/kg. [*3*]
THR = HIGH via iv route. Migrates to food from packaging materials. [*109*]
Fire Hazard: Slight, when exposed to heat or flame.

SODIUM OXALATE. White crystalline powder.
$Na_2C_2O_4$, mw: 136.0, d: 2.34.
Acute tox data: sc LD_{LO} (mice) = 100 mg/kg. [*3*]
THR = HIGH via sc and oral routes. See oxalates.

SODIUM OXIDE. See sodium monoxide.

SODIUM OZONIDE. NaO_3, mw: 71.
THR = Violent reaction with acids, water. [*19*]

SODIUM PALMITATE.
THR = U. A substance which migrates to food from packaging materials. [*109*]

SODIUM PECTINATE.
THR = No data. A general purpose food additive. [*109*]

SODIUM PENTABORATE. See sodium borate.

SODIUM PENTACHLOROPHENATE. Syn: *sodium pentachlorophenolate*. Tan powder. C_6Cl_5ONa, mp: 289.3.
Acute tox data: Oral LD_{50} (rat) = 210 mg/kg; sc LD_{50} (rat) = 72 mg/kg; dermal LD_{LO} (mice) = 164 mg/kg; dermal LD_{LO} (rabbit) = 270 mg/kg; it LD_{LO} (rat) = 146 mg/kg. [*3*]
THR = HIGH via oral, sc, it and dermal routes. See pentachlorophenol and chlorides. A fungicide.

SODIUM PENTACHLOROPHENOLATE. See sodium pentachlorophenate.

SODIUM PENTASULFIDE. Yellow crystals. Na_2S_5, mw: 206.32, mp: 251.8°.
THR = See sulfides.

SODIUM PENTABARBITAL. White powder.
$NaC_{11}H_{17}N_2O_3$, mw: 248.26.
Acute tox data: 0.4 mg/kg causes psychotropic effects on humans. Oral TD_{LO} (women) = 60 mg/kg causes CNS effects. Oral LD_{50} (rat) = 200 mg/kg; ip LD_{50} (rat) = 36 mg/kg; sc LD_{50} (rat) = 47 mg/kg; iv LD_{50} (rat) = 65 mg/kg; id LD_{50} (rat) = 39 mg/kg; oral LD_{50} (duck, wild birds) = 75 mg/kg. [*3*]
THR = HIGH via oral, ip, sc, iv, id routes. See also barbiturates.

SODIUM PERACETATE. $Na \cdot C_2H_4O_3$, mw: 99.1.
THR = An exper neo. [*3*]

SODIUM PERBORATE TETRAHYDRATE. Syns: *metaborate peroxyhydrate, sodium perborate tetrahydrate*. White crystals with saline taste, slightly water-sol. $NaBO_3 \cdot 4H_2O$, mw: 153.9, mp: 62°.
Acute tox data: ip LD_{50} (mice) = 538 mg/kg. [*3*]
THR = HIGH–MOD via ip route. See boron compounds and peroxides.
Fire Hazard: Slight, by chemical reaction. An oxidizer. Practically non-hazardous unless mixed with highly combustible or reactive organic compounds.

SODIUM PERBORSILICATE. White powder. Composition: sodium borate, sodium silicate and hydrogen peroxide.
THR = See silicates, boron compounds and peroxides.
Fire Hazard: Slight, by chemical reaction; can react with reducing materials.

SODIUM PERCARBONATE. Decomp in aqueous solution to hydrogen peroxide and sodium carbonate. $Na_2C_2O_6$, mw: 166.
THR = Probably toxic. An irr.
Fire Hazard: Dangerous; a powerful oxidizer.

SODIUM PERCHLORATE. Colorless deliquescent crystals. $NaClO_4$, mw: 122.45, mp: 482° (decomp).
Acute tox data: ip LD_{50} (mice) = 551 mg/kg. [*3*]
THR = MOD via ip route. See perchlorates. Forms an explosive with NH_4NO_3, CaH_2, charcoal, Mg, reducing agents, SrH_2. [*19*]

SODIUM PERIODATE. $NaIO_4$, mw: 213.9.
Acute tox data: ip LD_{50} (mice) = 58 mg/kg. [*3*]
THR = HIGH via ip route. A powerful oxidizer.

SODIUM PERMANGANATE. Purple to reddish-black crystals or powder. $NaMnO_4$, mw: 141.93, mp: decomp.
Acute tox data: iv LD_{LO} (rabbit) = 55 mg/kg. [*3*]
THR = HIGH via iv route. See also manganese compounds.
Fire Hazard: Mod, by chemical reaction; a strong oxidizer.
Disaster Hazard: Dangerous; will react vigorously with combustible materials.

SODIUM PEROXIDE. Syns: *sodium dioxide, sodium superoxide, sodium binoxide.* White powder, turning yellow when heated. Na_2O_2, mw: 77.99, mp: decomp @ 460°, bp: decomp, d: 2.805.
THR = HIGH irr to skin, eyes and mu mem. See sodium hydroxide and peroxides, inorganic.
Fire Hazard: Dangerous, by chemical reaction; a powerful oxidizing agent. See peroxides, inorganic.
Explosion Hazard: Reacts violently with water, acids, powdered metals, acetic acid, acetic anhydride, Al, ($Al + CO_2$), $(NH_4)_2S_2O_8$, aniline, Sb, As, benzene, BN, CaC_2, charcoal, Cu, (KNO_3 + dextrose), ethyl ether, H_2S, glycerine, hexamethylenetetramine, Mg, ($Mg + CO_2$), MnO_2, organic matter, P, K, Se_2Cl_2, (AgCl + charcoal), Na, SCl, Sn, Zn. [*19*]
Disaster Hazard: Dangerous; will react with water or steam to produce heat and toxic fumes; can react vigorously with reducing materials.
To Fight Fire: Carbon dioxide or dry chemical. Combustible materials ignited by contact with sodium peroxide should be smothered with soda ash, salt or dolomite mixtures. Chemical fire extinguishers should not be used. If the fire cannot be smothered, it should be flooded with large quantities of water from a hose.

SODIUM PEROXYCHROMATE. Orange plates. Na_3CrO_8, mw: 249.99, mp: decomp @ 115°.
THR = HIGH irr to skin, eyes and mu mem. A powerful oxidizer. See chromium compounds.
Fire Hazard: Mod, by chemical reaction; a strong oxidizer.
Disaster Hazard: Dangerous; will react vigorously with combustible materials.

SODIUM PERSULFATE. White crystalline powder, sol in water, decomp by alcohol. $Na_2S_2O_8$, mw: 238.13.
Acute tox data: iv LD_{LO} (rabbit) = 178 mg/kg; ip LD_{50} (mice) = 226 mg/kg. [*3*]
THR = HIGH via ip and iv routes.
Fire Hazard: Mod. An oxidizer.
Disaster Hazard: Dangerous; a powerful oxidizer. See sulfates.

SODIUM PHENOBARBITAL. White crystals. $NaC_{12}H_{11}N_2O_3$, mw: 254.22.
Acute tox data: Oral LD_{50} (rat) = 660 mg/kg; ip LD_{50} (rat) = 190 mg/kg; oral LD_{50} (mice) = 200 mg/kg; iv LD_{50} (mice) = 238 mg/kg; oral LD_{50} (rabbit) = 150 mg/kg. [*3*]
THR = HIGH via oral, ip and iv routes. An exper neo via oral route. [*3*] See also barbiturates.

SODIUM PHENOSULFONATE. $C_6H_6O_3S \cdot Na$, mw: 181.2.
Acute tox data: Oral LD_{50} (mice) = 3200 mg/kg. [*3*]
THR = MOD via oral route.
Disaster Hazard: Dangerous; see sulfonates.

SODIUM PHENOXIDE. White, deliquescent, crystalline needles. $NaOC_6H_5$, mw: 116.10.
THR = HIGH. See also phenol and sodium hydroxide. A powerful irr to skin, eyes and mu mem.
Disaster Hazard: Dangerous; when heated to decomp, or on contact with acid or acid fumes, emits highly toxic fumes.

SODIUM PHENYLACETATE. Dry powder. $C_6H_5CH_2COONa$, mw: 158.1.
THR = U. See also phenol.
Disaster Hazard: Dangerous; when heated to decomp, emits toxic fumes.

SODIUM PHENYL GLYCINAMINE-*p*-ARSONATE. See tryparsamide.

SODIUM-*o*-PHENYLPHENATE. Crystals or practically white flakes. $NaOC_6H_4C_6H_5$, mw: 192.2.
Acute tox data: Oral LD_{50} (rat) = 1160 mg/kg. [*3*]
THR = MOD via oral route. A fungicide.
Disaster Hazard: Dangerous; when heated to decomp, emits highly toxic fumes.

SODIUM-*o*-PHENYLPHENOLATE. See sodium-*o*-phenylphenate.

SODIUM PHOSPHATE, DIBASIC. Syns: *DSP, disodium phosphate, sodium orthophosphate (sec), disodium-o-phosphate, disodium hydrogen phosphate.* Colorless, translucent crystals or white powder, sol in

water, very slightly sol in alcohol. Na_2HPO_4, mw: 142.
THR = A nutrient and/or dietary food additive, a general purpose food additive, and a sequestrant food additive. [*109*] A MILD irr.
Acute tox data: ip LD_{LO} (rat) = 1000 mg/kg; sc LD_{LO} (rat) = 1000 mg/kg; im LD_{LO} (rat) = 1000 mg/kg; iv LD_{LO} (rabbit) = 986 mg/kg. [*3*]
THR = MOD via ip, sc, im and iv routes.
Disaster Hazard: Dangerous; see phosphates.

SODIUM PHOSPHATE, MONOBASIC. Syns: *sodium acid phosphate, sodium biphosphate, sodium-o-phosphate (primary), MSP, sodium dihydrogen phosphate.* (a) NaH_2PO_4, white crystalline powder, very sol in water, mw: 120; (b) $NaH_2PO_4 \cdot H_2O$, large translucent crystals, insol in alcohol, mw: 138, mp(b): loses water at 100°, d: 2.040.
Acute tox data: im LD_{50} (rat) = 250 mg/kg. [*3*]
THR = HIGH via im route. MILD irr. A sequestrant and a nutrient and/or dietary supplement food additive. [*109*]
Disaster Hazard: See phosphates.

SODIUM-*o*-PHOSPHATE. See trisodium phosphate.

SODIUM-*o*-PHOSPHATE, PRIMARY. See sodium phosphate, monobasic.

SODIUM-*o*-PHOSPHATE, SECONDARY. See sodium phosphate, dibasic.

SODIUM PHOSPHIDE. Red crystals. Na_3P, mw: 100.0, mp: decomp.
Acute tox data: Inhal LC_{LO} (rat) = 580 ppm for 1 hr; inhal LD_{LO} (cat) = 173 ppm for 2 hrs; inhal LC_{LO} (guinea pig) = 288 ppm for 2 hrs. [*3*]
THR = HIGH via inhal and oral routes. See phosphides. Reacts violently with water to yield phosphine. [*19*]

SODIUM PHOSPHINAMIDE. $NaNHPH_2$, mw: 71.
THR = Spont flam in air. [*19*]

SODIUM PHOSPHOALUMINATE. See sodium aluminum phosphate.

SODIUM PHOSPHOTUNGSTATE.
Acute tox data: Oral LD_{50} (rat) = 1600 mg/kg; oral LD_{50} (mouse) = 700 mg/kg. [*3*]
THR = MOD via oral route. See also tungsten compounds.

SODIUM PICRAMATE. Yellow, water-sol salt. $NaOC_6H_2[NO_2]_2NH_2$, mw: 221.2.
THR = See picric acid and explosives, high.

SODIUM PICRATE, DRY OR WET WITH LESS THAN 10% WATER. Needle-like crystals, mod sol in water and alcohol. $NaC_6H_2O_7 \cdot H_2O$, mw: 269.12, mp: $-H_2O$ @ 150°.
THR = Explodes @ 310°. See picric acid.

SODIUM PICRATE, WET WITH NOT LESS THAN 10% WATER. See sodium picrate.

SODIUM POLONIDE. Radioactive material. Na_2Po, mw: 256.
THR = See polonium.

SODIUM POLYMANNURATE. See sodium alginate.

SODIUM POTASSIUM ALLOYS. See NaK.

SODIUM POTASSIUM TARTRATE. Syns: *rochelle salt, potassium sodium tartrate.* Colorless, transparent, efflorescent crystals or white powder, sol in water, insol in alcohol. $KNaC_4H_4O_6 \cdot 4H_2O$, mw: 282. mp: 70°–80°, d: 1.77.
THR = U. A sequestrant and general-purpose food additive. [*109*]

SODIUM PROPIONATE. Transparent crystals or granules, almost odorless, very sol in water, slightly sol in alcohol. CH_3CH_2COONa, mw: 96 (anhydrous).
Acute tox data: sc LD_{50} (mice) = 2100 mg/kg; dermal LD_{50} (rabbit) = 1640 mg/kg. [*3*]
THR = MOD via sc and dermal routes. An allergen. A chemical preservative food additive. [*109*]

SODIUM-3-PYRIDINE SULFONATE. Yellowish white powder. $C_5H_4NO_2SNa$, mw: 181.2.
THR = U. Probably toxic.
Disaster Hazard: Dangerous; when heated to decomp, or on contact with acid or acid fumes; emits highly toxic fumes of SO_x and NO_x.

SODIUM PYROPHOSPHATE. See tetrasodium pyrophosphate.

SODIUM PYROPHOSPHATE, ACID. See sodium acid pyrophosphate.

SODIUM PYROPHOSPHATE PEROXIDE. White powder $Na_4P_2O_7 \cdot 2H_2O$, mw: 334.
THR = See sodium phosphate and hydrogen peroxide.
Fire Hazard: Mod, by chemical reaction; an oxidizing material; may ignite upon intimate contact with combustible matter.

SODIUM PYROVANADATE. Colorless hexagonal plates. $Na_4V_2O_7$, mw: 305.84, mp: 632°–654°.
Acute tox data: sc LD_{LO} (rat) = 67 mg/kg; iv LD_{LO} (rabbit) = 5 mg/kg. [*3*]
THR = HIGH via iv and sc routes. See vanadium compounds.

SODIUM RHODANATE. See sodium sulfocyanide.

SODIUM SACCHARIN. Syns: *sodium benzosulfimide, gluside, soluble saccharin.* White crystals or crystalline powder, odorless or faint aromatic odor. $C_7H_4NNaO_3S \cdot 2H_2O$, mw: 241.2.
Acute tox data: ip LD_{LO} (mice) = 512 mg/kg. [*3*]

THR = MOD via ip route. An exper carc via imp route. [*3*] See sweeteners. A non-nutrient sweetener food additive. [*109*]

SODIUM SALICYLATE. $C_7H_5O_3 \cdot Na$, mw: 160.1.
Acute tox data: Oral LD_{50} (rabbit) = 1700 mg/kg; oral LD_{50} (rat) = 1600 mg/kg; ip LD_{50} (rat) = 780 mg/kg; iv LD_{50} (mice) = 700 mg/kg. [*3*]
THR = MOD via oral, ip and iv routes. A powerful irr. Affects CNS. An exper teratogen via oral and pa routes. [*3*]

SODIUM SARCOSINATE. Syn: *sodium-N-methyl amine acetate.* Clear liquid. $HN(CH_3)CH_3COONa$, mw: 111.1, mp. d: 3.83.
THR = MOD irr to skin, eyes and mu mem.

SODIUM SARKOMYCIN. $C_7H_8O_3 \cdot Na$, mw: 163.1.
THR = An exper neo via sc route. [*3*]

SODIUM SELENATE. Colorless rhombic crystals. Na_2SeO_4, mw: 188.95, d: 3.098.
Acute tox data: ip LD_{LO} (rat) = 9 mg/kg; iv LD_{50} (rat) = 3 mg/kg; oral LD_{LO} (dog) = 4 mg/kg; pa LD_{LO} (dog) = 4 mg/kg; rec LD_{LO} (dog) = 4 mg/kg. [*3*]
THR = HIGH via rec, pa, oral, iv and ip routes. An exper (±) carc. [*3, 8*] Resembles arsenic in its effects, causes damage to liver and kidneys. See also selenium compounds.

SODIUM SELENIDE. White to red deliquescent crystals. Na_2Se, mw: 124.95, mp: >875°, d: 2.625 @ 10°.
Acute tox data: ip LD_{50} (mice) = 4 mg/kg. [*3*]
THR = HIGH via ip route. See selenium compounds.

SODIUM SELENITE. White crystals. $Na_2SeO_3 \cdot 5H_2O$, mw: 263.04.
Acute tox data: Oral LD_{50} (rat) = 7 mg/kg; oral LD_{50} (mouse) = 7 mg/kg; oral LD_{50} (dog) = 4 mg/kg. [*3*]
THR = An exper (±) carc. [*3, 8*] See selenium compounds.

SODIUM SESQUICARBONATE. Syn: *sesqui.* White needle-shaped crystals, sol in water.
$Na_2CO_3 \cdot NaHCO_3 \cdot 2H_2O$, mw: 226.1, d: 2.112, mp: decomp.
THR = Irr to skin and mu mem. A general-purpose food aditive. [*109*]

SODIUM SESQUISILICATE, ANHYDROUS. Crystals. $Na_6Si_2O_7$, mw: 306.1.
Acute tox data: Oral LD_{50} (rat) = 1300 mg/kg. [*3*]
THR = MOD via oral route See also sodium silicate.

SODIUM SESQUISILICATE, HYDRATED. Crystals. $Na_3HSiO_4 \cdot 5H_2O$, mw: 252.2.
Acute tox data: Oral LD_{50} (rat) = 1600 mg/kg. [*3*]
THR = MOD via oral route. See also sodium silicate.

SODIUM SILICATE. Na_2SiO_3, mw: 122.1.
Acute tox data: Oral LD_{50} (rat) = 1280 mg/kg; ip LD_{LO} (guinea pig) = 200 mg/kg. [*3*]
THR = HIGH via ip route and MOD via oral route. A caustic material, irr to skin and mu mem. Ingestion causes GI upset. A substance which migrates to food from packaging materials. [*109*] Violent reaction with F_2. [*19*]

SODIUM SILICIDE. Na_2Si_2, mw: 102.1.
THR = Self-ignites in air or water. [*19*] See also sodium and silicon.

SODIUM SILICOFLUORIDE. See sodium fluosilicate.

SODIUM SORBATE. $CH_3CH{:}CHCH{:}CHCOONa$, mw: 134.
Acute tox data: Oral LD_{LO} (rat) = 4000 mg/kg. [*3*]
THR = MOD via oral route. A chemical preservative food additive. It migrates to food from packaging materials. [*109*]

SODIUM STEARYL FUMARATE.
THR = U. A food additive permitted in food for human consumption. [*109*]

SODIUM SUCARYL.
Acute tox data: iv LD_{LO} (mice) = 1220 mg/kg; iv LD_{LO} (rabbit) = 4470 mg/kg. [*3*]
THR = MOD via iv route. See sodium cyclamate.

SODIUM SULFACHLOROPYRAZINE MONOHYDRATE.
THR = U. A food additive permitted in the food and drinking water of animals, and/or for the treatment of food-producing animals. [*109*]

SODIUM-N-(5-SULFAMOYL-1,3,4-THIADIAZOL-2-YL)-ACETAMIDE. $C_4H_6O_3N_4S_2 \cdot Na$, mw: 245.3.
THR = HIGH via oral route to hamsters. It affects pregnancy. A (S) teratogen. [*3*]

SODIUM SULFATE, ANHYDROUS. Syn: *sodium sulfate, exsiccated.* White crystals or powder, odorless, sol in water and glycerol, insol in alcohol. Na_2SO_4, mw: 142., d: 2.671, mp: 888°.
THR = A substance which migrates to food from packaging materials. [*109*] Reacts violently with Al. [*19*]
Disaster Hazard: See sulfates.

SODIUM SULFATE, EXSICCATED. See sodium sulfate, anhydrous.

SODIUM SULFHYDRATE. See sodium hydrosulfide.

SODIUM SULFIDE. Syn: *sodium monosulfide.* Amorphous, yellow-pink or white deliquescent crystals. Na_2S, mw: 78.06, mp: 1180°, d: 1.856 @ 14°.
Acute tox data: iv LD (rabbit) = 6 mg/kg. [*3*]
THR = HIGH via iv route. See also sulfides. Reacts violently with C, diazonium salts, *n,n*-dichloro-

methylamine, *o*-nitroaniline diazonium salt, water. [*19*] This material is unstable and can explode on rapid heating or percussion.

SODIUM SULFIDE NONAHYDRATE. $Na_2S \cdot 9H_2O$, mw: 240.2.

Acute tox data: ip LD_{50} (mice) = 53 mg/kg. [*3*]

THR = HIGH via ip and oral routes. See also sulfides.

SODIUM SULFITE. Hexagonal prisms or white powder. Na_2SO_3, mw: 126.06, bp: decomp, d: 2.633 @ 15.4°.

Acute tox data: iv LD_{50} (mice) = 175 mg/kg; oral LD_{LO} (rabbit) = 1181 mg/kg; in humans an oral TD_{LO} = 6 mg/kg causes CNS effects. [*3*]

THR = HIGH via iv and MOD via oral route. See sulfites. A chemical preservative food additive. [*109*]

SODIUM SULFOCYANATE. See sodium sulfocyanide.

SODIUM SULFOCYANIDE. Syns: *sodium sulfocyanate, sodium rhodanate, sodium rhodanide, sodium thiocyanate.* Colorless deliquescent crystals or white powder. NaCNS, mw: 81.08, mp: 287°.

Acute tox data: Oral LD_{50} (rat) = 764 mg/kg; ip LD_{50} (rat) = 540 mg/kg; iv LD_{50} (mice) = 484 mg/kg; sc LD_{LO} (rabbit) = 500 mg/kg. [*3*]

THR = MOD via oral, ip and sc; HIGH via iv routes. Large doses taken internally cause vomiting and convulsions. Chronic poisoning is manifested by weakness, confusion, diarrhea and skin rashes.

SODIUM SULFOVINATE. See sodium ethylsulfate.

SODIUM SUPEROXIDE. NaO_2, mw: 55.

THR = Decomp violently @ >250° to evolve O_2. Also reacts violently with water. See sodium peroxide for toxicity effects.

SODIUM TARTRATE. Syns: *sal tartar, disodium tartrate.* White crystals or granules, sol in water, insol in alcohol. $Na_2C_4H_4O_6 \cdot 2H_2O$, mw: 207.2, d: 1.82, loses $2H_2O$ @ 120°.

Acute tox data: Oral LD_{50} (rat) = 1290 mg/kg; sc LD_{LO} (cat) = 3000 mg/kg; oral LD_{50} (mice) = 4360 mg/kg. [*3*]

THR = MOD via oral and sc routes. A sequestrant food additive. [*109*]

SODIUM TELLURATE (VI). Hexagonal plates or white powder. $Na_2TeO_4 \cdot 2H_2O$, mw: 273.64, mp: decomp.

Acute tox data: Oral LD_{50} (rat) = 385 mg/kg; oral LD_{50} (mice) = 165 mg/kg; iv LD_{LO} (rat) = 31 mg/kg; ip LD_{LO} (rat) = 37 mg/kg. [*3*]

THR = HIGH via oral, ip and iv routes. See tellurium compounds.

SODIUM TELLURATE (IV). $Na_2H_2TeO_3$, mw: 223.6.

Acute tox data: Oral LD_{50} (rat) = 83 mg/kg; ip LD_{LO} (rat) = 4 mg/kg; iv LD_{LO} (rat) = 0.4 mg/kg. [*3*]

THR = HIGH via oral, ip and iv routes. See tellurium compounds.

SODIUM TETRAZOLYL-5-AZIDE. $NaCN_7$, mw: 133.1.

THR = Detonates by percussion. [*19*]

SODIUM TETRABORATE. Syn: *sodium borate.* White crystals. $Na_2B_4O_7$, mw: 201.27, mp: 741°, bp: 1575° (decomp), d: 2.367.

Acute tox data: Oral LD_{LO} (infant human) = 1700 mg/kg; oral LD_{LO} (human) = 214 mg/kg; oral LD_{50} (rat) = 2660 mg/kg. [*3*]

THR = MOD via oral route. See also boron compounds.

SODIUM-2,3,4,6-TETRACHLOROPHENATE. See sodium-2,3,4,6-tetrachlorophenol.

SODIUM-2,3,4,6-TETRACHLOROPHENOL. Syns: *sodium-2,3,4,6-tetrachlorophenate, sodium-2,3,4,6-tetrachlorophenolate.* Buff to light brown flakes. C_6HCl_4ONa, mw: 253.9, vap. d: 9.4.

THR = All chlorophenols are toxic. See also phenol and chlorinated phenols.

Disaster Hazard: Dangerous. See chlorophenols.

SODIUM-2,3,4,6-TETRACHLOROPHENOLATE. See sodium-2,3,4,6-tetrachlorophenol.

SODIUM TETRAHYDROALUMINATE. $NaAlH_4$, mw: 54.

THR = Violent reaction with water, tetrahydrofuran. [*19*]

SODIUM TETRAHYDROBORATE. See sodium borohydride.

SODIUM TETRAHYDROGEN-*o*-TELLURATE. See sodium tellurate (VI).

SODIUM TETRAPHENYL BORATE.

THR = U. See boron compounds.

SODIUM TETRASULFIDE. Yellow, cubic, hygroscopic crystals. Na_2S_4, mw: 174.26, mp: 275°, bp: decomp.

THR = HIGH. See sodium sulfide.

SODIUM THIOARSENATE. Monoclinic, yellow crystals. $Na_3AsS_4 \cdot 8H_2O$, mw: 416.29, mp: decomp.

THR = See arsenic compounds.

SODIUM THIOCYANATE. See sodium sulfocyanide.

SODIUM THIOGLYCOLATE. Hygroscopic crystals. $HSCH_2COONa$, mw: 114.1.

Acute tox data: ip LD_{50} (rat) = 140 mg/kg; iv LD_{LO} (rabbit) = 100 mg/kg; ip LD_{LO} (mice) = 200 mg/kg. [*3*]

THR = HIGH via ip and iv routes. See sulfides. This material yields hydrogen sulfide on decomp. The literature contains the report of a death attributed to the absorption of toxic decomp products from the use of this material in a permanent waving solution.

Disaster Hazard: Dangerous. See sulfides.

SODIUM THIOSULFATE PENTAHYDRATE. Syn: *hypo*. Monoclinic, colorless, odorless crystals. $Na_2S_2O_3 \cdot 5H_2O$, mw: 248.2, mp: 48° (rapid heating), d: 1.69.

Acute tox data: iv LD_{LO} (dog) = 3000 mg/kg; oral TD_{LO} (human) = 300 mg/kg for 7 days ⟶ blood effects. [*3*]

THR = HIGH via oral and mod via iv routes. See thiosulfates. Large doses internally have a cathartic action. A sequestrant food additive. It migrates to food from packaging materials. [*109*] Violent reaction with $NaNO_2$. [*19*]

Disaster Hazard: Dangerous. See thiosulfates.

SODIUM TIN CITRATE.

Acute tox data: Oral LD_{50} (mice) = 2700 mg/kg. [*3*]

THR = MOD via oral route. See also tin compounds.

SODIUM-*p*-TOLUENE SULFON CHLORAMINE. See chloramine-T.

SODIUM TRICHLOROACETATE. Crystals, water sol, $Cl_3CCOONa$, mw: 185.4.

Acute tox data: Oral LD_{50} (rat) = 3320 mg/kg; oral LD_{50} (mice) = 4970 mg/kg; iv LD_{50} (mice) = 2370 mg/kg. [*3*]

THR = MOD via oral and iv routes. Large doses cause CNS depression. An herbicide.

Disaster Hazard: Dangerous. See chlorides.

SODIUM-2,4,5-TRICHLOROPHENOLATE. Buff to brown flakes. $Cl_3C_6H_2ONa \cdot 1\frac{1}{2}H_2O$, mw: 246.4, vap. d: 8.5.

Acute tox data: Oral LD_{50} (rat) = 1620 mg/kg. [*3*]

THR = MOD via oral route. Used as a fungicide. See chlorophenols.

Disaster Hazard: Dangerous; when heated to decomp, emits highly toxic fumes of chlorides and oxides of sodium.

SODIUM TRIMETHOXY BOROHYDRIDE. Microcrystalline, fine white powder and white to grayish lumps. $NaBH(OCH_3)_3$, mw: 127.9, mp: 230° (decomp and evolves methyl borate), d: 1.24.

THR = See boron compounds and hydrides.

SODIUM TRIPERCHROMATE. $Na_3Cr(O_2)_4$, mw: 249.

THR = Explodes @ 115°. [*19*]

SODIUM TRI-*m*-PHOSPHATE. $(NaPHO_3)_3$, mw: 308.9.

Acute tox data: ip LD_{50} (rat) = 3650 mg/kg; iv LD_{LO} (rabbit) = 240 mg/kg. [*3*]

THR = HIGH via iv and MOD via ip routes.

SODIUM TRIPOLYPHOSPHATE. $P_3H_5O_{10}Na_5$, mw: 372.9.

Acute tox data: ip LD_{50} (rat) = 525 mg/kg. [*3*]

THR = MOD–HIGH via ip route. Ingestion of large doses of sodium phosphates causes catharsis. Sodium *m*- and pyrophosphates can cause hemorrhages from the intestine if taken internally in large doses. A sequestrant and general purpose food additive. It migrates to food from packaging materials. [*109*]

SODIUM TUNGSTATE. White rhombic crystals Na_2WO_4, mw: 293.91, mp: 698°, d: 4.179.

Acute tox data: Oral LD_{50} (rat) = 1190 mg/kg; oral LD_{50} (mice) = 240 mg/kg; im LD_{50} (rabbit) = 105 mg/kg; oral LD_{50} (rabbit) = 875 mg/kg; oral LD_{50} (guinea pig) = 1152 mg/kg. [*3*]

THR = HIGH via im and MOD via oral routes. See also tungsten compounds.

SODIUM URANATE. Syn: *sodium-m-uranate*. Gray, yellow or red plates or rhombic prisms or powder. Na_2UO_4, mw: 348.06.

THR = See uranium.

SODIUM URANYL ACETATE. See uranium sodium oxyacetate.

SODIUM URANYL CARBONATE. Yellow crystals. $2Na_2CO_3 \cdot UO_2CO_3$, mw: 542.09, mp: decomp @ 400°.

THR = See uranium.

SODIUM-*m*-VANADATE. Colorless, monoclinic, prismatic crystals or pale green crystalline powder. $NaVO_3 \cdot 4H_2O$, mw: 194.0, mp: 630°.

Acute tox data: Oral LD_{LO} (rat) = 200 mg/kg; iv LD_{LO} (rat) = 10 mg/kg. [*3*]

THR = HIGH via ip and oral routes. See vanadium compounds.

SODIUM-*o*-VANADATE. Colorless, hexagonal prisms. Na_3VO_4, mw: 183.94, mp: 850°–866°.

Acute tox data: sc LD_{LO} (rat) = 50 mg/kg; iv LD_{LO} (rabbit) = 2 mg/kg; sc LD_{LO} (mice) = 50 mg/kg. [*3*]

THR = HIGH via sc and iv routes. See vanadium compounds.

SODIUM XANTHOGENATE. See sodium ethyl xanthate.

SODIUM XENATE.

Acute tox data: iv LD_{50} (mice) = 15 mg/kg. [*3*]

THR = HIGH via iv route.

SODIUM ZINC URANYL ACETATE. Tablets or monoclinic crystals. $NaZn(UO_2)_3(C_2H_3O_2)_9 \cdot 9H_2O$, mw: 1447.6.

THR = See uranium.

SODIUM ZIRCONIUM LACTATE. See zirconium sodium lactate.

SOLAN. See 3′-chloro-2-methyl-*p*-valerotoluidide.

SOLDERS. Ductile, relatively low melting alloys. Pb + Sn (from 5–81.5% Pb). d: 7.5–10.2 (depending on composition), mp: 185°–232°.
THR = Fumes may be toxic on repeated inhal.

SOLUBLE SACCHARIN. See sodium saccharin.

SOLVENT EDM. See ethylene dichloride, carbon tetrachloride.

SOMAN. Syn: *fluoromethylpinacolyloxy phosphine oxide.* $(CH_3)_3CCH(CH_3)OP(CH_3)(O)(F)$, mw: 182.2, bp: 167°, fp: −70°, d: 1.026 @ 20°, vap. d: 6.33, vap. press: 0.207 mm @ 20°.
Acute tox data: Inhal LC_{LO} (human) = 70 mg/m^3; dermal LD_{LO} (human) = 18 mg/kg; sc LD_{50} (mouse) = 0.14 mg/kg; iv LD_{LO} (mouse) = 0.066 mg/kg. [*3*]
THR = VERY HIGH via all routes of exposure. Est LD for man = 0.01 mg/kg. Hydrofluoric acid is a product of hydrolysis. Very dangerous to eyes. Extremely toxic upon skin absorption; it penetrates the skin rapidly but does not injure it. Death may occur after 15 min of exposure. It acts like tabun, but faster and in lower conc. A very potent cholinesterase inhibitor. See also parathion.
Fire Hazard: Slight, when exposed to heat or flame.
Disaster Hazard: Highly dangerous; when heated to decomp, emits highly toxic fumes of PO_x and fluorides; will react with water, steam or acids to produce toxic and corrosive fumes; can react with oxidizing materials.
To Fight Fire: Foam, CO_2, dry chemical.

SOOT. A dark brown to black powdery material. Composition: carbon.
THR = Soot is an obstructive and irr dust, which is often carc. It has caused skin cancer. This property may be due to some coal tar product which adheres to the soot rather than the soot itself. The commercial product in the form of dust has caused cancer of nasal sinuses and lungs. [*14*] A common air contaminant.
Fire Hazard: Slight, when exposed to heat or flame.

SORBIC ACID. Syn: *2,4-hexadienoic acid.* Colorless needles. $CH_3CHCHCHCHCOOH$, mw: 112.12, bp: 228° (decomp), mp: 134.5°, flash p: 260°F (COC), vap. press: 0.01 mm @ 20°, vap. d: 3.87.
Acute tox data: Oral LD_{50} (rat) = 11,000 mg/kg; uk LD_{LO} (human) = 5000 mg/kg. [*3*]
THR = LOW via oral route. An exper carc via sc route. [*3, 23*] A chemical food preservative. [*109*]
Fire Hazard: Slight, when exposed to heat or flame; can react with oxidizing materials.
To Fight Fire: Water.

SORBITAN MONOLAURATE. A (S) carc. [*14*] See surfactants.

SORBITAN MONOLEATE. A (S) carc. [*14*] See surfactants.

SORBITAN MONOSTEARATE. Cream-colored waxy solid; slight odor. d: 1.0 @ 25°, mp: 54°.
THR = U. A food additive permitted in the feed and drinking water of animals and/or for the treatment of food-producing animals; also permitted in food for human consumption. [*109*]

***d*-SORBITE.** See sorbitol.

SORBITOL. Syns: *glucitol, d-sorbite, hexahydric alcohol.* White crystalline powder, odorless, sol in water, slightly sol in methanol, ethanol, acetic acid, phenol and acetamide, almost insol in other organic solvents. $C_6H_8(OH)_6$, mw: 182, d: 1.47 @ −5°, mp: 93° (metastable form); 97.5° (stable form), bp: 105°
THR = LOW. A nutrient and/or dietary supplement food additive; it migrates to food from packaging materials. [*109*]

SORBOSE. White crystalline powder, sol in water, slightly sol in ethyl or isopropyl alcohol, insol in ether, acetone, benzene and chloroform.
$HOCH_2CO(CHOH)_3CH_2OH$, mw: 180, mp: 165°.
THR = U. A substance which migrates to food from packaging materials. [*109*]

SORSAKA.
THR = An exper neo via sc route. [*3*]

SOTERENOL HYDROCHLORIDE.
$C_{12}H_{20}O_4N_2S \cdot HCl$, mw: 324.9.
THR = An exper neo and teratogen. [*3*]

SOUP. See nitroglycerin

SOYBEAN OIL. Syns: *soya bean oil, chinese bean oil.* Pale yellow to brownish-yellow liquid. Composition: glycerides of fatty acids. mp: 22.2°, flash p: 540°F (CC), autoign. temp.: 833°F, d: 0.925.
THR = No data. A substance which migrates to food from packaging materials. [*109*]
Fire Hazard: Slight, when exposed to heat or flame.
Spont Heating: Mod.
To Fight Fire: CO_2, dry chemical.

SPANISH FLY. See cantharides.

"SPAN." A co-carc. [*14, 23*]

SPARTEINE. Colorless oil. $C_{15}H_{26}N_2$, mw: 234.43, bp: 173° @ 8 mm; 180°–181° @ 20 mm in air, d: 1.020 @ 20°/4°.
Acute tox data: ip LD_{50} (rat) = 42 mg/kg; sc LD_{LO}

(mice) = 120 mg/kg; iv LD_{LO} (rabbit) = 30 mg/kg. [*3*]
THR = HIGH via iv, ip and sc routes.
Disaster Hazard: Dangerous; when heated, emits highly toxic fumes of NO_x.

SPEARMINT OIL. Syns: *oil of crisp mint*, *curled mint*. Colorless, yellow or greenish-yellow liquid.
Composition: 50% carvone, l-limonene, pinene. d: 0.917–0.934 @ 25°/25°.
THR = A MILD allergen.
Fire Hazard: Slight.

SPECIAL FIREWORKS. See explosives, low.

SPENT ACID. See mixed acid.

SPENT OXIDE. Syn: *gashouse tankage*. Composition: iron sponge + 5–10% nitrogen.
THR = U. Can contain toxic materials.
Fire Hazard: Mod, when exposed to heat or flame; can react with oxidizing materials.

SPENT SULFURIC ACID. See sulfuric acid.

SPERMACETI. A wax. mp: 42°-50°, d: 0.94.
THR = A MILD allergen.
Disaster Hazard: Slight, when heated, emits acrid fumes.

SPERM OIL. flash p: 428°F, d: 0.875–0.884 @ 25°/25°, autoign. temp.: 586°F, flash p(no. 2): 460°F.
THR = U. See thermic and oxidation products.
Fire Hazard: Slight, when exposed to heat or flame.
To Fight Fire: CO_2, dry chemical.

SPERRYLITE. See platinum arsenide.

SPHEROCOBALTITE. See cobaltous carbonate.

SPICES AND OTHER NATURAL SEASONINGS AND FLAVORINGS (LEAVES, ROOTS, BARKS, BERRIES, ETC.) See food additives. [*109*]

SPINDLE OIL. See lubricating oil.

SPIRIT OF GLYCERYL TRINITRATE. See spirits of nitroglycerin.

SPIRIT OF HARTSHORN. See aromatic spirits of ammonia.

SPIRIT OF NITER (SWEET) Syn: *ethyl nitrite spirit*. Pale straw-colored liquid, fragrant, pungent odor, burning taste.
Composition: alcohol solution: 3.5–4.5%. $C_2H_5NO_2$, d: not over 0.823 @ 25°.
THR = A MILD irr. See also nitrites.
Fire Hazard: Mod, when exposed to heat or flame.
Explosion Hazard: Mod, when vapors are exposed to heat or flame.
Disaster Hazard: Dangerous; when heated to decomp, emits highly toxic fumes of NO_x; can react with oxidizing materials.

SPIRIT OF NITROGLYCERIN. Syn: *spirit of glyceryl trinitrate*. Clear, colorless liquid. Composition: 1.0–1.1% glycerol trinitrate in alcoholic solution. d: 0.814–0.820 @ 25°.
THR = MOD via oral and inhal route. See also nitroglycerin.
Fire Hazard: Dangerous, when exposed to heat or flame. See also ethyl alcohol.
Explosion Hazard: See nitroglycerine and ethyl alcohol. If the alcohol evaporates, the residue is nitroglycerin.
Disaster Hazard: Dangerous; when dried out, shock will explode it; when heated to decomp, emits highly toxic fumes; on contact with oxidizing materials the mixture can react vigorously.

SPIRIT OF TURPENTINE. See turpentine oil.

SPIRIT OF WINE. See ethyl alcohol.

SPIROFULVIN. $C_{17}H_{17}O_6Cl$, mw: 352.8.
THR = An exper (+) carc, neo and teratogen via oral route. [*3, 7*]

SPONT COMBUSTION IN SOLID WASTES. See Section 6.

SPRAY COATING OF CONSTRUCTION AND INSTALLATION. See Section 2.

SPRAY COATING DESIGN. See Section 2.

SPRAY COATING FAN AND PIPING. See Section 2.

SPRAY COATING MAINTENANCE. See Section 2.

SPECIALTY SKILLS IN INDUSTRIAL HYGIENE. See Section 1.

SPRENGEL EXPLOSIVES. This type of explosive is a mixture of nitrobenzene and fuming nitric acid. It is a powerful and cheap explosive and would have many uses except that it is limited by practical disadvantages. The components have to be mixed in glass shortly before the explosive is used. This requires preparation and equipment not always available at the site of the explosion. This material can be destroyed by throwing it into large quantities of water, or possibly by burning in small quantities at a time. See explosives, high.

SQUILL RED. Syn: *sea onion*.
Acute tox data: Oral TD_{LO} (man) = 1414 mg/kg. [*3*]
THR = MOD via oral route. Large doses cause vomiting and cardiac depression.

STABILIZING FOOD ADDITIVE. [*109*]

STAGGER WEED. See delphinium.

STAINLESS STEEL SELENIUM ALLOYS. See selenium compounds.

STANNANE. See tin tetrahydride.

STANNIC-BIS-ACETYLACETONE DIBROMIDE. Crystals, sol in benzene, chloroform and acetone. $(C_5H_7O_2)_2SnBr_2$, mw: 476.8, bp: 187°.
THR = See tin compounds and bromides.

STANNIC-BIS-ACETYLACETONE DICHLORIDE. Crystals, water-sol. $(C_5H_7O_2)_2SnCl_2$, mw: 387.8, bp: 203°.
THR = See tin compounds and chlorides.

STANNIC-BIS-BENZOYL ACETONE DIBROMIDE. Yellow powder, slightly sol in organic solvents. $(C_{10}H_9O_2)_2SnBr_2$, mw: 608, bp: 214°.
THR = See tin compounds and bromides.

STANNIC-BIS-BENZOYL METHANE DIBROMIDE. Yellow crystals, insol in water. $(C_{15}H_{10}O_2)_2SnBr_2$, mw: 723.0, bp: 278°.
THR = See tin compounds and bromides.

STANNIC-BIS-3-ETHYL ACETYL ACETONE DIBROMIDE. Colorless crystals, sol in chloroform and benzene. $(C_7H_{11}O_2)_2SnBr_2$, mw: 532.9, mp: 166°.
THR = See tin compounds and bromides.

STANNIC BROMIDE. Syns: *tin bromide, tin tetrabromide.* White crystalline mass. $SnBr_4$, mw: 438.36, mp: 31°, bp: 202°, d(liquid): 3.340 @ 35°, vap. press: 10 mm @ 72.7°.
THR = See bromides and tin compounds. Violent reaction with ONOCl. [*19*]

STANNIC CHLORIDE. Syns: *tin chloride, tin tetrachloride.* Colorless, fuming caustic liquid or crystals. $SnCl_4$, mw: 260.53, mp: −33°, bp: 114.1°, d: 2.232, vap. press: 10 mm @ 10°.
Acute tox data: ip LD_{50} (mice) = 46 mg/kg. [*3*]
THR = HIGH via ip route. Corrosive. See hydrochloric acid.
Fire Hazard: Slight, by chemical reaction. Upon contact with moisture, considerable heat is generated. Violent reaction with K, Na, turpentine. [*19*]
Disaster Hazard: Dangerous; hydrochloric acid is liberated on contact with moisture or heat.

STANNIC CHROMATE. Brownish-yellow crystalline powder. $Sn(CrO_4)_2$, mw: 350.72, mp: decomp.
THR = See chromium compounds.

STANNIC IODIDE. Red cubic crystals. SnI_4, mw: 626.3. mp: 144.5°, bp: 364°, d: 4.473 @ 0°.
Acute tox data: iv LD_{LO} (rat) = 200 mg/kg. [*3*]
THR = HIGH via iv route. Strong reaction with NO_2Cl, (K + S), (Na + S). [*19*] See also tin compounds and iodides.

STANNIC OXIDE (CASSITERITE). See tin oxide.

STANNIC PHOSPHIDE. See tin monophosphide.

STANNIC SULFIDE. Syns: *artificial gold, tin bisulfide, mosaic gold, tin bronze, tin disulfide.* Yellow to brown powder, sol in conc hydrochloric acid, aqua regia, solutions of alkali hydroxides or sulfides, insol in water or dilute acids. SnS_2, mw: 183.82, d: 4.5, mp: decomp @ 600°.
THR = MILD irr. See also tin compounds and sulfides. Violent reaction with $Cd(ClO_3)_2$, $HClO_3$, ClO, $Mg(ClO_3)_2$, $Zn(ClO_3)_2$. [*19*]
Disaster Hazard: Dangerous; see sulfides.

STANNOUS ACETATE. White to yellowish powder, decomp by water. $Sn(C_2H_3O_2)_2$, mw: 236.79, mp: 182°.
THR = A MILD irr. See also tin compounds.

STANNOUS CHLORIDE. Syns: *tin crystals, tin salt, tin dichloride, tin protochloride.* Colorless crystals, sol in less than its own weight of water, very sol in hydrochloric acid (dilute or conc), sol in alcohol, ethyl acetate, glacial acetic acid and sodium hydroxide solution. $SnCl_2 \cdot 2H_2O$, mw: 225.65, d: 2.71, mp: 37°–38°.
Acute tox data: Oral LD_{50} (rat) = 700 mg/kg; oral LD_{50} (mice) = 1200 mg/kg; ip LD_{50} (mice) = 66 mg/kg; iv LD_{LO} (dog) = 20 mg/kg. [*3*]
THR = HIGH via oral route. A chemical preservative food additive. [*109*] Violent reactions with BrF_3, CaC_2, ethylene oxide, hydrazine hydrate, nitrates, K, Na. [*19*]
Disaster Hazard: See chlorides.

STANNOUS FLUORIDE. See tin fluoride.

STANNOUS NITRATE. $Sn(NO_3)_2$, mw: 242.7.
THR = May explode with flour dust. [*19*]

STANNOUS OXIDE. See tin oxide.

STANNOUS SULFIDE. SnS, mw: 150.8.
THR = HIGH. See sulfides. Very violent reactions with $Cd(ClO_3)_2$, $HClO_3$, ClO, $Mg(ClO_3)_2$, $Zn(ClO_3)_2$. [*19*]

STARCH DUST.
THR = A nuisance dust. An allergen.
Fire Hazard: Mod, when exposed to flame, can react with oxidizing materials.
Explosion Hazard: Mod, when exposed to flame.

STARCH GUM. See dextrin.

STEARAMIDO PROPYLIDIMETHYL-β-HYDROXY ETHYLAMMONIUM PHOSPHATE. Syn: "*Catanac*" *SP antistatic agent.*
Acute tox data: Oral LD_{50} (rat) = 2835 mg/kg. [*3*]
THR = MOD via oral route. A quaternary ammonium compound. Large doses by mouth cause vomiting and diarrhea. See also benzalkonium chloride.

STEARIC ACID. Syn: *octadecanoic acid.* White amorphous solid. $CH_3(CH_2)_{16}COOH$, mw: 284.47, mp: 69.3°, bp: 383°, flash p: 385°F (CC), d: 0.847, autoign. temp.: 743°F, vap. press: 1 mm @ 173.7°, vap. d: 9.80.

Acute tox data: iv LD_{50} (rat) = 22 mg/kg; iv LD_{50} (mouse) = 23 mg/kg. [*3*]

THR = HIGH via iv route. A substance which migrates to food from packaging materials. [*109*]

Fire Hazard: Slight, when exposed to heat or flame.

Spont Heating: Yes.

To Fight Fire: CO_2, dry chemical.

STEAROYL AZIRIDINE. $C_{20}H_{39}ON$, mw: 309.6.

THR = An exper neo via sc route. [*3, 23*]

STEAROYL PROPYLENE GLYCOL HYDROGEN SUCCINATE. See succistearin.

STEARYL ALCOHOL. Syn: *1-octadecanol.* Colorless solid or flakes. $C_{18}H_{37}OH$, mw: 270.5, mp: 58°, bp: 202° @ 10 mm, d: 0.8124 @ 59°/4°.

THR = Practically non-toxic. See alcohols.

Fire Hazard: Mod, when exposed to heat or flame; can react with oxidizing materials.

To Fight Fire: Foam, CO_2, dry chemical.

STEARYL CITRATE.

THR = A sequestrant food additive. [*109*]

STEATITE. See talc.

STEEL WOOL. It is composed of long curls of steel wire. Sometimes contains soap to aid in cleaning.

THR = See soap and iron. Reacts violently with H_2SO_4, $HClO_4$. [*19*]

STERIGMATOCYSTIN. $C_{18}H_{12}O_6$, mw: 324.3.

THR = An exper (+) carc via oral, dermal and sc routes. [*3, 9, 23*]

STIBENYL. See sodium-*p*-acetylamino phenyl antimonate.

STIBINE. See antimony hydride.

STIBIUM. See antimony.

4-STILBENAMINE. $C_{14}H_{13}N$, mw: 195.3.

THR = An exper carc and neo via sc and oral routes. [*3, 23*]

STILBENE. See diphenyl ethylene.

STILBESTROL. See diethyl stilbestrol.

STILBESTROL DIMETHYL ETHER. $C_{20}H_{24}O_2$, mw: 296.4.

THR = An exper neo. [*3*]

STILBESTROL DIETHYL DIPROPIONATE. See diethyl stilbestrol dipropionate.

STODDARD SOLVENT. Syns: *safety solvent, varnoline, cleaning solvents (kerosene class), naphtha safety solvent, white spirits.* Clear, colorless liquid. Composed of 85% nonane and 15% trimethyl benzene. bp: 220°–300°, flash p: 100°–110°F, lel = 1.1%, uel = 6%, autoign. temp.: 450°F, d: 1.0.

THR = See components as listed above. Probably a MILD irr and MOD via oral and inhal routes. See gasoline.

Fire Hazard: Mod, when exposed to heat or flame.

Explosion Hazard: Mod, when exposed to flame.

Disaster Hazard: Mod; when heated to decomp, emits acrid fumes and may explode; can react with oxidizing materials.

To Fight Fire: Foam, CO_2, dry chemical.

STORAGE AND HANDLING OF COMPRESSED GASES. See Section 7.

STORAGE AND HANDLING OF HAZARDOUS MATERIALS. See Section 7.

STORAGE OF TOXIC MATERIALS IN BODY. See Section 9.

STOVARSOL.

THR = A recog carc. [*14*] See arsenic compounds.

STOVE POLISH, LIQUID. Flash p: 80°F.

THR = Variable.

Fire Hazard: Dangerous, when exposed to heat, flame or powerful oxidizers.

Explosion Hazard: U.

Disaster Hazard: Dangerous, upon exposure to heat or flame; can react vigorously with oxidizing materials.

STRAW.

THR = A nuisance dust and an allergen.

Fire Hazard: Mod, when exposed to heat or flame.

Spont Heating: Yes, especially when damp.

Explosion Hazard: Mod, in the form of dust, when exposed to flame.

To Fight Fire: Water.

STRAWBERRY ALDEHYDE. See ethyl methyl phenyl glycidate.

STRAW OIL. See lubricating oil.

STREPTOMYCIN. An antibiotic, it is a base and readily forms salts with anions. $C_{21}H_{39}N_7O_{12}$, mw: 581.58.

Acute tox data: sc LD_{50} (mice) = 520 mg/kg; iv LD_{50} (mice) = 160 mg/kg. [*3*]

THR = HIGH via iv and MOD via sc routes. Toxic to kidneys and CNS. A food additive permitted in food for human consumption. [*109*] Has been implicated in aplastic anemia.

STREPTOZOTOCIN. $C_8H_{15}O_7N_3$, mw: 265.3, mp: 115°.

THR = HIGH via oral route. An exper (+) neo and carc [*3, 10*] via ip and iv routes. Also an exper mutagen. [*23*]

STREUNEX. See benzene hexachloride.

STROBANE. See octachlorocamphene.

STRONTIUM. A silvery, soft metal which in air rapidly turns yellow. Sr, atwt: 87.63, mp: 769°, bp: 1384°, d: 2.54, vap. press: 10 mm @ 898°.

THR = MOD irr via oral and inhal routes. See also specific compounds. It resembles calcium in its metabolism and behavior. The stable form has low toxicity.

Radiation Hazard: For permissible levels, see Section 5A, Table 5A.5. Artificial isotope ^{85}Sr, $T\frac{1}{2}$ = 64d, decays to stable ^{85}Rb by ec; emits γ's of 0.51 MeV and x-rays. Artificial isotope ^{89}Sr, $T\frac{1}{2}$ = 53d, decays to stable ^{89}Y via β's of 1.46 MeV. Artificial isotope ^{90}Sr, $T\frac{1}{2}$ = 28y, decays to radioactive ^{90}Y via β's of 0.54 MeV.

Fire Hazard: Dangerous, in the form of dust; also when exposed to flame. See also powdered metals.

Explosion Hazard: Mod, in the form of dust, by spont chemical reaction. Reacts with water to evolve hydrogen. Can be stored under liquid hydrocarbons.

Disaster Hazard: Highly dangerous, in form of radioactive isotopes; will react with water or steam to produce heat and hydrogen; on contact with oxidizing materials, can react vigorously.

STRONTIUM ACETATE. $Sr(C_2H_3O_2)_2$, mw: 205.7.

Acute tox data: iv LD_{LO} (rat) = 123 mg/kg. [*3*]

THR = HIGH via iv route.

STRONTIUM ACID-*o*-ARSENATE. Rhombic needles. $SrHAsO_4 \cdot H_2O$. mw: 245.56, mp: $-H_2O$ @125°, d: 3.606 @ 15°.

THR = A poison. See arsenic compounds.

STRONTIUM ALLOYS.

THR = May react violently with water, acids. [*19*]

STRONTIUM ARSENITE. Syn: *strontium-o-arsenite*. White powder. $Sr_3(AsO_3)_2 \cdot 4H_2O$, mw: 580.77.

THR = A poison. See arsenic compounds.

STRONTIUM BROMATE. Monoclinic colorless–yellowish hygroscopic crystals. $Sr(BrO_3)_2 \cdot H_2O$, mw: 361.48, mp: $-H_2O$ @ 120°, bp: decomp @ 240°.

THR = See bromates.

STRONTIUM BROMIDE. White hygroscopic needles. $SrBr_2$, mw: 247.46, mp: 643°, bp: decomp, d: 4.216 @ 24°.

Acute tox data: ip LD_{50} (rat) = 1000 mg/kg. [*3*]

THR = MOD via ip route.

STRONTIUM CARBIDE. SrC_2. mw: 111.6.

THR = Reacts violently with Se, S. [*19*]

STRONTIUM CHLORATE. White crystalline powder. $Sr(ClO_3)_2$, mw: 254.54, mp: 120° (decomp), d: 3.152.

THR = See chlorates. Can react violently with acids. [*19*]

STRONTIUM CHLORATES, WET. See chlorates.

STRONTIUM CHLORIDE. $SrCl_2$, mw: 158.4.

Acute tox data: ip LD_{50} (mice) = 908 mg/kg; iv LD_{50} (mice) = 148 mg/kg. [*3*]

THR = HIGH via iv route and MOD via ip route. See also strontium compounds.

STRONTIUM CHLORIDE FLUORIDE. Tetragonal crystals. $SrCl_2 \cdot SrF_2$, mw: 284.17, mp: 962°, d: 4.18.

THR = See fluorides and chlorides.

STRONTIUM CHROMATE. Monoclinic yellow crystals. $SrCrO_4$, mw: 203.64, d: 3.895 @ 15°.

THR = See chromium compounds. An exper (+) carc. [*3, 6*]

STRONTIUM COMPOUNDS.

THR = The strontium ion has a LOW order of toxicity. It is chemically and biologically similar to calcium. The oxides and hydroxides are mod caustic materials. As with other compounds, the toxicity may be a function of the anion.

STRONTIUM CYANIDE. White, rhombic, deliquescent crystals. $Sr(CN)_2 \cdot 4H_2O$, mw: 211.73, mp: decomp.

THR = See cyanides.

STRONTIUM DIOXIDE. See strontium peroxide.

STRONTIUM DIURANATE. See uranium strontium oxide.

STRONTIUM FLUOBORATE.

Acute tox data: Oral LD_{LO} (rat) = 500 mg/kg; inhal LC_{LO} (mice) = 650 mg/m^3 for 1/6 hr. [*3*]

THR = HIGH via oral and inhal routes. See also fluorides.

STRONTIUM FLUORIDE. Cubic, odorless crystals or white powder. SrF_2, mw: 125.63, mp: 1190°, d: 4.24.

Acute tox data: iv LD_{LO} (rat) = 625 mg/kg. [*3*]

THR = MOD via iv route. See fluorides.

STRONTIUM FLUOSILICATE. Monoclinic crystals. $SrSiF_6 \cdot 2H_2O$, mw: 265.72, mp: decomp, d: 2.99 @ 17.5°.

Acute tox data: Oral LD_{LO} (rat) = 250 mg/kg. [*3*]

THR = HIGH via oral route. See fluosilicates.

STRONTIUM HYDRIDE. White crystals, reacts with water. SrH_2, mw: 89.7, d: 3.27, mp: decomp at red heat.

THR = See strontium compounds and hydrides. Violent reaction with bromates, chlorates, perchlorates. [*19*]

STRONTIUM HYDROSULFIDE. Crystals. $Sr(HS)_2$, mw: 154.77, mp: decomp.

THR = See sulfides.

STRONTIUM HYDROXIDE. White deliquescent crystals. $Sr(OH)_2$, mw: 121.65, mp: 375°, d: 3.625.

THR = See strontium compounds.

STRONTIUM IODATE. Triclinic crystals. $Sr(IO_3)_2$, mw: 437.47, d: 5.045 @ 15°.
THR = See iodates.

STRONTIUM IODIDE. Colorless plates. SrI_2, mw: 341.47, bp: decomp, d: 4.549 @ 25°.
Acute tox data: ip LD_{50} (rat) = 800 mg/kg. [*3*]
THR = MOD via ip route. See iodides. Violent reaction with K. [*19*]

STRONTIUM, METALLIC, DISPERSION IN ORGANIC SOLVENT. See strontium.

STRONTIUM MONOSULFIDE. Cubic light gray crystals. SrS, mw: 119.70, d: 3.70 @ 15°.
THR = See sulfides.

STRONTIUM NITRATE. White powder. $Sr(NO_3)_2$, mw: 211.65, mp: 570°, d: 2.986.
Acute tox data: ip LD_{50} (rat) = 540 mg/kg. [*3*]
THR = MOD via ip route. See nitrates.

STRONTIUM NITRITE. Hexagonal crystals. $Sr(NO_2)_2 \cdot H_2O$, mw: 197.66, mp: $-H_2O$ @ >100°, bp: decomp @ 240°, d: 2.408 @ 0°/0°.
THR = See nitrites.

STRONTIUM OXALATE. White, odorless, colorless, crystalline powder. $SrC_2O_4 \cdot H_2O$, mw: 193.67, mp: $-H_2O$ @ 150°.
THR = See oxalates.

STRONTIUM PERCHLORATE. Colorless crystals. $Sr(ClO_4)_2$, mw: 286.54.
THR = See perchlorates.

STRONTIUM PERMANGANATE. Cubic purple crystals. $Sr(MnO_4)_2 \cdot 3H_2O$, mw: 379.54, mp: decomp @ 175°, d: 2.75.
THR = See manganese compounds.
Fire Hazard: Mod, a strong oxidizer.
Disaster Hazard: Mod; keep away from flam materials.

STRONTIUM PEROXIDE. White powder. SrO_2, mw: 119.63, mp: decomp. d: 4.56.
THR = See peroxides, inorganic.

STRONTIUM PHOSPHIDE. Sr_3P_2, mw: 324.9.
Acute tox data: inhal LC_{LO} (rat) = 580 ppm for 1 hr; inhal LC_{LO} (cat) = 173 ppm for 2 hrs. [*3*]
THR = HIGH via inhal route. See also phosphides. Violent reaction with halogens. [*19*]

STRONTIUM POTASSIUM CHLORATE. Syn: *potassium strontium chlorate.* White crystalline powder. $Sr(ClO_3)_2 \cdot 2KClO_3$, mw: 499.67.
THR = See chlorates.

STRONTIUM SALICYLATE. $C_{14}H_{10}O_6 \cdot Sr$, mw: 361.9.
Acute tox data: ip LD_{50} (rat) = 400 mg/kg. [*3*]
THR = HIGH via ip route.

STRONTIUM SELENATE. Rhombic crystals. $SrSeO_4$, mw: 230.59, d: 4.23.
THR = See selenium compounds.

STRONTIUM SULFIDE. See strontium monosulfide.

STRONTIUM SULFITE. Colorless crystals. $SrSO_3$, mw: 167.70, mp: decomp.
THR = See sulfites.

STRONTIUM TETRASULFIDE. Reddish crystals. $SrS_4 \cdot 6H_2O$, mw: 324.0, mp: 25°.
THR = See sulfides.

STROPHANTHIN. Syn: *K-strophanthin.* White or yellowish powder; very bitter taste. $C_{29}H_{44}O_{12} \cdot 8H_2O$.
Acute tox data: iv LD_{50} (mice) = 2.5 mg/kg; iv LD_{50} (frog) = 0.69 mg/kg. [*3*]
THR = HIGH via iv route.
Fire Hazard: Slight.

STRYCHINE AND COMPOUNDS. Hard, white, crystalline alkaloid, very bitter taste. $C_{21}H_{22}N_2O_2$, mw: 334.40, mp: 268°, bp: 270°, d: 1.359 @ 18°.
Acute tox data: Oral LD_{50} (human) = 30 mg/kg; oral LD_{50} (rat) = 16 mg/kg; ip LD_{50} (mice) = 0.98 mg/kg; iv LD_{50} (mice) = 0.41 mg/kg; oral LD_{50} (wild bird) = 4 mg/kg; oral LD_{50} (duck) = 3 mg/kg; sc LD_{LO} (dog) = 0.35 mg/kg. [*3*]
THR = VERY HIGH via oral, ip, iv and sc routes. An allergen. Lethal dose to man: 30–60 mg/kg.

A very poisonous alkaloid. If it is taken by mouth, the time of action depends upon the condition of the stomach, that is, whether empty or full, and the nature of the food present. If taken by subcutaneous injection, the place of administration of the injection will affect the time of action. The first symptoms are a feeling of uneasiness with a heightened reflex of irritability, followed by muscular twitching in some parts of the body. With larger doses, this is followed by a sense of impending suffocation. Convulsive movements begin which have the effect of mechanically causing the patient to cry out or to shriek; then follow the characteristic spasms which set in with violence. These are at first clonic and then tonic. There are successive attacks of spasms. With each successive attack, the symptoms become more violent, eventually resulting in death. A rodenticide.

Disaster Hazard: Dangerous; when heated, emits highly toxic fumes.
Treatment and Antidotes: Call a physician at once. The stomach should be washed out with potassium permanganate solution which is diluted to the color of port wine; give chloral hydrate per rectum to control convulsions. The use of apomorphine has been recommended and the use of phenobarbital on animals has proved successful. Barbiturates do not

act readily. One gram of Merck's carbo medicinolis will bind 580 mg of strychnine. When this is available the stomach should be washed out with it immediately. Sodium amytal has been used effectively as has sodium pentobarbital, particularly in animal exper.

STRYCHNINE SULFATE. $C_{21}H_{22}O_2N_2 \cdot \frac{1}{2}H_2SO_4$, mw: 383.5.
Acute tox data: Oral LD_{50} (rat) = 5 mg/kg; ip LD_{50} (rat) = 3 mg/kg; sc LD_{50} (rat) = 1.7 mg/kg; iv LD_{50} (rat) = 0.475 mg/kg; oral LD_{50} (wild birds) = 5 mg/kg. [*3*]
THR = HIGH via oral, ip, sc, iv routes. See also strychnine.

STYPHNIC ACID. See trinitroresorcinol.

STYPTICIN. See cotarnine chloride.

STYRALYL ALCOHOL. See phenyl methyl carbinol.

STYRENE (MONOMER). See phenyl ethylene.

STYRENE DIBROMIDE. See dibromoethyl benzene.

STYRENE OXIDE. Syn: *phenylethylene oxide.* Colorless liquid. $C_6H_5CHOCH_2$, mw: 120.1, bp: 194.2, flash p: 165° F (OC), fp: −36.7°, d: 1.0469 @ 25°/4°, vap. d: 4.14.
Acute tox data: dermal LD_{50} (rabbit) = 1060 mg/kg; oral LD_{50} (rat) = 4290 mg/kg; inhal LC_{LO} (rat) = 500 ppm for 4 hrs; ip LD_{50} (rat) = 460 mg/kg. [*3*]
THR = HIGH via inhal and ip routes. MOD via oral and dermal routes. An exper (±) carc via dermal route. [*13, 23, 3*]
Fire Hazard: Mod, when exposed to heat, flame or oxidizers.
Disaster Hazard: Mod dangerous; when heated, emits acrid fumes; can react with oxidizing materials.
To Fight Fire: Foam, CO_2, dry chemical.

STYRENE POLYMER. Syn: *polystyrene.* $(C_8H_8)_n$.
THR = An exper neo by implantation. [*3*]

STYRYL 430. $C_{27}H_{25}O_2N_4 \cdot C_2H_3O_2$, mw: 496.6.
THR = An exper neo. [*3*]

4′-STYRYL ACETANILIDE. $C_{16}H_{15}ON$, mw: 237.3.
THR = HIGH to animals. It is an exper carc via ip and sc routes. [*3, 23*]

6-STYRYL BENZO(a)PYRENE. $C_{28}H_{18}$, mw: 354.5.
THR = An exper neo. [*3, 23*]

***trans*-N-(*p*-STYRYL PHENYL)ACETOHYDROXAMIC ACID.** $C_{16}H_{15}O_2N$, mw: 253.3.
THR = An exper carc. [*3*]

***trans*-N-(*p*-STYRYL PHENYL)ACETOHYDROXAMIC ACID** (Copper ++ complex).
$C_{34}H_{28}O_6N_2 \cdot Cu$, mw: 624.2.
THR = An exper carc. [*3*]

SUBERANE. See cycloheptane.

SUBERONE. See cycloheptanone.

SUBGALLATE. See bismuth gallate, basic.

SUBLAMINE. Syn: *mercury ethylene diamine sulfate.* White crystalline powder, contains approx 43% mercury. $HgSO_4 \cdot 4CH_2NH_2 \cdot 2H_2O$, mw: 452.87.
THR = See mercury compounds, organic.

SUBSALICYLATE (COMMERCIAL). See bismuth salicylate, basic.

SUBSTANCES ON GRAS LIST. [*109*]

SUBTILISINS. See proteolytic enzymes.
THR = HIGH.

SUBTOSAN. See polyvinyl pyrrolidone.

SUCCINCHLORIMIDE. See N-chlorosuccinimide.

SUCCINDIALDEHYDE. See glutaraldehyde.

SUCCINIC ACID. Syns: *ethylene succinic acid, ethylene dicarboxylic acid, butanedioc acid.* Colorless crystals. $COOH(CH_2)_2COOH$, mw: 118.09, mp: 185°, bp: 235° (decomp), d: 1.564 @ 15°/4°.
Acute tox data: sc LD_{LO} (frog) = 2000 mg/kg. [*3*]
THR = MOD via sc route. A general purpose food additive. [*109*]
Fire Hazard: Slight, when exposed to heat or flame; can react with oxidizing materials.

SUCCINIC ACID ANHYDRIDE. See succinic anhydride.

SUCCINIC ACID PEROXIDE. Syns: *butanedioic peroxide, succinyl peroxide.* Fine white powder, odorless with tart taste, mod sol in water. $(HOOCCH_2CH_2CO)_2O_2$, mw: 234.2, mp: 125° (decomp).
THR = MOD irr. See also peroxides, organic.

SUCCINIC ANHYDRIDE. Syns: *butanedioic anhydride, succinic acid anhydride.* Colorless needles. $(CH_2CO)_2O$, mw: 100.07, mp: 261°, d: 1.104, vap. press: 1 mm @ 92.0°.
THR = An exper carc via sc route. [*3, 23, 81*]
Fire Hazard: Slight, when exposed to heat or flame.

SUCCINIC PEROXIDE. See succinic acid peroxide.

SUCCINIC CHLORIDE. $C_4H_4O_2Cl_2$, mw: 155.0.
Acute tox data: ip LD_{50} (mice) = 62 mg/kg. [*3*]
THR = HIGH irr to skin, eyes and mu mem and via ip route.

SUCCINONITRILE. Syns: *butanedinitrile, ethylene cyanide.* Colorless, odorless, waxy material. $CNCH_2CH_2CN$, mw: 80.09, mp: 58.1°, bp: 267°, flash p: 270° F (ASTM D92-46), d: 1.022 @ 25°, vap. press: 2 mm @ 100°, vap. d: 2.1.
Acute tox data: ip LD_{LO} (mice) = 100 mg/kg; sc LD_{LO} (frog) = 1000 mg/kg. [*3*]

THR = HIGH via ip and MOD via sc routes. See also nitriles.
Fire Hazard: Slight, when exposed to heat or flame.
Disaster Hazard: Dangerous; when heated to decomp or on contact with acid or acid fumes, emits highly toxic fumes of cyanides; can react with oxidizing materials.
To Fight Fire: Alcohol foam, CO_2, dry chemical.

4-SUCCINOYL AMINO-2,3-DIMETHYL AZOBENZENE. $C_{18}H_{19}O_3N_3$, mw: 325.4.
THR = An exper neo. [*3*]

SUCCINYL PEROXIDE. See succinic acid peroxide.

SUCCISTEARIN. Syn: *stearoyl propylene glycol hydrogen succinate.* The reaction product of succinic anhydride fully hydrogenated vegetable oil, and propylene glycol.
THR = U. A food additive permitted in food for human consumption. [*109*]

SUCROSE. Syn: *sugar.* White crystals. $C_{12}H_{22}O_{11}$, mw: 342.3, d: 1.587.
Acute tox data: Oral LD_{50} (rat) = 29,700 mg/kg. [*3*]
THR = NONE, via oral route. A non-toxic nuisance dust. See dust.

SUCROSE OCTAACETATE. Crystals. $C_{12}H_{14}O_3(OOCCH_3)_8$, mw: 678.58, mp: 89°, bp: 260° @ 1.0 mm, d: 1.28 @ 20°/20°.
THR = Probably LOW. No details.
Fire Hazard: Slight.

SUGAR OF LEAD. See lead acetate.

SULFAETHOXY PYRIDAZINE. Syn: *n-(6-ethoxy-3-pyridazinyl)sulfanilamide.* $C_{12}H_{14}N_4O_3S$, mw: 294. mp: 180°–186°.
THR = An antimicrobial food additive permitted in the feed and drinking water of animals and/or for the treatment of food-producing animals. Also permitted in food for human consumption. [*109*]
Disaster Hazard: Dangerous; see sulfates.

SULFAGUANIDINE. Needle-like crystals. $C_7H_{10}N_4O_2S \cdot H_2O$, mw: 232.3, mp: 190°–193°.
Acute tox data: ip LD_{100} (mice) = 1000 mg/kg; ip LD_{LO} (mice) = 500 mg/kg. [*3*]
THR = MOD via ip route.
Disaster Hazard: Dangerous; when heated to decomp, emits highly toxic fumes of SO_x and NO_x.

SULFAMETHAZINE. Syn: *N-(4,6-dimethyl-2-pyrimidyl)sulfanilamide.* White to yellow-to-white powder, almost odorless, sol in acetone, slightly sol in alcohol, very slightly sol in water and ether. $NH_2C_6H_4SO_2NHC_4N_2H(CH_3)_2$, mw: 278, mp: 197°–200°.
THR = An exper carc via imp route. [*3*] A food additive permitted in the feed and drinking water of animals and for the treatment of food-producing animals. Also permitted in food for human consumption. [*109*]
Disaster Hazard: Dangerous. See sulfates.

SULFAMETHYL THIADIAZOLE. $C_9H_{10}O_2N_4S_2$, mw: 270.4.
THR = An exper carc. [*3*]

SULFAMETHYL ISOXAZOLE. $C_{10}H_{11}O_3N_3S$, mw: 253.3.
THR = An exper neo. [*3*]

SULFAMIC ACID. Syn: *amidosulfonic acid.* White crystalline solid. H_2NSO_3H, mw: 97.09, mp: 200° (decomp), bp: decomp, d: 2.03 @ 12°.
Acute tox data: Oral LD_{50} (rat) = 1.6 mg/kg; ip LD_{LO} (rat) = 100 mg/kg. [*3*]
THR = HIGH via oral and ip route. A substance which migrates to food from packaging materials. [*109*] Violent reaction with Cl_2, fuming HNO_3. [*19*]
Disaster Hazard: Dangerous; see sulfonates.

SULFAMIDE. Syn: *sulfuryl amide.* Rhombic plates. $SO_2(NH_2)_2$, mw: 96.11, mp: 91.5°, bp: decomp @ 250°.
THR = Irr. See amides.
Disaster Hazard: Dangerous; when heated to decomp, emits highly toxic fumes of SO_x; will react with water or steam to produce toxic and corrosive fumes.

N-(5-SULFAMOYL-1,3,4-THIADIAZOL-2-YL) ACETAMIDE. Syn: *acetazolamide.* $C_4H_6O_3N_4S_2$, mw: 222.3.
THR = Mod via oral route to rats. It affects pregnancy. It is an exper teratogen. [*3*] Implicated in aplastic anemia.

SULFANILAMIDE. See sulfonamide.

SULFANILIC ACID. Syn: *o-amino benzenesulfonic acid.* Colorless crystals. $NH_2C_6H_4SO_3H \cdot H_2O$, mw: 191.2, mp: 288° (decomp).
THR = U. Animal exper suggest LOW toxicity and slight irr.
Disaster Hazard: Dangerous; when heated to decomp or on contact with acids, emits highly toxic fumes of NO_x and SO_x.

SULFANILIC ACID DIAZIDE. See *p*-diazobenzene sulfonic acid.

SULFANITRAN. See acetyl-(*p*-nitrophenyl)sulfanilamide.

SULFASAN. See 4,4-dithiomorpholine.

SULFATES.
THR = Variable. In general the toxic qualities of substances containing the sulfate radical is that of the material (cation) with which the sulfate (anion) is combined. See specific compound. Violent reaction with Al, Mg. [*19*]

Disaster Hazard: Dangerous. When heated to decomp, they emit highly toxic fumes of SO_x.

SULFATHIAZOLE. $C_9H_9O_2N_3S_2$, mw: 255.3.

THR = An exper carc. [*3*]

SULFIDES.

THR = Variable. The alkaline sulfides (potassium, calcium, ammonium and sodium) are similar in action to alkalies. They cause softening and irr of the skin. If taken by mouth they are corrosive and irr through the liberation of hydrogen sulfide and free alkali. Hydrogen sulfide is especially toxic. See hydrogen sulfide.

Sulfides of the heavy metals are generally insoluble and hence have little toxic action except through the liberation of hydrogen sulfide.

Sulfides are used as fungicides.

Fire Hazard: Mod, when exposed to flame or by spont chemical reaction. Many sulfides ignite easily in air at room temp. Others require a higher temp. or the presence of an oxidizer. Upon contact with moisture or acids, hydrogen sulfide is evolved. Many powerful oxidizers on contact with sulfides ignite violently. See also hydrogen sulfide.

Explosion Hazard: Many sulfides react violently and explosively on contact with powerful oxidizers. Hydrogen sulfide evolved can form explosive mixtures with air. See also hydrogen sulfide.

Disaster Hazard: Dangerous; when heated to decomp, they emit highly toxic fumes of SO_x; they react with water, steam or acids to produce toxic and flam vapors of hydrogen sulfide.

SULFITES.

THR = Fairly large doses of sulfites can be tolerated since they are rapidly oxidized to sulfates, although if swallowed they may cause irr of the stomach by liberating sulfurous acid. In exper animals, large doses of sodium sulfite have been shown to cause retarded growth, nerve irr, atrophy of bone marrow, depression and paralysis.

Disaster Hazard: Dangerous; when heated to decomp, they emit highly toxic fumes of sulfur dioxide; they will react with water, steam or acids to produce a toxic and corrosive material.

SULFOCARBOLIC ACID. See phenosulfonic acid.

1-SULFOCYANO-2,4-DINITROBENZENE. Powder. $CNSC_6H_3(NO_2)_2$, mw: 225.2.

THR = See nitrates and cyanides.

SULFOLANE. $C_4H_8O_2S$, mw: 120.2.

Acute tox data: Oral LD_{50} (rat) = 1540 mg/kg; iv LD_{50} (mice) = 1080 mg/kg; dermal LD_{50} (rabbit) = 3180 mg/kg. [*3*]

THR = Mod via oral, iv and dermal routes. Reacts violently with chlorosulfonic acid, oleum. [*19*]

SULFONATED CASTOR OIL. See turkey red oil.

SULFONAMIDE. Syn: *sulfanilamide.* $C_6H_8O_2N_2S$, mw: 172.2.

Acute tox data: Oral LD_{50} (rat) = 3900 mg/kg; iv LD_{50} (mice) = 621 mg/kg; oral LD_{50} (dog) = 2000 mg/kg; oral LD_{50} (rabbit) = 2000 mg/kg. [*3*]

THR = MOD via oral and iv routes. An exper (S) carc via sc and pa routes. [*3*] Implicated in aplastic anemia.

SULFONATES.

THR = Variable. See specific compounds. Usually irr.

Disaster Hazard: Dangerous; when heated to decomp, or on contact with acid or acid fumes, they emit highly toxic fumes of SO_x.

SULFONETHYL METHANE. See 2,2′-bis(ethyl sulfonyl)butane.

***p,p′*-SULFONYL BIS ACETANILIDE.** $C_{16}H_{16}O_4N_2S$, mw: 332.4.

THR = An exper carc. [*3*]

SULFONYL CHLORIDE. See sulfuryl chloride.

***p,p′*-SULFONYL DIANILINE.** See diamino diphenyl sulfone.

SULFOSALICYLIC ACID. Syns: *3-carboxy-4-hydroxy benzene sulfonic acid, 5-sulfosalicyl acid, 2-hydroxybenzoic sulfonic acid, salicylsulfonic acid.* White crystalline powder, very sol in water and alcohol, sol in ether. $C_7H_6O_6S$, mw: 218.2, mp: 120° (decomp at higher temp).

Acute tox data: Oral LD_{50} (rat) = 2450 mg/kg; oral LD_{LO} (rabbit) = 1300 mg/kg. [*3*]

THR = MOD via oral route.

Disaster Hazard: Dangerous; see sulfates.

SULFOTEPP. See tetraethyldithiopyrophosphate.

SULFOXIDE. Syn: *isosafrole-n-octyl sulfoxide.* Water-insol, slightly sol in petroleum oils, sol in most organic solvents. $C_{18}H_{28}O_3S$, mw: 324.5.

Acute tox data: Oral LD_{50} (rat) = 2000 mg/kg. [*3*]

THR = MOD via oral route. An exper carc and neo via oral and sc routes. [*3*] Reacts violently with $HClO_4$. [*19*]

Disaster Hazard: Dangerous; when heated to decomp, emits highly toxic fumes.

SULFUR. Syns: *brimstone, flowers of sulfur, sulfur flour.* Rhombic yellow crystals or yellow powder. S_8, mw: 256.48, mp: 119°, bp: 444.6°, flash p: 405°F (CC), d: 2.07; d(liquid): 1.803; autoign. temp.: 450°F, vap. press: 1 mm @ 183.8°.

THR = Very LOW. See nuisance dusts. A fungicide. Chronic inhal can cause irr of mu mem.

Radiation Hazard: For permissible levels, see Section 5A, Table 5A.5. Artificial isotope ^{35}S, $T_{\frac{1}{2}}$ = 88d, decays to stable ^{35}Cl via β's of 0.17 MeV.

Fire Hazard: Slight, when exposed to heat or flame, or by chemical reaction with oxidizers.

Spont Heating: No.

Explosion Hazard: In the form of dust, when exposed to flame. Can react violently with halogens, carbides, halogenates, halogenites, zinc, uranium, tin, sodium, lithium, nickel, palladium, phosphorus, potassium, indium, calcium, boron, aluminum, (aluminum + niobium pentoxide), ammonia, ammonium nitrate, ammonium perchlorate, BrF_5, BrF_3, $(Ca + VO + H_2O)$, $Ca(OCl)_2$, Ca_3P_2, Cs_3N, charcoal, (Cu + chlorates), ClO_2, ClO, ClF_3, CrO_3, $Cr(OCl)_2$, hydrocarbons, IF_5, IO_5, PbO_2, $Hg(NO_3)_2$, HgO, Hg_2O, NO_2, P_2O_3, $(KNO_3 + As_2S_3)$, K_3N, $KMnO_4$, $AgNO_3$, Ag_2O, NaH, $(NaNO_3$ + charcoal), $(Na + SnI_4)$, SCl_2, Tl_2O_3. [*19*]

Disaster Hazard: Dangerous; when heated it burns and emits highly toxic fumes of SO_x. Can react with oxidizing materials.

To Fight Fire: Water or special mixtures of dry chemical.

SULFUR BROMIDE. Syn: *sulfur monobromide.* S_2Br_2, mw: 223.96, mp: −40°, bp: 54° @ 0.2 mm, d: 2.635.

THR = HIGH irr to skin, eyes and mu mem. See also bromides.

Fire Hazard: Slight, when exposed to heat or flame.

Disaster Hazard: Dangerous; when heated to decomp, emits highly toxic fumes of SO_x and bromides; will react with water or steam to produce toxic and corrosive fumes.

SULFUR CHLORIDE. Syn: *sulfur monochloride.* Amber to yellowish-red, oily, fuming liquid, penetrating odor, decomp in water. S_2Cl_2, mw: 135.03, mp: −80°, bp: 138.0°, flash p: 245°F (CC), d: 1.6885 @ 15.5°/15.5°, autoign. temp.: 453°F, vap. press: 10 mm @ 27.5°, vap. d: 4.66.

Acute tox data: Inhal LC_{LO} (mice) = 150 ppm for 1 min; inhal LC_{LO} (cat) = 48 ppm for $\frac{1}{4}$ hr. [*3*]

THR = HIGH irr to skin, eyes and mu mem. HIGH via oral and inhal routes.

It is a fuming, corrosive liquid with a penetrating odor which is very irr to the eyes, lungs and mu mem. It decomp on contact with water to form hydrogen chloride, thiosulfuric acid, and sulfur. These decomp products are highly irr. Its toxic effects are irr of the upper respiratory tract, although the results of intoxication are usually transitory in nature. However, if hydrolysis is not complete in the upper respiratory tract, injury to the bronchioles and alveoli can result. The literature notes that conc of 2–9 ppm have been found in rubber factories and that these conc were observed to be mildly irr. A conc of 150 ppm has been stated to be fatal to mice after an exposure of only 1 min.

Fire Hazard: Contact with organic matter, P_2O_3, Na_2O_2, water, $Cr(OCl)_2$. [*19*] Low, via heat or flame.

Spont Heating: No.

Disaster Hazard: Dangerous; when heated to decomp, emits highly toxic fumes of chlorides and SO_x; will react with water or steam to produce heat and toxic and corrosive fumes; can react with oxidizing materials.

To Fight Fire: CO_2, dry chemical.

SULFUR CHLORIDE PENTAFLUORIDE. $SClF_5$, mw: 162.5.

Acute tox data: inhal LC_{LO} (rat) = 100 ppm for 1 hr. [*3*]

THR = HIGH via inhal route. HIGH irr to skin, eyes and mu mem. See also fluorides.

SULFUR COMPOUNDS. Variable formula.

THR = Variable. See specific material as listed. Common air contaminants.

Disaster Hazard: Dangerous; when heated to decomp, these materials can evolve highly toxic fumes containing SO_x. See sulfides.

SULFUR DECAFLUORIDE. S_2F_{10}, mw: 254.1.

Acute tox data: inhal LC_{LO} (rat) = 1780 ppm for 1 hr; iv LD_{LO} (dog) = 1 mg/kg; iv LD_{LO} (rabbit) = 5.79 mg/kg. [*3*]

THR = HIGH via iv and MOD via inhal route. See also fluorides.

SULFUR DIBROMIDE. SBr_2, mw: 191.9.

THR = HIGH irr to skin, eyes and mu mem. See also bromides. Violent reaction with Na, K. [*19*]

SULFUR DICHLORIDE. Reddish-brown liquid, pungent odor. SCl_2, mw: 103.0, mp: −78°, bp: 59°, d: 1.621 @ 15°/15°, vap. d: 3.55.

THR = HIGH irr to skin, eyes and mu mem. A corrosive to tissue.

Fire Hazard: Mod, when exposed to heat or flame. Violent reaction with Al, NH_3, K, Na. [*19*]

Disaster Hazard: Dangerous; when heated to decomp, emits highly toxic fumes of chlorides and SO_x; will react with water or steam to produce heat and toxic and corrosive fumes.

SULFUR DIOXIDE. Syn: *sulfurous acid anhydride.* Colorless gas or liquid, pungent odor. SO_2, mw: 64.06, mp: −75.5°, bp: −10.0°, d(liquid): 1.434 @ 0°, vap. d: 2.264 @ 0°, vap. press: 2538 mm @ 21.1°.

Acute tox data: Inhal TC_{LO} (human) = 3 ppm for 5 days ⟶ pulmonary effects; inhal TC_{LO} (man) = 4 ppm for 1 min ⟶ pulmonary effects; inhal LC_{LO} (rat) = 611 ppm for 5 hrs; inhal LC_{LO} (mice) = 764 ppm for ⅓ hr; inhal LC_{LO} (guinea pig) = 5000 ppm for 5 min. [*3*]

THR = HIGH irr via inhal route and to skin, eyes and mu mem.

This gas is dangerous to the eyes, as it causes irr and inflammation of the conjunctiva. It has a suffocating odor and is a corrosive and poisonous material. In moist air or fogs, it combines with water to form sulfurous acid, but is only very slowly oxidized to sulfuric acid. Conc of 6–12 ppm cause immediate irr of the nose and throat, while 0.3–1 ppm can be detected by the average individual possibly by taste rather than by sense of smell. 3 ppm has an easily noticeable odor and 20 ppm is the least amount which is irr to the eyes. 10,000 ppm is an irr to moist areas of the skin within a few minutes of exposure.

It chiefly affects the upper respiratory tract and the bronchi. It may cause edema of the lungs or glottis, and can produce respiratory paralysis. Conc of <1 ppm are believed to be injurious to plant foliage.

This material is so irr that it provides its own warning of toxic conc. 400–500 ppm is immediately dangerous to life and 50–100 ppm is considered to be the maximum permissible conc for exposures of 30–60 min. Excessive exposures to high enough conc of this material can be fatal. Its toxicity is comparable to that of hydrogen chloride. However, less than fatal conc can be borne for fair periods of time with no apparent permanent damage. It is used as a fumigant, insecticide and fungicide, and a chemical preservative food additive. [*109*] It is a common air contaminant. It reacts violently with acrolein, Al, $CsHC_2$, CsO, chlorates, ClF_3, Cr, FeO, F_2, Mn, KHC_2, $KClO_3$, Rb_2C_2, Na, Na_2C_2, SnO, lithium acetylene carbide diammino. [*19*]

Disaster Hazard: Dangerous; will react with water or steam to produce toxic and corrosive fumes.

Treatment and Antidotes: Personnel who have shown toxicity symptoms when exposed to this material should immediately be removed to fresh air. If the eyes are involved they should be irrigated with copious quantities of warm water. If the symptoms persist, call a physician.

SULFURETTED HYDROGEN. See hydrogen sulfide.

SULFUR FLOUR. See sulfur.

SULFUR FLUORIDE. Syn: *sulfur monofluoride.* Colorless gas. S_2F_2, mw: 102.12, mp: −104.5°, bp: −99°, d(liquid): 1.5 @ −100°.

THR = See fluorides and hydrofluoric acid.

SULFUR HEPTOXIDE. Syn: *persulfur heptoxide.* Viscous liquid or possibly needle-like crystals. S_2O_7, mw: 176.1, mp: 0°, bp: sublimes @ 10°.

THR = HIGH irr via oral and inhal to skin, eyes and mu mem.

Fire Hazard: Mod, when exposed to heat or flame or by chemical reaction. When heated, or in contact with water or alcohol, it liberates oxygen.

Disaster Hazard: Dangerous; when heated to decomp, emits highly toxic fumes of SO_x; can react with reducing materials.

To Fight Fire: CO_2, dry chemical.

SULFUR HEXAFLUORIDE. Colorless gas. SF_6, mw: 146.06, mp: −51° (sublimes @ −64°), vap. d: 6.602, d(liquid): 1.67 @ −100°.

THR = This material is chemically inert in the pure state and is considered to be physiologically inert as well. However, as it is ordinarily obtainable, it can contain variable quantities of the low sulfur fluorides. Some of these are toxic, very reactive chemically and corrosive in nature. These materials can hydrolyze on contact with water to yield hydrogen fluoride, which is highly toxic and very corrosive. In high conc and when pure it may act as a simple asphyxiant. Vigorous reaction with disilane. [*19*] May explode.

Disaster Hazard: Dangerous; when heated to decomp, emits highly toxic fumes of fluorides and SO_x.

SULFURIC ACID. Syns: *oil of vitriol, dipping acid.* Colorless, oily liquid. H_2SO_4, mw: 98.08, mp: 10.49°, bp: 330°, d: 1.834, vap. press 1 mm @ 145.8°.

Acute tox data: Oral LD_{50} (rat) = 2140 mg/kg. [*3*]

THR = MOD via oral route. Extremely irr, corrosive and toxic to tissue. Contact with the body results in rapid destruction of tissue, causing severe burns. No systemic effects due to continual ingestion of small amounts of this material have been noted. There are systemic effects secondary to tissue damage caused by contact with it. However, repeated contact with dilute solutions can cause a dermatitis, and repeated or prolonged inhal of a mist of sulfuric acid can cause an inflammation of the upper respiratory tract leading to chronic bronchitis. Sensitivity to sulfuric acid or mists or vapors varies with individuals. Normally 0.125–0.50 ppm may be mildly annoying and 1.5–2.5 ppm can be definitely unpleasant. 10–20 ppm is unbearable.

Workers exposed to low conc of the vapor gradually lose their sensitivity to its irr action. Inhal of conc vapor or mists from hot acid or oleum can cause rapid loss of consciousness with serious damage to lung tissue. In conc form it acts as a powerful caustic to the skin destroying the epidermis and penetrating some distance into the skin and subcutaneous tissues, in which it causes necrosis. This causes great pain, and, if much of the skin is involved, it is accompanied by shock, collapse and symptoms similar to those seen in severe burns. The fumes or mists of this material cause coughing and irr of the mu mem of the eyes and upper respiratory tract. Severe exposure may cause a chemical pneumonitis; erosion of the teeth due to exposure to strong acid fumes has been recog in industry. It is used as a general purpose food additive; it migrates to food from packaging materials. [*109*] A common air contaminant.

Fire Hazard: This is a very powerful, acidic oxidizer which can ignite or even explode on contact with many materials; i.e., acetic acid, acetone cyanhydrin, (acetone + HNO_3), (acetone + $K_2Cr_2O_7$), acetonitrile, acrolein, acrylonitrile, (acrylonitrile + H_2O), (alcohols + H_2O_2), allyl alcohol, allyl chloride, NH_4OH, 2-amino ethanol, NH_4triperchromate, aniline, (bromates + metals), BrF_5, *n*-butyraldehyde, carbides, $CoHC_2$, chlorates, (metals + chlorates), ClF_3, chlorosulfonic acid, Cu_3N, diisobutylene. (dimethyl benzylcarbinol + H_2O_2), epichlorohydrin, ethylene cyanhydrin, ethylene diamine, ethylene glycol, ethylene imine, fulminates, HCl, H_2, IF_7, (indene + HNO_3), Fe, isoprene, Li_6Si_2, Hg_3N_2, mesityl oxide, metals, (HNO_3 + glycerides), *p*-nitrotoluene, perchlorates, $HClO_4$, (C_6H_6 + permanganates), pentasilver trihydroxydiamino phosphate, (1-phenyl-2-methyl propyl alcohol + H_2O_2), P, $P(OCN)_3$, picrates, potassium-*tert*-butoxide, $KClO_3$, $KMnO_4$, ($KMnO_4$ + KCl), ($KMnO_4$ + H_2O), β-propiolactone, $RbHC_2$, propylene oxide, pyridine, Na, Na_2CO_3, NaOH, steel, styrene monomer, water, vinyl acetate, (HNO_3 + toluene). [*19*]

Disaster Hazard: Dangerous; when heated, emits highly toxic fumes; will react with water or steam to produce heat; can react with oxidizing or reducing materials.

Treatment and Antidotes: Speed in removing this material from contact with the body is of primary importance. Start first aid at once. In all cases of contact in any form, delay can result in serious injuries and all persons injured should be referred to a physician. However, immediately give prolonged applications of running water to wash the material off the body. Remove contaminated clothing. Subject patient to a deluge type of shower if this is available. Do not attempt to neutralize the acid in contact with the skin until all areas of contact have been thoroughly irrigated with running water. Then applications of mild alkaline solutions may be in order. Shock symptoms will often be noted in cases of severe or extensive burns. In such a case, put patient on his back, keep him warm but not hot until physician arrives. Do not apply oils or ointments to burned area without instructions from a physician. If eyes are involved, they should immediately be irrigated with copious quantities of warm water for at least 15 min.

If the material has been taken internally, it causes burns of the mu mem of the throat, esophagus, and stomach. Do not attempt to induce vomiting in patients who have swallowed strong solutions of sulfuric acid. Do not give anything by mouth to an unconscious patient. If he is conscious, encourage him to wash out his mouth with copious amounts of water, then have him drink milk mixed with whites of eggs. If this is not available, have him drink as much water as possible. Get medical help.

SULFURIC ACID, AROMATIC. Syn: *elixir of vitriol.* Clear, reddish-brown liquid, peculiar aromatic odor, pleasant acid taste when diluted.

THR = Corrosive. See sulfuric acid.

Fire Hazard: Mod, when exposed to heat or flame. See also ethyl alcohol and sulfuric acid.

Explosion Hazard: Mod, in the form of vapor (ethyl alcohol) when exposed to flame.

Disaster Hazard: Dangerous; see sulfuric acid and ethyl alcohol.

SURFURIC ACID, FUMING. See oleum.

SULFURIC ACID MIST. An airborne suspension of sulfuric acid in the form of droplets.

Acute tox data: Inhal TC_{LO} (human) = 3 $mg/m^3 \longrightarrow$ irr of the mouth, skin and eyes. Inhal TC_{LO} (human) = 0.35 mg/m^3 for 3 min $\longrightarrow$ pulmonary irr effects; inhal LC_{LO} (rat) = 178 ppm for 7 hrs; inhal LC_{LO} (mice) = 140 ppm for 3½ hrs. [*3*]

THR = HIGH irr to skin, eyes and mu mem.

SULFURIC ACID SLUDGE. See selenium compounds.

SULFURIC CHLORIDE. See sulfuryl chloride.

SULFURIC CHLOROHYDRIN. See chlorsulfonic acid.

SULFURIC ETHER. See ethyl ether.

SULFURIC OXYCHLORIDE. See sulfuryl chloride.

SULFURIC OXYFLUORIDE. See sulfuryl fluoride.

SULFUR MONOBROMIDE. See sulfur bromide.

SULFUR MONOCHLORIDE. See sulfur chloride.

SULFUR MONOFLUORIDE. See sulfur fluoride.

SULFUR MONOOXYTETRACHLORIDE. Dark red liquid. S_2OCl_4, mw: 221.96, bp: 60°–61°, d: 1.656 @ 0°.

THR = HIGH irr to skin, eyes and mu mem. See also hydrochloric acid.
Disaster Hazard: Dangerous; when heated to decomp, emits highly toxic fumes of SO_x and hydrochloric acid; will react with water or steam to produce toxic and corrosive fumes.

SULFUR NITRIDE. S_4N_4, mw: 184.3.
THR = Explodes upon impact or >30° or contact with $Ba(ClO_3)_2$. [*19*]

SULFUROUS ACID. Colorless liquid, suffocating sulfur odor (in solution only). H_2SO_3, mw: 82.08, d: about 1.03.
THR = HIGH irr to skin, eyes and mu mem and via oral and inhal routes. See also SO_x.
Disaster Hazard: Dangerous; when heated to decomp, emits highly toxic fumes.

SULFUROUS ACID ANHYDRIDE. See sulfur dioxide.

SULFUROUS ACID 2-(*p-tert*-BUTYL PHENOXY)-1-METHYL ETHYL-2-CHLOROETHYL ESTER. Syn: *aramite.* Liquid, miscible with many organic solvents, insol in water. $C_{15}H_{23}ClO_4S$, mw: 334.9, d: 1.145–1.1620, mp: −31.7°, bp: 175° @ 0.1 mm, vap. press: <10 mm @ 25°.
Acute tox data: Oral LD_{50} (rat) = 3.9 mg/kg; ip LD_{50} (mice) = 200 mg/kg. [*3*]
THR = HIGH via oral and ip routes. A pesticide. See also chlorinated hydrocarbons. An exper (+) carc via oral route. [*12, 3, 23*]
Disaster Hazard: Dangerous; when heated to decomp, emits highly toxic fumes of chlorides, etc.

SULFUROUS OXYCHLORIDE. See thionyl chloride.

SULFUR PENTAFLUORIDE. See sulfur decafluoride.

SULFUR SESQUIOXIDE. Blue green crystals. S_2O_3, mw: 112.13, mp: decomp @ 70°–95°.
THR = No data. See also sulfur dioxide and sulfur trioxide.
Disaster Hazard: Dangerous; when heated to decomp, emits highly toxic fumes of SO_x; can react with oxidizing materials.

SULFUR TETRACHLORIDE. Yellow-brown liquid or gas at ordinary temp. SCl_4, mw: 173.89, mp: −30°, bp: −15° (decomp).
THR = HIGH irr to skin, eyes and mu mem and via oral and inhal routes.
Disaster Hazard: Dangerous; when heated to decomp, emits highly toxic fumes of hydrochloric acid and SO_x; will react with water or steam to produce toxic and corrosive fumes.

SULFUR TETRAFLUORIDE. Gas. SF_4, mw: 108.06, bp: −40°, mp: −124°.
Acute tox data: Inhal LC_{LO} (rat) = 19 ppm for 4 hrs. [*3*]
THR = HIGH via inhal route. A powerful irr. See also fluorides.
Disaster Hazard: Dangerous; when heated to decomp, emits highly toxic fumes of fluorides; will react with water, steam or acids to produce toxic and corrosive fumes.

SULFUR TRIOXIDE. Syn: *sulfuric acid anhydride.* It exists in 3 forms; the most valuable commercially is the γ form (mp: 16.8°, bp: 44.8°) which has a strong tendency to polymerize to the straight chain β form (mp β: 32.5°) and subsequently to the x-linked α form (mp α: 62°). When the β or α forms are melted they tend to revert to the γ form liquid or ice-like crystals. SO_3 (β) $\longrightarrow$ asbestos-like crystals. vap. press: (β) 433 mm @ 250°, vap. press (α): 344 mm. [*73*] SO_3, mw: 80.07. vap. d: 2.76.
THR = HIGH irr to skin, eyes and mu mem. A corrosive poison. See also sulfuric acid. Violent reaction with O_2F_2, PbO, $NClO_2$, $HClO_4$, P, tetrafluorethylene. [*19*]
Disaster Hazard: Dangerous; when heated to decomp, emits highly toxic fumes; reacts with steam to form corrosive, toxic fumes of sulfuric acid.

SULFUR TRIOXIDE, STABILIZED. See sulfur trioxide.

SULFUR TRIOXYTETRACHLORIDE. White crystals. $S_2O_3Cl_4$, mw: 254.0, mp: 57° (decomp).
THR = U. This material readily decomp into toxic compounds and should be considered highly toxic.
Disaster Hazard: Dangerous; when heated to decomp, emits highly toxic fumes of SO_x and chlorides; will react with water or steam to produce toxic and corrosive fumes.

SULFURYL AMIDE. See sulfamide.

SULFURYL CHLORIDE. Syns: *chlorosulfuric acid, sulfonyl chloride*; *sulfuric chloride, sulfuric oxychloride.* Colorless liquid, pungent odor. SO_2Cl_2, mw: 135.0, mp: −54.1°, bp: 69.1°, d: 1.6674, vap. press: 100 mm @ 17.8°, vap. d: 4.65.
THR = HIGH irr to skin, eyes and mu mem. Corrosive. See sulfuric acid and hydrochloric acid, which are formed upon hydrolysis. Can explode with PbO_2. [*19*]
Disaster Hazard: Dangerous; when heated to decomp, emits highly toxic fumes of chlorides and SO_x; will react with water or steam to produce heat and toxic and corrosive fumes.

SULFURYL CHLORIDE FLUORIDE. Colorless gas. SO_2ClF, mw: 118.52, mp: −124.7°, bp: 7.1°.
Acute tox data: inhal LC_{LO} (mice) = 620 mg/m^3. [*3*]

THR = HIGH via inhal route. Very irr. See sulfuric acid, chlorides and fluorides.

SULFURYL FLUORIDE. Syns: *sulfuric oxyfluoride, vikane.* Colorless gas. SO_2F_2, mw: 102.07, mp: −137°, bp: −55°, d: 3.72 g/liter.

Acute tox data: Oral LD_{LO} (rat) = 100 mg/kg; oral LD_{50} (guinea pig) = 100 mg/kg. [*3*]

THR = HIGH via oral route. See also fluorides. Accidental exposure of a human resulted in nausea, vomiting, cramps and itching. May have narcotic action in high conc.

Disaster Hazard: Dangerous; when heated to decomp, emits highly toxic fumes of fluorides and SO_x; will react with water or steam to produce toxic and corrosive fumes.

SULPHENONE. Syn: *p-chlorophenyl phenyl sulfone.* Crystals, slight aromatic odor, no taste, insol in water. $C_{12}H_9ClO_2S$, mw: 252.7, mp: 90°–94°.

Acute tox data: Oral LD_{50} (rat) = 1400 mg/kg; LD_{LO} (rat) = 500 mg/kg. [*3*]

THR = HIGH via ip route and MOD via oral route.

Disaster Hazard: Dangerous; when heated to decomp, emits highly toxic fumes.

SUNLIGHT.

THR = A recog carc. [*14*] See radiation, ultraviolet.

SUPERPALITE. See trichloroacetyl chloride.

SUPINIDINE.

THR = An exper carc. [*23*]

SUPININE.

Acute tox data: uk LD_{50} (rat) = 400 mg/kg. [*3*]

THR = HIGH. An exper carc. [*23*]

SURFACTANTS.

THR = A (S) carc of the skin, lungs, alimentary canal, bladder. [*14*]

SUSCEPTIBILITY OF SURFACTANTS TO FISSION PRODUCT CONTAMINATION. See Section 5A.

SUSCEPTIBILITY TO SPONTANEOUS HEATING. See Section 7.

SWEET GUM (tannin-containing fraction).

THR = An exper neo to rats via sc routes. [*103*]

SWEET WAX.

THR = A recog carc. [*14*] See paraffins and petroleum waxes.

SWEET OIL. See olive oil.

SWEP. Syn: *methyl-n-(3,4-dichlorophenyl)carbamate.* A powder. $C_8H_7O_2NCl_2$, mw: 220.

Acute tox data: Oral LD_{50} (rat) = 522 mg/kg; dermal LD_{50} (rabbit) = 2480 mg/kg. [*3*]

THR = HIGH via oral route and MOD via dermal route. An herbicide. See also carbamates.

Disaster Hazard: Dangerous. See chlorides.

SYLVAN. See 2-methyl furan.

SYLVIC ACID. See abietic acid.

SYMPTOMATOLOGY AND CLINICAL TESTS FOR POISONING. See Section 9.

SYNTHETIC HYDROGENATED COAL.

THR = A recog carc. [*14*] See petroleum asphalt.

SYSTOX. See demeton.

t

2,4,5-T. See 2,4,5-trichlorophenoxyacetic acid.

TABUN. Syn: *cyanodimethylamino ethoxyphosphine oxide*. A colorless to brownish liquid.

$C_5H_{11}O_2N_2P$, mw: 162.15, bp: decomp @ 238°, fp: −49.4°, flash p: 172°F, d: 1.073 @ 25°, vap. press: 0.07 mm @ 25°, vap. d: 5.63.

Acute tox data: Inhal LC_{LO} (human) = 150 mg/m^3; inhal TC_{LO} (human) = 200 ppm ⟶ CNS effects; dermal LD_{LO} (human) = 23 mg/kg; iv LD_{LO} (human) = 0.014 mg/kg; oral LD_{50} (rat) = 3.7 mg/kg; iv LD_{50} (rat) = 0.066 mg/kg; dermal LD_{50} (mice) = 1 mg/kg; ip LD_{50} (mouse) = 0.6 mg/kg; sc LD_{50} (mice) = 0.34 mg/kg; iv LD_{50} (mice) = 0.15 mg/kg; dermal LD_{50} (monkey) = 9 mg/kg; oral LD_{50} (rabbit) = 16 mg/kg; dermal LD_{50} (rabbit) = 35 mg/kg; sc LD_{50} (rabbit) = 0.5 mg/kg; sc LD_{50} (guinea pig) = 0.2 mg/kg. [*3*]

THR = VERY HIGH via all routes. LD (man) = 0.01 mg/kg. A nerve gas. Vapor does not penetrate skin; liquid does so rapidly. The primary physiological action is on the sympathetic nervous system, causing a vasoparesis. Vapors when inhaled can cause nausea, vomiting and diarrhea, which can be followed by muscular twitchings and convulsions. See also parathion.

Fire Hazard: Mod, when exposed to heat or flame.

Disaster Hazard: Highly dangerous; emits highly toxic fumes; can react with oxidizing materials.

TABUTREX. See di-*n*-butyl succinate.

TAGAYASAN. See sawdust.

TALC. Syns: *talcum*, *French chalk*, *steatite*. White to grayish-white, fine, odorless powder. Powdered native hydrous magnesium silicate.

THR = The talc with $< 1\%$ asbestos is mainly regarded as a nuisance dust. Prolonged or repeated exposure can produce a form of pulmonary fibrosis (talc pneumoconiosis) which may be due to asbestos content. It is a substance which migrates to food from packaging materials. [*109*] It is a common air contaminant. See also asbestos.

TALC DUST. See talc.

TALCUM. See talc.

TALL OIL. Syns: *liquid rosin*, *tallol*. Flam liquid, dark brown, acrid odor. Composition: rosin acids, oleic and linoleic acids. d: 0.95, flash p: 360°F.

THR = A MILD allergen. A substance which migrates to food from packaging materials. [*109*]

Fire Hazard: Slight, when exposed to heat or flame; can react with oxidizing materials.

To Fight Fire: Dry chemical, CO_2.

TALLOL. See tall oil.

TALLOW. A solid fat. mp: 31°–37°, flash p: 509°F (CC), d: 0.895.

THR = LOW. A substance which migrates to food from packaging materials. [*109*]

Fire Hazard: Slight, when exposed to heat or flame; can react with oxidizing materials.

Spont Heating: Yes.

To Fight Fire: CO_2, dry chemical.

TALLOW OIL. Syn: *oleo oil*. mp: 42.8°, flash p: 492°F, d: 0.914, autoign. temp.: 980°F.

THR = U. Probably LOW.

Fire Hazard: Slight, when exposed to heat or flame; can react with oxidizing materials.

Spont Heating: Yes.

To Fight Fire: CO_2, dry chemical.

TAMARON. Syns: *o,s-dimethyl phosphoramidothioate; ortho—9006, chevron 9006*. $C_2H_8O_2PNS$, mw: 141.1, mp: 40°. Crystals slightly water sol; sol in alcohol.

Acute tox data: Oral LD_{50} (rats) = 7.5 mg/kg; dermal LD_{50} (rabbit) = 118 mg/kg. [*3*]

THR = HIGH via oral and dermal routes. An insecticide and acaricide. A cholinesterase inhibitor. See parathion.

Disaster Hazard: Dangerous; see parathion.

TANKAGE FERTILIZER.

THR = Details U; may contain ammonia, sulfides and other irr.

Fire Hazard: Mod, when exposed to heat or flame. Presence or absence of moisture can contribute to spont heating. Avoid storage before cooling, or extremes of moisture content; can react with oxidizing materials.

Spont Heating: Variable.

TANKAGE, ROUGH AMMONIATE. See tankage fertilizer.

TANK BOTTOM WAX.

THR = A recog carc. [*14*] See petroleum waxes and paraffins.

TANKS, EMPTY. See Section 7.

TANKS FOR STORAGE OF FLAMMABLE LIQUIDS. See Section 7.

TANNIC ACID. Syns: *gallotannin, gallotannic acid.* Yellowish-white or brown bulky powder or flakes. $C_{76}H_{52}O_{46}$, mw: 1701.2, mp: 200°, flash p: 390°F (OC), autoign. temp.: 980°F.
Acute tox data: Oral LD_{100} (rat) = 6000 mg/kg. [*2*] Oral LD_{50} (rabbit) = 5000 mg/kg; sc LD_{LO} (mice) = 75 mg/kg; iv LD_{LO} (mice) = 10 mg/kg. [*3*]
THR = HIGH via iv and sc; MOD via oral routes. An exper (+) carc [*3, 7, 14*] of the liver via sc route.
Fire Hazard: Slight, when exposed to heat or flame.
Spont Heating: No.
To Fight Fire: Water.

TANNIN. Extracted from bracken fern.
Acute tox data: Ip LD_{50} (mice) = 160 mg/kg. [*3*]
THR = HIGH via ip route. An exper (S) carc via imp route. [*3, 7*]

TANSY OIL. Syn: *oil of tansy.* Yellowish liquid, strong odor, d: 0.925–0.950 @ 15°/15°. Sol in alcohol, ether chloroform. Slightly sol in water.
THR = A poison.
Disaster Hazard: Dangerous; when heated, emits toxic fumes.

TANTALUM. A heavy, gray, dense, very hard metal. Ta, atwt: 180.948, mp: 2996°, bp: 5425°, d: 16.654.
THR = An exper neo by implantation. [*3*] See also tantalum compounds. Some industrial skin injuries from tantalum have been reported. However, systemic industrial poisoning is apparently not known.
Radiation Hazard: For permissible levels, see Section 5A, Table 5A.5. Artificial isotope ^{182}Ta, $T\frac{1}{2}$ = 115 d, decays to stable ^{182}W via β's of 0.36–0.51 MeV; emits γ's of 0.07–1.2 MeV.
Fire Hazard: The dry powder ignites spont in air; it reacts violently with BrF_3, F_2. [*19*] See also powdered metals.

TANTALUM BROMIDE. Yellow crystals. $TaBr_5$, mw: 580.46, mp: 240°, bp: 320°, d: 4.67.
THR = See bromides and tantalum compounds.

TANTALUM CHLORIDE. Light yellow crystalline powder. $TaCl_5$, mw: 358.17, mp: 221°, bp: 242°, d: 3.68 @ 27°.
Acute tox data: Oral LD_{50} (rat) = 1900 mg/kg; ip LD_{50} (rat) = 75 mg/kg. [*3*]
THR = HIGH via ip route and MOD via oral route. See chlorides and tantalum compounds.

TANTALUM COMPOUNDS.
THR = Some tantalum compounds have been suspected of causing skin irr and mild fibrosis of the lungs. See also tantalum.

TANTALUM FLUORIDE. Colorless crystals. TaF_5, mw: 275.95, mp: 96.8°, bp: 229.5°, d: 4.74, vap. press: 100 mm @ 130°.
Acute tox data: Iv LD_{50} (mice) = 110 mg/kg. [*3*]
THR = HIGH via iv route. See also fluorides and tantalum compounds.

TANTALUM PENTOXIDE. Syn: *tantalic acid anhydride.* Ta_2O_5, mw: 441.9.
THR = Reacts violently with BrF_3, ClF_3, Li. [*19*] See also tantalum.

TANTALUM POTASSIUM FLUORIDE. See potassium heptafluorotantalate.

TAPIOCA DEXTRIN. See dextrin.

TAR ACIDS (COAL). See phenol and cresol.

TARAPACAITE. See potassium chromate.

TAR BASES.
THR = A recog carc. [*14*] See coal tar and pitch.

TAR CAMPHOR. See naphthalene.

TAR, DEHYDRATED. See coal tar creosote.

TAR DUST.
THR = MOD irr to skin, eyes and mu mem.
Fire Hazard: Mod, when exposed to heat or flame; can react with oxidizing materials.
To Fight Fire: Dry chemical, CO_2, foam.

TAR, LIQUID. See coal tar creosote.

TAR OIL RECTIFIED. See pine tar oil.

TARTAN. Syn: *o,o-diethyl-S,N-(α-cyanoisopropyl)-carbamoyl methyl phosphorothioate.* A yellow liquid, sol in most organic solvents. $C_{10}H_{19}O_4PSN$, mw: 180.1, d: 1.200.
Acute tox data: Oral LD_{50} (rat) = 3.5 mg/kg; dermal LD_{50} (rat) = 105 mg/kg; oral LD_{50} (mice) = 12 mg/kg; oral LD_{50} (dog) = 20 mg/kg; oral LD_{50} (rabbit) = 8 mg/kg. [*3*]
THR = HIGH via oral and dermal routes. HIGH toxicity organophosphate acaricide and insecticide. See also parathion.
Disaster Hazard: Dangerous; see parathion.

TARTAR EMETIC. See antimony potassium tartrate.

TARTARIC ACID. Syn: *racemic acid.* White crystals. $HOOC(CHOH)_2COOH$, mw: 150.09, mp: 168°–170°, flash p: 410°F (OC), d: 1.76, autoign. temp.: 797°F.
Acute tox data: iv LD_{50} (mice) = 485 mg/kg; oral LD_{LO} (dog) = 5000 mg/kg. [*3*]
THR = HIGH via iv and MOD via oral routes. It is used as a sequestrant and general purpose food additive; it is a substance which migrates to food from packaging materials. [*109*]
Fire Hazard: Slight, when exposed to heat or flame. Violent reaction possible with Ag. [*19*]

Spont Heating: No.
To Fight Fire: Water.

TAR, WATER GAS.
THR = MOD irr.
Fire Hazard: Slight, when exposed to heat.
Disaster Hazard: Mod dangerous; when strongly heated, emits acrid fumes.

TAUROCHOLIC ACID. Syn: *cholaic acid, cholytaurine*. Occurs as a sodium salt in the bile, crystals, freely sol in water, sol in alcohol, insol in ether and ethyl acetate. $C_{26}H_{45}NO_7S$, mw: 515, mp: 125°.
Acute tox data: ip LD_{50} (rat) = 450 mg/kg; ip LD_{50} (mice) = 620 mg/kg. [*3*]
THR = HIGH–MOD via ip route. An emulsifying agent food additive. [*109*]
Disaster Hazard: Dangerous, see sulfides.

TCM. See trichloromelamine.

TCNE. See tetracyanoethylene.

2,4,5-TCPPA. See 2-(2,4,5-trichlorophenoxy)propionic acid.

TDE. See DDD.

TDI. See 2,4-tolylene diisocyanate.

TEAK. See sawdust.

TEAR GAS MATERIAL, LIQUID OR SOLID, N.O.S.
THR = HIGH irr to skin, eyes and mu mem.
Disaster Hazard: Dangerous; emits highly toxic fumes.

TECHNETIUM. Hexagonal crystal structure. A chemical element. Tc, atwt: 97.
THR = U.
Radiation Hazard: For permissible levels, see Section 5A, Table 5A.5. Artificial isotope ^{97}Tc, $T\frac{1}{2}$ = 2.6×10^6y, decays to stable ^{97}Mo by ec; emits x-rays. Artificial isotope ^{99}Tc, $T\frac{1}{2} = 2.1 \times 10^5$y, decays to stable ^{99}Ru via β's of 0.29 MeV.

TEDION. Syns: *duphar; tetrachlorodiphenyl sulfone, tetradifon*. Crystals, nearly water-insol. $C_{12}H_6Cl_4O_2S$, mw: 356.1, mp: 147°.
Acute tox data: Oral LD_{50} (rat) = 566 mg/kg. [*3*]
THR = MOD via oral route. A food additive permitted in food for human consumption. [*109*]
Disaster Hazard: Dangerous; when heated to decomp, emits highly toxic fumes of chlorides and SO_x.

TEDP. See tetraethyldithiopyrophosphate.

TEFLON. See polytetrafluoroethylene.

TELEVISION TUBE RADIATIONS.
THR = A recog carc. [*14*] See radiation, ionizing.

***o*-TELLURIC ACID.** Needles. $H_6TeO_6 \cdot 4H_2O$; mw: 301.72, mp: $-4H_2O$ @ 100°.
Acute tox data: iv LD_{LO} (rat) = 31 mg/kg; oral LD_{LO} (rabbit) = 56 mg/kg; iv LD_{LO} (rabbit) = 5.6 mg/kg. [*3*]
THR = HIGH via iv and oral routes. See tellurium compounds.

TELLURIUM. Silvery-white, metallic, lustrous element, quite brittle. Te, atwt: 127.6, mp: 449.5°, bp: 989.8°, d: 6.24 @ 20°, vap. press: 1 mm @ 520°.
Acute tox data: Sc LD_{LO} (dog) = 290 mg/kg. [*3*]
THR = HIGH via sc route. See tellurium compounds.
Radiation Hazard: For permissible levels, see Section 5A, Table 5A.5. Artificial isotope ^{125m}Te, $T\frac{1}{2}$ = 58d, decays to stable ^{125}Te via γ's of 0.04, 0.11 MeV. Artificial isotope ^{127m}Te, $T\frac{1}{2}$ = 109d, decays to radioactive ^{127}Te via γ's of 0.09 MeV. Artificial isotope ^{129m}Te, $T\frac{1}{2}$ = 34d, decays to radioactive ^{129}Te via γ's of 0.11 MeV.
Fire Hazard: Mod, in the form of dust when exposed to heat or flame or by chemical reaction with Cd, Cl, ClF, ClF_3, F_2, Zn, Li_6Si_2, $(K + H_2)$, $AgBrO_3$, Na. [*19*] See also powdered metals.

TELLURIUM COMPOUNDS.
THR = Elemental tellurium has relatively LOW toxicity. It is converted in the body to dimethyl telluride which imparts a garlic-like odor to the breath and sweat. Heavy exposures may, in addition, result in headache, drowsiness, metallic taste, loss of appetite and nausea. Various tellurium salts may also produce similar symptoms. Large doses can be fatal, as was the case following accidental administration of sodium tellurite.
Disaster Hazard: Dangerous; when heated or on contact with acid or acid fumes, they emit highly toxic fumes.

TELLURIUM DIBROMIDE. Needles. $TeBr_2$, mw: 287.44, mp: 210°, bp: 339°.
THR = See tellurium compounds and bromides.

TELLURIUM DICHLORIDE. Crystals or amorphous solid, unstable. $TeCl_2$, mw: 198.52, mp: 209 ± 5°, bp: 327°, d: 7.05.
THR = See tellurium compounds and chlorides.

TELLURIUM DIETHYL DITHIOCARBAMATE.
Syn: *ethyl tellurac*. Orange-yellow powder. $C_{20}H_{40}N_4S_8 \cdot Te$; mw: 720.6, d: 1.44, mp: 108°–118°.
THR = An exper (±) carc and neo [*12, 3*] via oral and sc routes.

TELLURIUM DIIODIDE. Crystals. TeI_2, mw: 381.45, mp: sublimes.
THR = See tellurium compounds and iodides.

TELLURIUM DIOXIDE. TeO_2, mw: 159.6.
Acute tox data: uk LD_{LO} (rabbit) = 324 mg/kg. [*3*]
THR = HIGH via uk route.

TELLURIUM HEXAFLUORIDE. Colorless gas, unpleasant odor. TeF_6, mw: 241.61, mp: −37.6°, bp: −38.9° (subl), d: 3.025 @ −35.5°.
Acute tox data: Inhal LC_{LO} (rat) = 5 ppm. [*3*]
THR = HIGH via inhal route. HIGH irr to skin, eyes and mu mem. See fluorides and tellurium compounds.

TELLURIUM HYDRIDE. Colorless gas. H_2Te, mw: 130, mp: −48.9°, bp: 2.2°.
THR = HIGH. See hydrides and tellurium.

TELLURIUM NITRIDE. Solid.
THR = HIGH. See tellurium compounds and nitrides.
Explosion Hazard: Severe, when shocked or exposed to heat.
Disaster Hazard: Dangerous; shock will explode it; when heated or on contact with acid or acid fumes, emits highly toxic fumes of tellurium and may explode; will react with water or steam to produce toxic fumes.

TELLURIUM SULFIDE. Amorphous powder. TeS_2, mw: 191.73.
THR = See sulfides and tellurium compounds.

TELLURIUM SULFITE. Amorphous, deep red solid. $TeSO_3$, mw: 207.67, mp: softens @ 30°, bp: decomp.
THR = See tellurium compounds and sulfites.

TELLURIUM TETRACHLORIDE HYDROCHLORIDE. $HTeCl_5$, mw: 305.9.
THR = When reacted with NH_3, an explosive may be formed. [*19*]

TELLURIUM TETRAFLUORIDE. White crystals. TeF_4, mw: 203.61, mp: sublimes, bp: >97°.
THR = See fluorides and tellurium compounds.
Acute tox data: Oral LD_{50} (rat) = 4.8 mg/kg; dermal LD_{50} (rat) = 5 mg/kg; ip LD_{50} (rat) = 3.56 mg/kg; oral LD_{50} (chicken) = 2 mg/kg; dermal LD_{50} (guinea pig and rabbit) = 2 mg/kg. [*3*]
THR = HIGH via oral, dermal and ip routes. See also fluorides and tellurium compounds.

TELODRIN. Syn: *isobenzan*. $C_9H_4OCl_8$, mw: 411.7.
THR = An exper neo via oral route. [*3*]

TELONE. See 1,3-dichloropropene.

TEM. See triethylene melamine.

TEMIK. Syn: *2-methyl-2-(methylthio)propionaldehyde-o-(methylcarbamoyl)oxime*. A solid material. $C_7H_{14}O_2N_2S$, mw: 190.3.
Acute tox data: Oral LD_{50} (rat) = 1 mg/kg; dermal LD_{50} (rat) = 2.5 mg/kg; oral LD_{50} (duck) = 4.44 mg/kg; oral LD_{50} (mice) = 0.3 mg/kg. [*3*]
THR = HIGH via oral and dermal routes. Very HIGH systemic insecticide, acaricide and nematocide.

TEMUR. See tetramethyl urea.

TENAMENE. $C_{20}H_{36}N_2$, mw: 304.6.
Acute tox data: Oral LD_{LO} (rat) = 800 mg/kg; ip LD_{LO} (rat) = 800 mg/kg. [*3*]
THR = MOD via oral and ip routes.

TENAMENE 31. $C_{22}H_{40}N_2$, mw: 332.6.
Acute tox data: Oral LD_{50} (rat) = 2400 mg/kg; dermal LD_{50} (rabbit) = 1800 mg/kg. [*3*]
THR = MOD via oral and dermal routes.

TENAMENE 2. $C_{14}H_{24}N_2$, mw: 220.4.
Acute tox data: Dermal LD_{50} (guinea pig) = 5000 mg/kg; oral LD_{LO} (rat) = 200 mg/kg; inhal LC_{LO} (rat) = 600 mg/m^3 for 6 hrs. [*3*]
THR = HIGH via oral route; MOD via inhal and dermal route.

"TENAMINE-1." Lemon-yellow, mobile liquid. Comp: 48% N, *n*-butyl-*p*-aminophenol in isopropyl alcohol. d: 0.89–0.91 @ 25°/25°, flash p: 61°F, fp: −33°.
THR = U. See individual components.
Fire Hazard: Dangerous, when exposed to heat, flame or oxidizers.
Explosion Hazard: U.
Disaster Hazard: Dangerous; when heated to decomp, emits toxic fumes; can react with oxidizing materials.
To Fight Fire: CO_2, dry chemical.

"TENAMINE 60." Liquid. Comp: 20% toluene and 80% disalicylal propylenediimine. fp: −18.3°, flash p: 19°F (CC), d: 1.07 @ 25°/25°.
THR = Details U. See individual components.
Fire Hazard: Dangerous, when exposed to heat or flame.
Explosion Hazard: U.
Disaster Hazard: Dangerous; keep away from open flame or heat; emits toxic fumes when heated; can react with oxidizing materials.
To Fight Fire: Water, foam, CO_2, dry chemical.

TEP. See tetraethyl pyrophosphate.

TEPA. See tris(1-aziridinyl)phospine oxide.

TEPP. See tetraethyl pyrophosphate.

TERATOGENESIS. See Sections 9 and 8.

TERBACH. See sinbar.

TERBIUM. A silvery-gray, soft, ductile, malleable metallic element. Tb, atwt: 158.9254, mp: 1360°, bp: 3041°, d: 8.234.
THR = U. As a lanthanon it may impair blood coagulation. See also lanthanum.
Fire Hazard: In the form of dust in air or on contact with halogen. [*19*] See also powdered metals.
Radiation Hazard: For permissible levels, see Section 5A, Table 5A.5. Artificial isotope ^{160}Tb, $T_{\frac{1}{2}}$ = 72d, decays to stable ^{190}Dy via β's of 0.25–0.86 MeV; emits γ's of 0.9–1.31 MeV.

TERBIUM CHLORIDE. $TbCl_3$, mw: 265.3.
Acute tox data: ip LD_{50} (mice) = 332 mg/kg. [*3*]
THR = HIGH via ip route. See also terbium.

TERBIUM CITRATE.
Acute tox data: ip LD_{50} (mice) = 121 mg/kg; ip LD_{50} (guinea pig) = 74 mg/kg. [*3*]
THR = HIGH via ip route. See also terbium.

TERBIUM NITRATE. Colorless monoclinic needles. $Tb(NO_3)_3 \cdot 6H_2O$, mw: 453.1, mp: 89.3°.
Acute tox data: ip LD_{50} (rat) = 260 mg/kg; ip LD_{50} (mice) = 480 mg/kg. [*3*]
THR = HIGH via ip route. See nitrates and terbium.

TERBIUM OXIDE.
Acute tox data: Oral LD_{LO} (rat) = 1000 mg/kg. [*3*]
THR = MOD via oral route. See also terbium.

TEREPHTHALIC ACID. Syns: *p-phthalic acid, TPA, benzene-p-dicarboxylic acid.* White crystals or powder, insol in water, chloroform, ether and acetic acid, slightly sol in alcohol, sol in alkalies. $C_6H_4(COOH)_2$, mw: 166.13, d: 1.51, sublimes @ 300°.
Acute tox data: Ip LD_{50} (mice) = 1430 mg/kg; iv LD_{LO} (dog) = 767 mg/kg. [*3*]
THR = MOD via ip and iv routes.

TEREPHTHALOYL CHLORIDE. Syn: *p-phthalyl dichloride.* White crystalline material, musty odor. $C_8H_4O_2Cl_2$, mw: 203.0, bp: 266°, flash p: 356°F (COC), fp: 81.4°.
Acute tox data: Oral LD_{50} (rat) = 2500 mg/kg. [*3*]
THR = MOD via oral route. An irr.
Disaster Hazard: Dangerous. When heated to decomp, emits highly toxic fumes of chlorides.

TERGITOL (TYPICAL).
THR = MOD via oral and dermal routes. See also 7-ethyl-2-methyl-4-hendecanol sulfate, sodium salt.

TERPENE POLYCHLORINATES. Chlorinated mixed terpenes.
Acute tox data: Oral LD_{50} (rat) = 200 mg/kg; oral LD_{50} (chicken) = 139 mg/kg. [*3*]
THR = HIGH via oral route. An exper (+) carc via oral route. [*3, 12, 10*]

TERPENES. See turpentine oil.

***m*-TERPHENYL.** Syns: *m-phenylbiphenyl, benzene-1,3-diphenyl.* Colorless needles. $(C_6H_5)_2C_6H_4$, mw: 230.3, mp: 86°–87°, bp: 363°, flash p: 375°F (OC), d: 1.164, vap. press: 7.95.
THR = MOD via oral and inhal routes. See also p-terphenyl. Animal feeding exper show injury to liver and kidneys. Toxicity of terphenyls is proportional to sol. See also diphenyl.
Fire Hazard: Slight, when exposed to heat or flame.
To Fight Fire: CO_2, dry chemical.

***o*-TERPHENYL.** Syn: *1,2-diphenylbenzene.* A liquid. $C_6H_4(C_6H_5)_2$, mw: 230.3, bp: 332°, flash p: 325°F (OC), d: 1.14, vap. d: 7.95.
THR = U. See *p*-terphenyl.
Fire Hazard: Slight, when exposed to heat or flame.
To Fight Fire: CO_2, dry chemical.

***p*-TERPHENYL.** Syn: *benzene-1,4-diphenyl.* $(C_6H_5)_2C_6H_4$, mw: 230.3, mp: 213°, bp: 405°, flash p: 405°F (OC), d: 1.236, vap. d: 7.95.
Fire Hazard: Slight, when exposed to heat or flame.
To Fight Fire: Water, CO_2, dry chemical.

TERPINOLENE. Syn: *1,4(8)-p-menthadiene.* Colorless liquid. $C_{10}H_{16}$, mw: 136.23, bp: 185°, d: 0.855, flash p: 100°F (CC).
THR = U.
Fire Hazard: Mod, when exposed to heat or flame.
Disaster Hazard: Mod dangerous; keep away from open flame; can react with oxidizing materials.
To Fight Fire: Foam, CO_2, dry chemical.

TERTIARY ALCOHOL. See alcohols, N.O.S.

TESTOSTERONE. Syn: *Δ^4-androsten-17-(a)-ol-3-one.* Crystals. $C_{19}H_{28}O_2$, mw: 288.4, mp: 155°.
Acute tox data: ip LD_{LO} (rat) = 326 mg/kg. [*3*]
THR = HIGH via ip route. An exper (+) carc [*15*] and an exper neo [*3*] via sc route. Workers engaged in manufacture and packaging have shown effects from this hormone. Enlargement of the breasts in male workers has been observed. ip LD_{100} (female rats) = 325 mg/kg.
Disaster Hazard: Dangerous; when heated to decomp, emits toxic fumes.

TESTOSTERONE PROPIONATE. White or creamy white crystals or crystalline powder, odorless, freely sol in alcohol, dioxane, ether and other organic solvents, sol in vegetable oils, insol in water. $C_{19}H_{27}O \cdot OOCC_2H_5$.
THR = An exper neo. [*3*] A food additive permitted in the feed and drinking water of animals and/or for the treatment of food-producing animals. Also permitted in food for human consumption. [*109*]

TETRAAMINE COPPER SULFATE. See copper sulfate, ammonia and amines.

TETRAAMYL BENZENE. Liquid. $C_6H_2(C_5H_{11})_4$, mw: 358.6, bp: 320°, flash p: 295°F, d: 0.89.
THR = Details U. Probably irr and narcotic in HIGH conc.
Fire Hazard: Slight, when exposed to heat or flame; can react with oxidizing materials.
To Fight Fire: CO_2, dry chemical.

TETRA-*n*-AMYLTHIOGERMANIUM. Colorless liquid, sol in benzene. $Ge[S(CH_2)_4CH_3]_4$, mw: 485.4, d: 1.0697 @ 25°, bp: 241° @ 4 mm.
THR = See germanium compounds and organometals.

TETRA-*n*-AMYLTIN. Colorless stable liquid. $(C_5H_{11})_4Sn$, mw: 403.3, d: 1.0206, bp: 181° @ 10 mm.
THR = See tin compounds.

TETRAAQUOSTANNIC BIS-ACETYL ACETONE STANNIC BROMIDE. Colorless crystals, sol in benzene. $(C_5H_7O_2)_2Sn(OH_2)_4SnBr_6$, mw: 987.2, mp: 107°.
THR = See tin compounds and bromides.

TETRABENZYL GERMANIUM. Colorless powder. $Ge(CH_2C_6H_4)_4$, mw: 437.1, mp: 108°.
THR = Details U. See germanium compounds.

TETRABENZYL SILICANE. Liquid, insol in water. $(C_6H_5CH)_2Si$, mw: 392.6, bp: 127.5°.
THR = U. See silanes.

TETRABENZYLTIN. Colorless crystals, insol in water, sol in common organic solvents. $(C_6H_5CH_2)_4Sn$, mw: 483.2, mp: 43°.
THR = U. See tin compounds.

TETRA-*p*-BIPHENYLYL GERMANIUM. White crystals, insol in water. $Ge(C_6H_4C_6H_5)_4$, mw: 685.4, mp: 271°.
THR = U. See germanium compounds.

TETRABORANE. See boron hydrides.

TETRABORIC ACID. White powder. $H_2B_4O_7$, mw: 157.30.
THR = See boron compounds.

TETRABORONDECAHYDRIDE. See dihydrotetraborane.

TETRABROMETHYLENE. See dicarbon tetrabromide.

3,4,6,7-TETRABROMO-*o*-CRESOL. White to buff crystals, insol in water. $C_7H_4Br_4O$, mw: 423.8, mp: 205°–208° (decomp).
THR = Details U. Animal exper show corrosive action on skin but low systemic toxicity. See also cresol.
Disaster Hazard: Dangerous; when heated to decomp, emits highly toxic fumes.

TETRABROMOETHANE. See acetylene tetrabromide.

TETRABROMOMETHANE. See carbon tetrabromide.

TETRA-*p*-BROMOPHENYL THIOGERMANIUM. Colorless crystals, sol in benzene. $Ge(SC_6H_4Br)_4$, mw: 824.9, mp: 196°.
THR = See sulfates, bromides, germanium compounds and organometals.

TETRABROMOSILANE. See silicon bromide.

TETRABROMOSILICANE. See silicon bromide.

TETRA-*n*-BUTYLGERMANIUM. Colorless oily liquid. $Ge(C_4H_9)_4$, mw: 301.1, bp: 180°.
THR = Details U. See germanium compounds.

TETRA-*p-tert*-BUTYLPHENYL THIOGERMANIUM. Colorless crystals, sol in petroleum ether, ether and acetone. $Ge[SC_6H_4C(CH_3)_3]_4$, mw: 733.7, mp: 156°.
THR = See germanium compounds and organometals.

TETRABUTYL THIODISUCCINATE. Liquid. $[CH_2(COOC_4H_9)CH(COOC_4H_9)]_2S$, mw: 490.64, mp: −45°, bp: 246° @ 5 mm, flash p: 430°F (OC), d: 1.0543 @ 20°/20°, vap. press: 0.24 mm @ 200°, vap. d: 16.9.
Acute tox data: Oral LD_{50} (rat) = 100000 mg/kg. [*3*]
THR = NONE via oral route.
Fire Hazard: Slight, when exposed to heat or flame.
Disaster Hazard: Dangerous; when heated to decomp, burns and emits highly toxic fumes of SO_x; can react on contact with oxidizing materials.
To Fight Fire: Water, foam, CO_2, dry chemical.

TETRA-*n*-BUTYL THIOGERMANIUM. A liquid. $Ge[S(CH_2)_3CH_3]_4$, mw: 429.3, bp: 222.5° @ 4.5 mm, d: 1.1072 @ 25°.
THR = See germanium compounds, organometals and sulfur compounds.

TETRA-*sec*-BUTYL THIOGERMANIUM. A liquid, sol in benzene. $Ge[SCH(CH_3)C_2H_5]_4$, mw: 429.3, d: 1.1119 @ 25°, bp: 200.5° @ 4 mm.
THR = See germanium compounds, sulfur compounds and organometals.

TETRA-*tert*-BUTYL THIOGERMANIUM. Crystals, sol in absolute alcohol. $Ge[SC(CH_3)_3]_4$, mw: 429.3, mp: 173°, bp: sublimes @ 170° @ 4 mm.
THR = See germanium compounds, sulfur compounds and organometals.

TETRA-*n*-BUTYLTIN. Colorless stable liquid. $(C_4H_9)_4Sn$, mw: 347.2, d: 1.0572, bp: 145° @ 10 mm.
THR = HIGH via oral and dermal routes. See tin compounds.

TETRA-*n*-BUTYL TITANATE. Colorless liquid. $(C_4H_9O)_4Ti$, mw: 340, mp: < −40°, bp: 206° @ 10 mm, d: 0.9951 @ 25°, vap. d: 11.7.
THR = U. See esters and titanium compounds.
Fire Hazard: Mod, when exposed to heat or flame.

TETRABUTYL UREA. Liquid. $(C_4H_9)_2NCON(C_4H_9)_2$, mw: 284.5, mp: −60°, bp: 300°–325°, flash p: 200°F, d: 0.876, vap. d: 9.83.
THR = U.
Fire Hazard: Mod, when exposed to heat, flame or oxidizers.
Disaster Hazard: Dangerous; when heated to decomp, emits toxic fumes; can react with oxidizing materials.
To Fight Fire: Foam, CO_2, dry chemical.

TETRACENE. Syn: *naphthacene*. Orange crystals. $C_{18}H_{12}$, mw: 228.3, mp: 341°, d: 1.35.
THR = U.
Explosion Hazard: Mod, when shocked.
Disaster Hazard: Dangerous; shock will explode it; when heated, burns and emits acrid fumes; can react on contact with oxidizing materials.

TETRACETYLENE DICARBONIC ACID. Solid.
THR = U.
Explosion Hazard: Severe, when shocked or exposed to heat or by chemical reaction.
Disaster Hazard: Highly dangerous; shock or heat will explode it.

TETRACETYL THIOGERMANIUM. White crystals, sol in organic solvents. $Ge[SCH_2(CH_2)_{14}CH_3]_4$, mw: 1102.6, mp: 51°.
THR = See germanium compounds, organometals and sulfur compounds.

1,2,4,5-TETRACHLOROBENZENE. Syn: *benzene tetrachloride*. $C_6H_2Cl_4$, mw: 215.9, mp: 138°, bp: 245°, flash p: 311°F (CC), d: 1.734, vap. press: < 0.1 mm @ 25°, vap. d: 7.4.
Acute tox data: Oral LD_{50} (rat) = 1500 mg/kg; oral LD_{50} (mice) = 1035 mg/kg. [*3*]
THR = MOD via oral route. See also chlorinated hydrocarbons, aromatic.
Fire Hazard: Slight, when exposed to heat or flame.
Disaster Hazard: Dangerous; when heated to decomp, emits highly toxic fumes of chlorides; can react vigorously with oxidizing materials. Violent reaction with ($NaOH + CH_3OH$). [*19*]
To Fight Fire: CO_2, dry chemical.

TETRACHLORO-*p*-BENZOQUINONE. See tetrachloroquinone.

TETRACHLORODIAMMINE PLATINUM (IV). Rhombic or hexagonal orange-yellow plates or needles. $[Pt(NH_3)_2Cl_4]$, mw: 371.12.
THR = See platinum compounds and chlorides.

2,3,7,8-TETRACHLORODIBENZO-*p*-DIOXIN. Syn: *TCDD*. Colorless needles. $C_{12}H_4O_2Cl_4$, mw; 322, mp: 305°.
Acute tox data: Oral LD_{50} (male and female guinea pigs) = 0.0006–0.0021 mg/kg; oral LD_{50} (male and female rats) = 0.022–0.045 mg/kg; oral LD_{50} (male and female rabbits) = 0.15 mg/kg; dermal LD_{50} (male and female rabbits) = 0.275 mg/kg; ip LD_{50} (male and female rabbits) = 0.252 mg/kg. [*81*]
THR = HIGH via oral, dermal and ip routes. It causes death in rats by hepatic cell necrosis. Death can follow a lethal dose by weeks. Acute and subacute exposure ⟶ hepatic necrosis, thymic atrophy, hemorrhage, lymphoid depletion, chloracne.
THR = An exper teratogen [*3*] and [*81*] mutagen and carc.

1,1,2,2-TETRACHLORODIFLUOROETHANE. Liquid. $C_2F_2Cl_4$, mw: 203.8, bp: 92.8°, d: 1.6447 @ 25°, vap. d: 7.03.
Acute tox data: Inhal LC_{LO} (rat) = 15,000 ppm for 4 hrs. [*3*]
THR = LOW via inhal route. An exper carc. [*23*]
Disaster Hazard: Dangerous; when decomp, emits highly toxic fumes of fluorides and chlorides.

1,1,1,2-TETRACHLORO-2,2-DIFLUOROETHANE.
Acute tox data: Inhal LC_{LO} (rat) = 20,000 ppm for 1 hr. [*3*]
THR = LOW via inhal route. An exper carc. [*23*]

TETRACHLORODINITROETHANE. Crystals. $(CCl_2NO_2)_2$, mw: 257.87, bp: decomp @ 130° ⟶ NO_x.
THR = HIGH irr to skin, eyes and mu mem.
Disaster Hazard: Dangerous; when heated, emits highly toxic fumes; can react vigorously with oxidizing materials.

TETRACHLORODIPHENYLETHANE. See 1,1-dichloro-2,2-bis(*p*-chlorophenyl)ethane.

2,4,5,4-TETRACHLORODIPHENYL SULFONE. See tedion.

1,1,2,2-TETRACHLOROETHANE. See acetylene tetrachloride.

TETRACHLOROETHYLENE. See perchloroethylene.

N-(1,1,2,2-TETRACHLOROETHYLTHIO)-4-CYCLOHEXENE-1,2-DICARBOXIMIDE. See difolatan.

TETRACHLOROHEPTANE. $C_7H_{12}Cl_4$, mw: 238.
THR = Limited animal exper suggest LOW via oral route. See chlorinated hydrocarbons, aliphatic.
Disaster Hazard: Dangerous. See chlorides.

TETRACHLOROMETHYL ETHER. Fuming liquid, pungent odor. $O(CHCl_2)_2$, mw: 183.87, bp: 145°, d: 1.6537 @ 18°.
THR = HIGH irr to skin, eyes and mu mem. Although there is no data on the subject, it may have carc potential.
Disaster Hazard: Dangerous; when heated to decomp, emits highly toxic fumes.

TETRACHLORONAPHTHALENE. Syn: *1,2,3,4-tetrachloro-1,2,3,4-tetrahydronaphthalene*. Crystals. $C_{10}H_8Cl_4$, mw: 265.9, mp: 182°.
Acute tox data: Inhal TC_{LO} (human) = 3 mg/m^3 ⟶ systemic effects. [*3*]
THR = HIGH via inhal and dermal routes. See also chlorinated naphthalenes and chlorinated diphenyls.
Disaster Hazard: Dangerous; when heated to decomp, emits highly toxic fumes.

TETRACHLORONAPHTHOQUINONE.
THR = Probably toxic. A fungicide. See also quinones.
Disaster Hazard: Dangerous; when heated to decomp, emits highly toxic fumes.

1,2,3,4-TETRACHLORO-5-NITROBENZENE.
$C_6HO_2NCl_4$, mw: 260.9.
THR = HIGH. An exper neo. [*3*]

1,2,3,5-TETRACHLORO-4-NITROBENZENE.
THR = An exper neo. [*3*]

1,2,4,5-TETRACHLORO-3-NITROBENZENE.
THR = An exper neo. [*3*]

TETRACHLORONONANE. $C_9H_{16}Cl_4$, mw: 266.
THR = Limited animal exper show MOD acute. See chlorinated HC, aliphatic.
Disaster Hazard: Dangerous; see chlorides.

TETRACHLORO PENTANE. $C_5H_8Cl_4$, mw: 210.
THR = Limited animal exper suggest MOD via oral route. See chlorinated HC, aliphatic.
Disaster Hazard: Dangerous; see chlorides.

***m*-TETRACHLOROPHTHALODINITRILE.**
Acute tox data: ip LD_{50} (mouse) = 2.5 mg/kg. [*3*]
THR = HIGH via ip route. See nitriles.
Disaster Hazard: Dangerous; see nitriles.

2,3,4,6-TETRACHLOROPHENOL. Light brown mass, strong odor. Cl_4C_6HOH, mw: 231.9, mp: 69°–70°, bp: 288° (decomp), vap. press: 1 mm @ 100.0°.
Acute tox data: Oral LD_{50} (rat) = 140 mg/kg; ip LD_{50} (rat) = 130 mg/kg. [*3*]
THR = HIGH via oral and ip routes. An exper neo via sc route. [*3*]
Disaster Hazard: Dangerous; when heated to decomp, emits highly toxic fumes.

2,2,5,6-TETRACHLOROPHENOL. Brown solid, phenol odor. C_6HCl_4OH, mw: 231.92, mp: 50°, d: 1.65 @ 60°.
Acute tox data: ip LD_{LO} (mice) = 500 mg/kg. [*3*]
THR = MOD via ip route.
Disaster Hazard: Dangerous; when heated to decomp, emits highly toxic fumes.

***o*-TETRACHLOROPHTHALO DINITRILE.**
$C_8Cl_4N_2$, mw: 265.9.
Acute tox data: ip LD_{50} (mice) = 66 mg/kg. [*3*]
THR = HIGH via ip route. See also nitriles.

***p*-TETRACHLOROPHTHALODINITRILE.**
Acute tox data: ip LD_{50} (mice) = 1581 mg/kg. [*3*]
THR = MOD via ip route. See also nitriles.

TETRACHLOROQUINONE. Syns: *chloranil, tetrachloro-p-benzoquinone*. Yellow crystals, insol in water. $C_6O_2Cl_4$, mw: 245.9, mp: 290°.
Acute tox data: Oral LD_{50} (rat) = 4000 mg/kg; ip LD_{LO} (rat) = 500 mg/kg. [*3*]
THR = HIGH via ip and MOD via oral route. An exper carc via oral route. [*3*] Can cause CNS depression. May be irr to skin and mu mem. See quinones. A fungicide.
Disaster Hazard: Dangerous; when heated to decomp, emits highly toxic fumes of chlorides.

TETRACHLORORESORCINOL. Crystals.
$C_6(OH)_2Cl_4$, mw: 247.9.
THR = Probably HIGH. A seed disinfectant. See resorcinol and chlorinated HC, aromatic.
Disaster Hazard: Dangerous; when heated to decomp, emits highly toxic fumes of chlorides.

TETRACHLOROSILANE. See silicon chloride.

TETRACHLOROTETRAFLUOROPROPANE. Liquid. $C_3Cl_4F_4$, mw: 253.9.
Acute tox data: Inhal LC_{LO} (rat) = 16 ppm for 4 hrs. [*3*]
THR = HIGH via inhal route. See halogenated HC, aliphatic.
Disaster Hazard: Dangerous; see fluorides and chlorides.

1,2,3,4-TETRACHLORO-1,2,3,4-TETRAHYDRONAPHTHALENE. See tetrachloronaphthalene.

TETRACHLOROTHIOPHENE. C_4SCl_4, mw: 221.9.
Acute tox data: Oral LD_{50} (rat) = 70 mg/kg; ip LD_{LO} (mice) = 64 mg/kg; dermal LD_{50} (rabbit) = 256 mg/kg. [*3*]
THR = HIGH via oral, dermal and ip routes.

TETRACHLORO UNDECANE. $C_{11}H_{20}Cl_4$, mw: 294.
THR = Data shows LOW via oral route. See chlorinated HC, aliphatic.
Disaster Hazard: Dangerous; see chlorides.

TETRACOBALT DODECACARBONYL. See cobalt carbonyl.

TETRACYANOETHYLENE. Syns: *ethane tetracarbonitrile, TCNE*. Colorless crystals, sublimes > 120°. $(CN)_2C{:}C(CN)_2$, mw: 128, mp: 198°–200°, bp: 223°.
Acute tox data: Oral LD_{50} (mice) = 29 mg/kg. [*3*]
THR = HIGH via oral route. See also nitriles.

TETRACYCLINE. $C_{22}H_{24}O_8N_2$, mw: 444.5.
THR = HIGH to women (acute) via several routes. [*3*] Has been implicated in aplastic anemia.

TETRACYCLOHEXYL THIOGERMANIUM. Crystals, insol in water. $Ge(SC_6H_{11})_4$, mw: 533.4, d: 1.259–1.270, mp: 84°–88°.
THR = See germanium compounds, sulfur compounds and organometals.

TETRACYCLOHEXYLTIN. White crystals, insol in water. $(C_6H_{11})_4Sn$, mw: 451.3, mp: 264°.
THR = See tin compounds.

TETRADECANE. Liquid. $CH_3(CH_2)_{12}CH_3$, mw: 198.38, mp: 5.5°, lel = 0.5%, bp: 252.5°, flash p: 212°F, d: 0.765, vap. press: 1 mm @ 76.4°, vap. d: 6.83, autoign. temp.: 396°F.
THR = U. Probably irr and narcotic in HIGH conc. See also decane.
Fire Hazard: Low, when exposed to heat or flame.
Spont Heating: No.
Explosion Hazard: Mod, in the form of vapor when exposed to flame.
Disaster Hazard: Mod dangerous; when heated, emits acrid fumes; can react with oxidizing materials.
To Fight Fire: Foam, CO_2, dry chemical.

TETRADECANOL. Syn: *tetradecyl alcohol.* Opaque leaflets. $C_{14}H_{29}OH$, mw: 214.38, mp: 37.62°, bp: 264.1°, flash p: 285°F (OC), d: 0.8355 @ 20°/20°, liquid: 0.8236 @ 38°/4°, vap. press: 0.01 mm @ 20°, vap. d: 7.39.
Acute tox data: Oral LD_{50} (rat) = 33000 mg/kg; dermal LD_{50} (rabbit) = 7130 mg/kg. [*3*]
THR = LOW via dermal route and NONE via oral route.
Fire Hazard: Slight, when exposed to heat or flame; can react with oxidizing materials.
Spont Heating: No.
To Fight Fire: CO_2, dry chemical.

TETRADECANOYL ETHYLENE IMINE.
THR = An exper neo via sc route [*3*] and [*23*]

1-TETRADECENE. Syn: *α-tetradecylene.* Colorless liquid. $CH_2CH(CH_2)_{11}CH_3$, mw: 196.36, mp: −12°, bp: 127° @ 15 mm, d: 0.7737 @ 20°/4°, vap. d: 6.78, flash p: 230°F.
THR = U. Probably irr and narcotic in high conc.
Fire Hazard: Low; can react with oxidizing materials.
To Fight Fire: CO_2, dry chemical.

TETRADECYL ALCOHOL. See tetradecanol.

TETRADECYLAMINE. Oil, amine odor. $C_{14}H_{29}NH_2$, mw: 213, fp: 38.2°, vap. press: 32 mm @ 180°.
THR = Details U. See amines.

α-TETRADECYLENE. See 1-tetradecene.

TETRADIFON. See tedion.

TETRADYMITE. See bismuth tritelluride.

TETRAETHANOL AMMONIUM HYDROXIDE. Crystals. $(HOCH_2CH_2)_4 \cdot NOH$, mw: 211.3, mp: 123°, vap. press: < 0.01 mm @ 20°, vap. d: 7.28.
Acute tox data: Oral LD_{50} (rat) = 2250 mg/kg; oral LD_{50} (guinea pig) = 3510 mg/kg. [*3*]
THR = MOD via oral route. See also ammonia.
Disaster Hazard: Mod; when heated to decomp, emits toxic fumes.

TETRAETHOXYL GERMANIUM. See tetraethyl germanate.

TETRAETHOXYPROPANE. Liquid. $(C_2H_5O)_2CHCH_2CH(C_2H_5O)_2$, mw: 220.3, mp: −90°, bp: 219.9°, flash p: 190°F (OC), d: 0.9197 @ 20°/20°, vap. d: 7.58.
Acute tox data: Oral LD_{50} (rat) = 1610 mg/kg; ip LD_{LO} (mice) = 200 mg/kg. [*3*]
THR = HIGH via ip and MOD via oral routes.
Fire Hazard: Mod, when exposed to heat or flame; can react with oxidizing materials.
To Fight Fire: Foam, CO_2, dry chemical.

TETRAETHOXY SILANE. See ethyl silicate.

TETRA(2-ETHYLBUTYL)SILICATE. Insol in water, slightly sol in methanol, miscible with most organic solvents. $[C_2H_5CH(C_2H_5)CH_2O]_4Si$, mw: 422, d: 0.8920–0.9018 @ 20°/20°, mp: <−100°, bp: 238° @ 50 mm, flash p: 335°F (OC).
Acute tox data: Oral LD_{50} (rat) = 20,000 mg/kg. [*3*]
THR = Practically NONE via oral route.
Fire Hazard: Low flam when exposed to heat or open flame.
To Fight Fire: Mist, spray, dry chemical.

TETRAETHYL DIARSINE. See arsenic diethyl.

TETRAETHYLDIARSYL. See arsenic diethyl.

TETRAETHYL DITHIOPYROPHOSPHATE. Syns: *sulfotepp, bladafume, TEDP.* A liquid almost insol in water. $(C_2H_5O_4)P_2OS_2$, mw: 322.
Acute tox data: Oral LD_{50} (rat) = 5 mg/kg; im LD_{50} (rat) = 0.055 mg/kg; ip LD_{50} (mice) = 0.94 mg/kg; sc LD_{50} (mice) = 8 mg/kg; dermal LD_{50} (rabbit) = 20 mg/kg; oral LD_{50} (wild birds) = 100 mg/kg. [*3*]
THR = HIGH via oral, im, ip, sc and dermal routes. A HIGH cholinesterase inhibitor. See parathion.
Disaster Hazard: Dangerous; see parathion.

TETRAETHYLENE GLYCOL. Colorless to pale straw-colored liquid. $HO(C_2H_4O)_3C_2H_4OH$, mw: 194.22, bp: 327.3°, fp: −6°, flash p: 360°F (OC), d: 1.1248 @ 20°/20°, vap. press: 1 mm @ 153.9°.
Acute tox data: Oral LD_{50} (rat) = 29000 mg/kg. [*3*]
THR = Practically NONE via oral route.
Fire Hazard: Slight, when exposed to heat or flame; can react with oxidizing materials.
Spont Heating: No.
To Fight Fire: Alcohol foam, water, CO_2, dry chemical.

TETRAETHYLENE GLYCOL DIMETHYL ETHER. See dimethoxytetraglycol.

TETRAETHYLENEPENTAMINE. Viscous, hygroscopic liquid. $NH_2(CH_2CH_2NH)_3CH_2NH_2$, mw: 189.30, bp: 333°, flash p: 325°F (OC), d: 0.9980 @ 20°/20°, vap. press: < 0.01 mm @ 20°.
Acute tox data: Oral LD_{50} (rat) = 3990 mg/kg; dermal LD_{50} (rabbit) = 660 mg/kg. [*3*]
THR = MOD via dermal and oral routes.

Fire Hazard: Slight, when exposed to heat or flame.
Disaster Hazard: Dangerous; when heated to decomp, emits toxic fumes of NO_x; can react with oxidizing materials.
To Fight Fire: CO_2, dry chemical.

TETRAETHYLENE PYROPHOSPHORAMIDE. $C_8H_{16}O_3N_4P_2$, mw: 278.2.
THR = An exper mutagen. [*3*]

TETRAETHYL GERMANATE. Syn: *tetraethoxyl germanium*. Colorless liquid. $Ge(OC_2H_5)_4$, mw: 252.8, mp: −81°, bp: 186°.
THR = Details U. See germanium compounds.

TETRAETHYLGERMANIUM. Syn: *germanium tetraethyl*. Colorless oil, decomp by water. $(C_2H_5)_4Ge$, mw: 188.8, d: 1.198 @ 0°, mp: −90°, bp: 163°.
Acute tox data: Oral LD_{LO} (rat) = 700 mg/kg; ip LD_{LO} (rat) = 590 mg/kg; oral LD_{LO} (mice) = 2870 mg/kg. [*3*]
THR = MOD via oral and ip routes. Animal exper show stimulation of blood formation. See also germanium compounds.

TETRA-(2-ETHYLHEXYL)SILICATE. Insol in water. $(C_4H_9CH(C_2H_5)CH_2O)_4Si$, mw: 544, d: 0.8838; bp: 350°–370°, fp: −90°, flash p: 390°F (OC).
THR = U. See also esters.
Fire Hazard: Slight, when exposed to heat or flame.
To Fight Fire: Spray, dry chemical, CO_2.

TETRA-2-ETHYLHEXYL TITANATE. Light yellow liquid. $Ti(OC_8H_{17})_4$, mw: 564, mp: $<-25°$, bp: 194° @ 0.25 mm; d: 1.0711 @ 25°, vap. d: 19.5.
THR = U. See titanium compounds.
Fire Hazard: Mod, when exposed to heat or flame; can react with oxidizing materials.
To Fight Fire: Water, foam, CO_2, dry chemical.

TETRAETHYL LEAD. See lead tetraethyl.

***0,0,0,0*-TETRAETHYL-*s,s*-METHYLENE DIPHOSPHORODITHIOATE.** See ethion.

TETRAETHYL PYPROPHOSPHATE. Syns: *TEP*, *fosvex*, *hesamite*, *nifos*, *TEPP*, *tetron*, *vaptone*, *others*. Water white to amber hygroscopic liquid. $(C_2H_5)_4P_2O_7$, mw: 290.20, d: 1.20.
Acute tox data: Oral LD_{LO} (human) = 2 mg/kg; oral TD_{LO} (human) = 0.432 mg/kg ⟶ CNS effects; im LD_{LO} (human) = 0.4 mg/kg; pa TD_{LO} (human) = 0.1 mg/kg ⟶ CNS effects; oral LD_{50} (rat) = 0.5 mg/kg; dermal LD_{50} (rat) = 2.4 mg/kg; ip LD_{50} (rat) = 0.85 mg/kg; oral LD_{50} (wild birds) = 1 mg/kg; dermal LD_{50} (rabbit) = 5 mg/kg. [*3*]
THR = VERY HIGH via all routes. The action is similar to that of parathion. Briefly, the action results in an irreversible inhibition of the cholinesterase molecules and the consequent accumulation of large amounts of acetylcholine. See also parathion.
Dangerous chronic dose: Exposure to any organic phosphorus insecticide lowers the cholinesterase level and, until that enzyme has been completely regenerated, the exposed organism remains susceptible to relatively small doses of tetraethyl pyrophosphate. In other words, small doses at frequent intervals are largely additive (see parathion for further details).
Signs and symptoms of poisoning: Findings are similar to those for parathion.
Treatment of poisoning: Same as for parathion.
Disaster Hazard: Dangerous, when heated, emits highly toxic fumes. See parathion.

TETRAETHYL-*o*-SILICATE. See ethyl silicate.

TETRAETHYL THIOGERMANIUM. A liquid, sol in benzene. $Ge(SC_2H_5)_4$, mw: 317.1, d: 1.2574 @ 25°, bp: 165° @ 5 mm.
THR = See germanium compounds and organometals.

TETRAETHYLTHIURAM SULFIDE. Crystals. $[(C_2H_5)_2NCS]_2S$, mw: 264, bp: 225°–240° @ 3 mm, d: 1.12 @ 20°/20°, mp: 31°.
Acute tox data: Ip LD_{50} (mice) = 750 mg/kg. [*3*]
THR = MOD via ip route. See also bis(diethylthiocarbamyl) disulfide.
Disaster Hazard: Dangerous; when heated to decomp, emits highly toxic fumes of SO_x; can react vigorously with oxidizing materials.

TETRAETHYLTIN. Colorless liquid, insol in water, sol in organic solvents. $(C_2H_5)_4Sn$, mw: 234.9, d: 1.187 @ 23°, mp: −112°, bp: 181°.
Acute tox data: Oral LD_{50} (rat) = 16 mg/kg; iv LD_{LO} (rat) = 25 mg/kg; ip LD_{LO} (mice) = 32 mg/kg; oral LD_{50} (rabbit) = 7 mg/kg. [*3*]
THR = HIGH via oral, iv, ip and dermal routes. See tin compounds.

TETRAFLUOROETHYLENE. Syn: *perfluoroethylene*. Colorless gas. CF_2CF_2, mw: 100.02, mp: −142.5°, bp: −78.4°.
Acute tox data: Inhal LC_{50} (rat) = 40,000 ppm for 4 hrs. [*3*]
THR = VERY LOW via inhal route. Can act as an asphyxiant and may have other toxic properties. See also compressed gases. Reacts violently with (IF_5 + limonene), SO_3. [*19*]
Disaster Hazard: Dangerous; when heated to decomp, emits highly toxic fumes of fluorides.

TETRAFLUOROETHYLENE, POLYMER. Syns: *teflon*, *fluon*.
THR = An exper neo to rats and mice via imp route. [*103*]

TETRAFLUOROHYDRAZINE. Colorless gas, colorless liquid or white solid when pure. N_2F_4, mw: 104.0, mp: −163°, bp: −73°, d(liquid): 1.5 @ −100°.
THR = Probably HIGH. See hydrofluoric acid.
Fire Hazard: Highly reactive with reducing agents.
Explosion Hazard: Can react explosively with reducing agents at normal temp. When ignited with hydrogen it can explode. At high pressures it can explode due to shock or blast. Violent reactions with NF_3, O_2. [*19*]
Disaster Hazard: Dangerous; when heated to decomp, emits highly toxic fumes. Heat, shock or blast can detonate it when under pressure. Can react explosively with reducing agents.

TETRAFLUOROMETHANE. See carbon tetrafluoride.

TETRAFLUOROSILANE. See silicon tetrafluoride.

TETRA-*n*-HEPTYLTIN. Liquid. $(C_7H_{15})_4Sn$, mw: 515, d: 0.9748, bp: 239° @ 10 mm.
THR = Details U. See tin compounds.

TETRA-*n*-HEXYLTIN. Liquid. $(C_6H_{13})_4Sn$, mw: 458.8, d: 0.9959, bp: 209° @ 10 mm.
THR = See tin compounds.

1,2,3,6-TETRAHYDROBENZALDEHYDE. Liquid. $CH_2CHCH(CH_2)_2CHCHO$, mw: 110.15, mp: −110°, bp: 164.5°, d: 0.9733 @ 20°/20°, vap. press: 1.6 mm @ 20°, vap. d: 3.80, flash p: 130°F (OC).
Acute tox data: Oral LD_{50} (rat) = 2460 mg/kg; dermal LD_{50} (rabbit) = 1770 mg/kg. [*3*]
THR = MOD via oral and dermal routes.
Disaster Hazard: Slightly dangerous; when heated emits acrid fumes.
Fire Hazard: Mod, when exposed to heat, flame or oxidizers.
To Fight Fire: Alcohol foam, mist, fog, dry chemical.

1,2,3,4-TETRAHYDROBENZENE. See cyclohexene.

1-*trans*-Δ-TETRAHYDROCANNABINOL. $C_{21}H_{30}O_2$, mw: 314.5.
THR = HIGH acute, oral. An exper teratogen. [*3*]

5,6,7,8-TETRAHYDROCARBAZOLE. $C_{12}H_{13}N$, mw: 171.2.
THR = MOD acute oral. [*3*] Reacts violently with acrylonitrile. [*19*]

1,2,3,4-TETRAHYDRODIBENZ(a,h)ANTHRACENE. $C_{22}H_{18}$, mw: 282.4.
THR = An exper carc. [*3*]

1,2,3,4-TETRAHYDRODIBENZ(a,i)ANTHRACENE.
THR = An exper carc. [*3*]

8,9,10,11-TETRAHYDRO-7,12-DIMETHYL BENZ(a)ANTHRACENE. $C_{19}H_{20}$, mw: 248.4.
THR = An exper carc. [*3*]

TETRAHYDRO DIMETHYLFURAN. Liquid. $C_6H_{12}O$, mw: 100.2.
THR = U. See tetrahydrofuran.
Fire Hazard: Mod, when exposed to heat or flame; can react with oxidizing materials.
To Fight Fire: Spray, foam, dry chemical, CO_2.

TETRAHYDRO-3,5-DIMETHYL-2H-THIADIAZINE-2-THIONE. See mylone.

TETRAHYDROFURAN. Syn: *cyclotetramethylene oxide-1,4-epoxy butane*. Colorless, mobile liquid, ether-like odor. $OCH_2CH_2CH_2CH_2$, mw: 72.10, bp: 65.4°, flash p: 6°F (TCC), lel = 2.3%, uel = 11.8%, fp: −108.5°, d: 0.888 @ 20°/4°, vap. press: 114 mm @ 15°, vap. d: 2.5, autoign. temp.: 610°F.
Acute tox data: Oral LD_{LO} (rat) = 3000 mg/kg; ip LD_{LO} (rat) = 500 mg/kg; inhal TC_{LO} (human) = 25000 ppm ⟶ CNS effects. [*3*]
THR = MOD via oral, ip and inhal routes. Irr to eyes and mu mem. Narcotic in high conc. Reported as causing injury to liver and kidneys. It is a food additive resulting from contact with containers or equipment. [*109*]
Fire Hazard: Dangerous, via heat, flames, oxidizers.
Explosion Hazard: In common with other ethers, unstabilized tetrahydrofuran forms thermally explosive peroxides on exposure to air. It must always be tested for peroxide prior to distillation. Peroxides can be removed by treatment with strong ferrous sulfate sol made slightly acidic with sodium bisulfate. Reacts violently with air on standing, $LiAlH_2$, KOH, $NaAlH_2$, NaOH. [*19*]
Disaster Hazard: Dangerous; when heated to decomp, emits toxic fumes; can react with oxidizing materials.
To Fight Fire: Foam, dry chemical, CO_2.

2,5-TETRAHYDROFURAN DIMETHANOL. Syn: *THF-glycol*. Hygroscopic liquid, faint odor, water-sol. $C_6H_{12}O_3$, mw: 132.2, d: 1.1719 @ 0°/4°, mp: <−50°, bp: 265°.
THR = Details U. A strong irr. Avoid contact with eyes. See also tetrahydrofuran.

TETRAHYDROFURFURYL ALCOHOL. Syns: *THFA, tetrahydrofuranmethanol*. A hygroscopic liquid, water-sol. $C_4H_7OCH_2OH$, mw: 102.13, mp: <−80°, lel = 1.5%, uel = 9.7% @ 72° to 122°F, bp: 178° @ 743 mm, flash p: 167°F (OC), d: 1.0485 @ 20°/4°, autoign. temp.: 540°F, vap. d: 3.5.
Acute tox data: Oral LD_{50} (rat) = 2500 mg/kg; oral LD_{50} (guinea pig) = 3000 mg/kg; oral LD_{50} (mice) = 2300 mg/kg. [*3*]
THR = MOD via oral route. An irr. See also alcohols.

Fire Hazard: Mod, when exposed to heat or flame; can react with oxidizing materials.
Explosion Hazard: Mod, in the form of vapor when exposed to flame.
To Fight Fire: Alcohol foam, water, CO_2, dry chemical.

TETRAHYDROFURFURYL OLEATE. A liquid. $C_{23}H_{42}O_3$, mw: 366.6, bp: 240° @ 5 mm, flash p: 390°F, d: 0.93, vap. d: 12.65.
THR = Details U. See tetrahydrofuran and esters.
Fire Hazard: Slight, when exposed to heat or flame; can react with oxidizing materials.
To Fight Fire: CO_2, dry chemical.

1,2,5,6-TETRAHYDRO-*o*-METHYLBENZOIC ACID. Crystals. $C_8H_{13}O_2$, mw: 141.2.
THR = Details U. A mite-repellant. See also methyl benzoate.
Disaster Hazard: Slightly dangerous; when heated, emits acrid fumes.

1,2,3,4-TETRAHYDRONAPHTHALENE. Syn: *tetralin*. Colorless liquid, menthol odor. $C_{10}H_{12}$, mw: 132.20, mp: −30°, bp: 207.2°, flash p: 160°F (OC), d: 0.981, vap. press: 1 mm @ 38.0°, vap. d: 4.55, autoign. temp.: 725°F, lel = 0.8% @ 212°F, uel = 5.0% @ 302°F.
Acute tox data: Oral LD_{50} (rat) = 2860 mg/kg; dermal LD_{50} (rabbit) = 17000 mg/kg; inhal LC_{LO} (guinea pig) = 275 ppm for 8 hrs/day for 17 days. [*3*]
THR = MOD via oral and inhal routes. VERY LOW via dermal route. An irr. Narcotic in high conc. Reported as causing cataracts and kidney injury in exper animals.
Fire Hazard: Mod, when exposed to heat or flame; can react with oxidizing materials.
Spont Heating: No.
To Fight Fire: Water, foam, CO_2, dry chemical.

1,2,3,4-TETRAHYDRONAPHTHOL. Syn: *α-naphthol-1,2,3,4-tetrahydride*. Colorless liquid, $C_{10}H_{12}O$, mw: 148.2, bp: 140° @ 17 mm, d: 1.090.
Acute tox data: Oral LD_{50} (rat) = 1620 mg/kg. [*3*]
THR = MOD via oral route. Limited animal exper show mod.
Fire Hazard: Mod, when exposed to heat or flame; can react with oxidizing materials.

1,2,3,4-TETRAHYDRO-2-NAPHTHOL.
Acute tox data: oral LD_{50} (rat) = 1000 mg/kg; oral LD_{50} (rabbit) = 2800 mg/kg; oral LD_{50} (guinea pig) = 1000 mg/kg. [*3*]
THR = MOD via oral route.

TETRAHYDRO-3-NITROSO-1,3-OXAZINE. $C_4H_8O_2N_2$, mw: 116.1.
THR = An exper carc. [*3, 23*]

TETRAHYDRO-2-NITROSO-2H-1,2-OXAZINE.
THR = An exper carc. [*3*]

TETRAHYDRO-1,4-OXAZINE. See morpholine.

TETRAHYDROPHTHALIC ANHYDRIDE. White powder. $C_8H_8O_3$, mw: 152.14, mp: 101.9°, bp: 195° @ 50 mm, flash p: 315°F (OC), d: 1.375 @ 25°/20°, vap. press: < 0.01 mm @ 20°, vap. d: 5.25.
Acute tox data: Oral LD_{LO} (rat) = 4590 mg/kg; ip LD_{LO} (mice) = 500 mg/kg. [*3*]
THR = HIGH via ip and MOD via oral routes. An irr.
Fire Hazard: Slight, when exposed to heat or flame.
Disaster Hazard: Slightly dangerous; when heated, emits acrid fumes; will react with water or steam to produce heat; can react with oxidizing materials.
To Fight Fire: Water, foam, CO_2, dry chemical.

TETRAHYDROPYRAN. Syn: *pentamethylene oxide*. Colorless, mobile liquid, ether-like odor. $(CH_2)_5O$, mw: 86.13, bp: 88°, flash p: −4°F, d: 0.8814 @ 20°/4°, vap. d: 4.0, mp: −49°.
THR = MOD irr via oral and inhal routes.
Fire Hazard: Dangerous, when exposed to heat, flame or oxidizers.
To Fight Fire: Alcohol foam.
Explosion Hazard: Mod, by chemical reaction. It can form explosive peroxides if stored in uninhibited condition.
Disaster Hazard: Dangerous, upon exposure to heat or flame; can react vigorously with oxidizing materials.

TETRAHYDROPYRAN-2-METHANOL. Liquid. $OCH_2CH_2CH_2CHCH_2OH$, mw: 116.16, fp: −70°, bp: 187°, d: 1.0272 @ 20°/20°, vap. d: 4.02, vap. press: 0.4 mm @ 20°, flash p: 200°F (CC).
Acute tox data: Oral LD_{50} (mice) = 2870 mg/kg; dermal LD_{50} (rabbit) = 4000 mg/kg. [*3*]
THR = MOD via oral and dermal routes.
Fire Hazard: Mod, when exposed to heat or flame; can react with oxidizing materials.
To Fight Fire: Alcohol foam, spray, mist, dry chemical.

1,2,5,6-TETRAHYDROPYRIDINE. Clear liquid. C_5H_9N, mw: 83.1, bp: 115°–120°, d: 0.913 @ 20°/4°, flash p: 61°F.
THR = Details U. See also pyridine.
Fire Hazard: HIGH when exposed to heat, sparks or flame.
To Fight Fire: Foam, spray, dry chemical.

TETRAHYDROSYLVAN. See 2-methyltetrahydrofuran.

TETRAHYDROURUSHIOL. See 3-pentadecyl catechol.

TETRAIODODISILANE ETHYLENE. See disilicon tetraiodide.

TETRAIODOETHYLENE. $C_2H_2I_4$, mw: 533.6.
THR = Violent reaction with IF_5. [*19*]

TETRAIODOMETHANE. See carbon tetraiodide.

TETRAIODOSILANE. See silicon tetraiodide.

TETRA IRIDIUM DODECACARBONYL. $Ir_4(CO)_{12}$, mw: 1105.0.
THR = HIGH via oral and inhal routes. See also carbon monoxide and iridium. An exper carc. [*23*]

TETRAISOAMYL GERMANIUM. Colorless oily liquid. $(C_5H_{11})_4Ge$, mw: 357.2, d: 0.9147 @ 20°/20°, bp: 164°.
THR = Details U. See germanium compounds.

TETRAISOAMYLTIN. Liquid. $(C_5H_{11})_4Sn$, mw: 403.3, d: 1.035, bp: 188° @ 24 mm.
THR = Details U. See tin compounds.

TETRAISOBUTYLLEAD. Crystals. $Pb[CH_2CH(CH_3)_2]_4$, mw: 435.7, d: 1.324, mp: 23°.
THR = HIGH. See lead compounds.

TETRAISOBUTYL THIOGERMANIUM. A liquid, sol in alcohol. $Ge[SCH_2CHCH(CH_3)_2]_4$, mw: 429.3, d: 1.0984 @ 25°, bp: 200° @ 5 mm.
THR = See organometals and germanium compounds.

TETRAISOBUTYLTIN. Colorless liquid, insol in water, sol in organic solvents. $(C_4H_9)_4Sn$, mw: 347.2, d: 1.054 @ 23°, mp: −13°, bp: 267°.
THR = See tin compounds.

TETRAISOPROPYLLEAD. Colorless liquid, insol in water, decomp in air. $Pb[CH(CH_3)_2]_4$, mw: 379.6, d: 1.4504, mp: −53.5°, bp: 120° @ 14 mm.
THR = HIGH. See lead compounds.

TETRAISOPROPYL THIOGERMANIUM. Liquid, sol in absolute alcohol. $Ge[SCH(CH_3)_2]_4$, mw: 373.2, d: 1.478 @ 25°, mp: 15°, bp: 163° @ 4 mm.
THR = See germanium compounds and organometals.

TETRAISOPROPYL TITANATE. Colorless liquid. $(C_3H_7O)_4Ti$, mw: 284, mp: 20°, bp: 232°, d: 0.955 @ 25°, vap. d: 9.8.
THR = U. See esters and titanium compounds.
Fire Hazard: Mod, when exposed to heat or flame; can react with oxidizing materials.

TETRAKIS(DIMETHYL DITHIOCARBAMATE) SELENIUM. Syn: *ethyl selenac.* $Se \cdot C_{12}H_{24}N_4S_8$, mw: 480.8.
THR = An exper carc to mice via oral route. [*103*]

TETRAKIS(HYDROXYMETHYL)PHOSPHONIUM CHLORIDE. $ClC_4H_{12}O_4P$, mw: 190.6.
THR = An exper mutagen. [*90*]

TETRALIN. See tetrahydronaphthalene.

TETRALITE. See tetryl.

TETRAM. Syn: *o,o-diethyl-3-(β-diethylamino)ethyl phosphorothiolate hydrogen oxalate.* Liquid. $(C_2H_5O)_2POSCH_3CH_2N(C_2H_5)_2$, mw: 269.4, bp: 110° @ 0.2 mm, mp: 98°.
Acute tox data: Oral LD_{50} (rat) = 9 mg/kg. [*94*] Oral LD_{50} (rat) = 5 mg/kg; oral LD_{50} (mouse) = 860 mg/kg; ip LD_{50} (mouse) = 0.5 mg/kg. [*3*]
THR = HIGH via oral and ip routes. An organic phosphate insecticide, hence a cholinesterase inhibitor. See parathion.
Disaster Hazard: Dangerous; when heated to decomp, emits highly toxic fumes.

TETRAMETHOXYDIBORANE. Colorless liquid, decomp by water. $(CH_3O)_4B_2$, mw: 145.8, mp: −24°, bp: 21° @ 44 mm.
THR = Details U. See also boron compounds and boron hydrides.

TETRAMETHYL AMMONIUM CHLORITE. $(CH_3)_4NClO_2$, mw: 141.6.
THR = A sensitive explosive by percussion. [*19*]

TETRAMETHYL AMMONIUM HYDROXIDE. A liquid. $(CH_3)_4NOH$, mw: 91, d: 1.
Acute tox data: sc LD_{LO} (mice) = 20 mg/kg. [*3*]
THR = HIGH irr to skin, eyes and mu mem and via inhal and oral routes. A powerful caustic.

7,8,9,11-TETRAMETHYL BENZ(c)CRIDINE. $C_{21}H_{19}N$, mw: 285.4.
THR = An exper neo. [*3, 23*]

7,8,9,12-TETRAMETHYL BENZ(a)ANTHRACENE. $C_{22}H_{20}$, mw: 284.4.
THR = An exper carc and neo via dermal and sc routes. [*3, 23*]

3,2,4,6-TETRAMETHYL-4-BIPHENYL AMINE.
THR = An exper carc. [*23*]

TETRAMETHYL DIALUMINUM. $(CH_3)_4(HAl)_2$, mw: 116.
THR = Spont flam in air. [*19*]

TETRAMETHYL DIGALLINE. $(CH_3)_4Ga_2$, mw: 200.
THR = Spont flam in air. [*19*] See gallium compounds.

TETRAMETHYLDISTIBINE. $C_4H_{12}Sb_2$, mw: 181.9.
THR = Self-ignites in air. [*19*] See antimony compounds.

2,2,4,4-TETRAMETHYL-1,3-CYCLOBUTANEDIOL. White crystalline solid, sol in methanol, slightly sol in water. $(CH_3)_2C(CHOH)_2C(CH_3)_2$, mw: 144.2, mp: 125°–135°, bp: 222°, vap. press: 144 mm @ 171°, flash p: 125°–135° F (COC).
THR = Considerably more toxic than cyclohexanol. Has a convulsant action on exper animals. Causes a mod skin irr.
Fire Hazard: Mod, via heat, flame or oxidizers.
To Fight Fire: Spray, fog, alcohol foam, dry chemical.

2,2,4,4-TETRAMETHYL-1,3-CYCLOBUTANEDIONE. White crystalline solid, insol in water, sol in alcohol and acetic acid. $(CH_3)_2C(CO)_2C(CH_3)_2$, mw:

140.2, mp: 116° (sublimes), bp: 159°, vap. press: 6 mm @ 52°, d: 1.11.
Acute tox data: sc TD_{LO} (mouse) = 210 mg/kg intermittant for 53 weeks ⟶ exper neo. [3]

TETRAMETHYL DIAMINOBENZOPHENONE. $C_{17}H_{20}N_2$, mw: 268.4.
THR = An exper neo. [3]

TETRAMETHYL DIARSINE. See cacodyl.

TETRAMETHYL DIARSYL. See cacodyl.

TETRAMETHYL DIBORANE. Colorless liquid, decomp by water. $B_2H_2(CH_3)_4$, mw: 83.8, mp: −72.5°, bp: 68.6°.
THR = See boron compounds and boron hydrides. Self-ignites in air. [19]

TETRAMETHYL DIPROPYLENE TRIAMINE. $H_2NCHCH_3CHCH_3CH_2NHCHCH_3$, mw: 189.
Acute tox data: Oral LD_{50} (rat) = 1620 mg/kg; dermal LD_{50} (rabbit) = 310 mg/kg. [3]
THR = HIGH via dermal and MOD via oral routes. Limited animal exper suggest HIGH via irr. See also amines.

TETRAMETHYLENE. See cyclobutane.

TETRAMETHYLENE CYANIDE. See adiponitrile.

TETRAMETHYL ETHYLENE DIAMINE. $(CH_3)_2NH_2CCNH_2(CH_3)_2$, mw: 116.2.
Acute tox data: Oral LD_{LO} (rat) = 1580 mg/kg; dermal LD_{50} (mice) = 7000 mg/kg. [3]
THR = MOD via oral and dermal routes. See also amines.
Fire Hazard: Mod, when exposed to heat or flame; can react with oxidizing materials.

TETRAMETHYLENE GLYCOL. $C_4H_{10}O_2$, mw: 90, autoign. temp.: 734°F, d: >1, bp: 110°.
THR = No data. See also glycols.
Fire Hazard: Low, via heat, flames.
To Fight Fire: Alcohol foam.

TETRAMETHYL GERMANIUM. Colorless liquid, sol in organic solvents. $Ge(CH_3)_4$, mw: 132.7, d: 1.006 @ 0°, mp: −88°, bp: 43.4°.
THR = Details U. See germanium compounds.
Fire Hazard: Probably dangerous.

TETRAMETHYL HYDRAZINE HYDROCHLORIDE. $C_4H_{12}N_2 \cdot HCl$, mw: 124.6.
THR = An exper oral neo to mice. [103]

TETRAMETHYLLEAD. See lead tetramethyl.

TETRAMETHYL LEAD MIXTURE. See compounds, motor fuel and antiknock.

TETRAMETHYL METHYLENE DIAMINE. Liquid. $(CH_3)_2NCH_2N(CH_3)_2$, mw: 102.2.
THR = Details U. See amines.
Fire Hazard: Dangerous, when exposed to powerful oxidizers, heat or open flame.

2,2,3,3-TETRAMETHYL PENTANE. C_9H_{20}, mw: 128, autoign. temp.: 806°F, lel = 0.8%, uel = 4.9%, d: 0.7, vap. d: 4.4, bp: 134°, flash p: <70°F.
THR = U.
Fire Hazard: Dangerous, via heat, flame, oxidizers.
To Fight Fire: Foam, spray, dry chemical, CO_2.
Explosion Hazard: Mod dangerous.

2,2,3,4-TETRAMETHYL PENTANE.
Fire Hazard and To Fight Fire: See 2,2,3,3-tetramethyl pentane.

1,2,3,4-TETRAMETHYL PHENANTHRENE. $C_{18}H_{18}$, mw: 234.4.
THR = An exper carc. [3, 23]

TETRAMETHYL PHENANTHRENE, RUTHENIUM SALT.
THR = An exper neo. [3]

TETRAMETHYL-*p*-PHENYLENEDIAMINE. Syn: *Wurster's reagent*. Leaflets, slightly sol in cold water, more sol in hot water, freely sol in alcohol, chloroform, ether, petroleum ether. $(CH_3)_2NC_6H_4N(CH_3)_2$, mw: 164.24, mp: 51°, bp: 260°.
Acute tox data: Inhal LC_{LO} (mice) = 1030 mg/m^3. [3]
THR = MOD via inhal route.

TETRAMETHYL PHOSPHORODIAMIDIC FLUORIDE. See hanane.

TETRAMETHYL SILICANE. Colorless liquid, sol in ether. $(CH_3)_4Si$, mw: 88.2, d: 0.651 @ 15°, bp: 26.5°.
THR = Details U. See silanes. Self-ignites when exposed to air or O_2. [19]

TETRAMETHYLSTANNANE. See tin tetramethyl.

TETRAMETHYL SUCCINONITRILE. Syn: *TSN*. Crystallizes in plates, almost no odor. $C_8H_{12}N_2$, mw: 136.0, mp: 169° (sublimes).
Acute tox data: Inhal LC_{LO} (rat) = 60 ppm. [3]
THR = HIGH via inhal and dermal routes. In the preparation of sponge rubber, an azo compound is used, which decomp to form tetramethylsuccinonitrile or TSN. Animal exper indicate that it is toxic. Rats exposed to a conc of 90 ppm exhibit their first convulsion after 1.5–2 hrs or less. Rats exposed to conc of 5.5 ppm exhibited their first convulsions in 27–31 hrs and were dead in from 31–46 hrs. HIGH toxicity. See nitriles. Absorbed via skin.
Disaster Hazard: Dangerous. See nitriles.
Treatment and Antidotes: It has been pointed out that this nitrile is different from other nitriles in that thiosulfate proved to be a poor antidote for intoxication. Barbiturates proved adequate for the control of convulsions. However, the barbiturates which have a short or medium period of action will

relieve the condition only for a time. Afterwards, the symptoms may reappear, and eventually the death of the animal may ensue. This indicates that TSN is but slowly detoxified by the body. The fatal dose is thought to be about 25 mg/kg of body weight. See also cyanides.

TETRAMETHYL THIOGERMANIUM. A liquid, sol in alcohol and benzene. $Ge(SCH_3)_4$, mw: 261.0, d: 1.4364 @ 25°, mp: −3°, bp: 140° @ 4 mm.
THR = See germanium compounds and organometals.

TETRAMETHYLTHIURAM DISULFIDE. See *bis*-(dimethylthiocarbamyl) disulfide.

TETRAMETHYLTHIURAM MONOSULFIDE. Syn: *bis(dimethyl thiocarbamyl)sulfide.* Yellow powder. $[(CH_3)_2NCS]_2S$, mw: 208.39.
Acute tox data: Ip LD_{50} (mice) = 250 mg/kg; oral LD_{LO} (rat) = 500 mg/kg; ip LD_{LO} (rat) = 25 mg/kg; oral LD_{LO} (cat) = 100 mg/kg. [*3*]
THR = HIGH via ip and oral routes. See also *bis*-(dimethyl thiocarbamyl)disulfide.
Disaster Hazard: Dangerous; when heated to decomp, emits highly toxic fumes of SO_x.

TETRAMETHYL TRIBORINE TRIAMINE. Colorless liquid, hydrolyzed by water. $(CH_3)_4B_3N_3H_3$, mw: 137.6, bp: 158°.
THR = U. See also boron compounds and amines.

1,1,3,3-TETRAMETHYLUREA. Liquid, fat odor. $[(CH_3)_2N]_2CO$, mw: 116.2, bp: 177°, mp: −1.2°, d: 0.969, flash p: 167°F.
Acute tox data: Oral LD_{50} (mice) = 2920 mg/kg; iv LD_{50} (mice) = 2230 mg/kg. [*3*]
THR = MOD via iv and oral routes.
Fire Hazard: Mod, via heat, flame and oxidizers.
To Fight Fire: Foam, mist, spray, dry chemicals.

TETRAMMINE CADMIUM PERRHENATE. Crystals. $[Cd(NH_3)_4](ReO_4)_2$, mw: 681.16, d: 3.714 @ 25°/4°.
THR = Details U. See also cadmium and rhenium compounds.
Fire Hazard: Slight, by chemical reaction with reducing agents.

TETRAMMINE COPPER II SULFATE. Rhombic blue crystals. $[Cu(NH_3)_4]SO_4 \cdot H_2O$, mw: 245.75, mp: decomp @ 150°, d: 1.81.
THR = See copper compounds and sulfates.

TETRAMMINE PLATINUM (II) CHLORIDE. Tetragonal colorless crystals. $[Pt(NH_3)_4]Cl_2 \cdot H_2O$, mw: 352.29, mp: 250°, $-H_2O$ @ 100°, d: 2.737.
THR = See platinum compounds and chlorides.

TETRAMMINE PLATINUM (II) CHLOROPLATINATE. Syn: *magnus salt*. Green or red crystals. $[Pt(NH_3)_4]PtCl_4$, mw: 600.42, mp: decomp, d: <4.1.
THR = See platinum compounds and chlorides.

TETRANAPHTHENE. Crystals. $C_6H_3C_4H_7C_2H_4$, mw: 158.2.
THR = MILD irr.
Fire Hazard: Slight, when exposed to heat or flame; can react with oxidizing materials.

TETRANAPHTHENOYL TRIETHYLENE TETRAMINE. Liquid. Flash p: 325°F, d: 1.01.
THR = U. See also amines.
Fire Hazard: Slight, when exposed to heat or flame; can react with oxidizing materials.
To Fight Fire: Foam, CO_2, dry chemical.

TETRANITROANILINE. Syn: *TNA*. Solid. $C_6H_3N_5O_8$, mw: 273.12, mp: 170°, bp: explodes @ 237°.
Acute tox data: sc LD_{LO} (dog) = 2500 mg/kg. [*3*]
THR = MOD via sc route.
Fire Hazard: See nitrates.
Explosion Hazard: Severe, when shocked or exposed to heat. Tetranitroaniline is a powerful and sensitive high explosive, similar to tetryl. It deteriorates in the presence of moisture. It is used as a booster for high explosive shells and in primer and detonating compositions. See also explosives, high, and nitrates.
Disaster Hazard: Dangerous; shock or heat will explode it; when heated to decomp, emits highly toxic fumes of NO_x; can react vigorously with reducing materials.

TETRANITROCARBAZOLE. Crystals. $(NO_2)_4(C_6H_2)_2NH$, mw: 347.2.
Acute tox data: uk LD_{50} (rat) = 250 mg/kg. [*3*]
THR = HIGH via uk route. An insecticide. See nitrates, organic.
Fire Hazard: See nitrates.
Explosion Hazard: Dangerous; shock will explode it; when heated to decomp, emits highly toxic fumes; can react vigorously with oxidizing materials.

TETRANITRODIGLYCERIN.
THR = See nitroglycerin.
Fire Hazard: See nitrates.
Explosion Hazard: Severe, when shocked or exposed to heat. Tetranitrodiglycerin resembles nitroglycerin, but is less sensitive. It is a component of low-freezing dynamites due to its own low-freezing point.
Disaster Hazard: Dangerous; shock will explode it; when heated to decomp, emits highly toxic fumes of NO_x; can react vigorously with oxidizing materials.

TETRANITROMETHANE. Colorless or yellow liquid. $C(NO_2)_4$, mw: 196.04, mp: 13°, bp: 125.7°, d: 1.650 @ 13°, vap. press: 10 mm @ 22.7°.
Acute tox data: Inhal LC_{LO} (rat) = 33 ppm for 6 hours; inhal LD_{LO} (mammal) = 500 mg/kg. [*3*]
THR = HIGH via inhal route. This material irr the eyes and respiratory passages and does serious damage to the liver. It occurs as an impurity in crude TNT, and is thought to be mainly responsible for the irr properties of that material. It can cause pulmonary edema, mild methemoglobinemia and fatty degeneration of the liver and kidneys. From animal exper it has been found that conc as low as 0.1 ppm have proved rapidly fatal and that conc of 3.3 to 25.2 ppm produced very rapid and marked irr of mu mem of the eyes, mouth and upper respiratory tract.
Fire Hazard: Dangerous. See nitrates and explosives, high.
Explosion Hazard: Severe, when shocked or exposed to heat. It can form very powerful explosives when mixed with other nitro high explosives which are somewhat oxygen-deficient, or hydrocarbons. [*19*] Its primary use is in blasting explosives and in detonating compositions.
Disaster Hazard: Highly dangerous; shock will explode it; when heated to decomp, emits highly toxic fumes of NO_x; can react vigorously with oxidizing materials.

TETRANITRONAPHTHALENE. Crystals. $C_{10}H_4(NO_2)_4$, mw: 308.2, mp: 200° (approx), bp: explodes.
THR = U. See nitrates, organic.
Fire Hazard: Dangerous; see nitrates.
Explosion Hazard: Severe, when shocked or exposed to heat. Tetranitronaphthalene is a much used high explosive equal to but somewhat less sensitive to impact than TNT. It is used for bursting charges. See also nitrates and explosives, high.
Disaster Hazard: Dangerous; shock or heat will explode it, when heated to decomp, emits highly toxic fumes of NO_x; can react vigorously with reducing materials.

TETRA-*n*-OCTYLTIN. Liquid. $(C_8H_{17})_4Sn$, mw: 571.6, d: 0.9605, bp: 268° @ 10 mm.
THR = See tin compounds.

TETRAPHENOXYGERMANIUM. Colorless oil, sol in benzene. $Ge(OC_6H_5)_4$, mw: 445.0, bp: 210°–220° @ 0.3 mm.
THR = Details U. See germanium compounds.

TETRAPHENYL ARSONIUM BROMIDE. Crystals. $(C_6H_5)_4AsBr \cdot 2H_2O$, mw: 499.3, mp: 282°.
THR = See arsenic compounds and bromides.

TETRAPHENYL ARSONIUM CHLORIDE. Crystals. $(C_6H_5)_4AsCl \cdot 2H_2O$, mw: 454.8, mp: 259°.
THR = HIGH. See arsenic compounds and chlorides.

TETRAPHENYL DIARSINE. $(C_6H_5)_4As_2$, mw: 458.2.
THR = HIGH via inhal and oral routes. See also arsenic compounds.

TETRA(2-PHENYLETHYL)GERMANIUM. Colorless crystals, sol in ether. $Ge(C_6H_5C_2H_4)_4$, mw: 493.2, mp: 57°.
THR = U. See germanium compounds.

TETRAPHENYL LEAD. White crystals, sol in benzene. $Pb(C_6H_5)_4$, mw: 515.6, mp: 228°.
THR = HIGH. See phenol and lead compounds.

TETRAPHENYL SILICANE. Colorless, solid, sol in acetic anhydride. $(C_6H_5)_4Si$, mw: 336.5.
THR = Details U. See silanes.

TETRAPHENYL THIOGERMANIUM. Colorless crystals, sol in organic solvents. $Ge(SC_6H_5)_4$, mw: 509.2, mp: 101.5°.
THR = See sulfides and phosphides.
Fire Hazard: Mod, when exposed to heat or flame or by chemical reaction. See also sulfides.
Explosion Hazard: Mod, see sulfides.
Disaster Hazard: Dangerous; when heated to decomp, emits highly toxic fumes of SO_x and PO_x; will react with water, steam or acids to produce toxic and flam vapors; can react with oxidizing materials.

TETRAPHOSPHORUS HEXASULFIDE. See phosphorus trisulfide.

TETRAPHOSPHORUS TRIIODIDE. P_4I_3, mw: 504.6.
THR = Reacts violently with HNO_3. [*19*] See also phosphides and iodides.

TETRAPHOSPHORUS TRISELENIDE. Orange-red crystals. P_4Se_3, mw: 360.80, mp: 242°, bp: 360°–400°, d: 1.31.
THR = See selenium compounds and phosphorus. Can self-ignite upon warming in air. [*19*]

TETRAPHOSPHORUS TRISULFIDE. See phosphorus sesquisulfide.

TETRAPROPIONYL GLYCOSYL PROPIONATE. See glucose pentapropionate.

TETRAPROPYLENE OXIDE.
Acute tox data: Oral LD_{50} (rat) = 1800 mg/kg. [*3*]
THR = MOD via oral route.

TETRA-*n*-PROPYLGERMANIUM. Colorless mobile liquid. $Ge(C_3H_7)_4$, mw: 245.0, d: 0.9539 @ 20°/20°, mp: −73°, bp: 225° @ 746 mm.
THR = Details U. See germanium compounds.

TETRA-*n*-PROPYLLEAD. Colorless liquid, sol in benzene. $Pb(C_3H_7)_4$, mw: 379.6, d: 1.44, bp: 126° @ 13 mm.

Acute tox data: Oral LD_{LO} (rat) = 395 mg/kg; pa LD_{50} (rat) = 200 mg/kg. [*3*]
THR = HIGH via oral and pa routes. See lead and lead compounds.

TETRAPROPYL THIOGERMANIUM. A liquid, sol in absolute alcohol. $Ge(SC_3H_7)_4$, mw: 373.2, d: 1.1662 @ 25°, bp: 192° @ 5 mm.
THR = See germanium compounds and organometals.

TETRAPROPYLTIN. Colorless liquid, insol in water, sol in organic solvents. $(C_3H_7)_4Sn$, mw: 291.1, d: 1.1065, bp: 225°.
THR = See tin compounds.

TETRAPYRIDINE CADMIUM FLUOSILICATE. White crystals. $Cd(C_5H_5N)_4SiF_6$, mw: 570.86, d: 2.282.
THR = See cadmium compounds and fluosilicates.

TETRAPYRIDINE COPPER II FLUOSILICATE. Rhombic purplish-blue crystals. $Cu(C_5H_5N)_4SiF_6$, mw: 521.99, d: 2.108.
THR = HIGH. See fluosilicates and copper compounds.

TETRAPYRIDINE NICKEL II FLUOSILICATE. Rhombic blue-green crystals. $Ni(C_5H_5N)_4SiF_6$, mw: 517.14, d: 2.307.
THR = HIGH. See silicofluorides and nickel compounds.

TETRA-N-PYRRYL GERMANIUM. Light yellow crystals, sol in chloroform. $Ge(C_4H_4N)_4$, mw: 336.9, mp: 202°.
THR = U. See germanium compounds.

TETRASILANE. Colorless liquid. Si_4H_{10}, mw: 122.44, mp: −93.5°, bp: 109°, d: 0.825 @ 0°, vap. press: 7.8 mm @ 0°.
THR = Details U. See also silanes.
Fire Hazard: Severe. Details U.
Explosion Hazard: Severe, by chemical reaction with oxygen; can detonate in air. Reacts violently with CCl_4. [*19*]
Disaster Hazard: Dangerous; on decomp, emits highly toxic fumes; can react vigorously with oxidizing materials, water, chlorinated aliphatics.

TETRASODIUM(ETHYLENEDINITRILO)ACETIC ACID. See sodium EDTA.

TETRASODIUM PYROPHOSPHATE. Syn: *TSPP.* White powder. $Na_4P_2O_7$, mw: 266, mp: 988°, d: 2.534.
Acute tox data: ip LD_{50} (rat) = 59 mg/kg; oral LD_{LO} (mice) = 40 mg/kg; iv LD_{LO} (rabbit) = 50 mg/kg. [*3*]
THR = HIGH via oral, ip and iv routes. It is not a cholinesterase inhibitor. It is a substance which migrates to food from packaging materials and a sequestrant food additive. [*109*]

TETRA-α-THIENYL GERMANIUM. White crystals, insol in water. $Ge(C_4H_3S)_4$, mw: 405.1, mp: 150°.
THR = See germanium compounds.

TETRA-*m*-TOLYL GERMANIUM. White crystals, insol in water. $Ge(C_6H_4CH_3)_4$, mw: 437.1, mp: 146°.
THR = Details U. See germanium compounds.

TETRA-*p*-TOLYL GERMANIUM. White crystals, insol in water. $Ge(C_6H_4CH_3)_4$, mw: 437.1, mp: 227°.
THR = Details U. See germanium compounds.

TETRA-*p*-TOLYL THIOGERMANIUM. Colorless crystals, sol in benzene. $Ge(SC_6H_4CH_3)_4$, mw: 565.4, mp: 111°.
THR = See sulfates and germanium compounds.

TETRA-*m*-TOLYLTIN. Colorless crystals, insol in water, sol in benzene. $(C_6H_4CH_3)_4Sn$, mw: 483.2, mp: 128.5°.
THR = See tin compounds.

TETRA-*o*-TOLYLTIN. White crystals, insol in water, sol in ether and benzene. $(C_6H_4CH_3)_4Sn$, mw: 483.2, mp: 159°.
THR = See tin compounds.

TETRA-*p*-TOLYLTIN. White crystals, insol in water, sol in organic solvents. $(C_6H_4CH_3)_4Sn$, mw: 483.2, mp: 233°.
THR = See tin compounds.

TETRA-*m*-XYLYLTIN. Crystals, slightly sol in organic solvents. $[C_6H_3(CH_3)_2]_4Sn$, mw: 539.3, mp: 219.5°, bp: 360° (decomp).
THR = See tin compounds.

TETRA-*p*-XYLYLTIN. Crystals, insol in water. $[C_6H_3(CH_3)_2]_4Sn$, mw: 539.3, mp: 273°, bp: 360° (decomp).
THR = See tin compounds.

TETRAZENE. Syn: *guanyl nitrosaminoguanyl tetrazene*. Crystals.
THR = See nitrates, organic. A HIGH explosive.
Fire Hazard: Dangerous, see nitrates and explosives, high.
Explosion Hazard: Severe, when shocked or exposed to heat. It is a high explosive which evolves much flame. It is used in priming compositions and sometimes in combination with lead azide to lower the flash point of the azide.
Disaster Hazard: Highly dangerous. Shock will explode it; when heated to decomp, emits highly toxic fumes of NO_x and explodes.

TETRON. See tetraethyl pyrophosphate.

TETRYL. Syns: *tetralite, trinitrophenylmethyl nitramine, nitramine.* Yellow monoclinic crystals. $(NO_2)_3C_6H_2N(NO_2)CH_3$, mw: 287.15, mp: 130°, bp: explodes @ 187°, d: 1.57 @ 19°.

Acute tox data: sc LD_{LO} (dog) = 5000 mg/kg. [3]
THR = MOD via sc and dermal routes. An irr, sensitizer and allergen. The chief effect produced by exposure to tetryl is the development of dermatitis. Conjunctivitis may be caused by rubbing the eyes with contaminated hands or through exposure to air-borne dust. Iridocyclitis and keratitis have developed as a sequel to the conjunctivitis. Some authorities consider that tetryl may be a cause of tracheitis and asthma. Sensitization which frequently occurs as a result of exposure to tetryl may play a part in all these conditions. Tetryl workers may develop gastrointestinal symptoms, though these complaints are more common among TNT workers. Anemia has been reported to occur frequently.
Fire Hazard: Dangerous; see nitrates.
Explosion Hazard: Severe, when shocked or exposed to heat or flame. It is a powerful explosive quite sensitive to percussion and more sensitive to shock and friction than TNT. It can be compressed into pellets for use as a booster explosive. It is used in reinforced detonators and is considered to be the standard booster charge for high explosive shells. A high explosive. It reacts violently with hydrazine. [19]
Disaster Hazard: Dangerous; see nitrates.

TEXTILE WASTE, WET.
Fire Hazard: Mod, when exposed to heat or by chemical reaction with oxidizing agents.
Spont Heating: Mod.

"TG-9." See triethylene glycol dipelargonate.

TGP. See triglycidyl phosphate.

THALIDOMIDE. $C_{13}H_{10}O_4N_2$, mw: 258.3.
Acute tox data: Oral LD_{50} (rat) = 113 mg/kg; dermal LD_{50} (rat) = 1550 mg/kg. [3]
THR = HIGH via oral and MOD via dermal routes. A recog human teratogen via oral, ip and sc routes. [3]

(−) THALIDOMIDE.
THR = MOD via oral route. An exper teratogen via oral route. [3]

(+) THALIDOMIDE.
THR = HIGH via oral route. An exper teratogen via oral route. [3]

(±) THALIDOMIDE.
THR = An exper teratogen via oral route. [3]

THALLIC NITRATE. Crystals. $Tl(NO_3)_3$, mw: 390.41.
THR = A poison. See thallium compounds and nitrates. The trihydrated form has reacted violently with (formic acid + vanillin). [19]

THALLIC OXIDE. Hexagonal black crystals, amorphous prisms. Tl_2O_3, mw: 456.78, mp: 717° ± 5°, bp: $-O_2$ @ 875°, d(amorphous): 9.65 @ 21°, d(hexagonal): 10.19 @ 22°.
Acute tox data: Oral LD_{50} (rat) = 22 mg/kg; ip LD_{LO} (rat) = 80 mg/kg; iv LD_{LO} (rabbit) = 44 mg/kg; oral LD_{LO} (dog) = 34 mg/kg. [3]
THR = HIGH via oral, ip and iv routes. A poison; see thallium compounds.
Fire Hazard: Slight, by chemical reaction. Evolves O_2 @ 875°. See thallium compounds.

THALLIC SULFATE. $TlSO_4$, mw: 300.4.
Acute tox data: Oral LD_{LO} (rat) = 23 mg/kg; oral LD_{50} (mice) = 24 mg/kg. [3]
THR = HIGH via oral route. See also thallium compounds.

THALLIC SULFIDE. Black amorphous powder. Tl_2S_3, mw: 504.98.
THR = See sulfides and thallium compounds.

THALLIUM. Bluish-white, soft, malleable metal. Tl, atwt: 204.37, mp: 303.5°, bp: 1457°, d: 11.85 @ 20°, vap. press: 1 mm @ 825°.
THR = See thallium compounds.
Radiation Hazard: For permissible levels, see Section 5A, Table 5A.5. Artificial isotope ^{204}Tl, $T_{\frac{1}{2}}$ = 3.8 y, decays to stable ^{204}Pb via β's of 0.77 MeV.
Fire Hazard: Mod, in the form of dust, when exposed to heat or flame. See also powdered metals. Violent reaction with F_2. [19]

THALLIUM ACETATE. Syn: *thallous acetate.* Silk-white crystals. $TlC_2H_3O_2$, mw: 263.43, mp: 110°, d: 3.68.
Acute tox data: ip LD_{50} (rat) = 30 mg/kg; oral LD_{50} (mice) = 35 mg/kg; oral LD_{LO} (dog) = 19 mg/kg; iv LD_{LO} (rabbit) = 26 mg/kg. [3]
THR = HIGH via ip, iv and oral routes. A poison. See thallium compounds.

THALLIUM AZIDE. Yellow crystals. TlN_3, mw: 246.41, mp: 334°.
THR = See azides and thallium compounds. Decomp @ 334°; is very unstable. [19]

THALLIUM BROMATE. Colorless crystals. $TlBrO_3$, mw: 332.31.
THR = See thallium compounds and bromates.

THALLIUM BROMIDE. Syn: *thallous bromide.* Yellowish-white powder. TlBr, mw: 284.31, mp: 460° (approx), bp: 815°, d: 7.557, vap. press: 10 mm @ 522°.
Acute tox data: Oral LD_{LO} (mice) = 29 mg/kg. [3]
THR = HIGH via oral route. A poison. See thallium compounds and bromides. Violent reaction with Na, K. [19]

THALLIUM CARBONATE. Monoclinic colorless crystals. Tl_2CO_3, mw: 468.79, mp: 273°, d: 7.11.
Acute tox data: Oral LD_{LO} (rat) = 23 mg/kg; oral LD_{50} (mice) = 21 mg/kg; sc LD_{LO} (rat) = 18 mg/kg. [*3*]
THR = HIGH via sc and oral routes. A poison. Toxic via dermal route. See thallium compounds.

THALLIUM CHLORATE. Syn: *thallous chlorate*. Solid. $TlClO_3$, mw: 287.85, d: 5.047 @ 90°.
THR = See chlorates and thallium compounds.

THALLIUM CHLORIDE. Syn: *thallous chloride*. Colorless or white powder. TlCl, mw: 239.8, mp: 430°, bp: 720°, d: 7.00, vap. press: 10 mm @ 517°.
Acute tox data: Oral LD_{50} (mice) = 24 mg/kg; ip LD_{50} (mice) = 24 mg/kg. [*3*]
THR = HIGH via oral and ip routes, also via dermal route. See thallium compounds and chlorides. Violent reaction with F_2, K. [*19*]

THALLIUM CHLOROPLATINATE. Pale orange crystals. Tl_2PtCl_6, mw: 816.75, d: 5.76 @ 17°.
THR = See thallium and platinum compounds.

THALLIUM CHROMATE. Yellow crystals. Tl_2CrO_4, mw: 524. 8.
THR = See chromium and thallium compounds.

THALLIUM COMPOUNDS.
THR = Acute poisoning usually follows the ingestion of toxic quantities of a thallium-bearing depilatory, or accidental or suicidal ingestion of rat poison. Children have been known to tolerate 8 mg of thallium acetate per kg of weight, but adults and adolescents have not. Acute poisoning results in swelling of the feet and legs, arthralgia, vomiting, insomnia, hyperesthesia and paresthesia of the hands and feet, mental confusion, polyneuritis with severe pains in the legs and loins, partial paralysis of the legs with reaction of degeneration, angina-like pains, nephritis, wasting and weakness, and lymphocytosis and eosinophilia. About the 18th day, complete loss of the hair of the body and head occurs. Fatal poisoning has been known to occur. Industrial poisoning is reported to have caused discoloration of the hair (which later falls out), joint pain, loss of appetite, fatigue, severe pain in the calves of the legs, albuminuria, eosinophilia, lymphocytosis and optic neuritis followed by atrophy. Cases of industrial poisoning are rare, however.
Disaster Hazard: Dangerous; when heated, they emit highly toxic fumes.

THALLIUM CYANIDE. Syn: *thallous cyanide*. Tablets. TlCN, mw: 230.41, mp: decomp.
THR = See cyanides and thallium compounds.

THALLIUM DICHROMATE. Syn: *thallous dichromate*. Red crystals. $Tl_2Cr_2O_7$, mw: 624.80.
THR = See chromium and thallium compounds.

THALLIUM DIETHYL DITHIOCARBAMATE.
Acute tox data: ip LD_{50} (rat) = 0.017 mg/kg. [*3*]
THR = VERY HIGH via ip route. A deadly poison.

THALLIUM DITHIONATE. Syn: *thallous dithionate*. Monoclinic crystals. $Tl_2S_2O_6$, mw: 568.91, mp: decomp, d: 5.57.
THR = See thallium compounds.

THALLIUM ETHOXIDE. Colorless liquid. $Tl(C_2H_5O)_3$, mw: 339.5, mp: −3°, bp: decomp @ 80°, d: 3.522.
THR = HIGH. See thallium compounds.
Fire Hazard: Mod, when exposed to heat or flame.
Disaster Hazard: Dangerous, when heated to decomp, emits highly toxic fumes; can react with oxidizing materials.

THALLIUM FERROCYANIDE. Triclinic yellow crystals. $Tl_4Fe(CN)_6 \cdot 2H_2O$, mw: 1065.55, d: 4.541.
THR = See thallium compounds and ferrocyanides.

THALLIUM IODIDE. Triclinic yellow crystals. TlI, mw: 331.27, d: 7.1, mp: 440°, bp: 824°, soluble in KI sol.
Acute tox data: Oral LD_{LO} (mice) = 28 mg/kg. [*3*]
THR = HIGH via oral route. See thallium compounds and iodides. A poison.

THALLIUM MONOFLUORIDE. Syn: *thallous fluoride*. Colorless cubic crystals. TlF, mw: 223.39, mp: 327°, d: 8.36.
Acute tox data: Oral LD_{LO} (rat) = 50 mg/kg. [*3*]
THR = HIGH via oral route. See thallium compounds and fluorides.

THALLIUM MONOIODIDE. See thallium iodide.

THALLIUM MONOSULFIDE. See thallium sulfide.

THALLIUM MONOXIDE. Syns: *thallium oxide, thallous oxide*. Black deliquescent crystals. Tl_2O, mw: 424.78, mp: 300°, bp: 1865°, $-O_2$ @ 1080°, d: 9.52 @ 16°.
THR = See thallium compounds.

THALLIUM NITRATE. Syn: *thallous nitrate*. Cubic crystals. $TlNO_3$, mw: 266.4, mp: 206°, bp: 430°, d: 5.55.
Acute tox data: sc LD_{LO} (rat) = 20 mg/kg; oral LD_{50} (mice) = 33 mg/kg; oral LD_{LO} (dog) = 45 mg/kg. [*3*]
THR = HIGH via oral and sc routes. See thallium compounds and nitrates.

THALLIUM NITRIDE. Tl_3N, mw: 627.1.
THR = HIGH; see thallium compounds. Violent reaction with water, dilute acids, shock or heat. [*19*]

THALLIUM OLEATE. Syn: *thallous oleate.* White crystalline clusters. $TlC_{18}H_{33}O_2$, mw: 485.83, mp: 131°-132°.
THR = See thallium compounds.

THALLIUM OXALATE. Syn: *thallous oxalate.* Monoclinic prisms. $Tl_2C_2O_4$, mw: 496.80, d: 6.31.
THR = See oxalates and thallium compounds.

THALLIUM OXIDE. See thallium monoxide.

THALLIUM PERCHLORATE. Syn: *thallous perchlorate.* Colorless crystals. $TlClO_4$, mw: 303.85, mp: 501°, bp: decomp, d: 4.89.
THR = See thallium compounds and perchlorates.

THALLIUM PEROXIDE. See thallic oxide.

THALLIUM PHENOXIDE. Syn: *thallous phenoxide.* White crystals. $TlOC_6H_5$, mw: 297.49, mp: 233°.
THR = See thallium compounds and phenol.

THALLIUM-*o*-PHOSPHATE. Syn: *thallous-o-phosphate.* Colorless needles. Tl_3PO_4, mw: 708.15, d: 6.89.
THR = See thallium compounds and phosphates.

THALLIUM PICRATE. Red or yellow crystals. $TlC_6H_2N_3O_7$, mw: 432.49, mp: explodes @ 273°-275°, d(red): 3.164 @ 17°, d(yellow): 2.993 @ 17°.
THR = See thallium compounds and picric acid.

THALLIUM PYROVANADATE. Solid. $Tl_4V_2O_7$, mw: 1031.46, mp: 454°, d: 8.21 @ 19°.
THR = See thallium compounds and vanadium.

THALLIUM SALTS, N.O.S. See thallium compounds.

THALLIUM SELENATE. Syn: *thallous selenate.* Rhombic needles. Tl_2SeO_4, mw: 551.74, mp: 400°, d: 6.875.
THR = See selenium and thallium compounds.

THALLIUM SELENIDE. Syn: *thallous selenide.* Gray leaf. Tl_2Se, mw: 487.74, mp: 340°.
THR = See selenium and thallium compounds.

THALLIUM SELENITE. TlSe, mw: 283.3.
Acute tox data: Oral LD_{LO} (rat) = 50 mg/kg. [*3*]
THR = HIGH via oral route. See also thallium and selenium compounds.

THALLIUM SESQUICHLORIDE. Hexagonal yellow crystals or yellow powder. Tl_2Cl_3, mw: 515.15, mp: 400°-500°, bp: decomp, d: 5.9.
THR = See thallium compounds and chlorides.

THALLIUM SILVER NITRATE. Syn: *thallous silver nitrate.* White crystalline powder. $TlNO_3 \cdot AgNO_3$, mw: 436.29, mp: 75°.
THR = See thallium compounds, nitrates, silver compounds.

THALLIUM STEARATE. Syn: *thallous stearate.* Needles. $TlC_{18}H_{35}O_2$, mw: 487.85, mp: 119°.
THR = See thallium compounds.

THALLIUM SULFATE. Syn: *thallous sulfate.* Colorless crystals. Tl_2SO_4, mw: 504.84, mp: 632°, bp: decomp, d: 6.77.
Acute tox data: Oral LD_{LO} (human) = 3 mg/kg; sc LD_{LO} (rat) = 13 mg/kg; oral LD_{50} (mice) = 29 mg/kg. [*3*] oral LD_{50} (brown rat) = 16 mg/kg.
THR = HIGH via oral and sc routes. This material is a dangerous, highly toxic metal salt which is used mainly as a rodenticide. Its main hazard is as a chronic poison due to its cumulation, especially in liver, brain and skeletal muscle; readily absorbed via GI tract and skin. A cellular toxicant like arsenic. Fatal human dose is about 500 mg of Tl. Intake of Tl causes depilation. Many reported fatalities. See thallium compounds and sulfates.

THALLIUM SULFIDE. Syns: *thallium monosulfide, thallous sulfide.* Blue-black powder. Tl_2S, mw: 440.84, mp: 443°, bp: decomp, d: 8.0.
THR = See thallium compounds and sulfides.

THALLIUM SULFITE. Syn: *thallous sulfite.* Crystals, Tl_2SO_3, mw: 488.85, d: 6.427.
THR = See thallium compounds and sulfites.

THALLIUM-*m*-TELLURATE. Syn: *thallous-m-tellurate.* Heavy white precipitate. Tl_2TeO_4, mw: 600.39, mp: red heat, d: 6.760 @ 17.6°.
THR = See thallium and tellurium compounds.

THALLIUM THIOCYANATE. Syn: *thallous thiocyanate.* Tetragonal colorless crystals. TlSCN, mw: 262.47.
THR = See thallium compounds and thiocyanates.

THALLIUM TRIFLUOROACETATE.
Acute tox data: sc LD_{50} (rat) = 20 mg/kg. [*3*]
THR = HIGH via sc route. A poison. See thallium compounds and fluoroacetates.

THALLIUM TRIIODIDE. Brown needles. TlI_3, mw: 585.15.
THR = See thallium compounds and iodides.

THALLOUS ACETATE. See thallium acetate.

THALLOUS AMIDE. $TlNH_2$, mw: 220.4.
THR = Explodes with dilute acids or water. [*19*]

THALLOUS BROMIDE. See thallium bromide.

THALLOUS CHLORATE. See thallium chlorate.

THALLOUS CHLORIDE. See thallium chloride.

THALLOUS CYANIDE. See thallium cyanide.

THALLOUS DICHROMATE. See thallium dichromate.

THALLOUS DITHIONATE. See thallium dithionate.

THALLOUS FLUORIDE. See thallium monofluoride.

THALLOUS HYDROGEN SULFATE. Syn: *thallium hydrogen sulfate.* Crystals. $TlHSO_4$, mw: 301.46, mp: 120°.

Acute tox data: Oral LD_{50} (rat) = 16 mg/kg; oral LD_{50} (wild birds) – 35 mg/kg. [*3*]
THR = HIGH via oral route. See thallium compounds and sulfuric acid.

THALLOUS HYDROXIDE. See thallium hydroxide.

THALLOUS NITRATE. See thallium nitrate.

THALLOUS OLEATE. See thallium oleate.

THALLOUS OXALATE. See thallium oxalate.

THALLOUS OXIDE. See thallium oxide.

THALLOUS PERCHLORATE. See thallium perchlorate.

THALLOUS PHENOXIDE. See thallium phenoxide.

THALLOUS-*o*-PHOSPHATE. See thallium-*o*-phosphate.

THALLOUS PHOSPHIDE. Tl_3P, mw: 644.1.
THR = Ignites when heated in air. [*19*] VERY poisonous. See also thallium compounds and phosphides.

THALLOUS SELENATE. See thallium selenate.

THALLOUS SELENIDE. See thallium selenide.

THALLOUS SILVER NITRATE. See thallium silver nitrate.

THALLOUS STEARATE. See thallium stearate.

THALLIUM SULFATE. See thallium sulfate.

THALLOUS SULFITE. See thallium sulfite.

THALLOUS-*m*-TELLURATE. See thallium-*m*-tellurate.

THALLOUS THIOCYANATE. See thallium thiocyanate.

THANITE. See isobornyl thiocyanoacetate.

THBP. See 2,4,5-trihydroxy butyrophenone.

THEBAINE. Syn: *p-morphine*. White to slightly yellowish, lustrous leaflets or prisms. $C_{19}H_{21}NO_3$, mw: 311.37, mp: 193°, d: 1.305.
Acute tox data: Sc LD_{50} (mice) = 31 mg/kg; sc LD_{50} (rabbit) = 13.9 mg/kg. [*3*]
THR = HIGH via oral and sc routes. Its action on humans resembles that of strychnine.
Disaster Hazard: Dangerous; when heated to decomp, emits highly toxic fumes.

THEINE. See caffeine.

THEOBROMINE. Syn: *3,7-dimethylxanthine*. White powder, bitter tasting alkaloid. $C_7H_8N_4O_2$, mw: 180.17, mp: 357°, bp: sublimes.
Acute tox data: Oral TD_{LO} (human) = 26 mg/kg ⟶ CNS effects; oral LD_{50} (cat) = 200 mg/kg. [*3*]
THR = HIGH via oral route.
Disaster Hazard: Dangerous; when heated to decomp, emits toxic fumes.

THEOBROMINE LITHIUM. See theobromose.

THEOBROMINE SODIUM ACETATE. See theobromine.

THEOBROMINE SODIUM FORMATE. See theophorin.

THEOBROMINE SODIUM SALICYLATE. See theobromine.

THEOBROMOSE. Syn: *theobromine lithium*. Needle-like crystals. $C_7H_7O_2N_4Li$, mw: 186.10.
THR = See also theobromine and lithium compounds.

THEOPHORIN. Syn: *theobromine-sodium formate*. White powder. $C_7H_7N_4O_2Na \cdot HCOONa \cdot H_2O$, mw: 288.13.
THR = HIGH irr to skin, eyes and mu mem.
Disaster Hazard: Dangerous; when heated, emits highly toxic fumes.

THEOPHYLLINE. Syn: *1,3-dimethylxanthine*. Monoclinic, odorless needles, bitter taste. $C_7H_8N_4O_2$, mw: 180.17, mp: 269.72°.
Acute tox data: Oral LD_{LO} (child) = 8.4 mg/kg; oral LD_{50} (mice) = 600 mg/kg; ip LD_{50} (mice) = 200 mg/kg; sc LD_{50} (mice) = 184 mg/kg; iv LD_{LO} (rat) = 240 mg/kg; iv TD_{LO} (human) = 10 mg/kg ⟶ CNS effects. [*3*]
THR = HIGH via oral, ip, sc and iv routes. An exper teratogen. [*3*]
Disaster Hazard: Dangerous; when heated to decomp, emits toxic fumes.

THEOPHYLLINE ETHYLENE DIAMINE. $C_{14}H_{16}O_4N_8$, mw: 420.5.
THR = HIGH acute oral with CNS effects. An exper teratogen. [*3*]

THERMIC AND OXIDATION PRODUCTS. In the case of vegetable and animal derived oils and fats and also waxes and greases, it is a susp carc of the lung, alimentary system and bladder. [*14*]

"THERMIT." Composition: Fe_2O_3 + Al.
THR = See aluminum and iron compounds.
Fire Hazard: Dangerous, when exposed to heat or flame. The reaction of Fe_2O_3 + Al is typical of a series of oxide-metal reactions. They are very dangerous in that once started they are very difficult to stop, as they supply their own oxygen. They may attain a temp. of about 2500°.
Disaster Hazard: Dangerous; keep away from combustible materials.

THIABENDAZOLE. Syn: *2-(4-thiazolyl)benzimidazole*. White to tan, odorless, insol in water, slightly sol in alcohol and acetone, very slightly sol in ether and chloroform. $C_{10}H_7N_3S$, mw: 201, mp: 304°.
Acute tox data: Oral LD_{50} (rat) = 3800 mg/kg; oral

LD_{50} (rat) = 3100 mg/kg; oral LD_{50} (mice) = 1395 mg/kg. [*3*]
THR = MOD via oral route. A food additive permitted in the feed and drinking water of animals and/or for the treatment of food-producing animals; also permitted in food for human consumption. [*109*]
Disaster Hazard: Dangerous; see sulfides.

THIACETIC ACID. See thioacetic acid.

THFA. See tetrahydrofurfuryl alcohol.

THF-GLYCOL. See 2,5-tetrahydrofuran dimethanol.

THIALDINE. Syn: *5,6-dihydro-2,3,6-trimethyl-1,3,5-dithiazine*. Powder. $SCH(CH_3)SCH(CH_3)NHCHCH_3$, mw: 163.29, mp: 44.4°, bp: decomp, flash p: 200°F (OC), d: 1.191, vap. d: 5.63.
THR = MOD acute. [*3*]
Fire Hazard: Mod, when exposed to heat, flame or oxidizers.
Disaster Hazard: Dangerous; when heated to decomp, emits toxic fumes.
To Fight Fire: CO_2, alcohol foam, dry chemical.

THIAMINE HYDROCHLORIDE. Syn: *vitamin* B_1. Small white crystals or crystalline powder, hygroscopic, nut-like odor, sol in water and glycerol, slightly sol in alcohol, insol in ether and benzene. $C_{12}H_{17}ClN_4OS \cdot HCl$, mw: 337, mp: 248° (decomp).
Acute tox data: Oral LD_{50} (mice) = 8224 mg/kg; ip LD_{50} (mice) = 200 mg/kg; iv LD_{50} (mice) = 89 mg/kg. [*3*]
THR = HIGH via iv and ip routes; LOW via oral route. A nutrient and/or dietary supplement food additive. [*109*]
Disaster Hazard: Dangerous; see chlorides.

THIAMINE MONONITRATE. White crystals or crystalline powder, non-hygroscopic, slightly sol in water, alcohol and chloroform. $C_{12}H_{17}N_5O_4S$, mw: 327, mp: 196°-200° (decomp).
Acute tox data: iv LD_{50} (rabbit) = 113 mg/kg; ip LD_{50} (mice) = 387 mg/kg; iv LD_{50} (mice) = 84 mg/kg. [*3*]
THR = HIGH via ip and iv routes. A nutrient and/or dietary supplement food additive. [*109*]
Fire Hazard: A powerful oxidizer. See nitrates.
Disaster Hazard: Dangerous; see nitrates.

THIANTHRENE PERCHLORATE. $C_{12}H_8S_2 \cdot HClO_4$, mw: 252.6.
THR = Very unstable and explosive. [*19*]

THIAZOSULFONE. See 2-amino-5-sulfanilylthiazole.

THICKENING FOOD ADDITIVES. [*109*]

THIENYL-α-PYRROLIDINE.
THR = U. An insecticide.
Disaster Hazard: Dangerous; when heated to decomp, emits highly toxic fumes.

THIIRANE. See ethylene sulfide.

THIMET. See phorate.

THINNING COMPOUNDS, PAINT, VARNISH, LACQUER, etc. See compounds, lacquer, paint, varnish, etc.

THIOACETAMIDE. Colorless leaflets, mercaptan odor. CH_3CSNH_2, mw: 75.20, mp: 113°.
Acute tox data: Oral LD_{LO} (rat) = 200 mg/kg; ip LD_{LO} (mice) = 300 mg/kg; sc LD_{LO} (mice) = 2000 mg/kg. [*3*]
THR = HIGH via oral and ip routes. MOD via sc route. Liver damage is reported. An exper (+) carc. [*3, 1*] via oral route.
Disaster Hazard: Dangerous; see sulfides.

THIOACETIC ACID. Syns: *ethanethiolic acid, methanecarbothiolic acid, thiacetic acid*. Colorless liquid, pungent, disagreeable odor. CH_3COSH, mw: 76.11, mp: $<-17°$, bp: 93°, d: 1.074 @ 10°/4°.
Acute tox data: ip LD_{LO} (mice) = 750 mg/kg. [*3*]
THR = MOD irr to skin, eyes and mu mem and via oral and inhal routes. See sulfides.

2,2′-THIOBIS(4-CHLOROPHENOL). See bis(2-hydroxy-5-chlorophenyl) sulfide.

2,2′-THIOBIS(4,6-DICHLOROPHENOL). See bithionol.

4,4′-THIOBIS(6-*tert*-BUTYL-*m*-CRESOL). Light gray to tan powder. mp: 150°, d: 1.10.
Acute tox data: ip LD_{LO} (mice) = 50 mg/kg. [*3*]
THR = HIGH via ip route and probably oral and inhal routes also.

THIOCARBAMIDE. See thiourea.

THIOCARBAMOSULFONAMIDES.
THR = Details U. Probably toxic. A fungicide. See also amides.
Disaster Hazard: Dangerous; when heated to decomp, emits highly toxic fumes of SO_x.

THIOCARBANIL. See phenyl mustard oil.

THIOCARBANILIDE. Syns: *N,N′-diphenyl thiourea, 1,3-diphenyl-2-thiourea*. White to faint gray powder. $C_6H_5NHC(S)NHC_6H_5$, mw: 228.3, mp: 154°, bp: decomp, d: 1.32 @ 25°.
Acute tox data: Oral LD_{50} (rabbit) = 1500 mg/kg; [*94*] ip LD_{LO} (mice) = 500 mg/kg; oral LD_{LO} (cat) = 720 mg/kg. [*3*]
THR = MOD via ip and oral routes.
Disaster Hazard: Dangerous; when heated to decomp, emits highly toxic fumes of SO_x.

THIOCARBONYL CHLORIDE. See thiophosgene.

THIOCARBONYL TETRACHLORIDE. See perchloromethyl mercaptan.

THIOCYANATES.
THR = Variable. Thiocyanates are not normally dissociated into cyanide; they have a low acute toxicity. Prolonged absorption may produce various skin eruptions, running nose, and occasionally dizziness, cramps, nausea, vomiting and mild or severe disturbances of the nervous system. Violent reactions have occurred when mixed with chlorates, nitrates, HNO_3, organic peroxides, peroxides, $KClO_3$, $NaClO_3$. [*19*]
Disaster Hazard: Dangerous; when heated to decomp, or on contact with acid or acid fumes; they emit highly toxic fumes of cyanides.

THIOCYANIC ACID. Colorless gas or white solid. HSCN, mw: 59.09, mp: 5°, bp: decomp.
THR = See thiocyanates.

***p*-THIOCYANOCHLOROBENZENE.** Crystals. ClC_6H_4SCN, mw: 169.6.
THR = Details U. A fumigant. See chlorobenzene and thiocyanates.
Disaster Hazard: Dangerous; when heated to decomp or on contact with acid or acid fumes, emits highly toxic fumes of cyanides and chlorides.

THIOCYANOGEN. Liquid or yellow solid. $(SCN)_2$, mw: 116.17, mp: −2° to −3° (decomp).
THR = Details U. Probably HIGH.
Disaster Hazard: Dangerous; when heated to decomp or on contact with acid or acid fumes, emits highly toxic fumes of SO_x and cyanides.

THIOCYANOPROPYL PHENYL ETHER.
THR = Details U. An insecticide.
Fire Hazard: Details U. See also ethers.
Disaster Hazard: Dangerous; when heated to decomp, emits highly toxic fumes.

THIODAN. See endosulfan.

4,4′-THIODIANILINE. $C_{12}H_{12}N_2S$, mw: 216.3, needles, mp: 108°.
THR = An exper carc via oral route [*3*] in rats. [*108*]

THIODIETHYLENE GLYCOL. Syn: *thiodiglycol.* Syrupy, colorless liquid, characteristic odor. $(CH_2CH_2OH)_2S$, mw: 122.2, mp: −11.2°, bp: 282°, flash p: 320°F (OC), d: 1.1847 @ 20°/20°, vap. d: 4.21.
Acute tox data: iv LD_{50} (rabbit) = 3000 mg/kg; oral LD_{50} (guinea pig) = 3960 mg/kg; sc LD_{50} (rat) = 4000 mg/kg. [*3*]
THR = MOD via iv, sc and oral routes. See also glycols.
Fire Hazard: Slight, when exposed to heat or flame. Reacts violently with (acetone + H_2O_2). [*19*]
Disaster Hazard: Dangerous; when heated to decomp, emits highly toxic fumes of SO_x; can react with oxidizing materials.
To Fight Fire: Alcohol foam, CO_2, dry chemical.

THIODIGLYCOL. See thiodiethylene glycol.

THIODIGLYCOLIC ACID. A white powder. $HOOCCH_2SCH_2COOH$, mw: 150, mp: 128°.
Acute tox data: Ip LD_{LO} (mice) = 300 mg/kg. [*3*]
THR = HIGH via ip route. See also thioglycolic acid.
Disaster Hazard: Mod dangerous; when heated to decomp or on contact with acid or acid fumes, emits toxic fumes.

THIODIPHENYLAMINE. See phenothiazine.

THIODIPROPIONIC ACID. $S(CH_2CH_2CH_2CO_2H)_2$, mw: 178, mp: 134°.
Acute tox data: Oral LD_{50} (rat) = 3980 mg/kg; ip LD_{50} (rat) = 500 mg/kg; oral LD_{50} (mice) = 2000 mg/kg; iv LD_{50} (mice) = 175 mg/kg. [*3*]
THR = HIGH via iv and ip routes; MOD via oral route. A chemical preservative food additive. [*109*]

β,β′-THIODIPROPIONITRILE. White crystals. $S(CH_2CH_2CN)_2$, mw: 140.20, mp(α): 28.65°, mp(β): 22.10°, d: 1.1095 @ 30°.
Acute tox data: Oral LD_{50} (rat) = 4210 mg/kg; oral LD_{50} (cat) = 4210 mg/kg; ip LD_{LO} (mouse) = 300 mg/kg. [*3*]
THR = HIGH via ip and MOD via oral routes. See cyanides and nitriles.

THIOETHYL ETHER. See ethyl sulfide.

THIOFURAN. See thiophene.

THIOGLYCOLIC ACID. Syns: *mercaptoacetic acid, 2-mercaptoethanoic acid, thiovanic acid.* Liquid, strong odor. $HSCH_2COOH$, mw: 92.11, mp: −16.5°, bp: 104°–106° @ 11 mm.
Acute tox data: Oral LD_{50} (rat) = 250 mg/kg; oral LD_{50} (rabbit) = 126 mg/kg; ip LD_{LO} (mice) = 100 mg/kg; sc LD_{LO} (mice) = 1000 mg/kg. [*3*]
THR = HIGH via oral and ip routes and MOD via sc route. For other properties, see hydrogen sulfide, which is readily evolved by this compound.

THIOGLYCOLLATES. Syn: *mercaptoacetates.*
THR = MOD irr to skin, eyes and mu mem and via oral route.
Disaster Hazard: Dangerous; when heated to decomp, they emit highly toxic fumes.

6-THIOGUANOSINE. $C_{10}H_{13}O_4N_5S$, mw: 299.3.
THR = An exper teratogen. [*3*]

2-THIO-4-KETOTHIAZOLIDINE. See rhodanine.

THIOMALIC ACID. Off-white powder. $HCO_2CH_2C(SH)HCO_2H$, mw: 150.2, mp: 150°.

Acute tox data: Oral LD_{50} (rat) = 800 mg/kg; ip LD_{50} (mice) = 500 mg/kg. [*3*]

THR = HIGH via ip and MOD via oral routes. Has been proposed as an antidote for heavy metal poisoning. Allergic dermatitis in humans has been reported.

Disaster Hazard: Dangerous; when heated to decomp, or on contact with acid or acid fumes, emits toxic fumes.

THIOMORPHOLIDOPHOSPHORIC DIETHYLENIMIDE. $C_8H_{16}ON_3SP$, mw: 233.3.

THR = An exper (+) carc. [*3, 8*]

THIONAZIN. See zinophos.

THIONYL BROMIDE. Yellow liquid. $SOBr_2$, mw: 207.90, mp: −52°, bp: 138° @ 773 mm, d: 2.68 @ 18°, vap. press: 10 mm @ 31.0°.

THR = HIGH irr to skin, eyes and mu mem and via oral and inhal routes.

Disaster Hazard: Dangerous; when heated, emits highly toxic fumes; will react with water, steam or acids to produce toxic and corrosive fumes.

THIONYL CHLORIDE. Syn: *sulfurous oxychloride*. Colorless to yellow to red liquid. $SOCl_2$, mw: 119.0, mp: −105°, bp: 78.8° @ 746 mm, d: 1.640 @ 15.5°/15.5°, vap. press: 100 mm @ 21.4°.

THR = This material has a pungent odor similar to that of sulfur dioxide; it fumes upon exposure to air. In the presence of moisture it decomp into hydrogen chloride and sulfur dioxide. Both these decomp products are very toxic and constitute serious toxicity hazards. The material itself is more toxic than sulfur dioxide. In exper with animals, it was found that an exposure of 20 min to a conc of 17.5 ppm was fatal to cats. It is classified as a corrosive liquid and can cause burns of the skin, eyes and mu mem wherever it comes in contact with the body. See also hydrogen chloride and sulfur dioxide. May react violently with dimethyl sulfoxide, 2,4-hexadiyn-1, 6-diol, *o*-nitrobenzoyl acetic acid, H_2O, *o*-nitrophenylacetic acid. [*19*]

Disaster Hazard: Corrosive. See hydrochloric acid and sulfur dioxide.

Treatment and Antidotes: Personnel exposed should follow the treatment outlined under hydrogen chloride and sulfur dioxide.

THIONYL CHLORIDE FLUORIDE. Gas. SOClF, mw: 102.52, mp: −139.5°, bp: 12.2°.

THR = HIGH irr to skin, eyes and mu mem, and via inhal route. See also hydrochloric acid and fluorides.

Disaster Hazard: Dangerous; when heated, emits highly toxic fumes; will react with water or steam to produce toxic and corrosive fumes.

THIONYL FLUORIDE. Colorless gas. Suffocating odor. SOF_2, mw: 86.07, bp: −44°, d: 2.93, mp: −130°.

Acute tox data: Inhal LC_{50} (rat) = 1920 mg/kg. [*3*]

THR = HIGH irr to skin, eyes and mu mem and via inhal route. See also fluorides.

Disaster Hazard: Dangerous; when heated, emits highly toxic fumes; will react with water or steam to produce toxic and corrosive fumes.

THIOPHANATE. Syns: *fungo 50, topsin M, enovit M*. Colorless plate-like crystals. $C_{12}H_{14}O_4N_4S_2$, mw: 342.4.

Acute tox data: Oral LD_{50} (rat) = 9700 mg/kg; ip LD_{50} (rat) = 1140 mg/kg; oral LD_{50} (mice) = 3400 mg/kg; oral LD_{50} (rabbit) = 2270 mg/kg. [*3*]

THR = MOD-LOW via oral; MOD via ip routes. A systemic fungicide of low to mod toxicity.

Disaster Hazard: Dangerous; see sulfur compounds.

THIOPHENE. Syn: *thiofuran*. Clear and colorless liquid. SCHCHCHCH, mw: 84.13, bp: 84.1°, fp: −38.3°, flash p: 30°F, d: 1.0583 @ 25°/4°, vap. press: 40 mm @ 12.5°, vap. d: 2.9.

Acute tox data: Inhal LC_{LO} (mice) = 8700 ppm; ip LD_{LO} (mice) = 100 mg/kg. [*3*]

THR = HIGH via ip; MOD via inhal routes.

Fire Hazard: Dangerous when exposed to heat or flame. May explode with HNO_3. [*19*]

Disaster Hazard: Dangerous; when heated to decomp, emits highly toxic fumes of SO_x; can react vigorously with oxidizing materials.

To Fight Fire: Foam, CO_2, dry chemical.

THIOPHENOL. See phenyl mercaptan.

α-THIOPHENYL BORIC ACID. Colorless crystals, sol in water. $(C_4H_3S)B(OH)_2$, mw: 128.0, mp: 134°.

THR = See boron compounds.

THIOPHOS. See parathion.

THIOPHOSGENE. Syns: *thiocarbonyl chloride, thiocarbon aldehyde*. Reddish liquid. $CSCl_2$, mw: 115, bp: 73.5°, d: 1.5085 @ 15°.

Acute tox data: Inhal LC_{50} (mice) = 370 mg/m^3; oral LD_{50} (rat) = 929 mg/kg. [*3*]

THR = HIGH irr to skin, eyes and mu mem and via inhal route. A very strong irr. See also phosgene.

Disaster Hazard: Dangerous; when heated to decomp or upon hydrolysis, emits highly toxic fumes.

THIOPHOSPHORAMIDE. Amorphous yellow-white powder. $PS(NH_2)_3$, mw: 111.12, mp: decomp @ 200°, d: 1.7 @ 13°.

THR = U. See also amides.

Disaster Hazard: Dangerous; when heated to decomp or on contact with acid or acid fumes, emits highly toxic fumes of SO_x and PO_x.

THIOPHOSPHORYL BROMIDE. Cubic yellow crystals. $PSBr_3$, mw: 302.8, mp: 38°, bp: decomp @ 175°, d: 2.85 @ 17°.
Acute tox data: Inhal LC_{50} (mice) = 2600 mg/m^3. [*3*]
THR = MOD via inhal route. See also bromides and thiophosphoryl chloride.
Disaster Hazard: Dangerous; when heated to decomp, emits highly toxic fumes of bromides, SO_x and PO_x; will react with water or steam to produce toxic and corrosive fumes.

THIOPHOSPHORYL CHLORIDE. Syn: *phosphorus sulfochloride*. Colorless, mobile liquid, pungent odor. $PSCl_3$, mw: 169.45, bp: 125°, fp: −35°, flash p: none, d: 1.63 @ 25°/4°, vap. press: 22 mm @ 25°, vap. d: 5.86.
THR = HIGH irr to skin, eyes and mu mem and via oral and inhal routes. A corrosive.
Disaster Hazard: Dangerous; when heated, emits highly toxic fumes; will react with water or steam to produce toxic and corrosive fumes.

THIOPHOSPHORYL DIBROMIDE CHLORIDE. Pale green, fuming liquid. $PSBr_2Cl$, mw: 258.34, mp: −60°, bp: 98° @ 60 mm, d(liquid): 2.48 @ 0°.
THR = HIGH corrosive. See thiophosphoryl chloride.
Disaster Hazard: Dangerous; when heated to decomp, emits highly toxic fumes; will react with water or steam to produce toxic and corrosive fumes.

THIOPHOSPHORYL FLUORIDE. Gas. PSF_3, mw: 120.05, mp: 3.8° @ 7.6 atm, bp: decomp.
THR = HIGH irr to skin, eyes and mu mem and via inhal route. See also fluorides. May ignite or explode spont. [*19*]
Disaster Hazard: Dangerous; will react with water or steam to produce toxic and corrosive fumes.

"THIOSEMICARBAZONE." Syn: *p-acetylamino benzaldehyde thiosemicarbazone*. Pale yellow crystals. $C_{10}H_{12}N_4OS$, mw: 236.3, mp: 207°.
Acute tox data: Oral LD_{LO} (mice) = 950 mg/kg; sc LD_{50} (mice) = 1000 mg/kg. [*3*]
THR = MOD via oral and sc routes. An allergen. Can cause nausea, vomiting, skin rashes, liver injury and bone marrow depression.
Disaster Hazard: Dangerous; when heated to decomp or on contact with acid or acid fumes, emits highly toxic fumes.

THIOSULFATES.
THR = Up to 12 g of sodium thiosulfate can be taken daily by mouth with no ill effects except catharsis. Most of the thiosulfates are LOW in acute toxicity.
Disaster Hazard: Dangerous; when heated to decomp, they emit highly toxic fumes of SO_x.

THIOSULFURIC ACID. $H_2S_2O_2$, mw: 114.15.
THR = HIGH via irr. Corrosive. See also sulfuric acid.
Disaster Hazard: Dangerous, when heated to decomp, emits highly toxic fumes of SO_x; will react with water or steam to produce heat.

THIOTEPA. See tris(1-aziridinyl)phosphine sulfide.

2-THIOURACIL. Syn: *2-mercapto-4-hydroxy pyrimidine*. Small crystals, bitter taste, practically insol in water, alcohol, ether and acids, sol in alkalis. $C_4H_4N_2OS$, mw: 128.2.
THR = An exper (+) carc [*1*] and neo. [*3*] A toxic material used in medicine.

THIOUREA. Syn: *thiocarbamide*. White powder or crystals. NH_2CSNH_2, mw: 76.1, mp: 177°, bp: decomp, d: 1.405.
Acute tox data: ip LD_{50} (rat) = 436 mg/kg; oral TD_{LO} (women) = 1600 mg/kg for 5 weeks ⟶ blood effects. [*3*]
THR = HIGH via ip route. An exper (+) carc. [*3, 1*] of the liver and thyroid via oral route. A poison. 1 mg has proved fatal to a rat. It is said to cause depression of bone marrow with anemia, leukopenia and thrombocytopenia. May also cause allergic skin eruptions. A goitrogen. A cause of hepatic tumors upon chronic administration. [*2*] May react violently with acrolein. [*19*]
Disaster Hazard: Dangerous; when heated to decomp, emits highly toxic fumes of SO_x.

THIOVANIC ACID. See thioglycolic acid.

THIOVANOL. See monothioglycerol.

1,4-THIOXANE. Water white, refractive mobile liquid, characteristic odor. $O(CH_2CH_2)S$, mw: 104.1, bp: 148.7°, fp: −17°, flash p: 108°F (CC), d: 1.117 @ 20°.
THR = HIGH via oral, sc and iv routes.
Fire Hazard: Mod, when exposed to heat, flame or oxidizers.
Disaster Hazard: Dangerous; when heated to decomp, emits highly toxic fumes of SO_x; can react with oxidizing materials.
To Fight Fire: Water, CO_2, dry chemical.

***p*-THIOXENE.** Syns: *2,4-thioxene; 2,4-dimethylthiophene*. Liquid. $(CH_3)_2C_4H_2S$, mw: 112.2, bp: 138°, d: 0.9956 @ 20°.
THR = MOD via inhal route. Possibly via dermal route.
Fire Hazard: Mod, when exposed to heat, flame or oxidizers.
Disaster Hazard: Dangerous; when heated to decomp, emits highly toxic fumes of sulfur; can react with oxidizing materials.
To Fight Fire: Foam, CO_2, dry chemical.

THIRAM. See bis(dimethylthiocarbamyl)disulfide.

THORAZINE. See chloropromazine.

THORAZINE HYDROCHLORIDE. $C_{17}H_{19}N_2SCl \cdot HCl$, mw: 355.4.
THR = HIGH acute oral. An exper teratogen. [*3*]

THORIUM. Silvery-white, air stable, soft, ductile metal. Th, atwt: 232.0381, mp: 1750° (approx), d: 11.72.
THR = On an acute basis it has caused dermatitis. However, taken internally, as ThO_2, it has proven to be carc due to its radioactivity.
Radiation Hazard: For permissible levels, see Section 5A, Table 5A.5. Natural isotope ^{228}Th (Radiothorium, Thorium Series), $T\frac{1}{2}$ = 1.9y, decays to radioactive ^{224}Ra via α's of 5.3–5.4 MeV. Natural isotope ^{230}Th (Ionium, Uranium Series), $T\frac{1}{2} = 8 \times 10^4$y, decays to radioactive ^{226}Ra via α's of 4.6–4.7 MeV. Natural isotope ^{232}Th (Thorium Series), $T\frac{1}{2} = 1.4 \times 10^{10}$y, decays to radioactive ^{228}Ra via α's of 4.0 MeV. Natural isotope ^{234}Th (UX-I, Uranium Series), $T\frac{1}{2}$ = 24d, decays to radioactive ^{234}Pa via β's of 0.10 (35%), 0.19 (65%) MeV.
Fire Hazard: Mod, in the form of dust, when exposed to heat or flame or by chemical reaction with oxidizers, such as Cl_2, P, S. [*19*]

THORIUM CARBIDE. ThC_2, mw: 256.1.
THR = For radioactivity, see thorium; see also carbides. Readily incandesces when heated in air. Can react violently with S, Se. [*19*]

THORIUM CHLORIDE. Syn: *thorium tetrachloride.* White odorless crystals, sol in water and alcohol. $ThCl_4$, mw: 373.8, d: 4.59, mp: 770°, bp: 921°.
Acute tox data: Iv LD_{LO} (rat) = 15 mg/kg. [*3*]
THR = HIGH via iv route. See also thorium.
Disaster Hazard: Dangerous; see chlorides.

THORIUM (IV) CHLORIDE HEPTAHYDRATE. $ThCl_4 \cdot 7H_2O$, mw: 499.9.
Acute tox data: Iv LD_{LO} (rabbit) = 30 mg/kg. [*3*]
THR = HIGH via iv route. A carc. [*14*] See also thorium.

THORIUM COMPOUNDS. For radiation hazard, see thorium. Toxicity other than due to radiation has not been described.

THORIUM DIHYDRIDE. Black metallic crystals, reacts with water. ThH_2, mw: 234.1, d: 8.24, mp: decomp.
THR = A recog carc. [*14*] See thorium compounds and hydrides.

THORIUM DIOXIDE. Heavy, white crystalline powder. ThO_2, mw: 264.1, d: 9.7, mp: > 2800°.
THR = A recog carc. [*14*] See also thorium.

THORIUM NITRATE. White crystalline mass, sol in water and alcohol. $Th(NO_3)_4 \cdot 4H_2O$, mw: 552.2.
Acute tox data: ip LD_{50} (rat) = 68 mg/kg; iv LD_{LO} (dog) = 8.4 mg/kg; ip LD_{LO} (rabbit) = 500 mg/kg. [*3*]
THR = HIGH via iv and ip routes. See also thorium. A recog carc. [*14*]
Disaster Hazard: Dangerous; oxidizing material; when in contact with readily combustible substances, will cause violent combustion or ignition.

THORIUM NITRIDE. Th_3N_4, mw: 669.1.
THR = Radiotoxic; see thorium and nitrides. It incandesces in air and hydrolyzes violently. [*19*]

THORIUM OXYSULFIDE. ThOS, mw: 280.1.
THR = Self-ignites in air. [*19*] See also thorium.

THORIUM PICRATE. Crystalline powder. $Th(C_6H_2N_2O_7)_4 \cdot 10H_2O$, mw: 1324.68.
THR = A recog carc. [*14*] See picric acid and thorium.

THORIUM PHOSPHIDE. Th_3P_4, mw: 737.1.
THR = Reacts with acids or acid fumes to $\longrightarrow PH_3$, which is spont flam. [*19*] Very toxic. See also phosphine and thorium.

THORIUM TETRACHLORIDE. See thorium chloride.

THORIUM TETRAHYDRIDE. Black metallic crystals, reacts with water. ThH_4, mw: 233.1, d: 8.24, mp: decomp explosively @ red heat.
THR = A recog carc. [*14*] See thorium and hydrides.

THORON. An inert gaseous element. Tn or ^{220}Rn, atwt: 220.
THR = See radiation hazard under thorium and Rn.

"THOROTRAST." See thorium dioxide.

THEREONINE. Syn: *2-amino-β-hydroxy butyric acid.* An essential amino acid, colorless crystals, sol in water, optically active. $CH_3CH(OH)CH(NH_2)COOH$, mw: 119, mp(*dl* form): 255°–257° with decomp.
THR = U. A nutrient and/or dietary supplement food additive. [*109*]

THRESHOLD LIMIT VALUES (TLV). See Section 1.

THULIUM. A bright, silvery-gray, lustrous, soft, malleable, ductile metallic element. Tm, atwt: 168.9342, mp: 1545°, bp: 1727°, d: 9.314 @ 25°.
THR = As a lanthanide it has probably at least a MOD degree of toxicity.
Radiation Hazard: For permissible levels, see Section 5A, Table 5A.5. Artificial isotope ^{171}Tm, $T\frac{1}{2}$ = 1.9y, decays to stable ^{171}Yb via β's of 0.10 MeV.
Fire Hazard: Mod, in the form of dust, when exposed to flame. See also powdered metals.
Explosion Hazard: In the form of dust, when exposed to flame or violent reaction with air, halogens. [*19*] See also powdered metals.

THULIUM CHLORIDE. $TmCl_3$, mw: 275.3.
Acute tox data: ip LD_{50} (mice) = 485 mg/kg; ip LD_{50} (guinea pig) = 144 mg/kg. [*3*]
THR = HIGH via ip route. See also thulium.

THULIUM CITRATE.
Acute tox data: ip LD_{50} (mice) = 80 mg/kg; ip LD_{50} (guinea pig) = 55 mg/kg. [*3*]
THR = HIGH via ip route. See also thulium.

THULIUM NITRATE HEXAHYDRATE. $Tm(NO_3)_3 \cdot 6H_2O$, mw: 463.1.
Acute tox data: ip LD_{50} (rat) = 285 mg/kg; ip LD_{50} (mice) = 255 mg/kg. [*3*]
THR = HIGH via ip route. See also thulium.

THYME CAMPHOR. See thymol.

THYME OIL. Colorless to reddish-brown liquid, pleasant odor, sharp taste. d: 0.930 @ 25°/25°.
Acute tox data: Oral LD_{50} (rat) = 2840 mg/kg. [*3*]
THR = MOD via oral route. An allergen and an irr.
Fire Hazard: Slight.

THYMOL. Syn: *thyme camphor*. Colorless, translucent crystals. $C_{10}H_{14}O$, mw: 150.2, mp: 51°, bp: 233°, d: 0.972, vap. press: 1 mm @ 64°.
Acute tox data: Oral LD_{50} (rat) = 980 mg/kg; oral LD_{50} (mice) = 1800 mg/kg; oral LD_{50} (guinea pig) = 880 mg/kg. [*3*]
THR = MOD via oral route. An allergen.
Disaster Hazard: Mod dangerous; when heated, emits toxic fumes.

THYMOL IODIDE. Red brown powder or crystals, slight aromatic odor, sol in ether, chloroform and fixed or volatile oils, slightly sol in alcohol, insol in water. Practically, dithymol diiodide; $C_{20}H_{24}(IO)_2$, mw: 550.2; > 100° iodine vapors.
THR = U. A trace mineral added to animal feeds. [*109*]

THYNON. Syns: *5,10-dihydroxy-5,10-dioxonaphtho-(2,3,6)-p-dithiin-2,3-dicarbonitrile, dithianone*. Gray-brown needle-like crystals. $C_{14}H_4N_2O_2S_2$, mw: 296.3, mp: 220°.
Acute tox data: Oral LD_{50} (rat) = 1000 mg/kg. [*3*]
THR = MOD via oral route. A fungicide.
Disaster Hazard: Dangerous; see nitriles.

TIBAL. See triisobutyl aluminum.

TIEMANNITE. See mercuric selenide.

TIGLIC ACID. Syn: *crotonalic acid*. Thick, syrupy liquid or colorless crystals, spicy odor. $CH_3CHC(CH_3)CO_2H$, mw: 100.11, mp: 65°, bp: 198.5°, d: 0.9641, vap. press: 1 mm @ 52.0°.
THR = A MILD irr. MOD via oral route.
Disaster Hazard: Mod dangerous; when heated to decomp, emits toxic fumes.

TIGLIUM OIL. See croton oil.

TIME FUZES. See explosives, high.

TIN(α). Cubic, gray, crystalline metallic element. Sn, atwt: 118.70, mp: 231.9°, stabilizes <18°, bp: 2260°, d: 7.31, vap. press: 1 mm @ 1492°.
THR = See tin compounds.
Radiation Hazard: For permissible levels, see Section 5, Table 5A.5. Artificial isotope ^{113}Sn, $T_{\frac{1}{2}}$ = 115d, decays to radioactive ^{113m}In by ec; emits γ's of 0.26 MeV and x-rays.
Fire Hazard: Slight, in the form of dust, when exposed to heat or by spont chemical reaction with Br_2, BrF_3, Cl_2, ClF_3, $Cu(NO_3)$, K_2O_2, S. [*19*] See also powdered metals.

TIN BIFLUORIDE. See tin fluorides.

TIN BROMIDE. See stannic bromide.

TINCAL. See borax.

TIN CHLORIDE. See stannic chloride.

TIN COMPOUNDS.
THR = Elemental tin is not generally considered toxic. Some inorganic tin salts are irr or can liberate toxic fumes on decomp. The latter is particularly true of tin halogens. Alkyl tin compounds may be highly toxic and produce skin rashes. Dust of tin oxides have caused a pneumoconiosis, which is relatively benign. Organic tin compounds are absorbed via the skin.

TINCTURE OF TABASCO PEPPER. See capsicum.

TIN FLUORIDE. Syns: *stannous fluoride, tin bifluoride*. White, lustrous crystalline powder. SnF_2, mw: 156.70.
THR = See tin compounds and fluorides. Violent reaction with Cl_2. [*19*]

TINKAL. See borax.

TIN MONOPHOSPHIDE. Syns: *tin phosphide, stannic phosphide*. Silver-white crystals. SnP, mw: 149.68, d: 6.56.
THR = See phosphides and tin compounds. A flam solid.

TIN ORE.
THR = A recog carc. [*14*] See also arsenic compounds.

TIN OXIDES. Syns: (*a*) *stannous oxide*, (*b*) *stannic oxide* (*cassiterite*). (a) tetragonal black powder, (b) white powder; (a) SnO, (b) SnO_2; mw(a): 134.70, mw(b): 150.7; mp(a): decomp @ 700°–950°, mp(b): decomp @ 1127°; d(a): 6.446 @ 0°, d(b): 6.95.
THR = See tin compounds. Violent reaction with ClF_3, H_2S_3, Mg, K, Na. [*19*]

TIN PHOSPHIDE. See tin monophosphide.

TIN TETRABROMIDE. See stannic bromide.

TIN TETRACHLORIDE. See stannic chloride.

TIN TETRACHLORIDE, ANHYDROUS.

TIN TETRAHYDRIDE. Syn: *stannane.* Gas. SnH_4, mw: 122.73, mp: −150°, bp: −52° (decomp).
THR = See hydrides and tin compounds.
Fire Hazard: Mod, when exposed to flamc. See also hydrides.
Explosion Hazard: Slight, when exposed to flame. See hydrides.
Disaster Hazard: Mod dangerous; when heated, emits toxic fumes and may explode; can react with oxidizing materials.

TIN TETRAMETHYL. Syn: *tetramethyl stannane.* Colorless liquid. $Sn(CH_3)_4$, mw: 178.84, bp: 78°, lel = 1.9%, d: 1.314 @ 0°/4°, vap. d: 6.2, flash p: <70°.
THR = HIGH via oral and possibly dermal routes. See tin compounds and organometals.
Fire Hazard: Mod, when exposed to heat, flame or oxidizers.
Spont Heating: No.
Explosion Hazard: Mod, when exposed to flame.
Disaster Hazard: Dangerous; when heated, emits acrid fumes and may explode; can react with oxidizing materials.
To Fight Fire: Water, foam, CO_2, dry chemical.

TIN TETRAPHENYL. Colorless crystals. $(C_6H_5)_4Sn$, mw: 427.1, mp: 226°, bp: 424°, flash p: 450°F (CC), d: 1.490.
THR = HIGH. See phenol and tin compounds.
Fire Hazard: Slight, when exposed to heat or flame.
Disaster Hazard: Dangerous; when heated, emits toxic fumes; can react with oxidizing materials.
To Fight Fire: CO_2, dry chemical.

TIN TRIPHOSPHIDE. Crystals. SnP_3, mw: 211.64, mp: decomp > 415° to Sn_4P_3, d: 4.10 @ 0°.
THR = See phosphides and tin compounds.

TITANIUM. Dark gray amorphous powder or lustrous white metal. Ti, atwt: 47.90, mp: 1720°, bp: >3000°, d: 4.5 @ 20°, autoign. temp.: 1200° for massive metal in air, 250° for powder.
THR = See titanium compounds.
Fire Hazard: Mod, in the form of dust, when exposed to heat or flame or by chemical reaction. See also powdered metals. Titanium can burn in an atmosphere of carbon dioxide, nitrogen or air. Also reacts violently with BrF_3, CuO, PbO, (Ni + $KClO_3$), HNO_3, O_2, $KClO_3$, KNO_3, $KMnO_4$, steam @ 704°, trichloroethylene, trichlorotrifluoroethane. [*19*] Ordinary extinguishers are often ineffective against titanium fires. Such fires require the special extinguishers designed for metal fires. See magnesium. In airtight enclosures, titanium fires can be controlled by the use of argon or helium. When titanium burns in the absence of moisture, it burns slowly but evolves much heat. The application of water to burning titanium can cause an explosion.
Explosion Hazard: Finely divided titanium dust and powders, like most metal powders, are potential explosion hazards when exposed to sparks, open flame or high heat sources. See also magnesium.

TITANIUM ALLOY.
THR = Exposure to red fuming HNO_3, O_2 or liquid O_2 may cause an explosion. [*19*]

TITANIUM BUTYLATE. See butyl titanate.

TITANIUM CARBIDE. TiC, mw: 59.9.
THR = See carbides. The dust can explode in air. [*19*]

TITANIUM COMPOUNDS.
THR = This material is considered to be physiologically inert. There are no reported cases in the literature where titanium as such has caused intoxication. The dusts of titanium or titanium compounds such as titanium oxide may be placed in the nuisance category. Titanium tetrachloride, however, is an irr and corrosive material, because when exposed to moisture, it hydrolyzes to hydrogen chloride. See hydrochloric acid.

TITANIUM DIBROMIDE. Black powder. $TiBr_2$, mw: 207.73, mp: decomp >500°.
THR = See bromides and titanium compounds.

TITANIUM DICHLORIDE. Light brown to black, deliquescent solid. $TiCl_2$, mw: 118.81, mp: sublimes in H_2, d: 3.13.
THR = See titanium compounds and hydrochloric acid. This material is highly flam in air @ room temp. In an inert atmosphere, $TiCl_4$ and Ti (which can ignite in air) are formed. [*19*]

TITANIUM DIOXIDE. Syn: *rutile.* Blue crystals. TiO_2, mw: 79.90, mp: 1860° (decomp), d: 4.26.
THR = See titanium compounds. A common air contaminant and nuisance dust. Violent reaction with Li. [*19*]

TITANIUM DISULFIDE. Yellow scales. TiS_2, mw: 112.03.
THR = See sulfides. Can detonate when heated with KNO_3. [*19*]

TITANIUM HYDRIDE. Dark gray metallic powder or crystals. TiH_2, mw: 49.9, d: 3.76.
THR = See hydrides and titanium compounds.
Fire Hazard: Mod, when exposed to heat or flame. Burns brilliantly in air. See hydrides.
Explosion Hazard: Mod, in the form of dust, by chemical reaction.
Disaster Hazard: See hydrides.

TITANIUM ISOPROPYLATE. See isopropyl titanate.

TITANIUM MONOSULFIDE. Reddish solid. TiS, mw: 79.97.
THR = See sulfides.

TITANIUM NITRIDE. Brassy crystals. TiN, mw: 61.9, mp: 2950°, d: 5.43.
THR = See titanium compounds and nitrides.

TITANIUM OXALATE. Yellow prisms. $Ti_2(C_2O_4)_3 \cdot 10H_2O$, mw: 540.02.
THR = See oxalates and titanium compounds.

TITANIUM OXIDE. See titanium dioxide.

TITANIUM PHOSPHIDE. Gray metallic solid. TiP, mw: 78.88, d: 3.95 @ 25°.
THR = See phosphides and titanium compounds.

TITANIUM POTASSIUM FLUORIDE. See potassium hexafluorotitanate.

TITANIUM SESQUISULFATE. See titanous sulfate.

TITANIUM SESQUISULFIDE. Grayish-black crystals. Ti_2S_3, mw: 192.00.
THR = See sulfides and titanium compounds.

TITANIUM SULFATE SOLUTION. See titanium compounds.

TITANIUM TETRACHLORIDE. Colorless to light yellow liquid, fumes in moist air. $TiCl_4$, mw: 189.73, mp: −30°, bp: 136.4°, d: 1.772 @ 25°/25°, vap. press: 10 mm @ 21.3°.
Acute tox data: Inhal LC_{LO} (mice) = 10 mg/m^3 for 2 hrs. [*3*]
THR = HIGH irr to skin, eyes and mu mem and via inhal route. HIGH corrosive because it liberates heat and hydrochloric acid upon contact with moisture. If spilled on skin, wipe off with dry cloth before applying water. Reacts violently with K. [*19*]
Disaster Hazard: See hydrochloric acid.

TITANIUM TETRAFLUORIDE. White powder. TiF_4, mw: 123.90, mp: >400°, d: 2.798 @ 20.5°.
THR = See fluorides.

TITANIUM TRIFLUORIDE. Purple-red or violet crystals. TiF_3, mw: 104.90.
THR = See fluorides.

TITANOCENE. $C_{10}H_{10}Ti$, mw: 178.1.
Acute tox data: im LD_{LO} (rat) = 50 mg/kg. [*3*]
THR = HIGH via im route. See also titanium.

TITANIUM DICHLORIDE. $C_{10}H_{10}Cl_2Ti$, mw: 248.
THR = An exper carc. [*3*]

TITANOUS SULFATE. Syn: *titanium sesquisulfate.* Green, deliquescent, crystalline powder. $Ti_2(SO_4)_3$, mw: 383.98.
THR = See titanium compounds and sulfates.

TNA. See tetranitroaniline.

TNT. See trinitrotoluene.

TNX. See 2,4,6-trinitroxylene.

TOBACCO. Dried leaves.
THR = A nicotine-containing dried leaf of the tobacco plant. The smoke produced by burning tobacco contains the highly toxic alkaloid, nicotine, tars and phenols, carbon monoxide, cyanides, nitrates, nitrites, carc, co-carc and perhaps 100 other chemicals, α-emitters, etc. Habitual inhal of tobacco smoke is considered a leading cause of lung cancer and circulatory problems, cardiac problems, etc. See also nicotine.
Fire Hazard: Slight, when exposed to heat or flame.

TOBACCO QUID.
THR = A susp carc. [*14*] See plant and fungal products.

TOBACCO WOOD. See witch hazel.

TOCHLORINE. See sodium-*p*-toluene sulfonchloramine.

α-TOCOPHEROL ACETATE. A form of vitamin E, yellow, nearly odorless, clear viscous oil, insol in water, freely sol in alcohol, miscible with acetone and vegetable oils. $C_{29}H_{49}O \cdot OOCCH_3$, mw: 472, d: 0.950–0.964.
THR = U. A nutrient and/or dietary supplement food additive. [*109*]

TOCOPHEROLS. Syn: *vitamin E.* A group of related substances, α,β,γ,δ-tocopherol, which constitute vitamin E. These vitamin constituents are viscous oils, sol in lipid solvents, insol in water.
THR = U. Used as a chemical preservative, nutrient, and/or as a dietary supplement food additive. [*109*]

TOE PUFFS. Toe puffs are box toe boards used in the manufacture of boots and shoes and may consist of several layers of fabric impregnated with celluloid solvent, rosin and dye.
THR = U.
Fire Hazard: Dangerous; they are liable to spont combustion.

TOFRANIL. $C_{19}H_{24}N_2$, mw: 280.5.
THR = An exper teratogen. HIGH via oral [*3*] route.

TOLAMINE. See sodium *p*-toluene sulfonchloramine.

TOLBUTAMIDE. Syn: *orinase.* $C_{12}H_{18}O_3N_2S$, mw: 270.4.
THR = MOD to HIGH acute to rats. An exper teratogen. [*3*] Has been implicated in aplastic anemia.

TOLAN. See diphenyl acetylene.

***o*-TOLIDINE.** Syn: *3,3'-dimethylbenzidine.* White to reddish crystals. $C_{14}H_{16}N_2$, mw: 212.3.
Acute tox data: Oral LD_{50} (rat) = 404 mg/kg; ip LD_{LO} (rat) = 125 mg/kg. [*3*]

THR = HIGH via ip and oral routes. An exper (+) carc via oral, sc and imp routes. [*3*, *9*]

***o*-TOLIDINE FLUOSILICATE.** Small white crystals, slightly sol in alcohol. $(C_6H_3NH_2CH_3)_2 \cdot H_2SiF_6$, mw: 356.4, mp: 269°.
THR = See fluosilicates and *o*-tolidine.

TOLUENE. Syns: *methylbenzene*, *phenylmethane*, *toluol*. Colorless liquid, benzol-like odor. $C_6H_5CH_3$, mw: 92.13, mp: −95° to −94.5°, bp: 110.4°, flash p: 40°F (CC), ulc: 75–80, lel = 1.27%, uel = 7%, d: 0.866 @ 20°/4°, autoign. temp.: 896°F, vap. press: 36.7 mm @ 30°, vap. d: 3.14.
Acute tox data: Inhal TC_{LO} (human) = 200 ppm ⟶ CNS effects; inhal TC_{LO} (man) = 100 ppm ⟶ psychotropic effects; oral LD_{50} (rat) = 5000 mg/kg; inhal LC_{LO} (rat) = 4000 ppm for 4 hrs; ip LD_{50} (rat) = 1640 mg/kg; inhal LC_{50} (mice) = 5300 ppm; dermal LD_{50} (rabbit) = 14000 mg/kg. [*3*]
THR = MOD via oral, inhal and ip routes; LOW via dermal route. Toluene is derived from coal tar, and commercial grades usually contain small amounts of benzene as an impurity. Acute poisoning, resulting from exposures to high conc of the vapors, are rare with toluene. Inhal of 200 ppm of toluene for 8 hrs may cause impairment of coordination and reaction time; with higher conc (up to 800 ppm) these effects are increased and are observed in a shorter time. In the few cases of acute toluene poisoning reported, the effect has been that of a narcotic, the workman passing through a stage of intoxication into one of coma. Recovery following removal from exposure has been the rule. An occasional report of chronic poisoning describes an anemia and leucopenia, with biopsy showing a bone marrow hypoplasia. These effects, however, are less common in people working with toluene, and they are not as severe.

Exposure to conc up to 200 ppm produces few symptoms. At 200–500 ppm, headache, nausea, loss of appetite, a bad taste, lassitude, impairment of coordination and reaction time are reported, but are not usually accompanied by any laboratory or physical findings of significance. With higher conc, the above complaints are increased and in addition, anemia, leucopenia and enlarged liver may be found in rare cases.

A common air contaminant.
Fire Hazard: Slight, when exposed to heat, flame or oxidizers.
Explosion Hazard: Mod, when exposed to flame or reacted with $(H_2SO_4 + HNO_3)$, N_2O_4, $AgClO_4$. [*19*]
Disaster Hazard: Mod dangerous; when heated, emits toxic fumes can react vigorously with oxidizing materials.
To Fight Fire: Foam, CO_2, dry chemical.

1,*o*-TOLUENE AZONAPHTHYLAMINE-2. See yellow OB.

2,4-TOLUENEDIAMINE. Syn: *tolylenediamine*. Prisms. $CH_3C_6H_3(NH_2)_2$, mw: 122.17, mp: 99°, bp: 280°, vap. press: 1 mm @ 106.5°.
Acute tox data: Oral LD_{LO} (rat) = 500 mg/kg; sc LD_{LO} (dog) = 200 mg/kg. [*3*]
THR = HIGH via sc and MOD via oral routes. An exper carc via oral route. [*3*] This material has a marked toxic action upon the liver and can cause fatty degeneration of that organ. It is also thought to be an irr. When solutions of it come in contact with the skin, it can cause irr and blisters, particularly to individuals who are sensitive to it.
Disaster Hazard: Mod dangerous; when heated, emits toxic fumes.

2,5-TOLUENDIAMINE. Syns: *2,5-tolylenediamine*, *2,5-diaminotoluene*. Colorless, crystalline tablets. $CH_3C_6H_3(NH_2)_2$, mw: 122.17, mp: 64°, bp: 274°.
Acute tox data: Oral LD_{LO} (mammal) = 3600 mg/kg. [*3*, *108*] A mutagen being studied for carc [*108*] properties.
THR = MOD via oral route. An irr. This material has a toxic action upon the liver and can cause fatty degeneration of that organ. Its total effect upon the body seems to take place 3 different ways. It is toxic to the CNS. It produces jaundice by action on the liver and spleen, and it produces anemia by destruction of the red blood cells. In this action it is quite similar to aniline, although by no means identical with it. Its high bp and the fact that the material is solid at room temp. makes it somewhat less hazardous than aniline, particularly at ordinary working temp. The literature contains a reference to a permanent injury to an eye due to the use of this material as an eyelash dye. It is considered to be an irr dye material. It can cause irr and blisters on the fingers of individuals whose skins are sensitive to it.
Disaster Hazard: Mod dangerous; when heated, emits toxic fumes.

TOLUENE DIISOCYANATE. See 2,4-tolylene diisocyanate.

TOLUENE SUBSTITUTE. Composed largely of octanes. bp: 100°, flash p: 30°F, d: 0.743.
THR = See octane.
Fire Hazard: Dangerous, when exposed to heat or flame.
Explosion Hazard: U.

Disaster Hazard: Dangerous, upon exposure to heat or flame; can react vigorously with oxidizing materials.

To Fight Fire: Foam, CO_2, dry chemical.

***o*-TOLUENESULFONIC ACID.** Syn: *methylbenzene-sulfonic acid.* Crystals. $CH_3C_6H_4SO_3H$, mw: 172.2.

THR = HIGH irr to skin, eyes and mu mem and via oral and inhal routes. See also *p*-toluene sulfonic acid.

Disaster Hazard: Dangerous; see sulfonate.

***p*-TOLUENESULFONIC ACID.** Syn: *p-toluene sulfonate.* Colorless leaflets, sol in alcohol, ether and water. $C_6H_4(SO_3H)(CH_3)$, mw: 172, mp: 107°, bp: 140° @ 20 mm.

Acute tox data: Oral LD_{50} (mice) = 400 mg/kg. [*3*]

THR = HIGH via oral route. Animal exper show MOD systemic toxicity and HIGH irr. See also *o*-toluene sulfonic acid.

Disaster Hazard: Dangerous; see sulfonates.

***p*-TOLUENE SULFONYLAMIDE.** $C_7H_9O_2NS$, mw: 171.2.

Acute tox data: Oral LD_{50} (wild birds) = 75 mg/kg; ip LD_{50} (mice) = 250 mg/kg. [*3*]

THR = HIGH via oral and ip routes. A fungicide. See also amides.

Disaster Hazard: Dangerous; see sulfonates.

TOLUENE-*p*-SULFONYL METHYL NITROSAMIDE.

THR = An exper carc. [*23*]

α-TOLUENE THIOL. See benzyl mercaptan.

TOLUENE TRICHLORIDE. See benzotrichloride.

α-TOLUIC ACID. See phenylacetic acid.

***m*-TOLUIDINE.** Syn: *m-methylaniline.* Colorless liquid. $CH_3C_6H_4NH_2$, mw: 107.2, mp: −50.5°, bp: 203.3°, d: 0.989 @ 20°/4°, vap. press: 1 mm @ 41°, vap. d: 3.90.

Acute tox data: Oral LD_{50} (rat) = 974 mg/kg; ip LD_{50} (mice) = 150 mg/kg. [*3*]

THR = HIGH via ip and MOD via oral route. An exper neo via oral route. [*3*]

Fire Hazard: Mod, when exposed to heat or flame.

Disaster Hazard: Dangerous; when heated, emits highly toxic fumes; can react vigorously on contact with oxidizing materials.

To Fight Fire: Foam, CO_2, dry chemical.

***o*-TOLUIDINE.** Syn: *o-methylaniline.* Colorless liquid. $CH_3C_6NH_2$, mw: 107.2, mp: −16.3°, bp: 199.7°, ulc: 20–25, flash p: 185° (CC), d: 1.004 @ 20°/4°, autoign. temp.: 900°F, vap. press: 1 mm @ 44°, vap. d: 3.69.

Acute tox data: Oral LD_{LO} (cat) = 300 mg/kg; oral LD_{50} (rat) = 900 mg/kg; ip LD_{50} (mice) = 150 mg/kg; dermal LD_{50} (rabbit) = 3250 mg/kg; oral LD_{LO} (frog) = 5 mg/kg. [*3*]

THR = HIGH via oral and ip routes. MOD via dermal route. An exper neo via oral route. [*3, 23*] This material can produce severe systemic disturbances. The main portal of entry into the body is the respiratory tract, particularly in cases of industrial exposure. The symptoms produced by intoxication due to this compound are headache, weakness, difficulty in breathing, air hunger, psychic disturbances, and marked irr of the kidneys and bladder. The literature does not yield any good data for comparing the toxicity of the *o*-, *m*- and *p*-isomers. Their behavior is generally comparable to that of aniline, and while the most frequent type of exposure is inhal, a certain amount of exposure occurs by skin contact. It has been determined exper that a conc of approx 100 ppm is the maximum endurable for 1 hr without serious consequences and that from 6–23 ppm is endurable for several hours without serious disturbances. See aniline. A recog carc. [*14*]

Fire Hazard: Mod, when exposed to heat or flame. Reacts violently with HNO_3. [*19*]

Spont Heating: No.

Disaster Hazard: Dangerous; when heated, emits highly toxic fumes; can react with oxidizing materials.

To Fight Fire: Foam, CO_2, dry chemical.

***p*-TOLUIDINE.** Syn: *p-methylaniline.* Colorless leaflets. $CH_3C_6H_4NH_2$, mw: 107.2, mp: 44.5°, bp: 200.4°, flash p: 188°F (CC), d: 1.046 @ 20°/4°, autoign. temp.: 900°F, vap. press: 1 mm @ 42°, vap. d: 3.90.

Acute tox data: Oral LD_{50} (rat) = 656 mg/kg; ip LD_{50} (mice) = 50 mg/kg; oral LD_{50} (wild birds) = 42 mg/kg. [*3*]

THR = HIGH via oral and ip routes. An exper carc via oral route. [*3*]

Fire Hazard: Mod, when exposed to heat or flame or oxidizers.

Spont Heating: No.

Disaster Hazard: Dangerous; when heated, emits highly toxic fumes; can react vigorously on contact with oxidizing materials.

To Fight Fire: Foam, CO_2, dry chemical.

***o*-TOLUIDINE HYDROCHLORIDE.** $C_7H_9N \cdot HCl$, mw: 143.6.

Acute tox data: Oral LD_{50} (rat) = 2951 mg/kg; ip LD_{50} (rat) = 150 mg/kg; ip LD_{50} (mice) = 113 mg/kg. [*3*]

THR = HIGH via ip; MOD via oral routes. An exper carc. [*23, 108*]

***p*-TOLUIDINE HYDROCHLORIDE.** See chlorotoluidine.

p-TOLUIDINE MUSTARD. Syn: *N,N-di(2-chloroethyl)-p-toluidine.*
THR = An exper carc. [*23*]

TOLUOL. See toluene.

O-TOLUOL SULFOACID. Liquid. $C_{14}H_{14}O_3S$, mw: 262.32, flash p: 363°F.
THR = U.
Fire Hazard: Slight, when exposed to heat or flame.
Disaster Hazard: Dangerous; when heated to decomp, emits highly toxic fumes of SO_x; can react with oxidizing materials.
To Fight Fire: Foam, CO_2, dry chemical.

TOLUYLENE BLUE. $C_{15}H_{19}N_4Cl$, mw: 290.1.
THR = An exper neo. [*3*]

TOLYLACETAMIDE. See *p*-acetotoluidide.

m-TOLYLAZOACETANILIDE. $C_{15}H_{16}ON_3$, mw: 254.3.
THR = An exper neo. [*3*]

p-(α-TOLYLAZO)-o-CRESOL. $C_{14}H_{14}ON_2$, mw: 226.3.
THR = An exper neo. [*3*]

4-(m-TOLYLAZO)-m-TOLUIDINE. See 4-amino-2,3-azotoluene.

4-(o-TOLYLAZO)-o-TOLUIDINE. See *o*-amino azotoluene.

4-(p-TOLYLAZO)-m-TOLUIDINE. See 4-amino-4,2-azotoluene.

4-(p-TOLYLAZO)-p-TOLUIDINE. See 4-amino-4,3-azotoluene.

2-(o-TOLYLAZO)-p-TOLUIDINE. See 2-amino-2,5-azotoluene.

o-TOLYL BIGUANIDE HYDROCHLORIDE. Colorless crystals. $CH_3C_6H_4NHC(NH)NHC(NH)NH_2 \cdot HCl$, mw: 227.5, mp: 227°, d: 1.264 @ 30°.
THR = U.
Disaster Hazard: Dangerous; when heated to decomp, emits highly toxic fumes.

p-TOLYL BROMIDE. See *p*-bromotoluene.

p-TOLYL-N-CARBAMOYL AZIRIDINE. $C_{10}H_{13}ON_2$, mw: 177.3.
THR = An exper neo. [*3*]

p-TOLYL DIETHANOLAMINE. Crystals. $(HOC_2H_4)_2NC_6H_4CH_3$, mw: 195.25, mp: 63.2°, flash p: 385°F (OC), vap. d: 6.73.
THR = Details U. See also amines.
Fire Hazard: Slight, when exposed to heat or flame; can react with oxidizing materials.
To Fight Fire: Foam, CO_2, dry chemical.

TOLYLENE DIISOCYANATE. Syns: *2,4-diisocyanotoluene, TDI.* Clear, faintly yellow liquid. $H_3CC_6H_3(NCO)_2$, mw: 174.16, bp: 118°–120° @ 10 mm, fp: 20°, d: 1.22 @ 20°/4°, flash p: 270°F (OC), vap. d: 6.0, lel = 0.9%, uel = 9.5%.
Acute tox data: Inhal TC_{LO} (human) = 0.5 ppm ⟶ irr effects; oral LD_{50} (rat) = 6170 mg/kg; inhal LC_{50} (rat) = 14 ppm for 4 hrs; inhal LC_{50} (rabbit) = 8 ppm for 4 hrs. [*3*]
THR = HIGH via inhal; MOD via oral routes. An irr. Capable of producing severe dermatitis and bronchial spasm. Following inhal (especially if severe), victim should be observed by a physician. Particularly irr to the eyes. A common air contaminant.
Fire Hazard: Combustible, when exposed to heat or flame.
Disaster Hazard: Dangerous; when heated to decomp, emits highly toxic fumes.
To Fight Fire: Dry chemical, CO_2.

o-TOLYLETHANOLAMINE. Liquid. $CH_3C_6H_4NHCH_2CH_2OH$, mw: 151.2, mp: −25°, bp: 297.1°, flash p: 290°F (OC), d: 1.0723 @ 20°/20°.
Acute tox data: Dermal LD_{50} (rabbit) = 1000 mg/kg. [*3*]
THR = MOD via dermal route. See also amines.
Fire Hazard: Slight, when exposed to heat or flame. Can react with oxidizing materials.
To Fight Fire: Foam, CO_2, dry chemical.

TOLYLGERMANIUM TRIBROMIDE. Colorless liquid, hydrolyzed by water. $(CH_3C_6H_4)GeBr_3$, mw: 403.5, mp: 156° @ 13 mm.
THR = See germanium compounds and bromides.

TOLYLGERMANIUM TRICHLORIDE. Colorless liquid, hydrolyzed by water. $(CH_3C_6H_4)GeCl_3$, mw: 270.1, bp: 116° @ 12 mm.
THR = See germanium compounds and chlorides.

TOLYLGERMANIUM TRIIODIDE. Colorless crystals, hydrolyzed by water. $(CH_3C_6H_4)GeI_3$, mw: 545.0, mp: 72°.
THR = See germanium compounds and iodides.

m-TOLYL HYDRAZINE. Oily liquid. $C_7H_{10}N_2$, mw: 122.2, bp: 243°, d: 1.061.
THR = U. See hydrazine.

TOLYL MALEIMIDE. $C_{11}H_9O_2N$, mw: 187.2.
Acute tox data: Oral LD_{LO} (rat) = 500 mg/kg; ip LD_{LO} (rat) = 50 mg/kg; dermal LD_{50} (rat) = 360 mg/kg. [*3*]
THR = HIGH via oral, ip and dermal routes.

p-TOLYL MERCURIC BROMIDE. Lustrous crystals, sol in organic solvents. $(C_7H_7)HgBr$, mw: 371.7, mp: 228°.
THR = HIGH. See mercury compounds, organic, and bromides.

***p*-TOLYL MERCURIC CHLORIDE.** Silky tablets, insol in water. $(C_7H_7)HgCl$, mw: 327.2, mp: 233°.
THR = HIGH. See mercury compounds organic, and chlorides.

TOLYL MERCURY SALICYLATE.
THR = HIGH. See mercury compounds, organic.

***m*-TOLYL-*m*-METHYL CARBAMATE.** See tsumacide.

***o*-TOLYL PHOSPHATE.** See tri-*o*-cresyl phosphate.

N,*m*-TOLYL PHTHALAMIC ACID. See duraset.

***p*-TOLYLPROPANOLAMINE HYDROCHLORIDE.** Liquid. $C_{10}H_{15}ON \cdot HCl$, mw: 201.7.
Acute tox data: ip LD_{LO} (rat) = 50 mg/kg; sc LD_{LO} (rat) = 80 mg/kg; iv LD_{LO} (rabbit) = 33 mg/kg. [*3*]
THR = HIGH via ip, sc and iv routes.
Fire Hazard: Mod, via heat, flame or oxidizers.
To Fight Fire: Spray, foam, dry chemical.

TOLYLTETRATHIO-*o*-STANNATE. Solid. $(SC_6H_4CH_3)_4Sn$, mw: 611.5, mp: 100°.
THR = See tin compounds and sulfates.

***m*-TOLYL THIOUREA.**
Acute tox data: Oral LD_{LO} (rat) = 5 mg/kg. [*3*]
THR = HIGH via oral route.

***o*-TOLYL THIOUREA.** $C_8H_{10}N_2S$, mw: 166.3.
Acute tox data: Oral LD_{LO} (rat) = 5 mg/kg; ip LD_{50} (mice) = 150 mg/kg. [*3*]
THR = HIGH via oral and ip routes.

***p*-TOLYL THIOUREA.**
Acute tox data: Oral LD_{LO} (rat) = 5 mg/kg. [*3*]
THR = HIGH via oral route.

***o*-TOLYLTIN TRICHLORIDE.** Liquid. $(C_6H_4CH_3)SnCl_3$, mw: 316.2, d: 1.7619, bp: 158° @ 20 mm.
THR = See tin compounds, hydrochloric acid and chlorides.

***m*-TOLYLTIN TRICHLORIDE.** Colorless liquid. $(C_6H_4CH_3)SnCl_3$, mw: 316.2, d: 1.7516, mp: $<-20°$, bp: 151° @ 23 mm.
THR = See tin compounds, hydrochloric acid and chlorides.

***p*-TOLYLTIN TRICHLORIDE.** Liquid, decomp by water. $(C_6H_4CH_3)SnCl_3$, mw: 316.2, d: 1.7522, bp: 157° @ 23 mm.
THR = See tin compounds, hydrochloric acid and chlorides.

***o*-TOLYL-*p*-TOLUENESULFONATE.** Liquid. $C_{14}H_{14}O_3S$, mw: 262.31, flash p: 363°F (CC).
THR = U. See also esters.
Fire Hazard: Slight, when exposed to heat or flame; can react with oxidizing materials.
Disaster Hazard: Dangerous; see sulfonates.
To Fight Fire: Dry chemical, CO_2.

TOMARIN. Syn: *3-(α-p-chlorophenyl-β-acetylethyl)-4-hydroxycoumarin*. Crystals. $C_{19}H_{15}ClO_4$, mw: 342.8, mp: 164°.
Acute tox data: Oral LD_{50} (rat) = 900 mg/kg. [*3*]
THR = MOD via oral route. See warfarin which closely resembles this material.
Disaster Hazard: Dangerous; when heated to decomp, emits highly toxic fumes of chlorides.

TOPPED TAR.
THR = A recog carc. [*14*] See coal tar and pitch.

"TORDON." Syn: *picloram*. Crystalline material. $C_6H_3Cl_3N_2O_2$, mw: 241.5, mp: 218°.
Acute tox data: Oral LD_{50} (rat) = 3750 mg/kg; oral LD_{50} (mouse) = 1500 mg/kg; oral LD_{50} (rabbit) = 2000 mg/kg. [*3*]
THR = MOD via oral route. An herbicide.

TOXADUST. See toxaphene.

TOXALBUMIN. See abrin.

TOXAPHENE. See octachlorocamphene.

TOXIC HAZARDS STORAGE AND HANDLING. See Section 7.

TOXICITY CRITERIA. See Sections 8, 9.

TOXICITY FACTORS. See Section 9.

TOXILIC ACID. See maleic acid.

TOXILIC ANHYDRIDE. See maleic anhydride.

TOXIC MATERIALS AND ENZYME INDUCTION. See Section 9.

TOXICOLOGY FOR INDUSTRIAL HYGIENE. See Section 1.

TOY CAPS. See explosives, low.

TRAGACANTH. Syn: *gum tragacanth*. Powder is white, pieces are white to pale yellow, translucent and horny.
THR = A MILD allergen, probably LOW toxicity. A stabilizer food additive. [*109*]
Fire Hazard: Slight, when exposed to heat or flame.

TRANSFER STATIONS AS DANGER POINTS IN SOLID WASTE HANDLING. See Section 6.

TRANSFORMER OIL. Syn: *transil oil*. Liquid. flash p: 295°F (OC), d: 0.9.
THR = U.
Fire Hazard: Slight, when exposed to heat or flame.
Spont Heating: No.
Disaster Hazard: Slightly dangerous; when heated, emits acrid fumes; can react with oxidizing materials.
To Fight Fire: CO_2, dry chemical.

TRANSIL OIL. See transformer oil.

"TRANSOTE." See creosote.

TREFLAN. See α,α,α-trifluoro-2,6-dinitro-N,N-dipropyl-*p*-toluidine.

TREMOLITE. A variety of asbestos, white to light green, vitreous to silky. $Ca_2Mg_5Si_8O_{22}(OH)_2$, mw: 710, d: 3.0–3.3.
THR = See asbestos.

TREMTONE. $C_{13}H_{14}O_2$, mw: 202.2, d: 1.080 @ 23°/4°. Liquid.
THR = HIGH via oral route. Found in richwheat or snake-root. Can occur in milk of cows that have eaten either of these.

TRENIMON. $C_{12}H_{13}O_2N_3$, mw: 231.3.
Acute tox data: iv LD_{50} (rat) = 0.47 mg/kg. [*3*]
THR = HIGH via iv route. An exper (+) carc via iv route [*3*, *8*] and mutagen [*3*] via ip route. [*23*]

TRETAMINE. See triethylene melamine.

TRIACETIN. See glyceryl triacetate.

TRIALLYLAMINE. Liquid. $(H_2C{:}CHCH_2)_3N$, mw: 137, d: 0.800 @ 20°/4°, mp: <−70°, bp: 150°–151°, flash p: 103°F (TOC).
Acute tox data: Inhal TC_{LO} (mice) = 13 ppm for 5 min ⟶ pulmonary effects; oral LD_{50} (rat) = 954 mg/kg; inhal LC_{LO} (rat) = 550 ppm for 4 hrs; oral LD_{50} (mice) = 492 mg/kg; ip LD_{50} (mice) = 187 mg/kg; dermal LD_{50} (rabbit) = 2250 mg/kg. [*3*]
THR = HIGH via ip, oral, inhal routes.
Fire Hazard: Mod, via heat, flame or oxidizers.
To Fight Fire: Foam, alcohol foam, fog.

TRIALLYL BORATE.
Acute tox data: Oral LD_{50} (mice) = 1800 mg/kg. [*3*]
THR = MOD via oral route. See also boron compounds.

TRIALLYL CYANURATE. Colorless liquid or solid. $C_{12}H_{15}O_3N_3$, mw: 249.26, bp: 120° @ 5 mm, fp: 27.3°, flash p: > 176°F (TOC), d: 1.1133 @ 30°, vap. press: 1 mm @ 100°.
THR = MOD irr to skin, eyes and mu mem and via oral and inhal routes.
Fire Hazard: Mod, when exposed to heat, flame or oxidizers.
Disaster Hazard: Dangerous, when heated to decomp, on contact with acid or acid fumes, emits highly toxic fumes of cyanides.
To Fight Fire: Spray, foam, dry chemical.

TRIALLYL PHOSPHATE. See allyl phosphate.

2,4,6-TRIAMINO-8-TRIAZINE. See melamine.

TRIAMYLAMINE. A clear, water white or pale yellow liquid, amine odor. $(C_5H_{11})_3N$, mw: 227.42, bp: 232°, vap. press: 7 mm @ 26°, flash p: 215°F (OC), d: 0.79–0.80 @ 20°/20°, vap. d: 7.83.
THR = U. See amines.
Fire Hazard: Low, when exposed to heat or flame; can react with oxidizing materials.
Disaster Hazard: Mod; when heated to decomp, emits toxic fumes.
To Fight Fire: Dry chemical, CO_2.

TRIAMYLBENZENE. A clear liquid. $(C_5H_{11})_3C_6H_3$, mw: 288.50, bp: 300°, flash p: 270°F (OC), d: 0.87.
THR = Details U. Probably irr and narcotic in high conc. See benzene.
Fire Hazard: Slight, when exposed to heat or flame; can react with oxidizing materials.
To Fight Fire: CO_2, dry chemical.

TRIAMYL BORATE. Syn: *tri-n-amyl borate*. A clear liquid, odor of *n*-amyl alcohol. $B(C_5H_{12}O)_3$, mw: 272.23, bp: 110°–114° @ 2 mm, flash p: 180°F (OC), d: 0.852 @ 27°, vap. d: 9.4.
THR = See boron compounds and esters.
Fire Hazard: Mod, when exposed to heat or flame; can react with oxidizing materials.
Spont Heating: No.
To Fight Fire: Foam, CO_2, dry chemical.

TRI-*n*-AMYLTIN BROMIDE. Liquid. $(C_5H_{11})_3SnBr$, mw: 412.0, d: 1.2678.
THR = See tin compounds and bromides.

TRI-*p*-ANISYL BORON. White crystals. $(C_6H_4OCH_3)_3B$, mw: 332.2, mp: 128°.
THR = Details U. See also boron compounds.

TRIANON. $C_{11}H_{11}O_2N_3S$, mw: 249.3.
THR = An exper neo via sc route. [*3*]

TRIAZINE. $C_3H_3N_3$, mw: 81.1.
THR = Violent reaction with HNO_3. [*19*]

***s*-TRIAZOBORANE.** See borazole.

***sym*-TRIAZINETRIOL.** See *n*-cyanuric acid.

TRIAZOBENZENE. See phenyl azoimide.

TRIBASIC COPPER SULFATE. Syns: *microgel*, *copper sulfate, tribasic*. Aqua-colored powder of extremely fine particle size, water-insol. $CuSO_4 \cdot 3Cu(OH)_2 \cdot H_2O$, mw: 284.
THR = U. A fungicide. See also copper compounds.
Disaster Hazard: Dangerous; see sulfates.

TRIBENZO(a,c,j)NAPHTHACENE.
THR = An exper carc. [*23*]

TRIBENZYL BORINE. Colorless crystals, insol in water. $B(C_6H_5CH_2)_3$, mw: 284.2, mp: 47°, bp: 230° @ 13 mm.
THR = U. See boron compounds.

TRIBENZYL ETHYL TIN. Colorless crystals. $(C_6H_5CH_2)_3(C_2H_5)Sn$, mw: 421.1, mp: 32°.
THR = U. See tin compounds.

TRIBENZYL GERMANIUM BROMIDE. Colorless crystals. $(C_6H_5CH_2)_3GeBr$, mw: 425.9, mp: 145°.
THR = See germanium compounds and bromides.

TRIBENZYL GERMANIUM CHLORIDE. Colorless crystals. $(CH_2C_6H_5)_2GeCl$, mw: 381.4, mp: 155°.
THR = See germanium compounds and chlorides.

TRIBENZYL GERMANIUM FLUORIDE. Colorless crystals. $(CH_2C_6H_5)_3GeF$.
THR = See fluorides and germanium compounds.

TRIBENZYL GERMANIUM IODIDE. Colorless crystals. $(CH_2C_6H_5)_3GeI$, mw: 472.9, mp: 141°.
THR = See germanium compounds and iodides.

TRIBENZYL GERMANIUM OXIDE. Solid, sol in petroleum ether. $[(C_6H_5CH_2)_3Ge]_2O$, mw: 708.0, mp: 135°.
THR = Details U. See germanium compounds.

TRIBENZYLTIN CHLORIDE. White crystals, insol in water, sol in organic solvents. $(C_6H_5CH_2)_3SnCl$, mw: 427.5, mp: 144°.
THR = See tin compounds and chlorides.

TRIBENZYLTIN HYDROXIDE. Colorless crystals, sol in hot organic solvents. $(C_6H_5CH_2)_3SnOH$, mw: 409.1, mp: 121°.
THR = Details U. See tin compounds.

TRIBENZYLTIN IODIDE. Crystals. $(C_6H_5CH_2)_3SnI$, mw: 519.0, mp: 103°.
THR = See tin compounds and iodides.

TRI-2-BIPHENYLYL PHOSPHATE. White granular solid. $(C_6H_5C_6H_4)_3PO_4$, mw: 554.6, mp: 113°–115°.
THR = See phosphorus compounds and phosphates.

TRIBORON SILICIDE. Rhombic black crystals. B_3Si, mw: 60.52, d: 2.52.
THR = Details U. See boron compounds.
Fire Hazard: Mod, by chemical reaction. Reacts with moisture to liberate H_2.
Explosion Hazard: Mod, in the form of dust by chemical reaction with oxidizers.
Disaster Hazard: Mod dangerous; will react with water or steam to produce heat and H_2.

TRIBROMOACETIC ACID. Lustrous leaflets, sol in water, alcohol and ether, slightly sol in petroleum ether. $CBr_3 \cdot COOH$, mw: 130°–131°, bp: 245° (decomp).
THR = U. See trichloroacetic acid. Probably very irr. See also bromides.
Disaster Hazard: Dangerous; see bromides.

2,4,6-TRIBROMOANILINE. Syn: *aniline tribromide*. Needles, insol in water, sol in hot alcohol, chloroform and ether, slightly sol in cold alcohol. $C_6H_2Br_3NH_2$, mw: 329.85, d: 2.35, mp: 120°–122°, bp: 300°.
Acute tox data: ip LD_{LO} (mice) = 500 mg/kg. [*3*]
THR = HIGH via ip and probably oral routes as well. See also bromides.
Disaster Hazard: Dangerous; see bromides and aniline.

TRIBROMOETHANOL. Syn: *avertin*. Crystals, ethereal odor, aromatic taste, slightly water-sol, sol in alcohol and organic solvents. CBr_3CH_2OH, mw: 282.8, mp: 70°–82°, bp: 92°–93° @ 10 mm.
Acute tox data: Oral LD_{LO} (rat) = 1000 mg/kg; sc LD_{LO} (rat) = 530 mg/kg; ip LD_{LO} (mice) = 600 mg/kg; iv LD_{LO} (rabbit) = 120 mg/kg; rec LD_{LO} (rabbit) = 700 mg/kg. [*3*]
THR = HIGH via iv; MOD via oral, sc, ip and rec routes.
Disaster Hazard: Dangerous; see bromides.

1,3,7-TRIBROMOFLUOREN-2-AMINE. $C_{13}H_8NBr_3$, mw: 418.
THR = An exper neo. [*3*]

TRIBROMOGERMANE. Syn: *germanium bromoform*. Colorless liquid. $GeHBr_3$, mw: 313.36, mp: −24.0, bp: decomp.
THR = See bromides and germanium compounds.

TRI-*n*-BROMOMELAMINE. White powder. $(C_3H_3N_6)Br_3$, mw: 362.9.
THR = Details U. Probably toxic. See also melamine.
Explosion Hazard: Slight, by chemical reaction or on contact with allyl alcohol at room temp.
Disaster Hazard: Mod dangerous; when heated to decomp, or on contact with acid fumes, emits toxic fumes of bromides.

TRIBROMOMETHANE. See bromoform.

1,1,1-TRIBROMO-2-METHYL-2-PROPANOL. Fine white crystals. $CBr_3C(CH_3)_2OH$, mw: 310.8, mp: 176°–177°.
THR = Details U.
Disaster Hazard: Dangerous, see bromides.

TRIBROMONEOPENTYL ALCOHOL. $C_4H_7OBr_3$, mw: 299.8.
THR = A mixture with (ethylacetoacetate + Zn) has exploded. [*19*]

TRIBROMONITROMETHANE. See bromopicrin.

TRIBROMOPHENOL-2,4,6. Syn: *bromol*. Long crystals, sol in 14,000 parts water @ 15°, sol in alcohol, chloroform, ether and glycerols. $C_6H_2Br_3OH$, mw: 330.83, d: 2.55, mp: 94°–96°, bp: 244°.
Acute tox data: HIGH via oral route. A powerful irr to skin, eyes and mu mem. May be absorbed dermally.
Disaster Hazard: Dangerous; see bromides and phenol.

1,2,3-TRIBROMOPROPANE. $C_3H_5Br_3$, mw: 280.8.
THR = An exper mutagen. [*90*] See also bromides.

TRIBROMO SILANE. Syn: *silicobromoform*. Mobile liquid, spont flam in air. $SiHBr_3$, mw: 268.9, d: 2.7 @ 17°/4°, mp: −73.5°, bp: 112°, vap. press: 8.8 mm @ 0°.
THR = Readily hydrolyzes to liberate hydrogen bromide, which is a powerful irr. See also hydrobromic acid.
Disaster Hazard: Dangerous; when heated to decomp or brought into contact with acid fumes or moisture, emits highly toxic fumes.

TRIBROMOTRIETHOXY DIALUMINUM. $Al_2Br_3C_6H_{15}O_3$, mw: 428.9.
THR = Reacts violently with air, ethanol, water. [*19*]

TRIBROMOTRIMETHYL DIALUMINUM. $C_3H_9Br_3Al_2$, mw: 338.8.
THR = Spont flam in air. Reacts violently with water. [*19*]

TRIBUTOXY BORANE. See tri-*n*-butyl borate.

TRIBUTOXY ETHYL PHOSPHATE. Light-colored liquid, butyl-like odor. $(C_4H_9OC_2H_4O)_3PO$, mw: 398.54, mp: −70°; bp: 200°–230° @ 4 mm, flash p: 435°, d: 1.02 @ 20°/20°, vap. press: 0.03 mm @ 150°, vap. d: 13.8.
Acute tox data: Oral LD_{50} (guinea pig) = 3000 mg/kg. [*3*]
THR = MOD via oral route. See esters.
Fire Hazard: Slight, when exposed to heat or flame.
Disaster Hazard: Dangerous; see phosphates; can react with oxidizing materials.
To Fight Fire: Water, foam, CO_2, dry chemical.

TRIBUTYL ALUMINUM. Colorless pyroforic liquid. $C_{12}H_{27}Al$, mw: 198, d: 0.823 @ 20°, fp: 26.7°.
THR = A dangerous fire hazard. May ignite spont.

TRIBUTYLAMINE. A colorless liquid. $(C_4H_9)_3N$, mw: 185.35, mp: −70°, bp: 213°, flash p: 187°F (OC), d: 0.78–0.79, vap. d: 6.38.
Acute tox data: Oral LD_{50} (rat) = 540 mg/kg; inhal LC_{LO} (rat) = 75 ppm for 4 hrs; dermal LD_{50} (rabbit) = 250 mg/kg. [*3*]
THR = HIGH via dermal and inhal routes; MOD via oral route. CNS stimulant, irr, sensitizer.
Fire Hazard: Mod when exposed to heat, flame or oxidizers.
Disaster Hazard: Mod; when heated to decomp, emits toxic fumes; can react with oxidizing materials.
To Fight Fire: Foam, CO_2, dry chemical.

TRI-*n*-BUTYL BORANE. Colorless pyroforic liquid, insol in water, sol in most organic solvents. $C_{12}H_{27}B$, mw: 182, mp: 34°, bp: 170° @ 222 mm, d: 0.747 @ 25°, vap. press: 1 mm @ 20°, flash p: −32°F.
THR = See boranes.
Fire Hazard: Dangerous; can ignite spont.

TRI-*n*-BUTYL BORATE. Colorless mobile liquid, odor like *n*-butyl alcohol. $B(OC_4H_9)_3$, mw: 230.16, bp: 230°, fp: < −70°, flash p: 200°F (COC), d: 0.847 @ 28°, vap. d: 7.95.
Acute tox data: Oral LD_{50} (mice) = 1740 mg/kg; ip LD_{LO} (mice) = 500 mg/kg. [*3*] See boron compounds and *n*-butanol.
Fire Hazard: Mod, when exposed to heat, open flame or oxidizers.
Disaster Hazard: Mod dangerous; when heated to decomp, or on contact with acid or acid fumes, can emit toxic fumes; on contact with oxidizing materials, can react vigorously.
To Fight Fire: Foam, CO_2, dry chemical.

TRI-*sec*-BUTYL BORATE. Colorless liquid, odor of *sec*-butanol. $[CH_3CH_2C(CH_3)OH]_3B$, mw: 230.16, bp: 184°–192°, flash p: 165°F (COC), d: 0.829 @ 24°.
Acute tox data: Oral LD_{50} (mice) = 2100 mg/kg. [*3*]
THR = MOD via oral route. See boron compounds and *sec*-butanol.
Fire Hazard: Mod, when exposed to heat or flame; can react with oxidizing materials.
To Fight Fire: Foam, CO_2, dry chemical.

TRI-*n*-BUTYLBORINE. See tri-butylborane.

TRI-*tert*-BUTYLBORINE. Colorless mobile liquid, not sol in water. $B(C_4H_9)_3$, mw: 182.2, bp: 71° @ 12 mm.
THR = Details U. See also boron compounds.

TRIBUTYL CITRATE. Syn: *butyl citrate*. Liquid. $(CH_2COOC_4H_9)_2COHCOOC_4H_9$, mw: 360.44, mp: −20°, bp: 232°, flash p: 315°F (COC), d: 1.042 @ 25°/25°, autoign. temp.: 695°F, vap. d: 12.41.
THR = See butyl alcohol and citric acid.
Fire Hazard: Slight, when exposed to heat or flame; can react with oxidizing materials.
To Fight Fire: CO_2, dry chemical.

TRIBUTYL-2,4-(DICHLOROBENZYL)PHOSPHONIUM CHLORIDE. See phosphon.

TRI-*n*-BUTYLGERMANIUM. $(C_4H_9)_3Ge$, mw: 243.6.
THR = U. Animal exper suggest LOW. See also germanium.

TRIBUTYL PHOSPHATE. Colorless, odorless liquid. $(C_4H_9)_3PO_4$, mw: 266.32, mp: < −80°, bp: 292°, flash p: 295°F (COC), d: 0.982 @ 20°, vap. d: 9.20.
Acute tox data: Oral LD_{50} (rat) = 3000 mg/kg; ip LD_{LO} (mice) = 63 mg/kg. [*3*]
THR = HIGH via ip and MOD via oral route. Causes stimulation of CNS.
Fire Hazard: Slight, when exposed to heat or flame.
Spont Heating: No.
Disaster Hazard: Dangerous; see phosphates.
To Fight Fire: CO_2, dry chemical, fog or mist.

TRIBUTYL PHOSPHINE. Colorless liquid, garlic odor, almost insol in water, miscible with ether, methanol, ethanol and benzene. $(C_4H_9)_3P$, mw: 202, d: 0.8100 @ 25°/4°, fp: −60° to −65°, bp: 240°, flash p: 104°F, autoign. temp.: 392°F.
Acute tox data: Oral LD_{50} (rat) = 750 mg/kg. [*3*]
THR = MOD via oral route. See esters.
Fire Hazard: Mod; spont flam in air. [*19*]
Disaster Hazard: Dangerous; when heated to decomp, emits highly toxic fumes.
To Fight Fire: Foam, alcohol foam, fog.

TRIBUTYL PHOSPHINE OXIDE. Crystals. $(C_4H_9)_3PO$, mw: 218.3.
Acute tox data: Oral LD_{50} (rat) = 217 mg/kg. [*3*]
THR = HIGH via oral route. Has been known to damage the eyes of exper animals.
Disaster Hazard: Dangerous; when heated to decomp, emits highly toxic fumes.

TRIBUTYL PHOSPHINE SULFIDE. $(C_4H_9)_3PS$, mw: 234.
Acute tox data: Oral LD_{50} (rat) = 930 mg/kg; dermal LD_{50} (rabbit) = 1000 mg/kg. [*3*]
THR = MOD via oral and dermal routes.
Disaster Hazard: Dangerous; when heated to decomp, emits highly toxic fumes of PO_x and SO_x.

TRIBUTYL PHOSPHITE. Liquid, decomp in water. $(C_4H_9)_3PO_3$, mw: 202.3, flash p: 248°F (OC), d: 0.9, bp: 120° @ 7 mm.
Acute tox data: Oral LD_{50} (rat) = 3000 mg/kg. [*3*]
THR = MOD via oral route. Limited animal exper show LOW via acute exposures. Has been known to damage the eyes of exper animals. See phosphorus acid and butanol.
Disaster Hazard: See tributyl phosphine oxide.
To Fight Fire: Dry chemical, CO_2.

TRIBUTYL PHOSPHOROTHIOITE. Colorless liquid, mild characteristic odor. $(C_4H_9O)_3PS$, mw: 282.3, bp: 142°–145° @ 4.5 mm, flash p: 295°F (COC), d: 0.987 @ 20°/4°.
Acute tox data: Oral LD_{50} (rat) = 84 mg/kg. [*3*]
THR = HIGH via oral route. A cholinesterase inhibitor, see parathion.
Fire Hazard: Slight, when exposed to heat or flame.
Disaster Hazard: Dangerous; when heated to decomp, emits highly toxic fumes of PO_x and SO_x; can react vigorously with oxidizing materials.

TRIBUTYL PHOSPHOROTRITHIOITE. Syn: "*DEF.*" Liquid, insol in water, sol in aliphatic, aromatic and chlorinated hydrocarbons. $(C_4H_9S)_3PO$, mw: 314, bp: 150° @ 0.3 mm.
Acute tox data: Oral LD_{50} (rat) = 150 mg/kg; dermal LD_{50} (rat) = 168 mg/kg; ip LD_{50} (rat) = 210 mg/kg; oral LD_{50} (guinea pig) = 260 mg/kg. [*3*]
THR = HIGH via oral, dermal and ip routes. Animal exper show anti-cholinesterase effect. See also parathion.
Disaster Hazard: Dangerous; see phosphates and sulfates.

TRI-*tert*-BUTYL THIOGERMANIUM CHLORIDE. Colorless crystals, sol in organic solvents. $Ge[SC(CH_3)_3]_3Cl$, mw: 375.6. mp: 67°, bp: 157° @ 4 mm.
THR = See germanium compounds and organometals.

TRIBUTYLTIN ACETATE. $C_{14}H_{30}O_2 \cdot Sn$, mw: 349.1.
Acute tox data: Oral LD_{50} (rat) = 99 mg/kg; ip LD_{LO} (rat) = 10 mg/kg; oral LD_{LO} (mouse) = 100 mg/kg. [*3*]
THR = HIGH via oral and ip routes. See also tin compounds.

TRI-*n*-BUTYLTIN BROMIDE. Liquid. $(C_4H_9)_3SnBr$, mw: 370.0, d: 1.3365.
Acute tox data: Inhal LC_{LO} (mice) = 1030 mg/m^3. [*3*]
THR = MOD via inhal route. See tin compounds and bromides.

TRI-*n*-BUTYLTIN IODIDE. $C_{12}H_{27}SnI$, mw: 417.0.
Acute tox data: Inhal LC_{LO} (mice) = 1340 mg/kg. [*3*]
THR = MOD via inhal route.

TRI-*n*-BUTYLTIN OXIDE. $C_{24}H_{54}O \cdot 2Sn$, mw: 596.2.
Acute tox data: Oral LD_{50} (rat) = 194 mg/kg; ip LD_{50} (rat) = 7 mg/kg; dermal LD_{LO} (rabbit) = 1170 mg/kg. [*3*]
THR = HIGH via oral and ip routes; MOD via dermal route. See also tin compounds.

TRIBUTYRIN. Syn: *glyceryl tributyrate.* $C_{15}H_{26}O_6$, mw: 302.4, mp: −75°, bp: 312°–315°, d: 1.0356 @ 20°/20°.
Acute tox data: Oral LD_{LO} (rat) = 3200 mg/kg; iv LD_{50} (mice) = 320 mg/kg. [*3*]
THR = HIGH via iv and MOD via oral routes. An exper carc via oral route. [*3, 23*] A synthetic flavoring substance and adjuvant. [*109*]
Fire Hazard: Slight, when exposed to heat or flame; can react with oxidizing materials.

TRICALCIUM-*o*-ARSENATE. See calcium arsenate.

TRICALCIUM CITRATE. See calcium citrate.

TRICALCIUM PHOSPHATE. See calcium phosphate, tribasic.

TRICALCIUM SILICATE. See cement, portland and other cement articles. Used as an anti-caking agent in foods. [*109*]

TRICAMBA. See 3,5,6-trichloro-*o*-anisic acid.

TRICAPRYLIN. $C_{30}H_{56}O_6$, mw: 512.9.
Acute tox data: iv LD_{50} (mice) = 3700 mg/kg. [*3*]
THR = MOD via iv route. An exper carc. [*23*]

TRICHLORACETIC ACID. Colorless, rhombic deliquescent crystals. CCl_3COOH, mw: 163.40, bp: 197.5°, fp: 57.5°, flash p: none, d: 1.6298 @ 61°/4°, vap. press: 1 mm @ 51.0°.
Acute tox data: Oral LD_{50} (rat) = 3320 mg/kg; ip LD_{LO} (mice) = 500 mg/kg. [*3*]
THR = MOD via ip and oral routes. Solutions are corrosive and HIGH irr to skin, eyes and mu mem.
Disaster Hazard: Dangerous; see chlorides.

TRICHLORFENSON. $C_{12}H_8O_3SCl_2$, mw: 303.2.
Acute tox data: Oral LD_{50} (rat) = 2000 mg/kg. [*3*]
THR = MOD via oral route. An exper neo. [*3*]

TRICHLORFON. See dylox.

TRICHLORMETHINE. $C_6H_{12}NCl_3 \cdot HCl$, mw: 241.
THR = An exper (S) carc [*8*] and neo. [*3*] Affects CNS and blood.

TRICHLOROACETANILIDE. $C_8H_6ONCl_3$, mw: 238.5.
Acute tox data: ip LD_{LO} (mice) = 1000 mg/kg. [*3*]
THR = MOD via ip route.

TRICHLOROACETALDEHYDE. See chloral.

TRICHLOROACETONITRILE. See chlorocyanohydrin.

TRICHLOROACETYL CHLORIDE. Syn: *superpalite*. Liquid. $CCl_3COCOCl$, mw: 181.86, bp: 118°, d: 1.629 @ 16°.
THR = HIGH irr to skin, eyes and mu mem and via oral and inhal routes. A military poison.
Disaster Hazard: Dangerous; see chlorides; will react with water or steam to produce toxic and corrosive fumes.

TRICHLOROACETYL CHLOROETHYLAMIDE.
THR = Details U. A mosquito-repellant. See also amides.
Disaster Hazard: Dangerous; sec chlorides.

TRICHLOROACETYL NITRILE. See chlorocyanohydrin.

3,5,6-TRICHLORO-*o*-ANISIC ACID. Syns: *banvel T, tricamba*. A powder, freely sol in alcohol, mod sol in xylene, its salts are freely sol in water. $C_7H_3O_2Cl_3$, mw: 225.5, mp: 138°.
Acute tox data: Oral LD_{50} (rat) = 970 mg/kg. [*94*] Oral LD_{50} (rat) = 300 mg/kg. [*3*]
THR = HIGH via oral route. An herbicide.
Disaster Hazard: Dangerous; see chlorides.

2,4,6-TRICHLOROANISOLE. Syn: *tyrene*. Crystals, faint odor, nearly water-insol. $C_7H_5Cl_3O$, mw: 211.5, mp: 59°, bp: 132° @ 27 mm.
THR = See anisole and chlorides.

1,2,3-TRICHLOROBENZENE. White crystals. $C_6H_3Cl_3$, mw: 181.5, mp: 52.6°, bp: 221°, flash p: 235°F (CC), d: 1.69 @ 25°/25°, vap. press: 1 mm @ 40.0°, vap. d: 6.26.
THR = MOD irr to skin, eyes and mu mem. Has caused exper loss of hair. Liver injury has been reported; see also chlorinated hydrocarbons, aromatic.
Fire Hazard: Slight, when exposed to heat, flame or oxidizers.
Disaster Hazard: Dangerous; see chlorides; can react with oxidizing materials.
To Fight Fire: Water, foam, CO_2, dry chemical.

1,2,4-TRICHLOROBENZENE. Syn: *uns-trichlorobenzene*. Colorless liquid. $C_6H_3Cl_3$, mw: 181.5, mp: 17°, bp: 213°, flash p: 230°F (CC), d: 1.454 @ 25°/25°, vap. press: 1 mm @ 38.4°, vap. d: 6.26.
Acute tox data: Oral LD_{50} (rat) = 756 mg/kg; oral LD_{50} (mice) = 766 mg/kg; ip LD_{LO} (mice) = 500 mg/kg. [*3*]
THR = MOD via oral and ip routes.
Fire Hazard: Slight, when exposed to heat or flame.
Disaster Hazard: Dangerous; see chlorides; can react vigorously with oxidizing materials.
To Fight Fire: Water, foam, CO_2, dry chemical.

1,3,5-TRICHLOROBENZENE. Syn: *sym-trichlorobenzene*. White crystals. $C_6H_3Cl_3$, mw: 181.5, mp: 63.4°, bp: 208.5°, flash p: 225°F (CC), vap. press: 10 mm @ 78.0°, vap. d: 6.26.
THR = MOD irr to skin, eyes and mu mem. [*2*] Causes an exper hair loss. MOD via oral and inhal routes. See also 1,2,3-trichlorobenzene and 1,2,4-trichlorobenzene.
Fire Hazard: Mod, when exposed to heat or flame.
Disaster Hazard: Dangerous; see chlorides; can react vigorously with oxidizing materials.
To Fight Fire: Water, foam, CO_2, dry chemical.

***sym*-TRICHLOROBENZENE.** See 1,3,5-trichlorobenzene.

***uns*-TRICHLOROBENZENE.** See 1,2,4-trichlorobenzene.

TRICHLOROBENZOQUINONIMINE. $C_6H_2ONCl_3$, mw: 210.41.
THR = Spont explosive. [*19*]

TRICHLOROBENZYL CHLORIDE. $C_6H_2Cl_3CH_2Cl$, mw: 229.9.
Acute tox data: Oral LD_{50} (rat) = 3075 mg/kg. [*3*]
THR = MOD via oral route. An irr. May cause skin irr.
Disaster Hazard: Dangerous; see chlorides.

1,1,1-TRICHLORO-2,2-BIS(*p*-BROMOPHENYL) ETHANE. Syn: *bromine analog of DDT*. Crystals. $(BrC_6H_4)_2CHCCl_3$, mw: 443.4.
THR = Details U. A pesticide. See DDT.

Disaster Hazard: Dangerous; see chlorides and bromides.

1,1,1-TRICHLORO-2,2-BIS(*p*-CHLOROPHENYL) ETHANE. See DDT.

1,1,1-TRICHLORO-2,2-BIS(*p*-FLUOROPHENYL) ETHANE. Syn: *fluorine analog of DDT*. Crystals. $(FC_6H_4)_2CHCCl_3$, mw: 321.6.

Acute tox data: Oral LD_{50} (rat) = 1120 mg/kg; oral LD_{LO} (mice) = 600 mg/kg. [*3*]

THR = MOD via oral route. See also fluorides. A pesticide. See also DDT.

Disaster Hazard: Dangerous; see chlorides and fluorides.

1,1,1-TRICHLORO-2,2-BIS(*p*-METHOXYPHENYL) ETHANE. See methoxychlor.

TRICHLOROBORINE DIMETHYL ETHERATE. Colorless crystals, decomp by water. $(CH_3)_2OBCl_3$, mw: 163.3, mp: 76° (decomp).

THR = See boron compounds, chlorides and ethers.

TRICHLOROBORINE TRIMETHYLAMMINE. Colorless crystals, sol in hot water and alcohol. $(CH_3)_3NBCl_3$, mw: 176.3, mp: 243°.

THR = See boron compounds and chlorides.

TRICHLOROBUTANE. Liquid. $Cl_3C_4H_7$, mw: 154.5, bp: 168°, flash p: 195°F.

THR = Details U. See chlorinated hydrocarbons, aliphatic.

Fire Hazard: Mod, when exposed to heat, flame or oxidizers.

Disaster Hazard: Dangerous; see chlorides; can react vigorously with oxidizing materials.

To Fight Fire: Foam, dry chemical.

α,α,β-TRICHLOROBUTYRAMIDE. Crystals. $CH_3CHClCCl_2CONH_2$, mw: 190.5.

THR = Details U. A pesticide and stomach insecticide; see also amides.

Disaster Hazard: Dangerous; see chlorides.

TRICHLOROCRESOLS. Compounds with the formula $OHC_6HCH_3Cl_3$ (there are a number of different isomers).

THR = See cresol and chlorides.

TRICHLOROCYANIDINE. See cyanuric chloride.

1,2,4-TRICHLORO-3,5-DINITROBENZENE. $Cl_3C_6H(NO_2)_2$, mw: 271.5.

Acute tox data: Oral LD_{50} (rat) = 425 mg/kg; dermal LD_{50} (rat) = 425 mg/kg. [*3*]

THR = HIGH via oral and dermal routes. See also benzene and nitrobenzene. A pesticide.

Fire Hazard: Mod, when exposed to heat or flame. See nitrates.

Explosion Hazard: See nitrates.

Disaster Hazard: Dangerous; when heated to decomp, emits highly toxic fumes of NO_x and chlorides; can react vigorously with reducing materials.

1,2,3-TRICHLORO-4,6-DINITROBENZENE. THR = An exper neo via oral and sc routes. [*3*]

α-TRICHLOROETHANE. Syns: *1,1,1-trichloroethane, methyl chloroform*. Colorless liquid. CH_3CCl_3, mw: 133.42, bp: 74.1°, fp: −32.5°, flash p: none, d: 1.3492 @ 20°/4°, vap. press: 100 mm @ 20.0°.

Acute tox data: ip LD_{50} (mice) = 4700 mg/kg; oral LD_{50} (dog) = 750 mg/kg; oral LD_{50} (rabbit) = 5660 mg/kg; [*3*] 920 ppm for 70 min ⟶ CNS effects in humans; inhal LC_{LO} (man) = 27000 mg/m^3 for 10 min. [*3*]

THR = MOD via ip and oral routes. Causes a proarrhythmic activity which sensitizes the heart to epinephrine-induced arrhythmias. This sometimes will cause a cardiac arrest particularly when this material is massively inhaled as in drug abuse for euphoria. [*115*] Reacts violently with acetone, N_2O_4, O_2, O_2 liquid, Na, NaOH, Na-K alloy. [*19*] Narcotic in HIGH conc.

Disaster Hazard: Dangerous; see chlorides.

β-TRICHLOROETHANE. Syns: *1,1,2-trichloroethane, vinyl trichloride*. Liquid, pleasant odor. $CH_2ClCHCl_2$, mw: 133.4, bp: 114°, fp: −35°, d: 1.4416 @ 20°/4°, vap. press: 40 mm @ 35.2°.

Acute tox data: Oral LD_{50} (rat) = 1140 mg/kg; inhal LC_{LO} (rat) = 500 ppm for 8 hrs; ip LD_{50} (mice) = 994 mg/kg; sc LD_{50} (mice) = 227 mg/kg; iv LD_{LO} (dog) = 95 mg/kg. [*3*]

THR = HIGH via iv and sc routes. MOD via ip, inhal, oral and probably dermal routes. Trichloroethane has narcotic properties and acts as a local irr to the eyes, nose and lungs. It may also be injurious to the liver and kidneys. A fumigant.

Disaster Hazard: Dangerous; see chlorides.

TRICHLOROETHANOL. Liquid. CCl_3CH_2OH, mw: 149.5, mp: 17.8°, bp: 150° @ 765 mm, d: 1.54 @ 25°/4°, vap. press: 1 mm @ 20°, vap. d: 5.16.

Acute tox data: Oral LD_{50} (rat) = 600 mg/kg; iv LD_{50} (mice) = 201 mg/kg; ip LD_{LO} (rat) = 300 mg/kg. [*3*]

THR = HIGH via iv and ip routes. MOD via oral route. An anesthetic.

Disaster Hazard: Dangerous; see chlorides.

1,1,3-TRI(2-CHLOROETHOXY)PROPANE. $OHCH_2CHClC_2H_4CH(OHCH_2CHCl)_2$, mw: 279.5.

THR = Details U. Limited animal exper suggest HIGH.

Disaster Hazard: Dangerous; see chlorides.

TRICHLOROETHYL CARBAMATE. $C_4H_4O_2NCl_3$, mw: 204.4.

THR = An exper carc. [*3*]

TRICHLOROETHYLENE. Syns: *ethinyl trichloride, ethylene trichloride*. Stable, colorless, heavy, mobile liquid, chloroform-like odor. $CHClCCl_2$, mw: 131.40, mp: −73°, bp: 87.1°, fp: −86.8°, d: 1.45560 @ 25°/4°, autoign. temp.: 788°F; vap. press: 100 mm @ 32°, vap. d: 4.53, flash p: none, lel = 12.5%, uel = 90%.

Acute tox data: Oral LD_{LO} (human) = 857 mg/kg; 160 ppm for 83 min ⟶ human CNS effects; 110 ppm for 8 hrs ⟶ inhal human irr effects; oral LD_{50} (rat) = 4920 mg/kg; inhal LC_{LO} (rat) = 8000 ppm for 4 hrs; ip LD_{50} (dog) = 1900 mg/kg; iv LD_{LO} (dog) = 150 mg/kg. [*3*]

THR = HIGH via iv; MOD via ip, inhal, oral routes. An exper (S) carc. [*3, 13*] Inhal of high conc causes narcosis and anesthesia. A form of addiction has been observed in exposed workers. Prolonged inhal of mod conc causes headache and drowsiness. Fatalities following severe, acute exposure have been attributed to ventricular fibrillation resulting in cardiac failure. There is damage to liver and other organs from chronic exposure. Cases have been reported but are of questionable validity. Determination of the metabolites trichloracetic acid and trichloroethanol in urine reflects the absorption of trichloroethylene. A food additive permitted in food for human consumption. [*109*] A common air contaminant.

Fire Hazard: Low, when exposed to heat or flame. High conc of trichloroethylene vapor in high-temp. air can be made to burn mildly if plied with a strong flame. Though such a condition is difficult to produce, flames or arcs should not be used in closed equipment which contains any solvent residue or vapor. Can react violently with Al, Ba, N_2O_4, Li, Mg, liquid O_2, O_2, KOH, KNO_3, Na, NaOH, Ti. [*19*]

Spont Heating: No.

Disaster Hazard: Dangerous; see chlorides.

TRICHLOROETHYL SILANE. $C_2H_5SiCl_3$, mw: 163.5.

THR = Reacts violently with water. [*19*]

TRICHLOROFLUOROGERMANE. Colorless liquid. $GeCl_3F$, mw: 197.97, mp: −49°, bp: 37.5°.

THR = See fluorides, germanium compounds and chlorides.

1,1,1-TRICHLOROFLUOROETHANE. $C_2H_2Cl_3F$, mw: 151.4.

THR = No data. See fluorides. Violent reaction with Ba. [*19*]

TRICHLOROFLUOROMETHANE. See fluorotrichloromethane.

TRICHLOROGERMANE. Syn: *germanium chloroform*. Colorless liquid. $GeHCl_3$, mw: 179.98, mp: −71.0°, bp: 75.2°, d: 1.93 @ 0°C.

THR = See hydrochloric acid and germanium compounds.

TRICHLOROISOCYANURIC ACID. White crystals, chlorine odor, mod sol in water. $(ClNCO)_3$, mw: 232.5, mp: 225°−230° (decomp).

Acute tox data: Oral LD_{50} (rat) = 700–800 mg/kg.

THR = MOD–HIGH via oral route. Toxicity symptoms include emaciation, lethargy, weakness and delayed death. Autopsy shows inflammation of GI tract, liver discoloration and kidney hyperemia. A powerful oxidizer.

Disaster Hazard: Dangerous; when heated to decomp, emits chloride and carbon monoxide fumes.

1,1,1-TRICHLOROISOPROPYL ALCOHOL. Syns: *isopral, 1,1,1-trichloro-2-propanol*. Crystals, camphor-like odor, pungent taste, water-sol. $C_3H_5Cl_3O$, mw: 163.4, mp: 50°, bp: 162°.

Acute tox data: Oral LD_{LO} (rat) = 1000 mg/kg. [*3*]

THR = MOD via oral route. See also chlorinated hydrocarbons, aliphatic.

Disaster Hazard: Dangerous; see chlorides.

TRICHLOROMELAMINE. Syn: *TCM*. White powder, slightly water-sol. $C_3H_3Cl_3N_6$, mw: 229.4, autoign. temp.: 320°F.

Acute tox data: Oral LD_{50} (mice) = 490 mg/kg. [*3*]

THR = HIGH via oral route.

Fire Hazard: Mod, in the pure state, when heated or ignited by spark or flame; reacts vigorously to evolve smoke and heat; reacts with acetone, NH_3, aniline, diphenylamine, turpentine. [*19*] Vendor can supply directions for handling.

Disaster Hazard: Dangerous; when heated to decomp, emits highly toxic chloride and NO_x fumes.

TRICHLOROMETHANE. See chloroform.

TRICHLOROMETHANE SULFENYL CHLORIDE. See perchloromethyl mercaptan.

TRICHLOROMETHYL CHLOROFORMATE. See diphosgene.

TRICHLOROMETHYL ETHER. A liquid of pungent odor. $CHCl_2OCH_2Cl$, mw: 149.42, bp: 130°−132°, d: 1.5066 @ 10°.

THR = HIGH irr to skin, eyes and mu mem and via oral, inhal routes. See also ethers.

Disaster Hazard: Dangerous; when heated to decomp, emits highly toxic fumes; will react with water or steam to produce toxic and corrosive fumes.

TRICHLOROMETHYL PERCHLORATE. Cl_3CClO_4, mw: 217.8.

THR = Detonates @ 40°.

***cis*-N-(TRICHLOROMETHYL)THIO-4-CYCLOHEXENE-1,2-DICARBOXIMIDE.** Syn: *captan.* Odorless crystals, insol in water, sol in benzene and chloroform. $C_9H_8O_2NSCl_3$, mw: 300.6.
Acute tox data: Oral LD_{LO} (human) = 1071 mg/kg; oral LD_{50} (rat) = 9000 mg/kg. [*3*]
THR = LOW via oral route. Large ingested doses may cause vomiting and diarrhea. A fungicide. A food additive permitted in food for human consumption. [*109*] An exper teratogen and neo; a mutagen via oral and ip routes. [*3*]
Disaster Hazard: Dangerous; see chlorides and sulfates.

N-(TRICHLOROMETHYLTHIO)PHTHALIMIDE. See *n*-trichloromethyl thiotetrahydrophthalimide.

TRICHLORONAPHTHALENE. A white solid. $C_{10}H_5Cl_3$, mw: 231.51.
Acute tox data: Inhal TC_{LO} (human) = 30 mg/m^3 ⟶ systemic effects. [*3*]
THR = HIGH via inhal route. Toxic via dermal route. See also chlorinated naphthalenes and chlorinated diphenyls.
Disaster Hazard: Dangerous; see chlorides.

TRICHLORONITROMETHANE. See chloropicrin.

3,4,6-TRICHLORO-2-NITROPHENOL. Syn: *dowlap.* Pale yellow crystals. $C_6H_2Cl_3NO_3$, mw: 230.5, mp: 93°.
THR = See chlorinated phenols.

TRICHLORONITROPROPANOL. Crystalline solid. $CCl_3CHOHCH_2NO_2$, mw: 208.4, mp: 40°, bp: 120° @ 5 mm, flash p: 352°F (OC), d: 1.605 @ 45°/4°, vap. press: 0.1 mm @ 20°.
THR = See alcohols and chlorinated hydrocarbons, aliphatic.

TRICHLORONITROSOMETHANE. Dark blue liquid, unpleasant odor. CCl_3NO, mw: 148.39, bp: 5° @70 mm, d: 1.5 @ 20°.
Acute tox data: Inhal LC_{LO} (mice) = 1650 mg/m^3. [*3*]
THR = MOD via inhal route. Very irr to skin, eyes and mu mem. A lachrymator type military poison.
Disaster Hazard: Dangerous; see chlorides and nitrates; will react with water or steam to produce toxic and corrosive fumes.

1,1,2-TRICHLORO-1,1,3,3,3-PENTAFLUOROPROPANE. $C_3F_5Cl_3$, mw: 237.4.
Acute tox data: Oral LD_{50} (rat) = 15,000 mg/kg; inhal LC_{LO} (rat) = 8000 ppm for 4 hrs. [*3*]
THR = VERY LOW via oral and LOW via inhal route.
Disaster Hazard: Dangerous; see chlorides and fluorides.

2,3,6-TRICHLOROPHENOL. Colorless needles. $Cl_3C_6H_2OH$, mw: 197.5, mp: 62°, bp: 253°.
Acute tox data: ip LD_{50} (rat) = 308 mg/kg. [*3*]
THR = HIGH via ip route. See chlorinated phenols.

2,4,5-TRICHLOROPHENOL. Colorless needles or gray flakes. $C_6H_2Cl_3OH$, mw: 197.5, bp: 252°, fp: 57.0°, d: 1.678 @ 25°/4°, vap. press: 1 mm @ 72.0°.
Acute tox data: Oral LD_{50} (rat) = 820 mg/kg; ip LD_{50} (rat) = 355 mg/kg; sc LD_{50} (rat) = 2260 mg/kg; oral LD_{50} (guinea pig) = 1000 mg/kg. [*3*]
THR = HIGH via ip and MOD via oral and sc routes. See chlorinated phenols.

2,4,6-TRICHLOROPHENOL. Colorless needles or yellow solid, strong phenolic odor. $Cl_3C_6H_2OH$, mw: 197.5, mp: 68°, bp: 244.5°, fp: 62°, d: 1.490 @75°/4°, vap. press: 1 mm @ 76.5°.
Acute tox data: Oral LD_{50} (rat) = 820 mg/kg; ip LD_{50} (rat) = 276 mg/kg. [*3*]
THR = HIGH via ip and MOD via oral route. An exper carc via oral route. [*3*] See chlorinated phenols.

2,4,5-TRICHLOROPHENOXY ACETIC ACID. Syn: *2,4,5-T.* Crystals, light tan solid. $Cl_3C_6H_2OCH_2COOH$, mw: 255.5, mp: 151°–153°.
Acute tox data: Oral LD_{50} (rat) = 300 mg/kg; oral LD_{50} (dog) = 100 mg/kg. [*3*]
THR = HIGH via oral routes. An exper teratogen via oral and sc routes. [*3*] A highly toxic chlorinated phenoxy acid herbicide; rapidly excreted after ingestion. Readily absorbed via inhal and ingestion, slowly via skin. Signs of intoxication include weakness, lethargy, anorexia, diarrhea, ventricular fibrillation and/or cardiac arrest and death. An exper (±) carc via oral route. [*81*]
Disaster Hazard: Dangerous; see chlorides.

2-(2,4,5-TRICHLOROPHENOXY)ETHYL-2,2-DICHLOROPROPIONATE. See erbon.

2-(2,4,5-TRICHLOROPHENOXY)PROPIONIC ACID. Syn: *2,4,5-TCPPA.* Crystals, slight water-sol. $C_9H_7Cl_3O_3$, mw: 269.5, mp: 182°.
Acute tox data: Oral LD_{50} (rat) = 650 mg/kg; oral LD_{50} (mammal) = 650 mg/kg. [*3*]
THR = MOD via oral route. Animal exper have shown liver and kidney injury. See also 2,4-D.
Disaster Hazard: Dangerous; see chlorides.

2-(2,4,5-TRICHLOROPHENOXY)PROPIONIC ACID PROPYLENE GLYCOL BUTYL ETHER ESTER. Syns: *kuron, silvex.* Liquid, insol in water. $C_9H_7Cl_3O_3 + CH_3CHOHCH_2OCH_2CH_2CH_3$.
Acute tox data: uk LD_{50} (rat) = 500 mg/kg. [*3*]
THR = HIGH via uk route. In exper animals, it has

caused damage to liver and kidneys. See also *2,4-dichlorophenoxy acetic acid.*
Disaster Hazard: Dangerous; when heated to decomp, emits highly toxic fumes.

2,4,5-TRICHLOROPHENYL ACETIC ACID. Syn: *fenac.* Colorless crystals. $Cl_3C_6H_2OOCH_3$, mw: 239.5, mp: 160°.
Acute tox data: Oral LD_{50} (rat) = 3000 mg/kg; oral LD_{50} (rat) = 1780 mg/kg. [*3*]
THR = MOD via oral route. A toxic fungicide.
Disaster Hazard: Dangerous; see chlorides.

TRI-*o*-CHLOROPHENYL BORATE. White solid, odor of *o*-chlorophenol. $(ClC_6H_4O)_3B$, mw: 399.15, mp: 47°–49°; bp: 264°–270° @ 14 mm.
THR = No data. See esters, boron compounds.

TRICHLOROPHENYL CHLOROACETATE.
Crystals. $Cl_3C_6H_2OOCCH_2Cl$, mw: 274.0.
Acute tox data: Oral LD_{50} (guinea pig) = 1000 mg/kg. [*3*]
THR = MOD via oral route.

1,1,1-TRICHLOROPROPANE.
Acute tox data: Inhal LC_{LO} (rat) = 8000 ppm for 4 hrs. [*3*]
THR = MOD via inhal route.

1,1,2-TRICHLOROPROPANE.
Acute tox data: Oral LD_{50} (rat) = 1230 mg/kg; inhal LC_{50} (rat) = 2000 ppm for 4 hrs. [*3*]
THR = MOD via oral and inhal routes.

1,2,2-TRICHLOROPROPANE.
Acute tox data: Oral LD_{50} (rat) = 1230 mg/kg. [*3*]
THR = MOD via oral route.

1,2,3-TRICHLOROPROPANE. Syns: *glycerol trichlorohydrin, allyl trichloride, trichlorohydrin.* Colorless liquid. $CH_2ClCHClCH_2Cl$, mw: 147.44, mp: −14.7°, bp: 156.17°, ulc: 20–25, lel = 3.2%, uel = 12.6% @ 150°, flash p: 180°F (TOC), d: 1.3888 @ 20°, autoign temp.: 579°F (commercial), vap. press: 10 mm @ 46°, vap. d: 5.0.
Acute tox data: Oral LD_{50} (rat) = 320 mg/kg; inhal LC_{LO} (rat) = 1000 ppm for 4 hrs; oral LD_{LO} (dog) = 200 mg/kg; dermal LD_{50} (rabbit) = 1770 mg/kg. [*3*]
THR = MOD irr via dermal routes. HIGH irr via oral, inhal routes. A lipoid solvent. Cumulative toxicity. See also chlorinated hydrocarbons, aliphatic.
Fire Hazard: Mod, when exposed to heat, flame or oxidizers.
Disaster Hazard: Dangerous; when heated to decomp, burns and emits highly toxic fumes of chlorides; can react vigorously with oxidizing materials.
To Fight Fire: Water, foam, CO_2, dry chemical.

1,1,1-TRICHLORO-2-PROPANOL. See 1,1,1-trichloroisopropyl alcohol.

1,2,3-TRICHLOROPROPENE. Syn: *allyltrichloride.* $ClCH_2CCl{:}CHCl$, mw: 145.4, bp: 142°, d: 1.414 @ 20°/20°, flash p: 180°F (OC).
Acute tox data: Oral LD_{50} (rat) = 616 mg/kg; inhal LC_{LO} (rat) = 500 ppm for 4 hrs; dermal LD_{50} (rabbit) = 640 mg/kg. [*3*]
THR = MOD via oral, dermal routes. HIGH via inhal route.
Disaster Hazard: High: when heated to decomp, yields highly toxic fumes.
Fire Hazard: Mod, when exposed to heat, flames (sparks) or powerful oxidizers.
To Fight Fire: Water (as a blanket), spray, mist, dry chemical.

2,2,3-TRICHLOROPROPIONALDEHYDE.
THR = No data. Limited animal exper suggest high toxicity and irr. See also aldehydes
Disaster Hazard: Dangerous; see chlorides.

2,2,3-TRICHLOROPROPIONIC ACID.
$ClCH_2CCl_2COOH$, mw: 177.
Acute tox data: Oral LD_{50} (rat) = 2460 mg/kg; dermal LD_{LO} (rabbit) = 1770 mg/kg. [*3*]
THR = MOD via oral and dermal routes.
Disaster Hazard: Dangerous; see chlorides.

TRICHLOROSILANE. Syn: *silicochloroform.* Colorless, very volatile liquid, decomp in water. $SiHCl_3$, mw: 135.44, mp: −134°, bp: 31.8°, flash p: 7°F (OC), d: 1.35 @ 0°, vap. press: 400 mm @ 14.5°, vap. d: 4.7.
Acute tox data: Oral LD_{50} (rat) = 1030 mg/kg; inhal LC_{LO} (rat) = 1000 ppm for 4 hrs. [*3*]
THR = MOD via oral and inhal routes.
Fire Hazard: Very dangerous; when exposed to flame or by chemical reaction. Spont flam in air.
Explosion Hazard: Unknown.
Disaster Hazard: Dangerous; when heated to decomp, emits highly toxic fumes of chlorides; will react with water or steam to produce heat and toxic and corrosive fumes; on contact with oxidizing materials, can react vigorously.
To Fight Fire: CO_2, dry chemical.

α-TRICHLOROTOLUENE. See benzotrichloride.

TRICHLORO-*s*-TRIAZINE. See cyanuric chloride.

1,1,2-TRICHLORO-1,2,2-TRIFLUOROETHANE. See trifluorotrichloroethane.

TRICK MATCHES AND TRICK NOISE-MAKERS, EXPLOSIVE. See explosives, low.

TRICOPPER ANTIMONIDE. Gray crystals, Cu_3Sb, mw: 312.38, mp: 687°, d: 8.51.
THR = See antimony and copper compounds.

Fire Hazard: Mod, by chemical reaction with moisture to evolve antimony hydride. See also stibine.
Explosion Hazard: Slight, by chemical reaction with powerful oxidizers. See also stibine.
Disaster Hazard: Dangerous; when heated to decomp, or on contact with acid or acid fumes, emits highly toxic fumes of antimony; will react with water or steam to produce toxic and flam vapors.

TRICOPPER ARSENIDE. Syn: *domeykite.* Hexagonal crystals. Cu_3As, mw: 265.53, mp: 830°, d: 8.0.
THR = See arsenic compounds and arsenides.

11-TRICOSENE. A liquid. $C_{10}H_{21}CHCHC_{11}H_{23}$, mw: 322, bp: 168°–170° @ 2.4 mm, flash p: 284°F, d: 0.80.
THR = No data.
Fire Hazard: Slight, when exposed to heat or flame; can react with oxidizing materials.
To Fight Fire: Foam, CO_2, dry chemical.

TRI-*m,p*-CRESYL BORATE. Straw-yellow, viscous liquid, characteristic odor. $(CH_3C_6H_4OH)_3B$, mw: 335.2, bp: 179°–210° @ 0.1 mm, flash p: 240°F (COC), d: 1.053 @ 27.6°, vap. d: 11.6.
THR = See cresol and boric acid.
Fire Hazard: Slight, when exposed to heat or flame.
Disaster Hazard: Dangerous; when heated to decomp, emits highly toxic fumes; can react vigorously with oxidizing materials.
To Fight Fire: Water, CO_2, dry chemical.

TRI-*o*-CRESYL BORATE. Straw-yellow liquid, odor of *o*-cresol. $(CH_3C_6H_4O)_3B$, mw: 332.2, bp: 189°–195° @ 2 mm, flash p: 345°F (COC), d: 1.079 @ 22°, vap. d: 11.4.
Acute tox data: Oral LD_{50} (mice) = 400 mg/kg. [3]
THR = HIGH via oral route.

TRI-*o*-CRESYL PHOSPHATE. Syn: *o-tolyl phosphate.* Colorless liquid. $(CH_3C_6H_4)PO_4$, mw: 368.36, mp: −25° to −30°, bp: 410° (slight decomp), flash p: 437°F, d: 1.17, autoign temp.: 725°F, vap. d: 12.7.
Acute tox data: Oral LD_{LO} (human) = 1000 mg/kg; oral TD_{LO} (human) = 6 mg/kg ⟶ CNS effects; oral LD_{50} (rat) = 3000 mg/kg; ip LD_{LO} (mice) = 50 mg/kg. [3]
THR = MOD via oral and HIGH via ip route. Most of the cases of tri-*o*-cresyl phosphate poisoning have followed its ingestion. In 1930, some 15,000 persons were affected in the United States, and of these, 10 died. The responsible material was found to be an alcoholic drink known as Jamaica ginger, or "jake." This beverage had been adulterated with about 2% of tri-*o*-cresyl phosphate. The affected persons developed a polyneuritis, which progressed, in many cases, with degeneration of the peripheral motor nerves, the anterior horn cells and the pyramidal tracts. Sensory changes were absent. Since 1930 they have been several other outbreaks of poisoning following ingestion of the material. Recently 3 cases of polyneuritis occurring in England in connection with the manufacture of the tri-*o*-cresyl phosphate have been reported. Absorption was probably through the respiratory tract, though there may have been some absorption through the skin. All three men made a good recovery.

From ingestion exper with cockerels, it appears that tri-*o*-cresyl phosphate is more toxic than the *m*-form, and much more so than tri-*p*-cresyl phosphate or triphenyl phosphate.

Irrespective of whether absorption has been by ingestion or by inhal or skin absorption, the history is usually one of early, transient gastro-intestinal upset, with nausea, vomiting, diarrhea and abdominal pain. These clear up, and are followed in 1–3 weeks by soreness of the lower leg muscles, "numbness" of the toes and fingers, and a few days later by weakness of the toes and bilateral foot-drop. After another week or so, weakness of the fingers and bilateral wrist-drop follow. There are no sensory changes. Recovery is slow, and the degree of residual paralysis depends upon the extent of damage to the nervous system. Many cases recover completely. In 1958 several thousand persons in Morocco were poisoned with this material which was present in lubricating oil which had been mixed with edible oils by dishonest merchants. Many of the victims suffered a permanent paralysis.
Fire Hazard: Slight, when exposed to heat or flame.
Spont Heating: No.
Disaster Hazard: Dangerous; when heated to decomp, emits highly toxic fumes of PO_x; can react with oxidizing materials.
To Fight Fire: CO_2, dry chemical.

TRICYANIC ACID. See *n*-cyanuric acid.

TRICYANOGEN CHLORIDE. See cyanuric chloride.

TRICYCLOHEXYL BORATE. Large, needle-like white crystals, nearly odorless. $[\underline{CH_2(CH_2)_4CHO}]_3B$, mw: 308.3, mp: 59°–61°, bp: 330°.
THR = See boron compounds and esters.
Fire Hazard: Slight, when exposed to heat or flame.
To Fight Fire: Spray, mist, dry chemical, foam.

TRICYCLOHEXYL BORINE. Syn: *boron tricyclohexyl.* Colorless crystals, insol in water, sol in ether. $(C_6H_{11})_3B$, mw: 260.3, mp: 100°, bp: 194° @ 15 mm.
THR = Details U. See also boron compounds.

TRI-(2-CYCLOHEXYLCYCLOHEXYL)BORATE. White solid, odor of 2-cyclohexylcyclohexanol,

$(C_{12}H_{21}O)_3B$, mw: 554.7, mp: 172°–175°, bp: 230°–250° @ 0.3 mm.

Acute tox data: Oral LD_{50} (mice) = 2050 mg/kg. [*3*]

THR = MOD via oral route. See boron compounds and esters.

Fire Hazard: Slight, when exposed to heat or flame; can react with oxidizing materials.

TRICYCLOHEXYL GERMANIUM BROMIDE. Colorless crystals, hydrolyzed in water. $(C_6H_{11})_3Br$, mw: 402.0, mp: 110°.

THR = See germanium compounds and chlorides.

TRICYCLOHEXYL GERMANIUM FLUORIDE. Colorless crystals, hydrolyzed in water. $(C_6H_{11})_3GeF$, mw: 341.1, mp: 92°.

THR = See fluorides.

TRICYCLOHEXYL GERMANIUM HYDROXIDE. Solid material, sol in organic solvents. $(C_6H_{11})_3GeOH$, mw: 339.1, mp: 177°.

THR = Details U. See germanium compounds.

TRICYCLOHEXYL GERMANIUM IODIDE. Colorless crystals, hydrolyzed by water. $(C_6H_{11})_3GeI$, mw: 449.0, mp: 100°.

THR = See germanium compounds and iodides.

TRICYCLOHEXYL TIN HYDROXIDE. Syn: *plictran.*

Acute tox data: Oral LD_{50} (rat) = 190 mg/kg; ip LD_{50} (rat) = 13 mg/kg. [*3*]

THR = HIGH via oral and ip routes. See tin compounds.

TRICYCLOQUINAZOLINE. $C_{21}H_{12}N_4$, mw: 320.4.

THR = An exper carc via sc and dermal routes. [*3*, *23*]

TRIDANE. See tridecyl benzene.

TRIDECANOL. Syn: *tridecyl alcohol.* General term for a commercial mixture of isomers of the formula $C_{12}H_{25}CH_2OH$, water-white liquid, pleasant odor. mw: 200, boiling range: 252°–272°, d: 0.845 @ 20°/20°, flash p: 250°F (COC), vap. d: 6.9.

Acute tox data: Oral LD_{LO} (rat) = 4750 mg/kg; dermal LD_{50} (rabbit) = 7070 mg/kg. [*3*]

THR = MOD via oral and LOW via dermal routes. See alcohols.

Fire Hazard: Mod to slight, when exposed to heat or flame.

To Fight Fire: Mist, spray, dry chemical, foam.

TRIDECYL ACRYLATE. Insol in water. $H_2CCHCOOC_{13}H_{27}$, mw: 254, d: 0.9, bp: 150° @ 10 mm, flash p: 270°F (OC).

THR = Details U. Limited animal exper suggest low toxicity. See also esters.

Fire Hazard: Slight, when exposed to heat or flame.

To Fight Fire: Spray, mist, dry chemical, foam.

TRIDECYL ALDEHYDE. $C_{13}H_{26}O$, mw: 198.3.

THR = Self-ignites in air. [*19*]

TRIDECYL BENZENE. Syns: *tridane, 1-phenyl tridecane.* Colorless liquid. $C_6H_5(CH_2)_{12}CH_3$, mw: 260, d: 0.85–0.86 @ 15°/15°, bp: 346°, fp: 10°.

THR = U. Vapors may have a narcotic effect. An irr.

1-TRIDECYL-2-BENZYL-2-HYDROXYETHYL IMIDAZOLIUM CHLORIDE. Syn: *alrosept MBC.*

THR = MOD irr to skin, eyes and mu mem and via oral and inhal routes. Effects similar to other quaternary amines. Exper has produced cholinesterase inhibition and muscular paralysis.

Disaster Hazard: When heated to decomp, emits highly toxic fumes.

1-TRIDECYL-2-METHYL-2-HYDROXYETHYL IMIDAZOLINIUM CHLORIDE. Syn: *alrosept MM.* $C_{19}H_{37}ON_2Cl$, mw: 345.0.

Acute tox data: Oral LD_{50} (rat) = 500 mg/kg. [*3*]

THR = HIGH via oral route.

TRIDECYL PHOSPHITE. Water-white liquid, decyl alcohol odor, insol in water. $(C_{10}H_{21}O)_3P$, mw: 502, d: 0.892 @ 25°/15.5°, mp: $<0°$, flash p: 455°F (OC).

THR = See esters and phosphites.

Fire Hazard: Slight, when exposed to heat or flame.

Disaster Hazard: Dangerous; when heated to decomp, emits highly toxic fumes.

To Fight Fire: Mist, CO_2, dry chemical.

TRIDEUTERO AMMONIA. See ammonia-d_3.

TRI-(DIISOBUTYL CARBINYL)BORATE. White crystals, odor of diisobutyl carbinol. $C_{27}H_{57}O_3B$, mw: 440.55, mp: 99°–100°, bp: 198°–209° @ 22 mm.

THR = See diisobutylcarbinol and boron compounds.

TRI-*n*-DODECYL BORATE. Light straw-yellow, oily liquid. $[CH_3(CH_2)_{10}CH_2O]_3B$, mw: 566.8, bp: 479°, flash p: 465°F (COC), d: 0.845 @ 26.8°, vap. d: 19.6.

THR = See boron compounds and esters.

Fire Hazard: Slight, when exposed to heat or flame; can react with oxidizing materials.

To Fight Fire: Foam, CO_2, dry chemical.

1,2,4,5,9,10-TRIEPOXY DECANE. $C_{10}H_{16}O_3$, mw: 184.3.

THR = An exper neo. [*3*]

TRIETHANOLAMINE. Pale yellow viscous liquid. $(CH_2OHCH_2)_3N$, mw: 149.19, mp: 21.2°, bp: 360°, flash p: 355°F (CC), d: 1.1258 @ 20°/20°, vap. press: 10 mm @ 205°, vap. d: 5.14.

Acute tox data: Oral LD_{50} (rat) = 8680 mg/kg; oral LD_{50} (guinea pig) = 8000 mg/kg. [*3*]

THR = LOW via oral route. Liver and kidney damage has been demonstrated in animals from chronic exposure.

Fire Hazard: Slight, when exposed to heat or flame.
Spont Heating: No.
Disaster Hazard: Dangerous; when heated to decomp, emits toxic fumes of NO_x; can react vigorously with oxidizing materials.
To Fight Fire: Alcohol foam, CO_2, dry chemical.

TRIETHANOLAMINE BORATE. White odorless solid. $C_6H_{15}O_3NB$, mw: 160, mp: 235.5°–238.5°.
THR = See triethanolamine.

TRIETHANOLAMINE-*o-sec*-BUTYL PHENATE.
THR = U. A fungicide. See triethanolamine and esters.

TRIETHANOLAMINE TITANATE. Yellow liquid. $Ti[(OCH_2CH_2)NCH_2CH_2OH]_2$, mw: 254.14, d: 1.05, vap. d: 8.78.
THR = See esters and triethanol amine.
Fire Hazard: Slight, when exposed to heat or flame; can react with oxidizing materials.

TRIETHOXY BORON. See ethyl borate.

1,1,3-TRIETHOXY BUTANE.
$CH(OC_2H_5)_2CH_2CH(OC_2H_5)CH_3$, mw: 190.
Acute tox data: Oral LD_{50} (rat) = 4920 mg/kg; inhal LC_{LO} (rat) = 2000 ppm for 4 hrs; dermal LD_{LO} (rabbit) = 1770 mg/kg. [*3*]
THR = MOD via oral, dermal and inhal routes.

1,1,3-TRIETHOXY HEXANE. Liquid, insol in water. $CH(OC_2H_5)CH_2CH(OC_2H_5)C_3H_7$, mw: 218, d: 0.8746 @ 20°/20°, bp: 133° @ 50 mm, fp: −100°, flash p: 210°F (OC), vap. d: 7.5.
Acute tox data: Oral LD_{50} (rat) = 17,000 mg/kg. [*3*]
THR = LOW via oral route.
Fire Hazard: Low, when exposed to heat or flame.
To Fight Fire: Foam, alcohol foam, fog.

TRIETHOXY METHANE. See triethyl-*o*-formate.

1,3,3-TRIETHOXY PROPENE-1.
$(C_2H_5O)HCCHCH(C_2H_5O)_2$, mw: 174.2.
Acute tox data: Oral LD_{50} (rat) = 2460 mg/kg; inhal LC_{LO} (rat) = 250 ppm for 4 hrs; dermal LD_{50} (rabbit) = 370 mg/kg. [*3*]
THR = HIGH via dermal and inhal routes; MOD via oral route.

TRIETHYL ALLYL GERMANE. $C_9H_{20}Ge$, mw: 200.9.
Acute tox data: Oral LD_{LO} (rat) = 330 mg/kg; ip LD_{LO} (rat) = 22 mg/kg; oral LD_{LO} (mice) = 770 mg/kg. [*3*]
THR = HIGH via oral and ip routes. See germanium compounds.

TRIETHYL ALUMINUM. Syn: *aluminum triethyl.* $(C_2H_5)_3Al$, mw: 114, fp: −52.5°, d: 0.837 @ 20°, vap. press: 4 mm @ 83°, flash p: −63°F, bp: 194°.
THR = HIGH. Extremely destructive to living tissue.
Fire Hazard: Dangerous; ignites spont in air.
To Fight Fire: CO_2, dry sand, dry chemical. See metal fires. Do not use water, foam or halogenated fire-fighting agents.

TRIETHYL ALUMINUM ETHERATE. Colorless liquid, ether odor. $4Al(C_2H_5)_3 \cdot 3(C_2H_5)_2O$, mw: 679.0, bp: 112° @ 16 mm.
THR = Details U. See triethyl aluminum and organometals.
Fire Hazard: Dangerous; see ethers; self-ignites in air. [*19*]
Explosion Hazard: Dangerous; see ethers; upon contact with moisture, explodes and evolves ethane.
Disaster Hazard: Dangerous; explodes upon contact with moisture. Warming can evolve copious fumes of ether. See ethers.
To Fight Fire: See ethers.

TRIETHYLAMINE. Colorless liquid, ammonia odor. $(C_2H_5)_3N$, mw: 101.19, mp: −114.8°, bp: 89.5°, flash p: 20°F (OC), d: 0.7255 @ 25°/4°, vap. d: 3.48, lel = 1.2%, uel = 8.0%.
Acute tox data: Oral LD_{50} (rat) = 460 mg/kg; inhal LC_{LO} (rat) = 1000 ppm for 4 hrs; dermal LD_{50} (rabbit) = 570 mg/kg; oral LD_{50} (mice) = 546 mg/kg; inhal LD_{50} (guinea pig) = 1000 ppm for 4 hrs. [*3*]
THR = HIGH irr via oral, inhal route; MOD via dermal routes. Exper animals have shown kidney and liver damage.
Fire Hazard: Dangerous, when exposed to heat, flame or oxidizers.
Explosion Hazard: U.
Disaster Hazard: Highly dangerous; keep away from heat or open flame; can react with oxidizing materials.
To Fight Fire: CO_2, dry chemical, alcohol foam.

TRIETHYL-*n*-AMYLTIN. Liquid. $(C_2H_5)_3Sn(C_5H_{11})$, mw: 277.0, bp: 102° @ 10 mm.
THR = Toxic. Details U. See tin compounds.

TRIETHYL ANTIMONITE. See antimony ethoxide.

TRIETHYL ANTIMONY SULFATE. $(C_2H_5)_3SbSO_4$, mw: 305.
THR = Spont flam in air. [*19*]

TRIETHYL ARSENIC. Syn: *arsenic triethyl.* Colorless liquid. $As(C_2H_5)_3$, mw: 162.1, bp: 140° @ 736 mm, d: 1.152, vap. d: 5.59.
THR = HIGH. See arsenic compounds.
Fire Hazard: Mod.
Disaster Hazard: Dangerous; when heated to decomp, or on contact with acid or acid fumes, emits highly

toxic fumes of arsenic; can react vigorously with oxidizing materials.

TRIETHYL ARSINE. $(C_2H_5)_3As$, mw: 162.2.
THR = HIGH. See also arsenic compounds. Spont flam in air. [*19*]

TRIETHYLBENZENE. Clear, colorless liquid. $C_6H_3(CH_2CH_3)_3$, mw: 162.3, mp: <−70°, bp: 218°–219°, flash p: 181°F (CC), d: .870 @ 25°/25°, vap. d: 5.6.
Acute tox data: Oral LD_{LO} (rat) = 5000 mg/kg. [*3*]
THR = MOD via oral route.
Fire Hazard: Mod, when exposed to heat or flame; can react with oxidizing materials.
To Fight Fire: Foam, CO_2, dry chemical.

TRIETHYL BISMUTHINE. Syn: *bismuth triethyl.* Liquid. $Bi(C_2H_5)_3$, mw: 296.18, bp: 107° @ 79 mm (can explode), d: 1.82.
THR = See bismuth compounds.
Fire Hazard: Dangerous, when exposed to heat or flame. Self-ignites in air. [*19*]
Explosion Hazard: Mod, when exposed to heat. Explodes at 302°F.
Disaster Hazard: Mod; when heated to decomp, emits toxic fumes and explodes; can react with oxidizing materials.

TRIETHYL BORANE. See boron triethyl.

TRIETHYL BORATE. See ethyl borate.

TRIETHYL BORON. See boron triethyl.

TRIETHYL CITRATE. $C_3H_5O(COOC_2H_5)_3$, mw: 276.3, bp: 294°, flash p: 303°F (COC), d: 1.136 @ 25°, vap. press: 1 mm @ 107.0°.
THR = Probably low. See citric acid. A general purpose food additive. [*109*] See esters.
Fire Hazard: Slight, when exposed to heat or flame.
To Fight Fire: Dry chemical, CO_2.

TRI-(1-ETHYL CYCLOHEXYL)BORATE. $B(C_8H_{15}O)_3$, mw: 392.5.
THR = Probably toxic. See boron compounds and esters.
Fire Hazard: A flam material, when exposed to heat or flame.

TRIETHYL DIBORANE. $(C_2H_5)_3B_2H_3$, mw: 111.8.
THR = See boranes. Self-ignites in air. [*19*]

TRIETHYL(*p*-DIMETHYL AMINOPHENYL)TIN. Liquid. $(C_2H_5)_3(CH_3)_2NC_6H_4Sn$, mw: 326.1, d: 1.2425, bp: 173° @ 3 mm.
THR = Details U. See tin compounds.

TRIETHYLENE DIAMINE. Hygroscopic crystals. $N(CH_2CH_2)_3N$, mw: 112, mp: 158°, bp: 174°.
THR = An irr, allergen, and skin-sensitizer. See also amines.

TRIETHYLENE GLYCOL. Syns: *2,2′-ethylene dioxydiethanol, glycol bis(hydroxyethyl)ether.* Colorless liquid, hygroscopic. $(CH_2OCH_2CH_2OH)_2$, mw: 150.17, bp: 291.2°; fp: −7.3°, flash p: 350°F, d: 1.122 @ 25°/25°, lel = 0.9%, uel = 9.2%, autoign. temp.: 700°F, vap. press: 1 mm @ 114°, vap. d: 5.17.
Acute tox data: Oral LD_{50} (mice) = 18,500 mg/kg; sc LD_{50} (mice) = 8750 mg/kg; iv LD_{50} (mice) = 6500 mg/kg. [*3*]
THR = MOD via iv route; LOW via sc route and practically NONE via oral route. See also glycols. A fungicide.
Fire Hazard: Slight, when exposed to heat or flame; can react with oxidizing materials.
Spont Heating: No.
Explosion Hazard: Mod, in the form of vapor when exposed to flame, spark or heat source.
To Fight Fire: Alcohol foam, dry chemical.

TRIETHYLENE GLYCOL CAPRYLATE. Syn: *plasticizer SC.* Clear liquid. $C_7H_{15}CO_2CH_2CH_2OCH_2CH_2OCH_2CH_2CO_2C_7H_{15}$, mw: 402.6, d: 0.973 @ 20°, mp: −3°, bp: 243° @ 5 mm.
THR = Details U. Animal exper suggest very low toxicity and mod irr effects on skin. See also glycols.

TRIETHYLENE GLYCOL DIBENZOATE. Crystals. $C_{20}H_{22}O_6$, mw: 358.3, bp: 210°–223°, fp: 46°, flash p: 457°F (TOC), d: 1.168 @ 25°/4°.
THR = Details U, but probably quite low. See also glycols.
Fire Hazard: Slight, when exposed to heat or flame; can react with oxidizing materials.
To Fight Fire: Water, foam, CO_2, dry chemical.

TRIETHYLENE GLYCOL DICHLORIDE. See triglycol dichloride.

TRIETHYLENE GLYCOL DI-2-ETHYL BUTYRATE. Colorless liquid. $C_{18}H_{34}O_6$, mw: 346.45, mp: −65°, bp: 197° @ 5 mm, flash p: 385°F (OC), d: 0.9945 @ 20°/20°, vap. press: 5.8 mm @ 200°, vap. d: 11.95.
Acute tox data: Oral LD_{50} (guinea pig) = 3110 mg/kg. [*3*]
THR = MOD via oral route. See glycols.
Fire Hazard: Slight, when exposed to heat or flame; can react with oxidizing materials.
To Fight Fire: Foam, CO_2, dry chemical.

TRIETHYLENE GLYCOL DI-2-ETHYLHEXOATE. Colorless liquid with mild odor. $C_{23}H_{42}O_5$, mw: 402.56, mp: −58°, bp: 218° @ 5 mm, flash p: 405°F, d: 0.9679 @ 20°/20°, vap. press: 1.9 mm @ 200°, vap. d: 13.9.
Acute tox data: Oral LD_{50} (rat) = 31,000 mg/kg; oral LD_{50} (guinea pig) = 21,000 mg/kg. [*3*]
THR = NONE via oral route. See glycols.

Fire Hazard: Slight, when exposed to heat or flame; can react with oxidizing materials.
To Fight Fire: Foam, CO_2, dry chemical.

TRIETHYLENE GLYCOL DIGLYCIDYL ETHER. $C_{12}H_{22}O_6$, mw: 262.3.
THR = An exper (+) carc [*13*] and neo [*3*, *23*] via ip route.

TRIETHYLENE GLYCOL DIMETHYL ETHER. Water-white liquid, mild ethereal odor. $C_8H_{18}O_4$, mw: 178.22, bp: 216°, fp: −46°, flash p: 232°F (OC), d: 0.982 @ 20°/20°, autoign. temp.: 1166°F, vap. press: 0.9 mm @ 20°.
THR = See glycols.
Fire Hazard: Slight, when exposed to heat or flame; can react with oxidizing materials.
To Fight Fire: CO_2, dry chemical.

TRIETHYLENE GLYCOL DINITRATE. $C_6H_{12}O_4 \cdot 2NO_2$, mw: 240.2.
Acute tox data: Oral LD_{50} (rat) = 1000 mg/kg; ip LD_{50} (rat) = 796 mg/kg; sc LD_{50} (rat) = 2520 mg/kg. [*3*]
THR = MOD via oral, ip and sc routes. See also glycols and nitrates.

TRIETHYLENE GLYCOL DIPELARGONATE. Syn: *TG-9*. Clear liquid, mild characteristic odor. $C_{24}H_{46}O_6$, mw: 438, fp: −4° to 1°, flash p: 420°F, bp: 251° @ 5 mm, d: 0.964 @ 20°/20°, vap. d: 15.1.
THR = Details U. See glycols.
Fire Hazard: Slight, when exposed to heat or flame; can react with oxidizing materials.
To Fight Fire: Foam, CO_2, dry chemical.

TRIETHYLENE GLYCOL METHYL ETHER ACETATE. See methoxytriglycol acetate.

TRIETHYLENE GLYCOL MONOBUTYL ETHER. Syn: *butoxy triglycol*. Liquid, completely sol in water. $C_{10}H_{22}O_4$, mw: 206, d: 1.0021 @ 20°/20°, bp: decomp, fp: −47.4°, flash p: 290°F.
Acute tox data: Oral LD_{50} (rat) = 6730 mg/kg; dermal LD_{50} (rabbit) = 3540 mg/kg. [*3*]
THR = MOD via oral and dermal routes. See also glycols.
Fire Hazard: Slight, when exposed to heat or flame.
To Fight Fire: Water, foam, fog.

TRIETHYLENE GLYCOL MONOMETHYL ETHER. See methoxy triglycol.

TRIETHYLENE MELAMINE. Syns: *TEM; 2,4-tris(1-aziridinyl)triazine, tetramine*. Small crystals, water-sol. $C_9H_{12}N_6$, mw: 204.2, decomp @ 139°.
Acute tox data: Oral LD_{50} (rat) = 1 mg/kg; ip LD_{50} (rat) = 1 mg/kg; im LD_{50} (rat) = 1.5 mg/kg; oral LD_{50} (mice) = 15 mg/kg. [*3*]
THR = VERY HIGH via oral, ip and im routes. An exper (+) carc, mutagen, neo and teratogen, via ip, sc, dermal routes. [*3*, *8*, *23*] Has been reported as capable of causing GI disturbances and bone marrow depression.
Disaster Hazard: Dangerous; when heated to decomp, emits highly toxic fumes of NO_x.

TRIETHYLENE PHOSPHORAMIDE. See tris-(1-aziridinyl)phosphine oxide.

TRIETHYLENE TETRAMINE. Moderately viscous, yellowish liquid. $H_{20}N_4C_7$, mw: 146.24, bp: 278°, flash p: 275°F, d: 0.982, vap press: $<$0.01 mm @ 20°, autoign. temp.: 640°F.
Acute tox data: Oral LD_{50} (rat) = 4340 mg/kg; dermal LD_{50} (rabbit) = 820 mg/kg. [*3*]
THR = MOD via dermal and oral routes. Causes skin irr and sensitization.
Fire Hazard: Slight, when exposed to heat or flame; can react with oxidizing materials.
Spont Heating: No.
To Fight Fire: CO_2, dry chemical, alcohol foam.

TRIETHYL ETHOXY DIPHOSPHINYL OXIDE. $(C_2H_5)_4O_2P_2$, mw: 210.2.
THR = Spont flam in air. [*19*]

TRIETHYL-*o*-FORMATE. Syns: *o-formic ester, triethoxymethane*. Clear liquid, pungent odor. $(C_2H_5O)_3CH$, mw: 148.20, bp: 145.9°, flash p: 86°F (CC), d: 0.895 @ 20°/20°, vap press: 10 mm @ 40.5°, vap. d: 5.11.
Acute tox data: Oral LD_{50} (rat) = 2920 mg/kg; inhal LC_{LO} (rat) = 8000 ppm for 4 hrs. [*3*]
THR = MOD via oral and inhal routes. An irr. See also esters.
Fire Hazard: Dangerous, when exposed to heat, flame or oxidizers.
Explosion Hazard: U.
Disaster Hazard: Mod dangerous; when exposed to heat or open flame, can react with oxidizing materials.
To Fight Fire: Foam, CO_2, dry chemical.

TRIETHYL GALLIUM. Colorless liquid, decomp by water. $(C_2H_5)_3Ga$, mw: 156.9, d: 1.0576 @ 30°, mp: 82.3°, bp: 142.6°.
THR = See gallium compounds. Spont flam in air. [*19*]

TRIETHYL GALLIUM MONAMMINE. Colorless liquid, decomp by water. $(C_2H_5)_3GaNH_3$, mw: 173.9.
THR = U. See gallium compounds.

TRIETHYL GALLIUM MONOETHERATE. Colorless liquid, decomp by water. $(C_2H_5)_3Ga \cdot O \cdot C_2H_5$, mw: 231.0.
THR = See gallium compounds and ethyl ether.

TRIETHYL GERMANIUM ACETATE.
$(C_2H_5)_3GeOOCH_3$, mw: 207.
THR = U. Animal exper show toxicity much lower than corresponding lead compound. See also germanium.

TRIETHYL GERMANIUM BROMIDE. Colorless liquid, hydrolyzed in water. $(C_2H_5)_3GeBr$, mw: 239.7, mp: −33°, bp: 191°.
THR = See germanium compounds and bromides.

TRIETHYL GERMANIUM CHLORIDE. Colorless liquid, hydrolyzed by water. $(C_2H_5)_3GeCl$, mw: 195.2, mp: < −50°, bp: 176°.
THR = See germanium compounds and chlorides.

TRIETHYL GERMANIUM FLUORIDE. Colorless liquid, hydrolyzed by water. $(C_2H_5)_3GeF$, mw: 178.8, bp: 149° @ 751 mm.
THR = See fluorides.

TRIETHYL GERMANIUM HYDRIDE. Colorless liquid, insol in water. $(C_2H_5)_3GeH$, mw: 160.8, bp: 124° @ 751 mm.
THR = See germanium compounds and hydrides.

TRIETHYL GERMANIUM IMINE. Colorless liquid, hydrolyzes in water. $[(C_2H_5)_3Ge)]_2NH$, mw: 334.6, bp: 100° @ 0.1 mm.
THR = U. See germanium compounds.

TRIETHYL GERMANIUM IODIDE. Colorless liquid, hydrolyzed by water. $(C_2H_5)_3GeI$, mw: 286.7, mp: < −50°, bp: 212°.
THR = See germanium compounds and iodides.

TRIETHYL GERMANIUM OXIDE. Colorless liquid, insol in water. $[(C_2H_5)_3Ge]_2O$, mw: 335.6, mp: < −50°, bp: 254°.
THR = Details U. See germanium compounds.

TRI(2-ETHYLHEXYL)BORATE. Colorless mobile liquid. Odor of 2-ethylhexanol.
$[C_4H_9(C_2H_5)CHCH_2O]_3B$, mw: 398.5, bp: 350°–354°, flash p: 350°F (COC), d: 0.857 @ 23.6°, vap. d: 13.8.
Acute tox data: Oral LD_{50} (rat) = 2830 mg/kg. [*3*]
THR = MOD via oral route. See esters.
Fire Hazard: Slight, when exposed to heat or flame; can react with oxidizing materials.
To Fight Fire: Foam, CO_2, dry chemical.

TRI(2-ETHYLHEXYL)PHOSPHATE. Light colored liquid. $(C_2H_5C_6H_{12})_3PO_4$, mw: 434.6, mp: −74°, bp: 216° @ 5 mm, flash p: 405°F, d: 0.9262 @ 20°/20°, vap press: 0.23 mm @ 150°.
Acute tox data: Oral LD_{50} (rat) = 37000 mg/kg; dermal LD_{LO} (rabbit) = 20000 mg/kg. [*3*]
THR = NONE via oral and dermal routes. See also phosphates.

TRIETHYL-*o*-HYDROXYPHENYLTIN. Liquid. $(C_2H_5)_3HOC_6H_4Sn$, mw: 299.0, d: 1.3229 @ 25°, bp: 200° @ 3 mm.
THR = Details U. See tin compounds.

TRIETHYL INDIUM. $(C_2H_5)_3In$, mw: 202.
THR = Spont flam in air. [*19*]

TRIETHYL ISOAMYLTIN. Liquid. $(C_2H_5)_3Sn(C_5H_{11})$, mw: 277.0, d: 1.1203, bp: 111° @ 18.5 mm.
THR = Details U. See tin compounds.

TRIETHYL ISOBUTYLTIN. Liquid. $(C_2H_5)_3Sn(C_4H_9)$, mw: 263.0, d: 1.139, bp: 96.5° @ 17 mm.
THR = Details U. See tin compounds.

TRIETHYL LEAD. Syn: *hexaethyldilead*. A liquid, insol in water. $Pb_2(C_2H_5)_6$, mw: 588.8, d: 1.471.
Acute tox data: Pa LD_{50} (rat) = 11 mg/kg. [*3*]
THR = HIGH via pa route. See lead compounds.

TRIETHYL LEAD CHLORIDE. $C_6H_{15}ClPb$, mw: 329.9.
Acute tox data: ip LD_{LO} (rat) = 5 mg/kg; sc LD_{LO} (rat) = 1 mg/kg; ip LD_{LO} (rabbit) = 13 mg/kg. [*3*]
THR = HIGH via ip and sc routes. See also lead compounds.

TRIETHYL LEAD FLUOROACETATE. $C_2H_2O_2F \cdot PbC_6H_{15}$, mw: 371.44.
Acute tox data: Inhal by humans of 1.7 mg/m^3 for $\frac{1}{6}$ hr ⟶ pulmonary effects; sc LD_{50} (mouse) = 15 mg/kg. [*3*]
THR = HIGH via inhal, oral and sc routes. See also lead compounds.

TRI(7-ETHYL-2-METHYL-4-UNDECYL)BORATE.
See tritetradecyl borate.

TRIETHYL PHENYL GERMANIUM. Colorless liquid, insol in water. $Ge(C_2H_5)_3(C_6H_5)$, mw: 236.9, bp: 117° @ 13 mm.
THR = Details U. See germanium compounds.

TRIETHYL PHENYL SILICANE. Solid. $(C_3H_5)_3(C_6H_5)Si$, mw: 192.3, mp: 148°, bp: 230°.
THR = Details U. See silanes.

TRIETHYL PHENYLTIN. Colorless liquid, insol in water, sol in organic solvents. $(C_2H_5)_3Sn(C_6H_5)$, mw: 283.0, d: 1.2639, bp: 254°.
THR = Details U. See tin compounds.

TRIETHYL PHOSPHATE. Syn: *TEP*. Liquid, sol in most organic solvents, insol in water. $(C_2H_5)_3PO_4$, mw: 182.2, mp: −56.5°, bp: 209°–218°, flash p: 240°F (OC), d: 1.067–1.072 @ 20°/20°, vap. press: 1 mm @ 39.6°, vap. d: 6.28.
Acute tox data: Oral LD_{LO} (rat) = 1600 mg/kg; ip LD_{LO} (rat) = 800 mg/kg; oral LD_{LO} (mice) = 1600 mg/kg; ip LD_{LO} (mice) = 800 mg/kg. [*3*]
THR = MOD via oral and ip routes. Causes cholines-

terase inhibition, but to a lesser extent than parathion. May be expected to cause nerve injury similar to that of other phosphate esters. See also tri-*o*-cresyl phosphate.
Fire Hazard: Slight, when exposed to heat or flame.
Disaster Hazard: Dangerous. See phosphates; can react vigorously with oxidizing materials.
To Fight Fire: CO_2, dry chemical, alcohol foam.

TRIETHYL PHOSPHINE. Colorless liquid. $(C_2H_5)_3P$, mw: 118.2, bp: 126°, d: 0.801 @ 20°/4°.
THR = Details U. Probably HIGH. See phosphides.
Disaster Hazard: Dangerous; see phosphates and phosphine.

TRIETHYL PHOSPHINE OXIDE. Colorless deliquescent crystals. $(C_2H_5)_3PO$, mw: 134.2, mp: 52.9°, bp: 242.9°.
THR = Details U. Probably HIGH.
Disaster Hazard: Dangerous; see phosphates.

TRIETHYL PHOSPHINE SULFIDE. Crystals. $(C_2H_5)_3PS$, mw: 150.2, mp: 94°, bp: ignites.
THR = Details U; probably high.
Fire Hazard: LOW, when exposed to heat, flame or oxidizers.
Disaster Hazard: Dangerous; see phosphates and sulfides; can react vigorously with oxidizing materials.
To Fight Fire: Spray, mist, dry chemical.

TRIETHYL PHOSPHOROTHIOATE. Colorless liquid, strong characteristic odor. $(C_2H_5O)_3PS$, mw: 198.2, bp: 93.2°–94° @ 10 mm, flash p: 225°F (COC), d: 1.074 @ 20°/4°.
Acute tox data: Inhal LC_{LO} (rat) = 41 ppm for 4 hrs. [*3*]
THR = HIGH via inhal and oral routes. A cholinesterase inhibitor. See parathion.
Fire Hazard: LOW, when exposed to heat, flame or oxidizers.
Disaster Hazard: Dangerous; see phosphates and sulfates; can react with oxidizing materials.

TRIETHYL PROPYL GERMANE. $C_9H_{22}Ge$, mw: 202.9.
Acute tox data: Oral LD_{LO} (rat) = 4700 mg/kg; ip LD_{LO} (rat) = 1430 mg/kg. [*3*]
THR = MOD via oral and ip routes.

TRIETHYL-*n*-PROPYL TIN. Liquid. $(C_2H_5)_3Sn(C_3H_7)$, mw: 249.0, d: 1.1780, bp: 82° @ 13 mm.
THR = Details U. See tin compounds.

TRIETHYL STIBINE. See antimony triethyl.

TRIETHYL THALLIUM. Yellow liquid. $(C_2H_5)_3Tl$, mw: 291.6, d: 1.957 @ 23°/23°, mp: −63°, bp: 192°.
THR = HIGH. See thallium compounds.

TRIETHYLTIN. Colorless liquid, insol in water. $(C_2H_5)_3Sn$, mw: 205.0, d: 1.3774, mp: < −75°, bp: 161° @ 23 mm.
THR = See tin compounds, organic.

TRIETHYLTIN ACETATE. $C_8H_{18}O_2Sn$, mw: 265.
Acute tox data: Oral LD_{50} (rat) = 4 mg/kg; iv LD_{50} (rat) = 4.2 mg/kg.[*3*]
THR = HIGH via oral and iv routes.

TRIETHYLTIN BROMIDE. Colorless liquid, sol in organic solvents. $(C_2H_5)_3SnBr$, mw: 285.8, d: 1.630, mp: −13.5°, bp: 224°.
Acute tox data: Inhal LC_{LO} (mice) = 1640 mg/m^3. [*3*]
THR = MOD via inhal route. See tin compounds and bromides.

TRIETHYLTIN CHLORIDE. Colorless liquid, insol in water, sol in organic solvents. $(C_2H_5)_3SnCl$, mw: 241.4, d: 1.428 @ 8°, mp: 10°, bp: 210°.
Acute tox data: ip LD_{50} (rat) = 5 mg/kg. [*3*]
THR = HIGH via ip and dermal routes. See tin compounds and chlorides.

TRIETHYLTIN ETHOXIDE. Colorless liquid, decomp by water, sol in organic solvents. $(C_2H_5)_3Sn(OC_2H_5)$, mw: 250.94, d: 1.2634, bp: 190°.
THR = Details U. See tin compounds and ethyl alcohol.

TRIETHYLTINHYDROXIDE. Colorless crystals, sol in water and organic solvents. $(C_2H_5)_3SnOH$, mw: 222.9, mp: 43°, bp: 271°.
THR = Details U. See tin compounds.

TRIETHYLTIN IODIDE. Colorless liquid, sol in organic solvents. $(C_2H_5)_3SnI$, mw: 332.8, d: 1.833, mp: 34.5°, bp: 225°.
THR = Toxic. See tin compounds and iodides.

TRIETHYLTIN SULFATE. $C_6H_{16}O_4SSn$, mw: 303.
Acute tox data: ip LD_{50} (rat) = 5.7 mg/kg; pa LD_{50} (rat) = 6 mg/kg; ip LD_{50} (guinea pig) = 3 mg/kg; iv LD_{LO} (bird) = 3 mg/kg; oral LD_{LO} (rat) = 10 mg/kg; sc LD_{LO} (rat) = 25 mg/kg. [*3*]
THR = HIGH via ip, pa, iv, oral, sc and dermal routes.

TRIETHYL-*p*-TOLYLGERMANIUM. Colorless liquid, insol in water. $(C_2H_5)_3Ge(C_6H_4CH_3)$, mw: 250.9, bp: 126° @ 12 mm.
THR = Details U. See germanium compounds.

2,2,2-TRIETHYL-1,1,1-TRIPHENYL DIGERMANE. Colorless crystals, insol in water. $(C_2H_5)_3Ge_2(C_6H_5)_3$, mw: 463.7, mp: 90°.
THR = Details U. See germanium compounds.

TRI-(1-ETHYNYLCYCLOHEXYL)BORATE. Pale-yellow liquid, odor of 1-ethynylcyclohexanol, mw: 380.34, bp: 150°–170° @ 0.5 mm, flash p: 190°F (COC), d: 1.006 @ 27°.
THR = See boron compounds and alcohols.

Fire Hazard: Mod, when exposed to heat or flame; can react with oxidizing materials.
To Fight Fire: Water, foam, CO_2, dry chemical.

TRIFLUOROACETIC ACID. Colorless liquid, strong pungent odor. CF_3COOH, mw: 114, mp: −15.25°, bp: 71.1° @ 734 mm, d: 1.535 @ 0°.
Acute tox data: Oral LD_{50} (rat) = 200 mg/kg; ip LD_{LO} (mice) = 150 mg/kg. [*3*]
THR = HIGH via oral and ip routes. A powerful irr. See also fluorides.

1,1,2-TRIFLUORO-4-BROMOBUTENE. $CF_2{:}CFCH_2CH_2Br$, mw: 189.
THR = Details U. Probably irr and narcotic in high conc.
Disaster Hazard: Dangerous; see fluorides and bromides.

TRIFLUOROBROMOMETHANE. See bromotrifluoromethane.

TRIFLUOROCHLOROETHENE. See trifluorochloroethylene.

TRIFLUOROCHLOROETHYLENE. See chlorotrifluoroethylene.

***α,α,α*-TRIFLUORO-2,6-DINITRO-*n,n*-DIPROPYL-*p*-TOLUIDINE.** Syn: *trifluralin*. Yellowish-orange solid, insol in water, sol in xylene, acetone and ethanol. $(CF_3)(NO_2)_2C_6H_2N(C_3H_7)_2$, mw: 335, mp: 48.5°–49°, bp: 139°–140° @ 4.2 mm.
Acute tox data: Oral LD_{50} (rat) = 500 mg/kg; oral LD_{50} (mice) = 5000 mg/kg. [*3*]
THR = MOD via oral route. An herbicide.
Disaster Hazard: Dangerous; see fluorides and nitrates.

TRIFLUOROMETHANE. See fluoroform.

TRIFLUOROMETHYLBENZENE. See benzotrifluoride.

TRIFLUOROMETHYL HEXACHLOROCYCLOHEXANE. $H_2C_6Cl_6F_3CH_3$, mw: 342.
THR = An insecticide. Specific toxicologic data not available. Toxicity may be similar to that of DDT and fluorides. See fluorides and chlorides.

TRIFLUOROMETHYL HYPOFLUORITE. A gas at normal temp. Syn: *fluorooxy trifluoromethane*. CF_3OF, mw: 104. An extremely toxic gas; see fluorine and fluorides.

TRIFLUOROMETHYL PHOSPHINE. F_3CPH_2, mw: 102.
THR = No data. Probably HIGH. Spont flam in air. [*19*]

TRIFLUOROMONOBROMOMETHANE. See bromotrifluoromethane.

TRIFLUORONITROANILINE. $C_6H_2O_3N_3F_3$, mw: 221.1.
THR = Has detonated on impact. [*19*]

TRIFLUORONITROSOMETHANE. Bright blue gas. CF_3NO, mw: 99.02, mp: −150°, bp: −80°.
THR = HIGH irr to skin, eyes and mu mem and via inhal route. A military poison.
Disaster Hazard: Dangerous; see fluorides and NO_x.

2,2,2-TRIFLUORO-N-(9-OXOFLUOREN-2-YL)-ACETAMIDE. $C_{15}H_{18}O_2NF_3$, mw: 291.2.
THR = An exper carc. [*3*]

TRIFLUOROPENTACHLOROPROPANE. Liquid. $C_3F_3Cl_5$, mw: 270.3, bp: 155°, flash p: 228°F.
THR = See chlorinated hydrocarbons, aliphatic.
Fire Hazard: When exposed to heat, flame or oxidizers.
Disaster Hazard: Dangerous; see fluorides and chlorides; can react vigorously with oxidizing materials.
To Fight Fire: Foam, CO_2, dry chemical.

1,1,2-TRIFLUORO-1,2,2-TRICHLOROETHANE. Syn: "*Freon 113*." Colorless gas. CCl_3CF_3, mw: 187.39, mp: 13.2°, bp: 45.8°, d: 1.5702, autoign. temp.: 1256°F.
Acute tox data: Oral LD_{LO} (rat) = 45 mg/kg; inhal LC_{LO} (rat) = 87000 ppm for 6 hrs; inhal of 4500 ppm by humans ⟶ CNS effects. [*3*]
THR = LOW via inhal route. An exper carc. [*23*]
Fire Hazard: Very slight; when exposed to heat or flame. Violent reaction with Al, Ba, Li, Sm, NaK alloy, Ti. [*19*]
Disaster Hazard: Dangerous; see chlorides and fluorides.

TRIFLUOROVINYL BROMIDE. A dense gas. $CF_2{:}CFBr$, mw: 161, bp: −2°.
THR = U.
Disaster Hazard: Dangerous; see fluorides and bromides.

TRIFLUPERDOL. $C_{22}H_{23}O_2NF_4$, mw: 409.5.
THR = HIGH acute oral toxicity. An exper teratogen. [*3*]

TRIFLURALIN. See α,α,α-trifluoro-2,6-dinitro-n,n-dipropyl-p-toluidine.

TRIGERMANE. Syn: *germanium hydride*. Colorless liquid. Ge_3H_8, mw: 225.86, mp: −105.6°, bp: 110.5°, d: 2.2.
THR = See hydrides and germanium compounds.

TRIGLYCIDYL PHOSPHATE. Syn: *TGP*.
THR = Probably HIGH. Details U. Animal exper show bone marrow depression. See also diglycidyl ether.

TRIGLYCOL DICHLORIDE. Syn: *triethylene glycol dichloride*. Colorless liquid. $Cl(C_2H_4O)_2C_2H_4Cl$, mw:

187.09, bp: 240°, fp: −31.5°, flash p: 250°F (OC), d: 1.197, vap. press: 0.03 mm @ 20°.
Acute tox data: Oral LD_{50} (rat) = 250 mg/kg; dermal LD_{50} (rabbit) = 1410 mg/kg; oral LD_{50} (guinea pig) = 120 mg/kg. [*3*]
THR = HIGH via oral route and MOD via dermal route.
Fire Hazard: Slight, when exposed to heat or flame.
Disaster Hazard: Dangerous; see chlorides; can react with oxidizing materials.
To Fight Fire: CO_2, dry chemical.

TRI-*n*-HEXYL BORATE. Colorless liquid, odor of *n*-hexanol. $C_{18}H_{39}O_3B$, mw: 314.4, bp: 140°-146° @ 2 mm, flash p: 300°F, d: 0.847 @ 28°, vap. d: 10.8.
Acute tox data: Oral LD_{50} (mice) = 1800 mg/kg. [*3*]
THR = MOD via oral route. See esters and boron compounds.
Fire Hazard: Slight, when exposed to heat or flame; can react with oxidizing materials.
To Fight Fire: Foam, CO_2, dry chemical.

TRIHEXYLENE GLYCOL BIBORATE. Colorless liquid, odor of hexylene glycol. $C_{18}H_{36}O_6B_2$, mw: 370.1, bp: 143°-149° @ 2 mm, flash p: 345°F, d: 0.982 @ 21°, vap. d: 12.8.
Acute tox data: ip LD_{LO} (mice) = 750 mg/kg. [*3*]
THR = MOD via ip route. See boron compounds and glycols.
Fire Hazard: Slight, when exposed to heat or flame; can react with oxidizing materials.
To Fight Fire: Foam, CO_2, dry chemical.

TRIHEXYL PHOSPHITE. Mobile colorless liquid, characteristic odor, decomp in water. $(C_6H_{13})_3PO_3$, mw: 334, d: 0.897, bp: 135°-141° @ 0.2 mm, flash p: 320°F (OC).
THR = See esters.
Fire Hazard: Slight, when exposed to heat or flame.
Disaster Hazard: Dangerous; see phosphorus compounds.
To Fight Fire: Foam, alcohol foam, mist.

TRIHYDROXYBENZENE. See pyrogallol.

3,4,5-TRIHYDROXYBENZOIC ACID. See gallic acid.

2,4,5-TRIHYDROXY BUTYROPHENONE. Syn: *THBP.* Yellow-tan crystals, very slightly sol in water, sol in alcohol and propylene glycol.
$C_6H_2(OH)_3COC_3H_7$, mw: 196.2, mp: 149°-153°, d: 6.0 lb/gal @ 20°.
Acute tox data: ip LD_{50} (mice) = 200 mg/kg. [*3*]
THR = HIGH via ip route. A food additive permitted in food for human consumption. [*109*]

TRIIODOMETHANE. See iodoform.

TRIIRON PHOSPHIDE. Gray crystals. Fe_3P, mw: 198.53, mp: 1100°, d: 6.74.
THR = See phosphides.

TRIISOAMYL BORATE. Liquid. $(C_5H_{11}O)_3B$, mw: 272.2, d: 0.872 @ 0°, bp: 255°.
THR = Details U. See also boron compounds and esters.

TRIISOAMYL BORON. Colorless mobile liquid, not sol in water, sol in ether. $(C_5H_{11})_3B$, mw: 224.2, d: 0.72, bp: 119° @ 14 mm.
THR = Details U. See also boron compounds and esters.

TRIISOAMYLTIN BROMIDE. Solid. $(C_5H_{11})_3SnBr$, mw: 412.0, d: 1.2613, mp: 21°, bp: 177° @ 15 mm.
THR = Toxic. See tin compounds and bromides.

TRIISOAMYLTIN CHLORIDE. Liquid. $(C_5H_{11})_3SnCl$, mw: 367.49.
THR = Toxic. See tin compounds and chlorides.
To Fight Fire: Foam, CO_2, dry chemical.

TRIISOAMYLTIN FLUORIDE. Crystals. $(C_5H_{11})_3SnF$, mw: 351.1, mp: 288°
THR = Toxic. See fluorides and tin compounds.

TRIISOAMYLTIN IODIDE. Liquid. $(C_5H_{11})_3SnI$, mw: 459.0, mp: −22°, d: 1.3777 @ 26.5°, bp: 182° @ 13 mm.
THR = Toxic. See tin compounds and iodides.

TRIISOBUTYL ALUMINUM. Clear colorless liquid. Ignites on exposure to air. $(C_4H_9)_3Al$, 198.3, d: 0.7859 @ 20°, vap. press: 1 mm @ 47°, flash p: < 4°, fp: 4.3°, bp: decomp.
THR = HIGH. Extremely destructive to living tissue.
Fire Hazard: Dangerous; pyrophoric in air. Reacts violently with moisture, acids, air, alcohols, amines and halogens. [*19*]
Disaster Hazard: Dangerous; in presence of air or moisture, reacts violently.
To Fight Fire: CO_2, dry sand, dry chemical. DO NOT use water, foam or halogenated extinguishing agents.

TRIISOBUTYL ALUMINUM CHLORIDE.
THR = HIGH irr to skin, eyes and mu mem. Action similar to but less intense than that of di-isobutyl aluminum chloride.
Disaster Hazard: Dangerous; see chlorides.

TRIISOBUTYL BORATE. Syn: *triisobutoxy borine.* Colorless liquid; odor of isobutyl alcohol. $[CH_3CH(CH_3)CH_2O]_3B$, mw: 230.16, bp: 212°, flash p: 185°F (COC), d: 0.843 @ 23°.
Acute tox data: Oral LD_{50} (mice) = 2020 mg/kg. [*3*]
THR = MOD via oral route.
Fire Hazard: Mod, via heat, flame or oxidizers.
To Fight Fire: Water spray, foam, dry chemical.

TRIISOBUTYL BORON. Colorless mobile liquid, insol in water. $(C_4H_9)_3B$, mw: 182.2, d: 0.74; bp: 188°. THR = Details U. See also boron compounds.

TRIISOBUTYLENE OXIDE. $(C_4H_8)_3O$, mw: 184.
Acute tox data: Oral LD_{50} (rat) = 6690 mg/kg; dermal LD_{50} (rabbit) = 14000 mg/kg. [3]
THR = MOD via oral and LOW via dermal route.

TRIISOBUTYL ETHYLTIN. Liquid. $(C_4H_9)_3Sn(C_2H_5)$, mw: 319.1, d: 1.0779 @ 21°, bp: 125° @ 16 mm.
THR = Details U. See tin compounds.

TRIISOBUTYL ISOAMYLTIN. Liquid. $(C_4H_9)_3Sn(C_5H_{11})$, mw: 361.2, d: 1.0356 @ 27°, bp: 152.9° @ 16.5 mm.
THR = Toxic. Details U. See tin compounds.

TRIISOBUTYLTIN BROMIDE. Liquid. $(C_4H_9)_3SnBr$, mw: 370.0, d: 1.3253, mp: −26.5°, bp: 148° @ 13 mm.
THR = Toxic. See tin compounds and bromides.

TRIISOBUTYLTIN CHLORIDE. Solid. $(C_4H_9)_3SnCl$, mw: 325.5, d: 1.1290 @ 34°, mp: 30.2°, bp: 174° @ 13 mm.
THR = Toxic. See tin compounds and chlorides.

TRIISOBUTYLTIN FLUORIDE. Crystals, slightly sol in organic solvents. $(C_4H_9)_3SnF$, mw: 309.0, mp: 244°.
THR = Toxic. See fluorides and tin compounds.

TRIISOBUTYLTIN IODIDE. Colorless liquid, insol in organic solvents. $(C_4H_9)_3SnI$, mw: 417.0, d: 1.378 @ 26.5°, mp: −22°, bp: 286°.
THR = Toxic. See tin compounds and iodides.

TRIISOOCTYL AMINE. $C_{24}H_{51}N$, mw: 353.8.
Acute tox data: Oral LD_{50} (rat) = 1620 mg/kg; dermal LD_{50} (rabbit) = 3180 mg/kg. [3]
THR = MOD via oral and dermal routes. See also amines.

TRIISOOCTYL PHOSPHINE. $(C_8H_{17})_3P$, mw: 370.2.
Acute tox data: Oral LD_{50} (rat) = 21000 mg/kg; dermal LD_{50} (rabbit) = 3970 mg/kg. [3]
THR = MOD via dermal and NONE via oral routes.

***o,o,o*-TRIISOOCTYL PHOSPHOROTHIOATE.** Colorless liquid, mild characteristic odor. $(C_8H_{17}O)_3PS$, mw: 450.7, bp: 160°–170° @ 0.2 mm, flash p: 410° F (COC), d: 0.933 @ 20°/4°.
THR = HIGH. A cholinesterase inhibitor. See parathion.
Fire Hazard: Slight, when exposed to heat or flame.
Disaster Hazard: Dangerous; see phosphates and sulfur compounds; can react vigorously with oxidizing materials.
To Fight Fire: Spray, alcohol foam, mist, dry chemical.

TRIISOPROPANOLAMINE. Pure crystalline white solid. $N(C_3H_6OH)_3$, mw: 191.27, mp: 45°, bp: 305°, flash p: 320° F (OC), d: 1.0200 @ 20°/20°, vap. press: < 0.01 mm @ 20°.
Acute tox data: Oral LD_{50} (guinea pig) = 1080 mg/kg; dermal LD_{LO} (rabbit) = 10000 mg/kg. [3]
THR = MOD via oral and LOW via dermal route. See also amines.
Fire Hazard: Slight, when exposed to heat or flame.
Spont Heating: No.
Disaster Hazard: Mod dangerous; when heated to decomp, emits toxic fumes.
To Fight Fire: Alcohol foam, water, CO_2, dry chemical.

TRIISOPROPYL BENZENE. Clear, colorless liquid. $C_6H_3[CH(CH_3)_2]_3$, mw: 204.3, mp: −15°, bp: 236°–237°, flash p: 205° F, d: 0.854 @ 25°/25°, vap. d: 7.0.
THR = Details U. See cumene.
Fire Hazard: Mod, when exposed to heat, flame or oxidizers.
Disaster Hazard: Mod dangerous; when heated, emits toxic fumes; can react with oxidizing materials.
To Fight Fire: Foam, CO_2, dry chemical.

TRIISOPROPYL BORATE. Colorless liquid. $(C_9H_{21}O_3)B$, mw: 188.08, mp: −59°, bp: 141.0°–142.4°, flash p: 82° F (TCC), d: 0.8138 @ 25°.
Acute tox data: Oral LD_{50} (mice) = 2500 mg/kg. [3]
THR = MOD via oral route. See esters and boron compounds.
Fire Hazard: Dangerous, when exposed to heat, flame or oxidizers.
Explosion Hazard: U.
Disaster Hazard: Dangerous; keep away from flame and heat; can react vigorously with oxidizing materials.
To Fight Fire: Foam, CO_2, dry chemical.

TRIISOPROPYLTIN ACETATE. $C_{11}H_{24}O_2Sn$, mw: 307.0.
Acute tox data: Oral LD_{50} (rat) = 44 mg/kg; iv LD_{50} (rat) = 12 mg/kg. [3]
THR = HIGH via oral and iv routes.

TRIISOPROPYLTIN BROMIDE. Liquid, sol in organic solvents. $(C_3H_7)_3SnBr$, mw: 327.9, d: 1.4263 @ 25°, mp: −49°, bp: 133° @ 12 mm.
THR = Toxic. See tin compounds and bromides.

TRIISOPROPYLTIN IODIDE. Liquid. $(C_3H_7)_3SnI$, mw: 374.9, d: 1.4378 @ 22°, bp: 151° @ 13 mm.
THR = Toxic. See tin compounds and iodides.

TRILAURYL TRITHIOPHOSPHITE. Pale yellow liquid. $[CH_3(CH_2)_{11}S]_3P$, mw: 634, d: 0.915 @ 25°/15°, mp: 20°, flash p: 398° F (OC).
THR = Details U. See esters.
Fire Hazard: Slight, when exposed to heat or flame.

Disaster Hazard: Dangerous; see phosphorus and sulfur compounds.
To Fight Fire: CO_2, dry chemical.

TRIMAGNESIUM PHOSPHATE. See magnesium phosphate, tribasic.

TRIMANGANESE DIPHOSPHIDE. Dark gray crystals. Mn_3P_2, mw: 226.75, mp: 1095°, d: 5.12 @ 18°.
THR = See phosphides and manganese compounds.

TRIMANGANESE TETROXIDE. Mn_3O_4, mw: 228.8.
THR = See also manganese compounds. Violent reaction @ < 100°. [*19*]

TRIMEDONE. See trimethadione.

TRIMELLITIC ANHYDRIDE. Syn: *TMA;* crystals, $C_9H_4O_5$, mw: 192.1, mp: 162°, bp: 240°–245° @ 14 mm, sol in acetone, ethyl acetate, dimethyl formamide.
Acute tox data: Oral LD_{50} (mice) = 2210 mg/kg. [*3*]
THR = To be regarded as HIGH tox. Has caused pulmonary edema from inhal. Irr to lungs and air passages. May be a powerful allergen. Typical attack consists of breathlessness, wheezing, cough, running nose, immunological sensitization and asthma symptoms. [*116*]

TRIMETHADIONE. Syn: *trimedone.* $C_6H_9O_3N$, mw: 143.2.
THR = Mod via oral route to rats. [*3*] Has been implicated in aplastic anemia.

TRIMETHOXY BORINE. See methyl borate.

3,4,5-TRIMETHOXY CINNAMALDEHYDE. $C_{12}H_{14}O_4$, mw: 222.3.
THR = An exper carc. [*3, 23*]

2′,4′,6′-TRIMETHYL ACETANILIDE. $C_{11}H_{15}ON$, mw: 177.3.
THR = An exper neo. [*3*]

TRIMETHYL ACETIC ACID. Syn: *pivalic acid.* Crystals. $(CH_3)_3CCOOH$, mw: 102.1, mp: 35.5°, bp: 164°, d: 0.91.
Acute tox data: Oral LD_{LO} (rat) = 5000 mg/kg. [*3*]
THR = MOD via oral route.

TRIMETHYL ACETYL CHLORIDE. C_5H_9OCl, mw: 120.6.
THR = No data. Probably at least mildly irr.

TRIMETHYL ADIPIC ACID. Powder. $C_9H_{16}O_4$, mw: 188.2.
THR = Probably LOW. An irr.
Disaster Hazard: Slightly dangerous; when heated, emits acrid fumes.

TRIMETHYL ALUMINUM. See aluminum methyl.

TRIMETHYL ALUMINUM DIETHYL ETHERATE. $(CH_3)_3Al(C_2H_5)_2O$, mw: 146.1.
THR = Spont flam in air. [*19*]

TRIMETHYL ALUMINUM DIMETHYL ETHERATE. $(CH_3)_3Al \cdot (CH_3)_2O$, mw: 118.1.
THR = Spont flam in air. [*19*]

TRIMETHYLAMINE. Colorless gas. $(CH_3)_3N$, mw: 59.11, bp: 2.87°, lel = 2%, uel = 11.6%, fp: −117.1°, d: 0.662 @ −5°, autoign. temp.: 374°F, vap d: 2.0, flash p: 20°F (CC).
Acute tox data: ip LD_{LO} (mice) = 75 mg/kg; iv LD_{50} (mice) = 90 mg/kg; sc LD_{LO} (mice) = 1000 mg/kg; sc LD_{LO} (rabbit) = 800 mg/kg. [*3*]
THR = HIGH via ip and iv routes; MOD via sc route.
Fire Hazard: Very dangerous, when exposed to flame.
Explosion Hazard: Mod, when exposed to spark or flame.
Disaster Hazard: Mod dangerous; when heated, emits toxic fumes; can react with oxidizing materials.
To Fight Fire: Stop flow of gas.

TRIMETHYL AMINE OXIDE PERCHLORATE. $(CH_3)_3NO \cdot HClO_4$, mw: 175.5.
A heat- and shock-sensitive explosive. [*19*]

TRIMETHYL AMINOBORANE. Colorless crystals, decomp by water, sol in ether. $(CH_3)_3NBH_3$, mw: 73.0, mp: 94°, bp: 172°.
Acute tox data: ip LD_{LO} (rat) = 970 mg/kg. [*3*]
THR = MOD via ip route. See also boron compounds.

TRIMETHYL AMINOMETHANE. See *tert*-butylamine.

TRI(METHYLAMYL)BORATE. Syn: *tri(methyl isobutylcarbinyl)borate.* Colorless mobile liquid, characteristic odor. $[(CH_3)_2CHCH_2CHOCH_3]_3B$, mw: 314.3, bp: 257°, flash p: 220°F (COC), d: 0.819 @ 29°, vap. d: 10.8.
THR = See boric acid and esters.
Fire Hazard: Slight, when exposed to heat or flame.
Disaster Hazard: Mod dangerous; when heated to decomp, emits toxic fumes; on contact with oxidizing materials, can react vigorously.
To Fight Fire: Foam, CO_2, dry chemical.

2,4,6-TRIMETHYL ANILINE. See mesidine.

2,4,5-TRIMETHYL ANILINE HYDROCHLORIDE. $C_9H_{13}N \cdot HCl$, mw: 171.7.
Acute tox data: Oral LD_{50} (rat) = 1585 mg/kg; ip LD_{50} (mice) = 340 mg/kg. [*3*]
THR = HIGH via ip and MOD via oral routes. An exper carc. [*23*]

TRIMETHYL ANTIMONY SULFATE. $(CH_3)_3SbSO_4$, mw: 262.9.
THR = See also antimony compounds. Spont flam in air. [*19*]

TRIMETHYL ARSENIC. Syn: *arsenic trimethyl.* Col-

orless liquid. $As(CH_3)_3$, mw: 120.0, bp: 70°, d: 1.124, vap. d: 4.14.
THR = A poison. See arsenic compounds.
Fire Hazard: Spont flam in air. [*19*]
Disaster Hazard: Dangerous; when heated to decomp, or on contact with acid fumes, emits highly toxic fumes of arsenic; can react vigorously with oxidizing materials.

9,10,12-TRIMETHYL BENZ(a)ACRIDINE. $C_{20}H_{17}N$, mw: 271.4.
THR = An exper neo. [*3, 23*]

3,8,12-TRIMETHYL BENZ(a)ACRIDINE.
THR = An exper neo. [*3, 23*]

5,7,11-TRIMETHYL BENZ(c)ACRIDINE.
THR = An exper neo. [*3, 23*]

7,8,11-TRIMETHYL BENZ(c)ACRIDINE.
THR = An exper neo. [*3, 23*]

7,9,10-TRIMETHYL BENZ(c)ACRIDINE.
THR = An exper neo. [*3, 23*]

7,9,11-TRIMETHYL BENZ(c)ACRIDINE.
THR = An exper neo. [*3, 23*]

8,10,12-TRIMETHYL BENZ(c)ACRIDINE.
THR = An exper neo. [*3, 23*]

4,5,10-TRIMETHYL BENZ(a)ANTHRACENE. $C_{21}H_{18}$, mw: 270.4.
THR = An exper carc. [*3, 23*]

4,7,12-TRIMETHYL BENZ(a)ANTHRACENE.
THR = An exper neo. [*3*]

6,7,8-TRIMETHYL BENZ(a)ANTHRACENE.
THR = An exper carc. [*3*]

6,7,12-TRIMETHYL BENZ(a)ANTHRACENE.
THR = An exper carc. [*3*]

6,8,12-TRIMETHYL BENZ(a)ANTHRACENE.
THR = An exper carc. [*3, 23*]

7,8,12-TRIMETHYL BENZ(a)ANTHRACENE.
THR = An exper neo and carc. [*3*]

7,9,12-TRIMETHYL BENZ(a)ANTHRACENE.
THR = An exper neo and carc. [*3*]

7,10,12-TRIMETHYL BENZ(a)ANTHRACENE.
THR = An exper carc. [*3*]

1,2,3-TRIMETHYL BENZENE. C_9H_{12}, mw: 120.2, autoign. temp.: 878° F, d: 0.89, vap. d: 4.15, bp: 176.1°.
Acute tox data: Oral LD_{LO} (rat) = 5000 mg/kg. [*3*]
THR = MOD via oral route.
Fire Hazard: Mod, via heat, flame and oxidizers.
To Fight Fire: Water spray, mist, dry chemical, CO_2, foam.

1,2,4-TRIMETHYL BENZENE. See pseudocumene.

1,3,5-TRIMETHYL BENZENE. See mesitylene.

1,3,6-TRIMETHYL BENZO(a)PYRENE. $C_{23}H_{18}$, mw: 294.4.
THR = An exper neo. [*3, 23*]

2,6,6-TRIMETHYL BICYCLO-(3,1,1)-2-HEPTENE. See pinene.

3,2′,5′-TRIMETHYL-4-BIPHENYL AMINE.
THR = An exper carc. [*23*]

TRIMETHYL BISMUTH. Syn: *bismuth trimethyl.* Liquid. $Bi(CH_3)_3$, mw: 254.10, bp: 110°, d: 2.300 @ 18°.
Acute tox data: sc LD_{50} (rabbit) = 182 mg/kg; oral LD_{50} (rabbit) = 484 mg/kg; iv LD_{50} (dog) = 12 mg/kg; dermal LD_{50} (dog) = 233 mg/kg; inhal LD_{LO} (dog) = 233 mg/kg. [*3*]
THR = HIGH via sc, oral, iv and dermal routes. Can cause narcosis and CNS depression. Prolonged exposure can cause encephalopathy similar to that of organic lead compounds. See also bismuth compounds.
Fire Hazard: Mod, when exposed to heat or flame.
Explosion Hazard: Spont flam in air. [*19*]
Disaster Hazard: Dangerous; when heated to decomp, emits toxic fumes of bismuth and reacts with oxidizing materials.

TRIMETHYL BORATE. See methyl borate.

TRIMETHYL BORINE. See boron trimethyl.

2,2,3-TRIMETHYL BUTANE. C_7H_{16}, mw: 100.2, flash p: <32° F, autoign. temp.: 842° F, d: 0.69, vap. d: 3.46, bp: 81.1.
THR = No data. See heptane.
Fire Hazard: Dangerous, via heat, flame or oxidizers.
To Fight Fire: Water spray, fog, foam, CO_2.

2,3,3-TRIMETHYL-1-BUTENE. C_7H_{14}, mw: 98.1, bp: 72.2°, flash p: <32° F, autoign. temp.: 707° F, d: 0.71, vap. d: 3.39.
THR = No data; probably LOW.
Fire Hazard: Dangerous, via heat, flame or oxidizers.
To Fight Fire: Water spray, mist, foam, CO_2.

TRIMETHYL CHLOROSILANE. Colorless liquid; sol in benzene, ether and perchloroethylene. $(CH_3)_3SiCl$, mw: 108.8, bp: 57°, d: 0.854 @ 25°/25°, flash p: −18° F.
Acute tox data: Inhal LC_{LO} (mice) = 500 mg/m^3 for 1/6 hr; ip LD_{LO} (mice) = 750 mg/kg. [*3*]
THR = MOD via inhal and ip routes. An exper neo via ip route. [*3*] See silanes.
Fire Hazard: Dangerous, when exposed to heat or flame. Violent reaction with water. [*19*]
Disaster Hazard: Dangerous; see chlorides.
To Fight Fire: Foam, alcohol foam and fog.

3,3,5-TRIMETHYL CYCLOHEXANOL. Syn: *trimethyl cyclohexanol.* Liquid. $C_9H_{18}O$, mw: 142.23,

fp: 37.0°, bp: 198°, flash p: 190°F (OC), d: 0.878 @ 40°/20°, vap. press: 0.1 mm @ 20°, vap. d: 4.91.
Acute tox data: Oral LD_{50} (rat) = 3250 mg/kg; dermal LD_{50} (rabbit) = 2800 mg/kg. [*3*]
THR = MOD via oral and dermal routes.
Fire Hazard: Mod, when exposed to heat, flame or oxidizers.
Disaster Hazard: Mod dangerous; when heated, emits toxic fumes; can react with oxidizing materials.
To Fight Fire: Alcohol, foam, CO_2, dry chemical.

TRIMETHYL CYCLOHEXANOL. Liquid. $C_9H_{18}O$, mw: 142.23, fp: 37.0°, bp: 198°, flash p: 165°F (OC), vap. d: 4.91, d: 0.878 @ 40°/20°, vap. press: 0.1 mm @ 20°.
Fire Hazard: See 3,3,5-trimethyl cyclohexanol.
To Fight Fire: See 3,3,5-trimethyl cyclohexanol.

TRIMETHYL CYCLOHEXANONE. Liquid. $C_9H_{16}O$, mw: 140.2.
THR = Probably LOW. A MILD irr.
Fire Hazard: Mod, when exposed to heat or flame; can react with oxidizing materials.

11,12,17-TRIMETHYL-15H-CYCLOPENTA(a) PHENANTHRENE.
THR = An exper carc. [*23*]

TRIMETHYL DIALUMINUM HYDRIDE. $(CH_3)_3Al_2H_3$, mw: 102.
Explosion Hazard: Reacts violently with water, which can lead to an explosion. [*19*]

1,1,2-TRIMETHYL DIBORANE. Colorless liquid, decomp by water. $B_2H_3(CH_3)_3$, mw: 69.8, mp: −123°, bp: 45.5°.
THR = See boron hydride and boron compounds. Spont flam in air. [*19*]

TRIMETHYL DIHYDROQUINOLINE POLYMER. Amber pellets. $(C_{12}H_{15}N)_3$, mw: 519.93 (approx), mp: softens @ 75°, d: 1.08.
Acute tox data: Oral LD_{50} (rat) = 2000 mg/kg; oral LD_{50} (mice) = 1450 mg/kg. [*3*]
THR = MOD via oral route. Animal exper suggest mod to low acute and chronic toxicity.
Disaster Hazard: Dangerous; when heated to decomp, emits highly toxic fumes of cyanides.

TRIMETHYLENE. See cyclopropane.

TRIMETHYLENE BROMIDE. Colorless liquid. $C_3H_6Br_2$, mw: 201.9, bp: 166.5°, fp: −33°, d: 1.977 @ 25°/25°, vap. d: 7.0.
Acute tox data: ip LD_{LO} (mice) = 750 mg/kg; rec LD_{LO} (rabbit) = 4250 mg/kg. [*3*]
THR = MOD via ip and rec routes. Also irr and narcotic in high conc.
Disaster Hazard: Dangerous; see bromides.

TRIMETHYLENE CHLORIDE. See 1,3-dichloropropane.

TRIMETHYLENE CHLOROBROMIDE. Colorless liquid. C_3H_6ClBr, mw: 157.5, bp: 143°–145°, flash p: none, d: 1.594 @ 25°/25°, vap. d: 5.5.
THR = U. Probably irr and narcotic in high conc.
Disaster Hazard: Dangerous; see chlorides and bromides.

TRIMETHYLENE DIMETHANE SULFONATE. $C_5H_{12}O_6S_2$, mw: 232.3.
THR = An exper teratogen. [*3*]

TRIMETHYLENE GLYCOL. Syns: *1,3-propanediol*, *1,3-propylene glycol*. Colorless, odorless liquid, sol in water, alcohol and ether. $CH_2OHCH_2CH_2OH$, mw: 76, d: 1.0536 @ 25°, bp: 210°–211°, vap. d: 2.6, autoign. temp.: 752°F.
THR = See glycols.
To Fight Fire: Alcohol foam.

TRIMETHYLENE GLYCOL DINITRATE.
THR = Details U. See ethylene glycol dinitrate.
Fire Hazard: See nitrates.
Explosion Hazard: See nitrates.
Explosive Range: Explodes at 225°F.
Disaster Hazard: Dangerous; see nitrates.

TRIMETHYLENE OXIDE. Syns: *1,3-epoxypropane*, *oxetane*. Oil, agreeable odor. C_3H_6O, mw: 58.1, d: 0.8930 @ 25°/4°, bp: 48° @ 750 mm.
Acute tox data: sc LD_{50} (rat) = 500 mg/kg. [*3*]
THR = HIGH via sc route. An exper carc via sc route. [*3, 23*] May be narcotic in high conc.

TRIMETHYLENE TRINITRAMINE. See cyclotrimethylene trinitramine.

TRIMETHYLETHANE. See liquified petroleum gas.

TRIMETHYLETHYLENE. See α,*n*-amylene.

TRIMETHYLETHYLTIN. Colorless liquid, insol in water, sol in organic solvents. $(CH_3)_3Sn(C_2H_5)$, mw: 192.9, bp: 108.2°.
THR = Details U. See tin compounds.

TRIMETHYL-*o*-FORMATE. Colorless liquid, pungent odor. $HC(OCH_3)_3$, mw: 106.12, vap. d: 3.67.
Acute tox data: Oral LD_{50} (rat) = 3130 mg/kg; inhal LC_{LO} (rat) = 5000 ppm for 4 hrs. [*3*]
THR = MOD via oral and inhal routes. An irr to skin, eyes and mu mem. See esters.
Fire Hazard: Slight, when exposed to heat or flame; can react with oxidizing materials.

TRIMETHYL GALLIUM. Colorless liquid, decomp by water. $(CH_3)_3Ga$, mw: 114.8, mp: −19°, bp: 55.7°.
THR = Details U. See gallium compounds.
Fire Hazard: Spont flam in air. [*19*]

TRIMETHYL GALLIUM MONAMMINE. White crystals, decomp by water. $Ga(CH_3)_3 \cdot NH_3$, mw: 131.9, mp: 31°.
THR = Details U. See gallium compounds.

TRIMETHYL GALLIUM MONOETHERATE. Colorless liquid, decomp by water. $(CH_3)_3GaO(C_2H_5)_2$, mw: 188.9, mp: $<-76°$, bp: 99°.
THR = Details U. See gallium compounds and ethyl ether. This material easily evolves ether. See ether.

TRIMETHYL GERMANIUM BROMIDE. Oily liquid, decomp by water. $(CH_3)_3GeBr$, mw: 197.6, d: 1.544 @ 18°/40°, mp: −25°, bp: 113.7°.
THR = See germanium compounds and bromides.

2,5,5-TRIMETHYL HEPTANE. $C_{10}H_{22}$, mw: 142.2, flash p: 131°F, autoign. temp.: 860°F, d: 0.73, vap d: 4.91, bp: 151.1°.
THR = No data. Probably LOW.
Fire Hazard: Mod, via heat, flame or oxidizers.
To Fight Fire: Water spray, fog, CO_2, foam, dry chemical.

2,2,5-TRIMETHYL HEXANE. A clear liquid. C_9H_{20}, mw: 128.25, bp: 125°, fp: −106°, flash p: 55°F (OC), d: 0.707 @ 20°/4°, vap. press: 12.9 mm @ 21°, vap. d: 4.7.
THR = U.
Fire Hazard: Dangerous, when exposed to heat, flame or oxidizers.
Explosion Hazard: U.
Disaster Hazard: Very dangerous; keep away from heat or open flame; can react vigorously with oxidizing materials.
To Fight Fire: Foam, CO_2, dry chemical.

2,3,3-TRIMETHYL HEXANE. Colorless liquid. C_9H_{20}, mw: 128.25, bp: 137.7°, fp: −116.8°, flash p: 79°F, d: 0.734 @ 25°/4°, vap. press: 10.1 mm @ 25°, vap. d: 4.43.
THR = U.
Fire Hazard: Dangerous, when exposed to heat, flame or oxidizers.
Explosion Hazard: U.
Disaster Hazard: Dangerous; keep away from heat or open flame; can react with oxidizing materials.
To Fight Fire: Foam, CO_2, dry chemical.

2,3,4-TRIMETHYL HEXANE. Colorless liquid. C_9H_{20}, mw: 128.25, bp: 139°, flash p: 80.8°F, d: 0.737 @ 25°/4°, vap. press: 9.1 mm @ 25°, vap. d: 4.43.
THR = U. Probably irr and narcotic in high conc.
Fire Hazard: Dangerous, when exposed to heat, flame or oxidizers.
Explosion Hazard: U.
Disaster Hazard: Dangerous; keep away from heat or open flame; can react vigorously with oxidizing materials.

3,3,4-TRIMETHYLHEXANE. Colorless liquid. C_9H_{20}, mw: 128.25, bp: 140.5°, fp: −101.2°, flash p: 79°F, d: 0.741 @ 25°/4°, vap. press: 8.6 mm @ 25°, vap. d: 4.43.
THR = U. Probably irr and narcotic in high conc.
Fire Hazard: Dangerous, when exposed to heat, flame or oxidizers.
Explosion Hazard: U.
Disaster Hazard: Dangerous; keep away from heat or open flame; can react with oxidizing materials.
To Fight Fire: Foam, CO_2, dry chemical.

3,5,5-TRIMETHYL HEXANOL. Colorless liquid. $C_9H_{20}O$, mw: 144.25, mp: −70°, bp: 195°, flash p: 200°F (OC), d: 0.824 @ 25°/4°, vap. d: 5.0.
THR = See alcohols.
Fire Hazard: Mod, when exposed to heat or flame; can react with oxidizing materials.
To Fight Fire: Foam, CO_2, dry chemical.

1,5-TRIMETHYL-4-HEXENYLAMINE. See isometheptene.

TRIMETHYL HYDRAZINE HYDROCHLORIDE. $C_3H_{10}N_2 \cdot HCl$, mw: 110.6.
THR = An exper oral carc to mice. [*103*]

TRI(METHYLISOBUTYL CARBONYL)BORATE. See tri(methylamyl)borate.

TRIMETHYL METHANE. See isobutane.

1,1,3-TRIMETHYL-3-NITROSOUREA. $C_4H_9O_2N_3$, mw: 131.2.
THR = An exper carc. [*3, 23*] High via oral route.

2,4,8-TRIMETHYL-6-NONANOL. Flash p: 199°F, bp: 255°.
THR = See 2,6,8-trimethyl-4-nonanol.
Fire Hazard: See 2,6,8-trimethyl-4-nonanol.
To Fight Fire: See 2,6,8-trimethyl-4-nonanol.

2,6,8-TRIMETHYL-4-NONANOL. Liquid. $C_{12}H_{26}O$, mw: 186.33, bp: 225.2, fp: −60°, flash p: 200°F (OC), vap. press: <0.01 mm @ 20°, d: 0.8193 @ 20°/20°, vap. d: 6.43.
Acute tox data: LD_{50} (rat) = 17,000 mg/kg; dermal LD_{50} (rabbit) = 11,000 mg/kg. [*3*]
THR = LOW via dermal and oral routes.
Fire Hazard: Mod, when exposed to heat or flame; can react with oxidizing materials.
To Fight Fire: Foam, CO_2, dry chemical.

2,6,8-TRIMETHYL-4-NONANE. $C_{12}H_{26}$, mw: 170.4.
THR = No data.
Fire Hazard: Mod, via heat, flame or oxidizers.
To Fight Fire: Water spray, fog, foam, CO_2, dry chemical.

2,6,8-TRIMETHYL NONANONE-4. Liquid. $C_{12}H_{24}O$, mw: 184.4, mp: −75°, bp: 211°–219°, flash p: 196°F (OC), d: 0.8165 @ 20°/20°, vap. d: 6.37.
Acute tox data: Oral LD_{50} (rat) = 8470 mg/kg; dermal LD_{50} (rabbit) = 11,000 mg/kg. [*3*]
THR = LOW via oral and dermal routes. See ketones.
Fire Hazard: Mod, when exposed to heat or flame; can react with oxidizing materials.
To Fight Fire: Foam, CO_2, dry chemical.

TRIMETHYLOL ETHANE. Syn: *2,2-dimethylol propanol-1.* White, odorless, crystalline powder. $CH_3C(CH_2OH)_3$, mw: 120.15, mp: about 200°.
THR = U. See alcohols.
Fire Hazard: Slight, when exposed to heat.

2,2,3-TRIMETHYL PENTANE. bp: 110°, flash p: <70°F, autoign. temp.: 806°F, d: 0.72, vap. d: 3.94.
Fire Hazard: Dangerous, via heat, flame or oxidizers.
To Fight Fire: Dry chemical, CO_2.

2,2,4-TRIMETHYL PENTANE. See isooctane.

2,3,3-TRIMETHYL PENTANE. bp: 115°, flash p: 70°F, autoign. temp.: 798°F.
Fire Hazard: See 2,2,3-trimethyl pentane.
To Fight Fire: Dry chemical, CO_2.

2,2,4-TRIMETHYL-1,3-PENTANEDIOL. White crystalline solid. $C_8H_{18}O_2$, mw: 146.2, mp: 49°–51°, bp: 109°–111° @ 4 mm, flash p: 235°.
Acute tox data: Oral LD_{LO} (rat) = 2000 mg/kg; ip LD_{LO} (rat) = 800 mg/kg; iv LD_{LO} (rat) = 145 mg/kg; oral LD_{LO} (mice) = 2200 mg/kg. [*3*]
THR = HIGH via iv and MOD via oral and ip routes. An insect-repellent.
Fire Hazard: Slight, when exposed to heat or flame; can react with oxidizing materials.

2,2,4-TRIMETHYL PENTANEDIOL DIISOBUTYRATE. $C_{16}H_{30}O_4$, mw: 286, bp: 536°F, d: 0.9, vap. d: 9.9, flash p: 250°F (OC).
THR = U.
Fire Hazard: Slight, when exposed to heat or flame.
To Fight Fire: Alcohol foam, spray, mist, dry chemical.

2,2,4-TRIMETHYL PENTANEDIOL MONOISOBUTYRATE BENZOATE. $C_{19}H_{28}O_4$, mw: 320, bp: 75° @ 10 mm, d: 1.0, flash p: 325°F (OC).
Acute tox data: Oral LD_{LO} (rat) = 3200 mg/kg; oral LD_{LO} (mice) = 3200 mg/kg. [*3*]
THR = MOD via oral route.
Fire Hazard: Slight, when exposed to heat or open flame.
To Fight Fire: Spray, dry chemical.

2,2,4-TRIMETHYL PENTANOL. $C_8H_{18}O$, mw: 130.
Acute tox data: Oral LD_{50} (rat) = 3730 mg/kg. [*3*]
THR = MOD via oral route.

2,4,4-TRIMETHYL PENTENE-1. See diisobutylene.

2,3,4-TRIMETHYL-1-PENTENE. bp: 101°, flash p: <70°F, autoign. temp.: 779°F.
Fire Hazard: Dangerous, via heat, flame or oxidizers.
To Fight Fire: Dry chemical, CO_2, foam.

2,4,4-TRIMETHYL PENTENE-2. A clear liquid. C_8H_{16}, mw: 112.2, bp: 104.5°, flash p: 35°F (TOC), fp: −106.4°, d: 0.724 @ 15.5°/15.5°, vap. press: 77.5 mm @ 38°, vap. d: 3.9, autoign. temp.: 581°F.
THR = U. Probably irr and narcotic in high conc. See also isooctene.
Fire Hazard: Dangerous, when exposed to heat or flame.
Explosion Hazard: U.
Disaster Hazard: Highly dangerous; keep away from heat or open flame; can react vigorously with oxidizing materials.
To Fight Fire: Foam, CO_2, dry chemical.

3,4,4-TRIMETHYL-2-PENTENE. bp: 112°, autoign. temp.: 617°F.
Fire Hazard: See 2,3,4-trimethyl-1-pentene.
To Fight Fire: See 2,3,4-trimethyl-1-pentene.

1,2,4-TRIMETHYL PHENANTHRENE.
THR = An exper carc. [*23*]

TRIMETHYL PHENYL GERMANIUM. Colorless liquid, insol in water. $(CH_3)_3Ge(C_6H_5)$, mw: 194.8, bp: 183°.
THR = U. See germanium compounds.

TRIMETHYL PHENYL METHYL CARBAMATE. See landrin.

TRIMETHYL PHOSPHATE. $C_3H_9O_4P$, mw: 140.1.
Acute tox data: Oral LD_{50} (rat) = 840 mg/kg; oral LD_{50} (mice) = 1470 mg/kg; dermal LD_{LO} (rabbit) = 2830 mg/kg. [*3*]
THR = MOD via oral and dermal routes. An exper mutagen via ip route. [*3*]

TRIMETHYL PHOSPHINE. Colorless liquid. $(CH_3)_3P$, mw: 76.1, bp: 42°, d: <1.
THR = Details U. Probably HIGH. See also phosphine.
Fire Hazard: Spont flam in air. [*19*] See also phosphine.
Explosion Hazard: Violent reaction in air. [*19*]
Disaster Hazard: Dangerous; when heated to decomp, or on contact with acid or acid fumes, emits highly toxic fumes of PO_x; can react vigorously with oxidizing materials.

TRIMETHYL PHOSPHITE. Colorless liquid, insol in water, sol in hexane, benzene, acetone, alcohol, ether, carbon tetrachloride and kerosene. $(CH_3)_3PO_3$, mw:

124, d: 1.046 @ 20°/4°, vap. d: 4.3, bp: 232°–234°F, flash p: 130°F (OC).
THR = Violent reaction with $Mg(ClO_4)_2$. [*19*]
Fire Hazard: Mod, when exposed to heat, flame or oxidizers.
Disaster Hazard: Dangerous; see phosphates.
To Fight Fire: Water, foam, fog, CO_2.

TRIMETHYL RHENIUM. Colorless oil. $Re(CH_3)_3$, mw: 231.4, bp: 60°.
THR = Details U. See rhenium compounds.

N-(TRIMETHYLSILYL)AMIDAZOLE. $C_6H_{12}N_2Si$, mw: 140.3.
THR = An exper neo. [*3*]

TRIMETHYL STANNYL TRIPHENYL GERMANIUM. White crystals, insol in water. $(CH_3)_3SnGe(C_6H_5)_3$, mw: 467.7, mp: 88°.
THR = Toxic. Details U. See tin and germanium compounds.

TRIMETHYL STIBINE. See antimony trimethyl.

2,4,6-TRIMETHYL-1,2,3,6-TETRAHYDROBENZALDEHYDE. Liquid. $C_{10}H_{16}O$, mw: 152.23, mp: −41°, bp: 204.5°, flash p: 185°F (OC), d: 0.9195 @ 20°/20°, vap. press: 0.3 mm @ 20°, vap. d: 5.25.
THR = Details U. See aldehydes.
Fire Hazard: Mod, when exposed to heat or flame; can react with oxidizing materials.
To Fight Fire: Foam, CO_2, dry chemical.

TRIMETHYL THALLIUM. $(CH_3)_3Tl$, mw: 249.5. Spont flam in air. [*19*]
THR = Toxic. See also thallium compounds.

TRIMETHYLTIN. Colorless liquid, insol in water. $(CH_3)_3Sn$, mw: 163.8, d: 1.570 @ 25°, mp: 23°, bp: 182°.
THR = See tin compounds.

TRIMETHYL TIN ACETATE. $C_5H_{12}O_2Sn$, mw: 222.9.
Acute tox data: Oral LD_{50} (rat) = 9 mg/kg. [*3*]
THR = VERY HIGH via oral route. See also tin compounds.

TRIMETHYL TIN BROMIDE. Colorless crystals, sol in water and organic solvents. $(CH_3)_3SnBr$, mw: 243.7, mp: 27°, bp: 165°.
THR = See tin compounds and bromides.

TRIMETHYL TIN CHLORIDE. Colorless crystals, sol in water and organic solvents. $(CH_3)_3SnCl$, mw: 199.3, mp: 37°.
THR = See tin compounds and chlorides.

TRIMETHYL TIN FLUORIDE. Colorless crystals, slightly sol in organic solvents. $(CH_3)_3SnF$, mw: 182.8, mp: 360°.
THR = See fluorides and tin compounds.

TRIMETHYL TIN HYDRIDE. Colorless oily liquid, sol in organic solvents. $(CH_3)_3SnH$, mw: 164.8, bp: 60°.
THR = See tin compounds and hydrides.

TRIMETHYL TIN HYDROXIDE. Colorless crystals, sol in water and many organic solvents. $(CH_3)_3SnOH$, mw: 180.8, mp: 118° (decomp).
Acute tox data: sc LD_{LO} (mice) = 1.8 mg/kg. [*3*]
THR = VERY HIGH via sc route. HIGH via dermal and oral routes. See tin compounds.

TRIMETHYL TIN IODIDE. Colorless liquid, sol in many organic solvents. $(CH_3)_3SnI$, mw: 290.7, d: 2.1432, mp: 3.4°, bp; 170°.
THR = See tin compounds and iodides.

TRIMETHYL TIN OXIDE. White powder, insol in water and organic solvents. $(CH_3)_3Sn\text{-}O\text{-}Sn(CH_3)_3$, mw: 343.6.
THR = Details U. See tin compounds.

TRIMETHYL TIN SULFATE. $C_3H_{10}O_4SSn$, mw: 260.9.
Acute tox data: Oral LD_{LO} (rat) = 30 mg/kg; ip LD_{LO} (rat) = 16 mg/kg. [*3*]
THR = HIGH via oral and ip routes. See also tin compounds.

TRIMETHYL TIN SULFIDE. Light yellow oil, insol in water, sol in organic solvents. $(CH_3)_3Sn\text{-}S\text{-}Sn(CH_3)_3$, mw: 359.7, d: 1.649 @ 25°, mp: 6°, bp: 233.5°.
THR = See tin compounds and sulfides.

TRIMETHYL TRIBORINE TRIAMINE(β). Colorless crystals or liquid, hydrolyzed by water. $B_3(CH_3)_3N_3H_3$, mw: 122.6, mp: 31.5°, bp: 129°.
THR = Details U. See also boron compounds and amines.

TRIMETHYL TRIBORINE TRIAMINE (N). Colorless liquid, hydrolyzed by water. $(CH_3)_3B_3N_3H_3$, mw: 122.6, bp: 134°.
THR = Details U. See also boron compounds and amines.

TRIMETHYL TRIBORINE TRIAMINE. Colorless liquid, hydrolyzed by water. $(CH_3)_3B_3N_3H_3$, mw: 122.6, bp: 139°.
THR = Details U. See also boron compounds and amines.

3,8,13-TRIMETHYL TRICYCLOQUINAZOLINE. $C_{24}H_{18}N_4$, mw: 362.5.
THR = An exper neo. [*3*]

α,α,α-TRIMETHYL TRIMETHYLENE GLYCOL. See 2-methyl-2,4-pentanediol.

2,4,6-TRIMETHYL-1,3,5-TRIOXANE. $C_6H_{12}O_3$, mw: 132.2.

Acute tox data: Dermal LD_{50} (rabbit) = 14,000 mg/kg; oral LD_{50} (rabbit) = 3304 mg/kg; oral LD_{50} (dog) = 3500 mg/kg; inhal LC_{LO} (rat) = 2000 ppm for 4 hrs; In humans 14 mg/kg (oral), 14 mg/kg (iv); 71 mg/kg (im) and 14 mg/kg (rec) ⟶ psychotropic effects. [*3*]

THR = MOD via oral and inhal; LOW via dermal routes.

TRI-β-NAPHTHYL BORATE. Colorless crystals, decomp by water, sol in benzene. $(C_{10}H_7O)_3B$, mw: 440.3, mp: 115°.

THR = Details U. See also β-naphthol and boron compounds.

TRI-α-NAPHTHYLBORINE. Colorless crystals, insol in water, sol in organic solvents. $B(C_{10}H_7)_3$, mw: 392.3, mp: 203°.

THR = Details U. See also α-naphthol and boron compounds.

TRINICKEL DIPHOSPHIDE. Dark green-black crystals. Ni_3P_2, mw: 238.03, d: 5.99.

THR = See phosphides and nickel compounds.

TRINITROACETONITRILE. Solid. $C_2N_4O_6$, mw: 176.05.

THR = See nitriles.

Explosion Hazard: Mod, when exposed to heat.

Explosive Range: Explodes @ 392°F.

Disaster Hazard: Dangerous; when heated to decomp, emits highly toxic fumes of cyanides and explodes; can react vigorously with oxidizing materials.

TRINITROANILINE. Syn: *picramide.* Crystals. $C_6H_4N_4O_6$, mw: 228.1, mp: 188°, bp: explodes, d: 1.762.

THR = See nitrates, aniline and 2,4-dinitrophenol.

Fire Hazard: See nitrates.

Explosion Hazard: Severe, when shocked or exposed to heat. A very sensitive high explosive.

Disaster Hazard: Highly dangerous; shock will explode it; when heated to decomp, emits highly toxic fumes of NO_x and explodes; can react vigorously with reducing materials.

TRINITROANISOLE. Syn: *methyl picrate.* Crystals. $CH_3OC_6H_2(NO_2)_3$, mw: 243.13, mp: 68.4°, d: 1.408 @ 20°/4°.

THR = MOD irr and allergen. HIGH via oral and inhal; MOD via dermal routes. See also picric acid.

Fire Hazard: See nitrates. A high explosive.

Explosion Hazard: Severe, when shocked or exposed to heat. Trinitroanisole resembles picric acid in its high explosive properties but does not attack metals provided it is protected from moisture. It has a lower mp than picric acid, which is an advantage in shell loading. It is used as a booster charge.

Disaster Hazard: Highly dangerous; shock will explode it; when heated to decomp, emits highly toxic fumes of NO_x and explodes; can react vigorously with reducing materials.

TRINITRO BENZALDEHYDE. Liquid. $C_6H_2CHO(NO_2)_3$, mw: 241.1, mp: 119°.

THR = See aldehydes and nitrates.

1,3,5-TRINITROBENZENE. Yellow crystals. $C_6H_3(NO_2)_3$, mw: 213.11, mp: 122°, bp: decomp, d: 1.760 @ 20°/4°.

Acute tox data: Oral LD_{50} (rat) = 505 mg/kg; oral LD_{50} (mice) = 572 mg/kg. [*3*]

THR = MOD–HIGH via oral route. See also nitro compounds of aromatic hydrocarbons.

Fire Hazard: See nitrates.

Explosion Hazard: Severe, when shocked or exposed to heat. Trinitrobenzene is considered a powerful high explosive and has more shattering power than TNT. Although it is less sensitive to impact than TNT, it is not used much because it is difficult to produce.

Disaster Hazard: Highly dangerous; shock will explode it; when heated to decomp, emits highly toxic fumes of NO_x and explodes; can react vigorously with reducing materials.

2,4,6-TRINITROBENZOIC ACID. Syn: *trinitrobenzoic acid.* Orthorhombic crystals, sol @ 25° (2.05% in water, 26.6% in alcohol, 14.7% in ether), sol in methanol, slightly sol in benzene. $C_6H_2(NO_2)_3COOH$, mw: 257.12, mp: 228.7°.

THR = No data. See trinitrotoluene.

TRINITROCHLORBENZENE. Solid. $C_6H_2Cl(NO_3)_3$, mw: 295.6.

THR = See nitro compounds and chlorinated compounds of aromatic hydrocarbons.

Fire Hazard: See nitrates.

Explosion Hazard: Severe when shocked or exposed to heat.

Disaster Hazard: Highly dangerous; shock will explode it; when heated to decomp, emits highly toxic fumes of phosgene and NO_x and explodes; can react vigorously with oxidizing materials.

2,4,6-TRINITRO-*m*-CRESOL. Syn: *cresolite.* Yellow crystals. $(NO_2)_3C_6H(CH_3)OH$, mw: 243.13, mp: 106°, bp: explodes @ 150°.

Acute tox data: ip LD_{LO} (mice) = 31 mg/kg; ip LD_{50} (mice) = 168 mg/kg. [*3*]

THR = HIGH via ip route. See nitro compounds of aromatic hydrocarbons.

Fire Hazard: See nitrates.

Explosion Hazard: Severe, when shocked or exposed to heat. Trinitrocresol is not as powerful a high explosive as TNT or picric acid. It has been used as a bursting charge and in combination with other high explosives.

Disaster Hazard: Highly dangerous; shock will explode it; when heated to decomp, emits highly toxic fumes of NO_x and explodes; can react vigorously with oxidizing materials.

2,4,4'-TRINITRODIPHENYLAMINE. Crystals. $C_{12}H_8N(NO_2)_2NO_2$, mw: 304.2.

THR = MOD via oral and inhal routes. See also amines.

Fire Hazard: See nitrates.

Explosion Hazard: Dangerous; see nitrates.

Disaster Hazard: Dangerous;, see nitrates.

TRINITRO ETHANOL. $C_2H_3O_7N_3$, mw: 181.1.

THR = Has exploded during distillation. [*19*]

2,4,7-TRINITROFLUOREN-9-ONE. $C_{13}H_5O_7N_3$, mw: 315.2.

THR = An exper neo. [*3*]

TRINITROMETHANE. Syn: *nitroform*. $CH(NO_2)_3$, mw: 151, mp: 15°, d: 1.469, bp: decomp $> 25°$.

THR = Somewhat irr to skin, eyes and mu mem. Inhal can cause headache and nausea. Causes mild narcosis.

Explosion Hazard: Dangerous; explodes when heated rapidly. Conc of >50% dissolve exothermally and can explode. [*19*] 90% concentration + 10% isopropyl alcohol in polyethylene bottles has exploded. Can explode during distillation.

TRINITRONAPHTHALENE(mixture of isomers). Syn: *naphthite*. White to yellow crystals, insol in water, sol in alcohol. $C_{10}H_5(NO_2)_3$, mw: 263.16, mp: 113°–247° (depending upon isomeric composition).

THR = See nitrates, N.O.S., and nitro compounds of aromatic hydrocarbons.

TRINITROPHENOL. See picric acid.

TRINITROPHENYL METHYL NITRAMINE. See tetryl.

TRINITROPHENYL NITRAMINE ETHYL NITRATE. Syn: *petryl*.

THR = Unknown. See nitro compounds of aromatic hydrocarbons.

Fire Hazard: See nitrates.

Explosion Hazard: Severe when shocked or exposed to heat. Explodes when heated; sol in nitroglycerin. Its high explosive sensitivity to impact and friction are about the same as that of tetryl, but its shattering power is much greater. It is used as a base charge in detonators.

Disaster Hazard: Extremely dangerous; shock will explode it; when heated to decomp, emits highly toxic fumes of NO_x and explodes, can react vigorously with oxidizing materials.

TRINITROPHLOROGLUCIN. Syn: *trinitrophloroglucinol*. Powder. $C_6(NO_2)_3(OH)_3$, mw: 261.1.

THR = See nitro compounds of aromatic hydrocarbons.

Fire Hazard: See nitrates.

Explosion Hazard: See nitrates.

Disaster Hazard: Dangerous; shock will explode it; when heated to decomp, emits highly toxic fumes of NO_x and explodes; can react vigorously with oxidizing or reducing materials.

TRINITRORESORCINOL. $C_6H_3O_8N_3$, mw: 245.

THR = U. See nitro compounds of aromatic hydrocarbons.

TRINITROTOLUENE. Syns: *TNT*, *sym-trinitrotoluol, triton*. Colorless monoclinic crystals. $(NO_2)_3C_6H_2CH_3$, mw: 227.13, mp: 80.7°, bp: 240° explodes, flash p: explodes, d: 1.654.

Acute tox data: Oral LD_{LO} (rat) = 700 mg/kg; oral LD_{LO} (cat) = 1850 mg/kg; sc LD_{LO} (cat) = 200 mg/kg. [*3*]

THR = HIGH via sc route and MOD via oral and dermal routes. Has been implicated in aplastic anemia.

Fire Hazard: See nitrates.

Explosion Hazard: Mod; will detonate under strong shock. See explosives, high. It detonates at around 240° C but can be distilled safely under reduced pressure. It is a comparatively insensitive explosive. In small quantities it will burn quietly if not confined. However, sudden heating of any quantity will cause it to detonate; the accumulation of heat when large quantities are burning will cause detonation. In other respects it is one of the most stable of all high explosives and there are but few restrictions to its handling. It is for this reason, from the military standpoint, that TNT is quantitatively the most used. It requires a fall of 130 cm for a 2 kg weight to detonate it. It is one of the most powerful high explosives. It can be detonated by the usual detonators and blasting caps (at least a No. 6). For full efficiency, the use of a high velocity initiator, such as tetryl, is required. TNT is one of those explosives containing an oxygen deficiency. In other words, the addition of products which are oxygen rich can enhance its explosive power. Also mono- and dinitrotoluene may be added for reduction of the temp. of the explosion and to make the explosion flashless. Various materials are added to TNT to make what is known as permissible explosives. TNT may be regarded as the equivalent of 40% dynamite and can be used under water. It is also used in the manufacture of detonator fuse known as Cordeau Detonant. For the military, TNT finds use in all types of bursting charges, including armor-piercing types, although it is somewhat too sensitive to be ideal for this purpose, and has since been replaced to a great extent by ammonium picrate. It is a relatively expensive explosive and does not compete seriously with dynamite for general commercial use.

Disaster Hazard: Highly dangerous; shock will explode it; when heated to decomp, emits highly toxic fumes of NO_x; can react vigorously with reducing materials.

2,4,6-TRINITROXYLENE. Syn: *TNX.* Rhombic crystals. $(NO_2)_3C_6H(CH_3)_2$, mw: 241.16, mp: 181.5°, d: 1.604 @ 19°.

THR = Details U. It is said to be less toxic than trinitrotoluene. See also nitro compounds of aromatic hydrocarbons.

Fire Hazard: See nitrates.

Explosion Hazard: Severe, when shocked or exposed to heat. This high explosive is not very powerful when used alone. However, the addition of picric acid or other nitro-type of high explosive serves to lower its mp and to reinforce its explosive power. It is also used in mixtures with ammonium nitrate; in detonating compositions, and mixed with other high explosives, as a bursting charge.

Disaster Hazard: Highly dangerous; shock will explode it; when heated to decomp, emits highly toxic fumes of NO_x and explodes; can react vigorously with oxidizing materials.

TRI-2-OCTYL BORATE. Colorless liquid, odor of 2-octanol. $[CH_3(CH_2)_5C(CH_3)HO]_3B$, mw: 398.47, bp: 340°–349°, flash p: 330°F (COC), d: 0.837 @ 24.5°, vap. d: 13.8.

THR = See boron compounds and esters.

Fire Hazard: Low, when exposed to heat or flame; can react with oxidizing materials.

To Fight Fire: Foam, CO_2, dry chemical.

TRI-*n*-OCTYL BORATE. Colorless liquid, odor of octyl alcohol. $[CH_3(CH_2)_7O]_3B$, mw: 398.5, bp: 192°–194° @ 2 mm, flash p: 370°F (COC), d: 0.846 @ 23°, vap. d: 13.7.

Acute tox data: Oral LD_{50} (mice) = 1290 mg/kg. [*3*]

THR = MOD via oral route. See esters and boron compounds.

Fire Hazard: Slight, when exposed to heat or flame; can react with oxidizing materials.

To Fight Fire: Foam, CO_2, dry chemical.

TRIOCTYL PHOSPHATE. Liquid. $[C_4H_9CH(C_2H_5)CH_2O]_3P{:}O$, mw: 434.63, mp: −74°, bp: 216° @ 5 mm, flash p: 405°F (OC), d: 0.9262 @ 20°/20°, vap. d: 14.95.

THR = Details U. See esters.

Fire Hazard: Slight, when exposed to heat or flame.

Disaster Hazard: Dangerous; see phosphates; can react with oxidizing materials.

To Fight Fire: Foam, CO_2, dry chemical.

TRIOCTYL PHOSPHITE. Syn: *tris(2-ethyl hexyl) phosphite.* Insol in water. $(C_8H_{17}O)_3P$, mw: 418, d: 0.9, bp: 212°F @ 0.01 mm, flash p: 340°F (OC).

THR = U. See esters.

Fire Hazard: Slight, when exposed to heat or flame.

Disaster Hazard: Dangerous; see phosphates.

To Fight Fire: CO_2, mist, dry chemical.

TRIOL-230. Clear, colorless liquid. $H_{22}O_4C_{10}$, mw: 206.28, bp: 196° @ 5 mm, flash p: 395°F (COC), fp: $<-5°$, d: 1.081 @ 20°/20°, vap. press: <0.01 mm @ 20°.

THR = U. See alcohols.

Fire Hazard: Slight, when exposed to heat or flame; can react with oxidizing materials.

To Fight Fire: Water, foam, CO_2, dry chemical.

TRIOLEYL BORATE. Pale yellow liquid, odor of oleyl alcohol. $[CH_3(CH_2)_7CHCH(CH_2)_7O]_3B$, mw: 813.20, bp: 300°–330° @ 0.5 mm, flash p: 495°F (COC), d: 0.860 @ 23.6°.

THR = See boron compounds and esters.

Fire Hazard: Slight, when exposed to heat or flame; can react with oxidizing materials.

To Fight Fire: Foam, CO_2, dry chemical.

TRIOSMIUM DODECACARBONYL. $Os_3(CO)_{12}$, mw: 906.7.

THR = HIGH via inhal route. See also osmium compounds and carbon monoxide. An exper carc. [*23*]

TRIOXANE. Syns: *sym-trioxane, α-trioxymethylene.* Colorless crystals, odor of ethyl alcohol. $(CH_2O)_3$, mw: 90.08, mp: 62°, lel = 3.6%, uel = 28.7%, bp: 114.5° (sublimes), flash p: 113°F (OC), d: 1.17 @ 65°/20°, autoign. temp.: 777°F, vap. press: 13 mm @ 25°, vap. d: 3.1.

Acute tox data: Oral LD_{50} (rat) = 800 mg/kg; dermal LD_{LO} (rabbit) = 10000 mg/kg. [*3*]

THR = MOD via oral and LOW via dermal routes. Principal action is that of primary irr. Used as a food additive permitted in food for human consumption. [*109*] See also formaldehyde. Can evolve formaldehyde when heated strongly or in contact with strong acids.

Fire Hazard: Mod, when exposed to heat, flame or oxidizers.

Explosion Hazard: Mod, in the form of vapor, when exposed to flame.

Disaster Hazard: Mod dangerous; can explode when heated; reacts with oxidizing materials; on contact with acid or acid fumes, can emit toxic fumes.

To Fight Fire: Foam, CO_2, dry chemical.

sym-TRIOXANE. See trioxane.

TRIOXIME.
THR = U.
Fire Hazard: U.
Explosion Hazard: Mod, when exposed to heat.
Explosive Range: Explodes @ 311°F.

α-TRIOXYMETHYLENE. See trioxane.

TRIPELENNAMINE. Syn: *pyribenzamine.* $C_{16}H_{21}N_3$, mw: 255.4.
Acute tox data: iv LD_{50} (rat) = 37 mg/kg; ip LD_{50} (mice) = 65 mg/kg; oral LD_{50} (mice) = 210 mg/kg; sc LD_{50} (mice) = 75 mg/kg. [*3*]
THR = HIGH via iv, ip, oral and sc routes. Addicts have added it to paregoric to make "blue velvet," which can cause a euphoria by injection. [*20*] Has been implicated in aplastic anemia.

TRIPENTAERYTHRITOL. White to ivory, odorless powder. $C_{15}H_{32}O_{10}$, mw: 372.4, mp: approx 240°, d: 1.30.
THR = See alcohols.
Fire Hazard: Slight.

TRIPENTYL BORATE. $C_{15}H_{33}O_3B$, mw: 272.3.
Acute tox data: Oral LD_{50} (mice) = 1600 mg/kg. [*3*]
THR = MOD via oral route. See also boron compounds.

2,3,3-TRIPHENYL ACRYLONITRILE. $C_{21}H_{15}N$, mw: 281.4.
Acute tox data: Oral LD_{50} (rat) = 284 mg/kg. [*3*]
THR = HIGH via oral route. See also nitriles. An exper carc via sc route. [*3*]

TRIPHENYL ALUMINUM. White crystals, decomp upon contact with moisture. $(C_6H_5)_3Al$, mw: 258.3, mp: 200°.
THR = See organometals and aluminum compounds. Explodes on contact with H_2O. [*19*]

TRIPHENYL AMINE. Crystals. $N(C_6H_5)_3$, mw: 245.3, d: 0.774 @ 0°/0°, mp: 127°, bp: 365°.
THR = No details. Probably HIGH. See also amines.

TRIPHENYL ANISYL GERMANIUM. White powder, sol in alcohol. $C_{25}H_{22}OGe$, mw: 411.0, mp: 159°.
THR = Details U. See germanium compounds.

TRIPHENYL ANTIMONY. See triphenyl stibine.

TRIPHENYL ARSENIC. Syn: *arsenic triphenyl.* White crystals. $As(C_6H_5)_3$, mw: 306.2, mp: 60°, d: 1.2225 @ 48°, bp: > 360° (in CO_2).
THR = See arsenic compounds.

TRIPHENYL BENZENE. $C_{24}H_{18}$, mw: 306.4.
THR = An exper neo via sc route. [*3*]

TRIPHENYL BENZYL TIN. Colorless crystals, sol in organic solvents except alcohol. $C_{25}H_{22}Sn$, mw: 441.1, mp: 90°, bp: 250° @ 3 mm.
THR = Details U. See tin compounds.

TRIPHENYL BISMUTHINE. Syn: *bismuth triphenyl.* Monoclinic crystals. $Bi(C_6H_5)_3$, mw: 440.30, mp: 78°, bp: 242° @ 14 mm, d: 1.585.
THR = See bismuth compounds.

TRIPHENYL BORATE. White to pink solid, odor of phenol, decomp by water. $(C_6H_5O)_3B$, mw: 290.12, mp: 35°, bp: > 360°.
Acute tox data: Oral LD_{50} (mice) = 200 mg/kg. [*3*]
THR = HIGH via oral route. See phenol and boron compounds.
Fire Hazard: Mod, when exposed to heat or flame. Hydrolyzes to phenol.
Disaster Hazard: Dangerous; when heated, emits highly toxic fumes; reacts with water or steam to form toxic fumes of phenol; in contact with oxidizing material, can react vigorously.
To Fight Fire: Foam, CO_2, dry chemical.

TRIPHENYL BORON. See boron triphenyl.

TRIPHENYL CYCLOHEXYL BORATE. $(C_{12}H_{15}O)_3B$, mw: 536.6.
Acute tox data: Oral LD_{50} (mice) = 1240 mg/kg. [*3*]
THR = MOD via oral route.

TRIPHENYL DIMETHYL AMINOPHENYL GERMANIUM. White crystals. $C_{26}H_{25}NGe$, mw: 424.1, mp: 141°.
THR = Details U. See germanium compounds and phenol.

TRIPHENYL-3,4-DICHLOROPHENYL PHOSPHONIUM CHLORIDE. Solid. $C_{24}H_{18}Cl_3P$, mw: 443.7.
THR = Details U. See germanium compounds and phenol.

TRIPHENYL-3,4-DICHLOROPHENYL PHOSPHONIUM CHLORIDE. Solid. $C_{24}H_{18}Cl_3P$, mw: 443.7.
THR = Details U; a moth-repellent. Probably toxic.
Disaster Hazard: Dangerous; see phosphates and chlorides.

TRIPHENYLETHYLTIN. White powder. $C_{20}H_{20}Sn$, mw: 379.1, d: 1.2953 @ 62°, mp: 56°.
THR = Details U. See tin compounds.

TRIPHENYL GERMANIUM AMIDE. White solid. $C_{18}H_{17}NGe$, mw: 319.9, mp: decomp to evolve NH_3.
THR = Details U. See amides, germanium compounds and phenol.

TRIPHENYL GERMANIUM BROMIDE. Crystals, hydrolyzed by hot water. $(C_6H_5)_3GeBr$, mw: 383.8, mp: 139°.
THR = See germanium compounds and bromides.

TRIPHENYL GERMANIUM CHLORIDE. White crystals, hydrolyzes in hot water. $(C_6H_5)_3GeCl$, mw: 339.4, mp: 118°, bp: 285° @ 12 mm.
THR = See germanium compounds and chlorides.

TRIPHENYL GERMANIUM FLUORIDE. White crystals, hydrolyzes in hot water. $(C_6H_5)_3GeF$, mw: 322.9, mp: 77°.
THR = See fluorides and phenol.

TRIPHENYL GERMANIUM HYDRIDE. White crystals, insol in water. $(C_6H_5)_3GeH$, mw: 304.9, mp:(α): 47°, mp(β): 27°.
THR = See germanium compounds and phenol.

TRIPHENYL GERMANIUM HYDROXIDE. White crystals, insol in water. $(C_6H_5)_3GeOH$, mw: 320, mp: 134°.
THR = Details U. See germanium compounds and phenol.

TRIPHENYL GERMANIUM IODIDE. White crystals, hydrolyzed by water. $(C_6H_5)_3GeI$, mw: 430.8, mp: 157°.
THR = See germanium compounds and iodides.

TRIPHENYL GERMANIUM OXIDE. Colorless crystals, insol in water $[(C_6H_5)_3Ge]_2O$, mw: 623.8, mp: 184°.
THR = See germanium compounds and phenol.

TRIPHENYL GERMANIUM SODIUM. Yellowish crystals, decomp by water. $(C_6H_5)_3GeNa$, mw: 326.9.
THR = HIGH. See sodium, germanium compounds and phenol.
Fire and Explosion Hazard: This material reacts with water to evolve hydrogen; see hydrogen.

TRIPHENYL GERMANIUM SODIUM OXIDE. White solid, decomp by water. $(C_6H_5)_3GeONa$, mw: 342.9.
THR = Details U. See germanium compounds.

TRIPHENYL GERMANIUM SODIUM TRIAMMINE. Yellow solid, decomp by water. $(C_6H_5)_3GeNa \cdot 3NH_3$, mw: 378.0, mp: decomp.
THR = Details U. See ammonia and germanium compounds.

TRIPHENYL METHANE. $C_{19}H_{16}$, mw: 244.3, flash p: 212°F, d: 1.01, vap. d: 8.43, bp: 359°.
Fire Hazard: Low.
To Fight Fire: Foam, CO_2, dry chemical, water spray.

TRIPHENYL METHANE DYES. Syn: *rosaniline dyes.*
THR = Details U; a fungicide; some are susp carc. [*14*] Can be strong allergen. [*20*]

TRIPHENYL METHYL TIN. Colorless crystals, sol in chloroform, benzene and ether. $(C_6H_5)_3Sn(CH_3)$, mw: 365.0, d: 1.3113 @ 64°, mp: 64°.
THR = Details U. See tin compounds and phenol.

TRIPHENYL-α-NAPHTHYL TIN. Colorless crystals, insol in benzene, ether and chloroform. $(C_6H_5)_3Sn(C_{10}H_7)$, mw: 477.2, mp: 125°.
THR = See tin compounds and phenol.

TRIPHENYL PHOSPHATE. Colorless, odorless, crystalline solid. $PO(OC_6H_5)_3$, mw: 326.28, mp: 48.5°, bp: 245° @ 11 mm, flash p: 428°F (CC), d: 1.268 @ 60°, vap. press: 1 mm @ 193.5°.
Acute tox data: Oral LD_{LO} (rat) = 3000 mg/kg; sc LD_{50} (cat) = 100 mg/kg. [*3*]
THR = HIGH via sc and MOD via oral routes. Absorbed (but slowly) particularly via skin. Not a potent cholinesterase inhibitor. [*20*] See also tricresyl-*o*-phosphate.
Fire Hazard: Slight, when exposed to heat or flame.
Spont Heating: No.
Disaster Hazard: Dangerous; see phosphates.
To Fight Fire: CO_2, dry chemical.

TRIPHENYL PHOSPHINE. Syn: *triphenyl phosphorus.* Crystals. $(C_6H_5)_3P$, mw: 262.3, mp: 79°, bp: > 360°, d: 1.194, flash p: 356°F (OC), vap. d: 9.0.
Acute tox data: Oral LD_{50} (rat) = 800 mg/kg; inhal LC_{50} (rat) = 1135 ppm for 4 hrs. [*3*]
THR = MOD via oral and inhal routes. See phosphine and phenol.
Fire Hazard: Low, when exposed to heat or flame.
Explosion Hazard: Slight, in the form of vapor when exposed to flame.
Disaster Hazard: Dangerous; when heated to decomp, emits highly toxic fumes of phosphine and PO_x; can react vigorously with oxidizing materials.
To Fight Fire: Dry chemical, fog, CO_2.

TRIPHENYL PHOSPHINE OXIDE. White crystals. $(C_6H_5)_3PO$, mw: 278.3, mp: 156°, bp: > 360°, d: 1.2124 @ 22.6°.
THR = Probably HIGH. See phosphine.
Explosive Hazard: U.
Disaster Hazard: Dangerous; see phosphates.

TRIPHENYL PHOSPHITE. Water white to pale yellow solid or oily liquid, clean and pleasant odor, insol in water. $(C_6H_5O)_3P$, mw: 310, d: 1.184 @ 25°/25°, mp: 22°–25°, bp: 155°–160° @ 0.1 mm, flash p: 425°F (OC).
Acute tox data: Oral LD_{50} (rat) = 1600 mg/kg; ip LD_{50} (rat) = 250 mg/kg; sc LD_{LO} (rat) = 2000 mg/kg. [*3*]
THR = HIGH via ip; MOD via oral and sc routes. Irr to skin. See phenol and phosphites.
Fire Hazard: Slight, when exposed to heat or flame.
Disaster Hazard: Dangerous; see phosphates.
To Fight Fire: CO_2, mist, dry chemical.

TRIPHENYL PHOSPHORUS. See triphenyl phosphine.

TRIPHENYL STANNYL METHANE. Crystals, insol in water, sol in organic solvents. $[(C_6H_5)_3Sn]_3CH$, mw: 1063.1, mp: 128°.
THR = Details U. See tin compounds.

TRIPHENYL STIBINE. Syn: *triphenyl antimony*. Crystals, water-insol, sol in organic solvents. $(C_6H_5)_3Sb$, mw: 353.1, mp: 50°, d: 1.4343 @ 25°, bp: > 360°.
Acute tox data: Oral LD_{50} (rat) = 187 mg/kg; ip LD_{50} (rat) = 168 mg/kg; oral LD_{50} (mice) = 650 mg/kg. [*3*]
THR = HIGH via oral and ip routes. See antimony compounds.
Fire Hazard: Mod, when exposed to heat or flame.
Disaster Hazard: Dangerous; see antimony compounds; can react vigorously with oxidizing materials.
To Fight Fire: Water, foam, mist.

TRIPHENYL TIN. White powder, insol in water. $(C_6H_5)_3Sn$, mw: 350.0, mp: 232.5°.
Acute tox data: uk LD_{50} (rat) = 125 mg/kg. [*3*]
THR = HIGH via uk route. See tin compounds.
Disaster Hazard: Dangerous; see phenol.

TRIPHENYL TIN ACETATE. See brestan.

TRIPHENYL TIN BROMIDE. Colorless crystals, insol in water, sol in organic solvents. $(C_6H_5)_3SnBr$, mw: 429.9, mp: 120.5°, bp: 249° @ 13.5 mm.
THR = See tin compounds and bromides.

TRIPHENYL TIN CHLORIDE. Colorless crystals, insol in water, sol in organic solvents. $(C_6H_5)_3SnCl$, mw: 385.5, mp: 106°, bp: 240° @ 13.5 mm.
THR = See tin compounds and chlorides.

TRIPHENYL TIN FLUORIDE. Crystals, slightly water-sol. $(C_6H_5)_3SnF$, mw: 369.0.
THR = See fluorides and phenol.

TRIPHENYL TIN HYDROXIDE. Solid. $(C_6H_5)_3SnOH$, mw: 367.0, mp: 122°.
Acute tox data: Oral LD_{50} (rat) = 46 mg/kg, ip LD_{LO} (rat) = 100 mg/kg; ip LD_{LO} (mice) = 8.5 mg/kg. [*3*]
THR = HIGH via oral and ip routes. See tin compounds and phenol.

TRIPHENYL TIN IODIDE. White crystals, insol in water, sol in organic solvents. $(C_6H_5)_3SnI$, mw: 476.9, mp: 121°, bp: 253 @ 13.5 mm.
THR = See tin compounds and iodides.

TRIPHENYL-*m*-TOLYLGERMANIUM. White crystals, insol in water. $(C_6H_5)_3Ge(C_6H_4CH_3)$, mw: 395.0, mp: 137°.
THR = Details U. See germanium compounds.

TRIPHENYL-*p*-TOLYLGERMANIUM. White crystals, sol in organic solvents. $(C_6H_5)_3Ge(C_6H_4CH_3)$, mw: 395.0, mp: 124°.
THR = Details U. See germanium compounds.

TRIPHENYL-*p*-TOLYLTIN. Crystals, sol in chloroform, benzene and ether. $(C_6H_5)_3Sn(C_7H_7)$, mw: 441.1, mp: 124°.
THR = Details U. See tin compounds.

TRI-*p*-PHENYLYL GERMANIUM BROMIDE. Syn: *Tri-p-biphenylyl germanium bromide*. White crystals, sol in benzene. $(C_6H_5C_6H_4)_3GeBr$, mw: 612.1, mp: 242°.
THR = See germanium compounds and bromides.

TRIPHENYL-*p*-XYLYLTIN. Crystals, sol in benzene, ether and chloroform. $(C_6H_5)_3Sn[C_6H_3(CH_3)_2]$, mw: 456.2, mp: 100.5°.
THR = Details U. See tin compounds.

TRIPHOSGENE. See hexachloromethyl carbonate.

TRIPOLI. Finely granulated white or gray siliceous rock. An amorphous form of SiO_2.
THR = A nuisance dust which may contain silica. See also silica. Effects due to silica content may be enough to produce pulmonary fibrosis.

TRIPROPOXY BORON. See tri-*n*-propyl borate.

TRIPROPYL ALUMINUM. See aluminum tripropyl.

TRIPROPYLAMINE. Liquid, very slightly sol in water. $N(C_3H_7)_3$, mw: 143.3, mp: −93°, bp: 156°, flash p: 105°F (OC), d: 0.75, vap. d: 4.9.
Acute tox data: Oral LD_{50} (rat) = 258 mg/kg; inhal LC_{LO} (rat) = 250 ppm for 4 hrs; dermal LD_{50} (rabbit) = 570 mg/kg. [*3*]
THR = HIGH via oral and inhal; MOD via dermal routes. See amines.
Fire Hazard: Mod, when exposed to heat, flame or oxidizers.
Disaster Hazard: Mod dangerous; when heated to decomp, emits toxic fumes; can react with oxidizing materials.
To Fight Fire: Foam, CO_2, dry chemical.

TRI-*n*-PROPYLBORATE. Syn: *propyl borate*. Colorless liquid, odor of *n*-propanol. $C_9H_{21}O_3B$, mw: 188.08, bp: 176°–170°, flash p: 155°F (COC), d: 0.856 @ 24°.
Acute tox data: Oral LD_{50} (mice) = 2008 mg/kg. [*3*]
THR = MOD via oral route. See esters and boron compounds.
Fire Hazard: Mod, when exposed to heat or flame; can react with oxidizing materials.

TRI-*n*-PROPYLBORON. Colorless liquid; insol in water, sol in ether. $(C_3H_7)_3B$, mw: 140.1, d: 0.725, bp: 156°.
THR = Details U. See boron compounds. Self ignites in air.

TRI-*sec*-PROPYLBORON. Colorless mobile liquid, insol in water, sol in ether. $B(C_3H_7)_3$, mw: 140.1, bp: 150°.
THR = Details U. See boron compounds. Self-ignites in air.

TRI-*n*-PROPYL-*n*-BUTYLTIN. Liquid. $(C_3H_7)_3Sn(C_4H_9)$, mw: 305.1, bp: 121° @ 10 mm.
THR = Details U. See tin compounds.

TRIPROPYLENE GLYCOL. Colorless liquid. $C_9H_{20}O_4$, mw: 192.3, mp: does not crystallize, bp: 267°, flash p: 285°F, d: 1.023 @ 25°/25°, vap. press: 1 mm @ 96.0°, vap. d: 6.63.
Acute tox data: Oral LD_{50} (rat) = 3000 mg/kg. [*3*]
THR = MOD via oral route. See glycols.
Fire Hazard: Slight, when exposed to heat or flame; can react with oxidizing materials.
To Fight Fire: Water, foam, CO_2, dry chemical.

TRIPROPYLENE GLYCOL METHYL ETHER. Syn: *tripropylene glycol monomethyl ether*. $C_{10}H_{22}O_4$, mw: 206.3, bp: 243°, flash p: 250°F, d: 0.967 @ 25°/25°, vap. d: 7.1.
Acute tox data: Oral LD_{50} (rat) = 3300 mg/kg; oral LD_{LO} (dog) = 5000 mg/kg. [*3*]
THR = MOD via oral route. See glycols.
Fire Hazard: Slight, when exposed to heat or flame; can react with oxidizing materials. See ethers.
To Fight Fire: Foam, CO_2, dry chemical.

TRIPROPYLENE GLYCOL MONOMETHYL ETHER. See tripropylene glycol methyl ether.

TRIPROPYL ETHYL TIN. Liquid. $(C_3H_7)_3Sn(C_2H_5)$, mw: 277.0, d: 1.1225 @ 22°, bp: 117.5° @ 23.3 mm.
THR = Details U. See tin compounds.

TRIPROPYL INDIUM. $(C_3H_7)_3In$, mw: 244.
THR = Spont flam in air. [*19*]

TRI-*n*-PROPYL ISOBUTYL TIN. Liquid. $(C_3H_7)_3Sn(C_4H_9)$, mw: 305.1, d: 1.0841 @ 24°, bp: 128° @ 18 mm.
THR = Details U. See tin compounds.

TRIPROPYL LEAD. $Pb(C_3H_7)_3$, mw: 336.5.
Acute tox data: pa LD_{LO} (rat) = 20 mg/kg. [*3*]
THR = HIGH via pa route. See also lead.

TRIPROPYL LEAD CHLORIDE. $ClPb(C_3H_7)_3$, mw: 371.9.
Acute tox data: Oral LD_{LO} (rat) = 40 mg/kg; ip LD_{LO} (rat) = 5.38 mg/kg; sc LD_{LO} (rat) = 11 mg/kg. [*3*]
THR = HIGH via oral, ip and sc routes. See also lead.

TRI-*n*-PROPYL PHOSPHATE. Liquid. $(C_3H_7O)_3PO$, mw: 224, bp: 97° @ 4 mm, d: 1.002 @ 25°/4°, vap. d: 7.72.
THR = U. See esters.
Fire Hazard: Slight, when exposed to heat or flame.
Disaster Hazard: Dangerous; see phosphates; can react with oxidizing materials.

TRIPROPYL TIN ACETATE. $O_2SnC_{11}H_{24}$, mw: 307.0.
Acute tox data: Oral LD_{50} (rat) = 118 mg/kg; iv LD_{LO} (rat) = 24 mg/kg. [*3*]
THR = HIGH via iv and oral routes. See also tin compounds.

TRIPROPYL TIN BROMIDE.
Acute tox data: Inhal LC_{LO} (mice) = 1650 mg/m^3. [*3*]
THR = MOD via inhal route. See also tin compounds and bromides.

TRIPROPYL TIN CHLORIDE. Colorless liquid, sol in organic solvents. $(C_3H_7)_3SnCl$, mw: 283.4, d: 1.2678 @ 28°, mp: −23.5°.
THR = See tin compounds and chlorides.

TRI-*n*-PROPYL TIN FLUORIDE. Crystals, sol in organic solvents. $(C_3H_7)_3SnF$, mw: 267.0, mp: 275°.
THR = See fluorides.

TRI-*n*-PROPYL TIN IODIDE. Colorless liquid, sol in organic solvents. $(C_3H_7)_3SnI$, mw: 374.9, d: 1.692 @ 16°, mp: −53°, bp: 262°.
THR = See tin compounds and iodides.

2,2′,2″-TRIPYRIDYL RHENICHLORIDE. Pale green crystals, insol in water. $(C_5H_4N)_3HReCl_6$, mw: 634.3.
THR = See rhenium compounds and chlorides.

TRIRUTHENIUM DODECACARBONYL. $Ru_3(CO)_{12}$, mw: 639.3.
THR = An exper carc. [*23*] See also ruthenium compounds and carbon monoxide.

TRIS. See Tris-BP.

TRIS-ACETYLACETONE GERMANIUM CUPRIBROMIDE. Green-black crystals. $[(C_5H_7O_2)_3Ge]CuBr_3$, mw: 673.2, mp: 139°.
THR = See copper compounds, germanium compounds and bromides.

TRIS-ACETYL ACETONE GERMANIUM CUPROBROMIDE. Colorless crystals, sol in chloroform. $[(C_6H_7O_2)_3Ge]CuBr_2$, mw: 593.3, mp: 166°.
THR = See copper compounds, bromides and germanium compounds.

TRIS-ACETYLACETONE GERMANIUM CUPROCHLORIDE. Colorless crystals, sol in chloroform. $[(C_5H_7O_2)_3Ge]CuCl_2$, mw: 504.4, mp: 148°.
THR = See copper compounds, germanium compounds and chlorides.

TRIS-ACETYLACETONE GERMANIUM DICUPROBROMIDE. Colorless crystals, sol in acetone. $[(C_5H_7O_2)_3Ge]Cu_2Br_3$, mw: 736.8, mp: 195° (decomp).
THR = See copper compounds, germanium compounds and bromides.

TRIS(*p*-AMINOPHENYL)CARBONIUM PAMOATE. $C_{19}H_8N_3 \cdot \frac{1}{2}C_{23}H_{14}O_6$, mw: 471.4.
THR = An exper carc. [*3*]

TRIS(1-AZIRIDINYL)PHOSPHINE OXIDE. Syns: *triethylene phosphoramide, tepa, APO, 1-aziridinyl phosphine oxide (tris)*. Colorless crystals, sol in water, alcohol and ether. $(NCH_2CH_2)_3PO$, mw: 173.2, mp: 41°, bp: 90° @ 23 mm.

Acute tox data: Oral LD_{50} (rat) = 37 mg/kg; dermal LD_{50} (rat) = 87 mg/kg; ip LD_{LO} (mice) = 0.156 mg/kg; iv LD_{LO} (dog) = 0.43 mg/kg; oral LD_{50} (chicken) = 151 mg/kg. [*3*]

THR = HIGH via oral, dermal, ip and iv routes. An exper (±) carc and mutagen. [*3, 8, 23*]

Disaster Hazard: Dangerous; see phosphorus compounds, organic.

TRIS(1-AZIRIDINYL)PHOSPHINE SULFIDE. Syn: *thiotepa*. $C_6H_{12}N_3SP$, mw: 189.2.

THR = An exper (+) carc and teratogen. [*3, 8, 23*] HIGH via oral route.

2,4-TRIS(1-AZIRIDINYL)-*s*-TRIAZINE. See triethylene melamine.

TRIS-BP. See tris(2,3-dibromopropyl)phosphate.

TRIS(2-CHLOROETHYL)AMINE. $C_6H_{12}NCl_3$, mw: 204.54.

Acute tox data: Dermal LD_{50} (rat) = 4.9 mg/kg; inhal LC_{LO} (rat) = 180 mg/m^3; iv LD_{50} (rat) = 0.7 mg/kg; dermal LD_{50} (mice) = 7 mg/kg; sc LD_{50} (mice) = 2 mg/kg; inhal LC_{50} (human) = 1000 mg/m^3. [*3*]

THR = HIGH irr to skin, eyes, and mu mem and via dermal, inhal, iv and sc routes. An exper carc. [*4, 23*]

TRIS(β-CHLOROETHYL)PHOSPHATE. Clear liquid. $(ClCH_2CH_2)_3PO_4$, mw: 285.5, flash p: 421°F (COC), boiling range: 210°–220° @ 20 mm, d: 1.425 @ 20°/20°, autoign. temp.: 1115°F, vap. press: 0.5 mm @ 145°.

Acute tox data: Oral LD_{50} (rat) = 1410 mg/kg; ip LD_{LO} (mice) = 250 mg/kg. [*3*]

THR = HIGH via ip and MOD via oral routes. An exper mutagen. [*90*]

Fire Hazard: Slight, when exposed to heat or flame.

Disaster Hazard: Dangerous; see phosphates and chlorides.

TRIS(β-CHLOROETHYL)PHOSPHITE. Water-white liquid. $(ClCH_2CH_2O)_3P$, mw: 269.51, bp: 125°–135° @ 7 mm, flash p: > 280°F (OC), d: 1.3348 @ 35°/4°, vap. press: > 1 mm @ 20°, vap. d: 9.32.

Acute tox data: ip LD_{LO} (mice) = 250 mg/kg. [*3*]

THR = HIGH via ip route. Vapors are highly toxic. A powerful irr to eyes, skin and mu mem. Absorbed via intact skin in tox amounts. [*20*]

Fire Hazard: Slight, when exposed to heat or flame.

Disaster Hazard: Dangerous; see phosphates and chlorides; can react vigorously with oxidizing materials; can isomerize vigorously.

To Fight Fire: Water, foam, CO_2, dry chemical.

1,2,3-TRIS(CHLOROMETHOXY)PROPANE. $C_6H_{11}O_3Cl_3$, mw: 237.5, bp: 155° @ 19 mm, d: 1.3575 @ 17.5°/4°.

THR = An exper (+) carc and neo via dermal, sc, ip routes. [*3, 81*]

TRIS(1-CHLOROMETHYL-2-CHLOROETHYL) PHOSPHATE. $C_9H_{11}Cl_6PO_4$, mw: 430.9.

THR = An exper mutagen. [*90*]

TRIS(2,3-DIBROMOPROPYL)PHOSPHATE. $C_9H_{15}O_4PBr_6$, mw: 697.67. Syns: *TRIS-BP, USAF DO-41*. Crystals, d: 2.24, flash p: > 112°.

Acute tox data: Oral LD_{50} (rat) = 1010 mg/kg; ip LD_{LO} (mouse) = 300 mg/kg. [*3*]

THR = HIGH via ip, MOD via oral routes. Used to control flam of cloth (children's sleepwear). Can be absorbed via human skin, or chewed or sucked off of sleepwear by infants. It is an exper carc, mutagen, teratogen. Can cause testicular atrophy and sterility. [*110*]

TRIS(2-ETHYL HEXYL)PHOSPHITE. See trioctyl phosphite.

1,1,3-TRIS(*p*-GLYCIDYLOXY PHENYL)PROPANE.

THR = MOD via dermal and oral routes. A powerful allergen.

TRIS(HYDROXYMETHYL)AMINOMETHANE. Crystals. $(CH_2OH)_3CNH_2$, mw: 121.14, mp: 171°–172°, bp: 219° @ 10 mm.

Acute tox data: iv LD_{50} (mice) = 1210 mg/kg. [*3*]

THR = MOD via iv route. An irr to skin and mu mem.

Fire Hazard: Slight, when exposed to heat or flame.

Disaster Hazard: Mod dangerous; when heated to decomp, emits toxic fumes; can react with oxidizing materials.

TRIS(HYDROXYMETHYL)NITROMETHANE. Crystalline. $(CH_2OH)_3CNO_2$, mw: 151.12, mp: 165°–170°, bp: decomp.

Acute tox data: Oral LD_{50} (rat) = 1900 mg/kg. [*3*]

THR = MOD via oral route. Probably an irr. See nitrates.

TRISILANE. Syns: *trisilicon octahydride, trisilicopropane*. Liquid. Si_3H_8, mw: 92.2, mp: −117.4°, bp: 52.9°, d: 0.743 @ 0°, vap. press: 95.5 mm @ 0°.

THR = Details U. See silanes.

Fire Hazard: Dangerous, by chemical reaction. Decomp in water.

Explosion Hazard: Severe, by spont chemical reaction. Detonates spont in air. Violent reaction with CCl_4, O_2. [*19*]

Disaster Hazard: Dangerous; when heated to decomp, emits toxic fumes and can explode; will react with water or steam to produce hydrogen and toxic fumes; can react vigorously with oxidizing materials.

TRISILICON OCTAHYDRIDE. See trisilane.

TRISILICOPROPANE. See trisilane.

TRISILICYLAMINE. Syn: *nitrilotrisilane*. Liquid $(SiH_3)_3N$, mw: 107.26, mp: −105.6°, bp: 52°, d: 0.895 @ −106°.
THR = Details U. See amines and silanes. Spont flam in air. [*19*]
Disaster Hazard: Dangerous; when heated to decomp, emits highly toxic fumes; will react with water or steam to produce toxic and flam vapors.

TRISILYL ARSINE. $As(SiH_3)_3$, mw: 168.2.
THR = Very toxic. See arsenic compounds. Spont flam in air. [*19*]

TRISILYL PHOSPHINE. $(SiH_3)_3P$, mw: 124.3. Liquid.
THR = Spont flam in air. [*19*] See also phosphine.

TRIS(1-(2-METHYL)AZIRIDINYL)PHOSPHINE OXIDE. Syns: *MAPO*, *metapa*. Amber-colored liquid, amine odor, miscible with water and all organic solvents. $[N(CH_2)_2CH_3]P{:}O$, mw: 215, bp: 118°–125° @ 1 mm, d: 1.079 @ 25°/25°.
Acute tox data: Oral LD_{50} (rat) = 136 mg/kg; dermal LD_{50} (rat) = 183 mg/kg; oral LD_{50} (chicken) = 329 mg/kg; sc LD_{50} (mice) = 140 mg/kg; ip LD_{LO} (mice) = 3.125 mg/kg. [*3*]
THR = HIGH via oral, dermal, sc and ip routes. An exper (S) carc, mutagen and teratogen via oral, ip routes. [*3*, *8*] Highly toxic by skin absorption as well as by ingestion. Animal exper suggest cholinesterase inhibition, possibly due to metabolic products of this material in body.

1,3,5-TRIS(NITROMETHYL)BENZENE. $C_6H_3(CH_2NO_2)_3$, mw: 255.1.
THR = Very possibly violently unstable. [*19*]

TRISODIUM ARSENATE HEPTAHYDRATE. $3NaAsO_4 \cdot 7H_2O$, mw: 334.
THR = Very highly toxic by ip route on mice. An exper (±) carc via ip route. [*3*, *6*]

TRISODIUM-1,3,6-NAPHTHALENE TRISULFONATE. Light tan to buff powder or crystals. $C_{10}H_5(SO_3Na)_3$, mw: 434.4.
THR = U. Probably irr.
Disaster Hazard: Dangerous; see sulfonates.

TRISODIUM NITROPHOSPHATE. White powder.
THR = Details U. See trisodium phosphate.
Disaster Hazard: Dangerous; see nitrates and phosphates.

TRISODIUM PHOSPHATE DODECAHYDRATE. Syn: *sodium-o-phosphate*. $NA_3PO_4 \cdot 12H_2O$, mw: 380.21, mp: 73.3°–76.7° (decomp), $-12H_2O$ @ 100°, d: 1.62 @ 20°.
Acute tox data: ip LD_{50} (mice) = 430 mg/kg; oral LD_{50} (rat) = 7400 mg/kg. [*3*]
THR = HIGH via ip route and LOW via oral route. A sequestrant, general purpose, and nutrient and/or dietary supplement food additive. [*109*] A strong, caustic material.
Disaster Hazard: Dangerous; see phosphates.

TRISOL-2-(METHYL)AZIRIDINYLOPHOSPHINE OXIDE. See tris-[1-(2-methyl) aziridinyl] phosphine oxide.

TRISTATE SPECIAL NO. 1. See explosives, high.

TRISTEARYL BORATE. White solid, odor of stearyl alcohol. $[CH_3(CH_2)_{16}CH_2O]_3B$, mw: 819.25, mp: 49.8°–54°, bp: 300°–331° @ 0.3 mm.
THR = See boron compounds and stearyl alcohol.

TRIS(TRIFLUOROMETHYL)PHOSPHINE. $(CF_3)_3P$, mw: 238.
THR = Spont flam in air. [*19*]

TRIS(TRIMETHYL SILYL)PHOSPHINE. $[(CH_3)_3Si]_3P$, mw: 250.5.
THR = Spont flam in air. [*19*]

TRITETRADECYL BORATE. Syn: *tri(7-ethyl-2-methyl-4-undecyl)borate*. Light straw-yellow, viscous liquid, slight odor. $(C_{14}H_{29}O)_3B$, mw: 1658.3, bp: 225° @ 0.6 mm, flash p: 395°F (COC), d: 0.846 @ 26°, vap. d: 22.5.
THR = See boron compounds and esters.
Fire Hazard: Slight, when exposed to heat or flame; can react with oxidizing materials.
To Fight Fire: Foam, CO_2, dry chemical.

TRI(TETRAHYDROFURFURYL)BORATE. Mobile yellow liquid, characteristic odor. $(C_5H_9O)_3BO_3$, mw: 314.2, bp: 321°, flash p: 295°F (COC), d: 1.103 @ 26.2°, vap. d: 10.8.
THR = See boron compounds and esters.
Fire Hazard: Slight, when exposed to heat or flame; can react with oxidizing materials.
To Fight Fire: Water, foam, CO_2, dry chemical.

TRITHION. See carbophenothion.

TRI-*n*-TIN BUTYL HYDRIDE. $C_{12}H_{27}SnH$, mw: 291.1.
Acute tox data: Inhal LC_{LO} (mouse) = 1460 mg/m^3. [*3*]
THR = MOD via inhal route. See also hydrides and tin compounds.

TRITIUM. Syn: *hydrogen-3*. A colorless, radioactive gaseous isotope of hydrogen. T_2, mw: 6.05.
THR = See hydrogen. For radiological information on tritium, see hydrogen.

TRITOLYL AMINE PERCHLORATE. $C_{15}H_{28}O_4NCl$, mw: 321.8.
THR = Very unstable; can detonate at >123°. [*19*]

TRITON WR-1339.
THR = An exper teratogen via ip and pa routes. [*3*]

TRI-*m*-TOLYL BORON. Colorless crystals, insol in water, sol in benzene and ether. $B(CH_3C_6H_4)_3$, mw: 284.2, mp: 175°, bp: 233° @ 12 mm.
THR = Details U. See boron compounds.

TRI-*m*-TOLYL GERMANIUM BROMIDE. White crystals, sol in organic solvents. $(C_6H_4CH_3)_3GeBr$, mw: 425.9, mp: 79°, bp: 223° @ 1 mm.
THR = See germanium compounds and bromides.

TRI-*o*-TOLYL GERMANIUM BROMIDE. Colorless oil. $BrGe(C_6H_4CH_3)_3$, mw: 425.9, bp: 205°–210° @ 1 mm.
THR = See germanium compounds and bromides.

TRI-*p*-TOLYL GERMANIUM BROMIDE. Colorless crystals, sol in petroleum ether. $(C_6H_4CH_3)_3GeBr$, mw: 425.9, mp: 129°.
THR = See germanium compounds and bromides.

TRI-*m*-TOLYL GERMANIUM CHLORIDE. Small crystals, sol in petroleum ether. $C_{21}H_{21}GeCl$, mw: 381.4, mp: 85°, bp: 221°–224° @ 4 mm.
THR = See germanium compounds and chlorides.

TRI-*o*-TOLYL GERMANIUM CHLORIDE. Colorless oil. $C_{21}H_{21}GeCl$, mw: 381.4.
THR = See germanium compounds and chlorides.

TRI-*p*-TOLYL GERMANIUM CHLORIDE. White crystals, sol in petroleum ether. $C_{21}H_{21}GeCl$, mw: 381.4, mp: 121°.
THR = See germanium compounds and chlorides.

TRI-*o*-TOLYL GERMANIUM HYDROXIDE. White powder. $C_{21}H_{21}GeOH$, mw: 363.0, bp: 212°–214° @ 1 mm.
THR = Details U. See germanium compounds.

TRI-*m*-TOLYL GERMANIUM OXIDE. White crystals, insol in water. $C_{42}H_{42}Ge_2O$, mw: 708.0, mp: 125°.
THR = Details U. See germanium compounds.

TRI-*p*-TOLYL GERMANIUM OXIDE. White crystals, insol in water. $C_{42}H_{42}Ge_2O$, mw: 708.0, mp: 150°.
THR = Details U. See germanium compounds.

TRI-*o*-TOLYL PHOSPHATE. See tri-*o*-cresyl phosphate.

TRI-*o*-TOLYLTIN BROMIDE. Crystals, sol in benzene and ether. $C_{21}H_{21}SnBr$, mw: 472.0, mp: 99.5°.
THR = See tin compounds and bromides.

TRI-*p*-TOLYLTIN BROMIDE. Crystals, sol in benzene and ether. $C_{21}H_{21}SnBr$, mw: 472.0, mp: 98.5°.
THR = See tin compounds and bromides.

TRI-*m*-TOLYLTIN CHLORIDE. Crystals. $C_{21}H_{21}SnCl$, mw: 427.5, mp: 108°.
THR = See tin compounds and chlorides.

TRI-*o*-TOLYLTIN CHLORIDE. Crystals, sol in benzene and ether. $C_{21}H_{21}SnCl$, mw: 427.5, mp: 99.5°.
THR = See tin compounds and chlorides.

TRI-*p*-TOLYLTIN CHLORIDE. Crystals, slightly sol in organic solvents. $C_{21}H_{21}SnCl$, mw: 427.5, mp: 97.5°.
THR = See tin compounds and chlorides.

TRI-*p*-TOLYLTIN HYDROXIDE. Solid. $C_{21}H_{21}SnOH$, mw: 409.1, mp: 109°.
THR = Details U. See tin compounds.

TRI-*o*-TOLYL TIN IODIDE. Crystals, sol in benzene and ether. $C_{21}H_{21}SnI$, mw: 519.0, mp: 119.5°.
THR = See tin compounds and iodides.

TRI-*m*-TOLYL-*p*-TOLYL GERMANIUM. White powder, sol in methyl alcohol. $C_{28}H_{28}Ge$, mw: 437.1, mp: 100°.
THR = Details U. See germanium compounds.

TRITOPINE. Syn: *laudanidine*. White crystalline alkaloid. $C_{20}H_{25}NO_4$, mw: 343.41, mp: 166°.
THR = A poison.
Disaster Hazard: Dangerous; when heated to decomp, emits toxic fumes.

TRI(TRIPHENYL GERMANIUM)NITRIDE. Colorless crystals, hydrolyzed by water. $C_{54}H_{45}Ge_3N$, mw: 925.7, mp: 164°.
THR = Details U. See ammonia and germanium compounds.

TRIVINYL ANTIMONY. C_6H_9Sb, mw: 202.9.
THR = Spont flam in air. [*19*] See also antimony.

TRIVINYL BISMUTHINE. $Bi(CH_2CH)_3$, mw: 290.1.
THR = No data. See bismuth compounds. Spont flam in air. [*19*]

TRI-*p*-XYLYLBORON. Colorless crystals, insol in water, sol in organic solvents. $BC_{24}H_{27}$, mw: 326.3, mp: 147°, bp: 221° @ 12 mm.
THR = Details U. See also boron compounds.

TRI-*p*-XYLYLTIN BROMIDE. Crystals, sol in benzene and ether. $C_{24}H_{27}SnBr$, mw: 514.1, mp: 151°.
THR = See tin compounds and bromides.

TRI-*p*-XYLYLTIN CHLORIDE. Crystals, sol in organic solvents. $C_{24}H_{27}SnCl$, mw: 469.6, mp: 141.5°.
THR = See tin compounds and chlorides.

TRI-*m*-XYLYLTIN FLUORIDE. Crystals, sol in benzene, ether and alcohol. $C_{24}H_{27}SnF$, mw: 453.2, mp: 205°.
THR = See fluorides and tin compounds.

TRI-*p*-XYLYLTIN FLUORIDE. Crystals, slightly sol in benzene. $C_{24}H_{27}SnF$, mw: 453.2, mp: 247°.
THR = See fluorides and tin compounds.

TRI-*p*-XYLYLTIN IODIDE. Crystals, sol in benzene, ether and chloroform. $C_{24}H_{27}SnI$, mw: 561.1, mp: 159.5°.
THR = See tin compounds and iodides.

TROILITE. See ferrous sulfide.

TROJAN COAL POWDERS. See explosives, high.

"TROLUOIL." Water-white liquid. bp: 90°–96°, flash p: 25°F, d: 0.741 @ 15.5°.
THR = U. See gasoline.
Fire Hazard: Dangerous; flam liquid; can react vigorously with oxidizing materials.
To Fight Fire: Foam, CO_2, dry chemical.

TROPACOCAINE HYDROCHLORIDE. Crystalline salt, very refractive. $C_{15}H_{19}NO_2 \cdot HCl$, mw: 281.78, mp: 271°.
Acute tox data: iv LD_{LO} (rat) = 20 mg/kg; sc LD_{LO} (rabbit) = 400 mg/kg; ip LD_{LO} (guinea pig) = 170 mg/kg. [*3*]
THR = HIGH via sc, iv and ip routes. An alkaloid poison.
Disaster Hazard: Dangerous; see nitrates and chlorides.

TROPINE. Syn: *N-methyltropoline*. White crystalline solid. $C_8H_{15}NO$, mw: 141.21, mp: 63°, bp: 233°, d: 1.039 @ 76°/4°.
THR = HIGH via inhal and oral routes. An alkaloid poison.
Disaster Hazard: Dangerous; see nitrates.

TROPINE CARBOXYLIC ACID. See ecgonine.

TROPINE PLATINUM HYDROCHLORIDE. Orange-red monoclinic tablets. $C_{16}H_{30}NO_2 \cdot HCl_2PtCl_4$, mw: 692.31, mp: 198°–200°.
THR = See tropine, platinum compounds and chlorides.

TROPYLIUM PERCHLORATE. $(C_7H_7)ClO_4$, mw: 190.8.
Explosion Hazard: Has exploded upon being touched by glass. [*19*]

TRUE ARSENIC ACID. See *o*-arsenic acid.

TRYPAFLAVIN.
THR = An exper carc via sc route. [*3*]

TRYPAN BLUE. Blue-gray powder, water-sol. $C_{34}H_{24}N_6Na_4O_{14}S_4$, mw: 960.8.
Acute tox data: ip LD_{LO} (rat) = 300 mg/kg; sc LD_{LO} (rat) = 300 mg/kg; iv LD_{LO} (rat) = 300 mg/kg; sc LD_{50} (mice) = 267 mg/kg. [*3*]
THR = HIGH via ip, sc, iv routes. An exper (+) carc and teratogen. [*3, 4*] A (S) carc of lymphoid tissue.

TRYPARSAMIDE. Syn: *sodium phenylglycinamine-p-arsonate*. White crystalline powder. $C_{16}H_{20}O_8N_2As_2Na_2 \cdot \frac{1}{2}H_2O$, mw: 305.1.
Acute tox data: iv LD_{50} (rabbit) = 700 mg/kg; oral LD_{LO} (rabbit) = 200 mg/kg; im LD_{LO} (rat) = 250 mg/kg. [*3*]
THR = HIGH via oral and im; MOD via iv routes. See arsenic compounds.

TRYPTOPHANE. Syns: *indole-α-amino propionic acid, 1,α-amino-3-indole propionic acid*. An essential amino acid, occurs in isomeric forms. We consider the *l* and *dl* forms. White crystals; *dl:* slightly sol in water; *l:* sol in water, hot alcohol and alkali hydroxides, insol in chloroform. $C_{11}H_{20}O_2N_2$, mw: 204.2, mp: decomp @ 289°.
THR = U. A nutrient and/or dietary supplement food additive. [*109*]

TRYPTOPHAN MUSTARD. $C_{15}H_{19}O_2N_3Cl_2$, mw: 344.3.
Acute tox data: ic LD_{50} (rat) = 0.004 mg/kg. [*3*]
THR = VERY, VERY HIGH via ic route.

TSN. See tetramethyl succinonitrile.

TSPP. See tetrasodium pyrophosphate.

T-STUFF. See hydrogen peroxide.

TSUMACIDE. Syn: *m-tolyl-N-methyl carbamate*.
Acute tox data: Dermal LD_{50} (rat) = 6000 mg/kg; oral LD_{50} (mice) = 268 mg/kg; dermal LD_{50} (mice) = 6 mg/kg; oral LD_{50} (wild birds) = 100 mg/kg. [*3*]
THR = HIGH via oral and dermal routes. An insecticide. See also carbamates.

TTD. See bis(diethylthiocarbamyl) disulfide.

TUBATOXIN. See rotenone.

β,β-TUBATOXYTHIO CYANODIETHYL ETHER. See β-butoxy thiocyanodiethyl ether.

TUNGATES. See driers, paint, varnish, etc.

TUNG NUT MEALS.
THR = Toxic via oral route. Contact causes dermatitis. Ingestion ⟶ nausea, vomiting, cramps, diarrhea and tenesmus, thirst, dizziness, lethargy and disorientation. Large doses can cause fever, tachycardia and respiratory effects. See also saponin.
Fire Hazard: Mod, in the form of dust when exposed to heat or flame; process material and cool thoroughly before storage so as not to over-dry; can react with oxidizing materials.

TUNG OIL. Syn: *china wood oil*. Thick yellowish liquid. mp: 31°, flash p: 552°F (CC), d: 0.94, autoign. temp.: 855°F.
THR = U. May act as a weak allergen.
Fire Hazard: Slight, when exposed to heat or flame;

can react with oxidizing materials. Avoid contact or leakage with combustibles. Keep cool and well ventilated.
Spont Heating: Mod.
To Fight Fire: CO_2, dry chemical.

TUNGSTEN. Syn: *wolfram.* A steely-gray to white, cuttable, forgeable and spinnable metal. W, atwt: 183.85, mp: 3410°, bp: 5660°, d: 19.3 @ 20°.
Acute tox data: ip LD_{50} (rat) = 5000 mg/kg. [*3*]
THR = MOD via ip route. See also tungsten compounds.
Radiation Hazard: For permissible levels, see Section 5A, Table 5A.5. Artificial isotope ^{181}W, $T\frac{1}{2} = 140d$, decays to stable ^{181}Ta by ec; emits γ's of 0.01 MeV and x-rays. Artificial isotope ^{185}W, $T\frac{1}{2} = 75d$, decays to stable ^{185}Re via β's of 0.43 MeV.
Fire Hazard: Mod, in the form of dust when exposed to flame. See also powdered metals.

TUNGSTEN CARBIDE. WC, mw: 195.9.
Explosion Hazard: Violent reaction with F_2, ClF_3, IF_5, PbO_2, NO_2 and N_2O. [*19*]

TUNGSTEN CARBONYL. Colorless rhombic crystals. $W(CO)_6$, mw: 351.98, mp: 50° (sublimes), bp: 175°, d: 2.65, vap. press: 1.2 mm @ 67°, vap. d: 12.1.
THR = See carbonyls and tungsten compounds.

TUNGSTEN COMPOUNDS.
THR = Tungsten compounds are considered somewhat more toxic than those of molybdenum. However, industrially, this element does not constitute an important health hazard. Exposure is related chiefly to the dust arising from the crushing and milling of the two chief ores of tungsten, namely, scheelite and wolframite. There is very little published with reference to its toxicity. The feeding of 2, 5, and 10% of diet as tungsten metal over a period of 70 days has been shown to be without marked effect upon the growth of rats, as measured in terms of gain in weight. Ammonium-*p*-tungstate has been found to be much less toxic to rats upon ingestion than either tungstic oxide or sodium tungstate. Recent studies have failed to indicate any serious toxic effect following the inhal or ingestion of various tungsten compounds, although heavy exposure to the dust or the ingestion of large amounts of the soluble compounds produces a certain rate of mortality in exper animals.

TUNGSTEN DIOXIDE. WO_2, mw: 215.9.
Explosion Hazard: Reacts violently with Cl_2. [*19*]

TUNGSTEN DISULFIDE. Dark-gray crystals. WS_2, mw: 248.5, d: 7.5 @ 10°.
THR = See sulfides and tungsten compounds.

TUNGSTEN HEXAFLUORIDE. Light yellow liquid or colorless gas. WF_6, mw: 297.9, mp: 2.5°, bp: 17.5°, d(gas): 12.9 g/liter, d(liquid): 3.44.
THR = See fluorides and tungsten compounds.

TUNGSTEN OXYCHLORIDE. Syn: *tungsten oxytetrachloride.* Red needles. $WOCl_4$, mw: 341.75, mp: 211°, bp: 227.5°.
THR = MOD irr to skin, eyes and mu mem.
Disaster Hazard: Dangerous; see chlorides.

TUNGSTEN OXYTETRACHLORIDE. See tungsten oxychloride.

TUNGSTEN PHOSPHIDE. Dark gray prisms. W_2P, mw: 398.82, d: 5.21.
THR = See phosphides and tungsten compounds.

TUNGSTEN TRIOXIDE. WO_3, mw: 231.9. Heavy yellow powder. Insol in water, sol in caustic alkalis.
Acute tox data: Oral LD_{LO} (rat) = 840 mg/kg. [*3*]
THR = MOD via oral route. See also tungsten compounds.
Explosion Hazard: Can react violently with ClF_3, Li. [*19*]

TUNGSTEN TRISULFIDE. Chocolate-brown powder. WS_3, mw: 280.12.
THR = See sulfides and tungsten compounds.

TUPERSAN. Syn: *1-(2-methylcyclohexyl)-3-phenyl urea.* White crystalline material, very slightly water-sol, sol in alcohol. $C_{14}H_{20}N_2O$, mw: 232.3, mp: 121°.
Acute tox data: Oral LD_{50} (rat) = >7500 mg/kg. Oral LD_{50} (rat) = 5000 mg/kg. [*3*]
THR = MOD via oral route.

TURKEY RED OIL. Syn: *sulfonated castor oil.* A reddish viscid liquid, characteristic odor. flash p: 476°F (CC), d: 0.95, autoign. temp.: 833°F.
THR = An irr.
Fire Hazard: Slight, when exposed to heat or flame.
Spont Heating: No.
To Fight Fire: Alcohol foam, CO_2, dry chemical.

TURMERIC.
THR = An allergen and irr.
Fire Hazard: Slight, when exposed to heat or flame.

TURPENTINE (GUM). Syn: *gum thus.* Opaque, sticky yellowish masses.
THR = A mild allergen. See also turpentine oil.
Fire Hazard: Mod.

TURPENTINE OIL. Syn: *spirit of turpentine.* Colorless liquid, characteristic odor. Principally $C_{10}H_{16}$. mw: 136, bp: 154°–170°, lel = 0.8%, flash p: 95°F (CC), d: 0.854–0.868 @ 25°/25°, autoign. temp.: 488°F, vap. d: 4.84, ulc: 40–50.
Acute tox data: Inhal TC_{LO} (human) = 175 ppm $\longrightarrow$ irr effects. [*3*] Aspiration into lungs causes a chem-

ical pneumonitis. Oral LD (man) = 150 ml. Oral exposure ⟶ injury to GI tract and kidneys. [*20*] Dermal contact ⟶ erythema and itching. [*20*]
THR = HIGH via inhal route. Can cause serious irr of kidneys. A common air contaminant.
Fire Hazard: Mod, when exposed to heat or flame. Avoid impregnation of leakage with combustibles. Keep cool and ventilated.
Spont Heating: Yes.
Explosion Hazard: Mod, in the form of vapor when exposed to flame; can react violently with $Ca(OCl)_2$, Cl_2, CrO_3, $Cr(OCl)_2$, $SnCl_4$, hexachloromelamine, trichloromelamine. [*19*]
Disaster Hazard: Mod; when heated, emits acrid fumes; can react with oxidizing materials.
To Fight Fire: Foam, CO_2, dry chemical.

TURPENTINE SUBSTITUTES. See mineral spirits #10.

TWEEN 80.
THR = An exper carc via sc route. [*14*, *3*, *23*] See also surfactants.

TWEEN 20-TRIS. A (S) carc. [*14*] See surfactants.

TUTANE. See butylamine.

TYLOSIN. An antibiotic; crystals. $C_{45}H_{77}NO_{17}$, mw: 904.2, mp: 128°–130°.
Acute tox data: ip LD_{50} (mice) = 594 mg/kg. [*3*]
THR = MOD via ip route. A food additive permitted in the feed and drinking water of animals and/or for the treatment of food-producing animals; also permitted in food for human consumption. [*109*]

TYPE-CLEANING COMPOUNDS, LIQUID.
THR = Variable; may contain benzene, carbon tetrachloride or other toxic materials. See specific components.

TYPES OF RESPIRATORS FOR HAZARD CONTROL. See Section 2.

TYPEWRITING RIBBONS.
THR = A mild allergen.

TYRAMINE. See ergot.

TYRENE. See 2,4,6-trichloroanisole.

TYROSINE. Syns: *β,p-hydroxyphenyl alanine*, *α-amino-β,p-hydroxy phenyl propionic acid.* A non-essential amino acid; white crystals, sol in water, slightly sol in alcohol, insol in ether, optically active. $C_9H_{11}O_3$, mw: 181.2.
THR = U. A nutrient and/or dietary supplement food additive. [*109*]

U

UC10854. Syn: *3-isopropyl phenyl methyl carbamate.* $C_{11}H_{15}O_2N$, mw: 193.3.
Acute tox data: Oral LD_{50} (rat) = 29 mg/kg; ip LD_{50} (rat) = 6 mg/kg; iv LD_{50} (rat) = 3.15 mg/kg; im LD_{50} (rat) = 14 mg/kg; oral LD_{50} (wild bird) = 3 mg/kg; oral LD_{50} (chicken) = 12 mg/kg. [*3*]
THR = HIGH via oral, ip, iv and im routes.

UINTAITE. See gilsonite.

ULTRASENE. Syn: *kerosene, deodorized.* Insol in water, flash p: 175°F.
THR = See kerosene.
Fire Hazard: Mod, when exposed to heat, flame or oxidizers.
To Fight Fire: Water, foam, mist.

ULTRAVIOLET RADIATION.
Radiation Hazard: Excessive ultraviolet radiation exposure can cause acute inflammation of the eyes as well as burns of the skin. A recog carc. [*14*] See ionizing radiation.

UNDECANAL. Syn: *hendecanal.* Colorless liquid. $[CH_3(CH_2)_9]CHO$, mw: 170.29, mp: −4°, bp: 117° @ 18 mm, flash p: 235°F (COC), d: 0.830 @ 20°/4°, vap. press: 0.04 mm @ 20°, vap. d: 5.94.
THR = MOD irr to skin, eyes and mu mem. See also aldehydes.
Fire Hazard: Slight, when exposed to heat or flame; can react with oxidizing materials.
To Fight Fire: CO_2, dry chemical.

UNDECANE. See hendecane.

UNDECANOL-2. Syn: *hendecanol-2.* Colorless liquid, insol in water, sol in alcohol and ether. $C_{11}H_{24}O$, mw: 172, d: 0.8363 @ 20°, mp: 12°, bp: 228°–229°, flash p: 235°F (OC).
Acute tox data: Oral LD_{50} (rat) = 3000 mg/kg; dermal LD_{50} (rabbit) = 4760 mg/kg. [*3*]
THR = MOD via oral and dermal routes. See alcohols.
Fire Hazard: Slight, when exposed to heat or flame.
To Fight Fire: Dry chemical, CO_2.

UNDECYLENIC ACID. Bright, clear, mobile liquid or crystals. $C_{11}H_{20}O_2$, mw: 184.27, mp: 24.5°, bp: 160° @ 10 mm, flash p: 295°F (COC), d: 0.910 @25°/25°.
Acute tox data: Oral LD_{50} (rat) = 2500 mg/kg; oral LD_{50} (mice) = 2300 mg/kg; ip LD_{50} (mice) = 960 mg/kg. [*3*]
THR = MOD via oral and ip routes. Ingestion may cause nausea, vomiting and urticaria.
Fire Hazard: Mod, when exposed to heat or flame; can react with oxidizing materials.
To Fight Fire: Foam, CO_2, dry chemical.

UNDECYL FLUOROACETATE. $C_{13}H_{25}O_2F$, mw: 232.4.
Acute tox data: ip LD_{50} (mice) = 60 mg/kg. [*3*]
THR = HIGH via ip route. See also fluorides.

UNDECYLIC ACID. $C_{11}H_{22}O_2$, mw: 186.3.
Acute tox data: iv LD_{50} (mice) = 140 mg/kg. [*3*]
THR = HIGH via iv route.

UNDERWRITERS LABORATORIES CLASSIFICATION (ULC). See Section 7.

UNITED NATIONS LABELING SYSTEM. See Section 11.

UNITS OF ENERGY. See Section 5A.

UNITS OF RADIATION DOSE. See Section 5A.

UNITS OF RADIATION INTENSITY. See Section 5A.

UNOXEPOXIDE 201. $C_{16}H_{24}O_4$, mw: 280.4.
THR = An exper (+) carc. [*3, 13*]

UNSLAKED LIME. See calcium oxide.

URANIUM. A heavy, silvery-white, malleable, ductile, softer-than-steel metal. U. atwt: 238.029, mp: 1132°, bp: 3818°, d: 18.95 (ca).
Radiation Hazard: For permissible levels, see Section 5A, Table 5A.5. Natural U consists of ^{238}U and ^{234}U in radioactive equilibrium, plus 0.7% of ^{235}U. Enriched U has increased percentages of the lighter isotopes, ^{234}U and ^{235}U. Its specific activity, and consequently its radiation hazard, are correspondingly increased. Natural isotope ^{232}U, $T\frac{1}{2}$ = 72y, decays to radioactive ^{228}Th via α's of 5.3 MeV. Artificial isotope ^{233}U, (neptunium series); $T\frac{1}{2} = 1.6 \times 10^5$y, decays to radioactive ^{229}Th via α's of 4.8 MeV. Natural isotope ^{224}U (U II; Uranium Series), $T\frac{1}{2} = 2.5 \times 10^5$y, decays to radioactive ^{230}Th via α's of 4.7–4.8 MeV. Natural (0.72%) isotope ^{235}U (actino-uranium; actinium series), $T\frac{1}{2} = 7 \times 10^8$y, decays to radioactive ^{231}Th via α's of 4.3–4.6 MeV. Artificial isotope ^{236}U, $T\frac{1}{2} = 2.4 \times 10^7$y, decays to radioactive ^{232}Th via α's of 4.5 MeV. Artificial isotope ^{237}U, $T\frac{1}{2}$ = 7d, decays to radioactive ^{237}Np via β's of 0.25 MeV. Natural isotope ^{238}U (U I; uranium series), $T\frac{1}{2} = 4.5 \times 10^9$y, decays to radioactive ^{234}Th via α's of 4.2 MeV.

THR = A recog carc [*14*] and a highly toxic element on an acute basis. The permissible levels for soluble compounds are based on chemical toxicity, while the permissible body level for insol compounds is based on radiotoxicity. The high chemical toxicity of U and its salts is largely shown in kidney damage, and acute necrotic arterial lesions. The rapid passage of sol U compounds through the body tends to allow relatively large amounts to be taken in. The high toxicity effect of insol compounds is largely due to lung irradiation by inhaled particles. This material is transferred from the lungs of animals quite slowly.
Fire Hazard: Dangerous, in the form of a solid or dust when exposed to heat or flame.
Explosion Hazard: It can react violently with air, Cl_2, F_2, HNO_3, NO, Se, S, water. [*19*]
Disaster Hazard: For further information, refer to AEC publications as listed in bibliography for Section 5A. See also Section 5A.

URANIUM AMMONIUM FLUORIDE. Syn: *ammonium uranium fluoride*. Greenish-yellow crystalline powder. $UO_2F_2 \cdot 3NH_4F$, mw: 419.2.
THR = A recog carc. [*14*] See uranium and fluorides.

URANIUM AMMONIUM PENTAFLUORIDE. Tetragonal crystals $(NH_4)_3UO_2F_5$, mw: 419.2, mp: decomp.
THR = See uranium and fluorides.

URANIUM BARIUM OXIDE. Syn: *barium diuranate*. Yellow or orange powder. BaU_2O_7, mw: 725.50.
THR = HIGH. A recog carc. [*14*] See uranium and barium compounds.

URANIUM BORIDE. Syn: *uranium diboride*. Hexagonal crystals. UB_2, mw: 248.9, mp: 2365°, d: 12.70.
THR = A recog carc. [*14*] See uranium and boron compounds.
Radiation Hazard: See uranium.
Fire Hazard: Mod; can react with moisture or acids to evolve boron hydride. See also boron hydrides.
Explosion Hazard: Mod; can react with moisture or acids to evolve boron hydride. See also boron hydrides.
Disaster Hazard: Mod; will react with water or steam to produce toxic and flam vapors; can react with oxidizing materials; on contact with acid or acid fumes, can emit toxic fumes.

URANIUM CARBIDE. Gray crystals. UC_2, mw: 262.09, mp: 2260°, bp: 4100°, d: 11.28 @ 18°.
THR = A recog carc. [*14*] See uranium and carbides.
Explosion Hazard: Can react violently with air, Br_2, Cl_2, F_2, H_2O. [*19*]

URANIUM DIBORIDE. See uranium boride.

URANIUM DIOXIDE. Rhombic or cubic brown-black crystals. UO_2, mw: 270.07, mp: 2176°, d: 10.9.
THR − A recog carc. [*14*] See uranium.
Fire Hazard: Self-ignites in air. [*19*]

URANIUM DISULFIDE. Tetragonal gray-black crystals. US_2, mw: 302.20, mp: > 1100°.
THR = A recog carc. [*14*] See uranium and sulfides.

URANIUM HEXAFLUORIDE. Monoclinic, colorless to pale yellow deliquescent crystals. Volatile. UF_6, mw: 352.07, mp: 69.2° @ 2 atm, bp: 56.2° @ 764.6 mm, d: 5.09 @ 20.7°, vap. press: 100 mm @ 18.2°.
THR = A radioactive and very corrosive material. HIGH. See uranium and fluorides.

URANIUM HYDRIDE. UH_3, mw: 241.1.
THR = HIGH. See uranium; also see hydrides.
Fire Hazard: Spont flam in air. [*19*]

URANIUM NEODYMIUM ALLOY.
Explosion Hazard: Has exploded during pickling with HNO_3. [*19*] See also uranium.

URANIUM NITRATE. See uranium and nitrates.

URANIUM NITRIDE. Brown-black crystals. UN, mw: 252, d: 14.31, mp: 2630 ± 50°. [*2*]
THR = A recog carc. [*14*] See uranium and nitrides.
Fire Hazard: Can self-ignite in air. [*19*]

URANIUM OXIDES. UO_2, UO_3, U_3O_8.
THR = HIGH. A recog carc. [*14, 2, 3*] Can react violently with BrF_3. [*19*]

URANIUM OXYACETATE. Syn: *uranyl acetate*. Rhombic yellow crystals. $UO_2(C_2H_3O_2)_2 \cdot 2H_2O$, mw: 424.19, mp: $-2H_2O$ @ 110°, bp: 275° (decomp), d: 2.893 @ 15°.
Acute tox data: ip LD_{50} (mice) = 400 mg/kg. [*3*]
THR = HIGH via ip route. A recog carc. [*14*] See uranium.

URANIUM OXYAMMONIUM CARBONATE. Syn: *uranyl ammonium carbonate*. Monoclinic yellow crystals. $2(NH_4)_2CO_3 \cdot UO_2CO_3 \cdot 2H_2O$, mw: 558.29, mp: decomp @ 100°, d: 2.773.
THR = A recog carc. [*14*] See uranium.

URANIUM OXYAMMONIUM CHLORIDE. Syn: *uranyl ammonium chloride*. Greenish-yellow deliquescent crystals. $UO_2Cl_2 \cdot 2NH_4Cl \cdot 2H_2O$, mw: 484.0.
THR = A recog carc. [*14*] See uranium and chlorides.

URANIUM OXYBENZOATE. Syn: *uranyl benzoate*. Yellow powder. $UO_2(C_7H_5O_2)_2$, mw: 512.29.
THR = A recog carc. [*14*] See uranium.

URANIUM OXYBROMIDE. Syn: *uranyl bromide*. Green-yellow, hygroscopic needles. UO_2Br_2, mw: 429.90.
THR = A recog carc. [*14*] See uranium and bromides.

URANIUM OXYCHLORIDE. Syn: *uranyl chloride.* Yellow deliquescent crystals. UO_2Cl_2, mw: 340.9, mp: < red heat.
Acute tox data: ip LD_{50} (mice) = 7 mg/kg. [3]
THR = HIGH via ip route. A recog carc. [14] See uranium and chlorides.

URANIUM OXYFLUORIDE. $U(OF)_2$, mw: 308.0.
Acute tox data: ip LD_{LO} (rat) = 40 mg/kg; iv LD_{50} (rat) = 40 mg/kg. [3]
THR = HIGH via ip and iv routes. See also fluorides and uranium.

URANIUM OXYFORMATE. Syn: *uranyl formate.* $UO_2(CHO_2)_2 \cdot H_2O$, mw: 378.12, mp: $-H_2O$ @ 110°, d: 3.695 @ 19°.
THR = A recog carc. [14] See uranium.

URANIUM OXYIODATE. Syn: *uranyl iodate.* Rhombic yellow crystals. $UO_2(IO_3)_2$, mw: 619.91, mp: decomp 250°, d: 5.2.
THR = A recog carc. [14] See uranium and iodates.

URANIUM OXYIODIDE. Syn: *uranyl iodide.* Red, deliquescent crystals. UO_2I_2, mw: 523.91, mp: decomp in air.
THR = A recog carc. [14] See uranium and iodides.

URANIUM OXYMONOHYDROGEN PHOSPHATE. Syn: *uranyl monohydrogen phosphate.* Tetragonal yellow plates. $UO_2HPO_4 \cdot 4H_2O$, mw: 438.12.
THR = A recog carc. [14] See uranium and phosphates.

URANIUM OXYNITRATE. Syn: *uranyl nitrate.* Rhombic, yellow deliquescent crystals. $UO_2(NO_3)_2 \cdot 6H_2$, mw: 502.18, mp: 60.2° (decomp) @ 100°), d: 2.807 @ 13°.
Acute tox data: ip LD_{50} (rat) = 135 mg/kg; sc LD_{50} (dog) = 4 mg/kg; iv LD_{LO} (rat) = 2 mg/kg; oral LD_{LO} (cat) = 238 mg/kg. [3]
THR = HIGH via ip, sc, iv and oral routes. A corrosive, irr material. A recog carc. [14] See uranium and nitrates.

URANIUM OXYOXALATE. Syn: *uranyl oxalate.* Yellow crystals. $UO_2C_2O_4 \cdot 3H_2O$, mw: 412.14, mp: $-H_2O$ @ 110°.
THR = A recog carc. [14] See uranium and oxalates.

URANIUM OXYPERCHLORATE. Syn: *uranyl perchlorate.* Yellow deliquescent crystals. $UO_2(ClO_4)_2 \cdot 6H_2O$, mw: 545.07, mp: 90° (decomp @ 110°).
THR = A recog carc. [14] See uranium and perchlorates. Can explode with ethanol. [19]

URANIUM OXYSULFATE. Syn: *uranyl sulfate.* Yellow-green crystals. $UO_2SO_4 \cdot 2H_2O$, mw: 420.18, mp: decomp @ 100°, d: 3.28 @ 16.5°.
THR = A recog carc. [14] See uranium and sulfates.

URANIUM OXYSULFIDE. Syn: *uranyl sulfide.* Brown-black tetragonal crystals. UO_2S, mw: 302.14, mp: decomp @ 40°–50°.
THR = A recog carc. [14] See uranium and sulfides.

URANIUM OXYSULFITE. Syn: *uranyl sulfite.* Pale green crystals. $UO_2SO_3 \cdot 4H_2O$, mw: 422.20.
THR = A recog carc. [14] See uranium and sulfites.

URANIUM PENTACHLORIDE. Dark green-gray needles, red by transparent light, deliquescent. UCl_5, mw: 415.36, mp: decomp @ 120°.
THR = A recog carc. [14] See uranium and chlorides.

URANIUM PEROXIDE. Pale yellow, hygroscopic crystals, $UO_4 \cdot 2H_2O$, mw: 338.10, mp: decomp @ 115°.
THR = A recog carc. [14] See uranium and peroxides, inorganic.

URANIUM PHOSPHATE. Syns: *mono-uranyl-o-phosphate, uranyl phosphate.* Tetragonal yellow plates, sol in acids. $UO_2HPO_4 \cdot 4H_2O$, mw: 438.1.
THR = A recog carc. [14] See uranium and phosphates.

URANIUM PHOSPHIDE. U_3P_4, mw: 838.1.
THR = Very toxic on an acute basis; see also phosphine. A recog carc. [14] In contact with HCl, liberates the self-igniting PH_3. [19]

URANIUM POTASSIUM NITRATE. Syns: *potassium uranium nitrate, uranyl potassium nitrate.* Yellow crystalline powder. $(KNO_3)_2 \cdot UO_2(NO_3)_2$, mw: 394.086.
THR = A recog carc. [14] See uranium and nitrates.

URANIUM POTASSIUM SULFATE. Syn: *potassium uranyl sulfate.* Monoclinic yellow crystals. $K_2SO_4 \cdot UO_2SO_4 \cdot 2H_2O$, mw: 576.41, mp: $-2H_2O$ @ 120°, d: 3.363 @ 19.1°.
THR = A recog carc. [14] See uranium and sulfates.

URANIUM SESQUISULFIDE. Gray-black needles. U_2S_9, mw: 572.34, mp: ignites in air.
THR = A recog carc. [14] See uranium and sulfides.

URANIUM SODIUM OXYACETATE. Syn: *sodium uranyl acetate.* Yellow crystals. $NaUO_2(C_2H_3O_2)_3$, mw: 470.20, d: 2.56.
THR = A recog carc. [14] See uranium.

URANIUM STRONTIUM OXIDE. Syn: *strontium diuranate.* Yellow powder. SrU_2O_7, mw: 675.77.
THR = A recog carc. [14] See uranium.

URANIUM TETRABROMIDE. Brown deliquescent leaflets. UBr_4, mw: 557.73, bp: volatilizes, d: 4.84 @ 21°/4°.
THR = A recog carc. [14] See uranium and bromides.

URANIUM TETRACHLORIDE. Cubic, dark green-gray deliquescent crystals. UCl_4, mw: 379.8, mp: 590°, bp: 791°, d: 4.725 @ 25°/4°.

Acute tox data: ip LD_{50} (rat) = 335 mg/kg; iv LD_{50} (rat) = 400 mg/kg. [*3*]
THR = HIGH via ip and iv routes. A recog carc. [*14*] See uranium and chlorides.

URANIUM TETRAFLUORIDE. Green amorphous powder. UF_4, mw: 314.07, mp: approx @ > 1100°.
THR = A recog carc. [*14*] See uranium and fluorides.

URANIUM TETRAHYDROBORATE. $U(BH_4)_3$, mw: 282.5.
THR = HIGH. A recog carc. [*14*] Self-ignites in air. [*19, 2*] See uranium compounds.

URANIUM TETRAIODIDE. Black needles. UI_4, mw: 745.75, mp: 500°, d: 5.6 @ 15°.
THR = A recog carc. [*14*] See uranium and iodides.

URANIUM TRIBROMIDE. Dark brown hygroscopic needles. UBr_3, mw: 477.82, bp: volatilizes.
THR = A recog carc. [*14*] See uranium and bromides.

URANIUM TRICHLORIDE. Dark red hygroscopic needles. UCl_3, mw: 344.44, d: 5.44 @ 25°/4°.
THR = A recog carc. [*14*] See uranium and chlorides.

URANIUM TRIHYDRIDE. Black metallic crystals, reacts with water. UH_3, mw: 241.1, d: 10.5, mp: decomp @ > 300°.
THR = A recog carc. [*14*] See uranium and hydrides.

URANIUM TRIOXIDE. Syn: *uranyl oxide*. Yellow-red powder, sol in acids. UO_3, mw: 286.07, mp: decomp, d: 7.29.
THR = A recog carc. [*14*] See uranium.

URANIUM X. See protactinium.

URANOUS SULFATE. Rhombic green crystals. $U(SO_4)_2 \cdot 4H_2O$, mw: 502.27, mp: $-4H_2O$ @ 300°.
THR = A recog carc. [*14*] See uranium and sulfates.

URANYL ACETATE. See uranium oxyacetate.

URANYL AMMONIUM CARBONATE. See uranium oxyammonium carbonate.

URANYL AMMONIUM CHLORIDE. See uranium oxyammonium chloride.

URANYL BENZOATE. See uranium oxybenzoate.

URANYL BROMIDE. See uranium oxybromide.

URANYL CHLORIDE. See uranium oxychloride.

URANYL FORMATE. See uranium oxyformate.

URANYL IODATE. See uranium oxyiodate.

URANYL IODIDE. See uranium oxyiodide.

URANYL MONOHYDROGEN PHOSPHATE. See uranium oxymonohydrogen phosphate.

URANYL NITRATE. See uranium oxynitrate.

URANYL OXALATE. See uranium oxyoxalate.

URANYL OXIDE. See uranium trioxide.

URANYL PERCHLORATE. See uranium oxyperchlorate.

URANYL PHOSPHATE. See uranium phosphate.

URANYL POTASSIUM NITRATE. See uranium potassium nitrate.

URANYL SULFATE. See uranium oxysulfate.

URANYL SULFIDE. See uranium oxysulfide.

URANYL SULFITE. See uranium oxysulfite.

***p*-URAZINE.** $C_2H_4O_2N_4$, mw: 116.1.
THR = Can explode, possibly due to an impurity. [*19*]

URBACID. Syn: *bis(dimethylthiocarbamoylthio)methylarsine*. Colorless crystals, not sol in water, sol in many organic solvents. $C_7H_{15}N_2S_4As$, mw: 330.3, mp: 144°.
Acute tox data: Oral LD_{50} (rat) = 100 mg/kg. [*3*]
THR = HIGH via oral route. A highly toxic organic arsenical fungicide. [*14*] See arsenic compounds.

UREA. See carbamide.

UREA HYDROGEN PEROXIDE. See urea peroxide.

UREA NITRATE.
$CO(NH_2)_2HNO_3$ (varying with moisture content), mw: 123.08.
THR = A mild irr. See also nitrates.
Fire Hazard: See nitrates.
Explosion Hazard: See nitrates.
Disaster Hazard: See nitrates.

UREA PEROXIDE. Syn: *urea hydrogen peroxide*. White crystals. $CO(NH_2)_2 \cdot H_2O_2$, mw: 94.08, mp: 75°–85° (decomp).
THR = A MOD irr to skin, eyes and mu mem. See also peroxides, organic.
Fire Hazard: See peroxides, organic.
Disaster Hazard: See peroxides, organic.

α-UREIDO ISOBUTYRIC ACID LACTAM. See dimethylhydantoin.

URETHAN ALDEHYDE CONDENSATE. A susp carc. [*14*] See carbamates.

URETHANE. See ethyl carbamate.

URIC ACID. Syns: *lithic acid, uric oxide*. White crystals. $CO(NH)_2COC_2CO(NH)_2$, mw: 168.11, mp: decomp to hydrogen cyanide, d: 1.855–1.893.
THR = Can evolve hydrogen cyanide when heated. See also cyanides.
Disaster Hazard: Dangerous; when heated to decomp, emits highly toxic fumes of cyanides.

URIC OXIDE. See uric acid.

URITONE. See hexamethylenctetramine.

UROTROPINE. See hexamethylenetetramine.

UROX. Syns: *3-(p-chlorophenyl)-1,1-dimethylurea trichloroacetate.* $CCl_3COOC_6H_4ClNCON(CH_3)_2$, mw: 360.1.
Acute tox data: Oral LD_{50} (rat) = 2300 mg/kg. [*3*]
THR = MOD via oral route.
Disaster Hazard: Dangerous; when heated to decomp, emits highly toxic fumes.

UROXIN. See alloxantin.

URSOL. See *p*-phenylenediamine.

URSOL P. See *p*-aminophenol.

URUSHIOL. Pale yellow liquid, sol in alcohol and ether. Main constituent of the irr oil of poison ivy. Composition: a mixture of several derivatives of catechol.
THR = A powerful allergen, particularly to sensitive individuals. Contact with the 3-pentadecyl catechol can result in typical poison ivy dermatitis in 24–36 hrs. The sensitizer is not volatile but can be transported about as smoke from burning leaves or the barks of the plants. This material is also contained by the cashew family, mango, pistachio and other rare tropical fruits. [*20*] Systemic effects are rare. The active irr of poison ivy. See also pentadecylcatechol.

UV RADIATION. See ultraviolet radiation.

UX_2. See protactinium.

V

VALERAL. See *n*-pentanal.

***n*-VALERALDEHYDE.** See pentanal.

VALERIC ACID. Syn: *pentanoic acid.* Colorless liquid, unpleasant odor, somewhat water-sol. $CH_3CH_2CH_2CH_2COOH$. mw: 102.1, d: 0.939 @ 20°/4°, mp: −34.5°, bp: 187°, flash p: 205°F (OC).
Acute tox data: iv LD_{50} (mice) = 1290 mg/kg, oral LD_{50} (mice) = 500 mg/kg, sc LD_{50} (mice) = 3590 mg/kg. [*3*]
THR = HIGH irr via oral route. MOD via iv and sc routes.
Fire Hazard: Low.
To Fight Fire: Dry chemical, foam, spray, CO_2.

γ-VALEROLACTONE. Colorless mobile liquid. ($C_5H_8O_2$), mw: 100.06, mp: −31°, bp: 205°–206.5°, flash p: 205°F (COC), d: 1.0518 at 25°/25°, vap. d: 3.45.
Acute tox data: Oral LD_{50} (rabbit) = 2604 mg/kg. [*3*]
THR = MOD via oral route. See also lactones.
Fire Hazard: Mod, when exposed to heat or flame; can react with oxidizing materials.
To Fight Fire: Water, foam, CO_2, dry chemical.

VALERYL CHLORIDE.
THR = A corrosive irr to skin, eyes and mu mem.

VALERYL DIETHYLAMIDE. See valyl.

VALINE. Syn: *α-amino isovaleric acid.* An essential amino acid, white crystalline solid, sol in water, very slightly sol in alcohol, insol in ether. $(CH_3)_2CHCH(NH_2)COOH$, mw: 117, mp(*dl*): 298° (decomp), mp(*l*): 315°, d(*l*): 1.230.
THR = A nutrient and/or dietary supplement food additive. [*109*]

VALIUM. Syn: *diazepam.* $C_{16}H_{13}ON_2Cl$, mw: 284.8.
Acute tox data: Oral LD_{50} (rat) = 710 mg/kg; oral LD_{50} (mice) = 758 mg/kg; ip LD_{50} (mice) = 220 mg/kg. [*3*] An exper teratogen [*3*] via oral route.
THR = HIGH via ip and MOD via oral routes.

VALONE. Syn: *2-isovaleryl-1,3-indandione.* Crystals. $C_{14}H_{14}O$, mw: 230.3, mp: 68°.
THR = Details U. In animal exper it has displayed fairly high toxicity. Acts as an anti-coagulant leading to hemorrhages. See also coumarin.

VALYL. Syns: *diethyl valeramide, valeryl diethylamide.* Colorless liquid, burning taste. $C_4H_9CON(C_2H_5)_2$, mw: 157.25, bp: 210°.
THR = MOD irr via oral route. May have a depressant action.
Fire Hazard: Slightly dangerous; when heated, will burn.

VANADIC SULFATE. See vanadyl sulfate.

VANADIUM. A bright, white, soft, ductile metal, slightly radioactive. V, atwt: 50.9414, mp: 1717°, bp: 3000°, d: 6.11 at 18.7°.
THR = See vanadium compounds.
Radiation Hazard: For permissible levels, see Section 5A, Table 5A.5. Artificial isotope ^{48}V, $T_{\frac{1}{2}}$ = 16d, decays to stable ^{48}Ti by ec (37%) via positrons of 0.70 MeV. Emits γ's of 0.99, 1.31 MeV and x-rays.
Fire Hazard: Mod, in the form of dust, when exposed to heat or flame. Can react violently with BrF_3, Cl_2, Li. [*19*]

VANADIUM BROMIDE. See vanadium tribromide.

VANADIUM CARBONYL. Syn: *vanadium hexacarbonyl.* A blue-green, air-sensitive powder. $V(CO)_6$, mw: 291.0, decomp @ 60°–70°.
THR = This is a toxic material, both from its V content and its ability to evolve CO. See vanadium compounds and carbon monoxide. This material should be stored under an inert gas such as argon.
Disaster Hazard: See carbonyls.

VANADIUM COMPOUNDS.
THR = Variable. Vanadium compounds act chiefly as irr to the conjunctivae and respiratory tract. Prolonged exposures may lead to pulmonary involvement. There is still some controversy as to the effects of industrial exposure on other systems of the body. Responses are acute, never chronic.

The first report of vanadium poisoning in humans described rather widespread systemic effects, consisting of polycythemia, followed by red blood cell destruction and anemia, loss of appetite, pallor and emaciation, albuminuria and hematuria, gastrointestinal disorders, nervous complaints and cough, sometimes severe enough to cause hemoptysis. More recent reports describe symptoms which, for the most part, are restricted to the conjunctivae and respiratory system, no evidence being found of disturbances of the gastrointestinal tract, kidneys, blood or CNS. Though certain workers believe that it is only the pentoxide which is harmful, other investigators have found that patronite dust (chiefly vanadium sulfide) is quite toxic to animals, causing acute pulmonary edema. The fumes are highly toxic.

Symptoms and signs of poisoning are pallor, greenish-black discoloration of the tongue, paroxysmal cough, conjunctivitis, dyspnea and pain in the chest, bronchitis, rales and rhonchi, bronchospasm, tremor of the fingers and arms, radiographic reticulation. See also specific compounds.

These are common air contaminants.

VANADIUM DIBORIDE. Hexagonal crystals. VB_2, mw: 72.59, d: 5.10.

THR = See vanadium compounds and boron compounds.

Fire Hazard: Mod; on contact with moisture or acids, hydrides can be formed. See also boron hydrides.

Explosion Hazard: Mod; on contact with moisture or acids, hydrides can be formed. See also born hydrides.

Disaster Hazard: Mod dangerous; will react with water or steam to produce toxic and flam vapors; can react with oxidizing materials; on contact with acid or acid fumes, can emit toxic fumes.

VANADIUM DICHLORIDE. Syn: *vanadous chloride*. Hexagonal green plates, deliquescent. VCl_2, mw: 121.86, d: 3.23 @ 18°.

Acute tox data: Oral LD_{50} (rat) = 540 mg/kg. [*3*]

THR = MOD via oral route. See hydrochloric acid and vanadium compounds.

Disaster Hazard: Dangerous; see chlorides; will react with water or steam to produce toxic and corrosive fumes.

VANADIUM DIOXIDE. See vanadium tetraoxide.

VANADIUM HEXACARBONYL. See vanadium carbonyl.

VANADIUM IODIDE. Green deliquescent crystals, $VI_3 \cdot 6H_2O$, mw: 539.81.

THR = See vanadium compounds and iodides.

VANADIUM MONOSULFIDE. Syn: *vanadium sulfide*. Black plates. VS, mw: 83.02, mp: decomp, d: 4.20.

THR = See sulfides and vanadium compounds.

VANADIUM MONOXIDE. Syn: *vanadium oxide*. Light gray crystals. VO, mw: 66.9, mp: ignites, d: 5.758 @ 14°.

THR = See vanadium compounds.

Fire Hazard: Mod, when heated.

VANADIUM OXIDE. See vanadium monoxide.

VANADIUM OXYBROMIDE. Violet crystals. VOBr, mw: 146.87, mp: decomp @ 480°, d: 4.00 @ 18°.

THR = See vanadium compounds and bromides.

VANADIUM OXYCHLORIDE. Yellow-brown powder. VOCl, mw: 102.41, bp: 127°, d: 3.64 @ 20°.

THR = See vanadium compounds and chlorides.

VANADIUM OXYDIBROMIDE. Brown deliquescent powder. $VOBr_2$, mw: 226.78, mp: decomp @ 180°.

THR = See vanadium compounds and bromides.

VANADIUM OXYDICHLORIDE. Syns: *hypovanadic hydrochloride, vanadyl chloride*. Dark green syrupy mass. $VOCl_2$, mw: 137.86, d: 2.88 @ 13°.

THR = See vanadium compounds and chlorides. Can react violently with K. [*19*]

VANADIUM OXYDIFLUORIDE. Yellow solid. VOF_2, mw: 104.95, mp: decomp, d: 3.396 @ 19°.

THR = See fluorides and vanadium compounds.

VANADIUM OXYTRIBROMIDE. Red liquid. $VOBr_3$, mw: 306.70, mp: decomp @ 180°, bp: 130° @ 100 mm, d: 2.933 @ 14.5.

THR = See vanadium compounds and bromides.

VANADIUM OXYTRICHLORIDE. Yellow deliquescent liquid. $VOCl_3$, mw: 173.32, mp: −77 ± 2°, bp: 126.7°, d: 1.811 @ 32°.

Acute tox data: Oral LD_{50} (rat) = 140 mg/kg. [*3*]

THR = HIGH via oral route. See vanadium compounds and hydrochloric acid. Can react violently with P. [*19*]

VANADIUM OXYTRIFLUORIDE. Yellow-white hygroscopic crystals. VOF_3, mw: 123.95, mp: 300°, bp: 480°, d: 2.459 @ 19°.

THR = See fluorides and vanadium compounds.

VANADIUM PENTACHLORIDE. VCl_5, mw: 228.2.

THR = A powerful irr to skin, eyes and mu mem. Can react violently with K, Na. [*19*] See also vanadium compounds and hydrochloric acid.

VANADIUM PENTAFLUORIDE. Liquid. VF_5, mw: 145.95, bp: 47.9°, d: 2.177 @ 19°, mp: 19°.

THR = See fluorides and vanadium compounds.

VANADIUM PENTASULFIDE. Black-green powder; V_2S_5, mw: 262.2, mp: decomp, d: 3.00.

THR = See sulfides and vanadium compounds.

VANADIUM PENTOXIDE DUST. Yellow to red crystalline powder, V_2O_5, mw: 181.90, mp: 690°, bp: decomp @ 1750°, d: 3.357 @ 18°.

Acute tox data: Oral LD_{50} (mice) = 23 mg/kg; inhal LC_{LO} (rat) = 70 mg/m^3 for 2 hrs; it LD_{LO} (rat) = 25 mg/kg; sc LD_{LO} (mice) = 88 mg/kg; iv LD_{LO} (rabbit) = 10 mg/kg; inhal TC_{LO} (human) = 1 mg/m^3 for 8 hrs ⟶ allergic symptoms; inhal TC_{LO} (human) = 0.1 mg/m^3 ⟶ eye effects. [*3*]

THR = HIGH via oral, inhal, it, sc and iv routes. See vanadium compounds. Can react violently with $(Ca + S + H_2O)$, ClF_3, Li. [*19*]

VANADIUM SESQUIOXIDE. Syn: *vanadium trioxide*. Black crystals. V_2O_3, mw: 149.9, mp: 1970°, d: 4.87 @ 18°.

Acute tox data: Oral LD_{50} (mice) = 130 mg/kg; it LD_{LO} (rat) = 125 mg/kg. [*3*]

THR = HIGH via it and oral routes. See vanadium compounds. Can self-ignite in air. [*19*]

VANADIUM SESQUISULFIDE. Syn: *vanadium trisulfide*. Green-black plates or powder. V_2S_3, mw: 198.1, mp: decomp, d: 4.7 @ 21°.

THR = See sulfides and vanadium compounds.

VANADIUM SULFATE. See vanadyl sulfate.

VANADIUM SULFIDE. See vanadium monosulfide.

VANADIUM TETRACHLORIDE. Reddish-brown liquid. VCl_4, mw: 192.78, mp: −28 ± 2°, bp: 148.5°, d: 1.816 @ 30°.

Acute tox data: Oral LD_{50} (rat) = 160 mg/kg. [*3*]

THR = HIGH via oral route. See hydrochloric acid and vanadium compounds.

VANADIUM TETRAFLUORIDE. Brown-yellow crystals, very hygroscopic. VF_4, mw: 127.0, mp: decomp @ 325°, d: 2.975 @ 23°.

THR = See fluorides and vanadium compounds.

VANADIUM TETRAOXIDE. Syn: *vanadium dioxide*. Black crystals. V_2O_4, mw: 165.90, mp: 1967°, d: 4.339.

THR = See vanadium compounds.

VANADIUM TRIBROMIDE. Green-black, deliquescent crystals. VBr_3, mw: 290.8, mp: decomp.

Acute tox data: sc LD_{LO} (rabbit) = 20 mg/kg. [*3*]

THR = HIGH via sc route. See vanadium compounds and bromides.

VANADIUM TRICHLORIDE. Pink crystals, VCl_3, mw: 157.32, mp: decomp, d: 3.00 @ 18°.

Acute tox data: Oral LD_{50} (rat) = 350 mg/kg; oral LD_{50} (mice) = 23 mg/kg. [*3*]

THR = HIGH via oral route. See vanadium compounds and hydrochloric acid.

VANADIUM TRIFLUORIDE. Rhombic green crystals. VF_3, mw: 108.0, mp: 800°, bp: sublimes, d: 3.363 @ 19°.

THR = HIGH. See fluorides and vanadium compounds.

VANADIUM TRIOXIDE. See vanadium sesquioxide.

VANADIUM TRISULFIDE. See vanadium sesquisulfide.

VANADOUS CHLORIDE. See vanadium dichloride.

VANADYL CHLORIDE. See vanadium oxydichloride.

VANADYL SULFATE. Syns: *vanadic sulfate*, *vanadium sulfate*. Blue crystals. $VOSO_4$, mw: 163.01.

Acute tox data: ip LD_{50} (mice) = 45 mg/kg; iv LD_{LO} (rabbit) = 16 mg/kg; sc LD_{LO} (rat) = 140 mg/kg. [*3*]

THR = HIGH via ip, iv and sc routes. See vanadium compounds and sulfates.

VANCIDE BN. $C_{12}H_4O_2Cl_4$, mw: 400.

Acute tox data: Oral LD_{50} (rat) = 492 mg/kg. [*3*]

THR = HIGH via oral route. An exper neo [*3*] via oral route.

VANCIDE TH. Syn: *hexahydro-1,3,5-triethyl-s-triazine*. A light yellow liquid, sol in water. $C_9H_{21}N_3$, mw: 171.2, d: 0.89 @ 25°.

Acute tox data: Oral LD_{50} (rat) = 0.32 mg/kg; dermal LD_{50} (rabbit) = 500 mg/kg. [*3*]

THR = VERY HIGH toxicity fungicide and bactericide via oral, HIGH via dermal.

VANILLA. Cured, full-grown unripe fruit of *vanilla planifolia*. Composition: 2–3% vanillin, 4% resin, 10% sugar, etc.

THR = Natural form of vanillin. A flavoring and perfume component. An allergen and a weak irr. See also vanillin.

VANILLIC ALDEHYDE. See vanillin.

VANILLIN. Syns: *3-methoxy-4-hydroxy benzaldehyde*, *vanillic aldehyde*. White crystalline needles, pleasant odor, sol in 125 parts water, 20 parts glycerol, 2 parts 95% alcohol, chloroform and ether. $(CH_3O)(OH)C_6H_3CHO$, mw: 152, d: 1.056, mp: 81°–83°, bp: 285°.

Acute tox data: Oral LD_{50} (rat) = 1580 mg/kg; ip LD_{50} (rat) = 1160 mg/kg; sc LD_{50} (rat) = 1500 mg/kg; iv LD_{LO} (dog) = 1320 mg/kg. [*3*]

THR = MOD via oral, ip, sc and iv routes. A synthetic flavoring substance and adjuvant. [*109*] Can react violently with Br_2, $HClO_4$, potassium-*tert*-butoxide, (*tert*-chlorobenzene + NaOH), (formic acid + $Tl(NO_3)_3$). [*19*]

VAPAM. See methyl dithiocarbamic acid, sodium salt.

VAPOPHOS. See parathion.

VAPOR PHASE DEGREASERS. See Section 2.

VAPOR PHASE DEGREASER DESIGN. See Section 2.

VAPOR PHASE DEGREASER LOCATION. See Section 2.

VAPOR PHASE DEGREASER OPERATION. See Section 2.

VAPORTONE. See tetraethyl pyrophosphate.

VARNISH. Usually contains oil, resins and solvents. flash p: 176°–248°F.

THR = Variable. An allergen. See specific components. A common air contaminant.

Fire Hazard: Dangerous, when exposed to heat or flame.
Explosion Hazard: U.
Disaster Hazard: Dangerous, keep away from heat and open flame; can react vigorously with oxidizing materials.
To Fight Fire: Foam, CO_2, dry chemical.

VARNISH MAKERS' NAPHTHA. See naphtha.

VARNISH SHELLAC. flash p: 40°–70° F.
THR = Details U. A variable allergen. See also specific components. A common air contaminant.
Fire Hazard: Dangerous, when exposed to heat or flame.
Explosion Hazard: U.
Disaster Hazard: Dangerous; keep away from heat or open flame; can react vigorously with oxidizing materials.
To Fight Fire: CO_2, dry chemical.

VARNOLINE. See Stoddard solvent.

V-C13 NEMACIDE. Syn: *o-2,4-dichlorophenyl-o,o-diethylphosphorothioate*. A non-volatile, residual organic phosphate nematocide and insecticide, insol in water, sol in most organic solvents. $C_{10}H_{13}O_3PSCl_2$, mw: 315.0, bp: 166° @ 0.1 mm, d: 1.3.
Acute tox data: Oral LD_{50} (rat) = 250 mg/kg; oral LD_{50} (chicken) = 148 mg/kg; oral LD_{50} (wild birds) = 14 mg/kg. [*3*]
THR = HIGH via oral route. HIGH toxicity insecticide and an organic phosphate. See parathion.
Disaster Hazard: Dangerous; see parathion and chlorides.

VEGEDEX. Syn: *2-chloroallyl diethyl dithiocarbamate*. Amber liquid. $C_8H_{14}ClNS_2$, mw: 233.8, bp: 129° @ 1 mm.
Acute tox data: Oral LD_{50} (rat) = 850 mg/kg. [*3*]
THR = MOD via oral route. A mild skin irr. See carbamates, see also disulfiram.
Disaster Hazard: Dangerous; see chlorides and sulfur compounds.

VEGETABLE GUM. See dextrin.

VEGETABLE OIL AND FAT. A susp carc. [*14*] See thermic and oxidation products.

VEGETABLE OIL, HYDROGENATED. flash p: 610°F (OC), d: < 1.
Fire Hazard: Slight, when exposed to heat or flame.
To Fight Fire: CO_2, dry chemical.

VENTILATION FOR RADIATION EXPOSURE CONTROL. Section 5.

VENZAR. Syn: *3-cyclohexyl-5,6-trimethylene uracil*. A solid, very slightly sol in water, somewhat sol in most organic solvents. $C_{13}H_{18}N_2O_2$, mw: 234.3, mp: 290°, d: 1.32.
Acute tox data: Oral LD_{50} (rat) = 11,000 mg/kg. [*94*]
THR = LOW via oral route. An herbicide with medium to low toxicity.

VERATRALDEHYDE. Syns: *3,4-dimethoxy benzaldehyde, veratric aldehyde, 3,4-dimethoxy benzene carbonal*. Needles, odor of vanilla beans, slightly sol in hot water, freely sol in alcohol and ether. $C_6H_3CHO(OCH_3)_2$, mw: 166.17, mp: 42°–43°, bp: 281°.
Acute tox data: iv LD_{50} (mice) = 220 mg/kg. [*3*]
THR = HIGH via iv route. See aldehydes.

VERATRIDINE. Yellow-white powder. $C_{36}H_{51}NO_{11}$, mw: 673.8, mp: 180°.
Acute tox data: ip LD_{50} (rat) = 3.5 mg/kg; iv LD_{50} (mice) = 0.42 mg/kg; ip LD_{50} (mice) = 1.35 mg/kg. [*3*]
THR = HIGH via ip and iv routes. An insecticide. See also veratrine.
Fire Hazard: Slight.

VERATRINE. See sabadilla.

VERATRINE SULFATE. See sabadilla.

VERATRYL CHLORIDE. See 3,4-dimethoxybenzyl chloride.

VERDIGRIS. See copper acetate, basic.

VERISAN. See 2-methoxyethylmercury silicate.

VERMICULITE. A hydrated magnesium-aluminum-iron silicate, monoclinic crystals, pseudo-hexagonal character, insol in water and organic solvents, dissolves in hot conc sulfuric acid. Composition: approx 39% SiO_2, approx 21% MgO, 15% Al_2O_3, 9% Fe_2O_3, 5–7% K_2O, 1% CaO, 5–9% H_2O, and small quantities of Cr, Mn, P, S, Cl.
THR = Can act as a nuisance dust.

VERMILION. See mercuric sulfide, red.

VERV-CA. See calcium stearyl-2-lactylate.

VETERINARY DRUG FOOD ADDITIVES. [*109*]

VETERINARY MEDICINE AFFECTED BY INDUSTRIAL TOXICOLOGY. See Section 9.

VIBRATION ISOLATION. See Section 3.

VIENNA GREEN. See copper acetoarsenite.

VIGORITE NO 5 L.F. See explosives, high.

VILLIAUMITE. See sodium fluoride.

VINCALEUKOBLASTINE. $C_{46}H_{58}O_9N_4$, mw: 811.1.
THR = An exper teratogen [*3*] via iv route.

VINCALEUKOBLASTINE SULFATE. $C_{46}H_{58}O_9N_4 \cdot H_2SO_4$, mw: 909.2.
THR = HIGH. An exper teratogen [*3*] via pa route.

VINCYCLOHEXENE DIOXIDE.
THR = An exper carc. [*14*] See aliphatic and aromatic epoxides.

VINEGAR. Clear to yellow liquid. Composition: approximately 6% sol of acetic acid.
THR = See acetic acid.

VINEGAR SALTS. See calcium acetate.

VINEGAR ACID. See acetic acid.

VINOL. See vinyl alcohol.

VINYL ACETATE. Colorless mobile liquid, polymerizes to solid on exposure to light. $CH_3COOCHCH_2$, mw: 86.05, mp: −100.2°, bp: 73°, flash p: 18°F, d: 0.9335 @ 20°, autoign. temp.: 800°F, vap. press: 100 mm @ 21.5°, lel = 2.6%, uel = 13.4%, vap. d: 3.0.
Acute tox data: inhal LC_{LO} (rat) = 4000 ppm for 4 hrs; ip LD_{LO} (rat) = 500 mg/kg. [*3*]
THR = MOD irr via inhal and ip routes. May act as a skin irr by its defatting action. High conc of vapor are narcotic but are formed only if an inhibitor is present.
Fire Hazard: Highly dangerous, when exposed to heat, flame or oxidizers.
Spont Heating: No.
Explosion Hazard: Can react violently with 2-amino ethanol, chlorosulfonic acid, ethylene diamine, ethylene imine, HCl, HF, HNO_3, oleum, peroxides, H_2SO_4. [*19*]
Disaster Hazard: Dangerous; when heated to decomp, burns and emits acrid fumes; can react with oxidizing materials.
To Fight Fire: Alcohol foam, CO_2, dry chemical.

VINYL ACETONITRILE. See allyl cyanide.

VINYL ALCOHOL. Syns: *ethenol*, *vinol*. An unstable liquid; isolated only in the form of its esters or the polymer, polyvinyl alcohol. CH_2CHOH, mw: 44.1.
THR = Details U. See alcohols.
Fire Hazard: Dangerous, when exposed to heat or flame.
Explosion Hazard: U.
Disaster Hazard: Dangerous, when exposed to heat or flame; can react vigorously with oxidizing materials.

VINYL ALLYL ETHER. See allyl vinyl ether.

VINYL AZIDE. $C_2H_3N_3$, mw: 69.1.
Explosion Hazard: Has exploded in glass. [*19*]

7-VINYL BENZ(a)ANTHRACENE. $C_{20}H_{14}$, mw: 254.3.
THR = An exper neo to rats via sc route. [*103*]

VINYL BENZENE. See phenylethylene.

VINYL BROMIDE. Syns: *bromoethylene*, *bromoethene*. A gas. CH_2CHBr, mw: 107.0, mp: −138°, bp: 15.6°, d: 1.51.
Acute tox data: Oral LD_{50} (rat) = 500 mg/kg.
THR = HIGH via oral route. An irr via inhal route.
Fire Hazard: Dangerous, when exposed to heat or flame.
Explosion Hazard: U.
Disaster Hazard: Dangerous; can react vigorously with oxidizing materials. See bromides.
To Fight Fire: CO_2, dry chemical or water spray.

VINYL BUTYL "CELLOSOLVE." Liquid. $C_8H_{16}O_2$, mw: 144.21, mp: −71°, bp: 88° @ 50 mm, d: 0.8654 @ 20°/20°, vap. d: 4.98.
Acute tox data: Oral LD_{50} (rat) = 3100 mg/kg; inhal LC_{LO} (rat) = 2000 ppm for 8 hrs; dermal LD_{50} (rabbit) = 3000 mg/kg. [*3*]
THR = MOD via oral, inhal and dermal routes. See glycols.
Fire Hazard: Mod, when exposed to heat or flame.
Disaster Hazard: Mod dangerous, when exposed to heat or flame; can react with oxidizing materials.

VINYL BUTYL ETHER. Liquid. $CH_2CHOC_4H_9$, mw: 100.16, mp: −112.7°, bp: 94.1°, flash p: 15°F (OC), d: 0.7803 @ 20°/20°, vap. d: 3.45.
Acute tox data: Oral LD_{50} (rat) = 10,000 mg/kg; dermal LD_{50} (rabbit) = 4240 mg/kg. [*3*]
THR = MOD via dermal route and LOW via oral route. See ethers.
Fire Hazard: Dangerous, when exposed to heat or flame.
Explosion Hazard: Details U. See ethers.
Disaster Hazard: Highly dangerous, when exposed to heat or flame; can react with oxidizing materials.
To Fight Fire: Foam, CO_2, dry chemical.

VINYL-S-(BUTYL MERCAPTOETHYL)ETHER. $C_8H_{16}OS$, mw: 160.3.
Acute tox data: Oral LD_{LO} (rat) = 2830 mg/kg. [*3*]
THR = MOD via oral route. Limited animal exper suggest mod toxicity.
Fire Hazard: See ethers.
Disaster Hazard: See ethers and mercaptans.

VINYL BUTYRATE. $CH_2CHOOCCH_2CH_2CH_3$, mw: 114.1, d: 0.9, vap. d: 4.0, bp: 116°, flash p: 68°F (OC), lel = 1.4%, uel = 8.8%.
Acute tox data: Inhal LC_{LO} (rat) = 4000 ppm for 4 hrs, oral LD_{50} (rat) = 8530 mg/kg. [*3*]
THR = LOW via inhal and oral routes. See also esters.
Fire Hazard: Dangerous, when exposed to heat, flame or oxidizers.
To Fight Fire: Alcohol foam.

VINYL CAPROATE. See vinyl-2-hexanoate.

VINYL CARBAZOLE. Liquid. $(C_6H_4)_2NCHCH_2$, mw: 193.2.
THR = An allergen. Probably irr to skin and mu mem. See also amines.
Fire Hazard: Mod, when exposed to heat or flame.

Explosion Hazard: U.
Disaster Hazard: Mod dangerous; when heated to decomp, emits toxic fumes; can react with oxidizing materials.

VINYL CARBINOL. See allyl alcohol.

VINYL-2-CHLOROETHYL ETHER. See 2-chloroethyl vinyl ether.

VINYL CHLORIDE. Syns: *chloroethylene, chloroethene*. Colorless liquid or gas (when inhibited), faintly sweet odor. CH_2CHCl, mw: 62.50, bp: −13.4°, lel = 3.6%, uel = 33%; flash p: −108°F (COC), fp: −159.7°, d(liquid): 0.9195 @ 15°/4°, vap. press: 2600 mm @ 25°, vap d: 2.15, autoign. temp.: 882°F.
Acute tox data: Inhal TC_{LO} (human) = 20 ppm ⟶ cardiovascular effects. [*3*] Inhal TC_{LO} (human) = 500 ppm intermitt ⟶ carc; oral LD_{50} (rat) = 500 mg/kg; inhal TC_{LO} (rat) = 6000 ppm for 4 hrs/day for 12–18 days ⟶ neo; inhal TC_{LO} (rat) = 250 ppm, intermitt for 4 hrs/day over 130 wks ⟶ carc. [*3*]
THR = HIGH irr via inhal route and to skin, eyes and mu mem. In high conc, it acts as an anesthetic. Causes skin burns by rapid evaporation and consequent freezing. Chronic exposure has shown liver injury in rats and rabbits. Circulatory and bone changes in the fingertips reported in workers handling unpolymerized materials. A recog human carc. [*3, 1, 23*] via inhal route. [*102*] May cause local irr.
Fire Hazard: Dangerous, when exposed to heat, flame or oxidizers. Large fires of this material are practically inextinguishable.
Spont Heating: No.
Explosion Hazard: Severe, in the form of vapor, when exposed to heat or flame. Also, on standing, forms peroxides in air and can then explode. [*19*]
Disaster Hazard: Very dangerous; when heated to decomp, emits highly toxic fumes of phosgene; can react vigorously with oxidizing materials. Before storing or handling this material, instructions for its use should be obtained from the supplier.
To Fight Fire: Stop flow of gas.

VINYL CROTONATE. Slightly sol in water. $CH_2CHOCH{:}CHCH_3$, mw: 112, d: 0.9, vap. d: 4.0, bp: 134°, flash p: 78°F (OC).
Acute tox data: Oral LD_{50} (rat) = 6500 mg/kg. [*3*]
THR = MOD via oral route. Probably very irr.
Fire Hazard: Dangerous, when exposed to heat, flame or oxidizers.
To Fight Fire: Alcohol foam.

VINYL CYANIDE. See acrylonitrile.

VINYL CYCLOHEXANE. $CH_3CH(HC{:}CH_2)C_4H_8$, mw: 110.2.
Acute tox data: Oral LD_{LO} (rat) = 4000 mg/kg; inhal LC_{LO} (rat) = 20 mg/m^3. [*3*]
THR = HIGH via inhal route and MOD via oral route. High conc produce narcosis, and lower levels on repeated exposure cause liver and kidney injury in animals.

4-VINYL CYCLOHEXENE-1. Liquid. C_8H_{12}, mw: 108.18, bp: 128°, fp: −109°, flash p: 60°F (TOC), d: 0.832 @ 20°/4°, autoign. temp.: 517°F, vap. press: 25.8 mm @ 38°, vap. d: 3.76.
THR = An exper carc and neo. [*3, 13*] via dermal route; MOD via oral route; LOW via inhal and dermal routes.
Fire Hazard: Dangerous, when exposed to heat, flame or oxidizers.
Disaster Hazard: Dangerous; when exposed to heat or open flame, can react with oxidizing materials.
To Fight Fire: Foam, CO_2, dry chemical.

VINYL CYCLOHEXENE DIOXIDE. Colorless liquid. $CH_2CHOC_6H_9O$, mw: 140, d: 1.098 @ 20°/20°, bp: 227°, flash p: 230°F.
Acute tox data: Oral LD_{50} (rat) = 2130 mg/kg; inhal LC_{50} (rat) = 800 ppm for 4 hrs; dermal LD_{50} (rabbit) = 620 mg/kg. [*3*]
THR = MOD via oral, inhal and dermal routes.
Fire Hazard: Slight, when exposed to heat or flame.
To Fight Fire: Water, foam, dry chemical.

4-VINYL CYCLOHEXENE MONOXIDE. Liquid, very slightly sol in water. $CH_2CHC_6H_9O$, mw: 136, d: 0.9598 @ 20°/20°, bp: 169°, flash p: 136°F, fp: −100°.
Acute tox data: Oral LD_{50} (rat) = 2000 mg/kg; dermal LD_{50} (rabbit) = 2830 mg/kg. [*3*]
THR = MOD via oral and dermal routes. An exper neo. [*3*]
Fire Hazard: Mod, when exposed to heat or flame.
To Fight Fire: Foam, alcohol foam, mist.

VINYL DECANOATE. $CH_3(CH_2)_8COOHC{:}CH_2$, mw: 198.
THR = Details U. Limited animal exper suggest low toxicity. See also esters.

VINYLENE CARBONATE. $C_3H_2O_3$, mw: 86.1.
THR = An exper carc. [*3, 23*]

VINYL ETHER. Syns: *divinyl ether, divinyl oxide*. Colorless liquid, very volatile. $(CH_2CH)_2O$, mw: 70.1, bp: 29°, ulc: 100, lel = 1.7%, uel = 27%, flash p: < −22°F (CC), d: 0.774 @ 20°/20°, autoign. temp.: 680°F, vap. d: 2.41.
Acute tox data: Inhal LC_{LO} (mice) = 51233 ppm. [*3*]
THR = LOW via inhal route. See also ethers. An inhal anesthetic. Prolonged exposure causes liver injury.

Fire Hazard: Dangerous, when exposed to heat, flame or oxidizers.
Explosion Hazard: Severe, in the form of vapor, when exposed to heat or flame. Violent reaction with air, O_2. [*19*] See also ethers.
Disaster Hazard: Highly dangerous; when exposed to heat or flame, can react vigorously with oxidizing materials.
To Fight Fire: CO_2, dry chemical.

VINYL ETHYLENE OXIDE. See butadiene monoxide.

VINYL ETHYL ETHER. Colorless liquid. $CH_2CHOC_2H_5$, mw: 72.104, bp: 35.6°, flash p: <−50°F, fp: −115°, d: 0.754, autoign. temp.: 395°F, vap. press: 428 mm @ 20°, lel = 1.7%, uel = 28%, vap. d: 2.5.
Acute tox data: Oral LD_{50} (rat) = 8160 mg/kg. [*3*]
THR = LOW via oral route. See also ethers.
Fire Hazard: Highly dangerous, when exposed to heat or flame; see ethers.
Explosion Hazard: Dangerous; see ethers.
Disaster Hazard: Severe; when heated or exposed to flame, can react vigorously with oxidizing materials. See ethers.
To Fight Fire: Alcohol foam, foam, CO_2, dry chemical. See also ethers.

VINYL-2-ETHYL HEXOATE. Liquid, insol in water. $CH_2CHOCOCH(C_2H_5)C_4H_9$, mw: 170, flash p: 165°F (OC), d: 0.8751, bp: 185.2°, fp: −90°, vap. d: 6.0.
Acute tox data: Oral LD_{50} (rat) = 4290 mg/kg. [*3*]
THR = MOD via oral route. See esters.
Fire Hazard: Mod, when exposed to heat, flame or oxidizers.
To Fight Fire: Foam, alcohol foam, mist.

VINYL-2-ETHYLHEXYL ETHER. Liquid. $CH_2CHO(C_2H_5)C_6H_{12}$, mw: 156.3, mp: −100°, bp: 177.5°, flash p: 135°F (OC), d: 0.810, autoign. temp.: 395°F, vap. d: 5.4.
Acute tox data: Oral LD_{50} (rat) = 1350 mg/kg. [*3*]
THR = MOD via oral route. See ethers.
Fire Hazard: Mod, when exposed to heat or flame; can react with oxidizing materials. See ethers.
Explosion Hazard: Details U. See ethers.
To Fight Fire: Alcohol foam, foam, CO_2, dry chemical. See also ethers.

VINYL-S-(ETHYL MERCAPTOETHYL)ETHER. $C_6H_{12}OS$, mw: 132.2.
Acute tox data: Oral LD_{50} (rat) = 240 mg/kg. [*3*]
THR = HIGH via oral route. See also ethers.
Disaster Hazard: Dangerous; see ethers and mercaptans.

2-VINYL-5-ETHYL PYRIDINE. Insol in water. $N:C(CH:CH_2)CH:CHC(C_2H_5):CH$, mw: 133, d: 0.9449 @ 20°/20°, bp: 138° @ 100 mm, vap. press: 0.2 mm @ 20°, fp: −50.9°, flash p: 200°F (COC).
THR = U. Probably toxic.
Fire Hazard: Mod, when exposed to heat, flame or oxidizers.
Disaster Hazard: Dangerous; see cyanides.
To Fight Fire: Foam, alcohol foam, mist.

VINYL FLUORIDE. Syn: *fluoroethylene*. Colorless gas, insol in water, sol in alcohol and ether. $CH_2:CHF$, mw: 46, bp: −72°; lel = 2.6%, uel = 21.7%.
THR = HIGH. See fluorides. No details.
Fire Hazard: Ignites in presence of heat or source of ignition. Highly dangerous.
Disaster Hazard: Dangerous; see fluorides.
To Fight Fire: Stop flow of gas.

VINYL FORMATE. See acrylic acid.

VINYL-2-HEXANOATE. Syn: *vinyl caproate*. $CH_2:CHCO_2H_{11}$, mw: 142.2.
THR = Details U. Animal exper suggest LOW toxicity.

VINYLIDENE CHLORIDE. Syn: *1,1-dichloroethylene*. Colorless volatile liquid. CH_2CCl_2, mw: 97.0, bp: 31.6°, lel = 7.3%, uel = 16.0%, fp: −122°, flash p: 0°F (OC), d: 1.213 @ 20°/4°, autoign. temp.: 1058°F.
THR = An exper carc [*3*] via inhal route. See vinyl chloride.
Fire Hazard: Highly dangerous, when exposed to heat or flame.
Explosion Hazard: Mod, in the form of gas, when exposed to heat or flame. Also can explode spont; reacts violently with chlorosulfonic acid, HNO_3, oleum. [*19*]
Disaster Hazard: Highly dangerous; see chlorides; can react vigorously with oxidizing materials.
To Fight Fire: Alcohol foam, CO_2, dry chemical.

VINYLIDENE FLUORIDE. See 1,1-difluoroethylene.

VINYL IODIDE. Syn: *iodethene*. Liquid, insol in water. CH_2CHI, mw: 154.0, d: 2.08 @ 0°, bp: 56°.
THR = Details U. Irr to mu mem. Narcotic in high conc.
Disaster Hazard: Dangerous; see iodides.

VINYL ISOBUTYL ETHER. See isobutyl vinyl ether.

VINYL ISOOCTYL ETHER. Insol in water. $CH_2CHO(CH_2)_5CH(CH_3)_2$, mw: 156, flash p: 140°F, d: 0.8, vap. d: 5.4, bp: 175°.
THR = U. See ethers.
Fire Hazard: Mod, when exposed to heat, flame or oxidizers. See ethers.
To Fight Fire: Water spray, foam, mist, dry chemical, CO_2.

VINYL ISOPROPYL ETHER. Liquid. $CH_2CHOC_3H_7$, mw: 86.2, flash p: −26°F (CC), autoign. temp.: 522°F.

THR = Details U. See ethers.
Fire Hazard: Highly dangerous, when exposed to heat or flame; see ethers.
Explosion Hazard: Details U. See ethers.
Disaster Hazard: Highly dangerous, keep away from heat or open flame; can react vigorously with oxidizing materials.
To Fight Fire: Alcohol foam.

VINYL-2-METHOXYETHYL ETHER. Liquid. $CH_2CHOCH_2CH_2OCH_3$, mw: 102.13, mp: −82.8°, bp: 108.8°, flash p: 65°F (OC), d: 0.8967, vap. d: 3.53.
Acute tox data: Oral LD_{50} (rat) = 3900 mg/kg; inhal LC_{LO} (rat) = 8000 ppm for 4 hrs. [*3*]
THR = MOD via oral and inhal routes. See ethers.
Fire Hazard: Dangerous, when exposed to heat or flame; can react with oxidizing materials.
Explosion Hazard: Details U. See ethers.
Disaster Hazard: Dangerous; keep away from heat and flame.
To Fight Fire: Foam, CO_2, dry chemical. See also ethers.

VINYL METHYL ETHER. Syn: *methyl vinyl ether.* Colorless, easily liquefied gas or colorless liquid. CH_2CHOCH_3, mw: 58.1, bp: 6.0°, flash p: −60°F, d: 0.7500, vap. d: 2.0, fp: −121.6°, vap. press: 1052 mm @ 20°.
Acute tox data: Oral LD_{50} (rat) = 4900 mg/kg. [*3*]
THR = MOD via oral route. See ethers.
Fire Hazard: Very dangerous, when exposed to heat, flame or oxidizers.
Explosion Hazard: Details U. See ethers.
Disaster Hazard: Very dangerous; can react vigorously with oxidizing materials.
To Fight Fire: Stop flow of gas. See also ethers.

VINYL METHYL KETONE. See methyl vinyl ketone.

VINYL NONYL ETHER. Liquid. $CH_2CHOC_9H_{19}$, mw: 170.3, bp: 161°, d: 0.81.
THR = Details U. See ethers.
Fire Hazard: Mod, when exposed to heat or flame; can react with oxidizing materials. See also ethers.
Explosion Hazard: Details U. See ethers.

VINYL-*n*-OCTADECYL ETHER. Insol in water. $CH_2{:}CHO(CH_2)_{17}CH_3$, mw: 308, mp: 28°, bp: 130°–186° @ 5 mm, d: 0.8, flash p: 350°F.
THR = U. See ethers.
Fire Hazard: Slight, when exposed to heat or flame.
To Fight Fire: Dry chemical, CO_2.

VINYL OCTANOATE. $CH_2CHOOC(CH_2)_6CH_3$, mw: 180.
THR = U. Limited animal exper suggest low toxicity. See also esters.

α-VINYL PIPERONYL ALCOHOL. $C_{10}H_{10}O_3$, mw: 220.2.
THR = An exper neo. [*3*]

VINYL PROPIONATE. Liquid, almost insol in water. $CH_2{:}CHOOCC_2H_5$, mw: 100, d: 0.9173 @ 20°/20°, bp: 95°, fp: −81.1°, flash p: 34°F (OC), vap. d: 3.3.
Acute tox data: Oral LD_{50} (rat) = 4760 mg/kg; inhal LC_{LO} (rat) = 4000 ppm for 4 hrs. [*3*]
THR = MOD via oral and inhal routes. See also esters.
Fire Hazard: Dangerous, when exposed to heat, flame or oxidizers.
To Fight Fire: Alcohol foam.

VINYL PYRIDINE. Liquid. C_7H_7N, mw: 105.1, bp: 159°, d: 0.9746 @ 20°.
THR = Details U. Animal exper suggest high toxicity. See pyridine. An irr to the skin, eyes and respiratory tract.
Fire Hazard: Mod, when exposed to heat or flame.
Explosion Hazard: U.
Disaster Hazard: Dangerous; when heated to decomp, emits highly toxic fumes of cyanide; can react with oxidizing materials.

***n*-VINYL-2-PYRROLIDONE.** Colorless liquid, water-sol. C_6H_9NO, mw: 111.1, bp: 148° @ 100 mm, fp: 13.5°, flash p: 209°F (OC), d: 1.04 @ 25°, autoign. temp.: 213°F, vap. d: 3.8.
Acute tox data: Oral LD_{50} (rat) = 1500 mg/kg. [*3*]
THR = MOD via oral route. Probably irr and narcotic in high conc.
Fire Hazard: Low when exposed to heat or flame.
Disaster Hazard: Dangerous; when heated to decomp, emits highly toxic fumes of NO_x; can react vigorously with oxidizing materials.
To Fight Fire: Alcohol foam, CO_2, dry chemical.

VINYL RESINS.
THR = Known allergens. No further details.
Fire Hazard: Slight, when exposed to heat or flame; can react with oxidizing materials.

VINYL STEARATE. White, waxy solid. $H_2CCHCO_2(CH_2)_{16}CH_3$, mw: 310.5, mp: 28°–30°, bp: 180° @ 2 mm, d: 0.881 @ 20°/20°.
THR = LOW. See esters.
Fire Hazard: Slight, when exposed to heat or flame; can react with oxidizing materials.
To Fight Fire: Foam, CO_2, dry chemical.

VINYL STYRENE. See *m*-divinylbenzene.

VINYL SULFONE. $C_4H_6O_2S$, mw: 118.2.
Acute tox data: Oral LD_{50} (rat) = 32 mg/kg; ip LD_{50} (rat) = 3 mg/kg; sc LD_{50} (rat) = 14 mg/kg; iv LD_{50} (rat) = 12 mg/kg; dermal LD_{50} (rabbit) = 22 mg/kg. [*3*]
THR = HIGH via dermal, iv, sc, ip and oral routes.
Disaster Hazard: Dangerous; see sulfonates.

VINYL TOLUENE. Syn: *methyl styrene.* Colorless liquid. $CH_2CHC_6H_4CH_3$, mw: 118.2, bp: 170°–171°, fp: −82.5°, flash p: 134°F (OC), d(monomer): 0.89 @ 25°/25°, d(polymer): 1.027 @ 25°/25°, vap. d: 4.08, autoign. temp.: 923°F, lel = 0.7%.
Acute tox data: Oral LD_{LO} (rat) = 4900 mg/kg; oral LD_{50} (mice) = 3160 mg/kg; inhal LC_{50} (mice) = 3020 mg/m^3. [*3*]
THR = MOD via oral and inhal routes. Chronic exposure damages kidneys and liver. Irr to eyes and mu mem. See also toluene. An exper teratogen. [*3*] via inhal route.
Fire Hazard: Mod, when exposed to heat or flame; can react with oxidizing materials.
To Fight Fire: Foam, CO_2, dry chemical.

VINYL TRICHLORIDE. See β-trichloroethane.

VINYL TRICHLOROSILANE. Fuming liquid. $CH_2CHSiCl_3$, mw: 162.5, bp: 90.6°, d: 1.265 @ 25°/25°, flash p: 16°F.
Acute tox data: Oral LD_{50} (rat) = 1280 mg/kg; inhal LC_{LO} (rat) = 500 ppm for 4 hrs; oral LD_{50} (mice) = 3160 mg/kg; dermal LD_{50} (rabbit) = 680 mg/kg; inhal LC_{50} (mice) = 3020 mg/m^3 for 4 hrs.
THR = MOD via oral, dermal and inhal routes.
Fire Hazard: Dangerous; reacts violently with water, moist air. [*19*]
Disaster Hazard: Dangerous; see chlorides; will react with water or steam to produce toxic and corrosive fumes.

VINYL TRIMETHYL NONYL ETHER. Liquid. $CH_2CHOCHCH_2(CH_3)_2$, mw: 212.36, mp: −90°, bp: 223.5°, flash p: 200°F (OC), d: 0.8075 @ 20°/20°, vap. d: 7.33.
Acute tox data: Oral LD_{50} (rat) = 1220 mg/kg. [*3*]
THR = MOD via oral route. See ethers.
Fire Hazard: Mod, when exposed to heat or flame; can react with oxidizing materials. See ethers.
To Fight Fire: Foam, CO_2, dry chemical.

VISHA. See *aconitum ferox.*

VISTARIL HYDROCHLORIDE. $C_{21}H_{27}O_2Cl \cdot HCl$, mw: 411.4.
THR = An exper teratogen. [*3*]

VITALLIUM. An alloy of chromium, cobalt and molybdenum.
THR = An exper neo [*3*] via imp route.

VITAMIN A. A suitable form of derivative of retinol. In liquid form it is a light yellow to red oil, very sol in chloroform and ether, sol in absolute alcohol and vegetable oils, insol in glycerin and water. $C_{20}H_{30}O$, mw: 286.5, mp: 62°.
Acute tox data: Oral LD_{50} (rat) = 2000 mg/kg; oral LD_{50} (mice) = 4000 mg/kg. [*3*]
THR = MOD via oral route. An exper teratogen via oral route. A nutrient and/or dietary supplement food additive. [*109*]

VITAMIN A ACETATE. Synthetic vitamin A acetic acid ester, finely divided, dry, light yellow crystalline powder, odorless. $C_{20}H_{29}OOCH_3$, mw: 328.5.
Acute tox data: Oral LD_{LO} (mice) = 1000 mg/kg. [*3*]
THR = MOD via oral route. An exper teratogen via oral route. [*3*] A nutrient and/or dietary supplement food additive. [*109*]

VITAMIN A PALMITATE. Synthetic vitamin A palmitic acid ester; yellow liquid, odorless. $C_{20}H_{29}OOC_{15}H_{31}$, mw: 525.
THR = An exper teratogen via oral route. [*3*] A nutrient and/or dietary supplement food additive. [*109*]

VITAMIN B_2. See riboflavin.

VITAMIN B_{12}. Syn: *cobalamin.* The anti-pernicious anemia vitamin; all vitamin B_{12} compounds contain the cobalt atom in its trivalent state. There are at least 3 active forms: cyanocobalamin, hydroxycobalamin and nitrocobalamin. Dark red crystals. $C_{63}H_{88}CoN_{14}P$, mw: 1355.4.
Acute tox data: ip LD_{LO} (mice) = 3 mg/kg. [*3*]
THR = HIGH via ip route. A nutrient and/or food supplement food additive. [*109*]

VITAMIN D_2. Syns: *ergocalciferol, calciferol.* White crystals, odorless, insol in water, sol in alcohol, chloroform, ether and fatty acids. $C_{28}H_{44}O$, mw: 396, mp: 115°–118°.
Acute tox data: Oral LD_{LO} (dog) = 4 mg/kg; ip LD_{LO} (dog) = 10 mg/kg; iv LD_{LO} (dog) = 5 mg/kg; im LD_{LO} (dog) = 5 mg/kg. [*3*]
THR = HIGH via oral, ip, iv and im routes. A nutrient and/or dietary supplement food additive. [*109*]

VITAMIN H. See biotin.

VM & P. See "dryolene."

VULCAN COAL POWDERS. See explosives, high.

WALL PLASTER.
THR = A mild allergen and irr. Used as an antidote: In spite of its slight adverse effects, plaster may be used in an emergency to neutralize a strong acid poison taken internally.

WALSH-HEALY ACT. See Section 1.

WARFARIN. Syns: *3-(α-acetonyl benzene)-4-hydroxy coumarin, compound 42, WARF-12.* Colorless, odorless, tasteless crystals. mp: 161°.
Acute tox data: Oral LD_{50} (rat) = 3 mg/kg; iv LD_{50} (rat) = 186 mg/kg; oral LD_{50} (dog) = 200 mg/kg; oral LD_{50} (rabbit) = 800 mg/kg; ip LD_{LO} (rat) = 420 mg/kg. [*3*]
THR = VERY HIGH to rats via oral route; HIGH via iv, oral and ip routes. A rodenticide. Warfarin has two actions—inhibition of prothrombin formation and capillary damage. There is evidence that these two actions are produced by the two moieties of the molecule. Thus, 4-hydroxycoumarin inhibits the formation of prothrombin and reduces the clotting power of the blood, while there is some evidence benzalacetone produces capillary damage and leads to bleeding upon the very slightest trauma. Significantly enough, vitamin K has an antidotal action against both actions of warfarin up to a certain point.

The action of warfarin is similar to that of the common drug dicoumarol, except that capillary damage (the so-called "toxic factor") is greatly increased.

Dangerous acute and chronic dose in man: Information on the toxicity of warfarin to man is available. Serious illness was induced by ingesting 1.7 mg of warfarin per kg. per day for 6 consecutive days with suicidal intent. This would correspond to eating almost 1 lb. of warfarin bait (0.025% warfarin) each day for 6 days. All signs and symptoms were caused by hemorrhage and, following multiple small transfusions and massive doses of vitamin K, recovery was complete.

Data from animal exper suggest that a single dose would be harmless. The rat is specifically susceptible to warfarin, yet mortalities are highly irregular and may be low following single doses of 50, 100, or even 150 mg/kg. To obtain a dose of 50 mg/kg the average 150 lb man would have to eat 0.7 kg or 1.5 lbs of the warfarin conc available on the market, although the amount of active ingredient would be only 3.5 g. To obtain the same dose with the strongest bait recommended, the same man would have to eat 14 kg or 30 lbs of the rat bait.

However, with repeated daily doses, the effective toxicity of the compound is greatly increased. For example, 5 daily doses of 1.0 mg/kg each (a total of 5.0 mg/kg) is sufficient to kill all rats which eat it. There is considerable species difference in susceptibility. For example, chickens may be raised to maturity on an adequate growing mash containing an effective rodenticidal concentration of warfarin.

The possibility of human poisoning by warfarin must be kept in mind, although the safety factors make it appear unlikely that poisoning will occur except with suicidal intent or as a result of gross carelessness and ignorance.

Signs and symptoms of poisoning in man: The initial symptoms in an attempted suicide using warfarin were back pain and abdominal pain. The onset occurred the first day after the sixth daily dose. A day after onset vomiting and attacks of nose bleeding occurred. On the second day of illness, when admitted to the hospital, the patient was observed to have a generalized petechial rash. The prothrombin time was greatly prolonged. The coagulation time appeared to be definitely increased by the Lee-White method of measurement. Bleeding time was normal. Urine was normal in appearance but contained many red cells on microscopic examination.

Treatment: After blood has been taken for prothrombin and other differential diagnostic tests, a blood transfusion should be given at once if there is reasonable assurance that warfarin poisoning has occurred, irrespective of whether or not signs and symptoms are present. Vitamin K in a dose of 65 mg repeated 3 times on the first day of treatment is suggested irrespective of symptoms. Smaller doses should be continued until the prothrombin time has reached normal. In a more seriously ill patient, small transfusions of carefully matched whole blood should be given daily until the patient has returned to normal. Should it ever be necessary to treat a patient in shock from blood loss resulting from warfarin poisoning, frequent small transfusions and a complete consideration of the blood chemistry would be in order. Any large hematomata should be the subject of surgical consultation, but any surgical action should be taken only after the clotting power of the blood is restored to normal.

WASTE BASKETS AS A DANGER POINT IN SOLID WASTE HANDLING. See Section 6.

WASTE CONTAINERS AS A DANGER POINT IN SOLID WASTE HANDLING. Section 6.

WASTE PAPER, WET.
Fire Hazard: Mod, when exposed to heat or by spont chemical reaction; can react with oxidizing materials. See Section 7.

WASTE RAGS, WET. See Section 7.

WASTE TEXTILE, WET.
Fire Hazard: Mod, when exposed to heat or by spont chemical reaction; can react with oxidizing materials. See Section 7.

WASTE WOOL, WET.
Fire Hazard: Mod, when exposed to heat or flame or by spont chemical reaction; can react with oxidizing materials. See Section 7.

WATAPANA SHIMARON. Syn: *acacia villosa.*
THR = An exper carc to rats via sc route and to hamsters via imp route. [*103*]

WATER GAS. See carbon monoxide and hydrogen. lel = 7.0%, uel = 72%.

WATER GLASS. See sodium silicate.

WATER OF AMMONIA. See ammonium hydroxide.

WATER-SENSITIVE FIRE AND EXPLOSION HAZARDS. Section 7.

WAX, MICROCRYSTALLINE. Derived from petroleum, white, amber or black solid, odorless. flash p: >400°F, d: 0.9.
THR = U. Probably VERY LOW on an acute basis.
Fire Hazard: Slight, when exposed to heat or flame.
To Fight Fire: Dry chemical, CO_2.

WAX MYRTLE. Tannin-containing fraction of bark.
THR = An exper neo to rats via sc route. [*103*]

WAX, OZOCERITE. flash p: 236°F, d: 0.9.
Fire Hazard: Low.
To Fight Fire: Dry chemical, CO_2.

WELDING AND FLAME CUTTING. See Section 2.

WELDING AND FLAME CUTTING VENTILATION CONTROL. See Section 2.

WELDING ARC RADIATION. A recog carc. [*14*] See ultraviolet radiation.

WELDING FUMES.
THR = When welding is done on a surface coated with cadmium tox fumes of cadmium are evolved. When zinc-coated surfaces are welded, toxic quantities of zinc oxide may be liberated. When painted surfaces are welded, lead or other pigment fumes may be liberated. And when fluoride fluxes are used in welding, very toxic fluoride fumes are evolved. When oily surfaces are welded, offensive and toxic fumes can be liberated, and when the welding torch is improperly ignited, carbon monoxide which is very toxic may be evolved. Also, NO_x may be formed. It is therefore considered hazardous to inhale excessive amounts of welding fumes. It is also possible to inhale sufficient quantities of iron oxide from welding to cause siderosis. Metal fume fever is a common reaction. It is characterized by chills, fever, sweating, and leucocytosis coming on several hours after exposure. Recovery is usually complete in 24–48 hrs. and there are no significant after-effects.

Safety goggles are required to protect against spatter. Light-filtering goggles are required to shield the eyes against the intense light from the arc.

WET HAIR. See hair.

WHALE OIL. d: 0.925, flash p: 446°F (CC), autoign temp.: 800°F.
THR = Details U. Probably VERY LOW.
Fire Hazard: Slight, when exposed to heat or flame.
Spont Heating: Yes.
To Fight Fire: CO_2, dry chemical.

WHEAT OIL.
THR = An allergen.
Disaster Hazard: Slightly dangerous; when heated, emits acrid fumes.

WHERE TO GO FOR LABELING INFORMATION. See Section 11.

WHISKEY. Light to deep amber liquid, characteristic odor and taste. Composition: approx 50% ethyl alcohol (by volume) + acetic acid and ethyl acetate. flash p: 82°F (CC), d: 0.935–0.923 @ 25°.
THR = MOD via oral route. See also ethanol and other components as listed. An exper carc, teratogen via oral, ip and rec routes. [*103*]
Fire Hazard: Mod, when exposed to heat or flame; can react vigorously with oxidizing materials.
Explosion Hazard: U.
To Fight Fire: CO_2, dry chemical.

WHITE ACID. A mixture of ammonium bifluoride and hydrofluoric acid; used for etching glass. See components as listed.

WHITE ARSENIC. See arsenic trioxide.

WHITE CAUSTIC. See sodium hydroxide.

WHITE FLAG. See orris.

WHITE LEAD. See lead compounds.

WHITE PRECIPITATE. See mercury, ammoniated.

WHITE SPIRIT. See turpentine substitutes.

WHITE TAR. See naphthalene.

WHITE WAX. See beeswax, white.

WINES, HIGH. Light golden to amber color, aromatic, alcoholic liquids. flash p: 60°–80°F.
THR = MOD via oral route. See also ethanol.
Fire Hazard: Dangerous, when exposed to heat or flame; can react with oxidizing materials.
Explosion Hazard: U.
To Fight Fire: CO_2, dry chemical.

WINES (SHERRY AND PORT). Light golden to amber color, aromatic, alcoholic liquids. flash p: 129°F.
THR = See ethanol.
Fire Hazard: Mod, when exposed to heat or flame; can react with oxidizing materials.
Explosion Hazard: U.
To Fight Fire: CO_2, dry chemical.

WINTERGREEN OIL. See methyl salicylate.

WITCH HAZEL. Syns: *hamamelis, tobacco wood.*
THR = A mild irr.
Fire Hazard: Slight, when exposed to heat or flame; can react with oxidizing materials.
Explosion Hazard: U.

WOLFRAM. See tungsten.

WOLFSBANE. See aconite.

WOOD ALCOHOL. See methanol or methyl alcohol.

WOOD AND PEAT CREOSOTE. A recog carc. [*14*] See creosote.

WOOD DUST. See sawdust.

WOOD GAS. See carbon monoxide.

WOOD TAR ACID.
THR = Details U. A fungicide, herbicide and insecticide.
Fire Hazard: U.
Disaster Hazard: Slightly dangerous; when heated, emits an irr smoke.

WOOD TAR OIL.
THR = Details U. A fungicide and insecticide.
Fire Hazard: Mod, when exposed to heat or flame; can react with oxidizing materials.

WOOD VINEGAR. See pyroligneous acid.

WOOL AND WOOL WASTES.
THR = A mild allergen.
Fire Hazard: Mod, when exposed to heat or flame; can react with oxidizing materials.
Spont Heating: Yes, particularly when wet.

WOOL FAT. A susp carc. See thermic and oxidation products. [*14*]

WOOL GREASE. See lanolin.

WORKMAN'S COMPENSATION LAWS. See Section 1.

WORKER HEALTH-RELATED LEGISLATION. See Section 1.

WORMWOOD. See absinthium.

WORMWOOD OIL. Syn: *absinthe.* Very strong odor and acrid taste.
THR = HIGH tox via oral route. Can cause GI symptoms, stupor, nervousness, convulsions, death. [*2*] This is the essence of absinthe, prolonged use of which can lead to mental deterioration.

WULFENITE. See lead molybdate.

WURTZITE. See zinc sulfide (α).

XANTHINE. $C_5H_4O_2N_4$, mw: 152.1.
Acute tox data: ip LD_{50} (mice) = 500 mg/kg. [*3*]
THR = HIGH via ip route. An exper neo. [*3*] via sc and imp routes.

XANTHINE-N-OXIDE. $C_5H_4O_3N_4$, mw: 168.1.
THR = An exper carc [*3*] via sc route.

XANTHOGEN DISULFIDE.
THR = Details U. See sulfides.

XANTHURIC ACID. $C_{10}H_7O_4N$, mw: 205.2.
THR = An exper neo [*3*] via imp route.

XENON. Colorless, gaseous element. Xe, atwt: 131.30, d(gas): 5.8878 g/liter, d(liquid): 3.52 @ −109°, mp: −112°, bp: −107°.
THR = A simple asphyxiant. For a discussion of toxicity effects, argon. A common air contaminant.
Radiation Hazard: For permissible levels, see Section 5A, Table 5A.5. Artificial isotope ^{133}Xe, $T_{\frac{1}{2}}$ = 5.3 d, decays to stable ^{133}Cs via β's of 0.35 MeV; emits γ's of 0.08 MeV. Artificial isotope ^{135}Xe, $T_{\frac{1}{2}}$ = 9h, decays to radioactive ^{135}Cs via β's of 0.91 MeV; emits γ's of 0.25 MeV. ^{133}Xe and ^{135}Xe are produced by neutron irradiation of stable Xe in air-cooled reactors.

XENON DIFLUORIDE. Colorless solid, stable at room temp., easily sublimed, has a significant vapor pressure at 20°, colorless, water-sol vapor with a powerful fluorine odor. XeF_2, mw: 169.3, mp: 140°, vap. press: 3.8 mm @ 25°, d(solid): 4.32.
THR = HIGH toxicity corrosive irr to skin, eyes and mu mem. See also fluorides. Its vapor hydrolyzes to form highly toxic hydrofluorides. Must be stored and handled as a very powerful fluorine type of oxidizer. In the presence of even traces of moisture, it will attack glass and metals. See hydrofluoric acid and fluorides.

XENON HEXAFLUORIDE. XeF_6, mw: 245.3.
THR = Has reacted explosively with water. [*19*] HIGH toxicity corrosive irr to skin, eyes and mu mem. See also fluorine and fluorides.

XENYLAMINE. See *p*-aminodiphenyl.

XENYLAMINE ANTI-OXIDANT.
THR = A recog carc. [*14*] See aromatic amines.

X-RAY FILM(NITROCELLULOSE BASE). See nitrates.

X-RAY FILM SCRAP(NITROCELLULOSE BASE). See nitrates.

X-RAY FILM(NITROCELLULOSE BASE), UNEXPOSED. See nitrates.

X-RAYS. Highly penetrating electromagnetic radiation produced by electrical means or by nuclear interaction.
Radiation Hazard: A recog carc. [*14*] See ionizing radiation and gamma rays.

X-RAY, PRIMARY PROTECTIVE BARRIER NEEDS. See Section 5A.

X-RAY, SECONDARY PROTECTIVE BARRIER NEEDS. See Section 5A.

***m*-XYLENE.** Syn: *m-xylol.* Colorless liquid.
$C_6H_4(CH_3)_2$, mw: 106.2, mp: −47.9°, bp: 139°, lel = 1.1%, uel = 7.0%, flash p: 84°F, d: 0.864 @ 20°/4°, vap press: 10 mm @ 28.3°, vap. d: 3.66, autoign temp.: 986°F.
Acute tox data: Oral LD_{50} (rat) = 5000 mg/kg; inhal LC_{LO} (rat) = 8000 ppm for 4 hrs; ip LD_{LO} (rat) = 2000 mg/kg; sc LD_{LO} (rat) = 5000 mg/kg. [*3*]
THR = MOD via oral, inhal, ip and sc routes. A common air contaminant.
Fire Hazard: Dangerous, when exposed to heat or flame; can react with oxidizing materials.
Explosion Hazard: Mod, in the form of vapor when exposed to heat or flame.
Disaster Hazard: Dangerous; keep away from open flame.
To Fight Fire: Foam, CO_2, dry chemical.

***o*-XYLENE.** Syn: *o-xylol.* Colorless liquid.
$C_6H_4(CH_3)_2$, mw: 106.2, bp: 144.4°, fp: −25.5°, ulc: 40-45, lel = 1.0%, uel = 6.0%, flash p: 90°F (CC), d: 0.880 @ 20°/4°, vap. press: 10 mm @ 32.1°, vap. d: 3.66, autoign. temp.: 869°F.
Acute tox data: Oral LD_{LO} (rat) = 5000 mg/kg; ip LD_{LO} (rat) = 1500 mg/kg; sc LD_{LO} (rat) = 2500 mg/kg; inhal LC_{LO} (mice) = 6920 ppm. [*3*]
THR = MOD via oral, ip, sc and inhal routes. A common air contaminant.
Fire Hazard: Dangerous, when exposed to heat or flame; can react with oxidizing materials.
Explosion Hazard: Slight, in the form of vapor, when exposed to heat or flame.
To Fight Fire: Foam, CO_2, dry chemical.

p-XYLENE. Syn: *p-xylol.* Clear liquid. $C_6H_4(CH_3)_2$, mw: 106.2, bp: 138.3°, lel = 1.1%, uel = 7.0%, fp: 13.2°, flash p: 81°F (CC), d: 0.8611 @ 20° / 4°, vap. press: 10 mm @ 27.3°, vap. d: 3.66, autoign. temp.: 986°F.

Acute tox data: Oral LD_{50} (rat) = 5000 mg/kg; ip LD_{LO} (rat) = 2000 mg/kg; sc LD_{LO} (rat) = 5000 mg/kg; inhal LC_{LO} (mice) = 3460 ppm. [*3*]

THR = MOD via oral, ip, sc and inhal routes. A common air contaminant.

Fire Hazard: Dangerous, when exposed to heat or flame; can react with oxidizing materials.

Explosion Hazard: Mod, in the form of vapor, when exposed to heat or flame.

To Fight Fire: Foam, CO_2, dry chemical.

m-XYLENE-α,α-DIAMINE. $C_8H_{12}N_2$, mw: 136.1.

Acute tox data: Oral LD_{50} (rat) = 930 mg/kg. [*3*]

THR = MOD via oral route.

XYLENE HEXAFLUORIDE. Syn: *bis(trifluoromethyl) benzene.* Clear, water white liquid. $C_6H_4(CF_3)_2$, mw: 214.11, mp: −40° to −50°, bp: 115°, d: 1.395 @ 20°/15.5°.

THR = See xylene and fluorides.

XYLENES(MIXED m- AND p-ISOMERS). A clear liquid. bp: 138.5°, flash p: 100°F (TOC), d: 0.864 @ 20°/4°, vap. press: 6.72 mm @ 21°.

Acute tox data: Inhal TC_{LO} (human) = 200 ppm → irr effects; oral LD_{50} (rat) = 4300 mg/kg; ip LD_{LO} (rat) = 2000 mg/kg. [*3*]

THR = MOD via inhal and oral routes. Some temporary corneal effects are noted, as well as some conjunctival irr by instillation. Very little dermal toxicity. [*69, 27, 51*]

Fire Hazard: Mod, in the presence of heat or flame; can react with oxidizing materials.

To Fight Fire: Foam, CO_2, dry chemical.

XYLENE SUBSTITUTES. Liquid. flash p: 45°F, d: 0.760.

THR = U.

Fire Hazard: Dangerous, when exposed to heat or flame.

Explosion Hazard: U.

Disaster Hazard: Dangerous; keep away from heat and flame; can react with oxidizing materials.

To Fight Fire: Foam, CO_2, dry chemical.

3,5-XYLENOL. Syns: *3,5-dimethyl phenol, 1-hydroxy-3,5-dimethyl benzene, 5-hydroxy-1,3-dimethyl benzene.* White crystals. $(CH_3)_2C_6H_3OH$, mw: 122.16, mp: 68°, bp: 219.5°, d: 1.0362, vap. press: 1 mm @ 62°.

Acute tox data: Oral LD_{50} (mice) = 477 mg/kg; oral LD_{50} (rat) = 608 mg/kg. An exper carc [*3*] via dermal route.

THR = HIGH–MOD via oral route.

XYLIDINE. Syn: *aminodimethylbenzene.* Usually liquid (except for *o*-4-xylidine). $(CH_3)_2C_6H_3NH_2$, mw: 121.2, bp: 213°–226°, flash p: 206° (CC), d: 0.97–0.99, vap. d: 4.17.

Acute tox data: Oral LD_{LO} (rat) = 610 mg/kg; inhal LC_{LO} (mouse) = 149 ppm; iv LD_{LO} (cat) = 120 mg/kg. [*3*]

THR = HIGH via iv, inhal and dermal routes, MOD via oral route. This material, which so closely resembles aniline in its toxic effects is actually twice as toxic as aniline, based on the determination of the LD_{50} for mice by inhal. It can cause injury to the blood and the liver. It does not necessarily give any alarm or warning, such as cyanosis, headache, and dizziness which characterizes aniline poisoning. Thus it may be considered a more insidious poison than aniline, and severe and possibly fatal intoxication may come about through skin absorption. From animal exper it has been further found that the minimum lethal dose for rabbits is 0.28 g/kg of body weight, and that the lethal dose intravenously is 240 mg/kg of body weight for rabbits. The signs of intoxication in animals are loss of weight, dyspnea, prostration, albuminuria, and occasional terminal convulsions. This compound pentrates the intact skin of rabbits in sufficient quantity to cause cyanosis and death. There are no local effects upon the skin. The application of 3.3 g or more of this material per kg of body weight on the intact skin of a rabbit for 1 hr or more always caused fatal results. A 2% solution of this material caused no harm to 3 rabbits in the course of 50 periods of cutaneous contact.

Fire Hazard: Low, when exposed to heat or flame.

Disaster Hazard: Dangerous; when heated to decomp, emits highly toxic fumes; can react vigorously with oxidizing materials.

To Fight Fire: Foam, CO_2, dry chemical.

Treatment and antidotes: In case of exposure, contact the medical department immediately. When it splashes or spills upon the person, the area of contact should be washed with copious quantities of warm water aided by soap. Remove all contaminated clothing and if possible wash the area affected under a deluge-type of shower. Protective clothing should be washed before reuse. This material must not be used without adequate ventilation as well as personal protective equipment. See also aniline. Personnel engaged working with it should do so under the direct supervision of a medical department.

2,3-XYLIDENE.

Acute tox data: Oral LD_{50} (rat) = 933 mg/kg. [*3*]

THR = MOD via oral route.

2,4-XYLIDENE. Syn: *1-amino-2,4-dimethylbenzene.* bp: 214°, mp: 16°.
Acute tox data: Oral LD_{50} (rat) = 467 mg/kg; oral LD_{50} (rat) = 467 mg/kg; oral LD_{50} (mouse) = 250 mg/kg. [*3*]
THR = HIGH via oral route. An exper carc via oral route in rats. [*108*]

2,5-XYLIDENE. Syn: *2,5-dimethyl aniline.* Colorless oil, bp: 214°, mp: 155°.
Acute tox data: Oral LD_{50} (rat) = 1297 mg/kg; oral LD_{50} (mouse) = 841 mg/kg. [*3*]
THR = MOD via oral route. An exper carc in rats via oral route. [*108*]

2,6-XYLIDENE. Syn: *2,6-dimethyl aniline.*
Acute tox data: Oral LD_{50} (rat) = 840 mg/kg; oral LD_{50} (mouse) = 707 mg/kg. [*3*]
THR = MOD to HIGH via oral route.

3,4-XYLIDENE. Syn: *3,4-dimethyl aniline.*
Acute tox data: Oral LD_{50} (rat) = 812 mg/kg; oral LD_{50} (wild birds) = 5 mg/kg. [*3*]
THR = MOD to HIGH via oral route, depending on species. Chronic toxicity. An exper carc. [*23*]

3,5-XYLIDENE.
Acute tox data: Oral LD_{50} (rat) = 707 mg/kg; oral LD_{50} (mouse) = 421 mg/kg. [*3*]
THR = MOD to HIGH via oral route.

XYLITOL.
THR = U. An additive permitted in food for human consumption. A sweetener. [*109*]

XYLOLS. See xylenes-*o,m,p.*

***m*-XYLYL BROMIDE.** Colorless liquid. C_8H_9Br, mw: 185.1, bp: 213°, d: 1.371 @ 23°.
Acute tox data: Inhal LC_{LO} (human) = 75 ppm for 1/6 hr. [*3*]
THR = HIGH via inhal route. A strong irr; see also bromides.
Fire Hazard: Mod, when exposed to heat or flame.
Disaster Hazard: Dangerous; see bromides; can react vigorously with oxidizing materials.

***o*-XYLYL BROMIDE.** Syn: *o-methyl benzyl bromide.* Crystals. C_8H_9Br, mw: 185.1, mp: 21°, bp: 223°, d: 1.381.
THR = See *m*-xylyl bromide.
Fire Hazard: Slight, when exposed to heat or flame.
Disaster Hazard: Dangerous; see bromides.

***p*-XYLYL BROMIDE.** White crystals. C_8H_9Br, mw: 185.1, mp: 36°, bp: 220° @ 740 mm, d: 1.324.
THR = See *m*-xylyl bromide.

***m*-XYLYL CHLORIDE.** Colorless liquid. $CH_3C_6H_4CH_2Cl$, mw: 140.61, bp: 196°, d: 1.064.
THR = HIGH irr to skin, eyes and mu mem and via oral and inhal routes.
Disaster Hazard: Dangerous; see chlorides.

***o*-XYLYL CHLORIDE.** Liquid. C_8H_9Cl, mw: 140.6, bp: 200°.
THR = *m*-xylyl chloride.

***p*-XYLYL CHLORIDE.** Fuming liquid, irr odor. C_8H_9Cl, mw: 140.6, bp: 200°–202°.
THR = See *m*-xylyl chloride.

XYLYL ETHANOL. $C_{10}H_{14}O$, mw: 150.2.
THR = Can react violently with ($H_2O_2 + H_2SO_4$). [*19*]

Y

YELLOW AB. Syn: *1-(phenylazo)-2-naphthylamine.* $C_{16}H_{13}N_3$, mw: 247.3.
THR = An exper (−) carc and neo [*3, 4, 23*] via oral route.

YELLOW BEESWAX. See beeswax.

YELLOW JASMINE. See *gelsemium sempervirens.*

YELLOW OB. $C_{17}H_{15}N_3$, mw: 261.4.
THR = An exper (+) carc and neo [*3, 4*] via sc route.

YELLOW PLUMBOUS OXIDE. See litharge.

YELLOW PRUSSIATE OF POTASH. See potassium ferrocyanide.

YELLOW RESIN. See rosin.

YEW. A wood.
THR = A mild allergen and irr in form of dust.
Fire Hazard: Mod, when exposed to heat or flame; can react with oxidizing materials.
Explosion Hazard: Mod, in the form of dust, when exposed to flame.

YOHIMBENE. Syn: *isomer of yohimbine.* Light-sensitive crystals. $C_{21}H_{26}N_2O_3$, mw: 354.4, mp: 278° (decomp).
THR = An alkaloid poison. See also yohimbine.
Disaster Hazard: Mod dangerous; when heated to decomp, emits toxic fumes.

YOHIMBINE. Syns: *corynine, aphrodine, quebrachine.* Colorless needles from water and alcohol. $C_{21}H_{26}N_2O_3$, mw: 354.4, mp: 235°.
Acute tox data: Oral LD_{LO} (mice) = 25 mg/kg; sc LD_{50} (frog) = 34 mg/kg; iv LD_{LO} (rabbit) = 11 mg/kg. [*3*]
THR = HIGH via oral, sc, iv and dermal routes. This material is a poison. Cases of poisoning have occurred from its use as an aphrodisiac. Upon local application, it produces anesthesia. However, absorption of it can give rise to toxic symptoms, such as salivation, increased respiration, and repeated defecation. With reference to the circulatory system, there may be a fall in blood pressure and sometimes myocardial damage, involving particularly the conduction system of the heart, with a resultant decrease in the efficiency of the heart.
Disaster Hazard: Mod dangerous; when heated to decomp, emits toxic fumes.

YOHIMBINE HYDROCHLORIDE. See yohimbine.

YPERITE. See mustard gas.

YTTERBIUM. A bright, silvery, lustrous, soft, malleable, ductile and fairly stable element. Yb, atwt: 173.04, mp: 824°, bp: 1193°, d: 6.972.
THR = U. As a lanthanon it may have an anticoagulant action on blood. See also lanthanum. An exper neo via imp route.
Radiation Hazard: For permissible levels, see Section 5A, Table 5A.5.
Fire Hazard: Mod, in the form of dust, when reacted with air, halogens. [*19*]

YTTERBIUM CHLORIDE. $YbCl_3$, mw: 279.4.
Acute tox data: ip LD_{50} (mice) = 395 mg/kg; ip LD_{50} (guinea pig) = 132 mg/kg. [*3*]
THR = HIGH via ip route.

YTTERBIUM CITRATE.
Acute tox data: ip LD_{50} (mice) = 143 mg/kg; ip LD_{50} (guinea pig) = 69 mg/kg. [*3*]
THR = HIGH via ip route.

YTTERBIUM NITRATE HEXAHYDRATE. $Yb(NO_3)_3 \cdot 6H_2O$, mw: 467.2.
Acute tox data: Oral LD_{50} (rat) = 3100 mg/kg; ip LD_{50} (rat) = 255 mg/kg; ip LD_{50} (mice) = 250 mg/kg. [*3*]
THR = HIGH via ip route and MOD via oral route.

YTTERBIUM SELENATE. Hexagonal plates. $Yb_2(SeO_4)_3 \cdot 8H_2O$, mw: 919.1, d: 3.30.
THR = See selenium and ytterbium.

YTTERBIUM SELENITE. $Yb_2(SeO_3)_3$, mw: 727.0.
THR = See selenium.

YTTRIA. See yttrium oxide.

YTTRIUM. Hexagonal, gray-black metallic element. Y, atwt: 88.9, mp: 1500°, bp: 3200°, d: 5.51.
THR = U. As a lanthanon, it may have an anticoagulant effect on the blood. See also lanthanum.
Radiation Hazard: For permissible levels, see Section 5A, Table 5A.5 Artificial isotope ^{90}Y, $T\frac{1}{2}$ = 64h, decays to stable ^{90}Zr via β's of 2.27 MeV. Yttrium exists in radioactive equilibrium with its parent, ^{90}Sr. Permissible levels are given for the equilibrium mixture. Artificial isotope ^{91}Y, $T\frac{1}{2}$ = 59d, decays to stable ^{91}Zr via β's of 1.54 MeV.
Fire Hazard: Mod, in the form of dust, when reacted with air, halogens. [*19*]

YTTRIUM BROMATE. Hexagonal, prismatic crystals. $Y(BrO_3)_3 \cdot 9H_2O$, mw: 634.8, mp: 74°, $-6H_2O$ @100°.
THR = See bromates and yttrium.

YTTRIUM BROMIDE. Deliquescent crystals. YBr_3, mw: 328.7.
THR = See bromides and yttrium.

YTTRIUM CHLORIDE. YCl_3, mw: 195.3.
Acute tox data: ip LD_{50} (rat) = 45 mg/kg; ip LD_{50} (mice) = 88 mg/kg. [3]
THR = HIGH via ip route. See also yttrium.

YTTRIUM FLUORIDE. Gelatinous material. $YF_3 \cdot \frac{1}{2}H_2O$, mw: 154.9.
THR = See fluorides and yttrium.

YTTRIUM NITRATE. Reddish-white prisms. $Y(NO_3)_3 \cdot 6H_2O$, mw: 383.0, d: 2.682.
Acute tox data: ip LD_{50} (rat) = 362 mg/kg; sc LD_{50} (mice) = 1662 mg/kg; iv LD_{50} (rabbit) = 500 mg/kg. [3] See nitrates and yttrium.
THR = HIGH via ip and iv routes. MOD via sc route. An exper neo via oral route. [3]

YTTRIUM OXIDE. Syn: *yttria.* White powder. Y_2O_3, mw: 225.8, d: 4.84.
Acute tox data: ip LD_{50} (rat) = 500 mg/kg. [3]
THR = HIGH via ip route. See yttrium.

YTTRIUM SULFIDE. Yellow-gray powder. Y_2S_3, mw: 274.04.
THR = See sulfides and yttrium.

Z

ZEARALENONE. $C_{18}H_{22}O_5$, mw: 318.4.
THR = An exper carc. [*23*]

ZEOLITE. A hydrated alkali aluminum silicate, capable of exchanging alkali for calcium and magnesium; used for softening water. There are a number of artificial zeolites now on the market. $Na_2O \cdot Al_2O(SiO_2)_4 \cdot (H_2O)_x$.
THR = A nuisance dust. See nuisance dusts. May be alkaline on contact with tissue moisture. LOW via oral route. [*20*]

ZEPHIRAN CHLORIDE. See benzalkonium chloride.

ZERO TOLERANCE CONCEPT. [*109*]

ZINC. A bluish-white lustrous metal. Zn, atwt: 65.38, mp: 419.58, bp: 907°, d: 7.133 @ 25°, vap. press: 1 mm @ 487°.
Acute tox data: ip LD_{50} (mice) = 15 mg/kg. [*3*]
THR = HIGH via ip route. See also zinc compounds.
Radiation Hazard: For permissible levels, see Section 5A, Table 5A.5. Artificial isotope ^{65}Zn, $T_{\frac{1}{2}}$ = 245d, decays to stable ^{65}Cu by ec; emits γ's of 1.12 MeV and x-rays.
Fire Hazard: Mod, in the form of dust when exposed to heat or flame.
Spont Heating: No.
Explosion Hazard: In the form of dust when reacted with acids, NH_4NO_3, BaO_2, $Ba(NO_3)_2$, Cd, CS_2, chlorates, Cl_2, ClF_3, CrO_3, (ethyl acetoacetate + tribromoneopentyl alcohol), F_2, hydrazine mononitrate, hydroxylamine, $Pb(N_3)_2$, (Mg + $Ba(NO_3)_2$ + BaO_2), $MnCl_2$, HNO_3, performic acid, $KClO_3$, KNO_3, K_2O_2, Se, $NaClO_3$, Na_2O_2, S, Te, H_2O. [*19*] See also powdered metals.
To Fight Fire: Special mixtures of dry chemical.

ZINC ACETATE. Monoclinic crystals. $Zn(C_2H_3O_2)$, mw: 183.47, mp: 242°, bp: sublimes in vacuum, d: 1.84.
Acute tox data: Oral LD_{50} (rat) = 2460 mg/kg. [*3*]
THR = MOD via oral route. See zinc compounds. A trace mineral added to animal feeds. [*109*]

ZINC ACETONYL ACETONATE.
Acute tox data: ip LD_{50} (rat) = 50 mg/kg. [*3*]
THR = HIGH via ip route. See also zinc compounds.

ZINC ALKYL AMINE *o*-PHENYL PHENATE.
THR = U. See phenols and zinc compounds. A fungicide.
Disaster Hazard: Dangerous; when heated to decomp, emits high toxic fumes.

ZINC ALLYL DITHIOCARBAMATE.
Acute tox data: Oral LD_{50} (rat) = 375 mg/kg; oral LD_{50} (mice) = 440 mg/kg. [*3*]
THR = HIGH via oral route. See also carbamates and zinc compounds.

ZINC AMIDE. Amorphous, white powder. $Zn(NH_2)_2$, mw: 97.43, mp: decomp @ 200°, d: 2.13 @ 25°.
THR = Can react violently with hydrazine. [*19*] See zinc compounds.

ZINC AMMONIUM NITRITE. Solid. $ZnNH_4(NO_2)_3$, mw: 221.5.
THR = See nitrites and zinc compounds.
Fire Hazard: Mod, by spont chemical reaction. A powerful oxidizing agent.
Explosion Hazard: U.
Disaster Hazard: See nitrites.

ZINC ARSENATE. White, odorless powder. Compositions: variable; approx 5ZnO, $2As_2O_5$.
THR = A poison. See arsenic and zinc compounds.

ZINC-*o*-ARSENATE. Monoclinic crystals. $Zn_3(AsO_4)_2 \cdot 8H_2O$, mw: 618.09, mp: decomp @ 100°, d: 3.309 @ 15°.
THR = See arsenic and zinc compounds.

ZINC ARSENIDE. Cubic crystals. Zn_3As_2, mw: 345.96.
THR = HIGH. See zinc and arsenic compounds.
Fire Hazard: Mod; can evolve arsine upon contact with moisture or acid; see also arsine.
Explosion Hazard: Mod; can evolve arsine upon contact with moisture or acid; see also arsine.
Disaster Hazard: Dangerous; when heated to decomp, or on contact with acid or acid fumes, emits highly toxic fumes of arsenic.

ZINC ARSENITE. See zinc-*m*-arsenite.

ZINC-*m*-ARSENITE. Syns: *ZMA, zinc arsenite*. White powder. $Zn(AsO_2)_2$, mw: 279.2.
THR = HIGH. A wood preservative, insecticide. See arsenic compounds; also see zinc compounds.

ZINC ASHES. See zinc.

ZINC BACITRACIN. Creamy white powder, slightly sol in water.
THR = U. An additive permitted in food for human consumption. Also permitted in the feed and drinking water of animals and/or for the treatment of food-producing animals. [*109*]

ZINC BENZENE DIAZONIUM CHLORIDE. $C_6H_5N_2Cl_3Zn$, mw: 276.8.
THR = Has exploded in storage. [*19*] See also zinc compounds.

ZINC BENZOATE. White powder, sol in water (40 parts). $Zn(C_7H_5O_2)_2$, mw: 307.60.
THR = A MILD respiratory irr. See also zinc compounds.

ZINC BERYLLIUM SILICATE. See beryllium zinc silicate.

ZINC BORATE. White amorphous powder or triclinic crystals. $3ZnO \cdot 2B_2O_3$, mw: 383.42, mp: 980°, d(amorphous): 3.64, d(crystal): 4.22.
THR = See boron and zinc compounds.

ZINC BROMATE. White, deliquescent powder. $Zn(BrO_3)_2 \cdot 6H_2O$, mw: 429.3, mp: 100°, bp: $-6H_2O$ @ 200°, d: 2.566.
THR = See bromates and zinc compounds. Can react violently with Al, As, C, Cu, metal sulfides, organic matter, P, S. [*19*]

ZINC BROMIDE. Rhombic, colorless, hygroscopic crystals. $ZnBr_2$, mw: 225.2, mp: 394°, bp: 650°, d: 4.219 @ 4°.
THR = See bromides and zinc compounds. Can react violently with K, Na. [*19*]

ZINC CAPRYLATE. Lustrous scales, slightly sol in boiling water, mod sol in boiling alcohol. $Zn(C_8H_{15}O_2)_2$, mw: 351.78, mp: 136°.
THR = U. On decomp, it releases irr fumes of caprylic acid. See also zinc compounds. A fungicide.

ZINC CARBONATE. White crystalline powder. $ZnCO_3$, mw: 125.4, mp: $-CO_2$ @ 300°, d: 4.42–4.45.
THR = See zinc compounds. A trace mineral added to animal feeds. [*109*]

ZINC CHELATE COMPLEX OF 8-QUINOLINOL. A susp carc. [*14*] See 8-hydroxyquinoline.

ZINC CHLORATE. Colorless, very deliquescent crystals. $Zn(ClO_3)_2 \cdot 4H_2O$, mw: 304.36, mp: decomp @ 60°, bp: decomp, d: 2.15.
THR = See chlorates and zinc compounds. Can react violently with Al, Sb_2S_3, As, As_2S_3, C, charcoal, Cu, CuS, MnO_2, metal sulfides, dibasic organic acids, organic matter, P, SnS, S, H_2SO_4. [*19*]

ZINC CHLORIDE. Syn: *butter of zinc.* Cubic, white, deliquescent crystals. $ZnCl_2$, mw: 136.30, mp: 290°, bp: 732°, d: 2.91 @ 25°, vap. press: 1 mm @ 428°.
Acute tox data: ip LD_{50} (mice) = 31 mg/kg; iv LD_{LO} (rat) = 30 mg/kg. Human inhal of 4800 mg/m^3 for $\frac{1}{2}$ hr $\rightarrow$ pulmonary effects.
THR = HIGH via ip, iv and inhal routes. See zinc compounds and chlorides. Used as a trace mineral added to animal feeds. Also as a nutrient and/or dietary supplement food additive. It is a substance that migrates to food from packaging materials. [*109*] The fumes are highly toxic. A poison. An exper neo. [*3*] Can react violently with K. [*19*]

ZINC CHLORIDE, CHROMATED. Solid.
THR = A recog carc. [*14*] See chromium compounds, chlorides and zinc compounds.

ZINC-5-CHLORO-2-MERCAPTOBENZOTHIAZOLE.
THR = U. See zinc compounds and chlorides.

ZINC CHROMATE. Lemon-yellow prisms. $ZnCrO_4$, mw: 181.4.
Acute tox data: iv LD_{LO} (mice) = 30 mg/kg. [*3*]
THR = HIGH via iv route. See also zinc compounds and chromium compounds.

ZINC CHROMATE HYDROXIDE.
$ZnCrO_4 \cdot Zn(OH)_2 \cdot H_2O$, mw: 299.
THR = An exper (+) carc [*3, 6*] via inhal route.

ZINC COMPOUNDS.
THR = Variable, generally of low toxicity. Zinc is not inherently a toxic element. However, when heated, it evolves a fume of zinc oxide which, when inhaled fresh, can cause a disease known as "brass founders' ague," or "brass chills." It is possible for people to become immune to it, but this immunity can be broken by cessation of exposure of only a few days. Zinc oxide dust which is not freshly formed is virtually innocuous. There is no cumulative effect to the inhal of zinc fumes. Fatalities, however have resulted from lung damage caused by the inhal of high conc of zinc chloride fumes. Sol salts of zinc have a harsh metallic taste; small doses can cause nausea and vomiting, while larger doses cause violent vomiting and purging. So far as can be determined, the continued administration of zinc salts in small doses has no effect in man except those of disordered digestion and constipation. Exposure to zinc chloride fumes can cause damage to the mu mem of the nasopharnyx and respiratory tract and give rise to a pale gray cyanosis. Workers in zinc refining have been reported as suffering from a variety of non-specific intestinal, respiratory and nervous symptoms. Ulceration of the nasal septum and eczematous dermatosis are also reported.

It has been stated that zinc oxide dust can block the ducts of the sebaceous glands and give rise to a papular, pustular eczema in men engaged in packing this compound into barrels. Sensitivity to zinc oxide in man is extremely rare. Zinc chloride, because of its caustic action, can cause ulceration of the fingers, hands and forearms of those who use it as a flux in soldering. This condition has even been observed in men who handle railway ties which have been impregnated with this material. It is the opinion of some who work with it that it is carc. Many Zn salts are known or susp carc. [*22*] Common air contaminants.

Treatment and antidotes: Personnel exposed to zinc chloride fumes should immediately wash the area of contact with copious quantities of warm water and soap. Remove all contaminated clothing at once and if the area of contact is large, subject patient to a deluge-type of shower as quickly as possible. If the eyes are involved in exposure to zinc chloride fumes, they should be irrigated for at least 15 min with warm water.

ZINC CYANIDE. Rhombic colorless crystals. $Zn(CN)_2$, mw: 117.4, mp: decomp @ 800°.

Acute tox data: ip LD_{LO} (rat) = 100 mg/kg. [*3*]

THR = HIGH via oral and ip routes. See cyanides and zinc compounds. Can react violently with Mg. [*19*]

ZINC DIBUTYL DITHIOCARBAMATE. White powder. $ZnC_{18}H_{38}N_2S_4$, mw: 476.2, mp: 104°–108°, d: 1.24 @ 20°/20°.

Acute tox data: ip LD_{50} (mice) = 100 mg/kg. [*3*]

THR = HIGH via ip route. An exper neo [*3*] via oral route. See carbamates.

Disaster Hazard: Dangerous; see sulfides and carbamates.

ZINC DICHROMATE. Orange-yellow powder; reddish-brown hygroscopic crystals. $ZnCr_2O_7 \cdot 3H_2O$; mw: 335.45.

THR = See chromium and zinc compounds.

ZINC DIETHYL. Syn: *zinc ethyl.* Liquid. $Zn(C_2H_5)_2$, mw: 123.5, mp: −28°, bp: 118°, d: 1.2065 @ 20°/4°.

THR = High. See zinc compounds.

Fire Hazard: Dangerous, by spont chemical reaction. Spont flam in air.

Explosion Hazard: Violent reaction with Cl_2, water, air, hydrazine. [*19*]

Disaster Hazard: Dangerous fire hazard; can react vigorously with oxidizing materials.

To Fight Fire: Do not use water, foam or halogenated extinguishing agents. Use dry materials, such as graphite, sand, etc.

ZINC DIETHYLDITHIOCARBAMATE. Syn: *ethyl ziram.* White powder. $C_{10}H_{22}S_4N_2Zn$, mw: 364.0, d: 1.47 @ 20°/20°.

Acute tox data: Oral LD_{50} (rat) = 3340 mg/kg; oral LD_{50} (rabbit) = 570 mg/kg. [*3*]

THR = HIGH–MOD via oral route. An exper neo [*3*] via oral route. See also carbamates and zinc compounds. See zinc ethylene bis(dithiocarbamate) and bis(diethyl thiocarbamyl) disulfide. This material is very irr to the eyes, nose and throat. Several hrs after the material may have gotten into the eyes, it causes an unbearable pain. A seed disinfectant, fungicide, rubber accelerator.

Disaster Hazard: Dangerous; see sulfides and cyanides.

ZINC DIMETHYL. Mobile liquid. $Zn(CH_3)_2$, mw: 95.5, mp: −40°, bp: 46°, d: 1.386 @ 10.5°/4°.

THR = See zinc compounds.

Fire Hazard: Dangerous, by chemical reaction. Spont flam in air.

Explosion Hazard: Can react violently with air, water, Cl_2, hydrazine. [*19*]

Disaster Hazard: Dangerous; when heated to decomp, burns and emits toxic fumes of zinc compounds; can react with oxidizing materials.

ZINC DIMETHYL DITHIOCARBAMATE. Syn: *ziram.* White powder. $C_6H_{14}S_4N_2Zn$, mw: 307.8, mp: 248°–250°, d: 1.65 @ 20°/20°.

Acute tox data: Oral LD_{50} (rat) = 1400 mg/kg; ip LD_{50} (rat) = 23 mg/kg; ip LD_{50} (mice) = 73 mg/kg; oral LD_{50} (wild birds) = 100 mg/kg. [*3*]

THR = HIGH via oral, ip routes. An exper (±) carc and neo [*12, 3*] via oral and imp routes. Some mutagenic activity. [*12*]

THR = Zinc dimethyl dithiocarbamate is very irr to the eyes, nose and throat. Several hrs after this material gets into the eyes, the pain becomes unbearable. A seed disinfectant and fungicide.

Disaster Hazard: Dangerous; see sulfides and cyanides.

ZINC DIMETHYL DITHIOCARBAMATE-CYCLOHEXYLAMINE COMPLEX. White powder.

THR = An exper carc. [*14*] A fungicide. See individual components.

Fire Hazard: Mod, when exposed to heat or flame.

Explosion Hazard: U.

Disaster Hazard: Dangerous; see sulfides and cyanides; can react with oxidizing materials.

ZINC DITHIOCARBAMATE.

THR = An exper carc. [*14*] See carbamates. See also zinc compounds.

Disaster Hazard: Dangerous; see sulfides and cyanides.

ZINC (II) EDTA COMPLEX.

Acute tox data: ip LD_{50} (mice) = 85 mg(Zn)/kg. [*3*]

THR = HIGH via ip route

ZINC ETHYL. See zinc diethyl.

ZINC ETHYLENE BIS(DITHIOCARBAMATE). Syn: *zineb.* Light colored powder, water-insol. $Zn(CS_2NHCH_2)_2$, mw: 275.8.

Acute tox data: uk LD_{50} (mammal) = 2000 mg/kg. [*3*]
THR = MOD via uk route. An irr via oral, inhal routes. An exper (±) carc and teratogen [*12, 3*] via oral and imp routes.
THR = A fungicide. Irr to the skin and mu mem. See also bis(diethyl thiocarbamyl)disulfide.
Disaster Hazard: Dangerous; when heated to decomp, emits highly toxic fumes of NO_x and SO_x.

ZINC FERROCYANIDE. White powder. $Zn_2Fe(CN)_6$, mw: 342.72.
Acute tox data: Oral LD_{50} (rat) = >5000 mg/kg. [*3*]
THR = MOD via oral route. See zinc compounds and ferrocyanides.

ZINC FLUOGALLATE. Colorless crystals. $(Zn \cdot 6H_2O) \cdot (GaF_5 \cdot 5H_2O)$, mw: 356.2, mp: $-5H_2O$ @ 110°, d: 2.33.
THR = See zinc compounds and fluorides.

ZINC FLUORIDE. White powder. ZnF_2, mw: 103.38, mp: 872°, bp: 1497°, d: 4.84 @ 15°, vap. press: 1 mm @ 970°.
Acute tox data: sc LD_{LO} (frog) = 280 mg/kg; oral LD_{LO} (guinea pig) = 200 mg/kg. [*3*]
THR = HIGH via sc and oral route. See fluorides and zinc compounds. Can react violently with K. [*19*]

ZINC FLUOSILICATE. Hexagonal colorless prisms. $ZnSiF_6 \cdot 6H_2O$, mw: 315.5, d: 2.104.
Acute tox data: Oral LD_{LO} (rat) = 100 mg/kg; sc LD_{LO} (guinea pig) = 200 mg/kg; sc LD_{LO} (frog) = 280 mg/kg. [*3*]
THR = HIGH via oral and sc routes. See also zinc compounds and fluosilicates.

ZINC FORMALDEHYDE SULFOXYLATE. Rhombic prisms. $Zn(HSO_2CH_2O)_2$, mw: 255.6, mp: decomp.
THR = See zinc compounds and formaldehyde.
Disaster Hazard: Dangerous; see sulfonates.

ZINC FORMATE. White crystals. $Zn(CHO_2)_2$, mw: 155.4, bp: decomp, d: 2.36.
THR = See zinc compounds and formic acid.

ZINC GALLATE. White crystals. $ZnGa_2O_4$, mw: 268.8, mp: <800°, d: 6.15 (theoretical).
THR = See zinc and gallium compounds.

ZINC GLUCONATE.
THR = U. A dietary supplement food additive. [*109*] See zinc compounds.

ZINC HEXAFLUOROSILICATE. See zinc fluosilicates.

ZINC HYDROSULFITE. Syn: *zinc dithionite*. White amorphous solid, sol in water. ZnS_2O_4, mw: 193.
THR = U. A substance which migrates to food from packaging materials. [*109*]
Disaster Hazard: Dangerous; See sulfites.

ZINC HYDROXIDE. White powder. $Zn(OH)_2$, mw: 99.4, mp: 125° (decomp), d: 3.053.
THR = See zinc compounds.

ZINC HYPOPHOSPHITE. Colorless, hygroscopic crystalline powder. $Zn(H_2PO_2)_2 \cdot H_2O$, mw: 213.4.
THR = See zinc compounds and hypophosphites.

ZINC IODATE. White crystalline powder. $Zn(IO_3)_2$, mw: 415.2, mp: decomp, d: 4.98.
THR = See iodates and zinc compounds. Can react violently with Al, As, C. [*19*]

ZINC IODIDE. Colorless cubic powder or white deliquescent powder. ZnI_2, mw: 319.22, mp: 446°, bp: 624°, d: 4.666 @ 14.2°.
THR = See iodides and zinc compounds. Can react violently with K.

ZINCITE. See zinc oxide.

ZINC LAURATE. White powder. $Zn(C_{12}H_{23}O_2)_2$, mw: 464.0, mp: 128°.
THR = See zinc compounds.
Fire Hazard: Slight, when exposed to heat or flame; can react with oxidizing materials.

ZINC MERCAPTOBENZOTHIAZOLE.
Acute tox data: Oral LD_{50} (rat) = 540 mg/kg; ip LD_{LO} (mice) = 200 mg/kg. [*3*]
THR = HIGH via oral and ip routes. A fungicide. See zinc compounds. An exper neo [*3*] via sc route.
Disaster Hazard: Dangerous; when heated to decomp, or on contact with acid or acid fumes, emits highly toxic fumes of SO_x.

ZINC MERCURY CHROMATE.
Acute tox data: Oral LD_{50} (rat) = 630 mg/kg. [*3*]
THR = MOD via oral route. See chromium compounds, zinc and mercury compounds.

ZINC NAPHTHENATE. A solid. $Zn(C_6H_5COO)_2$, mw: 319.7.
Acute tox data: Oral LD_{50} (rat) = 4920 mg/kg. [*3*]
THR = MOD via oral route. A fungicide and mildew preventive. See zinc compounds.
Fire Hazard: Slight, when exposed to heat or flame.

ZINC NITRATE. A: needles, B: tetragonal colorless crystals. A: $Zn(NO_3)_2 \cdot 3H_2O$, B: $Zn(NO_3)_2 \cdot 6H_2O$; mw(A): 243.33, mw(B): 297.49; d(B): 2.065 @ 14°; mp(A): 42.5°, mp(B): 36.4°; bp(B): $-6H_2O$ @ 105°–131°.
Acute tox data: Oral LD_{50} (rat) = 1190 mg/kg. [*3*]
THR = MOD via oral route. See nitrates and zinc compounds. Can react violently with C,Cu, metal sulfides, organic matter, P, S. [*19*]

ZINC NITRODITHIOACETATE.
THR = A fungicide. Details U. See zinc compounds and thioacetic acid.
Fire Hazard: U. An oxidizer.

Disaster Hazard: Dangerous; when heated to decomp, emits highly toxic fumes.

ZINC OXALATE. White powder. $ZnC_2O_4 \cdot 2H_2O$, mw: 189.43, mp: 100° (sublimes), d: 2.562 @ 24.5°.
THR = See oxalates and zinc compounds.

ZINC OXIDE. Syns: *zincite, chinese white, zinc white, flowers of zinc.* White or yellowish powder. ZnO, mw: 81.38, mp: >1800°, d: 5.47.
THR = A seed disinfectant. See zinc compounds. A fungicide; a trace mineral added to animal feeds; also a dietary supplement food additive. [*109*] Has exploded with chlorinated rubber. Violent reaction with Mg. [*19*]

ZINC OXYSULFATE. White powder. $ZnO \cdot ZnSO_4$, mw: 242.8.
THR = See zinc compounds and sulfates. A fungicide.

ZINC PERCHLORATE HEXAHYDRATE. $Zn(ClO_4)_2 \cdot 6H_2O$, mw: 372.4.
Acute tox data: ip LD_{LO} (mice) = 76 mg/kg. [*3*]
THR = HIGH via ip route. See also zinc compounds and perchlorates.

ZINC PERMANGANATE. Violet-brown or black hygroscopic crystals. $Zn(MnO_4)_2 \cdot 6H_2O$, mw: 411.34, mp: $-5H_2O$ @ 100°, d: 2.47.
THR = See manganese and zinc compounds.
Fire Hazard: Mod, by chemical reaction with reducing agents. A powerful oxidizing agent.

ZINC PEROXIDE. Yellow-white powder. ZnO_2 (theoretical), mw: 97.38, d: 1.571 (theoretical).
THR = Systemic toxicity is similar to zinc oxide. See peroxides and zinc compounds.
Fire Hazard: Mod, when exposed to heat or by chemical reaction with reducing materials. Finely divided powder is slightly sol in water, decomp rapidly at 150°C. It is not dangerous unless mixed with highly combustible materials (Section 7).
Explosion Hazard: Dangerous, when exposed to heat. Can react violently with Al, Zn. [*19*]
Explosive Range: Explodes at 212° for peroxide prepared from $ZnSO_4$, NH_3, H_2O_2; $4ZnO \cdot 3H_2O_2 \cdot H_2O$ explodes at 190°.
Disaster Hazard: Very dangerous; explodes when heated; will react with water or steam to produce heat; on contact with reducing material, can react vigorously.

ZINC PHOSPHIDE. Cubic, dark gray crystals or powder. Zn_3P_2, mw: 258.10, mp: 420°, bp: 1100°, d: 4.55 @ 13°.
Acute tox data: Oral LD_{50} (rat) = 40 mg/kg; oral exposure of women to 3600 mg/kg → CNS effects.
THR = HIGH via oral route. See also phosphides and zinc compounds. A highly toxic rodenticide. This material is stable while kept dry. In moist air, decomp slowly. It reacts violently with acids or acid fumes to emit the highly toxic and flam phosphine.

ZINC PICRATE. Yellow crystalline powder. $Zn(C_6H_2N_3O_7)_2 \cdot 8H_2O$, mw: 665.7, mp: explodes.
THR = See picric acid and zinc compounds.
Fire Hazard: Dangerous. See nitrates.
Explosion Hazard: Severe. See explosives, high and nitrates.
Disaster Hazard: Dangerous; see nitrates and explosives, high.

ZINC POLYACRYLATE.
THR = An exper neo via imp route. [*3*]

ZINC POWDER. See zinc.

ZINC(PROPYLENE BIS DITHIOCARBAMATE).
Acute tox data: Oral LD_{50} (rat) = 8500 mg/kg. [*3*]
THR = LOW via oral route. See also zinc compounds and carbamates.

ZINC-2-PYRIDINETHIOL-1-OXIDE.
Acute tox data: Oral LD_{50} (rat) = 309 mg/kg. [*3*]
THR = HIGH via oral route.

ZINC RICINOLEATE. Fine white powder. $Zn(C_{18}H_{33}O_3)_2$, mw: 659.9, mp: 92°, d: 1.10 @ 25°/25°.
THR = See zinc compounds.
Disaster Hazard: Slight; when heated, emits acrid fumes.

ZINC SELENATE. Triclinic crystals. $ZnSeO_4 \cdot 5H_2O$, mw: 298.42, mp: decomp @ 50°, d: 2.591.
THR = See selenium and zinc compounds.

ZINC SELENIDE. Cubic crystals. ZnSe, mw: 144.34, d: 5.42 @ 15°.
THR = See zinc and selenium compounds.
Fire Hazard: Mod; on contact with moisture or acids, can evolve selenium hydride; see hydrides.
Explosion Hazard: Mod; can evolve a hydride; see hydrides.
Disaster Hazard: See selenium.

ZINC SILICOFLUORIDE. See zinc fluosilicate.

ZINC SORBATE. $Zn(C_6H_7O_2)_2$, mw: 134.1.
Acute tox data: Oral LD_{LO} (rat) = 4000 mg/kg. [*3*]
THR = MOD via oral route. See also zinc compounds.

ZINC STEARATE. White powder. $Zn(C_{18}H_{35}O_2)_2$, mw: 632.30, mp: 130°, flash p: 530°F (OC), autoign. temp.: 790°F.
THR = Inhal of zinc stearate has been reported as causing pulmonary fibrosis. See zinc compounds. A nutrient and/or dietary supplement food additive. [*109*] Also a nuisance dust.
Fire Hazard: Slight, when exposed to heat or flame.
To Fight Fire: Water, foam, CO_2, dry chemical.

ZINC SULFATE. Syn: *zinkosite.* Rhombic colorless crystals. $ZnSO_4$, mw: 161.44, mp: decomp @ 740°, d: 3.74 @ 15°.
Acute tox data: ip LD_{50} (rat) = 40 mg/kg; ip LD_{50} (mice) = 29 mg/kg. [*3*]
THR = HIGH via ip route. See zinc compounds. A fungicide. A trace mineral added to animal feeds; a nutrient and/or dietary supplement food additive. [*109*]

ZINC SULFATE HEPTAHYDRATE. $ZnSO_4 \cdot 7H_2O$, mw: 287.6.
Acute tox data: Oral LD_{LO} (rat) = 2200 mg/kg; oral LD_{LO} (rabbit) = 1914 mg/kg; iv LD_{LO} (rat) = 49 mg/kg; sc LD_{LO} (rat) = 330 mg/kg. [*3*]
THR = HIGH via iv and sc routes. MOD via oral route. See also zinc compounds. An exper neo [*3*] via sc route.

ZINC SULFIDE(α). Syn: *wurtzite.* Hexagonal colorless crystals. ZnS, mw: 97.45, mp: 1850° @ 150 atm, bp: sublimes @ 1185°, d: 4.087.
THR = See sulfides and zinc compounds. A fungicide. Can react violently with ICl. [*19*]

ZINC SULFITE. White crystalline powder. $ZnSO_3 \cdot 2H_2O$, mw: 181.48.
THR = See sulfites and zinc compounds.

ZINC TELLURATE. Heavy, granular, white precipitate. Zn_3TeO_6, mw: 419.75.
THR = See tellurium and zinc compounds.

ZINC TELLURIDE. Cubic red crystals. ZnTe, mw: 192.99, mp: 1238.5°, d: 6.34 @ 15°.
THR = See tellurium and zinc compounds.
Fire Hazard: Mod; upon contact with moisture or acids, evolves tellurium hydride.
Explosion Hazard: Slight; can evolve the hydride of tellurium. See hydrides.
Disaster Hazard: Dangerous; see tellurium compounds.

ZINC TETRAMMINE CHLORATE. $Zn(NH_3)_4(ClO)_3)_2$, mw: 300.4.
THR = A shock-sensitive explosive. [*19*] See also zinc compounds and ammonia.

ZINC TETRAMMINE PERCHLORATE. $Zn(NH_3)_4 \cdot (ClO_4)_2$, mw: 332.1.
THR = A shock-sensitive explosive. [*19*] See also zinc compounds and ammonia.

ZINC THIOCYANATE. White powder. $Zn(SCN)_2$, mw: 181.55.
THR = See thiocyanates and zinc compounds.

ZINC-2,4,5-TRICHLOROPHENATE. Colorless crystals. $Zn(OC_6H_2Cl_3)_2$, mw: 458.3.
Acute tox data: Oral LD_{50} (rat) = 1000 mg/kg. [*3*]
THR = MOD via oral route. See also zinc. A fungicide and seed protectant. See chlorinated phenols.
Disaster Hazard: Dangerous; see chlorides.

ZINC WHITE. See zinc oxide.

ZINEB. See zinc ethylene bis(dithiocarbamate).

ZINGIBER. See ginger.

ZINKOSITE. See zinc sulfate.

ZINO PHOS. Syns: *o,o-diethyl-o-(2-pyrazinyl) phosphorothioate, thionazin.*
Acute tox data: Oral LD_{50} (rat) = 3.5 mg/kg; dermal LD_{50} (rat) = 11 mg/kg; dermal LD_{50} (guinea pig) = 10 mg/kg. [*3*]
THR = HIGH via oral and dermal routes. A cholinesterase inhibitor. See also parathion.
Disaster Hazard: Dangerous; when heated to decomp, emits highly toxic fumes.

ZIRAM. See zinc dimethyl dithiocarbamate.

ZIRCONIUM. A grayish-white lustrous metal, very slightly radioactive. Zr, atwt: 91.22, mp: 1852°, bp: 3577°, d: 6.506 @ 20°.
THR = See zirconium compounds.
Radiation Hazard: For permissible levels, see Section 5A, Table 5A.5. Artificial isotope ^{95}Zr, $T_{\frac{1}{2}}$ = 65d, decays to radioactive ^{95}Nb via β's of 0.36 (43%), 0.40 (55%) MeV; emits γ's of 0.72, 0.76 MeV.
Fire Hazard: Dangerous, in the form of dust, when exposed to heat or flame or by chemical reaction with oxidizers.
Spont Heating: No.
Explosion Hazard: Dangerous, in the form of dust, by chemical reaction with air, alkali hydroxides, alkali metal chromates, dichromates, molybdates, salts, sulfates, tungstates, borax, CCl_4, CuO, Pb, PbO, P, $KClO_3$, KNO_3. [*19*]
Explosive Range: 0.16 g/liter in air.
To Fight Fire: Special mixtures, dry chemical, salt or dry sand.

ZIRCONIUM ACETATE. $H_2ZrO_2(C_2H_3O_2)_2$, mw: 243.
THR = See zirconium compounds.

ZIRCONIUM ACETYL ACETONATE.
Acute tox data: ip LD_{LO} (mice) = 316 mg/kg. [*3*]
THR = HIGH via ip route.

ZIRCONIUM AMMONIUM FLUORIDE. A: Rhombic white crystals; B: colorless cubic crystals. A: $(NH_4)_2ZrF_6$, B: $(NH_4)_3ZrF_7$; mw(A): 243.0, mw(B): 278.34, d: 1.154.
THR = See fluorides.

ZIRCONIUM CARBIDE. Hard, gray, metallic crystals. ZrC, mw: 103.23, mp: 3540°, bp: 5100°, d: 6.73.
THR = See carbides and zirconium compounds. Spont flam in air. [*19*]

ZIRCONIUM CHLORIDE OXIDE OCTAHYDRATE. $ZrOCl_2 \cdot 8H_2O$, mw: 322.3.
THR = An exper neo [*3*] via id route.

ZIRCONIUM COMPOUNDS.
THR = Zirconium is not an important industrial poision. Deaths in rabbits have been caused by intravenous injection of 150 mg/kg of body weight. Most zirconium compounds in common use are insol and considered inert. Pulmonary granuloma in zirconium workers has been reported and sodium zirconium lactate has been held responsible for skin granulomas. Avoid inhal of Zr-containing aerosols, which can cause lung granulomas. Zr-containing drugs or cosmetic products are being controlled by the FDA. [*37*]

ZIRCONIUM DIBROMIDE. Black powder. $ZrBr_2$, mw: 251.05, mp: 350° (decomp).
THR = See bromides.
Fire Hazard: Dangerous, by spont chemical reaction; ignites spont in air.
Explosion Hazard: U.
Disaster Hazard: Dangerous; see bromides; can react vigorously with oxidizing materials.

ZIRCONIUM DICARBIDE. ZrC_2, mw: 115.2.
THR = Can react violently with Br_2, Cl_2, F_2, I. [*19*]

ZIRCONIUM DICHLORIDE. Black crystals. $ZrCl_2$, mw: 162.13, mp: decomp @ 350°.
THR = See hydrochloric acid and zirconium.
Disaster Hazard: Dangerous; see chlorides.

ZIRCONIUM DIOXIDE. Syn: *baddeleyite*. Colorless, yellow or brown monoclinic crystals. ZrO_2, mw: 123.22, mp: 2700°, bp: 4300°, d: 5.49.
THR = See zirconium compounds.

ZIRCONIUM FLUORIDE. See zirconium tetrafluoride.

ZIRCONIUM GLUCONATE.
Acute tox data: ip LD_{50} (rat) = 247 mg/kg. [*3*]
THR = HIGH via ip route. See also zirconium compounds.

ZIRCONIUM HYDRIDE. Metallic dark gray to black powder. ZrH_2, mw: 93.23, d: 5.6, autoign. temp.: 270° in air.
THR = See hydrides and zirconium compounds. Incandesces when heated in air. [*19*] Flam when wet.

ZIRCONIUM IODIDE.
THR = See zirconium and iodides.

ZIRCONIUM (IV)LACTATE. White, slightly moist pulp, very slightly sol in water and common organic solvents, sol in aqueous alkali solutions with the formation of salts. $C_3H_5O_3 \cdot Zr$, mw: 180.3.
Acute tox data: ip LD_{50} (rat) = 500 mg/kg. [*3*]
THR = HIGH via ip route. An exper neo [*3*] via id route. Prolonged inhal of dust caused interstitial pneumonia. See also zirconium compounds. Powerful skin allergen.

ZIRCONIUM(III)LACTATE. $H_4ZrO(CHO)_9$, mw: 372.4.
Acute tox data: ip LD_{50} (rat) = 670 mg/kg; dermal TD_{LO} (mice) = 0.02 mg/kg ⟶ skin effects. [*3*]
THR = MOD via ip route. Powerful skin allergen. See zirconium compounds.

ZIRCONIUM, METAL. See Zirconium.

ZIRCONIUM NITRATE. White crystals. $Zr(NO_3)_4 \cdot 5H_2O$, mw: 429.33, mp: decomp @ 100°.
Acute tox data: Oral LD_{50} (rat) = 3200 mg/kg; inhal LC_{LO} (rat) = 500 mg/m^3 for 1/2 hr. [*3*]
THR = HIGH via inhal route and MOD via oral route. See nitrates.

ZIRCONIUM NITRIDE. Brassy-colored powder, refractory. ZrN, mw: 105.22, mp: 2930°–1980°, d: 7.09.
THR = Details U. Probably LOW. See also nitrides.

ZIRCONIUM PHOSPHIDE. Gray crystals. ZrP_2, mw: 153.2, d: 4.77 @ 25°/4°.
THR = See phosphides.

ZIRCONIUM PICRAMATE, WET WITH 20% OF WATER.
THR = See nitrates.

ZIRCONIUM SCRAP (BORINGS, CLIPPINGS, SHAVINGS, SHEETS OR TURNINGS). See zirconium.

ZIRCONIUM SELENATE. Hexagonal transparent crystals. $Zr(SeO_4)_2 \cdot 4H_2O$, mw: 449.20, mp: $-3H_2O$ @ 100°, $-4H_2O$ @ 130°.
THR = See selenium and zirconium compounds.

ZIRCONIUM (IV)SILICATE. $ZrSiO_4$, mw: 183.3.
THR = No data. See also silicates and zirconium compounds.

ZIRCONIUM SODIUM LACTATE. Syn: *sodium zirconium lactate*. Straw-colored liquid. $NaH_3ZrO(CH_3CHOCOO)_3$, mw: 397, d: 1.28.
Acute tox data: id exposure to women of 0.000004 mg/kg ⟶ skin effects.
THR = HIGH via id route. Has caused skin granulomas in humans. Inhal exper on rabbits produced bronchiolar abscesses, lobar pneumonia and peribronchial granulomas. An exper neo [*3*] via id route.

ZIRCONIUM SULFATE. White crystalline powder, insol in alcohol. $Zr(SO_4)_2 \cdot 4H_2O$, mw: 355.42.
Acute tox data: Oral LD_{50} (rat) = 3500 mg/kg; ip LD_{50} (rat) = 175 mg/kg; sc LD_{LO} (rat) = 500 mg/kg. [*3*]

THR = HIGH via ip and sc routes. MOD via oral route. See zirconium compounds.

ZIRCONIUM SULFIDE. Steel-gray crystals. ZrS_2, mw: 155.35, d: 3.87.
THR = See sulfides.

ZIRCONIUM TARTRATE. White powder, slightly sol in water. $ZnC_4H_4O_6 \cdot H_2O$, mw: 231.47.
THR = See zirconium compounds and tartaric acid.

ZIRCONIUM TETRACHLORIDE. White lustrous crystals. $ZrCl_4$, mw: 233.05, mp: sublimes @ 300°, bp: 331°, d: 2.80, vap. press: 1 mm @ 190°.
Acute tox data: Oral LD_{50} (rat) = 1688 mg/kg; oral LD_{50} (mice) = 655 mg/kg. [*3*]
THR = MOD via oral route. See hydrochloric acid. Self-ignites in air. [*19*]

ZIRCONIUM TETRAFLUORIDE. Syn: *zirconium fluoride.* Refractive crystals, water-sol. ZrF_4, mw: 167.2, d: 4.6 @ 16°, sublimes @ 600°.
Acute tox data: iv LD_{50} (mice) = 98 mg/kg. [*3*]
THR = HIGH via iv route. See fluorides.

ZIRCONIUM TETRAHYDROBORATE. $Zr(BH_4)_4$, mw: 118.1.
THR = See zirconium and boron. Can react vigorously with HNO_3. [*19*]

ZIRCONIUM, WET OR SLUDGE. See zirconium.

ZIRCONOCENE DICHLORIDE. $Cl_2ZrC_{10}H_{10}$, mw: 292.3.
Acute tox data: ip LD_{50} (rat) = 30 mg/kg. [*3*]
THR = HIGH via ip route. See also zirconium compounds and hydrochloric acid.

ZIRCONYL ACETATE.
Acute tox data: Oral LD_{50} (rat) = 4100 mg/kg; ip LD_{50} (rat) = 300 mg/kg. [*3*]
THR = HIGH via ip and MOD via oral route. See also zirconium compounds.

ZIRCONYL BROMIDE. Brilliant deliquescent crystals. $ZrOBr_2 \cdot xH_2O$, mp: $-H_2O$ @ 120°.
THR = See bromides and zirconium compounds.

ZIRCONYL CHLORIDE. $ZrOCl_2$, mw: 178.1.
Acute tox data: Oral LD_{50} (rat) = 3500 mg/kg; ip LD_{50} (rat) = 400 mg/kg; id TD_{LO} (mice) = 1.2 mg/kg; sc LD_{LO} (rat) = 500 mg/kg. [*3*]
THR = HIGH via id, ip and sc routes, MOD via oral route. An exper neo via ip, in routes to mice. [*103*]

ZIRCONYL NITRATE.
Acute tox data: Oral LD_{50} (rat) = 2500 mg/kg; ip LD_{50} (rat) = 1250 mg/kg. [*3*]
THR = MOD via oral and ip routes. See also nitrates and zirconium compounds.

ZIRCONYL SULFATE.
Acute tox data: Oral LD_{50} (rat) = 3500 mg/kg; ip LD_{50} (rat) = 175 mg/kg. [*3*]
THR = HIGH via ip and MOD via oral routes. See also zirconium compounds.

ZIRCONYL SULFIDE. Yellow powder. ZrOS, mw: 139.29, d: 4.87.
THR = See sulfides and zirconium.

ZMA. See zinc-*m*-arsenite.

ZOALENE. Syn: *3,5-dinitro-o-toluamide.* Yellowish solid, very slightly sol in water, sol in acetone, acetonitrile and dimethyl formamide. $(O_2N)_2C_6H_2(CH_3)CONH_2$, mw: 225, mp: 177°.
Acute tox data: Oral LD_{50} (rat) = 600 mg/kg. [*3*]
THR = MOD via oral route. An additive permitted in the feed and drinking water of animals and/or for the treatment of food-processing animals. Also permitted in food for human consumption. [*109*]
Disaster Hazard: See nitrates.

BIBLIOGRAPHY

1. IARC Monographs on the Evaluation of Carcinogenic Risk of Chemicals to Man. Vol. 7, World Health Organization (1974).
2. The Merck Index, 9th edition, Merck & Co., Rahway, New Jersey (1976).
3. Registry of Toxic Effects of Chemical Substances NIOSH (1976).
4. IARC Monographs on the Evaluation of Carcinogenic Risk of Chemicals to Man, Vol. 8, World Health Organization (1975).
5. IARC Monographs on the Evaluation of Carcinogenic Risk of Chemicals to Man, Vol. 5, World Health Organization (1974).
6. IARC Monographs on the Evaluation of Carcinogenic Risk of Chemicals to Man, Vol. 2, World Health Organization (1973).
7. IARC Monographs on the Evaluation of Carcinogenic Risk of Chemicals to Man, Vol. 10, World Health Organization (1976).
8. IARC Monographs on the Evaluation of Carcinogenic Risk of Chemicals to Man, Vol. 9, World Health Organization (1975).
9. IARC Monographs on the Evaluation of Carcinogenic Risk of Chemicals to Man, Vol. 1, World Health Organization (1972).
10. IARC Monographs on the Evaluation of Carcinogenic Risk of Chemicals to Man, Vol. 4, World Health Organization (1974).
11. IARC Monographs on the Evaluation of Carcinogenic Risk of Chemicals to Man, Vol. 3, World Health Organization (1973).
12. IARC Monographs on the Evaluation of Carcinogenic Risk of Chemicals to Man, Vol. 12, World Health Organization (1976).
13. IARC Monographs on the Evaluation of Carcinogenic Risk of Chemicals to Man, Vol. 11, World Health Organization (1976).
14. Hueper, W. C., Medicolegal Considerations of Occupational and Non-occupational Environmental Cancers, in Lawyers' Medical Cyclopedia, Vol. 5B, C. J. Frankel and R. M. Patterson, (editors) The Allen Smith Co., Indianapolis, Indiana (1972).
15. IARC Monographs on the Evaluation of Carcinogenic Risk of Chemicals to Man, Vol. 6, World Health Organization (1974).
16. Goodman, L. S. and Gilman, A., The Pharmacological Basis of Therapeutics, 5th Edition. Macmillan Publishing Co., New York (1975).
17. Handbook of Chemistry and Physics, 56th Edition, Weast (1976).
18. Gosselin, R. E., Hodge, H. C., Smith, R. P., Gleason, M. N., Clinical Toxicology of Commercial Products, 4th Edition. Williams & Wilkins Co., Baltimore, Maryland (1976).
19. Chemistry Laboratory Safety Library, 491 M, 5th Edition, National Fire Protection Association, Boston, Massachusetts (1975).
20. Deichmann, W. B. and Gerarde, H. W., Toxicology of Drugs and Chemicals, Academic Press, New York (1969).
21. *Environment*, **19**:1, pp. 6–12 (January/February 1977).
22. Sunderman, F. W., Jr. and Maenza, R. M., "Comparisons of the Carcinogenicities of Nickel Subsulphide, Nickel Sulphide, Nickel dust, Iron dust, and Nickel-Iron Sulphide Following i.m. Administration in Fischer Rats." *Proc. Amer. Assn. Cancer Res.* (1976).
23. Searle, C. E. (editor) Chemical Carcinogens ACS Monograph 173 Washington, D.C. (1976).
24. Documentation of TLV's, 3rd Edition, ACGIH (1971).
25. "Occupational Carcinogenesis," Annals of New York Academy of Sciences, 271 (1976).
26. Browning, E., Toxicity of Industrial Metals, 2nd Edition, Appleton-Crofts, New York (1969).
27. Browning, E., Toxic Solvents, Edward Arnold & Co., London, England (1953).
28. *C&EN*, p. 22 (July 25, 1977).
29. Statement of Eula Bingham re DOW findings on news media.
30. *C&EN*, p. 11 (August 1, 1977).
31. *C&EN*, p. 151 (September 3, 1977).
32. *Environment*, **19**:5, 28 (June/July 1977).
33. *C&EN*, p. 30 (July 18, 1977).
34. *Occupational Health & Safety*, p. 8 (May/June 1977).
35. "Occupational Exposure to Benzoyl Peroxide," NIOSH (June 1977).
36. "Occupational Exposure to Inorganic Fluorides," NIOSH (1975).
37. *C&EN*, p. 14 (August 22, 1977).
38. *Aldrichimica Acta*, **10**:2 (1977).
39. "Canada's Moth War," May, E. E. *Environment*, **19**:6 (August/September 1977).
40. *Occupational Health & Safety*, p. 9 (July/August 1977).
41. *Occupational Health & Safety*, pp. 9–10 (July/August 1977).
42. *Occupational Health & Safety*, p. 12 (July/August 1977).
43. *American Industrial Hygiene Association Journal*, **38**, A-15–A-20 (July 1977).
44. *American Industrial Hygiene Association Journal*, **38**, 307–320 (July 1977).
45. "Trouble in the Air from Maryland Rock Quarry," *Science*, 197, 237-240 (July 15, 1977).
46. "Toxicity of Mild Prenatal Carbon Monoxide Exposure," *Science*, 197, 680–682 (August 12, 1977).
47. "Perilla Ketone: A Potent Lung Toxic from the Mint Plant, Perilla Frutesiens Britton," *Science*, 197 (August 5, 1977).
48. "Seveso: The Questions Persist where Dioxin Created a Wasteland," *Science*, 197 (September 9, 1977).

49. "Occupational Exposure to Boron Trifluoride," NIOSH (1976).
50. "Occupational Exposure to Cadmium," NIOSH (1976).
51. Fairhall, L. T., Industrial Toxicology, 2nd Edition, Williams & Wilkins, Baltimore, Maryland (1957).
52. Evaluation of Certain Food Additives and the Contaminants Mercury, Lead and Cadmium (Sixteenth Report of the Joint FAO/WHO Expert Committee on Food Additives) World Health Organication techn. Rep. Ser., #505 (1972).
53. Ferm, V. H., and Carpenter, S. J., "Teratogenic Effect of Cadmium and its Inhibition by Zinc," *Nature*, **216,** 1123–1124 (1967).
54. Fitzgerald, J. J. and Detwiler, C. G., "Optimum Particle Size for Penetration Through the Milliporefilter," *Arch. Ind. Health*, **15,** 3–8 (1957).
55. "Occupational Exposure to Carbaryl," NIOSH (September 1976).
56. Wolfe, H. R., Durham, W. J., and Armstrong, J. F., "Exposure of Workers to Pesticides," *Arch. Envir. Health*, 14 (1976).
57. Gaines, T. B., "Acute Toxicology of Pesticides to Rats," Toxicology of Applied Pharmacology 2 pp. 88–99 (1960).
58. "Occupational Exposure to Carbon Disulfide," NIOSH (May 1977).
59. Elkins, H. B., "Toxic Fumes," *Industrial Medicine*, **8,** 426–432 (1939).
60. Elkins, H. B. Chemistry of Industrial Toxicology, John Wiley & Sons, New York (1950).
61. Occupational Exposure to Chromium (VI) NIOSH (1975).
62. Hueper, W. C., Occupational and Environmental Cancers of the Respiratory Tract. Springer-Verlag, New York pp. 57–85 (1966).
63. Hueper, W. C. and Payne, W. W. "Experimental Cancers in Rats Produced by Chromium Compounds and Their Significance to Industry and Public Health," *AIHA Journal*, **20,** 274–280 (1959).
64. Mancuso, T. F., "Occupational Cancer and Other Health Hazards in a Chromate Plant:" A medical appraisal—II Clinic and Toxicological Aspects," *Industrial Medicine and Surgery*, **20,** 393–407 (1951).
65. "Occupational Exposure to Cotton Dust," NIOSH (1974).
66. Fox, A. J., *et al.*, "A Survey of Respiratory Disease in Cotton Operatives, Part I" *British Journal Industrial Medicine*, **30,** 42–47 (1973).
67. Fox, A. J., *et al.*, "A Survey of Respiratory Disease in Cotton Operatives, Part II," *British Journal Industrial Medicine*, **30,** 48–53 (1973).
68. "Occupational Exposure to Organotin Compounds," NIOSH (1976).
69. "Occupational Exposure to Xylene," NIOSH (1976).
70. "Catalytic Units," *Science News*, **112,** 171 (September 10, 1977).
71. "Occupational Exposure to Inorganic Nickel," NIOSH (May 1977).
72. "Occupational Exposure to Phosgene," NIOSH (1976).
73. Henderson, Y. and Haggard, H. W., Noxious Gases, 2nd Edition, Reinhold Publishing Corp., New York (Van Nostrand Reinhold) (1943).
74. Chlorine Manual, The Chlorine Institute, New York (1969).
75. "Occupational Exposure to Phenol," NIOSH (July 1976).
76. Patty, F., Industrial Hygiene & Toxicology, 2nd Edition, Interscience, New York (1963).
77. Sollman, A Manual of Pharmacology, 8th Edition W. B. Saunders Co., Philadelphia, Pennsylvania (1957).
78. "Occupational Exposure to Epichlorohydrin," NIOSH (September 1976).
79. "Occupational Exposure to Ethylene Dichloride," NIOSH (1976).
80. IARC Monographs on the Evaluation of Carcinogenic Risk of Chemicals to Man, Vol. 14, World Health Organization (1977).
81. IARC Monographs on the Evaluation of Carcinogenic Risk of Chemicals to Man, Vol. 15, World Health Organization (1977).
82. Kolness, A. W., "Selenium Linked to Nerve Disorder," *C&EN*, p. 19 (June 27, 1977).
83. "Occupational Exposure to Hydrogen Sulfide," NIOSH (May 1977).
84. "Occupational Exposure to Hydrogen Fluoride." NIOSH (March 1976).
85. Kleinfeld, M., "Acute Pulmonary Edema of Chemical Origin," *Arch. Environmental Health*, **10,** 942–946 (1965).
86. Stokinger, H. E., "Toxicity Following Inhalation of F_2 and HF" in Pharmacology and Toxicology of U Compounds Voegtlin and Hodge (Editors), McGraw-Hill Book Co., New York (1949).
87. "Occupational Exposure to Isopropyl Alcohol," NIOSH (March 1976).
88. "Occupational Exposure to Nitric Acid," NIOSH (March 1976).
89. "Occupational Exposure to Inorganic Arsenic," NIOSH (1975).
90. Blum, A. and Ames, B., "Flame Retardant Additions as Possible Cancer Hazards." *Science*, **195,** 76–78 (January 7, 1977).
91. Sviberly, J. L., U.S. Government Report (February 1, 1960).
92. *Bulletin of Environmental Toxicology*, **15,** 660 Springer-Verlag, New York (1976).
93. "Acrylonitrile Linked to Cancer in Workers." *C&EN* p. 6 (May 30, 1977).
94. Farm Chemicals Handbook. Meister Publishing Co., Willoughby, Ohio (1972).
95. Fishbein, L., "Environmental Metallic Carcinogens: An Overview of Exposure Levels." *Journal of Toxicology and Environmental Health*, **2,** 77–109 (1976).
96. "Occupational Exposure to Acetylene." NIOSH (July 1976).
97. "Occupational Exposure to Acrylamide," NIOSH (October 1976).
98. "Occupational Exposure to Alkanes," NIOSH (March 1977).
99. "Occupational Exposure to Allyl Chloride," NIOSH (September 1976).
100. "Occupational Exposure to Methyl Alcohol," NIOSH (March 1976).

101. "Occupational Exposure to Methyl Parathion," NIOSH (September 1976).
102. Occupational Carcinogenisis, Saffioti, U., and Wagoner, J. K. (Editors), New York Academy of Sciences, 271 (1976).
103. "Suspected Carcinogens," NIOSH (1977).
104. National Fire Codes 12. National Fire Protection Ass'n, 470 Atlantic Ave., Boston, Mass. (1977).
105. Chedekel, M. R., "Red-Brown Pigment May Cause Skin Cancer," *C&EN*, p. 5 (September 19, 1977).
106. Szara, S., "Interaction of Hallucinogens with Other Drugs." Annals of New York Academy of Sciences, 281 (1976).
107. National Cancer Institute Bulletin.
108. IARC Monographs on the Evaluation of Carcinogenic Risk of Chemicals to Man, Vol. 16, World Health Organization (June 1977).
109. Sax, N. I., Dangerous Properties of Industrial Materials, 4th Edition, Van Nostrand Reinhold, New York (1975).
110. *Science*, **201**, 1020–1023 (September 15, 1978).
111. NIOSH Current Intelligence Bulletin #25.
112. *Chemistry*, **51**:6, 21–23, (July/August 1978).
113. AIHA #39, pp. A-14, (June 1978).
114. Environmental Defense Letter (September/October 1977).
115. "Review of Inhalants: Euphoria to Dysfunction." Research Monograph #15 National Institute on Drug Abuse, pp. 171–184 (1977).
116. NIOSH, Current Intelligence Bulletin #21.
117. Registry of Toxic Effects of Chemical Substances, NIOSH (1977).
118. Carcinogen/Neoplastigen Subfile, November 15, 1978 of Registry of Toxic Effects of Chemical Substanccs.

ADDENDUM

Following is a list of over 400 materials for which the carc/neo/teratogen data have entered the literature since the publication of the 1976 *Registry of Toxic Effects of Chemical Substances by NIOSH*.

Because this material became available so late in the preparation of the 5th Edition, it has been included in this abbreviated form. All that is listed is the name of the material, its carc/neo/teratogen status, and the routes of exposure by which the data were obtained.

The reference for the information is no. 118 in the Bibliography which follows Section 12.

These items will be incorporated in the 6th Edition of the book.

***p*-ACETOPHENETIDIDE.** Oral carc to humans.

N-ACETOXY-N-MYRISTOYL-2-AMINO FLUORENE. Carc to rats via sc.

9-ACETYL-1,7,8-ANTHRACENE TRIOL. Carc to mice via dermal.

N-ACETYL-N′-(*p*-HYDROXYMETHYL)PHENYL HYDRAZINE. Neo to mice via oral.

N-ACETYL-N-MYRISTOYLOXY-2-AMINO FLUORENE. Carc to rats via sc.

ACRIFLAVINE + PROFLAVINE. Neo to rats via sc; mutagen to mice via ip.

AFLATOXIN G1 (56%) + AFLATOXIN B1 (38%). Neo to rats via oral; carc to rats via sc and it; carc to mice via sc.

AFLATOXIN G1. Carc to rats via sc.

***p*-ALLYL ANISOLE.** Carc to mice via sc.

ALLYL HYDRAZINE HYDROCHLORIDE. Carc to mice via oral.

ALLYL ISO THIOCYANATE. Neo to mice via dermal.

***o*-ALLYL PHENOL.** Neo to mice via dermal.

ALUMINUM OXIDE. Neo to rats via ipl.

ALUMINUM SILICATE. Neo to rats via ipl.

3-AMINO-2,5-DICHLORO BENZOIC ACID. Neo to mice via oral.

4-(2-AMINOETHYL)-1,2,3-BENZENE TRIOL HYDROCHLORIDE. Carc to mice via sc.

2-AMINO-4-(ETHYLTHIO)BUTYRIC ACID. Carc to rats via oral.

2-AMINO-4-(12-METHYL-7-BENZ(a)ANTHRYL METHYL)THIO BUTYRIC ACID. Carc to mice via iv.

AMPHIBOLE. See also Asbestos. Carc to rats via ipl.

AMSINCKIA INTERMEDIA. Carc and neo to rats via oral.

9-ANTHRONOL. Carc to mice via dermal.

ARECHOLINE BASE. Neo to hamsters via dermal.

L-ASPARAGINASE. Neo to mice via ip.

ASSAM TEA. Neo to rats via sc.

BARIUM SULFATE. Neo to rats via ipl.

BASIC RED 1. Neo to rats via sc.

BASORA CORRA. Neo to rats via sc.

BENLATE + SODIUM NITRATE. Carc to mice via oral.

BENZ(a)ANTHRACENE-1,2-DIHYDRODIOL. Neo to mice via dermal.

BENZ(a)ANTHRACENE-3,4-DIHYDRODIOL. Neo to mice via dermal.

BENZ(a)ANTHRACENE-5,6-DIHYDRODIOL. Neo to mice via dermal.

BENZ(a)ANTHRACENE-10,11-DIHYDRODIOL. Neo to mice via dermal.

BENZ(a)ANTHRACEN-7-YL TRICHLORO METHYL KETONE. Neo to rats and mice via sc.

BENZETHONIUM CHLORIDE. Neo to rats via sc.

BENZIMIDAZOLE METHYLENE MUSTARD. Carc to mice via ip.

BENZO(a)PYRENE-7,8-DIHYDRODIOL-9,10-EPOXIDE (anti). Neo to mice via dermal.

BENZO(a)PYRENE-4,5-EPOXIDE. Neo to mice via dermal.

BENZO(a)PYRENE-7,8-EPOXIDE. Neo to mice via dermal.

BENZO(a)PYREN-2-OL. Carc to mice via dermal.

BENZO(a)PYREN-6-OL. Neo to mice via dermal, sc and in.

BENZO(a)PYREN-11-OL. Carc and neo to mice via dermal.

N-BENZOYLOXY-4′-ETHYL-N-METHYL-4-AMINO AZOBENZENE. Neo to rats via sc.

BENZYL HYDRAZINE DIHYDROCHLORIDE. Neo to mice via oral.

BERYLLIUM FLUORIDE. Carc to rats via inhal.

BERYLLIUM SULFATE TETRAHYDRATE. Carc to rats via inhal.

BETEL NUT. Carc and neo to rats and mice via sc.

3,3′,4,4′-BIPHENYL TETRAMINE TETRAHYDROCHLORIDE. Neo to rats and mice via oral.

9,10-BIS(CHLOROMETHYL)ANTHRACENE. Carc to mice via iv.

4-BIS(2-HYDROXYETHYL)AMINO-2-(5-NITRO-2-FURYL)QUINAZOLINE. Carc to rats via oral.

3-BIS(2-HYDROXYETHYL)AMINO-2-(5-NITRO-2-THIENYL)QUINAZOLINE. Carc to rats vio oral.

BIS(TRIMETHYLSILYL)ACETAMIDE. Neo to mice via ip.

BP-9,10-DIHYDRODIOL. Neo to mice via dermal.

BRACKEN FERN TANNIN. Carc to mice via imp.

2-BROMOBUTANE. Neo to mice via ip.

2-BROMOERGOCRYPTINE. Carc to rats via oral.

9-BROMOMETHYL ANTHRACENE. Carc to mice via iv.

1-BROMO-2-METHYL PROPANE. Neo to mice via ip.

2-BROMO-2-METHYL PROPANE. Neo to mice via ip.

2-BUTENYL PHENOL. Neo to mice via dermal.

4-(BUTYL NITROSAMINO)-2-BUTANONE. Carc to rats via oral.

4-(BUTYL NITROSAMINO)-3-BUTANONE. Carc to rats via oral.

***o*-BUTYL PHENOL.** Neo to mice via dermal.

***p*-BUTYL PHENOL.** Neo to mice via dermal.

CADMIUM COMPOUNDS. Carc to humans via inhal.

CADMIUM SULFATE(1:1)HYDRATE(3:8). Neo to rats via sc.

CALAMUS OIL. Carc to rats via oral.

CALCIUM SULFATE DIHYDRATE. Neo to rats via ip.

CALOMEL + MAGNESIUM SULFATE(5:8). Neo to mice via oral.

CARBENDAZIM + SODIUM NITRITE(5:1). Carc to mice via oral.

CASEIN-SODIUM. Carc to mice via sc.

CHERRY BARK OAK. Neo to rats via sc.

CHESTNUT TANNIN. Neo to mice via sc.

CHLORDANE. Carc to mice via oral.

CHLORDIAZ EPOXIDE + SODIUM NITRITE(1:1). Neo to rats via oral.

5-β-CHOLESTAN-3-β-OL. Neo to mice via sc.

CHOLESTROL-α-OXIDE. Neo to mice via sc.

CHOLESTERYL-14-METHYL HEXADECANOATE. Carc to mice via ip.

***p*-CHLORO BENZENE THIOL.** Neo to mice via dermal.

2-CHLORO-4-BIPHENYLAMINE. Neo to rats via sc.

2-CHLOROBUTANE. Neo to mice via ip.

1-CHLORO-2,4-DINITRO NAPHTHALENE. Carc to rats via oral.

21-CHLORO-13-ETHYL-17-HYDROXY-18,19-DINOR-17-α-PREGN-4-EN-20-YN-ONE + 3-METHOXY-17-α-19-NORPREGNA-1,3,5(10)-TRIEN-2-YN-17-OL. Carc to dogs via oral.

(2-CHLOROETHYL)NITROSO CARBAMIC ACID, ETHYL ESTER. Carc to rats via oral.

7-CHLOROMETHYL BENZ(a)ANTHRACENE. Carc to mice via iv.

6-CHLOROMETHYL BENZO(a)PYRENE. Carc to rats via sc.

10-CHLOROMETHYL-9-CHLOROANTHRACENE. Carc to mice via iv.

10-CHLOROMETHYL-9-METHYL ANTHRACENE. Carc to mice via iv.

2-CHLORO-2-METHYL PROPANE. Neo to mice via ip.

1-CHLORO-2-NITRO BENZENE. Carc to mice via oral.

1-CHLORO-4-NITRO BENZENE. Neo to mice via oral.

***m*-CHLORO PHENOL.** Neo to mice via dermal.

***p*-CHLOROPHENOXY ANILINE.** Carc to rats and mice via oral.

3-CHLORO-4-STILBENAMINE. Neo to rats via sc.

4-CHLORO-*o*-TOLUIDINE HYDROCHLORIDE. Carc to mice via oral.

CHROMIUM ALLUMEN. Neo to rats via sc.

CHROMIUM CARBONYL. Neo to rats via it.

C.I. DIRECT BLACK 38. Carc(+) to rats via oral.

C.I. DIRECT VIOLET 1. Carc to rats via oral.

C.I. NATURAL BROWN 7. Carc to mice via dermal.

C.I. SOLVENT YELLOW 12. Neo to mice via sc.

CINNAMYL ANTHRANILATE. Neo to mice via ip.

COAL TAR. Carc to mice via dermal.

COAL TAR, AEROSOL. Neo to mice via inhal.

COAL TAR PITCH. Carc to mice via dermal.

COCOLODE. Neo to rats via sc.

COLTSFOOT. Carc to rats via oral.

CORONENE. Neo to mice via dermal.

CORUNDUM. Neo to rats via ip.

***m*-CRESOL.** Neo to mice via dermal.

CRISTOBALITE. Carc to rats via ip.

CYCLOHEXANE SULFAMIC ACID. Neo to humans via oral.

4H-CYCLOPENTA(def)PHENANTHRENE. Carc to rats via oral.

CYTOSTASAN. Carc to mice via oral and ip.

DECANE. Neo to mice via dermal.

DECYL ALCOHOL. Carc to mice via dermal.

7-DEHYDRO CHOLESTEROL. Neo to mice via sc.

DEHYDRO RETRONECINE. Neo to rats via sc.

DERRIS. Neo to rats via ip.

3,-β,17-β-DIACETOXY-17-α-ETHINYL-4-ESTRENE. Teratogen to mice via oral.

4,4′-DIAMINO-3,3′-BIPHENYLDIOL. Neo to rats via oral; carc to rats via sc; carc to rats via uk; neo to mice via oral, dermal and sc.

2,4-DIAMINOTOLUENE DIHYDROCHLORIDE. Carc to rats via oral; neo to mice via oral.

6,12-DIAZA ANTH ANTHRENE. Neo to rats via imp.

6,12-DIAZA ANTHANTHRENE SULFATE. Neo to rats via imp.

3,4-(DIBENZ(a,h)ANTHRYL-7,14-ENE)-2,5-DIOXO-1-PYRROLIDINE ACETIC ACID. Carc to mice via sc.

DIBENZO(def, mno)CHRYSENE. Neo to mice via dermal.

3,4-DIBROMO-1-NITROSO PIPERIDINE. Carc to rats via oral.

DICHLOROMETHYL METHYL ETHER. Neo to mice via dermal.

DIHYDROMETHYL PYRAN. Neo to rats via sc.

3,4-DICHLORO-1-NITROSO PIPERIDINE. Carc to rats via oral.

3,4-DICHLORO-*n*-NITROSO PYRROLIDINE. Carc to rats via oral.

(2,4-DICHLOROPHENOXY)ACETIC ACID. Teratogen to rats via oral; teratogen to mice via oral and sc.

2,2′-DICHLORO-4,4′-STILBENAMINE. Neo to rats via sc.

3,3′-DICHLORO-4,4′-STILBENDIAMINE. Neo to rats via sc.

N,N′-DIETHYL-N,N′-DINITROSO ETHYLENE DIAMINE. Carc to rats via oral.

1,2-DIETHYL HYDRAZINE DIHYDROCHLORIDE. Carc, teratogen to rats via iv.

2,10-DIFLUORO BENZO(rst)PENTAPHENE. Neo to mice via sc.

trans-4,5-DIHYDRO-4,5-DIHYDROXY BENZO(a)-PYRENE. Neo to mice via dermal.

7,8-DIHYDRO-7,8-DIHYDROXY BENZO(a)PYRENE. Carc to mice via dermal.

(+)trans-7,8-DIHYDRO-7,8-DIHYDROXY BENZO(a)PYRENE. Carc to mice via dermal.

9,10-DIHYDRO-9,10-DIHYDROXY BENZO(a)PYRENE. Neo to mice via dermal.

7,14-DIHYDRO-7,14-ETHANO DIBENZ(a,b)ANTHRACENE-15,16-DICARBOXYLIC ACID. Carc to rats via sc.

2,3-DIHYDRO-3-ETHYL-6-METHYL-1H-CYCLOPENTA(a)ANTHRACENE. Neo to rats via oral.

9,10-DIHYDRO-7-METHYL BENZO(a)PYRENE. Carc to mice via imp.

1,2-DIHYDRO-2-(5′-NITROFURYL)-4-HYDROXY QUINAZOLINE-3-OXIDE. Neo to dogs via oral.

1,2-DIHYDRO-2-(5-NITRO-2-THIENYL)QUINAZOLIN-4(3H)-ONE. Carc to rats via oral.

2,3-DIHYDROPHORBOL ACETATE MYRISTATE. Neo to mice via dermal.

5-β-DIHYDROTESTOSTERONE. Neo to mice via sc.

trans-8,9-DIHYDROXY-8,9-DIHYDROBENZ(a)ANTHRACENE. Neo to mice via dermal.

(+)-trans-7,8-DIHYDROXY-7,8-DIHYDROBENZO-(a)PYRENE. Neo to mice via dermal.

(−)-trans-7,8-DIHYDROXY-7,8-DIHYDROBENZO-(a)PYRENE. Neo to mice via dermal.

(+,−)-7,β,8-α-DIHYDROXY-9-β,10-β-EPOXY-7,8,9,10-TETRAHYDRO-BENZO(a)PYRENE. Neo to mice via dermal.

(+,−)-trans-7-β,8-α-DIHYDROXY-9,α,10-α-EPOXY-7,8,9,10-TETRAHYDROBENZO(a)PYRENE. Carc to mice via ip.

4-(2,3-DIHYDROXY PROPYLAMINO)-2-(5-NITRO-2-THIENYL)QUINAZOLINE. Carc to rats via oral.

2′,5′-DIMETHOXY STILBENAMINE. Carc to rats and mice via oral.

DIMETHYL AMINO SUCCINAMIC ACID. Carc to mice via oral.

9,10-DIMETHYL ANTHRACENE. Carc to mice via dermal.

N,N′-DIMETHYL-4,4′-AZODIACETANILIDE. Neo to mice via oral.

2,2′-DIMETHYL-4-BIPHENYLAMINE. Neo to rats via sc.

2,3′-DIMETHYL-4-BIPHENYLAMINE. Neo to rats via sc.

3,3′-DIMETHYL-4-BIPHENYLAMINE. Carc to rats via sc.

1,6-DIMETHYL-1,6-DINITROSO BIUREA. Neo to rats via sc.

DIMETHYL DODECYL AMINE HYDROCHLORIDE + SODIUM NITRITE(7:8). Carc to rats via oral.

7,14-DIMETHYL-7,14-ETHANODIBENZ(a,b)ANTHRACENE-15,16-DICARBOXYLIC ACID. Carc to rats via sc.

***α,α*-DIMETHYL-2-(6-METHOXY NAPHTHYL)-PROPIONIC ACID.** Carc to mice via oral.

N,*m*-DIMETHYL-N-NITROSO BENZYLAMINE. Carc to rats via oral.

***n,o*-DIMETHYL-N-NITROSO BENZYL AMINE.** Carc to rats via oral.

***n,p*-DIMETHYL-N-NITROSO BENZYL AMINE.** Carc to rats via oral.

***o,n*-DIMETHYL-N-NITROSO HYDROXYLAMINE.** Neo to rats via oral.

4,6-DIMETHYL-2-(5-NITRO-2-FURYL)-PYRIMIDINE. Carc to rats via oral.

2,5-DIMETHYL-N-NITROSO PYRROLIDINE. Neo to rats via oral.

2,3-DIMETHYL-4-PHENYL AZOANILINE. Carc to mice via oral.

N,N-DIMETHYL-4-(3,4,5-TRIMETHYLPHENYL)-AZO ANILINE. Carc to rats via oral.

2,4-DINITRO-1-FLUORO BENZENE. Neo to mice via dermal.

2,4-DINITRO TOLUENE. Neo to rats via oral.

N,N-DI-*n*-PROPYL ETHYL CARBAMATE. Neo to mice via ip.

DIRECT BROWN 95. Carc (+) to rats via oral.

DITOLYL ETHANE. Neo to mice via oral.

DODECANE. Carc to mice via dermal.

DIDECYL ALCOHOL. Carc to mice via dermal.

ELASIOMYCIN. Neo to rats via iv.

ENIDREL. Neo to mice via oral.

9,10-EPOXY-7,8,9,10-TETRA HYDRO BENZO(a)-PYRENE. Neo to mice via dermal.

ETHAMBUTOL + SODIUM NITRITE (1:1). Carc to mice via oral.

1-(*p*-ETHYL CARBOXYPHENYL)-3,3-DIMETHYL TRIAZENE. Carc to rats and mice via sc.

4-(ETHYL SULFONYL)-1-NAPHTHALENE SULFONAMIDE. Carc to mice via oral; neo to mice via im.

ETHYNERONE. Neo to dogs via oral.

ETHYNODIOL DIACETATE. Neo to mice via oral.

N-3-FLUORENYL ACETAMIDE. Carc to rats via im.

N-(2-FLUORENYL)-N-PHENYL HYDROXYLAMINE. Neo to rats via ip.

N-FLUOREN-2-YL-N-TETRADECANOYL HYDROXAMIC ACID. Carc to rats via sc.

3-FLUORO BENZO(rst)PENTAPHENE. Neo to mice via sc.

FOOD BLUE 3. Neo to rats via sc and im.

GOLD. Carc to rats and mice via imp.

GUANINE. Neo to rats via sc.

GUAVA. Neo to rats via sc.

HELIOTROPIUM SUPINUM L. Neo to rats via oral.

HEPTACHLOR. Carc to mice via oral.

17-*β*-HEPTANOYLOXY-19-NOR-17-*α*-PREGNEN-2-YNONE. Neo to rats via im.

HEXACHLORO-1,3-BUTADIENE. Carc to rats via oral.

4-(2-HYDROXY ETHYLAMINO)-2-(5-NITRO-2-THIENYL)QUINAZOLINE. Carc to rats via oral.

2-(7-HYDROXYFLUOREN-2-YL)ACETAMIDE. Neo to rats via oral.

7-HYDROXY GUANINE. Carc to rats via sc.

5-HYDROXY INDOLE ACETIC ACID. Neo to mice via sc.

17-HYDROXY-19-NOR-17-*α*-PREGN-4-EN-20-YN-3-ONE ACETATE + 19-NOR-17-*α*-PREGNA-1,3,5-(10)-TRIEN-2-YNE-3,17-DIOL. Neo to women via oral.

N-HYDROXY PHENACETIN. Carc to rats via oral.

***p*-HYDROXY PHENYL PYRUVIC ACID.** Carc to mice via sc.

1′-HYDROXY SAFROLE-2′,3′-OXIDE. Neo to mice via dermal.

ICR 10. Carc to mice via iv.

ICR 170. Carc to mice via iv.

ICR 292. Carc to mice via iv.

ICR 311. Carc to mice via iv.

ICR 340. Carc to mice via iv.

ICR 342. Carc to mice via iv.

ICR 394. Carc to mice via iv.

IH-INDOLE-3-ACETIC ACID. Carc to mice via sc; teratogen to mice and rats via oral.

INDOLE-3-ACRYLIC ACID. Carc to mice, hamsters, guinea pigs via sc.

IODOMETHANE. Neo to rats and mice via sc and ip.

2-IODOPROPANE. Neo to mice via ip.

2-IMIDAZOLIDINONE. Carc to mice via sc.

1-IODO BUTANE. Neo to mice via ip.

2-IODO BUTANE. Neo to mice via ip.

IRON DUST. Carc to rats via it.

IRON SODIUM GLUCONATE. Neo to mice via sc.

α-[(ISOPROPYLAMINO)METHYL]-2-NAPHTHALENE METHANOL. Carc to mice via unk.

α-[(ISOPROPYLAMINO)METHYL]-NAPHTHALENE METHANOL HYDROCHLORIDE. Neo to mice via oral.

ISOPROPYL OILS. Neo to mice via inhal and sc.

IVORY. Carc to rats and mice via imp.

LEAD-MOLYBDENUM CHROMATE. Neo to rats via sc.

LEUCO PARA FUCHSINE. Neo to rats and mice via oral, sc and dermal.

LIME OIL. Neo to mice via oral.

***d*-LIMONENE.** Neo to mice via oral.

LINOLEIC ACID + OLEIC ACID. Neo to mice via oral.

LUCANTHONE HYDROCHLORIDE. Carc to rats via oral.

MAGNESIUM OXIDE. Neo to hamsters via it.

MANGANESE. Carc to rats via im.

MANGANESE ACETYL ACETONATE. Neo to rats via im.

MANGANOUS CHLORIDE. Carc to mice via ip and sc.

MANGANOUS SULFATE, HYDRATE. Carc to mice via ip.

MARSH ROSEMARY. Neo to rats via sc.

MESTRANOL. Neo to mice via oral.

MESTRANOL + ANAGESTONE ACETATE (1:10). Carc to dogs via oral.

MESTRANOL + CHLORAMIDINONE ACETATE. Carc to women via oral; neo to mice via oral.

MESTRANOL + ETHYNODIOL DIACETATE. Neo to women via oral.

MESTRANOL + LYNESTRENOL. Neo to woman via oral.

METHAPYRILENE + SODIUM NITRITE (1:2). Carc to rats via oral.

METHOXY AZOXY METHANOL ACETATE. Carc to rats via pa.

2-METHOXY-4-BIPHENYLAMINE. Neo to rats via sc.

1-METHOXY ETHYL METHYL NITROSAMINE. Carc to rats via oral.

METHOXY METHYL ETHYL NITROSAMINE. Carc to rats via oral.

1-METHOXY-N-NITROSO DIETHYLAMINE. Carc to rats via oral.

1-METHOXY-N-NITROSO DIMETHYLAMINE. Carc to rats via oral.

1-METHOXY PROPYL PROPYL NITROSAMINE. Neo to hamsters via sc.

3-METHOXY-4-STILBENAMINE. Neo to rats via sc.

***p*-METHOXY-α-VINYL BENZYL ALCOHOL.** Carc to mice via sc.

N-METHYL ANILINE + SODIUM NITRITE (1.2:1). Carc to mice via oral.

N-METHYL ANILINE + SODIUM NITRITE (1:35). Carc to rats via oral.

3-(4-METHYL AMINO)-1-BUTENYL PYRIDINE. Neo to rats via it.

METHYL AZOXY OCTANE. Carc to mice via ip.

7-METHYL BENZ(a)ANTHRACEN-8-YL CARBAMIDE. Neo to mice via sc.

7-METHYL BENZ(a)ANTHRACENE-8-CARBONITRILE. Neo to mice via sc.

4′-METHYL-4-BIPHENYLAMINE. Neo to rats via sc.

4,4′-METHYLENE BIS(N-METHYL ANILINE). Neo to rats via sc.

***n*-METHYL-*n*-FORMYL HYDRAZINE.** Carc to mice via oral.

METHYL GAG. Neo to rats via oral.

METHYL GUANIDINE + SODIUM NITRITE (1:1). Carc to rats via oral.

METHYL HYDROXY OCTADECA DIENOATE. Neo to mice via dermal.

1-METHYL IMIDAZOLE-2-THIOL. Neo to rats via oral.

2-METHYL-1,4-NAPHTHOQUINONE. Neo to mice via dermal.

2-METHYL-1-NITRO ANTHRAQUINONE. Carc to rats and mice via oral.

1-METHYL-3-NITROGUANIDINE + SODIUM NITRITE (1:1). Carc to rats via oral.

4-(METHYL NITROSAMINO)-1-BUTANOL. Carc to rats via oral.

4-(N-METHYL-N-NITROSAMINO)-4-(3-PYRIDYL)-1-BUTANONE. Carc to mice via ip.

METHYL NITROSO CYANIDE. Carc and neo to rats via oral.

2-METHYL-4-STILBENAMINE. Neo to rats via sc.

3-METHYL-4-STILBENAMINE. Neo to rats via oral and sc.

MIMOSA TANNIN. Neo to rats and mice via sc.

MINERAL OIL. Neo to mice via ip.

MOLYBDENUM TRIOXIDE. Carc to mice via ip.

MORPHOLINE + SODIUM NITRITE (4:1). Carc to mice via oral.

4-MORPHOLINO-2-(5-NITRO-2-THIENYL)QUINAZOLINE. Carc to rats via oral.

MUSTARD OIL + 0.5% ARGEMONE OIL. Neo to mice via dermal.

N-MYRISTOYLOXY-N-MYRISTOYL-2-AMINO FLUORENE. Carc to rats via sc.

9-MYRISTOYL-1,7,8-ANTHRACENE TRIOL. Carc to mice via dermal.

MYRTAN TANNIN. Neo to mice via sc.

NAPHTHO(2,3-f)QUINOLINE. Carc to rats via sc.

NCI-C04137. Neo to rats and mice via oral.

NEMALITE DUST. Neo to rats via ip; neo to mice via ipl.

NEOCHROMIUM. Neo to rats via sc.

NICKEL-GALLIUM ALLOY (60:40). Carc to rats via imp.

NICKEL IRON SULFIDE. Carc and neo to rats via im.

NICKEL ISODECYL-O-PHOSPHATE (3:2). Neo to rats via im.

NICKELOUS CARBONATE. Neo to rats via imp.

NICKEL SULFATE (1:1). Neo to rats via imp.

NICKEL TITANIUM OXIDE. Carc to rats via im.

NITRILOTRIACETIC ACID. Neo to mice via oral.

3,3′,3″-NITRILOTRIS PROPIONAMIDE. Neo to rats via oral.

2-NITRONAPHTHALENE. Neo to dogs via oral.

1-NITRO-3-NITROSO-1-PROPYL GUANIDINE. Neo to rats via sc.

3-NITRO-1-NITROSO-1-PENTYL GUANIDINE. Neo to rats via sc.

2-NITROPROPANE. Carc to rats via inhal.

N-NITROSO DIALLYLAMINE. Carc to hamsters via sc.

***n*-NITROSO DIPENTYLAMINE.** Carc to rats via oral and sc.

***n*-NITROSO OXAZOLIDINE.** Carc to rats via oral.

1-NITROSO NAPHTHALENE. Neo to rats via ip.

2-NITROSO NAPHTHALENE. Neo to rats via dermal; neo to mice via sc.

N-NITROSO-4-PIPERIDINONE. Carc to rats via oral.

N-NITROSO-N-PROPYL-1-BUTANAMINE. Carc to rats via oral.

N-NITROSO-N-PROPYL PROPIONAMIDE. Neo to hamsters via sc.

NORGESTREL + MESTRANOL. Carc to women via oral.

N-NITROSO TETRAHYDRO-1,2-OXAZINE. Carc to rats via oral.

NSC-5265. Neo to mice via ip.

OIL OF ORANGE. Neo to mice via oral.

OVULEN. Carc to women via oral; neo to mice via oral.

7-OXIRANYL BENZ(a)ANTHRACENE. Neo to rats via sc.

6-OXIRANYL BENZO(a)PYRENE. Neo to rats via sc.

7-OXOCHOLESTEROL. Neo to mice via sc.

8-OXO-8H-ISOCHROMENO(4′,3′:4,5)PYRROLO-(2,3-f)QUINOLINE. Neo to mice via sc.

4-OXO STEARIC ACID. Neo to mice via sc.

PALYGORSCITE. Carc to rats via ip.

PANFURAN-S. Carc to rats via oral.

PERSIMMON. Neo to rats via sc.

4-(PENTYL NITROSAMINO)-1-BUTANOL. Neo to rats via oral.

***o*-(sec-PENTYL)PHENOL.** Neo to mice via dermal.

***p*-PENTYL PHENOL.** Neo to mice via dermal.

***p*-(sec-PENTYL)PHENOL.** Neo to mice via dermal.

PETASITENINE. Carc to rats via oral.

PHENYL BUTAZONE. Carc to women via oral; teratogen to rats via oral.

***m*-PHENYLENE DIAMINE.** Neo to rats via sc.

***p*-PHENYLENE DIAMINE.** Neo to rats via sc.

***m*-PHENYLENE DIAMINE DIHYDROCHLORIDE.** Neo to rats via sc.

***o*-PHENYLENE DIAMINE DIHYDROCHLORIDE.** Carc to rats and mice via oral.

N-PHENYL DIETHANOLAMINE MONO GLYCIDYL ETHER. Neo to mice via sc.

***n*-PHENYL-2-FLUORENAMINE.** Neo to rats via ip.

PHORBOL ACETATE MYRISTATE. Neo to mice via dermal.

PHTHALONITRILE. Neo to rats and mice via oral; carc to rats and mice via sc.

PIPERAZINE + SODIUM NITRITE (4:1). Carc to mice via oral.

PLATINUM. Carc to rats and mice via imp.

PODOPHYLLIN. Neo to mice via oral.

POLY BROMINATED BIPHENYL. Neo to rats via oral.

POLY ESTRADIOL PHOSPHATE. Carc to mice via sc.

POLYVINYL ACETATE CHLORIDE. Carc to mice via imp.

POTASSIUM ZINC CHROMATE (BASIC). Neo to mice via it.

POTATO (SAP FROM GREEN POTATO TOPS). Carc to rats via ip.

***n*-PROPYL IODIDE.** Neo to mice via ip.

PROPYL NITROSAMINO PROPYL ACETATE. Carc to hamsters via sc.

12H-PYRIDO(2,3-a)THIENO(2,3-i)CARBAZOLE. Neo to mice via sc.

PYROSET TKP. Neo to mice via dermal.

QUEBRACO TANNIN. Neo to rats and mice via sc.

REMAZOL BLACK B. Neo to rats via sc and oral.

RESORCINOL. Neo to mice via dermal.

RIFAMPICIN. Neo and teratogen to mice via oral; teratogen to rats via oral.

RUGULOSIN. Carc to mice via oral.

SAGRADO. Carc to rats via sc.

SASSAFRAS. Carc to rats via sc.

SELENIUM. Neo to mice via oral.

SHINING SUMAC. Carc to rats via sc.

SILVER NITRATE. Neo to mice via dermal.

SODIUM-7,14-DIHYDRO-7,14-ETHANODIBENZ-(a,h)ANTHRACENE-1,6-DICARBOXYLATE. Neo to rats via sc; carc to mice via sc; neo to mice via ip; carc to mice via sc.

SODIUM MAGNESIUM-HYDROXY AMPHIBOLE. Carc to rats via ipl.

SODIUM NITRITE + OXYTETRACYCLINE (1:1). Neo to rats via oral.

SORBITAN MONOLAURATE. Neo to mice via dermal.

STEARIC ACID. Neo to mice via imp.

4,4′-STILBENDIAMINE. Neo to rats via sc.

SULFACOMBIN. Carc to rats via pa.

N-SULFONOXY-N-ACETYL-2-AMINO FLUORENE. Neo to rats via sc.

SWEET GUM. Neo to rats via sc.

1,1,2,2-TETRACHLORO ETHANE. Carc (+) to mice via oral.

TETRACHLORO ETHYLENE. Carc to mice via oral.

TETRACHLORVINPHOS. Carc to mice via oral.

TETRADECANE. Carc to mice via dermal.

TETRAFLUORO-*m*-PHENYLENE DIAMINE DIHYDROCHLORIDE. Carc to mice via oral.

1,2,3,6-TETRAHYDRO-1-NITROSO PYRIDINE. Carc to rats via oral.

TETRAKIS(HYDROXYMETHYL)PHOSPHONIUM CHLORIDE. Neo to mice via dermal.

2,2′,4′,6-TETRAMETHYL-4-BIPHENYL AMINE. Neo to rats via sc.

***p*-(1,1,3,3-TETRAMETHYL BUTYL)PHENOL.** Neo to mice via dermal.

TETRAMETHYL HYDRAZINE HYDROCHLORIDE. Neo to mice via oral.

2,6,10,14-TETRAMETHYL PENTADECANE. Neo to mice via ip.

TIN. Carc to rats via imp; neo to mice via imp.

TITANIUM. Neo to rats via im.

TITANIUM ACETONYL ACETONATE. Carc to rats via im.

TITANIUM OXIDE. Carc to rats via im.

***m*-TOLUIDINE HYDROCHLORIDE.** Neo to mice via oral.

***o*-TOLUIDINE HYDROCHLORIDE.** Carc to mice via oral; neo to rats via oral.

***p*-TOLUIDINE HYDROCHLORIDE.** Carc to mice via oral.

***p*-TOLYL HYDRAZINE HYDROCHLORIDE.** Neo to mice via oral and sc.

TRIBROMO METHANE. Neo to mice via ip.

3,5,6-TRICHLORO-4-AMINOPICOLINIC ACID. Neo, carc (+) to rats via oral.

2,4,6-TRICHLORO ANILINE. Neo to mice via oral.

2,4,5-TRICHLORO PHENOL. Neo to mice via dermal.

TRIDYMITE. Neo to mice via it.

TRIFLURALIN. Carc and teratogen to mice via oral.

2,4,6-TRIMETHYL ANILINE. Neo, carc to rats via oral.

2,4,5-TRIMETHYL ANILINE HYDROCHLORIDE. Neo to mice via oral.

2,4,6-TRIMETHYL ANILINE HYDROCHLORIDE. Carc to rats and mice via oral.

2,2′,5′-TRIMETHYL-4-BIPHENYL AMINE. Neo to rats via sc.

TRIPHENYL ETHYLENE. Neo to mice via sc.

TRIMETHYL HYDRAZINE HYDROCHLORIDE. Carc to mice and hamsters via oral.

TRISODIUM NITRILO TRIACETIC ACID. Neo to rats via oral.

TWEEN 60. Carc to mice via dermal.

7-VINYL BENZ(a)ANTHRACENE. Neo to rats via sc.

DOT HAZARDOUS MATE

PLACARDING ANY QUANTITY—

MOTOR VEHICLES, FREIGHT CONTAINERS AND RAIL CARS

Placard motor vehicles, freight containers, and rail cars containing "any quantity" of hazardous materials listed in TABLE 1.

TABLE 1

HAZARDOUS MATERIAL CLASSED OR DESCRIBED AS	PLACARDS
Class A explosives	EXPLOSIVES A.
Class B explosives	EXPLOSIVES B.
Poison A	POISON GAS.
Flammable solid (DANGEROUS WHEN WET label only)	FLAMMABLE SOLID W.
Radioactive material	RADIOACTIVE.
Radioactive material:	
Uranium hexafluoride, fissile (containing more than 0.7 pct U^{235})	RADIOACTIVE AND CORROSIVE.
Uranium hexafluoride, low specific activity (containing 0.7 pct. or less U^{235})	RADIOACTIVE AND CORROSIVE.

YELLOW III labeled packagings only.

For Uranium Hexafluoride, see Sec. 172.504(a) and TABLE 1.

RAIL PLACARDS

SQUARE BACKGROUND FOR RAIL SHIPMENTS—Each "EXPLOSIVE A PLACARD," "POISON GAS PLACARD," and "POISON GAS—EMPTY PLACARD" affixed to a rail car must be placed on a square background measuring 14¼ inches on each side with a black border extending to 15½ inches on each side (illustrated in above chart). (See Sec. 172.510(a) and 172.527(a)).